W9-BBH-774

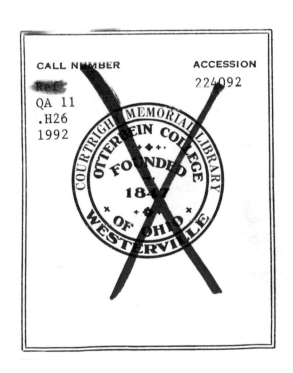

HANDBOOK OF RESEARCH ON MATHEMATICS TEACHING AND LEARNING

HANDBOOK OF RESEARCH ON MATHEMATICS TEACHING AND LEARNING

A Project of the
National Council of Teachers of Mathematics

DOUGLAS A. GROUWS
EDITOR

MACMILLAN PUBLISHING COMPANY
New York

Maxwell Macmillan Canada
Toronto

Maxwell Macmillan International
New York   Oxford   Singapore   Sydney

Macmillan Publishing Company   Maxwell Macmillan Canada, Inc.
866 Third Avenue   1200 Eglinton Avenue East, Suite 200
New York, NY 10022   Don Mills, Ontario M3C 3N1

Macmillan Publishing Company is part of the Maxwell Communication Group of Companies

Library of Congress Catalog Card Number: 91-37820

Printed in the United States of America

printing number
1 2 3 4 5 6 7 8 9 10

Library of Congress Cataloging-in-Publication Data
Handbook of research on mathematics teaching and learning / Douglas A.
Grouws, editor.
    p.   cm.
  "A project of the National Council of Teachers of Mathematics."
  Includes bibliographical references and index.
  ISBN 0-02-922381-4
  1. Mathematics–Study and teaching–Research.   I. Grouws, Douglas
A. II. National Council of Teachers of Mathematics.
QA11.H26 1992
510'.71–dc20   91-37820
                      CIP

■

CONTENTS

Part I

OVERVIEW

Part II

MATHEMATICS TEACHING

Part III
LEARNING FROM INSTRUCTION

Part IV
CRITICAL ISSUES

Part
V
PERSPECTIVES

PREFACE

Research on mathematics teaching and learning has flourished over the past two decades. There now exists a recognizable body of research that not only is conducted within the realm of mathematics but also takes the nature of the mathematics domain into account in all aspects of the work, including framing research questions, choosing a mode of investigation, designing instruments, collecting data, interpreting results, and suggesting implications. Certainly not all studies that are classified as mathematics education research give the mathematics involved the same attention, nor is each study equally sensitive to each of the aspects of investigation mentioned. However, there is sufficient progress in this direction to be able to characterize research on mathematics education as a research field and to identify those conducting these studies as the mathematics education research community. It is altogether fitting, therefore, that this *Handbook* has come into existence. The primary goals for the *Handbook* are to synthesize and reconceptualize past research, suggest areas of research most useful to advancing the field, and, where appropriate, provide implications of research for classroom practice.

AUDIENCE

A large number of scholars identify their research interest as mathematics education. Their productivity necessitates forums for sharing ideas and research results in order that the field can move forward based on collective wisdom and established findings, without wasted effort caused by uninformed or redundant investigations. The primary audience for the *Handbook* consists of mathematics education researchers and others doing scholarly work in mathematics education. This group includes college and university faculty, graduate students, investigators in research and development centers, and staff members at federal, state, and local agencies that conduct and use research within the discipline of mathematics. The intent of the authors of this volume was to provide useful perspectives as well as pertinent information for conducting investigations that are informed by previous work. The *Handbook* should also be a useful textbook for graduate research seminars.

Chapter authors were not directed to write specifically for curriculum developers, staff development coordinators, and teachers. The book should, however, be useful to all three

groups as they set policy and make decisions about curriculum and instruction for mathematics education in schools. Many researchers have advocated that mathematics teachers at all levels become involved in research because of the benefits that would accrue to the field, because of the teachers' keen insights and experience, and because the reflection involved and the experience gained would improve classroom instruction. I do not expect this volume to move many teachers in this direction. I do hope, however, that reading the *Handbook* will stimulate some teachers to conduct research in their own classrooms and that the book will be a useful tool for those teachers.

SCOPE

The *Handbook* was designed to be comprehensive in its coverage of research and research issues within mathematics education. Although the scope of the book is broad, some areas could undoubtedly have received more attention. Such areas will become increasingly obvious as research in mathematics education evolves. There is some overlap in coverage across chapters; this overlap helps to focus attention on the ideas and issues in the forefront of current research, provides a variety of perspectives on a particular issue as addressed by multiple authors, and demonstrates the interrelatedness of research in the field. Some chapters in the *Handbook* emphasize issues of learning mathematics, whereas others focus on teaching. Yet, interestingly, most of the authors explicitly address both of these perspectives to some degree. (In most cases this was done without much cajoling from the editor.) In the past these two dimensions have often defined two separate disciplines of scientific inquiry (Romberg & Carpenter, 1986). The attention here to both functions is significant because of the important relationship between teaching and learning.

STRUCTURE

Teaching and learning are considered in many *Handbook* chapters, but other ways that chapters can be naturally partitioned were used in developing the structure of the *Handbook*. The *Handbook* is composed of 29 chapters that are organized into five major sections. The first section, the Overview, comprises four chapters. It includes a history of research in mathematics education; an analysis of how the nature of mathematics as a discipline has been viewed over time and how this has influenced research in mathematics education; a perspective on scholarship in mathematics education, including the evolution of research methods and the role of theory; and a theoretical reconceptualization of one of the most crucial issues in mathematics education—the nature of understanding within the context of learning and teaching. The second section, Mathematics Teaching, includes chapters that attend to the culture of the mathematics classroom, teaching practices and their effects, teacher beliefs and conceptions, teacher knowledge, teaching with small groups, professionalism and mathematics teaching, and becoming a mathematics teacher as a process that spans preservice teacher education and teacher development once in the profession. The third section, Learning from Instruction, is composed of chapters organized about important mathematical domains that have been the focus of considerable research activity. Chapters deal with additive structures, multiplicative structures, rational numbers, problem solving and sense making in mathematics, estimation and number sense, algebra, geometry and spatial reasoning, probability and statistics, and advanced mathematical thinking including functions, limits, infinity, and proof. Whereas the authors for this section focus heavily on learning issues, reflecting the nature of much past research in these areas, they have thoughtfully considered teaching issues as an important part of their discussions. The fourth section, Critical Issues, includes teaching, learning, and content issues and deals with research and some of the most important and pervasive issues in the field today: technology, mathematics learning outside of school, affect, gender, race and ethnicity, and assessment. The final section, Perspectives, provides insightful views of three interesting and important areas. One chapter explores cross-national studies of mathematics achievement and their implications. Another chapter examines research in mathematics education from an international perspective with attention to commonalities and differences. The final chapter provides a philosophical look at where we are, the problems we face, and some thoughts on how we should proceed.

DEVELOPMENT OF THE *HANDBOOK*

The *Handbook* is a project of the National Council of Teachers of Mathematics. Lloyd Chilton, Executive Editor at Macmillan, suggested that a comprehensive handbook on the teaching and learning of mathematics would be a valuable addition to the field. John Dossey, president of the National Council of Teachers of Mathematics, enthusiastically endorsed the idea and took it to the board of directors for approval. They approved the idea in late 1986, and shortly thereafter I was asked to serve as editor. I accepted this commitment because I am convinced that such a book is needed and because it provides an opportunity to move research in mathematics education forward. I envisioned the editor's job as a large one and was never disappointed. My first task was to assemble an

editorial board to assist in providing direction and structure for the *Handbook* and to help identify authors. The members of the editorial board were Thomas P. Carpenter, University of Wisconsin; John Dossey, Illinois State University; James Hiebert, University of Delaware; Carolyn Kieran, University of Quebec at Montreal; Jeremy Kilpatrick, University of Georgia; and Douglas B. McLeod, San Diego State University. They are eminent scholars in mathematics education, and I was pleased that each accepted this responsibility without hesitation in spite of many other commitments. With the help of the editorial board, chapter topics and potential authors were identified. We sought to invite authors who were active and accomplished researchers with recognized reputations for excellence in their areas of specialization. We were pleased that nearly every potential author we approached accepted the challenge and, with one exception, fulfilled their obligations. It is simple mathematics to calculate the long development time necessary to complete the *Handbook*. Each chapter was reviewed by two outside experts who provided feedback in the form of criticism and commentary. The editor reviewed each chapter, as did a member of the editorial board in many cases.

The value of each chapter, and of the *Handbook,* was enhanced by each author taking seriously my charge to carefully and comprehensively review the research literature in his or her area and reconceptualize it in a way that would increase understanding and provide a base for productive future research. Noteworthy to me was the authors' sensitivity to the importance of work in other fields such as cognitive psychology, philosophy, and sociology and in other areas such as science education, reading, and writing. Their attention to the historical roots of many ideas in mathematics education as well as the evolving methodologies becoming prevalent provided a needed perspective and an element of interest. These insights and many others contribute to the quality of the book.

ACKNOWLEDGMENTS

Producing a handbook requires the diligent help of many people, and that certainly is the case with this *Handbook*. The quality of the authors' work and their work habits are the most important factors in producing a valuable handbook. Every one of the 40 authors involved in this endeavor produced high-quality work and did his or her best to meet production deadlines. I am grateful to each of them. The editorial panel provided needed scholarly advice and detailed help whenever I asked, for which I am indebted. Reviewers played an important part in the development of each chapter; they gave freely of their time and expertise, and I thank them for the vital role they played.

I want to thank Lloyd Chilton for initiating the idea for the *Handbook,* John Dossey for moving it through organizational channels, and the National Council of Teachers of Mathematics for endorsing it and asking me to serve as editor. I also want to express a special word of thanks to Thomas Cooney, University of Georgia, who provided reviews of several chapters and never turned down my many requests for advice and help, even on short notice.

Those at Publication Services worked hard to move this book quickly through to publication. I especially want to thank my liaison person, Maria Victoria Paras, for her hard work on this book and her pleasant and understanding manner in dealing with the difficulties that arose as the book was being assembled. I want to thank Bruce Biddle, Director of the Center for Research in Social Behavior, and Wayne Dumas, Chair of the Department of Curriculum and Instruction, for the support they gave me in this endeavor. I also want to thank Mary Beth Llorens and Linda Bolte for the editorial assistance and proofreading help they provided for many chapters, Kevin Evens for proofreading help, and my secretary, LeeAnn Debo, who provided efficient and cheerful help on all aspects of this project.

Finally, I want to acknowledge the understanding and support I received from my wife, Carol, especially during the times when things were stacking up and moving slowly. Thanks also to our sons, David and Michael, who frequently asked, "How is the *Handbook* coming, Dad?" Sometimes the comment was meant to prod and sometimes to encourage, but it was always intended to help.

Douglas A. Grouws, Editor
Columbia, Missouri

REFERENCE

Romberg, T. A., & Carpenter, T. P. (1986). Research on teaching and learning mathematics: Two disciplines of scientific inquiry. In M. C. Wittrock (Ed.), *Handbook of research on teaching* (3rd edition, pp. 850–873). New York: Macmillan.

OVERVIEW

A HISTORY OF RESEARCH IN MATHEMATICS EDUCATION

Jeremy Kilpatrick
UNIVERSITY OF GEORGIA

The history of research in mathematics education is part of the history of a field—mathematics education—that has developed over the last two centuries as mathematicians and educators have turned their attention to how and what mathematics is, or might be, taught and learned in school. From the outset, research in mathematics education has also been shaped by forces within the larger arena of educational research, which roughly a century ago abandoned philosophical speculation in favor of a more scientific approach. But like mathematics education itself, research in mathematics education has struggled to achieve its own identity. It has tried to formulate its own issues and its own ways of addressing them. It has tried to define itself and to develop a cadre of people who identify themselves as researchers in mathematics education.

During the past two decades, that task of self-definition has largely been accomplished. An international community of researchers exists that holds meetings, publishes journals and newsletters, promotes collaboration within and across disciplines in doing and critiquing research studies, and attempts to keep a research consciousness alive in the councils of the mathematics education organizations in which members of the research community participate. It is appropriate to look back at some of the people and events that have given form, direction, and substance to the field of research in mathematics education.

Any retrospective view is bound to be clouded by the difficulty of deciding what counts as research in the field. Studies that now seem to be exemplary cases of early research in mathematics education may not have been viewed that way by the investigators themselves and their contemporaries, who may have seen the studies as, for example, investigations in psychology, studies in history, or survey research on educational practice. In contrast, activities such as writing mathematics textbooks, which most people today would probably not consider as research, may have been thought by our forebears in mathematics education to epitomize research in the field. We do not have accounts of what people considered research to be as the field of mathematics education was taking shape. Researchers did not necessarily use the term *research in mathematics education* to describe what they did. In reviewing past work, therefore, we should be as inclusive as possible.

A useful broad definition of research is *disciplined inquiry* (Cronbach & Suppes, 1969). The term *inquiry* suggests that the work is aimed at answering a specific question; it is not idle speculation or scholarship for its own sake. The term *disciplined* suggests not only that the investigation may be guided by concepts and methods from disciplines such as psychology, history, philosophy, or anthropology but also that it is put on display so that the line of inquiry can be examined and verified. Disciplined inquiry need not be "scientific" in the sense of being based on empirically tested hypotheses, but like any good scientific work, it ought to be scholarly, public, and open to critique and possible refutation. Research in mathematics education, then, is disciplined inquiry into the teaching and learning of mathematics.

The purposes of research in mathematics education are manifold. If one views such research as akin to that of the natural sciences in following the empirical-analytic tradition (Popkewitz, 1984, pp. 35–40), one adopts the traditional aims of science: to explain, to predict, or to control (Carr & Kemmis,

I am grateful to the following people for answering queries and providing information: Rolf Biehler, Alan Bishop, Ken Clements, Geoffrey Howson, Jens Høyrup, Paul Hurd, David Johnson, Wan Kang, Christine Keitel, Neil Pateman, Hendrik Radatz, Gert Schubring, Yasuhiro Sekiguchi, Sharon Senk, Les Steffe, Pinchas Tamir, Zalman Usiskin, Sigrid Wagner, Heide Wiegel, and David Wheeler. My particular thanks to Doug McLeod, Jim Wilson, and most especially George Stanic for reading drafts and recognizing what I should have said.

1986, p. 83). If one views research in mathematics education as more like the interpretive understanding of a culture that an anthropologist might seek (Eisenhart, 1988), one tries instead to understand the meanings that the learning and teaching of mathematics have for those who are engaged in the activity. If one takes the approach of critical sociology and undertakes "action research" designed to help teachers and students gain greater freedom and autonomy in their work, the research is done both to improve practice and to involve the participants in that improvement (Carr & Kemmis, 1986, p. 165). A movement in mathematics education research over the past decade or so has led it away from the empirical-analytic tradition and, rather haltingly, toward interpretive and (to a lesser extent) critical approaches (Kilpatrick, 1988a, 1988b). Although the methods of the empirical-analytic tradition have dominated research in mathematics education for most of this century, the scientific aspirations of explanation, prediction, and control seem never to have been as predominant a motivation as the desire to understand and, especially, to improve both the learning and the teaching of mathematics. Despite that desire, however, understanding and improvement have not ordinarily meant adopting the participants' views or taking the instructional context as problematic. Research in mathematics education has dealt primarily with problems of learning and teaching as defined by researchers. They undertake research studies because they know of practices that ought to be better and have a vision of how that betterment might be accomplished. They attempt to do research that is applied rather than pure (Nisbet & Entwistle, 1973).

Of course, the noble purposes of understanding and improvement frequently mask other reasons for undertaking a research study. Ministries of education have been known to commission research studies in mathematics education for the sole purpose of justifying decisions already made or about to be made on other grounds. Moreover, no one can deny that the requirements for obtaining advanced degrees, tenure, and promotions in institutions of higher education in certain countries have led more than a few mathematics educators to conduct investigations they would not otherwise consider doing. The motives for doing research are often tangled and far from disinterested.

One cannot easily trace the various personal and political motivations that have yielded the enormous body of research in mathematics education that has accumulated during this century. In the controversies about what that research says, how it should be done, and what importance it has for practice, however, one can see a field in the process of defining itself. In other words, whatever an investigator's private reasons might have been for undertaking a certain study, that study may have helped determine what research in mathematics education is to be.

A Place in the University

Like mathematics education itself, research in mathematics education began primarily in the universities. The Protestant universities of Prussia at the beginning of the 19th century sparked a reform of higher education that eventually spread to other countries and that led to the differentiation and professionalization of the modern scientific disciplines (Grattan-Guinness, 1988; Jahnke, 1986; Pyenson, 1985; Schubring, 1988b;

Shils, 1978). With that reform came the expectation that university faculty would not only teach but also conduct research. In Europe, education began to be studied as a separate academic discipline. Progress was slow. The first chair in education was established at the University of Halle in 1779, yet by 1910 the number of staff members at German universities with teaching responsibilities in education numbered only 13 (Husén, 1983). The new professors of education tended to come from philosophy and history. In Sweden, lectures in education were given by a professor of philosophy at the University of Uppsala as early as 1804, but it was not until 1910 that the university established a chair in education. One of the first chairs in education in Great Britain was that of "Professor of the Theory, History, and Practice of Education" at the University of Edinburgh in 1876 (Cubberley, 1920, p. 826). In the United States, although New York University (in 1832), Brown University (from 1850 to 1855), and the University of Michigan (in 1860) began occasional courses in education, the first permanent professorship in education was not established until 1873, at the University of Iowa. As of 1890 there were fewer than a dozen chairs of education in the United States (Cubberly, 1920, p. 827).

All through the 19th century, universities graduated teachers of mathematics for secondary school, but instruction in the teaching of mathematics was at best a separate and minor part of the teacher's preparation. In Germany, pedagogical seminars were organized within the universities early in the century to prepare university students to teach in the gymnasiums. The seminars educated the students to do original research in mathematics (Pyenson, 1983, p. 23). Only at the end of the century, with the attempt to establish didactics as a discipline dealing with school knowledge as against a more general pedagogy (Gerner, 1968, p. 279), did German university students begin to receive practical training in mathematics teaching. One of the leaders in bringing methods courses into university education was Felix Klein, who not only began such courses at several universities but also supervised the first doctoral (*Habilitation*) degree in mathematics education, which was obtained at Göttingen in 1911 by Rudolf Schimmack (Schubring, 1988b).

In other countries, students preparing to teach mathematics studied mathematics, with perhaps an occasional lecture on classroom management or moral education as professional preparation. In France, for example, the École Normale Supérieure established by Napoleon in 1810 continued almost unchanged until the Second World War; its training of prospective mathematics teachers for the lycées consisted primarily of lectures in mathematics. (France's main strategy for contending with the shortage of well-trained teachers was to provide textbooks, which later became a mechanism for controlling instruction; Schubring, 1987, p. 47.) Elementary school teachers in these countries were typically prepared in separate institutions of pedagogy—termed colleges, institutes, seminaries, or normal schools—that were institutions of secondary rather than higher education. Japan, for example, adopted the American version of the normal school; by 1908, there were 75 Japanese normal schools (Fugisawa, 1912, p. 213). (The word *normal* came from the French word for a model, or rule—the prospective teachers were given rules for teaching; Butts, 1947, p. 493.) In the United States, meanwhile, the normal schools had begun to evolve into teachers colleges in a movement that accelerated after 1920

(Cremin, 1953). As countries began to establish national school systems, they found that they needed a larger supply of qualified teachers who had received a professional education. The specialized training in subject matter that might have sufficed to prepare teachers for elite schools—the gymnasium, the lycée, the English public school—was clearly insufficient for the cadres of teachers needed to staff the new secondary schools being founded. And demands for better qualified elementary school teachers led countries to upgrade the institutions in which they were prepared from secondary to higher education status.

Mathematics education as a field of study began to develop slowly at the end of the 19th century as universities in several countries, in response to the need for more and better prepared teachers, started expanding their programs in teacher education. By 1912, a survey by the International Commission on the Teaching of Mathematics reported that university lectures on mathematics education (to supplement mathematics lectures) were being offered in the United States, Great Britain, Germany, and Belgium (Schubring, 1988b). In some countries, new institutions of higher education were established to prepare teachers (such as the *pädogogische Hochschulen* in Germany in the 1920s) in which the emphasis was on practical preparation in how to teach. Research did not flourish in such institutions. But elsewhere (as in the United States, and later Germany, England, and Japan), some of the special schools for training teachers, whether elementary or secondary, were absorbed or enlarged into universities. An early example that became widely emulated around the world was the New York College for the Training of Teachers, established in 1887, which in 1890 was affiliated with Columbia University as its Teachers College. Reviewing the 1904-1905 catalog for Teachers College, Henri Fehr (1904) concluded that the college was not a professional school in the sense of preparing teachers as rapidly as possible to teach "the ABCs of mathematics" (p. 316), but rather that it gave future mathematics teachers, along with practical directions for their career, an excellent general culture in mathematics, the opportunity to pursue specialized studies, and an introduction to the historical development of their science.

In time, and somewhat differently in different countries, mathematics education came to be recognized as a university subject. The expectation that people engaged in the education of mathematics teachers in a university ought to be doing research and not just teaching led many of these people to undertake research in mathematics education.

Influences from the Disciplines

Two disciplines have had a seminal influence on research in mathematics education. The first is mathematics itself. Mathematicians have a long, if sporadic, history of interest in studying the teaching and learning of their subject. Disputes between pure and applied mathematicians over the appropriate preparation for advanced study in the discipline have led not only to periodic attempts to revise school and college curricula but at times to efforts to survey the extent of those revisions and to study how students respond to them. Concerns about inadequate preparation in the lower schools, falling enrollments in advanced courses, the potential erosion of mathe-

matics as a school subject, and threats to national status have from time to time prompted mathematicians to look into what the schools are doing and how it might be improved. Curiosity about how mathematics is created occasionally led to introspection by mathematicians regarding their own reasoning processes and to attempts to teach those processes. Observations of the results of their own children's or grandchildren's mathematical thinking encouraged some mathematicians to develop detailed analyses of that thinking or programs to improve it. Although these forays into mathematics education have not always been sustained, they have usually been given serious consideration, in part because of the mathematician's status in the society.

As mathematics education developed in the universities, it tended to attract people whose primary interest was in mathematical subject matter and who perhaps even thought of themselves as mathematicians. These mathematics educators conducted historical and philosophical studies, surveys, and eventually other types of empirical research. Their work and that of the mathematicians raised many of the issues that researchers are addressing today.

The second major influence on research in mathematics education is psychology. One of the preconditions for the development of mathematics education, according to Schubring (1988b), was the age-graded school in which the teacher could deal with a class as a group and begin to observe patterns of cognitive development. Near the turn of the 20th century, psychological institutes in Germany and psychology departments in the United States began to undertake empirical studies in education (Husén, 1983). Psychology became the "master science" of the school and consequently a central part of the normal school curriculum (Cubberley, 1920, p. 755). As a scientific discipline that originated and was developing within the university (Napoli, 1981, pp. 8–9), psychology provided professors in schools and departments of education with a science that might give their school a higher status and a set of methods they could use in the research they were expected to do. Furthermore, demographic pressures within psychology itself were pushing researchers into educational research (O'Donnell, 1985, p. 154). From the beginning of educational psychology, mathematics was a popular vehicle to use in investigating learning, probably because of perceptions regarding its important role in the school curriculum; its relative independence of nonschool influences; its cumulative, hierarchical structure as a school subject; its abstraction and arbitrariness; and the range of complexity and difficulty in the learning tasks it can provide. Mathematics educators have often been wary of psychological research because of what they have seen as an indifference to or ignorance of the academic discipline of mathematics, but they have never hesitated to borrow ideas and techniques freely from psychology.

In what follows, the roots of research in mathematics education are traced as they relate, first, to mathematics and, second, to psychology. As a convenient oversimplification, the roots in mathematics deal primarily with research into what mathematical content is taught and learned, and the roots in psychology deal primarily with research into how that content is taught and learned. As another oversimplification, mathematicians have tended to be interested in secondary and collegiate mathematics, and psychologists have tended to be interested in elementary mathematics. These oversimplifications have enough truth to make the categories useful for an overview and enough exceptions to make the story interesting.

After the roots have been sketched, the emergence of mathematics education as a field of study provides a framework for discussing the research produced during the first two-thirds of this century by the people who became known as mathematics educators. The formation over the next decade or so of an international community of researchers in mathematics education is then delineated, and the story is brought up to the beginning of the 1980s with an outline of some recent challenges and responses in the field.

ROOTS IN MATHEMATICS

On assuming the position of professor of mathematics at Erlangen in 1872 at the age of 23, Felix Klein published his famous "Erlanger Programm," which, by showing how a geometry could be characterized by the invariants of a transformation group, gave a productive new spin to research in geometry. At the same time, he delivered a less technical inaugural address on mathematics education (Rowe, 1983, 1985b) in which he propounded a rather traditional Prussian neohumanistic view of mathematics. Although he argued for more attention to the applications of mathematics to the other sciences, he stressed most of all the formal educational value that mathematics can have. He called for livelier instruction and a more spirited treatment of mathematics in the gymnasiums and in the universities. And he advocated that prospective teachers know mathematics to the point at which they could undertake independent research. (The Prussian teaching regulations of 1866 required candidates to publish an original study in their chosen field.) At the time, Klein was not concerned with the particular studies the prospective teachers undertook—the formal educational value of mathematics made one topic as good as any other. But later in life, he changed his mind and argued that teachers should concentrate on those studies that would be of fundamental importance for their profession (Rowe, 1985b, p. 128).

In his Erlangen address, Klein deplored the division between humanistic and scientific education. He saw mathematics as occupying a central position between the two. Klein's own education had given him a universalist approach both to fields within mathematics itself and to the connections of mathematics with the sciences. After arriving at Göttingen in 1885, he began to make it a center for applied mathematics and technological research (Rowe, 1985a, p. 281). In 1888, he proposed the unification of the Prussian technical institutes and the universities. That unification did not occur, however. By 1900, Klein had concluded that students coming from any of the three types of secondary schools (*Gymnasien, Realgymnasien,* and *Oberrealschulen*) should be prepared mathematically to study in either type of institution of higher education. The failure of his proposals for reforming higher education, according to Schubring (1988a), led Klein to undertake a more far-reaching reform that would begin with the secondary school. Analytic geometry and the calculus would be put back into the secondary school, and the banner for this reform would be the concept of *function.*

Like many mathematicians since, Klein believed that substantial changes in the school mathematics curriculum were essential if the demands of modern higher education were to be met.

The International Commission

At the Fourth International Congress of Mathematicians in Rome in 1908, a new organization was formed: the International Commission on the Teaching of Mathematics (for histories of the organization, see Howson, 1984; Kahane, 1988). David Eugene Smith, of Teachers College, Columbia University, had in 1905 first proposed the idea of an international commission to study the teaching of mathematics. He seems to have believed that an extensive cross-national study would help countries see the value of different ways of organizing the mathematics curriculum, but he did not argue for a reform agenda. Klein, although he did not attend the Rome congress, was able to use the new commission to help advance his reform goals. Klein was named president of the commission (serving until his death in 1925), with Henri Fehr of the University of Geneva as secretary. *L'Enseignement Mathématique,* the journal founded in 1899 by Fehr and Charles Laisant (from the École Polytechnique in Paris), became the commission's official journal.

The ostensible purpose of the international commission was to report on the state of mathematics teaching at all levels of schooling around the world. Countries were invited to participate in the commission's activities according to how active they had been in the previous International Congresses of Mathematicians (ICMs). National subcommittees were formed and invited to report back at the 1912 ICM in Cambridge, England. A tremendous flurry of activity ensued, with 17 subcommittees presenting national reports at the ICM (H. Fehr, 1913). The more lengthy reports included 11 volumes from the United States, 6 from Germany, 5 from France, and 2 large volumes from the United Kingdom. According to Schubring (1988a), of the 20 national subcommittees, 9 had by 1914 not only delivered national reports but also undertaken "significant reform activities" (p. 7). Those countries were Austria, Belgium, Denmark, France, Germany, Hungary, Sweden, the United Kingdom, and the United States.

The international comparisons based on these reports were confined to an overview from Hungary of the various reform efforts and four reports from the United States that made international comparisons of curricula and teacher training (Schubring, 1988a). The comparisons were descriptive rather than analytic; for example, the major curriculum comparison (J. C. Brown, 1915) consisted of little more than lists of syllabi, and the point seemed to be simply to show how the U.S. curriculum diverged from common European practice. Studies that attempted to probe more deeply into cross-national similarities and differences were undertaken by the commission itself. One of the topics chosen for study at the 1912 ICM, for example, was intuition and experiment in secondary mathematics teaching. David Eugene Smith (1913) reported on the commission's findings, which were based on questionnaire data obtained by Walther Lietzmann, of the Oberrealschule in Barmen, Germany. Smith found growing attention to descriptive geometry in several countries but was unable to determine whether the function concept was becoming more widely treated in an intuitive fashion. It should be noted that responses to the questionnaire came from one or two college or university professors in each of the six countries surveyed, so the adequacy of the report as a description of current practice was limited.

The reports from the International Commission on the Teaching of Mathematics (by 1920, it had produced 294 publications, according to Schubring, 1988a) marked the beginning of efforts by mathematicians and mathematics educators not only to reform school mathematics but also to gather information that could be used in that reform. Klein was far from the only mathematician to take an interest in school reform as the 20th century began. France reformed its geometry teaching by government decree in 1902, and mathematicians like Emile Borel, Jacques Hadamard, and Henri Lebesgue contributed articles and textbooks (Kahane, 1988). In England, John Perry, of the Royal College of Science, had developed a new syllabus in practical mathematics and in 1901 delivered to the British Association a sharp critique of contemporary mathematics teaching, arguing for a more intuitive and laboratory-based approach. Retiring in 1902 as president of the American Mathematical Society, Eliakim Hastings Moore (1903/1967) of the University of Chicago echoed Perry's critique and called for a curriculum in which the different branches of mathematics would be unified. He called on professional mathematicians to get involved in the reform of school mathematics. But involvement—often taking the form of textbook writing—was one thing; careful investigation was another. The data-gathering activities of the international commission were monumental, politically motivated, methodologically unsophisticated, and conceptually weak. The reports that resulted were more compilations of data than analyses or interpretations. They launched the process, however, of finding out what mathematics is being taught in the schools and how it is being taught.

The Study of Mathematical Thinking

Mathematicians have often been interested in the mysterious processes of mathematical creation. How do they do what they do when they are doing mathematics? They have used terms like *insight* and *intuition* to try to capture some of these processes. In the very first issue of *L'Enseignement Mathématique,* Henri Poincaré (1899) argued for more attention to intuition in mathematics instruction along with the attention given to logic: "It is by logic that one proves, but it is by intuition that one invents" (p. 161). Poincaré saw mathematical creation as a process of discernment, and he emphasized how a flash of insight might come after a period of intense concentration. In a later essay on mathematical discovery, he related the story of how he had put aside a difficult problem on which he had been working and suddenly had an insight into the solution as he was putting his foot on the step of a vehicle to go for a drive (Poincaré, 1952, p. 53). He claimed that such an insight had characteristic qualities of definiteness and certainty.

The editors of *L'Enseignement Mathématique,* Fehr and Laisant, surveyed over 100 mathematicians by questionnaire to learn how they did mathematics ("Enquête sur," 1902; H. Fehr, 1905; Fehr, Flournoy, & Claparède, 1908). The investigation followed a parallel inquiry by the mathematician E. Maillet, who

was especially interested in mathematical dreams (Hadamard, 1945/1954). Few of Maillet's respondents reported having had such dreams. Fehr and Laisant's investigation, in which they were assisted by the Genevese psychologists Edouard Claparède and Théodore Flournoy, covered a wider range of topics, from questions about recollections of early studies and interests ("At what time, as well as you can remember, and under what circumstances did you begin to be interested in mathematical sciences?") to work habits ("Does one work better standing, seated, or lying down?") to processes of inspiration ("Would you say that your principal discoveries have been the result of a deliberate endeavor in a definite direction, or have they arisen, so to speak, spontaneously in your mind?"). (For the complete set of questions, see Fehr or Fehr et al.; in English, Hadamard, 1945/1954, pp. 137–141.) Poincaré (1952, p. 46) saw the *Enseignement* inquiry as essentially confirming his conclusions, but Hadamard was disappointed because it asked only about successful discoveries, not about failures, and also because he found that almost all the responses came from "alleged mathematicians whose names are now completely unknown" (p. 10). The final report (published in 11 parts in *L'Enseignement Mathématique* from 1905 to 1908 and collected in Fehr et al.) is essentially a list of the verbatim responses to each group of questions accompanied by some general observations.

Hadamard (1945/1954) later undertook his own informal inquiry among mathematicians in America, querying such mathematicians as George Birkhoff, Norbert Wiener, George Pólya, and Albert Einstein about the mental images they used in doing mathematics. Davis and Hersh (1981, p. 308) use the term *cognitive style in mathematics* to characterize the difference in approach to solving problems that Hadamard identified. Indeed, Hadamard's investigation, casually conducted and incompletely reported as it was, did anticipate a subsequent body of research into cognitive styles and their relation to mathematical thinking, albeit the thinking of less gifted doers of mathematics.

In his book, Hadamard (1945/1954, pp. 1–2) spoke of *subjective* (introspective) and *objective* (behavioral) methods of psychological investigation. He argued for the use of subjective methods in the investigation of mathematical invention, claiming that the exceptionality of the phenomenon did not permit the comparison of numerous cases required by observation. He considered the behaviorist's rejection of attention to thought and consciousness to be "an unscientific attitude" (p. 1). The rise of behaviorism in the early part of this century, whether unscientific or not, greatly inhibited the study of mathematical thinking.

ROOTS IN PSYCHOLOGY

The same premiere issue of *L'Enseignement Mathématique* that carried the article on mathematics instruction by Poincaré also contained an article by Alfred Binet (1899) on *scientific pedagogy.* Binet was the director of the first French psychological laboratory, established 10 years before at the Sorbonne. In the article, he described a new movement in pedagogy that was appearing in several countries and that sought to replace a priori assertions by precise results based on data. Experimental pedagogy could not be undertaken in one's study or laboratory; it had to be done directly with children in schools and therefore primarily by teachers. Binet identified three principal methods of investigation: questionnaire, observation, and experiment. (He commented that questionnaires, although good for a preliminary inquiry, ought to address a theme whose importance was commensurate with the effort needed to fill them out, noting "*ces énormes questionnaires américains*," p. 32.) He argued for elementary school teachers to be given scientific training. Acknowledging that his readership was drawn from the secondary school teaching force, Binet asked for their interest and understanding, too.

Research on Thinking

Measuring Mental Ability. Today, Binet is hailed for his contributions to the psychology of thinking and as the originator of the intelligence test. Early in his career, he had followed the prevailing interest in looking for evidence of intelligence in the measurement of skulls (Gould, 1981, p. 146). Franz Josef Gall had initiated the "science" of *phrenology* (judging specific mental capacities by the configuration of the skull, particularly its "bumps"), and Paul Broca had developed phrenology into the study of *craniometry* (measuring the skull and its contents). The neurologist Paul Möbius (grandson of the mathematician August Möbius) studied the heads of eminent mathematicians—along with other characteristics—and concluded that mathematical talent is identifiable from the shape of the skull, although in a more complex fashion than Gall had suggested (Möbius, 1900, pp. 271–331). Binet's own careful studies of children and adolescents convinced him, however, that physical measurements could not yield good evidence of mental prowess. To study mental ability, one needed to pose tasks in which that ability could be demonstrated.

In England, Francis Galton (Charles Darwin's second cousin) had attempted to apply Darwin's theory of evolution to the study of psychology. He tried to find empirical evidence for the inheritance of mental ability, using a variety of physiological and psychological measures. Although his attempts to merge physical and mental phenomena, treating them as "faculties" and using statistics to justify his interpretations, were "mutually interpenetrating and logically self-confirming" (Hamilton, 1980, p. 157), Galton in the process began the science of mental testing. His successors Karl Pearson, Cyril Burt, and Charles Spearman extended his work, helping to establish differential psychology as a major branch of psychology. Galton's tests were limited in the responses they tested (reaction times, word associations, sensory discrimination). It was Binet who discovered how to assess mental ability by using "complex" tasks that were closer to what one might encounter in everyday life (Carroll, 1978, p. 7).

Had Binet's ideas about intelligence testing—the use of scores for diagnosis rather than ranking; the rejection of an innate, fixed quality known as "intelligence"—been preserved as his tests migrated across the Atlantic, "we would have been spared a major misuse of science in our century" (Gould, 1981, p. 155). Instead, American psychologists such as Henry Goddard, Lewis Terman, and Robert Yerkes developed out of Binet's tests a hereditarian theory of IQ that not only had some disastrous effects in its consequences for social policy

(Gould, 1981, ch. 5) but also colored the views of a generation of American researchers in mathematics education about the prospects for improving mathematical abilities.

Tracing Mental Development. Binet devised his IQ scale for the limited purpose of identifying those students whose performance suggested that they might be in need of some special education. He was less interested in the label a score might provide than in what help might be given a student with a particular pattern of performance. At Binet's laboratory in 1920, a recent recipient of the Ph.D. in natural sciences from Neuchâtel was given the task of standardizing with Parisian children some of the reasoning tests that Burt, another hereditarian, had developed in England. This new Ph.D., whose name was Jean Piaget, was intrigued by the processes the children used to obtain their answers—especially those answers that were incorrect. Adapting psychiatric examining procedures to the assessment of reasoning processes, Piaget developed the "clinical method" that later became his trademark (Flavell, 1963, p. 3). His ideas took longer than Binet's to migrate out of French-speaking Europe, but when they did eventually arrive on foreign shores they were to have a profound effect not only on the substance of research in mathematics education but also on its methodology.

Binet's psychological laboratory was one of dozens that were established in Europe, Asia, and North America from 1875, the date Wilhelm Wundt's laboratory was established at Leipzig and William James's at Harvard, until the end of the century (Madsen, 1988, p. 118). One of those laboratories was established at Johns Hopkins University in 1883 by G. Stanley Hall, who had studied with Wundt and who brought experimental pedagogy to America by launching a child study movement. Like many other American psychologists of the time, Hall was greatly influenced by Darwin's work. He contended that the child's development evolved in much the same fashion as human evolution and that therefore there was not much point in trying to influence intellectual development in childhood. Hall saw the teaching of elementary mathematics as a matter of instilling habits, and his elementary school curriculum devoted little time to arithmetic (Travers, 1983, p. 470). After World War I, Hall's ideas became part of an anti-intellectual movement in the United States that threatened the place of mathematics in the school curriculum (Stanic, 1983/1984, 1986a) and that influenced research in mathematics education primarily by stressing the difficulty of arithmetic and by emphasizing the primacy of the child's interests and the need for motivation (Buswell, 1930, p. 451). In France, experimental pedagogy did not flourish much beyond Binet's laboratory, but in a survey published in 1911, Claparède found lively activity in many parts of the rest of Europe and America, as well as in Russia and Japan.

Provoking Productive Thought. Psychologists at another psychological laboratory founded in 1896 at the University of Würzburg by another student of Wundt's, Oswald Külpe, broke away from Wundt's view that through introspection one could study the structure of consciousness. The Würzburg School claimed that abstract thoughts exist that are not accompanied by any image and that one needs to study thinking not through its contents but through its functions. One of Külpe's students, Max Wertheimer founded gestalt psychology, which, although it was primarily concerned with perception, also gave some attention to the processes of creativity and problem solving. Wertheimer (1959) later published a book on productive thinking that contained many examples from mathematics. The Würzberg school also included Otto Selz, whose work on problem solving influenced a generation of psychologists (Frijda & de Groot, 1982). Among them was Karl Duncker (1945), the author of an influential monograph that analyzed the processes of solving problems, some of which were complex mathematical problems. The work of the Würzberg and the gestalt psychologists in thinking and reasoning was of particular interest to mathematics educators in the United States after psychological research in their country came to be dominated by behaviorism (see H. F. Fehr, 1953, for an attempt to reconcile these disparate views and Resnick & Ford, 1981, ch. 6, for an introduction to gestalt psychology as applied to mathematics instruction).

Another psychologist who was influenced by Selz, although he deplored Selz's lack of attention to the relation of thought to speech, was Lev Semenovich Vygotsky (1934/1962, p. 122). Vygotsky formulated a theory of mental growth in which instruction guides development rather than following it. His use of a dynamic research methodology in which thinking processes are observed as they develop under the conditions of instruction has been adopted by Western researchers (see Kantowski, 1979, pp. 132–133, for a description of the methodology). And his concept of *the zone of proximal development*— the difference in level of difficulty between problems that one can solve alone and those one could solve with the help of others—is being used by researchers interested in the social mediation of cognitive change (Newman, Griffin, & Cole, 1989). Vygotsky was one of the developers of the Soviet theory of abilities, which was extended to the study of mathematical abilities by Vadim Andreevich Krutetskii (1968/1976). (See Goldberg, 1978, for a review of other relevant Soviet and East European psychological research.)

The studies of thinking begun by Galton and continued in various ways by Binet, Piaget, Wertheimer, Selz, Vygotsky, and others frequently touched on the mental abilities one uses in doing mathematics. Many of the tasks used in the studies were mathematical in nature, but even when they were not, the kinds of complex thought these psychologists identified often seemed to mathematics educators to be linked to mathematical performance. Psychological research into mathematical thinking has complemented and extended the insights mathematicians have had about how they work. It has not, however, had as pervasive an influence on research in mathematics education as studies of teaching and learning have had.

Research on thinking ordinarily examines individual or group differences in performance under "typical" conditions, although sometimes it attempts to optimize performance by providing hints or special instruction. It often looks at changes in performance over time. A major stream of research on thinking that was begun by Galton studies the intercorrelations among "factors" assumed to reflect stable (and perhaps inherent) mental characteristics (see Carroll, 1978, for a history of mental ability testing and test theory). A positive aspect of much of this research has been its willingness to examine naturally occurring relationships as a source of hypotheses to be tested. But a negative aspect, only partly redressed by the influence of

Vygotsky and Piaget, has been the heavy use of elaborate statistical procedures—such as correlation analysis, regression analysis, and factor analysis—which ordinarily require the questionable assumptions that relationships are linear and effects additive (Hamilton, 1980). Until recently, furthermore, this research has virtually ignored social and cultural influences on thinking.

Studies of Teaching and Learning

Treatments and Effects. A second tradition in psychology deals more directly with teaching and learning and has undoubtedly had more impact on the design of research in mathematics education. In this tradition, research undertakes to examine the "effects" of instructional "treatments." Teaching is taken as a treatment and learning as an effect. In an analogy to research in agriculture or pharmacology, the effects of various treatments are studied by systematic variation in the treatments followed by careful measurement of the presumed effects. Constructs such as knowledge about a mathematical topic are defined operationally as variables and, wherever possible, are quantified. Controls such as the random assignment of "subjects" to treatments allow the investigator to infer that a treatment has caused an effect (for more details on this approach, see Campbell & Stanley, 1963).

The paradigm for research into treatment effects is the field experiment. The basic technique for analyzing such effects within the compass of a single investigation is analysis of variance, which compares the variation between treatments with that within treatments. Analysis of variance was first developed by Ronald A. Fisher in the mid-1920s in connection with agricultural research. Fisher did not apply his technique outside of the biological sciences, but it gradually made its way into educational research anyway, especially after World War II. Acceptance of Fisher's ideas was facilitated by the same bias toward biological explanations for human actions that had given rise to behaviorism and that had helped make Galton's ideas about mental testing popular in education (Hamilton, 1980).

The Connectionists' Assault on Transfer. Before the refinement of analysis of variance was introduced into the analysis of experimental data, researchers simply compared the performance of groups given different treatments. A key notion is that of a control group, for without at least one point of reference or contrast, the description of an experimental group's performance must be quite limited. The use of a control group in educational research was popularized by Edward L. Thorndike, who in a series of experiments begun in 1900 with Robert S. Woodworth attempted to show the limitations of transfer of training (Clifford, 1968/1984, pp. 270–276). They found, for example, that practice in judging the sizes of rectangles of various dimensions did not improve one's ability to judge the size of a triangle.

Thorndike termed his psychology *connectionism*, and it was, along with Ivan Pavlov's conditioning, or reflexology, one of the forerunners of the behaviorism that dominated American psychology until around 1930, when it began to branch into the neobehaviorism of Edward Tolman and Clark L. Hull and the radical behaviorism of Burrhus Frederic Skinner (Madsen, 1988, p. 477). Thorndike emphasized the active, selective learning of satisfying responses. He argued that bonds between stimuli and responses are strengthened through exercise in which success is rewarded. He applied these principles to mathematics in a series of arithmetic textbooks published in 1917 and then in *The Psychology of Arithmetic* (1922) and *The Psychology of Algebra* (Thorndike et al., 1923). These books, in addition to having a profound effect on the teaching of arithmetic in the United States (Cronbach & Suppes, 1969, p. 97), gave rise to a substantial body of research on the effects of drill in learning arithmetic. (See H. F. Fehr, 1953, and Resnick & Ford, 1981, ch. 2, for discussions of connectionism applied to mathematics learning; Travers, 1983, pp. 280–285, for details about the books; and Suydam & Dessart, 1980, for a review of the literature on drill.)

Those educators who viewed mathematics as vital in the training of logical thinking saw Thorndike's research with Woodworth on transfer of training as a shot to the heart of their subject. Transfer of training was the psychological mechanism supporting the theory of mental discipline: the ancient idea that thinking can be trained in general through instruction in specific subjects. Mental discipline is allied both to formal discipline (the view that the form— for example, difficulty, abstraction—of a subject is more important than its content) and to faculty psychology (the view of the mind as a set of specific faculties requiring exercise for their development).

For centuries mathematics had had a secure place in the school curriculum as one of the liberal arts. Further, it had by the end of the 16th century become the foundation for mechanics and hence for the ensuing revolutions in science and technology (Keller, 1985). But by the end of the 19th century, it was competing for space in the secondary and collegiate curriculums with newer subjects such as history, the natural sciences, and modern foreign languages. Members of the fledgling field of mathematics education were arguing, in the face of claims that mathematics was not used much in everyday life, for the disciplinary value of their subject. For example, Jacob William Albert Young (1906/1925), of the University of Chicago, contended that "still more important than the subject matter of mathematics is the fact that it exemplifies most typically, clearly and simply certain modes of thought which are of the utmost importance to every one" (p. 17). Mathematicians, too, assumed that the study of mathematics had generative power: "The principal aim of mathematical education is to develop certain faculties of the mind" (Poincaré, 1952, p. 128). The argument that what one learns in the study of mathematics does not transfer to other domains engendered both a reconsideration of the justifications for teaching mathematics and an outpouring of research on transfer.

To be sure, Thorndike did not say that transfer of training was impossible "but only that transfer cannot be assumed to occur, that it is rarely automatic, and that direct teaching for desired outcomes is usually more efficient and economical than are hoped-for, spill-over effects" (Clifford, 1968/1984, p. 280). Thorndike's theory of transfer, however, was highly restrictive. He assumed that transfer occurs only if the transfer situation contains "identical elements" from the training situation. Later studies seemed to call that assumption into question (Travers,

1983, p. 267). Thorndike's identical-elements theory evolved into *cumulative learning theory* (Gagné, 1968, p. 186; Resnick & Ford, 1981, p. 39), a neobehaviorist approach in which instruction is based on the analysis of complex skills into so-called learning hierarchies.

An early study by Charles H. Judd (1908) proved to be especially powerful in arguing against Thorndike's theory. In Judd's study, fifth- and sixth-grade pupils practiced shooting darts at an underwater target. Those who had been taught the principle of refraction were more adaptable to changes in the depth of the target than those who had not been taught the principle. Judd argued that transfer occurs through generalization. He contended that although generalization might not occur at the lower levels of mental activity, it is typical of the kind of higher-order thinking the schools should cultivate:

The psychology which concludes that transfer is uncommon or of slight degree is the psychology of animal consciousness, the psychology of particular experiences. The psychology of the higher mental processes teaches that the end and goal of all education is the development of systems of ideas which can be carried over from the situations in which they were acquired to other situations. Systems of general ideas illuminate and clarify human experiences by raising them to the level of abstract, generalized, conceptual understanding. (Judd, 1936, p. 201)

Although Judd's research study appears to have had some methodological flaws (small nonequivalent groups) and was sketchily reported, several subsequent studies claimed to confirm his findings (Becker & McLeod, 1967; Orata, 1935). Nonetheless, American psychologists preferred to follow Thorndike's approach to transfer rather than Judd's (Travers, 1983, p. 333). Mathematics educators, however, seemed to find Judd's views more congenial (Rosskopf, 1953; Stanic, 1983/1984, pp. 145–155), even though both men's views have been criticized as overly mechanical and as inadequate to the requirements of education as the reconstruction of experience (Orata, 1937).

The Psychology of the School Subjects. Judd's efforts at the University of Chicago to develop the psychology of the school subjects also proved popular with mathematics educators. "No other school subject has even approached the level and frequency of studies conducted in the area of arithmetic" (Shulman, 1970, p. 25). Judd tried to relate the learning of complex subject matter to basic psychological processes in a way that would avoid Thorndike's reductionism. To Judd, each subject had its own facts and generalizations that needed to be understood before that knowledge could be used. Judd (1928) contended that Thorndike's view of arithmetic as a tool subject was false. Instead, arithmetic is a "general mode of thinking" (p. 6). "It is the business of the school to transmit to the pupils the intellectual methods of arrangement by which the complexities of the world may be unraveled and a new pattern made of experience. The most comprehensive and flexible patterns for the rearrangement of experiences are those supplied by the mathematical sciences" (p. 8). Although Judd often stressed the importance of studying the learning of secondary school subjects and not just those from the elementary school, his own research concentrated on simple concepts from arithmetic (Judd, 1927). Similarly, both his student Guy T. Buswell and his and Buswell's student William A. Brownell, developer of the "meaning" theory of arithmetic, confined their investigations of mathematics largely to arithmetic learning. (For an account of the development of the psychology of mathematics as a school subject, see Travers, 1983, pp. 370–398; see Shulman, 1974, for an updated portrayal of the psychology of the school subjects).

Judd's (1927) work on the concept of number—he investigated children's and adults' speed and errors in counting sequences of sounds or flashes of light—stands between the pioneering work of James A. McLellan and John Dewey (1895) and of D. E. Phillips (1897) and the efforts of Piaget (1941/1952) some years later. All of these studies attempted to show how the concept of number develops out of the child's activity. McLellan and Dewey argued that the child needs to relate parts to wholes and to make measurements as well as to count. Phillips, working under G. Stanley Hall and making extensive use of questionnaire data, claimed to have found that counting was much more fundamental than McLellan and Dewey believed. (Dewey, 1898, countered that Phillips had misinterpreted their work.) In contrast, Piaget attempted to embed the construction of the number concept in the development of logical thinking rather than in the operation of counting. For him, number was a consequence of reflection on one's actions; he claimed that there was a correspondence between the basic structures of modern mathematics and the mental structures formed by what he termed *reflective abstraction*. (Some of his interpretations were subsequently challenged by mathematicians; see, for example, Freudenthal, 1973, appendix 1; Rotman, 1977.) Judd took the intermediate position of considering counting to be a fundamental process but also emphasizing its complexity and its dependence on mental development. Brownell (1928) then extended Judd's work to the realm of dot patterns and showed the slow progression from a concrete to an abstract understanding of number. The concept of number, including the roles of counting and of numerical operations, has in the last half century been one of the most active research sites in mathematics education. (See Bergeron & Herscovics, 1990, and Fuson, this volume, for reviews of recent studies; Clason, 1969, for a history of number concepts in U.S. textbooks.)

Transfer as Touchstone and Catalyst. Judd's argument for the transfer of principles rather than elements, however, is probably his most enduring contribution to research in mathematics education. Thorndike may have demolished the concept of mental discipline—and thus narrowed the justification for teaching mathematics—as far as many psychologists and general educators were concerned, but mathematics educators with a humanistic outlook could not stand by and watch their subject elbowed out of the curriculum. D. E. Smith, W. F. Osgood, and J. W. A. Young, the American commissioners on the International Commission on the Teaching of Mathematics (1912), were not convinced by Thorndike's claims: "Few, if any, mathematicians who are conversant with the results of such experiments as have been made, and are sympathetic with their spirit, feel that aught has been as yet established which would require them to change their views of the value of the study of mathematics" (p. 32). Vevia Blair of the Horace Mann School in New York City, reporting to the National Committee on Mathematical Requirements in 1923, tried to make the

case for transfer, and consequently for the "disciplinary value" of mathematics, by surveying both the research literature and the opinions of several dozen educational psychologists, including Thorndike and Judd. She found support for the view that substantial transfer might occur, depending on the methods of teaching (although Thorndike balked at the use of the term *substantial*). Two decades later, the debate still raged. The Joint Commission of the Mathematical Association of America and the National Council of Teachers of Mathematics on "The Place of Mathematics in Secondary Schools," citing evidence in support of transfer, lambasted as unscientific and misleading the experimental investigations that had claimed to invalidate mental discipline (Reeve, 1940, pp. 217–222). Judd's research, and the research it spawned, helped keep the argument for transfer alive. The argument gradually came to be put as follows: Transfer is possible on a large scale, but one must teach for transfer. Judd phrased it in this way:

The real problem of transfer is a problem of so organizing training that it will carry over in the minds of students into other fields. There is a method of teaching a subject so that it will transfer, and there are other methods of teaching the subject so that the transfer will be very small. Mathematics as a subject cannot be described in my judgment as sure to transfer. All depends upon the way in which the subject is handled. (quoted in Young, 1925, p. 377)

The topic of transfer of training occupies a unique position in the relations between psychologists and mathematics educators. It has played a pivotal role in arguments for the place of mathematics in the curriculum that have sometimes found psychologists and mathematics educators in opposing camps (Stanic, 1983/1984, 1986b). As "the most important single concept in any educationally relevant theory of learning" (Shulman, 1970, p. 55), it has stimulated a large body of research, much of it in mathematics education. Like discovery learning (Shulman, 1970; Shulman & Keislar, 1966), it is a topic on which psychologists disagree among themselves and which mathematics educators tend to view positively and often optimistically. (Even pessimistic assessments by mathematics educators of the possibilities for discovery and transfer do not question either the reality or the desirability of those phenomena; e.g., Becker & McLeod, 1967.) And in North America at least, the topic epitomizes the ambivalence that researchers in mathematics education often felt during the first half of this century as they borrowed the research methods of behaviorist psychology while generally disdaining the behaviorist view of school mathematics. With the arrival of cognitive psychology in the 1950s and 1960s, marked by the availability of Piaget's work in English translation and the reinterpretation of that work by Jerome Bruner, researchers in mathematics education began to have a more judicious regard for psychological theory and to collaborate more frequently with psychologists. (See Shulman, 1970, for an analysis of Bruner's and Piaget's theories in contrast to those of Gagné and of David Ausubel.)

EMERGENCE OF A PROFESSION

The roots of mathematics education as a field of activity go back several millennia. The Sumerian scribes of 3000 B.C. had systematized applied mathematics in the schools and developed means of teaching place value, sexagesimal fractions, and the use of tables for calculation (Høyrup, 1980, pp. 19–20). In the fifth century B.C., Socrates could use adroit questioning to lead a slave boy to discover that the area of a square on the diagonal of another square is twice that of the smaller square, as related in Plato's *Meno*. The 16th-century scholar Robert Recorde not only taught mathematics but also wrote textbooks that used Socratic dialogue to deal with matters of proof, definition, and understanding (Fauvel, 1989; Howson, 1982, ch. 1). Early 19th-century educators such as Johann Pestalozzi, Friedrich Froebel, and Johann Friedrich Herbart proposed teaching methods based on concrete experience and educational aims concerned with the development of mental faculties that influenced the teaching of mathematics from kindergarten through the secondary school. In the United States, Warren Colburn published in 1821 an innovative arithmetic textbook based on Pestalozzi's pedagogy that gave students the opportunity to discover rules by induction from examples (Jones & Coxford, 1970, pp. 20–21). From 1816 to 1822 in Germany, Martin Ohm (brother of the physicist Georg Simon Ohm) published several works on mathematics education setting forth a view of mathematics as a system of relations between operations. He later wrote some influential textbooks in which the number concept was organized as a progressive extension from natural numbers to rational, negative, real, and complex numbers—the same organizing principle used today (Jahnke, 1986; Zerner, 1989). People have engaged in mathematics education, and in reflection on mathematics education, for centuries.

Early Mathematics Educators and Their Research

As noted in the introduction, however, it has only been since the last part of the 19th century, as national school systems were set up and the preparation of mathematics teachers began to be established in higher education, that mathematics education has emerged as a professional field. Only then did people begin to identify themselves as mathematics educators. In England, mathematics education was founded as a field of study in the early 20th century by educators such as Charles Godfrey, Benchara Branford, George St. Lawrence Carson, and Percy Nunn (Howson, 1982, ch. 8). In the United States, the major founders in that sense were D. E. Smith and J. W. A. Young (Jones & Coxford, 1970, p. 42).

These mathematics educators did not themselves undertake much research in mathematics education as currently defined. Smith did most of his research on the history of mathematics (some of it, to be sure, was on the history of mathematics education); Godfrey wrote on the examination system; all of them published major works on the teaching of mathematics; and most were noted textbook authors, including Smith and Carson, who coauthored algebra and geometry textbooks.

Smith (1895) did conduct a modest empirical research study while he was teaching mathematics at the Michigan State Normal School in Ypsilanti. (A few years later, Smith joined the faculty of Teachers College, Columbia, where he became well known for his work in the history of mathematics. Although he was eventually to serve as president of the Mathematical Association of America, he had studied mathematics only as an undergraduate, obtaining his doctorate in art history and becoming a lawyer before beginning his teaching career; Swetz, 1987, pp. 299–304; Travers, 1983, pp. 381–382.) Smith's study appears to have been one of the first to address a topic that was subsequently to become a major focus of research—differences between the sexes in learning and doing mathematics. Smith cited G. Stanley Hall's (1891) landmark study "The Contents of Children's Minds on Entering School," in which Hall reported, among many other things, that girls excelled in space concepts and boys in number. (For both Hall's study in Boston and the data Hall provided from Berlin, the results were more mixed and much less definitive than either Hall or Smith suggested.) Smith looked at the examination performance of students (mostly women) in his own classes when he had taught at Cortland (New York) State Normal School and in the school records of the classes at Ypsilanti. He found somewhat better performance by the men, particularly in arithmetic and geometry, but again the differences were small and the pattern of results mixed. Although his conclusions are open to challenge, Smith was astute enough to recognize some of the possible reasons for differential performance (including selection factors and sex of the examiner) that would not always be acknowledged by subsequent investigators. (See Friedman, 1989, and Leder, this volume, for recent reviews of sex differences in mathematics.)

An example that is probably more representative of how these early mathematics educators approached research is a chapter on "some experiments in teaching geometry to blind children" in Branford's (1908) *A Study of Mathematical Education*. The chapter contains little more than descriptions of two lessons, one taught by Branford to a class of eight blind children and the other taught by their regular teacher. As Howson (1988) observes, "The 'experiments' described would...have hardly satisfied today's best research criteria" (p. 267). It seems clear from Branford's comments in the preface and elsewhere in the book that he himself regarded his experiments as not so much organized attempts to find out or explain what happened as illustrations for teachers of activities they might wish to try.

Nonetheless, for all the limitations we can see in their work, these mathematics educators and their colleagues began the process of systematic inquiry in our field. Through their own activities and those of their students, they laid the foundations for research that would be done neither out of the mathematician's passing curiosity nor to serve the psychologist's need for subject matter but in response to a profession's questions about its practice.

Education as a Science

The scientific movement in education can be traced back to Alexander Bain's book *Education as a Science*, published in 1879. (See Selleck, 1968, ch. 8, for a history of the movement in England; Whipple, 1938, for an assessment of its contributions in the United States.) Out of the movement grew an interest in using scientific methods to study the techniques for teaching various school subjects. The first thesis in mathematics education to be written for a university degree in Britain was W. McClelland's 1910 B.Ed. thesis at the University of Edinburgh, entitled "An Experimental Study of the Different Methods of

Subtraction" (Frobisher & Joy, 1978, p. 1). Comparisons of (often sketchily specified) methods of teaching mathematical topics have long been popular as research studies (Breslich, 1938; Brownell, 1930; Buswell, 1938; Buswell & Judd, 1925, ch. 8; Smiler, 1970). Although often termed *experiments*, the studies done on methods during the first third of the 20th century, at the height of the scientific movement, seldom took the form of controlled comparisons of the effects of teaching students randomly assigned to different groups. Marilyn N. Suydam's (1967/1968, p. 456) evaluation of published research in elementary mathematics from 1900 through 1965 indicates that whereas one out of every three studies published in the years 1951–1965 could be classified as an experiment, in 1900–1929 only about one out of eight studies could be so classified. According to Suydam, over a quarter of the studies at that time did not include a control group. Furthermore, the sample size was likely to have been rather small, and the links to a strong theoretical rationale negligible. Ernst R. Breslich noted that "in most of these investigations the differences between the compared methods or conditions have been surprisingly low" (p. 134). Concluding his survey of the contribution that research on arithmetic had made as of 1938, Buswell sounded a wistful note of discouragement before drawing his confident conclusion: "While a consideration of individual researches in arithmetic leaves one with the feeling of frequent contradiction and confusion, a survey of the entire subject shows clearly that the combined results of research are exerting a rational directing influence upon the arithmetic program as a whole." (p. 128).

Child Study

The scientific movement had begun with the same sort of optimism, but its early efforts were in a different direction than the comparison of teaching methods. The field of child study had started late in the 18th century if one attributes its founding to Dieterich Tiedemann in Germany (Claparède, 1911, p. 13; Travers, 1983, p. 454), or somewhat later if one gives the credit to Charles Darwin for his observations of his son's behavior (Selleck, 1968, p. 277). In either case, child study seems to have lain dormant until it was taken up again in Germany and the United States near the end of the 19th century and became "the first of the concerns of the scientific educationists" (Selleck, 1968, p. 277) to be pursued systematically. Mention has already been made of Hall's establishment of child study in the United States. Several of the handful of doctoral dissertations in psychology completed at Clark (whose founder was Hall), Columbia, and Harvard in the early years of the 20th century dealt with the perception of number or space, in line with the child study movement's emphasis on naturalistic, descriptive studies (University Microfilms International, 1989). In a survey of research studies in mathematics education in the United States from 1880 through 1963, Helen B. Smiler (1970) found that of the 37 studies done through 1910, 13 could be classified as psychological studies of the development of concepts, which included studies of readiness, and an additional 4 were methods studies of teaching that concerned individual differences. Suydam's (1967, pp. 175–260, 299–353, 442) survey indicates that of 53 studies of elementary mathematics

published prior to 1925, 24 (by my count) were studies related to basic concepts and 6 were related to individual differences. Many of these studies were apparently inspired by the work of Hall and his followers. After 1910 other categories of studies became much more prominent. One can see in Smiler's data a clear decline in studies of concept development from 1880 to about 1925. In 1934, Elisabeth Williams published her pioneering work in England on the foundations of geometry learning, but it was not followed up at the time. Studies of the growth and development of mathematical concepts reemerged strongly again in the 1960s, when intellectual development along the lines charted by Piaget became a major research interest (Travers, 1983, p. 480; for a view from the U.K., see Peel, 1971).

Hall's influence, and that of another early 20th-century educator, Maria Montessori, on the use of activities and materials in teaching mathematics to young children should not be underrated. Many of the counting activities, puzzles, and pattern recognition apparatuses the advocates of child study developed are still in use today. Child study itself, however, yielded little in the way of worthwhile research in mathematics education. Child study advocates tended to do their best work with preschool children (Selleck, 1968, p. 281), and mathematics was seldom a focus of that work. Furthermore, much of the research was of low quality. The child study movement "investigated, fairly indiscriminately, a large number of questions, sometimes failing to enquire whether the questions were worth asking; its interviewing techniques were often slack, its observations loosely controlled, its use of the questionnaire method over-ambitious and naïve" (Selleck, 1968, p. 280).

As a movement, child study was part of a general uneasiness that began to emerge in the United States in the 1890s—a dissatisfaction with the formal methods of instruction employed in all school subjects but perhaps especially in elementary arithmetic, which sometimes occupied half the school day (Buswell, 1930, p. 123) and which was "the chief source of non-promotion in the elementary school" (Buswell & Judd, 1925, p. 7). Hall advocated "the postponement of all formal teaching of arithmetic to a much later stage in the school program than was common, holding that the earliest school years should be devoted to the accumulation of concrete experiences" (Buswell & Judd, 1925, p. 3). Hall's views on mathematics education were influential, but for several decades efforts to postpone formal instruction in arithmetic were largely overshadowed by two other phenomena in American education that became part of the scientific movement—the testing movement and the attempt to restrict the school curriculum to subjects having social utility. Each of these phenomena gave rise to a sizable body of research, and each had profound effects on school practice. Buswell and Judd observed in their survey of the research on arithmetic conducted before 1925:

Reorganization of mathematics is going on at the present time in many different lines and through the efforts of many experimenters as a direct result of the numerous current scientific studies....The leading characteristic of the present movement may be said to be its reliance for reform on the careful measurement of results of school work and on analytical studies of social demands and the intellectual processes of pupils. (p. 165)

The Testing Movement

The Advent of Written Tests. The testing movement had deep roots. In 1845, the Boston School Committee, proud of what their schools were accomplishing and under pressure from Horace Mann, Secretary of the State Board of Education in Massachusetts, to show that the schools were doing the job the state expected of them in view of generous financial appropriations, undertook a comprehensive school survey (Brigham, Coolidge, & Graves, 1845; Parsons, Howe, & Neale, 1845; see also Caldwell & Courtis, 1925, ch. 1). A subcommittee of the School Committee had determined that it would be impossible to give oral examinations to all the students (ages from roughly 7 to 14 years) in the system. Although the committee's charge had subsequently been limited to examining only the students in the highest grade in each school, they decided to use a battery of written tests. Each test was administered on a single day by members of the committee. The arithmetic test consisted of ten questions to be done in 70 minutes, which the committee considered insufficient time for most students but which meant they would be occupied the whole period. The other tests, on topics such as history, geography, and grammar, took an hour each. The arithmetic test was administered in 18 schools to 4 to 26 students in a school (Caldwell & Courtis, 1925, p. 28).

The examiners, and Mann as well, were disappointed in the generally low levels of performance. (See Caldwell & Courtis, 1925, for copies of the tests, the reports, and Mann's comments.) In arithmetic, for example, the average across students and questions was just under 35%, which prompted Mann (1845b) to complain: "Such a result repels comment. No friendly attempt at palliation can make it any better. No severity of just censure can make it any worse" (p. 363). Several questions were missed by everyone, and others were answered correctly by only a handful of students. One of the more difficult questions was the following:

The City of Boston has 120,000 inhabitants, half males, and its property liable to taxation is one hundred millions. It levies a poll tax of $\frac{2}{3}$ of a dollar each on one half of its male population. It taxes income to the amount of $50,000, and its whole tax is $770,000. What should a man pay whose taxable property amounts to $100,000?

A correct answer was given by only 1 of the 308 students tested, Miss Frances A. Lathrop of the Hawes School. Mann was scornful:

Who of all the boys in the Boston Grammar and Writing schools, shall hereafter be city assessors, when not one of them can tell what tax shall be levied on a hundred thousand dollars, when all the necessary conditions are given, with perfect precision and clearness? (p. 363)

The Boston examiners tried to offer some encouragement by pointing to the achievement of the best schools and exhorting the others to copy their practices. They recommended specific changes in instructional methods and school organization. They published a table giving the ranks of the schools in order of achievement, saying at the same time that the rankings were only approximate as measures of intellectual accomplishment and far from adequate as measures of the merit of a school (Caldwell & Courtis, 1925, pp. 180–181). Nonetheless, the ranking of schools by achievement test scores had begun.

Mann (1845a) was enthusiastic about the use of printed questions and written answers as a means of examination, saying that it would "constitute a new era in the history of our schools" (p. 330). He claimed that such examinations were common in Europe and found them to have many advantages over the traditional oral examinations given by school inspectors. Written examinations had indeed been in use for some time in Europe. Civil service testing had begun in China by 2200 B.C. and was well established there by 1115 B.C., when applicants were required to show proficiency in the five basic arts of music, archery, horsemanship, writing, and arithmetic (DuBois, 1965). The Chinese system of selecting candidates for state office by successive written competitive examinations (district, province, country) was subsequently imported to Europe and the United States in various forms. France began civil service testing in 1791, only to have it abolished by Napoleon. When the English set up examinations to select entrants for the Indian Civil Service in 1853, they used the Chinese system as a model. Competitive examinations to enter the military academies and external examinations for schools were also begun in England in the 1850s (Howson, 1982, p. 160). The use of written competitive examinations for admitting candidates to the civil service, certifying school achievement, awarding university scholarships, and evaluating schools began to raise questions of accuracy and fairness. Howson (1988) identified as "the first educational research, as we would tend to recognize it" (p. 267), the studies by Francis Y. Edgeworth (1888, 1890) of the sources of variation in marking competitive examination papers and how that error might be controlled. Whether the purpose was to evaluate how much a student had learned (and therefore his or her qualifications for advancement) or to judge the effectiveness of a school's program, written examinations and increasingly, especially in the United States, written short-answer tests became the medium of choice. Such tests were slow to catch on in other countries despite studies such as that by Philip B. Ballard (1923, pp. 180–203) in England claiming to demonstrate the superiority of tests with many questions to be answered quickly over examinations having only a few time-consuming questions.

Testing to Establish Standards. Returning in 1890 from visits to European schools and psychological laboratories, the reformer Joseph Mayer Rice began 2 years later to investigate the public school systems of various large cities in the United States. He was appalled by the rigid, mechanical, dehumanizing methods of instruction, by the harshness of manner he saw among the teachers, and by the passive responses from the students. When the muckraking articles he wrote failed to ignite the fires of reform, he decided to gather some data to support his claims. As an editor of the *Forum* in charge of its department of educational research, he undertook an investigation into arithmetic that was one of the first attempts at using empirical data to attack an educational problem. As Rice (1902) saw it, "the elementary schools had been conducted altogether on lines dictated by theories,... the ways and means of different schools had varied in accordance with different theories, ...and no attempt had been made to discover the comparative value of different processes by comparing the results" (p. 281).

Rice (1902) believed that facts ought to replace theories as the basis for guiding instruction. He developed an eight-item arithmetic test for each grade from 4 to 8, with some overlapping of items across grades (sample item at grades 6, 7, and 8: "A gentleman gave away $\frac{1}{7}$ of the books in his library, lent $\frac{1}{6}$ of the remainder, and sold $\frac{1}{5}$ of what was left. He then had 420 books remaining. How many had he at first?"). Rice supervised the administration of the tests to about 6,000 students from 18 schools in seven cities. His most striking finding was the large differences in achievement between cities, although there were some large differences between schools within cities as well. The average scores for the schools ranged from 25% to 80%, and the variation was even greater within some grades. Rice could not account for the differences between schools in terms of class size, average age of the students, or socioeconomic level of the neighborhood or city. Some schools in the slums had higher average scores than the "aristocratic" schools. Rice scored not only the number of problems the students got correct but also the number that would have been correct if a mechanical error had not been made. He found relatively few mechanical errors. The schools that had the highest number of problems correct in principle had the fewest mechanical errors, which prompted him to conclude that "the mechanical side of arithmetic has shown itself to be very closely related to the thought side" (pp. 291–292).

Rice thought it important to set standards for the schools to meet. In round percentages, the average school scores on his test for grades 4 to 8 had been 60, 70, 60, 40, and 50. Rice decided that a school's achievement was adequate if the overall average was 60% and four out of five grades reached the grade averages. He recognized that if another test were used, the level of difficulty might be different, but he was optimistic about the possibility of determining by research what percentage ought to be obtained on a given item by a given group of students: "In due course of time there ought to be no difficulty in establishing standards in arithmetic with mathematical precision" (Rice, 1902, p. 291). Rice did not attempt to suggest a standard for the mechanical side of arithmetic, but he did have something to say about the time that ought to be devoted to arithmetic instruction. He found that the average daily time of arithmetic instruction in the schools he surveyed varied from 30 to 60 minutes and concluded that, although there was no clear relationship between time and achievement, 45 minutes was probably sufficient.

In a second article, Rice (1903) examined other factors that might account for the differences between the schools in arithmetic achievement. He ruled out variations in homework assignments, finding that the school with the poorest results on his test was assigning the most homework. He also ruled out variations in teaching method because no special methods seemed to have been used in the schools that did well on his test. Other factors, such as amount of review and qualifications of the teacher, did not appear to be related to differences in achievement. Rice's method, which relied on inspection of the characteristics of high- and low-scoring schools, did not allow him to estimate the strength or shape of the relationship between some hypothesized factor and the level of achievement; it permitted him only to say that because not all the high-scoring schools had a characteristic that the low-scoring schools lacked, or vice versa, there must not be much of a relationship. Surprisingly, he found what he termed a *controlling factor*: the system of examination used to assess the teacher's performance.

As Rice saw it, teachers would supply what supervisors reasonably demanded. The schools in those systems in his study where teachers made up their own tests had relatively low levels of achievement, and the schools in the systems where tests were made up by the principals and supplemented by well-devised tests from the superintendent had relatively high levels of achievement. At the end of the article, Rice said that his aim since returning from his visits abroad had been to study not the results of instruction but what he termed "the spirit of the schools." A decade of inquiry and data gathering had led him to conclude that the superintendent and the supervisory staff of a school district were responsible for that spirit. By setting standards and administering examinations, they could get the results they wanted. The firebrand cum researcher was promoting the managerial approach to education. (See Travers, 1983, pp. 105–108, for a similar analysis.)

The Measurement Movement Begins. Buswell and Judd (1925) noted that although Rice's interpretations of his data and his conclusions were rejected by many educators, his objective methods of measurement attracted much attention. They contended that "it was not until 1908, however, when Stone published his results, that the measurement movement in arithmetic can be said to have really gained sanction" (p. 5). Cliff W. Stone, as his doctoral dissertation at Teachers College, Columbia University, under the guidance of a committee that included E. L. Thorndike and D. E. Smith, surveyed the arithmetical achievement of some 3,000 sixth graders from 26 school systems, administering the tests himself. Stone was critical of Rice's rather cavalier dismissal of the course of study as an influence on differences in performance across schools. (Rice had claimed, for example, that the effects of students not having learned to perform an operation necessary to solve some of the problems turned out to have an effect on the school average of no more than 2%, but he did not spell out how he arrived at that figure.) Stone also wanted to measure achievement both in reasoning (word problems similar to those used by Rice) and fundamentals (addition, subtraction, multiplication, and division of whole numbers); he wanted to standardize the administration and scoring of the tests; and he wanted to express relationships between achievement and various factors in terms of correlation coefficients.

What Stone (1908) found "is summed up in the one word, *diversity*" (p. 90). He found great variability across school systems in average test scores, errors, time spent each week on arithmetic (whether or not homework time was included), and the "average ratio of time to abilities." He observed "a difference in course of study excellence which can hardly be put in words" (p. 90). Unwilling to use Rice's seat-of-the-pants technique for ruling out factors that might be related to achievement, Stone ended up with a mound of correlation coefficients that pointed in various directions. Unlike Rice, Stone found some evidence that homework improved achievement. Because there seemed to be little relation between achievement and the time spent on instruction, he concluded that many systems must be wasting time on arithmetic.

Stone (1908) could see that no one factor could account for the variation he had observed in achievement, suggesting that "the course of study may be the most important single factor, but it does not produce abilities unless taught" (p. 91).

The extensive variation in achievement across systems obviously bothered him. He argued that "the greatest need shown by the research is standards of achievement" (p. 90). Such standards would make courses of study more uniform and reduce the variability in average test scores. Stone recommended his own tests as a start toward such standards. He also recommended that systems or schools examine *time expenditure* by calculating the ratio of the "time cost in week-minutes" (total time devoted to arithmetic each week) to the average test score and comparing it to the ratios found in his study. Stone's ratios were a way of linking achievement (product) to the time spent in learning (cost), a primitive measure of what later became known as the *efficiency* of instruction (see Osborne & Crosswhite, 1970, pp. 188–192; Stanic, 1983/1984, pp. 47–57).

Inspired by Stone's research, Stuart A. Courtis (1909–1911), of the Home and Day School in Detroit, wanted to know how well the students in his school, a private school for girls, would do on the same tests. He recognized that some of the test questions were within the capabilities of the third graders and that the length of the tests was sufficient to challenge the seniors (grade 13). By administering the tests to all students in grades 3 through 13, he discovered both a striking variability in scores within each grade and a pattern of growth in achievement across grades. To get a better picture of growth, he devised his own, more focused tests (including tests of speed as well as of reasoning and fundamentals) and administered them at the beginning and the end of the school year. Courtis saw the value of his efforts as demonstrating

that it is practicable to measure, not only the growth in ability and efficiency from grade to grade, the defects and needs of any one grade or individual, but the *effects of changes in method or procedure* as well. By a series of tests through a number of years, it ought to be possible to build up a real science of teaching and to determine by strict experimental methods the truth or falsity of any educational hypothesis. (p. 199)

Courtis went on to refine his tests into a carefully graded series that included all the different combinations of digits and, by collecting data from many school systems, to standardize the tests. He wanted what he termed "standard scores and standard growths for every grade" (Courtis, 1913, p. 10). His tests were used extensively by school systems (e.g., the first administration of standardized tests on a large scale, to over 33,000 students in New York City in 1913, was of Courtis's tests) and influenced both the construction of subsequent tests and the study of measurement issues such as reliability, validity, and speededness (Buswell & Judd, 1925, pp. 38–57).

In his study for the New York City Schools, Courtis (1913) compared the graphs of test scores across grades for a school in Michigan that had been using his standard tests for several years (very likely the Home and Day School) and one in New York that had not. The curve for the Michigan school was smoother, which suggested to Courtis that poor teachers in certain grades in the New York school might be the cause of the uneven performance there. He saw what he considered too much variation in average scores across classes within grades and schools in New York City—"much greater than efficient conditions would warrant" (p. 117)—and he attributed the problem to "inefficient teachers." He argued that although some low scores might be attributable to causes such as epidemics, with a knowledge of the conditions of schooling one could "determine exact cause" (p. 116) and thus compare the teachers. Courtis also used the data from New York to study the relation between the abilities assessed by his tests, the differences between the performance of boys and girls, and the effects of foreign parentage ("American" vs. "Jewish," "Italian," and "German"). (In 1915, Courtis, who had become the director of Detroit's new Bureau of School Efficiency, organized a meeting of seven directors of research in school systems into the National Association of Directors of Educational Research, the organization that in 1930 became the American Educational Research Association; Travers, 1983, pp. 125, 128.)

Testing to Improve Practice. Clifford B. Upton, in an investigation for the National Committee on Mathematical Requirements of the Mathematical Association of America published in 1923, described some of the dozens of tests in arithmetic, algebra, and geometry then being used in secondary schools, as well as the prognostic tests of mathematical ability published by Agnes L. Rogers in 1918 and Lewis Thurstone's Vocational Guidance Test of 1922. Upton ended his review by quoting at length from Rogers, who had observed, "We can scarcely grasp the full significance of the introduction of scientific measurement in education" (quoted in Upton, 1923, p. 416; see also Rogers, 1919, p. 162). She saw testing as a means of controlling "the human machine" so that it would be more efficient. And mathematics was the subject in which progress should be most rapid.

By 1925, Buswell and Judd could argue that the extensive use of tests such as those constructed by Courtis had been helpful not merely for the improvement of instructional practice but also for the development of educational science. The establishment and use of standards by means of the tests allowed "a critical comparison of different methods of teaching and at the same time…furnished the best possible material with which to perfect scientific measurement" (p. 45).

In the same year, Caldwell and Courtis, having adapted and distributed some of the questions from the Boston School Survey of 1845 to school districts around the United States in 1919, could claim that education in 1923 was superior to that of three quarters of a century earlier. As Travers (1983) observed, however, "the data do not bear close examination" (p. 156). The groups of children were not really comparable, and neither were the tests and the scoring procedures. The 40,000 children who took the test in 1919 did better on some of the test questions, notably in geography and history, and worse on others, notably in arithmetic, with the median score on all questions being somewhat higher in 1919 than in 1845. Caldwell and Courtis had chosen the five easiest arithmetic questions to repeat and had even supplied some additional information ("A rood is a measure of land no longer in use. There are 40 sq. rods in a rood and 4 roods in an acre."), but, for example, the question "What part of 100 acres is 63 acres, 2 roods, and 7 sq. rods?" was answered correctly by only 16% of the sample in 1919, whereas 92% of the Boston students surveyed in 1845 had answered it correctly without having to be reminded what a rood was. Part of Caldwell and Courtis's argument was that such test questions, dealing as they did with situations unlikely to arise in the children's experience, had been eliminated from modern textbooks and were therefore of little value for comparison purposes. This issue—the adequacy of comparisons based on test questions that might be more appropriate for one group of students than another—would arise again some decades later when national and international surveys of mathematics achievement were launched.

The Social Utility Movement

The other influential part of the scientific movement in American education, besides testing, was the effort to reorient the school curriculum around topics that were socially useful. That effort is hinted at by Caldwell and Courtis's (1925, pp. 68–70) comments to the effect that modern children, having no occasion outside of school to divide acres into roods and square rods, should not be expected to do such tasks in school. A question such as "How much is $\frac{1}{2}$ of $\frac{1}{3}$ of 9 hours and 18 minutes?" according to Caldwell and Courtis, had no "life value." The main reason to repeat the archaic arithmetic questions from 1845 in the 1919 survey had been to remind the public that "the subject matter of arithmetic is today being continuously modified to eliminate all elements of little direct use in the life of the child" (p. 68).

What Does Business Want? At the annual meeting of the National Education Association (NEA) held in Charleston, South Carolina, in July 1900, Edward W. Stitt, Principal of Public School No. 89 in New York City, reported the results of a survey he had made of the courses of study in arithmetic in 30 of the largest cities of the country. He was pleased to find that topics such as duodecimals, compound interest, and foreign exchange were being eliminated or reduced in importance but surprised and dismayed to discover many instances of what he termed "excessive requirements," including compound partnership, partial payments, compound proportion, and true discount. In response to complaints from parents of his students that their children were using methods that were different from those they themselves employed, he set out to determine how business practices varied from those in school and the degree to which the arithmetic curriculum might be reduced by limiting it to "those subjects which the business-men found most necessary" (p. 568). Stitt sent letters of inquiry, enclosing stamped return envelopes, to 600 businesses of various types in New York; the teachers and students of his school distributed additional copies, and various newspapers, including the *New York Times*, publicized the inquiry. Stitt did not give the total number surveyed or the number of replies, but he did report having received responses from most of the 57 types of business he had identified. By his account, 44% of the writers who replied

stated that there was absolutely no need for any arithmetic beyond the fundamental rules and common and decimal fractions. The great desiderata were accuracy and speed, and strong emphasis was laid on the fact that the pupils should fully understand the importance and logical sequence of every step of the processes involved. (p. 569)

He classified the information he received under the headings of "mechanical aids" (teach the multiplication table through

20×20, require legible figures, give little attention to fractions such as $\frac{8}{9}$ or $\frac{19}{23}$) and "processes of solution" (restrict interest calculations to finding the amount of interest, teach pupils to "think in percentage," insist on approximations of results before performing calculations). Stitt attributed the inaccuracy he saw in the children's work to a lack of drill in fundamental rules. He noted that "many merchants strongly urge the importance of mental arithmetic as a factor of business success" (p. 571), claiming that at least 60% of the arithmetic of the business world was done without pen or pencil. He faulted much arithmetic instruction for its lack of variety and for the heavy emphasis on what he termed the "pouring-in process"—"Occasionally the teacher should rest and give the children a chance to assimilate the instruction" (p. 572).

A Streamlined Curriculum. Stitt seems to have been the first to compare what business wanted with what the schools were doing in arithmetic, but he was not alone in thinking that the curriculum had become encumbered with arcane procedures and irrelevant concepts. Five years previously, William Torrey Harris, reporting to the NEA on behalf of their Committee of Fifteen on Elementary Education (Harris, Draper, & Tarbell, 1895), had asked for basic arithmetic instruction to begin in grade 2 and be completed by grade 6 so that some algebra could be taught in grades 7 and 8 and for a reduction in the amount of time each day spent on mathematics—in particular, an elimination of the practice of having two lessons a day in arithmetic, one "mental," the other "written." Too much time spent on arithmetic, according to Harris, was giving children "a bent or set in the direction of thinking quantitatively, with a corresponding neglect of the power to observe, and to reflect upon, qualitative and causal aspects" (Harris et al., p. 22); the result was "arrested development."

A consensus among educators in the NEA had begun to emerge that the schools were wasting time on unproductive instruction and that one needed to take a closer look at what the world outside the school did with ideas from various subjects, including mathematics. In 1904, Frank M. McMurry of Teachers College forcefully argued before the NEA convention for the elimination of useless subject matter. He "laid down principles for the selection of the subject-matter of the curriculum and gave them pointed application to arithmetic" (Wilson, 1919, p. 1). He offered a list of topics to be eliminated that included not only such perennials as true discount and cube root but also "all common fractions, except those of a very low denomination and customary in business,...[and] all algebra, except such simple use of the equation as is directly helpful in arithmetic" (McMurry, 1904, p. 198). The NEA did not have a department of mathematics, but several educators from other fields were willing to look at the elementary arithmetic curriculum.

What Arithmetic Do People Use? Walter S. Monroe (1917), professor of educational administration at the Kansas State Normal School in Emporia, classified the problems that appeared in four widely used arithmetic textbooks according to the type of human activity portrayed in the problem. He then compared the occupational activities to the distribution of occupations from the Federal Census. Of the 1,073 types of practical problems, he found that 71% might occur in an occupation. Most of these problems related to trade, but there were large numbers from manufacturing and agriculture as well. Monroe was concerned that the occupations of 55% of the working population were not represented, which he saw as limiting the value of arithmetic as a vocational subject. He was also disturbed by the great variation across textbooks in the types of problem they contained. Monroe's account of his work appeared in the second report of the NEA Department of Superintendence's Committee on Economy of Time in Public Education. Included in that same report was Guy M. Wilson's (1917) survey of the social and business use of arithmetic.

Wilson, a professor of agricultural education at the Iowa State College of Agriculture and Mechanic Arts, had been working for several years with a group of teachers in Connorsville, Indiana, to ascertain the arithmetic actually employed in everyday life. After surveying the business community, the teachers of Connorsville had recommended the elimination from the arithmetic course of some 14 topics, from the long method of finding the greatest common divisor to the calculation of cube roots. An investigation by the Iowa State Teachers' Association that asked several hundred citizens all over that state to check a list of 16 topics from the curriculum had found that 42% had no use for any of them. Wilson decided that a better way to get information on the actual adult use of arithmetic would be to use students to gather the data. He had superintendents and teachers ask sixth, seventh, and eighth graders in 16 Iowa towns and cities and one Minnesota city to collect over a 2-week period every arithmetic problem solved by their mother or father (a procedure the mathematics educator Raleigh Schorling, 1923, p. 239, termed "devious"). Wilson (1917) got data from 1,457 persons solving 5,036 problems. Not surprisingly, most of the problems involved rather simple arithmetic with numbers under 1,000. Calculations with money in buying or selling were frequent. Fractions seldom had numerators other than 1; $\frac{1}{2}$ was the most commonly used fraction. He concluded: "In actual usage, few problems of an abstract nature are encountered. The problems are concrete and relate to business situations. They require simple reasoning and a decision as to the processes to be employed" (p. 141). Clearly, "the necessary work in arithmetic can be mastered in much less time than is now being devoted to it" (p. 142). For his doctoral thesis, which used the same data, Wilson (1919) expanded the sample to 4,068 persons solving 14,583 problems. He offered a more detailed analysis of the problems and was more cautious in his conclusions. Nonetheless, he concluded that "many of the traditional processes of arithmetic should be entirely omitted" (p. 55), contending that "it is doubtful if arithmetic is entitled to the large time expenditure which is now allotted to it in the elementary grades" (p. 56).

Other studies followed in which investigators scanned advertisements, cookbooks, sales slips, and other artifacts of modern society for evidence as to which numbers were used in daily life and how (for summaries of 19 such studies, see Wilson, 1925, pp. 40–62). They found repeatedly that "the chief demand is for small numbers and very simple fractions" (Buswell & Judd, 1925, p. 16). Looking back in 1938 on what he termed the "reductionists," Buswell commented that the social utility idea had originally been applied in a negative fashion—to determine what should be dropped from the curriculum. At the outset, the casualties had been minor and unlamented, but as the research began to accumulate,

textbook-writers eliminated long addends in addition, fractions with large denominators, decimals beyond a certain number of places, and some entire processes that were found not to occur frequently. The influence of this kind of research reached its peak when one enterprising graduate student argued in his doctor's thesis that eighty-five percent of all arithmetic taught in the schools was useless. (p. 124)

Buswell saw the research of the 1930s as taking a more positive view of social utility and as looking beyond computation to the ways in which instruction in quantitative thinking might make the arithmetic learned in school more useful.

A Reaction to Reductionism

The watershed in the change from research aimed at trimming the curriculum to a more constructive attack on how arithmetic might be organized and taught was the Report of the National Society for the Study of Education's six-member Committee on Arithmetic (Whipple, 1930). Half the committee members were professors of educational psychology; the others were a lecturer in education, a director of research in a city system, and a state director of teacher training. They were identified more with psychology and the school subjects than most of the school administrators and curriculum theorists from the NEA who had launched the drive toward social utility for the curriculum. The first part of the report was essentially a defense of arithmetic as a school subject against its reductionist critics, led off by a forceful argument by Burdette R. Buckingham (1930) of Harvard on the social value of arithmetic. Among other points, Buckingham noted the folly of basing any curriculum on surveys of the frequency of adult usage: "Shall we say that 60 percent of the teaching of the schools in spelling and language should be devoted to the one hundred words of most frequent occurrence—to *the, and, but, to, he,* etc?" (p. 44). The second part of the report contained a classification by Brownell (1930) of the research techniques used in studies of arithmetic, a survey of the nature and findings of those studies by Buswell (1930), and 11 original studies selected to represent the best of current work.

The report was published as the 29th yearbook of the NSSE, which also had an appendix giving a critique by a panel of 15 reviewers. Leo J. Brueckner (1930) of the University of Minnesota contended in the critique that although the yearbook made a substantial contribution to the field, its authors showed an unfortunate preoccupation with the computational side of arithmetic to the neglect of its other aspects. Presumably the attacks of the reductionists had forced many of these researchers into a defensive posture, fighting the battle of computation rather than giving more attention to the informational, sociological, and psychological functions of arithmetic. The committee's chair, Frederic B. Knight (1930) of the State University of Iowa, had acknowledged in his introduction to the report that the psychology of learning assumed by the committee was "a behavioristic one, viewing skills and habits as fabrics of connections" (p. 5). Brueckner, speaking for the reviewers, took the committee to task for ignoring gestalt psychology, which Knight had claimed was not ready to be used as a basis for elementary education.

Perhaps the most notable feature of Knight's (1930) introduction, however, was its frank embrace of scientific (psychological) research as providing essential support for teaching. Psychology can help us with teaching methods: "How to teach the child can be separated, in discussion, from what to teach—and how to teach is fundamentally more a matter of psychology based on research and investigation than a matter of philosophy" (pp. 4–5). Moreover, efforts to base the curriculum on the "felt needs" and the interests of children are in error. Psychology can also help us with content: "What to teach should be decided by as wise adults as are available for the task, who will base their decisions as far as possible upon the available body of objective scientific data" (p. 6). These sentiments epitomize the idealistic views many researchers shared at the height of the scientific movement in education.

Buswell (1938) later claimed that the 29th yearbook "marked a turning point in the treatment of arithmetic" (p. 126), in part because of what he termed an "excellent" critique. Among other things, the critique had "laid stress upon the place of problem work in arithmetic. Since 1930, the drill theory of arithmetic has been subject to criticism from a number of quantitative investigations and more attention has been given to studies of problem-solving and quantitative thinking of a noncomputational variety" (p. 126).

Readiness and Incidental Learning

Drill theory was attacked severely, along with *incidental learning* theory—which held that children learn arithmetic better if it is not systematically taught to them—by Brownell (1935) in an article that set out his alternative, *meaning* theory. Throughout the 1930s and into the 1940s, advocates conducted research to support their position on incidental learning or its more moderate companion, *readiness* theory—the view that the teaching of a concept or skill should at least not occur until children are mature enough to learn it. That research was counteracted by other research that attempted to shift the ground away from mere computational proficiency or the child's need for arithmetic and toward higher levels of mathematical understanding.

The Grade Placement of Topics. An opening shot on the readiness front had been fired by Carleton W. Washburne (1930), superintendent of the Winnetka, Illinois, schools, writing for the Committee of Seven of the Northern Illinois Conference on Supervision in the 29th NSSE yearbook. The committee, under Washburne's leadership, had been experimenting with the teaching of problem solving since 1926, but this was their first major report on the grade placement of arithmetic topics. Washburne claimed to have found (by teaching topics at grades above and below the customary level and then testing for mental age, learning, and retention) the minimum mental age at which each of the topics should be taught. The Committee of Seven's investigations continued for the rest of the decade (Washburne, 1939) and had "an unquestioned influence on the organization of the arithmetic curriculum" (DeVault & Weaver, 1970, pp. 126–127), inducing curriculum makers to move topics higher in the grades and to separate topics that were related mathematically. Washburne, who seems to have been at least as interested in promot-

ing his innovations at Winnetka as in conducting rigorous research, soon tangled with his director of research, Lewis Raths, over the adequacy of the techniques and instrumentation being used. (After only a few years at Winnetka, Raths left for Ohio State University, where he subsequently played a major role in the Eight Year Study.) Then in 1938, Brownell weighed in with a strong critique of the committee's work, not only faulting the conception of maturation, the instruction, and the instruments used in the research but also chiding the committee for diverting research away from more fundamental issues. (See Washburne, 1939, for an account of Raths's and Brownell's criticisms and Washburne's rejoinders.)

The Committee of Seven was not alone in investigating the effects of postponing formal arithmetic instruction. As early as 1912, Ballard had found no ill effects from postponing arithmetic instruction until after age 7. Other studies followed that raised the grade level at which instruction began. (See Brueckner, 1939, for a review of these studies.) This line of research had an effect that persisted into the 1950s and was not confined to North America. The postponement of topics was the theme of a 1937 survey by the Scottish Council for Research in Education of head teachers in Scotland regarding current practice in their schools (Murray, 1939). In a 1957 review, Fred and Eleanor Schonell referred to "the lightening of the syllabus—so that children do less but understand what remains much better" (p. 37)—with no decline in "standards," that had occurred in British and Australian schools in response to the American research. Robert L. Morton, writing in 1953 on what research says to the classroom teacher about teaching arithmetic, claimed that the so-called stepped up program, in which topics were postponed to higher grades, was here to stay:

Some of the earlier findings on grade placement have been challenged on the ground that a better organized program and different teaching technics might well have produced very different results. Be that as it may, the fact remains that the upward movement of topics and phases of topics seems to be a permanent one. (p. 18)

The most dramatic study of grade placement began in 1929 when Louis P. Benezet (1935–1936), the superintendent of schools in Manchester, New Hampshire, arranged that no arithmetic would be taught until grade 7 beyond activities in estimating and using numbers in social situations. After a year's instruction, the students' test scores were at the level of those of comparable students who had undergone regular instruction. Benezet proudly summed up his accomplishments in his self-description for *Who's Who:* "Pioneer in curriculum revision; substituted social sciences and natural sciences for mathematics and ancient langs. in secondary curriculum, 1930, health studies, science readers, citizenship and vocabulary building for arithmetic in the five lower grades, 1929" (Marquis, 1936, p. 286). Benezet apparently saw the problem as one of compressing the study of mathematics into the smallest possible amount of time.

Eliminating Formal Instruction. The more extreme proponents of incidental learning proposed that formal instruction in mathematics be eliminated altogether rather than simply postponed. Many of these educators had been influenced by the progressive ideas of John Dewey and William Heard Kilpatrick and had concluded that all the mathematics anyone needed could be learned through experience. They usually concentrated on eliminating arithmetic as a separate subject from the elementary school curriculum (Jones & Coxford, 1970, p. 50). Children would learn number facts, for example, by engaging in activities in which number combinations occurred naturally. The research of Henry Harap and Charlotte E. Mapes (1934, 1936) was often cited. They studied a class of students who in grades 5 and 6 learned operations with fractions and decimals by engaging in activities such as organizing a candy sale, sewing a quilt, and making furniture polish. Brownell (1935) maintained that although incidental learning could help counteract the practice of teaching arithmetic as an isolated subject, it did not provide an organization in which "the meaningful concepts and the intelligent skills requisite to real arithmetical ability" (p. 16) could be developed.

Meaningful Arithmetic

From 1935 to 1949, Brownell undertook a series of studies in support of his meaning theory, several of which were later published in the series of *Duke University Research Studies in Education*. The last of these studies (Brownell & Moser, 1949) was later characterized as "one of the best-executed of all educational experiments" (Cronbach, 1965, p. 118), and Brownell's research on the teaching of arithmetic has been termed "brilliant" (Shulman, 1974, p. 335). A major reason for these accolades may have been Brownell's refusal to work within the confines of laboratory experimentation in which the effects of a single variable are sought by manipulating that variable and attempting to hold everything else constant. Brownell (1948) regarded such an approach as untenable when one is studying the learning situation in the school:

It is a mistake to believe, as some do, that all research which deserves the name must follow this [classic experimental] pattern and that investigations which do not are, ipso facto, untrustworthy and valueless. To take this extreme position would be to deny to botany and to astronomy (to cite but two examples) any validity as fields for research. In these fields, as in many others, systematic observation, competent analysis, and rigorous classification of data largely take the place of research in accordance with the Law of the Single Variable. (p. 495)

Brownell was noted for his use of a variety of techniques for gathering data, including extended interviews with individual children and teachers, and for his careful, extensive, and penetrating analyses of those data. (See Kilpatrick & Weaver, 1977; for a bibliography of Brownell's publications, see Weaver & Kilpatrick, 1972, pp. 329–337.)

According to one assessment, meaningful arithmetic programs "were initiated in the 1930s, grew in popularity through the 1940s, and were the commonly accepted programs of the 1950s" (DeVault & Weaver, 1970, p. 124). Although it is difficult to judge how well accepted the programs were, they appeared to survive early criticism of their failure to address adequately the child's interests and the adult's needs (DeVault & Weaver, 1970, p. 125). The extent to which the word *meaning* in Brownell's theory referred to mathematical meaning or to social meaning was not always clear (Jones & Coxford, 1970, p. 50), but the examples he used (Brownell, 1935, pp. 25–28; see also Brownell,

1945, p. 481) suggested that the emphasis was to be on mathematical relationships. In a sense, meaningful arithmetic anticipated the modern mathematics movement that began in the 1950s.

Research in Response to Curriculum Issues

The elementary school arithmetic curriculum in America in the first half of the 20th century can be likened to a battleground in which shifting combinations of school superintendents, principals, educational psychologists, teachers, curriculum specialists, and professors of education fought recurring skirmishes over what would be taught and how. (For a more genteel portrayal in terms of interest groups, see Stanic, 1983/1984, 1986a.) The great faith that most of these people repeatedly expressed in their writings was that scientific research studies would provide the means by which they could resolve their differences—or at least establish the superiority of their own position. Their continued wrangling generated a great number of research studies.

Mathematics in the Secondary School. Some of the issues associated with arithmetic teaching and learning spilled over into the secondary school, where the volume of empirical research was much less (though the volume of scholarly writing on mathematics teaching was at least as great). In a 1933 review, Harry E. Benz, while bemoaning the general lack of research on the teaching of high school mathematics, cited several studies of the uses of school mathematics, the effects of drill, the characteristics of textbooks, and the results of testing. A later summary by E. R. Breslich (1938) noted some research on these same topics. In both articles, however, most of the studies cited reflected the special character of the conflicts that were occurring over the secondary curriculum.

Although the elementary arithmetic curriculum was repeatedly being questioned by prominent educators, none of them proposed doing away with arithmetic as a subject for everyone to learn (as opposed to doing away with its study as a separate subject). People granted the importance of knowing arithmetic even when they saw it as best learned informally rather than formally. The secondary curriculum was different. There the question was primarily who should study mathematics, and only partly what form the mathematics should take. The requirement that all students study mathematics in high school was seriously challenged in the 1920s and 1930s; people spoke of the "crisis" that threatened the existence of mathematics as a required subject (Stanic, 1986a). Mathematics educators responded to the crisis in a variety of ways, including the formation in 1920 of the National Council of Teachers of Mathematics (NCTM) (Austin, 1921). Another response was to undertake research.

The attacks on the place of mathematics in the high school curriculum led some mathematics educators to reconsider the objectives of high school mathematics instruction. Breslich (1938) noted that the objectives had long been set by experts in mathematics, who wanted students prepared for the advanced mathematics they would encounter in college. Other objectives were possible, and investigators such as Raleigh Schorling (1925) set out to determine those objectives by surveying courses of study, textbooks, professional literature, students' test performance, and opinions of experts. Some researchers asked teachers to produce lists of objectives; others inferred objectives from classroom practice (see Benz, 1933, and Breslich, 1938, for discussions of this literature). The argument for the "disciplinary value," or transfer effect, of the study of high school mathematics was investigated extensively (Orata, 1935, 1937; Rosskopf, 1953). A dissertation that became a widely cited NCTM yearbook was Harold Fawcett's (1938) account of his attempt to base a high school course in demonstrative geometry on nonmathematical situations and student-developed terms, definitions, and assumptions, as a means of promoting reflective thinking.

Another set of studies dealt with individual differences in mathematical abilities. If not all students would study mathematics, or at least the same mathematics, then some means would have to be found to decide which students would profit from a particular course. A major issue was the question of ascertaining who was prepared for the study of algebra. Furthermore, students were increasingly being grouped according to ability on the assumption that instruction was more effective and more efficient if the students were "homogeneous" in ability than if they were "heterogenous." The testing movement provided the instruments for exploring questions about the prediction of success and the effects of grouping students by ability; both of these became popular research topics. An example is a study by Charles M. Austin (1924), who found that an intelligence test was better than an arithmetic reasoning test at predicting grades in an algebra course. Specially developed tests were also used to investigate the errors students made in algebra and geometry (Benz, 1933, pp. 43–46; Breslich, 1938, p. 134) and the correlation between mathematical abilities and other mental abilities (Benz, 1933, pp. 26–29). Attempts were made to analyze in detail the specific skills and concepts required in algebra (for example, Everett, 1928) and geometry (for example, Welte, 1926).

A final set of studies at the secondary school level dealt with whether subjects such as algebra, geometry, and trigonometry should be taught in separate courses or in courses of unified mathematics—also known as general or correlated mathematics, although the three terms eventually took on somewhat different meanings (Benz, 1933, pp. 24–25; Osborne & Crosswhite, 1970, pp. 173–179; Senk, 1981; Sigurdson, 1962). The idea of unifying various branches of mathematics in a single course emerged in France and Prussia in the 19th century. In England, the movement toward unification was slow. A course in unified mathematics was developed by David Mair in 1907, but unified examinations did not become available until 1921, and not until the 1960s did most Joint Matriculation Board candidates take the unified examinations (Howson, 1982, p. 163). In the United States, the movement was not simply slow; it was sidetracked. From the outset, traditionalists such as D. E. Smith opposed unified courses. E. H. Moore's (1903/1967) early call for unification was taken up most enthusiastically at the University of Chicago Laboratory School, where George W. Myers and E. R. Breslich developed textbooks and tried out unified courses from 1903 to 1923. To counter Smith's criticisms of extreme formalism and unteachability, Myers and Breslich obtained evidence on how well students were learning (Benz, 1933, p. 24; Senk, 1981, p. 57). Later, more controlled studies were done in which the achievement of students in unified courses was

compared with that of students in traditional courses (Benz, 1933, pp. 24–25). Although the preponderance of the research evidence favored the unified courses, that evidence was rather weak (Senk, 1981, p. 66). The supporters of traditional Euclidean geometry were numerous and firm in their belief that the subject afforded students an incomparable introduction to axiomatic reasoning and an opportunity to appreciate elegant mathematics. A year course of Euclidean geometry remained a mainstay of the American high school curriculum, and general mathematics was relegated to a course that served as an alternative to algebra for students not bound for college.

Evidence to Support Change. The lack of firm evidence on which to base school mathematics instruction was lamented by curriculum theorist Harold Rugg (1924) in a pointed review that contrasted Thorndike's work on algebra with the Mathematical Association of America's report on *The Reorganization of Mathematics in Secondary Education.* (Although not usually identified as a mathematics educator, Rugg chaired the committee that drew up the first constitution of the NCTM and served as its first vice-president; Stanic, 1983/1984, p. 157.) Rugg argued that no blue-ribbon committee should report in the future without conducting "scientific studies of the abilities of children, of the needs of society and of the learning process" (p. 19). He set forth the case for educational foundations to fund such studies:

Is there, in America today, one scientifically controlled classroom experiment in mathematics? It is my confident judgment that there is not. Yet there are more than fifty public school systems in each of which there is in working order a "bureau of research," directed by a person trained in the technique of measurement and in most cases of experimentation. It is practically impossible for an editor to discover for journal publication controlled classroom experimentation. Furthermore, it is practically impossible to persuade graduate students in our schools of education to undertake dissertation problems in experimentation. Why? Primarily because it costs too much; it takes two or three years to complete one controlled experiment. Our major "research" energies are still devoted to giving tests and to presenting "norms." (p. 18)

A major study, begun in 1932, that was well funded (by the Carnegie Corporation and the General Education Board) and that was conducted to assess the long-term effects of curriculum change was the Eight Year Study. The purpose was to allow secondary schools to experiment with innovative curricula without risking their graduates' chances of admission to college. An agreement was negotiated by the Progressive Education Association between 30 selected high schools and a large number of colleges that, for 5 years beginning in 1936, released the high schools from the usual requirement that college applicants complete a certain number of units of study in each of several subjects. The students from these high schools were to be admitted to college on the recommendations of principals and headmasters, together with a record of activities and test scores, rather than on the basis of grades and units (Chamberlin, Chamberlin, Drought, & Scott, 1942, p. 1). A comparison of the college careers of 1,475 students from the 30 participating schools with those of a similar number of students from conventional schools found that the students from the 30 schools did "a somewhat better job" (Chamberlin et al., 1942,

p. 208) than their peers on a variety of measures, from college grades, honors, and graduation rates to participation in organized activities and ratings of social concern (see Travers, 1983, pp. 144–153, for an analysis and critique of the study).

The Eight Year Study ended as World War II was underway and progressivism was fading as a movement in American schools. (The Progressive Education Association disbanded in 1955, but "progressivism, as its founders saw it, had been dead a long time by then"; Mayer, 1961, p. 64.) Although its modest findings had little impact, the study provided a training ground for a cadre of educational researchers who would later become eminent (Bruno Bettleheim, Oscar K. Buros, Chester W. Harris, and Hilda Taba, among others), and it set the pattern for evaluation studies for several decades. Mathematics was, of course, one of the subjects whose curriculum was revamped in the 30 schools, but, almost paradoxically, the effects of specific changes in school subjects were not studied. The evaluation model used for the study had been developed by its director, Ralph W. Tyler, and relied on the operational definition of objectives and the construction of instruments to measure their achievement. Although in view of the way such definitions have subsequently been used one might have expected the instruments to focus on the acquisition of specific skills and procedures, social adjustment and general aspects of thinking were central. Much effort went into the assessment of thinking skills (Smith, Tyler, & the Evaluation Staff, 1942), and the resulting tests "created more interest in their time than any other set of instruments produced by the entire program" (Travers, 1983, p. 150). Several mathematics educators were on the staff of the study. Among the group responsible for the thinking skills tests were Maurice L. Hartung and Harold C. Trimble; one of the tests the group constructed, on the nature of proof in nonmathematical situations, was used by Harold Fawcett (1938, appendix, part 1) in his dissertation on proof. Despite their involvement, however, mathematics educators did not learn from the study how variations in their subject's curriculum might affect students' learning. Although members of the Eight Year Study Committee served on the blue-ribbon Committee on the Function of Mathematics in General Education of the Progressive Education Association (1940), whose report *Mathematics in General Education* attempted to set forth principles for the mathematics curriculum, the results of the study could not be used to inform the report in the way Rugg had envisioned. For mathematics education, the greatest effect of the Eight Year Study undoubtedly came through subsequent national and international assessments of mathematics achievement that were influenced by Tyler and followed his evaluation model (Madaus & Stufflebeam, 1989, preface).

Reviewing the growing volume of (largely empirical) research in mathematics education during the first half of the 20th century, one cannot help being struck by the way curriculum issues—what mathematics should be taught and how—have dominated the research agenda from the outset. Popular topics such as whether students can transfer their knowledge of mathematics from one arena to another or how students of different ages learn the same mathematical skills were investigated not simply out of intellectual curiosity but primarily because reformers were proposing substantial changes in the school mathematics curriculum. Curriculum questions are ultimately questions of purpose and value; they concern what

ought to be rather than what is. Many researchers continued to believe, however, that scientific research could resolve these questions. Even in the Soviet Union, where research in mathematics education gradually emerged out of a more theoretically oriented instructional psychology (Menchinskaya, 1967/1969), issues such as whether students should learn mathematical problem solving by studying prototypical solutions or by learning steps in problem analysis (essentially a question of what transfers in learning but also a curriculum issue) occupied researchers from roughly the mid-1930s on (Goldberg, 1978, pp. 373–374; Krutetskii, 1968/1976, pp, 49–54). Although some of the resulting research appears to have been stimulated by the conflicting views of N. A. Menchinskaya and P. Y. Galperin on how knowledge is mastered, much of it appears to have been a response to complaints from teachers about the difficulties students encountered with the problems contained in the mathematics curriculum (Kilpatrick & Wirszup, 1972, pp. 1–5).

A Search for Identity

As mathematics educators attempted to define their field, they found themselves looking to mathematicians and psychologists for guidance as to what constituted scholarly work. Although they felt a strong allegiance to the prestigious discipline of mathematics—and indeed derived much of their status within colleges of education from that allegiance—these mathematics educators had ambivalent feelings about mathematicians' pronouncements on education. As early as 1905, D. E. Smith lamented that at the sessions on mathematics education at the Third International Congress on Mathematics, held the preceding year in Heidelberg, the participants spent their time debating mathematical minutiae rather than examining general problems of mathematics education. Reporting on changes and trends in the teaching of mathematics in the United States, William David Reeve (1929) pointed out the control exerted by college professors of pure mathematics, whose "opinions were law and gospel for all concerned" (p. 134), on the North Central Association of Colleges and Secondary Schools at the turn of the century. Although Reeve acknowledged that these professors had a legitimate interest in the improvement of mathematics teaching, he seemed to be saying that other educators needed to be heard. By the 1930s, people were calling for a truce between mathematicians and "educationists" (Havighurst, 1937; Hedrick, 1932). (The tension has not vanished entirely over the years. At a January 1990 meeting of the Executive Committee of the International Commission on Mathematical Instruction, the Secretary of the International Mathematical Union, the ICMI's parent body, took it as axiomatic that "the teaching of mathematics is so important it must be in the hands of mathematicians." See also Jackson, 1988.)

At the same time, the work of psychologists sometimes seemed irrelevant to the questions confronting mathematics educators. In lauding psychologists for helping mathematics educators adopt the learner's perspective, Reeve (1929) could not resist an opening dig at their indifference to the quality of the subject matter taught:

To be sure, the psychologists have shown us how to teach better some things which would better go untaught, but they have also helped us to organize our fundamental material along lines that are psychological rather than logical....[They] have discovered many useful facts and laws about how children learn most easily and most economically, how habits are formed, how abilities are developed and how they may be retained. (p. 148)

Reeve noted the minimal impact of research on mathematics instruction, attributing the changes that had occurred to "external social forces" (p. 185). By 1953, Howard F. Fehr could survey a somewhat richer psychological literature and conclude that the mathematics educator had to be pragmatic:

That these theories [of learning] conflict at a number of points should not concern us too much. If the application of one of the conflicting theories proves more useful for our purpose in a given situation, and application of another theory in another, we shall use each as it fits the occasion. Until psychology develops into a more significantly unified, scientific theory we must do this. (p. 8)

The Danish psychologist K. B. Madsen (1988) identified the years from 1933 until around 1960 as the period of integration for the psychological community, a period that he claims was completely dominated by American psychology (p. 336). International collaboration between psychologists had been shattered by World War II and only gradually resumed. Out of that period came a "mainstream psychology" (Madsen, 1988, ch. 12), whose most influential branch for school mathematics in many countries was the cognitive psychology of Jean Piaget and Jerome Bruner. Not all psychologists supported this integration, however, and by 1960 or so, several alternative schools were flourishing. In particular, humanistic psychology, Marxist psychology, and contemporary behaviorism (Madsen, 1988, chs. 9–11) began to spread, partly in response to the crisis in values raised by the social turmoil of the 1960s and partly because of dissatisfaction with different aspects of the paradigm followed by psychologists attempting to integrate their subject into one science.

The 1960s were also a time during which mathematicians "rediscovered the school." By the end of the 1950s, widening discontinuities between the mathematics taught in universities and that taught in the lower schools, as well as growing concern over declining enrollments in university mathematics courses, were beginning to give rise to a flood of curriculum reform projects in various countries that collectively became known as "the new math" (Cooper, 1985; Howson, Keitel, & Kilpatrick, 1981; Moon, 1986). International collaboration among mathematics educators had begun again with the reconstitution of the International Commission on Mathematical Instruction in 1952, followed by a series of special conferences and, in August 1969, the first International Congress on Mathematical Education (Howson, 1984). Once again, concerns about the school mathematics curriculum triggered research studies as people sought evidence to substantiate some of their proposals for change. This time around, however, mathematicians and psychologists were brought together in curriculum development projects, and studies were undertaken that drew upon both perspectives. A revival of interest in issues such as learning by discovery, readiness for learning, processes of learning, and aptitude for learning helped people from different disciplines see some common ground (Shulman, 1970). The ferment in the curriculum and in psychology itself during the 1960s had,

by the end of the decade, begun to accelerate the formation of a research community in mathematics education.

COALESCENCE OF A COMMUNITY

In December 1950, the *Mathematics Teacher* began a new department, "Research in Mathematics Education," edited by John J. Kinsella of New York University, that attempted to acquaint its readers with developments in research. Kinsella (1952) had observed that the burgeoning number of studies in mathematics education was making it increasingly difficult for people to keep abreast of the literature. The desire of practitioners to have research applied to issues in mathematics education, and not just interest in research itself, appeared to be growing. In 1953, Kenneth E. Brown of the U.S. Office of Education, in collaboration with a National Council of Teachers of Mathematics committee on research chaired at that time by Henry Van Engen of Iowa State Teachers College, published a list of research studies conducted in 1952. Another list was published the following year (K. E. Brown, 1954). Brown and the NCTM committee then began a series of biennial publications from the Office of Education that attempted not only to summarize but also to organize, and to some extent analyze, the recent literature (see, e.g., Brown & Abell, 1965).

Also in 1953, the National Education Association's Department of Classroom Teachers, in conjunction with the American Educational Research Association, published the second pamphlet in a series "What Research Says to the Teacher." (The NCTM had meanwhile become affiliated with the NEA in June 1950.) The pamphlet, entitled *Teaching Arithmetic*, by Robert L. Morton, gave teachers advice on teaching drawn from Morton's reading of the research literature, with relatively few references to specific studies. A companion in the series, *Teaching High-School Mathematics*, by Howard Fehr (1955), made only one reference to empirical research, Duncker's (1945) study of problem solving, relying instead on references to research summaries, mathematics textbooks, histories of mathematics, and pedagogical works. Herbert F. Spitzer's (1962) update of the Morton pamphlet focused on the influence of the new math movement: "the attempt to develop mathematical concepts, operations, and principles from a mathematical point of view with little or no reference to their use, even during the introductory consideration of a topic" (p. 12), and "a definite trend in 1962 to place arithmetic topics at a lower grade level than in 1952" (p. 6). Again, references to specific studies were few and accompanied by general references on teaching arithmetic. The series was continued by the NEA's Association for Supervision and Curriculum Development, and subsequent volumes (Callahan & Glennon, 1975; Glennon & Callahan, 1968) contained more comprehensive reviews of the research bearing on particular issues in elementary school mathematics. For these volumes, studies were classified according to whether they concerned the curriculum, the child, the learning environment, or the teaching method. A specific question would be raised ("What does the work of Piaget suggest about the cognitive development of the child?" "How can we best group children for learning mathematics?"), and a brief essay would follow in which opinions and findings from recent research were summarized.

These and other reviews that began to lengthen, proliferate, and focus on specific research studies during the 1950s and 1960s indicated that mathematics education was not only developing its own body of research literature but also moving beyond conventional psychological studies to incorporate other sources and models. In Kinsella's (1952) words, "The problem [of locating studies in mathematics education] is becoming more complex because investigators are becoming sensitive to the fact that learning and curriculum making in mathematics are related to many other fields" (p. 273). Beyond the problem of locating the many studies was the problem of finding studies that built upon previous work. In a talk to the AERA in February 1950, William Brownell reported that 1,413 research studies in arithmetic had been completed by the end of 1945. These studies were conducted by 778 different authors, 615 of whom never reported more than one such study and only 53 of whom reported more than three studies (Buswell, 1951, p. 283). Recognizing that studies conducted by individual investigators were not cumulating in a productive way, researchers began to issue calls for more coordination of research and for ways of bringing together people of various talents and specializations to carry out that research.

The Golden Age

Getting Acquainted. In the early to mid-1950s, reform of the school mathematics curriculum was being proposed from many sides. American schools were under attack from business and the military for graduating young adults who lacked basic computational skills, from colleges for failing to equip their entrants with a knowledge of mathematics adequate for college work, and from the public (agitated by a host of critics from Arthur Bestor to Hyman Rickover) for having watered down the curriculum in response to progressivism and life-adjustment education (for a brief early view of progressivism's effects, as well as some subsequent efforts to improve the mathematics curriculum, see Mayer, 1961, chs. 3 & 12). Late in 1954, the Carnegie Corporation asked the Educational Testing Service to propose a plan for research to address the improvement of mathematics courses and teaching. The resulting report, *A Survey of Mathematical Education: The Causes of Student Dropout, Failure, and Incompetence at the Elementary and Secondary Levels,* was released in October 1955 and published in abbreviated form the next year (Dyer, Kalin, & Lord, 1956). The authors of the report reviewed the research literature, surveyed the opinions of various authorities through questionnaire and interview, and observed and interviewed teachers and students. They concluded that the current and proposed efforts to reform the curriculum were generally to be applauded but observed that

curriculum revision strikes at only one side of the trouble in mathematical education. Much more needs to be known about the mental functioning of children who are expected to pass through the curriculum and about the teachers who are expected to operate it. (Dyer et al., p. 26)

The authors called for

a coordinated research effort which is feasible if those now expert in the several pieces of the problem can become intelligible to one an-

other and join forces long enough to map out and execute the interrelated lines of investigation. Mathematicians, psychologists, educationists, classroom teachers, and such consumers of mathematics as physicists, engineers, and economists must put their heads together, not for a day or a week, but for as long as is necessary to get experimental studies going and keep them going until hunches can be converted into verifiable conclusions based on actual data from classrooms and training centers. (Dyer et al., pp. 27–28)

The report outlined a series of 11 research projects, ranging from the development and evaluation of teacher-training courses to case studies of students in a remedial clinic. It recommended the formation of a committee on research in mathematics education that might either advise an agency that would carry out the research or serve as that agency itself (ETS, 1955, p. 90).

Although no such committee appeared, efforts at coordination continued. It became obvious that the gulfs between researchers having various professional specializations would not be easily bridged. As government support for educational research began to increase in the 1950s and 1960s, along with its support for curriculum development, sporadic meetings were organized of representatives from various constituencies concerned with research in mathematics education. The Cooperative Research Act of 1954 (Public Law 531) had empowered the U.S. Office of Education to fund colleges, universities, and state departments of education to conduct research, surveys, and demonstration projects. By 1965, some 30 projects had been funded in mathematics, and 7 had been completed (Brown & Abell, 1965, pp. 23–29). One project attempted to draw constituencies together by addressing psychological problems and research methods in the teaching of mathematics. It yielded a monograph reporting on a 1959 conference that reviewed the literature from 1948 to 1958, identified problem areas, and proposed new lines of research (DuBois & Feierabend, 1960). In 1962, the Committee on Intellective Processes Research of the Social Sciences Research Council held a conference that "brought together psychologists, mathematicians, and mathematics educators to discuss problems of mathematical learning and to try to learn how the work of one group was, or could be made, relevant to the work of the other" (Morrisett & Vinsonhaler, 1965, p. 2). The exchange did not take place easily:

Perhaps the most salient single feature of the conference was the difficulty that psychologists and mathematicians experienced in communicating with one another. It is obvious that the subject of mathematical learning is rich with ideas for psychological experimentation and theory. Much more research needs to be done on the subject, but it will probably be a rare psychologist who will be at home in this research until a continuing dialogue between mathematicians and psychologists produces a greater communality of understanding. (p. 3)

That was the view of the psychologist concerned about adapting ideas from mathematical learning and being at home in research on it.

Subsequent conferences focused more directly on the views of mathematicians and mathematics educators. They included three conferences on needed research in mathematics education. The first was a small meeting of mathematicians and educators in October 1965 sponsored by Teachers College, Columbia University (H. F. Fehr, 1966). In his remarks on resources for research, the mathematician Marshall Stone called for a service that would provide access at least nationally, but probably also internationally, to current information on problems of mathematics education. In the subsequent discussion, repository libraries and a research journal were proposed as means of access and dissemination. The last of the three conferences was a somewhat larger meeting of mathematicians and educators at Cornell University in the spring of 1968 (Long, Meltzer, & Hilton, 1970). The purpose of that conference was to identify topics for term projects, theses, and postdoctoral research in mathematics education; many potential topics were identified. Much of the discussion dealt with the criteria for a high quality doctoral program in mathematics education. Although psychology was seen as an important and potentially useful field for the mathematics educator to study,

there seemed to be a consensus that cognitive psychology at the present time is not oriented toward problems or approaches which are likely to be useful to mathematics education. . . . A substantial problem seems to be to find psychologists (mathematicians) who understand mathematics (psychology) in any but a shallow manner. (Long et al., p. 451)

The second of the three conferences was larger and more diverse in composition than either of the others. It was held at the University of Georgia in September 1967, and for many of the participants, it seemed to mark the beginnings of true interdisciplinarity and community among researchers in mathematics education (Hooten, 1967). The previous year, Vincent J. Glennon (1966) had asserted that "the Golden Age of educational research is now with us" (p. 367). J. Fred Weaver (1967) echoed that claim, arguing that it was equally true for research in mathematics education. The atmosphere at the conference suggested that many participants shared Weaver's opinion.

The conference was jointly sponsored by the National Council of Teachers of Mathematics, the University of Georgia, and the National Science Foundation. The NCTM had created an ad hoc Committee on Research in Mathematics Education in 1964. In April 1965, the committee structure was reorganized, and a new Research Advisory Committee was created as one of the curriculum subcommittees. The Research Advisory Committee began to take a more active role in promoting research activities within the council. Previous committees on research during the late 1950s had arranged some sessions on research at annual meetings and had promoted articles and departments on research in the *Mathematics Teacher* and in the new journal, the *Arithmetic Teacher*, that had begun in 1954. The reorganized Research Advisory Committee began several series of research-reporting and research-reviewing sessions at the NCTM annual meetings in 1967 and 1968 that brought together for an extended time people who were interested in research. The Philadephia meeting in 1968 also included a one-day presession on research. (The research-reporting sessions at the meeting itself had space for 8 papers and attracted 48 proposals; Romberg, 1968.) Such activities later became a regular feature of the annual meetings. The committee also developed plans for a research journal to be published by the council and set up an advisory committee to work with people at the University of Georgia in arranging the 1967 National Conference on Needed Research in Mathematics Education.

The conference was organized around three themes: the learning of mathematics, the teaching of mathematics, and the mathematics curriculum. Prepared papers and the ensuing discussions attempted to identify researchable problems, theoretical models, and research designs and methods under each theme. One of the major goals of the conference was the "encouragement of greater coordination of related activities" (Hooten & Wagner, 1967, p. 1). In his summary of the conference, Ralph Tyler (1967) noted how much time it had taken for the participants to develop a common language and get beyond their lamentations over the shortage of significant research so that they could arrive at a coherent perspective. The conference had illuminated the complexity of the research enterprise in mathematics education, but it had also suggested, at least to Tyler, that there was no need to avoid or bemoan that complexity. It ought to be addressed head on:

Within the research enterprise, it is desirable to encourage a variety of research interests rather than to emphasize a model of one superperson who has all the interests and desirable characteristics of good researchers. . . . Research is not a monolithic enterprise nor does it involve a single kind of person. It is a collective enterprise of considerable magnitude involving a variety of talents, and needing a variety of activities in order to be maximally effective. (p. 140)

The community of researchers in mathematics education would necessarily be a diverse community.

People began to consider ways in which researchers might meet together on a regular basis. The AERA had increased its membership considerably during the late 1960s and had begun to outgrow its divisional structure. It started to encourage the formation of special interest groups on various topics. At the AERA annual meeting in February 1968 in Chicago, a group met to organize the Special Interest Group for Research in Mathematics Education (SIG/RME). It was one of the first interest groups to organize, and it grew to some 75 members within a few months (J. W. Wilson, 1968). In addition to sponsoring sessions at the AERA annual meeting, the SIG/RME cooperated with the Research Advisory Committee in producing an NCTM newsletter on research, presessions at the NCTM annual meetings, and research sessions during those meetings. By 1981, the membership count was 300 from the United States and 45 from eight other countries, including Canada (SIG/RME, 1981).

Several other organizations for researchers in mathematics education appeared late in the 1970s. At the end of the fourth of a series of annual conferences, the Research Council for Diagnostic and Prescriptive Mathematics was formed in April 1977 with about 100 members (SIG/RME, 1977). Also in 1977, Ken Clements of Monash University and John Foyster of the Australian Council for Educational Research founded MERGA, the Mathematics Education Research Group of Australia (changed to "Australasia" in 1980), whose first conference in May attracted about 80 participants (Clements, 1989, p. 55). MERGA had been anticipated by the formation of a Research and Management subcommittee of the Australian Association of Mathematics Teachers in 1972 and by the subsequent creation of a research section in their journal, the *Australian Mathematics Teacher* (Jones, 1984). Soon to become the largest international organization, the International Group for the Psychology of Mathematics Education (PME) held its first meeting at Utrecht at the end of August and beginning of September 1977. The PME, which had been organized at the Third International Congress on Mathematical Education the previous summer in Karlsruhe, actually grew out of a round table at the first congress in 1969 and a workshop at the second in 1972 (Fischbein, 1990, p. 4). (As a Study Group of the International Commission on Mathematical Instruction, the PME currently has a membership of some 500 people from 39 countries. See Nesher & Kilpatrick, 1990, for accounts of PME research.) Some British members of the PME met at Chelsea College in March 1978 to form the British Society for the Psychology of Learning Mathematics, whose first newsletter appeared in July 1978; the North American branch of the PME held its first meeting at Northwestern University in September 1979.

Finding a Voice. Discussions of a research journal to be published by the NCTM had occurred sporadically in council committees at least since the formation of the 1956–1957 Research Committee, whose duties were characterized as follows: "To provide research section at convention; to prepare and provide means of collecting and publishing research. To consider establishment of supplement to *Mathematical Reviews* to contain World Review Research in mathematics education" (NCTM, 1956, p. 570). In his charge to the original Research Advisory Committee in 1965, NCTM President Bruce Meserve asked them to look into the question of a research journal. To demonstrate that there was both material and an audience for such a journal, the committee proposed a publication on research and named a special committee headed by Joseph M. Scandura of the University of Pennsylvania to develop it. The resulting booklet (Scandura, 1967) sold over 4,000 copies the first year (Johnson, 1968), which helped demonstrate to the NCTM Board of Directors that a journal would be financially feasible. Several papers in the booklet stressed the importance of *information-oriented* (basic) as well as *product-oriented* (applied) research in education, in part because basic research was seen as being neglected in favor of curriculum development efforts, in part because the distinction between the two kinds of research was on people's minds (Dyer et al., 1956, p. 29; Cronbach & Suppes, 1969, pp. 19–27), and in part because influential people within the NCTM were apparently skeptical about the value of theory-building efforts for teachers. One argument for publishing a research journal that had been raised in NCTM circles was that the articles on research in the council's other journals were too theoretical and too technical for teachers to read. Proponents of the research journal found themselves arguing that basic research was important to do and important for teachers to know about, but that a separate journal was needed to report that research in detail. Many members of the NCTM Board did not think it appropriate for an organization of teachers to publish such a journal.

At its meeting in Las Vegas in April 1967, the Research Advisory Committee decided to submit a formal proposal for a journal to the NCTM Executive Committee. In his report of the meeting, the chairman, Thomas A. Romberg (1967) of the University of Wisconsin, observed, "It is to be understood that getting this journal started has the highest priority of any of the activities of this committee" (p. 1). A proposal went forward, and with the deciding vote cast by NCTM President

Donovan A. Johnson, a divided NCTM Board in April 1968 approved the publication of a "self-supporting" research journal (Johnson, 1968). David C. Johnson of the University of Minnesota was appointed as its first editor. Despite some delays in making financial arrangements, an editorial office was set up and manuscripts were solicited, reviewed, and edited in time for Volume 1, Number 1, of the *Journal for Research in Mathematics Education* to appear in January 1970, the beginning of the NCTM's Golden Jubilee Year.

By that time, other journals that published articles on research had already appeared, although none was devoted to publishing reports of original research. In January 1969, the School Mathematics Study Group at Stanford University produced the first of a series of *Investigations in Mathematics Education*, an occasional journal of abstracts, together with annotated lists of references. In 1972, the journal was reorganized as a quarterly publication of the Center for Science and Mathematics Education at Ohio State University in cooperation with the ERIC Information Analysis Center there. At the Colloquium on Modern Curricula in Mathematical Education conducted by the reconstituted International Commission on Mathematical Instruction in December 1964 in Utrecht, the participants had expressed an "urgent need for more international information on national activities in mathematical education, which could be organized and spread by an active and accessible international center of information or by a high level periodical on mathematical education" (quoted in Freudenthal, 1978, p. 503). The need was met in part by the establishment of the Zentrum für Didaktik der Mathematik at the University of Karlsruhe and, beginning in June 1969, that center's publication of the *Zentralblatt für Didaktik der Mathematik*, a journal containing articles and documentation, in German and English, on the international literature in mathematics education. The need for a periodical was also met by Hans Freudenthal's establishment of *Educational Studies in Mathematics (ESM)*, which began publication in May 1968. The *ESM* publishes some research as part of its effort to report on didactical, methodological, and pedagogical issues and on new developments in the field.

The late 1960s and early 1970s were a fertile time not only for the establishment of journals and information centers but also for the founding of institutes and centers in which research might be carried out. In the United States, 21 research and development centers and 20 regional educational laboratories were established between 1965 and 1967 as a consequence of the Elementary and Secondary Education Act of 1965. Several centers and laboratories conducted research relating to mathematics education. In other countries, institutions appeared that were specifically designed to conduct or promote research in mathematics education (for accounts of the activities of some of these institutions, see Zweng, Green, Kilpatrick, Pollak, & Suydam, 1983, pp. 520–530). These included some 30 Instituts de Recherche pour l'Enseignement de Mathématiques in France (the first begun in 1969), the Shell Centres for Mathematical Education at Chelsea College and the University of Nottingham in England (the Nottingham Centre, begun in 1968 as an inservice education unit, was converted into an interdisciplinary research unit of the University of Nottingham in 1972), and the Institut für Didaktik der Mathematik (IDM), in Bielefeld, Germany (begun in 1973).

Research Takes Off

Introducing a set of articles reviewing 5 years of research in mathematics education Romberg (1969) claimed that over 1,000 studies of mathematics instruction had been reported. To organize the reviews, he used the following categories, which he said reflected "the stimulation of mathematics education by the new courses developed in the early 1960's and the growing interest and involvement of mathematics educators and psychologists in systematic studies of the learning and teaching of mathematics" (p. 473): (a) association learning, (b) activity learning, (c) problem solving and creative behavior, (d) teaching, (e) effectiveness of instructional programs, (f) learner characteristics associated with achievement, (g) attitudes, and (h) evaluation of achievement. He foresaw that research in mathematics education would become more programatic, with more attention given to basic research, the implementation of better curriculum evaluations, and the development of better achievement tests. He characterized the research under review as "large in quantity, poor but improving in quality, and diverse" (p. 473), a sentiment that found echoes in the articles he was introducing.

In the third volume of the series *New Trends in Mathematics Teaching* (Fehr & Glaymann, 1972), one of the chapters dealt with research. Recent trends in research activity were characterized as follows:

The volume of research in mathematics education has increased dramatically over the past two decades. In general, the quality of the published research has also increased, although perhaps not so dramatically. Much of the increase in research activity can be traced to the curriculum reform movement...but some of it [stems] from a more general movement in many countries toward greater expenditures on education and a correspondingly greater demand for evidence that such expenditures have been worthwhile. So-called "basic research" has been encouraged in some centers and research institutes, but the main thrust has been in the realm of "applied research"– chiefly the development and testing of new methods and materials. (pp. 129–130)

In the subsequent volume, Heinrich Bauersfeld (1979, pp. 203, 210), one of the founders of the IDM, remarked that he and his colleagues had located some 3,000 studies of the mathematical learning process in preparation for his report to the Third International Congress on Mathematical Education in Karlsruhe in 1976. He also noted that about 85% of the studies were conducted in the United States. He saw research interest as having shifted from the student and the curriculum to the teacher, with increased attention to real classroom situations and the social context of learning.

The Growing Magnitude of the Research Enterprise. The rapid acceleration in the volume of research that occurred from the mid-1950s to the mid-1970s was not confined to mathematics education, nor was it confined to the United States. Introducing a survey of educational research, policy, and practice around the world, John Nisbet (1985) observed:

It is only within the past 25 years that research in education has received public funding on any substantial scale. Prior to 1960, research was mainly a spare-time amateur affair, unorganized and often ignored until its findings had percolated through into generally accepted values.... In the years 1965–70 public funding of educational research and development expanded at an unprecedented rate. In Britain public expenditure on research in education multiplied ten-fold between 1964 and 1969, while in the USA expenditure doubled each year from 1964 through 1967. Never again can we expect to see such rapid growth. (p. 12)

We can get a picture of the growth in research in mathematics education over a century by combining data from several surveys (see Figure 1.1). Buswell and Judd (1925, p. 3) tallied 307 books, periodical reports, and bulletins on research in arithmetic from before 1892 to 1924. (They reported their data in frequencies over 4-year intervals. In the figure, the average yearly frequency for each interval has been assigned to the midpoint of that interval, beginning in 1890.) Smiler (1970, pp. 494–495) categorized 1,456 studies from 1880 to 1963 and reported categories in 5-year intervals (except for 1880 to 1900, which was put into one interval; data in the figure are average yearly frequencies plotted at the midpoint of the interval). Smiler reported her data as "some of the research in mathematics education which has been done in the United States" (p. 493), whereas Buswell and Judd had included work done outside the United States. Both tabulations, however, are fairly consistent in showing a gradual but steady rise in the number of studies reported up to 1925. That gradual rise continued until around the mid-1950s to mid-1960s, when it began a striking increase. The acceleration can be seen in the data from Suydam's (1967/1968, p. 442) tabulation of research reports on elementary school mathematics published in American journals each year from 1909 to 1965. It can also be seen in the yearly compilations by Suydam from 1955 to 1988 of journal-published articles and dissertations reporting studies at all grades from kindergarten to grade 12. These data are reported separately in Figure 1.1 to allow comparison of the rates of growth in the numbers of articles and dissertations; obviously, if they had been combined the increase would have been even more dramatic. (The data for 1955 to 1975 are taken from Suydam & Osborne, 1977, p. 21; the data from 1976 to 1988 are taken from Suydam's yearly research compilations in the July issues of the *Journal for Research in Mathematics Education*. Dissertations are those referenced in *Dissertation Abstracts* and, after 1969, *Dissertation Abstracts International*.) The growth in the number of articles lagged somewhat behind the growth in the number of dissertations, but both seemed to have reached a peak by 1980. Figure 1.2 shows the same data on dissertations from Suydam's yearly compilations compared with data from Frobisher and Joy's (1978) bibliography of theses and dissertations done in Britain and Ireland from 1910 to 1977. Although not all of Suydam's data are from North America, the overwhelming majority are. One can see the sharp rise occurring a little later in Frobisher and Joy's data than in Suydam's, but it occurs nonetheless.

Much of the growth in dissertations had come from the establishment of new doctoral programs in mathematics education, mostly in the United States but elsewhere as well. In some instances, these programs developed within departments of mathematics, but the great majority of U.S. programs were in schools or colleges of education (McIntosh & Crosswhite, 1973, p. 5). Estimates from a survey of programs suggested that the

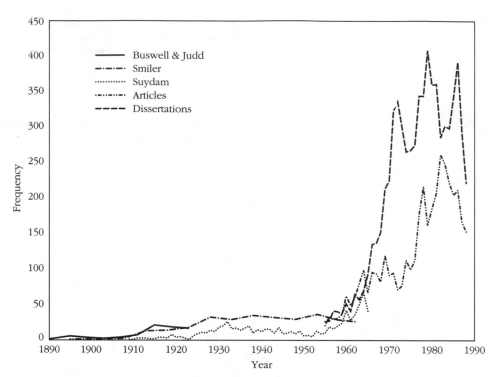

FIGURE 1-1. Research studies in mathematics education from 1890 to 1988.

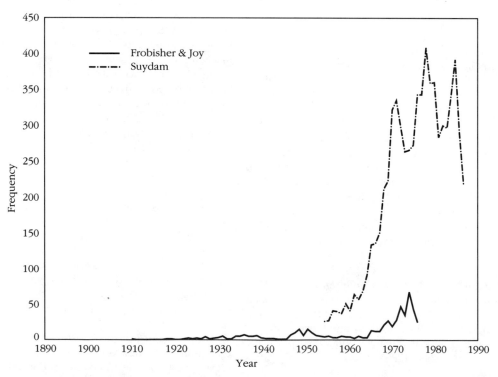

FIGURE 1-2. Theses and dissertations in mathematics education from 1910 to 1988.

number of doctorates in mathematics education awarded in the United States roughly doubled from 1967 to 1973, when about 1,000 students were pursuing studies and some 200 degrees were awarded (McIntosh & Crosswhite, 1973, p. 6). The establishment and expansion of doctoral programs created opportunities for mathematics educators with a interest in research to obtain faculty positions allowing them to not only teach mathematics and train teachers—as mathematics educators had done for years—but also conduct research studies with the aid of graduate students. A few people were able to obtain a full-time research position in a university or in one of the newly formed institutes and centers connected with a university; most, however, relied on special-purpose grants and contracts from foundations or the government to support their research. (For a similar report on the situation in the U.K., see Bishop, 1981.)

Expansion Around the World. In the 1960s and 1970s, research studies in mathematics education grew not only in number but also in scope as researchers increasingly moved across the boundaries of disciplines and countries. A visit to the United States in 1964 by Jean Piaget, with twin conferences in his honor at Cornell University and the University of California at Berkeley on cognitive studies and curriculum development (Ripple & Rockcastle, 1964), helped stimulate Piagetian research among North American mathematics educators. A subsequent conference in October 1970 at Columbia University (Rosskopf, Steffe, & Taback, 1981) brought Hermine Sinclair, one of Piaget's colleagues at Geneva, to the United States; researchers in mathematics education were brought face to face with developmental psychologists. At a joint U.S.–Japan seminar on mathematics education in April 1971 on the campus of the Japanese National Institute for Educational Research in Tokyo (Kawaguchi, 1971), several dozen mathematics educators from the two countries convened. Most of the reports and discussions concerned curriculum change, but a few dealt directly with research activities and the possibilities of collaborative studies. At the Second International Congress on Mathematical Education at Exeter, England in 1972, the working group on the psychology of mathematics learning proved to be "the most popular group in terms of numbers attending" (Howson, 1973, p. 15). The group's discussions led to a call for "periodical meetings (in the form of symposia, seminars, etc.) [to] be organized between psychologists, mathematicians and teachers" (p. 17).

In a paper prepared for a colloquium series at the Wisconsin Research and Development Center for Cognitive Learning, Edward G. Begle (1968) noted that the volume of research in mathematics education was increasing rapidly but also reported "the melancholy information that this large number of research efforts yields very little that can be used to improve mathematics education" (p. 44). Until the 1960s it had been impossible, because of insufficient financial support, to conduct the kind of research needed for comparing curriculum treatments: "either a large experiment or a large number of carefully correlated small ones" (p. 44). He then reported briefly on the National Longitudinal Study of Mathematical Abilities (NLSMA) he was directing that was examining the mathematics attainment of over a hundred thousand students. It had begun gathering data in September 1962 and had tested students in the autumn and spring each year for 5 years. The purpose was to ascertain the effects of the new-math curriculum revision efforts. (For a fuller description of the NLSMA, see Howson et al., 1981, pp. 189–195; for reviews of its findings, see Osborne, 1975.) Although the NLSMA was an exercise in curriculum evaluation and not, by some definitions, a research study, it had a substantial influence on the research enterprise. It brought together psychologists and mathematicians to develop instruments for assessing reasoning ability, the ability to apply mathematics in nonroutine contexts, and attitudes toward various facets of mathematics—instruments that were used in both subsidiary and ensuing research studies. It developed and refined new techniques for analyzing the multiple effects of complex treatments on nonrandomly chosen groups. And it trained quite a few researchers in mathematics education.

Subsequent surveys of mathematical attainment were undertaken not to compare curricula but to describe and contrast levels of performance. These included the mathematics assessments conducted in the United States every 4 or 5 years since 1972–1973 by the National Assessment of Educational Progress (NAEP, 1975; Dossey, Mullis, Lindquist, & Chambers, 1988); annual surveys, beginning in 1978, of mathematical performance by students in England, Wales, and Northern Ireland undertaken by the Assessment of Performance Unit (Foxman, 1985); and a series of tests developed in 1977 to assess students' performance in primary mathematics in Papua New Guinea (Roberts, 1983). (See Carss, 1986, pp. 182–183, for comments on some large-scale surveys and their effects.)

The most comprehensive studies of mathematical attainment have used mathematics as a vehicle for examining cross-national differences in educational systems. These have been studies undertaken by the International Association for the Evaluation of Educational Achievement (IEA), which has conducted studies of several school subjects. The first IEA mathematics study (Husén, 1967) began its data collection in 1964 in 12 countries and reported its findings in 1967. (For American views of the study, see Wilson & Peaker, 1971; for a strongly negative review, see Freudenthal, 1975.) The second IEA mathematics study (Burstein, in press; Robitaille & Garden, 1989; Travers & Westbury, 1989) was more oriented toward questions raised by mathematics educators and featured more detailed analyses of curricula and greater efforts to characterize instruction in the different countries. Data collection began in 1980 in 21 countries, and the international findings were reported in 1989.

The IEA studies, together with the various national surveys of attainment, represented a triumph of psychometrics and technology. Sophisticated procedures for sampling items and students and for equating data from disparate sources allowed researchers to make efficient use of testing time. Increased computer power permitted elaborate calculations with mammoth data bases. The large experiments and large sets of correlated small experiments that Begle wanted to see were not only affordable but feasible.

An Experimental Science

In his address to the First International Congress on Mathematical Education in Lyons in August 1969, Begle had argued that mathematics education needed to follow the pattern of observation and speculation employed by the physical and natural sciences:

We need to follow the procedures used by our colleagues in physics, chemistry, biology, etc. in order to build up a theory of mathematics education....We need to start with extensive, careful, empirical observations of mathematics teaching and mathematics learning. Any regularities noted in these observations will lead to the formulation of hypotheses. These hypotheses can then be checked against further observations, and refined and sharpened, and so on. To slight either the empirical observations or the theory building would be folly. They must be intertwined at all times. (p. 242)

He went on to note that empirical research need not involve large numbers. He cited clinical investigations being conducted in the Soviet Union as sources of hypotheses to be tested using wider selections of students and teachers. The complexity of the phenomena under study—the mind of the child, the mathematics classroom—however, meant that "to restrict ourselves to small scale observations would be to sacrifice the generality of our theories" (p. 243). Begle ended with a statement of his research credo: "Only by becoming more scientific can we achieve the humanitarian goal of improving education for our children and for everyone's children" (p. 243).

Over the next decade, Begle attempted to act on that credo by designing a series of structured teaching units to be used to control some of the variation that occurs when comparing teaching methods. He also began a series of reviews of the empirical literature that culminated in a book issued after his death in March 1978 (Begle, 1979). In that book he classified nearly 7,000 studies according to variables (characteristics of teachers, students, materials, instruction, and so on) that might affect the learning of mathematics and then reviewed the main findings of the studies.

Begle's book appeared at what seemed to be a propitious time for research in mathematics education. The same computers that were enabling the analysis of large sets of empirical data were also enabling cognitive scientists to develop models of thinking, including mathematical thinking. The 1972 publication of Allen Newell and Herbert Simon's *Human Problem Solving* had shifted the attention of researchers in mathematical problem solving to the mental "space" in which a solver represents the problem under consideration. Despite some reservations about the value of cognitive science for mathematics education (Long et al., 1970, p. 451), researchers in mathematics education were attempting to understand and use the cognitive scientist's ideas about mental representations and the processes for acting on those representations. The journal *Cognitive Science* had begun publication in January 1977, and the Cognitive Science Society was founded in 1979 (Gardner, 1985, p. 36). A rapprochement seemed to be building between researchers in mathematics education and cognitive psychologists (Resnick & Ford, 1981, esp. ch. 9). (For a history of the rise of cognitive science, see Gardner; for histories of cognitive psychology, see Madsen, 1988, pp. 530–548; Mandler, 1985; for an example of subsequent collaboration with mathematics education, see Schoenfeld, 1987.)

The Research in Science Education program initiated by the National Science Foundation in 1978 had begun to link together people from different disciplines who were working on research in mathematics education. A joint program of the NSF and the National Institute of Education was underway that aimed at bringing researchers from cognitive science together with their colleagues in mathematics and science education. The National Council of Teachers of Mathematics was about to publish a major reference work on research in mathematics education (Shumway, 1980). A special series on research in Begle's honor was being organized for the Fourth International Congress on Mathematics Education in Berkeley in 1980, and the number of research sections was being increased well beyond those of previous congresses (see Zweng et al., 1983, chs. 12 & 13).

In Europe, researchers were exploring the implications of the activity theory developed in the Soviet Union (Christiansen & Walther, 1986; Hedegaard, Hakkarainen, & Engeström, 1984; Leont'ev, 1981), which takes people's everyday activity in their societally and culturally mediated environment as both the locus of research and the unit for analyzing their mental functioning. European researchers were also pursuing what has been termed the *realistic approach* to learning and teaching (De Corte & Vershaffel, 1986), in which mathematics is viewed as a human activity arising out of real situations and in which students learn by investigating problems they have formulated (e.g., Bell, 1979). In both activity theory and the realistic approach, the social nature of the knowledge learned and of the teaching process used is central. Researchers were taking the social and cultural dimensions of mathematics education more seriously (Bauersfeld, 1980; Bishop, 1988).

Around the world, mathematics education was becoming a university discipline in many countries as teacher education became more professionalized, career patterns like those of the established disciplines began to emerge, the field began to reflect on its own activity, and communication among its members improved (Schubring, 1983). Research in mathematics education seemed poised to become the experimental science Begle and others envisioned. Yet as the 1970s ended, increasing doubts were being expressed about the contribution research was making to the educational process, and some people eventually began to wonder whether educational research could become a science at all.

REAPPRAISAL OF A FIELD

In the same article in which he heralded the golden age of research in mathematics education, Weaver (1967) also observed that questions were being raised about the future of educational research. Weaver was primarily concerned about such matters as insufficient collaboration, overly simple research designs, and inadequate reporting, but it later became clear that other issues were more threatening. In particular, educational research was in danger of having been oversold to governments and funding agencies. When the governments of the United States and other countries increased their financial support of educational research in the 1960s and early 1970s, "expectations were high among both educational research workers and the makers of government policy....Yet the 1970s culminated in sharp criticism, even disenchantment, about education in general and about educational research in particular" (Husén,

1983, p. 90). Educational researchers were viewed as unable to demonstrate that they had made advances that could be translated into practical benefits. In the words of one critic, "Educational research is growing but, in most cases, apparently is going nowhere" (Klitgaard, 1978, p. 353).

Realizing that their work was failing to meet expectations, and that it was being disregarded or disparaged, researchers began to ask whether it might somehow be wrongly directed. At a conference on the relation between theory and practice at the Institut für Didaktik der Mathematik in Bielefeld in December 1976, Decker F. Walker (1977) argued that education may not be capable of becoming a science. Educational research is plagued by uncertainty about its nature, fails to make continuous progress because of faddishness, and suffers from low prestige. "The fact of low prestige has led to a compulsive search for rigor through unassailable methodology that has virtually sterilized research in the field. . . . [It] has also led to a servile imitation of and dependence on established disciplines" (p. 60). Walker claimed that the search for academic respect had led educational researchers away from educational practitioners.

The complexity of educational phenomena that Begle had noted in 1969 appeared to be exacting a toll. Educational researchers were not satisfying the requests of practitioners that they provide useful information, they were not satisfying funding agencies and their colleagues in the scientific disciplines that their work was valuable, and they were not satisfying their own expectations for what research should be. Far from living in a golden age, they seemed to be entering a depression. Referring to the problems confronting American schools as the 1980s began, one commentator put it another way:

Because the problems are so serious and so much more complex than is generally admitted, there may be a certain utility in acknowledging how little we in fact know about schools, how much we may be living in the first days of educational research. (Graubard, 1981, p. v)

Speaking at the SIG/RME research presession to the NCTM annual meeting in Boston in April 1979, Bauersfeld (1980) suggested that the profession of mathematics education was undergoing a "fundamental change of paradigms" (p. 36). He identified three important theses for the transition stage: First, mathematics education lacks a theoretical orientation; second, research and practice are following different paradigms; and third, interdisciplinary approaches are promising, if not essential, for putting research knowledge into practice. These theses captured much of how the crisis in educational research was affecting research in mathematics education.

The absence of good theoretical frameworks may have arisen from pressures to get results quickly and to follow "acceptable" research procedures (Bauersfeld, 1980, p. 37). Certainly, many U.S. researchers felt pressures to publish exten-

sively, which often yielded elaborate studies of minor, sometimes trivial, research questions (Kilpatrick, 1981, p. 373). The courting of scientific respectability by mimicking the research designs of the natural sciences seemed to be leading investigators to conduct studies that were methodologically impeccable but conceptually barren. Whatever the cause, the good ideas needed to propel research in mathematics education forward appeared to be not so much unavailable as missing from the community's discourse.

The actions of a practitioner who interprets classroom events within their contexts could not be further removed from the inferences made by a researcher caught up in controlling variation, quantifying effects, and using statistical models. The end of the 1970s brought serious attempts to reconcile the different approaches of researchers and practitioners by abandoning classical models of hypothesis testing and adopting conceptual frameworks that encouraged researchers to take the practitioner's perspective (Sowder, 1989, pp. 12–19). More and more, research in mathematics education was moving out of the library and laboratory and into the classroom and school.

More importantly, practitioners were increasingly becoming key members of the interdisciplinary groups needed to help research link the complexity of practice to theoretical constructs. The techniques and concepts used by anthropologists, sociologists, linguists, and philosophers proved helpful in that task, but at the center was the teacher, whose experience provided the impulse for research and gave it validity. The 1980s began with the promise of a more fruitful integration of research and practice than at any previous time in the history of mathematics education.

Even though its history is not a long history, research in mathematics education is a conversation that began well before today's researchers appeared and that will continue long after they have gone. It is a conversation with thousands of voices speaking on hundreds of topics. Important issues are discussed, but the commentary is often difficult to understand. Listeners can get impatient and discouraged. It is easier to walk away and dismiss what is being said than to stay and listen and respond. American voices have often dominated the conversation (as they have in the present account), but voices from other places are being heard more frequently. Now and again psychologists and mathematicians have tried to contribute their ideas to the discourse. They have been welcomed into the circle, as have others, when they seemed to be responding to the common interest; they have been ignored when they have been seen as pursuing their exclusive concerns.

People often think the conversation has just begun or that only what they are saying today is worthwhile. They might think differently if they were to study the record of what has been thought and said. It is neither an empty slate nor a chronicle of failure. It is the impressive story of a community defining itself.

References

Austin, C. M. (1921). The National Council of Teachers of Mathematics. *Mathematics Teacher, 14,* 1–4.

Austin, C. M. (1924). An experiment in testing and classifying pupils in beginning algebra. *Mathematics Teacher, 17,* 46–56.

Bain, A. (1879). *Education as a science*. London: Kegan Paul.

Ballard, P. B. (1912). The teaching of mathematics in London public elementary schools. *The teaching of mathematics in the United Kingdom* (Board of Education Report on Educational Subjects, Vol. 26, Pt. 1, pp. 3–20). London: His Majesty's Stationery Office.

Ballard, P. B. (1923). *The new examiner*. London: Hodder & Stoughton.

Bauersfeld, H. (1979). Research related to the learning process. In H.-G. Steiner & B. Christiansen (Eds.), *New trends in mathematics teaching* (Vol. 4, pp. 199–213). Paris: Unesco.

Bauersfeld, H. (1980). Hidden dimensions in the so-called reality of a mathematics classroom. *Educational Studies in Mathematics, 11*, 23–41.

Becker, J. P., & McLeod, G. K. (1967). Teaching, discovery, and the problems of transfer of training in mathematics. In J. M. Scandura (Ed.), *Research in mathematics education* (pp. 93–107). Washington, DC: National Council of Teachers of Mathematics.

Begle, E. G. (1968). Curriculum research in mathematics. In H. J. Klausmeier & G. T. O'Hearn (Eds.), *Research and development toward the improvement of education* (pp. 44–48). Madison, WI: Dembar Educational Research Services.

Begle, E. G. (1969). The role of research in the improvement of mathematics education. *Educational Studies in Mathematics, 2*, 232–244.

Begle, E. G. (1979). *Critical variables in mathematics education: Findings from a survey of the empirical literature*. Washington, DC: Mathematical Association of America & National Council of Teachers of Mathematics.

Bell, A. W. (1979). The learning of process aspects of mathematics. *Educational Studies in Mathematics, 10*, 361–387.

Benezet, L. P. (1935–1936). The story of an experiment. *Journal of the National Education Association, 24*, 241–244, 301–303; *25*, 7–8.

Benz, H. E. (1933). A summary of some scientific investigations of the teaching of high school mathematics. In W. D. Reeve (Ed.), *The teaching of mathematics in the secondary school* (Eighth Yearbook of the National Council of Teachers of Mathematics, pp. 14–54). New York: Columbia University, Teachers College, Bureau of Publications.

Bergeron, J. C., & Herscovics, N. (1990). Psychological aspects of learning early arithmetic. In P. Nesher & J. Kilpatrick (Eds.), *Mathematics and cognition: A research synthesis by the International Group for the Psychology of Mathematics Education* (pp. 31–52). Cambridge: Cambridge University Press.

Binet, A. (1899). La pédagogie scientifique [Scientific pedagogy]. *L'Enseignement Mathématique, 1*, 29–38.

Bishop, A. (1981). Research in mathematics education in the UK. In B. Wilson (Ed.), *Mathematics education: The report of the first Anglo-Soviet Seminar held at St Antony's College, Oxford, September 8-16, 1981, under the auspices of the British Council and the Ministry of Education of the USSR* (pp. 88–91). London: British Council.

Bishop, A. (1988). *Mathematical enculturation: A cultural perspective on mathematics education*. Dordrecht, The Netherlands: Kluwer.

Blair, V. (1923). The present status of "disciplinary values" in education. In National Committee on Mathematical Requirements, *The reorganization of mathematics in secondary education* (pp. 89-104). Oberlin, OH: Mathematical Association of America.

Branford, B. (1908). *A study of mathematical education*. Oxford: Clarendon Press.

Breslich, E. R. (1938). Contributions to secondary-school mathematics. In G. M. Whipple (Ed.), *The scientific movement in education* (37th Yearbook of the National Society for the Study of Education, Pt. 2, pp. 128–134). Bloomington, IL: Public School Publishing Co.

Brigham, W., Coolidge, J. I. T., & Graves, H. A. (1845). [Extracts from the report on the writing schools]. *Common School Journal, 7*, 329–336.

Brown, J. C. (1915). *Curricula in mathematics: A comparison of courses in the countries represented in the International Commission on the Teaching of Mathematics* (U.S. Bureau of Education Bulletin, 1914, No. 45). Washington, DC: U.S. Government Printing Office.

Brown, K. E. (1953, July). *Mathematics education research studies—1952* (Circular No. 377). Washington, DC: U.S. Department of Health, Education, & Welfare, Office of Education.

Brown, K. E. (1954, May). *Mathematics education research studies—1953* (Circular No. 377-II). Washington, DC: U.S. Department of Health, Education, & Welfare, Office of Education.

Brown, K. E., & Abell, T. L. (1965). *Analysis of research in the teaching of mathematics* (OE Publication No. 29007-62). Washington, DC: U.S. Government Printing Office.

Brownell, W. A. (1928). *The development of children's number ideas in the primary grades* (Supplementary Educational Monographs No. 35). Chicago: University of Chicago.

Brownell, W. A. (1930). The techniques of research employed in arithmetic. In G. M. Whipple (Ed.), *Report of the society's committee on arithmetic* (29th Yearbook of the National Society for the Study of Education, pp. 415–443). Chicago: University of Chicago Press.

Brownell, W. A. (1935). Psychological considerations in the learning and the teaching of arithmetic. In W. D. Reeve (Ed.), *The teaching of arithmetic* (10th Yearbook of the National Council of Teachers of Mathematics, pp. 1–31). New York: Columbia University, Teachers College, Bureau of Publications.

Brownell, W. A. (1938). A critique of the Committee of Seven's investigations on the grade placement of arithmetic topics. *Elementary School Journal, 38*, 495–508.

Brownell, W. A. (1945). When is arithmetic meaningful? *Journal of Educational Research, 38*, 481–498.

Brownell, W. A. (1948). Learning theory and educational practice. *Journal of Educational Research, 41*, 481–497.

Brownell, W. A., & Moser, H. E. (1949). *Meaningful vs. mechanical learning: A study in Grade III subtraction* (Duke University Research Studies in Education No. 8). Durham, NC: Duke University Press.

Brueckner, L. J. (1930). A critique of the yearbook. In G. M. Whipple (Ed.), *Report of the society's committee on arithmetic* (29th Yearbook of the National Society for the Study of Education, pp. 681–709). Chicago: University of Chicago Press.

Brueckner, L. J. (1939). The development of ability in arithmetic. In G. M. Whipple (Ed.), *Child development and the curriculum* (38th Yearbook of the National Society for the Study of Education, Pt. 1, pp. 275–298). Bloomington, IL: Public School Publishing Co.

Buckingham, B. R. (1930). The social value of arithmetic. In G. M. Whipple (Ed.), *Report of the society's committee on arithmetic* (29th Yearbook of the National Society for the Study of Education, pp. 9–62). Chicago: University of Chicago Press.

Burstein, L. (Ed.). (in press). *The IEA Study of Mathematics III: Student growth and classroom processes in lower secondary school*. Oxford: Pergamon.

Buswell, G. T. (1930). A critical survey of previous research in arithmetic. In G. M. Whipple (Ed.), *Report of the society's committee on arithmetic* (29th Yearbook of the National Society for the Study of Education, pp. 445–470). Chicago: University of Chicago Press.

Buswell, G. T. (1938). Contributions to elementary-school mathematics. In G. M. Whipple (Ed.), *The scientific movement in education* (37th Yearbook of the National Society for the Study of Education, Pt. 2, pp. 123–128). Bloomington, IL: Public School Publishing Co.

Buswell, G. T. (1951). Needed research on arithmetic. In N. B. Henry (Ed.), *The teaching of arithmetic* (50th Yearbook of the National Society for the Study of Education, Pt. 2, pp. 282–297). Chicago: University of Chicago Press.

Buswell, G. T., & Judd, C. H. (1925). Summary of educational investigations relating to arithmetic (Supplementary Educational Monographs No. 27). Chicago: University of Chicago.

Butts, R. F. (1947). *A cultural history of education*. New York: McGraw-Hill.

Caldwell, O. W., & Courtis, S. A. (1925). *Then and now in education, 1845–1923: A message of encouragement from the past to the present*. Yonkers-on-Hudson, NY: World Book.

Callahan, L. G., & Glennon, V. J. (1975). *Elementary school mathematics: A guide to current research* (4th ed.). Washington, DC: Association for Supervision and Curriculum Development.

Campbell, D. T., & Stanley, J. C. (1963). *Experimental and quasi-experimental designs for research*. Chicago: Rand McNally.

Carr, W., & Kemmis, S. (1986). *Becoming critical: Education, knowledge and action research*. London: Falmer Press.

Carroll, J. B. (1978). On the theory-practice interface in the measurement of intellectual abilities. In P. Suppes (Ed.), *Impact of research on education: Some case studies* (pp. 1–105). Washington, DC: National Academy of Education.

Carss, M. (Ed.). (1986). *Proceedings of the Fifth International Congress on Mathematical Education*. Boston: Birkhäuser.

Chamberlin, D., Chamberlin, E., Drought, N. E., & Scott, W. E. (1942). *Did they succeed in college?* (Adventure in American Education Vol. 4). New York: Harper.

Christiansen, B., & Walther, G. (1986). Task and activity. In B. Christiansen, A. G. Howson, & M. Otte (Eds.), *Perspectives on mathematics education* (pp. 243–307). Dordrecht, The Netherlands: Reidel.

Claparède, E. (1911). *Experimental pedagogy and the psychology of the child* (M. Louch & H. Holman, Trans.). New York: Longmans, Green.

Clason, R. G. (1969). Number concepts in arithmetic texts of the United States from 1880 to 1966, with related psychological and mathematical developments. *Dissertation Abstracts International, 30*, 146A. (University Microfilms No. 69–12,075)

Clements, M. A. (1989). *Mathematics for the minority: Some historical perspectives of school mathematics in Victoria*. Geelong, Victoria: Deakin University Press.

Clifford, G. J. (1984). *Edward L. Thorndike: The sane positivist*. Middletown, CT: Wesleyan University Press. (Original work published 1968).

Cooper, B. (1985). *Renegotiating secondary school mathematics: A study of curriculum change and stability* (Studies in Curriculum History 3). London: Falmer.

Courtis, S. A. (1909–1911). Measurement of growth and efficiency in arithmetic. *Elementary School Teacher, 10*, 58–74, 177–199; *11*, 171–185, 360–370, 528–539.

Courtis, S. A. (1913). The Courtis tests in arithmetic. *Report on educational aspects of the public school system of the City of New York to the Committee on School Inquiry of the Board of Estimate and Apportionment* (Pt. 2, Subdiv. 1, Sect. D). New York: City of New York.

Cremin, L. A. (1953). The heritage of American teacher education. *Journal of Teacher Education, 4*, 163–170, 246–250.

Cronbach, L. J. (1965). Issues current in educational psychology. *Monographs of the Society for Research in Child Development, 30*(1, Serial No. 99), pp. 109–126.

Cronbach, L. J., & Suppes, P. (Eds.). (1969). *Research for tomorrow's schools: Disciplined inquiry for education*. New York: Macmillan.

Cubberley, E. P. (1920). *The history of education*. Boston: Houghton Mifflin.

Davis, P. J., & Hersh, R. (1981). *The mathematical experience*. Boston: Birkhäuser.

De Corte, E., & Vershaffel, L. (1986, April). *Research on the teaching and learning of mathematics: Some remarks from a European perspective*. Paper presented at the meeting of the American Educational Research Association, San Francisco. (ERIC Document Reproduction Service No. ED 275 535)

DeVault, M. V., & Weaver, J. F. (1970). Forces and issues related to curriculum and instruction, K–6. In P. S. Jones (Ed.), *A history of mathematics education in the United States and Canada* (32nd Yearbook of the National Council of Teachers of Mathematics, pp. 91–152). Washington, DC: National Council of Teachers of Mathematics.

Dewey, J. (1898). Some remarks on the psychology of number. *Pedagogical Seminary, 5*, 426–434.

Dossey, J. A., Mullis, I. V. S., Lindquist, M. M., & Chambers, D. L. (1988, June). *The mathematics report card: Are we measuring up?* (Report No. 17-M-01). Princeton, NJ: Educational Testing Service.

DuBois, P. H. (1965). A test-dominated society: China, 1115 B.C.–1905 A.D. In C. W. Harris (Ed.), *Proceedings of the 1964 Invitational Conference on Testing Problems* (pp. 3–11). Princeton, NJ: Educational Testing Service.

DuBois, P. H., & Feierabend, R. L. (Eds.). (1960). *Research problems in mathematics education* (Cooperative Research Monograph No. 3, OE-12008). Washington, DC: U.S. Government Printing Office.

Duncker, K. (1945). On problem-solving. *Psychological Monographs, 58*(5, Whole No. 270).

Dyer, H. S., Kalin, R., & Lord, F. M. (1956). *Problems in mathematical education*. Princeton, NJ: Educational Testing Service.

Edgeworth, F. Y. (1888). The statistics of examinations. *Journal of the Royal Statistical Society, 51*, 599–635.

Edgeworth, F. Y. (1890). The element of chance in competitive examinations. *Journal of the Royal Statistical Society, 53*, 460–475, 644–663.

Educational Testing Service. (1955, October). *A survey of mathematical education: The causes of student dropout, failure, and incompetence at the elementary and secondary levels*. Princeton, NJ: Author.

Eisenhart, M. A. (1988). The ethnographic research tradition and mathematics education research. *Journal for Research in Mathematics Education, 16*, 324–336.

Enquête sur la méthode de travail des mathématiciens [Inquiry into the working methods of mathematicians]. (1902). *L'Enseignement Mathématique, 4*, 208–211.

Everett, J. P. (1928). *The fundamental skills of algebra* (Teachers College Contributions to Education No. 324). New York: Columbia University, Teachers College.

Fauvel, J. (1989). Platonic rhetoric in distance learning: How Robert Record taught the home learner. *For the Learning of Mathematics, 9*(1), 2–6.

Fawcett, H. P. (1938). *The nature of proof* (13th Yearbook of the National Council of Teachers of Mathematics). Washington, DC: National Council of Teachers of Mathematics.

Fehr, H. (1904). Les études mathématiques à l'école normale de l'Université Columbia de New-York [Mathematical studies at the normal school of New York's Columbia University]. *L'Enseignement Mathématique, 6*, 313–316.

Fehr, H. (1905). L'enquête de "L'Enseignement Mathématique" sur la méthode de travail des mathématiciens ["L'Enseignement Mathématique"'s inquiry into the working methods of mathematicians]. In A. Krazer (Ed.), *Verhandlungen des Dritten Internationalen Mathematiker- Kongresses in Heidelberg vom 8. bis 13. August 1904* (pp. 603–607). Leipzig, Germany: B. G. Taubner.

Fehr, H. (1913). Liste des publications du comité central et des sous-commissions nationales [List of publications of the central committee and the national subcommissions]. In E. W. Hobson & A. E. H. Love (Eds.), *Proceedings of the Fifth International Congress of Mathematicians (Cambridge, 22–28 August 1912)* (Vol. 2, pp. 642–653). Cambridge: Cambridge University Press.

Fehr, H., Flournoy, T., & Claparède, E. (1908). [*Enquête de "L'Enseignement Mathématique" sur la méthode de travail des mathématiciens* ["L'Enseignement Mathématique"'s inquiry into the working methods of mathematicians]. Paris: Gauthier-Villars; Geneva: Georg & Cie.

Fehr, H. F. (1953). Theories of learning related to the field of mathematics. In H. F. Fehr (Ed.), *The learning of mathematics: Its theory and practice* (21st Yearbook of the National Council of Teachers of Mathematics, pp. 1–41). Washington, DC: National Council of Teachers of Mathematics.

Fehr, H. F. (1955). *Teaching high-school mathematics* (What research says to the teacher, No. 9). Washington, DC: National Education Association.

Fehr, H. F. (Ed.). (1966). *Needed research in mathematical education*. New York: Teachers College Press.

Fehr, H. F., & Glaymann, M. (Eds.). (1972). *New trends in mathematics teaching* (Vol. 3). Paris: Unesco.

Fischbein, E. (1990). Introduction. In P. Nesher & J. Kilpatrick (Eds.), *Mathematics and cognition: A research synthesis by the International Group for the Psychology of Mathematics Education* (pp. 1–13). Cambridge: Cambridge University Press.

Flavell, J. H. (1963). *The developmental psychology of Jean Piaget*. Princeton, NJ: D. Van Nostrand.

Foxman, D. (1985). Practical and oral mathematics assessments in monitoring surveys in the schools of England, Wales and Northern Ireland. In A. Bell, B. Low, & J. Kilpatrick (Eds.), *Theory, research and practice in mathematical education: Working group reports and collected papers* (pp. 423–434). Nottingham, UK: Shell Centre for Mathematical Education.

Freudenthal, H. (1973). *Mathematics as an educational task*. Dordrecht, The Netherlands: Reidel.

Freudenthal, H. (1975). Pupils' achievements internationally compared—The IEA. *Educational Studies in Mathematics, 6,* 127–186.

Freudenthal, H. (1978). Acknowledgment. *Educational Studies in Mathematics, 9,* 503–504.

Friedman, L. (1989). Mathematics and the gender gap: A meta-analysis of recent studies on sex differences in mathematical tasks. *Review of Educational Research, 59,* 185–213.

Frijda, N. H., & de Groot, A. D. (Eds.). (1982). *Otto Selz: His contribution to psychology*. The Hague: Mouton.

Frobisher, L. J., & Joy, R. R. (1978). *Mathematical education: A bibliography of theses and dissertations*. Leeds, England: Mathsed Press.

Fugisawa, R. (1912). *Summary report on the teaching of mathematics in Japan*. Tokyo: Ministry of Education.

Gagné, R. M. (1968). Contributions of learning to human development. *Psychological Review, 75,* 177–191.

Gardner, H. (1985). *The mind's new science: A history of the cognitive revolution*. New York: Basic Books.

Gerner, B. (1968). *Otto Willmann im alter* [Otto Willmann in old age]. Ratingen, Germany: A. Henn.

Glennon, V. J. (1966). Research needs in elementary school mathematics education. *Arithmetic Teacher, 13,* 363–368.

Glennon, V. J., & Callahan, L. G. (1968). *Elementary school mathematics: A guide to current research* (3rd ed.). Washington, DC: Association for Supervision and Curriculum Development.

Goldberg, J. G. (1978). Psychological research into mathematics learning and teaching in the U.S.S.R. and Eastern Europe. In F. J. Swetz (Ed.), *Socialist mathematics education* (pp. 371–406). Southampton, PA: Burgundy Press.

Gould, S. J. (1981). *The mismeasure of man*. New York: Norton.

Grattan-Guinness, I. (1988). Grandes écoles, petite université: Some puzzled remarks on higher education in mathematics in France, 1795-1840. *History of Universities, 7,* 197–225.

Graubard, S. R. (1981). Preface to the issue, "America's schools: Public and private." *Daedalus, 110*(3), v–xxiv.

Hadamard, J. (1954). *An essay on the psychology of invention in the mathematical field*. New York: Dover. (Original work published 1945)

Hall, G. S. (1891). The contents of children's minds on entering school. *Pedagogical Seminary, 1,* 139–173.

Hamilton, D. (1980). Educational research and the shadows of Francis Galton and Ronald Fisher. In W. B. Dockrell & D. Hamilton (Eds.), *Rethinking educational research* (pp. 153–168). London: Hodder & Stoughton.

Harap, H., & Mapes, C. E. (1934). The learning of fundamentals in an arithmetic activity program. *Elementary School Journal, 34,* 515–525.

Harap, H., & Mapes, C. E. (1936). The learning of decimals in an arithmetic activity program. *Journal of Educational Research, 29,* 686–693.

Harris, W. T., Draper, A. S., & Tarbell, H. S. (1895). *Report of the Committee of Fifteen on Elementary Education*. Boston: New England Publishing Co.

Havighurst, R. J. (1937). Can mathematicians and educationists cooperate? *Mathematics Teacher, 30,* 211–213.

Hedegaard, M., Hakkarainen, P., & Engeström, Y. (Eds.). (1984). *Learning and teaching on a scientific basis: Methodological and epistemological aspects of the activity theory of learning and teaching*. Aarhus, Denmark: Aarhus Universitet, Psykologisk Institut.

Hedrick, E. R. (1932). Desirable cooperation between educationists and mathematicians. *School and Society, 36,* 769–777.

Howson, A. G. (Ed.). (1973). *Developments in mathematical education: Proceedings of the Second International Congress on Mathematical Education*. Cambridge: Cambridge University Press.

Howson, A. G. (1982). *A history of mathematics education in England*. Cambridge: Cambridge University Press.

Howson, A. G. (1984). Seventy five years of the International Commission on Mathematical Instruction. *Educational Studies in Mathematics, 15,* 75–93.

Howson, A. G. (1988). Research in mathematics education. *Mathematical Gazette, 72,* 265–271.

Howson, G., Keitel, C., & Kilpatrick, J. (1981). *Curriculum development in mathematics*. Cambridge: Cambridge University Press.

Hooten, J. R., Jr. (Ed.). (1967). Proceedings of the National Conference on Needed Research in Mathematics Education [Special issue]. *Journal of Research and Development in Education, 1*(1).

Hooten, J. R., Jr., & Wagner, J. (1967). Preface. *Journal of Research and Development in Education, 1*(1), 1.

Høyrup, J. (1980). Influences of institutionalized mathematics teaching on the development and organization of mathematical thought in the pre-modern period. Investigations in an aspect of the anthropology of mathematics. *Studien zum Zusammenhang von Wissenschaft und Bildung* (Materialien und Studien, Band 20, pp. 7–137). Bielefeld, Germany: Universität Bielefeld, Institut für Didakdik der Mathematik.

Husén, T. (Ed.). (1967). *International Study of Achievement in Mathematics: A comparison of twelve countries* (Vols. 1 & 2). New York: Wiley.

Husén, T. (1983). Educational research and the making of policy in education: An international perspective. *Minerva, 21,*81–100.

International Commission on the Teaching of Mathematics. (1912). *Report of the American commissioners of the International Commission on the Teaching of Mathematics* (U.S. Bureau of Education Bulletin, 1912, No. 14). Washington, DC: U.S. Government Printing Office.

Jackson, A. (1988). Research mathematicians in mathematics education. *Notices of the American Mathematical Society, 35,* 790–794, 1123–1131.

Jahnke, H. N. (1986). Origins of school mathematics in early nineteenth-century Germany. *Journal of Curriculum Studies, 18,* 85–94.

Johnson, D. A. (1968). President's report: The state of the council, 1967/68. *Arithmetic Teacher, 15,* 571–575.

Jones, G. A. (1984, August). Research in mathematics education in Australia: MERGA provides a new awakening. In J. Briggs (Ed.), *Summary of research in mathematics education in Australia* (pp. 3–5). Brisbane: Brisbane College of Advanced Education.

Jones, P. S., & Coxford, A. F., Jr. (1970). Mathematics in the evolving schools. In P. S. Jones (Ed.), *A history of mathematics education in the United States and Canada* (32nd Yearbook of the National Council of Teachers of Mathematics, pp. 9–89). Washington, DC: National Council of Teachers of Mathematics.

Judd, C. H. (1908). The relation of special training to general intelligence. *Educational Review, 36,* 28–42.

Judd, C. H. (1927). *Psychological analysis of the fundamentals of arithmetic* (Supplementary Educational Monographs No. 32). Chicago: University of Chicago.

Judd, C. H. (1928). The fallacy of treating school subjects as "tool subjects." In J. R. Clark & W. D. Reeve (Eds.), *Selected topics in the teaching of mathematics* (Third Yearbook of the National Council of Teachers of Mathematics, pp. 1–10). New York: Columbia University, Teachers College, Bureau of Publications.

Judd, C. H. (1936). *Education as cultivation of the higher mental processes.* New York: Macmillan.

Kahane, J. P. (1988, December). ICMI and recent developments in mathematical education. In B. Winkelmann (Ed.), *Wissenschaftliches Kolloquium Hans-Georg Steiner zu Ehren* (Occasional Paper No. 116, pp. 13–23). Bielefeld, Germany: Universität Bielefeld, Institut für Didakdik der Mathematik.

Kantowski, M. G. (1979). Another view of the value of studying mathematics education research and development in the Soviet Union. In M. N. Suydam (Ed.), *An analysis of mathematics education in the Union of Soviet Socialist Republics* (pp. 130–147). Columbus, OH: ERIC Clearinghouse for Science, Mathematics, and Environmental Education.

Kawaguchi, T. (Ed.). (1971). Reports of U.S.–Japan Seminar on Mathematics Education [Special issue]. *Journal of Japan Society of Mathematical Education, 53*(Suppl. issue).

Keller, A. (1985). Mathematics, mechanics and the origin of the culture of mechanical invention. *Minerva, 23,* 348–361.

Kilpatrick, J. (1981). Research on mathematical learning and thinking in the United States. *Recherches en Didactique des Mathématiques, 2,* 363–379.

Kilpatrick, J. (1988a). Change and stability in research in mathematics education. *Zentralblatt für Didaktik der Mathematik, 20,* 202–204.

Kilpatrick, J. (1988b). Educational research: Scientific or political? *Australian Educational Researcher, 15,* 13–28.

Kilpatrick, J., & Weaver, J. F. (1977). The place of William A. Brownell in mathematics education. *Journal for Research in Mathematics Education, 8,* 382–384.

Kilpatrick, J., & Wirszup, I. (Eds.). (1972). *Instruction in problem solving* (Soviet Studies in the Psychology of Learning and Teaching Mathematics, Vol. 6). Stanford, CA: School Mathematics Study Group.

Kinsella, J. J. (1952). The problem of locating research studies in mathematics education. *Mathematics Teacher, 45,* 273–275.

Klitgaard, R. E. (1978). Justifying basic research in education [Review of *Fundamental research and the process of education*]. *Minerva, 16,* 348–353.

Knight, F. B. (1930). Introduction. In G. M. Whipple (Ed.), *Report of the society's committee on arithmetic* (29th Yearbook of the National Society for the Study of Education, pp. 1–7). Chicago: University of Chicago Press.

Krutetskii, V. A. (1976). *The psychology of mathematical abilities in schoolchildren* (J. Teller, Trans.). Chicago: University of Chicago Press. (Original work published 1968)

Leont'ev, A. N. (1981). The problem of activity in psychology. In J. V. Wertsch (Ed.), *The concept of activity in Soviet psychology* (pp. 37–71). Armonk, NY: M. E. Sharpe.

Long, R. S., Meltzer, N. S., & Hilton, P. J. (1970). Research in mathematics education. *Educational Studies in Mathematics, 2,* 446–468.

Madaus, G. F., & Stufflebeam, D. L. (Eds.). (1989). *Educational evaluation: Classic works of Ralph W. Tyler.* Boston: Kluwer.

Madsen, K. B. (1988). *A history of psychology in metascientific perspective.* Amsterdam: North-Holland.

Mandler, G. (1985). *Cognitive psychology: An essay in cognitive science.* Hillsdale, NJ: Lawrence Erlbaum.

Mann, H. (1845a). [Unsigned commentary on extracts from the report on the grammar schools]. *Common School Journal, 7,* 329–336.

Mann, H. (1845b). [Unsigned commentary on extracts from the report on the writing schools]. *Common School Journal, 7,* 344–368.

Marquis, A. N. (Ed.). (1936). *Who's who in America, Vol 19 (1936–1937).* Chicago: A. N. Marquis Co.

Mayer, M. (1961). *The schools.* New York: Harper & Bros.

McIntosh, J. A., & Crosswhite, F. J. (1973). *A survey of doctoral programs in mathematics education.* Columbus, OH: ERIC Information Analysis Center for Science, Mathematics and Environmental Education. (ERIC Document Reproduction Service No. ED 091 250)

McLellan, J. A., & Dewey, J. (1895). *The psychology of number and its applications to methods of teaching arithmetic.* New York: D. Appleton.

McMurry, F. M. (1904). What omissions are advisable in the present course of study, and what should be the basis for the same. *Proceedings and Addresses of the 43rd Annual Meeting of the National Education Association, 42,* 194–202.

Menchinskaya, N. A. (1969). Fifty years of Soviet instructional psychology. In J. Kilpatrick & I. Wirszup (Eds.), *The learning of mathematical concepts* (Soviet Studies in the Psychology of Learning and Teaching Mathematics, Vol. 1, pp. 5–27). Stanford, CA: School Mathematics Study Group. (Original work published 1967)

Möbius, P. J. (1900). *Über die unlage zur mathematik* [On mathematical ability]. Leipzig, Germany: J. A. Barth.

Monroe, W. S. (1917). A preliminary report of an investigation of the economy of time in arithmetic. In G. M. Whipple (Ed.), *Second report of the Committee on Minimal Essentials in Elementary-School Subjects* (16th Yearbook of the National Society for the Study of Education, Pt. 1, pp. 111–127). Bloomington, IL: Public School Publishing Co.

Moon, B. (1986). *The "new maths" curriculum controversy: An international story* (Studies in Curriculum History 5). London: Falmer.

Moore, E. H. (1967). On the foundations of mathematics. *Mathematics Teacher, 60,* 360–374. (Original work published 1903)

Morrisett, L. N., & Vinsonhaler, J. (Eds.). (1965). Mathematical learning [Special issue]. *Monographs of the Society for Research in Child Development, 30*(1, Serial No. 99).

Morton, R. L. (1953). *Teaching arithmetic* (What research says to the teacher, No. 2). Washington, DC: National Education Association.

Murray, J. (1939). Analysis of replies to a questionnaire in regard to certain processes and topics in the teaching of arithmetic. In *Studies in arithmetic* (Vol 1., pp. 11–78). London: University of London Press.

Napoli, D. S. (1981). *Architects of adjustment: The history of the psychological profession in the United States.* Port Washington, NY: Kennikat Press.

National Assessment of Educational Progress. (1975, October). *The first national assessment of mathematics* (Mathematics Report No. 04-MA-00). Washington, DC: U.S. Government Printing Office.

National Council of Teachers of Mathematics. (1956). NCTM committees (1956-57). *Mathematics Teacher, 49,* 567–570.

Nesher, P., & Kilpatrick, J. (Eds.). (1990). *Mathematics and cognition: A research synthesis by the International Group for the Psychology of Mathematics Education.* Cambridge: Cambridge University Press.

Newell, A., & Simon, H. A. (1972). *Human problem solving.* Englewood Cliffs, NJ: Prentice-Hall.

Newman, D., Griffin, P., & Cole, M. (1989). *The construction zone: Working for cognitive change in school.* Cambridge: Cambridge University Press.

Nisbet, J. (1985). Introduction. In J. Nisbet, J. Megarry, & S. Nisbet (Eds.), World yearbook of education 1985: Research, policy and practice (pp. 9–16). London: Kogan Page.

Nisbet, J. D., & Entwistle, N. J. (1973). The psychologist's contribution to educational research. In W. Taylor (Ed.), Research perspectives in education (pp. 113–120). London: Routledge & Kegan Paul.

O'Donnell, J. M. (1985). *The origins of behaviorism: American psychology, 1870–1920.* New York: New York University Press.

Orata, P. T. (1935). Transfer of training and educational pseudo-science. *Mathematics Teacher, 28,* 265–289.

Orata, P. T. (1937). Transfer of training and reconstruction of experience. *Mathematics Teacher, 30,* 99–109.

Osborne, A. R. (Ed.). (1975). Critical analyses of the NLSMA Reports [Special issue]. *Investigations in Mathematics Education, 8*(3).

Osborne, A. R., & Crosswhite, F. J. (1970). Forces and issues related to curriculum and instruction, 7-12. In P. S. Jones (Ed.), *A history of mathematics education in the United States and Canada* (32nd Yearbook of the National Council of Teachers of Mathematics, pp. 155–297). Washington, DC: National Council of Teachers of Mathematics.

Parsons, T., Howe, S. G., & Neale, R. H. (1845). Extracts from the report on the grammar schools. *Common School Journal, 7,* 290-326.

Peel, E. A. (1971). Psychological and educational research bearing on mathematics teaching. In W. Servais & T. Varga (Eds.), *Teaching school mathematics* (pp. 151–177). Harmondsworth, England: Penguin.

Perry, J. (Ed.). (1901). *Discussion on the teaching of mathematics.* London: Macmillan.

Phillips, D. E. (1897). Number and its application psychologically considered. *Pedagogical Seminary, 5,* 221–281.

Piaget, J. (1952). *The child's conception of number* (C. Gattegno & F. M. Hodgson, Trans.) London: Routledge & Kegan Paul. (Original work published 1941)

Poincaré, H. (1899). La logique et l'intuition dans la science mathématique et dans l'enseignement [Logic and intuition in the science of mathematics and in teaching]. *L'Enseignement Mathématique, 1,* 157–162.

Poincaré, H. (1952). *Science and method* (F. Maitland, Trans.). New York: Dover.

Popkewitz, T. S. (1984). *Paradigm and ideology in educational research: The social functions of the intellectual.* London: Falmer Press.

Progressive Education Association, Commission on the Secondary School Curriculum, Committee on the Function of Mathematics in General Education. (1940). *Mathematics in general education.* New York: Appleton-Century.

Pyenson, L. (1983). *Neohumanism and the persistence of pure mathematics in Wilhelmian Germany* (Memoirs Vol. 150). Philadelphia: American Philosophical Society.

Pyenson, L. (1985). *Cultural imperialism and exact sciences: German expansion overseas 1900–1930.* New York: Lang.

Reeve, W. D. (1929). United States. In W. D. Reeve (Ed.), *Significant changes and trends in the teaching of mathematics throughout the world since 1910* (Fourth Yearbook of the National Council of Teachers of Mathematics, pp. 131–186). New York: Columbia University, Teachers College, Bureau of Publications.

Reeve, W. D. (Ed.). (1940). *The place of mathematics in secondary education* (15th Yearbook of the National Council of Teachers of Mathematics). New York: Columbia University, Teachers College, Bureau of Publications.

Resnick, L. B., & Ford, W. W. (1981). *The psychology of mathematics for instruction.* Hillsdale, NJ: Lawrence Erlbaum.

Rice, J. M. (1902). Educational research: A test in arithmetic. *Forum, 34,* 281–295.

Rice, J. M. (1903). Educational research: Causes of success and failure in arithmetic. *Forum, 34,* 436–452.

Ripple, R. E., & Rockcastle, V. N. (Eds.). (1964). Piaget rediscovered: Selected papers from a report of the Conference on Cognitive Studies in Curriculum Development, March 1964 [Special issue]. *Journal of Research in Science Teaching, 2*(3).

Roberts, R. E. (1983). National assessment of mathematics achievement in Papua New Guinea. In M. Zweng, T. Green, J. Kilpatrick, H. Pollak, & M. Suydam (Eds.). *Proceedings of the Fourth International Congress on Mathematical Education* (pp. 553–556). Boston: Birkhäuser.

Robitaille, D. F., & Garden, R. A. (Eds.). (1989). *The IEA Study of Mathematics II: Contexts and outcomes of school mathematics.* Oxford: Pergamon.

Rogers, A. L. (1919). Tests of mathematical ability—Their scope and significance. *Mathematics Teacher, 11,* 145–164.

Romberg, T. A. (1967). *Report of the meeting of the Research Advisory Committee, April 21–22, 1967, Las Vegas, Nevada.* Unpublished manuscript.

Romberg, T. A. (1968). *Report of the meeting of the Research Advisory Committee, February 6–7, 1968, Chicago, Illinois.* Unpublished manuscript.

Romberg, T. A. (1969). Current research in mathematics education. *Review of Educational Research, 39,* 473–491.

Rosskopf, M. F. (1953). Transfer of training. In H. F. Fehr (Ed.), *The learning of mathematics: Its theory and practice* (21st Yearbook of the National Council of Teachers of Mathematics, pp. 205–227). Washington, DC: National Council of Teachers of Mathematics.

Rosskopf, M. F., Steffe, L. P., & Taback, S. (Eds.). (1981). *Piagetian cognitive-development research and mathematical education.* Washington, DC: National Council of Teachers of Mathematics.

Rotman, B. (1977). *Jean Piaget: Psychologist of the real.* Ithaca, NY: Cornell University Press.

Rowe, D. E. (1983). A forgotten chapter in the history of Felix Klein's *Erlanger Programm. Historia Mathematica, 10,* 448–457.

Rowe, D. E. (1985a). Essay review [Review of *Felix Klein; Universität, technische hochschule und industrie; & Neohumanism and the persistence of pure mathematics in Wilhelmian Germany*]. *Historica Mathematica, 12,* 278–291.

Rowe, D. E. (1985b). Felix Klein's *"Erlanger Antrittsrede"*: A transcription with English translation and commentary. *Historica Mathematica, 12,* 123–141.

Rugg, H. (1924). Curriculum making: What shall constitute the procedure of national committees? *Mathematics Teacher, 17,* 1–21.

Scandura, J. M. (Ed.). (1967). *Research in mathematics education.* Washington, DC: National Council of Teachers of Mathematics.

Schoenfeld, A. H. (Ed.). (1987). *Cognitive science and mathematics education.* Hillsdale, NJ: Lawrence Erlbaum.

Schonell, F. J., & Schonell, F. E. (1957). *Diagnosis and remedial teaching in arithmetic.* Edinburgh: Oliver & Boyd.

Schorling, R. (1923). Experimental courses in secondary school mathematics. In National Committee on Mathematical Requirements, *The reorganization of mathematics in secondary education* (pp. 177–278). Oberlin, OH: Mathematical Association of America.

Schorling, R. (1925). *A tentative list of objectives in the teaching of junior high school mathematics, with investigations for the determining of their validity.* Ann Arbor, MI: George Wahr.

Schubring, G. (1983). Comparative study of the development of mathematics education as a professional discipline in different countries. In M. Zweng, T. Green, J. Kilpatrick, H. Pollak, & M. Suydam (Eds.). *Proceedings of the Fourth International Congress on Mathematical Education* (pp. 482–484). Boston: Birkhäuser.

Schubring, G. (1987). On the methodology of analysing historical textbooks: Lacroix as textbook author. *For the Learning of Mathematics, 7*(3), 41–51.

Schubring, G. (1988a, February). *The cross-cultural "transmission" of concepts—The first international mathematics curricular reform around 1900, with an appendix on the biography of F. Klein* (Occasional Paper No. 92, corrected ed.). Bielefeld, Germany: Universität Bielefeld, Institut für Didakdik der Mathematik.

Schubring, G. (1988b). Factors determining theoretical developments of mathematics education as a discipline–Comparative historical studies of its institutional and social contexts. In H.-G. Steiner & A. Vermandel (Eds.), *Foundations and methodology of the discipline mathematics education (didactics of mathematics)* (Proceedings of the 2nd TME-Conference, pp. 161–173). Bielefeld, Germany & Antwerp, Belgium: University of Bielefeld & University of Antwerp.

Selleck, R. J. W. (1968). *The new education: 1870–1914*. London: Pitman.

Senk, S. L. (1981). *The development of a unified mathematics curriculum in the University High School: 1903–1923*. Unpublished manuscript, University of Chicago, Department of Education.

Shils, E. (1978). The order of learning in the United States from 1865 to 1920: The ascendancy of the universities. *Minerva, 16,* 159–195.

Shulman, L. S. (1970). Psychology and mathematics education. In E. G. Begle (Ed.), *Mathematics education* (69th Yearbook of the National Society for the Study of Education, Pt. 1, pp. 23–71). Chicago: University of Chicago Press.

Shulman, L. S. (1974). The psychology of school subjects: A premature obituary? *Journal of Research in Science Teaching, 11,* 319–339.

Shulman, L. S., & Keislar, E. R. (Eds.). (1966). *Learning by discovery: A critical appraisal.* Chicago: Rand McNally.

Shumway, R. J. (Ed.). (1980). *Research in mathematics education.* Reston, VA: National Council of Teachers of Mathematics.

Sigurdson, S. E. (1962). The development of the idea of unified mathematics in the secondary school curriculum: 1890–1930 (Doctoral dissertation, University of Wisconsin, 1962). *Dissertation Abstracts, 23,* 1997.

Smiler, H. B. (1970). A survey of research in mathematics education. In P. S. Jones (Ed.), *A history of mathematics education in the United States and Canada* (32nd Yearbook of the National Council of Teachers of Mathematics, pp. 493–499). Washington, DC: National Council of Teachers of Mathematics.

Smith, D. E. (1895). Sex in mathematics. *Educational Review, 10,* 84–88.

Smith, D. E. (1905). Réformes à accomplir dans l'enseignement des mathématiques: Opinion [Reforms to be accomplished in the teaching of mathematics: Opinion]. *L'Enseignement Mathématique, 7,* 469–471.

Smith, D. E. (1913). Intuition and experiment in mathematical teaching in the secondary schools. In E. W. Hobson & A. E. H. Love (Eds.), *Proceedings of the Fifth International Congress of Mathematicians (Cambridge, 22–28 August 1912)* (Vol. 2, pp. 611–632). Cambridge: Cambridge University Press.

Smith, E. R., Tyler, R. W., & the Evaluation Staff. (1942). *Appraising and recording student progress* (Adventure in American Education Vol. 3). New York: Harper.

Sowder, J. T. (Ed.). (1989). *Research agenda for mathematics education: Vol. 5. Setting a research agenda.* Reston, VA: National Council of Teachers of Mathematics; Hillsdale, NJ: Lawrence Erlbaum.

Special Interest Group for Research in Mathematics Education. (1977, May). *SIG/RME Newsletter,* p. 1.

Special Interest Group for Research in Mathematics Education. (1981, Spring). Membership & dues. *SIG/RME Newsletter,* p. 4.

Spitzer, H. F. (1962). *Teaching arithmetic* (What research says to the teacher, No. 2, rev. ed.). Washington, DC: National Education Association.

Stanic, G. M. A. (1984). Why teach mathematics? A historical study of the justification question (Doctoral dissertation, University of Wisconsin–Madison, 1983). *Dissertation Abstracts International, 44,* 2347A.

Stanic, G. M. A. (1986a). The growing crisis in mathematics education in the early twentieth century. *Journal for Research in Mathematics Education, 17,* 190–205.

Stanic, G. M. A. (1986b). Mental discipline theory and mathematics education. *For the Learning of Mathematics, 6*(1), 39–47.

Stitt, E. W. (1900). School and business arithmetic–Limitations and improvements. *Proceedings and Addresses of the 39th Annual Meeting of the National Education Association, 38,* 566–572.

Stone, C. W. (1908). *Arithmetical abilities and some factors determining them* (Teachers College Contributions to Education No. 19). New York: Columbia University, Teachers College.

Suydam, M. N. (1968). An evaluation of journal-published research reports on elementary school mathematics, 1900–1965 (Vols. 1 & 2) (Doctoral dissertation, Pennsylvania State University, 1967). *Dissertation Abstracts, 28,* 3387A–3388A.

Suydam, M. N., & Dessart, D. J. (1980). Skill learning. In R. J. Shumway (Ed.), *Research in mathematics education* (pp. 207–243). Reston, VA: National Council of Teachers of Mathematics.

Suydam, M. N., & Osborne, A. (1977). *The status of pre college science, mathematics, and social science education: 1955-1975: Vol. 2. Mathematics education* (SE-78-73). Washington, DC: U.S. Government Printing Office.

Swetz, F. J. (1987). *Capitalism and arithmetic.* La Salle, IL: Open Court.

Thorndike, E. L. (1922). *The psychology of arithmetic.* New York: Macmillan.

Thorndike, E. L., Cobb, M. V., Orleans, J. S., Symonds, P. M., Wald E., & Woodyard, E. (1923). *The psychology of algebra.* New York: Macmillan.

Travers, K. J., & Westbury, I. (Eds.). (1989). *The IEA Study of Mathematics I: Analysis of mathematics curricula.* Oxford: Pergamon.

Travers, R. M. W. (1983). *How research has changed American schools: A history from 1840 to the present.* Kalamazoo, MI: Mythos Press.

Tyler, R. (1967). Needed research in mathematics education: A summary of the conference. *Journal of Research and Development in Education, 1*(1), 133 141.

University Microfilms International. (1989). *Dissertation abstracts ondisc (1861–Jun 1980; DAI volumes 01/01–40/12)* [compact disc]. Ann Arbor, MI: Author.

Upton, C. B. (1923). Standardized tests in mathematics for secondary schools. In National Committee on Mathematical Requirements, *The reorganization of mathematics in secondary education* (pp. 279–428). Oberlin, OH: Mathematical Association of America.

Vygotsky, L. S. (1962). *Thought and language* (E. Hanfmann & G. Vakar, Trans.). Cambridge, MA: MIT Press. (Original work published 1934)

Walker, D. F. (1977). Thoughts on the current state of educational research. In M. Otte (Ed.), *Relating theory to practice in educational research* (Materialien und Studien, Band 6, pp. 57–65). Bielefeld, Germany: Universität Bielefeld, Institut für Didakdik der Mathematik.

Washburne, C. W. (1930). The grade placement of arithmetic topics. In G. M. Whipple (Ed.), *Report of the society's committee on arithmetic* (29th Yearbook of the National Society for the Study of Education, pp. 641–670). Chicago: University of Chicago Press.

Washburne, C. W. (1939). The work of the Committee of Seven on grade-placement in arithmetic. In G. M. Whipple (Ed.), *Child development and the curriculum* (38th Yearbook of the National Society for the Study of Education, Pt. 1, pp. 299–324). Bloomington, IL: Public School Publishing Co.

Weaver, J. F. (1967). Extending the impact of research on mathematics education. *Arithmetic Teacher, 14,* 314–318.

Weaver, J. F., & Kilpatrick, J. (Eds.). (1972). *The place of meaning in mathematics instruction: Selected research papers of William A. Brownell* (SMSG Studies in Mathematics, Vol. 22). Stanford, CA: School Mathematics Study Group.

Welte, H. D. (1926). *A psychological analysis of plane geometry* (University of Iowa Monographs in Education No. 1). Iowa City: University of Iowa, College of Education.

Wertheimer, M. (1959). *Productive thinking* (Enlarged ed.). New York: Harper & Row.

Whipple, G. M. (Ed.). (1930). *Report of the society's committee on arithmetic* (29th Yearbook of the National Society for the Study of Education, Pt. 1). Chicago: University of Chicago Press.

Whipple, G. M. (Ed.). (1938). *The scientific movement in education* (37th Yearbook of the National Society for the Study of Education, Pt. 2). Bloomington, IL: Public School Publishing Co.

Williams, E. (1934). The geometrical notions of young children. *Mathematical Gazette, 18,* 112–118.

Wilson, J. W. (1968, September 16). *Memorandum on plans and activities of the Special Interest Group for Research in Mathematics Education.* Unpublished manuscript.

Wilson, J. W., & Peaker, G. F. (Eds.). (1971). International Study of Achievement in Mathematics [Special issue]. *Journal for Research in Mathematics Education, 2*(2).

Wilson, G. M. (1917). A survey of the social and business use of arithmetic. In G. M. Whipple (Ed.), *Second report of the Committee on Minimal Essentials in Elementary-School Subjects* (16th Yearbook of the National Society for the Study of Education, Pt. 1, pp. 128–142). Bloomington, IL: Public School Publishing Co.

Wilson, G. M. (1919). *A survey of the social and business usage of arithmetic* (Teachers College Contributions to Education No. 100). New York: Columbia University, Teachers College.

Wilson, G. M. (1925). Arithmetic. In *Research in constructing the elementary school curriculum* (Third Yearbook of the Department of Superintendence, pp. 35–109). Washington, DC: National Education Association.

Young, J. W. A. (1925). *The teaching of mathematics in the elementary and the secondary school* (New ed. with supp.). New York: Longmans, Green. (Original work published 1906)

Zerner, M. (1989). [Review of *Martin Ohm (1792–1872): Universitäts und Schulmathematik in der neuhumanistischen Bildungs reform*]. *Educational Studies in Mathematics, 20,* 469–474.

Zweng, M., Green, T., Kilpatrick, J., Pollak, H., & Suydam, M. (Eds.). (1983). *Proceedings of the Fourth International Congress on Mathematical Education.* Boston: Birkhäuser.

▪2▪

THE NATURE OF MATHEMATICS:
ITS ROLE AND ITS INFLUENCE

John A. Dossey
ILLINOIS STATE UNIVERSITY

Perceptions of the nature and role of mathematics held by our society have a major influence on the development of school mathematics curriculum, instruction, and research. The understanding of different conceptions of mathematics is as important to the development and successful implementation of programs in school mathematics as it is to the conduct and interpretation of research studies. The literature of the reform movement in mathematics and science education (American Association for the Advancement of Science, 1989; Mathematical Sciences Education Board, 1989, 1990; National Council of Teachers of Mathematics, 1989) portrays mathematics as a dynamic, growing field of study. Other conceptions of the subject define mathematics as a static discipline, with a known set of concepts, principles, and skills (Fisher, 1990).

The rapid growth of mathematics and its applications over the past 50 years has led to a number of scholarly essays that examine its nature and its importance (Consortium for Mathematics and Its Applications, 1988; Committee on Support of Research in the Mathematical Sciences, 1969; Courant & Robbins, 1941; Davis & Hersh, 1980, 1986; Hardy, 1940; Hilton, 1984; Saaty & Weyl, 1969; Steen, 1978; Stewart, 1987; Wilder, 1968). This literature has woven a rich mosaic of conceptions of the nature of mathematics, ranging from axiomatic structures to generalized heuristics for solving problems. These diverse views of the nature of mathematics also have a pronounced impact on the ways in which our society conceives of mathematics and reacts to its ever-widening influence on our daily lives. Regarding this, Steen (1988) writes:

Many educated persons, especially scientists and engineers, harbor an image of mathematics as akin to a tree of knowledge: formulas, the-

orems, and results hang like ripe fruits to be plucked by passing scientists to nourish their theories. Mathematicians, in contrast, see their field as a rapidly growing rain forest, nourished and shaped by forces outside mathematics while contributing to human civilization a rich and ever-changing variety of intellectual flora and fauna. These differences in perception are due primarily to the steep and harsh terrain of abstract language that separates the mathematical rain forest from the domain of ordinary human activity. (p. 611)

Research shows that these differing conceptions have an influence on the ways in which both teachers and mathematicians approach the teaching and development of mathematics (Brown, 1985; Bush, 1982; Cooney, 1985; Good, Grouws, & Ebmeier, 1983; Kesler, 1985; McGalliard, 1983; Owens, 1987; Thompson, 1984). Some see mathematics as a static discipline developed abstractly. Others see mathematics as a dynamic discipline, constantly changing as a result of new discoveries from experimentation and application (Crosswhite et al., 1986). These contrasting views of the nature and source of mathematical knowledge have provided a continuum for conceptions of mathematics since the age of the Greeks. The lack of a common philosophy of mathematics has serious ramifications for both the practice and teaching of mathematics. This lack of consensus, some argue, is the reason that differing philosophies are not even discussed. Others conjecture that these views are transmitted to students and help shape their ideas about the nature of mathematics (Brown, Cooney, & Jones, 1990; Cooney, 1987). What follows is an overview of these conceptions of mathematics and their current and potential impact on the nature and course of mathematics education.

The author wishes to acknowledge the feedback and helpful suggestions made by Thomas Cooney, University of Georgia; Alan Osborne, Ohio State University; and Lynn Steen, St. Olaf College.

CONCEPTIONS OF MATHEMATICS

Historical

Discussions of the nature of mathematics date back to the fourth century B C. Among the first major contributors to the dialogue were Plato and his student, Aristotle. Plato took the position that the objects of mathematics had an existence of their own, beyond the mind, in the external world. In doing so, Plato drew clear distinctions between the ideas of the mind and their representations perceived in the world by the senses. This caused Plato to draw distinctions between arithmetic—the theory of numbers—and logistics—the techniques of computation required by businessmen. In the *Republic* (1952a), Plato argued that the study of arithmetic has a positive effect on individuals, compelling them to reason about abstract numbers. Plato consistently held to this view, showing indignation at technicians' use of physical arguments to "prove" results in applied settings. For Plato, mathematics came to "be identical with philosophy for modern thinkers, though they say that it should be studied for the sake of other things" (Aristotle, 1952, p. 510). This elevated position for mathematics as an abstract mental activity on externally existing objects that have only representations in the sensual world is also seen in Plato's discussion of the five regular solids in *Timaeus* (1952b) and his support and encouragement of the mathematical development of Athens (Boyer, 1968).

Aristotle, the student, viewed mathematics as one of three genera into which knowledge could be divided: the physical, the mathematical, and the theological:

[Mathematics is the one] which shows up quality with respect to forms and local motions, seeking figure, number, and magnitude, and also place, time, and similar things.... Such an essence falls, as it were, between the other two, not only because it can be conceived both through the senses and without the senses. (Ptolemy, 1952, p. 5)

This affirmation of the role of the senses as a source for abstracting ideas concerning mathematics was different from the view held by his teacher, Plato. Aristotle's view of mathematics was not based on a theory of an external, independent, unobservable body of knowledge. Rather it was based on experienced reality, where knowledge is obtained from experimentation, observation, and abstraction. This view supports the conception that one constructs the relations inherent in a given mathematical situation. In Aristotle's view, the construction of a mathematical idea comes through idealizations performed by the mathematician as a result of experience with objects. Thus, statements in applied mathematics are approximations of theorems in pure mathematics (Körner, 1960). Aristotle attempted to understand mathematical relationships through the collection and classification of empirical results derived from experiments and observations and then by deduction of a system to explain the inherent relationships in the data. Thus, the works and ideas of Plato and Aristotle molded two of the major contrasting themes concerning the nature of mathematics.

By the Middle Ages, Aristotle's work became known for its contributions to logic and its use in substantiating scientific claims. Although this was not contrary to the way in which Aristotle had employed his methods of logical reasoning, those who employed his principles often used them to argue against the derivation of evidence from empirical investigations. Aristotle drew clear lines between the ideal *forms* envisioned by Plato and their empirical realizations in worldly objects.

The distinctions between these two schools of mathematical thought were further commented upon by Francis Bacon in the early 1500s when he separated mathematics into pure and mixed mathematics:

To the pure mathematics are those sciences belonging which handle quantity determinate, merely severed from any axioms of natural philosophy.... For many parts of nature can neither be invented with sufficient subtlety, nor demonstrated with sufficient perspicuity, nor accommodated unto use with sufficient dexterity, without the aid and intervening of the mathematics. (1952, p. 46)

Similar discussions concerning the nature of mathematics were also echoed by Jean D'Alembert and other members of the French salon circle (Brown, 1988).

Descartes worked to move mathematics back to the path of deduction from accepted axioms. Though experimenting himself in biological matters, Descartes rejected input from experimentation and the senses in matters mathematical because it might possibly delude the perceiver. Descartes's consideration of mathematics worked to separate it from the senses:

For since the name "Mathematics" means exactly the same as "scientific study,"...we see that almost anyone who has had the slightest schooling, can easily distinguish what relates to Mathematics in any question from that which belongs to the other sciences....I saw consequently that there must be some general science to explain that element as a whole which gives rise to problems about order and measurement restricted as these are to no special subject matter. This, I perceived, was called "Universal Mathematics," not a far fetched designation, but one of long standing which has passed into current use, because in this science is contained everything on account of which the others are called parts of Mathematics. (1952, p. 7)

This struggle between the rationalists and the experimentalists affected all branches of science throughout the 17th and 18th centuries.

The German philosopher Immanuel Kant brought the discussion of the nature of mathematics, most notably the nature of geometry, back in to central focus with his *Critique of Pure Reason* (1952). Whereas he affirmed that all axioms and theorems of mathematics were truths, he held the view that the nature of perceptual space was Euclidean and that the contents of Euclidean geometry were a priori understandings of the human mind. This was in direct opposition to the emerging understandings of non-Euclidean geometry.

The establishment of the consistency of non-Euclidean geometry in the mid-1800s finally freed mathematics from the restrictive yoke of a single set of axioms thought to be the only model for the external world. The existence of consistent non-Euclidean geometries showed the power of man's mind to construct new mathematical structures, free from the bounds of an externally existing, controlling world (Eves, 1981; Kline, 1972, 1985; Körner, 1960). This discovery, exciting as it was, brought with it a new notion of "truth," one buried in the acceptance of an axiom or a set of axioms defining a model for an area of investigation. Mathematicians immediately began to apply this new freedom and axiomatic method to the study of mathematics.

Late 19th and Early 20th Century Views

New investigations in mathematics, freed from reliance on experimentation and perception, soon encountered new problems with the appearance of paradoxes in the real number system and the theory of sets. At this point, three new views of mathematics arose to deal with the perceived problems. The first was the school of logicism, founded by the German mathematician Gottlob Frege in 1884. This school, an outgrowth of the Platonic school, set out to show that ideas of mathematics could be viewed as a subset of the ideas of logic. The proponents of logicism set out to show that mathematical propositions could be expressed as completely general propositions whose truth followed from their form rather than from their interpretation in a specific contextual setting. A. N. Whitehead and Bertrand Russell (1910–13) set out to show this in their landmark work, *Principia Mathematica*. This attempt was equivalent to trying to establish classical mathematics from the terms of the axioms of the set theory developed by Zermelo and Frankel. This approach, as that of Frege, was built on the acceptance of an externally existing mathematics, and hence was a direct outgrowth of the Platonic school. Whitehead and Russell's approach failed through its inability to establish the axioms of infinity and choice in a state of complete generality devoid of context. This Platonic approach also failed because of the paradoxes in the system.

The followers of the Dutch mathematician L. E. J. Brouwer, on the other hand, did not accept the existence of any idea of classical mathematics unless it could be constructed via a combination of clear inductive steps from first principles. The members of Brouwer's school of thought, called the intuitionists, were greatly concerned with the appearance of paradoxes in set theory and their possible ramifications for all of classical mathematics. Unlike the logicists, who accepted the contents of classical mathematics, the intuitionists accepted only the mathematics that could be developed from the natural numbers forward through the mental activities of constructive proofs. This approach did not allow the use of the law of the excluded middle. This logical form asserts that the statement $p \lor \sim p$ is true and makes proof by contradiction possible.

In many ways, the ideas put forth by Brouwer were based on a foundation not unlike that professed by Kant. Brouwer did not argue for the "inspection of external objects, but [for] 'close introspection'" (Körner, 1960, p. 120). This conception portrayed mathematics as the objects resulting from "valid" demonstrations. Mathematical ideas existed only insofar as they were constructible by the human mind. The insistence on construction placed the mathematics of the intuitionists within the Aristotelian tradition. This view took logic to be a subset of mathematics. The intuitionists' labors resulted in a set of theorems and conceptions different from those of classical mathematics. Under their criteria for existence and validity, it is possible to show that every real-valued function defined for all real numbers is continuous. Needless to say, this and other differences from classical mathematics have not attracted a large number of converts to intuitionism.

The third conception of mathematics to emerge near the beginning of the 20th century was that of formalism. This school was molded by the German mathematician David Hilbert. Hilbert's views, like those of Brouwer, were more in line with the Aristotelian tradition than with Platonism. Hilbert did not accept the Kantian notion that the structure of arithmetic and geometry existed as descriptions of a priori knowledge to the same degree that Brouwer did. However, he did see mathematics as arising from intuition based on objects that could at least be considered as having concrete representations in the mind.

Formalism was grounded in the attempts to characterize mathematical ideas in terms of formal axiomatic systems. This attempt to free mathematics from contradictions was built around the construction of a set of axioms for a branch of mathematics that allowed for the topic to be discussed in a first-order language. Considerable progress was made in several areas under the aegis of formalism before its demise as a result of Kurt Gödel's 1931 landmark paper. Gödel (1931) established that it is impossible in axiomatic systems of the type Hilbert proposed to prove formally that the system is free of contradictions. Gödel also demonstrated that it is impossible to establish the consistency of a system employing the usual logic and number theory if one uses only the major concepts and methods from traditional number theory. These findings ended the attempt to so formalize all of mathematics, though the formalist school has continued to have a strong impact on the development of mathematics (Benacerraf & Putnam, 1964; van Heijenoort, 1967; Snapper, 1979a, 1979b).

The three major schools of thought created in the early 1900s to deal with the paradoxes discovered in the late 19th century advanced the discussion of the nature of mathematics, yet none of them provided a widely adopted foundation for the nature of mathematics. All three of them tended to view the contents of mathematics as products. In logicism, the contents were the elements of the body of classical mathematics, its definitions, its postulates, and its theorems. In intuitionism, the contents were the theorems that had been constructed from first principles via "valid" patterns of reasoning. In formalism, mathematics was made up of the formal axiomatic structures developed to rid classical mathematics of its shortcomings. The influence of the Platonic and Aristotelian notions still ran as a strong undercurrent through these theories. The origin of the "product"—either as a pre-existing external object or as an object created through experience from sense perceptions or experimentation—remained an issue.

Modern Views

The use of a product orientation to characterize the nature of mathematics is not a settled issue among mathematicians. They tend to carry strong Platonic views about the existence of mathematical concepts outside the human mind. When pushed to make clear their conceptions of mathematics, most retreat to a formalist, or Aristotelian, position of mathematics as a game played with symbol systems according to a fixed set of socially accepted rules (Davis & Hersh, 1980). In reality, however, most professional mathematicians think little about the nature of their subject as they work within it.

The formalist tradition retains a strong influence on the development of mathematics (Benacerraf & Putnam, 1964; Tymoczko, 1986). Hersh (1986) argues that the search for the foundations of mathematics is misguided. He suggests that the focus be shifted to the study of the contemporary practice of mathematics, with the notion that current practice is inherently fallible and, at the same time, a very public activity (Tymoczko, 1986). To do this, Hersh begins by describing the plight of the working mathematician. During the creation of new mathematics, the mathematician works as if the discipline describes an externally existing objective reality. But when discussing the nature of mathematics, the mathematician often rejects this notion and describes it as a meaningless game played with symbols. This lack of a commonly accepted view of the nature of mathematics among mathematicians has serious ramifications for the practice of mathematics education, as well as for mathematics itself.

The conception of mathematics held by the teacher has a strong impact on the way in which mathematics is approached in the classroom (Cooney, 1985). A teacher who has a formalist philosophy will present content in a structural format, calling on set theoretic language and conceptions (Hersh, 1986). Such a formalistic approach may be a good retreat for the individual who does not understand the material well enough to provide an insightful constructive view. Yet, if such formalism is *not* the notion carried by mathematicians, why should it dominate the presentation of mathematics in the classroom? To confront this issue, a discussion of the nature of mathematics must come to the foreground in mathematics education.

Tymoczko and Hersh argue that what is needed is a new philosophy of mathematics, one that will serve as a basis for the working mathematician and the working mathematics educator. According to Hersh, the working mathematician is not controlled by constant attention to validating every step with an accepted formal argument. Rather, the mathematician proceeds, guided by intuition, in exploring concepts and their interactions. Such a path places the focus on understanding as a guide, not long, formal derivations of carefully quantified results in a formal language.

This shift calls for a major change. Mathematics must be accepted as a human activity, an activity not strictly governed by any one school of thought (logicist, formalist, or constructivist). Such an approach would answer the question of what mathematics is by saying that:

Mathematics deals with ideas. Not pencil marks or chalk marks, not physical triangles or physical sets, but ideas (which may be represented or suggested by physical objects). What are the main properties of mathematical activity or mathematical knowledge, as known to all of us from daily experience?

1. Mathematical objects are invented or created by humans.
2. They are created, not arbitrarily, but arise from activity with already existing mathematical objects, and from the needs of science and daily life.
3. Once created, mathematical objects have properties which are well-determined, which we may have great difficulty in discovering, but which are possessed independently of our knowledge of them. (Hersh, 1986, p. 22)

The development and acceptance of a philosophy of mathematics carries with it challenges for mathematics and mathematics education. A philosophy should call for experiences that help mathematician, teacher, and student to experience the invention of mathematics. It should call for experiences that allow for the mathematization, or modeling, of ideas and events. Developing a new philosophy of mathematics requires discussion and communication of alternative views of mathematics to determine a valid and workable characterization of the discipline.

TEACHERS' CONCEPTIONS OF MATHEMATICS

The conception of mathematics held by the teacher may have a great deal to do with the way in which mathematics is characterized in classroom teaching. The subtle messages communicated to children about mathematics and its nature may, in turn, affect the way they grow to view mathematics and its role in their world.

Cooney (1987) has argued that substantive changes in the teaching of mathematics such as those suggested by the NCTM *Standards* (1989) will be slow in coming and difficult to achieve because of the basic beliefs teachers hold about the nature of mathematics. He notes that the most prevalent verb used by preservice teachers to describe their teaching is *present*. This conception of teaching embodies the notion of authority in that there is a presenter with a fixed message to send. Such a position assumes the external existence of a body of knowledge to be transmitted to the learners and is thus more Platonic than Aristotelian. The extension of this conception of how mathematics relates to education and its practice is an important one. The teacher's view of how teaching should take place in the classroom is strongly based on a teacher's understanding of the nature of mathematics, not on what he or she believes is the best way to teach (Hersh, 1986). To change the situation, one must construct alternative ways of conceptualizing the nature of mathematics and the implications of such conceptions for mathematics education.

Cooney (1987) used the work of Goffree (1985) and Perry (1970) in his analyses of the nature of mathematics portrayed in classrooms and concluded that school mathematics is bound up in a formal and external view of mathematics. Goffree presented a model for the way textbooks are developed and how teachers might employ them in the classroom to portray the nature of mathematics. The four textbook models were (a) the mechanistic, (b) the structuralist, (c) the empiricist, and (d) the realistic or applied. Each of these methods of textbook development portrays a view of the nature of mathematics. Goffree then crossed these textual characteristics with three ways in which teachers employ textbooks in the classroom:

Instrumental use—The teacher uses the textbook as an instrument, following its sequence and using its suggestions for dealing with the content.
Subjective use—The teacher uses the textbook as a guide, but provides a constructive overview of the materials, followed by a further

discussion of the concepts/principles/procedures based on the teachers' experience.

Fundamental use—At this level, the curriculum is developed from a constructive viewpoint. This approach is concerned with both the content and pedagogy involved in mathematics. (p. 26)

In many classrooms, the prevailing model is mechanistic-instrumental. Modern reform documents (NCTM, 1989) advocate a situation that is closer to realistic-fundamental. The enormous distance between these two models indicates the large role that the teacher's conception of the nature of mathematics can play in the teaching and learning process as it applies to school mathematics.

In related work, Cooney (1985) and his students (Brown, 1985; Bush, 1982; Kesler, 1985; McGalliard, 1983; Owens, 1987; Thompson, 1984) have also examined the nature of teachers' conceptions of mathematics using the levels of intellectual development created by William Perry (1970). Perry's model provides a means to describe the way in which humans view the world about them. Perry's hierarchical scheme sees individuals passing through stages from dualism to multiplistic perspectives to relativistic perspectives. In the dualistic stage, the individual assumes that one functions in a bipolar world with such choices as good or bad, right or wrong. At this stage, problems are resolved by an authority's ruling. The individual may grow to a stage where multiple perspectives are entertained; however, the perspectives are still viewed as discrete entities lacking structural relationships. Finally, a person may move to the stage of relativism, where a number of possible alternatives are considered relative to one another. At this stage, each of the alternatives is examined within its own frame of reference.

Kesler (1985) and McGalliard (1983) conducted studies of secondary school algebra and geometry teachers' conceptions of mathematics by analyzing their classroom teaching. Kesler found that algebra teachers differed greatly in their orientations. Some performed at the dualistic or multiplistic level of the hierarchy, whereas others showed signs of multiplistic-relativistic behaviors. McGalliard's study of geometry teachers showed that their view of mathematics was marked by dualism. These teachers viewed their task as one of presenting mathematics to their students. The teachers' main concern was in seeing that their students learned to perform easily the tasks required by their homework and tests. Thus, the learning of mathematics was reduced to knowing how rather than knowing why. The fact that fewer teachers in geometry exceeded the dualistic level might be a reflection of their lack of geometric experience. Cooney (1987) reflects on the predominance and implications of the presenting, or broadcast, mode for teaching. Presenting, by its very nature, involves authority. Such an orientation is not compatible with a style of classroom management and resource use that would promote student consideration of a number of perspectives on mathematics, its nature, and its use. These ideas, plus collaborating findings by Owens (1987) with preservice secondary teachers, suggest the great distance that must be covered to bring the classroom consideration of mathematics close to the fundamental-realistic combination envisioned by Goffree.

Owens's work, and that of Bush (1982), further indicated that many of the preservice teachers' dualistic or multiplistic views were strengthened by their experiences in upper-division mathematics content courses at the university level. There, they were exposed to teaching that strongly reflected the formalist view of mathematics as an externally developed axiom system. This influence only reinforces the conception that mathematics exists externally. Through direct intervention, Myerson (1977) was able to move some students to view mathematics on a somewhat higher level. But many still thought that there were specific, set methods to address each classroom question, reflecting the strong dualistic-multiplistic orientation of preservice teachers.

The reaction of students is a strong factor influencing a teacher's portrayal of the nature of mathematics in class. Brown (1985) and Cooney (1985) studied the reactions of a first-year teacher in the classroom. The teacher entered the classroom with an orientation that reflected both multiplistic and relativistic characteristics. He attempted to initiate a classroom style involving a good deal of problem solving and student activities aimed at providing a strong foundation for student learning. The students found these approaches threatening and their reactions led to his eventual return to a presenting mode. Cooney (1987) concludes: "I suspect that students gravitate toward a mechanistic curriculum and appreciate teachers whose interpretations of the text are quite predictable. If you believe the contrary, listen carefully to the negotiations that take place between students and teacher when test time arrives" (p. 27).

THE RELEVANCE OF CONCEPTIONS OF MATHEMATICS TO MATHEMATICS EDUCATION RESEARCH

The focus on mathematics education and the growth of research in mathematics education in the late 1970s and the 1980s reflects a renewed interest in the philosophy of mathematics and its relation to learning and teaching. At least five conceptions of mathematics can be identified in mathematics education literature (J. Sowder, 1989). These conceptions include two groups of studies from the external (Platonic) view of mathematics. The remaining three groups of studies take a more internal (Aristotelian) view.

External Conceptions

The work of two groups of researchers treats mathematics as an externally existing, established body of concepts, facts, principles, and skills available in syllabi and curricular materials. The work of the first group of researchers adopting the external view focuses on assisting teachers and schools to be more successful in conveying this knowledge to children. Their work takes a relatively fixed, static view of mathematics.

Studies investigating the role of teachers in mathematics classrooms commonly focus on the actions and instructional methods of the teachers rather than on the mathematics being taught or the methods by which that mathematics is being learned. Early studies of teacher actions by B. O. Smith and his coworkers (Smith, Meux, Coombs, Nuthall, & Precians, 1967) led to a number of studies of the relative efficacy of the use of logical discourse in the teaching of concepts and generalizations (Cooney, 1980; Cooney & Bradbard, 1976; L. Sowder, 1980). Later research on effective teaching selected mathematics classrooms as the site for data gathering (Brophy, 1986; Brophy & Evertson, 1981; Brophy & Good, 1986; Fisher & Berliner, 1985; Good, Grouws, & Ebmeier, 1983; Medley & Mitzel, 1963; Rosenshine & Furst, 1973; Slavin & Karweit, 1984, 1985) and focused on how teachers used domain-specific knowledge, and how they organized, sequenced, and presented it in attempts to promote different types of student performance in classroom settings (Berliner et al., 1988). Other studies have centered on efforts to delineate the differences in decision making between "novice" and "expert" teachers in planning for teaching and in instruction (Leinhardt, 1988; Leinhardt & Greeno, 1986). These studies often focus on the teaching acts that differentiate the performances of expert and novice teachers with classification based on student performance on standardized achievement tests, topic-specific tests, or student growth over a period of time. Leinhardt and co-workers have investigated the teaching of fractions, examining the role of teacher decision making, use of scripts, and the role and type of explanations. Good, Grouws, and Ebmeier (1983) examined the role of active teaching by expert teachers and then developed a prototype lesson organization that promoted student growth in mathematics.

Shavelson, Webb, Stasz, and McArthur (1988) provide warnings about the nature of findings from research based on the external conception point of view. First, the findings provide a picture of the existing situation, not a picture of what could be achieved under dramatically changed instruction. Second, the findings reflect the type of performance that was used to separate the teachers into the different categories initially. That is, when teachers were selected as experts on the basis of specific criteria, the results reflect the teaching patterns of instruction related to those criteria. The conduct of the studies and the external conception of the mathematics employed tend to direct the type of research questions asked, and those *not* asked. This research must include teachers with a wide variety of styles if findings generalizable to all teachers or all classrooms are desired.

The second group of researchers adopting the external view espouse a more dynamic view of mathematics, but they focus on adjusting the curriculum to reflect this growth of the discipline and to see how students acquire knowledge of the related content and skills. The underlying focus is, however, still on student mastery of the curriculum or on the application of recent advances in technology or instructional technology to mathematics instruction.

Thorpe (1989), in reviewing the nature of the teaching of algebra, states that "students have needed to learn pretty much the same algebra as did their parents and grandparents. But now something *has* changed. We have new tools" (p. 23). Kaput (1989) took this issue of the changing context for the teaching of algebra and provided a list of research questions concerning the role of linked representations in developing the symbol system of algebra. Thompson (1989) provided additional examples for the development of meanings for topics in numeration and quantity. The work of Wearne and Hiebert (1988) provides another example of such research in mathematics education.

Taking the concepts and skills related to fractions as given, Wearne and Hiebert looked at the ways in which students can come to understand and operate with decimal fractions and apply these learnings to situations calling for transfer of understanding to procedural skill. Each of these studies assumes the mathematics as given, but also allows for it to take on new meanings as time passes. The issue at heart is how teacher instruction or student understanding can be improved through research. The focus here is not on the creation of new content, but on the growth of individual knowledge of an existing portion of mathematics.

Internal Conceptions

The remaining three conceptions of mathematics found in mathematics education research focus on mathematics as a personally constructed, or internal, set of knowledge. In the first of these, mathematics is viewed as a process. Knowing mathematics is equated with doing mathematics. Research in this tradition focuses on examining the features of a given context that promotes the "doing." Almost everyone involved in the teaching and learning of mathematics holds that the learning of mathematics is a personal matter in which learners develop their own personalized notions of mathematics as a result of the activities in which they participate. Ernst von Glaserfeld (1987) described this conception of learning and teaching:

[As we] come to see knowledge and competence as products of the individual's conceptual organization of the individual's experience, the teacher's role will no longer be to dispense "truth" but rather to help and guide the student in the conceptual organization of certain areas of experience. Two things are required for the teacher to do this: on the one hand, an adequate idea of where the student is and, on the other, an adequate idea of the destination. Neither is accessible to direct observation. What the student says and does can be interpreted in terms of a hypothetical model—and this is one area of educational research that every *good* teacher since Socrates has done intuitively. Today, we are a good deal closer to providing the teacher with a set of relatively reliable diagnostic tools.

As for the helping and guiding, good teachers have always found ways and means of doing it because, consciously or unconsciously they realized that, although one can point the way with words and symbols, it is the student who has to do the conceptualizing and the operating. (p. 16)

This emphasis on students doing mathematics is the hallmark of this conceptualization of mathematics. It is the "doing"—the experimenting, abstracting, generalizing, and specializing—that constitutes mathematics, not a transmission of a well-formed communication. This approach to the learning of mathematics is reflected in the writing of Steffe (1988) and Romberg (1988), as well as in many of the emerging activity-oriented preschool and primary programs. This conception seems to be shared by George Polya as expressed in an address to the American Mathematical Society on his views on the learning of mathematics:

It has been said by many people in many ways that learning should be active, not merely passive or receptive; merely by reading books or listening to lectures or looking at moving pictures without adding some action of your own mind you can hardly learn anything and certainly you can not learn much.

There is another often expressed (and closely related) opinion: *The best way to learn anything is to discover it by yourself.* Lichtenberg… adds an interesting point: *What you have been obligated to discover by yourself leaves a path in your mind which you can use again when the need arises.* (Polya, 1965, pp. 102–103)

This personal construct approach to mathematics is a strong component of many of the recommendations of the NCTM *Curriculum and Evaluation Standards for School Mathematics* (1989) and has a strong history in mathematics education, including the work of Harold Fawcett (1938), the work of the Progressive Education Association (1938), and the NCTM *Agenda for Action* (1980).

A second personal, or internal, conceptualization of mathematics is based on the description of mathematical activities in terms of psychological models employing cognitive procedures and schemata. Larkin (1989) describes this approach in the following statement:

The central technique of cognitive science is modeling problem-solving behavior in the following way: A problem is considered as a *data* structure that includes whatever information is available about the problem. We then ask what kind of *program* could add information to that data structure to produce a solution to the problem. Because we want to have a model that explains human performance, we require that the model add information to the data structure in orders consistent with the orders in which humans are observed to add information. (p. 120)

This cognitive science approach to the study of mathematics can be found in the works and recommendations of Bransford, et al. (1988); Campione, Brown, and Connell (1988); Carpenter (1988); Chaiklin (1989); Hiebert (1986); Larkin (1989); Marshall (1988); Nesher (1988); Ohlsson (1988); Peterson (1988); Resnick (1987); and Wearne and Hiebert (1988). The diversity of this research adopting the cognitive modeling approach shows the apparent acceptance of it as a model for viewing the structure of mathematics learning. Its basic tenets are the identification of representations for mathematical knowledge, of operations individuals perform on that knowledge, and of the manner in which the human mind stores, transforms, and amalgamates that knowledge.

The third internal conception of mathematics that surfaces in mathematics education research is one that views mathematics knowledge as resulting from social interactions. Here the learning of mathematics is the acquiring of relevant facts, concepts, principles, and skills as a result of social interactions that rely heavily on context. The research describing this view (Bauersfeld, 1980; Bishop, 1985, 1988; Kieren, 1988; Lave, Smith, & Butler, 1988; Schoenfeld, 1988, 1989) focuses on building mathematics knowledge from learning in an apprentice mode, drawing on both the content and the context. Such an approach perhaps heightens the learner's ability to relate the mathematics to its applications and its potential use in problem-solving situations. The distance between the theoretical aspects of the content and the practical distinctions of applications is diminished. In social settings, the measurement of an individual's progress in mathematics is judged on the degree to which he or she has attained the content material transmitted. There is no measure of the cultural information transmitted or the relation of that material to the learner's position in life.

Schoenfeld (1988) argues that the nature of mathematics perceived by students is a result of an intricate interaction of cognitive and social factors existing in the context of schooling. If students are to learn and apply mathematics, they must come to see mathematics as having worth in social settings. Such "sense making" in the learning of mathematics calls for students to participate actively in "doing mathematics" to learn the skills of the discipline. Students must be called upon to participate aggressively in analyzing, conjecturing, structuring, and synthesizing numerical and spatial information in problem settings. These activities must also involve the students in seeing how the results of such activities relate to the solution of problems in the social setting from which the problems originated. Kieren (1988) similarly argues for the careful integration of mathematics learning with the features of the social context in which the mathematics has meaning.

Each of these three conceptions of the development and study of internal models for mathematics education provides important vantage points for research on the learning and teaching of mathematics. The election of one of these philosophies and its use in the design of research strongly influence the nature of the questions investigated, the manner in which relevant data are collected and analyzed, and the importance tied to the conclusions reached. Creators and users of research in mathematics education must pay closer attention to the role such philosophies play in the conduct of that research. To ignore this feature is to misinterpret findings and misapply the outcomes of such studies.

SUMMARY

The survey of the literature shows that conceptions of mathematics fall along an externally-internally developed continuum. Hersh's (1986) comments, along with others (Tymoczko, 1986), indicate that mathematicians behave like constructionalists until challenged. Similar findings may hold for mathematics teachers. The retreat to the external model to discuss their conceptions shows a strong predilection for Platonic views of mathematics. Such conceptions are strongly flavored by dualistic or multiplistic beliefs about mathematics, allowing few teachers to reject an authoritarian teaching style. Even so, the leaders and professional organizations in mathematics education are promoting a conception of mathematics that reflects a decidedly relativistic view of mathematics (Ernest, 1989). Steps to address the gaps between the philosophical bases for current mathematics instruction are important ones that must be addressed in the development and study of mathematics education at all levels.

The emergence of a process view of mathematics embedded in the NCTM *Standards* (1989) and in the works of modern mathematical philosophers (Tymoczko, 1986) presents many new and important challenges. Teacher educators and curriculum developers must become aware of the features and ramifications of the internal and external conceptions, and their ramifications for curricular development and teacher actions.

Further, all involved in applying mathematics education research must recognize the important influences of each conception of mathematics on both the findings cited and on the interpretation and application of such findings. Mathematics educators need to focus on the nature of mathematics in the development of research, curriculum, teacher training, instruction, and assessment as they strive to understand its impact on the learning and teaching of mathematics.

References

American Association for the Advancement of Science (AAAS). (1989). *Project 2061: Science for all Americans*. Washington, DC: Author.

Aristotle. (1952). Metaphysics. In R. M. Hutchins (Ed.), *Great books of the western world: Vol. 8. Aristotle 1* (pp. 495–626). Chicago: Encyclopaedia Britannica, Inc.

Bacon, F. (1952). *Advancement of learning*. In R. M. Hutchins (Ed.), *Great books of the western world: Vol. 30. Fancis Bacon* (pp. 1–101). Chicago: Encyclopaedia Britannica, Inc.

Bauersfeld, H. (1980). Hidden dimensions in the so-called reality of a mathematics classroom. *Educational Studies in Mathematics, 11*, 23–41.

Benacerraf, P., & Putnam, H. (1964). *Philosophy of mathematics*. Englewood Cliffs, NJ: Prentice Hall.

Berliner, D., Stein, P., Sabers, D., Clarridge, P., Cushing, K., & Pinnegar, S. (1988). Implications of research on pedagogical expertise and experience for mathematics teaching. In D. Grouws, T. Cooney, & D. Jones (Eds.), *Perspectives on research on effective mathematics teaching* (pp. 67–95). Reston, VA: National Council of Teachers of Mathematics.

Bishop, A. (1985). The social construction of meaning—a significant development for mathematics education? *For the Learning of Mathematics, 5*, 24–28.

Bishop, A. (1988). *Mathematical enculturation*. Boston: Kluwer Academic.

Boyer, C. (1968). *A history of mathematics*. New York: Wiley.

Bransford, J., Hasselbring, T., Barron, B., Kulewicz, S., Littlefield, ᴵ & Goin, L. (1988). The use of macro contexts to facilitate mathematical thinking. In R. Charles & E. Silver (Eds.), *The teaching and assessing of mathematical problem solving* (pp. 125–147). Reston, VA: National Council of Teachers of Mathematics.

Brophy, J. (1986). Teaching and learning mathematics: Where research should be going. *Journal for Research in Mathematics Education, 17*, 323–346.

Brophy, J., & Evertson, C. (1981). *Student characteristics and teaching*. New York: Longman.

Brophy, J., & Good, T. (1986). Teacher behavior and student achievement. In M. C. Wittrock (Ed.), *Handbook of research on teaching* (3rd ed., pp. 328–375). New York: Macmillan.

Brown, C. (1985). A study of the socialization to teaching of a beginning secondary mathematics teacher (Doctoral dissertation, University of Georgia). *Dissertation Abstracts International, 46A*, 2605.

Brown, G. (1988). Jean D'Alembert, mixed mathematics and the teaching of mathematics (Doctoral dissertation, Illinois State University). *Dissertation Abstracts International, 48A*, 2200–2201.

Brown, S., Cooney, T., & Jones, D. (1990). Mathematics teacher education. In W. R. Houston (Ed.), *Handbook of research on teacher education* (pp. 639–656). New York: Macmillan.

Bush, W. (1982). Preservice secondary mathematics teachers' knowledge about teaching mathematics and decision-making processes during teacher training (Doctoral dissertation, University of Georgia, 1982). *Dissertation Abstracts International, 43A*, 2264.

Campione, J., Brown, A., & Connell, M. (1989). Metacognition: On the importance of knowing what you are doing: In R. Charles and E. Silver (Eds.), *The teaching and assessing of mathematical problem solving* (pp. 93–114). Reston, VA: National Council of Teachers of Mathematics.

Carpenter, T. (1988). Teaching as problem solving. In R. Charles & E. Silver (Eds.), *The teaching and assessing of mathematical problem solving* (pp. 187–202). Reston, VA: National Council of Teachers of Mathematics.

Chaiklin, S. (1989). Cognitive studies of algebra problem solving and learning. In S. Wagner & C. Kieran (Eds.), *Research issues in the learning and teaching of algebra* (pp. 93–114). Reston, VA: National Council of Teachers of Mathematics.

Committee on Support of Research in the Mathematical Sciences (COSRIMS). (1969). *The mathematical sciences: A collection of essays*. Cambridge, MA: MIT. Press.

Consortium for Mathematics and Its Applications (COMAP). (1988). *For all practical purposes: Introduction to contemporary mathematics*. New York: W. H. Freeman.

Cooney, T. (1980). Research on teaching and teacher education. In R. Shumway (Ed.), *Research in mathematics education* (pp. 433–474). Reston, VA: National Council of Teachers of Education.

Cooney, T. (1985). A beginning teacher's view of problem solving. *Journal for Research in Mathematics Education, 16*, 324–336.

Cooney, T. (1987). The issue of reform: What have we learned from yesteryear? In Mathematical Sciences Education Board, *The teacher of mathematics: Issues for today and tomorrow* (pp. 17–35). Washington, DC: National Academy Press.

Cooney, T., & Bradbard, D. (1976). *Teaching strategies: Papers from a research workshop*. Columbus, OH: ERIC/SMEAC Center for Science, Mathematics, and Environmental Education.

Courant, R., & Robbins, H. (1941). *What is mathematics?* New York: Oxford University Press.

Crosswhite, F. J., Dossey, J. A., Cooney, T. J., Downs, F. L., Grouws, D. A., McKnight, C. C., Swafford, J. O., & Weinzweig, A. I. (1986). *Second international mathematics study detailed report for the United States*. Champaign, IL: Stipes.

Davis, P., & Hersh, R. (1980). *The mathematical experience*. Boston: Birkhauser.

Davis, P., & Hersh, R. (1986). *Descartes' dream: The world according to mathematics*. San Diego: Harcourt Brace Jovanovich.

Descartes, R. (1952). *Rules for the direction of the mind*. In R. M. Hutchins (Ed.), *Great books of the western world: Vol. 31. Descartes & Spinoza*: (pp. 1–40). Chicago: Encyclopaedia Britannica, Inc.

Ernest, P. (1989). Philosophy, mathematics and education. *International Journal of Mathematics Education in Science and Technology, 20*, 555–559.

Eves, H. (1981). *Great moments in mathematics (after 1650)*. Washington, DC: Mathematical Association of America.

Fawcett, H. (1938). *The nature of proof*. (Thirteenth Yearbook of the National Council of Teachers of Mathematics). New York: Bureau of Publications of Columbia University Teachers College.

Fisher, C. (1990). The Research Agenda Project as prologue. *Journal for Research in Mathematics Education, 21*, 81–89.

Fisher, C., & Berliner, D. (Eds.). (1985). *Perspectives on instructional time*. New York: Longman.

Gödel, K. (1931). Uberformal unentscheidbare Sätze der Principiä Mathematica und verwandter Systeme I. *Monatshefte für mathematik und physik, 38*, 173–198.

Goffree, F. (1985). The teacher and curriculum development. *For the Learning of Mathematics, 5*, 26–27.

Good, T., Grouws, D., & Ebmeier, H. (1983). *Active mathematics teaching*. New York: Longman.

Hardy, G. (1940). *A mathematician's apology*. Cambridge, England: Cambridge University Press.

Hersh, R. (1986). Some proposals for reviving the philosophy of mathematics. In T. Tymoczko (Ed.), *New directions in the philosophy of mathematics* (pp. 9–28). Boston: Birkhäuser.

Hiebert, J. (Ed.). (1986). *Conceptual and procedural knowledge: The case of mathematics*. Hillsdale, NJ: Lawrence Erlbaum.

Hilton, P. (1984). Current trends in mathematics and future trends in mathematics education. *For the Learning of Mathematics, 4*, 2–8.

Kant, I. (1952). The critique of pure reason. In R. M. Hutchins (Ed.), *Great books of the western world: Vol. 42. Kant* (pp. 1–250). Chicago: Encyclopedia Britannica, Inc.

Kaput, J. (1989). Linking representations in the symbol systems of algebra. In S. Wagner & C. Kieran (Eds.), *Research issues in the learning and teaching of algebra* (pp. 167–181). Reston, VA: National Council of Teachers of Mathematics.

Kesler, R. (1985). Teachers' instructional behavior related to their conceptions of teaching and mathematics and their level of dogmatism: Four case studies (Doctoral dissertation, University of Georgia, 1985). *Dissertation Abstracts International, 46*, 2606A.

Kieren, T. (1988). Personal knowledge of rational numbers: Its intuitive and formal development. In J. Hiebert & M. Behr (Eds.), *Number concepts and operations in the middle grades* (pp. 162–181). Reston, VA: National Council of Teachers of Mathematics.

Kline, M. (1972). *Mathematical thought from ancient to modern times*. New York: Oxford University Press.

Kline, M. (1985). *Mathematics and the search for knowledge*. New York: Oxford University Press.

Körner, S. (1960). *The philosophy of mathematics: An introduction*. New York: Harper & Row.

Larkin, J. (1989). Robust performance in algebra: The role of problem representation. In S. Wagner & C. Kieran (Eds.), *Research issues in the learning and teaching of algebra* (pp. 120–134). Reston, VA: National Council of Teachers of Mathematics.

Lave, J., Smith, S., & Butler, M. (1988). Problem solving as everyday practice. In R. Charles & E. Silver (Eds.), *The teaching and assessing of mathematical problem solving* (pp. 61–81). Reston, VA: National Council of Teachers of Mathematics.

Leinhardt, G. (1988). Expertise in instructional lessons: An example from fractions. In D. Grouws, T. Cooney, & D. Jones (Eds.), *Perspectives on research on effective mathematics teaching* (pp. 47–66). Reston, VA: National Council of Teachers of Mathematics.

Leinhardt, G., & Greeno, J. (1986). The cognitive skill of teaching. *Journal of Educational Psychology, 78*(2), 75–95.

Marshall, S. (1988). Assessing problem solving: A short-term remedy and a long-term solution. In R. Charles & E. Silver (Eds.), *The teaching and assessing of mathematical problem solving* (pp. 159–177). Reston, VA: National Council of Teachers of Mathematics.

Mathematical Sciences Education Board (MSEB). (1989). *Everybody counts*. Washington, DC: National Academy Press.

Mathematical Sciences Education Board (MSEB). (1990). *Reshaping school mathematics: A philosophy and framework for curriculum*. Washington, DC: National Academy Press.

McGalliard, W. (1983). Selected factors in the conceptual systems of geometry teachers: Four case studies (Doctoral dissertation, University of Georgia). *Dissertation Abstracts International, 44A*, 1364.

Medley, D., & Mitzel, H. (1963). Measuring classroom behavior by systematic observation. In N. L. Gage (Ed.), *Handbook of research on teaching* (pp. 247–328). Chicago: Rand McNally.

Myerson, L. (1977). Conception of knowledge in mathematics: Interaction with and applications for a teaching methods course (Doctoral dissertation, State University of New York at Buffalo). *Dissertation Abstracts International, 39A*, 2.

National Council of Teachers of Mathematics (NCTM). (1980). *An agenda for action*. Reston, VA: Author.

National Council of Teachers of Mathematics (NCTM). (1989). *Curriculum and evaluation standards for school mathematics*. Reston, VA: Author.

Nesher, P. (1988). Multiplicative school word problems: Theoretical approaches and empirical findings. In J. Hiebert & M. Behr (Eds.), *Number concepts and operations in the middle grades* (pp. 19–40). Reston, VA: National Council of Teachers of Mathematics.

Ohlsson, S. (1988). Mathematical meaning and application meaning in the semantics of fractions and related concepts. In J. Hiebert & M. Behr (Eds.), *Number concepts and operations in the middle grades* (pp. 53–92). Reston, VA: National Council of Teachers of Mathematics.

Owens, J. (1987). A study of four preservice secondary mathematics teachers' constructs of mathematics and mathematics teaching (Doctoral dissertation, University of Georgia). *Dissertation Abstracts International, 48A*, 588.

Perry, W. (1970). *Forms of intellectual and ethical development in the college years: A scheme*. New York: Holt, Rinehart and Winston.

Peterson, P. (1988). Teaching for higher-order thinking in mathematics: The challenge of the next decade. In D. Grouws, T. Cooney, & D. Jones (Eds.), *Perspectives on research on effective mathematics teaching* (pp. 2–26). Reston, VA: National Council of Teachers of Mathematics.

Plato. (1952a). The republic. In R. M. Hutchins (Ed.), *Great books of the western world: Vol. 7 Plato* (pp. 295–441). Chicago: Encyclopaedia Britannica, Inc.

Plato. (1952b). Timaeus. In R. M. Hutchins (Ed.), *Great books of the western world: Vol. 7. Plato* (pp. 442–477). Chicago: Encyclopaedia Britannica, Inc.

Polya, G. (1965). *Mathematical discovery: On understanding, learning, and teaching problem solving* (Vol. 2). New York: Wiley.

Progressive Education Association. (1938). *Mathematics in general education*. New York: Appleton-Century Company.

Ptolemy. (1952). The almagest. In R. M. Hutchins (Ed.), *Great books of the western world: Vol. 16. Ptolemy, Copernicus, & Kepler* (pp. 1–478). Chicago: Encyclopaedia Britannica, Inc.

Resnick, L. (1987). *Education and learning to think*. Washington, DC: National Academy Press.

Resnick, L. (1988). Treating mathematics as an ill-structured domain. In R. Charles, and E. Silver, (Eds.), *The teaching and assessing of mathematical problem solving* (pp. 32–60). Reston, VA: National Council of Teachers of Mathematics.

Romberg, T. (1988). Can teachers be professional? In D. Grouws, T. Cooney, & D. Jones (Eds.), *Perspectives on research on effective mathematics teaching* (pp. 224–244). Reston, VA: National Council of Teachers of Mathematics.

Rosenshine, B., & Furst, N. (1973). The use of direct observation to study teaching. In R. Travers (Ed.), *Second handbook of research on teaching* (2nd ed., pp. 122–183). Chicago, IL: Rand McNally.

Saaty, T., & Weyl, F. (1969). *The spirit and the uses of the mathematical sciences*. New York: McGraw-Hill.

Schoenfeld, A. (1988). Problem solving in context(s). In R. Charles & E. Silver, *The teaching and assessing of mathematical problem solving* (pp. 82–92). Reston, VA: National Council of Teachers of Mathematics.

Schoenfeld, A. (1989, March). *Reflections on a "practical philosophy."* Paper presented at the 1989 Annual Meeting of the American Educational Research Association, San Francisco, CA.

Shavelson, R., Webb, N., Stasz, C., & McArthur, D. (1988). Teaching mathematical problem solving: Insights from teachers and tutors. In R. Charles & E. Silver (Eds.), *The teaching and assessing of mathematical problem solving* (pp. 203–231). Reston, VA: National Council of Teachers of Mathematics.

Slavin, R., & Karweit, N. (1984). Mastery learning and student teams: A factorial experiment in urban general mathematics. *American Educational Research Journal, 21,* 725–736.

Slavin, R., & Karweit, N. (1985). Effects of whole-class, ability grouped, and individualized instruction on mathematics achievement. *American Educational Research Journal, 22,* 351–367.

Smith, B. O., Meux, M., Coombs, J., Nuthall, G., & Precians, R. (1967). *A study of the strategies of teaching.* Urbana, IL: Bureau of Educational Research, University of Illinois.

Snapper, E. (1979a). The three crises in mathematics: Logicism, intuitionism, and formalism. *Mathematics Magazine, 52,* 207–216.

Snapper, E. (1979b). What is mathematics? *American Mathematical Monthly, 86,* 551–557.

Sowder, J. (Ed.). (1989). *Setting a research agenda.* Reston, VA: National Council of Teachers of Mathematics.

Sowder, L. (1980). Concept and principle learning. In R. Shumway (Ed.), *Research in mathematics education* (pp. 244–285). Reston, VA: National Council of Teachers of Mathematics.

Steen, L. (1978). *Mathematics today: Twelve informal essays.* New York: Springer-Verlag.

Steen, L. (1988). The science of patterns. *Science, 240,* 611–616.

Steffe, L. (1988). Children's construction of number sequences and multiplying schemes. In J. Hiebert & M. Behr (Eds.), *Number concepts and operations in the middle grades* (pp. 119–140). Reston, VA: National Council of Teachers of Mathematics.

Stewart, I. (1987). *The problems of mathematics.* New York: Oxford University Press.

Thompson, A. (1984). The relationship of teachers' conception of mathematics and mathematics teaching to instructional practice. *Educational Studies in Mathematics, 15,* 105–127.

Thompson, P. (1989). Artificial intelligence, advanced technology, and learning and teaching algebra. In S. Wagner & C. Kieran (Eds.), *Research issues in the learning and teaching of algebra* (pp. 135–161). Reston, VA: National Council of Teachers of Mathematics.

Thorpe, J. (1989). Algebra: What should we be teaching and how should we teach it? In S. Wagner & C. Kieran (Eds.), *Research issues in the learning and teaching of algebra* (pp. 11–24). Reston, VA: National Council of Teachers of Mathematics.

Tymoczko, T. (1986). *New directions in the philosophy of mathematics.* Boston: Birkhauser.

van Heijenoort, J. (1967). *From Frege to Gödel.* Cambridge, MA: Harvard University Press.

von Glaserfeld, E. (1987). Learning as a constructive activity. In C. Janvier (Ed.), *Problems of representation in the teaching and learning of mathematics* (pp. 3–17). Hillsdale, NJ: Lawrence Erlbaum.

Wearne, D., & Hiebert, J. (1988). Constructing and using meaning for mathematical symbols. In J. Hiebert & M. Behr (Eds.), *Number concepts and operations in the middle grades* (pp. 220–235). Reston, VA: National Council of Teachers of Mathematics.

Whitehead, A. N., & Russell, B. (1910–13). *Principia mathematica* (3 vols.). New York: Cambridge University Press.

Wilder, R. (1968). *Evolution of mathematical concepts: An elementary study.* New York: Wiley.

·3·

PERSPECTIVES ON SCHOLARSHIP
AND RESEARCH METHODS

Thomas A. Romberg
UNIVERSITY OF WISCONSIN

A major reason why research methodology in education is such an exciting area is that education is not itself a discipline. Indeed, *education is a field of study*, a locus containing phenomena, events, institutions, problems, persons, and processes, which themselves constitute the raw material for inquiries of many kinds. (Shulman, 1988, p. 5)

During the past quarter century, scholars from many fields addressing the problems associated with the teaching and learning of mathematics have followed a variety of perspectives in carrying out their investigations. In this chapter I intend to identify in the social sciences the broad research trends that are related to the study of teaching and learning in school settings and to determine how those trends have influenced the study of mathematics in schools. As an aid for understanding the basis of these trends, I will (1) describe some features of mathematical sciences education as a field of study, (2) sketch the activities of researchers, and (3) outline the variety of research methods now in use. Five general research trends and their relationship to studies in mathematical sciences education are then identified and briefly described.

MATHEMATICAL SCIENCES
EDUCATION AS A FIELD OF STUDY

It is important to consider mathematical sciences education as a field of study because, as Shulman (1988) has argued,

school is complex; thus, the perspectives and procedures of scholarly inquiry of many disciplines have been used to investigate the questions arising from and inherent in the processes involved in the teaching and learning of mathematics in schools. E. G. Begle's (1961) diagram of school mathematics illustrates the interrelationship of the components in the schooling process and the need for multiple perspectives and procedures (see Figure 3.1). In this diagram, the enterprise of schooling is situated within a social context; the mathematical sciences curriculum involves a subset of mathematics; and instruction is carried out by a teacher with a group of students within a school classroom over time. This diagram was drawn to present a point of view regarding mathematics instruction by developing five basic points:

1. Schools have been *created by social groups* to prepare their young for membership in society.
2. *Sound* mathematics instruction is approached from a concern over what ideas from mathematics are taught and what uses are indicated.
3. Mathematics instruction can be *effective* if the learner is taken into consideration.

The author gratefully acknowledges the helpful feedback on the initial draft of this chapter provided by Richard Shumway, Ohio State University, and Richard Lesh, Educational Testing Service.

The research reported in this paper was supported by the Office of Educational Research and Improvement of the US Department of Education and by the Wisconsin Center for Education Research, School of Education, University of Wisconsin-Madison. The opinions expressed in this publication are those of the author(s) and do not necessarily reflect the view of the OERI or the Wisconsin Center for Education Research.

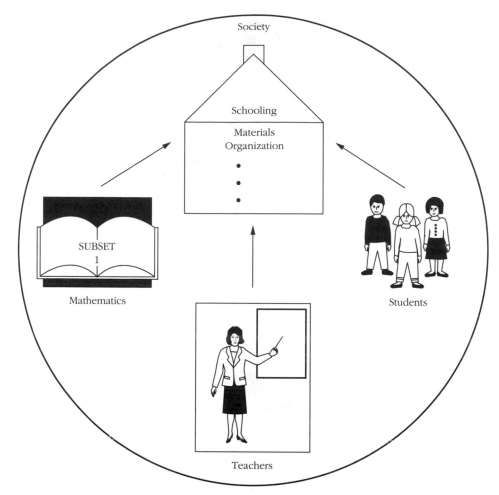

FIGURE 3–1. The relationship of society, mathematics, students, teachers, and schooling. (Adapted from E. G. Begle, Class notes for a seminar in mathematics education, 1961)

4. *Efficient* mathematics instruction can be accomplished through consideration of aspects of schooling.
5. Teachers are the *managers and guides* who make the instructional process work.

From these points, myriad questions can be raised, conjectures posed, and investigations then conducted. For example, such questions as the following could be raised:

- Who makes the decisions about what mathematics is included in a mathematics curriculum?
- Why is Euclidean geometry taught as a year-long course in American schools?
- What conceptions are held by teachers about how students learn to solve mathematical problems?
- How can instruction on rational numbers be sequenced and paced?
- What are students' conceptions about ratios and proportions?
- What impact on teaching does teacher isolation from the work of other mathematics teachers have?

Each of these questions deserves to be investigated. However, individual scholars might use different methods to study each question. Furthermore, scholars from different disciplines might study the same questions in quite different ways. Perspectives from mathematics, the sociology of knowledge, history, learning psychology, developmental psychology, agriculture, and anthropology have commonly been used to study such educational questions. As each of these disciplinary perspectives is brought to bear on the field of mathematical sciences education, it produces its own set of concepts, methods, and procedures. To understand the current trends in research in mathematical sciences education, one must be aware of these many perspectives and the principles upon which they are based. This is important because differences in methods do not merely comprise alternative ways of investigating the same questions. What distinguishes one method from another is not only the way in which information is gathered, analyzed, and reported, but also the very types of questions typically asked and the principles or paradigms upon which the methods to investigate such questions are based.

ACTIVITIES OF RESEARCHERS

The term *research* refers to processes—the things one does, not objects one can touch and see. Furthermore, doing research cannot be viewed as a mechanical performance or a set of activities that individuals follow in a prescribed or pre-determined fashion. The activities involved in doing research embody more characteristics of a craft than of a purely technical discipline. As in all crafts, there is agreement in a broad sense about what procedures are to be followed and what is considered acceptable work. These agreements arise from day-to-day intercourse among researchers. Given that doing research is a craft, what are its essential activities? For the purposes of this chapter, 10 activities are described (see Figure 3.2). There is nothing unique about this list; in fact, almost every research methods text outlines a similar set of activities. However, it is given here (1) to highlight some of the common problems that persons unfamiliar with research face in understanding the research process and (2) to provide a background for the discussion of research trends. Fur-

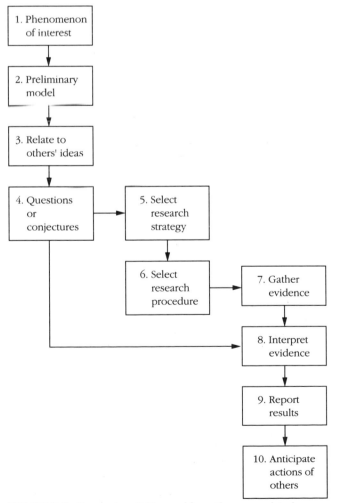

FIGURE 3–2. Research activities and how they are related.

thermore, although the activities are presented in a sequential order, they are not necessarily followed in that order. The interplay among factors such as the scholar's intent, assumptions, conjectures, availability of information, methods, and so forth cannot in practice be separated so neatly. The first four activities are the most important, for they are involved with situating one's ideas about a particular problem in the work of other scholars and deciding what to investigate. The next two activities involve making decisions about what kind of evidence to gather and how that is to be done. The next step is gathering data, and the last three have to do with making sense of the information gathered and reporting the results for others.

1. *Identify a phenomenon of interest.* All research begins with curiosity about a particular phenomenon in the real world. In mathematical sciences education, as suggested in Figure 3.1, the phenomenon involves teachers and students, how students learn, how students interact with mathematics, how students respond to teachers, how teachers plan instruction, and many other issues. Mathematics educators address a variety of areas, as one can see by examining any issue of the *Journal for Research in Mathematics Education*.

2. *Build a tentative model.* A researcher makes guesses about certain important aspects as variables of the phenomenon of interest and of how those aspects are related, then illustrates these in a model. For example, Thomas Carpenter and Elizabeth Fennema (1988) proposed the model in Figure 3.3 for integrating cognitive and instructional science to guide their current research. In this sense, a model is merely a set of descriptions of key variables and the implicit relationships among the variables. For most scholars, a model is merely a heuristic device to help clarify a complex phenomenon. Real situations are rarely well-defined and are often embedded in an environment that makes it hard to obtain a clear statement of the situation. Formulating a tentative model usually helps because doing so involves specifying the variables that one believes are operating in the real situation. Of course, the model is a simplification, since some features of reality will be significant and others irrelevant. Nevertheless, the model serves as a starting point or orientation to the situation of interest. Good researchers, like good artists in any field, as Jeremy Kilpatrick (1981) has suggested, are more creative in identifying variables and relationships that enable one to look afresh at familiar phenomena than are persons who are less imaginative.

3. *Relate the phenomenon and model to others' ideas.* An important activity is examining what other people think about the phenomenon and determining whether their ideas can be used to clarify, amplify, or modify the proposed model. A researcher interested in how children develop counting skills attempts to relate his or her ideas to other researchers' ideas about that phenomenon. To do this, the researcher must recognize that each investigator is a member of a particular scholarly group that holds a "world view." If one is to examine the potential contribution of others' ideas, one must relate those ideas to a particular world view. For example, a scholar who views the variety of children's understandings about fraction concepts from a constructivist viewpoint may argue that the typical experiences children have with fractions is impoverished. To build that argument, the researcher would have read and reflected

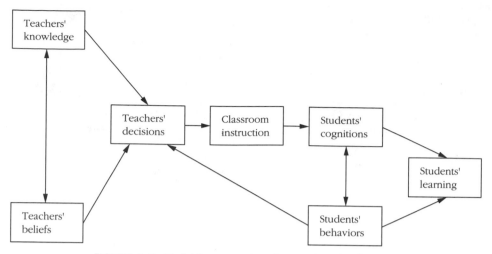

FIGURE 3–3. Model for research and curriculum development.

on the writings and studies of other constructivist scholars. On the other hand, if that researcher held a behaviorist view, his or her ideas would have to be assessed with respect to the works of other behaviorists.

4. *Ask specific questions or make a reasoned conjecture.* This is a key step in the research process because, as one examines a particular phenomenon, a number of potential questions inevitably arise. Deciding which questions to examine is not easy. John Platt (1964) argued that the choice of which question to examine is crucial. If the "critical" questions are asked, then "strong" inferences can be made; otherwise, a particular study may contribute little to a chain of inquiry. The notion of strong inferences leads to the important feature of most research programs—namely, the cumulative nature of a series of studies within a particular framework (Lakatos, 1976). In such a case, the constructivist scholar mentioned above might argue that a better understanding would occur if rich instructional activities were developed that allowed the child to construct meanings about part-whole relationships, whereas the behaviorist might argue for a better corrective feedback (reinforcement) schedule for a set of activities.

Questions usually take one of the following forms: How did things come to be this way? (past oriented), What is the status of things? (present oriented), or What will happen if I do the following? (future oriented). For example, another scholar studying fraction concepts may be interested in how the concepts have evolved in mathematics or possibly in how fractions have been defined and illustrated in textbooks in this century. Either interest could lead to specific past-oriented questions. Alternatively, one might be interested in teachers' current levels of knowledge about the domain of rational numbers, which would lead to specific present-oriented questions. Or one could argue that teaching calculator algorithms would contribute to students' understanding of fractions and thus raise specific future-oriented questions.

Of particular note is the fact that most past- and present-oriented studies are descriptive in character, whereas future-oriented studies are predictive. This distinction leads to a dispute as to whether one can make causal arguments from descriptive data. Experimentalists contend that only by manipulating variables under controlled situations can one reliably build causal arguments. Other scholars argue that one can build such arguments from descriptive data based on theoretical grounds.

Rather than simply raising interesting questions, researchers usually make one or more conjectures (reasoned guesses or predictions) about what it would take to answer the question(s). The conjectures are based on some relationship between the variables that characterize the phenomenon and on the ideas about those key variables and their relationships as outlined in the model.

5. *Select a general research strategy for gathering evidence.* The decision about what methods to use follows directly from the questions one selects, from the world view in which those questions are situated, from the tentative model one has built in order to explain the "phenomenon of interest," and from the conjecture that one has made about needed evidence. For example, if the questions to be answered are about the past, historiography would be appropriate. On the other hand, if the questions are present oriented, one may choose to do a survey or a case study, or to use one of several other data-gathering strategies.

6. *Select specific procedures.* To answer the specific questions that have been raised, evidence has to be gathered. It is at this step that the techniques usually taught in research methods courses are important: how to select a sample, how to gather information (interview, question, observe, test), how to organize information once it is collected, and so forth. There are a large number of specific procedures that one might follow for different types of questions. One must take care to select procedures that will shed light on the questions.

7. *Collect the information.* This step may be straightforward once one has decided to collect certain information to build an argument regarding the questions being asked. For example, if conducting a survey is appropriate, the procedures for gathering data, though often complex, can be planned. On

the other hand, if one is examining the culture of a classroom, the procedures for gathering information may expand or become more focused as one collects data.

8. *Interpret the information collected.* At this stage, one analyzes and interprets the information that has been collected. In many studies, the researcher scales information, aggregates it, and carries out appropriate statistical tests of significance on properties of the data. These are usually called *quantitative* methods, since assigning numbers to information is usual (scaling) and mathematical procedures are followed to aggregate and summarize the evidence. In other areas, such as a historical study, the researcher also categorizes, organizes, and interprets the relevant information that has been gathered. But if numbers are not used, the methods of analysis are called *qualitative*. It is important to realize, however, that in every investigation, more information is gathered than can be used to answer the question. Some of it is relevant, some of it is irrelevant, and some of it is not comprehensible. Teasing out the important information from all that is available is an art at which certain people are better than others.

9. *Transmit the results to others.* Being a member of a scholarly community implies a responsibility to inform other members about the completed investigation and to seek their comments and criticisms. Too often, researchers report only procedures and findings, not the model or world view. The findings of any particular study are interpretable only in terms of the world view. If it is not stated, readers will undoubtedly use their own notions to interpret the study. Significant differences between characteristics of distributions—such as the mean of the calculator algorithm group for understanding fractions and the mean of another group—are not important in themselves. Not just results, but also answers to questions that should have been embedded in a "normal science," must be transmitted to others.

10. *Anticipate the action of others.* Given the results of a particular investigation, every scholar is interested in what happens next and should anticipate later actions. Members of a scholarly community discuss ideas with each other, react to each other's ideas, and suggest new steps, modifications of previous studies, elaborations of procedures, and so forth. Scholars attempt to situate each study in a chain of inquiry. Things that came before and things that come after any particular study are important.

This outline suggests that three aspects of the research process need to be particularly addressed:

1. Researchers as members of a scholarly community
2. The ideology and paradigms of different research communities
3. The failure of many novice researchers to identify with a research community

Scholarly Communities

A scholar conducts research within a community. Scholarly communities involve commitments to certain lines of reasoning and premises for certifying knowledge. Each field of scholarship is characterized by particular constellations of questions, methods, and procedures. These constellations provide shared ways of "seeing" the world, of working, of testing each other's studies. Learning the exemplars of a field of inquiry involves more than learning the content of the field; it is also learning how to see, think about, and act toward the world.

Thomas Kuhn (1970) argues that within a discipline there is, at any one time, a dominant approach, with interrelated questions, procedures, and conceptual perspectives. This dominant approach, or "normal science," has well-established research problems, and the task of individual scholars is to fill in puzzles contained in the paradigm. "Normal science" is a product of a network of personal contacts that has been termed an "invisible college" (Crane, 1976). In each field of study there is often a small group of highly productive and influential scholars who communicate with each other informally about their work. These small groups set priorities for research and for the recruitment and training of new students; they also monitor the changing structure of their scholarly field. Even within the constraints of a particular community, researchers are rarely free to investigate any phenomenon that might strike their fancy. They are constrained by social pressure that dictates both what problems it is important to study and how resources are allocated, as well as by the expectations of other scholars about how problems are viewed, investigations carried out, and reports written.

In this sense all research communities operate within a cultural context. The social conditions in part dictate what it is fashionable to study and how scarce resources are to be allocated. In the 1960s, for example, the Civil Rights movement, protests against the Vietnam War, and changes in the social roles of women and men posed new themes for study. In the 1970s, the issue of gender differences in mathematics learning and then in teaching became an important issue. Today, the mandate is to produce a mathematically literate workforce composed of people who can reverse sagging industrial productivity. In this context, two interactive features between researchers and social problems should be mentioned. First, the way in which the social issues are conceived and examined is often based on

…the ability of particular theorists to create new themes for considering the world, thus creating new questions and leading [one] to search for different data to provide insights into how our world is constructed and changed. Contrary to prevailing belief, the potency of social science is not in the utility of its knowledge but in its ability to expand and to liberate the consciousness of people into considering the possibilities of their human conditions. (Popkewitz, 1984, p. 8)

Thus, scholars not only respond to social and political pressures, they help define the problems and create the political pressure for the resources needed to investigate those problems. Second, although the roots of what is studied lie in a social need, research by itself will never satisfy that need. As argued earlier, all research is carried out in order to understand some phenomenon. The assumption is that increased understanding will help to solve the problem or improve practice, but it must be recognized that a given study or combination of studies may not do so. For example, concern over ostensibly capable but "at-risk" students has spawned myriad studies on school alienation, on the sociological and psychological characteristics of these students, and on other relevant issues. The information gathered is useful and interesting, but will not by itself alleviate those concerns.

Finally, scholarly communities can restrict research. At certain times anomalies appear that seem unresolvable, and new concepts and tools must be created. These challenges to the community result in anxiety and debate that are not always resolved. The challenge to the existing disciplinary beliefs becomes not only a challenge to the objective questions held by the community but to basic premises about the organization of reality. For example, when competing notions of psychology (behavioral and gestalt) vied for acceptance in the 1930s, the issue of the debate was not evidence, but what should be considered legitimate psychological research. A similar conflict exists in the trend toward teacher empowerment in mathematics education, discussed later in this chapter. On one side, scholars such as Jere Brophy (1986) cite the importance of "process-product" teaching research, whereas others, such as Jere Confrey (1986), see that line of research as impoverished. Again the issue is not evidence, but what is important to study in the teaching of mathematics and how one views the job of teaching.

Ideology and Paradigms

To relate one's ideas to the work of other scholars, one must understand the philosophic perspectives that underlie one's work. Historically, the work of all social scientists has been grounded in the ideology and methods used to develop scientific theories in natural sciences such as physics or chemistry. During the last few centuries, the creation of such theories has led to what John Dewey called "authorized conviction" (1966, p. 189). These convictions about natural phenomena are propositions that are considered true as a consequence of systematic observations. Thus, the propositions are knowledge-claims about the world that have been uncovered by scholars. During the past century, this ideology regarding the nature of science, known as *positivism,* was adopted by social scientists in an attempt to build theories about social behaviors. In 1913, for example, J. B. Watson, the father of behaviorism, argued that "psychology, as the behaviorist views it, is a purely objective experimental branch of natural science. Its theoretical goal is the prediction and control of behavior" (1948), p. 457). This view about the scientific enterprise has dominated educational research until very recently.

Within the philosophy of science during the past half-century, there has been a rapid demise of the belief in positivism because, as W. P. Weimer has argued, "Our knowledge of the nature of science and its growth has increased in recent years, and traditional conceptions of science and its methodology have been examined, found wanting, and are in large part being abandoned" (1979, p. ix). Three aspects of this examination need to be mentioned. First, in large part because of the development of non-Euclidean geometries in the 19th century, it was demonstrated that one could develop equally consistent theories if one chose different assumptions. Thus, theories

do not reflect truth about the natural world; they reflect only mankind's assumptions about the world. Second, the position of philosopher Karl Popper (1968) on testing theoretical propositions—their truth can never be verified, only refuted—implies the fallibility of such propositions. The current view is that science is concerned with reliable knowledge, not with truth. A proposition may be useful for some predictions today because no one has as yet falsified it. However, it will quite possibly be shown to be false in the future. Finally, during this century, many social scientists and educators have rejected the physical sciences model as appropriate to their disciplines. Their nonnaturalistic ideological arguments (referred to as hermeneutic, interpretive, phenomenological, or critical) attempt to interpret the social interactions of humans within a culture, the rules that govern those interactions, and so forth.

Thus, although earlier in this century one could have assumed that most researchers held similar ideological views, one can no longer do so. Given this fact, it should not be surprising that during the past decade, inquiry into the learning and teaching of a variety of subjects (including mathematics) in school has been of many types for many reasons. First, scholars from a number of fields have chosen to examine different aspects of the general field of study. However, as illustrated earlier, the variety of perspectives has led to many disagreements between groups of scholars about the kinds of studies being conducted. The debates, voiced in educational journals and conventions, have led to a plethora of new journals and organizations. (For an analysis of the issues that underlie various research programs, the interested reader should see *Paradigm & Ideology in Educational Research* [Popkewitz, 1984]. The notions presented in this chapter draw heavily on that work.)

It has become clear that the differences in the conceptual lenses through which scholars view phenomena often represent deep-seated differences in assumptions about the nature of the world to be investigated. For the study of schooling, the different ideological perspectives are reflected in a group's assumptions about the knowledge that is to be taught, the work of students and how learning occurs, the work of teachers and professionalism, and the social organization and technology of the classroom and schools (Romberg, 1988). The implications of these different approaches can be illuminated by examining the particular world views associated with three paradigms that have emerged to give definition and structure to the practice of educational research: empirical-analytic, symbolic, and critical (Popkewitz, 1984).

The Empirical-Analytic Paradigm. The empirical-analytic approach to research, which has its origin in logical positivism, starts with the premise that one's goals involve explaining the relationship of humans to the natural world, and it uses those explanations to gain intellectual or technical control of the world. (Discussions of empirical-analytic science can be found in Easton [1971] and Homans [1967].) It is also assumed that what one knows can only be based on what can be observed or made observable (the empiric) and that observations are made to separate human behaviors into their constituent elements (the analytic). The product of such analyses are law-like theories of social behavior.

Popkewitz (1984) has argued that there are at least five other interrelated assumptions that give form to empirical-analytic research:

1. Theory is to be universal, not bound to a specific context or to actual circumstances in which generalizations are formulated. (p. 36)
2. There is a commitment to a disinterested science. The statements of science are believed to be independent of the goals and values which people may express within a situation. By eliminating contextual aspects, theory is only to describe the relationship of the "facts." (p. 37)
3. There is a belief that the social world exists as a system of variables. These variables are distinct and analytically separable parts of an interacting system.... The notion of a system of variables provides a specific meaning of causation within the empirical-analytic sciences. A cause is a relationship among empirical variables that can be explained or manipulated to produce conditionally predictable outcomes. (p. 37)
4. There is a belief in formalized knowledge. This involves making clear and precise the variables of inquiry prior to research. (p. 38)
5. The search for formal and disinterested knowledge creates a reliance upon mathematics in theory construction. Quantification of variables enables researchers to reduce or eliminate ambiguities and contradictions. It is also to elaborate the logical-deductive structure of knowledge by which hypotheses can be tested and improvements in theory made. (p. 38)

For education, these assumptions translate into the belief that the knowledge to be learned in any given area, such as mathematics, can be specified in terms of facts, concepts, procedures, and so forth; that the job of students is to master that knowledge; that the job of the teacher is to present that knowledge to pupils in an organized manner and to monitor their progress toward mastery; and that the organization and technology of the classroom and school are arranged to make the teaching and mastery of that knowledge as efficient as possible. The influence of these beliefs is apparent in many works on education, such as Bloom's *Taxonomy of Educational Objectives* (1956), *The Scientific Basis of the Art of Teaching* (Gage, 1978), and, in mathematics education, the work on "direct instruction" (such as that of Good & Grouws, 1981).

The Symbolic Paradigm. The goal of this approach to research is to understand how humans relate to the social world they have created. It is based on a belief that social life is created and sustained through symbolic interactions and patterns of conduct. Others have used the terms *hermeneutic, interpretive,* or *phenomenological* to describe this paradigm. (Symbolic sciences are discussed in Winch [1971], Schultz [1973], Luckmann [1977], and Cicourel [1969].) It is through the interactions of people that rules to govern social life are made and sustained. As Popkewitz (1984) has argued, "The ideas of 'rule-making' and 'rule-governed' can be contrasted to the law-like generalization of the empirical-analytic sciences. In the latter, it is methodologically assumed that there is an invariant nature to human behavior which can be discovered. The idea of 'rules' shifts attention from the invariant nature of behavior to the

field of human action, intent and communication" (p. 40). The goal is to develop theories about the social rules that underlie and govern social actions. Such theories are about the nature of discourse rather than behavior.

Thus, it assumed that the unique quality of being human is found in the symbols people invent to communicate meaning or to interpret the events of daily life. Language and its use within a culture—the interactions and negotiations in social situations—are assumed to define the possibilities of human existence. Like empirical-analytic science, however, the symbol paradigm is descriptive and neutral about social affairs. Thus, questions about actions to change social conditions, although informed by communication patterns, are subject to political or philosophical considerations.

This paradigm is common within such disciplines as sociology, political science, and anthropology. In education, this perspective translates into the belief that knowledge is situational and personal, that pupils learn by construction as a consequence of experiences, that the job of teaching is to create instructional experiences for students and negotiate with them intersubjective understandings gained from those experiences, and that the organization and technology of the classroom and school are arranged so that all of the experiences can be rich and meaningful. Examples of studies from this perspective include such works as *Inside a High School Student's World* (Cusick, 1973), *From Communication to Curriculum* (Barnes, 1975), and, from mathematics education, the variety of recent studies in "ethnomathematics" (such as that of D'Ambrosio, 1985).

The Critical Paradigm. The goal of the critical paradigm is to demystify the patterns of knowledge and social conditions that restrict our practical activities. The basic assumption of those holding this view is that humans, through thought and action, can improve the social world in which they live. This belief is rooted in the rapid social changes of the past century and in certain problems brought about by those changes. (Critical stances are discussed in Bernstein [1976] and Lukacs [1971].) For example, knowledge, for some, has become "professionalized." The consequence is that many individuals become dependent upon certain experts in society. The task of inquiry is to illuminate the assumptions and premises of social life so that individuals come to know themselves and their situation, understand the scope of and the boundaries placed upon their affairs, and offer arguments that are in opposition to the dominant culture and institutions. Thus, as Popkewitz (1984) has argued, "The function of critical theory is to understand the relations among value, interest, and action and, to paraphrase Marx, to change the world, not to describe it" (p. 45).

The impact of this paradigm on educational thought involves the beliefs that knowledge is gained by reflection on how humans can improve social conditions, that pupils learn through reflection and action, and that the job of teaching is to have students reflect on the social world in which they live and initiate actions to challenge current practices. Educational research from this perspective can be found in such works as *Education and Power* (Apple, 1982), *Learning to Labour* (Willis, 1977), and, in mathematics education, in the works of such scholars as Jean Anyon (1981).

Each of the three paradigms provides a different view of the nature and causes of our social situation. Each intellectual tradition provides a particular vantage point for considering the social actions involved in the learning and teaching of mathematics in school settings. It is important to realize that we live in a world of differing social visions. Such pluralism is a characteristic of the intellectual community in our society.

The Failure of Novices

This is merely a brief critical aside. Too many novices (usually graduate students) fail to see the importance of situating their study with the work of others. They often skip from a problem of interest to designing a study and gathering data. Failure to embed one's ideas within a community of scholars at best makes the results open to a variety of interpretations and most often leads to a study of little real value. Even if they review the literature related to their field of interest, they fail to appreciate the differences in perspectives of divergent authors (for example, supporting a position with quotes from both Skinner and Piaget, as if the two viewed the world through similar lenses). The consequences of such failures have been well documented by Jeremy Kilpatrick in his delightfully titled article, "The Reasonable Ineffectiveness of Research in Mathematics Education" (1981).

METHODS USED BY RESEARCHERS

In Figure 3.2, Activities 5 through 10 are those in which a scholar decides (1) what evidence is needed to address the questions or conjectures raised; (2) how to gather, analyze, and interpret that evidence; and (3) how to report the findings to others. It should be noted that researchers rarely begin an investigation with a fixed strategy for gathering evidence or with a specific method of analysis in mind. Only a novice would say "I just learned factor analysis, what can I study?" or "I just read an interesting ethnography, I would like to do an ethnographic investigation." The decisions about methods are made as a consequence of Activities 1 through 4. Given this caution, there are two aspects to the use of the term *research methods* that need to be understood. First, the specific methods discussed in the research literature may include the manner in which information is collected, the way it is aggregated and analyzed, or, sometimes, how it is reported. Second, the actual methods a researcher uses to gather evidence depend on at least five factors: world view, the time orientation of the questions being asked, whether the situation currently exists or not, the anticipated source of information, and product judgment.

World view, as discussed in the last section, situates the methods used within the beliefs of a particular scholarly community. *Time orientation* refers to whether the questions being raised are past, present, or future directed. *Situations* currently ex-

ist or need to be created. The *source* of evidence must be either artifacts (books, speeches, and the like), answers to questions, or observations of actions. *Judgment* refers to evaluation studies as a distinct category of research methods. An extensive number of specific methods exist in the literature that are based upon or use these five factors.

Methods Used with Existing Evidence

There are three methods in which researchers have no latitude for generating new data; that is, they must find what already exists and cannot alter the form in which the data appear. These methods are

- *Historiography.* In this approach, an effort is made to cast light on current conditions and problems through a deeper and fuller understanding of what has been done or occurred in the past.
- *Content analysis.* This method is used to investigate present-oriented questions when current artifacts can be examined.
- *Trend analyses.* This method is used to extrapolate from information about the past or present in order to make predictions about the future.

Methods Used When a Situation Exists and Evidence Must Be Developed

There are many different methods of investigation in which a situation exists and specific evidence must be gathered. In each method, the researcher has control over the form by which the information is gathered and aggregated:

- *Retrospective survey.* This method is used to study questions that are past oriented—the situation once existed, and individuals who were participants in the past situation can be interviewed using this method.
- *Mass-descriptive survey.* This very common method is used to study present-oriented questions. The procedure is to ask a carefully drawn sample of participants in the situation to answer a set of predetermined, structured questions.
- *Structured interviews.* This method is similar to the mass-descriptive survey except that it is assumed that by listening to (and coding) responses, researchers can find more illuminating information than by using mass surveys.
- *Clinical interviews.* This method of inquiry begins in a manner similar to structured interviews, but the sequence of questions varies with each respondent, depending on prior answers.
- *Projective surveys.* This method is used to make predictions about future events by asking a sample of persons to make judgments about trends.
- *Structured observations.* This method is used to study groups in which the actions of different members can be viewed. For example, to document the kinds of teacher and student interactions that take place during a mathematics lesson and to determine the extent to which students are engaged in learn-

ing, Charles Fischer and his colleagues (1978) developed a "time-on-task" observation procedure.

- *Clinical observations.* This method is used to study group behavior. The distinction between structured and clinical observations is similar to that made between structured and clinical interviews. The details of what one observes shift from predetermined categories to new categories, depending upon initial observations. Also, the observer is often a participant in the situation.

- *Longitudinal study.* To study change over time, the longitudinal approach is the most "natural" one. In fact, quite often its use is taken for granted. Studies of learning and memory, adaptation and habituation, fatigue, attitude change, and population change have inevitably resorted to a repeated-measurements approach as the most logical one for obtaining the desired information.

- *Cross-sectional designs.* Because the study of actual change over time is impractical in many instances, scholars have resorted to comparing "equivalent" samples of students at different ages and making inferences about growth or change.

- *Causal modeling.* This method is often used to investigate complex educational situations. The researcher begins by constructing a mathematical model that specifies the variables of interest and how they are related. Then a population is defined, a sample drawn, information gathered for each variable in the model and scaled, and relationships between variables studied. The information gathered usually involves a combination of data from surveys, interviews, and observations. This is a method commonly used by sociologists to build causal arguments without manipulating variables. Causality is claimed from the theoretical basis for the model.

- *Case studies.* This method is used to organize and report on information about the actions, perceptions, and beliefs of an individual or group under specific conditions or circumstances. The researcher is interested in telling an in-depth story about a particular case. What distinguishes a case study from other field study methods (ethnography, action research, and illuminative evaluation) is that researchers are writing a natural history of a particular situation. They are not interested in making judgments about a program or in testing a theoretical assumption.

- *Action research.* This term refers to a research strategy used to investigate schooling situations where the researcher assumes "wise practice" that needs to be documented and understood has evolved in schools or classrooms. Also, the documentation is often by the practioner.

- *Ethnography.* The term *ethnography* comes from anthropology and means "a picture of the 'way of life' of some identifiable group of people" (Wolcott, 1987, p. 188). The label is used to refer to both the combination of processes used when anthropologists do fieldwork and the product of that effort—the written ethnographic account. It is a method used to study the complex culture of schools. This approach usually requires a highly trained participant observer to examine artifacts, conduct clinical interviews, and do clinical observations. It differs from other methods in that the researcher

attempts to interpret the actions of people in terms of the culture in which they live.

Experiments

If one's conjecture involves predicting what will happen under conditions that do not now exist—that is, if it involves gathering evidence about the effects of a new and different product or program—one uses an *experimental approach.* In such studies, creating the new situation is a critical part of the effort. Through the study of the new situation and its effects, the researcher attempts to build a "causal" argument. In this category there are many possible distinctions among the variety of methods used by researchers. However, there are three general approaches used in education:

- *Teaching experiments.* This method of investigation is based on a common practice of good teachers. Periodically, most teachers try something new in their classroom and then judge the consequences of that action on student learning. However, the approach used by researchers is much more systematic in that hypotheses are first formed concerning the learning process, a teaching strategy is developed that involves systematic intervention and stimulation of the student's learning, and both the effectiveness of the teaching strategy and the reasons for its effectiveness are determined.

- *Comparative experiments.* This method of investigation is used to determine whether or not a specific set of actions "causes" a desired outcome. The determination is made by comparing the outcomes of a group treated by the set of actions with a similar group (the control group) that has not been so treated to see if there are predictable differences in outcomes. The design of a specific experiment involves attempts to isolate the treatment effects from other possible effects. For example, Campbell and Stanley (1963), in their now classic exposition on experimental designs, organized the variety of designs into three categories: pre-experimental, quasi-experimental, and true-experimental. The designs differ in terms of how potential sources of invalidity are controlled. True experiments, for example, control for internal invalidity by comparing treatment and control groups when units (students, classes) have been randomly assigned to the treatment groups.

- *Interrupted time-series experiments.* Quite often in education it is impractical or impossible to have a control group. For example, if a school district were to increase the number of computers in its computer lab, it would be impossible to determine the consequences of this increase (such as the number of students using the lab or average time spent on a computer) if the increase in access applied to some students (the treatment group) and not to others (the control group). In these circumstances, it is reasonable that the students be their own control. It is possible to compare outcomes prior to and after treatment. The key to using this strategy lies in gathering sufficient observations prior to treatment so that a trend can be determined and then comparing that trend with outcomes after treatment.

Evaluations

Finally, it is common, especially in mathematics education, that individuals or groups create new products intended to improve teaching and learning. The products may be new instructional materials, instructional techniques, or instructional programs. The development of a product involves an engineering process of inventing parts and putting them together to form something new. There are four stages to the development process: product design, product creation, product implementation, and product use. Complementing the production of new materials are four general methodologies that evaluators have developed to determine the product's quality at each stage.

- *Needs assessment.* To decide whether the design of a new product is "good," the researcher must answer three questions: (a) Is there a need for the product? (b) Is there a reasonable probability that the product being considered will fulfill that need? and (c) Among other products, what priority does this product have? To answer these questions, other existing or new evidence is collected and examined.

- *Formative evaluation.* At the creative stage of product development, an evaluator is interested in whether or not the product meets the design specifications. In order to determine whether the product is "good," evidence is gathered to answer four questions: (a) Is the content of the product of high quality? (b) Are intended performance outcomes reached? (c) Are unintended performance outcomes identified? and (d) Are necessary support services for installation provided?

- *Summative evaluation.* To determine whether or not a newly created product is ready for use, the researcher gathers evidence to answer four additional questions: (a) How is the content of the product different from its competitors'? (b) What performance differences exist between the product and its competitors? (c) What cost differences exist between it and its competitors? and (d) Have provisions been made for maintaining the use of the product?

- *Illuminative evaluation.* This term was coined by Malcolm Parlett and David Hamilton (1976) to characterize the study of "innovatory" programs in actual use. The procedure involves the application of field research methods (case study, ethnography, or action research) to the evaluation of new educational products. Note, however, that in this case the focus is on telling the story about the use of the product and making judgments about it.

This list of methods of gathering evidence is arbitrary. Many involve common procedures for gathering data or organizing it. Nevertheless, there are three important aspects associated with each method that need to be addressed: objectivity, quality of evidence, and generalizability.

Objectivity. Objectivity and its counterpart, bias, are of concern to all researchers as they consider problems, raise questions, gather evidence, organize evidence, and build arguments. On the one hand, a researcher must be both personally interested in the phenomenon and committed to investigating questions about it. In this sense all scholars are biased; in fact, a disinterested investigator should not be trusted. On the other hand, objectivity is expected in the manner in which a scholar gathers evidence, examines it, and reports findings. Everyone who has ever researched a problem admits that being objective is not always easy, for several reasons. For example, the kind of evidence one is likely to search for and examine is in part dictated by the community of scholars to which one belongs and its world view. Because of the concern about being objective, one feature of scholarly inquiry involves the presentation of the evidence gathered and the manner in which it has been examined so that others can judge or reexamine it.

Quality of Evidence. Two aspects of the evidence and the way in which it was gathered are usually considered when building an argument about the findings of an investigation: validity and reliability. In education, the *validity* of the evidence is often difficult to demonstrate because it refers to the degree to which the evidence being gathered is directly related to the phenomenon of interest. Too often, education investigators gather evidence on proxies (or indicators) that are only marginally related to the actual phenomenon. For example, because it is difficult to know a student's attitude about mathematics, an index derived from the student's answers to several questions is often used as an indicator of attitudes. The validity of such proxies is often questionable.

The *reliability* of the evidence concerns the accuracy of the evidence in light of the way in which it was gathered. For example, test scores are judged to be reliable if an index of consistency is high, and observations are considered reliable if one can demonstrate high agreement among judges. One problem in this regard involves the fact that no one wants to include in his or her analysis evidence that is in error or idiosyncratic. For this reason, researchers faced with the problem of "dirty," or unreliable, data attempt to "clean" the data before analysis.

The importance of having quality evidence cannot be overemphasized. This is true because, as Harold Larabee (1945) put it, "anyone who has surveyed the long history of man's claims about knowing is struck by the discrepancy between the pretentiousness of most knowledge-claims and the small amount of evidence actually available with which to back them up" (p. 82). The primary role of researchers is to provide reliable evidence to back up claims. Too many people are inclined to accept any evidence or statements that are first presented to them urgently, clearly, and repeatedly. Such statements are often dogmatic and assertive—that is, they are affirmations made about what is known with certainty (even if they are continually outrunning the evidence offered). Some persons have the intense conviction that they are absolutely right in their beliefs; this enables them to adopt and pursue bold policies with vigor and persistence. Knowing little or nothing is not always an obstacle to being confident; most people are slow to acknowledge that the true measure of their knowledge is their awareness of how much remains to be known. A researcher tries to be one whose claims of knowing go beyond a mere opinion, guess, or flight of fancy, to responsible claims with sufficient grounds for affirmation. In fact, Lee Cronbach and Pat Suppes (1969) claim that:

Disciplined inquiry has a quality that distinguishes it from other sources of opinion and belief. The disciplined inquiry is conducted and reported in such a way that the argument can be painstakingly examined. The report does not depend for its appeal on the eloquence of the writer or on any surface plausibility. (p. 15)

Unfortunately, as any journal editor can testify, there are too many research studies in education in which either the validity or the reliability of the evidence is questionable. Typical examples include the use of standardized test scores as a dependent variable in the study of problem solving, low return rates on surveys, the failure to control adequately for sources of invalidity in an experiment, and failure to triangulate evidence when building an argument in a case study. The gathering of evidence and the construction of an argument are the means by which researchers substantiate conjectures. This is an arduous process. Nor does it, once achieved, remain finished and complete; it has to be continually reachieved, since both what constitutes a reasonable argument and the given purposes for it are constantly changing.

Generalizability. The traditional view of generalizability (based on empirical-analytic assumptions) employs a procedural definition. According to this definition, the results of a study can be said to be generalizable because sampling procedures have been employed. This sense of generalizability has been called *horizontal generalizability* (Stephens, 1982). It enables one to attach a quantified estimate of likelihood to the repeatability of one's findings. Its essential feature is that it rests on notions of probability that enable one to predict how the elements of a sample, and the characteristics they bear, relate to a general population having the same characteristics.

The argument about generalizability could indeed rest on horizontal generalizability thus defined if researchers limited themselves only to making claims about other samples of the same population or about samples of other populations that bear the same characteristics as the one studied. However, many scholars look beyond the specific situation and link features of that situation to more abstract and general considerations. This other sense of generalizability might be called *vertical* (Stephens, 1982). In other words, on the basis of their evidence they make claims about theoretical propositions. They are not claiming replicability.

The contrast between vertical and horizontal generalizability can be illustrated by making a distinction between building interpretative theory and showing that these interpretations are likely to be useful in other, similar situations. For building interpretative theory, there may be value in selecting an instance that offers clearly delineated features and telling contrasts (vertical generalizability). However, once developed, this theory needs a more general application (horizontal generalizability), and an investigator will need to convince a skeptical audience that there are grounds for being confident that the features that gave rise to the interpretative theory are likely to be borne out across a range of groups or instances.

RESEARCH TRENDS

Given this background on research communities, their assumptions about the world, and their perspectives about research, it should not be surprising that the past quarter-century's research on teaching and learning in schools is difficult to summarize and its trends difficult to identify. Nevertheless, I believe that at least five broad trends in the social sciences can be described. By emphasizing broad trends, I am choosing not to discuss specific changes that have occurred in the kinds of research carried out in particular areas of mathematical sciences education. These are discussed in the other chapters of this book. Also, I want to point out that although each of the broad trends is discussed separately, they are interrelated and not truly independent.

Trend #1: Growth of Research

Although research on teaching and learning in school settings has a long history in the United States, there has been a dramatic growth in the past 30 years. This trend has been so obvious throughout the social sciences that it hardly needs to be mentioned. Since this growth promises to continue at an even faster pace in the next decade, however, four factors that have contributed to it need to be mentioned.

First, one of the reasons for growth has been the availability of research funds. Federal funds for educational research first became available in the 1960s from the newly created National Science Foundation and the Office of Education. In particular, the passage of the Elementary and Secondary Education Act (ESEA) in 1963 provided extensive funds to establish research centers, evaluate products, and so forth. Research on the teaching and learning of mathematics, though not always central to such concerns, grew accordingly. (For a more comprehensive examination of this trend in mathematical sciences education, see chapter 1 in this volume.)

Second, in part because of the availability of research funds and in part because of the increased social demands for the reform of schooling, there has been a dramatic increase in the involvement of scholars from fields other than education and mathematics in the study of the teaching and learning of mathematics. Psychologists, sociologists, anthropologists, economists, political scientists, mathematicians, and others, as well as mathematics educators, have contributed to the growth of research.

Third, since the 1960s there has been a proliferation of journals, research organizations, and meetings dedicated to educational research. These are necessary both for the creation of research communities and as a means of sharing work with colleagues. Prior to 1960, there were only three research journals in the United States that regularly published educational studies: *Psychometrica* (on testing issues, published by the Psychometric Society), the *Journal of Educational Psychology* (on human learning, published by the American Psychological Association), and the *Review of Educational Research* (which invited summary reviews on specific topics, published by the American Educational Research Association (this journal changed its practice in 1970 from the publication of invited reviews to acceptance of unsolicited manuscripts). The *American Educational Research Journal* was first published in 1963, and the two principle research journals in mathematics education, the *Journal for Research in Mathematics Education* and *Educational Studies in Mathematics*, were both started in 1969. Today, there are over fifty educational research journals published in the United States and a similar number in the rest of the world. There has been a concomitant growth in organizations dedicated to research. In mathematics education, AERA's Special Interest Group in Mathematics Education and the International Group for the Psychology of Mathematics Education have been organized since 1963 and have become particularly important groups nationally and internationally.

Finally, research centers dedicated to mathematics education have been created at universities in several countries. They are staffed by professors in mathematics education whose primary job is associated with research programs, not just teacher education. The development of such centers is critical if long-term research programs and communities are to develop. Of particular note are the following centers: The Freudenthal Institute at the University of Utrecht in The Netherlands, the Shell Centre for Mathematical Education at the University of Nottingham in Great Britain, the Institute für Didaktik der Mathematik (IDM) at the University of Bielefeldt in West Germany, and the National Center for Research in Mathematical Sciences Education (NCRMSE) at the University of Wisconsin in the United States.

Not surprisingly, it is clear that resources, vehicles for sharing information, and centers committed to long-term inquiry are major contributing factors to the dynamic growth of research in teaching and learning.

Trend #2: Growing Diversity in Research Methods

It should not be surprising that as a field of research grows, the variety of methods used by scholars grows. In 1963 Donald Campbell and Julian Stanley argued that conducting experiments was "the only means for settling disputes regarding educational practice, the only way of verifying educational improvements, and the only way of establishing a cumulative tradition in which improvements [could] be introduced without the danger of a faddish discard of old wisdom in favor of inferior novelties" (p. 173). They went on to chastise studies in one class (teaching experiments or case studies) as "well-nigh unethical" (p. 177).

By 1973, when experimentation was still seen as the ultimate method to be used in educational studies, other methods were nevertheless deemed appropriate as a part of a "descriptive-correlational-experimental loop" (Rosenshine & Furst, 1973). The argument was that, first, characteristics of a problem situation are decided upon and attempts are made to describe variability on those features. This usually involves the scaling of responses to questions or observations of behaviors. Second, the relationships between the variables are examined, via correlations or regression equations. Finally, one or more variables are manipulated and the effects of that manipulation examined through some experimental design.

The roots of the use of these methods in education are in the "empirical-analytic" traditions found in behavioral psychology, which in turn borrowed them from agriculture. However, it should be noted that Campbell and Stanley (1963) admitted that "because of the intransigency of the environment" (p. 171), experimenters lacked the kind of control that one could have in animal or agricultural experiments. Children cannot be controlled like pigeons, rats, or monkeys; alternative methods of instruction are not like different brands of fertilizer; and classrooms are not independent field plots.

Today, although correlational studies and experiments are commonly conducted, many scholars are using different strategies and methods. The methods (some were described in an earlier section of this chapter) have been adapted for the study of schooling from scholarly traditions in developmental psychology, sociology, anthropology, political science, and other social sciences. These traditions are based on different ideologies and reflect different intellectual histories. In addition, Americans have gradually become aware of the works of international scholars. Since these scholars view educational problems through their own cultural lenses, it is often difficult to accommodate such works to American purposes. In his 1966 review of theories of conceptual behavior, for example, Lyle Bourne dismissed Piaget's work because "it diverges widely from conventional American theory" (1966, p. 44). Today, however, scholars have become more tolerant of foreign perspectives.

Finally, as educational research has grown, so has membership in the research community. Much research is now done by research teams whose members bring different intellectual and practical knowledge to bear on a common set of problems. Such teams often include practicing teachers. One consequence of this diversity of perspectives is the conflict between those who believe the role of research is to make a stable institution (the school or classroom) more effective and efficient and those who see research as a force that can either shake up institutions or radically reform them.

Trend #3: A Shift in Epistemology

The question "What does it mean to know?" is currently being raised in all disciplines. In the past it was assumed that knowledge in any discipline was simply an accumulation of bits of information (concepts and skills) arranged in some sequential order. Today this assumption is being challenged. The alternative perspective involves a notion of *authentic* knowledge. Scholars in many fields are now engaged in attempts to articulate this notion as it applies to their discipline. Three aspects of such attempts need to be mentioned.

First, one must distinguish between *knowledge* and the *record of knowledge*. For example, when many nonmathematicians look at mathematics, they see a bounded and static set of concepts and skills to be mastered. This is perhaps a reflection of the mathematics they studied in school or college rather than an insight into the discipline itself. For many, *to know* means to identify the artifacts of the discipline—its basic concepts and procedures. For others more familiar with the discipline, *to know* mathematics is *to do* mathematics. A person gathers, discovers, or creates knowledge in the course of some activity having a purpose. Only if the emphasis is put on the process of *doing* is mathematics likely to make sense to students.

The distinction between knowledge and the record of knowledge can perhaps be made clear by an analogy with the discipline of music. Like mathematics, music has many branches categorized in a variety of ways (classical, jazz, rock; instrumental, vocal); it has a sparse notational system for preserving information (notes, time-signatures, clefs) and theories that describe the structure of compositions (scales, patterns). However, no matter how many of the artifacts of music one has learned, it is not the same as *doing* music. It is only when one performs that one knows music. Similarly, in mathematics one can learn the concepts about numbers, how to solve equations, and so on, but that is not *doing* mathematics. Doing mathematics involves solving problems, abstracting, inventing, proving, and so forth.

In a similar manner, James Greeno (1990) argues that to become competent in any field involves understanding that the field (for example, rational numbers or functions) should be seen as an environment in which a collection of resources for knowing, understanding, and reasoning in that domain is located. Learning to operate in that domain involves knowing what resources are available and how to access them when needed. Obviously, it is important to learn mathematical concepts and practice mathematical procedures (like learning to read musical notation and practice scales), but it is also important to have an opportunity to solve problems (perform) at the learner's level of capability. Too often, students find arithmetic, algebra, and calculus to be senseless, dull, and even intimidating. After all, who can enjoy routinely multiplying one four-digit number by another (analogous to playing scales on a piano) or solving a system of simultaneous linear equations? As a result of such limited experiences, many students are prejudiced against the broader, more interesting aspects of mathematics. In summary, the notion of authentic knowledge involves knowing how to do mathematics, sociology, music, or science.

Second, and contrary to popular opinion, mathematics is not a stable discipline. Lynn Steen (1990) has argued that the computer, calculator, and other new technological devices are changing what it means to do mathematics. He infers that "mathematics is a science of patterns" and that technology provides mathematicians with powerful tools to examine elaborate and complex patterns in a way never before possible. These tools make it possible to decouple calculation from mathematical investigations, in a manner similar to the way the typewriter decoupled penmanship from writing. Efficient computational skills, whether in arithmetic, algebra, or even calculus, are no longer prerequisites to being able to do mathematics. The technology allows everyone to carry out common calculations more efficiently, to perform new calculations that could not be done by hand, to illustrate information graphically in a multitude of ways, and simultaneously to see the relationships between representations.

Third is the issue of what is unique in the learning and teaching of any subject, such as mathematics. From Begle's perspective (see Figure 3.1), only the term *mathematics* would distinguish the relationships and possible questions to be studied from those questions raised about the teaching and learning of science, or social studies, or other disciplines. The concepts of mathematics are similar to those in other subjects; its processes, though often algorithmic, are little different from those employed in learning to conjugate regular verbs or solve chemical equations; nor can it be argued that problem solving is unique to mathematics. Thus, it should not be surprising that, when approaching research on teaching and learning in schools, many scholars from the social sciences have failed to see anything unique about the learning of mathematics. In fact, many researchers have chosen to study the learning of mathematics only because they felt it was the best organized, most sequential, most intact school subject.

During the past decade some scholars recognized the dynamic, changing nature of the discipline. This realization has led to at least two activities. First is a response to the difficulty that, according to Edward Barbeau (1989), "most of the population perceive mathematics as a fixed body of knowledge long set into final form. Its subject matter is the manipulation of numbers and the proving of geometrical deductions. It is a cold and austere discipline which provides no scope for judgement or creativity" (p. 2). These views are undoubtedly a reflection of the mathematics studied in school rather than an insight into the discipline itself. There has been a growing awareness by mathematicians of the need to represent better what mathematics is about, to illustrate what mathematicians do, and to popularize the discipline. This does not mean that good books about the discipline did not exist before the 1980s. Classics such as *The World of Mathematics* (Newman, 1956), *Mathematics: Its Content, Methods, and Meaning* (Aleksandrov, Kolmogorov, & Lavrent'ev, 1956), and *Mathematical Thought From Ancient to Modern Times* (Kline, 1972) have been available. Unfortunately, these books were written for readers well-trained in mathematics and science. However, a large number of books about mathematics have been published during the 1980s for a nontechnically educated audience. These include *The Mathematical Experience* (Davis & Hersh, 1981), *Descartes' Dream* (Davis & Hersh, 1986), *Mathematics and the Unexpected* (Ekeland, 1988), *Speaking Mathematically* (Pimm, 1987), *Innumeracy: Mathematical Illiteracy and Its Consequences* (Paulos, 1988), *Capitalism & Arithmetic: The New Math of the 15th Century* (Swetz, 1987), *Mathematics Counts* (Committee of Inquiry into the Teaching of Mathematics, 1982), *Renewing United States Mathematics: Critical Resource for the Future* (Commission on Physical Sciences, Mathematics, and Resources, 1984), *Perspectives on Mathematics Education* (Christiansen, Howson, & Otte, 1986), *Mathematics, Insight, and Meaning* (de Lange, 1987), *Cognitive Science and Mathematics Education* (Schoenfeld, 1985), *Mathematics Tomorrow* (Steen, 1981), *Everybody Counts* (Mathematical Sciences Education Board, 1989), *Curriculum and Evaluation Standards for School Mathematics* (National Council of Teachers of Mathematics, 1989), *Reshaping School Mathematics* (Mathematical Sciences Education Board, 1990), and *On the Shoulders of Giants* (Steen, 1990). These works present mathematics as a discipline with many facets. One can define mathematics "as a language, as a particular kind of logical structure, as a body of knowledge about numbers and space, as a series of methods for deriving conclusions, as the essence of our knowledge of the physical world, or as an amusing intellectual activity" (Kline, 1962, p. 2). Understanding these variations and the dynamic nature of mathematics is important because different features have been emphasized in school mathematics programs at different times by different authors, and because the reform arguments now being carried out hinge on the participants' view of mathematics.

Second, there has been a growing interest in the teacher's understanding of what it means to do mathematics. A number of questions are being addressed by scholars: Are teachers aware that mathematics is more than the compilation of discrete bits of propositional knowledge? (Ball & McDiarmid, 1989), Is the knowledge teachers are exposed to during training useful in teaching students to do mathematics? (Lappan & Even, 1989; McDiarmid, Ball, & Anderson, 1988), and Can a new pedagogy be created to prepare teachers for authentic instruction? (Lampert, 1988; Lampert & Ball, 1990). The argument is that mathematics is a unique domain—a sparse set of signs and symbols that can be used to model a wide variety of situations, a number of strategies (heuristics) used to examine features of that domain, and specific methods of reasoning. Teachers must be intimately aware of this if they are to represent mathematics adequately to their students.

Trend #4: A Shift in Learning Psychology

A new view of learning—cognitive science—is an outgrowth of the revolution in psychology that has become dominant during the past decade. The following seven notions of how the mind works characterize this view of learning.

First, processing begins with an experience. Information from the experience is filtered, organized, and stored in memory. This critical aspect of cognitive theory distinguishes among three types of memory: working memory, short-term memory, and long-term memory. *Working memory* involves the limited amount of information that one can process at a particular point in time. The units of information may be individual pieces or may be related pieces (chunked information). Information is stored in *short-term memory* to be used in a relatively short period of time. Then it is either dumped (forgotten) or stored in *long-term memory.* Second, although humans are capable of remembering a great deal, they have an extremely limited capacity to think about a number of different things at one time. As a consequence of this limited capacity, information stored in long-term memory must be well organized. The mind naturally organizes repeated similar experiences in long-term memory into what psychologists call *schemata.* These are complex networks of concepts, rules, and strategies, not isolated facts or algorithms. Having information stored in this way helps an individual cope with new experiences. Such schemata develop over long periods of time and by continual exposure to related contextual events. Third, new experiences either use one's existing schemata (called assimilation) or force a change in a particular schema (called accommodation). Fourth, naturally occurring

schemata are idiosyncratic to the individual, who is usually unaware of the organization. Fifth, although most learning occurs through natural assimilation and accommodation of experiences, it can also occur via preorganized and structured experiences. For example, learning the common names and features for geometric figures (for example, isosceles, parallel, perpendicular) on the basis of natural experience is unlikely. Learning the formal language, signs and symbols, properties, and principles associated with any mathematical domain is no easy matter. Instructional activities should serve as the means of connecting a student's informal, natural experiences with the formal aspects of mathematics. The assumption is that students will reorganize their informal quantitative and spatial schemata as a consequence of interacting in such activities. Sixth, schemata are never fixed. They continually change as the individual grows and has additional experiences. In fact, there is general agreement that the evolution of one's thinking goes through several cyclic stages. School-age children do not think like adults. They are more likely to respond in a concrete-symbolic manner to new information. Only in later grades do they begin to build formal relationships between disparate experiences, and, even then, they do not yet think in terms of composite, abstract concepts. Finally, students who have developed initial but not well-organized schema activity search for experiences that will provide them structure.

In summary, cognitive psychologists have provided the concept of "well-organized schemata" to explain how people impose order on experiential information. Assimilation, accommodation, and mode of functioning in response to new information are important in the enterprise of schooling. Without schemata into which new information can be assimilated, experience is incomprehensible, and little can therefore be learned from it. But the schema by which a student assimilates a lesson may not be that assumed by teachers or mathematicians. This mismatch can easily escape detection because the student will often be able to repeat segments of the text and lecture even though he or she understands them in terms of an incorrect, incomplete, or inconsistent framework. Indeed, students may develop specialized frameworks for maintaining the particular identity of lesson material in order to cope with the demand for veridical reproduction. Schema use must be a dynamic, constructive process, for people do not have a schema stored to fit every conceivable situation. In this view, acquisition of knowledge implies changes in schemata, not just the aggregation of information.

Trend #5: Growth of Political Awareness

It has long been assumed that scientific research offers tested, reliable knowledge that individuals can use to inform their thinking and practice, and that the search for such knowledge is independent of political debates and decisions about education. Today many scholars are becoming aware that educational research is both scientific and political. As Kilpatrick (1987) notes, however, "a cynic might say that educational research as we know it is neither scientific nor political. It pretends to a level of insight, precision, and clarity that it has not yet begun to attain, and it is routinely disparaged and ignored by decision makers in education" (p. 1).

There is no question that education is political in the United States. Our schools are governed by some 16,000 local school districts, as well as by state regulations and federal mandates. Public policy debates about how teaching and learning are to be organized, licensed, monitored, and so forth are ongoing in our society. Moreover, with the publication of such documents as *A Nation at Risk* (National Commission on Excellence in Education, 1983) and *Educating Americans for the 21st Century* (National Science Board Commission on Precollege Education in Mathematics, Science, and Technology, 1983), both critical of current practices particularly in mathematics and science education, many educators have felt compelled to enter into the debates about educational reform.

This growing awareness of the political importance of research is reflected in the emphasis on policy research and dissemination of research findings in the specifications for new research and development centers in education to be supported by the U.S. Department of Education. Also, the growing political awareness of the mathematical sciences education community is reflected in the formation of the Mathematical Sciences Education Board (MSEB) of the National Research Council of the National Academy of Sciences. This board was created to provide a national voice for mathematics education in Washington, DC. Finally, MSEB is in the process of encouraging the creation of similar coalitions in each of the 50 states.

In today's climate for educational reform, researchers cannot be dispassionate observers on the sideline. They must realize that policy decisions are being and will be made, and that they must contribute to the debates.

References

Aleksandrov, A. D., Kolmogorov, A. N., & Lavrent'ev, M. A. (1956). *Mathematics: Its content, methods, and meaning.* Cambridge, MA: MIT Press.

Anyon, J. (1981). Elementary schooling and distinctions of social class. *Interchange, 12,* 118–132.

Apple, M. (1982). *Education and power.* Boston: Routledge and Kegan Paul.

Ball, D. L., & McDiarmid, G. W. (1989). *The subject matter preparation of teachers* (Issue Paper 89–4). East Lansing, MI: Michigan State University, National Center for Research on Teacher Education.

Barbeau, E. J. (1989, September). *Mathematics for the public.* Paper presented at the meeting of the International Commission on Mathematical Instruction, Leeds University, England.

Barnes, D. (1975). *From communication to curriculum.* Harmondsworth, Middlesex: Penguin Books.

Begle, E. G. (1961). Seminar in mathematics education: Class notes. Unpublished.

Bernstein, R. (1976). *The restructuring of social and political theory.* New York: Harcourt Brace Jovanovich.

Bloom, B. S. (Ed.) (1956). *Taxonomy of educational objectives: The classification of educational goals. Handbook I: Cognitive domain.* New York: David McKay Co.

Bourne, L. E., Jr. (1966). *Human conceptual behavior.* Boston: Allyn and Bacon.

Brophy, J. (1986). Teaching and learning mathematics: Where research should be going. *Journal for Research in Mathematics Education, 17*(5), 323–346.

Campbell, D., & Stanley, J. (1963). Experimental and quasi-experimental designs for research on teaching. In N. L. Gage (Ed.), *Handbook of research on teaching* (pp. 171–246). Chicago: Rand McNally.

Carpenter, T., & Fennema, E. (1988). Research and cognitively guided instruction. In E. Fennema, T. P. Carpenter, & S. J. Lamon (Eds.), *Integrating research on teaching and learning mathematics* (pp. 2–19). Madison: National Center for Research in Mathematical Sciences Education, University of Wisconsin.

Christiansen, B., Howson, A. G., & Otte, M. (1986). *Perspectives on mathematics education.* Holland: D. Reidel Publishing Co.

Cicourel, A. (1969). *Method and measurement in sociology.* New York: Holt, Rinehart and Winston.

Commission on Physical Sciences, Mathematics, and Resources. (1984). *Reviewing United States mathematics: Critical resource for the future. Report of the Ad Hoc Committee on Resources for the Mathematical Sciences.* Washington, DC: National Academy Press.

Committee of Inquiry into the Teaching of Mathematics in Schools (CITMS). (1982). *Mathematics counts.* London: Her Majesty's Stationery Office.

Confrey, J. (1986). A critique of teacher effectiveness research in mathematics education. *Journal for Research in Mathematics Education, 17*(5), 347–360.

Crane, D. (1976). *Invisible colleges: Diffusion of knowledge in scientific communities.* Chicago: University of Chicago Press.

Cronbach, L. J., & Suppes, P. (Eds.) (1969). *Research for tomorrow's schools: Disciplined inquiry for education.* New York: Macmillan.

Cusick, P. (1973). *Inside a high school student's world.* New York: Holt, Rinehart and Winston.

D'Ambrosio, U. (1985). *Socio-cultural bases for mathematics education.* Cumpinas, Brazil: UNICAMP Centro de Producoes.

Davis, P. J., & Hersh, R. (1981). *The mathematical experience.* Boston: Birkhäuser.

Davis, P. J., & Hersh, R. (1986). *Descartes' dream.* New York: Harcourt Brace Jovanovich.

de Lange, J. (1987). *Mathematics, insight, and meaning.* The Netherlands: University of Utrecht.

Dewey, J. (1966). *Democracy and education.* New York: The Free Press.

Easton, D. (1971). *The political system: An inquiry into the state of political science.* New York: Alfred A. Knopf.

Ekeland, I. (1988). *Mathematics and the unexpected.* Chicago: University of Chicago Press.

Fischer, C., Berliner, D., Filby, N., Marliave, R., Cohen, L., Deshaw, M., & Moore, J. (1978). *Teaching and learning in elementary schools: A summary of the beginning teacher evaluation study.* San Francisco: Far West Laboratory for Educational Research and Development.

Gage, N. L. (1978). *The scientific basis of the art of teaching.* New York: Teachers College Press.

Good, T. L., & Grouws, D. A. (1981). *Experimental research in secondary mathematics classrooms: Working with teachers* (Contract No. 79–0103). Washington, DC: National Institute of Education.

Greeno, J. (1990). *Number sense as situated knowing in a conceptual domain* (IRL Report No. IRL90–0014). Palo Alto, CA: Stanford University, Institute for Research on Learning.

Homans, G. C. (1967). *The nature of social science.* New York: Harcourt, Brace and World.

Kilpatrick, J. (1981). The reasonable ineffectiveness of research in mathematics education. *For the Learning of Mathematics, 2*(2), 22–28.

Kilpatrick, J. (1987, December). *Educational research: Scientific or political?* Paper presented at the First Joint AARE/NZARE Conference, University of Canterbury, Christchurch, New Zealand.

Kline, M. (1962). *Mathematics: A cultural approach.* Reading, MA: Addison-Wesley.

Kline, M. (1972). *Mathematical thought from ancient to modern times.* New York: Oxford University Press.

Kuhn, T. (1970). *The structure of scientific revolutions.* Chicago: University of Chicago Press.

Lakatos, I. (1976). Falsification and the methodology of scientific research programs. In I. Lakatos & A. Musgrave (Eds.), *Criticism and the growth of knowledge* (pp. 91–196). Cambridge, England: Cambridge University Press.

Lampert, M. (1988). The teacher's role in reinventing the meaning of mathematical knowing in the classroom. *Proceedings of the Tenth Annual Meeting of the North American Chapter of the International Group for the Psychology of Mathematics Education* (pp. 433–480). DeKalb, IL: Northern Illinois University.

Lampert, M., & Ball, D. L. (1990). *Using hypermedia technology to support a new pedagogy of teacher education* (Issue Paper 90–5). East Lansing, MI: Michigan State University, National Center for Research on Teacher Education.

Lappan, G., & Even, R. (1989). *Learning to teach: Constructing meaningful understanding of mathematical content* (Craft Paper 89–3). East Lansing, MI: Michigan State University, National Center for Research on Teacher Education.

Larabee, H. (1945). *Reliable knowledge.* Cambridge, MA: Houghton Mifflin.

Luckmann, T. (Ed.) (1977). *Phenomenology and sociology.* Harmondsworth, Middlesex: Penguin Books.

Lukacs, G. (1971). *History and class consciousness: Studies in Marxist dialectics,* R. Livingstone (Trans). Cambridge, MA: MIT Press.

Mathematical Sciences Education Board (MSEB). (1989). *Everybody counts: A report to the nation on the future of mathematics education.* Washington, DC: National Academy Press.

Mathematical Sciences Education Board (1990). *Reshaping school mathematics.* Washington, DC: National Academy Press.

McDiarmid, G. W., Ball, D. L., & Anderson, C. W. (1988). *Why staying one chapter ahead doesn't really work: Subject-specific pedagogy* (Issue Paper 88–6). East Lansing, MI: Michigan State University, National Center for Research on Teacher Education.

National Commission on Excellence in Education. (1983). *A nation at risk: The imperative for educational reform.* Washington, DC: US Government Printing Office.

National Council of Teachers of Mathematics. (1989). *Curriculum and evaluation standards for school mathematics.* Reston, VA: Author.

National Science Board Commission on Precollege Education in Mathematics, Science, and Technology. (1983). *Educating Americans for the twenty-first century: A plan of action for improving the mathematics, science, and technology education for all American elementary and secondary students so that their achievement is the best in the world by 1995.* Washington, DC: National Science Foundation.

Newman, J. R. (1956). *The world of mathematics.* New York: Simon & Schuster.

Parlett, M., & Hamilton, D. (1976). Evaluation as illumination: A new approach to the study of innovatory programs. In G. V. Glass (Ed.), *Evaluation Studies: Review Annual, Vol. 1* (pp. 140–157). Beverly Hills, CA: SAGE Publications.

Paulos, J. A. (1988). *Innumeracy: Mathematical illiteracy and its consequences.* New York: Hill and Wang.

Pimm, D. (1987). *Speaking mathematically: Communication in mathematics classrooms.* London: Routledge and Kegan Paul.

Platt, J. (1964). Strong inference. *Science, 146*(3642), pp. 347–352.

Popkewitz, T. (1984). *Paradigm and ideology in educational research: The social functions of the intellectual.* London: The Falmer Press.

Popper, K. (1968). *Conjectures and refutations.* New York: Harper Torchbooks.

Romberg, T. A. (1988). Can teachers be professionals? In D. A. Grouws, T. J. Cooney, & D. Jones (Eds.), *Perspectives on research on effective mathematics teaching* (pp. 224–244). Reston, VA: The National Council of Teachers of Mathematics.

Rosenshine, B., & Furst, N. (1973). The use of direct observation to study teaching. In R. M. Travers (Ed.), *Second handbook of research on teaching* (pp. 122–183). Chicago: Rand McNally.

Schoenfeld, A. H. (1985). *Cognitive science and mathematics education.* Hillsdale, NJ: Lawrence Erlbaum.

Schultz, A. (1973). The problem of social reality. In M. Nathanson (Ed.), *Collected Papers I.* The Hague: Martinus Nijhoff.

Shulman, L. S. (1988). Disciplines of inquiry in education: An overview. In R. M. Jaeger (Ed.), *Complementary methods for research in education* (pp. 3–20). Washington, DC: American Educational Research Association.

Steen, L. A. (1981). *Mathematics tomorrow.* New York: Springer-Verlag.

Steen, L. A. (1990). *On the shoulders of giants: New approaches to numeracy.* Washington, DC: National Academy Press.

Stephens, M. (1982). A question of generalizability. *Theory and Research in Social Education, 9*(4), 75–89.

Swetz, F. J. (1987). *Capitalism and arithmetic: The new math of the 15th century.* La Salle, IL: Open Court.

Watson, J. B. (1948). Psychology as the behaviorist views it. In W. Dennis (Ed.), *Readings in the history of psychology* (pp. 457–471). New York: Appleton-Century-Crofts.

Weimer, W. B. (1979). *Notes on the methodology of scientific research.* Hillsdale, NJ: Lawrence Erlbaum.

Willis, P. (1977). *Learning to labour.* Farnborough, UK: Saxon House.

Winch, P. (1971). *The idea of a social science and its relation to philosophy.* New York: Humanities Press.

Wolcott, H. (1987). Ethnographic research in education. In R. M. Jaeger (Ed.), *Complementary methods for research in education* (pp. 187–206). Washington, DC: American Educational Research Association.

· 4 ·

LEARNING AND TEACHING
WITH UNDERSTANDING

James Hiebert
UNIVERSITY OF DELAWARE

Thomas P. Carpenter
UNIVERSITY OF WISCONSIN

One of the most widely accepted ideas within the mathematics education community is the idea that students should understand mathematics. The goal of many research and implementation efforts in mathematics education has been to promote learning with understanding. But achieving this goal has been like searching for the Holy Grail. There is a persistent belief in the merits of the goal, but designing school learning environments that successfully promote understanding has been difficult.

Learning with understanding is an issue that extends beyond the boundaries of mathematics education. Many general theories of learning, including those with differing paradigmatic origins, wrestle with the notion of understanding. For example, a number of recent efforts within the cognitive science tradition, with its emphasis on modeling internal representations with considerable precision (Gardner, 1985), can be interpreted as a recognition that understanding is a fundamental aspect of learning and that models of learning must attend to issues of understanding (Mayer, 1989; Ohlsson & Rees, 1988; Perkins & Simmons, 1988). At the same time, a number of theories have emerged that draw heavily from anthropology and sociology,

in part because of a dissatisfaction with the explications of understanding provided by cognitive science approaches (Brown, Collins, & Duguid, 1989; Lave, 1988). A major goal of these theories, with their emphasis on situated knowledge and social cognition, is to explain the apparent understanding that comes with learning in everyday contexts and the lack of such understanding that accompanies learning in formal school settings. To the benefit of mathematics education, much of the recent discussion is occurring in the context of learning mathematics.

The purpose of this chapter is to consider, again, learning mathematics with understanding. Drawing from old and new work in the psychology of learning, we present a framework for examining issues of understanding. The framework is applied first to questions of learning and then to questions of teaching. The questions of interest are those related to learning with understanding and teaching for understanding. We conclude the chapter with a shift of the person who needs to understand, from the student and the teacher to the researcher: What can be learned from students' efforts to understand that might inform researchers' efforts to understand understanding?

We would like to thank P. Cobb, Purdue University; D. Grouws, University of Missouri; R. Putnam, Michigan State University; and L. Resnick, University of Pittsburgh, for their comments on an earlier draft of this paper and the National Science Foundation (Grant Nos. MDR 8651552 and MDR 8855627) and the Office of Education Research and Improvement of the U.S. Department of Education for partial support while writing it. The opinions expressed in the paper are those of the authors and not necessarily those of the reviewers, NSF or OERI.

REPRESENTING AND CONNECTING: A FRAMEWORK FOR THINKING ABOUT UNDERSTANDING

The framework we propose for reconsidering understanding is based on the assumption that knowledge is represented internally, and that these internal representations are structured. A useful way of describing understanding is in terms of the way an individual's internal representations are structured. We point to some alternative ways of characterizing understanding but argue that the structure of represented knowledge provides an especially coherent framework for analyzing a range of issues related to understanding mathematics.

External and Internal Representation

To think about and communicate mathematical ideas, we need to represent them in some way. Communication requires that the representations be external, taking the form of spoken language, written symbols, pictures, or physical objects (cf. Lesh, Post, & Behr, 1987). A particular mathematical idea can often be represented in any one form or in all forms of representation.

To think about mathematical ideas we need to represent them internally, in a way that allows the mind to operate on them. Because mental representations are not observable, discussions of how ideas are represented inside the head are based on high degrees of inference. For years, the associationist perspective in psychology (Thorndike, 1914; Skinner, 1953) ruled out the discussion of mental representations because they cannot be observed. However, work in cognitive science has restored mental representations as a legitimate field of study. Indeed, the notion of mental representations is a central idea that brings together work on cognition from a variety of fields, including psychology, computer science, linguistics, and others (Gardner, 1985).

Our aim in this chapter is not to extend basic findings in cognitive science. Rather, we draw quite heavily from insights provided by work in cognitive science to deal with questions of learning and teaching mathematics. In particular, we build directly on two assumptions from cognitive science regarding mental or internal representations. First, we assume some relationship exists between external and internal representations. Second, we assume that internal representations can be related or connected to one another in useful ways.

Assuming a relationship between external and internal representations is consistent with much of the work in cognitive science, but it is an assumption not uniformly held. There is an ongoing debate, for example, about whether the form of a mental representation mimics in some way the external object or event being represented (Shepard, 1982; Shepard & Metzler, 1971) or whether there is a common form used to represent all information (Newell, 1980; Pylyshyn, 1980). Although the debate is not resolved, we believe it reasonable to assume that the nature of the internal representation is influenced and constrained by the external situation being represented (Kosslyn & Hatfield, 1984). We apply this assumption to mathematical

situations by assuming that the nature of external mathematical representations influences the nature of internal mathematical representations (Greeno, 1988a; Kaput, 1988). Evidence from a variety of task situations suggests that this is a reasonable assumption (Gonzalez & Kolers, 1982; Stigler, 1984). The important point here is that, when considering issues of representation in mathematics, we must consider both external and internal representations. That is, the form of an external representation (physical materials, pictures, symbols, etc.) with which a student interacts makes a difference in the way the student represents the quantity or relationship internally. Conversely, the way in which a student deals with or generates an external representation reveals something of how the student has represented that information internally.

Despite our assumptions about the existence of relationships between external and internal representations, we make no claims about the precise nature of internal representations. We do not presume, for example, that if second graders work with bundled sticks when dealing with two-digit numbers, they represent internally all quantities more than nine as mental images of sticks. We do presume, though, that students who interact with bundled sticks represent these quantities differently for themselves than students who work only with written symbols.

Connecting Representations

The second assumption drawn from work in cognitive science is that internal representations can be connected. Although these connections can only be inferred, we assume, just as with representations themselves, that they are influenced by external activity; connections between internal representations can be stimulated by building connections between corresponding external representations.

External Connections. Connections between external representations of mathematical information can be constructed by the learner between different representation forms of the same mathematical idea or between related ideas within the same representation form. Connections between different representations are often based on relationships of similarity ("these are alike in the following ways") and relationships of difference ("these are different in the following ways"). For example, whole numbers can be represented externally with spoken language, written symbols, and base-10 blocks. Connections between these forms can be built by examining how they are the same and how they are different. Connections within the same representation form are often made by noticing patterns or regularities in the system. For example, the spoken vocabulary for whole numbers contains a predictable regularity (after the teens). Connections between ones, tens, and hundreds, as words in the system, can be made by detecting these regularities. Connections between representation forms and within representation forms play a role in learning mathematics with understanding. Both will be elaborated later in the chapter.

Internal Connections. We propose that when relationships between internal representations of ideas are constructed, they produce networks of knowledge. It is not currently possible to specify the exact nature of the networks, but we believe it is useful to think about the networks in terms of two metaphors: Networks may be structured like vertical hierarchies, or they may be structured like webs. When networks are structured like hierarchies, some representations subsume other representations; representations fit as details underneath or within more general representations. Generalizations are examples of overarching or umbrella representations, whereas special cases are examples of details.

In the second metaphor a network may be structured like a spider's web. The junctures, or nodes, can be thought of as the pieces of represented information, and the threads between them as the connections or relationships. All nodes in the web are ultimately connected, making it possible to travel between them by following established connections. Some nodes, however, are connected more directly than others. The webs may be very simple, resembling linear chains, or they may be extremely complex, with many connections emanating from each node.

Both hierarchies and webs appear in the rather extensive work on knowledge structures and semantic nets (Chi, 1978; Geeslin & Shavelson, 1975; Greeno, 1978; Leinhardt & Smith, 1985; Quillian, 1968). An interesting illustration of both kinds of networks is contained in Chi's and Koeske's (1983) report of a four-year-old's knowledge of dinosaurs. From the child's responses to a variety of tasks involving knowledge of dinosaurs, the investigators plotted the information the child produced about dinosaurs, including the relationships between the pieces of information, as suggested by the child. Some pieces of information were grouped together under a higher order construct, as in a hierarchy. For example, a subset of dinosaurs were seen to be related because they ate the same kinds of things. In other cases, an individual piece of information was seen as related to another, but no hierarchy was suggested. Such a network appeared to be more like a web than a hierarchy.

There is no reason that the two metaphors of hierarchies and webs cannot be mixed, resulting in many additional forms of networks. The levels of the hierarchies, for example, can themselves be thought of as webs. Chi's and Koeske's (1983) four-year-old appears to have structured dinosaur knowledge in this way. Perhaps it is safe to assume that students' internal representations, whatever their nature, are extremely complex and that useful metaphors will need to capture this complexity.

Usefulness of the Framework

There is considerable current debate about whether understanding can be fully described in terms of internal knowledge structures. Perhaps a complete description will need to include analyses of situated activity and social interaction as well (Lave, 1988; Greeno, 1989; Resnick, 1987b). Whatever the outcome of this debate, we believe that the notion of connected representations of knowledge will continue to provide a useful way to think about understanding mathematics, for several reasons. First, it provides a level of analysis that makes contact with both theoretical cognitive issues and practical educational issues. The discussion of understanding can proceed in a way

that places it in a larger theoretical context, while at the same time providing sufficient specificity to draw substantive instructional implications. Second, it generates a coherent framework for connecting a variety of issues in mathematics teaching and learning, both past and present. The framework suggests a reinterpretation of some past issues in mathematics education and also reveals similarities and differences between our approach and other current efforts to understand mathematical understanding (Cobb, 1990; Herscovics, 1989; Pirie & Kieren, 1989; Putnam, Lampert, & Peterson, 1990). Third, it suggests interpretations of students' learning that help to explain their successes and failures, both in and out of school. That is, thinking about connections between internal representations is a useful way of thinking about the very data currently calling into question the descriptions of cognition based on structural mental models. Our aim is to deliver on these claims by applying the framework to the complex issues that surround learning and teaching mathematics with understanding.

LEARNING MATHEMATICS WITH UNDERSTANDING

Defining Understanding

We begin by defining understanding in terms of the way information is represented and structured. A mathematical idea or procedure or fact is understood if it is part of an internal network. More specifically, the mathematics is understood if its mental representation is part of a network of representations. The degree of understanding is determined by the number and the strength of the connections. A mathematical idea, procedure, or fact is understood thoroughly if it is linked to existing networks with stronger or more numerous connections.

The idea that understanding in mathematics is making connections between ideas, facts, or procedures is not new. It is a theme that runs through some of the classic works within mathematics education literature (Brownell, 1935; Fehr, 1955; McLellan & Dewey, 1895; Polya, 1957; Van Engen, 1949; Wertheimer, 1959) and emerges frequently in more recent discussions of representation and understanding in mathematics (Davis, 1984; Greeno, 1977; Hiebert, 1986; Janvier, 1987; Michener, 1978). Many of those who study mathematics learning agree that understanding involves recognizing relationships between pieces of information. The contribution of this chapter does not lie in presenting this basic idea but in describing it with more precision, in applying it to a range of learning and teaching situations and in pushing the ideas further to help us understand understanding.

There are several kinds of connections that learners construct to create mental networks. The kinds of relationships that might be constructed have been hinted at in suggesting metaphors for the networks; one kind of relationship is based on similarities and differences and a second kind is based on inclusion.

Relationships of similarity and difference may be created by noting correspondences both between different external representation forms and within the same form. As an example of building relationships between different representation forms, consider students' activities with base-10 blocks and the standard written notation for whole numbers. Once a particular size block is assigned a value of 1, the remaining blocks necessarily take on values of 10 to some nonzero power. Because each block is bigger or smaller by 10 than adjacent size blocks, the size of the block tells its value. Correspondences can be established between particular size blocks and particular positions in the written notation. Connections then can be created between the relative size of the block and the relative values of the positions.

The presumed benefit to the learners is that they bring with them a great deal of useful information already connected to the blocks. From their prior experiences they are likely to have built up ideas of relative size and weight (larger things have more), of counting objects, maybe even of trading things that are equal and putting together things that are alike. In other words, they are likely to have an internal network connected to their mental representation of the physical blocks. If the written symbols are linked to the block representation, they get linked to the entire block network. The written symbols are informed by the multiple associations the learners have already constructed for the blocks.

Connections need not be limited to the representations of quantity but can also be created between actions on quantity. Students are likely to add and subtract blocks by combining blocks of the same size. Parallel actions can be taken on the written symbols, and connections can be established between each action on the blocks and a corresponding action on the symbols. Making such connections explicit for students is the basis for a number of instructional suggestions on teaching multidigit addition and subtraction (Bell, Fuson, & Lesh, 1976; Merseth, 1978; Resnick, 1982).

It should be made clear that the instructional goal in connecting representations of quantity or actions on quantity is not for students to crystallize connections between particular blocks and particular positions. Rather, the goal is for students to build bridges from their knowledge of working with blocks (grouping by 10, combining like blocks, and so on) to the written symbols. Once the relative values of numerals are established, and they are seen as groupings of 10, the connections to the blocks become unnecessary.

We must also make clear that we are not suggesting that connections between base-10 blocks and written symbols are straightforward or easy for students to construct (see Schoenfeld, 1986). Only by thinking and talking about the differences and similarities between the symbol and block representation of whole numbers can students construct relationships between representations.

Another example of building connections between different representation forms is given by Nesher (1989) in her theoretical analysis of using Cuisenaire™ rods for adding and subtracting single-digit whole numbers. Because many young children have available associations for the salient features of the rods—length and color—these mental networks of information can be used to guide addition and subtraction with rods. Then, as students construct connections between rods and numerals, their knowledge of rods can stimulate and inform their actions with the written symbols. Although Nesher casts her argument in a slightly different way than we do, the theoretical principles she invokes are similar to those we express here.

Similarities and differences between different representation forms are the basis for relationships that reappear again and again throughout a student's mathematical career. For example, an understanding of the written epsilon definition of limit is presumed to be enriched if it is connected to the picture of an asymptote on a graph. Similarities and differences between alternate representations of the same information are relationships that can stimulate the construction of useful connections at all levels of expertise.

Relationships based on similarities and differences can also be established within a representation form. Consider, for example, the written notation for whole numbers along with the four arithmetic operations. Standard algorithms are ordinarily taught for each operation, but there are also numerous alternative procedures that can be used. All of these procedures are related in some way. For example, the standard division algorithm is the same as the subtraction algorithm at many points but it also differs in that, among other things, the subtrahend must be supplied through multiplication. As another example, two different procedures commonly are employed for subtracting multidigit numerals: the regrouping method and the equal additions method. The methods are the same in that they subtract individual columns of digits, but they are different in that the regrouping method retains the value of the minuend but changes the values in individual columns. The equal additions method retains the size of the difference but changes the values of the minuend and subtrahend.

By thinking and talking about similarities and differences between arithmetic procedures, students can construct relationships between them. In this case, the instructional goal is not necessarily to inform one procedure by the other but, rather, to help students build a coherent mental network in which all pieces are joined to others with multiple links. Constructing relationships within a representation form often increases the cohesion and structure of the network.

Relationships based on similarities and differences, whether between or within representation forms, are likely to be found in networks that resemble webs, since similarities and differences need not imply higher order relationships but, rather, may simply join pieces of information represented at about the same level of generality. Another kind of relationship is suggested when one mathematical fact or procedure is seen as a special case of another. This kind of relationship is based on notions of inclusion or notions of general cases and specific cases. It is likely that these kinds of relationships are found in hierarchical networks.

An example of connections built on inclusion relationships is taken from work on early addition and subtraction. It is well known that many young children solve addition and subtraction word problems using counting strategies that mirror the semantic structure of the problems (Carpenter & Moser, 1984). Riley, Greeno, and Heller (1983) propose a series of models that depict children's knowledge structures—their networks of representations—as they become increasingly proficient problem solvers. All of the models are based on problem schemata, which are representations of the basic types of semantic structures commonly appearing in word problems. Presumably, children interpret individual word problems as special cases of the general problem types and apply counting strategies that they have connected to these types. In the most advanced model, Riley et al. propose a problem schema of the most general kind. They suggest that children create an internal network built on relationships of parts to whole. Because all addition and subtraction problems can be treated as special cases of part-whole situations, a general part-whole schema provides a set of higher order connections that can subsume connections at the level of individual problems or even at the more abstract level of problem types. At this point, children's internal networks apparently look like well structured hierarchies, with part-whole relationships serving as umbrellas for more specific relationships.

The use of problem schemas in the models of Riley et al. (1983) represents an application of schemata or frames as descriptions of knowledge structures (Davis, 1984; Schank & Abelson, 1977). Generally, schemata are relatively stable internal networks that are constructed at a relatively high level of abstraction or generality. Schemata serve as templates that are used to interpret specific events. That is, schemata are abstract representations to which specific situations are connected as special cases. For our purposes, the important point is that connections which tie internal representations into networks can be based on relationships of subsumption and inclusion.

A second example of connections built on inclusion relationships, but an example somewhat different than schemata, is taken from Kieren's (1988) portrayal of rational number knowledge. Kieren proposes a model of how experts might structure their knowledge of rational numbers, a model Kieren describes as "an image of ideal personal rational number" (p. 162). At the lowest level, closest to interactions with the external environment, are mental representations of empirical data. These local representations are connected to more general principles or constructs which, in turn, are connected to higher-level constructs. Kieren suggests intuitive ideas of partitioning, of making quantities equivalent, and of forming units (which can be subdivided) as constructs at a medium level of generality, with a more formal notion of multiplicative structures as a highly general construct. Many of the relationships that tie the pieces together are inclusion relationships, interpreting one idea, fact, or procedure as a special case of another.

To summarize, we have argued that it is useful to think of students' knowledge of mathematics as internal networks of representations. Understanding occurs as representations get connected into increasingly structured and cohesive networks. The connections which create networks form several kinds of relationships, including similarities, differences, and inclusion and subsumption.

Building Understanding

Using our definition of understanding, we can now describe, at least in a metaphoric way, the structuring process that produces understanding. Networks of mental representations are built gradually as new information is connected to existing networks or as new relationships are constructed between previously disconnected information. Understanding grows as the networks become larger and more organized. Thus, understanding is not an all or none phenomenon. Understanding can be rather limited if only some of the mental representations of potentially related ideas are connected or if the connections are weak. Connections that are weak and fragile may be useless in the face of conflicting or nonsupportive situations. Understanding increases as networks grow and as relationships become strengthened with reinforcing experiences and tighter network structuring.

Growth of networks may occur in several ways. Perhaps the easiest to envision is adjoining a representation for a new fact or procedure to an existing network. Consider fourth-grade students who have already constructed an internal network for place value with whole numbers. Part of this network is the procedure for adding and subtracting by writing the numerals vertically with digits aligned on the right. Suppose the students have represented the procedure in such a way that they have connected the mechanics of aligning digits with the combining of quantities measured with the same unit (ones, tens, hundreds, and so on). When these students encounter addition and subtraction with decimal fractions, they are in a good position to connect the frequently taught procedure—line up the decimal points—with combining quantities measured with the same unit. If they build the connection, the addition and subtraction procedure becomes part of the existing network, the network becomes enriched, and adding and subtracting decimals is understood.

Although the image of adding to existing networks in a smooth cumulative way is appealing in its simplicity, it may turn out that the image is too simple. From a distance, the linear incremental growth of networks of understanding may appear plausible. However, up close analyses that catch students in the act of building understandings reveal a much more chaotic process (Hiebert, Wearne, & Taber, 1991; Schoenfeld, Smith, & Arcavi, in press; Steffe & Cobb, 1988). Growth can be characterized as changes in networks as well as additions to networks. The changes sometimes are manifested as temporary regressions as well as progressions. The changes appear to be intermittent and somewhat unpredictable; students seem to build understanding sporadically, rather than through smooth, monotonic increases.

Changes in networks might best be described as reorganizations. Representations are rearranged; new connections are formed, and old connections may be modified or abandoned. The construction of new relationships may force a reconfiguration of affected networks. The reorganizations may be local or widespread and dramatic, reverberating across numerous related networks. Reorganizations are manifested both as new insights, local or global, and as temporary confusions. Ultimately, understanding increases as the reorganizations yield more richly connected, cohesive networks.

For an example of network reorganization, consider first year algebra students who are observing that the sign of a number changes as it moves from one side of an equation to the other. If the students have represented their previous ideas about the equal sign in a way that connects them only with actions of writing the answer—-separating the problem statement on the left with the answer on the right—then they are unlikely to possess internal networks with which the new actions can connect. Because of their limited representations of the equal sign, the students are unlikely to represent the new information in a way that links it to their existing networks. But suppose the students engage in activities that focus on equivalence, encouraging them, for example, to treat the equal sign as the fulcrum of a balance. The new representations of equivalence and the new relationships that students construct may prompt them to replace existing connections in their relevant networks with new connections. Networks concerned with the equal sign are reorganized, and the actions of reading the problem statement on the left and writing the answer on the right, and changing the sign of a number as it moves from one side of the equation to the other, can both be related to equivalence within a single, cohesive network.

One consequence of the apparent pervasiveness of reorganization (Hiebert et al., 1991; Schoenfeld et al., in press; Steffe & Cobb, 1988) is that internal networks are better thought of as dynamic instead of as static. Networks are constantly undergoing realignment and reconfiguration as new relationships are constructed.

The processes of reorganizing networks and adjoining new representations to existing networks both depend, to some degree, on the networks that have already been created. Past experiences create mental networks that the learner uses to interpret and understand new experiences and information (Ausubel, 1968; Wittrock, 1974). "People continually try to understand and think about the new in terms of what they already know" (Glaser, 1984, p. 100). In other words, existing networks influence the relationships that are constructed, thereby helping to shape the new networks that are formed.

The degree to which existing networks influence the nature of new relationships probably varies widely. If the learner tries hard to fit a new idea, fact, or procedure into a current way of thinking, existing networks constrain the relationships that are created. At the other extreme, a learner may represent new information in a way that does not connect it with existing networks. The representation might be held in isolation and, if it does not decay, may be connected eventually with other compartmentalized representations or with a connected network (cf. Noddings, 1985; Nolan, 1973). But even here, the eventual relationships are influenced by those that already exist. From our perspective, the most likely scenarios for building understanding involve increases in either the size or the structure of networks, these processes both building on existing networks.

The notion of building understanding by constructing relationships that yield larger, more cohesive internal networks is useful in analyzing a number of issues related to understanding mathematics. In the sections that follow we apply our framework to several important issues of mathematical representation currently receiving attention in the mathematics education community.

Using Alternate Representations in Classrooms. For years mathematics educators have advocated using a variety of forms to represent mathematical ideas for students. Physical three-dimensional objects are often suggested to be especially useful. The simple rationale usually provided is that children understand better when ideas are presented with concrete materials.

Despite the intuitive appeal of using materials, investigations of the effectiveness of concrete materials in classrooms have yielded mixed results (Fennema, 1972; Raphael & Wahlstrom, 1989; Sowell, 1989; Suydam & Higgins, 1977). From our perspective, there are several explanations for the equivocal effects of interacting with concrete materials.

To say that children understand when using concrete materials carries a definite meaning within our framework and provides a way of thinking about how alternate representations may (or may not) support learning with understanding. From our perspective, to say that children understand ideas when they are presented with concrete materials is to say that children construct relationships which yield a connected network containing representations of the materials and their interactions with them. Children might do this by either representing the materials in a way that connects them with existing networks or constructing relationships that prompt a reorganization of networks. Consequently, it is important to consider both the internal networks that students carry with them and the classroom activities that promote construction of relationships between internal representations.

Many students bring with them representations of experiences with quantities from outside of school. For example, even young children have created a network of associations for attributes such as numerosity, length, volume, weight, and so on. The networks may be of varying degrees of scope and cohesion, but most children bring with them some knowledge of quantities. Children's existing associations of quantity may be exactly what is needed to interpret the concrete materials appropriately and to connect relevant features of the materials to an existing network (Nesher, 1989). For example, children's notions of equivalence—of trading equal amounts—and of combining like quantities are needed to support construction of appropriate relationships with base-10 blocks. The existence of prior representations of quantities and actions on quantities may account, at least in part, for the many positive results reported when using physical materials in the classroom.

But what about the negative results: Why are concrete materials sometimes ineffective? First, there is a flipside to the previous discussion: If students do not bring with them the kind of knowledge of quantities that teachers expect, it is not easy for students to relate their interactions with the materials to existing networks. They do not interpret the materials in the way that the teacher expects, and the use of concrete materials is then likely to generate only haphazard connections.

Second, negative results with concrete materials may be traced to two features of the activities in which students engage. The first concerns the distance between the concrete material and the mathematical relationships that we intend them to represent; the closer the match between salient features of the materials and the mathematical relationships, the more contextual support there is for students to construct intended connections. The match between targeted relationships in the situation and concrete materials can range from a very close, contextually reinforced match to a very distant one with few supporting cues. In the latter case, the concrete material takes on the features of an arbitrary symbol rather than a natural embodiment.

As an example of contextual distance, consider again the place value principle in the whole number notation system. Several different materials are commonly used with primary-grade students to represent the place value principle, including base-10 blocks, colored chips, and money. Imagine a situation in which 136 objects (cows, candies, pennies) are combined with 57 similar objects. If the situation is represented with base-10 blocks with the small cube standing for 1, the objects could be represented by one set of 136 small cubes and another set of 57 small cubes. However, without a special grouping, this representation does not capture the place value principle. On the other hand, the situation could be represented by taking advantage of the grouping built into the materials. The set of 136 objects could be represented with 6 small cubes, 3 long (tens) blocks, and 1 flat (hundreds) block. Similarly, 57 objects could be represented with 7 small cubes and 5 long blocks. It is important to note that, in both sets, the total number of objects is still visible; they have simply been grouped in a special way. Also, as noted earlier, there are several physical cues such as size and weight, that project the relative values of the blocks so that the grouping is represented in a reasonably salient, nonarbitrary way. That is, each larger block is 10 times bigger than the preceding block. For all these reasons, the blocks contain considerable contextual support for the place value principle, for there is a relatively close match between the features of the material and the mathematical features of place value.

Suppose the situation is represented with colored chips. We need to begin by arbitrarily assigning the unit value to one of the colors, say yellow. The 136 and the 57 objects could be represented by 157 yellow chips in one set and 57 yellow chips in another. To capture the place value principle, however, we need to assign the values of ten and hundred to other colors. In contrast to the blocks, these assignments are arbitrary as well. Once the values are assigned, the objects can be represented by the chips. But the individual objects are not visible in the chip representation. The 136 objects are represented by, say 1 red chip, 3 green chips, and 6 yellow chips. These chips can be operated on much the same as the blocks—like pieces can be combined, 10 of one piece can be traded for 1 of another and so on. But the chip material itself provides no physical clues about its value; the features of the colored chips representing the mathematical features of place value have been assigned arbitrarily. Compared with the blocks, the chips contain less contextual support for place value, thus the contextual distance is greater.

Now suppose the situation is represented with money. Coins and bills come with values assigned by convention. Pennies represent 1, dimes represent 10, and dollar bills represent 100. If students know these values from previous experiences, they can use their knowledge to represent 136 as 1 dollar bill, 3 dimes, and 6 pennies. Similarly, 57 can be represented as 5 dimes and 7 pennies. The money can then be manipulated as the blocks or chips. The material itself, however, achieves value in much the same way as the colored chips—by arbitrary assignment. There are no physical features of money that suggest the values of the pieces. So, in this sense, money is similar to colored chips in its contextual distance from place value ideas.

But money, as a physical representation of place value, raises a number of additional issues. Although the values of particular pieces of money are assigned arbitrarily, students often bring with them a variety of money associations, including knowledge of the values of each piece. That is, students often carry an internal network for money that has been constructed through a variety of experiences, usually out of school. Knowledge represented in the network may support features of money that capture place value principles, such as the appropriate values of pennies, dimes, and dollars, and the 10-for-1 trading that can be done between these pieces. Consequently, for most children money provides a much different representation for place value than colored chips, despite the fact that values for both are assigned arbitrarily.

Children's familiarity with money suggests that it may be an especially useful representation of place value. However, this conclusion may need to be tempered somewhat. Children's experiences with money have occurred in special cultural settings. Money takes on meaning in settings that involve purchasing or exchanging goods. The special role that money plays in our culture is evidenced by the fact that it is unusual to let money stand for other objects, such as cows or candies. The consequence of its special role is that children may not recognize immediately the way in which money represents place value. Children's associations for money may remain tied to its use outside of class and fail to connect with place value ideas that the teacher is emphasizing in class. This means that, although money is potentially a useful material, teachers should not assume students will make the appropriate place value connections immediately. Activities that aim to make the intended connections explicit for students may be needed with money just as with other concrete materials.

It is important to note that simply because one material is more distant from an intended relationship than another does not mean it is necessarily less useful. It may be productive to imagine the distance between the quantities or relationships that are of interest and the external form in which the quantities or relationships are ultimately to be represented. In most instruction programs, written symbols are the representation of choice, but are contextually more distant from quantities or relationships than many concrete materials. It may be helpful for students to fill this gap between quantities and written symbols with physical materials, beginning with contextually close materials that make contact in salient ways with the quantitative relationships of interest and, then, moving to contextually more distant materials that, in turn, make contact with written symbols (cf. Hart, 1989).

A second feature of classroom activities that might help explain the uneven effectiveness of using concrete materials in the classroom is a potential interaction between the materials and the social situation in which the materials are used. To set the stage for this argument, it is necessary to point out that all representations of a mathematical situation are not entirely faithful to the quantities or relationships in the situation that are of interest. That is, each representation of a quantity or relationship captures some of its features, but not others. To make the intended connections, students must focus on features of the representation that capture relationships of interest.

How can instruction help to focus students' attention? Social interaction in the classroom provides a powerful way of collecting students' attention and focusing it on shared experiences (Cobb, 1990; Lampert, 1989). Language has a distinctive orienting function (Maturana & Varela, 1980), and classroom discourse can exploit this function to orient students' attention to mathematical relationships of interest. Concrete materials provide a public representation to which attention can be drawn and a focus for discussion. Because the materials are public, they can be shared by all the students in the class. To the extent that the same mathematical features are perceived by the students, the language in the classroom can be about the same things. Communication among the students and teachers is enhanced because all participants can focus their attention on the same entities and relationships between entities. This is nontrivial because students' varied backgrounds and goals often make it difficult to share a discussion about a common event or idea. Concrete materials, by providing a comparatively unambiguous target, thus make such shared discussions possible.

Focusing students' attention on particular relationships is important because not all relationships are of equal interest. Certain materials are selected for classroom activities because they presumably capture relationships adults believe to be especially important. There is no guarantee, though, that students see the same relationships in the materials that we do. Through class discussions, teachers and students can talk about possible relationships, drawing attention to the relationships of interest. By interacting with the materials and with others about those materials, students are more likely to construct the relationships that the teachers intend. In fact, the language used to talk with others about materials may be crucial for students in constructing relationships (Greeno, 1988b; Resnick & Omanson, 1987). The social context in which materials are used may account, in part, for their effectiveness (or ineffectiveness) in helping students understand.

Understanding Written Symbols. The standard written symbol systems of mathematics play an especially important part in children's learning experiences in school, forming a primary representation system in the development of expertise. The way in which written symbols take on meaning and can be used deserves special consideration.

Meaning of written symbols can develop in the two ways that understanding develops for any representation form — connecting with other forms of representation or establishing connections within the representation. When written symbols are connected with other forms, such as physical objects, pictures, and spoken language, the source of meaning is presumed to be the internal networks that already have been created for these forms. Internal representations of written symbols can benefit from these networks by connecting with them. Meaning can also be established by building relationships within the symbol system; often, this occurs by recognizing patterns within the symbol system.

Meaning is drawn from outside the symbol systems as learners create connections between written symbols and other forms of representation, such as concrete materials. From earlier analyses, it is not simply the presence of concrete materials that provides meaning for symbols, nor is it simply the juxtaposition of materials and symbols. In order for symbols to acquire meaning, learners must connect their mental representations of written symbols with their mental representations of concrete materials. The potential for these connections to create understanding is complicated by the fact that the concrete materials themselves are representations of mathematical relationships and quantities. Thus, the usefulness of concrete materials as referents for symbols depends both on their embodiments of mathematical relationships and on their connections to written symbols.

Connections between written symbols and other representations can be made both for individual symbols and for actions on symbols. Most mathematical symbol systems that students encounter in school have two kinds of individual symbols: symbols that stand for quantities and symbols that stand for relationships between quantities. Numeric symbols, such as -3, $\frac{1}{2}$, and 2.8, take on meaning as they become connected with other representations of quantity, such as concrete materials or everyday experiences. Operation and relation symbols, such as $+$, $-$, and $=$, take on meaning as they become connected with actions and relations between the same representations of quantity that served as referents for the numeric symbols. For example, consider the $\times$ symbol in the sentence $3.2 \times 4.5 = \_\_$. Suppose that students have used base-10 blocks as referents for decimal numerals. The $\times$ symbol acquires meaning for students as they connect it with the anticipated action of creating 3 groups of the blocks representing 4.5 and 2-tenths of a group more. The meaning for $\times$ does not indicate how to manipulate the symbols; rather, it suggests what the manipulations will do for you when they are learned and executed. Note that the meaning for $\times$ illustrated here is only one meaning and, by itself, does not provide a full understanding of multiplication (Greer, this volume; Nesher, 1988; Schwartz, 1988; Vergnaud, 1988). Students increase their understanding of $\times$ as they connect it with a widening set of referents, each backed by an internal network of representations. The idea of a growing meaning for the written symbol applies not only to $\times$, but to all written symbols.

Once meanings are established for individual symbols, it is possible to think about creating meaning for rules and procedures that govern actions on these symbols. Meanings can be created for rules by connecting the actions on symbols that make up the rules with parallel actions on the referents for the numeric symbols. The degree of explicitness in the step-by-step correspondence between actions on referents and actions on symbols can vary (Bebout, 1990; Carpenter, Moser, & Bebout, 1988; Fuson, 1986; Resnick, 1982; Wearne & Hiebert, 1988). It has been shown that students can create connections under either implicit or explicit conditions, but there are important differences in instructional approaches. These are elaborated later in the chapter.

Understanding for written symbols also can be generated as connections are built within the system, itself. As noted earlier, connections within the system give it structure by relating its elements to one another. There are at least two ingredients in the process of building connections within a written symbol system. First, written symbols must be represented internally as mathematical objects rather than just as marks on paper that stand for other things. In a sense, symbols must change from being transparent, only revealing their referents, to being opaque, thus achieving an identity in their own right (Bruner, 1973; Davis, 1984; Mason, 1987). Students must recognize symbols as entities that can be elements of a system, related in various ways, and operated on. The fact that students are capable of treating symbols as entities is revealed in the metaphoric language they use to talk about symbols when using their intuitions in developing procedures (Trafton, 1989): When students say they are chopping numbers in half and knocking off or tacking on parts of numbers, they are suggesting that the symbols are, themselves, meaningful objects.

A second ingredient that contributes to the connecting process is the set of patterns and regularities in the symbol systems. Because the standard written symbol systems in mathematics are so tightly structured, such patterns abound (Steen, 1988). As with any specially structured external representation, though, the patterns are not inevitably transferred into students' internal representations; students must construct the relevant relationships for themselves.

As with concrete materials, it is likely that social interaction in the classroom influences the kinds of relationships that students construct. Sharing discussions about regularities and patterns in a written symbol system may support the personal construction of relevant relationships. It is also likely that building relationships requires a reflective frame of mind (Hiebert, 1990; Kilpatrick, 1985). Because the written symbols must be seen as mathematical objects, it is reasonable to think that students must consciously reflect on the symbols as elements of a system rather than simply moving them around on paper according to memorized rules. Of course, symbols can be manipulated without reflection, but manipulation without reflection is unlikely to stimulate construction of the relationships that lead to understanding.

Examples of relationships built within a symbol system can be seen at all levels of mathematics, although such relationships become increasingly predominant as mathematics becomes more advanced. As an example at the elementary level, consider again the notation system for whole numbers. A numeral such as 3,824 is usually treated as three thousand eight hundred twenty-four units, where the units are ones. Because of the structure built into the notation, however, the numeral also expresses how many units of any power of ten can be formed: three hundred eighty-two tens, thirty-eight hundreds, and three thousands. The same pattern extends to decimal fractions. The number of units of a particular kind can always be determined from the numeral by treating the unit of interest as the final digit. Such relationships within the notation, although perhaps obvious to adults, must be constructed by students.

For a second example, consider the process of changing a common fraction to a decimal. A significant consequence of this activity is that the decimal either terminates or repeats. Why could the string of digits not continue randomly? A clue lies in the pattern that turns up in the workspace when carrying out the long division. Suppose, for instance, that the fraction is $\frac{8}{17}$.

The subtraction steps in the procedure can yield a maximum of 17 different differences (0, 1, 2, ..., 16). If the difference is 0, the division (and the decimal representation) terminates. If the difference is not zero, it must repeat after 16 subtractions, at the most. Once the subtraction yields a difference obtained earlier, the entire computation, and the decimal quotient, repeats itself. The significance of this analysis is that it can be carried out entirely within the symbol system, with no appeals to outside referents.

A third example is drawn from high school algebra. Algebra students acquire a variety of procedures for simplifying expressions and solving equations. Understanding for these procedures can be derived from connections within the symbol system, specifically by relating them to basic properties of the number system. For example, simplification of an expression like $3x + 5x$ is based on the distributive property. If students can recognize that the distributive property applies to this expression, they have the knowledge necessary to simplify the expression without learning a new procedure for combining like terms. Connections with the distributive property help students recognize the situations to which the property applies so that they avoid such errors as $3x + 5y = 8xy$. Connections that relate basic properties of the number system to algebraic procedures bring about meaning for the procedure and for the property.

It is important to remember that recognizing patterns and constructing connections within a system require a shift in the activities' goal from one of proficient execution to one of reflective analysis of structure. The structure of the symbol system is a source of meaning for symbols, but the source is tapped only as students recognize the structure and represent it internally for themselves.

The argument that meaning can be derived through building relationships within the symbol systems of mathematics is not new. For example, many of Brownell's writings (1935, 1947) can be interpreted as essays in support of building meaning from mathematical relationships within the symbol systems. Brownell's argument is slightly different from ours, for he was contrasting meaning derived from mathematical relationships with meaning derived from real life contexts in which mathematics is often embedded. Furthermore, Brownell did not make the distinctions we have made between connections *within* representation forms versus connections *between* representation forms. Nevertheless, most of the examples Brownell used to illustrate mathematical relationships are examples within the system of written symbols and based on the structure of these systems.

To summarize and extend this analysis, meaning for written symbols can develop through connections between symbols and other representation forms, such as physical objects, or through connections within the system of symbols. In other words, written symbols can be thought of as intellectual tools with two primary functions: (1) a public function that involves recording what is already known in order to share and communicate it, and (2) a personal function that involves organizing and manipulating ideas. When symbols function as a record of what is already known, they stand for something; the learner has connected them with a mathematical situation represented in another way. The symbols, then, are used to re-present the ideas or actions that they represent. When the symbols are used to organize and manipulate ideas, the interrelationships or structure of the symbol system are most important. The symbols take the place of the ideas as the objects of thought. Both functions of symbols require connections; the public function usually requires connections to other representation forms, and the personal function is enhanced by additional connections within the symbolic form.

Consequences of Understanding Mathematics

The issue of understanding mathematics achieves its significance through the numerous claims made on its behalf. For example, proponents of emphasizing understanding in school mathematics have argued that those students who understand will retain what they learn and transfer it to novel situations. Some of these claims have strong theoretical support, some are backed by empirical data as well, and some have little of either. Claims that fit the first two categories are of primary interest and will be examined here in the context of representing and connecting knowledge.

Understanding Is Generative. It is now well accepted that students construct their own mathematical knowledge rather than receiving it in finished form from the teacher or a textbook. Within our framework, this means students create their own internal representations of their interactions with the world and build their own networks of representations. A crucial aspect of students' constructive processes is their inventiveness (Piaget, 1973; Resnick, 1980; Wittrock, 1974). Children continually invent ways of dealing with the world. Many of the errors they make, such as linguistic overgeneralizations, can be interpreted as the result of inventions. Similarly, in school mathematics, students rely on invented strategies to solve a variety of problems.

But students' inventions do not necessarily lead to productive mathematics. If students are working with written symbols unconnected to richer networks of knowledge, their inventions often produce flawed algorithms (Brown & Van-Lehn, 1982; Cauley, 1988; Hiebert & Wearne, 1985; Matz, 1980). For example, Cauley (1988) found that third-graders lacking conceptual knowledge of the multidigit subtraction procedure were more likely to make performance errors, such as subtracting the top smaller digit from the bottom larger digit. However, if the arguments of students' inventions are parts of well-connected networks, the resulting mathematics can be productive (Behr, Wachsmuth, Post, & Lesh, 1984; Carpenter & Moser, 1984; Heid, 1988; Payne, 1976; Wearne & Hiebert, 1988). Fourth-grade students, for example, who had connected decimal fraction numerals with physical representations of decimal quantities were more likely to invent appropriate procedures for dealing with problems they had not seen before, such as ordering decimals by size and changing between decimal and common fraction forms, than were those who had not made the same connections (Wearne & Hiebert, 1988).

The distinguishing feature of productive inventions is the nature of the mental representation on which the invention op-

erates (Resnick, 1987a). It seems reasonable that, if the mental representations are enriched by being connected within a network, then inventions are stimulated, guided, and monitored by much related knowledge. On the other hand, if inventions are operating on representations not connected with related knowledge, then the inventions are more likely to be flawed and counterproductive.

The significance of this analysis is that if understanding is built initially, then the ongoing inventive process can operate on mental representations with rich associations. The results of such inventions remain connected to the network of knowledge. Thus, inventions push students' current understanding. Inventions that operate on understandings can generate new understandings, suggesting a kind of snowball effect. As networks grow and become more structured they increase the potential for invention and proliferate their potential points of contact with newly represented information. Richer networks are more likely to connect with new representations than are impoverished networks; there are more ways the new representations can be related because there is simply more to which they can relate.

The theoretical case for the generation of understanding has significant implications for the growth of students' mathematical knowledge. If the argument is correct, it points to the importance of building understanding—of creating rich networks of knowledge—when a topic is first encountered. The emphasis should be placed initially on supporting students efforts to build relationships rather than encouraging them to become proficient executors of procedures. This hypothetical process stands in contrast to alternatives, such as storing pieces of information separately, perhaps through drill and practice, and then creating a network at some later point by connecting the previously stored pieces of information (Nolan, 1973).

Emphasizing relationships does not imply that students should be asked to memorize connections. Rather, the implication of generative understanding is that students can be placed in settings in which they can construct useful connections (Mayer, 1989). Understanding is more accurately viewed as generated by individual students than as provided by the teacher.

As noted earlier, the generative process is not smooth and predictable (Hiebert et al., 1991; Schoenfeld et al., in press), but the available evidence suggests that, over time, students do construct relationships, create productive inventions, and build their understanding (Bednarz & Janvier, 1988; Carpenter & Moser, 1984; Siegler & Jenkins, 1989; Steffe & Cobb, 1988).

Understanding Promotes Remembering. Since Frederic Bartlett (1932) presented his work on *Remembering,* it has become increasingly clear that memory is a constructive or reconstructive process, rather than a passive activity of storage. If the information to be remembered is more complex than nonsense syllables, people often structure it in such a way as to impose some meaning on it; by doing so, they often modify the information they remember. Evidence from verbal learning and comprehension suggests that these modifications are made to bring the information in line with the person's current knowledge (Rumelhart, 1975). That way, the information is represented by students in a way that fits with their existing network of knowledge. Making connections between new information and existing knowledge already represented in networks is one way of characterizing Bartlett's (1932) observation of a natural "effort after meaning."

One advantage of the inclination to create connections between new and existing knowledge is that well-connected knowledge is remembered better (Baddeley, 1976; Bruner, 1960; Hilgard, 1957). There are probably two explanations for this. First, an entire network of knowledge is less likely to deteriorate than an isolated piece of information. Second, retrieval of information is enhanced if it is connected to a larger network. There simply are more routes of recall. For example, think back to the students who are learning to add decimal numbers. Suppose their teacher says, "When you add decimals, line up the decimal points." If students store this information as a separate piece, retrieval depends on retracing a single link between the procedure and the perception of an addition problem as one involving decimals. However, if students connect this information with a network of knowledge of adding other quantities by recognizing the relation of adding like items together, they are in a much better position to retrieve the information. The retrieval process can be triggered by several external and internal cues, such as knowledge of place value, common denominators, or sizes of quantities. In fact, the information might be viewed as a special case of a more general principle, and could be reconstructed extemporaneously.

It appears, then, that the reason understanding promotes remembering is theoretically straightforward. Memory, if viewed as a reconstructive process, involves the same cognitive activity as understanding: constructing connections between representations of new knowledge and existing knowledge. Assuming the connections are appropriate mathematically, understanding and memory are increased concurrently.

Understanding Reduces the Amount that Must Be Remembered.

A consequence of understanding related to enhanced memory pertains to what must be remembered. If something is understood, it is represented in a way that connects it to a network. The more structured the network, the fewer individual pieces need to be retrieved separately. Memory for any single part of the network comes with memory for the network as a whole, reducing the number of items that must be remembered.

Some networks are so tightly structured that they are accessed and applied as a whole, as a single chunk. Accessing any part of the chunk means accessing the entire network. Recalling the discussion of schemata from earlier in the chapter, imagine students constructing a schema for equivalent fractions. Such a schema would include representations for the equivalence relation, for procedures used to generate fractions equivalent to a given fraction and to check whether two fractions are equivalent, and perhaps for some facts about equivalent fraction pairs (such as $\frac{1}{2} = \frac{2}{4}$). Now consider the procedures that commonly are taught for adding and subtracting fractions, reducing fractions to simplest form, and changing between improper and mixed fractions. Often, the procedures are taught as separate algorithms and students are implicitly encouraged to store and retrieve each algorithm separately. But a major component of each procedure involves the construction of equivalent fractions. If students possess a schema for equivalent fractions and recognize the appropriateness of applying the schema in each case, then each procedure does not need to be stored and retrieved as a separate algorithm. A single core procedure, constructing equivalent fractions, can be retrieved and adapted to each of the problem constraints.

Understanding Enhances Transfer.

It is obvious that transfer is essential for mathematical competence and that transfer occurs frequently in mathematics. Transfer is essential because new problems need to be solved using previously learned strategies. It would be impossible to become competent if a separate strategy would need to be learned for every problem. Transfer occurs frequently because many students improve their performance on some problems by learning to solve related ones. However, once the analysis moves beyond these global observations, difficulties arise. For example, to what degree must the problems be related before solving one enhances one's performance on the other?

Although transfer has been difficult to assess precisely and teach effectively, it has had a long history in psychology and mathematics education (Overman, 1935; Thorndike & Woodworth, 1901). It is receiving renewed attention as the field of cognitive science influences our views of learning (Campione, 1988; Cormier & Hagman, 1987; DeCorte, 1987; Larkin, 1989; Perkins & Salomon, 1989; Singley & Anderson, 1989) and as competing theories emerge (Brown et al., 1989; Lave, 1988). We briefly review the theoretical positions on transfer and then place the questions within our framework of representations and connections.

One of the earliest systematic explanations for transfer of learning can be found in Thorndike's (1914) view of shared components. According to Thorndike, the degree of positive transfer from one task to another is a function of the number of elements that the tasks share. If many of the elements or components that make up the tasks are alike, then the transfer from one task to the other is high.

Much of the work on transfer since Thorndike has attempted to improve the conceptualization of common elements between tasks. Debates have centered around whether the shared components between tasks are better conceived as specific elements or as more general principles. Some argue that recent work in cognitive science has made it possible to formulate hypotheses of more general transfer, based on shared principles or strategies, for example, while maintaining the same precision with which Thorndike formulated his hypotheses of specific transfer (Gray & Orasnu, 1987). In general, the work in cognitive science has encouraged a shift from viewing common elements as inherent in the external tasks to viewing them as features of internal representations of the tasks.

The empirical results show that transfer is usually quite specific and contextually bounded. That is, transfer between tasks is most apparent if there are specific external or internal elements in common and if the tasks are situated in common contexts. If the situational features of the task are considerably different, and if the tasks are similar only in general ways, transfer is unlikely. In response to these findings, there is a renewed recognition that transfer occurs only if the learner recognizes the similarity of the tasks; further, the likelihood of this recognition depends largely on the context in which the tasks are embedded (Gick & Holyoak, 1987; Pea, 1987; Voss, 1987).

The boundaries that the situation seems to place on transfer are severe enough that Lave (1988) suggests a complete reformulation of the theories of learning and transfer. According to Lave, theories are no longer tenable if they assume that the learning of cognitive skills can be disembedded from their original context and transferred to a range of tasks. Rather, theories should postulate situation specificity and situation discontinuity as fundamental and an inherent characteristic of cognition. Cognitive analysis should shift, Lave argues, from focusing on the abstract, disembedded character of internal representations to describing cognitive actions within a situation.

The framework we propose retains the notion of internal representations. In fact, we propose that the way internal representations are connected helps to explain the potential for transfer. But, we also suggest that the problem situations in which students engage influence the nature of the internal representations and their connections to other representations. In other words, the situation or context influences the amount of transfer that actually occurs.

To elaborate, consider an intriguing paradox present in the learning literature, particularly in the research on mathematics learning. On the one hand, there seems to exist in students a natural "effort after meaning" (Bartlett, 1932). Evidence of such an effort often can be found in the errors they make (Byers & Erlwanger, 1985). Many misconceptions seem to result from overgeneralizing or from combining pieces of information in inappropriate ways. For example, when students are first learning algebra, many of their errors can be understood as overgeneralizations of previously learned procedures (Matz, 1980). Many of the procedures they learn involve treating expressions as decomposable. Expressions such as $(ab)^n$ and $\sqrt{ab}$ can be written as $a^n b^n$ and $\sqrt{a}\,\sqrt{b}$, respectively. Perhaps it is not surprising that many students also write $(a + b)^n = a^n + b^n$ and $\sqrt{a + b} = \sqrt{a} + \sqrt{b}$. In our terms, students are making connections but the connections are inappropriate.

On the other hand, it seems that often students fail to make connections. They represent information as isolated pieces, compartmentalize their knowledge, and fail to recognize or create connections in mathematical situations that appear obvious to adults. These failures are at the heart of many recent analyses of students' mathematics learning (Davis, 1984; Hiebert, 1986; Hughes, 1986). As an example, Davis and McKnight (1980) explored third and fourth graders' treatment of the problem $7,002 - 25$. They identified six kinds of knowledge that students might connect with this problem: approximate size of quantities, making change with money, representations of quantities with base-10 blocks, strategies for solving similar problems with smaller numbers, mental computation strategies, and the standard written procedure. During their interviews, they found that not all students possessed all kinds of knowledge, but more importantly, when the knowledge was available it was not connected. In particular, knowledge of the written procedure was not connected to knowledge of any other kind. This was apparent when students demonstrated knowledge of making change,

but did not use this knowledge to correct a flawed written procedure or to question an unreasonable written response.

How is it that students seem to search for connections on the one hand but fail to recognize and create many important connections on the other? We believe the paradox can be explained, at least in part, by considering constraints that students are inadvertently encouraged to impose on their internal representations. Students represent for themselves the information and material with which they interact. The richness of the information and material influences the richness of their internal representations. If the mathematical activities in which students are engaged are overly restrictive, their internal representations are severely constrained, and consequently the networks that are built are bounded by these constraints. Connections between these restricted networks are difficult to establish. Students are forced to search for meaning within these relatively small bounded networks, and because meaning in mathematics often comes by relating ideas, facts, and procedures across a range of situations, a search for meaning within an overly restricted domain is bound to be deficient.

As an example of mathematical activities that might lead to overly restricted internal representations, consider the introduction of addition and subtraction situations in the primary grades. Most curricula in the United States present only a few of the many different problem situations (Stigler, Fuson, Ham, & Kim, 1986). Typical word problems in US textbooks describe the joining of two sets or the separating of one set from another and ask for the result. Few problems present comparison or equalizing situations, or vary the unknown in the problem, even though young children can interpret a variety of these problem types (Carpenter & Moser, 1984). Presenting only a restricted set of problems is likely to constrain students' internal representation of addition and subtraction.

Other examples of overly restricted activities in mathematics instruction have been suggested recently with increased research attention to mathematics learning in the middle grades. Multiplication, for example, is often viewed only as repeated-addition (Fischbein, Deri, Nello, & Marino, 1985). Such a representation is inadequate to deal with the variety of multiplicative situations (Bell, Greer, Grimison, & Mangan, 1989; Greer, this volume; Nesher, 1988, Schwartz, 1988). Indeed, simply changing the numbers in the problem from whole numbers to fractions creates situations that students are unable to handle with a repeated addition view of multiplication. By varying the problem situations in which students engage, it appears that instruction can encourage richer interpretations of multiplication (Nesher, 1988). Similarly, fractions are often introduced only through visual models showing parts of regions. Internal representations that students are likely to build from such a restricted set of activities are inadequate to deal with the wide range of fraction situations (Behr, Harel, Post & Lesh, this volume; Kieren, 1988; Lesh, Post, & Behr, 1988; Ohlsson, 1988).

The potential for students to build restricted representations is exacerbated by the fact that, in school, problem situations often are printed symbols on textbook pages or worksheets. These problems encourage students to think primarily about symbols. Internal representations are formed for written symbols and manipulations on symbols. If the problems do not promote connections to other representations, the internal networks of symbols that are formed remain severely bounded and unconnected to other knowledge. For example, many students do not connect the decimal fraction symbol system with that of common fractions (Hiebert & Wearne, 1986). So, the search for meaning—the attempt to make connections—must be carried out in confined networks that represent particular sets of symbols.

The concept of students searching for connections within overly restricted domains is analogous to the earlier descriptions of inventions using impoverished resources (Resnick, 1987a). In both descriptions, students are constructing knowledge by making connections, but the connections are overly restricted and the knowledge is of limited use.

The hypothesis that problem situations constrain the nature of internal representations is proposed here because we believe this phenomenon has a direct bearing on the issues of understanding and transfer. Problem situations influence the way in which internal representations are formed and structured. In turn, the structure of internal networks of knowledge defines the potential for transfer. According to our earlier analysis, understanding means more connections within and between networks, implying fewer boundaries and allowing the search for similarities between tasks within larger domains of knowledge. Recognizing similarities and differences between tasks increases the likelihood of transferring a strategy used on one task to a related task.

If our analysis is correct and problem situations help to place boundaries on internal representations, then learning environments should be designed to extend the limits of these boundaries. Several suggestions for the design of such environments are mentioned here and then elaborated later in the chapter. First, problem situations should either vary along critical dimensions or encourage engagement with central ideas that promote later connections rather than restrict them (Case & Sandieson, 1988; Kieren, 1988; Vergnaud, 1988). Second, construction of connections between different forms of external representations should be encouraged (Hiebert & Lindquist, 1990). Third, classroom discourse and social interactions that focus on mathematical relationships should be promoted to support recognition of connections, breakdown of mental boundaries, and reorganization of internal networks (Lampert, 1986, 1989).

Understanding Influences Beliefs. The discussion to this point has focused on cognitive consequences of understanding. It is important to note, though, that understanding yields affective consequences as well. Students' beliefs about mathematics influence the growth of understanding in important yet subtle ways (Schoenfeld, 1985). It also is plausible that the process of building understanding influences students' beliefs about mathematics.

A bridge between the cognitive process of understanding and the affective process of forming beliefs is provided by Doyle's (1983, 1988) analyses of classroom activities. Doyle argued that the kind of work students do determines how they think about a particular domain and what they believe about the nature of the subject. Work is defined by the tasks students complete and the cognitive processes they use. Because students in mathematics classes often are asked to memorize procedures and rules for manipulating symbols as individual pieces of information (Burns & Lash, 1988; Porter, 1989; Stodolsky, 1988), it is not surprising that many students believe that mathematics is mainly a matter of following rules (National Assessment of Educational Progress, 1983), that it consists mostly of symbols on paper (Mason, 1987), and that the symbols and rules are disconnected from other things they know about mathematics (Carpenter, Hiebert, & Moser, 1983; Carraher & Schliemann, 1985; Davis, 1984; Hiebert & Wearne, 1986; Schoenfeld, 1985). On the other hand, if students would be asked to construct connections between pieces of information—within a representation system or between different representations—one might expect that students would believe that mathematics is a cohesive body of knowledge, that information acquired in one setting will connect with information acquired in another, and that there are consistencies within representation systems and predictable correspondences between representation systems. Such beliefs would, in turn, support the further growth of mathematical knowledge. However, instructional programs that encourage students to construct connections over an extended period of time are rare, and empirical data that address the link between cognitive connecting processes and productive beliefs about mathematics are just becoming available (Nicholls, Cobb, Wood, Yackel, & Patashnick, 1990; Schoenfeld, 1985). Thus, although the theoretical argument is persuasive, the final analysis awaits the collection of additional data.

Past and Current Issues Related to Understanding

The framework for thinking about understanding proposed in this chapter is useful for rethinking a number of persistent issues in mathematics learning research. The ideas of representing mathematical quantities and relationships externally and internally and connecting representations to form internal networks are useful because they provide a common perspective from which to consider issues traditionally treated separately. This provides an opportunity to recognize relationships not usually considered and to see things in a new way. Some of the issues have been alluded to already and some have been discussed briefly; we now turn our full attention to them.

Conceptual and Procedural Knowledge. One of the longstanding debates in mathematics education concerns the relative importance of conceptual knowledge versus procedural knowledge or of understanding versus skill (Brownell, 1935; Bruner, 1960; Gagne, 1977; McLellan & Dewey, 1895; Thorndike, 1922). The debate has often been carried out in the context of proposing instructional programs emphasizing one kind of knowledge over the other. The prevailing view has seesawed back and forth, weighted by the persuasiveness of the spokesperson for each particular position. Although the arguments may have been convincing at times within the mathematics education community, we have not made great progress in our understanding of the issue.

Perhaps a first step in understanding how procedural and conceptual knowledge contribute to mathematical expertise is to agree that the question of which kind of knowledge is most important is the wrong question to ask. Both kinds of knowledge are crucial. A better question to ask is how conceptual and procedural knowledge are related (Glaser, 1979).

A second step toward a better understanding of the issue is to define *conceptual* and *procedural knowledge* precisely enough to permit descriptions of specific relationships between them. Useful definitions of these constructs often suggest relationships. We define conceptual knowledge in a way that identifies it with knowledge that is understood: Conceptual knowledge is equated with connected networks. In other words, conceptual knowledge is knowledge that is rich in relationships (Hiebert & Lefevre, 1986). A unit of conceptual knowledge is not stored as an isolated piece of information; it is conceptual knowledge only if it is part of a network. On the other hand, we define procedural knowledge as a sequence of actions. The minimal connections needed to create internal representations of a procedure are connections between succeeding actions in the procedure. Examples of procedures are standard computation algorithms in arithmetic. Often, in school mathematics, procedures prescribe the manipulation of written symbols in a step-by-step sequence.

Before considering potential relationships between conceptual and procedural knowledge, it is important to emphasize that both kinds of knowledge are required for mathematical expertise. Procedures allow mathematical tasks to be completed efficiently. Procedures that have been practiced and memorized can be executed quickly and with relatively little mental effort. Well-rehearsed procedures capture a kind of mathematical power because they exploit the consistency and patterns in mathematical systems and guide the seemingly effortless solution of routine problems.

Conceptual knowledge also is required for mathematical expertise. Some of its benefits already have been identified in the advantages proposed for understanding. But, in addition to the earlier claims made for understanding, conceptual knowledge contributes to mathematical expertise through its relationships with procedural knowledge. Given our framework in this chapter, we suggest that the relationships between conceptual and procedural knowledge depend upon the connections learners construct between their representations. From an expert's point of view, procedures in mathematics always depend upon principles represented conceptually. In other words, all mathematical procedures are potentially associated with connected networks of information. For example, the procedure for subtracting multidigit numerals is a step-by-step sequence of actions; but associated with the actions are mathematical principles involving equivalence and regrouping, sometimes represented as richly connected networks. Information that might be included in these networks is described well by Davis and McKnight's (1980) conjectures about the kinds of knowledge that might have informed their students' procedures: knowledge of making change, of working with special concrete materials, and of finding the approximate size of numbers. If the learner connects the procedure with some of the conceptual knowledge on which it is based, then the procedure becomes part of a larger network, closely related to conceptual knowledge.

The advantage of relating procedures to conceptual knowledge is, in part, the flexibility that accrues (Hatano, 1988;

Hiebert & Lefevre, 1986). Procedures connected to networks gain access to all information in the network. When encountering problems that differ from those for which a procedure was initially learned, the related conceptual knowledge may detect useful similarities and differences between problems, and subsequently, inform the procedure regarding appropriate adjustments. In this way, conceptual knowledge extends the procedure's range of applicability.

Relationships between conceptual and procedural knowledge may range from no relationship to a relationship so close that they become difficult to distinguish. While the degree of relationship is captured by considering the number and strength of the established connections, the nature of the relationship is not well described by considering only the theoretical notion of connections between mental representations. Fleshing out the ways in which conceptual and procedural knowledge interact and describing the forms these relationships can take moves us beyond the boundaries of this chapter. These issues are pursued further in Byers and Erlwanger (1984), Davis (1984), Gelman and Gallistel (1978), Greeno (1983), Hatano, (1988), Hiebert (1986), Ohlsson and Rees (1988), Resnick and Ford (1981), and Skemp (1978).

Even if we change our primary question from "Which kind of knowledge is more important?" to "How do conceptual and procedural knowledge interact?" teachers are still left with the question of whether to be concerned with conceptual relationships or procedural proficiency first. Educators are left with questions of how to design learning environments, and different designs are likely to be suggested by different answers to these questions. For purposes of discussion we focus the question a bit further on an especially important feature of school mathematics programs: Should we first help students develop efficient symbol manipulation procedures, or should we help them build relationships between symbol procedures and conceptual networks?

The bulk of the theoretical arguments supports building meaning for written mathematical symbols and rules before practicing the rules for efficient execution (Brownell, 1954; Brueckner, 1939; Fehr, 1955; Goldin, 1987; Hiebert, 1988; Kaput, 1987), although other possibilities have been proposed (Noddings, 1985; Nolan, 1973). Although the empirical data have not kept pace with theoretical speculations, the data that address the question seem to be consistent with the theories that argue for meaning before efficiency (Brownell & Chazal, 1935; Mack, 1990; Resnick & Omanson, 1987; Wearne & Hiebert, 1988). The evidence suggests that learners who possess well-practiced, automatized rules for manipulating symbols are reluctant to connect the rules with other representations that might give them meaning. For example, Wearne and Hiebert (1988) reported that fifth and sixth graders who had already practiced rules for adding and subtracting decimal fractions by lining up the decimal points were less likely than fourth graders with no such experience to acquire conceptual knowledge about decimals from instruction using base-10 blocks to represent decimal quantities and actions on quantities.

The tendency to persist in using procedures once they are well-rehearsed, without reflecting on them or examining them further, has been noted for some time in a variety of domains. Kuhn and Phelps (1982) observed students solving relatively complex chemistry problems and noted that the most difficult task for some students in becoming better problem solvers is abandoning old strategies. They suggest that students are reluctant to give up familiar strategies, especially if they do not understand them and thus do not recognize their inadequacies. Some years ago, Gestalt psychologists labeled what may be a related phenomenon *functional fixedness* (Duncker, 1945; Luchins, 1942; Werthheimer, 1959). When a particular approach or procedure is practiced it can become fixed, making it difficult to think of the problem situation in another way.

Within the cognitive science tradition, these related phenomena regarding the tendency of well-practiced procedures to become insulated can be explained by the relative inaccessibility of tightly connected procedural knowledge. Anderson (1983) suggested that knowledge is stored initially as declarative knowledge. As problems are solved, some pieces of knowledge are connected as steps in a procedure. As the procedure is practiced repeatedly, the individual pieces of knowledge lose their identity and become parts of a single procedure, making it difficult to reflect on those individual steps. As Hatano (1988) noted, connecting a procedure with conceptual knowledge requires the separation of a procedure into individual steps in order to reflect on its makeup. When analyzing the kind of knowledge acquired by abacus experts, Hatano (1988) observed that "This process of acceleration of calculation speed results in a sacrifice of understanding and of the construction of conceptual knowledge. It is hard to unpack a merged specific rule to find the meaning of any given step." (p. 64).

In summary, research efforts now being directed toward uncovering relationships between conceptual and procedural knowledge appear to be more useful than earlier attempts to establish the importance of one over the other. At this point, both theory and available data favor stressing understanding before skill proficiency. Within the framework for this chapter, teaching environments should be designed to help students build internal representations of procedures that become part of larger conceptual networks before encouraging the repeated practice of procedures.

Street Mathematics and School Mathematics. A good deal of interest has been generated recently by evidence that unschooled persons solve everyday mathematics problems successfully using invented strategies and that many schooled persons solve everyday mathematics problems using strategies different from those learned in school (Carraher, Carraher, & Schliemann, 1985, 1987; Ginsburg, 1978; Lave, 1988; Lave, Murtaugh, & de la Rocha, 1984; Saxe, 1988, 1991; Scribner, 1984). Often, these findings are interpreted to mean that considerable mathematical competence is developed outside of school and that school learning is not useful for solving real problems. It appears that mathematical behavior exhibited on everyday problems contains many features associated with good mathematical problem solving but it is worth considering further the relationship between street mathematics and school mathematics within the framework of representing and connecting knowledge.

A frequent criticism of school mathematics is that procedures for solving problems are learned in an overly mechan-

ical way. That is, school-learned procedures often cannot be used flexibly to solve problems other than those on which they were practiced and, thus, do not transfer well. We have argued that the problem situations with which students interact may set boundaries on their internal representations of the tasks. If the school program presents tasks only as written symbols and does not support connections with other external representation forms and problem situations, the knowledge acquired has limited transfer potential because the internal representation is severely constrained.

The nature of the learning process outside of school appears to be a bit more difficult to describe. But some interesting parallels may exist between learning mathematics in school and out of school. A hallmark of street mathematics is its contextualized nature: Everyday mathematical tasks are embedded in a natural, familiar setting. However, the mathematical skills that are developed through completing the tasks may remain deeply embedded in the context. In other words, like skills learned in school, skills learned outside of school may be constrained by contextual factors. The data from research on this hypothesis are mixed. For example, Saxe (1991) reported that young Brazilian street vendors recognized and worked with numerals only if they were written on familiar bills or coins. The procedures, themselves, seemed to be bound by the familiar context. Carraher, Schliemann, and Carraher (1988) found that some construction foremen and fishermen used proportion strategies flexibly to solve novel problems whereas others did not. The reasons for the differences are not clear, but it may be important that the successful workmen apparently had years of experience in using specific proportion strategies. A plausible conclusion at this point is that initial learning of both school mathematics and street mathematics is likely to be context bound and show limited transfer (Lave, 1988; Stigler & Baranes, 1988).

Given the previous argument of the relationship between problem situations and internal representations, the explanations for the lack of transfer of both street mathematics and school mathematics are rather straightforward. It is likely that students acquire mathematics in both settings through limited and specialized experiences. They represent mathematical ideas and procedures in ways that form small, specialized internal networks, not seen by their owners as related to other information and not connected to other networks. One would predict that if children possessed internal networks constructed both in and out of school, and if they recognized the connections between them, their understanding and performance in both settings would improve.

The evidence does not yet shed a great deal of light on the relationship between in-school and out-of-school learning; some learners seem to make connections between in-school and out-of-school mathematics whereas others do not, and the nature of the connections made is still unclear. There is a wealth of evidence, cited earlier, that many people fail to make connections. But, in some cases, school-learned mathematics seems to support a more flexible use of informal strategies. Acioly and Schliemann (1986) reported that lottery game bookies who had been to school were better able to solve new problems involving placing bets than bookies who had received no school training. Saxe (1991) found that school-learned computational procedures influenced the sophistication of pricing strategies used by Brazilian candy sellers. Conversely, some evidence suggests informal strategies acquired outside of school can facilitate the acquisition of formal school mathematics. Saxe (1991) reported that candy sellers outperformed nonsellers on some school tasks by applying the computational strategies they used for selling candy. Bebout (1990) and Carpenter et al. (1988) found that first and second graders in the United States could make sense of symbolic representations for addition and subtraction situations by using counting strategies for solving simple addition and subtraction stories. These studies indicate that mathematics learning is not doomed to compartmentalization; connections between internal representations are possible. But the data do not yet provide much guidance on how to structure learning experiences so out-of-school and in-school learnings become connected.

In concluding this section, it may be useful to consider again the notion of representation. Mathematics used in out-of-school, informal settings is mathematics represented externally with everyday experiences or scripts, physical objects, or pictures. Ordinarily, street mathematics is not represented extensively with written symbols. In contrast, school mathematics depends heavily on representation with written symbols. In each case, competence develops with an individual form of representation, but the problem situations do not encourage building connections between different representations. A major theme of the first part of this chapter is that mathematical knowledge grows and is reorganized as connections are made between representations. Perhaps the lesson learned thus far from the research on street and school mathematics is that both kinds of mathematics are frequently limited, for both often exploit only certain representation systems and create restricted networks of knowledge. Mathematical knowledge would increase and become more coherent for learners if they could establish connections between these networks. Internal networks that organize mathematical information from out-of-school experiences would become integrated with networks of in-school mathematics. Problem solving in one setting could then be informed by strategies acquired in other settings (Ginsburg, 1982).

Prior Knowledge and Current Learning. One observation that assumes near axiomatic status in cognitive science is that students' prior knowledge influences what they learn and how they perform. In the epigraph to his 1968 book, D. Ausubel says, "If I had to reduce all of educational psychology to just one principle, I would say this: The most important single factor influencing learning is what the learner already knows" (p. vi). Students interpret and respond to new situations in terms of what they know. In our framework of knowledge as networks of internal representations, the simple amount of knowledge a student has constructed can be thought of as the number of internal networks, while the degree of understanding associated with the knowledge can be thought of as the structure or coherence of these networks.

Consistent with our previous arguments, we propose that not only the amount of prior knowledge influences current learning but, also, the way that knowledge is structured. In

other words, prior knowledge that is well understood influences learning differently than prior knowledge that is less understood. Theoretical support for this position rests squarely on the arguments advanced earlier. For example, because memory for information that is well structured internally is better than memory for loosely connected information, students are more likely to retrieve well understood prior knowledge when they encounter new, but related, problems. Also, because understanding is generative, prior knowledge that has been understood is more likely to generate new understandings in new situations; relationships between prior knowledge and new material are more likely to be built.

Considering potential connections between students' existing internal networks and new information is one way of viewing the issue of readiness. If students possess little knowledge in a particular domain, or if their knowledge is poorly organized, they may have a difficult time learning advanced ideas in the domain. Students are ready to learn only if they have built internal representations to which the new information can be connected. Of course, connections to related ideas can be made by students through metaphors, analogies using alternate external representations, and so on, but making connections remains essential.

A great deal of work in developmental psychology over the years (Case, 1985; Piaget, 1970; Werner, 1957) indicates that our discussion of readiness is not complete. The number and coherence of existing internal networks may not be the only important factor in determining readiness; developmental constraints also seem to impose limitations on learning new material.

Case's (1985) theory is particularly relevant because it rests on the assumption that the ability to integrate separate pieces of information and build internal connections is a crucial developmental constraint. Younger children are not able to connect information in the same way as older children. Development enables the formation of larger, more integrated networks of knowledge. We certainly acknowledge the existence of developmental constraints on the ability of students to construct connections, even though they have not been emphasized here.

Summary

The growth of mathematical knowledge can be viewed as a process of constructing internal representations of information and, in turn, connecting the representations to form organized networks. An issue of special importance for mathematics educators is the issue of acquiring knowledge with understanding. Within the context of representing and connecting knowledge, understanding can be viewed as a process of making connections, or establishing relationships, either between knowledge already internally represented or between existing networks and new information. If one accepts this description, it is no wonder understanding has special significance in mathematics learning. Even with all of the arguments advanced in favor of understanding in the past, we are only beginning to fully appreciate its role in the growth of mathematical knowledge.

In addition to providing a way to understand understanding, the notions of internal representations and connections provides a framework for thinking about a variety of issues of interest to the mathematics education community. Issues of conceptual and procedural knowledge, street mathematics and school mathematics (or informal and formal knowledge), and the influence of prior knowledge on learning all can be discussed within the same context. More than that, the discussion suggests that these issues are not separate phenomena; rather, they are different ways of labeling cognitive activities that arise from the same fundamental mechanisms.

In this part of the chapter we have considered the ways in which this framework helps us understand some critical issues of learning. In the next part of the chapter we retain the framework and use it to consider our understanding of teaching.

TEACHING MATHEMATICS FOR UNDERSTANDING

Starting from the premise that the development of understanding should be a basic goal of mathematics instruction, an obvious implication of our characterization of understanding is that instruction should be designed so that students build connections. How should instruction be designed to accomplish this goal? It seems evident that procedures and concepts should not be taught as isolated bits of information, but it is less clear what connections are most important or what kind of instruction is most effective for promoting these connections. It is not feasible to address all possible connections, and different programs of research and instruction have focused on different kinds of connections. We consider first the kinds of connections that may be the focus of instruction and then the question of how instruction might help students form connections.

Alternative Frameworks for Structuring Connections

Different perspectives suggest different choices regarding what connections should be paramount when planning instruction in mathematics. One way of distinguishing between the perspectives is to ask whether to start with the knowledge students already have and focus on how that knowledge can be extended, or whether to start with an analysis of how knowledge should be connected once it is acquired and work backwards to decide how those connections might be formed. Although any thorough analysis ultimately must be concerned both with how knowledge is organized once it is acquired and how new knowledge is connected to students' prior knowledge, different programs of instruction may be more consistent with one particular approach.

Furthermore, some basic research on understanding inherently lends itself to one approach or the other. For example, there is a growing body of research documenting that children and adults acquire substantial knowledge of mathematics outside of school they can apply to solving a variety of problems in everyday situations (Carpenter & Moser, 1983; Carraher et al., 1987; Ginsburg, 1982; Lave, 1988; Saxe, 1991; Scribner, 1984).

This research provides a potential starting place for instruction to build on students' prior knowledge, but it does not suggest how that knowledge ultimately is integrated into a fully developed network including the concepts, procedures, and symbols of in-school mathematics. Applications of this research to instruction methods tend to focus on connections with prior knowledge. This can be labeled a *bottom-up* approach.

On the other hand, expert-novice studies generally provide a model of how knowledge might be connected once it is acquired (Chi, Feltovich, & Glaser, 1981). Many of these studies, though, tend to focus on limitations of the knowledge held by novices; thus, attaining expertise is more of a question of abandoning novice conceptions rather than of building upon them. The most straightforward applications of this research focus on how knowledge is potentially structured as a result of particular types of instruction. The goal is to attempt to teach students to make the same kinds of connections observed in experts. This can be labeled a *top-down* approach.

The bottom-up/top-down distinction is not only reflected in the models of students' thinking that may serve as a basis for research on instruction; it is also fundamental in the selection of the mathematics content that is emphasized. Although the issue has not been cast in black and white, there has been a continual discussion about whether instruction should focus on connections derived from problem contexts that give particular mathematical concepts and procedures meaning, or whether instruction should be based on analysis of mathematical structure. The distinction has a long history and is central to many of the debates over the goals of the mathematics curriculum. It is prominent in the work of Dewey and Bruner on the one hand, and Ausubel and Gagne on the other. The current mathematics curriculum reform movement reflected in the recommendations of the *Curriculum and Evaluation Standards for School Mathematics* (Commission on Standards for School Mathematics, 1989) and *Everybody Counts* (National Research Council, 1989), emphasizes the use of problem contexts to develop meaning, whereas the curriculum reforms of the 1960s were based on an analysis of mathematical structure and rigor. Currently, most curriculum reform recommendations assign a significant role to problem contexts in developing meaning for mathematics; however, some approaches reflect a more bottom-up approach to the analysis of problem contexts and others reflect a more top-down analysis.

The current discussion of situated cognition (Brown et al., 1989) reflects the former perspective. Brown and associates argue that instruction should not immediately attempt to separate abstract mathematical concepts and procedures from the contexts that initially give them meaning. They propose that concepts and procedures are linked inexorably to the contexts in which they are situated. The implication is that learning situations should be embedded in authentic problem situations that have meaning for the students. By learning concepts and procedures in problem solving contexts, it is presumed that the knowledge is connected so that it is accessible for problem solving.

The notion of conceptual field defined by Vergnaud (1983, 1988) also is concerned with the problem situations that underlie mathematical abstractions. However, conceptual fields are defined through a top-down analysis of the properties of problems and situations represented by an interconnected set of mathematical abstractions. This emphasis on the structure of the conceptual field is similar in many ways to an analysis based on mathematical structure. Although conceptual fields are defined in terms of problems, the emphasis is on fundamental semantic properties that define similarities and differences among problems rather than on the particular context of the problems.

The principles of situated cognition and conceptual fields do not necessarily represent conflicting perspectives of how knowledge is organized or how it should be structured. An analysis of how problem situations fit in a conceptual field may help to identify appropriate contexts in which to situate particular concepts and procedures. Furthermore, although conceptual fields may provide a basis for identifying how knowledge should be connected as a result of instruction, they also provide a basis for analyzing students' prior knowledge. However, fundamentally different programs of instruction may result, even though the constructs themselves are not in conflict, depending on whether the goals of instruction are driven by the organizing structures of the conceptual field or by the problem contexts that initially have meaning for students.

Example 1: Acquiring Knowledge of Written Symbols.
Carpenter, Fennema, and Peterson (Carpenter, Fennema, Peterson, Chiang, & Loef, 1989; Fennema, Carpenter, & Peterson, 1989; Peterson, Fennema, & Carpenter, 1991) have been studying how instruction on basic addition and subtraction concepts and skills may build upon the informal number concepts and problem solving skills that children have acquired before they enter school. Research on children's early number concepts has provided a highly structured analysis of the development of addition and subtraction concepts and skills as reflected in children's solutions of different types of word problems. The research provides a taxonomy of problems that can be solved by many children when they enter first-grade, a detailed analysis of the processes children use to solve different problems, and a map of how these processes evolve over time. For reviews of this research see Carpenter (1985), Carpenter and Moser (1983), or Riley et al. (1983).

In the study reported by Carpenter et al. (1989), the teachers built upon students' informal knowledge by starting with problems students could solve. Students spent a good deal of time solving and talking about these problems, using a variety of informal counting and modeling procedures. The explicit analysis of problems and children's solutions allowed teachers to select and adapt problems so that different students were able to deal with reasonably challenging problems.

By talking about how they solved problems, children learned to reflect on their solution methods and to articulate their solutions. The modeling and counting strategies that children used to solve different problems became more accessible. Thus they could readily be applied to a variety of problems and children could relate the strategies they used to different problems. Once children's informal strategies were readily accessible and were objects of discussion, symbols were introduced as ways of representing knowledge the children already had. In

this way, symbols were linked to the children's intuitive knowledge about addition and subtraction.

One implication of designing instruction to connect new concepts and procedures to students' prior knowledge is that the content and sequence of what is taught is selected so that the new knowledge can be related directly to what is already known. Introducing symbols for addition and subtraction means introducing symbols that children can use to represent problems consistent with children's informal strategies involving counting or modeling with physical objects or fingers. Consider, for example, the following problem:

Carlos has 4 dollars. How much more money does he have to earn to have enough money to buy a puppy that costs 11 dollars?

Most first-grade children solve the above problem by modeling the action described in the problem. If they use counters, they make a set of 4 counters, increment the set until there are a total of 11, and count the 7 counters added to find the answer. A somewhat more sophisticated strategy involves counting up from 4 to 11, keeping track of the number of counts in the counting sequence. For example, the child counts 4, pauses, then counts 5, 6, 7, 8, 9, 10, 11. With each count starting with 5, a finger is extended. The answer is found by counting the number of fingers extended. These two strategies reflect the joining action in the problem and correspond most directly to the noncanonical open-number sentence $4 + \_\_ = 11$. Studies by Bebout (1990), Carey (1991), and Carpenter, Moser, & Bebout (1988) indicate that children readily learn to write open number sentences that correspond directly to the counting and modeling strategies they use to solve problems. This work suggests that open-number sentences can be used to help children develop internal representations of early arithmetic symbols, connecting, in turn, with their intuitive knowledge about number operations.

An alternative approach is to start with a top-down analysis based on mathematical structure. A powerful aspect of mathematics resides in the fact that a number of seemingly diverse situations can be represented externally by the same mathematical symbols. Ultimately students need to learn that the different addition and subtraction problems can be represented by canonical number sentences of the form $a + b = \_\_$ or $a - b = \_\_$. Some researchers argue children initially should learn to represent problems with one of these two number sentences; for example the problem above would be represented by the number sentence $11 - 4 = \_\_$ (Rathmell & Huinker, 1989; Wilson, 1967). Children learn to relate the problem to the number sentence by analyzing the problem in terms of parts and wholes. The 4 dollars is part of the 11 dollars; when you have a part and a whole, you subtract to find the other part. The part-whole analysis provides a unifying framework for connecting numbers of different additive and subtractive situations; it is based on an analysis of skilled performance (Riley et al., 1983), however, and does not correspond to the way most young children initially think about such problems.

The goal of establishing internal connected networks does not resolve the question of whether to first introduce a part-whole analysis early in the curriculum or, instead, to start with a variety of open-number sentences. Both approaches have a goal of connecting knowledge, but they operate from different perspectives. The part-whole analysis provides the unifying schema of part-whole, into which almost all addition and subtraction problems are mapped. This approach provides a schema for relating seemingly diverse problems to one another. This analysis is based on a top-down perspective; it is an efficient way that skilled problem solvers can analyze addition and subtraction situations. The instructional approach, essentially, is to teach directly the schemata that skilled performers use to organize their knowledge. From the start, then, instruction focuses on long range goals.

The open-number sentence approach starts from the other end; it attempts to link children's emerging internal representations of symbols with the concepts and procedures for which children already have internal networks. The concern is that if links are not initially drawn by children between their informal knowledge and the written symbols, children may develop separate systems of arithmetic, one that operates in school and one that operates in the real world, and they will not readily see the connections between them (Carraher et al., 1987; Cobb, 1988; Ginsburg, 1982; Lave, 1988; Lawler, 1981). Both approaches intend to help students acquire flexible schemata for relating problems to each other, but they start from very different perspectives.

Example 2: Sequencing Topics. Helping students connect their intuitive knowledge of mathematics with written symbols is one way of helping them connect new knowledge with prior knowledge. Another factor to consider in helping students build on existing knowledge is the sequence in which topics are taught. Sequences for instruction frequently are specified based on a top-down analysis of the knowledge and skills required to learn a particular concept. By starting with students' existing knowledge rather than the objectives of instruction, different sequences of instruction may emerge.

Our second example illustrates how students' prior knowledge may influence the sequence of instruction. Mack (1990) taught sixth-grade students basic fraction concepts by relating fraction symbols and operations to the students' informal knowledge of fractions in problem contexts meaningful to the students. The contexts were structured so that students initially dealt with problem situations involving fractions by partitioning units into parts and treating the parts as whole numbers. Thus, problem contexts provided a mechanism for students to connect fraction concepts and procedures to their knowledge of whole number concepts and procedures. One of the types of problems children could solve before acquiring detailed knowledge of fraction algorithms was subtraction with regrouping, generally regarded as one of the more difficult fraction topics to master. (Kerslake, 1986; Kouba, Brown, Carpenter, Lindquist, Silver, & Swafford, 1988). Mack found, however, that students could relate fraction problems requiring regrouping to knowledge acquired in real-world contexts, and could use this knowledge to solve problems even though they lacked formal knowledge of regrouping concepts. Sixth-grade students could solve such problems as $4\frac{1}{8} - 1\frac{5}{8}$ by relating the problem to situations like sawing $1\frac{5}{8}$ feet off a board that was $4\frac{1}{8}$ feet long. Students solved a variety of problems like this before they were introduced formally to concepts like equivalence. When equivalence was taught, it could be related to the informal procedures that children had invented to solve different problems. Equivalence of fractions provided a context and became a unifying construct that connected a variety of different problem solving procedures.

In contrast, a top-down analysis may suggest other connections that provide an integrated perspective of rational number. A complete concept of rational number involves the integration of a number of subconstructs, including measure, ratio, quotient, and multiplicative operator (Kieren, 1988), all embedded in the broader construct of a multiplicative conceptual field (Behr et al., this volume). Kieren (1988) proposed a hierarchial structure in which the basic subconstructs are integrated in stages, leading to a unified concept of rational number. Behr, Harel, Post & Lesh (in press) identify fundamental principles like invariance of arithmetic operations and composition, decomposition, and conversion of units, connecting the multiplicative field to other conceptual fields. These analyses specify explicit connections that could serve as the focus of instruction. For example, Behr et al. (in press) observe that in most current curricula, equivalence of fractions is treated as an isolated topic. They propose that equivalence should be dealt with as a specific case of a more general notion—invariance of the result of arithmetic operations under different transformations of the operands. They recommend further that children in early grades be given tasks in which they are asked to consider the effects of changes in operands on the results of arithmetic operations.

One of the dangers in attempting to build upon students' knowledge is that students' informal conceptions may be limiting. The problem situations that initially are most meaningful for students may not provide a sufficiently rich context in which to develop a full understanding of a given construct. For example, children as young as first-graders can begin to develop an initial understanding of multiplication as a group of sets containing the same number of objects; by the third-grade, these same students relate multiplication to repeated addition (Kouba, 1989). This narrow conception of multiplication does not extend well to fractions and decimals, and results in such misconceptions as "multiplication always makes bigger" (Fischbein et al., 1985; Bell et al., 1989; Greer, this volume). Although a careful analysis of students' thinking was the driving force for Mack's (1990) instruction, she was able to avoid these pitfalls by keeping in sight the broader set of connections that were needed to be established in order for students to have a complete understanding of rational number. Although she initially gave students partitioning problems that they could relate to whole number partitioning and grouping situations, she gradually extended the problem situations to encompass most of the problem situations represented in the multiplicative conceptual field.

A danger in starting with a top-down analysis is that instruction will not make contact with what students already know. Students often develop separate systems of arithmetic—an informal system that they use to solve problems that are meaningful to them and a school arithmetic consisting of procedures that they apply to symbols or artificial story problems they are given in school—that operate independently of one another (Carraher et al., 1987; Cobb, 1988; Ginsburg, 1982). The curricular reforms of the 1960s emphasized structure to develop understanding, but this top-down analysis did not make contact with what students already knew: Thus, students had difficulty relating to the formal mathematical structure.

Using the more top-down analysis based on conceptual fields for planning curriculum, however, does not imply that initial instruction ignores students' entering knowledge. The analysis of a conceptual field can provide a framework for analyzing the strengths and limits of students' initial knowledge in the given domain. Although Behr et al. (in press) recommend early emphasis on fundamental principles that provide structure to the multiplicative conceptual field, they do not recommend these structures be imposed on students in situations they do not understand. Rather, they recommend situations be selected that are meaningful to students to embody the fundamental principles of the conceptual field. A thorough understanding of the conceptual field is necessary to select situations encompassing the critical relations that characterize it.

There is not a great deal of difference between the two approaches in the ultimate goals of instruction or in the concern that instruction makes contact with students' existing conceptions. However, the bottom-up approach, illustrated by Mack (1990), is probably more flexible in the sequence in which problem situations are encountered and in how connections may be constructed in the early phases of instruction. Students are allowed to select contexts and interpretations of problems that make sense to them, and they are given some latitude in how they make connections. The top-down approach, illustrated by Behr et al. (in press), is more concerned that students' understanding reflect the critical parameters of the conceptual field at each stage of instruction.

Example 3: Acquiring Computation Procedures. A third example of the different perspectives concerns the teaching of multidigit algorithms. Given that students acquire a great deal of informal mathematics for solving real world problems, it is not surprising that many examples of instruction attempting to build on students' prior knowledge appeal to implicit knowledge situated in real world contexts. We believe the critical issue regarding prior knowledge, however, is not whether it is based on implicit knowledge of real world problems but whether it has meaning for the students. Students are capable of extending more abstract, symbolic knowledge, provided that the symbolic knowledge, itself, is rich in connections. One example of the kinds of extensions students make is illustrated by procedures students invent to solve different computation problems before they are taught formal algorithms (Carpenter & Fennema, 1989; Cobb & Merkel, 1989; Kamii & Joseph, 1988; Madell, 1985; Weiland, 1985). The following is an example of how one first-grade student added 246 + 178:

Well, 2 plus 1 is 3, so I know it's two hundred and one hundred, so now it's somewhere in the three hundreds. And then you have to add the tens on. And the tens are 4 and 7 . . . well, um, if you started at 70,

80, 90, 100. Right? And that's four hundreds. So now you're already in the three hundreds because of that [100 + 200], but now you're in the four hundreds because of that [40 + 70]. But you've still got one more ten. So if you're doing it 300 plus 40 plus 70 you'd have four hundred and ten. But you're not doing that. So what you need to do then is add 6 more onto 10, which is 16. And then 8 more: 17, 18, 19, 20, 21, 22, 23, 24. So that's 124. I mean 424 (Carpenter & Fennema, 1989, p. 40).

By starting out inventing procedures for operating on multidigit numerals, children must draw on their knowledge of place value, because they have no algorithms on which to fall back. This encourages connections between place-value concepts and operations with multidigit numbers. Once students are reasonably proficient in constructing their own procedures for solving multidigit problems, standard algorithms can be introduced as more efficient procedures that keep track of steps in a more systematic way. Thus, the standard algorithms are connected to place value concepts through the invented procedures.

An alternative method for giving meaning to algorithms is to help students connect the symbolic algorithm to other external representations that have meaning for students, usually physical materials that embody the principles of place value. One of the most commonly used materials for representing place value is some variant of base-10 blocks (Bell et al., 1976; Fuson, 1986; Resnick, 1982). As we noted earlier, the blocks provide physical representations directly corresponding to written numerals in our base-10 system. Using the blocks combines a bottom-up approach based on students' prior knowledge about joining and separating objects with a top-down analysis of the principles of place value and the steps involved in standard algorithmic procedures. Because the blocks are simply counters assembled in a special way, they can be joined and separated to solve problems just like any other counters. The blocks, though, are not materials that students naturally encounter outside of school or generally invent for themselves. They are specially designed to embody the fundamental principles of place value and can be manipulated in a way that corresponds directly to the steps in standard addition and subtraction algorithms. Meaning can be given to algorithmic procedures by connecting each step in the algorithm to a corresponding action on the blocks (See Figure 4.1). Because the connections are made explicitly for the students using specially designed materials, this approach contrasts with encouraging invented procedures and allowing students' attention to be distributed across a range of potential connections.

Extensions to Advanced Topics. The three examples we have given of instruction building on students' prior knowledge have involved arithmetic concepts and operations. Much of the research on informal mathematics involves arithmetic concepts and procedures, and most of the initial instructional research focusing on students' prior knowledge has reflected this emphasis. The question arises, however, as to whether this approach is limited to elementary mathematical concepts and procedures. It could be argued that the structure of advanced mathematical topics is derived through formal analysis, and stu-

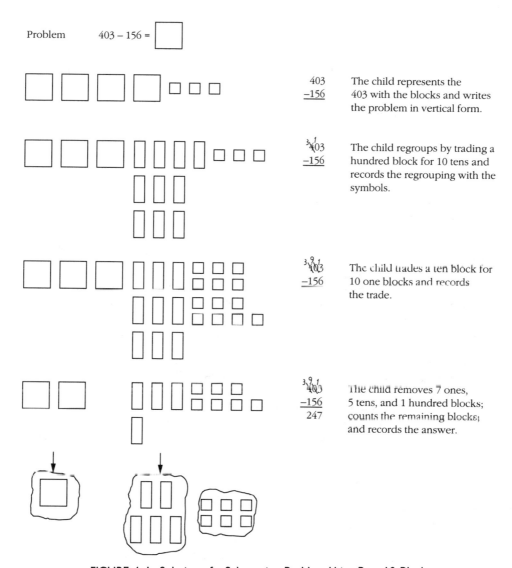

Problem   403 − 156 = ☐

403   The child represents the
−156   403 with the blocks and writes
          the problem in vertical form.

³₄̸03   The child regroups by trading a
−156   hundred block for 10 tens and
          records the regrouping with the
          symbols.

³₄̸⁹₀̸3   The child trades a ten block for
−156   10 one blocks and records
          the trade.

³₄̸⁹₀̸¹3   The child removes 7 ones,
−156   5 tens, and 1 hundred blocks;
247   counts the remaining blocks;
          and records the answer.

FIGURE 4–1. Solution of a Subtraction Problem Using Base 10 Blocks

dents have such a limited knowledge base that it is difficult to build more complex mathematical concepts on that knowledge.

Much research on the effect of prior knowledge on learning advanced mathematical topics like proportional reasoning and algebra has analyzed how prior knowledge from arithmetic leads to misconceptions when generalized to the more advanced topics (Hart, 1988; Matz, 1980). However, work by Davis (1964) as part of the *Madison Project* illustrates how a number of advanced principles of algebra can be developed by extending students' knowledge of arithmetic.

Most stage theories, like the work of van Hiele (1986; Clements & Battista, this volume; Fuys, Geddes, & Tischler, 1988) and Case (1985), are based on the premise that the learning of advanced topics builds on students' prior knowledge in important ways. More advanced stages are characterized by the ability to recognize and construct increasingly complex connec-

tions. Case's research provides another example of a bottom-up approach for constructing connections. He proposes that effective programs of instruction may be developed by recapitulating natural developmental stages in the instructional sequence. Instruction starts with problems to be solved at earlier stages and successively introduces problems to be solved in later stages. For example, he developed a program of instruction on ratio and proportion that begins with problems involving integral ratios readily solved by students using whole number operations. Problems that require more elaborate procedures are introduced in stages before the general operating procedures for ratio and proportion problems are taught. More research certainly is required with more advanced topics, but it is too early to discount the hypothesis that effective programs of instruction in advanced mathematics can be developed by building systematically upon students' existing knowledge.

Making Connections Explicit

Starting with the premise that instruction should foster connections does not provide clear guidelines about how instruction should help students make those connections any more than it resolves the question of which connections should be the focus of instruction. A central issue is how explicit should instruction be in specifying connections. Two questions need to be addressed: Whether connections should be made explicit for students and whether there are specific connections that should be taught. On the first question there is general consensus that connections between mathematical ideas should be discussed and that students should be encouraged to reflect on them. This does not imply that a teacher must have specific connections in mind; the connections can be generated by students. Several of the previously described examples of programs building on students' prior knowledge began by having students talk about their informal strategies in order to help students become aware of their implicit informal knowledge (Carpenter et al., 1989; Lampert, 1986; Mack, 1990). Cooperative groups and whole class discussions may provide the opportunity for students to describe and explain connections they have formed.

Our conception of understanding as connections may provide a framework for studying the effects of different instructional and grouping practices on the development of students' understanding. A great deal of research has been conducted recently on the effects of interaction within small groups on student learning (Good, Mulryan, & McCaslin, this volume; Slavin, 1989; Webb, 1989). Although it appears that achievement can be enhanced through task related interaction, it is not clear what factors account for learning. More detailed observation and assessment measures are needed to tease out the important cognitive features of student interactions. Since instruction provides opportunities for students to make connections explicit, studying what connections become explicit during student-teacher interactions and assessing what connections students make as a result of instructional activities may help us to understand the relation between specific programs of instruction and specific learning outcomes.

There is less agreement on the second of the two questions: Can we identify a priori specific connections that should be the object of instruction? Two assumptions underlie attempts to define specific connections as the focus of instruction: (a) connections that should be made explicit are the critical connections made by individuals proficient in the subject and (b) connections that are made explicit are internalized by students. Earlier we discussed several conceptions about how knowledge may be connected internally, one involving a hierarchical structure, the other analogous to a web. Insofar as connections are hierarchical, it may be possible to identify key constructs that provide structure to a number of related concepts; these key constructs could provide a coherent focus for organizing instruction. Students' attention could be drawn to relationships between the key constructs and special cases or instantiations of them. For example, the part-whole schema discussed earlier may integrate a number of alternative conceptions of addition and subtraction. The web analogy, on the other hand, portrays

a complex network of connections that can be so profuse and idiosyncratic that such connections are unlikely candidates for explicit instruction. Although we can never hope to make all possible fruitful connections explicit during instruction, expert-novice studies and theoretical analyses of the structure of different content domains have identified key constructs, or ways of chunking major concepts and procedures, that could provide an explicit structure for instruction.

It is more problematic to assume that the connections taught explicitly are internalized by students. The potential danger in teaching connections explicitly is that the information required to make the connections explicit will be internalized as one more piece of isolated knowledge rather than supporting the construction of useful connections. For example, instruction in place value often includes the use of expanded notation ($500 + 70 + 6$ to represent 576). One rationale for this approach is that expanded notation helps to provide a link between the numeral and physical representations like base-10 blocks (see Figure 4.1). Understanding the expanded notation, however, requires most of the same knowledge required to understand the standard numeral (Fuson, Chapter 12, this volume), and the expanded notation includes a number of additional complexities. Young children do not initially find expanded notation easy to comprehend. Unless a significant effort is devoted to establishing the connections between the different symbolic and physical representations, there is a risk that expanded notation will become one more isolated concept to learn, and it may actually interfere with the development of understanding.

On the other hand, it is possible to argue that students can benefit from explicit direction as they build connections. If students are not given sufficient guidance, they may focus on superficial connections that do not correspond to the connections required for skilled performance and understanding the central concepts. As noted earlier, students often construct connections in overly restricted domains and can construct connections that are inappropriate or unproductive.

The issue does not boil down to whether learning with understanding requires that specific connections be made explicit. Researchers must be sensitive to the assumptions underlying a particular approach and the potential benefits and misconceptions that can arise by adopting it. By carefully examining how instruction is designed to foster connections and carefully assessing the connections that students make as a result, we may begin to develop a principled basis for understanding the effects of different patterns of instruction.

Specific Cases. The issue of explicitly teaching specific connections is not new; it is reflected in the debates over discovery learning and meaningful expository learning of the 1960s (Ausubel, 1961; Bruner, 1961; Shulman, 1970; Shulman & Keislar, 1966). A central premise of Ausubel's work is that instruction should start by specifying how major ideas are interconnected and related to what students already know, whereas Bruner argued that by discovering critical relationships by solving problems, students, themselves, construct connections so that relevant knowledge is available to solve problems.

The focus has shifted somewhat so that researchers no longer are comparing instructional programs based on the level of guidance provided to students in learning mathematics. Rather, different programs of instruction and research are often based on different assumptions about whether it is appropriate or possible to specify connections for students. An example of specifying connections for students is implied by work on how physics experts' knowledge is connected. Larkin, McDermott, Simon, and Simon (1980) found that physics experts organized their knowledge of physics principles in chunks based on higher-order principles not corresponding to the organization of the principles in standard physics textbooks. The experts, for example, did not represent internally all principles related to the inclined plane in a single network and all principles related to pulleys in another network. Rather, they organized principles internally in ways that took into account relationships between different physical contexts. The research suggests that an alternative organization of the physics curriculum corresponding more directly to the way physics experts organize their knowledge may be appropriate. Principles that are chunked together by experts would be taught in some proximity and the nature of the connections would be made explicit for students.

It is important to note that the question of whether to design instruction to encourage specific connections is not identical to the distinction between building connections from the bottom-up or the top-down, even though studies of experts' knowledge were discussed earlier as examples of top-down approaches. Instruction building on students' prior knowledge may also be structured to foster explicit connections. One way of making explicit the connections between informal knowledge and the symbols and procedures of mathematics is for the teacher to allow and suggest symbolic representations that correspond directly to the informal strategies children used to solve problems. The use of noncanonical and canonical open-number sentences to represent addition and subtraction problems illustrates this approach (Bebout, 1990; Carey, 1991; Carpenter, Moser, & Bebout, 1988).

An alternative to connecting symbolic representations directly with students' informal knowledge is to generate physical representations that have salient connections with abstract symbols and procedures that are the objects of instruction. Base-10 blocks are an example of such representations, but instruction using base-10 blocks is not all of one kind in terms of how explicitly connections are presented. For example, base-10 blocks may be used explicitly to illustrate correspondences between manipulations of the physical materials and steps in computation algorithms. Consider the approach for teaching the multidigit subtraction algorithm for whole numbers illustrated in Figure 4.1. Blocks are used first to represent the larger number. To subtract, 10 for 1 exchanges are required so that there are enough units to remove six of them. The exchange corresponds to regrouping or borrowing in the subtraction algorithm, and when the exchange is made with blocks, the corresponding regrouping marks with the written symbols are noted. Thus, exchanges with the blocks are mapped directly to steps in the symbol algorithm. This process, which for obvious reasons is called mapping instruction, frequently is recommended for teaching algorithms (Bell et al., 1976; Merseth, 1978) and is illustrated in studies by Fuson (1986) and Resnick (1982; Resnick & Omanson, 1987). For this approach, it is critical that

the materials be constructed so the manipulations necessary to solve the problem correspond to steps in the algorithm being learned.

An alternative method of helping students connect symbolic algorithms with actions on base-10 blocks is proposed by Wearne and Hiebert (1988; Hiebert, 1988). They separate the connection of symbols with referents and the development and elaboration of symbol manipulation procedures. They advocate helping students establish meaning for symbols by building connections with referents, and then encouraging students to develop procedures for manipulating symbols by using the meanings of the symbols. More complicated algorithms are then worked out by elaborating previously constructed symbolic procedures. In other words, algorithmic procedures are learned by connecting them to the meanings of symbols and to the established simpler symbolic procedures, rather than relating them directly to physical referents. In applying this approach, Wearne and Hiebert (1988, 1989) taught students initial decimal place value concepts using base-10 blocks. Students were encouraged to add and subtract decimals, using the values of the digits as guides. The blocks were not used as manipulative devices to show step-by-step correspondences between actions on the blocks and symbolic addition and subtraction manipulations; rather, the students were prompted to relate their symbolic procedures to the place value concepts they had already acquired.

The approach used by Wearne and Hiebert requires that meaning for symbols and initial procedures be established at a level able to support the development of more complex procedures. In contrast, for some mapping instruction studies, basic symbol meanings are learned only tentatively at first and connections between symbols and referents are firmly established concurrently with the learning of complex symbol manipulation procedures as part of the mapping instruction (Fuson, 1988; 1990; Fuson & Briars, 1990).

Physical materials other than base-10 blocks can be used to help students create meaning for computation algorithms, and the way the materials are used provide additional contrasts with regard to explicitness of connections. For example, the solution illustrated in Figure 4.2 shows a common way that young children use Unifix cubes to solve the given problem. Often children break the tens apart rather than exchanging a ten for 10 ones. The approach is very different from mapping instruction because there is not a direct connection between manipulations of the cubes and the steps in the standard algorithm. The Unifix cubes allow a variety of different solutions, whereas the base-10 blocks cannot be broken apart and are more constrained. These constraints match those of mapping instruction, in which there is a specific way that the blocks are to be used so that the connections to the algorithm can be made explicit. The same manipulations are possible with the Unifix cubes, but teachers are less likely to focus on explicit connections between actions on the cubes and steps in the symbolic algorithms when multiple solution strategies are possible.

A number of basic questions need to be addressed regarding outcomes of different approaches. For example, are the connections that students form as a result of mapping instruc-

Problem   42 – 17

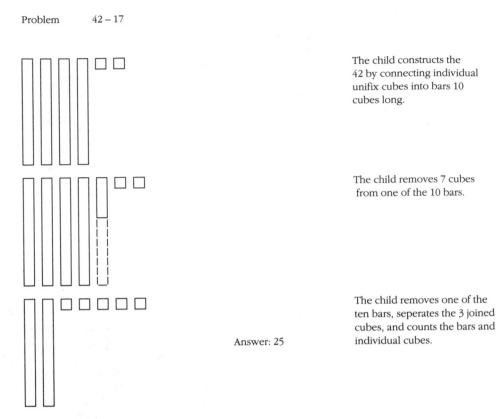

The child constructs the 42 by connecting individual unifix cubes into bars 10 cubes long.

The child removes 7 cubes from one of the 10 bars.

Answer: 25

The child removes one of the ten bars, seperates the 3 joined cubes, and counts the bars and individual cubes.

FIGURE 4–2. Solution of a Subtraction Problem Using Unifix Cubes

tion limited to the explicit connections specified during instruction? Can the connections be applied flexibly, or do they simply represent procedures that must be performed in a step-by-step manner? One might be concerned whether students can relate their operations on the Unifix cubes to specific symbolic procedures that correspond to their physical manipulations. In both cases, we should be concerned with how students generalize the connections they have made and extend them to unfamiliar situations. For example, if students have learned addition by mapping instruction or some alternative procedure, how readily can they adapt their knowledge to subtraction problems? Can they extend the symbolic procedures they have learned to larger numbers for which physical representations are not available? What beliefs have students formed about symbolic procedures and their ability to understand them?

Differences in how physical materials are used to develop understanding are not only grounded in the nature of those physical materials: There also are important differences in how materials may be used to make connections explicit. As noted earlier, the effects of using materials on students' understanding may have as much to do with the context in which they are used and the way in which students interact with materials as with the materials themselves. We have argued that an important variable to consider is the explicitness with which the connections are drawn for students between features of the materials and symbolic representations.

Misconceptions. Although our discussion to this point has assumed the desirability of connecting knowledge, students' errors and misconceptions also can be understood in terms of connections that have been formed. The analysis of students' errors in arithmetic and algebra suggests that students' errors frequently are caused by extending learned procedures to new problem situations incorrectly (Brown & VanLehn, 1982; Matz, 1980). In fact Resnick, Nesher, Leonard, Magone, Omanson, & Peled (1989) argue that errors are a natural consequence of attempting to integrate new procedures with prior knowledge. Thus, the problem is not that incorrect procedures are isolated from other knowledge, but critical connections that would make clear the nature of the errors are absent.

Much of the research on students' errors attempts to explain the errors by identifying specific ways knowledge may or may not be connected. For example, research on students' errors with decimal fractions indicates that some errors are caused because students apply rules for whole numbers to the decimal fractions; other errors appear because students attempt to apply rules for common fraction symbols to decimal symbols (Resnick et al., 1989). The two different sets of errors result from different sets of connections, and it could be argued that different instructional approaches are required to correct the errors.

One of the potential implications of research on students' errors is that instruction might be designed to address directly the specific deficits that the error analysis helps us to diagnose. In other words, research on students' errors makes it possible to identify specific deficits in the way students' knowledge is connected so that instruction can be designed to address the specific connections students lack or to point out why certain connections are inappropriate. The particular connections of interest would be made explicit for students.

Researchers in science education argue that it is necessary to address students' specific misconceptions directly or they will continue to persist even though students have learned scientific principles in direct conflict with the misconceptions (Hewson & Hewson, 1988; Posner, Strike, Hewson, & Gertzog, 1982). Unless students are forced to confront explicitly the conflict between their misconceptions and the scientific principles they have learned, the connections may never be made; the misconceptions and the scientific principles may coexist as separate islands of knowledge.

Earlier we argued that many errors in mathematics may result from students' attempts to build connections within overly restricted domains. For example, the misapplication of whole number rules to decimal fractions might be interpreted as students' attempts to connect decimal computation with something they already know, and the primary domain within which they search is whole number symbol procedures. The potentially relevant domain of meaningful referents for decimal symbols may be unavailable. This explanation suggests a somewhat different instructional approach. Rather than focusing on specific flaws in students' procedures, instruction would aim to push back the boundaries of students' search spaces. The goal would be to help students link the entire set of decimal procedures, instead of a particular procedure, with other domains, and to encourage them to build connections to a variety of networks, rather than to whole number procedures alone. This approach may be less explicit about specific connections because of the assumptions that a variety of connections are important and that different sets of connections may be equally appropriate.

Assessment

Our characterization of understanding not only has implications for instruction; it also has implications for issues related to instruction, such as how the outcomes of instruction should be assessed. One of the most pressing problems in education is the development of procedures for assessing higher-order thinking. Although some progress is evident (Kulm, 1990), both instructional practice and research on instruction have been limited by the narrow measures of achievement available (Romberg & Carpenter, 1986). Significant progress in research on teaching for understanding depends on good measures of understanding, so that the specific outcomes of instruction can be assessed. Progress in achieving widespread implementation of curriculum programs stressing understanding depends on being able to document the outcomes of such programs.

One reason understanding in the mathematics classroom has been such an elusive goal may be that assessing understanding is not straightforward. Because most theories or models of understanding, including the framework used in this chapter,

are built on internal operations of the mind, they cannot be observed directly. As a consequence, measuring understanding becomes a highly inferential task. That is, since understanding cannot be measured directly, it must be inferred by the measurer. Further, understanding usually cannot be inferred from a single response on a single task; any individual task can be performed correctly without understanding. A variety of tasks, then, are needed to generate a profile of behavioral evidence. All of these complications contribute to the difficulty of assessing understanding.

In a limited way, the analysis of students' errors provides one perspective on understanding. Theoretical analyses of students' errors may provide a basis for identifying specific limitations in students' understanding (Brown & VanLehn, 1982; Matz, 1980; Resnick et al., 1989). Frequently students' errors can be analyzed in terms of connections that have or have not been formed, so an analysis of procedural errors can provide a perspective on specific limits of understanding. For example, many computational errors with decimal fractions can be explained as the linking of whole number rules to decimal fraction symbols (Hiebert & Wearne, 1986; Resnick et al., 1989). Brown and VanLehn (1982) provide a fine-grained analysis of specific rules that account for specific errors in children's subtraction, and Matz (1980) connects a wide range of algebra errors to an overgeneralization of the distributive property.

Assessing the specific limits of understanding by analyzing students' errors requires errors to be treated differently than they are on most standard tests. Rather than focusing on an overall score, patterns of individual errors become important; it is critical to consider the nature of the error, not just whether a student missed a particular item. Although items need to be constructed and aggregated carefully to assess the range of possible misconceptions, the individual items used for error analysis often are similar to items measuring routine procedures on most standard tests. As a consequence, there are several good examples of tests that can accurately diagnose the nature of students' errors and identify the specific limits of their understanding (Brown & VanLehn, 1982). Such tests have implications for both instruction and research. Resnick's (1982) study illustrates that a careful diagnosis of students' errors provides a fine-grained picture of students' knowledge before instruction and how it changes as a result of instruction.

The limitation of error analysis is that it only provides part of the picture. Errors may imply a lack of understanding, but an absence of errors on the type of items used on most diagnostic instruments does not imply understanding is present. The assumption that understanding implies well-connected knowledge suggests that we should assess understanding by attempting to determine how knowledge *is* connected. There are several connections that might be assessed, and these generally correspond to the types of connections that may be addressed as part of instruction. In fact, assessment of understanding may depend on instruction and the language developed to talk about connections.

One type of connection that could be assessed is the connection between symbols and symbolic procedures and corresponding referents. For example, students can be asked to validate symbolic procedures using appropriate physical representations, and then to explain how their symbolic procedures correspond to actions on physical referents. There are several difficulties inherent in assessing these connections. Students generally must have some familiarity with the physical referents to have a reasonable probability of success on a given assessment task. Many potential referents, such as counters, coins, and the like, are familiar to all students, but specially constructed representations may not be familiar. Furthermore, unless students have experience using and talking about materials, they may not know what is expected of them or have the language to talk about it. They may be able to construct appropriate connections if the task is made clear to them, but they may not do so readily in a testing situation. This is not a problem for a teacher assessing students' understanding in an instructional situation, but it may be a problem for more standardized tests.

A second difficulty in assessing connections between symbols and referents is that it is not always easy to tell whether meaningful connections exist or whether students are simply applying a procedure they have learned mechanically from instruction. Students can learn a sequence of steps to manipulate physical objects and perform corresponding symbolic manipulations. To ensure that these manipulations reflect some level of understanding, it is necessary to ensure that the students have some flexibility in establishing appropriate correspondences.

Another kind of connection that might assess understanding is the connection between symbolic procedures and informal problem-solving procedures. Cobb (1988), Erlwanger (1975), Lawler (1981) and others have documented how students fail to acknowledge the relationship of different solutions to identical problems presented in different contexts. Connections between written symbols and intuitive knowledge may be assessed by presenting problems in contexts likely to elicit different solutions and asking students to discuss the relationships between the solutions. For example, a second-grade child might be asked to calculate 38 + 6 written in vertical notation and to solve a story problem involving the same addition. The first situation is likely to elicit a solution based on school procedures, whereas the second problem is likely to be solved by counting up from 38. A discussion of how a child perceives the relation between these two solutions can provide insights about how students connect symbolic procedures and intuitive counting and modeling procedures (Cobb, 1988).

A third type of connection that might be assessed are connections within symbol systems. One way of getting at these types of connections is to assess how students extend procedures they have learned to contexts beyond the ones they have studied directly. For example, students who have established meanings for decimal notation through hundredths might be assessed on their ability to extend this knowledge to thousandths, to compare and order decimals, or to represent decimals as common fractions (Hiebert & Wearne, 1988).

We do not presume to offer solutions to the problems of assessment but, rather, to suggest that the framework we propose, in particular the idea of connections between representations, provides a way of thinking about assessing understanding and provides a general criterion for constructing useful tasks. As, we proposed early in this chapter, it is useful to imagine some correspondence between the form of the external represen-

tations with which students interact and the internal representations they create. In developing understanding, recognizing connections between external representations may stimulate the construction of connections between corresponding internal representations. Perhaps it is useful to think of assessment as reversing this process. Internal connections may be tapped by having students reflect them back into connections between external representations. It seems clear that tasks which measure understanding will need to measure, in some way, connections that students have constructed. Examples of this kind of assessment work are in progress. Such work includes Sowder's (Sowder, 1988; Sowder & Wheeler, 1989) investigations into number sense, Marshall's (1988a, 1988b) work on assessing knowledge schemata, and Collis, Romberg, and Jurdak's (1986) discussion of a technique for assessing mathematical problem solving. In each case, the assessments build directly on specific cognitive research and theory that takes into account both content and the processes students use to solve problems in the domain.

These assessments provide a different picture of student achievement than most current standard tests. Rather than measuring understanding with a single score, these tests provide a profile of understanding along dimensions perceived as critical for the theory underlying the assessment. Items are not sampled from a broad pool of items with similar psychometric characteristics; instead, they are selected from a multifaceted domain defined by critical features required for understanding. A discussion of the issues involved in the construction of such assessment instruments can be found in Haertel (1985).

Finally, it is not only important that students have constructed critical connections for understanding the material that they have already learned. It is important that students develop the ability to form new connections and recognize the benefits of striving to understand new material (Bereiter & Scardamalia, 1989). These important goals also need to be taken into account as we plan assessment.

Teaching for Understanding and Teachers' Understanding

Our analysis of understanding extends to teachers as well as students, encompassing their pedagogical knowledge as well as their knowledge of mathematics. This implies that we should be concerned how teachers' knowledge is connected. The same basic issues that we identified discussing teaching students for understanding apply to the problem of helping teachers implement programs of instruction designed to develop understanding in students.

Teachers possess a great deal of intuitive knowledge about students and about teaching, but that knowledge is not always well connected (Carpenter, Fennema, Peterson, & Carey, 1988; Fennema & Loef, this volume). The research of Carpenter, Fennema, and Peterson described earlier (Carpenter et al., 1989; Fennema et al., 1989; Peterson et al., 1991) considers the effects of helping teachers connect their knowledge and build upon it. There are direct parallels between the ways teachers are taught and the instruction they implement in their classrooms as a result. Using video tapes of children solving problems, teachers are given the opportunity to discuss and make explicit their knowledge about students' thinking. The research on the development of addition and subtraction concepts in children provides a basis for structuring this knowledge and making it coherent. The research provides a framework for distinguishing among problems in terms of a few basic principles, and children's strategies for solving problems are related directly to this classification scheme. Thus, teachers develop a principled framework for integrating their knowledge of mathematics and their knowledge of children's thinking.

Cobb, Yackel, and Wood (1988) and Cobb et al. (1991) take a similar approach to building on teachers' knowledge of students' thinking. However, they develop specific instructional materials for teachers to use with children in specified ways. By interacting with children using these materials, teachers are forced to examine their own knowledge of children's thinking.

The question of whether to make connections explicit for students also has a parallel in how our knowledge of students' understanding is translated into instructional programs provided to teachers. One possibility is to provide teachers with explicit instruction programs based on principles derived from studying student and/or expert performance. For example, in the research program investigating the teaching of multidigit subtraction algorithms through mapping instruction described by Fuson and Briars (1990), specific instructional activities were provided to the teachers and in-service programs were conducted specifying how the materials were to be used. Although the teachers were helped to understand the principles underlying mapping instruction, the nature of the program requires materials to be used in a specified way.

Although Cobb et al. (1988, 1991) have based their instructional program on constructivist principles, they also provided teachers with explicit instructional materials and specific guidelines for grouping students in pairs. In this case, however, the teachers were given a great deal of latitude in how they used the materials, and the materials and activities served as a mechanism to help teachers understand students' thinking.

At the other end of the continuum from providing teachers with explicit instructional activities is the approach illustrated by the research of Carpenter, Fennema, and Peterson. Based on the research on teachers' thinking and decision making (Clark & Peterson, 1986), this program starts with the assumption that teaching is problem-solving (Carpenter, 1988). Rather than providing teachers with specific activities designed to develop understanding, Carpenter, Fennema, and Peterson encourage teachers to design their own instructional activities and adapt instruction to their own students' knowledge and abilities based on the general knowledge about children's thinking they acquire in the program.

The distinctions illustrated in the above examples are not distinctions between rote and meaningful implementation. Our fundamental assumption is that skilled performance in any complex domain requires understanding. Research and analyses of curriculum implementation efforts have documented the problems that occur when teachers attempt to implement instructional programs they do not understand (National Advisory Committee on Mathematical Education, 1975; Porter, 1988; Stephens, 1982). Any program of instruction requires the teacher's understanding of the mathematics and the pedagogy for it to be successful in the long run, and all of the approaches

discussed here can be implemented in a thoughtful way with the objective of assisting teachers in understanding alternative instructional programs.

IMPLICATIONS FOR RESEARCH ON UNDERSTANDING

Our premise that understanding is critical for both students and teachers extends to the research process. The extension applies in two ways. First, just as we believe it is important for students to connect things they are learning and for teachers to develop connected, coherent structures of knowledge about teaching, we also believe it is important for those engaged in research on mathematics learning and teaching to consider connections between the phenomena they study. Investigators need to continually search for relationships between different aspects of their own work and between their work and that of others. Working out connections or relationships between pieces of the domain means building theories. Second, accepting the premise that understanding is important for students and teachers should influence the content of research efforts and shape the way in which research is carried out.

Building Theories of Mathematics Learning and Teaching

Theories, the way they are used here, are our explanations for how things work. They are the products of making explicit our hypotheses and hunches about how mathematics is learned and taught, of making public our private intuitions about the important features and relationships in the teaching and learning process. They are statements of the connections we have constructed.

Given this definition, all investigators work from theories, even though the theories may be quite local and implicit. All mathematics researchers work from assumptions, intuitions, and hypotheses about how students learn and how teachers teach mathematics. It is not a question of whether or not we develop theories. As N. L. Gage put it in the first *Handbook of Research on Teaching* (1963), everyone has theories and "differ not in whether they use theory, but in the degree to which they are aware of the theory they use" (p. 94). The process of building theories is a process of making explicit the implicit hypotheses that guide our work.

There are compelling arguments from the history and philosophy of science for developing explicit theories, whether or not the theories are right, whether or not the truth of the theories can be established or agreed upon (Lakatos, 1978; Popper, 1963). The reasons that theories are so essential are suggested in several intriguing parallels between the development and use of theories by students as they learn and the development and use of theories by those engaged in investigating this learning. It is these parallels that we want to highlight, two of which seem especially interesting: The first has to do with the way in which we view theories and the second with the level of explicitness needed to create useful theories.

When investigating whether students understand the mathematics they are performing, we often ask students to explain their reasons for doing what they do. We look for evidence in their explanations that they have connected the pieces of knowledge that support their performance. Students' explanations are their theories of how things work. Asking students to verbalize their theories allows us to interact with them about their thinking. Similarly, the extent of our understanding of the teaching and learning process can be measured by how well we explain students' connections and relationships. Theories serve as our explanations; making the theories public allows researchers to interact, to establish links between individual research projects, to connect results from separate projects in meaningful ways, and to guide future collaborative efforts.

A second parallel between theories of students and theories of researchers is suggested by Karmiloff-Smith and Inhelder's (1975) article "If You Want to Get Ahead, Get a Theory." They describe young children's progress in understanding a game activity with blocks in terms of the theories the children form and the hypotheses they test. For children's theories to be useful for them, the theories must be explicit enough to generate predictions that are testable. Perhaps the same holds true for our theories of the teaching-learning process. The theories may be relatively primitive and local, but to be useful they need to be sufficiently explicit to generate predictions that can be tested.

A final comment about theories can also be related to children's learning. Karmiloff-Smith and Inhelder (1975) note that theories are useful for children even if they are wrong. But theories become an obstacle for children if they are held too long in the face of disconfirming evidence. Our theories are always tentative, serving to make explicit our explanations for ourselves and others but always in need of modification or replacement.

Setting Goals for Research

If students' and teachers' understandings are important, then research efforts should be directed toward describing and explaining these understandings. The specific research questions to be addressed will follow from making our theories explicit and our predictions testable. However, the theories may completely determine neither the research design nor other particulars of the research process.

We propose that research which focuses on students' and teachers' understanding be designed to provide a fine-grained analysis of how the elements of the teaching-learning process interact. We would argue that there is little value in research that pits one instructional treatment against another without providing a detailed account of the effects of instruction on students' learning and thinking. Without such accounts, the information does little to increase our own understanding of the teaching-learning process. The outcomes may be connected to specific features of the instruction or of the students in unknown ways. Research that contributes to our understanding of students' and teachers' understanding will be research that reveals students' and teachers' thinking as it occurs in classroom settings and as it changes over the course of instruction. Such research will take us well beyond earlier investigations based on simple achievement measures and will feed the development of increasingly useful theories of mathematics learning and teaching.

References

Acioly, N. M., & Schliemann, A. D. (1986). *Intuitive mathematics and schooling in understanding a lottery game.* Paper presented at the tenth international meeting of the Psychology of Mathematics Education, London.

Anderson, J. R. (1983). *The architecture of cognition.* Cambridge, MA: Harvard University Press.

Ausubel, D. P. (1961). In defense of verbal learning. *Educational Theory, 11,* 15–25.

Ausubel, D. P. (1968). *Educational psychology: A cognitive view.* New York: Holt, Rinehart & Winston.

Baddeley, A. D. (1976). *The psychology of memory.* New York: Basic Books.

Bartlett, F. C. (1932). *Remembering.* Cambridge: Cambridge University Press.

Bebout, H. C. (1990). Children's symbolic representation of addition and subtraction word problems. *Journal for Research in Mathematics Education, 21,* 123–131.

Bednarz, N., & Janvier, B. (1988). A constructivist approach to numeration in primary school: Results of a three year intervention with the same group of children. *Educational Studies in Mathematics, 19,* 299–331.

Behr, M., Harel, G., Post, T. & Lesh, R. (in press). Rational Numbers: Toward a semantic analysis—Emphasis on the operator construct. In T. P. Carpenter, E. Fennema, & T. A. Romberg (Eds.), *Rational numbers: An integration of research.* Hillsdale, NJ: Lawrence Erlbaum.

Behr, M. J., Wachsmuth, I., Post, T. R., & Lesh, R. (1984). Order and equivalence of rational numbers: A clinical teaching experiment. *Journal for Research in Mathematics Education, 15,* 323–341.

Bell, A., Greer, B., Grimison, L., & Mangan, C. (1989). Children's performance on multiplicative word problems: Elements of a descriptive theory. *Journal for Research in Mathematics Education, 20,* 434–449.

Bell, M. S., Fuson, K. C., & Lesh, R. A. (1976). *Algebraic and arithmetic structures: A concrete approach for elementary school teachers.* New York: Free Press.

Bereiter, C., & Scardamalia, M. (1989). Intentional learning as a goal of instruction. In L. B. Resnick (Ed.), *Knowing, learning, and instruction: Essays in honor of Robert Glaser* (pp. 361–392). Hillsdale, NJ: Lawrence Erlbaum.

Brown, J. S., Collins, A., & Duguid, P. (1989). Situated cognition and the culture of learning. *Educational Researcher, 18*(1), 32–42.

Brown, J. S., & VanLehn, K. (1982). Towards a generative theory of "bugs." In T. P. Carpenter, J. M. Moser, & T. A. Romberg (Eds.), *Addition and subtraction: A cognitive perspective* (pp. 117–135). Hillsdale, NJ: Lawrence Erlbaum.

Brownell, W. A. (1935). Psychological considerations in the learning and teaching of arithmetic. In W. D. Reeve (Ed.), *The teaching of arithmetic. Tenth yearbook of the National Council of Teachers of Mathematics* (pp. 1–31). New York: Teachers College, Columbia University.

Brownell, W. A. (1947). The place of meaning in the teaching of arithmetic. *Elementary School Journal, 47,* 256–265.

Brownell, W. A. (1954). The revolution in arithmetic. *Arithmetic Teacher, 1,* 1–5.

Brownell, W. A., & Chazal, C. B. (1935). The effects of premature drill in third-grade arithmetic. *Journal of Educational Research, 29,* 17–28.

Brueckner, L. J. (1939). The development of ability in arithmetic. In G. M. Whipple (Ed.), *Thirty-eighth yearbook of the National Society for the Study of Education: Part 1. Child development and the curriculum* (pp. 275–298). Bloomington, IL: Public School Publishing Co.

Bruner, J. S. (1960). *The process of education.* New York: Vintage Books.

Bruner, J. S. (1961). The act of discovery. *Harvard Education Review, 31,* 21–32.

Bruner, J. S. (1973). *Beyond the information given.* New York: Norton.

Burns, R. B. & Lash, A. A. (1988). Nine seventh-grade teachers' knowledge and planning of problem-solving instruction. *Elementary School Journal, 88,* 369–386.

Byers, V., & Erlwanger, S. (1984). Content and form in mathematics. *Educational Studies in Mathematics, 15,* 259–275.

Byers, V., & Erlwanger, S. (1985). Memory in mathematical understanding. *Educational Studies in Mathematics, 16,* 259–281.

Campione, J. C. (Chair). (1988, April). *Constraints and chances in transfer of knowledge.* Symposium conducted at the annual meeting of the American Educational Research Association, New Orleans.

Carey, D. A. (1991). Number sentences: Linking addition and subtraction word problems and symbols. *Journal of Research in Mathematics Education, 22* 266–280.

Carpenter, T. P. (1985). Learning to add and subtract: An exercise in problem solving. In E. A. Silver (Ed.), *Teaching and learning mathematical problem solving: Multiple research perspectives* (pp. 17–40). Hillsdale, NJ: Lawrence Erlbaum.

Carpenter, T. P. (1988). Teaching as problem solving. In R. I. Charles & E. A. Silver (Eds.), *The teaching and assessing of mathematical problem solving* (pp. 187–202). Reston, VA: National Council of Teachers of Mathematics.

Carpenter, T. P., & Fennema, E. (1989). *Building on the knowledge of students and teachers.* Paper presented at the thirteenth international meeting of the Psychology of Mathematics Education, Paris.

Carpenter, T. P., Fennema, E., Peterson, P. L., & Carey, D. A. (1988). Teachers' pedagogical content knowledge of students' problem solving in elementary arithmetic. *Journal for Research in Mathematics Education, 19,* 385–401.

Carpenter, T. P., Fennema, E., Peterson, P. L., Chiang, C. P., & Loef, M. (1989). Using knowledge of children's mathematics thinking in classroom teaching: An experimental study. *American Educational Research Journal, 26,* 499–531.

Carpenter, T. P., Hiebert, J., & Moser, J. M. (1983). The effect of instruction on children's solutions of addition and subtraction word problems. *Educational Studies in Mathematics, 14,* 55–72.

Carpenter, T. P., & Moser, J. M. (1983). The acquisition of addition and subtraction concepts. In R. Lesh & M. Landau (Eds.), *The acquisition of mathematical concepts and processes* (pp. 7–40). New York: Academic Press.

Carpenter, T. P., & Moser, J. M. (1984). The acquisition of addition and subtraction concepts in grades one through three. *Journal for Research in Mathematics Education, 15,* 179–202.

Carpenter, T. P., Moser, J. M., & Bebout, H. (1988). The representation of basic addition and subtraction word problems. *Journal for Research in Mathematics Education, 19,* 345–357.

Carraher, T. N., Carraher, D. W., & Schliemann, A. D. (1985). Mathematics in the streets and in schools. *British Journal of Development Psychology, 3,* 21–29.

Carraher, T. N., Carraher, D. W., & Schliemann, A. D. (1987). Written and oral mathematics. *Journal for Research in Mathematics Education, 18,* 83–97.

Carraher, T. N., & Schliemann, A. D. (1985). Computation routines prescribed by schools: Help or hindrance? *Journal for Research in Mathematics Education, 16,* 37–44.

Carraher, T. N., Schliemann, A. D., & Carraher, D. W. (1988). Mathematical concepts in everyday life. In G. B. Saxe & M. Gearhart (Eds.), *Children's mathematics* (pp. 71–87). San Francisco: Jossey-Bass.

Case, R. (1985). *Intellectual development: Birth to adulthood.* New York: Academic Press.

Case, R., & Sandieson, R. (1988). A developmental approach to the identification and teaching of central conceptual structures in mathematics and science in the middle grades. In J. Hiebert & M. Behr (Eds.), *Number concepts and operations in the middle grades* (pp. 236–259). Reston, VA: National Council of Teachers of Mathematics.

Cauley, K. M. (1988). Construction of logical knowledge: Study of borrowing in subtraction. *Journal of Educational Psychology, 80,* 202–205.

Chi, M. (1978). Knowledge structures and memory development. In R. Siegler (Ed.), *Children's thinking: What develops?* (pp. 73–96). Hillsdale, NJ: Lawrence Erlbaum.

Chi, M., Feltovich, P., & Glaser, R. (1981). Categorization and representation of physics problems by experts and novices. *Cognitive Science, 5,* 121–152.

Chi, M. T. H., & Koeske, R. D. (1983). Network representation of a child's dinosaur knowledge. *Developmental Psychology, 19,* 29–39.

Clark, C. M., & Peterson, P. L. (1986). Teachers' thought processes. In M. C. Wittrock (Ed.), *Handbook of research on teaching* (3rd ed.) (pp. 255–296). New York: Macmillan.

Cobb, P. (1988). The tension between theories of learning and instruction in mathematics education. *Educational Psychologist, 23,* 87–103.

Cobb, P. (1990). A constructivist perspective on information-processing theories of mathematical activity. *International Journal of Educational Research, 14,* 67–92.

Cobb, P., & Merkel, G. (1989). Thinking strategies as an example of teaching arithmetic through problem solving. In P. Trafton (Ed.) *New directions for elementary school mathematics* (pp. 70–81). Reston, VA: National Council of Teachers of Mathematics.

Cobb, P., Wood, T., Yackel, E., Nicholls, J., Wheatley, G., Trigatti, B., & Perlwitz, M. (1991). Assessment of a problem-centered second-grade mathematics project. *Journal for Research in Mathematics Education, 22,* 3–29.

Cobb, P., Yackel, E., & Wood, T. (1988). Curriculum and teacher development: Psychological and anthropological perspectives. In E. Fennema, T. Carpenter, & S. J. Lamon (Eds.), *Integrating research on teaching and learning mathematics* (pp. 92–131). Madison: Wisconsin Center for Education Research.

Collis, K. F., Romberg, T. A., & Jurdak, M. E. (1986). A technique for assessing mathematical problem-solving ability. *Journal for Research in Mathematics Education, 17,* 206–221.

Commission on Standards for School Mathematics (1989). *Curriculum and evaluation standards for school mathematics.* Reston, VA: National Council of Teachers of Mathematics.

Cormier, S. M., & Hagman, J. D. (Eds.). (1987). *Transfer of learning: Contemporary research and applications.* New York: Academic Press.

Davis, R. B. (1964). *Discovery in mathematics: A text for teachers.* Reading, MA: Addison-Wesley.

Davis, R. B. (1984). *Learning mathematics: The cognitive science approach to mathematics education.* Norwood, NJ: Ablex.

Davis, R. B., & McKnight, C. (1980). The influence of semantic content on algorithmic behavior. *Journal of Mathematical Behavior, 3*(1), 39–87.

DeCorte, E. (Ed.). (1987). Acquisition and transfer of knowledge and cognitive skills [Special issue]. *International Journal of Educational Research, 11*(6).

Doyle, W. (1983). Academic work. *Review of Educational Research, 53,* 159–199.

Doyle, W. (1988). Work in mathematics classes: The context of students' thinking during instruction. *Educational Psychologist, 23,* 167–180.

Duncker, K. (1945). On problem solving. *Psychological Monographs, 58* (5, Whole No. 270).

Erlwanger, S. H. (1975). Case studies of children's conceptions of mathematics—Part I. *Journal of Children's Mathematical Behavior, 1*(3), 157–183.

Fehr, H. (1955). A philosophy of arithmetic instruction. *Arithmetic Teacher, 2,* 27–32.

Fennema, E. H. (1972). Models and mathematics. *Arithmetic Teacher, 65,* 635–640.

Fennema, E., Carpenter, T. P., & Peterson, P. L. (1989). Learning mathematics with understanding: Cognitively guided instruction. In J. E. Brophy (Ed.), *Advances in research on teaching* (Vol. 1, pp. 195–221). Greenwich, CT: JAI Press.

Fischbein, E., Deri, M., Nello, M. S., Marino, M. S. (1985). The role of implicit models in solving verbal problems in multiplication and division. *Journal for Research in Mathematics Education, 6,* 3–17.

Fuson, K. C. (1986). Roles of representation and verbalization in the teaching of multi-digit addition and subtraction. *European Journal of Psychology of Education, 1,* 35–56.

Fuson, K. C. (1990). Issues in place-value and multi-digit addition and subtraction learning and teaching. *Journal for Research in Mathematics Education, 21,* 273–280.

Fuson, K. C., & Briars, D. J. (1990). Base-ten blocks as a first- and second-grade learning/teaching setting for multi-digit addition and subtraction place-value concepts. *Journal for Research in Mathematics Education 21,* 180–206.

Fuys, D., Geddes, D., & Tischler, R. (1988). The Van Hiele model of thinking in geometry among adolescents. *Journal for Research in Mathematics Education Monographs, 3.*

Gage, N. L. (1963). Paradigms for research on teaching. In N. L. Gage (Ed.), *Handbook of research on teaching* (pp. 94–141). Chicago: Rand McNally.

Gagne, R. M. (1977). *The conditions of learning* (3rd ed.). New York: Holt, Rinehart, & Winston.

Gardner, H. (1985). *The mind's new science.* New York: Basic.

Geeslin, W. E., & Shavelson, R. (1975). Comparison of content structure and cognitive structure in high school students' learning of probability. *Journal for Research in Mathematics Education, 6,* 109–120.

Gelman, R., & Gallistel, C. R. (1978). *The child's understanding of number.* Cambridge, MA: Harvard University Press.

Gick, M. L., & Holyoak, K. J. (1987). The cognitive basis of knowledge transfer. In S. M. Cormier & J. D. Hagman (Eds.), *Transfer of learning: Contemporary research and applications* (pp. 9–46). New York: Academic Press.

Ginsburg, H. P. (1982). *Children's arithmetic.* Austin, TX: Pro-Ed.

Ginsburg, H. P. (1978). Poor children, African mathematics, and the problem of schooling. *Educational Research Quarterly, 2, 26–43.*

Glaser, R. (1979). Trends and research questions in psychological research on learning and schooling. *Educational Researcher, 8*(10), 6–13.

Glaser, R. (1984). Education and thinking: The role of knowledge. *American Psychologist, 39,* 93–104.

Goldin, G. A. (1987). Cognitive representational systems for mathematical problem solving. In C. Janvier (Ed.), *Problems of representation in the teaching and learning of mathematics* (pp. 125–145). Hillsdale, NJ: Lawrence Erlbaum.

Gonzalez, E. G., & Kolers, P. A. (1982). Mental manipulation of arithmetic symbols. *Journal of Experimental Psychology: Learning, Memory, and Cognition, 8,* 308–319.

Gray, W. D., & Orasnu, J. M. (1987). Transfer of cognitive skills. In S. M. Cormier & J. D. Hagman (Eds.), *Transfer of learning: Contemporary research and applications* (pp. 183–215). New York: Academic Press.

Greeno, J. G. (1977). Process of understanding in problem solving. In N. J. Castellan, Jr., D. B. Pisoni, & G. R. Potts (Eds.), *Cognitive theory* (Vol. 2, pp. 43–83). Hillsdale, NJ: Lawrence Erlbaum.

Greeno, J. G. (1978). A study of problem solving. In R. Glaser (Ed.), *Advances in instructional psychology* (Vol. 1, pp. 13–75). Hillsdale, NJ: Lawrence Erlbaum.

Greeno, J. G. (1983). Conceptual entities. In D. Gentner & A. L. Stevens (Eds.), *Mental models* (pp. 227–252). Hillsdale, NJ: Lawrence Erlbaum.

Greeno, J. G. (1988a). *Situations, mental models, and generative knowledge* (Report No. IRL 88-0005). Palo Alto: Institute for Research on Learning.

Greeno, J. G. (1988b). The situated activities of learning and knowing mathematics. In M. J. Behr, C. G. Lacampagne, & M. M. Wheeler (Eds.), *Proceedings of the Tenth Annual Meeting of PME-NA* (pp. 481–521). DeKalb, IL: Northern Illinois University.

Greeno, J. G. (1989). A perspective on thinking. *American Psychologist, 44,* 134–141.

Haertel, E. (1985). Construct validity and criterion-referenced testing. *Review of Educational Research, 55,* 23–46.

Hart, K. (1988). Ratio and proportion. In J. Hiebert & M. Behr (Eds.), *Number concepts and operations in the middle grades* (pp. 198–219). Reston, VA: National Council of Teachers of Mathematics.

Hart, K. (1989). There is little connection. In P. Ernest (Ed.), *Mathematics teaching: The state of the art* (pp. 138–142). New York: Falmer Press.

Hatano, G. (1988). Social and motivational bases for mathematical understanding. In G. B. Saxe & M. Gearhart (Eds.), *Children's mathematics* (pp. 55–70). San Francisco: Jossey-Bass.

Heid, M. K. (1988). Resequencing skills and concepts in applied calculus using the computer as a tool. *Journal for Research in Mathematics Education, 19,* 3–25.

Herscovics, N., (1989). The description and analysis of mathematical processes. In C. A. Mayer, G. A. Goldin, & R. B. Davis (Eds.), *Proceedings of the eleventh annual meeting of the North American Chapter of the International Group for the Psychology of Mathematics Education* (Vol. 2, pp. 1–28). New Brunswick, NJ: Rutgers University.

Hewson, P. W., & Hewson, M. (1988). An appropriate conception of teaching science: A view from studies of science learning. *Science Education, 72,* 597–614.

Hiebert, J. (Ed.) (1986). *Conceptual and procedural knowledge: The case of mathematics.* Hillsdale, NJ: Lawrence Erlbaum.

Hiebert, J. (1988). A theory of developing competence with written mathematical symbols. *Educational Studies in Mathematics, 19,* 333–355.

Hiebert, J. (1990). The role of routine procedures in the development of mathematical competence. In T. J. Cooney (Ed.), *Teaching and learning mathematics in the 1990's: 1990 Yearbook* (pp. 31–40). Reston, VA: National Council of Teachers of Mathematics.

Hiebert, J., & Lefevre, P. (1986). Conceptual and procedural knowledge in mathematics: An introductory analysis. In J. Hiebert (Ed.), *Conceptual and procedural knowledge: The case of mathematics* (pp. 1–27). Hillsdale, NJ: Lawrence Erlbaum.

Hiebert, J., & Lindquist, M. M. (1990). Developing mathematical knowledge in the young child. In J. Payne (Ed.), *Mathematics for the young child* (pp. 17–36). Reston, VA: National Council of Teachers of Mathematics.

Hiebert, J., & Wearne, D. (1985). A model of students' decimal computation procedures. *Cognition & Instruction, 2,* 175–205.

Hiebert, J., & Wearne, D. (1986). Procedures over concepts: The acquisition of decimal number knowledge. In J. Hiebert (Ed.), *Conceptual and procedural knowledge: The case of mathematics* (pp. 199–223). Hillsdale, NJ: Lawrence Erlbaum.

Hiebert, J., & Wearne, D. (1988). Instruction and cognitive change in mathematics. *Educational Psychologist, 23,* 105–117.

Hiebert, J., Wearne, D., & Taber, S. (1991). Fourth grader's gradual construction of decimal fractions during instruction using different physical representations. *Elementary School Journal, 91,* 321–341.

Hilgard, E. R. (1957). *Introduction to psychology* (2nd ed.). New York: Harcourt Brace.

Hughes, M. (1986). *Children and number.* Oxford, Blackwell.

Janvier, C. (Ed.). (1987). *Problems of representation in the teaching and learning of mathematics.* Hillsdale, NJ: Lawrence Erlbaum.

Kamii, C., & Joseph, L. (1988). Teaching place value and double column addition. *Arithmetic Teacher, 35*(6), 48–52.

Kaput, J. J. (1987). Toward a theory of symbol use in mathematics. In C. Janvier (Ed.), *Problems of representation in the teaching and learning of mathematics* (pp. 159–195). Hillsdale, NJ: Lawrence Erlbaum.

Kaput, J. J. (1988). *Notations and representations as mediators of constructive processes.* Unpublished manuscript. Educational Technology Center, Harvard University, Cambridge.

Karmiloff-Smith, A., & Inhelder, B. (1975). "If you want to get ahead, get a theory." *Cognition, 3,* 192–212.

Kerslake, D. (1986). *Fractions: Children's strategies and errors. A report of the strategies and errors in secondary school mathematics project.* Windsor, England: NFER-Nelson.

Kieren, T. E. (1988). Personal knowledge of rational number: Its intuitive and formal development. In J. Hiebert & M. Behr (Eds.), *Number concepts and operations in the middle grades* (pp. 162–181). Reston, VA: National Council of Teachers of Mathematics.

Kilpatrick, J. (1985). Reflection and recursion. *Educational Studies in Mathematics, 16,* 1–26.

Kosslyn, S. M., & Hatfield, G. (1984). Representation without symbol systems. *Social Research, 51,* 1019–1045.

Kouba, V. L. (1989). Children's solution strategies for equivalent set multiplication and division word problems. *Journal for Research in Mathematics Education, 20,* 147–158.

Kouba, V. L., Brown, C. A., Carpenter, T. P., Lindquist, M. M., Silver, E. A., & Swafford, J. O. (1988). Results of the fourth NAEP assessment of mathematics: Number, operations, and word problems. *Arithmetic Teacher, 35*(8), 14–19.

Kuhn, D., & Phelps, E. (1982). The development of problem-solving strategies. In H. Reese (Ed.), *Advances in child development and behavior* (Vol. 17, pp. 1–44). New York: Academic Press.

Kulm, G. (Ed.) (in press). *Assessing higher order thinking in mathematics.* Washington, DC: American Association for the Advancement of Science.

Lakatos, I. (1978). *The methodology of scientific research programmes* (J. Worrall & G. Currie, Eds.). New York: Cambridge University Press.

Lampert, M. (1986). Knowing, doing, and teaching multiplication. *Cognition and Instruction, 3,* 305–342.

Lampert, M. (1989). Choosing and using mathematical tools in classroom discourse. In J. E. Brophy (Ed.), *Advances in research on teaching* (Vol. 1, pp. 223–264). Greenwich, CT: JAI Press.

Larkin, J. H. (1989). What kind of knowledge transfers? In L. B. Resnick (Ed.), *Knowing, learning, and instruction: Essays in honor of Robert Glaser* (pp. 283–305). Hillsdale, NJ: Lawrence Erlbaum.

Larkin, J. H., McDermott, J., Simon, D. P., & Simon, H. A. (1980). Models of competence in solving physics problems. *Cognitive Science, 4,* 317–345.

Lave, J. (1988). *Cognition in practice.* Cambridge: Cambridge University Press.

Lave, J., Murtaugh, M., & de la Rocha, O. (1984). The dialectical construction of arithmetic in grocery shopping. In B. Rogoff & J. Lave (Eds.), *Everyday cognition: Its development in social context* (pp. 67–94). Cambridge, MA: Harvard University.

Lawler, R. W. (1981). The progressive construction of mind. *Cognitive Science, 5,* 1–30.

Leinhardt, G., & Smith, D. A. (1985). Expertise in mathematics instruction: Subject matter knowledge. *Journal of Educational Psychology, 77,* 247–271.

Lesh, R., Post, T., & Behr, M. (1987). Representations and translations among representations in mathematics learning and problem solving. In C. Janvier (Ed.), *Problems of representation in the teaching and learning of mathematics* (pp. 33–40). Hillsdale, NJ: Lawrence Erlbaum.

Lesh, R., Post, T., & Behr, M. (1988). Proportional reasoning. In J. Hiebert & M. Behr (Eds.), *Number concepts and operations in the middle grades* (pp. 93–118). Reston, VA: National Council of Teachers of Mathematics.

Luchins, A. S. (1942). Mechanization in problem solving: The effect of Einstellung. *Psychological Monographs, 54*(6, Whole No. 248).

Mack, N. K. (1990). Learning fractions with understanding: Building on informal knowledge. *Journal for Research in Mathematics Education, 21,* 16–32.

Madell, R. (1985). Children's natural processes. *Arithmetic Teacher, 32*(7), 20–22.

Marshall, S. P. (1988a). Assessing problem solving: A short-term remedy and a long-term solution. In R. I. Charles & E. A. Silver (Eds.), *The teaching and assessing of mathematical problem solving* (pp. 159–177). Reston, VA: National Council of Teachers of Mathematics.

Marshall, S. P. (1988b, April). *Assessing schema knowledge.* Paper presented at the annual meeting of the American Educational Research Association, New Orleans.

Mason, J. H. (1987). What do symbols represent? In C. Janvier (Ed.), *Problems of representation in the teaching and learning of mathematics* (pp. 73–81). Hillsdale, NJ: Lawrence Erlbaum.

Maturana, H., & Varela, F. (1980). *Boston University Philosophy of Science Series: Vol. 42. Autopoiesis and cognition.* Dordrecht: D. Riedel.

Matz, M. (1980). Towards a computational theory of algebraic competence. *Journal of Mathematical Behavior, 3*(1), 93–166.

Mayer, R. E. (1989). Models for understanding. *Review of Educational Research, 59,* 43–64.

McLellan, J. A., & Dewey, J. (1895). *The psychology of number and its applications to methods of teaching arithmetic.* New York: D. Appleton.

Merseth, K. K. (1978). Using materials and activities in teaching addition and subtraction concepts. In M. N. Suydam (Ed.), *Developing computational skills: 1978 Yearbook* (pp. 61–77). Reston, VA: National Council of Teachers of Mathematics.

Michener, E. R. (1978). Understanding understanding mathematics. *Cognitive Science, 2,* 361–383.

National Advisory Committee on Mathematical Education. (1975). *Overview and analysis of school mathematics, grades K-12.* Washington, DC: Conference Board of the Mathematical Sciences.

National Assessment of Educational Progress. (1983). *The Third National Mathematics Assessment: Results, trends and issues.* Denver, CO: Education Commission of the States.

National Research Council. (1989). *Everybody counts: A report to the nation on the future of mathematics education.* Washington, DC: National Academy of Sciences.

Nesher, P. (1988). Multiplicative school word problems: Theoretical approaches and empirical findings. In J. Hiebert & M. Behr (Eds.), *Number concepts and operations in the middle grades* (pp. 19–40). Reston, VA: National Council of Teachers of Mathematics.

Nesher, P. (1989). Microworlds in mathematical education: A pedagogical realism. In L. B. Resnick (Ed.), *Knowing, learning, and instruction* (pp. 187–215). Hillsdale, NJ: Lawrence Erlbaum.

Newell, A. (1980). Physical symbol systems. *Cognitive Science, 4,* 135–183.

Nicholls, J. G., Cobb, P., Wood, T., Yackel, E., & Patashnick, M. (1990). Assessing students' theories of success in mathematics: Individual and classroom differences. *Journal for Research in Mathematics Education, 21,* 109–122.

Noddings, N. (1985). Formal modes of knowing. In E. Eisner (Ed.), *Learning and teaching the ways of knowing. Eighty-fourth yearbook of the National Society for the Study of Education, Part II* (pp. 116–132). Chicago: University of Chicago Press.

Nolan, J. D. (1973). Conceptual and rote learning in children. *Teachers College Record, 75,* 251–258.

Ohlsson, S. (1988). Mathematical meaning and applicational meaning in the semantics of fractions and related concepts. In J. Hiebert & M. Behr (Eds.), *Number concepts and operations in the middle grades* (pp. 53–92). Reston, VA: National Council of Teachers of Mathematics.

Ohlsson, S., & Rees, E. (1988). *An information processing analysis of the function of conceptual understanding in the learning of arithmetic procedures* (Tech. Rep. No. KUL 88-03). Pittsburgh: University of Pittsburgh, Learning Research and Development Center.

Overman, J. R. (1935). The problem of transfer in arithmetic. In W. D. Reeve (Ed.), *The teaching of arithmetic. Tenth yearbook of the National Council of Teachers of Mathematics* (pp. 173–185). New York: Teachers College, Columbia University.

Payne, J. N. (1976). Review of research on fractions. In R. A. Lesh (Ed.), *Number and measurement* (pp. 145–187). Columbus, OH: ERIC/SMEAC.

Pea, R. D. (1987). Socializing the knowledge transfer problem. *International Journal of Educational Research, 11,* 639–663.

Perkins, D. N., & Salomon, G. (1989). Are cognitive skills context bound? *Educational Researcher, 18*(1), 16–25.

Perkins, D. W., & Simmons, R. (1988). Patterns of misunderstandings: An integrative model for science, math, and programming. *Review of Educational Research, 58,* 303–326.

Peterson, P. L., Fennema, E., & Carpenter, T. P. (1991). Teachers' knowledge of students' mathematical problem-solving knowledge. In J. E. Brophy (Ed.), *Advances in research on teaching* (Vol. 2, pp. 49–86). Greenwich, CT: JAI Press.

Piaget, J. (1970). Piaget's theory. In P. Mussen (Ed.), *Carmichael's manual of child psychology* (Vol. 1). New York: Wiley.

Piaget, J. (1973). *To understand is to invent.* New York: Grossman.

Pirie, S. E. B., & Kieren, T. E., (1989). A recursive theory for mathematical understanding. *For the Learning of Mathematics, 9*(3), 7–11.

Polya, G. (1957). *How to solve it* (2nd ed.). Garden City, NY: Doubleday Anchor Books.

Popper, K. R. (1963). *Conjectures and refutations: The growth of scientific knowledge.* New York: Harper & Row.

Porter, A. C. (1988). *External standards and good teaching: The pros and cons of telling teachers what to do* (Occasional Paper No. 126). East Lansing: Michigan State University, Institute for Research on Teaching.

Porter, A. C. (1989). A curriculum out of balance: The case of elementary school mathematics. *Educational Researcher, 18*(5), 9–15.

Posner, G. J., Strike, K. A., Hewson, P. W., & Gertzog, W. A. (1982). Accommodation of a scientific conception: Towards a theory of conceptual change. *Science Education, 66,* 211–227.

Putnam, R. T., Lampert, M., & Peterson, P. L. (1990). Alternative perspectives on knowing mathematics in elementary schools. In C. Cazden (Ed.), *Review of research in education* (Vol. 16, pp. 57–150). Washington, D.C.: American Educational Research Association.

Pylyshyn, Z. W. (1980). Cognitive representation and the process architecture distinction. *The Behavioral and Brain Sciences, 3,* 154–169.

Quillian, M. R. (1968). Semantic memory. In M. Minsky (Ed.), *Semantic information processing* (pp. 227–270). Cambridge: MIT Press.

Raphael, D., & Wahlstrom, M. (1989). The influence of instructional aids on mathematics achievement. *Journal for Research in Mathematics Education, 20,* 173–190.

Rathmell, E. C., & Huinker, D. M. (1989). Using "part-whole" language to help children represent and solve word problems. In P. R. Trafton (Ed.), *New directions for elementary school mathematics* (pp. 99–110). Reston, VA: National Council of Teachers of Mathematics.

Resnick, L. B. (1980). The role of invention in the development of mathematical competence. In R. H. Kluwe & H. Spada (Eds.), *Developmental models of thinking* (pp. 213–244). New York: Academic Press.

Resnick, L. B. (1982). Syntax and semantics in learning to subtract. In T. P. Carpenter, J. M. Moser, & T. A. Romberg (Eds.), *Addition and subtraction: A cognitive perspective* (pp. 136–155). Hillsdale, NJ: Lawrence Erlbaum.

Resnick, L. B. (1987a). Constructing knowledge in school. In L. S. Liben (Ed.), *Development and learning: Conflict or congruence?* (pp. 19–50). Hillsdale, NJ: Lawrence Erlbaum.

Resnick, L. B. (1987b). Learning in school and out. *Educational Researcher, 16*(6), 13–20.

Resnick, L. B., & Ford, W. W. (1981). *The psychology of mathematics for instruction.* Hillsdale, NJ: Lawrence Erlbaum.

Resnick, L. B., Nesher, P., Leonard, F., Magone, M., Omanson, S., & Peled, I. (1989). Conceptual bases of arithmetic errors: The case of decimal fractions. *Journal for Research in Mathematics Education, 20,* 8–27.

Resnick, L. B., & Omanson, S. F. (1987). Learning to understand arithmetic. In R. Glaser (Ed.), *Advances in instructional psychology* (Vol. 3, pp. 41–95). Hillsdale, NJ: Lawrence Erlbaum.

Riley, M. S., Greeno, J., & Heller, J. (1983). The development of children's problem-solving ability in arithmetic. In H. P. Ginsburg (Ed.), *The development of mathematical thinking* (pp. 153–196). New York: Academic Press.

Romberg, T. A., & Carpenter, T. P. (1986). Research on teaching and learning mathematics: Two disciplines of scientific inquiry. In M. C. Wittrock (Ed.), *Handbook of research on teaching* (3rd edition, pp. 850–873). New York: Macmillan.

Rumelhart, D. E. (1975). Notes on a schema for stories. In D. G. Bobrow & A. M. Collins (Eds.), *Representation and understanding* (pp. 211–236). New York: Academic Press.

Saxe, G. B. (1988). Candy selling and math learning. *Educational Researcher, 17*(6), 14–21.

Saxe, G. B. (1991). *Culture and cognitive development: Studies in mathematical understanding.* Hillsdale, NJ: Lawrence Erlbaum.

Schank, R. C., & Abelson, R. (1977). *Scripts, plans, goals, and understanding.* Hillsdale, NJ: Lawrence Erlbaum.

Schoenfeld, A. H. (1985). *Mathematical problem solving.* New York: Academic Press.

Schoenfeld, A. H. (1986). On having and using geometric knowledge. In J. Hiebert (Ed.), *Conceptual and procedural knowledge: The case of mathematics* (pp. 225–264). Hillsdale, NJ: Lawrence Erlbaum.

Schoenfeld, A. H., Smith, J. P., & Arcavi, A. (in press). Learning: The microgenetic analysis of one student's evolving understanding of a complex subject matter domain. In R. Glaser (Ed.), *Advances in instructional psychology* (Vol. 4). Hillsdale, NJ: Lawrence Erlbaum.

Schwartz, J. (1988). Intensive quantity and referent transforming arithmetic operations. In J. Hiebert & M. Behr (Eds.), *Number concepts and operations in the middle grades* (pp. 41–52). Reston, VA: National Council of Teachers of Mathematics.

Scribner, S. (1984). Studying working intelligence. In B. Rogoff & J. Lave (Eds.), *Everyday cognition: Its development in social context* (pp. 9–40). Cambridge, MA: Harvard University.

Shepard, R. N. (1982). Perceptual and analogical bases of cognition. In J. Mehler, E. C. T. Walker, & M. Garrett (Eds.), *Perspectives on mental representation: Experimental and theoretical studies of cognitive processes and capacities* (pp. 49–67). Hillsdale, NJ: Lawrence Erlbaum.

Shepard, R. N., & Metzler, J. (1971). Mental rotation of three dimensional objects. *Science, 171,* 701–703.

Shulman, L. S. (1970). Psychology and mathematics education. In E. G. Begle (Ed.), *Mathematics education. Sixty-Ninth Yearbook of the Society for the Study of Education* (pp. 23–71). Chicago: University of Chicago Press.

Shulman, L. S., & Keisler, E. R. (1966). *Learning by discovery: A critical appraisal.* Chicago: Rand McNally.

Siegler, R. S., & Jenkins, E. (1989). *How children discover new strategies.* Hillsdale, NJ: Lawrence Erlbaum.

Singley, M. K., & Anderson, J. R. (1989). *The transfer of cognitive skill.* Cambridge, MA: Harvard University Press.

Skemp, R. R. (1978). Relational understanding and instrumental understanding. *Arithmetic Teacher, 26*(3), 9–15.

Skinner, B. F. (1953). *Science and human behavior.* New York: Macmillan.

Slavin, R. E. (1989). Cooperative learning and student achievement. In R. E. Slavin (Ed.), *School and classroom organization.* (pp. 129–156). Hillsdale, NJ: Lawrence Erlbaum.

Sowder, J. T. (1988). Mental computation and number comparison: Their roles in the development of number sense and computational estimation. In J. Hiebert & M. Behr (Eds.), *Number concepts and operations in the middle grades* (pp. 182–197). Reston, VA: National Council of Teachers of Mathematics.

Sowder, J. T., & Wheeler, M. M. (1989). The development of concepts and strategies used in computational estimation. *Journal for Research in Mathematics Education, 20,* 130–146.

Sowell, E. J. (1989). Effects of manipulative materials in mathematics instruction. *Journal for Research in Mathematics Education, 20,* 498–505.

Steen, L. A. (1988). The science of patterns. *Science, 240,* 611–616.

Steffe, L. P., & Cobb, P. (1988). *Construction of arithmetical meanings and strategies.* New York: Springer-Verlag.

Stephens, W. (1982). *Mathematical knowledge and school work: A case study of the teaching of Developing Mathematical Processes.* Madison: Wisconsin Center for Education Research.

Stigler, J. W. (1984). "Mental abacus": The effect of abacus training on Chinese children's mental calculation. *Cognitive Psychology, 16,* 145–176.

Stigler, J., & Baranes, R. (1988). Culture and mathematics learning. *Review of Research in Education, 15,* 253–306.

Stigler, J. W., Fuson, K. C., Ham, M., & Kim, M. S. (1986). An analysis of addition and subtraction word problems in American and Soviet elementary mathematics textbooks. *Cognition and Instruction, 3,* 153–171.

Stodolsky, S. S. (1988). *The subject matters: Classroom activity in math and social studies.* Chicago: University of Chicago Press.

Suydam, M. N., & Higgins, J. L. (1977). *Activity-based learning in elementary school mathematics: Recommendations from research.* Columbus, OH: ERIC/SMEAC.

Thorndike, E. L. (1914). *The psychology of learning.* New York: Teachers College.

Thorndike, E. L. (1922). *The psychology of arithmetic.* New York: Macmillan.

Thorndike, E. L., & Woodworth, R. S. (1901). The influence of improvement in one mental function upon the efficiency of other functions. *Psychological Review, 8,* 247–261, 384–395, 553–564.

Trafton, P. R. (1989). Reflections on the number sense conference. In J. T. Sowder & B. P. Schappelle (Eds.), *Establishing foundations for research on number sense and related topics: Report of a conference* (pp. 74–77). San Diego: San Diego State University, Center for Research in Mathematics and Science Education.

Van Engen, H. (1949). An analysis of meaning in arithmetic. *Elementary School Journal, 49,* 321–329; 395–400.

Van Hiele, P. M. (1986). *Structure and insight.* Orlando, FL: Academic Press.

Vergnaud, G. (1983). Multiplicative structures. In R. Lesh & M. Landau (Eds.), *Acquisition of mathematical concepts and processes* (pp. 127–174). New York: Academic Press.

Vergnaud, G. (1988). Multiplicative structures. In J. Hiebert & M. Behr (Eds.), *Number concepts and operations in the middle grades* (pp. 141–161). Reston, VA: National Council of Teachers of Mathematics.

Voss, J. F. (1987). Learning and transfer in subject-matter learning: A problem-solving model. *International Journal of Educational Research, 11,* 607–622.

Wearne, D., & Hiebert, J. (1988). A cognitive approach to meaningful mathematics instruction: Testing a local theory using decimal numbers. *Journal for Research in Mathematics Education, 19,* 371–384.

Wearne, D., & Hiebert, J. (1989). Cognitive changes during conceptually based instruction on decimal fractions. *Journal of Educational Psychology, 81,* 507–513.

Webb, N. M. (Ed.). (1989). Peer interaction, problem-solving, and cognition: Multidisciplinary perspectives [special issue]. *International Journal of Educational Research, 13*(1).

Weiland, L. (1985). Matching instruction to children's thinking about division. *Arithmetic Teacher, 33*(4), 34–35.

Werner, H. (1957). The concept of development from a comparative and organismic point of view. In D. B. Harris (Ed.), *The concept of development* (pp. 125–148). Minneapolis: University of Minnesota Press.

Wertheimer, M. (1959). *Productive thinking.* New York: Harper & Row.

Wilson, J. W. (1967). The role of structure in verbal problem solving. *Arithmetic Teacher, 14,* 486–497.

Wittrock, M. C. (1974). A generative model of mathematics learning. *Journal for Research in Mathematics Education, 5,* 181–196.

Part

II

MATHEMATICS TEACHING

·5·

THE CULTURE OF THE MATHEMATICS CLASSROOM: AN UNKNOWN QUANTITY?

Marilyn Nickson

UNIVERSITY OF CAMBRIDGE
LOCAL EXAMINATIONS SYNDICATE

A consideration of the culture of the mathematics classroom implies an acceptance of the idea that mathematics exerts a unique influence on the context in the classrooms in which it is being taught and learned. This is a relatively recent departure from traditional concerns in mathematics education and is indicative of shifts in perspectives within the field that have, in turn, given rise to some change of emphasis in related research. This chapter will explore these changes from three broad perspectives. The first will consider those related to the nature of mathematics as a discipline, the second will be concerned with research about teachers and the teaching of mathematics, and the third will consider pupil perspectives.

The intention here is to consider aspects of culture within the microcosm of the mathematics classroom in order to build a picture of what contributes to the establishment of a culture at this level. The much wider issues that relate to mathematics education at the macro level will not be discussed. The choice to restrict this examination to the classroom alone is a deliberate one. There is a tendency in educational research to adopt global perspectives and popular catchwords without taking the time to clarify what we mean by them and to consider properly their relevance. Culture is just such a phenomenon. It appears increasingly in considerations at broader levels as, for example, with respect to ethnicity (D'Ambrosio, 1985), politics (Fasheh, 1982; Joseph, 1990; Mellin-Olsen, 1987; Pimm, 1990), technology (Noss, 1988), cross-cultural issues (Zaslavsky, 1989), and multicultural mathematics (Nickson, 1988b). If, however, we are concerned with the classroom as our unit of study, we are forced to come to some understanding and agreement about the meaning of culture at that level rather than taking

for granted that culture is shared. Having achieved that, it becomes more meaningful to go beyond the classroom to consider wider aspects of culture and to explore, in turn, how they impinge on that of the mathematics classroom. This chapter is concerned with taking the first steps in that direction.

What is the Culture of the Mathematics Classroom?

A reasonably accessible view of culture from the wealth of interpretations available is offered by Levitas (1974) when he states:

Every child in every society has to learn from adults the meanings given to life by his society; but every society possesses with a greater or lesser degree of difference, meanings to be learned. In short, every society has a culture to be learned though cultures are different. (p. 3)

Socialization is seen as a universal process, and culture is seen as the content of the socialization process that differs from one society to another. The shared meanings that come to be accepted by a society form its content. In discussing culture in the context of the school curriculum, Smith, Stanley, and Shores (1971) note that, in curricula as a whole, emphasis tends to be upon the "more fundamental universals, or cultural core, such as the values, sentiments, knowledges, and skills that provide society with stability and vitality and individuals with the motivations and deep-lying controls of conduct" (p. 17). This is echoed by Feiman-Nemser and Floden (1986) when they state that a "focus on *culture* implies inferences about knowledge, values, and norms for action, none of which can be directly

The author gratefully acknowledges the helpful comments provided by Peter Hall, University of Missouri, and Robert Underhill, Virginia Tech University.

observed" (p. 506). In reviewing research related to cultures of teaching, they note that the idea of culture in this context offers a new way of examining educational practice suggestive of an anthropological perspective, although, as they point out, neither their approach nor the research they report is necessarily anthropological in nature.

For our purposes, what we shall be considering here are the invisible and apparently shared meanings that teachers and pupils bring to the mathematics classroom and that govern their interaction in it. A single chapter cannot deal with all aspects of the knowledge, beliefs, and values that are held by the actors in this setting. What can be done by exploring the kinds of questions addressed by research in mathematics education is to identify where and how some of the matters of a cultural nature affect what goes on in the classroom.

A Variety of Classroom Cultures. A danger in associating culture with the mathematics classroom is to assume there is only one such culture. However, since key aspects of culture are concerned with unseen beliefs and values, the culture of a mathematics classroom will depend to a very large extent on these hidden perspectives of teachers and pupils in relation to the subject. As a result, just as there is a multiplicity of teacher cultures (Feiman-Nemser & Floden, 1986), there are many variations in the culture of classrooms in which mathematics is taught. It is tempting to assume that a concern with a single discipline will in some way act as a unifying agent, but no two mathematics classrooms are exactly alike. Nevertheless, by focusing on culture, we can learn more about how the "invisible" components in the teaching and learning situation can contribute to or detract from the quality of the mathematical learning that takes place. An exploration of such issues as the influence of differing perceptions of mathematics as a subject, of teacher beliefs and actions, and of pupil perspectives may help clarify how some of these components contribute to the cultural context of the mathematics classroom.

The Anthropological Dimension. Culture is being considered here in its anthropological sense in that we are focusing on an aspect of the study of human culture; that is, we are concerned with what people do within a mathematics classroom. Scheffler (1976) draws attention to the organic metaphor of culture sometimes employed within an educational context, where education is largely interpreted in terms of growth. His suggestion is that culture in the anthropological sense may too easily be taken as similar to its use in the organic sense, where culture is interpreted in terms of renewal and adaptation, or of growth and sustaining equilibrium within an environment. The dangers in using the metaphor in the organic sense in a classroom environment is that educational processes are essentially likened to "the processes by which individuals take on the environing culture" in order to ensure the continuity of that culture (Scheffler, 1976, p. 53). The importance in distinguishing between the two here is to draw attention to the facts that education *mediates* between the individual and his or her culture and that the invisible aspects of the cultural core are brought into play, as it were, by the teacher. The pupils being taught do not merely "take on" mathematics. In the context of the mathematics classroom, teachers act as agents of a particular culture, and in this role they make judgments and choices about aspects of that culture to which their pupils will be introduced—in this case, what mathematics will be taught, to whom, and how. These are the invisibles of the cultural core for which teachers have a responsibility.

Cooney, Goffree, Stephens, and Nickson (1985) have noted the importance of metaphors used in mathematics education and their potential effect on what we do. They suggest that such metaphors can reveal much about our basic beliefs, and by considering them, we can "increase our sensitivity to the possible roles researchers and teachers play" (p. 25). It is important in exploring the mathematics classroom from the perspective of the culture it generates to remember that we are concerned with the people in that setting and what they bring to it. We must increase our sensitivity to the importance of their hidden knowledge, beliefs, and values for mathematics education.

FROM CLOSED TO OPEN

As noted at the beginning of this chapter, subject specificity in studying the culture of a classroom is a relatively new phenomenon. Reiss (1978) states that, "Empirical research in education ... is only hesitantly approaching the problem as a subject-matter specificity of teaching and learning processes. Research on the relationship between subject matter and patterns of social interaction in the classroom has been barely tentative, despite the fact that demands for it are repeatedly raised in related research resumés" (p. 400). Some years later, this situation is little changed and a plea for more classroom research related to the teaching and learning of mathematics is still being made (Good & Biddle, 1988). An encouraging example of an increase of interest in this area is Eisenhart's (1988) examination of the relationship between the ethnographic research tradition and research in mathematics education, which points to a greater concern with a study of mathematics classrooms and to a widening of the research base for such studies. Before exploring new directions in research methodology, the importance of a change in beliefs and values about the nature of mathematical knowledge will be discussed because such beliefs and values related to mathematics as a subject are in a sense responsible for much of what goes on in the classroom context.

Differing Views of Mathematics

One of the major shifts in thinking in relation to the teaching and learning of mathematics in recent years has been with respect to the adoption of differing views of the nature of mathematics as a discipline. Thom (1972) suggests that all mathematical pedagogy rests on a philosophy of mathematics, however poorly defined or articulated it might be. Even bearing in mind constraints imposed by being compelled to teach particular content, the way in which it is approached can be seen as a manifestation of a particular philosophy, a point implicitly acknowledged in the Nonstatutory Guidelines (National Curriculum Council, 1989) of the recently developed National Curriculum in the United Kingdom. Some idea of the range of perceptions of the nature of mathematical knowledge can be gained by considering two traditions of mathematical thought—one of long-standing and one that has more recently come to the fore.

The "Formalist" Tradition. The view of mathematics that has informed and historically transfixed most mathematics curricula has been what Lakatos (1976) refers to as the formalist view, based on the epistemology of logical positivism. Hamlyn (1970) describes a central thesis of logical positivism that "all propositions *other than those about mathematics or logic* are verifiable by reference to experience" (p. 37). In other words, the foundations of mathematical knowledge are not seen to be in any sense social in origin, but lie outside human action in what Lakatos (1976) calls a "formalist heaven" (p. 2). Mathematics is considered to consist of immutable truths and unquestionable certainty. This in turn means that much of what we know as mathematics, such as problem solving, and what is commonly understood as mathematics in "the full spectrum of its relationships to science, to technology, to the humanities, and to human life" (Wittman, 1989, p. 298) is not recognized. Equally important is the fact that such a view does not take into account how mathematics changes and grows—it is waiting "out there" to be discovered. In short, mathematics as a way of knowing and interpreting our experience is discounted and becomes removed from human activity and the context of everyday life.

Hersh (1979) notes that "the criticism of formalism in the high school has been primarily on pedagogic grounds." He goes on to state that "all such arguments are inconclusive if they leave unquestioned the dogma that real mathematics is precisely formal derivations from stated axioms....The issue then, is not, what is the best way to teach, but what is mathematics really all about" (p. 18). Others have considered the implications of such a view (Bloor, 1976; Lerman, 1990; Nickson, 1981; Plunkett, 1981), some of which will be discussed below in relation to how this view affects the classroom context and the culture engendered.

A "Growth and Change" View of Mathematics. Lakatos (1976) has been responsible for much of the new direction in thought in recent years about the nature of mathematical knowledge. He is a proponent of Popper's (1972) view of objective knowledge, where knowledge is seen as resulting from competing theories that are proposed, made public, and tested against other theories and held to be true until "falsified" by a better theory. Kuhn (1970) goes beyond this by identifying the importance of the subjective aspect involved in the selection of theories in the first place. He emphasizes the important part that judgment and choice have to play in such a situation and hence the role of different value systems. He suggests that "in many concrete situations, different values, though all constitutive of good reasons, dictate different conclusions, different choices" (p. 70). Thus, how knowledge comes into being, and also how it comes to be superseded and changed, is not only a social phenomenon but a cultural one.

Mathematics educators in recent years have increasingly been drawn to consider a view of mathematical knowledge seen in these terms. Researchers such as Confrey (1980), Wolfson (1981), Nickson (1981), Pimm (1982), Lerman (1986), and Cobb (1986) have been among those to study the implications of such a view for the mathematics classroom. Orton (1988) explores the issue of how such theories can contribute to the study of fundamental issues in mathematics education. We shall now go on to consider how this view and a formalist view may manifest themselves in the culture of the mathematics classroom.

Shared Views of Mathematics in the Classroom

Referring to the work connected with teaching cultures, Feiman-Nemser and Floden (1986) state that "teaching cultures are embodied in the work-related beliefs and knowledge teachers share—beliefs about appropriate ways of acting on the job and rewarding aspects of teaching, and knowledge that enables teachers to do their work" (p. 508). Whether or not teachers can identify the particular nature of the subject, they must hold beliefs and values with respect to mathematics that influence how they teach. These will affect what content they select, whether they consider it accessible to all pupils, and how they choose to make it accessible to them. It is reasonable to assume that their actions throughout these processes will reflect their personal perceptions and beliefs related to the subject and pedagogy. This has been confirmed to some extent by Thompson (1984), who investigated high school teachers' beliefs and their classroom teaching. She found that "the observed consistency between the teachers' professed conceptions of mathematics and the manner in which they typically presented the content strongly suggests that the teachers' views, beliefs, and preferences about mathematics do influence their instructional practice" (p. 125).

Evidence of Practice in the Mathematics Classroom. An analysis of classroom practice in the United Kingdom carried out over the past 15 years (Ashton, Kneen, Davies, & Holley, 1975; Department of Education and Science, 1978, 1979; Galton, Simon, & Croll, 1980; National Committee of Inquiry, 1982; Ward, 1979) suggests that the view of mathematics being projected was that of a linear subject, mainly concerned with mechanistically teaching facts and skills predominantly related to number, and generally characterized by paper-and-pencil activity, even where commercially published schemes were not in use. In the United States, Porter, Floden, Freeman, Schmidt, and Schwille (1988) have found a reliance on mathematics textbooks in elementary schools from which teachers choose to place an emphasis on computational skills; time spent on teaching these skills takes up approximately 75% of class mathematics time. They note that "the lack of balance in teacher attention to conceptual understanding, skills, and applications is problematic and should be addressed" (p. 106).

Although this may be limited evidence upon which to make judgments about shared views of mathematics and the classroom cultures they generate, some inferences can be made. Whether or not overtly identified, it appears that the shared view of mathematics reflected in these particular classrooms is close to a formalist perspective insofar as it tends to concentrate on the formalities of the number system and the abstract manipulation of number. The evidence does not suggest a classroom environment that reflects mathematics as "an integral part of human culture" (Wittman, 1989, p. 298). A heavy reliance on text-books and teaching number facts is not conducive to teacher-pupil exchange or to interaction among pupils themselves; it is more conducive to an environment dominated by a concern with right or wrong answers. It could be argued that teachers in these classrooms adopt particular teaching strategies because of their supposed effectiveness, regardless of personal teaching philosophy. However, the lack of success in achieving desirable results that stems largely from the adoption of teaching methods reflecting this formalist perspective has led to recent inquiries in the United States and the United Kingdom to study the failure of mathematics curricula in meeting the needs of these respective societies (National Committee of Excellence in Education, 1983; National Committee of Inquiry, 1982).

Where mathematics as a discipline has been perceived in formalist terms, it has on the whole remained inaccessible to teachers and hence to pupils. The traditional detachment of mathematical content from shared activity and experience, so that it remains at an abstract and formal level, erects a barrier around the subject that removes it from other spheres of social behavior. The messages conveyed in approaching the subject in this way are based upon assumptions that have been accepted unquestioningly and from which no deviation is permissible. The classroom culture that evolves will inevitably mirror this unquestioning acceptance. Brown and Cooney (1982) give an interesting example of this through an incident in which a geometry teacher was struggling with a below-average class. When it is suggested that the teacher downplay the importance of proof and engage in a more active approach to doing geometry, she resisted the idea because she had developed the strong belief that proof is necessary to learning geometry for all students, even though her students would not be tested on it. Brown and Cooney conclude that the intensity of the teacher's beliefs precluded considerations other than proof-oriented ones. The visibility and acceptance of what is right and wrong in mathematics and of what is done and not done are potent factors in stopping teachers from engaging in activities that they may instinctively feel are appropriate but that might challenge the supposedly inviolable essence of mathematics as they themselves were taught it.

New Directions. In recent years following the Cockcroft Report (National Committee of Inquiry, 1982) in the United Kingdom and building on previous good practice, new methodologies have been developed and their adoption encouraged in mathematics curricula in an attempt to improve a situation where past success had been limited. Teachers have been encouraged to ensure that as much mathematics teaching as possible is based upon shared activity and that the subject is presented as open to discussion, investigation, and hypothesis (Department of Education and Science, 1985). This is an approach that more closely reflects a growth-and-change view of the nature of mathematics, where mathematical knowledge (even at classroom level) is not held to be exempt from interpretations that require "reconsideration, revision and refinement" (Toulmin, 1972, p. 50). Teachers (especially at primary level) are coming to value the importance of discussion and problem solving and are allowing children to generate theories and test them. Teachers are even generating theories themselves (Alderson, 1988;

McGrath, 1988). The decisions that are made in implementing this kind of curriculum, however, rest upon teachers' confidence in the appropriateness of doing so. It is clearly important that they should be able to give reasons for their choices and to identify criteria for judging the worthiness of what they do. The beliefs and values underlying this perspective have to be made more clearly identifiable to them to sustain their confidence in their professional judgment as teachers of mathematics. A point has been reached where a rationale is emerging that can offer them the theoretical support they need in sustaining that confidence. The fundamentals of such a rationale already exist, but need to be clearly articulated and made explicit and accessible.

Social Contexts. The implementation of curricula that reflect a growth-and-change perspective of mathematical knowledge will affect the social context of the mathematics classroom by implicitly encouraging the active participation of all concerned. If mathematics is characterized in terms of openness and of questioning and testing ideas, what is necessarily involved is a sharing of ideas and problems among pupils and between pupils and teacher. Although it is clear that specific concepts and skills have to be learned, the learning and application can become more purposeful because of this shared nature. The context becomes more purposeful rather than artificially contrived, and it gives rise to the possibility for increased discussion that is considered to be an important aspect of the mathematical learning process, particularly in the early years (Desforges & Cockburn, 1987). More will be said about teachers and pupils in the classroom later in this chapter, but the point being established here is how differing perceptions of the nature of mathematics (that is, the subject itself) can influence the culture of the mathematics classroom and why these perceptions must be taken into account in order to understand fully that culture.

Reiss (1978) notes that "in mathematical education, it is not only important what is known, but how it is known, and how this knowledge is reproduced" (p. 394). She suggests that in order to explore issues of this nature, what needs to be studied are the different forms and levels this interaction takes in classrooms across different subject areas. She notes research that suggests that "different behaviors are socially rewarded" in different subjects at each level (p. 405). These levels of interaction are characterized by degrees of negotiation between the participants about the rules for interpreting and evaluating actions in the classroom. The first level is where interaction occurs most frequently, and this takes place in humanities and social sciences classes. The second level is where participants move within an established and predetermined system of rules, and it is at this level that "social exchange in instructional situations in mathematics and natural sciences will predominantly take place" (Reiss, 1978, p. 405).

This analytical structure helps clarify the important effect specific subjects have on the interactive social processes within classrooms. The fact that levels of interaction within the mathematics classroom have been found to be constrained by an established structure of rules also suggests that this is a reflection of mathematics perceived in somewhat rigid and traditional formalist terms. However, with a shift in perceptions about the nature of the subject implicit in new methodologies and approaches within curricula, it is arguable that we are moving from the second level of social interaction in mathematics classrooms to the first. It is likely that when mathematics instruction is taken beyond establishing facts and practicing skills to an approach using more openness, investigation, problem-solving, and critical discussion, there will be more social interactions, more negotiation, and more emphasis upon shared interpretation and evaluation of what goes on in mathematics classrooms.

Summary

The differing views held by teachers and pupils in relation to the nature of mathematical knowledge are an important component in the culture of a mathematics classroom, since they are linked with the way mathematics is taught and received. One perspective may result in a classroom context which could be described as "asocial" insofar as it emphasizes the abstractness of mathematics to be done individually and more or less in silence by pupils in the classroom. Another view emphasizes the social aspect of the foundations of mathematical knowledge that precipitates a different kind of teaching and learning context in which the mathematics is sometimes done by individuals but can also be shared and open to question, challenged, discussed, explored, and tested. These examples may represent the two extremes, but they serve to indicate the potential power exerted by beliefs about mathematics on the teaching of mathematics.

FROM TRANSMISSION TO PARTICIPATION

In recent years, educational research has become increasingly concerned with the study of classroom processes and the complexity of interactions within them. Lortie's (1975) sociological study of the schoolteacher, as well as other work such as that of Nash (1974) and Rutter, Maughan, Mortimore, & Ouston (1979), has helped to establish a background against which to consider issues of this nature. This increased interest in social and interpersonal aspects of the classroom has led to a shift in the nature of research undertaken, and it recognizes the importance of social interactions within classrooms and of what pupils and teachers bring to them.

Shifting Research Perspectives

Kilpatrick (1988) offers a succinct description of three approaches to educational research that include the behaviorist tradition, the interpretivist view, and action research. Where the behaviorist tradition is concerned, he states that "the goal is to uncover law-like regularities in educational phenomena; the methods are aimed at specifying behavior and analyzing it into components. The world is a system of interacting variables whose variation can be controlled experimentally and modeled mathematically." (p. 98). Brown, Cooney, and Jones (1990) refer to a broader category of methodologies of this type as positivist.

In contrast to this, as Kilpatrick (1988) points out, the interpretivist view sets out to "capture and share the understanding that participants in an educational encounter have of what they are teaching and learning." Action research, on the other hand, "adopts the so-called critical approach" and seeks "not merely to understand the meanings participants bring to the educational process but to change those meanings that have been distorted by ideology." (p. 98) Cohen and Manion (1980) point out that is difficult to ascribe a "comprehensive definition" to action research "because usage varies with time, place and setting" (p. 174). Their interpretation is that action research is essentially small-scale and has to do with intervening in a real-world situation and closely examining the results of such intervention, but no specific link with ideology is made. Kilpatrick (1988) sums up the three approaches to research in the following way:

The behaviorist stands apart from an educational encounter, aiming at general laws that will transcend time, place and circumstance. The interpretivist moves into that encounter, attempting to describe and explain it from a nonjudgmental stance. The action researcher enters the encounter with an eye toward obtaining greater freedom and autonomy for the participants. (p. 98)

He concludes that it would be unwise to view any of these approaches as existing in entirely separate compartments or as representing a shift away from "hard science." Trying to categorize research in mathematics education in line with these approaches would be extremely difficult and thus illustrates Kilpatrick's point. However, the application of new research techniques in the mathematics classroom has important consequences, in the long term, for the practicalities of how mathematics is taught and learned. Research is moving along a spectrum of activity from a position where the individual *acts* of teachers and pupils are studied, to a holistic approach where all teacher-pupil and pupil-pupil *interactions* are scrutinized together with the values, beliefs, and attitudes they bring to the situation. A brief consideration of each of the research paradigms helps to identify their potential to contribute to our knowledge of the culture of the mathematics classroom.

The Positivist Perspective in Mathematics Education Research. From the preceding interpretation of the behaviorist research paradigm, it can be seen that rather than focusing on the "whole" of a teaching and learning situation, research methodology of this type is concerned with how teachers and students function within the classroom. The purpose in offering the following examples related to the mathematics classroom is to identify the kinds of questions the methodology addresses. The intention is not to imply that researchers cited do not recognize the importance of social or cultural phenomena within the mathematics classroom, or that their research falls neatly into a single category. However, such studies do entail a choice with respect to those topics for research that may prove most fruitful. As such, they are offered as examples of research priorities and, at the same time, of the kinds of messages they may convey to teachers about what is important in mathematics education.

Many such studies have been concerned with exploring how students learn particular mathematical concepts such as conservation or operations with numbers (Bell, Fischbein, &

Greer, 1984; Carpenter & Moser, 1983; Gelman, 1969) the stages of children's learning (Donaldson, 1978), the characteristics of individual learners (Trown & Leith, 1975), time spent on task (Peterson, Swing, Stark, & Waas, 1984), and seatwork (Anderson, 1981). The search has been for evidence of an optimum way to teach particular mathematics to particular children, with the aim to achieve transfer of these findings to the teaching of the same mathematics to similar pupils in similar situations. Other research of this nature has set out to identify learning styles, such as studies undertaken by Bennett (1976) McLeod, Carpenter, McCornack, and Skvarcius (1978), and those reported in Biggs (1967) and Shulman (1970). In general, by focusing on individuals, the search is for clusters of common characteristics from which to generalize about particular types of teachers or learners. Cooney et al. (1985) state that in employing such a paradigm for research, "importance is placed on the revelation of general truths and principles," and it is judged according to its power to "yield predictions and useful prescriptions" (p. 25). Hence the value of research of this nature for the classroom teacher is seen to lie in its predictive nature and its potential for offering insights into successful ways of teaching mathematics.

Begle (1979) casts doubt on the relevance of much of this early work in the literature survey in which he set out to identify critical variables in mathematics education. He found that there was little correlation between the many teacher characteristics and variables identified and the effectiveness of teaching mathematics as measured by higher pupil achievement. He indicated that "the very concept of the effectiveness of a teacher may not be valid" (p. 37). This has not deterred further concern with the effectiveness of the individual teacher, however, as studies by Good, Grouws, and Ebmeier (1983), Berliner (1986), Leinhart (1988), and others have shown. Many studies of this kind have involved the observation and analysis of the behavior of individual teachers who have been successful or "expert," with a view to identifying behaviors that may be adopted by others. Although the aim of all such studies is to produce useful insights into successful teaching practice, the message conveyed to teachers may be that there is some ideal way of teaching mathematical content. As Lampert (1985) puts it, teachers who do look to this kind of research for guidance may hold the view that "the teacher's work is to find out what researchers and policymakers say should be done with or to students and then to do it.... If the teacher does what she is told, students will learn" (p. 191).

Further exploration has focused on the effects of teachers' personal mathematical knowledge (Bassham, 1962; Malle, 1988), teachers' attitudes and expectations (Bishop & Nickson, 1983; Good & Brophy, 1978), and teachers' perceptions of and beliefs about mathematics (Brown & Cooney, 1982; Nickson, 1981; Thompson, 1984) upon what they do in the classroom. Some of this research has been observational in nature and has sometimes drawn upon work done in classrooms in general and been applied to the context of the mathematics classroom in particular.

From the point of view of the culture of the mathematics classroom, research on the actions of teachers and how these actions relate to pupil outcomes is likely to tell us little about the shared invisibles such as values, beliefs, and meanings that combine to create that culture. Clearly, if results of this type of study are made accessible to teachers, they may select from the conclusions and improve their classroom practice. However, their *reasons* for making a selection in the first place—that is, what informs that judgment and what leads them to make particular choices—are important in understanding the classroom as a whole. It is these "work-related beliefs and knowledge teachers share," to which Feiman-Nemser and Floden (1986, p. 508) refer, that make a strong contribution to establishing the culture of a mathematics classroom.

A Constructivist Approach. The recent increased interest in constructivism as a theoretical perspective in mathematical pedagogy has contributed to the shift in the focus of research towards interactive processes and away from a behaviorist paradigm. Steffe and Killion (1986) state that, from the constructivist perspective, "*mathematics teaching* consists primarily of the *mathematical interactions* between a teacher and children" (p. 207). The teacher has an intended meaning to impart to the children, who interpret it and adjust it to their personal mathematics schemes, thereby constructing their own knowledge. This model of mathematical learning has formed the basis for a considerable body of research that includes work related to assessment (Brown, 1986) as well as to the learning of specific mathematical concepts (Herscovitz & Vinner, 1984; von Glaserfeld, 1981). In spite of the interactive character of this perspective, the focus remains on the researcher or teacher, the learner, and specific mathematical concepts, and usually does not extend to the whole classroom context in which teaching and learning take place.

Desforges (1985) suggests that research concerned with matching pupils with tasks to achieve improved outcomes is based on prescriptive models of teaching derived by learning theories and "entirely ignore[s] the constraints on the teachers and the social conditions of the classroom" (p. 96). Valuable detailed knowledge can be gained by studying how a pupil receives and accommodates some mathematical input, but the perceived relevance of such knowledge to teachers and the problem of its accessibility to them remains. The broad acceptance of children as constructors of their own mathematical knowledge has not been accompanied by a similar view of the teacher's research knowledge.

An Interpretivist Paradigm

It has been argued elsewhere (Nickson, 1988a) that bringing teachers into the arena of research activity can be an important step in increasing their understanding of research processes and results and their relation to classroom practice. It is the possibility of this type of collaboration that has been a factor in attracting researchers to the use of descriptive or naturalistic methodologies. Eisenhart's (1988) discussion of the ethnographic tradition of research in relation to mathematics education is particularly relevant here because it highlights the anthropological nature of the questions addressed in studying the culture of the mathematics classroom and it provides a valuable framework against which to discuss the interpretivist approaches to research increasingly being adopted in the field.

Ethnographic Research Methodology and Mathematics Education. Eisenhart (1988) makes the point that "few educational researchers have actually undertaken ethnographic research" (p. 99), which is defined as "the disciplined study of what the world is like for people who have learned to see, hear, speak, think, and act in ways that are different" (Spradley, 1980, p. 3). Spradley holds that ethnographic research has to do with learning from people, not from studying them.

The four methodologies used for data collection within this paradigm are identified as participant observation, ethnographic interviewing, a search for artifacts (available written or graphic materials related to the topic of study), and researcher introspection. These can be supplemented with others such as surveys and questionnaires. Data collection and analysis are carried out together throughout the study so that what is being tested is "an emergent theory of culture or social organization." The ultimate goal of the research is "a theoretical explanation that encompasses all the data and thus provides a comprehensive picture of the complex meanings and social activity" (Eisenhart, 1988, p. 107).

Eisenhart acknowledges that researchers in mathematics education are identifying questions for which an ethnographic approach is appropriate, but suggests that the tendency for these researchers is "to use case studies, in-depth interviews, or in-classroom observations without doing what most educational anthropologists would call ethnographic research" as described above (p. 99). She goes on to show how, when "ethnography is placed within the context of interpretivism and cultural anthropology and then compared to traditional educational research and psychology," it becomes clear why ethnography as a methodology has been difficult to adapt to research in mathematics education. There is a long tradition of other well-established research paradigms in the educational field with which it has to compete. Nevertheless, as the questions being addressed are seen to be increasingly appropriate for study using ethnographic methodology, stronger links will be forged between the fields of educational anthropology and mathematics education.

Interpretivist Studies. If we accept what Feiman-Nemser and Floden suggest when they say that the "complexity and immediacy of classroom events" (1986, p. 515-516) are what provide the link between teachers' beliefs and behavior patterns, then these events must be studied in order to gain any insight into the nature of the culture of a mathematics classroom. Eisenhart (1988) points out that the mixture of methodologies usually employed in interpretivist studies effectively limits them to a case study of a single class or of a small number of pupils or teachers. The inherent danger of the approach is the temptation to extrapolate from the local to the global (Davies, 1983). Davies describes some of the earlier work of this type, which he suggests has been detrimental to the field because researchers have jumped from "their little bits of the here-and-now to their big story" and in the process of so doing, have left "some of their actors (usually teachers) pretty trampled" (p. 133). If such approaches are to provide a way of bridging the gap between the mathematics teacher and the researcher, they must be used with sensitivity and care. They have the potential to involve teachers as partners in research, keeping in mind that it is the teacher's domain being studied and about which judgments will be made.

An example of the kind of public soul-searching teachers are often asked to engage in comes from Desforges and Cockburn's (1987) study of mathematics teaching in first schools in the United Kingdom. In this instance, the teacher sets out to establish the names of three-dimensional shapes. Following several acceptable student examples of sphere-shaped objects, Samantha suggests a circle. "Mrs G. thought, 'Help! This is going to be a lot more difficult than I thought. How am I going to explain the difference between a circle and a sphere?' " (p. 113) Later on, "head" is suggested as an example of a sphere. This time, Mrs. G. says:

At first I was pleased. That was a nice idea. It was quick to relate it to spheres. Then I thought that a head is not a perfect sphere....I did not want them to get the wrong ideas at the beginning—they are really capable of that. At the same time I did not want to bog them down with pernickety bits and pieces. It was hard to sort of weigh that out. I was really struggling. (p. 113–114)

This brief extract conveys both the immediacy of the dilemma that faced this teacher and her openness in admitting her dilemma. Similar dilemmas have been explored by Lampert (1985). This example shows that the teacher is trying to share what went on in her mind and to articulate her reason for actions in resolving the knowledge-practice conflict in which she finds herself. Lortie (1975) refers to "teachers' recurrent doubts about the value of their work with students" and states that in asking teachers to assess their own performance in his study, "no aspect of the teacher's work evoked as much emotion as this issue of assessing outcomes" (p. 142). Clearly the intention of research is not to add to teachers' doubts and anxiety. This kind of situation, where the teacher is almost justifying her choices, highlights the importance of sensitivity on the part of the researcher in interpretivist methodologies.

In Delamont's (1981) view, familiarity is the greatest danger in descriptive methodologies. In discussing their wider use in educational settings, she suggests that one way to enhance the chances of obtaining meaningful information is to study "unusual, bizarre, or 'different' classrooms" (p. 74). This approach contrasts with mathematics classroom studies with a more specific focus, for example, on the expert or the novice teacher (Berliner et al., 1988), where the focus is predetermined. In the latter kind of study, the expert teachers have been singled out according to the degree of achievement of pupils in their charge, and the actions of these teachers are studied to search for commonalities that could account for their pedagogic success. If Delamont's (1981) ideas were implemented, the situation could be turned around, and a class might be identified as "unusual" because of the generally high level of mathematical achievement of the pupils. The question to be explored would be, "What is there about *this classroom* that produces these results?" The whole context, including the pupils' perspectives and what they bring to the situation in terms of beliefs and expectations, would be studied, not just the perspectives and actions of the teacher. By taking these phenomena into account and exploring the interactive processes within the class as a whole and at different levels, new meanings and insights can emerge that have particular relevance to the cultural aspects of the classroom. For example, it may be that the success of the class has as much to do with the fact that the teacher and pupils share a common perception of what mathematics is about as it has to do with specific actions on the part of the teacher.

A good example of a valuable insight gained within an interpretivist study is provided by Yates (1978) when she describes her intervention in a classroom situation in which Linda is exploring the making of cubes from six-square nets. Yates asks questions to find out how Linda is approaching the problem and realizes that she has upset the girl's equilibrium. In her reflections on the situation, Yates asks herself why she intervened and concludes that it was because Linda's approach to the problem was not typical; she was approaching the problem differently than others. Yates identifies the uncertainty as her own and poses the question, "How often do we project our uncertainties onto our pupils?" (p. 62) This is a powerful question that many teachers of mathematics could benefit from asking themselves at frequent intervals. It is immediately accessible to the teacher, and the situation from which it was drawn is identifiable as one which teachers meet every day. Important phenomena such as the projection of our own uncertainties onto our pupils often become embedded in the exchanges and structures within the institutionalized classroom setting and then become lost as part of our pedagogical knowledge. They are an important part of the classroom processes that need to be brought to the surface. The self-examination to which Yates submits herself in this extract is a good example of "researcher introspection" that Eisenhart (1988) identifies as a component of the ethnographic tradition, and which in this particular situation clearly produced a valuable insight. The nature of teacher intervention conveys powerful messages to pupils. In this case, the pupil sensed disapproval on Yates' part that suggested unhappiness with the way she was proceeding. Yates was concerned that Linda would fail to achieve a satisfactory solution to the problem. Linda in turn was beginning to lose faith in her own problem-solving strategy, and her confidence in doing mathematics was undermined.

Good and Biddle (1988) suggest that "observational research has not been utilized sufficiently to improve mathematics education" and that it has the potential to "yield better theories for understanding the learning of mathematics and other subjects and can produce more adequate models for improving teaching" (p. 114). The foregoing examples provide some indication of how observation can also lead to a more reflective approach on the part of teachers and researchers and can impart a deeper understanding of the processes that contribute to the culture of mathematics classrooms and theories about them.

Action Research

Two definitions of action research were offered earlier (Cohen & Manion, 1980; Kilpatrick, 1988). Kilpatrick identifies its current popularity in Australia and it has also appeared in curriculum development research in the United Kingdom over the past few decades (Shipman, 1974). Although action research does not appear to figure strongly in educational research in the United States at the moment, it has not been ignored.

Cohen and Manion (1980) state that "the scene for its appearance began to be set in [the United States] in the 1920s with the application of the scientific method to the study of educational problems, growing interest in group interaction and group processes, and the emerging progressive movement" (p. 176). It is still characterized by a concern with groups and their interaction.

The importance of action research in the context of the culture of the mathematics classroom is the potential it holds for the involvement of the teacher in the research process. Cohen and Manion (1980) suggest five broad categories of purpose for action research in schools and classrooms: (1) diagnosing and remedying problems in specific situations; (2) providing inservice training to sharpen analytical powers and self-awareness of teachers; (3) injecting innovative approaches to teaching and learning into an ongoing system; (4) improving poor communication between the practicing teacher and the researcher and remedying the lack of clear prescriptions provided by traditional research; and (5) providing an alternative to more impressionistic and subjective approaches to solving problems in the classroom. It is clear that teachers can play a central part in addressing these kinds of problems in the classroom when they act as "both practitioner and researcher in one" (p. 177). Action research differs from interpretivist research methodologies in that it relies both on observational and behavioral data and "is therefore empirical," so that over a period of time "information is collected, shared, discussed, recorded in some way, evaluated and acted upon" (p. 179–180). The overlap of methodologies emphasizes the danger noted earlier of trying to be too specific in typifying research activities. Action research combines characteristics of both positivist and interpretivest paradigms.

The participant nature of action research with its direct involvement of teachers means that studies of this kind can be carried out by teachers researching in their own classrooms and schools, sometimes undertaking an inservice route towards a higher degree. An example of such a study with mathematics as a curriculum focus is an investigation of the effectiveness of parental involvement in a program of mathematics homework over a period of time and with specific teaching or learning objectives in mind (Alderson, 1988). Another teacher explores underachievers in mathematics in her own school (McGrath, 1988). These studies have involved the interaction of the teacher with other teachers, with pupils and parents, and with college researchers. The analytical and reflective skills and habits developed by teachers engaging directly in the research process have great potential for positively influencing their attitudes towards research generally. By developing self-confidence in the field, teachers are likely to be ready to question and evaluate other research rather than to ignore it, or worse, to accept it unquestioningly.

Summary

As long as the main expectations of research in mathematics education are linked with its potential prescriptive power, the values or the essentially descriptive nature of other perspectives will remain questionable to many. This overview of approaches to research related to mathematics teachers shows how increased interest in the adoption of an interpretivist paradigm could prove fruitful in giving us insight into the culture of the mathematics classroom. Brown, Cooney, and Jones (1990) suggest that interpretivist research also has the potential to influence mathematics teacher education in several ways. It can provide a deeper understanding of meaning-making processes of teachers and thereby provide a basis for constructing teacher education programs that are responsive to teachers' beliefs and needs. To respond to the hidden beliefs and meanings that all classroom participants bring to the teaching and learning situation, they must first of all be identified. The interpretivist research paradigm seems particularly suited to the task.

FROM ACCEPTING TO QUESTIONING

Feiman-Nemser and Floden (1986) point out in their explorations of teaching cultures that it is important to be aware of norms "that govern social interactions between teachers and other role groups" (p. 508). The most immediate and significant "other role group" for teachers are their pupils, who exert "the most important external constraint upon them" (Bishop & Nickson, 1983, p. 15). The developments described in the previous two sections suggest that pupils are increasingly playing a more active role in the mathematics classroom and potentially contribute much more overtly to the processes that combine to make up its culture. Judgments and choices are made by pupils as well as teachers.

Pupils' Perceptions of Mathematics: Accepting

Desforges and Cockburn (1987) provide an example of how the context in which mathematics is done conveys messages about the subject itself. As part of their study, they probed children's perceptions of mathematics. The children were working from commercially published schemes, and the researchers found that in spite of the fact that teachers encouraged work at a steady pace, the children were "very keen to race on through the scheme's workbooks" (p. 101). The following is an extract from an interview with a pupil engaged in using such a workbook:

Interviewer: Why did you measure these things?
    Kelly: For the book.
Interviewer: Why do you think it's in the book?
    Kelly: It's our work. (p. 100–101)

Mathematics is seen by Kelly not as the practicality of measuring to gain any relevant or useful information or to solve a real problem; it is what the book "tells her to do" and it is part of her "work." The children interviewed could recall very few instances when they had seen adults engaged in measuring. Desforges and Cockburn (1987) conclude that "they never saw adults doing mathematics and in any event they were far more concerned with the reality of their own world which, mathematically speaking, was the self-imposed business of getting on through the scheme" (p. 102). The children had been found not to rate discussion sessions in mathematics very highly, and Desforges and Cockburn (1987) suggest that "perhaps the children saw them as distraction from the main agenda." They go on to say:

The content of the work was dominated by the commercial schemes used, in part because the structure of the schemes looked like a race track to the children and in part because, forming the only overtly assessed part of the mathematics curriculum, they could be taken to be the whole of what was meant by mathematics. (p. 102)

Carraher (1989) makes a similar point when she writes:

Although the teacher may introduce a topic by making reference to everyday situations, the situations are usually stripped of their basic sense. The loss of meaning does not result from the pretend situation, from the role-playing of a situation which is not part of school life, but from the stripping off of central elements of the situation, which are viewed as extraneous to mathematics. (p. 639)

A further example offered by Cobb (1986) shows how learning mathematics in a particular context affects a pupil's beliefs about mathematics. He suggests that if a pupil asks a teacher for confirmation about the rightness or wrongness of a piece of work as soon as it is completed, this is evidence of the fact that the teacher is seen by the pupil as an authority. However, it is also evidence of the pupil's perceptions of the nature of mathematics, that is, that mathematics is a kind of knowledge that can always be assessed in terms of correctness or incorrectness. This is similar to the pupils in the science example offered by Lampert (1985), where they were confused by the possibility of two right answers, although in this case it probably had more to do with the way in which the knowledge was being assessed. Cobb (1986) states that "this inference would be further substantiated if the teacher responds by attempting to initiate a Socratic dialogue and the student shows irritation or frustration" (p. 4). Mathematics is not seen to be a subject for exploration or discussion where the possibility of negotiation about methodology or content may exist. The expectation is that alternatives do not exist.

The first of the preceding examples shows how children's views are in the process of being shaped by the way mathematics is being presented, while the second is an illustration of the long-term effects of this kind of approach, where expectations become narrowly focused on the criterion of the "right or wrong" nature of mathematical knowledge. Cobb (1987) also reports a study related to classroom contexts in which arithmetic is being taught by teachers who have declared their expectations that the children will produce correct answers, and, in most cases, the teachers have prescribed the methods they expect students to use. Cobb refers to the fact that the children were faced with problems, but says that "these problems were not purely arithmetical in origin. They derived from the children's social interactions in classrooms characterized by imposition," and further suggests that "the children had the responsibility of adjusting their goals and activities to fit their teachers' relatively rigid expectations" (Cobb, 1987, p. 121). The view of mathematics being developed is rigidly determined by the specific requirements of the teacher, and rather than the context being one of activity, negotiation, and discussion, it is one of repression, where the use of the pupils' own strategies and potential for solving problems is discouraged. There is no room for maneuver, and should the pupil instinctively wish to approach a problem differently, the teacher's clear expectations produce in the student an immediate internal conflict. This is reminiscent of Yates' (1978) dilemma identified earlier, when she realized that her interference with her pupil's problem-solving strategy upset the pupil's equilibrium.

Pupils' Perceptions of Mathematics: Questioning

Jaworski (1989) has studied mathematics classrooms in which an investigative approach has been adopted, which she links with a constructivist view of mathematics teaching and learning. In one case study, she describes a lesson in which a class of 14- and 15-year-olds were exploring shapes that had the same area and perimeter. The class worked in groups of their own choice after negotiating with each other about which shapes they wished to study. The teacher, over a period of time with the class, had established the importance of looking for pattern in their mathematical work, which symbolized his "belief in mathematics being about expressing generality" (Jaworski, 1989, p. 151). In her observations of what was going on in the classroom, Jaworski (1989) notes that "implicit in the interchange here was an emphasis on conjecturing, on trying out special cases, and on seeking for generality" (p. 151). Two boys remained after the lesson to discuss some aspects of it, and the following comment about the teaching methodology was made by one of them to the researcher with the teacher present:

To tell you the truth…I mean, Mr xxxx's…a different kind of teacher completely. Before, you've had sums that you've been set…At first, to tell you the truth, I didn't like him as a teacher. I thought, 'No. Pathetic!,—you know, 'this isn't maths—what's this got to do with maths?' And as I've come along, I've realized that it's got a lot to do with maths. To have to learn rather than just have to sit and [say] 'Oh, I've done 50 sums today', 'I've done a hundred.' You don't bother about that now, just concentrate, and at the end of a lesson you've learned something….I've really progressed. (p. 153–154)

This extract indicates how the beliefs about mathematics that the members of this class share with their teacher reflect a classroom context that differs considerably from those projected by pupils in the examples given earlier. The interactions between pupils and teacher and among pupils are based on a perception of mathematics that does not exclude problems that can be shared. However, what the pupil has to say also hints at the difficulties faced by teachers who hold constructivist views of mathematics and try to teach mathematics to pupils who have developed quite different perceptions from years of prescriptive teaching. If mathematics is seen by pupils to be solely about accuracy in doing number work or any other kind of mathematical work, it is bound to be difficult for teachers to alter such views. Pupils' expectations built up over the years can provide a considerable block to effecting such change. It is clear that such a conflict of views would manifest itself in some way in the culture of the classroom in which it exists. Teachers should be more aware of what the symptoms of such conflict might be and how to act upon them where they arise.

Pirie (1988) reports research that focuses on discussion as an aid to mathematical understanding. She describes how the teacher starts a lesson by promoting a whole class discussion by putting forward a mathematical idea or problem and keeping the discussion open so that "pupils are encouraged to offer their thinking to the rest of the class." The teacher remains as unobtrusive as possible and deflects questions back to the pupils. The intention is "that the pupils shall form their own intuitions about the structures of mathematics" (p. 2). The pupils then form small groups to pursue a solution to the problem. Pirie (1988) states that "the bedrock on which this style is built is the belief that knowledge must be actively constructed and not passively received if the pupils are truly to learn mathematics" (p. 2). She describes how one pair of girls share their discovery of how to divide fractions; she also discusses the strategies they evolve to cope with that problem. She suggests that concepts are understood not just by experiencing them, but by interpreting those experiences to each other and also, in this case, to the researcher.

In a classroom such as this, it is made clear to the pupils through the teaching practice used that their views and conjectures are as valuable as anyone else's. They are developing their perceptions of mathematics from those of their teachers and peers, and they are learning to articulate their own ideas and beliefs about them. They are actively engaged in doing mathematics and in the process, they are receiving messages that mathematics is about questioning, conjecturing, and trial and error. The tools necessary for engaging in these processes are learned as pupils advance and are in a context that is meaningful either because it is a problem that they have made their own through exploration and discussion, or because it is a real-life problem to be solved.

Summary

The studies referred to in this section show how the pupil's role in the mathematics classroom can vary dramatically according to the view of mathematics projected by the teacher. The linearity and formality associated with most teaching of mathematics from published schemes or textbooks tend to produce a passive acceptance of mathematics in the abstract, with little connection being made by pupils between their work and real life. The visibility of mathematics in terms of a "right or wrong" nature is accepted by pupils, and their main concerns seem to be with the quantity of mathematics done and its correctness. The culture of such a classroom seems to be fairly well defined, since the pupils' expectations are based on their acceptance of what appear to be clear messages from the scheme or the book and the teacher. The nature of the social interaction within the classroom is dominated by these perceptions.

Where beliefs about mathematics differ and where views of mathematics as socially constructed knowledge prevail, pupils take on quite a different role. The messages they receive are that they are expected to contribute their own ideas, to try their own solutions, and even to challenge the teacher. In such situations the culture of the mathematics classroom is likely to be more variable, since there exists the possibility for greater divergence of views. There is also greater potential for conflict in terms of what is agreed and shared, since the development of these views of mathematics is fairly recent and because not all pupils will necessarily accept them in the light of their previous educational experience in mathematics.

CONCLUSIONS

What has emerged in this study is that the culture of the mathematics classroom will vary according to the actors within it. The unique culture of each classroom is the product of what the teacher and pupils bring to it in terms of knowledge, beliefs, and values, and how these affect the social interactions within that context. It is all too easy to assume that these invisibles of the cultural core are shared by all the participants and that there is a harmony of views about the goals being pursued and the values related to them. The research suggests that there may once have been a fair degree of predictability about what could be expected, as well as uniformity in the views held and the methodologies they gave rise to. This resulted from a dominant perception of the nature of mathematical knowledge and the way it affected pedagogy, which in turn affected the social interaction within the classroom and, ultimately, the culture of that classroom. However, the concurrent changes over the three broad fronts considered here mean that an increased potential for variation now exists within the context of the mathematics classroom with respect to what is taught, as well as how and to whom it is taught. There is more possibility for choice and more possibility that those choices will be guided by different beliefs and values. Consequently, there will be greater variation in the cultures of mathematics classrooms. With this increased breadth of possibilities, however, comes increased potential for lack of consensus, in particular for mismatches between teachers' views and goals and those of pupils. This can result in what we might call productive classroom cultures and nonproductive classroom cultures. If we are to gain anything from the study of the culture of the mathematics classrooms, it will come from an understanding of the factors that contribute to their productivity in this sense.

What Needs to be Done?

Several issues have emerged in this study that can be explored in order to help to achieve this understanding. They are closely linked with the changes that have been discussed and that need to be addressed because of their potential for militating against a productive classroom culture and hence their potential for obstructing the effective teaching and learning of mathematics.

A New Rationale. There is a need for a clearly articulated rationale to provide a theoretical foundation for new methodologies and approaches that are developing concurrently with changing views about the nature of mathematical knowledge. This rationale should be accessible to teachers in order to avoid the conflict of values that can arise when they are asked to adopt these new methodologies without understanding their foundations.

Potential Areas of Conflict. It is important to be aware of the potential for conflicting views and their effects within the classroom where, for example, a pupil may be expected to assume a changed role within the context of the mathematics lesson. Teachers' expectations differ depending on their beliefs and values, yet pupils are expected to adjust and accept them as valid because the teacher is seen as an authority. We need to develop strategies for coping with such situations and to consider ways of negotiating shared meanings and goals.

Messages Conveyed by Research. An increase in the adoption of new research methods that study the reality of the classroom is needed. These methods can provide a rich opportunity for teachers to become drawn into research as equal partners and can allow them to explore questions about mathematics in their classrooms that *they* see as important. Topics for study in the past have conveyed messages that have placed a premium on the mechanics of maximizing mathematical learning. With teachers as partners in research, the message conveyed will be that their ideas and expertise have credence and value. Through their participation, teachers will come to appreciate the full effect that their personal beliefs and values have on their mathematics classroom.

Repression of Pupils' Ideas in the Mathematics Classroom. We need to be more aware of the subtle effects of social interaction in the classroom and how actions and comments by teachers can repress the pupil's individual mathematical thinking. New methodologies allow teaching and learning situations where the pupil is encouraged to challenge and question the teacher as well as other pupils. Implicit in this situation is the need to acknowledge and value what the pupil offers.

Accessibility. It has been clear throughout this chapter that matters related to the culture of the mathematics classroom all have to do, one way or another, with the accessibility of mathematics to teachers and pupils. The assumption is that all members of a given society have access to mathematical knowledge at some level. It is important to remember how vital that accessibility is to individuals and how the extent to which mathematical knowledge can add to the quality of their lives by helping to inform the judgments and decisions they make.

The importance of the culture of the mathematics classroom is implicit in the statement by Reiss (1978) that "individual acquisition of mathematical concepts and skills in school will always be accompanied by an acquisition of orientations about the subject matter related and the social significance of the things learned" (p. 394–395). She is hinting at the hidden messages that form a main part of the culture of the mathematics classroom. Popkewitz (1988) writes in a similar vein when he states that "school mathematics involves not only acquiring content; it involves participating in a social world that contains standards of reason, rules of practice and conceptions of knowledge" (p. 221). As mathematics educators, we need to be more aware of the hidden social messages in what we do and the power of their influence on the young people we teach.

References

Alderson, J. (1988). *Parental involvement in children's mathematics.* Unpublished master's thesis, Essex Institute of Higher Education, Brentwood, U.K.

Anderson, L. M. (1981). *Student responses to seatwork: Implications for the study of student's cognitive processes* (Research Series No. 102). East Lansing, MI: Institute for Research on Teaching, Michigan State University.

Ashton, P., Kneen, P., Davies, F., & Holley, B. (1975). *The aims of primary education: A study of teachers' opinions.* London: Macmillan Education Ltd.

Bassham, H. (1962). Teacher understanding and pupil efficiency in mathematics—a study of relationship. *Arithmetic Teacher, 9,* 383–387.

Begle, E. G. (1979). *Critical variables in mathematics education.* Washington, DC: Mathematics Association of America and the National Council of Teachers of Mathematics.

Bell, A., Fischbein, E., & Greer, B. (1984). Choice of operation in arithmetic problems: The effects of number size, problem structure and context. *Educational Studies in Mathematics, 15,* 129–147.

Bennett, N. (1976). *Teaching styles and pupil progress.* London: Open Books.

Berliner, D. C. (1986). In pursuit of the expert pedagogue. *Educational Researcher, 15,* 5–13.

Berliner, D. C., Stein, P., Sabers, D., Clarridge, P., Cushing, K., & Pinnegar, S. (1988). Implications of research on pedagogical expertise and experience for mathematics teaching. In D. A. Grouws, T. J. Cooney, & D. Jones, (Eds.), *Effective perspectives on research on teaching* (pp. 67–95). Reston, VA: National Council of Teachers of Mathematics.

Biggs, J. B. (1967). *Mathematics and conditions of learning.* Slough, U.K.: National Foundation for Educational Research (NFER).

Bishop, A. J., & Nickson, M. (1983). *The social context of mathematics education: A review of research in mathematical education* (Part B). Windsor, U.K.: NFER-Nelson.

Bloor, D. (1976). *Knowledge and social imagery.* London: Routledge & Kegan Paul.

Brown, C. A., & Cooney, T. J. (1982). Research on teacher education: A philosophical orientation. *Journal of Research and Development in Education, 15*(4), 13–18.

Brown, M. (1986, July). Developing a model to describe the mathematical progress of secondary school students (11–16 years): Findings of the graded assessment in mathematics project. In *Proceedings of the Tenth International Conference: Psychology of Mathematics Education* (pp. 135–140). London, U.K.

Brown, S. I., Cooney, T. J., & Jones, D. (1990). Mathematics teacher education. In W. R. Houston (Ed.), *Handbook of Research on Teacher Education* (pp. 639–656). New York: Macmillan.

Carpenter, T. P., & Moser, J. M. (1983). The acquisition of addition and subtraction: A cognitive perspective. In R. Lesh & M. Landau (Eds.), *Acquisition of mathematics concepts and processes* (pp. 7–44). New York: Academic Press.

Carraher, T. N. (1989). Negotiating the results of mathematical computations. *International Journal of Educational Research, 13*(6), 637–646.

Cobb, P. (1986). Contexts, goals, beliefs and learning mathematics. *For the Learning of Mathematics, 6*(2), 2–9.

Cobb, P. (1987). An investigation of young children's academic arithmetic contexts. *Educational Studies in Mathematics, 18,* 109–124.

Cohen, L., & Manion, L. (1980). *Research methods in education.* London: Croom Helm.

Confrey, J. (1980). *Conceptual change analysis: Implications for mathematics and curriculum inquiry.* East Lansing, MI: Institute for Research on Teaching, Science-Mathematics Teaching Center, Michigan State University.

Cooney, T. J., Goffree, F., Stephens, M., & Nickson, M. (1985). The professional life of teachers. *For the Learning of Mathematics, 5*(2), 24–30.

D'Ambrosio, U. (1985). Ethnomathematics and its place in the history and pedagogy for mathematics. *For the Learning of Mathematics, 5*(1), 44–48.

Davies, W. B. (1983). The sociology of education. In P. Hirst (Ed.), *Educational theory and its foundation disciplines* (pp. 100–137). London: Routledge & Kegan Paul.

Delamont, S. (1981). All too familiar? A decade of classroom research. *Educational Analysis, 3*(1), 47–68.

Department of Education and Science. (1978). *Primary education in England: A survey by H M Inspectors of Schools.* London: Her Majesty's Stationary Office.

Department of Education and Science. (1979). *Aspects of secondary education in England.* London: Her Majesty's Stationary Office.

Department of Education and Science. (1985). *GCSE: The national criteria for mathematics.* London: Her Majesty's Stationary Office.

Desforges, C. (1985). Matching tasks for children. In S. N. Bennett & C. Desforges, (Eds.) *Recent advances in classroom research* (pp. 92–104). London: Constable.

Desforges, C., & Cockburn, A. (1987). *Understanding the mathematics teacher: A study of practice in first schools.* London: The Falmer Press.

Donaldson, M. C. (1978). *Children's minds.* London: Croom Helm.

Eisenhart, M. A. (1988). The ethnographic research tradition and mathematics education research. *Journal for Research in Mathematics Education, 19*(2), 99–114.

Fasheh, M. (1982). Mathematics, culture and authority. *For the Learning of Mathematics, 3*(2), 2–8.

Feiman-Nemser, S., & Floden, R. E. (1986). The cultures of teaching. In M. C. Wittrock (Ed.), *Handbook of research in teaching* (3rd edition, pp. 505–526). London: Collier-Macmillan.

Galton, M., Simon, B., & Croll, P. (1980). *Inside the primary classroom.* London: Routledge & Kegan Paul.

Gelman, R. (1969). Conservation acquisition: A problem of learning to attend to relevant attributes. *Journal of Experimental Child Psychology, 7,* 167–187.

Good, T., & Biddle, B. J. (1988). Research and improvement of mathematics instruction: The need for observational resources. In D. A. Grouws, T. J. Cooney, & D. Jones (Eds.), *Perspectives on research on effective mathematics teaching* (Vol. 1, pp. 114–142). Reston, VA: National Council of Teachers of Mathematics.

Good, T., & Brophy, J. (1978). *Looking in classrooms.* New York: Harper & Row.

Good, T. L., Grouws, D., & Ebmeier, H. (1983). *Active mathematics teaching.* New York: Longman.

Hamlyn, D. W. (1970). *The theory of knowledge.* London: Macmillan Press.

Herscovitz, R., & Vinner, S. (1984). Children's concepts in elementary geometry—a reflection of teacher's concepts. In B. Southwell et al. (Eds.), *Proceedings of the Eighth International Conference for the Psychology of Mathematics Education* (pp. 28–34). Darlinghurst, Australia: International Group for the Psychology of Mathematics Education.

Hersh, R. (1979). Some proposals for revising the philosophy of mathematics. *Advances in Mathematics, 31*(1), 31–50.

Jaworski, B. (1989, July). *To inculcate versus to elicit knowledge.* Actes de la 13e conference internationale, Psychology of Mathematics Education, Paris.

Joseph, G. (1990, April). *The politics of anti-racist mathematics.* Paper presented at a conference on the Political Dimensions of Mathematics Education, Institute of Education, University of London.

Kilpatrick, J. (1988). Editorial, *Journal for Research in Mathematics Education, 19*(2), 98.

Kuhn, T. (1970). *The structure of scientific revolutions* (2nd ed.). Chicago: The University of Chicago Press.

Lakatos, I. (1976). *Proofs and refutations.* Cambridge, U.K.: Cambridge University Press.

Lampert, M. (1985). How do teachers manage to teach? *Harvard Educational Review, 55*(2), 178–194.

Leinhart, G. (1988). Expertise in instructional lessons: An example from fractions. In D. A. Grouws, T. J. Cooney, & D. Jones (Eds.), *Perspectives on Research on Effective Mathematics Teaching* (pp. 47–66). Reston, VA: National Council of Teachers of Mathematics; Hillsdale, NJ: Lawrence Erlbaum.

Lerman, S. (1986). *Alternative views of the nature of mathematics and their possible influence on the teaching of mathematics.* Unpublished doctoral dissertation, Center for Educational Studies, King's College, University of London.

Lerman, S. (1990). Alternative perspectives of the nature of mathematics and their influence on the teaching of mathematics. *British Educational Research Journal, 16*(1), 15–61.

Levitas, M. (1974). *Marxist perspectives in the sociology of education.* London: Routledge & Kegan Paul.

Lortie, D. C. (1975). *Schoolteacher: A sociological study.* Chicago: The University of Chicago Press.

Malle, G. (1988). *The question of meaning in teacher education.* Paper presented at the Sixth International Congress on Mathematical Education, Budapest, Hungary.

McGrath, J. E. (1988). *Looking for a common denominator: A study of underachievement in the primary mathematics classroom.* Unpublished master's thesis, Essex Institute of Higher Education, Brentwood, U.K.

McLeod, D. B., Carpenter, T. P., McCornack, R. L., & Skvarcius, R. (1978). Cognitive style and mathematics learning: The interaction of field independence and instructional treatment in numeration systems. *Journal of Research in Mathematical Education, 9,* 163–174.

Mellin-Olsen, S. (1987). *The politics of mathematics education.* Dordrecht, Holland: D. Reidel.

Nash, R. (1974, November). Pupils' expectations of their teachers. *Research in Education, 47*–61.

National Committee of Excellence in Education. (1983). *A nation at risk: The imperative for educational reform.* Washington, DC: National Institute of Education.

National Committee of Inquiry into the Teaching of Mathematics in Schools. (1982). *Mathematics counts* (The Cockcroft report). London: Her Majesty's Stationary Office.

National Curriculum Council. (1989). *Mathematics in the National Curriculum.* London: Her Majesty's Stationary Office.

Nickson, M. (1981). *Social foundations of the mathematics curriculum: A rationale for change.* Unpublished doctoral dissertation, Institute of Education, University of London.

Nickson, M. (1988a). Pervasive themes and some departure points for research into effective mathematics teaching. In D. A. Grouws, T. J. Cooney, & D. Jones (Eds.), *Perspectives on research on effective mathematics teaching* (Vol. 1, pp. 245–252). Reston, VA: National Council of Teachers of Mathematics.

Nickson, M. (1988b). What is multicultural mathematics? In P. Ernest (Ed.), *Mathematics teaching: The state of the art* (pp. 236–241). Lewes, Sussex: Falmer Press.

Noss, R. (1988). The computer as a cultural influence in mathematical learning. *Educational Studies in Mathematics, 19,* 251–268.

Orton, R. E. (1988). Two theories of 'theory' in mathematics education: Using Kuhn and Lakatos to examine four foundational and issues. *For the Learning of Mathematics, 8*(1), 36–43.

Peterson, P. L., Swing, S. R., Stark, K. D., & Waas, G. A. (1984). Students' cognition and time on task during mathematics instruction. *American Educational Research Journal, 21*(3), 487–515.

Pimm, D. (1982). Why the history and philosophy of mathematics should not be rated X. *For the Learning of Mathematics, 3*(1), 12–15.

Pimm, D. (1990, April). *Mathematical versus political awareness: Some political dangers inherent in the teaching of mathematics.* Paper presented at a conference on the Political Dimensions of Mathematics Education, Institute of Education, University of London.

Pirie, S. E. B. (1988). Understanding: Instrumental, relational, intuition, constructed, formalized…? How can we know? *For the Learning of Mathematics, 8*(3), 2–6.

Plunkett, S. (1981). Fundamental questions for teachers. *For the Learning of Mathematics, 2*(2), 46–48.

Popkewitz, T. S. (1988). Institutional issues in the study of school mathematics: Curriculum research. In A. J. Bishop (Ed.), *Mathematics education and culture* (pp. 221–249). Dordrecht, Holland: Kluwer Academic Publishers.

Popper, K. (1972). *Objective knowledge—an evolutionary approach.* Oxford: Oxford University Press.

Porter, A., Floden, R., Freeman, D., Schmidt, W., & Schwille, J. (1988). Content determinants in elementary school mathematics. In D. Grouws, T. L. Cooney, & D. Jones (Eds.), *Perspectives on research on effective mathematics teaching* (Vol. 1, pp. 96–114). Reston, VA: National Council of Teachers of Mathematics.

Reiss, V. (1978, September). Socialization phenomena in the mathematics classroom: Their significance for interdisciplinary teaching. In *Cooperation between science teachers and mathematics teachers* (pp. 392–417). IDM Bielefeld Materialen und Studien Ban 16, University of Bielefeld.

Rutter, M., Maughan, B., Mortimore, P., & Ouston, J. (1979). *Fifteen thousand hours: Secondary schools and their effects on children.* London: Open Books.

Scheffler, I. (1976). *The language of education.* Springfield, IL: Charles C. Thomas.

Shipman, M. D. (1974). *Inside a curriculum project.* London: Methuen.

Shulman, L. S. (1970). Psychology and mathematics education. In E. G. Begle (Ed.), *Mathematics Education* (pp. 23–71), 69th Yearbook of the National Society for the Study of Education. Chicago: University of Chicago Press.

Smith, B. O., Stanley, W. O., & Shores, J. H. (1971). Cultural roots of the curriculum. In R. Hooper (Ed.), *The curriculum: Context, design and development* (pp. 16–19). Edinburgh: Oliver & Boyd.

Spradley, J. (1980). *Participant observation.* New York: Holt, Rinehart & Winston.

Steffe, L. P., & Killion, K. (1986, July). Mathematics teaching: A specification in a constructionist frame of reference. In L. Burton and C. Hoyles (Eds.), *Proceedings of the Tenth International Conference, Psychology of Mathematics Education* (pp. 207–216). London: University of London Institute of Education.

Thom, R. (1972). Modern mathematics: Does it really exist? In A. G. Howson (Ed.), *Developments in mathematical education* (pp. 194–209). Cambridge, U.K.: Cambridge University Press.

Thompson, A. G. (1984). The relationship of teachers' conceptions of mathematics and mathematics teaching to instructional practice. *Educational Studies in Mathematics, 15*(2), 105–127.

Toulmin, S. (1972). *Human understanding.* Oxford: Clarendon Press.

Trown, E. A., & Leith, G. O. M. (1975). Decision rules for teaching strategies in primary schools: Personality—treatment interactions. *British Journal of Educational Psychology, 45,* 130–140.

von Glaserfeld, E. (1981). An attentional model for the construction of units and number. *Journal of Research in Mathematical Education 12*(2), 83–94.

Ward, M. (1979). *Mathematics and the 10-year-old.* Schools Council Working Paper No. 61. London: Evans-Methuen Educational.

Wittman, E. C. (1989). The mathematical training of teachers from the point of view of education. *Journal für Mathematik—Didaktik, 10*(4), 291–308.

Wolfson, P. (1981). Philosophy enters the classroom. *For the Learning of Mathematics, 2*(1), 22–26.

Yates, J. (1978). *Four mathematical classrooms: An inquiry into teaching method.* Southampton, U.K.: Faculty of Mathematical Studies, University of Southampton.

Zaslavsky, C. (1989). Integrating mathematics with the study of cultural traditions. In C. Keitel, P. Damerow, A. J. Bishop, & P. Gerdes (Eds.), *Mathematics, education and society* (pp. 14–15). Paris: United Nations Educational, Scientific and Cultural Organization (UNESCO).

▪6▪

MATHEMATICS TEACHING PRACTICES
AND THEIR EFFECTS

Mary Schatz Koehler
SAN DIEGO STATE UNIVERSITY

Douglas A. Grouws
UNIVERSITY OF MISSOURI-COLUMBIA

The main purpose of this chapter is to examine the current status of research on mathematics teaching practices. First, a brief review of research on teaching in the content domain of mathematics is given. This is followed by an exploration of current research programs that focus on the combined issues of the teaching and learning of mathematics. Finally, important and often neglected dimensions of quality of instruction and meaningful learning are discussed.

LEVEL OF COMPLEXITY IN TEACHER RESEARCH

In order to understand current research fully, we need to know some history of research on mathematics teaching. Research on teaching has gone through several periods of reform. Rosenshine (1979) and Medley (1979) have reviewed research on teaching in terms of several phases or cycles. Rosenshine (1979) sees three cycles to this point. The first cycle focused on teacher personality and characteristics, the second cycle focused on teacher-student interaction, and the third cycle focused on student attention and the content students mastered. Medley (1979) presents a similar view, noting that research focused on identifying effective teachers first through their characteristics, then through the methods they used, next through

teacher behaviors and classroom climate, and finally through their command of a repertoire of competencies. Both of these reviews look at research on teaching from a chronological or historical point of view. This chapter, however, examines research on teaching from the perspective of complexity. Four levels of complexity and representative models are presented that reflect changes and progress in research on teaching.

Level I

Research at the first level represents work on teacher effectiveness that uses a fairly simplistic model. A particular component of teaching or a specific teacher characteristic is studied in isolation; little or no attention is given to other factors or to the quality of teaching. Early examples of this type of research are studies (mostly done before 1950) that examine teacher characteristics (for example, years of teaching experience, number of mathematics courses taken, and so forth) and personality traits (for example, enthusiasm). The model in Figure 6.1 illustrates this type of research. The dashed line between teacher characteristics and pupil outcomes indicates the assumption that teacher characteristics directly affect pupil outcomes, although pupil outcomes were usually not measured. Instead, effective teachers were identified on the basis of opinions of supervi-

The authors gratefully acknowledge the helpful comments and feedback provided by Barbara Dougherty, University of Hawaii, and Barbara Pence, San Jose State University. Time to prepare this chapter was partially provided by the National Science Foundation and the National Center for Research in Mathematics Education, Madison, WI. The ideas expressed are those of the authors. We also wish to thank Mary Beth Llorens for editorial help and LeeAnn Debo and Pat Shanks for typing the manuscript.

FIGURE 6–1. Basic Level 1 Research Model

sors, principals, and occasionally students. Based on this information, conclusions were drawn about whether a particular characteristic or trait was associated with successful teachers. This research emphasized *teachers* rather than *teaching*. These studies include those done by Charters and Waples (1929) and Barr and Emans (1930), which identified traits such as good judgment, considerateness, enthusiasm, personal magnetism, personal appearance, and loyalty as important characteristics of effective teachers.

Later examples of Level 1 research include studies that examine only one component in the whole teaching-learning process. Although student outcomes were considered, the narrow focus of these studies and their lack of attention to the quality of teaching was limiting. Sometimes these studies were experiments in which only one aspect of the teaching process was changed in some way. Two groups of studies provide examples of this type of Level 1 research. Studies such as Shipp and Deer (1960), Shuster and Pigge (1965), Zahn (1966), Sindelar, Gartland, and Wilson (1984), and Seifert and Beck (1984) all examined time allocation within a mathematics class period. These studies examined whether spending more or less time on development, practice, or seatwork would lead to greater student achievement. Smith (1977), Smith and Cotten (1980), and Hines, Cruickshank, and Kennedy (1985) examined the issue of teacher clarity, and they contrasted lessons that had varying degrees of vagueness and discontinuities to determine which lessons were best for students. The model in Figure 6.2 represents this later phase of Level 1 research. It shows the clear acknowledgement of the influence teacher behaviors have on pupil outcomes.

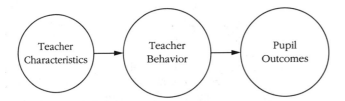

FIGURE 6–2. Elaborated Level 1 Research Model

Level 2

Studies at this level are more complex than studies at Level 1. They usually involve multiple classroom observations that provide extensive detail concerning instruction in mathematics. One form of Level 2 research is often referred to as process-

product research. Classroom processes (usually defined as what is going on in the classroom) are observed, and the frequency of particular teacher and student behaviors are noted. Figure 6.3 illustrates this type of research. The assumptions that teacher behavior influences student behavior and that pupil behavior influences teacher behavior are important ones in this model. A central feature of these studies is careful documentation of what teachers and students *do* during mathematics instruction. This usually involves the development of quite elaborate coding schemes for recording classroom events during observation. Often the influence of teachers and students on each other's behavior is observed through analysis of teacher-student interactions. Examples of behavior coded in some studies include types of questions asked, length of responses to questions, number and type of examples used, amount of instructional time devoted to practice activities, frequency of use of manipulative materials, amount of time allocated to developing new concepts, and amount of time spent on review activities. A theoretical basis for using particular observation categories is not often provided; instead, a logical argument is usually given. Student outcomes (usually achievement test scores) are then correlated with the frequency of the observed behaviors to determine what behaviors are associated with large gains in learning. Again, not much attention is paid to the quality of the teaching under observation. Brophy and Good (1986) provide a comprehensive review of process-product studies.

Some later studies that fit this level were more carefully designed and contributed significantly to the knowledge base of mathematics teaching. Some examples of these large studies are the Texas Teacher Effectiveness Study (Evertson, Anderson, Anderson, & Brophy, 1980; Evertson, Emmer, & Brophy, 1980) and the Missouri Mathematics Program (Good, Grouws, & Ebmeier, 1983). The Texas studies provided useful information about differences between mathematics teaching and the teaching of other disciplines, as well as specific information about successful classroom management techniques. The Missouri Mathematics Program followed a descriptive–correlational–experimental loop and resulted in a model for whole-class mathematics teaching, referred to as Active Mathematics Teaching, that resulted in large student achievement gains.

Level 3

The complexity of the view of teaching increases again as we move to Level 3. As can be seen in Figure 6.4, a primary distinction between this level and the previous level is the inclusion of the category of pupil characteristics and the broadening of the category of pupil outcomes to include attitudes as well as achievement. Pupil characteristics such as gender, race, and confidence level can affect teacher practice and pupil actions.

The Autonomous Learning Behavior study (Fennema & Peterson, 1986; Peterson & Fennema, 1985) carried out in fourth-grade classrooms is an example of a Level 3 study. Some process-product ideas were used in the sense that it was a large-scale observational study that correlated classroom processes with student achievement gains. However, it went beyond previous process-product research in that it considered both high

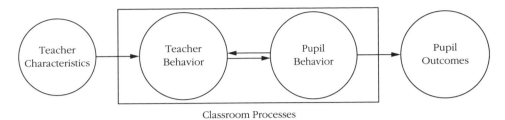

FIGURE 6–3. Level 2 Research Model

and low cognitive-level gains and considered differential gains for females and males.

Other studies at this level began to use a blending of methodologies along with consideration of various student characteristics. Hart (1989) used classroom observations as well as field notes, and she considered the confidence level and gender of the student in her research with seventh-grade mathematics classes. She found that there were gender differences in teacher-student interaction patterns and that these seemed to be linked to particular teacher patterns or styles. Koehler (1986, 1990) combined observations of teacher-student interactions with field notes to examine the differential effectiveness of certain classroom processes on the high and low cognitive-level achievement of females and males in algebra classes. She found evidence indicating that females performed better in classes where they were given less teacher help than in other classes, thus encouraging them either directly or indirectly to become more independent learners of mathematics. Stanic and Reyes (1986), in order to examine the question of differential achievement by race and gender, conducted an in-depth case study of teaching in the seventh grade that involved audio recordings, classroom observations, and student and teacher interviews. They concluded that a full understanding of teacher-student interactions must take into account the teachers' intentions. They noted that differential outcomes can result from differential teacher treatment of students, but also from equal treatment of students, since not all students respond in the same way to a particular teaching style.

Level 4

Teaching is indeed very complex. Not only is teaching more complex than was originally realized, this complexity is increasing. Teachers have a wider range of ability levels to reach, a broader range of mathematics topics to teach (National Council of Teachers of Mathematics [NCTM], 1989), and a greater assortment of teaching methodologies that they can choose from (for example, manipulatives, small groups). In addition, research techniques and data analysis techniques have become more advanced, allowing researchers to examine more complex questions. The most substantial change in the view of research on teaching is the awareness of the need to pair research on teaching with research on learning. Romberg and Carpenter (1986) argue quite forcefully in favor of integration of the two domains. Fennema, Carpenter, and Lamon (1988) and Grouws (1988) elaborate on the theme. Putnam, Lampert, and Peterson (1990) discuss how research on learning has progressed in recent years, and they present new views on learning from the perspectives of cognitive psychology, mathematics, and the classroom.

Currently, research questions in teaching and learning are being approached from several perspectives. Research at Level 4 might be defined as research that has a strong theoretical foundation and is based on a model that involves many factors. Level 4 research might be based on the model in Figure 6.5 or a similar model.

In this model, the outcomes of learning are based on the students' own actions or behaviors. In most cases, these actions

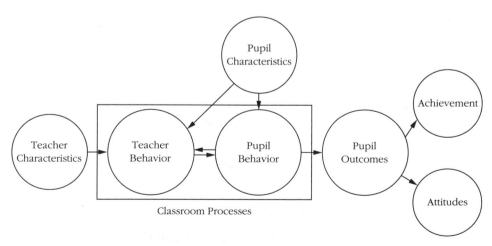

FIGURE 6–4. Level 3 Research Model

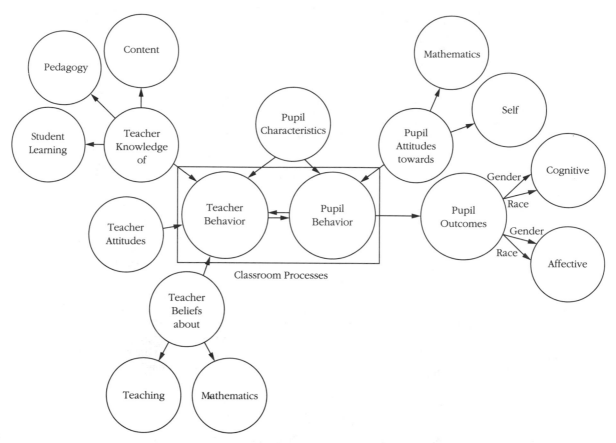

FIGURE 6–5. Level 4 Research Model

are influenced largely by what the teacher does or says within the classroom. Also influencing the students' actions are the students' attitudes or beliefs about themselves as learners of mathematics and their beliefs about mathematics as a discipline. Students' confidence in their ability to learn mathematics, their belief that mathematics will be useful to them, and their feelings about whether they can "discover" mathematics are some examples of the components that influence students' actions. Teacher behavior is influenced by the teacher's knowledge of (a) the mathematics content being taught, (b) how students might learn or understand that particular content, and (c) methods of teaching that particular content. Also influencing teacher behavior are the teachers' attitudes and beliefs about teaching and mathematics. For example, teachers who believe that students learn by explicit example and repetition or by extensive practice, and who see their role as dispensers of information, would behave differently in the classroom than teachers who believe students learn by discovery or investigation and who see their role as "co-explorers" with the student. These latter teachers might ask more open-ended questions, engage in more problem posing, and be less tied to the textbook. As in Levels 2 and 3, there is an important two-way connection between pupil behavior and teacher behavior, and pupil characteristics influence both the teacher's and the pupil's behavior.

The rest of this chapter describes several Level 4 studies— those that emphasize an examination of the teaching act per se with attention to student learning. Studies involving teacher knowledge, beliefs, and decision-making, although integral to the teaching act, will not be considered directly as they are covered elsewhere in this volume.

MULTIPLE RESEARCH PERSPECTIVES

There are many perspectives from which to examine the teaching act. This section will explore various perspectives that current research has taken.

Constructivist Approach

One way in which research on teaching has been linked with research on learning is through a constructivist perspective. Cobb et al. (1991) explain that "from the constructivist perspective, mathematical learning is not a process of internalizing carefully packaged knowledge but is instead a matter of reorganizing activity, where activity is interpreted broadly to include conceptual activity or thought" (p. 5). Cobb, Yackel, and Wood (1988) have used a constructivist point of view to create a framework in which to discuss both teaching and learning. This is not an intuitively obvious concept, however. Cobb (1988) points out some of the difficulties encountered. He notes that, "Although constructivist theory is attractive when the issue of learning is considered, deep-rooted problems arise when attempts are made to apply it to instruction" (p. 87). This is especially true if one subscribes to the "transmission of knowledge" view of teaching, rather than to the view that "students construct knowledge for themselves by restructuring their internal cognitive structures" (p. 87). The constructivist assumptions about how students learn changes the assumptions about what teacher actions or behaviors might be desirable. If one subscribes to a constructivist view of learning, then the goal is no longer one of developing pedagogical strategies to help students receive or acquire mathematical knowledge, but rather to structure, monitor, and adjust activities for students to engage in. A researcher in this situation needs to view observations of teaching quite differently. For example, rather than seeing teachers' actions as having direct influence on students' learning, the view would be more indirect. Cobb notes that "teachers' actions do influence the problems that students attempt to solve and thus the knowledge they construct" (p. 92).

Cobb and Steffe (1983) explain a constructivist view of teaching by noting that "in the constructivist view, teachers should continually make a conscious attempt to 'see' both their own and the children's actions from the children's point of view" (p. 85). A constructivist teacher would initiate activities, and the child would reflect on and abstract patterns or regularities from these.

In a further elaboration, Cobb (1988) notes that teaching can be perceived as a continuum with negotiation and imposition as extremes. One who views teaching as the transmission of knowledge would follow a teaching-by-imposition model, and those who view teaching as the facilitation of the construction of knowledge would follow a teaching-by-negotiation model. Cobb et al. (1991) explain that

The teacher's role in initiating and guiding mathematical negotiations is a highly complex activity that includes highlighting conflicts between alternative interpretations or solutions, helping students develop productive small-group collaborative relationships, facilitating mathematical dialogue between students, implicitly legitimizing selected aspects of contributions to discussion in light of their potential fruitfulness for further mathematical constructions, redescribing students' explanations in more sophisticated terms that are nonetheless comprehensible to students, and guiding the development of taken-to-be-shared interpretations when particular representational systems are established. (p. 7)

Yackel, Cobb, Wood, Wheatley, and Merkel (1990) point out that, in addition to giving students problems to resolve, much learning or construction of knowledge takes place through social interactions, with the teacher and peers as part of problem solving. When children are given the opportunity to interact with each other and the teacher, they can "verbalize their thinking, explain or justify their solutions, and ask for clarifications. Attempts to resolve conflicts lead to opportunities for children to reconceptualize a problem and to extend their conceptual framework to incorporate the alternative solution methods" (p. 19).

One of the research techniques used to examine teaching from a constructivist perspective is to study a *teaching episode*—usually with an individual student—in detail. In one such detailed description, Cobb (1988) notes that since learning is an active problem-solving process "one of the teachers' primary responsibilities should be to engage students in activities that give rise to genuine mathematical problems for them" (p. 95). Often these teaching episodes are videotaped and then discussed with the teacher. The researchers might make suggestions about initial activities to use that could foster the child's conception of the mathematics being taught.

Another research technique is the *teaching experiment*. The context for one teaching experiment was children's acquisition of early number concepts. A second-grade class was used in a year-long experiment (Cobb, Yackel, & Wood, 1988). Researchers worked closely with an experienced but traditional teacher to develop instructional activities that allowed students to construct their knowledge of important mathematical ideas and concepts. Some activities had been developed in the year preceding the experimental year, but many more were constructed and modified during the experimental year partly as a response to differences in the new students' mathematical conceptions. Data included videotapes of every lesson as well as interviews with the children and audiotaped interviews with the teacher. The experiment was considered successful by both the researchers and the district personnel so, as a consequence, an increased pool of teachers was able to implement the instructional procedures in a third year of the project (Cobb et al., 1991).

In addition to considering different instructional procedures based on a particular perspective of how students learn, this line of research also considered noncognitive outcomes for the students. Nicholls, Cobb, Wood, Yackel, and Patashnick (1990) examined students' beliefs about the causes of success in mathematics, using the sample of second-grade students described above, who were in classes that engaged in collaborative problem solving and social interactions. The topic of this chapter is not affective issues, so details will not be included here, but we feel it is important to point out that, by examining noncognitive outcomes, this particular chain of inquiry incorporates more fully some of the elements of the model presented in Figure 6.5.

Cognitively Guided Instruction

The underlying philosophy of Cognitively Guided Instruction (CGI) is that teachers need to make instructional decisions based on knowledge from cognitive science about how students learn particular content. With the view that understanding involves linking new knowledge to existing knowledge, Fennema, Carpenter, and Peterson (1989) explain that teachers need to be aware of the knowledge their students have at various stages so they can provide appropriate instruction. This philosophy has been the framework for a research program in which the "major focus has been to study the effects of programs designed to teach teachers about learners' thinking and how to use that information to design and implement instruction" (Carpenter & Fennema, 1988, p. 11). The model illustrat-

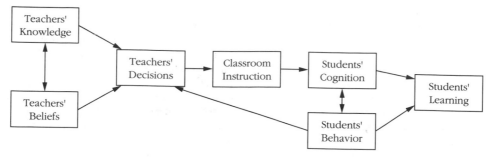

FIGURE 6–6. Cognitively Guided Instruction (CGI) Research Model (Fennema, Carpenter, & Peterson, 1989, p. 204)

ing CGI is given in Figure 6.6. The model shows that classroom instruction is based on "teachers' decisions [which] are presumed to be based on their own knowledge and beliefs as well as their assessment of students' knowledge through their observation of students' behaviors" (Carpenter & Fennema, 1988, p. 9).

Fennema, Carpenter, and Peterson (1989) explain further that "the major tenets of CGI are: (1) Instruction must be based on what each learner knows, (2) Instruction should take into consideration how children's mathematical ideas develop naturally, and (3) Children must be mentally active as they learn mathematics" (p. 203). Carpenter, Fennema, Peterson, Chiang, and Loef (1989) conducted a study to test some of these hypotheses regarding CGI. Their study involved 40 first-grade teachers, half of whom were randomly assigned to a treatment group. These treatment teachers attended a one–month summer workshop on research regarding the learning of addition and subtraction concepts. All 40 teachers and their students were observed for a minimum of 16 days from November through April. Various assessments were made, including a pretest and post-test of student achievement with various measures of problem-solving ability, measures of student confidence and beliefs, and student interviews. Results indicate that students in the experimental classes performed more favorably on measures of problem solving and also on recall of number facts.

The group of experimental teachers spent more time on word problems than control teachers, and control teachers spent more time on number facts problems. Experimental teachers focused more often on the process students used to solve problems, and control teachers focused more frequently on the answer to the problem. In addition, experimental teachers allowed students a wider variety of strategies to solve problems. The following excerpt illustrates a difference in teaching behavior exhibited by the two groups of teachers:

A typical activity that was observed in CGI classes was for a teacher to pose a problem to a group of students. After providing some time for students to solve the problem, the teacher would ask one student to describe how he or she solved the problem. The emphasis was on the process for solving the problem, rather than on the answer. After this student explained his or her problem-solving process, the teacher would ask whether anyone had solved the problem in a different way and give another student a chance to explain her or his solution. The teacher would continue calling on students until no student would report a

way of solving the problem that had not already been described.... In contrast to CGI teachers, control teachers less often (a) posed problems, (b) listened to students' strategies, and (c) encouraged the use of multiple strategies to solve problems. They spent more time reviewing material covered previously, such as drilling on number facts, and more time giving feedback to students' answers. (Carpenter, Fennema, Peterson, Chiang, & Loef, 1989, p. 528)

Several vignettes describing teaching practices of CGI teachers in more detail are given in Fennema, Carpenter, and Peterson (1989).

Expert-Novice Paradigm

Another approach in examining teaching behaviors is the expert-novice paradigm. Leinhardt (1989) explains that the focus of this approach is to consider

teaching as one of the more interesting and complex cognitive processes in which human beings engage and to analyze the tasks, resources, and constraints in which teachers are enmeshed. With this perspective, one stands figuratively behind the shoulder of the teacher and watches as the teacher juggles the multiple goals of script completion, tactical information processing, decision-making, problem solving and planning. (p. 52)

In this type of research the underlying philosophy is actually similar to the process-product paradigm in that two categories of teachers are observed (in this case experts are contrasted with novices; in the process-product research, effective teachers are contrasted with less effective ones) with the intention of identifying the qualities or behaviors necessary for successful teaching. "The ultimate goal is to understand expertise well enough to develop instruction for novice teachers that will make their early performance more expert-like, and, eventually, to move the performance of all teachers into the expert range" (Leinhardt, 1989, p. 53). The expert-novice paradigm differs from the process-product paradigm in terms of what is observed and how success or effectiveness is measured.

In one study that exemplifies the expert-novice tradition, four expert teachers and two novice teachers were observed, interviewed, and videotaped. Their lessons were analyzed in great detail, and differences were found between the experts and the novices. The ability to construct and teach lessons, the "crafting" of lessons, is one dimension on which experts and novices differed. Leinhardt (1989) explains that a lesson has three components: an agenda, a lesson segment, and explanation. "An agenda is a unique operational plan that a teacher uses to teach a mathematics lesson. It includes both the objectives or goals for lesson segments and the actions that can be used to achieve them" (p. 55). Lesson segments, also called activity structures, are "segments of social events in which various actors assume specific roles to accomplish specific tasks" (p. 56). Other researchers (Good, Grouws, & Ebmeier, 1983) have also identified lesson segments and have investigated their sequencing and effects. The third lesson component investigated by Leinhardt is the explanation or "system of goals and actions involved in the actual transmission of subject matter content" (p. 56).

In the analysis of the data it was found that experts had richer agendas—their plans contained more detailed information, explicitly referenced student actions, and planned instructional actions. In terms of lesson segments, it was found that experts spent less time in transitions, and more consistently distributed their time among other lesson components. Novices were much more variable in the amount of time spent in each lesson segment—giving the overall impression of instability. The explanation portion of the lesson showed that experts gave better explanations of new material in that they contained more critical features and fewer errors. Often novices did not complete their explanations.

In summary, it was found that

Expert mathematics teachers weave a series of lessons together to form an instructional topic in ways that consistently build upon and advance material introduced in prior lessons. Experts also construct lessons that display a highly efficient internal structure, one that is characterized by fluid movement from one type of activity to another, by minimal student confusion during instruction, and by a transparent system of goals.... Novice teachers' lessons, on the other hand, are characterized by fragmented lesson structures with long transitions between lesson segments, by frequent confusion caused by missent signals, and by an ambiguous system of goals that often appear to be abandoned rather than achieved. (p. 73)

The preceding line of research also looked at how students learn, but the perspective was quite different from other research on learning. Leinhardt and Putnam (1987) examined lessons from a student's point of view and hypothesized how students actually learn from a lesson. They note that, in order for a student to learn from a mathematics lesson, the student needs certain cognitive capabilities. These include (a) *an action system,* which enables a student to act appropriately in school; (b) *a lesson parser,* which enables the student to recognize and anticipate lesson components; (c) *an information gatherer,* which absorbs information from a lesson and incorporates it into existing knowledge; (d) *a knowledge generator,* which seeks new knowledge and acts as a motivator; and (e) *an evaluator,* which assesses the meaningfulness of the new material. This research on learning was done in parallel with the research on teaching, and the two are interconnected in the sense that both address aspects of existing lessons.

Livingston and Borko (1990) have also considered the question of differences in expert-novice teaching, and they present two conceptual frameworks for examining teaching. Their first framework views teaching as a *complex cognitive skill.* Central to this framework are the ideas of pedagogical reasoning, pedagogical content knowledge, and schemata. They explain that

A schema is an abstract knowledge structure that summarizes information about many particular cases and the relationships among them. ...Pedagogical reasoning is the process by which teachers transform subject matter knowledge into forms comprehensible to their particular students.... Pedagogical content knowledge is the domain specific knowledge of teaching that integrates content and pedagogy. (p. 373–374)

They have found that expert teachers have more elaborate and more interconnected schemata than novice teachers. In addition, pedagogical reasoning is at an early stage of development in novice teachers.

The second conceptual framework deals with teaching as an *improvisational performance.* The teacher has an overall plan or outline but does not follow an exact script. Rather, the teacher relies on a repertoire of routines and instructional moves in response to the needs or actions of the students. In comparing student teachers with their more experienced cooperating teachers, the researchers (Borko & Livingston, 1989; Livingston & Borko, 1990) found that using the two conceptual frameworks enabled them to examine the developmental differences between experts and novices. Novice teachers had limited pedagogical content knowledge about student learning. They had little knowledge of student misconceptions. Their schemata was adequate for their own understanding, but was insufficiently developed, interconnected, and accessible to enable them to be responsive, flexible teachers. These limitations resulted in the novices being less skilled at improvisation. Borko and Livingston (1989) point out further that experts became in some ways like novices when teaching new content, which underscores the important influence of knowledge of the content being taught. It is believed that one cannot acquire pedagogical reasoning or pedagogical content knowledge without actually teaching the specific content. Borko and Livingston recommend the necessity of longitudinal studies of the development of expertise in teaching.

Sociological and Epistemological View

Lampert (1988b, 1989, 1990) has combined research on teaching with research on learning through a sociological and epistemological view. By considering how new knowledge is established in the discipline of mathematics, and by applying this process to the teaching-learning process of the classroom, she brings a fresh perspective to the field.

Lampert (1990) believes that new mathematics is brought about through "a process of 'conscious guessing' about relationships among quantities and shapes, with proof following a 'zig-zag' path starting from conjectures and moving to the examination of premises through the use of counterexamples or 'refutations' " (p. 30). She points out further that currently how one comes to know mathematics within the discipline of mathematics is quite different than how one comes to know mathematics in school, where "*doing* mathematics means following the rules laid down by the teacher, *knowing* mathematics means remembering and applying the correct rule when the teacher asks a question; and mathematical *truth is determined* when the answer is ratified by the teacher" (p. 32). Ideally, "new knowledge is constructed as a joint venture in the class rather than as a communication from teacher to student" (1989, p. 257).

The research methodology Lampert has employed is unique. She has engaged herself in the role of "teacher-scholar," and as such she has been teaching mathematics in fourth– and fifth–grade classes while collecting data on her own teaching. Some of the data has consisted of audiotapes and videotapes of the lessons, classroom observations, samples of students' work in the form of notebooks and homework, and field notes on the planning and implementation of lessons. Lampert (1990) points out that her methods are a blending of action research and interpretive social science. The method of analysis of the data is also unique in the sense that it is interpreted by educational psychologists, sociolinguists, and mathematicians. Lampert refers to this method of analysis as "textual exegesis" (1988b, p. 134) in the sense that teaching is the managing of multiple and conflicting goals and that any teaching action can be interpreted in multiple ways. She notes that the purpose of the analysis is "not to determine whether general propositions about learning or teaching are true or false but to further our understanding of the character of these particular kinds of human activity" (1990, p. 37).

Lampert believes that the ultimate goal of teaching is to "encourage conjecturing and arguing and to make the environment safe for students to express their thinking" (1988a, p. 16). She sees the goal of her research as using disciplinary knowledge to "describe the classroom interaction between teachers and students in school mathematics lessons" (1988b, p. 136). Her premise regarding teaching is that it "will involve getting students to reveal and examine the assumptions they are making about mathematical structures, and it will involve presenting new material in a way that enables them to consider the reasonability of their own and teacher's assertions" (p. 136). This belief about the nature of teaching then leads to two tasks for teachers: (a) choosing and posing problems, including raising questions and asking for clarification in order to engage students in mathematical discourse, and (b) finding language and symbols that students and teachers can use to enable them to talk about the same mathematics content.

Lampert's research has given her insight into characteristics of problems that would be good to pose for lesson construction. The first characteristic is that the domain of the problem is familiar to the student. This will allow all of the students in the class to state and test mathematical guesses. A second characteristic is that the mathematical problems have the "potential to lead students into unfamiliar and important mathematical territory, and in particular, to lead them into territory that relates to the curricular agenda" (p. 137).

Lampert believes that students construct their own knowledge and that teachers are responsible for educating students in the tools of their culture. She attempted, through discourse based on problem posing, to make "knowing mathematics in the classroom more like knowing mathematics in the discipline" (1990, p. 59). Lampert points out that although she was successful in engaging students in an "authentic mathematical activity... the problem of defining what knowledge students have acquired remains" (1990, p. 59).

Mathematics Content View

This perspective of research on teaching and learning has as its primary goal generating "information on the relationships between acquiring key cognitive processes in a domain and the instructional events that support or hinder such acquisitions" (Hiebert & Wearne, 1988b, p. 180). This perspective studies how a particular topic or content strand within the domain of mathematics is learned, and it applies that knowledge directly to the teaching of that specific content. For this reason it is also referred to as the Cognitive Instructional Approach and is described as "studying learning to inform teaching" (p. 179).

Hiebert and Wearne (1988a) describe more fully the methodology used in research on teaching and learning from this perspective. First, a well-defined content domain is selected. In their early research, they used the content domain of addition and subtraction of decimal fractions. Next the key cognitive processes for successful performance in that domain are identified. In terms of competence with the decimal fraction system, Wearne and Hiebert (1988) have identified four key cognitive processes: "(a) *connecting* individual symbols with referents, (b) *developing* symbol-manipulation procedures, (c) *elaborating and routinizing* the rules for symbols, and (d) using the symbols and rules as referents for a more *abstract* symbol system" (p. 372). The third component of the methodology is finding existing instruction or designing special instruction that promotes the use of the key cognitive processes. In the Hiebert and Wearne research, new instructional strategies were developed. The last component of the methodology is to evaluate the instruction, both in terms of direct assessment of performance in the content domain and in terms of transfer tasks.

Wearne and Hiebert (1988) report an example of research using this methodology. A sample of fourth, fifth, and sixth graders were given a series of nine lessons incorporating the first two cognitive processes identified in the acquisition process of learning how to compute with decimals. It was found that students who were able to use the "semantic process"—the process involving the key cognitive processes—outperformed students who could not. They also found that previous instruction (as in the case of some fifth and most sixth graders) interfered with the acquisition of these key cognitive concepts. This led them to conclude that it is better to develop meanings for symbols before practicing syntactic routines.

In this research view, the teacher's behaviors are somewhat prescribed. The goal of the work is to use research on how learners acquire expertise with very specific mathematical concepts and then to carefully structure instruction that leads students through the key cognitive processes.

In a continuation of their line of research, Hiebert, Wearne, and Taber (1991) focused on how students construct understandings. A class of low-achieving fourth graders were given instruction on decimal fractions that was specifically deigned to promote understanding. Student understanding was assessed several times during the 11-day unit, using written tests and interviews. It was found that understanding comes gradually, with small erratic-looking changes, rather than with a sudden flash of insight. It is not the case that complete understanding can be characterized as the piecing together of partial understandings into an ever larger network. Their analysis has shown that "the appropriate model for the development of understanding may be one of change and flux and reorganization rather than steady monotonic growth.... Disconnecting, connecting and reorganizing appear to be the rule rather than gradual addition to a stable structure" (p. 339).

Mack (1990) has also examined teaching and learning from the perspective of mathematical content. Like the Hiebert group, she has examined the development of students' understanding about fractions. In particular, she has focused on how students' informal knowledge of fractions influences their understanding of instruction on addition and subtraction of fractions in an attempt to see how students build on this knowledge to give meaning to fraction symbols and procedures. Mack's research involved eight sixth-grade students of average ability who were given individualized instruction that incorporated clinical interviews. Since the instruction was tailored to each student's own needs based on their informal knowledge, misconceptions, and responses to questions, this line of research is closely related to the CGI principles.

Mack found that students had a rich store of informal knowledge about fractions that they could build on to develop understanding of formal mathematical symbols and procedures. Mack points out that it is not yet clear whether students can then expand on this core of knowledge to develop a broad conception of rational numbers. She notes finally that "the results add more evidence to the argument in favor of teaching concepts prior to procedures and suggest that students can construct meaningful algorithms by building upon informal knowledge" (p. 30).

SUMMARY AND COMPARISON

In the preceding pages, several different research programs or paradigms were presented that examine research on teaching and learning. The following is our attempt to summarize and contrast these paradigms.

In the *constructivist approach,* teaching behavior is examined from the viewpoint of how much it encourages or facilitates learner construction of knowledge. Teaching is viewed on a continuum between negotiation and imposition, and the teacher's role is to find and adjust activities for students. Social interactions are seen as a critical part of knowledge construction. Research methodologies include studying a teaching episode in detail and conducting a teaching experiment.

In *cognitively guided instruction,* the primary notion is that teachers make decisions about their teaching based on their knowledge and beliefs about how children learn. Since it is believed that students learn by linking new knowledge to existing knowledge, the role of the teacher is to provide instruction appropriate for each student. Listening to students is critical. Research involves informing teachers about theories of how

children learn and research on how children's mathematical ideas develop on particular topics, and then monitoring how this new knowledge might change the teachers' behavior in the classroom.

The goal behind the research on the *expert-novice paradigm* is more than just identifying the behaviors of expert teachers in their "crafting" of lessons. It is to identify the developmental process that teachers go through as they move from novice to expert. Teaching is viewed as both a complex cognitive skill and an improvisational performance. The research methodology involves extensive observations and in-depth analysis of the implementation of lessons.

In the *sociological* or *epistemological view* of mathematics teaching and learning, it is believed that students will only come to know mathematics if they learn mathematics in the way it is developed in the discipline. In that regard, teaching is viewed as helping students to construct knowledge in the discipline through problem posing and engaging students in mathematical discourse so that they might examine their own assumptions about mathematics. The research technique used thus far is a unique one and involves the researcher becoming a "teacher-scholar."

Finally, in the *mathematics content view,* teaching is regarded as an agent of cognitive change for the learner. The goal is to design instructional sequences and develop instructional techniques that would readily facilitate this cognitive growth and change. The research methodology has involved examining the key cognitive processes used in acquiring a particular concept and then designing instructional methods to help students develop these processes.

The perspectives of recent research on teaching and learning in mathematics are varied, and comparing them reveals important differences among the models. First, at the risk of oversimplification, let us mention some basic points of agreement. All these perspectives accept the premise that students are not passive absorbers of information, but rather have an active part in the acquisition of knowledge and strategies. All the perspectives basically view the teacher as an informed and reflective decision maker. Beyond these similarities, there seem to be important differences among the perspectives. For example, Lampert believes that students construct knowledge in much the same way that knowledge is constructed within the discipline of mathematics. Leinhardt hypothesizes that students construct knowledge through their interpretation of the lesson. Hiebert and Wearne believe that knowledge is constructed through acquisition of key cognitive processes. Mack believes that students' informal knowledge is the basis for understanding. Cobb et al. (1991) and the CGI group believe students construct knowledge idiosyncratically. Putnam, Lampert, and Peterson (1990) elaborate on different views of what it means to know and understand mathematics, some of which are similar to the views taken by the preceding research perspectives. They discuss understanding in terms of (a) representation, (b) knowledge structures, (c) connections among types of knowledge, (d) active construction of knowledge, and (e) situated cognition. These different views of the acquisition or construction of knowledge lead to different views of teaching. Hiebert and Wearne seem to provide more detail and structure in their attempt to identify a sequence of lessons that teachers can use to facilitate student learning in specific content domains. CGI seems most open and idealistic in suggesting that teachers know for each student precisely their stage of cognitive development with respect to a given content area and that they modify instruction continually to meet individual needs. Cobb et al. (1991) believe that teachers can work from lessons that have been developed with cognitive strategies in mind and can modify those lessons as necessary. Lampert believes that teachers need to focus on selecting and posing appropriate problems, and Leinhardt (1989) focuses attention less on specific content and more on structure of lessons. Borko and Livingston (1989) explain that teachers improvise as they teach, using a rich repertoire of instructional moves. There is obviously considerable room for variation in the interpretation of effective teaching—even with agreement on the basic premises that research on teaching should be combined with research on learning, that students construct their knowledge, and that teachers are thoughtful decision makers.

CRUCIAL DIMENSIONS: QUALITY OF INSTRUCTION, MEANING, AND DEVELOPMENT

Although there are multiple perspectives from which research on teaching can be approached (see, for example, Grouws, Cooney, & Jones, 1988) and multiple interpretations of the teaching act, one underlying theme that needs to be more adequately addressed in all research on mathematics teaching, regardless of the philosophical perspective brought to the work, is the notion of quality of instruction. Although there is general agreement that quality of mathematics instruction is important, it is generally not directly addressed in most research studies. It seems to be a variable that researchers have been reluctant to tackle head-on.

Quality of instruction can be considered analytically or holistically. For instance, an analytic analysis could be based on teaching actions and a holistic analysis might focus on teaching episodes or a global examination of mathematics lessons. Grouws (1988) points out that many teaching actions—such as asking a question, giving an example, and drawing a diagram—have a quality dimension associated with them. He notes that "it is relatively easy to identify actions that would fall at the extreme ends of a quality continuum: making diagrams that cannot be read, asking ambiguous questions, using examples that do not fit the conditions of a definition, and so on" (p. 231). Consideration of instruction quality should not be limited, however, to evaluating specific teaching behaviors. It should include examining "the way classroom events fit together to form a meaningful learning situation" (p. 232). Future research based on any of the five research perspectives previously discussed (even though their underlying philosophies are varied) would benefit from greater consideration of instruction quality.

There are many ways to move forward in addressing the issues of instruction quality and meaningful learning. One approach is to study the part of a mathematics lesson referred to in some research as "development." Research on development as a specific part of a lesson began with some of the early time-allocation studies previously discussed (for example, Shipp & Deer, 1960; Shuster & Pigge, 1965). These studies experi-

mentally contrasted the amount of time allocated to development with the amount of time devoted to other parts of a lesson, such as seatwork or practice. Research on the topic of development has advanced considerably since these early studies were done, as has research on teaching in general.

In early studies on development the value of increasing development time was clearly shown, but no consideration was given to the quality of development. Often in these studies "development" was taken to be synonymous with "teacher lecture." Development as it has been more recently defined, however, actually encompasses much more than teacher lecture. A more current definition of development is given by Good, Grouws, and Ebmeier (1983) as "the process whereby a teacher facilitates the meaningful acquisition of an idea by a learner" (p. 206). This process includes *whatever* the teacher may do to facilitate learning, whether it be structuring small-group work, providing guided-discovery activities, leading a class discussion, organizing individual investigations, or presenting a lecture.

It is important to note that the emphasis in recent definitions of development is on *meaningful* acquisition of ideas. In his far-reaching research, Brownell (1935) elaborated on the importance of meaning and understanding in the teaching and learning of mathematics. Some of his observations are as pertinent today as they were over 50 years ago. As Pirie (1988) points out, "If we, as mathematics educationalists, are to devise effective teaching strategies, make sense of pupils' actions, provide experiences which enable children to construct their own mathematical concepts, we must first have a viable model of understanding on which to build" (p. 2). While advocating continued attention to advancing our comprehension of understanding and developing models of it, he also suggests that in actuality we will probably never fully comprehend "understanding" itself. More recent scholarly investigations of the notion of meaning and understanding have yielded important concepts, such as relational and instrumental understanding (Skemp, 1978), symbolic understanding (Backhouse, 1978), and a four-level hierarchy of understanding (Herscovics & Bergeron, 1983). More details and reconceptualization of ideas and investigations of understanding from the student perspective are presented in Chapter 4 of this volume by Hiebert and Carpenter.

The importance of meaning and understanding has also been recognized in the work of Good, Grouws, and Ebmeier (1983) who identified five components of development as part of their study of successful mathematics teachers: attending to prerequisites, attending to relationships, attending to representation, attending to perceptions, and generality of concepts. They emphasize that "development is a very complex phenomenon and the particular elements that constitute successful development may vary from one teaching context to the next," and suggest that "it is also important to realize that effective development may not be composed of the same combination of behaviors, even in similar classroom environments" (p. 207).

The components of development that they identified were not mutually exclusive and did not necessarily span all the dimensions of development that could be identified. They may, however, provide a useful starting point for a study of the quality of mathematics instruction that takes account of the student but also provides a focus on the teacher's role in the process. This latter emphasis goes beyond many programs of research that often end at the point of conceptualizing some aspect of student understanding of mathematics.

CONCLUSIONS AND FUTURE DIRECTIONS

Brophy (1986) points out that "although progress in the last twenty years has been remarkable, research on classroom teaching, including research on school mathematics instruction, is still in its infancy" (p. 328). It seems safe to say that research has progressed considerably since Brophy made his observation. The research has become more sophisticated in the sense that every aspect of the teaching act is looked at in great detail, and more attention is being paid to the content being taught. Some areas need more emphasis, however. For example, in the area of student outcomes more attention should be given to better assessment of the cognitive level of the learning that takes place, to noncognitive outcomes, and also to possible differential outcomes for particular groups of students.

In addition, the majority of the research has been done, for a variety of reasons, on the elementary and middle-level grades. It is imperative that we look at the secondary level to see whether the theories of learning and teaching discussed thus far would be viable in a more complex mathematical setting, with older students, and with teachers who have (usually) a more thorough mathematics background. Brophy (1989) indicates that research is also needed on the postsecondary level.

Researchers of mathematics teaching have become more adept at using many skills (research methodologies), more willing to consider two things at once (research on teaching and learning), and somewhat more tolerant of different points of view, and, although they do not always agree, researchers are at least willing to attempt to communicate. With the current calls for reform in mathematics classrooms and the potential for increasing our understanding of mathematics teaching demonstrated by the varied approaches outlined in this chapter, it is more likely than ever that research will contribute in important ways to increasing student mathematics learning in important ways.

References

Buulthouse, J. K. (1978). Understanding school mathematics—a comment. *Mathematics Teaching, 82,* 39–41.

Barr, A. S., & Emans, L. (1930). What qualities are prerequisites to success in teaching? *Nation's Schools, 6,* 60–64.

Borko, H., & Livingston, C. (1989). Cognition and improvisation: Differences in mathematics instruction by expert and novice teachers. *American Educational Research Journal, 26*(4), 473–498.

Brophy, J. (1986). Teaching and learning mathematics: Where research should be going. *Journal for Research in Mathematics Education, 17*(5), 323–346.

Brophy, J. (1989). Toward a theory of learning. In J. Brophy (Ed.), *Advances in research on teaching* (pp. 345–355). Greenwich, CT: JAI Press.

Brophy, J., & Good, T. L. (1986). Teacher behavior and student achievement. In M. Wittrock (Ed.), *Handbook of research on teaching* (3rd edition, pp. 328–375). New York: Macmillan.

Brownell, W. A. (1935). Psychological considerations in the learning and teaching of arithmetic. In W. D. Reeve (Ed.), *The teaching of arithmetic: The tenth yearbook of the National Council of Teachers of Mathematics* (pp. 1–31). New York: Teachers College Press.

Carpenter, T. P., & Fennema, E. (1988). Research and cognitively guided instruction. In E. Fennema, T. P. Carpenter, & S. J. Lamon (Eds.), *Integrating research on teaching and learning mathematics* (pp. 2–19). Madison, WI: University of Wisconsin, Wisconsin Center for Education Research.

Carpenter, T. P., Fennema, E., Peterson, P. L., Chiang, C., & Loef, M. (1989). Using knowledge of children's mathematics thinking in classroom teaching: An experimental study. *American Educational Research Journal, 26*(4), 499–531.

Charters, W. W., & Waples, D. (1929). *The commonwealth teacher-training study.* Chicago: University of Chicago Press.

Cobb, P. (1988). The tension between theories of learning and instruction in mathematics education. *Educational Psychologist, 23*(2), 87–103.

Cobb, P., & Steffe, L. P. (1983). The constructivist researcher as teacher and model builder. *Journal for Research in Mathematics Education, 14*(2), 83–94.

Cobb, P., Wood, T., Yackel, E. Nicholls, J., Wheatley, G., Trigatti, B., & Perlwitz, M. (1991). Assessment of a problem-centered second-grade mathematics project. *Journal for Research in Mathematics Education, 22*(1), 3–29.

Cobb, P., Yackel, E., & Wood, T. (1988). Curriculum and teacher development: Psychological and anthropological perspectives. In E. Fennema, T. P. Carpenter, & S. J. Lamon (Eds.), *Integrating research on teaching and learning mathematics* (pp. 92–131). Madison, WI: University of Wisconsin, Wisconsin Center for Education Research.

Evertson, C. M., Anderson, C. W., Anderson, L. M., & Brophy, J. E. (1980). Relationships between classroom behaviors and student outcomes in junior high mathematics and English classes. *American Educational Research Journal, 17*(1), 43–60.

Evertson, C. M., Emmer, E. T., & Brophy, J. E. (1980). Predictors of effective teaching in junior high mathematics classrooms. *Journal for Research in Mathematics Education, 11*(3), 167–178.

Fennema, E., Carpenter, T. P., & Lamon, S. J. (Eds.). (1988). *Integrating research on teaching and learning mathematics.* Madison, WI: University of Wisconsin, Wisconsin Center for Education Research.

Fennema, E., Carpenter, T. P., & Peterson, P. L. (1989). Learning mathematics with understanding: Cognitively guided instruction. In J. Brophy (Ed.), *Advances in research on teaching* (pp. 195–221). Greenwich, CT: JAI Press.

Fennema, E., & Peterson, P. L. (1986). Teacher-student interactions and sex-related differences in learning mathematics. *Teaching and Teacher Education, 2*(1), 19–42.

Good, T. L., Grouws, D. A., & Ebmeier, H. (1983). *Active mathematics teaching.* New York: Longman.

Grouws, D. A. (1988). Improving research in mathematics classroom instruction. In E. Fennema, T. P. Carpenter, & S. J. Lamon (Eds.), *Integrating research on teaching and learning mathematics* (pp. 220–237). Madison, WI: University of Wisconsin, Wisconsin Center for Education Research.

Grouws, D. A., Cooney, T. J., & Jones, D. (Eds.). (1988). *Perspectives on research on effective mathematics teaching.* Hillsdale, NJ: Lawrence Erlbaum.

Hart, L. E. (1989). Classroom processes, sex of student, and confidence in learning mathematics. *Journal for Research in Mathematics Education, 20*(3), 242–260.

Herscovics, N., & Bergeron, J. C. (1983, February). Models of understanding. *Zentralblatt für Didaktik der Mathematik,* 75–83.

Hiebert, J., & Wearne, D. (1988a). Instruction and cognitive change in mathematics. *Educational Psychologist, 23*(2), 105–117.

Hiebert, J., & Wearne, D. (1988b). Methodologies for studying learning to inform teaching. In E. Fennema, T. P. Carpenter, & S. J. Lamon (Eds.), *Integrating research on teaching and learning mathematics* (pp. 168–193). Madison, WI: University of Wisconsin, Wisconsin Center for Education Research.

Hiebert, J., Wearne, D., & Taber, S. (1991). Fourth graders' gradual construction of decimal fractions during instruction using different physical representations. *Elementary School Journal, 91*(4), 321–341.

Hines, C. V., Cruickshank, D. R., & Kennedy, J. J. (1985). Teacher clarity and its relationship to student achievement and satisfaction. *American Educational Research Journal, 22*(1), 87–99.

Koehler, M. S. (1986). Effective mathematics teaching and sex-related differences in algebra one classes. (Doctoral dissertation, University of Wisconsin, 1985). *Dissertation Abstracts International, 46,* 2953A.

Koehler, M. S. (1990). Classrooms, teachers, and gender differences in mathematics. In E. Fennema & G. Leder (Eds.), *Mathematics and gender* (pp. 128–148). New York: Teachers College Press.

Lampert, M. (1988a, April). *Cognition in mathematical practice: A response to Jean Lave.* Paper presented at the American Educational Research Association Annual Meeting, New Orleans, LA.

Lampert, M. (1988b). Connecting mathematical teaching and learning. In E. Fennema, T. P. Carpenter, & S. J. Lamon (Eds.), *Integrating research on teaching and learning mathematics* (pp. 132–167). Madison, WI: University of Wisconsin, Wisconsin Center for Education Research.

Lampert, M. (1989). Choosing and using mathematical tools in classroom discourse. In J. Brophy (Ed.), *Advances in research on teaching* (pp. 223–264). Greenwich, CT: JAI Press.

Lampert, M. (1990). When the problem is not the question and the solution is not the answer: Mathematical knowing and teaching. *American Educational Research Journal, 27*(1), 29–63.

Leinhardt, G. (1989). Math lessons: A contrast of novice and expert competence. *Journal for Research in Mathematics Education, 20*(1), 52–75.

Leinhardt, G., & Putnam, R. T. (1987). The skill of learning from classroom lessons. *American Educational Research Journal, 24*(4), 557–588.

Livingston, C., & Borko, H. (1990). High school mathematics review lessons: Expert-novice distinctions. *Journal for Research in Mathematics Education, 21*(5), 372–387.

Mack, N. (1990). Learning fractions with understanding: Building on informal knowledge. *Journal for Research in Mathematics Education, 21*(1), 16–32.

Medley, D. M. (1979). The effectiveness of teachers. In P. L. Peterson & H. J. Walberg (Eds.), *Research on teaching: Concepts, findings, and implications* (pp. 11–27). Berkeley, CA: McCutchan Publishing Corporation.

National Council of Teachers of Mathematics. (1989). *Curriculum and evaluation standards for school mathematics.* Reston, VA: Author.

Nicholls, J. G., Cobb, P., Wood, T., Yackel, E., & Patashnick, M. (1990). Assessing students' theories of success in mathematics: Individual and classroom differences. *Journal for Research in Mathematics Education, 21*(2), 109–122.

Peterson, P. L., & Fennema, E. (1985). Effective teaching, student engagement in classroom activities, and sex-related differences in learning mathematics. *American Educational Research Journal, 22*(3), 309–335.

Pirie, S. E. B. (1988). Understanding: Instrumental, relational, intuitive, constructed, formalised...? How can we know? *For the Learning of Mathematics, 8*(3), 2–6.

Putnam, R. T., Lampert, M., & Peterson, P. L. (1990). Alternative perspectives on knowing mathematics in elementary schools. In C. B. Cazden (Ed.), *Review of research in education* (pp. 57–150). Washington, DC: American Educational Research Association.

Romberg, T. A., & Carpenter, T. P. (1986). Research on teaching and learning mathematics: Two disciplines of scientific inquiry. In M. Wittrock (Ed.), *Handbook of research on teaching* (3rd ed., pp. 850–873). New York: Macmillan.

Rosenshine, B. V. (1979). Content, time and direct instruction. In P. L. Peterson & H. J. Walberg (Eds.), *Research on teaching: Concepts, findings, and implications* (pp. 28–56). Berkeley, CA: McCutchan Publishing Corporation.

Seifert, E. H., & Beck, J. J. (1984). Relationships between task time and learning gains in secondary schools. *Journal of Educational Research, 78*(1), 5–10.

Shipp, D. E., & Deer, G. H. (1960). The use of class time in arithmetic. *The Arithmetic Teacher, 7*(3), 117–121.

Shuster, A., & Pigge, F. (1965). Retention efficiency of meaningful teaching. *The Arithmetic Teacher, 12,* 24–31.

Sindelar, P. T., Gartland, D., & Wilson, R. J. (1984). The effects of lesson format on the acquisition of mathematical concepts of fourth graders. *Journal of Educational Research, 78*(1), 40–44.

Skemp, R. R. (1978). Relational understanding and instrumental understanding. *The Arithmetic Teacher, 3,* 9–15.

Smith, L. R. (1977). Aspects of teacher discourse and student achievement in mathematics. *Journal for Research in Mathematics Education, 8*(3), 195–204.

Smith, L. R., & Cotten, M. L. (1980). Effects of lesson vagueness and discontinuity on student achievement and attitudes. *Journal of Educational Psychology, 72*(5), 670–675.

Stanic, G. M. A., & Reyes, L. H. (1986, April). *Gender and race differences in mathematics: A case study of a seventh-grade classroom.* Paper presented at the annual meeting of the American Educational Research Association, San Francisco, CA.

Wearne, D., & Hiebert, J. (1988). A cognitive approach to meaningful mathematics instruction: Testing a local theory using decimal numbers. *Journal for Research in Mathematics Education, 19*(5), 371–384.

Yackel, E., Cobb, P., Wood, T., Wheatley, G., & Merkel, G. (1990). The importance of social interaction in children's construction of mathematical knowledge. In T. J. Cooney & C. R. Hirsch (Eds.), *Teaching and learning mathematics in the 1990s* (pp. 12–21). Reston, VA: National Council of Teachers of Mathematics.

Zahn, K. G. (1966). Use of class time in eighth-grade arithmetic. *The Arithmetic Teacher, 13*(2), 113–120.

·7·

TEACHERS' BELIEFS AND CONCEPTIONS: A SYNTHESIS OF THE RESEARCH

Alba G. Thompson

SAN DIEGO STATE UNIVERSITY

One's conceptions of what mathematics *is* affects one's conception of how it should be presented. One's manner of presenting it is an indication of what one believes to be most essential in it.... The issue, then, is not, What is the best way to teach? but, What is mathematics really all about? (Hersh, 1986, p. 13)

There is no universal agreement on what constitutes "good mathematics teaching." The opening quote indicates what one considers to be desirable ways of teaching and learning mathematics is influenced by one's conception of mathematics. René Thom (1973) noted this when he stated that "all mathematical pedagogy, even if scarcely coherent, rests on a philosophy of mathematics" (p. 204). It is unlikely that disagreements about what constitutes good mathematics teaching can be resolved without addressing important issues about the nature of mathematics.

For many educated persons, mathematics is a discipline characterized by accurate results and infallible procedures, whose basic elements are arithmetic operations, algebraic procedures, and geometric terms and theorems. For them, knowing mathematics is equivalent to being skillful in performing procedures and being able to identify the basic concepts of the discipline. The conception of mathematics teaching that follows from this view is one in which concepts and procedures are presented in a clear way and opportunities are afforded the students to practice identifying concepts and performing procedures. Such a conception of mathematics teaching, though, can lead to instruction that places undue emphasis on the manipulation of symbols whose meanings are rarely addressed, as documented in the research literature (Thompson, 1982; 1984).

An alternative account of the meaning and nature of mathematics emerges from a sociological analysis of mathematical knowledge based on the ongoing practice of mathematicians. Such an account can be found in a recently published collection of works on the philosophy of mathematics (Tymoczko, 1986b). In it, mathematicians and philosophers of mathematics depict mathematics as a kind of mental activity, a social construction involving conjectures, proofs, and refutations, whose results are subject to revolutionary change and whose validity, therefore, must be judged in relation to a social and cultural setting.

In response to the question "What is mathematics?" Hersh (1986) offered the following as "the most straightforward, natural answer" (p. 22):

I am deeply indebted to Doug Jones, University of Kentucky, for the valuable suggestions he offered during the planning stage of the chapter. I wish to thank him and Penelope Peterson, Michigan State University, for their thoughtful and thorough reactions to an early draft. Finally, I wish to thank Pat Thompson, San Diego State University, for his helpful comments during our numerous discussions while the chapter was in progress. Preparation of this chapter was supported in part by National Science Foundation Grant No. MDR 89-50311. Any opinions or conclusions expressed are those of the author and do not represent an official position of NSF.

Mathematics deals with ideas. Not pencil marks or chalk marks, not physical triangles or physical sets, but ideas (which may be represented or suggested by physical objects). What are the main properties of mathematical activity or mathematical knowledge, as known to all of us from daily experience? (1) Mathematical objects are invented or created by humans. (2) They are created, not arbitrarily, but arise from activity with already existing mathematical objects, and from the needs of science and daily life. (3) Once created, mathematical objects have properties which are well-determined, which we may have great difficulty discovering, but which are possessed independently of our knowledge of them. (p. 22–23)

Adopting the point of view of the practicing mathematician, Hersh (1986), Lakatos (1986), and Putnam (1986) challenged the basic assumption that mathematical knowledge is a priori and infallible. They argued that mathematical knowledge is, in fact, fallible and, in this respect, is similar to knowledge in the natural sciences. "Our inherited *and unexamined* philosophical dogma is that mathematical truth should possess absolute certainty. Our actual experience in mathematical work offers uncertainty in plenty" (Hersh, 1986, p. 17). The position that real mathematics is constituted by formal derivations from formally stated axioms is also challenged by Hersh.

Anyone who has even been in the least interested in mathematics, or has even observed other people who were interested in it, is aware that mathematical work is work with ideas. Symbols are used as aids to thinking just as musical scores are used as aids to music. The music comes first, the score comes later. Moreover, the score can never be a full embodiment of the musical thoughts of the composer. Just so, we know that a set of axioms and definitions is an attempt to describe the main properties of a mathematical idea. But there may always remain an aspect of the idea which we use implicitly, which we have not formalized because we have not yet seen the counterexample that would make us aware of the possibility of doubting it. (Hersh, 1986, p. 18–19)

An assumption underlying Hersh's view of mathematics is that *knowing* mathematics is *making* mathematics. What characterizes mathematics is its making, its creative activities or generative processes.

The view of mathematics as "in the making" is consistent with the conception of mathematics teaching held by prominent mathematicians (Halmos, 1975; Polya, 1963; Steen, 1988; Thom, 1973) and many in mathematics education, a conception reflected in documents such as The Cockcroft Report (Committee of Inquiry into the Teaching of Mathematics in Schools, 1983), the *Curriculum and Evaluation Standards for School Mathematics* (National Council of Teachers of Mathematics [NCTM], 1989), and *Everybody Counts* (National Research Council, 1989). The conception of mathematics teaching that can be gleaned from these documents is one in which students engage in purposeful activities that grow out of problem situations, requiring reasoning and creative thinking, gathering and applying information, discovering, inventing, and communicating ideas, and testing those ideas through critical reflection and argumentation. This view of mathematics teaching is in sharp contrast to alternative views in which the mastery of concepts and procedures is the ultimate goal of instruction. However, it does not deny the value and place of concepts and procedures in the mathematics curriculum.

We do not assert that informational knowledge has no value, only that its value lies in the extent to which it is useful in the course of some purposeful activity. It is clear that the fundamental concepts and procedures from some branches of mathematics should be known by all students.... But instruction should persistently emphasize "doing" rather than "knowing that." (NCTM, 1989, p. 7)

For years there has been much discourse in mathematics education about the need to think of mathematics teaching as not just explaining its content, but also engaging students in the processes of doing mathematics. Yet, as noted in the *Standards* (NCTM, 1989), traditional teaching emphases have been on the mastery of symbols and procedures, largely ignoring the processes of mathematics and the fact that mathematical knowledge often emerges from dealing with problem situations. Indeed, the converse of Hersh's statement can be used to characterize typical school mathematics—first comes the score, but the music never follows.

Those who denounce current school mathematics instruction argue that not only does it misrepresent mathematics to the students, but also accounts in large part for their poor performance in national and international assessments. Such arguments have led to numerous calls in recent years for reform in mathematics education in the United States.

Explanations for the state of mathematics instruction in schools are plentiful and diverse, but few seem to be informed by research. One would caution against planning remedies on the basis of superficial, albeit seemingly plausible, explanations; they can lead to misguided, futile efforts. For example, there was a time when educators naively thought that producing a "teacher-proof" curriculum would go a long way in solving the problems of mathematics instruction. Thanks to studies of teachers' thinking and decision-making, educators now recognize that how teachers interpret and implement curricula is influenced significantly by their knowledge and beliefs (Clark & Peterson, 1986; Romberg & Carpenter, 1986). By recognizing that bringing about changes in what goes on in mathematics classrooms depends on individual teachers changing their approaches to teaching and that these approaches, in turn, are influenced by teachers' conceptions, mathematics educators have acknowledged the importance of this line of research.

This chapter examines the literature on teachers' conceptions pertaining to mathematics education. Some studies of teachers' knowledge of mathematics are discussed, but only insofar as they are relevant to the topic of teachers' conceptions; research on teachers' knowledge of mathematics is discussed in another chapter (Fennema & Loef, this volume). The chapter starts with a brief historical overview of the study of beliefs in this century, followed by a discussion of some philosophical distinctions between beliefs and knowledge. It then goes on to discuss the research that has been done, including theoretical models, methodology, and findings. Finally, contributions and implications are discussed with recommendations for future studies.

THE STUDY OF BELIEFS: A BRIEF HISTORY

Around the beginning of this century and well into the 1920s, there was considerable interest among social psychologists in the study of the nature of beliefs and their influence on people's actions. In the decades that followed, that interest faded and nearly disappeared as a topic in psychological literature, due in part, to the difficulty in accessing these beliefs for study, and, in part, to the emergence in the 1930s of associationism and, subsequently, behaviorism. In the 1960s, interest in the study of beliefs was somewhat renewed, but was quite varied among psychologists (Abelson & Carroll, 1965; Rokeach, 1960). The advent of cognitive science in the 1970s created "a place for the study of belief systems in relation to other aspects of human cognition and human affect" (Abelson, 1979, p. 355). The 1980s witnessed a resurgence of interest in beliefs and belief systems among scholars from disciplines as diverse as psychology, political science, anthropology, and education.

Among educators, interest in the study of teachers' beliefs and conceptions was fueled by a shift in paradigms for research on teaching. Prompted in part by information processing theory and other developments in cognitive science, research on teaching began a shift in the 1970s from a process-product paradigm, in which the object of study was teachers' behaviors, to a focus on teachers' thinking and decision-making processes (see Clark & Peterson, 1986; Shavelson & Stern, 1981; Shulman & Elstein, 1975). The shift of focus to teachers' cognition, in turn, led to an interest in identifying and understanding the composition and structure of "belief systems and conceptions," "action mind frames" (Shavelson, 1988), and "implicit theories" (Clark, 1988) underlying teachers' thoughts and decisions.

Still under the influence of a legacy of behaviorism, there were occasional studies in the decades of the 1960s and 1970s, conducted mainly by attitude researchers, that either directly or indirectly addressed teachers' beliefs and conceptions (Harvey, Prather, White, & Hoffmeister, 1968; Kerlinger, 1967). However, very few studies were related specifically to mathematics education. Since 1980, however, many studies in mathematics education have focused on teachers' beliefs about mathematics and mathematics teaching and learning. For the most part, these researchers have worked from the premise that "to understand teaching from teachers' perspectives we have to understand the beliefs with which they define their work" (Nespor, 1987, p. 323).

BELIEFS AND KNOWLEDGE

Despite the current popularity of teachers' beliefs as a topic of study, the concept of belief has not been dealt with in a substantial way in the educational research literature. For the most part, researchers have assumed that readers know what beliefs are. One explanation for the scarcity of reasoned discourses on beliefs in the educational literature is the difficulty of distinguishing between beliefs and knowledge. Because of the close connection that exists between beliefs and knowledge, distinctions between them are fuzzy (Scheffler, 1965). Researchers have noted it is frequently the case that teachers treat their beliefs as knowledge; an observation that has led many who initially set out to investigate teachers' knowledge to also consider teachers' beliefs (Grossman, Wilson, & Shulman, 1989).

Another explanation for the lack of discussions in educational research literature on the nature of beliefs and the distinctions between beliefs and knowledge is that the value of searching for definitive characterizations of the two concepts is arguable for educational research. Having suffered extension and abuse for well over one thousand years, the two concepts—and the words associated with them—are so broad that to search for a definitive characterization of either may prove to be futile (Needham, 1972; Wolgast, 1977). Some educators have argued that it is not useful for researchers to search for distinctions between knowledge and belief, but, rather, to search for whether and how, if at all, teachers' beliefs—or what they may take to be knowledge—affect their experience.

These arguments notwithstanding, researchers interested in studying teachers' beliefs should give careful consideration to the concept, both from a philosophical as well as a psychological perspective. Philosophical works can be helpful in clarifying the nature of beliefs. Psychological studies may prove useful in interpreting the nature of the relationship between beliefs and behavior as well as in understanding the function and structure of beliefs. The intent of the next section is to provide some common ground for the discussions that follow; it is not to offer a complete treatise on beliefs. A thorough treatment of the concept can be found in philosophical and psychological works (Abelson, 1979; Green, 1971; Hintikka, 1962; Russell, 1948; Scheffler, 1965).

Distinctions Between Beliefs and Knowledge

Beliefs have been distinguished from knowledge in a number of ways. For the purpose of this discussion, only a few distinctive features of beliefs and belief systems—to be elaborated in the next section—that seem most pertinent to the study of teachers' beliefs shall be considered. One feature of beliefs is that they can be held with varying degrees of conviction.

The believer can be passionately committed to a point of view, or at the other extreme could regard a state of affairs as more probable than not, as in "I believe that micro-organisms will be found on Mars." This dimension of variation is absent from knowledge systems. One would not say that one knew a fact strongly. (Abelson, 1979, p. 360)

Another distinctive feature of beliefs is that they are not consensual. "Semantically, 'belief' as distinct from knowledge carries the connotation of disputability—the believer is aware that others may think differently" (Abelson, 1979, p. 356). A common stance among philosophers is that disputability is associated with beliefs; truth or certainty is associated with knowledge. Scheffler (1965) argued that a claim to knowledge must satisfy a truth condition, whereas beliefs are independent of their validity.

In general, if you think I am mistaken in my belief, you will deny that I know, no matter how sincere you judge me to be and no matter how strong you consider my conviction. For X [an individual] to be judged mistaken is sufficient basis for rejecting the claim that he knows. It follows that if X is admitted to know, he must be judged not to be mistaken, and this is the point of the truth condition. ... Knowing, it would appear, is incompatible with being wrong or mistaken, and when I describe someone as knowing, I commit myself to his not being mistaken.... knowing unlike believing, has independent factual reference. (p. 23–24)

From a traditional epistemological perspective, a characteristic of knowledge is general agreement about procedures for evaluating and judging its validity; knowledge must meet criteria involving canons of evidence. Beliefs, on the other hand, are often held or justified for reasons that do not meet those criteria, and, thus, are characterized by a lack of agreement over how they are to be evaluated or judged.

Belief systems often include affective feelings and evaluations, vivid memories of personal experiences, and assumptions about the existence of entities and alternative worlds, all of which are simply not open to outside evaluation or critical examination in the same sense that the components of knowledge systems are. (Nespor, 1987, p. 321)

It should be noted that the evidence against which a claim to knowing is evaluated may change over time as old theories are replaced by new ones. Indeed, it is commonly accepted within the philosophy of science that even what one takes to be factual knowledge is dependent upon current theories (Feyerabend, 1975; Kuhn, 1962; Lakatos, 1976). Therefore, what may have been rightfully claimed as knowledge at one time may, in light of later theories, be judged as belief. Inversely, a once-held belief, with time, may be accepted as knowledge in light of new supporting theories. Thus, Scheffler's statement about the incompatibility of *knowing* and *being mistaken* fails to acknowledge the temporal quality of theories as canons of evidence.

The issue of consensuality seems particularly relevant to research on teachers' beliefs. In education, more so than in the sciences, it is common for alternative theories to coexist, even when aspects of one theory contradict the other, which may help explain the difficulty of sorting out teachers' knowledge from beliefs. Therefore, it is important that researchers make explicit, to themselves as well as others, the theory or theories of teaching and learning and the conceptualizations of the nature of mathematics with which they are approaching the study of mathematics teachers' beliefs. Without explicit attention to them, the significance of a study may be obscured, making it easy for readers to dismiss the research as inconsequential, albeit interesting.

Belief Systems

Just as the concept of belief has been used freely by researchers, so has the concept of a belief system been used without explication. The notion of a belief system is a metaphor for examining and describing how an individual's beliefs are organized (Green, 1971; Rokeach, 1960). It seems appropriate, at least from a structural point of view, to conceive of a belief system much in the same way that we think of a cognitive structure in a particular conceptual domain. As such, belief systems are dynamic in nature, undergoing change and restructuring as individuals evaluate their beliefs against their experiences.

Green (1971) identified three dimensions of belief systems, having to do not with the content of the beliefs themselves, but with the way in which they are related to one another within the system. The first of these dimensions has to do with the observation that a belief is never held in total independence of all other beliefs, and that some beliefs are related to others in the way that reasons are related to conclusions. Thus, belief systems have a quasi-logical structure, with some *primary* beliefs and some *derivative* beliefs. To illustrate, consider a teacher who believes it important to present mathematics "clearly" to the students; this is a primary belief. To this end, the teacher believes it important (a) to prepare lessons thoroughly, to ensure a clear, sequential presentation, and (b) to be prepared to answer readily any question posed by the students; these are both derivative beliefs.

Green's (1971) second dimension is related to the degree of conviction with which beliefs are held or to their psychological strength. According to Green, the beliefs in the system can be viewed as either *central* or *peripheral*—the central ones being the most strongly held beliefs, and the peripheral ones those most susceptible to change or examination. He noted that logical primacy and psychological centrality are orthogonal dimensions, arguing that they are two different features or properties of a belief. "A belief may be logically derivative and yet be psychologically central, or it may be logically primary and psychologically peripheral" (p. 46). In the example given earlier, the derivative belief in the importance of being prepared to answer student questions may be more important or psychologically central to the teacher for reasons of maintaining authority and credibility ("Teachers are supposed to know their stuff.") than for clarifying the subject to the students.

The third of Green's (1971) dimensions has to do with the claim that "beliefs are held in clusters, more or less in isolation from other clusters and protected from any relationship with other sets of beliefs" (p. 48). This clustering prevents crossfertilization among clusters of beliefs or confrontations between them, and makes it possible to hold conflicting sets of beliefs. This clustering property may help explain some of the inconsistencies among the beliefs professed by teachers, documented in several studies (Brown, 1985; Cooney, 1985; Thompson, 1982, 1984).

In addition to the notion of a belief system, this chapter will refer to teachers' "conceptions," viewed as a more general mental structure, encompassing beliefs, meanings, concepts, propositions, rules, mental images, preferences, and the like. Though the distinction may not be a terribly important one, it will be more natural at times to refer to a teachers' conception of mathematics as a discipline than to simply speak of the teachers' beliefs about mathematics.

Beliefs and Mathematics Teaching and Learning

The nature of teachers' beliefs about the subject matter and about its teaching and learning, as well as the influence of those beliefs on teachers' instructional practice, are relatively new topics of study. As such, these topics constitute largely uncharted areas of research on teaching. Nevertheless, a number of studies in mathematics education (Dougherty, 1990; Grant, 1984; Kesler, 1985; Kuhs, 1980; Lerman, 1983; Marks, 1987; McGalliard, 1983; Shroyer, 1978; Steinberg, Haymore, & Marks, 1985; Thompson, 1984), have indicated that teachers' beliefs about mathematics and its teaching play a significant role in shaping the teachers' characteristic patterns of instructional behavior.

In a theoretical paper based partly on empirical findings of studies of mathematics teachers' beliefs, Ernest (1988) noted that among the many key elements that influence the practice of mathematics teaching, three are most notable:

1. The teacher's mental contents or schemas, particularly the system of beliefs concerning mathematics and its teaching and learning;
2. The social context of the teaching situation, particularly the constraints and opportunities it provides; and,
3. The teacher's level of thought processes and reflection. (p. 1)

Part of teachers' mental "contents" or schemas is their knowledge of mathematics. Ernest (1988) contended that, although important, knowledge of mathematics alone does not account for differences in practice across mathematics teachers. According to Ernest, the research literature on mathematics teachers' beliefs, although scant, indicates that teachers' approaches to mathematics teaching depend fundamentally on their systems of beliefs, in particular on their conceptions of the nature and meaning of mathematics, and on their mental models of teaching and learning mathematics.

The influence of belief systems on performance has not been restricted to teachers and studies of teaching. A number of cognitive researchers in mathematics education have made similar observations with regard to students' mathematical performance. A well-known example is Schoenfeld's (1983) analysis of the role played by students' conceptions of mathematics in their interpretation of problem tasks and their problem-solving performance. Likewise, De Corte and Verschaffel (1985) spoke of young students' *word problem schema,* defining it to include "knowledge of a number of implicit rules, suppositions, and agreements inherent in the 'word problem game' that enables students to interpret ambiguities and obscurities correctly and to compensate for insufficiencies" (p. 7–8). Yet another example is that of Mellin-Olsen's (1981) discussion of the S-Rationale and I-Rationale, and how these are manifested in children's approaches to learning mathematics.

Curiously, an awareness of the significant role beliefs play in cognitive behavior and ensuing interest in belief systems as a topic of study appear to have developed concurrently, yet independently, among mathematics education researchers interested in teachers' cognitions and those interested in students' cognition. For both groups, the study of beliefs has emerged in recent years as an important, legitimate line of research. Its potential for making significant contributions to the field is becoming widely recognized in mathematics education.

MATHEMATICS TEACHERS' BELIEFS AND CONCEPTIONS

This section attempts to provide a synthesis of the empirical literature on mathematics teachers' beliefs and conceptions. Before discussing specific sets of studies, themes, or results, a general overview of the literature will be helpful to gain a sense

of the scope of the research and of the diversity of purpose, design, and technique that can be gleaned across studies.

Overview

Studies of mathematics teachers' beliefs and conceptions have focused on beliefs about mathematics, beliefs about mathematics teaching and learning, or both. Some studies have examined the relationship between teachers' beliefs and their instructional practices. The beliefs of both elementary and secondary teachers have been studied, but investigations of junior and senior high mathematics teachers' beliefs appear to be more common than those of elementary teachers. Moreover, some studies have involved pre-service teachers, others in-service teachers. A search of the literature in mathematics education revealed no single study specific to the topic of beliefs involving both pre- and in-service teachers, or a mix of teachers from the elementary, middle, and high school levels.

Most research on teachers' beliefs and conceptions is interpretive in nature and employs qualitative methods of analysis. Numerous techniques for obtaining data have been used, including Likert-scale questionnaires, interviews, classroom observations, stimulated recall interviews, linguistic analysis of teacher talk, paragraph completion tests, responses to simulation materials such as vignettes describing hypothetical students or classroom situations, and concept generation and mapping exercises such as the Kelly Repertory Grid Technique (Fransella & Bannister, 1977). Most studies have employed a combination of these techniques, rather than a single technique.

Research designs have also varied considerably, depending on the purpose of the study, from ethnographic case studies of one or two teachers (Brown, 1985; Cooney, 1985) to standardized administration of a belief inventory (Peterson, Fennema, & Carpenter, 1987). The purpose of some studies has been to describe or document the substance of teachers' beliefs (Helms, 1989; Owens, 1987; Stonewater & Oprea, 1988). The intent of other studies has been to examine the relationship between teachers' conceptions and instructional practice (Cooney, 1985; Dougherty, 1990; Grant, 1984; Kesler, 1985; McGalliard, 1983; Shaw, 1989; Thompson, 1984). Investigation of the phenomenon of changing teachers' conceptions has been the purpose of yet another set of studies (Meyerson, 1978; Schram & Wilcox, 1988; Schram, Wilcox, Lanier, & Lappan, 1988; Thompson, 1988). The diversity of purposes, methods, designs, and analytical frameworks used by researchers has led to great variability in how teachers' conceptions have been described.

It should be noted that to look at research on mathematics teachers' beliefs and conceptions in isolation from research on mathematics teachers' knowledge will necessarily result in an incomplete picture. Therefore, studies of pre-service and in-service teachers' subject matter knowledge in mathematics are cited, where needed.

The discussion that follows is organized into five sections or subsections. The first section discusses findings regarding teachers' conceptions of mathematics. Following that discussion is a subsection on the relationship between teachers' conceptions of mathematics and their instructional practice. Teachers' conceptions of mathematics teaching and learning are examined next, followed by a discussion of the relationship of conceptions of mathematics teaching and learning to instructional practice. Finally studies regarding the issue of changing teachers' conceptions are addressed. Comments on methodology and descriptions of theoretical frameworks are integrated throughout the discussion.

Teachers' Conceptions of Mathematics

A teacher's conception of the nature of mathematics may be viewed as that teacher's conscious or subconscious beliefs, concepts, meanings, rules, mental images, and preferences concerning the discipline of mathematics. Those beliefs, concepts, views, and preferences constitute the rudiments of a philosophy of mathematics, although for some teachers they may not be developed and articulated into a coherent philosophy (Ernest, 1988; Jones, Henderson, & Cooney, 1986). The significance for teaching of teachers' conceptions of subject matter has been widely recognized, both across a range of curricular areas (Clark & Peterson, 1986; Feiman-Nemser & Floden, 1986; Grossman, Wilson, & Shulman, 1989) and, as noted earlier, in mathematics (e.g., Ernest, 1985; Hersh, 1986; Lerman, 1983; Thom, 1973; Thompson, 1982, 1984).

Out of a number of possible variations, Ernest (1988) distinguished three conceptions of mathematics because of their significance in the philosophy of mathematics (Benacerraf & Putnam, 1964; Davis & Hersh, 1980; Lakatos, 1976), and also because they have been documented in empirical studies of mathematics teaching. He summarized them as follows:

First of all, there is a dynamic, problem-driven view of mathematics as a continually expanding field of human creation and invention, in which patterns are generated and then distilled into knowledge. Thus mathematics is a process of enquiry and coming to know, adding to the sum of knowledge. Mathematics is not a finished product, for its results remain open to revision (the problem-solving view).

Secondly, there is the view of mathematics as a static but unified body of knowledge, a crystalline realm of interconnecting structures and truths, bound together by filaments of logic and meaning. Thus mathematics is a monolith, a static immutable product. Mathematics is discovered, not created (the Platonist view).

Thirdly, there is the view that mathematics, like a bag of tools, is made up of an accumulation of facts, rules and skills to be used by the trained artisan skillfully in the pursuance of some external end. Thus mathematics is a set of unrelated but utilitarian rules and facts (the instrumentalist view). (p. 10)

It is quite conceivable, indeed probable, for an individual teacher's conception of mathematics to include aspects of more than one of the above—even seemingly conflicting aspects. As noted earlier, the clustering quality of belief systems may help explain the occurrence of conflicting beliefs. Thompson (1984) referred to the integratedness of conceptual systems to describe the absence of conflicting beliefs held in isolated clusters. It may appear that only to the extent that an individual's conceptual system is integrated can the conceptions of mathematics described above be used to characterize it. However, assessments of the psychological strength or logical primacy (Green, 1971) of an individual's professed beliefs can be helpful in deciding how best to characterize the individual's conceptions.

Thompson (1984), for example, found remarkable consistency—indicative of a well-integrated system—in the views of mathematics expressed by two of the three junior high mathematics teachers she studied. Kay, one of the teachers in her study, held a problem-solving view of mathematics, while Jeanne's professed beliefs and instructional practice were more aligned with a Platonist view. Although the third teacher, Lynn, expressed somewhat inconsistent views, it was clear from her instructional practice and from most of her professed beliefs that she held an instrumentalist view of the subject.

Lerman (1983) identified two alternative conceptions of the nature of mathematics, which he called absolutist and fallibilist views. He argued that the absolutist and fallibilist views correspond to two competing schools of thought in the philosophy of mathematics: Euclidean and Quasi-empirical (Lakatos, 1978). According to Lerman, from an absolutist perspective, all of mathematics is based on universal, absolute foundations, and, as such, it is "the paradigm of knowledge, certain, absolute, value-free and abstract, with its connections to the real world perhaps of a platonic nature." From a fallibilist perspective, mathematics develops through conjectures, proofs, and refutations, and uncertainty is accepted as inherent in the discipline. In a theoretical discussion of the relationship between philosophy of mathematics and teaching mathematics, Tymoczko (1986a) argued that the quasi-empirical view of mathematics—what Lerman called the fallibilist view—is the only one appropriate for teachers. The parallelism between Lerman's absolutist and fallibilist views and Ernest's Platonic and problem-solving views is readily observable.

Lerman (1983) offered a theoretical discussion of the connections of these two views with the teaching of mathematics, explaining how each can lead to very different models of mathematics teaching. He obtained data in support of the hypothesized correspondence between the two conceptions of mathematics and alternative views of teaching in a study involving pre-service secondary teachers, using an instrument designed to assess views ranging from absolutist to fallibilist. Four of the preservice teachers, two found to be at the absolutist extreme of the dimension and two at the fallibilist, were individually asked to react to a video recording of a segment of a mathematics lesson. The reactions of the teachers were consistent with their assessed views about the nature of mathematics. The absolutist teachers were critical of the teacher in the video for "not directing the students enough" with the content of the lesson. The fallibilist teachers, in contrast, were critical of the teacher in the video for being too directive.

A number of researchers (Copes, 1979, 1982; Dougherty, 1990; Helms, 1989; Kesler, 1985; McGalliard, 1983; Meyerson, 1978; Owens, 1987; Stonewater & Oprea, 1988) have used Perry's (1970) scheme of intellectual and ethical development, or adaptations of it, as a framework for analyzing and characterizing teachers' conceptions of mathematics. The scheme emerged from patterns in interview data obtained from college students by William G. Perry, Jr., while he was counselor at Harvard University during the 1950s and 1960s. The scheme consists of nine stages or "positions" that describe the intellectual and ethical development of college students from the viewpoint of their conception of knowledge. The scheme addresses conceptions of knowledge in general.

An adaptation of Perry's scheme for the study of conceptions of mathematical knowledge was made by Copes (1979), who proposed four types of conceptions: absolutism, multiplism, relativism, and dynamism. Copes described each type as corresponding to a conception of mathematical knowledge prevailing at different periods of its historical development. For example, absolutism prevailed from the time of the Egyptians and Babylonians until the middle of the nineteenth century. From an absolutist perspective, mathematics was viewed as a collection of facts whose truth is verifiable in the physical world. Multiplism, according to Copes, emerged with the advent of non-Euclidean geometries. Mathematical facts no longer needed to be verified by observable physical phenomena. Multiplism was characterized by the coexistence of different mathematical systems that might contradict each other. The advent of relativism was marked by the abandonment of efforts to prove the logical consistency of the different systems and the concomitant acceptance of their coexistence as equally valid systems. Dynamism is characterized by a commitment to a particular system or approach within the context of relativism.

Copes (1979) discussed applications of his framework to the teaching of mathematics, and suggested ways in which different teaching styles can communicate different conceptions. For example, a teaching style that emphasizes the transmission of mathematical facts, right versus wrong answers and procedures, and single approaches to the solutions of problems may communicate an absolutist or dualist view of mathematics.

Although one would expect any given teacher's professed beliefs and views about the nature of mathematics, when considered as a system, to exhibit some inconsistencies rendering classification into one category difficult, internal consistency in teachers' professed mathematical beliefs has been documented in studies employing Copes adaptation of Perry's scheme. For example, in a study of the mathematical beliefs of three high school teachers, Stonewater and Oprea (1988) reported remarkable internal consistency in the beliefs professed by each teacher about the nature of truth and the role of authority in mathematics—two themes central to Perry's (1970, 1981) theory. None of the teachers' beliefs were contradictory to the thought structures predicted by Perry's theory on the basis of their assessed position along the scheme. This observation raises the question of whether teachers' mathematical beliefs can be predicted by their level of intellectual development, a question of methodological significance for this line of research.

No discussion of teachers' conceptions of mathematics would be complete without Richard Skemp's (1978) frequently referenced theoretical discussion of conceptions of mathematics. Skemp proposed that two different conceptions of what constitutes mathematics account for sharp differences in instructional approaches and emphases. He credited Mellin-Olsen for drawing his attention to a distinction between two different meanings generally associated with "understanding" as it relates to mathematics: "relational understanding" and "instrumental understanding." Skemp described the former as "knowing both what to do and why." With respect to instrumental understanding, he noted:

Instrumental understanding I would until recently not have regarded as understanding at all. It is what I have in the past described as "rules without reasons," without realising that for many pupils *and their teachers* the possession of such a rule, and ability to use it, was what they meant by "understanding." (p. 9)

Skemp (1978) proposed a corresponding distinction between "instrumental mathematics" and "relational mathematics"; the distinction resides in the type of knowledge each reflects. According to Skemp, instrumental knowledge of mathematics is knowledge of a set of "fixed plans" for performing mathematical tasks. The characteristic of these plans is that they prescribe a step-by-step procedure to be followed in performing a given task, with each step determining the next. "The kind of learning that leads to instrumental [knowledge of] mathematics consists of the learning of an increased number of fixed plans, by which pupils can find their way from particular starting points to required finishing points" (p. 14). In contrast, relational knowledge of mathematics is characterized by the possession of conceptual structures that enable the possessor to construct several plans for performing a given task. In learning relational mathematics, "the means become independent of particular ends to be reached thereby" (p. 14); that is, the learner acquires knowledge of inclusive principles adequate to accommodate a multitude of events or tasks. With regard to the distinction between instrumental and relational mathematics, Skemp noted that

We are not talking about better and worse teaching of the same kind of mathematics.... It has taken me some time to realize that this is not the case. I used to think that math teachers were all teaching the same subject, some doing it better than others. I now believe that *there are two effectively different subjects being taught under the same name "mathematics."* (p. 11)

Skemp (1978) added that the difference in these two conceptions of what constitutes mathematical understanding and mathematical knowledge is at the root of many of the difficulties we have experienced in mathematics education. Specifically, this difference in conceptions is at the root of disagreements about what constitutes "sound" approaches to the teaching of mathematics and what constitutes "sound" student assessment practices.

Absent from Skemp's discussion is explicit reference to empirical data to validate his theoretical distinctions. However, the obvious correspondence between Skemp's instrumental view and Ernest's (1988) instrumentalist view can be regarded as a measure of validation. Moreover, the pervasiveness of the instrumentalist view among pre-service and in-service teachers stands out from a perusal of the research literature. As for "relational mathematics," it may be viewed as analogous to mathematics from a Platonist view, although it is not necessarily in conflict with Ernest's (1988) description of a problem-solving view, both of which have been documented in studies.

Readily observable from the literature is the small number of cases of teachers with informed historical and philosophical perspectives of mathematics (Steinberg et al., 1985; Thompson, 1984). For the majority of the teachers whose mathematical beliefs have been reported in studies done in the United States, mathematics is the mathematics of the school curriculum: arithmetic, algebra, geometry, and so on. This is true even for teachers who have completed undergraduate majors in mathematics (see Steinberg et al., 1985; Owens, 1987). This narrow, static view of the discipline, based on school mathematics, may help explain the preponderance of the absolutist/dualistic view of the discipline among studies using Perry's scheme, and of the instrumentalist and Platonist views among other studies.

Another noteworthy finding from studies focusing on teachers' conceptions of mathematics is the internal consistency of individual teachers' professed beliefs. Whereas diversity of conceptions has been documented across teachers, the beliefs professed by individual teachers concerning the nature of mathematics have been found to be generally consistent. This consistency was explicitly reported for all three of the teachers in Stonewater and Oprea's (1988) study, for two of the three teachers in Thompson's (1984) study, and can be inferred from Lerman's (1983) study in which, reportedly, two of the four teachers were at the absolutist extreme of the dimension, with the other two at the fallibilist extreme.

Insofar as the practical implications of this research are a concern, the internal consistency or integration of teachers' belief systems about the nature of mathematics, by itself, may appear not to be as relevant an issue as consistency between a teacher's conceptions of mathematics and his or her instructional practice. The next section looks at the literature on the relationship between teachers' conceptions of mathematics and their patterns of instructional behavior.

The Relationship Between Conceptions of Mathematics and Instructional Practice

Researchers have reported varying degrees of consistency between teachers' professed beliefs about the nature of mathematics and the teachers' instructional practices. Thompson (1984), for example, observed a high degree of consistency. She noted: "Although the complexity of the relationship between conceptions and practice defies the simplicity of cause and effect, much of the contrast in the teachers' instructional emphases may be explained by differences in their prevailing views of mathematics" (p. 119). For example, Lynn, whose view of mathematics was best characterized as instrumentalist, taught in a prescriptive manner emphasizing teacher demonstrations of rules and procedures. Jeanne, on the other hand, viewed mathematics primarily as a coherent subject consisting of logically interrelated topics and, accordingly, emphasized the mathematical meaning of concepts and the logic of mathematical procedures. Finally, Kay, who held a problem-solving view of mathematics, emphasized activities aimed at engaging students in the generative processes of mathematics.

Strong relationships between the knowledge base of novice teachers and their instructional practice were also reported by Steinberg et al. (1985). For example, Joe, one of the novice teachers studied who had completed doctoral level coursework in mathematics, had an elaborate and integrated conceptual map of the discipline. His teaching was conceptually oriented, stressing the "whys" of mathematical procedures and providing students with problems of his own design. He would point out to students how the topic at hand related to other topics they had studied and how all these fit into the larger scheme of mathematics. In contrast, Laura, with a much narrower experi-

ential base in the study of mathematics and an instrumentalist view of the subject, emphasized drill on procedures, rarely justifying them. "While Joe would allow students to generate their own algorithms for working problems and then discuss why they did or did not work, Laura was reluctant to allow students to use algorithms not included in the text" (Grossman et al., 1989, p. 27).

On the other hand, in a study of four senior high school mathematics teachers, Kesler (1985) reported some variability in the degree of consistency between teachers' conceptions and their teaching practice. He observed that the teachers' conceptions of mathematics ranged from dualistic/absolutistic to multiplistic/relativistic conceptions. Although the two teachers with dualistic conceptions taught in a manner consistent with their conceptions, the instructional practice of the two teachers with multiplistic conceptions ranged from "strict authoritarian to an inquiry mode of instruction" (Kesler, 1985).

McGalliard (1983) observed a high degree of consistency between the mathematics conceptions of four senior high mathematics teachers and their instructional practice while teaching geometry. He reported that based on their dualistic conceptions of mathematics, the teachers acted in "authoritative" ways regarding the content of the lessons, adopting a "right vs. wrong" stance and emphasizing the use of rules without explanations or justifications. The teachers emphasized to the students the importance of memorizing answers and taking notes in class, thus ostensibly engendering in the students the belief that external authority is the source of mathematical justification. Interestingly, McGalliard reported the teachers had much to say about how mathematics, particularly geometry, helps promote students' logical thought processes.

Inconsistencies between professed beliefs and instructional practice, such as those reported by McGalliard (1983), alert us to an important methodological consideration. Any serious attempt to characterize a teacher's conception of the discipline he or she teaches should not be limited to an analysis of the teacher's professed views. It should also include an examination of the instructional setting, the practices characteristic of that teacher, and the relationship between the teacher's professed views and actual practice. Scheffler (1965) underscored this point when he noted that

with independent knowledge of the social context, we may judge belief as revealed in word and deed. Where these latter two diverge, we may need to decide whether to postulate weakness of will, or irrationality, or deviant purpose, or ignorance, or bizarre belief, or insincerity, and the choice may often be difficult. . . . It will, in any case, never be reasonable to take belief simply as a matter of verbal response: belief is rather a "theoretical" state characterizing, in subtle ways, the orientation of the person in the world. (p. 89–90)

That the investigation of teachers' beliefs and conceptions is fundamentally problematic was underscored by Munby (1982).

The immediate response to the question "How shall we determine what teachers believe?" ought to be "Ask them!" But, for the following reasons, this alluringly simple approach is unsuited. First, the fact that the question is worth asking implies a commitment to the view that people have different beliefs and, thus, perspectives. To honor this is to comprehend the awkwardness of asking a question which gives no hint of the perspective from which it might be answered. To be sure, the perspective ought to be that of the teacher, but it is difficult to grasp this perspective before asking a question about what it is. Next, while subscribing to the view that our beliefs construct our experience, it is necessary to recognize that individually we may not be the best people to clearly enunciate our beliefs and perspectives since some of these may lurk beyond ready articulation. (p. 217)

Clearly, these are considerations with important implications for designing and conducting studies of teachers' beliefs. Such considerations alert us against taking matters of research design and methodology lightly.

At the very least, investigations of teachers' mathematical beliefs should examine teachers' verbal data along with observational data of their instructional practice or mathematical behavior; it will not suffice to rely solely on verbal data. In the case of pre-service teachers, data about their mathematical behavior as they encounter tasks in training content courses would be useful. Information of this kind would be valuable to reform efforts in mathematics teacher education. Furthermore, the examination and interpretation of verbal and observational data must be done in light of independently obtained information of the social context.

In the case of observed discrepancies between professed mathematical beliefs and practice, one must question the extent to which teachers are aware of such discrepancies and, if so, how they explain them. In some cases, such inconsistencies may be explained by the presence of unexamined clusters of conflicting beliefs. In other cases, the explanations offered by teachers may reveal various sources of influence on their instructional practice, causing them to subordinate their beliefs. In the case where consistency is observed, information about how teachers arrived at current beliefs and practices can be valuable. It is not until we have a clearer picture of how teachers modify and reorganize their beliefs in the presence of classroom demands and problems, and, conversely, how their practice is influenced by their conceptions of mathematics, that we can claim to understand the relationship between beliefs and practice.

Teachers' Conceptions of Mathematics Teaching and Learning

What a teacher considers to be desirable goals of the mathematics program, his or her own role in teaching, the students' role, appropriate classroom activities, desirable instructional approaches and emphases, legitimate mathematical procedures, and acceptable outcomes of instruction are all part of the teacher's conception of mathematics teaching. Differences in teachers' conceptions of mathematics appear to be related to differences in their views about mathematics teaching (Copes, 1979; Lerman, 1983; Thompson, 1984). For example, Thompson found that differences in the teachers' prevailing views of mathematics were related both to differences in their views about the appropriate locus of control in teaching and of what constituted evidence of mathematical understanding in their students, and to differences in their perceptions of the purpose of planning lessons.

Teachers' conceptions of mathematics teaching are also likely to reflect their views, though tacit, of students' mathematical knowledge, of how they learn mathematics, and of the roles and purposes of schools in general. A strong relationship has been observed between teachers' conceptions of teaching and their conceptions of students' mathematical knowledge (Cobb, Wood, & Yackel, in press; Carpenter, Fennema, Peterson, & Carey, 1988).

It is difficult to conceive of teaching models without some underlying theory of how students learn mathematics, even if the theory is incomplete and implicit. There seems to be a logical, natural connection between the two. Certainly, the connections have been drawn in theoretical discussions of alternative models of mathematics teaching. A classical example of this is Brownell's (1935) discussion of *drill, incidental,* and *meaning* theories of instruction. Although it seems reasonable to expect a model of mathematics teaching to be somehow related to or derived from some model of mathematics learning, for most teachers it is unlikely that the two have been developed and articulated into a coherent theory of instruction. Rather, conceptions of teaching and learning tend to be eclectic collections of beliefs and views that appear to be more the result of their years of experience in the classroom than of any type of formal or informal study. Clark (1988) underscored this point when he noted:

Research on teacher thinking has documented the fact that teachers develop and hold implicit theories about their students (Bussis, Chittenden, & Amarel, 1976), about the subject matter that they teach (Ball, 1986, see 1988; Duffy, 1977; Elbaz, 1981; Kuhs, 1980) and about their roles and responsibilities and how they should act (Ignatovich, Cusick, & Ray, 1979; Olson, 1981). These implicit theories are not neat and complete reproductions of the educational psychology found in textbooks or lecture notes. Rather, teachers' implicit theories tend to be eclectic aggregations of cause-effect propositions from many sources, rules of thumb, generalizations drawn from personal experience, beliefs, values, biases, and prejudices. (p. 6)

In studying the source of pre-service teachers' beliefs about mathematics teaching and learning, researchers have noted that those beliefs, for the most part, are formed during the teachers' schooling years and are shaped by their own experience as students of mathematics (Ball, 1988; Bush, 1983; Owens, 1987). The task of modifying long-held, deeply rooted conceptions of mathematics and its teaching in the short period of a course in methods of teaching remains a major problem in mathematics teacher education.

Models of Mathematics Teaching. Mindful of the fact that due to the eclectic nature of teachers' conceptions of mathematics teaching, a given teacher's conception is unlikely to fit any given teaching model; nevertheless, one should consider the predominant models of mathematics teaching that can be gleaned from the literature. The conceptions of the nature of mathematics, and the models of mathematics learning implicit in each, are briefly discussed.

Based on a review of the literature in mathematics education, teacher education, the philosophy of mathematics, the philosophy of education, and research on teaching and learning, Kuhs and Ball (1986) identified "at least four dominant and distinctive views of how mathematics should be taught:"

1. *Learner-focused:* mathematics teaching that focuses on the learner's personal construction of mathematical knowledge;
2. *Content-focused with an emphasis on conceptual understanding:* mathematics teaching that is driven by the content itself but emphasizes conceptual understanding;
3. *Content-focused with an emphasis on performance:* mathematics teaching that emphasizes student performance and mastery of mathematical rules and procedures; and
4. *Classroom-focused:* mathematics teaching based on knowledge about effective classrooms. (p. 2)

According to Kuhs and Ball (1986), a constructivist view of mathematics learning (Cobb & Steffe, 1983; Confrey, 1985; Thompson, 1985; von Glasersfeld, 1987) typically underlies the *learner-focused* view of mathematics teaching. Because the learner-focused view centers around the students' active involvement in doing mathematics—in exploring and formalizing ideas—it is the instructional model most likely to be advocated by those who have a problem-solving view of mathematics, who view mathematics as a dynamic discipline, dealing with self-generated ideas and involving methods of inquiry (Ernest, 1988; see earlier discussion). From a learner-focused perspective of teaching, the teacher is viewed as facilitator and stimulator of student learning, posing interesting questions and situations for investigation, challenging students to think, and helping them uncover inadequacies in their own thinking (Kuhs & Ball, 1986). Students are viewed as ultimately responsible for judging the adequacy of their own ideas. Knowledge is assessed in terms of the consistency between the students' constructed ideas and the shared meaning of the idea in the discipline, as well as in terms of their ability to validate conjectures and support or defend their conclusions.

The second view discussed by Kuhs and Ball (1986), *the content-focused with emphasis on understanding,* is the view of teaching that would follow naturally from the conception of the nature of mathematics that Ernest (1988) labeled Platonist. Kuhs and Ball characterized this view as one in which instruction makes mathematical content the focus of classroom activity while emphasizing students' understanding of ideas and processes. This view of teaching is akin to Brownell's (1935) "meaning theory of instruction" in that it emphasizes students' understanding of the logical relations among various mathematical ideas and the concepts and logic underlying mathematical procedures. Criteria for judging student knowledge are similar to those of the learner-focused model.

Kuhs and Ball (1986) distinguished between the first two views of teaching by the way subject matter is organized. Unlike the learner-focused model, in which students' ideas and interests are primary considerations, content is organized in the content-focused model according to the structure of mathematics, following some notion of scope and sequence the teacher may have. Kuhs and Ball indicated that what distinguishes the content-focused view emphasizing conceptual understanding from the other three views is "the dual influence of content and learner. On one hand, content is focal, but on the other, understanding is viewed as constructed by the individual" (p. 15).

The third view, *the content-focused view with emphasis on performance,* also makes mathematical content its focal point. Yet, underlying this view are conceptions of the nature of mathematics, of mathematics learning, and of schooling in general that are very different from those underlying the first two views (Kuhs & Ball, 1986). The content-performance view of teaching is analogous to what Brownell (1935) described as "drill theory." It is the view of teaching that would follow naturally from the instrumentalist view of the nature of mathematics. Kuhs and Ball stated some of the central premises of this view:

- Rules are the basic building blocks of all mathematical knowledge and all mathematical behavior is rule-governed.
- Knowledge of mathematics is being able to get answers and do problems using the rules that have been learned.
- Computational procedures should be "automatized."
- It is not necessary to understand the source or reason for student errors; further instruction on the correct way to do things will result in appropriate learning.
- In school, knowing mathematics means being able to demonstrate mastery of the skills described by instructional objectives. (1986, p. 22)

In the instrumentalist view of teaching, the content is organized according to a hierarchy of skills and concepts; it is presented sequentially to the whole class, to small groups, or to an individual, following a pre-assessment of students' mastery of prerequisite skills. Many of the self-paced instructional programs of the early 1970s, such as the Individually Prescribed Instruction (IPI) program and the more recent Team-Assisted Individualization (TAI) program (Slavin, 1987), are patterned after an instrumentalist view of mathematics teaching. From an instrumentalist perspective, the role of the teacher is to demonstrate, explain, and define the material, presenting it in an expository style. Accordingly, the role of the students is to "listen, participate in didactic interactions (for example, responding to teacher questions) and do exercises or problems using procedures that have been modeled by the teacher or text" (Kuhs & Ball, 1986, p. 23).

None of the proposed models of mathematics teaching has been the object of more criticism by mathematics educators than the model following most naturally from an instrumentalist perspective. Critics of this model object to taking a student's ability to obtain correct answers, perform algorithms, and state definitions as evidence of their "knowing" mathematics. These critics often base their objections on reports of studies (Erlwanger, 1975; Leinhardt, 1985; Schoenfeld, 1985) documenting that students who perform adequately on routine mathematical tasks often have impoverished conceptions and significant misunderstandings of the mathematical ideas in those tasks. Results of national (Lindquist, 1989) and international assessments (McKnight et al., 1987) are often cited to substantiate the inappropriateness of interpreting computational proficiency as evidence of knowledge of mathematics. Some critics, however, object to instrumentalism because of the "bond" theory of learning implicit in it, and also on the grounds that such instruction does not help students understand the structure of mathematics (Steffe & Blake, 1983). Another criticism that has been voiced, particularly by those with a problem-solving view of mathematics, is that instrumentalism does not actively involve the students in the processes of exploring and investigating ideas; therefore, it not only denies students the opportunity to do "real" mathematics, it also misrepresents mathematics to the students.

The fourth and last of the distinctive views of how mathematics should be taught, identified by Kuhs and Ball (1986), is the *classroom-focused* view of teaching. Central to this view is the notion that classroom activity must be well structured and efficiently organized according to effective teacher behaviors identified in process-product studies of teaching effectiveness. Kuhs and Ball noted that unlike the other models of mathematics teaching, this model, in its purest form, does not address questions about the content of instruction. Rather, it assumes that content is established by the school curriculum. In addition, this model is not necessarily grounded on any particular theory of learning: "The assumption is that students learn best when classroom lessons are clearly structured and follow principles of effective instruction (for example, maintaining high expectations, insuring a task-focused environment)" (Kuhs & Ball, 1986, p. 27).

In the classroom-focused model of teaching, the teacher is viewed as playing an active role directing all classroom activities, clearly presenting the material of the lesson to the whole class or to subgroups thereof, and providing opportunities for students to practice individually. From this perspective, effective teachers are those who "skillfully explain, assign tasks, monitor student work, provide feedback to students, and manage the classroom environment, preventing, or eliminating, disruptions that might interfere with the flow of planned activity" (Kuhs & Ball, 1986, p. 26). Accordingly, the students' role is to listen attentively to the teacher and cooperate by following directions, answering questions, and completing the tasks assigned by the teacher.

As an example of the generic, classroom-focused model of teaching, Kuhs and Ball (1986) referred to Madeline Hunter's popular method, with its focus on the structure of lessons and general pedagogical skills. In contrast to the Hunter model of teaching, which ignores subject matter content, Kuhs and Ball cited the Missouri Mathematics Program (Good, Grouws, & Ebmeier, 1983) as patterned after the classroom-focused model, but with provisions and discussions specific to mathematics that closely parallel the views of the content-focused model with emphasis on understanding.

The four models of mathematics teaching identified by Kuhs and Ball (1986) are useful in describing major differences among current views of mathematics teaching. As in the case of conceptions of mathematics, a given teacher's conception of mathematics teaching is more likely to include various aspects of several models than it is to fit perfectly into the description of a single model. Nevertheless, as discussed earlier, a teacher's conception of teaching can be characterized on the basis of the more psychologically central or logically primary (Green, 1971) beliefs the teacher holds.

Because they are eclectic aggregations of beliefs, values, propositions and principles, teachers' models of mathematics teaching may often reflect inconsistencies. Such inconsistencies have been documented in a number of studies (Shaw, 1989;

Thompson, 1984). When inconsistencies are observed among the beliefs professed by teachers, it can be inferred that the inconsistent beliefs are held in isolation of one another (Green, 1971). However, if it appears a teacher has modified his or her beliefs to resolve or avoid inconsistencies—either in thought or deed—then it is legitimate to infer that the teacher is holding these beliefs in relation to one another; thus, the teacher has integrated them into an apparently coherent system. Support for this statement can be found in the case of *Kay* (Thompson, 1984). Therefore, it is reasonable to assume that the degree to which a teacher's conception of teaching constitutes an integrated, coherent system depends on the extent to which the teacher has reflected on and made explicit to himself or herself the beliefs and values he or she holds as well as the propositions and principles he or she has abstracted from experience.

The fact that the models of teaching described in this section cannot be used to neatly categorize teachers according to their beliefs might lead one to wonder why researchers bother with them at all. The reason for including them was discussed in an earlier part of this chapter, but is worth restating; it is important that researchers interested in examining teachers' beliefs make explicit, to themselves and to others, the perspectives from which they are approaching their work. This is particularly important because of the interpretive nature of most of the studies in this line of research. Without information of the perspective from which the analysis was done, the reader may have difficulty understanding the significance of some findings and dismiss this line of research as simply interesting but inconsequential. Furthermore, the models of mathematics teaching identified by Kuhs and Ball (1986) can be viewed as constituting a consensual knowledge base regarding models of teaching. Indeed, they bear close correspondence to the models of mathematics teaching that Ernest (1988) proposed after surveying the empirical and theoretical literature in mathematics education.

Relationship Between Beliefs about Teaching and Instructional Practice. For the most part, studies of the relationship between teachers' beliefs about teaching and instructional practice have examined the congruence between teachers' professed beliefs and their observed practice. The findings have not been as consistent across studies, or across teachers, as findings on the relationship between conceptions of the nature of mathematics and instructional practice. Some researchers have reported a high degree of agreement (Grant, 1984; Shirk, 1973) between teachers' professed views of mathematics teaching and their instructional practice, whereas others have reported sharp contrasts (for example, Cooney, 1985; Shaw, 1989; Thompson, 1982).

Shirk, for example, examined the conceptual frameworks of four pre-service elementary teachers and their relation to the teachers' behavior when teaching mathematics to small groups of junior high school students. He described the teachers' conceptual frameworks in two parts: the teachers' conceptions of mathematics teaching and their conceptions of their roles as teachers. He observed that although the teachers' conceptions had elements in common, the unique combination of elements in each case accounted for their different teaching behaviors. He noted that the teachers' conceptions appeared to be activitated in teaching situations, resulting in the teachers behaving in ways that were consistent with their conceptions. Likewise, Grant (1984) reported congruence of professed beliefs and instructional practice in the case of three senior high mathematics teachers.

Other studies, however, have found discrepancies between teachers' professed beliefs about teaching mathematics and their practice (Brown, 1985; Cooney, 1985). Within a single study, some teachers reportedly professed beliefs about mathematics teaching that were largely consistent with their instructional practices, whereas other teachers in the same study showed a great disparity (Thompson, 1984).

The inconsistencies reported in these studies indicate that teachers' conceptions of teaching and learning mathematics are not related in a simple cause-and-effect way to their instructional practices. Instead, they suggest a complex relationship, with many sources of influence at work; one such source is the social context in which mathematics teaching takes place, with all the constraints it imposes and the opportunities it offers. Embedded in this context are the values, beliefs, and expectations of students, parents, fellow teachers, and administrators; the adopted curriculum; the assessment practices; and the values and philosophical leanings of the educational system at large.

The influence of the social context on a teacher's instructional practice was documented by Brown (1985) in her study of the socialization to teaching of Fred, a beginning secondary mathematics teacher (see also Cooney, 1985). Cooney described the tensions and conflicts Fred experienced between his strong views of mathematics teaching, favoring an emphasis on what he conceived as problem solving—non-standard, recreational problems that would serve more a motivational purpose than a mathematical objective—and his perceptions of the realities of his teaching situation, which he described as imposing numerous obstacles to actualizing his views. Thus, when faced with pressures to cover subject matter and maintain class control, Fred readily compromised his belief in problem-solving as an instructional goal.

Addressing the effect of the social context on teachers' instructional decisions and actions, Ernest (1988) noted that

These sources lead the teacher to internalise a powerful set of constraints affecting the enactment of the models of teaching and learning mathematics. The socialization effect of the context is so powerful that despite having differing beliefs about mathematics and its teaching, teachers in the same school are often observed to adopt similar classroom practices. (p. 4)

It appears that as teachers interact with their environment, some experience no conflict between their beliefs and their practice and some learn to live with unresolved conflicts, as Fred did. Other teachers, however, seem to reorganize their beliefs in response to the pressures encountered in teaching.

Reported inconsistencies between professed beliefs and observed practice can also be explained in part by the way teachers' beliefs have been measured. Reliance on verbal responses to questions posed at an abstract level of thought as the only source of data is problematic. The fact that some of the beliefs professed by teachers are more a manifestation of a verbal commitment to abstract ideas about teaching than of an operative

theory of instruction may account for some of the inconsistencies reported in the literature (Shaw, 1989). Such inconsistencies may be fewer among experienced teachers than among novice teachers who have not had many occasions to operationalize and test those ideas and modify their views accordingly. In any case, it will not be appropriate, from a methodological point of view, to take verbal expression alone as evidence of belief. Scheffler (1965) warned about this when he noted that "it seems particularly important to avoid mistaking verbal dispositions for belief. To this end, it is crucial that we recognize not only the ramifications of belief in conduct but also the influence of motivation and social climate on verbal expression" (p. 90).

The political climate may also account for some of the observed discrepancies between teachers' professed beliefs and their instructional practice. For example, state-level policies, such as those being implemented in California, may influence teachers' practices without necessarily affecting their views. Also, historical events, such as the current reform movement in mathematics education, and the publication of documents such as the NCTM's *Curriculum and Evaluation Standards for School Mathematics,* may have an influence on teachers' verbal expressions regarding their views of mathematics teaching and learning.

Finally, one must not ignore that there is a great deal of knowledge essential for successfully implementing certain models of mathematics teaching (Ball, 1988, 1990; Dewey, 1964; Hawkins, 1973; Shulman, 1986; Steinberg, Haymore, & Marks, 1985). Thus, some inconsistencies between teachers' professed beliefs and practices may also be manifestations of espoused teaching ideals that cannot be realized because the teachers do not possess the skills and knowledge necessary to implement them. For example, the learner-focused model of teaching identified by Kuhs and Ball (1986) requires that the teacher possess a broad knowledge base in mathematics in order to recognize and capitalize on opportunities that arise naturally in the mathematics class or in other classes to examine or apply mathematical ideas and procedures. Ball (1988) reported that pre-service teachers, themselves, recognized the need for broader, deeper understandings of mathematics in order to teach conceptually.

It should be clear from the foregoing discussion that the relationship between teachers' conceptions of teaching and their practice is not a simple one. Yet, an assumption that appears to underlie most investigations is that the relationship is one of linear causality, where first come the beliefs and then follows the practice. The literature, however, suggests that the relationship is more complex, involving a give and take between beliefs and experience and, thus, is dialectical in nature. In this regard, Cobb, Wood, and Yackel (1990) noted:

In our view, arguments about the direction of the assumed causality miss the point; *the very nature of the relationship needs to be reconceptualized.* Our current work with teachers is based on the alternative assumption that beliefs and practice are dialectically related (italics added). (p. 145)

There is support in the literature for the claim that beliefs influence classroom practice; teachers' beliefs appear to act as filters through which teachers interpret and ascribe meanings to their experiences as they interact with children and the subject matter. But, at the same time, many of a teacher's beliefs and views seem to originate in and be shaped by experiences in the classroom. By interacting with their environment, with all its demands and problems, teachers appear to evaluate and reorganize their beliefs through reflective acts, some more than others.

Based on her data, Thompson (1984) observed that the extent to which experienced teachers' conceptions are consistent with their practice depends in large measure on the teachers' tendency to reflect on their actions—to think about their actions vis-á-vis their beliefs, their students, the subject matter, and the specific context of instruction. This is not to suggest, however, that, upon reflection, all tensions and conflicts between beliefs and practice will be resolved. However, it is by reflecting on their views and actions that teachers gain an awareness of their tacit assumptions, beliefs, and views, and how these relate to their practice. It is through reflection that teachers develop coherent rationales for their views, assumptions, and actions, and become aware of viable alternatives. Ernest (1988) also recognized the central role reflection plays on teaching when he noted that by reflecting on the effect of their actions on students, teachers develop a sensitivity for context that enables them to select and implement situationally appropriate instruction in accordance with their own views and models.

Changing Teachers' Conceptions

A growing awareness of the role that teachers' beliefs play in teaching has led researchers to address a number of related questions: How do these conceptions form? How do they evolve? Particularly, how can they be affected? The latter question has guided a number of investigations of how teachers' conceptions of mathematics and mathematics teaching and learning can be influenced and enriched. Although, as a research topic, effecting change in teachers' conceptions seems to have gained popularity in recent years, a few studies in mathematics education predate the flurry of the 1980s.

Collier (1972) used Likert scales to measure pre-service elementary teachers' beliefs about mathematics and mathematics teaching along a formal-informal dimension. The formal end of the dimension was characterized by items depicting mathematics as rigid and exact, free of ambiguity and contradiction, and consisting of rules and formulas for solving problems. A formal view of mathematics instruction was defined in terms of items that emphasized teacher demonstration, memorization of facts and procedures, and single approaches to the solution of problems. In contrast, the informal pole of the dimension was characterized by items depicting mathematics as aesthetic, creative, and investigative in nature and as allowing for a multiplicity of approaches to the solution of problems. An informal view of mathematics instruction was characterized by an emphasis on student discovery, experimentation, and inventiveness, the use of trial-and-error methods, and the encouragement of original thinking.

Collier (1972) defined a quotient of ambivalence and used it, as well as the formal-informal dimension, to describe the beliefs of prospective teachers at different stages of the preparation program. Prospective teachers nearing the end of the program had more informal and less ambivalent views about mathematics and mathematics teaching than teachers beginning the program. Also, prospective teachers who had been identified as high-achievers viewed mathematics as less formal and had less ambivalent views of mathematics instruction than the low-achievers. However, most scores reflected a neutral position along the formal-informal dimension. Collier concluded that, allowing for the cross-sectional nature of the samples, the results indicated a slight progression in the beliefs of the teachers toward an informal view of mathematics and mathematics instruction as they went through the program.

In his study of four pre-service elementary teachers enrolled in a mathematics methods course, Shirk (1973), unlike Collier, found no discernable change in the teachers' conceptions. Shirk observed some changes in instructional behavior, but indicated those changes were consistent with the teachers' conceptions.

Taken together, the results of Collier (1972) and Shirk (1973) suggest that prospective teachers' conceptions are not easily altered, and that one should not expect noteworthy changes to come about over the period of a single training course. Still, one might wonder if it would be possible to obtain more encouraging results than those reported by either Collier or Shirk through intervention of comparable duration, but specifically designed to induce change in the teachers' conceptions.

A case in point was a study conducted by Meyerson (1978), who reported the results of one such intervention in a pre-service methods course for secondary mathematics teachers. The course was designed to affect change in the participants' conception of knowledge with respect to mathematics and mathematics teaching. The teachers' conceptions were diagnosed according to their position on Perry's scheme applied to knowledge of mathematics and mathematics teaching. During the course, the teachers engaged in exercises focusing on seven themes: mathematical mistakes, surprise, doubt, reexamination of pedagogical truisms, feelings, individual differences, and problem solving. Meyerson reported some success in moving teachers along the scheme, noting that the key factor affecting change was doubt. "Doubting one's relationship with authority and reexamining one's beliefs" (p. 137) were essential in moving from one stage to the next. Doubt was generally aroused in problem-posing situations that caused confusion and created controversy.

A more recent study examining the effect of courses on pre-service elementary teachers' mathematical conceptions was carried out by Schram, Wilcox, Lanier, and Lappan (1988). Schram et al. set out to examine changes in undergraduate education majors' knowledge about mathematics, mathematics learning, and mathematics teaching as they progressed through a sequence of three innovative mathematics courses. The courses emphasized conceptual development, group work, and problem-solving activities. Changes in students' thinking about mathematics were attributed to their participation in one of the courses in the sequence. At the end of the 10-week course, changes were reported in the participants' conceptions of the nature of mathematics, of the structure of mathematics classes, and of the process of learning mathematics.

Schram and Wilcox (1988) also conducted case studies of two prospective elementary teachers enrolled in the first of the three courses. The case studies focused specifically on the students' views about how mathematics is learned and what it means to know mathematics. The students' views were examined against a framework developed by the researchers, consisting of three levels that reflected different orientations to mathematics teaching and learning. Whereas one student changed his original views of what it means to know mathematics, the other student appeared to take in the new experiences and conceptual ideas by modifying them to fit into her original conceptions.

This phenomenon of assimilation without accommodation has been observed among British mathematics teachers as well. For example, Lerman (1987, cited in Ernest, 1988) noted that when teachers were faced with a new external examination requiring students to carry out mathematical investigations and projects for assessment, they treated such inquiries in a didactical manner.

The phenomenon of teachers modifying new ideas to fit their existing schemas is not well understood. Yet, understanding why teachers do this instead of restructuring their current schemas is central to effecting change. Skemp (1978) offered the following as one of four factors contributing to the difficulty of teachers changing their instructional practices:

"The great psychological difficulty for teachers of accommodating (restructuring) their existing and longstanding schemas, even for the minority who know they need to, want to do so, and have time for study." (p. 13)

Generally, though, studies that have dealt with change in teachers' beliefs have not provided the detailed analyses necessary to shed light on the question of why it is so difficult for teachers to accommodate their schemas and internalize new ideas. A better understanding of the sources of this difficulty is critical to the design of strong, successful teacher education and enhancement programs, programs that go beyond simply raising the level of enthusiasm of participating teachers. Such detailed analyses should seek to explain, at least in part, why it is that out of a group of teachers participating in an in-service program, only a few walk away to implement or try out the new ideas with some measure of success—something to which all those who work closely with teachers regretfully can attest. Unfortunately, the literature on teacher change, though rich with tips, does not offer explanations for this phenomenon. Furthermore, those tips are based on assumptions that may be at odds with the goals of many intervention programs.

A set of studies reporting a high degree of success in changing teachers' beliefs and practices was conducted by Carpenter, Fennema, and Peterson at the University of Wisconsin (Carpenter, Fennema, Peterson, Chiang, & Loef, 1989). The studies were designed to investigate the effect information about children's thinking in solving simple addition and subtraction word problems would have on primary school teachers' instructional practice. Carpenter et al. observed important changes in the instructional decisions of the teachers; reportedly, the teachers also spent more time during class listening to their students'

explanations of problem-solving strategies and less time engaging students in rote activities.

If one agrees that an important goal of studies focusing on change in teachers' conceptions and practice is to shed light on what happens cognitively to teachers as they participate in those programs, then it would seem necessary to study individual teachers in depth and to provide detailed analyses of their cognitive processes. One such investigation was a case study of a teacher conducted by Cobb, Wood, and Yackel of Purdue University (1990). Insights gained from their study led the researchers to state: "The crucial point in our development of a collaborative relationship with the project teacher occurred when she began to realize her current practice might be problematic" (Cobb, Wood, & Yackel, 1990, p. 131–132). Cobb et al. underscored the importance of teachers seeing their current practice as problematic as some sort of prerequisite mental state necessary for beneficial collaboration between teachers and researchers or staff developers. Cobb et al. used the teacher's classroom as an environment where the teacher learned by doing and by reflecting on her actions. The researchers' role was to help the teacher "develop personal, experientially-based reasons and motivations for reorganizing her classroom practice" (Cobb et al., 1990, p. 144) and to assist the teacher in doing so, rather than to show the teacher how to teach in a specified way.

Insightful analyses and detailed accounts of how teachers internalize new ideas and develop new instructional practices can contribute to our understanding of the cognitive processes involved in teachers changing their conceptions and practices. From a practical viewpoint, such detailed accounts should prove particularly valuable at a time when federal, state, and local agencies are investing considerable funds in teacher enhancement and teacher preparation programs in the United States.

Summary and Future Directions

Studies of the relationship between teachers' beliefs and practice lead us to question the adequacy of two related assumptions underlying a number of studies. One of them is that belief systems are static entities to be uncovered. The second assumption is that the relationship between beliefs and practice is a simple linear-causal one. Thoughtful analyses of the nature of the relationship between beliefs and practice suggest that belief systems are dynamic, permeable mental structures, susceptible to change in light of experience. The research also strongly suggests that the relationship between beliefs and practice is a dialectic, not a simple cause-and-effect relationship. Thus, future studies, particularly those having to do with effecting change, should seek to elucidate the dialectic between teachers' beliefs and practice, rather than try to determine whether and how changes in beliefs result in changes in practice.

Early in this chapter I noted that some educators have argued that because of the close conceptual connections between beliefs and knowledge, it is not useful for researchers to distinguish between teachers' knowledge and teachers' beliefs. Instead, they should search for whether and how teachers' beliefs, or what they may take to be knowledge, relate to their experience. Because of this argument, it seems more helpful for researchers to focus their studies on teachers' conceptions—mental structures, encompassing both beliefs and any aspect of the teachers' knowledge that bears on their experience, such as meanings, concepts, propositions, rules, mental images, and the like—instead of simply teachers' beliefs.

Finally, a question that has received virtually no attention from researchers on teachers' conceptions is the extent to which teachers' and students' conceptions interact during instruction. Documented research shows that teachers' conceptions, for the most part, are reflected in their instructional practices. But we know little about how instructional practices, in turn, communicate those conceptions or others to students, if they do so at all. Insofar as one observes congruence between the mathematical beliefs of students (Schoenfeld, 1983) and those of teachers (Thompson, 1988), it is natural to infer that some communication is effected. Furthermore, since teachers are the primary mediators between the subject matter of mathematics and the students, it is also natural to infer that the teachers' conceptions are indeed communicated to students through practices in the classroom. This chain of inferences, however, remains to be empirically validated. There is a great deal that we can learn from insightful analyses of the nature of that interaction.

CONTRIBUTIONS AND IMPLICATIONS

There is little doubt that the study of mathematics teachers' beliefs and conceptions has established a place for itself within the mathematics education research enterprise. What remains unclear, however, is what this line of research has to contribute to mathematics education.

After examining the literature, the contributions may appear to be more tenuous than obvious. It may be that much of what this line of research has to contribute is yet unrealized. Be that as it may, there are several areas of mathematics education to which research on teachers' beliefs and conceptions has made important contributions: They are mathematics teacher education and research on teacher education, and research on mathematics teaching and learning.

Contributions to Mathematics Teacher Education and Research on Teacher Education

Research on teachers' conceptions of mathematics and its teaching has produced information that can be used as an impetus for mathematics teacher educators and staff developers to reexamine aspects of their work. Indeed, some teacher educators have already begun to raise thoughtful and important questions such as: What conceptions of mathematics and of mathematics teaching and learning do teachers (pre- and inservice) bring to teacher education and staff development programs? What can those programs offer to support or challenge those conceptions? Although these questions are not answered by research on teachers' beliefs and conceptions, they are nevertheless important questions that might not have been asked in the absence of this research.

In 1980, Fenstermacher argued that transforming teachers' beliefs requires knowledge of current beliefs, and he called for more emphasis on descriptive studies of teaching that include attention to teachers' mental states and cognitive processes. Over a decade later, a knowledge base of teachers' beliefs about mathematics and its teaching and learning is beginning to develop. How we use this information to help teachers reflect on their own beliefs and practice is an issue that deserves attention. Indeed, the use of this information to help teachers at all levels reflect on their teaching was one of the directions for future research called for at the Conference on Effective Mathematics Teaching of the Research Agenda for Mathematics Education Project (Cooney, Grouws, & Jones, 1988). Certainly case studies resulting from this research can be and have been used in teacher education programs, much in the same manner as case studies are used in law and medical schools.

In discussing the contributions of research on teacher thinking to teacher education Clark (1988) noted:

Research on teacher thinking does not constitute the ground for radical revision of the form and content of teacher preparation. Some of the most important contributions to teacher education may take the form of rationalizing, justifying, and understanding practices that have long been in place in teacher education. Furthermore, many contributions of research on teacher thinking will not make teacher education easier, but they may make teacher preparation more interesting. (p. 6)

By underscoring the influence of teachers' conceptions on their practice, this line of research can help end the debate over whether teachers should be well educated or highly trained. It certainly cautions against reducing teacher preparation to training on specific, well-defined skills and competencies—a practice rooted in the process-product research of the 1960s and 1970s.

As with research on teacher thinking in general, research on teachers' mathematical conceptions may not offer a blueprint for revision of mathematics teacher education programs, but it certainly alerts us to question our current practices. As observed earlier, very few cases of teachers with an informed historical and philosophical perspective of mathematics have been documented in the literature. This observation may suggest the need to revise curricula to include courses in the history and philosophy of mathematics. Yet it will probably require more than the inclusion of two more courses in teachers' undergraduate coursework to overcome the apparently pervasive influence of their pre-college mathematical experiences on their conceptions. It appears that much about the nature of the discipline is effectively conveyed by the very manner in which instruction in the content of mathematics is conducted.

Research on teachers' beliefs and conceptions has stimulated reflection on our role as teacher educators. What do we believe about mathematics teaching and learning? To what extent are our methods of teaching and evaluating students' work consistent with our professed beliefs about mathematics and mathematics teaching and learning? What challenges do we face in changing our practices to better reflect our conceptions? What effect, if any, do our methods of teaching mathematics and the materials we use have on our students' conceptions of mathematics? Although research on teachers' mathematical conceptions does not provide specific guidelines on how to educate teachers, it has provided us with examples of concepts, methods, and ideas on which to reflect. It certainly has alerted us to the influence our conceptions might have on how our students interpret and internalize the experiences we offer them in teacher education programs.

That research on teachers' conceptions and beliefs has influenced research and practice in education is evidenced in two projects currently underway. The projects are designed to investigate the impact of teacher education programs on prospective teachers. One of these projects is of national scope and is housed at the National Center for Research on Teacher Education at Michigan State University. The other project is *The Learning to Teach Mathematics* project at Virginia Polytechnic Institute and State University. Rather than simply assessing prospective teachers' growth in terms of knowledge and skills as has been traditionally done, both projects include an examination of the teachers' beliefs about mathematics and about the teaching and learning of mathematics. Had these projects been implemented 15 years earlier, it is unlikely that they would have included teachers' beliefs in their analyses of teachers' growth.

Contributions to Research on Mathematics Teaching and Learning

As part of the broader area of research on teachers' cognitions, studies of teachers' conceptions have contributed to a conceptual shift in the field of research on teaching, moving away from a behavioral conception of teaching towards a conception that takes account of teachers as rational beings. This conceptual shift has caused researchers to ask very different kinds of questions about teaching than they asked in the 1960s and early 1970s. The reconceptualization of teaching has led to a change in the research agenda—from determining the basic skills or competencies of teaching associated with student growth to understanding teaching from teachers' perspectives. An understanding of teaching from teachers' perspectives complements our growing understanding of learning from learners' perspectives, which, in turn, enriches the idea of schooling as the negotiation of norms, practices, and meanings (Cobb, 1988).

With a more informed view of teaching and learning, the mathematics education community now knows, for example, that it was naive to try to measure teachers' knowledge of mathematics in terms of number of completed college level mathematics courses and to look for connections between knowledge of mathematics, thus measured, and teaching effectiveness—measured in terms of student growth scores. It is also understood now that it was inappropriate to attempt to define teacher effectiveness solely in terms of behaviors associated with student gain scores. Researchers understand why such simple designs for studying mathematics teaching in classrooms were inadequate. A more informed view, although still incomplete, of the role that teachers' conceptions play in teaching has led to the acknowledgement that no description of mathematics teaching and learning is adequate and complete unless it includes consideration of the beliefs and intentions of teachers and students (Fenstermacher, 1980).

Thoughtful analyses of the relationship between teachers' beliefs and practice have increased our awareness of the strong influence the social context of the classroom has on teachers' intentions and actions, complementing what is known to be similarly the case with students. Along with an increased awareness of that influence has come an increased appreciation of the need to view the mathematics classroom as a socially organized unit where social situations are "constituted at every moment through the interaction of reflective subjects" (Bauersfeld, 1980, p. 30). In short, research on teachers' beliefs has made clearer to us that no simple model of teaching and learning can be used to account for teachers' and students' actions in the classroom.

Finally, a contribution of this line of research to research on teaching has been to highlight the importance of teacher reflection as a vehicle for knowledge growth. It is curious that while the research literature on teacher reflection is rapidly growing, the contribution of mathematics educators to that research has been mainly through studies of mathematics teachers' beliefs.

Much of the unrealized potential of research on teachers' beliefs and conceptions lies in the study of the relationship between teachers' conceptions and students' conceptions. Plausible as it may seem to assume that teachers' conceptions influence students' conceptions, virtually nothing is known about the extent to which teachers communicate their mathematical conceptions to students, and how this occurs. Likewise, virtually nothing is known about whether students' views of the subject matter influence teachers' instructional decisions and actions as well as their views of the subject. Findings from such investigations would be rich in practical implications for mathematics education.

Practical Implications

The studies addressed in this chapter indicate that, as in the case of children, teachers' conceptions entail more than just knowledge of specific mathematical content and pedagogical skills. Indeed, they entail knowledge and beliefs about mathematics, about the rules in effect while teaching and learning the subject, and more. Much of this is not taught explicitly in schools or in teacher preparation programs; hence, much of it must be learned from experience in the classroom.

If there are certain mathematical beliefs and conceptions that we as teachers want to cultivate in our students, be they education or engineering majors, then we must examine the extent to which the structure and conduct of our mathematics classes are conducive to cultivating them. We must carefully examine the ways in which we and the materials we use portray mathematics to our students. We must develop a sensitivity for the many subtle ways in which unintended messages and meanings might be communicated to our students.

At the same time that we as teachers reflect on our own practices, as teacher educators we must investigate ways of helping teachers reflect on theirs. We must find ways of helping teachers become aware of the implied rules and beliefs that operate in their classrooms and help them to examine their consequences. For example, teachers' often unexamined assumptions or beliefs about what children are capable or not capable of learning can render them impervious to matters of children's cognitions; we must find ways of helping teachers examine those assumptions.

As teacher educators or staff developers, we must consider and test alternatives to presenting ourselves as pedagogical authorities who possess all the answers and teach teachers what they ought to believe and how they ought to teach. We must explore ways to help teachers examine their beliefs and practices, develop intrinsic motivations for considering alternatives to their current practices, and develop personal reasons for justifying their actions. For teachers, intrinsic motivation for considering alternatives must come from their own experience in the classroom. Thus, we may consider helping them develop awareness of shortcomings in their current practices that will prompt them to consider alternatives. Only then, it would seem, true collaboration between teacher and staff developers can begin (see Cobb et al., 1990, for specifics of such an approach to staff development efforts).

We should not take lightly the task of helping teachers change their practices and conceptions. Attempts to increase teachers' knowledge by demonstrating and presenting information about pedagogical techniques have not produced the desired results. Indeed, the research reviewed here suggests that teachers' conceptions of mathematics, of how it should be taught, and of how children learn it are deeply rooted. Research would caution us against underestimating the robustness of those conceptions and practices. The tendency of teachers to interpret new ideas and techniques through old mindsets— even when the ideas have been enthusiastically embraced— should alert us against measuring the fruitfulness of our work in superficial ways. We should regard change as a long-term process resulting from the teacher testing alternatives in the class-

room, reflecting on their relative merits vis-á-vis the teacher's goals, and making a commitment to one or more alternatives.

It would seem that some research could be useful in helping teachers examine their beliefs and practices. Research can provide food for thought; certainly, case studies of teachers can be used intentionally to prompt teachers to reflect upon and examine their own beliefs and practices. We should be cautiously optimistic, however, about the potential of using research to bring about desirable changes in teachers. Changes in teachers' practices are unlikely to result from presenting, examining, or discussing research studies alone. Whether such efforts have an effect depends on the extent to which teachers accommodate their existing conceptions to the new ideas and how, if at all, those ideas are translated into action. Schon (1987) noted that

In the terrain of professional practice, applied science and research-based technique occupy a critically important though limited territory, bounded on several sides by artistry. There [is] an art of problem framing, an art of implementation, and an art of improvisation—all necessary to mediate the use in practice of applied science and technique. (p. 13)

Finally, as researchers interested in teachers' cognition as it relates to the teaching of mathematics, we can begin to investigate how teachers learn from their experiences in the classroom as they interact with the students and the subject matter, how they might assimilate new information about mathematics, its teaching, and its learning, and how that information is internalized. Detailed accounts of these processes in the form of explanatory models can be most useful to mathematics teacher educators and staff developers.

References

Abelson, R. (1979). Differences between belief systems and knowledge systems. *Cognitive Science, 3*, 355–366.

Abelson, R. P. & Carroll, J. D. (1965). Computer simulation of individual belief systems. *American Behavioral Scientist, 8*(9), 24–30.

Ball, D. L. (1986, September). *Unlearning to teach mathematics.* Paper presented at the meeting of the North American Chapter of the International Group for the Psychology of Mathematics Education, East Lansing, MI.

Ball, D. L. (1988). Unlearning to teach mathematics. *For the Learning of Mathematics, 8*(1), 40–48.

Ball, D. L. (1990). Prospective elementary and secondary teachers' understanding of division. *Journal for Research in Mathematics Education, 21*(3), 132–144.

Bauersfeld, H. (1980). Hidden dimensions in the so-called reality of a mathematics classroom. *Educational Studies in Mathematics, 11*, 23–41.

Benacerraf, P. & Putnam, H. (1964). *Philosophy of mathematics: Selected readings.* Oxford: Basil Blackwell.

Brown, C. A. (1985). *A study of the socialization to teaching of a beginning mathematics teacher.* Unpublished doctoral dissertation, University of Georgia, Athens.

Brownell, W. (1935). Psychological considerations in the learning and teaching of arithmetic. In *The teaching of arithmetic, NCTM Tenth Yearbook.* (pp. 1–31). New York: Bureau of Publications, Teachers College, Columbia University.

Bush, W. (1983). Preservice secondary mathematics teachers' knowledge about teaching mathematics and decision-making during teacher training (Doctoral Dissertation, University of Georgia, 1982). *Dissertation Abstracts International, 43*, 2264A.

Bussis, A. M., Chittenden, F., & Amarel, M. (1976). *Beyond surface curriculum.* Boulder, CO: Westview Press.

Carpenter, T. P., Fennema, E., Peterson, P. L., & Carey, D. A. (1988). Teachers' pedagogical content knowledge in mathematics. *Journal for Research in Mathematics Education, 19*, 385–401.

Carpenter, T. P., Fennema, E., Peterson, P. L., Chiang, C. P., & Loef, M. (1989). Using knowledge of children's mathematics thinking in classroom teaching: An experimental study. *American Educational Research Journal, 26*(4), 499–532.

Clark, C. (1988). Asking the right questions about teacher preparation: Contributions of research on teacher thinking. *Educational Researcher, 17*(2), 5–12.

Clark, C. M., & Peterson, P. L. (1986). Teachers' thought processes. In M. C. Wittrock (Ed.), *Handbook of research on teaching* (3rd ed., pp. 255–296). New York: Macmillan.

Cobb, P. (1988). The tension between theories of learning and theories of instruction in mathematics education. *Educational Psychologist, 23,* 87–104.

Cobb, P., & Steffe, L. P. (1983). The constructivist researcher as teacher and model builder. *Journal for Research in Mathematics Education, 14,* 83–94.

Cobb, P., Wood, T., & Yackel, E. (1990). Classrooms as learning environments for teachers and researcher. In R. Davis, C. Maher, & N. Noddings, (Eds.), Constructivist views on the teaching and learning of mathematics. *Journal for Research in Mathematics Education Monograph* (pp. 125–146). Reston, VA: National Council of Teachers of Mathematics.

Cobb, P., Wood, T., & Yackel, E. (in press). A constructivist approach to second grade mathematics. In E. von Glasersfeld (Ed.), *Constructivism in mathematics education.* Dordrecht, Holland: Reidel.

Collier, C. P. (1972). Prospective elementary teachers' intensity and ambivalence of beliefs about mathematics and mathematics instruction. *Journal for Research in Mathematics Education, 3,* 155–163.

Committee of Inquiry into the Teaching of Mathematics in Schools. (1983). *Mathematics Counts* (The Cockcroft Report). London: Her Majesty's Stationery Office.

Confrey, J. (1985, April). *A constructivist view of mathematics instruction. Part I: A theoretical perspective.* Paper presented at the annual meeting of the American Educational Research Association, Chicago, IL.

Cooney, T. J. (1985). A beginning teacher's view of problem solving. *Journal for Research in Mathematics Education, 16,* 324–336.

Cooney, T. J., Grouws, D. A., & Jones, D. (1988). An agenda for research on teaching mathematics. In D. A. Grouws, T. J. Cooney, & D. Jones (Eds.), *Perspectives on research on effective mathematics teaching* (pp. 253–261). Reston, VA: National Council of Teachers of Mathematics.

Copes, L. (1979). The Perry development scheme and the teaching of mathematics. Paper presented at the annual meeting of the International Group for the Psychology of Mathematics Education, Warwick, England.

Copes, L. (1982). The Perry development scheme: A metaphor for learning and teaching mathematics. *For the Learning of Mathematics, 3,* 38–44.

Davis, P. J., & Hersh, R. (1980). *The Mathematical Experience.* Boston: Birkhäuser.

De Corte, E., & Verschaffel, L. (1985). Beginning first graders' initial representation of arithmetic word problems. *The Journal of Mathematical Behavior, 4,* 3–21.

Dewey, J. (1964). The nature of subject matter. In R. R. Archambault (Ed.), *John Dewey on education* (pp. 359–372). Chicago: University of Chicago Press. (Original work published 1916)

Dougherty, B. J. (1990). Influences of teacher cognitive/conceptual levels on problem-solving instruction. In G. Booker et al. (Eds.), *Proceedings of the Fourteenth International Conference for the Psychology of Mathematics Education* (pp. 119–126). Oaxtepec, Mexico: International Group for the Psychology of Mathematics Education.

Duffy, G. (1977, December). *A study of teacher conceptions of reading.* Paper presented at the National Reading Conference, New Orleans.

Elbaz, F. (1981). The teacher's "practical knowledge": Report of a case study. *Curriculum Inquiry, 11,* 43–71.

Erlwanger, S. (1975). Benny's conceptions of rules and answers in IPI mathematics. *Journal of Mathematical Behavior, 1*(2), 7–25.

Ernest, P. (1985). The philosophy of mathematics and mathematics education. *International Journal of Science and Technology, 16*(5), 603–612.

Ernest, P. (1988, July). *The impact of beliefs on the teaching of mathematics.* Paper prepared for ICME VI, Budapest, Hungary.

Feiman-Nemser, S., & Floden, R. E. (1986). The cultures of teaching. In M. C. Wittrock (Ed.), *Handbook of research on teaching* (3rd ed., pp. 505–526). New York: Macmillan.

Fenstermacher, G. D. (1980). On learning to teach effectively from research on teacher effectiveness. In C. Denham & A. Lieberman (Eds.), *Time to learn* (pp. 127–138). Washington, DC: National Institute of Education.

Feyerabend, P. K. (1975). *Against method.* Atlantic Highlands: Humanities Press.

Fransella, F., & Bannister, J. (1977). *A manual for repertory grid technique.* London: Academic Press.

Good, T. L., Grouws, D. A., & Ebmeier, H. (1983). *Active mathematics teaching.* New York: Longman.

Grant, C. E. (1984). A study of the relationship between secondary mathematics teachers' beliefs about the teaching-learning process and their observed classroom behaviors (Doctoral dissertation, University of North Dakota, 1984). *Dissertation Abstracts International, 46,* DA8507627.

Green, T. F. (1971). *The activities of teaching.* New York: McGraw-Hill.

Grossman, P., Wilson, S., & Shulman, L. (1989). Teachers of substance: Subject matter knowledge for teaching. In M. C. Reynolds (Ed.), *Knowledge base for the beginning teacher* (pp. 23–34). Oxford: Pergamon Press.

Halmos, P. R. (1975). The teaching of problem solving. *American Mathematical Monthly, 82*(5), 446–470.

Harvey, O., Prather, M., White, B., & Hoffmeister, J. (1968). Teachers' beliefs, classroom atmosphere, and student behavior. *American Educational Research Journal, 5,* 151–166.

Hawkins, D. (1973). Nature, man, and mathematics. In A. G. Howson (Ed.), *Developments in mathematical education* (pp. 115–135). Cambridge: Cambridge University Press.

Helms, J. M. (1989). *Preservice secondary mathematics teachers' beliefs about mathematics and the teaching of mathematics: Two case studies.* Unpublished doctoral dissertation, University of Georgia, Athens.

Hersh, R. (1986). Some proposals for revising the philosophy of mathematics. In T. Tymoczko (Ed.), *New directions in the philosophy of mathematics* (pp. 9–28). Boston: Birkhauser.

Hintikka, J. (1962). *Knowledge and belief.* Ithaca, NY: Cornell University Press.

Ignatovich, F. R., Cusick, P. A., & Ray, J. E. (1979). *Value/beliefs patterns of teachers and those administrators engaged in attempts to influence teaching* (Research Series No. 33). East Lansing: Institute for Research on Teaching, Michigan State University.

Jones, D., Henderson, E., & Cooney, T. (1986). Mathematics teachers beliefs about mathematics and about teaching mathematics. In G. Lappan & R. Even (Eds.), *Proceedings of the eighth annual meeting of the North American chapter of the International Group for the Psychology of Mathematics Education* (pp. 274–279). East Lansing, MI: Michigan State University.

Kerlinger, F. N. (1967). Social attitudes and their criterial referents: A structural theory. *Psychological Review, 74,* 110–122.

Kesler, R., Jr. (1985). *Teachers' instructional behavior related to their conceptions of teaching and mathematics and their level of dogmatism: Four case studies.* Unpublished doctoral dissertation, University of Georgia, Athens.

Kuhn, T. (1962). *The structure of scientific revolutions.* Chicago: University of Chicago Press.

Kuhs, T. (1980). *Teachers' conceptions of mathematics.* Unpublished doctoral dissertation, Michigan State University, East Lansing.

Kuhs, T. M., & Ball, D. L. (1986). *Approaches to teaching mathematics: Mapping the domains of knowledge, skills, and dispositions.* East Lansing: Michigan State University, Center on Teacher Education.

Lakatos, I. (1976). *Proofs and refutations.* Cambridge, U.K.: Cambridge University Press.

Lakatos, I. (1978). *Mathematics science and epistemology: Philosophical papers* (Vol. 2). Cambridge, U.K.: Cambridge University Press.

Lakatos, I. (1986). A renaissance of empiricism in the recent philosophy of mathematics? In T. Tymoczko (Ed.), *New directions in the philosophy of mathematics* (pp. 29–48). Boston: Birkhäuser.

Leinhardt, G. (1985). *Getting to know: Tracing students' mathematical knowledge from intuition to competence.* Pittsburgh: University of Pittsburgh, Learning Research and Development Center.

Lerman, S. (1983). Problem solving or knowledge centered: The influence of philosophy on mathematics teaching. *International Journal of Mathematical Education in Science and Technology, 14*(1), 59–66.

Lerman, S. (1987). Investigations: Where to now. In P. Ernest (Ed.), *Teaching and learning mathematics, Part 1 (Perspectives 33)* (pp. 47–56). Exeter: University of Exeter School of Education.

Lindquist, M. M. (Ed.). (1989). *Results from the fourth mathematics assessment.* Reston, VA: National Council of Teachers of Mathematics.

Marks, R. (1987). Those who appreciate: The mathematician as secondary teacher. A case study of Joe, a beginning mathematics teacher. *Knowledge growth in a profession series.* Stanford, CA: Stanford University, School of Education.

McGalliard, W. A., Jr. (1983). Selected factors in the conceptual systems of geometry teachers: Four case studies (Doctoral dissertation, University of Georgia, 1982). *Dissertation Abstracts International, 44,* 1364A.

McKnight, C. C., Crosswhite, F. J., Dossey, J. A., Kifer, E., Swafford, J. O., Travers, K. J., & Cooney, T. J. (1987). *The underachieving curriculum: Assessing U.S. school mathematics from an international perspective.* Champaign, IL: Stipes Publishing Company.

Mellin-Olsen, S. (1981). Instrumentalism as an educational concept. *Educational Studies in Mathematics, 12,* 351–367.

Meyerson, L. N. (1978). Conception of knowledge in mathematics: Interaction with and applications to a teaching methods course (Doctoral dissertation, State University of New York, Buffalo, 1977). *Dissertation Abstracts International, 38,* 733A.

Munby, H. (1982). The place of teachers' beliefs in research on teacher thinking and decision making, and an alternative methodology. *Instructional Science, 11,* 201–225.

National Council of Teachers of Mathematics. (1989). *Curriculum and evaluation standards for school mathematics.* Reston, VA: Author.

National Research Council. (1989). *Everybody counts: A report to the nation on the future of mathematics education.* Washington, DC: National Academy Press.

Needham, R. (1972). *Belief, language, and experience.* Oxford: Basil Blackwell.

Nespor, J. (1987). The role of beliefs in the practice of teaching. *Journal of Curriculum Studies, 19,* 317–328.

Olson, J. K. (1981). Teacher influence in the classroom. *Instructional Science, 10,* 259–275.

Owens, J. E. (1987). *A study of four preservice secondary mathematics teachers' constructs of mathematics and mathematics teaching.* Unpublished doctoral dissertation, University of Georgia, Athens.

Perry, W. G., Jr. (1970). *Forms of intellectual and ethical development in the college years: A scheme.* New York: Holt, Rinehart, and Winston.

Perry, W. G., Jr. (1981). Cognitive and ethical growth: The making of meaning. In A. Chickering (Ed.), *The modern American college* (pp. 76–116). New York: Jossey-Bass.

Peterson, P. L., Fennema, E., & Carpenter, T. P. (1987). *Teachers' pedagogical content beliefs in mathematics.* Paper presented at the AERA Conference, Washington, DC.

Polya, G. (1963). On learning, teaching, and learning teaching. *American Mathematical Monthly, 70,* 605–619.

Putnam, H. (1986). What is mathematical truth? In T. Tymoczko (Ed.), *New directions in the philosophy of mathematics* (pp. 49–65). Boston: Birkhäuser.

Rokeach, M. (1960). *The open and closed mind.* New York: Basic Books.

Romberg, T. A., & Carpenter, T. P. (1986). Research on teaching and learning mathematics: Two disciplines of scientific inquiry. In M. C. Wittrock (Ed.), *Handbook of research on teaching,* (3rd ed., pp. 850–873). New York: Macmillan.

Russell, B. (1948). *Human knowledge: Its scope and limits.* New York: Simon and Schuster.

Scheffler, I. (1965). *Conditions of knowledge: An introduction to epistemology and education.* Glenview, IL: Scott, Foresman, and Company.

Schoenfeld, A. (1983). Beyond the purely cognitive: Belief systems, social cognitions, and metacognitions as driving forces in intellectual performance. *Cognitive Science, 7,* 329–363.

Schoenfeld, A. (1985). *Mathematical problem solving.* San Diego, CA: Academic Press, Inc.

Schon, D. A. (1987). *Educating the reflective practitioner.* San Francisco: Jossey-Bass.

Schram, P., & Wilcox, S. K. (1988, November) Changing preservice teachers' conceptions of mathematics learning. In M. J. Behr, C. B. Lacampagne, & M. M. Wheeler (Eds.), *PME-NA: Proceedings of the tenth annual meeting* (pp. 349–355). DeKalb, IL: Northern University.

Schram, P., Wilcox, S., Lanier, P., & Lappan, G. (1988). *Changing mathematical conceptions of preservice teachers: A content and pedagogical intervention* (Research Report No. 1988-4). East Lansing, MI: National Center for Research on Teacher Education.

Shavelson, R. J. (1988). Contributions of educational research to policy and practice: Constructing, challenging, changing cognition. *Educational Researcher, 17*(7), 4–11.

Shavelson, R. J., & Stern, P. (1981). Research on teachers' pedagogical thoughts, judgments, decisions, and behavior. *Review of Educational Research, 51,* 455–498.

Shaw, K. (1989). *Contrasts of teacher ideal and actual beliefs about mathematics understanding: Three case studies.* Unpublished doctoral dissertation, University of Georgia, Athens.

Shirk, G. B. (1973). *An examination of conceptual frameworks of beginning mathematics teachers.* Unpublished doctoral dissertation, University of Illinois at Urbana-Champaign.

Shroyer, J. C. (1978, March). *Critical moments in the teaching of mathematics.* Paper presented at the annual meeting of the American Educational Research Association, Toronto.

Shulman, L. S. (1986). Those who understand: Knowledge growth in teaching. *Educational Researcher, 15*(2), 4–14.

Shulman, L. S., & Elstein, A. S. (1975). Studies of problem solving judgment and decision making. In F. N. Kerlinger (Ed.), *Review of research in education* (Vol. 3, pp. 3–42). Itasca, IL: F. E. Peacock.

Skemp, R. R. (1978). Relational understanding and instrumental understanding. *Arithmetic Teacher, 26*(3), 9–15.

Slavin, R. E. (1987). Cooperative learning and individualized instruction. *Arithmetic Teacher, 35,* 14–16.

Steen, L. (1988). The science of patterns. *Science, 240,* 611–616.

Steffe, L. P., & Blake, R. N. (1983). Seeking meaning in mathematics instruction: A response to Gagne. *Journal for Research in Mathematics Education, 14,* 210–213.

Steinberg, R., Haymore, J., & Marks, R. (1985, April). *Teachers' knowledge and structuring content in mathematics.* Paper presented at the annual meeting of the American Educational Research Association, San Francisco.

Stonewater, J. K., & Oprea, J. M. (1988). An analysis of in-service teachers' mathematical beliefs: A cognitive development perspective. In M. J. Behr, C. B. Lacampagne, & M. M. Wheeler (Eds.), *PME-NA: Proceedings of the tenth annual meeting* (pp. 356–363). DeKalb, IL: Northern University.

Thom, R. (1973). Modern mathematics: Does it exist?. In A. G. Howson (Ed.), *Developments in mathematical education: Proceedings of the Second International Congress on Mathematics Education* (pp. 194–209). Cambridge: Cambridge University Press.

Thompson, A. (1982). *Teachers' conceptions of mathematics: Three case studies*. Unpublished doctoral dissertation, University of Georgia, Athens.

Thompson, A. (1984). The relationship of teachers' conceptions of mathematics teaching to instructional practice. *Educational Studies in Mathematics, 15,* 105–127.

Thompson, A. (1988). Learning to teach mathematical problem solving: Changes in teachers' conceptions and beliefs. In R. I. Charles & E. Silver (Eds.), *The teaching and assessing of mathematical problem solving* (pp. 232–243). Reston, VA: National Council of Teachers of Mathematics.

Thompson, P. W. (1985). Experience, problem solving, and learning mathematics: Considerations in developing mathematics curricula. In E. A. Silver (Ed.), *Teaching and learning mathematical problem solving: Multiple research perspectives* (pp. 189–236). Hillsdale, NJ: Lawrence Erlbaum.

Tymoczko, T. (1986a). Making room for mathematicians in the philosophy of mathematics. *The Mathematical Intelligencer, 8,* 44–50.

Tymoczko, T. (Ed.). (1986b). *New directions in the philosophy of mathematics.* Boston: Birkhauser.

von Glasersfeld, E. (1987). Learning as a constructive activity. In C. Janvier (Ed.), *Problems of representation in the teaching and learning of mathematics* (pp. 3–17). Hillsdale, NJ: Lawrence Erlbaum.

Wolgast, E. H. (1977). *Paradoxes of knowledge.* Ithaca, NY: Cornell University Press.

·8·

TEACHERS' KNOWLEDGE AND ITS IMPACT

Elizabeth Fennema
Megan Loef Franke

UNIVERSITY OF WISCONSIN–MADISON

To be a teacher requires extensive and highly organized bodies of knowledge.
(Shulman, 1985, p. 47)

The single factor which seems to have the greatest power to carry forward
our understanding of the teachers' role is the phenomenon of teachers'
knowledge. (Elbaz, 1983, p. 45)

No one questions the idea that what a teacher knows is one of the most important influences on what is done in classrooms and ultimately on what students learn. However, there is no consensus on what critical knowledge is necessary to ensure that students learn mathematics. Many components of teachers' knowledge have been identified. Some scholars suggest that since one cannot teach what one does not know, teachers must have in-depth knowledge not only of the specific mathematics they teach, but also of the mathematics that their students are to learn in the future. Only with this intensive knowledge of mathematics can a teacher know how to structure her or his own mathematics teaching so that students continue to learn. Others suggest that knowledge of cultural and ethnic diversity is essential for effective teaching. Since the United States is becoming increasingly multicultural, and since a student's culture is a major determinant of how the student learns, before a teacher can be effective, he or she must understand the cultural diversity of the students. Still other scholars suggest that knowledge of how students think and learn is vital knowledge for teachers, while others believe that knowledge of general pedagogical principles is a necessary component of teachers' knowledge.

Is there critical information that can be gained from research that will help us to identify whether or not these are critical components of teachers' knowledge? Do these components affect what teachers do in classrooms and what students learn? Can past research indicate directions for future research about teachers' knowledge that would help in the improvement of mathematics education?

This chapter will not answer the above questions precisely, but will provide a critical review of the scholarly literature that has dealt with these questions about teacher knowledge. Instead of including a total review of the literature, we have identified and discussed some important research themes, presented summaries of reviews, and provided examples of well-conceived and implemented studies that have addressed those themes. We speculate on how knowledge gained from past research addresses the questions, and we suggest how future research can be designed to provide useful information about teacher knowledge, classroom instruction, and pupil learning. The focus of this chapter is on teacher knowledge. Although we believe it is impossible to separate beliefs and knowledge, beliefs are covered by Thompson, Chapter 7, this volume. We will omit most discussion of beliefs in this chapter.

We would like to thank Thomas Cooney, University of Georgia, and Deborah Ball, Michigan State University, for their thoughtful comments on an earlier draft of this chapter and the National Science Foundation and the Office of Education Research and Improvement of the Department of Education for partial support while writing it. The opinions expressed are those of the authors and not necessarily those of the reviewers, NSF, or OERI.

We have attempted to examine literature from many different viewpoints, with mathematics education literature as our starting point. We have gone into other curriculum areas (such as history or science) when there was informative scholarship to consider. In addition, we looked at general scholarship on teaching as well as theories of knowledge acquisition and organization posited by cognitive scientists. The questions dealing with teacher knowledge are receiving an increasing amount of attention by researchers. Even so, the area is complex, ill defined, and often poorly studied. This chapter's goal is to present the area with all of its difficulties in the hope that future research will provide the important information about teacher knowledge that is critically needed.

COMPONENTS OF TEACHER KNOWLEDGE

Common sense suggests that teacher knowledge is not monolithic. It is a large, integrated, functioning system with each part difficult to isolate. While many have speculated on the components of teacher knowledge, only a few components have received major attention from researchers: content knowledge, knowledge of learning, knowledge of mathematical representations, and pedagogical knowledge. Some have studied knowledge as integrated, but most have not. Because of the emphasis of many researchers, we have been forced to violate our premise of the importance of the integration of knowledge and start with a consideration of some individual components of teacher knowledge. When possible, we also report on how some of the components have been studied as integrated. After examining the research that has been done with individual components, we shall move on to some hypothesized frameworks and models of integrated knowledge. We conclude by proposing a research model for examining the integration of teacher knowledge.

Knowledge of Mathematics

When one addresses the issue of teacher knowledge, mathematics immediately presents itself. Within mathematics education, there exists much rhetoric that reflects strong beliefs about the importance of mathematical knowledge to teachers. For example, one of the most widely offered explanations of why students do not learn mathematics is the inadequacy of their teachers' knowledge of mathematics. Another example is certification requirements for secondary school teachers, which almost always list the number and type of mathematics courses that must be completed before a person is allowed to teach. Underlying these beliefs is an assumption that mathematical knowledge is critical for teachers before they can help students learn.

The belief in the importance of mathematical knowledge is shared by scholars in the field. Consider these statements: "Knowledge of mathematics is obviously fundamental to being able to help someone else learn it" (Ball, 1988a, p. 12). "A firm grasp of the underlying concepts is an important and necessary framework for the elementary teacher to possess...[when] teaching related concepts to children....[and] many teachers simply do not know enough mathematics" (Post, Harel, Behr, & Lesh, 1988, pp. 210, 213).

There is some evidence that the mathematical knowledge of teachers, particularly elementary and middle school teachers, is not very good. Consider recent reports: after reviewing the studies about elementary preservice teachers' knowledge of content, Brown, Cooney, and Jones (1990) concluded that "research of this type leaves the distinct impression that preservice elementary teachers do not possess a level of mathematical understanding that is necessary to teach elementary school mathematics as recommended in various proclamations from professional organizations such as NCTM" (p. 643). Post et al. (1988) investigated 218 intermediate grade teachers' knowledge about the conceptual underpinnings of rational numbers. They summarized their results by saying:

Regardless of which item category is selected, a significant percentage of teachers were missing one half to two thirds of the items. This percentage varied by category, but in general, 20 to 30 percent of the teachers scored less than 50 percent on the overall instrument (p. 191).

In spite of the beliefs in the importance of mathematical knowledge and the evidence that some teachers do not have adequate knowledge of mathematics, research has provided little support for a direct relationship between teachers' knowledge of mathematics and student learning. Consider one set of studies that investigated this relationship. The relationship of teacher knowledge and student learning was reported in the National Longitudinal Study of Mathematical Abilities (NLSMA) (School Mathematics Study Group, 1972). The NLSMA investigators carefully ascertained how many mathematics courses teachers had taken and then computed correlations between that number and student learning. No important relationships were found. Five years later, Eisenberg (1977) replicated this study and reported the same results. It is important to note that these studies defined teachers' knowledge as the number of university-level mathematics courses successfully completed. No attempt was made to measure what the teachers knew about mathematics or to ascertain accurately the mathematics covered in the various courses completed.

Some studies reported on the relationship between student learning and teacher knowledge as measured by achievement tests such as the National Teachers Examination. Other studies investigated the relationship between teachers' acquisition of knowledge and their students' learning. Neither type of study indicated much of a relationship between teachers' knowledge and the learning of their students (General Accounting Office, 1984). These results were so discouraging, Begle concluded that "the effects of a teacher's subject matter knowledge... seem to be far *less* (emphasis added) powerful than most of us had realized....Our attempts to improve mathematics education would *not* profit from further studies of teachers." (1979, pp. 54–55).

Thus, it appears that while we can document that teachers' knowledge (at least elementary school teachers) is not very good, there has been little evidence until very recently that this lack of knowledge critically influenced learning. However, before one totally dismisses the belief that teacher knowledge of mathematics is a valuable area of study, consider how most of these studies measured knowledge. For example, in the NLSMA studies, the number of university courses taken was often used as a proxy measure for knowledge, and little evidence was presented about how teacher knowledge was integrated or whether a relationship existed between university courses taken and classroom teaching. Other studies used some form of a standardized test to identify teacher knowledge, and once again, no attempt was made to measure the complexity of teacher knowledge or the relationship between the formal mathematics that teachers knew and what they taught. Furthermore, correlational techniques were often used to assess the relationships between teacher knowledge and student performance so that little is known about the directionality of any existing relationships. Perhaps the inadequate measures of knowledge and relatively limited research methodology concealed any relationships that did exist between teachers' knowledge and student learning.

Recently, some studies have reported somewhat different results, partly because of a change in research methodology. Instead of using correlational techniques to measure the relationship between some measure of teacher knowledge and their students' learning, scholars have been looking at teaching itself. What teachers do in classrooms has been studied as a mediator between teachers' content knowledge and their students' learning. These studies have been, for the most part, conducted within the interpretive tradition (Erickson, 1986), and have concentrated on providing rich descriptions of teachers in action in their own classrooms. Usually a small number of teachers are described and inferences are drawn about the relationship between teachers' subject-matter knowledge and various aspects of the classroom. Often, expert teachers are described; more expert teachers are compared with less expert teachers; or a teacher is compared with him or herself in domains where he or she has more or less knowledge.

Three examples are described in depth in order to illustrate the ongoing methodology, the nature of the results, and what can and cannot be concluded from studies of this type. All three studies suggest that content knowledge does influence the decisions teachers make about classroom instruction. Consider the work of a first grade teacher and two university professors. These are very different teachers who teach students of different ages and experiences. From the descriptions of their teaching, it is clear that the teachers' knowledge of mathematics had an impact on their instruction. The nature of that impact and the impact on their students' learning is less clear.

An Expert First Grade Teacher. The first illustration involves an elementary school teacher who was observed over a two-year period teaching in content areas in which her content knowledge differed. Ms. Jackson is a teacher who had been working intensively with the Cognitively Guided Instruction (CGI) project (Carpenter & Fennema, in press; Fennema, Carpenter, & Peterson, 1989). After the first year of the project, the CGI staff identified her as an expert CGI teacher based on observations done in her classroom of both her teaching and her students' learning. During the first two years of the project, Ms. Jackson taught first grade. During the third year of the project, she moved with her class of first graders into the second grade.

We did intensive observations of her as she worked with her students in many content areas, but shall report here on observations done in addition, subtraction, and fractions during the year in which she was teaching second grade. As part of the CGI project, Ms. Jackson had studied and learned a great deal about the domain of addition and subtraction. In particular, she learned about the variety of problem types that can be solved by addition or subtraction, the relationships between addition and subtraction problems, and the strategies children use to solve them (Carpenter, Fennema, Peterson, & Carey, 1988). She received no inservice training in fractions. Not only had Ms. Jackson not studied in the area of fractions for a number of years, but she also reported that her knowledge of fractions was inadequate and that she wished she knew as much about fractions as she did about addition and subtraction. When her knowledge of fractions was measured, it was found to be limited in content, but rich in pedagogy (Lehrer & Loef, in press). It was clear that her knowledge of addition and subtraction content was much richer than her knowledge of fractions.

Ms. Jackson's methods of instruction for addition and subtraction were different than those for fractions. While the central activity of her class was always problem solving, when addition and subtraction was the focus, a wide variety of problem types were used: join, separate, part–part-whole, and compare. When fractions were taught, one basic type of problem was used: part-whole. The classroom discourse was largely directed by the students during addition and subtraction. They often wrote their own problems and discussed at length different solution strategies. If Ms. Jackson tried to move on to another problem before the students were ready, they would refuse to go on until they had resolved the original problem. Ms. Jackson let them direct the discourse and took most of her cues for pacing and selection of activities from the students themselves. Ms. Jackson was thrilled when an unexpected response from the children was given, and that often provided the direction for the rest of the class period or week. She would talk at great length to us about a certain child's response and how it helped determine what the child should do next.

The classroom discourse was much different when fractions were the context of problem solving; Ms. Jackson directed the students' interactions to a much greater extent. When an unexpected (to Ms. Jackson) response was given during addition or subtraction, it often formed the basis of more problem solving or extended discussion about the appropriateness of the response. Ms. Jackson was able to take the response and use it to question children or to present a different type of problem. When an unanticipated response was given during a lesson on fractions, Ms. Jackson listened carefully to the child, but did not act on what the child had said. She usually did not follow up on student responses by asking specific clarification questions, nor did she attempt to pose another problem that would build on the response (Lehrer & Loef, in press). Overall, there was less discussion and less mathematics occuring during lessons involving fractions than during problem solving involving addition and subtraction. Children did not discuss problem solutions as well or as comfortably in fractions as in addition and subtraction. Thus, the amount of variability in the context of the problems, as well as the amount of classroom discourse and follow-up activities differed depending on whether addition and subtraction or fractions were the context of the lesson. It seems reasonable to conclude that this difference in classroom behavior of Ms. Jackson was at least partly created by the difference in her knowledge of the two content areas.

The children in Ms. Jackson's second-grade class were tested at the end of first grade and at the end of second grade. Their ability to solve addition and subtraction problems increased dramatically during the year, while their fraction problem-solving skills increased less dramatically. In no way do we wish to indicate that what happened in Ms. Jackson's class with fractions was inadequate. The work with fractions performed by the second-grade children went beyond what most second-grade classes do, and the activities they did with fractions enabled the children to develop basic concepts about fractions as well as learn about multiplication and division. However, it is clear that in the area in which Ms. Jackson was more knowledgeable, instruction and subsequent learning was richer.

A Fourth-Fifth Grade Teacher with High Mathematical Knowledge.

Magdalene Lampert has taught elementary school for 10 years. She is currently a professor in the Institute for Research on Teacher Learning at Michigan State University, and, as part of her scholarly activities, teaches mathematics daily to one group of children. Although never explicitly measured, her background in mathematics, as well as education, is thorough. From reading Lampert's descriptions of her own teaching (Lampert, 1989) and a case study of Lampert by one of her colleagues (Ball, 1991), it is possible to identify some important aspects of her teaching. Lampert's overall goals of teaching mathematics include "learning what mathematics is and how one engages in it." She states that "these goals are purposefully coequal and interconnected with acquiring the stuff—concepts and procedures—of mathematics" (Ball, 1991, p. 31). She wants her students to learn about mathematical thinking and to discover that they have the power to do mathematics. "The goal is to help students develop mathematical power and to become active participants in mathematics as a system of human thought" (Ball, 1991, p. 35).

Before instruction begins, Lampert carefully thinks about the mathematics she wants to teach. During instruction, she engages the children in dialogue and continually responds to them by saying "How do you know that?" or "Why do you think that?" She listens carefully to the children's responses and keeps the lessons focused on the content she wants the children to learn. She is able to use almost any response given by a child as a mechanism to push the discussion further. Children appear to be actively engaged in mathematics and to be constructing their own ways to understand mathematics, which is congruent with Lampert's analysis of the mathematical ideas. "Lampert's pedagogy subtly blends goal and process.... She models mathematical thinking and activity, and asks questions that push students to examine and articulate their ideas.... Lampert draws the strategy and rationale for her approach from the discipline of mathematics itself" (Ball, 1991, p. 32).

Clearly, Lampert is using her knowledge of mathematics to structure what is done in her classroom. She recognizes that

mathematics consists of concepts and processes as well as procedures, and she integrates all three into her instruction. While we have rich descriptions of her teaching, she and her colleague unfortunately have not reported the effect of this instruction on her students' mathematics learning, so we have no evidence that they learn more than in a traditional classroom. We can only assume that such a rich environment must positively influence what her students learn. The impact of rich environments created by a teacher who knows mathematics is a topic that needs to be investigated.

A Knowledgeable University Teacher.

Alan Schoenfeld, a professor of mathematics, conducts research concerned with mathematical problem solving (Schoenfeld, 1985). As part of this work, he has taught and written about a number of intensive, university-level classes in heuristics of problem solving. There are somewhat different emphases in each class he has described, depending upon the specific research question under investigation. However, there are also similarities, as noted by Schoenfeld:

As might be expected, virtually all of our class time was devoted to solving problems and to discussing problem solutions. The outline of a typical day's work was as follows. Class usually began with a discussion of homework problems, the solutions to which were often presented by students. After the homework had been discussed, a sheet of problems would be distributed. The class then broke into small groups (three or four students in each) to work on the problems. I traveled around the room as a roving consultant while they worked. When they had made a reasonable amount of progress, or had run out of steam, we reconvened to see what they had accomplished and to tackle the hard problems as a group.

There were, of course, lectures on the problem-solving strategies in which sample solutions were given. Decision-making processes were explicitly role modeled when sample solutions were presented. When I arrived at a critical point in a solution, I raised three or four possible options and evaluated. On the basis of those evaluations I chose to pursue one option (perhaps the wrong one).

When the class convened as a whole to work problems, I served as orchestrator of the students' suggestions. My role was not to lead the students to a predetermined solution, although there were obviously times when it was appropriate to demonstrate particular mathematical ideas.... (I asked questions of the type): What do you think we should do? Are you all sure you understand the problem? Which suggestion should we try to implement and why?

Solving problems in small groups provided the opportunity to work with the students individually as they grappled with the problems. During this part of the instruction, my role, in essence, was that of a problem-solving coach. (Schoenfeld, 1985, pp. 221–222)

Once again, the classroom discourse is partly directed by the students, and the emphasis of the classroom is on the students' understanding of the processes and interrelationships in mathematics. Schoenfeld has not reported on what his students learned, so again, we can assume, but not verify, that their learning was affected by Schoenfeld's classroom behaviors.

Although we have different levels of information regarding each teacher's knowledge of the content, it seems that the mathematical knowledge of these three teachers had an impact on instruction. While the three situations had many components in common, some of the most important are the richness of the mathematics available for the learners and the structure of the classroom, which enabled the learners to participate in authentic mathematical activity. The emphasis of the mathematics in each class was on problem solving and the power of the learners to understand and to do mathematics. Learners in all three teachers' classes were doing the mathematics themselves, were thinking about the mathematics that they had done, and were carrying on mathematical discourse with their peers and teachers. The students in these classes experienced mathematics in a way that is fundamentally different than the procedural mathematics often experienced in classrooms.

Knowledge obtained from studies of this type is valuable. These studies provide rich descriptions of how a teacher who knows mathematics can structure classrooms so that students can interact with mathematics in a way that has been widely hypothesized to result in positive outcomes, both affective and cognitive (NCTM, 1989). Most researchers believe that students in such classrooms develop positive belief systems about themselves and mathematics, as well as learn a rich set of interrelated mathematical ideas. However, such descriptions are not or should not be ends in and of themselves. Until more information is provided about the impact of such classrooms on student outcomes, the research is incomplete. Certainly, a fertile area for research is to study the impact on students of rich classroom environments taught by knowledgeable teachers.

Content Knowledge and Other Disciplines. There are studies in other disciplines which reinforce the idea that teacher knowledge of content has an impact on students. Hashweh (1986) examined the effects of teachers' subject-matter knowledge on instruction in biology and physics. He measured teachers' knowledge in a well-defined domain in which they specialized and also in another related domain where they possessed less knowledge. Within their specialty, teachers had more detailed topic knowledge, more knowledge of other discipline concepts, more knowledge of higher-order principles, and more knowledge of ways to connect the topic entities to each other. In other words, in a topic in which teachers knew more, aspects of their knowledge were more related to other aspects and these pieces were integrated. Hashweh also gathered information about a number of teaching behaviors. In the areas in which they had less knowledge, teachers wrote exam questions focusing on recall, while questions asking for application, transfer, and drawing relationships between concepts were written in areas in which teachers had more knowledge. When teachers were not working within their specialty, they followed the text chapters more closely and failed to reject inappropriate sections. Furthermore, in the areas in which they were more knowledgeable, teachers dealt with students' misconceptions, while in the less knowledgeable areas, they either did not recognize the misconceptions, agreed with the students, or chose not to deal with the misconceptions. Hashweh concluded that knowledge of subject matter greatly contributed to the transformation of written curriculum into an active curriculum.

Wineburg and Wilson (1991) described the instructional effects of two teachers' knowledge of history. Both of these teachers had been identified as expert teachers by university professors, schools administrators, and other teachers. Data were collected through interviews, self reports, and classroom observations. Reported are protocols that illustrate extreme differences in teaching styles between the two teachers. One was a showman who captivated his class by his theatrical performance. The other had established an environment where her students were deeply involved in studying an assigned point of view so that they could debate with other students. Each teacher exhibited a strong grasp of the major historical concepts being taught. Wineburg and Wilson conclude that students in both classes learned a great deal of history because of their teachers' subject-matter knowledge. In trying to understand how this knowledge dictated their teaching methods, Wineburg and Wilson concluded that:

The role played by subject-matter knowledge in teaching is complicated: A teacher's knowledge of American history does not dictate in any algorithmic way what he or she chooses to teach. Teaching history melds subject-matter knowledge with a host of other understandings—knowledge of students, of learning and teaching, of schools, of curriculum and of educational aims.... What teachers know about U.S. history influences not only what they choose to teach but also how they choose to teach it. (p. 310)

The studies of the mathematics, science, and history teachers possessing a rich, integrated knowledge of subject matter lend credence to the belief that teacher knowledge can influence instruction. In these studies, the content of instruction appeared to be at least partially dependent on teacher knowledge, as did the discourse of the class. Knowledge did not dictate precisely what was done, as was clearly evident in the examples presented of the history teachers. However, the richness of the material being taught appeared to be directly related to the subject-matter knowledge of the teacher.

Certainly these studies, done in the interpretive research mode, do not produce definitive results about the impact of teacher knowledge of subject matter on instruction and, ultimately, on affective and cognitive outcomes. What is needed is work that explicitly and systematically measures teacher knowledge of content and that focuses on what students learn in environments where teachers differ in subject-matter knowledge. These studies do, however, point out the need to study knowledge of subject matter as it interacts with other aspects of teacher knowledge. The results of the studies are intriguing and lend credence to the belief that content knowledge of teachers is positively related to the structure of their classrooms.

The Organization of Teacher Content Knowledge. What is there about teachers' knowledge of mathematics that might have an impact on how they structure learning activities? The evidence is beginning to accumulate to support the idea that when a teacher has a conceptual understanding of mathematics, it influences classroom instruction in a positive way. In an attempt to understand what a conceptual understanding of mathematics might be, two constructs should be considered: (1) the nature of mathematics itself, and (2) teachers' mental organization of the knowledge they have about mathematics. Even though these two constructs appear to be quite divergent, in reality they are not, because interrelationships are vital in both instances. It is useful to consider each of these in turn to fully understand their similarities and what they might mean for the study of teacher knowledge.

THE NATURE OF MATHEMATICS. First consider the nature of mathematics. In recent years, many have written about this (see Romberg, 1983; Schoenfeld, in press; Steen, 1988), and some of the common themes are the ever-growing content of mathematics, the interrelationships of its major structural elements, its ability to represent the world, its use in communication, and its creative use in solving problems of many kinds. The researchers stress two kinds of relationships: those within mathematics and those that relate mathematics to the real world. While the consideration of the nature of mathematics has resulted in demands that the mathematics curriculum be changed to emphasize the processes of mathematics, the interrelations of mathematical ideas, and problem solving in the real world, the nature of mathematics has not been adequately considered in many studies of teacher knowledge of content. The early studies tended to use inadequate measures, and the interpretive studies have seldom documented teachers' knowledge. Measuring an individual's knowledge of mathematics using a complex definition that adequately addresses the nature of mathematic is not easy, and most measurement instruments used have not been based on such definitions. Needed are procedures that measure the interrelations of ideas, the applications of ideas, and the processes of mathematics (see Webb, Chapter 26, this volume). However, such procedures are now being developed for students, and researchers who are concerned about teachers' knowledge will have to modify these procedures or develop their own so that the complex nature of mathematics can be considered as teachers' knowledge of mathematics is measured.

THE MENTAL ORGANIZATION OF TEACHER KNOWLEDGE. Another way to understand the conceptual nature of teachers' mathematical knowledge is to turn to the literature in cognitive science that has dealt with the mental organization of knowledge. This literature has provided new theories and new methodologies for studying the organization and intricacies of teacher knowledge: how knowledge is acquired, structured, and retrieved. Cognitive researchers attempt to elucidate an individual's mental representations (Gardner, 1986). These researchers then explore how these representations are developed, related to one another, and retrieved. Information thus obtained provides a basis for understanding more fully the effects of teachers' knowledge on their instruction and on student outcomes. This information gives us insight into and provides hypotheses about how teachers represent their plans, organize their routines, evaluate their teaching in view of student feedback, and determine when they have been successful.

Although there exists within the field of cognitive science a variety of competing models for describing the general cognitive system, consistent information is available concerning aspects of an individual's knowledge structure. For instance, given appropriate experience, individuals are able to "chunk" information and organize it in a hierarchical fashion (where specific knowledge is embedded within more generally applicable knowledge). This hierarchical organization of knowledge allows efficient access to information and increased memory capacity.

One approach to understanding hierarchical knowledge structures has been to study experts and novices as they solve problems. Researchers consistently report that experts are more efficient and have different solution paths than do novices. Experts efficiently search problem space, possess meta-statements to aid in decision-making, and organize their knowledge based on properties rather than on individual rules or instances (Chi, Feltovich, & Glaser, 1981; Chi, Glaser, & Rees, 1982; Larkin, 1980; Voss, Greene, Post, & Penner, 1983). Thus, expert knowledge is better integrated and more accessible than is the knowledge of novices. For experts, knowledge is organized in specific ways: connections exist between ideas, the relationship between ideas can be specified, the links can differ among ideas, and the manner in which the knowledge is organized is relevant to understanding and application.

Researchers have also begun to investigate how information about the mental organization of knowledge can be used to facilitate acquisition of new knowledge. Eylon and Reif (1984) report that effective use of hierarchical organization depends on explicit mental connections between different levels of the hierarchy and other units, and some relationship between the organization of the material presented and the organization of the internal knowledge of an individual. They also report that individuals with high levels of knowledge organization are better able to recall information. The organization of an individual's knowledge influences success in accessing that information, in acquiring new information, and thus, in solving problems (Rumelhart & Norman, 1978).

These ideas are not foreign to research within mathematics education. There has been a history of research that supports the idea that knowledge is retained or useful only when mathematics learning is coupled with an understanding of mathematical processes, or when learners are able to build relationships between what they already know and new ideas (Hiebert, 1986; Resnick & Ford, 1981). In the past, this definition of understanding has drawn heavily on an analysis of the structure of mathematics (see, for example, the work of Brownell, 1935), but new work within the cognitive tradition has expanded what it means to learn with understanding (Hiebert and Carpenter, Chapter 4, this volume). A common element appears to be the interrelationships of ideas, both within the structure of mathematics and within the mental structure of the learner. However, most of this research has been done in relation to students' learning of mathematics, and only recently is the construct of understanding being applied to the study of teacher knowledge. What has been done with teachers often refers to the conceptual knowledge of teachers. Careful examination of this work reveals that the construct used to study teacher knowledge is very similar to that used by those who study learners.

A number of studies have examined teachers' knowledge in a way that indicates that the conceptual or interrelated nature of such knowledge was being considered. These studies almost uniformly conclude that when teacher knowledge of content has been defined in a way that is congruent with the nature of mathematics and/or when a conceptual organization of knowledge was considered, a positive relationship was found between content knowledge of teachers and their instruction.

Leinhardt and Smith (1985) gathered data about teachers' understanding of fractions. Their subjects were novices (student teachers) and experts (teachers whose students had showed unusual and consistent growth scores over a five-year period). Leinhardt and Smith developed semantic nets of the teachers' knowledge by interviewing them, observing them as they taught, and having them complete various card sorts. After comparing the semantic nets of experienced and novice teachers, they reported that while there was variation in the knowledge among the expert teachers, "the more experienced and competent teachers exhibited a more refined hierarchial structure to their knowledge" (p. 252). In other words, these expert teachers appeared to know not only the procedural rules of solving fractions problems, but understood the interrelationships of the procedures.

Steinberg, Haymore, and Marks (1985) not only investigated the interrelationships of teachers' knowledge, they also examined the impact of interrelated knowledge on teaching. They reported a relationship between teachers' formal education and background, their actual knowledge of the specific subset of mathematics they were teaching, their knowledge of how that subset fit in the "big picture" of mathematics, and their approach used in teaching. The teachers whose mathematical knowledge appeared to be connected and conceptual were also more conceptual in their teaching, while those without this type of knowledge were more rule-based.

The relationship of teachers' conceptual knowledge of subject matter to student learning and classroom behavior has also been investigated in other disciplines. Carlsen (1991) reported a study dealing with science and described the subject-matter knowledge of four teachers. For each teacher, he identified topics in which the teacher had either high or low knowledge, then collected one year's lesson plans, transcripts of eight science lessons divided between low and high knowledge topics, and teacher and student comments. He concluded that when teachers had low conceptual knowledge in an area, they tended to use seatwork and nonlaboratory projects. The same teachers tended to use lectures and laboratory activities when they had high knowledge of a topic. In low-knowledge areas, teachers discouraged student questions and participation, whereas teachers with high knowledge actively encouraged student questions. When teachers had low knowledge in an area, they may have communicated the content, but an instructional context was built which was antithetical to the nature of science. "Science was communicated to learners, but it was an incomplete view of science" (Carlsen, 1991, p. 137).

Hashweh (1986), whose study was reported earlier in this paper, also described teachers' knowledge in the domain in which they specialized and in another related domain about which they had less knowledge. Within their field of expertise, teachers had more detailed topic knowledge, more knowledge of other discipline concepts, more knowledge of higher-order principles, and more knowledge of ways to connect the topic entities to each other. In other words, their knowledge was related and integrated.

No study cited above is particularly convincing by itself. However, the similarity of the conclusions does lend credence to the belief that when teachers have an integrated, conceptual understanding of specific subject matter, they structure their classrooms so that students are able to interact with the conceptual nature of the subject. In the various studies, the depth of the material being taught appeared to be directly related to the amount of subject-matter knowledge of the teacher.

The evidence available indicates that for teacher knowledge of content to positively influence classroom instruction, the knowledge must be organized in a particular way. It is reasonable to believe that the positive relationship currently being reported between teachers' knowledge and classroom behavior is due partly to the fact that the interrelationships of the mathematical ideas of the teachers were being investigated. The important factor in a positive relationship between content knowledge and classroom instruction appears to be the mental organization of the knowledge that the teacher possesses. As stated by Brophy,

Where [teachers'] knowledge is more explicit, better connected, and more integrated, they will tend to teach the subject more dynamically, represent it in more varied ways, and encourage and respond fully to student comments and questions. Where their knowledge is limited, they will tend to depend on the text for content, de-emphasize interactive discourse in favor of seatwork assignments, and in general, portray the subject as a collection of static factual knowledge (Brophy, 1991, p. 352)

Knowledge of Mathematical Representations

Another type of knowledge, not totally disparate from content, is how mathematics should be represented in instruction. This involves taking complex subject matter and translating into representations that can be understood by students. This translation of mathematics into understandable representations is what distinguishes a mathematics teacher from a mathematician. Consider Wineburg and Wilson's (1991) discussion of history teachers:

Their aim is not to create new knowledge in the discipline but to create understanding in the minds of learners. Unlike the historian, who only has to face inward toward the discipline, the teacher of history must face inward and outward, being at once deeply familiar with the content of the discipline while never forgetting that the goal of this understanding is to foster it in others.... It is precisely in this meeting of subject matter and pedagogy,... that we see the expertise of...teachers most clearly. (p. 335–336)

What is true for history teachers is also true for teachers of mathematics. Even though many describe the ideal mathematics teacher as being one who lets students "do" mathematics (Lampert, 1986), the mathematics that students are taught to "do" must be put into a framework that is understandable by the learners. The mathematics must be translated for them so that they can see the relationship between their knowledge and the new knowledge that they are to learn. Mathematics is composed of a large set of highly related abstractions, and if teachers do not know how to translate those abstractions into a form that enables learners to relate the mathematics to what they already know, they will not learn with understanding.

One context in which most school mathematics can be represented is to use real-world situations and problems. Each mathematical idea can be represented with almost an infinite number of such situations. For example, within simple addition and subtraction one can solve joining problems, separating problems, change problems, and comparison problems. Within each of these categories there are also an infinite number of real-world problems that can be understood and solved by very young children. Not only are there real-world situations that can be represented by mathematics, there is a highly sophisticated set of concrete and pictorial representations by which to model mathematical ideas. For example, Cuisenaire™ rods can be used to represent all whole-number mathematical ideas, and arrays can be used to represent operations on fractions. In cases involving representations of this type, the teacher models a mathematical idea that students should learn. The work by Hiebert and Wearne (1986) illustrates this type of representation; it is concerned with the teaching of decimals and how Dienes base-ten blocks can facilitate learning.

It is believed that use of both real-world situations and concrete or pictorial representations help students learn the abstract ideas of mathematics with understanding. Thus, for teachers to facilitate learning with understanding, they must know how to interpret or represent the mathematical ideas they wish their students to learn. Some studies have investigated teachers' knowledge of mathematical representation and most have indicated that many teachers are lacking this kind of knowledge.

Ball (1988b) reported on 19 preservice teachers' abilities to develop a representation of $1\frac{3}{4}$ divided by $\frac{1}{2}$, either as a story problem or other kind of a model. While all but two of the teachers could accurately calculate the answer to the problem, none of the preservice elementary teachers and only about half of the preservice secondary teachers could develop an appropriate representation of the problem. These preservice teachers knew the rules or procedures for calculating the answer, but apparently were unable to translate this procedural knowledge into a form that could help students understand the concept. Orton (1988) also investigated teachers' knowledge of representations of fraction concepts. He asked 29 inservice elementary teachers how they would teach a fraction concept to a hypothetical student who had a specific misconception about fractions. Most of the teachers relied heavily on procedural and symbolic representations rather than on a representation that would promote conceptual understanding.

It is somewhat puzzling that the teachers in these two studies did not demonstrate knowledge of representations of fractions. Most elementary preservice mathematics methods courses emphasize the use of representations, although they are usually called manipulatives. Would different results have been obtained if whole-number operations had been the topic under consideration? Do teachers know the representations of the content they ordinarily teach? Does knowing these representations make any difference in how teachers teach or what students learn? Where does knowledge of such representations fit into the knowledge structure of teachers? Such questions remain unanswered.

Teachers' Knowledge of Students

Another component of the complex knowledge structure of teachers is their knowledge about learners. The idea that teachers need to have knowledge about learners is not new. It has been recognized for decades that teachers need to know how students acquire knowledge and develop positive self-images. It has been a major tenet of professional education that if teachers had this knowledge, they would use it in their instructional decision-making, so that learning would be improved. Helping teachers acquire this knowledge has been the responsibility of many educational psychology classes that have included general principles of learning ranging from behaviorism to the developmental psychology of Piaget. It was assumed that these principles were applicable to all students and to all content domains and if teachers knew them, they would transfer their knowledge as they planned lessons and implemented them. However, there is little evidence indicating whether or not this knowledge is useful to teachers in making decisions about teaching. Many teachers appear to be unable to use such knowledge except in procedural ways.

Certainly one reason why learning theories have not been useful to teachers is because no one considered teachers and their cognitions when curricula and teacher education programs were developed. It was considered adequate if curriculum materials were thoughtfully developed and based on experts' knowledge of how students learn. It was assumed that knowledge of general principles by the teacher would ensure that the teacher would use the materials in the way in which they were designed. Not considered was the filtering of these materials through the cognitions of individual teachers before they could have impact on learners.

Consider the case of *Developing Mathematical Processes* (DMP) (Romberg, Harvey, Moser, & Montgomery, 1974), an elementary school mathematics program developed with large expenditures of federal funds. This program was organized around two basic themes: the structure of mathematics and how children learn mathematics. Learning activities were carefully designed to represent important mathematical ideas in such a way that the children themselves might learn their interrelationships. In addition, there was an attempt to provide teachers with all the directions and aids they would need to make instructional decisions. In fact, the program was designed to make most of the important decisions for the teacher about the selection of content, how that content should be taught, and how it should be assessed. It was believed that most if not all of the knowledge necessary for effective teaching was provided.

However, DMP was never implemented as the authors had planned. Stephens (1982) studied the implementation of DMP and concluded that teachers who used the program became managers of an "efficient transfer of a body of subject matter to students and as such subverted the major instructional objectives held by the authors.... DMP had, in effect, been presented to teachers as a complete mathematics curriculum, which they were to implement and manage" (pp. 241, 244). Teachers were asked to assume a predetermined set of beliefs and knowledge, which had been selected and articulated by the authors of DMP.

Because the DMP authors' beliefs, knowledge, and objectives were not necessarily shared by the teachers, most of the teachers did not use the program as planned. The knowledge of how children learn, although implemented into the program in the best manner possible by curriculum developers, was not useful to teachers as they actually made instructional decisions about mathematics teaching.

It might be argued that understanding about how learners acquire knowledge is irrelevant to teachers. The argument could be made that any learning theory only reflects what has been known about the acquisition of knowledge in the past and does not reflect what an innovative teacher can or should do in the classroom. As it could also be argued that learning is based on what happens in the classroom, and thus, learning is dependent on how the teacher structures the learning environment, not on what the student does. As it appears that such scholars believe there are no general principles of learning that can be formulated to have impact on instructional decisions that influence learning; perhaps considering specific principles of learning would be useful.

Knowledge of Students' Cognitions. Research on learning from the perspective of cognitive science has renewed the discussion about knowledge of learners' thinking and how that knowledge should be used in teachers' decision-making. Many believe that principles derived through cognitive science research have more potential than other learning theories for influencing instructional decisions as these principles are integrated with subject matter in a way that makes them of direct use to teachers. The organization of cognitive principles into a unified whole appears to be congruent with what we know about how the organization of teachers' knowledge affects their classroom behavior.

It would appear that this type of knowledge should be useful to teachers as they plan instruction for individuals within their classrooms. However, some have suggested that it would be impossible for teachers to remember and use knowledge about each child's thinking process in a way that would enable them to make instructional decisions. For example, Putnam and Leinhardt (1986) report that they saw no evidence that expert teachers were aware of children's thinking processes during instruction. They concluded that knowing the thinking processes of learners was not an important component of expert teaching. However, there was no evidence reported in this study that the expert teachers had been given access to integrated knowledge about children's thinking.

One series of studies done by the authors of this chapter, along with Thomas Carpenter and Penelope Peterson, illustrates the methodology of research that uses knowledge of children's thinking to study teachers' knowledge. These studies suggest that teachers' knowledge of children's thinking might have an important influence on classroom learning (Carpenter, Fennema, Peterson, Chiang, & Loef, 1989; Fennema et al., 1989). These studies built on the research investigating young children's learning of addition and subtraction (Carpenter & Moser, 1983). Carpenter and Moser did a conceptual analysis of the content domain of addition and subtraction. This resulted in the identification of 11 types of addition and subtraction word problems that could be solved by young children.

Using individual interviews with children, they identified hierarchies of difficulty for specific types of problems and the solution strategies used by young children to solve these problems. As children matured, they could solve more difficult problem types with increasingly mature strategies. Many others have confirmed the findings of Carpenter and Moser (Fuson, 1988).

Thus, knowledge of addition and subtraction is an integrated whole made up of a conceptual content analysis combined with children's thinking about the various components of the analysis. It is specific to important content taught in schools; it gives a readily understandable picture of children's cognitions; and it is robust. It contains elements of developmental psychology in that it reflects children's maturing strategies and abilities to solve more difficult problem types.

As a part of a National Science Foundation-sponsored project called Cognitively Guided Instruction (CGI), a number of studies have been conducted to determine if the knowledge about addition and subtraction gathered through research would make a difference in the instructional decisions of teachers. To start, we found that teachers did know quite a bit about addition and subtraction content, even before they were formally taught about it. However, that knowledge was not particularly well organized and teachers often did not recognize the relationships between problem types, solution strategies, and the difficulty of the various problem types. The knowledge they possessed appeared to be fragmented and was not organized in a way that would enable them to understand children's thinking and to apply it to instruction (Carpenter, Fennema, Peterson, & Carey, 1988).

In the next study, we compared what experimental teachers and control teachers knew about the thinking of the children in their classrooms by asking children to solve problems and then asking teachers to predict how the children would solve those problems. The experimental (CGI) teachers were then given access to the knowledge about children's thinking in addition and subtraction, and for a year we compared their instructional decisions (as evidenced by classroom behavior) to the classroom behavior of the control teachers (Carpenter, et al., 1989). At the end of the instructional year, we once again measured the knowledge that the experimental teachers and control teachers had about their students. The experimental teachers knew more than did the control teachers about the mental processes used by children as they solved problems and worked with number facts. Experimental teachers' instructional decisions differed in some important aspects from those of the control teachers. Experimental teachers spent less time during class on drill activities and more time on problem-solving activities. They also spent more time listening to children explain the mental processes they used while solving word problems and expected and accepted a larger variety of problem-solving strategies from the children than did the control teachers. Children in the experimental classes also learned more than did children in the control classes, both in computational skills and in problem solving.

This set of studies illustrates that knowledge derived from research on learners' thinking can be used by teachers in a way that has impact on educational outcomes. The studies done with CGI were built to investigate the idea that knowledge about childrens' thinking in addition and subtraction would enable teachers to assess children's knowledge and to adapt instruction for each child depending upon what that child knew. In many cases, what actually happened in classrooms was not what we had anticipated, but we believe that the knowledge about children's thinking did influence what teachers did in their classrooms; they were able in varying degrees to attend to individual children, and the learning of children improved as a result of what the teachers had learned about children's thinking.

The CGI studies provide evidence that teachers can attend to individual students when they have appropriate and well-organized knowledge. The knowledge to which the CGI teachers were given access was both very specific and well organized. We speculate that the reason this knowledge was so helpful was because it was specific in regard to children's problem solving in a particular domain that was already part of the teachers' curriculum. It appears that this knowledge empowered teachers to understand the children's thinking in a way they had been unable to do before. As one of the experimental teachers commented: "I have always known that it was important to listen to kids, but before I never knew what questions to ask or what to listen for." We also speculate that one reason that Putnam and Leinhardt's (1986) expert teachers did not attend to individuals was that they did not have the knowledge about children's thinking processes that CGI teachers had.

It should be pointed out that there is not a robust, integrated set of knowledge available about most content areas within the mathematics curriculum (Hiebert and Wearne, 1991). As yet, we have little information regarding how much knowledge of children's thinking is essential in order to have an impact on teaching. Do we have to wait for cognitive scientists to provide the type of detailed knowledge that we have in addition and subtraction, or would detailed knowledge in one area be sufficient to enable teachers to understand children's thinking in other areas? Is it even possible for teachers to have such a complete body of knowledge for each content area? There are many such unanswered questions, and we are currently investigating the extension of CGI to other content areas of the primary school. However, the CGI studies support the idea that knowledge of children's thinking, when it is integrated, robust, and a part of the known curriculum, can influence the teaching and learning of mathematics.

Teachers' General Knowledge of Teaching and Decision-Making

During the last decade, researchers have increasingly studied teachers and how the decisions they make influence student learning. Most of this work has been done outside of the area of mathematics education and has been concerned with general principles of teaching. Much of it has been done from the perspective that teachers are thoughtful professionals with unique human abilities to plan, implement, evaluate, and reflect on their instruction. Teachers have been regarded as not only those who engage in observable classroom behaviors, but also as active processors of information before, during, and after classroom instruction. Comprehensive reviews of this research have been done by Shavelson and Stern (1981) and by Clark and Peterson (1986). This work documents that teachers have theories and belief systems that influence their perceptions, plans, and actions in the classroom. Teachers are reflective, thoughtful individuals, engaging in a complex and cognitively demanding human process. Teachers' beliefs, knowledge, judgments, and thoughts have a profound effect on the decisions they make, which in turn determine to a large extent what students learn in their classrooms.

Research has clearly documented that the decisions made by teachers before, during, and after instruction are a dominant influence on what is learned by students. While decisions made during these times are not and cannot be totally separated, it is useful to examine these decisions individually. The four steps of curriculum planning taught to many teachers (specify objectives, select activities, organize activities, and specify evaluation procedures) do not appear to be used often by teachers (Clark & Peterson, 1986), but teachers do spend time planning; they often make year-long plans, semester plans, unit plans, weekly plans, and daily plans. These plans are nested within each other, and modification of one results in modification of others. These plans vary in length, detail, and specificity. In fact, some plans are never written, but are held in mind by the teacher.

During the preactive or planning phase, teachers make decisions that affect instruction dramatically. They decide what to teach, how they are going to teach, how to organize the classroom, what routines to use, and how to adapt instruction for individuals. After thoroughly reviewing the literature on decision-making, Bush (1982) concluded that the majority of instructional decisions were preactive decisions. One should not interpret this as being in conflict with what teachers did in the CGI studies or in studies that report more knowledgeable teachers' responses during instruction to students. This response to students is, in fact, part of the planning of these teachers.

Teachers also make many decisions during instruction; these decisions are often called interactive decisions. Among the choices that teachers make are decisions to modify their plans, to respond to a child in a particular way, to call on a given child for an answer, to reward or reject certain answers, to discipline an unruly child, to encourage a shy child, and to speed up or slow down a lesson. Teachers make these decisions partly by assessing whether their plans are working and how children are responding. Teachers reflect about the instruction after instruction. While the reflection may be informal, teachers often refer to specific information they have gathered. For example, when worksheets are checked, a teacher gains information about what children learned that day. She or he may use this information to evaluate the day's teaching and to plan for the next day's class.

One major influence on teacher decision-making is the knowledge that drives teachers' actions and decisions and provides them with the flexibility that enables them to reason, to judge, to weigh alternatives, to reflect, and to act (Clark & Yinger, 1978). Ultimately, how a teacher behaves and the potential effectiveness of a teacher rest on the knowledge that a teacher possesses. Most of the work with decision-making and how it is influenced by knowledge has to do with personal and practical (general) knowledge of teachers. It usually includes knowledge of specifics in the organization and running of classrooms, such as how to maintain discipline and how to establish efficient routines. It includes the structure of various lessons taught in the past and how various learners have reacted to such lessons.

While this knowledge builds up through a variety of experience, it is usually growing and changing. Clandinin (1985) states that "personal practical knowledge is viewed as tentative, subject to change, and transient rather than something fixed, objective, and unchanging" (p. 364). The purpose of practical knowledge is to inform wise action, to express purposes, and to give shape and meaning to teachers' experiences. It is situation-specific knowledge that is personally compelling and oriented toward action (Elbaz, 1983). In other words, it is an important basis for teachers' instructional decisions. The basis for this perspective stems from the belief that professional and personal experiences shape teachers' knowledge (Elbaz, 1983). Teachers are perceived as reflective beings whose knowledge is based on experiences as interpreted through beliefs, insights, and habits. (See Feiman-Nemser & Floden, 1986, for a complete review of the research based on this perspective.)

Frameworks and Cognitive Models of Teacher Knowledge

We now turn to various ways in which knowledge as an integrated phenomenon has been studied and consider some proposed frameworks and cognitive models that indicate how the various components might be organized. We have chosen to report briefly the frameworks of Shulman and Peterson, and to discuss three models in depth: one that has been studied in relation to mathematics teaching and two others that we speculate have potential for study.

Shulman (1986) proposed a framework for analyzing teachers' knowledge that distinguished between different categories of knowledge: subject-matter knowledge, pedagogical content knowledge, and curricular knowledge. *Subject-matter* knowledge is the "amount and organization of the knowledge per se in the mind of the teacher" (p. 9). This is the mathematical knowledge of teachers.

Pedagogical content knowledge, according to Shulman, includes "the most useful forms of representation of those [content] ideas, the most powerful analogies, illustrations, examples, explanations, and demonstrations—in a word, the ways of representing and formulating the subject that make it comprehensible to others." It also includes "an understanding of what makes the learning of specific topics easy or difficult, the conceptions and preconceptions that students of different ages and backgrounds bring with them to the learning of those most frequently taught topics and lessons" (Shulman, 1986, p. 9). Included in this category would be the specifics of how mathematics might be interpreted for learners, such as how to represent a fraction so that children would be able to understand the numerator and denominator (as a geometric region or as a set with parts shaded to represent the fraction). Also included in pedagogical content knowledge is knowing how students think within specific mathematical domains. Work within the cognitive science tradition completed in the last few years fits readily into this category; the work of Carpenter and Moser (1983), for example, as it deals with young children's problem solving in addition and subtraction.

Curricular knowledge is knowledge of instructional materials available for teaching various topics and the "set of characteristics that serve as both the indications and contraindications for the use of particular curriculum or program materials in particular circumstances" (Shulman, 1986, p. 10). Knowledge of various program materials, such as sets of manipulatives that represent mathematical ideas or computer software is included in this category. Thus, Shulman suggests that the important components of teachers' knowledge are what they know about mathematics, about how learners think about mathematical topics, and about instructional materials developed to teach mathematics.

Peterson (1988) builds on and modifies Shulman's framework. She argues that in order to be effective, teachers of mathematics need three kinds of knowledge: how students think in specific content areas, how to facilitate growth in students' thinking, and self-awareness of their own cognitive processes. Peterson does not ignore the content knowledge necessary to teach, but argues convincingly that this knowledge must be held in relation to the three categories she has identified. Mathematics knowledge isolated from children's cognition and teachers' metacognition does not appear important to her. According to Peterson, unless a teacher can understand his or her own thinking in mathematics, knowledge of content will not be useful in structuring the classroom so that students can learn.

Cognitive Models of the Skills of Teaching. In formulating models of teachers' cognition, some psychologists, including Shavelson, have attempted to describe mental schemata that provide the mechanism for decision-making (cited in Livingston & Borko, 1990). While these researchers have provided evidence that such schemata exist and are used in teacher decision-making, most of the work has been done independently of any specific content. One notable exception to the content-independent work has been the line of work by Leinhardt and her colleagues (Leinhardt & Greeno, 1986; Leinhardt, Putnam, Stein, & Baxter, 1991; Leinhardt & Smith, 1985; Putnam & Leinhardt, 1986).

Leinhardt and her colleagues' goal has been to describe in depth the mental structures of skilled teachers. This work is based on the belief that teaching is "a complex cognitive skill amenable to analysis in a manner similar to other skills described by cognitive psychology" (Leinhardt & Greeno, 1986, p. 75). Using cognitive science methodology, they first hypothesized a model of mental structures. Next, they examined and contrasted the teaching behavior of expert and novice teachers to see if their behavior was congruent with the hypothesized model.

THE MODEL. The skill of teaching, according to Leinhardt and her colleagues, is determined by at least two fundamental, related systems of knowledge: subject matter (content knowledge) and lesson structure (practical knowledge). The structuring of a lesson takes priority and is both supported and constrained by the teacher's knowledge of the content to be taught.

Knowledge of lesson structure is organized into agendas, scripts, and routines. While the distinction between agendas and scripts is not absolute and has changed as the line of inquiry has progressed, the agenda can be characterized as a dynamic master plan that includes the overall goals and activities of a lesson that can be modified as the lesson proceeds. A script "consists of a loosely ordered set of goals and actions that a teacher has built up over time for teaching a particular topic" (Leinhardt et al., 1991, p. 89). A script is what the teacher has foremost in his or her mind as the lesson is started, and it includes the various activities such as review, presentation, and supervised practice that will take place during the instructional period. It is the major determinant of the lesson and is rarely changed during instruction. These scripts are developed and polished over time and are used whenever appropriate.

Routines are activities that are performed frequently by both teacher and students. They "allow relatively low-level activities to be carried out efficiently, without diverting significant mental resources from the more general and substantive activities and goals of teaching" (Leinhardt & Greeno, 1986, p. 76).

Thus, they hypothesized that as a teacher starts a lesson, she or he has an overall goal, some procedures or activities that will be used in an ordered way to achieve the goal, and some routines that enable the class to function smoothly. While input from students may slightly modify what happens, any change in the planned lesson is only modest at best.

Knowledge of subject-matter content, which is defined as "the knowledge that a teacher needs to have or uses in the course of teaching a particular school-level curriculum" (Leinhardt et al., 1991, p. 88), supports the agendas and scripts. Thus, subject-matter knowledge includes not only knowledge of mathematics, but also includes knowledge of curriculum activities, effective methods of presentation, and assessment procedures. Subject-matter knowledge aids the teacher in establishing the agendas and scripts, particularly in deciding what representations of the mathematical topic to use and/or how to present the topic. However, according to the model proposed by Leinhardt, subject matter knowledge itself is not a primary determinant of teaching behavior. Instead, the rich repertoire of agendas and scripts built up over time determine instructional methods.

TESTING THE MODEL. To examine the validity of the cognitive model of teaching hypothesized, Leinhardt and her colleagues have compared and contrasted experts (teachers whose students have high gains on standardized tests, and/or peer or supervisor nomination) to novices (student teachers). They observed these teachers in action, interviewed them, and gathered data in a variety of other ways.

After data gathering, the researchers developed semantic nets that represented the mental structures of the teachers as reflected in their teaching behaviors. Scripts and agendas were identified. In particular, the script concept has proven to be powerful in both characterizing teaching and in differentiating expert and novice teachers. The researchers report that scripts allowed stable and cohesive lessons to take place. While scripts did not change much during the course of a lesson, they were modified as teachers gained more experience and enabled the teacher to maintain focus on important mathematical ideas. Leinhardt et al. (1991) conclude that "teachers' cur-

riculum scripts provide an especially fruitful site for exploring teacher subject-matter knowledge because they represent the transformation of teachers' knowledge of content into a form that is accessed during the teaching of lessons" (p. 111).

Expert teachers' agendas, scripts, and routines were systematically different from those of novices. Expert teachers used more varied and explicit routines. The segments of the lessons of experts were more precise. In fact, as one reads the descriptions of these expert teachers, they appear very similar to an "active teacher" as described by Good, Grouws, and Ebmeier (1983). Novice teachers appeared to have less elaborate and interconnected schemata than did expert teachers. Also, the schemata information was less accessible to novice teachers than to expert teachers.

CRITIQUE OF THE MODEL. This work by Leinhardt and her associates is significant and can be used as a basis for future work in identifying teacher knowledge. Seldom within the field of mathematics education can we point to a true chain of inquiry where a researcher publishes a series of investigations built on a carefully hypothesized model. This work is one such chain. The methodology is consistently thorough and well reported. However, as is often the case with researchers whose main interest is not mathematics education, the mathematics studied has been limited. The emphasis of the teaching studied has been on the learning of procedures [such as reducing a fraction to lowest terms (Leinhardt and Smith, 1985)], or the structure placed on the observed lessons by the researchers has reflected procedures rather than understanding.

The methodology involved observing teachers and then preparing detailed semantic nets of the mathematics presented (Leinhardt & Smith, 1985). This methodology lends itself to breaking mathematical ideas down into small component parts and studying these relationships. Although the procedures for obtaining these semantic nets (developed from observation of teachers) are very different than the procedures used by Gagne (1970) to develop his content hierarchies, the end results are very similar, particularly for expert teachers.

Teacher knowledge, to Leinhardt and her associates (Leinhardt & Smith, 1985), is made up of the skills and abilities needed to run classrooms well and to adequately interpret and explain certain procedural mathematical ideas so that the students are able to acquire the mathematical skills that the teacher feels are important. Currently, both cognitive psychologists and mathematics education researchers question whether attention to the details of mathematics achieves the important goals of mathematics. Mathematics is more than the sum of a number of small ideas, and overall understanding of mathematics has not been achieved by attention to such details.

Additional work using mathematical understanding as the focus of instruction is needed to test this model. One has to wonder whether the attention to predeveloped routines, scripts, and agendas by teachers will enable students to learn and understand mathematics in all of its richness. Leinhardt suggests that it is impossible for teachers to attend to individuals as they learn (Putnam & Leinhardt, 1986), yet only when individuals interact with mathematics appropriate for them that they learn (see Hiebert and Carpenter, Chapter 4, this volume). Therefore, on at least two points is this model lacking: the mathematics that students are being asked to learn and the lack of attention to individuals.

Knowledge as Practical and Personal. One model, although an incomplete one because of its limited attention to content concerns, has some potential for understanding practical knowledge of mathematics teachers. Elbaz (1983) suggests that the practical, personal conception of teacher knowledge posits knowledge as dynamic, context driven, and related across past, present, and future. (Once again, the division between knowledge and beliefs becomes blurred, if not invisible.) This knowledge is structured into a set of rules and principles that develop over time, mainly through interaction with students in the classroom environment, although other professional and personal experiences can contribute. It is this knowledge that directs the teachers' instructional decision-making and that the teacher uses to justify his or her actions. Elbaz believes that the structure of teachers' knowledge includes three dimensions: rules of practice, practical principles, and images. The rules and principles embody the instructional knowledge, while the images direct the decision-making. In images, which are unique to each teacher, all aspects of teacher knowledge come together. Images are temporal, interactive, and nonneutral; they evolve and are dependent on all aspects of knowledge. Images involve emotions and morality and take into account existing knowledge and allow for the development of new knowledge (Clandinin, 1985). Elbaz (1981) states that a teachers' "feelings, values, needs and beliefs combine as she formulates brief metaphoric statements of how teaching should be and marshals experience, theoretical knowledge, and school folklore to give substance to these images" (p. 61).

Images order all aspects of practical knowledge and are used in making judgments. Rules and principles used to direct teaching are chosen through images. If the rules and principles happen to conflict with the image a teacher holds about a particular situation, different rules are chosen. Images can also extend knowledge by generating new rules and principles. Elbaz (1983) posits that a teacher's practical knowledge is oriented toward a particular practical context and these orientations can be examined on a number of dimensions: social, personal, experiential, theoretical, and situational. Orientation addresses how a teachers' knowledge is held; it provides a sense of order to an otherwise complex set of information. The examination of orientation allows those attempting to understand teacher knowledge to attend to a broad range of detail and to take note of the variety and comprehensiveness of teacher knowledge.

While Elbaz's ideas have not been studied in relation to mathematics teachers' knowledge, it is interesting to speculate about what they might mean. Certainly, the idea of orientation can be seen in the literature about teacher beliefs. Teachers whose orientation includes the belief that mathematics is procedural might use rules and principles that are in line with their beliefs to guide their decision-making. The decisions, then, are oriented to ensuring that mathematics is presented clearly and that students have ample opportunity to practice the procedures presented. (see Thompson, Chapter 7, this volume.) One might also speculate that if a teacher's knowledge of content is poorly integrated, then his or her image of mathematics would probably reflect this image and the content of instruction in the classroom would also be poorly integrated. Theoretical models such as this provide fertile ground for investigation. They speak to the complexity of teachers' knowledge and the necessity for clearly identifying and studying this complexity.

The models of Leinhardt (Leinhardt & Greeno, 1986; Leinhardt & Smith, 1985; Leinhardt et al., 1991; Putnam & Leinhardt, 1986) and Elbaz (1981; 1983) each include a system for explaining the knowledge a teacher possesses, its interacting nature, and how a teacher decides what to do in the classroom context. Leinhardt discusses the relationship between agendas, scripts, and routines and proposes an interactive system consisting of lesson structure and subject-matter knowledge. Teacher behavior is driven by agendas that build on scripts (which are dependent on subject-matter knowledge) and routines. Within the model, the agendas work as a goal structure for determining which scripts and agendas are applied in the classroom. As long as the goals remain constant, so can the scripts and routines, even given a different context. Elbaz proposes that rules and principles are driven by images, all of which are governed by a teacher's orientation.

There appear to be some specific similarities in the models proposed by Elbaz and Leinhardt. Compare scripts with rules and routines with principles. Agendas and images also appear somewhat similar, although agendas appear to be more limited in scope than do images. The case is not made for the inclusion of beliefs, emotions, or selection of goals in agendas as it is in images. Agendas differ from images in their impact on decision-making; images tend to incorporate many forms of knowledge and beliefs and to guide classroom decisions, but do not determine particular behaviors, as do agendas.

Teachers' Knowledge as Situated. Another framework that can be applied to the understanding of teacher knowledge is that of situated knowledge. While the complexity and richness of the work on situated knowledge is beyond the scope of this chapter, the construct is worth considering as we attempt to understand teachers' knowledge and its impact on teacher behavior and student learning. We will briefly review the construct of situated knowledge and then speculate about what it might mean as we consider teachers' knowledge.

When discussing situated knowledge and its acquisition, several have pointed out the contrast of knowledge gained by students in school and out of school (Resnick, 1987; Brown, Collins, & Duguid, 1988; 1989). These scholars have theorized that in-school knowledge is acquired by working alone to memorize laws and rigid concepts and by solving well-defined problems. Such knowledge is explicit and difficult to transfer. Out-of-school knowledge is acquired by working in a social situation to decide on the causes of events, to solve ill-defined problems, and to construct one's own understanding. This knowledge is personal and transferable because it can be retrieved and used. These scholars go on to explicate the idea that knowledge acquired either in or out of school is not independent of the situation in which it is learned and used, but is dependent on the situation in which it is acquired (Brown et al., 1989). All knowledge is situated and is partly a result of the activity, context, and culture in which it is developed (Brown et al., 1989). Not only is knowledge a partial result of these three interrelated components, but the component also remains a referent by which knowledge is retrieved, interpreted, and used.

Knowledge acquired in school is not situated in the broader life of an individual because the activities, contexts, and culture of the school are not related by the learner to his or her out-of-school culture. School knowledge is not particularly useful because it is fragmented, isolated from reality, too explicit to be transferred, and quickly forgotten.

Knowledge acquired through activities set in a context which enable a learner to connect the knowledge to his or her broader culture is generalizable and useful. Even when the knowledge is originally quite specific, through what Brown and his colleagues (Brown et al., 1989) call authentic activity, it becomes situated across multiple contexts. Such knowledge can be used to solve many diverse problems and to gain new knowledge. The learning environments described by Lampert (1986) and Schoenfeld (in press) are often characterized as classrooms where mathematical knowledge can be learned in ways congruent with these ideas.

It appears that most teachers' knowledge of mathematics may be similar to what has been described as in-school acquired knowledge. Teachers' knowledge of mathematical content is learned by studying mathematics in schools. While some elementary school mathematics might be taught in such a way that would enable a future teacher to situate it partly in the broader context of his or her culture, it is not unreasonable to conclude that most teachers' knowledge of advanced mathematics has been learned in typical classrooms where mathematics has been explained, and practice and/or homework assigned. Students tend to study the assigned mathematics alone and to solve well defined problems that illustrate the mathematical topics of the day. Mathematics thus becomes situated only in well-defined problems and in relation to the culture of the school. Because most teachers have learned their mathematics in this way, it seems logical to believe that their knowledge of only the most rudimentary mathematics is situated outside the classroom. Since most teachers have no opportunity to learn mathematics in any other way, either before or after they start teaching, their knowledge of mathematics remains knowledge that is useful only in schools.

Teachers' knowledge of pedagogical procedures has also been learned predominantly in classrooms where they have been students. While preservice education experiences have given them some opportunity to observe and practice some procedures, it would be naive to think that all knowledge of pedagogy is learned in preservice education. Being a student in many diverse classrooms for 12 to 14 years before one enters preservice education undoubtedly provides much knowledge that is situated in the reality of classrooms. However, the role of the teacher is different from the role of the student, and a beginning teachers' knowledge of pedagogical procedures is only from the student's perspective. Preservice experiences somewhat refine this pedagogical knowledge so that it is more nearly congruent with the teachers' perspective. Thus, a beginning teachers' knowledge of pedagogical procedures is situated within his or her own school experiences. As the teacher gains experience in classrooms, this knowledge grows and matures.

Certain things work or don't work and thus her or his repertoire of pedagogical procedures grows.

Another type of teacher knowledge that we have talked about is knowledge of students. From various preservice courses, novice teachers have often gained general information about how learners acquire knowledge. While such knowledge has not proved to be useful to teachers, there is evidence that as they gain experience teachers acquire knowledge that could be characterized as situated knowledge. Consider the knowledge of first-grade teachers as reported by Carpenter et al. (1989). These experienced teachers, before they participated in the experimental treatment, were asked about the relative difficulty and solution strategies of several types of addition and subtraction word problems for first-grade children in general and for six specific children in their classrooms. Most of the teachers were familiar with frequently used strategies for solving the problems and could identify relatively accurately the success that their own students would have in solving the problems. While this knowledge was not well integrated, these teachers had the knowledge needed to predict their own students' problem-solving performance. Thus, these teachers had knowledge that was specific to the children that they were teaching.

What would situated knowledge for mathematics teachers be? It is the interaction of knowledge of mathematics, pedagogical procedures, and students. A teacher must take his or her knowledge of mathematics, pedagogical procedures, and learners in general and apply that to the structuring of his or her classroom learning activities for specific learners. This knowledge is dynamic. Starting with the rudimentary knowledge of beginning teachers, it grows and matures as it interacts with specific learners in a classroom.

The entire construct of situated knowledge is so new that a research paradigm to substantiate it has yet to develop or to be applied to the study of teachers. However, it holds great promise for increasing our understanding of learners' knowledge and perhaps even greater potential for increasing our understanding of teachers' knowledge. This model has many implications for teacher education, both pre- and inservice. The mathematics that teachers learn must be learned in a context that is much broader than traditional in-school learning so that the teachers' knowledge is more similar to what we have called the nature of mathematics. Teachers' pedagogical knowledge must also be considered so that they can learn about classrooms in which students are able to participate in mathematics. The idea of situated knowledge and how it is acquired offers credence to the many statements of the necessity of total reorganization of both schools and teacher education programs.

The impact of understanding teachers' situated knowledge on curriculum and instructional change could be profound. How does one provide situated knowledge of classrooms that meet the goals of the *Standards* (NCTM, 1989), if classrooms where knowledge is situated are as rare as is believed? Certainly, the impact of the knowledge teachers have gained as both learners and teachers of mathematics in traditional classrooms cannot be ignored as changes in mathematics classrooms are proposed.

A Synthesis of the Models. We have reported several frameworks and models of teachers' knowledge. Each has provided new insight into defining teacher knowledge and providing new ways of thinking about it. The approaches taken and the conceptualizations of teacher knowledge proposed are not inconsistent; rather they build on each other. Identification of their consistencies, inconsistencies, and relationships can help researchers identify useful conceptions of teacher knowledge that take into account the various aspects of the models proposed.

Shulman (1986) initiated much new thinking about teacher knowledge by providing a categorization of teacher knowledge that included more than a single dimension. Peterson (1988) built on the work of Shulman by identifying the importance of the cognitions of the teacher in her discussion of metacognition. In their initial conceptions, Shulman and Peterson described many of the same components of teacher knowledge and both stress the importance of considering content to be taught as an important part of teachers' knowledge, its organization, and how it should be studied. In addition, each made a strong case for the importance of considering knowledge of learner's cognitions about the subject matter.

Although the definition of subject matter was inadequate, knowledge of content was important in the work of Leinhardt and her colleagues. The knowledge of mathematical content was different within the expert teacher group and there were also major differences between expert and novice teachers. The expert teachers' knowledge tended to be organized into a hierarchical structure. They used richer systems of representations, and they tended to present more detailed conceptual and procedural knowledge. This knowledge of subject matter had an impact in several ways: (1) on agendas, because teachers with more knowledge had richer mental plans than did teachers with less knowledge; (2) on scripts, because more knowledgeable teachers were able to use more representations and richer explanations; and (3) on teachers' response to students' comments and questions during instruction (Leinhardt et al., 1991). While Elbaz (1983) did not provide an adequate place for the impact of content in her model, her ideas stress the growing dynamic nature of knowledge. She provided insight into the necessity of recognizing that knowledge is ever changing and never static. In a narrow sense, the work of Elbaz relates to that of situated knowledge. Combining the two approaches could enrich personal practical knowledge and provide a framework for continuing to conceptualize and measure teachers' situated knowledge.

The idea of situated knowledge provides insight into the isolation of school-based knowledge and suggests that school knowledge is different from real-world knowledge. We suggested that the components of teachers' knowledge are also situated and appear to be situated in the narrow environment of the school. This situating of teacher knowledge influences the way teachers make instructional decisions and what their students learn.

Knowledge as a Process of Development. One consistency across all conceptions of teacher knowledge is the view that teachers' knowledge is continually changing and developing. Knowledge is not static. Aspects of teacher knowledge grow through interactions with mathematics, in the classroom environment, with students, and through professional experiences,

to name but a few. Knowledge acquisition is a process; the crucial aspect is not what knowledge exists, but rather that the process continues. The challenge then is to understand knowledge as it grows and changes and to discover what experiences contribute to this growth and change.

This view of knowledge as a process has ramifications for measuring teacher knowledge, for developing models that describe how teacher knowledge is acquired, and for making decisions regarding the knowledge teachers need before teaching. This perspective promotes the idea that a teacher's knowledge is constantly changing and that the knowledge he or she possesses about teaching (while dependent on the knowledge that is possessed about the content and the pedagogy before entering the classroom), continues to change. In order to change knowledge, teachers have to redevelop their knowledge in terms of new knowledge. Not only is this cognitively demanding, but it is also time consuming. It is no wonder that change takes time or that teachers often do not change following an inservice program.

The implications for methodology of the belief that knowledge is dynamic are profound. Understanding developing knowledge is limited when measures of static knowledge are used. Furthermore, if beliefs or orientations are as important to determining knowledge as many propose, then measures need to be developed that can account for teachers' differing orientations. Researchers must also be aware of their own orientations, as they may influence the measures they choose and the results that they gain.

A MODEL FOR RESEARCH ON TEACHERS' KNOWLEDGE

Understanding teacher knowledge of mathematics and its influence on teaching and learning is difficult and has become increasingly more difficult as the complexity and dynamic nature of knowledge has become recognized. Researching teacher knowledge means more than investigating the number of mathematics courses teachers have taken or the procedural knowledge of mathematics they possess. Knowledge of mathematics teaching includes knowledge of pedagogy, as well as understanding the underlying processes of the mathematical concepts, knowing the relationship between different aspects of mathematical knowledge, being able to interpret that knowledge for teaching, knowing and understanding students' thinking, and being able to assess student knowledge to make instructional decisions. Teacher knowledge is no longer viewed as an isolated construct in its effects on teachers' classroom behavior and student learning. Teacher knowledge cannot be separated from the subject matter being investigated, from how that subject matter can be represented for learners, from what we know about students' thinking in specific domains, or from teacher beliefs.

Now that we are gaining an understanding of the components of teacher knowledge that can make a difference in classroom practices and student learning, each component requires further study in terms of definition, parameters, and relationships with other components. Only by clearly defining the components and their relationships can an understanding of teacher knowledge and a consensus about terminology be gained. The study of these components cannot be performed out of context or in isolation. Not only must components be studied in their interrelationships, the dynamic nature of teacher knowledge must be considered. Measuring knowledge at one point in time does not do justice to how knowledge changes as teachers participate in various experiences, both planned and unplanned.

The model that we put forth for examination and discussion centers on teachers' knowledge as it occurs in the context of the classroom (see Figure 8.1). The model, which shows both the interactive and dynamic nature of teacher knowledge, includes the components of teacher knowledge of the content of mathematics, knowledge of pedagogy, knowledge of students' cognitions, and teachers' beliefs. It also shows each component in context. Each component shall be described briefly and a discussion will follow regarding the role of context in relation to the various components.

The *content of mathematics* component includes teacher knowledge of the concepts, procedures, and problem-solving processes within the domain in which they teach, as well as in related content domains. It includes knowledge of the concepts underlying the procedures, the interrelatedness of these concepts, and how these concepts and procedures are used in various types of problem solving. Crucial also to teacher knowledge of content is the manner in which the knowledge is organized, indicating teacher knowledge of the relationships between mathematical ideas.

Pedagogical knowledge includes teachers' knowledge of teaching procedures such as effective strategies for planning, classroom routines, behavior management techniques, classroom organizational procedures, and motivational techniques.

Learners' cognitions includes knowledge of how students think and learn and, in particular, how this occurs within specific mathematics content. This includes knowledge of how students acquire the knowledge of the mathematics content being addressed, as well as understanding the processes the students will use and the difficulties and successes that are likely to occur.

The center triangle of our model indicates the teachers' knowledge and beliefs in context or as situated. The context is the structure that defines the components of knowledge and beliefs that come into play. Within a given context, teachers' knowledge of content interacts with knowledge of pedagogy and students' cognitions and combines with beliefs to create a unique set of knowledge that drives classroom behavior. In order to understand the way in which we define context and the role that it plays in our model, consider this example.

A teacher gains knowledge from an inservice program that indicates that knowing the mental processes used by her students to solve addition problems would help to improve her instruction and enhance her students' learning. The teacher makes a decision to try to use this knowledge in instruction. The teacher's knowledge of the content of addition allows her to choose problem types within addition that are connected to what the students already know about numeration. However, in structuring the lesson, the teacher does not have any specific teaching techniques or activities that will help her find out what the children are thinking. Furthermore, she has no contingencies for what happens if she is unable to elicit from the children what they are thinking. The teacher draws upon other pedagogical skills in her repertoire, such as questioning skills or a routine used in reading comprehension, to elicit understanding. During instruction, the knowledge about children's thinking processes and the corresponding pedagogy is slowly adapted so that it is successful in the current situation. New knowledge is gained about how to get at what the students are thinking, a new routine of question asking is developed in relation to addition problems, and instruction is adapted to what the teacher has learned about the students.

What we have done in our example is show that some components of teacher knowledge evolve through teaching. This has been corroborated by other researchers as well (Orton, 1989; Shulman, 1987). Teaching is a process within which new knowledge is created.

Knowledge often develops based on the teachers' pedagogical knowledge and through classroom interactions with the subject matter and the students in the classroom. Shulman (1987) states:

The key to distinguishing the knowledge base of teaching lies at the intersection of content and pedagogy, in the capacity of a teacher to *transform* the content knowledge he or she possesses into forms that are pedagogically powerful and yet adaptive to the variations in ability and background presented by the students. (p. 15)

The critical word here is *transform*. Teachers have to take their complex knowledge and somehow change it so that their students are able to interact with the material and learn. This transformation is not simple, nor does it occur at one point in time. Instead, it is continuous and must change as the students who are being taught change. In other words, teachers' use of their knowledge must change as the context in which they work changes.

Examination of teachers is beginning to indicate that knowledge can be and is transformed through classroom interaction. The example above indicates that when knowledge is transformed during instruction, that knowledge becomes tied to

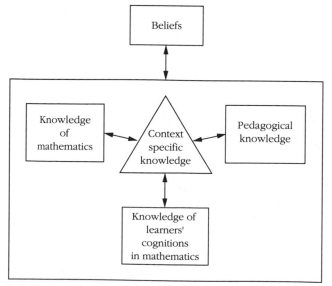

FIGURE 8–1. Teachers' Knowledge: Developing in Context

the context in which it was developed. If the context in which the teacher is a part was to change (different content being covered, different classroom structure, or different students), the knowledge drawn upon by the teacher will also change. The degree of change in the knowledge drawn upon will depend on the knowledge base of the teacher involved, the complexity of the knowledge base, its interconnections, and the specificity of the knowledge available. If the knowledge specific to the given context is unavailable, then the teacher must either use more general knowledge relevant to a variety of situations or use knowledge that closely matches the situation. When knowledge is brought to a new situation, the knowledge can be adapted and stored as new knowledge now relevant to the current situation.

The transforming of knowledge in action is understandably complex. Little research is available that explains the relationship between the components of knowledge as new knowledge develops in teaching, nor is information available regarding the parameters of knowledge being transformed through teacher implementation. Here all aspects of teacher knowledge and beliefs come together and all must be considered to understand the whole. The challenge is to develop methodology and systematic studies that will provide information to enlighten our thinking in this area. The future lies in understanding the dynamic interaction between components of teacher knowledge and beliefs, the roles they play, and how the roles differ as teachers differ in the knowledge and beliefs they possess.

References

Ball, D. L. (1988a). *Unlearning to teach mathematics* (Issue Paper 88-1). East Lansing: Michigan State University, National Center for Research on Teacher Education.

Ball, D. L. (1988b). *Examining the subject matter knowledge of prospective mathematics teachers*. Unpublished manuscript.

Ball, D. L. (1991). Research on teaching mathematics: Making subject matter knowledge part of the equation. In J. E. Brophy (Ed.), *Advances in research on teaching: Teachers' subject matter knowledge and classroom instruction* (Vol. 2, pp. 1–48). Greenwich, CT: JAI Press.

Begle, E. G. (1979). *Critical variables in mathematics education: Findings from a survey of empirical literature.* Washington, DC: Mathematics Association of America and the National Council of Teachers of Mathematics.

Brophy, J. E. (1991). Conclusion to advances in research on teaching, Vol. 2: Teachers' knowledge of subject matter as it relates to their teaching practice. In J. E. Brophy (Ed.), *Advances in research on teaching: Teachers' subject matter knowledge and classroom instruction* (Vol. 2, pp. 347–362). Greenwich, CT: JAI Press.

Brown, J. S., Collins, A., & Duguid, P. (1988). *Situated cognition and the culture of learning* (Research Report No. IRL 88-0008). Palo Alto, CA: Xerox Palo Alto Research Center, Institute for Research on Learning.

Brown, J. S., Collins, A., & Duguid, P. (1989). Situated cognition and the culture of learning. *Educational Researcher, 18*(1), 32–42.

Brown, S. I., Cooney, T. J., & Jones, D. (1990). Mathematics teacher education. In W. R. Houston (Ed.), *Handbook of Research on Teacher Education* (pp. 639–656). New York: Macmillan.

Brownell, W. A. (1935). Psychological considerations in the learning and teaching of arithmetic. In W. D. Reeve (Ed.), *The teaching of arithmetic. Tenth yearbook of the National Council of Teachers of Mathematics* (pp. 1–31). New York: Teachers College, Columbia University.

Bush, W. S. (1982). *Preservice secondary mathematics teachers' knowledge about teaching mathematics and decision-making process during teacher training.* Unpublished doctoral dissertation, University of Georgia.

Carlsen, W. (1991). Subject-matter knowledge and science teaching: A pragmatic perspective. In J. E. Brophy (Ed.), *Advances in research on teaching: Teachers' subject matter knowledge and classroom instruction* (Vol. 2, pp. 115–143). Greenwich, CT: JAI Press.

Carpenter, T. P., & Fennema, E. (in press) Cognitively Guided Instruction: Building on the knowledge of students and teachers. In R.

Glaser & L. Bond (Eds.), *International Journal of Educational Research* [special issue].

Carpenter, T. P., Fennema, E., Peterson, P. L., & Carey, D. A. (1988). Teachers' pedagogical content knowledge of students' problem solving in elementary arithmetic. *Journal for Research in Mathematics Education, 19*(5), 385–401.

Carpenter, T. P., Fennema, E., Peterson, P. L., Chiang, C. P., & Loef, M. (1989). Using knowledge of children's mathematics thinking in classroom teaching: An experimental study. *American Educational Research Journal, 26*(4), 499–532.

Carpenter, T. P., & Moser, J. M. (1983). The acquisition of addition and subtraction concepts. In R. Lesh, & M. Landau (Eds.), *The acquisition of mathematical concepts and processes* (p. 7–14). New York: Academic Press.

Chi, M. T. H., Feltovich, P. J., & Glaser, R. (1981). Categorization and representation of physics problems by experts and novices. *Cognitive Science, 5*, 121–152.

Chi, M. T. H., Glaser, R., & Rees, E. (1982). Expertise in problem solving. In R. Sternberg (Ed.), *Advances in the psychology of human intelligence* (pp. 7–75). Hillsdale, NJ: Lawrence Erlbaum.

Clandinin, D. J. (1985). Personal practical knowledge: A study of teachers' classroom images. *Curriculum Inquiry, 15*, 361–385.

Clark, C. M., & Peterson, P. L. (1986). Teachers' thought processes. In M. C. Wittrock (Ed.), *Handbook of research on teaching* (3rd ed., pp. 255–296). New York: Macmillan.

Clark, C. M., & Yinger, R. J. (1978). *Research on teaching thinking* (Research Series No. 12). East Lansing, MI: Institute for Research on Teaching.

Eisenberg, T. A. (1977). Begle revisited: Teacher knowledge and student achievement in algebra. *Journal for Research in Mathematics Education, 8*, 216–222.

Elbaz, F. (1981). The teacher's "practical knowledge": Report of a case study. *Curriculum Inquiry, 11*, 43–71.

Elbaz, F. (1983). *Teacher thinking: A study of practical knowledge.* New York: Nichols.

Erickson, F. (1986). Qualitative methods in research on teaching. In M. C. Wittrock (Ed.), *Handbook of research on teaching* (3rd ed., pp. 119–161). New York: Macmillan.

Eylon, B., & Reif, F. (1984). Effects of knowledge organization on task performance. *Cognition and Instruction, 1*, 5–44.

Feiman-Nemser, S., & Floden, R. E. (1986). The cultures of teaching. In M. C. Wittrock (Ed.), *Handbook of research on teaching* (3rd ed., pp. 505–526). New York: Macmillan.

Fennema, E., Carpenter, T. P., & Peterson, P. L. (1989). Teachers' decision making and cognitively guided instruction: A new paradigm for curriculum development. In K. Clements & N. F. Ellerton (Eds.), *Facilitating change in mathematics education*. Geelong, Victoria, Australia: Deakin University Press.

Fuson, K. C. (1988). *Children's counting and concepts of number*. New York: Springer-Verlag.

Gagne, R. M. (1970). *The conditions of learning* (2nd ed.). New York: Holt, Rinehart, & Winston.

Gardner, H. (1986). *The mind's new science: A history of cognitive revolution*. New York: Basic Books.

General Accounting Office. (1984). *New directions for federal programs to aid math and science teaching* (GAO/PEMO-85-5). Washington, DC: Author.

Good, T., Grouws, D., & Ebmeier, H. (1983). *Active mathematics teaching*. New York: Longman.

Hashweh, M. Z. (1986). *Effects of subject-matter knowledge in the teaching of biology and physics*. Paper presented at the annual meeting of the American Educational Research Association, San Francisco.

Hiebert, J. (Ed.). (1986). *Conceptual and procedural knowledge: The case of mathematics*. Hillsdale, NJ: Lawrence Erlbaum.

Hiebert, J., & Wearne, D. (1986). Procedures over concepts: The acquisition of decimal number knowledge. In J. Hiebert (Ed.), *Conceptual and procedural knowledge: The case of mathematics* (pp. 199–223). Hillsdale, NJ: Lawrence Erlbaum.

Hiebert, J., & Wearne, D. (1991). Methodologies for studying learning to inform teaching. In E. Fennema, T. P. Carpenter, & S. J. Lamon (Eds.), *Integrating research on teaching and learning mathematics* (pp. 153–176). Albany, NY: SUNY Press.

Lampert, M. (1989). Choosing and using mathematical tools in classroom discourse. In J. E. Brophy (Ed.), *Advances in research on teaching (Vol. I)*. Greenwich, CT: JAI Press.

Lampert, M. (1986). Knowing, doing, and teaching multiplication. *Cognition and Instruction, 3,* 305–342.

Larkin, J. G., McDermott, J., Simon, D. P., & Simon, H. A. (1980). Models of competence in solving physics problems. *Cognitive Science, 4,* 317–345.

Leinhardt, G., & Greeno, J. C. (1986). The cognitive skill of teaching. *Journal of Educational Psychology, 2,* 75–95.

Leinhardt, G., Putnam, R. T., Stein, M. K., & Baxter, J. (1991). Where subject knowledge matters. In J. E. Brophy (Ed.), *Advances in research on teaching: Teachers' subject matter knowledge and classroom instruction* (Vol. 2, pp. 87–113). Greenwich, CT: JAI Press.

Leinhardt, G., & Smith, D. A. (1985). Expertise in mathematics instruction: Subject matter knowledge. *Journal of Educational Psychology, 3,* 247–271.

Lehrer, R., & Loef, M. (in press). Understanding teachers' knowledge of fractions. *Journal for Research in Mathematics Education*.

Livingston, C., & Borko, H. (1990). High school mathematics review lessons: Expert-novice distinctions. *Journal for Research in Mathematics Education 21*(5), 372–387.

National Council of Teachers of Mathematics. (1989). *Curriculum and evaluation standards for school mathematics*. Reston, VA: Author.

Orton, R. E. (1988). *Using representations to conceptualize teachers' knowledge*. Paper presented at the Psychology of Mathematics Education-North America, DeKalb, IL.

Orton, R. E. (1989). *A study of the foundations of mathematics teachers' knowledge (a preliminary investigation)*. Paper presented at the annual meeting of the American Educational Research Association, San Francisco.

Peterson, P. L. (1988). Teachers' and students' cognitional knowledge for classroom teaching and learning. *Educational Researcher, 17*(5), 5–14.

Post, T. R., Harel, G., Behr, M. J., & Lesh, R. (1991). Intermediate teachers' knowledge of rational number concepts. In E. Fennema, T. P. Carpenter, & S. J. Lamon (Eds.), *Integrating research on teaching and learning mathematics* (pp. 177–198). Albany, NY: SUNY Press.

Putnam, R. T., & Leinhardt, G. (1986). *Curriculum scripts and the adjustment of content in mathematics lessons*. Paper presented at the annual meeting of the Educational Research Association, San Francisco.

Resnick, L. B. (1987). Learning in school and out. *Educational Researcher, 4,* 13–20.

Resnick, L. B., & Ford, W. W. (1981). *The psychology of mathematics for instruction*. Hillsdale, NJ: Lawrence Erlbaum.

Romberg, T. A. (1983). A common curriculum for mathematics. In G. I. Fenstermacher & J. I. Goodlad (Eds.), *Individual differences and the common curriculum* (pp. 121–159). Chicago: NSSE.

Romberg, T. A., Harvey, J., Moser, J., & Montgomery, M. (1974). *Developing mathematical processes*. Chicago: Rand McNally.

Rumelhart, D. E., & Norman, D. A. (1978). Accretion, tuning, and restructuring: Three modes of learning. In J. W. Cotton & R. Klatzky (Eds.), *Semantic factors in cognition*. Hillsdale, NJ: Lawrence Erlbaum.

Schoenfeld, A. H. (1985). *Mathematical problem solving*. Orlando, FL: Academic Press.

Schoenfeld, A. H. (in press). On mathematics as sense-making: An informal attack on the unfortunate divorce of formal and informal mathematics. In D. N. Perkins, J. Segal, & J. Voss (Eds.), *Informal reasoning and education*. Hillsdale, NJ: Laurence Erlbaum.

School Mathematics Study Group. (1972). Correlates of mathematics achievement: Teacher background and opinion variables. In J. W. Wilson & E. A. Begle (Eds.), *NLSMA Reports* (No. 23, Part A). Palo Alto, CA: Author.

Shavelson, R. J., & Stern, P. (1981). Research on teachers' pedagogical thoughts, judgments, decisions, and behavior. *Review of Educational Research, 51,* 455–498.

Shulman, L. S. (1985). On teaching problem solving and solving the problems of teaching. In E. A. Silver (Ed.), *Teaching and learning mathematical problem solving: Multiple research perspectives* (pp. 439–450). Hillsdale, NJ: Laurence Erlbaum.

Shulman, L. S. (1986). Those who understand: Knowledge growth in teaching. *Educational Researcher, 15*(2), 4–14.

Shulman, L. (1987). Knowledge and teaching: Foundations of the new reform. *Harvard Educational Review, 57*(1), 1–22.

Steen, L. A. (1988). Out from achievement. *Issues in Science and Technology, 5*(1), 88–93.

Steinberg, T., Haymore, J., & Marks, R. (1985). *Teachers' knowledge and structuring content in mathematics*. Paper presented at the annual meeting of the American Educational Research Association, Chicago.

Stephens, W. M. (1982). *Mathematical knowledge and school work: A case study of the teaching of developmental processes*. Unpublished doctoral dissertation, University of Wisconsin-Madison.

Voss, J. F., Green, T. R., Post, T. A., & Penner, B. C. (1983). Problem-solving skill in the social sciences. In G. H. Bower (Ed.), *The psychology of learning and motivation* (pp. 165–213). New York: Academic Press.

Wineburg, S., & Wilson, S. M. (1991). Subject matter knowledge in the teaching of history. In J. E. Brophy (Ed.), *Advances in research on teaching: Teachers' subject matter knowledge and classroom instruction* (Vol. 2, pp. 303–345). Greenwich, CT: JAI Press.

GROUPING FOR INSTRUCTION IN MATHEMATICS: A CALL FOR PROGRAMMATIC RESEARCH ON SMALL-GROUP PROCESSES

Thomas L. Good
UNIVERSITY OF MISSOURI–COLUMBIA

Catherine Mulryan
ST. PATRICK'S COLLEGE

Mary McCaslin
UNIVERSITY OF MISSOURI–COLUMBIA

My best card ever! I looked forward with considerable enthusiasm to my core seventh-grade class (a two-period combined English and American History class). My favorable attitude toward the course was not due to the subject matter but rather to the opportunity it provided for writing and for peer exchange. My colleagues and I—four of us shared the same square table on the dark side of the room (in the far corner away from the windows)—had developed a passion for writing and exchanging baseball cards. These were not just any old cards; they described in rich detail our fantasy baseball careers. It started simply with only a few statistics (for example, minor league career statistics), but soon our cards began to describe extraordinary exploits (huge salary contracts, our being traded for five other players, civic honors no less!). The cards reflected our expanding sense of baseball "fame," and in our writing we generated outrageous but enjoyable baseball histories, histories that were rewritten (shamelessly) daily.

I would be hard pressed to defend what we learned during our off-task baseball-card writing activities. I remember with considerable detail many of the exchanges with peers, although I recall little of the English and American history that we studied in that class. I would venture to say, however, that the attempt to write clearly and vividly and to use statistical information convincingly (with peers' feedback and under their close scrutiny) led to many satisfying and enjoyable learning activities. Indeed, no other junior high class stands out in my mind as much as this one does.

Upon further reflection, I recall that in my K–12 schooling I spent no more than 5% of my time learning from and with peers. Indeed, my most memorable experience (I still recall the best card I ever wrote!) was in an activity not intended to involve peer learning. Considering the recent advocacy for "cooperative learning," I often wonder what opportunities today's youth have for peer learning.

We acknowledge the support provided by the Center for Research in Social Behavior at the University of Missouri–Columbia, and we especially thank Teresa Hjellming for typing the manuscript and Gail Hinkel for editing it. The authors acknowledge the helpful feedback from Jere Brophy, Michigan State University; Doug Grouws, University of Missouri; and Bob Slavin, Johns Hopkins University, and wish to especially thank Mary Lindquist for reading the manuscript on two occasions and providing helpful feedback. We also gratefully acknowledge the support of the National Science Foundation (NSF Grant MDR 8550619 and NSF Grant TPE-8955171), which made this work possible. However, the ideas presented here are the authors', and no endorsement from NSF should be inferred.

On a recent walk I asked my daughter (who has just completed her freshman year in college) about her experiences in small-group work. She suggested that in grades K–12 she spent about 20% of her time in group work with peers and speculated that perhaps 10% of that work was highly cooperative. About 75% of group work involved learning that was active and enjoyable. I asked her *why* some group experiences were better than others, and she said "It depends! It varies with countless factors, including the work, how it would be graded, who you worked with, whether you have worked with group members before, and so on." She continued to name 20 or so other factors. I proceeded to inform her that my interest in her small-group experience in school was motivated by the fact that two colleagues and I were in the midst of reviewing the literature on use of small-group models in mathematics instruction. I mentioned that much of the literature only reported achievement findings and that it ignored the "it depends" factors (who actually works together, how well they work together, and what task they work on). She was absolutely amazed and responded, "How can you know about the value of small-group instruction if you don't *observe* students or *ask* them?"

Although one student's reflections on her school experiences do not constitute evidence about the use of cooperative procedures in schools, her comments remind us that cooperative groups take many forms, serve various purposes, and vary widely in quality. This rich, complex perspective stands in stark contrast to some of the writing on cooperative learning that suggests cooperative methods virtually ensure student learning.

In writing the chapter, we have learned much from exchanging ideas with each other. Our active and cooperative collaboration has altered our individual views considerably. As members of a group comprised of a researcher who explores teacher communication in whole-class settings, a teacher educator and former classroom teacher who made extensive use of grouping strategies, and an interview researcher who focuses on the perceptions of individual learners through interviews, our individual experiences have been integrated as we have examined the use of grouping for mathematics instruction.

We have a great interest in small-group instruction and believe that it can facilitate students' cognitive and affective growth. The quality of instruction, however, is much more important than the form, and small-group instruction can be used in inappropriate as well as appropriate ways. Moreover, most students also benefit from instruction in whole-class and individual settings. Thus, more use of cooperative techniques should occur, but the techniques must be integrated with other organizational models.

Problems with Mathematics Curriculum and Instruction

Major problems with the mathematics curriculum, mathematics instruction, and student performance in mathematics have been identified in recent times (McKnight et al., 1987). Although our society has changed considerably over the last 50 years, the mathematics curriculum has remained fairly constant. Advancing technology offers new methods of accomplishing mathematical tasks and of communicating mathematical ideas. The curriculum of today, as reflected in textbooks and assessment tools, however, emphasizes arithmetic. According to the *Standards* (NCTM, 1989), "instruction has emphasized computational facility at the expense of a broad, integrated view of mathematics and has reflected neither the vitality of the subject nor the characteristics of the student" (p. 65). Not only do students' experiences in mathematics involve outmoded methods of calculating, but there is evidence that considerable time in elementary classrooms is spent on review of previously taught topics (Barr, 1988; Flanders, 1987; Perry, 1988). The emphasis on arithmetic and drill leaves little time for important topics such as geometry and statistics or for activities that encourage students to think logically and discover mathematical principles and ideas.

In a comprehensive set of naturalistic and experimental studies, Good, Grouws, and Ebmeier (1983) found that many teachers did not provide students with a meaningful orientation to the mathematics being presented. In a study of mathematics as taught in first- and fifth-grade American, Japanese, and Chinese classrooms, Stigler and Perry (1988) found that there were more *coherence* (students were better able to construct a coherent account of the sequence of the mathematics lesson) and *reflection* (thinking and verbalization rather than performance and practice) in Asian classes than in U.S. classes.

In integrating results of two studies of mathematics content in third-, fourth-, and fifth-grade classes, Porter (1988) concluded that too little attention was given to conceptualization of concepts and to problem solving and that extensive time was spent on skill development. A second problem was teachers' tendency to teach for exposure rather than for understanding (that is, teaching many topics briefly). Porter concluded that such teaching may inadvertently communicate to students that speed and accuracy in mathematics are more important than understanding and application.

Some of these issues have been addressed by the National Council of Teachers of Mathematics. The NCTM's *Curriculum and Evaluation Standards for School Mathematics* (1989), for example, recommends a broadened mathematics curriculum and specifies that student activities should evolve from problem situations and that learning is enhanced through active involvement of students with mathematics. There is a need for better mathematics instruction, more teaching for understanding (that is, meaningful development, coherence, reflection, and complex problem solving), and less time on fragmented practice and review in elementary classrooms. This is a complex problem that involves texts, curriculum materials generally, tests, and instruction.

Recent theory and research illustrate that students learn more when learning tasks require them to think about what is being learned (Nickerson, 1988). Although some educators view basic skills acquisition and higher-order thinking as mutually exclusive, recent research in cognitive science indicates that the two types of learning go hand in hand. There is growing recognition that a key to meaningful learning involves helping students to relate new information or knowledge to what they already know (Duffy & Roehler, 1988; Fennema, Carpenter, & Peterson, 1989). In a recent study, Fennema et al. (1989) illustrated that when teachers' understanding of children's cognition about addition and subtraction was enhanced, the teachers were able to design more meaningful instruction. In part, Fennema et al. argued that teachers were more successful because they allowed students more time to solve word problems and more opportunity to explain how they solved them. In turn, this gave teachers access to students' cognition and enabled them to base subsequent instruction on it.

Much research suggests that students are too passive and need to become more involved intellectually in classroom activities (for example, Good, Slavings, Harel, & Emerson, 1987; Goodlad, 1983). However, many students have developed negative views of themselves as learners and engage in invidious social comparisons that make it difficult to predict the conditions under which they will become mentally more active (see Newman, 1990). Although many writers interpret recent cognitive science research as suggesting the need for less teacher support and more learner independence (see Nickerson, 1988), what this means in practice is far from clear. For example, does heavy reliance on more talented peers mean less dependency than moderate reliance on the teacher? Do different subject matter concepts require various amounts of expert modeling and coaching so that simple statements about appropriate practice are highly misleading?

Still, in too many classrooms students are required to do little more than listen passively. They need more teacher explanations and modeling of problem solving that help them relate what they already know to present instruction. Students need to see teachers think and apply concepts (make predictions, use multiple approaches). Further, students also need extensive opportunity to act on concepts themselves and to share their thinking with teachers and peers.

Small-Group Instruction as a Possible Solution

Some educators believe that small-group instruction can be a useful strategy for responding to problems discussed in the previous section. That is, they argue that use of small groups will lead to more meaningful assignments and less time spent on needless review and individualized seatwork. However, simply increasing the amount of small-group instruction is unlikely to make the discussion or practice of mathematics more meaningful (Good & Biddle, 1988). In some classrooms small-group work may offer students more opportunities for interaction but may lessen coherence and increase drill and review (for example, the teacher supplies students with abundant easy work so as not to be interrupted when working with another group). In other classes small-group teaching may increase both verbalization and coherence, and in some classrooms neither will be affected.

Others have raised issues about the quality of curriculum tasks assigned for small-group work. "Simply because a lesson is implemented cooperatively does not assure its value. Using cooperative techniques to have students cover the same boring, inconsequential, or biased material or to have them 'get through' worksheets with more efficiency doesn't demonstrate the approach's full potential for changing what goes on in schools" (Sapon-Shevin & Schniedewind, 1989–90, p. 64).

Good and Biddle (1988) argue that small-group instruction is not a panacea but an attractive instructional format that, when properly implemented (for example, careful organization, appropriate tasks), could enable teachers to achieve certain goals: practicing meaningfully on mathematical topics of appropriate difficulty and interest, learning prosocial skills, taking different approaches to problem solving, verbalizing thoughts about mathematics, and enhancing social intelligence (Noddings, 1989). However, the advantages of small-group instruction await verification. Current research on small-group instruction illustrates that under certain conditions the motivational components of small-group work can promote students' mastery of mathematical skills and concepts (Johnson, Johnson, & Maruyame, 1983; Slavin, 1983, 1989). Indeed, the general consistency of these gains is impressive. Yet to be demonstrated is whether small-group instruction influences the development of critical thinking, students' views of mathematics content and problem solving, students' ability to reason mathematically, and their developing social intelligence.

Good (1983) and Good and Biddle (1988) argue that most educational reform movements in the United States have addressed single variables or clusters of variables focused on only one problem of schooling and have not been based on research. The general assumption appears to be that there is a common problem; therefore, there ought to be a common, simple solution. At various times, educators in this century have advocated as answers large-group instruction, small-group teaching, and individualized teaching (Good & Biddle, 1988). Similarly, both direct instruction and discovery learning have been lauded at different times as means for improving education. Unfortunately, this logic defies experience as well as research. The problems of U.S. schooling vary from school to school (Good & Brophy, 1986; Good & Weinstein, 1986), and research has shown that even teachers at the same grade level in the same school may have different problems. Simple characteristics of instruction have never predicted student achievement, although many reform efforts have focused on such characteristics (Good, 1983). The important issue is not whether individualized instruction, small-group instruction, or discovery learning is emphasized but rather the *quality* of planning and instruction. If important qualitative changes are to occur, increased research that relates processes to student outcomes, as well as research on new curriculum materials that facilitate group processes, will be necessary.

Purpose and Organization of the Chapter

In this chapter we review some of the literature on small-group instruction, although our review is not comprehensive. Several summaries of this literature can be found elsewhere (for example, Bossert, 1988–1989; Johnson et al., 1983; Johnson, Maruyame, Johnson, Nelson, & Skon, 1981; Slavin, 1983; and Slavin, 1989b). There is clear and compelling evidence that small-group instructional models can facilitate student achievement (most notably in basic skills) as well as more favorable attitudes toward peers and subject matter. However, there is a paucity of process data to indicate what happens during small-group instruction and which dimensions of the model are important if small-group instruction is to achieve these goals. That is, the "it depends" factors have received little research attention. Thus our major task in this chapter is to argue the need for more process-observational and interview research and to detail how some of this research could be conducted in the 1990s. Much of the literature on cooperative learning is about *cooperating to learn*. In this chapter, however, we focus on *learning to cooperate* as well as cooperating to learn (Rohrkemper, 1986). We also stress that much of the learning that occurs during student-student exchanges involves informal knowledge and incidental learning—as in the baseball card example.

At present, important new expectations are being announced in various publications (for example, the NCTM *Standards,* 1989). For example, learning to take risks in making mathematical judgments and developing confidence in number sense and mathematical reasoning are now seen as at least as important as learning to multiply. Small-group instruction is a strategy that is frequently recommended for making learners more active. However, some research shows that students do not necessarily become active learners just because they are placed in small groups. Thus, if instruction is to be improved, educators must understand better how students learn to cooperate during small-group instruction. A decade of programmatic research on small-group instructional processes could yield information that would provide practitioners with concepts, findings, and theories that could help them to design better small-group instructional activities.

We hope that readers will leave the chapter with many questions about how to improve group instruction in mathematics. Further, we hope that readers will be encouraged to conduct research and/or to engage in creative clinical practices that might indicate how grouping can stimulate creative, productive mathematical thought. The chapter will not yield a synthesis of "what works," because the relative absence of information about small-group processes (lack of classroom observation and student interviews) prohibits simple statements, especially about how small-group instruction facilitates learning of higher-order content. The chapter should, however, stimulate thinking about productive future research. We must have more basic research on small-group processes if we are to understand its potential more fully. Further, we believe that opportunities for substantial curriculum/instructional improvements require extensive basic research and cooperation between teachers and researchers. Unfortunately, curriculum reform seldom allows this type of research collaboration to develop.

In this chapter, we first discuss some research on the typical use of grouping (for example, ability groups) for mathematics instruction. Second, we describe some of our research in small-group contexts. Third, we illustrate a range of ways in which small-group processes can be examined and some of the differences among theoretical conceptions of cooperative small-group work (for example, Piaget's sense of "co-operation" versus Vygotsky's [1978] sense of socially situated learning). This part of the chapter demonstrates the complexity of student learning in social settings. Finally, we discuss research questions/issues that have resulted from our work. Many additional research topics could be examined; however, the diverse questions we propose are reasonably comprehensive and serve to illustrate why a decade of research on small-group process would be an important investment.

LITERATURE REVIEW

This literature review is highly selective. There are hundreds of studies on grouping, and numerous studies have explored the relation between grouping and mathematics achievement. We summarize this general literature quickly because comparatively little information can be derived from studies that do not examine classroom processes (see also Bossert, 1988–1989; Brophy & Good, 1974; Dunkin & Biddle, 1974; Good & Biddle, 1988). In short, we argue that the form of grouping (whole class, two groups, three groups) is less important than the quality of instruction. However, research that involves observation of group processes consistently shows that, in comparison to teachers who teach the whole class, teachers who use two and three groups for instruction do not generally teach more new content or present mathematics in a more coherent and understandable way. Hence, grouping does not in itself improve lessons. We discuss studies that illustrate the value of observing group practices in order to clarify how process dimensions either support or hinder learning.

In the next section of the literature review we describe research on student interaction during multiple small-group (3–4 students) instruction. Again, we do not emphasize outcomes because there are several excellent reviews in this area. We focus instead on knowledge of group processes. Few studies describe what takes place during small-group instruction (see also Bossert, 1988–1989). However, what research we have shows that some students have difficulty participating in groups and that gains from participation are far from automatic. Hence, although many studies show favorable outcomes of small-group instruction, other studies indicate problems associated with the implementation of the model.

Grouping and Organization for Instruction in Mathematics

The question of how to assign students to classrooms for instruction has long been debated. For example, should all three third-grade teachers at a school receive about equal numbers of high-, middle-, and low-achieving students (heterogeneous assignment), or should one teacher receive the top third of the students, another the middle third, and the other the lowest one-third of students? After students have been assigned to classrooms, there is again the issue of how to organize them for instruction. Should students be taught as a whole class or divided into two, three, or more groups for instruction?

Mason (1990) documented the need to consider organizational structure (assignment of students to classrooms within schools), macro formats (how teachers usually organize classrooms for instruction), and micro formats (how teachers organize classes for a given instructional topic or lesson). He argued that these distinctions must be kept in mind if we are to understand how classroom structures influence classroom processes and ultimately student learning.

The relationship between assignment of students to classes and organization of students within classrooms has been a much discussed and researched topic. For example, in 1932 Billet reviewed over 140 articles, and more recently Slavin

(1987a) reviewed several hundred studies. Despite the extensive number of studies on this topic, educators continue to debate whether students of similar ability should be taught together or whether classes of mixed ability or mixed groups facilitate achievement.

Part of the reason this literature is so difficult to learn from is that most researchers failed to observe instructional processes (Bossert, 1988–1989; Good & Marshall, 1984; Wilkinson, 1988). Mason (1990) indicated that several grouping formats have yet to be studied and that studies are often confounded due to imprecise descriptions about how students were grouped. For example, it is possible that teachers who share students across grade levels in order to obtain groups of students who are similar in achievement may in fact be teaching groups that are more diverse than teachers (in other schools) who do not regroup within a grade or across grades.

What is most problematic and more often debated in contemporary circles is the extent to which teachers should use within-classroom grouping in order to narrow the range of students they teach to make it easier to plan instruction. Although the literature shows mixed results, studies that include observation of instruction consistently suggest that low-achieving students who are grouped within classrooms tend to receive less appropriate instruction than higher achievers (Wilkinson, 1988). Moreover, there is little evidence that groups of low achievers receive appropriately interesting and challenging instruction.

It is interesting that reading experts are currently pressuring teachers to reduce their heavy reliance on within-class homogeneous grouping and to increase their use of small heterogeneous groups or whole classes (see *Becoming a Nation of Readers*). In contrast, in mathematics there are two conflicting contemporary trends. On the one hand, there is increased emphasis on small heterogeneous groups (Johnson et al., 1983; Slavin, 1989a, 1989b); on the other hand, some experts also advocate within-classroom ability grouping (for example, Slavin, 1987a, 1987b). We return to the issue of cooperative heterogeneous groups later in the chapter because this movement is our central interest. Now we want to discuss teachers' use of within-classroom ability grouping for instruction in mathematics.

Descriptive Studies of Within-Classroom Grouping

SURVEY DATA. To obtain information about how teachers use small-group instruction, Good, Grouws, and Mason (1990) surveyed 1,509 teachers from elementary schools in 10 districts in three states. Although 23% of the sample indicated that they occasionally used small-group instruction, only 13% primarily taught students in groups, and only 5% indicated that they frequently allowed students to work cooperatively with peers. These results suggest that teachers infrequently use small-group instruction and that such instruction serves a relatively narrow range of functions. The size of groups that most teachers form (e.g., breaking the class into two large groups) and the limited time that teachers allocate for group work do not seem conducive to students reasoning and thinking critically. The picture of grouping that emerges from the survey is one in which grouping is used to allow students in one group to practice while the teacher works with another group.

OBSERVATIONAL DATA. An observational study was conducted to describe what actually occurs during small-group instruction when the class is divided into two or three groups (see Good, Grouws, Mason, Slavings, & Cramer, 1990, for details). How do teachers actually employ these formats? How do teachers differ in their behavior when they use grouping procedures and when they teach mathematics to the class as a whole? Thirty-three teachers who reported grouping students for mathematics instruction took part in the study. Study findings showed that teachers who grouped also used whole-class teaching for a variety of purposes.

Students in two- and three-group classrooms received less development and spent more time in transition and in non-mathematical activities than those in whole-class, ad hoc classes. Whole-class, ad hoc teachers allocated more time for development per lesson than two- and three-group teachers (whole class, 15.9; two group, 9.6; and three group, 9.0 minutes). When considering the amount of time that a group of students received development (that is, theoretically, students in a three-group setting received only one-third of the total amount of time spent on development), it was clear that teachers in whole-class settings provided substantially more time for development. The quality of development was also higher in the whole-class, ad hoc classes than in group classes.

It is interesting that the whole-class structure seemed to allow teachers more time to individualize instruction. That is, after teaching the lesson, teachers provided enrichment or remediation to a few students or talked to individual students about their work. In contrast, when two- and three-group teachers finished with one group, they needed to deal with the next group.

Finally, it is noteworthy that teachers who used the three-group model had the *most* time on task but the *least* student-student interaction (when students worked without the teacher) during mathematics lessons. Two- and three-group classes were seen as high in promoting group self-management. Unfortunately, the self-management skills were confined to "doing one's work without disturbing other students" rather than self-management in the broader sense of knowing when and how to use peers or other resources to obtain information. Hence, a change in group structure did not lead to a change in the nature of the learning community.

This study suggests that simply using more grouping does not necessarily lead to more student verbalization, critical thinking, or increased collaboration on mathematics projects. Indeed, in some classrooms grouping meant less opportunity for these important activities and more time on review.

GERLEMAN'S STUDY. The findings of the survey and observational studies described in the preceding sections are consistent with those obtained by Gerleman (1987). She reported that few teachers utilized small-group work in fourth-grade mathematics. She studied small-group instruction in 11 fourth-grade mathematics classrooms drawn from three large urban school districts. Observational data indicated that instruction was primarily teacher-directed and differed from whole-class instruction only in that the teacher taught two or three groups separately. According to Gerleman, teachers used one of two general formats. In Type 1 grouping (two of 11 classrooms), groups covered the same content separately but simultaneously (more advanced students often had an enrichment assignment while other students caught up). In Type 2 grouping (eight classrooms), each group received different content and moved at its own pace. In the other classroom, a more flexible, individualized approach was used.

Gerleman found that the basic instructional pattern in both Type 1 and Type 2 classes involved the teacher presenting the lesson and then students doing practice work individually. Gerleman described the small-group teaching as disappointing in that it failed to yield some of the potential benefits of whole-class teaching (for example, extended discussion of problems; a focus on conceptual understanding) or of small-group instruction (students can collaborate). Both the time spent on development and the quality of development were low. Teachers tended to use only textbook examples and did not emphasize student understanding. They spent over *50% of available time on review* because of perceived needs to "manage" students. According to Gerleman, the small-group teaching did not allow academic interaction among students. Furthermore, virtually no curriculum tasks required students to use higher cognitive processes (fo rexample, explain answers by citing rules; solve problems by using principles; collect data and make inferences about them). Gerleman concluded that the apparent purpose of grouping was to accommodate differences in student achievement, not to introduce new content or to change the nature of mathematical tasks assigned.

Outcome Studies of Within-Classroom Grouping. However, despite these descriptive process studies that question the value of within-class ability grouping in mathematics, several studies have focused on achievement outcomes and suggest possible advantages of within-classroom ability grouping (Slavin 1987a).

Slavin (1987a) organized the literature on grouping and imposed rigorous, explicit criteria in an attempt to find the best studies before conducting a meta-analysis. He then looked for consistency of findings across comparable studies. Slavin was optimistic about the use of within-grade-level regrouping for instruction in mathematics because, of the seven studies found, five favored regrouped classes over heterogeneous, self-contained classes (that is, in terms of enhanced mathematics achievement). Although the findings appear to be considerably stronger in reading than in mathematics, Slavin did not comment on this in the review.

Slavin also examined research in which the teacher divided the mathematics classroom into two or three groups. In the area of elementary school mathematics instruction, his review yielded eight "adequate" (that is, met his criteria) studies. Six of these studies indicated that teachers who grouped achieved better results than teachers who taught the class as a whole. On the basis of these findings, Slavin recommends within-classroom ability grouping for mathematics instruction.

Despite the consistency of these "adequate" studies, several critics (for example, Gamoran, 1987; Hiebert, 1987) have questioned the validity of Slavin's recommendations because there were no observational data and hence there is no way of knowing why or how grouping patterns influenced achievement.

Mason (1990) criticized Slavin's recommendations because the studies still blurred the issue of group composition. He ar-

gued that even though studies are organized by context, they may still vary to a large extent (for example, because of teachers' use of different macro- and micro-formats). Mason noted that Slavin combined studies of teachers who used two and three groups in a within-classroom grouping format, although the instructional processes of these two formats may differ drastically (see Good, Grouws, Mason, Slavings, & Cramer, 1990).

A Study Comparing Process and Outcome Dimensions of Two-Group and Whole-Class Settings. Mason (1990) was intrigued by research showing advantages of whole-class instruction under certain conditions (for example, Good et al., 1983) and by Slavin's (1987a) support of within-classroom ability grouping (Slavin, 1987a; Slavin & Karweit, 1985). He conducted an experimental, observational study to test the efficacy of two grouping strategies in sixth-grade mathematics classrooms to which students had been assigned in a homogeneous, cross-classroom structure (within–grade-level assignment of students to classes to reduce variability in achievement). The critical question was how organization of the classroom for instruction influenced achievement.

There were three conditions in Mason's study. Teachers in control classrooms used the two-group instructional model that they had been using in previous years. Mason wanted to determine if a whole-class, ad hoc instructional model (teacher teaches whole class—then makes decisions about practice, remediation, or enrichment) would result in better achievement than the two-group, within-classroom teaching model. Although control-group teachers received no training, both experimental groups (whole-class, ad hoc and two-group) received extensive training to ensure that they were aware of research on active teaching and learning and National Council of Teachers of Mathematics (1989) guidelines concerning instruction and curriculum (for example, attention to problem solving, estimation, and mental arithmetic).

In the whole-class model teachers responded to students' learning needs through *situational adaptation*. That is, they taught the lesson to the whole class and then assessed student understanding after the development phase of the lesson. Whole-class teachers then decided on subsequent remedial or enrichment activities.

In contrast, the two-group model addresses the reduced heterogeneity of students through within-classroom ability grouping. This *structural adaptation* is based on the assumption that differential instruction, pace, and curriculum content will more effectively meet the needs of students. Thus, in the preactive stage of teaching, teachers could plan mathematics lessons for students who were comparable in mathematical ability, although they had to plan for two groups of students. Whereas adaptation to student needs in the whole-class, ad hoc model occurred during the interactive stage of teaching, the adaptation in the two-group model was more likely to occur in the preactive stage of teaching. Although two-group teachers were free to revise plans during the instructional day, the constraints of the two-group model made it difficult for them to make these adaptations spontaneously.

Following the in-service training for experimental teachers, observations were made in all classrooms (whole-class, ad hoc; two-group experimental; and two-group control) over the next several months. Observational data provided clear evidence that treatment implementation was highest in the whole-class, ad-hoc classes. Moreover, achievement was consistently higher in these classes than in classes using two-group procedures (both treatment and control). The performance of students in the whole-class, ad hoc classes surpassed the performance of students in the classes of teachers who used the other two models in the following areas: computation ($p < .0001$); concepts ($p < .01$); and mental arithmetic and estimation ($p < .001$). Whole-class students' performance on a problem-solving test exceeded the performance of students in the other two conditions but did not reach statistical significance ($p < .08$). What is significant is not that the whole-class model was more effective than the two- group model (we suspect that under certain conditions two-group strategies might be more effective) but rather the convincing evidence that instruction was more easily adjusted and more meaningful to students in some classrooms than in others. Although the quality of teaching in most classrooms was reasonably high, students in the whole-class, ad hoc classes received more and better development (that is, instruction designed to facilitate the meaningful acquisition of mathematical ideas), and teachers using this model were better able to assess students' understanding (that is, they paid more attention to student cognition and its implications for subsequent instruction). Thus in whole-class, ad hoc classes the focus on mathematics appeared to be more coherent and meaningful (for example, Stigler & Perry, 1988), as well as more sensitive to student understanding (for example, Fennema et al., 1989). Previous research has shown that these conditions facilitate student understanding of mathematics.

On-the-spot adaptation—the provision of enrichment or remediation—occurred considerably more frequently in whole-class, ad hoc classrooms than in two-group classrooms, indicating that the situational model associated with whole-class teaching facilitated adjustment of instruction to the needs of individual students or groups of students in comparison to the structural adaptation model of two-group teaching.

Mason also found that whole-class, ad hoc teachers extended more individualized time to students than did two-group teachers (both experimental and control). This extra time for individuals occurred in various ways (for example, assessing understanding of new content, conducting reviews or reteaching, etc.).

Even though two-group teachers taught students that in theory should have benefitted from a given lesson, observers rated the quality (and amount) of development as lower in two-group classes than in others. Further, managing two groups at the same time (for example, giving directions so that the two groups can begin work at the same time, monitoring one group while teaching the other, etc.) makes it difficult for teachers to engage in spontaneous adaptations during instruction. Other teachers or teachers with additional or different training might make two-group instruction work well. For two-group models to work, the lesson quality must be high. Simply reducing the ability range of students who receive instruction does not in itself enhance teaching/learning. Teachers' perception that a whole-class model may not meet the needs of all students simultaneously may stimulate more careful assessment of students' understanding. In contrast, two-group teachers, under the assumption that their model better addresses students' instruc-

tional needs, may not feel as compelled to assess student understanding.

To reiterate, the issue is how instruction can be improved, not which model is best. That is, how do different models for organizing classrooms interact with classroom processes? As we have argued previously, Mason's research shows the value of observation in helping researchers to understand why variation in teaching may be associated with differential learning outcomes. The whole-class, ad hoc model was associated with enhanced student learning because the treatment appeared to improve teachers' explanations (for example, Duffy & Roehler, 1989) by increasing the meaningfulness (development) of the mathematics presented (for example, Good et al., 1983; Stevenson, Lee, & Stigler, 1986) and teachers' attention to student understanding (for example, Fennema et al., 1989).

Our review of research on grouping in which teachers divide the class into two or three groups based on student achievement convinces us that this practice often does not lead to a productive learning environment—especially for students who are placed into low-achievement groups. We next turn attention to the consideration of the use of smaller individualized groups—especially grouping procedures that allow students to work cooperatively in heterogeneous groups.

Cooperative Small-Group Instruction

The literature includes increasing discussion of small-group mathematics instruction in elementary schools. Noddings (1989) notes that articles on this topic abound and that at least some policymakers are advocating more small-group instruction. For example, California policymakers want some emphasis on small-group instruction built into textbooks used in the state (Mathematics Framework for California Public Schools, 1985). Ample research illustrates that small-group methods can under certain conditions facilitate achievement and affective gains (Slavin, 1989b).

There are now several excellent reviews of the literature on the use of cooperative groups to facilitate student achievement (Davidson, 1985; Johnson & Johnson, 1974; Sharan, 1980; Slavin, 1989b; Slavin et al., 1985). These studies provide consistent evidence that students' achievement is facilitated by cooperative learning techniques.

Slavin (1989a) notes that in the past 15 years there have been several models for organizing classrooms so that students can learn academic material from one another in small groups. Most cooperative learning methods have students work in heterogeneous groups of about four members. Group members are encouraged to help one another learn; however, program strategies vary considerably (see Slavin, 1989b, for descriptions of some popular cooperative learning methods and their defining characteristics).

Reviews of studies of the effects of cooperative learning have generally yielded positive findings. Slavin (1989a) contends that research has shown that these programs enhance various affective outcomes, including intergroup relations, acceptance of mainstream academically handicapped students by their classmates, self-esteem, enjoyment of class or subject, and general acceptance of others. Further, achievement effects of cooperative learning are generally positive. In his "best-evidence" review (1987a), Slavin found that 72% of the 68 "adequate" studies showed higher achievement for cooperative learning than for control conditions. He contends that achievement effects appear to be due primarily to combined cooperative and competitive incentives that motivate students to encourage each other to learn. However, these effects are not automatic, and reviewers differ sharply about the conditions necessary to obtain achievement effects.

Slavin (1989b) does conclude that two factors necessary for achievement gain (skill acquisition) are (1) use of group goals and (2) individual accountability. These factors provide *incentives* for students to help one another to learn. According to Slavin, if these two factors are not present, students have no real interest and/or have a counterproductive interest in one another's success. Thus, if these two factors are not present, most students do not provide one another with rich explanations needed to understand material and to achieve. When there are group goals but no individual accountability, as in some cooperative learning models, students may view interacting with other students as a waste of time and may be reluctant to stop and explain concepts to other group members who are having problems. Slavin argues that under these conditions, the most able students may simply do the work for the rest of the group; the less able may be exposed to less instruction.

The literature indicates that cooperative learning can be an effective means of increasing student achievement and that group goals and individual accountability are important issues to examine. However, group goals can take many forms and, in terms of certain theoretical perspectives and student outcomes (for example, higher-order thinking), there is comparatively little evidence with which to assess cooperative learning.

Empirical Issues. Davidson (1985) provides a comprehensive review of small-group learning and teaching in mathematics. He summarizes various approaches that have been taken in the area of mathematics and notes that, in addition to cooperative programs, several texts have been designed for small-group learning in mathematics. Davidson notes that the utility of small-group models in mathematics is generally supported in the literature. However, he argues that achievement comparisons have been made chiefly in reference to computational skills, simple concepts, and simple application problems. Thus, the fact that cooperative group work is reported as beneficial should be qualified by the narrow range of dependent measures that has been examined. More information is needed about how higher-order skills are affected by small-group mathematics learning.

Davidson also contends that the tendency for reasonable achievement to occur under cooperative-learning conditions typically holds at a general, main-effect level, but that when aptitude-treatment interaction considerations are taken into account, the pattern becomes more complex. For example, he notes that Peterson and Janicki (1979) and Peterson, Janicki, and Swing (1981) found that students of high mathematics ability achieved significantly more in small-group than in large-group approaches. Further, as he notes, students' attitudes toward subject matter and the method of instruction are a complex topic and show variation from study to study. For example, in some studies (for example, Gilmer, 1978), students reported a preference for small-group instruction; however, in other studies (for example, Peterson & Janicki, 1979) they preferred instruction in a large-group setting. According to Davidson, the interaction between student ability and attitude toward instruction is not consistent across studies. For example, in one study (Peterson & Janicki, 1979), students learned less under instruction using the method that was their initial preference.

Theoretical Issues. Pepitone (1985) argues that the cooperative-learning models that have been used in the classroom have tended to equate group reward and group goals. In essence, most of the early theoretical work on cooperation and competition was based on the notion of goals toward which individuals would work cooperatively in groups—goals they had developed and integrated on their own. Pepitone argues that group goals, as used in group dynamics theory, suggest that the focus of the activity is what the group *wishes to accomplish*. In most schools, however, group goals are assigned by teachers rather than discussed and decided on by students.

Pepitone notes that tasks have been characterized in many different ways and that the difficulty of a task and the extent to which the task can be subdivided are perhaps the most salient ways in which tasks have been differentiated. Building upon group dynamics theory, she distinguishes three major sources of the skills and related dispositions that are needed in competitive and cooperative situations. The first consideration concerns task-activity requirements, that is, demands that stem directly from the properties of an assigned task (for example, numbers have to be manipulated, measurements have to be made, etc.). Second, task role requirements are the interpersonal relationships that are dictated by the demands of a task: Who will do which measurements? What will be done after the measurements have been completed? How will people do the work together? The third consideration involves group role requirements. These demands do not stem directly from the properties of a task but rather from the internal needs of the group to maintain itself for task performance (for example, Who will initiate discussion? Who will help to mediate conflict?, etc.).

Pepitone argues that the concept of group roles has largely been neglected in cooperative research. That is, the internal needs of the group to maintain itself in order to accomplish its work, how this is played out in various settings, and the consequences of different ways in which groups work have not been carefully researched. She contends that the importance of distinguishing between task roles and group roles has long been considered (for example, Benne & Sheats, 1948); however, classroom research on group roles has been neglected. According to Pepitone, no group can make much progress unless its members are willing to break the ice, mediate disputes, and generally work to maintain the group itself. Obviously, the use of small groups as a way of actively learning the intended curriculum is an important activity, but it should not be confused with another possible goal—the use of small groups to allow students to construct goals and pursue self-defined goals. Thus, from certain theoretical perspectives, extant research on cooperative grouping has ignored key distinctions between group and individual goals and between the goals of the school curriculum and those of individual learners.

Some theorists have argued persuasively that cooperative groups may not lead to enhanced participation and on-task involvement for low-status students. Cohen provides a theoretical framework based on expectation-states theory (Berger, Cohen, & Zelditch, 1972) that suggests that cooperative small-group structures make status differences among students more salient. Hence, because students work together closely during small-group activities, achievement differences are likely to become more evident (especially on certain types of tasks), and status differences may increase. Because of these status differences, high achievers will dominate and low-status students will remain relatively passive. Indeed, empirical research on student interaction in cooperative small groups has shown that high-status students tended to dominate small-group discussion because they were expected to be more competent, and low-status members were more passive because they were not expected to make contributions (for example, Cohen, 1982; Humphreys & Berger, 1981; Rosenholtz, 1985; Stulac, 1975).

Cohen argues that achievement differences are likely to be most salient in reading and mathematics. Assignments like those typically given to individual students (for example, involve one correct solution) are the ones most unsuitable for small-group work. According to Cohen, only tasks that require multiple abilities and contributions for task completion are likely to promote cooperative activity and collaboration by all students in a group. Hence, although efforts to help students develop skills for working with others may assist small groups to work productively, the most important factor is the development of appropriate and meaningful tasks that encourage cooperative interaction. Some theorists would therefore question the extent to which curriculum materials used in much cooperative research actually represent the opportunity for meaningful collaboration (as opposed to "tutorial" interaction).

Problem Solving as Adaptive Learning in Cooperative Groups. A special theoretical concern that we want to argue is the conceptualization of problem solving as adaptive learning. Of the various potential uses of small-group instruction, we believe that perhaps its major strength is to help students develop problem-solving skills and dispositions. However, if students are to become successful mathematical problem solvers, they must first be adaptive learners. Other chapters in this volume discuss problem solving more directly in terms of the mathematical problem (see, for example, Schoenfeld). Here we talk about the *problem solver*, and our concern in this section is with a *psychological* definition of problem solving that fuses the affective with the intellectual. In this section, we deal with the adaptive learning of individual students as well as the special problems of adaptive learning in small-group settings.

Noddings (1989) offers different frameworks that support the use of small-group teaching. Two of the frameworks she has proposed are outcome and development. Outcome theorists emphasize learning outcomes more than learning processes. They focus on how small groups may be used to supplement current instructional methods—how to help students learn the content of the traditional curriculum. Developmental theorists focus on cognitive, social, and moral development. Theorists like Dewey (1902) and Vygotsky (1978) have argued that cognitive growth occurs through social interaction. In mathematics the developmental view often translates into experiential learning, the active development of knowledge through concrete exploration and social conversation.

In our use of the term *outcome perspective* in this section, we are implying a narrow definition of outcome (knowledge of isolated concepts, collections of facts to be memorized, and processes to be practiced) because this is primarily what extant small-group research has studied. We recognize that researchers could emphasize outcomes that are process dimensions requiring considerable thinking—finding a counterexample, detecting a pattern, summarizing a relationship among variables. We do not suggest that a focus on outcomes per se is improper, but we do express concern for narrow definitions of outcomes (heavy emphasis on fast, correct answers). Thus, our contrast between outcome and development should be seen as a relative contrast between algorithmic responding and problem solving.

These views lead to some important value differences. For example, to some outcome theorists the issue of competitive versus cooperative reward structures would be an interesting question (that is, Do cooperative groups function better when rewarded in certain ways?), but developmental theorists would more likely question the interplay of competition and cooperation. The research reviewed previously in this chapter has emphasized an outcome perspective. In this section the focus placed on small groups is a developmental-social perspective: problem solving as adaptive learning in a social setting. Here we emphasize a distinct theoretical perspective of learning to cooperate rather than cooperating to learn (Rohrkemper, 1986).

We turn now to a psychological definition of problem solving that incorporates three elements: (1) maintaining the intention to learn (2) while enacting alternative task strategies (3) in the face of uncertainty. This conception implies an integration of the affective and the intellectual in the confrontation of difficult or novel tasks. It is consistent with Vygotskian theory (Vygotsky, 1962, 1978) and present-day extensions as represented in Soviet "activity theory" (for example, Davydov & Radzikhovskii, 1985). This conception of problem solving merges the spheres of motivation, emotion, cognitive strategy, and metacognitive awareness—typically compartmentalized in traditional problem-solving research. Vygotsky (1962) argued that "... their separation [intellect and affect] as subjects of study is a major weakness of traditional psychology since it makes the thought process appear as an autonomous flow of 'thoughts thinking themselves,' segregated from the fullness of life, from the personal needs and interests, the inclinations and impulses of the thinker" (p. 8). Similarly, most research on student learning in small groups has ignored how students think and learn while interacting with peers on a cooperative task.

In this section of the chapter, we explore the interplay of these intrapsychological processes during the goal-directed activity of problem solving—as that interplay is revealed through the mechanism of "inner speech," defined as speech for oneself that is self-directive and guides thought in nonautomatic cognition; truncated and dense, it turns words into thoughts (Vygotsky, 1962, p. 131). We discuss students' reported inner speech when they are engaged in individualized work on tasks of varying demands. We outline groundwork on the nature and function of students' reported inner speech when they are confronted by tasks in the context of small groups. Attention is given to hypothesized socialization processes, the potential tension between the interpersonal demands of group membership and intrapersonal hardiness (for example, willingness to present one's own view), and the generalization and transfer of strategies relevant to the tasks of problem solving in the social and physical realms.

Effective problem-solving behavior, or adaptive learning in this problem-solving context, refers to the processes students enact as they attempt to solve problems; it does not refer to the objective correctness of the outcome of those processes. In this definition, effective problem-solving behavior involves the recognition, marshaling, and enactment of dual personal resources: those that transform the task and those that transform the self. This definition also assumes, indeed *requires*, that the problem-solving tasks encourage, or at least allow, the student to take a flexible approach. Not all classroom tasks are like this. As Rohrkemper and Corno (1988) argue, teachers often engineer tasks to enhance success and, in so doing, design tasks that are so predictable that they impede student opportunity to explore alternative task- or self-regulation strategies. Restriction of transformation opportunities limits students' ability to take charge of their learning: Students learn little about the potential or limits of their knowledge and personal resources and even less about how to mediate (Feurstein, Rand, Hoffman, & Miller, 1980) or compensate for those limitations. One wonders if the parallel in the social domain—the restriction of social interaction to individualized settings or noninteractive large groups—has similar costs in the development and flexibility of strategies to learn from, with, and in the social world.

Intrapersonal and Interpersonal Processes in Problem Solving

FUNCTIONAL LANGUAGE. Language is the hypothesized link between the interpersonal and intrapersonal spheres (Vygotsky, 1962, 1978). It is the bond between two essential and mutually interdependent processes: (1) internalization of the social/instructional environment and (2) the capacity for self-direction or self-regulation through inner speech. Understanding the relation between—for Vygotsky, the dialectical integration of—these processes is the critical task of the Vygotskian perspective. The dynamics of internalization, termed "emergent interaction" (Wertsch & Stone, 1985), are at the very heart of a Vygotskian perspective because Vygotsky "...conceives of mind as the product of social life and treats it as a form of activity which was earlier shared by two people (originated in communication), and which only later, as a result of mental development, becomes a form of behavior in one person" (Luria, 1979, p. 143). Thus the social environment and its transformation and internalization are pivotal influences on individual thought; the individual is a product of personal social experiences. And the capacity for self-regulation through adaptive inner speech begins in the social world.

Language is social in origin and acquires two distinct functions: communication with others and self-direction. Self-directive inner speech implies a dialogue rather than a monologue (see also Meichenbaum, 1977); it is the mechanism for

self-regulation in "effortful cognition" (Posner, 1973). It merges the affective with the intellectual in the pursuit of motivated thinking and learning. Two types of inner speech have been identified that reflect this integration: self-involved and task-involved (Rohrkemper, 1986; Rohrkemper & Bershon, 1984; Rohrkemper, Slavin, & McCauley, 1983). Self-involved inner speech reflects control over the self through motivational and affective statements. Task-involved inner speech reflects control over the task through problem-solving, strategic instructional statements afforded by the task, and modification of the task if necessary and possible.

TASK AFFORDANCE. Together, self-involved and task-involved inner speech differentially mediate tasks. And tasks differentially mediate inner speech. For example, some tasks senselessly overwhelm, fail to stimulate, or are so prescriptive and narrow that they do not allow transformation of self or task. These tasks do not afford the enactment and, hence, the empowerment of inner speech. Instead such tasks paralyze—whether through sheer terror or numbing boredom, the effect is the same—students do not engage or refine adaptive, self-regulating inner speech (McCaslin Rohrkemper, 1989).

Tasks afford and promote the enactment of inner speech in the first place. Mathematics educators are familiar with this discussion: Task definition is the recurring theme in curriculum debate. A Vygotskian perspective calls for informative conceptual and problem-solving tasks—those that optimally challenge and thereby allow initiation of enhancing self-involved and strategic task-involved inner speech. There is some evidence that tasks subjectively experienced as moderately difficult most facilitate that development (D'Amico, 1986; McCaslin Rohrkemper, 1989; Rohrkemper, 1986; Rohrkemper & Bershon, 1984).

Recent work on students' reported inner speech while engaged in a group task, although outside of classroom learning (and accountability) systems, may be informative. For example, Bershon (1987) integrated work by Sharan (1984) with a Vygotskian perspective on functional inner speech by having students (grades 3–6) in four-person groups engage in a cooperative Lego construction task. Student groups (single gender, but varied in ability composition) were filmed, and students were subsequently interviewed individually about their inner speech with either a fixed-process-tracing or open-recall procedure. Bershon found that students' general sense of their problem-solving behavior consisted primarily of task-involved inner speech (72% of all students reported inner speech). When in the cooperative group and engaged in a group task, however, students' reported inner speech changed so that the majority of inner-speech reports concerned self-involved inner speech (51%). Task-involved inner speech when engaged in a group task consisted of only 35% of the total inner-speech reports. Group tasks were also associated with reported inner speech considered "off-task" and "about group members."

Bershon's pattern of results supports Pepitone's (1985) concerns and also sparks interesting questions about the tension between intrapersonal awareness and interpersonal dynamics. Students in this study evidenced an increase in *self*-awareness as participants in a group task. The reduction in reported task-involved inner speech points to the possibility that learning to work on group tasks is a learning task in and of itself, and that the increased focus on oneself is part of coping with and meeting those learning demands. One wonders about the extent to which the focus away from the assigned group task increases and enhances self-involved inner speech and the personal efficacy that individual students bring to subsequent group tasks. Or is the shift in focus detrimental in that it indulges a narcissism that does little to engender hardiness in the individual learner or the group? Bershon found no developmental, gender, or ability differences in student reports. Replication to school tasks with concern for personality and social-cognition variables with students beyond middle childhood years may increase our understanding of the emergent interaction of students engaged in group tasks in small-group settings.

EMERGENT INTERACTION AND SMALL-GROUP LEARNING. In conjunction with task opportunities, inner speech also serves as an interface between the social/instructional environment of the classroom and the internal resources of the student. As such, inner speech is the hypothesized vehicle for both the acquisition and enactment of self-regulation. Moreover, students differ in the fluidity of their reported inner speech, the sophistication of the task-involved strategies they employ, and the types of motivational and affective configurations that allow them to persevere (McCaslin & Rohrkemper, 1989). Research on students' self-regulation indicates that students are able directly to teach and model for their peers what we here term adaptive inner speech: strategic self- and task-control (e.g., Schunk, 1989). Indeed, Schunk (1989) has found that in some situations, peer modeling is a more powerful social/instructional tool than is modeling by an adult.

How do the developmental dynamics of social-comparison processes (Nicholls, 1984; Stipek, 1984) interact with exposure to multiple social contexts that vary in size, configuration, and activity? McCaslin (in progress) found that students' (grades 3–6) self-comparisons to classmates varied with participation in cooperative groups. In general, students in cooperative groups reported lowered ability self-comparisons with their classmates than did their peers who did not participate in cooperative groups. This ordering held for grades 3, 4, and 5. Lowered self-evaluation is *not* typically associated with younger students. It may be that the opportunity to compare oneself to classmates and teammates and to rate one's team relative to others' results in more social-comparison information than younger children are used to having and handling.

Equally important, however, is that lowered ability self-comparison *is* expected by sixth grade, when it is considered more "realistic"; yet cooperative groups appeared to mediate detrimental evaluation for these students. It may be that the emerging concern with social cognitive skills and a general increase in the importance of peer relationships for sixth graders expand these students' definition of *able*. Their participation in small groups likely provided varied yet overlapping cues about their expanding sense of ability.

McCaslin also asked students in cooperative groups to rank their team (of eight teams) and their relative standing within their team (of four students). These social-comparison rankings indicated that, for all grades, self-comparison to classmates-in-general was the most positive self-appraisal provided (even

though these appraisals were more negative for students in cooperative groups than were those provided by nonparticipating peers in grades 3–5). Self-comparisons to classmates-in-general are also the least accessible to disconfirming and personally damaging information. These patterns point to the importance of continued research and thinking about the dynamic tension between interpersonal and intrapersonal constructions and their emergent interaction with the developing learner.

Thus, we argue that intrapersonal dynamics are unique constructions that are a function of the emergent interaction between the developing individual and the multiple social/instructional environments in her life. Their interplay within a specific learning event and as they unfold over time promises to be an exciting area of inquiry into the dynamic tension between interpersonal and intrapersonal processes and the enhancement of self-directive inner speech.

Some Barriers. One impediment to initiating research on small-group learning dynamics is our lack of conceptual knowledge about the processes involved in such learning contexts. We have seen that a Vygotskian framework provides a dialectical model between self and other and the language of emergent interaction. Other models and constructs include locus assumptions and dimensional analyses (for example, group adoption of *self* vs. *task* goals [Homans, 1950]; *reward* or *goal* structure imposed on groups vs. the *task* structure they confront [Ames, 1984; Slavin et al., 1985]), and conceptions.

The student group as "team" is well represented in the literature (for example, Slavin et al., 1985). Much concern has been expressed about the applicability of the team concept to group participation and learning (for example, Pepitone, 1985). One could also ask if the team concept is more appropriate for males than for females, even in today's climate of expanding gender roles and opportunities. Similarly, a second concept, group as "family," leaves much to be desired. The prevalence of single-parent, blended, and extended families among public school children, even as family size approaches a bimodal distribution, seriously challenges the usefulness of the traditional family construct. There does not seem to be a shared experience of *family* on which to construct and elaborate students' understanding of *small group*.

Alternative conceptions and models of the constructive and interactive quality of student learning groups have not been forthcoming. The recent explosion of models for teaching (for example, adaptive teaching [Corno & Snow, 1985]; coaching [Palincsar & Brown, 1989]; guided apprenticeship [Rogoff, 1989]; and mediation [Feurstein et al., 1980]) and for classrooms (for example, places for learning [Marshall, 1988]; workplaces [Doyle, 1983]) has not inspired the small-group instruction researcher.

Competing models and conceptions can generate more informative research questions. Some hesitation, however, may be beneficial. The power of concepts such as size, composition, task, and goal structure (cooperative, competitive, and individualistic) in conjunction with interactive-process constructs and individual-difference variables seems to offer much to explore. The proliferation of models, however, may encourage research on model building rather than on the phenomenon of interest. Even so, the culture of individualism in the United

States makes it unlikely that taxonomies of group processes and new conceptions of "group" will emerge without careful attention to—observation of and interviews with—groups in process and groups as reconstructed and internalized by participants. A Vygotskian perspective would add as well that we cannot understand the intrapersonal workings of the individual without an understanding of former and ongoing interpersonal processes. Hence, there can be no real knowledge of the individual student without an understanding of the groups in which he participates. As we will see in the next section, there is comparatively little research that combines attention to the process, individual students' interpretation and mediation of the process, and the social and task setting in which the learning occurs.

Research on Small-Group Instructional Processes. According to Bossert (1988–1989), researchers of cooperative learning have developed several interesting explanations of why cooperative models work. Building on the research of Johnson and Johnson (1985), he notes that four major mediating (process) explanations could account for the success of cooperative methods: (1) reasoning strategies—exchange in cooperative groups may stimulate students to engage in more higher-order thinking; (2) constructive controversy—heterogeneous cooperative groups force the accommodation of the opinions of various members, and students must therefore search, engage in problem-solving, and take another's perspective; (3) cognitive processing—cooperative methods increase opportunities for students to rehearse information orally and to integrate information, especially explanations of how to approach a particular task; and (4) peer encouragement and involvement in learning—students help one another during group work. These positive interactions increase friendship, acceptance, and cognitive information processing. Unfortunately, these potential mediating explanations have not received systematic research attention.

Bossert notes that most cooperative learning researchers have employed a "black box" approach. That is, students are assigned to one of two treatments, outcomes are measured, and the treatments are compared in terms of their effects. When effects are found, post hoc rationales are used to explain the results. According to Bossert, mediating factors that explain why cooperative procedures work must be examined in observational research, and assumptions about desirable learning processes during small-group work must be verified and modified on the basis of such research.

Problems in Cooperative Small-Group Learning. Bossert points out that researchers often fail to verify whether students have even engaged in cooperative interactions. As an example of the need to observe student-student interaction, he cites a study by Johnson, Johnson, and Stanne (1986) in which students assigned to a competitive group were explicitly instructed not to work together, and students in the cooperative group were given explicit instructions to work together. However, observation showed that boys in the competitive situation had more task-related exchanges than girls in the cooperative group. Hence, assumptions about what takes place during group interaction can often be erroneous, and the need for ob-

servational research, at least to verify experimental conditions (if not to explain differences in learning), is obvious.

According to Bossert, when researchers observe instructional processes directly, the results have not always supported theories of cooperative learning. For example, Slavin (1987b) found that students who were asked to work together and to stay on-task were off-task more—and their achievement was lower—than students who were asked to work individually. As we argued earlier, there is a notable difference between studies that show achievement effects (using black box designs) and observational studies that examine classroom processes and report more problematic findings.

King (1989) examined the covert behavior of third graders in one classroom as they learned mathematics in a small-group cooperative-learning format called groups-of-four. He analyzed four mathematics lessons and used a stimulated-recall methodology to collect covert process data from two groups of four students each (two high achievers and two low achievers). King observed student performance during group interaction and interviewed students after their participation in groups-of-four learning activities.

Interviews with individual students about their thoughts during group work revealed notable variance between students in the two groups. For example, one group reported more self-oriented thoughts, and the other reported more task-oriented thoughts. However, the interviews also showed differences between high and low achievers in both groups. Highs reported more interactive thoughts, especially task-oriented interactive thoughts. Thus highs were thinking more during the group work task than were low achievers. Interviews also revealed differences in mathematical (content) understanding and in understanding of the work (task) requirements. High achievers could explain tasks clearly, but low achievers had great difficulty explaining group tasks. Low achievers may be less active in groups partly because they do not understand the tasks (see also Mulryan, 1989).

King observed group work and found that high-achieving students dominated group thinking and task initiations and generally played leadership roles. In contrast, low achievers were relatively passive during group work, although they enjoyed working in small groups (especially the social aspects) and looked forward to more opportunities for such work. Such affective reports need to be considered by researchers who use black box approaches to understanding cooperative interactions. That is, students' reports of satisfaction may mask important problems in group functioning.

Other observational research also illustrates that small groups do not provide an environment that is equally advantageous for all students (Lindow, Wilkinson, & Peterson, 1985). Although these researchers found that second and third graders could handle controversy that arose during group discussions, the means of resolving conflict varied with gender and achievement. Boys and high achievers gave more demonstrations (attempts to explain one's position) and had more prevailing answers (answers accepted by other participants) than did girls and low-ability students. High-ability students dominated the discussion associated with task disagreement and provided the greatest number of sophisticated process demonstrations. Of 56 process demonstrations coded, 30 were provided by high-ability students, 15 by medium-ability students, and 11 by students of lower ability (and eight of these were provided by a single student).

Wilkinson (1988) reviewed several studies of small-group processes and concluded that heterogeneous groups may only maintain the status quo concerning students' participation levels, since high achievers tend to become dominant. Low-ability students may lack skills for requesting and providing needed information during group interaction. According to Wilkinson, teachers will have to develop strategies for changing group interactions that occur naturally if all students are to benefit from small-group learning. Similar concerns have been expressed by other researchers (for example, Cazden, 1986; Cohen, 1984, 1986).

Facilitating Student Performance in Small Groups. As we have seen, several studies have suggested that certain students do not benefit from small-group processes. Other research shows that although the processes can be problematic, under certain conditions favorable outcomes may be obtained. Webb (1985, 1989) reviewed the literature on small groups and concluded that giving explanations is related to student achievement but that information that is presented without explanations is not associated with achievement. Moreover, receiving explanations tends to be associated with achievement; however, receiving information without explanation and receiving no help are detrimental to achievement. Hence, one rich framework for examining work groups involves studying the extent to which students are encouraged to provide explanations to other students. It is important to explore how individual students perceive cooperative behavior and the extent to which they can differentiate between providing answers and providing explanations.

Webb's review also suggests that group composition is important. She concludes that mixed-ability groups that include high-, medium-, and low-ability students are beneficial to relatively high- and low-achieving students but not to moderate-achieving students. However, she concludes that two other forms of group composition appear to be beneficial to most group members: mixed-ability groups with high-ability and medium-ability students or medium- ability and low-ability students, as well as groups with all medium-ability students.

Although Webb's insights are useful, she calls for additional work if we are to understand how students influence one another in small-group settings. In addition to advocating the examination of variables other than giving and receiving help, Webb notes that even within these categories, our information is still incomplete. She suggests that subsequent work needs to differentiate types of help and notes that research with a focus on giving and receiving help but without distinctions made among different types of help is likely to be uninformative and may even be misleading. Webb cautions that because detailed information about peer interaction is needed—and because to make certain distinctions it may be necessary to listen to group interaction on several occasions—verbatim audio or video records are likely to be essential research tools. Unfortunately, most process examinations of students' work in groups (that is, those studies that Webb examines in her review) have not included detailed audio recordings; hence, the data base for understanding how students learn in groups is fragmentary at present.

Other research has qualified Webb's findings (Lindow, Wilkinson, & Peterson, 1985; Peterson et al., 1981). In a study of small-group learning in second- and third-grade classrooms, Lindow et al. (1985) did not obtain the relationship between students' provision of explanations and their achievement that Webb (1984) found. They suggest that their results might be due partly to the younger age of students in their sample or perhaps to the mathematics content being studied (process explanations may be more necessary for certain types of content). As we will argue later, much more research is needed to explore how "desirable" process factors are mediated by student characteristics and content studied.

Still, several studies (Cazden, 1986; Cohen, 1984; King, 1989) indicate that students who provide active instruction and task leadership during group work are most likely to benefit from grouping. Webb (1989) reviewed 19 published studies exploring small-group processes and student achievement in mathematics and computer science classes. She asserts that whether help is beneficial in small groups depends on: (1) whether the help given is relevant to the student's particular misunderstanding; (2) whether the degree and type of elaboration correspond to the level of help needed; (3) whether the timing is appropriate, that is, in close proximity to student error or question; (4) whether the student understands the explanation; and (5) whether the target student has the chance to use the explanation to solve the problem. Webb notes that few studies have examined factors related to whether students benefit from help they receive.

This review also qualifies Webb's previous position that giving help facilitates the achievement of the help-giver. She notes that helpers must clarify and organize their thinking and often give explanations in new or different ways. However, when helpers only give right answers (no explanations) there is less reason for them to reorganize their thinking. Further, she concludes that students who give explanations show high achievement in part because they are more knowledgeable to begin with. According to Webb, help-giving that involves explanation is likely to be done by the most able students, and these students enhance their understanding of material by giving help. Thus, under certain conditions, small groups may allow more able students to become even more competent. We agree with Webb (1989) that it is important to study whether low achievers' achievement will increase if they give explanations to peers.

Other researchers have also argued the importance of peer interaction as students attempt to construct mathematical knowledge. Yackel, Cobb, Wood, Wheatley, and Merkel (1990) describe their efforts to build and implement problem-centered instruction in second-grade classrooms. In their approach all mathematical concepts and skills are taught in dyadic problem-solving activity (two students working together) followed by whole-class discussion in which diversity and thinking (rather than correct answers) are emphasized. Data suggest that their approach has been successful in that students in experimental classes (in comparison to control students) performed higher on state-mandated tests of mathematical concepts and application. Further, project students reported valuing cooperation more and were more willing to think for themselves and to persevere in the problem solving. This program will likely yield information about techniques teachers can use to help students resolve conflict and about the differential effects of various strategies. Further, this research is likely to yield valuable information about how teachers can intervene in group problem solving without giving too much direction, and about how teachers can create class norms that emphasize cooperative learning and the value of others' ideas.

Another study of students working in groups of two (Phelps & Damon, 1989) showed that fourth graders of equal abilities working in pairs (with minimal adult assistance) can make significantly more progress in understanding basic mathematics and spatial-reasoning concepts than students in control groups. For two years they studied 152 students who were divided into four groups: mathematics, spatial reasoning, and two control groups. Two students worked in pairs to solve problems or tasks that neither could do previously.

Results showed that the effectiveness of peer collaboration varied with the task. In the mathematics domain, peer sessions had little effect on performance during missing-addends and multiplication sets. However, peer collaboration resulted in significant improvement for ratio and proportion tasks. In the spatial domain, similar differential effects were obtained. Model-copying performance was not enhanced by peer collaboration, but collaboration improved performance on spatial-perspective problems.

On the basis of their study, Phelps and Damon argue that peer collaboration is a good method for promoting conceptual development but not for enhancing rote learning. They note that tasks least affected by peer collaboration tend to involve formulas and procedures (basic skills) that can be presented directly. However, collaboration facilitates learning of concepts that are not as amenable to direct instruction (for example, proportionality).

A Study of Work Groups in Mathematics. Good, Reys, Grouws, and Mulryan (1989–1990) use the term *work groups* to refer to heterogeneous small groups of three to five students in which teachers want students to work cooperatively. The term *work groups* is used to distinguish these groups from other types of small groups, including the achievement groups discussed earlier.

Good et al. (1989–1990) administered to over 400 teachers a survey about how often they used work groups during mathematics. Since the researchers were interested in what teachers and students actually did during work-group instruction, teachers who reported using work groups more than once a week and for half of the mathematics period or more were recruited. Also, an attempt was made to obtain a sample of teachers across various grade levels.

The observational sample was drawn from three school districts in a large, urban metropolitan area. Fifteen teachers in nine elementary schools participated. A total of 63 observations were conducted (15 in primary grades and 48 in intermediate grades). This was a naturalistic study of how teachers implemented the work-group format. The intent in this research was to study general strengths and weaknesses—not to study any one group or classroom in detail.

Although there was considerable variation among teachers (that is, work groups appeared more effective in some classes than others), the following strengths generally appeared across the lessons (for more details see Good et al., 1989–1990):

Active learning. Most (but not all) students exchanged mathematical ideas when they were in small, heterogeneous work groups.

Opportunities for peer interaction. Observers' comments about students' interaction within groups suggest that many students were developing social and communication skills—the ability to work with and build on strengths that others bring to a task—as well as becoming more aware of the needs and interests of others.

More interesting mathematics. Many lessons were designed to develop higher-order thinking skills (discovery, concepts, problem solving) rather than to emphasize computational skills and the quick processing of mathematical information.

Enhanced opportunity for mathematical thinking. Work groups frequently provided students an opportunity to explore diverse and, in some cases, more advanced mathematics.

Teacher as curriculum developer. Lessons that teachers had developed themselves or had found in supplementary books were most prevalent; only during a few observations did a teacher use a lesson from a textbook or teacher's manual. Some teachers demonstrated skill in writing curriculum.

The study also indicated that work-group lessons had some apparent disadvantages. We now turn to discussion of some of the *weaknesses* discovered in the lessons we observed.

Inadequate curriculum. The paucity of curriculum materials (designed explicitly for small-group instruction) was the greatest hindrance to the effective use of work-group instruction, and it forced teachers to rely on textbooks or to develop their own materials. Unfortunately, most problems in mathematics textbooks are designed for students to work alone.

Curriculum discontinuity. When teachers were left to their own devices to develop small-group tasks (taking examples from in-service programs, buying supplementary books, writing their own curricula), there often was no continuity in content difficulty across grades. Sometimes more sophisticated content was introduced at lower grades than at higher grades (for example, multiples and probability).

Designing appropriate tasks. Some teachers attempted to force too many topics and activities into the work-group model, in spite of the fact that some mathematical content is probably best taught in other formats or in various work-group formats (two students rather than four working together). Work-group tasks should be group-dependent; that is, there should be some reason for the group to work together rather than as individuals.

Implementing new tasks. Some teachers allowed too little time for groups to work on assigned tasks. Time allowed for group work varied considerably, from 30 seconds to 28 minutes. Some students were given so little time to complete tasks that they did not even get organized. Other students did not take tasks seriously because they knew (as they indicated during interviews) the teacher would soon intervene and provide assistance and direction.

Assigning student roles. Although some in-service programs apparently recommend that students be assigned roles to fulfill within groups, our observations led us to conclude that this practice is highly questionable. Designation of students as leaders, recorders, or materials managers seemed artificial in many cases (that is, students fought over roles or ignored role assignments).

Student passivity. Although we have stressed the positive social climate that existed for many students, a subset of students was passive during group work. If teachers did not address this problem (for example, calling on passive students to summarize group work for the entire class; emphasizing diversity rather than speed of responding), passive students were content to allow other students to do the work.

Lesson structure and accountability. Unfortunately, many groups we observed were not held accountable for completing tasks. Students need to discuss what they have learned with the teacher and other students. Too many lessons ended abruptly without sufficient time for students to summarize what they had learned.

Student Behavior in and Perceptions of Small Cooperative Groups.

A study by Mulryan (1989) focused on fifth and sixth graders' attending and interactive behavior in cooperative small groups working on mathematics. It attempted to identify group processes that influenced students' involvement (for example, why some students withdraw while others are more active). The study built on the work by Good et al. (1989–1990) described in the preceding section, particularly their findings concerning differential responding of students in small groups. Recent research on student mediation of instruction (Wittrock, 1986) also influenced the design and methodology of the study.

Mulryan investigated students' attending and interactive behavior in three classroom contexts: cooperative small-group mathematics, whole-class mathematics, and reading group. Students' interactive behavior in the classroom was compared with their interactive behavior on the school playground. She also investigated students' and teachers' perceptions about cooperative small-group instruction in relation to the other two classroom contexts. A major focus of this research study was the generation of concepts and understandings about students' responses to cooperative small-group learning opportunities, concepts and understandings that would provide a basis for future cooperative small-group studies.

Data collected in the study made it possible to compare the on-task behavior of students in whole-class mathematics and group reading contexts with their on-task behavior in the cooperative small groups in mathematics. Mulryan also compared, across the various classroom contexts, the attending behavior of high- and low-achieving students, boys and girls, students with internal and external locus of control, and students with high and low perceived competence in mathematics. Comparisons were made between students' and teachers' perceptions of the nature and purpose of cooperative small-group work in mathematics, about appropriate student behavior in this context, and about a range of task variables relating to cooperative activity in small groups.

METHOD. Five sixth-grade classes and one fifth-grade class in which students worked in cooperative small groups in mathematics at least once a week were chosen for the study. Eight students from each classroom were chosen as target students

for observation and interviews. Target students were chosen on the basis of teachers' within-class ranking of their mathematics and reading achievement and according to perceptions of their ability and willingness to cooperate and work well in cooperative small groups. Students' attending and interactive behavior was studied through classroom observation. Interviews were conducted to examine students' and teachers' perceptions and expectations in relation to cooperative small-group instruction.

Teachers agreed to arrange for their students to work on the same seven cooperative problem-solving tasks during mathematics observations over the period of the study: the consecutive sums problem (Burns, 1987), tangrams, pentominoes (Burns, 1987), the popcorn lesson (Burns, 1987), palindromes (Burns, 1987), word problems (Haenisch & Hill, 1985), and Cuisenaire™ rod patterns (Davidson & Willcutt, 1983). This allowed useful comparisons across classes. Tasks were chosen for their suitability for mixed-ability cooperative groups and for their variety and challenge level.

For each observed mathematics lesson, teachers agreed to organize their instruction according to the following model: (a) approximately 30 minutes of teacher presentation to the class as a whole (for example, review of previously taught content, introduction and/or teaching of new content); (b) approximately 30 minutes of students working in small groups (from three to five students) on cooperative problem-solving tasks; (c) approximately 5 to 15 minutes summarizing the cooperative work in which students have been engaged. Teachers were free to change the sequencing of (a) and (b), but all three parts of the model were to be included in each observed lesson.

An observational instrument and accompanying codes distinguished (a) the type of attending behavior of students, (b) the nature of attending behavior, and (c) the nature and extent of off-task behavior. Three types of on-task behavior, from on-task engrossed to minimally on-task, were discriminated. Six categories of attending were identified to indicate whether on-task behavior was cooperative or individual, and to indicate the type of student engagement in cooperative or individual work. The nature of the activity a student was engaged in was coded, and six categories of off-task behavior were identified.

Aspects of student interactive behavior coded included the nature of student initiating behavior and the mode of student interaction. The categories of student interacting behavior made it possible to collect data on students' successful and unsuccessful attempts to initiate interaction with the teacher, student(s), or group, and students' willingness to respond to the initiations of others. The content of student-student and student-teacher interactions could also be coded in one or more of 10 categories (for example, gives opinion, evaluation, or analysis; gives suggestions or directions; asks for suggestions or directions). Two categories of student socio-emotional behavior (that is, positive and negative) were included on the observational form. These categories were from Bales's (1951) Interaction Process Analysis System.

In addition to data on student attending and interactive behavior, other information, including high-inference measures, was recorded during observations: group composition and group processes; role acquisition or assignment in the group; length of time that the group had been working together; group and individual cooperation; group interaction; on-task involvement; and group and individual accountability. Data on students' perceived competence in mathematics and reading were obtained by having students rank themselves in comparison with their peers. Locus of control data was obtained using the Intellectual Achievement Responsibility Scale (Crandall, Katkovsky, & Crandall, 1965).

Teachers and target students were interviewed at the beginning and end of the study. Target students were also interviewed after each mathematics observation. The initial teacher interview focused on several dimensions, including (a) perception of students' attitudes toward mathematics and reading, (b) expectations for appropriate student behavior, (c) perceptions of students' behavior in the various learning contexts, and (d) the nature and degree of student-student interaction and student-teacher interaction that were permitted in different learning contexts.

The initial student interview measured several dimensions, including: (a) the nature of mathematics, the purpose of mathematics learning, and general attitudes toward mathematics, including perception of self as a mathematics learner; (b) perceptions of teachers' expectations for appropriate student behavior during whole-class mathematics presentations, seatwork, reading group work, and cooperative small-group work in mathematics; (c) students' perceptions of their own attending and interactive behavior and that of their peers; and (d) attitudes towards cooperative small-group work in mathematics. The final teacher and student interview investigated teachers' and students' perceptions of a range of issues relating to the group problem- solving tasks used in the study, and the responses of students and groups to these tasks.

After each observation, target students were interviewed about a range of issues relating to their perceptions of group tasks and group processes. Questions about particular occurrences within the observed group were included in these interviews when relevant (for example, conflicts among group members, the splitting of the group, passivity or dominant behavior by one or more students).

The study was carried out over 9 weeks during the spring. Teacher and student interviews were conducted during the first and final weeks. During the middle 7 weeks of the study, each classroom was visited for one mathematics period a week, and teachers organized students to work on one of the cooperative small-group tasks during each visit. Four of the eight target students were observed in a whole-class mathematics context and in the cooperative small-group context during each observation. After each mathematics observation, the target students who had been the focus of observation were interviewed. Details of group composition and group work style were also recorded.

Each target student was observed for a minimum of 20 minutes in the cooperative small-group context, and 20 minutes in the whole-class mathematics context during the study. Target students were also observed on three occasions during reading-group work. Reading observations varied from 6 to 12 minutes each. Observational data were collected using a script-taping method in which details of student attending and interactive behavior were recorded in 20-second blocks.

SELECTED FINDINGS. The discussion provided here briefly highlights some of the key process and perceptual differences that Mulryan (1989) obtained. We focus on differences between what high- and low-achieving students did during small-group work and some of their attitudes about small-group processes. However, the study also provided important information about gender differences and between-classroom variation (for example, teacher effects on small-group processes, illustrating that small-group interaction may provide different experiences for various types of students).

Attending behavior: Students manifested more attending behavior in the cooperative small-group context than in the whole-class mathematics and reading-group contexts. These results are consistent with previous research (for example, Ziegler, 1981): Students are more active during small-group cooperative work than in other instructional settings.

Mulryan, however, found that not all students benefitted in the same way from increased involvement during small-group work. She identified three types of student attending behavior: Type 1, on-task engrossed; Type 2, on-task but not engrossed; and Type 3, minimally on-task. Her results indicated that high achievers engaged in significantly more Type 1 behavior than did low achievers in cooperative small groups. Further, a comparison of the combined amount of Type 1 and Type 2 attending behavior for high and low achievers showed that high achievers manifested significantly more quality attending behavior in cooperative small groups than did low achievers. In addition, there was no significant difference between the quality of attending behavior of low achievers in whole-class versus small-group mathematics lessons, whereas for high achievers a significant difference was found between these two contexts.

That involvement in small groups led to differential opportunities for students as a function of achievement can be seen in other ways as well. For example, the data indicate that high achievers spent 5% of their time in cooperative small groups engaged in off-task behavior; low achievers were off-task 13% of the time.

High achievers also engaged in more interaction than low achievers during cooperative small-group work. However, differences in interactive behavior reached significance only in the case of suggestion and direction-giving, with high achievers providing significantly more suggestions and directions than did low achievers. In contrast, low-achieving students asked more questions than high achievers in cooperative small groups. This finding is consistent with Webb's (1982) finding that question-asking in cooperative small groups was associated with low achievement.

Students' attitudes: Student interview responses showed that high- and low-achieving students perceived mathematics in somewhat different ways. Most students—especially low achievers—reported in interviews that mathematics was work with numbers and number operations. Students' conceptions of mathematics may be related to their differential behavior during small-group work—especially given the types of tasks used during small-group work. The interview data also indicated a narrower and less differentiated view of cooperative small-group work among low achievers than among high achievers.

Because high-achieving students see more varied reasons for engaging in small-group work, they may be more adaptable than low achievers.

In expressing the purpose of cooperative small-group work in mathematics, low achievers tended to focus more on the non-task-related aspects of work. For example, most low achievers considered "to learn more and better," and/or "to learn to work with others" as the main purpose(s) of cooperative small-group instruction. More high achievers focused on task demands and the cognitive processing required by cooperative tasks (for example, "to learn different approaches to tasks," "to combine skills for harder work").

A notable finding was that, whereas at the beginning of the study most students viewed passive behavior in small groups as beyond the control of students (shy, don't understand), at the end of the study all students were more likely to view passivity as intentional. There were minor differences between high and low achievers in this area (for example, high achievers were more likely to cite wanting to work alone as a reason for passivity than were low achievers).

Students' views of students who were active and dominating in groups showed strong differences by achievement. Interview responses indicated that high and low achievers viewed verbal domination differently. High achievers tended to explain active behavior with phrases such as because they "like the people in the group," "understand the task well," and "are depended upon." In contrast, low achievers more frequently reported that students dominate because they "think they know more than others," "want to control the group," or "don't want to give others a chance." Such differential student perceptions may lead to differential behavior during small-group work, since students may interpret the same behavior in various ways (as a helpful suggestion versus an attempt to take over).

Students' interview responses usually indicated that they preferred the cooperative small-group section of the lesson to the whole-class teacher presentation. Students' preferences for cooperative work showed no gender or achievement effects. However, the preference for small-group work occurred partly because the format provided task and social stimulation without requiring students to deal with other students (learn how to cooperate). Thirty-eight percent of students' time was spent working individually during small-group sessions, and this appeared to be due partly to students' preferences. Although students realized that teachers expected them to work together, subdivision of tasks was common, with consequent individualized work, and many students regarded subdivisibility as a positive feature of group tasks. Moreover, interaction in small groups often took place between two individuals—rather than one member addressing the three other students.

In general, the work conditions in small groups provided a context in which students could remain relatively passive if they wished. High tolerance among students appeared to exist, provided students did not frustrate the goals of other group members through consistent disruptive behavior. Target students, for example, reported that students who withdrew from group work (that is, remained passive) were not usually encouraged to become involved. The knowledge that group membership was temporary is likely to have encouraged students

not to deal with problems presented by other students (passivity, disruption, etc.). In the student interviews, discussions of uncooperative behavior initiated by other students were rarely accompanied by descriptions of action taken to solve problems.

DISCUSSION OF FINDINGS. Although Mulryan found that students' involvement in curriculum tasks increased during cooperative small-group work, her study also uncovered important differences between low- and high-achieving students in attention and participation as well as in perceptions of and reactions to peers' behavior. These findings support Davidson's (1985) argument that the results of experimental studies of cooperative small-group instruction do not always hold when consideration is given to aptitude-treatment interactions. That is, the opportunity provided during small-group instruction varies for different students. Mulryan's study also supports the argument by Good and Biddle (1988) that student gains from involvement in cooperative small groups are not automatic and that more attention needs to be paid to the conditions under which gains are most likely to occur. In combination, Mulryan's findings on student attention, participation, and perceptions of appropriate and expected student behavior in cooperative small groups indicate that, although most students had some problems adapting to the group context, low achievers had the most difficulty.

Mulryan argues that the data suggest that many low achievers may not have perceived the changes in work style and engagement that cooperative small-group learning requires. For many students, especially low achievers, the legitimization of peer interaction and talking about work may have been the only major perceived changes. This may have led to the failure by some students to engage in the more active, on-task cognitive processing and sharing usually involved in cooperative small-group work. It is also possible that, just as low achievers generally experience more varied expectations and treatment from teachers (Good et al., 1987), they may be subject to more varied expectations and treatment from peers in the cooperative small-group context. Such experiences would likely result in considerable uncertainty for low achievers in this context and would reduce their involvement in group tasks, especially if the situation were ambiguous. In the interviews, some high achievers expressed dislike and avoidance of giving help to low achievers, while others saw their role as a help-giving one. This indicates that low achievers probably received varied treatment from high-achieving peers.

The motivation issue is further complicated by the factor of speed of task completion. In some cooperative small groups, low achievers may be discouraged from active participation in tasks because the more academically competent group members are concerned with getting tasks completed as quickly as possible. Speed of task completion was mentioned as a criterion of group success by both high and low achievers in interviews.

During interviews, both students and teachers mentioned help-giving and/or help-seeking as a purpose of and expected behavior in small cooperative groups. Observations showed that students often did ask one another questions and check information. Although question-asking in cooperative small groups was relatively infrequent, low achievers asked the most questions, and high achievers showed proportionately more information-giving behavior. The roles of "helper" and "helped" that emerged in many small groups may not have been useful, especially when the same students played these roles consistently. In such circumstances a "caste" system that discourages the "helped" from active initiation or information-giving may develop during small-group work. In the interviews following the observed lessons, some students indicated that they perceived low achievers as a burden to other members of cooperative groups.

It is important to note that not all low achievers in this study asked questions or asked for help from peers in cooperative small groups. Research on student help-seeking may help explain this finding. Some students might have been reluctant to seek help because they wanted to avoid appearing stupid, receiving negative sanctions from group members, or receiving inadequate help (Newman & Goldin, 1990; van der Meij, 1988). Some students in Mulryan's study (usually high achievers) reacted negatively to low achievers' questions, especially if these questions were frequent. In the interviews, some high achievers expressed frustration about being requested to explain and to give help. Many low achievers appeared to depend on the good will of other, more able students in order to be involved productively in cooperative small-group work or to obtain needed information.

Yet other factors may explain participants' behavior during small-group work. For example, students' feelings of self-efficacy, especially in relation to mathematics, may also affect behavior. Low achievers' failure experiences in mathematics may result in negative self-efficacy feelings that transfer to cooperative small-group work. These feelings (that is, the conviction that success on a task is unlikely or impossible) are likely to impede students' participation, especially on relatively demanding and unfamiliar tasks (Corno & Mandinach, 1983; Rohrkemper & Corno, 1988), as was the case in most of the cooperative-learning tasks explored in Mulryan's study.

The behavior of some low-achieving students in the cooperative small groups may be explained at least in part by the literature on social loafing (for example, Latané, Williams, & Harkins, 1979). Some students may exert less effort in groups because they do not perceive themselves as personally accountable for the group product, and they consider their individual contributions not easily identifiable by the teacher. Conditions that support social loafing may have existed in the Mulryan study, as evidenced from the finding that most students believed the teacher did not know how much work individual students did in cooperative small groups.

In summary, it is clear that the passive behavior of some students during small-group work is a significant concern. The problems that certain students experience in whole-class settings appear again when these students are asked to play somewhat different roles in small-group settings. Teachers who hope to increase student involvement will have to develop strategies that allow all students to play active roles during small-group work.

Mulryan identified several pervasive problems of small-group work that will require additional study and reflection. The problems are not simple and probably do not have simple solutions. That is, the relatively passive behavior of low achievers in cooperative small groups found in this study is likely due to *many* factors. The preceding discussion indicates some possible reasons for the lower involvement of these students in cooperative small-group work. A combination of these and other factors likely contributes to passive student behavior.

FUTURE RESEARCH

We have described some of the available research literature and have presented our own research experience in this area. We have illustrated that research on cooperative small-group learning yields intriguing and useful findings that shed light on a range of complex issues. Although this research has increased our understanding of many aspects of cooperative small-group learning, it is merely a beginning, and much relating to this area remains to be investigated. Current research may be more usefully perceived as raising new questions and hypotheses for future research rather than providing findings that can be applied to learning contexts. Likewise, research provides only suggestions for practice. What is needed at this time—to examine and interpret extant and new research and to serve as a basis for planning subsequent research—is a structure or overall view that identifies important variables that need investigation.

The Conduct of Research

Here we emphasize the need for many studies that explore different variables and different combinations of variables in order to yield theories about what group processes facilitate various learner outcomes. We suspect that different *process* and *group composition* variables are associated with both successful basic skills acquisition and higher-order learning. Considering the limited number of studies that have involved specific outcome variables (especially higher-order variables) and interesting mathematical topics (for example, probability, measurement), attempts to characterize the relation between grouping and higher-order learning seem premature. Slavin (1989–1990), who has contributed much to the field of cooperative learning, agrees that there is considerable debate on whether and how cooperative methods might facilitate higher-order conceptual learning. Because of (1) the need for more higher-order mathematical thinking in the curriculum, and (2) questions about the effects of cooperative learning methods on higher-order learning, most new research should focus on this topic.

We agree with Piaget (1970), that theory is constructed out of meaningful activity—interaction and involvement. Although the ultimate goal is to produce theoretical knowledge, researchers must first use multiple frameworks to examine the role of cooperative group work. As basic knowledge accumulates (for example, how students' conceptions of mathematics and cooperation vary with age, structure, and topic), results of studies can be used to develop propositional knowledge and can be applied in classrooms by creative teachers. However, knowledge will always have to be applied in terms of teachers' understanding of their students and their particular context.

In particular, we call for researchers from various disciplines to study cooperative groups. For example, an economist might explore how groups make differential use of resources (time, materials, individual members), a sociologist might study resource allocation across groups, and a clinical psychologist might ask how individual members' decisions and perceptions affect how they use resources available either in the group or the classroom. Research by subject-matter experts is also vital if we are to learn how students' understanding of mathematics is affected by learning in a group setting. For what mathematical topics are group and individual work most appropriate? Researchers from various disciplines who perceive issues differently should offer a more rigorous examination of grouping.

Furthermore, classroom teachers should be involved in the study of small cooperative groups. If researchers are to build theoretical models that are relevant to practice, they must more systematically collaborate with teachers about factors that inhibit students' ability to think in social settings and about what can be done to overcome these problems. Recent research in sociology indicates that important scientific knowledge and theory can be advanced by using participants (from the social setting) in more appropriate, collegial ways (see Whyte, 1989).

Variables that Merit Research

In this section we identify and discuss many variables that in our view merit research in the context of cooperative small-group learning. This list is not exhaustive but provides an adequate orientation to subsequent research. The framework includes these major categories of variables: *task variables* (including content variables), *teacher and instructional variables,* and *individual* (including developmental differences) and *group variables.* These variables have been identified in the light of our previous research in this area. Although many of the variables interact with one another (for example, ability of students and task difficulty), a comprehensive discussion of this interaction is beyond the scope of this chapter.

Task Variables

NATURE OF LEARNING TASKS. The nature of learning tasks exerts an important influence on the way cooperative small groups work together and on the learning outcomes expected and achieved. Learning tasks can differ in several ways, and research is needed to investigate the way in which students in cooperative groups respond to and work on various types of tasks, as well as to determine task characteristics that lead to desired outcomes.

Learning tasks, for example, might involve skill development, concept development, and/or problem solving. They might be organized to allow students to engage in initial exploration of a mathematics topic or area; to review, consolidate, or extend previous learning and/or apply it in a new context; or tasks may involve a combination of some or all of these. Tasks might involve half a mathematics period or less, an en-

tire mathematics period, or several mathematics periods. Students may work with concrete materials or manipulatives on some tasks, whereas other tasks may involve pictorial representation or paper-and-pencil computation. Tasks may be simple or coordinated; that is, students may work on some group tasks individually, and some tasks may require subdivision and/or coordination of activity by group members. Cooperative learning tasks can also vary in difficulty. Research needs to consider the responses of students in a group when they are assigned one main task (that is, a task that is sufficiently complex to allow for multiple inputs and viewpoints) as compared with several related tasks or subparts of a task. Are tasks that can be easily subdivided more —or less— likely to promote cooperative activity in groups than tasks that are less easily subdivided? Is subdivision a possible source of conflict and/or inequity among students in a group?

SUBJECT MATTER. The issue of what, in general, constitutes appropriate mathematical knowledge (for teachers and students) is of course complex, and considerable interest abounds in increasing emphasis on appropriate mathematics (see, for example, Ball, 1991). The type of mathematical content to be assigned for small-group work also merits analysis and research. Considering problems with the current mathematics curriculum and students' learning, small-group learning would be an excellent means of emphasizing problem solving, measurement, estimation, statistics, and extended opportunities for mathematical reasoning. Recent recommendations from the NCTM concerning such content also support this view. We suspect that small groups working on tasks that emphasize problems, measurement, and inferential thinking present a particularly viable way for students to learn mathematical concepts and reasoning in ways that are congruent with real-world tasks (Lampert, 1986; Schoenfeld, 1983). Applying mathematical reasoning to social problems and issues is another effective use for small groups. How to promote sustained mathematical dialogue in such settings is an interesting conceptual and empirical issue.

DIALOGUE AND INQUIRY. One putative advantage of small groups is that they allow students to use the resources of other students and to apply mathematical reasoning to complex issues and problems that benefit from cooperative interaction. Unfortunately, some teachers appear to assign relatively easy and sometimes trivial assignments in order to reduce management issues and controversy. Thus, how to use small groups to promote sustained mathematical dialogue and inquiry appears to be an interesting conceptual and empirical issue. Newman (1990) notes that certain beliefs work against thoughtful inquiry in schools.

Most knowledge is certain rather than problematic; knowledge is created primarily by outside authority, not within one's self; knowledge is to be comprehended and expressed in small, fragmented chunks; knowledge is to be learned as quickly as possible, rather than pondered. Knowledge may seem counterintuitive or mysterious with respect to one's experience, but should be believed anyway; arguments and conflict about the nature of knowledge are personally risky because winners are favored over losers. (p. 12)

Whether small-group work can increase students' willingness to explore, ponder, and resist the urge to go faster awaits experimental verification. According to Newman, even in very supportive environments individuals have great difficulty subjecting themselves to continuous scrutiny, and at times they are unwilling to resolve ambiguity and contradiction. Yet, as Newman argues, higher-order reasoning requires recognition and resolution of conflicting views and tolerance for uncertainty and ambiguity. Research on small-group processes that require students to engage in sustained higher-order thinking and dialogue is needed.

TEACHERS' ALLOCATION OF TIME AND MATERIALS. Other factors related to cooperative learning tasks include the amount of time allocated for students to work on a task and the time that students actually spend working on the task. For example, do students respond differently to cooperative learning tasks requiring some or all of one mathematics period than to tasks requiring several days or weeks? Also, what work times maximize small-group cooperation and learning for various types of cooperative tasks?

Another issue that needs to be investigated is materials allocation during small-group activities. Is the sharing of scarce resources in the cooperative small-group setting a potential source of conflict among students (Pepitone, 1985)? What considerations should guide resource allocation in order for productive cooperative work to occur?

Teacher and Instructional Variables

TEACHER PERCEPTIONS AND EXPECTATIONS. It is generally recognized that teachers' expectations, beliefs, and knowledge play an important part in determining their classroom behavior and the nature of the learning opportunities they provide (Good & Brophy, 1987; Peterson, Fennema, & Carpenter, 1991). Studies are needed of the relation between teachers' expectations and beliefs about cooperative small-group learning and the nature, quality, and amount of cooperative learning in their classrooms.

Researchers also need to examine teachers' perceptions of the nature of mathematics and their perceptions of the skills and competencies that students need to develop in this area. For example, do teachers who perceive mathematics as mainly computation and the mechanical manipulation of numbers regard cooperative mathematics work differently from teachers who view mathematics as more open-ended? How do teachers' perceptions of the nature of mathematics and the purpose and nature of cooperative small-group work compare to their students' perceptions? Do teachers' perceptions and expectations influence students' perceptions and behavior differentially according to type of cooperative-learning activity and grade level? For example, are some activities likely to be more self-motivating for small groups than others and therefore less influenced by teacher variables? Mulryan (1989), for instance, found that cooperative small-group tasks that involved the use of manipulatives tended to be more motivating for fifth- and sixth-grade students than pencil-and-paper tasks. Also, can work groups help students become less dependent on the teacher and more self-regulating? If so, do peer expectations become more important than teacher expectations?

CLASSROOM CHARACTERISTICS. Studies might also investigate the relation between general classroom atmosphere (for example, teachers' respect for individual differences) and student behavior in cooperative small groups. For example, it is likely that teacher-student relationships and the way in which students individually and as groups are treated by the teacher in both learning and non-learning contexts influence the way in which students relate to one another (Bandura, 1982; Good & Brophy, 1987). Do students in cooperative small groups respond to and work with low achievers and students with poor social skills in a more positive way in classrooms where teachers show more positive attitudes and greater acceptance of these students and of individual differences generally? Are performance expectations for low achievers likely to increase in this context? Research also needs to consider the effects of peer modeling (Schunk, Hanson, & Cox, 1987) in the context of cooperative small-group learning. Are high achievers who witness help-giving behavior on the part of other high achievers in groups more likely to manifest help-giving behavior themselves in this context? Are low achievers more likely to ask for help when they witness other students asking for help and obtaining it? Research needs to examine the extent to which these modeling effects operate and identify other characteristics of classrooms in which cooperative small-group learning in mathematics is being successfully implemented.

TEACHER ROLE. Teachers' preparation of students for cooperative small-group activities and the follow-up instruction they provide are also likely to affect students' engagement in cooperative work, as well as the amount and quality of student learning. More information is needed about the type of preparation that will stimulate students' interest and enthusiasm, provide adequate information and instruction, and yet not reduce the challenge of the task. What sort of immediate preparation is most appropriate for different task types and students in various grades? What sort of advance preparation should students receive in order to learn the skills necessary to work effectively in small cooperative groups?

Researchers need to examine what teacher follow-up and feedback are most appropriate for particular types of tasks and students. Should follow-up and feedback to cooperative groups be done in a whole-class context or with groups individually? Which models encourage student autonomy? Which models best allow teachers to obtain information about students' cognitions and subsequent instructional needs?

Another aspect of the teacher's role during cooperative small-group activity concerns what the teacher should be doing while students are engaged in cooperative-learning tasks. In what context(s) should teachers interact with students while groups are working? What teacher interventions are likely to be most appropriate in different contexts? How can teachers assist students without providing solutions or strategies that students might benefit from discovering for themselves? How can students be taught to seek help when misunderstanding, confusion, or frustration halt productive group activity?

CLASSROOM MANAGEMENT. More information is needed on classroom management in the context of cooperative small-group learning. What management variables are relevant to cooperative small-group work in math? Cooperative small-group work frequently differs from other learning contexts in that student-student interaction is usually encouraged. How can teachers and students be helped to cope with these differences? How can teachers best organize and manage the classroom during cooperative group work so that discipline problems do not arise, interaction between students primarily involves tasks, and pupils still have sufficient freedom to contribute to and participate in the group discussions that sometimes may involve disagreements between group members?

PREPARATION OF TEACHERS. What forms of preparation are most appropriate for teachers? How can teachers be helped to cope with the ambiguity for themselves and their students that can result during small-group cooperative work (for example, when students in a group adopt different approaches to a learning task and/or come up with a range of equally plausible alternative solutions to a problem)? How much ambiguity can students and teachers be expected to tolerate?

What kind of support do teachers need as they plan cooperative group work? Lack of support by administrators, especially principals, is likely to result in teacher avoidance of cooperative small-group work or in teachers using it at a cost to themselves. For example, if teaching success and teacher competence are determined by administrators and others, including parents, on the basis of student achievement on a narrow range of curriculum objectives that include few skills and competencies that are achieved during cooperative group work, teachers' commitment to cooperative group work is likely to be low. At worst, cooperative activity may be excluded altogether. At best, teachers may regard cooperative small-group tasks as ancillary and supplemental to "more important work" or as optional tasks to be used when students have completed other work. Influences on teachers' attitudes toward cooperative small-group work and the amount and nature of group work in their classes should be examined.

ASSESSMENT AND ACCOUNTABILITY. Assessment and evaluation of cooperative small-group work have received considerable attention from researchers. However, new constructs, variables, and dependent measures are needed in this area, especially regarding attempts to understand more about small-group processes and student motivation. What reward systems, if any, most affect motivation during small-group work, and what are the effects of the presence or absence of competition within and between groups? Some work has been done in this area (for example, Ames, 1984), but much remains to be done. Are grades or other credit systems appropriate motivators, and in what contexts? What is the relation between achievement and motivation? Should students' work in cooperative groups be evaluated on a group or individual basis? How can teachers monitor and evaluate the work of individuals and groups in ways that are nondisruptive to the group process? Are some students likely to invest more effort in cooperative-learning tasks if grades or other rewards are offered on an individual or group basis? Is competition more likely to promote or impede cooperative activity in some or all contexts, and with certain students or among students generally? Do some students work better and cooperate more when they engage in an activity

for its intrinsic value and enjoyment rather than for extrinsic motivators?

In what ways do developmental variables interact with reward structures in cooperative small-group work? For example, do younger children generally regard rewards as informative whereas older children perceive rewards as attempts at control by the teacher (Deci, 1980)?

Research on cooperative learning concerned with *skill development* has consistently illustrated that two factors are associated with effective performance: group goals and individual accountability (Slavin, 1989–1990). However, there is evidence (Mulryan, 1989) that some students see such accountability devices as superficial. Further, there are good theoretical reasons to argue that cooperative learning itself can be enjoyable and satisfying. Is it possible that close individual accountability is needed only when the task involves drill rather than meaningful learning?

How can teachers promote in-group accountability (i.e., students holding each other accountable for group on-task participation), and is this form of accountability likely to increase students' engagement in cooperative tasks? Are various motivational and accountability measures appropriate for different types of activities and for students who vary in ability, age, and prior experiences?

Individual Student and Group Variables

AGE/GRADE. Research should focus on the appropriateness of cooperative group work in mathematics for students at different ages or grades. Can students in the primary grades be expected to work cooperatively on mathematics tasks to the same degree and in the same way as students in the middle and intermediate grades? How does the meaning of cooperation differ among students in various grades, and what are the implications of these differences for the planning and implementation of cooperative group methods in schools?

Teachers must construct tasks that are consistent with students' development, and research is needed to guide this process. For example, first-grade teachers would likely have difficulty explaining to students the differences between the teacher's role during small-group versus whole-class instruction. That is, because students have learned that the teacher dominates during whole-class instruction, they might not understand that they are to learn from peers as well as from the teacher during group work. In contrast, the issue of teacher authority would pose relatively few problems in fifth-grade classrooms. Because fifth graders are preoccupied by questions such as who they are and how they relate to their peers, they enjoy exchanging ideas with peers and would be less concerned about attempting to please the teacher.

Another potential problem with small-group work—how students react to group members who perform poorly—would also take different forms at various grades. First graders are concrete and immediate in their thinking. They would not pay much attention to repeated failures or difficulties of other students. Rather, they probably believe that students who have difficulty will understand content or tasks next week. In contrast, this issue would probably be difficult for fifth graders to deal with because of their sense of fair play and their belief that effort should lead to success. Fifth graders who are having trouble might engage in face-saving strategies (make excuses such as a headache or lost book, behave as class clown, project disinterest in the project, etc.). Students who are trying to work with them make it easy for these students to engage in face-saving behavior by accepting their explanations and not challenging them to think or to engage in tasks. In other cases, better students might contribute to the process by slowing down the speed at which they work or by withholding solutions because they do not want to embarrass slower students. In the middle grades, because of students' emerging conceptions of ability and their strong sense of fair play, dealing constructively with a variety of learning capacities poses difficulties that will require teachers to select tasks for group work carefully and to help students develop skills for dealing with individual differences.

Age also affects students' attention spans and energy for academic work, and teachers should vary tasks accordingly. In first-grade classrooms students have high energy but short attention spans. In comparison, fifth graders have moderate energy and moderate attention spans for academic work. However, fifth graders have high energy and interest in peer work.

ACHIEVEMENT. Other important student characteristics include achievement (achievement perceived as the interaction between ability/knowledge, motivation, and task). What kinds of cooperative-learning tasks and approaches are most appropriate for individuals and groups of different mathematics achievement levels? How is mathematics competency—generally as well as in specific areas—related to students' responses to cooperative learning? Can students be more easily assigned to cooperative groups in grade 1 than in grade 6 because their mathematical abilities are less varied?

Webb (1982) found that giving and receiving explanations in cooperative small groups related positively to student achievement. How can students be encouraged to give and seek information from group members? Do some students need more support and help than others? For example, do students with poor social skills and/or students with lower ability require different preparation from more confident or high-achieving students? Do some high-ability students have difficulty working cooperatively with less competent students and need special guidance by the teacher to promote sharing and cooperation?

GENDER AND SES. Other student characteristics (for example, gender and socioeconomic status) might also be considered in this context (Erickson, 1986). Are girls and boys likely to benefit to the same extent from cooperative activity during mathematics learning? Fennema and Peterson (1987), for example, found that girls appear to gain more than boys from participation in cooperative small-group work, at least in terms of achievement. However, this potential is not realized in all small groups (Mulryan, 1989).

Is socioeconomic status an important variable in the context of cooperative small-group learning? If so, what is its influence? Related to socioeconomic status is the notion of culture. The interface between home and school and between school and the broader culture affects how mathematics is conceptualized and learned (see Stigler & Baranes, 1988). Research is needed concerning whether the experiences of students in different cultures affect their behavior in cooperative groups, and if so, the nature of these effects and their implications for implementing cooperative small-group learning approaches should be studied (Erickson, 1986).

PREVIOUS EXPERIENCES IN SCHOOL. Students' previous experiences in school, including their internalization of appropriate and required expectations for classroom behavior, are likely to influence their behavior in cooperative group settings, especially when cooperative group work is being introduced. Research is needed to investigate the relation between students' perceptions of appropriate classroom behavior and their responses to cooperative group settings. How can students who have had little experience relating to peers and interacting productively in learning contexts learn to cooperate in small groups?

PERSONALITY VARIABLES. Certain personality types may be associated with different kinds of responding in cooperative groups, and researchers should identify personality factors that relate to domination, passivity, or effective cooperative behavior in groups. What effects do dominant and/or aggressive students have on small-group processes? Do more (or less) independent (and dependent) students behave differently in cooperative groups? How do students with different personality characteristics respond to the freedom and self-regulation that characterize cooperative small-group work?

SOCIALIZATION OF INNER SPEECH. One promising area of research on small-group instruction concerns the dynamics of student enactment and internalization of group members' inner speech, both speech that is directly taught or intentionally modeled and speech that is spontaneous and spoken aloud in situations of difficulty. It may well be that teachers and peers have quite different strengths to bring to students whose inner speech is deficient. One hypothesis concerns the relative efficacy of self-involved compared with task-involved inner speech modeling. For instance, teachers may be a relatively more potent source of influence on task-involved inner speech strategies. There is some indication that students ascertain the extent of their misunderstanding in mathematics before seeking help: The more lost, the greater is the reliance on the teacher rather than a group member or other classmate (Rohrkemper & Bershon, 1984). Similarly, students who work in cooperative teams for rewards, yet work on individualized tasks within the group, claim to assess their own knowledge before venturing to help another; this is especially the case with relatively low-achieving students (Rohrkemper & Bershon, 1984). In comparison, parents and peers may model more convincing and facilitative self-involved inner speech strategies. As a result, self-involved inner speech is likely much more varied and idiosyncratic than is task-involved inner speech (McCaslin Rohrkemper, 1989).

Within a group, "who models whom" is yet another promising area of research on the internalization of inner speech in small-group learning. Small groups likely confound and transform personality, learning, and motivational constructs that researchers have carefully delineated and kept isolated.

For instance, what are the implications for internalization by group members when the learner with the most powerful task-involved inner speech also displays guilt-laden self-involved inner speech? What are the dynamics of modeling processes when efficacious task-involved inner speech is not self-reinforced? Do peers selectively attend and model? If so, which aspect is internalized and by whom? How would these dynamics differ as a function of grade level?

Do some group configurations encourage a too-ready reliance on others' inner speech as scaffolding tools so that *self-*direction is replaced by *other* instruction? There is some evidence that learners identified as helpless by their teachers, even though they possess the needed ability and skills, consistently display negative self-involved inner speech and engage in directive task-involved inner speech only after individual failure and with social supports (D'Amico, 1986). Would these kinds of learners be rendered more active or passive in their initiation of self-directive inner speech in small-group contexts? If these learners were more apt to engage in enhancing self-involved inner speech because of the known task-involved inner speech resources available to them, would this predict an increasingly positive sense of self—and subsequently influence increased task-involved initiatives or, in its absence, increase the price and fragility of unrealistic self-appraisal? Alternatively, would exposure to others' self-directive inner speech exacerbate already negative social-comparison processes so that each group success reflected negatively on helpless learners' personal resources and efficacy? That is, can others' effective task-involved inner speech increase a sense of personal unworthiness? How would these social-comparison processes inform the design of groups and the structure of the tasks in which they engage?

We know that social dominance is not perfectly correlated with relative ability (see Mulryan, 1989). The variation among personality factors, cognitive facility, and inner-speech configurations provides an interesting set of research puzzles. For instance, what happens to the intrapersonal dynamics of the less-confident learner whose hold on effective task-involved inner speech is tenuous and too readily dominated by the more confident but less able group participant? Alternatively, if a learner is confident that she knows fractions but is less sure of her ability to cooperate, what would be the appropriate task to challenge and enhance abilities: one ensuring that the group obtains correct answers, or one engaging participant behaviors that allow others to assume instructional leadership roles, perhaps at the cost of incorrect answers? Would the latter decision result in detrimental self-involved inner speech as the less-than-perfect outcome—either because group members have not learned as much as needed or because one's own outcome does not accurately reflect his understanding of fractions?

STUDENT PERCEPTIONS. Students' perceptions—for example, of their own ability and the ability of peers—also influence their behavior in small groups. What expectations do students have for themselves and others? Do some lower-achieving students expect to be less involved and less able to make a useful contribution to group activity than other students whom they perceive as more able? On the other hand, do some middle and high achievers expect low achievers to make few useful con-

tributions to task completion? How do students perceive the cohesiveness of the group in which they work and the degree to which the group values them as co-workers? Mulryan (1989) found that low achievers, especially those attending remedial instruction outside regular classrooms, are frequently excluded from ongoing group activity by other group members. Are low achievers aware of their rejection, and if so how does this affect their perception of cooperative group learning, their peers, and their own self-worth? Do some low-achieving students welcome this exclusion?

The relation between students' perceptions of mathematics and mathematics learning and their responses to cooperative settings is also important. How will students who perceive mathematics as primarily involving the manipulation of numbers and the finding of predetermined solutions to exercises and textbook problems respond to cooperative small-group mathematics tasks that may involve exploration, experimentation, discovery, multiple steps, and possibly multiple solutions?

GROUP COMPOSITION. Clearly, the dynamics of a particular group, and consequently the degree to which the group works together productively and cooperatively, will be determined largely by the mix of students within the group. Also, group composition is an important variable in relation to task selection. Research needs to investigate what group composition(s) work best for different types of tasks and at various grades. Although we know from Webb's work (which did not allow for division of labor among group members) that middle achievers do poorly when placed in groups with high and low achievers, it would be interesting to see if these findings hold when tasks are more complex. Which kinds of cooperative group work does group composition really affect? How can group composition be best determined? For example, in some contexts is random selection of group members more appropriate than grouping students who are likely to work well together?

Will students with poor social skills benefit more in groups with some empathic and supportive students than in groups of more aggressive, task-oriented, and competitive students? Under what conditions do some students dominate small groups, either academically or socially? Do students manifest similar behavior regardless of the composition of their groups, or does the presence of certain students in the group or certain task variables cause them to modify their behavior?

Research also needs to investigate the extent to which various combinations of students work well together toward the accomplishment of different group learning goals and whether specific criteria for group composition are necessary or appropriate for various goals. For example, are there some contexts in which small-group composition should be determined by student achievement (that is, having students of similar ability working together)? For what learning goals is mixed-ability grouping most appropriate?

Another variable that needs to be addressed is whether groups should be homogeneous or heterogeneous. What are the benefits of having students with different mathematics abilities work together in groups versus students with similar abilities? In heterogeneous groups, which students are most likely to gain? How do high-, low-, and middle-ability students respond to each other and to learning activities in heterogeneous groups? How do the dynamics of heterogeneous groups compare with those of homogeneous groups? What students will gain most in homogeneous groups? Are low-achieving students likely to be less challenged in homogeneous cooperative groups and consequently likely to be more passive and learn less in this setting than they would in heterogeneous groups? Will high achievers gain more and move at a faster pace when working with students of similar ability? In what ways other than mathematics ability might homogeneity or heterogeneity be determined for cooperative group work, and in what contexts? In sum, what are the trade-offs of different bases for group member selection? The research suggested here would extend work already carried out by Webb (1989).

STABILITY OF GROUP MEMBERSHIP. In our initial study of work groups (Good et al., 1989–1990), we were impressed by the variability among classrooms of group membership stability. Some teachers changed membership every time they used the work-group model, other teachers allowed groups to remain intact for several days, and yet other teachers allowed group membership to remain stable for a month or so. If the goal is simply to use cooperative methods as a way for students to learn academic content or improve social behavior (for example, listening, turn-taking), stability of group membership may not be important. However, if teachers use work groups so that students will learn to cooperate (Rohrkemper, 1986) or develop more sophisticated dispositions about mathematics, then it seems advantageous to have more time for students to learn how to use one another as resources. For example, if during a lesson one student dominates a group (and if group membership changes frequently), it is unlikely that other group members would attempt to alter the behavior of the dominating student unless it were irritating or disrupted the group's work. Similarly, if a talented student is passive and does not contribute to group work in a flexibly grouped classroom, the group members have little need to develop the social skills that would allow them to draw the student into the group.

For various reasons, teachers might change group membership frequently. For example, if two students are truly difficult to interact with (for example, a class bully or a student who is extremely hyperactive or withdrawn), a teacher may want to rotate these students in and out of various groups to minimize other students' contact with them and to keep all groups functioning more smoothly. By changing group membership frequently (for these students in particular), the teacher is able to maintain satisfactory learning conditions for most students.

Thus, there are advantages and disadvantages to having students work together in the same group for a relatively long period, say for five or more cooperative activities or tasks. Group stability allows students to become familiar with the work styles, competencies, and personal characteristics of other group members, and it allows group norms for student behavior to develop. As argued above, students who work in relatively permanent groups might be more likely to work out their differences than students in groups whose membership changed frequently. In contrast, changing group membership

after one or two periods may occasionally benefit individual students and groups by allowing them to work with a wider range of students, perhaps to avoid the negative consequences of personal conflicts and disagreement that could characterize more stable groups. Needed is research to explore how the types of tasks, age of students, and other relevant variables interact with stability of group membership.

RESEARCH RECOMMENDATIONS

In the previous section we discussed variables that merit additional reflection and research. In this section, we distinguish between two general paradigms or approaches to studying student groups: the *aptitude-treatment interaction (ATI)* approach, which focuses on the individual in the group; and the *socially situated paradigm,* which focuses on the meaning for the group. Although the ATI paradigm has generated much useful information about individual students' reactions to small-group learning, at times learning in small groups requires a second type of research paradigm, one that allows examination of the social context of learning, or how students *collectively* and individually construct and reconstruct meaning in social settings. We call this interactive construction of meaning the *socially situated learning paradigm,* or the dialectic model. In this paradigm the learning is located *in the cooperation.*

We integrate the discussion of the two research approaches with the discussion of proposed research in order to illustrate the potential of a multiple-perspective approach for the study of cooperative groups. Finally, in the discussion of group interviews we point out that at times it may be useful to interview groups of students in order to stimulate students' thinking about the social context of group work. We also illustrate one way of blending social and individual approaches by suggesting that thoughtful selection of students for group interviews may yield much useful information for both students and researchers.

Definitions of "Students"

Consideration of students in small-group contexts invites an expansion of educators' concepts of what it means to be a student. Students in whole-class learning settings are perhaps more appropriately conceived of as generic students who potentially differ in age, gender, ability, motivation, and prior knowledge than are students who work in small groups. As much of the previous discussion underscores, such conceptions of "student-as-learner" likely leave teachers at a loss to interpret and influence *as intended* students' interpersonal and intrapersonal processes in small groups. We have noted, for example, the role of students' individual personality constructs, social-cognitive skills, and interpersonal presence as relevant concerns in the construction of groups so that teacher and student goals may be realized.

Student-characteristic variables lend support to an aptitude-treatment interaction (ATI) model (Cronbach & Snow, 1977) for understanding the functional relationship between individuals and settings—here, social/instructional environments. That is, the characteristics that a student brings to the classroom setting are understood as potentially interactive with other setting or treatment variables (here, for example, group members, task criteria, etc.). ATI models rely on internal and relatively stable "defining" characteristics of students; interest is in how best to enhance, mediate, compensate, or otherwise modify those intrapersonal factors through social/instructional environmental manipulation.

An alternative conception of the relationship between the individual and the social/instructional environment would maintain that the individual is inextricably a part of the context within which he functions. In this contextual model, the individual and social/instructional setting are in emergent, dialectic tension, so that neither is understood in isolation from the other as in an ATI model, where factors may be seen as independent. Hence, in a contextual model it is not as easy to locate learning or motivation; each may well reside in the socially constructed setting and, indeed, precede learner internalization. Concern is with the process—the dialectic tension—between self and other, the intrapersonal and the interpersonal. Co-constructions between self and other are the foci of interest rather than internal characteristics as distinct from external conditions. The dialectic model has long been represented in the literature on development of self (for example, Baldwin, 1897) and consciousness or mind (for example, Marx & Engels, 1888; Vygotsky, 1962, 1978). Recently, it has been rejuvenated with much enthusiasm in work on socially shared cognition (for example, Lave, 1988; Rogoff, 1989; Wertsch, 1989).

An Illustration. Each model of student learning differs considerably from others in its implications for classroom practice. Consider a case of an individual learner, a student confronting new learning in the classroom who, initially with strategic teacher questions (for example, "What is the first question you should ask yourself?") and subsequently with more cryptic cues (looking at teacher), is able to apply reasonable, task-appropriate strategies and persist at in-class practice tasks. The same student, however, is unable to complete the review work at home; without the props and cues of the classroom context, the student is unable to organize the task or self to initiate, let alone transform, the task.

In an ATI perspective, a deficit would be located either within the student (for example, lacks organizational skills, short attention span, immature, unmotivated) or within the environment (for example, uneven lesson development, insufficient time for practice, inappropriate task difficulty) or some combination. The point of locating the deficit would be to develop a better match between student need and the environment to maximize student outcome.

In a dialectic model, the student's learning is socially situated; it is not yet located within (internalized by) the learner. This is not due to a mismatch in instruction. Quite the reverse: The teacher has achieved an appropriate "bandwidth" (Palincsar & Brown, 1989) or zone of proximal development (Vygotsky, 1962) for the student, who is being socially supported in learning that he is currently unable to do for himself. This perspective maintains that, with appropriate continued "scaffolding," the student will internalize the structure of the new learning. He does not yet own the knowledge; from a motiva-

tional and learning perspective, the learner's recognition that he does not yet "know" is an important and integral part of the process of learning (see Fridman, 1986). In the contextual model, student adaptation to the task of difficult learning does not imply a deficiency; rather, it is an important part of the process of motivated learning and the process of coming to know oneself as a learner. The social/instructional environment in a contextual perspective, then, does not seek to obviate the student's difficulty as it likely would in an ATI model.

How would each model inform the construction and interpretation of student learning in small groups? It seems likely that an ATI perspective would, as we have in much of this chapter, delineate student characteristics that are considered relevant to social processes and that are potentially interactive with task demands and instructional styles. An ATI perspective would likely locate the goal of group learning in an outcome, be it student achievement, cooperative behavior, or what have you. Group composition and duration, task selection, and instructional decisions would seek to enhance the attainment of these goals. The generative power of this model has been amply demonstrated throughout this chapter.

A socially situated perspective examines small-group learning in a different manner, however. It is more concerned with the tension between the social/instructional environment—including peers, tasks, and instructional supports—and the individual student than it is with a specific outcome. This is partially a function of the focus on reciprocal processes: An "outcome" is continuously reconstructed; it is not relatively stable, as posited in an ATI approach. A contextual perspective is also an historical/cultural one: What a student believes herself to be in relation to others—her self and cultural identity—informs her understanding and internalization of present experiences.

In a socially situated perspective, then, the individual is always engaged in relation to others; learning, motivation, and socialization are not distinct processes. Rather, because of the focus on the social processes in an individual's coming to know, "socialization" encompasses learning, motivation, and socialization processes. A contextual approach to small-group learning rests on a construct of socialization that involves dynamic change among evolving systems; it is concerned with the emergent interaction of self, task, and other in an historical and ongoing sense.

Instructional Contexts

Thus far, we have considered learner characteristics and learning processes in relation to individualized and small-group contexts and how these conceptions differ in locus and stability in two models: ATI and socially situated (or contextual). We now consider the dimension of stability as it organizes instructional contexts and informs the relative appropriateness of the ATI and contextual models of student learning.

Whole-class learning formats tend to be relatively stable events: Not only do lessons usually involve standardized procedures and thus become predictable for students (for example, review, overview, development, practice, homework expectations), but they are of sufficient duration (40–50 minutes) that subcomponents are readily identifiable. Whole-class learning formats are a constant in U.S. classrooms; not only do they occur in Los Angeles and Lancaster, but across generations: One author's grandmother understands perfectly what the author describes and adds examples from her childhood in a one-room school in the Copper Country of Upper Peninsula, Michigan. These formats are part of our culture, our heritage. In an important sense they provide a common experience across time and place and thus function as a source of shared identity across generations and locations (see also Anderson-Levitt, Sirota, & Mazurier, 1991). From the socio-historical perspective of the contextual approach, whole-class learning is part of our cultural and individual identity.

In contrast, small-group learning settings are much more varied in design, implementation, duration, and perhaps more importantly, by experience. It is doubtful that there is a common shared experience of small-group learning—not across individuals on the same task, not within individuals across differing tasks, and not across groups or even within the same group over time. Groups evolve, tasks affect group evolution, individuals influence as they are influenced by tasks and groups and time. An individual internalizes and aggregates some experiences and not others as he constructs a sense of self as learner, group member, and participant.

Experiences in small-group instruction are better thought of as probabilistic than as predictable. Moreover, because small-group learning has not been consistently implemented in classrooms, there is not even the commonality of confusion that might stabilize the experience through oral history. So small-group experiences are fluid and transitory, difficult to compartmentalize, and distinct from a shared cultural identity of schooling. Indeed, most small-group experiences likely occur outside the classroom (in teams, birthday parties, and family reunions) and probably do not seem relevant to classroom experiences.

Research Constructions

Implementation of a research program involves not only decisions about the model of the learner, the dynamics of the teaching and learning process, and the structure of the learning context; it also involves a choice about what the researcher wants to study. Given the relative stability and discrete stages of whole-class learning formats, for example, an ATI perspective has potential for exploring differential student learning within a whole-class setting. An ATI perspective may well generate alternative strategies for improving whole-class instruction for different types of learners. This perspective also has potential for exploring small-group instruction when the anticipated outcomes of instruction are associated with traditional achievement measures (concepts, skills, and individual problem-solving) and normative social behavior (turn-taking, listening, sharing, respecting individual differences). Each of the authors has participated in research on student learning and/or social behavior within an ATI framework that ascribed independent variables: internal to students, and external to settings (Good & Power, 1976; Mulryan, 1989; Rohrkemper, 1984, 1985). Research on small-group processes, however, presents an opportunity to explore ATI conceptions and to consider the power of alternative models.

The small-group learning format, as we have noted, lacks some of the qualities of whole-class learning formats that readily afford an ATI model (process stability, known and discrete stages) and thus expands research-decision rules to include decisions about the kind of understanding one wishes students to attain. This format also affords the consideration of alternative conceptions of classroom learning processes. Changing our conceptions of what we wish students to learn and the processes by which that learning is to occur also affects our definition of *student*.

One model we have posited as useful for framing and mediating changing conceptions of classroom learning goals that may result from the use of small-group instruction is the socially situated model. This model allows the examination of socialization processes broadly defined as students learning about themselves and others, and themselves in relation to others. Socially situated models appear to provide useful heuristics for examining nonlinear change processes and nontraditional educational goals. As such, they may increase our understanding of how students learn mathematics in a social-interactive environment. Such understanding is crucial if students are to use complex mathematical skills and understandings in a complex social world that demands identifying problems, conceptualizing solution paths, and using multiple resources (including others) to solve problems.

We stress that choice of research perspective is not an either/or issue. Some research will necessarily require a focus on individual learners in group settings and, accordingly, the ATI approach, which emphasizes variation among individuals, will be most helpful. However, other questions concern individualized processes of internalization and/or group questions. Researchers need to recognize that such questions require a contextual or social approach.

Combining Models

In this section we will illustrate how a research program might combine both an ATI perspective and a socially situated model in the same study. In doing so we are not arguing that the models need to be combined, because the research question should dictate the choice of model. By discussing the use of group interviews we hope to illustrate the value of new data collection procedures (that is, use of group interviews) as well as to clarify more fully the implications of the two models that we have discussed.

Recent research on classroom teaching and learning has increasingly involved the study of teacher and student cognition (for example, Clark & Peterson, 1986; Shulman, 1986; Wittrock, 1986) and the use of interviews (for example, Rohrkemper & Bershon, 1984) to understand perceptions that influence classroom behavior. It is interesting that almost all researchers interview students (or teachers) individually rather than question two or more participants together. Group interview techniques, however, could provide improved data as well as an important learning context in which valuable socialization could occur.

The decision to interview an individual implicitly suggests that the importance of learning about a group (or classroom) process depends on each individual's memory and construction of the learning task. That is, if a researcher wants to understand *individual learning,* or if interest is confined to traditional learning or social behavior outcomes (in a small-group setting), then a focus on the individual might be advantageous. However, if the intent is to study group processes and how groups vary in their construction of a task and in their productivity, then interviewing two or more group members together might be a better strategy than trying to learn from one participant who views and reports group work from an idiosyncratic perspective. Also, as we argued earlier, the socially situated paradigm suggests that group learning is located in the cooperation among members; thus social/cooperative interviews might provide a window into how students construct meaning in social settings. Group interviews could be constructed to take advantage of both the ATI and socially situated perspectives.

Various measures could be used for assigning students to work groups (gender, previous ability in mathematics, etc.). Here we use cognitive style as an example of the type of research that could be conducted. One cognitive-style measure that has been researched is the extent to which individuals are field dependent or field independent. Persons low in psychological differentiation (that is, field dependent) have difficulty differentiating stimuli from the contexts in which they are embedded; thus their perceptions are easily influenced by social contexts. In contrast, individuals who are high in psychological differentiation (field independent) perceive situations more analytically. These students can separate stimuli from background so that they change their perceptions less when changes in context are introduced. Field-dependent students' views (perceptions and memories) of group problem solving would more likely be affected by a particular task or group composition than would the views of more field-independent students. Interviews with two or three field-dependent students would therefore probably show that these students' reactions to group problem solving were more unstable than other students' and that they varied more from individual to individual. Since memory is socially constructed, researchers could probably obtain more comprehensive information about group processes by interviewing students whose collective perceptions are likely to mirror the group experience. In addition, students (whether field dependent or independent) are likely to benefit from participating jointly in interviews and otherwise reflecting on their group experiences. Field-independent students could learn much from field-dependent students, who are likely to be attentive to and make more use of the prevailing social frames of reference within groups. For example, field-dependent students would more frequently look at the faces of other students to obtain cues about what others are thinking and feeling and would attend more to group discussion. They would be much more likely to consider how other members view a task and which students are most cooperative at a particular time.

Conversely, field-dependent students can also learn from field-independent students. Field-independent students might help other students understand how small-group work differs from whole-class work. They are also much more likely to attend to the academic demands of a task, although they are less introspective than field-dependent students about the social demands of tasks. If researchers want to learn how groups of students work together (as opposed to how individuals learn in a group setting), the group interview method has much to offer. It can yield information about students' thinking during learning activities, may simulate the emergent group process as the interview unfolds, and might help students to learn from one another.

Interviews could yield information that might enable students to learn new strategies for working cooperatively. Whole-class instruction allows little incidental learning to occur (except for bright students, who can listen to the teacher and also reflect on the material). However, during interviews students could reflect jointly on tasks (that is, listen to peers discuss and enact group participation, conflict, and cooperation during group work). This might help some students consider how they can assume different and perhaps more productive roles in small groups. Five half-hour reflective conversations about what happened during small-group instruction may have more powerful effects on how students *think* (for examlpe, about cooperation, the importance of initiative, and the need for giving rich explanations) than would 20 hours of direct instruction by the teacher or 20 hours of group work. Too much of the socialization teachers use to prepare students for roles in groups appears to involve abstract imperatives (share, give explanations) rather than students' reflection on experiences they have had recently (that is, sharing materials, division of labor). Discussions about strategies for resolving conflict and for listening to members who felt left out (and were thus not viewed by other members as resources) might be important sources of insight as students further develop skills for learning how to cooperate, as opposed to simply cooperating to learn (Rohrkemper, 1986).

Consider the following example from a recent interview with four students who were discussing their reactions to seven mathematics lessons taught in a groups-of-four framework. The comments below occurred at the beginning of the interview (this was the first time that students had reflected on their group experiences).

(I = interviewer   F = female student   M = male student)
I: Now, did you enjoy working on the activities a little, a lot, or somewhere in the middle?
(chorus of "yeah")
F: I enjoyed working on the activities when the group would allow me to help. Some of them wanted their ideas, and they were sort of picky about how they wanted help.
I: Okay, so they wouldn't let you get your ideas in, was that it?
F: Sometimes they did.
I: How do the rest of you feel about it?
M: I like the rod patterns the best, I think they were fun.
I: And _____, what was your overall impression?
F: I liked the popcorn.
M: I liked the ones where you had to find ...where they reversed, take the reverse and you'd say numbers.
I: So you're talking about the palindrome activity, basically. Okay, now my next question—some people have already begun to answer this question. Which of all the seven activities you worked on was your favorite one?

M: I liked the popcorn because you could ...see which ones would come out and predicting which ones wouldn't. It's just a fun activity because it made you think of what color you're going to get. Then sometimes you kept getting the same color and you're asking, "Why I'm always getting this color?" Or, try and get another color and it's, "well, we can't because we can't see the colors."

I: So you really enjoy that one, I can see that. What were the other activities—what other activities did the rest of you enjoy?

F: I liked the popcorn and the consecutive numbers.

I: Okay, can you tell me why?

F: Well, the popcorn was fun because you got to guess and the consecu—

I: Consecutive numbers, okay.

F: That was just fun because it was sort of challenging.

I: Okay, how about you, _____?

F: I liked the rod patterns and the popcorn.

I: Why was that?

F: I thought the popcorn thing was fun and then the rod patterns. It was kind of hard, but it was fun.

I: Okay, _____, how about you?

F: I liked the word problems because they were challenging and you had to be creative during it.

I: Okay. Now the least preferred. What one did you not care about too much, of all of those seven? Or the one that you liked least of all?

M: I would say word problems 'cause it was really easy for me and our group it was ...in our group it was ...doing the word problems, it was real easy for us. It was just sorta boring 'cause it was like doing normal math or something like that.

I: How did the rest of you feel about the word problems?

F: I didn't like 'em 'cause I'm not that good at 'em.

I: Okay. And it makes a difference?

F: Yeah.

I: Okay. How about you,_____? Word problems?

F: They're okay, but it was ...I don't know. I just don't like doing the problems.

I: In groups or just generally?

F: I just don't like doing them.

I: At all. Okay, and _____ ...the word problems that you shared _____'s ...you had said word problems were your favorite, didn't you?

F: Yes.

(later in the interview)

I: Can you tell me, what sort of activities work best in cooperative groups? You've had different kinds of activities there. You've had activities like ones that involved writing numbers, computing, like the consecutive numbers and the palindromes, and the word problems. Then you had ones that you were kind of moving shapes around, paper shapes and the pentominoes and the tangrams, and the rods and the popcorn in which you were handling materials. Now, do all of those kinds of activities work pretty well in groups, or are there some kinds that work better than others?

In this interview students discuss and in a limited way defend what they liked and disliked about particular group problem-solving tasks. Such discussions would likely help students to recognize that peers have various reactions to different activities and to become more aware of their own attitudes toward learning tasks. It is instructive that the interviewer begins with an individual question (that is, "What did *you* enjoy?") rather than a social question (for example, "When did the *group* work best?") and that students take turns answering. Still, a group norm directs the individual contributions—students consistently explain their fun and interest as due in part to the "challenge," language that the students had introduced first. Obviously, in a group-interview setting the responses of individuals are not independent but rather are informed by the social context. When the interviewer introduces "least preferred," the students again adopt a common criterion—task difficulty, which is transformed into individual affect.

At the end of the interview excerpt the interviewer raises a question that may produce more social interaction—that is, an attempt for the group to discuss shared meaning. For example, discussion of what makes a good group may help less thoughtful and plan-oriented students to become more articulate about how and why groups function successfully. It also might generate strategies to help students understand what other group members do when they are uncertain about what others have said or when they are concerned about how others in the group are behaving. Such reflective interviews might enable students to think more systematically about their own *groups* as well as their developing skills for changing those situations.

SUMMARY

In this chapter we have argued that there are major problems in the mathematics curriculum with content, the way content is presented, how students are tested for mathematical thinking abilities, and students' understanding of mathematics. Among other things, there is too much emphasis on drill of facts and low-level concepts, and too little instruction that provides rich teacher explanations that help students relate their knowledge to current instruction. Students receive too few chances to solve problems in diverse ways, discuss the application of mathematics, and explain their thinking.

Various educators have suggested that one way to enhance mathematics instruction is by increased use of small-group instruction. In this chapter we have stressed that increased use of small-group instruction per se is unlikely to be part of the solution. This is because the important issue is not whether individualized instruction, small-group instruction, whole-class instruction, or discovery learning is stressed; the important issue is the *quality* of planning and instruction (Good, 1983). For example, we have reviewed literature demonstrating that for many students, increased attempts to understand mathematics are not automatic simply because they are placed in small groups (Bossert, 1988–1989). Indeed, there is evidence that some students may become passive and develop inappropriate perceptions as a result of their learning experiences in small groups. Mulryan (1989) found in one study that, during their initial experiences with small-group learning, most students perceived other students' passivity as beyond these students' control. However, after more experience with small-group work, students came to perceive passivity as deliberate and intentional—thus students may be learning to blame peers for their inactivity.

Much has been learned about how small groups can be used to facilitate students' acquisition of skills. Information is needed, however, about how groups can facilitate certain student attitudes and problem-solving abilities, including the knowledge of when to work with others, when to work alone, and how to vary these two approaches. Problems of learning are complex, and we need more process studies that illustrate how groups of students attempt to reduce ambiguity and risk when faced with difficult problems requiring creative thought, as well as studies that illustrate how and what students learn from failure. Research in whole-class settings shows that these issues are difficult to resolve (Doyle, 1977; McCaslin Rohrkemper, 1989), and we suspect the same is true in group settings.

We believe that programmatic research on small-group instructional processes could yield information that is extremely helpful to practitioners by providing them with concepts, findings, and theories that could lead them to design better small-group instructional activities. In calling for new research, we have attended to variables (for example, stability and composition of group membership) as well as research paradigms. We have stressed that traditional research models are useful for exploring how individuals learn in a group. However, we have noted that new models may be needed for certain questions. For example, we noted that a socially situated model appears to provide useful heuristics for examining nonlinear change processes and nontraditional educational goals.

Despite some of the problems identified in this chapter, we feel that increased use of *appropriate* small-group instruction can make mathematics learning more meaningful. Indeed, at present, two of the authors are engaged in cooperative research with classroom teachers in an attempt to develop curriculum materials and instructional models that support small-group learning (Good, McCaslin, & Reys, in press). There are many potential advantages of utilizing small-group instructional models. For example, Lindquist (1989) suggests that the appropriate use of small groups for mathematics instruction and learning can encourage students' verbalization; increase students' responsibility for their own learning; encourage students to work cooperatively in ways that build social skills; add variety to the routine of mathematics classes; enable teachers to individualize instruction and accommodate students' needs, interests, and abilities; and increase the possibility of students solving certain problems or looking at problems in a variety of ways.

Small-group learning will not replace whole-class or large-group teaching, because there is ample evidence that students can meaningfully construct mathematical knowledge during whole-class activities (see Fennema et al., 1989; Good, Grouws, & Ebmeier, 1983; Lampert, 1988). Still, increased use of small-group learning, better use of computers, better curriculum activities, better whole-class teaching, and more selective use of content units can improve practice.

The percentage of time that students spend in a particular format will vary with the topic, students' development, and many other variables. We agree with Slavin (1989–1990), that research supports the increased use of cooperative learning. However, it is important to recognize the diverse conceptualizations of "cooperation," and the field needs to accommodate this diversity with a richer definition. In addition, some forms of cooperative learning are more important than other forms. Further topics that await research are how much learning should be cooperative, and what group processes facilitate students' ability to reason mathematically.

Many educational problems cannot be reduced by improved knowledge of cooperative group processes in mathematics (state use of low-level tests; too little time for teachers to reflect on instruction and develop new curriculum materials); however, systematic and large investments in research in this area in the 1990s could provide invaluable process data for building better theories and ultimately transforming how we think about schooling in the 21st century.

References

Ames, C. (1984). Competitive, cooperative, and individualistic goal structures: A cognitive-motivational analysis. In R. Ames & C. Ames (Eds.), *Research on motivation in education, Vol. 1: Student motivation* (pp. 177–207). Orlando, FL: Academic Press.

Anderson-Levitt, K. M., Sirota, R., & Mazurier, M. (1991). Elementary education in France. *Elementary School Journal, 92,* 79–93.

Baldwin, J. M. (1897). *Social and ethical interpretations in mental development: A study in social psychology.* New York, NY: Macmillan.

Bales, R. F. (1951). *Interaction process analysis: A method for the study of small groups.* Reading, MA: Addison-Wesley.

Ball, D. (1991). Research on teaching mathematics: Making subject-matter knowledge part of the equation. In J. Brophy (Ed.), *Advances in research on teaching, Vol. 2: Teachers' subject-matter knowledge* (pp. 1–48). Greenwich, CT: JAI Press.

Bandura, A. (1982). Self-efficacy mechanism in human agency. *American Psychologist, 37,* 122–148.

Barr, R. (1988). Conditions influencing content taught in nine fourth-grade mathematics classrooms. *Elementary School Journal, 88,* 387–412.

Benne, K., & Sheats, P. (1948). Functional roles of group members. *Journal of Social Issues, 4,* 41–49.

Berger, J., Cohen, B., & Zelditch, M., Jr. (1972). Status characteristics and social interaction. *American Sociological Review, 6,* 479–508.

Bershon, B. L. (1987). *Elementary students' reported inner speech during a cooperative problem-solving task.* Unpublished doctoral dissertation, University of Maryland, College Park.

Billet, R. (1932). *The administration and supervision of homogeneous grouping.* Columbus, OH: Ohio State University Press.

Bossert, S. (1988–1989). Cooperative activities in the classroom. In E. Rothkopf (Ed.), *Review of research in education,* (Vol. 15, pp. 225–250). Washington, DC: American Educational Research Association.

Brophy, J., & Good, T. (1974). *Teacher-student relationships: Causes and consequences.* New York, NY: Holt, Rinehart, & Winston.

Burns, M. (1987). *A collection of math lessons from grades 3 through 6.* New York, NY: The Math Solution Publications.

Cazden, C. (1986). Classroom discourse. In M. C. Wittrock (Ed.), *Handbook of research on teaching* (3rd ed., pp. 432–463). New York, NY: Macmillan.

Clark, C., & Peterson, P. (1986). Teachers' thought processes. In M. C. Wittrock (Ed.), *Handbook of research on teaching* (3rd ed., pp. 255–298). New York, NY: Macmillan.

Cohen, E. (1982). Expectation states and interracial interaction in school settings. *Annual Review of Sociology, 8,* 209–235.

Cohen, E. (1984). Talking and working together: Status interaction and learning. In P. Peterson, L. Wilkinson, & M. Hallinan (Eds.), *The social context of instruction: Group organization and group process* (pp. 171–186). New York, NY: Academic Press.

Cohen, E. (1986). *Designing group work: Strategies for the heterogeneous classroom.* New York, NY: Teachers College, Columbia University.

Corno, L., & Mandinach, E. (1983). The role of cognitive engagement in classroom learning and motivation. *Educational Psychologist, 18,* 88–108.

Corno, L., & Snow, R. (1985). Adapting teaching to individual differences among learners. In M. C. Wittrock (Ed.), *Handbook of research on teaching* (3rd ed., pp. 605–629). New York, NY: Macmillan.

Crandall, V. C., Katkovsky, W., & Crandall, V. J. (1965). Children's beliefs in their own control of reinforcement in intellectual-academic achievement situations. *Child Development, 36,* 91–109.

Cronbach, L. J., & Snow, R. (1977). *Aptitudes and instructional methods.* New York, NY: Irvington/Nailburg.

D'Amico, A. J. (1986). *Individual differences in adolescents' classroom behavior and reported problem solving inner speech.* Unpublished doctoral dissertation, Bryn Mawr College, Bryn Mawr, PA.

Davidson, N. (1985). Small-group learning and teaching in mathematics: A selective review of the literature. In R. Slavin, S. Sharan, S. Kagan, R. Lazarowitz, C. Webb, & R. Schmuck (Eds.), *Learning to cooperate, cooperating to learn* (pp. 211–230). New York, NY: Plenum.

Davidson, P. S., & Willcutt, R. E. (1983). *Spatial problem solving with Cuisenaire rods.* New York, NY: Cuisenaire Company of America.

Davydov, V. V., & Radzikhovskii, L. A. (1985). Vygotsky's theory and the activity-oriented approach in psychology. In J. Wertsch (Ed.), *Culture, communication, and cognition: Vygotskian perspectives* (pp. 35–65). New York, NY: Cambridge University Press.

Deci, E. (1980). *The psychology of self-determination.* Lexington, MA: Heath.

Dewey, J. (1902). *The school and society.* Chicago, IL: University of Chicago Press.

Doyle, W. (1977). Paradigm for research on teacher effectiveness. In L. Shulman (Ed.), *Review of research in education* (Vol. 5, pp. 163–199). Itasca, IL: Peacock.

Doyle, W. (1983). Academic work. *Review of Educational Research, 53,* 159–200.

Duffy, G., & Roehler, L. (1989). Tension between information giving and mediation: Perspectives on instructional explanations. In J. Brophy (Ed.), *Advances in research on teaching* (Vol. 1, pp. 1–33). Greenwich, CT: JAI Press.

Dunkin, M. J., & Biddle, B. J. (1974). *The study of teaching.* New York, NY: Holt, Rinehart, & Winston.

Erickson, F. (1986). Qualitative methods in research on teaching. In M. Wittrock (Ed.), *Handbook of research on teaching* (3rd ed., pp. 119–161). New York, NY: Macmillan.

Fennema, E., Carpenter, T., & Peterson, P. (1989). Learning mathematics with understanding: Cognitively-guided instruction. In J. Brophy (Ed.), *Advances in research on teaching* (Vol. 1, pp. 195–221). Greenwich, CT: JAI Press.

Fennema, E., & Peterson, P. (1987). Effective teaching of boys and girls: The same or different? In D. C. Berliner & B. V. Rosenshine (Eds.), *Talks to teachers: A festschrift for N.L. Gage* (pp. 111–125). New York, NY: Random House.

Feurstein, R., Rand, Y., Hoffman, M., & Miller, R. (1980). *Instrumental enrichment: An intervention program for cognitive modifiability.* Baltimore, MD: University Park Press.

Flanders, J. (1987). How much of the content in mathematics textbooks is new? *Arithmetic Teacher, 35*(1), 18–23.

Fridman, L. M. (1986). Shaping the motivation to learn. *Soviet Education, 28,* 60–86.

Gamoran, A. (1987). Organization, instruction, and the effects of ability grouping: Comment on Slavin's "best-evidence synthesis." *Review of Educational Research, 57,* 341–345.

Gerleman, S. (1987). An observational study of small-group instruction in fourth-grade mathematics classrooms. *Elementary School Journal, 88,* 3–28.

Gilmer, G. (1978). *Effects of small discussion groups on self-paced instruction in a developmental algebra course.* Unpublished doctoral dissertation, Marquette University, Milwaukee, WI.

Good, T. (1983). Classroom research: A decade of progress. *Educational Psychologist, 18,* 127–144.

Good, T., & Biddle, B. (1988). Research and the improvement of mathematics instruction: The need for observational resources. In D. Grouws & T. Cooney (Eds.), *Perspectives on research on effective mathematics teaching* (pp. 114–142). Hillsdale, NJ: Lawrence Erlbaum.

Good, T., & Brophy, J. (1986). *Educational psychology: A realistic approach* (3rd ed.). New York, NY: Longman.

Good, T., & Brophy, J. (1987). *Looking in classrooms* (4th ed.). New York, NY: Harper & Row.

Good, T., Grouws, D., & Ebmeier, H. (1983). *Active mathematics teaching.* New York, NY: Longman.

Good, T., Grouws, D., & Mason, D. (1990). Teachers' beliefs about small-group instruction in elementary school mathematics. *Journal for Research in Mathematics Education, 21,* 2–15.

Good, T., Grouws, D., Mason, D., Slavings, R., & Cramer, K. (1990). An observational study of small-group mathematics instruction in elementary schools. *American Educational Research Journal, 27* (4), 755–782.

Good, T., McCaslin, M., & Reys, B. (in press). Investigating work groups to promote problem solving in mathematics. In J. Brophy (Ed.), *Advances in research on teaching* (Vol. 3). Greenwich, CT: JAI Press.

Good, T., & Marshall, S. (1984). Do students learn more in heterogeneous or homogeneous groups? In P. Peterson, L. Wilkinson, & M. Hallinan (Eds.), *The social context of instruction* (pp. 15–38). New York, NY: Academic Press.

Good, T., & Power, C. (1976). Designing successful environments for different types of students. *Journal of Curriculum Studies, 8,* 45–60.

Good, T., Reys, B., Grouws, D., & Mulryan, C. (1989–1990). Using work groups in mathematics instruction. *Educational Leadership, 47,* 56–62.

Good, T., Slavings, R., Harel, K., & Emerson, H. (1987). Student passivity: A study of question asking in K-12 classrooms. *Sociology of Education, 60,* 181–199.

Good, T., & Weinstein, R. (1986). Schools make a difference: Evidence, criticisms, and new directions. *American Psychologist, 41,* 1090–1097.

Goodlad, J. (1983). A study of schooling: Some findings and hypotheses. *Phi Delta Kappan, 64,* 464–470.

Haenisch, S., & Hill, M. (1985). *Riverside mathematics: Problem-solving resource book (grade 6).* Chicago, IL: Riverside.

Hiebert, E. (1987). A context of instruction and student learning: An examination of Slavin's assumptions. *Review of Educational Research, 57,* 337–340.

Homans, G. C. (1950). *The human group.* New York, NY: Harcourt, Brace, & World.

Humphreys, P., & Berger, J. (1981). Theoretical consequences of the status characteristic formulation. *American Journal of Sociology, 86,* 953–983.

Johnson, D., & Johnson, R. (1974). Instructional goal structure: Cooperative, competitive, or individualistic. *Review of Educational Research, 44,* 213–240.

Johnson, D., & Johnson, R. (1985). The internal dynamics of cooperative learning groups. In R. Slavin, S. Sharan, S. Kagan, R. Hertz-Lazarowitz, C. Webb, & R. Schmuck (Eds.), *Learning to cooperate, cooperating to learn* (pp. 103–124). New York, NY: Plenum.

Johnson, D., Johnson, R., & Maruyame, G. (1983). Interdependence and interpersonal attraction among heterogeneous and homogeneous individuals: A theoretical formulation and a meta-analysis of the research. *Review of Educational Research, 53,* 5–54.

Johnson, D., Maruyame, G., Johnson, R., Nelson, D., & Skon, L. (1981). Effects of cooperative, competitive, and individualistic goal structures on achievement: A meta-analysis. *Psychological Bulletin, 89,* 47–62.

Johnson, R., Johnson, D., & Stanne, M. (1986). Comparison of computer-assisted cooperative, competitive, and individualistic learning. *American Educational Research Journal, 23,* 382–392.

King, L. (1989). Student classroom perceptions and cooperative learning in small groups (Technical Report 475). Columbia: University of Missouri, Center for Research in Social Behavior.

Lampert, M. (1986). Teaching multiplication. *Journal of Mathematical Behavior, 5,* 241–280.

Lampert, M. (1988). Connecting mathematics teaching and learning. In E. Fennema, T. Carpenter, & S. Lamon (Eds.), *Integrating research on teaching and learning of mathematics* (pp. 132–167). Madison: University of Wisconsin, National Center for Research in Mathematical Sciences Education.

Latané, B., Williams, K., & Harkins, S. (1979). Many hands make light the work: The causes and consequences of social loafing. *Journal of Personality and Social Psychology, 37,* 822–832.

Lave, J. (1988). *Cognition and practice: Mind, mathematics and culture in everyday life.* Cambridge, England: Cambridge University Press.

Lindow, J., Wilkinson, L., & Peterson, P. (1985). Antecedents and consequences of school-age children's verbal disagreements during small-group learning. *Journal of Educational Psychology, 77,* 658–667.

Lindquist, M. (1989). Mathematics content and small-group instruction in grades 4 through 6. *Elementary School Journal, 89,* 625–632.

Luria, A. R. (1979). *The making of mind: A personal account of Soviet psychology.* M. Cole & S. Cole (Eds.). Cambridge, MA: Harvard University Press.

Marshall, H. (1988). Work or learning implications of classroom metaphors. *Educational Researcher, 17,* 9–16.

Marx, K. (1972). Theses on Feuerbach. In R.C. Tucker (Ed.), *The Marx-Engels reader* (pp. 107–109). New York, NY: Norton. (Original work published 1888.)

Mason, D. (1990). *The effects of two small-group models of active teaching and active learning on elementary school mathematics achievement.* Unpublished doctoral dissertation, University of Missouri–Columbia.

Mathematics Framework for California Public Schools. (1985). Sacramento, CA: California State Department of Public Instruction.

McCaslin Rohrkemper, M. (1989). Self-regulated learning and academic achievement: A Vygotskian view. In B. Zimmerman & D. Schunk (Eds.), *Self-regulated learning and academic achievement: Theory, research, and practice* (pp. 143–167). New York, NY: Springer-Verlag.

McKnight, C., Crosswhite, M., Dossey, J., Kifer, E., Swafford, J., Travers, K., & Cooney, T. (1987). *The underachieving curriculum: Assessing U.S. views of mathematics from an international perspective.* Champaign, IL: Stipes Publishing Company.

Meichenbaum, D. (1977). *Cognitive behavior modification.* New York, NY: Norton.

Mulryan, C. (1989). *A study of intermediate-grade students' involvement and participation in cooperative small groups in mathematics.* Unpublished doctoral dissertation, University of Missouri–Columbia.

National Council of Teachers of Mathematics (NCTM). (1989). *Curriculum and evaluation standards for school mathematics.* Reston, VA: Author.

Newman, R. (1990). Children's help-seeking in the classroom: The role of motivational factors and attitudes. *Journal of Educational Psychology, 82,* 71–80.

Newman, R., & Goldin, L. (1990). Children's reluctance to seek help with schoolwork. *Journal of Educational Psychology, 82,* 92–100.

Nicholls, J. G. (1984). Conceptions of ability and achievement motivation. In R. Ames & C. Ames (Eds.), *Research on motivation in education: Student motivation* (Vol. 1, pp. 39–73). Orlando, FL: Academic Press.

Nickerson, R. (1988). On improving thinking through instruction. In E. Rothkopf (Ed.), *Review of research in education* (Vol. 15, pp. 3–57). Washington, DC: American Educational Research Association.

Noddings, N. (1989). Theoretical and practical concerns about small groups in mathematics. *Elementary School Journal, 89,* 607–623.

Palincsar, A., & Brown, A. (1989). Classroom dialogues to promote self-regulated comprehension. In J. Brophy (Ed.), *Advances in research on teaching* (Vol. 1, pp. 35–71). Greenwich, CT: JAI Press.

Pepitone, E. (1985). Children in cooperation and competition: Antecedents and consequences of self-orientation. In R. Slavin, S. Sharan, S. Kagan, R. Lazarowitz, C. Webb, & R. Schmuck (Eds.), *Learning to cooperate, cooperating to learn* (pp. 17–67). New York, NY: Plenum.

Perry, M. (1988). Problem assignment and learning outcomes in nine fourth-grade mathematics classes. *Elementary School Journal, 88,* 413–426.

Peterson, P., Fennema, E., & Carpenter, T. (1991). Teachers' knowledge of students' mathematics problem-solving knowledge. In J. Brophy (Ed.), *Advances in research on teaching, Vol. 2: Teachers' subject-matter knowledge* (pp. 49–86). Greenwich, CT: JAI Press.

Peterson, P., & Janicki, T. (1979). Individual characteristics and children's learning in large-group and small-group approaches. *Journal of Educational Psychology, 71,* 677–687.

Peterson, P., Janicki, T., & Swing, S. (1981). Ability × treatment interaction effects on children's learning in large-group and small-group approaches. *American Educational Research Journal, 18,* 452–474.

Phelps, E., & Damon, W. (1989). Problem solving with equals: Peer collaboration as a context for learning mathematics and spatial concepts. *Journal of Educational Psychology, 81,* 639–646.

Piaget, J. (1970). Piaget's theory. In P. Mussen (Ed.), *Carmichael's manual of child psychology* (3rd ed., Vol. 1, pp. 703–732). New York, NY: John Wiley & Sons, Inc.

Porter, A. (1988). *A curriculum out of balance* (Report Series No. 191). East Lansing, MI: Institute for Research on Teaching, Michigan State University.

Posner, M. (1973). *Cognition: An introduction.* Glenview, IL: Scott, Foresman & Company.

Rogoff, B. (1989, February). *Social interaction as apprenticeship to thinking.* Presentation at the Conference on Socially Shared Cognition, Pittsburgh.

Rohrkemper, M. (1984). The influence of teacher socialization style on students' social cognition and reported interpersonal classroom behavior. *Elementary School Journal, 85,* 245–275.

Rohrkemper, M. (1985). Individual differences in students' perceptions of routine classroom events. *Journal of Educational Psychology, 77,* 29–44.

Rohrkemper, M. (1986). Education and cooperation. *Review of Education, 12,* 19–22.

Rohrkemper, M., & Bershon, B. L. (1984). The quality of student task engagement: Elementary school students' reports of the causes and effects of problem difficulty. *Elementary School Journal, 85,* 127–147.

Rohrkemper, M., & Corno, L. (1988). Success and failure on classroom tasks: Adaptive learning and classroom teaching. *Elementary School Journal, 88,* 299–312.

Rohrkemper, M., Slavin, R., & McCauley, K. (1983, April). *Investigating students' perceptions of cognitive strategies as learning tools.* Paper presented at the annual meeting of the American Educational Research Association, Montreal.

Rosenholtz, S. (1985). Effective schools: Interpreting the evidence. *American Journal of Education, 93,* 352–388.

Sapon-Shevin, M., & Schniedewind, N. (1989–1990). Selling cooperative learning without selling it short. *Educational Leadership, 47,* 63–65.

Schoenfeld, A. (1983). Episodes and executive decisions in mathematical problem solving. In R. Lesh & M. Landau (Eds.), *Acquisition of mathematics concepts and processes* (pp. 345–395). New York, NY: Academic Press.

Schunk, D. (1989). Self-regulated learning and academic achievement: A social cognitive view. In B. Zimmerman & D. Schunk (Eds.), *Self-regulated learning and academic achievement: Theory, research, and practice* (pp. 83–110). New York, NY: Springer-Verlag.

Schunk, D. H., Hanson, A. R., & Cox, P. D. (1987). Peer-model attributes and children's achievement behaviors. *Journal of Educational Psychology, 79,* 54–61.

Sharan, S. (1980). Cooperative learning in small groups: Recent methods and effects on achievement, attitudes, and ethnic relations. *Review of Educational Research, 50,* 241–271.

Sharan, S. (1984). *Cooperative learning in the classroom: Research in desegregated schools.* Hillsdale, NJ: Lawrence Erlbaum.

Shulman, L. (1986). Paradigms in research programs in the study of teaching: A contemporary perspective. In M. Wittrock (Ed.), *Handbook of research on teaching* (3rd ed., pp. 3-36). New York, NY: Macmillan.

Slavin, R. (1983). *Cooperative learning.* New York, NY: Longman.

Slavin, R. (1987a). Ability grouping and student achievement in elementary schools: A best-evidence synthesis. *Review of Educational Research, 57,* 293–336.

Slavin, R. (1987b). Cooperative learning and the cooperative school. *Educational Leadership, 45,* 7–13.

Slavin, R. (1989a). Cooperative learning and student achievement. In R. Slavin (Ed.), *School and classroom organization* (pp. 129–156). Hillsdale, NJ: Lawrence Erlbaum.

Slavin, R. (Ed.). (1989b). *School and classroom organization.* Hillsdale, NJ: Lawrence Erlbaum.

Slavin, R. (1989–1990). Research on cooperative learning: Consensus and controversy. *Educational Leadership, 47,* 52–54.

Slavin, R., & Karweit, N. (1985). Effects of whole-class, ability-group, and individualized instruction on mathematics achievement. *American Educational Research Journal, 22,* 351–367.

Slavin, R., Sharan, S., Kagan, S., Lazarowitz, R., Webb, C., & Schmuck, R. (Eds.). (1985). *Learning to cooperate, cooperating to learn.* New York, NY: Plenum.

Stigler, J., & Baranes, R. (1988). Culture and mathematics learning. In E. Rothkopf (Ed.), *Review of research in education* (Vol. 15, pp. 253–306). Washington, DC: American Educational Research Association.

Stigler, J., & Perry, M. (1988). Cross-cultural studies of mathematics teaching and learning: Recent findings and new directions. In D. Grouws & T. Cooney (Eds.), *Perspectives on research on effective mathematics teaching* (Vol. 1, pp. 194–223). Hillsdale, NJ: Lawrence Erlbaum.

Stipek, D. J. (1984). The development of achievement motivation. In R. Ames & C. Ames (Eds.), *Research on motivation in education: Student motivation* (Vol. 1, pp. 145–174). Orlando, FL: Academic Press.

Stulac, J. (1975). *The self-fulfilling prophecy: Modifying the effects of academic competence in task-oriented groups.* Unpublished doctoral dissertation, Stanford University, Stanford, CA.

van der Meij, H. (1988). Constraints on question-asking in classrooms. *Journal of Educational Psychology, 80,* 401–405.

Vygotsky, L. S. (1962). *Thought and language.* Cambridge, MA: MIT Press.

Vygotsky, L. S. (1978). *Mind in society: The development of higher psychological processes.* Cambridge, MA: Harvard University Press.

Webb, N. (1982). Peer interaction and learning in cooperative small groups. *Journal of Educational Psychology, 74,* 642–655.

Webb, N. (1984). Sex differences in interaction and achievement in cooperative small groups. *Journal of Educational Psychology, 76,* 33–44.

Webb, N. (1985). Student interaction and learning in small groups: A research summary. In R. Slavin, S. Sharan, S. Kagan, R. Lazarowitz, C. Webb, & R. Schmuck (Eds.), *Learning to cooperate, cooperating to learn* (pp. 147–176). New York, NY: Plenum.

Webb, N. (1989). Peer interaction and learning in small groups. *International Journal of Educational Research, 13,* 21–39.

Wertsch, J. (1989, February). *Problems of meaning in communication in a sociocultural approach to mind.* Paper presented at the Conference on Socially Shared Cognition, Pittsburgh.

Wertsch, J., & Stone, C. (1985). The concept of internalization in Vygotsky's account of the genesis of higher mental functions. In J. Wertsch (Ed.), *Culture, communication, and cognition: Vygotskian perspectives* (pp. 162–179). New York, NY: Cambridge University Press.

Whyte, W. (1989). Advancing scientific knowledge through participatory action research. *Sociological Forum, 4,* 367–385.

Wilkinson, L. C. (1988). Grouping children for learning: Implications for kindergarten education. In E. Rothkopf (Ed.), *Review of research in education* (Vol. 15, pp. 203–223). Washington, DC: American Educational Research Association.

Wittrock, M. (1986). Students' thought processes. In M. C. Wittrock (Ed.), *Handbook of research on teaching* (3rd ed., pp. 297–314). New York, NY: Macmillan.

Yackel, E., Cobb, P., Wood, T., Wheatley, G., & Merkel, G. (1990). Teaching and learning mathematics in the 1990s. In T. Cooney (Ed.), *1990 yearbook of the National Council of Teachers of Mathematics.* Reston, VA: NCTM.

Ziegler, S. (1981). The effectiveness of cooperative learning teams for increasing cross-ethnic friendship: Additional evidence. *Human Organization, 40,* 264–268.

·10·

PROFESSIONALIZATION AND
MATHEMATICS TEACHING

Nel Noddings
STANFORD UNIVERSITY

Current efforts to reform education include recommendations for the professionalization of teaching (Carnegie Task Force, 1986; Holmes Group, 1986). Using law and medicine as exemplars of the professions, these reform groups have suggested that preparation for teaching be conducted at the graduate level, a professional teaching hierarchy be created, and that a board of professionals should oversee the testing and licensing of teaching candidates. All of these recommendations can be challenged both from within a professional framework and from perspectives that criticize the very idea of professionalization as it is now defined. Further, the sheer number of teachers required by our schools acts against the formation of a profession modeled after law or medicine.

In this chapter I will discuss professionalism in terms of its definitions, history, growth, and conflicts. Secondly, I will examine the various aspects of professionalism with the purpose of analyzing what mathematics teachers would need to do in order to be professionals. Finally, I will summarize the issues, dangers, and possibilities that present themselves in the struggle toward professional status.

GROWTH AND CONFLICT
IN THE PROFESSIONS

What does it mean to be a professional? In recent decades, professionalism, once greatly admired, has become associated with self-interest and power-seeking. To avoid what Metzger (1987) calls the "spectre of professionism," it might be well to distinguish between *professionalism* and *professionalization*.

The former refers to a set of standards and practices approved by a profession; in a global sense, it refers to a highly skillful and ethically admirable way of performing in an occupation. For example, we often say of a proficient mechanic or plumber, "He is a real professional." When we use *professionalism* in this way, we refer to a person or group's adherence to a set of high standards internal to the practice under consideration.

Professionalization, in contrast, usually refers to the status characteristics of an occupation. For example, when we speak of the professionalization of teaching or nursing, we suggest certain changes in those occupations that will make them more like established professions. We cite such features as extensive, highly specialized preparation, high prestige, better-than-average remuneration, internal control over occupational practices and admission (self-regulation), and a relatively high level of autonomy for individual practitioners. Both professionalism and professionalization suggest the possession and organization of specialized knowledge, but the latter refers specifically to a process of seizing and isolating knowledge—making it the particular domain of "professionals." As the sociologist Dorothy Smith (1987) has pointed out, "The organization of professional knowledge is more than a guarantee of standards, more than a monopoly of knowledge and skill, it is a monopolization of control within a dominant class" (p. 217). Professionalism and professionalization are, of course, not discrete concepts; they are two aspects of establishing and maintaining a profession, but they refer to different perspectives and give rise to different emphases.

Moves to professionalize an occupation may be characterized by more or less attention to professionalism—that is, to

My thanks to David Berliner, Thomas Cooney, Douglas Grouws, Denis Phillips, and David Tyack for helpful comments on the first draft of this chapter.

the internal or intrinsic excellence of the practice (MacIntyre, 1981). In the early nineteenth century, professions emphasized such aspects of professionalism as altruism and service. Commitment to service lay at the heart of professionalism, and even today significant vestiges of that emphasis remain in various codes and oaths (Greenwood, 1957; Larson, 1977). But this century has seen an overt shift to power and privilege as the earmarks of professions (Larson, 1977); that is, the emphasis has shifted from professionalism to professionalization. With the pursuit of power and prestige, professions have taken on characteristics that may be inimical to service: hierarchies of control, less and less direct contact between professional and client, highly specialized language underscoring the powerlessness of clients, great monetary investments in preparatory education, an increase in internal talk as contrasted to interaction with the larger community, and an overall exclusivity marked by racism, sexism, and classism (Sykes, 1987).

Conflicts regularly arise between the self-interests of professionals and their commitment to the well being of clients and the larger community. Indeed, current debates over the professionalization of schoolteaching and nursing can be viewed, in part, as just such a conflict. A basic question for those deeply concerned about students is this: What effect will the professionalization movement have on students? Critics of the Holmes and Carnegie reports are not necessarily opposed to professionalization in general (see the issue of *Teachers College Record*, 1987, devoted to the Holmes recommendations), but they urge attention to the professional-client relationship. How will students benefit from the higher status of their teachers? Will they benefit at all?

Changing an occupation or semi-profession (Etzioni, 1969) into a full profession is not just difficult (Romberg, 1988), it is ethically and socially problematic as well. Issues have already been located in the tension between professionalism and professionalization; more questions will arise as we look at the growth of professions in the United States.

In the last half of the nineteenth century and the early decades of the twentieth, professions began to take on the characteristics identified in our discussion of professionalization. Their growth was closely connected to the rise of American universities and the transformation of higher education into an enterprise concentrating on the creation of knowledge (Light, 1983). This close connection induced a conflict between two professional spheres: the academic profession responsible for educating both its own members and practitioners and the world of professional practice. The professor of medicine, for example, would owe some allegiance to both the academic profession and the medical profession. Conflicts of loyalty arose; hierarchies were established, attacked, overthrown, and rearranged.

Mathematics teachers are dominated by two related professions that share responsibility for their preparation: professors of education and academic mathematicians. The pervasive lack of genuine cooperation between these two groups further weakens the professional status of mathematics teachers. A professional, by twentieth century standards of professionalization, should possess highly specialized knowledge; mathematics teachers cannot usually claim to have the knowledge of academic mathematicians or of professional educators in the academy.

The new emphasis on pedagogical content knowledge (Shulman, 1987) is meant, in part, to address this lack of specialized knowledge. However, at the present time, the expression "pedagogical content knowledge" is more a political rallying cry than a label for an actual body of knowledge. First, we need to identify and describe this body of knowledge as it appears in actual practitioners. Second, we need to decide whether the knowledge presently manifested in practitioners is sufficient; that decision entails an answer to the question "sufficient for what?" Should we insist that pedagogical content knowledge be demonstrably linked to better teacher performance? If other professions serve as models, the practical effects of having pedagogical content knowledge may not be as important as the status that accompanies it. Some of the difficulties involved in establishing a claim to such specialized knowledge will be discussed later in this chapter.

The growth of highly specialized knowledge supplies one theoretical explanation for the rise of a profession, but another set of theories suggests that power and politics are more important than knowledge. Academics can create huge bodies of knowledge (parts of which are not even useful to practitioners) and use this knowledge as means to exclude intruders and monopolize a set of services (Light, 1983). From this perspective, several of the Holmes and Carnegie recommendations are clearly attempts to gain control and place knowledge firmly in the hands of "professionals." In other words, a natural growth in knowledge may enhance professional status, as it did in medicine, or an artificial growth—deliberate creation of knowledge of doubtful worth—can be used to construct an important component of professional status. Who creates this knowledge and evaluates it? Will it be constructed by theoreticians or practitioners? Contemporary critics of professionalization want to know who these professionals will be and which groups will stand at the top of the professional hierarchy (Freedman, 1988; Noddings, 1990).

Both kinds of theoretical explanations have been used to describe the growth of medicine. The growth of medical knowledge helped to establish medical schools as the appropriate domain for medical training; the old apprenticeship model was clearly inferior in its capacity to transmit the latest knowledge. But cliques of wealthy, powerful physicians worked to establish standards that would both squeeze out the incompetent and, simultaneously, enhance their own prestige and wealth. The reasons given for exclusionary practices were tenets of professionalism but the moves actually made showed signs of professionalization.

The history of professionalization in medicine can be instructive for educators. Physicians highly trained in the best early medical schools were, as Donald Light (1983) points out, often less effective than their field-trained counterparts. "In fact, given their bold, confident practice of prescribing mercury and draining large quantities of blood, they were often more dangerous" (p. 360). This is something to keep in mind when considering the recommendation of the Holmes Group that the undergraduate education major be abolished. This recommendation is rationalized as a necessary part of professionalization;

that is, all truly professional education is given at the graduate level, and, therefore, teaching cannot be a true profession if its practitioners are prepared as undergraduates. Present evidence, though, suggests that elementary teaching—for which teachers are usually prepared in undergraduate education—is somewhat better on several counts than most high school teaching, which generally requires graduate level preparation. Holmes' writers even admit that most observers evaluate teaching in elementary schools as "more lively, imaginative, and considerate of students" (Holmes Group, 1986, p. 16) than that observed in high schools, but they suggest that weaknesses at both levels can be remedied by strengthening the liberal arts component of undergraduate education and locating most professional education at the graduate level. This decision to brush aside the judgment of "most observers" can only be seen as a political move, an emphasis on professionalization rather than professionalism.

Another set of issues can be located by considering the attempts of nursing to achieve professional status. Nursing, like teaching, has long had the status of a semi-profession (Etzioni, 1969). Overshadowed and dominated by medicine and staffed almost entirely by women, nursing has been unable to break into the ranks of full professions. Nursing's struggle is further complicated by its courageous emphasis on professionalism—on its own internal definition of nursing and criteria of excellence. Whereas medicine is marked by increasing distance between professional and client, nursing insists that caring is central to its practice and that "caring can be effectively demonstrated and practiced only interpersonally" (Watson, 1985, p. 8). A major conflict for nursing is its strong affirmation of values, such as direct caring, that are devalued in the world of professions (Reverby, 1987). Nurse theorists ask, "Must nursing sell its soul to be a true profession?"

Although caring, construed as a set of activities characterized by an attitude, has always been a defining feature of nursing (Reverby, 1987), the new wave of theory on caring draws directly from feminist analyses in psychology (Gilligan, 1982), philosophy (Grimshaw, 1986; Vetterling-Braggin, Elliston, & English, 1977), ethics (Hoagland, 1987; Noddings, 1984), political theory (Eisenstein, 1979; Ferguson, 1984; French, 1985; Jaggar, 1983), and theology (Daly, 1973, 1978; Ruether, 1983). At a recent conference on caring and nursing, Delores Gaut called on the conferees as *women and nurses* to build not only a new profession but a new world. This is a bold, exciting challenge, but there is little in the history of professions to predict a successful outcome.

Nurse theorists like Watson (1985), Gaut (1979), and Leininger (1984) emphasize the importance of relatedness and responsiveness in caring. They do not reject the medical model entirely; on the contrary, they endorse technical competence and the appropriate use of medical technology. However, they strongly deny the completeness of the medical model and its centrality to nursing. At the same time, they, like their nursing predecessors, have adopted a professional model of education (Reverby, 1987), and nursing will soon require more and more graduate education. It is hard to say whether this move is motivated by the internal demands of professionalism or the external pressure to professionalize; probably both sets of

demands are operating. What nursing holds out to teaching is the challenge to define or redefine a profession in the long tradition of service, close collaboration with clients, and shared power. It may be a hard idea to sell in today's world of self-interest.

Before turning to an analysis of the components of professionalism and professionalization, and their direct bearing on mathematics teaching, it may be instructive to consider the struggles of one other profession: the ministry. Seminaries for the preparation of ministers were established as graduate schools before the rise of modern universities. Andover Theological Seminary was started in the early 1800s and served as the model for many later seminaries (Light, 1983). When specialized research universities were developed, many of the earlier seminaries affiliated loosely with them. In this setting, people training for the ministry found themselves in a predicament similar to that of teachers in training. their professional training was often dominated by academic departments, and the religious and pastoral functions lost emphasis. Further, professionally trained ministers often had to compete with lay preachers, charismatic personalities who had great appeal among the masses. The ministry lost prestige as the country became more highly educated. Some people simply gave less attention to religion; others turned to more soul-satisfying (if less "professional") preaching. Graduate education—even at major universities—could not save a declining profession.

Preaching, like teaching, has often been considered a gift, and many lay people have questioned the need for professional training. Although the mainline Protestant and Catholic churches have always employed highly trained professionals, many sects rely on lay preachers and evangelical zeal (Stark & Bainbridge, 1985). Similarly, many people believe that anyone who knows a subject matter fairly well can teach it to high school students; anyone who loves children and can read, write, and spell can teach elementary school. In both cases, many people doubt that there is a substantial body of knowledge essential to the practice.

These beliefs work against both professionalization and professionalism—the former because they make it seem unnecessary and undesirable, the latter because qualities that many think should be at the heart of professionalism become its whole being and, therefore, lead professionals to downplay the complexity of their calling. Nursing, for example, was long thought to be mere "women's work;" "call a nurse" meant "call a woman." In their long struggle toward professional status, nurses freed themselves from cooking, washing sheets, and scrubbing floors—work "any woman" could do. It is still often suggested that "anyone" can do the direct caring that nurses identify as the heart of their profession (Noddings, 1990, Reverby, 1987). This attitude rests on a profound misunderstanding and an unwarranted devaluation of the interpersonal skills as well as the caring attitude that facilitates their effective use. Just as nurses have been tempted to deny the centrality of caring in order to be regarded as true professionals, highly trained ministers are likely to soft-pedal evangelical rhetoric and techniques of healing through touch (even though many regard both as important) so as to escape identification with lay preachers. Now many teachers fear that they, too, are being pressed to abandon their central mission—devotion to students as whole persons—for highly specialized work as professionals (Biklen, 1987; Boston Women's Teachers Group, 1983; Freedman, 1988).

With the problems identified in this background discussion, we can see that professionalization is, at best, a partial good. Mathematics teachers seeking professional status must ask sophisticated, normative questions as well as instrumental and strategic ones. How to become a professional is too simple a question—even if the process is anything but simple. How to define one's occupation as a profession, how to resist pressures to adopt the appearances of professionalism, and how to balance self-interest with concern for students and colleagues are the tough issues that teachers face as they consider contemporary moves toward professionalization.

THE FEATURES OF PROFESSIONS

A distinction between professionalism and professionalization was useful in locating issues and raising questions about benefits and harms that may accompany a drive toward professional status. Now I will present an analysis of the specific features of professions.

There are many models and descriptions of professions and professional occupations (Bledstein, 1976; Flexner, 1915; Hall, 1968; Larson, 1977). All of them mention, in one form or another, the following features of professions: selection and regulation, specialized knowledge, altruism or service, privilege and status hierarchies, collegiality, and autonomy.

Selection and Regulation

All professions strive for self-regulation; that is, professions seek to control the processes by which people are admitted to the profession and the standards to which they will be held as they practice. Sometimes standard-setting is so rigorous that shortages are created in the ranks of professionals, and the end result for those who survive may be an increase in both prestige and remuneration. The story of medicine's deliberate attempts to alleviate overcrowding in the field and simultaneously improve practice through tough standard-setting is well documented (Markowitz & Rosner, 1979; Starr, 1982).

The recommendation for a hierarchy in teaching (Carnegie Task Force, 1986; Holmes Group, 1986) is clearly a move in this direction. Although the overall number of teachers cannot be controlled by the profession, the number admitted to the upper echelon *can* be limited. Such control is designed in part to increase the power and prestige of teaching by demonstrating that at least some of its members bear the mark of true professionals. They will have achieved years of advanced study, successful completion of board examinations, recognition in a subfield of specialization, and respectable salary levels.

Control of an upper echelon is a first step in gaining a greater share in overall control. State regulation now dominates teacher assessment and accreditation, and it is often considered a yoke to be thrown off in the pursuit of professional status. However, regulation by the state is, in fact, a major first step

in professionalization (Sykes, 1986). Any work that evokes the state's compelling interest is important work, and professional elites have often cooperated with political elites to initiate state regulation (Deegan, 1989). The state's legitimate interest is to protect the public; in the best interest of the public and itself, a profession provides the content for standards established by the state. The state may, for example, require advanced board examinations, and the profession then contributes the substance of the examination.

This necessary partnership of state and profession induces considerable tension. Not only is there a continual struggle for dominance, but control itself becomes a measure or test of professional status; as we saw earlier, the push for control can distract members of a group from professionalism toward professionalization. The *appearances* of professional status become ends in themselves.

The present emphasis on teacher assessment is part of the struggle for control and for professionalization. Tom Bird (1988) writes: "Assessment for certification is not solely a conceptual or technical matter. It is also part of a strategy to build a professional organization that is respected by the public and that provides support, recognition, and leadership to its members" (p. 19). A bit later he writes, "The Board's certificate should confer some advantage or standing on the teachers who earn it. Otherwise, there is no point to the Board" (p. 19). Bird means, of course, that there is no point from the perspective of teachers interested in advancement, but his statement, along with many others, illustrates the shift from public service to self-interest that often marks professionalization. Critics are right to ask what the point of such a Board is from the perspective of parents and students. As Sykes (1986) points out, the reform of medicine "produced a mix of costs and benefits whose overall impact is difficult to evaluate" (p. 12). It is predictable that, at least in the short run, part of the population will suffer from lack of expert services, lower access to the profession, and a general feeling of discrimination.

The situation for mathematics teachers is complicated (as it is for most secondary school teachers) by their connections to other professional groups that exercise considerable influence over their work. Professional mathematicians and scientists periodically take interest in the mathematics curriculum (American Association for the Advancement of Science, 1987; Conference Board of the Mathematical Sciences, 1984; National Science Board Commission on Precollege Education in Mathematics, Science, and Technology, 1983). The recommendations of these groups reflect current trends in mathematics, but they are sometimes naive with respect to the psychology of learning and pedagogy. These groups are unlikely to serve as reference groups for mathematics teachers. When mathematics teachers identify a national reference group, it is usually the National Council of Teachers of Mathematics (NCTM) (Yamashita, 1987), and the interpretations of this group are crucial in their impact on teachers. In effect, classroom teachers are at the mercy of competition or cooperation between powerful professional groups. One group controls the undergraduate subject matter preparation of teachers; the other strongly affects both their graduate professional training and, to a lesser degree, the content of what they will actually teach. It is not surprising that award-winning teachers often cite participation in the activities of NCTM as highly significant in their professional lives (Yamashita, 1987).

Although academic mathematicians and educators are not in an overt battle for control over the standards for mathematics teachers, the traditional separation between the groups is worth exploring in connection with the new reform movement. When many teachers were produced by state teachers colleges, mathematics professors were professional educators, and most of the course work in mathematics had a pedagogical flavor. As teachers colleges were expanded into multi purpose state schools to accommodate the great influx of veterans after World War II, the primary reference group of college mathematics teachers may have changed. Mathematics courses now had to prepare everyone who needed mathematics, regardless of his or her future occupation. It is questionable whether this system has served precollegiate education well. Now influential advisors like the Holmes Group want to put even greater emphasis on liberal education. To their credit, they recognize that the educational advantage of such a move depends heavily on genuine cooperation between liberal arts and professional education groups; however, there is little historical evidence that such cooperation will become a reality. More will be said on this in the discussion about the knowledge aspect of professions.

Before we leave this section, a concern raised in the introductory analysis needs reiteration. Standard-setting tends to bring with it exclusivity and, sometimes, outright discrimination (Sykes, 1986, 1987). As more study and examinations are required, women and minorities tend to be squeezed out of the upper ranks. This concern may not be quite so relevant as it was decades ago, since more women and minorities are entering the professions, but it remains a matter for serious consideration. In particular, teaching has long been a profession peculiarly attractive to women raising children. Many of these women define "professional" in terms of caring for children; that is, they see a "professional" as a "good teacher"—one who cares deeply about the growth and development of children as whole persons. Their home and school lives have a seamless quality (Biklen, 1987; Boston Women's Teachers' Group, 1983; Freedman, 1987). As the demand grows for longer on-site hours, for more non-student-oriented activities—more of the trappings of professionalization—these women may become second-class citizens, even though they continue to do much of the direct contact work with children.

In conclusion, standard-setting in a profession effects costs in the communities it serves. Educators hope, of course, that higher standards will serve both professionals and clients better, but the results of standard-setting in other professions suggest that the results will be mixed and hard to evaluate.

Specialized Knowledge

One of the hardest problems facing the professionalization movement is the definition and description of the knowledge needed by teachers (Holmes Group, 1986). Clearly, secondary school teachers need to know their subject, and they need to know how to teach it, but of what does the latter knowledge consist? Are there other bodies of knowledge that mathematics

teachers should have? What does it mean to "know mathematics" today?

As discussed earlier, mathematics teachers are not in a position to define or create mathematics. This means, of course, that they are not in a political position to define what mathematics is or to create mathematics that will be recognized by a society of professional mathematicians. In a cognitive sense, everyone working with mathematics, teachers and students alike, creates or constructs mathematics. (See Davis, Maher, & Noddings, 1990). But professionalization puts far greater emphasis on political issues than on cognitive or intellectual matters. What happens in the field of mathematics, itself, affects the undergraduate mathematical education of teachers, as it does that of all mathematics students, but undergraduate mathematics education often lags behind mathematical development (National Science Board Task Committee on Undergraduate Science and Engineering Education, 1986, 1987). Undergraduate mathematics preparation for teachers is, thus, likely to be deficient in several ways: First, the whole field suffers educational neglect as academic mathematicians are often loathe to take time away from the creation of mathematics to construct new curricula or to experiment with modes of pedagogy suitable for various groups. Second, when catch-up recommendations are made, they rarely take into consideration the needs of prospective teachers; mathematics courses, especially for teachers, have usually been perceived or even designed as watered-down versions of "real" mathematics courses. Third, not only is the mathematical content needed by teachers unconsidered, but related knowledge of great importance to teachers is also neglected: the history and epistemology of mathematics, its connection to other fields, and the psychology of learning mathematics (Holmes Group, 1986; Romberg, 1988). Finally, even though students who plan to teach often do as well or better than other mathematics majors, persistent rumor has it that the poorest students go into teaching, a rumor sustained by the unwillingness of many departments to provide rigorous courses specially designed to meet the needs of prospective teachers. If such courses are perceived as "pushovers," the students in them will be seen as inferior. Working partnerships between mathematics and education departments are essential if the mathematics preparation of teachers is to be improved. This is an important function, increasingly well filled by university professors of mathematics education.

There is some debate, however, on just how important subject matter knowledge is to teaching. Although one would suppose such knowledge a necessary, though not sufficient, condition for adequate teaching, research supplies little evidence to back such an intuitively plausible claim (Druva & Anderson, 1983). Certainly mathematics teachers must know some mathematics, but how much and of what sort? Simple correlations between teachers' mathematical knowledge and student outcomes are generally low. However, this observation should not lead us to suppose that mathematical knowledge is unimportant in mathematics teaching. Indeed, correlations between knowledge and outcome measures are surprisingly low in most other professions as well, but professional opinion has not, for that reason, devalued knowledge. Perhaps the best approach is to continue exploration and to analyze and refine the undergraduate program as discussed above.

202 • MATHEMATICS TEACHING

In any case, knowledge of mathematics cannot be sufficient to describe the professional knowledge of teachers. What does a mathematics *teacher* know that someone with similar mathematical preparation does not? What specialized knowledge does the teacher have? Several investigators are now concentrating on these and related questions (Buchmann, 1983; Conroy, 1987; Romberg, 1985, 1988). Lee Shulman (1986, 1987, 1989) is studying pedagogical content knowledge—knowledge specific to the teaching of mathematics and other subjects. Although this line of research is intuitively significant, it creates many new issues; for example, how are content knowledge and pedagogical content knowledge related? If two teachers, A and B, both know three methods for teaching X, what else affects their performance? Perhaps A chooses a particular method because she has psychological knowledge about her students that B lacks, and, therefore, she proves more effective in teaching X. On the other hand, perhaps B gets along so well with his students that he teaches X effectively no matter which method he chooses. Maybe a third teacher, C, turns the whole matter of learning X over to the students, and they do even better!

Continuing research on teacher knowledge is crucial not only for the conduct of teaching itself but also for teacher preparation. At this stage we do not know how much pedagogical content knowledge can be imparted prior to actual teaching, but evidence suggests that most of it is learned on the job. The results of research in this area are important for both traditional and nontraditional teacher preparation programs. Among recruits to the latter, "making educational coursework more rigorous, more specific to subject matter pedagogical needs, and more practically informative" (Darling-Hammond, Hudson, & Kirby, 1989) is a high-priority recommendation. The old-fashioned methods course, done well, is a basic necessity for mathematics teachers.

In discussing pedagogical content knowledge, we cannot avoid considering cognitive science and psychology, for each has something to contribute to the specialized knowledge that makes teachers professionals. To constructivists, for example, knowledge of how children learn mathematics is essential in teaching (Steffe, Cobb, & von Glasersfeld, 1988; Davis, Maher, & Noddings, 1990). Important work is being done on metacognition and the role it plays in problem solving (Schoenfeld, 1987). Work related to both constructivist perspectives and metacognition concentrates on the creation of meaning in mathematics classrooms (Lampert, 1985, 1988). Further, studies in cognitive anthropology are beginning to shed light on mathematical behavior in everyday life in a variety of cultures (Carraher, Carraher, & Schliemann, 1985; Lave, 1988; Saxe, 1988). Familiarity with work of this sort marks the professional teacher, educated at a major research university. It is knowledge that teachers prepared through alternative methods are unlikely to acquire.

A question remains, of course, as to whether the knowledge just mentioned facilitates good teaching; surely it is not *necessary* for teaching. Because it is so difficult to establish connections between teacher knowledge and student performance—or even teacher performance—it seems likely that, if such knowledge becomes part of the licensing expectations for Board certified teachers, it will do so on the basis of expert recommendation (Shulman, 1989).

Current views on teacher knowledge often put great emphasis on liberal education (Carnegie Task Force, 1986; Holmes Group, 1986; Shulman, 1989). This recommendation implies that a broad liberal education might be better preparation for teaching than, say, an engineering or technical education. Plausible as the recommendation sounds, there is little evidence to support it. Indeed critical theorists (Giroux, 1988), liberation educators (Freire, 1970; Illich, 1971), and feminist theorists (Martin, 1985; Noddings, 1990; Rich, 1979) have launched powerful attacks on liberal education as a mechanism of exclusivity. From a critical perspective, the current call for greater emphasis on liberal education is politically—not epistemologically or psychologically—motivated. To be liberally educated is to achieve a certain status; the connection between such education and effectiveness in the classroom remains unclear. Thus, a traditional liberal education for teachers might be seen as a greater contribution to professionalization than to professionalism.

Finally, one may consider general pedagogical skills as part of specialized teacher knowledge. All teachers need to know how to manage a classroom, discipline unruly students, evaluate students, report to parents, fill out forms, and fulfill other administrative expectations. Which of these functions are properly associated with teaching and which are artifacts of a misdirected educational system? This question is fundamental for teacher education programs. On the one hand, teacher educators want their graduates to be prepared for the real world of the classroom; on the other, they hope that their graduates will help to change the system in ways characteristic of true professionalism. Teacher education programs are constantly trying to balance such demands. Is it a mark of true professionalism to keep order with a baseball bat? At the opposite extreme, is it appropriate for professionals to avoid the worst of present conditions and find more congenial situations for their practice? These questions lead logically into the next aspect of professional life.

Altruism and Service

Altruism and service have always been part of the description of professional life. It is expected that a professional will work toward the best interests of his or her clients. Indeed, professions grew out of an earlier notion of *vocation,* a lifework to which people were "called." Although the idea of public service is still formally encoded in all professions, the semi-professions (for example, nursing and teaching) are often recognized as more altruistic than their fully professional counterparts in medicine and law.

It has never been considered good professional taste to seek one's own financial betterment overtly. For example, most physicians and lawyers do not advertise, and their professional associations are supposedly devoted first to the advancement of their respective sciences and only secondarily to the welfare of their own members. When teachers joined labor unions, this was taken as a sign of anti-professionalism. From the accounts discussed earlier, we know that professionals have often acted in their own best interest—sometimes to the detriment of clients—but that such activities cannot be overt moves for more money or better working conditions. To admit publicly that one's salary or working conditions are unsatisfactory is to confess a likeness to the uneducated masses, to labor. It should be noted, however, that even unionism strives for a professional image; the union label has become a proud mark of quality, denoting a product that the public can trust. Likewise, teachers' unions have continually stressed professional aspects of teaching. The perceived connection between public trust and professionalism is a strong one.

Altruism and service, as components of professional life, are usually defined as expert application of theoretical knowledge to the problems of clients (Argyris & Schon, 1974). The emphasis on theory tends to outweigh the importance of service, and occupations that involve continuous and direct interpersonal relations with clients are generally devalued. As shown in the opening discussion, nursing has failed to achieve full professional status, in part because of its insistence that caring can be effectively accomplished only interpersonally (Watson, 1985). The current reform movement in teaching rarely says anything about students directly, concentrating, instead, on transforming the appearance of teaching so that it more nearly resembles other professions. It is assumed, but not demonstrated, that more professional teachers will use their enhanced theoretical knowledge for the betterment of students.

Some of the current arguments over professionalization focus on potential effects on students (*Teachers College Record*, 1987), and a few of these analyze the movement from a feminist perspective (Freedman, 1988; Noddings, 1990). A main complaint against both the professionalization movement and many of its critics is that they both ignore the experience of women. They take for granted that people enter professions with an eye on advancement and power over others, but this orientation has not been one historically associated with women. Even women who aspire to be administrators often display a lack of interest in the status and prestige of administrative positions (Edson, 1988, p. 10).

Although neither the Holmes nor Carnegie groups can be accused of accepting the old stereotypes of women as passive, lacking drive, uninterested in intellectual work, and lukewarm in professional commitment (Biklen, 1987), they do accept the stereotypical-masculine notions of professional life. In contrast, there is considerable evidence that women teachers put great emphasis on the altruistic component of teaching. They are often deeply committed to their students (Biklen, 1987; Boston Women's Teachers' Group, 1983; Edson, 1988; Gilligan, 1982). Further, they do not see a sharp separation between their work as homemakers and teachers, accepting responsibility for the growth of others as central to their professionalism and deriving deep satisfaction from their nurturing roles. The picture of deeply committed nurturers is very different from the earlier stereotype used as an excuse for women's absence in administrative positions, but it is still a picture at odds with the current description of professional life.

Closely connected to the emphasis on altruism in professional life is a concern with teaching as a moral enterprise. Many writers have noted the centrality of moral considerations in teaching, and some critics have expressed concern about the neglect of moral matters in the professionalization movement (Sockett, 1987). Every profession has moral aspects—indeed, every human interaction has a moral component—but no other profession bears quite the moral responsibility that teaching does (Sichel, 1988). Even the ministry, by nature deeply concerned with morality, does not have the power to direct young people's moral development in daily life to the extent that teaching does. This awesome responsibility and the need to share it has led some to call for the *deprofessionalization* of teaching (Noddings, 1984); that is, a redefinition of professionalism with an emphasis on the altruistic and service aspects.

Mathematics teachers may need to give more attention to the moral conduct of their teaching. In a time when " 'They don't care,' is the number one complaint heard from student dropouts" (Comer, 1988, p. 35), it may be necessary to cultivate trusting relationships with students in order to teach them mathematics and guide them toward more caring lives (Noddings & Greeno, in press). Furthermore, the special problems of women and minorities in mathematics add another dimension to the array of moral problems in mathematics education. As Sockett (1987) asserts, these problems are not mere components of teaching; rather, the moral nature of teaching is fundamental to the enterprise and underlies all other dimensions. It is this observation that critics often have in mind when they call for greater attention to altruism and service in professional life.

Privilege and Status Hierarchies

People enter teaching for a variety of reasons, among them: "The desire to teach in general or to teach a subject in particular (22%); the idea of teaching as a good and worthy profession (18%); and a desire to be of service to others (17%)" (Goodlad, 1984). Seen as a "good and worthy profession," teaching should provide the privileges and status generally accorded to professions.

We have already discussed several impediments to achieving full professional status for teachers. The sheer number of teachers is surely another. Medicine advanced in part by creating a noticeable shortage of physicians; such a tactic could not work in the ministry because congregations could turn to talented lay preachers. Similarly, it cannot work for schoolteachers because, first, children need to be supervised and, second, it is believed that many reasonably well-educated adults can do an adequate job of teaching with or without specialized training. This state of affairs makes it difficult to elevate the status of *all* teachers substantially.

There is another way, however, to make teaching more attractive as a profession. A hierarchy can be created in which the top echelon enjoys the status of a true profession. Calling this group *Career Professionals*, the Holmes report (1986) says:

But such a cadre could only be small—we estimate roughly one-fifth of all teachers. A majority of the teaching force should be *Professional Teachers:* people who have proven their competence at work, in rigorous professional qualification examinations, and in their own education. Their jobs would differ from those of today's teachers in their more serious educational requirements, and in the stiff standards for entry into and continuance in teaching. These would be jobs to which only bright, highly qualified people could gain entry. They also would be jobs in which teachers could continue to learn and improve—among other ways, through their work with one another and with Career Professionals. They would be jobs that could lead to a Career Professional position, if the incumbent were sufficiently gifted and willing to invest the time in advanced study and examinations. (pp. 8–9)

Clearly the prospect of achieving Career Professional status is an attractive one, and the possibility of such advancement might well enhance the status of teaching. But deep concerns have properly been raised: one such worry is that success in teaching—in the important sense of affecting students' lives and futures—might not play a large role in securing advancement. The people who invest their time in actual contact with students may not have time enough for the advanced study required. Further, many of these people may regard advanced study as irrelevant if its usefulness in the classroom cannot be demonstrated. Finally, because the work of Career Professionals takes them out of the classroom for a significant part of the day, teachers may conclude that the greatest rewards come to those who have least direct contact with students.

In addition to concerns about the devaluation of people who continue to do the daily work of teaching, there are concerns about the elitism and exclusivity inherent in hierarchical arrangements. As we saw earlier, standard-setting often leads to the exclusion of racial and ethnic minorities, women, and the poor. The decision to limit the ranks of Career Professionals to one-fifth of the teaching force underscores this concern. On the one hand, it is an eminently practical decision: one-fifth of the teaching force is about equal to the number of physicians in this country. Thus, the move seems economically feasible. On the other hand, though, it just happens that the major research universities train about one-fifth of the nation's teachers, and surely it is these people who will find it easiest to become Career Professionals, since their institutions will define what knowledge is required. The threat of exclusivity is, therefore, very real.

This discussion can be extended to a full-blown critique of professions and professionalization, as suggested in the opening section of this chapter. The worry expressed 50 years ago by Virginia Woolf is salient today; expressing considerable ambivalence about women's participation in professions, she wrote

Do we wish to join that [academic] procession, or don't we? On what terms shall we join the procession? Above all, where is it leading us, the procession of educated men? ...What is this 'civilization'? What are the ceremonies and why should we take part in them? What are these professions and why should we make money out of them? Where in short is it leading us, the procession of the sons of educated men? (pp. 62–63)

Questions about the prevalent use of hierarchies and challenges to their legitimacy mark every form of feminism (Offen, 1988, p. 152). Feminist legal scholars, among others, have launched critical attacks on professional hierarchies (MacKinnon, 1987; Menkel-Meadow, 1988), and feminist scholars in education are now awakening to the threat of new hierarchies in schoolteaching (Freedman, 1987).

There may be a compromise that will satisfy both the opponents of hierarchies and those who want to advance the professional status of teachers by providing extended opportunities. The Holmes Group and other reformers are correct in pointing out the short-sightedness and wastefulness of failing to develop teachers' specialized talents in pedagogy, curriculum development, community leadership, action research, and other functions rarely delegated to teachers. They are right again when they deplore the shameful failure to use the experience of expert teachers in guiding novices. But recognizing this state of affairs does not entail the introduction of hierarchies of prestige and rewards. Options of the sort just mentioned could be provided as choices within teaching, and one choice that should stand on a level at least equal to all others is the choice to specialize in working with students—*to teach*.

Probably no human being, certainly not one expected to fulfill the obligation of a professional, should be forced to teach five classes and meet 150 or more students every day, as most mathematics teachers must under present conditions. Plans for true professionalism must consider the daily lives of all teachers and make careful provision for the welfare of students. The professional-client relationship should remain central in any discussion of professionalism.

Collegiality

Professionals identify with a reference group that protects their interests, guides their behavior with respect to clients, and increases their professional knowledge. Virtually every commentator on teaching mentions the lack of opportunity for teachers to engage in professional activities with colleagues, and most reform groups emphasize the centrality of collegiality in professionalization (Holmes Group, 1986, p. 66).

In a recent review of the literature on collegiality, Romberg (1988) found its importance well documented. Within-school collegial relationships are highly correlated with satisfying school climate and general effectiveness (Little, 1982). Teachers who have opportunities to plan together, observe each other, and diagnose and evaluate students together are apparently happier with teaching as a profession than those who do not have such opportunities.

In an interesting comparison of presidential award-winning mathematics teachers with other professional math teachers (as identified by NCTM membership), Yamashita (1987) reports several findings that could be useful for professional development. Awardees reported more collegial connections outside their own schools: Whereas none in the comparison group mentioned collegial relations with professors, district, state, or county administrators, or former in-school colleagues, 12 to 15 percent of awardees reported such connections. Within their own schools, awardees tended to interact less often with their own departmental colleagues, but far more often with their department chair. Yamashita's most striking result is the tremendous emphasis awardees placed on participation in the NCTM; for these teachers, the NCTM provides the single most important professional forum in their careers. Similarly, they work much more frequently with math teachers in other schools than their counterparts in the comparison group.

In contrast, Yamashita's comparison group reported spending more time than the awardees with nonmath teachers at their own schools, and they described much of their interaction as social rather than professional. The awardees, the most highly respected math teachers, have found a reference group in *mathematics educators*. They regard themselves, together with professors, administrators, and other teachers in math education, as a focal professional group. This does not mean, of course, that such teachers are more professional than others but, rather, that they have resolved ambiguities of allegiance in a particular way. It may be that other teachers locate their primary reference group in organizations like National Education Association (NEA) or the American Federation of Teachers (AFT).

Although the concept of professionalization is rife with controversial issues, there is rarely a negative comment about collegiality. The notion is interpreted differently, as we have seen, but it is hard to imagine anyone saying that collegiality is a bad thing. There are, however, phenomena to be watched. As collegiality grows or changes, an occupation may be transformed (Bucher & Strauss, 1961; Romberg, 1988); for example, increased collegiality could mean less time spent on teacher-student relationships and a greater gap between teachers and lay members of the community. Thus, even this highly regarded attribute of professions cannot be approached uncritically.

Autonomy

Autonomy can be viewed in several ways. A *profession* may secure control over its own preparation, admission to its ranks, and regulation of member behavior. A *professional subgroup*, such as a mathematics department, may have control over the content of instruction; another larger subgroup, a collective bargaining unit, for example, may exercise considerable control over the working conditions of teachers. Finally, an *individual professional* is a person who has major control in professional-client relations. For mathematics teachers, this last autonomy would imply freedom from interference by administrators or lay people in matters such as the choice of specific curricular materials, instructional methods, student diagnosis and evaluation, and classroom management. In all of these senses of autonomy, teachers fall so far short that Romberg (1988) concludes: "teachers are not now professionals" (p. 233).

The first sense of autonomy is almost synonymous with professionalization, and the current reform movement is aimed in this direction. As we have seen, a first step toward professionalization is often public control. However, if this goes too far and the profession does not secure considerable autonomy for itself, the result can be a severe loss of autonomy for the individual. As Darling-Hammond and Berry (1988) observed, "increasing public regulation of teaching has decreased the control of teachers over what is taught and how it is taught, lessening their professional responsibility and autonomy" (p. vi).

Darling-Hammond and Berry (1988) and McNeil (1987) both discuss "first" and "second wave" educational reforms in the 1980s. The two movements represent a tension between public and professional control. The first wave is associated with standardization, teacher competency testing, and a heavy emphasis on learning outcomes that are standardized. The second is represented by the reform groups already discussed and, although a host of issues have arisen in connection with their recommendations, the vision of teaching is creative, engaged, exciting, and open-ended. Such views of teaching have been advanced several times before in this century, each time to be displaced by a public effort to control and standardize (Darling-Hammond & Berry, 1988, p. vi).

The present contest is riddled with ambiguities: For example, some states that have implemented more specific standards for teacher education and licensure have also created alternative routes to certification that bypass the requirements. States have assumed responsibility for decisions formerly made by institutions of higher education. On the other hand, they have also delegated previously state-controlled decisions to local employers—for example, letting school district personnel decide whether to award continuing state licenses to beginning teachers. (Darling-Hammond & Berry, 1988, p. xiv)

Concurrently states require tougher standards for teacher education programs and allow people to enter the profession with almost no formal preparation. This ambiguity points to a deep public distrust of professional educators; the second wave is designed in part to overcome this distrust. Playing to the public admiration of traditional liberal education, the demand for higher test grades for admission, and the general faith in board-regulated credentials, the new reformers have followed earlier professional movements in aligning themselves with powerful public opinion to gain control of their own profession (Deegan, 1989; Light, 1983). Whether this can be done in a time of impending teacher shortages and massive social problems remains a serious question.

The second kind of autonomy is often developed and sustained by individual faculties. Sometimes with the encouragement of administration and sometimes in direct opposition to it, faculty cultures emerge and sustain professional working groups; mathematics departments and grade-level planning groups are among the subsets usually supported by administration. These groups can play a substantial role in curriculum policy, student evaluation, and instructional methods. When appropriate control rests at this level, individual teachers usually feel autonomous because they share in making the decisions that guide their daily practice.

When such control is removed from teachers, however, the effects can be devastating. In one study, administrative pressure for standardization led teachers "to take their resistance out on the students" (McNeil, 1987, p. 9). These teachers deliberately used the methods pushed on them to reduce their own work in retaliation for their loss of control. Everything they taught was "easily conveyed and tested" (McNeil, p. 10), but both teachers and students were, as a result, "deskilled." Even when a faculty culture is strong enough to resist initial attempts to control what goes on in classrooms, anti-professional administrators can tighten restrictions so that teachers feel thoroughly unprofessional (Carnegie Task Force, 1986; McNeil, 1987; Romberg, 1988). McNeil (1987) reports the case of faculties at magnet schools in Texas, so demoralized by standardization that, while many continue to resist through minimal accommodation, some have withdrawn from collective action into their own private classroom world while others have left teaching entirely.

Besides the sense of autonomy gained by participation in departments and grade-level groups, teachers sometimes feel more autonomous through union activities. (It is fair to say that some teachers also object to union restrictions on their teaching and feel less autonomous as a result of strong union activity.) Evidence now available suggests that union success in securing certain bread and butter items is strongly correlated with "greater professionalism" (McDonnell & Pascal, 1988, p. vi). The study just cited lists 15 "professional teaching conditions" (p. 7) including items such as controls on administration of standardized tests to students (a condition hardly ever attained), mandated class size, controls on numbers of classroom interruptions, a ban on administrators intervening to change grades, and the right of teachers to refuse assignment outside their subject or grade-level expertise. As several researchers have noted (Kerchner & Mitchell, 1986; Sykes, 1985), the professional teaching conditions are, in a sense, preliminary or foundational to professionalism; that is, they are necessary but not sufficient conditions for professionalism.

The six contract provisions so far associated with high professional teaching conditions scores are (McDonnell & Pascal, 1988, p. 9):

- Duration of school day is specified.
- Teachers are guaranteed preparation periods.
- Maximum class sizes are specified.
- Involuntary teacher transfers are permitted only under specific conditions.
- Teachers can eject disruptive students.
- Reduction-in-force (RIF) procedures are specified.

The need to negotiate such items is, itself, a sign of the pervasive lack of teachers' professional autonomy.

The final form of autonomy—that of the individual teacher in the classroom—is greatly affected by the first and second forms, an observation that underscores the importance of collegiality discussed earlier. Teachers feel relatively autonomous when they have a strong national reference group, when they participate in decision making with competent and respected departmental colleagues, and when they have control over events in their own classrooms (Conference Board of the Mathematical Sciences, 1984).

Dictation of specific learning objectives and models of instruction are particularly galling to teachers' professional identity (McNeil, 1987). Such control over their core enterprise suggests to teachers that they are not "really teachers, just workers" (McNeil, 1987, p. 101). Piles of paper work, deadly routines, check-list evaluations, and rules that must be followed rigidly all tend to make teaching less of a profession.

Given the discontent many teachers feel with current working conditions, one would suppose they would heartily endorse the new reforms. Unfortunately, here, too, many teachers feel that they have been and will be left out of making crucial decisions and that the future of teaching—their future—is in the hands of an elite minority.

SUMMARY AND CONCLUSION

Whether we look at *professionalism* or *professionalization*, mathematics teachers—in fact, all teachers—fall short of professional status. At present, teacher-led organizations have little control over standard-setting for the profession. There is no consensus on the knowledge that teachers must have, and control over teacher-knowledge does not rest with teachers. Devotion to service, to the lives and well-being of students, has become a mark of semi-professional rather than true professional life. There is little prestige or status attached to teaching. Teachers still labor in isolation, lacking the collegiality necessary for rich professional life. Finally, external regulation has severely constrained individual teacher autonomy.

There are bright spots, however, in this bleak picture. Reform movements are pressing for changes designed to professionalize teaching. It seems likely that new standards for teaching will emerge, that teacher-knowledge will be described and transformed for instruction, that new patterns of collegiality will be encouraged, and that at least some teachers will achieve a reasonable level of autonomy in their work. But the bright side of the picture has its own tarnished spots. As any occupation moves toward professionalization, important social and professional issues arise: standard-setting tends to induce exclusivity in the forms of racism, classism, and sexism; the teaching hierarchy recommended by the Holmes and Carnegie groups provides true professional status for some teachers while those who work directly with students remain beneath the professionals, the new emphasis on collegiality may lead to an increasing separation from students and lay members of the community and de-emphasis of the altruistic component of professional life is regarded by some as "selling the soul" of the profession.

What recommendations can be made? It is reasonable to suggest genuine cooperation among groups that control teacher knowledge; in the case of mathematics teachers, cooperation among mathematicians, mathematics users, and educators is essential. Courses in mathematics should be designed especially for teachers. From the evidence at hand, it is prudent for teacher organizations to continue their press for better working conditions; teachers should familiarize themselves with the correlations between essential items and professional working conditions. Given the variety of skeptical analyses available, teacher educators and others should reexamine recommendations for reform and try to lessen or remove the predicted effects that many thoughtful critics find pernicious. In particular, a great deal more attention should go to teacher-student relationships as central to the very concept of professionalism. Finally, administrators and teachers should work together to develop faculty cultures that exemplify professional relations and attributes. Professionals depend on the trust of their clients. Teachers, then, need opportunities to earn that trust, and such opportunities can only arise as teachers are entrusted by those now controlling their work.

References

American Association for the Advancement of Science. (1987). *What science is most worth knowing?* Draft report of Phase I, Project 2061. Washington, DC: Author.

Argyris, C. & Schon, D. A. (1974). *Theory in practice: Increasing professional effectiveness.* San Francisco: Jossey-Bass.

Biklen, S. K. (1987). Women in American elementary school teaching: A case study. In P. A. Schmuck (Ed.), *Women educators: Employees of schools in Western countries* (pp. 223–242). Albany: State University of New York Press.

Bird, T. (1988). *Teacher assessment and professionalization.* Paper prepared for the Task Force on Teaching as a Profession. New York: Carnegie Forum on Education and the Economy.

Bledstein, B. (1976). *The culture of professionalism.* New York: W. W. Norton.

Boston Women's Teachers Group (Freedman, S., Jackson, J., & Boles, C.). (1983). Teaching: An imperiled "profession." In L. S. Shulman & G. Sykes (Eds.), *Handbook of teaching and policy* (pp. 261–299). New York: Longman.

Bucher, R. & Strauss, A. (1961). Professions in process. *American Journal of Sociology, 66,* 325–334.

Buchmann, M. (1983). *The priority of knowledge and understanding in teaching.* Occasional paper no. 61. East Lansing, MI: Institute for Research on Teaching, Michigan State University.

Carnegie Task Force on Teaching as a Profession. (1986). *A nation prepared.* New York: Carnegie Forum on Education and the Economy.

Carraher, T. N., Carraher, D. W., & Schliemann, A. D. (1985). Mathematics in the streets and in schools. *British Journal of Developmental Psychology, 3,* 21–29.

Comer, J. P. (1988). Is "parenting" essential to good teaching? *NEA Today, 6,* 34–40.

Conference Board of the Mathematical Sciences. (1984). *New goals for mathematical sciences education.* Washington, DC: Author.

Conroy, J. (1987). Monitoring school mathematics: Teachers' perceptions of mathematics. In T. A. Romberg & D. M. Stewart (Eds.) *The monitoring of school mathematics: Background papers. Volume 3: Schooling, teachers and teaching* (pp. 75–92). Madison, WI: Wisconsin Center for Education Research.

Daly, M. (1973). *Beyond God the father.* Boston: Beacon Press.

Daly, M. (1978). *Gyn/ecology: The metaethics of radical feminism.* Boston: Beacon Press.

Darling-Hammond, L. & Berry, B. (1988). *The evolution of teacher policy.* Santa Monica, CA: Rand.

Darling-Hammond, L., Hudson, L., & Kirby, S. N., (1989). *Redesigning teacher education: Opening the door for new recruits to science and mathematics teaching.* Santa Monica, CA: Rand.

Davis, R. B., Maher, C. A., & Noddings, N. (Eds.). (1990). *Constructivist views on the teaching and learning of mathematics* (JRME Monograph Series, Vol. 4). Reston, VA: National Council of Teachers of Mathematics.

Deegan, P. (1989). *Professionalism in the curriculum field.* Unpublished doctoral dissertation, Stanford University.

Druva, C. A. & Anderson, R. D. (1983). Science teacher characteristics by teacher behaviour and by student outcome: A meta-analysis of research. *Journal of Research in Science Teaching, 20,* 467–479.

Edson, S. K. (1988). *Pushing the limits: The female administrative aspirant.* Albany: State University of New York Press.

Eisenstein, Z. (Ed.) (1979). *Capitalist patriarchy and the case for socialist feminism.* New York: Monthly Review Press.

Etzioni, A. (Ed.) (1969). *The semi-professions and their organization: Teachers, nurses, and social workers.* New York: Free Press.

Ferguson, K. (1984). *The feminist case against bureaucracy.* Philadelphia: Temple University Press.

Flexner, A. (1915). Is social work a profession? *School and Society, 1*(26), 901–911.

Freedman, S. (1988, May). *Weeding woman out of "woman's true profession"—A critical look at professionalizing teaching.* Paper presented at the annual meeting of Professors of Curriculum, Boston.

Freire, P. (1970). *Pedagogy of the oppressed* (Myra Bergman Ramos, Trans.) New York: Herder & Herder.

French, M. (1985). *Beyond power.* New York: Summit Books.

Gaut, D. (1979). *An application of the Kerr-Soltis model to the concept of caring in nursing education.* Unpublished doctoral dissertation, University of Washington, Seattle.

Gilligan, C. J. (1982). *In a different voice.* Cambridge: Harvard University Press.

Giroux, H. (1988). *Teachers as intellectuals.* Granby, MA: Bergin & Garvey.

Goodlad, J. I. (1984). *A place called school.* New York: McGraw-Hill.

Greenwood, E. (1957). Attributes of a profession. *Social Work, 2,* 45–55.

Grimshaw, J. (1986). *Philosophy and feminist thinking.* Minneapolis: University of Minnesota Press.

Hall, R. H. (1968). Professionalization and bureaucratization. *American Sociological Review, 33*(1), 92–104

Hoagland, S. (1987). Moral agency under oppression: Beyond praise and blame. *Trivia, 10,* 24–40

Holmes Group. (1986). *Tomorrow's teachers.* East Lansing: Author.

Illich, I. (1971). *Deschooling society.* New York: Harper & Row.

Jaggar, A. M. (1983). *Feminist politics and human nature.* Totowa, NJ: Rowman & Allanheld.

Kerchner, C. T. & Mitchell, D. E. (1986). Teaching reform and union reform. *The Elementary School Journal, 86*(4), 448–470.

Lampert, M. (1985). Mathematics learning in context: The voyage of the Mimi. *Journal of Mathematical Behavior, 4,* 157–167.

Lampert, M. (1988). The teacher's role in reinventing the meaning of mathematical knowing in the classroom. In M. J. Behr, C. B. Lacampagne, & M. M. Wheeler, (Eds.), *Proceedings of the Tenth Annual Conference of the North American Chapter of the Psychology of Mathematics Education Group* (PME) (pp. 433–480). DeKalb: Northern Illinois University.

Larson, M. S. (1977). *The rise of professionalism.* Berkeley: University of California Press.

Lave, J. (1988). *Cognition in practice.* New York: Cambridge University Press.

Leininger, M. (1984). *Care: The essence of nursing and health.* Thorofare, NJ: Slack.

Light, D. W. (1983). The development of professional schools in America. In K. H. Jarausch (Ed.), *The transformation of higher learning 1860–1930* (pp. 345–365). Stuttgart, Germany: Klett-Cotta.

Little, J. W. (1982). Norms of collegiality and experimentation: Workplace conditions of school success. *American Educational Research Journal, 19,* 325–340.

MacIntyre, A. (1981). *After virtue.* Notre Dame, IN: University of Notre Dame Press.

MacKinnon, C. A. (1987). *Feminism unmodified.* Cambridge: Harvard University Press.

Markowitz, G. & Rosner, D. (1979). Doctors in crisis. In S. Reverby & D. Rosner (Eds.), *Health care in America: Essays in social history* (pp. 185–205). Philadelphia: Temple University Press.

Martin, J. R. (1985) *Reclaiming a conversation*. New Haven, CT: Yale University Press.

McDonnell, L. M. & Pascal, A. (1988). *Teacher unions and educational reform*. Santa Monica, CA: Rand.

McNeil, L. M. (1987). Exit, voice, and community: Magnet teachers' responses to standardization. *Educational Policy, 1*(1), 93–113.

Menkel-Meadow, C. J. (1988). Feminist legal theory, critical legal studies, and legal education or "The fem-crits go to law school." *Journal of Legal Education, 38*(1 & 2), 61–86.

Metzger, W. P. (1987). A spectre is haunting American scholars: The spectre of "professionism." *Educational Researcher, 16*(6), 10–21.

National Science Board Commission on Precollege Education in Mathematics, Science and Technology. (1983). *Educating Americans for the 21st century*. Washington, DC: National Science Foundation.

National Science Board Task Committee on Undergraduate Science and Engineering Education. (1986). *Undergraduate science, mathematics and engineering education*. Washington, DC: National Science Foundation.

National Science Board Task Committee on Undergraduate Science and Engineering Education. (1987). *Undergraduate science, mathematics and engineering education, Volume II: Source materials*. Washington, DC: National Science Foundation.

Noddings, N. (1984). *Caring: A feminine approach to ethics and moral education*. Berkeley and Los Angeles, CA: University of California Press.

Noddings, N. (1990). Feminist critiques in the professions. In C. B. Cazden (Ed.), *Review of Research in Education*, (Vol. 16, pp. 393–424). Washington, DC: American Educational Research Association.

Noddings, N. & Greeno, J. (in press). Accelerating the mathematics performance of educationally at-risk students. In H. Levin (Ed.), *Accelerating the education of at-risk students*. New York and London: Falmer.

Offen, K. (1988). Defining feminism: A comparative historical approach. *Signs, 14*(1), 119–157.

Reverby, S. (1987). *Ordered to care*. Cambridge, England: Cambridge University Press.

Rich, A. (1979). *On lies, secrets, and silence*. New York and London: W. W. Norton.

Romberg, T. A. (Ed.) (1985). *Using research in the professional life of mathematics teachers*. Madison, WI: Wisconsin Center for Education Research.

Romberg, T. A. (1988). Can teachers be professionals? In D. A. Grouws, T. J. Cooney, & D. Jones (Eds.), *Perspectives on research on effective mathematics teaching* (Vol. I, pp. 224–244). Reston, VA: National Council of Teachers of Mathematics; Hillsdale, NJ: Lawrence Erlbaum.

Ruether, R. R. (1983) The feminist critique in religious studies. In E. Langland & W. Gove (Eds.), *A feminist perspective in the academy* (pp. 52–66). Chicago: University of Chicago Press.

Saxe, G. B. (1988). Candy selling and math learning. *Educational Researcher, 17*(6), 14–21.

Schoenfeld, A. H. (Ed.). (1987). *Cognitive science and mathematics education*. Hillsdale, NJ: Lawrence Erlbaum.

Shulman, L. S. (1986). Those who understand: Knowledge growth in teaching. *Educational Researcher, 15*(2), 4–14.

Shulman, L. S. (1987). Knowledge and teaching: Foundations of the new reform. *Harvard Educational Review, 56*, 1–22.

Shulman, L. S. (1989). The pedagogy of a discipline is the highest stage of learning. *The Holmes Group Forum, 3*(3), 4–5.

Sichel, B. A. (1988). *Moral education*. Philadelphia: Temple University Press.

Smith, D. E. (1987). *The everyday world as problematic: A feminist sociology*. Boston: Northeastern University Press.

Sockett, H. T. (1987). Has Shulman got the strategy right? *Harvard Educational Review, 57*, 208–219.

Stark, R. & Bainbridge, W. S. (1985). *The future of religion*. Berkeley, CA: University of California Press.

Starr, P. (1982). *The social transformation of American medicine*. New York: Basic Books.

Steffe, L. P., Cobb, P., & von Glasersfeld, E. (1988). *Construction of arithmetical meanings and strategies*. New York: Springer-Verlag.

Sykes, G. (1985, March). *Prospects for the reform of teaching through policy*. Paper presented at the American Educational Research Association annual meeting, Chicago.

Sykes, G. (1986). *The social consequences of standard-setting in the professions*. Paper prepared for the Task Force on Teaching as a Profession, Carnegie Forum on Education and the Economy.

Sykes, G. (1987). Teaching and professionalism: A cautionary perspective. Paper presented at Michigan State University, East Lansing.

Teachers College Record. (1987). *88*(3).

Vetterling-Braggin, M., Ellison, F., & English, J. (Eds.). (1977). *Feminism and Philosophy*. Boston: Little, Brown.

Watson, J. (1985). *Nursing: Human science and human care*. Norwalk, CT: Appleton-Century-Crofts.

Woolf, V. (1966). *Three guineas*. New York: Harcourt Brace. (Original work published 1938)

Yamashita, J. (1987). *Outstanding mathematics teachers: Their perspectives on professional development, affiliation, and rewards*. Unpublished doctoral dissertation, Stanford University.

BECOMING A MATHEMATICS TEACHER

Catherine A. Brown

VIRGINIA POLYTECHNIC INSTITUTE AND STATE UNIVERSITY

Hilda Borko

UNIVERSITY OF MARYLAND

This chapter examines the process of becoming a mathematics teacher, focusing on the individual and on changes he or she undergoes in assuming the role of a professional teacher. The goals of this chapter are to review the research related to becoming a mathematics teacher and to discuss implications of this research both for mathematics teacher education as well as for future research.

Interest in the process of becoming a teacher has increased in recent years, motivated, at least in part, by such national reports as *Tomorrow's Teachers: A Report of the Holmes Group* (The Holmes Group, 1986) and *A Nation Prepared: Teachers for the 21st Century* (Carnegie Task Force, 1986). Both of those reports indicate that existing teacher education programs are inadequate to fulfill our country's need for teachers prepared to educate students in the 21st century.

Within the mathematics education community, documents such as the National Council of Teachers of Mathematics' *Curriculum and Evaluation Standards for School Mathematics* (NCTM, 1989a), *Guidelines for the Post-Baccalaureate Education of Teachers of Mathematics* (NCTM, 1989b), and *Professional Standards for Teaching Mathematics* (NCTM, 1990) highlight the importance of mathematics teacher education and echo the concerns about the adequacy of existing teacher preparation programs. Collectively, these documents suggest current practices in teacher education will not produce teachers able to teach mathematics in the manner envisioned by the community.

Given the renewed interest in teacher preparation—in mathematics teacher preparation in particular—the need is clear for the educational research community to take stock of what we know and don't know about the process of becoming

a teacher. Thus, it should not be surprising that two recent research handbooks contain chapters addressing topics related to the focus of this chapter. The chapter by Lanier and Little in the *Third Handbook of Research on Teaching* (Wittrock, 1986) examines research on teacher education using a framework based on Schwab's (1978) four commonplaces of teaching: the teacher, the student, the curriculum, and the milieu. Lanier and Little reviewed research related to teacher educators, pre-service and in-service teachers, the teacher education curriculum, and the context of teacher education. The focus of their chapter, in contrast to this one, is on programmatic aspects of teacher education rather than on the characteristics of individuals in the process of becoming teachers.

The chapter by Brown, Cooney, and Jones (1990) in the recently published *Handbook of Research on Teacher Education* specifically addresses mathematics teacher education. Brown et al. chose to write a chapter focusing on philosophical issues in research on mathematics teacher education, rather than a more traditional literature review, in part because of the limited body of research on *mathematics* teacher education or becoming a *mathematics* teacher.

We have chosen to handle limitations in existing research differently than Brown et al. This chapter is primarily a literature review. However, it is not limited to research with a mathematical focus; instead this chapter will examine research related to becoming a teacher, whether or not it addresses mathematics specifically. When appropriate, we attempt to draw implications for becoming a mathematics teacher from the findings and conclusions of research with a generic focus.

The organization of this chapter reflects the three research traditions within which most research on becoming a teacher

The authors gratefully acknowledge the helpful comments and suggestions provided by Deborah Ball, Michigan State University, and Carolyn Maher, Rutgers University.

has been conducted: learning to teach, socialization, and adult development. Within the learning-to-teach perspective, we include research on teacher knowledge, beliefs, thinking and actions, with major emphasis on research conducted within the discipline of psychology and grounded in the assumptions of cognitive psychology. Research on teacher socialization views the teacher as a member of a professional society and examines the process of entry into that society. The discussion focuses on three major traditions in teacher socialization research—functionalist, interpretive, and critical—all of which address the question of how teachers acquire or construct their culture-specific beliefs, values, and attitudes. Research on teacher development draws upon theories of adult development and teacher development to guide the study of the professional growth of teachers. Becoming a teacher is characterized by a series of internally guided changes that can be described by predictable sequences of growth, adaptation, and transformation.

In the section of the chapter devoted to each research tradition, we present the assumptions and theoretical frameworks underlying that tradition's perspective on becoming a teacher, a number of key studies, and patterns of findings in the research. The next section contains a brief description of a study—focusing on becoming a mathematics teacher—that draws upon multiple research traditions. We end the chapter with a discussion of implications for teacher education research and practice. Before beginning the review, we present three assumptions about becoming a mathematics teacher that guided all aspects of our work on the chapter.

ASSUMPTIONS ABOUT BECOMING A MATHEMATICS TEACHER

Throughout the writing of this chapter, our work has been guided by three assumptions, the first of which is that good mathematics teaching can be identified and described. As criteria for "good mathematics teaching," we adopted the guidelines set forth in NCTM's (1990) *Professional Standards for Teaching Mathematics,* which presents a vision of what a mathematics teacher must know and be able to do in order to teach mathematics as envisioned by the NCTM (1989a) *Curriculum and Evaluation Standards.* The *Professional Standards for Teaching Mathematics* defines the mathematics teacher's major roles as:

1. creating a classroom *environment* to support teaching and learning mathematics;
2. setting goals and selecting or creating mathematical *tasks* to help students achieve these goals;
3. stimulating and managing classroom *discourse* so that students and teachers are clearer about what is being learned;
4. *analyzing* student learning, the mathematical tasks, and the environment in order to make ongoing instructional decisions (NCTM, 1990, p. 4).

Being a good mathematics teacher means fulfilling these roles in the classroom so that students learn mathematics as it is described in NCTM's (1989a) *Curriculum and Evaluation Standards for School Mathematics.* According to these standards, computational algorithms, manipulations of symbols, and mem-

orization of rules must no longer dominate school mathematics; rather, mathematical reasoning, problem solving, communication, and connections must be central. At the heart of the *Curriculum and Evaluation Standards* is the commitment to developing the mathematical literacy and power of all students. Mathematical literacy includes having an appreciation of the value and beauty of mathematics as well as being able to appraise and use mathematical information; mathematical power encompasses the ability to "explore, conjecture, and reason logically, as well as the ability to use a variety of mathematical methods effectively to solve nonroutine problems" (p. 5) and the self-confidence and disposition to do so (NCTM, 1989a).

The *Curriculum and Evaluation Standards* not only set goals for school mathematics, but also implied a significant departure from the traditional practice of mathematics teaching. The *Professional Teaching Standards* set explicit goals for teaching that are compatible with the curricular goals. For example, teachers are to provide and structure the time necessary for students to explore mathematics and grapple with ideas and problems. Teachers are to pose tasks that involve sound, significant mathematics and stimulate students to make mathematical connections. Teachers should use technology and engage in ongoing analysis of teaching and learning in order to inform their short- and long-term planning. These professional standards serve as measures of quality by which we often compare mathematics teachers in our analysis of the process of becoming a mathematics teacher.

The second assumption on which this chapter is based is that the process of becoming a mathematics teacher includes both generic and subject matter specific components. Most research on becoming a teacher has investigated generic pedagogical issues; very little has focused on mathematics-specific concerns. Nevertheless, we believe that results and conclusions from research with a generic focus can tell us a great deal about certain aspects of becoming a mathematics teacher. For example, the research results suggest reasons that novice teachers might be hesitant to implement the NCTM *Curriculum and Evaluation Standards* or that they might be reluctant to relinquish the role of "authority" in the classroom, as suggested in the NCTM *Professional Standards for Teaching Mathematics.* The chapter, therefore, includes research that addresses both generic and mathematics-specific components and characteristics of becoming a mathematics teacher.

Our third assumption is that becoming a teacher is a life-long process; that is, teachers begin to learn about teaching long before their formal teacher education begins (Wright & Tuska, 1968) and continue to learn and change throughout their careers (Sprinthall & Thies-Sprinthall, 1983). While we believe that becoming a teacher is a life-long process, our emphasis in this chapter is on research related to the student teaching and induction phases of that process. Student teaching, or the clinical field experience, is a component of virtually every pre-service teacher preparation program and is commonly considered to be an essential—if not *the* essential—element of these programs (Conant, 1963; Feiman-Nemser, 1983). Student teaching is also the most widely studied aspect of pre-service teacher education programs (Feiman-Nemser, 1983), although there continues to be much disagreement over the degree and nature of its influences on the process of learning to teach

(Zeichner, 1985a). The induction years constitute the time during which a teacher makes decisions about whether to stay in the teaching profession; during these years, novice teachers develop many of the beliefs and actions that they retain throughout their careers (Borko, 1986; McDonald, 1980).

LEARNING TO TEACH

In this section of the chapter, we examine the question of becoming a mathematics teacher from the perspective of learning to teach. Research on learning to teach has traditionally been conducted within the discipline of psychology; in recent years the underlying framework for the research has shifted from one grounded in behavioral psychology to one grounded in cognitive psychology. In keeping with this trend, our major emphasis will be on cognitive psychological research on learning to teach.

This section of the chapter is organized into three parts. In the first part, we examine the assumptions and theoretical frameworks underlying cognitive psychological research on learning to teach. The second part presents research on learning to teach. In keeping with the overall focus of the chapter, we emphasize research on the pre-service phase of becoming a teacher; however, we also briefly address studies of expert novice differences in teaching and investigations of in-service teachers' learning to teach. The section concludes with a discussion of patterns of findings from these sets of learning-to-teach studies.

A Theoretical Framework from Cognitive Psychology

Cognitive psychology is the scientific study of mental events, primarily concerned with the contents of the human mind (knowledge, beliefs) and the mental processes in which people engage (thinking, problem solving, planning). Most cognitive psychologists hold the view that knowledge and thinking are completely internal to the human mind (Putnam, Lampert, & Peterson, 1990; however, see these authors' discussion as a growing movement within cognitive science to consider cognition to be interactively situated in physical and social contexts). As Resnick (1985) notes, "The heart of cognitive psychology is the centrality given to the human mind and the treatment of thinking processes as concrete phenomena that can be studied scientifically" (p. 124).

One fundamental assumption underlying cognitive psychological theory and research is that knowledge is organized and stored in structures in the human mind. Cognitive psychologists agree that "the essence of knowledge is structure. Knowledge is not a 'basket of facts' " (R. Anderson, 1984, p. 5). The centrality of this assumption led Putnam, Lampert, and Peterson (1990) to suggest that "a basic, though overly simplified, definition of knowledge in cognitive theories" is that "knowledge *is* the cognitive structures of the individual knower" (p. 67).

A second fundamental assumption of cognitive psychology is that an individual's knowledge structures and mental representations of the world play a central role in that individual's perceptions, thoughts, and actions (Putnam, Lampert, & Peterson, 1990; Shuell, 1986). Unlike behavioral psychologists, cognitive psychologists assume that teachers' thinking is directly influenced by their knowledge. Their thinking, in turn, guides their actions in the classroom. People "continually try to understand and think about the new in terms of what they already know" (Glaser, 1984, p. 100). Learning involves making connections between new information and existing systems of knowledge; teaching should facilitate making these connections by helping students to relate new knowledge to knowledge they have already developed.

Learning to teach entails the acquisition of knowledge systems or schemata, cognitive skills such as pedagogical problem solving and decision-making, and a set of observable teaching behaviors. To understand learning to teach, one must study how these systems—and the relationships among them—develop and change with experience, as well as identify the factors that influence this change process. Given this perspective, it is not surprising that two major areas of focus in cognitive psychological literature on learning to teach are the content and processes of teacher thinking, teachers' knowledge systems and cognitive skills. In practice, these two aspects of teacher thinking are difficult to distinguish. Indeed, researchers typically study both of them simultaneously. However, at a conceptual level it is possible to make the distinction between the two. Separate theoretical models have been developed to represent teachers' knowledge systems and their cognitive processes. Below, we examine several of the models that have been developed, which underlie a number of programs of research on learning to teach. (For further information on these and related issues, see Chapter 7, Thompson, this volume, and Chapter 2, Fennema and Loef, this volume.)

Teacher Knowledge. Cognitive psychological research on teacher knowledge begins with the assumption that human knowledge is organized and structured. Schema theory (R. Anderson, 1984) provides one model for the representation and organization of knowledge: Schema is an abstract knowledge structure that summarizes information about many particular cases and the relationships among them. People store knowledge about objects and events in their experiences in schemata representing those experiences.

Shulman's (Shulman & Grossman, 1988; see also Wilson, Shulman, & Richert, 1987) theoretical model of domains of teachers' professional knowledge is particularly relevant to research on learning to teach. Shulman et al. hypothesized that teachers draw from seven domains of knowledge—sets of cognitive schemata—as they plan and implement instruction: knowledge of subject matter, pedagogical content knowledge, knowledge of other content, knowledge of the curriculum, knowledge of learners, knowledge of educational aims, and general pedagogical knowledge. Their research program (described later in the chapter) focused primarily on the first two domains. It provided the educational research community with elaborated definitions of subject matter knowledge and pedagogical content knowledge and findings about their relationship to classroom practice.

Subject matter knowledge consists of an understanding of the key facts, concepts, principles and explanatory frameworks in a discipline, known as *substantive knowledge*, as well as the rules of evidence and proof within that discipline, known as *syntactic knowledge* (Shulman & Grossman, 1988). In mathematics, for example, substantive knowledge includes mathematical facts, concepts, and computational algorithms; syntactic knowledge encompasses an understanding of the methods of mathematical proof and other forms of argument used by mathematicians.

Pedagogical content knowledge, or knowledge of a subject matter for teaching, consists of an understanding of how to represent specific subject matter topics and issues in ways appropriate to the diverse abilities and interests of learners. It includes,

...for the most regularly taught topics in one's subject area, the most useful forms of representation of those ideas, the most powerful analogies, illustrations, examples, explanations, and demonstrations—in a word, the ways of representing the subject that make it comprehensible to others.... [It] also includes an understanding of what makes the learning of specific topics easy or difficult: the conceptions and preconceptions that students of different ages and backgrounds bring with them to learning." (Shulman, 1986, p.9)

Examples within the discipline of mathematics are unlimited, but one topic that has received explicit attention by researchers is that of addition and subtraction of whole numbers (Carpenter, Fennema, Peterson, & Carey, 1988). Research on students' learning of addition and subtraction concepts (Carpenter & Moser, 1983) provided an explicit framework for analyzing word problems and the processes children use to solve them. Pedagogical content knowledge, in this case, consisted of knowledge of this framework and knowledge of how to use this framework to help first-grade students learn addition and subtraction. Pedagogical content knowledge is the one domain of knowledge unique to the teaching profession, distinguishing teachers from other content specialists such as theoretical mathematicians, research scientists, and journalists. Not surprisingly, it is relatively undeveloped in novice teachers.

Ball (1988a) developed a conceptual framework for exploring teachers' subject matter knowledge specifically in the area of mathematics. She claimed that understanding mathematics for teaching entails both knowledge *of* mathematics and knowledge *about* mathematics. Knowledge *of* mathematics is closely related to Shulman's dimension of substantive knowledge, and it includes both propositional and procedural knowledge. To teach mathematics effectively, Ball argued, individuals must have knowledge of mathematics characterized by an explicit conceptual understanding of the principles and meaning underlying mathematical procedures and by connectedness—rather than compartmentalization—of mathematical topics, rules, and definitions.

Knowledge *about* mathematics is related to Shulman's dimension of syntactic knowledge, and it includes an understanding of the nature of knowledge in the discipline—where it comes from, how it changes, how truth is established, and what it means to know and do mathematics. Ball's research related to this conceptual framework (described later in the chapter) examined the knowledge of the mathematics that prospective teachers bring to teacher education.

Teachers' Cognitive Processes. Researchers on teachers' thinking traditionally have drawn a distinction between planning, which occurs in the empty classroom, and the thought pro-

cesses in which teachers engage during classroom interaction. This distinction follows from the work of Jackson (1968), who identified three temporally distinct phases of teaching—preactive, interactive, and postactive—and is based on the assumption that the thinking teachers do during classroom interaction is qualitatively different from the thinking they do away from students (Crist, Marx, & Peterson, 1974). (Readers are referred to recent reviews by Borko and Shavelson, 1990, and Clark and Peterson, 1986, for further discussion of research on teachers' cognitive processes and for research-based models of teacher planning and interactive decision making.)

Shavelson (1986) suggested that the distinction between preactive and interactive thinking, although previously useful, now seems inappropriate. Preactive and interactive decision-making are not conceptually distinct processes; they are interrelated components of a process of developing and enacting agendas based on teaching schemata. Research on learning to teach must examine how teachers learn to translate the knowledge stored in their teaching schemata into operational plans or agendas for classroom action, and how they learn to carry out these agendas in the classroom. To research one component of this process, such as planning, and not the other, such as interactive teaching, is impossible.

Wilson and colleagues' model of pedagogical thinking and action (Wilson, Shulman, & Richert, 1987) provides support for Shavelson's view that preactive and interactive thinking are not conceptually distinct. The model describes six common components of teaching: comprehension, transformation, instruction, evaluation, reflection, and new comprehension. *Comprehension* is the process of critically understanding a set of ideas to be taught. During *transformation,* the teacher moves from a personal comprehension of the ideas to be taught to an understanding of how to facilitate students' comprehension of these ideas. *Instruction,* the process of facilitating students' comprehension, consists of a variety of teaching acts, such as organizing and managing the classroom, presenting clear explanations, and providing for student practice. *Evaluation,* or checking for student understanding, encompasses both on-line assessments of student understanding and more formal testing and evaluation procedures. *Reflection* entails evaluating one's own teaching; it is the set of processes that enables a professional to learn from experience. As a result of engaging in these processes, the teacher develops a *new comprehension* of the subject matter.

Central to this model of pedagogical thinking and action is the concept of pedagogical reasoning, the process of transforming subject matter "into forms that are pedagogically powerful and yet adaptive to the variations in ability and background presented by the students" (Shulman, 1987, p.15). Pedagogical reasoning includes the identification and selection of strategies for representing key ideas in the lesson, and the adaptation of these strategies to the characteristics of the learners. Like pedagogical content knowledge, it is unique to the profession of teaching and is relatively undeveloped in novice teachers.

The conceptual frameworks of teacher knowledge and cognitive processes described above provide lenses through which one can examine research on learning to teach. As shown in the following section of the chapter, the concepts of content knowledge, pedagogical content knowledge, and pedagogical reasoning are particularly relevant to this task.

Acquiring the Knowledge and Cognitive Processes of Teaching

Although the distinction between knowledge and thinking is fairly clear conceptually, the content and processes of teacher cognition are intertwined. Teachers draw upon their knowledge and thinking skills simultaneously, as they engage in the planning and interactive activities of teaching. Further, many investigators address questions of knowledge and thinking within a single program of research on teaching or learning to teach. Similarly, we do not separate the two sets of topics in our discussion of the research; instead, we categorize studies by the level of the participants' experience. Our major focus is on studies of learning to teach in the pre-service phase of teacher education. However, this section begins with a brief discussion of expert-novice studies of teaching and concludes with an examination of studies of in-service teachers' learning to teach. This section of the chapter highlights investigations of learning to teach mathematics, although the research reported here is not limited to these studies alone.

Studies of Expert and Novice Teachers. Some of the most extensive research on teacher cognition has been conducted within the expert-novice research tradition; these studies provide a fairly consistent set of findings and conclusions about differences in the knowledge, thinking, and actions of expert and novice teachers. The findings and conclusions of this research help us to understand the outcomes of learning to teach, although they do not completely illuminate the process. Given the indirect relevance to the central theme of this chapter, we summarize patterns in the research rather than extensively reporting individual studies.

Before summarizing patterns, however, one should identify the expert-novice studies relevant here, studies that focused specifically on mathematics teaching. It is important to note that although these studies used mathematics teaching and learning as a research site, they addressed questions about generic teaching strategies, and their findings and conclusions were not intended to be specific to this subject matter.

Berliner, Carter, and their colleagues conducted a research program to identify differences in the perception, memory, and problem solving of secondary mathematics and science teachers, varying in experience and expertise (Berliner, 1986, 1987, 1988; Carter, Cushing, Sabers, Stein, & Berliner, 1988; Carter, Sabers, Cushing, Pinnegar, & Berliner, 1987). They developed a series of laboratory tasks to simulate classroom events, and then compared the performance on these tasks of three groups of subjects: experts (secondary mathematics and science teachers with over 5 years of experience nominated by school superintendents or principals and observed by the project research staff), novices (secondary mathematics and science student teachers or first-year teachers), and postulants (individuals employed in business or industry who expressed an interest in teaching and had educational backgrounds in mathematics or science but no formal pedagogical training).

Leinhardt and colleagues worked closely with a small number of expert and novice elementary teachers in a multi-year program of research investigating expertise in mathematics teaching (Leinhardt, 1986; Leinhardt & Greeno, 1986; Lein-

hardt & Smith, 1985). Expert teachers were selected on the basis of unusual and consistent growth scores of their students in mathematics over a five-year period; novices were in their last year of pre-service teacher preparation and were identified as the best student teachers by their university supervisors. Data collection consisted primarily of classroom observations and interviews with participants about their teaching and their knowledge of mathematics. Based on these data, the researchers developed theoretical models of the cognitive processes and knowledge that provide the basis of effective teaching.

Borko and colleagues investigated the nature of pedagogical expertise by comparing the planning, teaching, and post-lesson reflections of a small number of expert and novice mathematics and science teachers (Borko & Livingston, 1989; Borko, Bellamy, Sanders, & Sanford, 1989; Livingston & Borko, 1990), "Novices" consisted of student teachers, selected on the basis of strong academic records in mathematics or science and performance in subject matter methods courses. "Experts," the cooperating teachers with whom they were placed, were recommended by building administrators and county teacher center coordinators. Participants were observed teaching for one week of instruction; they were interviewed about their instructional planning prior to each observed lesson and about their after-teaching reflections following the lesson.

Findings from these and other studies of pedagogical expertise suggest that characteristics of expertise identified in other cognitively complex domains such as physics problem solving (Chi, Feltovich, & Glaser, 1981; Larkin, McDermott, Simon, D.P., & Simon, H.A., 1980) and chess (de Groot, 1965) are shared by expert teachers as well. To summarize briefly, expert teachers displayed more pedagogical knowledge, content knowledge, and pedagogical content knowledge than did novices. Further, their conceptual systems, or cognitive schemata, for organizing and storing this knowledge are more elaborate, interconnected, and accessible than the novices' schemata. As a result, expert teachers are more efficient than novices in their processing of information during both the planning and the interactive phases of teaching (Borko & Shavelson, 1990).

Expert teachers plan more in their heads than do novices, using self-created mental scripts to guide the direction of their lessons. They also plan more quickly and efficiently than novices, because they are able to combine information from existing schemata to fit the particulars of a given lesson. Novices, in contrast, often have to develop, or at least modify and elaborate, their available schemata. Novices' schemata for pedagogical content knowledge seem particularly limited. Whereas experts' schemata include stores of powerful explanations, demonstrations, and examples for representing the subject matter to students, novices must develop these representations as part of the planning process for each lesson. Further, because their pedagogical reasoning skills are less developed than experts', this planning, itself, is often inefficiently carried out.

Experts' interactive teaching shows a greater use of instructional and management routines than novices'. These routines are mutually known by teacher and students, and their implementation requires little or no explanation or monitoring. Because they can be carried out with little conscious attention, routines reduce the cognitive demands on the teachers. Further, experts are able to reduce the cognitive load during interactive teaching by being more selective in the information to which they attend, focusing only on aspects of the classroom that are relevant to instructional decision-making. By reducing cognitive demands in these ways, experts free themselves both to focus on the important and/or dynamic features of the lesson content and to focus on information from students about how the lesson is progressing. Further, when such information suggests that a lesson is not going well, or when student questions or comments lead them away from their mental scripts, experts are better than novices at successfully changing the direction of their lesson, that is, at improvising (Borko & Livingston, 1989). Their greater success at improvisation can be explained by experts' more extensive networks of interconnected, easily accessible cognitive schemata; novices do not have as many potentially appropriate scripts for instructional and management routines to draw upon in classroom situations, nor do they have schemata for pedagogical content knowledge sufficiently developed to enable the construction of explanations or examples on the spot.

As we noted earlier, these differences in the thinking and actions of expert and novice teachers help us to understand the outcomes, but not the processes, of learning to teach. Berliner (1988) provided an initial step toward understanding these processes by suggesting a theory of pedagogical expertise development. Borrowing from a general model of skill development (Dreyfus & Dreyfus, 1986), he theorized that teachers progress through five stages in the journey toward expertise: novice, advanced beginner, competent, proficient, and expert. However, as he noted, the theory is "derived from speculations about how expert systems in the field of artificial intelligence can be created. Its application to pedagogy is still unknown" (p.8). Studies focusing on learning to teach, as discussed in the next section of this chapter, provide empirical evidence that may be used to assess the usefulness of Berliner's heuristic model.

Studies of Novice Teachers' Learning to Teach. Five research programs serve as the basis for our discussion of research on the pre-service phase of learning to teach, each selected on the following criteria: The program employs a longitudinal (multi-year) examination of learning to teach; the program specifically addresses the content and/or processes of teacher cognition; the program is conducted in an actual university or public school setting; and the program uses a combination of data-collection strategies. Only two of the studies that met these criteria focus on learning to teach mathematics, and only preliminary data are available for those investigations.

The research programs which we discuss in detail include: the National Center for Research on Teacher Education's (NCRTE) investigation of the role of teacher education in learning to teach; a longitudinal study of innovative mathematics and methods courses also conducted at the NCRTE; Borko's work, with colleagues at Virginia Tech and The University of Maryland, examining novice teachers' planning, implementation and evaluation of instruction; a project coordinated by Feiman-Nemser at Michigan State University that explored pre-service teachers' transitions to pedagogical thinking; and Shulman and colleagues' program of research on the development of subject matter knowledge, conducted at Stanford University. The discussion begins with an examination of the two programs that focus explicitly on learning to teach mathematics.

TEACHER EDUCATION AND LEARNING TO TEACH STUDY. The first study, conducted by the National Center for Research on Teacher Education at Michigan State University, investigates the role of teacher education in learning to teach (Ball, 1988b; Feiman-Nemser, 1990; National Center for Research on Teacher Education, 1988). The overall goals of the project are to identify the reasoning behind different ways of helping teachers learn to teach and to describe the impact of the different approaches on teachers' learning. The research design combines case studies of teacher education programs with longitudinal studies of teachers' learning, enabling the Center to describe the purpose and character of different teacher education programs, determine whether and how participating teachers' ideas and practices change throughout the programs, and explore the relationship between opportunities to learn and learning outcomes.

Eleven programs, each representing different types of learning opportunities, serve as the settings for the research. This sample includes pre-service, induction, in-service, and alternative route programs, thus allowing the Center to explore teachers' learning at different stages in their careers. At each site, researchers follow a sample of teachers over time, tracking changes in their knowledge, skills, and dispositions as they move through teacher education and into independent teaching. They focus on learning to teach as it relates to two subject areas—mathematics and writing—in the belief that the contrast between the two areas will increase understanding of the relationship between subject matter and teaching.

Three types of data-collection instruments are used at each site: written questionnaires, interviews, and classroom observation guides. The questionnaire is designed to measure respondents' beliefs and knowledge about teaching and learning, learners, the nature of knowledge, learning to teach, and teacher education, as well as to collect demographic data. Interviews, conducted with a smaller number of teachers and students at each site, elicit respondents' ideas and ways of thinking about teaching and learning mathematics and writing. Classroom observations serve as the primary source of data on teachers' dispositions and interactive skills in teaching mathematics and writing, as well as a source from which to make inferences about their knowledge and beliefs.

Ball (1988a, 1990) analyzed a portion of the baseline data from questionnaires and interviews of 19 elementary and secondary mathematics teaching candidates to describe the mathematical understandings they brought to teacher education as well as their ideas about teaching, learning, students, and teachers' roles. She analyzed the candidates' understanding of mathematics along the dimensions of truth value, legitimacy, and connectedness. In general, more participants were able to give correct answers than were able to explain the reasons that their answers were correct. The results showed almost no evidence in their responses of connectedness of topics in mathematics; for example, when presented with 3 division problems, all but one prospective teacher responded to each question in terms of the specific mathematical knowledge entailed—division of fractions, division by zero, or solving algebraic equations involving division. Their knowledge of division seemed founded more on memorization than conceptual understanding. They tended to search their memory for particular rules rather than focusing on the underlying meanings of the problems or relying on general meanings of division.

Based on these findings, Ball argued that prospective teachers do not have an understanding of the principles underlying mathematical procedures adequate for teaching. Also, their knowledge of mathematics is not sufficiently connected to enable them to break away from the common approach to teaching and learning math by compartmentalizing topics. She concluded that subject matter knowledge should be a central focus of teacher education programs and that much more knowledge is needed about how teachers can be helped to increase and develop their understandings of mathematics in order to teach mathematics effectively.

LEARNING TO TEACH IN INNOVATIVE MATHEMATICS AND METHODS COURSES. The second longitudinal study, also conducted at Michigan State University's NCRTE, examines the implementation of a sequence of three innovative mathematics courses and a coordinated mathematics methods course and curriculum seminar for undergraduate elementary majors. Researchers are attempting to trace the nature and extent of changes in the beliefs and knowledge about mathematics, mathematics learning, and mathematics teaching among pre-service teachers as a result of the innovative courses (Schram, Wilcox, Lanier, & Lappan, 1988; Schram & Wilcox, 1988; Schram, Wilcox, Lappan, & Lanier, 1989; Wilcox, Schram, Lappan, & Lanier, 1990).

The participants consist of 24 pre-service teachers who entered Michigan State University's Academic Learning Program in Fall, 1987. Data were collected during their participation in the two-year teacher preparation program and their first year of teaching. Researchers recorded fieldnotes and videotaped class sessions for the mathematics and mathematics methods courses, and participants completed questionnaires at several points throughout the program. In addition, a small number of participants were selected as an intensive sample—6 in the first year; 4 of the 6 in the second and third years. Data for the intensive sample include tape-recorded interviews, written assignments, exams, observations of student teaching, tape-recorded conferences with mentor teachers and fieldwork instructors, and observations and interviews during their first year of teaching.

Three research questions serve as the basis for the analysis of year 1 and year 2 data: (1) What does it mean to know mathematics? (2) How is mathematics learned? (3) What is the teacher's role in creating effective mathematical experiences for children? The researchers developed an analytic framework describing three levels corresponding to different orientations to teaching and learning mathematics. These levels provide a way to analyze changes in participants' beliefs and knowledge about mathematics, mathematics learning, and mathematics teaching as they progress through the teacher education program.

Preliminary analyses of data from year 1 and year 2 have been completed. During year 1, the prospective teachers participated in the first mathematics course in the sequence. From the first day of class, the instructor began to create a community in which students could collectively and cooperatively engage in mathematical inquiry and learn new ways to understand mathematics. Problem situations were used consistently to introduce mathematical concepts. Students worked in groups to explore and validate possible solutions for problem situations, used multiple representations to examine parallel relationships, and identified connections among mathematical ideas; the students were then encouraged to generalize their solutions and communicate results from their exploration of mathematical ideas.

Students entered the course with a traditional view of mathematics as an abstract, mechanical, and meaningless series of symbols and rules, but by the end of the course, the majority were beginning to question this view and develop a more conceptual understanding of mathematics. For their own learning, they were beginning to value a learning community organized around problem solving, group work, and discussion about mathematics. However, although the course seemed to be a powerful intervention into how they thought about mathematics for themselves, this thinking did not carry over to how they thought about mathematics for young children. Nearly half the participants still associated elementary mathematics with basics—number facts and whole-number operations, hierarchically ordered content, mastery of computational skills before problem solving, and a constrained, sequential view of proceeding from one level of learning to the next. The researchers attributed the change in participants' thinking about their own learning of mathematics to characteristics of the course. In contrast, they concluded that a single, 10-week course is insufficient to persuade teachers to resist the contextual constraints impeding conceptual approaches to teaching in elementary classrooms.

Results reported after the second year of data collection are based on the first two mathematics courses, the mathematics methods course, and student teaching. Researchers continued to see significant changes in the prospective teachers' beliefs about what it means to know mathematics and how mathematics is learned, as participants showed growing conceptual orientations to the study of mathematics. However, in the context of student teaching, researchers discovered tension between participants' views of themselves as adult learners of mathematics and their practice with young children. Further, participants differed with respect to the mathematical goals they set for children, the learning opportunities they provided, and the degree to which conceptual understanding or algorithmic thinking focused their instructional efforts. These findings provide further evidence regarding the complexity of the process of changing prospective teachers' beliefs and knowledge about mathematics, mathematics learning, and mathematics teaching.

NOVICE TEACHERS' PREACTIVE AND INTERACTIVE TEACHING. The third program of study, Borko's research with colleagues at Virginia Tech and the University of Maryland (Borko, 1985; Borko, Lalik, & Tomchin, 1987; Borko, Livingston, McCaleb, & Mauro, 1988), focused on novice teachers' thinking during the preactive and interactive phases of teaching. The research program is based on the assumption that decision-making is a central skill of teaching (Shavelson, 1976). To understand teaching, one must understand the decisions teachers make during the planning and interactive stages, as well as the factors influencing those decisions.

In one investigation (Borko, 1985; Borko, Lalik, & Tomchin, 1987), researchers followed novice elementary teachers through their final year of an undergraduate teacher preparation program. They analyzed journals kept by the student teachers throughout their year-long professional field experience, in which they described and evaluated their most successful and least successful lessons for the week. Four student teachers, two weaker and two stronger, also participated in an in-depth study of the planning and evaluation of reading lessons. Each participant was observed and videotaped teaching a reading lesson four times during her field experience. Before each observation, the student teacher was interviewed about her planning for the lesson. Following the observation, she was asked to evaluate the lesson.

Results of the study showed stronger and weaker student teachers both had similar conceptions of successful lessons. Further, these conceptions did not change over the course of the year-long field experience. All student teachers recorded similar descriptions of successful lessons in their journal entries. And, the four who participated in the observational study adopted similar approaches to modifying and supplementing basal reading lessons. However, weaker student teachers taught more lessons that they considered to be unsuccessful, and their unsuccessful lessons seemed to be more discrepant from the shared conception of successful teaching. Stronger student teachers planned in more detail, considered more aspects of the lessons in their planning, and used a problem-solving approach by which they anticipated problems and made plans to lessen or circumvent them.

In a subsequent study conducted at the University of Maryland, Borko and colleagues (Borko, Livingston, McCaleb, & Mauro, 1988) examined the planning and post-lesson reflections of 12 elementary and secondary student teachers by observing and interviewing each participant for a period of two consecutive days. Patterns in the data revealed four factors related to participants' planning and post-lesson reflections: subject matter and pedagogical knowledge, content area influence, teaching multiple sections of the same course, and student teacher responsibility and control. The importance of strong content preparation was illustrated by the relationship between subject matter knowledge and characteristics of planning and teaching. For example, participants with stronger content preparation required less time and effort for daily planning, focused more attention on planning instructional strategies and less on learning content, were more flexible in their planning and teaching, and were more confident in their teaching. Content area influenced the relative impact of the textbook, curriculum guides, and personal knowledge and experience on classroom instruction. Textbooks and curriculum guides played a more central role in planning for mathematics instruction than for instruction in other subject areas; in contrast, student teachers in theater arts and literature classes relied on personal knowledge, experience, and interests when designing classroom instruction much more than did mathematics student teachers. Secondary students made differential use of the opportunity created by teaching multiple sections of the same course, with maximum benefit being derived from using the opportunity to obtain the cooperating teacher's feedback and revise instruction between sections. Finally, student teachers who perceived themselves as responsible for, and in control of, classroom events more frequently took active steps to correct problem situations and improve their teaching.

Several factors associated with success in learning to teach were evident across the studies within this research program:

careful, detailed planning which incorporates strategies for minimizing potential problems; strong subject matter preparation to enable novice teachers to focus planning energies productively; and a perception—shared by the novice teacher, colleagues, and administrators—that the novice teacher is responsible for, and in control of, classroom events.

KNOWLEDGE USE IN LEARNING TO TEACH. The fourth research program, coordinated by Feiman-Nemser at Michigan State University, examined pre-service teachers' learning experiences during formal preparation, and the manner in which these experiences helped and hindered the transition to "pedagogical thinking" (Feiman-Nemser & Buchmann, 1986, 1987; Feiman-Nemser, 1990). Feiman-Nemser and colleagues identified the transition to pedagogical thinking as a major component of learning to teach, defining pedagogical thinking as thinking about teaching that focuses on the students' needs rather than on oneself as the teacher or on the subject matter, alone. To understand this transition, they described and analyzed what pre-service teachers learn in relation to what they are taught, both in coursework and in fieldwork. They also attempted to determine if and how the experiences that comprise pre-service teacher education add up to preparation for teaching.

Between 1982 and 1984, members of the research team followed six elementary education students through two years of undergraduate teacher preparation. The students were in two contrasting programs: the Academic Program, which emphasized theoretical and subject matter knowledge, teaching for understanding, and conceptual change, and the Decision-Making Program, which emphasized generic teaching methods and research-based decision making. Participants were interviewed each semester about what they learned in their courses and field placements and how they thought these experiences would help them in their teaching. Researchers also observed a core course in each program, recording fieldnotes about content, activities, and interactions. During student teaching, which occurred in the second year of the teacher preparation programs, researchers visited their assigned classrooms weekly to observe and document the student teacher's activities and to talk informally with the student teacher, the cooperating teacher, and the university supervisor. Formal interviews also were conducted with each participant before and after student teaching. Reports of the research focus primarily on two cases from the first year and two cases from the second.

University coursework dominated the first year of formal teacher preparation. Both participants experienced difficulty making the transition to pedagogical thinking due, at least in part, to their limited knowledge of both subject matter and pedagogy. Coursework in the teacher preparation programs did not remedy the participants' subject matter knowledge deficits. Nor did it provide enough "teacher education"—instruction, supervision, practice, and reflection—to sufficiently enhance their pedagogical knowledge. Both participants attempted to compensate for their limited knowledge by relying on their own schooling, textbooks, and practical experience in learning to teach.

The theme of transition to pedagogical thinking emerged as a significant factor in the second year case studies, which focused on two participants' student teaching experiences. Both student teachers experienced difficulty in recognizing the differences between going through the motions of teaching, such as checking homework, talking at the board, and giving assignments, and connecting these activities to what students should be learning over time. Often they did not capitalize on potential learning opportunities present in the classroom activities they orchestrated.

Based on their research findings, Feiman-Nemser and colleagues concluded that, without guidance, pre-service teachers find it difficult to make the transition to pedagogical thinking. By themselves, they can rarely see beyond what they want or need to do, or what the setting requires. They cannot be expected to analyze the knowledge and beliefs they draw upon in making instructional decisions, or their reasons for these decisions, while trying to cope with the demands of the classroom. The researchers suggested that teacher educators—in this study the university personnel and cooperating teachers—take an active role in guiding pre-service teachers' pedagogical thinking and actions, perhaps by demonstrating teaching actions and verbalizing pedagogical thinking, or by stimulating pre-service teachers to analyze and discuss their actions and decisions.

KNOWLEDGE GROWTH IN TEACHING. The fifth research project, conducted by Shulman and colleagues at Stanford University, focused primarily on "how teachers learn to transform their own understanding of subject matter into representations and forms of presentation that make sense to students" (Shulman & Grossman, 1988, p. 1). The participants were 20 student teachers enrolled in 3 fifth-year teacher preparation programs in California during the 1984-85 academic year, representing the subject areas of English, social studies, biology, and mathematics. Twelve of the participants were followed into their first year of full-time teaching.

During the first year of the investigation, researchers conducted a series of interviews that focused on participants' intellectual histories, general knowledge of their subject area, knowledge of the specific courses they were assigned to teach, general pedagogical knowledge, and pedagogical knowledge of their content area. In some of these interviews, they used structured tasks to elicit knowledge about subject matter and its relationship to the knowledge and practice of teaching. They also conducted a series of planning-observation-reflection data collection cycles. Each cycle included a planning think-aloud and a planning interview focusing on knowledge of content and on what the participant wanted students to learn about that content, and an interview about the observed teaching to detect changes in the participant's knowledge of subject matter and pedagogy as well as the perceived sources of those changes. Participants followed into their first year of teaching were observed during teaching and interviewed about the observed lessons. Interviews prior to the observations stressed the teachers' knowledge of content to be taught, their preparation for teaching the content, their expectations regarding potential student difficulties, and their ideas about adapting content for different types of learners. After each observation, teachers were asked to reflect about the lesson, student performance, and their own teaching.

Based on the first-year data, Shulman and colleagues developed the two theoretical frameworks presented earlier in this chapter to describe the domains of teachers' professional knowledge and the process of pedagogical reasoning (Shulman, 1987; Shulman & Grossman, 1988; Wilson, Shulman, & Richert, 1987). They used these frameworks to interpret patterns in participants' teaching and learning to teach.

Teachers' subject matter knowledge affected classroom instruction in a number of ways. For example, "prior subject matter knowledge and background in a content area affect the ways in which teachers select and structure content for teaching, choose activities and assignments for students, and use textbooks and other curriculum materials" (Shulman & Grossman, 1988, p. 12). The depth and character of participants' subject matter knowledge influenced both the style and substance of instruction. Greater subject matter knowledge enabled teachers to connect topics within a subject and to provide conceptual explanations, as opposed to purely algorithmic ones. In mathematics, participants with greater subject matter knowledge were more likely to see problem solving as central to mathematics instruction and to emphasize a conceptual approach to teaching. In comparison to teachers with much less mathematical knowledge, they gave more explanations about why certain procedures work or do not work; conveyed to students the nature of mathematics by addressing the relationships among concepts and showing applications of the material studied; presented material in a more abstract form; and engaged the students in more problem solving activities (Steinberg, Haymore, & Marks, 1985). Similarly, science teachers with greater subject matter knowledge were more likely to stress the importance of scientific inquiry in their teaching (Baxter, Richert, & Saylor, 1985). Across the subject areas, participants were more likely to welcome student questions and engage in open-ended discussions when teaching familiar topics (Grossman, 1987; see also Carlsen, 1987).

Almost all of the participants experienced growth in their subject matter knowledge as a result of teaching and preparing to teach, as they often needed to review content as they prepared to teach. Further, because the match between a college major and the secondary school curriculum is far from perfect in fields such as science and social studies, participants in those certification areas expanded their subject matter knowledge as they prepared to teach material they had never studied.

Participants also began to develop pedagogical content knowledge—knowledge of pedagogically powerful ways of representing the content to students—as they attempted to communicate their own understanding to students. As Shulman and Grossman (1988) explained, "Their knowledge of subject matter became infused with their knowledge of students, their knowledge of teaching, and their knowledge of curriculum, and resulted in knowledge of their subjects that was specific to the task of teaching" (p.19). For example, participants developed alternative frameworks for thinking about teaching particular subjects (for example, teaching mathematics as problem solving). They also acquired knowledge about students' understanding and misconceptions of a subject as well as curricular guidelines and materials available for teaching particular subjects. As they searched for ways to present the content of their disciplines for students, they sometimes generated representations or transformations to facilitate students' developing understandings. These representations took a variety of forms, including analogies, metaphors, demonstrations, examples, and illustrations.

In addition to developing this new type of knowledge, the student teachers developed *pedagogical reasoning skills*, new ways of thinking that helped them to generate subject matter transformations. Shulman and colleagues characterized the student teachers' struggles to find ways of presenting content to their students as attempts to use their developing pedagogical reasoning skills and pedagogical knowledge to transform their own knowledge into "learnable" forms.

Participants acquired their pedagogical content knowledge and pedagogical reasoning skills from a number of different sources. Their initial ideas about teaching were based on recollections of the ways in which their own high school teachers taught their subjects. In their teacher preparation programs, they learned about teaching through both coursework and fieldwork. Participants credited coursework more frequently as influencing their general pedagogical knowledge and conceptions of the subject matter; they mentioned fieldwork more frequently though as influencing their curriculum knowledge and knowledge of student understanding (Grossman & Richert, 1986).

PATTERNS IN PRE-SERVICE STUDIES OF LEARNING TO TEACH MATHEMATICS. Several patterns are apparent across this diverse set of studies of pre-service teachers' learning-to-teach experiences. The studies provide evidence of limitations in pre-service teachers' pedagogical content knowledge and pedagogical reasoning skills, the components of knowledge and thinking that are unique to the teaching profession. In addition, they illustrate the interconnections among knowledge, thinking, and classroom actions. For example, limitations in subject matter knowledge and pedagogical content knowledge were associated with a difficulty in making the transition to pedagogical thinking, an inability to connect topics during classroom instruction, and a focus on procedural rather than conceptual understanding. These patterns in novice teachers' learning-to-teach experiences will be discussed more fully later in the chapter.

In-service Studies of Learning to Teach. Learning to teach is a life-long process, of which pre-service preparation is just one phase (Feiman-Nemser, 1983). Ideally, teachers emerge from this phase as strong novices, equipped with the skills and dispositions to facilitate continuation of the learning process. In-service educational programs should help them to become "competent" teachers and should help some to attain the status of expert (Berliner, 1988).

In this part of the chapter, we describe investigations of two in-service programs designed to foster continued teacher learning. Because in-service programs are not the major focus of the chapter, the discussion is limited to research on programs that specifically address learning to teach mathematics and emphasize teacher knowledge and/or thinking skills.

COGNITIVELY GUIDED INSTRUCTION. The Cognitively Guided Instruction (CGI) project is a multi-year, multi-phased program of curriculum development and research directed by Carpenter, Fennema, and Peterson. The project was designed to provide primary-grade teachers with recent findings from cognitive science research in children's mathematics learning so that the teachers might alter their mathematics instruction to encourage children's problem solving. In the first phase of the project, the researchers attempted to understand and describe the relationships among first-grade teachers' pedagogical content beliefs, pedagogical content knowledge, their reported approaches to teaching, and students' achievements in mathematics (Carpenter, Fennema, Peterson, & Carey, 1988; Peterson, Fennema, Carpenter, & Loef, 1989). The second phase was an experimental investigation of the relationships among teachers' knowledge and beliefs about students' mathematical knowledge, teachers' mathematics instruction in the classroom, and their students' achievement in mathematics computation and problem solving (Carpenter, Fennema, Peterson, Chiang, & Loef, 1989; Peterson, Carpenter, & Fennema, 1989). Because the in-service workshop for teachers was part of the second phase, that phase will be described in detail.

The major question addressed in the second phase of the project was whether knowledge derived from classroom-based research on teaching and from laboratory-based research on learning would improve teachers' classroom instruction and children's achievement. Forty first-grade teachers participated. Twenty were randomly assigned to the experimental group and participated in a four-week summer workshop on Cognitively Guided Instruction; the remaining 20 teachers—the control group—participated in a half-day workshop on problem-solving. Teachers and students were observed for 16 days throughout the school year. Teachers' knowledge and their beliefs about student knowledge were assessed using questionnaires and interviews. Students completed standardized and researcher-constructed tests of achievement in computation and problem solving, as well as interviews to assess their problem-solving strategies, confidence, and reported understanding of mathematics.

The workshop for teachers in the experimental group was designed in accord with assumptions about teacher cognition and student cognition. The content of the workshop was based on research on children's solutions of addition and subtraction problems. The format was based on the assumption that teachers are thoughtful professionals who construct their own knowledge and understanding. Teachers were provided with access to knowledge about types of addition and subtraction problems and the learning and development of addition and subtraction concepts and problem-solving skills in young children. They worked both together and separately to design programs of instruction based on that information. However, they were not *trained* in specific techniques for altering their teaching or their curriculum.

Despite the fact that instructional practices were not prescribed in the CGI workshop, experimental teachers taught problem-solving significantly more and number facts significantly less than control teachers. Experimental teachers also posed problems to students more often, encouraged students to use more of a variety of problem-solving strategies, and listened more frequently to the processes used by students to solve problems. They believed more than did control teachers that instruction should be built on existing knowledge, and they knew more about students' strategies for both number facts and problem solving. Experimental students exceeded control students in number fact knowledge and some tests of problem-solving. They were also more confident of their abilities to solve mathematics problems and reported a significantly greater understanding of the mathematics than did control students.

Some aspects of teaching and learning did not fit these general patterns, though. For example, CGI and control teachers did not differ significantly in their knowledge of students' problem-solving abilities, as assessed by their predictions of students' performance on complex addition and subtraction word problems and on advanced problems. Also, both groups of teachers increased substantially in their agreement with the idea that children construct mathematical knowledge, and, at post-test, the two groups did not differ significantly in their agreement with this view. With respect to student achievement, CGI and control classes showed similar achievement on the computation test and solving of advanced problems.

A correlational analysis, using only data from the experimental teachers, provided additional insights into the relationships among teachers' pedagogical content knowledge, mathematics instruction, and students' achievement. Teachers' knowledge of students' problem-solving ability showed a significant positive relationship to students' mathematics problem-solving achievement as well as to teachers' questioning of students about their problem-solving processes and listening to students solving problems. This research suggests that experienced teachers' pedagogical content knowledge and pedagogical content beliefs can be affected by in-service workshops, and that such changes are associated with changes in their classroom instruction and increased student understanding and problem solving in mathematics.

THE SECOND GRADE CLASSROOM TEACHING PROJECT. The Second Grade Classroom Teaching project conducted by Cobb, Wood, and Yackel, examined second graders' construction of mathematical knowledge in classroom instructional settings. A major goal of the project was to develop a form of educational practice compatible with constructivist theory and viable within the constraints of a public school system. One component of the project consisted of a teaching experiment in which the researchers worked with one teacher to implement cognitively-based instructional activities in her classroom (Cobb, Yackel, & Wood, 1988; Cobb, Wood, & Yackel, 1990; Wood, Cobb, & Yackel, 1990). In a subsequent component, the researchers designed and implemented an in-service program, based in part on what they learned from the classroom teaching experiment, to help teachers develop classroom practices compatible with constructivism (Cobb, Wood, & Yackel, 1990; Cobb et al., 1991).

In the spring prior to the classroom teaching experiment the researchers met with the project teacher to discuss the cognitive models of children's mathematical learning. However, they soon realized that the teacher was not attempting to understand children's mathematical activities or thinking processes. Only when she began to see her current instructional practices as problematic did she develop a genuine collaborative relationship with the researchers and become motivated to modify her classroom practice. At that point, the teacher began trying to develop ways of understanding her interactions with students, grounded in her own pragmatic knowledge rather than in the researchers' formal cognitive models. As a result, her teaching,

as well as her conceptualization of her role as a teacher and the children's roles as learners, changed in ways that were more compatible with constructivist theory.

Based on this experience, the researchers drew an analogy between teachers' construction of pedagogical knowledge and students' construction of mathematical knowledge. They reasoned that to help teachers develop classroom practices compatible with a constructivist approach to mathematics learning, they must help them to see aspects of their text-based instruction as problematic and, then, encourage them to resolve these problematic elements by reorganizing their pedagogical content knowledge and beliefs and modifying their classroom practice. Guided by this awareness, the researchers developed an in-service program designed to encourage teacher autonomy and help teachers develop forms of practice that they could rationalize and justify. The program consists of several components, including a one-week summer institute, researcher visits to participating teachers' classrooms at least once every two weeks during the first year of their participation, weekly small group meetings in which teachers discuss their experiences, and four after-school work sessions during the school year. The goal of the summer institute is to create situations in which teachers begin to question their assumptions about both the mathematical knowledge children can construct and their own current instructional practices. During classroom visits, project staff address teachers' pragmatic concerns and help them to see certain aspects of their practice outside of their awareness as problematic. Working sessions, initiated at the teachers' request, focus on methods children use to solve arithmetic problems.

Eighteen teachers participated in the first round of the in-service program; a subset of 10 participated in an assessment of the project. Teacher and student data from the 10 classrooms were compared to data from eight non-project classrooms at the same schools. Project teachers' pedagogical content beliefs were more compatible with constructivism than were those of their non-project colleagues. Project students were superior in conceptual understanding at the end of the school year, although computational proficiency was comparable across groups. Project students also held stronger beliefs about the importance of understanding and collaborating, and they attributed less importance to conforming to the solution methods of others, being competitive, and looking for task-extrinsic reasons for success.

PATTERNS IN IN-SERVICE TEACHERS' LEARNING TO TEACH. Findings from both research projects indicate that in-service programs can affect teachers' pedagogical content knowledge and beliefs, and that these changes in knowledge and beliefs are associated with changes in classroom practices and student achievement. Although such findings must certainly be viewed with optimism, some caution is also in order: Conclusions about teacher change in the second-grade study are limited by the fact that systematic data were not collected about teachers' knowledge, their thinking processes, or the nature of their classroom practices. Further, since all teachers in the experimental group were volunteers, initial beliefs of project and non-project teachers may have differed. In the CGI study, experimental teachers and students did not outperform control teachers and students on several measures of belief and knowledge.

These patterns of mixed findings are supported by data from the Missouri Mathematics Effectiveness Project as well as by investigations in other content domains. The Missouri Mathematics Effectiveness Project was a multi-year, multi-phase program of research in which Good, Grouws, and colleagues developed and tested a model for whole-class instruction in mathematics (Good & Grouws, 1979; Good, Grouws, & Ebmeier, 1983). The model and associated research focused on effective mathematics teaching practices, instead of on teacher knowledge or thought processes. In an experimental investigation, teachers in the experimental group learned about the model via a manual and two 90-minute workshops. These teachers implemented more of the program's instructional elements than did control teachers, and achievement gains of the experimental students were substantially higher than gains of control students. The one major exception to this pattern was the difficulty teachers in the experimental group had in implementing the requested allocation of time to the development portion of the lesson. Good and Grouws (1979) speculated one reason for the low level of implementation might be that "teachers might not have had the knowledge base necessary to focus on development for relatively long periods of time" (p. 358). The manual and workshops that comprised the in-service program might not have been extensive enough to enable the growth in content knowledge or pedagogical content knowledge necessary to alter significantly teachers' presentations of new skills and concepts.

Two research programs in science education revealed similar limitations in teachers' knowledge growth following participation in in-service programs. Krajcik and Layman (1989) found that middle school science teachers who participated in a three-week summer workshop on the teaching of heat energy and temperature concepts showed some improvement in their understanding of these concepts (content knowledge) and ideas about how to teach them (pedagogical content knowledge). However, many participants still exhibited alternative or weak concepts, particularly regarding how to teach the concept of heat energy. Similarly, Smith and Neale (1989) examined the subject matter knowledge and pedagogical content knowledge of participants in a four-week summer program on teaching for conceptual change in primary science classes. Improvements in teachers' content knowledge were fairly dramatic; however, changes in their pedagogical content knowledge were more varied. Although most of the teachers increased their use of conceptual teaching strategies, powerful examples, metaphors, and other ways of representing content were still missing from most of their lessons.

Overall, these studies suggest that experienced teachers can increase their content knowledge and pedagogical content knowledge of mathematical and scientific concepts through participation in in-service programs. However, the process of helping teachers improve in these areas requires careful attention over an extended period of time.

Patterns and Implications in Learning to Teach

This section of the chapter concludes with a discussion of patterns in the learning-to-teach process, evident in the diverse set of research programs presented. In keeping with the focus of the chapter, we emphasize patterns derived from research on pre-service teachers' learning to teach, noting how they are supported and/or extended by research both on expert-novice distinctions and on in-service teachers' learning experiences. The patterns we discuss are organized around three issues: 1) the impact of content knowledge on teaching; 2) novices' learning of pedagogical content knowledge; and 3) difficulties in learning pedagogical reasoning skills. These issues relate to the theoretical framework from cognitive psychology, in which content knowledge and pedagogical content knowledge are identified as key domains of teachers' professional knowledge, while pedagogical reasoning skills are identified as a central process in pedagogical thinking and action. In addition, pedagogical content knowledge and pedagogical reasoning are seen as aspects of cognition unique to the teaching profession. The fact that these issues emerge in an analysis of research on learning to teach attests to the power of the theoretical framework as a tool for helping us to understand the process of learning to teach.

Content Knowledge. Findings from several of the research programs confirm the importance of strong preparation in one's content area prior to student teaching. Without adequate content knowledge, student teachers spend much of their limited planning time learning content, rather than planning how to present the content to facilitate the student's understanding. Student teachers with strong content preparation are more likely to be flexible in their teaching and responsive to students' needs, and to provide conceptual explanations, instead of purely procedural ones. They also tend to place greater emphasis on the organization and connectedness of knowledge within the discipline and less on the provision of specific information. Student teachers without adequate content knowledge are likely to lack confidence in their ability to teach well. Clearly, a teacher must have strong content knowledge to teach in a manner consistent with the conception of good teaching proposed by NCTM and endorsed in this chapter.

Unfortunately, the research also suggests that prospective teachers often do not have adequate content knowledge when they begin student teaching. What constitutes adequate knowledge? C. Anderson (1989) uses the concepts of *high* and *low* literacy, borrowed from Bereiter and Scardamalia (1987) to address this question. He contrasts "high literacy, necessary for sustained analytical efforts or the production of original work, with low literacy, the skills and knowledge necessary for basic functioning in our society" (p. 89), and suggests that teachers should be "adequately prepared to help their students attain an achievable form of high literacy" (p. 91). To accomplish this goal, they, themselves, must be highly literate in the subjects they teach. They must be able to "think deeply and flexibly about the relationships among facts, concepts, and procedures that constitute the structure of knowledge in the discipline, about the many functions that the content to be taught might have in the classroom and outside, and about the many different forms or levels of understanding that students exhibit as they develop disciplinary knowledge" (p. 103). He suggests that most student teachers have "highly developed low literacy," and that even the strongest have only "latent high literacy." This analysis is supported by Ball's (1990) findings regarding prospective teachers' mathematical understandings and her conclusions about the inadequacy of their knowledge for teaching.

Anderson further suggests that neither current academic nor current professional educational coursework are particularly good at helping prospective teachers develop high literacy in their content areas. Not surprisingly, he, like Ball, recommends that improvements in both should be a focus of teacher education reforms; we concur.

Pedagogical Content Knowledge. Because pedagogical content knowledge is unique to the profession of teaching, we expect it to be relatively undeveloped in novice teachers, and thus to be a primary focus of their educational experiences. The research examined confirms these expectations: In a number of studies, novice teachers showed evidence of growth in pedagogical content knowledge as a result of teaching and preparing to teach. Observations and interviews also indicated that the growth process was not always an easy one. Student teachers struggled, and sometimes failed, to come up with powerful means of representing subject areas to students. Further, their efforts were often time-consuming and inefficient.

Pedagogical content knowledge also emerged as an important explanatory construct in studies of expert-novice distinctions and in-service teachers' learning experiences. Many differences in the teaching skills of experts and novices can be explained by the assumption that novices' schemata for pedagogical content knowledge are less elaborate, interconnected, and accessible than those of experts. Further, studies of in-service programs for experienced teachers revealed that these teachers, too, are likely to encounter difficulty in their attempts to expand their pedagogical content knowledge base. In fact, growth in content knowledge appears to be easier for experienced teachers than growth in pedagogical content knowledge. These patterns of findings support the recommendation that acquisition of pedagogical content knowledge be a central priority in pre-service teacher education programs, and that it continue to receive attention in in-service programs, available to teachers throughout their careers.

Pedagogical Reasoning. Pedagogical reasoning, the process of transforming content knowledge into forms that are pedagogically powerful and adaptive to particular groups of students, is at the core of successful teaching. Like pedagogical content knowledge, it is relatively undeveloped in novice teachers. Further, research evidence suggests that one of the most difficult aspects of learning to teach is making the transition from a personal orientation to a discipline to thinking about how to organize and represent the content of that discipline to facilitate student understanding. Pre-service teachers in several of the research programs experienced difficulty in making this transition to pedagogical reasoning. Perhaps even more troublesome, evidence suggests that, in several instances, these teachers did not even realize they were having problems.

Expert-novice research reveals that differences in pedagogical reasoning skills constitute another factor distinguishing expert from novice teachers. The work by Cobb and his colleagues suggests that for teachers to develop classroom practices compatible with the NCTM conception of good teaching, teacher education programs must help teachers to see current practice as problematic and to reason in new ways about mathematics instruction. It seems clear, therefore, that pedagogical reasoning should be a central focus of teacher preparation programs.

TEACHER SOCIALIZATION

This section of the chapter discusses research on the socialization of teachers. We first review the three perspectives from which most of the research on teacher socialization has been conducted. We then present studies representing each of these traditions. This section ends with a summary of patterns found in the results of teacher socialization research.

Research on teacher socialization views the teacher as a member of a professional culture; the process of becoming a teacher, then, is viewed as becoming a member of that culture. Thus, research on teacher socialization attempts to answer the question "How do teachers acquire or construct the beliefs, values, and attitudes of the culture of teachers?" Studies of teacher socialization posit external forces, persons, and mechanisms that influence teachers as they become participants in this culture. The three main traditions in teacher socialization research have been identified by Zeichner and Gore (1990): functionalist, interpretive, and critical theory.

Many studies of teacher socialization assume a functionalist model, stressing the notion that socialization fits the individual to society. The functionalist model assumes a socialization process with a particular end, that end being a person fully matured and capable of taking his or her place in the society and continuing the status quo. Functionalist accounts of teacher socialization portray novice teachers as relatively passive entities constantly giving way to socializing forces in the schools (Lacey, 1977).

Recently, studies of teacher socialization have begun to challenge the deterministic character of the functionalist view and to emphasize the role of the individual teacher in the construction of professional identity. From this interpretive perspective, teacher socialization is seen as involving a constant interplay between choice and constraint, between individual and institutional factors. Individual teachers are not merely passive recipients in the process of socialization; rather, they play active roles in their own socialization into the profession.

An emergent view of socialization is that of critical theory. The critical approach seeks to bring issues such as class, gender, race, justice, and equality to the forefront. The goal of research conducted within the critical approach is social transformation. Even more than from an interpretive perspective, "People must be considered as both the creators and the products of the social situations in which they live" (Bolster, 1983, p. 303).

In this section we discuss selected studies conducted within each of these three research traditions. Our discussion is not exhaustive; rather we have chosen studies that represent the questions asked, methodologies used, and results found in each perspective. In keeping with our interest in the individual becoming a teacher, we review studies focusing on teachers' experiences of socialization, rather than on the institutions or cultures into which they are being socialized. Our interest in the individual, coupled with a belief that teachers are active participants in the process of becoming a teacher, led us to emphasize socialization research conducted within the interpretive and critical perspectives, although our review of teacher socialization from the critical perspective is limited due to the scarcity of studies.

Teacher Socialization from a Functionalist Perspective

Studies of teacher socialization from a functionalist perspective attempt to explain the status quo in predictive, generalizable ways. Because teachers are viewed as passive, and external influences as deterministic, research from this perspective often has focused on describing external factors and their influence on novice teachers. We discuss studies that represent two primary areas of concern to researchers in this tradition: (1) the influence of experiences prior to formal teacher education; and (2) the influences of elements of school practice on novice teachers. These studies were selected because, although they attempt to generalize to the population of teachers without much regard for context or individual differences, they focus on experiences of the novice teachers rather than on those of the institutions acting on the novices.

Experiences Prior to Formal Teacher Education. It is widely accepted that pre-service teachers do not enter a teacher education program *tabulae rasae*, but that they bring with them prior knowledge, beliefs, and attitudes about mathematics and teaching they develop during their years as students. Lortie (1975) conducted a socialization study of teachers in six school districts that documented this influence of teachers' years as students. Analyzing data gathered by intensive interviews, observations, and surveys in these six school districts, and drawing on data from other studies, he found that one of the most dominant influences on teachers was what he termed an *apprenticeship of observation.* He argued that "there are ways in which being a student is like serving an apprenticeship in teaching; students have protracted face-to-face and consequential interactions with established teachers" (p. 61). During this apprenticeship of observation, Lortie suggested, teaching models are internalized and the student learns to "take the role" of the teacher.

In their responses to a question asking them to describe an outstanding teacher they had, teachers in Lortie's study volunteered information about how their current work was affected by the teaching they received. Forty-two percent of the teachers interviewed went beyond a simple description and connected their own teaching practices with those of the teacher they were describing. This research confirms that those planning to teach form definite conceptions about the nature of teaching long before they actually enter the role themselves, or even before they formally begin preparation for it. However, Lortie argued, the apprenticeship-of-observation does not instill a sense of the problematics of teaching because of the limits of the vantage points of students. Thus, teachers enter the profession with simplistic expectations of their roles.

Petty and Hogben (1980) conducted a two-year study of experienced elementary school teachers, student teachers, and students not majoring in education. They used a semantic dif-

ferential instrument to gather data about the bureaucratic and professional orientations of those studied and to allow the subjects to describe the meaning of teaching in a fairly comprehensive manner. Their results indicated "factor structures for groups of respondents were similar, dominated by immediate practical concerns, and remained largely uniform throughout the two years of this study" (Petty & Hogben, p. 54). These results, they argued, suggest common attitudes toward teaching held by the three groups studied. This supports Lortie's contention that a great deal of socialization occurs before formal teacher education and is largely unaffected by teacher education or teaching experience.

Wright and Tuska (1968) explored the impact of some of a child's significant others (mother, father, teacher) on the child's decision to become a teacher and on actual teaching practice. Data were collected on 508 women using a Teaching Attitudes Questionnaire administered three times during the women's professional development: before student teaching, after student teaching, and toward the end of the first year of teaching. The questionnaire was designed to record each subject's conception of self and her role as teacher. The results of the study indicate that early experiences exert a powerful influence on the images future teachers have of teachers and teaching, and that these images continue to influence teachers even once they have assumed the role of teacher themselves. The research suggests that unless formal education can change these preexisting images, teachers will employ methods similar to the methods their own teachers used.

Findings such as these are troublesome when we consider current efforts to reform mathematics teaching in directions radically different from past practices. The research reviewed here suggests that novice mathematics teachers who have experienced years of traditional instruction may be socialized to this manner of teaching; therefore, it may be very difficult to design experiences for novice—and experienced—teachers that will help them develop different conceptualizations of mathematics teaching.

Influences of Elements of School Practice. A number of studies from the functionalist perspective have examined the impact of experiences in classrooms on the socialization process. Often these studies investigated changes in attitudes of student teachers and beginning teachers; these changes constitute a shift from the progressive, liberal attitudes formed during university or college education toward the more conservative (realistic) attitudes of practicing teachers.

Hoy's (1969) study of Pupil Control Ideology (PCI) is typical of this research. *Control ideology* is conceptualized along a continuum, ranging from *humanistic* to *custodial.* A humanistic orientation is indicated by "an accepting trustful view of students and confidence in their ability to be self-disciplining and responsible." A custodial orientation is characterized by "stress on the maintenance of order; distrust of students; and a punitive, moralistic orientation" (Hoy, 1969, p. 258). One hundred seventy-five elementary and secondary teachers completed a 20-item instrument, the Pupil Control Ideology Form (PCI Form) at four points in time: prior to student teaching, at the conclusion of eight weeks of student teaching, one year after student teaching, and two years after student teaching.

Analysis of the data indicated that

regardless of teaching level, elementary or secondary, the pupil control ideology of beginning teachers became progressively and significantly more custodial, both after student teaching and after the first year of teaching experience. However, pupil control ideology remained virtually unchanged as teachers acquired their second year of teaching experience. (p. 260)

Other studies of attitude change (McArthur, 1978; Iannaccone, 1963) also indicate that teachers shift toward more traditional, utilitarian views of teaching as they move out of teacher education and into the classroom.

From a functionalist perspective, changes in novice teachers are assumed to be the result of some external influence. Most often, cooperating teachers and other colleagues have been viewed as the primary source of influence, although this is not well documented by research. A number of studies (Copeland, 1978, 1980; Iannaccone, 1963) suggest that the attitudes and behaviors of student teachers do, in fact, shift toward those of their cooperating teachers in the course of the student teaching experience. However, the causes of the shift and the extent to which this shift occurs are unclear.

Copeland (1980) conducted an experimental study exploring the processes which mediate the relationship between cooperating teacher and student teacher classroom actions. He designed the study to test the effect of the classroom ecological system on student teachers. He identified cooperating teachers who were either "high" or "low" in the skill of "asking probing questions" and assigned student teachers to these cooperating teachers in a rather complicated way. Eight student teachers were assigned to observe a "high" cooperating teacher and eight student teachers were assigned to observe a "low" cooperating teacher for three weeks. At the end of these three weeks, four of the student teachers in each group were assigned to teach in the classroom of a cooperating teacher with an opposite skill level rating. The other four were assigned to cooperating teachers with the same skill level as the teacher they had observed.

Copeland found that not only did the behavior of the cooperating teacher relate to student teachers' use of particular skills, but the degree of skill usage was also a function of the classroom ecosystem. That is, if students in a classroom had been exposed to, and were receptive to, a particular level of skill of asking probing questions, they would reinforce the student teacher's use of this skill. Copeland argued that the cooperating teachers' use of particular techniques in the classroom may also be influenced by the ecology of the classroom—that is, molded by students' responses. He concluded that the shifts by student teachers toward teaching styles similar to those of cooperating teachers were due largely to the ecology of the classroom, not to the direct influence of the cooperating teachers (see Doyle & Ponder, 1975, for a discussion of the ecology of the classroom). Other studies have documented the influence of student behavior on novice teachers' behavior (Fiedler, 1975; Klein, 1971), supporting Copeland's findings that student behavior is a strong socializing agent.

Much of the research on changes in novice teachers holds the underlying assumption that formal teacher education influences pre-service teachers in liberal ways. This assumption has been challenged by studies that suggest that teacher education programs, in fact, act as conservative forces (Tabachnick, Popkewitz, and Zeichner, 1979–1980; Bartholomew, 1976). Bartholomew's (1976) analysis of teacher education in the United Kingdom led him to conclude that the view of the college and university as a liberal influence is produced by comparing *attitudes* in teacher education with *practice* in the classroom:

If we compare attitude or theory found in the college with practice in the college we find the same gap between liberal theory and conservative practice that can be found between the college and the school. Because the college is only seen in terms of its theory or attitudes, and not its practices, it is made to appear liberal, but the appearance masks a different reality. (p. 117)

The emphasis in both the schools and the universities is on the transmission of knowledge, where knowledge is separate from teacher and learners. Students are expected to demonstrate mastery of this knowledge, essentially divorced from practice: "The key is that as a student…[the teacher] never experiences in practice the liberalism which he [sic] is so freely allowed to express in theory. The change to conservative attitudes merely expresses what was the position in practice all the time" (p. 123). These results further contribute to the functionalist explanation of the tendency of novice teachers to teach as they were taught, in the tradition dominant in schools today.

Patterns in Functionalist Research. Research on teacher socialization from a functionalist perspective posits external forces that mold novice teachers to fit the teaching society existant in the schools. Although each of the research studies reviewed here focused on different socializing agents encountered in the process of becoming a mathematics teacher, the results of these studies and other studies done within the functionalist perspective are consistent. They suggest that novice teachers' actions and attitudes change in the direction of the actions and attitudes of currently practicing teachers. Furthermore, they indicate that the major factors influencing novice teachers—biography, teacher education programs, and experiences in the classroom—consistently encourage beginning teachers to conform to the status quo.

Researchers viewing socialization from a functionalist perspective interpret the results of these studies as describing shifts in novice teachers determined by the influences of socializing agents on them. Because these studies reported general tendencies in novice teachers and did not investigate the details of individual experiences of socialization, the research does not provide information to indicate the reasons behind novice changes. Functionalist studies also do not consider the differences in the context into which novices are socialized. In spite of the limitations of these studies, they do indicate that there is a tendency for beginning teachers to develop attitudes and behaviors that are dominant in the existing culture of the schools. Such findings are troublesome to people who hope to see mathematics instruction reformed, for they suggest that novice teachers most likely will not implement innovations in either the mathematics curriculum or teaching practice unless those innovations are also part of the culture of the schools where they learned and where they teach and learned to teach.

Teacher Socialization from the Interpretive Perspective

Interpretive studies of teacher socialization, unlike those from a functionalist perspective, are typically "aimed at developing a model of the socialization process that would encompass the possibility of autonomous change emanating from the choices and strategies adopted by individuals" (Lacey, 1985, p. 4076). Much of the recent interpretive research has been inspired by the work of Lacey (1977), who utilized both participant observation and questionnaire data to understand the experiences of student teachers from the perspectives of the student teachers. Lacey developed the concept of *social strategy* which is key to his explanation of beginning teachers' socialization: "A social strategy involves...the selection of ideas and actions and working out their complex interrelationships (action-idea systems) in a given situation. The selection of these action-idea systems as a student [teacher] moves from situation to situation, need not be consistent" (p.68). Lacey claimed beginning teachers employ three distinct social strategies in dealing with institutional constraints they face in their roles as teachers.

Internalized adjustment refers to the first strategy, in which the individual complies with the constraints imposed by a situation and believes that the constraints are for the best. The individual using this strategy takes on the characteristics expected of persons in that setting, conforming their behavior and making a value commitment.

The second strategy, *strategic compliance,* refers to a response in which an individual teacher complies with the constraints of a situation but retains private reservations about doing so. Individuals act in a way inconsistent with their personal beliefs; conformity of behavior is simply an adaptive response and does not include a corresponding change in values.

The third strategy, *strategic redefinition,* is a response in which an individual is successful in an attempt to change the situation, even though he or she possesses no formal power to do so. This change is achieved by enabling or causing those with formal power to change their definitions of what is appropriate in the situation. Using Lacey's framework, one must take into account the constraints of the situation into which a teacher is being socialized as well as the individual teacher's purposes within that situation. His theory implies that the ideas and actions of a teacher are interpretable only in the context of specific situations.

Another study investigating teacher socialization from an interpretive perspective, conducted by Zeichner and his colleagues (Zeichner & Tabachnick, 1985; Zeichner, Tabachnick, & Densmore, 1987), examined socialization to teaching as manifested in changes in beginning teachers' *teaching perspectives.* The program was designed as a two-phase study conducted over a two-year period. The first phase, consisting of 13 student teachers in a university teacher education program, attempted to discover ways in which the student teaching experience influenced the development of teaching perspectives and the factors that influenced these changes. The second phase followed four of the original 13 participants into their first year of teaching, with the intent of discovering how the particular characteristics, dispositions, and abilities of the beginning teachers and the various people and institutional characteristics in their schools influenced the development of teaching perspectives during that first year. *Perspective* is used here as Becker, Geer, Hughes, and Strauss (1961) defined it: a coordinated set of ideas and actions a person uses in dealing with some problematic situation. It is assumed that teacher behavior and teacher thinking are inseparable and that both reflect perspectives toward teaching.

The 13 participants in the first phase of the study were selected to create a group of pre-service teachers who appeared to have quite different beliefs within each category measured by the *Teacher Belief Inventory* (TBI), a 47-item instrument that assesses student teacher beliefs related to six specific categories: (1) the teacher's role; (2) teacher-pupil relationships; (3) knowledge and curriculum; (4) student diversity; (5) the role of the community in school affairs; and (6) the role of the school in society (the last two categories proved not to be useful in the study). The participants were interviewed and observed during their student teaching experience; university supervisors and cooperating teachers were interviewed in order to establish the substance and dimensions of the student teachers' perspectives and the manner and degree to which these perspectives changed during the course of the semester.

The dominant trend among subjects was for teaching perspectives to grow in a direction consistent with the latent culture student teachers brought to the experience. As Zeichner and colleagues (1987) reported, "data clearly indicate that student teaching did not result in a homogenization of teacher perspectives.... On the contrary, with the exception of three students, teaching perspectives solidified but did not change fundamentally over the course of the 15-week semester" (p.36).

Although there were no significant shifts in the substance of student teachers' perspectives in their study, several kinds of changes did occur for most participants. Student teachers appeared to gain a more realistic perception of the job of teaching and the teacher's role; they became increasingly more confident in their abilities to handle a classroom in their preferred styles, and increasingly less anxious about observations and evaluations of their teaching.

Three students who did not develop in ways consistent with the perspectives they brought to student teaching employed what Lacey (1977) termed "strategic compliance." These student teachers experienced severe constraints in their school placements against which they reacted strongly. Because of their position as student teachers and the strength of the institutional constraints, they acted publicly in ways demanded by their situations, but privately continued to hold strong reservations about their actions. Teaching perspectives of these three student teachers did not develop or change over the semester of the study.

Most student teachers in the study had purposefully selected themselves into situations that corresponded with their ideas about teaching, where they would be able to act in certain preferred ways. It is therefore not surprising that the perspectives of 10 teachers did not radically change. It is also true, however, that even when placed in the same school and faced with common institutional constraints, student teachers each reacted somewhat uniquely to their situations, but still consistent with the perspectives they brought to student teaching. These results seem to contradict the functionalist perspective of passive responses to institutional forces.

The four participants Zeichner and colleagues followed through their experiences as first-year teachers were relatively alike in their apparent perspectives toward teaching when they began their first teaching jobs. Three of the four teachers worked in very different situations as first-year teachers than as student teachers, yet two of these three attempted to implement a style of pedagogy similar to that evidenced during student teaching. Only one teacher significantly changed her perspective in response to differing institutional demands. The study suggests that under some conditions beginning teachers are able to maintain a perspective that is in conflict with the dominant institutional cultures in their schools. One explanation for beginning teachers' persistence in the face of institutional pressure is that school cultures are often diverse. Zeichner and colleagues found that subcultures exist in schools and that these subcultures sometimes attempt to influence beginning teachers in contradictory ways. These contradictions seemed to provide novices with opportunities to establish individual expressions of teaching. The researchers also suggested that a more careful examination of the informal cultures present in school settings might provide further explanation of individual differences in beginning teachers' socialization experiences.

Other studies have supported the notion that understanding the informal cultures of teachers, pupils, and schools is important to understanding the socialization of both student teachers and beginning teachers. For example, informal cultures in the school setting were explored in Brown's (1985) study of one secondary mathematics teacher's socialization to teaching. This study investigated changes in the teacher's *conception of mathematics teaching* during his first year of teaching, defined as having three components: a conception of mathematics; beliefs about appropriate goals and tasks for the mathematics classroom; and beliefs about the relative responsibilities of teacher and students concerning motivation, discipline, and evaluation.

The study was based in the symbolic interactionist perspective, which assumes teachers and students together define the situation of the classroom, constrained by their backgrounds and the physical, temporal, and organizational context in which the classroom is embedded (Delamont, 1983). Observational and interview data were analyzed to determine the extent of change in the novice teacher and the factors that influenced change.

The influence of the pupils' conceptions of mathematics teaching, as manifested in their actions and their talk, was one of the most significant factors in the development of the novice's conception of mathematics teaching. For example, the novice teacher repeatedly justified his classroom actions by describing what he believed his students were capable of doing, or were willing to do. His perceptions of his students were incongruent with the conception of mathematics teaching he had developed through his own school and university experiences. Brown described how the persistence of student behavior through the year created a situation in which this novice teacher felt compelled to change the way he thought about teaching and acted in the mathematics classroom.

Although these perceptions motivated the novice teacher to change his thinking and actions in that situation, he did not change some of his most basic general conceptions of mathe-

matics and teaching. This pattern fits Lacey's notion of strategic compliance; the teacher conformed his behavior to the students' expectations, yet retained ideals for teaching mathematics which he construed as being impossible to implement in his present situation.

Patterns in Research from the Interpretivist Perspective. Studies of teacher socialization from an interpretivist perspective indicate the multitude of influences that present choices and constraints to the beginning teacher. These investigations are predominately case studies and provide detailed looks at the experiences of individual teachers as opposed to central tendencies in populations of teachers. The studies reveal that individual differences in novice teachers mediate the ways they are influenced by elements in their environments, such as students, colleagues, and institutional or organizational factors. Although there are differences in the effects on individual teachers, these elements nonetheless are shown to be important influences in the socialization process of teachers.

The research presented in this section does not provide information about why different novice teachers respond differently to elements in the socialization process. It is not known why one novice mathematics teacher maintains a problem-solving focus in her teaching while another abandons her ideals in the face of institutional pressures to teach "basic skills." Is depth of understanding of mathematics a factor? Does strength of commitment to the belief that problem-solving is important make a difference? These are questions unanswered by interpretive research.

Although the studies reviewed in this section describe the interactions of individuals with elements in the school culture, the emphasis has been on how teachers have been influenced—not on how the culture has been changed—in these studies and in others from this perspective. The interpretive view of socialization as an interactive process should lead researchers to pay attention not only to the individual's experience of socialization but also to how institutions, colleagues, and pupils are influenced by novice teachers.

Teacher Socialization from the Critical Perspective

Research on teacher socialization from the critical perspective is conducted with two primary goals: (1) understanding the socialization process within a broad social context; and (2) transforming the socialization process to rid it of inequities in race, gender, and class relations. Most of the published writing from this perspective has been theoretical or reflective (Popkewitz, 1987). There are currently very few empirical studies of socialization from the critical perspective (Zeichner & Gore, 1990).

Ginsburg's (1988) two-year participant observation study of pre-service teacher socialization is one of the few good examples of empirical research within the critical tradition. Ginsburg enrolled in classes with pre-service teachers in order to study closely "teacher education, its organizational structure, curricular content and lived experience" (p. 1). More specifically, he wanted to better understand the relationship between teacher education and the reproduction of inequalities in wealth, power, and status as well as gender, race, and social class. Data for the study were compiled from field notes, interviews with students and instructors, and evidence such as class handouts and assignments. Ginsburg described the many ways in which the status quo was perpetuated in the teacher education program he studied. For example, he argued that pre-service teachers were being prepared to "deliver what was seen as given curriculum knowledge" (p. 119). As they progressed through the program, some students became less and less willing to challenge the existing curriculum guides or textbooks. He suggested that "it may not be coincidental that a predominantly female population is being 'anticipatorily deskilled' " (p. 121). Their teacher preparation programs seemed to prepare these female pre-service teachers for a life of accepting curricula most likely developed and administered by males.

In his last chapter, a discussion of critical praxis by teacher educators, Ginsburg argued that all university faculty who encounter pre-service teachers have a responsibility to more actively address issues of social importance in their instruction. Typical of critical theorists, he placed the process of teacher education within a broader political and social context, with the purpose of criticizing those things that are simply taken for granted in teacher education, the status quo.

Densmore (1987) investigated the development of teachers' understandings of professionalism in teaching in a study of two first-year female teachers' socialization. The two teachers, Beth and Sarah, were observed and interviewed over a period of a year. Observations of and interviews with principals, other teachers, parents, and students also were used as sources of data in this study. Several organizational and ideological features of teachers' work functioned as controls on the two teachers: architectural design, curricular programs, school staff norms, elements of the professionalist ideology, meeting frequency and agenda, parental pressure, the principal's role, the absence of opportunities for genuine teacher collectivity, and a heavy work volume (p. 152). Densmore argued that the teachers' understandings of professionalism in teaching were responses to these factors in their working conditions.

For Beth, professionalism came to mean adherence to accepted institutional norms. By midyear, because her responsibilities became more than she could handle well, teaching became a "job," in which she invested little of herself. Beth became focused on getting her job done; being professional came to mean meeting the expectations of the administration, students, and parents rather than making her own decisions about what to teach and how.

For Sarah, professionalism meant, in part, being a school representative, most often to parents. Many responsibilities, skills, and knowledge had been removed from Sarah's work by the organization of teaching in her school. She used her status as a professional, and, thus, an expert, to keep parents from removing even more control from her hands. As the year progressed, Sarah felt compelled by her sense of professionalism—and parental pressure—to supplement the curriculum with additional creative lessons. In doing so, she increased the quantity of her work, which was already considerable.

Beth and Sarah found little autonomy in teaching, gave over control of the curriculum to others, and were overloaded with work. They seemed to define teacher professionalism in ways

that more closely reflected teachers as workers than as professionals. Based on this analysis, Densmore argued for a conceptualization of teachers as workers rather than teachers as professionals. She suggested that the ideology of professionalism in teaching is in conflict with the reality of the teaching situation, and noted that teacher education programs and schools refer to the rhetoric of professionalism but still treat teachers as workers.

In a discussion typical of critical studies of teacher socialization, Densmore placed teaching in the context of the more general labor process in a capitalistic society. She called for a transformation of schools, but in the context of broader efforts to transform exploitative social relations. Teachers, she argued, must organize and act to gain more power in the workplace, but also to help transform our educational programs to be less supportive of social inequalities.

Patterns in Critical Studies of Socialization. Critical studies are clearly intent on revealing the role of class, race, and gender relations in the socialization of teachers. They interpret teacher socialization in the broader context of society, seeking to encourage transformation of the status quo as they challenge the values of our current educational system by exposing inequities within it. For example, they argue the inequity of male teacher educators and school administrators having power over groups of predominately female novice teachers. They also point out that while teacher education institutions and schools may talk about teacher autonomy, research suggests that this autonomy is limited. Investigations from a critical perspective reveal the complexity of the structures and ideologies of the institutions in which novice teachers learn and work, and indicate ways in which novices are controlled by those structures and ideologies.

Although critical studies are grounded in an assumption that people are both creators and products of the social situations in which they work, there is much more emphasis on describing the effects of social situations on novice teachers than the effects of novice teachers on their social situations. Neither Ginsburg nor Densmore report changes in teacher education or schools as a result of novice teachers' participation. It seems important that researchers document the ways in which novice or experienced teachers influence teacher education programs and schools, so that we can better understand processes by which teachers can begin to transform these institutions.

Patterns and Implications in Teacher Socialization

Socialization studies suppose that external forces influence teachers as they become members of the teaching culture. The functionalist perspective views these influences as determining the "final product" that a novice teacher becomes. The interpretive and critical perspectives view the teacher as taking an active role in interpreting these influences and, to some extent, modifying them. Taken as a whole, these studies suggest that several aspects of the novice teacher's experiences are important in the process of becoming a teacher.

The socialization literature confirms that pre-service teachers come to formal teacher education with many already-established notions about teaching. From their experiences, as children and as students, they have formed ideas about the role of teacher, and these are the ideas to which many novice teachers aspire. There is debate about the extent to which formal teacher education programs can encourage teachers to modify the beliefs that they bring with them.

Cooperating teachers and other colleagues are potential sources of influence on novice teachers' patterns of thinking and action. However, it appears that there is not one unique culture of teaching in a school, but, rather, different groups of teachers sharing different subcultures. This lack of unity, the brevity of the student teaching experience, and the relative isolation of novices from their colleagues may all serve to limit the effects of colleagues in the socialization of beginning teachers. Unfortunately, because few studies provide data sensitive to informal and often subtle influences of colleagues, existing research does not reveal how extensive and lasting this influence is.

There is evidence that novice teachers' classroom actions are also influenced by the behaviors of students and the novices' perceptions of student abilities and motivations. These factors not only are major constraints on teachers' classroom actions, but appear also to influence novices' conceptions of what is possible in the classroom. Most research on socialization has not been sensitive to the role of the student; thus, there is not a clear picture of the extent and magnitude of the influence of expert students on novice teachers.

The conclusions of socialization research are potentially troublesome to those mathematics educators and other teacher educators who subscribe to the vision of mathematics teaching described in the NCTM standards, as most of the mathematics teaching experienced by novice teachers when they were students would not meet these standards. Twelve to fourteen years of experience as students in classrooms serve as a strong socializing influence not easily overcome later in formal teacher education programs or even as teachers in classrooms.

It may be the case, though, that the influence of teacher education programs and early classroom teaching experiences can be substantially increased if these experiences present a common culture of teaching to novices. If, for example, university teacher education programs, cooperating teachers for student teachers, mentors for beginning teachers, and colleagues *all* support good mathematics teaching as described in the NCTM documents, novice teachers would more likely be socialized into that culture of teaching. However, unless novice teachers experience good mathematics teaching as students, see it modeled by teachers they respect, and are situated in a culture of teaching that accepts and practices good teaching, it will be difficult for them to implement and maintain good teaching in their own classrooms. This line of reasoning argues for the potential value of professional development schools as locations for teacher education programs.

Whether one accepts a functionalist, interpretive, or critical perspective on teacher socialization, there is little argument that there are currently strong, competing sources of influence in the environment of beginning teachers. There is also a growing consensus about the interactive nature of the socialization process and the role of novice teachers in making choices and influencing the culture into which they are being

socialized. There are few studies, however, that take into account influences on teacher socialization emanating from beyond the school context. Race, social class, gender and other broader social issues have not been well integrated into studies of teacher socialization. As society begins to recognize the importance of mathematics education for all students, regardless of race, gender, or social class, it seems important that researchers pay attention to the role socialization may play in the reproduction of inequality in these areas.

TEACHER DEVELOPMENT

Research on teacher development is based on a view of the teacher as an adult learner whose development results from changes in cognitive structures; these cognitive structures, it is assumed, are the thinking patterns by which a person relates to the environment. Several different theories of developmental stages provide frameworks for studying how teachers organize their worlds: (1) cognitive development, based on Piaget (1972); (2) moral decision-making, based on Kohlberg (1969); (3) ego development, based on Loevinger (1976); (4) conceptual development based on Hunt (1971); (5) intellectual and ethical development based on Perry (1970); and (6) stages of concerns, based on Fuller (1969). All of these theories assume the existence of stages of development, suggesting these stages occur in a hierarchical, sequential, and invariable order. From the perspective of teacher development, becoming a teacher involves the process of moving from the less complex stages to the more complex ones. (For a more complete discussion of these theories see Oja, 1980, and Sprinthall & Thies-Sprinthall, 1983.)

This section of the chapter examines three programs of study based on developmental stage theory: (1) research on mathematics teachers' development, based on Perry's (1970) theory of epistemological and ethical development; (2) research based on Fuller's (1969) conception of teacher development as a process of passing through stages of concerns; and (3) research on teacher education or staff development efforts grounded in Loevinger's theory of ego development. These three research programs were selected because they are most in harmony with the focus on the process of individuals becoming mathematics teachers. They also represent major programs of study within the developmental perspective that facilitate understanding of novice teachers' development. Perry's scheme has been utilized in a number of studies of mathematics teachers. Fuller's theory of teacher concerns is frequently used in investigations of beginning teachers. Loevinger's theory, one of the most comprehensive of the developmental theories (Cummings & Murray, 1989), has been used by a number of researchers to investigate factors mediating the effect of teacher education or staff development efforts on teachers.

Perry's Scheme

Perry's theory of epistemological and ethical development (1970) is probably the only developmental stage theory utilized in studies of mathematics teacher education. Perry's scheme is a description of an evolution in students' interpretations of their lives, obtained from their accounts of experiences during their college years. It grew out of patterns which emerged from analysis of long, open-ended interviews conducted with male students in the context of counseling sessions at Harvard in the 1950s and 1960s. The scheme attempts to capture the developmental nature of the worldviews of college students, although it appears to be applicable to the worldviews of other populations as well.

The full scheme consists of nine positions from which persons view their worlds. Perry describes the college students he studied as moving through these positions in order, although some of them regressed—something most developmental theories would frown upon. The nine positions are most frequently collapsed into four categories: simple dualism (dualism), complex dualism (multiplicity), relativism, and commitment.

In a *dualistic* position, a person believes that every question has an answer, that there is a solution to every problem, and that an authority or expert will know and can deliver these answers. An authority who doesn't deliver the answers is either incompetent or trying to get the students to learn the answers themselves. In mathematics, this perspective conforms to the classic view that there are right and wrong answers in mathematics and that the teacher should know the difference.

Persons with a *multiplistic* view of the world believe that all views are equally legitimate and that the role of authority is to make students think for themselves. No one knows "the truth," therefore, anything is worth thinking and believing. This view is manifested in mathematics as formalism—the perspective that everyone has a right to an axiom system, and that each axiom system is equally valid since mathematics is only a systematic use of symbols.

Persons operating from a *relativistic* view of the world have come to believe that not all opinions are equally good. Quality counts, and there are criteria upon which beliefs or actions can be evaluated depending on the context of the evaluation. Solutions to mathematical problems may be elegant but abstract, and other solutions may be better for exemplifying techniques that are more understandable.

Finally, persons who have reached the stage of *commitment* realize that decisions can only be made on the basis of uncertainty and that there is risk involved. At this stage, persons are willing to accept alternative viewpoints and recognize that knowledge is a personal structure for interpreting experience. Mathematics teachers at this stage would be most open to a constructivist perspective on learning and teaching.

Several studies have used Perry's scheme as a theoretical perspective from which to explore both pre-service and in-service mathematics teachers' views of mathematics and mathematics teaching. Meyerson (1977) sought to answer the questions "Is it possible to locate a person's conception of mathematical knowledge and conception of mathematics teaching along the Perry continuum?" and "Is it possible to create a classroom atmosphere that encourages development with respect to Perry's scheme?" His study investigated the effect that a mathematics methods course designed to foster students' development along Perry's scheme had on pre-service teachers' conception of mathematics and mathematics teaching. Students first were diagnosed with regard to their position on the conception of mathematical knowledge and mathematics teaching

continua, and then were assigned exercises designed to encourage growth along the continua. These exercises involved the use of mathematical errors, surprise, doubt, reexaminations of common beliefs about teaching, feelings, individual differences, and problem-posing. Case studies were developed to describe the diagnostic and growth processes based on data collected from student diaries, formal interviews, informal discussions, classroom exercises, homework assignments, final projects, and exams.

Meyerson reported that doubt was a key factor for movement through Perry's scheme. Doubt was elicited by confronting students with mathematical situations that caused confusion and created controversy in the class; as they were forced to face their confusion and resolve the controversy, some students changed their views of what constituted authority in mathematics. Meyerson reported mixed results in his attempt to encourage students' growth along Perry's scheme. He hypothesized that students' views of mathematics had been developed over many years of mathematics classes, and he suggested that one course could not be expected to encourage great changes in these long-held views.

In his study of four experienced high school geometry teachers, McGalliard (1983) investigated the conception of geometrical knowledge communicated through instruction, aims in teaching geometry, and evaluative assessments of students. Observation of classroom episodes, interviews, and written responses were used to develop case studies of individual teachers' classroom activities. McGalliard found that teachers' classroom practices could be generally described as representing a dualistic view of authority. However, when the classes observed dealt with geometric proof there was some indication of a slight degree of multiplism in the instruction. Unfortunately, the research did not attempt to directly relate the teachers' instructional practices with their own levels of development on Perry's scheme.

In a more recent study, Owens (1987) used Kelly's Personal Construct Theory (Kelly, 1955), together with Perry's scheme, to investigate the constructs and worldviews that influence student teachers' conceptions of the subject matter of secondary mathematics and of the roles associated with teaching and becoming a teacher. Each of four pre-service secondary mathematics teachers completed a series of seven one-hour interviews eliciting constructs and elements (Kelly, 1955). These interviews provided the data for exploring the experiential, mathematical, and pedagogical frameworks from which the participants viewed their undergraduate experiences and anticipated their roles as mathematics teachers. Owens used Perry's theory to help explain the ways in which student teachers thought about mathematics and mathematics teaching. He hypothesized that differences in mathematics and teaching understanding would depend on individuals' perceived relationship with "authority" (p. 52).

Perry's (1970) own work suggested that few undergraduates reach the level of *commitment* by the end of their college careers. Similarly, only one of the teachers Owens studied had reached the level of *relativism*. This teacher, Susan, seemed to be caught between a realization of inner authority and a continuing need for external structure in her life. With respect to mathematics, the need for external structure seemed to be fairly strong. The other three participants in Owen's study were at the level of *multiplism* on Perry's scheme. Laura, for example, often talked about different ways of learning or teaching, although she could seldom supply examples to clarify her point and consistently avoided choosing between competing ideas (p. 138); this realization of alternatives, but avoidance of commitment to a specific option, is characteristic of the multiplistic stage. Laura seemed to want an authority to choose for her; she wanted her principal to "set up exactly what (he) wants done," her advisor to "tell me what courses to take," her professors to "teach us how to teach," and the text to decide on the appropriate method of solution (p. 139). Ellen and Tim, the other two subjects in the study, were also at the level of *multiplicity,* for they indicated no inclination to reflect on or justify their actions during student teaching.

Patterns in Research Based on Perry's Scheme. Research based on Perry's scheme indicates that both experienced and novice teachers tend to rely on external authority in making decisions regarding course content and pedagogical strategies. Teachers tend to be at the low levels of Perry's scheme, and it is difficult to move teachers to more advanced levels on the scheme. This research suggests that novice mathematics teachers may, therefore, have difficulty embracing a constructivist conception of mathematics teaching and learning that places authority for mathematical knowledge within all individuals, including students.

There have been no studies attempting to relate teachers' positions on Perry's scheme with their acceptance or implementation of innovative programs. It is therefore difficult to anticipate how teachers low or high on Perry's scheme might react to NCTM recommendations for curriculum, evaluation, and teaching. It seems that teachers low on Perry's scheme—those who seek outside experts as authority—would depend on their perceptions regarding the authority of the National Council of Teachers of Mathematics. If it is perceived as more "expert" than the more traditional sources of curriculum and teaching strategies, such as textbooks, state curriculum guides, and school mandates about lesson formats, then teachers' acceptance and implementation might be high. On the other hand, teachers who are high on Perry's scheme—who accept the existence of multiple viewpoints and recognize knowledge as being within individuals—may be reluctant to accept the authority of NCTM unless they view it as an organization that represents their own views. If NCTM is viewed as an abstract entity attempting to exert control over teachers' classrooms, teachers high on Perry's scheme may choose to ignore NCTM's recommendations. However, teachers who have reached Perry's level of relativism or commitment, and who find NCTM's recommendations congruent with their own ideas, should be attracted to the overall message of the recommendations of NCTM regarding the teaching of mathematics.

Teacher Concerns

Fuller's (1969, 1970; Fuller & Bown, 1975) theory of stages of teachers' concerns was motivated by her convictions regarding pre-service teachers. Fuller noted pre-service teachers had needs not being met by teacher education, as teacher education programs had not taken into account characteristics of pre-service teachers as learners.

The initial theory (Fuller, 1969) identified and described novice teachers' *concerns* about teaching, based on data collected from student teachers about their perceptions of teacher education experiences, their foundations for reports of dissatisfaction, and their motivations for learning what they brought to their professional experiences. Data were collected by taping group counseling sessions with student teachers and by having student teachers write about their concerns. Fuller also reanalyzed data from other studies of perceived problems of novice teachers.

The more fully developed theory, as described in Fuller & Bown (1975), points out four stages of development in learning to teach, each with its own set of characteristic concerns. The first stage is that of *preteaching concerns,* in which students who have never taught are not yet concerned about teaching, but, instead, are concerned about themselves as students. In the second stage teachers are primarily occupied with *self-concerns*—concerns about their survival as a teacher and their adequacy in the classroom, about class control, about being liked by pupils, and about being observed, evaluated, praised, and criticized. Fuller claimed that these self-concerns seem to be more prevalent in pre-service than in-service teachers. The third phase is characterized by *task concerns*—concerns about the teaching situation, time pressures, inflexible situations, too many students, too few materials, too many noninstructional duties, and the like. Frustrations about tasks are generally felt by people in actual teaching situations, that is, in-service rather than pre-service teachers. The fourth phase consists of *pupil concerns*—concerns about the social and emotional needs of pupils, fairness, and the appropriateness of curriculum and instruction for individuals in the classroom.

From a developmental perspective, the concerns related to self are considered less mature and less desirable than the concerns related to pupils. The progression of these concerns is explained by the general human tendency to be preoccupied with basic needs (survival) until they are satisfied. Thus, it is argued that teachers cannot reach the more mature, more desirable stage of concerns about pupils until they have resolved concerns about their own survival and teaching situation (Fuller & Bown, 1975). Fuller suggested that concerns move through a series of steps of arousal and resolution; arousal of concerns seemed to result from affective experiences, whereas resolution of concerns was described as more of a cognitive event, requiring teachers to develop more knowledge and skill.

Marso & Pigge (1989) conducted a study to investigate and further validate Fuller's theory of concerns, also hoping to determine whether any identified changes in teaching concerns might be related to selected teaching field—elementary, secondary, special education, and specialized areas such as art, music, and physical education—or gender. Six groups of pre-service and in-service teachers participated in the study. The teacher education students (pre-service teachers) consisted of a group of 559 students in a required educational orientation course, a group of 151 students about to begin student teaching, and a group of 162 students who had just completed student teaching. The in-service teachers consisted of 94 teachers completing their first year of teaching, 104 teachers completing their third year, and 123 teachers completing their fifth year. The subjects completed a Teacher Concern Questionnaire, an Attitude Toward Teaching as a Career Scale, and a questionnaire requesting demographic information, including gender and teaching field.

The researchers identified three patterns of concerns that occurred during the pre-service and in-service years: low job-specific concerns early in the pre-service years, an increase in job-specific concerns as the prospective teachers anticipated instructional tasks in university experiences, and an additional increase when they actually experienced the complex tasks of instruction in full-time teaching. Furthermore, there were decreases in self-concerns with additional teaching experience (p. 38). In general, these changes tended to be relatively positive, predictable, and consistent with Fuller's theory.

However, the nature of the changes varied considerably for particular subgroups of individuals. For example, secondary teachers at every level were more likely than other teachers to be concerned about the teaching setting being too routine and too inflexible. Consistent with Fuller's model, total concerns about teaching and self-concerns decreased with additional education and early teaching experience, while task-related concerns were initially low but increased as the individuals began to teach. However, the study indicated secondary teachers' level of task concerns and self-concerns changed from the lowest among the four teaching fields in the pre-service years to the highest among these groups during the in-service years. There was also a decrease in secondary teachers' concerns about their impact upon pupils as they moved from pre-service education to in-service teaching.

Contrary to the Fuller model, pupil concerns were stable and highest among the three types of concerns for all teachers in the Marso and Pigge study. This pattern is consistent with other studies investigating teacher concerns and rooted in Fuller's model (Adams, Hutchinson, & Martray, 1980; Adams & Martray, 1981), which, over a five-year pre-service teacher education program, have reported changes in self-concerns but little or no change in either task or pupil concerns.

In a longitudinal study, Adams (1982) investigated the change in teachers' perceived problems, concerns, and classroom behavior over a six-year period using data collected as part of Western Kentucky University's Teacher Preparations Evaluation Program (TPEP). Adams administered two instruments, the Teacher Preparation Evaluation Inventory and the Teacher Concerns Checklist, and conducted classroom observations at four points in time: during student teaching and near the end of the first, third, and fifth years of teaching experience. Data from this study indicate that instructional impact concerns were highest for all teachers, although they were more prominent concerns for elementary in-service teachers than for secondary in-service teachers. In general, self-concerns decreased from student teaching through the fifth-year, and task concerns related to instruction increased with experience, both supporting Fuller's theory. However, concerns related to impact on students were the highest of all concerns and did not change across levels of experience, contradicting Fuller's theory that impact concerns are primarily found in mature teachers.

Patterns in Teacher Concern Research. Though the results of studies based on Fuller's model of teachers' concerns are not consistent and do not completely support the model, they do indicate that there are changes in teacher's concerns as they progress through teacher education and into the initial years of teaching. Fuller's framework provides us with a global picture of changes in teacher concerns. However, it is apparent that this framework can not account for the complex networks of concerns that novice teachers have. The cross-sectional and longitudinal studies reviewed in this section also suggest that changes in teacher concerns are not fixed in sequence, as Fuller originally proposed. Teachers have concerns about self, task, and pupils in all phases of becoming a teacher; the strength of these concerns is what changes over time. Self-concerns appear to be strongest during the pre-service and initial in-service years, decreasing with experience. All teachers, no matter what their level of experience, seem to be more concerned about their impact on pupils than they are about classroom tasks. The studies also suggest that there are differences in the concerns of secondary and elementary teachers, and, perhaps, male and female teachers. However, the data are insufficient to indicate the complexity of these differences or their causes.

The available information on changes in teacher concerns is derived primarily from large scale statistical studies from which we can learn about general tendencies in the populations studied. It seems necessary now to investigate more closely individual teachers' concerns and the changes in these concerns as the teachers move through their pre-service experience and initial years of teaching. Perhaps more intensive, more focused studies of individuals will help us to understand the concerns of novice mathematics teachers and the roles these concerns play in the process of becoming a mathematics teacher. Current studies do not address elementary teachers' concerns about their knowledge of mathematics or secondary teachers' concerns about getting through the mandated mathematics curriculum. Such concerns may interfere with teachers' implementation of reform in mathematics teaching. Research results currently available also do not suggest methods of moving teachers through the lower stages of concerns to more mature concerns or ways of arousing concerns and helping teachers resolve them.

Loevinger's Theory of Ego Development and Teacher Education/Staff Development. Loevinger (1976, 1980) conceived of ego development as holistic, encompassing moral and personality development, cognitive complexity, and interpersonal style. The beginning stages of the model, the presocial (I-1) and impulsive (I-2) are rarely found in adults. Later stages, more common in adults, include the following:

conformist (I-3), characterized by a conventional, stereotypical view of the world; self-aware (I-3/4), characterized by increased awareness of inner feelings and multiplicity of thinking; conscientious (I-4), characterized by differentiated thinking about self and others; individualistic (I-4/5), characterized by increased awareness of inner conflict and toleration of paradox; autonomous (I-5), characterized by cherishing individual differences and toleration of ambiguity; and integrated (I-6) characterized by integration of inner conflict. (Cummings & Murray, 1989).

These ego stages are thought to occur in an invariant, irreversible sequence. Loevinger proposes that it is through interaction with the environment that one grows or moves to higher levels of ego development.

Several studies have investigated relationships between teachers' level of ego development and their responses to teacher education or staff development efforts, including those by Spatig, Ginsburg, & Liberman (1982), Cummings & Murray (1989), and Oja and colleagues (Oja, 1980, 1989; Oja & Ham, 1984). Although only Spatig et al. focused on novice teachers, the other studies help us build a more complete picture of the issues addressed, methodologies used, and results found in research conducted within the developmental perspective.

Spatig, Ginsburg, and Liberman (1982) used ego development to examine the extent to which pre-service teachers' orientation to teaching with respect to discipline and control (Conform D), pupil involvement in classroom decision-making (Conform P), and facts—versus pupil's interest—oriented instructional objectives (Conform F) were influenced by their instructors. The researchers hypothesized that students at the Conformist stage of ego development would be more likely to accept the ideas and attitudes of the instructor than those at the Conscientious stage. Participants in the study were 57 students in a course entitled "Introduction to the Profession of Teaching." A sentence-completion test was used to classify their level of ego development. Students' and instructors' orientations toward teaching were measured using seventeen, five-point Likert scale items.

Students at the conformist stage developed less varied orientations to teaching, closer to the orientations of their instructors than students at the conscientious stage—a pattern that provides some support for the researchers' hypothesis. However, this pattern was the case only for the teachers' orientations toward classroom control (Conform D). Spatig, Ginsburg, and Liberman also interpreted these results within a socialization framework and suggested that both a functionalist and an interpretivist view of socialization may have validity; that is, although some pre-service teachers appear to be molded by their instructors, other pre-service teachers "seem to creatively construct their occupational identities, or at least not have such identities determined or largely influenced by their instructors" (p. 323).

Cummings and Murray (1989) studied the relationship between ego development and teachers' views of various aspects of education, measuring both constructs by sentence-completion tests. The study also examined whether ego level could predict the participation and achievement of teachers in a university professional education course—measured by grades in the course and teacher ratings of students with respect to participation—and on a set of scales based on Bloom's Taxonomy of Educational Objectives. The participants were 58 teachers enrolled in a summer university program on guidance counseling, with teaching experience ranging from 0 to 24 years.

Very weak, though statistically significant, correlations were found between ego development and cognitive and affective achievement as measured by the Bloom rating scale, and between ego development and views of education. Contrary to the researchers' expectations, ego level did not correlate significantly with grades or student participation. However, participation correlated significantly with both grades and Bloom rating.

Although these results do not indicate much predictive power for ego development, the authors argued that they suggest potentially fruitful avenues for future investigations. For example, completions of the stem "A good teacher" by teachers at the conformist (I-3) level of ego development indicated that they conceptualized the role of a good teacher only as a presenter of information or as someone who cares for children. Not until teachers were at the conscientious (I-4) level of ego development did concerns for variety in teaching methods, individualizing instruction, and new approaches to teaching appear.

Oja's (1980, 1989; Oja & Ham, 1984) program of research on staff development, using the developmental stage theories of Loevinger, Kohlberg, and Hunt, not only contributes to a better understanding of how teachers' developmental levels relate to their responses to staff development opportunities, but also suggests ways to assist teachers to grow developmentally. Oja reported on a staff development project called Action Research on Change in Schools (ARCS) and its relationship to the life periods, career cycles, and developmental stages of the teachers who participated (for a description of life periods, career cycles, and specific developmental theories used, see Oja, 1989). The ARCS staff development program consisted of three phases over the course of a summer and fall.

Phase I. Building supportive interpersonal relationships within small groups to create an environment necessary for developmental growth.
Phase II. Learning new skills appropriate for more complex role-taking. For example: skills in interpersonal effectiveness, indirect teaching, individualizing instruction, and supervision. Familiarizing teachers with the theory of developmental stages of growth.
Phase III. Applying the newly acquired skills and theory to the teachers' own classroom setting with consistent on-going supervision in small groups and advising in individual conferences. (Oja, 1980, p. 37)

In the study, teachers were asked to respond to three tests designed to measure stages of moral judgment, ego development, and conceptual complexity. A teacher's levels of development were related to the individual teacher's participation in and perception of activities of the ARCS staff development project.

Oja described teachers representing different developmental stages and their participation in the ARCS program. For example, Elliot exhibited characteristics of the Individualistic stage of development; he was able to keep his own needs and goals in mind while fulfilling obligations to the group. Oja contrasted Elliot with a group member operating at the Conscientious stage of development, who had difficulty maintaining her obligation to the group without sacrificing her own personal goals.

Florence, at the Conscientious stage, had experienced frustration about a school-mandated staff development program. Florence had already studied in the area of the mandated program and felt quite skillful in that area. The school program provided no alternatives for teachers like Florence, but rather forced all teachers to participate in the same workshop. This, Florence felt, was alienating her and causing her to resist a program she might otherwise have supported enthusiastically.

Oja concluded that age, life period, and teaching experience can help explain key issues in a teachers' life and career and can often explain why a teacher will choose to become involved in certain staff development activities. However, teachers' performance, thoughts, problem solving, and group behavior while participating in a particular staff development activity appear to be related to their cognitive-developmental stages (Oja, 1989). Oja argued that teachers' performances are not simply idiosyncratic and grounded in the individual. Rather, cognitive-development stage characteristics help explain how certain teachers think and perform in staff development. Based on this analysis, Oja argued that staff development programs attending to the developmental characteristics of participating teachers will be more successful in achieving their goals than programs that ignore teachers' development. Thus, according to Oja, staff development programs should be designed to involve teachers at different developmental levels and also to create safe environments that allow for teachers to development further. The importance of helping teachers to reach higher levels of development is supported by classroom-based research that provides evidence that states, in general "...persons judged at higher stages of development function more complexly, possess a wider repertoire of behavioral skills, perceive problems more broadly, and can respond more accurately and empathically to the needs of others" (Sprinthall & Thies-Sprinthall, 1983). For example, in Hunt's (1971) studies of teachers in their own classrooms, teachers exhibiting higher stages of development were more adaptive, more flexible, and more tolerant in their teaching style. These teachers also used a variety of teaching methods, such as lectures, small group discussions, inquiry, and role playing.

Patterns in Ego Development Research. There are few patterns in the studies discussed in this section. Although researchers found relationships between teachers' levels of ego development and their responses to pre-service and in-service education programs, most of the relationships were weak. Novice teachers at low levels of ego development are more likely to conform to teaching orientations held by course instructors than are novices at high levels of ego development. However, teachers at low levels of development also were found to conceptualize the role of teacher in much narrower ways than teachers at higher levels.

Given these findings, it is difficult to predict how teachers' developmental levels may be related to teachers' being receptive to learning about and implementing good mathematics instruction as described in NCTM's *Professional Standards* and the *Standards for Curriculum and Evaluation.* Teachers at lower levels of ego development may not be able to accept teachers using a variety of approaches to mathematics instruction or students working independently in the classroom. It may, therefore, be important to provide experiences for mathematics teachers that encourage teachers to grow developmentally, concurrent with or even before providing experiences more directly related to implementing NCTM standards. This means that staff development programs with the goal of improving mathematics teaching should not only address mathematics and pedagogy, but also create safe environments that allow for teacher development in ego maturity, principled moral and ethical reasoning, and increased conceptual complexity.

Staff development programs designed to serve teachers of differing developmental levels will be more successful in achieving their goals than programs that ignore them. Attempting to train all teachers to implement new curricula may be counterproductive if the teachers are at differing developmental levels; teachers who have been using problem-solving or small group work successfully for ten years most likely will resent the in-service education that might be new and exciting to teachers unaware of these teaching techniques.

Patterns and Implications in Teacher Development

This section concludes with an examination of patterns in teachers' developmental stages apparent across the three research programs discussed. The theories in which this research is grounded presuppose stages of development that occur in a hierarchical, sequential, and invariant order, ranging from less to more complex or mature. In general, the studies reviewed suggest that there are, indeed, differences in teachers in terms of their developmental stages, but that these differences are not necessarily based on age or amount of experience.

There is also evidence that the progression through developmental stages is not as linear as the theories suggest. Individual teachers were found to have characteristics of several stages simultaneously. For example, a teacher's beliefs about authority differed depending on what aspect of his or her teaching the researcher probed. Also, an individual teacher could concurrently have concerns related to self, task, and impact on students, and, thus be dealing with competing concerns. Novice teachers were found to have concerns about survival in the classroom, while still holding concerns about the impact they have on student learning.

The results of the studies reviewed in this section suggest that novice teachers are sometimes not developmentally ready to assume the roles required of them as good mathematics teachers. They look outside themselves for authorities on pedagogy and mathematics, but may not be aware of "experts" other than the textbook or curriculum guide. The NCTM documents call for teachers to work with their students, encouraging students to construct mathematical meaning for themselves in the classroom. Currently, however, the more traditional external authorities—textbooks and state-mandated assessments—encourage teachers to teach in ways inconsistent with NCTM's documents. Teachers at low levels with respect to either Perry's or Loevinger's developmental theories appear to have difficulty accepting varieties of approaches to teaching or viewpoints that deviate from the traditional; this implies that these teachers will tend to reject innovation and hold to traditional modes of mathematics teaching. An important topic for research in this area is how experiences can be designed to help teachers attain the higher developmental stages that would enable them to teach mathematics well.

We urge caution in expecting too much from such experiences, however. Most of the changes in development found in the studies presented were merely statistically significant, and neither indicated movement through many developmental levels nor showed many teachers who had attained high levels of development. This lack of major change may be due, in part, to the fact that, in general, this research has not been longitudinal. Longer interventions or more experience in the classroom may lead to greater developmental changes than have been noted to date. Also, there is little evidence to indicate whether the changes reported in these studies are long-lasting or only temporary. Investigations of more longitudinal interventions would address some of these deficiencies.

THE LEARNING TO TEACH MATHEMATICS PROJECT

Before concluding this chapter with a discussion of implications of the research for practice and research, we present a brief description of a study we are currently conducting. The design of this study takes into account several of the limitations found in the studies previously discussed in this chapter. It draws upon multiple research perspectives; follows teachers for a period of two years, through their final year of teacher preparation and first year of teaching; and examines changes in the teachers' classroom actions as well as in their cognitions. Thus, we believe it is an example of research that has the potential to improve our understanding of the complexity of the process of becoming a mathematics teacher.

The Learning to Teach Mathematics project (Borko, et al., 1990) brings together perspectives from anthropology, cognitive psychology, and mathematics education to study the process of becoming a middle-school mathematics teacher, as it occurs for a small number of novice teachers during their final year in a university teacher education program and first year of teaching. We consider both generic and mathematics-specific components of becoming a mathematics teacher as we attempt to describe and explain changes in the teachers' knowledge, beliefs, thinking, and actions. Central to our explanations are the influences of the participants' university and public school experiences, including the teaching cultures and social organization of mathematics instruction in the two kinds of settings and key university and public school personnel's expectations and actions related to learning to teach.

Eight seniors in an elementary teacher education program participated in the first year of the investigation. All eight are women who began their final year of teacher preparation with the intention to teach middle school mathematics, and who had mathematics as an "area of concentration" in their certification program. The design of the program called for students to participate in four different student teaching placements (7 weeks each, 2 per semester) over the course of the year. During the first 3 placements, the cohort taught half-days and took methods courses taught by University faculty at a central location.

Data collection instruments were designed to explore the process of learning to teach mathematics in numerous ways and from numerous perspectives. During the first year of the study, for example, questionnaires and semi-structured interviews were administered at three points during the school year to assess participants' knowledge and beliefs about mathematics, mathematics pedagogy, general pedagogy, and learning to teach. During each of three observation cycles designed to provide information about participants' thinking and actions in the classroom, we observed each participant teaching for a week of instruction and interviewed her before and after each observed lesson. We also observed conferences between the participants and their respective cooperating teachers and university supervisors, and interviewed each person about interactions during the conferences. Interviews designed to gather information concerning the sociocultural environment of the participants' school division, schools, and classrooms were conducted with the student teachers and a variety of people, including principals, cooperating teachers, and university supervisors. Parallel interviews were conducted with the student teachers and various university personnel, including the model (program) director, instructors, and university supervisors, to gather information about the sociocultural environment of the university program. In addition, we observed all class sessions of the mathematics methods course and interviewed participants and the course instructor about course goals, objectives, and content. A similar data collection plan was used to follow four of the participants through their first year of teaching middle-grades mathematics.

We are analyzing these data using a framework which allows us to examine our four major areas of interest: participants' knowledge and beliefs, participants' thinking and actions in the classroom, the sociocultural environment of the public school setting; and the sociocultural environment of the university teacher education program. In the first phase of analysis, we examine each component of the framework separately. We then examine the four sets of analyses together in order to understand both the process of learning to teach and the influences of university and public school experiences on that process. Our analysis to date covers the experiences of two of the participants—Ms. Daniels and Ms. Jenkins—during their first semester of student teaching and their mathematics methods course. (See Borko, et al., 1990; Borko, et al., in press; Eisenhardt, Behm, & Romagnano, 1991.)

Our analysis of interviews about participants' knowledge and beliefs reveals that Ms. Daniels and Ms. Jenkins both had strong procedural understanding of middle-grades mathematics, but lacked strong conceptual understanding; their understanding of techniques typically employed to facilitate teaching for conceptual understanding was also limited. Ms. Daniels's and Ms. Jenkins's goals for teaching mathematics included helping students to understand why mathematics works, as well as how to do mathematics, and they were aware their lack of conceptual understanding would cause difficulties in achieving these goals.

Another consistent theme in these data is their desire to make mathematics class fun for their students, viewing the use of games as a potential solution to this desire. They talked about games and activities not in terms of how much mathematics their students might learn, but, rather, in terms of how much more fun and exciting the games would make the mathematics class.

These issues of conceptual versus procedural understanding and making mathematics fun were also apparent in our analysis of observation cycle data. However, these data showed that both student teachers acknowledged and demonstrated a limited understanding of techniques typically employed in teaching for conceptual understanding. Further, almost all the explanations they provided while teaching were procedural.

In keeping with their desire to make mathematics fun. Ms. Daniels and Ms. Jenkins placed a priority on planning lessons that students would enjoy and that would help them to develop positive attitudes toward math. Their teaching reflected this priority; observed lessons frequently included games and other hands-on activities.

Participants looked to the mathematics methods course for ideas about how to teach math, paying particular attention to strategies for incorporating concrete and semi-concrete representations into classroom activities. Whereas the mathematics methods course instructor presented these ideas as strategies for illustrating conceptual underpinnings of mathematical procedures, though, the student teachers used them primarily as ways to motivate and actively engage pupils. Unfortunately, they sometimes did not help pupils to see the connection between the concrete or semi-concrete representations and the mathematical concepts they were meant to illustrate; instead, their presentations focused on the procedural aspects of the activities, such as how to fold paper and shade areas to represent multiplication of fractions.

Our analysis of the university experience suggests that the teacher education program rather abruptly separates the student teachers from their familiar roles as university students, requiring them to balance the dual roles of university student and public school teacher. For example, student teachers must attend classes and take university course work seriously, while at the same time being competent as a classroom teacher. The teacher education program suggests it is the student teachers' responsibility to negotiate these demanding, and sometimes competing, sets of expectations.

In addition, in the methods courses taken simultaneously with student teaching, content includes both theoretical and applied components. Both aspects are considered by the university to be important; however, theoretical work is given priority. In the public school setting, on the other hand, almost all of the student teachers and their cooperating teachers give priority to learning ideas that work in the classroom—the applied aspect of their university coursework—and express impatience with what they consider to be the irrelevant contribution of theory.

Not surprisingly, the student teachers viewed themselves as straddling the two worlds of the university and the public school. Faced with a tangle of competing expectations, they demanded to be taught ideas that work in the classroom; such as techniques that hold students' attention while focusing on procedural content. They found the university experience to be of little practical value and argued most of the information in the methods courses was too theoretical to help in their classroom teaching. Also, participants noted that the work required for the university courses often conflicted with the time needed to prepare lessons for their student teaching.

In a number of important ways, then, the university program did not work. First, it did not work if its purpose was to produce beginning teachers who taught according to the mathematics education community's conception of good mathematics instruction. Second, it did not work if its purpose was to turn out inquisitive, informed, and confident beginning teachers. We turn to perspectives from cognitive psychology and anthropology to examine possible reasons for this lack of success.

In the section on Learning to Teach, we noted that three concepts are central to understanding learning to teach from a cognitive psychological perspective: subject matter knowledge, pedagogical content knowledge, and pedagogical reasoning skills. These concepts, and conclusions reported in that section of the chapter, help to explain patterns in Ms. Daniels's and Ms. Jenkins's teaching. Like other student teachers who lack adequate subject matter knowledge, Ms. Daniels's and Ms. Jenkins's teaching was characterized by procedural rather than conceptual explanations, and they lacked confidence in their ability to teach well. Further, Ms. Daniels and Ms. Jenkins often were unable to come up with powerful ways of presenting the discipline of mathematics to their students. Given that they spent their limited planning time designing fun activities, and that they lacked conceptually-based subject matter knowledge, it is not surprising that they encountered difficulty in expanding their pedagogical content knowledge base.

White's (1989) model of student teaching as a "rite of passage"—a program staged by the teaching profession only to reproduce itself—helps us to understand the program's lack of success in terms of the student teachers' reactions to the competing demands of the university and public school. According to her model, the rite-of-passage ritual includes three stages: *separation* (the disruption of old social networks and erosion of existing social identity), *transition* (an attempt to reorganize novices' behavior, appearance, and ways of thinking to bring them in line with conventions within the profession), and *incorporation* (the conferring of the label and credentials of their new status). Viewed from this perspective, the university was only partially successful in facilitating transition from student to teacher. It did orchestrate the disruption of old social networks to set a need for the transitional stage. However, to a large extent, it turned the task of enculturation over to the public schools, which stressed the acquisition of practical, rather than theoretical knowledge. Thus, the university created a situation in which student teachers received conflicting messages about how to reorganize their ways of thinking and acting as professionals, and it gave the message that student teachers were expected to resolve the competing demands of the two institutions on their own, failing to become a source of assistance in the student teachers' transition.

The constraints imposed by this situation seemed to set the student teachers up to need and demand "quick fixes" for the learning-to-teach activities they faced. When was there time to develop a conceptual explanation of a mathematics problem if it must be taught tomorrow and one's own understanding was only procedural? When was there time to reflect on today's lesson if there were only a few hours left to prepare tomorrow's? Facing these problems, it is not surprising that student teachers demanded to be taught ideas that work in the classroom or that they selectively focused on practically relevant ideas, ignoring other, conceptually oriented aspects of their university coursework. (For a more extended presentation of this analysis, see Eisenhart, Behm, and Romagnano, 1991.)

IMPLICATIONS FOR PRACTICE AND RESEARCH

The research presented in this chapter is diverse in both content and perspective. The studies included provide snapshots (pictures) of novice teachers through various lenses (perspectives). There are patterns that can be seen through each individual lens, but there are also patterns common to all lenses. Whether we view becoming a teacher as a process of learning, socialization, or development, it seems clear that the process is a complex one and that it is influenced by multiple factors. Further, experiences during early phases of becoming a teacher—professional coursework, student teaching, and the initial years in the classroom—are critical. Important changes occur as a result of these experiences, as novice teachers begin their long journeys toward expertise.

The photography metaphor used here is appropriate, not only because the research uses different lenses (or theoretical frameworks), but also because existing research provides very little data on the *process* of becoming a teacher. In preparing this chapter, we searched for longitudinal programs of research and investigations that examined the impact of teacher education interventions. We found that longitudinal studies following novice teachers for more than two years are almost nonexistent, despite agreement among teacher educators and researchers that becoming a teacher is a lifelong process. Also, we found few research programs that included intervention components. More is known about the process of becoming a mathematics teacher today than was a decade ago, and the implications for practice that follow reflect these new understandings; however, understanding is still quite limited. The final section of this chapter suggests directions for research designed to remedy this limitation.

Implications for Practice

It has long been assumed that teachers need to know how children learn mathematics in order to design effective instructional programs. Similarly, the research reviewed in this chapter suggests that the designs of effective teacher education programs, pre-service or in-service, are dependent upon designers understanding teachers' learning, development, and socialization. What does the research on becoming a teacher suggest for the design and implementation of teacher education programs?

The literature on learning to teach indicates that teacher education programs should provide opportunities for growth in content knowledge, pedagogical content knowledge, and pedagogical reasoning. The NCTM *Curriculum and Evaluation Standards for School Mathematics* (NCTM, 1989b) and *Professional Standards for Teaching Mathematics* (NCTM, 1990) give some guidance as to the content and pedagogy teachers must understand in order to teach appropriate mathematics well. Both sets of standards provide the direction, but not the mechanism, for reform in school mathematics; however, neither NCTM document is well grounded in systematic research, although both are widely accepted as guidelines representing the mathematics education community's "best thinking" on these topics. Research on learning to teach provides initial ideas for designing teacher education programs conforming to these guidelines. For example, the research program by Schram and colleagues (Schram et al., 1988; Schram et al., 1989) illustrates the kinds of experiences in mathematics and methods courses that can enhance teachers' conceptual understandings of mathematics.

The socialization research, together with Ball's (1988) analysis of the knowledge, beliefs, and dispositions held by prospective teachers when they enter teacher education, suggest that teacher education programs must sometimes help participants to "unlearn" as well as to learn. That is, mathematics teacher education must help teachers discard some of the knowledge, beliefs, and dispositions regarding mathematics and pedagogy they bring to the pre-service and in-service programs.

Research on teacher socialization also indicates that, in order for school experiences to influence novice teachers' socialization in positive ways, a school must exhibit a cohesive and coherent teaching culture. If a student teaching placement is in a school where several cultures of teaching are represented, it is unlikely that the placement will have much of a socializing effect. It follows, then, that student teaching placements should be carefully chosen to ensure that the knowledge, beliefs, and attitudes present in a cooperating teacher's classroom, and in the school in general, are in agreement with the goals of the teacher education program. Schools and universities should work together to determine a teaching culture congruent with the goals for teacher education, and should cooperate to establish experiences for teachers that will encourage their socialization into such a culture.

Finally, the research on teacher development indicates that participants in teacher education programs may be at different developmental stages and have very different needs for assistance. For example, teachers at the dualistic position on Perry's scheme may respond very differently to instruction about problem solving as a focus of school mathematics than teachers at the multiplistic position. Similarly, teachers with poorly developed egos may be reluctant to see cooperative learning as a viable way of teaching mathematics, while preferring to learn more teacher-centered styles of classroom instruction.

The developmental perspective argues that for teacher education programs to be most effective—that is, to help students achieve higher levels of development—they should assess participants' levels of development and then provide experiences either at the teachers' level or one stage higher in the developmental hierarchy. However, although there is some evidence that teachers at higher levels of development are more flexible, indirect, and empathic, there is little evidence that they are, in fact, more effective teachers of mathematics.

Unfortunately, currently available research is not sufficient to produce designs of teacher education programs that will address all of the issues raised by the research. Too often, research informs us about what is wrong, but does not offer concrete suggestions on how to fix it. Further, most investigations examine becoming a teacher using a single framework, thus providing limited insights into the complexity of the process. The suggestions for research that follow address these limitations.

Implications for Research

Most existing research on becoming a mathematics teacher is descriptive in nature and, therefore, provides limited evidence about the design and implementation of good mathematics teacher education programs. However, as we pointed out in the previous section, this research does suggest that teacher education programs with certain elements may facilitate the process of becoming a good mathematics teacher. A logical next step for researchers is to design, implement, and study programs for teachers that incorporate these elements. Careful documentation of the experiences of teachers in such programs, and the resulting changes in their knowledge, beliefs, dispositions, thinking, and actions, will provide further insights into the process of becoming a mathematics teacher.

The research reviewed confirms that learning, socialization, and development are long and complex processes. Yet, most of the studies investigated very short periods within the process of becoming a teacher. The longest studies followed novices for five years, and these focused only on changes in teacher concerns; longitudinal studies of longer duration are necessary to increase understanding of the process of becoming a teacher. Researchers must not only study interventions for the life of the intervention, but continue to study the effects of the intervention for several years. This is the only means by which we can learn if the effects of innovations in teacher education are long-lasting and can withstand the pressures of the classroom and the school cultures.

In designing and implementing intervention studies, it is also important that researchers focus more on teachers' classroom actions. Most of the studies presented here described teachers' beliefs, knowledge, dispositions, or thinking; there was much less emphasis on teachers' classroom actions. It is clear from the research on teacher knowledge and beliefs that cognitions are not always reflected in action in the classroom (see Thompson, Chapter 7, this volume; Fennema & Loef, Chapter 8, this volume). If a goal of research on becoming a mathematics teacher is to help teachers become better able to teach mathematics as envisioned in NCTM documents, then it is important to study teacher actions as well as cognitions, as well as to identify conditions under which changes in teacher cognition are likely to be accompanied by compatible changes in classroom actions.

Intervention studies also need to address the complexity of the process of becoming a mathematics teacher. The perspectives of learning, socialization, and development were used to organize this chapter; like most researchers on teaching, these perspectives were treated as distinct conceptual frameworks, when in reality, development, socialization, and learning processes occur simultaneously. In order to best understand the complexity of the process of becoming a mathematics teacher, studies must incorporate multiple perspectives, and must attempt to untangle the relationships between and among the various aspects of becoming a teacher. Such studies would draw together researchers from different research traditions, such as mathematics, psychology, anthropology, and sociology, each with different frameworks within which to view becoming a teacher.

Finally, in designing intervention studies, researchers and teacher educators should both incorporate NCTM's guidelines regarding good mathematics teaching, and draw upon research evidence on factors that facilitate becoming a mathematics teacher. The NCTM documents describe the mathematics and the pedagogy that should be the goals of the learning, socialization, and development experiences within teacher education programs. As we stated in the assumptions underlying this chapter, becoming a mathematics teacher includes both generic and mathematics-specific aspects, and both must be the subject of investigations.

As researchers and teacher educators continue to seek ways to understand and improve the process of becoming a mathematics teacher, they should try to cooperate in designing programs of teacher education that are grounded in research. Teachers in these programs should be studied as they participate in the experiences over time. Researchers should collaborate to bring a variety of perspectives to the study of the complex process of becoming a mathematics teacher.

References

Adams, R. D. (1982). A look at changes in teacher perceptions and behavior across time. *Journal of Teacher Education, 33*(4), 40–43.

Adams, R. D., Hutchinson, S., & Martray, C. (1980, April). *A developmental study of teacher concerns across time.* Paper presented at the annual meeting of the American Educational Research Association, Boston.

Adams, R. D., & Martray, C. (1981, April). *Teacher development: A study of factors related to teacher concerns for pre, beginning, and experienced teachers.* Paper presented at the annual meeting of the American Educational Research Association, Los Angeles.

Anderson, C. (1989). The role of education in the academic disciplines in teacher preparation. In A. E. Woolfolk (Ed.), *Research perspectives on the graduate preparation of teachers* (pp. 88–107). Englewood Cliffs, NJ: Prentice Hall.

Anderson, R. C. (1984). Some reflections on the acquisition of knowledge. *Educational Researcher, 13*(10), 5–10.

Ball, D. L. (1988a, April). *Prospective teachers' understandings of mathematics: What do they bring with them to teacher education?* Paper presented at the annual meeting of the American Educational Research Association, New Orleans.

Ball, D. L. (1988b). Research on teacher learning: Studying how teachers' knowledge changes. *Action in Teacher Education, 10*(2), 17–24.

Ball, D. L. (1990). Prospective elementary and secondary teachers' understanding of division. *Journal for Research in Mathematics Education, 21,* 132–144.

Bartholomew, J. (1976). Schooling teachers: The myth of the liberal college. In G. Witty & M. Young (Eds.), *Explorations in the politics of school knowledge* (pp. 114–124). Driffield: Nafferton Books.

Baxter, J., Richert, A., & Saylor, C. (1985, April). *Science group: Content and process of biology.* Paper presented at the Annual Meeting of the American Educational Research Association, Chicago.

Becker, H., Geer, B., Hughes, E., & Strauss, A. (1961). *Boys in White.* Chicago, IL: University of Chicago Press.

Bereiter, C., & Scardamalia, M. (1987). An attainable version of high literacy: Approaches to teaching higher-order skills in reading and writing. *Curriculum Inquiry, 17*(1), 9–30.

Berliner, D. C. (1986). In search of the expert pedagogue. *Educational Researcher, 15*(7), 5–13.

Berliner, D. C. (1987). Ways of thinking about students and classrooms by more and less experienced teachers. In J. Calderhead (Ed.), *Exploring teachers' thinking* (pp. 60–83). London: Cassell Educational Limited.

Berliner, D. C. (1988, February). *The development of expertise in pedagogy.* Paper presented at the annual meeting of the American Association of Colleges for Teacher Education, New Orleans.

Bolster, A. S. (1983). Toward a more effective model of research on teaching. *Harvard Educational Review*, 53(3), 294–308.

Borko, H. (1985). Student teachers' planning and evaluation of reading lessons. In J. A. Niles & R. Lalik (Eds.) *Issues in literacy: A research persective* (Thirty-fourth yearbook of the National Reading Conference, pp. 263–71). New York, NY: The National Reading Conference, Inc.

Borko, H. (1986). Clinical teacher education: the induction years. In J. V. Hoffman & S. A. Edwards (Eds.), *Reality and reform in clinical teacher education* (pp. 45–63). New York, NY: Random House.

Borko, H., Bellamy, M. L., Sanders, L., & Sanford, J. (1989, March). *Experienced teachers' and student teachers' planning and instruction*. Paper presented at the annual meeting of the American Educational Research Association, San Francisco, CA.

Borko, H., Brown, C. A., Underhill, R. G., Eisenhart, M., Jones, D., & Agard, P. C. (1990). Learning to teach mathematics (Year 2 Progress Report submitted to the National Science Foundation). Blacksburg, VA: Virginia Polytechnic Institute and State University.

Borko, H., Eisenhardt, M., Underhill, R. G., Brown, C. A., Jones, D., & Agard, P. C. (in press). To teach mathematics for conceptual or procedural knowledge?: A dilemma of learning to teach in the 'New World Order' of mathematics education reform. *Journal for Research in Mathematics Education*.

Borko, H., Lalik, R., & Tomchin, E. (1987). Student teachers' understanding of successful teaching. *Teaching and Teacher Education*, 3, 77–90.

Borko, H., & Livingston, C. (1989). Cognition and improvisation: Differences in mathematics instruction by expert and novice teachers. *American Educational Research Journal*, 26, 473–498.

Borko, H., Livingston, C., McCaleb, J., & Mauro, L. (1988). Student teachers' planning and post-lesson reflections. Patterns and implications for teacher preparation. In J. Calderhead (Ed.), *Teachers' professional learning* (pp. 65–83). London: Falmer Press.

Borko, H., & Shavelson, R. J. (1990). Teachers' decision making. In B. Jones & L. Idols (Eds.), *Dimensions of thinking and cognitive instruction* (pp. 311–346). Hillsdale, NJ: Lawrence Erlbaum.

Brown, C. A. (1985). *A study of the socialization to teaching of a beginning secondary mathematics teacher*. Unpublished doctoral dissertation, University of Georgia, Athens.

Brown, S. I., Cooney, T. J., & Jones, D. (1990). Mathematics teacher education. In W. R. Houston (Ed.), *Handbook of research on teacher education* (pp. 639–656). New York, NY: Macmillan.

Carlsen, W. S. (1987, April). *Why do you ask? The effects of science teacher subject-matter knowledge on teacher questioning and classroom discourse*. Paper presented at the Annual Meeting of the American Educational Research Association, Washington, D.C.

Carnegie Task Force on Teaching as a Profession. (1986). *A Nation Prepared: Teachers for the 21st Century*. Hyattsville, MD: Carnegie Forum on Education and the Economy.

Carpenter, T. P., Fennema, E., Peterson, P. L., & Carey, D. A. (1988). Teachers' pedagogical content knowledge of students' problem solving in elementary arithmetic. *Journal for Research in Mathematics Education*, 19, 385–401.

Carpenter, T. P., Fennema, E., Peterson, P. L., Chiang, C., & Loef, M. (1989). Using knowledge of children's mathematics thinking in classroom teaching: An Experimental study. *American Educational Research Journal*, 26, 499–531.

Carpenter, T. P., & Moser, J. M. (1983). The acquisition of addition and subtraction concepts. In R. Lesh & M. Landau (Eds.), *The acquisition of mathematics concepts and processes* (pp. 7–44). New York, NY: Academic Press.

Carpenter, T. P. & Peterson, P. L. (1988). Learning through instruction: The study of students' thinking during instruction in mathematics. *Educational Psychologist*, 23, 79–85.

Carter, K., Cushing, K., Sabers, D. Stein, P., & Berliner, D. (1988). Expert-novice differences in perceiving and processing visual classroom stimuli. *Journal of Teacher Education*, 39(3), 25–31.

Carter, K., Sabers, D., Cushing, K., Pinnegar, S., & Berliner, D. (1987). Processing and using information about students: A study of expert, novice, and postulant teachers. *Teaching and Teacher Education*, 3, 147–157.

Chi, M., Feltovich, P., & Glaser, R. (1981). Categorization and representation of physics problems by experts and novices. *Cognitive Science*, 5(2), 121–152.

Clark, C. M., & Peterson, P. L. (1986). Teachers' thought processes. In M. C. Wittrock (Ed.), *Handbook of research on teaching* (3rd ed., pp. 255–296). New York, NY: Macmillan.

Cobb, P., Wood, T., & Yackel, E. (1990). Classrooms as learning environments for teachers and researchers. In R. Davis, C. Maher, & N. Noddings (Eds), *Constructivist views on the teaching and learning of mathematics* (Number 4, pp. 125–146). Reston, VA: NCTM.

Cobb, P., Wood, T., Yackel, E., Nicholls, J., Wheatley, G., Trigatti, B., & Perlwitz, M. (1991). Assessment of a problem-centered second-grade mathematics project. *Journal for Research in Mathematics Education*, 22, 3–29.

Cobb, P., Yackel, E., & Wood, T. (1988). Curriculum and teacher development: Psychological and anthropological perspectives. In E. Fennema, T. P. Carpenter, & S. J. Lamon (Eds.), *Integrating research on teaching and learning mathematics* (pp. 92–131). Madison, WI: Wisconsin Center for Education Research, University of Wisconsin–Madison.

Conant, J. (1963). *The education of American teachers*. New York, NY: McGraw Hill.

Copeland, W. (1978). Processes mediating the relationship between cooperating teacher behavior and student-teacher classroom performance. *Journal of Educational Psychology*, 70, 95–100.

Copeland, W. (1980). Student teachers and cooperating teachers: An ecological relationship. *Theory into Practice*, 18, 194–199.

Crist, J., Marx, R. W., & Peterson, P. L. (1974). *Teacher behavior in the organizational domain* (Report submitted to the National Institute of Education). Stanford, CA: Stanford Center for R & D in Teaching.

Cummings, A. L., & Murray, H. G. (1989). Ego development and its relation to teacher education. *Teaching and Teacher Education*, 5, 21–32.

deGroot, A. D. (1965). *Thought and choice in chess*. The Hague: Mouton.

Delamont, S. (1983). *Interaction in the classroom*. London: Methuen.

Densmore, K. (1987). Professionalism, proletarianization and teacher work. In T. Popkewitz (Ed.), *Critical Studies in Teacher Education: Its Folklore, Theory and Practice* (pp. 130 160). New York, NY: Falmer Press.

Doyle, W., & Ponder, G. (1975). Classroom ecology: Some concerns about a neglected dimension of research on teaching. *Contemporary Education*, 46, (3), 183–188.

Doyle, W. (1978). Paradigms for research on teacher effectiveness. In L. S. Shulman (Ed.), *Review of research in education* (Vol. 5, pp. 163–198). Itasca, IL: Peacock.

Dreyfus, H. L., & Dreyfus, S. E. (1986). *Mind over machine*. New York, NY: Free Press.

Eisenhart, M., Behm, L., & Romagnano, L. (1991). Learning to teach: Developing expertise or rite of passage? *Journal of Education for Teaching*, 17 (1), 51–71.

Feiman-Nemser, S. (1983). Learning to teach. In L. Shulman and G. Sykes (Eds.), *Handbook on teaching and policy* (pp. 150–170). New York, NY: Longman.

Feiman-Nemser, S. (1990). Teacher preparation: Structural and conceptual alternatives. In W. R. Houston (Ed.), *Handbook for research on teacher education* (pp. 212–233). New York, NY: Macmillan.

Feiman-Nemser, S., & Buchmann, M. (1986). The first year of teacher preparation: Transition to pedagogical thinking? *Journal of Curricular Studies*, 18, 239–256.

Feiman-Nemser, S. & Buchmann, M. (1987). When is student teaching teacher education? *Teaching and Teacher Education*, 3, 255–273.

Fiedler, M. (1975). Bidirectionality of influence in classroom interaction. *Journal of Educational Psychology, 67,* 735–744.

Fuller, F. F. (1969). Concerns of teachers: A developmental conceptualization. *American Educational Research Journal, 6,* 207–226.

Fuller, F. F. (1970). *Personalized Education for Teachers. An Introduction for Teacher Educators.* Austin, TX: Research and Development Center for Teacher Education, The University of Texas. (ERIC Document Reproduction Service No. ED 048 105)

Fuller, F., & Bown, O. (1975). On becoming a teacher. In K. Ryan (Ed.), *Teacher education* (The 74th yearbook of the National Society for the Study of Education, pp. 25–52). Chicago, IL: University of Chicago Press.

Ginsburg, M. (1988). *Contradictions in teacher education and society: A critical analysis.* New York, NY: Falmer Press.

Glaser, R. (1984). Education and thinking: The role of knowledge. *American Psychologist, 39,* 93–104.

Good, T. L., & Grouws, D. A. (1979). The Missouri Mathematics Effectiveness Project: An experimental study in fourth-grade classrooms. *Journal of Educational Psychology, 71,* 355–362.

Good, T. L., Grouws, D. A., & Ebmeier, H. (1983). *Active mathematics teaching.* New York, NY: Longman.

Grossman, P. L. (1987, April). *A tale of two teachers: The role of subject matter orientation in teaching.* Paper presented at the annual meeting of the American Educational Research Association, Washington, DC.

Grossman, P. L., & Richert, A. E. (1986, April). *Unacknowledged knowledge growth: A re-examination of the effects of teacher education.* Paper presented at the annual meeting of the American Educational Research Association, San Francisco.

Holmes Group (1986). *Tomorrow's Teachers.* East Lansing, MI: The Holmes Group.

Hoy, W. (1969). Pupil control ideology and organizational socialization: A further examination of the influence of experience on the beginning teacher. *School Review, 77,* 257–265.

Hunt, D. E. (1971). *Matching models in education.* Toronto: Ontario Institute for Studies in Education.

Iannaccone, L. (1963). Student teaching: A transitional stage in the making of a teacher. *Theory into Practice, 2,* 73–80.

Jackson, P. W. (1968). *Life in classrooms.* New York, NY: Holt, Rinehart, & Winston.

Kelly, G. A. (1955). *The psychology of personal constructs. Vol. 1. A theory of personality.* New York, NY: Norton.

Klein, S. (1971). Student influence on teacher behavior. *American Educational Research Journal, 8,* 403–421.

Kohlberg, L. (1969). Stage and sequence: The cognitive-developmental approach to socialization. In D. A. Goslin (Ed.), *Handbook of socialization theory and research* (pp. 347–380). Chicago, IL: Rand-McNally.

Krajcik, J. S., & Layman, J. W. (1989, March). *Middle school teachers' conceptions of heat and temperature: Personal and teaching knowledge.* Paper presented at the Annual Meeting of the National Association for Research in Science Teaching, San Francisco.

Lacey, C. (1977). *The socialization of teachers.* London: Methuen.

Lacey, C. (1985). Professional socialization of teachers. In T. Husen & T. N. Postlethwaite (Eds.), *The international encyclopedia of education* (pp. 4073–4084). Oxford: Pergamon Press.

Lanier, J. E. & Little, J. W. (1986). Research on teacher education. In M. C. Wittrock (Ed.), *Handbook of research on teaching* (3rd ed., pp. 527–569). New York, NY: Macmillan.

Larkin, J., McDermott, J., Simon, D. P., & Simon, H. A. (1980). Expert and novice performance in solving physics problems. *Science, 208,* 1135–1142.

Leinhardt, G. (1986, April). *Math lessons: A contrast of novice and expert competence.* Paper presented at the annual meeting of the American Educational Research Association, San Francisco.

Leinhardt, G., & Greeno, J. G. (1986). The cognitive skill of teaching. *Journal of Educational Psychology, 78,* 75–95.

Leinhardt, G., & Smith, D. (1985). Expertise in mathematics instruction: Subject matter knowledge. *Journal of Educational Psychology, 77,* 247–271.

Livingston, C., & Borko, H. (1990). High school mathematics review lessons: Expert-novice distinctions. *Journal for Research in Mathematics Education, 21,* 372–387.

Loevinger, J. (1976). *Ego development.* San Francisco, CA: Jossey-Bass.

Loevinger, J. (1980). Some thoughts on ego development and counseling. *Personnel and Guidance Journal, 58,* 389–390.

Lortie, D. (1975). *Schoolteacher.* Chicago, IL: The University of Chicago Press.

Marso, R. N., & Pigge, F. L. (1989). The influence of preservice training and teaching experience upon attitude and concerns about teaching. *Teaching & Teacher Education, 5*(1), 33–41.

McArthur, J. (1978). What does teaching do to teachers? *Educational Administration Quarterly, 14,* 89–103.

McDonald, F. (1980). The problems of beginning teachers: A crisis in training (Vol. 1). *Study of induction programs for beginning teachers* (N.I.E. Contract No. 400-78-0069). Princeton, NJ: Educational Testing Service.

McGalliard, W. A. (1983). *Selected factors in the conceptual systems of geometry teachers: Four case studies.* Unpublished doctoral dissertation, University of Georgia, Athens.

Meyerson, L. N. (1977). *Conception of knowledge in mathematics: Interaction with and application to a teaching methods course.* Unpublished doctoral dissertation, State University of New York at Buffalo.

Mitroff, I., & Kilmann, R. (1978). *Methodological approaches to social sciences.* San Francisco, CA: Jossey-Bass.

National Center for Research on Teacher Education. (1988). Teacher education and learning to teach: A research agenda. *Journal of Teacher Education, 39*(6), 27–32.

National Council of Teachers of Mathematics. (1989a). *Curriculum and Evaluation Standards for School Mathematics.* Reston, VA: Author.

National Council of Teachers of Mathematics. (1989b). *Guidelines for the Post-Baccalaureate Education of Teachers of Mathematics.* Reston, VA: Author.

National Council of Teachers of Mathematics. (1990). *Professional Standards For Teaching Mathematics.* Reston, VA: Author.

Oja, S. N. (1980). Adult development is implicit in staff development. *Journal of Staff Development, 1*(2), 8–55.

Oja, S. N. (1989). Teachers: Ages and stages of adult development. In M. L. Holly & C. S. McLoughlin (Eds.), *Perspectives on teacher professional development* (pp. 119–154). London: Falmer Press.

Oja, S. N., & Ham, M. C. (1984). A cognitive-developmental approach to collaborative action research with teachers. *Teachers College Record, 86*(1), 171–92.

Owens, J. E. (1987). *A study of four preservice secondary mathematics teachers' constructs of mathematics and mathematics teaching.* Unpublished doctoral dissertation, University of Georgia, Athens.

Perry, W. G. (1970). *Forms of intellectual and ethical development in the college years: A scheme.* New York, NY: Holt, Rinehart, and Winston.

Peterson, P. L., Carpenter, T. P., & Fennema, E. (1989). Teachers' knowledge of students' knowledge and cognition in mathematics problem solving. *Journal of Educational Psychology, 81,* 558–569.

Peterson, P. L., Fennema, E., Carpenter, T. P., & Loef, M. (1989). Teachers' pedagogical content beliefs in mathematics. *Cognition and Instruction, 6,* 1–40.

Petty, M., & Hogben, D. (1980). Explorations of semantic space with beginning teachers: A study of socialization into teaching. *British Journal of Teacher Education, 6,* 51–61.

Piaget, J. (1972). Intellectual evolution from adolescence to adulthood. *Human Development, 15*(1), 1–12.

Popkewitz, T. (Ed.). (1987). *Critical Studies in Teacher Education: Its Folklore, Theory and Practice.* New York, NY: Falmer Press.

Putnam, R. T., Lampert, M., & Peterson, P. L. (1990). Alternative perspectives on knowing mathematics in elementary schools. In C. Cazden (Ed.), *Review of Research in Education* (Vol. 16, pp. 57–150). Washington, D.C.: American Educational Research Association.

Resnick, L. B. (1985). Cognition and instruction: Recent theories of human competence. In B. L. Hammonds (Ed.), *Master lecture series: Vol. 4 Psychology and learning* (pp. 123–186). Washington, D. C.: American Psychological Association.

Schram, P., & Wilcox, S. K. (1988). Changing preservice teachers' conception of mathematics learning. In M. Behr & C. Lacampagne (Eds.), *Proceedings of the Tenth Annual Meeting of the North American Chapter of the International Group for the Psychology of Mathematics Education.* DeKalb, IL: Northern Illinois University.

Schram, P., Wilcox, S., Lanier, P., & Lappan, G. (1988). *Changing mathematical conceptions of preservice teachers: A content and pedagogical intervention* (pp. 296–302). Paper presented at the annual meeting of the American Educational Research Association, New Orleans.

Schram, P., Wilcox, S. K., Lappan, G., & Lanier, P. (1989). Changing preservice teachers' beliefs about mathematics education. In C. Maher, G. Goldin, & R. Davis (Eds.), *Proceeding of the Eleventh Annual Meeting of the North American Chapter of the International Group for the Psychology of Mathematics Education.* (pp 296–302). New Brunswick, NJ: Rutgers University.

Schwab, J. (1978). *Science, curriculum, and liberal education: Selected essays.* Chicago, IL: University of Chicago Press.

Shavelson, R. J. (1976). Teachers' decision making. In N. L. Gage (Ed.), *The psychology of teaching methods* (Yearbook of National Society for the Study of Education, pp. 372–414). Chicago, IL: University of Chicago Press.

Shavelson, R. J. (1986). *Interactive decision making: Some thoughts on teacher cognition.* Invited address, I Congreso Internacional, "Pensamientos de los Profesores y Toma de Decisiones," Seville, Spain.

Shuell, T. J. (1986). Cognitive conceptions of learning. *Review of Educational Research, 56,* 411–436.

Shulman, L. S. (1986). Those who understand: Knowledge growth in teaching. *Educational Researcher, 15,* 4–14.

Shulman, L. S. (1987). Knowledge and teaching: Foundations of the new reform. *Harvard Educational Review, 57*(1), 1–22.

Shulman, L. S., & Grossman, P. L. (1988). *Knowledge growth in teaching: A final report to the Spencer Foundation.* Stanford, CA: Stanford University.

Smith, D. C., & Neale, D. C. (1989). The construction of subject matter knowledge in primary science teaching. *Teaching and teacher education. 5,* 1–20.

Spatig, L., Ginsburg, M. B., & Liberman, D. (1982). Ego development as an explanation of passive and active models of teacher socialization. *College Student Journal, 16,* 315–325.

Sprinthall, N. A., & Thies-Sprinthall, L. (1983). The teacher as an adult learner: A cognitive-developmental view. In G. A. Griffin (Ed.), *Staff Development* (82nd yearbook of the National Society for the Study of Education, pp. 13–35). Chicago, IL: University of Chicago Press.

Steinberg, R., Haymore, J., & Marks, R. (1985). *Teachers' knowledge and structuring content in mathematics.* Paper presented at the annual meeting of the American Educational Research Association, Chicago.

Tabachnick, B. R., Popkewitz, T., & Zeichner, K. (1979–80). Teacher education and the professional perspectives of student teachers. *Interchange, 10,* 12–29.

White, J. (1989). Student teaching as a rite of passage. *Anthropology and Education Quarterly, 20,* 177–195.

Wilcox, S., Schram, P., Lappan, G., & Lanier, P. (1990, April). *The role of a learning community in changing preservice teachers' knowledge and beliefs about mathematics education.* Paper presented at the annual meeting of the American Educational Research Association, Boston.

Wilson, S. M., Shulman, L. S., & Richert, A. E. (1987). "150 different ways" of knowing: Representations of knowledge in teaching. In J. Calderhead (Ed.), *Exploring teachers' thinking* (pp. 104–124). London: Cassell Educational Limited.

Wittrock, M. C. (Ed.) (1986). *Handbook of research on teaching* (3rd ed.). New York, NY: Macmillan.

Wood, T., Cobb, P., & Yackel, E. (1990). The contextual nature of teaching: Change in mathematics but stability in reading. *Elementary School Journal, 90,* 497–513.

Wright, B., & Tuska, S. (1968). From dream to life in the psychology of becoming a teacher. *School Review, 26,* 183–193.

Zeichner, K. (1985). The ecology of field experience: Toward an understanding of the role of field experiences on teacher development. *Journal of Research and Development in Teacher Education, 18,* 44–52.

Zeichner, K., & Gore, J. (1990). Teacher socialization. In W. R. Houston (Ed.), *Handbook of research on teacher education* (pp. 329–348). New York, NY: Macmillan.

Zeichner, K., & Tabachnick, B. R. (1985, January). The development of teacher perspectives: Social strategies and institutional control in the socialization of beginning teachers. *Journal of Education for Teaching, 11,* 1–25.

Zeichner, K., Tabachnick, B., and Densmore, K. (1987). Individual, institutional, and cultural influences on the development of teachers' craft knowledge. In J. Calderhead (Ed.), *Exploring teachers' thinking* (pp. 21–59). London: Cassell.

Part

·III·

LEARNING FROM INSTRUCTION

▪12▪

RESEARCH ON WHOLE NUMBER
ADDITION AND SUBTRACTION

Karen C. Fuson

NORTHWESTERN UNIVERSITY

The 1980s saw many calls for changes in mathematics education in the United States. Many of these are summarized in the *Curriculum and Evaluation Standards for School Mathematics* (1989). These calls stemmed from at least three sources: (a) evidence of shortcomings of our present school mathematics efforts as evaluated by the National Assessment of Educational Progress (Brown et al., 1989; Kouba et al., 1988) and by international comparisons (McKnight et al., 1987; Rohlen, 1983; Song & Ginsburg, 1987; Stevenson, Lee, and Stigler, 1986; Stigler, Lee, Lucker, & Stevenson, 1982), (b) evidence that children use a variety of different conceptual structures that result in different solution procedures for the same mathematical problem, which indicates that mathematics education must explicitly consider how a given child is thinking about a given problem, and (c) an increasing awareness that we are in the process of changing from an industrial to an informational society, which implies radical changes for school mathematics. Research on whole number addition and subtraction contributed considerably to all three sources, providing evidence that U.S. children demonstrated inadequate place-value and multidigit subtraction knowledge in third grade (the earliest grade tested) and showed less knowledge of addition and subtraction even in first grade than children in several Asian countries. Many studies uncovered a variety of patterns of children's thinking in addition and subtraction situations. Some studies have begun to explore teaching addition and subtraction in ways more consistent with this changing society.

School mathematics education now needs to undergo a shift from teaching children to be calculators for known adult problem situations to helping children learn to use calculators—and other future tools—to solve problem situations not yet imag-

ined by adults. Clearly, it is still important for children to learn to add and subtract; however, the focus needs to change from one of children rapidly producing accurate solutions to pages of stereotypical numeral problems to one of children discussing in the classroom alternative solution procedures for a variety of addition and subtraction situations. The major task for the future is to create a new vision of the mathematics classroom that both reflects new knowledge about children's thinking and is consistent with the new educational goals.

Children's understanding of whole number addition and subtraction has been the focus of a great deal of research in the past 15 years. Researchers in mathematics education, developmental psychology, and cognitive psychology have made considerable headway in understanding the variety of ways children think about single-digit addition and subtraction situations, and they are beginning to understand some important aspects of children's thinking about multidigit addition and subtraction situations. There has been considerable interaction among researchers from these areas, initiated by the Wingspread conference (see Carpenter, Moser, & Romberg, 1982). Separate islands of research remain, however, that reflect different paradigms, models, and results; future research could profit from efforts to bridge and even integrate these islands.

This chapter will first outline the real world domain of whole number addition and subtraction situations. It will then describe the developmental progression of conceptual structures that children between the ages of 2 and 8 construct to interpret and solve these situations. The unitary conceptual structures built for numbers up to one hundred will be considered first, and then the multiunit conceptual structures built for

The author gratefully acknowledges the helpful comments and suggestions provided by Art Baraody, University of Illinois; Tom Carpenter, University of Wisconsin; and Paul Cobb, Purdue University.

multi-digit whole numbers will be discussed. A vision of what might be possible in preschool and primary school classrooms will then be summarized. The treatment of all topics presented here is necessarily brief; a longer review of many of these topics is contained in Fuson (in press), and recent reviews of some topic areas discussed in this chapter are available in Baroody and Ginsburg (1986), Carpenter and Moser (1984), Fuson (1988a, 1990a), Bergeron & Herscovics (1990), Labinowicz (1985), and Riley, Greeno, & Heller (1983).

In this chapter the word "mark" is used to describe what is written down about addition and subtraction situations (including numerals, +, −, and =) instead of the more commonly used word "symbol," because the latter may imply that the mark itself contains meaning, whereas "mark" clearly reminds us that the viewer must interpret that mark and give it meaning. This reminder is especially important in thinking about very simple mathematics, such as whole number addition and subtraction, for which most adults have established meanings for marks that are so routine and automatic that it is difficult for them even to conceive of others not understanding these meanings.

THE MATHEMATICAL AND REAL WORLD WHOLE NUMBER ADDITION AND SUBTRACTION DOMAIN

Addition and Subtraction Situations

The kinds of whole number addition and subtraction situations that exist in the real world have been the focus of a considerable amount of analytical and empirical research. Many different terms have been given to the identified situations, but there is considerable overlap in the situations used in most category systems; the category system here in Figure 12.1 and Table 12.1 draws on all of these earlier efforts and their crucial features (see earlier reviews by Carpenter & Moser, 1983, and Riley et al., 1983). Examples have been taken from many studies to indicate the range of problem examples given even within problem types. The implications these different problem types have for children's solutions are discussed in the section concerning children's unitary conceptual structures.

There are four basic addition and subtraction situations: Compare, Combine, Change Add To, and Change Take From. When there are *two* quantities, one can compare them or combine them; the Compare and Combine situations are *binary* operations, in which two numbers are operated on to produce a unique third number. When there is only *one* quantity, one can add to that quantity or take from that quantity; the Change Add To and Change Take From situations are *unary* operations, in which one number is operated on to produce a unique third number. When objects are used to make each of these real world situations, and the operations are carried out on these objects, the initial situation disappears in all but the Compare situation.

The distinction between *static* situations—of related quantities that do not change—and *active* situations—in which quantities do change—has been made in almost all category systems. Most category systems collapse this static/active distinc-

tion into the binary/unary distinction, yielding only the static binary Combine and Compare categories and the active unary Change Add To and Change Take From categories. However, active binary forms of Combine and Compare problems can be constructed; the actions in these problems cue solution procedures, so they frequently are easier than static forms. Active binary Compare problems, called *Equalize* problems, are combinations of Compare and Change problems in which the difference between two quantities is expressed as unary Change Add To or Change Take From actions (see the examples in Table 12.1 and the drawings in Figure 12.1) rather than as a static state as in Compare problems. One can also make active binary Combine problems, in which the combining is done explicitly in the problem rather than implicitly, using class inclusion terms or conceptually with words like "altogether," as in static Combine problems. Children's solution procedures for Combine situations show both active and static conceptual combining, but this distinction may have little importance except for immature or inexperienced problem solvers—it is included here in Table 12.1 and Figure 12.1 for the sake of completeness. Piaget's (1941/65) conceptually advanced definition of what constitutes children's understanding of addition and subtraction (see the later discussion in "The Increasing Integration of Sequence, Count, and Cardinal Meanings of Number Word") may be relevant to the class inclusion conceptual Combine problems. However, the existence of many Combine problem situations not requiring class inclusion may be one reason that failure on Piagetian concrete-operational tasks has not been found to limit children's ability to solve a range of addition and subtraction situations (Bisanz, 1989; Lemoyne & Favreau, 1981; Underhill, 1986). In particular, the use of the word "altogether" with class terms such as "children" (boys, girls, children) or "candy" (lollipops, Tootsie rolls, candy) may enable children who do not yet show class inclusion to understand a Combine situation just as the use of "all the children" enables children to solve class inclusion tasks (Fuson, Lyons, Pergament, Hall, & Kwon, 1988). It may be preferable to refer to the result of the putting-together combining operation as "all" rather than "whole" for two reasons. First, in real world situations, a "whole" is rarely formed but one can consider "all the entities." Second, "all" seems to be a more natural response for children—for example, kindergarteners and first graders asked to justify correct class-inclusion responses frequently referred to "all the pigs" but never to "the whole" (Fuson, Lyons, Pergament, Hall, & Kwon, 1988).

Each addition and subtraction situation involves three quantities, any one of which can be unknown. Thus, three problem subtypes exist for each situation in Table 12.1. These are presented in order of the difficulty (least to greatest) children have in solving the problems. In the Combine situation, one could distinguish a missing first part from a missing second part. However, because these parts usually differ little conceptually in actual situations, and because De Corte and Verschaffel (1987a) reported that children commute these parts fairly readily, the usual practice of not distinguishing these subtypes is followed here, resulting in only two kinds of Combine problems.

Addition is an operation that makes a sum out of two known addends, and subtraction is an operation that makes an addend out of a known sum and a known addend. Thus,

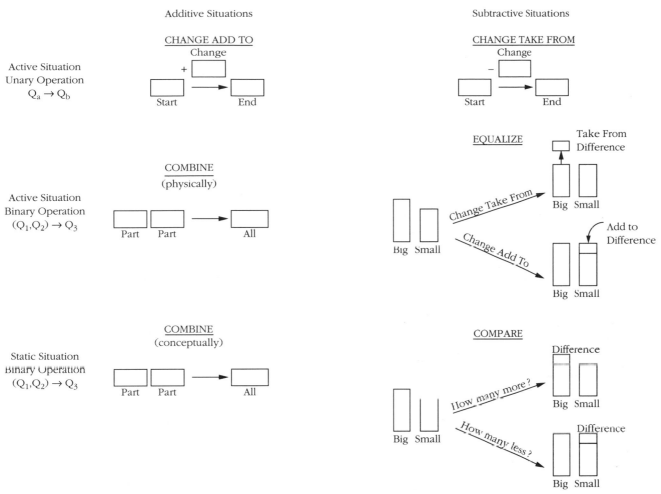

FIGURE 12–1. Real World Addition and Subtraction Situations

two of the four major situations are addition situations (Combine and Change Add To) while the remaining two are subtraction situations (Compare/Equalize and Change Take From). However, within each of the four situations, one of the three problem subtypes requires an addition operation to find the answer (the problem subtype in which the two addend quantities are known) and the other two problem subtypes require a subtraction operation to find the answer (those in which the sum and one addend are known). Thus, there is an important distinction to be made between the problem situation and the operation (addition or subtraction) required to find the unknown quantity; that is, a distinction exists between the problem situation and the solution procedure. In the canonical forms of word problems—those in which the difference, the end, or the all are unknown—these are identical, but they differ in many of the problem subtypes.

The different problem situations present different meanings for the +, −, and = marks used in number sentences. For example, the − and the = in the sentence 14 − 6 = 8 can have various meanings: "14 marbles take away 6 marbles becomes (or results in) 8 marbles" (the − has the Change Take From meaning and the = means "becomes" or "results in"), "compare 6 to 14 to find how much more or less" (the − has

the Compare meaning and the = means "is the same number as"), and "all 14 entities minus the 6 that are part of the 14 is the other part of the 14" (the − has the Combine "unknown part" meaning and the = means "is identical to" because only one set of entities is involved in the Combine situation—the entities that present the 6 and the 8 also present the 14). The Compare meaning for = is closest to the mathematical meaning of "equals" or "is equivalent to." However, textbooks rarely provide children an opportunity to consider different meanings for the +, −, and = marks, and the meanings ordinarily given in textbooks are the Change meanings (Fuson, in press). Perhaps for this reason many children's interpretations of the = mark are limited to a "results in," unidirectional, Change interpretation in which such forms as 6 = 6 and 5 = 2 + 3 are not seen as permissible (Baroody & Ginsburg, 1983; Behr, Erlwanger, & Nichols, 1980). This "do something" interpretation of the = mark, in which the meaning of alternative forms is not accessed, may also result from the manner in which textbooks organize and teachers use pages of addition and subtraction mark problems for rote practice (Cobb, 1987c).

Consideration of the full range of addition and subtraction situations requires an extension to the integers (including negative as well as positive whole numbers), which necessitates

TABLE 12–1. Whole Number Addition and Subtraction Word Problems

Additive Situations	Subtractive Situations

Change Add To

Missing End

Pete had 3 apples. Ann gave Pete 5 more apples. How many apples does Pete have now?

Missing Change

Kathy had 5 pencils. How many more pencils does she have to put with them so she has 7 pencils altogether?

Missing Start

Bob got 2 cookies. Now he has 5 cookies. How many cookies did Bob have in the beginning?

Change Take From

Missing End

Joe had 8 marbles. Then he gave 5 marbles to Tom. How many marbles does Joe have now?

Missing Change

Fred had 11 pieces of candy. He lost some of the pieces. Now he has 4 pieces of candy. How many pieces of candy did Fred lose?

Missing Start

Karen had some word problems. She used 22 of them in this table. She still has 79 word problems. How many word problems did she have to start with?

Combine physically

Missing All

Sara has 6 sugar donuts and 9 plain donuts. Then she puts them all on a plate. How many donuts are there on the plate?

Missing Part

Joe and Tom have 8 marbles when they put all their marbles together. Joe has 3 marbles. How many marbles does Tom have?

Equalize

Add To

Difference Unknown

Susan has 8 marbles. Fred has 5 marbles. How many more marbles does Fred have to get to have as many marbles as Susan has?

Difference Sentence Cues Solution

There were 6 boys on the soccer team. Two more boys joined the team. Now there is the same number of boys as girls on the team. How many girls are on the team?

Difference Sentence Cues Opposite Solution Procedure

Connie has 13 marbles. If Jim wins 5 marbles, he will have the same number of marbles as Connie. How many marbles does Jim have?

Take From

Difference Unknown

Jane has 7 dolls. Ann has 3 dolls. How many dolls does Jane have to lose to have as many as Ann?

Difference Sentence Cues Solution

There were 11 glasses on the table. I put 4 of them away so there would be the same number of glasses as plates on the table. How many plates were on the table?

Difference Sentence Cues Opposite Solution Procedure

There were some girls in the dancing group. Four of them sat down so each boy would have a partner. There are 7 boys in the dancing group. How many girls are in the dancing group?

Combine conceptually

Missing All

There are 6 boys and 8 girls on the soccer team. How many children are on the team?

Missing Part

Brian has 14 flowers. Eight of them are red and the rest are yellow. How many yellow flowers does Brian have?

Compare

Difference Unknown

Joe has 3 balloons. His sister Connie has 5 balloons. How many more balloons does Connie have than Joe?

Difference Sentence Cues Solution

Luis has 6 pet fish. Carla has 2 more fish than Luis. How many fish does Carla have?

Difference Sentence Cues Opposite Solution Procedure

Maxine has 9 sweaters. She has 5 sweaters more than Sue. How many sweaters does Sue have?

Difference Unknown

Janice has 8 sticks of gum. Tom has 2 sticks of gum. Tom has how many sticks less than Janice?

Difference Sentence Cues Solution

The milkman brought on Sunday 11 bottles of milk and on Monday he brought 4 bottles less. How many bottles did he bring on Monday?

Difference Sentence Cues Opposite Solution Procedure

Jim has 5 marbles. He has 8 fewer marbles than Connie. How many marbles does Connie have?

Note: Most of the above problems were taken from Briars & Larkin (1984), Carpenter & Moser (1982, 1983), De Corte & Verschaffel (1987a), Fuson & Willis (1986), Hendrickson & Thompson (1982), Nesher & Teubal (1975), and Riley & Greeno (1988).

246

an avoidance of terminology or educational practices in the lower grades that interfere with later comprehension of these integers. This category system will easily and accurately extend to all three kinds of integer meanings identified by Vergnaud (1982), and in fact two integer meanings are already involved in the system. Vergnaud's relationship meaning (R) is used in the Compare situations to refer to the difference as an integer state: how much more one person has $(+x)$ or how much less the other person has $(-x)$. This integer quality of the difference is one reason difference statements are difficult for children to decode: They contain information about who has more or less as well as information about the quantity of this difference, because the difference is always relative.

The Equalize problem situations use Vergnaud's transformation (T) meaning of integers: they refer to the difference as a transformation—how much more one person needs to get $(+x)$ or how much less the other person needs to have, that is, how much that person needs to give away $(-x)$. These transformation meanings of integers arise in the Change situations also. A Change Add To situation for whole numbers always results in a larger whole number, but a Change Take From situation for whole numbers only results in a whole number if the start number is larger than the change taken away. Taking a larger from a smaller number yields a negative number, thus requiring integers, and so − is not commutative in the whole numbers. However, many young children ignore the order of written take-away marks, causing $3 − 8$ to function as $|3 − 8|$, and therefore yielding the answer 5 just as $8 − 3$ does; whether this is because of a general left-to-right order difficulty, belief in commutativity, or a last-ditch attempt to provide an answer (Baroody & Ginsburg, 1986, p. 102) is not clear.

Vergnaud's third integer meaning is a state meaning (S) of integer measures that are a record of such transformations, such as checks and bills; these integer measures can describe any two opposite states that cancel each other out. Although Vergnaud only discusses the addition and subtraction of certain combinations of these meanings, all three integer meanings— R, T, and S—can be compared, put together, added to, and taken away, thus yielding a wide range of integer addition and subtraction situations. These capture the usual range of integer situations if both count (discrete countable objects) and measure (continuous entities requiring units of measure) meanings are allowed for each situation. Little work has been done on children's understanding of integers (a notable exception is Vergnaud, 1982, and see Fayol, Abdi, & Gombert, 1987, and Peled, Resnick, & Mukhopadhyay, 1988), but so much is now known about children's understanding of whole number addition and subtraction situations that useful research could well begin on their various extensions to negative numbers.

The category system of whole number addition and subtraction situations—and its extension to integers—is intended to capture the range of addition and subtraction situations that occur in the real world, but much research, instead, uses word problems, restricted school versions of the real world. Research has only begun to cope with the added complexities of out-of-school situations, in which identification and posing of problems complicate the problem solving. However, the way tradi-

tional school mathematics treats word problems as exercises to practice marks meanings to be learned in an algorithmic, rote fashion may be more of a problem than the gap between word problems and real world situations. Posing, solving, and discussing word problems can be productive mathematical situations for children.

Differences Between Spoken Number Words and Written Number Marks

There are differences between most systems of spoken number words and the system of written number marks used worldwide (see Fuson, 1990a; Fuson & Kwon, in press-a, in press-b). These differing features have implications for single-digit and multidigit addition and subtraction that will be discussed in the following sections. English and most European languages are irregular variants of a regular named-value system, in which words from one through nine are paired with words that name larger values, for example, five *thousand* eight *hundred* six *ten* four. Although regular for hundreds and thousands, English has many irregularities in the words between ten and one hundred including special words, different pronunciations, reversals in the teen words, and two different modifications of "ten," "teen" and "ty," neither of which clearly says ten. These irregularities have serious consequences for English-speaking children's learning in contrast to children who speak a regular named-value language; these consequences are described as the relevant conceptual structures are discussed in the following sections.

The system of written number marks is a positional base-ten system, in which larger values are indicated by placing marks in different relative positions to the left of the units position. Values that are missing are not just omitted, as they are with words; instead, a zero mark (0) is required for each missing value, to keep other marks in their correct relative positions. Written marks can be used to express numbers of arbitrarily large value, while English words must name a new value for each larger value—for example, the 47th place does not have a commonly known name. English meets this requirement by being a base-thousand system with a subbase of ten. The regularity of the marks enables regular repetitive procedures for adding and subtracting large whole numbers to be developed; however, these procedures can easily be learned without quantitative understanding because the value meanings for the multidigit marks are neither explicit nor salient.

CHILDREN'S UNITARY CONCEPTUAL STRUCTURES FOR ADDITION AND SUBTRACTION

Precursors to Addition and Subtraction

Perceptual Quantity. Children's first cardinal meanings of number words are as labels for small sets of perceived objects. These perceptual quantity meanings seem to be biologically prepared: Research on human infants demonstrated their ability to differentiate sets of one, two, and possibly three objects from each other (Cooper, 1984; Starkey & Cooper, 1980; Strauss

& Curtis, 1984). This visually immediate apprehension and la-belling of small numbers has been termed *subitizing* (Kaufman, Lord, Reese, & Volkmann, 1949). The basis for this differentiating ability is still a source of dispute, with patterns and attentional mechanisms the main explanations (Chi & Klahr, 1975; Cooper, 1984; Klahr & Wallace, 1973; Mandler & Shebo, 1982; von Glasersfeld, 1982). Lower animal species seem to have some perceptual number abilities, but only birds and primates have shown the additional ability to connect a subitized number with a written mark or auditory label (see the review by Davis & Perusse, 1988). Perceptual quantity provides an early basis for addition, as children "see" the addends and the sum as in "two olives and two olives make four olives." Perceptual quantity continues to play a role in the more advanced conceptual levels of addition and subtraction, for children frequently first demonstrate some conceptual advance by using perceptual methods. Thus, they may perform at a higher level of addition and subtraction with very small numbers than they would with larger numbers. For example, a child may add by counting on one or two (solve 4 + 2 by saying "4, 5, 6") but not be able to count on five or more (cannot solve 4 + 5 by counting "4, 5, 6, 7, 8, 9").

Counted Quantity. Counting is the method used in all cultures to differentiate and label quantities not easily or accurately differentiated by perceptual means. The culture determines a standard sequence of number words, body positions (Lancy, 1983; Saxe, 1982), or finger or hand gestures (Mukhopadhyay, personal communication, June, 1989; Secada, 1985; Zaslavsky, 1973) that are then used with an indicating act linking each number label with a to-be-counted entity. The last counted label then shifts from its counting meaning, as paired with a single object, to a cardinal meaning that refers to the whole group of entities and tells the number of entities. In developing their counting ability children in each culture must (a) learn the number sequence of their own culture, (b) learn the indicating act of their own culture (usually pointing), (c) learn to use the indicating act to connect one number label to one entity (make local correspondences), (d) learn methods to remember the already-counted entities so that entities are not recounted (make a global correspondence), and (e) learn the cardinal significance of counting. Research related to all of these aspects of counting is presented and reviewed in Fuson (1988a), including the considerable controversy about how children first relate counting and cardinality meanings of number words. Counting is a powerful general method limited only by the size of the number-label sequence and the ability to carry out the counting accurately. Most preschool children in the U.S. have considerable experience with counting, and by age 5 show competence in counting up to ten, 20, or even 30 entities (for example, Fuson, 1988a; Gelman & Gallistel, 1978; Ginsburg & Russell, 1981; Saxe, Guberman, & Gearhart, 1988).

The Increasing Integration of Sequence, Count, and Cardinal Meanings of Number Words

Children's first uses of number words occur in sequence uses (saying number words in a sequence), counting uses (say-ing number words in correspondence with indicating objects), and cardinal uses (referring to the number of one or more objects by a single number word). Some children may begin such number word uses before the age of two (Durkin, Shire, Riem, Crowther, & Rutter; 1986; Fuson, 1988a; Wagner & Walters, 1982); over the next six years of development, these three uses of number words become increasingly interrelated.

The meanings for the number words said in the number-word sequence reflect these changing relationships in a series of developmental levels (see Table 12.2) identified by Fuson, Richards, and Briars (1982). The String Level involves only the sequence, which may not yet even be unitized into separate number words. At the Unbreakable List Level, the sequence is first unitized into words, which, in turn, are used in counting by relating each word to a perceptual unit item (Steffe, von Glasersfeld, Richards, & Cobb, 1983), that is, an object taken as a single, to-be-counted item. Initially this counting has no cardinal meaning. Then children gradually begin to relate the last counted word to cardinal meanings for the group of counted objects, acting in accord with a cardinality principle (Gelman & Gallistel, 1978; Schaeffer, Eggleston, & Scott, 1974). Although their first use of this principle may reflect only a last-word rule (say the last counted word in response to a how-many question) (Fuson, 1988a), they soon learn to make a count-to-cardinal transition in word meaning in which that word has cardinal meaning (Fuson, 1988a; Fuson & Hall, 1983). In order to count out a given number of objects, children must later learn to shift from the given cardinal meaning of the number word to its counting meaning as the last counted word; in other words, children must make the opposite cardinal-to-count transition in word meaning and remember the word while they are counting (Fuson, 1988a). With this knowledge, a child can add two given number words by first counting out objects for one number word, then counting out objects for the other number word, and finally counting all of the objects.

At the Breakable Chain Level, children can start saying the sequence from an arbitrary number word. They eventually use this ability in combination with an embedded cardinal-to-count transition in word meaning (in Table 12.2 see the horizontal arrow from the cardinal four to the count four) to add by a more efficient *counting-on method,* in which the final sum counting begins with the first addend number word ("three, four, five, six") instead of starting with one ("one, two, ..., six"). At the Numerable Chain Level, it is not objects that present addition and subtraction situations. Rather the child uses sequence words themselves to present these situations. Counting on as a sequence procedure sounds just like the earlier counting on with objects ("three, four, five, six"). However, objects do not present the addends at this level, some method of keeping track of the words said for the second addend must be used so that the sum count will be accurate. Children use auditory or visual patterns, sequentially extended fingers, and double counting to keep track of this second addend (Fuson, 1982; Steffe et al., 1983; Steinberg, 1984).

In the final Bidirectional Chain/Truly Numerical Counting Level, each word of the sequence becomes an identical iterable one that is both a sequence word and a cardinal word (Fuson, 1988a). Each word refers to all of the words up to and including itself (see the first picture at this level in Table 12.2),

TABLE 12–2. Developmental Levels Within the Number-Word Sequence

Sequence level	Meanings related	Conceptual structures within the sequence and relationships among different number-word meanings	
String	Sequence	onetwothreefourfivesixseven	Words may not be differentiated
Unbreakable list	Sequence	one-two-three-four-five-six-seven-	Words are differentiated
	Sequence-Count	one-two-three-four-five-six-seven	Words are paired with objects
		o o o o o o o	
	Sequence-Count-Cardinal	one-two-three-four-five-six-seven→[seven]	Counting objects has a cardinal result
		o o o o o o o	
Breakable Chain	Sequence-Count-Cardinal	[four]⟶four-five-six-seven → [seven]	The addends are embedded within the sum count; the embedded first addend count is abbreviated via a cardinal-to-count transition in word meaning
Numerable Chain	Sequence-Count-Cardinal	[four]⟶four (five) (six) (seven) PI keeping-track method for the second addend	The sequence words become cardinal entities; a correspondence is made between the embedded second addend and some other presentation of the second addend
Bidirectional Chain/Truly Numerical Counting	Sequence-Count-Cardinal / Sequence-Count-Cardinal	① ② ③ ④ ⑤ ⑥ ⑦; 7 6 / ±1 = ±1 / 6 6 7 + 6 = 12 + 1 = 13 because 6 + 6 = 12; (8)(5) → (8)(2)(3) → (10)(3) → (13); (n−1) cardinal (n) ordinal; Know each number as all combinations (5) (1)(4) (2)(3) (3)(2) (4)(1); (5)(7) / +1 −1 = / (6)(6) 5 + 7 = 12 because 6 + 6 = 12	The sequence becomes a unitized seriated embedded numerical sequence: both addends exist outside of and equivalent to the sum: relationships between two different addend/addend/sum structures can be partitioned

Note: A rectangle drawn around related meanings indicates meanings that have become integrated. A number word alone has sequence meaning, a number word enclosed by a bracket has a cardinal meaning.

and each next word is one larger than the previous word in both the cardinal and the sequence meanings. Thus, this cardinalized sequence displays both class inclusion—each cardinal number is embedded within the next—and seriation, meeting Piaget's (1965/41) requirement for a truly operational cardinal number. An ordinal-to-count transition, or the seriated embedded sequence, permits understanding that the nth ordinal word is preceded by a cardinal group of $n - 1$ objects, another Piagetian operational aspect of number. Both addends can now exist outside of the sum and are equivalent to it. Any given small number can be broken down into all of its possible addend pairs. The triangular relationship of the addends and the sum allows addition and subtraction to be seen as inverses. There is relatively little work on children's conceptions at this level (except see Baroody, Ginsburg, & Waxman, 1983; Piaget, 1965/41; Steffe & Cobb, 1988; Steffe et al., 1983); it may be that there are sublevels within this level and that all of these relationships are not constructed simultaneously (see Fuson, 1988a, and Steffe & Cobb, 1988) .

Conceptual Structures for Addition and Subtraction

Children in the United States display a progression of successively more complex, abstract, efficient, and general conceptual structures for addition and subtraction. Each successive level demonstrates cognitive advances and requires new conceptual understandings. Number-word sequence abilities are prerequisites for these new conceptual understandings, although the sequence abilities may precede conceptual advances by a considerable period of time. Children become capable of solving more complex word problems at each level. Much of the research on these conceptual and word problem advances are summarized in Table 12.3, which is an integration, extension, and simplification of several tables in Fuson (1988a). There is not space to credit or discuss many table entries; see Fuson (1988a, Chapter 8) and Fuson (in press) for a fuller discussion of much of this research.

The work on counting and cardinal conceptual units (entities that can present the numbers in the addition and subtraction situation), cardinal conceptual operations (operations that conceptually integrate entities into a single set having a cardinal quantity), and cardinal conceptual structures (the relationships among the counting and cardinal conceptual units forming the addends and sum) was stimulated by Steffe and is a summary of his work (Steffe & Cobb, 1988; Steffe et al., 1983) and of my own related work (Fuson, 1988a). The table omits the new relationships between counting and cardinality meanings of number words that must be understood at each level (see Fuson, 1988a, Chapter 8, for a discussion of these relationships); some of these were discussed with respect to Table 12.2. Major empirical work on levels of children's word problem solution procedures comes from a longitudinal study by Carpenter and Moser (1982; 1983; 1984) and from several studies by De Corte and Verschaffel (1985a, 1985b, 1986, 1987a, 1987b, 1987c; Verschaffel & De Corte, 1988, 1990), and by Nesher (Nesher, Greeno, & Riley, 1982; Nesher & Katriel, 1977). Empirical work, coordinated with the development of a computer model describing successive levels of word problem solving has been carried out by Greeno, Riley, and colleagues (Kintsch & Greeno, 1985; Riley & Greeno, 1988; Riley, Greeno, & Heller, 1983) and by Briars and Larkin (1984). Many researchers have also been influenced by Brownell's early work on children's meaningful learning (1928, 1935; Brownell & Stretch, 1931). Questions have been raised about various aspects of proposed models; reviews of several of the major books in this research area can be found in Bryant (1989), Carpenter (1985), Clements (1989), Cobb (1987a), Hiebert (1984), and Rathmell (1987).

Three levels of development are distinguished in Table 12.3. In the first level, children must present addition or subtraction situations to themselves using objects, or, *perceptual unit items*. These objects are used to model directly the addition or subtraction operation given in the situation. At a given moment each object can only present an addend or a sum, though the role can change over time (an object can first be an addend entity and can later be considered as an entity in the sum). In Level II, all three quantities in an addition or subtraction situation can be presented simultaneously by embedding entities for the addends within the sum and considering them simultaneously as the addend entities and as the sum, a process known as *embedded integration*. Children now can carry out the solution procedures having an abbreviated counting of the sum by starting or ending that count at the first addend. At this level, sequence words no longer are used to count entities presenting the quantities; the words, themselves, present the quantities in addition or subtraction situations. In all of these sequence solution procedures, some method of keeping track of the second addend words must be used, requiring a paired integration of the second addend words with some other representation of the second addend generated by the child (such as fingers successively extended or double counting). In Level III, quantities can be presented by *ideal unit items,* items that can be combined and separated in flexible ways. Ideal unit items enable children to carry out derived fact solutions using a known fact to find a related, but unknown, answer. For quantities at this level, the addends are not embedded within the sum but are outside and can be compared to the sum. Numbers, themselves, become units that comprise numerical triads—two known addends and a known sum. The conceptual advances in each level allow children to present to themselves problem situations not presentable at lower levels; they also allow children to present given problem situations in a new way.

Present data concerning the placement of entries in Table 12.3 are incomplete or inadequate in several ways. First, there are many fewer data on Compare and Equalize problems than on other problems, and the available data often show Compare problems to be particularly difficult. However, several instructional studies (Fuson, 1988b; Fuson & Willis, 1986, 1989) found that children could rapidly learn the meaning of Compare problems, especially if Equalizing problems were given to explain them. Most of the placements of Compare and Equalize problems in Table 12.3 should be considered as tentative (they are largely based on partially analyzed data), and future work about these types should consider at least minimal interventions in helping children understand the meaning of the Compare words and Compare situation. More generally, few studies have used even minimal demonstrations of possible problem solutions (as done, for example, in Briars & Larkin, 1981), so we have little notion how many of the table entries are influenced by lack of previous experience with a given problem type. Second, De Corte and Verschaffel (1987a) reported that both derived fact procedures and recalled fact solutions changed with the word problem to model the problem situation as additive or subtractive (see Level III examples in Table 12.3). However, in other studies, neither derived facts nor recalled facts were differentiated into additive or subtractive categories, so, at this time, data are not available concerning the extent of direct modeling with facts for problem types other than those from the De Corte and Verschaffel (1987a) study shown in Level III. Third, almost all studies only ask the child to solve a given problem once, so the available data indicate children's preferred solution procedures but not the range of solutions that they can carry out for a given problem. Fourth, some problems can be solved by trial and error (Change Take From with an unknown Start), and most studies do not differentiate between a trial and error solution and one carried out directly. Fifth, because time-consuming interviews are required in order to identify solution procedures, usual experimental controls, such as counterbalancing problem types and giving multiple problems of each type, often have been abandoned. We consequently have little information about how these variables affect performance and, thus, affect the placement of problems at different levels.

There are four issues that complicate the placement of problems within levels. First, children show considerable variability in the solution procedures they use for a given kind of word problem or for an oral or numeral problem ("four plus two" or $4 + 2$), even within the same session or on the next day (Carpenter & Moser, 1984; Siegler, 1987b; Steinberg, 1984). New solution procedures ordinarily enter a child's repertoire of available strategies and are used gradually to replace old solution procedures; this is true both for word problems and for oral or written marks problems (Baroody & Ginsburg, 1986; Carpenter & Moser, 1984; Siegler & Jenkins, 1989; Steinberg, 1984). Responses are affected by the presence of objects, with children solving at Level 1 if objects are provided but solving at Level II if objects are not available, necessitating for Carpenter and Moser (1984) a mixed-use category between direct modeling with objects and the more efficient, embedded procedures. Placing a child at a given level means that the child is able to use the solution procedures at that level, but he or she may not always do so.

Second, variations of problem phrasing affect the percentages of children able to solve many problem types and sometimes the solution procedures used. Problem situations that contain cues directing particular actions are solved earlier than forms of the same situation lacking such cues (Carpenter & Moser, 1984; De Corte, Verschaffel, & De Win, 1985; Fuson & Willis, 1986; Hudson, 1983; Nesher & Teubal, 1975). Forms of situations that provide fuller descriptions of given situations also facilitate performance (Carpenter et al. 1981; De Corte, Verschaffel, & De Win, 1985; Lindvall & Ibarra, 1980), especially during children's early exposure to word problems, when they may not yet possess what Nesher (Nesher, 1980; Nesher & Katriel,1977) called SCH-PROBs knowledge and De Corte & Verschaffel (1985a) called a "word problem schema" for inferring unstated information. The order of the problem quantities can affect the solution procedure, but this variable affects the solution of some problem types more than others (for example, Combine unknown Part more than Change Add To unknown Change problems, De Corte & Verschaffel, 1987a; Verschaffel & DeCorte, 1990).

Third, solution procedures may vary with the particular numbers in a problem. Because children learn small number facts before larger number facts, a recalled fact solution is more likely with problems containing small numbers. Knowledge of

a particular number triad (7, 5, 12) may enable children to solve a more difficult problem type than they can solve without knowing the number triad (Briars & Larkin, 1984; Carpenter & Moser, 1984). The sequence counting procedures are done earlier for problems that have one small addend, because perceptual methods of keeping track of the words counted on, up, or back can be used.

Finally, instruction can also affect performance and the solution procedure used on particular problem types. Japanese first graders learn both the Change Take From and Compare meanings for subtraction and in contrast to U.S. children, they do much better on Compare unknown Difference problems than on Change Add To unknown Change problems, which they do not learn until second grade (Ishida, 1988). U.S. first graders who received instruction focused on regarding all problem types as part-whole situations, and then writing an open sentence to solve the problem, moved from direct modeling of various situations requiring subtraction to solving all such situations by separating; on some problem types, children had more wrong operation errors after this instruction than before it (Carpenter, Hiebert, & Moser, 1983). In contrast, U.S. first and second graders taught counting up for subtraction marks problems counted up to solve various word problems requiring subtraction (Fuson, 1986b; Fuson & Willis, 1988).

Level I: The Single Representation of an Addend or the Sum. At Level I, children can solve those problem types (in Table 12.1) that require a set of entities to present only an addend or a sum at a given time. The types for which there is evidence of child solutions with such single entity uses are given in Level 1 of Table 3. In all cases, children directly model the problem situations with entities of some kind: the number of entities is determined by the numbers in the problem and what is done with the entities is determined by the problem situation. Thus, children do not seem at this level to go through three separate problem-solving phases: they do not first represent (present to themselves) the problem situation, then decide on a solution procedure (decide to add or subtract), and finally, carry out the solution procedure, perhaps with entities. Rather, the child's construction of the situation, numbers, and solution procedure is a complex, interrelated whole in which the addition or subtraction meaning is taken directly from the problem situation and modeled with entities. Thus, it is no wonder children solve such problems correctly without first writing a correct solution procedure sentence $(7 - 2)$, as frequently required by school instruction (Carpenter, Hiebert, & Moser, 1983; Thompson & Hendrickson, 1983); such sentences are redundant and unnecessary for problems solved by direct modeling. Children often carry out direct modeling solution procedures with larger numbers such as sums in the teens and even the twenties. However, much of the work on word problems and on number word and number marks problems—"three plus two" or $3 + 2$—has used only addends of five and less. Perceptual pattern procedures are available for these smaller numbers, but these procedures do not easily extend to addends of more than five (for example, it is considerably more difficult to use the fingers to show addends of more than five).

There are many variations of counting all used for word problems and for addition situations given in spoken num-

TABLE 12-3. Developmental Levels of Conceptual Structures and Solution Procedures for Word Problems

Level	Counting and Cardinal Conceptual Units	Cardinal Conceptual Operation	Cardinal Conceptual Structures	Additive (Forward) Solution Procedures	Subtractive (Backward) Solution Procedures
				addend + addend = [sum]	sum − addend = [addend]
				addend + [addend] = sum	sum − [addend] = addend
I	Perceptual unit items: Single presentation of the addend or the sum	Cardinal integration		**Count all (A or PT)** ChAdd: Start A Change → [End] PT: Part PT Part → [All] EqAdd: Small/Start A Diff/Ch → [Big/End] Cm^m: Small A Diff → [Big] **Add-on-up-to a (AOUT or PTUT)** ChAdd: Start AOUT [Change] → End PT: Part PTUT [Part] → All EqAdd: Small/Start AOUT [Diff/Ch] = Big/End Cm: Small AOUT [Diff] → Big	**Take-away a (TA)** ChTake: Start TA Change → [End] PT: All TA Part → [Part] EqTake: Big/Start TA Diff/Ch → [Small/End] Cm: Big TA Small → [Diff] **Separate-to a (ST)** ChTake: Start ST [Change] → End EqTake: Big/Start ST [Diff/Ch] = Small/End **Match (M)** Cm: Big M Small → [Diff] Eq: Big M Small → [Diff]
II	Sequence unit items: Simultaneous presentation of each addend within the sum Unproduced first addend sequence unit items	Embedded integration for both addends First addend is embedded and abbreviated by a cardinal-to-sequence transition		**Sequence-count-on a (CO)** ChAdd: Start# CO Change → [End] PT: Part# CO Part → [All] EqAdd: Small/St# CO Diff/Ch → [Big/End] Cm^m: Small# CO Diff → [Big] *Opp Cm^l: Small# CO Diff → [Big] *Rev EqTake: Small/End# CO Diff/Ch → [Big/St] *Rev ChTake: End# CO Change → [Start]	**Sequence-count-down a (CD)** ChTake: Start# CD Change → [End] PT: All# CD Part → [Part] EqTake: Big/Start# CD Diff/Ch → [Small/End] Cm: Big# CD Small → [Diff] Cm^l: Big# CD Diff → [Small] *Opp Cm^m: Big# CD Diff → [Small] *Rev ChA: End# CD Change → [Start] *SE ChTake: Start# CD End → [Change]

Sequence-count-down-to a (CDT)

ChTake: Start# CDT [Change] → End
EqTake: Big/Start# CDT [Diff/Ch] → Small/End

Direct-subtractive derived and known facts

ChTake [End]: to find 13 − 7 use
down over ten: 13 − 3 − 4 = 10 − 4 = 6
or the known fact 13 − 7 = 6

Sequence-count-up-to a (CU)

ChAdd: Start# CU [Change] → End
PT: Part# CU [Part] → [All]
EqAdc: Small/St# CU [Diff/Ch] = Big/End
Cm: Small# CU [Diff] → Big
Cm': Diff# CU [Small] → Big
*Opp Cm^m: Diff# CU [Small] → Big
*Opp EqAdc: Diff/Ch# CU [Small/St] → Big/End
*Rev ChTake: End# CU [Change] → Start
*SE ChAdd: Change# CU [Start] → End
*R,SE ChTake: Change# CU [End] → Start

Direct-additive derived and known facts

PT[End]: to find 7 + 5 use
up over ten: 7 + 3 + 2 = 10 + 2 = 12
or the known fact 7 + 5 = 12

Indirect-additive derived and known facts

ChAdd [Change]: to find 5 knowing 7 and 12 use
up over ten: 7 + 3 [is ten] + 2 more is 12
so 3 and 2 is 5
or the known fact 7 + 5 = 12

Second addend:
Paired integrations of the second addend and keeping-track unit items

Keeping-track unit items: Second addend entities are generated to correspond with the second addend words

III

Ideal chunkable unit items: Simultaneous nonembedded mental presentation of both addends and the sum

Numerical equivalence

Nonembedded simultaneous addends and sum

Numbers as units

Triad addend/addend/sum structure

253

Notes. Dotted lines enclose unknown quantities, and solid lines enclose known quantities. PI means paired integration. Brackets enclose the unknown, word problem quantities written out (e.g., Start, Big) are presented by entities (except for Diff which should be read as if it said Difference), underlined quantities require paired integration with entities in a keeping-track procedure, quantity labels followed by # denote an abbreviated first addend (a single sequence number word produced for that quantity). Problem quantities newly solved at a given level are marked with *. Opp denotes a solution procedure opposite to the difference sentence. Rev or R denotes reversibility and SE denotes subset equivalence used on the original problem situation. Problem types given in Level III are exemplary only. Solution procedures are abbreviated as follows: A is add to, PT is put together; TA is take away, AOUT is add on up to, PTUT is put together up to, ST is separate to, M is match, CO is count on, CD is count down, CU is count up to, CDT is count down to. The superscripts ^{m} and ^{i} refer to the word "more" or "less" used in a compare problem.

bers or in written marks. For word problems, the counting-all solution procedures sometimes model the problem situation; children tend to add objects to the first addend objects for Change Add To problems, and make separate sets and put them together for Combine problems (De Corte & Verschaffel, 1987a). Children also sometimes use a conceptual putting-together in which the addend objects are not moved but are simply counted all together in the final sum count. Thus, children's solution procedures show both active physical and static conceptual combining of two sets and active unary addition of one set to an initial set. Because these variations have seldom been differentiated in the literature (except see Bergeron & Herscovics, 1990), it is not clear how long and to what extent solution procedures show these distinctions. Counting all solutions for small number problems (addends of five or less) given in number words or written number marks display a developmental sequence using fingers (in most of these studies children are not provided with objects for counting as they usually are in the word problem studies). Children initially extend each finger sequentially while counting; later they learn patterns of fingers and extend all the necessary fingers at once (Baroody, 1987; Siegler & Robinson, 1982; Siegler & Shrager, 1984). Children initially count all of the extended fingers, but eventually they may learn to recognize the sum as a pattern (Baroody, 1987; Siegler & Shrager, 1984). Later, fingers may not first be sequentially extended for each addend but may only be extended for the final sum (Baroody, 1987). This requires some method of keeping track of the second addend fingers so that the sum count can be stopped correctly: thus, this method is easier for number pairs with at least one small addend. Finally, counting all may be done without any observable objects, as the final sum counting may all be said out loud (Baroody, 1984; Baroody & Gannon, 1984; Baroody & Ginsburg, 1986; Siegler & Robinson, 1982; Siegler & Shrager, 1984). In this case, it is not clear what is being counted nor how the child keeps track of the second addend except by a pattern.

There is less information available about direct modeling subtraction procedures than about modeling addition procedures. In word problem situations children carry out subtraction by the three unknown addend procedures given in Level 1 of Table 12.3 and by matching. In take-away *a*, children make the given sum and then take the given addend entities away from the sum, leaving the unknown addend entities to be counted. In add-on-up-to *s*, they make the known addend, add on more entities until the sum has been reached, and then count the added on entities to get the unknown addend. In separate-to-*a*, children make the given sum, count the known addend entities and push the rest away, then count the rest to find the unknown addend. In matching, they make the known sum and the known addend, match these two sets, and then count the extra unmatched entities to find the unknown addend. Thus, these alternative meanings of subtraction clearly are available and usable by kindergarten and first-grade children. However, in textbooks and in studies of subtraction outside the word problem literature, subtraction and the − mark almost always are only given the take-away meaning, thus limiting children's subtraction experiences and the kinds of Level II and III procedures they may invent.

When using a take-away definition for oral number word problems with addends of five or less, Siegler (1987b) found 5- and 6-year-olds' take-away solution procedures were analogous to the count-all procedures reported for addition. Siegler found that the factors that predicted subtraction performance on these small number combinations were (1) how well the corresponding addition problem was learned (fewer errors were made if the addition fact was known), (2) the amount of counting down required (more errors were made with larger known addends counted down), and (3) whether the known addend was five (these problems were particularly easy because a whole hand of fingers was taken away).

There may actually be two sublevels within Level I. Many, but not all, studies have found that more children could solve the canonical Change and Combine problems—the Change problems with the unknown End quantity and Combine unknown All problem—than the other problems listed in Level I (except most studies have not included Equalize problems); these canonical problem types are solvable by many kindergarten children. The other problems in Level I clearly are solvable by many first graders' direct modeling, but their solution may lag behind that of the canonical problems. The add-on procedure requires that the objects added on be kept separate from the initial objects; children sometimes fail to do this, and then give the sum number as the answer (Dean & Malik, 1986). This separation could occur intentionally, requiring knowledge of the successively different use of those objects, or it could occur serendipitously. Such separation of the unknown addend objects is accomplished serindipitously as part of the separate-to solution procedure; thus, many studies show children solving the Change Take From missing Change problem (by separating to) before they solve the Change Add To missing Change problem (by adding up to). Both the Riley/Greeno and the Briars/Larkin word problem models place the latter problem at a higher level than the canonical Change Add To and Change Take Away problems and propose that it is solvable only when counting on occurs. However, both the Carpenter and Moser (1984) longitudinal data and the results of De Corte and Verschaffel (1987a) contradict this prediction; instead, they find that many first graders solve this Change Add To missing Change problem by adding on before they use counting on. Thus this problem seems to be in an advanced sublevel beyond the canonical Level I problems but below the Level II problems. Even a third level might be differentiated for direct modeling with objects within Level I, a trial and error level in which still more difficult problems, such as the Change Take From missing Start, are solved, but only by trial and error (Briars & Larkin, 1981, and Dean & Malik, 1986, report such solutions).

Level II: Abbreviated Sequence Counting Procedures. Most children in the United States spontaneously invent efficient, abbreviated sequence counting procedures, in which number words present both addends embedded within the sum. Children do not count out the addends first; rather, they only make a final count of the sum, but each addend is embedded within this sum as a recognizable, separate section of number words. The conceptual operation forming these addends is embedded integration, and each addend undergoes its own conceptual development to result in the advanced procedures placed at Level II. The first addend sequence unit items are not actually produced; instead, a cardinal-to-count transition in word meaning allows the production of the sum words to begin with the first addend word. The second addend becomes paired with some kind of units for keeping track of how many words are said for the second addend. These conceptual developments result in the four sequence procedures listed in Table 12.3, the unknown in each case in dotted lines and the known quantities in solid lines.

The forward sequence procedures, sequence-count-on and sequence-count-up-to a, are abbreviations of the Level I object counting procedures count all and add-on-up-to a. In both sequence procedures the counting of the first addend is abbreviated to saying the first addend word; such as, "8, 9, 10, 11, 12, 13, 14." Saying these words is sequence counting on to find the solution to $8 + 6$ when the words are accompanied by a method of keeping track of the 6 words counted on past 8 (for example putting out successive fingers until 6 fingers have been extended). The same words are sequence counting up to find $14 - 8$ when a tracking method is used for the words said after 8 and up through 14 (e.g., successively extending fingers until 14 is said and then looking at the fingers to find the missing second addend—here, 6).

The backward sequence counting procedures, sequence-count-down a and sequence-count-down to a, are not just abbreviations of their Level I object-counting predecessors take-away a and separate-to a. In those object procedures all of the counting was forward counting; the sum was counted before the known addend was counted out from the sum objects. The backward sequence counting procedures are reversals of the forward sequence counting procedures counting on and counting up to a. Sequence-count-down in order to solve $14 - 6$ is "14, 13, 12, 11, 10, 9, 8," where some method of keeping track of 6 words counted down from 14 is used and the last word tells the unknown first addend—here, 8. Sequence-count-down-to a in order to solve the same problem is "14, 13, 12, 11, 10, 9, 8, 7, 6," where the words stop when the 6 is said and some method of keeping track of how many words are between 14 and 6 is used to find the unknown second addend, 8.

The conceptual progress that leads to the sequence counting procedures can be made for each addend separately, but the data at this point are not clear about which advance is ordinarily made first. These advances may, in fact, be independent, and individual children may take separate paths to the abbreviated sequence procedures. The abbreviation of the first addend can occur with object-counting versions of count-on, count-up-to, count-down, and count-down-to. In these solution procedures, objects present the addends within the sum; keeping-track methods for the second addend are not required because objects are present to be counted for the second addend. Conversely, keeping-track methods can be used for the second addend without abbreviating the first; that is, all of the sequence words can be said for the first addend. Thus, in sequence-count-all, a child would add $5 + 4$ by saying "1, 2, 3, 4, 5, 6, 7, 8, 9," but would need a keeping-track method for the second addend to be sure that exactly 4 words are said past 5. The data concerning embedded abbreviated object-counting procedures come mainly from situations in which addition or subtraction situa-

tions are presented by objects or hidden objects (for example, Steffe et al., 1983), and the data concerning sequence-count all come from addition problems presented by numerals or stated number words ($5 + 4$ or "five and four") (for example Baroody, 1987).

Conclusions concerning the developmental order of these transitional procedures—object-count-on and sequence-count all—are clouded by the failure of most studies to distinguish between keeping-track methods that result in object-counting-on (those in which objects, including fingers, are put out for the second addend and are then counted) and keeping-track methods that result in sequence-counting all or sequence-counting-on (those in which the keeping-track method is carried out concurrently with saying the sequence words, such as fingers raised successively). Abbreviated object-counting procedures were placed at Level II of an analysis similar to Table 12.3 in Fuson (1988a, in press), and abbreviated sequence procedures were placed at Level III. Here the abbreviated object-counting procedures are viewed as transitional processes that may be used by some children and that can be facilitated by particular classroom experiences. Understanding the range of circumstances under which children construct these abbreviated object counting procedures, and the frequency of their use, must await future research in which observations of keeping-track methods differentiate between abbreviated object and sequence procedures.

Knowledge about the developmental order of the conceptual progress made concerning the first addend (abbreviated procedures) and the second addend (keeping-track methods) is also limited by the fact that children can count one, two, or three words past the first addend, simply by subitizing those words, at a time when they have not yet constructed general methods of keeping-track, to be used for larger numbers (Fuson, 1982; Steffe et al., 1983). Thus, studies that use only small addends may overestimate children's abilities and attribute sequence procedures to them before they really possess general sequence procedures. The size of the first addend may also affect whether a child sequence-counts-all or sequence-counts-on; research has shown that children are more likely to sequence-count-on when the first addend is large (Baroody & Ginsburg, 1986; Siegler & Campbell, 1989).

Object-counting-on is not just a rote abbreviation of object-counting-all. It requires number-word sequence competence—being able to start counting from an arbitrary number word (Fuson et al., 1982; Table 12.2)—and three kinds of conceptual understanding, all of which require the child to consider objects as belonging simultaneously to an addend and to the sum; that is, the perceptual unit items children construct from the objects embed the addends within the sum (Fuson, 1988a; Steffe & Cobb, 1988; Steffe et al., 1983). Two of these kinds of conceptual understanding were identified by Secada, Fuson, and Hall (1983), and all three are discussed in Fuson (1988a). Fuson and Secada (1986) reported success with classroom tasks operationalizing these understandings; teachers in this study went on to help children move on to sequence-counting-on using one-handed finger patterns to keep track of the second addend.

The circumstances under which children spontaneously make the transition from counting all to object-counting-on or to sequence-counting-on are still not clear. Certainly, considerations of efficiency play a role (Baroody & Ginsburg, 1986; Resnick & Neches, 1984), but it is still not fully understood what prompts a child to see the first addend count within the final sum count as unnecessary. Baroody (1987), Carpenter and Moser (1984), and Siegler and Jenkins (1989) reported longitudinal observations that indicated this transition did not come all at once; children varied widely in their recognition of the nature and value of the new procedure (Siegler & Jenkins, 1989). Children are more likely to count on when at least one addend is large (Baroody and Ginsburg, 1986; Siegler & Jenkins, 1989). Two transitional strategies seem to play a role in the shift from counting-all to counting-on. Siegler and Jenkins (1989) reported that children used a short-cut to the counting-all approach in which they did not generate objects for the addends before counting all; they carried out an embedded counting all procedure in which objects simultaneously presented the addends and the sum. Baroody (1984), Davydov and Andronov (1981), Fuson (1982), and Steffe (personal communication, 1980) all reported a transitional procedure used by some children: children said the number words for the first addend very rapidly and then said the second addend words as with usual counting. These children seemed to need to present the first addend to themselves in some way, as if they did not yet trust the cardinal-to-count transition in word meaning used in the abbreviated procedures; these rapid run throughs frequently occurred with large first addends.

The relationship between counting-on from the first given addend and counting-on from the larger addend has been a source of controversy (the literature discussed in the remainder of this section does not differentiate between abbreviated object and sequence procedures). Groen and Resnick (1977) proposed that for marks problems counting-on from first preceded counting-on from larger (termed "min"), and Resnick and Neches (1984) implemented this proposal in a running computer model. Both the Riley/Greeno and the Briars/Larkin word problem models place counting-on from larger at a higher level than counting-on from first, because the former requires knowing somehow that the addends can be exchanged. However, more recent evidence indicates that children are fairly cavalier about their starting addend for marks problems. Children may verbally count all beginning with the larger number without necessarily understanding commutativity, and they may count all from the larger number on some problems but not on others (see Baroody & Gannon, 1984, and Baroody & Ginsburg, 1986, for discussions). Baroody (1987) and Siegler and Jenkins (1989) conducted longitudinal studies that found counting-on from first was done by only a small number of subjects; most used object or sequence procedures beginning with the larger number. For word problems, the counting-on solution procedure may depend on the problem structure; children were much more likely to commute the addends and count-on from larger for a Put Together problem, with its similar parts, than they were for a Change Add To problem, where they more often counted on from the first given addend (De Corte & Verschaffel, 1987a). Carpenter and Moser (1984) reported little support for placement by computer models of counting-on from first as a distinct level preceding counting-on from larger. Rather, they found most chil-

dren using both of these procedures, with no strong order of which procedure was used first by a given child.

The developmental relationship among the four abbreviated counting procedures is clear neither for the sequence procedures nor for the earlier abbreviated object-counting procedures. There is some evidence that counting-on for addition precedes the three unknown addend procedures with hidden object procedures (Secada, 1982), but the evidence is not so conclusive for word problems (Carpenter & Moser, 1984).

The evidence concerning the developmental relationships among the three unknown addend procedures, counting-up-to, counting-down, and counting-down-to, is conflicting and complicated by the usual failures to differentiate object from sequence procedures and to ascertain children's meanings for the unknown addend (subtraction) problems they were given. All three procedures can be used to find any subtraction answer, but the procedures available to a child are obviously constrained by the child's meanings for the unknown addend situation, such as the available meanings for −. This issue has important instructional ramifications, for there is considerable evidence that counting down is more difficult and error-prone than counting up (Baroody, 1984; Baroody & Ginsburg, 1983; Blume, 1981; Carraher & Schliemann, 1985; Fuson, 1984; Fuson et al., 1982; Siegler, 1987b). First, counting backwards is simply much more difficult than counting forwards (for example, Fuson et al., 1982). Second, there are two different counting-down procedures that children tend to confuse. To solve $13 - 5$, one can say (and think) "13, 12, 11, 10, 9 (five words taken away from the thirteen words), 8 (words left as the answer)" or "12 (one taken away), 11, 10, 9, 8 (five taken away, so 8 is the answer)." Children often combine parts of these two procedures, thereby getting an answer one too many or one too few (Carraher & Schliemann, 1985; Fuson, 1984; Steinberg, 1984), suggesting that their counting down lacks sufficient conceptual basis.

Children use both counting-up and counting-down on word problems (Carpenter & Moser, 1984; De Corte & Verschaffel, 1987a). With word problems, the relative ease of counting-up leads many children to choose it in preference to counting-down when their conceptual structures are sufficient to free them from direct modeling; children even use counting-up on take-away problems (Burghardt & Fuson, 1989; Carpenter & Moser, 1984; Fuson & Willis, 1988; Verschaffel & De Corte, 1990). First graders can learn to give the − mark a counting-up meaning and, thus, can count up to solve subtraction marks problems (Fuson, 1986b; Fuson & Willis, 1988; Wynroth, 1980, reported in Baroody, 1984). Steffe & Cobb (1988) found in a teaching experiment with hidden object situations that counting-up and counting-down occurred at the same developmental level. However, for subtraction marks problems, some researchers have reported that children counted down first and only later counted up (Baroody, 1984; Cobb, 1985; Woods, Resnick, & Groen, 1975); some researchers have also found that older children used a choice model for subtraction marks problems in which they counted up when the known addend was large and counted down when the known addend was small (Woods et al., 1975). The differences among these results seem to rest on the meaning of subtraction used by a given child. If a child knew or used only a take-away meaning for the word "subtraction" or for the − mark, or if the researcher provided the take-away meaning as the only meaning given for the problems, the child would count down for subtraction, only beginning to count up later. Because of the considerable difficulties children experience in counting down accurately, learning a counting-up meaning for subtraction marks problems would seem to be advantageous. Such learning enabled first graders to solve written marks problems for all subtraction combinations up to 18 and to be as rapid and accurate at subtracting as at adding by counting on (Fuson, 1986b; Fuson & Willis, 1988). These results contrast sharply with the usual findings of considerable delay in mastering subtraction as compared to addition (Kouba et al., 1989; Stigler, Lee, & Stevenson, 1990; Thornton & Smith, 1988).

Some addition and subtraction situations cannot be solved by direct modeling alone; they must be operated upon conceptually in order to be solved. These are the new, starred problem subtypes in Level II. How children solve these most difficult problems is still not clear. The Briars and Larkin model proposes that children use *Subset Equivalence* and/or *Reversibility* knowledge (SE and R in Table 12.3) to solve these difficult problems. Subset equivalence knowledge is knowing that one can start addition with either addend; so, for example, a Change Add To problem with an unknown Start quantity is solved by switching the Change and Start quantities and then counting up from the known Change quantity. Reversibility knowledge enables one to think about a situation in reverse, such as solving the same problem by counting down the known Change quantity from the known End quantity. Steffe and Cobb (1988) reported that second graders in a teaching experiment used advanced levels of cardinal conceptual structures, like reversibility and subset equivalence, to solve object and open-number sentences. The Riley/Greeno model proposes different conceptual operations—using a Part-Part-Whole schema—for the difficult Change and Compare problems; evidence concerning this position is discussed in the section on the knowledge structures used for word problems. The difficult Equalize and Compare problems in which the change or the comparing more/less statement is opposite to the required solution procedure (marked "Opp" in Table 12.3) also can be solved by reversibility or subset equivalence. The abbreviated sequence counting procedures emphasize *both* addends embedded within the sum and require conceptual progress for both addends, in contrast to the transitional procedures that involve conceptual progress in only one addend; this emphasis of both addends would seem to support reversibility and subset equivalence thinking.

Another issue that is still unresolved for the abbreviated counting procedures is the conditions under which these solution procedures model the problem situation directly. It is clear that at some point children free themselves from direct modeling and use abbreviated counting procedures in flexible ways. Carpenter and Moser (1984) reported a pronounced tendency for children to count up for all problem subtypes requiring subtraction (having one unknown addend), and Fuson and Willis (1988) and Burghardt and Fuson (1989) found that many children who learned to sequence count up to solve subtraction marks problems such as $14 - 8$ counted up to solve different types of subtraction word problems as well, but directly modeled these problems when asked to solve the problems

using objects. Whether this potential freedom from the problem situation accompanies the first use of Level II abbreviated sequence counting procedures or only follows at some later time is not clear.

Level III: Derived Fact and Known Fact Procedures. At Level III the abbreviated sequence procedures are chunked into derived fact procedures in which the numbers in the given problem are redistributed to become numbers whose sum or difference is already known. For example, instead of solving $7 + 5$ by counting 5 words on from 7, the 5 is chunked into 3 (to make 10 with the 7) and 2 more to make the simpler $10 + 2 = 12$. Redistributed derived fact analogues for counting down and counting up to are given in Table 12.3. Children also use derived fact strategies in which one addend and the sum are added to or, more rarely, subtracted from in order to change those numbers into a known fact: the sum $8 + 6$ is $6 + 6 (+2) = 12 + 2 = 14$ (using the known doubles fact $6 + 6 = 12$).

There has been relatively little research done on either the conceptual structures required by different derived facts or on the developmental relationships between sequence counting strategies and derived facts (except see Cobb, 1985; Putnam, deBettencourt, & Leinhardt, 1990; and Steffe & Cobb, 1988). Few studies even differentiate the particular derived facts used by children. Steinberg (1984) found that middle-class second graders could learn derived facts and that this learning did not depend upon already knowing advanced counting strategies. However, learning subtraction derived facts was considerably more difficult than learning addition derived facts, and a considerable part of the second-grade class did not learn them. Thornton (1990; Thornton & Smith, 1988) taught average and above-average first graders addition and subtraction strategies, including some derived facts. The children did much better than the control group (who were using the textbook), but most did not get to the more difficult subtraction combinations in contrast to all achievement levels of first graders receiving sequence counting-up instruction (Fuson, 1986b; Fuson and Willis, 1988), who solved all single-digit subtraction combinations to 18 and did so as easily as the related addition combinations. Almost all the children in Thornton's (1990) experimental group could solve large single-digit addition marks problems by counting on and large subtraction marks problems by counting up, even though they had not been taught these procedures; thus it appears these sequence procedures may well precede most or all derived fact solution procedures. These issues for derived facts are complicated, as usual, by the special perceptual processes and number-word sequence abilities that may enable a child to carry out a derived fact procedure involving 1 (or another small number) before that procedure can be carried out in general; thus, some simple derived fact procedures may be available before Level III. Full resolution of these issues will require a classification of derived fact strategies that enable them to be compared to the sequence counting strategies.

Certain derived fact procedures are quite important in Japan, Korea, mainland China, and Taiwan, where they are taught to all children in first grade. These procedures are all related to ten and are easier to carry out in these languages, which are all regular named-value systems that explicitly name ten: twelve is said as "ten two" and fifteen is said as "ten five." Addition is taught as "up over ten" in which one addend (usually the smaller) is broken into (a) the number that will make ten with the other addend and (b) the left-over number: eight plus six, for example, is thought of as "eight plus two from the six is ten plus the four left over from the six is ten four." The answer can be found just by saying ten and the part of the second addend that is left after making ten. In English this method is more difficult because the English teen words are not automatically given by finding the left-over part, and many U.S. second graders do not know what ten plus any number is; they must count on to find that ten plus four is fourteen (Steinberg, 1984). Many U.S. first-grade children also do not learn another prerequisite for this method: giving for a given number the number that makes ten (that is, giving 3 for 7 and 6 for 4). This is especially true in classes receiving standard textbook instruction but also occurs in experimental classes given visual support for such combinations to ten (Thornton, 1990). The Asian children are taught two different approaches for subtraction with minuends over ten. One is a down-over-ten method that is the reverse of the up-over-ten method; this down-over-ten method is exemplified in Level III of Table 12.3 as the direct subtractive derived fact method (this was De Corte and Verschaffel's [1987a] term for it). The other Asian method is a subtract-from-ten method that becomes an additive method: for "ten four minus eight" (14 − 8), take the eight from the ten leaving two which is then added to the four to make six. These derived fact procedures are taught in first grade in all these countries (Fuson & Kwon, in press-a; Fuson, Stigler, & Bartsch, 1988). First graders there reach high levels of addition and subtraction performance on such problems, much higher than do U.S. children receiving regular instruction (Fuson & Kwon, in press-a; Song & Ginsburg, 1987; Stigler, Lee, & Stevenson, 1990) or instruction on derived facts (Steinberg, 1985; Thornton & Smith, 1988).

Because children clearly can free themselves from direct modeling when they use the Level II abbreviated sequence procedures, and, indeed, the most difficult word problems require such freeing for their solution, one would think that this freedom would be reflected in derived fact solutions. However, De Corte and Verschaffel (1987a) have recently reported a considerable amount of direct modeling with derived fact procedures and even with recalled facts. Examples of these results are given in Table 12.3. Whether this reflects only a preference of solution procedure by the child—almost all studies elicit only the child's first choice of solution—or a limited ability to solve certain problem types in certain ways is not clear at this point. Few studies have differentiated derived or known fact solution procedures according to whether they model the problem situation, so it is also not clear how typical the De Corte and Verschaffel results are. A prolonged dependence upon direct modeling approaches, even with conceptually advanced solution procedures, raises concerns about the ability of such children to solve word problems with large, multidigit numbers, as it may be more difficult to conceptualize direct modeling solution procedures with multidigit numbers than with smaller numbers, and to solve two-step word problems, about which virtually nothing is known (except that Soviet children are exposed to many such problems while children in the United States see almost none, Stigler et al., 1986).

Most children eventually recall the sum or the difference of two given numbers without any counting or derived fact procedure. Recalled facts for some numbers coexist with other procedures at all levels. A child may know 1 + 1 = 2 or 2 + 2 = 4 while still counting all to find the sum of most other pairs of numbers. At present, no developmental account or model describes adequately the range of different procedures including recall that are used at a given time by a given child nor describes how this range changes over time as children move through the levels in Table 12.3. Stage accounts, such as those given in Table 12.3, describe the best performance of which a child is capable at a given time, but do not capture the variability in actual performance of a given child on a given set of problems on a given day (variability due to a range of factors including the size of numbers, the nature of the addition or subtraction situations in the given problems, the changing mood, or other internal variables of the child).

Most of the research on fact recall has involved the chronometric approach, in which reaction times for different number combinations are used to infer the nature of the processing involved and/or attributes of the organization of these combinations in memory (Ashcraft, 1982, 1985; Ashcraft & Battaglia, 1978; Ashcraft & Fierman, 1982; Ashcraft & Stazyk, 1981; Groen & Parkman, 1972; Kaye, Post, Hall, & Dineen, 1986; McCloskey, Sokol, & Goodman, 1986; Miller, Perlmutter, & Keating, 1984; Resnick, 1983; Siegler & Campbell, 1989; Siegler & Shrager, 1984; Woods, Resnick, & Groen, 1975). Such studies have shown that doubles ($a + a$ or $2a - a$) are much easier to learn than most other combinations, that giving sums for larger addends takes more time than doing so for smaller addends, and that children through about third grade add and subtract many combinations by using counting procedures but children in higher grades and adults primarily use fact recall (though they may occasionally use other procedures). The rate at which children recall certain combinations does seem to be affected by their opportunity to try various combinations. Siegler and Shrager (1984) found that parents presented doubles to preschool children particularly frequently and reported presenting $x + 1$ problems more frequently than $1 + x$ problems (the former are easier), and Hamann and Ashcraft (1986) reported that textbooks contain more examples of easy combinations than of difficult ones (and have done so for a long time). Bisanz and LeFevre (1990) argued that these frequencies may reflect cognitive limitations in children that result in parents or textbooks presenting problems that they have found children can solve. How facts are stored and accessed in memory is still an open issue; most treatments postulate a semantic network in which activation of addends automatically activates the sum (Ashcraft, 1983, 1987; Bisanz & LeFevre, 1990). The extent to which children and adults use relational knowledge such as rules or patterns (for example, 0 + any number = that number) is not clear at present (Ashcraft, 1985; Baroody, 1985, 1987; Cobb, 1983), though even kindergarten and mentally handicapped children seem able to use a rule for 0 (Baroody, 1988, 1989). The chronometric approaches often oversimplify the developmental picture by assuming within-child and across-child homogeneity at a given age. Data pooled across children and across different trials by the same child frequently yield patterns that seem to support a single, nonrecall procedure called "min," which functions like counting on but uses a metaphorical "counter in the head" (Groen & Parkman, 1972). Siegler (1987a, 1989) has recently pointed out the problems resulting from averaging chronometric data, and instead suggests the use of a microgenetic approach, implementing intensive trial-by-trial, longitudinal observations on individual children (Siegler & Jenkins, 1989). Most chronometric studies also have not attempted to determine conceptual structures that might be required for memorizing certain facts, although there is some evidence that this might be a useful approach. Cobb (1986), for example, reported that only children who counted on to find sums knew all four $n + 1$ facts, even though all the children studied could count on one word as a sequence skill removed from an addition context.

The most ambitious attempt to model the progression of orally stated number combinations (for example "two and three") from the earliest counting all solution procedures to the final fact state has been two models developed by Siegler and colleagues (Siegler & Jenkins, 1989; Siegler & Schrager, 1984; Siegler & Shipley, 1987). In the early distributions-of-associations model, a child first tries to generate an answer to a given combination from memory and sets a threshold of certainty for accepting a generated answer. If no stored answer meets this threshold, the child uses one of the counting-all solution procedures described for Level I. Over time and with practice, associative strengths for particular answers gradually begin to build up for particular number pairs. Particular elaborated visual or kinesthetic problem representations (such as a figural and kinesthetic image of 2+3 fingers making 5 fingers) and particular solution procedures also become associated with particular number pairs (see Geary & Burlingham-Dubree, 1989, for evidence concerning this aspect of the model). How rapidly a distribution of associations for a particular pair of numbers converges on a single correct answer depends both on the accuracy of the solution procedures carried out and on the threshold set by a given child—and presumably also on the accuracy and strength of any elaborated problem representations. Accurate solution procedures and a high threshold for an answer both lead more rapidly to a distribution of possible answers peaked at the correct one. A computer simulation of the model using the associations of answers obtained from preschool children has learned over many trials the small facts to 10; the distributions become increasingly peaked at the correct answers. This model emphasizes how important it is for teachers to facilitate children's construction and use of accurate solution procedures by creating classroom environments where accuracy and meaningful solutions are more important than rapidly produced answers, for stress on the latter leads to increased associations with incorrect answers for good and not-so-good students who set low thresholds for generating an answer without solving the problem (Siegler, 1988).

A more recent strategy-choice model is currently being developed. In this model a solution procedure is chosen from among all available strategies, and the associations between a particular number pair and a particular strategy reflect the past speed and accuracy of that strategy on that particular pair, on other pairs sharing certain features with that pair (for example smaller addends of particular sizes), and on all pairs on which

that strategy has been used (Siegler & Jenkins, 1989). The issue of how a new strategy is chosen over reliable, old strategies is presently addressed by giving new strategies novelty points that decrease with use. A computer simulation of this model, run on all single-digit sums, generalized its knowledge of a strategy to new pairs. This strategy-choice model obviously is closer to the kind of complex cognitive behavior required for children who use a range of solution procedures across different number pairs and different addition and subtraction situations. This strategy-choice approach also may better fit the behavior of young children asked to give answers but not allowed to carry out addition and subtraction counting procedures; kindergarteners seem to use a range of strategies in such a situation rather than just generating an answer from a distribution of associations, as in the early Siegler model (Baroody, 1989).

A full model of children's strategies for addition and subtraction will need to be extended in several ways. It will eventually have to include the whole range of solution procedures used by children, from the strategies for small numbers presently in the model through to sequence counting strategies and derived facts for larger addends. Data on the model at present are restricted to the canonical Change definitions of addition and subtraction; extensions to the whole range of addition and especially subtraction situations will eventually be required. The model does not yet address how children first construct a new strategy, nor does it say anything about children's conceptual structures; such extensions are obviously important. The microgenetic approach (Siegler & Jenkins, 1989) and the teaching experiment (Cobb & Steffe, 1983; Steffe, 1983) are methodologies better suited than usual chronometric methods for studying such new learning and the conceptual structures involved in various strategies. Ideally, future work by many researchers will yield a better understanding of the conceptual structures children use in carrying out various strategies and the kinds of experiences that help them construct more advanced conceptual structures.

A final issue just beginning to be addressed is the extent to which the developmental sequences of solution procedures identified for children in the United States apply to children in other cultures. It is quite clear that experiences within the classroom, such as the meaning given to the − mark, influence the course of these solution procedures within children in the United States and elsewhere. Just beginning to emerge are data concerning the possibility that the structure of the sequence of number words used by the child and the culturally specific way in which children learn to show numbers on their fingers can influence the nature of the solution procedures children devise and/or learn (see discussion in Fuson & Kwon, 1991b). Neuman (1987) reported interview data with school-entering Swedish children—who are one year older than children in the United States upon school entry—indicating a developmental sequence of solution procedures in those children opposite to that of children in the United States. These Swedish children put the second addend on fingers contiguous to the fingers used for the first addend, rather than on separate hands as is typical in the United States, resulting in an eventual assignment by these children of particular number names to particular fingers. In more advanced solution procedures, the fingers present the sum and the second addend is estimated (or it could be counted with words); this, too, is opposite to the solution procedures for children in the United States, in which the number words count or present the sum and fingers present the second addend. The number line is structured in the Swedish fashion, with the number line presenting the sum and number words counting the second addend. Therefore, in addition to the usual confusion between length and count presentations of number (for example, is a six the length from 0 to 6—six intervals—or is it the little 6 mark?), this reversal of the roles of number words may be confusing to some children in the U.S. A third pattern of folding and unfolding fingers was identified in Korean children (Fuson & Kwon, 1991b, in press-a), whose method involved reusing fingers to show numbers over ten. In combination with Korean number words that name the ten, this finger use supports all aspects of the addition and subtraction methods structured around ten. Swedish and English number-word sequences are similar in structure, so solution procedures clearly can depend both on the structure of the sequence and on how fingers show numbers.

Future cross-cultural and subcultural research may uncover more dependencies and may suggest particularly helpful procedures that might be shared with children. Teachers should be aware of the solution procedures supported in children's homes in order to understand and reduce any possible interference with solution procedures supported within the classroom. Increased attention both to effects of the structure of the number-word sequence spoken in the home and to ways in which the home presents and conceptualizes numbers may help increase mathematics learning for cultural minorities within this country.

Unresolved Issues Concerning the Knowledge Structures Used for Word Problems

The role played by a part-part-whole conceptual schema in children's comprehension and solution of word problems is controversial. Resnick (1983) proposed that children present to themselves and solve all the varieties of word problems by using a part-part-whole schema, but gave no data supporting the use of such a schema. The Riley/Greeno word problem computer model assumes that the more difficult Change and Compare problems are solved by re-representing them with a part-part-whole schema (Riley & Greeno, 1988; Riley et al., 1983). No evidence has been reported to support this re-representation hypothesis, except for overall percentages of problems solved by children and by the computer model, and this hypothesis may be particularly inadequate for Compare problems (Riley & Greeno, 1988). But these views are so influential that, for example, Dean and Malik (1986) throughout their paper called solving a Change Add To unknown Start problem "having part-whole knowledge" even though there was no evidence of children's use of a part-part-whole conceptual schema. Many discussions of such a schema use the term "part-whole," which also can refer to class inclusion, multidigit numbers, and multiplication; the term "part-part-whole" is used here instead in order to restrict the meaning to addition/subtraction situations.

The empirical evidence currently does not support widespread use of a general part-part-whole conceptual schema for Change and Compare problems. Rathmell (1986) reported considerable success with a 2-year intervention in which all problem types were analyzed verbally within a part-part-whole instructional context; the actual conceptual structures constructed by children, though, were not clear because the problems were all solved by using Change meanings (putting the parts together or taking the known part from the whole). Other evidence more directly supports use of separate Change and Compare schemata for these problems and contradicts use of a part-part-whole schema for problem types other than Combine problems. De Corte & Verschaffel (1987a) found direct modeling of solution procedures, even through Level III, that demonstrated major differences between solution procedures for Combine and Change Add To problems. Wolters (1983) found that teaching the part-part-whole schema had a negative effect on solving Change and Compare word problems, and De Corte and Verschaffel (1985b) reported that first graders who used a single part-part-whole drawing had less success solving word problems than did first graders who used different drawings for Change, Combine, and Compare problems. Willis and Fuson (1988) and Fuson and Willis (1989), in a study of second graders with three different schematic drawings available to solve Change, Combine, and Compare problems, reported three findings. First, the study showed no evidence, either in worksheets during learning or in post-tests, of more attempts to use the part-part-whole Combine drawing for Change and Compare problems than to use Change and Compare drawings for Combine problems. Second, when a Combine drawing was used for Change or Compare problems, it tended to be associated with fewer correct strategies and answers than did the Change or Compare drawing. Finally, the largest and most consistent intrusion of drawings was the use by many children over several classes of a Compare drawing for the Combine unknown Part problems.

Another unresolved issue is the relationship between children's presentations to themselves of the numbers involved in word problems and their presentations of the word problem situation. Numbers are very salient features of word problems, with some children going so far as to read only the numbers in a problem and not the words (De Corte & Verschaffel, 1986, 1987b). This emphasis on numbers is increased by the common U.S. textbook feature of having whole pages of a single kind of word problem, in contrast to the textbook used in the Soviet Union in which word problems on a page are quite varied in type (Stigler, Fuson, Ham, & Kim, 1986). Children's conceptual structures for numbers go through the developmental levels in Table 12.3, but this evolution ordinarily may have little to do with experience with word problems, given the extremely restricted and simple kinds of word problems U.S. children ordinarily see (Stigler et al., 1986). Much of the research on the conceptual structures involved in these levels uses addition and subtraction tasks involving objects and hidden sets or subsets, rather than using word problems at all (Fuson, 1988a; Steffe & Cobb, 1988; Steffe, von Glasersfeld, Richards, & Cobb, 1983). It is not at all clear that children have problem-type schemata with which they present to themselves problem situations with unspecified quantities, to which a specified numerosity is then appended to be used in the solution procedure. Instead, the numbers in the problems may play a central role from the beginning interpretation of the problem situation, as, for example, with the children who argued vociferously in class that Combine unknown Part problems compared the big number (the known all) to the small number (the known part) and so chose a Compare rather than a Combine drawing for these problems (Fuson & Willis, 1989).

There are still controversies concerning the related, more general issue of the nature of the knowledge structures that are useful for children to comprehend and solve word problem situations. Major approaches to this issue include the following: model reasonable possible knowledge structures in computer programs, ascertaining whether they are sufficient for problem solutions and predicting developmental levels and answers consistent with empirical data (Briars & Larkin, 1984; Riley & Greeno, 1988; Riley et al., 1983); ask children to retell word problems as a measure of word problem structure (Cummins, Kintsch, Reusser, & Weimer, 1988; De Corte & Verschaffel, 1987a, 1987c; Verschaffel & De Corte, 1990); ask children to sort problems that go together (Morales, Shute, & Pelligrino, 1985); ask children to give the final question in a Change problem (Dean & Malik, 1986); see whether children choose Compare, Combine, or Change drawings for given problems and fill numbers correctly into these drawings (Fuson & Willis, 1989; Willis & Fuson, 1988); and see whether children can write number sentences that model the problem situation (Bebout, 1990; Carpenter, Moser, & Bebout, 1988).

The two word problem computer models address this issue better in combination than they do separately. The Briars and Larkin model is more appropriate for the earlier solution procedures in Table 12.3 because its knowledge structures contain knowledge of words that enable the computer to model directly with objects each sentence of a word problem. The separation of problem schemata from action-based, solution procedure schemata, as in the Riley and Greeno (1988) model, is more appropriate for later developmental levels, when there is evidence of a separation between the child's presentations of the problem situation and the solution procedure and there has been time for experience with problems of a given type to have built up a problem-type schema able to be retrieved from memory (Stigler et al., 1986). The retelling data locate most problem-solving difficulties with children's presentations of problem situations and not with their solution of a successfully presented problem. Children retell more difficult problems as easier problems (Cummins et al., 1988) and also may mispresent the question (De Corte & Verschaffel, 1987c; Verschaffel & DeCorte, 1988). Some of these mispresentations are consistent with those generated by the computer models, but others are not (De Corte & Verschaffel, 1987c; Verschaffel & DeCorte, 1988). Children may also assimilate a problem directly into a correct but different presentation that leads to a correct solution (Fuson & Willis, 1989).

For some children, there is a tension between presenting the problem situation and presenting the solution procedure to themselves, and these functions of objects and numerical sentences are often confused in present instruction. Children asked to use schematic drawings or numerical sentences to show the problem situation show evidence of confusion of these two

functions (Fuson & Willis, 1989; Ishida, 1988; Willis & Fuson, 1988), because they sometimes, instead, seem to show the solution procedure. Alternatively, children also solve some word problems correctly for which they do not write the correct solution procedure equation (Carpenter, Hiebert, & Moser, 1983; Thompson & Hendrickson, 1983), indicating that the common school practice of writing the solution procedure equation is unnecessary for these problem types. Children can write open sentences (for example, $\_\_ + 7 = 15$) that show the problem situation for all forms of Change problems (Bebout, 1987; Carpenter, Moser, & Bebout, 1988). Children also seem to interpret some open sentences as Compare and Equalize situations (Lemoyne & Favreau, 1981), though this has not been true for U.S. children who seem only to use the Change interpretations (Lindvall & Ibarra, 1980). Children's ability to solve at least some word problems is closely related to their ability to solve the related situational sentence (Grouws, 1972). Both teaching and research will benefit if the function of objects or written marks as presenting problem structure is clearly differentiated from the function of presenting the solution procedure.

A final issue concerning knowledge structures for word problems concerns the separate islands of research that exist in the field. To date most investigations have used only one or two possible ways to pose addition and subtraction settings to children. Thus, we know too little about relationships among children's understanding of real world or classroom addition and subtraction situations, of addition and subtraction word problems, of written number pairs ($4 + 2$), of sets of entities with or without hidden sets, and of open sentences (for example, $\_\_ - 6 = 8$). Hopefully, researchers in the future will begin to use a wider range of tasks, thus beginning to integrate the knowledge coming from many different research paradigms.

It will also be important to attend to two kinds of extensions of present research—to larger numbers and to integers. Direct modeling procedures would seem to be much more difficult for problems with multidigit numbers, though direct-modeling thinking about these larger numbers might still be important in deciding whether to add or subtract. With such larger numbers, a conscious choice of solution procedure is necessary, making it even more important that the distinction between addition and subtraction situations and the solution procedures required to solve particular situations be clear to children and to teachers. The extension to integers and to general algebraic methods will require that the conceptual structures built by children for whole numbers support, rather than interfere, with these extensions.

Conclusions and Implications for the Classroom

Many unresolved issues and some of the obstacles to resolving them have been discussed. However, it is important to emphasize the large amount we do now know about how children present to themselves and solve a range of addition and subtraction situations. There is a great deal of direct modeling of the situation, even at the highest levels, but children also show flexible solutions to many problems. The evidence is more than enough to indicate that the usual school textbook approach to single-digit addition and subtraction (discussed in

Fuson, in press) is woefully inadequate. In most text series, this moves from addition and subtraction with pictures to marks problems $(8-3)$ without pictures; children are expected to go from primitive counting all and separating to memorized facts. Teachers, and textbooks, overestimate children's use of modeling with objects and recall and underestimate the use of counting strategies (Carpenter, Fennema, Peterson, Carey, 1988). In too many classrooms, the abbreviated object and sequence solution procedures, and even derived fact procedures, are socially unacceptable, secret underground children's inventions.

U.S. children experience a very restricted range of addition and subtraction situations and meanings. The only meaning given to subtraction initially, and for a year or even several years in many text series, is a take-away meaning. Most of the word problems in texts are only of the simplest canonical forms in which the solution procedure is the same as the underlying situation, word problems follow marks problems as applications and not as initial providers of meaning, word problems are concentrated on particular pages, in many series only problems of the same type are given on a page, practically no two-step problems are given, and extraneous information is rarely included (Stigler et al., 1986). These features result both in children showing huge drops in performance on two-step problems and problems containing extraneous information as well as in a weak understanding of mathematical expressions— for example, 52% of the third graders chose $49 = 83 + 132$ as the answer to the question "If $49 + 83 = 132$ is true, which of the following is true?" (Kouba et al., 1989; see also Cobb, 1987c).

The effort to understand developmental changes in children's presentations of addition and subtraction situations to themselves has been hampered by the extremely restricted set of situations to which most U.S. and Belgian (De Corte & Verschaffel's subjects) children are exposed in traditional instruction. If, instead, from the beginning of kindergarten, children were exposed to a rich range of addition and subtraction situations, if understanding and solving these situations in several different ways were emphasized, and if materials to support alternative solutions were provided and discussed, children might be able to understand considerably more than they do in the present impoverished, narrow classroom environment. This opportunity is given in the Soviet Union, where addition and subtraction word problems are distributed fairly equally over the whole range of problem types, problems of different kinds appear on many pages, and 37% of first grade and 53% of second-grade addition and subtraction word problems are two-step problems involving two different problem types from Table 12.1 (Stigler et al., 1986). In a year-long teaching study, in which children in nine first- and second-grade classrooms were helped through the developmental levels of solution procedures (Table 12.3) and were given opportunity to solve all kinds of word problems using schematic drawings as conceptual aids, I found that the above-average second graders reached ceiling with almost all problem types using three-digit numbers requiring trading, the second graders of all achievement levels and the first graders of above-average achievement reached ceiling on almost all problem types with sums to 18, and the first graders of average achievement reached ceiling on many problem types with sums to 10 (Fuson, 1988b; Fuson

& Willis, 1989). Clearly children in this country can do much better than present research indicates if they are given experience with a wide range of problems and support for moving through the developmental levels of solution procedures.

CHILDREN'S MULTIUNIT CONCEPTUAL STRUCTURES FOR MULTIDIGIT ADDITION AND SUBTRACTION

U. S. Children's Understanding of Multidigit Numeration, Addition, and Subtraction

Children in the United States receiving ordinary classroom instruction have considerable difficulty constructing concepts of multidigit numeration, addition, and subtraction. The National Assessment of Educational Progress reported that only 64% of third graders could identify the digit in the tens place in a four-digit number and less than half identified the hundreds or thousands digit; only 72% of the seventh graders correctly gave the number that is 100 more than 498 (Brown et al., 1989; Kouba et al., 1988). First through fifth graders also show considerable difficulty in relating ungrouped objects to the tens digit and ones digit, in relating objects that show tens and ones groupings to the tens and ones digits, and in focusing on the size of each grouping as well as on the number of groups (Kamii & Joseph, 1988; Ross, 1986, 1989).

This inability to understand or use multidigit concepts affects multidigit addition and subtraction. Many third graders align numbers on the left instead of by their positional values (i.e., on the right) in order to add or subtract them (Ginsburg, 1977; Labinowicz, 1985; Tougher, 1981). Many third graders identify the 1 traded over to the tens or hundreds column as a one and not as a ten or a hundred (Labinowicz, 1985; Resnick, 1983; Resnick & Omanson, 1987; Silvern, 1989). On the National Assessment of Educational Progress, a third of the third graders did not correctly do two-digit subtraction problems requiring a trade, and only half did the three-digit problem correctly (Brown et al., 1989; Kouba et al., 1988). In a heterogeneous sample of fifth graders from the Chicago area, only 69% solved correctly a three-digit subtraction requiring two trades (Stigler, Lee, & Stevenson, 1990). Davis and McKnight (1980) found, in interviewing third and fourth graders from several schools with above-average students, that not a single child solved $7,000-2$ correctly.

Many errors U.S. children make on numeration and on multidigit addition and subtraction tasks indicate that they interpret and treat multidigit numbers as single-digit numbers placed adjacent to each other, rather than using multiunit meanings for the digits in different positions. Thus, they seem to be using a concatenated single-digit conceptual structure for multidigit numbers (see Fuson, 1990a, for discussion of the evidence concerning this incorrect conceptual structure and the errors it generates). For many children, this concatenated single-digit conception is so strong that they do not even question it if they get a different answer by other means. Instead, solving vertical multidigit addition or subtraction marks problems seems to occur within a separate written marks world in which it is

(a) acceptable to get different answers for the same marks problem (because no meanings other than the concatenated single-digit meaning are available to check any marks procedures) and (b) acceptable to get different answers using a written marks procedure (for example, finding 37 + 8 to be 315 or 117 when 37 + 8 is written vertically as a marks problem) and using a unitary representation (for example, finding 37 + 8 to be 45 by counting on 8 from 37) (Cobb & Wheatley, 1988; Davis, 1984; Davis & McKnight, 1980; Resnick 1982, 1983; Resnick & Omanson, 1987).

The use of this concatenated single-digit meaning for multi-digit numbers may stem from classroom experiences that do not sufficiently support children's construction of multiunit meanings, do require children to add and subtract multidigit numbers in a procedural, rule-directed fashion, and do set expectations that school mathematics activities do not require one to think or to access meanings. School textbooks present multidigit numeration, addition, and subtraction in ways that interfere with children's ability to make generalizations. (textbook evidence is discussed in Fuson, in press, and refers to texts published in 1987 and 1988, except for one published in 1985). Work on multidigit addition and subtraction is distributed over four or five years, while in Asian countries and the Soviet Union such work is completed by the third grade (Fuson, 1990b; Fuson, Stigler, & Bartsch, 1988). In U.S. texts, two-digit multiunit place-value and addition and subtraction problems with no trading (for example, 32 + 45) precede the more difficult single-digit sums and differences to 18, even though the single-digit sums and differences require only the unitary conceptual structures in Table 12.3 while efficient solutions for two-digit problems require the construction of multiunit conceptual structures for tens and ones. Work on multidigit addition and subtraction is fragmented, with many months or years separating problems with no trading from problems with trading, problems with one trade from problems with two trades, and problems with different numbers of digits (2-, 3-, 4-, and 5-digits) from each other. The approach to problems in most texts is rule-based: add the ones, next, add the tens, and, finally, add the hundreds. Although pictures of different-sized objects for each place may be used to explain or justify the procedures, these are abandoned in most texts long before most children can construct multiunit conceptual structures that link these different-sized quantities to each mark's position. In some texts the layout of these pictures and the written marks is so confusing that it is difficult to see how the pictures relate to the marks. Many texts use expanded notation, possibly to provide a named-value meaning to the positional base-ten marks. However, this notation is not used well even by fifth grade, where only 45% of the children could even imitate a four-digit example (Stigler et al., 1990). Expanded notation may instead support children's use of incorrect named-value forms of written marks such as 5000700803 (for 5783) or 10032 (for 132) (Bell and Bell, 1988; Stigler et al., 1990).

Children's Multiunit Conceptual Structures for Multidigit Numbers

The unitary conceptual structures for number discussed in earlier parts of this chapter can be used to solve addition and subtraction of two-digit numbers; children can count all or count on and can separate, count up, or count down with numbers larger than ten. But these unitary procedures rapidly become error-prone and time-consuming as the two-digit numbers get larger. Large numbers in all systems of spoken number words and written number marks are formed by making larger and larger multiunits, composed of a certain number of units or smaller multiunits. To understand their culture's number words or written number marks, children must construct conceptual structures that reflect the kinds of multiunits used in their culture's usual interpretations of their number words and marks. English-speaking children, therefore, need to construct multiunit conceptual structures that enable them to understand the differing features of both named-value English number words and positional base-ten written marks (see Fuson, 1990a, and the second section of this chapter for differences between the English systems of number words and number marks), and allow them to relate these two symbol systems to each other.

Table 12.4 shows the conceptual structures required to understand the surface features of the written marks and the surface features of the spoken words, and describes six multiunit structures that are referents for the marks positions and the named values of the words. The conceptual structures in the first six rows of the table are sufficient for understanding multidigit numeration and multiunit addition and subtraction, if all six of these conceptual structures are related to each other (see the discussion in Fuson, 1990b, from which the table is taken). The last four multiunit conceptual structures listed in the table are more advanced understandings, useful in understanding multidigit multiplication and division and exponential and scientific notation.

Translating between written marks and spoken words is complicated by two differences between the marks and words. First, the values of the spoken words are explicitly named, but the values of the marks are implicit in the positions. Thus, children hearing "five hundred sixty-two" want to write the named values "five hundred" and then "sixty" and then "two" (500602) rather than write what looks like "five six two" (562). Second, the positions in the written marks do not have absolute values like those in the named-value words, but have only relative values with respect to the rightmost position. Therefore, to say the words for a written multidigit numeral, one must count or subitize backwards from the right in order to find the name of the first (leftmost) numeral; only after this backward procedure can one read the marks in a forward way, inserting the named value of each position after one says the numeral in that position. This relative positional feature also means that a value cannot just be omitted if there is none of that value, as values are omitted with words ("five hundred two"); some written mark must be used for that missing value to keep all of the other digits in their correct relative position (502, not 52).

Although written marks and spoken words have these differing features, the marks and words do use the same multiunit quantities: the quantities ten, hundred, thousand, and so on, that are the increasing multiples of ten forming the referents for the named values in the words and the relative positions in the written marks. The multiunit quantities conceptual structure contains knowledge about the quantities that are the referents for the multiunit names. Children need perceptual support for their construction of these multiunit items, just as they required perceptual support for the earlier construction of single unit items for smaller numbers. Materials that present collections of tens, hundreds, and thousands (size embodiments) enable children to construct multiunit quantities conceptual structures and link them to written marks and to multidigit words (Fuson, 1986a; Fuson & Briars, 1990; Labinowicz, 1985; Resnick, 1983; Resnick & Omanson, 1987; Steffe & Cobb, 1988; Thompson, 1982). Relative advantages of different kinds of size embodiments—such as those in which children actively make the larger multiunits out of the smaller multiunits versus those in which the larger multiunits already exist—are not yet clearly established by research. The availability of multiunits in particular materials (for example, base-ten blocks) also does not mean that a given child uses conceptual multiunits for those materials. For example, a child may not see a ten-unit as composed of ten single unit items even if the child uses the verbal label

TABLE 12–4. Conceptual Structures for Multiunit Numbers

Name of the conceptual structure	Nature of the conceptual structure				
Features of the marks					
Visual layout	———	———	———	———	———
Positions ordered in increasing value from the right	Fifth	Fourth	Third	Second	First
Features of the words					
Multiunit names	*Wan*	*Qian*	*Bai*	*Shi*	*Yi*
Words ordered in decreasing values as they are said	Ten-thousand	Thousand	Hundred	Ten	Ones
Multiunit structures					
Multiunit quantities					

Regular ten-for-one and one-for-ten trades	Ten thousands one ↔ ten	Ten hundreds one ↔ ten	Ten tens one ↔ ten	Ten ones one ↔ ten	
Positions/values as cumulative trades	Four trades	Three trades	Two trades	One trade	No trades
Positions/values as cumulative multiples of ten	Four multiples of ten $t \times t \times t \times t$	Three multiples of ten $t \times t \times t$	Two multiples of ten $t \times t$	One multiple of ten t	No multiples of ten
Positions/values as exponential words for multiples of ten	Ten to the fourth power	Ten to the third power	Ten to the second power	Ten to the first power	Ten to the zero power
Positions/values as exponential marks for multiples of ten	10^4	10^3	10^2	10^1	10^0

Note. The Chinese multiunit names are used to indicate the learning task for the child.

"ten" for that block (Cobb & Wheatley, 1988; Labinowicz, 1985, 1989; Ross, 1986, 1989; Steffe & Cobb, 1988). For this reason, Fuson (1990b) differentiated between the potentially "collectible" multiunits presented by size embodiments and the "collected" conceptual multiunits formed by children who saw and used the presented multiunits. Children can also make larger and larger multiunit quantities by seeing the ten-for-one trade rule in the first several quantities and then making objects that show the size of the multiunits for the fifth, sixth, and following places (for example, Joslyn, 1990). This relationship between contiguous values/positions is the focus of the "regular ten-for-one and one-for-ten trades" conceptual structures.

Addition and subtraction of multidigit numbers can be understood and carried out correctly using the first six conceptual structures in Table 12.4. The first crucial notion for multidigit addition and subtraction is that one must add or subtract like multiunits. One cannot combine different multiunits: 3 hundreds and 2 tens are not 5 hundreds or 5 tens, they are only 3 hundreds and 2 tens. Second, because multidigit marks can have only nine or less of any multiunit, multidigit addition and subtraction are carried out as single-digit addition and subtraction of like multiunits: to add $5862 + 2574$, the 5 thousand-unit-items must be added to the 2 thousand-unit-items (making seven thousand-unit-items), the 8 hundreds must be added to the 5 hundreds (making thirteen hundreds), and so on. The multiunit quantities conceptual structure is sufficient to support both of these aspects of multiunit addition and subtraction. The third crucial aspect requires the one/ten trades conceptual structures; in cases where the sum of a given kind of multiunit exceeds nine, one must recognize and solve the problem of how to write that multiunit. One can say "thirteen hundreds," but if one writes 13 in the hundreds position, the thousands digit gets moved over to the fifth place. The ten-for-one trade relationship for adjacent positions moving to the left (and for larger named values) suggests trading ten of the thirteen hundreds for one thousand, leaving three hundreds to be written in the hundreds position. Knowledge of the one-for-ten trades in the left-to-right direction is required for multidigit subtraction problems in which any named value or position has fewer of that multiunit in the minuend than in the subtrahend; one must trade one next-larger multiunit to get ten of that multiunit so that one can subtract that kind of multiunit. If one thinks of subtraction as the inverse of addition, then the subtraction problems that require a one-for-ten trade in order to subtract a given kind of multiunit will come from exactly those addition problems in which a given kind of multiunit exceeds nine and thus require a ten-for-one trade for the sum of that multiunit to be written. Multiunit subtraction, as does single-digit subtraction, also requires the understanding that subtraction is not commutative but must be carried out in the direction specified. Children's understanding of subtraction directionality is complicated by the opposite order of different English phrases for subtraction: $5 - 2$ can be said "take 2 from 5" and "5 take away 2" and "5 minus 2" (Fuson & Burghardt, 1991). The lack of features in multidigit written marks that connect the single digits of a given multiunit number together, and the fact that one subtracts only one kind of multiunit at a time, also makes it easy for children to reverse the directionality of subtraction across different multiunits (Fuson & Burghardt, 1991; Fuson, 1990b; VanLehn, 1986).

Children in the United States seem to use at least two different kinds of conceptual multiunit items for multiunit quantities, collected multiunits and sequence multiunits, leading to different methods of multidigit addition and subtraction, especially for two-digit numbers (evidence concerning these items is reviewed in Fuson, 1990b). Collected multiunits are based on collections of objects, on collectible multiunits; a ten-unit item is made by conceptually collecting ten single unit items, a hundred-unit item is made by collecting 100 single unit items, a thousand-unit item is made by collecting 1000 single unit items, and so on. A multidigit number word is then conceptually composed of a certain number of each of these collected multiunit items: For example, five thousand six hundred eighty nine is just five of the thousand units, six of the hundred units, eight of the ten units, and nine of the single unit items. Originally these collected multiunits are dependent upon perceptual support: children can only construct and use them in the presence of collectible objects (for example, base-ten blocks, in which the ten block is a single block that is ten units long, the hundred block is a flat that is ten units by ten units, and so on). Later, children do not need the collectible objects to be present in order to use their conceptual collected multiunits (Fuson, 1986a; Thompson, 1982).

Sequence multiunits are multiunit chunks within the number-word sequence. These multiunits require the sequence skills of skip counting by tens (10, 20, 30, or 14, 24, 34, 44), by hundreds, and by thousands. Such counting may initially be based only on patterns of the number words; for it to reflect the use of conceptual sequence multiunits, such counting must be connected to quantities. Sequence multiunits involve the recognition that a count by ten (for example, 26, 36) is equivalent to a count of ten sequence single-unit items (for example, 26, 27, 28, 29, 30, 31, 32, 33, 34, 35, 36). Thus, a count forward by ten increases the starting sequence number by ten, and a count backward by ten decreases the starting sequence number by ten (Steffe & Cobb, 1988; Thompson, 1982). Similarly, a sequence hundred-unit item involves the recognition that a count by one hundred (for example, 248, 348) is actually just a short-cut for a count forward of 100 sequence single-unit items (one would say 100 words if one actually counted from 248 to 348 by ones). With these conceptual sequence multiunits, a multidigit number is something that is built up by counting within the number-word sequence using these sequence multiunits. Rather than being the result of saying every number word beginning from one, five thousand six hundred eighty nine is thought of as being the result of making five thousand-unit jumps (one thousand, two thousand, … , five thousand), six hundred-unit jumps, eight ten-unit jumps, and nine single-unit item steps.

Multidigit addition and subtraction with conceptual collected multiunits has the three crucial components of multiunit addition and subtraction discussed above. Children's comprehension of these components is facilitated by thinking about each of these components in the physical presence of collectible multiunit objects. When children add tens or hundreds, they first actually carry out and later potentially see in their mind collected tens being added to more collected tens and collected hundreds being added to other collected hundreds. Such an approach can lead to second graders achieving high levels of accuracy and understanding of four-digit addition and subtraction (Fuson, 1986a; Fuson & Briars, 1990) and can enable children to self-correct errors, including subtraction with zeros in the minuend, by thinking about collectible multiunits (Fuson, 1986a).

Addition and subtraction of multidigit numbers thought of as sequence multiunit items are extensions of the unitary sequence counting procedures in which the counting-on, counting-up-to, or counting-down is done using sequence multiunit jumps of ten, hundred, and thousand. Such counting procedures can be carried out within positions (or named values) or can involve the whole first addend. Thus, for $5,862 + 2,574$ one could count on the two thousands (five thousand, six thousand, seven thousand), then the five hundreds while remembering the seven thousand (eight hundred, nine hundred, ten hundred, eleven hundred, twelve hundred, thirteen hundred, or eight hundred, nine hundred, one thousand, one thousand one hundred, one thousand two hundred, one thousand three hundred), then the tens while remembering the thousands and the hundreds (sixty, seventy, eighty, ninety, one hundred, one hundred ten, one hundred twenty, one hundred thirty). Whenever the sum of a given multiunit is over ten, the count goes up over (in addition) or down over (in subtraction) a decade change, a hundred change, or a thousand change (see the hundreds and tens counting-on above). Multiunit counting procedures that involve the whole first addend—or the whole sum, in subtraction—make multiunit jumps only within a given value but carry along the whole number in the sequence counting. For two-digit numbers, this procedure is more difficult than the former ($62 + 74$ is "sixty-two, seventy-two, eighty-two, ninety-two, one hundred two, one hundred twelve, one hundred twenty-two, one hundred thirty-two" rather than "sixty, seventy, eighty, …"). For larger numbers it is quite difficult (for example, five thousand eight hundred sixty-two, six thousand eight hundred sixty-two, seven thousand eight hundred sixty-two).

The evidence at this time seems to indicate that for hundreds and thousands, and perhaps for tens, the developmental sequence for using these collected and sequence multiunit items in multidigit addition and subtraction follows the order for the earlier unitary single-digit numbers: Children present multiunit numbers to themselves by counting and reflecting upon collections of objects before presenting such numbers by counting one's own counting words and reflecting upon these words (as in sequence counting on or back by hundreds or thousands). For two-digit numbers, second graders may construct sequence ten-unit items or collected ten-unit items without constructing the other (Cobb and Wheatley, 1988), but constructing sequence multiunits for numbers over one hundred seems to present more of a challenge to many second and third graders. Many children show considerable difficulties in even carrying out the sequence skills required for sequence counting on by tens and hundreds, especially when such counting involves multiunit sums over ten that require counting over hundred or thousand changes (Bell & Burns, 1981; Cobb & Wheatley, 1988; Labinowicz, 1985; Miller & Stigler, 1987; Resnick, 1983; Steffe & Cobb, 1988; Thompson, 1982). Thompson (1982) found that children could do tasks involving collected multiunits before they could do similar tasks requiring sequence multiunits. When taught using base-ten blocks linked to written addition and subtraction, second graders of all achievement levels

and high-achieving first graders were able to construct perceptual and mental collected multiunits for four-digit numbers and use these multiunits to add and subtract such numbers more accurately and meaningfully than third graders receiving ordinary U.S. instruction (Fuson, 1986a; Fuson & Briars, 1990). Intensive instructional efforts at helping children construct sequence multiunits have not yet been tried, so it is not clear when children could build such multiunits with support. The research described in the last section indicates that at present few U.S. classrooms provide sufficient support for children to construct either of these conceptual multiunit quantities; thus, children are largely limited to the incorrect concatenated single-digit conception of multidigit addition and subtraction, especially when such problems are presented in vertical format.

U.S. Children's Linguistic and Cultural Disadvantages in Constructing and Using Multiunit Conceptual Structures

The lack of explicit naming of the tens in English makes it more difficult for English-speaking children to construct and use multiunit conceptual structures and to add and subtract multidigit numbers meaningfully, than for Asian children, who speak a language that is a totally regular named-value system including the tens (12 is "ten two," 58 is "five ten eight," and so on). Most English-speaking U.S. first graders, given base-ten longs (tens) and units (ones), make unitary models of two-digit numbers using only the units, while most Chinese, Japanese, and Korean first graders use the tens and ones to make multiunit models (Miura, 1987; Miura, Kim, Chang, & Okamoto, 1988); this is true even for Japanese first graders before school work on tens and ones and U.S. first graders after school work on tens and ones (Miura & Okamoto, 1989). Multidigit items on curriculum-fair written and interview tasks given to a large sample of first and fifth graders in the U.S. (in the Chicago area), in Japan (in Sendai), and in Taiwan (in Taipei) indicated considerably lower scores by the U.S. children at both grades (Stigler, Lee, & Stevenson, 1990). Korean second and third graders explained the trading for tens and hundreds better and calculated more accurately than U.S. third graders (Fuson & Kwon, in press-b; Song & Ginsburg, 1987).

This difference begins with addition and subtraction of single-digit sums and differences over ten. Such sums are given in multiunit form in Asian languages (for example, $8 + 4$ is "ten two") but only in unitary form in English ($8 + 4$ is "twelve," whose initial meaning is a unitary cardinal or sequence meaning with no connotation of a ten and a two). Asian children are taught derived fact strategies for adding and subtracting that use the multiunit of ten, while U.S. children are taught only the Level I unitary counting strategies for adding and subtracting and go on to invent the developmental sequence of unitary strategies described earlier. Thus, much of the time spent by Asian children learning single-digit sums and differences over ten involves constructing multiunit conceptual structures for two-digit numbers while that time for U.S. children involves only using unitary conceptual structures. Under ordinary classroom conditions, most U.S. children who invent the over-ten strategy for addition only do so at the end-point of the whole sequence of unitary conceptual structures, so this invention is relatively delayed. How easily U.S. children could learn single-digit strategies structured around ten, the nature of the linguistic, perceptual, or instructional support necessary for such learning, and the extent to which such supports could replace the usual sequence of unitary conceptual structures are all unclear at this time.

Consequently, when traditionally instructed U.S. children begin to add and subtract multidigit numbers, they not only have to construct new multiunit conceptual structures, but they also have highly automatized unitary conceptual structures that can interfere with their construction of multiunit tens and ones. For example, Madell (1985) found that many U.S. 6-year-olds using base-ten blocks solved $48 - 14$ by trading a long from 48 to get eighteen units from which they could take fourteen units rather than conceptualizing fourteen as a multiunit one ten (one long) and four ones which they then took from the four longs and eight ones. The extended experience in finding single-digit sums and differences using unitary conceptual structures means that U.S. children, when adding and subtracting multidigit numbers that require trading, must shift back and forth between a unitary meaning for a single-digit sum or difference and a multiunit meaning required for meaningful trading. In English, $8 + 4$ is twelve, and ten of these twelve units must be traded for one ten-unit that is then put with the other tens in the second position. In Asian languages, single-digit sums and differences are already given in multiunit form, so multidigit addition and subtraction can be carried out completely within a multiunit conception. In these languages, $8 + 4$ is the multiunit "ten two," which actually suggests the multidigit procedure: putting the ten with the tens (see Fuson & Kwon, 1991a, in press b, for further discussion of relative advantages of English and Asian regular named value systems for single-digit and multidigit calculation).

There are few other cultural supports in the United States for constructing multiunits based on ten. Almost all other countries in the world use the metric system, which provides many intuitive experiences for multiunits based on ten; the United States does not use the metric system. Our monetary system has intrusions of five (nickels, quarters, and half dollars; five, twenty, and fifty dollar bills) within the tens structure (pennies, dimes, dollars, ten dollars, one hundred dollars). Fortunately, we so far have not experienced the run-away inflation of some other countries that supply children daily experience with large numbers in the thousands in order to carry out transactions involving candy bars (Carrahar & Schliemann, 1985; Saxe, 1987). We have no culturally entrenched folk calculator like the abacus that is based on ten, compared to Asian countries and the Soviet Union. Even the usual way we teach children to show numbers on their fingers (putting each addend on a separate hand) leads to the sequence of unitary structures discussed above, while Korean children learn methods of showing numbers with fingers that are more easily structured around ten (Fuson & Kwon, 1991b, in press-a).

Helping Children Construct Multiunit Conceptual Structures

Research concerning these linguistic and cultural disadvantages and how to compensate for them is just beginning. At least two alternative strategies exist for overcoming these disadvantages. From the beginning, one could try to support children's construction of concepts of number structured around ten and, thus, cut short the present long-term dependence upon unitary structures for single-digit numbers (discussed above and in Baroody, 1990). This is the approach taken by the present curriculum as manifested in textbooks, and the present implementation of the approach clearly is a failure. Textbooks show two-digit numerals as tens and ones before doing single-digit addition and subtraction above ten, but they do not support addition and subtraction methods structured around ten. Therefore, children continue to solve such problems with unitary conceptual structures. The extent to which this approach can succeed with different learning supports is an empirical question.

A second strategy for overcoming linguistic and cultural disadvantages is to use the support of the regular English named-value hundreds and thousands, bypass the interference of the unitary conceptual structures for two digit numbers, and begin the task of constructing multiunit conceptual structures by focusing on four-digit numbers. This strategy was tested in a series of studies and resulted in second graders having much higher levels of understanding of multidigit addition and subtraction and of place value than those students receiving traditional school instruction (Fuson, 1986a; Fuson & Briars, 1990). An embodiment showing collectible multiunits (base-ten blocks) and an embodiment showing the positional marks (digit cards) were both used to help children construct meanings for English number words and written four-digit number marks and for addition and subtraction of four-digit numbers. Beginning the construction of multiunit conceptual structures with four-digit (rather than with two-digit) numbers avoids interference from unitary two-digit conceptual structures and provides children with an opportunity to construct a more general view of multidigit addition and subtraction as adding or subtracting similar multiunits—they can see the repetitive ten/one trades and similar addition and subtraction over several positions. This approach rectifies several drawbacks of the extended distribution of multidigit addition and subtraction in present textbooks (Fuson, 1990b, in press).

Unknown at this time, and central to the choice between these two strategies, is the question of whether English-speaking children can construct multiunit conceptual structures only after constructing the developmental sequence of unitary conceptual structures that children display now (shown in Table 12.3). Children not given special support for structuring addition and subtraction around ten show this limitation (Steffe & Cobb, 1988). The base-ten block studies made this assumption and provided support for children to move through the developmental sequence of unitary conceptual structures up to sequence counting on for single-digit addition and sequence counting up for single-digit subtraction, which were then used for finding single-digit sums and differences not known by a given child. These studies indicate that it is not necessary for English-speaking children to construct multiunits of ten for adding and subtracting single-digit or two-digit numbers before constructing multiunits for ten, hundred, and thousand for adding and subtracting four-digit numbers meaningfully and accurately.

With the second strategy children have to switch from unitary conceptual structures for finding single-digit sums and differences to multiunit meanings for trading multiunits, at least for the length of time it takes to construct robust ten-unit items for the teens. If future research on the first strategy results in English-speaking children constructing strategies structured around ten, the best approach might be a combination of both strategies in which supports for ten-structured strategies for adding and subtracting single-digit numbers are followed by supported work with three- or four-digit numbers, to help children construct several different conceptual multiunits. This combination would eliminate the need to switch from unitary to multiunit conceptual structures, but would maintain the advantages of constructing a more general understanding of multidigit addition and subtraction than is possible when such problems are spread over three or four years, as occurs at present (Fuson, 1990a, 1990b).

There is little current research concerning experiences in mathematics classrooms that help children construct multiunit conceptual structures for multidigit addition and subtraction. There is sufficient research to indicate that size embodiments (such as base-ten blocks) that present collectible multiunits are helpful to many children in understanding multidigit numbers. But there is little research or agreement about just how size embodiments should be used or which size embodiments might be best for which children (see Baroody, 1990; Cobb, 1987b; Davis, 1984; Fuson, 1990a, 1990b; Hiebert, 1984; Kamii & Joseph, 1988; Labinowicz, 1985; and Ross, 1988, for discussions of some of the issues). Much of the research examines only sequence multiunits, and, thus, may underestimate children's ability to solve the given problem if collected multiunits were examined or supported instead. Research is needed to identify multidigit addition and subtraction procedures that children invent for three-digit and larger numbers and to explore contexts that support such inventions. At present, almost all reports of invented procedures involve only two-digit numbers (see Labinowicz, 1985, for a summary), and most of these procedures do not readily generalize to larger numbers. Some methods of classroom support of invented procedures seem to work much better for addition than for subtraction, where only 34% of the third graders who had reinvented two-digit addition without traditional instruction solved $43 - 16$ correctly (Kamii, 1989). Future research on embodiments and on children's solution procedures would be improved by making distinctions between collected multiunits and sequence multiunits and between features of the named-value English words and the unnamed value positional written marks and by supporting children's construction of related conceptual structures for the first six rows in Table 12.4. The comparative ease of counting-up and other forward interpretations over taking away and other backward interpretations of subtraction and - sign suggest that forward interpretations also be tried with multidigit subtraction.

A VISION OF FUTURE MATHEMATICS CLASSROOMS

Three lines of argument and evidence converge on similar visions of mathematics classrooms as places where children construct meanings for mathematical concepts, words, and written marks and carry out, discuss, and justify solution procedures

for mathematical situations. One is the considerable evidence discussed above that children do construct meanings and do possess a range of different solution procedures. A second is the future for which children need to be prepared, a future in which technological, computational resources will continue to increase rapidly, creating mathematical needs impossible to envision now; this rate of technological and workplace change will continue to accelerate. Both of these require citizens and workers who can confidently attack and analyze complex situations and evaluate alternative solution procedures before they are carried out. A third is the empirical evidence, now emerging, concerning the mathematical competence of children in certain countries and the classroom environments producing this competence. This evidence can help to clarify our vision of mathematics classrooms focused on meaning.

Recent evidence indicates that the superiority of mathematical performance of elementary school children in Japan and Taiwan over U. S. children is not limited to computation but extends to applications in real world and word problems and to mathematics as an abstract system and ranges over all topics in the mathematics curriculum (Stigler, Lee, & Stevenson, 1990). Classroom observations in the three countries indicated that most U.S. class time was spent on individual seatwork, teacher talk was focused on rote procedures, and explanations or discussions were infrequent. In contrast, Japanese and Taiwanese teachers used real world problems and concrete, manipulable objects much more than did U.S. teachers: Every child in Japan and every classroom in Taiwan had a mathematics set of colorful, interesting materials used extensively to illustrate and model mathematical concepts. Japanese teachers spent a long time on exploration and explanation of one carefully chosen problem, emphasized alternative solution methods, and discussed errors made by children in a nonthreatening way, thus helping all children making that error. U.S. teachers on the other hand, posed many problems in a single class and frequently moved from topic to topic, seldom demonstrated or elicited more than one solution method, and avoided discussing errors for fear of embarrassing the child making the error (Stigler, 1988). Illinois mathematics teachers visiting a range of Japanese schools came to similar conclusions and also noted that students frequently worked for substantial periods of time in pairs or cooperative groups on a few problems before presenting individual or group solutions to the class (*Mathematics Teaching in Japanese Elementary and Secondary Schools*, 1989). Elementary school mathematics education in Japan has also been described as helping children master a multiplicity of conceptual structures for mathematical ideas by using concrete objects and situations from which such conceptual structures can be constructed (Kroll & Yabe, 1987). Japanese teachers, parents, and children attribute successful mathematics learning much more to effort, while their U.S. counterparts attribute such learning much more to ability (Stevenson, Lummis, Lee, & Stigler, 1990). It is socially acceptable in this country for a child to stop working at learning mathematics because of a lack of sufficient "math ability," while in Japan and Taiwan failure in mathematics is attributed chiefly to lack of hard work and can be repaired by working harder. Japanese and Taiwanese children spend many more hours on mathematics than do U.S. children, because there are more school days, more weekly time in mathematics classes, and more children working on mathematics during mathematics classes in Japan and Taiwan (Stigler, Lee, & Stevenson, 1987). However, research does exist on some of the nontime features of these cross-cultural differences; it indicates that when U.S. teachers spend more time on the developmental portion of the lesson, achievement increases dramatically (Good, Grouws, & Ebmeier, 1983). Research needs to examine the effectiveness with U.S. children of the other nontime features of the Asian classrooms.

Our envisioned school mathematics classrooms thus are places where (a) children engage in mathematical situations that are meaningful and interesting to them, (b) the emphasis is on sustained engagement in mathematical situations, not on rapidly obtaining answers, (c) alternative solution procedures are accepted, discussed, and justified, and (d) errors are just expected way stations on the road to solutions and should be analyzed in order to increase everyone's understanding. The whole range of addition and subtraction situations given in Table 12.1 needs to be explored in at least three settings—real world situations, verbal situations (including those with extraneous information and those requiring two-step solutions), and numerical and mathematical mark situations—and these settings need to be related to each other. Children need to pose problems as well as solve them. The use of written mathematical marks for additive and subtractive structures needs always to be able to be accompanied by a description of those marks' meaning. Children need to be provided with opportunities to reflect on the solution procedures they are using so that they can move through the developmental progression of conceptual structures for additive and subtractive situations; these opportunities may arise from concrete situations or from discussions. This developmental progression should be known to teachers, so they can facilitate children's movement through it (Carpenter, Fennema, Peterson, Chiang, & Loef, 1989), and/or be reflected in learning materials used in the classroom (Cobb, Wood, & Yackel, in press; Cobb et al., 1991; Fuson & Secada, 1986; Fuson & Willis, 1988). Teachers will need to create new classroom norms in order for children to function in this new way so that (a) children feel free to make and correct their own errors, (b) sustained effort and progress, not the number of problems completed, is rewarded, and (c) children figure out their own solution and explain it rather than searching for or remembering the "right" answer (Cobb, Yackel, Wood, Wheatley, & Merkel, 1988). Such a shift positively affects children's attitudes toward mathematics and feelings of confidence about mathematics (Cobb et al., 1991; Nicholls et al., 1990). For multidigit additive and subtractive situations, experiences will need to be provided that enable children to construct multiunit conceptual structures.

Fortunately, the research on children's understanding of addition and subtraction indicates that children enter school ready to function in such a classroom. Kindergarteners already have a considerable amount of knowledge of addition and subtraction situations, and they approach these situations in ways meaningful both to them and to an observer. It is only after exposure to school instruction that emphasizes rote procedures for addition and subtraction that children stop eliciting meanings in additive and subtractive situations, make errors that may be incomprehensible, and are not concerned about answers

that do not make sense. Thus, if we can create classrooms in which children expect and demand that addition and subtraction be meaningful, this new classroom vision seems able to become a reality.

CONCLUSION

Children's competence in addition and subtraction of whole numbers has been an exceedingly active and productive area of research. Research has yielded a description of a developmental progression of successively more abstract and efficient conceptual structures that children in the U.S. construct for addition and subtraction of numbers up to about one hundred. We know less about how and why children move through this progression and about how parents and teachers can best help them to do so. A great deal of progress has also been made in understanding the different kinds of addition and subtraction situations that exist in the real world, and there is considerable agreement concerning a categorical system of these situations. We know much about how children solve many of these situations when they are presented in word problem form and about how these solutions are affected by the developmental progression of conceptual structures for addition and subtraction. Questions still remain, though, about solutions of some kinds of problem situations, relationships between children's conceptual structures for problem situations and their structures for solution procedures, and the best ways to help children understand the most complex addition and subtraction situations. We know a great deal about errors children make in multidigit addition and subtraction calculations and about their lack of understanding of multidigit addition and subtraction and of place-value concepts. We know something about conceptual structures children have for multidigit addition and subtraction and how to help children construct accurate and useful conceptual structures. However, there also are many unanswered questions in this area. Our knowledge of children's understandings of multidigit addition and subtraction lags far behind our knowledge of children's understandings of addition and subtraction of numbers less than twenty.

This chapter has discussed the shape and extent of our present knowledge and has identified issues that need attention. Many of these issues arise from gaps in our present knowledge, and resolving them would fill these gaps. However, most of our present knowledge about U.S. children's understanding of addition and subtraction is based on research done with children who have received traditional mathematics school instruction; many of the limitations of this instruction have been pointed out in this chapter. We have very little knowledge about how much or in what ways children could learn about addition and subtraction if they had different experiences in the classroom. Thus, it seems more important now to begin to develop and test ways to help children overcome the present limitations of their school mathematics experience—including adapting and testing approaches from cultures that are more successful in mathematics teaching—than it does to fill in all the gaps in our knowledge about children's learning or understanding under the present far-from-maximum learning and teaching conditions. It is clear that we need to progress to different kinds of mathematics classrooms if children's understanding of addition and subtraction is to improve. The probability that future classroom changes will actually reflect progress rather than regress will be increased if research can effectively address questions concerning how to improve children's learning about addition and subtraction.

References

Ashcraft, M. H. (1982). The development of mental arithmetic: A chronometric approach. *Developmental Review, 2*, 213–236.

Ashcraft, M. H. (1983). Procedural knowledge versus fact retrieval in mental arithmetic: A reply to Baroody. *Developmental Review, 3*, 231–235.

Ashcraft, M. H. (1985). Is it farfetched that some of us remember our arithmetic facts? *Journal for Research in Mathematics Education, 16*, 99–105.

Ashcraft, M. H. (1987). Children's knowledge of simple arithmetic: A developmental model and simulation. In J. Bisanz, C. J. Brainerd, & R. Kail (Eds.), *Formal methods in developmental psychology* (pp. 302–338). New York: Springer-Verlag.

Ashcraft, M. H. & Battaglia, J. (1978). Cognitive arithmetic: Evidence for retrieval and decision processes in mental addition. *Journal of Experimental Psychology: Human Learning and Memory, 4*, 527–538.

Ashcraft, M. H., & Fierman, B. A. (1982). Mental addition in third, fourth, and sixth graders. *Journal of Experimental Child Psychology, 33*, 216–234.

Ashcraft, M. H., & Stazyk, E. H. (1981). Mental addition: A test of three verification models. *Memory & Cognition, 9*, 185–196.

Baroody, A. J. (1984). Children's difficulties in subtraction: Some causes and questions. *Journal for Research in Mathematics Education, 15*, 203–213.

Baroody, A. J. (1985). Mastery of the basic number combinations: Internalization of relationships or facts? *Journal for Research in Mathematics Education, 16*, 83–98.

Baroody, A. J. (1987). The development of counting strategies for single-digit addition. *Journal for Research in Mathematics Education, 18*, 141–157.

Baroody, A. J. (1988). Mental-addition development of children classified as mentally handicapped. *Educational Studies in Mathematics, 19*, 369–388.

Baroody, A. J. (1989). Kindergartners' mental addition with single-digit combinations. *Journal for Research in Mathematics Education, 20*, 159–172.

Baroody, A. J. (1990). How and when should place-value concepts and skills be taught? *Journal for Research in Mathematics Education, 21*, 281–286.

Baroody, A. J., & Gannon, K.E. (1984). The development of the commutativity principle and economical addition strategies. *Cognition and Instruction, 1*, 321–339.

Baroody, A. J. & Ginsburg, H. P. (1983). The effects of instruction on children's understanding of the "equals" sign. *The Elementary School Journal, 84*, 199–212.

Baroody, A. J., & Ginsburg, H. P. (1986). The relationship between initial meaningful and mechanical knowledge of arithmetic. In J. Hiebert (Ed.), *Conceptual and procedural knowledge: The case of mathematics* (pp. 75–112). Hillsdale, NJ: Lawrence Erlbaum.

Baroody, A. J., Ginsburg, H. P., & Waxman, B. (1983). Children's use of mathematical structure. *Journal of Research in Mathematics Education, 14*, 156–168.

Bebout, H. C. (1990). Children's symbolic representations of addition and subtraction word problems. *Journal for Research in Mathematics Education, 21*, 123–131.

Behr, M., Erlwanger, S., & Nichols, E. (1980). How children view the equals sign. *Mathematics Teaching, 92*, 13–15.

Bell, M. S., & Bell, J. B. (1988). *Assessing and enhancing the counting and numeration capabilities and basic operation concepts of primary school children*. University of Chicago, unpublished manuscript.

Bell, M. & Burns, J. (1981). Counting and numeration capabilities of primary school children: A preliminary report. In T. R. Post & M. P. Roberts (Eds.). *Proceedings of the Third Annual Meeting of the North American Chapter of the International Group for the Psychology of Mathematics Education* (pp. 17–23). Minneapolis, Minn: University of Minnesota.

Bergeron, J. C., & Herscovics, N. (1990). Psychological aspects of learning early arithmetic. In P. Nesher & J. Kilpatrick (Eds.), *Mathematics and cognition: A research synthesis by the International Group for the Psychology of Mathematics Education* (pp. 31–52). Cambridge: Cambridge University Press.

Bisanz, J. (1989, March). *Development of arithmetic computation and number conservation skills*. Paper presented at the annual meeting of the American Educational Research Association, San Francisco.

Bisanz, J. & LeFevre, J. (1990). Strategic and nonstrategic processing in the development of mathematical cognition. In D. Bjorklund (Ed.), *Children's strategies: Contemporary views of cognitive development* (pp. 213–244). Hillsdale, NJ: Lawrence Erlbaum.

Blume, G. W. (1981). *Kindergarten and first-grade children's strategies solving addition and subtraction problems in abstract and verbal problem contexts* (Technical Report No. 583). Madison: Wisconsin Center for Educational Research.

Briars, D. J., & Larkin, J. H. (1981). Young children's best efforts in solving word problems: Strategies and performance of individuals. In T. R. Post & M. P. Roberts (Eds.), *Psychology of Mathematics Education: Proceedings of the Third Annual Meeting of the North American Chapter of the International Group for the Psychology of Mathematics Education* (pp. 30–36). Minneapolis, Minn.

Briars, D. J., & Larkin, J. H. (1984). An integrated model of skills in solving elementary word problems. *Cognition and Instruction, 1*, 245–296.

Brown, C. A., Carpenter, T. P., Kouba, V. L., Lindquist, M. M., Silver, E. A., & Swafford, J. O. (Eds.) (1989). *Results of the fourth mathematics assessment: National assessment of educational progress*. Reston, VA: National Council of Teachers of Mathematics.

Brownell, W. A. (1928). *The development of children's number ideas in the primary grades*. Chicago: The University of Chicago.

Brownell, W. A. (1935). Psychological considerations in the learning and the teaching of arithmetic. In D. W. Reeve (Ed.), *The teaching of arithmetic. Tenth yearbook, National Council of Teachers of Mathematics* (pp. 1–31). New York: Teachers College, Columbia University.

Brownell, W. A. & Stretch, L. B. (1931). *The effect of unfamiliar settings on problem solving*. Durham, NC: Duke University.

Bryant, P. (1989). [Review of *Children's counting and concepts of number*]. *British Journal of Developmental Psychology, 6*, 395–397.

Burghardt, B., & Fuson, K. (1989). [Relationships between sequence counting and derived facts]. Unpublished raw data.

Carpenter, T. P. (1985). Toward a theory of construction [Review of *Children's counting types: Philosophy, theory, and application*] *Journal for Research in Mathematics Education, 16*, 70–76.

Carpenter, T. P., Fennema, E., Peterson, P. L., & Carey, D. A. (1988). Teachers' pedagogical content knowledge of students' problem solving in elementary arithmetic. *Journal for Research in Mathematics Education, 19,* 385–401.

Carpenter, T. P., Fennema, E., Peterson, P. L., Chiang, C., & Loef, M. (1989). Using knowledge of children's mathematics thinking in classroom teaching: An experimental study. *American Educational Research Journal, 26,* 499–531.

Carpenter, T. P., Hiebert, J., & Moser, J. M. (1981). Problem structure and first-grade children's initial solution processes for simple addition and subtraction problems. *Journal for Research in Mathematics Education, 12,* 27–39.

Carpenter, T. P., Hiebert, J., & Moser, J. M. (1983). The effect of instruction on children's solutions of addition and subtraction word problems. *Educational Studies in Mathematics, 14,* 55–72.

Carpenter, T. P., & Moser, J. M. (1982). The development of addition and subtraction problem-solving skills. In T. P. Carpenter, J. M. Moser, & T. Romberg (Eds.), *Addition and subtraction: A cognitive perspective* (pp. 9–24). Hillsdale, NJ: Lawrence Erlbaum.

Carpenter, T. P., & Moser, J. M. (1983). The acquisition of addition and subtraction concepts. In R. Lesh & M. Landau (Eds.), *Acquisition of Mathematics: Concepts and Processes* (pp. 7–44). New York: Academic Press.

Carpenter, T. P., & Moser, J. M. (1984). The acquisition of addition and subtraction concepts in grades one through three. *Journal for Research in Mathematics Education, 15,* 179–202.

Carpenter, T. P., Moser, J. M., & Bebout, H. C. (1988). Representation of addition and subtraction word problems. *Journal for Research in Mathematics Education, 19,* 345–357.

Carpenter, T. P., Moser, J. M., & Romberg, T. A. (Eds.) (1982). *Addition and subtraction: A cognitive perspective.* Hillsdale, NJ: Lawrence Erlbaum.

Carraher, T. N., & Schliemann, A. D. (1985). Computation routines prescribed by schools: Help or hindrance? *Journal for Research in Mathematics Education, 16,* 37–44.

Chi, M. T. C., & Klahr, D. (1975). Span and rate of apprehension in children and adults. *Journal of Experimental Child Psychology, 19,* 434–439.

Clements, D. H. (1989). Consensus, more or less. [Review of *Construction of arithmetical meanings and strategies and Children's counting and concepts of number*]. *Journal for Research in Mathematics Education, 20,* 111–119.

Cobb, P. (1981). *Children's strategies for finding sums and differences.* (Doctoral dissertation, University of Georgia, 1983). *Dissertation Abstracts International, 44,* 2396.

Cobb, P. (1985, April). *Children's concepts of addition and subtraction: From number to part-whole.* Paper presented at the annual meeting of the American Educational Research Association, Chicago.

Cobb, P. (1986). An Investigation into the sensory-motor and conceptual origins of the basic addition facts. *Proceedings of the Tenth International Conference of the International Group for the Psychology of Mathematics Education* (pp. 141–146). London: University of London Institute of Education.

Cobb, P. (1987a). An analysis of three models of early number development. *Journal for Research in Mathematics Education, 18,* 163–179.

Cobb, P. (1987b). Information-processing psychology and mathematics education-A constructivist perspective. *The Journal of Mathematical Behavior, 6,* 3–40.

Cobb, P. (1987c). An investigation of young children's academic arithmetic contexts. *Educational Studies in Mathematics, 18,* 109–124.

Cobb, P., & Steffe, L. P. (1983). The constructivist researcher as teacher and model builder. *Journal for Research in Mathematics Education, 14,* 83–94.

Cobb, P., & Wheatley, G. (1988). Children's initial understandings of ten. *Focus on Learning Problems in Mathematics, 10,* 1–28.

Cobb, P., Wood, T., & Yackel, E. (in press). A constructivist approach to second grade mathematics. In E. von Glasersfeld (Ed.), *Constructivism in mathematics education.* Dordrecht: Reidel.

Cobb, P., Wood, T., Yackel, E., Nicholls, J. G., Wheatley, G., Trigatti, B., & Perlwitz, M. (1991). Assessment of a problem-centered second-grade mathematics project. *Journal for Research in Mathematics Education, 22,* 3–29.

Cobb, P., Yackel, E., Wood, T. Wheatley, G., & Merkel, G. (1988). Creating a problem-solving atmosphere. *Arithmetic Teacher 36*(1), 46–47.

Cooper, R. G. (1984). Early number development: Discovering number space with addition and subtraction. In C. Sophian (Ed.), *Origins of cognitive skills* (pp. 157–192). Hillsdale, NJ: Lawrence Erlbaum.

Cummins, D. D., Kintsch, W., Reusser, K., & Weimer, R. (1988). The role of understanding in solving word problems. *Cognitive Psychology, 20,* 405–438.

Davis, R. B. (1984). *Learning mathematics: The cognitive science approach to mathematics education.* Norwood, NJ: Ablex.

Davis, R. B., & McKnight, C. C. (1980). The influence of semantic content on algorithmic behavior. *Journal of Children's Mathematical Behavior, 3,* 39–87.

Davis, R. B., & Perusse, R. (1988). Numerical competence in animals: Definitional issues, current evidence, and a new research agenda. *Behavioral and Brain Sciences, 11,* 561–579.

Davydov, V. V., & Andronov, V. P. (1981). *Psychological conditions of the origination of ideal actions* (Project Paper 81–2), (English translation), Madison: Wisconsin Research and Development Center for Individualized Schooling, The University of Wisconsin.

De Corte, E., & Verschaffel, L. (1985a). Beginning first graders' initial representation of arithmetic word problems. *Journal of Mathematical Behavior, 1,* 3–21.

De Corte, E., & Verschaffel, L. (1985b). Working with simple word problems in early mathematics instruction. In L. Streefland (Ed.), *Proceedings of the Ninth International Conference for the Psychology of Mathematics Education* (pp. 304–309). Utrecht, The Netherlands: Research Group on Mathematics Education and Educational Computer Center.

De Corte, E., & Verschaffel, L. (1986, April). *Eye-movement data as access to solution processes of elementary addition and subtraction problems.* Paper presented at the meeting of the American Educational Research Association, San Francisco.

De Corte, E., & Verschaffel, L. (1987a). The effect of semantic structure on first-graders' strategies for solving addition and subtraction word problems. *Journal for Research in Mathematics Education, 18,* 363–381.

De Corte, E., & Verschaffel, L. (1987b). First graders' eye movements during elementary addition and subtraction word problem solving. In G. Luez & V. Lass (Eds.), *Fourth European Conference on Eye Movements, Vol. 1: Proceedings* (pp. 148–150). Gottingen: Hogzefe.

De Corte, E., & Verschaffel, L. (1987c). Using retelling data to study young children's word problem-solving. In J. A. Sloboda & D. Rogers (Eds.), *Cognitive Processes in Mathematics* (pp. 42–59). Oxford: Clarendon Press.

De Corte, E., & Verschaffel, L. (1988). Computer simulation as a tool in research on problem solving in subject-matter domains. *International Journal of Educational Research, 12,* 49–69.

De Corte, E., Verschaffel, L., & De Win, L. (1985). The influence of rewording verbal problems on childrens' problem representations and solutions. *Journal of Eductional Psychology, 77,* 460–470.

Dean, A. L., & Malik, M. M. (1986). Representing and solving arithmetic word problems: A study of developmental interaction. *Cognition and Instruction, 3,* 221–228.

Durkin, K., Shire, B., Riem, R., Crowther, R. D., & Rutter, D. R. (1986). The social and linguistic context of early number word use. *British Journal of Developmental Psychology, 4,* 269–288.

Fayol, M., Abdi, H., & Gombert, J. E. (1987). Arithmetic problems formulation and working memory load. *Cognition and Instruction, 4,* 187–202.

Fuson, K. C. (1982). An analysis of the counting-on solution procedure in addition. In T. P. Carpenter, J. M. Moser, & T. A. Romberg (Eds.), *Addition and subtraction: A cognitive perspective* (pp. 67–81). Hillsdale, NJ: Lawrence Erlbaum.

Fuson, K. C. (1984). More complexities in subtraction. *Journal for Research in Mathematics Education, 15,* 214–225.

Fuson, K. C. (1986a). Roles of representation and verbalization in the teaching of multi-digit addition and subtraction. *European Journal of Psychology of Education, 1,* 35–56.

Fuson, K. C. (1986b). Teaching children to subtract by counting up. *Journal for Research in Mathematics Education, 17,* 172–189.

Fuson, K. C. (1988a). *Children's counting and concepts of number.* New York: Springer-Verlag.

Fuson, K. C. (1988b). First and second graders' ability to use schematic drawings in solving twelve kinds of addition and subtraction word problems. In M. J. Behr, C. B. Lacampagne, M. M. Wheeler (Eds.), *Proceedings of the Tenth Annual Meeting of the North American Chapter of the International Group for the Psychology of Mathematics Education* (pp. 364–370). DeKalb, IL: Northern Illinois University.

Fuson, K. C. (1990a). Conceptual structures for multiunit numbers: Implications for learning and teaching multidigit addition, subtraction, and place value. *Cognition and Instruction, 7,* 343–404.

Fuson, K. C. (1990b). Issues in place-value and multidigit addition and subtraction learning. *Journal for Research in Mathematics Education, 21,* 273–280.

Fuson, K. C. (in press). Research on learning and teaching addition and subtraction of whole numbers. In G. Leinhardt, R. T. Putnam, & R. Hattrup (Eds.), *Analysis of arithmetic for mathematics teaching.* Hillsdale, NJ: Lawrence Erlbaum.

Fuson, K. C., & Briars, D. J. (1990). Using a base-ten blocks learning/teaching approach for first- and second-grade place-value and multidigit addition and subtraction. *Journal for Research in Mathematics Education, 21,* 180–206.

Fuson, K. C., & Burghardt, B. (1991). *Second graders' construction of multidigit addition and subtraction in small cooperative groups.* Manuscript in preparation.

Fuson, K. C., & Hall, J. W. (1983). The acquisition of early number word meanings. In H. Ginsburg (Ed.), *The development of children's mathematical thinking* (pp. 49–107). New York: Academic Press.

Fuson, K. C., & Kwon, Y. (1991a). Chinese-based regular and European irregular systems of number words: The disadvantages for English-speaking children. In K. Durkin & B. Shire (Eds.), *Language and mathematical education* (pp. 211–226). Milton Keynes, G.B.: Open University Press.

Fuson, K. C., & Kwon, Y. (1991b). Systemes de mots nombres et autres outils culturels: Effets sur les premiers calculs de l'enfant (Learning addition and subtraction: Effects of number words and other cultural tools). In J. Bideaud, C. Meljac, & J. P. Fischer (Eds.), *Les chemins du nombre (Pathways to number)* (pp. 351–374). Villeneuve d'Ascq, France: Presses Universitaires de Lille and Hillsdale: NJ: Lawrence Erlbaum.

Fuson, K. C., & Kwon, Y. (in press-a). Korean children's single-digit addition and subtraction: Numbers structured by ten. *Journal for Research in Mathematics Education.*

Fuson, K. C., & Kwon, Y. (in press-b). Korean children's understanding of multidigit addition and subtraction. *Child Development.*

Fuson, K. C., Lyons, B. G., Pergament, G. G., Hall, J. W., & Kwon, Y. (1988). Effects of collection terms on class inclusion and on number tasks. *Cognitive Psychology, 20,* 96–120.

Fuson, K. C., Richards, J., & Briars, D. J. (1982). The acquisition and elaboration of the number word sequence. In C. Brainerd (Ed.),

Progress in cognitive development research: Vol 1. Children's logical and mathematical cognition (pp. 33–92). New York: Springer-Verlag.

Fuson, K. C., & Secada, W. G. (1986). Teaching children to add by counting on with finger patterns. *Cognition and Instruction, 3,* 229–260.

Fuson, K. C., Stigler, J. W., & Bartsch, K. (1988). Grade placement of addition and subtraction topics in China, Japan, the Soviet Union, Taiwan, and the United States. *Journal for Research in Mathematics Education, 19,* 449–458.

Fuson, K. C., & Willis, G. B. (1986). First and second graders' performance on compare and equalize word problems. *In Proceedings of the Tenth International Conference on the Psychology of Mathematics Education* (pp. 19–24). London: University of London Institute of Education.

Fuson, K. C., & Willis, G. B. (1988). Subtracting by counting up: More evidence. *Journal for Research in Mathematics Education, 19,* 402–420.

Fuson, K. C., & Willis, G. B. (1989). Second graders' use of schematic drawings in solving addition and subtraction word problems. *Journal of Educational Psychology, 81,* 514–520.

Geary, D. C., & Burlingham-Dubree, M. (1989). External validation of the strategy choice model for addition. *Journal of Experimental Child Psychology, 47,* 175–192.

Gelman, R., & Gallistel, C. R. (1978). *The child's understanding of number.* Cambridge, MA: Harvard University Press.

Ginsburg, H. P. (1977). *Children's arithmetic: The learning process.* New York: D. Van Nostrand.

Ginsburg, H. P., & Russell, R. L. (1981). Social class and racial influences on early mathematical thinking. *Monographs of the Society for Research in Child Development, 46,* (6, Serial No. 193).

Good, T. L., Grouws, D. A., & Ebmeier, H. (1983). *Active mathematics teaching.* New York: Longman.

Groen, G. J., & Parkman, J. M. (1972). A chronometric analysis of simple addition. *Psychological Review, 79,* 329–343.

Groen, G. J., & Resnick, L. B. (1977). Can preschool children invent addition algorithms? *Journal of Educational Psychology, 69,* 645–652.

Grouws, D. A. (1972). Open sentences: Some instructional considerations from research. *The Arithmetic Teacher, 19,* 595–599.

Hamann, M. S., & Ashcraft, M. H. (1986). Textbook presentations of the basic addition facts. *Cognition and Instruction, 3,* 173–192.

Hiebert, J. (1984). Complementary perspectives [Review of *Acquisition of mathematics concepts and processes; Children's logical and mathematical cognition: Progress in cognitive development research; The development of mathematical thinking*]. *Journal for Research in Mathematics Education, 15,* 229–234.

Hudson, T. (1983). Correspondences and numerical differences between joint sets. *Child Development, 54,* 84–90.

Illinois Council of Teachers of Mathematics (1989). *Mathematics teaching in Japanese elementary and secondary schools: A report of the ICTM Japan mathematics delegation (1988).* Carbondale, IL: Southern Illinois University.

Ishida, J. (1988, July/August). *An analysis of first grade children's writing number sentences in solving word problems.* Paper given at the International Congress of Mathematics Education, Budapest.

Joslyn, R. E. (1990). Using concrete models to teach large-number concepts. *Arithmetic Teacher, 38*(3), 6–9.

Kamii, C. (1989). *Young children continue to reinvent arithmetic-2nd grade: Implications of Piaget's theory.* New York: Teachers College Press.

Kamii, C. & Joseph, L. (1988). Teaching place value and double-column addition. *Arithmetic Teacher, 35*(6), 48–52.

Kaufman, E. L., Lord, M. W., Reese, T. W., & Volkmann, J. (1949). The discrimination of visual number. *American Journal of Psychology, 62,* 498–525.

Kaye, D. B., Post, T. A., Hall, V. C., & Dineen, J. T. (1986). Emergence of information-retrieval strategies in numerical cognition: A developmental study. *Cognition and Instruction, 3,* 127–150.

Kintsch, W., & Greeno, J. G. (1985). Understanding and solving word arithmetic problems. *Psychological Review, 92,* 109–129.

Klahr, D., & Wallace, J. G. (1973). The role of quantification operators in the development of conservation of quantity. *Cognitive Psychology, 4,* 301–327.

Kouba, V. L., Brown, C. A., Carpenter, T. P., Lindquist, M. M., Silver, E. A., & Swafford, J. O. (1988). Results of the fourth NAEP assessment of mathematics: Number, operations, and word problems. *Arithmetic Teacher, 35*(8), 14–19.

Kouba, V. L., Carpenter, T. P., & Swafford, J. O. (1989). Number and operations. In C. A. Brown, T. P. Carpenter, V. L. Kouba, M. M. Lindquist, E. A. Silver, & J. O. Swafford (Eds.), *Results of the fourth mathematics assessment: National assessment of educational progress.* Reston, VA: National Council of Teachers of Mathematics.

Kroll, D. L., & Yabe, T. (1987). A Japanese educator's perspective on teaching mathematics in the elementary school. *Arithmetic Teacher, 35*(2), 36–43.

Labinowicz, E. (1985). *Learning from children: New beginnings for teaching numerical thinking.* Menlo Park, CA: Addison-Wesley.

Labinowicz, E. (1989, March). *Tens as numerical building blocks.* Paper presented at the annual meeting of the American Educational Research Association. San Francisco.

Lancy, D. F. (1983). *Cross-cultural studies in cognition and mathematics.* New York: Academic Press.

Lemoyne, G., & Favreau, M. (1981). Piaget's concept of number development. Its relevance to mathematics learning. *Journal for Research in Mathematics Education, 12,* 179–196.

Lindvall, C. M., & Ibarra, C. G. (1980). Incorrect procedures used by primary grade pupils in solving open addition and subtraction sentences. *Journal for Research in Mathematics Education, 11,* 50–62.

Madell, R. (1985). Children's natural processes. *Arithmetic Teacher, 32*(7), 20–22.

Mandler, G., & Shebo, B. J. (1982). Subitizing: An analysis of its component processes. *Journal of Experimental Psychology: General, 111,* 1–22.

McCloskey, M., Sokol, S. M., & Goodman, R. A. (1986). Cognitive processes in verbal-number production: Inferences from the performance of brain-damaged subjects. *Journal of Experimental Psychology: General, 115,* 307–330.

McKnight, C. C.; Crosswhite, F. J.; Dossey, J. A.; Kifer, E.; Swafford, J. O.; Travers, K. J.; & Cooney, T. J. (1987). *The underachieving curriculum: Assessing U. S. school mathematics from an international perspective.* (A national report on the Second International Mathematics Study). Champaign, IL: Stipes.

Miller, K. F., Perlmutter, M., & Keating, D. (1984). Cognitive arithmetic: Comparison of operations. *Journal of Experimental Psychology: Learning, Memory, and Cognition, 10,* 46–60.

Miller, K., & Stigler, J. W. (1987). Counting in Chinese: Cultural variation in a basic cognitive skill. *Cognitive Development, 2,* 279–305.

Miura, I. T. (1987). Mathematics achievement as a function of language. *Journal of Educational Psychology, 79,* 79–82.

Miura, I. Kim, C. C., Chang, C., & Okamoto, Y. (1988). Effects of language characteristics on children's cognitive representation of number: Cross-national comparisons. *Child Development, 59,* 1445–1450.

Miura, I. & Okamoto, Y., (1989). Comparisons of American and Japanese first graders' cognitive representation of number and understanding of place value. *Journal of Educational Psychology, 81,* 109–113.

Morales, R. V., Shute, V. J. & Pelligrino, J. W. (1985). Developmental differences in understanding and solving simple mathematics word problems. *Cognition and Instruction, 2,* 41–57.

Nesher, P. (1980). The stereotyped nature of school word problems. *For the Learning of Mathematics, 1,* 41–48.

Nesher, P., Greeno, J. G., & Riley, M. S. (1982). The development of semantic categories for addition and subtraction. *Educational Studies in Mathematics, 13,* 373–394.

Nesher, P., & Katriel, T. (1977). A semantic analysis of addition and subtraction word problems in arithmetic. *Educational Studies in Mathematics, 8,* 251–269.

Nesher, P., & Teubal, E. (1975). Verbal cues as an interfering factor in verbal problem solving. *Educational Studies in Mathematics, 6,* 41–51.

Neuman, D. (1987). *The origin of arithmetic skills: A phenomenographic approach.* (No. 62, 5–24, Goteburg Studies in Educational Sciences). Goteberg, Sweden: ACTA Universitatis Gothoburgensis.

Nicholls, J. G., Cobb, P., Wood, T., Yackel, E., & Patashnick, M. (1990). Assessing students' theories of success in mathematics: Individual and classroom differences. *Journal for Research in Mathematics Education, 21,* 109–122.

Peled, I., Resnick, L. B., & Mukhopadhyay, S. (1988, November). *Formal and informal sources of mental models for directed number.* Paper presented at the annual meeting of the Psychonomic Society, Chicago, IL.

Piaget, J. (1965/1941). *The child's conception of number.* New York: W. W. Norton & Company, Inc. (Translated and published in English, New York: Humanities, 1952, from original publication with A. Szemiska, La Genese Du Nombre Chez l'Infant, 1941.)

Putnam, R. T., deBettencourt, L. V., & Leinhardt, G. (1990). Understanding of derived-fact strategies in addition and subtraction. *Cognition and Instruction, 7,* 245–285.

Rathmell, E. C. (1986). Helping children learn to solve story problems. In A. Zollman, W. Speer, & J. Meyer (Eds.), *The fifth mathematics methods conference papers.* Bowling Green, Ohio. Bowling Green State University.

Rathmell, E. C. (1987). Only connect. [Review of *Learning from children*]. *Journal for Research in Mathematics Education, 18,* 408–410.

Resnick, L. B. (1982). Syntax and semantics in learning to subtract. In T. P. Carpenter, J. M. Moser, & T. A. Romberg (Eds.), *Addition and subtraction: A cognitive perspective* (pp. 136–155). Hillsdale, NJ: Lawrence Erlbaum.

Resnick, L. B. (1983). A developmental theory of number understanding. In H. P. Ginsburg (Ed.), *The development of mathematical thinking* (pp. 109–151). New York: Academic Press.

Resnick, L. B. & Neches, R. (1984). Factors affecting individual differences in learning ability. In R. Sternberg (Ed.), *Advances in the psychology of human intelligence* (Vol. 2, pp. 275–323). Hillsdale, NJ: Lawrence Erlbaum.

Resnick, L. B., & Omanson, S. F. (1987). Learning to understand arithmetic. In R. Glaser (Ed.), *Advances in instructional psychology* (Vol. 3) (pp. 41–95). Hillsdale, NJ: Lawrence Erlbaum.

Riley, M. S., & Greeno, J. G., (1988). Developmental analysis of understanding language about quantities and of solving problems. *Cognition and instruction, 5* (1), 49–101.

Riley, M. S., Greeno, J. G., & Heller, J. I. (1983). Development of children's problem-solving ability in arithmetic. In H. Ginsburg (Ed.), *The development of mathematical thinking* (pp. 153–196). New York: Academic Press.

Rohlen, T. P. (1983). *Japan's high schools.* Berkeley: University of California Press.

Ross, S. H. (1986, April). *The development of children's place-value numeration concepts in grades two through five.* Paper presented at the annual meeting of the American Educational Research Association, San Francisco.

Ross, S. H. (1988, April). *The roles of cognitive development and instruction in children's acquisition of place-value numeration concepts.* Paper presented at the annual meeting of the National Council of Teachers of Mathematics, Chicago.

Ross, S. H. (1989). Parts, wholes, and place value: A developmental view. *Arithmetic Teacher, 36*(6), 47–51

Saxe, G. B. (1982). Culture and the development of numerical cognition: Studies among the Oksapmin of Papua New Guinea. In C. J. Brainerd (Ed.), *Progress in cognitive development research: Vol. 1. Children's logical and mathematical cognition* (pp. 157–176). New York: Springer-Verlag.

Saxe, G. B. (1987, April). *Cognition in context: Studies with Brazilian candy sellers.* Paper presented at the Biennial Meeting of the Society for Research in Child Development, Baltimore, Maryland.

Saxe, G. B., Guberman, S. R., & Gearhart, M. (1988). Social and developmental processes in children's understanding of number. *Monographs of the Society for Research in Child Development, 52* (2, Serial No. 216).

Schaeffer, B., Eggleston, V. H., & Scott, J. L. (1974). Number development in young children. *Cognitive Psychology, 6,* 357–379.

Secada, W. G. (1982, March). *The use of counting for subtraction.* Paper presented at the annual meeting of the American Educational Research Association, New York.

Secada, W. G. (1985). Counting in sign: The number string, accuracy and use. (Doctoral dissertation, Northwestern University, 1984). *Dissertation Abstracts International, 45,* 3571A–3572A.

Secada, W. G., Fuson, K. C., & Hall, J. W. (1983). The transition from counting-all to counting-on in addition. *Journal for Research in Mathematics Education, 14,* 47–57.

Siegler, R. S. (1987a). The perils of averaging data over strategies: An example from children's addition. *Journal of Experimental Psychology: General, 116,* 250–264.

Siegler, R. S. (1987b). Strategy choices in subtraction. In J. A. Sloboda & D. Rogers (Eds.), *Cognitive processes in mathematics* (pp. 81–106). New York: Oxford University Press.

Siegler, R. S. (1988). Individual differences in strategy choices: Good students, not-so-good students, and perfectionists. *Child Development, 59,* 833–851.

Siegler, R. S. (1989). The hazards of mental chronometry: An example from children's subtraction. *Journal of Educational Psychology, 81,* 497–506.

Siegler, R. S., & Campbell, J. (1990). Diagnosing individual differences in strategy choice procedures. In N. Frederiksen, R. Glaser, A. Lesgold, & M. G. Shafto (Eds.), *Diagnostic monitoring of knowledge and skill acquisition* (pp. 113–139). Hillsdale, NJ: Lawrence Erlbaum.

Siegler, R. S., & Jenkins, E. (1989). *How children discover new strategies.* Hillsdale, NJ: Lawrence Erlbaum.

Siegler, R. S., & Robinson, M. (1982). The development of numerical understandings. In H. W. Reese and L. P. Lipsitt (Eds.), *Advances in child development and behaviour, Vol. 16* (pp. 242–312). New York: Academic Press.

Siegler, R. S., & Shipley, C. (1987). The role of learning in children's strategy choices. In L.S. Liben (Ed.), *Development & learning: Conflict or congruence* (pp. 71–107). Hillsdale, NJ: Lawrence Erlbaum.

Siegler, R. S., & Shrager, J. (1984). Strategy choices in addition and subtraction: How do children know what to do. In C. Sophian (Ed.), *Origins of cognitive skills* (pp. 229–293). Hillsdale, NJ: Lawrence Erlbaum.

Silvern, S. B. (1989). *Children's understanding of the double-column addition algorithm.* Paper presented at the Annual Meeting of the American Educational Research Association, San Francisco, March, 1989.

Song, M. J., & Ginsburg, H. P. (1987). The development of informal and formal mathematical thinking in Korean and U.S. children. *Child Development, 58,* 1286–1296.

Starkey, P., & Cooper, R. G. (1980). Perception of numbers by human infants. *Science, 210,* 1033–1035.

Steffe, L. P. (1983). The teaching experiment methodology in a constructivist research program. In M. Zweng, T. Green, J. Kilpatrick,

H. Pollack, & M. Suydam (Eds.), *Proceedings of the Fourth International Congress on Mathematical Education* (pp. 469–471). Boston: Birkhauser.

Steffe, L. P., & Cobb, P. (1988). *Construction of arithmetical meanings and strategies.* New York: Springer-Verlag.

Steffe, L. P., von Glasersfeld, E., Richards, J., & Cobb, P. (1983). *Children's counting types: Philosophy, theory, and application.* New York: Praeger Scientific.

Steinberg, R. (1984). A teaching experiment of the learning of addition and subtraction facts (Doctoral dissertation, University of Wisconsin-Madison, 1983). *Dissertation Abstracts International, 44,* 3313A.

Steinberg, R. M. (1985). Instruction on derived facts strategies in addition and subtraction. *Journal for Research in Mathematics Education, 16,* 337–355.

Stevenson, H. W., Lee, S. Y., & Stigler, J. W. (1986). Mathematics achievement of Chinese, Japanese, and American children. *Science, 231,* 693–699.

Stevenson, H. W., Lummis, M., Lee, S. Y., & Stigler, J. W. (1990). *Making the grade in mathematics: Elementary school mathematics in the Unites States, Taiwan, and Japan.* Reston, VA: National Council of Teachers of Mathematics.

Stigler, J. W. (1988). The use of verbal explanation in Japanese and American classrooms. *Arithmetic Teacher, 36*(2), 27–29.

Stigler, J. W., Fuson, K. C., Ham, M., & Kim, M. S. (1986). An analysis of addition and subtraction word problems in American and Soviet elementary mathematics textbooks. *Cognition and Instruction, 3,* 153–171.

Stigler, J. W., Lee, S. Y., Lucker, G. W., & Stevenson, H. W. (1982). Curriculum and achievement in mathematics: A study of elementary school children in Japan, Taiwan, and the United States. *Journal of Educational Psychology, 74,* 315–322.

Stigler, J. W., Lee., S. Y. & Stevenson, H. W. (1987). Mathematics classrooms in Japan, Taiwan, and the United States. *Child Development, 58,* 1272–1285.

Stigler, J. W., Lee, S. Y., & Stevenson, H. W. (1990). *The mathematical knowledge of Japanese, Chinese, and American elementary school children.* Reston, VA: National Council of Teachers of Mathematics.

Strauss, M. S., & Curtis, L. E. (1984). Development of numerical concepts in infancy. In C. Sophian (Ed.), *Origins of cognitive skills* (pp. 131–155). Hillsdale, NJ: Lawrence Erlbaum.

Thompson, C., & Hendrickson, A. D. (1983). Verbal addition and subtraction problems: New research focusing on levels of difficulty of the problems and of the related number sentences. *Focus on Learning Problems in Mathematics, 5,* 33–45.

Thompson, P. W. (1982). A theoretical framework for understanding young children's concepts of whole number numeration. (Doctoral dissertation, University of Georgia, 1982). *Dissertation Abstracts International, 43,* 1868A.

Thornton, C. A., & Smith P. J. (1988). Action research: Strategies for learning subtraction facts. *Arithmetic Teacher, 35*(8), 8–12.

Thornton, C. A. (1990). Solution strategies: Subtraction number facts. *Educational Studies in Mathematics, 21,* 241–263.

Tougher, H. E. (1981). Too many blanks! What workbooks don't teach. *Arithmetic Teacher, 28*(6), 67.

Underhill, R. (1986). Early addition story problem performance: How does it relate to schooling and conservation? In L. Streefland (Ed.), *Proceedings of the Tenth International Conference, Psychology of Mathematics Education* (pp. 63–68). Utrech, the Netherlands: PME.

VanLehn, K. (1986). Arithmetic procedures are induced from examples. In J. Hiebert (Ed.), *Conceptual and procedural knowledge: The case of mathematics* (pp. 133–179). Hillsdale, NJ: Lawrence Erlbaum.

Vergnaud, G. (1982). A classification of cognitive tasks and operations of thought involved in addition and subtraction problems. In T. P.

Carpenter, J. M. Moser, & T. A. Romberg (Eds.), *Addition and subtraction: a cognitive perspective* (pp. 39–59). Hillsdale, NJ: Lawrence Erlbaum.

Verschaffel, L., & De Corte, E. (1990). Do non-semantic factors also influence the solution process of addition and subtraction word problems? In H. Mandl, E. De Corte, N. Bennett, & H. F. Friedrich (Eds.), *Learning and instruction, European research in an international context. Volume 2.2: Analysis of complex skills and complex knowledge domains* (pp. 415–429). Oxford: Pergamon Press.

von Glasersfeld, E. (1982). Subitizing: The role of figural patterns in the development of numerical concepts. *Archives de Psychologie, 50*, 191–218.

Wagner, S., & Walters, J. A. (1982). A longitudinal analysis of early number concepts: From numbers to number. In G. Forman (Ed.), *Action and thought* (pp. 137–161). New York: Academic Press.

Willis, G. B., & Fuson, K. C. (1988). Teaching children to use schematic drawings to solve addition and subtraction word problems. *Journal of Educational Psychology, 80*, 192–201.

Wolters, M. A. D. (1983). The part-whole schema and arithmetical problems. *Educational Studies in Mathematics, 14*, 127–138.

Woods, S. S., Resnick, L. B., & Groen, G. J. (1975). An experimental test of five process models for subtraction. *Journal of Educational Psychology, 67*, 17–21.

Working Groups of the Commission on Standards for School Mathematics of the National Council of Teachers of Mathematics (1989). *Curriculum and evaluation standards for school mathematics.* Reston, VA: National Council of Teachers of Mathematics.

Wynroth, L. (1980). *Wynroth math program—The natural numbers sequence.* Ithaca, NY: Author.

Zaslavsky, C. (1973). *Africa counts: Number and pattern in African culture.* Boston: Prindle, Weber, & Schmidt.

·13·

MULTIPLICATION AND DIVISION
AS MODELS OF SITUATIONS

Brian Greer

QUEEN'S UNIVERSITY, BELFAST

Multiplication and division of positive integers and rational numbers may be considered relatively simple from a mathematical point of view. The research reviewed in this chapter, however, reveals the psychological complexity behind the mathematical simplicity. In particular, this complexity is manifested when the operations are considered not just from the computational point of view, but in terms of how they model situations.

The chapter is in four parts. In the first, the range of applications of multiplication and division is set out, and the corresponding variety of external representations for the operations is illustrated. In the second part, the complexity is demonstrated further by reviewing current theoretical frameworks that treat the topic from several different perspectives. There follows an outline of a broader framework, the key points of which are that (a) multiplication and division model many distinguishable classes of situations, and (b) a fundamental conceptual restructuring is necessary when multiplication and division are extended beyond the domain of positive integers. The final part of the chapter points to the future by considering the potential of computer representations, putting forward suggestions for the improvement of teaching about multiplication and division, and setting out an agenda for further research.

PSYCHOLOGICAL COMPLEXITY OF MULTIPLICATION AND DIVISION

Applications for Multiplication and Division of Integers

The most basic way in which numbers are derived from the environment is through counting acts, giving rise to the pos-itive integers—the so-called natural numbers. (Negative numbers are not dealt with in this chapter; for convenience, therefore, *integers* will be used throughout to mean positive integers). The most important classes of situations involving multiplication and division of integers include: equal groups; multiplicative comparison; Cartesian product; rectangular area.

A situation in which there is a number of groups of objects having the same number in each group normally constitutes a child's earliest encounter with an application for multiplication. For example,

3 children have 4 cookies each. How many cookies do they have altogether?

Within this conceptualization, the two numbers play clearly different roles. The number of children is the *multiplier* that operates on the number of cookies, the *multiplicand*, to produce the answer. A consequence of this asymmetry is that two types of division may be distinguished. Dividing the total by the number of groups to find the number in each group is called *partitive* division, which corresponds to the familiar practice of equal sharing (with social connotations of equity). Dividing the total by the number in each group to find the number of groups is called *quotitive* division (sometimes termed *measurement* division, reflecting its conceptual links with the operations of measurement).

The equal-groups situation can arise in a variety of ways. Some examples are the mathematization of cases of natural replication (for example, n people have $5n$ fingers), repetition of a sequence of actions (for example, taking three steps four times), and human practices such as giving the same number of objects to a number of people.

The author gratefully acknowledges the helpful comments provided by Guershon Harel, Purdue University, James Hiebert, University of Delaware, and Larry Sowder, San Diego State University.

An alternative way of conceptualizing the equal-groups situation is in terms of a rate. From this point of view, the initial example may be expressed thus:

If there are 4 cookies per child, how many cookies do 3 children have?

In general, the number per group is multiplied by the number of groups to find the total number. There is implicit in this conceptualization an invariant relationship linking number of children and number of cookies; the situation described in the example is the *particular* instantiation of this relationship when the number of children is 3.

A different type of application is multiplicative comparison, verbally expressed by "n times as many as," as in the following example.

John has 3 times as many apples as Mary. Mary has 4 apples. How many apples has John?

Here the multiplicative factor may be conceived of as the multiplier. However, it is also possible to view the situation in terms of a many-one correspondence (3 apples of John's for every 1 apple of Mary's), which makes 3 the multiplier.

Cartesian products provide a quite different context for multiplication of natural numbers. An example of such a problem is

If 4 boys and 3 girls are dancing, how many different partnerships are possible?

This class of situations corresponds to the formal definition of $m \times n$ in terms of the number of distinct ordered pairs that can be formed when the first member of each pair belongs to a set with m elements and the second to a set with n elements. This sophisticated way of defining multiplication of integers was formalized relatively recently in historical terms.

There is a symmetry between the roles of the two numbers here, and hence only one type of division problem. Given that there are 12 possible partnerships, there is no essential difference between (a) being told that there are 4 boys and asked how many girls there are and (b) being told that there are 3 girls and asked how many boys. (In fact, it would be unusual to pose division problems of this type).

The final situation to be considered is rectangular area, where the sides of the rectangle are integral, say 4 cm by 3 cm. In this case, the rectangle can be partitioned into squares of side 1 cm so that the area may be found by *counting* these squares—it is *literally* 12 square cms. Such a diagram bears an obvious similarity to the physical arrangement of mn objects in a rectangular array with m rows and n columns. (Just as we still use the term *square* with a numerical meaning, Plato and Aristotle used the word *oblong* to mean any number that could be represented by a rectangular array). Either the rectangular array or the rectangle with integer sides provides a useful representation for making certain properties of multiplication as a binary operation, such as commutativity, intuitively acceptable. As with Cartesian products, the two numbers multiplied play equivalent roles, so they are not distinguishable as multiplicand

and multiplier; consequently, there are not two distinct types of division problems.

Extension to (Positive) Rational Numbers

Numbers arise, in the first instance, as mathematizations of reality—through counting procedures applied to discrete objects and through measuring procedures applied to continuous quantities. Other numbers are generated by successive applications of mathematical operations to existing numbers (for example, by taking the average of a set of measures). In this section, the extension of multiplication and division is considered for (a) numbers that can be traced back to counting procedures but that go outside the natural numbers through application of division at some point, and (b) numbers as measures. Numbers of the former sort are represented in the form a/b where a and b are integers; for convenience, the term *fractions* will be used for these throughout the chapter. Numbers of the latter sort are generally represented using decimal notation; for convenience, the term *decimals* will be used for these throughout the chapter. (As with integers, it is implicit that only positive numbers are being considered).

Extensions of the integers, and operations on them, have both a formal and an applicational aspect in terms of the wider range of problems that can be solved thereby. Formally, fractions are constructed as a reaction to the lack of closure of the integers under division. For integers a and b, $a \div b$ is defined (as c) if there exists an integer c such that $b \times c = a$. However, $14 \div 3$, for example, is defined only as 4, remainder 2. By extending the concept of number to fractions, and extending the definitions of multiplication and division accordingly, closure is achieved for division as well, that is, any fraction can be divided by any nonzero fraction and the result is a fraction.

This extension is paralleled within applications. For example, 14 pizzas cannot be equally divided among 3 children, as long as a pizza is considered an indivisible whole; each child can be given 4 pizzas, with 2 left over. By a shift of perspective, whereby a pizza is considered as something that can be cut into fractions (fractured), a solution becomes possible.

Thus, one use of the notation a/b (where a and b are integers) is to represent the result of partitive division, as in the sharing of a pizzas equally among b people, the result being a/b pizzas per person. A problem such as this can be solved by the following conceptual maneuver. If the a pizzas are reconceptualized as $ab(1/b$ pizzas) then the division can be carried out within the integer domain yielding $a(1/b$ pizzas) per person, which can then be converted back to a/b pizzas per person. Behr, Harel, Post, and Lesh, Chapter 14, this volume, present a comprehensive working out of such maneuvers.

Bell, Fischbein, and Greer (1984, p. 138) reported the following word problem generated by a student for the calculation $4 \div 24$:

Mary has 4 lb apples. She has got to share them between her and her 23 friends. How many apples do they get each? (12 apples to a lb.)

This student's stratagem for turning an uncongenial calculation into a congenial one could be seen as an intermediate step toward the general method that would substitute for the final clause in her problem "12 ($\frac{1}{12}$ lb) to a lb"—(or "6 ($\frac{1}{6}$ lb) to a lb").

The characterization of equal groups as a rate, as in 4 pizzas per child, generalizes to what might be termed a rational rate, such as 14/3 pizzas per child, which can be expressed in terms of integers as the equivalent rate of 14 pizzas for every 3 children.

A different use of the notation a/b is to represent a multiplicative part-whole relationship, as in the following:

There are 36 children in a class, of whom $\frac{2}{3}$ are girls. How many girls are there in the class?

Part-whole relationships are common within social practices—decimation, tithes, tax percentages, for example. The probability of an event corresponding to *a* out of *b* equally likely outcomes is a related application. The inverse task is to find a quantity *x* given that *y* is *a/b* of *x*. This task was termed the "construct-the-unit problem" by Behr and Post (1988, p. 198), who commented that it is almost never included in elementary curricula.

One can very conveniently represent *a/b* by a circle (or some other shape) divided into *b* equal sectors, of which *a* are differentiated in some way; concentration on this particular representation can result in it dominating students' conceptualizations of fractions to an unhealthy extent. Silver (1986) found that, among college students he tested, the fact that "almost universally, the students had only one available model for a fraction—the part/whole model expressed as sectors of a circle" (p. 189) contributed to fundamental conceptual errors.

Multiplicative comparison can be generalized beyond integer multipliers. In particular, a multiplicative comparison such as

John has 3 times as many apples as Mary.

can be inverted:

Mary has $\frac{1}{3}$ as many apples as John.

In general, if John has *a* apples and Mary has *b* apples, or if John has *a* apples for every *b* apples Mary has, then John has *a/b* times as many apples as Mary (and Mary has *b/a* times as many apples as John). .

Rectangular area can be generalized to rectangles the measures of whose sides are fractional. This context is a traditional way of making intuitively acceptable the definition of multiplication for fractions (De Morgan, 1910—[originally 1831], Chapter 4).

The further extension to quantities derived by measurement increases the range and complexity of situations modeled. Measurement is a matter of indefinitely subdividing units so that the quantity concerned is represented by a fraction to the desired degree of accuracy. When a fixed subdivision factor of 10 is used, the resulting measures are expressed as decimals. The extension of the decimal system from integers to nonintegers might appear relatively straightforward. However, there are hidden complexities, and it is well known that students experience considerable difficulty both with word problems in-

volving decimals and with purely computational tasks (Hiebert, in press).

Classes of situations already covered generalize to situations involving measures. The equal-groups class generalizes straightforwardly to the class of situations in which a measure is replicated some number of times, for example "three lots of 4.2 liters." An integral multiplier is retained in this generalization.

Problems within a great variety of contexts can be classified as rate problems involving decimals—such as speed problems, price problems, and mixture problems, as illustrated by these examples:

If a boat moves at a steady speed of 4.2 meters per second, how far does it move in 3.3 seconds?
How much does it cost for 3.3 gallons of paint at $4.20 per gallon?
A recipe calls for 3.3 pounds of butter for each pound of flour. How much butter is needed with 4.2 pounds of flour?

The general structure may be symbolized as

$$x \text{ [measure}_1\text{]} \times y \text{ [measure}_2 \text{ per measure}_1\text{]} = xy \text{ [measure}_2\text{]}$$

Measure conversion problems, such as

An inch is about 2.54 centimeters. About how long is 3.3 inches in centimeters?

may be considered a special case, having the particular property that measure$_1$ and measure$_2$ are alternative measures of the same quantity. (A further variation is when the measures are within the same system, as in conversion from feet to inches).

Multiplicative comparisons generalize straightforwardly to situations where the multiplicative factor is a decimal, as in this example:

Iron weighs 0.88 times as much as copper. A piece of copper weighs 4.2 gms. How much would a piece of iron the same size weigh?

A related type of situation is where some quantity undergoes a multiplicative change—for example, the expansion of a metal rod when heated. Such a situation may be construed as a particular form of multiplicative comparison where the comparison is between the same quantity before and after a transformation.

Another class of situations arises through the product of measures. For example,

If an appliance uses 3.3 kilowatts for 4.2 hours, how many kilowatt-hours of electricity does it consume?

(Sometimes, as in this case, the unit used in measuring the product, like a double-barreled surname, shows its parentage). This class of situations may be considered a generalization of Cartesian product since, in a sense, each part of the 3.3 kilowatts is combined with each part of the 4.2 hours.

The Range of Applications: Summary

The classes of situations distinguished in the foregoing analysis are summarized in Table 13.1, with examples of multiplication and division word problems in each case. Wherever multiplicand and multiplier can be distinguished—that is to say, one of the quantities involved in multiplication is conceptualized as operating on the other to produce the result—two types of division problem have been distinguished, corresponding to division by the multiplier and division by the multiplicand.

This classification is by no means exhaustive. Moreover, the extension of the concepts of multiplication and division can be continued indefinitely to encompass directed numbers, complex numbers, matrices, and so on. Of particular interest, both historically and psychologically, is the extension of multiplication and division to include negative numbers (Fischbein, 1987, Chapter 8), but this and other extensions lie beyond the scope of this chapter.

The distinctions between classes of situations are important pedagogically and provide an analytical framework useful for guiding research. However, it is important to realize that the way in which a situation is interpreted is not inherent in the situation, but depends on the student's construal of it. For example, a problem that would be classified as involving a Cartesian product, such as

How many different pizzas can be made with 3 different types of base and 4 different types of topping?

could be transposed into an equal-groups problem by thinking of it in the following terms:

4 different toppings (with base 1) + 4 different toppings (with base 2) + 4 different toppings (with base 3)

Moreover, apparently minor changes in the verbal formulation could influence the interpretation one way or another. Compare, for example,

(a) A painter mixes a color by using 3.2 times as much red as yellow. How much red does he need with 4.6 pints of yellow?
(b) A painter mixes a color using 3.2 pints of red for each pint of yellow. How much red should he use with 4.6 pints of yellow?

These two descriptions of the same scenario are biased toward different construals—multiplicative comparison and rate, respectively.

External Representations

The diversity of situations reviewed so far is reflected in the variety of external representations used both for teaching purposes and in the course of research, a selection of which is presented in Figure 13.1 (for further examples, see the references cited).

Figure 13.1a represents partitive division of 12 objects by dealing out, as in card games (Anghileri & Johnson, 1988, p. 147).

Figure 13.1b illustrates the number-line model for representing either multiplication (Vest, 1975, p. 24) or quotitive division (Anghileri & Johnson, 1988, p. 161).

Figure 13.1c is one of several ways to represent the 12 possible combinations of elements from a set with 3 elements and a set with 4 elements (cf. Freudenthal, 1983, p. 65).

Figure 13.1d illustrates a traditional representation for the product of two fractions ($\frac{1}{3} \times \frac{1}{4} = \frac{1}{12}$).

Figure 13.1e is a pictorial representation of Cartesian product (cf. Freudenthal, 1983, p. 62).

Figure 13.1f is a pictorial representation of an equal-groups situation (Beattys & Maher, 1989, p. 111; see also Freudenthal, 1983, p. 111).

Figures 13.1g (Beattys & Maher, 1989, p. 111) and 13.1h (Peled & Nesher, 1988, p. 257) are examples of representations used in research investigating children's ability to ignore irrelevant distortions of canonical representations.

Figures 13.1i (Sowder, 1988b, p. 75) and 13.1j (Greer, 1984, p. 296) schematically represent operations with large and indefinite numbers respectively.

Three general points may be made about such representations. First, it is easier to represent sets of objects and measured quantities than to represent derived quantities (such as miles per hour). Second, a diagram cannot adequately convey the essentially dynamic nature of a process such as "dealing out" (Figure 13.1a) or systematically generating combinations (Figures 13.1c, 13.1e). Dynamic computer versions of these representations would be much more powerful than any static representation. Third, one representation can be metamorphosed into another. For example 3 rows of 4 symbols in a line (representing 3 groups of 4) can be changed into a rectangular array by arranging the rows below each other (Freudenthal, 1983, p. 109); such metamorphoses could be particularly effectively shown as continuous transformations on a computer screen.

CURRENT THEORETICAL PERSPECTIVES

The outline in the first part of this chapter of the most salient circumstances giving rise to a need for multiplication and division of numbers within different numerical domains serves as a background to a wide range of theories and analyses prominent in current research, a selection of which is summarized here.

Vergnaud's Analysis of Multiplicative Structures

Vergnaud (1983, 1988) sets multiplication and division within the larger context of what he terms "the conceptual field of multiplicative structures," which he defines as consisting of

all situations that can be analyzed as simple and multiple proportion problems and for which one usually needs to multiply or divide. Several kinds of mathematical concepts are tied to those situations and the thinking needed to master them. Among these concepts are linear and n-linear functions, vector spaces, dimensional analysis, fraction, ratio, rate, rational number, and multiplication and division. (Vergnaud, 1988, p. 141).

TABLE 13–1. Situations Modelled by Multiplication and Division

Class	Multiplication problem	Division (by multiplier)	Division (by multiplicand)
Equal groups	3 children each have 4 oranges. How many oranges do they have altogether?	12 oranges are shared equally among 3 children. How many does each get?	If you have 12 oranges, how many children can you give 4 oranges?
Equal measures	3 children each have 4.2 liters of orange juice. How much orange juice do they have altogether?	12.6 liters of orange juice is shared equally among 3 children. How much does each get?	If you have 12.6 liters of orange juice, to how many children can you give 4.2 liters?
Rate	A boat moves at a steady speed of 4.2 meters per second. How far does it move in 3.3 seconds?	A boat moves 13.9 meters in 3.3 seconds. What is its average speed in meters per second?	How long does it take a boat to move 13.9 meters at a speed of 4.2 meters per second?
Measure conversion	An inch is about 2.54 centimeters. About how long is 3.1 inches in centimeters?	3.1 inches is about 7.84 centimeters. About how many centimeters are there in an inch?	An inch is about 2.54 centimeters. About how long in inches is 7.84 centimeters?
Multiplicative comparison	Iron is 0.88 times as heavy as copper. If a piece of copper weighs 4.2 kg how much does a piece of iron the same size weigh?	Iron is 0.88 times as heavy as copper. If a piece of iron weighs 3.7 kg, how much does a piece of copper the same size weigh?	If equally sized pieces of iron and copper weigh 3.7 kg and 4.2 kg respectively, how heavy is iron relative to copper?
Part/whole	A college passed the top 3/5 of its students in an exam. If 80 students did the exam, how many passed?	A college passed the top 3/5 of its students in an exam. If 48 passed, how many students sat the exam?	A college passed the top 48 out of 80 students who sat an exam. What fraction of the students passed?
Multiplicative change	A piece of elastic can be stretched to 3.3 times its original length. What is the length of a piece 4.2 meters long when fully stretched?	A piece of elastic can be stretched to 3.3 times its original length. When fully stretched it is 13.9 meters long. What was its original length? Division	A piece of elastic 4.2 meters long can be stretched to 13.9 meters. By what factor is it lengthened?
Cartesian product	If there are 3 routes from A to B, and 4 routes from B to C, how many different ways are there of going from A to C via B?	If there are 12 different routes from A to C via B, and 3 routes from A to B, how many routes are there from B to C?	If there are 12 different routes from A to C via B, and 3 routes from A to B, how many routes are there from B to C?
Rectangular area	What is the area of a rectangle 3.3 meters long by 4.2 meters wide?	If the area of a rectangle is 13.9 m² and the length is 3.3 m what is the width?	
Product of measures	If a heater uses 3.3 kilowatts of electricity for 4.2 hours, how many kilowatt-hours is that?	A heater uses 3.3 kilowatts per hours. For how long can it be used on 13.9 kilowatt-hour of electricity?	A heater uses 3.3 kilowatts per hours. For how long can it be used on 13.9 kilowatt-hour of electricity?

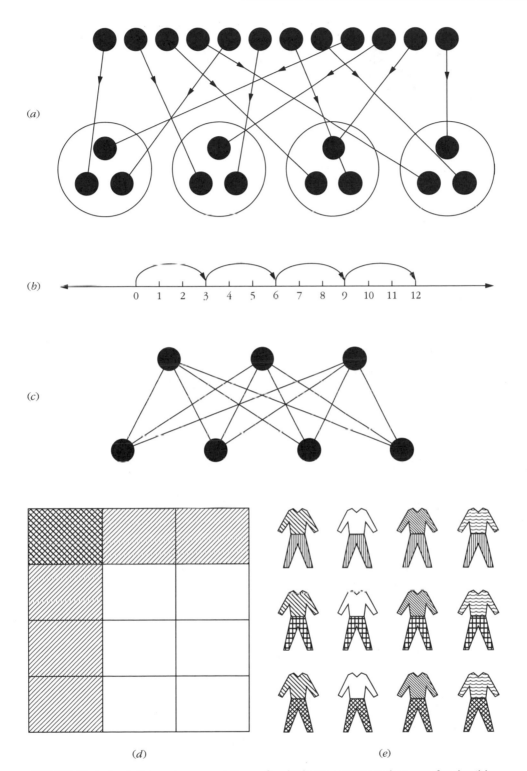

FIGURE 13–1. (a-e) External representations of multiplicative situations (see text for details).

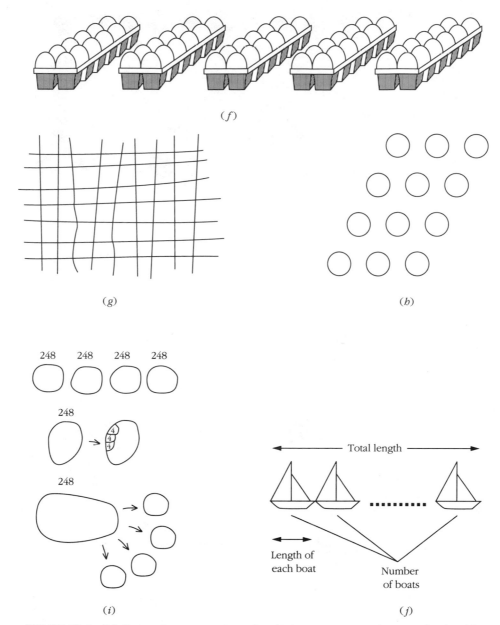

FIGURE 13–1. *(f-i)* External representations of multiplicative situations (see text for details).

The development of understanding of this conceptual field, he stresses, takes a long time—from age 7 to 18 at least. Moreover, there are important links with the conceptual field of additive structures (see Fuson, Chapter 12, this volume).

Three main classes of problems within multiplicative structures are identified (Vergnaud, 1983), termed *isomorphism of measures, product of measures,* and *multiple proportions.*

Isomorphism of measures covers all situations where there is a direct proportion between two measure spaces M_1 and M_2; Vergnaud's corresponding schematic diagram is shown in Figure 13.2*a*. No distinctions are made as to what types of numbers are involved—the quantities within each measure space may be integers, fractions, or decimals. The first seven classes in Table 13.1 are all subsumable within this category; the examples that follow are schematically illustrated in Figure 13.2.

Multiplication: 3 children each have 4 oranges. How many oranges do they have altogether? (Figure 13.2*b*).

Division by multiplier: A boat moves 13.9 meters in 3.3 seconds. What is its average speed in meters per second? (Figure 13.2*c*).

Division by multiplicand: An inch is about 2.54 centimeters. About how long in inches is 7.84 centimeters? (Figure 13.2*d*).

M_1	M_2	Children	Oranges	Seconds	Meters	Inches	Centimeters
1	a	1	4	1	?	1	2.54
b	c	3	?	3.3	13.9	?	7.84
(a)		(b)		(c)		(d)	

FIGURE 13–2. Schematic representation of isomorphism of measures (Vergnaud, 1983, 1988).

Multiplication and division problems can therefore be seen as special cases of *rule-of-three* problems (i.e., problems involving two equal ratios), in which one of the terms is 1; depending on which of the three remaining terms is unknown, the problem is solved by multiplication or by division of one type or the other. This interpretation demonstrates the closeness of the link between multiplication/division and proportional reasoning as components of the conceptual field.

The tabular representation has been used explicitly to help children solve word problems (Mechmandarov, 1987; Nesher, 1988; Sellke, Behr, & Voelker, 1988). Indeed, its pedagogical application goes back at least to the 16th century, when it was used by Robert Recorde and "the older English arithmeticians" (Cajori, 1917, p. 194).

The second major category identified by Vergnaud, termed *product of measures,* covers situations where two measure spaces, M_1 and M_2, are mapped onto a third, M_3; the corresponding schematic diagram is shown in Figure 13.3a. (Note that *product of measures* is used in a more restrictive sense in this chapter; in Vergnaud's scheme, this category subsumes both Cartesian products and rectangular area). Examples of Vergnaud's schematic representation of problems within this category (taken from Table 13.1) are as follows:

Multiplication: What is the area of a rectangle 3.3 meters long by 4.2 meters wide? (Figure 13.3b).
Division: A heater uses 3.3 kilowatts per hour. For how long can it be used on 13.9 kilowatt-hours of electricity? (Figure 13.3c).

Note that there is no clear distinction here between two types of division problem as there is in the case of isomorphism of measures.

The following is an example of a multiple-proportion problem cited by Vergnaud (1983, p. 139):

A family of 4 persons wants to spend 13 days at a resort. The cost per person per day is $35. What will the total cost of the holiday be?

However, this problem can be decomposed into simpler problems falling within the classes already defined—in the following way, for example,

$$4 \text{ people} \times 13 \text{ days} = 52 \text{ person-days}$$

$$\$35 \text{ per person-day} \times 52 \text{ person-days} = \$1820$$

Such problems are not included for separate analysis in this chapter. They are important, especially in physics, where formulas often require multiple proportions involving several quantities—more generally integral or fractional powers of several quantities.

The contrast is very striking between the variety of situations identified earlier in the chapter, as listed in Table 13.1, and the broad classes identified by Vergnaud on the basis of common mathematical structure. Moreover, Vergnaud's classification does not differentiate between situations dealing with sets of discrete objects and those dealing with measures. He does recognize, however, that specific contexts have peculiar characteristics and need to be subjected to separate analysis; volume is one context that he has investigated in some detail (Vergnaud, 1983).

Considerable attention has been paid by Vergnaud to the variety of methods used by children to solve problems. A finding of importance and generality is that relationships *within* measure spaces are usually more easily handled than relationships *between* measure spaces. Thus, the problem depicted in Figure 13.2b is more likely to be solved by the reasoning that 3 children have 3 times as many oranges as 1

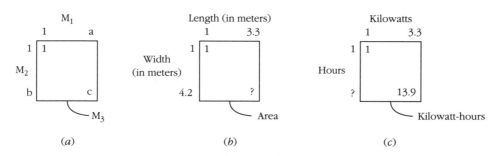

(a)   (b)   (c)

FIGURE 13–3. Schematic representation of product of measures (Vergnaud, 1983, 1988).

child than by the reasoning that the number of oranges is 4 times the number of children.

The notion of *theorem-in-action* is introduced to describe "mathematical relationships that are taken into account by students when they choose an operation or a sequence of operations to solve a problem" (Vergnaud, 1988, p. 144). In the example above, the first pattern of reasoning corresponds to the mathematical property of a linear function (a function of the form $f(x) = kx$, where k is a constant) that $f(\lambda \cdot 1) = \lambda \cdot f(1)$, which is a special case of the more general property that $f(\lambda \cdot x) = \lambda \cdot f(x)$. (For the technical details, the reader is referred to Vergnaud [1983, 1988]). The key point is that Vergnaud is pointing to the *implicit* use of structural properties in the course of problem-solving; such uses he terms *theorems-in-action*. Typically, they are limited in scope (generally to "easy" numbers), and they are not always correct. Nevertheless, he suggests that they offer access to students' intuitive grasp of multiplicative structures that teachers could exploit by making them explicit and general, with the aim of turning them into what might be called "theorems, in action," the *explicit* application of structural properties.

Schwartz and Kaput: The Role of Intensive Quantities

Focusing on mathematics as a modeling activity, Schwartz (1988) draws attention to the importance of the links between numbers and their referents. Numbers arise

1. In quantifying aspects of the world, either by counting (discrete quantities) or by measuring (continuous quantities). These are *extensive* quantities.
2. Through the application of arithmetical operations to already defined quantities. Such operations may be complex—for example, the calculation of a correlation coefficient.

(It should perhaps be added that there is a further way in which quantities are derived, namely through complex interactions within human practices—for example, conventions governing what percentage should be given as a tip in various circumstances.)

Among quantities derived by the application of mathematical operations to other quantities, *intensive* quantities constitute a particularly important kind. To use an example from Schwartz (1988), consider a pile of coffee beans with the following associated quantities:

Weight of coffee	5.0 lb
Cost of coffee	$15.00
Price of coffee	$3.00 per lb

The third quantity differs in kind from the first two. Whereas the weight and cost (extensive quantities) are properties of the *whole* pile, the price per lb is a property of *any part* of the pile. If the pile is added to a second pile of coffee, the weights and costs can be added, but not the price per lb. The essential characteristic of an intensive quantity is that it numerically expresses a constant multiplicative relationship between two other quantities (not necessarily extensive). Intensive quantities—such as

constant speed, acceleration, and mass density—are very important in physics.

The following classification of situations is based on this fundamental distinction between extensive (E) and intensive (I) quantities (see Kaput, 1985, for details and for examples of other cases).

1. Problems with structure $I \times E = E'$. Such problems correspond to Vergnaud's *isomorphism of measures* problems. For example: "3 children each have 4.2 liters of orange juice. How much do they have altogether?" fits into this pattern thus:

$$4.2 \text{ liters per child} \times 3 \text{ children} = 12.6 \text{ liters}$$

There are two associated types of division problem (as in Table 1): $E'/E = I$, $E'/I = E$. This class, according to Schwartz (1988, p. 50) "accounts for the vast majority of the multiplication and division problems we ask students to undertake."
2. Problems with structure $E \times E' = E''$. Such problems correspond to Vergnaud's *product of measures,* namely situations such as Cartesian product or area.

Kaput (1985, p. 13) asserts as a basic principle that, "the elementary mathematics of school should not be, as tacitly assumed, exclusively the mathematics of number with applications regarded as separate, but rather should begin with the mathematics of quantity, so that the mathematics and its 'applications' are of a piece from the very beginning." Thus, attention should be paid not simply to the numbers in a problem but also to the *referents* of the numbers. An important difference contributing to the greater difficulty of multiplication and division problems as compared to addition and subtraction problems is that the latter are unidimensional, whereas the former have dimensional complexity—an intensive quantity such as miles per hour may be seen as transforming a quantity with referent "hours" to a quantity with referent "miles."

A formal treatment of referents is outlined by Schwartz (1988) and embodied in his *Semantic Calculator* (Schwartz, 1982). Software developed by Shalin and Bee (1985) and by Thompson (1989), referred to in a later section titled "The Potential of Computer Software," also explicitly draws attention to the referents as well as the numbers in the solution of word problems. Software specifically developed to build understanding of intensive quantities (Kaput, 1985, 1990) is also mentioned in that section.

Like Vergnaud, Schwartz (1988) and Kaput (1985) point out that a word problem, such as

A boat moves at a steady speed of 4.2 meters per second. How far does it go in 3.3 seconds?

has as background an implicit functional relationship: Distance (in meters) = 4.2 × time (in seconds). Links can be made with more general functions. In particular, as Schwartz (1988, p. 48) has pointed out, "the graphical representations of the relationships between the three quantities of interest are in fact

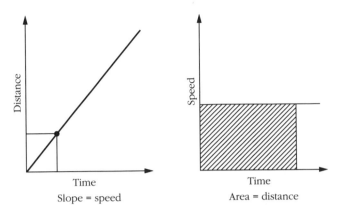

FIGURE 13–4. Links between multiplication, division and calculus (Schwartz, 1988).

representations of the essential ideas of the differential and integral calculus respectively"—namely slopes and areas under curves (Figure 13.4).

Nesher's Semantic Analysis

Nesher (1988) has analyzed the propositional structures of three classes of situations. The first, which she terms *mapping rule*, corresponds to Vergnaud's isomorphism of measures and Schwartz and Kaput's $I \times E = E'$ category. The second, which she terms *Cartesian multiplication*, is included under product of measures by Vergnaud and within the $E \times E' = E''$ category by Schwartz and Kaput.

Multiplicative comparison, referred to in the first part of this chapter but not distinguished as a separate class by either Vergnaud (who subsumes it under isomorphism of measures) or Schwartz and Kaput (who subsume it under the $I \times E' = E''$ category), is exemplified by the following problem (Nesher, 1988, p. 22):

> *Dan has 5 marbles. Ruth has 4 times as many marbles as Dan. How many marbles does Ruth have?*

Nesher suggests further that the compared objects need not be of the same kind—the number of Ruth's stamps could be compared to the number of Dan's marbles. This comparison seems rather unnatural; within the classification proposed in this chapter, such a problem would be treated as a rate problem.

Nesher's distinction of multiplicative comparison as a separate class may well be related to the simplicity of expressing multiplicative comparison in Hebrew (in which language there is an expression P such that P-5, for example, means *5 times as many as*). Israeli children aged 10–12 were asked (Nesher, 1988) to

a. Write a multiplication word problem corresponding to 3×4.
b. Write down how they would explain to another child how to distinguish between a multiplication and an addition word problem.

The most common responses to task (a) were multiplicative-comparison problems (41%) and mapping-rule problems (34%). For task (b), responses referred most commonly to

multiplicative-comparison characteristics (53%) or mapping rule characteristics (14%).

The finding that the preponderance of word problems were the multiplicative-comparison type contrasts with studies of English-speaking students in which multiplicative-comparison word problems have not been found (Bell et al., 1984; Kaput, 1985; Mangan, 1986). Taken together, these results have a number of implications.

1. They point to the importance of education in the formation of young children's conceptualizations of multiplication (the P formulation tends to be introduced by teachers in Israel at an early stage).
2. They raise the question of cross-cultural differences, which have not been extensively studied in research on multiplication and division.
3. They are relevant to the general question of the relation between thought and language in young children. Nesher's summarization (p. 34) that "children found it easier to give a verbal sign for the [multiplicative-comparison] problems ...but could imagine the [mapping-rule] problems better" could be interpreted as broadly in line with the Piagetian view that language alone does not produce full understanding unless the underlying logical structure is grasped.

By far the most common method of confronting students with situations to be modeled is through word problems. These can be analyzed as texts constituting a distinct genre. From this point of view the task may be seen as one of translating from the natural-language representation of the word problem to the mathematical-language representation of the model (and back). For competent performance on word problems it is generally agreed that, in the course of constructing an appropriate mathematical formulation, an intermediate representation of the situation being verbally described is needed. For example, De Corte and Verschaffel (1985, p. 4), referred to "the construction of a global, mental representation of the problem situation." Similarly, Kintsch (1986) distinguished between the *textbase* and a *situation model*, and Reusser (1987) has developed a computer simulation that elaborates on this idea. While De Corte and Verschaffel, Kintsch, and Reusser have applied their analyses only to addition and subtraction problems, they seem equally applicable to multiplication and division problems, although these would be more complex to analyze in detail due to the wider range of situations and the dimensional complexity.

As Nesher (1980) points out, it is clear from many findings that students very frequently bypass any such intermediate representation and move directly to a mathematical expression on the basis of syntactical, surface clues. Students develop strategies such as those identified by Sowder (1988a), which include:

Look at the numbers; they will tell you which operation to use.
Try all the operations and choose the most reasonable answer.
Look for key words or phrases to tell which operation to use.

The stereotyped nature of word problems presented to students (Nesher, 1980) reinforces such behavior since it produces a high degree of superficial success. Reusser (1988) concluded that "whoever observes students in classroom and homework situations can find again and again how few common textbook problems force students to do an in-depth semantic analysis."

In describing the linguistic analysis of word problems through propositional description and textual analysis, Nesher (1988) pointed out that

The child's first glance shows a text consisting of natural language with some numerical data. It is clear that the numerical data do not provide any information regarding the kind of operation to be performed; rather *all* the decisive information is embedded in the verbal formulation of the text. (p. 24)

This statement is logically true, but data summarized in discussing the two theoretical approaches remaining to be covered show that for many students (and also teachers) the numerical data *are* a major factor in determining the ease with which the correct operation is identified for a problem; consequently, the types of numbers cannot be dismissed as "a trivial variable" (Nesher, 1988, p. 27).

Fischbein's Theory of Primitive Models

Fischbein, Deri, Nello, and Marino (1985) proposed a theory to account for children's performance on word problems. Specifically, they tested students' ability in *choice of operation,* in which the task is to nominate the correct operations for single-operation word problems without being required to carry out any calculation. The theory proposed by Fischbein et al. was stated as follows:

Each fundamental operation of arithmetic generally remains linked to an implicit, unconscious, and primitive intuitive model. Identification of the operation needed to solve a problem with two items of numerical data takes place not directly but as mediated by the model. The model imposes its own constraints on the search process. (p. 4)

For multiplication, they proposed that the primitive model is repeated addition. In an equal-groups situation, such as 3 children having 4 oranges each, the situation can be conceptualized as 4 oranges + 4 oranges + 4 oranges, and the answer can then be calculated by repeated addition. This representation generalizes naturally to a situation such as 3 children having 4.2 liters of orange juice each, which can be conceptualized as 4.2 liters + 4.2 liters + 4.2 liters. For a situation to be assimilable to this model, the crucial factor is that the multiplier must be an integer; no restriction applies to the multiplicand. Moreover, this model of multiplication carries the implication that the result is always larger than the multiplicand.

For equal groups or equal measures, this condition is met by definition. However, the multiplicand/multiplier distinction applies in other classes of situation (see Table 13.1) and, in general, multiplicand and multiplier may be integers, fractions, or decimals. For example, consider the following contrasting pair:

A rocket travels at a speed of 16 miles per second. How far does it travel in 0.85 seconds?
A rocket travels at a speed of 0.85 miles per second. How far does it travel in 16 seconds?

From a purely computational point of view both problems involve the multiplication of 16 and 0.85, but the former is more difficult to envisage as requiring multiplication for solution; many children, indeed, judge that the answer would be given by $16 \div 0.85$ (Greer, 1988).

Results from several experiments using problems from a variety of situation classes consistently show the *multiplier effect* (De Corte, Verschaffel, & Van Coillie, 1988, p. 203), namely that the difficulty of recognizing multiplication as the appropriate operation for the solution of a problem depends on whether the multiplier is an integer, a decimal greater than 1, or a decimal less than 1 (Bell et al., 1984; De Corte et al., 1988; Fischbein et al., 1985; Luke, 1988; Mangan, 1986). The size of the effect, in terms of the difference in percentage of correct choices, is of the order of 10–15% for the difference between integer and decimal greater than 1 as multiplier. For the difference between integer and decimal less than 1, the size of the effect is of the order of 40–50%. When the multiplier is less than 1, there is the added difficulty that the result is smaller than the multiplicand, which is incompatible with the repeated addition model. By contrast, the findings from these experiments show that it makes no appreciable difference what type of number appears as the multiplicand. Thus, these results for the interpretation of word problems modeled by multiplication show a clear pattern that is consistent with the theory advanced by Fischbein et al.

For division, Fischbein et al. proposed two primitive models, namely partition (sharing into equal subcollections or subquantities) and quotition (determining how many subcollections or subquantities of a given size are contained in a collection or quantity). If the operation in a problem is to be readily perceivable as partitive division, then the divisor must be an integer and smaller than the dividend. For the operation in a problem to be readily perceivable as quotitive division, the divisor must be smaller than the dividend.

Partitive and quotitive division were defined earlier for the equal-groups situation, but the definitions can be generalized to all situations in which multiplicand and multiplier can be distinguished (Table 13.1) by defining partition as division by the multiplier, and quotition as division by the multiplicand. On this basis, the following problems both involve partitive division:

The cost of 3 liters of orange juice is $6. What is the price per liter?
The cost of $\frac{2}{3}$ of a liter of orange juice is $6. What is the price per liter?

It is much easier to recognize division as the appropriate operation in the former case than in the latter. Likewise, each of the following problems involves quotitive division:

An inch is about 2.54 centimeters. About how long in inches is 7.84 centimeters?
An inch is about 2.54 centimeters. About how long in inches is 1.84 centimeters?

Again, the former but not the latter conforms to the constraint identified by Fischbein et al. for quotitive division and is easier to construe as requiring division.

Fischbein et al. (1985, p. 14) suggest that partition is the original intuitive primitive model for division and that the quotitive model is acquired later through instruction. In studies in which subjects have been asked to write a word problem corresponding to a division such as $12 \div 3$, partitive division problems have been produced in the overwhelming majority of cases (Af Ekenstam & Greger, 1983; Bell et al., 1984; Kaput, 1985; Mangan, 1986). Graeber and Tirosh (1988) reported a predominance of partitive over quotitive word problems when similar tasks were given to preservice elementary school teachers.

In general, the data for division problems present a much more complicated picture than is the case for multiplication problems. Fischbein's explanation in terms of constraints bearing on partitive and quotitive division does not account for experimental findings (Bell, Greer, Grimison, & Mangan, 1989). As discussed below, it is necessary to augment his explanation by reference to numerical and computational considerations.

Fischbein et al. (1985, p. 15) suggest that the primitive models reflect both "features of human mental behavior that are primary, natural, and basic" *and* "the way in which the corresponding concept or operation was initially taught in school." Moreover, as pointed out by Graeber and Tirosh (1988, p. 272), it is difficult to disentangle the effects of early conceptualizations from the effects of having an extended period of working with multiplication and division within the restricted domain of the integers. The general point is that children's early experience with multiplication and division, both in and out of school, is grounded in behavior largely limited to simple situations involving discrete objects and mathematically restricted within the integer domain. When it becomes necessary to broaden the domain beyond the integers and to deal with new classes of situations, problems are caused by the deep-rootedness of the early conceptualizations. This phenomenon within the context of multiplication and division concepts is paradigmatic of the growth of mathematical concepts in general, both historically and within the individual (Fischbein, 1987).

Numerical and Computational Effects (Bell and Greer)

Bell, Greer, and associated researchers have drawn particular attention to effects associated with the types and properties of numbers appearing in word problems and to the interfering effects of students' thoughts about calculations, even when no calculation is required (Bell et al.,1984; Bell et al.,1989; Greer, 1987, 1988; Mangan, 1986).

The most obvious manifestation of numerical/computational aspects is the influence of familiar misconceptions, notably that multiplication always makes bigger, and division smaller, and that division is always division of the larger number by the smaller. These misconceptions show up in purely numerical items (Greer, 1989) as well as in word-problem tasks, and their prevalence has also been shown among preservice and elementary teachers (Graeber & Tirosh, 1988; Graeber, Tirosh, & Glover, 1989; Harel, Behr, Post, & Lesh, in preparation; Tirosh & Graeber, 1989; Vinner & Linchevski, 1988). A frequently observed pattern is that students are aware of the size of the result relative to the operand and choose multiplication or division on that basis; when the operand is less than 1 this strategy leads to division being chosen instead of multiplication, and vice versa.

When children are presented with single-step word problems and asked to state which operation would be necessary to find the solution (without having to carry out the calculation) the numbers have a very marked effect on the difficulty of the task. For multiplication, as mentioned earlier, the effect of changing the multiplier from an integer to a decimal less than 1 (and leaving the problem otherwise unchanged) is to reduce correct answers by about 40–50%; this effect occurs in adults as well as children (Bell et al., 1989; Greer & Mangan, 1986; Mangan, 1986). For division problems, changing the numbers has an even greater effect, of the order of 60–70% in the extreme cases (Bell et al., 1989; Greer & Mangan, 1986; Mangan, 1986). Bell et al. (1989) argue that in order to account for the complex findings on division problems, it is necessary to consider the student's perception of the difficulty of the calculation that would be required, even though the choice of operation task *does not require the calculation to be carried out.*

Another way in which students' thinking about calculations interferes with their conceptualization of problems is illustrated by the following excerpt from an interview of a preservice elementary teacher who had given $15 \div 5$ as the operation required to solve a problem about 15 friends sharing 5 pounds of cookies (Graeber & Tirosh, 1988).

(I = interviewer, S = subject)
I: Why did you write it that way [$15 \div 5$]?
S: I did it automatically. You know *usually people will write the question in a way that the larger number will come first* . . .
I: Did someone tell you about the larger number coming first?
S: It is just *something that's in all problems and all examples you are given,* I bet.
I: But perhaps someone pointed this out to you?
S: I guess I realized it by myself.
(p. 272, emphasis added)

Nesher (1987) commented that multiplication and division word problems in textbooks rarely offer counterexamples to the misconceptions that multiplication always makes bigger and division smaller, and that division is always of the larger number by the smaller. Answers may be determined by these misconceptions, as in the example above, rather than by a logical analysis of the situation described.

Several investigators have reported observations of children changing their choice of operation from multiplication to division or vice versa when presented *successively* with problems differing essentially only in terms of the numbers (Bell et al., 1981). For example, Af Ekenstam and Greger (1983) presented subjects 12 to 13 years old with the following two problems:

A cheese weighs 5 kg. 1 kg costs 28 kr. Find out the price of the cheese. Which operation would you have to perform?

$$28 \div 5 \qquad 5 \times 28 \qquad 5 + 28 \qquad 28 + 28 + 28 + 28$$

A piece of cheese weighs 0.923 kg. 1 kg costs 27.50 kr. Find out the price of the cheese. Which operation would you have to do?

$$27.50 + 0.923 \qquad 27.50 \div 0.923 \qquad 0.923 \times 27.50 \qquad 27.50 - 0.923$$

They reported that when these two problems were presented in interviews *one after the other,* students would frequently choose multiplication for the first and division for the second, *even when the interviewer drew their attention to the similarity between the two problems.* It seems clear that the choice of division for the second problem is based on the realization that the answer will be less than 27.50, combined with the belief that multiplication *always* makes bigger, and division smaller.

This phenomenon of changing choice of operation between successively presented problems when only the numbers in the problem are changed, labeled *nonconservation of operations,* has been further studied and analyzed by Greer (1987, 1988) as an indication of underlying weakness in children's understanding of multiplication and division when they are extended beyond the domain of the integers.

Multiple Perspectives

As the foregoing sample shows, multiplication and division have been analyzed from a variety of angles. Moreover, the coverage is incomplete—in particular, there is a considerable body of work exploring in detail the developmental origins of multiplicative concepts (Anghileri, 1989; Kouba, 1989; Nantais & Herscovics, 1989; Steffe, 1988).

Within the range covered, there are some marked contrasts to be drawn. Lesh, Post, and Behr (1988) argued that the approaches of Vergnaud and of Schwartz and Kaput represent different traditions concerned, respectively, with the generalization of integer-based concepts into the domain of rational numbers, and with a mathematics of quantity rather than a mathematics of number. In a situation such as traveling at 4 miles per hour for 3 hours, Schwartz characterizes the intensive quantity, 4 miles per hour, as operating in a referent-transforming way on the extensive quantity, 3 hours, to produce another extensive quantity, 12 miles. This identification of the intensive quantity as the operator is shown in the labeling of the class of situations to which the example belongs as $I \times E = E'$. Vergnaud's interpretation in terms of an isomorphism between two measure spaces, hours and miles, is quite different. The problem can be solved *within* measure spaces by deriving a scalar operator in one measure space and then applying it in the other:

$$3 \text{ hours} = 3 \times 1 \text{ hour}$$
$$\updownarrow$$
$$3 \times 4 \text{ miles} = 12 \text{ miles}$$

Within this conceptualization of the solution process, the number attached to the extensive quantity is the multiplier. This identification of what subjects conceptualize as multiplicand and multiplier agrees with that of Fischbein and of Bell and Greer (though the theoretical bases for the identification are different) and has been assumed in the analysis presented in this chapter (Table 13.1). Support for this assumption is provided by the experimental findings summarized as the multiplier effect (p. 286).

The central theme of this chapter, the generalization of integer-based concepts of multiplication and division, is addressed in contrasting ways within the theoretical perspectives reviewed so far—through the analysis of proportionality (Vergnaud), intensive quantities (Schwartz and Kaput), the

semantic structure of word problems (Nesher), effects of intuitive integer-based models (Fischbein), and lack of awareness of invariances (nonconservation of multiplication and division) combined with numerical effects on conceptualization (Bell and Greer). Common to all these approaches is a realization that the generalization of multiplication and division beyond the integer domain is difficult and requires a major conceptual restructuring.

Currently, therefore, a wide range of theoretical frameworks exists (Harel & Confrey, in press). These frameworks emphasize different, often complementary facets, with some areas of disagreement. In the next section, a broader perspective related to very general characteristics of mathematics is presented.

TOWARD A MORE COMPREHENSIVE FRAMEWORK

In this section, a framework is proposed for the analysis of multiplication and division of (positive) integers, fractions, and decimals as models of situations. The two central themes of this framework are

1. The range of situations modeled, and the variation across this range in the nature of modeling.
2. The long course of development of multiplicative concepts; in particular, the radical conceptual restructuring required to extend these concepts beyond the integer domain (Greer, in press).

These themes, fully generalized, reflect fundamental characteristics of mathematics—its role in modeling situations and thereby enabling problems to be solved, and its intrinsically developmental nature, whereby concepts arising and defined "naturally" within a limited domain become extended, both as a response to more complex problems and as the result of the application of formal analysis (either of which may precede the other, as can be illustrated by historical examples).

A Differentiated Classification of Situations Modeled by Multiplication and Division

At the start of the chapter, an analysis of situations was presented, dealing in turn with (positive) integers, fractions, and decimals (Table 13.1). Table 13.2 summarizes this classification, incorporating differentiation on the basis of the types of quantities involved. It also shows the correspondence with the broad categories defined by Vergnaud, Schwartz and Kaput, and Nesher.

It is characteristic of mathematics to identify structural unity underlying situations that appear on the surface to be very different; this characteristic is a source of great power, as Freudenthal (1983) has pointed out.

Mathematics is powerful thanks to its universality. One can count all sets by the same sort of numbers, as one can measure all magnitudes by the same sort of numbers.... Multiplication has the aspect of repeated addition and of pair formation, and likewise division has its own variety of aspects. But in spite of this wealth of aspects, it is always the same operation—a fact that expresses itself by algorithmics. As a calculator one may forget about the origin of one's numbers and the origin of one's *arithmetical* problem in some *word* problem. But at the same time one must be able to return from the algorithmic simplicity to the phenomenal variety in order to discover the simplicity in the variety. (pp. 116–117)

From a structural point of view, the situations labeled in this chapter as equal groups and rates may be considered the same; likewise for Cartesian product and product of measures. From the psychological and educational points of view, however, it is not clear to what extent a focus on common structure is useful in characterizing children's thinking or as a guide to pedagogical strategies. (A parallel could be drawn here with the debate over the generality of the cognitive structures proposed by Piaget in light of experimental evidence showing the importance of contextual and situational factors).

Davis and Hersh (1981, p. 74) declared that "there can be no comprehensive systematization of all the situations in which it is appropriate to add" and doubtless a similar statement applies to multiplying and dividing. The classification proposed in Table 13.2 therefore represents a minimal set of distinctions (covering some of the most important classes of situations that students up to the age of about 15 are likely to encounter in and out of school), which is open to both extension and refinement. More complex combinations could be added, such as Vergnaud's multiple proportions (generalizing to products of powers of quantities, common in physics). Also, many applications of multiplication involve products of more than two quantities, as in exponentiation, which has links with very general phenomena of growth (Confrey, in press). Indefinite refinement is possible; within classes of situations, further distinctions can be made—some very subtle but psychologically significant. For example, within equal-groups situations, the nature of the group membership varies (Kaput, 1985, p. 23) from physical containment (for example, candies per bag) to metaphorical inclusion (for example, children per family).

It must always be remembered that a structural analysis provides only a framework for research and cannot be assumed to provide a description of students' thinking. As discussed earlier, construals of the situations will not necessarily conform to any a priori classification. Considerable ingenuity has been deployed in devising methodologies that make it possible to infer students' solution strategies and their underlying representations.

The Modeling Role of Multiplication and Division

The focus of concern is mathematics as a modeling activity. (Schwartz, 1988, p. 41)
Mathematical concepts are rooted in situations and problems. (Vergnaud, 1988, p. 142)

The role of multiplication and division in modeling aspects of reality (and thereby enabling problems to be solved) constitutes yet another basis for differentiation, since the nature of the relationships between situations and corresponding calculations varies widely. This variety can be illustrated (although not exhaustively) by considering a number of examples in the style of the Davis and Hersh (1981, p. 70 et seq.) discussion of addition problems.

1. The result of a calculation may be exact or approximate. If an employee is paid \$15.20 per hour, then for 4 hours, the exact pay is \$60.80. However, if a piece of timber is 15.20 m long (measured to an accuracy of two decimal places) then 4 such pieces placed end to end may measure 15.78, 15.79, 15.80, 15.81 or 15.82 m to two decimal places, since the actual length of each piece lies within the interval from 15.195 m to 15.205 m. The calculation that leads to the answer 15.80 m must be interpreted in light of this characteristic of measurement. As Hilton (1984, p.8) put it, "$2 + 2 = 4$ in counting arithmetic; but $2 + 2 = 4$ with a probability of $3/4$ if we are dealing with measurement."

2. "An army bus holds 36 soldiers. If 1,128 soldiers are being bused to their training site, how many buses are needed?" This question is the notorious item from the National Assessment of Educational Progress (1983), to which 29% of 13-year-olds gave the answer "31, remainder 12," and 18% the answer "31." Such results underline Hilton's (1984, p. 8) judgment that "the separation of division from its context is an appalling feature of traditional arithmetic." Students' attention should be drawn to variations in the interpretation of division, as in this example given by Streefland (1988, p. 81):

$6394 \div 12$, invent stories belonging to this sum such that the result is, respectively

532	*532.84 rem. 4*
533	*532.833333*
532 rem. 10	*about 530*

3. "One can of tuna fish costs \$1.05. How much do two cans of tuna fish cost?" (Davis & Hersh, 1981, p. 71). Your grocer, like the one used by Davis and Hersh, may offer two cans for \$2.00. The point is that setting cost proportional to quantity is a convention that is usually reasonable but is not inviolable and may be departed from for various reasons.

4. "A man can run a mile in 4 minutes. How long will it take him to run 3 miles?." An expectation that 12 minutes should be given as the answer illustrates the unrealistic nature of many textbook problems (the precision of the 4 minutes is itself spurious). It *can* make sense to calculate 12 minutes here, but taking 4 minutes as an approximation, and 12 minutes as an approximation on the high side. Further analysis could lead to a refined model that accounts for the effects of fatigue.

5. "How long is 30 cms in inches?" The answer here depends on the degree of accuracy required.

6. "The exchange rate is \$1.82 dollars per pound. How many dollars will you get for £100?" (In real life, the answer partly depends on what you are charged for the transaction.) Unlike measure conversions (example 5), exchange rates vary in time, and from bank to bank. Such quantities are the results of very complex sets of mathematical operations.

7. "If a die is thrown 600 times, about how many 6s will be obtained?" Probabilistic situations such as this represent a very special sort of relationship between the result of the calculation, 100, as the *expected value*, and reality.

The central point here is that instead of the sanitized view of the world generally presented in the "word problem game" (De Corte & Verschaffel, 1985, p. 8), students should experience a wide range of examples in which the nature of the relationship between reality and calculation—and the appropriate interpretation of the result of the calculation—vary and in which implicit modeling assumptions (such as that of proportionality) are made explicit. Such an approach could meet widespread criticisms of the use of stereotyped word problems in schools (Nesher, 1980, 1988; Reusser, 1988).

The Extended Development of Multiplicative Concepts

The range of situations covered in Table 13.2 in terms of semantic structure and numerical types also relates to different stages in the conceptual development of children as mediated by educational experience. Vergnaud (1988) and many others have emphasized the extended span of time over which children's understanding of the conceptual field of multiplicative structures is built up. A comprehensive elaboration of this theme would extend beyond the scope of this chapter to negative numbers, complex numbers, and so on, and to links (Schwartz, 1988; Vergnaud, 1988) with more advanced mathematical topics and concepts from physics.

Multiplication and division are conceptually complex even within the restricted domain of whole numbers, both in terms of the range of situations modeled (Anghileri, 1989; Kouba, 1989; Vest, 1971) and in terms of basic conceptual understanding (Nantais & Herscovics, 1989; Steffe, 1988). However, the main focus of attention in this chapter has been the extension of multiplicative concepts beyond the integers to fractions and decimals. This extension is paradigmatic of the intrinsically developmental nature of mathematics whereby operations, relations, functions, and the like are extended by formal means beyond their intuitive, behaviorally based origins.

The problems children have in restructuring their conceptualizations of multiplication and division beyond the integer domain have been shown by a range of experimental findings, which may be summarized as follows:

1. There are well-known common misconceptions that can be attributed to overgeneralization of experience within the integer domain, notably that multiplication always makes bigger and division always makes smaller, and that division is always of the larger number by the smaller. Such misconceptions are inferrable from performance on both purely numerical tasks and word problems and are often explicitly stated by subjects (Af Ekenstam & Greger, 1983; Bell et al., 1981; Bell et al., 1984; Greer, 1987, 1989; Sowder, 1988); they have also been shown to be quite common among preservice and elementary teachers (Graeber & Tirosh, 1988; Graeber et al., 1989; Harel et al., in preparation; Tirosh & Graeber, 1989; Vinner & Linchevski, 1988).

2. The added difficulty of moving from multiplication problems involving only integers to problems involving decimals is not simply a matter of introducing "harder" numbers. A clear pattern has been found, namely the *multiplier effect*, for classes of situation in which multiplicand and multiplier can be differentiated (De Corte et al., 1988; Greer & Mangan, 1986; Luke, 1988). For multiplication to be intuitively seen as the appropriate operation, the key factor is that the multiplier must be an integer; as mentioned earlier, this pattern of findings is consistent with the theory advanced by Fischbein et al. (1985).

3. For division problems, the picture is much more complex. There are some indications that the generalized partitive/quotitive distinction has some explanatory power, but it seems clear that many factors, including numerical and computational considerations, interact in a complex way when students attempt to interpret problems involving division (Bell et al., 1989).

4. When children are given a calculation and asked to write word problems for which that calculation would be appro-

TABLE 13-2. Classifications of Situations Modeled by Multiplication and Division

Proposed classification				Other classifications		
Integers	Integer multiplier	Fractions	Decimals	Vergnaud	Schwartz & Kaput	Nesher
Equal groups	Equal measures	Rational rate	Rate			
			Measure conversion	Isomorphism between measure spaces spaces	$I \times E = E'$	Mapping rule
Multiplicative comparison		Multiplicative comparison Part/whole	Multiplicative comparison Part/whole Multiplicative change			Multiplicative comparison
Rectangular array Rectangular area		Rectangular area	Rectangular area	Product of measures	$E \times E' = E''$	Cartesian product
Cartesian product			Product of measures			

priate, they show a limited ability to generate appropriate problems except for the simplest of cases (Af Ekenstam & Greger, 1983; Bell et al., 1984; Mangan, 1986).

5. Another difficulty is the move away from quantities derived from counting to those derived by measuring or by derivation of new quantities through mathematical operations applied to existing ones. In particular, children have difficulty with intensive quantities (Bell & Onslow, 1987; Kaput, 1985; Schwartz, 1988).

6. A number of observations and experiments have shown that children often do not conserve operations; that is, they are unaware of, and fail to exploit, the invariance of operations over the numbers involved in word problems (Greer, 1987, 1988).

7. There are specific differences associated with specific contexts, such as volume, which Vergnaud (1983) has investigated in some detail. Contexts differ in the extent to which they draw on knowledge from everyday life (for example, price problems) or depend more on education (for example, density). Relatively little systematic comparison across contexts has been done, although some comparative data can be found in Bell et al. (1989) and Vergnaud (1983, 1988).

It can be illuminating to consider historical parallels. Any developmentalist or educator must keep in mind that we are attempting to help children construct concepts that culturally took centuries to evolve. The conceptual obstacles faced by children have historical echoes; for example, Pacioli, an Italian mathematician of the 15th century (according to Cajori, 1917, pp. 182–183) "was greatly embarrassed by the use of the term 'multiplication' in case of fractions, where the product is less than the multiplicand," quoting from the Bible ("Be fruitful and multiply") to prove that *multiply* means *increase*. Cajori continues,

That, in the historical development, multiplication and division should have been considered primarily in connection with integers, is very natural. The same course must be adopted in teaching the young. First come the easy but restricted meanings of multiplication and division, applicable to whole numbers. In due time the successful teacher causes students to see the necessity of modifying and broadening the meanings assigned to the terms. (p. 183)

A number of suggestions along the lines of the last sentence in this quotation have been made. It is common, for example, to introduce multiplication of fractions in the context of area (De Morgan, 1910). Semadeni (1984) proposes that the formal considerations (notably invariance of properties) guiding mathematicians in the extension of concepts should be paralleled by what he calls the "concretization permanence principle" whereby the extension is motivated and justified within an appropriate context. Greer (1988) has suggested as an example of this approach a situation in which a bucket is being winched up from a well (Figure 13.5). Within this context, there is a natural extension from integer multipliers (how far is the bucket raised by some number of complete turns of the winch) to noninteger multipliers (incomplete turns).

Another important theme with developmental implications is the shift in perspective from viewing a given problem as an isolated relationship between three quantities to viewing that problem against the background of an implicit functional rela-

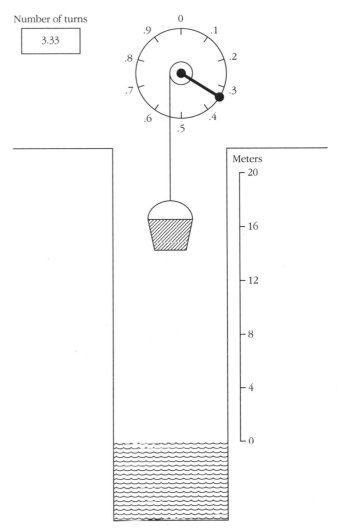

FIGURE 13–5. Raising the bucket: A context for the extension from discrete to continuous multiplication.

tionship. Consider a typical word problem, such as

A man walks 12 miles in 4 hours. At what speed is he walking?

Behind the specific quantities mentioned is an implicit general relationship between elapsed time and distance traveled.

In Vergnaud's analysis, the problem would be set in the context of an isomorphism between two measure spaces (Figure 13.2a). In Thompson's (1989) terms, the problem is to generalize the ratio, 12 miles : 4 hours, to a rate. In Kaput's (1985, p. 54) terms, we are dealing with a small sample of data from the domain of a linear function.

The complexity of the radical conceptual restructurings implied by the foregoing analysis stands in stark contrast to the apparent simplicity of the extra rules needed to extend multiplication and division calculations. As Schwartz (1988) concluded,

The commonly held view among mathematics educators and in the mathematics education research community is that children's early number knowledge and number strategies lead in a more or less continuous way to an understanding of multiplicative structures and rational numbers. The present analysis, coupled with the empirical observation of the difficulties children have with these latter concepts, would suggest that this commonly held view is faulty. (pp. 51–52)

LOOKING FORWARD

The Potential of Computer Software

The ability of computers to open "new representational windows" (Kaput, 1986, p. 187) can be harnessed to strengthen students' understanding of multiplicative concepts and can be expected to prove of increasing importance in the future. One approach is to use representations that make explicit the types of quantity and their referents (as in Schwartz's *Semantic Calculator* [1982]), and the relationships between these quantities. A further development along these lines is the representational system devised by Shalin and Bee (1985), exemplified for a simple problem in Figure 13.6 (taken from Greeno's (1987) summary) but capable of extension to multistep problems. Thompson (1989) has developed this approach further in *Word Problem Assistant,* which is based on an elaborated theory of quantity-based reasoning and extends naturally from arithmetic to algebraic problems.

Marshall has developed a different approach to word problems (covering addition and subtraction as well as multiplication and division), based on five very general "semantic profiles" (Marshall, 1988; Marshall, Barthuli, Brewer, & Rose, 1989). Students are taught to match situations to the corresponding schemas, which are iconically represented on the computer.

Kaput (1990, in press) has developed software designed to represent intensive quantities, with an emphasis on multiple linked representations. For example, an intensive quantity such as 3 umbrellas for every 2 animals may be represented (for different instantiations of the relationship) in terms of boxes, each of which contains 3 umbrella icons and 2 animal icons; a table with two columns listing numbers of umbrellas and animals as related pairs; and a coordinate graph plotting number

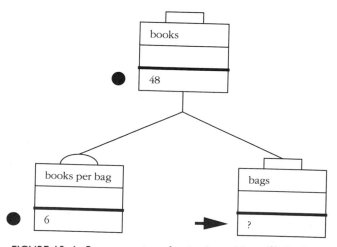

FIGURE 13–6. Representation of a simple problem. (Shalin & Bee, 1985)

of umbrellas against number of animals. Extensions to continuous intensive quantities by replacing numerosity of icons by lengths of line segments are being developed (Kaput, in preparation). Greer (1988) has suggested using computers to represent dynamically situations in which quantities vary with a constant ratio, such as the number of turns of the winch and the change in height of the bucket in the well (Figure 13.5).

Among many other natural applications for software within the domain of multiplicative structures, the use of geometrically similar figures as a means of tackling misconceptions in proportional reasoning deserves mention (Hoyles, Noss, & Sutherland, 1989). See also the comments made in this first section of this chapter on the advantages of dynamic versions of the static representations shown in Figure 13.1.

Educational Recommendations

On the basis of research and analysis, numerous suggestions have been made to improve teaching of multiplicative concepts (for example, Greer, 1989; Mangan, 1989); these are briefly summarized here.

A straightforward first step would be to improve students' experiential bases in various ways. At present, for example, as Nesher (1987, p. 37) has pointed out, "the probability of the occurrence of multiplication and division word problems in the textbooks that detect...misconceptions is low." This is a clear case where the findings of research could be directly beneficial. A related recommendation is the routine use of "harder" numbers; the use of unrealistically "easy" numbers has many detectable effects, such as that students often judge whether they are proceeding along the right lines on the basis of whether the result is coming out "cleanly" (Reusser, 1988). The predominant use of easy numbers is presumably due to the well-intentioned belief expressed by De Morgan (1910, pp. 21-22): "The power of the mind cannot be directed to two things at once: if the complexity of the numbers used requires all the student's attention, he cannot observe the principle of the rule which he is following." The problem with this strategy is that it assumes that the principle will be unproblematically generalized, yet there is plenty of evidence that this is not a justifiable assumption. In an age of ready access to calculators, realistic hard numbers should be used routinely.

More fundamentally, there is agreement among analysts and researchers that students need to learn about a much wider range of situations modeled by the operations than is currently the case, and that the operations and extensions of them should be introduced within problem-posing contexts.

There are many ways in which the impoverished diet of word problems currently offered should be enriched. Problems should be varied so that students cannot assume that if there are two numbers in the problem, the answer will be found by adding, subtracting, multiplying or dividing the two numbers. More problems involving more than one step should be used; at a stroke, this would cut out many of the superficial solution strategies (Sowder, 1988b). Problems should be included with insufficient and/or superfluous information. More attention should be paid to realistic modeling, rather than "easy, quasi-realistic situations" (Nesher, 1988, p. 38). Relationships between quantities should be analyzed qualitatively (cf. Behr, Harel, Post,

& Lesh, Chapter 14, this volume). In general, the aim may be summed up as forcing students to think and to construct conceptual models of the situations described (Reusser, 1988).

Various conceptual obstacles to the development of multiplicative concepts (particularly at the point of extension beyond the integer domain) have been identified, and suggestions have been made to help students overcome these (Greer, 1989). Obstacles include (a) misconceptions due to overgeneralization of rules that are valid in the integer domain, (b) linguistic aspects—natural language both reflects and reinforces intuitive conceptualizations (as in the example of Pacioli cited earlier), and (c) failure to dissociate concepts from algorithms. Pedagogical strategies for tackling these obstacles include the use of cognitive conflict (Bell, Swan, Onslow, Pratt, & Purdy, 1985; Hart, 1988). Fischbein (1987, 1990) advocates a broadly metacognitive approach to acquaint the student with the limitations and traps of intuition, the nature and causes of fallacious patterns of thinking mediated by primitive models, and the usefulness of resources exploited by experts, such as "alarm devices" (Fischbein, 1987, p. 40) and "critics" (Davis, 1984, p. 43).

Agenda for Further Research

A considerable range of empirical studies and theoretical analyses has been reviewed in this chapter. Such is the complexity of the field, however, that there is no shortage of tasks to be tackled; the level of activity in the field characteristic of the 1980s is likely to increase in the 1990s.

A general priority should be to complement *ascertaining* experiments, which diagnose existing ills, with *teaching* experiments designed to test whether recommended remedies (such as those outlined above) can be effectively administered in the classroom. A related recent development has been the study of teachers' conceptual problems, and such investigations are likely to continue. An important new contribution would be systematic analysis of textbook treatments of multiplication and division.

A second suggested priority is detailed, longitudinal case studies to trace the development in individuals of multiplicative concepts. Such studies have been carried out at the stage of initial construction of multiplicative concepts (Steffe, 1988), but investigations in similar detail are needed at the point where the concepts extend beyond the integer domain.

Cognitive scientists have contributed extensively in the last decade to the study of addition and subtraction word problems. It is to be hoped that they will increasingly turn their attention to the more complex field of multiplication and division word problems.

Analysis of the conceptual field of multiplicative structures will doubtless continue. There is a need for synthesis of hitherto rather separated bodies of research on multiplication and division word problems, proportional reasoning, and rational number concepts. Beyond this, a longer-term objective is analysis of the relationships between the conceptual fields of additive and multiplicative structures.

References

Af Ekenstam, A., & Greger, K. (1983). Some aspects of children's ability to solve mathematical problems. *Educational Studies in Mathematics, 14,* 369–384.

Anghileri, J. (1989). An investigation of young children's understanding of multiplication. *Educational Studies in Mathematics, 20,* 367–385.

Anghileri, J., & Johnson, D. C. (1988). Arithmetic operations on whole numbers: Multiplication and Division. In T. R. Post (Ed.), *Teaching mathematics in grades K-8* (pp. 146–189). Boston, MA: Allyn and Bacon.

Beattys, C., & Maher, C. (1989). Exploring children's perceptions of multiplication through pictorial representations. In G. Vergnaud, J. Rogalski, & M. Artigue (Eds.), *Proceedings of the Thirteenth Annual Conference of the International Group for the Psychology of Mathematics Education* (Vol. 1, pp.109–116). Paris: G. R. Didactique.

Behr, M., & Post, T. (1988). Teaching rational number and decimal concepts. In T. R. Post (Ed.), *Teaching mathematics in grades K-8* (pp. 190–230). Boston, MA: Allyn and Bacon.

Bell, A., Fischbein, E., & Greer, B. (1984). Choice of operation in verbal arithmetic problems: The effects of number size, problem structure and context. *Educational Studies in Mathematics, 15,* 129–147.

Bell, A., Greer, B., Grimison, L., & Mangan, C. (1989). Children's performance on multiplicative word problems: Elements of a descriptive theory. *Journal for Research in Mathematics Education, 20,* 434–449.

Bell, A., & Onslow, B. (1987). Multiplicative structures: Development of the understanding of rates (intensive quantities). In J. C. Bergeron, N. Herscovics, & C. Kieran (Eds.), *Proceedings of the Eleventh Annual Conference of the International Group for the Psychology of Mathematics Education* (Vol. 2, pp. 275–281). Montreal: University of Montreal.

Bell, A., Swan, M., Onslow, B., Pratt, K., & Purdy, D. (1985). *Diagnostic teaching: Teaching for long term learning.* Nottingham: University of Nottingham, Shell Centre for Mathematical Education.

Bell, A., Swan, M., & Taylor, G. (1981). Choice of operations in verbal problems with decimal numbers. *Educational Studies in Mathematics, 12,* 399–420.

Cajori, F. (1917). *A history of elementary mathematics.* New York: Macmillan.

Confrey, J. (in press). Splitting, similarity and rate of change: A new approach to multiplication and exponential functions. In G. Harel & J. Confrey (Eds.), *The development of multiplicative reasoning in the learning of mathematics.* Albany, NY: SUNY Press.

Davis, P. J., & Hersh, R. (1981). *The mathematical experience.* Boston, MA: Birkhäuser.

Davis, R. B. (1984). *Learning mathematics: The cognitive science approach to mathematics education.* London: Croom Helm.

De Corte, E., & Verschaffel, L. (1985). Beginning first graders' initial representation of arithmetic word problems. *Journal of Mathematical Behavior, 4,* 3–21.

De Corte, E., Verschaffel, L., & Van Coillie, V. (1988). Influence of number size, problem structure and response mode on children's solutions of multiplication word problems. *Journal of Mathematical Behavior, 7,* 197–216.

De Morgan, A. (1910). *Study and difficulties of mathematics.* Chicago, IL: University of Chicago Press.

Fischbein, E. (1987). *Intuition in science and mathematics.* Dordrecht, The Netherlands: Reidel.

Fischbein, E. (1990). Intuition and information processing in mathematical activity. In B. Greer & L. Verschaffel (Eds.). Mathematics education as a proving-ground for information-processing theories

[Special issue]. *International Journal of Educational Research, 14* (1), pp. 31–50.

Fischbein, E., Deri, M., Nello, M. S., & Marino, M. S. (1985). The role of implicit models in solving verbal problems in multiplication and division. *Journal for Research in Mathematics Education, 16,* 3–17.

Freudenthal, H. (1983). *Didactical phenomenology of mathematical structures.* Dordrecht, The Netherlands: Reidel.

Graeber, A., & Tirosh, D. (1988). Multiplication and division involving decimals: Preservice elementary teachers' performance and beliefs. *Journal of Mathematical Behavior, 7,* 263–280.

Graeber, A. O., Tirosh, D., & Glover, R. (1989). Preservice teachers' misconceptions in solving verbal problems in multiplication and division. *Journal for Research in Mathematics Education, 20,* 95–102.

Greeno, J. G. (1987). Instructional representations based on research about understanding. In A. H. Schoenfeld (Ed.), *Cognitive science and mathematics education* (pp. 61–88). Hillsdale, NJ: Lawrence Erlbaum.

Greer, B. (1984). Cognitive psychology and mathematics education. *Journal of Structural Learning, 8,* 291–300.

Greer, B. (1987). Nonconservation of multiplication and division involving decimals. *Journal for Research in Mathematics Education, 18,* 37–45.

Greer, B. (1988). Nonconservation of multiplication and division: Analysis of a symptom. *Journal of Mathematical Behavior, 7,* 281–298.

Greer, B. (1989). Conceptual obstacles to the development of the concepts of multiplication and division. In H. Mandl, E. de Corte, S. N. Bennett, & H. F. Friedrich (Eds.), *Learning and instruction: European research in an international context.* (Vol. 2.2, pp. 461–476). Oxford: Pergamon.

Greer, B., & Mangan, C. (1986). Understanding multiplication and division: From 10-year-olds to student teachers. In L. Burton & C. Hoyles (Eds.), *Proceedings of the Tenth International Conference for the Psychology of Mathematics Education* (pp. 25–30). London: London Institute of Education.

Greer, B. (in press). Extending the meaning of multiplication and division. In G. Harel & J. Confrey (Eds.), *The development of multiplicative reasoning in the learning of mathematics.* Albany, NY: SUNY Press.

Harel, G., Behr, M., Post, T., & Lesh, R. (in preparation). Teachers' knowledge of multiplication and division concepts.

Harel, G. & Confrey, J. (Eds.) (in press). *The development of multiplicative reasoning in the learning of mathematics.* Albany, NY: SUNY Press.

Hart, K. (1988). Ratio and proportion. In J. Hiebert & M. Behr (Eds.), *Number concepts and operations in the middle grades* (pp.198–219). Reston, VA: National Council of Teachers of Mathematics; Hillsdale, NJ: Lawrence Erlbaum.

Hiebert, J. (in press). Mathematical, cognitive, and instructional analyses of decimal fractions. In G. Leinhardt, & R. T. Putnam (Eds.), *Cognitive research: Mathematics learning and instruction.*

Hilton, P. (1984). Current trends in mathematics and future trends in mathematics education. *For the Learning of Mathematics, 4* (1), 2–8.

Hoyles, C., Noss, R., & Sutherland, R. (1989). A LOGO-based microworld for ratio and proportion. In G. Vergnaud, J. Rogalski, & M. Artigue (Eds.), *Proceedings of the Thirteenth Annual Conference of the International Group for the Psychology of Mathematics Education* (Vol. 2, pp.115–122). Paris: G. R. Didactique.

Kaput, J. (1985). *Multiplicative word problems and intensive quantities: An integrated software response* (Technical Report 85-19). Cambridge, MA: Harvard University, Educational Technology Center.

Kaput, J. (1986). Information technology and mathematics: Opening new representational windows. *Journal of Mathematical Behavior, 5,* 187–208.

Kaput, J. (1990). Applying the computer's representational plasticity to create bridging notations to ramp from the concrete to the abstract. In B. Bowen (Ed.), *Designing for learning* (pp. 63–70). Cupertino, CA: Apple Computers.

Kaput, J. & West, M. (in press). Missing value proportional reasoning problems: Factors affecting informal reasoning patterns. In G. Harel & J. Confrey (Eds.), *The development of multiplicative reasoning in the learning of mathematics.* Albany, NY: SUNY Press.

Kintsch, W. (1986). Learning from text. *Cognition and Instruction, 3,* 87–108.

Kouba, V. L. (1989). Children's solution strategies for equivalent set multiplication and division word problems. *Journal for Research in Mathematics Education, 20,* 147–158.

Lesh, R., Post, T., & Behr, M. (1988). Proportional reasoning. In J. Hiebert & M. Behr (Eds.), *Number concepts and operations in the middle grades* (pp. 93–118). Reston, VA: National Council of Teachers of Mathematics; Hillsdale, NJ: Lawrence Erlbaum.

Luke, C. (1988). The repeated addition model of multiplication and children's performance on mathematical word problems. *Journal of Mathematical Behavior, 7,* 217–226.

Mangan, C. (1986). *Choice of operation in multiplication and division word problems.* Unpublished doctoral dissertation, Queen's University, Belfast.

Mangan, C. (1989). Multiplication and division as models of situations: What research has to say to the teacher. In B. Greer, & G. Mulhern (Eds.), *New directions in mathematics education* (pp. 107–127). London: Routledge.

Marshall, S. P. (1988, April). *Assessing schema knowledge.* Paper presented at the Annual Meeting of the American Educational Research Association, New Orleans, LA.

Marshall, S. P., Barthuli, K. E., Brewer, M. A., & Rose, F. E. (1989). *Story problem solver: A schema-based system of instruction* (Tech. Rep. No. 89-01). San Diego: San Diego State University, Center for Research in Mathematics and Science Education.

Mechmandarov, I. (1987). *The role of dimensional analysis in teaching multiplicative word problems.* Unpublished manuscript, Tel-Aviv, Israel, Center for Educational Technology.

Nantais, N., & Herscovics, N. (1989). Epistemological analysis of early multiplication. In G. Vergnaud, J. Rogalski, & M. Artigue (Eds.), *Proceedings of the Thirteenth Annual Conference of the International Group for the Psychology of Mathematics Education (Vol. 3),* pp.18–24. Paris: G. R. Didactique.

National Assessment of Educational Progress (1983). *The third national mathematics assessment: Results, trends and issues.* Denver: Education Commission of the States.

Nesher, P. (1980). The stereotyped nature of school word problems. *For the Learning of Mathematics, 1*(1), 41–48.

Nesher, P. (1987). Towards an instructional theory: The role of student's misconceptions. *For the Learning of Mathematics, 7*(3), 33–40.

Nesher, P. (1988). Multiplicative school word problems: Theoretical approaches and empirical findings. In J. Hiebert & M. Behr (Eds.), *Number concepts and operations in the middle grades* (pp. 19–40). Reston, VA: National Council of Teachers of Mathematics; Hillsdale, NJ: Lawrence Erlbaum.

Peled, I., & Nesher, P. (1988). What children tell us about multiplication word problems. *Journal of Mathematical Behavior, 7,* 239–262.

Reusser, K. (1987, September). *The computer's effort after meaning: Word problem solving as a process of text comprehension and mathematization.* Paper presented at the Second European Conference for Research on Learning and Instruction, Tubingen, Germany.

Reusser, K. (1988). Problem solving beyond the logic of things: Contextual effects on understanding and solving word problems. *Instructional Science, 17* (4), 309–338.

Schwartz, J. (1982). The semantic calculator. *Classroom Computer News, 2*, 22–24.

Schwartz, J. (1988). Intensive quantity and referent transforming arithmetic operations. In J. Hiebert & M. Behr (Eds.), *Number concepts and operations in the middle grades* (pp. 41–52). Reston, VA: National Council of Teachers of Mathematics; Hillsdale, NJ: Lawrence Erlbaum.

Sellke, D. H., Behr, M. J., & Voelker, A. M. (1988). Representing and solving multiplicative story problems as proportion problems with a unit ratio–a teaching experiment. In M. Behr, C. B. Lacampagne, & M. M. Wheeler (Eds.), *Proceedings of the Tenth Annual Meeting of the North American Chapter of the International Group for the Psychology of Mathematics Education* (pp. 121–126). DeKalb, IL: Northern Illinois University.

Semadeni, Z. (1984). A principle of concretization permanence for the formation of arithmetical concepts. *Educational Studies in Mathematics, 15*, 379–395.

Shalin, V., & Bee, N. V. (1985). *Analysis of the semantic structure of a domain of word problems.* Pittsburgh: University of Pittsburgh, Learning Research and Development Center.

Silver, E. A. (1986). Using conceptual and procedural knowledge: A focus on relationships. In J. Hiebert (Ed.), *Conceptual and procedural knowledge: The case of mathematics* (pp. 181–198). Hillsdale, NJ: Lawrence Erlbaum.

Sowder, L. (1988a). Children's solutions of story problems. *Journal of Mathematical Behavior, 7*, 227–238.

Sowder, L. (1988b). *Concept-driven strategies for solving story problems in mathematics.* (National Science Foundation Project Report). San Diego, CA: San Diego State University

Steffe, L. (1988). Children's construction of number sequences and multiplying schemes. In J. Hiebert & M. Behr (Eds.), *Number con-*

cepts and operations in the middle grades (pp. 119–140). Reston, VA: National Council of Teachers of Mathematics; Hillsdale, NJ: Lawrence Erlbaum.

Streefland, L. (1988). Reconstructive learning. In A. Borbas (Ed.), *Proceedings of the Twelfth International Conference for the Psychology of Mathematics Education, (Vol. 1*, pp. 75–91). Veszprem, Hungary: OOK Printing House.

Thompson, P. (1989, March). *A cognitive model of quantity-based algebraic reasoning.* Paper presented at the Annual Meeting of the American Educational Research Association, San Francisco, CA.

Tirosh, D., & Graeber, A.O. (1989). Preservice elementary teachers' explicit beliefs about multiplication and division. *Educational Studies in Mathematics, 20*, 79–96.

Vergnaud, G. (1983). Multiplicative structures. In R. Lesh & M. Landau (Eds.), *Acquisition of mathematics concepts and processes* (pp. 127–174). New York: Academic Press.

Vergnaud, G. (1988). Multiplicative structures. In J. Hiebert & M. Behr (Eds.), *Number concepts and operations in the middle grades* (pp. 141–161). Reston, VA: National Council of Teachers of Mathematics; Hillsdale, NJ: Lawrence Erlbaum.

Vest, F. R. (1971). A catalog of models for multiplication and division of whole numbers. *Educational Studies in Mathematics, 3*, 220–228.

Vest, F. R. (1975). Episodes with several models of multiplication. *Mathematics in School, 14*, 24–26.

Vinner, S. & Linchevski, L. (1988). Is there any relation between division and multiplication? Elementary teachers' ideas about division. In A. Borbas (Ed.), *Proceedings of the Twelfth International Conference for the Psychology of Mathematics Education, (Vol. 2*, pp. 625–632). Veszprem, Hungary: OOK Printing House.

·14·

RATIONAL NUMBER, RATIO, AND PROPORTION

Merlyn J. Behr

NORTHERN ILLINOIS UNIVERSITY

Guershon Harel

PURDUE UNIVERSITY

Thomas Post

UNIVERSITY OF MINNESOTA

Richard Lesh

EDUCATIONAL TESTING SERVICE

There is a great deal of agreement that learning rational number concepts remains a serious obstacle in the mathematical development of children. This consensus is manifested in the similarity of the opening remarks of a number of recent papers on the topic of children's construction of rational number knowledge (Bigelow, Davis, & Hunting, 1989; Freudenthal, 1983; Kieren, 1988; Ohlsson, 1987, 1988). In stark contrast, there is no clear agreement about how to facilitate learning of rational number concepts. Numerous questions about how to facilitate children's construction of rational number knowledge remain unanswered even if clearly formulated. For one thing, we must find out what types of experiences children need in order to develop their rational number knowledge. We also have to agree on the concepts of fraction and rational number. We are able to give clear and precise mathematical definitions of rational numbers and fractions: Rational numbers are elements of an infinite quotient field consisting of infinite equivalence classes, and the elements of the equivalence classes are fractions. However, when fractions and rational numbers as applied to real-world problems are looked at from a pedagogical point of view, they take on numerous "personalities." From the perspective of research and curriculum development, the problem is to describe these personalities in sufficient depth and clarity so that the organization of learning experiences for children will have a firm theoretical foundation.

In this chapter, we describe some of the "personalities" of rational numbers from two perspectives: the mathematics of quantity (Schwartz, 1988) and the rational number as an element of the multiplicative conceptual field (Vergnaud, 1983, 1988). Kieren (1976) first introduced the idea that rational numbers consist of several constructs and that understanding the concept of a rational number depends on gaining an under-

The development of this paper was in part supported with funds from the National Science Foundation under Grant No. DPE 84-70077 (The Rational Number Project). Any opinions, findings, and conclusions expressed are those of the authors and do not necessarily reflect the views of the National Science Foundation.

The authors wish to thank Thomas E. Kieren, University of Alberta, and Leslie Steffe, University of Georgia, for their helpful reviews of two earlier drafts of this chapter.

standing of the confluence of these constructs. Behr, Lesh, Post, and Silver (1983) restated these constructs in somewhat different terms, and Nesher's (1985 [cited in Ohlsson, 1987]) analysis of rational numbers distinguished among the following concepts: fraction as a part-whole relationship; rational number as the result of the division of two numbers, as a ratio, as an operator, and as a probability. More recently, Kieren (1988) indicated that he believes a fully developed rational number concept comprises four subconstructs, namely measure, quotient, ratio number, and multiplicative operator. Vergnaud (1983) and Freudenthal (1983) seem to agree. A factor analysis study reported by Rahim and Kieren (1987) indicates that these are separate constructs.

THE RATIONAL NUMBER CONSTRUCT THEORY

In his seminal work, Kieren (1976) argues that exposure to numerous rational number constructs is necessary to gain a full understanding of rational number. With no claim that it is exhaustive, his list of rational number constructs consists of fractions; decimal fractions; equivalence classes of fractions; numbers of the form p/q, where p and q are integers and $q \neq 0$ (in which form rational numbers are called ratio numbers); multiplicative operators; and elements of an infinite ordered quotient field. Freudenthal (1983) considers fractions to be the phenomenological source of rational number concepts and speaks of four aspects of fractions. First, he uses the term fraction as *fracturer* to refer to the experiential aspects of fractions that are based on activities such as comparing quantities and magnitudes by sight and feel, folding, and weighing parts in one's hands or on a balance. The idea behind such experiences is to involve the child in primitive measuring activity before formal measurement is used. Second, he adds that the most concrete way fractions present themselves is as wholes being split into equal parts through activities such as splitting, slicing, cutting, or coloring. Freudenthal's third construct of fractions is as *comparers*. This notion extends the part-of-a-whole concept to one in which parts of different wholes are compared; it also extends the fraction concept to include fractions greater than one. Freudenthal observes that a fourth concept, fraction as an *operator,* comes up in the other three fraction constructs as well.

Vergnaud (1983) places the concepts of fraction and ratio in the broader context of the multiplicative conceptual field. He considers these concepts in the framework of three problem types: isomorphism-of-measures, product-of-measures, and multiple proportions. Of greatest interest to the issues addressed in this chapter is the isomorphism-of-measures class of problems. Isomorphism-of-measures is a structure consisting of simple direct proportion between two measure spaces M_1 and M_2. Four different types are identified:

Schema 1		Schema 2		Schema 3		Schema 4	
M_1	M_2	M_1	M_2	M_1	M_2	M_1	M_2
1	a	1	$x = f(1)$	1	$a = f(1)$	a	b
b	x	a	$b = f(a)$	x	$b = f(x)$	c	x

The first schema is called multiplication, in which the problem is to find x given a and b. The second and third schemata are

commonly called partitive and quotitive division, respectively, and the problems are to find $f(1)$ and x, respectively, given a and b. The fourth schema is referred to as the rule of three problems, in which the problem is to find x (given a, b, and c), where x can appear in any one of the four positions. In Vergnaud's formulation, multiplication and division problems appear as special cases of direct proportion problems.

Ohlsson (1987) criticizes these lists because they are not exhaustive and they include things that should not be included. Ohlsson goes on to criticize his own earlier attempt to address the range of interpretations of the rational number concept, an attempt in which he considered rational numbers from the perspective of a semantic field. The idea is to assign a structure to the field that explains the meanings of concepts in the field and brings out the semantic relationships between them. In his analysis, Ohlsson began by considering an ordered pair of integers and placing constraints on the domain of referents for these integers. Because each pairing of constraints would lead to a potential interpretation of rational number, it was hoped that this approach would limit and exhaust the possible interpretations. Since both components of the pair are integers that can be interpreted as quantities or as parameters for operations (that is, the number of times it can be repeated), there would be four potential interpretations of fractions. While Ohlsson considered this analysis an advancement over previous ones, he found serious weaknesses and attempted yet another approach.

In the new approach, he attempted to substantially advance the construct theory of rational numbers; because of this we give a more extensive overview of his analysis. He suggested that the symbol of the form x/y (that is, the fraction bar) has come to denote four mathematical constructs, which he identifies as the quotient function, a rational number, a binary vector, and a particular kind of composite function. The development proceeds by considering both a mathematical theory for each of these four constructs and corresponding applications of the mathematical theory. In each of the four cases the various applications of the theory turn out to be individual interpretations of an entity symbolized by x/y. The analysis procedure is somewhat similar to his earlier one in that certain constraints are placed on the numerator and the denominator of x/y to define the applications for a given interpretation of x/y.

The first interpretation, the quotient function, identifies four applications of x/y: partitioning, extracting, shrinking, and educing. In the first and third applications, x and y represent a quantity and a parameter, respectively, and in the second and fourth both represent quantities. A difficulty with the first and third applications in terms of their definitions is that the concept of quotient does not seem to be well defined by operands of two different types, quantity and parameter. The quotient interpretation of x/y as rational number is identified with two applications—fractions and measures. Here the fraction application does not seem different from the usual part-whole notion of fractional part and, as defined, the measure application does not relate to the measure concept of rational number in the sense of Kieren (1976) or Behr, et al. (1983). Moreover, as defined, it is very close to the part-whole concept of fractions and is limited to the use of standard units of measure.

Applications of the binary vector interpretation of x/y are ratio, intensive quantity, rate, and proportion. The notion of ratio is that of comparison of two quantities; intensive quantity refers to a ratio of quantities from different measure spaces to which a common name, such as density, is applicable; proportion fits the part-whole concept of fractions when the targeted parts are unitized into a composite (for example, $\frac{3}{4}$ of a pie is considered to be one piece equal in size to three $\frac{1}{4}$ pieces); and rate is considered to be a ratio in which the reference quantity is time. These interpretations seem not to add anything new to our understanding of rational numbers because they are redundant with concepts described by earlier analyses or, as with his concept of intensive quantity, they are too limited in scope. Finally, the interpretation of x/y as a composite function leads to the usual interpretation of rational number as operator (Kieren, 1976). It would appear, then, that five subconstructs of rational number—part-whole, quotient, ratio number, operator, and measure—which have to some extent stood the test of time, still suffice to clarify the meaning of rational number. A major thrust of this chapter is to apply the concepts of mathematics of quantity to the five subconstructs to provide a deeper semantic understanding of them. This analysis is given in section 2, "Rational Number Construct Theory: Toward a Semantic Analysis," of this chapter.

Questions such as how rational number knowledge is acquired and organized were recently addressed by Kieren (1988) and Pirie (1988). Resnick (1986) addressed mathematical knowledge from a more global perspective with emphasis on intuitive mathematical knowledge. Their contributions are summarized here.

Acquisition of Rational Number Knowledge

Kieren (1988) presents a theoretical model of mathematical knowledge-building and relates it specifically to rational number knowledge. One aspect of his theory is the notion of an ideal network of personal rational number knowledge. This network consists of six levels of knowledge and can be thought of as an image of ideal rational number knowledge. The first (most primary level) contains constructs that are very local and close to the fact level. The next level comprises the constructs of partitioning, equivalencing, and forming dividable units. The four constructs of rational numbers—measure, quotient, ratio number, and operator—form the third level. Moving up to the fourth level, Kieren posits knowledge of the scalar and functional relationships upon which the more formal construct of fraction and rational-number equivalence depend. The penultimate level synthesizes the constructs of rational number and related concepts to produce the general construct of the multiplicative conceptual field. This network of knowledge structures is capped by knowledge of rational numbers as an element of an infinite quotient field. It is emphasized that the ideal network of knowledge is very interactive in that knowing rational numbers as elements of a quotient field—the highest knowledge level—permits one not only to prove theorems about the structure of the mathematical system but to explain various phenomena at all lower levels of the network as well. Kieren (1988) also presents a model of rational number knowledge represented by four concentric rings. The inner ring consists of the basic knowledge that one acquires as a result of living in a particular environment; it is what d'Ambrosio (1985

[cited in Kieren, 1988]) calls ethnomathematical knowledge. Moving outward, the next ring suggests a level of intuitive knowledge; Kieren considers this to be schooled knowledge built from and related back to one's everyday experience. The third ring represents technical symbolic language that involves the use of standard language, symbols, and algorithms. The outer ring represents axiomatic knowledge of the system. An important observation about this model of rational number knowledge is that it is thought to be dynamic, organic, and interactive; that is, a mature rational number knower must be able to engage in the whole range of thought and action and interrelate thought and action at one level with thought and action at other levels. Aspects of this model are further elaborated by Pirie (1988) in an essay on mathematical thinking as a recursive function. Pirie points out that any level of rational number thinking can become the input into rational number thinking at any other level.

The notion of the existence and development of intuitive mathematical knowledge, its interaction with the development of formal mathematical structures, and its facilitation of that development is of considerable interest in research on the acquisition of mathematical knowledge. Resnick (1986) presents an analysis of the intuitive knowledge that children bring with them on entering school and intuitive knowledge that they exhibit on more advanced but early school tasks. At this level, intuitive knowledge is characterized by the fact that children are able to relate their knowledge of the additive composition property of numbers to the place-value system of numeration. That is, in the examples Resnick was able to give, the symbol system functions as part of the child's intuitive understanding of mathematics. She goes on to conjecture that many fundamental concepts—including proportions, ratios, and other multiplicative relationships—can only be developed when formal notations are well established and incorporated into the learner's intuitive mathematical system. An important implication of this is that we investigate instructional situations that help children develop intuitive mathematical knowledge and then, as characterized by the work of Mack (1990), investigate ways of tying instruction to children's informal or intuitive knowledge. Resnick (1986) offers two characteristics of intuitive knowledge, as she uses the term:

First, intuitive knowledge is self-evident and obvious to the person who has it; it does not, phenomenologically, require justification in terms of prior premises.... Second, intuitive knowledge is easily accessible and linked in memory to a variety of specific situations. It thereby provides the basis for highly flexible application of well-known concepts, notations, and transformational rules. It thus frees people from excessive reliance on fixed algorithms, and allows them to invent procedures for problems not previously encountered and to work ahead of formal instruction in constructing mathematical knowledge (Resnick, 1986, p. 188).

The Scope and Organization of this Work. During recent years, a number of books and conferences have focused on reviews of literature related to children's rational number concepts and proportional reasoning. For example, relevant chapters are contained in Lesh and Landau (1983), Post (1988), and Hiebert and Behr (1989). However, these publications tend to summarize past research rather than identify priorities (or provide tools) to facilitate future research. Because comprehensive literature reviews are already available, this chapter emphasizes future-oriented factors that appear to be especially critical to future research and development initiatives.

A recurring theme emphasized in recent analysis-and-synthesis conferences and publications has been the need to develop a precise language and notation system to facilitate communication among related but relatively unconnected strands of research bearing on the topic of rational numbers and proportional reasoning. For example,

- In previous publications, researchers mapping out the conceptual terrain of rational numbers have approached the topic from the following diverse perspectives: linguistics (Nesher), computer science (Ohlsson), cognitive science (Greeno), science (Karplus, Schwartz), developmental psychology (Vergnaud), and curriculum development in mathematics education (Hart). This multidisciplinary approach has many strengths, but diversity has also led to communication difficulties. It is sometimes difficult to determine points of agreement or disagreement among apparently related studies.
- Much productive mathematics education research has been in topics surrounding the area of rational numbers and proportional reasoning. Examples include *whole-number arithmetic concepts* (Carpenter & Moser, 1982; and Steffe, Cobb, & von Glasersfeld, 1988), *algebra concepts* (Wagner & Kieran, 1989), and *problem solving* (Charles & Silver, 1989).

However, research in each of these areas has proceeded in a parallel and independent fashion so that, even though we know many cases where these strands of research develop simultaneously, developmental interdependencies are usually not clear. For example, we know that certain ideas about fractions, ratios, or proportions are developing at the same time as some concepts related to (a) whole number multiplication and "composite units" (Steffe, von Glasersfeld, Richards, & Cobb, 1983), (b) exponentiation and "unit splitting operations" (Confrey, 1988), or (c) rates or "intensive measures" (Hart, 1981; Kaput, 1985; Schwartz, 1988). Yet, in general little is known about developmental interrelationships.

To improve dialogue among related research endeavors, this chapter will (a) describe a semantic/mathematical analysis of rational number concepts, and (b) introduce a mathematical notation system that is sufficiently nontechnical to be useful to nonmathematicians but is also consistent with:

- *formal mathematics*—the usual symbol systems that mathematicians use to describe structural similarities among problem situations characterized by related (for example, homomorphic) systems of rational number relations and operations.
- *children's mathematics*—the reasoning patterns influenced by problem characteristics such as those related to the types of "units" or quantities that are involved in various situations (for example, composite versus nondecomposable units, continuous versus discrete quantities, or intensive versus extensive quantities).

The analysis and notation system introduced in this chapter will focus on the notion of "unit types" and "composite units"

(Steffe, 1989), because

- These are constructs that have been neglected in past analyses of rational number concepts.
- This is a construct that we regard as critical for establishing links between research on rational numbers and research on other conceptual areas, such as whole-number arithmetic (Steffe, 1989), exponentiation (Confrey, 1988), and algebra (Thompson, 1989).

The notation system described in this chapter has been designed to allow integration into the system described in Lesh, Post, and Behr (1988), which was used to define the logic driver for a computer-based "problem transformer tool" (PAT). As with PAT, the notational system introduced in this chapter attempts to describe the reasoning processes of children. The analysis described by this notational system suggests uses of manipulative materials to help children develop certain concepts about rational numbers. It also suggests hypotheses for further research about mathematical behavior.

This first component of the chapter consists of a presentation of our analysis of several rational number constructs. This appears in the section "Rational Number Construct Theory: Toward a Semantic Analysis."

The second component of the chapter considers children's development of intuitive, or qualitative, knowledge about rational number and proportion situations. Our view is that this intuitive knowledge can be developed by children through appropriate instructional situations constructed to exemplify reasoning *principles* for rational number and proportion situations. An extensive analysis was conducted by Harel, Behr, Post, and Lesh (in press) of the problem representations and solution strategies seventh grade children use in solving proportion situations based on the blocks task. The blocks task was designed to involve qualitative reasoning rather than quantitative reasoning (Chi, Feltovich, & Glaser, 1981). The children's problem representations and solution strategies suggested implicit or intuitive knowledge of *qualitative reasoning principles* for rational number and proportion situations. These principles were inferred from the children's problem-solving protocols and abstracted into precise principles. These principles can be considered qualitative reasoning principles for proportion situations.

Simple proportion problems (Harel et al., in press) can be classified according to whether they involve a pair of ratios or a pair of products. The qualitative reasoning principles for proportion situations are classified in a similar manner.

Qualitative reasoning in a proportion situation then involves qualitative reasoning about a rational number or a ratio a/b or a product $a \cdot b$. Of concern is the qualitative question of the *direction* of change, or no change, in a/b or $a \cdot b$ as a result of combinations of qualitative changes of increase, decrease, or no change in a and b.

Qualitative questions about the order or equivalence of ratios, products, or rational numbers relate to the important concept of mathematical variability. This involves the question of whether a change occurs in a relation or operation under a transformation (increase, decrease, or no change) on the components of the relation or operation. If a change does take place, the question of whether to increase or decrease the outcome in order to compensate for the change becomes the issue.

Our analyses suggest principles that seem foundational for the development of intuitive knowledge about the order and equivalence of rational numbers, ratios, and products. These principles are presented in the section "Principles for Qualitative Reasoning in Fraction, Ratio, and Product Comparison." A subsection, "The Semantic Analysis of Rational Numbers: Some Implications for Curriculum," addresses two issues: (a) How different interpretations of rational numbers, which we present in our analysis, provide the basis for alternate problem representations and solution strategies of some division problems. The sample problems presented were written to *illustrate this point* and may therefore not be entirely realistic. (b) How looking at mathematics from the perspective of units of quantity provides a link between additive and multiplicative structures.

A third section, "Research on Teaching," (that is, research on teaching rational numbers) was pressed upon us by the initial charge for the development of this chapter. Unfortunately, we were unable to find much research that specifically targeted teaching rational number concepts. We present some aspects of research on teaching that bear on the teaching of rational numbers and mathematics more generally in the final section.

RATIONAL NUMBER CONSTRUCT THEORY: TOWARD A SEMANTIC ANALYSIS

Based on our own and others' research in the content domain of multiplicative structures, we have come to the realization that many of the limited, alternative (or mis-) conceptions that children and some adults (middle grades teachers, for example) have about many multiplicative concepts result from deficiencies in the curricular experiences provided in school. The well-known intuitive rules children form about multiplication and division (for example, multiplication makes bigger and division makes smaller) apparently result from a curricular overemphasis on multiplication and division of whole numbers. In the absence of counterexperience or "high-level" mathematical education, these limited conceptions remain into adult life.

Some Curriculum Concerns

We believe that the elementary school curriculum is deficient by failing to include the basic concepts and principles relating to multiplicative structures necessary for later learning in the intermediate grades. A second deficiency is that multiplicative concepts are presented in the middle grades in such a way that they remain cognitively isolated rather than interconnected. Both of these deficiencies arise from lack of research or analytical understanding of how multiplicative concepts interrelate from theoretical, mathematical, and cognitive perspectives.

There are five broad deficiencies:

1. Lack of problem situations that provide experience with composition, decomposition, and conversion of conceptual units (Steffe et al., 1988; Steffe, 1988).

2. Lack of consideration of arithmetic operations for both whole and rational numbers from the perspective of the mathematics of quantity (Schwartz, 1988).
3. Lack of problem situations that provide a wide range of experience so children develop less-constrained models of multiplication and division (Bell, Fischbein, & Greer, 1984; Graeber, Tirosh, & Glover, 1989; Greer, 1987).
4. Lack of experience with qualitative reasoning about number size, order relations, and the outcome of operations.
5. Lack of problem and computational situations that exemplify the invariance or variance of arithmetic operations and that exemplify variability principles fundamental to both qualitative and quantitative proportional reasoning (Harel et al., in press).

The remainder of this chapter will elaborate each of these five deficiencies.

Conceptual Units. Steffe (1988) discusses the general issue of children's abilities to form units of quantity. He identifies four types of units germane to our discussion: counting units, composite units, measurement units, and units of units. Extensive results about how children's formation of the whole-number concept and addition and subtraction concepts depends on formation of such units is given in Steffe et al. (1988). More recent results (Steffe, 1988) suggest that concepts of multiplication and division also depend on formation and reformation of these four unit types.

The Mathematics of Quantity. Schwartz (1976, 1988) and Kaput (1985) address the importance of developing arithmetic through the mathematics of quantity. This means that *units of measure* and magnitude of quantities are both significant to understanding number relations and operations (Schwartz, 1976, 1988). In particular, we showed that the units of measure and magnitude of quantities affect the problem representations that a solver of proportion problems makes (Harel & Behr, 1988). An analysis we have begun, emphasizing units analysis and mathematics of quantity, is yielding new insight into the subconstructs of rational numbers as well as the meaning of operations on whole and rational numbers. Based on this perspective, the remainder of this section presents findings from our investigation of the rational-number subconstructs of part-whole, quotient, and operator.

Components of the Analysis

The analysis we have underway concerning the multiplicative conceptual field incorporates the notions of units and the concepts of mathematics of quantity (Schwartz, 1988). We have employed two forms of analysis: drawing diagrams to represent the physical manipulation of objects, and using the notation of mathematics of quantity. Our aim is to present a semantic representation of the concepts analyzed with the diagrams, and a mathematical analysis with a mathematics-of-quantity model that has a "close" step-by-step relationship with the diagram.

The Bridging Notation—Motivation. In the course of conducting this analysis we have developed a notational system that might bridge the gap between commonly used representations—(a) contextualized pictorial or physically manipulative representations and (b) symbolic mathematical representations—of the entities, relationships, and operations that are involved. In one sense, the bridging notation can be thought of as a generic, noncontextualized, pictorial system representing the manipulation of objects at the concrete level. In this sense, the entities in the system can be replaced by physical representations of the real objects, with the sequence of diagrams in the bridging notation suggesting appropriate manipulations of the real objects. We hope the analysis, using this notation, suggests the cognition involved in understanding the mathematics. From the perspective of the symbolic mathematical representation, we believe the notation is sufficiently rich to afford nearly one-for-one matching between representations in the bridging notation and the mathematical symbolism.

Measure Unit for Discrete Quantity. To motivate the components of the bridging symbol system we start with a contextual realistic scenario: A mother, in preparing for a birthday party, puts four party favors in each cup. What are some of the tacit or implicit understandings we have about the quantity 4 favors per cup? A first observation is that this defines *1-cup* as *the unit* for party favors. In directing her helper to distribute the favors after they have been organized as 4 favors per cup, the mother would say "give each child one cup" in preference to saying "give each child four favors." This is because *the unit* for *measuring* party favors is *one cup*. In this particular context, the individual party favors lose their identity, and the concern is on the 1-cup unit. The identity of the individual favors is restored when the children each receive a cup of favors, look in, and count or compare (for sake of illustration) the number of favors they received with the number that others received. What might the child do who mistakenly received three instead of four favors in a cup? One scenario is that the child would go to the hostess and report, "I didn't get a cupful." Upon investigating the hostess could respond, "Oh my, you got only *three favors*," or thinking in terms of the 1-cup unit she might think, "I filled the cup only three-fourths full." She and the child might see the party favors in (or removed from) the particular cup, and those in a prototypical cup, in different ways: for example, in the particular cup, three singleton favors or one composite of three favors; in the prototypical cup, four singleton favors or one composite of four favors. We take the position that the child's and the hostess's construction of the fractional "cupfulness" is based on the type of mental objects each constructs from the party favors and the numerical relationship each sees between the amounts of these objects. For our purpose, we consider two ways in which the fractional cupfulness (that is, the measure with respect to the 4-favors-per-cup unit) of one party favor (or a cup with one favor in it) *could be* cognized and thus described as one-fourth of a 1-cup unit or as one $\frac{1}{4}$ cup unit. A cup with three party favors could be described in one of three ways: three-fourths of a one-cup unit, as one $\frac{3}{4}$ cup unit, or as three ($\frac{1}{4}$ cup unit)s. Similar descriptions hold for a cup with x favors compared to a unit of n favors per cup: x/n of a 1-cup unit, one x/n cup unit, $x(1/n$ cup unit)s, or $c(d/n$ cup unit)s, where $cd = x$ and x, c, and d are natural numbers.

Whether in a cup, taken out of a cup, or not yet put into a cup, the favors can also be cognized independently of their relationship to a prototype 1-cup unit. We identify three different conceptualizations: (a) If several physical attributes of the favors—exclusive of, or in addition to, their attribute of oneness—such as size, color, or number of spots are attended to, then our interpretation is that they are being cognized as *physical objects*. That is, they are not cognized as units for counting. (b) If the favors are counted by the child or matched one-to-one with another child's favors so that the attribute of oneness is particularized as the attribute of concern, then our interpretation is that the favors are conceptualized as *singleton units*—one favor as 1(1-unit), two favors as 2(1-unit)s, ..., and n favors as n(1-unit)s. This corresponds to Steffe's (1986) concept of a counting unit. (c) The child's attention might be drawn to a group of favors because they have some attribute (in addition to oneness) in common, such as all being red. If the child counted these red favors or in some other way (such as subitizing) determined the cardinality, then our interpretation would be that the child has determined a *composite unit*. An even stronger indicator of a composite unit formation would be if the child counted 5 cup units of favors and then multiplied 5×4 or repeatedly added 4 to determine the total number of favors. There could be other forms of evidence that a child has formed a composite unit. We will denote a conceptual composite of two objects or two singleton units (that is, 2(1-unit)s) as 1(2-unit), n conceptual replicates or physical duplicates of a (2 unit) as n(2-unit)s, a composite of three objects or singleton units as 1(3-unit), and n replications or duplications as n(3 unit)s. In general, 1(x-unit) and n(x-unit)s have analogous interpretations.

Measurement of Discrete Quantity. The next notion that we will illustrate in a realistic context is the process of using a unit such as 4 oranges per bag as a measurement unit to determine the measure of a set of less than 4 apples. In this illustration, the quantity to be measured (set of apples) is disjoint from, and of smaller cardinality than, the measurement quantity. An illustration to determine the measure when the quantity to be measured is contained in or contains the measurement quantity would be similar. An alternative statement of how a quantity such as 4 oranges per bag is used as a measurement unit is how it is used in a physical context as a divisor for quotitive division. Distinguishing characteristics of measure units (versus singleton and composite units) are that they are iterable (Steffe, 1986) and intensive rather than extensive quantity. We wish to emphasize the fact that there is more than one way for an individual to cognize the sets of apples and oranges in terms of units. Three apples could be thought of as just a group of apples, as 3 singleton units of one apple—3(1-unit)s—or as one composite 3-apple unit—1(3-unit). Similar statements hold for the four oranges. In addition, the oranges—any singleton orange, any subset, the whole collection, or singleton or composite units—must in some instances during the measurement process be cognized in relation to the bag as the unit of measure. As above, 1, 2, 3 or 4 oranges (taken as one composite unit) defines the fractional "bagfulness" of that number of oranges or any set of matching apples. Moreover, in order to fully understand this measurement process, the cognizer must be able to reunitize rather freely from one form of unit to another. For example, the ability to reunitize 3 singleton-unit apples—3(1-unit)s—as a 3-apple composite unit—1(3-unit)—

or 3 oranges as 1(3-unit) or as 1($\frac{3}{4}$-unit) is necessary before a measurement can be made of the apples.

ILLUSTRATION: 3 APPLES ÷ 4 ORANGES PER BAG. To introduce the bridging notation, which we will discuss more formally after this illustration, we will do parallel illustrations. One illustration uses contextualized pictures of apples and oranges per bag, and the other the bridging notation. With the bridging notation, 0s represent apples, *s oranges, and grouping symbols ( ), { }, or [ ] indicate unitization. For example, (0) (0) (0) represents three singleton apple units.

1.

3 apples

$\frac{4 \text{ oranges}}{\text{bag}}$

We consider in this example that the apples and oranges are conceptualized as singleton apples and oranges, that is, as (1-unit)s and {1-unit}s, respectively.

2.

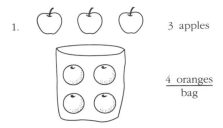

Measurement of discrete quantity is accomplished by matching (1-unit)s of the to-be-measured quantity with {1-unit}s of the measurement quantity. Here one apple is matched with one orange. With considerable additional cognition, one could determine at this point that the measure of this one apple with respect to the measure unit of 1-bag is one $\frac{1}{4}$ bag (or 1[$\frac{1}{4}$-unit]). In a similar sense we can now determine that the quotitive division, 1 apple divided by 4 oranges per bag is one $\frac{1}{4}$ bag. Where illustrations of measurement are involved, we take the point of view throughout that the cognition necessary to determine the measure is done only when the set of objects to be measured through the matching process has been exhausted.

3.

A second apple, or (1-unit), of the quantity to be measured is matched with a second orange, or {1-unit}, of the measurement quantity; in the meantime, the first matched pair is dropped from attention.

4.

The matchings are accumulated; that is, both matched pairings are brought into cognitive attention.

5.

Matching the third, and last apple, with a third orange is accomplished. In the meantime the first two matched pairs are dropped from attention. One can now observe why it is necessary to cognize the apples and oranges as individual singleton units, since matching is accomplished by cognizing and acting upon individual objects as single cognitive entities.

6.

The matchings are accumulated.

7.

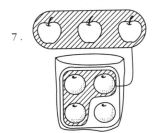

Measurement of the discrete set of apples using the 4 oranges per bag indirectly establishes the cardinality of the set of apples by establishing the cardinality of a subset of the oranges. Since cardinality is an attribute of a composite of objects and not of the individual objects, it is necessary to unitize the 3 apples into a composite 3-apple unit, the 4 oranges into a composite 4-orange unit, and the 3 matched oranges into a composite 3-orange subunit. Finally, a single matching is established between the composite 3-apple unit and the 3-orange unit, and the measure of the 3-apple unit is established by finding the fractional bagfulness determined by the 3-orange composite unit. The individual apples and oranges could now lose their identity and could be cognized generically as units of fruit. The "bagfulness" of the 3-orange unit is one $\frac{3}{4}$ bag, that is, $1(\frac{3}{4}$ unit$)$. The subtlety of why we call this one $\frac{3}{4}$ bag rather than $\frac{3}{4}$ of a 1-bag is discussed later in the general discussion of the bridging notation.

We now offer a mathematics of quantity representation that fits both the contextual picture model and the bridging model, using the grouping-symbol unit notation along with the words *apple* and *orange* to aid the reader. For symbolic processing of 3 apples ÷ 4 oranges, it is necessary at some point in the process to transform apple units and orange units into a common unit, such as fruit. Numbers in parentheses correspond to step numbers in the illustration above.

The Bridging Notation. As suggested, we will use any mark, especially, 0, *, and # to denote real objects, as in the example

0 to denote an apple, * to denote an orange, and # to denote a fruit. When an object (denoted by any of 0, *, or #) is conceptualized not just as an object but as a 1-object unit (such as a 1-apple unit, 1-orange unit, or 1-fruit unit), we enclose the symbol for the object within grouping symbols, (), [], or { }. Thus, (0), [*], and {#} could denote units of 1-apple, 1-orange, and 1-fruit in our previous example. When different grouping symbols are used to denote a unit, it is to be assumed that the units come from different measure spaces. Groups of 1-object units will be represented as (0) (0) (0) (0) or [*] [*] [*] and using the notation (1-unit) for a 1-object unit we will denote the preceding groups as 4 (1-unit)s and 3 [1-unit]s, respectively.

We will account for two types of composite units with our notation. A composite unit of 3 *objects* will be represented as

$$1. \quad 3(1\text{-apple})s \div \frac{4\{1\text{ orange}\}s}{[1\text{-bag}]} \tag{1}$$

$$2. \quad = (1(1\text{-apple}) + 1(1\text{-apple}) + 1(1\text{-apple})) \div \frac{4\{1\text{-orange}\}s}{[1\text{-bag}]} \tag{2}$$

$$3. \quad = 1(1\text{-apple}) \div \frac{4\{1\text{-orange}\}s}{[1\text{-bag}]} + (1(1\text{-apple}) + 1(1\text{-apple})) \div \frac{4\{1\text{-orange}\}s}{[1\text{-bag}]} \tag{3}$$

$$4. \quad = \left(1(1\text{-apple}) \div \frac{4\{1\text{-orange}\}s}{[1\text{-bag}]} + 1(1\text{-apple}) \div \frac{4\{1\text{-orange}\}s}{[1\text{-bag}]}\right) + 1(1\text{-apple}) \div \frac{4\{1\text{-orange}\}s}{[1\text{-bag}]} \tag{4}$$

$$5. \quad = 2(1\text{-apple})s \div \frac{4\{1\text{-orange}\}s}{[1\text{-bag}]} + 1(1\text{-apple}) \div \frac{4\{1\text{-orange}\}s}{[1\text{-bag}]} \tag{5}$$

$$6. \quad = 3(1\text{-apple})s \div \frac{4\{1\text{-orange}\}s}{[1\text{-bag}]} \tag{6}$$

$$7. \quad = 3(1\text{-fruit})s \div \frac{4(1\text{-fruit})s^*}{[1\text{-bag}]} \tag{7}$$

$$8. \quad = 1(3\text{-fruit})s \div \frac{\frac{4}{3}(3\text{-fruit})s}{\frac{4}{3}[\frac{3}{4}\text{-bag}]} \tag{7}$$

$$9. \quad = 1(3\text{-fruit}) \div \frac{1(3\text{-fruit})}{1[\frac{3}{4}\text{-bag}]} \tag{7}$$

$$10. \quad = 1[\tfrac{3}{4}\text{-bag}].^\dagger \tag{7}$$

*This is a convention we adopt to show a unit conversion to a superunit. In this case, apple units and orange units are sub-measure space units of the fruit-unit measure space. In general, if ( ) and {} denote units from two measure spaces and there is a third measure of which each is a subspace, we denote the superspace unit with a boldface overprint of the two grouping symbols used for the subspaces; in this case, overprints in boldface type of (and {, and) and } for left and right parentheses, respectively. Thus the grouping symbol used to denote the superspace units for ( ) and {} is ( ).

†It may at first seem strange to the reader to think of division of 3 apples by 4 oranges per bag as being equal to a $\frac{3}{4}$-bag. But consider an analogous situation:

$$5 \text{ cm} \div \frac{10 \text{ dec}}{\text{meter}} = \tfrac{1}{20} \text{ meter.}$$

The two situations are essentially the same except for the familiarity of the context. To determine the measure of a 5 cm length with a meter stick graduated into decimeters *only*, may not be an everyday occurrence, but nothing seems strange about the symbolic statement. This is likely because we immediately recognize that decimeters and centimeters can be changed to a common unit and can therefore be canceled as the symbolic processing is carried out. But this concern for a common unit is necessary for *symbolic* processing of the division, not for processing the division at the level of manipulative aids. At the level of manipulative aids one need not be concerned about apples and oranges per se, but just that the quantity 4 objects per unit is being used to measure 3 objects and the the 3 objects are not contained in the 4 objects (as a subset relationship).

(0 0 0), and a composite of 3 *1-object units*—that is, a composite of 3 (1-unit)s—as ((0) (0) (0)). Both are notated algebraically as 1 (3-unit). (More correctly ((0) (0) (0)) is a unit-of-units, a concept that we introduce shortly.) Composite units of n objects and of n 1-object units will be represented similarly. We observe in passing that unitization (and reunitization) is a cognitive process that we have attempted to represent externally, both with the bridging notation and the algebraic notation of mathematics of quantity.

We turn next to the question of how a collection of discrete objects serves as a measure unit. From our perspective, a measure unit is an intensive quantity such as 4 oranges per bag. We need to consider two types of measure units (we illustrate using 4 discrete objects): One type of measure unit is composed of 1 (4-unit) per [1-unit] and another of 4 (1-unit)s per [1-unit], which are notated respectively as [(* * * *)] and [(*) (*) (*) (*)]. Conceptually the 1-unit denoted by the [ ] plays the same role as the bag or the cup in our earlier examples. On occasion we will need to represent and interpret a fractional part of a composite unit and of the two types of measure units. We will use the traditional means of shading to designate a fractional part of a unit of discrete objects. For example, (●●●○) could denote $\frac{3}{4}$ of a (4-unit). Using units with 5 "objects," we illustrate the bridging notation and the corresponding algebraic notations: (● 0 0 0 0) and ((●) 0 0 0 0) both are interpreted as $\frac{1}{5}$(5-unit), [(● 0 0 0 0)] and [((●) 0 0 0 0)] as $\frac{1}{5}$[1-unit], and [(●) (0) (0) (0) (0)] as 1[$\frac{1}{5}$-unit]. The same interpretations hold for n discrete objects, where n is any natural number. Some subtleties arise for fractional parts other than unit fractional parts: (● ● 0 0 0) is interpreted as two $\frac{1}{5}$(5-unit)s, ((● ●) 0 0 0) as one $\frac{2}{5}$(5-unit), [(● ● 0 0 0)] as two $\frac{1}{5}$[1-unit]s, [((● ●) 0 0 0)] as one $\frac{2}{5}$[1-unit], [(●) (●) (0) (0) (0)] as 2[$\frac{1}{5}$-unit]s, and [((●) (●)) (0) (0) (0)] as 1 [$\frac{2}{5}$-unit]. We make one final observation about a relationship between various units and unit types that follows from the interpretations above: $\frac{1}{5}$(5-unit) = 1(1-unit), $\frac{2}{5}$(5-unit) = 1(2-unit), or in general $1/n$(n-unit) = 1(1-unit) and x/n(n-unit) = 1(x-unit), where x and n are any nonzero natural numbers.

Although it has only one application in this chapter, we introduce a final concept of composite units, namely units-of-units. The need to conceptualize units-of-units arises for the learner and knower of fraction concepts in a situation where an (8-unit) is used in a part-whole representation of $\frac{3}{4}$. The (8-unit)—(0 0 0 0 0 0 0 0)—would first be partitioned into 4 groups of 2—(0 0 / 0 0 / 0 0 / 0 0)—and then the groups of two are unitized—((0 0)(0 0)(0 0) (0 0)). When the 4 (2-unit)s are unitized into a composite unit, the result is a composite unit-of-units, in this case we denote it as 1(4(2-unit)s-unit). Similarly, ((000) (000) (000) (000) (000)) is a (5-unit) composition of (3-unit)s; that is, 1(5(3-unit)s-unit). We will apply part-whole fraction designation to units-of-units; ((●●●) (●●●) (●●●) (0 0 0) (0 0 0)) is interpreted as $\frac{3}{5}$(5(3-unit)s-unit), or an alternate unitization could be 3($\frac{1}{5}$(5(3-unit)s-unit))s. Finally, it is possible to unitize the 3($\frac{1}{5}$-unit)s, which would be pictured as (((●●●) (●●●) (●●●)) (0 0 0) (0 0 0)) and denoted as 1($\frac{3}{5}$(5(3-unit)s-unit)-unit).

We note that composite units-of-units may have several interpretations; that is, they can be reunitized in various ways. When a child counts five longs (Dienes's blocks) as "1, 2, 3, 4,

5; 50," there are two implicit recognitions about the composite 5-unit of (10-unit)s or 1(5(10-unit)s-unit). The counting 1, 2, 3, 4, 5 indicates awareness that the 1(5(10-unit)s-unit) constitutes a (5-<u>unit</u>) (underlining here is to suggest that the cognizer is aware that the unit is a composite). Announcement that the cardinality of the composite unit-of-units in terms of units of one demonstrates recognition that the 1(5(10-unit)s-unit) represents the same amount of quantity as 1(50-unit).

THE FUNCTION OF UNIT TYPES. We have suggested that x discrete objects can be cognitively unitized as follows: x(1-units), 1(x-unit), x(1-units)/ [1-unit], and 1(x-unit)/[1-unit]; in addition, n(x-unit)s can be unitized as a composite unit-of-units, 1(n(x-unit)s-unit). The same is true of a continuous object visually segmented into x continuous subsections of equal measure (Steffe, 1989). Each of these unit types can function in certain capacities as follows:

1. Any x(1-units) can be *put into* m[1-units] to give n(1-units)/ [1-unit], where n, m, and x are natural numbers and $n = x \div m$ where $m|x$.
2. Any x(1-units) can be distributed among m[1-units] to give n(1-units)/[1-unit], where n, m, and x are natural numbers and $n = x \div m$ where $m|x$.
3. One (x-unit) can be equi-partitioned into n parts of cardinality m, and the parts can be unitized to give n (m-unit)s as subunits of the (x-unit), where n, m, and x are natural numbers and $n = x \div m$ where $m|x$.
4a. Any (x-unit) can be separated into a part-part-whole relationship, and the parts can be unitized to give subunits of 1(m-unit) and 1(n-unit), where n, m, and x are natural numbers and $n + m = x$. Conversely, if 1(m-unit) and 1(n-unit) are composite units from the same measure space, they can be combined to form 1(y-unit) so that a part-part-whole relationship exists among the 1(m-unit), 1(n-unit), and 1(y-unit), where n, m, and y are natural numbers and $y = m + n$.
4b. Any (x-unit) can be separated into a part-part-whole relationship, and the parts can be unitized to give subunits of m/x(x-unit) and n/x(x-unit), where n, m, and x are natural numbers and $n + m = x$. Conversely, if m/x(x-unit) and n/x(x-unit) are composite units from the same measure space, they can be combined to form an y/x(x-unit) so that a part-part-whole relationship exists among the m/x(x-unit), n/x(x-unit), and y/x(x-unit), where n, m, y, and x are natural numbers and $y = m + n$.
5a. For any x(1-unit)s/[1-unit] quantity, the x(1-unit)s can be separated into parts of n(1-unit)s and m(1-unit)s, and the parts can be unitized to have measure n[1/x-unit]s and m[1/x-unit]s, respectively, where n, m, and x are natural numbers and $n + m = x$. Conversely, if q(1-unit)s and r(1-unit)s are disjoint parts (perhaps not exhaustive) of the x(1-units) in the quantity x(1-unit)s/[1-unit], then their measures q[1/x-unit]s and r[1/x-unit]s can be combined to give the measure of the q(1-unit)s and r(1-unit)s together to be y[1/x-unit]s, where q, r, y, and x are natural numbers and $y = q + r$.
5b. For any x(1-unit)s/[1-unit] quantity, the x(1-unit)s can be separated into parts of n(1-unit)s and m(1-unit)s, and the

parts can be unitized to have measure 1[n/x–unit] and 1[m/x-unit], respectively, where n, m, and x are natural numbers and $n + m = x$. Conversely, if q(1-unit)s and r(1-unit)s are disjoint parts (perhaps not exhaustive) of the x(1-units) in the quantity x(1-unit)s/[1-unit], then their measures 1[q/x-unit] and 1[r/x-unit] can be combined to give the measure of the q(1-unit)s and r(1-unit)s together to be 1[y/x-unit], where q, r, y, and x are natural numbers and $y = q + r$.

6a. For any (x-unit)/[1-unit] quantity, the (x-unit) can be separated into parts of 1(n-unit) and 1(m-unit), and the parts can be unitized to have measure n[1/x-unit]s and m[1/x-unit]s, respectively, where n, m, and x are natural numbers and $n + m = x$. Conversely, if 1(q-unit) and 1(r-unit) are disjoint parts (perhaps not exhaustive) of the (x-unit) in the quantity (x-unit)/[1-unit], then their measures q[1/x-unit]s and r[1/x-unit]s can be combined to give the measure of the (q-unit) and (r-unit) together to be y[1/x-unit]s, where q, r, y, and x are natural numbers and $y = q + r$.

6b. For any (x-unit)/[1-unit] quantity, the (x-unit) can be separated into parts of 1(n-unit) and 1(m-unit), and the parts can be unitized to have measure 1[n/x-unit] and 1[m/x-unit], respectively, where n, m, and x are natural numbers and $n + m = x$. Conversely, if 1(q-unit) and 1(r unit) are disjoint parts (perhaps not exhaustive) of the (x-unit) in the quantity (x–unit)/[1-unit], then their measures 1[q/x-unit] and 1[r/x-unit] can be combined to give the measure of the 1(q-unit) and 1(r-unit) together to be 1[y/x-unit], where q, r, y, and x are natural numbers and $y = q + r$.

7. Any x(1-unit)s can be measured by m(1-unit)s/[1-unit] to give a measurement of x[1/m–unit]s, where m and x are natural numbers.

8. Any x(1-unit)s can be measured by 1(m-unit)/[1-unit] to give a measurement of x1/m[1-unit]s, where m and x are natural numbers.

9. Any 1(x-unit) can be measured by m(1-unit)s/[1-unit] to give a measurement of 1[x/m-unit], where m and x are natural numbers.

10. Any 1(x-unit) can be measured by 1(m-unit)/[1-unit] to give a measurement of x/m[1-unit], where m and x are natural numbers.

An important question is how a child forms units and constructs relationships among the units suggested in the above list. The analyses given later in this chapter (see also Behr, Harel, Post, & Lesh, 1990) based on the bridging notation suggests how manipulation of objects would facilitate children's construction of such units. It remains for curriculum development and research to create problem situations whose solutions would involve these manipulations and to determine whether such solutions can be created by children. In addition, teachers may need help with this perspective on rational numbers and rational number problem situations.

Various authors take the view that one model of 4×3 is an expression such as $4 \times$ (3 apples). This model is to indicate to children that 4, the multiplier, tells the number of times to replicate the set of 3 apples. This is called *scalar multiplication*.

In our view, however, scalar multiplication is an advanced mathematical construct that has no basis in experience. It is not the set of 3 apples that is replicated but a *relationship* of 3 apples to one-unit. Thus the model for 4×3 is 4[1-units] with 3(1-units) per each [1-unit]. This formulation agrees with a units analysis approach in which one thinks of "canceling the units" in the expression 4[1-units] $\times$ 3(1-units)/[1-unit]; the indicated multiplication then has the meaning of forming the 3(1-units) for [1-unit] relationship for each of the 4[1-units]. This way, the cancelation of the unit labels corresponds to a reunitization of the 4 3(1-unit)s to [1-unit] relationship to 12 (1-unit)s. According to this thinking, the interpretation of $4 \times$ (3 apples) is as 4 sets $\times$ 3 apples per set.

This model for 4×3 is also different from Schwartz's (1988) model, which proposes thinking of $4 \times$ (3 apples) as 4 apples per apple times 3 apples. If one considers problems as origins of mathematical models, then both our formulation and Schwartz's are necessary. Schwartz's formulation is the appropriate model for the problem:

Jane has 3 apples. Bill has 4 times as many as Jane. How many apples does Bill have?

The interpretation of 4×3 that we have given is, however, the better model for the problem:

There are 3 apples in a bag and there are 4 bags. How many apples are there in all?

Both of these formulations have been used in our analysis of the operator concept of rational number.

UNIT-CONVERSION PRINCIPLES. We have identified several basic unit-conversion principles for solving whole and rational-number word and computation problems.

1. 1(a-unit) = a(1-unit)s.
2. a(b-unit)s = b(a-unit)s.
3. a(b-unit)s = c(d-unit)s, where $ab = cd$ (or $a/c = d/b$),
4. $\frac{(a\text{-unit})}{[a\text{-unit}]} = \frac{(c\text{-unit})}{[c\text{-unit}]}$, a, b, c, and d are any rational numbers.
5. 1(x-unit) = 1(n(m-unit)s-unit), where $m \times n = x$, and m, n, and x are natural numbers.

A note is necessary about what interpretation can be given to the equal sign in this context. Since any of the variables used above denote real numbers, and any string of symbols within a pair of grouping symbols denotes a unit of measure, any symbol string such as c(d-unit)s must denote a quantity. Thus the equal sign in this context, as in any mathematical context, indicates that the symbol strings on each side of it are names for the same quantity. No indications are given of the syntactic or semantic transformations necessary to derive one member of an equality from the other. These equations are external representations of the equivalence of a quantity under different unitizations. While we do capture some of the external manifestations of the cognition involved in these reunitizations with the bridging notation, we make no such claims for the algebraic notation.

A View of the Literature

Children as young as 6 and 11 years old can deal with problems that depend on these principles. Steffe (1988) reports a child named Zackery who was able to abstract 4×10 from a set of cards arranged in order within suits, but who was unable to reconceptualize the cards arranged by denomination across suits as 10×4—an application of the second unit-conversion principle. Pirie (1988) tells of an 11-year-old named Katie who was working on the task $6\frac{1}{3} \div \frac{2}{3}$ and commented that the actual drawing of pictures and the $\frac{2}{3}$ pieces was "boring." She suggested that "6 wholes would have 6 times 3 . . . 18 thirds and half that number of two-thirds." So, $6\frac{1}{3}$ has 19 thirds and $9\frac{1}{2}$ two-thirds.

This thinking is easily modeled by using unit labels, and unit conversion principles as follows:

$$
\begin{aligned}
6\tfrac{1}{3} &= 6\tfrac{1}{3}(1\text{-units}) \div 1(\tfrac{2}{3}\text{-unit}) \\
&= 19(\tfrac{1}{3}\text{-units}) \div 1(\tfrac{2}{3}\text{-unit}) \\
&= \tfrac{19}{2}(\tfrac{2}{3}\text{-units}) \div 1(\tfrac{2}{3}\text{-unit}) \\
&= \frac{9\tfrac{1}{2}(\tfrac{2}{3}\text{-unit})\text{s}}{(\tfrac{2}{3}\text{-unit})} \\
&= \frac{9\tfrac{1}{2}(1\text{-units})\text{s}}{1(1\text{-unit})} \\
&= 9\tfrac{1}{2}.
\end{aligned}
$$

Hunting (1981) investigated the performance of students in grades 4, 6, and 8 on numerous sharing tasks, which he interpreted in terms of formation and reformation of composite units. Examples of these tasks follow, with interpretations in terms of units notation. One situation involved showing 4 dolls and 12 cookies to the child and asking how many cookies each doll would get if the cookies were shared equally. This was analyzed by Hunting (represented in our notation) as the problem

$$1(12\text{cookie-unit}) = 4(x\text{cookie-unit})\text{s, find } x$$

A second task was to take 25 cookies and share them among some dolls so that each doll gets five cookies. As a task in reunitization, this has the form

$$1(25\text{cookie-unit}) = x(5\text{cookie-unit})\text{s, find } x$$

Each of these two tasks involve unit-conversion principle 3, with $a = 1$. Another task with $a, b, c,$ and $d \neq 1$, which Hunting used and interpreted as a units task, asked the child to take 8 piles of 3 cookies and rearrange them so that 4 dolls could share them equally. This problem has the form $8(3\text{cookie-unit})\text{s} = 4(x\text{cookie-unit})\text{s}$. Hunting found his subjects (grades 4, 6, and 8) to be quite successful with these kinds of problems. The strategy most frequently used by these subjects involved anticipating the solution based on whole-number knowledge rather than experimenting with solutions by manipulating objects. Can much younger children construct the knowledge necessary to solve such problems through experiences with manipulative aids?

We believe curriculum development should provide children with problem situations that give them an experiential base for internalizing the unit-conversion principles they will apply to the concepts of fraction, rational number, rate, ratio, and proportion, as well as to multiplication and division problems that are more sophisticated than those traditionally given. Moreover, we maintain that research needs to determine the extent to which problem situations in both the whole and rational number domains could be provided as early as first grade.

An unpublished study by the Rational Number Project focused on the knowledge structures that children use in their attempts to solve problems of the form x is a/b of y where x and a/b are given and y is the unit whole; problems of the form $y = c/b$ with c greater than b were also investigated. The concern of this study is the flexibility of the concept of unit, where units vary across a wide domain of contexts and complexity of numerical relationships. The tasks presented fifth grade students with a fractional part of a unit whole and asked them to reconstruct the unit, some part of the unit, or some multiple of the unit. These problems were posed in both continuous and discrete situations. Student responses fall into one of five response categories.

Category 1: unit fraction decomposition and composition. Students initially decomposed a given fractional part into unit fractions of the form $1/m$. They then regenerated the unit whole by iterating this unit fraction.

Category 2: unit parts decomposition and composition. Here students decomposed a given fractional part into parts corresponding to the number in the numerator of the given fraction. The unit whole was then composed of the requisite number of parts. (This is thought to be a somewhat less sophisticated response than category 1 and was therefore separated from it.)

Category 3: no unit part of unit fraction decomposition. Subjects showed no awareness that the fractional part is composed of, or decomposable to, unit parts or unit fractions equal in number to the numerator.

Category 4: given fractional part used as a unit. Here the subject used the given fractional part as the unit whole.

Category 5: given fractional part used as the unit fraction or unit part. Here subjects used the given fractional part as the unit fractional part.

Categories 1 and 2 generally led to correct solutions while categories 3, 4, and 5 did not. Some conclusions that we made from these results follow.

Reconstruction of the whole, given a fractional part of the whole, is more difficult than the more common task of partitioning to find a specific fractional part of a unit whole. This was explained in terms of the additional m-space (Case & Sandieson, 1988) required for the former task.

The data suggest that the discrete context is easier for children initially, probably because discrete interpretations are amenable to already well-developed counting procedures, where continuous formats are not. Continous formats must invoke concepts of rational number partitioning that are not readily derived from previously internalized counting procedures.

Usual instruction on fractions involves providing a process for finding and perceiving a fraction as part of a whole; $\frac{3}{5}$, for example, is perceived as embedded in the whole. A deeper perception of $\frac{3}{5}$ (Novillis, 1976) is to conceive a separate entity that is $\frac{3}{5}$ as much (as long, as big, as sour) as the given unit. This concept of fraction likely helps develop the notion of $\frac{3}{5}$ as an entity in addition to being 3 of 5 parts. Instruction on fractions is incomplete if it does not distinguish a fraction from the whole of which it is part.

Among the low-scoring subjects in our study, only 28% of the responses were in categories 1 and 2, suggesting again that fraction concepts are difficult for many children even into grade 5. Children who were subjects in this teaching experiment had instruction and review of the process of solving these problems. What can be expected from school children in regular classrooms in grades 5 and above, even into high school? Although not addressed directly by our research, the question of when it is most appropriate to begin relating these two reversible processes should receive future research attention.

Typical instruction on fractions in schools emphasizes the part-whole construct. This reinforces the perception of $\frac{3}{5}$ as 3 of 5 parts, but not as the iteration of 3 abstract units of size one-fifth—that is, as $3(\frac{1}{5}$-unit)s. We propose that rational number instruction limited to the part-whole construct is inadequate to develop a complete understanding of rational numbers and should be extended to include other constructs in the context of mathematics of quantity.

Semantic Analysis—Overview. We have been conducting analyses of whole number and rational number concepts and operations. One of our current analyses is of the subconstructs of rational numbers, and we present partial results in this section. The rational-number subconstruct analysis employs two notations, the bridging notation and the notation of mathematics of quantity, and relies heavily on the conceptualizations of unit formation, reunitization, and principles of units conversion. The thrust of the analysis is to use the bridging notation to represent the physical manipulations that would be embodied in real world problem situations, in order to facilitate learners' development of the construct. We believe that this bridging notation strongly exhibits the semantics of the mathematics involved in the rational-number construct. During the analysis, representation of the physical manipulations by the bridging notation interacts with the mathematical analysis. The result is nearly a one-for-one matching between a sequence of manipulations of physical objects and a sequence of manipulations of mathematical symbols. We claim that this matching indicates that the manipulation of objects, from which children's understanding arises, is also a true representation of the mathematics involved. Thus, we claim to be able to offer situations from which children *can* (not necessarily *will*) construct knowledge for understanding the mathematics. The analysis considers the part-whole, quotient, and operator subconstructs of rational numbers. We report our analysis of the part-whole and quotient subconstructs in some depth after giving an overview of where our analysis has progressed to, some of the general considerations involved in the analysis, and some of the resulting interpretations of rational numbers.

Researchers of rational numbers, as well as teachers of the concepts, normally think of using both discrete and continuous quantity representations for rational numbers. In this chapter we limit our report to analysis based on discrete quantity, with several exceptions. Our choice of discrete quantity over continuous quantity is based on the assertion by Bigelow et al. (1989) that children's first concepts of rational numbers should be based on experience with discrete quantities. Moreover, our analysis suggests that the concepts children need to work with continuous quantity have their basis in situations involving discrete quantity. The choice was necessary because a description of the totality of the analysis that we have underway would go well beyond the scope of this chapter.

We present in Figure 14.1 a flowchart to illustrate some of the considerations that went into the analysis and interpretations of rational numbers.

Interpretations of Rational Number. We next consider some demonstrations to give information about the analysis of rational-number constructs of part-whole and quotient. Following the demonstrations on the part-whole and quotient constructs, we give a brief overview of the considerations that have gone into the analysis of the operator construct, but a full report of this analysis goes beyond the scope of this chapter.

The Part-Whole Construct. Using a units analysis, the part-whole subconstruct leads to two interpretations of rational numbers; we illustrate with $\frac{3}{4}$.

THREE-FOURTHS AS PARTS OF A WHOLE. The units interpretation of three-fourths as *parts* of a whole is $3(\frac{1}{4}$-unit)s for continuous quantity and $3(\frac{1}{4}(4(n$-unit)s)-unit)s for discrete quantity. (A note on reading the notation for units-of-units: To determine the basic type of unit expressed, look for the right-most symbol string, "-unit)"; then look for the first numeral (or variable) inside the left-most parenthesis. The combination of these define the basic unit type. Thus, $(\frac{1}{4}(4(2$-unit)s)-unit) is basically a $(\frac{1}{4}$-unit)—it could also be thought of as $(\frac{1}{4}(4$-unit)-unit)—while $(4(2$-unit)s-unit) is basically a $(4$-unit). Because of the difference in the notation for the continuous-quantity unit and the discrete-quantity unit, it appears at first that the interpretation of three-fourths differs across the two types of quantity. However, the conceptual outcome in each case is that three-fourths is three $\frac{1}{4}$ units. We illustrate the physical manipulations to arrive at these interpretations with both a discrete-quantity model and a continuous-quantity model in Figures 14.2 and 14.3, respectively. Similar demonstrations hold for any $(4n$-unit); where $n = 1, 2, 3, \ldots$, each would lead to the interpretation of three-fourths as $3(\frac{1}{4}(4(n$-unit)s-unit)-unit)s.

THREE-FOURTHS AS A COMPOSITE PART OF A WHOLE. The units interpretation of three-fourths as a composite unit is $1(\frac{3}{4}$-unit); a presentation with the bridging notation is given in Figure 14.4. We omit a continuous quantity for this construct. Similar demonstrations hold for any $(4n$-unit); where $n = 1, 2, 3, \ldots$, each would lead to the interpretation of

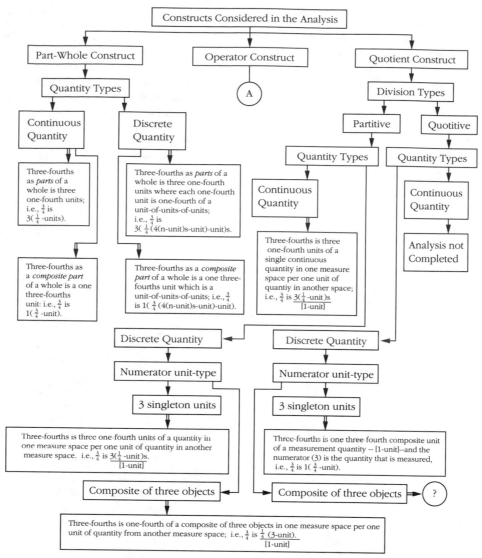

FIGURE 14–1. The construct theory of rational numbers, overview of semantic analysis—interpretations of three-fourths.

three-fourths as ($\frac{3}{4}$(4(*n*-unit)s)-unit). Evidence that children make these interpretations is given by Behr, et al. (1983).

The Quotient Construct. We consider the quotient construct of rational numbers from perspectives of both quotitive and partitive division. To get a complete picture of a rational number as a quotient one needs to consider several variables. Consideration should be given to both continuous and discrete quantity and to different possible unitizations of the numerator and denominator. In the rational number x/y the numerator can be interpreted as x(1-unit)s or as 1(x-unit). The denominator can be similarly interpreted as y(1-unit)s or 1(y-unit). To date, our analysis for partitive division has considered the two cases for the numerator but only the case for y(1-unit)s for the denominator.

PARTITIVE DIVISION. Two interpretations of a rational number result from a partitive interpretation of division. One interpre-

tation leads to the concept that three-fourths is 3($\frac{1}{4}$-unit)s per [1-unit]; the other, that three-fourths is $\frac{1}{4}$(3-unit) per [1-unit]. We illustrate the first interpretation using the bridging notation based on discrete (Figure 14.5) and continuous (Figure 14.6) quantity and follow this with a corresponding mathematics-of-quantity representation.

In the mathematics-of-quantity model that follows (Figure 14.7), we demonstrate the related nature of the continuous, discrete, and mathematical models by noting in parentheses, following each step in the mathematical derivation, the corresponding steps in the discrete (Figure 14.5) and continuous (Figure 14.6) models.

The notation used in the continuous model (Figure 14.6) represents a true model of how *some* children are known to solve such problems through sharing (other partitions are given in the literature and can be modeled by this notation equally well). They do this by first equi-partitioning each of the 3(1-units) into four parts and then distributing these parts eq-

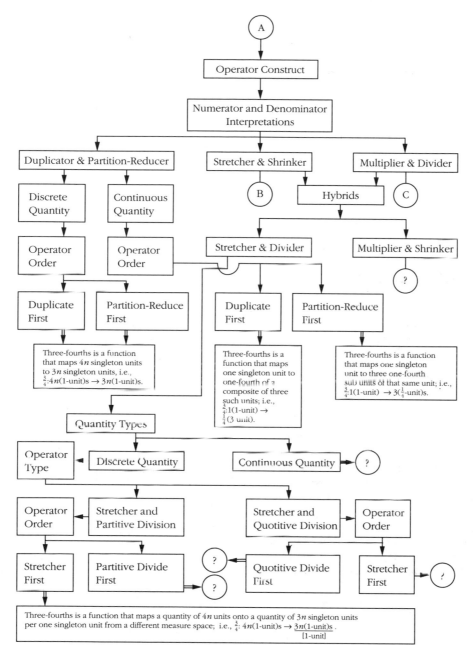

FIGURE 14–1. (Continued).

uitably among the 4[1-units] (Kieren, Nelson, & Smith, 1985). On the other hand, if 3 is interpreted as a composite 1(3-unit) instead of as 3(1-units) as above, this gives $\frac{3}{4}$ as $\frac{1}{4}$(3-unit) per [1-unit], or for the discrete case (Figure 14.8), as $\frac{1}{4}$(3(4-unit)s-unit) per [1-unit].

A note to the interested reader: The model for the continuous case can be constructed by choosing for the numerator quantity one (3-unit) in the form of three circular regions and then carefully following the type of partitioning that is done with the (3-unit)s of four discrete objects. In an investigation of partitioning behavior of children in grades 6, 7, and 8, Kieren and Nelson (1981) presented the task of shading the amount one child gets if 3 candy bars are shared equally by 4 chil-

dren. The candy bars were presented as three rectangles partitioned into 8 parts. Of 196 responses, 12.5% interpreted the three candy bars as one composite unit—one (3-unit)—and (apparently) gave the following left to right partition and solution (Figure 14.9). This solution suggests that three-fourths is $\frac{1}{4}$(3-unit). This partition given represents a lower level of partitioning performance (Bigelow, et al., 1989) than one based on successive halving as we illustrated for the discrete case above (Figure 14.8). However, we suggest that children need the experience of interacting with other children and teachers in making partitions. Figure 14.10 presents another partition of a (3-unit) that children can accomplish with urging from a teacher.

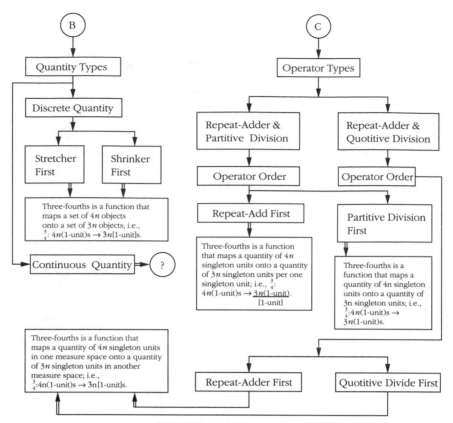

FIGURE 14–1. (*Continued*).

QUOTITIVE DIVISION. Each of the two representations that arise from the partitive-division interpretation involves the quotient of two extensive quantities, and the result is a representation of the rational number $\frac{3}{4}$ as intensive quantity. Two additional representations for the quotient subconstruct of rational numbers result from the quotitive (measurement) meaning of division. We find four ways to look at this division according to two different interpretations of the numerator,

as 3(1-units) or as a composite 1(3-unit), and according to two similar interpretations for the denominator. Because of space considerations, we give demonstrations for the 3(1-unit)s interpretation of the numerator with the two interpretations of the denominator (Figures 14.11 and 14.13) respectively.

Figure 14.12 presents the mathematics-of-quantity model that corresponds to the demonstration in Figure 14.11. In this case, we are able to match statements in the mathematics-of-

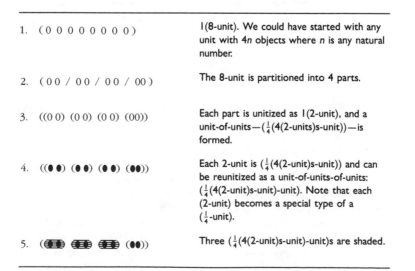

1. $(\,0\ 0\ 0\ 0\ 0\ 0\ 0\ 0\,)$ — 1(8-unit). We could have started with any unit with 4*n* objects where *n* is any natural number.

2. $(\,0\ 0\ /\ 0\ 0\ /\ 0\ 0\ /\ 0\ 0\,)$ — The 8-unit is partitioned into 4 parts.

3. $(\,(0\ 0)\ (0\ 0)\ (0\ 0)\ (00)\,)$ — Each part is unitized as 1(2-unit), and a unit-of-units—$(\frac{1}{4}(4(2\text{-units})s\text{-unit}))$—is formed.

4. $(\,(\bullet\ \bullet)\ (\bullet\ \bullet)\ (\bullet\ \bullet)\ (\bullet\bullet)\,)$ — Each 2-unit is $(\frac{1}{4}(4(2\text{-unit})s\text{-unit}))$ and can be reunitized as a unit-of-units-of-units: $(\frac{1}{4}(4(2\text{-unit})s\text{-unit})\text{-unit})$. Note that each (2-unit) becomes a special type of a $(\frac{1}{4}\text{-unit})$.

5. $(\,(\text{⬤} \text{⬤} \text{⬤})\ (\bullet\bullet)\,)$ — Three $(\frac{1}{4}(4(2\text{-unit})s\text{-unit})\text{-unit})$s are shaded.

FIGURE 14–2. Three-fourths as *parts* of a discrete unit-whole leads to the interpretation that three-fourths is $3(\frac{1}{4}\text{-unit})$s.

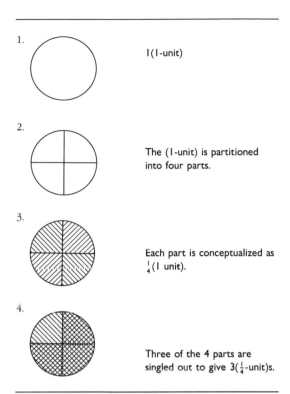

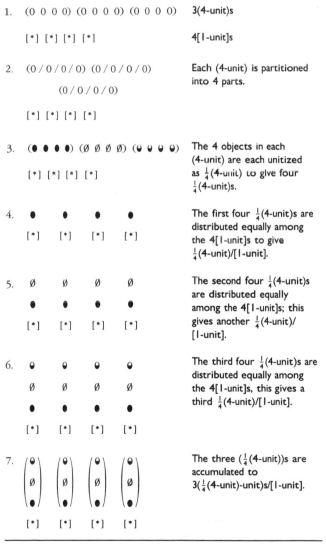

responding symbolic process deserves comment. We hypothesize first (step 7, Figure 14.11) that in order to assign a magnitude to the quantity to be measured—conceptualized as 3(1-unit)s—it needs to be reconceptualized as the composite unit, or 1(3-unit). To be sure, one can think of the measure of a quantity as being the sum of the measures of its parts, as in finding the total area of some irregular geometric regions. But we hypothesize that even in this case the assignment of a single number and unit as the measure of a quantity implies conceptualization of the quantity as a single cognitive entity. This is consistent with the interpretation of the difference in children's responses to the directions of "count these objects" as compared to "tell me how many things there are here." In the first instance, most children count and say "1, 2, 3, 4, …, n"; in the second case they will count in the same

1. ◯　　　　　　1(1-unit)

2. ⊕　　　　　The (1-unit) is partitioned into four parts.

3. ⊕ (shaded)　　Each part is conceptualized as $\frac{1}{4}$(1 unit).

4. ⊕ (fully shaded)　Three of the 4 parts are singled out to give 3($\frac{1}{4}$-unit)s.

FIGURE 14–3. Three-fourths as *parts* of a continuous unit-whole leads to the interpretation that three-fourths is 3($\frac{1}{4}$-units).

quantity model almost one for one with steps in the diagram, so numbers from the diagram are indicated in parentheses at the end of each algebraic statement to note the correspondence.

The cognition we hypothesize as necessary to go from step 7 to step 8 in Figure 14.11 and Figure 14.13 and the cor-

1. (0 0 0 0 0 0 0 0)　　1(8-unit). We could have started with any unit with 4n objects where *n* is any natural number.

2. (0 0 / 0 0 / 0 0 / 0 0)　The (8-unit) is partitioned into 4 parts.

3. ((0 0) (0 0) (0 0) (0 0))　Each part is unitized as 1(2-unit), and a unit-of-units—1(4(2-unit)s-unit)—is formed.

4. ((●●) (●●) (●●) (●●))　Each 2-unit is $\frac{1}{4}$ of a (4(2-unit)s-unit) and can be reunitized as a unit-of-units-of-units: ($\frac{1}{4}$(4(2-unit)s-unit)-unit). Note that each (2-unit) becomes a special type of a ($\frac{1}{4}$-unit).

5. (⬭⬭⬭ (●●))　Three ($\frac{1}{4}$(4(2-unit)s-unit)-unit)s unitized to ($\frac{3}{4}$(4(2-unit)s-unit)-unit.).

FIGURE 14–4. Three-fourths as a *composite part* of a discrete unit-whole leads to the interpretation of three-fourths as ($\frac{3}{4}$(4(n-unit)s)-unit).

1. (0 0 0 0) (0 0 0 0) (0 0 0 0)　3(4-unit)s

[*] [*] [*] [*]　4[1-unit]s

2. (0 / 0 / 0 / 0) (0 / 0 / 0 / 0)　Each (4-unit) is partitioned into 4 parts.
(0 / 0 / 0 / 0)

[*] [*] [*] [*]

3. (● ● ● ●) (Ø Ø Ø Ø) (◓ ◓ ◓ ◓)　The 4 objects in each (4-unit) are each unitized as $\frac{1}{4}$(4-unit) to give four $\frac{1}{4}$(4-unit)s.

[*] [*] [*] [*]

4. ●　　●　　●　　●　The first four $\frac{1}{4}$(4-unit)s are distributed equally among the 4[1-unit]s to give $\frac{1}{4}$(4-unit)/[1-unit].
[*]　[*]　[*]　[*]

5. Ø　Ø　Ø　Ø　The second four $\frac{1}{4}$(4-unit)s are distributed equally among the 4[1-unit]s; this gives another $\frac{1}{4}$(4-unit)/[1-unit].
●　●　●　●
[*]　[*]　[*]　[*]

6. ◓　◓　◓　◓　The third four $\frac{1}{4}$(4-unit)s are distributed equally among the 4[1-unit]s, this gives a third $\frac{1}{4}$(4-unit)/[1-unit].
Ø　Ø　Ø　Ø
●　●　●　●
[*]　[*]　[*]　[*]

7. (◓ Ø ●) (◓ Ø ●) (◓ Ø ●) (◓ Ø ●)　The three ($\frac{1}{4}$(4-unit))s are accumulated to 3($\frac{1}{4}$(4-unit)-unit)s/[1-unit].
[*]　[*]　[*]　[*]

FIGURE 14–5. The partitive division of 3 ÷ 4, based on discrete quantity, leads to the interpretation that three-fourths is 3($\frac{1}{4}$-units) in one measure space per one unit of quantity from another measure space.

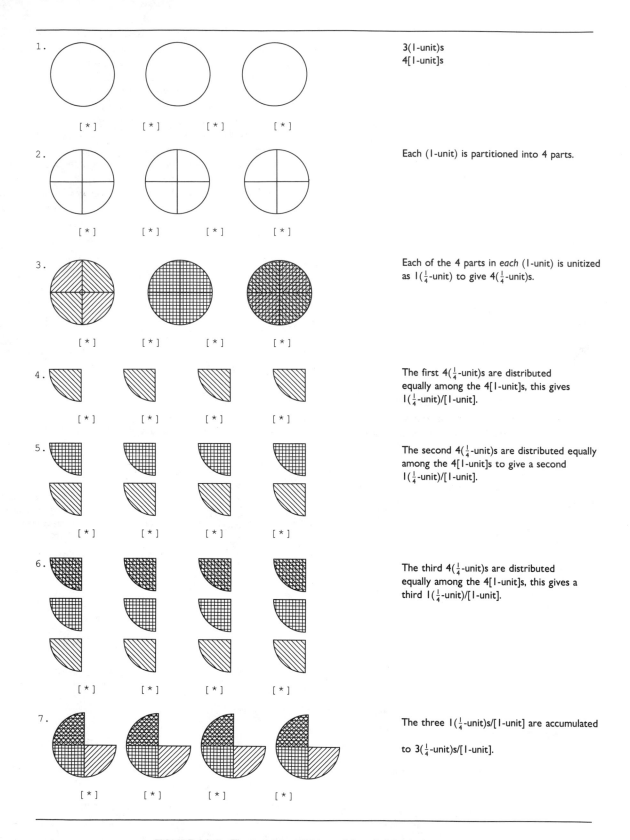

1.
$3(1\text{-unit})$s
$4[1\text{-unit}]$s

2.
Each (1-unit) is partitioned into 4 parts.

3.
Each of the 4 parts in *each* (1-unit) is unitized as $1(\frac{1}{4}\text{-unit})$ to give $4(\frac{1}{4}\text{-unit})$s.

4.
The first $4(\frac{1}{4}\text{-unit})$s are distributed equally among the $4[1\text{-unit}]$s, this gives $1(\frac{1}{4}\text{-unit})/[1\text{-unit}]$.

5.
The second $4(\frac{1}{4}\text{-unit})$s are distributed equally among the $4[1\text{-unit}]$s to give a second $1(\frac{1}{4}\text{-unit})/[1\text{-unit}]$.

6.
The third $4(\frac{1}{4}\text{-unit})$s are distributed equally among the $4[1\text{-unit}]$s, this gives a third $1(\frac{1}{4}\text{-unit})/[1\text{-unit}]$.

7.
The three $1(\frac{1}{4}\text{-unit})s/[1\text{-unit}]$ are accumulated to $3(\frac{1}{4}\text{-unit})s/[1\text{-unit}]$.

FIGURE 14–6. The partitive division of 3 ÷ 4, based on continuous quantity leads to the interpretation that three-fourths is $3(\frac{1}{4}\text{-units})$ of quantity in one measure space per one unit of quantity in another measure space.

1. $\frac{3}{4} = 3(\text{1-unit})s \div 4[\text{1-unit}]s$ (1)

2. $= 3 \times (4(\tfrac{1}{4}\text{-unit})s) \div 4[\text{1-unit}]s$ (2)

3. $= (4(\tfrac{1}{4}\text{-u})s + 4(\tfrac{1}{4}\text{-u})s + 4(\tfrac{1}{4}\text{-u})s \div 4[\text{1-unit}]s$ (3)

4. $= 4(\tfrac{1}{4}\text{-u})s \div 4[\tfrac{1}{4}\text{-u}]s + \left(4(\tfrac{1}{4}\text{-u})s + 4(\tfrac{1}{4}\text{-u})s\right) \div 4[\text{1-unit}]s$

5. $= \dfrac{1(\tfrac{1}{4}\text{-u})}{4[\text{1-unit}]s} + \left(4(\tfrac{1}{4}\text{-u})s + 4(\tfrac{1}{4}\text{-u})s\right) \div 4[\tfrac{1}{4}\text{-u}]s$ (4)

6. $= \dfrac{1(\tfrac{1}{4}\text{-u})}{[\text{1-unit}]} + 4(\tfrac{1}{4}\text{-u})s \div 4[\tfrac{1}{4}\text{-u}]s + \cdots$ (5)

7. $= \dfrac{1(\tfrac{1}{4}\text{-u})}{[\text{1-unit}]} + \dfrac{1(\tfrac{1}{4}\text{-u})}{[\text{1-unit}]} + 4(\tfrac{1}{4}\text{-u})s \div 4[\tfrac{1}{4}\text{-u}]s$ (5)

8. $= \dfrac{1(\tfrac{1}{4}\text{-u})}{[\text{1-unit}]} + \dfrac{1(\tfrac{1}{4}\text{-u})}{[\text{1-unit}]} + \dfrac{1(\tfrac{1}{4}\text{-u})}{[\text{1-unit}]}$ (6)

9. $= \dfrac{3(\tfrac{1}{4}\text{-u})s}{[\text{1-unit}]}$

FIGURE I4–7. The partitive division of $3 \div 4$ that leads to the interpretation that three-fourths is $3(\tfrac{1}{4}$-units) in one measure space per one unit of quantity in another measure space.

way, pause, and repeat "*n*." Saying "*n*" a second time in the latter case is interpreted to mean that the child distinguishes between counting as a process of establishing a one-to-one correspondence and giving the cardinality (the measure) of the collection of things. It is assumed (Markman, 1979) that cardinality is an attribute of a *collection,* not of the individual elements in the collection. Similarly, we assume measure of a quantity is an attribute of the quantity conceptualized as an entity and not an attribute of its constituent parts.

The assumption that the quantity to be measured is cognized as a composite unit introduces an additional requirement that the quantity to which it is matched in the measurement process is also conceptualized as a composite unit. Thus, in step 8 of Figure 14.11, it is suggested that the three $[\tfrac{1}{4}$-unit]s of step 7 are reunitized as $1[\tfrac{3}{4}$-unit] composite unit.

These hypotheses refer to the essential cognitive structures needed to complete the measurement process *at the level of manipulative materials.* The corresponding symbolic processing (steps 7–10, Figure 14.12) responds to two sets of constraints—to model the manipulative processing as closely as possible, and to obey the syntax rules of the symbol system. The syntax rules for handling symbolic quantity units, in particular the syntax operation of canceling units, requires that the units be the same; this is represented in step 7 of Figure 14.12. Here it is assumed that the unit represented by the ( )—that is, the print-over of (on { and of) on }—is a common unit to the units represented by ( ) and { }. This is analogous to the situation in our motivational example in which units of "apple" and "orange" were changed to the common unit of "fruit." If our motive in Figure 14.12 had been to give the briefest symbolic

1. $((0\,0\,0\,0)\ (0\,0\,0\,0)\ (0\,0\,0\,0))$
    $[^{*}]\ \ [^{*}]\ \ [^{*}]\ \ [^{*}]$
        One (3(4-unit)s-unit) 4[1-unit]s.

2. $((0\,0\,0\,0)\ (0\,0\,/\,0\,0)\ (0\,0\,0\,0))$
    $[^{*}]\ \ [^{*}]\ \ [^{*}]\ \ [^{*}]$
        The (3(4-unit)s-unit) is partitioned into two halves, each denoted by $\tfrac{1}{2}$(3(4-unit)s-unit).

3. $((0\,0\,0\,/\,0)\ (0\,0\,/\,0\,0)\ (0\,0\,0\,0))$
    $[^{*}]\ \ [^{*}]\ \ [^{*}]\ \ [^{*}]$
        Each $\tfrac{1}{2}$(3(4-unit)s-unit) is partitioned into halves, each denoted by $\tfrac{1}{4}$(3(4-unit)s-unit).

4. $((\bullet\bullet\bullet\,/\,\emptyset)\ (\emptyset\,\emptyset\,/\,\emptyset\emptyset)\ (\emptyset\,/\,\emptyset\emptyset\emptyset)$
    $[^{*}]\quad [^{*}]\quad [^{*}]\quad [^{*}]$
        Each of the four $\tfrac{1}{4}$(3(4-unit)s-unit) is matched with one [1-unit]. This results in four instances of $\tfrac{1}{4}$(3(4-unit)s-unit)/[1-unit]…

    or

5.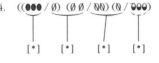
    $[^{*}]\quad [^{*}]\quad [^{*}]\quad [^{*}]$
        Each of the four $\tfrac{1}{4}$(3(4-unit)s-unit) are distributed to one of the [1-unit]s. This also results in four instances of $\tfrac{1}{4}$(3(4-unit)s-unit)/[1-unit].

FIGURE I4–0. The partitive division of $3 \div 4$ based on discrete quantity with 3 interpreted as 1(3-unit) to give an interpretation of three-fourths as one-fourth of a (3-unit) (a unit-of-units) in one measure space per one unit of quantity in another measure space: $\frac{3}{4} = \tfrac{1}{4}(3(4\text{-unit})\text{s-unit})/[\text{1-unit}]$.

sequence, step 7 could have been transformed to the following alternate step 8:

8. $3\{\text{1-unit}\}s \div \dfrac{3\{\text{1-unit}\}s}{3[\tfrac{1}{4}\text{-unit}]s}$

and then to

9. $3\{\text{1-unit}\}s \div \dfrac{3\{\text{1-unit}\}s}{[\tfrac{3}{4}\text{-unit}]}$

and then to

10. $1[\tfrac{3}{4}\text{-unit}]$.

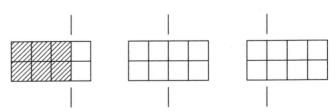

FIGURE I4–9. Partitioning of three candy bars as a (3-unit), which shows $\frac{3}{4}$ as $\tfrac{1}{4}$(3-unit).

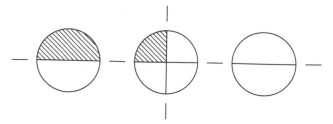

FIGURE 14–10 A partition of three circles conceptualized as 1 (3-unit) based on successive halving of units.

Note that the reconceptualization of 3{1-unit}s as 1{3-unit} is not modeled in this alternative symbolic sequence. It is in the interest of responding to both sets of constraints that the more complicated steps 8 and 9 are given in Figure 14.12.

On the Operator Construct. The operator concept of rational numbers suggests that the rational number $\frac{3}{4}$ is thought of as a function applied to some number, object, or set. As such we can think of an application of the numerator quantity to the object, followed by the denominator quantity applied to this result, or vice versa. The basic notion is that the natural-number numerator causes an extension of the quantity, while the denominator causes a contraction. The question of the nature of the extension and contraction leads to interesting variations of this construct of rational number. Accordingly, we give to the numerator and denominator the following paired interpretations:

1. Duplicator and partition reducer,
2. Stretcher and shrinker,
3. Multiplier and divisor.

The analyses of interpretations 2 and 3 that we have conducted suggest that certain conditions lead to at least two hybrid interpretations:

4. Stretcher and divisor,
5. Multiplier and shrinker.

These hybrid pairings arise because of different ways that a learner might cognize units following the application of a stretcher in interpretation 2, or following a multiplier in interpretation 3.

THE DUPLICATOR/PARTITION-REDUCER INTERPRETATION. The duplicator/partition-reducer interpretation seems the most basic and closest to the application of the concept in the domain of manipulative materials or real objects. Our ongoing analysis has given attention to both discrete and continuous quantities and to the question of the order in which the duplication and partition-reducer operators are applied. Issues of units composition and recomposition become very significant in these analyses.

THE STRETCHER/SHRINKER CONSTRUCT. There is an important conceptual and mathematical difference between a stretch of some

continuous unit, or a set of discrete objects conceptualized as a unit, and a repeat-add or duplicate of either of these units. A repeat-add or duplication are iterative actions on the entire conceptual unit. A stretch or shrink, on the other hand, acts uniformly on any subset of discrete objects or on a continuous subset of points of a continuous object to transform it into one that measures n times the original subset. This raises important considerations for providing experiences to help children conceive the stretcher/shrinker operator as one interpretation of rational numbers.

To interpret rational numbers by way of the stretcher/shrinker operator, we consider the numerator to be a stretcher and the denominator a shrinker. With symbolic representation in terms of mathematics of quantity, the outcome of applying a rational

1.	(0) (0) (0)	3(1-unit)s
	[{*} {*} {*} {*}]	$\frac{4\{1\text{-unit}\}s}{[1\text{-unit}]}$
2.	(0) (0) (0)	The measurement quantity is subunitized as $4[\frac{1}{4}\text{-units}]$.
	[{*} {*} {*} {*}]	
3.	(0) (0) (0) ↑	The "first" (1-unit) of the to-be-measured quantity is matched with the "first" sub measurement-unit (i.e., iteration of the measurement-unit is begun).
	[{*} {*} {*} {*}]	
4.	(0) (0) (0) ↗	The "second" (1-unit) of the to-be-measured quantity is matched with the "second" sub measurement-unit. Meanwhile the first matched pair is dropped from attention.
	[{*} {*} {*} {*}]	
5.	(0) (0) (0) ↑ ↗	The matchings are accumulated; i.e., both matched pairs are brought into cognitive attention.
	[{*} {*} {*} {*}]	
6.	(0) (0) (0) ↗	The "third" (1-unit) of the to-be-measured quantity is matched with the "third" sub measurement-unit. Meanwhile the first two matched pairs are dropped from attention.
	[{*} {*} {*} {*}]	
7.	(0) (0) (0) ↑ ↑ ↗	The matchings are accumulated.
	[{*} {*} {*} {*}]	
8.	((0) (0) (0)) ↑	The 3(1-unit)s of the measured quantity are-unitized to a 1(3-unit) and the 3 matched [$\frac{1}{4}$-unit] of the measurement quantity are unitized as 1[$\frac{3}{4}$-unit]. The measure of the quantity that was to be measured is 1[$\frac{3}{4}$-unit].
	[{{*} {*} {*}} {*}]	

FIGURE 14–11. Quotitive division of 3(1-units) by 4{1-units}/ [1-unit], which leads to the interpretation that three-fourths is 1[$\frac{3}{4}$-unit].

1. $\dfrac{3}{4} = 3(\text{1-unit})s \div \dfrac{4\{\text{1-unit}\}s}{[\text{1-unit}]}$ (1)

2. $= (1(\text{1-unit}) + 1(\text{1-unit}) + 1(\text{1-unit})) \div \dfrac{4\{\text{1-unit}\}s}{4[\frac{1}{4}\text{-unit}]s}$ (2)

3. $= 1(\text{1-u}) \div \dfrac{4\{\text{1-u}\}s}{4[\frac{1}{4}\text{-u}]s} + (1(\text{1-u}) + 1(\text{1-u})) \div \dfrac{4\{\text{1-u}\}s}{4[\frac{1}{4}\text{-u}]s}$ (3)

4. $= 1(\text{1-u}) \div \dfrac{4\{\text{1-u}\}s}{4[\frac{1}{4}\text{-u}]s} + 1(\text{1-u}) \div \dfrac{4\{\text{1-u}\}s}{[\frac{1}{4}\text{-u}]} + \left(1(\text{1-u}) \div \dfrac{4\{\text{1-u}\}s}{4[\frac{1}{4}\text{-u}]s}\right)$ (4)

5. $= 2(\text{1-u})s \div \dfrac{4\{\text{1-u}\}s}{4[\frac{1}{4}\text{-u}]s} + \left(1(\text{1-u}) \div \dfrac{4\{\text{1-u}\}s}{4[\frac{1}{4}\text{-u}]s}\right)$ (5)

6. $= 2(\text{1-u})s \div \dfrac{4\{\text{1-u}\}s}{4[\frac{1}{4}\text{-u}]s} + 1(\text{1-u}) \div \dfrac{4\{\text{1-u}\}s}{4[\frac{1}{4}\text{-u}]s}$ (6)

7. $= 3(\text{1-u})s \div \dfrac{4\{\text{1-u}\}s}{4[\frac{1}{4}\text{-u}]s}$ (7)

8. $= 1(\text{3-unit}) \div \dfrac{4/3(\text{3-unit})s}{4/3[\frac{3}{4}\text{-unit}]s}$ (8)

9. $= 1(\text{3-unit}) \div \dfrac{\{\text{3-unit}\}}{[\frac{3}{4}\text{-unit}]}$ (8)

10. $= 1[3/4\text{-unit}]$ (8)

FIGURE 14–12. The mathematics-of-quantity model of the quotitive division of $3 \div 4$, which leads to the interpretation that three-fourths is one ($\frac{3}{4}$-unit).

number to some unit does not change regardless of the order of applying the stretcher and shrinker; moreover, the process does not change substantially. In the domain of physical representations, on the other hand, the process does vary considerably.

THE MULTIPLIER-DIVISOR INTERPRETATION. For this interpretation we consider various meanings of *multiplier*—repeat-adder, times-as-many (or greater-than) factor, first factor in a cross-product—and also various meanings of *divisor*—partitive divisor, quotitive divisor, times-as-few (smaller-than) factor—and finally two types of quantity—discrete and continuous. Our thinking is that the times-as-many (greater-than) factor and times-fewer (less) combination is exactly the same as the stretcher/shrinker interpretation, and that the first component of a cross-product is not a possible (at least not a reasonable) interpretation. Thus, the analysis we have under way considers numerator as repeat-adder, denominator as partitive and quotitive divisor, numerator operator applied first, and denominator operator applied first.

Section Summary

This section of the chapter had several purposes. One was to call attention to critical deficiencies in the mathematics curric-

1. (0) (0) (0)
    [{* * * *}]

    3(1-unit)s
    $\dfrac{1\{\text{4-unit}\}}{[\text{1-unit}]}$

2. (0) (0) (0)
    [{* * * *}]

    The measurement quantity is subunitized as four $\frac{1}{4}$[1-units].

3. (0) (0) (0)
    [{* * * *}]

    The first (1-unit) of the quantity to be measured is matched with the first submeasurement unit (i.e., iteration of the measurement unit is begun).

4. (0) (0) (0)
    [{* * * *}]

    The second (1-unit) of the quantity to be measured is matched with the second submeasurement unit. Meanwhile, the first matched pair is dropped from attention.

5. (0) (0) (0)
    [{* * * *}]

    The matchings are accumulated; i.e., both matched pairs are brought into cognitive attention.

6. (0) (0) (0)
    [{* * * *}]

    The third (1-unit) of the quantity to be measured is matched with the third sub measurement unit. Meanwhile, the first two matched pairs are dropped from attention.

7. (0) (0) (0)
    [{* * * *}]

    The matchings are accumulated.

8. ((0) (0) (0))
    [{{* * *} *}]

    The 3(1-unit)s of the measured quantity are unitized to a 1(3-unit) and the 3 matched $\frac{1}{4}$[1-unit]s of the measurement quantity are unitized as $\frac{3}{4}$[1-unit]. The measure of the quantity that was to be measured is $\frac{3}{4}$[1-unit].

FIGURE 14–13. Quotitive division of 3(1-units) by 1{4-units}/[1-unit], which leads to the interpretation that $\frac{3}{4}$ is $\frac{3}{4}$[1-unit].

ula of elementary and middle schools. We identified five areas where the teaching of multiplicative structures is deficient:

1. Composition, decomposition, and conversion of units.
2. Operations on numbers from the perspective of mathematics of quantity.
3. Constrained models.
4. Qualitative reasoning.
5. Variability principles.

A second purpose was to suggest that research and development leading to curricular reform should be guided by an extensive content/semantic analysis of the domain of multiplicative structures or, more specifically, the portion of this domain concerned with rational number, ratio, and proportion.

A goal of our analysis is to provide a theoretical context that will guide research into the cognition underlying students' manipulations in transforming physical representations. A second goal is to associate these manipulations with the abstract representation of mathematics of quantity. We claim that our bridging notation is essentially a one-to-one map between learners' cognitive structures and the mathematics-of-quantity representation.

The initial results from this analysis suggest that the concepts in the multiplicative structures domain are inextricably interrelated and exceedingly complex. The purpose of a content/semantic analysis is to make use of the knowledge gained from research into the knowledge structures that children form in cognizing concepts in the content domain and to make assumptions about the necessary cognitive structures where research is lacking. These assumptions suggest further research issues and questions. For example, numerous assumptions were made in the demonstrations about unit formation and reformation. Can these formations and reformations be conceptualized by children? Research needs to address the issue of whether situations can be developed that will help children construct implicit and intuitive knowledge about unit conversion principles first, and then explicit knowledge of them. Analyses of rational number constructs suggest that their understanding depends on a grasp of these unit conversion principles as well as rather deep knowledge about concepts of measurement.

We argue that future research and curriculum development should be based on and improve our analysis of multiplicative structures. A great deal of effort will be necessary to develop situations from which children can construct knowledge about these ideas. For example, research has given us some information about children's ability to partition both discrete and continuous quantity (Hunting, 1986; Kieren & Nelson, 1981; Pothier & Sawada, 1983), but little is known about instructional situations that might facilitate children's ability to partition.

PRINCIPLES FOR QUALITATIVE REASONING IN FRACTION AND RATIO COMPARISON

Mathematical Variability—A Fundamental Issue

The flexibility of thought that Resnick (1986) mentions as a characteristic of intuitive knowledge is readily observable in children's protocols involving intuitive reasoning. (Harel & Behr, 1988; Kieren, 1988; Pirie, 1988; Post, Wachsmuth, Lesh, & Behr, 1985; Resnick, 1982, 1986). One can observe a flexible interaction among different levels of representation or understanding (Kieren, 1988)—for example, between thought directed at actual manipulative aids, or mental images of the manipulation, and an oral or written symbolic representation of the quantities involved. Moreover, while one does not often see explicit reference to mathematical principles (for example, principles of place value [Resnick, 1986]) in these protocols, consistent behavior and uniform solution strategies suggest understandings that can be abstracted from the protocols in the form of mathematical principles. Germane to this section are the numerous observations that suggest an implicit awareness of the invariance, or a compensation for variation, of the value of a quantity under certain transformations. Awareness of invariance (or variation) or the search for invariance under certain transformations is a central concept, almost a defining concept, of mathematical thinking. Invariance and compensation for variation are basic to many areas of elementary mathematics—the development of basic fact strategies (Carpenter & Moser, 1982), children's invention and use of alternative computation algorithms (Harel & Behr, 1988; Pirie, 1988; Resnick, 1986), as well as fraction equivalence and proportion problems (Lesh et al., 1988).

Research on rational-number learning suggests the importance of fraction order and equivalence to the understanding of a rational number as an entity (that is, as a single number) and to the understanding of the size of the number (Post et al., 1985; Smith, 1988). Fundamentally, the question of whether two rational numbers are equivalent or which is less is a question of invariance or variation of a multiplicative relation (Lesh, et al., 1988). Two rational numbers a/b and c/d, can be compared in terms of equivalence or nonequivalence by investigating whether there is a transformation of a/b to c/d, defined as changes from a to c and from b to d, under which the multiplicative relationship between a and b is or is not invariant. In the current mathematics curriculum, the issue of fraction order and equivalence is treated as an isolated topic rather than as a special instance of mathematical variability. We view the concepts of fraction order and equivalence and proportionality as one component of this very significant and global mathematical concept. Fraction equivalence can be viewed in the context of the invariance of a multiplicative relation between the numerator and denominator, or as the invariance of a quotient.

Research and development must address the need to provide children in early elementary grades with situations that involve variability. There are two issues here. First, children require adequate experience to understand what is meant by the concept of change, or difference. For example, the question of what change in 4 will result in 8 is one that very young children can deal with. Moreover, as early as possible, children should be brought to understand that the change in 4 to get 8 (or the difference between 4 and 8) can be defined in two ways: additively (with an addition or a subtraction rule) or multiplicatively (with a multiplication or a division rule). The additive rule for changing 4 to 8 is to "add 4," while the multiplicative rule is to "multiply by 2." The additive-change rule to get 8 from 4 is the same as to get 17 from 13; this is not true for multiplication. The second issue is the investigation of change and the direction of change in additive and multiplicative relationships and operations under transformation on the components in the relation or operation. While the additive relation between 4 and 8 is invariant under the transformation of adding 9 to both 4 and 8, this is not true for the multiplicative relation between the same two numbers. The ability to represent change (or difference) in both additive and multiplicative terms and to understand their behavior under transformation is fundamental to understanding fraction and ratio equivalence.

The mathematics curriculum needs to develop learning situations, problems, and computation that will help children develop at least an implicit understanding of the principles that underlie mathematical invariance. That is, the curriculum should provide school experience to help children construct intuitive knowledge about fraction and ratio equivalence. Situations must be developed in which children systematically build their understanding of principles that underlie the invariance and the compensation for variation within additive, subtractive, multiplicative and divisive relations and operations. The goal is to help children construct these principles as "theorems in action" (Vergnaud, 1988).

Qualitative Reasoning. In our research on children's thinking strategies applied to fraction and proportion tasks, we became interested in children's ability to reason qualitatively about the order relation between fractions and between ratios. Our interest was in children's use of qualitative reasoning about situations modeled by $a/b = c$, including reasoning such as, if a stays the same and b increases then a/b decreases, or if both a and b increase, then qualitative methods of reasoning are no longer adequate to determine whether a/b increases, decreases, or stays the same in value, and quantitative procedures are necessary.

The concern in qualitative reasoning about a situation modeled by the equation $a/b = c$ is to determine the direction, as opposed to the amount, of change (or no change) in c as a result of information about only the direction of change (or no change) in a and b, or to establish that the direction of change is indeterminate based only on qualitative reasoning. Our thinking about the role of qualitative reasoning was influenced by the work of Chi and Glaser (1982), who found that expert problem solvers are known to reason qualitatively about problem components and relationships among them before attempting to describe these components and relationships in quantitative terms. Consensus is that experts' reasoning about a problem leads to a superior problem representation that enables the expert to know when qualitative reasoning is inadequate and quantitative reasoning is necessary.

It is not the case that experts use qualitative reasoning and novices do not; rather, experts' qualitative reasoning is based on scientific principles and involves formation of relationships among problem components based on these principles. Novices, on the other hand, reason about the surface structure of the problem. It appears that the qualitative reasoning to which Chi and Glaser refer is not unlike the schooled intuitive reasoning to which Kieren (1988) refers.

The issue in this section is to present basic principles upon which to base intuitive knowledge about fraction order and equivalence and proportional reasoning. They are principles that need to become self-evident to children through experience. These principles should then provide the basis for flexible application of additive and multiplicative notation and transformation rules to the solution of problems involving fraction or ratio equivalence (Resnick, 1986).

There are two aspects to our analysis of qualitative reasoning as applied to a problem involving questions of fraction order or equivalence or the proportionality of two ratios. One aspect of the problem is *determinability*: Can the order relation requested in the problem be determined through qualitative reasoning? The second aspect concerns *determination*: What is the order relation requested in the problem, if it can be determined? These questions are discussed in the next two sections. The discussion is a recapitulation of an analysis in Harel et al. (in press).

Principles for Solving Multiplicative Tasks

Two Categories of Multiplicative Tasks. In Harel et al., (in press) we showed that, from the perspective of invariance of relations, proportion tasks can be classified into two broad categories: invariance of ratio and invariance of product. The orange juice task used by Noelting (1980) and the balance beam task used by Siegler (1976) are representatives of these two categories, respectively. In the orange juice task, one of two pairs of ratios—amount of water to amount of orange concentrate, or amount of water to total amount of mixture—are compared across two mixtures to determine which one tastes the more orangy or if they taste the same. In the balance task, the expected solution procedure is to compare the products of the values of the distance and the weight of objects on each side of the fulcrum to determine which side of the balance beam goes down. Strictly speaking, a task would be classified according to the way a given subject solved it. For the sake of discussion we classify these two tasks according to expected solution procedures; most subjects do solve the tasks as expected.

Quite different mathematical principles, and thus different reasoning patterns, are involved in the solution of the invariance-of-ratio and the invariance-of-product tasks. To describe principles forming the basis for qualitative reasoning about these two types of tasks, we introduce a refinement of each type. We identify two subcategories of the invariance-of-product category and illustrate the subcategories with hypothetical balance beam or orange juice tasks.

FIND-PRODUCT-ORDER SUBCATEGORY. This subcategory consists of problems in which order relations between values of corresponding task quantities are given, and these order relations form the factors of the two products. These problems ask about the relation between two values of the quantity represented by the product. For example, let a_1 and a_2 denote the weights of objects placed on each side of the fulcrum of a balance beam, and let b_1 and b_2 represent their respective distances from the fulcrum. Further, suppose it is given that $a_1 < a_2$ and that $b_1 = b_2$. Which side of the fulcrum will go down? That is, the task is to determine the relationship between the products $k_1 = a_1 \times b_1$ and $k_2 = a_2 \times b_2$. We call attention to the fact that the information given in this task is about the directionality of the order relations, and it need not include specific numerical values for the weights and distances. This is characteristic of a qualitative proportional reasoning task. Different forms of the task can be formulated by taking all possible combinations of the three order relations between a_1 and a_2 and the three possible order relations between b_1 and b_2. The order relation between k_1 and k_2 will in some cases be indeterminate through qualitative reasoning alone.

FIND-FACTOR-ORDER SUBCATEGORY. This category consists of problems in which an order relation is given between values of quantities represented by two products and an order relation is given between two corresponding factors in the two products. The problem asks about an order relation between the other two factors. If in the example above it were given that $a_1 < a_2$ and $k_1 = k_2$, where $k_1 = a_1 \times b_1$ and $k_2 = a_2 \times b_2$ and the question was about the order relation between b_1 and b_2, then this would be an exemplar of this subcategory.

Likewise, there are two subcategories of the invariance-of-ratio category.

FIND-RATE-ORDER SUBCATEGORY. This subcategory consists of problems that give two order relations between values of corresponding quantities in two rate pairs and ask about the order relation between the values of the quantities represented by the two rates.

An example can be formulated from the orange juice context of Noelting (1980). If two orange juice mixtures (1 and 2) are made from amounts of water a_1 and a_2 with $a_1 < a_2$ and amounts of orange concentrate b_1 and b_2 with $b_1 = b_2$, which of the two mixtures, 1 or 2, tastes the more orangy, or do they taste the same? The decision of which is more orangy would be based on the order relation between the two ratios a_1/b_1 and a_2/b_2, or between $a_1/(a_1 + b_1)$ and $a_2/(a_2 + b_2)$.

FIND-RATE-QUANTITY SUBCATEGORY. This subcategory consists of problems that give an order relationship between the value of quantities represented by two rates and an order relation between the values of two corresponding quantities in the two rate pairs, the problem asks about the other order relation between the values of the two corresponding quantities in the two rate pairs.

Again, an example can be formulated from the orange juice context. If two orange juice mixtures (1 and 2) were made so that mixture 1 tastes more orangy than mixture 2 (that is, $a_1/b_2 > a_2/b_2$) and the amount of orange concentrate in mixture 1 is greater than the amount of orange concentrate in mixture 2 (that is, $a_1 > a_2$), then which of the mixtures has more water, or do they both have the same amount?

Multiplicative Determinability and Determination Principles. Each of the four subcategories of problems involves two number pairs, *a* and *b*, *c* and *d*, and either a pair of products $a \times b$ and $c \times d$ or a pair of ratios a/b and c/d. For each of the three pairs in a given problem there are three possible order relations; the structure of the problems in the subcategories above is that the order relations between two of the three pairs are given and the problem is to (a) decide if the third order relation is determinable from the given information and if so (b) to ascertain what that order relation is. The knowledge required to solve these kinds of tasks relies on principles that we have placed into two categories—*multiplicative determinability principles* and *multiplicative determination principles*. The multiplicative determinability category consists of principles that specify the conditions under which order relations between factors in the product of the values of two quantities can lead to declaring whether the order relation between the values of these quantities is determinate or indeterminate (for example, if *a* and *b* are equal but *c* and *d* are unequal, then the order relation between the products $a \times c$ and $b \times d$ is determinate). The multiplicative determination category consists of principles that specify the conditions under which order relations between factors of two quantities can lead to declaring that the relation between the two quantities is less than, greater than,

or equal to (for example, if *a* and *b* are equal but *c* is greater than *d*, then the relation that holds between a/c and b/d is that $a/c < b/d$).

The principles in the first class of multiplicative determinability category give information about the determinability of the order relation between products. When order information is given about corresponding factors within each product, we refer to these as *product composition* (PC) principles.

PC1. The order relation between the products $a \times c$ and $b \times d$ is *determinate* if the order relation between *a* and *b* is the same as between *c* and *d* or if one of them is the equal-to relation.

PC2. The order relation between the products $a \times c$ and $b \times d$ is *indeterminate* if the order relation between *a* and *b* conflicts with (is in the opposite direction of) the order relation between *c* and *d*.

We offer some examples, in this case from the balance beam context, to illustrate problems in which these principles are the mathematical formulation of the knowledge needed to solve the problem.

A classmate knows that Billy is heavier than Jane, and he tells Billy to sit farther from the center of a teeter-totter than Jane. Before he has Billy and Jane exert the force of their weight onto the teeter-totter, he tells the class about the relationship between their weights and then asks the class to tell which end of the teeter-totter—Billy's or Jane's—will go down. In this case the direction of movement of the teeter-totter, off horizontal is indeterminate from the information given. To be able to determine the order relation of the two angular moments, numerical data on Billy and Jane's weight and distance would be needed and the moments would need to be calculated.

Another example: A classmate tells the class that Billy weighs 90 lbs and Jane 80 lbs. He has Billy sit 4 ft from the center of the teeter-totter and Jane 3 ft; he gives everyone in the class this information and asks which side of the teeter-totter will go down. In this case (a) the direction of the order relation between moments (that is, the order relation between the weight and distance products) is determinate and (b) Billy's side will go down. Notice that the directionality of the order relations of the respective weights and distances is sufficient to determine the order relation between the moments; there was not a need to compute the exact moments to determine this order relation. This is what characterizes the solution as qualitative or intuitive. If for some reason, say by teacher direction, calculation of the moments was required, and if the solver actually compared these computed values to determine the order relation, then the solution would be characterized as quantitative. Nevertheless, one can see how a prior qualitative analysis of the problem could guide the quantitative calculations and serve as a check on them.

We also state principles on which knowledge is based to decide the determinability of the order of one pair of factors of a product when the order relation between the other two factors and the product are given. We refer to these principles as *product decomposition* (PD) principles.

PD1. The order relation between the factors *a* and *b* in the products $a \times c$ and $b \times d$ is *determinate* if the order relation between the factors c and d is in the opposite direction from the order relation between the products $a \times c$ and $b \times d$ or if one of the order relations is the equal-to relation.

PD2. The order relation between the factors a and b in the products $a \times c$ and $b \times d$ is *indeterminate* if the order relations between the other two factors c and d, and between the products $a \times c$ and $b \times d$ are the same but neither is the equal-to relation.

An example where PD1 can be applied, constructed from the balance beam context: A classmate has Billy and Jane sit on a teeter-totter so that Jane's side would go down. Billy is heavier than Jane. Is Jane's distance from the center equal to, less than, or greater than Billy's distance from the center? In this case the requested order relationship is (a) determinate and (b) Jane's distance from the center is greater than Billy's. The structure of this problem in terms of the stated principles is that the order relationship between the product of weight and distance from the center is given for both Billy and Jane and the order relation between their weights is given. The question is about the relationship between the other corresponding factors—distances from the center—in the two products. Note that, if the problem is changed so that Billy's weight is less than Jane's, then the order relationship between the distances is indeterminate based on the given qualitative data.

Two *ratio composition* (RC) principles make up the third category of determinability principles.

RC1. The order relation between the ratios a/c and b/d is *determinate* if the order relations between a and b is in the opposite direction from the order relation between c and d, or if one of them is the equal-to relation.

RC2. The order relation between the ratios a/c and b/d is *indeterminate* if the order relation between a and b is the same as the order relation between c and d, but neither of them is the equal-to relation.

We offer an example constructed from the orange juice context: If the amount of water in mixture 1 is less than in mixture 2 and the amount of orange concentrate in mixture 1 is less than in mixture 2, then the relationship between the orangy tastes of mixtures 1 and 2 is indeterminate (that is, the relationship between the two water-to-concentrate ratios is indeterminate).

On the other hand, if the amount of water in mixture 1 is less than in mixture 2 and the amount of orange concentrate in mixture 1 is greater than in mixture 2, then (a) the order relation between the two water-to-concentrate ratios is determinate and (b) the water-to-concentrate ratio for mixture 1 is greater than for mixture 2, so mixture 1 tastes more watery and less orangy.

Once a determinability principle is applied and the requested order relation is found to be determinable, then a determination principle can be applied to ascertain whether that relation is the less-than, equal-to, or greater-than relation. There is a determination principle that corresponds to the determinability principles; we denote them by [PC1], [PD1], [RC 1], and [RD1]. For example, [PD1] is as follows: if $c \geq d$ and $a \times c < b \times d$ then the order relation between a and b is $a < b$.

TABLE 14-1. Effect on k, the change between $k_1 = a_1/b_1$ ($k_1 = a_1 \times b_1$) and $k_2 = a_2/b_2$ ($k_2 = a_2 \times b_2$) from change between a_1 and a_2 (Δa) and between b_1 and b_2 (Δb).

	Δa		
Δb	+	0	−
+	? (+)	−(+)	− (?)
0	+(+)	0(0)	−(−)
−	+ (?)	+(−)	? (−)

Note. The entries in each of the 18 cells are qualitative values of Δk. The ? means the value of Δk is indeterminate with qualitative reasoning; more precisely the ? means that the value of Δk might be any one of +, −, or 0. The entries in the parentheses refer to invariance of products; the others refer to invariance of ratio.

The entire set of determinability and determination principles can be easily summarized into a concise table as shown in Table 14.1. Success on problems from the find-rate and find-product subcategories can be achieved by reasoning as to how a qualitative change in a_1 to get a_2 and b_1 to get b_2 affects the size and thus the comparison of k_1 and k_2, where $k_1 = a_1 \times b_1$ (or a_1/b_1) and $k_2 = a_2 \times b_2$ (or a_2/b_2). The changes in a_1 to get a_2 and b_1 to get b_2, and k_1 to get k_2 can be denoted by Δa, Δb, and Δk, respectively, and the qualitative value (or directionality) of these changes can be denoted by +, 0, or − according to whether the change is an increase, no change, or decrease, respectively. In Table 14.1, selecting a pair of values, one from the vertical and one from the horizontal axis, and locating the corresponding value in the body of the table gives information about how qualitative changes in rate- or fraction-quantities (factors of a product, parentheses in text here correspond to parentheses in Table 14.1) affect the qualitative value of the rate or fraction (product). Note that the question marks in the body of the table indicate that the value of Δk is indeterminate, which means that instances can be found in which Δk is +, others for which it is −, and still others for which it is 0. Thus, each of the question marks could be replaced either mentally—or physically in the table—by a disjunctive listing of the three possibilities (+, −, or 0). In this way, Table 14.1 describes the knowledge needed to solve find-rate and find-product problems based on the qualitative relationships among the problem quantities.

Selecting a pair of values, one from an axis and another within the body of the table along the row or column of the first, and then locating the corresponding value on the other axis, Table 14.1 describes the knowledge about the qualitative relationships among problem quantities needed to solve problems from the find-rate-quantity and find-product subcategories. We give two examples to illustrate the information in the Table 14.1.

We have two fractions and we know that the numerator of the first is greater than the numerator of the second (that is, the change from the first to the second is − indicating a decrease) and that the two fractions are the same size (the change in value of fraction 1 to get fraction 2 is 0, that is, no change). Is the denominator of the first fraction less than, greater than, or equal to the denominator of the second fraction? We can analyze the problem in this way: The change from numerator 1 to numerator 2 is − (a decrease in the numerator), find − on the horizontal axis of Table 14.1. The change from fraction 1 to fraction 2 is 0 (no change), so we look for 0 in the body of the table in the column under −. Since the *only* 0 in this column appears implicitly, under the guise of ?, we move along that row to the vertical axis and conclude that the direction of change from denominator 1 to denominator 2 is a decrease. Again we note that it is only the direction of change that we determine from the table, not the amount of change.

Next consider the situation of having two products, we know that one factor decreases and that the product decreases. What happens to the other factor? Using the table, we notice that under the column for a decrease in the first factor, a minus sign (−), indicating a decrease in the product appears three times: twice explicitly, and once under the guise of an indeterminate (?). Thus the change in the second factor can be any one of +, −, or 0 because any one of three (horizontal axis entry, table entry, vertical axis deduction) is possible—(−, −, +), (−, −, 0), and (−, −, −).

Now suppose we have two products, product 1 and product 2, and that the change in the first factor of product 1 to the first factor of product 2 is an increase, and the same for factor 2. What is the order relation between product 1 and product 2? We find + on the left of the table, and + on the top; the qualitative change in the product, +, is given in the body of the table at the intersection of the row and column in parentheses, (+). If both factors of a product increase, then the product increases as well.

Section Conclusion

Qualitative reasoning is known to be a significant variable in problem-solving performance. Expert problem-solvers are known to reason qualitatively about problem components and relationships among them before attempting to describe these components and relationships in quantitative terms (Chi & Glaser, 1982). The consensus of research is that an expert's reasoning about a problem leads to a superior problem representation because it contains numerous qualitative considerations about problem components and their interactions (Chi et al., 1981; Chi & Glaser, 1982; Chi, Glaser, & Rees, 1982). While some tasks used in traditional studies of the proportion concept make it possible for subjects to solve the task using qualitative reasoning (for example, Siegler & Vago, 1978; Siegler, 1976; Karplus, Pulos, & Stage ,1983; Noelting, 1980), it has only been very recently that qualitative reasoning has become an object of study in this area of research.

Studies on Qualitative Rational-Number and Ratio Reasoning. Harel, et al. (in press) compared the tasks used by the researchers cited in the conclusion above using several criteria, including whether the tasks were solvable by qualitative reasoning alone. They found that some variations of the balance scale task (Siegler, 1976) were solvable by qualitative reasoning; for example when there is more weight on one side of the fulcrum and the weight on the other side is further from the fulcrum. Noelting (1980) gave 23 orange concentrate and water mixture tasks, of which only two were solvable by qual-

itative reasoning; for example, those where the differences in amounts of orange concentrate and water between two mixtures could be defined as (+, −) and (−, +) according to Table 14.1. All of the other tasks were of the form (+, +) or (−, −). The fullness tasks given by Siegler and Vago (1978) require quantitative reasoning.

The question of whether children can use qualitative reasoning in solving fraction-equivalence and proportion problems has been investigated by the Rational Number Project. A study reported by Heller, Ahlgren, Post, Behr, and Lesh (1989) investigated seventh-grade children's performance on numerical (quantitative) and qualitative problems, using both missing-value and comparison-type problems. Examples of qualitative missing-value and comparison-type problems, respectively, follow.

If Cathy ran less laps in more time than she did yesterday, her running speed would be: faster, exactly the same, slower, there is not enough information to tell?

Bill ran the same number of laps as Greg. Bill ran longer than Greg. Who was the faster runner: Bill, Greg, they ran exactly the same speed, there is not enough information to tell?

A conclusion drawn from their work was that qualitative reasoning is helpful (not completely necessary) but certainly not sufficient for successful performance on quantitative proportional-reasoning problems. They found that some subjects' performance on the qualitative problems was low, while their performance on quantitative proportion problems was high. They concluded that quantitative proportional-reasoning problems can be solved without good qualitative proportional reasoning by applying memorized procedures. They suggest that intuitive understanding of the direction of change (qualitative understanding) in the value of a ratio or fraction should precede quantitative exercises.

Another study reported by Larson, Behr, Harel, Post, Lesh (1989) directly investigated qualitative-reasoning ability among seventh-grade children. The tasks required the child to determine whether the value of a fraction or ratio would increase, decrease, or stay the same under given changes in the numerator or denominator as follows: Increase(I)/Decrease(D), D/I, D/D, Same(S)/I, and I/S. Based on the protocol data collected, numerous unsuccessful qualitative-reasoning strategies were identified, and several successful ones as well. We call attention to some of the more interesting strategies. One category of strategies, observed on tasks in which the changes in the numerator and denominator were in the same direction, suggest that some children believe the value of a fraction or ratio changes in the same direction as the changes in its two components. Another category of responses suggests that the child believes a greater change in one of the two components will result in some change in the value of the fraction or ratio. A refinement of this strategy by some children is that the direction of change in the value of the fraction or ratio is in the same direction as the directional effect of the component that has the greater amount of change. For example, if it is given that both the numerator and denominator increase, the child may reply that it depends on which one changes most; if the denominator increases most, then the value of the fraction or ratio will decrease. Some children mulled over a change in the value of the fraction or ratio in a sort of composition-of-effects strategy. For example, given that the numerator increases and the denominator decreases, they would reason that increasing the numerator causes an increase in the value of the fraction, decreasing the denominator causes an increase in the value of the fraction, the two increases together result in an increase. These results were obtained at different times during a teaching experiment, and they suggest that instructional situations can lead children to reason qualitatively about the size of fractions and ratios. It should be noted that these children, having attained seventh grade, had already learned other quantitative strategies for comparing fractions and ratios; a certain amount of unlearning of existing incorrect strategies was necessary before the new strategies could be successfully applied. Greater success might be possible with a younger child whose thinking about fractions and ratios was less affected by prior knowledge of quantitative strategies.

In a third study (Harel et al., in press) a "blocks task" was developed specifically to investigate children's qualitative reasoning in a proportion context. The task involved two pairs of composite blocks (A, B) and (C, D) with A and C constructed from the same kind of smaller (unit) blocks, which were nevertheless larger than those used to construct B and D. The number of unit blocks in A was less than the number of unit blocks in C, and these numbers remained constant across all variations of the blocks tasks used in the study. Three instances of blocks B and D were used in which the number of unit blocks was one less, the same, or one more, compared to A and C, respectively. Given information about the weight relationship between and A and B (<, =, >) in the context of a visually observable number relationship among the four composite blocks, seventh-grade children were asked to determine the weight relationship between C and D. Of 27 possible task variations, 9 were used in the study. Of particular interest in this study were the type of problem representations that children formed in response to the problem presentation, the solution strategies that were used, and the relationship between the sophistication of the problem representation and the solution strategy. Hierarchies of three problem representations and three categories of solution strategies with two solution strategies in the high category, three in the middle, and one in the low category were identified. A high correlation was found between the level of the problem representation and the level of solution strategy. By matching a type of problem representation with its most highly correlated solution strategy, a hierarchy of six solution *processes* was identified. A high correlation was found between the level of solution process and the level of the students' mathematics ability.

Van den Brink and Streefland (1979) give evidence that children as young as 6 and 7 years old have intuitive knowledge about ratios and proportions that suggests implicit knowledge of the principles given in this section. They tell of a child who, during a discussion with his father about how the propeller of a ship works, looked at a toy boat, referred to a picture in his room of a large sea-going ship with a man standing near the propeller, and inquired of his father how big the propeller on such a ship really is. "It wouldn't fit into your room," an-

swered the father. After some moments of reflection the child responded,

It is true. In my book on energy is a propeller like this (shows a distance of about 3 cm between thumb and forefinger) with a little man like that (about 1 cm).

The authors indicate that the child compared the relationship between the seagoing ship's propeller (bigger than the boy's room) and his father to the picture of the big ship with a man beside the propeller. By qualitatively maintaining an invariant relationship between man and propeller, the child confirms his father's statement that the ship's propeller would be bigger than his room.

In her dissertation, Lamon (1989) identified 16 proportion problem types by crossing four problem dimensions with four semantic categories. The problem dimensions were relative/absolute change, recognization of ratio-preserving mappings, covariance/invariance, and construction of ratio-preserving mappings; the semantic categories were well-chunked measures, part-part-whole, unrelated sets, and stretchers/shrinkers. She notes that recognizing relative change is likely an important prerequisite to moving beyond additive relationships into multiplicative structures. Multiplicative relationships arise from an evaluation of the size of a change by considering its relationship to the starting values and not merely in terms of its absolute amount. The problem dimensions of relative/absolute change and covariance/invariance are particularly germane to the issues raised in this section. She found that the performance of sixth-grade students on relative/absolute change across the semantic categories (in the order given above) was 6.8%, 57.4%, 83.3%, and 18.9%. Performance of sixth-grade students on the covariance/invariance problem dimension across the four semantic types was 72.3%, 79.5%, 60.5%, and 50.0%. Tasks on this dimension involve ability to recognize variability or invariance of the relationship between the two components of a ratio (or fraction) under change in each component.

Teaching for Qualitative Reasoning. While the evidence to date is sketchy, there is a trend in the direction of supporting the notion that qualitative knowledge can be constructed through school situations. Moreover, though the evidence is again slight, a reasonable hypothesis seems to be that the qualitative knowledge an individual has about a situation is similar to what Resnick and Kieren have referred to as intuitive knowledge. It is knowledge that "belongs" to the individual, is constructed from real experience, and provides for considerable flexibility in thought. The work of Chi and her colleagues gives an important reason for stating principles on which to base qualitative reasoning about proportional situations. A characteristic of experts' qualitative reasoning as compared to the qualitative reasoning of novices is that it proceeds at a semantically deep level and incorporates principles of the content domain. The qualitative reasoning of novices, on the other hand, is at the surface level and is directed at comparison of formulas and procedures for attempting to isolate the problem unknown. The aim of initial instruction in rational numbers and proportions should be to put children in situations where they are able to construct principles to apply qualitatively to questions of order, equivalence, and size of fractions and ratios. The objective of helping children construct principles for qualitative reasoning is based on the belief that this knowledge can guide their quantitative thinking, as it does for experts, in a content domain.

Research needs to determine how knowledge of principles for qualitative thinking and the ability to think qualitatively will help children make connections between this intuitive, informal knowledge and the symbol system that is the basis for quantitative methods. While work in this area has advanced in the domain of early numbers, very little has been done in the area of rational numbers and proportions. Recent work (Mack, 1990) represents a beginning. Mack points out that numerous studies demonstrate that children possess a store of informal knowledge about fractions. The issue she addresses is whether instruction can build on this informal knowledge in a way that extends it to, or connects with, the formal system of fraction notation. An instructional move in her work that seems to help children make this connection is to give a child a problem in symbolic form and ask the child to reason about it in terms of a real situation. Mack (1990) reports that while children were able to build on their informal knowledge, the results also suggest that knowledge of rote procedures interferes with their attempts to construct procedures that are meaningful to them. Van den Brink and Streefland (1979) give suggestions for the type of instruction that would help children develop intuitive or qualitative knowledge about the principles stated in this section. We give an example of an instructional situation that they suggest. A story is told to the class about Liz Thumb, who once upon a time became as small as a thumb. Liz Thumb is pictured on a worksheet, and students are asked to draw common objects *into the picture*—a flower, a stone, one of the child's own shoes, and the like. These drawings will demonstrate children's conceptions of ratio, and discussion among the students and the teacher can help the class to come to agreement about how big the drawn objects should be and why.

The Semantic Analysis of Rational Number: Some Implications for Curriculum

In section 2 of this chapter, "Rational Number Construct Theory: Toward a Semantic Analysis," we showed that any rational number x/y can be interpreted in any one of four ways x/y(1-unit), x(1/y-unit), 1/y(x-unit), and 1(x/y-unit). Another analysis we have underway deals with operations on rational numbers from the perspective of mathematics of quantity. Each of the four different interpretations of rational numbers can be shown to be embedded in real world, or at least in conceivable textbook word problems. Our analysis of rational-number operations from the perspective of mathematics of quantity has progressed most with the operation of division. We will use problem examples in this domain to illustrate the richness of problem situations and alternative problem representations, as well as solution procedures that arise from the analysis.

Division Problem Examples. We will present one problem with some analysis of its solution from the perspectives of the embedded rational-number interpretations and the mathematics of quantity involved. In addition, we will present other problems to represent their possible range, but we give only the mathematical model that could be used to solve them and the interpretations of rational number that are represented. The numbers are the same in each of the example problems to make comparison among models and rational-number interpretations easier. We will give two solutions of a problem, based on different problem representations. We call attention to the fact that solving problems such as these *depends on an understanding of the different interpretations of rational numbers and on the different unit-conversion principles;* these problems are not expected to facilitate initial learning of these notions. They would provide situations in which they can be applied in order to deepen the understanding, but knowing the rational-number interpretations listed in the opening sentence of this section and the unit-conversion principles listed earlier are considered prerequisite knowledge to problem interpretation and representation.

The following problem is presented to illustrate.

Bob mixed 6 tubes of paint. He used $\frac{1}{8}$ of the mixture to paint 2 pieces of wood, each having an area of $\frac{1}{8}m^2$. How many tubes would he need to paint $1m^2$?

We can interpret this problem in more than one way. Traditionally, it would be interpreted as a multistep problem involving multiplication and division. Using this interpretation and notation, which is analogous to that which we used in section 2 for the mathematics of quantity, the solution of the problem could be as follows: First, identify the problem quantities as 6(1-tube)s of paint in 1(1-mixture), 2(1-piece)s of wood, and $\frac{1}{8}$(1-m²) measure of each piece of wood. The question is to find the number of (1-tube)s per each (1-m²), then needed computations are carried out,

1. $\dfrac{6(1\text{-tube})s}{(1\text{-mixture})} \times \frac{1}{8}(1\text{-mixture}) = \frac{6}{8}(1\text{-tube})$

2. $2(1\text{-piece})s \times \dfrac{\frac{1}{8}(1\text{-m}^2)}{(1\text{-piece})} \quad = \frac{2}{8}(1\text{-m}^2)$

3. $\frac{6}{8}(1\text{-tube}) \div \frac{2}{8}(1\text{-m}^2) \quad = \dfrac{(\frac{6}{8} \div \frac{2}{8})(1\text{-tube})}{(1\text{-m}^2)}$

4. $\quad = \dfrac{3(1\text{-tube})s}{(1\text{-m}^2)}$

We offer some observations about this problem representation and solution. Rather than there being a holistic problem representation, the components of the problem are represented separately and, later in the third step of the solution, the solver must remember or determine (depending on whether he has a problem plan) how these two components go together. Moreover, the solver must remember or determine that the quantities computed in steps 1 and 2 must be divided and must decide which is the dividend and divisor, respectively. In the tree diagram that we give later for this problem, the solution would be considered bottom-up rather than top-down. This solution represents all the problem quantities in terms

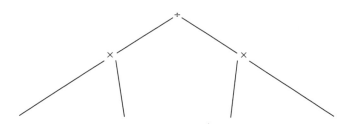

FIGURE 14–14. A hierarchical tree-structure for problem I.

of singleton units. Does this give a relatively strong cognitive representation of the problem, or would other unitizations of the problem quantities lead to a more powerful representation? The solution process illustrated might be characterized as first interpreting units, then converting units to units of one, followed by partitive division.

With an alternative interpretation of the units, it is a partitive division followed by units conversion. That is, if we identify the given quantities as $\frac{1}{8}$(6-tube) and $2(\frac{1}{8}$-m^2)s, it remains to find the number of (1-tube)s per each (1-m^2). The steps in the solution based on this problem representation would be as follows:

$$1. \quad \tfrac{1}{8}\text{(6-tube)} \div 2(\tfrac{1}{8}\text{-m}^2) = \frac{(\tfrac{1}{8} \div 2)\text{(6-tube)}}{(\tfrac{1}{8}\text{-m}^2)}$$

$$2. \qquad\qquad = \frac{\tfrac{1}{16}\text{(6-tube)}}{(\tfrac{1}{8}\text{-m}^2)}$$

$$3. \qquad\qquad = \frac{\tfrac{48}{16}(\tfrac{1}{8}\text{-tube})\text{s}}{(\tfrac{1}{8}\text{-m}^2)}$$

$$4. \qquad\qquad = \frac{3(\tfrac{1}{8}\text{-tube})\text{s}}{(\tfrac{1}{8}\text{-m}^2)}$$

$$5. \qquad\qquad = \frac{3(\text{1-tube})}{(\text{1-m}^2)}.$$

In contrast to the first solution, this one proceeds from a holistic problem representation and each subsequent step of the problem can be derived from the one before it using previously learned principles of units formation and conversion. This solution could be described as top-down rather than bottom-up. We see several interpretations of fractions in this solution: $\frac{6}{8}$ as $\frac{1}{8}$(6-unit), $\frac{2}{8}$ as $2(\frac{1}{8}$-unit)s, and $\frac{3}{8}$ as $3(\frac{1}{8}$-unit)s. Thus, while one might consider this a higher-level problem representation, it appears that the problem representation and solution also requires a higher level of thinking. But the development and use of higher-order thinking is something that we advocate for mathematics curricula in middle school and earlier grades.

Another possible advantage to our second interpretation is that it leads to uniformity in problem representation across division problems. In terms of a hierarchical tree-structure for the problem (see Figure 14.14), our interpretation represents the division (the central operation in the problem) at the top level of the tree. The two multiplications are at lower levels. The traditional approach (our first interpretation) to solving the problem would perform the multiplications first and then use these results as operands in the division, a bottom-up solution.

A top-down solution would perform the division first and then do a units conversion. The issue of problem representation in this context is similar to the one Larkin (1989) discussed for algebraic equations. Important research questions about problem representation lurk here.

Our analysis of partitive division problems has led us to identify three stages in the solution of partitive division word or computational problems: (a) units interpretation, (b) distributing and counting units (or putting in and counting units), and (c) units conversion. It appears that the units interpretation is the first step, while (b) and (c) are interchangeable. In the first solution, (c) was done before (b); in the second, the order was reversed.

Other Division Problem Examples.

1. One-eighth of the heat consumed for the house provides $\frac{1}{8}$ of what is needed to heat the basement. If the house consumes 6 units of heat and the basement has two equal-size rooms, how much heat is needed to heat each of the basement rooms?

The top-down representation using a mathematics-of-quantity interpretation for this is as follows: $\frac{1}{8}$(6-heat-unit) $\div \frac{1}{8}$(2-room-unit)s = ? (1-heat-unit)/(1-room-unit). In this representation $\frac{6}{8}$ and $\frac{2}{8}$, respectively, have the interpretation of $\frac{1}{8}$(6-unit) and $\frac{1}{8}$(2-unit).

2. David can save $\frac{1}{8}$ of his monthly income. He found that 6 months saving is enough for 2 payments, each of which is $\frac{1}{8}$ of the price of the car he wants to buy. In how many months can David buy the car he wants?

The mathematics of quantity interpretation: $6(\frac{1}{8}$-monthly-income) $\div 2(\frac{1}{8}$-car payment)s = ? (1-monthly-income)/(1-car payment). In this case, $\frac{6}{8}$ and $\frac{2}{8}$ have the following interpretations, respectively — $6(\frac{1}{8}$-unit)s and $2(\frac{1}{8}$-unit)s.

3. In 6 hours, $\frac{1}{8}$ block of snow is melted into an amount of water that fills $\frac{1}{8}$ of 2 cans of equal size. How many blocks are needed to fill 1 can of water?

Interpretation: $6(\frac{1}{8}$-block)s $\div \frac{1}{8}$(2-can)s = ? (1-block)/(1-can); $\frac{6}{8}$ and $\frac{2}{8}$, respectively, have the interpretation $6(\frac{1}{8}$ unit)s and $\frac{1}{8}$(2-unit).

4. If the value of a function at the point $\frac{3}{4}$ is 2, what is the slope of the function?

Interpretation: 1(2-unit) $\div 1[\frac{3}{4}$-unit] = ? (1-unit)/[1-unit]; the fraction $\frac{3}{4}$ has the interpretation of $1[\frac{3}{4}$-unit].

A Wider Set of Problem Situations.
Restricted interpretations of arithmetic operations and prescribed problem interpretations and representations in the curriculum has led to a limited range of problem situations and thus, in children, to constrained cognitive models for these operations. Because to date our analysis has concentrated on division, the thrust of our remarks on this issue will concern division, with some references to addition.

CHILDREN'S MODELS FOR DIVISION. Fischbein, Deri, Nello, and Marino (1985) indicate that the dominant models children and (Graeber et al., 1989) teachers use to solve multiplication and division problems have a very limited range of applicability. Work by Kouba (1986), Fischbein et al. (1985), and Greer (1987) clearly suggests that distribution is by far the dominant model. The distribution model and the problem types to which it is applicable leads to conceptions such as "the divisor must be a whole number." Our analysis suggests the existence of another model for partitive division, which we call the put-in model. This model is characterized by physically, or conceptually, putting the objects represented by the dividend *into* the object(s) represented by the divisor.

MORE APPROPRIATE MODELS FOR DIVISION. Units interpretation interacts in an important way with appropriate models or representations for division problems. There are four question types for a division word problem with the given quantities $x(a$-unit)s and $y[b$-units]:

1. How many (a-unit)s for each [b-unit]?
2. How many (1-unit)s for each [b-unit]?
3. How many (a-unit)s for each [1-unit]?
4. How many (1-unit)s for each [1-unit]?

Mathematics of quantity suggests at least two strategies to answer each of these questions. One strategy is to compute the quotient of x divided by y and then do appropriate conversion of units; that is, use the problem representation $x(a$-unit)s $\div$ $y[b$-unit]s and proceed in a top-down order for the solution. The second is to convert units as appropriate, so that the (given) units in the problem data statement correspond to the (target) units in the problem question. The former might be a more powerful problem representation; it provides for a common problem representation for any one of the four questions, and it avoids the matter of classification of problems as multistep (question 4 above) or as extraneous data (questions 1, 2, and 3). Still a third method is to convert all the problem quantities to units of one and then proceed. This representation seems to be the least efficient and least accurate representation of some of the problem forms. The necessary numerical computation resulting from either strategy is essentially the same; the efficiency comes through the understanding exemplified in the holistic problem representation (Larkin, 1989). Important research issues about problem representation are implicit in this discussion. Some examples of research issues that can be investigated are:

1. If a problem solver first changes the units of the problem's given quantities to be the same as the units of the quantities in the problem question, will problem-solving performance be improved?
2. Do children who are aware of the different interpretations of rational numbers perform better on problem solving than those who don't?
3. Do children who exhibit knowledge of the several units-conversion principles exhibit better problem-solving performance than those who don't?

Links Between Additive and Multiplicative Structures. Considering the arithmetic of whole and rational numbers from the joint perspectives of units—composition, decomposition, and conversion—and the mathematics of quantity provides an essential link between whole-number concepts (the additive conceptual field [Vergnaud, 1983]) and multiplicative concepts, including rational-number concepts, (the multiplicative conceptual field).

First-grade children use a concrete-quantity representation for $2 + 5$, such as 2 apples plus 5 apples gives 7 apples. Abstractly, this has the form 2(1-unit)s + 5(1-unit)s = 7(1-unit)s; an analogous model holds for $\frac{2}{8} + \frac{5}{8}$: 2(1-eighth-unit)s + 5(1-eighth-unit)s equals 7(1-eighth-unit)s. A situation of 2 stones plus 5 boys is cause for pause because stones and boys do not add in the same way as apples and apples, unless a *common counting unit* is found for stones and boys—objects, for example. Similarly, for $\frac{2}{4} + \frac{5}{6}$ there is the same need for a common counting unit. A firm understanding of the need for a common counting unit for addition situations, along with the recognition that $\frac{2}{4}$ and $\frac{5}{6}$ can be considered to be 2($\frac{1}{4}$-unit)s and 5($\frac{1}{6}$-unit)s, should help to lay the conceptual base for avoiding the addition of $\frac{2}{4}$ and $\frac{5}{6}$ as $\frac{7}{10}$.

Another type of whole-number problem that will help develop conceptual understanding of the need for common counting units is the following (the problem form is more important than its context):

> *Jane has 2 bags with 4 candies in each, and 5 bags with 6 candies in each. How many bags with 2 candies in each can she make?*

This, again, is traditionally a multistep problem; the first step, multiplication, changes everything to units of one in spite of the fact that the problem question asks about composite units of two. To our knowledge, how children might solve this problem before school instruction forces this solution model on them is not known; some recent but limited pilot work suggests a likely solution to be to change the bags of 4 candies and 6 candies to bags of 2 candies, that is, to convert to the unit requested in the problem question and then count or add to find the total number of bags of 2 candies. In terms of our notation, this problem solution is as follows:

$$2(4\text{-unit})s + 5(6\text{-unit})s = 4(2\text{-unit})s + 15 (2\text{-unit})s$$
$$= 19(2\text{-unit})s.$$

We call attention to the conceptual similarity between this problem and $\frac{2}{4} + \frac{5}{6}$: (a) units interpretation indicates a need for a common unit, (b) the magnitude of the common unit is a common divisor of the two given units, (c) units conversion is needed before counting units (that is, addition) can take place. After this, the strategy for solving the whole-number problem and the fraction addition problem is conceptually and procedurally exactly the same.

We consider the following questions to be of fundamental importance to research and development in the area of multiplicative structures: (a) Do firm cognitive links exist between the two systems? (b) If cognitive links do exist, what are they?

(c) If cognitive links exist between the two structures, how can we help children develop these links? (d) Does the acquisition of these links facilitate the transition to the field of multiplicative structures so that learning of concepts, operations, and relationships in this complex domain would be easier than it currently is for children? (e) If such links are found to exist, can multiplicative concepts be learned earlier in the curriculum and be somewhat concurrent with learning about additive structures?

We believe that some important links do exist, and part of our current analysis seeks to identify them. Unfortunately, we have not progressed sufficiently far to be able to elaborate at this point in time.

TEACHING STUDIES AND RATIONAL NUMBER CONCEPTS

In the following sections we examine current research on teaching and the implications of these studies on the teaching of rational numbers in classroom settings. We begin with programs that have not directly involved rational numbers, and move progressively toward considering investigations that give explicit attention to rational numbers.

Cognitively Guided Instruction

The Cognitively Guided Instruction (CGI) Research Paradigm acknowledges four fundamental assumptions that appear to underlie much contemporary cognitive research on childrens' learning (Peterson, Fennema, & Carpenter, in press; Cobb, Yackel, & Wood, 1988).

1. Children construct their own mathematical knowledge.
2. Mathematics instruction should be organized to facilitate children's construction of knowledge.
3. Children's development of mathematical ideas should provide the basis for sequencing topics of instruction.
4. Mathematical skills should be taught in relation to understanding and problem-solving.

The CGI model basically assumes research-based knowledge of children's learning within specific content domains. To date, the CGI model has been used only with primary children's addition and subtraction concepts, although attempts are currently under way to provide implications for other situations (Carpenter & Fennema, personal communication, November 1989). In a recent study relating teachers' knowledge to student problem-solving behavior with one-step addition and subtraction word problems, Peterson et al. (in press) suggested four CGI principles for applying research-based knowledge of children's learning from classroom instruction.

1. Teachers should assess not only whether a child can solve a particular word problem, but also how the child solves the problem. Teachers should analyze children's thinking by asking appropriate questions and listening to children's responses.
2. Teachers should use the knowledge that they derive from assessment and diagnosis of the children to plan appropriate instruction.
3. Teachers should organize instruction to involve children so that they actively construct their own knowledge with understanding.
4. Teachers should ensure that elementary mathematics instruction stresses relationships between mathematics concepts, skills, and problem solving, with greater emphasis on problem solving than exists in most instructional programs.

The primary Wisconsin CGI study was conducted with 40 first-grade teachers, half of whom participated a four-week summer CGI program. Participants were urged to develop instructional sequences based on their interpretations of the research literature on children's learning of addition and subtraction. Many of the successful CGI teachers adapted a loosely structured discussion format where students were encouraged to solve problems their own way and to look for alternative solution strategies. Students of CGI teachers had encouraging achievement results, and the CGI teachers appeared to have a better grasp of students' capabilities and solution strategies.

It is important to understand here that the CGI model attempted to capitalize on children's previous knowledge of addition and subtraction, which they had essentially acquired prior to formal instruction. In fact, didactic formal instruction in the traditional sense of the word, was not an element in the Cognitively Guided Instructional model. The basic research into children's understanding of addition and subtraction concepts involves the join, separate, combine, and compare types of problems that are discussed frequently in the literature (Moser, 1988; Carpenter, Heibert, & Moser, 1981; Carpenter & Moser, 1982).

What implications might the CGI model have for research into teaching and learning rational number concepts? It is not at all clear that the basic tenets of the CGI model are directly generalizable to the more complex mathematical structures embedded in rational-number usage. The subtle complexities within the domain are recurring themes in research papers. With the introduction of the domain of rational-number concepts comes a cognitively complex multivariate system requiring relativistic thinking, a system where counting strategies and their variations no longer form the basis of successful solution strategies.

The following differences between addition-and-subtraction studies and rational number studies highlight difficulties involved in generalizing the CGI model:

1. The addition-and-subtraction studies depended on children's informal concepts of those operations and attempted to develop these informal notions through informal discussion techniques. As yet, research has not established that children have the same degree of informal rational-number concepts. It is our position that these concepts must be developed in classroom environments. Such environments do not, however, preclude children's construction of meaningful rational-number concepts or content-organization plans that build upon children's intuitions about mathematics.
2. Primary teachers basically understand the content of addition and subtraction and their variations. The same cannot be said for teachers' understanding of rational-number con-

cepts. In our recent survey of over 200 intermediate-level teachers in Minnesota and Illinois, one-quarter to one-third did not appear to understand the mathematics they were teaching (Post, Harel, Behr, & Lesh, 1988). Although there is no indication that teachers cannot learn these concepts, large-scale in-service in these areas becomes a logical necessity. It is our position that teachers must be generally well-informed about a content domain in order to provide appropriate instruction for children.

3. First- and second-grade classes have traditionally spent a major part of their time dealing with addition and subtraction concepts revolving around basic facts, and their application in problem-solving settings. This was precisely the context within which the CGI studies were conducted. This will not be the case for rational-number instruction. The content advocated for the intermediate grades will be very different from what is currently in the mainstream curriculum, with far less attention to symbolic operations and far more attention to the underlying conceptual structure—including order, equivalence, concept of unit, and so on.

4. The impact of standardized tests on the curriculum are far more complex in the rational-number domain. The issues transcend those concerned with instructional paradigms, but they must be reconciled nonetheless in any attempts to substantively change the nature of school curricula.

It appears, then, that research on teaching rational numbers will be more complex than research on teaching additive structures. We suspect that research models will be firmly "situated" (Grenno, 1983) in specific topics within the domains of rational numbers, proportional reasoning, and other multiplicative structures. We envision a more direct approach to instruction, one with less emphasis on traditional objectives and more attention to complex conceptual underpinnings for the domain. Attention to children's knowledge construction will, of course, be an essential element. Teachers must always be encouraged to learn about student constructions and thinking strategies. It does not follow, however, that students cannot or should not receive mathematical knowledge from teachers, that mathematics instruction should not be organized to facilitate the teachers' clear presentation of knowledge, that the structure of mathematics should not provide a basis for sequencing topics of instruction, or that mathematical skills cannot be integrated and taught along with student understanding and problem solving.

As we look to the decade ahead, we are reminded of Gage's (1989) admonition to once again make use of a variety of research paradigms in our attempts to provide more viable and more informed research on teaching. The Rational Number Project has employed one paradigm that appears to hold promise for research on teaching rational-number concepts and operation, proportions, geometry, and the like. Its theory base is discussed and its implications for research on teaching is presented in the next section.

Rational Number Project Teaching Experiments

The Rational Number Project (RNP) has been researching children's learning of rational-number concepts (part-whole, ratio, decimal, operator-and-quotient, and proportional rela-

tionships) since 1979. The project has conducted experimental studies (Cramer, Post, & Behr, 1989), surveys (Heller, Post, Behr, & Lesh, 1990), and teaching experiments (Behr, Wachsmuth, Post, and Lesh, 1984; Post et al., 1985).

The primary source of RNP data has been four different teaching experiments conducted with students in grades 4, 5, and 7. Our teaching experiments focused on the process of mathematical concept development rather than on achievement as measured by written tests. They were conducted with 6–9 students and involved observation of the instructional process by persons other than the instructor. Instruction was controlled by detailed lesson plans (in some cases scripted), activities, written tests, and student interviews. As in most teaching experiments, our interest was to observe the learning process as it occurred and to gauge the depth and direction of student understandings resulting from interaction with carefully constructed, theory-based, instructional materials.

The interview was the primary source of data. The interview is ideally suited to obtain detailed information about an individual's acquisition of new mathematical concepts. Our interviews began as structured sets of questions but quickly became tailored to the specific responses given by students. Consequently, we were able to probe student interest and appeal, while at the same time assessing the depth of student understandings and misunderstandings. Careful study of transcribed interviews (protocols) resulted in insights about students' evolving cognitions. The RNP conducted teaching experiments that were 12, 18, 30, and 17 weeks in duration and that were conducted simultaneously in the Twin Cities area and in DeKalb, Illinois. Weekly interviews given to each student were transcribed and analyzed. In this way, each individual's progress was charted. Since some of the questions were repeated from session to session, it was possible to be quite precise in documenting individual development, the stability of conceptual attainment, and areas of need. It was also possible to contrast each student's progress with the others', although that was not a primary concern. This contrast was conducted not for grading purposes but to identify patterns of growth and understanding across students.

Since statistical assumptions of significantly large numbers of subjects were not met, alternative analytic strategies were employed. These included protocol and videotape analysis and the use of descriptive statistics. Teaching experiments are not easy to generalize, but this shortcoming is compensated for by the richness of the information provided. In one case, the understanding unearthed in one of the teaching experiments was tested in an experimental setting with well-defined experimental and control conditions (Cramer et al., 1989).

Rational Number Project Teaching Experiments: Theoretical Model

The Rational Number Project has relied on two basic theoretical models for the development and execution of its four teaching experiments. We must state initially that our position is squarely within the cognitive psychological camp. We have been influenced by the work of Piaget (1965), Dienes (Dienes & Golding, 1971), Bruner (1966) and a host of more contemporary researchers. We, like Dienes (1960), believe that learning mathematics can ultimately be integrated into one's per-

Perceptual Variates

Mathematical Variates	Fraction Circles	Cuisenaire Rods	Number Lines	Paper Folding	Chips
Part–whole					
Measure					
Ratio					
Decimal					
Operator					

FIGURE 14–15. Matrix representation of Dienes' mathematical and variability principles applied to rational-number concepts.

sonality and become a means of genuine personal fulfillment. We have embraced the four basic components of his theory of mathematics learning (the dynamic principle, perceptual variability, mathematical variability, and the constructivity principle) and have tried to embed their substance and spirit within our student materials. In our materials we have utilized the play stage of student development and have made provision for the transformation of play into more-structured stages of fuller awareness (the dynamic principle). In addition, we have actualized the notion that construction precedes analysis (constructivity principle) by focusing heavily on individuals' interaction with their environment. We have also provided opportunities for students to talk about mathematics with their peers. In more contemporary terms, we approached instruction from a constructivist perspective. The two variability principles were used to guide the construction of the teaching experiment lessons. The model employed is essentially a two-dimensional matrix with one of the variability principles defining each dimension. The model as originally suggested involved numeration systems, with various manipulative materials contrasted to several number bases (Reys & Post, 1973; Post, 1974). It was quickly realized that the mathematical and perceptional variability principles applied to a wide array of mathematical entities, rational numbers included. In our initial model, the five rational-number subconstructs identified by Kieren (1976) constituted the mathematical variability dimension, while a wide array of manipulated materials made up the perceptual variability dimension. This original model appears in Figure 14.15. At this point we had a "helicopter" perspective of the teaching experiments. What remained was to develop sequences of lessons within appropriate cells in the matrix. Dienes contended that, psychologically, the perceptual variability provides the opportunity for mathematical abstraction, while the mathematical variability is concerned with the generalization of the concept(s) under consideration. Certainly, both are important aspects of mathematical conceptual development. Additionally, the variability principles provide for some attention to individualized learning rates and learning styles. The lessons would require very active physical and mental involvement on the part of

the learner. The scope and sequence of the rational-number lessons appear in Table 14.2 (Behr et al., 1984).

Having now the basic orientation as to the broad parameters of our instructional development, attention must be paid to the specific nature of the ways in which individuals would interact with these mathematical concepts. We found Bruner's (1966) notion of modes of representation useful in this regard. In his early work, Bruner suggested that an idea might exist at three levels — or modes — of representation (inactive, iconic, and symbolic). Although never specifically stated by Bruner, these modes were interpreted to occur in a linear and sequential order. Literally generations of curricula were developed suggesting first inactive, then iconic, and then symbolic involvement on the part of the student; a misinterpretation of Bruner's (1966) original intent (Lesh, personal communication, 1975). Realizing the artificial nature of such linearity, Lesh (1979) extended the model to two additional modes (spoken symbols and real-world situations), eliminated the linearity, and stressed the interactive nature of these modes of representation. Various analyses have shown that manipulative aids are just one part of the development of mathematical concepts. Other modes of representation — for example, pictorial, verbal, symbolic, and real-world situations — also play a role (Lesh, Landau, & Hamilton, 1983). The model suggests, and it has been our contention, that the translations within and between modes of representation make ideas meaningful for children. The Lesh translation model appears in Figure 14.16.

The reader will note the inclusion of Bruner's three modes of representation as the central triangle in this model. Arrows denote translations between modes and the concurrent ability to reconceptualize a given idea in a different mode. For example, asking a student to draw a picture of $\frac{1}{2}$ plus $\frac{1}{3}$ (first written on the blackboard) would be a translation from written symbols to pictures, a translation between modes. Similarly, asking a child to demonstrate $\frac{2}{3}$ with Cuisenaire rods, given a display of $\frac{2}{3}$ with fraction circles, represents a translation within modes — in this case, an instance of Dienes' perceptual-variability principle. Post (1988) elaborates on the cognitive function of these translations.

TABLE 14–2. Design of Part-Whole Instructional Materials

Lesson	Embodiment	Activity
1	Color-coded circular pieces	Name pieces Compare sizes Observe that as size decreases, number to make whole increases Observe equivalence
2	Color-coded rectangular pieces	Name pieces Compare sizes Observe that as size decreases, number to make whole increases Observe equivalence
3	Color-coded circular and rectangular pieces	Observe similarities and differences between circular and rectangular pieces Translate between circular and rectangular pieces
4	Color-coded circular and rectangular pieces	Attach unit fraction names (one-fourth, one-fifth) to parts of whole Work with equivalent sets of fractions
5	Paper folding with circles and rectangles	Attach fraction names to shaded parts of folded regions Learn names for unit and proper fractions Associate names with color-coded parts and with shaded parts of folded regions Note similarities and differences
6	Cuisenaire rods	Attach fraction names to display of rods Note fractions as sums of unit fractions Compare display with colored pieces and paper folding Investigate real-world problems
7	All materials from Lessons 1 to 6	Review, using all four embodiments Identify proper fractions orally, in written form, symbolically, and pictorially Translate from one mode to another

(continued)

A bit of reflection will indicate that these translations cannot be made unless a child understands the concept under consideration in the given mode. Further, the child must reinterpret the concept in order to display it in another mode with another material in the same mode. This understanding and reinterpretation are important cognitive (intellectual) processes and need to be encouraged in the teaching-learning process. For this reason, the Lesh model has been a powerful tool for us in the design of our teaching experiments, but it can also be a powerful tool for the classroom teacher. Future research will need to determine which of the many paths through the model are crucial, necessary, or important to mathematical learning, although we hypothesize that all are important. We have been particularly interested in the triad of manipulative aids, spoken symbols, and written symbols. And the triad of real-world situations, manipulative aids, and written symbols. When implemented they dictate relatively intense levels of student involvement, discussion, use of manipulative aids, and use of intellectual processes not

TABLE 14–2. Design of Part-Whole Instructional Materials *(Continued)*

Lesson	Embodiment	Activity
8	Chips	Review division as partitioning ($18 \div 3 = 6$ represents 18 chips, 3 groups, 6 in each group) Represent fractions by covering equal-sized groups of chips Associate fractions with amount covered (3 of 4 groups covered is $\frac{3}{4}$)
9	All materials from Lessons 1 to 7	Translate to any mode, using any embodiment, a fraction represented with physical objects, orally, or with a written symbol
10	Color-coded pieces Paper folding Chips	Represent improper fractions using pieces Translate between improper and mixed number notation orally and in writing Translate representation to paper folding Translate representation to chips
11	Number line	Associate whole numbers, fractions, and mixed numbers with points on the number line Convert improper fractions to whole or mixed numbers Determine equivalence Add fractions with same denominators
12	All materials from Lessons 1 to 11	Using pieces, chips, rods, or pictures of parts of a unit, construct the unit
13	Chips Paper folding	Represent a model for multiplication of fractions Generalize to algorithm with product of numerators and product of denominators

commonly found in mathematics instruction. Levels of student achievement and understanding have been gratifying and have been documented elsewhere. The translation issue suggested in this model should not be taken lightly. Behr (1976) found a significant gap between manipulative aids and symbols and suggested that the mental bridge to cross this gap is complex. Specific attention needs to be directed toward helping students implement the various translations (Post, 1988, pp. 15–16).

Implications for Teaching

Within the domain of mathematics learning, perceptual variability is hypothesized by Dienes (1960) to promote mathematical abstraction, while mathematical variability provides for generalization and the opportunity for expanded understanding of broader perspectives of the issues under consideration. In a similar fashion, teachers need to be exposed to various aspects of teaching in a wide variety of conditions or contexts. For this reason, it is important to focus on a broad spectrum of teacher roles (for example, as an instructor of large and small groups, as a tutor, as a student, as an interviewer, as diagnostician, as a confidant, and so forth.) and to relate these roles to specific tasks teachers are expected to perform (Leinhardt & Greeno, 1986). Just as mathematical abstractions are not themselves contained in the materials which children use, abstractions and generalizations relating to the teaching profession are not necessarily embedded in any single role which the teacher might assume. Such abstractions and generalizations can only be extracted from consideration of a variety of situational, contextual, and model activities, roles, and tasks. In the same way that children are encouraged to discuss similarities and differences between various isomorphs of mathematical concepts, teachers should be encouraged to discuss similarities and differences between pedagogically related actions in various mathematical contexts. A wide variety of avenues should be exploited to provide the foundation for these discussions. Clinically based experiences,

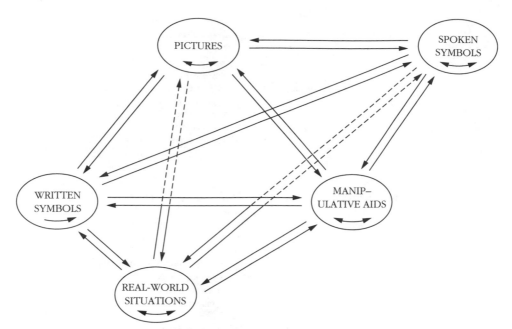

FIGURE 14–16. Lesh's model for translations between modes of representation (adapted from Bruner).

Child 1
↓
Child 4 → Models or Situations ← Child 2
↑
Child 3

Diagram A

Cooperative Groups

Model 1
↑
Model 4 ←(Children)→ Model 2
↓
Model 3

Diagram B

Perceptual Variability

FIGURE 14–17. Diagrams to represent two interpretations of the perceptual-variability principles in an instructional context.

videotapes, demonstration lessons, and other types of sharing experiences come immediately to mind. We hypothesize that it is the opportunity to examine a variety of situations from a number of perspectives and to simultaneously gain the perspectives of other individuals that fosters the development of higher-order understanding and processes of teaching. This is directly parallel to our belief that it is the students' ability to make translations within and between modes of representation that makes ideas meaningful for them (Post, Behr, & Lesh, 1986).

In the Rational Number Project teaching experiments and the related Applied Mathematical Problem Solving Project (AMPS) (Lesh, 1980), cooperative groups of intermediate-level children were asked to focus on a variety of mathematical models, concepts, and problem situations and then to discuss and come to agreement as to the intended meaning(s) (Figure 14.17, Diagram A). Individual students were also asked to focus on several models or embodiments of a single mathematical idea and to indicate similarities and differences in the different interpretations (Figure 14.17, Diagram B). Later the group task was to reconcile these

interpretations in a way so as to arrive at the most probable and widely agreed upon meaning(s). We believe that teachers can also profit from discussing single pedagogical incidents and attempting to reconcile the most probable meanings.

The models in Figure 14.17 can be used directly with teachers and can also be extended to include teacher-education instructional settings as depicted in Figure 14.18. Notice that each of these situations is in fact a variation of the perceptual-variability or multiple-embodiment principle as applied to various patterns of human interaction.

In our early work with children, we continually attempted to stress higher-order thinking and processing, defining higher-order thinking in part as the ability to make these translations. It was important to us to encourage children to go beyond a single incident and to reflect about general meanings. This invariably involved intellectual processes called metacognition: We were encouraging children to think about their own thinking. In similar fashion, it seems reasonable to encourage teachers to think seriously about their and others' teaching acts. The RNP and the AMPS projects determined that successful

Teacher 1

↓

Teacher 4 → Situation Instructional ← Teacher 2

↑

Teacher 3

FIGURE 14–18. A diagram of how Dienes' variability principles can be extended to an instructional setting for teacher education.

problem solvers tend to think at more than one level. They think about the problem at hand, but are also aware of their own thinking. The best problem solvers also generalize problem approaches and problem and solution types as described by Krutetskii (1976). The ability to be simultaneously the "doer" and the "observer" is critical to the solution of many multistage problems. Likewise, it is important that teachers be able to identify and describe their own thoughts about teaching at a number of levels.

Teachers not only teach content; they also implicitly transmit attitudes, beliefs, and understanding of mathematics. Whether desirable or not, students think of teachers as models of correct problem-solving behavior. As teachers act out or demonstrate solutions to problems, it is especially important that they reflect on their own problem-solving behavior so as to help students identify their own metacognitive processes. The ability to accurately and insightfully observe one's own problem-solving behavior is probably related to the ability to accurately observe, describe, and critique the problem-solving behavior of others (Post, et al., 1988).

References

Behr, M. (1976) *The effect of manipulatives in second graders' learning of mathematics.* (Report No. 11, Vol 1) Tallahassee, FL (ERIC Document Reproduction Service No. ED 144 809).

Behr, M., Harel, G., Post, T., & Lesh, R. (1990, April). *On the operator concept of rational numbers: Towards a semantic analysis* Paper presented at the Annual Meeting of the American Educational Research Association. Boston, MA.

Behr, M., Lesh, R., Post, T. R., & Silver, E. A. (1983). Rational number concepts. In R. Lesh & M. Landau (Eds.), *Acquisition of mathematical concepts and processes* (pp. 91–126). New York: Academic Press.

Behr, M., Wachsmuth, I., Post, T., & Lesh, R. (1984). Order and equivalence of rational numbers A clinical teaching experiment. *Journal for Research in Mathematics Education, 14,* 323–341.

Bell, A., Fischbein, E., & Greer, B. (1984). Choice of operation in verbal arithmetic problems: The effects of number size, problem structure and context. *Educational Studies in Mathematics, 15,* 129–147.

Bigelow, J. C., Davis, G. E., & Hunting, R. P. (1989, April). *Some Remarks on the Homology and Dynamics of Rational Number Learning.* Paper presented at the research presession of the National Council of Teachers of Mathematics Annual Meeting, Orlando, FL.

Bruner, J. (1966). *Toward a theory of instruction.* New York: W. W. Norton & Company, Inc.

Carpenter, T., Hiebert, J., & Moser, J. (1981). Problem structure and first grade children's initial solution process for simple addition and subtraction problems. *Journal for Research in Mathematics Education, 12,* 27–39.

Carpenter, T. P., & Moser, J. M. (1982). The development of addition and subtraction problem-solving skills. In T. P. Carpenter, J. M. Moser, & T. A. Romberg (Eds.), *Addition and subtraction: A cognitive perspective.* New York: Academic Press.

Case, R., & Sandieson, R. (1988). A developmental approach to the identification and teaching of central conceptual structures. In J. Hiebert & M. Behr (Eds.), *Research agenda for mathematics education: Number concepts and operations in the middle grades* (pp. 236–259). Reston, VA: National Council of Teachers of Mathematics.

Charles, R., & Silver, E. (1989). *The teaching and assessing of mathematical problem solving.* Reston, VA: National Council of Teachers of Mathematics.

Chi, M. T. H., Feltovich, P. J., & Glaser, R. (1981). Categorization and representation of physics problems by experts and novices. *Cognitive Science, 5*(2), 121–152.

Chi, M. T. H., & Glaser, R. (1982). *Final report: Knowledge and skill difference in novice and experts.* (Contract NO. N00014- 78-C-0375). Washington, DC: Office of Naval Research.

Chi, M. T. H., Glaser, R., & Rees, E. (1982) Expertise in problem solving. In R. J. Sternberg (Ed.), *Advances in the psychology of human intelligence* (Vol. 1, pp. 7–75). Hillsdale, NJ: Lawrence Erlbaum.

Cobb, P., Yackel, E., & Wood, T. (1988). Curriculum and teacher development: Psychological and anthropological perspectives. In E. Fennema, T. Carpenter, & S. J. Lamon (Eds.), *Integrating research on teaching and learning mathematics* (pp. 92–131) Madison: University of Wisconsin, National Center for Research in Mathematical Sciences Education.

Confrey, J. (1988). Multiplication and splitting: Their role in understanding exponential functions. *Proceedings of the Tenth Annual Meeting of the North American Chapter of the International Group for the Psychology of Mathematics Education.* (PME-NA) DeKalb, IL.

Cramer, K., Post, T., & Behr, M. (1989). Cognitive restructuring ability, teacher guidance, and perceptual distractor tasks: An aptitude treatment interaction study. *Journal for Research In Mathematics Education, 20,* 103–110.

d'Ambrosio, U. (1985). Ethnomathematics and its place in the history and pedagogy of mathematics. *For the Learning of Mathematics, 5*(1), 44–48.

Dienes, Z. (1960) *Building up mathematics.* London: Hutchinson Educational Ltd.

Dienes, Z., & Golding, E. (1971). *Approach to modern mathematics.* New York: Herder and Herder.

Fischbein, E., Deri, M., Nello, M., & Marino, M. (1985). The role of implicit models in solving verbal problems in multiplication and division. *Journal for Research in Mathematics Education, 16,* 3–17.

Freudenthal, H. (1983). *Didactical phenomenology of mathematical structures.* Boston: D. Reidel.

Gage, N. L. (1989). The paradigm wars and their aftermath, a "historical" sketch of research on teaching since 1989. *Educational Researcher, 18*(7), 4–10.

Graeber, A., Tirosh, D., & Glover, R. (1989). Preservice teachers misconceptions in solving verbal problems in multiplication and division. *Journal for Research in Mathematics Education, 20,* 95–102.

Greer, B. (1987). Understanding of arithmetical operations as models of situations. In J. A. Sloboda & D. Rogers (Eds.), *Cognitive processes in mathematics* (pp. 660–680). New York: Oxford University Press.

Grenno, J. G. (1983). Conceptual entities. In D. Gentner & A. Stevens (Eds.), *Mental Models*. Hillsdale, NJ: Lawrence Erlbaum.

Harel, G., & Behr, M. (1988). Structure and hierarchy of missing value proportion problems and their representations. The *Journal of Mathematical Behavior, 8,* 77–119.

Harel, G., Behr, M., Post, T., & Lesh, R. (in press). The blocks task and qualitative reasoning skills of 7th grade children in solving the task. *Cognition and Instruction.*

Hart, K. (1981). *Children's understanding of mathematics: 11–16.* London: Murray.

Heller, P., Ahlgren, A., Post, T., Behr, M., & Lesh, R. (1989). Proportional reasoning: The effect of two context variables, rate type and problem setting. *Journal for Research in Science Teaching, 26*(3), 205–220.

Heller, P., Post, T., Behr, M., & Lesh, R. (1990). Qualitative and numerical reasoning about fractions and ratios by seventh and eighth grade students. *Journal for Research in Mathematics Education, 21,* 388–402.

Hiebert, J., & Behr, M. (Eds.). (1989). *Number concepts and operations in the middle grades.* Reston, VA: National Council of Teachers of Mathematics.

Hunting, R. P. (1981). *The role of discrete quantity partition knowledge in the child's construction of fractional number.* (Doctoral dissertation, University of Georgia, 1981). *Dissertation Abstracts International, 41,* 4380A-4381A. (University Microfilms No. 8107919).

Hunting, R. P. (1986). Rachel's schemes for constructing fraction knowledge. *Educational Studies in Mathematics. 17,* 49–66.

Kaput, J. (1985). *Multiplicative word problems and intensive quantities: An integrated software response* (Tech. Rep.). Harvard Graduate School of Education, Educational Technology Center.

Karplus, R., Pulos, S., & Stage, E. K. (1983). Early adolescents proportional reasoning on "rate" problems. *Educational Studies in Mathematics, 14,* 219–233.

Kieren, T. (1976). On the mathematical, cognitive, and instructional foundations of rational numbers. In R. Lesh (Ed.), *Number and measurement: Papers from a research workshop* (pp. 101–144). Columbus, OH: ERIC/SMEAC.

Kieren, T. (1988). Personal knowledge of rational numbers: Its intuitive and formal development. In J. Hiebert & M. Behr (Eds.), *Number concepts and operations in the middle grades* (pp. 53–92). Reston, VA: National Council of Teachers of Mathematics.

Kieren, T. E., & Nelson, D. (1981). Partitioning and unit recognition in performances on rational number tasks. In T. Post & M. Roberts (Eds.), *The proceedings of the third annual meeting of the International Group for the Psychology of Mathematics Education—North American Chapter* (pp. 91–102).

Kieren, T., Nelson, D., & Smith, G. (1985). Graphical algorithms in partitioning tasks. *The Journal of Mathematical Behavior, 4,* 25–36.

Kouba, V. L. (1986). *How young children solve multiplication and division word problems.* Paper presented at the research presession of the National Council of Teachers of Mathematics, Washington, DC.

Krutetskii, V. (1976). *The psychology of mathematical abilities in school children.* Chicago: The University of Chicago Press.

Lamon, S. J. (1989). *Ratio and proportion: Preinstructional cognitions.* Unpublished doctoral dissertation, University of Wisconsin-Madison.

Larkin, J. H. (1989). Robust performance in algebra: The role of the problem. In S. Wagner & C. Kieran (Eds.), *Research issues in the learning of algebra* (pp. 120–134). Reston, VA: National Council of Teachers of Mathematics.

Larson, S., Behr, M., Harel, G., Post, T., & Lesh, R. (1989). Proportional reasoning in young adolescents: An analysis of strategies. In C. A. Maher, G. A. Goldin, & R. B. Davis (Eds.), *Proceedings of the Eleventh Annual Meeting of the American Chapter of the International Group for the Psychology of Mathematics Education* (pp. 181–197). New Brunswick, NJ.

Leinhardt, G., & Greeno, J. (1986). The cognitive skill of teaching. *Journal of Educational Psychology, 78*(2), 75–95.

Lesh, R. (1979). Mathematical learning disabilities: Consideration for identification, diagnosis, and remediation. In R. Lesh, D. Mierkiewicz, & M. G. Kantowski (Eds.), *Applied Mathematical Problem Solving.* Columbus, OH: ERIC/SMEAC.

Lesh, R. (1980). Applied problem solving in Middle School Mathematics. Proposal submitted to the National Science Foundation.

Lesh, R., & Landau, M. (Eds.). (1983). *Acquisition of mathematical concepts and processes.* New York: Academic Press.

Lesh, R., Landau, M., & Hamilton, E. (1983). Conceptual models in applied mathematical problem solving. In R. Lesh & M. Landau (Eds.), *Acquisition of mathematics concepts and processes.* (pp. 263–343). New York: Academic Press.

Lesh, R., Post, T., & Behr, M. (1988). Proportional reasoning. In J. Hiebert & M. Behr (Eds.), *Number concepts and operations in the middle grades* (pp. 93–118). Reston, VA: National Council of Teachers of Mathematics.

Mack, N. K. (1990). Learning fractions with understanding: building on informal knowledge. *Journal for Research in Mathematics Education, 21,* 16–32.

Markman, E. M. (1979). Class and collections: Conceptual organization and numerical abilities. *Cognitive Psychology, 11,* 395–411.

Moser, J. (1988). Arithmetic operation on whole numbers; addition and subtraction. In T. Post (Ed.), *Teaching mathematics in grades K-8: Research based methods* (pp. 111–144.) Boston: Allyn and Bacon.

Nesher, P. (1985). *An outline for a tutorial on rational numbers.* Unpublished manuscript.

Noelting, G. (1980). The development of proportional reasoning and the ratio concept, part I—Differentiation of stages. *Educational Studies in Mathematics, 11,* 217–254.

Novillis, C. G. (1976). An analysis of the fraction concept into a hierarchy of selected subconcepts and the testing of the hierarchical dependencies. *Journal for Research in Mathematics Education, 7,* 131–144.

Ohlsson, S. (1987). Sense and reference in the design of iterative illustrations for rational numbers. In R. W. Lawler & M. Yazdani (Eds.), *Artificial intelligence and education* (pp. 307–344). Norwood, NJ: Ablex.

Ohlsson, S. (1988). Mathematical meaning and applicational meaning in the semantics of fractions and related concepts. In J. Hiebert & M. Behr (Eds.), *Number concepts and operations in the middle grades* (pp. 53–92). Reston, VA: National Council of Teachers of Mathematics.

Peterson, P. L., Fennema, E., & Carpenter, T. P. (in press). Teachers' Knowledge of Students' Mathematics Problem Solving Knowledge. In J. Brophy (Ed.), *Advances in research on Teaching: Vol. 2. Teachers' subject matter knowledge.* Greenwich, CN: JAI Press.

Piaget, J. (1965). *The child's conception of number.* England: W. W. Norton.

Pirie, S. E. B. (1988). Understanding: Instrumental, relational, intuitive, constructed, formalized . . . ? How can we know? *For the Learning of Mathematics. 8*(3), 2–6.

Post, T. (1974) A model for the construction and sequencing of laboratory activities. *The Arithmetic Teacher, 21*(7), pp. 616–622.

Post, T. (1988). Some notes on the nature of mathematics learning. In T. Post (Ed.), *Teaching Mathematics in grades K-9: Research-based methods* (pp. 1–19). Boston: Allyn and Bacon.

Post, T., Behr, M., & Lesh, R. (1986). Research-based observations about children's learning of rational number concepts. *Focus on Learning Problems in Mathematics. 8*(1), 39–48.

Post, T., Harel, G., Behr, M., & Lesh, R. (1988). Intermediate teacher's knowledge of rational number concepts. In L. Fennema, T. Carpenter, & S. Lamon (Eds.), *Integrating research on teaching and learning mathematics* (pp. 194–219). Madison, WI: Center for Educational Research.

Post, T., Wachsmuth, E., Lesh, R., & Behr, M. (1985). Order and equivalence of rational number: A cognitive analysis. *Journal for Research in Mathematics Education, 15*, 18–36.

Pothier, Y., & Sawada, D. (1983). Partitioning: The emergence of rational number ideas in young children, *Journal for Research in Mathematics Education, 14*, 307–317.

Rahim, M., & Kieren, T. (1987). A preliminary report on the reliability and factorial validity of the rational number thinking test in the republic of Trinidad and Tobago. In M. Behr, C. Lacampagne, & M. Wheeler (Eds.), *Proceedings of the tenth annual conference of the International Group for the Psychology of Mathematics Education—North American Chapter* (pp. 114–120). DeKalb, IL: Northern Illinois University.

Resnick, L. B. (1982). A developmental theory of number understanding. In H. P. Ginsburg (Ed.), *Children's Knowledge of Arithmetic.* New York: Academic Press, Inc.

Resnick, L. B. (1986). The development of mathematical intuition. In M. Perlmutter (Ed.), *Perspectives on intellectual development: The Minnesota Symposia on Child Psychology* (Vol.19, pp. 159–194). Hillsdale, NJ: Lawrence Erlbaum.

Reys, R., & Post, T. (1973). *The mathematics laboratory: Theory to practice.* Boston: Prindle, Webber & Schmidt, Inc.

Schwartz, J. L. (1976). *Semantic aspects of quantity.* Unpublished manuscript, MIT, Cambridge.

Schwartz, J. L. (1988). Intensive quantity and referent transforming arithmetic operations. In J. Hiebert & M. Behr (Eds.), *Number concepts and operations in the middle grades* (pp. 41–52). Reston, VA: National Council of Teachers of Mathematics.

Siegler, R. S. (1976). Three aspects of cognitive development. *Cognitive Psychology, 8*, 481–520.

Siegler, R. S., & Vago, S. (1978). The development of a proportionality concept: Judging relative fullness. *Journal of Experimental Child Psychology, 25*, 311–395.

Smith, J. (1988). *Classroom tasks for learning fractions: A new perspective.* Paper presented at the annual meeting of the American Educational Research Association, New Orleans, LA.

Steffe, L. (1986). *Composite units and their constitutive operations.* Paper presented at the Research Presession to the Annual Meeting of the National Council of Teachers of Mathematics, Washington, DC.

Steffe, L. (1988). Children's construction of number sequences and multiplying schemes. In J. Hiebert & M. Behr (Eds.), *Number concepts and operations in the middle grades* (pp. 119–140). Reston, VA: National Council of Teachers of Mathematics.

Steffe, L. (1989). *Children's construction of the rational numbers of arithmetic.* Proposal to the National Science Foundation. Washington, DC: National Science Foundation.

Steffe, L., Cobb, P., & von Glasersfeld, E. (1988). *Construction of arithmetical meanings and strategies.* New York, NY: Springer-Verlag.

Steffe, L. P., von Glasersfeld, E., Richards, J., & Cobb, P. (1983). *Children's counting types: Philosophy, theory, and application.* New York: Praeger.

Thompson, P. W. (1989). Artificial intelligence, advanced technology, and learning and teaching algebra. In S. Wagner & C. Kieran (Eds.), *Research issues in the learning and teaching of algebra.* Reston, VA: National Council of Teachers of Mathematics.

Van den Brink, J., & Streefland, L. (1979). Young children (6-8), ratio and proportion. *Educational Studies in Mathematics, 10*, 403–420.

Vergnaud, G. (1983). Multiplicative structures. In R. Lesh & M. Landau (Eds.), *Acquisition of mathematical concepts and processes* (pp. 127–174). New York: Academic Press.

Vergnaud, G. (1988). Multiplicative structures. In J. Hiebert & M. Behr (Eds.), *Number concepts and operations in the middle grades* (pp. 141–161). Reston, VA: National Council of Teachers of Mathematics.

Wagner, S., & Kieran, C. (Eds.) (1989). *Research issues in the learning and teaching of algebra.* Reston, VA: National Council of Teachers of Mathematics.

LEARNING TO THINK MATHEMATICALLY: PROBLEM SOLVING, METACOGNITION, AND SENSE MAKING IN MATHEMATICS

Alan H. Schoenfeld

THE UNIVERSITY OF CALIFORNIA—BERKELEY

The goals of this chapter are (1) to outline and substantiate a broad conceptualization of what it means to think mathematically, (2) to summarize the literature relevant to understanding mathematical thinking and problem solving, and (3) to point to new directions in research, development, and assessment consonant with an emerging understanding of mathematical thinking and the goals for instruction outlined here.

The use of the phrase "learning to think mathematically" in this chapter's title is deliberately broad. Although the original charter for this chapter was to review the literature on problem solving and metacognition, the literature itself is somewhat ill defined and poorly grounded. As the literature summary will make clear, *problem solving* has been used with multiple meanings that range from "working rote exercises" to "doing mathematics as a professional"; *metacognition* has multiple and almost disjoint meanings (from knowledge about one's thought processes to self-regulation during problem solving) that make it difficult to use as a concept. This chapter outlines the various meanings that have been ascribed to these terms and discusses their role in mathematical thinking. The discussion will not have the character of a classic literature review, which is typically encyclopedic in its references and telegraphic in its discussions of individual papers or results. It will, instead, be selective and illustrative, with main points illustrated by extended discussions of pertinent examples.

Problem solving has, as predicted in the 1980 *Yearbook* of the National Council of Teachers of Mathematics (Krulik, 1980, p. xiv), been the theme of the 1980s. The decade began with NCTM's widely heralded statement, in its *Agenda for Action*, that "problem solving must be the focus of school mathematics" (NCTM, 1980, p. 1). It concluded with the publication of *Everybody Counts* (National Research Council, 1989) and the *Curriculum and Evaluation Standards for School Mathematics* (NCTM, 1989), both of which emphasize problem solving. One might infer, then, that there is general acceptance of the idea that the primary goal of mathematics instruction should be to have students become competent problem solvers. Yet, given the multiple interpretations of the term, the goal is hardly clear. Equally unclear is the role that problem solving, once adequately characterized, should play in the larger context of school mathematics. What are the goals for mathematics instruction, and how does problem solving fit within those goals?

Such questions are complex. Goals for mathematics instruction depend on one's conceptualization of what mathematics is, and what it means to understand mathematics. Such conceptualizations vary widely. At one end of the spectrum, mathematical knowledge is seen as a body of facts and procedures dealing with quantities, magnitudes, and forms, and the relationships among them; knowing mathematics is seen as having mastered these facts and procedures. At the other end of

The author thanks Frank Lester, Indiana University, and Jim Greeno, Stanford University, for their insightful comments on a draft of the manuscript. The current version is much improved for their help. The writing of this chapter was supported in part by the U.S. National Science Foundation through NSF grants MDR-8550332 and BNS-8711342. The Foundation's support does not necessarily imply its endorsement of the ideas or opinions expressed in the paper.

the spectrum, mathematics is conceptualized as the "science of patterns," an (almost) empirical discipline closely akin to the sciences in its emphasis on pattern-seeking on the basis of empirical evidence.

The author's view is that the former perspective trivializes mathematics; that a curriculum based on mastering a corpus of mathematical facts and procedures is severely impoverished—in much the same way that an English curriculum would be considered impoverished if it focused largely, if not exclusively, on issues of grammar. The author characterizes the mathematical enterprise as follows:

Mathematics is an inherently social activity, in which a community of trained practitioners (mathematical scientists) engages in the science of patterns—systematic attempts, based on observation, study, and experimentation, to determine the nature or principles of regularities in systems defined axiomatically or theoretically ("pure mathematics") or models of systems abstracted from real world objects ("applied mathematics"). The tools of mathematics are abstraction, symbolic representation, and symbolic manipulation. However, being trained in the use of these tools no more means that one thinks mathematically than knowing how to use shop tools makes one a craftsman. Learning to think mathematically means (a) developing a mathematical point of view—valuing the processes of mathematization and abstraction and having the predilection to apply them, and (b) developing competence with the tools of the trade, and using those tools in the service of the goal of understanding structure—mathematical sense-making. (Schoenfeld, forthcoming)

This notion of mathematics has gained increasing currency as the mathematical community has grappled, in recent years, with issues of what it means to know mathematics and to be mathematically prepared for an increasingly technological world. The following quotation from *Everybody Counts* typifies the view, echoing themes in the NCTM *Standards* (NCTM, 1989) and *Reshaping School Mathematics* (National Research Council, 1990a).

Mathematics is a living subject which seeks to understand patterns that permeate both the world around us and the mind within us. Although the language of mathematics is based on rules that must be learned, it is important for motivation that students move beyond rules to be able to express things in the language of mathematics. This transformation suggests changes both in curricular content and instructional style. It involves renewed effort to focus on

- seeking solutions, not just memorizing procedures;
- exploring patterns, not just memorizing formulas;
- formulating conjectures, not just doing exercises.

As teaching begins to reflect these emphases, students will have opportunities to study mathematics as an exploratory, dynamic, evolving discipline rather than as a rigid, absolute, closed body of laws to be memorized. They will be encouraged to see mathematics as a science, not as a canon, and to recognize that mathematics is really about patterns and not merely about numbers. (National Research Council, 1989, p. 84)

From this perspective, learning mathematics is empowering. Mathematically powerful students are quantitatively literate. They are capable of interpreting the vast amounts of quantitative data they encounter on a daily basis and of making balanced judgments on the basis of those interpretations. They use mathematics in practical ways, from simple applications such as using proportional reasoning for recipes or scale models, to complex budget projections, statistical analyses, and computer modeling. They are flexible thinkers with a broad repertoire of techniques and perspectives for dealing with novel problems and situations. They are analytical, both in thinking through issues themselves and in examining the arguments put forth by others.

This chapter is divided into three main parts, the first two of which constitute the bulk of the review. The first part, "Toward an Understanding of 'Mathematical Thinking,' " is largely historical and theoretical, having as its goals the clarification of terms like *problem, problem solving,* and *doing mathematics.* It begins with "Immediate Background: Curricular Trends in the Latter 20th Century," a brief recapitulation of the curricular trends and social imperatives that produced the focus on problem solving as the major goal of mathematics instruction in the 1980s. The next section, "On Problems and Problem Solving: Conflicting Definitions," explores contrasting ways in which the terms *problem* and *problem solving* have been used in the literature and the contradictions that have resulted from the multiple definitions and the epistemological stances underlying them. "Enculturation and Cognition" outlines recent findings suggesting the large role of cultural factors in the development of individual understanding. "Epistemology, Ontology, and Pedagogy Intertwined" describes current explorations into the nature of mathematical thinking and knowing and the implications of these explorations for mathematical instruction. The first part concludes with "Goals for Instruction and a Pedagogical Imperative."

The second part, "A Framework for Understanding Mathematical Cognition," provides more of a classical empirical literature review. "The Framework" briefly describes an overarching structure for the examination of mathematical thinking that has evolved over the past decade. It will be argued that all of these categories—core knowledge, problem-solving strategies, effective use of one's resources, having a mathematical perspective, and engagement in mathematical practices—are fundamental aspects of thinking mathematically. The sections that follow elaborate on empirical research within the categories of the framework. "Resources" describes our current understanding of cognitive structures: the constructive nature of cognition, cognitive architecture, memory, and access to it. "Heuristics" describes the literature on mathematical problem-solving strategies. "Monitoring and Control" describes research related to the aspect of metacognition known as self-regulation. "Beliefs and Affects" considers individuals' relationships to the mathematical situations they find themselves in and the effects of individual perspectives on mathematical behavior and performance. Finally, "Practices" focuses on the practical side of the issue of socialization discussed in the first part, describing instructional attempts to foster mathematical thinking by creating microcosms of mathematical practice.

The third part, "Issues," raises some practical and theoretical points of concern as it looks to the future. It begins with a discussion of issues and terms that need clarification and then points to the need for an understanding of methodological tools for inquiry into problem solving. It continues with a discussion of unresolved issues in each of the categories of the framework discussed in the second part, and concludes with a brief commentary on important issues in program design, implementation, and assessment. The specification of new goals for mathematics instruction consonant with current understandings of what it means to think mathematically carries with it an obligation to specify assessment techniques—methods for determining whether students are achieving those goals. Some preliminary steps in those directions are considered.

TOWARD AN UNDERSTANDING OF "MATHEMATICAL THINKING"

Immediate Background: Curricular Trends in the Latter 20th Century

The American mathematics education enterprise is now undergoing extensive scrutiny, with an eye toward reform. The reasons for the reexamination, and for a major overhaul of the current mathematics instruction system, are many and deep. Among them are the following:

- Poor American showings on international comparisons of student competence. On objective tests of mathematical "basics," students in the United States score consistently near the bottom, often grouped with third-world countries (International Association for the Evaluation of Educational Achievement, 1987; National Commission on Excellence in Education, 1983). Moreover, the mathematics education infrastructure in the United States differs substantially from those of its Asian counterparts, whose students score at the top. Asian students take more mathematics and have to meet much higher standards both at school and at home (Stevenson, Lee, & Stigler, 1986).

- Mathematics dropout rates. From grade 8 on, America loses roughly half of the student pool taking mathematics courses. Of the 3.6 million ninth graders taking mathematics in 1972, for example, fewer than 300,000 survived to take a college freshman mathematics class in 1976; 11,000 earned bachelor's degrees in 1980, 2700 earned master's degrees in 1982, and only 400 earned doctorates in mathematics by 1986. (National Research Council, 1989, 1990a).

- Equity issues. Of those who drop out of mathematics, there is a disproportionately high percentage of women and minorities. The effect, in our increasingly technological society, is that women and minorities are disproportionately blocked access to lucrative and productive careers (National Research Council, 1989, 1990b; National Center of Educational Statistics, 1988a).

- Demographics. "Currently, 8% of the labor force consists of scientists or engineers; the overwhelming majority are White males. But by the end of the century, only 15% of the net new labor force will be White males. Changing demographics have raised the stake for all Americans" (National Research Council, 1989, p. 19). The educational and technological requirements for the work force are increasing, while prospects for more students in mathematics-based areas are not good (National Center of Educational Statistics, 1988b).

The 1980s, of course, is not the first time that the American mathematics enterprise has been declared "in crisis." A major renewal of mathematics and science curricula in the United States was precipitated on October 4, 1957, by the Soviet Union's successful launch of the space satellite *Sputnik*. In response to fears of impending Soviet technological and military supremacy, scientists and mathematicians became heavily involved in the creation of new educational materials, often referred to collectively as the alphabet curricula (SMSG in mathematics, BSCS in biology, PSSC in physics). In mathematics, the "new math" flourished briefly in the 1960s and then came to be perceived as a failure. The general perception was that students had not only failed to master the abstract ideas they were being asked to grapple with in the new math, but they had also failed to master the basic skills that the generations of students who preceded them in the schools had managed to learn successfully. In a dramatic pendulum swing, the new math was replaced by the back-to-basics movement. The idea, simply put, was that the fancy theoretical notions underlying the new math had not worked and that we as a nation should make sure that our students had mastered the basics—the foundation upon which higher-order thinking skills were to rest.

By the end of the 1970s it became clear that the back-to-basics movement was a failure. A decade of curricula that focused on rote mechanical skills produced a generation of students who, for lack of exposure and experience, performed dismally on measures of thinking and problem solving. Even more disturbing, they were no better at the basics than the students who had studied the alphabet curricula. The pendulum began to swing in the opposite direction, toward "problem solving." The first major call in that direction was issued by the National Council of Supervisors of Mathematics in 1977. It was followed by the National Council of Teachers of Mathematics' (1980) *Agenda for Action*, which had as its first recommendation that problem solving be the focus of school mathematics (p. 1). Just as back-to-basics was declared to be the theme of the 1970s, problem solving was declared to be the theme of the 1980s (see, for example, Krulik, 1980). Here is one simple measure of the turnaround: In the 1978 draft program for the 1980 International Congress on Mathematics Education (ICME IV, Berkeley, California, 1980; see Zweng, Green, Kilpatrick, Pollak, & Suydam, 1983), only one session on problem solving was planned, and it was listed under "unusual aspects of the curriculum." Four years later, problem solving was one of the seven main themes of the next International Congress (ICME V, Adelaide, Australia; see Burkhardt, Groves, Schoenfeld, & Stacey, 1988; Carss, 1986). Similarly, "metacognition," coined in the late 1970s, appeared occasionally in the mathematics education literature of the early 1980s, and then with ever-increasing frequency through the decade. Problem solving and metacognition, the lead terms in this article's title, are perhaps the two most overworked and least understood buzzwords of the 1980s.

This chapter suggests that, on the one hand, much of what passed under the name of problem solving during the 1980s has been superficial, and that were it not for the current "crisis," a reverse pendulum swing might well be on its way. On the other hand, it documents that we now know much more about mathematical thinking, learning, and problem solving than during the immediate post-*Sputnik* years, and that a reconceptualization of both problem solving and mathematics curricula that do justice to it is now possible. Such a reconceptualization will in large part be based in part on advances made in the past decade: detailed understandings of the nature of thinking and learning and of problem-solving strategies and metacognition; evolving conceptions of mathematics as the "science of patterns" and of doing mathematics as an act of sense-making; and cognitive apprenticeship and "cultures of learning."

On Problems and Problem Solving: Conflicting Definitions

In a historical review focusing on the role of problem solving in the mathematics curriculum, Stanic and Kilpatrick (1988) provide the following brief summary:

Problems have occupied a central place in the school mathematics curriculum since antiquity, but problem solving has not. Only recently have mathematics educators accepted the idea that the development of problem solving ability deserves special attention. With this focus on problem solving has come confusion. The term *problem solving* has become a slogan encompassing different views of what education is, of what schooling is, of what mathematics is, and of why we should teach mathematics in general and problem solving in particular. (p. 1)

Indeed, "problems" and "problem solving" have had multiple and often contradictory meanings through the years—a fact that makes interpretation of the literature difficult. For example, a 1983 survey of college mathematics departments (Schoenfeld, 1983) revealed the following categories of goals for courses that were identified by respondents as "problem solving" courses:

- to train students to "think creatively" and/or "develop their problem-solving ability" (usually with a focus on heuristic strategies);
- to prepare students for problem competitions such as the Putnam examinations or national or international Olympiads;
- to provide potential teachers with instruction in a narrow band of heuristic strategies;
- to learn standard techniques in particular domains, most frequently in mathematical modeling;
- to provide a new approach to remedial mathematics (basic skills) or to try to induce "critical thinking" or "analytical reasoning" skills.

The two poles of meaning indicated in the survey are nicely illustrated in two of Webster's (1979, p. 1434) definitions for the term *problem*:

Definition 1: "In mathematics, anything required to be done, or requiring the doing of something."
Definition 2: "A question...that is perplexing or difficult."

Problems as Routine Exercises. Webster's first definition, cited immediately above, captures the sense of the term *problem* as it has traditionally been used in mathematics instruction. For nearly as long as we have written records of mathematics, sets of mathematics tasks have been with us—as vehicles of instruction, as means of practice, and as yardsticks for the acquisition of mathematical skills. Often such collections of tasks are anything but problems in the sense of the second definition. They are, rather, routine exercises organized to provide practice on a particular mathematical technique that, typically, has just been demonstrated to the student. We begin this section with a detailed examination of such problems, focusing on their nature, the assumptions underlying their structure and presentation, and the consequences of instruction based largely, if not exclusively, on such problem sets. That discussion sets the context for a possible alternative view.

A generic example of a mathematics problem set, with antecedents that Stanic and Kilpatrick (1988) trace to antiquity, is the following excerpt from a late 19th century text, W. J. Milne's *A Mental Arithmetic* (1897). The reader may wish to obtain an answer to problem 52 by virtue of mental arithmetic before reading the solution Milne provides.

52. How much will it cost to plow 32 acres of land at $3.75 per acre?

    SOLUTION: $3.75 is $\frac{3}{8}$ of $10. At $10 per acre the plowing would cost $320, but since $3.75 is $\frac{3}{8}$ of $10, it will cost $\frac{3}{8}$ of $320, which is $120. Therefore, etc.
53. How much will 72 sheep cost at $6.25 per head?
54. A baker bought 88 barrels of flour at $3.75 per barrel. How much did it all cost?
55. How much will 18 cords of wood cost at $6.66 $\frac{2}{3}$ per cord?

[These exercises continue down the page and beyond.]

(Milne, 1897, page 7; cited in Stanic & Kilpatrick, 1988)

The particular technique students are intended to learn from this body of text is illustrated in the solution of problem 52. In all of the exercises, the student is asked to find the product $(A \times B)$, where A is given as a two-digit decimal that corresponds to a price in dollars and cents. The decimal values have been chosen so that a simple ratio is implicit in the decimal form of A. That is, $A = r \times C$, where r is a simple fraction and C is a power of 10. Hence, $(A \times B)$ can be computed as $r \times (C \times B)$. Thus, working from the template provided in the solution to problem 52, the student is expected to solve problem 53 as follows:

$$(6.25 \times 72) = ([\tfrac{5}{8} \times 10] \times 72) = (\tfrac{5}{8} \times [10 \times 72])$$
$$= (\tfrac{5}{8} \times 720) = 5 \times 90 = 450.$$

The student can obtain the solutions to all the problems in this section of the text by applying this algorithm. When the conditions of the problem are changed ever so slightly (in problems 52 to 60 the number C is 10, but in problem 61 it changes from 10 to 100), students are given a suggestion to help extend the procedure they have learned:

61. The porter on a sleeping car was paid $37.50 per month for 16 months. How much did he earn?

    SUGGESTION: $37.50 is $\frac{3}{8}$ of $100.

Later in this section we will examine, in detail, the assumptions underlying the structure of this problem set and the effects on students of repeated exposure to such problem sets. For now, we simply note the general structure of the section and the basic pedagogical and epistemological assumption underlying its design.

STRUCTURE:

1. A task is used to introduce a technique.
2. The technique is illustrated.
3. More tasks are provided so that the student may practice the illustrated skills.

BASIC ASSUMPTION:

Having worked this cluster of exercises, the students will have a new technique in their mathematical tool kit. Presumably, the sum total of such techniques (the curriculum) reflects the corpus of mathematics the student is expected to master; the set of techniques the student has mastered comprises the student's mathematical knowledge and understanding.

Traditional Uses of "Problem Solving" (in the Sense of Tasks Required To Be Done): Means to a Focused End. In their historical review of problem solving, Stanic and Kilpatrick (1988) identify three main themes regarding its usage. In the first theme, which they call "problem solving as context," problems are employed as vehicles in the service of other curricular goals. They identify five such roles that problems play:

1. *As a justification for teaching mathematics.* "Historically, problem solving has been included in the mathematics curriculum in part because the problems provide justification for teaching mathematics at all. Presumably, at least some problems related in some way to real-world experiences were included in the curriculum to convince students and teachers of the value of mathematics" (p. 13).
2. *To provide specific motivation for subject topics.* Problems are often used to introduce topics with the implicit or explicit understanding that once you have learned the lesson that follows, you will be able to solve problems of this type.
3. *As recreation.* Recreational problems are intended to be motivational, in a broader sense than in number 2, above. They show that "math can be fun" and that there are entertaining uses for the skills students have mastered.
4. *As a means of developing new skills.* Carefully sequenced problems can introduce students to new subject matter and provide a context for discussions of subject-matter techniques.
5. *As practice.* Milne's exercises, and the vast majority of school mathematics tasks, fall into this category. Students are shown a technique and then given problems to practice until they have mastered the technique.

In all five of these roles, problems are seen as rather prosaic entities (recall Webster's first definition) and are used as a means to one of the ends listed above. That is, problem solving is not usually seen as a goal in itself, but solving problems is seen as facilitating the achievement of other goals. Problem solving has a minimal interpretation: working the tasks that have been presented.

The second theme identified by Stanic and Kilpatrick (1988) is "problem solving as skill." This theme has its roots in a reaction to Thorndike's work (Thorndike & Woodworth, 1901). Thorndike's research debunked the simple notion of "mental exercise," in which it was assumed that learning reasoning skills in domains such as mathematics would result in generally improved reasoning performance in other domains. Hence, if mathematical problem solving was to be important, it was not because it made one a better problem solver in general, but because solving mathematical problems was valuable in its own right. This led to the notion of problem solving as skill—a skill still rather narrowly defined (that is, being able to obtain solutions to the problems assigned), but worthy of instruction in its own right. Though there might be some dispute on the matter, this author's perspective is that the vast majority of curricular development and implementation that went on under the name of "problem solving" in the 1980s was of this type.

> Problem solving is often seen as one of a number of skills to be taught in the school curriculum. According to this view, problem solving is not necessarily seen as a unitary skill, but there is a clear skill orientation....
> Putting problem solving in a hierarchy of skills to be acquired by students leads to certain consequences for the role of problem solving in the curriculum.... [D]istinctions are made between solving routine and nonroutine problems. That is, nonroutine problem solving is characterized as a higher level skill to be acquired after skill at solving routine problems (which, in turn, is to be acquired after students learn basic mathematical concepts and skills). (Stanic & Kilpatrick, 1988, p. 15)

It is important to note that, even though in this second interpretation problem solving is seen as a skill in its own right, the basic underlying pedagogical and epistemological assumptions in this theme are precisely the same as those outlined for Milne's examples in the discussion above. Typically, problem-solving techniques (such as drawing diagrams, looking for patterns when $n = 1, 2, 3, 4, \ldots$) are taught as subject matter, with practice problems assigned so that the techniques can be mastered. After receiving this kind of problem-solving instruction (often a separate part of the curriculum), the students' "mathematical tool kit" is presumed to contain problem-solving skills as well as the facts and procedures they have studied. This expanded body of knowledge presumably comprises the students' mathematical knowledge and understanding.

The third theme identified by Stanic and Kilpatrick (1988) is "problem solving as art." This view, in strong contrast to the previous two, holds that real problem solving (that is, working problems of the "perplexing" kind) is the heart of mathematics, if not mathematics itself. We now turn to that view, as expressed by some notable mathematicians and philosophers.

On Problems That Are Problematic: Mathematicians' Perspectives. As noted earlier, mathematicians are hardly unanimous in their conceptions of problem solving. Courses in problem solving at the university level have goals that range from remediation to critical thinking to developing creativity. Nonetheless, there is a particularly mathematical point of view regarding the role that problems play in the lives of those who do mathematics.

The unifying theme is that the work of mathematicians, on an ongoing basis, is solving problems—problems of the "perplexing or difficult" kind, that is. Halmos makes the claim, quite simply, that solving problems is "the heart of mathematics."

What does mathematics *really* consist of? Axioms (such as the parallel postulate)? Theorems (such as the fundamental theorem of algebra)? Proofs (such as Gödel's proof of undecidability)? Definitions (such as the Menger definition of dimension)? Theories (such as category theory)? Formulas (such as Cauchy's integral formula)? Methods (such as the method of successive approximations)?

Mathematics could surely not exist without these ingredients; they are all essential. It is nevertheless a tenable point of view that none of them is at the heart of the subject, that the mathematician's main reason for existence is to solve problems, and that, therefore, what mathematics *really* consists of is problems and solutions. (1980, p. 519)

Some famous mathematical problems are named as such, for example, the four-color problem (which when solved, became the four-color theorem). Others go under the name of hypothesis (such as the Riemann hypothesis) or conjecture (Goldbach's conjecture, that every even number greater than 2 can be written as the sum of two odd primes). Some problems are motivated by practical or theoretical concerns oriented in the real world (applied problems), and others by abstract concerns (for example, what is the distribution of twin primes?). The ones mentioned above are the "big" problems that have been unsolved for decades and whose solution earns the solvers significant notice. But they differ only in scale from the problems encountered in the day-to-day activity of mathematicians. Whether pure or applied, the challenges that ultimately advance our understanding take weeks, months, and often years to solve. This being the case, Halmos argues, students' mathematical experiences should prepare them for tackling such challenges. That is, students should engage in "real" problem solving, learning during their academic careers to work problems of significant difficulty and complexity.

I do believe that problems are the heart of mathematics, and I hope that as teachers, in the classroom, in seminars, and in the books and articles we write, we will emphasize them more and more, and that we will train our students to be better problem posers and problem solvers than we are. (1980, p. 524)

The mathematician best known for his conceptualization of mathematics as problem solving and for his work in making problem solving the focus of mathematics instruction is Pólya. Indeed, the edifice of problem-solving work erected in the past two decades stands largely on the foundations of his work. The mathematics education community is most familiar with Pólya's work through his (1945/1957) introductory volume *How to Solve It,* in which he introduced the term "modern heuristic" to describe the art of problem solving, and through his subsequent elaborations on the theme in the two-volume sets, *Mathematics and Plausible Reasoning* (1954) and *Mathematical Discovery* (1962, 1965/1981). In fact, Pólya's work on problem solving and "method" was apparent as early as the publication of his and Szegö's (1925) *Problems and Theorems in Analysis.* In this section we focus on the broad mathematical and philosophical themes woven through Pólya's work on problem solv-

ing. Details regarding the implementation of heuristic strategies are pursued in the research review.

It is essential to understand Pólya's conception of mathematics as an activity. As early as the 1920s, Pólya had an interest in mathematical heuristics, and he and Szegö included some heuristics (in the form of aphorisms) as suggestions for guiding students' work through the difficult problem sets in *Aufgaben und Lehrsätze aus der Analysis I* (1925). Yet the role of mathematical engagement—of hands-on mathematics, if you will—was central in Pólya's view.

General rules which could prescribe in detail the most useful discipline of thought are not known to us. Even if such rules could be formulated, they could not be very useful...[for] one must have them assimilated into one's flesh and blood and ready for instant use....The independent solving of challenging problems will aid the reader far more than the aphorisms which follow, although as a start these can do him no harm. (p. vii)

Part of that engagement, according to Pólya, was the active engagement of discovery, one which takes place in large measure by guessing. Eschewing the notion of mathematics as a formal and formalistic deductive discipline, Pólya argued that mathematics is akin to the physical sciences in its dependence on guessing, insight, and discovery.

To a mathematician, who is active in research, mathematics may appear sometimes as a guessing game; you have to guess a mathematical theorem before you prove it, you have to guess the idea of the proof before you carry through all the details.

To a philosopher with a somewhat open mind all intelligent acquisition of knowledge should appear sometimes as a guessing game, I think. In science as in everyday life, when faced with a new situation, we start out with some guess. Our first guess may fall short of the mark, but we try it and, according to the degree of success, we modify it more or less. Eventually, after several trials and several modifications, pushed by observations and led by analogy, we may arrive at a more satisfactory guess. The layman does not find it surprising that the naturalist works this way....And the layman is not surprised to hear that the naturalist is guessing like himself. It may appear a little more surprising to the layman that the mathematician is also guessing. The result of the mathematician's creative work is demonstrative reasoning, a proof, but the proof is discovered by plausible reasoning, by guessing....

Mathematical facts are first guessed and then proved, and almost every passage in this book endeavors to show that such is the normal procedure. If the learning of mathematics has anything to do with the discovery of mathematics, the student must be given some opportunity to do problems in which he first guesses and then proves some mathematical fact on an appropriate level. (1954, pp. 158–160)

For Pólya, mathematical epistemology and mathematical pedagogy are deeply intertwined. Pólya takes it as given that for students to gain a sense of the mathematical enterprise, their experience with mathematics must be consistent with the way mathematics is done. The linkage of epistemology and pedagogy is, as well, the major theme of this chapter. The next section of this chapter elaborates a particular view of mathematical thinking—discussing mathematics as an act of sensemaking that is socially constructed and socially transmitted. It argues that students develop their sense of mathematics—and thus how they use mathematics—from their experiences with mathematics (largely in the classroom). It follows that classroom mathematics must mirror this sense of mathematics as a sense-making activity, if students are to come to understand and use mathematics in meaningful ways.

Enculturation and Cognition

An emerging body of literature (Bauersfeld, 1979; Brown, Collins, & Duguid, 1989; Collins, Brown, & Newman, 1989; Greeno, 1989; Lampert, 1990; Lave, 1988; Lave, Smith, & Butler, 1988; Resnick, 1988; Rogoff & Lave, 1984; Schoenfeld, 1989a, 1990b; see especially Nunes, Chapter 22, this volume) conceives of mathematics learning as an inherently social (as well as cognitive) activity, and an essentially constructive activity instead of an absorptive one.

By the mid-1980s, the constructivist perspective, with roots in Piaget's work (1954) and with contemporary research manifestations such as the misconceptions literature (Brown & Burton, 1978; diSessa, 1983; Novak, 1987), was widely accepted in the research community as being well grounded. Romberg and Carpenter (1986) stated the fact bluntly: "The research shows that learning proceeds through *construction, not absorption*" (p. 868). The constructivist perspective pervades this handbook as well. However, the work cited in the previous paragraph extends the notion of constructivism from the purely cognitive sphere, where much of the research has been done, to the social sphere. As such, it blends with some theoretical notions from the social literature. Resnick, tracing contemporary work to antecedents in the work of George Herbert Mead (1934) and Lev Vygotsky (1978), states that "several lines of cognitive theory and research point toward the hypothesis that we develop habits and skills of interpretation and meaning construction though a process more usefully conceived of as *socialization* than *instruction*" (1988, p. 39).

The notion of socialization as identified by Resnick (also called enculturation—entering and picking up the values of a community or culture) is central, in that it highlights the importance of perspective and point of view as core aspects of knowledge. The case can be made that a fundamental component of thinking mathematically is having a mathematical point of view, that is, seeing the world in the ways mathematicians do.

[T]he reconceptualization of thinking and learning that is emerging from the body of recent work on the nature of cognition suggests that becoming a good mathematical problem solver—becoming a good thinker in any domain—may be as much a matter of acquiring the habits and dispositions of interpretation and sense-making as of acquiring any particular set of skills, strategies, or knowledge. If this is so, we may do well to conceive of mathematics education less as an instructional process (in the traditional sense of teaching specific, well-defined skills or items of knowledge), than as a socialization process. In this conception, people develop points of view and behavior patterns associated with gender roles, ethnic and familial cultures, and other socially defined traits. When we describe the processes by which children are socialized into these patterns of thought, affect, and action, we describe long-term patterns of interaction and engagement in a social environment. (1988, p. 58)

This cultural perspective is well grounded anthropologically, but it is relatively new to the mathematics education literature. The main idea, that point of view is a fundamental determinant of cognition, and that the community to which one belongs shapes the development of one's point of view, is made eloquently by Clifford Geertz.

Consider…Evans-Pritchard's famous discussion of Azande witchcraft. He is, as he explicitly says but no one seems much to have noticed, concerned with common-sense thought—Zande common-sense thought—as the general background against which the notion of witchcraft is developed.…

Take a Zande boy, he says, who has stubbed his foot on a tree stump and developed an infection. The boy says it's witchcraft. Nonsense, says Evans-Pritchard, out of his own common-sense tradition: you were merely bloody careless; you should have looked where you were going. I did look where I was going; you have to with so many stumps about, says the boy—*and if I hadn't been witched I would have seen it*. Furthermore, all cuts do not take days to heal, but on the contrary, close quickly, for that is the nature of cuts. But this one festered, thus witchcraft must be involved.

Or take a Zande potter, a very skilled one, who, when now and again one of his pots cracks in the making, cries "witchcraft!" Nonsense! says Evans-Pritchard, who, like all good ethnographers, seems never to learn: of course sometimes pots crack in the making; it's the way of the world. But, says the potter, I chose the clay carefully, I took pains to remove all the pebbles and dirt, I built up the clay slowly and with care, and I abstained from sexual intercourse the night before. And *still* it broke. What else can it be but witchcraft? (1983, p. 78)

Geertz's point is that Evans-Pritchard and the African tribesmen agree on the data (the incidents they are trying to explain), but that their interpretations of what the incidents mean are radically different. Each person's interpretation is derived from his or her own culture and seems common-sensical. The anthropologist in the West and the Africans on their home turf have each developed points of view consonant with the mainstream perspectives of their societies. And those culturally determined (socially mediated) views determine what sense they make of what they see.

The same, it is argued, is true of members of *communities of practice*, groups of people engaged in common endeavors within their own culture. Three such groups include the community of tailors in "Tailors' Alley" in Monrovia, Liberia, studied by Jean Lave (in preparation), the community of practicing mathematicians, and the community that spends its daytime hours in schools. In each case, the "habits and dispositions" (see the quotation from Resnick, above) of community members are culturally defined and have great weight in shaping individual behavior. We discuss the first two here: the third is discussed in the next section. First, Lave's study (which largely inspired the work on cognitive apprenticeship discussed below) examined the apprenticeship system by which Monrovian tailors learn their skills. Schoenfeld summarized Lave's perspective on what "learning to be a tailor" means, as follows:

Being a tailor is more than having a set of tailoring skills. It includes a way of thinking, a way of seeing, and having a set of values and perspectives. In Tailors' Alley, learning the curriculum of tailoring and learning to *be a tailor* are inseparable: the learning takes place in the context of doing real tailors' work, in the community of tailors. Apprentices are surrounded by journeymen and master tailors, from whom they learn their skills—and among whom they live, picking up their values and perspectives as well. These values and perspectives are not part of the formal curriculum of tailoring, but they are a central defining feature of the environment, and of what the apprentices learn. The apprentice tailors are apprenticing themselves into a *community*, and when they have succeeded in doing so, they have adopted a point of view as well as a set of skills—both of which define them as tailors. [If this notion seems a bit farfetched, think of groups of people such as lawyers, doctors, automobile salesmen, or university professors in our own society. That there are political (and other) stereotypes of these groups indicates that there is more to membership in any of these communities than simply possessing the relevant credentials or skills.] (1989c, pp. 85–86)

Second, there is what might be called "seeing the world through the lens of the mathematician." As illustrations, here are two comments made by the applied mathematician Henry Pollak.

How many saguarro cacti more than 6 feet high are in the state of Arizona? I read that the saguarro is an endangered species. Developers tear them down when they put up new condominiums. So when I visited Arizona two or three years ago I decided to try an estimate. I came up with 10^8. Let me tell you how I arrived at that answer. In the areas where they appear, saguarros seem to be fairly regularly spaced, approximately 50 feet apart. That approximation gave me 10^2 to a linear mile, which implied 10^4 in each square mile. The region where the saguarros grow is at least 50 by 200 miles. I therefore multiplied $10^4 \times 10^4$ to arrive at my final answer. I asked a group of teachers in Arizona for their estimate, and they were at a loss as to how to begin.

If you go into a supermarket, you will typically see a number of checkout counters, one of which is labeled "Express Lane" for x packages or fewer. If you make observations on x, you'll find it varies a good deal. In my home town, the A&P allow six items; the Shop-Rite, eight, and Kings, 10. I've seen numbers vary from 5 to 15 across the country. If the numbers vary that much, then we obviously don't understand what the correct number should be. How many packages should be allowed in an express line? (1987, pp. 260–261)

Both of these excerpts exemplify the habits and dispositions of the mathematician. Hearing that the saguarro is endangered, Pollak almost reflexively asks how many saguarro there might be; he then works out a crude estimate on the basis of available data. This predilection to quantify and model is certainly a part of the mathematical disposition and is not typical of those outside mathematically oriented communities. (Indeed, Pollak notes that neither the question nor the mathematical tools to deal with it seemed natural to the teachers with whom he discussed it.) That disposition is even clearer in the second example, thanks to Pollak's language. Note that Pollak perceives the supermarket as a mathematical context—again, hardly a typical perspective. For most people, the number of items allowed in the express line is simply a matter of the supermarket's prerogative. For Pollak, the number is a variable, and the task of determining the "right" value of that variable is an optimization problem. The habit of seeing phenomena in mathematical terms is also part of the mathematical disposition.

In short, Pollak sees the world from a mathematical point of view. Situations that others might not attend to at all serve for him as the contexts for interesting mathematical problems. The issues he raises in what to most people would be non-mathematical contexts—supermarket check-out lines and desert fields—are inherently mathematical in character. His language ("for x packages or fewer") is that of the mathematician, and his approaches to conceptualizing the problems (optimization for the supermarket problem, estimation regarding the number of cacti) employ typical patterns of mathematical reasoning. There are, of course, multiple mathematical points of view. For a charming and lucid elaboration of many of these, see Davis and Hersh (1981).

Epistemology, Ontology, and Pedagogy Intertwined

In short, the point of the literature discussed in the previous section is that learning is culturally shaped and defined; people develop their understandings of any enterprise from their participation in the "community of practice" within which that enterprise is practiced. The lessons students learn about mathematics in our current classrooms are broadly cultural, extending far beyond the scope of the mathematical facts and procedures (the explicit curriculum) that they study. As Hoffman (1989) points out, this understanding gives added importance to a discussion of epistemological issues. Whether or not one is explicit about one's epistemological stance, he observes, what one thinks mathematics is will shape the kinds of mathematical environments one creates, and thus the kinds of mathematical understandings that one's students will develop.

Here we pursue the epistemological-to-pedagogical link in two ways. First, we perform a detailed exegesis of the selection of "mental arithmetic" exercises from Milne (1897), elaborating the assumptions that underlie it, and the consequences of curricula based on such assumptions. That exegesis is not derived from the literature, although it is consistent with it. The author's intention in performing the analysis is to help establish the context for the literature review, particularly the sections on beliefs and context. Second, we examine some issues in mathematical epistemology and ontology. As Hoffman observes, it is important to understand what doing mathematics is, if one hopes to establish classroom practices that will help students develop the right mathematical point of view. The epistemological explorations in this section establish the basis for the pedagogical suggestions that follow later in the chapter.

On Problems as Practice: An Exegesis of Milne's Problem Set. The selection of exercises from Milne's *Mental Arithmetic* introduced earlier in this chapter has the virtue that it is both antiquated and modern: One can examine it at a distance because of its age, but one will also find its counterparts in almost every classroom around the country. We shall examine it at length.

Recall the first problem posed by Milne: "How much will it cost to plow 32 acres of land at $3.75 per acre?" His solution was to convert $3.75 into a fraction of $10, as follows. "$3.75 is $\frac{3}{8}$ of $10. At $10 per acre the plowing would cost $320, but since $3.75 is $\frac{3}{8}$ of $10, it will cost $\frac{3}{8}$ of $320, which is $120." This solution method was then intended to be applied to all of the problems that followed.

It is perfectly reasonable, and useful, to devote instructional time to the technique Milne illustrates. The technique is plausible from a practical point of view, in that there might well be circumstances where a student could most easily do computations of the type demonstrated. It is also quite reasonable from a mathematical point of view. Being able to perceive $A \times B$ as $(r \times C) \times B = r \times (BC)$ when the latter is easier to compute, and carrying out the computation, is a sign that one has developed some understanding of fractions and of multiplicative structures; one would hope that students would develop such understandings in their mathematics instruction. The critique that follows is not based in an objection to the potential value or utility of the mathematics Milne presents, but in the ways in which the topic is treated.

ISSUE 1: FACE VALIDITY. At first glance, the technique illustrated in Milne's problem 52 seems useful, and the solutions to the subsequent problems appear appropriate. As noted above, one hopes that students will have enough number sense to be able to compute $32 \times \$3.75$ in the absence of paper and pencil. However, there is the serious question as to whether one would really expect students to work the problems the way Milne suggests. In a quick survey as this chapter was being written, the author asked four colleagues to solve problem 52 mentally. Three of the four solutions did convert the ".75" in $3.75 to a fractional equivalent, but none of the four employed fractions in the way suggested by Milne. The fourth avoided fractions altogether, but also avoided the standard algorithm. Here is what the four did:

- Two of the people converted 3.75 into $3\frac{3}{4}$, and then applied the distributive law to obtain

$$(3\tfrac{3}{4})(32) = (3 + \tfrac{3}{4})(32) = 96 + (\tfrac{3}{4})(32) = 96 + 24 = 120.$$

- One expressed 3.75 as $(4 - \tfrac{1}{4})$, and then distributed as follows:

$$(4 - \tfrac{1}{4})(32) = 128 - (\tfrac{1}{4})(32) = 128 - 8 = 120.$$

- One noted that 32 is a power of 2. He divided and multiplied by 2's until the arithmetic became trivial:

$$(32)(3.75) = (16)(7.5) = (8)(15) = (4)(30) = 120.$$

In terms of mental economy, we note that each of the methods used is as easy to employ as the one presented by Milne.

ISSUE 2: THE EXAMPLES ARE CONTRIVED TO ILLUSTRATE THE MATHEMATICAL TECHNIQUE AT HAND. In real life one rarely if ever encounters unit prices such as $\$6.66\frac{2}{3}$. (But we commonly see prices such as "3 for $20.00.") The numbers used in problem 55 and others were clearly selected so that students could successfully perform the algorithm taught in this lesson. On the one hand, choosing numbers in this way makes it easy for students practice the technique. On the other hand, the choice makes the problem itself implausible. Moreover, the problem settings (cords of wood, price of sheep, and so on) are soon seen to be window dressing designed to make the problems appear relevant, but which in fact have no real role in the problem. As such, the artificiality of the examples moves the corpus of exercises from the realm of the practical and plausible to the realm of the artificial.

ISSUE 3: THE EPISTEMOLOGICAL STANCE UNDERLYING THE USE OF SUCH EXERCISE SETS. In introducing Milne's examples, we discussed the pedagogical assumptions underlying the use of such structured problem sets in the curriculum. Here we pursue the ramifications of those assumptions.

Almost all of Western education, particularly mathematics education and instruction, is based on a traditional philosophical perspective regarding epistemology, which is defined as "the theory or science of the method or grounds of knowledge" (*Oxford English Dictionary,* page 884). The fundamental concerns of epistemology regard the nature of knowing and knowledge. "*Know,* in its most general sense, has been defined by some as 'to hold for true or real with assurance and on (what is held to be) an adequate objective foundation' " (*Oxford English Dictionary,* page 1549). In more colloquial terms, the generally held view—often unstated or implicit, but nonetheless powerful—is that what we *know* is what we can justifiably demonstrate to be true; our *knowledge* is the sum total of what we know. That is, one's mathematical knowledge is the set of mathematical facts and procedures one can reliably and correctly use. Jim Greeno pointed out in his review of this chapter that most instruction gives short shrift to the "justifiably demonstrate" part of mathematical knowledge—that it focuses on using techniques, with minimal attention to having students justify the procedures in a deep way. He suggests that if demonstrating is taken in a deep sense, it might be an important curricular objective.

A consequence of the perspective described above is that instruction has traditionally focused on the content aspect of knowledge. Traditionally, one defines what students ought to know in terms of chunks of subject matter and characterizes what a student knows in terms of the amount of content that has been mastered. (The longevity of Bloom's (1956) taxonomies and the presence of standardized curricula and examinations provide clear evidence of the pervasiveness of this perspective.) As natural and innocuous as this view of "knowledge as substance" may seem, it has serious entailments (see Issue 4). From this perspective, "learning mathematics" is defined as mastering, in some coherent order, the set of facts and procedures that comprise the body of mathematics. The route to learning consists of delineating the desired subject-matter content as clearly as possible, carving it into bite-sized pieces, and providing explicit instruction and practice on each of those pieces so that students master them. From the content perspective, the whole of a student's mathematical understanding is precisely the sum of these parts.

Commonly, mathematics is associated with certainty; knowing it, with being able to get the right answer, quickly (Ball, 1988; Schoenfeld, 1985b; Stodolsky, 1985). These cultural assumptions are shaped by school experience, in which *doing* mathematics means following the rules laid down by the teacher; *knowing* mathematics means remembering and applying the correct rule when the teacher asks a question; and mathematical truth *is determined* when the answer is ratified by the teacher. Beliefs about how to do mathematics and what it means to know it in school are acquired through years of watching, listening, and practicing. (Lampert, 1990, p. 31)

These assumptions play out clearly in the selection from Milne. The topic to be mastered is a particular, rather narrow technique. The domain of applicability of the technique is made clear: Initially it applies to decimals that can be written as $(a/b) \times 10$, and then the technique is extended to apply to decimals that can be written as $(a/b) \times 100$. Students are constrained to use this technique, and when they master it, they move on to the next. For many students, experience with problem sets of this type is their sole encounter with mathematics.

ISSUE 4: THE CUMULATIVE EFFECTS OF SUCH EXERCISE SETS. As Lampert notes, students' primary experience with mathematics—the grounds upon which they build their understanding of the discipline—is their exposure to mathematics in the classroom. The impression given by this set of exercises, and thousands like it that students work in school, is that there is one right way to solve the given set of problems—the method provided by the text or instructor. As indicated in the discussion of Issue 1, this is emphatically not the case; there are numerous ways to arrive at the answer. However, in the given instructional context only one method appears legitimate. There are numerous consequences to repeated experiences of this type.

One consequence of experiencing the curriculum in bite-size pieces is that students learn that answers and methods to problems will be provided to them; the students are not expected to figure out the methods by themselves. Over time, most students come to accept their passive role and to think of mathematics as "handed down" by experts for them to memorize (Carpenter, Lindquist, Matthews, & Silver, 1983; National Assessment of Educational Progress, 1983).

A second consequence of the nonproblematic nature of these "problems" is that students come to believe that in mathematics, (1) one should have a ready method for the solution of a given problem, and (2) the method should produce an answer to the problem in short order (Carpenter et al., 1983; National Assessment of Educational Progress, 1983; Schoenfeld, 1988, 1989b). In the 1983 National Assessment, about half of the students surveyed agreed with the statement, "Learning mathematics is mostly memorizing." Three quarters of the students agreed with the statement, "Doing mathematics requires lots of practice in following rules," and nine students out of ten agreed with the statement, "There is always a rule to follow in solving mathematics problems" (NAEP, 1983, pp. 27–28). As a result of holding such beliefs, students may not even attempt problems for which they have no ready method, or may curtail their efforts after only a few minutes without success.

More importantly, the methods imposed on students by teacher and texts may appear arbitrary and may contradict the alternative methods that the students have tried to develop for themselves. For example, all of the problems given by Milne, and more generally, in most mathematics, can be solved in a variety of ways. However, only one method was sanctioned in Milne's text. Recall also that some of the problems were clearly artificial, negating the practical virtues of the mathematics. After consistent experiences of this type, students may simply give up trying to make sense of the mathematics. They may consider the problems to be exercises of little meaning, despite their applied cover stories; they may come to believe that mathematics is not something they can make sense of, but rather something almost completely arbitrary (or at least whose meaningfulness is inaccessible to them) and which must thus be memorized

without looking for meaning—if they can cope with it at all (Lampert, 1990; Stipek & Weisz, 1981; Tobias, 1978). More detail is given in the section on belief systems.

The Mathematical Enterprise. Over the past two decades, there has been a significant change in the face of mathematics (its scope and the very means by which it is carried out) and in the community's understanding of what it is to know and do mathematics. A series of recent articles and reports (Hoffman, 1989; National Research Council, 1989; Steen, 1988) attempts to characterize the nature of contemporary mathematics and to point to changes in instructions that follow from the suggested reconceptualization. The main thrust of this reconceptualization is to think of mathematics, broadly, as "the science of patterns."

MATHEMATICS . . . *searching for patterns*
Mathematics reveals hidden patterns that help us understand the world around us. Now much more than arithmetic and geometry, mathematics today is a diverse discipline that deals with data, measurements, and observations from science; with inference, deduction, and proof; and with mathematical models of natural phenomena, of human behavior, and of social systems.
The cycle from data to deduction to application recurs everywhere mathematics is used, from everyday household tasks such as planning a long automobile trip to major management problems such as scheduling airline traffic or managing investment portfolios. The process of "doing" mathematics is far more than just calculation or deduction; it involves observation of patterns, testing of conjectures, and estimation of results.
As a practical matter, mathematics is a science of pattern and order. Its domain is not molecules or cells, but numbers, chance, form, algorithms, and change. As a science of abstract objects, mathematics relies on logic rather than observation as its standard of truth, yet employs observation, simulation, and even experimentation as a means of discovering truth. (National Research Council, 1989, p. 31)

In this quotation there is a major shift from the traditional focus on the content aspect of mathematics discussed above (where attention is focused primarily on the mathematics one "knows"), to the process aspects of mathematics (what is known as "doing mathematics"). Indeed, content is mentioned only in passing, while modes of thought are specifically highlighted.

In addition to theorems and theories, mathematics offers distinctive modes of thought which are both versatile and powerful, including modeling, abstraction, optimization, logical analysis, inference from data, and use of symbols. Experience with mathematical modes of thought builds mathematical power—a capacity of mind of increasing value in this technological age that enables one to read critically, to identify fallacies, to detect bias, to assess risk, and to suggest alternatives. Mathematics empowers us to understand better the information-laden world in which we live. (National Research Council, 1989, pp. 31–32)

One main change, then, is that there is a large focus on process rather than on mathematical content in describing both what mathematics is and what one hopes students will learn from studying it. In this sense, mathematics appears much more like science than it would if one focused solely on the subject matter. Indeed, the "science of patterns" may seem so broad a definition as to obscure the mathematical core contained therein. What makes it mathematical is the domain over which the abstracting or patterning is done and the choice of tools and methods typically employed. To repeat from the introductory definition, mathematics consists of "systematic attempts, based on observation, study, and experimentation, to determine the nature or principles of regularities in systems defined axiomatically or theoretically (pure mathematics) or models of systems abstracted from real world objects (applied mathematics). The tools of mathematics are abstraction, symbolic representation, and symbolic manipulation" (Schoenfeld, in preparation).

A second main change, reflected in the statement that "mathematics relies on logic rather than observation as its standard of truth, yet employs observation, simulation, and even experimentation as a means of discovering truth" (National Research Council, 1989, p. 32) reflects a growing understanding of mathematics as an empirical discipline of sorts, one in which mathematical practitioners gather data in the same ways that scientists do. This theme is seen in the writings of Lakatos (1977, 1978), who argued that mathematics does not, as it often appears, proceed inexorably and inevitably by deduction from a small set of axioms; rather that the community of mathematicians decides what is axiomatic, in effect making new definitions if the ones that have been used turn out to have untoward consequences. A third change is that doing mathematics is increasingly coming to be seen as a social and collaborative act. Steen's (1988) examples of major progress in mathematics—in number theory (the factorization of huge numbers and prime testing, requiring collaborative networks of computers), in the Nobel Prize-winning application of the Radon Transform to provide the mathematics underlying the technology for computer-assisted tomography (CAT) scans, and in the solution of some recent mathematical conjectures such as the four-color theorem—are all highly collaborative efforts. Collaboration, on the individual level, is discussed with greater frequency in the "near mathematical" literature, as in these two excerpts from Albers and Alexanderson's (1985) *Mathematical People: Profiles and Interviews.* Peter Hilton lays out the benefits of collaboration as follows:

First I must say that I do enjoy it. I very much enjoy collaborating with friends. Second, I think it is an efficient thing to do because... if you are just working on your own [you may] run out of steam.... But with two of you, what tends to happen is that when one person begins to feel a flagging interest, the other one provides the stimulus.... The third thing is, if you choose people to collaborate with who somewhat complement rather than duplicate the contribution that you are able to make, probably a better product results. (p. 141)

Persi Diaconis says the following:

There is a great advantage in working with a great co-author. There is excitement and fun, and it's something I notice happening more and more in mathematics. Mathematical people enjoy talking to each other.... Collaboration forces you to work beyond your normal level. Ron Graham has a nice way to put it. He says that when you've done a joint paper, both co-authors do 75% of the work, and that's about right.... Collaboration for me means enjoying talking and explaining, false starts, and the interaction of personalities. It's a great, great joy to me. (pp. 74–75)

For these individuals, and for those engaged in the kinds of collaborative efforts discussed by Steen, membership in the mathematical community is without question an important part of their mathematical lives. However, there is an emerging epistemological argument suggesting that mathematical collaboration and communication have a much more important role than indicated by the quotes above. According to that argument, *membership in a community of mathematical practice is part of what constitutes mathematical thinking and knowing.* Greeno notes that this idea takes some getting used to.

The idea of a [collaborative] practice contrasts with our standard ways of thinking about knowledge. We generally think of knowledge as some content in someone's mind, including mental structures and procedures. In contrast, a practice is an everyday activity, carried out in a socially meaningful context in which activity depends on communication and collaboration with others and knowing how to use the resources that are available in the situation...

An important [philosophical and historical] example has been contributed by Kitcher (1984). Kitcher's goal was to develop an epistemology of mathematics. The key concept in his epistemology is an idea of a mathematical practice, and mathematical knowledge is to be understood as knowledge of mathematical practice. A mathematical practice includes understanding of the language of mathematical practice, and the results that are currently accepted as established. It also includes knowledge of the currently important questions in the field, the methods of reasoning that are taken as valid ways of establishing new results, and metamathematical views that include knowledge of general goals of mathematical research and appreciation of criteria of significance and elegance. (1988, pp. 24–25)

That is, "having a mathematical point of view" and "being a member of the mathematical community" are central aspects of having mathematical knowledge. Schoenfeld makes the case as follows:

I remember discussing with some colleagues, early in our careers, what it was like to be a mathematician. Despite obvious individual differences, we had all developed what might be called the mathematician's point of view—a certain way of thinking about mathematics, of its value, of how it is done, etc. What we had picked up was much more than a set of skills; it was a way of viewing the world, and our work. We came to realize that we had undergone a process of acculturation, in which we had become members of, and had accepted the values of, a particular community. As the result of a protracted apprenticeship into mathematics, we had become mathematicians in a deep sense (by dint of world view) as well as by definition (what we were trained in, and did for a living). (1987a, p. 213)

The epistemological perspective discussed here dovetails closely with the enculturation perspective discussed earlier in this chapter. Recall Resnick's (1989) observation that "becoming a good mathematical problem solver—becoming a good thinker in any domain—may be as much a matter of acquiring the habits and dispositions of interpretation and sense-making as of acquiring any particular set of skills, strategies, or knowledge" (p. 58). The critical observation in both the mathematical and the school contexts is that one develops one's point of view by the process of acculturation, by becoming a member of the particular community of practice.

Goals for Instruction, and a Pedagogical Imperative

The Mathematical Association of America's Committee on the Teaching of Undergraduate Mathematics recently issued a *Source Book for College Mathematic Teaching* (Schoenfeld, 1990a). The *Source Book* begins with a statement of goals for instruction, which seem appropriate for discussion here.

Goals for Mathematics Instruction

Mathematics instruction should provide students with a sense of the discipline—a sense of its scope, power, uses, and history. It should give them a sense of what mathematics is and how it is done, at a level appropriate for the students to experience and understand. As a result of their instructional experiences, students should learn to value mathematics and to feel confident in their ability to do mathematics.

Mathematics instruction should develop students' understanding of important concepts in the appropriate core content....Instruction should be aimed at conceptual understanding rather than at mere mechanical skills, and at developing in students the ability to apply the subject matter they have studied with flexibility and resourcefulness.

Mathematics instruction should provide students the opportunity to explore a broad range of problems and problem situations, ranging from exercises to open-ended problems and exploratory situations. It should provide students with a broad range of approaches and techniques (ranging from the straightforward application of the appropriate algorithmic methods to the use of approximation methods, various modeling techniques, and the use of heuristic problem-solving strategies) for dealing with such problems.

Mathematics instruction should help students to develop what might be called a "mathematical point of view"—a predilection to analyze and understand, to perceive structure and structural relationships, to see how things fit together. (Note that those connections may be either pure or applied.) It should help students develop their analytical skills, and the ability to reason in extended chains of argument.

Mathematics instruction should help students to develop precision in both written and oral presentation. It should help students learn to present their analyses in clear and coherent arguments reflecting the mathematical style and sophistication appropriate to their mathematical levels. Students should learn to communicate with us and with each other, using the language of mathematics.

Mathematics instruction should help students develop the ability to read and use text and other mathematical materials. It should prepare students to become, as much as possible, independent learners, interpreters, and users of mathematics. (Schoenfeld, 1990a, p. 2)

In light of the discussion from *Everybody Counts,* we would add the following to the second goal: Mathematics instruction should help students develop mathematical power, including the use of specific mathematical modes of thought that are both versatile and powerful, including modeling, abstraction, optimization, logical analysis, inference from data, and use of symbols.

If these are plausible goals for instruction, one must ask what kinds of instruction might succeed at producing them. The literature reviewed in this part of the chapter, in particular the literature on socialization and epistemology, produces what is in essence a pedagogical imperative:

If one hopes for students to achieve the goals specified here—in particular, to develop the appropriate mathematical habits and dispositions of interpretation and sense-making as well as the appropriately mathemati-

cal modes of thought—then the communities of practice in which they learn mathematics must reflect and support those ways of thinking. That is, classrooms must be communities in which mathematical sense-making, of the kind we hope to have students develop, is practiced.

A FRAMEWORK FOR EXPLORING MATHEMATICAL COGNITION

The Framework

The first part of this chapter focused on the mathematical enterprise—what *Everybody Counts* calls "doing" mathematics. Here we focus on the processes involved in thinking mathematically—the psychological support structure for mathematical behavior. The main focus of our discussion is on developments over the past quarter century. It would seem short-sighted to ignore the past 2,000 years of philosophy and psychology related to mathematical thinking and problem solving, so we begin with a brief historical introduction (see Peters, 1962, or Watson, 1978, for detail) to establish the context for the discussion of contemporary work and explain why the focus, essentially de novo, is on the past few decades. For ease of reference, we refer to the enterprise under the umbrella label "psychological studies," including contributions from educational researchers, psychologists, social scientists, philosophers, and cognitive scientists, among others. General trends are discussed here, with details regarding mathematical thinking given in the subsequent sections.

The roots of contemporary studies in thinking and learning can be traced to the philosophical works of Plato and Aristotle. More directly, Descartes's (1952) *Rules for the Direction of the Mind* can be seen as the direct antecedent of Pólya's (1945, 1954, 1981) prescriptive attempts at problem solving. However, the study of the mind and its workings did not turn into an empirical discipline until the late 19th century. The origins of that discipline are usually traced to the opening of Wundt's laboratory in Leipzig, Germany, in 1879. "Wundt was the first modern psychologist—the first person to conceive of experimental psychology as a science.... The methodological prescriptive allegiances of Wilhelm Wundt are similar to those of the physiologists from whom he drew inspiration....[H]e subscribed to methodological objectivism in that he attempted to quantify experience so that others could repeat his procedures... Since the combination of introspection and experiment was the method of choice, Wundt fostered empiricism" (Watson, 1978, p. 292). Wundt (1904) and colleagues employed the methods of experimentation and introspection (self-reports of intellectual processes) to gather data about the workings of mind. These methods may have gotten psychology off to an empirical start, but they soon led to difficulties; members of different laboratories reported different kinds of introspections (corresponding to the theories held in those laboratories), and there were significant problems with both reliability and replicability of the research findings.

American psychology's origins at the the turn of the century were more philosophical, tied to pragmatism and functionalism. William James is generally considered the first major American psychologist, and his (1890) *Principles of Psychology* is an exemplar of the American approach. James's student, E. L. Thorndike, began with animal studies and moved to studies of human cognition. Thorndike's work, in particular, had great impact on theories of mathematical cognition.

One of the major rationales for the teaching of mathematics, dating back to Plato, was the notion of mental discipline. Simply put, the idea is that those who are good at mathematics tend to be good thinkers; those who are trained in mathematics learn to be good thinkers. As exercise and discipline train the body, the theory went, the mental discipline associated with doing mathematics trains the mind, making one a better thinker. Thorndike's work challenged this hypothesis. He offered experimental evidence that transfer of the type suggested by the notion of mental discipline was minimal (Thorndike & Woodworth, 1901) and argued that the benefits attributed to the study of mathematics were correlational: Students with better reasoning skills tended to take mathematics courses (Thorndike, 1924). His research, based in animal and human studies, put forth the "law of effect," which says in essence "you get good at what you practice, and there isn't much transfer." His "law of exercise" gave details of the ways (recency and frequency effects) learning took place as a function of practice. As Peters notes, "Few would object to the first, at any rate, of these two laws, as a statement of a necessary condition of learning; it is when they come to be regarded as sufficient conditions that uneasiness starts" (1963, p. 695).

Unfortunately, that sufficiency criterion grew and held sway for quite some time. On the continent, Wundt's introspectionist techniques were shown to be methodologically unreliable, and the concept of mentalism came under increasing attack. In Russia, Pavlov (1924) achieved stunning results with conditioned reflexes, his experimental work requiring no concept of mind at all. Finally, mind, consciousness, and all related phenomena were banished altogether by the behaviorists. John Watson (1930) was the main exponent of the behaviorist stance; and B. F. Skinner (1974) was a zealous adherent. The behaviorists were vehement in their attacks on mentalism and provoked equally strong counter-reactions.

John Watson and other behaviorists led a fierce attack, not only on introspectionism, but also on any attempt to develop a theory of mental operations. Psychology, according to the behaviorists, was to be entirely concerned with external behavior and not to try to analyze the workings of the mind that underlay this behavior:

> Behaviorism claims that consciousness is neither a definite nor a usable concept. The behaviorist, who has been trained always as an experimentalist, holds further that belief in the existence of consciousness goes back to the ancient days of superstition and magic. (Watson, 1930, p. 2)

> ...The behaviorist began his own conception of the problem of psychology by sweeping aside all medieval conceptions. He dropped from his scientific vocabulary all subjective terms such as sensation, perception, image, desire, purpose, and even thinking and emotion as they were subjectively defined. (Watson, 1930, p. 5)

The behaviorist program and the issues it spawned all but eliminated any serious research in cognitive psychology for 40 years. The rat

supplanted the human as the principal laboratory subject, and psychology turned to finding out what could be learned by studying animal learning and motivation. (Anderson, 1985, p. 7)

While behaviorism held center stage, alternate perspectives were in the wings. Piaget's work (1928, 1930, 1971), while rejected by his American contemporaries as being unrigorous, established the basis for the constructivist perspective, the now well-established position that individuals do not perceive the world directly, but that they perceive interpretations of it mediated by the interpretive frameworks they have developed. The Gestaltists, particularly Duncker, Hadamard, and Wertheimer, were interested in higher-order thinking and problem solving. The year 1945 was a banner year for the Gestaltists. Duncker's (1945) monograph *On Problem Solving* appeared in English, as did Hadamard's (1945) *Essay on the Psychology of Invention in the Mathematical Field* [which provides a detailed exegesis of Poincare's (1913) description of his discovery of the structure of Fuchsian functions], and Wertheimer's (1945/1959) *Productive Thinking,* which includes Wertheimer's famous discussion of the "parallelogram problem" and an interview with Einstein on the origins of relativity theory. These works all continued the spirit of Graham Wallas's (1926) *The Art of Thought,* in which Wallas codified the four-step Gestalt model of problem solving: saturation, incubation, inspiration, and verification. The Gestaltists, especially Wertheimer, were concerned with structure and deep understanding. Unfortunately their primary methodological tool was introspection, and they were vulnerable to attack on the basis of the methodology's lack of reliability and validity. (They were also vulnerable because they had no plausible theory of mental mechanism, while the behaviorists could claim that stimulus-response chains were modeled on neuronal connections.) To cap off the year 1945, Pólya's *How to Solve It*—compatible with the Gestaltists' work, but more prescriptive, à la Descartes, in flavor—appeared as well.

The downfall of behaviorism and the renewed advent of mentalism, in the form of the information processing approach to cognition, began in the mid-1950s (see Newell & Simon, 1972, pp. 873 ff. for detail). The development of artificial intelligence programs to solve problems, such as Newell & Simon's (1972) "General Problem Solver," hoist the behaviorists by their own petard.

> The simulation models of the 1950s were offspring of the marriage between ideas that had emerged from symbolic logic and cybernetics, on the one hand, and Würzburg and Gestalt psychology, on the other. From logic and cybernetics was inherited the idea that information transformation and transmission can be described in terms of the behavior of formally described symbol manipulation systems. From Würzburg and Gestalt psychology were inherited the ideas that long-term memory is an organization of directed associations and that problem solving is a process of directed goal-oriented search. (Simon, 1979, pp. 364–365)

Note that the information-processing work discussed by Simon met the behaviorists' standards in an absolutely incontrovertible way: Problem-solving programs (simulation models and artificial intelligence programs) produced problem-solving behavior, and all the workings of the program were out in the open for inspection. At the same time, the theories and methodologies of the information-processing school were fundamentally mentalistic—grounded in the theories of mentalistic psychology, and using observations of humans engaged in problem solving to infer the structure of their (mental) problem-solving strategies. Although it was at least a decade before such work had an impact on mainstream experimental psychology (Simon, 1979), and it was as late as 1980 that Simon and colleagues (Ericsson & Simon, 1980) were writing review articles hoping to legitimize the use of out-loud problem-solving protocols, an emphasis on cognitive processes emerged, stabilized, and began to predominate in psychological studies of mind.

Early work in the information-processing (IP) tradition was extremely narrow in focus, partly because of the wish to have clean, scientific results. For many, the only acceptable test of a theory was a running computer program that did what its author said it should. Early IP work often focused on puzzle domains (such as the Tower of Hanoi problem and its analogues), with the rationale that in such simple domains one could focus on the development of strategies, and then later move to semantically rich domains. As the tools were developed, studies moved from puzzles and games (logic, cryptarithmetic, and chess, for example) to more open-ended tasks, focusing on text-book tasks in domains such as physics and mathematics (and later, in developing expert systems in medical diagnosis, mass spectroscopy, and other areas). Nonetheless, work in the IP tradition remained quite narrow for some time. The focus was on the "architecture of cognition" (and machines): the structure of memory, of knowledge representations, of knowledge retrieval mechanisms, and of problem solving rules.

During the same time period (the first paper on metamemory by Flavell, Friedrichs, & Hoyt appeared in 1970; the topic peaked in the mid-to-late 1980s) metacognition became a major research topic. Here too, the literature is quite confused. In an early paper, Flavell characterized the term as follows:

Metacognition refers to one's knowledge concerning one's own cognitive processes or anything related to them, e.g. the learning-relevant properties of information or data. For example, I am engaging in metacognition...if I notice that I am having more trouble learning A than B; if it strikes me that I should double-check C before accepting it as a fact; if it occurs to me that I should scrutinize each and every alternative in a multiple-choice task before deciding which is the best one....Metacognition refers, among other things, to the active monitoring and consequent regulation and orchestration of those processes in relation to the cognitive objects or data on which they bear, usually in the service of some concrete [problem solving] goal or objective. (1976, p. 232)

This kitchen-sink definition includes a number of categories which have since been separated into more functional categories for exploration: (1) individuals' declarative knowledge about their cognitive processes, (2) self-regulatory procedures, including monitoring and "on-line" decision-making, and (3) beliefs and affects and their effects on performance. (Through the early 1980s, the cognitive and affective literatures were separate and unequal. The mid-1980s saw a rapprochement, with the notion of beliefs extending the scope of the cognitive inquiries to be at least compatible with those of the affective domain. Since then, the enculturation perspective discussed earlier has moved the two a bit closer.) These subcategories are considered in the framework elaborated below.

Finally, the tail end of the 1980s saw a potential unification of aspects of what might be called the cognitive and social perspectives on human behavior, in the theme of enculturation. The minimalist version of this perspective is that learning is a social act, taking place in a social context; that one must consider learning environments as cultural contexts and learning as a cultural act. (The maximal version, yet to be realized theoretically, is a unification that allows one to see what goes on "inside the individual head" and "distributed cognition" as aspects of the same thing.) Motivated by Lave's work on apprenticeship (1988, in preparation), Collins et al. (1989) abstracted common elements from productive learning environments in reading (Palincsar & Brown, 1984), writing (Scardamalia & Bereiter, 1983) and mathematics (Schoenfeld, 1985a). Across the case studies they found a common, broad conceptualization of domain knowledge that included not only the specifics of domain knowledge, but also an understanding of strategies and aspects of metacognitive behavior. In addition, they found that all three programs had aspects of "the culture of expert practice," in that the environments were designed to take advantage of social interactions to have students experience the gestalt of the discipline in ways comparable to the ways that practitioners do.

In general, research in mathematics education followed a similar progression of ideas and methodologies. Through the 1960s and 1970s, correlational, factor-analytic, and statistical "treatment A vs. treatment B" comparison studies predominated in the "scientific" study of thinking, learning, and problem solving. By the mid-1970s, however, researchers expressed frustration at the limitations of the kinds of contributions that could, in principle, be made by such studies of mathematical behavior. For example, Kilpatrick (1978) compared the research methods prevalent in the United States at the time with the kinds of qualitative research being done in the Soviet Union by Krutetskii (1976) and his colleagues. The American research, he claimed, was "rigorous" but somewhat sterile. In the search for experimental rigor, researchers had lost touch with truly meaningful mathematical behavior. In contrast, the Soviet studies of mathematical abilities were decidedly unrigorous, if not unscientific, but they focused on behavior and abilities that had face validity as important aspects of mathematical thinking. Kilpatrick suggested that the research community might do well to broaden the scope of its inquiries and methods.

Indeed, researchers in mathematics education turned increasingly to process-oriented studies in the late 1970s and 1980s. Much of the process-oriented research was influenced by the trends in psychological work described above, but it also had its own special character. As noted above, psychological research tended to focus on "cognitive architecture"—studies of the structure of memory, of representations, and so forth. From a psychological point of view, mathematical tasks were attractive as settings for such research because of their (ostensibly) formal, context-independent nature. That is, topics from literature or history might be "contaminated" by real-world knowledge, a fact that would make it difficult to control precisely what students brought to, or learned in, experimental settings. But purely formal topics from mathematics (for example, the algorithm for base 10 addition and subtraction or the rules for solving linear equations in one variable) could be taught as purely formal manipulations, and thus one could avoid the difficulties of "contamination." In an early information-processing study of problem solving, for example, Newell and Simon (1972) analyzed the behavior of students solving problems in symbolic logic. From their observations, they abstracted successful patterns of symbol manipulation and wrote them as computer programs. However, Newell and Simon's sample explicitly excluded any subjects who knew the meanings of the symbols (for example, that "P → Q" means "if P is true, then Q is true"), because their goal was to find productive modes of symbol manipulation without understanding the symbols, since the computer programs they intended to write wouldn't be able to reason on the basis of those meanings. In contrast, of course, the bottom line for most mathematics educators is to have students develop an understanding of the procedures and their meanings. Hence, the IP work took on a somewhat different character when adapted for the purposes of mathematics educators.

The state of the art in the early and late 1980s, respectively, can be seen in two excellent summary volumes—Silver's (1985) *Teaching and Learning Mathematical Problem Solving: Multiple Research Perspectives* and Charles and Silver's (1988) *The Teaching and Assessing of Mathematical Problem Solving.* Silver's volume was derived from a conference held in 1983, which brought together researchers from numerous disciplines to discuss results and directions for research in problem solving. Some confusion, a great deal of diversity, and a flowering of potentially valuable perspectives are evident in the volume. There was confusion, for example, about baseline definitions of "problem solving." Kilpatrick (1985), for example, gave a range of definitions and examples that covered the spectrum discussed in the first part of this chapter. And either explicitly or implicitly, that range of definitions was exemplified in the chapters of the book. At one end of the spectrum, Carpenter (1985) began his chapter with a discussion of the following problem: "James had 13 marbles. He lost 8 of them. How many marbles does he have left?" Carpenter notes that "such problems frequently are not included in discussions of problem solving because they can be solved by the routine application of a single arithmetic operation. A central premise of this paper is that the solutions of these problems, particularly the solutions of young children, do in fact involve real problem-solving behavior" (p. 17). Heller and Hungate (1985) implicitly take their definition of problem solving to mean being able to solve the exercises at the end of a standard textbook chapter, as does Mayer. At the other end of the spectrum, "the fundamental importance of epistemological issues (for example, beliefs, conceptions, and misconceptions) is reflected in the papers by Jim Kaput, Richard Lesh, Alan Schoenfeld, and Mike Shaughnessy (p. ix.)". Those chapters took a rather broad view of problem solving and mathematical thinking. Similarly, the chapters reveal a great diversity of methods and their productive application to issues related to problem solving. Carpenter's chapter presents detailed cross-sectional data on children's use of various strategies for solving word problems of the type discussed above. Heller and Hungate worked within the "expert-novice" paradigm for identifying the productive behavior of competent problem solvers and using such behavior as a guide for instruction for novices. Mayer discussed the application of schema theory, again within the expert-novice paradigm. Kaput

discussed fundamental issues of representation and their role in understanding, Shaughnessy discussed misconceptions, and Schoenfeld explained the roles of metacognition and beliefs. Alba Thompson (1985) studied teacher beliefs and their effects on instruction. There was also diversity in methodology: experimental methods, expert-novice studies, clinical interviews, protocol analyses, and classroom observations, among others. The field had clearly flowered, and there was a wide range of new work.

The Charles and Silver volume (1988) reflects a maturing of and continued progress in the field. By the end of the decade, most of the methodologies and perspectives tentatively explored in the Silver volume had been studied at some length, with the result that they had been contextualized in terms of just what they could offer in terms of explaining mathematical thinking. For example, the role of information-processing approaches and the expert-novice paradigm could be seen as providing certain kinds of information about the organization and growth of individual knowledge, but as illuminating only one aspect of a much larger and more complex set of issues. With more of the methodological tools in place, it became possible to take a broad view once again, focusing, for example, on history (the Stanic and Kilpatrick chapter in the Charles and Silver volume) and epistemology as grounding contexts for explorations into mathematical thinking. In the Charles and Silver volume, one sees the theme of social interactions and enculturation emerging as central concerns, while in the earlier Silver volume such themes were noted but put aside as "things we aren't really ready to deal with." What one sees is the evolution of overarching frameworks, such as cognitive apprenticeship, that deal with individual learning in a social context. That social theme is explored in the work of Greeno (1988), Lave et al. (1988), and Resnick (1988), among others. There is not at present anything resembling a coherent explanatory frame—that is, a principled explanation of how the varied aspects of mathematical thinking and problem solving fit together. However, there does appear to be an emerging consensus about the necessary scope of inquiries into mathematical thinking and problem solving. Although the fine detail varies [Collins et al. (1989) subsume the last two categories under a general discussion of "culture"; Lester, Garofalo, & Kroll (1989) subsume problem-solving strategies under the knowledge base, while maintaining separate categories for belief and affect], there appears to be general agreement on the importance of these five aspects of cognition:

- The knowledge base
- Problem-solving strategies
- Monitoring and control
- Beliefs and affects
- Practices

These five categories provide the framework employed in the balance of the review.

The Knowledge Base

Research on human cognitive processes over the past quarter century has focused on the organization of, and access to, information contained in memory. In the crudest terms, the underlying issues have been how information is organized and stored in the head; what comprises understanding; and how individuals have access to relevant information. The mainstream idea is that humans are information processors and that in their minds humans construct symbolic representations of the world. According to this view, thinking about and acting in the world consist, respectively, of operating mentally on those representations and taking actions externally that correspond to the results of our minds' internal workings. While these are the mainstream positions—and the ones elaborated below—it should be noted that all of them are controversial. There is, for example, a theoretical stance regarding distributed cognition (Pea, 1989), which argues that it is inappropriate to locate knowledge "in the head"—that knowledge resides in communities and their artifacts and in interactions between individuals and their environments (which include other people). The related concept of situated cognition (see, for example, Barwise & Perry, 1983; Brown et al., 1989; Lave & Wenger, 1989) is based on the underlying assumption that mental representations are not complete and that thinking exploits the features of the world in which one is embedded, rather than operating on abstractions of it. Moreover, even if one accepts the notion of internal cognitive representation, there are multiple perspectives regarding the nature and function of representations (see, for example, Janvier, 1987, for a collection of papers regarding perspectives on representations in mathematical thinking; for a detailed elaboration of such issues within the domain of algebra, see Wagner & Kieran, 1989, especially the chapter by Kaput), or of what "understanding" might be. (For a detailed elaboration of such themes with regard to elementary mathematics, see Putnam, Lampert, & Peterson, 1989.) Hence, the sequel presents what might be considered "largely agreed upon" perspectives.

Suppose a person finds him or herself in a situation that calls for the use of mathematics, either for purposes of interpretation (mathematizing) or problem solving. In order to understand the individual's behavior—which options are pursued, in which ways—one needs to know what mathematical tools the individual has at his or her disposal. Simply put, the issues related to the individual's knowledge base are: What information relevant to the mathematical situation or problem at hand does he or she possess? And how is that information accessed and used?

Although these two questions appear closely related, they are, in a sense, almost independent. By way of analogy, consider the parallel questions with regard to the contents of a library: What's in it, and how do you gain access to the contents? The answer to the first question is contained in the catalogue: a list of books, records, tapes, and other materials the library possesses. The contents are what interest you if you have a particular problem or need particular resources. How the books are catalogued or how you gain access to them is somewhat irrelevant (especially if the ones you want aren't in the catalogue). On the other hand, once you are interested in finding and using something listed in the catalogue, the situation changes. How the library actually works becomes critically important: Procedures for locating a book on the shelves, taking it to the desk, and checking it out must be understood.

Note, incidentally, that these procedures are largely independent of the contents of the library. One would follow the same set of procedures for accessing any two books in the general collection.

The same holds for assessing the knowledge base an individual brings to a problem-solving situation. In analyses of problem-solving performance, for example, the central issues most frequently deal with what individuals know (the contents of memory) and how that knowledge is deployed. In assessing decision-making during problem solving, for instance, one needs to know what options problem solvers had available. Did they fail to pursue particular options because they overlooked them or because they didn't know of their existence? In the former case, the difficulty might be metacognitive or not seeing the right "connections"; in the latter case, it is a matter of not having the right tools. From the point of view of the observer or experimenter trying to understand problem-solving behavior, then, a major task is the delineation of the knowledge base of individuals who confront the given problem-solving tasks. It is important to note that in this context, the knowledge base may contain things that are not true. Individuals bring misconceptions and misunderstood facts to problem situations; it is essential to understand that *those* are the tools they work with.

The Knowledge Inventory (Memory Contents). Broadly speaking, aspects of the knowledge base relevant for competent performance in a domain include informal and intuitive knowledge about the domain, facts, definitions, and the like; algorithmic procedures; routine procedures; relevant competencies; and knowledge about the rules of discourse in the domain. (This discussion is abstracted from pages 54–61 of Schoenfeld, 1985a.) Consider, for example, the resources an individual might bring to the following problem:

Problem

You are given two intersecting straight lines and a point P marked on one of them, as in the figure below. Show how to construct, using a straightedge and compass, a circle that is tangent to both lines and that has the point P as its point of tangency to one of the lines.

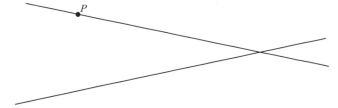

Informal knowledge an individual might bring to the problem includes general intuitions about circles and tangents and notions about "fitting tightly" that correspond to tangency. It also includes perceptual biases, such as a strong predilection to observe the symmetry between the points of tangency on the two lines. (This particular feature tends to become less salient, and ultimately negligible, as the vertex angle is made larger.) Of course, Euclidean geometry is a formal game; these informal understandings must be exploited within the context of the rules for constructions. As noted above, the facts, definitions, and algorithmic procedures the individual brings to the

TABLE 15–1. Part of the Inventory of an Individual's Resources for Working the Construction Problem

Degree of Knowledge	of	facts	and	procedures
Does the student:				
a. know nothing about b. know about the existence of, but nothing about the details of c. partially recall or suspect the details, but with little certainty d. confidently believe		the tangent to a circle is perpendicular to the radius drawn to the point of tangency (true) any two constructible loci suffice to determine the location of a point (true with qualifications) the center of an inscribed circle in a triangle lies at the intersection of the medians (false)		a (correct) procedure for bisecting an angle a (correct) procedure for dropping a perpendicular to a line from a point an (incorrect) procedure for erecting a perpendicular to a line through a given point on that line

problem situation may or may not be correct; they may be held with any degree of confidence from absolute (but possibly incorrect) certainty to great uncertainty. Part of this aspect of the knowledge inventory is outlined in Table 15.1.

Routine procedures and relevant competencies differ from facts, definitions, and algorithmic procedures in that they are somewhat less cut-and-dried. Facts are right or wrong. Algorithms, when applied correctly, are guaranteed to work. Routine procedures are likely to work, but with no guarantees. For example, the problem above, although stated as a construction problem, is intimately tied to a proof problem. One needs to know what properties the desired circle has; the most direct way of determining the properties is to prove that in a figure including the circle (see Figure 15.1), PV and QV are the same length, and CV bisects angle PVQ.

The relevant proof techniques are not algorithmic, but they are somewhat routine. People experienced in the domain know that one should seek congruent triangles, and that it is appropriate to draw in the line segments CV, CP, and CQ; that one of the standard methods for proving triangles congruent (SSS, ASA, AAS, or hypotenuse-leg) should probably be used; and that this knowledge should drive the search process. We note that all of the comments made in the discussion of Table 15.1 regarding the correctness of resources and the degree of certainty with which they are held apply to relevant procedures and routine competencies—what counts is what the individual

holds to be true. Finally, we note the importance of understanding the rules of discourse in the domain. As noted above, Euclidean geometry is a formal game; one has to play by certain rules. For example, we can't "line up" a tangent by eye or determine the diameter of a circle by sliding a ruler along until the largest chord is obtained. While such procedures may produce the right values empirically, they are proscribed in the formal domain. People who understand this will behave very differently from those who don't.

Access to Resources (the Structure of Memory). We now turn to the issue of how the contents of memory are organized, accessed, and processed. Figure 15.2, taken from Silver (1987), provides the overarching structure for the discussion. See Norman (1970) or Anderson (1983) for general discussions.

Here, in brief, are some of the main issues brought to center stage by Figure 15.2. First is the notion that human beings are information processors, acting on the basis of their coding of stimuli experienced in the world. That is, one's experiences—visual, auditory, tactile—are registered in sensory buffers and then (if they are not ignored) converted into the forms in which they are employed in working and long-term memory. The sensory buffer (also called iconic memory, for much of its content is in the forms of images) can register a great deal of information, but hold it only briefly. Some of that information will be lost, and some will be transmitted to working memory (one can take in a broad scene perceptually, but only reproduce a small percentage of it). Speaking loosely, working or short-term memory is where "thinking gets done." Working memory receives its contents from two sources—the sensory buffer and long-term memory.

The most important aspect of working or short-term memory (STM) is its limited capacity. Pioneering research by Miller (1956) indicated that, despite the huge amount of information humans can remember in general, they can only keep and operate on about seven "chunks" of information in short-term memory. For example, a person, unless specially trained, will find it

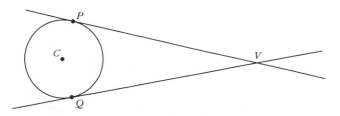

FIGURE 15–1. The Desired Configuration

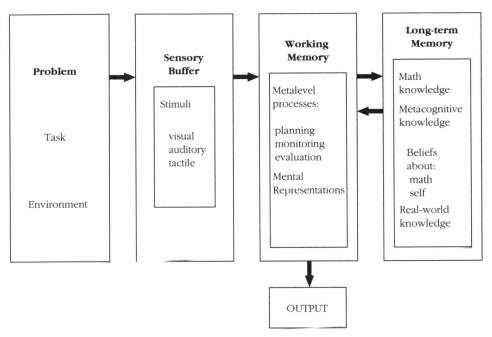

FIGURE 15–2. The Structure of Memory

nearly impossible to find the product of 637 and 829 mentally; the number of subtotals one must keep track of is too large for STM to hold. In this arithmetic example, the pieces of information in STM are relatively simple. Each of the 7 ± 2 chunks in STM can, however, be quite complex. As Simon (1980) points out:

A chunk is any perceptual configuration (visual, auditory, or what not) that is familiar and recognizable. For those of us who know the English language, spoken and printed words are chunks...For a person educated in Japanese schools, any one of several thousand Chinese ideograms is a single chunk (and not just a complex collection of lines and squiggles), and even many pairs of such ideograms constitute single chunks. For an experienced chess player, a "fianchettoed castled Black King's position" is a chunk, describing the respective locations of six or seven of the Black pieces. (1980, p. 83)

In short, the architecture of STM imposes severe constraints on the kinds and amount of mental processing people can perform. The operation of chunking—by which one can have compound entities in the STM slots—only eases the constraints somewhat. "Working-memory load" is indeed a serious problem when people have to keep multiple ideas in mind during problem solving. The limits on working memory also suggest that for "knowledge-rich" domains [chess a generic example (see below), but mathematics certainly one as well], there are severe limitations to the amount of "thinking things out" that one can do; the contents of the knowledge base are critically important.

Long-term memory (LTM) is an individual's permanent knowledge repository. Details of its workings are still very much open to question and too fine-grained for this discussion, but a general consensus appears to be that some sort of "neural network" representation—graphs whose vertices (nodes) represent chunks in memory and whose links represent connections between those chunks—is appropriate. Independent of these architectural issues, the fundamental issues have to do with the nature of knowledge and the organization of knowledge for access (that is, to be brought into STM) and use.

Before turning to issues of organization and access, one should note a long-standing distinction between two types of knowledge, characterized by Ryle (1949) respectively, as "knowing that" and "knowing how." More modern terminology, employed by Anderson (1976), is that of "declarative" and "procedural" knowledge, respectively. The relationship between the two is not clear-cut; see Hiebert (1985) for a set of contemporary studies exploring the connections between them.

One of the domains in which the contents of memory has been best elaborated is chess. De Groot (1965) explored chess masters' competence, looking for explanations such as "spatial ability" to explain their ability to "size up" a board rapidly and play numerous simultaneous games of chess. He briefly showed experts and novices typical midgame positions and asked them to recreate the positions on nearby chess boards. The masters' performance was nearly flawless; the novices' quite poor. However, when the two groups were asked to replicate positions where pieces had been randomly placed on the chess boards, experts did no better than novices, and when they were asked to replicate positions that were *almost* standard chess positions, the masters often replicated the standard positions—the ones they expected to see. That is, the experts had "vocabularies" of chess positions, some 50,000 well-recognized configurations, which they recognized and to which they responded automatically. These vocabularies formed the base (but not the whole) of their competence.

The same, it is argued, holds in all domains, including mathematics. Depending on the knowledge architecture invoked, the knowledge chunks may be referred to as scripts (Schank & Abelson, 1977), frames (Minsky, 1975), or schemata (Hinsley, Hayes, & Simon, 1977). Nonetheless, the basic underlying notion is the same: People abstract and codify their experiences, and the codifications of those experiences shape what people see and how they behave when they encounter new situations related to the ones they have abstracted and codified. The Hinsley, Hayes, and Simon study is generic in that regard. In one part of their work, for example, they read the first few words of a problem statement to subjects and asked the subjects to categorize the problem—to say what information they expected the problem to provide and what they were likely to be asked.

[A]fter hearing the three words "a river steamer" from a river current problem, one subject said, "It's going to be one of those river things with upstream, downstream, and still water. You are going to compare times upstream and downstream—or if the time is constant, it will be distance."...After hearing five words of a triangle problem, one subject said, "This may be something about 'how far is he from his goal?' using the Pythagorean theorem." (1977, p. 97)

The Hinsley, Hayes, and Simon findings were summed up as follows:

1. People can categorize problems into types.
2. People can categorize problems without completely formulating them for solution. If the category is to be used to cue a schema for formulating a problem, the schema must be retrieved before formulation is complete.
3. People have a body of information about each problem type which is potentially useful in formulating problems of that type for solutions,... directing attention to important problem elements, making relevance judgments, retrieving information concerning relevant equations, etc.
4. People use category identifications to formulate problems in the course of actually solving them (Hinsley et al., 1977, p. 92).

In sum, the findings of work in domains such as chess and mathematics point strongly to the importance and influence of the knowledge base. First, it is argued that expertise in various domains depends of having access to some 50,000 chunks of knowledge in LTM. Since it takes some time (perhaps 10 seconds of rehearsal for the simplest items) for each chunk to become embedded in LTM, and longer for knowledge connections to be made, that is one reason expertise takes as long as it does to develop. Second, a lot of what appears to be strategy use is in fact reliance on well-developed knowledge chunks of the type, "in this well-recognized situation, do the following." Nonetheless, it is important not to overplay the roles of these knowledge schemata, for they do play the role of vocabulary—the basis for routine performance in familiar territory. Chess players, when playing at the limit of their own abilities, do rely automatically on their vocabularies of chess positions, but also do significant strategizing. Similarly, mathematicians have immediate access to large amounts

of knowledge, but also employ a wide range of strategies when confronted with problems beyond the routine (and those, of course, are the problems mathematicians care about.) However, the straightforward suggestion that mathematics instruction focus on problem schemata does not sit well with the mathematics education community, for good reason. As noted in the historical review, IP work has tended to focus on performance but not necessarily on the basic understandings that support it. Hence, a reliance on schemata in crude form—"When you see these features in a problem, use this procedure"—may produce surface manifestations of competent behavior. However, that performance may, if not grounded in an understanding of the principles that led to the procedure, be error prone and easily forgotten. Thus, many educators would suggest caution when applying research findings from schema theory. For an elaboration of the underlying psychological ideas and the reaction from mathematics education, see the papers by Mayer (1985) and Sowder (1985).

Problem-solving Strategies (Heuristics). Discussions of problem-solving strategies in mathematics, or heuristics, must begin with Pólya. Simply put, *How to Solve It* (1945) planted the seeds of the problem-solving "movement" that flowered in the 1980s: Open the 1980 NCTM *Yearbook* (Krulik, 1980) to any page, and you are likely to find Pólya invoked, either directly or by inference in the discussion of problem-solving examples. The *Yearbook* begins by reproducing the *How to Solve It* problem-solving plan on its flyleaf and continues with numerous discussions of how to implement Pólya-like strategies in the classroom. Nor has Pólya's influence been limited to mathematics education. A cursory literature review found his work on problem solving cited in *American Political Science Review*, *Annual Review of Psychology*, *Artificial Intelligence*, *Computers and Chemistry*, *Computers and Education*, *Discourse Processes*, *Educational Leadership*, *Higher Education*, and *Human Learning*, to name just a few. Nonetheless, a close examination reveals that while his name is frequently invoked, his ideas are often trivialized. Little that goes in the name of Pólya also goes in the spirit of his work. Here we briefly follow two tracks: research exploring the efficacy of heuristics, or problem-solving strategies, and the "real world" implementation of problem-solving instruction.

MAKING HEURISTICS WORK. The scientific status of heuristic strategies such as those discussed by Pólya in *How to Solve It*—strategies in his "short dictionary of heuristics" such as (exploiting) analogy, auxiliary elements, decomposing and recombining, induction, specialization, variation, and working backwards—has been problematic, although the evidence appears to have turned in Pólya's favor over the past decade.

There is no doubt that Pólya's accounts of problem solving have face validity, in that they ring true to people with mathematical sophistication. Nonetheless, through the 1970s there was little empirical evidence to back up the sense that heuristics could be used as a means to enhance problem solving. For example, Wilson (1967) and Smith (1973) found that the heuristics that students were taught did not, despite their ostensible generality, transfer to new domains. Studies of problem-solving behaviors by Kantowski (1977), Kilpatrick (1967), and Lucas (1974) did indicate that students' use of heuristic strategies was positively correlated with performance on ability tests and on specially constructed problem-solving tests; however, the effects were relatively small. Harvey and Romberg (1980), in a compilation of dissertation studies in problem solving over the 1970s, indicated that the teaching of problem-solving strategies was "promising" but had yet to pan out. Begle had the following pessimistic assessment of the state of the art as of 1979:

A substantial amount of effort has gone into attempts to find out what strategies students use in attempting to solve mathematical problems. …No clear-cut directions for mathematics education are provided by the findings of these studies. In fact, there are enough indications that problem-solving strategies are both problem- and student-specific often enough to suggest that finding one (or few) strategies which should be taught to all (or most) students are far too simplistic. (1979, pp. 145-146)

In other fields such as artificial intelligence, where significant attention was given to heuristic strategies, strategies of the type described by Pólya were generally ignored (see, e.g., Groner, Groner, & Bischof, 1983; Simon, 1980). Newell, in summing up Pólya's influence, states the case as follows.

This chapter is an inquiry into the relationship of George Pólya's work on heuristic to the field of artificial intelligence (hereafter, AI). A neat phrasing of its theme would be *Pólya revered and Pólya ignored*. Pólya revered, because he is recognized in AI as the person who put heuristic back on the map of intellectual concerns. But Pólya ignored, because no one in AI has seriously built on his work.…

Everyone in AI, at least that part within hailing distance of problem solving and general reasoning, knows about Pólya. They take his ideas as provocative and wise. As Minsky (1961) states, "And everyone should know the work of Pólya on how to solve problems." But they also see his work as being too informal to build upon. Hunt (1975) has said "Analogical reasoning is potentially a very powerful device. In fact, Pólya [1954] devoted one entire volume of his two volume work to the discussion of the use of analogy and induction in mathematics. Unfortunately, he presents ad hoc examples but no general rules [p. 221]."

The 1980s have been kinder to heuristics à la Pólya. In short, the critique of the strategies listed in *How to Solve It* and its successors is that the characterizations of them were descriptive rather than prescriptive. That is, the characterizations allowed one to recognize the strategies when they were being used. However, Pólya's characterizations did not provide the amount of detail that would enable people who were not already familiar with the strategies to be able to implement them. Consider, for example, an ostensibly simple strategy such as "examining special cases." (This discussion is taken from pp. 288–290 of Schoenfeld [1987b, December].)

To better understand an unfamiliar problem, you may wish to exemplify the problem by considering various special cases. This may suggest the direction of, or perhaps the plausibility of, a solution.

Now consider the solutions to the following three problems.

Problem 1. Determine a formula in closed form for the series

$$\sum_{k=1}^{n} \frac{k}{(k+1)!}$$

Problem 2. Let P(x) and Q(x) be polynomials whose coefficients are the same but in "backwards order":

$$P(x) = a_0 + a_1 x + a_2 x^2 + \ldots a_n x^n, \text{ and}$$
$$Q(x) = a_n + a_{n-1} x + a_{n-2} x^2 + \ldots a_0 x^n$$

What is the relationship between the roots of P(x) and Q(x)? Prove your answer.

Problem 3 Let the real numbers a_0 and a_1 be given. Define the sequence $\{a_n\}$ by

$$a_n = \tfrac{1}{2}(a_{n-2} + a_{n-1}) \text{ for each } n \geq 2.$$

Does the sequence $\{a_n\}$ converge? If so, to what value?

Details of the solutions will not be given here. However, the following observations are important. For Problem 1, the special cases that help are examining what happens when the integer parameter n takes on the values 1, 2, 3, … in sequence; this suggests a general pattern that can be confirmed by induction. Yet trying to use special cases in the same way on the second problem may get one into trouble: Looking at values $n = 1, 2, 3, \ldots$ can lead to a wild goose chase. In Problem 2, the "right" special cases of P(x) and Q(x) to look at are easily factorable polynomials. Considering $P(x) = (2x + 1)(x + 4)(3x - 2)$, for example, leads to the discovery that its "reverse" can be factored without difficulty. The roots of P and Q are easy to compare, and the result (which is best proved another way) becomes obvious. Again, the special cases that simplify the third problem are different in nature. Choosing the values $a_0 = 0$ and $a_1 = 1$ allows one to see what happens for the sequence that those two values generate. The pattern in that case suggests what happens in general and (especially if one draws the right picture!) leads to a solution of the original problem.

Each of these problems typifies a large class of problems and exemplifies a different special-cases strategy. We have

Strategy 1. When dealing with problems in which an integer parameter n plays a prominent role, it may be of use to examine values of $n = 1, 2, 3, \ldots$ in sequence, in search of a pattern.

Strategy 2. When dealing with problems that concern the roots of polynomials, it may be of use to look at easily factorable polynomials.

Strategy 3. When dealing with problems that concern sequences or series that are constructed recursively, it may be of use to try initial values of 0 and 1—if such choices don't destroy the generality of the processes under investigation.

Needless to say, these three strategies hardly exhaust "special cases." At this level of analysis—the level necessary for implementing the strategies—one could find a dozen more. This is the case for almost all of Pólya's strategies. The indications are that students can learn to use these more carefully delineated strategies (see, for example, Schoenfeld, 1985a).

Generally speaking, studies of comparable detail have yielded similar findings. Silver (1979, 1981), for example, showed that "exploiting related problems" is much more complex than it first appears. Heller and Hungate (1985), in discussing the solution of (routine) problems in mathematics and science, indicate that attention to fine-grained detail, of the type suggested in the AI work discussed by Newell (1983), does allow for the delineation of learnable and usable problem-solving strategies. Their recommendations, derived from detailed studies of cognition are (1) make tacit processes explicit; (2) get students talking about processes; (3) provide guided practice; (4) ensure that component procedures are well learned; and (5) emphasize both qualitative understanding and specific procedures. These recommendations appear to apply to heuristic strategies as well as to the more routine techniques Heller and Hungate (1985) discuss. Similarly, Rissland's (1985) "tutorial" on AI and mathematics education points to parallels and to the kinds of advances that can be made with detailed analyses of problem-solving performance. There now exists the base knowledge for the careful, prescriptive characterization of problem-solving strategies.

"PROBLEM SOLVING" IN SCHOOL CURRICULA. In classroom practice, unfortunately, the rhetoric of problem solving has been seen more frequently than its substance. The following are some summary statements from the Dossey, Mullis, Lindquist, and Chambers (1988) *Mathematics Report Card*.

Instruction in mathematics classes is characterized by teachers explaining material, working problems at the board, and having students work mathematics problems on their own—a characterization that has not changed across the eight-year period from 1978 to 1986.

Considering the prevalence of research suggesting that there may be better ways for students to learn mathematics than listening to their teachers and then practicing what they have heard in rote fashion, the rarity of innovative approaches is a matter for true concern. Students need to learn to apply their newly acquired mathematics skills by involvement in investigative situations, and their responses indicate very few opportunities to engage in such activities. (p. 76)

According to the *Mathematics Report Card*, there is a predominance of textbooks, workbooks, and ditto sheets in mathematics classrooms; lessons are generically of the type Burkhardt (1988) calls the "exposition, examples, exercises" mode. Much the same is true of lessons that are supposedly about problem solving. In virtually all mainstream texts, "problem solving" is a separate activity and highlighted as such. Problem solving is usually included in the texts in one of two ways. First, there may be occasional "problem-solving" tasks sprinkled through the text (and delineated as such) as rewards or recreations. The implicit message contained in this format is, "You may take a breather from the real business of doing mathematics, and enjoy yourself for a while." Second, many texts contain "problem-solving" sections in which students are given drill-and-practice on simple versions of the strategies discussed in the previous

section. In generic textbook fashion, students are shown a strategy (say "finding patterns" by trying values of $n = 1, 2, 3, 4$ in sequence and guessing the result in general), given practice exercises using the strategy, given homework using the strategy, and tested on the strategy. Note that when the strategies are taught this way, they are no longer heuristics in Pólya's sense; they are mere algorithms. Problem solving, in the spirit of Pólya, is learning to grapple with new and unfamiliar tasks when the relevant solution methods (even if only partly mastered) are not known. When students are drilled in solution procedures as described here, they are not developing the broad set of skills Pólya and other mathematicians who cherish mathematical thinking have in mind.

Even with good materials (and more problem sources are becoming available; see, e.g., Groves & Stacey, 1984; Mason, Burton, & Stacey, 1982; Shell Centre, 1984), the task of teaching heuristics with the goal of developing the kinds of flexible skills Pólya describes is a sometimes daunting task. As Burkhardt notes, teaching problem solving is harder for the teacher in these ways:

Mathematically: The teachers must perceive the implications of the students' different approaches, whether they may be fruitful and, if not, what might make them so.

Pedagogically: The teacher must decide when to intervene, and what suggestions will help the students while leaving the solution essentially in their hands, and carry this through for each student, or group of students, in the class.

Personally: The teacher will often be in the position, unusual for mathematics teachers and uncomfortable for many, of *not knowing*; to work well without knowing all the answers requires experience, confidence, and self-awareness. (1988, p. 18)

That is, true problem solving is as demanding on the teacher as it is on the students—and far more rewarding, when achieved, than the pale imitations of it in most of today's curricula.

Self-regulation, or Monitoring and Control. Self-regulation, or monitoring and control, is one of three broad arenas encompassed under the umbrella term *metacognition*. For a broad historical review of the concept, see Brown (1987). In brief, the issue is one of resource allocation during cognitive activity and problem solving. We introduce the notion with some generic examples.

As you read some expository text, you may reach a point at which your understanding becomes fuzzy; you decide to either reread the text or stop and work out some illustrative examples to make sure you've gotten the point. In the midst of writing an article, you may notice that you've wandered from your intended outline. You may scrap the past few paragraphs and return to the original outline, or you may decide to modify it on the basis of what you've just written. Or, as you work a mathematical problem you may realize that the problem is more complex than you had thought at first. Perhaps the best thing to do is start over and make sure that you've fully understood it. Note that at this level of description, the actions in all three domains—the reading, writing, and mathematics—is much the same. In the midst of intellectual activity ("problem solving," broadly construed), you kept tabs on how well things were going. If things appeared to be proceeding well, you continued along the same path; if they appeared to be problematic, you took stock and considered other options. Monitoring and assessing progress "on line," and acting in response to the assessments of on-line progress, are the core components of self-regulation.

During the 1970s, research in at least three different domains—the developmental literature, artificial intelligence, and mathematics education—converged on self-regulation as a topic of importance. In general, the developmental literature shows that as children get older, they get better at planning for the tasks they are asked to perform and are better at making corrective judgments in response to feedback from their attempts. (Note: Such findings are generally cross-sectional, comparing the performance of groups of children at different age levels; studies rarely follow individual students or cohort groups through time.) A mainstream example of such findings is Karmiloff-Smith's (1979) study of children, ages four through nine, working on the task of constructing a railroad track. The children were given pieces of cardboard representing sections of a railroad track and told that they needed to put all of the pieces together to make a complete loop, so that the train could go around their completed track without ever leaving the track. They were rehearsed on the problem conditions until it was clear that they knew all of the constraints they had to satisfy in putting the tracks together. Typically the four- and five-year old children in the study jumped right into the task, picking up sections of the track more or less at random and lining them up in the order in which they picked them up. They showed no evidence of systematic planning for the task or in its execution. The older children in the study, ages eight and nine, engaged in a large amount of planning before engaging in the task. They sorted the track sections into piles (straight and curved track sections) and chose systematically from the piles (alternating curved and straight sections, or two straight and two curved in sequence) to build the track loops. They were, in general, more effective and efficient at getting the task done. In short, the ability and predilection to plan, act according to plan, and take on-line feedback into account in carrying out a plan seem to develop with age.

Over roughly the same time period, researchers in artificial intelligence came to recognize the necessity for "executive control" in their own work. As problem-solving programs (and expert systems) became increasingly complex, it became clear to researchers in AI that "resource management" was an issue. Solutions to the resource allocation problem varied widely, often dependent on the specifics of the domain in which planning or problem solving was being done. Sacerdoti (1974), for example, was concerned with the time sequence in which plans are executed—an obvious concern if one tries to follow the instructions, "Put your shoes and socks on" or "Paint the ladder and paint the ceiling" in literal order. His architecture, NOAH (for Nets of Action Hierarchies), was designed to help make efficient planning decisions that would avoid execution roadblocks. NOAH's plan execution was top-down, fleshing out

plans from the most general level downward, and only filling in specifics when necessary. Alternate models, corresponding to different domains were bottom-up; and still others, most notably the "Opportunistic Planning Model," or OPM, of Hayes-Roth and Hayes-Roth (1979), was heterarchical—somewhat top-down in approach, but also working at the local level when appropriate. In many ways, the work of Hayes-Roth and Hayes-Roth paralleled emerging work in mathematical problem solving. The task they gave subjects was to prioritize and plan a day's errands. Subjects were given a schematic map of a (hypothetical) city and list of tasks that should, if possible, be achieved that day. The tasks ranged from trivial and easily postponed (ordering a book) to essential (picking up medicine at the druggist). There were too many tasks to be accomplished, so the problem solver had to both prioritize the tasks and find reasonably efficient ways of sequencing and achieving them. The following paragraph summarizes those findings and stands in contrast to the generically clean and hierarchical models typifying the AI literature.

[P]eople's planning activity is largely *opportunistic*. That is, at each point in the process, the planner's current decisions and observations suggest various opportunities for plan development. The planner's subsequent decisions follow up on selected opportunities. Sometimes these decision processes follow an orderly path and produce a neat top-down expansion.... However, some decisions and observations might suggest less orderly opportunities for plan development. For example, a decision about how to conduct initial planned activities might illuminate certain constraints on the planning of later activities and cause the planner to refocus attention on that phase of the plan. Similarly, certain low-level refinements of a previous, abstract plan might suggest an alternative abstract plan to replace the original one. (Hayes-Roth & Hayes-Roth, 1979, p. 276)

Analogous findings were accumulating in the mathematics education literature. In the early 1980s, Silver (1982), Silver, Branca, and Adams (1980), and Garofalo and Lester (1985) pointed out the usefulness of the construct for mathematics educators; Lesh (1983, 1985) focused on the instability of students' conceptualizations of problems and problem situations and of the consequences of such difficulties. Speaking loosely, all of these studies dealt with the same set of issues regarding effective and resourceful problem-solving behavior. Their results can be summed up as follows: It's not just what you know; it's how, when, and whether you use it. The focus here is on two sets of studies designed to help students develop self-regulatory skills during mathematical problem solving. The studies were chosen for discussion because of (1) the explicit focus on self-regulation in both studies, (2) the amount of time each devoted to helping students develop such skills, and (3) the detailed reflections on success and failure in each.

Schoenfeld's (1985a, 1987a) problem-solving courses at the college level have as one of their major goals the development of executive or control skills. Here is a brief summary, adapted from Schoenfeld (1989d.)

The major issues are illustrated in Figures 15.3 and 15.4. Figure 15.3 shows the graph of a problem-solving attempt by a pair of students working as a team. The students read the problem, quickly chose an approach to it, and pursued that

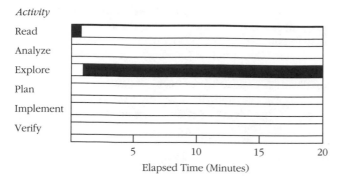

FIGURE 15–3. Time-line graph of a typical student attempting to solve a non-standard problem.

approach. They kept working on it, despite clear evidence that they were not making progress, for the full 20 minutes allocated for the problem session. At the end of the 20 minutes, they were asked how that approach would have helped them to solve the original problem. They couldn't say.

The reader may not have seen this kind of behavior too often. Such behavior does not generally appear when students work routine exercises, since the problem context in that case tells the students which techniques to use. (In a unit test on quadratic equations, for example, students know that they'll be using the quadratic formula.) But when students are doing *real* problem solving, working on unfamiliar problems out of context, such behavior more reflects the norm than not. In Schoenfeld's collection of (more than a hundred) videotapes of college and high-school students working unfamiliar problems, roughly 60% of the solution attempts are of the "read, make a decision quickly, and pursue that direction come hell or high water" variety. And that first, quick, wrong decision, if not reconsidered and reversed, *guarantees* failure.

Figure 15.4, which stands in stark contrast to Figure 15.3, traces a mathematics faculty member's attempt to solve a difficult two-part problem. The first thing to note is that the mathematician spent more than half of his allotted time trying to make sense of the problem. Rather than committing himself to any one particular direction, he did a significant amount of analyzing and (structured) exploring—not spending time in

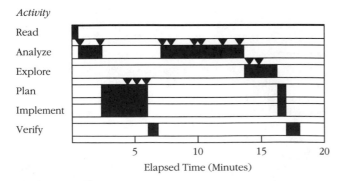

FIGURE 15–4. Time-line graph of a mathematician working a difficult problem.

unstructured exploration or moving into implementation until he was sure he was working in the right direction. Second, note that each of the small inverted triangles in Figure 15.4 represents an explicit comment on the state of his problem solution, for example, "Hmm. I don't know exactly where to start here" (followed by two minutes of analyzing the problem) or "OK. All I need to be able to do is [a particular technique] and I'm done" (followed by the straightforward implementation of his problem solution). It is interesting that when this faculty member began working the problem, he had fewer of the facts and procedures required to solve the problem readily accessible to him than did most of the students who were recorded working the problem. And, as he worked through the problem, the mathematician generated enough potential wild goose chases to keep an army of problem solvers busy. But he didn't get deflected by them. By monitoring his solution with care—pursuing interesting leads and abandoning paths that didn't seem to bear fruit—he managed to solve the problem, while the vast majority of students did not.

The general claim is that these two illustrations are relatively typical of adult student and "expert" behavior on unfamiliar problems. For the most part, students are unaware of or fail to use the executive skills demonstrated by the expert. However, it is the case that such skills can be learned as a result of explicit instruction that focuses on metacognitive aspects of mathematical thinking. That instruction takes the form of "coaching," with active interventions as students work on problems.

Roughly one third of the time in Schoenfeld's problem-solving classes is spent with the students working problems in small groups. The class divides into groups of three or four students and works on problems that have been distributed, while the instructor circulates through the room as "roving consultant." As he moves through the room, he reserves the right to ask the following three questions at any time:

What (exactly) are you doing? (Can you describe it precisely?)
Why are you doing it? (How does it fit into the solution?)
How does it help you? (What will you do with the outcome when you obtain it?)

He begins asking these questions early in the term. When he does so, the students are generally at a loss regarding how to answer them. With the recognition that, despite their uncomfortableness, he is going to continue asking those questions, the students begin to defend themselves against them by discussing the answers to them in advance. By the end of the term, this behavior has become habitual. (Note, however, that the better part of a semester is necessary to obtain such changes.)

The results of these interventions are best illustrated in Figure 15.5, which summarizes a pair of students' problem-solving attempt after taking the course. After reading the problem, they jumped into one solution attempt which, unfortunately, was based on an unfounded assumption. They realized this a few minutes later and decided to try something else. That choice too was a bad one, and they got involved in complicated computations that kept them occupied for $8\frac{1}{2}$ minutes. But at that point they stopped once again. One of the students said, "No, we aren't getting anything here [What we're doing isn't justi-

Activity

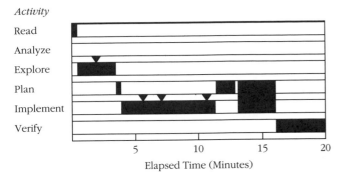

FIGURE 15–5. Time-line graph of a solution attempt after explicit training in monitoring and control.

fied]...Let's start all over and forget about this." They did, and found a solution in short order.

The students' solution is hardly expert-like in the standard sense, since they found the "right" approach quite late in the problem session. Yet in many ways their work resembled the mathematician's behavior illustrated in Figure 15.4 far more than the typical student behavior illustrated in Figure 15.3. The point here is not that the students managed to solve the problem, for to a significant degree solving nonstandard problems is a matter of luck and prior knowledge. The point is, that by virtue of good self-regulation, the students gave themselves the opportunity to solve the problem. They curtailed one possible wild goose chase shortly after beginning to work on the problem and truncated extensive computations halfway through the solution. Had they failed to do so (and they and the majority of their peers did fail to do so prior to the course), they never would have had the opportunity to pursue the correct solution they did find. In this, the students' behavior *was* expert-like. And in this, their solution was also typical of post-instruction attempts by the students. In contrast to the 60% of the "jump into a solution attempt and pursue it no matter what" attempts prior to the course, fewer than 20% of the post-instruction solution attempts were of that type. There was a concomitant increase in problem-solving success.

Lester et al. (1989, June) recently completed a major research and intervention study at the middle school-level, "designed to study the role of metacognition (that is, the knowledge and control of cognition) in seventh graders' mathematical problem solving" (p. v). The goal of the instruction, which took place in one "regular" and one "advanced" seventh-grade mathematics class, was to foster students' metacognitive development. Ways of achieving this goal were to have the teacher (1) serve as external monitor during problem solving, (2) encourage discussion of behaviors considered important for the internalization of metacognitive skills, and (3) model good executive behavior. Table 15.2 delineates the teacher behaviors stressed in the instruction. The total instruction time focusing on metacognition in the experiment was 16.1 hours spread over 12 weeks of instruction, averaging slightly more than $\frac{1}{3}$ (35.7%) of the mathematics classroom time during the instructional period.

The instruction included both "routine" and "nonroutine" problems. An example of a routine problem designed to give students experience in translating verbal statements into mathematical expressions was as follows.

Laura and Beth started reading the same book on Monday. Laura read 19 pages a day and Beth read 4 pages a day. What page was Beth on when Laura was on page 133?

The nonroutine problems used in the study included "process problems" (problems for which there is no standard algorithm for extracting or representing the given information) and problems with either superfluous or insufficient information. The instruction focused on problems amenable to particular strategies (guess-and-check, work backwards, look for patterns) and included games for whole-group activities. Assessment data and tools employed before, during, and after the instruction included written tests, clinical interviews, observations of individual and pair problem-solving sessions, and videotapes of the classroom instruction. Some of the main conclusions drawn by Lester et al. were as follows:

- There is a dynamic interaction between the mathematical concepts and processes (including metacognitive ones) used to solve problems using those concepts. That is, control processes and awareness of cognitive processes develop concurrently with an understanding of mathematical concepts.
- In order for students' problem-solving performance to improve, they must attempt to solve a variety of types of problems on a regular basis and over a prolonged period of time.
- Metacognition instruction is most effective when it takes place in a domain-specific context.
- Problem-solving instruction, metacognition instruction in particular, is likely to be most effective when it is provided in a systematically organized manner under the direction of the teacher.
- It is difficult for the teacher to maintain the roles of monitor, facilitator, and model in the face of classroom reality, especially when the students are having trouble with basic subject matter.
- Classroom dynamics regarding small-group activities are not as well understood as one would like, and facile assumptions that "small-group interactions are best" may not be warranted. The issue of "ideal" class configurations for problem-solving lessons needs more thought and experimentation.
- Assessment practices must reward and encourage the kinds of behaviors we wish students to demonstrate (1989, pp. 88–95).

Briefly, the findings discussed in this section are that developing self-regulatory skills in complex subject-matter domains is difficult and often involves behavior modification—"unlearning" inappropriate control behaviors developed through prior instruction. Such change can be catalyzed, but it requires a long period of time, with sustained attention to both cognitive and metacognitive processes. The task of creating the "right" instructional context, and providing the appropriate kinds of modeling and guidance, is challenging and subtle for

TABLE 15–2. Teaching Actions for Problem-Solving

Teaching Action	Purpose
BEFORE	
1. Read the problem—discuss words or phrases students may not understand	Illustrate the importance of reading carefully; focus on special vocabulary
2. Use whole-class discussion to focus on importance of understanding the problem	Focus on important data, clarification process
3. (Optional) Whole-class discussion of possible strategies to solve a problem	Elicit ideas for possible ways to solve the problem
DURING	
4. Observe and question students to determine where they are	Diagnose strengths and weaknesses
5. Provide hints as needed	Help students past blockages
6. Provide problem extensions as needed	Challenge early finishers to generalize
7. Require students who obtain a solution to "answer the question"	Require students to look over their work and make sure it makes sense
AFTER	
8. Show and discuss solutions	Show and name different strategies
9. Relate to previously solved problems or have students solve extensions	Demonstrate general applicability of problem solving strategies
10. Discuss special features, e.g. pictures	Show how features may influence approach

(Adapted from Lester et al., 1989, p. 26)

the teacher. The two studies cited (Lester et al., 1989; Schoenfeld, 1989d) point to some effective teacher behaviors and classroom practices that foster the development of self-regulatory skills. However, these represent only a beginning. They document the teaching efforts of established researchers who have the luxury to reflect on such issues and prepare instruction devoted to them. Making the move from such "existence proofs" (problematic as they are) to standard classrooms will require a substantial amount of conceptualizing and pedagogical engineering.

Beliefs and Affects

Once upon a time there was a sharply delineated distinction between the cognitive and affective domains, as reflected in the two volumes of Bloom's (1956) *Taxonomy of Educational Objectives*. Concepts such as mathematics anxiety, for example, clearly resided in the affective domain and were measured by questionnaires dealing with how the individual feels about mathematics (see, for example, Suinn, Edie, Nicoletti, & Spinelli, 1972). Concepts such as mathematics achievement and problem solving resided within the cognitive domain and were assessed by

tests focusing on subject-matter knowledge alone. As our vision gets clearer, however, the boundaries between those two domains become increasingly blurred.

Given the space constraints, to review the relevant literature or even try to give a sense of it would be an impossibility. Fortunately, one can point to McLeod, Chapter 23, this volume, and to books such as McLeod and Adams's (1989) *Affect and Mathematical Problem Solving: A New Perspective* as authoritative starting points for a discussion of affect. Beliefs—to be interpreted as an individual's understandings and feelings that shape the ways that the individual conceptualizes and engages in mathematical behavior—will receive a telegraphic discussion. The discussion will take place in three parts: student beliefs, teacher beliefs, and general societal beliefs about doing mathematics. There is a fairly extensive literature on the first, a moderate but growing literature on the second, and a small literature on the third. Hence, length of discussion does not correlate with the size of the literature base.

Student Beliefs. As an introduction to the topic, we recall Lampert's commentary:

Commonly, mathematics is associated with certainty; knowing it, with being able to get the right answer, quickly (Ball, 1988; Schoenfeld, 1985b; Stodolsky, 1985). These cultural assumptions are shaped by school experience, in which *doing* mathematics means following the rules laid down by the teacher; *knowing* mathematics means remembering and applying the correct rule when the teacher asks a question; and mathematical *truth is determined* when the answer is ratified by the teacher. Beliefs about how to do mathematics and what it means to know it in school are acquired through years of watching, listening, and practicing. (1990, p. 31)

An extension of Lampert's list, including other student beliefs delineated in the sources she cites, is given in Table 15.3.

The basic arguments regarding student beliefs were made in the first part of this chapter. As an illustration, consider the genesis and consequences of one belief regarding the amount of time students think is appropriate to spend working mathematics problems. The data come from year-long observations of high-school geometry classes.

Over the period of a full school year, none of the students in any of the dozen classes we observed worked mathematical tasks that could seriously be called problems. What the students worked were exercises: tasks designed to indicate mastery of relatively small chunks of subject matter, and to be completed in a short amount of time. In a typical five-day sequence, for example, students were given homework assignments that consisted of 28, 45, 18, 27, and 30 "problems" respectively. ...[A particular] teacher's practice was to have students present solutions to as many of the homework problems as possible at the board. Given the length of his assignments, that means that he expected the students to be able to work twenty or more "problems" in a fifty-four minute class period. Indeed, the unit test on locus and construction problems (a uniform exam in Math 10 classes at the school) contained twenty-five problems—giving students an average two minutes and ten seconds to work each problem. The teacher's advice to the students summed things up in a nutshell: "You'll have to know all your constructions cold *so you don't spend a lot of time thinking about them.*" [emphasis added]...

In sum, students who have finished a full twelve years of mathematics have worked thousands upon thousands of "problems"—virtually none of which were expected to take the students more than a few minutes to complete. The presumption underlying the assignments was as follows: If you understand the material, you can work the exercises. If you can't work the exercises within a reasonable amount of time, then you don't understand the material. That's a sign that you should seek help.

TABLE 15–3. Typical Student Beliefs about the Nature of Mathematics

- Mathematics problems have one and only one right answer.
- There is only one correct way to solve any mathematics problem—usually the rule the teacher has most recently demonstrated to the class.
- Ordinary students cannot expect to understand mathematics; they expect simply to memorize it and apply what they have learned mechanically and without understanding.
- Mathematics is a solitary activity, done by individuals in isolation.
- Students who have understood the mathematics they have studied will be able to solve any assigned problem in five minutes or less.
- The mathematics learned in school has little or nothing to do with the real world.
- Formal proof is irrelevant to processes of discovery or invention.

Whether or not the message is intended, students get it. One of the open-ended items on our questionnaire, administered to students in twelve high school mathematics classes in grades 9 through 12, read as follows: "If you understand the material, how long should it take to answer a typical homework problem? What is a reasonable amount of time to work on a problem before you know it's impossible?" Means for the two parts of the question were 2.2 minutes ($n = 221$) and 11.7 minutes ($n = 227$), respectively. (Schoenfeld, 1988, pp. 159–160)

Unfortunately, this belief has a serious behavioral corollary. Students with the belief will give up on a problem after a few minutes of unsuccessful attempts, even though they might have solved it had they persevered.

There are parallel arguments regarding the genesis and consequences of the each of the beliefs listed in Table 15.3. Recall, for example, the discussion of the artificial nature of Milne's mental arithmetic problems in the first part of this chapter. It was argued that, after extended experience with "cover stories" for problems that are essentially algorithmic exercises, students come to ignore the cover stories and focus on the "bottom line": performing the algorithm and writing down the answer. That kind of behavior produced an astonishing and widely quoted result in the third National Assessment of Educational Progress (NAEP, 1983), when a plurality of students who performed the correct numerical procedure on a problem ignored the cover story for the problem and wrote that the number of buses required for a given task was "31 remainder 12." In short:

1. Students abstract their beliefs about formal mathematics—their sense of their discipline—in large measure from their experiences in the classroom.
2. Students' beliefs shape their behavior in ways that have extraordinarily powerful (and often negative) consequences.

Teacher Beliefs. Belief structures are important not only for students, but for teachers as well. Simply put, a teacher's sense of the mathematical enterprise determines the nature of the classroom environment that the teacher creates. That environment, in turn, shapes students' beliefs about the nature of mathematics. We briefly cite two studies that provide clear documentation of this point. Cooney (1985) discussed the classroom behavior of a beginning teacher who professed a belief in "problem solving." At bottom, however, this teacher felt that giving students "fun" or nonstandard problems to work—that teacher's conception of problem solving—was, although recreational and motivational, ultimately subordinate to the goal of having students master the subject matter to be covered. Under the pressures of content coverage, the teacher sacrificed (essentially superficial) problem-solving goals for the more immediate goals of drilling students on the things they would be held accountable for.

Thompson (1985) presents two case studies demonstrating the ways that teacher beliefs play out in the classroom. One of her informants was named Jeanne.

Jeanne's remarks revealed a view of the content of mathematics as fixed and predetermined, as dictated by the physical world. At no time during either the lessons [Thompson observed] or the interviews did she allude to the generative processes of mathematics. It seemed apparent that she regarded mathematics as a finished product to be assimilated.... Jeanne's conception of mathematics teaching can be characterized in terms of her view of her role in teaching the subject matter and the students' role in learning it. Those were, in gross terms, that she was to disseminate information, and that her students were to receive it. (p. 286)

These beliefs played out in Jeanne's instruction. The teacher's task, as she saw it, was to present the lesson planned, without digressions or inefficient changes. Her students experienced the kind of rigid instruction that leads to the development of some of the student beliefs described above.

Thompson's second informant was named Kay. Among Kay's beliefs about mathematics and pedagogy were the following:

- Mathematics is more a subject of ideas and mental processes than a subject of facts.
- Mathematics can be best understood by rediscovering its ideas.
- Discovery and verification are essential processes in mathematics.
- The main objective of the study of mathematics is to develop reasoning skills that are necessary for solving problems.
- The teacher must create and maintain an open and informal classroom atmosphere to insure the students' freedom to ask questions and explore their ideas.
- The teacher should encourage students to guess and conjecture and should allow them to reason things on their own rather than show them how to reach a solution or an answer.
- The teacher should appeal to students' intuition and experiences when presenting the material in order to make it meaningful (pp. 288–290).

Kay's pedagogy was consistent with her beliefs and resulted in a classroom atmosphere that was at least potentially supportive of the development of her students' problem-solving abilities.

One may ask, of course, where teachers obtain their notions regarding the nature of mathematics and of the appropriate pedagogy for mathematics instruction. Not surprisingly, Thompson notes: "There is research evidence that teachers' conceptions and practices, particularly those of beginning teachers, are largely influenced by their schooling experience prior to entering methods of teaching courses" (p. 292). Hence, teacher beliefs tend to come home to roost in successive generations of teachers, in what may for the most part be a vicious pedagogical/epistemological circle.

Societal Beliefs. Stigler & Perry (1989) report on a series of cross-cultural studies that serve to highlight some of the societal beliefs in the United States, Japan, and China regarding mathematics.

There are large cultural differences in the beliefs held by parents, teachers, and children about the nature of mathematics learning. These beliefs can be organized into three broad categories: beliefs about what is *possible,* (i.e., what children are able to learn about mathematics at different ages); beliefs about what is *desirable* (i.e., what children should learn); and beliefs about what is the best *method* for teaching mathematics (i.e., how children should be taught). (p. 196)

Regarding what is possible, the studies indicate that people in the United States are much more likely than the Japanese to believe that innate ability (as opposed to effort) underlies children's success in mathematics. Such beliefs play out in important ways. First, parents and students who believe "either you have it or you don't" are much less likely to encourage students to work hard on mathematics than those who believe "you can do it if you try." Second, our nation's textbooks reflect our uniformly low expectations of students. "U.S. elementary textbooks introduce large numbers at a slower pace than do Japanese, Chinese, or Soviet textbooks, and delay the introduction of regrouping in addition and subtraction considerably longer than do books in other countries" (Stigler & Perry, 1989, p. 196). Regarding what is desirable, the studies indicate that, despite the international comparison studies, parents in the United States believe that reading, not mathematics, needs more emphasis in the curriculum. And finally, on methods:

Those in the U.S., particularly with respect to mathematics, tend to assume that understanding is equivalent to sudden insight. With mathematics, one often hears teachers tell children that they "either know it or they don't," implying that mathematics problems can either be solved quickly or not at all.... In Japan and China, understanding is conceived of as a more gradual process, where the more one struggles the more one comes to understand. Perhaps for this reason, one sees teachers in Japan and China pose more difficult problems, sometimes so difficult that the children will probably not be able to solve them within a single class period. (Stigler & Perry, 1989, p. 197)

In sum, whether acknowledged or not, whether conscious or not, beliefs shape mathematical behavior. Beliefs are abstracted from one's experiences and from the culture in which one is embedded. This leads to the consideration of mathematical practice.

Practices

As an introduction to this section, we recall Resnick's comments regarding mathematics instruction:

Becoming a good mathematical problem solver—becoming a good thinker in any domain—may be as much a matter of acquiring the habits and dispositions of interpretation and sense-making as of acquiring any particular set of skills, strategies, or knowledge. If this is so, we may do well to conceive of mathematics education less as an instructional process (in the traditional sense of teaching specific, well-defined skills or items of knowledge), than as a socialization process. (1989, p. 58)

The preceding section on beliefs and affects described some of the unfortunate consequences of entering the wrong kind of mathematical practice—the practice of "school mathematics." Here we examine some positive examples. These classroom environments, designed to reflect selected aspects of the mathematical community, have students interact (with each other and the mathematics) in ways that promote mathematical thinking. We take them in increasing grade order.

Lampert (1990) explicitly invokes a Pólya-Lakatosian epistemological backdrop for her fifth-grade lessons on exponentiation, deriving pedagogical practice from that epistemological stance. She describes

a research and development project in teaching designed to examine whether and how it might be possible to bring the practice of knowing mathematics in school closer to what it means to know mathematics within the discipline by deliberately altering the roles and responsibilities of teacher and students in classroom discourse....A [representative] case of teaching and learning about exponents derived from lessons taught in the project is described and interpreted from mathematical, pedagogical, and sociolinguistic perspectives. To change the meaning of knowing and learning in school, the teacher initiated and supported social interactions appropriate to making mathematical arguments in response to students' conjectures. The activities in which students engaged as they asserted and examined hypotheses about the mathematical structures that underlie their solutions to problems are contrasted with the conventional activities that characterize school mathematics. (p. 1)

Lampert describes a series of lessons on exponents in which students first found patterns of the last digits in the squares of natural numbers and then explored the last digits of large numbers (e.g., what is the last digit of 7^5?). In the process of classroom discussion, students found patterns, made definitions, reasoned about their claims, and ultimately defended their claims on mathematical grounds. At one point, for example, a student named Sam asserted flatly that the last digit of 7^5 is a 7, while others claimed that it was 1 or 9.

[Lampert] said: "You must have a proof in mind, Sam, to be so sure," and then [she] asked, "Arthur, why do you think it's a 1?"

The students attempted to resolve the problem of having more than one conjecture about what the last digit in seven to the fifth power might be. [The discussion] was a zig-zag between proofs that the last digit must be 7 and refutations of Arthur's and Sarah's alternative conjectures. The discussion ranged between observations of particular answers and generalizations about how exponents—and numbers more generally—work. Students examined their own assumptions and those of their classmates. [Lampert] assumed the role of manager of the discussion and sometimes participated in the argument, refuting a student's assertion. (p. 47)

At the end of the lesson, in which the class explored simple ways of looking at the last digits of 7^8 and 7^{16},

some students were verging on declaring an important law of exponents: $(n^a)(n^b) = n^{a+b}$, which they would articulate more fully, and prove the legitimacy of, in the next few classes. They were also beginning to develop a modular arithmetic of "last digits" to go with different base numbers, leading them into further generalizations about the properties of exponents. (pp. 54-55)

Note that Lampert did not "reveal truth," but entered the dialogue as a knowledgeable participant—a representative of the mathematical community who was not an all-knowing authority but rather one who could ask pointed questions to help students arrive at the correct mathematical judgments. Her pedagogical practice, in deflecting undue authority from the teacher, placed the burden of mathematical judgment (with constraints) on the shoulders of the students.

Balacheff (1987) exploits social interactions in a different way, but with similar epistemological goals. He describes a series of lessons for seventh graders, concerned with the theorem that "the sum of the angles of a triangle is 180°." The lessons begin with the class divided into small groups. Each group is given a worksheet with a copy of the same triangle and asked to compute the sum of its angles. The groups then report their answers, which vary widely—often from as little as 100° to as much as 300°! Since the students know they had all measured the same triangle, this causes a tension that must be resolved; they work on it until all students agree on a value. Balacheff then hands out a different triangle to each group, and has the group conjecture the sum of the angles before measuring it. Groups compare and contrast their results, and repeat the process with each other's triangles. This causes conflicts within and across groups, and the discussions that result in the resolutions of those conflicts make the relevant mathematical issues salient and meaningful to the students, so that they are intellectually prepared for the theoretical discussions (of a similar dialectical nature) that follow.

In a classic study that is strikingly contemporary in its spirit, Fawcett (1938) describes a two-year course in plane geometry he taught at the Ohio State University Laboratory School in the 1930s. Fawcett's goals were that students develop a good understanding of the subject matter of geometry, the right epistemological sense about the mathematics, and a sense of the applicability of the reasoning procedures that they had learned in geometry to situations outside the mathematics classroom. In order for this to happen, he believed, (1) the students had to engage in doing mathematics in a way consistent with his mathematical epistemology, (2) the connections between mathematical reasoning in the formal context of the classroom and mathematical reasoning outside it would have to be made explicit, and (3) the students would need to reflect both on their doing of mathematics and on the connections between the reasoning in both contexts.

For example, the issue of definition is important in mathematics. Fawcett pointed out that definitions have consequences: In his school, for example, there was an award for the "best teacher." Many students favored the librarian—but was the librarian a teacher? He also used sports as an analogy. In baseball, for example, there might be varying definitions of "foul ball" (Is a fly ball that hits the foul pole fair or foul?), but once the rules are set, the game can be played with consistency. After such discussions, Fawcett notes "no difficulty was met in leading the pupils to realize that these rules were nothing more than agreements which a group of interested people had made and that they implied certain conclusions" (p. 33). In the mathematical domain, he had his students debate the nature and usefulness of various definitions. Rather than provide the definition of "adjacent angle," for example, he asked the class to propose and defend various definitions. The first was "angles that share a common side," which was ruled out by Figure 15.6(*a*). A second suggestion, "angles that share a common vertex," was ruled out by Figure 15.6(*b*). "Angles that share a common side and a common vertex" had a good deal

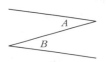 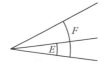

a. two angles that share a common side

b. two angles that share a common vertex

c. two angles that share a common side and a common vertex

FIGURE 15–6. Student definitions of "adjacent angles" discussed in Fawcett (1938).

of support until it was ruled out by Figure 15.6(c). Finally the class agreed upon a mathematically correct definition.

To recall a statement on the nature of mathematical *doing* by Pólya, "To a mathematician who is active in research, mathematics may appear sometimes as a guessing game; you have to guess a mathematical theorem before you prove it, you have to guess the idea of a proof before you carry through all the details" (1954, p. 158). Fawcett's class was engineered along these lines. He never gave assignments of the following form:

> *Prove that the diagonals of a parallelogram bisect each other but are not necessarily mutually perpendicular; also prove that the diagonals of a rhombus are mutually perpendicular.*

Instead, he would pose the problems in the following form:

1. Consider the parallelogram ABCD in Figure 15.7(*a*), with diagonals AC and BD. State all the properties of the figure that you are willing to accept. Then, give a complete argument justifying why you believe your assertions to be correct.
2. Suppose you also assume that AB = BC, so that the quadrilateral ABCD is a rhombus [Figure 15.7(*b*)]. State all the additional properties of the figure that you are willing to accept. Then, give a complete argument justifying why you believe your additional assertions to be correct.

Needless to say, different students had different opinions regarding what they would accept as properties of the figures. Fawcett had students supporting the different positions argue their conclusions—that is, a claim about a property of either figure had to be defended mathematically. The class (with Fawcett serving as an "especially knowledgeable member," but not as sole authority) served as "jury." Class discussions included not

only what was right and wrong (that is, Does a figure have a given property?), but also reflections on the nature of argumentation itself: Are inductive proofs always valid? Are converses always true? and so on. In short, Fawcett's students were acting like mathematicians, at the limits of their own community's (the classroom's) knowledge.

We continue with two examples at the college level. Alibert and his colleagues (1988) have developed a calculus course at Grenoble based on the following principles:

1. Coming to grips with uncertainty is a major part of the learning process.
2. A major role of proofs (the product of "scientific debate") is to convince first oneself, and then others, of the truth of a conjecture.
3. Mathematical tools can evolve meaningfully from the solution of complex problems, often taken from the physical sciences.
4. Students should be induced to reflect on their own thought processes.

Their course, based on these premises, introduces major mathematics topics with significant problems from the physical sciences (for example, the Riemann integral is introduced and motivated by a problem asking students to determine the gravitational attraction exerted by a stick on a marble). While in typical calculus classes the historical example would soon be abandoned and the subject matter would be presented in cut-and-dried fashion, the Grenoble course is true to its principles. The class, in a debate resembling that discussed in the examples from Lampert and Fawcett, formulates the mathematical problem and resolves it (in the sense of the term used by Mason et al., 1982) by a discussion in which ideas spring from the class and are nurtured by the instructor, who plays a facilitating and critical, rather than show-and-tell, role.

According to Alibert, experiences of this type result in the students' coming to grips with some fundamental mathematical notions. After the course, notes Alibert, "their conceptions of mathematics are interesting—and important for their learning. A large majority of the students answer the...question 'What does mathematics mean to you?' at an epistemological level; their school epistemology has almost disappeared" (p. 35).

Finally, Schoenfeld's problem-solving courses at the college level have many of the same attributes. As in Fawcett's case, no problems are posed in the "prove that" format; all are "What do you think is true, and why?" questions. Schoenfeld (forthcoming) explicitly deflects teacher authority to the student community, both in withholding his own understandings of problem solutions (many problems the class works on for days or weeks are problems for which he could present a 10-minute lecture solution) and developing in the class the critical sense of mathematical argumentation that leads it, as a community, to accept or reject on appropriate mathematical grounds the proposals made by class members.

For example, in a discussion of the Pythagorean theorem Schoenfeld (1990b) posed the problem of finding all solutions in integers to the equation $a^2 + b^2 = c^2$. There is a known

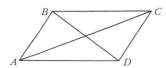 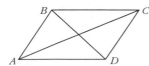

a. ABCD is a parallelogram. What do you think must be true?

b. ABCD is a rhombus. What else do you think must be true?

FIGURE 15–7. Problems posed in the style of Fawcett (1938).

solution, which he did not present. The class made a series of observations, among them:

1. Multiples of known solutions (e.g., the $\{6, 8, 10\}$ right triangle as a multiple of the $\{3, 4, 5\}$) are easy to obtain, but of no real interest. The class would focus on triangles whose sides were relatively prime.
2. The class observed, conjectured, and proved that in a relatively prime solution, the value of c is always odd.
3. Students observed that in all the cases of relatively prime solutions they knew—$\{3, 4, 5\}$, $\{5, 12, 13\}$, $\{7, 24, 25\}$, $\{8, 15, 17\}$, $\{12, 35, 37\}$—the larger leg (b) and the hypotenuse (c) differed by either 1 or 2. They conjectured that there are infinitely many triples in which b and c differ by 1 and by 2, and no others.
4. They proved that there are infinitely many solutions where b and c differ by 1 and where b and c differ by 2; they proved there are no solutions where b and c differ by 3. At that point a student asked if, should the pattern continue (i.e., if they could prove their conjecture), they would have a publishable theorem.

Of course, the answer to the student's question was no. First, the conjecture was wrong: There is, for example, the $\{20, 21, 29\}$ triple. Second, the definitive result—all Pythagorean triples are of the form $\{M^2 - N^2, 2MN, M^2 + N^2\}$—is well known and long established within the mathematical community. But to dismiss the students' results is to do them a grave injustice. In fact, all three of the results proved by the students in number 4 above were new to the instructor. The students were *doing* mathematics, at the frontiers of their community's knowledge.

In all of the examples discussed in this section, classroom environments were designed to be consonant with the instructors' epistemological sense of mathematics as an ongoing, dynamic discipline of sense making through the dialectic of conjecture and argumentation. In all, the authors provide some anecdotal and some empirically "objective" documentations of success. Yet, the existence of these positive cases raises far more questions that it answers. The issues raised here, and in general by the research discussed in this chapter, are the focus of discussion in the next section.

ISSUES

We conclude with an assessment of the state of the art in each of the areas discussed in this paper, pointing to both theoretical and practical issues that need attention and clarification. Caveat lector: The comments made here reflect the opinions of the author and may be shared to various degrees by the research community at large.

This chapter has focused on an emerging conceptualization of mathematical thinking based on an alternative epistemology in which the traditional conception of domain knowledge plays an altered and diminished role, even when it is expanded to include problem-solving strategies. In this emerging view, metacognition, belief, and mathematical practices are considered critical aspects of thinking mathematically. But there is

more. The person who thinks mathematically has a particular way of seeing the world, of representing it, of analyzing it. Only within that overarching context do the pieces—the knowledge base, strategies, control, beliefs, and practices—fit together coherently. We begin the discussion with comments on what it might mean for the pieces to fit together.

A useful idea to help analyze and understand complex systems is that of a nearly decomposable system. The idea is that one can make progress in understanding a large and complex system by carefully abstracting from it subsystems for analysis and then combining the analyses of the subsystems into an analysis of the whole. The study of human physiology provides a familiar example. Significant progress in our understanding of physiology has been made by conducting analyses of the circulatory system, the respiratory system, the digestive system, and so on. Such analyses yield tremendous insights and help move us forward in understanding human physiology as a whole. However, insights at the subsystem level alone are insufficient: Interactions among the subsystems must be considered, and the whole is obviously much more than the sum of its parts.

One can argue, convincingly I think, that the categories in the framework identified and discussed in the second part of this chapter provide a coherent and relatively comprehensive near decomposition of mathematical thinking (or at least, mathematical behavior). The individual categories cohere, and within them (to varying degrees of success) research has produced some ideas regarding underlying mechanisms. But the research community understands little about the interactions among the categories, and less about how they come to cohere—in particular how an individual's learning in those categories fits together to give the individual a sense of the mathematical enterprise, his or her "mathematical point of view." My own bias is that the key to this problem lies in the study of enculturation, of entry into the mathematical community. For the most part, people develop their sense of any serious endeavor—be it their religious beliefs, their attitude toward music, their identities as professionals or workers, their sense of themselves as readers (or nonreaders), or their sense of mathematics—from interactions with others. And if we are to understand how people develop their mathematical perspectives, we must look at the issue in terms of the mathematical communities in which students live and the practices that underlie those communities. The role of interactions with others will be central in understanding learning, whether it be understanding how individuals come to grips with the specifics of the domain (Moschkovich, 1989; Newman, Griffin, & Cole, 1989; Schoenfeld, Smith, & Arcavi, in press) or more broad issues about developing perspectives and values (Lave & Wenger, 1989; Schoenfeld, 1989c, forthcoming). This theme will be explored a bit more in the section on practices. We now proceed with a discussion of issues related to research, instruction, and assessment.

Fundamental issues remain unaddressed or unresolved in the general area of problem solving and in each of the particular areas addressed in the second part of this chapter. To begin, the field needs much greater clarity of the meanings of the term "problem solving." The term has served as an umbrella under which radically different types of research have been conducted. At minimum there should be a de facto requirement (now the exception rather than the rule) that every study or discussion of problem solving be accompanied by an operational definition of the term and examples of what the author means—whether it be working the exercises at the end of the chapter, scoring well on the Putnam exam, or "developing a mathematical point of view and the tools to go with it" as discussed in this chapter. Although one is loath to make recommendations that may result in jargon proliferation, it seems that the time is overdue for researchers to form some consensus on definitions about various aspects of problem solving. Great confusion arises when the same term refers to a multitude of sometimes contradictory and typically underspecified behaviors.

Along the same general lines, much greater clarity is necessary with regard to research methods. It is generally accepted that all research methodologies (1) address only particular aspects of problem solving behavior, leaving others unaddressed; (2) cast some behaviors into high relief, allowing for a close analysis of those; and (3) either obscure or distort other behaviors. The researchers' tool kit is expanding from the collection of mostly statistical and experimental techniques largely employed through the 1970s (comparison studies, regression analyses, and so on) to the broad range of clinical, protocol analysis, simulation, and computer modeling methods used today. Such methods are often ill- or inappropriately used. Those we understand well should, perhaps, come with "user's guides" of the following type: "This method is suited for explorations of A, B, and C, with the following caveats; or, It has not proven reliable for explorations of D, E, and F." Here is one example, as a case in point:

The protocol parsing scheme, that produced Figures 15.3, 15.4, and 15.5, analyzed protocols gathered in non-interventive problem solving sessions, is appropriate for documenting the presence or absence of executive decisions in problem solving, and demonstrating the consequences of those executive decisions. However, it is likely to be useful only on problems of Webster's type 2—"perplexing or difficult" problems, in which individuals must make difficult choices about resource allocation. (Control behavior is unlikely to be necessary or relevant when individuals are working routine or algorithmic exercises.) Moreover, the method reveals little or nothing about the mechanisms underlying successful or unsuccessful monitoring and assessment. More interventive methods will almost certainly be necessary to probe, on the spot, why individuals did or did not pursue particular options during problem solving. These, of course, will disturb the flow of problem solutions; hence the parsing method will no longer be appropriate for analyzing those protocols.

Indeed, a contemporary guide to research methods would be a useful tool for the field.

With regard to resources (domain knowledge), the two main issues that require attention are (1) finding adequate descriptions and representations of cognitive structures, and (2) elaborating the dynamic interaction between resources and other aspects of problem-solving behavior as people engage in mathematics. Over the past decade, researchers have developed some careful and fine-grained representations of mathematical structures, but the field still has a way to go before there is a strong congruence between the ways we describe knowledge structures and our sense of how such structures work phenomenologically. And we still lack a good sense of how the pieces fit together. How do resources interact with strategies, control, beliefs, and practices?

Much of the theoretical work with regard to problem-solving strategies has already been done; the remaining issues are more on the practical and implementational levels. The spadework for the elaboration of problem-solving strategies exists, in that there is a blueprint for elaborating strategies. It has been shown that problem-solving strategies can be described, in detail, at a level that is learnable. Following up on such studies, we now need carefully controlled data on the nature and amount of training and over what kinds of problems, that result in the acquisition of particular strategies (and how far strategy acquisition transfers). That is a demanding task, but not a theoretically difficult one.

We have made far less progress with regard to control. The importance of the idea has been identified and some methodological tools have been developed for charting control behaviors during problem solving. Moreover, research indicates that students (at least at the advanced secondary and college level) can be taught to develop productive control behaviors, although only in extended instruction that, in effect, amounts to behavior modification. However, there remain some fundamental issues to be resolved.

The first issue is mechanism. We lack an adequate characterization of control. That is, we do not have good theoretical models of what control is, and how it works. We do not know, for example, whether control is domain-independent or domain-dependent; nor do we know what the mechanisms might be for tying control decisions to domain knowledge. The second issue is development. We know that in some domains, children can demonstrate astonishingly subtle self-regulatory behaviors, in social situations, for example, where they pick up behavioral and conversational cues regarding whether and how to pursue particular topics of conversation with their parents. How and when do children develop such skills in the social domain? How and when do they develop (or fail to develop) the analogous skills in the domain of mathematics? Are the similarities merely apparent, or do they have a common base in some way? We have barely a clue regarding the answers to these questions.

The arena of beliefs and affects is reemerging as a focus of research, and it needs concentrated attention. It is basically underconceptualized, and it stands in need of new methodologies and new explanatory frames. The older measurement tools and concepts found in the affective literature are simply inadequate; they are not at a level of mechanism and most often tell us that *something* happens without offering good suggestions as to how and why. Recent work on beliefs points to issues of importance that straddle the cognitive and affective domains, but much of that work is still at the "telling good stories" level rather than the level of providing solid explanations. Despite some theoretical advances in recent years and increasing interest in the topic, we are still a long way from a unified perspective that allows for the meaningful integration of cognition and affect or, if such unification is not possible, from understanding why it is not.

Issues regarding practices and the means by which they are learned—enculturation—may be even more problematic. Here, in what may ultimately turn out to be one of the most important arenas for understanding the development of mathematical thinking, we seem to know the least. The importance of enculturation has now been recognized, but the best we can offer thus far in explication of it is a small number of well-described case studies. Those studies, however, give only the barest hints at underlying mechanisms. On the one hand, the tools available to cognitivists have yet to encompass the kinds of social issues clearly relevant for studies of enculturation—such as how one picks up the biases and perspectives common to members of a particular subculture. On the other hand, extant theoretical means for discussing phenomena such as enculturation do not yet operate at the detailed level that results in productive discussions of what people learn (for example, about mathematics) and why. There are hints regarding theoretical means for looking at the issue, such as Lave and Wenger's (1989) concept of "legitimate peripheral participation." Roughly, the idea is that by sitting on the fringe of a community, one gets a sense of the enterprise; as one interacts with members of the community and becomes more deeply embedded in it, one learns its language and picks up its perspectives as well. It remains to be seen, however, how such means will be developed and whether they will be up to the task.

Turning to practical issues, one notes that there is a host of unsolved and largely unaddressed questions dealing with instruction and assessment. It appears that as a nation we will be moving rapidly in the direction of new curricula, some of them very much along the lines suggested in this chapter. At the national level, *Everybody Counts* (National Research Council, 1989) represents the Mathematical Sciences Education Board's attempt to focus discussion on issues of mathematics education. *Everybody Counts* makes the case quite clearly that a perpetuation of the status quo is a recipe for disaster, and it calls for sweeping changes. The NCTM *Standards* (National Council of Teachers of Mathematics, 1989) reflects an emerging national consensus that all students should study a common core of material for (a minimum of) three years in secondary school. *Reshaping School Mathematics* (National Research Council, 1990a) supports the notion of a three-year common core and provides a philosophical rationale for a curriculum focusing on developing students' mathematical power. With such national statements as a backdrop, some states are moving rapidly toward the implementation of such curricula. In California, for example, the 1985 *Mathematics Framework* (California State Department of Education, 1985) claimed that "mathematical power, which involves the ability to discern mathematical relationships, reason logically, and use mathematical techniques effectively, must be the central concern of mathematics education" (p. 1). Its classroom recommendations for teachers are as follows:

- Model problem-solving behavior whenever possible, exploring and experimenting along with students.
- Create a classroom atmosphere in which all students feel comfortable trying out ideas.
- Invite students to explain their thinking at all stages of problem solving.

- Allow for the fact that more than one strategy may be needed to solve a given problem and that problems may require original approaches.
- Present problem situations that closely resemble real situations in their richness and complexity so that the experience that students gain in the classroom will be transferable (p. 14).

The 1991 *Mathematics Framework* (California State Department of Education, forthcoming), currently in draft form, builds on this foundation and moves significantly further in the directions suggested in this chapter. It recommends that lessons come in large, coherent chunks; that curricular units be anywhere from two to six weeks in length, be motivated by meaningful problems, and be integrated with regard to subject matter (for example, containing problems calling for the simultaneous use of algebra and geometry, rather than having geometry taught as a separate subject, as if algebra did not exist); and that students engage in collaborative work, often on projects that take days and weeks to complete. Pilot projects for a radically new secondary curriculum, implementing these ideas for grades 9–11, began in selected California schools in September 1989.

The presence of such projects, and their potential dissemination, raises significant practical and theoretical issues. For example, what kinds of teacher knowledge and behavior are necessary to implement such curricula on a large scale? One sees glimmers of ideas in the research (see Grouws & Cooney, 1989, for an overview), but in general, conceptions of how to teach for mathematical thinking have of necessity lagged behind our evolving concept of what it is to think mathematically. There are some signs of progress. For example, a small body of research (Peterson, Fennema, Carpenter, & Loef, 1989) suggests that with the appropriate in-service experiences (weeks of intensive study, not one-day workshops), teachers can learn enough about student learning to change classroom behavior. Much more research on teacher beliefs—how they are formed, how they can be made to evolve—is necessary. Also needed is research at the systemic level: What changes in school and district structures are likely to provide teachers with the support they need to make the desired changes in the classroom?

Offered as a conclusion is a brief discussion of what may be the single most potent systemic force in motivating change: assessment. *Everybody Counts* (National Research Council, 1989) states the case succinctly: "What is tested is what gets taught. Tests must measure what is most important" (p. 69). To state the case bluntly, current assessment measures (especially the standardized multiple-choice tests favored by many administrators for "accountability") deal with only a minuscule portion of the skills and perspectives encompassed by the phrase *mathematical power* and discussed in this chapter. The development of appropriate assessment measures, at both the individual and the school or district levels, will be a very challenging practical and theoretical task. Here are a few of the relevant questions:

What kinds of information can be gleaned from "open-ended questions," and what kinds of scoring procedures are reliable and informative both to those who do the assessing and those who are being tested? Here is one example of an interesting question type, taken from *A Question of Thinking* (California State Department of Education, 1989).

Imagine you are talking to a student in your class on the telephone and want the student to draw some figures. [They might be part of a homework assignment, for example]. The other student cannot see the figures. Write a set of directions so that the other student can draw the figures exactly as shown below.

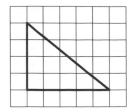 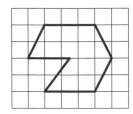

To answer this question adequately, one must understand the geometric representation of the figures *and* be able to communicate using mathematical language. Such questions, while still rather constrained, clearly focus on goals other than simple subject matter "mastery." A large collection of such items would, at minimum, push the boundaries of what is typically assessed. But such approaches are only a first step. Two other approaches currently being explored (by the California Assessment Program, among others) are discussed next.

Suppose the student is asked to put together a portfolio representing his or her best work in mathematics. How can such portfolios be structured to give the best sense of what the student has learned? What kind of entries should be included ("the problem I am proudest of having solved," a record of a group collaborative project, a description of the student's role in a class project) and how can they be evaluated fairly?

Next, how can one determine the kinds of collaborative skills learned by students in a mathematics program? Suppose one picks four students at random from a mathematics class toward the end of the school year, gives them a difficult open-ended problem to work, and videotapes what the students do as they work on the problem for an hour. What kinds of inferences can one make, reliably, from the videotape? One claim is that a trained observer can determine within the first few minutes of watching the tape whether the students have had extensive experience in collaborative work in mathematics. Students who have not had such experience will most likely find it difficult to coordinate their efforts, while those who have often worked collaboratively will (one hopes) readily fall into certain kinds of cooperative behaviors. Are such claims justified? How can one develop reliable methods for testing them? Another claim is that students' fluency at generating a range of approaches to deal with difficult problems will provide information about the kinds of instruction they have received, and about their success at the strategic and executive aspects of mathematical behavior. But what kinds of information, and how reliable the information might be, is very much open to question.

In sum, the imminent implementation of curricula with ambitious pedagogical and philosophical goals will raise a host of unavoidable and fundamentally difficult theoretical and practical issues. It is clear that we have our work cut out for us, but it is also clear that progress over the past decade gives us at least a fighting chance for success.

References

Albers, D., & Alexanderson, J. (1985). *Mathematical people: Profiles and interviews*. Chicago, IL: Contemporary books.

Alibert, D. (1988, June). Towards new customs in the classroom. *For the learning of mathematics, 8*(2), 31–35, 43.

Anderson, J. R. (1976). *Language, memory, and thought*. Hillsdale, NJ: Lawrence Erlbaum.

Anderson, J. R. (1983). *The architecture of cognition*. Cambridge, MA: Harvard University Press.

Anderson, J. R. (1985). *Cognitive psychology and its implications* (2nd ed.). New York, NY: Freeman.

Balacheff, N. (1987). *Devolution d'un probleme et construction d'une conjecture: Le cas de "la somme des angles d'un triangle."* Cahier de didactique des mathematiques No. 39. Paris: IREM Universite Paris VII.

Ball, D. (1988). *Knowledge and reasoning in mathematical pedagogy: Examining what prospective teachers bring to teacher education*. Unpublished doctoral dissertation, Michigan State University.

Barwise, K. J., & Perry, J. (1983). *Situations and attitudes*. Cambridge, MA: MIT Press.

Bauersfeld, H. (1979). Hidden dimensions in the so-called reality of a mathematics classroom. In R. Lesh & W. Secada (Eds.), *Some theoretical issues in mathematics education: Papers from a research presession* (pp. 13–32). Columbus, OH: ERIC.

Begle, E. (1979). *Critical variables in mathematics education*. Washington, DC.: Mathematical Association of America and National Council of Teachers of Mathematics.

Bloom, B. S. (1956). *Taxonomy of educational objectives. Handbook I: Cognitive domain. Handbook II: Affective domain*. New York, NY: David McKay.

Brown, A. (1987). Metacognition, executive control, self-regulation, and other more mysterious mechanisms. In F. Reiner & R. Kluwe (Eds.), *Metacognition, motivation, and understanding* (pp. 65–116). Hillsdale, NJ: Lawrence Erlbaum.

Brown, J. S., & Burton, R. R. (1978). Diagnostic models for procedural bugs in basic mathematical skills. *Cognitive Science, 2,* 155–192.

Brown, J. S., Collins, A., & Duguid, P. (1989, January-February). Situated cognition and the culture of learning. *Educational Researcher, 18*(1), 32–42.

Burkhardt, H. (1988). Teaching problem solving. In H. Burkhardt, S. Groves, A. Schoenfeld, & K. Stacey (Eds.), *Problem solving—A world view (Proceedings of the problem solving theme group, ICME 5)* (pp. 17–42). Nottingham, England: Shell Centre.

Burkhardt, H., Groves, S., Schoenfeld, A., & Stacey, K. (Eds.). (1988). *Problem solving—A world view. (Proceedings of the problem solving theme group, ICME 5)*. Nottingham, England: Shell Centre.

California State Department of Education. (1985). *Mathematics framework for California public schools kindergarten through grade twelve*. Sacramento, CA: California State Department of Education.

California State Department of Education. (1989). *A question of thinking*. Sacramento, CA: State Department.

California State Department of Education. (forthcoming). *Mathematics framework for California public schools kindergarten through grade twelve*. Sacramento, CA: California State Department of Education.

Carpenter, T. P. (1985). Learning to add and subtract: An exercise in problem solving. In E. A. Silver (Ed.), *Teaching and learning mathematical problem solving: Multiple research perspectives* (pp. 17–40). Hillsdale, NJ: Lawrence Erlbaum.

Carpenter, T. P., Lindquist, M. M., Matthews, W., & Silver, E. A. (1983). Results of the third NAEP mathematics assessment: Secondary school. *Mathematics Teacher, 76*, 652–659.

Carss, M. (Ed.). (1986). *Proceedings of the Fifth International Congress on Mathematics Education*. Boston, MA: Birkhäuser.

Charles, R., & Silver, E. A. (Eds.). (1988). *The teaching and assessing of mathematical problem solving*. Hillsdale, NJ: Lawrence Erlbaum.

Collins, A., Brown, J. S., & Newman, S. (1989). Cognitive apprenticeship: Teaching the craft of reading, writing, and mathematics. In L. B. Resnick (Ed.), *Knowing, learning, and instruction: Essays in honor of Robert Glaser*. Hillsdale, NJ: Lawrence Erlbaum.

Cooney. T. (1985). A beginning teacher's view of problem solving. *Journal for Research in Mathematics Education, 16*, 324–336.

Davis, P., & Hersh, R. (1981). *The mathematical experience*. Boston, MA: Houghton-Mifflin.

deGroot, A. (1965). *Thought and choice in chess*. The Hague: Mouton

Descartes, R. (1952). *Rules for the direction of the mind* (E. S. Haldane and G. R. I. Ross, Trans.). In *Great Books of the Western World* (Vol. 31). Chicago, IL: Encyclopedia Brittanica, Inc.

diSessa, A. (1983). Phenomenology and the evolution of intuition. In D. Gentner & A. Stevens (Eds.), *Mental Models* (pp. 15–33). Hillsdale, NJ: Lawrence Erlbaum.

Dossey, J., Mullis, I., Lindquist, M., & Chambers, D. (1988). *The mathematics report card: Are we measuring up? Trends and achievement based on the 1986 National Assessment*. Princeton, NJ: Educational Testing Service.

Duncker, K. (1945). *On problem solving*. Psychological Monographs 58, No. 5. (Whole # 270.) Washington, DC: American Psychological Association.

Ericsson, K., & Simon, H. (1980). Verbal reports as data. *Psychological Review, 87*, 215–251.

Fawcett, H. P. (1938). *The nature of proof* (Yearbook of the National Council of Teachers of Mathematics). New York, NY: Columbia University Teachers College Bureau of Publications.

Flavell, J. (1976). Metacognitive aspects of problem solving. In L. Resnick (Ed.), *The nature of intelligence* (pp. 231–236). Hillsdale, NJ: Lawrence Erlbaum.

Flavell, J. H., Friedrichs, A. G., & Hoyt, J. D. (1970). Developmental changes in memorizations processes. *Cognitive Psychology, 1*, 323–340.

Garofalo, J., & Lester, F. (1985). Metacognition, cognitive monitoring, and mathematical performance. *Journal for Research in Mathematics Education, 16*, 163–176.

Geertz, C. (1983). *Local knowledge*. New York, NY: Basic Books.

Greeno, J. (1988). For the study of mathematics epistemology. In R. Charles & E. Silver (Eds.), *The teaching and assessing of mathematical problem solving* (pp. 23–31). Reston, VA: National Council of Teachers of Mathematics.

Groner, R., Groner, M., & Bischof, W. (Eds.). (1983). *Methods of heuristics*. Hillsdale, NJ: Lawrence Erlbaum.

Grouws, D., & Cooney, T. (Eds.) (1988). *Effective mathematics teaching*. Hillsdale, NJ: Lawrence Erlbaum.

Groves, S., & Stacey, K. (1984). *The burwood box*. Melbourne, Australia: Victoria College, Burwood.

Hadamard, J. (1945). *An essay on the psychology of invention in the mathematical field*. Princeton, NJ: Princeton University Press.

Halmos, P. (1980). The heart of mathematics. *American Mathematical Monthly, 87*, 519–524.

Harvey, J. G., & Romberg, T. A. (1980). *Problem-solving studies in mathematics*. Madison, WI: Wisconsin Research and Development Center Monograph Series.

Hayes-Roth, B., & Hayes-Roth, F. (1979). A cognitive model of planning. *Cognitive Science, 3*, 275–31.

Heller, J., & Hungate, H. (1985). Implications for mathematics instruction of research on scientific problem solving. In E. A. Silver (Ed.), *Teaching and learning mathematical problem solving: Multiple research perspectives* (pp. 83–112). Hillsdale, NJ: Lawrence Erlbaum.

Hiebert, J. (1985). (Ed.) *Conceptual and procedural knowledge: The case of mathematics*. Hillsdale, NJ: Lawrence Erlbaum.

Hinsley, D.A., Hayes, J. R., & Simon, H. A. (1977). From words to equations: meaning and representation in algebra word problems. In M. Just & P. Carpenter (Eds.), *Cognitive processes in comprehension* (pp. 89–106). Hillsdale, NJ: Lawrence Erlbaum.

Hoffman, K. (1989, March). *The science of patterns: A practical philosophy of mathematics education*. Paper presented at the Special Interest Group for Research in Mathematics Education at the 1989 Annual Meeting of the American Educational Research Association, San Francisco.

Hunt, E. (1975). *Artificial intelligence*. New York, NY: Academic Press.

International Association for the Evaluation of Educational Achievement. (1987). *The underachieving curriculum: Assessing U.S. school mathematics from an international perspective*. Champaign, IL: Stipes Publishing Company.

James, W. (1890). *Principles of psychology (2 volumes)*. New York, NY: Holt.

Janvier, C. (Ed.). (1897). *Problems of representation in the teaching and learning of mathematics*. Hillsdale, NJ: Lawrence Erlbaum.

Kantowski, M. G. (1977). Processes involved in mathematical problem solving. *Journal for Research in Mathematics Education, 8*, 163–180.

Kaput, J. (1985). Representation and problem solving: issues related to modeling. In E. A. Silver (Ed.), *Teaching and learning mathematical problem solving: Multiple research perspectives* (pp. 381–398). Hillsdale, NJ: Lawrence Erlbaum.

Kaput, J. (1989). Linking representations in the symbol system of algebra. In S. Wagner & C. Kieran (Eds.), *Research issues in the learning and teaching of algebra* (pp. 167–194). Hillsdale, NJ: Lawrence Erlbaum.

Karmiloff-Smith, A. (1979) Problem solving construction and representations of closed railway circuits. *Archives of psychology, 47*, 37–59.

Kilpatrick, J. (1967). Analyzing the solution of word problems in mathematics: An exploratory study. (Unpublished doctoral dissertation, Stanford University). *Dissertation Abstracts International, 28*, 4380A. (University Microfilms 68-5, 442.)

Kilpatrick, J. (1978). Variables and methodologies in research on problem solving. In L. Hatfield (Ed.), *Mathematical problem solving* (pp. 7–20). Columbus, OH: ERIC.

Kilpatrick, J. (1985). A retrospective account of the past twenty-five years of research on teaching mathematical problem solving. In E. A. Silver, *Teaching and learning mathematical problem solving: Multiple research perspectives* (pp. 1–16). Hillsdale, NJ: Lawrence Erlbaum.

Kitcher, P. (1984). *The nature of mathematical knowledge*. New York, NY: Oxford University Press.

Krulik, S. (Ed.). (1980). *Problem solving in school mathematics*. (Yearbook of the National Council of Teachers of Mathematics). Reston, VA: NCTM.

Krutetskii, V. A. (1976). *The psychology of mathematical abilities in school children* (J. Teller, Trans; J. Kilpatrick, & I. Wirszup, Eds.). Chicago, IL: University of Chicago Press.

Lakatos, I. (1977). *Proofs and refutations* (revised ed.). Cambridge, MA: Cambridge University Press.

Lakatos, I. (1978). *Mathematics, science, and epistemology.* Cambridge, U.K.: Cambridge University Press.

Lampert, M. (1990). When the problem is not the question and the solution is not the answer: Mathematical knowing and teaching. *American Educational Research Journal, 27,* 29–63.

Lave, J. (1988). *Cognition in practice.* Boston, MA: Cambridge University Press.

Lave, J. (in preparation). *Tailored learning: Apprenticeship and everyday practice among craftsmen in West Africa.*

Lave, J., Smith, S., & Butler, M. (1988). Problem solving as everyday practice. In R. Charles & E. Silver (Eds.), *The teaching and assessing of mathematical problem solving* (pp. 61–81). Reston, VA: National Council of Teachers of Mathematics.

Lave, J., & Wenger, E. (1989). *Situated learning: Legitimate peripheral participation.* (Manuscript available from author, School of Education, University of California, Berkeley.)

Lesh, R. (1983). *Metacognition in mathematical problem solving.* Unpublished manuscript. Available from author, Educational Testing Service, Rosedale Road, Princeton, NJ.

Lesh, R. (1985). Conceptual analyses of problem solving performance. In E. A. Silver (Ed.), *Teaching and learning mathematical problem solving: Multiple research perspectives* (pp. 309–330). Hillsdale, NJ: Lawrence Erlbaum.

Lester, F., Garofalo, J., & Kroll, D. (1989). *The role of metacognition in mathematical problem solving: A study of two grade seven classes.* Final report to the National Science Foundation of NSF project MDR 85-50346.

Lucas, J. (1974). The teaching of heuristic problem-solving strategies in elementary calculus. *Journal for Reseach in Mathematics Education, 5,* 36–46.

Mason, J., Burton, L., & Stacey, K. (1982) *Thinking mathematically.* New York: Addison Wesley.

Mayer, R. (1985). Implications of cognitive psychology for instruction in mathematical problem solving. In E. A. Silver (Ed.), *Teaching and learning mathematical problem solving: Mulitiple research perspectives* (pp. 123–138). Hillsdale, NJ: Lawrence Erlbaum.

McLeod, D., & Adams, V. (1989). *Affect and mathematical problem solving: A new perspective.* New York, NY: Springer-Verlag.

Mead, G. H. (1934). *Mind, self, and society.* Chicago, IL: University of Chicago Press.

Miller, G. A. (1956). The magical number seven, plus or minus two: Some limits on our capacity for processing information. *Psychological Review, 63,* 81–97.

Milne, W. J. (1897). *A mental arithmetic.* New York, NY: American Book.

Minsky, M. (1961). Steps toward artificial intelligence. *Proceedings of the Institute of Radio Engineers, 49,* 8–30.

Minsky, M. (1975). A framework for representing knowledge. (1977). In P. Winston (Ed.), *The psychology of computer vision* (pp. 170–195). New York, NY: McGraw-Hill.

Moschkovich, J. (1989, March). *Constructing a problem space through appropriation: A case study of tutoring during computer exploration.* Paper presented at the 1989 annual meetings of the American Educational Research Association, San Francisco.

National Assessment of Educational Progress. (1983). *The third national mathematics assessment: Results, trends, and issues* (Report No. 13-MA-01). Denver, CO: Educational Commission of the States.

National Center of Educational Statistics. (1988a). *Trends in minority enrollment in higher education, Fall 1986-Fall 1988.* Washington, DC: U.S. Department of Education.

National Center of Educational Statistics. (1988b). *Digest of education statistics, 1988.* Washington, DC: U.S. Department of Education.

National Commission on Excellence in Education. (1983). *A nation at risk: The imperative for educational reform.* Washington, DC: U.S. Government Printing Office.

National Council of Supervisors of Mathematics. (1977, October). Position paper on basic mathematical skills. *Arithmetic Teacher 25,* (1) 19–22.

National Council of Teachers of Mathematics. (1980). *An agenda for action.* Reston, VA: NCTM.

National Council of Teachers of Mathematics. (1989). *Curriculum and evaluation standards for school mathematics.* Reston, VA: NCTM.

National Research Council. (1989). *Everybody counts: A report to the nation on the future of mathematics education.* Washington, DC: National Academy Press.

National Research Council. (1990a). *Reshaping school mathematics: A philosophy and framework for curriculum.* Washington, DC: National Academy Press.

National Research Council. (1990b). *A challenge of numbers.* Washington, DC: National Academy Press.

Newell, A. (1983) The heuristic of George Pólya and its relation to artificial intelligence. In R. Groner, M. Groner, & W. Bischof (Eds.), *Methods of heuristics* (pp. 195–243). Hillsdale, NJ: Lawrence Erlbaum.

Newell, A., & Simon, H. (1972). *Human problem solving.* Englewood Cliffs, NJ: Prentice-Hall.

Newman, D., Griffin, P., & Cole, M. (1989). *The construction zone: Working for cognitive change in school.* Cambridge, MA: Cambridge University Press.

Norman, D. (Ed.). (1970). *Models of human memory.* New York, NY: Academic Press.

Novak, J. (Ed.). (1987). *Proceedings of the Second International Seminar on Misconceptions and Educational Strategies in Science and Mathematics.* Ithaca, NY: Cornell University.

Oxford University Press. *Oxford English Dictionary* (compact ed.). Oxford: Author.

Palincsar, A., & Brown, A. (1984). Reciprocal teaching of comprehension-fostering and comprehension-monitoring activities. *Cognition and Instruction, 1,* pp. 117–175.

Pavlov, I. P. (1928). *Lectures on conditioned reflexes* (3rd ed.). (W. H. Gantt, Trans.) New York, NY: International Publishers.

Pea, R. (1989). *Socializing the knowledge transfer problem.* (IRL report IRL 89-0009). Palo Alto, CA: Institute for Research on Learning.

Peters, R. S. (1962) *Brett's history of psychology* (edited and abridged by R. S. Peters). Cambridge, MA: The M.I.T. Press.

Peterson, P., Fennema, E., Carpenter, T., & Loef, M. (1989). Teachers' pedagogical content beliefs in mathematics. *Cognition and Instruction 6,* 1–40.

Piaget, J. (1928). *The language and thought of the child.* New York, NY: Harcourt Brace.

Piaget, J. (1930). *The child's conception of physical causality.* New York, NY: Harcourt Brace.

Piaget, J. (1954). *The construction of reality in the child* (M. Cook, trans.) New York, NY: Ballantine Books.

Piaget, J. (1971). *The child's conception of time.* (original French version published 1927). New York, NY: Ballantine Books.

Poincaré, H. (1913). *The foundations of science* (G. H. Halstead, Trans.). New York, NY: Science Press.

Pollak, H. (1987). Cognitive science and mathematics education: A mathematician's perspective. In A. H. Schoenfeld (Ed.), *Cognitive science and mathematics education* (pp. 253–264). Hillsdale, NJ: Lawrence Erlbaum.

Pólya, G. (1945; 2nd edition, 1957). *How to solve it.* Princeton, NJ: Princeton University Press.

Pólya, G. (1954). *Mathematics and plausible reasoning; Vol. 1. Induction and analogy in mathematics; Vol. 2. Patterns of plausible inference.* Princeton, NJ: Princeton University Press.

Pólya, G. (1962,1965/1981). *Mathematical discovery* (Volume 1, 1962; Volume 2, 1965). Princeton: Princeton University Press. (Combined paperback edition, 1981). New York, NY: Wiley.

Pólya, G., & Szego G. (1925) *Aufgaben und Lehrsätze aus der Analysis I*. Berlin, Germany: Springer. An English version, *Problems and theorems in analysis I* (D. Aeppli, Trans.), was published by Springer (New York) in 1972.

Putnam, R. T., Lampert, M., & Peterson, P. (1989). Alternative perspectives on knowing mathematics in elementary schools. Elementary subjects center series number 11. Michigan State University: Center for Learning and Teaching Elementary Subjects.

Resnick, L. (1988). Treating mathematics as an ill-structured discipline. In R. Charles & E. Silver (Eds.), *The teaching and assessing of mathematical problem solving* (pp. 32–60). Reston, VA: National Council of Teachers of Mathematics.

Rissland, E. (1985). Artificial intelligence and the learning of mathematics: A tutorial sampling. In E. A. Silver (Ed.), *Teaching and learning mathematical problem solving: Multiple research perspectives* (pp. 147–176). Hillsdale, NJ: Lawrence Erlbaum.

Rogoff, B., & Lave, J. (Eds.). (1984). *Everyday cognition: Its development in social context*. Cambridge, MA: Harvard University Press.

Romberg, T., & Carpenter, T. (1986). Research in teaching and learning mathematics: Two disciplines of scientific inquiry. In M. Wittrock (Ed.), *Handbook of research on teaching* (3rd ed. pp. 850–873). New York, NY: Macmillan.

Ryle, G. (1949). *The concept of mind*. London: Hutchinson.

Sacerdoti, E. (1974) Planning in a hierarchy of abstraction spaces. *Artificial Intelligence 5*, 115–136.

Scardamalia, M., & Bereiter, C. (1983). Child as co-investigator: Helping children to gain insight into their own mental processes. In S. G. Paris, M. Olson, & H. W. Stevenson (Eds.), *Learning and motivation in the classroom* (pp. 61–82). Hillsdale, NJ: Lawrence Erlbaum.

Schank, R., & Abelson, R. (1977). *Scripts, plans, goals, and understanding*. Hillsdale, NJ: Lawrence Erlbaum.

Schoenfeld, A. (1983). *Problem solving in the mathematics curriculum: A report, recommendations, and an annotated bibliography*. Washington, D.C.: Mathematical Association of America.

Schoenfeld, A. (1985a). *Mathematical problem solving*. New York, NY: Academic Press.

Schoenfeld, A. (1985b). Metacognitive and epistemological issues in mathematical understanding. In E. A. Silver (Ed.), *Teaching and learning mathematical problem solving: Multiple research perspectives* (pp. 361–380). Hillsdale, NJ: Lawrence Erlbaum.

Schoenfeld, A. (1987a). What's all the fuss about metacognition? In A. Schoenfeld (Ed.), *Cognitive science and mathematics education* (pp. 189–215). Hillsdale, NJ: Lawrence Erlbaum.

Schoenfeld, A. (1987b, December). Pólya, problem solving, and education. *Mathematics Magazine, 60*(5), 283–291.

Schoenfeld, A. (1988). When good teaching leads to bad results: The disasters of "well taught" mathematics classes. *Educational Psychologist, 23*, 145–166.

Schoenfeld, A. (1989a). Problem solving in context(s). In R. Charles & E. Silver (Eds.), *The teaching and assessing of mathematical problem solving*, (pp. 82–92). Reston, VA: National Council of Teachers of Mathematics.

Schoenfeld, A. (1989b). Explorations of students' mathematical beliefs and behavior. *Journal for Research in Mathematics Education, 20*, 338–355.

Schoenfeld, A. (1989c). Ideas in the air: Speculations on small group learning, environmental and cultural influences on cognition, and epistemology. *International Journal of Educational Research, 13*, 71–88.

Schoenfeld, A. (1989d). Teaching mathematical thinking and problem solving. In L. B. Resnick & B. L. Klopfer (Eds.), *Toward the thinking curriculum: Current cognitive research* (pp. 83–103). (1989 Yearbook of the American Society for Curriculum Development). Washington, DC: ASCD.

Schoenfeld, A. (Ed.) (1990a) *A source book for college mathematics teaching*. Washington, DC: Mathematical Association of America.

Schoenfeld, A. (1990b). On mathematics as sense-making: An informal attack on the unfortunate divorce of formal and informal mathematics. In D. N. Perkins, J. Segal, & J. Voss (Eds.), *Informal reasoning and education* (pp. 281–300). Hillsdale, NJ: Lawrence Erlbaum.

Schoenfeld, A. (in preparation). Reflections on doing and teaching mathematics. In A. Schoenfeld (Ed.), *Mathematical thinking and problem solving*.

Schoenfeld, A., Smith, J., & Arcavi, A. (in press). Learning: The microgenetic analysis of one student's understanding of a complex subject matter domain. In R. Glaser (Ed.), *Advances in instructional psychology* (Vol. 4). Hillsdale, NJ: Lawrence Erlbaum.

Shaughnessy, M. (1985). Problem-solving derailers: The influence of misconceptions on problem solving performance. In E. A. Silver (Ed.), *Teaching and learning mathematical problem solving: Multiple research perspectives* (pp. 399–416). Hillsdale, NJ: Lawrence Erlbaum.

Shell Centre for Mathematical Education. (1984). *Problems with patterns and numbers*. Nottingham, England: Shell Centre.

Silver, E. A. (1979). Student perceptions of relatedness among mathematical verbal problems. *Journal for Research in Mathematics Education, 10*, 195–210.

Silver, E. A. (1981). Recall of mathematical problem information. Solving related problems. *Journal for Research in Mathematics Education, 12*, 54–64.

Silver, E. A. (1982). *Thinking about problem solving: Toward an understanding of metacognitive aspects of problem solving*. Paper presented at the Conference on Thinking, Suva, Fiji, January.

Silver, E. A. (Ed.). (1985). *Teaching and learning mathematical problem solving: Multiple research perspectives*. Hillsdale, NJ: Lawrence Erlbaum.

Silver, E. A. (1987). Foundations of cognitive theory and research for mathematics problem solving instruction. In A. Schoenfeld (Ed.), *Cognitive science and mathematics education* (pp. 33–60). Hillsdale, NJ: Lawrence Erlbaum.

Silver, E. A., Branca, N., & Adams, V. (1980). Metacognition: The missing link in problem solving? In R. Karplus (Ed.), *Proceedings of the Fourth International Congress on Mathematical Education* (pp. 429–433). Boston, MA: Birkhäuser.

Simon, H. (1979). Information processing models of cognition. *Annual Review of Psychology, 30*, 363–96.

Simon, H. (1980). Problem solving and education. In D. Tuma & F. Reif (Eds.), *Problem solving and education: Issues in teaching and research* (pp. 81–96). Hillsdale, NJ: Lawrence Erlbaum.

Skinner, B. F. (1974). *About behaviorism*. New York, NY: Knopf.

Smith, J. P. (1973). The effect of general versus specific heuristics in mathematical problem solving tasks. *Dissertation Abstracts International, 34*, 2400A. (University Microfilms 73-26, 637.)

Sowder, L. (1985). Cognitive psychology and mathematical problem solving: A discussion of Mayer's paper. In E. A. Silver (Ed.), *Teaching and learning mathematical problem solving: Multiple research perspectives* (pp. 139–145). Hillsdale, NJ: Lawrence Erlbaum.

Stanic, G., & Kilpatrick, J. (1988). Historical perspectives on problem solving in the mathematics curriculum. In R. Charles & E. Silver (Eds.), *The teaching and assessing of mathematical problem solving* (pp. 1–22). Reston, VA: National Council of Teachers of Mathematics.

Steen, L. (1988). The science of patterns. *Science, 240,* 611–616.

Stevenson, H. W., Lee, S-Y., & Stigler, J. W. (1986). Mathematics achievement of Chinese, Japanese, and American children. *Science, 231,* 693–698.

Stigler, J., & Perry, M. (1989). Cross cultural studies of mathematics teaching and learning: Recent findings and new directions. In D. Grouws & T. Cooney (Eds.), *Effective mathematics teaching* (pp. 194–223). Hillsdale, NJ: Lawrence Erlbaum.

Stipek, D. J., & Weisz, J. R. (1981). Perceived personal control and academic achievement. *Review of Educational Research 51,* 101–137.

Stodolsky, S. S. (1985). Telling math: Origins of math aversion and anxiety. *Educational Psychologist, 20,* 125–133.

Suinn, R. M., Edie, C. A., Nicoletti, J., & Spinelli, P. R. (1972). The MARS, a measure of mathematics anxiety: Psychometric data. *Journal of Clinical Psychology, 28,* 373–375.

Thompson, A. (1985). Teachers' conceptions of mathematics and the teaching of problem solving. In E. A. Silver, *Teaching and learning mathematical problem solving: Multiple research perspectives* (pp. 281–294). Hillsdale, NJ: Lawrence Erlbaum.

Thorndike, E. L. (1924). Mental discipline in high school studies. *Journal of Educational Psychology, 15,* 1–22, 83–98.

Thorndike, E. L., & Woodworth, R. S. (1901). The influence of improvement in one mental function on the efficiency of other mental functions. *Psychological Review, 8,* 247–261.

Tobias, S. (1978). *Overcoming math anxiety.* New York, NY: W. W. Norton.

Vygotsky, L. (1978). *Mind in society.* Cambridge, MA: Cambridge University Press.

Wagner, S., & Kieran, C. (Eds.). (1989). *Research issues in the learning and teaching of algebra.* Hillsdale, NJ: Lawrence Erlbaum.

Wallas, G. (1926). *The art of thought.* Selections in P. E. Vernon (Ed., 1970), *Creativity,* Middlesex, England: Penguin, p. 91–97.

Watson, J. (1930). *Behaviorism* (2nd ed.). New York, NY: Norton.

Watson, R. I. (1978). *The great psychologists.* Philadelphia, PA: Lippincott.

Webster's. (1979) *New universal unabridged dictionary.* Second edition. New York, NY: Simon & Schuster.

Wertheimer, M. (1945/1959). *Productive thinking.* New York, NY: Harper and Row.

Wilson, J. (1967) Generality of heuristics as an instructional variable. *Dissertation Abstracts International, 28,* 2575A. (University Microfilms 67-17, 526).

Wundt, W. (1904). *Principles of physiological psychology* (5th German ed., Vol. 1). (E. B. Titchener, Trans.). New York, NY: Macmillan.

Zweng, M., Green, T., Kilpatrick, J., Pollak, H., & Suydam, M. (Eds.). (1983). *Proceedings of the Fourth International Congress on Mathematics Education.* Boston, MA: Birkhäuser.

·16·

ESTIMATION AND NUMBER SENSE

Judith Sowder

SAN DIEGO STATE UNIVERSITY

The Many Facets of Estimation

Many types of questions and situations call for producing an estimate: Do I have enough cash to pay for these books? How much paint do I need for this room? How many people are in the stadium? How much do I spend on groceries each week? How long will it take me to drive to the dentist's office? Answering these questions involves estimating results of computations, estimating measures, and estimating numerosity. Each requires different kinds of understandings and different sets of skills.

The most common type of computational estimation problem requires estimating the result of a computation by performing some mental computation on approximations of the original numbers. To be correct, the answer must fall within a certain interval, as determined by the problem itself or some outside source, such as a teacher. Rubenstein (1983, 1985) also includes in this category tasks calling for decisions on whether or not a computational answer is reasonable, whether a number given is larger or smaller than the exact answer, whether the answer is larger or smaller than a given reference number, and whether or not an estimate is of the correct order of magnitude. The bulk of estimation research falls within the category of computational estimation, and a major portion of this chapter is devoted to this research.

Estimating measures and estimating numerosity call upon some of the same skills. To find the length of a room, one might estimate the length of a tile, count the number of tiles, and multiply. Estimating the length of the tile, however, calls for a very different type of skill than estimating numerosity, since the estimate required is of a continuous quantity rather than of a discrete quantity. A mental reference unit acquired through experience with measuring rather than through counting is required. Bright (1976) has provided a very thorough analysis of what is involved in estimating measurements. He defines estimation as the process of arriving at a measurement without using measuring tools, whereas measuring is the process of comparing an attribute of a physical object to some preselected unit. Measurement sets up a correspondence between the attribute and the real numbers. In order to form an estimate then, one must have a mental reference unit, that is, a mental "picture" or "feel" for the size of the unit. Bright notes that good estimators have a well-developed set of these mental reference units, which can be used in different ways. For example, an estimate of area might be found in one instance by using linear estimates in an area formula and in another instance by estimating the number of square units that would fit within the boundaries.

"How many" is usually a question of numerosity and asks that the number of items in a set be found. In many cases an estimate is sufficient and perhaps all that is even possible. As adults, we sometimes try to estimate the number of people in a theater, the number of cars in a parking lot, the number of strawberries in a container. The typical procedure used in numerosity estimation is to take a count of a sample, then multiply it by the number of such samples estimated to be present. So to estimate the number of people at a baseball game, we might count or estimate the people in a small section, estimate the number of such sections, and use the product as our estimate of the total. Variations on this procedure could use cross-sections (perhaps in finding the number of jelly beans in a jar), ratios (how many of the jelly beans in the jar are green), and other techniques. (Clayton, 1988a, has noted that this type of problem requiring an estimate of the number of

The preparation of this paper was supported in part by a grant from the National Science Foundation (Grant No. MDR-8751373). The opinions expressed in this paper do not necessarily reflect the position, policy, or endorsement of the National Science Foundation. I wish to thank Sandra Marshall, San Diego State University, Jack Hope, University of Saskatchewan, and Larry Sowder, San Diego State University, for their helpful comments on an earlier version of this chapter.

items in a container leads to what he calls the "jelly bean" effect in which estimates are particular rather than general, such as "583" rather than "600"). Maxfield (1976) has described how a parametric model, using areas or volumes known in terms of parameters such as the length of a side, can be used to estimate the number of individuals in a rapidly shifting crowd. In one example she used a wedge shape, together with triangular numbers (1, 3, 6, 10, … where $3 = 1 + 2$, $6 = 1 + 2 + 3$, $10 = 1 + 2 + 3 + 4$, and so on), to estimate the number of fish in a school, 1 fish deep, with about 8 rows, as approximately 36 fish ($1 + 2 + … + 8$).

Numerosity problems can become quite complex. Enrico Fermi liked to pose problems such as "How many piano tuners are there in Chicago?" This problem, in fact, has become so well known that it serves to illustrate the genre of problems called "Fermi problems." Another such problem well known in the problem-solving research literature is credited to Schoenfeld and asks for an estimate of the number of cells in an average-sized adult human body (Schoenfeld, 1985). Ross and Ross (1986) have pointed out that such problems are recursive in the sense that once some formulation of the problem is made, the quantities identified as necessary to the solution process are often themselves Fermi subproblems. The Rosses identify the distinguishing characteristic of a Fermi problem as the total reliance on information stored away in the head of the problem solver. Such problems usually *must* be answered with an estimate, since the exact answer is not available.

These three categories of estimation are the most commonly considered types of estimation (O'Daffer, 1979) because they are frequently used in daily life, and they are beginning to appear in elementary and middle school curricula. But there are other forms of estimation. Smart (1982) has described estimating trigonometric functions, estimating numerical values of the derivative for a graph of a function, and estimating with a calculator (for example, $6.159^{2.317}$) as other examples. Statistics and probability are other areas where estimation skills would be useful and might lead to better understanding. Although estimating the mean of a set of data might be undertaken as a computational estimation task, a statistician can often estimate not only the mean but also the standard deviation of the data, an estimate based more on previous experience than on specific computation. Estimating the probability of an event, such as whether school will be closed for a snow day within the next month, may or may not call for calculations, depending upon information such as location and time of year. As with other types of information used in daily life, an estimate of a statistical value or of a probability is often all that is needed and in some cases all that can be found. Development of the ability to make such estimates has not been investigated.

Background for Research on Estimation

Until recently, curriculum developers have given estimation short shrift. Before 1980, only a few voices spoke in support of its role in the mathematics curriculum. The World Book series of mathematics texts, with John Clark as principal author, gave estimation a prominent role throughout the grades, but this was not the usual state of affairs in mathematics texts; in fact, estimation disappeared from later editions of World Books texts. It

seems that computational estimation, at least, did not fit the objectives of the "new math" movement. A few scattered articles on estimation appeared in the *Arithmetic Teacher* (Corle, 1960, 1963; Faulk, 1962; O'Daffer, 1979; Sauble, 1955) and in *Science Education* (Swan & Jones, 1971, 1980) and some doctoral dissertations considered aspects of computational estimation (Fesharaki, 1979; Hall, 1977; Ibe, 1973; Nelson, 1967; Paull, 1972), but by and large the topic was ignored. Nevertheless, the role of estimation in mathematics, particularly in the mathematics curriculum, was recognized by Bell (1980) in his list of topics "everyman" should know and by Skvarcius (1973) in his paper at the Cape Ann Conference on Junior High Mathematics.

Then in 1977, the National Council of Supervisors of Mathematics, using the Euclid Conference Report (National Institute of Education, 1975) as a basis for developing a position paper on basic skills, included estimation and approximation within the 10 basic skill areas in need of development. This was followed by the *Agenda for Action* (National Council of Teachers of Mathematics, 1980), which called for incorporating estimation activities "into all areas of the program on a regular and sustaining basis" (p. 7), and the National Science Board Commission's recommendation (1982) that the development of estimation skills receive substantially more emphasis.

Very little research attention had been given to estimation (Buchanan, 1980) before these reports and recommendations. However, NAEP data (Carpenter, Coburn, Reys, & Wilson, 1976; Carpenter, Corbitt, Kepner, Lindquist, & Reys, 1980) offered ample evidence that students lacked basic and necessary estimation skills for carrying out these recommendations. It is not surprising then that most of the estimation research, particularly research on computational estimation, has been published since 1980.

A Chapter Overview

This chapter will focus on topics in estimation and related areas that have proved to be interesting to researchers. While other topics, such as the role of estimation in calculus, might also be interesting and educational, the lack of research associated with other areas would force discussions to be highly speculative. Even on topics that *have* received research attention, there will be more speculation than some might wish, since there simply is not a rich research base on estimation.

Computational estimation has received the most research attention, and the bulk of this chapter attends to studies of how people estimate computations and what abilities are related to estimation ability, how computational estimation concepts develop, the effects of instruction on computational estimation ability, and affective factors related to computational estimation. Mental computation, which is closely allied with computational estimation, is discussed next. The chapter also includes a brief discussion of some recent thinking about number sense and its importance for estimation.

The discussion of number sense is followed by the section on measurement estimation, with ability to estimate and instruction on estimation as the two major topics. Numerosity estimation has received the least research attention, and because the only two studies located combine it with measurement estimation, it is so combined here.

Assessment issues related to estimation are in some sense unique and different from other assessment issues, and therefore a section on assessment is also included. Finally, some psychological models for continuing research on these topics are examined.

But before beginning the major topics of this chapter, we need a brief digression to clarify two commonly misunderstood terms.

Estimation and Approximation

Some authors do not care to make clear distinctions between the terms *estimation* and *approximation*. Smart (1982), for example, defines estimation as "forming an approximate opinion of size, amount, or number that is sufficiently exact for a specified purpose" (p. 642). But Hall (1984) claims that estimation and approximation are not synonymous and writes that "whereas estimation is usually a mental exercise, approximating usually requires a tool of some kind" (p. 517). His example is that an estimate of the decimal representation of 4/17 to the closest thousandth could be 0.238, since 4/17 is slightly less than 1/4. But to approximate the decimal representation to the closest thousandth requires a calculator or paper and pencil to carry out the long division. Thompson (1979) calls an estimate an educated guess, usually made in the context of the number of objects in a collection, the result of a numerical computation, or the measure of an object. It is often difficult to assess the magnitude of error in an estimate. Approximating, on the other hand, is attempting to close in on a target value. It is often possible to get as close as one desires without ever reaching the exact value.

This particular distinction between estimation and approximation is not universal. Siegel, Goldsmith, and Madson (1982) consider estimation to be a process leading to a solution to a problem of counting or measurement; what we refer to as computational estimation they refer to as approximation. However, the Hall and Thompson distinction is the one mathematicians usually make and is best exemplified in Bakst's (1937) classical treatise on approximate computation, for which he distinguishes three approaches: the analysis of errors of approximate numbers, the writing of approximate numbers with one digit to the right of the last significant digit corresponding to the apparent error, and the probability of occurrences of error.

Approximation plays an important role in measurement. Every measurement of a continuous quantity is actually an approximation, since the "estimate then measure" activities often used in classrooms call for finding an estimate and then finding an approximate value of the measure. The existence or development of approximation skills has not been a focus of research. Yet it is an extremely important skill to have, particularly when mathematics is applied to real-world situations, as in science. Measures of level of accuracy of measurement are widely misunderstood. Approximation has its own "arithmetic" and rules. Further development on this topic is beyond the scope of this chapter, but a very readable treatment on approximate numbers by Hilton and Pedersen can be found in the NCTM year book *Estimation and Mental Computation* (Schoen & Zweng, 1986).

COMPUTATIONAL ESTIMATION

Studies of Estimation Ability

The NAEP data cited above, coupled with the recommendations to include more estimation in the curriculum, prompted several investigations of student performance on computational estimation. The most comprehensive of these was a study by Reys, Rybolt, Bestgen, and Wyatt (1980, 1982), undertaken to identify and characterize the computational processes used in estimation by in-school pupils and out-of-school adults. In the first part of the study, the researchers identified a group of good estimators by administering a 55-item computational estimation test to over 1,200 students in upper-track classes from grades 7 to 12 and to selected adults. A portion of those scoring in the top 10% of each age group were then interviewed in order to determine what strategies they used in solving estimation problems.

From the 59 interviews, the investigators identified three key processes characteristically used by good estimators. The first they called *reformulation*, or "the process of altering numerical data to produce a more mentally manageable form while leaving the structure of the problem intact" (p. 187). Rounding, truncating, substituting more compatible numbers (for example, $[6 \times 347] \div 43$ changed to $[6 \times 350] \div 42$), or using equivalent (or nearly equivalent) forms of numbers (for example, 30% changed to 1/3) would be examples of reformulation. The second process identified was called *translation*, or "changing the mathematical structure of the problem to a more mentally manageable form" (p. 188). Examples given were the recasting of $8,946 + 7,212 + 7,814$ to $8,000 \times 3$ or changing $(347 \times 6) \div 43$ to $347 \times (6 \div 43)$ and then to $350 \div 7$. The third process consistently used by these good estimators was *compensation*: "Adjustments made to reflect numerical variation that come about as a result of translation or reformulation of the problem. These adjustments were typically a function of the amount of time available to make a response but were also influenced by the manageability of the numerical data, context of the problem, and the individual's tolerance for error" (p. 189). Compensations were made both during the estimation process and after an initial estimate had been found. In an example of compensation after an initial estimate, a problem requiring an estimate of $\$21,319,908 \div 26$ was found by dividing 26,000,000 by 26, yielding 1,000,000, then compensating downward to \$850,000.

The investigators also noted that good estimators used a variety of strategies within these three key processes. Perhaps the most useful strategy was one of focusing on the left-most digit or digits, commonly known as the front-end strategy. This strategy took one of four forms: operating with rounded numbers with the same number of digits ($3,852 \to 4,000$), operating with the extracted portion of the rounded numbers ($3,852 \to 4$), truncating and filling in with zeros ($3,852 \to 3,000$), or truncating alone ($3,852 \to 3$). Of the four variations, only the first was based on formal instruction. However, good estimators often ignored school-taught rules of rounding and rounded to numbers more suitable for solving the problem at hand. In the problem given earlier ($[6 \times 347] \div 43$), 43 was rounded to 42 because 42 was a convenient multiple of 7. The distinction between rounding and truncation is important, because rounding is often a difficult skill for students to learn. Furthermore, as Trafton (1978) has pointed out, when students are working mentally with the approximated numbers, rounding hides the original numbers, whereas truncation keeps the significant digits in sight.

The investigators characterized good estimators as individuals who, besides having the ability to use these three processes, had a good grasp of basic facts, place value, and arithmetic properties; were skilled at mental computation; were self-confident and tolerant of error; and could use a variety of strategies and switch easily between strategies.

While the information obtained from investigating the processes used by good estimators has been helpful in guiding recent curriculum planning and other research studies, research literature has shown us many times that we also have much to gain by looking at the ways in which students of lower abilities approach particular kinds of problems. Other studies of performance on estimation tasks do this. Levine (1982) interviewed 89 college students of varying mathematical backgrounds and abilities in her study of how people estimate. Each student was asked to estimate answers for 10 multiplication and 10 division problems, with both whole numbers and decimal numbers involved. The students also took a quantitative ability test. Not surprisingly, Levine found that students who scored higher on the quantitative ability test used more estimation strategies, used different estimation strategies, and were better estimators than students who scored lower on the quantitative ability test. The most frequent strategy (for some, the only strategy) used by students of lower quantitative ability was to try to calculate, estimate, and then combine all partial products or quotients. For example, an estimate of 64.6×0.16 was found by taking 6×646, obtaining 3,876, rounding it to 3,000, then finding 10×646, or 6,460, rounding this to 6,000, adding 3,000 and 6,000 to obtain 9,000, and then placing a decimal point for a final estimate of 9. Levine noted that this strategy did not "require the individual to sense any relationships . . . or to have any 'number sense' to carry it out" (p. 358). Students of higher quantitative ability often used fractional relationships (for example, change 424×0.76 to $424 \times 3/4$) together with rounding one or both numbers.

These findings are consistent with the Reys et al. (1982) results, even though the designs of the studies and the methods of analysis are quite different. Good estimators in Levine's study used more strategies and appeared to be more flexible in their thinking than poor estimators, but her types of estimation items and her categories of strategies were more limited than those in the Reys et al. (1982) study, making it impossible to note whether some of the more sophisticated strategies found in the Reys et al. (1982) study were within the capabilities of the good students in her study. For example, Levine's better estimators may well have engaged in some sort of compensation, but since this was not one of the strategies studied by Levine, it is unclear whether or not compensation was attempted by either the good or the poor estimators.

Another investigation prompted by the poor performance by students on items from the NAEP tests was undertaken by Threadgill-Sowder (1984). She selected 12 representative NAEP or NAEP-like items and studied responses given by 26 students in sixth through ninth grade who were given the items individually and asked to work them aloud. Both multiple-choice and open-ended problems were included, with some context-embedded and some context-free problems in each category. Multiple-choice items were frequently answered correctly for the wrong reason, which implies that NAEP results might be even worse than reported. For example, one item read: "A quart of water weighs about 2.1 lbs. About how much do 13.8 quarts of water weigh? The answer is closest to (a) 6 1/2 (b) 7 (c) 26 (d) 28." Of the 16 students who selected 28 as their answer, 4 of them selected it for a wrong reason, such as looking for a number ending in 8 since $1 \times 8 = 8$.

Students who gave acceptable responses consistently demonstrated a quantitative intuition, or "number sense." Unacceptable responses went hand in hand with glaring examples of inability to grasp the meaning of the numbers in the problem. For example, over a third of the students gave a decimal answer for the estimated sum of 148.72 and 51.351, most with 200 as the whole number part of the answer.

Although giving a context for the estimation seemed to help in some cases, context can also make problems more difficult. Using the unfamiliar word *assessed* in the context of a tax problem in this study made the problem extremely difficult for some students. Aside from the obvious difficulty caused by using unfamiliar words in story problems, it seems reasonable to expect that any computation or estimation problem set within a context will make the problem more comprehensible to most students. Problems in verbal form were found to be easier than purely numerical problems in the study by Reys et al. (1980). This was also the case in the Carraher, Carraher, and Schliemann study (1987) in which arithmetic problems given as computation problems were significantly less likely to be solved than word problems or simulated store problems of the same arithmetic difficulty.

The use of context in estimation was a major factor in a study by Morgan (1988). She presented parallel questions as word problems and as computations and found that context helped in two ways. Difficulties in conceptualizing operations, such as multiplying by a number less than 1, were easier to overcome in problems set in a context. Also, the presence of a context seemed to discourage an algorithmic approach. For example, students were more likely to recognize digits after the decimal point as relatively insignificant when the decimal numbers were linked to a context.

Some abilities and skills seem to be very naturally related to computational estimation. In the Levine study just discussed, it was noted that quantitative ability is related to estimation ability. Hall (1977) found that estimation ability is related to problem-solving ability. And Paull (1972) concluded that the ability to estimate answers to arithmetic problems was related to mathematical and verbal ability.

Rubenstein (1985) identified eight factors that seemed likely to be related to computational estimation and devised 10-item scales on each factor, which she then gave to 309 eighth graders, together with a 64-item estimation test. In a regression analysis, the three factors contributing most to predicting estimation performance were the abilities to multiply and divide by powers of 10, to choose the larger or smaller of two numbers, and to choose which of two problems would yield the larger answer. This result was somewhat surprising, particularly since place value, knowledge of number facts, and operating with multiples of 10 explained little of the variance in the stepwise regression. Rubenstein noted that the factors that predicted estimation performance are not emphasized in textbooks to the extent that the less-predictive skills are.

These studies provide a good deal of information on processes and strategies used by good estimators, on the algorithmic-bound strategies used by poor estimators, and on some of the factors related to estimation ability. The picture is far from complete, however, and sometimes the data are not consistent. As is usually true of any new field of research, results are somewhat piecemeal. In the studies reviewed thus far, just as in the majority of those yet to be discussed here, the researchers began their work without established definitions of the phenomena being studied, without agreed-upon methodologies to guide their investigations, and without strong theories upon which to interpret their results. A "call to action" in the form of nationally made recommendations, together with beliefs born of classroom experience and knowledge of curricula in other areas of mathematical learning, led to these studies, to their designs, and to their results. These points are not made in a spirit of negative criticism, since it takes a body of researchers working on the same topic to reach a point where there can be any consensus on such matters as theory, definitions, or methodology, and any long-range research planning. Rather, these points are made to allow the reader to understand the limitations of the research, the reasons why some research questions were addressed rather than others, and some reasons for what may appear to be discrepancies or inconsistencies in the data.

For example, Rubenstein's data appear to show that place value and knowledge of basic facts are not good predictors of estimation performance, and yet these were the first characteristics noted by Reys et al. (1980) in their description of good estimators. It could be that other factors Rubenstein found to be good predictors were too closely related to place value and knowledge of basic facts for all to be significant in a stepwise regression analysis, or it could be that the items used to test these factors were not appropriate or that the students in her study differed little in their ability to recall basic facts. A less palatable interpretation might be that just because these two factors were found to be characteristic of good estimators in no way implies that they are directly related to estimation. And yet another reason for the inconsistency might be that data obtained from multiple-choice items, completion items, and scales sometimes lead to different results from those obtained from analyzing interview data. The two methodologies uncover very different aspects of any particular phenomenon. In the case of place value, it is easy to imagine test items that a majority of students could answer correctly, while interview data might show that only a few of the students truly understand place-value concepts.

There is much yet to be learned about estimation and related skills. Indeed, this information may in some ways become even more difficult to obtain, because future studies will also need to account for the effects of instruction on estimation. Such instruction is now taking place, but in very different ways and to varying degrees, and will make some research more difficult to interpret.

Yet there is one clear message that comes through several of these studies. *Good estimators are flexible in their thinking, and they use a variety of strategies. They demonstrate a deep understanding of numbers and operations, and they continually draw upon that understanding.* This point was dramatically made in a study by Dowker (1988), who gave Levine's (1982) 20 multiplication and division estimation items to 35 professional mathematicians at several universities. Not only were the mathematicians very accurate estimators but they also used a very large number of strategies, 22 different ones on $546 \div 33.5$ alone. Despite the fact that several items were designed to elicit a particular strategy, as many as 10 or more strategies were actually used on those items. The versatility and flexibility were even more striking when seven of the mathematicians were retested several months later: On about half of the items they used strategies different from those they had used when they were first tested.

Although, of course, estimation ability exists on some sort of a continuum, the picture that emerges from these studies tends to accentuate the extremes. Not only are very good estimators extremely versatile but poor estimators seem to be bound, with only slight variations, to one strategy—that of applying algorithms more suitable for finding the exact answer. Poor estimators have only a vague notion of the nature and purpose of estimation; they believe it to be inferior to exact calculation (Morgan, 1988) and equate it with guessing (Threadgill-Sowder, 1984).

There are several instructional questions at issue here. Can the particular strategies used by good estimators be taught to less skilled estimators? Or is it the flexibility of thinking and the inventing of strategies that should be our instructional focus? The research on experts and novices in other areas of mathematical and scientific endeavors has shown that experts organize subject matter very differently than novices, and novices become experts not simply by learning more facts and procedures but through a long process of mentally reorganizing their cognitive structure of the subject matter.

Hatano (1988) argues that there are, in fact, two quite different forms of expertise. Routine experts can solve routine problems quickly and accurately, using procedures which they have automatized. They have a body of procedural knowledge that is easily accessible. Such expertise is not without merit, and in some situations it is all that is necessary to accomplish whatever tasks are to be completed. An individual with adaptive expertise, on the other hand, understands how and why procedures work and can modify procedures as the constraints of the problem change. In other words, the adaptive expert can solve novel problems, while the routine expert cannot. This distinction seems particularly relevant when thinking about instruction on computational estimation. Many estimation problems are novel in nature, making it difficult if not impossible to delineate a list of procedures which, if learned, would constitute an adequate knowledge base for estimation. The adaptive expert who is not limited to particular rules and procedures but who has access to several ways of finding solutions is more likely to be successful at estimation. This model should influence the way we think about planning instruction on estimation.

Development of Computational Estimation Concepts and Skills

As has already been noted, estimation has not been a regular topic in the school curriculum, and students are only now being exposed to any instruction on estimation. Yet from the Reys et al. (1980) study, for example, we know that some people *do* become good estimators, and most people learn to estimate at least to some degree. A natural question to ask then is how estimation abilities develop in the absence of instruction.

In all of the studies discussed thus far, the subjects were middle-grade and secondary-grade students or adults. Do younger children estimate? And if so, what strategies do they use? Baroody (1989) investigated this question by analyzing the ways in which young children determine sums before they have learned the correct answers. His study was based on a schema-knowledge view that mental addition tasks, attempted before the answers are known, are actually estimation exercises for the children. The kindergartners in his study were not allowed to compute. Several strategies were noted for finding sums. Mechanical ones included constructing a teen from an addend (for example, $8 + 5$ is 18); adding 1 to the last addend ($8 + 5$ is 6); and adding 1 to the largest addend ($8 + 5 = 9$). As children became developmentally more prepared for addition, their invented estimation strategies yielded more plausible answers. Rather than adding 1 to the largest addend, some larger number was given, so $8 + 5$ might be 10, 11, 12, or 13, or else a teen *not* ending in one of the addends would be stated.

In a study of estimation strategies used by 5- to 9-year-olds, Dowker (1989) divided children into four levels according to their performance on a set of addition tasks. Depending on the level in which they were placed, students were asked to complete a set of estimation problems. For example, a Level 1 item was $5 + 2$, while a Level 4 item was $217 + 285$. Children were introduced to the idea of telling *about* how much the sum was through examples of answers by hypothetical students, answers that were classified as "good guesses" and "silly guesses." The students were then asked to make "good guesses" to additional problems. Children at Level 1 performed much worse than children at higher levels, even though the problems they received were easier. Many of their estimates were less than one of the addends or more than twice the exact answer. Several students used Baroody's (1989) mechanical strategies, as described in the preceding paragraph. At Levels 3 and 4 some students rounded or used a front-end strategy. The most striking finding in this study was that children frequently used appropriate strategies on estimation problems slightly above their level, but not on more difficult tasks. Dowker believed that very difficult tasks demoralized the children and encouraged wild guesses.

In a study by Sowder and Wheeler (1987) with slightly older children, students in grades 2, 4, 6, 8, and 10 were given estimation tasks all dealing with money in a "store" setting. Each task was given to students over several grade levels, and in each case, the percentage of students able to complete the problem increased by grade level. For example, students in grades 4, 6, 8, and 10 were shown a picture of four tires, each marked $43.97, with four answers for the total. The percentage of students choosing the correct "most likely" answer (without paper or pencil available) increased from 42% to 57% to 63% to 76% over the four grade levels. Interviews with four students at each of these grade levels revealed that before sixth grade, students had very few estimation skills and rarely even understood what they had to do in order to solve an estimation task such as telling whether a $5 bill on the table was enough to buy three price-marked items also on the table. An analysis of the interview responses in this study, together with results from the studies on estimation performance reviewed above, led to the formulation of a breakdown of factors involved in computational estimation, which appears in Table 16.1.

In a follow-up study, Sowder and Wheeler (1989) investigated the development of the understanding of the concepts and processes most directly associated with computational estimation (see Sections I and II of Table 16.1.) Twelve students of average ability were interviewed at each of grades 3, 5, 7, and 9. Most of the interview tasks consisted of descriptions

TABLE 16–1. An Analysis of the Components Involved in Computational Estimation (from Sowder and Wheeler, 1989, p. 132)

I. Conceptual Components
  A. Role of Approximate Numbers
    1. Recognition that approximate numbers are used to compute
    2. Recognition that an estimate is an approximation
  B. Multiple Processes/Multiple Outcomes
    1. Acceptance of more than one process for obtaining an estimate
    2. Acceptance of more than one value as an estimate
  C. Appropriateness
    1. Recognition that appropriateness of process depends on context
    2. Recognition that appropriateness of estimate depends on desired accuracy
II. Skill Components
  A. Processes
    1. Reformulation: Changing the numbers used to compute
      a. Rounding
      b. Truncating
      c. Averaging
      d. Changing the form of a number
    2. Compensation: Making adjustments during or after computing
    3. Translation: Changing the structure of the problem
  B. Outcomes
    1. Determination of correct order of magnitude of the estimate
    2. Determination of the range of acceptable estimates
III. Related Concepts and Skills
  A. Ability to work with powers of ten
  B. Knowledge of place value of numbers
  C. Ability to compare numbers by size
  D. Ability to compute mentally
  E. Knowledge of basic facts
  F. Knowledge of properties of operations and their appropriate use
  G. Recognition that modifying numbers can change outcome of computation
IV. Affective Components
  A. Confidence in ability to do mathematics
  B. Confidence in ability to estimate
  C. Tolerance for error
  D. Recognition of estimation as useful

of situations calling for computational estimation followed by solutions and explanations given by hypothetical students. The students being interviewed were asked whether they found the hypothetical estimation processes and answers acceptable and appropriate. Other tasks were open-ended.

There was very little change found over grade levels in children's willingness to accept the use of approximate numbers (in this case obtained by rounding) in finding an estimate. However, when students were shown a hypothetical estimate obtained by computing and then rounding, the numbers of students who found this process *preferable* to rounding and then computing *increased* by grade level. By ninth grade, only 4 of the 12 students interviewed preferred the round-then-compute strategy over the compute-then-round strategy. Perhaps the older students had a smaller tolerance for error, resulting from years of instruction where only the exact answer was acceptable.

There was a strong developmental trend in the willingness of students to accept more than one estimation process and more than one estimation result. Still, students were reluctant to accept two "right answers." This result also seems to be related to the usual emphasis on "one right answer" in mathematics classes, beginning in the lower grades.

As children in succeeding grades became more proficient in rounding, they became less willing to deviate from the rules of rounding learned in mathematics classes. For example, students in grades 5 and 7 objected to rounding 267 to 250 rather than to 300. Students seemed to find compensation rather complicated except in one very straightforward question. (Only seventh and ninth graders were tested on compensation.) Not until grade 9 did students attempt to compensate on open-response tasks, even when the compensation was straightforward and would obviously have led to a closer estimate.

On open-response tasks given to students after the hypothetical-answer tasks described above, third and fifth graders and a few seventh and ninth graders consistently attempted exact computation. Third graders did not demonstrate any estimation skills on open-response tasks. A few fifth graders attempted estimation on only one very simple task. Seventh graders demonstrated a limited repertoire of estimation skills but were generally unsuccessful at tasks such as estimating sums with three addends of four digits each. The performance of ninth graders was only slightly better. Thus, although they had just seen hypothetical estimates obtained by rounding and then computing, they chose exact computation, even on tasks where this was extremely difficult to do mentally.

While these two studies tracked over several grade levels the changes in children's strategies and understandings about estimation, they reveal little about the relationship between cognitive development and computational estimation. In a third study, Case and Sowder (1990) first analyzed computational estimation in terms of the two qualitatively different components involved in estimating: converting from exact to approximate numbers and computing (usually mentally) with those numbers. Then, based on Case's theory of cognitive development (Case, 1985), they predicted what types of tasks could be successfully undertaken at each level of development.

In Case's theory, cognitive growth during school years proceeds through two major stages, each having three substages.

Children in the *dimensional* stage (approximately ages 5–10) are able to focus on only one component, having one or more dimensions, at any one time. Children in the *vectorial* stage (approximately ages 11–18) can coordinate two or more qualitatively different components of a task at one time. Since computational estimation has two complex, multidimensional components—namely, approximation and mental computation—it theoretically belongs in grades where children have reached the vectorial stage of cognitive development.

In the Case and Sowder study, the theoretically appropriate problems for each substage of the dimensional and vectorial stages were tested in an interview setting with 12 children at each of grades K, 2, 4, 7, 9, and 11/12. The tasks predicted to be appropriate for each particular substage were also given to children at the preceding substage (except, of course, for the grade K tasks). Tasks for grades K, 2, and 4 involved mental computation and approximation tasks. Grade 4 tasks, for example, included finding 23 + 45 mentally and choosing which was closer to 32: 28 or 62. Children in grades 7, 9, and 11/12 were given successively more difficult computational estimation tasks, where difficulty depended upon such factors as the number of digits per addend, difference in levels of significance of addends, and whether or not any compensation was expected, for example, in estimating 4 × $8.27. Predictions of level at which success was possible were remarkably accurate, not only in terms of the majority of students at each level being successful with the tasks selected for that level, but also in terms of students at the previous level *not* being successful with the same tasks. The authors point out that a critical feature of this study is that the children had *not* been taught a set of formal algorithms for estimation. It was this feature that enabled the researchers to predict performance by the students on the basis of general intellectual characteristics.

There are several implications that follow from these studies on the development of understanding of computational estimation. The major one is that educators should not be in too much of a rush to teach estimation, so that we do not make the same mistakes that have been made with instruction on other skills, namely, teaching specific procedures to children too young to master them in anything but a superficial manner. We should take advantage of the natural development of estimation in adolescents and encourage them to use estimation in many contexts in order to counteract the "rote" tendency they acquire on other tasks. Instructional programs on estimation should take advantage of spontaneous mathematical intuitions and their development. The planning of estimation as a curricular topic must be done with care. There is a great need for research on the appropriate grade-level placement of estimation topics and how placement and instruction influence one another. There have been only a few studies so far on the effects of instruction on estimation. They are examined in the following section.

Instruction on Computational Estimation

Previous to work on identification of strategies used by good estimators, research on the effects of instruction on estimation focused on the strategy of rounding. Nelson (1967) taught a rounding procedure to 12 fourth-grade and 12 sixth-grade classes. The instruction led to significant improvement on an estimation test. Schoen, Friesen, Jarrett, and Urbatsch (1981) used a variety of contexts (computer-assisted instruction, calculator-aided worksheets, teacher instruction) to teach rounding and front-end estimation as strategies in fourth through sixth grades. They found the instruction effective in all the contexts, both on immediate and retention measures. There was also some evidence that meaningful estimation instruction transferred to estimation in verbal problems.

Rounding was also a key strategy taught to pre-service teachers in a study by Bestgen, Reys, Rybolt, and Wyatt (1980). One group of pre-service teachers received 5 minutes of instruction per week for 10 weeks, a weekly quiz, and after each quiz a short discussion of the four quiz problems for which the strategy of that week was appropriate. Another treatment group received only the weekly quizzes. While the improvement in both groups was more than that for a control group, the post-test scores of the two treatment groups were not noticeably different. This is perhaps not so surprising, since instruction was brief. What *is* important is the message that even frequent practice alone can lead to better estimation skills.

As a follow-up to the study that characterized the strategies used by good estimators (Reys et al. 1980, 1982), Reys, Trafton, Reys, and Zawojewski (1984) developed instructional materials on estimation for average ability students in grades 6, 7, and 8. Instruction now included not only rounding and front-end strategies but also extensive work on compatible numbers and on averaging, strategies that were found to be hallmarks of successful estimators in the earlier study. The materials were field-tested in 24 classrooms matched with 24 control classrooms. The instructional units consisted of 10 full-period lessons and several shorter (5–10 minute) lessons, spread throughout the school year. Although differences between treatment and control groups on the post-tests were not statistically significant, individual class mean growths were higher in the treatment group, and item analyses revealed that on some items in particular, the difference between pretest and post-test scores were quite dramatic. Interviews (12 per grade) indicated that the students in the treatment groups understood that estimation called for computing on approximate numbers rather than on exact numbers. They also had an improved understanding of number concepts.

Based on the Case and Sowder (1990) analysis of estimation as consisting of the coordination of mental computation and approximating numbers, Sowder and Markovits (1990) preceded an instructional unit on computational estimation in seventh grade with instruction on mental computation and on comparing and ordering fractional and decimal numbers. The estimation unit focused on making reasonable estimates, particularly with traditionally difficult topics such as multiplying and dividing by fractions or decimals. For example, the question "About how much is 0.52 × 789?" usually prompts rounding 0.52 to 1 or to 0 and also might call up the "multiplication makes bigger" rule. The estimation unit also included consideration of absolute and relative error. Even though the instruction was brief (10 days on estimation), children became much more flexible in their approach to estimation problems and were quite successful later working problems of the type included in the instruction.

In a study limited to estimating with fractions, Mack (1988) worked with eight sixth-grade students on a one-to-one basis for approximately 10 half-hour periods. Working with the students individually allowed her to focus on and draw upon each student's knowledge from both in-school and out-of-school experiences. While no student could estimate 7/8 + 5/6 at the beginning of instruction, all became quite proficient at estimating sums and differences of fractions and could use this knowledge to give meaning to computational algorithms and to avoid common misconceptions about operations on fractions. Similar results were found in work by Tierney (1988). Mack suggests that all instruction on computational algorithms with fractions be postponed until students are able to form estimates for the operations. This suggestion has also been made by Owens (1987) in the context of learning operations on decimal numbers.

Estimating has also been taught as a heuristic strategy for verbal problem solving. DeCorte and Somers (1982), in a teaching experiment with 12-year-olds, directed students through a solution procedure that consisted of reading and understanding the task, estimating the solution, representing the estimated solution, working out the problem, then verifying the solution by comparing it to the estimate. Although some design problems prevented the investigators from claiming conclusive proof of the effectiveness of this strategy, there is certainly enough support to warrant consideration of teaching estimation as a heuristic strategy in problem solving.

Perhaps the most interesting experiment with instruction on estimation is one for which no "hard" results are available. The interest is, rather, in the basic premise of the instruction and in its implementation. Edwards (1984) and his colleagues, working with Papua New Guinean business development officers, recognized that with the advent of the calculator, the standard "commercial mathematics" course taken by these people was no longer appropriate, and it was dropped from the curriculum. The mathematical skills the students still needed were identified. Basically, they needed to be able to make sense of numbers, but this was not possible without first being able to estimate. A course on estimation was developed, which is described in detail in Edwards's article. An easier course for those who did not yet know basic multiplication facts was also designed, as was a follow-up course in which estimation was applied to the types of problems these business people faced. These courses appear to have worked well for the people served by the courses. Could this plan work for other groups and in other cultures? It would seem so.

Affective Factors

An individual who feels that estimation is a valuable strategy for obtaining needed information, whether in the form of an answer to a problem that requires only an estimate or in the form of assurance that an exact answer is in the right ballpark, is more likely to use estimation frequently and to be proficient in its use. This person believes that estimation is useful. This belief is not universally held, particularly among school children. In the Sowder and Wheeler study (1989), students in grades 7 and 9 were willing to accept as estimates values found by rounding and then computing with the approximate numbers, but they preferred estimates found by performing an exact computation and then rounding off the result and calling that the estimate. This second method of finding an estimate allowed them, they said, to see how "far off" the estimate was from the real answer. Their concern was with the accuracy of the estimate, rather than with the accuracy of the exact answer.

The need for finding an exact answer is instilled in children from the early grades on. Silver (in Sowder & Schappelle, 1989) has noted that it is unreasonable to expect children, during a lesson on estimation, to put aside their belief that every problem has one right answer and one right method of obtaining that answer, particularly if they know that tomorrow and the next day and on most days following, one answer and one method will again be the rule of the day.

Believing that estimation is useful depends on understanding what estimation is all about. Yet many children don't really understand this. In a study with English 12- to 14-year-olds, Morgan (1988) found that most of the children she interviewed did not have a clear conception of the purpose or the nature of estimation. They seemed to consider estimation to be an inferior alternative to exact computation, and they described estimation as "guessing." In fact, some of their estimates were given in a "nonrounded" form because they believed that the estimate should "look" as though it could be the exact answer. Other investigators (Clayton, 1988b; Sowder & Wheeler, 1909) have also found indications that young children do not understand what estimation means.

There is, as one might expect, a relationship between one's self-concept as a "doer" of mathematics and one's ability to estimate. "Confident students tend to learn more, feel better about themselves, and be more interested in pursuing mathematical ideas than students who lack confidence" (Reys, 1984, p. 560). Bestgen et al. (1980) found that better estimators had favorable attitudes toward estimation and perceived doing estimation as more understandable than did poor estimators. In the extensive study carried out by Reys et al. (1982) in grades 7–12 and with adults, good estimators were found to be confident about their ability to estimate, except in some cases where they were confronted with conflicting evidence.

In a study with prospective elementary teachers, Sowder (1989) also found this relationship to hold in most cases. Sowder explored the relationship in terms of Weiner's (1972) theory of attribution. From this theory she predicted that students who were good estimators would be confident of their mathematical ability and would be more likely to attribute success to ability and failure to an external cause such as task difficulty or lack of effort. On the other hand, poor estimators would lack confidence in their ability to do mathematics, and they would attribute success to causes such as effort or the assistance of an outsider and failure to lack of ability. It seemed that this second group of students would be unlikely to do well on estimation tasks because they would view estimation as "risky." They would believe it safer to stay with algorithmic processes that they knew would lead to success. Using data obtained from the attribution items and from other written and interview tasks, Sowder was able to generate a profile of good estimators as people who

have strong self-concepts in mathematics, attribute success to ability, have too little experience with failure to attribute it to any cause, and value mental computation and estimation. Poor estimators, on the other hand, have low self-concepts in mathematics, attribute success to time and effort, attribute failure to task difficulty and lack of time and effort, and do not place much value on mental computation or estimation. The profiles were not universally true, and the exceptions provided some interesting insights into estimation ability. One good estimator clearly distinguished between everyday mathematics, for which her self-concept was high, and school mathematics, for which her self-concept was low. In fact, she used estimation to help cope with school mathematics. Another good estimator had made a fairly recent, conscious decision that, because of the convergence of several stressful events in her life, she had to "let go" of some of the things in her life that added to her stress; being accurate was one of those things. Before that decision, she claimed that she hardly ever used estimation because she preferred to have the exact answer. A third student, a poor estimator, had a very good self-concept regarding her mathematical ability. She was very skilled at carrying out long algorithmic procedures, and she enjoyed doing so. Her positive self-concept was related to her belief that what she was good at doing *was* mathematics.

Another factor that might have some influence on the ability to estimate is an individual's tolerance for error. Reys et al. (1980) believed that it was the understanding of the concept of an estimate that led good estimators to be comfortable with some error, and they noted that tolerance for error was one of the hallmarks of good estimators. This point received some support in a study by Wyatt (1986), in which high-performing estimators were much less concerned about precision than the low-performing estimators. Tolerance for error is an elusive phenomenon, however, and one that is difficult to measure with any degree of confidence. Sowder (1989) attempted to find items measuring tolerance for error in her study with prospective teachers. She gave five items to students and asked them to score each item on a five-point scale ranging from strongly disagree to strongly agree. Examples of items include the following: When I shop and the clerk makes a mistake that is more than a few cents off, I would notice the amount I'm being charged is wrong; when I do math problems, I feel better having the exact answer rather than a ball park estimate; when I balance a check book and I'm just a few cents off, I let it go rather than spend a long time trying to locate the error. None of the five items differentiated between good and poor estimators.

Efforts to emphasize computational estimation in the mathematics curricula must take affective issues into account if they are to be effective. Students must see the relevance of learning estimation and must believe they are capable of doing it. They will need to overcome the belief that there is always one right answer and one right procedure for obtaining it. But even more fundamentally, students (and teachers) need to develop a disposition toward making sense out of numbers. Silver (1989) has warned that focusing on cognitive competence while failing to address children's disposition toward numerical activity is unlikely to be successful.

MENTAL COMPUTATION

Computational estimation and mental computation are frequently grouped together as curriculum topics. There are good reasons for this linkage. One is that computational estimation requires a certain facility with mental computation. Another is that both have the potential of increasing students' understanding of the number system, particularly when carried out by "inventing" procedures that are idiosyncratic but appropriate for the particular problem. This view of mental computation, although frequently advanced in the literature (for example, Beberman, 1959; Markovits & Sowder, 1988; Plunkett, 1979), is contrary to the more popular view that mental computation is doing problems quickly in your head. When the second view is taken, a teacher may be satisfied with chain problems, such as "$3 + 12, -6, -7, \times 4, +18, -6, \div 2, +14$"; "tricks" such as the one for multiplying by 11; and drill. The opportunity to provide students with insight into the workings of the number system and the power to use it to their advantage will be lost.

There is evidence that skill in mental computation is associated with understanding the structure of the number system. Some evidence comes from studies of people skilled at mental computation. Hope and Sherrill (1987) compared the mental computation procedures of skilled and unskilled mental calculators among secondary students. Unskilled students almost exclusively favored the mental analogue of the pencil-and-paper algorithm, and they ignored even the most obvious number properties. Skilled mental calculators, on the other hand, used a variety of strategies, involving primarily different forms of distributivity and factoring. Their methods were considerably more efficient that those of the unskilled mental calculators. They avoided "carrying," frequently worked from left to right, and reduced memory demands by continually incorporating interim calculations.

While it might seem that the better mental calculators were more skilled because they had benefited more from schooling, this was clearly not the case in studies comparing the mental calculation strategies of schooled and unschooled children and adults (Ginsburg, Posner, & Russell, 1981; Petitto & Ginsburg, 1982). The studies were conducted with the Diola people of the Ivory Coast. As members of a mercantile society, the unschooled children developed superior mental computation skills that took advantage of the structure and properties of the whole number system and relations between operations. These children began by counting to solve addition problems, and the repetition of counting acts led to knowledge of simple number facts. In the marketplace, they frequently set out merchandise in groups of five and ten, leading to construction of a model for regrouping. In contrast, the schooled Diola (and schooled Americans also included in the study) relied heavily on paper-and-pencil algorithms and were not nearly as proficient at mental calculation.

Additional evidence that skill at mental computation and number understanding develop together can be found in a study by Markovits and Sowder (1988) in which fourth-grade and sixth-grade students received frequent instruction on mental computation over approximately 3 months. The instruction emphasized solving problems in many ways, and no rules were taught. Between the preinstructional interview and the final interview, fourth-grade students selecting standard algorithmic methods decreased from 72.5% to 35%, while their selection of nonstandard methods increased from 7.5% to 51%. (Methods were classified as standard, transitional, or nonstandard.) Sixth-grade statistics were similar. The nonstandard methods favored were left-to-right procedures, changing the order of operations, and decomposing and recomposing numbers. Although there were no tests of number understanding given, interview transcripts indicated that instructional time spent on exploring different strategies led to better understanding of place value, number decomposition, order of operations, and number properties.

Number knowledge underlying different computational strategies was analyzed in a study of mental computation ability by Dworkin (1988). Dworkin listed 10 mental computation strategies (other than the use of the pencil-and-paper algorithm) used by the fifth graders she interviewed. She then analyzed each strategy for the number knowledge needed to carry out the strategy: basic facts, place-value knowledge, knowledge of arithmetic operations, number properties, and number proximity. Dworkin also gave number sense problems to the students, where number sense was defined in terms of number magnitude, number proximity, order of numbers, properties of operations on numbers, and multiples of 10. She found a substantial positive relationship between mental computation and number sense.

The role of number knowledge in carrying out mental computation has also been considered by Sowder (1988). She theorized that carrying out a mental computation calls for addressing two fundamental questions: (1) How can I express the numbers to obtain basic fact questions? and (2) How will the operational sequence proceed as a result of the way the numbers are expressed?

The first question is a result of the need to reexpress the problem in a manner so that prior knowledge of number combinations can be used. Answering this question often requires not only knowledge of basic facts but also understanding of place value, the ability to decompose numbers, and the ability to operate with multiples and powers of 10. The second question also requires the ability to work with powers of 10, in addition to the ability to apply the commutative, associative, and distributive properties. For example, one might solve $83 - 26$ by changing it to $83 + 3 - 26 - 3$, using operation knowledge, because obtaining $86 - 26$ will allow the use of the familiar fact that 8 (tens) $-$ 2 (tens) is 6 (tens). In addition, one must know the basic fact $3 + 3$ to form the 86, the place-value knowledge that $86 - 26$ is $(80 - 20) + (6 - 6)$, and order of operation knowledge to carry out $86 - 26 - 3$. Since there are usually several ways to answer the first question (in this example, one could use the fact that $6 + 4$ is 10 in order to begin a counting-up procedure), choices can be made based on speed and ease of carrying out the operations. Exploration of choices leads to greater flexibility in mental computation.

We have emphasized the role of mental computation in bringing about a better understanding of the number system. Mental computation is also useful in its own right. "In the everyday world of the consumer and worker...there is more need for an exact or a reasonably accurate mental calculation than for a pencil-and-paper calculation" (Hope, 1987, p. 331). These two objectives are not contradictory, and both can be realized through appropriate instruction. How that instruction should take place is being questioned (Sowder & Schappelle, 1989). Some believe that instruction would be more effective if it were integrated into regular work on calculations rather than relegated to 10 minutes a day, as is often suggested. This question deserves long-term research attention.

NUMBER SENSE

Carpenter and his colleagues (1976), after analyzing NAEP data on estimation, concluded that before students can estimate well, they must develop a quantitative intuition, a feel for quantities represented by numbers. In more recent years, this quantitative intuition has come to be referred to as number sense. In the *Curriculum and Evaluation Standards for School Mathematics* (NCTM, 1989), number sense is "an intuition about numbers that is drawn from all the varied meanings of number" (p. 39). The authors of the *Standards* proposed that children with number sense understand numbers and their multiple relationships, recognize relative magnitude of numbers and the effect of operating on numbers, and have developed referents for quantities and measures. More broadly stated, number sense refers to a well-organized conceptual network that enables one to relate number and operation properties and to solve number problems in flexible and creative ways (Sowder, 1988).

These attempts at defining number sense do not characterize it to a degree desirable for guiding instructional and assessment efforts. In attempting to explain why defining number sense is so difficult, Resnick (1989) categorized number sense with other nondeterministic, open-ended forms of reasoning and thinking. She characterized number sense by substituting *number sense* for *higher order thinking* in her depiction of thinking skill in her book on *Education and Learning to Think* (1987, p. 3). It is instructive to consider this:

[Number sense] resists the precise forms we have come to associate with the setting of specified objectives for schooling. Nevertheless, it is relatively easy to list some key features of [number sense] when it occurs. Consider the following:

- [Number sense] is *nonalgorithmic*. That is, the path of action is not fully specified in advance.
- [Number sense] tends to be *complex*. The total path is not "visible" (mentally speaking) from any single vantage point.
- [Number sense] often yields *multiple solutions,* each with costs and benefits, rather than unique solutions.
- [Number sense] involves *nuanced judgment* and interpretation.
- [Number sense] involves the application of *multiple criteria,* which sometimes conflict with one another.
- [Number sense] often involves *uncertainty.* Not everything that bears on the task at hand is known.
- [Number sense] involves *self-regulation* of the thinking process. We do not recognize it in an individual when someone else "calls the plays" at every stop.

- [Number sense] involves *imposing meaning,* finding structure in apparent disorder.
- [Number sense] thinking is *effortful.* There is considerable mental work involved in the kinds of elaborations and judgments required.

How *can* we think about number sense in a way that will guide instruction? Silver (see Sowder & Schappelle, 1989) compared discussions on number sense to discussions on metacognition. People in the field of metacognition have argued that it is better to use words like *regulation* and *monitoring* that actually describe what they are talking about. Silver suggests that number sense is also a paralyzingly large phenomenon that we don't quite know how to define operationally and that perhaps we should focus on "pieces" of number sense for now, until we understand better how they all fit together. In a similar vein, Trafton (1989) speaks of number sense as evoking a certain approach to teaching, rather than being a highly defined construct, and suggests that pursuing number sense qua number sense might not be as useful as pursuing those aspects of number sense that have direct relation to how children process numbers in computational situations. This might include such things as recognizing the relative magnitude of numbers, being able to describe a quantity in terms of other quantities, and making reasonable quantitative judgments when solving problems or performing computations.

A thesis has been advanced in this chapter that instruction on estimation and mental computation can provide an avenue for developing number sense, or quantitative intuition. In order to explore this further, it is helpful to take a closer look at mathematical intuition. The development of mathematical intuition is a topic of special interest to Resnick (1986), and her thinking about this topic is relevant to our discussion here. Resnick characterizes mathematical intuition in two ways: It is self-evident to the person who has it, and it is easily accessible because it is linked in memory to specific situations. Both characteristics can be found in the responses of children successful on tasks such as comparing the sizes of two numbers. When children are asked why they say 17 is larger than 13, they respond that "it just is." They are unable, when asked, to give any further justification (Sowder & Wheeler, 1987).

How does mathematical intuition develop? Resnick proposed that concepts based on additive composition are acquired early and universally. Work on early number learning (for example, Carpenter & Moser, 1983) provides evidence that children come to school with the ability to solve addition and subtraction problems through counting and informal modeling. Resnick further argues, however, that the school focus on symbol manipulation "discourages children from bringing their developed intuitions to bear on school learning" (Resnick, p. 162). Furthermore, good mathematics learners are successful in linking school-learned symbols to the correct concepts, while for weaker students, the symbols become disassociated from the referents. Good students expect to make sense of mathematical rules and are willing to spend time and energy on making sense of these rules.

In the research discussed in this chapter, students who were good at estimation and mental computation were easily able to link symbols to concepts. Interview protocols with these students (Reys et al., 1980; Hope, 1987) provide many examples of their sense-making efforts. If possessing mathematical intuition "provides the basis for highly flexible application of well-known concepts, notation, and transformational rules,... frees people from excessive reliance on fixed algorithms, and allows them to invent procedures for problems not previously encountered and to work ahead of formal instruction in constructing mathematical knowledge" (Resnick, p. 188), then these exceptional students exemplified the power of mathematical intuition in the ways in which they solved problems.

And what about those who performed poorly on estimation and mental computation tasks? If Resnick is correct, then we must focus instruction on making sense of the symbols we use in mathematics. Instruction cannot help but develop quantitative intuition if it allows and encourages the invention of algorithms, promotes questioning of how numbers can be decomposed and recomposed and of how place-value concepts can be applied, admits to multiple answers and procedures, and demands reflection on reasonableness. Estimation and mental computation are not only useful tools in everyday life but they can also lead to better number sense.

MEASUREMENT ESTIMATION

Estimating measurements calls upon quite different abilities than estimating computations does. It is certainly more contextually bound than computational estimation usually is. A measurement estimate can be made without using any arithmetic operations, although some simple ones are frequently involved. In fact, the two types of estimation are often intertwined. For example, if tiles are estimated to be 10 inches per side, and there are about 18.5 tiles down the length of a room and about 13.5 tiles across, the area of a ceiling could be estimated to be 185 inches by 135 inches, or 15 feet by 11 feet, which is 165 square feet.

Bright (1976) has defined measurement estimation as the "process of arriving at a measurement or a measure without the aid of measuring tools. It is a mental process, though there are often visual or manipulative aspects to it" (p. 89). Bright described eight basic conditions under which this type of estimate can be made. These conditions are the possible triples formed by considering whether the object to be measured is present or absent, whether the unit of measure is present or absent, and whether the measurement or the object is specified. Of course, these eight conditions can be crossed with each of the different types of attributes that can be measured. A classroom problem such as "name something in the room that is about one meter long or high" is a measurement estimation situation that deals with length and for which the object is present, the unit of measure is not present (presumably), and the measurement is specified. Bright predicted that there would be more transfer among the eight conditions, given one attribute, than there would be across attributes. This analysis of measurement estimation should be useful in designing both research questions and instructional units on this topic. I will refer to these conditions when reviewing the research on measurement estimation.

Ability to Estimate Measurements

Research on measurement estimation is sparse. The few studies conducted before 1980 focused on how well children and adults estimated. Crawford and Zylstra (1952) found that high school seniors were poor at making measurement estimates, but the report of their study did not indicate whether attributes besides height and length were included in the test items. One interesting aspect of this study was that the authors claimed that problems on their "Estimating Test" had to be estimated and could not be computed. Yet an example problem given in the report was to estimate the length of the diagonal line on "this 8 1/2 by 11 page." Even more interesting is that of the choices given for the answer—8, 11, 14, 17, and 20—only 8 and 20 were considered "incorrect responses" after allowing for near estimates. It would seem then that the scores might be somewhat inflated, making performance even worse than reported. The investigators also reported that measurement estimation had low correlations with tests of mental and computational ability and mathematics course grades.

Two other early studies on measurement estimation were reported by Corle (1960, 1963). The first study examined performance by fifth and sixth graders, while the second focused on performance by elementary teachers and pre-service teachers. Items given to children included finding estimates for weight, length, thickness, diameter (of a ball), temperature, time, and liquid capacity. Adults were given problems on estimating measures of the same attributes, plus a problem on volume (of the ball) and one on barometric pressure. Sixth graders were more accurate in their estimates than fifth graders were, and boys were more accurate than girls, although no group performed well. The average percent of estimate error was over 50% for all groups, where estimate error was the ratio of error to actual measure, with 100% considered the maximum in order for the error scores not to be unduly influenced by extreme guesses. The two adult groups performed better than the children did, but teachers were not more accurate than the pre-service college students. Low correlations were found between estimation scores and arithmetic achievement. There was no indication given of *how* estimates were made. This raises some interesting questions about some of the results. For instance, less than 4% of the teachers could estimate the "cubic contents" of a basketball. Of the few who could, did they actually try to guess the volume, or did they estimate the diameter, call up a formula, and then calculate? If they used the formula, then the data from this item could be interpreted in terms of knowledge of the formula for the volume of a sphere rather than simply ability to estimate volume.

A particularly interesting aspect of the study with young students was that the students, after giving estimates, were then shown a table with measuring apparatus and asked to measure all the attributes just estimated. Measurement errors were approximately half as frequent as estimating errors at each grade level. Some pupil errors, reported as anecdotes, indicated that some students did not believe a light object such as a blackboard eraser weighed anything. Other errors included measuring the thickness of a pencil by wrapping a tape measure around it, measuring temperature with a spring balance, and using a ruler to measure the number of quarts of water in a bucket. If children cannot measure, they surely cannot be expected to estimate measures. It was not clear whether these students had had any instruction on measurement. It would seem they had not. The authors also pointed out that if children have so little sense of measure, they cannot be expected to recognize unreasonable results to verbal problems with measurement answers.

Two later studies by Swan and Jones (1971, 1980) were similar to the Corle studies in that both children and adults were involved and performance on estimates of several types of attributes was considered. The problems given, however, included finding many "outdoor" estimates, such as the height of a tree, the distance between two buildings, and acreage of a rectangular area. The investigators found a progression in estimating ability from elementary students through junior and senior high students, with adults performing considerably better than students. Males estimated distance and height better than females, but this difference did not hold for weight and temperature. The best estimates were made for temperature, and as one might expect, the worst estimates were for acreage. Shorter distances were easier to estimate than long ones, and both were easier than estimating weight. But overall, the results were the same as in the Corle studies and the Crawford and Zylstra study: School children and adults are not good estimators. More than 50% of the estimates in the Swan and Jones studies were considered to be poor estimates. As opposed to the 147 students included in the Corle studies, the Swan and Jones (1980) study included almost 1,300 students, from both rural and urban areas. The schools had been placing an emphasis on using metric units, so it seems almost certain that students had received instruction on measurement, although perhaps not on measurement estimation.

One could easily argue that many of the poor estimates in the Corle and the Swan and Jones studies were due to inexperience with the types of measures being estimated. A grocery shopper who can give quite good estimates of the weight of a package of meat might give extremely poor estimates of barometric pressure or the acreage of a field. Moreover, one could question the value of knowing how to estimate measures that are not part of one's daily life.

A more recent study of estimation skills (Siegel, Goldsmith, & Madson, 1982) was based on a competence model of estimation processes. The model contained two related procedures, *benchmark* estimation (application of a known standard) and *decomposition/recomposition* estimation (decomposing into small samples in order to apply a benchmark, then recomposing to obtain a final estimate). Both were applied to measurement situations and to numerosity situations. The model, presented as a flowchart, checked to see if a benchmark could be applied and, if not, whether a decomposition strategy would work. The study was designed to test the model and to assess developmental differences in estimation skills of children in grades 2 through 8. Tasks included estimating the length of a bat (benchmark measurement) and estimating the number of names on a page of a phone book (regular decomposition, numerosity). The investigators found marked age differences for the two types of problems given here as examples and for numeracy and length estimates calling for decomposition into irregular parts. There was only a weak relationship between accuracy in estimation and reported use of strategies. The authors suggested a revision of their model to account for the diversity of children's performances and the more hierarchical nature of estimation indicated by the responses. Curiously, the revised model makes no distinction between numerosity and linear measurement, a fact that might make such a model difficult to apply in instructional situations since numerosity is discrete and linear measurement is continuous.

Clayton (1988b) also studied estimates of measurement (how high is a double decker bus) and numerosity (how many sugar cubes would fill this matchbox). Questions such as these were given to over 1,000 schoolchildren. Clayton found that when answers were less than 100, children's responses were within 20% of the actual answer, but performance deteriorated as sizes of answers increased. He interviewed 37 secondary students, since he was curious to know whether students used any computational estimation to complete such tasks. They did not. Answers were found by guessing or by using a calculator and attempting to find an exact answer.

Instruction on Measurement Estimation. How effective is instruction on estimation? In a study by Ibe (1973), estimation before measuring was one of the instructional treatments given to sixth graders; the other treatment involved measuring without estimating first. The estimation treatment was superior to the measurement-only treatment on transfer, estimation, and achievement. Pretest scores in this study did show significant correlations with general ability and number facility, as well as with spatial ability and flexibility of perceptual closure. These results need to be interpreted in light of the topic of the test, angular measurement, which is a quite specialized area of estimation.

In a later study by Bright (1979), middle school teachers enrolled in a summer workshop improved their linear estimation skills through practice. Two other instructional studies used computer-based instructional units. Immers (1983) found that a 10-day computer-based instructional unit on estimating length was effective. The control group received no instruction on estimation, so it is difficult to know whether the fact that the instruction was computer-based made any difference. Morgan (1986) used a computer-based instructional unit on estimating area and linear measurement. Her second treatment group received paper-and-pencil instruction paralleling the computer treatment. The computer group did not outperform the non-computer group. Both groups learned some measurement estimation. Instructional studies such as these three do not offer any general prescriptive advice for combating the poor performance found in the earlier cited studies. Perhaps part of the reason is that these studies focused on outcomes rather than on processes.

A recent study by Hildreth (1980, 1983) did focus on measurement estimation processes. He found several strategies that led to success in estimating length and area. They include (1) successive mental application of the length or area unit together with counting (unit iteration); (2) subdividing the length using information given or known (using subdivision clues); (3) using prior information about the object or unit (prior knowledge); (4) comparing the target object with another object, either in or out of sight (comparison); (5) subdividing the object so that when comparing the unit to one of the segments, the ratio of unit to target object is smaller (chunking); (6) making estimates that are a little more and a little less than the largest objects, then narrowing in on the measurement (squeezing); (7) using length estimates to estimate area dimensions and then using a formula (length times width); and (8) rearranging part of the target figure to make it easier to estimate its area (rearrangement). Inappropriate strategies included (1) finding area by adding length and width (length plus width); (2) using a nonsquare rectangular unit and estimating the area of another rectangle using one side of the original unit for estimating width and the other side of the original unit for estimating length of the other rectangle (count around); (3) reporting area by considering only one of the dimensions (centering); (4) using an inappropriate unit to measure (use of inappropriate unit); (5) estimating area by first estimating perimeter, then "throwing in" some for the "middle" (count around plus some for the middle); and (6) wild guessing. The appropriate strategies generally fit the model for estimating physical measurement put forth by Moskol (1981). In her model, students use some type of direct process, including recall, intuition, or formulation of a one-to-one correspondence, or an indirect method where the estimate is broken into subquantities on which the direct method is used, followed by some sort of arithmetic operation.

Hildreth designed an instructional treatment based on his appropriate strategies, and compared its efficacy with an estimate-then-measure treatment. The strategy treatment certainly evoked more and better strategy use on the post-test, but it is not clear whether the first group performed better quantitatively than the second group, since the focus of both the pre- and post-test was on strategy use.

Although previously cited studies indicate that adults are better estimators than children, Hildreth showed that the variety of successful estimation strategies used by the two groups are not qualitatively different. Estimation ability was found to be related to perceptual ability, but not to mathematical ability, for the fifth and sixth graders.

A recent intervention study that included measurement estimation items was done by Markovits (1987) with Israeli sixth and seventh graders, pre-service teachers, and experienced teachers, much as in the Corle and the Swan and Jones studies. A student treatment group received an instructional unit of 8 to 10 lessons, covering several aspects of estimation. Post-test scores for the treatment group were compared to scores on the same test given to a control group of students and to the two adult groups. Instruction and test items included the following measurement topics: length and area estimation with drawings present; height, width, and area without drawings; liquid capacity; distance and speed of described objects; and absolute and relative error of measurements. In each of the areas tested, the experienced teachers performed much better than the other groups. Although this result is not in accord with the Corle finding that performance of experienced teachers does not differ from that of in-service teachers, the items in this study, more so than in the Corle studies, seemed to be the type that would yield better scores for those individuals with more maturity and experience. The instructional treatment was quite effective in all areas, and the treatment group scores were usually at or above the scores of pre-service teachers. The Markovits study provides evidence that a well-designed instructional unit can be quite effective in bringing about increased skill in performing measurement estimation.

ASSESSMENT ISSUES

Both researchers and curriculum developers have a stake in the development of appropriate assessment procedures for estimation, mental computation, and number size. In all of the research we have reported thus far, there has been some type of assessment made of understanding, or of school learning, or of treatment effects. Part of the difficulty in making sense of the research reports is due to the variety of techniques used for assessment, particularly for determining "success." For example, Clayton (1988a) examined the criteria for success on estimation tasks used by different investigators and found that in one instance an investigator considered answers to be "accurate" when they were within 50% of the actual value, while in another instance an investigator used an accuracy criteria of within 15% of the actual value.

The problem is further compounded by the fact that "reasonableness" can vary with the size of the numbers so that a straight percentage criterion does not seem valid. Clayton's example is that estimating a group of 10 to be 5 seems a greater error than estimating a crowd of 100,000 to be 50,000. In order to deal with this problem, at least in the area of numeracy estimation, he proposed a "Criteria of Reasonableness" (COR) scale for judging reasonableness of estimates of quantities larger than four. The COR is logarithmic in nature so that the requirements for an acceptable estimate are more exacting for small numbers than for large numbers. This model for reasonableness is not directly extendable to other types of estimation, but it does suggest that other models could be developed that might help standardize the meaning of reasonableness of estimates in research studies.

Concern on the part of curriculum developers for appropriate assessments is at least equal to that of researchers, since topics that cannot be adequately tested have slight chance of being in the taught curriculum, regardless of whether or not they are in the intended curriculum. Moreover, testing must measure the processes and concepts that are the goals of instruction if it is to promote meaningful instruction (Schoen, Blume, & Hart, 1987).

Assessment of number sense and related topics was a topic of discussion at a recent conference (Sowder & Schappelle, 1989). At that conference, Resnick pointed out that it is possible to assign numbers, if that is what is wanted, by using the model of judging Olympic performance. There might be several dimensions that are judged for events such as diving or figure skating, and judges rate each dimension. But deciding on what the dimensions are for number sense or estimation is a major research undertaking, particularly if we want to do this not only for individuals but for pairs or groups or whole classes. Another assessment technique that might be appropriate here is the use of portfolios, in which work over an extended period of time is collected and submitted to a judge or panel of judges. Assessment by portfolios is, however, a new and little-tried technique, still surrounded by some amount of controversy. Consistency and validity of this type of assessment are particularly troublesome.

While the appropriateness of portfolios as a means of assessment might be a research agenda item, it will not have much influence on testing over the short haul. What then do we have to guide us? Only assessment of computational estimation has so far received any attention. Reys and Bestgen (1981) have pointed out some psychometric problems related to assessing computational estimation. The first is the matter of timing. If students are given enough time to carry out the computation, they are less likely to estimate. There are two issues involving timing. One is the matter of how to time items. In research studies timing has frequently been done in group settings via a time-controlled carousel slide projector (for example, Reys et al., 1982) or with an overhead projector and timer (for example, Rubenstein, 1983). The second issue is one of deciding how much time should be allocated per item. Reys and Bestgen (1981) described a situation in which there was a 10–20% change upward in the percentage of exact answers recorded when the time on an exercise was increased by only five seconds. The implication was that this allowed students time to compute an exact answer. However, Sowder (1989) found that when students were individually allowed whatever time they needed to estimate problems previously incorrect on a group test, some students simply needed extra time to decide upon an estimation strategy and carry it out. And Schoen et al. (1987) found that on a test with a total time limit rather than an item time limit nearly all students did try to estimate. Whether students estimate or attempt exact computation seems to depend on whether they understand that they cannot finish all items by exact computation and whether they are willing to try to estimate, either because instruction on estimation has led them to place some value on estimation or because they realize (as participants in a research study, for example) that getting wrong answers by at least attempting to estimate is not going to affect them personally. (Adults are more likely to understand this point.) Exact computations are also mildly discouraged by using darkened paper with room only for answers in light-colored answer spaces or by distributing paper with room only for answers.

Question format has received more research attention than any other assessment issue. Assessment instruments are usually either open-ended or multiple-choice. While an open-ended format might seem to be a better choice in terms of allowing maximum freedom for responses (Reys & Bestgen, 1981), this is not always the case. Schoen et al. (1987) found that an open-ended test failed to measure students' ability with respect to any strategy but rounding.

Multiple-choice items can employ several kinds of formats. Rubenstein (1985) used scales asking subjects to decide whether an answer was reasonable or not, whether the answer was larger or smaller than a reference number, and which order of magnitude was appropriate for an answer. The reasonable versus unreasonable scale was not reliable, and of the remaining three scales, the open-ended items were considerably more difficult than the other items and the order of magnitude items were the easiest. Schoen and his colleagues (1987, 1988) were interested in how stems of multiple-choice questions could

influence performance. They used standard multiple-choice items (for example, the closest estimate of $1,926 + 851 + 3,273$ is $5,000, 6,000, 7,000,$ or $1,300$), foils containing the operation (for example, the correct answer for the closest estimate was $2,000 + 1,000 + 3,000$), foils containing ranges (for example, the correct answer was the range 5,500–6,500), benchmarks (is the above more than or less than 7,000? More than because..., less than because...), and order of magnitude (the closest estimate is 600, 6,000, 60,000, 600,000). The estimation items were designed to elicit processes identified in the Reys et al. (1982) study as those used by good estimators. That is, the foils were constructed in a manner that required or encouraged specific processes. The investigators found that all formats except the benchmark quite accurately measured aspects of estimation. Range items were most difficult, followed by standard multiple-choice items.

Some educators would argue that even though we have not yet developed appropriate assessment instruments for estimation and mental computation, we should be able to convince teachers of their value in the curriculum. Reys (see Sowder & Schappelle, 1989) believes that we need to convince teachers that students who have acquired these skills will perform better on standardized tests even without items targeting estimation and mental computation. A study by Rea and French (1972) offers some support for this belief. Their sixth graders, who received mental computation instruction for 15 minutes a day over 24 days, advanced almost a full year on the SRA mathematics achievement test. Unfortunately, without more evidence it may be difficult to convince teachers of Reys's belief, and yet one wonders whether obtaining more evidence is really worthwhile if it means conducting more research in which scores for existing standardized tests are used and therefore made to seem an important criterion.

CONSIDERATIONS FOR CONTINUING RESEARCH ON ESTIMATION AND RELATED TOPICS

At a recent conference held to investigate the theoretical foundations for research on number sense and estimation (Sowder & Schappelle, 1989), mathematics educators and psychologists found that although they agreed on the importance of these topics for learning mathematics, they did not agree on how best to incorporate the topics into school curricula, nor on how research on number sense and estimation ought to proceed. Case (1989) has suggested that the lack of agreement was at least partly due to the influence of three different epistemological traditions upon the participants' beliefs about the nature of knowledge acquisition. An investigator working within the *empiricist* tradition, for example, would think of knowledge as originating in the external world. Much of the current work in information processing stems from earlier work completed within this tradition. Case noted also that this epistemology sets the stage for the notion that it is possible, through a rationally designed curriculum, to "take children from where they are to where the curriculum designer feels they should be, in an optimally efficient fashion" (p. 59).

For the *rationalist,* on the other hand, knowledge originates in the mind. Among those who would be considered rationalists are investigators working within the Piagetian tradition, including Case himself. Within this tradition, educational methods tend to be more child-centered, and there is concern that curriculum materials are developmentally appropriate.

In the *sociocultural* tradition, knowledge is thought to be "essentially a social construction, with certain particular forms of knowledge constituting the intellectual tools for generating new knowledge and being passed on from one generation to the next" (p. 60). Current research on situated knowledge fits within this tradition. Case believes that all three traditions are legitimate ways to study the acquisition of knowledge, mathematical knowledge in particular. The different models of number sense and estimation advanced by conference participants clearly showed the influence of these different beliefs about how knowledge is acquired.

Marshall (1989), whose work is based on a schema theory model, has suggested that we think of number sense in terms of the "connectedness" of a vast network of nodes of mathematical knowledge. In this model, for example, students may have knowledge about whole number properties and about addition of whole numbers, but these topics may be encoded in memory without being connected. Thinking of number sense as the connectedness of number knowledge implies that number sense is not a body of knowledge to be taught, but rather that instruction should focus on the development of links between nodes of number knowledge. The type of instruction that would bring about these rich connections needs to be investigated, and this research would be dependent upon finding ways to measure the strength of such connections.

The schema theory model can also be a useful way of thinking about instruction on estimation and mental computation. Sweller and his colleagues (for example, Tarmizi & Sweller, 1988), for example, have shown that in some areas of problem-solving students can learn better when they work problems only *after* studying the worked examples of an expert. The rationale for this success is that the traditional means-end strategy of problem solving exhausts students' cognitive resources and obscures important structural aspects of the problems. Worked examples lighten the students' cognitive load, particularly in terms of holding material in working memory and processing it. Schema acquisition is thus more successful. Past research on estimation and mental computation (Reys et al., 1982; Hope & Sherrill, 1987) has yielded many good examples of expert thinking in these domains. Could it be that seeing these examples, perhaps with instruction that helps highlight the salient points, is an effective way to produce better performance in these domains, particularly for certain populations?

Within the rationalist tradition, the work of Case and Sowder (1990) offers a model of cognitive development applied to the development of understanding and skill in computational estimation. Using a top-down approach, these investigators first used the Case (1985) model of cognitive development to predict performance, then checked their predictions against data collected on students' performance. The predictions were remarkably accurate, giving validity to the model while aiding the investigators in making curricular recommendations based on the developmental levels of the students. That research fo-

cused on additive structures. There is a need to examine more closely the development of these abilities within the field of multiplicative structures.

Greeno (1989) has suggested an interesting metaphor for exploring learning within the domain of number sense, estimation, and mental computation, a metaphor that fits well with the view that cognition is situated in contexts. He characterizes this domain as an environment, with the collection of resources needed for knowing, understanding, and reasoning all at different places within this environment. "Learning in the domain, in this view, is analogous to learning to get around in an environment and to use the resources there in conducting one's activities productively and enjoyably" (p. 45). Numbers and quantities, together with their structure of relations and operations, form the resources within the environment of numbers. People with number sense can move around easily within this environment because of their access to the necessary resources. Teaching becomes the act of "indicating what resources the environment has, where they can be found, what some of the easy routes are, and where interesting sites are that are worth visiting" (p. 48).

This metaphor provides a rationale for emphasizing active exploration by students, particularly through cooperative group learning. According to this model, social groups play an important role in learning one's way around an environment, and understandings based on shared experiences tend to be richer and more valuable. Greeno cautions, however, that skills such as automatic recognition and generation of relations and patterns are still necessary for living in such an environment and that the development of these skills will require a significant amount of practice.

What is notable about all of these models is that when investigators apply them to the question of how number sense is acquired, the generally unanimous view is that instruction must be very different from what is currently the norm. There is consensus on the fact that number sense should permeate the curriculum beginning in the early grades rather than being relegated to "special lessons" designed to "teach number sense." And although some may disagree on how to approach instruction on estimation and mental computation—for example, whether direct or indirect instruction is more appropriate—these topics should be part and parcel of any instruction on arithmetic operations.

SUMMARY

Three categories of estimation were considered in this chapter: computational estimation, measurement estimation, and numerosity estimation.

Computational estimation has been the focus of a considerable amount of research in recent years. Results from several investigations of how individuals estimate showed that good estimators are flexible in their thinking, use a variety of estimation strategies, and demonstrate a deep understanding of numbers and operations. Poor estimators are bound to applying algorithms, find it difficult to think of a problem as having more than one right answer or solution procedure, do not value estimation, and often equate estimation with guessing.

Computational estimation is a complex process that develops slowly. In fact, development of estimation strategies for work with rational numbers is not universal, and many individuals never develop this ability on their own. Instruction should build on mathematical intuitions as they develop. A disposition toward making sense out of numbers should be an instructional goal. Other affective factors, such as valuing estimation, need to be considered when planning instruction.

Mental computation is an important component of computational estimation, but it is also an important skill in its own right. Like computational estimation, skill in mental computation is also associated with understanding the structure of the number system. Individuals skilled at mental computation use this understanding to their advantage while those poor at mental calculation tend to try to use mental analogues of paper-and-pencil algorithms.

Both computational estimation and mental computation are closely related to number sense. Although there is currently a great deal of interest in number sense, number sense has not been a focus in instruction. Like higher-order thinking, it is difficult to define and therefore difficult to assess. Part of the reason children lack number sense is that they have come to disassociate symbols and their referent concepts.

Estimating measures is highly dependent upon ability to measure, which is often weak in school-age children. Adults are usually better than children at measurement estimation, although the strategies they use are fundamentally the same as those children use. There is some evidence that the teaching of particular strategies for measurement estimation can be useful. Research on measurement estimation and numerosity is sparse, and these topics continue to be neglected by researchers.

Assessing number sense, mental computation, and all three kinds of estimation presents many difficulties. For example, with estimation there is the problem of multiple correct answers. With mental computation there is the difficulty of determining whether or not the computation was indeed carried out mentally, particularly in group settings where students have pencil and paper to write down answers. With number sense there is the lack of an operational definition on which to base assessment items. New types of assessment are needed to measure success in these areas. If the learning of these topics is not evaluated, they are unlikely to find their way into school curricula.

Investigators interested in studying number sense, computational estimation, and mental computation agree on the importance of these topics, but do not necessarily agree on which are the most important research issues to pursue, how research ought to proceed, or how these topics should be incorporated into the curriculum. This lack of agreement is primarily due to the different epistemological viewpoints of the investigators. All do agree, however, that number sense ought to permeate the curriculum and that computational estimation and mental computation should be incorporated into all instruction on computation.

References

Bakst, A. (1937). *Approximate computation: Twelfth yearbook of the National Council of Teachers of Mathematics*. New York: Bureau of Publications, Teachers College, Columbia University.

Baroody, A. J. (1989). Kindergartners' mental addition with single-digit combinations. *Journal for Research in Mathematics Education, 20*, 159–172.

Beberman, M. (1959). Introduction to C. H. Shutter & R. L. Spreckelmeyer, *Teaching the Third R*. Washington: Council for Basic Education. Cited in S. Josephine, Mental arithmetic in today's classroom, *Arithmetic Teacher, 1*, 199–200, 207.

Bell, M. S. (1980). Early teaching for effective numeracy. *Arithmetic Teacher, 28* (4), 2.

Bestgen, B. J., Reys, R. E., Rybolt, J. F., & Wyatt, J. W. (1980). Effectiveness of systematic instruction on attitudes and computational estimation skills of preservice elementary teachers. *Journal for Research in Mathematics Education, 11*, 124–136.

Bright, G. W. (1976). Estimating as part of learning to measure. In D. Nelson & R. E. Reys (Eds.), *Measurement in school mathematics: 1976 yearbook* (pp. 87–104). Reston, VA: NCTM.

Bright, G. W. (1979). Measuring experienced teachers' linear estimation skills at two levels of abstraction. *School Science and Mathematics, 79*, 161–164.

Buchanan, A. D. (1980). *Estimation as an essential mathematical skill* (Position paper 39). Los Alamitos, CA: Southwest Regional Laboratory for Educational Research and Development. (ERIC Document Reproduction Service No. ED 167 385)

Carpenter, T. P., Coburn, T. G., Reys, R. E., & Wilson, J. W. (1976). Notes from national assessment: Estimation. *Arithmetic Teacher, 23* (4), 296–302.

Carpenter, T. P., Corbitt, M. K., Kepner, H. S., Jr., Lindquist, M. M., & Reys, R. E. (1980). National assessment: A perspective of students' mastery of basic mathematics skills. In M. M. Lindquist (Ed.), *Selected issues in mathematics education* (pp. 215–257). Chicago: National Society for the Study of Education, and Reston, VA: NCTM.

Carpenter, T. P., & Moser, J. M. (1983). The acquisition of addition and subtraction concepts. In R. Lesh & M. Landau (Eds.), *The acquisition of mathematics concepts and processes* (pp. 7–44). New York: Academic Press.

Carraher, T. N., Carraher, D. W., & Schliemann, A. D. (1987). Written and oral mathematics. *Journal for Research in Mathematics Education, 18*, 83–97.

Case, R. (1985). *Intellectual development*. Orlando, FL: Academic Press.

Case, R. (1989). Fostering the development of children's number sense. In J. T. Sowder & B. P. Schappelle (Eds.), *Establishing foundations for research on number sense and related topics: Report of a conference* (pp. 57–64). San Diego: San Diego State University Center for Research in Mathematics and Science Education.

Case, R., & Sowder, J. T. (1990). The development of computational estimation: A neo-Piagetian analysis. *Cognition and Instruction, 7*, 79–104.

Clayton, J. G. (1988a). *How can teachers encourage children to estimate?* Unpublished manuscript.

Clayton, J. G. (1988b). Estimation. *Mathematics Teaching, MT125* (Dec), 18–19.

Corle, C. G. (1960). A study of the quantitative values of fifth and sixth grade pupils. *Arithmetic Teacher, 7* (3), 333–340.

Corle, C. G. (1963). Estimates of quantity by elementary teachers and college juniors. *Arithmetic Teacher, 10* (2), 347–353.

Crawford, B. M., & Zylstra, E. W. (1952). A study of high school seniors [sic] ability to estimate quantitative measurements. *Journal of Educational Research, 46,* 241–248.

DeCorte, E., & Somers, R. (1982). Estimating the outcome of a task as a heuristic strategy in arithmetic problem solving: A teaching experiment with sixth-graders. *Human Learning, 1,* 105–121.

Dowker, A. D. (1988, March). Computational estimation strategies of professional mathematicians. Paper presented at the Conference of the British Society for Research into Learning Mathematics, Brighton Polytechnic, Oxford, England.

Dowker, A. D. (1989, May). *Computational estimation by young children.* Paper presented at the Conference of the British Society for Research into Learning Mathematics, Brighton Polytechnic, Brighton, England.

Dworkin, L. (1988). *A study of mental computation ability and its relationship to whole number sense in fifth-grade students.* Unpublished master's thesis, University of California, Berkeley, CA.

Edwards, A. (1984). Computational estimation for numeracy. *Educational Studies in Mathematics, 15,* 59–73.

Faulk, C. J. (1962). How well do people estimate answers? *Arithmetic Teacher, 9,* 436–440.

Fesharaki, M. (1979). A study of the effect of hand-calculators on achievement, estimation and retention of seventh and eighth graders on decimals and percent (Doctoral dissertation, University of Missouri-Columbia, 1978). *Dissertation Abstracts International, 39,* 6004A.

Ginsburg, H. P., Posner, J. K., & Russell, R. L. (1981). The development of mental addition as a function of schooling and culture. *Journal of Cross-Cultural Psychology, 12* (2), 163–178.

Greeno, J. G. (1989). Some conjectures about number sense. In J. T. Sowder & B. P. Schappelle (Eds.), *Establishing foundations for research on number sense and related topics: Report of a conference* (pp. 43–56). San Diego: San Diego State University Center for Research in Mathematics and Science Education.

Hall, L. T., Jr. (1984). Estimation and approximation—not synonyms. *Mathematics Teacher, 77,* 515–517.

Hall, W. D. (1977). A study of the relationship between estimation and mathematical problem solving among fifth-grade students (Doctoral dissertation, University of Illinois, 1976). *Dissertation Abstracts International, 37,* 6324A–6324B.

Hatano, G. (1988). Social and motivational bases for mathematical understanding. In G. B. Saxe & M. Gearhart (Eds.) *Children's Mathematics* (pp. 55–70). San Francisco: Jossey-Bass.

Hildreth, D. J. (1980). Estimation strategy uses in length and area measurement tasks by fifth- and seventh-grade students (Doctoral dissertation, The Ohio State University, 1980). *Dissertation Abstracts International, 41,* 4319A–4320A.

Hildreth, D. J. (1983). The use of strategies in estimating measurements. *Arithmetic Teacher, 30* (5), 50–54.

Hope, J. A. (1987). A case study of a highly skilled mental calculator. *Journal for Research in Mathematics Education, 18,* 331–342.

Hope, J. A., & Sherrill, J. M. (1987). Characteristics of unskilled and skilled mental calculators. *Journal for Research in Mathematics Education, 18,* 98–111.

Ibe, M. D. (1973). The effects of using estimation in learning a unit of sixth-grade mathematics (Doctoral dissertation, University of Toronto, 1971). *Dissertation Abstracts International, 33,* 5036A.

Immers, R. C. (1983). Linear estimation ability and strategy use by students in grades two through five (Doctoral dissertation, University of Michigan, 1983). *Dissertation Abstracts International, 44,* 416A.

Levine, D. R. (1982). Strategy use and estimation ability of college students. *Journal for Research in Mathematics Education, 13,* 350–359.

Mack, N. K. (1988, April). *Using estimation to learn fractions with understanding.* Paper presented at the meeting of the American Educational Research Association, New Orleans, LA.

Markovits, Z. (1987). *Estimation—research and curriculum development.* Unpublished doctoral dissertation, Weizmann Institute, Rehovot, Israel.

Markovits, Z., & Sowder, J. (1988). Mental computation and number sense. In M. J. Behr, C. B. Lacampagne, M. M. Wheeler (Eds.), *PME-NA: Proceedings of the Tenth Annual Meeting* (pp. 58–64). DeKalb, IL: Northern Illinois University.

Marshall, S. P. (1989). Retrospective paper: Number sense conference. In J. T. Sowder & B. P. Schappelle (Eds.), *Establishing foundations for research on number sense and related topics: Report of a conference* (pp. 40–42). San Diego: San Diego State University Center for Research in Mathematics and Science Education.

Maxfield, M. W. (1976). Estimating a shaped crowd—Parametric models. *International Journal of Mathematical Education in Science and Technology, 7,* 71–73.

Morgan, C. (1988). *A study of estimation by secondary school children.* Unpublished Masters' dissertation. Institute of Education, London.

Morgan, V. R. L. (1986). A comparison of an instructional strategy oriented toward mathematical computer simulations to the traditional teacher-directed instruction on measurement estimation. (Doctoral dissertation, Boston University, 1986). *Dissertation Abstracts International, 47,* 456A.

Moskol, A. E. (1981). An exploratory study of the processes that college mathematics students use to solve real-world problems (Doctoral dissertation, University of Maryland, 1980). *Dissertation Abstracts International, 41,* 4320A.

National Council of Teachers of Mathematics. (1980). *An agenda for action: Recommendations for school mathematics of the 1980s.* Reston, VA: Author.

National Council of Teachers of Mathematics. (1989). *Curriculum and evaluation standards for school mathematics.* Reston, VA: Author.

National Institute of Education. (1975). *Conference on basic mathematical skills (Euclid conference).* Los Alamitos, CA: SWRL Educational Research and Development.

National Science Board Commission: Report of the Conference Board of the Mathematical Sciences. (1982). *The mathematical science curriculum K-12: What is still fundamental and what is not.* Washington, DC: National Science Foundation.

Nelson, N. Z. (1967). The effect of the teaching of estimation on arithmetic achievement in the fourth and sixth grades (Doctoral dissertation, University of Pittsburgh, 1966). *Dissertation Abstracts International, 27,* 4172A.

O'Daffer, P. (1979). A case and techniques for estimation: Estimation experiences in elementary school mathematics—essential, not extra! *Arithmetic Teacher, 26* (6), 46–51.

Owens, D. T. (July, 1987). Decimal multiplication in grade seven. In J. C. Bergeron, N. Herscovics, & C. Kieran (Eds.), *Proceedings of the Eleventh International Conference: Psychology of Mathematics Education.* (Vol. 2, pp. 423–429). Montreal.

Paull, D. R. (1972). The ability to estimate in mathematics (Doctoral dissertation, Columbia University, 1971). *Dissertation Abstracts International, 32,* 3567A.

Petitto, A. L., & Ginsburg, H. P. (1982). Mental arithmetic in Africa and America: Strategies, principles, and explanations. *International Journal of Psychology, 17,* 81–102.

Plunkett, S. (1979). Decomposition and all that rot. *Mathematics in Schools, 8,* (3), 2–5.

Rea, R. E., & French, J. (1972). Payoff in increased instructional time and enrichment activities. *Arithmetic Teacher, 19* (8), 663–668.

Resnick, L. B. (1986). The development of mathematical intuition. In M. Perlmutter (Ed.), *Perspectives on intellectual development: The Minnesota Symposia on Child Psychology* (Vol. 19, pp. 159–194). Hillsdale, NJ: Lawrence Erlbaum.

Resnick, L. B. (1987). *Education and learning to think.* Washington, DC: National Academy Press.

Resnick, L. B. (1989). Defining, assessing, and teaching number sense. In J. T. Sowder & B. P. Schappelle (Eds.), *Establishing foundations for research on number sense and related topics: Report of a conference* (pp. 35–39). San Diego: San Diego State University Center for Research in Mathematics and Science Education.

Reys, R. E., & Bestgen, B. J. (1981). Teaching and assessing computational estimation skills. *Elementary School Journal, 82*(2), 117–127.

Reys, R. E., Rybolt, J. F., Bestgen, B. J., & Wyatt, J. W. (1980). *Identification and characterization of computational estimation processes used by in-school pupils and out-of-school adults.* Final Report, Grant #NIE 79-0088. (ERIC Document Reproduction Service No. 197 963)

Reys, R. E., Rybolt, J. F., Bestgen, B. J., & Wyatt, J. W. (1982). Processes used by good computational estimators. *Journal for Research in Mathematics Education, 13*, 183–201.

Reys, R. E., Trafton, P. R., Reys, B. B., & Zawojewski, J. (1984). *Developing computational estimation materials for the middle grades.* Final Report of NSF Grant No. NSF 81/13601.

Ross, J., & Ross, M. (1986). Fermi problems, or how to make the most of what you already know In H. L. Schoen & M. J. Zweng (Eds.), *Estimation and mental computation: 1986 yearbook* (pp. 175–181). Reston, VA: NCTM.

Rubenstein, R. N. (1983). Mathematical variables related to computational estimation. *Dissertation Abstracts International, 44*, 695A. (University Microfilms No. 83–06, 935)

Rubenstein, R. N. (1985). Computational estimation and related mathematical skills. *Journal for Research in Mathematics Education, 16*, 106–119.

Sauble, I. (1955). Development of ability to estimate and to compute mentally. *Arithmetic Teacher, 2* (2), 33–39.

Schoen, H. L., Blume, G., & Hart, E. (1987, April,). *Measuring computational estimation processes.* Paper presented at the meeting of the American Educational Research Association, Washington, DC.

Schoen, H. L., Blume, G., & Hoover, H. D. (1988). *Outcomes and processes on estimation test items in different formats.* Unpublished manuscript.

Schoen, H. L., Friesen, C. D., Jarrett, J. A., & Urbatsch, T. D. (1981). Instruction in estimating solutions of whole number computations. *Journal for Research in Mathematics Education, 12*, 165–178.

Schoen, H. L., & Zweng, M. J. (Eds.) (1986). *Estimation and mental computation: 1986 yearbook.* Reston, VA: NCTM.

Schoenfeld, A. H. (1985). *Mathematical problem solving.* Orlando, FL: Academic Press.

Siegel, A. W., Goldsmith, L. T., & Madson, C. R. (1982). Skill in estimation problems of extent and numerosity. *Journal for Research in Mathematics Education, 13*, 211–232.

Silver, E. A. (1989). On making sense of number sense. In J. T. Sowder & B. P. Schappelle (Eds.), *Establishing foundations for research on number sense and related topics: Report of a conference* (pp. 92–96). San Diego: San Diego State University Center for Research in Mathematics and Science Education.

Skvarcius, R. (1973). The place of estimation in the mathematics curriculum of the junior high school. *The Cape Ann Conference on Junior High School Mathematics.* Boston: Physical Sciences Group. (ERIC Document Reproduction Service No. ED 085 257)

Smart, J. R. (1982). Estimation skills in arithmetic. *School Science and Mathematics, 82*, 642–649.

Sowder, J. T. (1988). Mental computation and number comparison: Their roles in the development of number sense and computational estimation. In J. Hiebert & M. Behr (Eds.), *Number concepts and operations in the middle grades* (pp. 182–197). Hillsdale, NJ: Lawrence Erlbaum; Reston, VA: NCTM.

Sowder, J. T. (1989). Affective factors and computational estimation ability. In D. B. McLeod & V. M. Adams (Eds.), *Affect and mathematical problem solving* (pp. 177–191). New York: Springer-Verlag.

Sowder, J. T., & Markovits, Z. (1990). Relative and absolute error in computational estimation. In G. Booker, P. Cobb, & T. N. deMendicuti (Eds.), *Proceedings of the Fourteenth Psychology of Mathematics Education Conference* (pp. 321–328). Mexico.

Sowder, J. T., & Schappelle, B. P. (Eds.). (1989). *Establishing foundations for research on number sense and related topics: Report of a conference.* San Diego, CA: San Diego State University Center for Research in Mathematics and Science Education.

Sowder, J. T., & Wheeler, M. M. (1987). *The development of computational estimation and number sense: Two exploratory studies* (Research report). San Diego, CA: San Diego State University Center for Research in Mathematics and Science Education.

Sowder, J. T., & Wheeler, M. M. (1989). The development of concepts and strategies used in computational estimation. *Journal for Research in Mathematics Education, 20*, 130–146.

Swan, M., & Jones, O. (1971). Distance, weight, height, area, and temperature percepts of university students. *Science Education, 55* (3), 353–360.

Swan, M., & Jones, O. E. (1980). Comparison of students' percepts of distance, weight, height, area, and temperature. *Science Education, 64* (3), 297–307.

Tarmizi, R. A., & Sweller, J. (1988). Guidance during mathematical problem solving. *Journal of Educational Psychology, 80*, 424–436.

Thompson, A. G. (1979). Estimating and approximating. *School Science and Mathematics, 79* (8), 575–580.

Threadgill-Sowder, J. (1984). Computational estimation procedures of school children. *Journal of Educational Research, 77* (6), 332–336.

Tierney, C. C. (1988, April). *Challenging students to generate representations for fractions.* Presentation at the Annual Conference of the New England Educational Research Association, Rockland, Maine.

Trafton, P. R. (1978). Estimation and mental arithmetic: Important components of computation. In M. N. Suydam & R. E. Reys (Eds.), *Developing computational skills: 1978 Yearbook* (pp. 196–213). Reston, VA: NCTM.

Trafton, P. R. (1989). Reflections on the number sense conference. In J. T. Sowder & B. P. Schappelle (Eds.), *Establishing foundations for research on number sense and related topics: Report of a conference.* (pp. 74–77). San Diego: San Diego State University Center for Research in Mathematics and Science Education.

Weiner, B. (1972). Attribution theory, achievement motivation, and the educational process. *Review of Educational Research, 42*, 203–215.

Wyatt, J. W. (1986). A case-study survey of computational estimation processes and notions of reasonableness among ninth-grade students (Doctoral dissertation, University of Missouri–Columbia, 1985). *Dissertation Abstracts International, 46*, 3280A.

·17·

THE LEARNING AND TEACHING
OF SCHOOL ALGEBRA

Carolyn Kieran

UNIVERSITÉ du QUÉBEC à MONTRÉAL

In reporting the results of the fourth mathematics assessment of seventh- and eleventh-grade U.S. students by the National Assessment of Educational Progress (NAEP), Brown et al. (1988) concluded that

Secondary school students generally seem to have some knowledge of basic algebraic and geometric concepts and skills. However, the results of this assessment indicate, as the results of past assessments have, that students often are not able to apply this knowledge in problem-solving situations, nor do they appear to understand many of the structures underlying these mathematical concepts and skills. (pp. 346–347)

However, to cover their lack of understanding, it appears that students resort to memorizing rules and procedures and they eventually come to believe that this activity represents the essence of algebra. Brown et al. reported that a large majority of the students in the NAEP study felt that mathematics is rule-based, and about half considered that learning mathematics is mostly memorizing. These findings are not restricted to the NAEP evaluation. They have been reported in countless studies conducted in other countries.

What compels many students to resort to memorizing the rules of algebra? What makes the comprehension of school algebra a difficult task for the majority? Is the *content* of algebra the source of the problem? Or is it the way it is *taught* that causes students not to be able to make sense of the subject? Or do students approach algebraic tasks in a way that is inappropriate for *learning* the subject material?

This chapter reviews the research literature that bears on these questions and shows how each one of the three points above (content, teaching, learning) contributes to the difficulties that students have in learning algebra. The chapter begins with an historical analysis of the development of algebra, followed by a description of the content of school algebra and a discussion of the psychological demands made on the algebra learner by the mathematical content. Then a brief overview of the teaching perspective is provided. Major findings from research on the learning and teaching of algebra are discussed in relation to the historical-psychological framework of the prior sections. The chapter concludes with a synthesis based on the three points above and includes some suggestions for further research.

HISTORICAL ANALYSIS

The following historical account of the development of algebraic symbolism and its transformational rules highlights the distinction between using letters to represent *unknowns* in equation solving and using letters to represent *givens* in expressing general solutions and as a tool for proving rules governing numerical relations. It also underscores the gradual loss in meaning inherent in moving from ordinary language descriptions of problem situations and their solutions to symbolic representations and procedures. Thus, the main emphasis in this account is on issues surrounding the development

I appreciate the helpful comments of Seth Chaiklin, Columbia University, and Sidney Rachlin, University of Hawaii, who reviewed several drafts of this chapter. I also wish to thank Anna Sfard, Hebrew University of Jerusalem, for her insightful contributions to the "process-object" discussions. This work was supported by the Québec Ministry of Education, FCAR Grant #89EQ4159, and by the Social Sciences and Humanities Research Council of Canada, Grant #410-88-0798. Any opinions, findings, conclusions, or recommendations expressed herein are those of the author and do not necessarily reflect the views of the Québec Ministry of Education or the Social Sciences and Humanities Research Council of Canada.

of algebraic symbolism. However, the historical summary suggests that the development of algebraic symbolism permitted a change from a procedural to a structural perspective on algebra. This procedural-structural distinction is a theme that will not only be carried through these introductory sections but also be used to interpret many of the research findings that are represented in the principal part of the chapter.

Some of the cognitive processes involved in learning school algebra find their roots in the historical development of algebra as a symbol system. I summarize the three stages through which algebra has evolved (see Harper, 1979, 1987, for an informative discussion of these stages). The rhetorical stage, which belongs to the period before Diophantus (c. 250 A.D.), was characterized by the use of ordinary language descriptions for solving particular types of problems and lacked the use of symbols or special signs to represent unknowns. The second stage, syncopated algebra, was initiated by Diophantus, who introduced the use of letters for *unknown* quantities. Harper has pointed out that the primary concern of algebraists during the 3rd to 16th centuries was discovering the identity of the letter(s), rather than attempting to express the general. Diophantus had no general methods; each of the 189 problems of his *Arithmetica* was solved by a different method. Kline (1972) has remarked that the work of Diophantus "reads like the procedural texts of the Egyptians and Babylonians, which tell us how to do things" (p. 144). It is to be noted that the syncopated symbolism introduced by Diophantus did not develop to any great extent until the turn of the 17th century. After the Moslem conquest of the 7th century, Arab mathematicians kept alive Greek and Hindu mathematics, but their algebra was primarily rhetorical. This style continued in the centuries that followed, during which time Arabic algebra was transmitted to Europe. Some notational modifications were sporadically adopted during the Renaissance, such as using abbreviations of normal words (for example, *p* for plus, *m* for minus) and other special terms, but no major symbolic advances occurred.

In the 1500s Diophantus's work was translated into Latin and began to circulate in printed form among European mathematical scholars. Vieta (1540–1603) read this work and was directly influenced by it. This led to Vieta's innovative use of a letter to stand for the *given,* as well as the unknown quantity, and thereby ushered in the crucial third stage in the development of algebraic symbolism, symbolic algebra. At this point it became possible to express general solutions and to use algebra as a tool for proving rules governing numerical relations. During the centuries that followed, the growth of symbolism and the paring away of superfluous detail went hand-in-hand with, and in fact facilitated, the development of other mathematical concepts, such as the concept of function.

Kleiner (1989) points out that even though one of the events that was essential to the development of the concept of function was the creation of a symbolic algebra (another was the wedding of algebra and geometry), the crucial stage in the development of the concept of function involved the notion of independent and dependent variables—a notion that was elaborated by Euler (1707–1783) in 1755. The early concept of function as an input–output procedural notion was soon replaced by more structural conceptions. Euler's concept of func-

tion was modified in the 1830s by Dirichlet (1805–1859), who viewed a function as an arbitrary correspondence between real numbers, and was generalized another hundred years later by Bourbaki, who defined function as a relation between two sets.

Implicit in the development of algebraic symbolism is the gradual evolution of structural conceptions. Until Vieta's invention of a truly symbolic algebra, the essence of rhetorical and syncopated algebra was the solving of certain kinds of problems by means of verbal prescriptions that involved a mixture of natural language and special characters; in other words, these prescriptions were basically descriptions of computational processes. Vieta's invention of an extremely condensed notation permitted algebra to be more than merely a procedural tool; it allowed the symbolic forms to be used structurally as objects. As we have seen, this led eventually to the development of the concept of function which subsequently went through its own procedural structural cycle. The structural developments in algebra during the last 150 years have had a considerable impact not only on the way algebra is perceived by algebraists but also on the way it is presented in school textbooks. (For another perspective on the relation between algebra and school algebra, see Davydov, 1975.)

SCHOOL ALGEBRA

A study of the historical development of algebra suggests that currently algebra is conceived as the branch of mathematics that deals with symbolizing general numerical relationships and mathematical structures and with operating on those structures. The way that algebra is presented in most American mathematics textbooks echoes this structural perspective. (Currently, very little algebra finds its way into grade school textbooks [Davis, 1985]—save for the occasional letter as a placeholder in "missing-addend sentences." Exceptions to this state of affairs are presented in a later section. Thus, the remarks in this particular section apply primarily to secondary school textbooks.) Typical topics include (a) the properties of real and complex numbers; (b) the forming and solving of first- and second-degree equations in one unknown; (c) the simplification of polynomial and rational expressions; (d) the symbolic representation of linear, quadratic, exponential, logarithmic, and trigonometric functions, along with their graphs; and (e) sequences and series. Not all students study all these topics; the more advanced themes are usually addressed only by students going on to college. The study of algebra usually begins in grade 8 or 9, with those students who are preparing for college generally finishing the equivalent of two years of algebra.

The content of high school algebra has changed very little over the years. At the beginning of this century, the topics covered in beginning algebra courses generally included the simplification of literal expressions, the forming and solving of both linear and quadratic equations, the use of these techniques to find answers to problems, and practice with ratios, proportions, powers and roots. In the decades that followed, there were efforts to include some practical aspects and the use of graphical methods. In the early 1960s the large and widening gap between the mathematics being taught in school and the subject as required for post-college jobs in such fields as nuclear physics, space exploration, communications, and computer technology became obvious. This observation led to the development of the New Mathematics movement. Algebra in the revised curriculum was to incorporate new topics like inequalities, emphasize unifying concepts like set and function, and be taught so that its structure and deductive character were apparent. However, those changes that have persisted into today's algebra curricula have been more cosmetic than substantial.

Nevertheless, the structural character that was evident at the beginning of the century has been maintained. Examples of the structural aspects of traditional high school algebra curricula include simplifying and factoring expressions, solving equations by performing the same operation on both sides, and manipulating parameters of functional equations such as $y = v + (x - b)^3$ to yield families of functions. The introductory chapter of most textbooks emphasizes links to arithmetic. Algebraic representations are treated as generalized statements of the operations carried out in arithmetic; that is, treated in procedural terms whereby numerical values are substituted into algebraic expressions to yield specific output values. However, once this relatively smooth introduction is completed, algebraic representations begin immediately to be treated as mathematical objects on which certain structural operations can be carried out, such as combining literal terms, factoring, or subtracting the same term from both sides of an equation.

It is important to distinguish here the way in which the terms *procedural* and *structural* are being used in this chapter. *Procedural* refers to arithmetic operations carried out on numbers to yield numbers. For example, if we take the algebraic expression, $3x + y$, and replace x and y by 4 and 5 respectively, the result is 17. Another example involves the solving of $2x + 5 = 11$ by substituting various values for x until the correct one is found. In both these ostensibly algebraic examples, the objects that are operated on are not the algebraic expressions but their numerical instantiations. Furthermore, the operations that are carried out on these numbers are computational—they yield a numerical result. Thus, both of these examples illustrate a procedural perspective in algebra.

The term *structural*, on the other hand, refers to a different set of operations that are carried out, not on numbers, but on algebraic expressions. For example, if we take the algebraic expression $3x + y + 8x$, this can be simplified to yield $11x + y$ or divided by z to yield $(11x + y)/z$. Equations such as $5x + 5 = 2x - 4$ can be solved by subtracting $2x$ from both sides to yield $5x - 2x + 5 = 2x - 2x - 4$, which can subsequently be simplified to $3x + 5 = -4$. In both of these examples, the objects that are operated on are the algebraic expressions, not some numerical instantiation. The operations that are carried out are not computational. Furthermore, the results are yet algebraic expressions.

As I have pointed out, most algebra textbooks attach a façade of procedural approaches onto their introduction to algebraic objects by providing a few exercises involving numerical substitution in algebraic expressions and various arithmetical techniques for solving algebraic equations—techniques that allow the students, in a sense, to bypass the algebraic symbolism. However, this pretense is soon dropped when expressions are to be simplified and equations are to be solved by formal methods. The implicit objectives of school algebra are structural. That students may attempt to circumvent or may not be prepared to handle the structural intent of the curriculum is discussed in the section immediately following. The cognitive demands involved in operating on algebraic expressions as objects with operations that are quite unlike the operations of arithmetic are clearly reminiscent of the intellectual struggles that occurred during the historical development of algebra, when procedural interpretations made way for structural ones.

PSYCHOLOGICAL CONSIDERATIONS

Sfard (1991) has suggested that abstract mathematical notions can be conceived in two fundamentally different ways: structurally (as objects) or operationally (as processes). She claims that the operational conception is, for most people, the first step in the acquisition of new mathematical notions. The transition from a "process" conception to an "object" conception is accomplished neither quickly nor without great difficulty. After they are fully developed, both conceptions are said to play important roles in mathematical activity. (Note that, in the context of school algebra, my use of the term *procedural* is intended to mean the same as Sfard's use of the term *operational*.)

Sfard contrasts the distinctions between the two conceptions in the following way:

> There is a deep ontological gap between operational and structural conceptions....Seeing a mathematical entity as an object means being capable of referring to it as if it was a real thing—a static structure, existing somewhere in space and time. It also means being able to recognize the idea "at a glance" and to manipulate it as a whole, without going into details....In contrast, interpreting a notion as a process implies regarding it as a potential rather than actual entity, which comes into existence upon request in a sequence of actions. Thus, whereas the structural conception is static, instantaneous, and integrative, the operational is dynamic, sequential, and detailed. (p. 4)

The existence of historical stages during which various mathematical concepts such as number and function evolved from operational to structural has led Sfard to create a parallel three-phase model of conceptual development that, as we shall see shortly, is supported by the findings of several research studies. During the first phase, called interiorization, some process is performed on already familiar mathematical objects. The second phase, called condensation, is one in which the operation or process is squeezed into more manageable units. The condensation phase lasts as long as a new entity is conceived only procedurally, or operationally. The third phase, reification, involves the sudden ability to see something familiar in a new light. Whereas interiorization and condensation are lengthy sequences of gradual, quantitative rather than qualitative changes, reification seems to be a leap: A process solidifies into an object, into a static structure. The new entity is detached from the process that produced it. For example, Sfard states, "In the case of function, reification may be evidenced by proficiency in solving equations in which 'unknowns' are functions (differential and functional equations, equations with parameters), by ability to talk about general properties of different processes performed on functions (such as composition and inversion), and by ultimate recognition that computability is not a necessary characteristic of the sets of ordered pairs which are to be regarded as functions" (p. 20).

The historical analysis above allows us to view the development of algebra as a cycle of procedural-structural evolution. In a similar way, the study of school algebra can be interpreted as a series of process-object (i.e., procedural-structural) adjustments that students must make in coming to understand the structural aspect of algebra. I have already briefly discussed the adaptations that students must make when they begin the study of algebraic expressions and equations—that students cannot long interpret these entities as arithmetic operations upon some number, but rather must very quickly learn to view them as objects in their own right upon which higher level processes (that is, operations) are carried out. In other words, students must soon realize that the objects that are operated upon are algebraic expressions rather than numbers; furthermore, the operations that are carried out are those of simplifying, factoring, rationalizing the denominator, solving, differentiating, and so on, rather than simply adding, subtracting, multiplying, and dividing as was the case in arithmetic. Tall (1989) has pointed out that, as a process, $2(a + b)$ is perceived quite differently from $2a + 2b$. Until a student is able to conceive of an algebraic expression as a mathematical object rather than as a process, algebraic manipulation can, according to Tall, be a source of conflict.

Another adjustment to be made by novice algebra students learning to deal with the structure of algebra, in particular the symbolic representation of numerical relationships, concerns the translation of problem situations into equations. Algebraic equations are structural representations that involve a non-arithmetic perspective on both the use of the equal sign and the nature of the operations that are depicted. Note in the following examples that the required change in perspective from an arithmetic to an algebraic one can be interpreted as a movement from a procedural to a structural conception.

In elementary school the equal sign is used more to announce a result than to express a symmetric and transitive relation. In attempting to solve the problem:

Daniel went to visit his grandmother, who gave him $1.50. Then he bought a book costing $3.20. If he has $2.30 left, how much money did he have before visiting his grandmother?

sixth graders will often write $2.30 + 3.20 = 5.50 - 1.50 = 4.00$ (Vergnaud, Benhadj, & Dussouet, 1979). The symmetry and the transitivity of the equal sign are violated. The equal sign is read as "it gives," that is, as a left-to-right directional signal. The interpretational shift that must occur in algebra with respect to the equal sign is precisely that of respecting the symmetric and transitive character of equality (Vergnaud, 1984, 1986).

The above example also illustrates how elementary school students obtain their answers, by working backwards using a linear, sequential approach involving a string of inverse operations but usually without any formalizing of either the problem situation or method of solution. In arithmetic, the goal is to find the answer, and this goal is usually accomplished by carrying out some sequence of arithmetic operations on either the given numbers of the problem or on the intermediate values derived therefrom.

In contrast to arithmetic, the answer-driven, working-backwards approach is often inapplicable in algebra. Consider the following problem:

The Westmount Video Shop offers two rental plans. The first plan costs $22.50 per year plus $2.00 per video rented. The second plan offers a free membership for one year but charges $3.25 per video rented. For what number of rental videos will these two plans cost exactly the same?

With problems of this type, students can usually no longer rely on the approaches they used in arithmetic. Filloy and Rojano (1984) have emphasized that a rupture occurs with problems such as these which can be modeled by equations of the type $ax \pm b = cx \pm d$. Students must not only begin to think in terms of forward operations in order to model these problems by equations, but also use a solving procedure that operates on both sides of the equation, that is, a process that operates on an algebraic object. Lesh, Post, and Behr (1987) have distinguished algebra problem solving from arithmetic problem solving by pointing out that, in algebra, the problem requires "first *describing* and then calculating" [italics added] (p. 657).

But even the activity of "describing" in algebra problem solving can be done in either procedural or structural terms. A good example of the difference between the two is provided by the set of studies carried out with university students by Clement and his colleagues (for example, Clement, 1982; Soloway, Lochhead, & Clement, 1982). The structural interpretation with respect to the now-classic Student-Professor problem (that is, "Write an equation using the variables S and P to represent the following statement, 'There are six times as many students as professors at this university'; use S for the number of students and P for the number of professors.") requires generating an equation in which the letters are treated as variables and the equal sign is used to express an equivalence. A variation of this question, which allows for a procedural interpretation, uses the following form: "Given the following statement, 'There are six times as many students as professors at this university,' write a computer program in BASIC which will output the number of students when supplied (via user input at the terminal) with the number of professors. Use S for the number of students and P for the number of professors." One of the characteristics of the procedural approach for this situation is an input-output interpretation that involves numerical inputs and outputs. Thus, even though the computer program involves literal terms, students are able to interpret them as numbers upon which a specific arithmetic operation is performed in order to produce a numerical output. That students have been found to be considerably more successful with a procedural approach, which specifies an algorithm for computing one magnitude by means of another, than with a structural approach, which specifies an equality relation among variables, is discussed in a later section.

The example above suggests that even university-level algebra students may find more meaning in procedural representations based on numerical interpretations than in structural representations. We have seen how, historically, procedural representations endured for several centuries. Wheeler (1989) has pointed out that a considerable amount of the underlying meaning disappeared when a specialized symbolic language was developed. Rhetorical and syncopated algebra were both fairly easy to follow and understand until the 16th century, when the notation began to be too complex to be understood in words. But the step to a symbolic system eliminated the meanings of individual items and even of the operations acting on them. Symbolic language is powerful because it removes many of the distinctions that the vernacular preserves and vastly expands its applicability. However, Wheeler emphasizes that symbolic language is semantically extremely weak, introducing the difficulty for the learner that, by suiting several contexts, the language appears to belong to none.

Thus, the cognitive demands placed on algebra students include, on the one hand, treating symbolic representations, which have little or no semantic content, as mathematical objects and operating upon these objects with processes that usually do not yield numerical solutions, and, on the other hand, modifying their former interpretations of certain symbols and beginning to represent the relationships of word-problem situations with operations that are often the inverses of those that they used almost automatically for solving similar problems in arithmetic. It took centuries for the field of algebra to undergo these developments. Yet students beginning their first algebra course are expected to reify (Sfard, 1991) algebraic representations almost immediately. The ways in which students attempt to cope with the cognitive demands made on them throughout their algebra courses are revealed in several of the studies described below.

TEACHING PERSPECTIVES

Having just considered the learning side of the coin, we now turn to the teaching side. Unfortunately, there is a grave scarcity not only of models of the teaching of algebra but also of literature dealing with the beliefs and attitudes of algebra teachers (Clark & Peterson, 1986; Grouws, Cooney, & Jones, 1988; Wagner & Kieran, 1989). For the most part, the literature on mathematics teaching does not describe the ways in which the teaching of algebra ought to be considered in a different light from, say, the teaching of geometry or arithmetic. This body of research tends to focus not on the distinctions to be made according to the various subject matters, but rather on the commonalities in the teaching of mathematics classes, such as time spent on whole-group instruction versus seat work, teaching for rote learning versus teaching for understanding, the role

of reviewing, constructivist approaches to teaching, motivation, social dynamics of the classroom, and so on. In much of the research literature on teaching mathematics, the actual content to be delivered is generally treated as a variable. From the few reports available that deal specifically with algebra teachers (for example, Slovin, 1990), one is led to conclude that algebra teachers, like teachers of other mathematics subjects, seem primarily concerned with managing their classrooms and covering the material. They tend to follow the chapters in the textbook, emphasizing with explanations some of the given examples and then assigning homework exercises. (This tendency is also documented in summary reports of the Second International Mathematics Study—for example, in McKnight, Travers, Crosswhite, & Swafford, 1985.) Many algebra teachers are also inclined to view themselves primarily as providers of information:

When I teach a classical, traditional Algebra I class I often get the feeling, got the feeling, that I could hire anyone, a graduate math student, somebody with a science background, to give a 15 minute lecture on the lesson of the day, say, quadratic equations by factoring, and you could do the same thing that I could, just walk around the class for the next 15 minutes making sure everybody gets started and that was it. There was no conversation, no dialogue; I lectured. (Interview, 3/21/89.) (Slovin, 1990, p. 7)

In another study involving a beginning teacher who professed to believe that problem solving is the principal activity of mathematics, Cooney (1985) described this teacher's attempts to integrate a problem-solving approach into his teaching of algebra. However, his view of problem solving was limited to using recreational problems [problems chosen primarily for the enjoyment solving them provides] as a motivational device before getting "down to brass tacks" (p. 331). Ultimately, "the textbook was the primary factor in determining his curriculum and, for the most part, his method of presentation" (p. 330).

If algebra teachers tend to teach what is in the textbooks, then a seemingly obvious first step toward changing the teaching of algebra—assuming that one wanted to change the way in which algebra is taught—would appear to be that of modifying the way algebra is presented in textbooks. Despite reform movements in the past century that have emphasized certain variations (for example, meaningful learning theory) in the teaching of a curriculum that has remained fairly stable, these variations have never, according to Weaver and Suydam (1972), been clearly interpreted at the secondary level as they have for elementary school mathematics. Furthermore, up until very recently, curriculum decisions made by textbook authors were not based on research results. However, there are signs that this situation is changing and that a few innovative algebra curricula are being developed that reflect findings from research on the teaching and learning of algebra (for example, Rachlin, 1987; Rachlin, Matsumoto, & Wada, 1988). Nevertheless, Rachlin (1989) emphasizes that "regardless of what content society ascribes to algebra, there is a need for research on the learning and teaching of the curriculum at two levels—that of the students and that of the teachers" (p. 259). Rachlin points out further that it would seem insufficient merely to change the textbook if one wanted to change substantially the teaching of algebra.

He suggests that "we must understand the nature of teachers' beliefs and cognitions and the roles these beliefs and cognitions play in the decisions teachers make as they present the new curriculum to their students" (p. 261). Unfortunately, very little literature exists on algebra teachers' beliefs and cognitions.

Another factor that has a bearing on how teachers interpret and adapt the material in a textbook is their understanding of both their students' cognitions and the role of students' behaviors. In a study aimed at developing a teacher's awareness of the effectiveness of his own teaching, Rachlin (1982) utilized a technique that was found to be quite beneficial. Rachlin first asked the teacher to predict the solution processes that the teacher's students would use for a given set of tasks. The teacher was then requested to review videotaped interviews conducted by the researcher with the teacher's students, to comment on students' errors and procedures, to indicate anything unexpected in reviewing the protocols, and to offer suggestions that might be useful for other teachers. This technique was an excellent means of conveying to the teacher the realization that certain notions and processes, which supposedly had been adequately taught in class, had not really been treated with enough depth. The teacher was inspired to comment that "sometimes (maybe very often) what we teach is not what they learn" (p. 144).

Even though the research community knows very little about how algebra teachers teach algebra and what their conceptions are of their own students' learning, this is not to suggest that there has not been considerable research on new approaches to the teaching of algebra—there has. However, the analysis of results has usually been directed toward detailing learning phenomena, not toward presenting teaching variables. (See Kieran, 1985a, for a description of the Soviet Teaching Experiment, a methodology that is often used in one form or another in many of these studies of learning phenomena.) In view of the scarcity of research emphasizing the role of the classroom teacher in algebra instruction, the findings of the few studies that do exist are discussed in the upcoming sections according to the particular algebraic content that is included in the given research.

COGNITIVE STUDIES IN THE LEARNING AND TEACHING OF ALGEBRA

The previous sections have brought into focus the historical-psychological factors that I believe are critical in making sense of the research studies that have investigated the learning and teaching of school algebra. The historical evolution of algebraic symbolism from verbal prescriptions to the representation of unknowns to the expression of general relationships—an evolution that can be described in procedural-structural terms—and the development of a psychological model by Sfard (1991) that postulates the need for a lengthy transition period in moving from operational (procedural) to structural conceptions suggest an organization of the research findings according to the order in which algebra topics are typically presented to students. In this way, we can follow the development of algebra learning and look for the evolution of thinking according to procedural-structural considerations.

Many first-year algebra courses begin with literal terms and their relation to numerical referents within the context of, first, algebraic expressions and, then, equations. After a brief period involving numerical substitution in both expressions and equations, the course generally continues with the properties of the different number systems, the simplification of expressions, and the solving of equations by formal methods. The manipulation and factoring of polynomial and rational expressions of varying degrees of complexity soon become a regular feature. Interspersed among the various chapters are word problems, thinly disguised as "real world" applications of whatever algebraic technique has just been learned. Students eventually encounter functions and their algebraic, tabular, and graphical representations. The functions covered usually include linear, quadratic, cubic, exponential, logarithmic, and trigonometric.

The above topics, which are generally accepted as the core of "school algebra," serve as an organizational framework for generating the various sections of this part of the chapter: literal terms and expressions, simplifying expressions, equations, solving equations, word problems, and functions and graphs. In addition to describing research findings according to the topic focused upon, this chapter considers whether there was some kind of instructional intervention. One last organizational note is that studies involving the use of computers, as well as those dealing with teachers and their perceptions, are not discussed separately but rather are integrated throughout the chapter.

Literal Terms and Expressions

School algebra is sometimes referred to as generalized arithmetic (Booth, 1984), that is, the writing of general statements representing given arithmetical rules and operations. In line with this characterization of school algebra, Booth has suggested that if elementary school students (ages 6-12) do not recognize that the total number of items in two sets containing, say, five and eight items, respectively, can be written as $5 + 8$ (rather than 13), it is highly unlikely that they will recognize that $a + b$ represents the total number of items in the sets containing a and b items. In other words, being able to treat $a + b$ as an object in algebra has some intuitive precursor in arithmetic. Algebra demands that students recognize, for example, that $a + b - c$ is not the same as $a - b + c$ (unless $b = c$).

Do students recognize these same structural constraints in arithmetic? There is some evidence to suggest that students are not aware of the underlying structure of arithmetic operations and of their properties (Kieran, 1989). Chaiklin and Lesgold (1984) worked with five sixth graders who were given the task of judging the equivalence (without computing the totals) of three-term arithmetic expressions with a subtraction and an addition operation (for example, $685 - 492 + 947$, $947 + 492 - 685$, $947 - 685 + 492$, $947 - 492 + 685$). They found that students used several different methods to combine numerical terms, even within the same expression, depending on the expression with which it was being compared. They also noted that students preferred to calculate in order to decide whether expressions were equivalent. This suggests that the pupils were not in a position to be able to judge equivalence without computing. Similar findings have been reported by Cauzinille-Marmeche, Mathieu, and Resnick (1984).

Collis (1974, 1975) used a task that involved finding the value of □ in, for example, $(235 + □) + (679 - 122) = 235 + 679$ to investigate whether or not students could recognize the relations among the various operations or would seek recourse to calculation. He used three formats: small numbers, large numbers (as shown), and letters. Collis found that younger children succeeded only with the small-number items since their only available method was to calculate. Collis described the ability to work with expressions without reducing them by calculating as "Acceptance of Lack of Closure."

Other studies investigating students' intuitive bases for the symbolism and structure of algebra have been carried out by Booth (1981a, 1981b, 1984, 1988). In one of her studies, Booth (1984) used a written test that included some items from the Concepts in Secondary Mathematics and Science (CSMS) assessment such as finding the area of the rectangle shown in Figure 17.1—an item on which 42% percent of the 13-year-olds tested in the CSMS survey responded with $7f3$ or $f21$ or $f + 21$. Booth then carried out interviews with students in her study who had made the same notational errors as had the students who had taken the CSMS test in order to investigate whether they were aware of the underlying procedure of finding the area of a rectangle, and if so, whether their difficulties were related to understanding the conventions of algebraic symbolism. Booth concluded that the ability to describe a method verbally does not necessarily entail the ability to recognize the correct symbolization of that method.

Discriminating among the various ways in which letters can be used in algebra can present difficulties to students. Küchemann (1978, 1981) carried out a large-scale study of students' interpretations of literal terms in 1976 as part of the CSMS project. He administered a 51-item paper-and-pencil test to 3000 British high school students, 13 to 15 years of age. Using a classification originally developed by Collis (1975), Küchemann categorized each item into one of the following six levels of interpretation of letters according to the minimum level required for successful performance:

(a) Letter evaluated: The letter is assigned a numerical value from the outset;

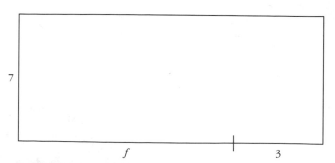

FIGURE 17–1. Adapted from the CSMS test item asking students to determine the area of the indicated rectangle (from Küchemann, 1981, p. 115).

(b) Letter not considered: The letter is ignored or its existence is acknowledged without giving it a meaning;
(c) Letter considered as a concrete object: The letter is regarded as a shorthand for a concrete object or as a concrete object in its own right;
(d) Letter considered as a specific unknown: The letter is regarded as a specific but unknown number;
(e) Letter considered as a generalized number: The letter is seen as representing, or at least as being able to take on, several values rather than just one;
(f) Letter considered as a variable: The letter is seen as representing a range of unspecified values and a systematic relationship is seen to exist between two such sets of values.

Küchemann found that, even though the interpretation that students chose to use depended in part on the nature and complexity of the question, only a very small percentage of the 13- to 15-year-old pupils were able to consider the letter as a generalized number—despite classroom experience in representing number patterns as generalized statements. Even fewer were able to interpret letters as variables, in the sense of Küchemann's classification. A greater number of pupils were able to interpret letters as specific unknowns than as generalized numbers. Nevertheless, the majority of students (73% of 13-year-olds, 59% of 14-year-olds, 53% of 15-year-olds) either treated letters as concrete objects or ignored them. In terms of the procedural-structural model of Sfard (1991), Küchemann's findings also suggest that many of the students tested had not yet begun to interpret literal expressions as numerical input-output procedures—the first phase in Sfard's hypothesized evolutionary process of developing a structural conception of algebraic expressions.

There is some evidence that long-term experience in Logo programming can assist students in developing such an understanding of variables and of algebraic expressions. One of the aims of the Logo Maths Project (Hoyles, Sutherland, & Evans, 1985; Noss, 1986; Sutherland & Hoyles, 1986)—a British study involving four pairs of students engaged in Logo programming activities during their normal mathematics class throughout a three-year period—was to develop and test materials designed to help students relate their understanding of the concept of variable in Logo to their understanding of variable in paper-and-pencil algebra. The researchers were responsible for the various Logo interventions throughout the study. Longitudinal case studies of the students (11 years old at the beginning of the study) were followed by a structured interview when the project terminated. During this interview pupils were asked to (a) make a generalization and formalize it in an algebra context, (b) make a generalization and formalize it in a Logo context, (c) answer algebra questions related to the meaning of letters (taken from the CSMS questionnaire), (d) answer Logo questions related to the meaning of variable names, and (e) represent a function in both Logo and algebra. The results showed that, when compared with a class of non-Logo students and also with the students tested in the CSMS project, the case-study students performed substantially better (Sutherland, 1987, 1988). A deeper understanding of the concept of variable also results from the activity of programming in BASIC among 12-year-olds (Oprea, 1988; Thomas & Tall, 1986). The positive results of the Logo Maths Project with respect to students' learning to relate their Logo experience to their paper-and-pencil algebra contrast with the findings of Roberts, Carter, Davis, and Feurzeig (1989). They report that the sixth graders with whom they worked throughout half a school year—a time far shorter than that in which the Logo Maths Project students were involved—were generally unable to transfer the algebra they learned in a Logo environment to the more traditional algebra.

It is to be noted that the use of variable terms in Logo programming includes (a) statements of explicit operations to be carried out on the input variable(s), (b) functional input-output representations of procedures, as well as (c) generalized statements expressing relationships between quantities. Two aspects of the Logo Maths Project seem critical: One is the considerable length of time spent on the programming activities; the other is the nature of the activities, that is, many were devoted to developing programs that represent input output relationships. Sutherland (1987) reports, however, that, even though the case-study subjects did considerably better on the CSMS test than did the subjects evaluated by Küchemann (1981) on questions involving the use of letters as specific unknowns and generalized numbers, they were not able to succeed on the highest-level questions in which the letter represented a range of unspecified values and for which a systematic relationship existed between two such sets of values. In other words, long term programming experience did not seem to equip the case-study students with the cognitive tools necessary to handle the CSMS questions that required a structural conception of literal terms and expressions.

A study with an instructional component that focused on introducing algebraic expressions to a group of grade 6 and 7 students was carried out by Chalouh and Herscovics (1988). The problems involved in the teaching experiment included rectangular arrays of dots, lines divided into segments, and areas of rectangular plots—all of the problems having one of the dimensions either hidden or expressed as an unknown quantity. For example, one of the questions presented to the subjects is the following: "Can you write down the area of this rectangle?" (see Figure 17-2). The teaching sequence allowed students to construct a procedural meaning for algebraic expressions such as $4x + 4y$. However, the students believed that these expressions were somehow incomplete unless they expressed them as part of an equality, such as "Area $= 4x + 4y$" or as "$4x + 4y = something$"—suggesting that a procedural in-

terpretation of an algebraic expression requires that part of the representation indicate the result of carrying out the procedure.

Similarly, Wagner, Rachlin, and Jensen (1984) found that many algebra students tried to add "$= 0$" to any expressions they were asked to simplify. The need to transform expressions into equations was also illustrated by the results of a study by Kieran (1983) who found that some of the students could not assign any meaning to a in the expression $a + 3$ because the expression lacked an equal sign and right-hand member.

Simplifying Expressions

The research results presented in the previous section have described some of the initial procedural conceptions of students with respect to literal terms and expressions. The absence of structural conceptions has been evident. Nevertheless, students are asked fairly soon in their algebra classes to simplify expressions—an activity that, for simple expressions, can be related initially to a numerical, procedural conception of expressions but which cannot for very long remain at that level. The complexity of the expressions, as well as the nature of the simplifications called for, quickly make such tasks undoable unless the student is able to develop a sense of operating on the algebraic expression as a mathematical object in its own right. Such a structural conception involves applying properties not to numbers but to expressions. The difficulties that students encounter in operating on algebraic expressions have been documented in studies ranging from those dealing with novices to those involving the more experienced student.

Greeno (1980) has suggested that the process of solving problems involves apprehending the structure of relations in the problem. To test this idea, Greeno (1982) carried out a study with beginning algebra students on tasks involving algebraic expressions. He found that student performance appeared to be quite haphazard, for a while at least. Their procedures were fraught with unsystematic errors, indicating an absence of knowledge of the structural features of algebra. Their confusion was evident from the way that they partitioned algebraic expressions into component parts. According to Greeno, beginning algebra students are consistent neither in their approach to the testing of conditions before performing some operation nor in their process of performing the operations. For example, they might simplify $4(6x - 3y) + 5x$ as $4(6x - 3y + 5x)$ on one occasion, but do something else on another. Wenger (1987) has described some of the poor strategic decisions made by students with extensive algebra experience—decisions that result in their "going round in circles" while carrying out simplification transformations because they cannot seem to "see" the right things in algebraic expressions. Even students who have successfully mastered the techniques of simplifying one type of expression, say, polynomials, have been found to be unable to transfer what they have learned to the next kind of simplification task involving, say, radicals; furthermore, they appear to perceive the two topics as separate (Rachlin, 1981). Eisenberg and Dreyfus (1988) suggest that because a recent trend in algebra is to emphasize procedures rather than the underlying structures (Coopersmith, 1984), students consequently consider mathematics as simply a compendium of algorithms, thus

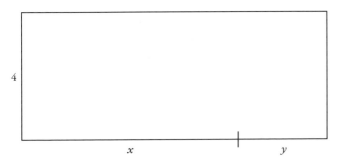

FIGURE 17-2. Adapted from the rectangular figure presented to subjects in the Chalouh and Herscovics (1988, p. 40) study.

making it difficult for them to generalize and apply what they have learned.

Other studies have also documented students' errors in parsing algebraic expressions (for example, Davis, 1975; Davis, Jockusch, & McKnight, 1978; Firth, 1975; Lewis, 1981; Matz, 1979; Sleeman, 1984). An example of an error that has frequently been seen is simplifying, say, $39x - 4$ to $35x$ or $2yz - 2y$ to z. That these errors are not restricted to novice algebra students is indicated by the research of Carry, Lewis, and Bernard (1980). In their study of the equation-solving processes used by college students, they found that this type of error, which they called the "deletion" error, was the most prevalent one that students made when simplifying expressions at various steps in the equation-solving process. In discussing this error, Carry, Lewis, and Bernard suggested that some students are overgeneralizing certain mathematically valid operations, arriving at a single generic deletion operation that often produces incorrect results.

An extension of this suggestion is that, because many students continue to view letters as labels for concrete objects, they simplify algebraic expressions by computing according to the rules of arithmetic and then tack on the letters. Matz (1979) has pointed out that in working with algebraic expressions, students tend to "slap a veneer of names on an arithmetic base, but all the work remains in the arithmetic" (p. 4).

A recent study with relevance to teaching has shown that instruction can improve students' ability to recognize and to operate on the structure of algebraic expressions. Thompson and Thompson (1987) worked with 8 seventh-graders over an 8-day period using a computer program that involved expressions and equations in two formats: usual symbolic form and expression trees (see Figure 17.3). To change an expression by the use of a field property or other transformation, students first had to choose the transformation and then place a pointer on top of the operation to be transformed in the tree representation of the expression (see also Thompson, 1989). The computer would not carry out any transformation containing an error. An integral part of this teaching experiment was a set of worksheets involving numeric transformation and algebraic identity problems—problems designed to direct students' attention to structure, that is, to the application of properties when operating on expressions. Thompson and Thompson report the rapid development of these novices' understanding of the structure of expressions and equations in this environment.

They also describe how students came to see variables as a replacement not only for numbers, but also for subexpressions. The potential value of this sort of environment for developing students' structural conceptions of algebraic expressions and for the operations carried out on algebraic expressions would certainly seem to suggest that more research in this area would be productive. When much of what was proposed during the New Math movement fell into disfavor, the field properties were one of the topics that came to receive less explicit attention and space in textbooks. The results of the Thompson and Thompson study illustrate the advantages of spending more time in algebra classes on this kind of activity.

Another aspect of learning the structure of algebraic expressions involves an awareness of the conventions of algebraic syntax. Bell, Malone, and Taylor (1987) report that beginning algebra students are often perplexed at being permitted to combine $2a + a + 15$ to $3a + 15$ but not $a + a + a \times 2$ to $3a \times 2$. Freudenthal (1973) points out that, if in ab the a is replaced by $-a$, it becomes $-ab$; but if b is replaced by $-b$, it does not become $a - b$ but $a(-b)$. The student must learn where to add brackets and where not. By conscious bracketing, the text is structured. Kieran (1979) has found that beginning algebra students tend to read expressions from left to right and do not see the need for brackets. A related issue concerns the punctuation of algebra. According to Freudenthal, the oldest means of syntactic structuring of algebraic expressions is the connecting force of certain algebraic operations. That raising to a power takes precedence over multiplication which takes priority over addition has been—according to Freudenthal—a visual characteristic of the way expressions have been punctuated, space-wise, for several centuries. Kirshner (1987, 1989) conducted a study of 400 students drawn from grades 9 and 11 and a first-year college calculus class to assess their reliance on visual cues in algebraic syntax. He found that for some students the surface features of ordinary notation provide a necessary cue to successful syntactic decision, but that the majority of students relied on propositional knowledge embodied in the rules for order of operations. Unfortunately for the learner who depends on visual cues, there is not always a strong correlation between the propositional rules and the surface features of algebraic notation.

Equations

As pointed out earlier, one of the requirements for generating and adequately interpreting structural representations such as equations is a conception of the symmetric and transitive character of equality—sometimes referred to as the "left-right equivalence" of the equal sign. The notion among beginning algebra students that the equal sign is a "do something signal" (Behr, Erlwanger, & Nichols, 1976) rather than a symbol of the equivalence between left and right sides of an equation is indicated by their initial reluctance to accept statements such as $4 + 3 = 6 + 1$ or $3 = 3$. They think that the right side should indicate the answer, that is, $4 + 3 = 7$. That older algebra students continue to view the equal sign as a separator symbol rather than as a sign for left-right equivalence is seen in their abbreviation of steps in equation solving and in their alternations when "adding the same thing to both sides" (Byers &

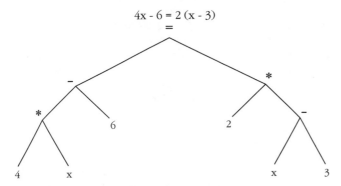

$$4x - 6 = 2 (x - 3)$$

FIGURE 17–3. Expression-tree format for the equation $4x - 6 = 2(x - 3)$ (adapted from Thompson, 1989, p. 152).

Herscovics, 1977):

$$\text{Solve for } x : 2x + 3 = 5 + x$$
$$2x + 3 - 3 = 5 + x$$
$$2x = 5 + x - x - 3$$
$$2x - x = 5 - 3$$
$$x = 2$$

This misuse of the equal sign persists well into college, as documented by Mevarech and Yitschak (1983), who showed that the students they tested had a poor understanding of the meaning of the equal sign, despite their ability to solve successfully different types of single-variable equations.

A teaching experiment (Kieran, 1981) carried out with six 12- and 13-year-olds was designed to enlarge novice students' understanding of the equal sign and to aid them in constructing meaning for algebraic equations of the type $ax \pm b = cx \pm d$. During individual interviews, students were first asked what the equal sign meant to them, followed by the request for an example showing the use of the equal sign. It is telling that most of them described the equal sign in terms of the *answer* and limited themselves to examples involving an operation on the left side and the result on the right. The ensuing teaching sessions focused on extending subjects' use of the equal sign to include multiple operations on both sides. This was done by having them construct arithmetic equalities, initially with one operation on each side, for example, $2 \times 6 = 4 \times 3$ and $2 \times 6 = 10 + 2$. They then went on to construct equalities with two operations on each side and then with multiple operations on each side, for example, $7 \times 2 + 3 - 2 = 5 \times 2 - 1 + 6$.

These equalities were given the name "arithmetic identities" in order to reserve the term "equation" for equalities with letters. Despite initial insistence from one subject on inserting the "answer" between both sides (i.e., $5 \times 3 = 15 = 10 + 5$), subjects seemed in general to be quite comfortable with equality statements containing multiple operations on both sides. They justified them in terms of both sides being equal because they had the same value. The comparisons that subjects were eventually able to make between left and right sides of the equal sign suggest that the equality symbol was being seen at this stage more as a relational symbol than as a "do something signal." The right side, by this time, did not have to contain the answer, but rather could be some expression that had the same value as the left side.

The reason for extending the notion of the equal sign to include multiple operations on both sides was to provide a foundation for the later construction of meaning for algebraic equations having multiple operations on both sides. If this extension were not done first, the idea that the result is always on the right side of the equal sign would accompany the student into the study of algebraic equations. Thus, equations such as $3x + 5 = 26$ might fit the student's existing notions, but others such as $3x + 5 = 2x + 12$ might not. Kieran also considered that cognitive strain would be eased considerably if the notion of the equal sign were first extended within the framework of arithmetic equalities.

The next step, introducing the concept of equation, involved taking one of the student's arithmetic identities and hiding any one of the numbers. The hiding was done first by a finger, then by a box, and finally by a letter (for example, $7 \times a - 3 = 5 \times 2 + 1$). Thus, an equation was defined as an *arithmetic identity with a hidden number.* Eventually the students hid the same number twice—one occurrence on the left side of the equal sign and the other occurrence on the right—as, for example, with $2 \times 3 + 7 = 5 \times 3 - 2$ being transformed into $2 \times c + 7 = 5 \times c - 2$. Just as with arithmetic identities, the right side of an algebraic equation did not have to contain the answer, but rather could be some expression that had the same value as the left side. For example, equations such as $2x + 3 = 4x + 1$ were described by the students: "If you know what number x is, then 2 times that number plus 3 has the same value as 4 times that number plus 1." This study showed that it is possible to change beginning algebra students' uni-directional and answer-on-the-right-side perception of the equal sign and of arithmetic equalities into a procedural view of algebraic equations that includes (a) letters standing for numbers, (b) an equal sign representing the equivalence of left and right sides, and (c) a right-hand member not necessarily consisting of a single numerical term, but rather an algebraic expression—an awareness considered crucial in the development of a structural conception of equations.

Another approach to developing meaning for equations is described by Bell, Malone, and Taylor (1987), who have provided in their report detailed lesson-by-lesson observations of student work. Their approach was a problem-solving one in which three classes of 14-year-old students were led to construct equations for problems such as the following one, which was presented in Lesson 5:

Students were given the problem of 3 piles of rocks: A, B, C where B has 2 more than A and C has 4 times as many rocks as A. The total number of rocks is 14. Their task was to find the number of rocks in each pile using x and to do the problem in 3 "different ways"—i.e. using the x in three different positions. (p. 108)

Bell, Malone, and Taylor reported that all students started with pile A as x, giving $x + 2$ and $4x$ for the other two piles. With pile B as x, students wrote $x - 2$ and $4x - 2$ for the remaining two; none used brackets for $4(x - 2)$. The resulting equation, $x - 2 + x + 4x - 2 = 14$, did not provide the same solution as before and consequently led to a discussion on the need for brackets. The final approach, with pile C as x, gave $1/4x$ and $2 + 1/4x$ for the other piles, which most students wrote as $x \div 4 + 2 + x \div 4 + x = 14$. After collecting $3x$ together, they wondered what to do with the numbers. Intervention helped the students learn certain techniques for dealing with the mechanics of forming equations. The researchers noted that the initial conceptual obstacle of how to express word-problem statements (for example, "15 more than x") were overcome; however, the second-order difficulties, that is, treating an algebraic expression as an object (for example, coping with "15 more than $[x - 30]$"), were less fully resolved.

Solving Equations

Operating on an equation as a mathematical object involves the formal procedure of performing the same operation on both sides of the equation. However, this is generally not the first method that is taught to students. "Guess and test" methods involving numerical substitution, as well as other informal techniques such as the cover-up method and working backwards, are often used as introductory approaches to equation solving (Bernard & Cohen, 1988). These are approaches that would appear to link well with procedural conceptions of equations. However, very soon afterwards, students are taught the formal method. Several studies have investigated student equation solving—in fact, this topic seems to have received major research attention.

The various solving methods used by algebra students have generally been classified according to the following types:

(a) use of number facts,
(b) use of counting techniques,
(c) cover-up,
(d) undoing (or working backwards)
(e) trial-and-error substitution,
(f) transposing (that is, Change Side–Change Sign),
(g) performing the same operation on both sides.

The last two approaches are often referred to as formal methods—transposing being considered an abbreviation of performing the same operation on both sides. Algebra students are usually not taught the first two of these approaches; they bring them along from their elementary school experience with "missing-addend sentences" (for example, $2 + \square = 5$). For example, solving $5 + n = 8$ by recalling the addition number fact that 5 plus 3 is 8 would be a use of known number facts. Solving the same equation by counting 5, 6, 7, 8 and noting that three numbers were named after the 5 in order to arrive at 8 would be an example of solving by counting techniques. Booth (1983) has reported the use of both methods among novice algebra students.

Bell, O'Brien, and Shiu (1980) have seen students use the cover-up method to solve equations such as $2x + 9 = 5x$: "Since $2x + 9$ totals $5x$, the 9 must be the same as $3x$ because $2x + 3x$ also equals $5x$; so x is 3." Whitman (1976) researched the relationship between the cover-up method and the formal procedure of performing the same operation on both sides of the equal sign in a teaching experiment involving six intact classes of seventh graders. She found that students who learned to solve equations by means of only the cover-up method performed better than those who learned both ways in close proximity, whereas students who learned to solve equations only formally performed worse than those who learned both techniques. These findings suggest that the students who had been taught to solve equations by the formal method alone were not conceptually prepared to operate on equations as mathematical objects with formal, structural operations.

The undoing method is analogous to the working-backwards approach used in arithmetic problem solving. For example, to solve $2x + 4 = 18$, the student takes the numerical result on the right side and, proceeding in a right-to-left order, undoes each operation as he/she comes to it by replacing the given operation with its inverse; thus, the student is able to operate exclusively with numbers and avoid dealing with the equivalence structure of this mathematical object. A computer environment embodying this approach is the Marble Bag Microworld developed by Feurzeig (1986). According to Feurzeig, an advantage of this approach, which has been used with sixth graders in a context of creating and solving simple marble bag stories, is that standard algebraic notation can be introduced as a rapid way of writing these stories.

The use of trial-and-error substitution as an equation-solving method (for example, solving $2x + 5 = 13$ by trying different values such as 2, 6, and then, possibly, 4) is very time-consuming and places a heavy burden on working memory, unless all trials are systematically recorded. As soon as algebra students learn to handle a formal method of equation solving, they tend to drop the use of substitution as an equation-solving technique (Kieran, 1985b). Unfortunately, students also seem to drop it as a device for verifying the correctness of their solution to an equation (Lewis, 1980); furthermore, they have been found not to use substitution as a means of testing the validity of their algebraic simplifications (Davis et al., 1978; Lee & Wheeler, 1989). Nevertheless, there is evidence that students who use substitution as an early equation-solving device—and not all of them do—possess a more developed notion of the balance between left and right sides of an equation and of the equivalence role of the equal sign than do students who never use substitution as an equation-solving method (Kieran, 1988). As we shall see, this awareness is helpful in successfully making the transition to the formal method of equation solving that involves performing the same operation on both sides of the equal sign.

Petitto (1979) has remarked that the first five methods above, which she refers to as intuitive, often do not generalize. In her study of 7 ninth-grade algebra students who were solving equations, she found that students who used a combination of formal and intuitive processes were more successful than those who used only one of these methods.

Another approach to equation solving that does not require a structural conception of equations but rather capitalizes on students' procedural input-output view of equations is one that uses computer-generated tables (Heid & Kunkle, 1988). Solving single-variable (one or more occurrences) equations in this environment means setting up and manipulating a table having a column for each of the expressions on the two sides of an equation. Students then input trial values for the variable term until the two expressions are equal. This study illustrates one of the ways in which computers can be used to advantage as equation-solving tools by making explicit the numerical output of both sides of an equation for each numerical input that is provided.

Formal methods of equation solving include transposing and performing the same operation on both sides of an equation. Although many algebra teachers consider transposing to be a shortened version of the procedure of performing the same operation on both sides, these two solving methods appear to be perceived quite differently by beginning algebra students (Kieran, 1988, 1989). Performing the same operation on both sides of an equation emphasizes the symmetry of an equation; this emphasis is absent in the procedure of transposing. As we shall see, there is some evidence to suggest that many students who use transposing are not operating on the equation as a mathematical object but rather are blindly applying the Change Side–Change Sign rule.

In a teaching experiment designed to aid grade 7 students in constructing meaning for the procedure of performing the same operation on both sides of an equation, Kieran (1988) found that at the outset of the study the subjects showed one of two preferences in solving simple one-operation equations: Some used trial-and-error substitution and the others used undoing. For two-operation equations such as $2x - 5 = 11$, the latter group of subjects extended their right-to-left undoing technique: Take 11, add 5 to it, then divide by 2. For multi-operation equations such as $3x + 4 - 2x = 8$, they generalized their method and simply undid each operation as they came to it—for this example, taking 8, dividing it by 2, adding 4, and then subtracting 3 (they had to ignore the last operation of multiplication because they had run out of operands). The preference that these subjects had for the undoing method of equation solving seemed to work against them in their efforts to make sense of the procedure that was being taught. The subjects who had begun the study with a preference for the undoing method were found, in general, to be unable to make sense of "performing the same operation on both sides." The instructional sequence seemed to have its greatest impact on those students who had begun the study with an initial preference for the substitution solving method and who viewed the equation as a balance between left and right sides. This observation suggests that learning to operate on the structure of an equation by performing the same operation on both sides may be easier for students who already view equations as entities with symmetric balance. Trial-and-error substitution is an elementary solving method that may provide an intuitive basis for the more structural solving methods. In contrast, although the use of a sequential, right-to-left undoing approach in equation solving seems much closer to the problem-solving methods used in arithmetic, it is clearly limited to equations having a single occurrence of the unknown term in one of a few restricted locations and unfortunately appears to encourage the learner to continue to bypass the algebraic symbolism rather than deal directly with the equation as a structural object.

In a previous section, it was noted that students often make poor strategic decisions when attempting to simplify algebraic expressions. Similar findings have been reported in studies that investigated the solving of multi-operation equations (for example, Carry, Lewis, & Bernard, 1980). Students have generally been found to lack the ability to generate and maintain a global overview of the features of an equation that should be attended to in deciding upon the next algebraic transformation to be carried out. Researchers working on the development of intelligent computer tutors for algebra are addressing this issue by including in their preliminary analyses systematic study of the approaches used by expert teachers while instructing in complex equation solving. For example, Shavelson, Webb, Stasz, and McArthur (1988) have described how a teacher they observed focused explicitly on the higher-level reasoning processes required in equation solving rather than on the lower-level transformations to be executed. She organized classroom discussions around goal structures in a top-down fashion—for example, beginning with the goal of eliminating fractions. Once the initial goal had been implemented, the teacher did not simply go on to the next step, but reminded the students of the reason for the low-level simplifications that had been necessitated by the goal.

McArthur (1985) and his colleagues (McArthur, Stacz, & Hotta, 1987) have incorporated the teaching of these global strategies into their intelligent tutor for solving algebraic equations. The present version of the system has an "inspectable expert" that can show a student the details of its reasoning. The student enters the initial equation to be solved and then can continue, should the student wish, to type additional steps in the equation-solving chain. If the student afterward wishes to edit a line, the program creates a branch from that step to indicate that any new equations constitute a different solution path from the one the student was originally on. The student can ask the expert if a particular step is correct; the expert will respond, as well as comment on the appropriateness of the step. The expert can also be asked to supply a next step or to elaborate on a step already provided. The use of this system has been found to develop a sense of heuristic search with respect to formal methods of equation solving and "to teach the student that an important part of learning a cognitive skill is *learning to study your own reasoning processes*" (McArthur et al., 1987, p. 317).

Another equation-solving, computer environment which has been found to facilitate acquiring a global view of a procedure is Algebraland (Brown, 1985). Brown suggests that the screen record of the student's solution steps presents a structure that "reifies the solution path" (p. 197) and consequently permits the student to examine it as a meta-object ("an object of study in its own right" [p. 197]). One of the features of this environment is that the system performs all of the algebraic operations for the student. According to Brown, the student need only decide which operator (such as isolate, collect, group, split, or simplify) to apply and what to apply it to, "thus removing any chance of making a syntactic error in a skill that she is expected already to have mastered" (Brown, 1985, p. 199)—an expectation that other research findings suggest may be completely unwarranted.

Other intelligent tutoring systems that permit students to focus on strategies for solving equations include the Algebra Workbench (Roberts et al., 1989) and the Algebra Tutor (Lewis, Milson, & Anderson, 1987). As of this writing, the Algebra Workbench is an operating system in prototype stage that has been used only briefly in a study investigating how much algebra can be taught to classes of sixth graders using such courseware. The Algebra Tutor likewise has had only limited use in actual classroom situations. Laboratory observations indicate that after students work with the Tutor they know more about algebra than before they started, but Anderson and his colleagues (Anderson, Boyle, Corbett, & Lewis, 1990) point out that they have no evidence comparing the value of time spent with the tutor and time spent in other algebra learning environments.

The effectiveness of concrete models in the teaching of formal equation-solving procedures has been researched by Filloy and Rojano (1984, 1985a, 1985b). In their teaching experiments they have aimed at helping students create meaning for equations of the types $ax \pm b = cx$ and $ax \pm b = cx \pm d$ and for the algebraic operations used in solving these equations. Their principal approach has been a geometric one, although they have also used the balance model in some of their studies. The geometric approach involves cover stories such as the

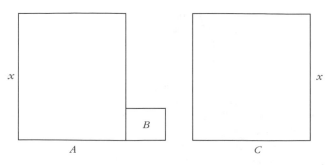

FIGURE 17–4. Adapted from the pictorial representation of a geometric situation used to model equations of the form $Ax + B = Cx$ in the Filloy and Rojano (1985a, p. 156) study.

following, accompanied by drawings similar to the one in Figure 17.4:

> *A person has a plot of land of dimensions A by x. Next (s)he buys an adjacent plot with an area of B square meters. A second person proposes to exchange this plot for another on the same street having the same area overall and the same depth as the first plot of land, but a better shape. How much should the depth measure so that the deal is a fair one?*

Filloy and Rojano carried out teaching interviews with three classes of 12- and 13-year-olds who already knew how to solve equations of the types $x \pm a = b$ and $ax \pm b = c$ (which Filloy and Rojano refer to as "arithmetical" equations), but who had not yet seen equations of the types $ax \pm b = cx$ and $ax \pm b = cx \pm d$ (which they refer to as "algebraic" equations). They designed their interview questions to uncover some of the obstacles experienced by students during the period of transition from arithmetical to algebraic equations, that is, during the period of the "didactical cut"—as Filloy and Rojano refer to this transition. The interviews revealed that the use of these two concrete models (the balance and the area models) did not significantly increase most students' ability to operate at the symbolic level with equations having two occurrences of the unknown. The well-known equation-solving error of combining constants and coefficients was also seen in this study, in particular with the use of the geometric model. Filloy and Rojano report that many students tended to fixate on the models and seemed unable to see the links between the operations performed with the model and the corresponding algebraic operations. As a result, the students remained dependent on the model even when it was no longer useful. In fact, students tried to use the model for simple equations that could have been solved more easily by the intuitive equation-solving methods that they had used before being taught the new method. They were so focused on the concretely-modeled procedure being taught that they seemed to forget their previously-used methods. This study provides further evidence that the transition from procedural to structural conceptions of algebraic equations is a difficult one for students to achieve.

In solving equations, many students try to apply the working-backwards approach that they used in arithmetic problem solving—operating on numbers rather than letters. As we have seen, the procedure of performing the same operation on both sides of the equal sign, whereby students operate explicitly on the structure of the equation, is not easily acquired. Consequently, many students, according to Filloy and Rojano (1984), are not really functioning in an algebraic mode. This is not to say that several of them do not become quite expert at solving equations. Studies (for example, Brown et al., 1988; Kieran, 1984; Mevarech & Yitschak, 1983) have shown that many students do learn to manipulate equations in a kind of non-thinking, automatic fashion, usually by using the Change Side–Change Sign procedure. However, these studies also demonstrate that the same students are generally not aware of the structure underlying the manipulations they perform.

Additional evidence of students' lack of a structural conception of equations is shown by some of the results of a study carried out by Kieran (1982, 1984). Most algebra teachers would expect that beginning students know that, say, $3 + 4 = 7$ can be expressed as $3 = 7 - 4$ and can generalize this knowledge to algebraic equations such as $x + 4 = 7$ and $x = 7 - 4$; thus, the teachers would expect students to consider the equations to be equivalent and thereby have the same solution. Kieran found that 12-year-old beginning algebra students have difficulty in judging equivalent expressions of the addition/subtraction relation. Two of the errors these students committed—the "Switching Addends" error and the "Redistribution" error—indicated their confusion with regard to recognizing the equivalent forms of addition and subtraction when the representations involved literal terms. In the "Switching Addends" error, $x + 37 = 150$ was judged to have the same solution as $x = 37 + 150$; in the "Redistribution" error, $x + 37 = 150$ was judged to have the same solution as $x + 37 - 10 = 150 + 10$. These errors suggest that beginning algebra students may be somewhat unsure of the structural relationships between addition and subtraction or, at the least, unsure of the written forms of these relationships when the written forms involve a literal term.

Still more evidence of the inability of students to distinguish structural features of equations is provided by Wagner, Rachlin, and Jensen (1984). Several tasks in their study were designed to probe students' awareness that the solution to an equation is determined by the structure of the equation and not by the particular letters used to represent the variable. This study drew on the research of Wagner (1977, 1981) on conservation of equation in which 29 students from middle and high school were individually interviewed and asked questions that included the following: "For the equations, $7 \times W + 22 = 109$ and $7 \times N + 22 = 109$, which would be larger, W or N?" Only 38% of those interviewed by Wagner answered correctly. The remainder believed either that the two equations would have to be solved, providing evidence of a procedural conception of equations, or that W is larger because of its position in the alphabet, evidence of an even more primitive view that does not include the use of letters to represent numbers.

Wagner, Rachlin, and Jensen (1984) asked ninth-grade students to solve the equation, $s/8 - 3 = 14$, and then to find the solution to the same equation after an alphabetic transformation of the variable had been effected—from s to t. In contrast to the subjects of the earlier Wagner study, most students knew immediately that the solution to the equation would not change. In the next task, the literal term t of the equation was changed to $t + 1$, and students were asked for the value of

$t + 1$. The majority re-solved the equation, some solving for $t + 1$ directly, but most solving for t and then figuring out the value of $t + 1$. In a later task, the students were asked to solve for $(2r + 1)$ in $4(2r + 1) + 7 = 35$. Only one student solved directly for $2r + 1$. The findings of this study show that most algebra students have trouble dealing with multiterm expressions as a single unit and suggest that students do not perceive that the basic surface structure of, for example, $4(2r + 1) + 7 = 35$ is the same as, say, $4x + 7 = 35$. Linchevski and Vinner (1990) refer to this as the hidden structure of an equation and have found that the ability of students to recognize such structure depends upon the recency of their experience with this kind of activity.

Greeno (1982) has pointed out that algebra novices lack knowledge of the constraints that determine whether transformations are permissible. For example, they do not know how to show that an incorrect solution is wrong, except to re-solve the given equation. They do not seem to be aware that an incorrect solution, when substituted into the original equation, will yield different values for the left and right sides of the equation. Nor do they realize that it is only the correct solution that will yield equivalent values for the two expressions in any equation of the equation-solving chain. However, not only novice equation solvers lack knowledge of these equivalence constraints. Kieran (1984) found that a group of nine experienced, competent high school solvers also lacked this knowledge.

Word Problems

The research literature on algebra word problems is divided into three subsections. The first part deals with traditional word problems that are found in many textbooks, such as age problems, distance-rate-time problems, and so on. The second part deals with problems that have been approached from a functional perspective. The third part deals with open-ended generalization problems seldom found in American algebra textbooks.

Traditional Word Problems. Generating equations to represent the relationships found in typical word problems is well known to be one of the major areas of difficulty for high school algebra students. In elementary school, children hardly ever write equations to represent arithmetic problems (Carpenter & Moser, 1982). In fact, if an equation in the form of an open sentence (for example, $4 + ? = 7$) is requested, children will solve the problem first and then attempt to provide the equation afterwards (Briars & Larkin, 1984). If children do write an equation, it usually represents the operations they carried out in arriving at the final answer to the problem (for example, $7 - 4 = 3$). I have already noted in the Psychological Considerations section that a major turnaround must occur in the thought processes of beginning algebra students when they are asked to think in terms of the forward operations that represent the structure of the problem rather than in terms of the solving operations. An example of a traditional algebra word problem is the following:

Ben is 4 years older than Juan. In 2 years, the sum of their ages will be 50. How old is each person now?

Much of the research that has addressed the issue of the problem representation processes used by algebra students has found that students tend to use either a direct-translation or a principle-driven approach (Chaiklin, 1989).

The direct-translation approach involves a phrase-by-phrase translation of the word problem into an equation containing numbers, variables, and operations. Some semantic knowledge is often required to formulate these equations; but solvers typically use nothing more than syntactic rules (Hinsley, Hayes, & Simon, 1977; Paige & Simon, 1966; Reed, 1984).

The principle-driven approach uses a mathematical principle to organize the variables and constants of a problem. For example, when some students read a problem that begins, "A canoeist paddles upstream...," they immediately realize that this is a problem involving distance, rate, and time. They are also aware of the effect of the current, depending on whether the canoeist is going upstream or downstream. These interpretive operations have been categorized into *schemata* to describe the principles used by algebra problem solvers (Hinsley et al., 1977; Mayer, 1980).

In general, it has been found that students have difficulty noticing structural similarities among problems with different cover stories (Reed, 1987). Often they resort to syntactic translation approaches and sometimes substitute various numerical values in order to verify the adequacy of their equations (Reed, Dempster, & Ettinger, 1985). They also attempt to use tables of relations as an intermediate step in generating equations for problems such as the distance-rate-time variety; however, they are generally unable to represent the relations correctly in these tables (Hoz & Harel, 1989). Chaiklin (1989) points out that evidence from the majority of cognitive studies in algebra problem solving clearly suggests that students have considerable difficulty in specifying relations among variables. Seemingly slight differences in problems can have a large effect on students' ability to construct correct equations. However, cognitive studies in algebra problem solving have, up to now, been unable to explain why certain methods of instruction in the learning of schematic relations for solving word problems are more effective with certain students than with others.

The approach most commonly used in teaching algebra classes how to solve word problems is to formulate an equation (or system of equations) involving unknowns and operations (usually forward operations) according to some mathematical relation and then, by a process of algebraic manipulation, to isolate the unknown term to find the solution. An alternate approach to representing and solving word problems by means of algebraic equations has been tested in the Soviet Union with elementary school children (Davydov, 1962; Freudenthal, 1974). To eliminate the difficulties encountered by children in having to decide whether to use forward operations or backward operations (called *direct* and *indirect* methods by the Soviets) according to the type of problem being solved, Davydov designed and experimented with an approach based on extensive teaching of part-whole relations. For example, in the first phase of the study involving classes of 8-year-olds, strips of paper cut into parts, volumes of water, weights of finely-crushed rock, and so on, were used to show children the meaning of wholes and parts. From the beginning, the wholes and parts were indicated by letters on drawings—numbers were never used. Later on, the wholes and parts were related in equations involving plus, minus, and equal signs. Children learned to make drawings corresponding to formulas such as $k = a - c - b - f$. After about three dozen lessons on the relation of wholes and parts, the second phase on problem solving began. Texts such as, "There were a red and b blue pencils in a box, and together there were c pencils," were to be translated into a drawing, a scheme, and three formulas. Then the children had to invent texts to correspond to given drawings and, later on, to given formulas. Eventually numerical values were introduced. An example of a problem-solving scenario is the following:

The teacher shows a graduated glass where the water level (k) is marked by an elastic band. Then he takes another graduated glass with c water and pours both into a drinking glass, which now contains b water. The pupils make a drawing, representing the connection between the magnitudes k, c, b and note down the formulas $c = b - k$, $b = c + k, k = b - c$.

Teacher: We indicated the volume of water by letters, but we can measure it also in numbers. With which magnitude shall we start measuring?
Pupil: With magnitude k, the water of the first graduated glass.
Teacher: Can you determine the volume of water in this graduated glass?
Pupil (going to the table and looking at the graduations): This water was 30 grams.
Teacher: Well, let us write $k = 30$.
Pupil: Now, the other graduated glass. This was 70 grams. $c = 70$.
Teacher: How much water is there in the drinking glass now? There is no graduation on it. How can we know how much is b? (The pupils are embarrassed. Then hands go up.)
Ljuda B.: One should pour it into a graduated glass and see how much there is.
Sereza S.: No need. b is our whole, isn't it? And k and c are parts. k is 70 and c is 30. To get the whole, one has to add the parts k and c, and 100 results.
(Freudenthal, 1974, p. 399)

It is notable that students wrote down all of the part-whole formulas and, when the question was asked, chose the formula in which the unknown was isolated. This activity, which is contrary to current practice in the West, demands that the student mentally generate a representation, which can be rather complex, involving several inverse operations. Sample problems that the 8-year-olds could handle with no apparent difficulty included: "In the morning a tractors worked on the land. In the course of the day some joined them. Then there were b of them working. How many had joined them?"

The positive results obtained from the studies with the 8-year-olds were again seen with the 9- to 11-year-old children who were able to solve even more difficult problems, still having only literal data. An example is the following:

Teacher (proposing the problem): In a basket are k apples, and in another one, so many that, if n apples are taken out, there are twice as much left as there are in the first one. How many are there in both altogether?
Tolja N. (noting immediately down the schema: $x =$ 1st basket + 2nd basket): In the first basket there were k ... (hesitating, thinking).
Teacher: What is said about the second basket?
Pupil: There is twice as much in it ...
Teacher: In the whole?
Pupil: No ... twice as much after they took away n apples.... (Pause.) This means: there is twice as much in it *and n more* apples. (Notes down the formula part $k \cdot 2 + n$.)
Teacher: Has the problem now been solved?
Pupil: Not yet. (Writes the whole formula and reads): $x = k + (k \cdot 2 + n)$. (Freudenthal, 1974, pp. 403–404)

These methods were then extended to include problems with numerical data.

Freudenthal (1974) points out that the technique used for these more complex problems is one of unfolding: The principal dependency is first established; components of this dependency are then successively unpacked until the final formula is derived. For example, for the problem: "A workbench weighed k kg. After an improvement in the construction it became m times lighter. How much metal is saved with d such workbenches?" the teacher emphasizes that the total saving is the product of the saving per workbench and their number: $I \cdot d$. Since I must be computed from the difference between the old and new weight, it is unfolded to be replaced by $k - new\ weight$. The new weight is, in turn, unfolded to k/m. It now becomes possible to express the final formula, $x = [k - (k/m)] \cdot d$.

These Soviet studies show that children can effectively be taught general methods for both representing and solving problems at a far younger age than is currently seen in non-Soviet algebra classes and suggest that the way in which we presently attempt to teach the translation of word problems into equations could be reconsidered. The reliance on part-whole relations and the use of problems involving only literal data are practices that are rarely emphasized in our standard approaches to the teaching of algebra.

Problems Approached from a Functional Perspective. The problem-solving research discussed in this section includes problems that are not unlike the traditional algebra word problems of several of the studies above; however, their mode of presentation and the solving approach that has been encouraged are usually different from those used in more standard problem-solving environments. In general, some functional relationship between two variables is established before the particular problem is solved. Often, the representation used for the expression of the functional relationship helps make explicit a procedural interpretation.

In adopting a functional approach to problem solving in their studies, researchers have attempted to provide an alternative avenue for students to come to understand unknowns and variables. Past studies have illustrated how impoverished students' conceptions of variable can be. For example, the results of the CSMS study reported by Küchemann (1978, 1981) showed that the majority of the 13- to 15-year-old students tested viewed letters as concrete objects or as labels for concrete objects. That older algebra students continue to use this label interpretation of literal terms is indicated by the research of Clement and his colleagues (for example, Clement, 1982; Clement, Lochhead, & Monk, 1981). They found that 37% of the engineering students they tested with the "Student-Professor" problem incorrectly expressed the functional relationship between the number of students and the number of professors; of these, two-thirds expressed the relationship as $6S = P$. That is, they wrote an "equation" to represent the idea that "there are six students for every professor" in the same way that they might express the relationship 3 feet = 1 yard. A related phenomenon was reported by Mevarech and Yitschak (1983), who found that 38% of the 150 college students they questioned answered that, in the equation $3k = m$, k is greater than m. These studies illustrate that many students continue to consider letters as labels rather than as numbers in an equivalence relation. This interpretation of letters precludes interpreting an equation not only as a structured mathematical object but also as a procedure since a procedural interpretation requires minimally that the student consider, for the examples above, k (or S) as a number, which when multiplied by 3 (or 6), yields the value m (or P).

Attempts by Rosnick and Clement (1980) to remedy this impoverished conception of algebraic variable within the context of the original version of the "Student-Professor" problem met with only limited success. However, when the demands of the task were changed from generating an equation to generating a computer program to model the same functional relationship described in an earlier section, the success rate was much higher (Soloway et al., 1982). A programming environment allowed students to inject into their reasoning a dynamic approach to representing the problem based on an input-output mode of reasoning (that is, "To calculate the value of the variable S, I must multiply the value of P by 6.").

Other studies provide further evidence of the power of a computer environment to support a procedural approach that involves specifying an algorithm for computing one magnitude by means of another. Kieran, Boileau, and Garançon (1989), as part of a research program on the use of computers in the learning of algebra, conducted a teaching experiment with 12 average-ability, seventh graders who participated in pairs in hourly sessions twice a week for four months. The environment was a problem-solving one that emphasized a functional interpretation of the relationships in problem situations. Students would enter into the computer a kind of natural-language program, one operation per line, for calculating the values of the variables in the program. For example, a problem such as:

The concessions manager at the Montreal Forum offered two pay plans for people willing to sell peanuts in the stands for the Canadiens hockey games. The first plan pays $28.68 plus $0.17 per bag sold. The second plan pays $11.00 plus $0.38 per bag sold. For what number of bags sold will these two pay plans give exactly the same pay?

might be represented as:

Input: number of bags
Program: number of bags × .17 gives first partial pay
                    first partial pay + 28.68 gives first pay plan
                    number of bags × .38 gives second partial pay
                    second partial pay + 11.00 gives second pay plan
Output: first pay plan
                    second pay plan

After entering their program for this problem, the students would then input various trial values for *number of bags* until the two output variables had the same value. One advantage of this kind of representation, as opposed to the standard equation representation, $.17x + 28.68 = .38x + 11$, is its closeness to natural language. Recall that the use of ordinary language descriptions for representing and solving problems characterized the pre-Diophantine stage in the development of algebraic symbolism. Another benefit of this computer-supported functional approach to problem representation is that it helps students to think in terms of forward operations—a step that is crucial in the later development of the ability to represent word problems by equations. The specific way in which students were moved toward thinking in terms of forward operations rather than inverse operations was by *separating the functional situation from the actual problem question* for a period of a few weeks (see also Owen & Sweller, 1989). For example, a problem such as the one above would be given in two parts. After the details of the two pay plans were presented, students would be asked questions such as, "If you sold 50 bags, how much would the first plan pay?...How much would the second plan pay?" and so on with varying numbers of bags sold. After students had set up a program to calculate the payments for a variable number of bags, they would then be given the actual problem question (that is, "For what number of bags sold will these two pay plans give exactly the same pay?"). After a few weeks of problem posing in two parts, students were presented with the entire problem, which they continued to represent by means of a program involving several one-operation lines. The teaching approach used in this study was effective in helping students develop a problem-solving method that they could formalize with apparent ease. In addition, problems such as the one above appeared to be no more difficult to represent and solve than traditional algebra word problems involving only one occurrence of the variable. This is in contrast to findings such as those of Filloy and Rojano (1985a) which showed students experienced considerable difficulty not only in setting up a single equation involving two occurrences of the variable but also in solving it by formal methods.

Fey (1989a) and Heid (Heid, 1988; Heid, Sheets, Matras, & Menasian, 1988) have developed a computer-intensive, "functional approach to problem solving," algebra curriculum that has been tested with entire classes of first-year algebra students. It includes the use of many different kinds of software, for example, curve-fitting programs, generators of tables of values, symbolic manipulators, and function graphers. The curriculum centers on "the use of these computer tools to (a) develop students' understanding of algebra concepts, and their ability to solve problems requiring algebra, before they master symbol manipulation techniques, and (b) make the concept of function a central organizing theme for theory, problem solving, and technique in algebra" (Heid et al., 1988, p. 2). An adaptation of a sample problem from their curriculum (Fey, 1989a) and the functional approach used in presenting it to students is the following:

> *Carla is planning a one-week vacation in the Poconos with her cousin, Kate. She borrows $195 from her mother to purchase a lawn mower so that she can earn money for the trip by cutting lawns. Let's say that she decides to charge $10 per lawn.*

1. Make a table charting her profit for 0, 5, 10, 15, 20, 25, and 30 lawns.
2. Now how many lawns must Carla mow in order to "break even"?
3. Write the calculations needed to compute Carla's net worth after having mowed 35 lawns.
4. Write a rule that explains how to calculate Carla's profit as a function of the number of lawns mowed.
   Profit = _____.
5. How many lawns does Carla have to mow in order to make a profit of $500 for her trip to the Poconos?

The above problem exemplifies one of several types used by Fey and Heid in their curriculum project. Samples of other types, as well as the ways in which various computer tools are used by students in their problem solving, can be found in Heid (1990). Fey and Heid have been testing their computer-intensive approach to the teaching of algebra for the past five years. Their evaluations have taken many forms and all of them have shown positive results. For example, one of their earlier evaluations (Heid, 1988) involved end-of-year interviews with both project and control pupils. In the interview tasks, the project pupils outperformed their counterparts on such mathematical modeling goals as constructing, interpreting, and linking representations; they also surpassed the pupils from the conventional classes in improvement of problem-solving abilities and did as well on a department final examination—a test requiring traditional algebraic manipulations.

Problems like the one from the Fey and Heid algebra curriculum have also been used by Demana and Leitzel (1988) in their instructional research with middle school students. A typical scenario from the Demana and Leitzel study might begin with a problem situation such as, "For some rectangle, the length is four centimeters more than the width." The researchers then work with the students in helping them to complete a table of numerical values that includes these headings: width, length, perimeter, and area (see Figure 17.5). The last line of the table contains a variable that describes the general case. Using the algebraic expression that has been formulated as a basis, Demana and Leitzel then go on to ask questions such as, "Find w if $w^2 + 4w = 45$" and "Find w if $4w + 8 = 41.6$." Since these numerical values are already in the table, students have all the information they need to "solve" the equations. The next step is to ask the students to solve for numerical values not in the table. For example, "Find w, the width of the rectangle, if $4w + 8 = 72$." Demana and Leitzel report that many

For some rectangles, the length of the rectangle is four centimeters more than the width. Complete the following table:			
Width (cm)	Length (cm)	Perimeter (cm)	Area (cm²)
1	5	12	5
5	9	28	45
8.4	12.4	41.6	104.16
12	16	56	192
w	$w + 4$	$4w + 8$	$w^2 + 4w$

FIGURE 17–5. Adapted from the table of values task used by Demana and Leitzel (1988, p. 65).

students develop their own solving methods, such as substituting different values of w into the equation until the left and right sides balance.

Open-ended Generalization Problems. The earlier historical account of the development of algebraic symbolism indicated that the use of letters to represent *unknowns* in equations occurred much earlier than the use of letters both to represent *givens* and to prove rules governing numerical relations. Harper (1987) offers some support for the statement that algebra students pass through the same stages in the development of their ability to handle algebraic symbolism as were evidenced in the history of algebraic symbolism. The kinds of problems Harper used were open-ended generalization problems. From his interviews of 144 secondary school pupils from Years 1 to 6, Harper found evidence of the three types of solutions identified in the history of mathematics by asking questions such as:

> *If you are given the sum and difference of any two numbers, show that you can always find out what the numbers are.*

With the rhetorical method, the pupil does not use algebraic symbolism but nevertheless specifies a procedure that is general (for example, "You divide the sum by 2 then divide the difference by 2; then to get the first number add the sum divided by 2 to the difference divided by 2; to get the second number take the difference divided by 2 away from the sum divided by 2." [Harper, 1987, p. 81]). With the Diophantine method, the pupil uses a letter (or letters) to represent an unknown quantity (for example, $x - y = 2$ and $x + y = 8$, solving for x and y) and states that the method can be applied to any numbers but does not use symbols for a general "given" quantity. With the Vietan method, the pupil uses letters for both unknown and given quantities:

$$
\begin{aligned}
\text{Let nos. } &= x \text{ and } y \\
m &= \text{sum of } x \text{ and } y \\
n &= \text{difference of } x \text{ and } y \\
\text{General equations}: \quad m &= x + y \\
n &= x - y \\
\text{Add together}: \quad m + n &= 2x \\
\ldots \text{Find } x \text{ and substitute back for } y.
\end{aligned}
$$

It is important to note that the Diophantine solution assumes that the same process can be carried on no matter which sum and difference are chosen and, thus, x is an unknown whose value is to be found. The Vietan solution, on the other hand, has a means of expressing any sum and any difference and of specifying the solution: The two numbers are $(m + n)/2$ and $(m - n)/2$. Not only is this solution general, but also it uses letters rather than conventional numerals. The Vietan approach provides a clear example of the use of algebraic equations as mathematical objects (on the other hand, it is not obvious that Harper's Diophantine method of solution requires operating on the equations $x + y = 8$ and $x - y = 2$ as objects).

In Year 1, all of the correct solutions were rhetorical. In Years 2 and 3, the rhetorical solutions continued to outnumber the other two types. It was only from Year 4 onward that the balance shifted in favor of, first, Diophantine and, then, Vietan solutions. Harper points out that these phenomena do not merely reflect pupils' classroom experiences and the normal structuring of algebra curricula because:

(i) pupils in the school were not encouraged to provide rhetorical type reponses in any of their work;
(ii) pupils were introduced to "letters for unknowns" and were expected to use these in problem solving activities during Year 1 and onwards;
(iii) pupils were using letters as "givens" in the context of functions and [as a tool] to make generalisations as early as Year 2;
(iv) simultaneous equations were introduced in Year 2. (Harper, 1987, p. 84)

Apparently students can verbalize a general solution to a problem prior to being able to generate a symbolic representation involving letters. Furthermore, Harper's study suggests that the use of the letter as a Diophantine "unknown" is more cognitively accessible than is the use of the letter as a "given," and that this latter usage is adopted by only a minority of more able pupils. A mere 28 of the 144 subjects used a Vietan type of response. The use of this approach rises dramatically in Year 6 of the secondary school, but mostly among the more mathematically-mature students—20 of the 28 who used a Vietan response were in Year 6. Harper notes that these findings accord well with the 8% success rate among Year 4 students of the CSMS study (Küchemann, 1978) on the question: "Which is larger, $2n$ or $n + 2$; why?" He suggests the existence of stages in the understanding of a literal term as variable and emphasizes that students use literal terms far earlier than they are able to conceptualize them as variables—that is, before they are able to perceive the general in the particular. These findings can also be interpreted to mean that procedural conceptions of literal terms precede structural ones.

Another study focusing on the use of letters to express the general was carried out by Chevallard and Conne (1984); they describe students' use of algebraic symbolism as a tool for proving rules governing numerical relations. As part of their study, the researchers presented an eighth-grade student with a sequence of questions that included the following:

Take three consecutive numbers. Now calculate the square of the middle one, subtract from it the product

of the other two.... Now do it with another three consecutive numbers.... Can you explain it with numbers?... Can you use algebra to explain it?

The student began with the three consecutive numbers 3, 4, and 5, which led to the calculation of $16 - 15$ to yield the result of 1. He then tried the numbers 10, 11, and 12, which led to the same result. When asked to explain what was happening, using algebra, he at first wrote $x^2 - yz = 1$, simply replacing all of the "given" numbers by letters. Having then realized that the use of only one letter would be better, the student proved the rule governing this numerical relation with the formulation, $x^2 - [(x + 1)(x - 1)] = 1$. Chevallard and Conne point out that this student, although only in the eighth grade, was one who had unusual facility with structural representations and their use as thinking tools.

Many students, however, have been found to be less successful in using algebraic symbolism as a tool with which to think about and to express general numerical relationships. Lee and Wheeler (1987), in their study of students' conceptions of generalization and justification, tested 354 grade 10 students and then interviewed 25 of them. One of the questions they presented to some of the students was the following (based on an example from Bell, 1976):

A girl multiplies a number by 5 and then adds 12. She then subtracts the original number and divides the result by 4. She notices that the answer she gets is 3 more than the number she started with. She says, "I think that would happen, whatever number I started with." Using algebra, show that she is right.

Of the 118 students who were given this problem, only nine set up the expression $(5x + 12 - x)/4$ and then algebraically worked it down to $x + 3$. Four of these nine students then went on to "demonstrate further" by substituting a couple of numerical values for x, suggesting that these students might still have been in transition between a procedural conception, which derives support from numerical operations, and a structural conception. Thirty-four others set up $(5x + 12 - x)/4 = x + 3$ and then proceeded to simplify the left side, yet did not base their conclusions on their algebraic work. Instead, they worked numerical examples and concluded from these examples. The interviews provided further evidence that students ignored their algebra and based all of their arguments on numerical evidence. Lee and Wheeler's findings suggest that the majority of these students lacked the kind of structural conception of algebraic expressions and equations that would permit them to use these objects as a notational tool for proving mathematical relations.

Another problem used in the Lee and Wheeler study was the following one: "Show, using algebra, that the sum of two consecutive numbers is always an odd number." Although the way the problem was formulated is different from Chevallard and Conne's, in that the latter initially asked students to work with numerical examples, it makes the same demands as were eventually required by the task of the Chevallard and Conne study. Lee and Wheeler report that only 7% of their subjects succeeded with this problem. Although the interviews showed that students *did* appreciate an algebraic demonstration when they or someone else produced it, they seemed far happier with their own numerical examples.

Rizzuti and Confrey (1988), in their report of interviews with a second-year university student majoring in nutritional sciences, describe a somewhat similar experience that illustrates the force of numerical examples. The environment was a problem-solving one involving exponential functions; the student was trying to recognize a pattern in her successive computations for calculating depreciation. Generalizing and factoring were found to be less useful than a numerical example in her final realization of the solution to the exponential problem.

Lee and Wheeler point out that "formulating the algebraic generalisation was not a major problem for the students who chose to do so; using it and appreciating it as a general statement was where these students failed" (Lee & Wheeler, 1987, p. 149). However, most studies that have investigated students' ability to generate symbolic representations suggest that algebra students generally experience severe difficulty not only in "using it and appreciating it as a general statement," as Lee and Wheeler have remarked, but also in generating an appropriate algebraic generalization. In a teaching experiment designed specifically to encourage the use of a letter as generalized number, Booth (1982, 1983) found strong resistance on the part of students. She suggests that "the attainment of this level of conceptualization is related to the development of 'higher-order' cognitive structures" (Booth, 1984, p. 88). Booth (1983) also pointed out that students may respond correctly to items requiring the use of certain notation or conventions and yet be unable to discriminate between correct and incorrect representations. This suggests, according to Booth, that the understanding of notation may proceed in stages—a finding echoed by Harper (1987) in his study of students' use of letters in representing certain kinds of problem situations. It is noteworthy that, in many of the above studies, the proportion of secondary school students who demonstrate evidence of a structural conception of expressions and equations tends to fall between 7% and 10%. The challenge for algebra instructors is to find a means of making the structural aspects of algebra accessible to a greater percentage of students.

A study carried out by Peck and Jencks (1988) with fifth graders illustrates an approach whereby algebra can "arise as a byproduct of making arithmetic sensible" (p. 85). The research involved a series of successful classroom interventions on the teaching of multiplication. Rather than simply have the children compute the answers to various multiplications, the teacher asked them to use graph paper and small masking strips to show, say, "2 times 3" (three squares two times). Later, special graph paper heavily marked on every tenth line was used to model multiplications involving two-digit numbers, such as 25×27. Experience with working out the product of these numbers using graph paper led eventually to written statements such as $(\Box + 5) \cdot (\Box + 7) = \Box^2 + 12\Box + 35$. Peck and Jencks emphasize that "if arithmetic becomes completely sensible to children and becomes a tool for their thinking, the decisions which make algebra sensible flow naturally from it" (p. 85). Davis (1985) has also described approaches used successfully with elementary school children in which they (a) think about the numerical relations of a situation, (b) discuss them explicitly in simple everyday language, and (c) eventually learn to represent them with letters or other nonmisleading notation.

Functions and Graphs

When Freudenthal (1973, 1982) characterized functions, he emphasized the notion of dependency: "Our world is not a calcified relational system but a realm of change, a realm of variable objects depending on each other; functions is a special kind of dependences, that is, between variables which are distinguished as dependent and independent" (Freudenthal, 1982, p. 12). Unfortunately, as Shuard and Neill (1977) point out, the idea of functional dependence has been totally eliminated from the current definition of function. In almost all algebra textbooks a function is now defined as a relation between members of two sets (not necessarily numerical) or members of the same set, such that each member of the domain has only one image. Some modern definitions do include mention of a rule; however, the notion of dependency is gone. Thus, as the research cited in this section shows, the teaching of functions in algebra classes tends to emphasize structural rather than procedural interpretations

Some of the effects of using set-theoretic definitions in the teaching of functions are illustrated by the findings of a study by Even (1988). She sent a questionnaire on the relation between functions and equations to prospective secondary mathematics teachers in their last year at eight Midwestern American universities. Respondents were asked to write a definition of function, to indicate how functions and equations are related to each other, and to find the number of real solutions to a quadratic equation (given a positive value and a negative value of the quadratic expression). The instruction that these prospective teachers had received in both their secondary school and college mathematics classes had been based on material found in standard, modern algebra textbooks. From the 152 responses received, Even concluded that (a) these prospective teachers hold a limited view of functions as equations only, (b) they do not know how to make sense of the modern definition of function, and (c) they cannot relate solutions of equations to values of corresponding functions in a graphical representation.

Verstappen (1982) distinguishes three categories for recording functional relations using mathematical language: (a) geometric—schemes, diagrams, histograms, graphs, drawings; (b) arithmetical—numbers, tables, ordered pairs; and (c) algebraic—letter symbols, formulas, mappings. Since functions are usually introduced in algebra classes by means of a formal set-theoretic definition, that is, as a many-to-one correspondence between elements of a domain and range, the representations that are generally invoked initially are mapping diagrams, equations, and ordered pairs. These representations are then usually extended to include tables of values and Cartesian graphs. Thus, the teaching of functions includes representations denoting various degrees of procedurality and structure. However, as studies discussed below show, students tend to bypass the more formal structural definitions and representations and interpret functions as procedures for computing one magnitude by means of another.

Sfard (1987) administered a questionnaire to sixty 16- and 18-year-olds, who were well-acquainted with the notion of function and with its formal structural definition, to find out whether they conceived of functions procedurally or structurally. The majority of the pupils questioned conceived of functions as a process rather than as a static construct. In a second phase of the study involving ninety-six 14- to 17-year-olds, students were asked to translate four simple word problems into equations and also to provide verbal prescriptions (algorithms) for calculating the solutions to similar problems. They succeeded much better with the verbal prescriptions than with the construction of equations. These findings support the results of the study by Soloway, Lochhead, and Clement (1982) that showed that students can cope with translating a word problem into an "equation" when that equation is in the form of a short computer program (algorithm) specifying how to compute the value of one variable based on another.

Even though students are formally introduced to functions in their algebra classes, they have already had considerable intuitive experience with function machines and other input-output representations in their arithmetic classes. Their earlier work with simple formulas, such as $P = 4 \times S$ for the perimeter of a square, also provides a basis for understanding functions in their algebra classes. Furthermore, functions are taught in science classes too—but as dependency relations between variables. However, the teaching of functions in algebra courses does not appear to capitalize on any of this prior experience. That students are exposed to two different definitions of functions in their science and algebra classes, with a possible interaction between the two of them, was of interest to Markovits, Eylon, and Bruckheimer (1983, 1986).

Markovits et al. (1983) investigated the effect of context (mathematics vs. science) on the results from tasks requiring students to draw the graph of a function that would pass through given noncollinear points in a Cartesian plane. Two groups of ninth graders—a high-ability group and a low-ability group—were tested. Both groups had already had experience in both their mathematics and science classes with graphing functions. Results indicated that most students provided a linear response, that is, almost all the graphs were composed of straight-line segments. Furthermore, the context had an effect on the success rate. Of those students who provided nonlinear responses, high-ability students were more successful with the pure mathematics problems than they were with the problems embedded in a scientific context. This trend was reversed for the low-ability group. These findings suggest that, for the high-ability group, the transition to function as a structural object may have been more advanced than it was for the low-ability group for whom function was still being perceived as a process.

Other results of the Markovits et al. (1986) study are also worth noting. They found that:

i. Whatever the particular nature of the question, three types of function caused difficulty: the constant function, a function defined piecewise, and a function represented by a discrete set of points.
ii. There was a general neglect of domain and range, whether attention to them was explicitly required by the question or not.
iii. In both the algebraic and the graphical form, the concept and representation of images and preimages was only partially understood.
iv. The variety of examples in the students' repertoire of functions was limited in both the graphical and algebraic form, but more especially in the latter.
v. Transfer from graphical to algebraic form was more difficult than vice versa, and both were conditioned by the limited repertoire noted in the previous point.
vi. "Complexity" of technical manipulations inhibited success.
vii. When examples of functions were required, there was an excessive adherence to linearity.
viii. Many of the above difficulties were clearly in evidence in the questions on functions defined by constraints. (p. 24)

Their finding that many ninth graders "hold a linear prototypical image of functions" (Markovits et al., 1983, p. 276) is of pedagogical interest. With the advent of technology, it is no longer necessary to restrict instruction in graphing to simple, linear functions.

Another study illustrating procedural-structural preferences in the representation of functional relations was carried out by Dreyfus and Eisenberg (1981, 1982). They investigated the intuitive bases for concepts of function among 440 sixth- to ninth-grade students. They asked questions in both concrete and abstract contexts on image, preimage, growth, extrema, and slope in three representational settings—graph, diagram, and table of ordered pairs. They found that high-ability students preferred the graphical setting for all questions, whereas low-ability students preferred the tabular setting. Though neither the graphical setting nor the tabular setting specifies directly how to compute one magnitude by means of another, the findings of this study suggest that low-ability students may be able to derive this information more easily from tabular settings than from graphical settings.

Further evidence that may support this suggestion is provided by Kaput (1988) from a study that involved a computer game modeled after the W. W. Sawyer/Robert Davis activity *Guess My Rule* (Barclay, 1985). In this activity, the computer has a "secret" function that the student is to guess on the basis of the computer's response to the student's numerical inputs. The student can choose the form of the computer's feedback—either numerical or graphical. Kaput found that the students, who were in first- and second-year algebra courses, preferred a numerical form of feedback (that is, a table of values) rather than a graphical one; significant teaching and prompting were needed to get them to use the graphically represented information.

Yerushalmy (1988) has investigated whether the graphical feedback provided by a computer can be used as a means of improving students' ability to operate with algebraic expressions. In an exploratory study that involved observations of seven mixed-ability students from seventh to ninth grade, she made use of the software program RESOLVER—a combined symbolic and graphical environment allowing the user to transform algebraic expressions and providing feedback on the correctness of the transformation. For each algebraic transformation carried out with RESOLVER, a parallel (simultaneous) display of three graphs is available: (a) a graph of the original expression, (b) a graph of the current transformed expression, and (c) a graph of the difference of the two expressions. Since any legitimate transformation of an algebraic expression does not affect the graph of the expression, the difference graph is intended to provide both qualitative and quantitative information about the correctness of each step. Although the graphical feedback was convincing to the students who were already familiar with graphs, Yerushalmy found it was much less effective than she had expected. She suggests that students' lack of prior experience with Cartesian graphs prevented them from interpreting the information that was contained in the graphs. She concluded that prior learning of graphs and functions is a vital prerequisite to using graphical feedback as a means of evaluating the correctness of algebraic transformations. Yerushalmy also emphasizes that, since the students tended to misinterpret the information presented by the graphs—a phenomenon that has also been well documented by Goldenberg (1988)—further research with very carefully planned teacher intervention is needed in this area of computer-supported learning.

The learning of graphs presents many difficulties to algebra students. Swan (1982) points out that students are routinely asked to generate tables of values satisfying algebraic equations in two variables, plot points on a suitably-scaled Cartesian graph, and read the coordinates of points off a graph, sometimes with the aim of solving an equation or system of equations. According to Swan, the consequences of emphasizing exclusively these skills are that students lose sight of the meaning of the task, rarely meet graphs other than those of straight lines, and get little practice at interpreting graphs in terms of realistic situations. The findings of several studies (for example, Bell & Janvier, 1981; Clement, 1985; Herscovics, 1980, 1989; Kerslake, 1977, 1981; Leinhardt, Zaslavsky, & Stein, 1990; Ponte, 1984) that have investigated students' understanding of the graphs of functions support this statement.

For example, Kerslake (1981) presented the 13-, 14-, and 15-year-olds of the CSMS assessment with a task of describing which of the graphs of Figure 17.6 represent journeys. Only 14% of the 13-year-olds and 25% of the 15-year-olds succeeded—despite classroom experience with travel graphs. Many described the third graph of Figure 17.6 in terms of "climbing up a mountain" or "going up, down, and then up again," illustrating students' confusion of the graph with a "picture" of the situation. (See also the classic race-track example of Janvier, 1978.) Kerslake found evidence to suggest that graphs related to real-world situations are no easier for students to interpret than graphs that are related merely to symbolic, decontextualized equations. Many students seemed unable to relate numerical data to the coordinate points and axes of a Cartesian plane. Similar difficulty has been identified regarding the number line, notably in dealing with scales (Vergnaud & Errecalde, 1980); nevertheless, Rogalski (1985) has reported that screen feedback in a computer environment can be very helpful in developing an understanding of scales and intervals for both the number line and the coordinate plane.

Further indication of students' difficulties with graphs is provided by another British study, the Assessment of Performance Unit (1980), which tested a random sample of 14,000 sixteen-year-old students in England, Wales, and Northern Ireland. This study reported that only 22% of its sample responded correctly to the question, "Which one of the following could be the graph of $y = (x - 1)(x + 4)$?" and only 9%, to the question, "The graph shown is a representation of the function $f(x)$ where $f(x) = x(a - x)$; what is the numerical value of a?" The computer and graphing calculator have great potential to help develop students' skills in this area by providing an opportunity to discover not only what parameters are but also the effects of changing them, as well as experience in drawing whole families of curves on the same set of axes (for example, Dreyfus & Halevi, 1988; Waits & Demana, 1988). However, computer graphing programs and the graphing calculator are not to be viewed simply as instructional panaceas. As shown by the research of Schoenfeld, Smith, and Arcavi (in press), who described in detail the learning of their case-study subject in the computer-graphing environment, GRAPHER, and by the related research of Moschkovich (1990), who observed two classes of 9th and 10th graders using not only graphing calculators but also Superplot and Green Globs, technology-enhanced learning environments do not of themselves help students decide which features of lines or equations are the relevant ones to focus on nor how to describe their observations or conclusions. These two studies point out, nevertheless, that the use of such technology, in conjunction with the necessary instructional support of a capable teacher, can help students to objectify graphs and their related equations by encouraging students to operate with these mathematical entities and to talk about them.

Other studies have also reported that students encounter obstacles with formal, functional notation. One of the tasks of the second mathematics assessment of the NAEP (Carpenter, Corbitt, Kepner, Lindquist, & Reys, 1981) asked students to evaluate $a + 7$ when $a = 5$ and then $f(5)$ when $f(a) = a + 7$. Although nearly all students (98%) could correctly answer the first question, only 65% were successful with the second one. Other studies (for example, Markovits et al., 1983; Thomas, 1969; Vinner, 1983) have also shown that formal notation for functions introduces additional difficulties for the algebra student.

Studies discussed above pointed out that students derive information from tabular representations more easily than from graphical ones. However, if the pupil must translate this information into an expression or equation, the task is considerably more complex. For example, an NAEP task involved completing the table shown in Figure 17.7 (Carpenter et al., 1981). Most of the students with one or two years of algebra could recognize the pattern—adding 7—from the given numerical values (success rates of 69% and 81%, respectively). However, they were less successful when asked to derive from the same table the value of y when $x = n$ (success rates of 41% and 58%, respectively).

The generation of algebraic representations in a functions environment has also been researched by Greeno (1988) and

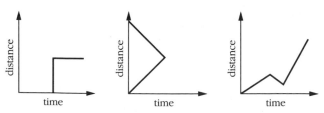

FIGURE 17–6. Adaptation of the CSMS graph task asking subjects to indicate which of the graphs represent journeys and to describe what happens in each case (from Kerslake, 1981, p. 128).

X	1	3	4	7	n
Y	8		11	14	

FIGURE 17–7. Adaptation of the NAEP task asking subjects to give the values of y when x = 3 and when x = n (from Carpenter et al., 1981, p. 68).

colleagues. For example, Gurtner (1989) investigated how and when general algebraic expressions are spontaneously used by students in a context of situated learning involving a winch box. He worked with small groups of 7th, 9th, and 11th graders. He found that all of the students comprehended the key features of linear functions in the situation, but that only the 11th graders understood that algebraic expressions involving variables could be given as solutions for problems where lack of specific information made numerical answers impossible.

Recently Schwarz (1988, 1989) has been developing and evaluating the Triple Representation Model curriculum (TRM—the three representations implemented are the algebraic, graphical, and tabular), a computer-based environment that includes most of the topics taught in introductory units on functions. In addition, TRM has been designed to help students avoid some of the conceptual obstacles usually encountered in traditional first-year algebra courses, such as their exclusive attachment to linearity (Markovits et al., 1986), their difficulty in transferring information between representations, their lack of a dynamic conception of functions, and their inability to view a function as a mathematical object. The curriculum was tested in a ninth-grade class in which computers were available in a ratio of one per pair of students. An aspect of student learning that Schwarz and his colleagues (Schwarz, Dreyfus, & Bruckheimer, 1990) have reported in detail concerns the issue of student attachment to linearity. They compare their findings with those reported by Markovits et al. (1986) and Karplus (1979). Schwarz, Dreyfus, and Bruckheimer found that 69% of their subjects reached partial or full curved-line reasoning (that is, an infinite number of curves can be drawn to pass through two given points), as opposed to 18% in the Karplus study and 17% in the Markovits study. According to Schwarz and his colleagues, full curved-line reasoning entails a complete integration of structural and procedural knowledge. They attribute their successful results to the TRM curriculum, which includes extensive experience in the qualitative representation of functions by means of graphs. However, they also point out that the teachers were continuously guided during their teaching in didactical and technological aspects of TRM and suggest that "a great effort has to be put into pedagogical engineering in order to enable teachers to master composite systems which contain group work techniques, computer sessions, and conventional lessons" (p. 20).

In contrast to the positive findings of the Schwarz study, much of the research on functions has, as we have seen, found that students are ill-equipped to interpret graphical representa-

tions from a structural perspective. Furthermore, their understanding of functions with respect to algebraic forms of representation tends to be clearly biased in favor of procedural conceptions. That teachers may not be sensitive to the seeming need of students to treat functions as procedures for a considerable period of time is shown by the results of a study by Dreyfus and Vinner (Dreyfus & Vinner, 1982; Vinner & Dreyfus, 1989). Five groups of students were tested: (a) college students taking a low-level course in mathematics, (b) college students taking intermediate-level mathematics courses, (c) college students in high-level mathematics courses, (d) college mathematics majors, and (e) junior high school mathematics teachers. One of the questions asked was, "What is a function, in your opinion?" Responses to this question indicated that, despite the emphasis on a set-theoretic definition of the function concept taught in their mathematics courses, the great majority of students did not accept it. The first three groups of students overwhelmingly (73%, 70%, and 63%, respectively) produced variations on a *procedural* definition (a function is a dependence relation, a rule, an operation, a formula, and so on). Even the mathematics majors were about equally split between a *procedural* definition and a formal, *structural* definition. Only the teachers showed a strong preference (73%) for a formal interpretation. If we can infer from this finding that most high school mathematics teachers have a structural conception of function, and that this is what they attempt to teach, the research cited in this section suggests that more instructional effort should be placed into helping students bridge the gap between procedural and structural conceptions of functions. Moreover, the research of Slovin (1990) on teachers' beliefs and attitudes suggests that a formal, structural approach to the teaching of algebra begins long before instructors teach the concept of function. From interviews with teachers prior to an in-service course devoted to teaching algebra through a problem-solving approach (Rachlin, Matsumoto, & Wada, 1988), she observed that some of the algebra teachers seemed to feel the need to push students quickly toward "efficient" (that is, structural) methods. In fact, when the in-service instructor began to describe the advantages for students of the guess-and-test method, one teacher confessed to thinking that the method was "pretty primitive."

In light of the emphasis throughout this chapter on the procedural-structural model of algebra and of algebra learning developed by Sfard (1991), it is appropriate that the final study reviewed in this section is one that she carried out. Her goal was to help students make the transition from procedural to structural conceptions of function. Previous studies cited in this chapter have shown that, despite an "object-oriented" way of teaching, a fully developed structural conception of algebraic objects seems rather rare in high school students; students' conceptions tend to be more procedural than structural. Sfard (1989) designed a study on functions in which students would be taught initially by an operational (the term used by Sfard, which corresponds to my use of the term *procedural*) approach that would gradually be transformed into a structural approach. The principal goal was to see whether a structural conception could be provoked in students by teaching that adhered faithfully to an operational-structural sequence.

The concept of function was taught to a class of young adults as part of their 60-hour programming course on algorithms and computability. Initially the term *function* was used almost synonymously with *algorithm*; later, *function* was used to name the *product* of an algorithm. The first attempt at separating functions from algorithms was made only after the set of already-known algorithms and the resulting set of functions had been broadened several times to include recursive and "split domain" calculations, among others. Different methods of constructing functions from other functions were discussed; thus, the view of a function as a self-contained entity that could serve as a building block for other entities was gradually promoted. The input-output description of function was replaced by the abstract Bourbaki definition only after a long period during which students' attention was focused on the static products of different algorithms rather than on the algorithms themselves. This final generalization led to the question of a noncomputable (that is, non-algorithmic) function. The last problem was expected to be the ultimate trigger for reification.

By observing the classroom during the entire course, Sfard found that the first attempts at transition to the structural approach met with resistance and lack of understanding: Many students could not cope either with sets of functions or with general definitions of operations performed on functions. The difficulty diminished with time but it did not disappear completely. Sfard notes that when the students were asked to describe the set of recursive functions, almost half the group indicated a difficulty with treating functions as building blocks for other functions. Not surprisingly, according to Sfard, the idea of a noncomputable function, when mentioned explicitly, evoked astonishment and opposition.

At the end of the course, Sfard administered a questionnaire. Responses showed that substantial progress toward structural conceptions had been achieved. Nevertheless, she notes that "our attempt to promote the structural conception cannot be regarded as fully successful" (Sfard, 1989, p. 158) and conjectures that "reification is inherently so difficult that there may be students for whom the structural conception will remain practically out of reach whatever the teaching method" (p. 158). Despite Sfard's seeming disappointment at the size of the gap between the efforts invested and the progress made, the results of the study do suggest that an approach to functions that is based primarily on programming activities, though valuable for emphasizing procedural aspects, may be too closely tied to computability to permit a full-blown structural conception of functions.

CONCLUDING REMARKS

In this chapter, I have adopted an historical-epistemological perspective and have attempted to reconceptualize much of the existing algebra research in terms of a procedural-structural model of algebra and of algebra learning and teaching. In so doing, I hope to have created a framework that permits better understanding of the difficulties students have in learning algebra. In my concluding remarks, I take up again the three factors that I suggested in the introduction were potential contributors to these difficulties—content, teaching, and learning. However, the order in which I synthesize these points is the reverse of that followed in the chapter. My reason for this is primarily that

the bulk of the research that was reviewed bears upon learning. This research, in turn, has implications for teaching; and when the research on teaching is taken into consideration, the findings from both of these areas lead to my suggestions regarding issues of content. While discussing each of these points, I briefly mention a few areas where further research is warranted.

Learning

Learning is, in a sense, the easiest factor to deal with, for most of the research in school algebra has focused on this issue. From the findings of the research presented in this chapter, two overriding themes emerge: the accessibility of procedural over structural interpretations and the difficulty in acquiring a structural conception of algebra. (The ways in which the terms *procedural* and *structural* have been used in this chapter are explained on page 392.)

The accessibility of procedural interpretations was seen in, for example, the greater ease with which students generate verbal prescriptions or computer programs for calculating solutions than equations to represent the same relations, the effortlessness with which they create functional input-output representations of problem situations, the way that certain programming environments seem to facilitate a view of letters as numbers, students' preference for numerical justifications when generalizing relationships, various indications of concept images of functions that reflect notions of dependence, and so on.

The difficulty that students experience with understanding the structure of algebra, even its most elementary aspects such as are found in high school textbooks, was exemplified by their early attempts to convert expressions into equations in order to have a representation that includes a result, the unsystematic and strategic errors they committed while simplifying expressions, their resistance to operating on an equation as an object as shown by their not using the solving procedure of "doing the same thing to both sides," their not treating the equal sign as a symbol of symmetry, their widespread inability both to consider a letter as a variable or as a given and to translate word problems into equations, their difficulty in seeing the "hidden" structure of equations, their non-use of algebra as a tool for proving numerical relations, their general inability to convert "process" interpretations of algebraic entities into "object" interpretations, and so on.

The overall conclusion that emerges from an examination of the findings of algebra learning research is that the majority of students do not acquire any real sense of the structural aspects of algebra. The research findings in this chapter, which are organized from the early topics that are taught to the later topics, suggest that most students never reach the structural part of the procedural-structural cycle. At worst, they memorize a pseudo-structural content; at best, they develop and continue to rely on procedural conceptions. However, there are signs of promise in this otherwise bleak picture of algebra learning. Some studies with a teaching component have shown that students can develop structural conceptions of certain aspects of algebra if they are provided with experiences involving, for example, the field properties in both arithmetic and algebraic settings, an introduction to equations containing algebraic expressions on both sides, the translation of word problems into not just one equation but several possible equations, the recording of numerical generalizations using unambiguous notation (even in elementary school), the solving of problems that contain only literal data, the manipulation of parameters of functional equations with the aid of computer graphing software that can easily display the graphs of families of functions, and so on.

The evidence from these latter studies suggests that, first, greater effort ought to be invested in classroom instruction toward, creating a solid base for developing the structural conceptions of algebra by spending considerably more time with procedural conceptions. Sfard (1991) emphasizes that a lengthy period of experience is required before procedural conceptions can be transformed into structural conceptions. Research indicates that procedural, input-output, conceptions are accessible and that strength in this area could serve to make algebraic activity more understandable and more meaningful. Second, the transition from procedural to structural conceptions would seem to require more explicit attention than it currently receives in most textbooks. Only a few studies have addressed the issue of how to facilitate this transition. It should be noted, however, that the acquisition of structural conceptions by which expressions, equations, and functions are conceived as objects and are operated on as objects does not eliminate the continued need for the procedural conception which served as a foundation for the construction of the structural conception; as Sfard points out, both play important roles in mathematical activity. But even fewer studies have addressed this last issue of the role and interaction of both conceptions in doing algebra. Lee and Wheeler (1989) have shown that students behave as if arithmetic and algebra were two closed systems. The challenge to classroom instruction is not only to build upon the arithmetic-to-algebra connection but also to keep alive the algebra-to-arithmetic connection, that is, to develop the abilities to move back and forth between the procedural and structural conceptions and to see the advantages of being able to choose one perspective or the other—depending on the task at hand.

Of all the areas where further research could profitably be carried out, the one that seems to stand out as being clearly in need of attention is that of finding ways to develop structural conceptions in students. An example of such research is the last study reported in this review, the one in which Sfard (1989) attempted to provoke a structural conception of function in students by means of a faithful adherence to a procedural-structural sequence of activities. Even though she was only partly successful as far as achieving the desired learning in her students, this type of study potentially can advance our knowledge of how students can be helped to create and to operate upon those entities that we perceive to be the objects of algebra.

Teaching

As we have seen, the amount of research that has been carried out with algebra teachers is minimal. Findings from these few studies suggest that their conceptions are structural and that this is the approach that they favor in their teaching. Despite the nature of algebra teaching, students seem rarely able to develop full-fledged structural conceptions.

Teachers who would like to consider changing their structural teaching approaches and not to deliver the material as it is currently developed and sequenced in most textbooks are obliged to look elsewhere for guidance. However, such guidance is not easily available. First, there is no readily accessible form of communication by which teachers can learn of research findings and how they might be applied to classroom instruction. Some of the journals of mathematics education research have as specific goals to disseminate cognitive findings and discuss their instructional implications; however, many of the published articles appear to be written primarily for other researchers. Second, teachers have very little time to hunt down research results in nearby university libraries. Yet this is what must be done in most cases. In-service courses on algebra teaching and collaboration with researchers, though effective in raising teachers' awareness levels of other approaches worth trying in their classrooms, are often even more difficult to find.

Since many teachers report teaching primarily what is in the textbook, a step toward ensuring that procedural conceptions are both developed and carefully crafted into structural conceptions would be to have textbooks reflect this stance. Nevertheless, the remarks made earlier in the section on Teaching Perspectives still apply. No matter what content is prescribed by the textbooks, there is still an enormous gap in the existing research literature on teaching regarding how algebra teachers interpret and deliver that content. This gap is possibly one of the areas most in need of research attention.

Content

If students experience difficulty with the algebra that is taught by their teachers and teachers teach the algebra that is presented in the textbooks, then a major factor contributing to the difficulty with algebra would appear to lie, by default, with the content of algebra as laid out in the textbooks. Presently, the content of most algebra textbooks does not incorporate a procedural-structural perspective on student learning of mathematics; nor does it appear to reflect how algebra evolved historically. A recently published series of secondary school algebra texts in England (NMP, 1987), to which both Harper and Küchemann contributed, illustrates how an algebra course can be based on the idea of successively developing the notions of letters as specific unknowns and then as givens in a gradually evolving sequence from the procedural to the structural. The approach adopted in this series suggests how algebraic content can be presented so as to avoid some of the pitfalls associated with introducing algebra according to its most recently developed formal definitions and structural perspectives (Boas, 1981). However, to my knowledge, the long-term cognitive impact of these textbooks has not yet begun to be researched. Nevertheless, research that has been carried out with certain elements of this approach has produced some promising results (for example, Bell, 1988; Bell et al., 1987).

Another aspect to be considered by algebra curriculum decision-makers and textbook authors at this time, and one that I have not discussed directly, is the role of technology. Proponents of reforming the algebra curriculum in such a way as to take advantage of current technology tend to de-emphasize symbol manipulation and stress procedural interpretations (for example, Fey, 1989a, 1989b). One of the rationales used to defend the excising of major portions of the current content devoted to the simplification of algebraic expressions is the presence of computerized symbol-manipulation tools that can do the same work. However, as Maurer (1987) points out:

You can use those systems to transform expressions from one form to another in some cases, but you have to be clear and persistent about what you want to do.... Perhaps students can learn without a great deal of specific pencil and paper work. Perhaps they can learn by seeing a lot of examples solved for them by a computer, but it's not clear to me how. (p. 248)

In other words, it is not obvious how the use of symbol manipulators in the early stages of learning algebra can help students develop a structural conception of algebraic expressions. This is a question for future research.

I have argued for extending the content of algebra by adding more emphasis to activities that both promote the development of procedural interpretations and make explicit the transition from procedural to structural conceptions. Increasing the prominence given to procedural notions is in accord with proposals coming from those who would like to see technology used more extensively in algebra courses. Note that proposals of this nature have also been put forward by the National Council of Teachers of Mathematics (1989) in its recent *Curriculum and Evaluation Standards for School Mathematics*. Certainly the evidence from studies that have involved the use of the computer as tool in a functional, input-output, approach to problem solving is a compelling indication of how far students can go with procedural interpretations. However, with the exception of studies such as the one carried out by Schwarz (1989), the potential of evolving a procedural conception into a structural one in a technologically-supported curriculum is still far from evident. Kaput (1989) has described in detail how technology can be used to assist students in forging links among the various two-variable notational systems: tables of data, coordinate graphs, and equations. But, except for graphs, which as with most geometric objects permit holistic and structural perceptions, most of the cognitive activity involved in working with these notations remains at the procedural level (Kieran, in press). The transition between the two conceptions is still largely implicit.

Both procedural and structural conceptions are important in the teaching and learning of algebra. As stated earlier, even though mathematicians profess to rely on structural conceptions, they also use procedural ones (Lampert, 1988). Furthermore, we have seen how algebra developed historically along a procedural-structural cycle. There is some strong evidence to support the notion that students pass through the same cycle in their learning of algebra. This review of the research literature has shown that both the teaching and the content of algebra emphasize structural considerations. However, research has shown that most students do not reach this goal. Thus, it has been suggested that both the content and the teaching of algebra ought to be reconsidered in the light of a procedural-structural dynamic of learning mathematics. However, the research base on which these suggestions rest is still rather slender. Nevertheless, recent findings indicate that we appear to be heading in the right direction. Continued research on the perspective taken in this chapter will provide data on which to base the eventual decision.

References

Anderson, J. R., Boyle, C. F., Corbett, A. T., & Lewis, M. W. (1990). Cognitive modeling and intelligent tutoring. *Artificial Intelligence, 42,* 7–49.

Assessment of Performance Unit. (1980). *Mathematical development, secondary survey, report no. 1.* London: HMSO.

Barclay, T. (1985). *Guess my rule* [Computer program]. Pleasantville, NY: HRM Software.

Behr, M., Erlwanger, S., & Nichols, E. (1976). *How children view equality sentences* (PMDC Technical Report No. 3). Tallahassee: Florida State University. (ERIC Document Reproduction Service No. ED144802)

Bell, A. (1976). A study of pupils' proof-explanations in mathematical situations. *Educational Studies in Mathematics, 7,* 23–40.

Bell, A. (1988). Algebra—choices in curriculum design. In A. Borbas (Ed.), *Proceedings of the Twelfth International Conference for the Psychology of Mathematics Education* (Vol. I, pp. 147–153). Veszprem, Hungary: OOK.

Bell, A., & Janvier, C. (1981). The interpretation of graphs representing situations. *For the Learning of Mathematics, 2*(1), 34-42.

Bell, A., Malone, J. A., & Taylor, P. C. (1987). *Algebra—an exploratory teaching experiment.* Nottingham, England: Shell Centre for Mathematical Education.

Bell, A., O'Brien, D., & Shiu, C. (1980). Designing teaching in the light of research on understanding. In R. Karplus (Ed.), *Proceedings of the Fourth International Conference for the Psychology of Mathematics Education* (pp. 119–125). Berkeley: University of California.

Bernard, J. E., & Cohen, M. P. (1988). An integration of equation-solving methods into a developmental learning sequence. In A. F. Coxford (Ed.), *The ideas of algebra, K–12* (1988 Yearbook, pp. 97–111). Reston, VA: National Council of Teachers of Mathematics.

Boas, R. P. (1981). Can we make mathematics intelligible? *American Mathematical Monthly, 88,* 727–731.

Booth, L. R. (1981a). Strategies and errors in generalised arithmetic. In C. Comiti & G. Vergnaud (Eds.), *Proceedings of the Fifth International Conference for the Psychology of Mathematics Education* (pp. 140–146). Grenoble, France: Laboratoire I.M.A.G.

Booth, L. R. (1981b). Child methods in secondary mathematics. *Educational Studies in Mathematics, 12,* 29–41.

Booth, L. R. (1982). Developing a teaching module in beginning algebra. In A. Vermandel (Ed.), *Proceedings of the Sixth International Conference for the Psychology of Mathematics Education* (pp. 280–285). Antwerp, Belgium: Universitaire Instelling Antwerpen.

Booth, L. R. (1983). A diagnostic teaching programme in elementary algebra: Results and implications. In R. Hershkowitz (Ed.), *Proceedings of the Seventh International Conference for the Psychology of Mathematics Education* (pp. 307–312). Rehovot, Israel: Weizmann Institute of Science.

Booth, L. R. (1984). *Algebra: Children's strategies and errors.* Windsor, UK: NFER-Nelson.

Booth, L. R. (1988). Children's difficulties in beginning algebra. In A. F. Coxford (Ed.), *The ideas of algebra, K–12* (1988 Yearbook, pp. 20–32). Reston, VA: National Council of Teachers of Mathematics.

Briars, D. J., & Larkin, J. H. (1984). An integrated model of skill in solving elementary word problems. *Cognition and Instruction, 1,* 245–296.

Brown, C. A., Carpenter, T. P., Kouba, V. L., Lindquist, M. M., Silver, E. A., & Swafford, J. O. (1988). Secondary school results for the fourth NAEP mathematics assessment: Algebra, geometry, mathematical methods, and attitudes. *Mathematics Teacher, 81,* 337–347.

Brown, J. S. (1985). Process versus product: A perspective on tools for communal and informal electronic learning. *Journal of Educational Computing Research, 1,* 179–201.

Byers, V., & Herscovics, N. (1977). Understanding school mathematics. *Mathematics Teaching, 81,* 24–27.

Carpenter, T. P., Corbitt, M. K., Kepner, H. S., Jr., Lindquist, M. M., & Reys, R. E. (1981). *Results from the second mathematics assessment of the National Assessment of Educational Progress.* Reston, VA: National Council of Teachers of Mathematics.

Carpenter, T. P., & Moser, J. M (1982). The development of addition and subtraction problem-solving skills. In T. P. Carpenter, J. M. Moser, & T. A. Romberg (Eds.), *Addition and subtraction: A cognitive perspective* (pp. 9–24). Hillsdale, NJ: Lawrence Erlbaum.

Carry, L. R., Lewis, C., & Bernard, J. (1980). *Psychology of equation solving: An information processing study* (Final Technical Report). Austin: University of Texas at Austin, Department of Curriculum and Instruction.

Cauzinille-Marmeche, E., Mathieu, J., & Resnick, L. B. (1984, April). *Children's understanding of algebraic and arithmetic expressions.* Paper presented at the annual meeting of the American Educational Research Association, New Orleans, LA.

Chaiklin, S. (1989). Cognitive studies of algebra problem solving and learning. In S. Wagner & C. Kieran (Eds.), *Research issues in the learning and teaching of algebra* (pp. 93–114). Reston, VA: National Council of Teachers of Mathematics; Hillsdale, NJ: Lawrence Erlbaum.

Chaiklin, S., & Lesgold, S. (1984, April). *Prealgebra students' knowledge of algebraic tasks with arithmetic expressions.* Paper presented at the annual meeting of the American Educational Research Association, New Orleans, LA.

Chalouh, L., & Herscovics, N. (1988). Teaching algebraic expressions in a meaningful way. In A. F. Coxford (Ed.), *The ideas of algebra, K–12* (1988 Yearbook, pp. 33–42). Reston, VA: National Council of Teachers of Mathematics.

Chevallard, Y., & Conne, F. (1984). Jalons à propos d'algèbre. *Interactions Didactiques, 3,* 1–54 (Universités de Genève et de Neuchâtel).

Clark, C. M., & Peterson, P. L. (1986). Teachers' thought processes. In M. C. Wittrock (Ed.), *Handbook of research on teaching* (3rd ed., pp. 255–296). New York: Macmillan.

Clement, J. (1982). Algebra word problem solutions: Thought processes underlying a common misconception. *Journal for Research in Mathematics Education, 13,* 16–30.

Clement, J. (1985). Misconceptions in graphing. In L. Streefland (Ed.), *Proceedings of the Ninth International Conference for the Psychology of Mathematics Education* (Vol. 1, pp. 369–375). Utrecht, The Netherlands: State University of Utrecht.

Clement, J., Lochhead, J., & Monk, G. (1981). Translation difficulties in learning mathematics. *American Mathematical Monthly, 88,* 286–290.

Collis, K. F. (1974, June). *Cognitive development and mathematics learning.* Paper presented at the Psychology of Mathematics Workshop, Centre for Science Education, Chelsea College, London.

Collis, K. F. (1975). *The development of formal reasoning.* Newcastle, Australia: University of Newcastle.

Cooney, T. J. (1985). A beginning teacher's view of problem solving. *Journal for Research in Mathematics Education, 16,* 324–336.

Coopersmith, A. (1984). Factoring trinomials: Trial and error? Hardly ever! *Mathematics Teacher, 77,* 194–195.

Davis, R. B. (1975). Cognitive processes involved in solving simple algebraic equations. *Journal of Children's Mathematical Behavior, 1*(3), 7–35.

Davis, R. B. (1985). ICME-5 report: Algebraic thinking in the early grades. *Journal of Mathematical Behavior, 4,* 195–208.

Davis, R. B., Jockusch, E., & McKnight, C. (1978). Cognitive processes in learning algebra. *Journal of Children's Mathematical Behavior, 2*(1), 10–320.

Davydov, V. V. (1962). An experiment in introducing elements of algebra in elementary school. *Soviet Education, V*(1), 27–37.

Davydov, V. V. (1975). The psychological characteristics of the "prenumerical" period of mathematics instruction. In L. P. Steffe (Ed.), *Children's capacity for learning mathematics* (Vol. VII of *Soviet studies in the psychology of learning and teaching mathematics,* pp. 109–205). Chicago, IL: University of Chicago, Survey of Recent East European Mathematical Literature.

Demana, F., & Leitzel, J. (1988). Establishing fundamental concepts through numerical problem solving. In A. F. Coxford (Ed.), *The ideas of algebra, K–12* (1988 Yearbook, pp. 61–68). Reston, VA: National Council of Teachers of Mathematics.

Dreyfus, T., & Eisenberg, T. (1981). Function concepts: Intuitive baseline. In C. Comiti & G. Vergnaud (Eds.), *Proceedings of the Fifth International Conference for the Psychology of Mathematics Education* (pp. 183–188). Grenoble, France: Laboratoire I.M.A.G.

Dreyfus, T., & Eisenberg, T. (1982). Intuitive functional concepts: A baseline study on intuitions. *Journal for Research in Mathematics Education, 13,* 360–380.

Dreyfus, T., & Halevi, T. (1988, July–August). Quadfun—a case study of pupil computer interaction. Paper presented to the theme group on Microcomputers and the Teaching of Mathematics at the Sixth International Congress on Mathematical Education, Budapest, Hungary.

Dreyfus, T., & Vinner, S. (1982). Some aspects of the function concept in college students and junior high school teachers. In A. Vermandel (Ed.), *Proceedings of the Sixth International Conference for the Psychology of Mathematics Education* (pp. 12–17). Antwerp, Belgium: Universitaire Instelling.

Eisenberg, T., & Dreyfus, T. (1988). Polynomials in the school curriculum. In A. F. Coxford (Ed.), *The ideas of algebra, K–12* (1988 Yearbook, pp. 112–118). Reston, VA: National Council of Teachers of Mathematics.

Even, R. (1988). Pre-service teachers conceptions of the relationships between functions and equations. In A. Borbas (Ed.), *Proceedings of the Twelfth International Conference for the Psychology of Mathematics Education* (Vol. I, pp. 304–311). Veszprem, Hungary: OOK.

Feurzeig, W. (1986). Algebra slaves and agents in a Logo-based mathematics curriculum. *Instructional Science, 14,* 229–254.

Fey, J. T. (Ed.). (1989a). *Computer-intensive algebra.* College Park: University of Maryland.

Fey, J. T. (1989b). School algebra for the year 2000. In S. Wagner & C. Kieran (Eds.), *Research issues in the learning and teaching of algebra* (pp. 199-213). Reston, VA: National Council of Teachers of Mathematics; Hillsdale, NJ: Lawrence Erlbaum.

Filloy, E., & Rojano, T. (1984). From an arithmetical to an algebraic thought. In J. M. Moser (Ed.), *Proceedings of the Sixth Annual Meeting of PME-NA* (pp. 51–56). Madison: University of Wisconsin.

Filloy, E., & Rojano, T. (1985a). Obstructions to the acquisition of elemental algebraic concepts and teaching strategies. In L. Streefland

(Ed.), *Proceedings of the Ninth International Conference for the Psychology of Mathematics Education* (Vol. I, pp. 154–158). Utrecht, The Netherlands: State University of Utrecht.

Filloy, E., & Rojano, T. (1985b). Operating on the unknown and models of teaching. In S. K. Damarin & M. Shelton (Eds.), *Proceedings of the Seventh Annual Meeting of PME-NA* (pp. 75–79). Columbus, OH: Ohio State University.

Firth, D. E. (1975). *A study of rule dependence in elementary algebra*. Unpublished master's thesis, University of Nottingham, England.

Freudenthal, H. (1973). *Mathematics as an educational task*. Dordrecht, The Netherlands: Reidel.

Freudenthal, H. (1974). Soviet research on teaching algebra at the lower grades of the elementary school. *Educational Studies in Mathematics, 5,* 391–412.

Freudenthal, H. (1982). Variables and functions. In G. van Barneveld & H. Krabbendam (Eds.), *Proceedings of Conference on Functions* (pp. 7–20). Enschede, The Netherlands: National Institute for Curriculum Development.

Goldenberg, E. P. (1988). Mathematics, metaphors, and human factors: Mathematical, technical, and pedagogical challenges in the educational use of graphical representation of functions. *Journal of Mathematical Behavior, 7,* 135–173.

Greeno, J. G. (1980). Trends in the theory of knowledge for problem solving. In D. T. Tuma & F. Reif (Eds.), *Problem solving and education: Issues in teaching and research* (pp. 9–23). Hillsdale, NJ: Lawrence Erlbaum.

Greeno, J. G. (1982, March). *A cognitive learning analysis of algebra*. Paper presented at the annual meeting of the American Educational Research Association, Boston, MA.

Greeno, J. G. (1988). The situated activities of learning and knowing mathematics. In M. J. Behr, C. B. Lacampagne, & M. M. Wheeler (Eds.), *Proceedings of Tenth Annual Meeting of the North American Chapter of the International Group for the Psychology of Mathematics Education* (plenary paper, pp. 481–521). DeKalb, IL: Northern Illinois University.

Grouws, D. A., Cooney, T. J., & Jones, D. (Eds.). (1988). *Perspectives on research on effective mathematics teaching*. Reston, VA: National Council of Teachers of Mathematics; Hillsdale, NJ: Lawrence Erlbaum.

Gurtner, J.-L. (1989). Understanding and discussing linear functions in situations: A developmental study. In G. Vergnaud, J. Rogalski, & M. Artigue (Eds.), *Proceedings of the Thirteenth International Conference for the Psychology of Mathematics Education* (Vol. II, pp. 31–38). Paris: G. R. Didactique, CNRS.

Harper, E. (1979). *The child's interpretation of a numerical variable*. Unpublished doctoral dissertation, University of Bath, England.

Harper, E. (1987). Ghosts of Diophantus. *Educational Studies in Mathematics, 18,* 75–90.

Heid, M. K. (1988). *"Algebra with Computers": A description and an evaluation of student performance and attitudes* (Report submitted to the State College Area School District Board of Education). State College: Pennsylvania State University.

Heid, M. K. (1990). Uses of technology in prealgebra and beginning algebra. *Mathematics Teacher, 84,* 194–198.

Heid, M. K., & Kunkle, D. (1988). Computer-generated tables: Tools for concept development in elementary algebra. In A. F. Coxford (Ed.), *The ideas of algebra, K–12* (1988 Yearbook, pp. 170–177). Reston, VA: National Council of Teachers of Mathematics.

Heid, M. K., Sheets, C., Matras, M. A., & Menasian, J. (1988, April). *Classroom and computer lab interaction in a computer-intensive algebra curriculum*. Paper presented at the annual meeting of the American Educational Research Association, New Orleans, LA.

Herscovics, N. (1980). Constructing meaning for linear equations: A problem of representation. *Recherches en Didactique des Mathématiques, 1,* 351–385.

Herscovics, N. (1989). Cognitive obstacles encountered in the learning of algebra. In S. Wagner & C. Kieran (Eds.), *Research issues in the learning and teaching of algebra* (pp. 60–86). Reston, VA: National Council of Teachers of Mathematics; Hillsdale, NJ: Lawrence Erlbaum.

Hinsley, D. A., Hayes, J. R., & Simon, H. A. (1977). From words to equations: Meaning and representation in algebra word problems. In M. A. Just & P. Carpenter (Eds.), *Comprehension and cognition* (pp. 89–106). Hillsdale, NJ: Lawrence Erlbaum.

Hoyles, C., Sutherland, R., & Evans, J. (1985). *The Logo Maths Project: A preliminary investigation of the pupil-centred approach to the learning of Logo in the secondary mathematics classroom, 1983–4*. London: University of London, Institute of Education.

Hoz, R., & Harel, G. (1989). The facilitating role of table forms in solving algebra speed problems: Real or imaginary? In G. Vergnaud, J. Rogalski, & M. Artigue (Eds.), *Proceedings of the Thirteenth International Conference for the Psychology of Mathematics Education* (Vol. II, pp. 123–130). Paris: G. R. Didactique, CNRS.

Janvier, C. (1978). *The interpretation of complex Cartesian graphs—studies and teaching experiments*. Unpublished doctoral dissertation, University of Nottingham, England.

Kaput, J. J. (1988, April). *Translations from numerical and graphical to algebraic representations of elementary functions*. Paper presented at the annual meeting of the American Educational Research Association, New Orleans, LA.

Kaput, J. J. (1989). Linking representations in the symbol systems of algebra. In S. Wagner & C. Kieran (Eds.), *Research issues in the learning and teaching of algebra* (pp. 167-194). Reston, VA: National Council of Teachers of Mathematics; Hillsdale, NJ: Lawrence Erlbaum.

Karplus, R. (1979). Continuous functions: Students' viewpoints. *European Journal of Science Education, 1*(3), 397–415.

Kerslake, D. (1977). The understanding of graphs. *Mathematics in School, 6*(2), 22–25.

Kerslake, D. (1981). Graphs. In K. M. Hart (Ed.), *Children's understanding of mathematics: 11–16* (pp. 120–136). London: John Murray.

Kieran, C. (1979). Children's operational thinking within the context of bracketing and the order of operations. In D. Tall (Ed.), *Proceedings of the Third International Conference for the Psychology of Mathematics Education* (pp. 128–133). Coventry, England: Warwick University, Mathematics Education Research Centre.

Kieran, C. (1981). Concepts associated with the equality symbol. *Educational Studies in Mathematics, 12,* 317–326.

Kieran, C. (1982, March). *The learning of algebra: A teaching experiment*. Paper presented at the annual meeting of the American Educational Research Association, New York. (ERIC Document Reproduction Service No. ED 216 884)

Kieran, C. (1983). Relationships between novices' views of algebraic letters and their use of symmetric and asymmetric equation-solving procedures. In J. C. Bergeron & N. Herscovics (Eds.), *Proceedings of the Fifth Annual Meeting of PME-NA,* (Vol. 1, pp. 161–168). Montréal, Canada: Université de Montréal.

Kieran, C. (1984). A comparison between novice and more-expert algebra students on tasks dealing with the equivalence of equations. In J. M. Moser (Ed.), *Proceedings of the Sixth Annual Meeting of PME-NA* (pp. 83–91). Madison: University of Wisconsin.

Kieran, C. (1985a). The Soviet teaching experiment. In T. A. Romberg (Ed.), *Research methods for studies in mathematics education: Some considerations and alternatives*. Madison: Wisconsin Education Research Center.

Kieran, C. (1985b). Use of substitution procedure in learning algebraic equation-solving. In S. K. Damarin & M. Shelton (Eds.), *Proceedings of the Seventh Annual Meeting of PME-NA* (pp. 145–152). Columbus, OH: Ohio State University.

Kieran, C. (1988). Two different approaches among algebra learners. In A. F. Coxford (Ed.), *The ideas of algebra, K–12* (1988 Yearbook, pp. 91–96). Reston, VA: National Council of Teachers of Mathematics.

Kieran, C. (1989). The early learning of algebra: A structural perspective. In S. Wagner & C. Kieran (Eds.), *Research issues in the learning and teaching of algebra* (pp. 33–56). Reston, VA: National Council of Teachers of Mathematics; Hillsdale, NJ: Lawrence Erlbaum.

Kieran, C. (in press). Functions, graphing, and technology: Integrating research on learning and instruction. In T. A. Romberg, E. Fennema, & T. P. Carpenter (Eds.), *Integrating research on the graphical representation of function*. Hillsdale, NJ: Lawrence Erlbaum.

Kieran, C., Boileau, A., & Garançon, M. (1989). Processes of mathematization in algebra problem solving within a computer environment: A functional approach. In C. A. Maher, G. A. Goldin, & R. B. Davis (Eds.), *Proceedings of the Eleventh Annual Meeting of PME-NA* (pp. 26–34). New Brunswick, NJ: Rutgers University.

Kirshner, D. (1987). *The grammar of symbolic elementary algebra*. Unpublished doctoral dissertation, University of British Columbia, Vancouver.

Kirshner, D. (1989). The visual syntax of algebra. *Journal for Research in Mathematics Education, 20*, 274–287.

Kleiner, I. (1989). Evolution of the function concept: A brief survey. *College Mathematics Journal, 20*(4), 282–300.

Kline, M. (1972). *Mathematical thought from ancient to modern times*. New York: Oxford University Press.

Küchemann, D. (1978). Children's understanding of numerical variables. *Mathematics in School, 7*(4), 23–26.

Küchemann, D. (1981). Algebra. In K. M. Hart (Ed.), *Children's understanding of mathematics 11–16* (pp. 102–119). London: John Murray.

Lampert, M. (1988, March). *Cognition in mathematical practice*. Invited address presented at the annual meeting of the American Educational Research Association, New Orleans, LA.

Lee, L., & Wheeler, D. (1987). *Algebraic thinking in high school students: Their conceptions of generalisation and justification* (Research Report). Montreal: Concordia University, Department of Mathematics.

Lee, L., & Wheeler, D. (1989). The arithmetic connection. *Educational Studies in Mathematics, 20*, 41–54.

Leinhardt, G., Zaslavsky, O., & Stein, M. K. (1990). Functions, graphs, and graphing: Tasks, learning, and teaching. *Review of Educational Research, 60*, 1–64.

Lesh, R., Post, T., & Behr, M. (1987). Dienes revisited: Multiple embodiments in computer environments. In I. Wirszup & R. Streit (Eds.), *Developments in school mathematics education around the world* (pp. 647–680). Reston, VA: National Council of Teachers of Mathematics.

Lewis, C. (1980, April). *Kinds of knowledge in algebra*. Paper presented at the annual meeting of the American Educational Research Association, Boston, MA.

Lewis, C. (1981). Skill in algebra. In J. R. Anderson (Ed.), *Cognitive skills and their acquisition*. Hillsdale, NJ: Lawrence Erlbaum.

Lewis, M. W., Milson, R., & Anderson, J. R. (1987). Designing an intelligent authoring system for high school mathematics ICAI: The TEACHER'S APPRENTICE Project. In G. P. Kearsley (Ed.), *Artificial intelligence and instruction: Applications and methods*. Reading, MA: Addison-Wesley.

Linchevski, L., & Vinner, S. (1990). Embedded figures and structures of algebraic expressions. In G. Booker, P. Cobb, & T. N. de Mendicuti (Eds.), *Proceedings of the Fourteenth International Conference for the Psychology of Mathematics Education* (Vol. II, pp. 85–92). Mexico City, Mexico: Centro de Investigacion y de Estudios Avanzados del I.P.N.

Markovits, Z., Eylon, B.-S., & Bruckheimer, M. (1983). Functions—linearity unconstrained. In R. Hershkowitz (Ed.), *Proceedings of the Seventh International Conference for the Psychology of Mathematics Education* (pp. 271–277). Rehovot, Israel: Weizmann Institute of Science.

Markovits, Z., Eylon, B.-S., & Bruckheimer, M. (1986). Functions today and yesterday. *For the Learning of Mathematics, 6*(2), 18–24.

Matz, M. (1979). *Towards a process model for high school algebra errors* (Working Paper 181). Cambridge: Massachusetts Institute of Technology, Artificial Intelligence Laboratory.

Maurer, S. B. (1987). [Remarks following the Wenger chapter]. In A. H. Schoenfeld (Ed.), *Cognitive science and mathematics education* (p. 248). Hillsdale, NJ: Lawrence Erlbaum.

Mayer, R. E. (1980). *Schemas for algebra story problems* (Report No. 80-3). Santa Barbara: University of California, Department of Psychology, Series in Learning and Cognition.

McArthur, D. (1985). Developing computer tools to support performing and learning complex skills. In D. Berger, K. Pedzek, & W. Ganks (Eds.), *Applications of cognitive psychology*. Hillsdale, NJ: Lawrence Erlbaum.

McArthur, D., Stasz, C., & Hotta, J. Y. (1987). Learning problem-solving skills in algebra. *Journal of Educational Technology Systems, 15*, 303–324.

McKnight, C. C., Travers, K. J., Crosswhite, F. J., & Swafford, J. O. (1985). Eighth-grade mathematics in U.S. schools: A report from the Second International Mathematics Study. *Arithmetic Teacher, 32*(8), 20–26.

Mevarech, Z. R., & Yitschak, D. (1983). Students' misconceptions of the equivalence relationship. In R. Hershkowitz (Ed.), *Proceedings of the Seventh International Conference for the Psychology of Mathematics Education* (pp. 313–320). Rehovot, Israel: Weizmann Institute of Science.

Moschkovich, J. (1990). Students' interpretations of linear equations and their graphs. In G. Booker, P. Cobb, & T. N. de Mendicuti (Eds.), *Proceedings of the Fourteenth International Conference for the Psychology of Mathematics Education* (Vol. II, pp. 109–116). Mexico City, Mexico: Centro de Investigacion y de Estudios Avanzados del I.P.N.

National Council of Teachers of Mathematics. (1989). *Curriculum and evaluation standards for school mathematics*. Reston, VA: NCTM.

NMP. (1987). *National Mathematics Project*. London: Longman.

Noss, R. (1986). Constructing a conceptual framework for elementary algebra through Logo programming. *Educational Studies in Mathematics, 17*, 335–357.

Oprea, J. M. (1988). Computer programming and mathematical thinking. *Journal of Mathematical Behavior, 7*, 175–190.

Owen, E., & Sweller, J. (1989). Should problem solving be used as a learning device in mathematics? *Journal for Research in Mathematics Education, 20*, 322–328.

Paige, J. M., & Simon, H. A. (1966). Cognitive processes in solving algebra word problems. In B. Kleinmuntz (Ed.), *Problem solving: Research, method, and theory* (pp. 51–119). New York, NY: Wiley.

Peck, D. M., & Jencks, S. M. (1988). Reality, arithmetic, algebra. *Journal of Mathematical Behavior, 7*, 85–91.

Petitto, A. (1979). The role of formal and non-formal thinking in doing algebra. *Journal of Children's Mathematical Behavior, 2*(2), 69–82.

Ponte, J. P. M. (1984). Functional reasoning and the interpretation of Cartesian graphs (Doctoral dissertation, University of Georgia, 1984). *Dissertation Abstracts International, 45*(6), 1675A. (University Microfilms No. 8421144)

Rachlin, S. L. (1981). *Processes used by college students in understanding basic algebra*. Unpublished doctoral dissertation, University of Georgia, Athens.

Rachlin, S. L. (1982). A teacher's analysis of students' problem-solving processes in algebra. In S. Wagner (Ed.), *Proceedings of the Fourth Annual Meeting of PME-NA* (pp. 140–147). Athens: University of Georgia, Department of Mathematics Education.

Rachlin, S. L. (1987). Using research to design a problem-solving approach for teaching algebra. In Sit-Tui Ong (Ed.), *Proceedings of the Fourth Southeast Asian Conference on Mathematical Education (ICMI-SEAMS)* (pp. 156–161). Singapore: Singapore Institute of Education.

Rachlin, S. L. (1989). The research agenda in algebra: A curriculum development perspective. In S. Wagner & C. Kieran (Eds.), *Research issues in the learning and teaching of algebra* (pp. 257–265). Reston, VA: National Council of Teachers of Mathematics; Hillsdale, NJ: Lawrence Erlbaum.

Rachlin, S. L., Matsumoto, A. N., & Wada, L. T. (1988). *Algebra 1: A process approach*. Honolulu: University of Hawaii, Curriculum Research & Development Group.

Reed, S. K. (1984). Estimating answers to algebra word problems. *Journal of Experimental Psychology: Learning, Memory and Cognition, 10,* 778–790.

Reed, S. K. (1987). A structure-mapping model for word problems. *Journal of Experimental Psychology: Learning, Memory and Cognition, 13,* 124–139.

Reed, S. K., Dempster, A., & Ettinger, M. (1985). Usefulness of analogous solutions for solving algebra word problems. *Journal of Experimental Psychology: Learning, Memory and Cognition, 11,* 106–125.

Rizzuti, J., & Confrey, J. (1988). A construction of the concept of exponential functions. In M. J. Behr, C. B. Lacampagne, & M. M. Wheeler (Eds.), *Proceedings of the Tenth Annual Meeting of PME-NA* (pp. 260–267). DeKalb: Northern Illinois University.

Roberts, N., Carter, R., Davis, F., & Feurzeig, W. (1989). Power tools for algebra problem solving. *Journal of Mathematical Behavior, 8,* 251–265.

Rogalski, J. (1985). Acquisition of number-space relationships: Using educational and research programs. In L. Streefland (Ed.), *Proceedings of the Ninth International Conference for the Psychology of Mathematics Education* (Vol. 1, pp. 71–76). Utrecht, The Netherlands: State University of Utrecht.

Rosnick, P., & Clement, J. (1980). Learning without understanding: The effect of tutoring strategies on algebra misconceptions. *Journal of Mathematical Behavior, 3*(1), 3–27.

Schoenfeld, A. H., Smith, J. P., & Arcavi, A. (in press). Learning. In R. Glaser (Ed.), *Advances in instructional psychology* (Vol. 4). Hillsdale, NJ: Lawrence Erlbaum.

Schwarz, B. (1988, July–August). *The Triple Representation Model curriculum for the function concept*. Paper presented to the theme group on Microcomputers and the Teaching of Mathematics at the Sixth International Congress on Mathematical Education, Budapest, Hungary.

Schwarz, B. (1989). *The use of a microworld to improve the concept image of a function: The Triple Representation Model curriculum*. Unpublished doctoral dissertation, Weizmann Institute of Science, Israel.

Schwarz, B., Dreyfus, T., & Bruckheimer, M. (1990). A model of the function concept in a three-fold representation. *Computers & Education, 14,* 249–262.

Sfard, A. (1987). Two conceptions of mathematical notions: Operational and structural. In J. C. Bergeron, N. Herscovics, & C. Kieran (Eds.), *Proceedings of the Eleventh International Conference for the Psychology of Mathematics Education* (Vol. III, pp. 162–169). Montréal, Canada: Université de Montréal.

Sfard, A. (1989). Transition from operational to structural conception: The notion of function revisited. In G. Vergnaud, J. Rogalski, & M. Artigue (Eds.), *Proceedings of the Thirteenth International Conference for the Psychology of Mathematics Education* (Vol. 3, pp. 151–158). Paris: G. R. Didactique, CNRS.

Sfard, A. (1991). On the dual nature of mathematical conceptions: Reflections on processes and objects as different sides of the same coin. *Educational Studies in Mathematics, 22,* 1–36.

Shavelson, R. J., Webb, N. M., Stasz, C., & McArthur, D. (1988). Teaching mathematical problem solving: Insights from teachers and tutors. In R. I. Charles & E. A. Silver (Eds.), *The teaching and assessing of mathematical problem solving* (pp. 203–231). Reston, VA: National Council of Teachers of Mathematics; Hillsdale, NJ: Lawrence Erlbaum.

Shuard, H., & Neill, H. (1977). *From graphs to calculus*. Glasgow: Blackie.

Sleeman, D. H. (1984). An attempt to understand students' understanding of basic algebra. *Cognitive Science, 8,* 387–412.

Slovin, H. (1990, April). *A study of the effect of an in-service course in teaching algebra through a problem-solving process*. Paper presented at the research presession of the annual meeting of the National Council of Teachers of Mathematics, Salt Lake City, UT.

Soloway, E., Lochhead, J., & Clement, J. (1982). Does computer programming enhance problem solving ability? Some positive evidence on algebra word problems. In R. J. Seidel, R. E. Anderson, & B. Hunter (Eds.), *Computer literacy*. New York, NY: Academic Press.

Sutherland, R. (1987). A study of the use and understanding of algebra related concepts within a Logo environment. In J. C. Bergeron, N. Herscovics, & C. Kieran, (Eds.), *Proceedings of the Eleventh International Conference for the Psychology of Mathematics Education* (Vol. I, pp. 241–247). Montréal, Canada: Université de Montréal.

Sutherland, R. (1988). *A longitudinal study of the development of pupils' algebraic thinking in a Logo environment*. Unpublished doctoral dissertation, University of London, Institute of Education, England.

Sutherland, R., & Hoyles, C. (1986). Logo as a context for learning about variable. In *Proceedings of the Tenth International Conference for the Psychology of Mathematics Education* (pp. 301–306). London: University of London, Institute of Education.

Swan, M. (1982). The teaching of functions and graphs. In G. van Barneveld & H. Krabbendam (Eds.), *Proceedings of Conference on Functions* (pp. 151–165). Enschede, The Netherlands: National Institute for Curriculum Development.

Tall, D. (1989, April). *Concept image, computers, and curriculum change*. Invited address presented at the research presession of the annual meeting of the National Council of Teachers of Mathematics, Orlando, FL.

Thomas, H. L. (1969). *An analysis of stages in the attainment of a concept of function*. Unpublished doctoral dissertation, Columbia University, New York.

Thomas, M., & Tall, D. (1986). The value of the computer in learning algebra concepts. In *Proceedings of the Tenth International Conference for the Psychology of Mathematics Education* (pp. 313–318). London: University of London, Institute of Education.

Thompson, P. W. (1989). Artificial intelligence, advanced technology, and learning and teaching algebra. In S. Wagner & C. Kieran (Eds.), *Research issues in the learning and teaching of algebra* (pp. 135–161). Reston, VA: National Council of Teachers of Mathematics; Hillsdale, NJ: Lawrence Erlbaum.

Thompson, P. W., & Thompson, A. G. (1987). Computer presentations of structure in algebra. In J. C. Bergeron, N. Herscovics, & C. Kieran, (Eds.), *Proceedings of the Eleventh International Conference for the Psychology of Mathematics Education* (Vol. I, pp. 248–254). Montréal, Canada: Université de Montréal.

Vergnaud, G. (1984). Understanding mathematics at the secondary-school level. In A. Bell, B. Low, & J. Kilpatrick (Eds.), *Theory, research & practice in mathematical education* (Report of ICME5 Working Group on Research in Mathematics Education, pp. 27–35). Nottingham, UK: Shell Centre for Mathematical Education.

Vergnaud, G. (1986, November). *Long terme et court terme dans l'apprentissage de l'algèbre*. Paper presented at the Colloque Franco-Allemand de didactique des mathématiques et de l'informatique, Marseilles, France.

Vergnaud, G., Benhadj, J., & Dussouet, A. (1979). *La coordination de l'enseignement des mathématiques entre le cours moyen 2e année et la classe de sixième*. Paris: Institut National de Recherche Pédagogique.

Vergnaud, G., & Errecalde, P. (1980). Some steps in the understanding and the use of scales and axis by 10–13 year-old students. In R. Karplus (Ed.), *Proceedings of the Fourth International Conference for the Psychology of Mathematics Education* (pp. 285–291). Berkeley, CA: University of California.

Verstappen, P. (1982). Some reflections on the introduction of relations and functions. In G. van Barneveld & H. Krabbendam (Eds.), *Proceedings of Conference on Functions* (pp. 166–184). Enschede, The Netherlands: National Institute for Curriculum Development.

Vinner, S. (1983). Concept definition, concept image and the notion of function. *International Journal of Mathematical Education in Science and Technology, 14*, 293–305.

Vinner, S., & Dreyfus, T. (1989). Images and definitions for the concept of function. *Journal for Research in Mathematics Education, 20*, 356–366.

Wagner, S. (1977, April). *Conservation of equation and function and its relationship to formal operational thought*. Paper presented at the annual meeting of the American Educational Research Association, New York.

Wagner, S. (1981). Conservation of equation and function under transformations of variable. *Journal for Research in Mathematics Education, 12*, 107–118.

Wagner, S., & Kieran, C. (Eds.). (1989). *Research issues in the learning and teaching of algebra* Reston, VA: National Council of Teachers of Mathematics; Hillsdale, NJ: Lawrence Erlbaum

Wagner, S., Rachlin, S. L., & Jensen, R. J. (1984). *Algebra Learning Project: Final report*. Athens: University of Georgia, Department of Mathematics Education.

Waits, B. K., & Demana, F. (1988, July–August). *New models for teaching and learning mathematics through technology*. Paper presented to the theme group on Microcomputers and the Teaching of Mathematics at the Sixth International Congress on Mathematical Education, Budapest, Hungary.

Weaver, J. F., & Suydam, M. N. (1972). *Meaningful instruction in mathematics education*. Columbus, OH: ERIC Information Analysis Center for Science, Mathematics, and Environmental Education.

Wenger, R. H. (1987). Cognitive science and algebra learning. In A. H. Schoenfeld (Ed.), *Cognitive science and mathematics education* (pp. 217–251). Hillsdale, NJ: Lawrence Erlbaum.

Wheeler, D. (1989). Contexts for research on the teaching and learning of algebra. In S. Wagner & C. Kieran (Eds.), *Research issues in the learning and teaching of algebra* (pp. 278–287). Reston, VA: National Council of Teachers of Mathematics; Hillsdale, NJ: Lawrence Erlbaum.

Whitman, B. S. (1976). Intuitive equation solving skills and the effects on them of formal techniques of equation solving (Doctoral dissertation, Florida State University, 1975). *Dissertation Abstracts International, 36*, 5180A. (University Microfilms No. 76-2720)

Yerushalmy, M. (1988). *Effects of graphic feedback on the ability to transform algebraic expressions when using computers* (Interim report to the Spencer Fellowship Program of the National Academy of Education) Haifa, Israel. The University of Haifa.

▪18▪

GEOMETRY AND SPATIAL REASONING

Douglas H. Clements
STATE UNIVERSITY OF NEW YORK AT BUFFALO

Michael T. Battista
KENT STATE UNIVERSITY

Spatial understandings are necessary for interpreting, understanding, and appreciating our inherently geometric world. (National Council of Teachers of Mathematics, 1989, p. 48)

Geometry is grasping space...that space in which the child lives, breathes and moves. The space that the child must learn to know, explore, conquer, in order to live, breathe, and move better in it. (Freudenthal, in National Council of Teachers of Mathematics, 1989, p. 48)

Arising out of practical activity and man's need to describe his surroundings, geometric forms were slowly conceptualized until they took on an abstract meaning of their own. Thus from a practical theory of earth measure, there developed a growing set of relations or theorems that culminated in Euclid's Elements, the collection, synthesis, and elaboration of all this knowledge. (Fehr, 1973, p. 370)

Equations are just the boring part of mathematics. I attempt to see things in terms of geometry. (Hawking, National Research Council, 1989, p. 35)

School geometry is the study of those spatial objects, relationships, and transformations that have been formalized (or mathematized) and the axiomatic mathematical systems that have been constructed to represent them. Spatial reasoning, on the other hand, consists of the set of cognitive processes by which mental representations for spatial objects, relationships, and transformations are constructed and manipulated. Clearly, geometry and spatial reasoning are strongly interrelated, and most mathematics educators seem to include spatial reasoning as part of the geometry curriculum. Usiskin (1987), for instance, has described four dimensions of geometry: (a) visualization, drawing, and construction of figures; (b) study of the spatial aspects of the physical world; (c) use as a vehicle for representing nonvisual mathematical concepts and relationships; and (d) representation as a formal mathematical system. The first three of these dimensions require the use of spatial reasoning.

When the term "school geometry" is used, it almost universally refers to Euclidean geometry, even though there are numerous approaches to the study of the topic (for example, synthetic, analytic, transformational, and vector). The traditional, secondary school version of geometry is axiomatic in nature; elementary school geometry traditionally has emphasized

The authors gratefully acknowledge the helpful comments provided by David Fuys, Brooklyn College, and Sharon Senk, Michigan State University. Time to prepare this material was partially provided by the National Science Foundation under Grant No. MDR-8651668. Any opinions, findings, and conclusions or recommendations expressed in this publication are those of the authors and do not necessarily reflect the views of the National Science Foundation.

measurement and informal development of those basic concepts needed in high school. According to Suydam (1985) there is a great deal of agreement that the goals of geometry instruction should be to

- develop logical thinking abilities;
- develop spatial intuition about the real world;
- impart the knowledge needed to study more mathematics; and
- teach the reading and interpretation of mathematical arguments (p. 481).

The National Council of Teachers of Mathematics (1989) *Curriculum Standards* calls for all students to

- identify, describe, compare, model, draw, and classify geometric figures in two and three dimensions;
- develop spatial sense;
- explore the effects of transforming, combining, subdividing, and changing geometric figures;
- understand, apply, and deduce properties of and relationships between geometric figures, including congruence and similarity;
- develop an appreciation of geometry as a means of describing and modeling the physical world;
- explore synthetic, transformational, and coordinate approaches to geometry, with college-bound students also required to develop an understanding of an axiomatic system through investigating and comparing various geometric systems; and
- explore a vector approach to certain aspects of geometry.

This chapter contains seven major sections. First, students' performance in geometry is briefly summarized as a background to the entire research corpus. Second, research on three major theoretical perspectives on the development of geometric thinking—Piaget, the van Hieles, and cognitive science—is reviewed. Third, the establishment of truth in geometry is discussed, highlighting both theoretical and empirical work. Fourth, the relationship between spatial thinking and mathematics, the nature of spatial reasoning and imagery, and attempts to teach spatial abilities are considered. The fifth section, representations of geometric ideas, includes issues related to concepts, diagrams, manipulatives, and computers. Sixth, we examine group and cross-cultural differences. Finally, broad conclusions are drawn from this research corpus.

STUDENTS' PERFORMANCE IN GEOMETRY

According to extensive evaluations of mathematics learning, elementary and middle school students in the United States are failing to learn basic geometric concepts and geometric problem solving; they are woefully underprepared for the study of more sophisticated geometric concepts and proof, especially when compared to students from other nations (Carpenter, Corbitt, Kepner, Lindquist, & Reys, 1980; Fey et al., 1984; Kouba et al., 1988; Stevenson, Lee, & Stigler, 1986; Stigler, Lee, & Steven-

son, 1990). For instance, fifth graders from Japan and Taiwan scored more than twice as high as U.S. students on a geometry test (Stigler et al., 1990). Japanese students in both first and fifth grades also scored much higher (and Taiwanese students only slightly higher) than U.S. students on tests of visualization and paper-folding. Stigler et al. (1990) postulate that the latter results may be due both to Japanese classrooms' heavy reliance on visual representations for concepts and to expectations that Japanese students become competent at drawing. Data from the Second International Mathematics Study (SIMS) showed that, in geometry, U.S. 8th and 12th graders scored at the 25th international percentile or below (McKnight, Travers, Crosswhite, & Swafford, 1985; McKnight, Travers, & Dossey, 1985).

Usiskin (1987), citing data from the 1982 U.S. National Assessment of Educational Progress (NAEP), reported that fewer than 10% of 13-year-olds could find the measure of the third angle for a triangle, given the measure of the other two angles; only 20% could find the length of the hypotenuse of a right triangle given its legs. (He concluded that a greater number of students could do the more difficult computation because it is taught to more students.) In the 1986 NAEP, Kouba et al. (1988) reported students' performance at identifying common geometric figures, such as parallel lines and the diameter of a circle, acceptable, but students' performance with figures not frequently encountered in everyday life, such as perpendicular lines and the radius of a circle, were reported as deficient. Performance dealing with properties of figures, visualization, and applications was poor. For example, only 60% of seventh grade students could identify the image of an object reflected through a line; only about 10% of seventh graders could find the area of a square, given the length of one of its sides (56% found the area of a rectangle, given the lengths of its sides); and less than 10% of seventh graders could identify which set of numbers could be the lengths of the sides of a triangle (even though 66% could do it if segments were given). Apparently, students can handle some problems much better if the problem is presented visually rather than verbally (Carpenter et al., 1980; Driscoll, 1983b; Kouba et al., 1988).

The situation is even worse at the high school level. First, only about half of all high school students enroll in a geometry course. Of those enrolled at the beginning of the school year only 63% were able to correctly identify triangles that were presented along with distractors (Usiskin, 1987). According to the 1978 NAEP in mathematics, only 64% of the 17-year-olds knew that a rectangle is a parallelogram, only 16% could find the area of a region made up of two rectangles, and just 9% could solve the problem "How many cubic feet of concrete would be needed to pave an area 30 feet long and 20 feet wide with a layer 4 inches thick?" Of 17-year-olds that had a full year of high school geometry, only 57% could calculate the volume of a rectangular solid, 54% could find the hypotenuse of a right triangle whose legs were multiples of 3 and 4, and 34% could find the area of a right triangle. Only 52% of entering secondary students could state the area of a square when its sides were given (Usiskin, 1982). On the 1986 assessment, 11th-grade students who had not taken high school geometry scored at about the same level as seventh graders (Lindquist & Kouba, 1989). There were few performance differences in visualization between those students who had taken geometry and those who had not, although there were large differences on items requiring knowledge of geometric properties and on applications. Less than 25% of 11th-grade students correctly identified which figures had lines of symmetry, whether they had taken geometry or not (even though symmetry is studied throughout elementary and middle school). Even more incriminating of the curriculum is the fact that only about 30% of high school geometry students enrolled in a course for which proof was a goal were able to write proofs or exhibit any understanding of the meaning of proof (Senk, 1985; Suydam, 1985). It is no wonder that doing proofs was the least liked mathematics topic by 17-year-olds on the 1982 NAEP and that less than 50% of the students rated the topic as important.

This depressing picture of students' knowledge of geometry is elaborated through a consideration of students' misconceptions. Here are some examples (Clements & Battista, 1989; Fuys, Geddes, & Tischler, 1988; Hoffer, 1983):

- an angle must have one horizontal ray
- a right angle is an angle that points to the right
- to be a side of a figure a segment must be vertical
- a segment is not a diagonal if it is vertical or horizontal
- a square is not a square if its base is not horizontal
- the only way a figure can be a triangle is if it is equilateral
- the height of a triangle or parallelogram is a side adjacent to the base
- the angle sum of a quadrilateral is the same as its area
- the Pythagorean theorem can be used to calculate the area of a rectangle
- if a shape has four sides, then it is a square
- the area of a quadrilateral can be obtained by transforming it into a rectangle with the same perimeter

Apparently, much learning of geometric concepts has been rote. Properties, class inclusions, relationships, and implications are frequently not perceived (Mayberry, 1983).

A primary cause of this poor performance may be the curriculum, both in what topics are treated and how they are treated. The major focus of standard elementary and middle school curricula is on recognizing and naming geometric shapes, writing the proper symbolism for simple geometric concepts, developing skill with measurement and construction tools such as a compass and protractor, and using formulas in geometric measurement (Porter, 1989; Thomas, 1982). These curricula consist of a hodgepodge of unrelated concepts with no systematic progression to higher levels of thought—levels requisite for sophisticated concept development and substantive geometric problem solving. In addition, teachers often do not teach even the impoverished geometry curriculum that is available to them. Porter, for instance, reported whole districts in which fourth- and fifth-grade teachers spent "virtually no time teaching geometry" (Porter, 1989, p. 11). Even when taught, geometry was the topic most frequently identified as being taught merely for "exposure"; that is, geometry was given only brief, cursory coverage. The SIMS data for the eighth-grade level indicated that teachers rated the opportunity to learn geometry much lower than the opportunity to learn any other topic (McKnight, Travers, Crosswhite, & Swafford, 1985). At the secondary level, the traditional emphasis has been on formal proof, despite the fact the students are unprepared to deal with it. Indeed, as Usiskin (1987) summarizes

There is no geometry curriculum at the elementary school level. As a result, students enter high school not knowing enough geometry to succeed. There is a geometry curriculum at the secondary level, but only about half of the students encounter it, and only about a third of these students understand it. (p. 29)

THE DEVELOPMENT OF GEOMETRIC THINKING

Piaget and Inhelder: The Child's Conception of Space

Two Major Themes. Two major themes of Piaget and Inhelder's (1967) influential theory on children's conception of space will be discussed. First, representations of space are constructed through the progressive organization of the child's motor and internalized actions, resulting in operational systems. Therefore, the representation of space is not a perceptual "reading off" of the spatial environment, but is the build-up from prior active manipulation of that environment. Second, the progressive organization of geometric ideas follows a definite order, and this order is more logical than historical in that initially topological relations (for example, connectedness, enclosure, and continuity) are constructed, and later projective (rectilinearity) and Euclidean (angularity, parallelism, and distance) relations. This has been termed the topological primacy thesis.

It is important to reiterate that Piaget and Inhelder were discussing the child's ability to represent space. They maintain that perceptual space is constructed early in the sensorimotor period. Nevertheless, perceptual space precludes the development of representational, or conceptual, space, in that its development also embodies the topological primacy thesis; it, too, is constructed rather than existent from the outset of development. Representational space, in addition, reflects properties of logical operational thought.

Topological Primacy and Constructivism

HAPTIC EVIDENCE. Piaget and Inhelder's first experiments provide evidence supporting both the themes of constructivism and topological primacy. Children were asked to explore hidden objects tactilely (haptic perception) and to either draw these objects or match them with duplicates. Preschool children were reported initially to discriminate objects on the basis of topological features, such as being closed or otherwise topologically equivalent. Only later could they discriminate rectilinear from curvilinear forms and, finally, among rectilinear closed shapes, such as squares and diamonds.

Piaget and Inhelder claim that the development of more sophisticated spatial concepts involves increasingly systematic and coordinated action. During the first stages of development, children are basically passive in their explorations. For example, children may touch one part of a shape, and this action results in a tactile perception; touching another part involves another action and perception. When children regulate such actions by establishing relations among them, an accurate representation of the shape can be built. For example, the systematic return to each movement's starting point allows the parts of the figure to be synthesized. It is when each mental action becomes reversible that it can be distinct yet coordinated with every other action into a coherent whole.

From this perspective, then, abstraction of shape is not a perceptual abstraction of a physical property, but is the result of a coordination of children's actions. Children "can only 'abstract' the idea of such a relation as equality on the basis of an action of equalization, the idea of a straight line from the action of following by hand or eye without changing direction, and the idea of an angle from two intersecting movements" (p. 43).

DRAWING EVIDENCE. Because making a drawing is an act of representation, not of perception, Piaget and Inhelder claim that inaccurate drawings reflect the inadequacy of mental tools for spatial representation. Indeed, the inability of young children to draw a copy of even simple shapes is taken as an indication that coordination of actions, rather than passive perception, lies at the foundation of the conceptual development of space. They also claim that children's drawn copies of geometric shapes represent topological features first. For example, at Stage 0 (before the age of 3) no purpose or aim can be discerned; children simply scribble. At substage IA (up to about age 3 years 11 months), a circle is drawn as an irregular closed curve, and squares and triangles are not distinguished from circles. While children do not distinguish between straight-sided and curved figures, there is a correct rendering of topological properties (for example, closed paths with small closed paths inside, on, or outside them). An obvious objection to such arguments is that inaccuracies in drawing might be attributable to motor difficulties. However, Piaget and Inhelder do not accept such arguments, providing supportive examples like the child who could draw a pine tree with branches at right angles but not a square.

At stage II (about age 4), there is a progressive differentiation of Euclidean shapes. The criterion for this stage is the successful reproduction of the square or rectangle. Euclidean relationships, such as angle and inclination, develop only slowly. Only at stage III (about 6–7 years) are all problems overcome; for example, Piaget and Inhelder state that at least two years of work is required to pass from copying the square to copying the rhombus, demonstrating that construction of a "Euclidean shape" requires more than a correct visual impression. Such a task involves a complex interplay of actions. To Piaget and Inhelder, then, topological relationships develop first because they represent the simplest organization of those actions from which shape is abstracted (for example, the dissociated elements of primitive motor rhythms in scribbling). Other relationships develop over long periods of time.

Projective Space. To Piaget and Inhelder, the difference between topological and projective or Euclidean relations concerns the way in which the different figures or objects are related to one another. The former are internal to a particular figure; the latter involve relations between figure and subject (projective) or between figures themselves (Euclidean). Projective relations begin psychologically, at the point when the figure is no longer viewed in isolation but begins to be considered in relation to a point of view. For example, the concept of the straight line results from the child's act of taking aim, or sighting. Children perceive a straight line since the earliest years, of course, but they cannot place objects along a straight path not parallel to the edges of a table. Instead, they tend to follow the edges of the table or curve the line toward such a path. This is not a perceptual problem; they realize that the line is not straight, but they cannot construct an adequate representation to make it so. They possess only an intuitive, spatial representation, an internalized imitation of previous perceptions that can be altered by distracting perceptual configurations (e.g., the edges of a table). Internal representation is based on operations and can, therefore, limit the influence of perceptual configurations. Thus, at about 7 years of age, straight paths are constructed by children spontaneously aiming or sighting along a trajectory, putting themselves in line with the two posts to be linked by the straight line.

Such findings are confirmed by experiments such as the "three mountains" task, in which children had to construct a scene from the perspective of a doll. For each new position of the doll, young children methodically went about their task of re-creating the appropriate viewpoint, but it always turned out to be from the same perspective: their own. Thus, Piaget and Inhelder infer that children must construct systems of reference not from familiarity born of experience, but rather, from operational linking and coordination of all possible viewpoints, each of which they are conscious. They conclude that such global coordination of viewpoints is the basic prerequisite in constructing simple projective relations. For although such relations are dependent upon a given viewpoint, nevertheless a single point of view cannot exist in an isolated fashion, but necessarily entails the construction of a complete system linking together all points of view.

Euclidean Space. Piaget and Inhelder next investigate the development of notions presumed to be intermediate between the projective and Euclidean spaces, for example, the idea of constructing similar figures. These experiments illustrate the gradual procurement of angle and parallelism concepts during middle childhood. Finally, in the development of Euclidean space, children come to "see" objects as located in a two-dimensional frame of reference. That is, Piaget and Inhelder challenge the claim that that there is an innate tendency or ability to organize objects in a two- or three-dimensional reference frame. Spatial awareness does not begin with such an organization; rather, the frame itself is a culminating point of the development of Euclidean space.

To test this hypothesis in the case of horizontality, children were shown jars half-filled with colored water and asked to predict the spatial orientation of the water level when the jar was tilted. For verticality, a plumb line was suspended inside an empty jar, which was similarly tilted, or children were asked to draw trees on a hillside. Children initially were incapable of representing planes; the water, for example, was represented by a scribble. At the next stage, the level of the water was always drawn perpendicular to the sides of the jar, regardless of tilt. Satisfaction with such drawings was in no way undermined even when an actual water-filled tilted jar was placed next to the drawing. It is, then, quite striking "how poorly commonly perceived events are recorded in the absence of a schema within which they may be organized" (p. 388). Sometimes, sensing that the water moves towards the mouth of the jar, children raised the level of the water, still keeping the surface perpendicular to the sides. Only at the final stage (at about 9 years of age) did children ostensibly draw upon the larger spatial frame of reference—the tabletop—in order to ascertain the horizontal.

Ultimately, the frame of reference constituting Euclidean space is analogous to a container made up of a network of sites or positions. Objects within this container may be mobile, but the positions are stationary. From the simultaneous organization of all possible positions in three dimensions emerges the Euclidean coordinate system. This organization is rooted in the preceding construction of the concept of straight line (as the maintenance of a constant direction of travel), parallels, and angles, followed by the coordination of their orientations and inclinations. This leads to a gradual replacement of relations of order and distance between objects with similar relations between the positions themselves. It is as if a space were emptied of objects so as to organize the space itself. Thus, intuition of space is not a "reading" or innate apprehension of the properties of objects, but a system of relationships born in actions performed on these objects.

Criticisms of Piaget and Inhelder's Work. As with much of the Genevan work, Piaget and Inhelder's theory of children's concept of space has been widely influential and widely criticized. One criticism has been that Piaget and Inhelder's use of terms such as topological, separation, proximity, and Euclidean, as well as the application of these and related concepts to the design of their studies, are not mathematically accurate (Darke, 1982; Kapadia, 1974; Martin, 1976a).

Related to this criticism is the problem of classifying figures as topological or Euclidean. Every figure possesses both these characteristics to an equivalent degree (that is, one cannot discuss an exclusively Euclidean figure), but Piaget and Inhelder's experiments depend on a mutually exclusive classification of figures into these two categories. Furthermore, because many of the figures they used were topologically equivalent (see Figure 18.1), one cannot be certain whether young children's choices were made on the basis of topological characteristics (Martin, 1976a). It is not clear why some anomalies such as these are dismissed as lack of drawing skill, whereas others, such as the inability to draw a square's straight sides, are not. Given such problems, replicative research is critical.

Other Research on the Theory. Researchers closely replicating Piaget and Inhelder's experiments have generally confirmed their findings (Laurendeau & Pinard, 1970; Lovell, 1959; Page, 1959; Peel, 1959). However, even within these studies ambiguities arise. For example, several researchers reported that even at the earliest ages (2–3 years) children can distinguish between curvilinear and rectilinear shapes, contrary to the theory (Lovell, 1959; Page, 1959).

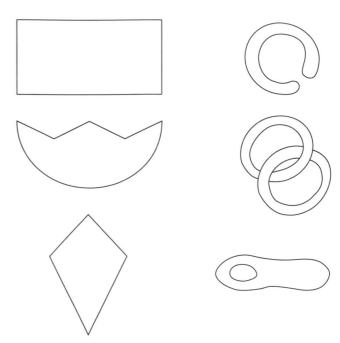

FIGURE 18–1. Shapes such as those on the left were considered Euclidean by Piaget and Inhelder in their haptic perception experiments; those on the right were considered to be topological forms.

However, it is still possible that children show a bias toward topological versus Euclidean characteristics. Here, too, basically corroborative results have been mixed with those contrary to predictions. For example, curvilinear shapes were identified at least as easily as purportedly topological ones (Laurendeau & Pinard, 1970; Lovell, 1959).

Several studies have attempted to ameliorate the problem of ambiguity by showing children a test shape and then, after its removal, asking children to identify a shape most like it. Esty (1970; cited in Darke, 1982) found that 4-year-olds classified the topologically equivalent shapes to be most like the original. However, older groups of children chose them as least like the original. Overall, the author claimed Piaget's thesis was supported with the important condition that distortions from the original were not substantial; this, however, is a metric concept as it involves measurement (see also Cousins & Abravanel, 1971; Jahoda, Deregowski, & Sinha, 1974). Martin (1976b) used a set of shapes and three variants: A was topologically equivalent to the model, while B and C, though not strictly equivalent to the model in a Euclidean sense, preserved as many Euclidean properties of the model as consistent with the fact that they had been altered to eliminate a particular topological property. Thus, B and C preserved properties such as straightness, curvature, line segment length, and angle size that A failed to preserve. But B failed to preserve connectedness and C varied closedness. Four-year-olds tended to choose the topologically equivalent copy as the worst copy of the model less often than older children did. But, the worst scores were at or above chance levels and, thus, did not support Piaget and Inhelder's theory. In addition, 4-year-olds sacrificed topological properties in their selections as freely as did 8-year-olds.

A difficulty in designing such experiments is in quantifying the degree of equivalence of the shapes. Geeslin and Shar (1979) modeled figures via a finite set of points on a grid. Degree of distortion was defined as the sum of displacements of these points. The authors postulate that children compare two figures in terms of the amount of "distortion" necessary to transform one figure into another, after an attempt at superimposition using rigid motions and dilations. This model received strong support. In agreement with other research, children preschool to grade 4 were cognizant of both topological and Euclidean properties and of how these properties distinguished variants. A small number of students at each level favored either topological or Euclidean properties. Note that, as the authors admit, these studies dealt with perception, while Piaget and Inhelder specifically address representation. Further, the model is more predictive than it is explanatory.

In sum, results of many of the Piagetian studies may be an artifact of the particular shapes chosen and the abilities of young children to identify and name these shapes (Fisher, 1965). If true, however, this does not support a strong version of the topological primacy thesis. It may not be topological properties as a class which enable young children to identify certain shapes. Visually salient properties (such as holes, curves, and corners), simplicity, and familiarity—rather than topological versus Euclidean properties—may underlie children's discrimination.

Similar problems have been found in drawing experiments (Dodwell, 1963; Lovell, 1959). Martin (1976b) reported that the drawings of 4-year-olds did not reflect predominantly topological features. He suggested that it may not be attention to topological properties that enables children to draw homeomorphic copies; rather, it may be their increasing coordination of Euclidean or projective properties, given that the coordination of such properties automatically preserves topological properties. Thus, despite evidence that many young children produce a circle when drawing a square, results do not confirm a strong topological primacy position. Research is needed that uses other techniques to infer children's internal representations and that more closely examines children's actions and thoughts in the process of drawing shapes. For example, one research program has confirmed a hierarchical developmental sequence of (a) reproduction of geometric figures requiring only encoding (that is, building a matching configuration of shapes with the original constantly in sight), (b) reproduction requiring memory (building a matching configuration from recall), and (c) transformation involving rotation and visual perspective-taking (building a matching configuration either from recall after a rotation or from another's perspective); preschool children are able to perform at only the first two levels (Rosser, Lane, & Mazzeo, 1988).

Research on children's construction of a frame of reference for Euclidean space—a coordinate system—similarly has revealed alterations and elaborations of Piaget and Inhelder's theory. For example, young children are more competent, and adolescents and adults are less competent, than the theory might suggest. Regarding the latter, not all high school seniors or college students are successful on Piaget's tasks designed to as-

sess an underlying Euclidean conceptual system (Liben, 1978; Mackay, Brazendale, & Wilson, 1972; Thomas & Jamison, 1975). On the other hand, it appears that young children's grasp of Euclidean spatial relationships is more adequate than the theory posits. Very young children can orient a horizontal or vertical line in space (Rosser, Horan, Mattson, & Mazzeo, 1984). Similarly, 4- to 6-year-old children can extrapolate lines from positions on both axes and determine where they intersect (Somerville & Bryant, 1985). Piagetian theory seems correct in postulating that the coordination of relations develops after such early abilities. Young children fail on double-axis orientation tasks, even when misleading perceptual cues are eliminated (Rosser et al., 1984). Similarly, the greatest difficulty is in coordinating two extrapolations, developing at the 3- to 4-year-old level, with the ability to extrapolate those lines developing as much as a year earlier (Somerville, Bryant, Mazzocco, & Johnson, 1987).

These results suggest an initial inability to utilize a conceptual coordinate system as an organizing spatial framework. Nevertheless, it is instructionally significant that, by the time they enter school, children can use coordinates when these are provided for them, even if, in facing traditional tasks, they are not yet able or predisposed spontaneously to construct coordinates for themselves.

However, performance on coordinate tasks is influenced by a variety of factors at all ages. Performance on horizontality and verticality tasks may reflect bias toward the perpendicular in copying angles, possibly because this reference is learned early (Ibbotson & Bryant, 1976). Representations of figures are also distorted, either locally by angle bisection or by increasing symmetry of the figure as a whole (Bremner & Taylor, 1982). Finally, performance on these Piagetian spatial tasks correlates with disembedding as well as with general spatial abilities (Liben, 1978). Such results indicate a general tendency to produce symmetry or simplicity in constructions which confound the traditional Piagetian interpretation (Bremner & Taylor, 1982; Mackay et al., 1972).

Such multiple determination of performance also characterizes the literature on perspective-taking abilities: Piaget and Inhelder's projective space. For example, perspective-taking tasks are easier if the children move around the objects or are provided a model of the room, suggesting the locations of the objects are coded individually with respect to an external framework of landmarks. Thus, coding the location of small objects may develop from association (coincidence) with a single external landmark, to proximity to a single landmark, to distance from several landmarks. By age 5, and possibly as early as 3, children encode the location of small objects with respect to a framework of landmarks. Such encoding continues into adulthood (Newcombe, 1989), implying that the development of projective space may involve not just the coordination of viewpoints but also the establishment of an external framework. For both perspective-taking and coordinate system abilities, the key may be the construction and selection of increasingly coordinated reference systems as frameworks for spatial organization.

Conclusion. Overall, while not totally disproven, the topological primacy theory is not supported. It may be that children do not construct first topological and later projective and

Euclidean ideas. Rather, it may be that ideas of all types develop over time, becoming increasingly integrated and synthesized. These ideas are originally intuitions grounded in action—building, drawing, and perceiving. Thus, research is needed to identify the specific, original intuitions and ideas that develop and the order in which they develop. For example, children might learn to coordinate certain actions that produce curvilinear shapes before coordinating those that produce rectilinear ones (Martin, 1976a), although theoretical explanations of such sequences are lacking. Observed lack of synchrony between perceptual and conceptual abilities supports Piaget's constructivist position (Rosser et al., 1984), but specific cognitive constructions generally have not been identified.

Such research should also explore the deformations that children do accept in their representation of figures. For example, some children equate an "almost closed" figure with a closed variant. Previously unexplored factors such as language, schooling, and the immediate social culture also demand attention (Darke, 1982). Piagetian interpretations of students' performance on tasks too often emphasize logical failures at the expense of uncovering the development of ideas not yet differentiated and integrated; thus, new approaches are warranted. In any case, it appears that certain Euclidean notions are present at an early age (Rosser et al., 1984; Rosser et al., 1988) and, contrary to Piaget and Inhelder and interpreters (Peel, 1959), even preschool children may be able to work with certain Euclidean ideas.

Similarly, results regarding Piaget and Inhelder's constructs of projective and Euclidean space reveal that young children have basic competencies in establishing spatial frameworks that could be effectively built upon in the classroom. However, we should probably expect, in students of all ages, a general Gestalt tendency toward symmetry and simplicity, for example, in matching and reproduction tasks. Research is needed to identify instruction facilitating the construction and selection of increasingly sophisticated reference systems for organizing spatial information.

Researchers have tended not to discuss Piaget and Inhelder's second major theme: Children's representation of space is not a perceptual "reading off" of their spatial environment, but is constructed from prior active manipulation of that environment. This is surprising, in that most of Piaget and Inhelder's results do support this hypothesis at least implicitly (results from one study, Wheatley & Cobb, 1990, provide direct support, as discussed in a succeeding section). More articulated research from a constructivist position is needed. In this regard, we turn briefly to other work consistent with this critical idea.

Fischbein (1987) argues that people's intuition of space is not innate and not reducible to a conglomerate of sensorial images. Space representations constitute a complex system of conceptions—although not necessarily formulated explicitly—which exceed the data at hand and the domain of perception in general. Subjective space is an interpretation of reality, not a reproduction of it. It is shaped by and exceeds experience. Consistent with Piaget's constructivism, Fischbein's theory further elaborates the nature of the intuition of space. To begin, intuition is not merely a reflection of objectively given space properties; rather, it is a "highly complex system of expecta-

tions, and programs of action, related to the movements of our body and its parts, which constitutes the intuition of space" (p. 87). Thus, intuitions consist of sensorimotor and intellectual skills organized into a system of beliefs and expectations that constitute an implicit theory of space. Most important, intuitions thus constructed are enactively meaningful; they are subjectively self-evident because they express the direct behavioral meaningfulness of an idea.

For example, the notion of straight line seems self-evident. A sophisticated adult is convinced that one may go on extending the line indefinitely, or that by following the straight line one uses the shortest path to reach a given point. These appear unequivocal "facts," properties of the "object" called a straight line. However, the straight line is an abstraction, not a perceptual object. It is a convention based on axioms which could be changed. It is through extrapolation from a behavioral meaning that one tends to believe in the absoluteness of the conception. People know that they can draw a straight line, recognize a straight line, and run along a straight line to reach a goal in minimum distance. They imbue the notion with the qualities of unequivocal evidence and credibility because it is behaviorally meaningful.

However, building intuition based on experience cuts both ways. The limitations of human experience account not only for the adaptive and organizing functions of intuitions, but also for distorted or erroneous representations of reality. Thus, space intuitions, like other intuitions, do not develop inevitably into increasing correspondence with pure logic or mathematics, as a reading of Piaget may suggest. Intuitive representations of space are non-homogeneous and anisotropic (exhibiting properties with different values when measured along axes in different directions). For example, people tend to attribute absolutely privileged directions to space, such as "up" and "down." They view space as centered (for example, at one's home) and having increasing density as one approaches the centration zones, with the effect that distances are increasingly amplified upon approach. Thus, our intuitive representation of space is a mixture of possibly contradictory properties, all related to our terrestrial life and our behavioral adaptive constraints (Fischbein, 1987).

The van Hieles: Levels of Geometric Thinking and Phases of Instruction

Levels of Geometric Thought. According to the theory of Pierre and Dina van Hiele, students progress through levels of thought in geometry (van Hiele, 1959; van Hiele, 1986; van Hiele-Geldof, 1984), from a Gestalt-like visual level through increasingly sophisticated levels of description, analysis, abstraction, and proof. The theory has the following defining characteristics:

- Learning is a discontinuous process. That is, there are "jumps" in the learning curve which reveal the presence of discrete, qualitatively different levels of thinking.
- The levels are sequential and hierarchical. For students to function adequately at one of the advanced levels in the van Hiele hierarchy, they must have mastered large portions of

the lower levels (Hoffer, 1981). Progress from one level to the next is more dependent upon instruction than on age or biological maturation. Teachers can reduce subject matter to a lower level, leading to rote memorization, but students cannot bypass levels and still achieve understanding (memorization is not an important feature of any level). Acquiring understanding requires working through certain phases of instruction.

- Concepts implicitly understood at one level become explicitly understood at the next level. "At each level there appears in an extrinsic way that which was intrinsic at the preceding level. At the base level, figures were in fact also determined by their properties, but someone thinking at this level is not aware of these properties" (van Hiele, 1959/1985, p. 246).

- Each level has its own language. "Each level has its own linguistic symbols and its own system of relations connecting these symbols. A relation which is 'correct' at one level can reveal itself to be incorrect at another. Think, for example, of a relation between a square and a rectangle. Two people who reason at different levels cannot understand each other. Neither can manage to follow the thought processes of the other" (van Hiele, 1959/1985, p. 246). Language structure is a critical factor in the movement through the levels.

Both the number and numbering of the levels have been variable. We shall initially describe the van Hieles' original five levels and later discuss a sixth. Our rationale is twofold. First, although van Hiele's recent works have described three rather than the original five levels, both empirical evidence (reviewed in succeeding sections) and the need for precision in psychologically oriented models of learning argue for maintaining finer delineations. Second, the empirical evidence also suggests a level that is more basic than van Hiele's "visual" level. There have been several different numbering systems used for the levels; we have adopted one system and have transposed those of each researcher to this scheme for consistency's sake.

LEVEL 1: VISUAL. Initially, students identify and operate on shapes and other geometric configurations according to their appearance. They recognize figures as visual gestalts, and, thus, they are able to mentally represent these figures as visual images. In identifying figures, they often use visual prototypes; students say that a given figure is a rectangle, for instance, because "it looks like a door." They do not, however, attend to geometric properties or to characteristic traits of the class of figures represented. That is, although figures are determined by their properties, students at this level are not conscious of the properties. At this level, students' reasoning is dominated by perception. For example, they might distinguish one figure from another without being able to name a single property of either figure, or they might judge that two figures are congruent because they look the same; "There is no why, one just sees it" (van Hiele, 1986, p. 83). During students' transition from the visual to the descriptive level, classes of visual objects begin to be associated with their characteristic properties.

At the visual level, the objects about which students reason are classes of figures recognized visually as "the same shape." For instance, by the statement "This figure is a rhombus," the

student means "This figure has the shape I have learned to call 'rhombus'" (van Hiele, 1986, p. 109). The end product of this reasoning is the creation of conceptualizations of figures that are based on the explicit recognition of their properties (that is, after this conceptual construction, the student is at Level 2).

LEVEL 2: DESCRIPTIVE/ANALYTIC. Upon reaching the second level, students recognize and can characterize shapes by their properties. For instance, a student might think of a rhombus as a figure with four equal sides; so the term "rhombus" refers to a collection of "properties that he has learned to call 'rhombus'" (van Hiele, 1986, p. 109). Students see figures as wholes, but now as collections of properties rather than as visual gestalts; the image begins to fall into the background. Properties are established experimentally by observing, measuring, drawing, and modeling. Students discover that some combinations of properties signal a class of figures and some do not; thus, the seeds of geometric implication are planted. Students at this level do not, however, see relationships between classes of figures (for example, a student might contend that a figure is not a rectangle because it is a square).

At this level, the objects about which students reason are classes of figures, thought about in terms of the sets of properties that the students associate with those figures. The product of this reasoning is the establishment of relationships between and the ordering of properties and classes of figures.

LEVEL 3: ABSTRACT/RELATIONAL. At Level 3, students can form abstract definitions, distinguish between necessary and sufficient sets of conditions for a concept, and understand and sometimes even provide logical arguments in the geometric domain. They can classify figures hierarchically (by ordering their properties) and give informal arguments to justify their classifications; a square, for example, is identified as a rhombus because it can be thought of as a "rhombus with some extra properties." They can discover properties of classes of figures by informal deduction. For example, they might deduce that in any quadrilateral the sum of the angles must be 360° because any quadrilateral can be decomposed into two triangles, each of whose angles sum to 180°.

As students discover properties of various shapes, they feel a need to organize the properties. One property can signal other properties, so definitions can be seen not merely as descriptions but as a method of logical organization. It becomes clear why, for example, a square is a rectangle. This logical organization of ideas is the first manifestation of true deduction. However, the students still do not understand that logical deduction is the method for establishing geometric truths.

At this level, the objects about which students reason are properties of classes of figures. Thus, for instance, the "properties are ordered, and the person will know that the figure is a rhombus if it satisfies the definition of quadrangle with four equal sides" (van Hiele, 1986, p. 109). The product of this reasoning is the reorganization of ideas achieved by interrelating properties of figures and classes of figures.

LEVEL 4: FORMAL DEDUCTION. Students establish theorems within an axiomatic system when they reach Level 4. They recognize the difference among undefined terms, definitions, axioms, and theorems. They are capable of constructing original proofs; that is, they can produce a sequence of statements that logically justifies a conclusion as a consequence of the "givens."

At this level, students can reason formally by logically interpreting geometric statements such as axioms, definitions, and theorems. The objects of their reasoning are relationships between properties of classes of figures. The product of their reasoning is the establishment of second-order relationships—relationships between relationships—expressed in terms of logical chains within a geometric system.

LEVEL 5: RIGOR/METAMATHEMATICAL. At the fifth level students reason formally about mathematical systems. They can study geometry in the absence of reference models, and they can reason by formally manipulating geometric statements such as axioms, definitions, and theorems. The objects of this reasoning are relationships between formal constructs. The product of their reasoning is the establishment, elaboration, and comparison of axiomatic systems of geometry.

Research on the Levels. As with the work of Piaget and Inhelder, van Hiele's theory has been influential and extensively studied. Research results will be discussed under the rubric of several critical questions.

DO THE VAN HIELE LEVELS ACCURATELY DESCRIBE STUDENTS' GEOMETRIC THINKING? Generally, empirical research, from both the U.S. and abroad, has confirmed that the van Hiele levels are useful in describing students' geometric concept development, from elementary school to college (Burger & Shaughnessy, 1986; Fuys et al., 1988; Han, 1986; Hoffer, 1983; Wirszup, 1976). For example, Usiskin (1982) found that about 75% of secondary students fit the van Hiele model (it should be noted that the percentage classifiable at a level varies with the instrument and scoring scheme). Burger and Shaughnessy (1986) administered clinical interviews to students from kindergarten to college. They reported that students' behaviors were generally consistent with the van Hieles' original general description of the levels. For example, students were to identify and describe all the squares, rectangles, parallelograms, and rhombuses in a set of quadrilaterals similar to those in Figure 18.2. Students who included imprecise visual qualities and irrelevant attributes (for example, orientation) in describing the shapes while omitting relevant attributes were assigned to Level 1. References to visual prototypes ("a rectangle looks like a door") were frequent among students assigned to this level. Students who contrasted shapes and identified them by means of their properties were assigned Level 2. One girl, for example, said that rectangles have "two sides equal and parallel to each other. Two longer sides are equal and parallel to each other, and they connect at 90 degrees" (p. 39). Squares were not included. Students who gave minimal characterization of shapes by referencing other shapes were assigned Level 3 (for example, a square is a parallelogram that has all the properties of a rhombus and a rectangle). One student frequently made conjectures and attempted to verify these conjectures by means of formal proof, indicating Level 4 thinking.

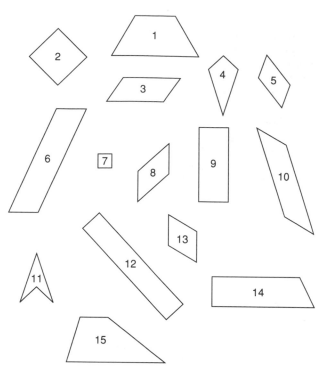

FIGURE 18–2. Quadrilaterals to be identified.

The existence of unique linguistic structures at each level has been supported in that, for example, "rectangle" means different things to students at different levels (for example, a visual gestalt vs. a "bearer of properties") (Burger & Shaughnessy, 1986; Fuys et al., 1988; Mayberry, 1983). In sum, the levels appear to exist and describe students' geometric development, validated through both interviews and written assessments.

ARE THE LEVELS DISCRETE? IS THERE A DISCONTINUITY BETWEEN LEVELS? Soviet research seems to indicate a positive answer (Hoffer, 1983; Wirszup, 1976). On the whole, however, results are mixed. First, several researchers have reported that students in transition are difficult to classify reliably (Fuys et al., 1988; Usiskin, 1982); this is especially true for Levels 2 and 3 (Burger & Shaughnessy, 1986). Difficulties in deciding between levels were considered by these researchers as evidence questioning the discrete nature of the levels.

Fuys et al. (1988) developed and documented a working model of the van Hiele levels and characterized the geometric thinking of sixth and ninth graders. The researchers used six to eight 45-minute instructional-assessment interviews, which allowed them to chart students' ability to progress within and between levels as a result of instruction. They determined both an entry level and a potential level (the level demonstrated after instruction). Whereas some of the students appeared to be on a plateau, there were also those who moved flexibly to different levels during the teaching episodes; an entry level assessment alone might have underestimated their abilities (Vygotsky, 1934/1986). Further, there was instability and oscillation between the levels in several cases. Similar results were observed in a teaching experiment on polyhedra (Lunkenbein,

1983). Continuity rather than jumps in learning was frequently observed.

DO STUDENTS REASON AT THE SAME VAN HIELE LEVELS ACROSS TOPICS? This question is also relevant to the issue of the discreteness of the levels, and evidence on this question is similarly mixed. A test of consensus revealed that pre-service elementary teachers were on different levels for different concepts (Mayberry, 1983), as were middle school (Mason, 1989) and secondary students (Denis, 1987). Similarly, Burger and Shaughnessy (1986) reported that students exhibited different preferred levels on different tasks. Some even oscillated from one level to another on the same task under probing. The researchers characterized the levels as dynamic rather than static and of a more continuous nature than their discrete descriptions would lead one to believe. Gutiérrez and Jaime (1988) compared the level of reasoning of pre-service teachers on three geometric topics: plane geometry, spatial geometry (polyhedra), and measurement. The levels reached across topics were not independent, but the data did not support the theorized global nature of the levels. The researchers hypothesized that as students develop, the degree of the globality of the levels is not constant, but increases with level. That is, as children develop, they grasp increasingly large "localities" of mathematical content and thus understand larger areas of mathematics.

Fuys et al. (1988) agreed that when first studying a new concept, students frequently lapsed to Level 1 thinking. They maintained, however, that students were quickly able to move to the higher level of thinking they had reached on prior concepts. The researchers, therefore, claim that these results support the contention that a student's *potential* level of thinking remains stable across concepts. The question is still open, but there has been the suggestion that assessment instruments must be topic specific (Senk, 1989).

DO THE LEVELS FORM A HIERARCHY? Research more consistently indicates that the levels are hierarchical, although here too there are exceptions (Mason, 1989). For example, Mayberry (1983) employed Gutman's scalogram analysis to show that her tasks representing the levels formed a hierarchy for pre-service teachers. These results were replicated by Denis (1987) for Puerto Rican secondary students. Gutiérrez and Jaime (1988) reported similar analysis and results, but only for Levels 1 to 4; Level 5 was found to be different in nature from the other levels. Most other researchers did not test the hypothesis in a similarly analytic manner; nevertheless, they interpret their results as supporting this hypothesis (Burger & Shaughnessy, 1986; Fuys et al., 1988; Usiskin, 1982).

Thus, the levels appear to be hierarchical, although there remains a need to submit this hypothesis to rigorous tests. As the van Hieles posited, however, this does not imply a maturational foundation. First, assignments to levels does not seem to be strictly related to age or grade (Burger & Shaughnessy, 1986; Mayberry, 1983). Second, development through the hierarchy appears to proceed under the influence of a teaching/learning process (Wirszup, 1976).

WHAT IS THE MOST BASIC LEVEL; THAT IS, DOES A LEVEL 0 EXIST? As previously mentioned, there is evidence for the existence of a Level 0 more basic than the van Hiele's "visual" level. For example, 9–34% of secondary students have failed to demonstrate thinking characteristic of even the visual level; 26% of the students who began the year at Level 0 remained at Level 0 at the end of the year (Usiskin, 1982). Such stability argues for the existence of level 0 (Senk, 1989). Likewise, 13% of the response patterns of pre-service teachers do not meet the criterion for Level 1 (Mayberry, 1983). Finally, students who enter a geometry course at Level 1 perform significantly better at writing proofs than those who enter at Level 0 (Senk, 1989).

This issue is not resolved, however. Fuys et al. (1988) specified that to be "on a level" students had to consistently exhibit behaviors indicative of that level. They also state, however, that Level 1 is different from the other levels, in that students may not be able to exhibit the corresponding behaviors (that is, they may not be able to name shapes). According to the researchers, these students should not be described as "not yet at Level 1." Nonetheless, they do remark such behaviors as "weak Level 1." In a later work, Fuys (1988) hypothesizes that students with weak Level 1 thinking are using one of Reif's (1987) case-based models as a foundation for their concepts, rather than a more sophisticated, rule-based model. Whether this actually argues for a separate level or for sublevels is as yet an open question.

However, the bulk of the evidence from van Hiele-based research, along with research from the Piagetian perspective, indicates the existence of thinking more primitive than, and probably prerequisite to, van Hiele's Level 1. Therefore, we postulate the following additional level:

LEVEL 0: PRE-RECOGNITION. At the pre-recognition level, children perceive geometric shapes, but perhaps because of a deficiency in perceptual activity, may attend to only a subset of a shape's visual characteristics. They are unable to identify many common shapes. They may distinguish between figures that are curvilinear and those that are rectilinear but not among figures in the same class; that is, they may differentiate between a square and a circle, but not between a square and a triangle. According to Piaget, "it is one thing to perceive a circle or a square and quite another to reconstruct a visual image of it to the point where it can be picked out from a group of models, or drawn after a purely tactile exploration" (Piaget & Inhelder, 1967, p. 37). Thus, students at this level may be unable to identify common shapes because they lack the ability to form requisite visual images. These images presuppose mental representations constructed from the child's own actions. That is, "The image is at first no more than an internal imitation of previously performed actions, then later, of actions capable of being performed" (Piaget & Inhelder, 1967, p. 449).

At this level, the "objects" about which students reason are specific visual or tactile stimuli; the product of this reasoning is a group of figures recognized visually as "the same shape."

SHOULD OTHER CHARACTERISTICS OF THE LEVELS BE CONSIDERED? Levels are complex structures involving the development of both concepts and reasoning processes (Burger & Shaughnessy, 1986). In addition, researchers have emphasized the importance of several interrelated notions: intent, belief systems, and metacognition. Fuys et al. (1988) posit that at each level students must become aware of what is expected, intentionally thinking in a certain way. For example, students at higher levels used such language as "explain," "provide it," "clinch it," and "be technical" in justifying their reasoning. Students at lower levels believed that they should respond to a task on paper exactly as it appeared (for example, changing its orientation is not allowed). More of these students labeled an oblique obtuse triangle a "triangle" when a manipulative triangle was used (Fuys et al., 1988).

Actually, metacognitive knowledge always has been an implicit part of the van Hiele model, in its emphasis on intent and insight, or understanding. According to Hoffer (1983), students show such understanding when they perform competently and intentionally a method that resolves an unfamiliar problem. They understand what they are doing, why they are doing it, and when it should be done. If certain beliefs, intentions, and the related metacognitive and even epistemic knowledge characterize each level, they need to be further articulated and incorporated into the model.

WHAT LEVELS OF THINKING ARE EVINCED GIVEN "TRADITIONAL" INSTRUCTIONAL PARADIGMS? Once the basic characteristics of the model have been generally validated, the question arises: What levels of geometric thinking are achieved by students in their present educational environment? Studies by Pyshkalo and Stolyar indicated a significant number of Soviet students were perceiving shapes only as wholes. Students stayed at Level 1 for a considerable time; by the end of grade 5, only 10–15% reached Level 2 (note that Soviet students enter grade 1 at age 7, compared to age 6 in the U.S.). This delay was even greater with respect to solids, for which there was no noticeable leap until the seventh grade (Pyshkalo, 1968; Wirszup, 1976).

Of the 16 sixth graders they studied, Fuys (1988) found that 19% were uniformly at Level 1, focusing on shapes as a whole without analyzing shapes in terms of their properties, even after instruction. They could identify familiar shapes singly, but not in complex configurations and sometimes not in different orientations. They had great difficulty with the concept of angle. They gained only a little Level 1 knowledge—visual thinking about shapes and parallelism—from work with manipulatives. The authors described these students as "geometry deprived." Another 31% made progress within Level 1 and were progressing toward Level 2. The final 50% began with Level 2 thinking and progressed toward Level 3. Nevertheless, they had to review some Level 1 knowledge and firm up ideas at Level 2. Some made deductive arguments, but most equated "proof" with generalization by examples (inductive reasoning).

The ninth graders similarly fell into three groups. As with the lowest group of sixth graders, about 12% of the ninth graders' little school experience with geometry, coupled with language and memory difficulties, resulted in Level 1 performance. They seldom realized that they could figure things out in mathematics by thinking about them, and progress within Level 1 was limited. The 44% in the middle group functioned at Level 2, with lapses to Level 1. They knew *familiar* shapes in terms of their properties, but had no knowledge of parallelograms and trapezoids. Another 44% performed consistently at Level 2, with progress to Level 3, working more rapidly and confidently. They were not only thoughtful and inventive about the problems they were doing but were also reflective about their own thinking.

Secondary students do not fair much better. Many students who have studied geometry formally are nonetheless on Levels 0 to 2, not Level 3 or 4; almost 40% finish high school geometry below Level 2 (Burger & Shaughnessy, 1986; Suydam, 1985; Usiskin, 1982). In fact, because many students have not developed Level 3 thought processes, they may not benefit from additional work in formal geometry because their knowledge and the information presented in the textbook will be organized differently.

WHAT LEVELS OF THINKING DO TRADITIONAL TEXTBOOKS REFLECT? Given the sorry state of students' level of geometric thinking, it is natural to ask what levels are promoted by textbooks. Fuys et al. (1988) analyzed several current geometry curricula as evidenced by American text series (grades K–8) in light of the van Hiele model. Four components of geometry lessons were analyzed: the aim, expository material, exercises, and related test and review questions. Not surprisingly, textbook series were found to be deficient in this aspect. Most work involved naming shapes and relations like parallelism, and students were only infrequently asked to reason with the figures.

Most questions were answerable at Level 1. There was little Level 2 or above thinking required in the lessons or tests, starting only slightly in grades 7–8. Average students would not need to think above Level 1 for almost all of their geometry experiences through grade 8. There were some jumps across levels; for example, exposition might occur at a higher level than the exercises. Topics were repeated across grades at the same level; the researchers termed this a "circular" rather than a "spiral" curriculum. Worse, perhaps, properties and relationships among polygons were sometimes not taught clearly or correctly.

Similar analyses of older Soviet textbooks (those written before several major reforms) revealed the absence of any systematic choice of geometric material, large gaps in its study, and a markedly late, one-sided acquaintance with many of the most important geometric concepts (Wirszup, 1976). Only about 1% of all problems dealt with geometry. This left sixth-grade students, from the very first lessons, doing work corresponding to the first three levels of geometric development simultaneously.

Phases of Instruction. The van Hiele model includes more than levels of geometric thinking. According to the van Hieles, progress from one level to the next depends little on biological maturation or development; instead, it proceeds under the influence of a teaching/learning process. The teacher plays a special role in facilitating this progress, especially in providing guidance about expectations (Fuys et al., 1988). Given that van Hiele level and achievement account for 40% to 60% of the variance in writing proofs, much of a student's achievement in this area is directly controlled by the teacher and the curriculum (Senk, 1989).

The van Hiele theory, though, does not support an "absorption theory" model of learning and teaching. The van Hieles claim that higher levels are achieved not via direct teacher telling, but through a suitable choice of exercises. Also, "children themselves will determine when the moment to go to the higher level has come" (P. van Hiele, personal communication, Sept. 27, 1988). Nevertheless, without the teacher, no progress would be made. For each phase, we will describe the goal for student learning and the teacher's role in providing instruction that enables this learning.

PHASE 1: INFORMATION. The students become acquainted with the content domain. The teacher discusses materials clarifying this content, placing them at the child's disposal. Through this discussion, the teacher learns how students interpret the language and provides information to bring students to purposeful action and perception.

PHASE 2: GUIDED ORIENTATION. In this phase, students become acquainted with the objects from which geometric ideas are abstracted. The goal of instruction during this phase is for students to be actively engaged in exploring objects (for example, folding, measuring) so as to encounter the principal connections of the network of relations that is to be formed. The teachers' role is to direct students' activity by guiding them in appropriate explorations—carefully structured, sequenced tasks (often one-step, eliciting specific responses) in which students manipulate objects so as to encounter specific concepts and procedures of geometry. Teachers should choose materials and tasks in which the targeted concepts and procedures are salient.

PHASE 3: EXPLICITATION. Students become conscious of the relations and begin to elaborate on their intuitive knowledge. Thus, in this phase, children become explicitly aware of their geometric conceptualizations, describe these conceptualizations in their own language, and learn some of the traditional mathematical language for the subject matter. The teacher's role is to bring the objects of study (geometric objects and ideas, relationships, patterns, and so on) to an explicit level of awareness by leading students' discussion of them in their own language. Once students have demonstrated their awareness of an object of study and have discussed it in their own words, the teacher introduces the relevant mathematical terminology.

PHASE 4: FREE ORIENTATION. Children solve problems whose solution requires the synthesis and utilization of those concepts and relations previously elaborated. They learn to orient themselves within the "network of relations" and to apply the relationships to solving problems. The teacher's role is to select appropriate materials and geometric problems (with multiple solution paths), to give instructions to permit various performances and to encourage students to reflect and elaborate on these problems and their solutions, and to introduce terms, concepts, and relevant problem-solving processes as needed.

PHASE 5: INTEGRATION. Students build a summary of all they have learned about the objects of study, integrating their knowledge into a coherent network that can easily be described and applied. The language and conceptualizations of mathematics are used to describe this network. The teacher's role is to encourage students to reflect on and consolidate their geometric knowledge, increasing emphasis on the use of mathematical structures as a framework for consolidation. Finally, the consolidated ideas are summarized by embedding them in the structural organization of formal mathematics. At the completion of Phase 5, a new level of thought is attained for the topic studied.

Critical Issues

ISSUES REGARDING LEVELS OF THINKING. There are problems with the research on the veracity of the theorized levels. For example, Fuys et al. (1988) interpret their results as supporting their validity. However, they also claim support for van Hiele's recent characterization of the model in terms of three levels: visual (previously Level 1, according to these researchers), analytic (previously 2), and theoretical (previously 3–5). They state that van Hiele agrees with this interpretation, but they caution that the three-level model may not be sufficiently refined to characterize thinking, especially considering their findings that students progressed toward Level 3 with no sign of axiomatic thinking. There are two additional problems with the three-level model, however. First, it seems that van Hiele describes the new visual level as combining aspects of the previous levels 1 and 2; therefore, the mapping from one model to the other is not unambiguous. Furthermore, if levels can be changed and combined, their hypothesized discrete, hierarchical psychological nature must be questioned. In a related vein, we have seen reports of both stronger and weaker performances at certain levels, and overlaps between levels; the question is, how wide a band can be permitted before the notion of hierarchical dependency disintegrates?

It is not even clear when a student is "at" a level. What does it mean for students to think of shapes in terms of their properties? Do students achieve Level 2 when they evince cognizance of the characteristics of shapes, or must they identify specific properties? When do students think *primarily* in terms of properties? Do they have to identify properties of specific shapes or classes of shapes?

Further, should students' thinking be characterized as "at" a single level? For example, Gutiérrez, Jaime, and Fortuny (1991) attempt to take into account students' capacity to use each van Hiele level rather than assign a single level. They use a vector with four components to represent the degree of acquisition of van Hiele Levels 1 through 4 (for example, one student might have a grade component for Level 1 of 96.67%; Level 2, 82.50%; Level 3, 50.00%; and Level 4, 3.75%; the researchers could not measure Level 5 to their satisfaction). They found many students who are apparently developing two consecutive levels of reasoning simultaneously, and hypothesized these results from mathematics instruction that leads students to begin the acquisition of level $n + 1$ before level n had been completely acquired (Gutiérrez et al., 1991). Such alternate conceptualizations of levels of thinking need to be explored, as they may bring into question the very nature of the levels; that is, the levels seem to have face validity, but if the number of levels is malleable and if performance is spread across levels and determined by what is taught, then it is unclear whether the levels are more logical or psychological.

Another problem with the levels is the observed lack of "discontinuities in learning." Some have defended the theory in the face of this evidence, claiming that the observations may reflect continuity not in learning but rather in teaching (Fuys et al., 1988; Hoffer, 1983). This is an open question, but a problem with the defense is that it makes it virtually impossible to disprove the theory, a criticism frequently waged against Piaget's theories. If there is a great deal of "transition," then (also like Piaget) this brings into question a strict stage interpretation.

Questions also arise concerning observations of reasoning at different levels across topics. Some, attempting to make minimal elaborations to the theory, have hypothesized that these students can "quickly move to the higher level of thinking" regarding the lower-level topic. It is not clear exactly what this means. Would others move reliably more slowly, and is this speed not attributable to other factors such as learning potential not directly tied to levels of thinking?

Thus, it is critical that research be conducted on valid assessment of van Hiele levels. Paper and pencil testing should be further refined and evaluated (for a recent discussion of this issue see Crowley, 1990; Usiskin & Senk, 1990; and Wilson, 1990). Different interview techniques, possibly less dependent on specific educational experiences, should be developed. For example, we have created a triad polygon sorting task designed to determine the level of geometric thinking for polygons. These tasks were created by the authors and Richard Lehrer. Students are presented with three polygons and asked, "Which two are most alike? Why?" For example, one student, presented with the following shapes, A □ B △ C △ chose B and C, saying that they "looked the same, except that B is bent in." She was attending to the visual aspects of the shapes, a Level 1 response. After working with our Logo-based geometry curriculum (Battista & Clements, 1990), the student chose A and B, saying that they both had four sides. Thus, she tended to let the overall visual aspect of the figures fade into the background, attending instead to the shapes' properties, a Level 2 response. Finally, research is needed on the relative usefulness of static and dynamic approaches to assessment. Dynamic approaches, which can assess "potential" level of thinking and the amount of instruction students need to achieve that level, may be more illuminating than the more typical static, or "snapshot," approach (Vygotsky, 1934/1986).

The way in which students, especially young students, learn geometric concepts has also been questioned. First, research demonstrates that young children can discriminate some of the characteristics of shapes, and often think of two-dimensional figures in terms of paths and motions used to construct them (Battista & Clements, 1987; Clements & Battista, 1989, 1990; Kay, 1987). This is inconsistent with the levels as presently conceived. Secondly, while young children are currently taught by a "template" (visual prototype) approach to recognizing geometric patterns, Kay (1987) maintains that this is appropriate if there is only one such template for each class (this does not apply to hierarchical-based classes). In contrast, Kay provided first graders with instruction that began with the more general case,

quadrilaterals, proceeded to rectangles, and then to squares. It addressed the relevant characteristics of each class and the hierarchical relationships among classes, using terms that embody these relationships: quadrilateral, rectangle-quadrilateral, and square-rectangle. At the end of instruction, most students identified characteristics of quadrilaterals, rectangles, and squares, and about half identified hierarchical relationships among these classes, though none had done so previously. Thus, Kay maintains that the van Hiele theory does not capture the full complexity of how young children come to understand geometric concepts. Some concepts always will be initially understood through inductive processes if the definition of the concept involves a complex deductive argument yet can be represented by a small number of visual templates (for example, "circle"). If the definition of the concept involves a relatively simple deductive argument and the concept cannot be represented easily by a template, then initial understanding will be deductive (for example, "quadrilateral"). This dichotomy is similar to Vygotsky's (1934/1986) formulation of spontaneous vs. scientific concepts. While both the depth of these first graders' understanding (especially of hierarchical relations) and the generalizations made on the basis of the empirical results must be questioned, such alternate hypotheses deserve further investigation. Future investigations should ensure that students are not simply mirroring repetitive verbal training; "Direct teaching of concepts is impossible and fruitless. A teacher who tries to do this usually accomplishes nothing but empty verbalism, a parrotlike repetition of words by the child, simulating a knowledge of the corresponding concepts but actually covering up a vacuum" (Vygotsky, 1934/1986, p. 150).

In a similar vein, de Villiers (1987) concluded that, contrary to van Hiele's theory, hierarchical class inclusion and deductive thinking develop independently and depend more on teaching strategy than on van Hiele level. He then describes a successful teaching strategy in which eighth- and ninth-grade students were taught first about quadrilaterals, and how special quadrilaterals could be obtained by specifying properties. This approach was contrasted with the traditional approach in which students associate names of figures with visual prototypes.

The defining quality associated with the name is therefore determined by the visual perception of the figure....We believe that the observation that children think of shapes as a whole without explicit reference to their components, is the direct result of our actually teaching children from the start to think of shapes as a whole and in terms of visual prototypes, and with no reference to their components." (p. 19)

However, the author makes this claim based on experiments with students intellectually capable of attending to properties. Our research, for instance, indicated that after being taught about the properties of squares and rectangles, many first graders explain why they say that squares are special kinds of rectangles by simply saying "because the teacher told us" (Battista & Clements, 1990). However, the criticism of the van Hiele levels raised by de Villiers, that the levels are very dependent on the curriculum, is certainly worthy of further research. He concluded that further research on the level at which both hierarchical classification and deduction occur is needed.

Such questions lead to the conclusion that, while van Hiele research has added to our knowledge considerably, the corpus has not yet been structured so as to simultaneously test alternate hypotheses (for example, finding students whose behavior seems to support a characteristic of the levels does not provide a strong test of competing hypotheses for the given behavior). In addition, the van Hiele theory describes students' behaviors; we also need to account for them. Research needs to address such questions as:

- How specifically is students' knowledge represented and structured at each level? Are new operations and concepts always constructed out of those that came before, as in Piaget's theory?

- Do levels represent discrete stages of major knowledge reorganization? That is, can the levels properly be described as stages? For instance, do they satisfy the following criteria as described by Steffe and Cobb (1988):

    1. *Constancy:* some property, state, or activity remains constant throughout each stage;
    2. *Incorporation:* the earlier stage must become incorporated in the next;
    3. *Order invariance:* the stages must emerge developmentally in a constant order; and
    4. *Integration:* the structural properties that define a given stage must form an integrated whole.

- Can we operationalize the levels? Most studies have used different testing instruments, some of which are content-oriented, while others are process-oriented. In addition, Fuys (1988) has suggested that the mode of presentation—verbal, pictorial, or concrete—might influence students' performance on such tasks. Because the levels clearly depend on instruction, we must be especially careful to consider the relationship between instruction and levels in all future research.

- Exactly what ideas do students construct and what mental operations must be attained in learning geometry? How does this development occur in the early years?

- Does a transition from one level to the next depend on the acquisition of certain types of knowledge, a restructuring of knowledge, or both? Does this vary by topic, especially topics outside of plane geometry?

- How can level of thinking be related to, yet distinguished from achievement (Senk, 1989)?

- What curriculum factors help facilitate transitions from one level to the next? This brings us to the next set of critical issues.

ISSUES REGARDING IMPLICATIONS FOR TEACHING. Theory and research from the van Hiele perspective has strong implications for instruction; most of these implications have not been adequately addressed in the research literature.

- *Educational goals for levels of thinking.* Opportunities for the construction of geometric ideas should be offered early. Students do not reach the descriptive level of geometry in part because they are not offered geometric problems in their early years (van Hiele, 1987). This "prolonged period of geo-

metric inactivity" (Wirszup, 1976, p. 85) in the early grades leads to "geometricly deprived" children (Fuys et al., 1988).

- Van Hiele suggested that the initial focus of the study of geometry must have as its goal the attainment of the second level of thought: "Geometric figures must become the bearer of their properties" (Wirszup, 1976, p. 88). He said that the subsequent focus of this study should be the attainment of the third level of thought. Students should understand the relations that connect properties of figures and begin to logically order the properties of shapes. Most researchers agree that achieving Level 2 and 3 thinking is an important goal of presecondary geometry instruction. At what age van Hiele believed students should attain these levels, however, remains in doubt. In certain writings, he indicates that students in grades 1 to 5 should concern themselves with deepening thinking at Level 1 and that higher levels should not be valued more highly than lower levels ("There are no arguments to push towards a descriptive level; the visual level is so extensive that the subjects there will last for years," P. van Hiele, personal communication, Sept. 27, 1988). However, in other writings, both van Hiele and other researchers emphasize the goal of Level 2 thinking earlier—for instance, by the end of the primary grades (Wirszup, 1976). Such a goal may be attainable. The familiarity of an experimental class of second graders with the geometry of solids enabled them to reach Level 2, surpassing seventh graders in the traditional curriculum (Wirszup, 1976). The Russian researchers also claim that the period of accumulating facts inductively should not be extended too long; they urge that simple deductions be encouraged in elementary school. It is important to continue study of these issues, because research consistently indicates that learning is rote if levels are skipped (Wirszup, 1976).

- *Language.* Imprecise language plagues students' work in geometry and is a critical factor in progressing through the levels (Burger & Shaughnessy, 1986; Fuys et al., 1988; Mayberry, 1983). Instruction should carefully draw distinctions between common usage and mathematical usage (Battista & Clements, 1990; Clements & Battista, 1990; Fuys et al., 1988). Teachers need to constantly remember that children's concepts underlying language may be vastly different than teachers think (Burger & Shaughnessy, 1986; Clements & Battista, 1989). Thus, when mathematical language is used too early and when the teacher does not use everyday speech as a point of reference, mathematical language is learned without concomitant mathematical understanding (van Hiele-Geldof, 1984).

- *Manipulatives and "real world" objects.* Language, of course, rests on a foundation of real-world experiences; beginning with such experiences is strongly indicated by research.

The deductive system of Euclid from which a few things have been omitted cannot produce an elementary geometry. In order to be elementary, one will have to start from the world as perceived and as already partially globally known by the children. The objective should be to analyze these phenomena and to establish a logical relationship. Only through an approach modified in that way can a geometry evolve that may be called elementary according to psychological principles. (van Hiele-Geldof, 1984, p. 16)

Students should manipulate concrete geometric shapes and materials so that they can "work out geometric shapes on their own" (p. 88). Further research has concurred that students respond favorably to initial introduction of concepts in real world settings and that manipulatives are important and helpful, especially at the Levels 0 and 1. The visual approach seemed not only to maintain student interest but also to assist students in creating definitions and conjectures and in gaining insight into relationships (Fuys et al., 1988). Considering this information, along with the deficiencies noted in textbooks, it is imperative that teachers not rely solely on the text.

PHASES OF INSTRUCTION. The phases of instruction are inextricably connected with the levels of thinking, and potentially more important for education; therefore, it is surprising and unfortunate that little research other than the van Hieles' has examined the phases directly. One study indicated that 20 days of phase-based instruction significantly raised high school students' van Hiele level of thought (more so from Level 1 to Level 2 than for any other levels), but it did not result in greater achievement in standard content or proof writing (Bobango, 1988). Additional studies are sorely needed, especially given unresolved questions and concerns regarding the phases; for example:

- How are the phases of instruction related to the levels of thinking? Hypothetically, students must be led by the teacher through all five phases to reach each new level. However, certain phases (for example, 2 and 3) appear to require of students types of thinking that are bound to a given level (for example, level 2). Van Hiele (1959) criticized Piaget and Inhelder (1967) for attaching his "stages of development" to one (pre-operational) period, but van Hieles' phases may make the opposite mistake in being too flexible and iteratable across levels.

- Should the teacher attempt to proceed linearly through the phases, or approach them as recursive within each level?

- Should the teacher introduce many concepts and guide students through the levels on each of them in parallel, or work through the levels (say to level 2 or 3) with a single concept and then use this as scaffolding to develop higher levels of thinking for other concepts?

- Is there a need for differentiation between pedagogical approaches for different types of learning outcomes, such as concepts, skills, or problem-solving abilities?

- What is the role of automatization and practice?

- The final phases would seem to enhance transfer; must transfer also be aided through the provision of systematic spaced reviews, which include a variety of problems?

COGNITIVE SCIENCE: PRECISE MODELS OF GEOMETRIC KNOWLEDGE AND PROCESSES

A third major theoretical perspective that has been applied to understanding students' learning of geometry is that of cognitive science. This field attempts to integrate research and theoretical work from psychology, philosophy, linguistics, and artificial intelligence.

Anderson's Model of Cognition. One cognitive science model, Anderson's (1983) ACT*, postulates two types of knowledge: *declarative* and *procedural*. Declarative knowledge is "knowing that"; for example, postulates and theorems would be stored in schemas along with knowledge about their function, form, and preconditions. Procedural knowledge, "knowing how," is stored in the form of production systems, or sets of condition-action pairs. If the condition, or cognitive contingency that specifies the circumstances under which the production can apply, matches some existing patterns of declarative knowledge, the action is performed (usually adding new elements to working memory; that is, the store of information the system can currently access).

According to the ACT* model, all knowledge initially comes in declarative form and must be interpreted by general procedures (for example, one uses general, recipe-following procedures to cook a new dish using the declarative knowledge read in a cookbook). Thus, procedural learning occurs only in executing a skill; one learns by doing. When declarative information is in the form of direct instructions, step-by-step interpretation is straightforward. However, information is usually not that direct. In the case of high school geometry, students may use declarative information to provide data required by general problem-solving operators, such as general search, sequential decomposition of problems, means-ends analysis, inferential reasoning, or making analogies between worked examples and new problems. Importantly, geometry textbooks assume student facility with such operators and virtually never directly specify which procedures should be applied. This assumption is sometimes mistaken. For example, several students studied by Anderson all had misunderstandings about how one determines whether a statement is implied by a rule. In general, the behaviors of these students on beginning proof problems was captured accurately by this model.

In performing the task, proceduralization gradually replaces the original interpretive application with productions that perform the behavior directly. For example, instead of verbally rehearsing the side-angle-side rule in geometry and figuring out how it applies, students build a production that directly recognizes the application. In "English," such a production might be:

IF   the goal is to prove $\triangle XYZ \cong \triangle UVW$
          and $XY \cong UV$
          and $YZ \cong VW$
THEN   set a subgoal to prove $\angle XYZ \cong \angle UVW$ so SAS can be used

Proceduralization is complemented by a composition process combining sequences of productions. Together, proceduralization and composition are called *knowledge compilation,* the creation of task-specific productions through practice. One form of support for the knowledge compilation theory lies in protocols. One, for example, illustrates the protracted, tediously incremental process initially followed by a student in recognizing the application of the SAS postulate to a problem, compared to the following recognition on a new task five problems later: "Right off the top of my head I am going to take a guess at what I am supposed to do . . . the side-angle-side postulate is what they are getting to" (Anderson, 1983, p. 234). Three differences are noted: The application of the postulate is faster, the statement of the postulate is no longer verbally rehearsed (that is, evoking a declarative representation into working memory, replete with failures leading to inaccuracies), and the original piecemeal application of the postulate is replaced by a single step of recognition. In sum, learning in this theory involves: (1) acquisition of declarative knowledge, (2) application of declarative knowledge to new situations by means of search and analogy, (3) compilation of domain-specific productions, and (4) strengthening of declarative and procedural knowledge (Anderson, Boyle, Corbett, & Lewis, undated).

Thus, an important key to success in proof-oriented geometry problem solving is the development of data-driven rules. These rules respond to configurations of information and result in further development of the problem. For example, experts quickly perceive relations such as triangle congruence, even without recognizing at that time how this will figure in the proof. How might such achievement be facilitated? In a succeeding section, the efforts of Anderson and his colleagues in building an AI (artificial intelligence) tutor based on this theory will help constitute their answer to this important question.

Greeno's Model of Geometry Problem Solving. Greeno's (1980) model of geometry problem solving is similar to Anderson's model of cognition. Based on think-aloud protocols obtained from six ninth-grade students, a computer simulation was designed that could solve the same problems these students were able to solve, and in the same general ways the students solved them. The simulation is a production system within three types of productions, reflecting the following three domains of geometry required for students to solve the problems they are given. First, propositions are used in making inferences (familiar statements about geometric relations, such as "Corresponding angles formed by parallel lines and a transversal are congruent") that constitute the main steps in geometry problem solving. Second, perceptual concepts are used to recognize patterns mentioned in the antecedents of many propositions (for example, the corresponding angles). Third, strategic principles are used in setting goals and planning (for example, when solution requires showing that two angles are congruent, one approach is to use relations such as corresponding angles; another is to prove that triangles containing the angles are congruent).

Of these three domains, the first two are included explicitly in instructional materials; however, strategic knowledge is not. References to that knowledge in the materials is indirect at best, and most teachers do not explicitly identify principles of strategy in their teaching. Students must acquire this knowledge through induction from sequences of steps observed in example solutions. Thus, the induced strategic principles are in the form of tacit procedural knowledge, involving processes the student can perform but cannot describe or analyze. These strategic principles are quite specific to the domain of problems; should they be taught directly? Greeno suggests that they should because it is unlikely that unguided discovery is more effective than a more explicit form of instruction. If direct teaching is interpreted as the teacher imposition of prescribed steps on students, it contrasts with van Hiele's characterization of students finding their own way in the network of relations; if it is interpreted as teacher facilitation of students' construction and development of explicit awareness of strategies, the two positions are not disparate.

Parallel Distributed Processing (PDP) Networks. Other cognitive science research suggests models with even more low-level detail. For example, a PDP (parallel distributed processing) network model might explain the holistic template representations of the lower levels in the van Hiele hierarchy. Such a network possesses processing *units*, representing conceptual objects such as features, words, or concepts, and *connections*, with activation weights between these units. It is the pattern of interconnections among the units that constitutes the processing system's knowledge structure in the domain—what it knows and how it responds (McClelland, Rumelhart, & The PDP Research Group, 1986).

How might such PDP networks more precisely represent students' knowledge structures at different van Hiele levels? During the pre-recognition level, neural network units that recognize certain commonly-occurring visual features are formed; thus, these features become recognizable. Shapes are "recognized" when certain patterns of links among features become established and enable the child to respond to any of a class of visual stimuli.

When a sufficient number of visual features become recognizable and their detectors interconnected in patterns that correspond to common shapes, the child progresses to the visual level. At this level, networks of detector units serve as "shape recognizers" with patterns of activation representing initial schema for figures. Figures that match visual prototypes closely enough cause certain patterns to be activated and, in turn, the figures to be recognized. Properties of figures are not recognized explicitly; the visual features that embody these properties simply activate the prototype recognizers. These representations are not usually reflected upon by the child. However, when some reflection is necessitated (usually by external requests, such as copying a figure), a pattern might be activated, but this pattern may be inadequate. For example, students often encode the basic configuration of a polygon rather than the number of sides, describing a nonconvex quadrilateral as a "triangle with a notch" or a "triangle with a side bent in" (Clements & Battista, 1989). With appropriate instruction, property recognition units begin to form; that is, visual features become sentient in isolation and are linked to a verbal label. The student becomes capable of reflecting on the visual features and, thus, recognizing the shapes' properties, eventually leading to Level 2 thought. While this is all conjecture, it is meant to illustrate a possible cognitive science interpretation of the current van Hiele theory. There is a frustrating lack of progress in further explicating such notions as "network of relations" and "find his way about in the field of symbols." These ideas are interesting and provocative, but they have not progressed to any greater degree of theoretical specificity than that which they attained at their inception (for a possible path for elaboration, see Minsky, 1986, p. 131).

Research has also substantiated the PDP-postulated existence of multiple schemas. That is, students may possess several different visual subschemas for figures (for example, a vertically- and a horizontally-oriented rectangle) without accepting the "average" case (for example, an obliquely-oriented rectangle). Neumann (1977) studied geometric patterns consisting of large rectangles, each with two rectangles within it. The patterns varied on three dimensions: the size of the outer rectangle, the size of the lower rectangle, and the number of stripes in the upper rectangle. Subjects studied a preponderance of stimuli with extreme values and few intermediate values of these three variables. They were presented with test patterns and asked to rate how confident they were that they had already studied that pattern. Subjects did not average what they had studied; instead, they rated the extremes much higher than the mean value patterns, showing that they could extract out of multiple foci of centrality in a stimulus set. While not addressing this question directly, studies on the van Hiele theory are consistent with this finding (Burger & Shaughnessy, 1986; Fuys et al., 1988).

As a final example, such a perspective helps explain people's recognition of a two-dimensional representation of a three-dimensional cube. First, why do we see a line as an edge, or a region as an area of a three-dimensional object? "Our vision-systems seem virtually compelled to group the outputs of our sensors into entities" like these (Minsky, 1986, p. 254). What enables us to see those features as grouped together to form larger objects? "Our vision-systems again are virtually compelled to represent each of those features, be it a corner, edge, or area, as belonging to one and only one larger object at a time.... Our vision-systems are born equipped, at each of several different levels, with some sort of 'locking-in' machinery that at every moment permits each 'part,' at each level, to be assigned to one and only one 'whole' at the next level" (Minsky, 1986, p. 254). How do we recognize these objects as cubes? Our "memory-frame machinery also uses 'locking-in' machinery that permits each object to be attached only to one frame [that is, schema] at a time. The end result is that in every region of the picture, the frames must compete with each other to account for each feature" (Minsky, 1986, p. 254). A PDP model might postulate units representing competing hypotheses concerning each vertex of a Necker cube—for example, the lower left vertex may either be in the front or back of the cube. The network consists of two interconnected subnetworks, one corresponding to each of the two global interpretations of the cube. They are mutually exclusive, and, thus, the spread of activation through the network forces one pattern or the other— but not both—to be activated at any given moment (McClelland et al., 1986).

The Three Theoretical Perspectives

The cognitive science models bring a precision to models of geometric thinking not always present in the theories of Piaget and van Hiele. For example, Anderson's and Greeno's models identify knowledge structures and processes in detail, and the PDP models bring explicitness to certain specific aspects of students' representations at lower levels. How-

ever, limitations of the cognitive science models also must be noted. With the small number of subjects usually involved, generalizability is a concern. These models tend not to explain the unsuccessful student, processes such as conjecturing and problem-finding, and the mechanisms of knowledge restructuring. Also, it is not clear that the genesis of all procedural knowledge lies in the compilation of previously-learned declarative knowledge. In fact, it would seem that many students in the current curriculum acquire mathematical ideas *only* procedurally, without connecting procedural to conceptual knowledge. That is, students often perform sequences of mathematical processes without being able to describe what they are doing or why, perhaps as visually moderated sequences as described by Davis (1984). Most of these models do not address students' development of qualitatively different levels of thinking and representation, belief systems, motivation, and meaningful interpretation of subject matter, and they de-emphasize the roles of sensorimotor activity, intuition, and culture in mathematical thinking (Cobb, 1989; Fischbein, 1987). In fact, similarities between computer simulations and student performance may be a reflection of paucity of situations in which learning and teaching are meaningful. Nevertheless, the theories provide insights and useful metaphors, as well as specific explications missing from most other perspectives.

The theories of Piaget and the van Hieles share certain important characteristics. Both, for example, emphasize the role of the student in actively constructing their own knowledge, as well as the nonverbal development of knowledge that is organized into complex systems. For example, van Hiele emphasizes that successful students do not learn facts, names, or rules, but networks of relationships that link geometric concepts and processes and are eventually organized into schemata (van Hiele, 1959). Thus, students must abstract mathematics from their own systematic patterns of activities. Teachers cannot successfully provide direct help to students who have not yet attained a certain level. "If you want to know how far children have made progress, do not wait for their imitation of your argumentation, but listen to them for what they have found out themselves" (P. van Hiele, personal communication, Sept. 27, 1988). Thus, both Piaget and van Hiele strongly disagree with the belief that good teachers merely explain clearly to children to teach them. Some mechanisms of development are also similar. Piaget stresses the role of disequilibrium and resolution of conflicts. Van Hiele implores teachers to recognize students' difficulties, but not avoid "crises of thinking," because these facilitate the transition to the higher level. In addition, both tend to avoid two positions: (1) the goal of education defined as unabatedly accelerating development ("we've identified levels; now how fast can we get children through them"), and (2) the complete devaluation of thinking at a lower level once a higher level is achieved. Espousal of these positions, however, can be heard in discussions of some who apply their theories to practice; the wisdom of either stance is unknown.

There are also important differences. As previously discussed, van Hiele emphasizes that the course of development is strongly influenced by the teaching/learning process. More significantly, van Hiele (1959) criticizes Piaget's belief in logic as a basis of thinking, claiming that logic can only develop on the foundation of earlier levels of thinking, levels Piaget supposedly "missed" because he already discovered stages of a different nature. By Piaget's "stages," van Hiele meant the stages of transition (for example, from preoperational to concrete operational thought). In moving through these three stages, the child first does not understand a certain idea, then moves into transition, and finally understands. Van Hiele states that the "stages and periods described by Piaget are not essentially connected with a particular age, but are characteristic for very many learning processes irrespective of the age at which they take place" (1959, p. 14). While intriguing, it should be noted that these three stages constitute but a small part of Piaget's developmental theory.

There are also problems with van Hiele's claim that it "escaped Piaget that the object of thought is quite different at the different levels, so that there can be no question of reasoning at the first level being based on a mastery of logical relations which belong to the third level and therefore cannot yet be known" (1959, p. 14). First, van Hiele makes a similar mistake that he attributes to Piaget: defining logic in his own image, or from his constrained perspective. Second, and more importantly, Piaget did hypothesize that objects of thought differed at different developmental stages. Both theorists believe that a critical instructional dilemma is teaching about objects that are not yet objects of reflection for students.

Little research has been conducted on the issues of similarities, differences, and potential syntheses of Piaget's and van Hiele's theories. They appear connected; Denis (1987) indicated that, for high school students, the van Hiele levels appear to be hierarchical across concrete and formal operational Piagetian stages. She reported that only 36% of students who had taken high school geometry had reached formal operational stage, and that most of them attained only Level 3 in the van Hiele hierarchy. She also found a significant difference in van Hiele level between students at the concrete and formal operational stages (although the nature of this relationship is equivocal).

Recall that the argument presented for the existence of a Level 0 is based on a partial synthesis of the theories of Piaget and van Hiele. Greater synthesis might be possible, though, for example, it may be that van Hiele's Level 2 represents a reconstruction on the abstract/conscious/verbal plane of those geometric conceptualizations that Piaget and Inhelder (1967) hypothesized were first constructed on the perceptual plane and then reconstructed on the representational/imaginal plane. Thus, Level 2 may depend in many ways on what Piaget termed the construction of "articulated mental imagery." There is a potential and a need for more detailed work in this area. For example, investigations need to consider how visual thinking is manifested when higher levels are achieved. As our discussions will indicate, it is doubtful that it is untransformed and merely "pushed into the background" by more sophisticated ways of thinking. The same psychological process, visual thinking, probably has a number of psychological layers, from primitive to sophisticated and interconnected with other ways of thinking, all of which play different roles in thinking depending on which layer is activated. Study of the development must go beyond investigating only the growth of increasingly sophisticated levels

of geometric thinking to investigating the continual development of processes such as visual thinking that appear initially well developed (Vygotsky, 1934/1986).

In general, research that builds on the strengths of all three theoretical perspectives might have potential. For example, Piaget's schemes, van Hiele's network of relations, and cognitive science's more explicit declarative networks certainly possess commonalities in their views of knowledge structure, and it is possible that a synthesis of these would yield a richer, more veridical model. Ideally, such a model would have the explication of the cognitive science perspective and the developmental aspects of the Piagetian and van Hiele perspectives.

ESTABLISHING TRUTH IN GEOMETRY

How do mathematicians establish truth? They use *proof*, logical, deductive reasoning based on axioms. How do they find truth? Most frequently by methods intuitive or empirical in nature (Eves, 1972). In fact, the process by which new mathematics is established is belied by the deductive format in which it is recorded (Lakatos, 1976). In creating mathematics, problems are posed, conjectures made, counterexamples offered, and conjectures revised; a theorem results when this refinement of ideas is judged to have answered a significant question. Bell (1976) distinguishes three functions for proof in mathematics: *verification*, which is concerned with establishing the truth of a proposition; *illumination*, which is concerned with conveying insight into why a proposition is true; and *systematisation*, which is organization of propositions into a deductive system.

In geometry, as in other areas of mathematics, empirical and deductive methods should interact and reinforce each other. For instance, often when one is stymied in taking a deductive approach, empirical investigations can generate explorable possibilities. However, for most students in geometry, deduction and empirical methods are separate domains with different ways to establish correctness (Schoenfeld, 1986). In fact, the use of formal deduction among students who are taking or have taken secondary school geometry is nearly absent (Burger & Shaughnessy, 1986; Usiskin, 1982). According to Schoenfeld, most students who have had a year of high school geometry are "naive empiricists whose approach to straightedge and compass constructions is an empirical guess-and-test loop" (Schoenfeld, 1986, p. 243). Students make a conjecture, then test it by examining their construction. If the construction looks sufficiently accurate, the student is satisfied that the conjecture has been verified. "In various problem sessions students have rejected correct solutions because they did not look sufficiently accurate and have accepted incorrect solutions because they looked good" (p. 243). In one series of interviews, college students were asked to solve a construction problem after having solved a proof problem that provided a solution to the construction problem. "Nearly a third of the students began the second problem by making conjectures that flatly violated the results they had just proved!" (Schoenfeld, 1988, p. 150). Evidently, the students' proof activity either had not really established knowledge for the students—or the knowledge was established but was compartmentalized in such a way that it was not accessible in the domain of constructions. According to Schoenfeld, instructional strategies used in the high school classrooms might be the cause for this compartmentalization. Although theorems and deduction were used to introduce and validate constructions, the emphasis was on constructions as procedures—that is, on skill acquisition.

It should be noted that Schoenfeld's investigation of "empirical methods" was restricted to students' use of constructions, which are usually taught as procedures with value in and of themselves. Other empirical approaches might produce somewhat different results. For instance, constructions on a computer might be better for students for two reasons. First, computers require more precise specification than those done with paper and pencil. Second, because the computer performs the constructions, the teacher can treat the topic less as a set of procedures to be learned, thus focusing more on concept development. Even with computer constructions, however, we might need to worry about the pitfalls of promoting an empiricist approach. For instance, in commenting on the computer software *Geometric Supposer,* Schoenfeld (1986) wondered if the ability of the program to repeat constructions automatically would lead students to be overly empirical. Judah Schwartz, the software's author, replied by recounting an instance where the opposite seemed to have happened. One student tried to convince another student that something was true by appealing to a large collection of confirming examples. Another student countered that the class had seen lots of constructions that worked for many examples but later had turned out to be incorrect. Thus, maybe because deduction arises from empirical approaches, the latter student had, through experience with empiricism, discovered its limitations. This was certainly an important step toward appreciating a need for the deductive approach to establishing truth in geometry. Apparently this is not an isolated case: As we will discuss below, research indicates that the *Geometric Supposer*'s empirical measurement approach does not negatively affect students' ability to learn proof.

Other studies have confirmed the existence of student confusion concerning methods of justifying mathematical statements. In a study by Martin and Harel (1989), pre-service elementary teachers were asked to judge the mathematical correctness of inductive and deductive verifications of statements. For each statement, more than 50% of the students accepted an inductive argument and more than 60% accepted a deductive argument as a valid mathematical proof. Fifty-two percent accepted an incorrect deductive argument as valid for an unfamiliar statement. Fischbein and Kedem (1982) found that high school students, after finding or learning a correct proof for a statement, maintained that surprises are still possible and that further checks are desirable. In a study of students age 12 to 15, Galbraith (1981) found that over a third of the students did not understand that counterexamples must satisfy the conditions of a conjecture but violate the conclusion; 18% felt one counterexample was not sufficient to disprove a statement.

According to Martin and Harel, in everyday life people consider "proof" essentially to be "what convinces me." Bell (1976) suggested that "conviction arrives most frequently as the result

of the mental scanning of a range of items which bear on the point in question, this resulting eventually in an integration of the ideas into a judgment" (p. 24). According to Bell, proof grows out of internal testing and the resultant acceptance or rejection of a generalization. Later, one subjects the generalization to criticism by others, first through oral and then through written statements that present not only the generalization but evidence for its validity in the form of a proof. Thus, students will not appreciate the purpose of formal proof until they recognize the public status of knowledge and the resultant need for public verification.

Of crucial importance here is the idea of internal testing. For a mathematician, internal tests eventually take the form of proof, as one attempts to perform the socially accepted criticism of one's argument oneself. However, do students perform this testing, and, if so, how? When quizzed about their justifications for ideas, students may refer to general propositions, specific instantiations of general propositions, diagrams, and isolated results unconnected to justifications (Talyzina, 1971). Of course, the latter two justifications are problematic, the first of these because students often make unwarranted assumptions based on diagrams, and the second because students are ignoring the need for one's reasoning to be laid out in detail so that it can be evaluated. Talyzina also found that younger students much more often than older students recalled a general proposition rather than a specific, relevant instantiation of that general proposition. It was claimed that the latter behavior "is characteristic of the higher stages of mastering the ability to solve geometry problems" (p. 98). That is, a characteristic of more accomplished thinking in geometry is a curtailment of step-by-step, deductive thought.

Students' Proof Performance

About one month before the end of the school year, Senk (1985) tested the proof-writing ability of 1,520 students in geometry classes that had studied the topic. A proof was considered correct if all the steps followed logically, even if there were minor errors in notation, vocabulary, or names of theorems. Seventy percent of students were correct on a simple six-step proof in which they were to supply either the reason or the statement. Fifty-one percent of students were correct on a simple proof requiring an auxiliary line and in which the students had to write both statements and reasons themselves. Thirty-two percent of the students could prove that the diagonals of a rectangle are congruent, but a mere 6% proved a somewhat more difficult theorem that did not follow directly from the triangle-congruence postulates and theorems. Just 3% of the students received perfect scores on the test. On only three of the twelve problems which required a full proof were at least half of the students successful. Senk concluded that only about 30% of all students in full-year geometry courses that teach proof reach a 75% mastery level in proof writing.

Brumfield (1973) queried 52 high school students who had taken an accelerated geometry course the previous year and were headed for an advanced placement calculus course about geometry. When asked to list as many postulates as they could remember, 50% of the students listed nothing at all and 31% listed only statements that were not postulates. Forty percent of the students could not list a single theorem, with many mixing theorems with axioms, definitions, and false statements. When asked to choose one interesting theorem and prove it, 81% of students did not attempt a proof, and only one of the 10 students who did was correct. Apparently, even bright students get very little meaningful mathematics out of the traditional, proof-oriented high school geometry course.

The Development of Proof Skills

Given students' poor performance on proof writing, it is imperative to investigate the development of this important skill. What are its components? What are the prerequisites for proof writing? When do children first attempt to justify their conclusions? How do they go about it?

Several components for understanding and constructing proofs have been suggested by Galbraith (1981): (a) understanding and being able to perform an exhaustive check of the set of possibilities required to verify a statement, (b) detecting and utilizing a relevant pattern or principle in the data, (c) utilizing a chain of inferences without needing to establish intermediate steps with concrete referents, (d) recognizing the domain of validity of a generalization, (e) correctly interpreting statements and definitions, and (f) understanding the formal structure of proof. Van Hiele, Piaget, and other theorists have offered several different perspectives on the development of proof skills. Piaget (Piaget, 1928; Piaget, 1987; Piaget, Inhelder, & Szeminska, 1960) for example, described levels for the notions of justification and proof. We will give a broad overview of these levels (the numbers used for the levels are a synthesis, so they do not necessarily correspond to those of Piaget).

Level 1 (up to age 7–8). In decisions about the truth of ideas, there is a lack of integration of observations and local conclusions. Each piece of data collected or example examined is treated as a separate event not integrated with others. Exploration proceeds randomly, without a plan, and local conclusions may be contradictory. This lack of direction in thinking is due to the fact that "there is nothing here which tends to make thought conscious of itself and consequently to systematize or 'direct' its successive judgments" (Piaget, 1928, p. 15). Being egocentric, the child attempts neither to see the point of view of another nor to think about making his or her viewpoint understood by others.

At the end of this level, there is some degree of integration and more purposeful exploration. Thought is more directed; students begin to understand that several clues or examples must be integrated to draw conclusions. Although patterns are established empirically, this is done without attempting to understand why the patterns occur. For instance, when students were putting together angles of a triangle, they were shown what happened for one triangle and asked to predict what would happen for others. The students counted the angles and predicted that shapes with three angles would produce semicircles and shapes with four would produce circles.

However, they ignored the size of angles; they did not attempt to determine why the pattern occurred.

The child at this level is capable of the most elementary form of deduction. It "consists either of foreseeing what will happen when such and such conditions are given, or in reconstructing what has happened when such and such results are given" (Piaget, 1928, p. 66).

Level 2 (ages 7–8 through 11–12). As students begin this level, they not only make predictions based on empirical results, they begin trying to justify their predictions. Induction and deduction often conflict. In the angles task, they attempted to analyze the angles for each new example. Because they were unable to see the sizes of the three angles as interdependent, though, they were often misled by the appearance of the angles, and often seemed less advanced than the students at Level 1. Children's incorrect predictions seemed to lead them to analyze interrelationships between the angles, but they were unable to establish a general relationship. There is also an anticipatory character and purposefulness to searches for information. For example, students might use information to establish classes of possibilities and nonpossibilities in a search task.

During the latter part of this level, inductive generalizations take place more quickly and often immediately. Each instance is compared with previous instances. On the angles task, children were able to establish a relationship between the three angles of a triangle. Also, there is no longer a contradiction between the analysis of the angles in individual triangles and the inductive generalization about their sum. "On the contrary, the induction itself which leads the subject to believe that the angles of any triangle will yield a semicircle provides an anticipatory schema which guides the composition of the angles of new triangles" (p. 204). The discovery becomes universal: "They always do [form a semicircle]" (p. 204). Students at this level, however, do not establish logical necessity. Even after empirically eliminating all possible answers but one, one student said "I prefer opening other ones [clues], you never know" (p. 114).

There is the capability of implication for these children "when reasoning rests upon beliefs and not upon assumptions, in other words, when it is founded on actual observation. But such deduction is still realistic, which means that the child cannot reason from premises without believing in them. Or even if he reasons implicitly from assumptions which he makes on his own, he cannot do so from those which are proposed to him" (pp. 251–52). Thus, although thought is logical, it is empirical in nature.

Level 3 (ages 11–12 and beyond). The child is capable of formal, deductive reasoning based on *any* assumptions. The soundness of the assumptions does not affect the validity of the argument. Logical necessity is established by the method employed. Students integrate information that has been revealed by various actions and decide what information must be obtained from further actions. At Level 2, "the deficiency that remains is the failure to recognize exhaustivity, which prevents subjects from considering their proofs as sufficient even when they are. Only at Level 3 does progress in integration lead to the conviction that the conditions established as being necessary, when taken together, are also sufficient" (p. 116).

For the angle task, students progressed from simply believing that the angles will always make a semicircle to a belief, based on logical reasoning, that this must necessarily be so. For example, they might state that the three angles make a semicircle because "the angles (at the base of an elongated isosceles triangle) aren't quite right angles, and the point makes up the difference" (p. 205). However, the logical reasoning that is used by the students is not based on formal mathematics. So the students may have believed that the three angles must necessarily form a semicircle, but they usually could not provide a formal reason.

In summary, at Level 1, the child's thinking is nonreflective and unsystematic, and therefore, not logical. At Level 2, thought is logical, but restricted to being empirical. Only at Level 3 is the child capable of logical deduction and of consciously operating within a mathematical system. What causes progress through the levels? From where does the need for verification arise? "Surely it must be the shock of our thought coming into contact with that of others, which produces doubt and the desire to prove.... Proof is the outcome of argument" (Piaget, 1928, p. 204). Due to contact with others, the child becomes ever more aware of his or her own thought, becoming "conscious of the definitions of the concepts he is using" and acquiring "a partial aptitude for introspecting his own mental experiments" (p. 243). The child becomes increasingly able to take the perspective of others. Finally, with the onset of formal thought, mental experiments, in which reality is constructed by reproducing in thought sequences of events as they have happened or were imagined to have happened, are replaced by logical experiments, in which the actual mechanism for construction is reflected upon. At this time, arguments can, in a real sense, be internalized; "Logical experiment is therefore an experiment carried out on oneself for the detection of contradiction" (p. 237).

Van Hiele's View

According to van Hiele, the *reasoning* of students at the visual and descriptive/analytic levels is quite different when they identify a figure. For the student at the visual level, the judgment is "based on an observation" (van Hiele, 1986, p. 110), "There is no why, one just sees it" (p. 83). For the student at the descriptive/analytic level, the judgment results "from a network of relations" (p. 110). The thinking of students at the descriptive/analytic level may involve observation; it may be that they see an image whenever they consider a given figure. The image is not the basis for judgment; the network of relations is. Even if a figure was imperfectly drawn (or distorted on a computer screen), such students' thinking would not be swayed if they were assured it was the intention of the drawer to make all sides equal. Van Hiele continues that it is this network of relations that distinguishes between the two levels. At the beginning, one does not possess the network.

Once a class of shapes is thought of as a collection of properties (at Level 2), the relationship between a figure and other figures is determined, and can be reflected upon. An elementary form of associational implication can take place. However,

"What it means to say that some property 'follows' from another cannot be explained" (p. 111). At Level 3, in contrast, the contents of statements A and B are not important. "The only things of importance for the further train of thought are the links existing between A and B. With these links the new network of relations is constructed.... When this second network of relations is present in so perfect a form that its structure can be read from it, when the pupil is able to speak to others about this structure, then the building blocks are present for the network of the third level" (p. 112). A technical language develops that makes it possible to communicate with others about the essential ideas in the network of relations (that is, to reason). "Without the network of relations, reasoning is impossible" (p. 110). But with the power of communication that results from the technical language comes an obligation to "stick to the network of relations." That is, with formal reasoning come constraints. Therein lies the difficulty for many students; they do not know what the constraints are nor do they understand why they apply.

According to van Hiele, the intuitive foundation of proof "begins with a pupil's statement that belief in the truth of some assertion is connected with belief in the truth of other assertions. The notion of this connection is intuitive: The laws of such a connection can only be learned by analysis" (1986, p. 124). Logic is created by analyzing and abstracting these laws, that is, by operating on the network of links between statements. De Villiers (1987) concurs that deductive reasoning first occurs at Level 3, when the network of logical relations between properties of concepts is established. He continues that because students at Levels 1 and 2 do not doubt the validity of their empirical observations, proof is meaningless to them; they see it as justifying the obvious.

Van Dormolen (1977) describes three levels of proof performance and relates them to the van Hiele levels. In the first, justifications are made for single cases; conclusions are restricted to the specific example for which the justification is given (for example, a particular rectangle). In the second, justifications and conclusions may be for specific cases, but refer to *collections* of similar objects (for example, the class of rectangles); several examples will be considered to illustrate a pattern, with students capable of generating further examples. In the third, students justify statements by forming arguments that conform to accepted norms; that is, they are capable of giving formal proofs. Van Dormolen relates his first level to van Hiele's visual level of thinking, his second to van Hiele's descriptive/analytic, and his third to van Hiele's level of formal deduction in which students attend to the properties of arguments. It should be observed that although students in Van Dormolen's second level have made progress, their method is fraught with potential for error. For instance, as students reason about a class of shapes by examining specific cases, they often attend to properties of the particular instances, thus making mistakes about the class in general.

Van Hiele Levels and the Ability to Construct Proofs.

As can be seen from the above descriptions, a proof-oriented geometry course requires thinking at least at Level 3 in the van Hiele hierarchy. However, over 70% of students begin high school geometry at Levels 0 or 1, and only those students who enter at level 2 (or higher) have a good chance of becoming competent with proof by the end of the course (Shaughnessy & Burger, 1985). It follows, therefore, that instruction should help students attain higher levels of geometric thought *before* they begin a proof-oriented study of geometry.

Senk (1989) investigated the relationship between van Hiele levels, writing geometry proofs, and achievement in nonproof geometry. Students enrolled in full-year geometry classes were tested in the fall for van Hiele level and entry level geometry knowledge, and in the spring for van Hiele level, knowledge of geometry, and proof-writing ability. It was found that achievement in writing geometry proofs was positively correlated with van Hiele level (.50 in the fall, .60 in the spring) and to achievement on nonproof content (.70 in the spring). Senk argued that students who start geometry at Level 0 have little chance of learning to write proofs, students at Level 1 have less than a one-in-three chance, and students at Level 2 have a 50-50 chance. Level 2 is the critical entry level. Senk noted that at the end of the school year, students at Level 3 or above significantly outperformed on proof students at Level 2 or below, but students at Levels 4 and 5 did not score significantly better than students at Level 3. (Possibly because of the low number of students in the upper two levels.) Indeed, 4%, 13%, and 22% of students at Levels 0, 1, and 2, but 57%, 85%, and 100% at Levels 3, 4, and 5, respectively, were classified as having mastered proof writing. These data seem to support that van Hiele Level 4 is the level at which students master proof, with Level 3 being a transitional level. One might conjecture that students at Level 3 probably are not able to do substantial proofs that they have not seen before (or actually understand what the proofs entail or accomplish). However, van Hiele's hierarchical theory that only students at Levels 4 or 5 should be expected to consistently write formal proofs was not strictly supported by the research.

Spring achievement on nonproof content accounted for 57% of the variance in proof scores, but van Hiele test scores accounted for only an additional 3%. Although Han (1986) also found that van Hiele level predicted performance on a proof-writing test, the relatively small contribution of van Hiele level to prediction above that of standard achievement test scores must be explained (Senk, 1989). It may be due to the difficulty of separating level of thinking from content. (The correlation between the van Hiele test and content was .6.)

A Conflict

There seems to be a great difference in emphasis between van Hiele and Piaget on how geometric reasoning and proof develop. For van Hiele, the emphasis seems to be on content; one progresses to higher levels of thought in geometry when the network of relations becomes sufficiently built up. The ability to reason logically in geometry is dependent on the amount and organization of content-specific knowledge. According to Piaget, however, certain logical operations develop in students independent of the content to which they are applied. These operations can be applied in a variety of contexts; it is through these operations that new mathematical knowledge is estab-

lished. For instance, if a student knows that a rectangle is a figure that has opposite sides equal and four right angles, and a square is a figure that has all sides equal and four right angles, then the student may deduce and internalize the fact that all squares are rectangles. The conclusion is newly created knowledge that has, by virtue of deduction, been integrated into the student's current cognitive structure.

In support of the Piagetian perspective, Driscoll (1983b) has emphasized the role of cognitive development in the acquisition of the ability to construct proofs, claiming that students need to be formal operational thinkers to completely understand and construct proofs. As evidence, he noted that on logic items from the 1978 NAEP (on which even the 17-year-olds did poorly), there was a much greater jump in performance between the 7- and 13-year-olds than between the 13- and 17-year-olds. Furthermore, Gardner (1983) suggests that only during the formal operational period can individuals deal with the idea of abstract spaces or with formal rules governing space. That is, formal geometry can be constructed only by individuals who can integrate logico-mathematical and spatial intelligence into a scientific system. McDonald (1989) provides support for the notion that the nature of thought of which a student is capable affects the student's construction of knowledge in geometry. Twenty secondary students classified as formal operational and 20 as concrete operational made judgments about the similarity of 13 geometric concepts from the area of similarity and congruence. Multidimensional scaling techniques indicated that prototypical cognitive maps could be drawn for both the formal and concrete operational students. Furthermore, formal operational students' structure of the content was significantly more like that of subject matter experts than that of the concrete operational students. Finally, in contrast to van Hiele's theory, Mason (1989) reported that the reasoning ability of fourth to eighth graders was far beyond what may have been anticipated, given their low van Hiele level of geometric thinking. These students evinced logical thinking indicative of Level 2, but without knowledge of specific definitions or geometric content corresponding to that level.

We conclude, then, that this is not a "chicken or egg" type of problem; there exists a dynamic interplay between level of reasoning and organization of knowledge. An important research issue is the elaboration of this interplay: How does the organization of knowledge depend on the stage of operational thinking, and how does the stage of thinking depend on knowledge organization?

Proof and Instruction

There have been numerous attempts to improve students' proof skills by teaching *formal* proof in novel ways, almost all of which have been unsuccessful (Harbeck, 1981; Ireland, 1974; Martin, 1971; Summa, 1982; Van Akin, 1972). An alternate approach claims that for students to develop an ability with proof, they must understand its nature. For example, Driscoll (1983b) reported a study by Greeno and Magone in which college students who had had high school geometry but were not very good at it were given a two-hour training program on proof-checking. The instructional program not only taught students specific steps to follow in checking a proof, it provided students the opportunity to analyze the nature of proofs. The researchers found that the experimental students were not only more effective than control students at checking proofs, but they did better at constructing proofs as well. They hypothesized that students needed to understand the nature of proof and how it differed from everyday argumentation.

Fuys claimed that some sixth and ninth graders in their teaching experiment made progress toward Level 3 thinking "by following and summarizing deductive explanations, and giving deductive arguments" (Fuys, 1988, p. 9). There was indication that some students' inability to do proofs was attributable to students believing that justification is something that others do for them; "this is true because it is a theorem or procedure I learned in class."

Bell (1976) suggested that success in proof could be promoted through cooperative investigations by students in which conjectures were made and conflicts were resolved by students presenting arguments and evidence. Fawcett (1938) conducted a 2-year experiment in geometry with the results supporting this contention. Students were challenged to develop their own axioms, definitions, and theorems and to examine, debate, and justify their conjectures. At the end of the two years, the experimental students scored higher than traditional students in geometry, and both experimental students and their parents claimed that the students' deductive thinking had improved. Fawcett also observed student behaviors that indicated an understanding of proof, behaviors such as asking that significant words and phrases in statements be carefully defined; requiring evidence to support conclusions; analyzing evidence; recognizing, analyzing, and re-evaluating stated and unstated assumptions; and evaluating arguments.

In a similar vein, Human and Nel (1989) reported success in a geometry course, based on the van Hiele model, that attempted a "gentle introduction" to deductive and axiomatic thinking. The intent was to make proof as meaningful as possible by initially including proofs of nonobvious statements and by having students formulate their own hypotheses to prove. The materials began to develop deductive skill by reviewing some familiar statements about triangles, and then asking students to solve problems by *thinking* with the aid of these statements. For example: "Given triangle PQR, $PQ = PR$, and the exterior angle at R being 100°, supply as much additional information about the triangle as you can and explain how you obtained this information."

The results of studies by Human and Nel, and especially by Fawcett, are consistent with an analysis of proof by Hanna (1989c). She argues that because mathematical results are presented formally by mathematicians in the form of theorems and proofs, this rigorous practice is mistakenly seen by many as the core of mathematical practice. It is then assumed that "learning mathematics must involve training in the ability to create this form" (pp. 22–23). Instructional treatments that have been based on this view have generally failed to accomplish their goals, probably because the students are attempting to follow formal rules unconnected to any activity they find meaningful. On the other hand, studies that have attempted to involve students in the crucial elements of mathematical discovery and discourse—conjecturing, careful reasoning, and the building of

validating arguments that can be scrutinized by others—have shown more positive effects.

In summary, we have seen that students are extremely unsuccessful with formal proof in geometry. This is disappointing, given the amount of time in the curriculum devoted to this goal. However, our analysis of students' proof-making abilities reveals a far more devastating finding: Students are deficient in their ability to establish truth in geometry, and, indeed, in all of mathematics. They have not developed those beliefs and schemas that motivate and allow them to establish mathematical truths. Indeed, if we adopt a constructivist perspective on mathematics learning, as students construct mathematical meaning and build a network of knowledge in mathematics, the process by which they establish mathematical truth for themselves becomes vitally important. For, in the process of constructing and restructuring mathematical knowledge, students must decide what they believe to be mathematical truths. Each newly encountered idea is accepted as true or rejected as false based on current knowledge and reasoning structures. Each of these decisions, in turn, either buttresses the current structures or causes them to be reorganized.

Obviously, more research attention must be devoted to how students establish truth and how they come to understand and utilize proof in their mathematical thinking. Work such as Piaget's, which documents the development of knowledge verification skills, is important; much of this work, though, was done in a context divorced from formal mathematics. Are the same results obtained in situations where students are exploring mathematics in classroom situations? Can we elaborate on the levels of verification? What types of tasks and environments encourage students to progress to higher levels of verification? How is students' knowledge organization affected by their methods of establishing truth?

SPATIAL REASONING

Gardner (1983, p. 8) argues that spatial ability is one of the several "relatively autonomous human intellectual competences" which he calls "human intelligences." Spatial thinking is essential to scientific thought; it is used to represent and manipulate information in learning and problem solving. The "metaphoric ability to discern similarities across diverse domains derives in many instances from a manifestation of spatial intelligence" (Gardner, 1983, p. 176). An example is when scientists draw analogies between human society and microorganisms or brain function. According to Harris (1981), the U.S. Employment Service estimates that most technical and scientific occupations, such as draftsman, airplane designer, architect, chemist, engineer, physicist, and mathematician, require persons having spatial ability at or above the 90th percentile.

The Relationship Between Spatial Thinking and Mathematics

Hadamard argued that much of the thinking required in higher mathematics is spatial in nature. Einstein commented that his elements of thought were not words, but "certain signs and more or less clear images which can be voluntarily repro duced or combined" (Gardner, 1983, p. 190). Numerous mathematicians and mathematics educators have suggested that spatial ability and visual imagery play vital roles in mathematical thinking (Lean & Clements, 1981; Wheatley, 1990). Perhaps underlying this position is the recognition that different modes of thought are used in mathematics.

Krutetskii (1976), for instance, refers to two different modes of thought: verbal-logical and visual-pictorial. He argues that the balance between these two modes of thought allows for different "mathematical casts of mind" which determine how an individual operates on mathematical ideas. He classified as *analytic,* those students who prefer verbal-logical modes of thought in mathematical problem solving, even for problems that would yield to a relatively simple visual approach; *geometric,* those who prefer visual-pictorial schemes even on problems more easily solved with analytic means; and *harmonic,* those who have no specific preference for either verbal-logical or visual-pictorial thinking. The theory of hemispheric specialization of the brain corroborates the existence of two modes of thought (Springer & Deutsch, 1981). A great deal of physiological evidence indicates that the two hemispheres of the brain are specialized for different modes of thought processing. In general, the left hemisphere is specialized for analytic/logical thinking in both verbal and numerical operations; it excels in sequential tasks, logical reasoning, and analysis of the components of a stimulus. Language is processed in the left hemisphere. The right hemisphere, on the other hand, predominates for spatial tasks, artistic endeavor, and body image and seems specialized for holistic thinking. "Although the left hemisphere seems to be as competent as the right in identifying the Euclidean (and nameable) properties of objects (that is, points, lines, and planes), it is much less capable than the right in identifying the less nameable topological properties such as transformations involving changes in lengths, angles, and shapes" (Franco & Sperry, 1977, p. 108).

Positive correlations have been found between spatial ability and mathematics achievement at all grade levels (Fennema & Sherman, 1977; Fennema & Sherman, 1978; Guay & McDaniel, 1977). It is not difficult to see why this relationship exists for there are numerous concepts in mathematics that have an obvious visual dimension. Davis (1986), for example, describes what "cognitive building blocks" are needed to determine the area of a rotated square on a geoboard. In addition to mental images of squares and triangles, he cites mental representations of the acts of rotating and translating triangles, of putting them together to make other shapes, and even of cutting apart squares to get triangles. Similarly, Soviet researchers have emphasized the importance of spatial thinking in geometry, "Visualizations are used as a basis for assimilating abstract [geometric] knowledge and individual concepts" (Yakimanskaya, 1971, p. 145). For instance, understanding the concept of rectangle and its properties requires that students analyze the spatial relationship of the sides of a rectangle—that is, understand "opposite" sides and distinguish them from "adjacent" sides. It was argued that teachers should provide activities for developing students' spatial imagination because assimilation would be "formalistic" if the teacher did not develop students' spatial

images, but provided verbal information about the properties of figures instead.

Hershkowitz (1989) outlined the role of visualization in the development of a student's conceptualization of a geometric idea and related this development to the van Hiele levels. First, a prototypical example is used as a reference to which possible exemplars are compared visually (van Hiele Level 1). Second, the prototypical visual example is used to derive the critical attributes of the concept (transition from Level 1 to Level 2), which are then applied in judging other figures. Finally, the critical attributes or properties of the concept are used to judge whether figures are instances of the concept (Level 2). Battista and Clements (1990) found a similar developmental sequence among students doing geometry in a Logo environment.

Moreover, visual thinking is utilized by many students in representing and operating on concepts that do not inherently contain a spatial aspect (Krutetskii, 1976; Lean & Clements, 1981). Because of instruction, for instance, students may think of fractions and operations on fractions in visual terms (Clements & Del Campo, 1989). In fact, heavy reliance on visual representations of mathematical ideas might be especially important at the elementary school level (Stigler et al., 1990) because young children rely more heavily on imagery than do adults (Kosslyn, 1983).

Indeed, Johnson (1987) argues that imagery is what enables us to utilize our bodily experiences to structure all thought, not just mathematics. There are two mechanisms by which this process occurs. The first is the image schema, "a recurring, dynamic pattern of our perceptual interactions and motor programs that gives coherence and structure to our experience" (p. xiv). For example, the vertical schema is the abstract cognitive structure that emerges from our natural tendency to employ an up-down orientation in structuring our experience. We encounter this structure repeatedly as we perceive objects and maneuver about the world.

The second concept useful for understanding the role of bodily experience in thinking is the metaphor. According to Johnson, a metaphor is "a pervasive mode of understanding by which we project patterns from one domain of experience in order to structure another domain of a different kind" (p. xv). It is one of the primary cognitive mechanisms by which we structure and make sense of experiences. Because physical experience is so fundamental to intellectual development, image schema become a primary source for metaphors. "Through metaphor, we make use of patterns that obtain in our physical experience to organize our more abstract understanding" (p. xv). For example, the idea that "more is up," as part of the vertical schema, is used to help us understand the abstract notions of more/less and change in quantity. Johnson argues that the use of image schema as metaphors for understanding abstract notions is pervasive and natural in human understanding. Even the idea of deduction derives from the spatial concept of "following a path."

Despite the claims for the importance of imagery and spatial thinking in mathematics, the relationship between spatial thinking and learning nongeometric concepts is not straightforward. For instance, Fennema and Tartre (1985) present somewhat conflicting results. Students high in spatial ability and low in verbal ability (high/low students) tended to translate problems into pictures more completely than low/high students, and there was some indication that low/high students were less able to draw and to use pictorial representations than high/low students. However, when using a problem-solving process that emphasized the use of spatial visualization, students who had high spatial visualization skill solved no more problems than students who had low visualization skill. Lean and Clements (1981) concluded that students who process mathematical information by verbal-logical means outperform students who process this information visually. Similarly, Hershkowitz (1989) claimed that the use of imagery in mathematical thinking can cause difficulties. For example, if a concept is tied too closely to a single image, its critical attributes might not be recognized or use of the concept in problem-solving situations might be limited because of over reliance on this image. On the other hand, Brown and Wheatley (1989) reported that, although the fifth-grade girls with low spatial ability that they interviewed performed well in school mathematics, their understanding of multiplication and division was instrumental, whereas the high spatial girls' understanding was more relational. One high spatial girl, although evincing an excellent grasp of mathematical ideas and problem solving in the interviews, performed poorly in school mathematics. Similarly, Tartre (1990a) suggested that 10th grade students who scored high on spatial orientation were better able to understand nongeometric problems and to link them to previous work than were students who scored low in spatial orientation.

Thus, there is reason to believe that spatial ability is important in students' construction and use of mathematical concepts—even nongeometric ones. But the role that such thinking plays in this construction is elusive and, even in geometry, multifaceted. For example, in van Hiele Level 1, one relies on and is restricted to visual processing. At van Hiele Level 2 and higher, one's use of visual images is constrained by one's verbal/propositional knowledge. Images and transformations of images incorporate this knowledge and, as a result, might behave differently at the different levels. Finally, there is the use of imagery in thinking about nongeometric concepts. One might be manipulating mental entities that have neither a visual nor a verbal format but are nonetheless operated on by visual-like transformations. Future research should attempt to elaborate these different types of uses of imagery and take care in distinguishing between the qualitatively different types of visual thinking.

The Nature of Spatial Abilities

Gardner (1983, p. 173) states that "Central to spatial intelligence are the capacities to perceive the visual world accurately, to perform transformations and modifications upon one's initial perceptions, and to be able to re-create aspects of one's visual experience, even in the absence of relevant physical stimuli." Two major components or factors of spatial tasks have been identified (Bishop, 1980; Harris, 1981; McGee, 1979). Spatial *orientation* is understanding and operating on the relationships between the positions of objects in space with respect to one's own position—for instance, finding one's way in a building. Spatial *visualization* is comprehension and performance of imagined movements of objects in two- and three-dimensional

space. Others have debated the characterization into these two factors (Clements, 1979; Eliot, 1987), and indeed, there is debate about the nature of spatial ability and its measurement. For instance, Bishop (1983) has suggested two spatial components that he believes are especially relevant for mathematics learning. The first is the ability to interpret figural information and involves understanding visual representations and vocabulary. The second is the ability for visual processing, involving manipulation and transformation of visual representations and images and translation of abstract relationships into visual representations. Other authors (Guay, McDaniel, & Angelo, 1978) believe that the essence of true spatial ability is the formation and transformation of visual images as organized wholes. They argue that many so-called spatial tests can be done effectively with analytic processing and, thus, are not good measures of spatial ability. Furthermore, there is evidence that different groups of individuals use different processes on spatial tasks. Some represent problems visually; others represent them verbally. Some attend to the whole stimulus at once; others attend to parts of it at a time. Some individuals use processing aids, such as marks on paper, object manipulation, and body movement.

Imagery

Several researchers have suggested that the most important determinant of spatial visualization ability is maintenance and manipulation of a high quality image of the stimulus, whereas others argue that performance on most spatial tests is best understood not in terms of imagery but rather in terms of reasoning and problem solving (Eliot, 1987). However, most factor analysts and developmentalists agree that "persons with well-developed spatial skills should be capable of imagining spatial arrangements of objects from different points of view and of manipulating visual images" (Clements, 1979, p. 15). Thus, the concept of image plays a central role in our study of spatial ability.

Images are internally perceived, wholistic representations of objects or scenes that are isomorphic to their referents. They are mentally changed by continuous transformations corresponding to physical transformations. Kosslyn (1983) defines four classes of image processes: generating an image, inspecting an image to answer questions about it, transforming and operating on an image, and maintaining an image in the service of some other mental operation.

According to Eliot (1987), there is consensus among spatial researchers about the following four points:

1. The mental processes that underlie the experience of an image are similar to those that underlie the perception of objects or pictures.
2. An image is a coherent, integrated representation of a scene or object from a particular viewpoint and is open to a perceptual-like process of scanning.
3. An image can be subject to apparently continuous mental transformations, such as rotations, in which intermediate states correspond to intermediate views of an actual object undergoing the corresponding physical transformation.

4. Images represent not only objects, but also interrelationships between an object's component parts and other objects. That is, "the functional relations among objects as imagined must to some degree mirror the functional relations among those same objects as actually perceived" (Shepard, 1978, p. 131).

While there has been much debate about the physiological nature of imagery, one noted researcher in mental imagery concludes the following:

Although the brain processes that underlie a mental image need not themselves be like any sort of a picture, they must necessarily contain the information that could in principle permit the reconstruction of a picture with a high degree of isomorphism to the external object imagined.... What is behind the common tendency to think of a mental image as some sort of a picture may be the fact that the brain process that underlies a purely mental image is very much like the brain process that is produced by looking at a corresponding picture. (Shepard, 1978, p. 128)

Supporting this view, Shepard and his colleagues found that the amount of time it takes individuals to judge whether two spatial forms (two-dimensional depictions of three-dimensional objects made of congruent cubes) are the same is dependent on the number of degrees that one form must be rotated to match the other. Kosslyn's research also supports this point of view. He and his colleagues found, for instance, that more time was required to scan greater distances on a mental image and that larger images required more time to scan than smaller ones. They concluded that their experiments support the claim that "portions of images depict corresponding portions of the represented object(s) and that the spatial relations between portions of the imaged object(s) are preserved by the spatial relations between the corresponding portions of the image" (Kosslyn, Reiser, & Ball, 1978, p. 59).

Shepard (1978, p. 128) also noted that the existence of ambiguous pictures such as optical illusions indicates that "not all of what is perceived or imagined is contained in the concrete picture that is externally presented or reconstructed" (p. 129). A mental image, perceptual or imagined, has a "deep as well as a surface structure." But how are mental images formed and how is imagery related to perception? Let us return to Piaget:

Perception is the knowledge of objects resulting from direct contact with them. As against this, representation or imagination involves the evocation of objects in their absence or, when it runs parallel to perception, in their presence. It completes perceptual knowledge by reference to objects not actually perceived.... Now in all probability the image is an internalized imitation..., and is consequently derived from motor activity, even though its final form is that of a figural pattern traced on the sensory data. (Piaget & Inhelder, 1967, p. 17)

As discussed previously, one of Piaget's experimental tasks was to have children first examine flat geometrical shapes without seeing them and then identify the shapes by drawing, naming, or pointing them out. Of this task, Piaget and Inhelder say, "Clearly, the reaction involves translating tactile-kinesthetic impressions from an invisible object into a spatial image of a visual kind" (Piaget & Inhelder, 1967, p. 18). To gain insight

into how images are constructed, let us examine Piaget and Inhelder's description of stages of performance on this task.

In Stage 1, (2–4 yrs), the child cannot construct a complete image of the geometric figure, "and, according to whether he has felt a curved or straight side or a point, he likens the shape touched to a visual shape possessing the same characteristic, not bothering about the rest of the object or attempting to put together the total structure.... It is obvious that these errors are due to inadequate exploration of the objects.... to recognize geometrical shapes the child has to explore the whole contour" (p. 23). Thus, if a child is unable to take in a whole shape with a single tactile centration, he or she is compelled to move his or her hands across the object to produce a series of centrations. "The perceptual recognition of the shape is consequently a result of the co-ordination of these centrations" (p. 38). "The lack of exploration on the part of children at this stage may therefore be explained as the result of a general deficiency in perceptual activity itself" (p. 24). Children of this stage could not draw a copy of even the simplest shapes because perceptual *activity*, not perception, is the source of imitation. That is, since these children failed to explore the edges of the surfaces, they could not draw the surfaces.

In Stage 2 (4–7 yrs), perceptual activity becomes apparent.

A general distinction is drawn between two major classes of shape, curvilinear or without angles, and rectilinear or with angles.... It is not the straight line itself which the child contrasts with round shapes, but rather the conjunction of straight lines which go to form an angle. (p. 30)

Furthermore, the child constructs his representation of angle not as two intersecting lines, but rather as the "outcome of a pair of movements (of eye and hand) which conjoin" (p. 31). In fact, "Euclidean shapes...are at least as much abstracted from particular actions as they are from the object to which the actions relate" (p. 31). That is, the image the child extracts from an object is what can be constructed from his own actions performed on the object. Even drawings express an object not so much as visually or tactilely perceived, but in terms of the related perceptual activity. However, although perceptual activity is capable of being carried to completion during Stage 2, the analysis remains empirical; thus, for complex shapes, it fails to achieve a synthesis and coordination of perceptual data based on reasoning.

During Stage 3 (7–8 yrs), exploration of shapes is still performed by means of the same type of perceptual activity as in the earlier stages. However, this activity is now directed by an "operational method which consists of grouping the elements perceived in terms of a general plan, and starting from a fixed point of reference to which the child can always return." At this level, the construction of the image of a shape assembles data into an anticipatory schema that includes possible features such as straight or curved lines, angles, parallels, order, and equal or unequal lengths. "In other words, every perceived shape is assimilated to the schema of the actions required to construct it" (p. 37).

Most of the discussion above relates to "haptic perception." The difference between a haptic and visual centration is that a visual centration can take in more elements at the same time. Thus, simple shapes can be visually apprehended in a single centration.

According to Piaget and Inhelder, "it is one thing to perceive a circle or a square and quite another to reconstruct a visual image of it to the point where it can be picked out from a group of models, or draw it after a purely tactile exploration" (p. 37). But a visual image of a shape presupposes a mental representation, and thus "the image is not a direct outcome of perception" (p. 38). Furthermore, "the power to imagine the shapes visually when they are perceived through the sense of touch alone, is an expression of the sensorimotor schema involved in their perception" (p. 41). Piaget concludes that in all three stages, "children are able to recognize, and especially to represent, only those shapes which they can actually reconstruct through their own actions" (p.43).

Thus, the mental representation of a figure—its image—is seen by Piaget as an internal imitation of actions. The reconstruction of shapes as visual images (so that they can be drawn, for instance) "is not just a matter of isolating various perceptual qualities, nor is it a question of extracting shapes from the objects without more ado. The reconstruction of shapes rests upon an active process of *putting in relation,* and it therefore implies that the abstraction is based on the child's own actions and comes about through their gradual co-ordination" (p. 78).

Although evidence from experimental psychologists has not been gathered to support Piaget's claim for the importance of action, it is at least consistent with it. For instance, according to Kosslyn (1983), images of objects are built out of separately stored parts, such as line segments or common geometric shapes, along with information (verbal or nonverbal) specifying how the parts are to be arranged relative to one another. Images of figures never before seen could be generated by amalgamating parts that have been previously generated from experience. Using this theory to support Piaget only requires us to equate Kosslyn's component parts of images to memory traces of a person's actions, physical or perceptual.

Also taking a constructivist approach to spatial thinking, Wheatley and Cobb (1990) claim that individuals give meaning and structure to spatial patterns based on their experiences, conceptual structures, intentions, and ongoing social interactions. "We do not conceive of the pattern [of shapes] as being 'out there' for the child to capture visually and store *as is* in her head but believe that each child constructs through her actions an image of the pattern which may later be represented and transformed" (p. 3). These "actions" can be physical or perceptual, conscious or unconscious. Furthermore, construction is not necessarily a conscious process; for example, when a person gives meaning to a diagram, they may be unaware of the "meaning-making" process (G. Wheatley, personal communication, August 11, 1989). In Wheatley and Cobb's study, first- and second-grade students were given five tangram pieces—a medium triangle, two small triangles, a square, and a parallelogram—and asked to make a pictured square. To help the students, a line diagram showing the placement of the three triangles was shown to students for three seconds initially and at any other times the students requested. They found the students at different levels of performance on several aspects of the task. Some interpreted the diagram as a set of triangular regions and others as a set of lines forming a design. Only the most advanced students not only interpreted the diagram as a set of triangular regions but also constructed the relative placement of these triangles. Some students were able to rotate images of the tangram pieces mentally and, thus, anticipate their positions; others determined the correct orientations by physical trial and error.

In summary, we have seen that images are internal, wholistic representations of objects that are isomorphic to their referents and can be inspected and transformed. The construction of images is certainly affected by existing cognitive structure, but it would be helpful to know more about how this actually occurs and whether it can be controlled. If we accept that images are based on actions, by what mechanism are images derived from these actions? Is the image of an object simply a replay of the sequence of actions involved in perceiving it? How, then, are images generated in the absence of objects; that is, what psychological mechanisms support the representation of an image?

Improvement in Spatial Ability

Numerous studies have indicated that spatial ability can be improved through training (Bishop, 1980). Ben-Chaim, Lappan, and Houang (1988) reported that a three-week instructional training program significantly increased the spatial visualization ability for all students in grades 5–8, with no sex differences in the gains. They suggested that seventh grade may be the optimal time for spatial visualization training. Bishop (1980) found that students taught in primary schools where the use of manipulative materials was prevalent performed better on tests of spatial ability than students who were in schools lacking use of such materials. Battista, Wheatley, and Talsma (1982) found that students' spatial skills improved during the course of an informal geometry course. However, there are reports that no improvement in spatial ability results from a standard course in high school geometry (Bishop, 1980). There is also evidence that performance on spatial tasks increases with grade level (Ben-Chaim et al., 1988; Johnson & Meade, 1987).

Given the obvious connection between spatial thinking and transformational geometry, one might hypothesize that work with the latter would improve skills in the former. As Fey stated, the "transformation approach makes geometry an appealing, dynamic subject that will develop spatial visualization ability and also the ability to reason" (Fey et al., 1984, p. 44). In agreement with this hypothesis, Del Grande (1986) found that such a geometry unit improved spatial perception of grade 2 students. Williford (1972) also investigated the effects of teaching the concepts of rigid motions and congruence. He concluded that second and third graders learned manual procedures for producing transformation images, but not how to perform such transformations mentally. Kidder's (1976) results extend Williford's in that they indicate students' ability to mentally perform isometries without training is limited at the middle and junior high level. He conjectured that "the ability to perform transformations at the representational level derives from formal-operational thought (in a Piagetian sense), and that the thirteen-year-old subjects of the present study were not in the formal operational stage" (p. 50). Ali Shah (1969) found that performance on these transformations increased greatly from age 7–8 to age 10–11, but that only about 50% of the latter age group mastered these topics. Moyer (1978) investigated whether, for children of ages 4 to 8, understanding of a two-dimensional isometry is dependent upon an explicit awareness of the physical motion related to the transformation. There were no significant effects for slides and flips (with the trend being a negative effect for younger children), but there was a dramatic, beneficial effect for turns. He also found that a slide task was at least as easy as a flip, and turns were most difficult. Finally, Usiskin (1972) compared the effects of a transformational approach to high school geometry to the traditional approach. Both the experimental and control groups showed significant increases on a spatial/perceptive test, although not differentially so. Both groups showed a decline in attitudes towards mathematics, but only the control group's was significant. It was also found that girls' attitudes towards mathematics declined more than boys' for both the transformational and traditional approaches.

REPRESENTATIONS OF GEOMETRIC IDEAS

Concept Images

Vinner and Hershkowitz (1980) claim that in thinking, people do not use definitions of concepts, but rather concept images, combinations of all the mental pictures and properties that have been associated with the concept. Their research validated not only that these concept images existed for a number of geometric concepts, but that such images could be adversely affected by inappropriate instruction. For example, the fact that, for many students, the concept image of an obtuse angle has a horizontal ray might result from the limited set of examples they see in texts and a "gravitational factor" (that is, a figure is "stable" only if it has one horizontal side, with the other side ascending). (The findings of other studies agree that students limit concepts to studied exemplars and consider inessential but common features as essential to the concept; Burger & Shaughnessy, 1986; Fisher, 1978; Fuys et al., 1988; Kabanova-Meller, 1970; Zykova, 1969). Components of concept images were also identified; for example, students' concept image for a right triangle were most likely to include a right triangle with a horizontal and a vertical side, less likely to include a similar triangle rotated slightly, and least likely to include a right isosceles triangle with a horizontal hypotenuse. Study of such concept images may provide useful information about errors that students make. For example, students who know a correct verbal description of a concept, but also have a specific visual image or prototype associated tightly with that concept, may have difficulty applying the verbal description correctly (Clements & Battista, 1989; Hershkowitz et al., 1990; Vinner & Hershkowitz, 1980).

Fischbein (1987) relates concept images to intuition. Subjects attach a particular presentation to the concept, which has a strong impact on their cognitive decisions. Even when the definition is explicitly mentioned, most subjects are not able to respond correctly. "The manner in which a concept functions in a reasoning process is highly dependent on its paradigmatic connections. The fact of knowing explicitly the definition does not eliminate the constraints imposed by the tacitly intervening paradigm" (p. 146). This also helps explain students' resistance to hierarchical relationships among quadrilaterals. The images attached to each figure function cognitively, not as particular cases but as general models. Thus, students have to learn the decisive role of explicitly defining concepts to avoid errors in using the terms that signify them. They have to construct a meaningful synthesis of this definition with a range of exemplars. Employing such a synthesis of analytic and verbal processes to construct robust concepts is possible, especially for students in grade 5 and beyond (Hershkowitz et al., 1990).

This formulation is highly consistent with Reif's (1987) "ideal" model for interpreting mathematical concepts reliably and efficiently. In familiar situations, this model first applies nonformal, case-based knowledge (see Fischbein's [1987] paradigmatic models), then checks doubtful conclusions with explicit formal knowledge. In unfamiliar situations or whenever inconsistencies or needs to make general inferences arise, the ideal model turns directly to formal knowledge. Nonformal knowledge, then, is still useful in providing checkpoints for more abstract arguments. Reif found the processes of effective experts fit this ideal model closely and recommends activities that encourage its development in students, such as teaching students concept-interpretation procedures and letting them implement those procedures in various typical and error-prone cases, thus compiling repertoires of knowledge about special cases and common errors.

This line of inquiry is relevant to both theory and practice; it is unfortunate, then, that it has neither adequately acknowledged nor built upon earlier work. For example, Vygotsky's (1934/1986) construct of "word sense" as the sum of all the psychological events aroused in a person's consciousness by the word is in some ways a more elaborate formulation. Word sense is a dynamic, complex, fluid whole, which has several zones of unequal stability (the word's definition is only one of the zones of word sense, albeit the most stable and precise zone). A word acquires its sense from the context in which it appears; in different contexts, it changes its sense. "The primordial word by no means could be reduced to a mere sign of the concept. Such a word is rather a picture, image, mental sketch of the concept. It is a work of art indeed. That is why such a word has a 'complex' character and may denote a number of objects belonging to one complex" (Vygotsky, 1934/1986, p. 133).

Using Diagrams

The Soviet researcher Kabanova-Meller states "Mastery of geometric theorems is characteristically accomplished through the perception of diagrams and is intimately connected with the development of spatial images" (1970, p. 7). To be successful with proof problems in which a diagram is used, students must establish semantic connections in the diagram. The solution process must be "constructive"; that is, students must form conclusions from the conditions ("expand" the condition) and find new relationships in the diagram ("transform" the diagram). Sometimes students derive the solution by expanding the condition step by step until it includes the solution; other times students represent the solution through a visual image and then determine the logical steps required to derive it. In addition, when perceiving a diagram for a problem, a student must focus on what is essential and dismiss what is nonessential. Kabanova-Meller found that some pupils could not do so, and she attributed this inability to the original learning of the relevant theorems. When learning theorems, students often incorporated information contained in a specific diagram as part of a theorem (for instance, thinking that the exterior angle of a triangle must be obtuse because the diagram given with the theorem pictured an obtuse exterior angle). This information later constrained the application of the theorem. That is, the student did not recognize the theorem was relevant for another problem because the diagram contained an acute exterior angle.

As students use a diagram to interpret a theorem, they must alter the corresponding mental image "by distinguishing its essential aspects during the abstraction process" (p. 46). The image becomes a guide for thinking about and applying the theorem. In a successful teaching experiment using multiple drawings to illustrate theorems, students learned how the placement, direction, and magnitude of geometric elements in a diagram might vary within the conditions set forth in a theorem. It was concluded that

> The process of abstracting the essential aspects of a concept or theorem must be inseparably linked with an awareness of how the features accompanying a diagram may vary. The pupil, therefore, must learn to formulate verbally the principle by which the features as well as the geometric elements and their relationships, may vary in different diagrams without destroying the sense of the concept or theorem. Unless the diagrams are used in this manner, varying the visual material will be ineffective. (p. 48)

In a similar vein, Vladimirskii concluded that the diagram accompanying the discussion of a geometric statement is not always helpful in reasoning (Vladimirskii, 1971). Students might mistake features of the diagram as essential features of the geometric relationship being considered, thus introducing irrelevant ideas into the concept. Alternately, a theorem might be linked to only the example diagram given with its statement; the "visual images corresponding to the relationship being studied become inert and inflexible" (p. 81). Thus, students will be unable to effectively use the theorem in problem solving. The author found that an instructional treatment designed to have students analyze the "system of features of a concept" in the context of diagrams was successful in helping students' knowledge become unlinked from specific diagrams.

Although these authors discussed how diagrams affect students' representation of concepts and theorems, it should be clear that similar comments about misconceptions and instructional remedies could be made about students' representation of geometric problems. Indeed, there are numerous accounts of persons arriving at incorrect solutions to problems due to improper problem representations (Davis, 1984).

While we have been focusing on how diagrams can affect students' representation of concepts, theorems, and problems, we should not overlook misconceptions that could arise out of students' interpretations of the diagrams themselves. For example, Parzysz (1988) reports that students often attribute characteristics of a drawing to the geometric object it represents, fail to understand that drawings do not necessarily represent all known information about the object represented, and commonly attempt to draw figures so that they preserve both viewing perspective and student knowledge about the properties of the object being drawn (for example, drawing the base of a regular pyramid as a square instead of a parallelogram).

An interview with a bright fourth-grade student, referred to as KL, at the end of the school year suggests caution in making even the most straightforward assumptions about how students interpret diagrams (Clements & Battista, 1990). The student was asked to identify all the rectangles out of a set of quadrilaterals consisting of rectangles, squares, rhombuses, parallelograms, and trapezoids of various sizes and orientations. He said that two parallelograms were "sorta like" rectangles, "like you're looking at it a different way. Like you'd look side-ways or something." When the interviewer asked if one of these parallelograms was really a rectangle, KL said "Yeah, it makes it if you're looking at it, like if it's like a piece of paper and you're looking below it...." The interviewer went on to ask what if it was a piece of paper that was cut that way. KL: "Then it wouldn't be a rectangle." For KL, the figures drawn on paper represented *pictures* of geometric objects; they were not considered as geometric objects themselves. In fact, it is true that a parallelogram might very well be a picture of a rectangle drawn in perspective. But the fact that KL was almost willing to name such a picture a rectangle is most instructive. Initially for children, pictures are meant to be depictions of real objects. In mathematics, however, pictures are meant to be symbols for concepts. KL seemed to be in transition between these two interpretations and was, thus, vacillating between them. This example is especially illustrative because it demonstrates the wide discrepancy that can exist between a child's and a teacher's or textbook's interpretations of a simple diagram.

Driscoll (1983b) relates other difficulties that students have with drawings to the notion of concept images discussed above. For instance, middle school students had a great deal of difficulty identifying right triangles when they were drawn with a nonstandard orientation (legs not vertical and horizontal). If students' concept images do not include relevant and irrelevant attributes of the concept, they become too restricted for proper identification and use of future examples.

Along with the previously-discussed PDP models, this research corpus on geometric representations has strong instructional implications: There is a dire need to provide variety in exemplars and, later, to help students construct a meaningful verbal synthesis or definition from these exemplars (including dealing with special cases and common errors). "By analyzing the paradigm in the light of the concept, by learning to find correct examples and counter-examples corresponding to the concept, the student may reach this stage of grasping a concept which is not void, related to exemplars which are not misused" (Fischbein, 1987, p. 152).

Results of research have additional implications for the presentation of examples and nonexamples. Psychological research often implies that positive instances are more useful than mixed negative-and-positive instances, whereas educational research suggests the usefulness of negative instances (Gibson, 1985). Wilson's results (1986) suggest a resolution: When every feature of every irrelevant dimension was equally likely, positive instances were more helpful in learning. When certain irrelevant features predominate, mixed positive-and-negative instances were more helpful. Note that the former is unlikely in the classroom context (especially considering typical prototypes in textbooks), and, thus, a mixture will be superior in most instances, especially for more difficult concepts (Charles, 1980; Gibson, 1985). Further, strategies for sequencing examples and nonexamples are not equally effective. Rational sequences (nonexamples of a concept matched with divergent examples to focus attention of the critical attributes) are superior to random sequences and can lead to a high level of achievement (Petty & Jansson, 1987). Pre-service teachers can be successfully trained to use such strategies, including exemplification moves (presenting example or nonexample) and characterization moves (statements about attribute, relevant or not) (Charles, 1980).

The Role of Action

Mental action is deemed important to the learning of geometry by each major theoretical perspective. Recall that Piaget and Inhelder claimed children's representation of space is not a perceptual reading off of their spatial environment, but is the result of prior active manipulation of that environment. The second teaching phase in the van Hiele scheme centers around students' manipulation of objects. Johnson provides another perspective on this important issue by arguing for the importance of image schema as metaphors.

The Use of Manipulatives. Although there are exceptions (Anderson, 1957), the majority of studies verify that the use of manipulatives should facilitate the construction of sound representations of geometric concepts. Exposure to a greater variety of stimuli positively affects achievement in geometry (Greabell, 1978). Such tactile-kinesthetic experiences as body movement and manipulating geometric solids help children, especially young children, learn geometric concepts (Gerhardt, 1973; Prigge, 1978). Children also fare better with solid cutouts than printed forms, the former encouraging the use of more senses (Stevenson & McBee, 1958).

There is empirical support that even for older students, especially those at lower levels in the van Hiele hierarchy, manipulatives are an essential aid in learning geometry (Fuys et al., 1988). Use of manipulatives seemed to allow students to try out their ideas, examine and reflect on them, and modify them. This physical approach seemed to maintain student interest, to assist students in creating definitions and new conjectures, and to aid them in gaining insight into new relationships. Similarly, sixth-grade students learn concepts of vector space better with concrete materials (Lamon & Huber, 1971). Fourth graders benefited more from an advanced organizer on motion geometry using concrete models than one using applications, whereas seventh graders benefited equally; the younger students may have been less able to impose the relevant relationships (Lesh & Johnson, 1976). Overall, the benefits of manipulatives hold across grade level, ability level, and topic, given that use of a manipulative "makes sense" for that topic (Driscoll, 1983a; Sowell, 1989). However, U.S. textbooks only infrequently suggest the use of manipulatives in geometry, and even when they do the suggested uses are not aimed at developing higher levels of thinking (Fuys et al., 1988; Stigler et al., 1990). In contrast, the evidently more successful Japanese instruction and instructional materials feature far greater use of manipulatives (Stigler et al., 1990). Similarly, Mitchelmore (1980) found that British students were about 3 years ahead of U.S. children in both spatial and three-dimensional drawing ability. He suggested that differences are attributable to different teaching approaches, in that British teachers tend to take a more informal approach to geometry and use more manipulative materials at the elementary level and more diagrams at higher levels. Another, complementary possibility is that the curricula reflect each country's attitude towards the use of spatial models.

Unfortunately, nearly half of K–6 teachers report that their students use manipulatives less than once a week, or not at all (Driscoll, 1983a). This may be important, because manipulative use for a school year or longer results in significant differences of moderate to large size in favor of the manipulative groups; use of shorter duration often does not produce significant results (Sowell, 1989). Further, use of manipulatives is not sufficient; simply using manipulatives does not guarantee meaningful learning (Raphael & Wahlstrom, 1989). Students must be guided to reflect on their use of manipulatives and to relate manipulative models to their informal concepts.

If manipulatives are accepted as important, what of pictures? Pictures can be important; even children as young as 5 or 6 years (but not younger) can use information in pictures to build a pyramid (for example, see Murphy and Wood, 1981). Thus, pictures can give students an immediate, intuitive grasp of certain geometric ideas. However, pictures need to be varied so that students aren't led to form incorrect concepts (see the previous discussions of concept images and diagrams). However, research indicates that it is rare for pictures to be superior to manipulatives. In fact, in some cases, pictures may not differ in effectiveness from instruction with symbols (Sowell, 1989). But the reason may not lie in the "nonconcrete" nature of the pictures as much as it lies in their "nonmanipulability"; that is, children cannot act on them as flexibly and extensively. This suggests investigation of manipulatable pictures such as graphic computer representations, the subject to which we turn.

The Promise of Computers. Computers' graphic capabilities may also facilitate the construction of geometric representations. As with other topics, students instructed in geometry with computers often score significantly higher than those having just classroom instruction (Austin, 1984; Morris, 1983). Even with children as young as preschool, computer-based programs are as effective in teaching about shapes as teacher-directed programs (von Stein, 1982) and more effective at teaching spatial relational concepts than television (Brawer, cited in Lieberman, 1985). Computer games have been found to be marginally effective at promoting learning of angle estimation skills (Bright, 1985) and significantly more effective than traditional instruction in facilitating achievement in coordinate geometry (Morris, 1983). However, optimism about the use of computers must be tempered by consideration of the findings of comparative media research. Decades of pre-computer research comparing the effects of different media on achievement show basically the same result: No significant difference (Clements, 1984). A change of curriculum or teaching strategy may explain the positive results of many comparative studies of Computer-Assisted Instruction, or CAI (Clark, 1983). However, there are certain functions computers can perform that cannot be duplicated in other situations.

LOGO. Logo represents such an application. We have seen that children's initial representations of space are based on action (Piaget & Inhelder, 1967). One implication is that Logo activities designed to help children abstract the notion of path may provide a very fertile environment for developing their conceptualizations of simple two-dimensional shapes. Logo environments are, in fact, action-based. These actions are both perceptual—watching the turtle's movements—and physical—interpreting the turtle's movement as physical motion that could be performed by oneself. By first having children form paths by walking, then using Logo, children can learn to think of the turtle's actions as ones that they can perform; that is, the turtle's actions become "body syntonic" (Battista & Clements, 1988; Papert, 1980). Because the mathematical concept of path can be thought of as a record of movement, the path concept may constitute a particularly good starting point for the study of geometry. Having students visually scan the side of a building, run their hands along the edge of a rectangular table, or walk a straight path will give students experience with the concept of straightness. However, this concept can be brought to a more explicit level of awareness with path activities in Logo; it is easy to have students use the turtle to discover that a straight path is one that has no turning. Thus, such experiences can help students develop a description or formalization of the concept of straightness.

These and other types of Logo activities might be used to encourage students to progress to Levels 2 (descriptive/analytic) and 3 (abstract/relational) in the van Hiele hierarchy. For instance, with the concept of rectangle, students initially are able only to identify visually presented examples, a Level 1 (visual) activity in the van Hiele hierarchy. In Logo, however, students can be asked to construct a sequence of commands (a procedure) to draw a rectangle. This "...allows, or obliges, the child to externalize intuitive expectations. When the intuition is translated into a program it becomes more obtrusive and more accessible to reflection" (Papert, 1980, p. 145). That is, in constructing a rectangle procedure, the students must analyze the visual aspects of the rectangle and reflect on how its component parts are put together, an activity that encourages Level 2 thinking. Furthermore, if asked to design a rectangle procedure that takes the length and width as inputs, students must construct a form of definition for a rectangle, one that the computer understands. Thus, they begin to build intuitive knowledge about the concept of defining a rectangle, knowledge that can later be integrated and formalized into an abstract definition (a Level 3 activity). Asking students if a square or a parallelogram can be drawn by their rectangle procedure if given the proper inputs encourages students to start ordering figures logically (another Level 3 activity).

Research suggests that these theoretical predictions are valid. Grade 7 students' work in Logo relates closely to their level of geometric thinking (Olson, Kieren, & Ludwig, 1987). In addition, appropriate use of Logo helps elementary students begin to make the transition from the Levels 0 and 1 to Level 2 of geometric thought. For example, Logo experience has been shown to have a significantly positive effect on elementary school children's plane figure concepts (Clements, 1987; Clements & Battista, 1989; Hughes & Macleod, 1986). This may be because, as recommended by van Hiele researchers, Logo incorporates implicitly the types of properties which will be developed by Level 1 thinkers explicitly, something that textbooks often fail to do (Battista & Clements, 1987, 1988, 1991; Fuys et al., 1988). Logo experience encourages students to view and describe geometric objects in terms of the actions or procedures used to construct them (Clements & Battista, 1989). When asked to describe geometric shapes, children with Logo experience proffer not only more statements overall, but also more statements that explicitly mention components and geometric properties of shapes, an indication of Level 2 thinking (Clements & Battista, 1989, 1990; Lehrer & Smith, 1986).

Similar results have emerged in the area of symmetry and motion geometry. Working with a Logo unit on motion geometry, students' movement to van Hiele levels was slow, but there was definite evidence of a beginning awareness of the properties of transformations (Olson et al., 1987). Similarly, intermediate grade students in the United States were engaged in symmetry and motion geometry activities using either Logo or paper and pencil (Johnson-Gentile, Clements, & Battista, 1990). Interviews conducted with a subsample revealed that both treatment groups performed at a higher level of geometric thinking than did the control group; Logo students performed at a higher level than the noncomputer students on four of the six interview tasks, noncomputer students performed at a higher level on one. Both Logo and non-Logo groups outperformed the control group on immediate and delayed post-tests; in addition, though the two treatment groups did not significantly differ on the immediate post-test, the Logo group outperformed the non-Logo group on the delayed post-test. Thus, there was support for the notion that the Logo-based version enhanced conceptual reconstruction of previously-learned ideas. Compared to students using paper and pencil, students using Logo worked with more precision and exactness (Gallou-Dumiel, 1989; Johnson-Gentile et al., 1990). Thus, there is evidence in support of the hypothesis that Logo experiences can help elementary to middle school students become cognizant of their mathematical intuitions and facilitate the transition from visual to descriptive/analytic geometric thinking in the domains of shapes, symmetry, and motions (Clements & Battista, 1990).

Several research projects have investigated the effects of Logo experience on students' conceptualizations of angle, angle measure, and rotation. In one study, responses of intermediate grade control students were more likely to reflect little knowledge of angle or common language usage, whereas the responses of the Logo students indicated more generalized, mathematically-oriented conceptualizations (including angle as rotation and as a union of two lines/segments/rays) (Clements & Battista, 1989). Other researchers studied how Logo might provide experiences at the second and third van Hiele levels for ninth-grade students (Olive, Lankenau, & Scally, 1986). Logo students gained more than the control students on interviews that operationalized the van Hiele levels for the concept of angle. Several other researchers have similarly reported a positive effect of Logo on students' angle concepts (Kieran, 1986a; Olive et al., 1986), although in some situations, benefits did not emerge until after more than a year of Logo experience (Kelly, Kelly, & Miller, 1986-1987).

This line of research also indicates that students hold many different schemas regarding not only angle concept but also angle measure. Third graders frequently relate the size of an angle to the length of the line segments that formed its sides, the tilt of the top line segment, the area enclosed by the triangular region defined by the drawn sides, the length between the sides (from points sometimes, but not always, equidistant from the vertex), the proximity of the two sides, or the turn at the vertex (Clements & Battista, 1989). Intermediate grade students often possess one of two schemas. In the "45-90 schema," slanted lines are associated with 45° turns; horizontal and vertical lines with 90° turns. In the "protractor schema," inputs to turns are based on usage of a protractor in "standard" position (thus, to have a turtle at home position turn left 45°, students might use an input of 135°, which corresponds to a protractor's reading when its base is horizontal) (Kieran, Hillel, & Erlwanger, 1986). Logo experiences may foster some misconceptions of angle measure, including viewing it as the angle of rotation along the path or the degree of rotation from the vertical (Clements & Battista, 1989). In addition, such experiences do not replace previous misconceptions of angle measure (Davis, 1984). For example, students' misconceptions about angle measure and difficulties coordinating the relationships between the turtle's rotation and the constructed angle have persisted for years, especially if not properly guided by their teachers (Hershkowitz et al., 1990; Hoyles & Sutherland, 1986; Kieran, 1986a, 1986b; Kieran et al., 1986). In general, however, Logo experience appears to facilitate angle measure understanding. Logo children's conceptualizations of larger angles are more likely to reflect mathematically correct, coherent, abstract ideas (Clements & Battista, 1989; Findlayson, 1984; Kieran, 1986b; Noss, 1987) and show a progression from van Hiele Level 0 to Level 2 in the span of the treatment (Clements & Battista, 1990). If Logo experiences emphasize the difference between the angle of rotation and the angle formed as the turtle traced a path, misconceptions regarding the measure of rotation and the measure of the angle may be avoided (Battista & Clements, 1991; Clements & Battista, 1989; Kieran, 1986b).

There is some evidence that Logo experiences affect measurement competencies beyond the measure of rotation and angle. Observations show that first graders invent their own standard units of measure to make Logo drawings, such as a rectangle with a width of 44 and a length of 88 via Forward 44 Forward 44 (Kull, 1986). Research indicates that Logo can help young children learn about measurement and aid researchers in learning more about what young children learn about measurement. Logo provides an arena in which young children may use units of varying size, define and create their own units, maintain or predict unit size, and create length rather than end point representations through either iterative or numeric distance commands. Further, Logo permits the child to manipulate units and to explore transformations of unit size and number of units without the distracting dexterity demands of measuring instruments and physical quantity (Campbell, 1987). Working with kindergartners and first graders, Campbell found that:

1. Children have difficulty adjusting to changes in unit size, especially when the unit size is halved. They may be using perceptual/spatial strategies rather than numerical/mathematical ones to solve these problems.
2. Children are remarkably accurate when estimating halved distances, even if the midpoint of the total distance has to be imagined.
3. Children understand that a distance traversed with a smaller unit requires a greater number of units than that same distance traversed with a larger unit. However, they consistently underestimate the strength of the inverse relationship between unit size and numerosity of units.
4. Contrary to expectations, estimation of lines with oblique orientations is usually not more difficult than lines with horizontal or vertical orientations. There were no differences across grades; a rather modest amount of Logo experience (two hours) may have helped the kindergarten children ignore the problem-irrelevant variables in the spatial field.

In almost all comparisons, Logo children were more accurate than control children. The control children were more likely to underestimate distances, particularly the longest distances; have difficulty compensating for the halved unit size; and underestimate the inverse relationship between unit size and unit numerosity. The Logo experience may have contributed to estimation accuracy. Nevertheless, even inexperienced children knew that the smaller numbers were to be associated with the shorter lengths and that progressively larger numbers would be assigned to progressively longer lengths. This basic principle of measurement seems to be acquired early in life and may not be dependent on specific measurement experiences.

Not all research has been positive. First, it should be noted that none of the studies have reported students' mastery of the concepts investigated, but merely a facilitative effect. Second, some studies show no significant differences between Logo and control groups (Johnson, 1986). Third, some studies have shown limited transfer; for example, students from two ninth-grade Logo classes did not differ significantly from control students on subsequent high school geometry grades and tests (Olive et al., 1986).

One problem is that students do not always think mathematically, even when the Logo environment invites such thinking. For example, some students rely excessively on visual cues and eschew analytical work, such as looking for exact mathematical and programming relations within the geometry of the figure (Hillel & Kieran, 1988). The visual approach is not correlated with the ability to visualize, but, instead, refers to reliance on the visual "data" of a geometric figure in determining students' Logo constructions (for example, "this last side looks like 60...try FORWARD 60"). In terms of van Hiele levels, the visual approach to solving Logo problems involves reasoning that lies between Levels 1 and 2. Although important in beginning phases of learning, continued use of a visual approach inhibits children from arriving at mathematical generalizations related to their Logo activity. There may be little reason for students to abandon visual approaches unless they are presented with tasks whose resolution requires an analytical approach. In addition, dialog between teacher and students is essential for encouraging higher-level reasoning. Care must be taken to help students establish and reflect upon path-command correspondences; that is, connections between geometric paths drawn by the turtle and the Logo commands that produce these paths (Battista & Clements, 1987, 1991; Clements, 1987; Clements & Battista, 1990).

In sum, studies that have shown the most positive effects involve carefully planned sequences of Logo activities and teacher mediation of students' work with those activities. It would appear that Logo's potential to develop geometric ideas will be fulfilled to the extent that teachers and instructional materials properly guide students' Logo experiences. This should include encouraging students to reflect on and forge links between Logo-based procedural knowledge and more traditional conceptual knowledge (Clements & Battista, 1989; Lehrer & Smith, 1986).

GRAPHIC TOOLS AND CONSTRUCTION PROGRAMS (GEOMETRIC SUPPOSER). Logo provides a powerful and flexible environment for students' representation and exploration of geometric ideas; other computer drawing and construction programs provide less flexibility but no less viable learning environments. For example, use of a computer "boxes" function, which allowed children to draw rectangles by stretching an electronic "rubber band," gave children a different perspective on geometric figures (Forman, 1986). The area fill function, which fills closed regions with color, prompted children to reflect on such topological features as closure. The potential of such drawing tools lies in the possibility that children will internalize such functions, thus constructing new mental tools; research has not yet adequately addressed this issue. Interaction with certain computer environments may help students build less restricted concept images. For example, producing random examples of isosceles and right triangles varying in shape and orientation helped 2–6-year-old children generalize their concepts of triangles to include a greater variety of triangular shapes and orientations (Shelton, 1985).

Computer graphics tools may positively affect spatial skills, with special benefits for girls. Although boys may outperform girls in computer games dealing with spatial ability (Pepin, Beaulieu, Matte, & Leroux, 1985), specially designed computer graphics modules have been shown to increase the spatial skills of high school girls to a level significantly beyond that of boys; although girls started with a lower composite mathematics score, after this training they had a significantly higher score than boys (Luchins, Rogers, & Voytuk, 1983).

Present-day media use two-dimensional representations to present most three-dimensional information. Research indicates that people find this difficult (Ben-Chaim et al., 1988).

Computers allow the dynamic manipulation of two-dimensional representations of three-dimensional figures. Osta (cited in Hershkowitz et al., 1990) studied this potential using two commercial programs, one in which operations could be performed on three-dimensional objects represented on the screen and the other, a "paint" program, in which operations could be performed only on two-dimensional figurative designs. The programs have constraints that necessitate the use of geometric properties rather than just visual information. Osta created problem situations in which students modified figures to move between two- and three-dimensional representations. Solution strategies of students in grades 8–9 were studied. At first, their work was local, dealing with small parts of figures through only perceptual strategies; with experience, though, students considered more global criteria and replaced inefficient perceptual strategies with strategies based on geometric properties.

The focus of construction programs, such as the *Geometric Supposer* software series, is to facilitate students making and testing conjectures. The Supposer programs allow students to choose a primitive shape, such as a triangle or quadrilateral (depending on the specific program), and to perform measurement operations and geometric constructions on it. The programs record the sequence of constructions and can automatically perform it again on other triangles or quadrilaterals. Thus, students can explore the generality of the consequences of constructions. Reports indicate that Supposer can be used effectively. In one evaluation, Supposer students performed as well as or better than their non-Supposer counterparts on geometry exams (Yerushalmy, Chazan, & Gordon, 1987). In addition, students' learning went beyond standard geometry content—for example, reinventing definitions, making conjectures, posing and solving significant problems, and devising original proofs. Making conjectures did not come easily to students and there was much frustration at the beginning of the year, but by the end nearly all students were making conjectures on large scale projects and felt the need to justify their generalizations. On specially-designed tests, Supposer students produced the same or higher level generalizations than the comparison group. There is conflicting evidence, however. For example, Bobango (1988) found that while instruction based on the van Hiele phases using Supposer significantly raised students' van Hiele level of thought, more so from Level 1 to Level 2 than for any other levels, it did not result in greater achievement in standard content or proof writing.

We have already seen that students' interpretation of diagrams is complex. A promising finding is that students make gains in understanding of diagrams and their limitations using Supposer. After such experience, they approached diagrams flexibly, treated a single diagram as a model for a class of diagrams, became aware that this model contains characteristics not representative of the class, and added auxiliary lines to diagrams (Yerushalmy & Chazan, 1988; Yerushalmy et al., 1987).

We have also seen that students have difficulty understanding proof; for instance, they often do not distinguish between two sources of knowledge about geometrical statements, measurement evidence, and deductive proofs. They believe that measuring examples "proves" a statement true for all members of an infinite set and/or that deductive proof pertains to one example only. Many students who held both beliefs still preferred the deductive proofs, not due to the influence of their teachers as much as to the explanatory characteristic of the proofs (Chazan, 1989). Supposer-based activities designed to change these beliefs engendered a movement away from considering measurement evidence as proof (Chazan, 1989; Wiske & Houde, 1988), although some students still thought there might be counterexamples to deductively proven results.

Considering post-test performance, percentages of students from experimental and control groups who produced informal and formal proofs were about the same in five comparisons conducted in one evaluation, and greater for the Supposer group in the sixth (Yerushalmy et al., 1987). The researchers noted, however, that teachers too frequently promoted a linear ordering of data collection and analysis, conjecture derivation, and proof, which led to nonreflective data gathering and obscured the differences between representations of specific instances during data collection phases and representations during conjecture and proof phases. Students, then, did not appreciate the different levels of generality these phases represented. The researchers recommend starting Supposer activities as investigations of a geometric relationship or concept (Yerushalmy et al., 1987).

Unlike textbook theorems, students believed Supposer-generated theorems needed to be proved before they could be accepted as true, leading to ownership of the theorems. They seemed to engage in van Hiele's Phase 4 learning, orienting themselves within the network of geometric relations (Lampert, 1988). In sum, it appears that, with proper support from the teacher, students using Supposer can come to understand the importance of formal proof as a way of establishing mathematical truth, although this seems to be related more to Bell's illumination function of proof than the verification and systematization functions (Yerushalmy, 1986). Chazan (1989), arguing from a philosophy of mathematics perspective, suggests that differences between deductive arguments and arguments based on measurement evidence are not as distinct as often supposed. He offers suggestions for teaching the differences, including developing proof as a social process and as an exploratory and explanatory process, rather than as an endpoint.

To implement the Supposer's guided inquiry approach successfully, research suggests the need for teaching strategies that connect students' inquiry with the curriculum and encourage inquiry as a way to learn successfully what needs to be known. In general, teachers in regular classrooms face multiple difficulties using Supposer; however, such struggles seem to have the potential to change teachers' practice, and thus change their beliefs about the meaning of knowing geometry and the acquisition of this knowledge in classrooms. Overall, while use of Supposer was found to demand hard work from and cause some frustration for both teachers and students, benefits were evident (Lampert, 1988; Wiske & Houde, 1988; Yerushalmy et al., 1987).

INTELLIGENT TUTORING: THE GEOMETRY TUTOR. Another field in which computers may make a unique contribution is that of intelligent tutors. Anderson, Boyle, and Reiser (1985) claim that it is feasible to build computer systems that are as effective as intelligent human tutors. Their Geometry Tutor was based on a set of pedagogical principles derived from the ACT* theory. First, productions in this theory always include a goal in the condition; thus, the goal structure of each problem is made explicit. Second, students are helped to cope with working memory demands by placing on the computer screen much of the information that the student might forget in the form of a proof graph. Third, because knowledge compilation occurs only during problem solving, formal instruction is made part of the problem-solving process. Students are given immediate feedback on their errors to make it easier for them to integrate this instruction into new productions they form.

The Geometry Tutor presents the statement to be proved at the top of the screen and the given statements at the bottom. The student adds to a developing "proof graph" by pointing to statements on the screen and by typing information. Each logical inference involves a set of premises, a reason, and a conclusion. Reasoning forward, the student points to the premises, types the reason, points to relevant geometric points in the diagram, and points to the conclusion (if already on the screen) or types it. These are connected in the proof graph with arrows. The proof is completed when there is a set of logical inferences connecting the given statements to the statements to be proved. The resultant proof graph makes concrete two abstract characteristics of proof problem solving: logical relations among the premises and conclusions and the search process used to find a correct proof. As the student works, the tutor infers which rule the student applied by determining which one matches the student's response. If correct, the tutor is silent; otherwise, instruction is given. All instruction, thus, is in the context of solving problems.

Three students of varying levels of ability learned geometry with the Geometry Tutor. According to the researchers, all learned geometry successfully and were solving problems more complex than those usually assigned in classrooms. All had positive attitudes (Anderson et al., 1985). Since this initial test, the Geometry Tutor has been used successfully in a public high school with four classes, from regular academic track to gifted (Anderson et al., n.d.). It was reported that students were enthusiastic. For the first year, all groups showed statistically significant improvement from pre- to post-test. A second year's test utilized a control group and found similar gains with a significant difference of the Geometry Tutor students over control group students, with improvements of about one standard deviation. Students unanimously reported that they prefer the proof-graph structure to traditional two-column proof formats. Further, such use has led to important revisions in the underlying ACT* theory.

Results of other evaluations of the program are more complex. Kafai (1989) used the Geometry Tutor to help students do congruence proofs in three classes. Generally, the Geometry Tutor was a success. All students had positive attitudes toward the program and recommended it for other students, all completed the minimal problem set, and most of them did all of the additional problems—a teacher stated that this had never happened before. However, when students used the proof-path method for noncomputer work, their performance decreased (most also used the statement-reason method). When using the proof-path method, students tended to reverse the order of statements and reasons and confuse the sequence of steps. This may indicate a serious problem with the program, for confusing the sequence of steps in a proof is an indication that students do not understand the nature of proof. As with many students in a standard geometry course, these students may have been learning merely how to superficially operate within a set of rules. Another observed problem involved the rigidity of the Tutor. It is not certain whether changes in the program or length of students' exposure to the program would ameliorate these difficulties. In another study, experimental students spent 30% of their time on the Geometry Tutor (Wertheimer, 1990). Here, there was no reported difficulty shifting from proof-graph to two-column format. Two experimental classes averaged 79% on a post-test, whereas the control class averaged 69% (about a letter grade in difference). The teacher/researcher commented that an increase in his individualization of instruction was the greatest impact of the program's use. However, the Tutor still makes considerable demands on the teacher; in fact, one report stated that 2–3 teachers were needed to keep the laboratory running smoothly (Schofield & Verban, 1988, note that software was not in its final form). These researchers proffered one additional warning: black and female students were more likely to express hostility toward or displeasure with the computer tutor. Although the results are positive for the most part, it is not clear that the evaluations were adequate to test students understanding of proof in either environment. More research that investigates students' thinking in this environment is needed.

Implications of this research program are not limited to computer tutors. Anderson claims that one of the reasons why traditional instruction is often so inadequate is that the teacher has an inadequate conception of the flow of control in the student. His model may illuminate these processes. Similarly, the structures of both problems and processes by which a proof is generated might be more explicitly represented and discussed.

IMPLICATIONS ACROSS COMPUTER ENVIRONMENTS. Several findings are intriguingly consistent across studies using different computer environments. First, researchers and teachers consistently report that in such contexts students cannot hide what they do not understand. That is, difficulties and misconceptions that are easily masked by traditional approaches emerge in computer environments and must be dealt with, leading to some frustration on the part of both teachers and students but also to greater development of mathematics abilities (Clements & Battista, 1989; Schofield & Verban, 1988; Yerushalmy et al., 1987). Second, at least at the high school level, students can become confused regarding the purpose of different components of a course; a single location for computer work, discussion, and lecture may alleviate this confusion. A monitor or projector for group discussions is noted as essential for all three environments. (These may seem to be lower-level concerns, but another general finding is that the importance and problems of arranging and managing hardware and software must not be underestimated.) Third, evaluation of learning in such environ-

ments must be reconsidered, as traditional approaches did not assess the full spectrum of what was learned; in some cases, these approaches made little sense (for example, when students worked on self-selected inquiries). On at least one issue, the various environments differed: With the Geometry Tutor, putting more than one student at a computer was deemed particularly unsuccessful (Wertheimer, 1990), but in contrast, two students working cooperatively at a computer seemed ideal in the more exploratory environments of *Geometric Supposer* and Logo. Indeed, one of the strengths of such environments is the spontaneous generation of cooperative learning and teaching (Clements & Battista, 1989; Lampert, 1988; Yerushalmy et al., 1987).

In summary, appropriately-designed software can engender higher-level interaction with geometric ideas. Certain computer environments allow the manipulation of screen objects in ways that assist students in viewing them as representatives of a class of geometric objects. This apparently develops students' ability to reflect on the properties of the class of objects and to think in a more general and abstract manner. Thoughtful sequences of computer activities and teacher mediation of students' work with those activities appear to be critical components of an efficacious educational environment. Perhaps even more fundamental, inquiry environments such as the Supposer- and Logo-based environments appear to have the potential to serve as catalysts both in promoting teachers' and students' reconceptualization of what it means to learn and understand geometry and in promoting the growth of students' autonomy in mathematical thinking. Fundamental changes demand considerable effort on the part of teachers and call for extensive support from teacher educators or advisors, from peers, and, ultimately, from the greater school system and culture.

GROUP AND CROSS-CULTURAL DIFFERENCES

We have already reported several cross-cultural differences in geometry achievement. There have also been studies of cross-cultural differences in spatial reasoning. For instance, Mitchelmore (1976) reported that native Africans of all nationalities had lower perceptual development when compared to Europeans with the same age and length of schooling as well as when compared to illiterate Eskimos and North American Indians. He also reported that the spatial thinking of students in Kingston, Jamaica was about three years behind that of students in Columbus, Ohio, which, in turn, was three years behind that of students in Bristol, England (Mitchelmore, 1980). He attributed these cross-cultural differences to various factors such as physical environment, social/cultural environment, and the school mathematics curriculum. Results from the fourth NAEP in the U.S. indicated that blacks and Hispanics in grades 7 and 11 had more difficulty than whites with graphically presented information, measurement, and geometry (Johnson, 1989). Johnson and Meade (1987) found that whites' spatial scores were higher than blacks'. However, the factors that contribute to these latter two differences have not been adequately investigated. On the other hand, researchers have investigated differences in performance between males and females extensively.

Gender Differences

In Geometry. In an examination of the mathematics items from the 1978 NAEP, Fennema and Carpenter (1981) found that males significantly outperformed females in the areas of geometry and measurement. Females' performance was lower than males' for geometry and measurement for all cognitive levels (knowledge, skill, understanding, application), and all ages (9, 13, 17 years), with the difference increasing with age. In the 1985 assessment, males outperformed females at all grade levels, but the differences were significant only at grade 11 for geometry and grades 3 and 11 for measurement (Meyer, 1989). Fennema and Carpenter observed large differences between males' and females' performance both for geometry items that were presented spatially and those that were presented verbally. Also, consistent with the hypothesis that differences in spatial ability somehow contribute to differences in mathematics achievement (Fennema & Carpenter, 1981), there was some indication from the sample items discussed in the 1985 data that the male advantage was greater on items presented with an accompanying diagram (Meyer, 1989). Smith and Walker (1988) found significant gender differences on the 10th grade geometry New York Regents exams, in favor of males. They rated the difference as small, and equated it to one-half of a question.

Hanna (1989a, 1989b) reported gender differences for eighth graders studied in the Second International Mathematics Study (SIMS). Overall, boys scored higher than girls on 75% of the geometry items; girls scored higher than boys on 25%. Boys' mean score in geometry was higher than girls' for 18 of the 20 countries—significantly so for 10, including the U.S. The significant differences in mean percentages ranged from 2% to 6%. Ethington (1990), analyzing a subset of this same data, found that, overall, males did better than females on the geometry items.

On the other hand, in their study of over 1,300 high school students, Senk and Usiskin found equal geometry proof-writing skills among males and females, when examining all students and just high-achieving students (Senk & Usiskin, 1983). They also found that males significantly outperformed females on the entering and end-of-year geometry tests, but attributed the latter difference to the former. In contrast, they found no gender differences in van Hiele level at the beginning of the year, with males, nevertheless, surpassing females in this respect at the end of the year (Usiskin, 1982). They conjectured that the occurrence of gender differences in geometric problem solving and the lack of gender differences in proof writing could be explained by the difference in formal educational experiences relating to these two types of tasks. Boys and girls do equally well on mathematical tasks for which in-class and out-of-class experiences are equivalent. Boys might have more spatial and problem-solving experiences than girls, and so they do better at these tasks. An alternate explanation is that girls do better at proof writing because it is a task that requires strict adherence to a classroom-mandated, formal set of rules—a "grammar," in a sense. This explanation is consistent with conjectures by Badger suggesting that girls tend to "keep to specific methods that have been approved by their teachers" (1981, p. 12) and by Linn and Hyde (1989) that "females may be more likely than

males to use the techniques they learned in school" (p. 22). Closely following prescribed procedures can certainly be helpful in situations where true understanding is lacking (as is the case with most students and proof).

In a comparison on various measures of geometry achievement, students in a standard geometry curriculum were compared to students studying the geometry curriculum that was created by the University of Chicago School Mathematics Project (UCSMP). Flores (1990) reported that the UCSMP treatment either reduced or reversed the usual gender gap favoring males. However, because the analyses were performed on scores adjusted for pre-treatment performance on a geometry readiness test, we do not know how males and females compared in achievement, only on their gains in achievement during the school year. Despite this caveat, the finding indicates that further investigation of gender differences in the UCSMP is warranted. It is noteworthy that females seemed to do better in a curriculum that placed heavy emphasis on reading (an activity in which females have traditionally excelled) and applications (in which females have usually done more poorly).

In Spatial Skills. Fennema and Carpenter (1981) hypothesized that one possible factor that may underlie gender differences in geometry achievement is spatial visualization. Mathematics achievement generally correlates with spatial visualization in the range of .30 to .60, and males' spatial scores have consistently been found to be higher than females' (Battista & Clements, 1990; Ben-Chaim et al., 1988; Fennema & Tartre, 1985; Tartre, 1990b). In her review of gender-related spatial differences, Harris states that gender differences in favor of males appear in a broad spectrum of tasks and situations and concludes that gender differences on spatial tasks are "real, not illusory" (Harris, 1981, p. 90). Linn and Petersen stated "Differences between males and females in spatial ability are widely acknowledged, yet considerable dispute surrounds the magnitude, nature, and age of first occurrence of these differences" (Linn & Petersen, 1985, p. 1479). For instance, they reported that spatial perception and rotation tasks were easier for males than for females but that tasks characterized by an analytic combination of visual and nonvisual strategies were equally difficult for males and females.

Johnson and Meade (1987) administered a battery of spatial tests to over 1,800 public school students, grades K–12. They found that males' spatial ability exceeded females' by fourth grade, with evidence that this superiority also existed at the earlier grades but was masked by females' superiority in verbal skills. (In fact, males' spatial scores were significantly higher than females in grades 1–4 when reading scores were used as a covariate). Their data also suggests that the male/female difference almost doubled starting at the tenth grade. They found a male spatial advantage for both blacks and whites.

While it has been argued that the commonly reported male superiority in spatial ability is of little consequence, Johnson and Meade suggest that the "effect sizes are substantial enough to create a sizable male majority if spatial ability were actually to influence performance levels on important variables or selection rates for certain occupations or training programs" (1987, p. 738). For instance, if the mean were used as a selection cutoff, their data for grades 4–12 indicate that 62% of the boys, but only 38% of the girls, would be selected.

DIFFERENCES IN STRATEGIES. There are several reports that indicate the relationship between spatial ability and mathematics achievement differs for males and females. Liben and Golbeck (1980) found that scores on Piagetian horizontality (water-level) and verticality tasks (plumb-line) were significantly and highly related to spatial ability for boys, but not for girls (grades 4-12). Kyllonen, Lohman, and Snow (1984) found that females' verbal aptitude and males' spatial aptitude each correlated with performance on a paper folding task. And Tartre (1990b) found that, although male and female high school students did not differ on mathematics (geometric and nongeometric) achievement or spatial orientation (SO) skill, low SO girls scored much lower in mathematics than did high SO girls and high and low SO boys. These findings are consistent with the hypothesis that there are differences in the processes underlying spatial and geometric thinking in males and females.

Indeed, there have been reports that males prefer nonverbal modes and females verbal modes of thought (Clements, 1983). In fact, Tartre (1990b) reported that females, more than males, tended to keep a written record of information during problem-solving. Also, Clements (1983) found that a much greater percentage of females than males used an inefficient visual-whole strategy on the Differential Aptitude Space Relations test, and that females used less effective concrete strategies and males used more efficient abstract strategies on spatial tasks. In fact, gender differences may "result from the propensity of females to select and consistently use less efficient or less accurate strategies for these tasks" (Linn & Petersen, 1985, p. 1492). Battista, Wheatley, and Talsma (1989) reported a tendency for female pre-service elementary teachers enrolled in geometry not to use the strategies that they used most effectively.

Battista (1990) found that males outperformed females in spatial visualization and in high school geometry, but that there were no gender differences in logical reasoning ability or use of geometric problem-solving strategies. When predicting geometry performance, spatial visualization was the most important factor for females, whereas logical reasoning and, as a secondary factor, the discrepancy between spatial visualization and logical reasoning were most important for males. It was suggested that there is a fundamental difference in the role that spatial visualization and logical reasoning play in males' and females' learning of geometric ideas. It was also found that the more spatial ability males, but not females, possessed in relation to their logical reasoning ability, the more likely they were to use a visualization strategy and the less likely to use a drawing strategy. This finding seems to indicate a fundamental difference in the way that the males and females represent geometric problems. Males with relatively higher spatial ability seemed to forego the use of the drawing strategy in preference for visualization, perhaps because their high visualization skills made them feel drawing is unnecessary. Females with relatively higher spatial ability, on the other hand, were more likely to use a diagram and less likely to use visualization. Similarly, Tartre (1989) found that high SO females were more likely than low SO females to use a drawing, but there were no differences for high and low SO males.

Another promising area of investigation is the interaction of teachers' style along a visual/nonvisual continuum and students' representational preference or visualization ability. Battista (1990) found that spatial visualization was more highly correlated with geometry learning for students in the classes of a teacher who emphasized and felt more confident about the role of visualization in geometry than a teacher who de-emphasized and felt less confident about this role. Males scored slightly higher than females in the former teacher's classes, whereas males scored much higher than females in the latter teacher's classes. Presmeg (1986) found that students who were visualizers seemed to do better in classes of teachers who neither over- nor underemphasized visual presentations. Bishop (1989) suggested that the latter study called into question the usefulness of simplistic ATI studies in this area. More thoughtful research is needed that investigates the interaction between teachers' instructional emphases and students' preference for and ability with visualization.

Thus, there is evidence suggesting that males and females, or at least subgroups thereof, may differ in the processes they use to solve mathematics, particularly geometry problems. There seem to be differences along the spatial dimension. However, these differences have not yet been adequately investigated. More research is needed in which students are carefully observed and interviewed as they solve problems.

BRAIN ORGANIZATION AND GENDER DIFFERENCES. It has been suggested by researchers in the physiology of the brain that "sex differences in verbal and spatial abilities may be related to differences in the way that those functions are distributed between the cerebral hemispheres in males and females" (Springer & Deutsch, 1981, p. 121). For instance, it has been conjectured that "both language abilities and spatial abilities are represented more bilaterally in females than in males" (Springer & Deutsch, 1981, p. 123). Further, this hypothesis posits that greater lateralization of function (that is, specialization to one side of the brain) may be essential for high spatial performance but less lateralization more important for verbal performance, so males should be superior in spatial tasks and females in verbal tasks. Thus, gender differences in geometry performance — performance which involves both the spatial and logical modes of thought — might productively be examined in terms of these two types of thought.

For example, drawings by left-side, brain-injured patients tend to be deficient in internal features and in organization of planning but adequate in overall spatial configuration, with the reverse true for patients with injury on the opposite side. These differences were mirrored by 5–13-year-old boys and girls asked to draw a complex figure. "At the youngest age, girls drew more internal details and more of the discrete parts, whereas boys concentrated more on the external configuration. At 11 years, boys drew their designs in 'long, sweeping, continuous lines,' whereas girls 'drew theirs part by part' (Waber, 1979, p. 173). In other words, where stylistic differences appeared, the boys' style tended to be characteristic of right-hemisphere processing, the girls' of left-hemisphere processing" (Harris, 1981, p. 103).

It is important to keep in mind the complexity of the research on gender differences. Such differences are observed in some areas of mathematics but not others, on some spatial tests but not others, in some cultures but not others (Hanna, 1989a). For instance, while males have performed better than females on various spatial tasks, the largest and most consistently observed male advantage appears on items requiring the rapid mental rotation of figures (Halpern, 1989), with the source of the difference being identified as the rate of mental rotation, not accuracy (Alderton, 1989). It has also been found that training tends to reduce these differences (Alderton, 1989; Linn & Hyde, 1989). Gender differences have been attributed to both biological and cultural factors and to a combination of the two (Geary, 1989). In a meta-analysis, Baenninger and Newcombe (1989) found (a) a weak relationship between participation in spatial activities and spatial ability, and (b) that spatial performance can be improved by training. Neither the participation/activity relationship nor the increases in performance differed for males and females. Even when there are no gender differences on a task, it should not be assumed that males and females are using the same strategies (Newcombe, Dubas, & Baenninger, 1989; Tartre, 1990b). Finally, Linn and Hyde (1989) and Feingold (1988) have argued that cognitive gender differences, including those in mathematics and spatial ability, are declining. Halpern (1989), on the other hand, has argued that these trends are artifacts of the testing instruments and changes in test populations.

CONCLUSIONS AND IMPLICATIONS

Without doubt, geometry is important. It offers us a way to interpret and reflect on our physical environment. It can serve as a tool for the study of other topics in mathematics and science. More importantly, however, spatial thinking, which obviously undergirds geometry, has been suggested by famous mathematicians such as Hadamard and Einstein to be essential to creative thought in all high level mathematics. Given their importance, therefore, it is essential that geometry and spatial reasoning receive greater attention in instruction and in research.

As we have seen, and belying its obvious importance in the curriculum, students' performance in geometry is woefully lacking. Neither what students learn in geometry nor the methods by which they learn it are satisfactory. There has been too little instructional attention given in the United States to spatial reasoning.

Evidence supports a constructivist position on how children learn spatial and geometric ideas. It appears that there is a progressive construction of geometric concepts from the perceptual to the conceptual plane, as well as developmental sequences in which children build increasingly integrated and synthesized geometric schemata. Research is needed to identify the specific cognitive constructions that children make at all age levels, especially in the context of supportive environments (for example, those including manipulatives, computer tools, and engaging tasks). Previously unexplored factors such as language, schooling, and the immediate social culture deserve attention in any such research program. Research on geometric representations from the cognitive science perspective should also be integrated in this quest. Ideally, such research should inform

us how to build on the strength of children's existing intuitions while ultimately aiding them in transcending the shortcomings of these intuitions.

Similarly, research that elaborates and extends the van Hiele theory is crucial. This perspective appears to hold much promise for the improvement of both research and instruction, and further elaboration and explication of such notions as "levels of geometric thinking" can help realize this promise. In addition, many of the implications for instruction previously described can and should be directly assessed in future studies. Especially crucial is research on the phases of instruction, which has not received adequate attention.

Perhaps the greatest strengths and weaknesses of the cognitive science approach lie in its extreme degree of specification. It provides much-needed details on cognitive processes (and thus forces explication on notions too often left vague in other theories, such as "networks of relations"). It may in particular provide explicitness to limited aspects of students' representations at lower van Hiele levels (for example, visual representations via PDP representations) and of concept formation. However, in a quest for machine-codable formats, some theories eschew such important constructs as belief systems, motivation, and meaningful interpretation of subject matter, while de-emphasizing the roles of intuition and culture. Thus, a synthesis of cognitive science, Piagetian, van Hiele, and other constructivist theories may yield particular riches.

Empowering students with methods by which they can establish for themselves mathematical truth and, thus, helping students develop intellectual autonomy, is a critical goal of geometry instruction, and indeed of all mathematics instruction. Currently, this goal is treated in a formal sense only in geometry, and, unfortunately, our instructional attempts at achieving this goal are failing. While we may dismiss formal proof as an instructional objective that only the best students need to master—thus we need not be overly concerned with our inability to teach it—we cannot so dismiss the process of establishing mathematical truth. It is the essence of mathematics. Without it, students cannot do mathematics; they can only examine noncritically what others have done. Thus, much more attention should be focused on research aimed at discovering how this important process develops, its effect on students' beliefs about and structuring of geometry knowledge, and how, for students, it can culminate in giving formal proofs.

Future investigations need to consider how visual thinking is manifest in higher levels of geometric thinking. It has been postulated here that visual thinking has a number of psychological layers, from primitive to sophisticated, each of which plays a different role in thinking, depending on which layer is activated.

We know a substantial amount about students' learning of geometric concepts. We need teaching/learning research that leads students to construct robust concepts through a meaningful synthesis of diagrams and visual images on the one hand, and through verbal definitions and analyses on the other. Such research should address the interrelationships between verbal and visual processing. It might also study how to help students build on, strengthen, and elaborate their existing intuitions about space and, ultimately, develop second-order, geometric intuitions.

Computers can help establish fecund environments for the study of students' geometric thinking. Research indicates that appropriately designed software, such as Logo and the *Geometric Supposer,* can engender high levels of geometric thinking. We need to learn more about the design of engaging tasks and teacher mediation, especially with an eye to using such software as a catalyst for the development of classroom cultures in which both teachers and students expand their beliefs about learning and understanding geometry.

We have reported gender differences in both geometry and spatial reasoning favoring males. If researchers focus not on whether such differences exist but rather on their exact nature, such differences can provide one perspective for investigating the mental processes of all students. That is, by investigating how males and females reason differently when dealing with spatial ideas and geometry, we may be able to better understand the development of geometric and spatial thinking for all students. However, because there is obviously much more variability in performance and processing *within* genders than between them, we should eventually be able to move beyond studying gender differences to the study of the different cognitive profiles that underlie successful performance in geom-

etry. More importantly, however, gender differences represent a cause for concern. Why do gender differences arise? Are there current instructional practices that tend to exacerbate the differences? What instructional practices can help ameliorate them?

Finally, geometry learning is an area rich with possibilities for future research. Given students' poor performance in this area, such research is sorely needed. Given a constructivist view of learning, research that describes the development of geometric concepts and thinking in various instructional environments is certainly required. Indeed, we believe that qualitatively different and improved environments for education in geometry will not emerge without the presence of (a) the theoretically cognizant teacher and (b) the student armed with a full array of tools for geometric investigations, including manipulatives and—perhaps most important—a computer replete with appropriate software. However, these agents cannot evolve without research that investigates the use of innovative materials, examines how students' knowledge develops within different instructional environments, and discovers how teachers can utilize both these environments and this knowledge about students' learning.

References

Alderton, D. L. (1989, March). *The fleeting nature of sex differences in spatial ability.* Paper presented at the meeting of the Annual Meeting of the American Educational Research Association, San Francisco, CA.

Anderson, G. R. (1957). Visual-tactual devices and their efficacy: An experiment in grade eight. *Arithmetic Teacher, 4,* 196–203.

Anderson, J. R. (1983). *The architecture of cognition.* Cambridge, MA: Harvard University Press.

Anderson, J. R., Boyle, C. F., Corbett, A., & Lewis, M. (n.d.). *Cognitive modelling and intelligent tutoring.* Unpublished manuscript, Carnegie-Mellon University, Pittsburgh, PA.

Anderson, J. R., Boyle, C. F., & Reiser, B. J. (1985). Intelligent tutoring systems. *Science, 228,* 456–462.

Austin, R. A. (1984). Teaching concepts and properties of parallelograms by a computer assisted instruction program and a traditional classroom setting. *Dissertation Abstracts International, 44,* 2075A. (University Microfilms No. DA8324939.)

Badger, M. E. (1981). Why aren't girls better at maths? A review of research. *Educational Research, 24*(1), 11–23.

Baenninger, M., & Newcombe, N. (1989). The role of experience in spatial test performance: A meta-analysis. *Sex Roles, 20,* 327–344.

Battista, M. T. (1990). Spatial visualization and gender differences in high school geometry. *Journal for Research in Mathematics Education, 21,* 47–60.

Battista, M. T., & Clements, D. H. (1987, June). *Logo-based geometry: Rationale and curriculum.* Paper presented at Learning and teaching geometry: Issues for research and practice working conference, Syracuse, NY: Syracuse University.

Battista, M. T., & Clements, D. H. (1988). A case for a Logo-based elementary school geometry curriculum. *Arithmetic Teacher, 36,* 11–17.

Battista, M. T., & Clements, D. H. (1990, April). *Logo environments and geometric learning.* Paper presented at the meeting of the National Council of Teachers of Mathematics, Salt Lake City, UT.

Battista, M. T., & Clements, D. H. (1991). *Logo geometry.* Morristown, NJ: Silver, Burdett & Ginn.

Battista, M. T., Wheatley, G. H., & Talsma, G. (1982). The importance of spatial visualization and cognitive development for geometry learning of preservice elementary teachers. *Journal for Research in Mathematics Education, 13,* 332–340.

Battista, M. T., Wheatley, G. W., & Talsma, G. (1989). Spatial visualization, formal reasoning, and geometric problem-solving strategies of preservice elementary teachers. *Focus on Learning Problems in Mathematics, 11,* 17–30.

Bell, A. W. (1976). A study of pupils' proof-explanations in mathematical situations. *Educational Studies in Mathematics, 7,* 23–40.

Ben-Chaim, D., Lappan, G., & Houang, R. T. (1988). The effect of instruction on spatial visualization skills of middle school boys and girls. *American Educational Research Journal, 25,* 51–71.

Bishop, A. J. (1980). Spatial abilities and mathematics achievement—A review. *Educational Studies in Mathematics, 11,* 257–269.

Bishop, A. J. (1983). Space and geometry. In R. Lesh & M. Landau (Eds.), *Acquisition of mathematics concepts and processes.* New York, NY: Academic Press.

Bishop, A. J. (1989). Review of research on visualization in mathematics education. *Focus on Learning Problems in Mathematics, 11*(1), 7–16.

Bobango, J. C. (1988). Van Hiele levels of geometric thought and student achievement in standard content and proof writing: The effect of phase-based instruction. *Dissertation Abstracts International, 48,* 2566A. (University Microfilms No. DA8727983.)

Bremner, J. G., & Taylor, A. J. (1982). Children's errors in copying angles: Perpendicular error or bisection error? *Perception, 11,* 163–171.

Bright, G. (1985). What research says: Teaching probability and estimation of length and angle measurements through microcomputer instructional games. *School Science and Mathematics, 85,* 513–522.

Brown, D. L., & Wheatley, G. W. (1989, September). *Relationship between spatial ability and mathematical knowledge.* Paper presented at the meeting of the North American Chapter of the International Group for the Psychology of Mathematics Education, New Brunswick, NJ.

Brumfield, C. (1973). Conventional approaches using synthetic Euclidean geometry. In K. B. Henderson (Ed.), *Geometry in the mathematics curriculum: 1973 Yearbook* (pp. 95–115). Reston, VA: National Council of Teachers of Mathematics.

Burger, W., & Shaughnessy, J. M. (1986). Characterizing the van Hiele levels of development in geometry. *Journal for Research in Mathematics Education, 17,* 31–48.

Campbell, P. F. (1987). *Measuring distance: Children's use of number and unit.* Final report submitted to the National Institute of Mental Health Under the ADAMHA Small Grant Award Program. Grant No. MSMA 1 R03 MH423435-01. University of Maryland, College Park.

Carpenter, T. P., Corbitt, M. K., Kepner, H. S., Lindquist, M. M., & Reys, R. E. (1980). National assessment. In E. Fennema (Ed.), *Mathematics education research: Implications for the 80s* (pp. 22–38). Alexandria, VA: Association for Supervision and Curriculum Development.

Charles, R. I. (1980). Exemplification and characterization moves in the classroom teaching of geometry concepts. *Journal for Research in Mathematics Education, 11,* 10–21.

Chazan, D. (1989). Instructional implications of a research project on students' understandings of the differences between empirical verification and mathematical proof. In D. Hergert (Ed.), *Proceedings of the First International Conference on the History and Philosophy of Science in Science Teaching.* Tallahassee, FL: Florida State University Science Education and Philosophy Department.

Clark, R. E. (1983). Reconsidering research on learning from media. *Review of Educational Research, 53,* 445–459.

Clements, D. H. (1984, November). Implications of media research for the instructional application of computers with young children. *Educational Technology,* pp. 7–16.

Clements, D. H. (1987). Longitudinal study of the effects of Logo programming on cognitive abilities and achievement. *Journal of Educational Computing Research, 3,* 73–94.

Clements, D. H., & Battista, M. T. (1989). Learning of geometric concepts in a Logo environment. *Journal for Research in Mathematics Education, 20,* 450–467.

Clements, D. H., & Battista, M. T. (1990). The effects of Logo on children's conceptualizations of angle and polygons. *Journal for Research in Mathematics Education, 21,* 356–371.

Clements, M. A. (1979). Sex differences in mathematical performance: An historical perspective. *Educational Studies in Mathematics, 10,* 305–322.

Clements, M. A. (1983). The question of how spatial ability is defined and its relevance to mathematics education. *Zentralblatt für Didaktik der Mathematik, 15,* 8–20.

Clements, M. A., & Del Campo, G. (1989). Linking verbal knowledge, visual images, and episodes for mathematical learning. *Focus on Learning Problems in Mathematics, 11*(1), 25–33.

Cobb, P. (1989). *A constructivist perspective on information-processing theories of mathematical activity.* Unpublished manuscript, Purdue University, West Lafayette, IN.

Cousins, D., & Abravanel, E. (1971). Some findings relevant to the hypothesis that topological spatial features are differentiated prior to euclidean features during growth. *British Journal of Psychology, 62,* 475–479.

Crowley, M. L. (1990). Criterion referenced reliability indices associated with the van Hiele geometry test. *Journal for Research in Mathematics Education, 21,* 238–241.

Darke, I. (1982). A review of research related to the topological primacy thesis. *Educational Studies in Mathematics, 13,* 119–142.

Davis, R. B. (1984). *Learning mathematics: The cognitive science approach to mathematics education.* Norwood, NJ: Ablex.

Davis, R. B. (1986). Conceptual and procedural knowledge in mathematics: A summary analysis. In J. Hiebert (Ed.), *Conceptual and procedural knowledge: The case of mathematics* (pp. 265–300). Hillsdale, NJ: Lawrence Erlbaum.

Del Grande, J. J. (1986). Can grade two children's spatial perception be improved by inserting a transformation geometry component into their mathematics program? *Dissertation Abstracts International, 47,* 3689A.

Denis, L. P. (1987). Relationships between stage of cognitive development and van Hiele level of geometric thought among Puerto Rican adolescents. *Dissertation Abstracts International, 48,* 859A. (University Microfilms No. DA8715795.)

de Villiers, M. D. (1987, June). *Research evidence on hierarchical thinking, teaching strategies, and the van Hiele theory: Some critical comments.* Paper presented at Learning and teaching geometry: Issues for research and practice working conference, Syracuse, NY, Syracuse University.

Dodwell, P. C. (1963). Children's understanding of spatial concepts. *The Canadian Journal of Psychology, 17,* 141–161.

Driscoll, M. J. (1983a). *Research within reach: Elementary school mathematics and reading.* St. Louis, MO: CEMREL, Inc.

Driscoll, M. J. (1983b). *Research within reach: Secondary school mathematics.* St. Louis, MO: CEMREL, Inc.

Eliot, J. (1987). *Models of psychological space.* New York, NY: Springer-Verlag.

Ethington, C. A. (1990). Gender differences in mathematics: An international perspective. *Journal for Research in Mathematics Education, 21,* 74–80.

Eves, H. (1972). *A survey of geometry.* Boston: Allyn and Bacon, Inc.

Fawcett, H. P. (1938). *The nature of proof.* New York, NY: Teachers College.

Fehr, H. F. (1973). Geometry as a secondary school subject. In K. B. Henderson (Ed.), *Geometry in the mathematics curriculum: 1973 Yearbook* (pp. 369-380). Reston, VA: National Council of Teachers of Mathematics.

Feingold, A. (1988). Cognitive gender differences are disappearing. *American Psychologist, 43,* 95–103.

Fennema, E., & Carpenter, T. P. (1981). Sex-related differences in mathematics: Results from National Assessment. *Mathematics Teacher, 74,* 554–559.

Fennema, E., & Sherman, J. (1977). Sex-related differences in mathematics achievement, spatial visualization and affective factors. *American Educational Research Journal, 14,* 51–71.

Fennema, E. H., & Sherman, J. A. (1978). Sex-related differences in mathematics achievement and related factors. *Journal for Research in Mathematics Education, 9,* 189–203.

Fennema, E., & Tartre, L. A. (1985). The use of spatial visualization in mathematics by girls and boys. *Journal for Research in Mathematics Education, 16,* 184–206.

Fey, J., Atchison, W. F., Good, R. A., Heid, M. K., Johnson, J., Kantowski, M. G., & Rosen, L. P. (1984). *Computing and mathematics: The impact on secondary school curricula.* College Park, MD: The University of Maryland.

Findlayson, H. M. (1984, September). *What do children learn through using Logo? D.A.I. Research Paper No. 237.* Paper presented at the meeting of the British Logo Users Group Conference, Loughborough, U.K.

Fischbein, E. (1987). *Intuition in science and mathematics.* Dordrecht, The Netherlands: Reidel.

Fischbein, E., & Kedem, I. (1982). Proof and certitude in the development of mathematical thinking. In A. Vermandel (Ed.), *Proceedings of the Sixth International Conference for the Psychology of Mathematics Education* (pp. 128–131). Antwerp, Belgium: Universitaire Instelling Antwerpen.

Fisher, G. H. (1965). Developmental features of behaviour and perception. *British Journal of Educational Psychology, 35,* 69–78.

Fisher, N. D. (1978). Visual influences of figure orientation on concept formation in geometry. *Dissertation Abstracts International, 38,* 4639A. (University Microfilms No. 7732300.)

Flores, P. V. (1990, April). *How Dick and Jane perform differently in geometry: Test results on reasoning, visualization, transformation, applications, and coordinates.* Paper presented at the meeting of the American Educational Research Association., Boston.

Forman, G. (1986). Observations of young children solving problems with microcomputers and robots. *Journal of Research in Childhood Education, 1*(2), 60–74.

Franco, L., & Sperry, R. W. (1977). Hemispheric lateralization for cognitive processing of geometry. *Neuropsychologia, 15,* 107–114.

Fuys, D. (1988, November). *Cognition, metacognition, and the van Hiele model.* Paper presented at the meeting of the International Group for the Psychology in Mathematics Education—North American Chapter, DeKalb, IL.

Fuys, D., Geddes, D., & Tischler, R. (1988). The van Hiele model of thinking in geometry among adolescents. *Journal for Research in Mathematics Education Monograph, 3.*

Galbraith, P. L. (1981). Aspects of proving: A clinical investigation of process. *Educational Studies in Mathematics, 12,* 1–29.

Gallou-Dumiel, E. (1989). Reflections, point symmetry and Logo. In C. A. Maher, G. A. Goldin, & R. B. Davis (Eds.), *Proceedings of the Eleventh Annual Meeting, North American Chapter of the International Group for the Psychology of Mathematics Education* (pp. 149–157). New Brunswick, NJ: Rutgers University.

Gardner, H. (1983). *Frames of mind: The theory of multiple intelligences.* New York, NY: Basic Books.

Geary, D. C. (1989). A model for representing gender differences in the pattern of cognitive abilities. *American Psychologist, 44,* 1155–1156.

Geeslin, W. E., & Shar, A. O. (1979). An alternative model describing children's spatial preferences. *Journal for Research in Mathematics Education, 10,* 57–68.

Gerhardt, L. A. (1973). *Moving and knowing: The young child orients himself in space.* Englewood Cliffs, NJ: Prentice Hall.

Gibson, S. (1985). The effects of position of counterexamples on the learning of algebraic and geometric conjunctive conceptions. *Dissertation Abstracts International, 46,* 378A. (University Microfilms No. DA8507804)

Greabell, L. C. (1978). The effect of stimuli input on the acquisition of introductory geometric concepts by elementary school children. *School Science and Mathematics, 78,* 320–326.

Greeno, J. G. (1980). Some examples of cognitive task analysis with instructional implications. In R. E. Snow, P. Federico, & W. E. Montague (Eds.), *Aptitude, learning, and instruction, Volume 2: Cognitive process analyses of learning and problem solving* (pp. 1–21). Hillsdale, NJ: Lawrence Erlbaum.

Guay, R. B., & McDaniel, E. (1977). The relationship between mathematics achievement and spatial abilities among elementary school children. *Journal for Research in Mathematics Education, 8,* 211–215.

Guay, R. B., McDaniel, E. D., & Angelo, S. (1978). *Analytic factor confounding spatial ability measurement.* West Lafayette, IN: Purdue University: U. S. Army Research Institute for the Behavioral Sciences.

Gutiérrez, A., & Jaime, A. (1988). *Globality versus locality of the van Hiele levels of geometric reasoning.* Unpublished manuscript, Universitat De Valencia, Valencia, Spain.

Gutiérrez, A., Jaime, A., & Fortuny, J. M. (1991). An alternative paradigm to evaluate the acquisition of the van Hiele levels. *Journal for Research in Mathematics Education, 22,* 237–251.

Halpern, D. F. (1989). The disappearance of cognitive gender differences: What you see depends on where you look. *American Psychologist, 44,* 1156–1157.

Han, T. (1986). The effects on achievement and attitude of a standard geometry textbook and a textbook consistent with the van Hiele theory. *Dissertation Abstracts International, 47,* 3690A. (University Microfilms No. DA8628106.)

Hanna, G. (1989a, July-August). Gender differences in mathematics achievement among eighth graders: Results from twenty countries. *Newsletter of the Association for Women in Mathematics,* pp. 11–17.

Hanna, G. (1989b, July-August). Gender differences in mathematics achievement among eighth graders: Results from twenty countries. Part 2. *Newsletter of the Association for Women in Mathematics,* pp. 11–17.

Hanna, G. (1989c). More than formal proof. *For the Learning of Mathematics, 9,* 20–23.

Harbeck, S. C. A. (1981). Experimental study of the effect of two proof formats in high school geometry on critical thinking and selected student attitudes. *Dissertation Abstracts International, 33,* 4243A.

Harris, L. J. (1981). Sex-related variations in spatial skill. In L. S. Liben, A. H. Patterson, & N. Newcombe (Eds.), *Spatial representation and behavior across the life span* (pp. 83–125). New York, NY: Academic Press.

Hershkowitz, R. (1989). Visualization in geometry—Two sides of the coin. *Focus on Learning Problems in Mathematics, 11,* 61–76.

Hershkowitz, R., Ben-Chaim, D., Hoyles, C., Lappan, G., Mitchelmore, M., & Vinner, S. (1990). Psychological aspects of learning geometry. In P. Nesher & J. Kilpatrick (Ed.), *Mathematics and cognition: A research synthesis by the International Group for the Psychology of Mathematics Education* (pp. 70–95). Cambridge, MA: Cambridge University Press.

Hillel, J., & Kieran, C. (1988). Schemas used by 12-year-olds in solving selected turtle geometry tasks. *Recherches en Didactique des Mathématiques, 8*(1.2), 61–103.

Hoffer, A. (1981). Geometry is more than proof. *Mathematics Teacher, 74,* 11–18.

Hoffer, A. (1983). Van Hiele—based research. In R. Lesh & M. Landau (Eds.), *Acquisition of mathematics concepts and processes* (pp. 205–227). New York, NY: Academic Press.

Hoyles, C., & Sutherland, R. (1986). *When 45 equals 60.* London, England, University of London Institute of Education, Microworlds Project.

Hughes, M., & Macleod, H. (1986). Part II: Using Logo with very young children. In R. Lawler, B. du Boulay, M. Hughes, & H. Macleod (Eds.), *Cognition and computers: Studies in learning* (pp. 179–219). Chichester, England: Ellis Horwood Limited.

Human, P. G., & Nel, J. H. (1989). *Alternative teaching strategies for geometry education: A theoretical and empirical study. RUMEUS curriculum materials series No. 11.* University of Stellenbosch.

Ibbotson, A., & Bryant, P. E. (1976). The perpendicular error and the vertical effect in children's drawing. *Perception, 5,* 319–326.

Ireland, S. H. (1974). The effects of a one-semester geometry course which emphasizes the nature of proof on student comprehension of deductive processes. *Dissertation Abstracts International, 35,* 102A–103A.

Jahoda, G., Deregowski, J. B., & Sinha, D. (1974). Topological and Euclidean spatial features noted by children. A cross-cultural study. *International Journal of Psychology, 9,* 159–172.

Johnson, E. S., & Meade, A. C. (1987). Developmental patterns of spatial ability: An early sex difference. *Child Development, 58,* 725–740.

Johnson, M. (1987). *The body in the mind.* Chicago, IL: The University of Chicago Press.

Johnson, M. L. (1989). Minority differences in mathematics. In M. M. Lindquist (Ed.), *Results from the fourth mathematics assessment of the National Assessment of Educational Progress* (pp. 135–148). Reston, VA: NCTM.

Johnson, P. A. (1986). *Effects of computer-assisted instruction compared to teacher-directed instruction on comprehension of abstract concepts by the deaf.* Unpublished doctoral dissertation, Northern Illinois University.

Johnson-Gentile, K., Clements, D. H., & Battista, M. T. (1990). *The effects of computer and noncomputer environment on students' conceptualizations of geometric motions.* Manuscript submitted for publication.

Kabanova-Meller, E. N. (1970). The role of the diagram in the application of geometric theorems. In J. Kilpatrick & I. Wirszup (Eds.), *Soviet studies in the psychology of learning and teaching mathematics (Vol. 4)* (pp. 7–49). Chicago, IL: University of Chicago.

Kafai, Y. (1989). What happens if you introduce an intelligent tutoring system in the classroom: A case study of the Geometry Tutor. In W. C. Ryan (Ed.), *Proceedings of the National Educational Computing Conference* (pp. 46–51). Eugene, OR: International Council on Computers for Education.

Kapadia, R. (1974). A critical examination of Piaget-Inhelder's view on topology. *Educational Studies in Mathematics, 5,* 419–424.

Kay, C. S. (1987). Is a square a rectangle? The development of first-grade students' understanding of quadrilaterals with implications for the van Hiele theory of the development of geometric thought. *Dissertations Abstracts International, 47,* 2934A. (University Microfilms No. DA8626590.)

Kelly, G. N., Kelly, J. T., & Miller, R. B. (1986–87). Working with Logo: Do 5th and 6th graders develop a basic understanding of angles and distances? *Journal of Computers in Mathematics and Science Teaching, 6,* 23–27.

Kidder, F. R. (1976). Elementary and middle school children's comprehension of Euclidean transformations. *Journal for Research in Mathematics Education, 7,* 40–52.

Kieran, C. (1986a). Logo and the notion of angle among fourth and sixth grade children. In *Proceedings of PME 10* (pp. 99–104). London, England: City University.

Kieran, C. (1986b). Turns and angles: What develops in Logo? In G. Lappan (Ed.), *Proceedings of the Eighth Annual PME-NA.* Lansing, MI: Michigan State University.

Kieran, C., Hillel, J., & Erlwanger, S. (1986). Perceptual and analytical schemas in solving structured turtle-geometry tasks. In C. Hoyles, R. Noss, & R. Sutherland (Eds.), *Proceedings of the Second Logo and Mathematics Educators Conference* (pp. 154–161). London, England: University of London.

Kosslyn, S. M. (1983). *Ghosts in the mind's machine.* New York, NY: W. W. Norton.

Kosslyn, S. M., Reiser, B. J., & Ball, T. M. (1978). Visual images preserve metric spatial information: Evidence from studies of image scanning. *Journal of Experimental Psychology: Human Perception and Performance, 4*(1), 47–60.

Kouba, V. L., Brown, C. A., Carpenter, T. P., Lindquist, M. M., Silver, E. A., & Swafford, J. O. (1988). Results of the fourth NAEP assessment of mathematics: Measurement, geometry, data interpretation, attitudes, and other topics. *Arithmetic Teacher, 33,* 10–16.

Krutetskii, V. A. (1976). *The psychology of mathematical abilities in schoolchildren.* Chicago, IL: University of Chicago Press.

Kull, J. A. (1986). Learning and Logo. In P. F. Campbell & G. G. Fein (Eds.), *Young children and microcomputers* (pp. 103–130). Englewood Cliffs, NJ: Prentice Hall.

Kyllonen, P. C., Lohman, D. F., & Snow, R. E. (1984). Effects of aptitudes, strategy training, and task facets on spatial task performance. *Journal of Educational Psychology, 76,* 130–145.

Lakatos, I. (1976). *Proofs and refutations: The logic of mathematical discovery.* New York, NY: Cambridge University Press.

Lamon, W. E., & Huber, L. E. (1971). The learning of the vector space structure by sixth grade students. *Educational Studies in Mathematics, 4,* 166–181.

Lampert, M. (1988). *Teachers' thinking about students' thinking about geometry: The effects of new teaching tools.* Technical Report. Cambridge, MA: Educational Technology Center, Harvard Graduate School of Education.

Laurendeau, M., & Pinard, A. (1970). *The development of the concept of space in the child.* New York, NY: International Universities Press.

Lean, G., & Clements, M. A. (1981). Spatial ability, visual imagery, and mathematical performance. *Educational Studies in Mathematics, 12,* 267–299.

Lehrer, R., & Smith, P. C. (1986, April). *Logo learning: Are two heads better than one?* Paper presented at the meeting of the American Educational Research Association, San Francisco.

Lesh, R. A., & Johnson, H. (1976). Models and applications as advanced organizers. *Journal for Research in Mathematics Education, 7,* 75–81.

Liben, L. S. (1978). Performance on Piagetian spatial tasks as a function of sex, field dependence, and training. *Merrill-Palmer Quarterly, 24,* 97–110.

Liben, L. S., & Golbeck, S. L. (1980). Sex differences in performance on Piagetian spatial tasks: Differences on competence or performance? *Child Development, 51,* 594–597.

Lieberman, D. (1985). Research on children and microcomputers: A review of utilization and effects studies. In M. Chen & W. Paisley (Eds.), *Children and microcomputers: Research on the newest medium* (pp. 59–83). Beverly Hills, CA: Sage.

Lindquist, M. M., & Kouba, V. L. (1989). Geometry. In M. M. Lindquist (Ed.), *Results from the fourth mathematics assessment of the National Assessment of Educational Progress* (pp. 35–43). Reston, VA: NCTM.

Linn, M. C., & Hyde, J. S. (1989). Gender, mathematics, and science. *Educational Researcher, 18*(8), 17–27.

Linn, M. C., & Peterson, A. C. (1985). Emergence and characterization of sex differences: A meta-analysis. *Child Development, 56,* 1479–1498.

Lovell, K. (1959). A follow-up study of some aspects of the work of Piaget and Inhelder on the child's conception of space. *British Journal of Educational Psychology, 29,* 104–117.

Luchins, E. H., Rogers, E. H., & Voytuk, J. A. (1983). *Improving spatial skills in pre-college mathematics through computer graphics. Final report, Grant SED 80-12633 from the National Science Foundation.* Unpublished manuscript, Rensselaer Polytechnic Institute, Troy, NY.

Lunkenbein, D. (1983). Observations concerning the child's concept of space and its consequences for the teaching of geometry to younger children [Summary]. In *Proceedings of the Fourth International Congress on Mathematical Education* (pp. 172–174). Boston, MA: Birkhäuser.

Mackay, C. K., Brazendale, A. H., & Wilson, L. F. (1972). Concepts of horizontal and vertical: A methodological note. *Developmental Psychology, 7,* 232–237.

Martin, J. L. (1976a). An analysis of some of Piaget's topological tasks from a mathematical point of view. *Journal for Research in Mathematics Education, 7,* 8–24.

Martin, J. L. (1976b). A test with selected topological properties of Piaget's hypothesis concerning the spatial representation of the young child. *Journal for Research in Mathematics Education, 7,* 26–38.

Martin, R. C. (1971). A study of methods of structuring a proof as an aid to the development of critical thinking skills in high school geometry. *Dissertation Abstracts International, 31,* 5875A.

Martin, W. G., & Harel, G. (1989). Proof frames of preservice elementary teachers. *Journal for Research in Mathematics Education, 20,* 41–51.

Mason, M. M. (1989, March). *Geometric understanding and misconceptions among gifted fourth-eighth graders.* Paper presented at the meeting of the American Educational Research Association, San Francisco.

Mayberry, J. (1983). The van Hiele levels of geometric thought in undergraduate preservice teachers. *Journal for Research in Mathematics Education, 14,* 58–69.

McClelland, J. L., Rumelhart, D. E., & The PDP Research Group (1986). *Parallel distributed processing: Explorations in the microstructure of cognition; Vol. 2. Psychological and biological models.* Cambridge, MA: MIT Press.

McDonald, J. (1989). Cognitive development and the structuring of geometric knowledge. *Journal for Research in Mathematics Education, 20,* 76–94.

McGee, M. G. (1979). Human spatial abilities: Psychometric studies and environmental, genetic, hormonal, and neurological influences. *Psychological Bulletin, 86,* 889–918.

McKnight, C. C., Travers, K. J., Crosswhite, F. J., & Swafford, J. O. (1985). Eighth-grade mathematics in U.S. schools: A report from the Secondary International Mathematics Study. *Arithmetic Teacher, 32*(8), 20-26.

McKnight, C. C., Travers, K. J., & Dossey, J. A. (1985). Twelfth-grade mathematics in U.S. high schools: A report from the Secondary International Mathematics Study. *Mathematics Teacher, 78*(4), 292–300.

Meyer, M. L. (1989). Gender differences in mathematics. In M. M. Lindquist (Ed.), *Results from the fourth mathematics assessment of the National Assessment of Educational Progress* (pp. 149–159). Reston, VA: NCTM.

Minsky, M. (1986). *The society of mind.* New York: Simon and Schuster.

Mitchelmore, M. C. (1976). Cross-cultural research on concepts of space and geometry. In J. L. Martin & D. A. Bradbard (Eds.), *Space and geometry. Papers from a research workshop* (pp. 143–184). Athens, GA: University of Georgia, Georgia Center for the Study of Learning and Teaching Mathematics. (ERIC Document Reproduction Service No. ED 132 033).

Mitchelmore, M. C. (1980). Three-dimensional geometrical drawing in three cultures. *Educational Studies in Mathematics, 11,* 205–216.

Morris, J. P. (1983). Microcomputers in a sixth-grade classroom. *Arithmetic Teacher, 31*(2), 22–24.

Moyer, J. C. (1978). The relationship between the mathematical structure of Euclidean transformations and the spontaneously developed cognitive structures of young children. *Journal for Research in Mathematics Education, 9,* 83–92.

Murphy, C. M., & Wood, D. J. (1981). Learning from pictures: The use of pictorial information by young children. *Journal of Experimental Child Psychology, 32,* 279–297.

National Council of Teachers of Mathematics. (1989). *Curriculum and evaluation standards for school mathematics.* Reston, VA: Author.

National Research Council. (1989). *Everybody counts.* Washington, DC: National Academy Press.

Neumann, P. G. (1977). Visual prototype formation with discontinuous representation of dimensions of variability. *Memory and Cognition, 5,* 187–197.

Newcombe, N. (1989). The development of spatial perspective taking. In H. W. Reese (Ed.), *Advances in Child Development and Behavior* (pp. 203–247). New York, NY: Academic Press.

Newcombe, N., Dubas, J. S., & Baenninger, M. (1989). Associations of timing of puberty, spatial ability, and lateralization in adult women. *Child Development, 60,* 246-254.

Noss, R. (1987). Children's learning of geometrical concepts through Logo. *Journal for Research in Mathematics Education ,18,* 343–362.

Olive, J., Lankenau, C. A., & Scally, S. P. (1986). *Teaching and understanding geometric relationships through Logo: Phase II. Interim Report: The Atlanta–Emory Logo Project.* Atlanta, GA: Emory University.

Olson, A. T., Kieren, T. E., & Ludwig, S. (1987). Linking Logo, levels, and language in mathematics. *Educational Studies in Mathematics, 18,* 359–370.

Page, E. I. (1959). Haptic perception: A consideration of one of the investigations of Piaget and Inhelder. *Educational Review, 11,* 115–124.

Papert, S. (1980). *Mindstorms: Children, computers, and powerful ideas.* New York: Basic Books.

Parzysz, B. (1988). Knowing vs. seeing. Problems of the plane representation of space geometry figures. *Educational Studies in Mathematics, 19,* 79–92.

Peel, E. A. (1959). Experimental examination of some of Piaget's schemata concerning children's perception and thinking, and a discussion of their educational significance. *British Journal of Educational Psychology, 29,* 89–103.

Pepin, M., Beaulieu, R., Matte, R., & Leroux, Y. (1985). Microcomputer games and sex-related differences: Spatial, verbal, and mathematical abilities. *Psychological Reports, 56,* 783–786.

Petty, O. S., & Jansson, L. C. (1987). Sequencing examples and nonexamples to facilitate concept attainment. *Journal for Research in Mathematics Education, 18,* 112–125.

Piaget, J. (1928). *Judgment and reasoning in the child.* New York: Harcourt, Brace, and Co.

Piaget, J. (1987). *Possibility and necessity. Vol. 2. The role of necessity in cognitive development.* Minneapolis: University of Minnesota Press.

Piaget, J., & Inhelder, B. (1967). *The child's conception of space.* New York: W. W. Norton & Co.

Piaget, J., Inhelder, B., & Szeminska, A. (1960). *The child's conception of geometry.* London: Routledge and Kegan Paul.

Porter, A. (1989). A curriculum out of balance: The case of elementary school mathematics. *Educational Researcher, 18,* 9–15.

Presmeg, N. C. (1986). Visualization in high school mathematics. *For the Learning of Mathematics, 6,* 42–46.

Prigge, G. R. (1978). The differential effects of the use of manipulative aids on the learning of geometric concepts by elementary school children. *Journal for Research in Mathematics Education, 9,* 361–367.

Pyshkalo, A. M. (1968). *Geometry in grades 1–4: Problems in the formation of geometric conceptions in pupils in the primary grades.* Moscow, USSR: Prosveshchenie Publishing House.

Raphael, D., & Wahlstrom, M. (1989). The influence of instructional aids on mathematics achievement. *Journal for Research in Mathematics Education, 20,* 173–190.

Reif, F. (1987). Interpretation of scientific or mathematical concepts: Cognitive issues and instructional implications. *Cognitive Science, 11,* 395–416.

Rosser, R. A., Horan, P. F., Mattson, S. L., & Mazzeo, J. (1984). Comprehension of Euclidean space in young children: The early emergence of understanding and its limits. *Genetic Psychology Monographs, 110,* 21–41.

Rosser, R. A., Lane, S., & Mazzeo, J. (1988). Order of acquisition of related geometric competencies in young children. *Child Study Journal, 18,* 75–90.

Schoenfeld, A. H. (1986). On having and using geometric knowledge. In J. Hiebert (Ed.), *Conceptual and procedural knowledge: The case of mathematics* (pp. 225–264). Hillsdale, NJ: Lawrence Erlbaum.

Schoenfeld, A. H. (1988). When good teaching leads to bad results: The disasters of well-taught mathematics courses. *Educational Psychologist, 23,* 145–166.

Schofield, J. W., & Verban, D. (1988). Computer usage in the teaching of mathematics: Issues that need answers. In D. A. Grouws, T. J. Cooney, & D. Jones (Eds.), *Perspectives on research on effective mathematics teaching: Vol. 1* (pp. 169–193). Hillsdale, NJ: Lawrence Erlbaum.

Senk, S. L. (1985). How well do students write geometry proofs? *Mathematics Teacher, 78,* 448–456.

Senk, S. L. (1989). Van Hiele levels and achievement in writing geometry proofs. *Journal for Research in Mathematics Education, 20,* 309–321.

Senk, S., & Usiskin, Z. (1983). Geometry proof writing: A new view of sex differences in mathematics ability. *American Journal of Education, 91,* 187–201.

Shah, S. A. (1969). Selected geometric concepts taught to children ages seven to eleven. *Arithmetic Teacher, 16,* 119–128.

Shaughnessy, J. M., & Burger, W. F. (1985). Spadework prior to deduction in geometry. *Mathematics Teacher, 78,* 419–428.

Shelton, M. (1985). Geometry, spatial development and computers: Young children and triangle concept development. In S. K. Damarin & M. Shelton (Eds.), *Proceedings of the Seventh Annual Meeting of the North American Branch of the International Group for the Psychology of Mathematics Education* (pp. 256–261). Columbus OH: Ohio State University.

Shepard, R. N. (1978). The mental image. *American Psychologist,* 125–137.

Smith, S. E., & Walker, W. J. (1988). Sex differences on New York State Regents examinations: Support for the differential course-taking hypothesis. *Journal for Research in Mathematics Education, 19,* 81–85.

Somerville, S. C., & Bryant, P. E. (1985). Young children's use of spatial coordinates. *Child Development, 56,* 604–613.

Somerville, S. C., Bryant, P. E., Mazzocco, M. M. M., & Johnson, S. P. (1987, April). *The early development of children's use of spatial coordinates.* Paper presented at the meeting of the Society for Research in Child Development, Baltimore, MD.

Sowell, E. J. (1989). Effects of manipulative materials in mathematics instruction. *Journal for Research in Mathematics Education, 20,* 498–505.

Springer, S. P., & Deutsch, G. (1981). *Left brain, right brain.* New York, NY: W. H. Freeman and Company.

Steffe, L., & Cobb, P. (1988). *Construction of arithmetical meanings and strategies.* New York, NY: Springer-Verlag.

Stevenson, H. W., Lee, S., & Stigler, J. W. (1986). Mathematics achievement of Chinese, Japanese, and American children. *Science, 231,* 693–699.

Stevenson, H. W., & McBee, G. (1958). The learning of object and pattern discrimination by children. *Journal of Comparative and Psychological Psychology, 51,* 752–754.

Stigler, J. W., Lee, S. Y., & Stevenson, H. W. (1990). *Mathematical knowledge of Japanese, Chinese, and American elementary school children.* Reston, VA: National Council of Teachers of Mathematics.

Summa, D. F. (1982). The effects of proof format, problem structure, and the type of given information on achievement and efficiency in geometric proof. *Dissertation Abstracts International, 42,* 3084A.

Suydam, M. N. (1985). The shape of instruction in geometry: Some highlights from research. *Mathematics Teacher, 78,* 481–86.

Talyzina, N. F. (1971). Properties of deductions in solving geometry problems. In J. Kilpatrick & I. Wirszup (Eds.), *Soviet studies in the psychology of learning and teaching mathematics* (Vol. 4, pp. 51–101). Chicago, IL: University of Chicago Press.

Tartre, L. A. (1990a). Spatial orientation skill and mathematical problem solving. *Journal for Research in Mathematics Education, 21,* 216–229.

Tartre, L. A. (1990b). Spatial skills, gender, and mathematics. In E. Fenneman & G. Leder (Eds.), *Mathematics and Gender: Influences on teachers and students* (pp. 27–59). New York, NY: Teachers College Press.

Thomas, B. (1982). *An abstract of kindergarten teachers' elicitation and utilization of children's prior knowledge in the teaching of shape concepts.* Unpublished manuscript, School of Education, Health, Nursing, and Arts Professions, New York University.

Thomas, H., & Jamison, W. (1975). On the acquisition of understanding that still water is horizontal. *Merrill-Palmer Quarterly, 21,* 31–44.

Usiskin, Z. (1972). The effects of teaching Euclidean geometry via transformations on student achievement and attitudes in tenth-grade geometry. *Journal for Research in Mathematics Education, 3,* 249–259.

Usiskin, Z. (1982). *Van Hiele levels and achievement in secondary school geometry (Final report of the Cognitive Development and Achievement in Secondary School Geometry Project).* Chicago, IL: University of Chicago, Department of Education.

Usiskin, Z. (1987). Resolving the continuing dilemmas in school geometry. In M. M. Lindquist & A. P. Shulte (Eds.), *Learning and Teaching Geometry, K-12: 1987 Yearbook* (pp. 17–31). Reston, VA: National Council of Teachers of Mathematics.

Usiskin, Z., & Senk, S. (1990). Evaluating a test of van Hiele levels: A response to Crowley and Wilson. *Journal for Research in Mathematics Education, 21,* 242–245.

Van Akin, F. E. (1972). An experimental evaluation of structure in proof in high school geometry. *Dissertation Abstracts International, 33,* 1425A.

Van Dormolen, J. (1977). Learning to understand what giving a proof really means. *Educational Studies in Mathematics, 8,* 27–34.

van Hiele, P. M. (1959). Development and learning process. *Acta Paedagogica Ultrajectina, 17.*

van Hiele, P. M. (1959/1985). The child's thought and geometry. In D. Fuys, D. Geddes, & R. Tischler (Eds.), *English translation of selected writings of Dina van Hiele-Geldof and Pierre M. van Hiele* (pp. 243–252). Brooklyn, NY: Brooklyn College, School of Education. (ERIC Document Reproduction Service No. 289 697).

van Hiele, P. M. (1986). *Structure and insight.* Orlando,: Academic Press.

van Hiele, P. M. (1987, June). *A method to facilitate the finding of levels of thinking in geometry by using the levels in arithmetic.* Paper presented at Learning and teaching geometry: Issues for research and practice working conference, Syracuse, NY: Syracuse University.

van Hiele-Geldof, D. (1984). The didactics of geometry in the lowest class of secondary school. In D. Fuys, D. Geddes, & R. Tischler (Eds.), *English translation of selected writings of Dina van Hiele-Geldof and Pierre M. van Hiele* (pp. 1–214). Brooklyn, NY: Brooklyn College, School of Education. (ERIC Document Reproduction Service No. 289 697.)

Vinner, S., & Hershkowitz, R. (1980). Concept images and common cognitive paths in the development of some simple geometrical concepts. In R. Karplus (Ed.), *Proceedings of the Fourth International Conference for the Psychology of Mathematics Education* (pp. 177–184). Berkeley, CA: Lawrence Hall of Science, University of California.

Vladimirskii, G. A. (1971). An experimental verification of a method and system of exercises for developing spatial imagination. In J. Kilpatrick & I. Wirszup (Eds.), *Soviet studies in the psychology of learning and teaching mathematics,* Vol. 5, (pp. 57–117). Chicago, IL: University of Chicago Press.

von Stein, J. H. (1982). An evaluation of the microcomputer as a facilitator of indirect learning for the kindergarten child. *Dissertation Abstractions International, 43,* 72A. (University Microfilms No. DA8214463.)

Vygotsky, L. S. (1934/1986). *Thought and language.* Cambridge, MA: MIT Press.

Waber, D. T. (1979). Cognitive abilities and sex-related variations in the maturation of cortical functions. In M. A. Wittig & A. C. Petersen (Ed.), *Sex-related differences in cognitive functioning* (pp. 161–186). New York, NY: Academic Press.

Wertheimer, R. (1990). The geometry proof tutor: An "intelligent" computer-based tutor in the classroom. *Mathematics Teacher, 83,* 308–317.

Wheatley, G. W. (1990). Spatial sense and mathematics learning. *Arithmetic Teacher, 37*(6), 10–11.

Wheatley, G., & Cobb, P. (1990). Analysis of young children's spatial constructions. In L. P. Steffe & T. Wood (Eds.), *Transforming early childhood mathematics education: International perspectives* (pp. 161–173). Hillsdale, NJ: Lawrence Erlbaum.

Williford, H. J. (1972). A study of transformational geometry instruction in the primary grades. *Journal for Research in Mathematics Education, 3,* 260–271.

Wilson, M. (1990). Measuring a van Hiele geometry sequence: A reanalysis, *Journal for Research in Mathematics Education, 21*, 230–237.

Wilson, P. S. (1986). Feature frequency and the use of negative instances in a geometry task. *Journal for Research in Mathematics Education, 17,* 130–139.

Wirszup, I. (1976). Breakthroughs in the psychology of learning and teaching geometry. In J. L. Martin & D. A. Bradbard (Eds.), *Space and geometry. Papers from a research workshop* (pp. 75–97). Athens, GA: University of Georgia, Georgia Center for the Study of Learning and Teaching Mathematics. (ERIC Document Reproduction Service No. ED 132 033.)

Wiske, M. S., & Houde, R. (1988). *From recitation to construction: Teachers change with new technologies. Technical Report.* Cambridge, MA: Educational Technology Center, Harvard Graduate School of Education.

Yakimanskaya, I. S. (1971). The development of spatial concepts and their role in the mastery of elementary geometric knowledge. In J. Kilpatrick & I. Wirszup (Eds.), *Soviet studies in the psychology of learning and teaching mathematics* (Vol. 5, pp. 145–168). Chicago, IL: University of Chicago Press.

Yerushalmy, M., & Chazan, D. (1988). *Overcoming visual obstacles with the aid of the Supposer.* Cambridge, MA: Educational Technology Center, Harvard Graduate School of Education.

Yerushalmy, M., Chazan, D., & Gordon, M. (1987). *Guided inquiry and technology: A year long study of children and teachers using the Geometric Supposer: ETC Final Report.* Newton, MA: Educational Development Center.

Zykova, V. I. (1969). Operating with concepts when solving geometry problems. In J. Kilpatrick & I. Wirszup (Eds.), *Soviet studies in the psychology of learning and teaching mathematics* (Vol. 1, pp. 93–148). Chicago, IL: University of Chicago.

·19·

RESEARCH IN PROBABILITY AND STATISTICS: REFLECTIONS AND DIRECTIONS

J. Michael Shaughnessy
OREGON STATE UNIVERSITY

Some years ago, I happened to be teaching an introductory college level course on probability, combinatorics, and elementary statistical concepts. At the same time I was attending an interdisciplinary graduate seminar on research in problem solving, judgment, and decision making. Among the research we addressed in the seminar were some then-recent papers on judgment under uncertainty. What fascinated me about those papers was that the subjects in those studies *could* have used elementary probability and statistics concepts to estimate the likelihood of events in the research tasks they were given, but they didn't. The mathematical concepts I was teaching to college freshman would have served them well, if only they had used them. Some of the subjects in those studies of judgment were even quite well schooled in statistics, and yet they still were caught using certain judgmental heuristics, rather than employing their hard-earned statistical knowledge. What was even more unnerving, from my point of view, was that I found myself falling prey to some of the same misconceptions that snared the subjects in the studies. Of course, my initial reaction as a mathematics teacher was that since most of the subjects in these studies were naive about probability and statistics, their nonmathematical ways of estimating likelihoods would disappear if only we could find the "right way" to teach them. So, I embarked on what has been a continually fascinating journey into research on the teaching and learning of probability and statistics. I soon found that the problem of getting people to use stochastics in their judgments and decisions, when appropriate, is not easily remedied just by "teaching them the right way." (Stochastics is the common European term to include "probability and statistics." This convenient abbreviation will be used throughout this chapter.) Intuitions, preconceptions, misconceptions, misunderstandings, nonnormative explanations—whatever one might call them—abound in the research on learning probability and statistics.

The cross fertilization of research traditions and methodologies in probability and statistics makes it one of the theoretically richest branches of research in mathematics education. Psychologists—cognitive, developmental, and even behaviorist—have played a major role in building theory in this field. Mathematics and statistics educators from all over the world have made major contributions, particularly in the last 10 years. Research in this field has been truly interdisciplinary, and perhaps that is how it should be, as probability and statistics are certainly not the sole domain of mathematics educators. In fact much of the research in this area has not been done by mathematics or statistics educators at all, and the contributions of North American mathematics educators have been all too few. Most of the contributions have been made either by cognitive psychologists or by European mathematics and statistics educators.

It is not all that surprising that there has not been much involvement by North American mathematics educators in research on the teaching and learning of stochastics. Much of what is researched in mathematics education is driven by what is taught in schools, or by what is required to enter college, or perhaps by what a document recommends that we should be teaching (See *An Agenda for Action,* NCTM, 1980, for example). Since very little probability or statistics has been systematically taught in our schools in the past, there has been little impetus to carry out research on the problems that students have in learning it. However, over the past ten years, there have been

I would like to thank the many colleagues who helped by sending me research from all over the world. It has made my task all the more enjoyable to see that there is much ongoing activity in research in stochastics. I would especially like to thank Joan Garfield, University of Minnesota, and Cliff Konold, University of Massachusetts—Amherst, for their comments on and reactions to earlier versions of this chapter. I would also like to thank Ruma Falk, Heinz Steinbring, and Anne Hawkins for their timely suggestions.

an increasing number of cries for the introduction of stochastics topics in our schools, starting in the middle grades and continuing throughout secondary school. (Davidson & Swift, 1988; Rade, 1983; Shulte, 1981). These pleas have finally started to take root. The recent *Standards* document published by NCTM (1989) recommends the introduction of a significant number of stochastics concepts throughout the school years. Thus, the need for a stepped-up, ongoing research program in the area of probability and statistics has never been greater, and it will become acute as more and more teachers and curriculum developers begin to implement NCTM's intentions for probability and statistics as presented in the Standards report.

I have two main goals in writing this chapter. First, I would like to provide an overview, interpretation, and focus of the research in learning and teaching stochastics. In no way do I claim to be exhaustive in my review of the existing research, but I do hope to be representative. The current state of the research in this area is far too eclectic to admit a complete synthesis; however, I do hope to provide a framework for researchers who are new to this area, as well as to provide some reconceptualization of the literature for those who have already been involved in research in stochastics. Second, I would like to point out several concerns that I have about research in stochastics, and to suggest some research directions that I believe need immediate attention.

The Current State of Stochastics in Schools and Universities

Stochastics appears to have been an integral part of the mathematics curriculum in many European countries for some time. However, in the United States, stochastics has not yet made it into the mainstream of school mathematics. The evidence at the secondary level indicates that very few secondary schools actually offer a separate, systematic course in probability and statistics. A recent study found that only 2% of college bound high school students had taken a course in probability and statistics in the United States, though 90% of these students had taken a course in algebra (De Beres, 1988). Many secondary schools that do include probability and statistics in their curriculum treat it as a 6–9 week unit inside of another course. As such, most students will not take it, and teachers may even be tempted to skip it altogether. This is a very unfortunate situation, since there is perhaps no other branch of the mathematical sciences that is as important for *all* students, college bound or not, as probability and statistics.

No comprehensive surveys of how much probability and statistics is taught in grades K–4, or 5–8 is currently available, but until very recently one could confidently say practically none. However, in the last 4 or 5 years, projects such as the Quantitative Literacy Project (Gnanadesikan, Scheaffer, & Swift, 1987; Landewehr & Watkins, 1986; Landewehr, Watkins & Swift, 1987; Newman, Obremski, & Scheaffer, 1987) and the Middle Grades Mathematics Project (Phillips, Lappan, Winter, & Fitzgerald, 1986) have begun to make some inroads into the teaching of probability and statistics in the middle grades.

For the college bound, the lack of secondary level preparation in stochastics for our students stands in sharp contrast to the demand for statistical literacy and reasoning that lie ahead

for them. A recent survey at a major university discovered that 160 different statistics courses were being taught in 13 different departments (Garfield & Ahlgren, 1988a). All of our students eventually become consumers and citizens at some level, and will enter a society in which the use of data and graphs to communicate information and influence decisions is ever increasing. In order to turn out informed citizens and responsible consumers, our schools will have to attend more keenly to the teaching of probability and statistics than they have in the past, as indicated by the NCTM *Standards* document (NCTM, 1989).

Most of the courses in probability and statistics that are offered at the university level continue to be either rule-bound, recipe-type courses for calculating statistics, or overly mathematized introductions to statistical probability that were the norm a decade ago (Shaughnessy, 1977). Thus, college level students, with all their prior beliefs and conceptual misunderstandings about stochastics, rarely get the opportunity to improve their statistical intuition or to see the applicability of the subject as undergraduates. University courses may, therefore, only make a bad situation worse, by masking conceptual and psychological complexities in the subject. Suggestions for improving the teaching of stochastics at the college and university level abound in the proceedings of the First and Second International Conference on Teaching Statistics (ICOTS I and II; Davidson and Swift, 1988; Rade, 1983). Fair and unfair games, simulations, interactive computer packages, statistics on spreadsheets, and student statistical consulting projects are a few of the many recommendations that can be found in the proceedings from these two conferences. (See, for example, Barnett, 1988; Bentz, 1983; Biehler, 1988; Piazza, 1988; Shaughnessy, 1983a). However, colleges and universities, like secondary and middle schools, have been slow to change the way stochastics is taught, despite these continual recommendations.

Garfield (1988) cites four issues that hinder the effective teaching of stochastics: (a) the role of probability and statistics in the curriculum, (b) links between research and instruction, (c) the preparation of mathematics teachers, and (d) the way learning is currently being assessed. It is interesting to note that these same four concerns are raised by a number of authors throughout the recent NCTM publication *The Teaching and Assessing of Mathematical Problem Solving* (NCTM, 1989). The place of mathematical problem solving in the *curriculum* is examined historically by Kilpatrick and Stannic (1989) and contextually by Schoenfeld (1989). Difficulties with *assessment* of mathematical problem-solving are discussed by Marshall (1989). Thompson (1989) talks about the difficulties in changing *teachers' beliefs and conceptions* about mathematics and about problem-solving. She refers to teaching problem solving as a "double art": one art for teaching and an extra one for problem solving. Charles (1989) discusses systematic attempts to develop effective in-service programs to enable teachers to incorporate problem solving into their classrooms. Carpenter (1989) and Lester (1989) highlight the importance of making connections and *links between research and instruction*.

It is not surprising that the impediments to effective teaching of probability and statistics in our schools are the same ones that hinder effective implementation of problem solving in our schools. The close ties between these two areas of research have been pointed out previously (Shaughnessy, 1985). For example, the teaching and learning of stochastics involves building models of physical phenomena, development and use of strategies (such as simulation strategies and counting strategies), and comparison and evaluation of several different approaches to problems in order to monitor possible misconceptions or misrepresentations. In these respects, *teaching stochastics is teaching problem solving,* albeit in a particular content domain. In addition, teachers' backgrounds are weak or nonexistent in both stochastics and in problem solving. This is not their fault, as historically our teacher preparation programs have not systematically included either stochastics or problem solving for prospective mathematics teachers. Finally, problem solving and stochastics are both, in some sense, new kids on the mathematical block, at least from the standpoint of inclusion in the secondary curriculum. We are asking teachers to make room for both problem solving and stochastics in the curriculum that they teach.

Garfield's four issues may really be symptomatic of the times in mathematics education. Under the new NCTM *Standards* recommendations, every content branch of mathematics education must wrestle with the issues of place in the curriculum, teacher background and implementation strategies, a research agenda that informs and collaborates with instruction, and creative assessment methodologies.

In the section above we discussed at some length the traditional complete absence of stochastics from the school curriculum. School curricula haven't required stochastics, and teachers aren't prepared to teach it anyway, so it hasn't been taught. Up until very recently, another reason for the absence of stochastics in secondary schools was the concurrent absence of appropriate materials; this is no longer the case. Curriculum materials such as the Quantitative Literacy Series (Gnanadesikan, Scheaffer, & Swift, 1987; Landewehr, Swift, & Watkins, 1987; Landewehr & Watkins, 1985; Newman, Obremski, & Scheaffer, 1987), the Probability unit of the Middle Grades Mathematics Project (Phillips et al., 1986), and *Using Statistics* (Travers, Stout, Swift, & Sextro, 1985) provide excellent opportunities to introduce students from a wide range of grade levels, 5–12, to stochastics concepts. There are interesting materials on stochastics available from Europe, such as *L'Enseignement des Probabilites* (Engel, 1975) and *Azar y Probabilidad* (Godino, Batanero, & Canizares, 1987). In addition, technology has enhanced our ability to develop fine interactive curricula on computers, for example in *Hands-on Statistics* (Weissglass, Thies, & Finzer, 1986) and in the Stretchy Histograms and Shifty Lines materials from the Reasoning Under Uncertainty Project (Rubin, Roseberry, & Bruce, 1988). There is even an entire stochastics curriculum just devoted to improving students' thinking under uncertainty (Beyth-Marom & Dekel, 1983).

Thus, the real barriers for improvement of stochastics teaching in North America are fundamentally (a) getting stochastics into the mainstream of the mathematical science school curriculum at all, (b) enhancing teachers' background and conceptions of probability and statistics, and (c) confronting students' and teachers' beliefs about probability and statistics. If it seems a bit odd to spend time in a chapter, supposedly about research, addressing concerns of curriculum and teaching, I have done so because research is usually caried out in those areas that

either researchers themselves or the public welfare designate as areas of concern. Unless stochastics starts to enter the mainstream of our schools on a wide basis, the number of people who do research in this area will continue to be small but vocal. As we shall see, the research in stochastics thus far highlights just how important it is that we actually include stochastics in our schools, because people are going to use it, and abuse it—perhaps more than any other branch of mathematics—whether or not we teach it to them.

Some Historical, Philosophical, and Epistemological Considerations

The historical and philosophical setting for the development of what we mean today by "probability" provides an essential backdrop for an overview of the research in stochastics learning. How we obtain stochastical knowledge, or even what it means to "know" something in a stochastic setting, are potential stumbling stones for researchers. We are saddled with considerable baggage, both philosophical and historical, which can provide obstacles not only to our research in learning probability and statistics, but also to our ability to communicate results to other researchers.

Ripples from the 17–18th-century epistemological schism between the continental rationalist tradition and the British empiricists can still be seen in the way some Europeans conduct research on the teaching and learning of stochastics. Seventeenth-century rationalist philosophers on the continent, such as Descartes, Leibnitz, and Spinoza, saw the attainment of knowledge as a process of pure reason, discovering or deducing absolute truths that were a priori in the mind (Copleston, 1963). The rationalists were much in the tradition of Plato; according to his theory of Forms, all knowledge is the process of rediscovering things already true and universal (Copleston, 1962). Knowledge for the rationalists was, therefore, deductive, involved uncovering a priori truths, and was not based on sense perception. At the same time that the rationalist movement was blossoming on the European continent, the British empiricist tradition in England, led by Locke, Berkeley, and Hume, fundamentally doubted the existence of absolute truths. They held that all knowledge is based upon inferences made from sensory observations of things. The empiricists claimed that we do not really *know* things, we only *know about* things. In this respect, they followed more closely in the tradition of Plato's pupil, Aristotle, who was more concerned with the concrete and the particular, while his mentor was concerned with the general and the abstract. For Aristotle, sense perception plays a more important role, though not the only role, in how we come to know things (Bambrough, 1963). The rise of empiricism as a philosophy was stimulated by reliance upon factual data and hypothesis testing, such as in the work of Sir Isaac Newton. In order to obtain information about the world around us, one needed to rely on sense perception, rather than discovering a priori truths. Therefore, for the empiricists, knowledge began with sense perception, and theory was built inductively.

The process of conducting research, and what it means to know that something is true, has continued to be influenced by this 17th-century epistemological schism. There are still traces of empiricism versus rationalism in the research on learning probability and statistics, if one compares the work, say, of Green (1983a, 1983b, 1987, 1988) in England with some of the work done by German and Austrian researchers (Bentz & Borovcnik, 1985, in press; Steinbring, in press). Green has conducted huge surveys of thousands of British 11–16 year olds to find out how they respond to probability tasks. Bentz, Borovcnik, and Steinbring have provided insightful theoretical perspectives on the teaching and learning of probability. Steinbring's research, in direct contrast to Green's, does not often report actual tasks administered or interviews given to students to see how they are thinking. Rather, Steinbring provides us with a fundamentally theoretical perspective to help change and improve the teaching of stochastics. Also, although Bentz and Borovcnik do undertake an analysis of empirical tasks that have been carried out by others, their primary purpose seems to be to point out all the pitfalls that can occur when one attempts to do empirical research (Bentz & Borovcnik, in press). One gets the impression that Bentz and Borovcnik are rather skeptical of the attempts made by others to investigate stochastics learning empirically. While it is beneficial to point out the pitfalls of empirical research, it is also necessary to move ahead in our assessment of students' notions of probability and statistics; this is not possible to do from a purely theoretical perspective.

The important point to be made here is that there are, indeed, different epistemological traditions in philosophy, and they have led to different research traditions in stochastics. As a result, there is not a common point of philosophical reference in the research community for doing research on stochastics, and there are, consequently, opportunities for researchers to misunderstand each other.

In addition to philosophical influences on the epistemology of stochastics, there are some historical developments in the probability concept, itself, and in the notion of what constitutes "acceptable evidence for truth", that have influenced research in stochastics. Perhaps the best critical account of the historical and philosophical development of the probability concept is found in Ian Hacking's book *The Emergence of Probability* (Hacking, 1975). The development of mathematical probability and statistics is a rather recent phenomenon, evolving from solutions to gaming problems and counting problems that were of interest in the mid-17th century. Thus, the field of probability and statistics is barely a mathematical adolescent when compared to geometry or to algebra, or even to the roots of the calculus traceable back to Eudoxes and Archimedes. Hacking reviews a number of possible reasons why stochastics has been a late bloomer, and then, one-by-one, dismisses each possible reason as irrelevant or insufficient. In Hacking's view, neither a preoccupation with determinism and first causes, nor religious doctrine, nor lack of an economic theory, nor the lack of "technology" (random devices) could, in itself, explain the historically slow emergence of mathematical probability. Hacking claims that this slow emergence was primarily due to the dual meaning that has historically been attached to the word *probability,* and to the respective definitions of scientific evidence that accompanied each of these two meanings.

The word *probability* has long been used to indicate both "degree of belief" and "calculations of stable frequencies for random events." Hacking notes that at the time that mathematical probability theory suddenly appeared in the mid-17th and early 18th-century writings of Pascal, Huygens, Leibnitz, and Bernoulli, the term probability still held both statistical (random events) and empirical (degree of belief) connotations. The empirical hurdles for a mathematized version of probability turned out to be formidable. The scientific investigations of scholars, such as Galileo and Francis Bacon, were considered the realm of *scientia,* or knowledge by demonstration; demonstration of absolute truths from first principles was the goal of such scholars. The task was to put forth a proper set of axioms, and then deduce the truths from these axioms. The notion of experimental evidence from samples that we hold today was not championed by Galileo and Bacon, who felt that such arguments could be mere opinions, and could be the result of bad experiments. Even though Galileo and Bacon are often credited with being the founders of the experimental method in science, they preferred to argue from First Causes in the tradition of Aristotle, and to present their theories in a deductive fashion (Hacking, pp. 27–28). In no way did Galileo or Bacon believe that one could infer absolute truth from experimental evidence. Thus, a duality arose between *scientia* (knowledge), and *opinio* (belief). There were "high sciences" (such as mathematics, mechanics, astronomy, and philosophy), which sought absolute truths, and then there were "low sciences" (such as medicine and astrology), which rendered opinions based on empirical evidence (Hacking, 1975). For Galileo and Bacon, probability was considered the low life of opinion, because at that time, anyone could render an opinion on any topic by basing it on a particular instance. One opinion could carry as much weight as any other opinion. In fact, one opinion was often considered more probable than another if it happened to have some authority behind it, such as the church.

The notion of what constitutes acceptable evidence is interwoven with these two meanings of probability. Hacking distinguishes between the evidence of testimony which is extrinsic, and the evidence of things, which is intrinsic. The evidence of testimony was tied to the tradition of *scientia,* arguing effects from first causes. In contrast, the evidence of things was long considered the realm of *opinio,* what we would call inductive evidence from many similar cases. As Hacking points out and Konold (1989a) reiterates, in order for our modern mathematical version of probability to emerge, the whole notion of what was acceptable evidence had to change. The tradition of *opinio* was deeply rooted in the 15th and 16th centuries, and had support of the Jesuits, who were quite powerful at that time. Opinions on scripture and church teachings frequently had to be rendered in the case of conflicting views. In essence, all one needed to get an opinion at that time was a bit of previous writing and a lot of authority. (One can see the roots of our modern legal system in this notion of *opinio.*) Probability was considered a matter of "approval" rather than a mathematical calculation, prior to the 17th century. The idea of experimental evidence finally began to gain some respectability in the 17th century under the theoretical nurturing of Pascal and through the work of Huygens (1657). Whereas *scientia* attempted to deduce effects from First Causes, this new science would attempt to induce causes from observed effects. Herein lie the seeds of our statistical decision theory, based on stable frequencies of random events.

Unfortunately, the dualistic tradition of the notion of probability is still with us, and it has crept into our research debates in a rather insidious way. The literature discusses the merits of different kinds of probability: *classical, frequentist,* and *subjective* (Kapadia, 1988; Konold, 1989a). Classical probability refers to the assignment of probabilities in an experiment with a random device where all outcomes are equally likely. Mathematicians would call this a uniform probability distribution. Frequentist theory considers probabilities to be assigned based on the long run behavior of random outcomes. Mathematically, this involves the theory of limits and convergence. Subjective probability is the 20th century term for *opinio,* or degree of belief. It is even possible to mathematize subjective probability, with a heavy reliance upon Bayes Theorem, and a theory of revision of probabilities based on accessible information.

The fact that these different conceptions of probability are discussed is not, in and of itself, a problem. What is a problem is that some researchers talk about it as if it were a battle to be won. Kapadia (1988) claims that the subjectivist view is winning out over the other two views. I see no evidence for this in the research literature of mathematics education. What does seem to be happening is that researchers are much more aware of the existence of subjectivist viewpoints in the students they teach, and in the teachers that use the materials they develop (Falk, 1988, in press; Garfield & Ahlgren, 1988a; Konold, 1989a, 1991; Pollatsek, Konold, Well, & Lima, 1984; Rubin & Roseberry, 1988; Shaughnessy, 1977, 1981, 1983a). This is all to the good, as it will force us to deal with subjective notions of probability in our teaching, research, and curriculum development efforts. However, the notion that there is one correct view of probability seems to me to be an empty debate, at best. (I will never forget one psychologist who, after meeting me, glared and quipped, "you must be a damn frequentist!") Mathematics education is concerned with the teaching and learning of probability and statistics, and as such, a *modeling* point of view is a much healthier perspective from which to discuss different views of probability.

There are, indeed, certain normative ideas of stochastics that we want our students to understand and to be able to use. However, while some probability experiments can best be modeled by a uniform probability space, others can best be modeled from a frequentist perspective. In particular the latter category includes examples of simulations which can be carried out by computers. There are problems in which a "marriage" between experimental frequencies and classical theory is desirable (such as tossing dice or using equal area spinners). There are other problems in which either a theoretical solution does not yet exist, or is not readily available to our students (see Problems for Stimulation by Simulation, Shaughnessy, 1983b); in such cases, a frequentist approach has great merit. There are also probability problems in which the conflict between subjectivist tendencies and classical theory can be resolved by taking a frequentist point of view and running a simulation (See Monty's Dilemma, p. 475). Thus, I would advocate a pragmatic approach which in-

volves modeling several conceptions of probability. The model of probability that we employ in a particular situation should be determined by the task we are asking our students to investigate, and by the types of problems we wish to solve. And if as we encounter new stochastic challenges, either mathematical or educational, our current set of stochastic models proves inadequate, a new paradigm for thinking about probability will have to evolve (Kuhn, 1962).

Psychologists and Mathematics Educators: Two Different Perspectives

A few years ago I wrote in detail about the differences in research in probability and statistics from the perspectives of psychologists and mathematics educators (Shaughnessy, 1983b). The phenomena of errors in stochastical reasoning continues to be of interest to researchers in both fields of investigation, but for different reasons. Most of the research concerns of the psychologists understandably focus on how reasoning occurs in situations of uncertainty. Their motivation for investigating judgment and decision-making under uncertainty comes in part from a concern about how doctors, judges, financial advisors, military experts, political advisors, and others make crucial decisions (perhaps with lives at stake) in situations where the information is probabilistic, at best. The psychologists are, therefore, mainly *observers* and *describers* of what happens when subjects wrestle with cognitive judgmental tasks. Customarily they attempt to explain what they observe on the basis of some theoretical model (Kahneman, Slovic, & Tversky 1982; Nisbett & Ross, 1980). Researchers in mathematics and statistics education are, however, natural *interveners.* Since their task is to improve students' knowledge of stochastics, educators are not usually content just to observe the troubles that people have with reasoning under uncertainty; rather, they wish to change students' conceptions and beliefs about probability and statistics.

The perspectives of the observer and the intervener provide an excellent basis for cooperation and cross fertilization of theoretical models and research methodologies. Researchers in mathematics and statistics education have thus far been greatly influenced by the investigations of psychologists. However, so far most of the sharing has been in one direction. It continues to amaze me that cognitive psychologists lament the depth and tenacity of certain nonnormative conceptions of probability, like the conjunction fallacy (Tversky & Kahneman, 1983), and yet make little attempt to team up with mathematics educators to see if the misconceptions can be diminished under instruction. Previous authors Schrage (1983), Shaughnessy (1977), Konold (1991), and delMas and Bart (1987) have pursued research questions that should be of great interest to psychologists. Perhaps psychologists are concerned about muddying the primeval waters of cognition with murky educational experiments that may distort their observation paradigm. This is rather an enigma, since many psychologists are in the business of teaching statistics, also (Nisbett, Krantz, Jepson, & Kunda, 1983). In any case, it is worth noting that research in the area of learning stochastics has come from two distinctly different research traditions, with two apparently different research agenda.

RESEARCH IN STOCHASTICS: WHAT RESEARCH TELLS US THUS FAR

Recent reviews and summaries of the literature, both by psychologists (Kahneman et al., 1982; Nisbett & Ross, 1980) and by mathematics and statistics educators (Bentz & Borovcnik, in press; Biehler, in press; Garfield & Ahlgren, 1988; Hawkins & Kapadia, 1984; Nisbett & Ross, 1980; Scholz, in press; Steinbring, in press) provide us with in-depth research perspectives from a number of viewpoints. Thus, rather than attempt a complete re-review of all the literature pertinent to the teaching and learning of stochastics, I chose to explore what I see as the main themes in stochastics research of interest to mathematics education, while reflecting and commenting on the perspectives of previous reviewers.

Two main types of studies are reported in the research literature. The first type *describes* how people think; the second type is concerned with *influencing* how people think. The first type investigates primitive conceptions or intuitions of probability and statistics, misconceptions, fallacies in thinking, judgmental biases, and so forth. The second type is concerned with influencing beliefs or conceptions, even changing them if possible. It is true that the first type has been carried out primarily by psychologists, and the second type primarily by mathematics/statistics educators. However, this is not entirely the case, as psychologists try to influence how people think by manipulating the tasks they present. In fact, some psychologists have recently been interested in the issues of specific training in stochastic reasoning (Nisbett, Fong, Lehman, & Cheng, 1987; Well, Pollatsek, & Boyce, 1990). Also, mathematics educators have discovered they must become familiar with students' preexisting stochastic conceptions before they try to teach the mathematical concepts of probability and statistics. In the following sections, we will examine some of the principal results from research in stochastics from both disciplines.

Judgmental Heuristics and Biases

The research of psychologists Daniel Kahneman and Amos Tversky, and many of their colleagues, has provided mathematics educators with a theoretical framework for researching learning in probability and statistics. Their work has also stimulated much of the ongoing psychological research and debate on reasoning under uncertainty. Kahneman and Tversky's original thesis was that people who are statistically naive make estimates for the likelihood of events by using certain judgmental heuristics, such as *representativeness* and *availability* (Kahneman & Tversky, 1972, 1973a, 1973b; Tversky & Kahneman, 1974). Although they have broadened their initial perspective to include other types of explanations for reasoning under uncertainty on some types of tasks (Tversky & Kahneman, 1982a; Tversky & Kahneman, 1983), much of their subsequent research appears to substantiate their original hypotheses. During the past twenty years, evidence in support of the use of judgmental heuristics has continually been found (Bar-Hillel, 1980; Bar-Hillel & Falk, 1982; Kahneman et al., 1982; Pollatsek et al., 1984; Shaughnessy, 1977; Taylor, 1982; Tversky &

Kahneman, 1983). Some authors take issue with the sweeping generalizations that Kahneman and Tversky occasionally make for reasoning under uncertainty via these heuristics (Konold, 1991; Scholz & Bentrup, 1984). However, there is little doubt of the importance of their perspective for diagnosing the psychological bases of subjects' misconceptions of probability and statistics. Thus, it is worthwhile to explore research on judgmental heuristics in detail.

Representativeness. According to the representativeness heuristic, people estimate likelihoods for events based on how well an outcome represents some aspect of its parent population (Kahneman & Tversky, 1972). People believe that a sample should either reflect the distribution of the parent population, or that a sample—perhaps even a single outcome—should mirror the process by which random events are generated. For example, many people believe that in a family of six children, the sequence BGGBGB is more likely to occur for having the children than either BBBBGB or BBBGGG (Kahneman & Tversky, 1972, Shaughnessy, 1977). In the first case, the sequence BGGBGB may appear more representative of the near 50-50 distribution of boys and girls than the sequence BBBBGB. In the second case, the sequence BBBGGG does not appear representative of the random process of having children. From a normative point of view, all 64 such sequences are equally likely to occur. ("Normative" is used in the literature to refer to some theoretical model for assigning probabilities or likelihoods to events.)

Another example of reliance on representativeness occurs with neglect of sample size. People may believe that the chance of getting at least 7 white balls in 10 draws from a population of 50% black and 50% white is the same as the chance of getting at least 70 whites in 100 draws (Schrage, 1983). In the extreme case, they may see no difference between the chances of getting at least 2 heads in 3 tosses of a coin, and at least 200 heads in 300 tosses. Thus, the effect of sample size on probability and variation is not a factor for people who are statistically naive. It is not apparent to them that extreme events (such as all heads in a given number of coin tosses) are more likely to occur within smaller sample sizes than in larger. The phenomenon of representativeness runs deeper, however, than just in naive subjects. Kahneman and Tversky report that even social science researchers trained in the use of statistics are prone to the same types of judgmental errors as the naive subjects (Tversky & Kahneman, 1971). Researchers may even be willing to gamble away their research hypotheses on studies in which the power of their statistical analysis is suspect.

Representativeness has also been used to explain the *negative recency effect,* or "gambler's fallacy." Many subjects tend to believe that after a run of heads, tails should be more likely to come up. Cohen (1957, 1960) did experiments drawing blue and yellow beads out of a container. Subjects were first asked to predict what color they thought would be drawn, then a bead was drawn, shown to the subject, and replaced in the container. The beads were pre-mixed in a given ratio in the container, say a ratio of 3:1, blue over yellow, but the subjects did not know the population proportions of blue vs. yellow. Cohen found evidence that adults tended to first predict the color that was appearing less often (a negative recency strategy in order to "balance things off"), and then after a small number of trials, they would switch to predicting the color that was appearing more often (a positive recency strategy). It appears that until subjects obtain information about the distribution of outcomes in a binomial experiment, they believe that things should balance out, and be representative of a 50-50 distribution. Once subjects do have some information about the distribution, even from small sample sizes, they tend to put too much faith in that information, predicting the more likely outcome much more than would be warranted by chance. Thus, people may believe that even small samples must be representative of the parent population. The phenomenon of negative recency can be found in the late innings of any baseball game, when the hometown hero comes up to bat in a crucial situation. If the ballplayer hasn't gotten a hit all day, the crowd murmurs "he's due!" The fans believe that a batting average of .300 should be reflected in the daily box scores, and that an 0 for 4 day shouldn't occur.

Another phenomenon that is often attributed to representativeness is the *base-rate fallacy* (Bar-Hillel, 1980; Tversky & Kahneman, 1974). A frequent technique in the literature is to provide a description of a person, and then ask subjects to rank some statements about the person, most probable to least probable (Kahneman & Tversky, 1973a; Tversky & Kahneman, 1982a). For example, subjects may be told that the person is male, 45, conservative, ambitious, and has no interest in political issues. Then they are asked which is more likely the case: (a) the person is a lawyer, or (b) the person is an engineer. The responses are overwhelmingly in favor of the engineer in such instances, because the description appears "more representative" of our stereotype of an engineer. What is a bit surprising, though, is that if subjects are also told that the person in the description was randomly drawn from a population 30% engineers and 70% lawyers, this base-rate information does not have much effect on their predictions for the person's occupation. The response rates for the engineer are still far higher than would be warranted under any normative calculation. People tend to put too much faith in the description, believe it to be an accurate representation of perceived population attributes, while neglecting the reliable quantitative information in the base rates.

There is one particular task involving base rates which, together with its variants, has perhaps been the single most thoroughly investigated task in all the research on judgment under uncertainty. This task was first reported in the literature by Tversky and Kahneman (1980) and by Bar-Hillel (1980), and was referred to again in *Judgment Under Uncertainty* (Tversky & Kahneman, 1982b). The problem, which has come to be known as the Taxi Problem, usually is stated something like this.

A cab was involved in a hit and run accident at night. There are two cab companies that operate in the city, a Blue Cab company, and a Green Cab company. It is known that 85% of the cabs in the city are Green and 15% are Blue. A witness at the scene identified the cab involved in the accident as a Blue Cab. This witness was tested under similar visibility conditions, and made correct color identifications in 80% of the trial instances. What is the probability that the cab involved in the accident was a Blue Cab rather than a Green one?

The cognitive psychologists have found that people tend to neglect the base-rate information (15% Blue Cabs) which should suggest that a Blue Cab is an unlikely event, and place their faith in the reliability of the witness. One explanation for this is the representativeness heuristic. People may feel that the single instance of the accident should be representative of the witness' 80% reliability data. Indeed, the modal probability prediction that it was a Blue Cab is at or near 80% in many of the instances when I have tried this problem with students and teachers. There are, however, a wide range of response estimates given, and many of these responses cannot be easily explained by simply appealing to a representativeness argument. For example, subjects who say that the probability that the cab was Blue is 100% clearly are not using representativeness. As a result, the Taxi Problem has been partially responsible for the current search by researchers for alternatives to representativeness to explain people's flawed probability estimates. One alternative explanation discussed by Ajzen (1977), and also by Well, Pollatsek, and Konold (1983), is causality. Konold suggests another explanation, which he calls the "outcome approach," to explain some of the responses to the Taxi Problem. Scholz and Bentrup (1984) investigate a great number of variants of the original Taxi Problem, and conclude that responses to the problem are much more complicated than could be explained solely by use of representativeness. We shall return in more detail to some of these alternative explanations later in the chapter.

From a mathematics education point of view, the Taxi Problem is a difficult problem, even for students who have studied probability. There is a good deal of cognitive strain involved in reading the problem, and keeping everything straight; it is difficult for the students to interpret exactly what they are being asked to do. If we actually expect some sort of normative computation from the subjects who are given the Taxi Problem, rather than a gut level reaction, the subjects would have to recognize that the situation calls for computing *conditional* probabilities. Contingency tables, tree diagrams, and Bayes theorem have to be used to model the conditional probabilities in the Taxi Problem (See Figure 19.1). As we shall see, conditional probability and the notion of independent events are particularly difficult concepts for students to grasp (Bar-Hillel & Falk, 1982; Borovcnik, 1984, 1988; Falk, 1979, 1983, 1988; Falk & Bar-Hillel, 1983; Kelly & Zwiers, 1988). The Taxi Problem is a very complex task, and it is not likely that use of the

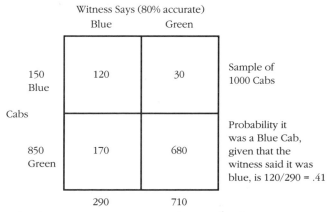

FIGURE 19–1. Contingency Table for the Taxi Problem

representativeness heuristic alone will suffice to explain all the difficulties people have with it, or account for all the reasons that the base-rate information is neglected.

On the other hand, many of the "occupation scenario" tasks and personality descriptions in the literature are rather straightforward, and raise less cognitive strain than the Taxi Problem. The neglect of base-rate data may occur because certain attributes are perceived to be intrinsic to a particular occupation, or to a particular personality. For example, the reason people select the engineer as the more likely candidate in the engineer-lawyer scenario above may be because the attributes of "ambitious," "conservative," and having "little interest in politics" are believed to be inherent in the very nature of the personality of an engineer; this is called the *fundamental attribution error* by Nisbett and Borgida (1975) and by Ross (1975). Ross and Anderson (1982) claim that the roots of attribution theory can be found in a book by Heider (1958). According to attribution theory, there is a tendency for attributors—judges in this case—to underestimate the impact of situational variables and overestimate the significance of dispositional factors. Thus, people estimating the likelihood that attributes describe an engineer rather than a lawyer would attend to the personality descriptions and ignore the situational base-rate data. Similarly, people would tend to believe the witness in the Taxi Problem, and ignore the base-rate data, thus attributing undue accuracy to the witness.

Whether one explains these phenomena by the use of representativeness, by the fundamental attribution error, or by some sort of causal reasoning, the fact remains that people are very susceptible to making these types of judgments and probability estimates. This poses a major problem for the teaching and learning of probability and statistics concepts. Our students are not *tabulae rasae*, waiting for the normative theory of probability to descend from our lips. Students already have their own built-in heuristics, biases, and beliefs about probability and statistics.

Availability. When people estimate the likelihood of events based on how easy it is for them to call to mind particular instances of the event, they are employing the availability heuristic. This judgmental heuristic can induce significant bias based on one's own narrow experience and personal perspective. For example, if you have driven into a town and been hit by a car running a stop sign, you are much more likely to give a higher estimate for the frequency of accidents in that town than someone who has driven there accident-free for several years. In fact, both of these individual perspectives may be far from the truth. If several of your friends have recently divorced their spouses, you may be led to believe that the local incidence of divorce is on the rise, when in fact it has not changed. We all have egocentric impressions of the frequency of events Often these impressions are biased, because even a single occurrence of an event can take on inflated significance when it happens to us. We don't perceive events that happen to us as just one more tally in a big objective frequency distribution. In a paper on the judgment of likely or unlikely coincidences, Falk (1989) found that subjects thought their own coincidences were more surprising than coincidences that happened to others. For example, it would be much more surprising to me if *I* bumped

into a friend from the States in the Budapest train station, than if someone else told a story of how *they* had such a chance meeting in the train station. Self-coincident events are more likely to be remembered or recalled than similar coincidences of others. Thus, self-meaningful coincidences were rated more surprising than self-meaningless coincidences in Falk's study, which, in turn, were rated more surprising than coincidences meaningful to others. "When an event (coincidence) happens to someone else, it is easy to view that person as one of many individuals. However, it is harder to see oneself as a replaceable object. One fails to consider alternative possibilities after one's own coincidence has occurred (Falk, 1989)." Coincidences available to us are easily recalled events, but in this case we are liable to underestimate the likelihood of a coincident event occurring, because it is so unique in our experience.

Kahneman and Tversky report a number of tasks in which subjects tend to give likelihood estimates based on how easy it is to construct instances, or on the ease of memory recall (Kahneman & Tversky, 1973b). They claim that subjects use the strength of certain associations as a basis for judging frequency. For example, subjects were asked to rank the relative frequency of words in the English language that had R in the first position or R in the third position. In which position would R be more likely to occur? Most subjects responded that R is more likely in the first position. In fact R, and several other consonants (K, L, N, V) are much more likely to appear in the third than in the first position in the English language. In similar tasks, college subjects were asked to produce words ending in _____ ing and words ending in _____ n _. (Kahneman & Tversky, 1983). They were able to produce many more words ending in _____ ing." Evidently, it is much easier to construct examples of "ing" words than of _ n _ words, even though the former is merely a special case of the latter.

Availability also comes into play when people are asked to give estimates for complex combinatorial tasks. If subjects who are naive about counting techniques are asked if they think it is possible, from a group of 10, to make up more different committees of 2 people or of 8 people, they choose committees of 2 people by a wide margin (Kahneman & Tversky, 1973b; Shaughnessy, 1977). One explanation for this phenomenon is the availability heuristic; it is just easier to construct examples of committees of 2 than to construct examples of committees of 8. Thus, availability is a heuristic which affects or reflects our own perception of relative frequency. The psychologists, indeed, have cause for concern in how availability may affect judgment. In diagnosing a disease, a physician draws on his or her past experience. It may be easier to recall instances where some symptoms and the disease occur together than to remember counterexamples, where either the symptom was present without the disease or vice versa. This can lead to errors or delays in obtaining a correct diagnosis of the disease. Similarly, financial investors and legal judges may also be susceptible to this same type of overreliance on particularly salient memories, thus biasing their legal or investment decisions.

The Conjunction Fallacy. Recent research suggests that naive subjects, and even those trained in probability and statistics, are prone to rate certain types of conjunctive events as much more likely to occur than their parent stem events. For example, Kahneman and Tversky (1983) found that naive college subjects rated the percentage of people that were 55 *and* had a heart attack higher than the percentage of people that just had a heart attack. Similarly, when primed with a description of a woman who is "bright, single, 31 years old, outspoken, and concerned with issues of social justice," subjects rated the statement "She is a bank teller *and* a feminist" more likely than the statement "She is a bank teller." It didn't matter whether subjects were statistically naive, knowledgeable, or even expert in statistics and decision theory. In every case, between 85% and 90% of them gave estimates that violated the conjunction rule of probability.

Kahneman and Tversky give several reasons why people do this. People may see that the personality description as representative of a feminist, and therefore put much more weight on the conjunction than it should receive because they overemphasize the feminist variable, and, in this case, pair up two rather weakly linked variables—bank teller and feminist. This type of decision is very similar to the fundamental attribution error discussed earlier. In the heart attack scenario, however, the two variables of age and incidence of heart attacks may be strongly (though perhaps falsely) linked in people's minds. This may be either because subjects believe age is a cause of heart attacks, or perhaps because in their own experience most of the people they know who have had heart attacks are older. Thus, they may either be using the availability heuristic or relying on some causal mechanism to give their estimates. There is another possibility here, too: The language "had a heart attack and is over 55" may be interpreted by some people as "had a heart attack *given that* they are over 55." Thus, the conjunction and the conditional probabilities can be confused because of the way "is over 55" can be interpreted by subjects, whether they are statistical novices or statistical literates. If we simply turn the order around to "is over 55 and had a heart attack," someone might reasonably understand it to mean "this person is known to be over 55, now what is the chance they have had a heart attack?"

Kahneman and Tversky claim that their subjects were not confusing the conditional and conjunctional versions of the heart attack task, because likelihood estimates were higher for a more explicit version of the conditional task. However, just because the percentage was higher doesn't necessarily mean there was no confusion between the two. As usual in Kahneman and Tversky's research, we have no information from the subjects other than their raw percentage estimates. There was no attempt made to probe the thinking of any of these subjects. Thus, their claim that subjects were not confusing conjunction with conditional is mostly speculation. A replication study with a better research design is needed. In fact, in a study on conditional probabilities Pollatsek, Well, Konold, & Hardiman (1987) contradict Kahneman and Tversky, as they found evidence that students were, indeed, confusing the conditional P(A | B) with the conjunction P(A ∩ B). The problem seems to occur primarily with the students' translation of conditional probability tasks, which then affects their understanding of the problem. Unfortunately, the Pollatsek study also relies primarily on a forced-choice task methodology, and so they were unable to obtain detailed insight into why the students were confusing conditionals with conjunctions.

It would be interesting to ask statistically naive subjects this question.

Imagine that there are two buckets of marbles in front of you, a green bucket and a red bucket. Each of the two buckets has 100 black marbles and 100 white marbles in it, all mixed up. You will flip a coin to decide which bucket to draw from, then you will pull a marble out of that bucket. What do you think is more likely to happen?
a) You will pull a white marble or,
b) You will choose the red bucket, and pull a white marble? Why?

This is the type of question we would expect to be rather easy for students who had seen a little probability. They might set up a tree diagram, and multiply probabilities. Or, they might realize that $P(W \cap R) < P(R)$ because the conjunction $W \cap R$ is a subset of either of the outcomes W or R.

The reason I suggest this particular question is that it is relatively context free. Although there has already been research on how naive subjects compare the likelihood of conjunctive statements to the likelihood of their supersets (Tversky & Kahneman, 1983), most of the conjunctive statements that have been investigated are laced with contextual traps. This version of the conjunction fallacy with the marbles is relatively context-free, and may be helpful in determining just how much of the conjunction fallacy is due to the context and how much is due to mathematical misconceptions.

Although the concepts of conjunction, conditional probability, and independent events are difficult enough to teach, Kahneman and Tversky's research suggests that there is psychological interference that makes the teaching of these ideas all the more complex: Most people, even if they know probability theory, are not going to take the time to mathematize the heart attack or bank teller scenarios. Their subjective responses will be based on representativeness, availability, or the fundamental attribution error, rather than on a normative model of probability of events.

Research on Conditional Probability and Independence. The difficulties students have with conditional probabilities and with concepts of independent events have been written about by a number of mathematics and statistics educators (Bar-Hillel & Falk, 1982; Borovcnik, 1984, 1988; Falk, 1979, 1981, 1982, 1983, 1986, 1988, 1989, in press; Falk & Bar-Hillel 1983; Kelly and Zwiers, 1988; Konold, 1989b; Pollatsek et al., 1987). One of the most prominent misconceptions of conditional probabilities arises when a conditioning event occurs after the event that it conditions. Falk (1983, 1988, in press) and Borovcnik (1988) discuss the following problem, which is now referred to as the "Falk phenomenon" in the literature.

An urn has two white balls and two black balls in it. Two balls are drawn out without replacing the first ball. 1) What is the probability that the second ball is white, given that the first ball was white? 2) What is the probability that the first ball was white given that the second ball is white?

In mathematical notation, the first question asks us to compute the conditional probability $P(W_2 | W_1)$, and the second question asks us to compute $P(W_1 | W_2)$. It is easy for students to see that the first probability is 1/3, because when we go to draw the second ball, there are one white and two black balls left in the jar. What is not so easy for students to see, or even to believe, is that the second probability is also 1/3. Students often feel that the answer to the second question is 1/2 because they cannot think of the event W_1 as contingent on W_2. In fact, students sometimes believe that you cannot even calculate the second probability, because the outcome of the first draw cannot, in their eyes, depend on the outcome of the second draw. Falk (in press) suggests that the reason this problem is so difficult for students is that they can explain the first conditional probability by a causality argument, while it is hard for students to infer cause for an event that is conditioned on an a posteriori event, as in the second conditional. Furthermore, the usual method of calculating such probabilities, Bayes theorem, gives little or no intuitive feel as to why the calculation yields 1/3 for the second problem as well. Falk suggests a teaching technique for the urn problem in which one ball is drawn and put aside, then a second ball is drawn. The students are shown that the second ball is white, and then asked what the probability that the first ball (still hidden) was also white. In other words, a physical simulation of the problem, acted out, may help students see that the second event can, indeed, be the conditioning event.

In other problems, Falk shows how difficulties in selecting the event to be the conditioning event can lead to misconceptions of conditional probabilities. This version of an old probability problem is a nice example.

> *There are three cards in a bag. One card has both sides green, one card has both sides blue, and the third card has a green side and a blue side. You pull a card out, and see that one side is blue. What is the probability that the other side is also blue?*

People often mistakenly believe that the probability that the second side is blue is 1/2. They argue that there are two possible cards to consider after the first one is shown to have a blue side, the BB card and the BG card. However, the conditioning event is not *the card*, it is *the side of a card*. Thus, the BB card has *two* ways to give us a second side that is blue. The probability that the other side is blue is thus 2/3.

Another type of misconception of conditional probabilities is the confusion between a conditional and its inverse. For example, there is a difference between the probability that I have measles given that I have a rash, and the probability that I have a rash, given that I have measles. The latter situation has a much higher probability than the former. There is a good deal of confusion about such things among the public, particularly where diagnosing disease is concerned (Eddy, 1982). If a person tests positive for AIDS, P(AIDS | + Test) does not equal P(+Test|AIDS). However, the public often sees no difference between these two conditional situations. Surely a person will have many of the symptoms of AIDS if they have the disease. On the other hand, the symptoms provide diagnostic information, not causal information. Falk (1986, 1988) points out

that the confusion between a conditional and its inverse often occurs in the interpretation of what it means to reject the null hypothesis (H_0) in a research study. If we obtain an event A (a statistical result, say) from a sample under a null hypothesis H_0, it is easy to confuse $P(A|H_0)$, and $P(H_0|A)$. If $P(A|H_0)$ is small enough, we traditionally reject the null hypothesis, since the probability of A occurring under the null hypothesis is so small. However, Falk notes that the question "what is the chance that we erroneously rejected the null hypothesis" is answered by computing the inverse conditional, $P(H_0|A)$, and that that is the question we would really like to answer. Even researchers trained in the use of statistics may confuse these two conditionals when interpreting their results.

Falk (in press) mentions a number of reasons why situations involving conditional probabilities are difficult for our students.

1. Students may have difficulty determining the conditioning event (like the colored card problem);
2. may confuse conditionality with causality, and thus investigate $P(B | A)$ when they are asked to investigate $P(A | B)$;
3. may believe the "time axis" prevents an event from being a conditioning event (as in the case with the white-black balls problem);
4. may be confused about what they are given to work with, because of the wording or framing of a conditional probability problem (like the Taxi Problem).

All the researchers of misconceptions of conditional probabilities advocate the use of real world examples to help students understand conditionals. For example, Falk notes that a posteriori events on the time axis are often used as conditioning events in other sciences, such as archeology or astronomy, where inferences about what happened long ago are based on what is happening now. Kelly and Zwiers (1988) stress the importance of examples that will help students to "bridge the gap between the fuzzy distinctions apparent in nature and the very rigid distinctions made in mathematics" (p. 97). Dependence does not imply causality, for example, for oxygen does not cause fires, but fires depend on oxygen to keep burning. Kelly and Zwiers also mention the difference between "real world" and "mathematical" notions of the concept of mutually exclusive events. In the real world, mutually exclusive events are not necessarily complementary events. They suggest that students be shown the difference between *contrary* and *contradictory* events. Events are contradictory when they cannot both be true and cannot both be false. For example the events "I won the tennis match" (A) and "I lost the tennis match" (B) are contradictory events. However, the events "all days in Oregon are rainy" (C) and "no days in Oregon are rainy" (D) are contrary events, because both of these statements can be false; in fact, a third proposition, different from each of C and D, can be true, namely "many days in Oregon are rainy."

Although there has been a lot written about misconceptions of conditional probabilities, there is almost no actual empirical research reported on students' beliefs and intuitions that deals specifically with conditional probabilities. One exception is the study by Pollatsek et al. (1987). However, that study also used written responses of students to forced-choice items. Although the researchers encouraged students to write reasons for their choices, the students did not provide useful information from what little they wrote. A clinical methodology is essential to get at what students are actually thinking. To my knowledge, there has not been any research reported that has attempted to change students' misconceptions or beliefs specifically on conditional tasks; the literature is long on excellent didactical suggestions, but short on hard research in the area. It is clear that many of the didactical suggestions are actually being utilized, but what is needed is a *systematic* investigation of students' conceptions and beliefs about conditional probabilities, with long-term documentation about what happens when one tries some of the didactical techniques suggested in the literature.

One problem I have used extensively with students to point out some of the difficulties with conditional information is a situation called Monty's Dilemma (Shaughnessy & Dick, 1991).

Monty's Dilemma. *During a certain game show, contestants are shown three closed doors. One of the doors has a big prize behind it, and the other two have gag gifts behind them. The contestants are asked to pick a door. Then the game show host, Monty, opens one of the remaining closed doors and shows it to the contestant, always revealing a gag gift. The contestants are then given the option to stick with their original choice, or to switch to the other unopened door. What should they do?*

To highlight the mathematics of Monty's Dilemma, we usually impose the following problem on the puzzle situation above.

The Problem. *If you were the contestant, which strategy below would you choose, and why?*
Strategy 1 (Stick)—Stick with the original door
Strategy 2 (Flip)—Flip a coin, stick if it showed heads, switch if it showed tails.
Strategy 3 (Switch)—Switch to the other door

This problem highlights the difficulty with *what* is actually known, *when* it is known, and *how* the new information obtained is used. We have administered this task to students from secondary school level through graduate school, including pre-service and in-service teachers. Most of the students believe that as soon as Monty opens one of the junk doors, the chance of winning the big prize automatically increases from 1/3 to 1/2. They reason that "now you know that one of the doors is no good." However, if no action is taken by the contestant to improve the chance of winning by actually using this new information, the probability of winning the big prize is still 1/3. The "flip" strategy is introduced to emphasize the difference between actively (re)choosing one of the remaining two doors, and passively sticking to the original door. The flip strategy increases the probability of winning to 1/2, while the stick strategy is still only 1/3. A dramatic change occurs in students' understanding of this problem when they actually carry out a simulation of *each* of the three strategies 100 times with an equal area spinner (3 equal parts) and a coin. Then they realize that rechoosing really does make a difference. Furthermore, they discover that it is best for the contestants to *always* switch, for the only way to lose in the switch strategy is to have picked the prize door the first time. There is only a 1/3 chance of picking the correct door at first, so there is a probability of 2/3 of winning the big prize in the switch strategy.

The success of a simulation strategy with Monty's dilemma provides further evidence that the suggestions of Falk (in press), Borovcnik (1983), Bentz (1983) and Shaughnessy (1981) to model probability problems via simulations is a promising technique for confronting and overcoming misconceptions. However, this is an area where much more research is needed. It is an area ripe for further empirical investigations, and for long-term teaching experiments.

Misconceptions of conditional probability may be closely related to students' understanding of independent events and of randomness in general. Some mathematical presentations of independent events even prefer to define "A is independent of B" if $P(A|B) = P(A)$ (in other words B doesn't change the likelihood of A), rather than that $P(A \cap B) = P(A) \times P(B)$. These are equivalent definitions, but Kelly and Zwiers (1988), among others, advocate introducing the concept of independence via the conditional probability definition, as they believe this is more intuitive for students.

With respect to independence, we have mentioned the research of Cohen (1957, 1960) in which subjects relied on either negative recency or positive recency heuristics, and demonstrated little recognition of independence in a binomial task involving drawing balls sequentially from a bin. In a related type of experiment in which students were asked to judge whether outcomes were random, Falk (1981) found that we humans are poor judges when it comes to recognizing and constructing random events.

In one study Falk showed secondary students sequences of 21 green and yellow cards, and asked them to decide how random the shuffle was. For example, how random do each of these sequences appear?

A. Y Y Y G G Y Y G G Y Y Y G G G G G G Y Y Y
B. G Y G Y G Y G Y Y G Y G Y G Y Y G Y G G Y G G

Students tended to pick the sequences where more "switches" between the colors occur, like B above. In fact, sequences with longer runs of one color are more likely to be the random ones. The second sequence appears more mixed to students, and thus is judged more random, when exactly the opposite is true. Falk found the same results when she asked students to generate their own sequences of 21 cards. Similar results occurred when two dimensional "chess boards" of two colored squares were presented. Students were not very reliable at either recognizing truly random arrangements, or at generating them. Falk concludes that we tend to see patterns in random events when there are none, and thus we reject the notion that certain outcomes are random. Also, we tend to infer randomness when it is not really present, for example, when we rate a sequence with more color switches than chance would predict as the most random sequence. The message from Falk's study is very clear: human beings should never be responsible for trying to generate "random" choices.

Konold (1989b) has evidence that even though our students appear to be reasoning on the basis of independence, if we dig a little deeper, we may find they are inconsistent in their responses. An alarming number of subjects in Konold's study initially rated four sequences of five tosses of a coin (THTHH, HHHTT, THTTT, and so forth) to be equally likely when they were asked "Which of these sequences is most likely to occur," and then turned around and picked a particular sequence when they were asked "Which sequence is least likely to occur?" From an analysis of the reasons students gave for their choices on this task, Konold concludes that students were not basing their responses on either the gambler's fallacy (the negative/positive recency phenomenon described by Cohen) or the representativeness heuristic, as claimed by Kahneman and Tversky on similar tasks. Konold suggests that many people make decisions about probability tasks based on what he calls the *outcome approach*. We explore the outcome approach in more detail in the next section.

Other Decision Schema in Stochastic Situations. Some responses of subjects to probability estimation tasks cannot easily be categorized as misconceptions based on judgmental heuristics. We have already noted the causal nature of some predictions in the section on conditional probability. Studies by Ajzen (1977), Nisbett and Ross (1980), and Tversky and Kahneman (1980) emphasize the differences between the relative impact of causal data and diagnostic data on likelihood estimates. For example, people have greater confidence in predicting weight from height than they do height from weight. Similarly, people predict that blue eyes in a child of a blue-eyed mother is more likely to occur than the event that the mother has blue eyes, given that the child has blue eyes. The mother appears to be the "cause" of the child's blue eyes, while the child is not perceived as causing the mother's blue eyes. Tversky and Kahneman (1980) found that subjects tend to rate the causal conditional (child blue given mother blue) more likely than the diagnostic conditional (mother blue given child blue). The terms "causal" and "diagnostic" come from medical scenarios, where in the one case the disease causes certain symptoms, whereas in the other case the symptoms help to diagnose the disease. The greater the perceived causal effect of one event upon the other, the more likely subjects will overestimate the probability of the caused event. This is a very reasonable strategy, when there really is cause and effect taking place. More often than not, however, there is no basis for this causal schema because the perceived causal relationship is illusory. Thus, reliance on causal schemas can distort probability estimates away from normative values.

Even though judgmental heuristics (such as availability and representativeness) and causal schemas can both lead to misconceptions of stochastics, both at least seem to involve an overview of stochastics as a process. Judgmental heuristics are based in perceptions of *random* processes, whereas causal schemas are elicited by *deterministic* processes. There is, however, evidence that some students might not have any process model for stochastic experiments. These students do not see the results of a single trial of an experiment as embedded in a sample of many such trials. Konold (1983, 1989b) reports that some subjects perceive each trial of an experiment as a separate, individual phenomenon. Konold calls this schema the *outcome approach*.

Outcome-oriented people may believe their task is to correctly decide for certain what the next outcome will be, rather than to estimate what is likely to occur. During a course in elementary statistics Konold presented college students with a number of stochastic scenarios, and let them reason aloud while he probed their thought processes. According to Konold (1989b), an outcome-oriented student uses a 50% chance as a guide to deciding between a "certain yes" and a "certain no" for an outcome. Events with significantly higher than a 50% chance (say 70%) *will* happen, and events with lower than a 50% chance (say 30%) *won't* happen. In the case of exactly 50%, students will say they "don't know" or "can't decide." On several of his tasks, Konold obtained very strong evidence for his outcome hypothesis. In one task, called the bone problem, subjects were asked to predict which side of an irregularly shaped die would land face down. Several subjects tenaciously maintained their original predictions even when confronted with frequency data from many rolls of this die, data which predicted otherwise. Konold also obtained responses to the Taxi Problem, indicating strong belief in the witness and total disregard for the base-rate information. Unlike Kahneman and Tversky, who, at least in their early research, would attribute the disregard of base-rate information to the fact that a Blue Cab is representative of the witness' reliability, Konold contends that some of his subjects were using the single outcome approach, and produces protocols that support his claims. Konold's subjects just believed the witness was right. Outcome-oriented decision-makers see the taxi scenario as a single accident, rather than one occurrence in a repeatable experiment. Konold concludes that subjects are likely to use an outcome approach for predicting events if either (1) the outcomes are not all equally likely to occur, as in the bone problem or (2) the experiment does not appear to be repeatable, as in the Taxi Problem. "As long as students believe there is some way that they can 'know for sure' whether a hypothesis is correct, the better part of statistical logic and all of probability theory will evade them" (Konold, 1989a, p. 92).

We have already discussed the cognitive complexity of the Taxi Problem. It may be very difficult to clearly distinguish reliance on representativeness from reliance on an outcome-oriented prediction on the taxi task. However, Konold's study is one of the few in the area of misconceptions of stochastics that has used a clinical methodology. Many of the tasks that have been administered by psychologists, such as Kahneman and Tversky and others, have involved forced-choice responses to particular item stems. A forced-choice methodology may be self-fulfilling in that alternative responses to such items do not get a chance to surface. Konold's methodology provides an opportunity to explore the full range of student responses to stochastic tasks. Clinical methodologies seem most appropriate for mathematics educators interested in exploring students' cognitive and affective processes on stochastic tasks.

Other Stochastical Misconceptions

We have seen that even researchers trained in the use of statistics hold statistical misconceptions. They may feel that even smaller sample sizes are sufficient in a replication study, since the small samples should be representative of the population (Tversky & Kahneman, 1971). If trained researchers have trouble with statistics concepts, it should not surprise us that students have misconceptions of some of the most elementary concepts, such as mean and variance. In an attempt to see if students have a mechanistic approach to computing the mean, such as an active balancing strategy, Pollatsek and his colleagues (Pollatsek, Konold, Well, & Lima, 1984; Pollatsek, Lima, & Well, 1981) gave statistically naive college students several tasks similar to one that Tversky and Kahneman had used. One version of their task is stated as follows.

The average SAT score for all high school students in a district is known to be 400. You pick a random sample of 10 students. The first student you pick had an SAT of 250. What would you expect the average SAT to be for the entire sample of 10?

Many students given this task will respond that the average should be 400. These students may be using representativeness as a guide, because they believe that even such a small sample should mirror the whole population. On the other hand, they may be confusing the population mean with the sample mean. The population mean doesn't change, but the expected sample mean is affected by knowledge of a score. Once we know the low score is in our sample, the normative calculation for this problem says that we should revise our estimate for the sample mean by *weighting* the 9 unknown scores by the mean, and tossing in the known score; thus, the revised mean is

$$(9 \times 400 + 250)/10 = 385$$

Pollatsek et al. (1981) discuss students' confusion of the sample mean in situations where it must be calculated as a weighted average. Most students' understanding of the mean is the "add-them-all-up-and-divide" algorithm, because that computation is all they were ever taught. When the outcomes are not equally likely, or the scores need to be weighted, estimates of the expected value are wrong. This is as much a mathematical deficiency as a psychological phenomenon. Students may not have been exposed to the general concept of the mean. They have only an instrumental understanding of the mean, and a partial one at that. When asked what the expected mean was for the remaining nine scores, the researchers did not find as much evidence for an active balancing strategy as they expected. An active balancer would predict that the average of the remaining nine scores should be higher than 400, to "fix up" that strangely low score. Instead, students tended to either predict 400 again, or to predict *lower* than 400, because they didn't believe the population mean was really 400 after they got a score of 250. Pollatsek et al. (1984) conclude that students rely on representativeness more heavily than on any active balancing mechanism. What is most striking about this research is that even after these students were shown alternative solutions to the problem in an interview setting, they stuck to their guns when asked what the best solution was; not very many changed their minds. The sample of subjects interviewed was rather small, but, nevertheless, this result showed a tenacious reliance upon the representativeness heuristic, even when confronted with reasonable alternatives.

Pollatsek and his colleagues note that their subjects were very uncomfortable estimating a point value for the mean in problems like the SAT score problem. This suggests that perhaps we would obtain quite different results if we changed the estimation tasks to be confidence interval estimates rather than point value estimates. It may even be better from a didactical point of view to put the emphasis on estimating confidence intervals for events, rather than on estimating one fixed probability value, because this may be part of the conceptual block of what Konold calls the single-outcome believer. Students who are asked to give a single number for the probability of an outcome may not understand that that outcome can occur repeatedly; it may be easier to teach repeatability via confidence intervals. What would the results of the SAT task be if we used several confidence interval versions such as the following?

The average SAT score for all high school students in a district is known to be 400. You pick a random sample of 10 students. The first student you pick had an SAT of 250.

A) Which interval is most likely to "catch" the mean of the sample?
   i) 250–375
   ii) 375–425
   iii) 425–550
   iv) An interval different than these, _____ (write in the bounds).
B) Which interval is most likely to "catch" the mean of the sample?
   i) 200–300
   ii) 300–400
   iii) 400–500
   iv) An interval different than these, _____ (write in the bounds).

Of course, questions might arise with these confidence interval versions of the SAT problem, too: "Where does one pick the cutoffs," "are there several possible confidence intervals," and so forth. It would be useful to pilot such items accompanied by interviews, so that the whole range of student responses could be explored. Once again, it appears that the task, itself, may have a good deal of impact on the interpretation of our research results.

In another study on statistical misconceptions, (Mevarech, 1983) found that college students mistakenly attributed group structure properties like associativity and closure to the operations of computing means and variances. In particular, these college students thought it possible to "average averages" (closure) by the "add-them-up-and-divide" algorithm. The "average-the-averages" misconception is a familiar one to anyone who has taught introductory statistics at the high-school or college level. Once again we see profound misunderstanding of the mean in situations where the averages need to be weighted by sample size. Mevarech's results are especially troubling for statistics teachers, because they occurred *after* these students had an elementary course on descriptive statistics. Mevarech subsequently attempted to teach a course in descriptive statistics that confronted students' misconceptions of the mean and variance. The research reports greater achievement at the end of a two-term course in statistics for those students who had explicit mastery learning instruction designed to confront their misconceptions. Unfortunately, there is no data reported by Mevarech on how the students actually did on overcoming their initial misconceptions.

Early studies by Kahneman and Tversky (1972) found that statistically naive college students do not attend to the effects of sample size when making estimates for the likelihood of events. These results suggested such students do not have an operational understanding of the law of large numbers, because they rely upon representativeness (rather than on sample size) to estimate likelihoods on tasks. More recently, though, Nisbett et al. (1983) claim that even people with no formal training in statistics use the law of large numbers when solving problems in everyday life, and, furthermore, one's ability to apply the law of large numbers in a wide variety of situations actually can be enhanced by training (Nisbett et al., 1987).

In an attempt to resolve some of this contradictory evidence about students' understanding of the law of large numbers, Well et al. (1990) conducted a series of experiments. Different versions of the problems were presented to students. In one version, called the "accuracy" version, students were just asked which would be closer to the population average, the average in a large sample or the average in a small sample. In the "tail" version of this problem, students were asked to estimate how likely it was that the sample average was a certain distance from the population average. Students tended to do well on the accuracy version, but poorly on the tail version. Thus, the emerging picture of students' intuitive understanding of the law of large numbers is not a simple one; task variables affect student performance a great deal. Well et al. carried out a follow-up experiment in which students received training about sampling distributions from a computer simulation experience. Students generated and observed a computer graphics display of the distribution of 100 samples of size 10 and 100 samples of size 100. After all the training, Well and his colleagues still found that many students believed the variability in the sampling distributions for size 10 and size 100 would be the same. This looks ominously like a return to representativeness. My own interpretation of the results of this study are that without explicit teaching on the concepts of variability in sampling distributions, and on how sample size affects this variability, students will not improve in their understanding of the law of large numbers. Mere exposure to the graphics displays of sampling distributions is probably not enough. Someone must explicitly point out the patterns in variability in sampling distributions, and the quantitative relationships that are involved. The fact that there are numbers involved in the tail version in the Well study makes it significantly more difficult than the accuracy version. The inability of the students to deal with the numerical version of this task reminds me of some of the examples in Paulos' book, *Innumeracy* (Paulos, 1989). Another problem that students have with sampling distribution problems is their continual confusion between individual scores and means of

scores. Students need direct, hands-on experience calculating families of means and creating distributions of means before they are given computer simulations.

I would like to close this section on other stochastic misconceptions by sharing a list of common statistical misunderstandings that was recently presented by Landewehr (1989).

1. People have the misconception that any difference in the means between two groups is significant.
2. People inappropriately believe there is no variability in the "real world."
3. People have unwarranted confidence in small samples.
4. People have insufficient respect for small differences in large samples.
5. People mistakenly believe that an appropriate size for a random sample is independent of overall population size.

To this list I would add one more shortcoming.

6. People are unaware of regression to the mean in their everyday lives.

The "real world" for many people is a world of deterministic causes. There is no such thing as variability for them, because they do not believe in random events or chance. We have seen plenty of examples of unwarranted confidence in small samples. In fact, television advertisements play off this misconception all the time, with phrases like "two out of three doctors say." On the other hand, Landewehr also implies that most people wouldn't recognize a statistically significant difference between nice, large samples if it stared them in the face, nor do they realize that a carefully drawn sample of a few hundred instances can tell us much about a very large population. People do not apply representativeness in those instances where it is really appropriate do to so. Finally, people tend to look for some reason that their golf score soared after several good rounds, or that the quality of food at a restaurant seems to be worse over time. Often a good deal of this variability can be accounted for just by regression to the mean. Somehow to admit this, though, we would seem to lose control of our fate, but maybe that is because we are entranced with single outcomes, rather than concerned with long-term results.

Heuristics are Often Helpful

Although there are circumstances where reliance upon heuristics such as representativeness and availability can result in biased, nonnormative probability estimates, in some contexts these heuristics are very useful and often give us good information. If you have stored in your memory several graphic scenarios of traffic accidents or near misses at a particular intersection, you will be more cautious at that intersection. Similarly, the doctor's available diagnostic frequency experience may be correct in a majority of cases. Availability is not always such a bad thing: In fact, it is often a very useful organizer for decision-making. Likewise, representativeness is often at the very heart of much of statistics. Borovcnik (1986) points out that representativeness is a statistical idea which should allow transfer of traits of samples to underlying populations. The very reason we try to draw a random sample from a population is so that it will be representative of the population. We wish to infer to the population any traits we discover in the random sample. Representativeness is, therefore, fundamental to the epistemology of statistical events; it is the way we claim to know something about a population with a certain degree of confidence. Our task as mathematics educators is to point out circumstances in which judgmental heuristics can adversely affect peoples' decisions, and to distinguish these from situations in which such heuristics are helpful. We are obliged to point out this difference to our students; it is not that there is something wrong with the way our students think, just that they—and we—can carry the usefulness of heuristics too far. (Cliff Konold has noted that psychologists actually have to search for situations that will lead people astray.)

Developmental Psychology and the Probability Concept

In their book, *Le Genese de l'idee de Hazard chez l'Enfant,* Piaget and Inhelder present evidence from clinical interviews with children that the probability concept is acquired in stages, in accordance with their theory of development (Piaget & Inhelder, 1951, 1975). Like most Piagetian research, the methodology requires a high degree of verbalization from the children, and the task selection and task administration can be accused of begging the research question it proposes to answer—namely, that there are these levels. However, Piaget's descriptions of what children do and know at various ages, are quite in line with the results of more recent research by cognitive psychologists.

Briefly, Piaget claims that the development of probability occurs in three stages. In the first stage, generally characteristic of children under 7 years of age, the child is unable to distinguish between necessary events and possible events. There is no evidence for a concept of uncertainty. According to Piaget, children at this age will try to find order in a random mixture. If they are shown instances of events A and B, and if A appears more frequently, they will predict B because "it was skipped too often." This is reminiscent of the gambler's fallacy, though the children may not be choosing B for quite the same reasons. They may have been using Konold's outcome approach. However, problems in understanding exactly what young children are thinking arise in their verbalization attempts, which are often not very clear. Piaget and Inhelder also noticed that some children predict the most frequently observed event, with total disregard for the population proportion. For example, in a task I have used, if young children are shown a box with 3 black marbles and 1 white marble, and another box with 6 black and 2 white marbles, and then asked whether there would be the same chance or a different chance of picking a black marble from each box, they frequently say there is a better chance of picking a black marble from the box with 6 marbles. These children do not understand probability as ratio (see Figure 19.2).

In the second stage, up to 14 years, the child recognizes the distinction between necessary and possible events, but has no systematic approach to generating a list of possibilities. A child at this stage supposedly does not possess the combinatorial

FIGURE 19–2 A Piagetian-Type Marble Task

skills or mathematical maturity to make an abstract model of a probability experiment. In the third stage, over age 14, the child begins to develop facility with combinatorial analyses, and understands probability as the limit of relative frequency. Thus, the concept of ratio would appear crucial to the development of the probability concept, from Piaget's point of view.

In what has been the largest study yet undertaken to investigate young adolescents' concepts of probability, Green (1979, 1983a, 1983b, 1987, 1988) surveyed over 3,000 students in England, age 11–16 to determine their level of Piagetian development and to find out what they knew about probability concepts and the language of uncertainty. Green's tasks included tree diagrams, visual representations of randomness, spinners with area models of probability, and Piagetian marble tasks like the one above. Green found that most students had not attained the stage of formal operations by their 16th year. Furthermore, Green found that tree diagrams and the multiplication principle were poorly understood, and that noncontiguous regions acted as distractors for the area model spinners. Items that involved counting outcomes were done much better than those that involved ratios. Nearly 50% of Green's subjects chose the 6B 2W box as more likely than the 3B 1W box to yield a B marble. Green's subjects were especially poor at distinguishing randomly distributed snowflakes in a rectangular grid from nonrandom distributions, or in picking out the random sequences of zeroes and 1s from the hand-manufactured sequences. A concept of randomness was almost totally lacking in these students, supporting the findings of Falk (1981) discussed above.

Green concludes that: (a) the ratio concept is crucial to a conceptual understanding of probability; (b) students are weak when it comes to understanding and using the common language of probability, such as "at least" or "certain," or "impossible"; and (c) only an extensive, systematic program of stochastics in the schools is likely to eliminate this fallacious thinking in children. We are not given information in Green's papers about the stochastic background of these students (what probability and statistics, if any, to which they may have been exposed). Most of Green's items are multiple choice, and although he asked his subjects to give reasons for their choices on some of the items, most of the information from such attempts was very sketchy and incomplete, and, therefore, not very helpful in clarifying students' thinking. Green paints a bleak picture of the stochastics situation in the schools in England. Because of the lack of attention to stochastics in American schools, we might expect the results to be just as bad, if not worse, in the United States, if such a giant survey were to be administered in the U.S.

The only large sample survey data on probability concepts in the U.S. comes from a very few items on the National Assessment of Educational Progress (NAEP). Carpenter, Corbitt, Kepner, Lindquist, and Reys (1981) report that students seem to have some intuitive notions of probabilities in very elementary situations, and that these notions grow with age. However, many students do not know how to describe their intuitions mathematically. This really is not surprising, since they probably were not taught how to build a probability model. When items get as complicated as predicting the probability of all heads in three tosses of a coin, only 5% of U.S. 17-year-olds made the correct computation. The NAEP items are much easier than most of Green's items, and seem to deal more with computations, and less with the concepts and stochastical understanding researched by Green.

The subject of students' probabilistic intuition, mentioned somewhat in passing by the NAEP reviewers, has been the major focus of several researchers' investigations for a number of years. Fischbein and others have written extensively about children's intuitions of probability and combinatorial concepts (Fischbein, 1975; Fischbein & Gazit, 1984; Fischbein, Pampu, & Manzat, 1970a; Fischbein, Pampu, & Manzat, 1970b) and the influence of instruction upon intuitive notions of probability. We will postpone the discussion of the influence of instruction to the next section, while taking up the matter of intuition in more detail here.

For Fischbein, an intuition is a cognitive belief. Intuitions are immediate, wholistic, and obvious to the believer, and there are many examples of intuitions in both mathematics and science; someone may have an intuitive notion of gravity, acceleration, whole number, or relative frequency. For Fischbein, these intuitions are adaptable, and, therefore, can be influenced by systematic instruction. Thus, Fischbein differs with Piaget on the influence that instruction can have on moving children from concrete operations to more formal operations. In his recent book, Fischbein distinguishes between *primary* and *secondary* intuitions (Fischbein, 1987). Primary intuitions are the ideas and beliefs that we have before instructional intervention; secondary intuitions are restructured cognitive beliefs that we accept and use as a result of instruction or experience within a particular cultural community. In probability, for example, many people have an intuition that they would need 180 people or more to have an even chance of getting a match of birthdays. However, after instruction, students may replace this intuition with a secondary intuition that can be applied to all such matching problems. They do not lose their primary intuitions, but rather have just built another intuition for this special context of matching problems. For Fischbein, the process of replacing a primary intuition by a secondary one is not a gradual process; it takes place as a whole, all at once. This is very much like the "Aha" experience in gestalt psychology—the moment of discovery or insight in the problem-solving process.

While intuitions are adaptable, they are also subject to overextension. There are numerous examples in mathematics and science where initial intuitive representations of ideas are limited. For example, the intuition of number as a measure of length holds up for rational and real values, but breaks down when we consider complex numbers. A new secondary intuition of a number, as a solution to an equation, can be built in its place to accommodate the complex numbers. In his review of Fischbein's book, Cobb (1989) calls this the "double edged

sword" of intuitions. Intuitions can mislead and promote misconceptions of scientific reality, as well as provide simplifying cognitions of that reality. For this reason, Fischbein speaks of the importance of developing an intuition of the "non-intuitive." This is particularly important in some branches of mathematics (such as probability and statistics) in which many phenomena conflict with our initial cognitive beliefs.

Fischbein does not refer to the judgmental heuristics of Kahneman and Tversky (representativeness and availability) as intuitions; however, there are some similarities. Judgmental heuristics also serve as double-edged swords, sometimes helping and sometimes hindering. Both heuristics and intuitions are beliefs from the perspective of those who use them, and both model external reality. On the other hand, intuitions appear in a much wider range of scientific contexts than just stochastics, and judgmental heuristics are process strategies rather than personal representations of scientific reality. It would seem that neither heuristics nor intuitions is a subset of the other.

The fact that there is an external, empirical mathematic and scientific reality to be discovered is fundamental to Fischbein's theory of intuition. This belief—that there actually is something outside of ourselves to be learned—drives the theory of primary and secondary intuitions and takes us back to our epistemological roots.

The Effects of Intervention on Stochastic Conceptions and Beliefs

In the last 15 years, mathematics and statistics educators have become very interested in the cognitive processes involved in the teaching and learning of stochastics. As we shall see, the research of cognitive psychologists has had an effect on the types of questions that mathematics educators have explored in probability and statistics. Prior to the strong influence of cognitive psychology, research conducted by mathematics educators fell primarily into three types of studies:

1. Feasibility studies undertaken to determine whether topics in probability and statistics could be taught to elementary or middle school age children (Doherty, 1965; Jones, 1974; Leake, 1962; Leffin, 1971; Mullenex, 1968);
2. Experimental or correlational studies attempting to look for the effects of teaching probability on other variables, or to find relationships between other variables (such as attitudes, symbol use, computation) and success in probability (Gipson, 1971; McKinley, 1960; Shepler, 1970; Shepler & Romberg, 1973; Shulte, 1968; White, 1974);
3. Studies which compared the effects of several approaches to teaching a unit on probability (Austin, 1974; Barz, 1970; Geeslin, 1974; Kipp, 1975; McLeod, 1972; and Moyer, 1974).

Most of the feasibility studies indicate that upper-elementary- and middle-school-age children already possess some idea of probability before instruction, and that it is possible to teach stochastic concepts to children in this age range (Doherty, 1965; Leake, 1962; Mullenex, 1968). Leffin (1971) and Jones (1974) found that IQ was an accurate predictor of children's performance on a probability test. Leffin (1971), Shepler (1970), and Gipson (1971) all found evidence that students did not always successfully apply information from a correctly listed sample space to calculate probabilities. They exhibited, therefore, only partial understanding of the concept of a probability model. Leffin also found that the concept of combinations was a difficult one for children in the grade 4–7 range. Leffin's study is really part stochastics and part combinatorics. In this chapter, I have taken the position that combinatorics (counting techniques) is a separate set of concepts from stochastics. Granted, applied combinatorics can be helpful in calculating probabilities; however, the fundamental concepts of randomness, independence, samples, distributions, and population parameters, do not have any conceptual ties to counting techniques. Too often people only think of combinatorics when they think of probability, perhaps because they had a very formal set-theoretic course in probability.

Jones's study (1974) was done with younger children (first–third graders) and was one of the first by a mathematics educator to use a clinical methodology in research on stochastics. Jones used taped interviews with children to gain insight into the types of errors they were making. Jones found that color biased children's choices of what outcome they thought would be most likely to occur on an area spinner, because the children often had a favorite color and would choose it over the color with the most area. So, in some cases, manipulatives might actually interfere with children's thinking about probability. A paper by Smock and Belovicz (1968) warns educators not to assume too much when they attempt to implement probability in the elementary schools, since the extent and generalizability of children's stochastic knowledge at this age is very limited. The general picture from most of this research, however, is that probability concepts can and should be introduced into the school at a fairly early age.

White (1974), McKinley (1960), and Shulte (1968) each found that students' achievement in probability concepts increased significantly after a unit on probability. Achievement in probability was found to correlate positively with computational ability, concept attainment, reading comprehension, language skills, and general mathematics achievement. Shulte found no significant effects of a probability unit on either attitudes or computational skill, although he did find significant effects upon students' ability to use and interpret mathematical formulas.

Moyer's (1974) experimental group learned a lot of probability, but other than that there were no significant differences between treatment groups on computation skills, reasoning ability, or attitudes. Barz (1970) found evidence that a historical practical-involvement course in probability tended to show higher student achievement results than a traditional, set-theoretic approach to probability. For some groups, the difference was significant.

Many of these earlier studies that taught a unit on probability may have involved too short a duration for significant changes in attitudes and skills to actually show up. What was needed—at all age levels—was longer, more intense exposure to stochastics, and greater use of clinical methodologies to investigate students' thinking processes.

Fischbein and his colleagues were among the first to report extensive changes in intuitions and conceptions of probability over the course of instruction. When children, even preschool children, received instruction on Piagetian probability ratio tasks, they improved in their predictions of outcomes (Fischbein et al., 1970a). The improvement was most pronounced among 9–10 year olds, and, according to Fischbein, this contradicts Piaget's theory of stages. Similar improvement was also noted when instruction was given on combinatorial estimation tasks (Fischbein et al., 1970b). Unfortunately, neither of these studies appears to have assessed the children's understanding prior to instruction, or to have provided a control group for comparison. In a later study, Fischbein and Gazit (1984) looked at the effects that teaching twelve lessons about probability and relative frequency had on both children's conceptions and intuitions. They found gains in conceptions, but obtained mixed results on improving probabilistic intuitions when experimental and control groups were compared. There may even have been a negative effect of the instruction on the Piagetian ratio tasks; however, this study also did not assess student's conceptions or intuitions prior to the instruction, so a negative effect cannot be inferred. It is clear that Fischbein believes that instruction can improve students' intuitive ideas of probability.

In an attempt to look at the effects of instruction on stochastically naive college students' use of judgmental heuristics, I conducted a 12-week, intensive teaching experiment (Shaughnessy, 1976, 1977). Students' knowledge of probability and use of heuristics were assessed both by written responses and by taped interviews before and after the instruction. Tasks similar to those of Kahneman and Tversky were used to assess reliance upon the judgmental heuristics, although the "either-or" forced response format was relaxed, and students were encouraged to write out or verbalize a reason for their answers. Significant differences between two control classes and two experimental classes were found at the end of the instructional intervention, with greatly reduced reliance upon the heuristics in the experimental groups.

There are several things about this study that are very clear to me, in retrospect. First, the experimental course was very intensive, meeting five days a week for an hour each day. Students carried out many experiments and simulations while working entirely in small groups. I could watch and record the changes in their conceptions: the day they realized there were "actually 64 outcomes" when you flipped 6 coins, the day they tossed tacks and realized that not all probability experiments had equally likely outcomes, and so on. The students kept a daily journal of all their class experiments and homework problems, with personal comments on how they felt or what they had learned. As a class, we were totally immersed in the activities of the course.

Second, even though the course was highly successful from a class standpoint in overcoming misconceptions of probability and statistics, there were still some students who did not change their responses or beliefs on those judgment tasks, even after the intense instructional experience. It is very difficult to replace a misconception with a normative conception, a primary intuition with a secondary intuition, or a judgmental heuristic with a mathematical model. Beliefs and conceptions are slow to change.

Third, I attribute most of the success in overcoming misconceptions to the instructional model for the activities. Students first had to "buy into the task" by *making a guess* for the outcome of the experiment. This made it their problem, set the hook, so to speak. Next, they had to carry out a structured task, *gathering and organizing* their data. Then students answered questions solely based on their data, after which we compared their experimental results to their initial guesses. Misconceptions were *explicitly confronted* with experimental evidence. Finally, we *built a theoretical probability model* that might account for the experimental data they had collected. Students then compared all three pieces of information: their initial guesses, the experimental empirical results, and the results predicted by the model. Throughout this whole instructional process, students were constantly placed in a position of having to reconcile the dissonance between their stochastic misconceptions and their empirical observations.

The potential that guessing and comparing outcomes has for helping students to confront misconceptions has also been reported by delMas and Bart (1987). These researchers conjectured that unless students are forced to record their predictions and then compare these predictions with actual experimental results, they will tend to look for and, subsequently, find confirming evidence for such deeply entrenched misconceptions as availability, the law of averages, and the law of small numbers. DelMas and Bart prescribed an evaluative task in which students in an elementary statistics class listed their predictions for a series of coin tosses, one at a time. The students' written list of predictions was listed alongside the randomly generated results. Pre/post comparisons on tasks similar to the coin toss experiment showed a marked increase in use of a normative, frequentist explanation among the students who did the prediction task, along with a marked decrease of frequentist explanations among control students. DelMas and Bart conclude that forcing students to explicitly compare their predictions to empirical results helps them to employ the frequentist model they are being taught. On the other hand, their results also suggest that students who are not given an opportunity to confront their stochastic misconceptions will end up relying even more heavily on judgmental heuristics after instruction than before.

For some time researchers in probability and statistics have been saying that mere instruction in stochastics is not sufficient to overcome the misconceptions and prior intuitions that students already have. However, delMas and Bart claim that not only is traditional instruction insufficient, it may even have negative effects on students' understanding of stochastics. More research of this type is needed to confirm delMas and Bart's results, but the implications of their study pose a big challenge to mathematics and statistics educators. One of the main goals of stochastics instruction should be to provide students with examples of how statistics is misused and abused, and how misconceptions of probability can lead to erroneous decisions (Shaughnessy, 1981, 1983a). Schrage (1983) provides some real-life examples of erroneous, stochastically-based decisions by members of the legal and medical professions. In one court case, a man was actually on trial for murder because of an erroneous probability argument. In another instance, Schrage discovered a faulty question involving probabilities on the German medical school entrance examination. Both these real-life situations involved several misconceptions, including disregard for base-rate data. Schrage was successful at turning the court's attention to the erroneous probability calculations, and the man

was acquitted. However, despite his best efforts, the group of "30 medical experts" who constructed the entrance exam failed to see their mistake. Schrage advocates the inclusion of just such real-life mistakes in courses on probability and statistics. Initially students may reason much like the doctors did, but only persistent attention to fallacies in reasoning and their social implications will help turn students' attention to the importance of normative thinking in stochastics.

Both Konold (1989b) and Garfield and delMas (1989) provide further evidence on the difficulty in changing student's stochastic conceptions. Much of Konold's research has been concerned with uncovering people's interpretation of probabilistic situations, rather than with confronting misconceptions. For example, Konold investigated people's explanations for what it means to say "there is a 40% chance of rain today" in research that led to his postulating the outcome phenomena (Konold, 1983). However, Konold (1989b) does describe an attempt to influence misconceptions with a computer modeling intervention. He found that some students changed their interpretation of probability tasks after instruction, while some of the students also persisted in maintaining nonnormative interpretations of the tasks, particularly the outcome phenomenon. Garfield and delMas (1989) used a computer tutorial program they developed called Coin Toss to examine the effects of a computer-based experience on statistically naive college students' misconceptions of probability. The computer software deals with the concepts of variability in samples, effect of sample size on variability, and independence and randomness. Their results were mixed; some students showed positive changes in their stochastic thinking after using Coin Toss, but a large number of students persisted in their misconceptions about sample size and variation after the experience. These results are consistent with those of Well et al. (1990) who found that many students still did not understand the effect of sample size on variability after considerable computer experience with modeling sampling distributions. The message in all these studies is consistent and clear: Stochastic misconceptions are very difficult to overcome and change in some students.

Konold provides us with some very salient reflections on the role of instruction in a situation fraught with students' prior stochastic misconceptions.

Long before their formal introduction to probability, students have dealt with countless situations involving uncertainty and have learned to use words such as probable, random, independent, lucky, chance, fair, unlikely. They have a coherent understanding that permits them to utter sentences using these words that are comprehensible to others in everyday situations. It is into this web of meanings that students attempt to integrate and thus make sense of their classroom experience.... My assumption is that students have intuitions about probability, and that they can't check these in at the classroom door. The success of the teacher depends on how these notions are treated in relation to those the teacher would like the student to acquire.... How students think about probability before and during instruction can facilitate communication between the student and the teacher. (Konold, 1991, p. 144).

Konold goes on to suggest several instructional techniques for confronting and overcoming stochastic misconceptions which have been used by physics educators who also must deal with students' prior misconceptions. Physics educators have designed instruction specifically to confront misconceptions (Clement, 1987; Hake, 1987). Konold categorizes the techniques from physics education as follows:

Students are encouraged to

1. test whether their beliefs coincide with the beliefs of others,
2. test whether their beliefs are consistent with their own beliefs about other related things, and
3. test whether their beliefs are born out with empirical evidence.

The instructional model I used (Shaughnessy, 1977) explicitly addressed (1) and (3); there was also a great deal of evidence for the students' search for internal consistency (2) in their journals. I agree with Konold that these three suggestions are among the necessary instructional steps we must take to begin to replace our students' primary intuitions with normative secondary intuitions of probability and statistics. Unfortunately, these steps may not be sufficient, as some students tenaciously maintain misconceptions even after such instruction.

Konold is a bit of an exception among psychologists, who as a group have not been particularly concerned with the influence of instruction on misconceptions of stochastics. For example, Kahneman and Tversky have not been particularly interested in intervening in their subjects' thought processes in most of their psychological research. However, they did manipulate tasks in their research on the conjunction fallacy in an attempt to make the tasks more transparent and, thereby, to improve the rate of correct responses (Tversky & Kahneman, 1983). They found that "cueing" the subjects—first asking them to estimate the frequency of men who were 55 and over—decreased subjects' tendency to rate the conjunction "is over 55 *and* had a heart attack" more likely than the simple event "had a heart attack." The cueing acted as a form of instruction in their experiment, or at least as an attention-grabber. They also found that subjects made fewer conjunction errors when probability information was given in the form of frequencies rather than percentages. The influence of the task itself and the task environment is evident in the research of Kahneman and Tversky, as it is in studies conducted by Well et al. (1990). The presentation of the task, the wording of the task, the subject's understanding of the task and the subjects' previous exposure to such tasks all can have a major influence on the subject's response. Although the influence of task variables has received considerable attention in other areas of mathematics education (see Goldin & McClintock, 1984), there has been little consistent concern about task variables in stochastic research. Some exceptions are the work of Scholz and Bentrup (1984) on variations of the Taxi Problem, the Well et al. (1990) study on the law of large numbers, and the review of empirical research by Bentz and Borovcnik (in press).

Among psychologists, Nisbett and his colleagues have probably done the most work exploring the effects statistics training has on subjects (Fong, Krantz, & Nisbett, 1986; Nisbett, Krantz, Jepson, & Fong, 1982; Nisbett et al., 1983; Nisbett et al., 1987). They have found that both the frequency and quality of statistical answers improved markedly with the amount of training in statistics (Fong et al., 1986). For example, given a scenario of a woman who is disappointed because the quality of the food in a restaurant seems to have gone downhill since her first visit, students with more statistical training gave more answers based on regression to the mean than their statistically naive colleagues (regression was indicated by 20% of beginning statistics students, 40% of those students with 1 to 3 statistics courses, and 80% of doctoral candidates). Nisbett is also interested in the transfer of statistical reasoning to everyday situations. Nisbett et al. (1987) describes the effects of specific training in the law of large numbers and how subjects were able to generalize applications of the law to a wide variety of situations after training. Nisbett is really interested in whether it is possible to teach general reasoning skills transferable to many areas, concluding it is possible to transfer training on the law of large numbers. I have already mentioned that other researchers contradict the findings of Nisbett; both Kahneman and Tversky (1972) and Well et al. (1990) have contrary evidence indicating that subjects still have misconceptions of the law of large numbers and sampling distributions, even after training. My general impression of Nisbett's tasks is that they do not elicit misconceptions because the tasks are relatively easy and straightforward, and they avoid numbers, numerical estimates, and any kind of quantification that may trigger the use of heuristics.

Other studies involving instructional interventions to influence stochastical conceptions have been done by Mevarech (1983) and Beyth-Marom and Dekel (1983). The Mevarech study successfully decreased misuses of computational operations on the mean and variance, such as "averaging averages." Beyth-Marom and Dekel created an entire curriculum directed at improving thinking under uncertainty. The emphasis in their course is on probabilistic thinking and on discussing and reacting to probabilistic scenarios, rather than on teaching the mathematics of statistics. This course appears to have arisen almost entirely out of the research from cognitive psychology, as it makes little or no reference to either the research or to didactical ideas from the mathematics or statistics education literature. Although it provides an interesting introduction to thinking under uncertainty, this course is yet another example of the relatively mutual isolation in which psychologists and mathematics educators have pursued their attempts to change people's stochastic conceptions. Beyth-Marom and Dekel mention that the classroom teachers who worked with them teaching the new curricula "themselves had difficulties comprehending some of the curriculum topics." The problem of dealing with teachers' misconceptions when we attempt instructional intervention is also cited by Rubin and Roseberry (1988).

Rubin et al. (1988) report on the implementation of innovative computer-based statistics curricula developed at BBN Laboratories in Boston in the Reasoning Under Uncertainty project. Interactive Macintosh software packages—Stretchy Histograms, Shifty Lines, and Sampler—were developed and field tested with three teachers. They found that teachers' prior knowledge of statistics is not necessarily structured to facilitate reasoning beyond simple formulas. The software demands an understanding of the conceptual structure of stochastics concepts with which teachers might not be familiar. One of the teachers did not understand that the median can remain unchanged when data is added to one of the ends of a distribution, feeling that something must be wrong with the computer software. In another instance, both the teacher and students totally misunderstood the meaning and implications of randomization in an experiment. The in-service problems encountered by Beyth-Marom and Dekel, and by Rubin and Roseberry dramatize the importance of the teacher (Garfield, 1988) in any curricular innovations we might attempt to overcome stochastic misconceptions. We have to first deal with teachers' misconceptions before we can expect them to be competent at helping their students to overcome misconceptions.

Research on Computer-Based Interventions

To date, there have only been a few studies on the effects that computers or computer simulations have on students' learning and understanding of probability and statistics. Among them are the studies by Garfield and delMas (1989) and Well et al. (1990) that we have already discussed. This lack of computer research results is only a temporary situation, as there is a lot of ongoing activity in the area; for example, there is a project at the Technical Education Research Center (TERC) in Cambridge, Massachusetts concentrating on creating a software environment in which students can create their own data models (Hancock, 1988). The TERC project operates on the premise that data-modeling activities, such as constructing, manipulating, and interpreting data, consist of skills and concepts that come before the process of actually doing statistics. Thus, the TERC software environment, called Tablemaker, has focused on representations of category and attribute data, with which students can form their own data structures. The step of constructing data is often left out of teaching statistics, as we tend to bring prestructured data into our classrooms in order to teach statistics concepts. In the TERC project, people create their own data. In another project, software for a Probability Simulator and a Statistical Analyzer are being developed as part of a curriculum project, and a companion research component of the project will investigate the influences and effects of the computer-based materials on students' misconceptions of probability and statistics (Konold, Sutherland, & Lockhead, 1988). Thus, in the near future, there will be more research results on the effects of using computer environments in stochastics.

There is almost universal agreement among stochastics researchers that computer simulations, computer spreadsheets, and the use of computers to conduct Exploratory Data Analysis (EDA) are the directions in which stochastics education should be headed. Many presentations at both the ICOTS I and ICOTS II meetings emphasized the benefits of computers in stochastics education (see Biehler, in press; Piazza, 1988; Rade, 1983). Articles by Biehler (in press) and Biehler, Rachs, and Winklemann (1988) emphasize the ability of computers to provide multiple representations for stochastic information, particularly the graphical abilities of computers. Methods from EDA, such as stem-and-leaf plots, box plots, scatter plots, and data smoothing, can be investigated graphically and dynamically on computers. Thus, computers provide both an exploratory aspect and a representational aspect. One of the best examples of this potentially dual nature of computer software may be the Stretchy Histograms and Shifty Lines materials of the Reasoning Under Uncertainty project (Rubin, Roseberry, & Bruce,

1988). According to the authors, field testing showed the computer features of visualization and features of interactive, linked representations in these materials help to aid learning. However, there has not yet been any large scale implementation of these materials, nor has there been any systematic research on how students' concepts of probability and statistics evolve while working with them.

In a review of the uses of computers in probability education Biehler et al. (1988) identify two "targets of difficulty" where computers can help: lack of experience and what they call the "concept-tool gap." Addressing the first concern, Biehler el al. claim that computers can provide much more experience with data manipulation and data representation than would ever be possible by hand in the same time frame. By the concept-tool gap, Biehler and colleagues mean that many problems in school mathematics traditionally depend on students' computational skills and teachers tend to avoid them. The computer provides us with simulation as an alternative problem-solving strategy, enabling us to investigate more realistic situations than were previously possible. Arguments for the combined benefits of computer simulations and Exploratory Data Analysis can be found throughout Biehler's papers. He believes that EDA may serve as the missing link between strictly deterministic situations and strictly random situations. When using EDA, it is not obvious at the start which parts of the data can or should be interpreted as strictly random. Biehler, then, sees the use of EDA as an opportunity to connect the two extremes of determinism and randomness (Biehler, in press).

The writings on uses of computers in probability and statistics are long on didactical suggestions but short on research suggestions. There are seemingly excellent curricular materials being developed for use in the classroom, but little accompanying research and evaluation work has been published. Many of these materials are so new that the research and evaluation work is still in progress. In their review, Biehler et al. (1988) say very little about empirical research on the use of computers in probability. They do mention several studies which they say demonstrate the American penchant for comparing analytic and simulation methods. Among these is a study by Atkinson (1975), who found strong evidence for the superiority of simulation methods over analytic methods in a course on probability. However, one gets the impression that Biehler does not believe that such comparative studies "in the American tradition" are very beneficial. Biehler appears to encourage studies from what he calls a more cognitively oriented research tradition, citing a microworlds study by Dreyfus (1986) as an example. The themes of constructivism and cognitive science are rather recent in mathematics education, and many of the studies that Biehler lumps in the American tradition were done prior to the emergence of cognitive perspectives on research. I agree with Biehler that case studies of how students' conceptions of stochastics change over a period of time when they are exposed to computer environments are an essential direction for research. On the other hand, it is also important that both the effects of computer-based instruction in stochastics and the evaluation of new curricular materials be included in our research programs. It *is* important for us to examine the effects of different approaches on groups as well as on individuals, because in the real world, *teachers teach groups*.

In closing this section, let me mention two concerns I have with uses of computers in stochastics education. First, what is the role of the teacher in computer environments? This question is certainly not unique to stochastics; however, I see precious little attention given to concerns of teaching in the development of new computer approaches to probability and statistics. After all, the teacher is our delivery system, more so in a computer age when the teacher is more the "guide on the side" rather than the "sage on the stage." The workshops conducted by the Quantitative Literacy Project (QLP) assist teachers with computer materials once they have been developed. However, I am more concerned with involving teachers early in projects, at both the research and the development ends. We can tap teachers' expertise on how kids learn and think by enlisting them in longer-term teaching experiments, where teacher and researcher take turns teaching, recording, and interviewing in the classroom.

Second, I caution researchers who use computers in teaching probability and statistics not to completely abandon other, more concrete representations of stochastics experiments. I have frequently used the computer in the teaching of probability and statistics throughout the past decade. Although I have no formal research evidence, it seems to be very important for many students to have the experience of actually generating and gathering their own data *physically*, with random devices such as dice or spinners, before they can understand or accept computer simulations. I have consistently found this to be the case when considering problems which may elicit misconceptions, like Monty's Dilemma (discussed previously). It is as though the students must proceed concretely through the same steps as the computer in order to believe the results. It is important for us to continue developing connections between concrete simulations and computer simulations in our teaching and investigating the effects of the transition between the two in our research.

A MICRO MODEL OF STOCHASTIC CONCEPTUAL DEVELOPMENT

There are very few formal models of conceptual development in probability and statistics. In fact, the only detailed model of stochastic thinking I am aware of is the Structure-Process model of thinking (SPT), outlined by Scholz (in press). (We will return to Scholz's information-processing model for stochastic thinking in the next section.) The difficulty with building models in research on stochastics is that if a model were to try to incorporate the results of all the different types of studies by mathematics educators and psychologists, that model would run the risk of being so complicated that it may be of no practical use either to researchers or to teachers. On the other hand, if one focuses on a piece of the picture in stochastics and builds a simple model, one runs the risk of oversimplifying and leaving out important points of view. The history of this discipline seems to have been to oversimplify the matter first, making adjustments later. I will attempt to follow somewhat in this tradition and err in the direction of a simpler model, while hoping that the model might be of some use to researchers and

teachers as they attempt to understand what we find out from research in stochastics.

When researchers have asked subjects to estimate likelihoods, predict outcomes, or make judgments or decisions under conditions of uncertainty, a variety of responses (and reasons for these responses) have been found. The different types of reasoning exhibited by our students indicate various levels of conceptual sophistication, ranging from a total lack of understanding of random events to a tolerance for comparing and contrasting several mathematized versions of random events. People's conceptions of random events are determined, in part, by their own experiences (primary intuitions), but they can also be influenced by instruction (secondary intuitions). Here is a characterization of people's stochastic conceptions which I have found helpful for describing research results and for planning instruction. In no way, though, do I claim these are exhaustive.

Types of Conceptions of Stochastics

1. *Non-statistical.* Indicators: responses based on beliefs, deterministic models, causality, or single outcome expectations; no attention to or awareness of chance or random events.
2. *Naive-statistical.* Indicators: use of judgmental heuristics, such as representativeness, availability, anchoring, balancing; mostly experientially based and nonnormative responses; some understanding of chance and random events.
3. *Emergent-statistical.* Indicators: ability to apply normative models to simple problems; recognition that there is a difference between intuitive beliefs and a mathematized model; perhaps some training in probability and statistics; beginning to understand that there are multiple mathematical representations of chance, such as classical and frequentist.
4. *Pragmatic-statistical.* Indicators: an in-depth understanding of mathematical models of chance (i.e. frequentist, classical, Bayesian); ability to compare and contrast various models of chance; ability to select and apply a normative model when confronted with choices under uncertainty; considerable training in stochastics; recognition of the limitations of and assumptions of various models.

The results of some research studies, especially ones with clinical methodologies that try to probe subjects' thought processes, have shown that some people do not have a conception of a random event as one outcome among many possible outcomes within a repeatable experiment. Konold's subjects that exhibited the outcome approach were trying to guess the results of the very next trial, rather than trying to estimate the frequency of a particular outcome over the course of many trials. Investigations into the Taxi Problem have shown a tendency to believe the witness, and disregard the base-rate data. The dominance of deterministic models with algorithmic presentations in our science and mathematics teaching precludes much exposure to models of chance and uncertainty for many of our students. Thus, they may look for causal influences to make decisions under uncertainty. There are clearly many people who do not operate in a stochastic setting; we will refer to these conceptions as *non-statistical*.

Other research—particularly with college students who have not been taught probability and statistics—has uncovered a number of systematic, nonnormative, conceptions of probability and statistics which are based in experience. Judgmental heuristics, such as anchoring, balancing, representativeness, and availability, characterize the conceptions of probability and statistics in the *naive-statistical* stage. These heuristics are usually triggered by analogy, because they work and are useful in other settings. Such heuristics are similar to the intuitions referred to by Fischbein, but they are neither purely primary nor purely secondary intuitions. College students have had many more cognitive experiences than the young children Fischbein studied, so their primary intuitions of probability have been distorted and influenced by experiences other than instruction. College students have been found to exhibit both non-statistical and naive-statistical conceptions of stochastics. We really do not know when or why these judgmental heuristics begin to show up in students' reasoning.

Nisbett et al. (1983) claim that if the sample space and the sampling process are clear, and if the role of chance in producing the events is clear (causal type tasks are suppressed), people really can learn to apply normative stochastic models. In other words, if things are simple, people are more likely to apply the principles of probability and statistics rather than resort to other explanations for uncertain events. Many of the tasks that have been used in research on judgment and decision-making are anything but simple. These tasks often involve scenarios in which the role of chance is not immediately obvious, and the sample space is not apparent. However, research has shown that people can be taught to translate complex scenarios of uncertainty into mathematical problems. It is possible to teach people to solve problems in stochastic settings. Students in this *emergent-statistical* stage are under the influence of didactical interventions, and their conceptions are changing. They are beginning to see the difference between degree of belief, and a mathematical model of a sample space. They are becoming able to apply normative models to an ever widening range of stochastic settings, but initially they may still be subject to falling back upon the familiar causal or heuristic explanations when confronted with an unfamiliar type of task. Their normative conceptions are developing, but still unstable.

The ability to function at the *pragmatic-statistical* stage is not reached by very many of our students, or even by many of our teachers. In this conceptual stage, people realize there are various models of uncertainty which attempt to represent chance experiments. They are able to compare and contrast these models—a frequentist model with a Bayesian model, say—and apply the models to calculate probabilities. More importantly, they are able to modify or extend these models and adapt them to unfamiliar situations, remaining aware that each model has its strengths and its limitations.

I do not wish to imply that these types of conceptions are necessarily linear, or that they are mutually exclusive. That is, I do not think it is necessary for a person to first be a "naive statistician" in order to become an "emergent statistician." Also, it is possible for people to function with several of these conceptions operative. For example, when psychologists find that graduate students and fellow researchers in psychology use heuristics to estimate likelihoods, they have found subjects who are naive-statistical in some settings, while emergent-statistical in others. The nature of the task can affect the type of conception we employ.

The students in our beginning courses in probability and statistics are usually in one of the first two categories, non-statistical or naive-statistical. Since the entire teaching-learning process occurs in the emergent-statistical realm, it is important for us to begin our instruction by confronting both our students' deterministically entrenched paradigms, and their statistically naive heuristics. We must create some dissonance within our students' past belief systems if we are to have a chance of replacing them with mathematical models. The role of the teacher is paramount in this; I am convinced that the latter two conceptual stages will not occur without carefully guided learning experiences under the tutelage of a well-trained teacher who is mathematically and statistically competent as well as sensitive to the types of beliefs and misconceptions that students have about stochastics. Even this is no guarantee, though, as research has shown our students stochastic tenets to be amazingly robust.

RESEARCH ISSUES: CURRENT AND FUTURE

There have been a number of recent critical reviews and commentaries on the teaching and learning of probability and statistics; Bentz and Borovcnik (in press), Scholz (in press), Steinbring (in press), Garfield and Ahlgren (1988a) and Hawkins and Kapadia (1984) have all contributed different perspectives on research in stochastics. A number of issues that concern researchers in stochastics were raised in these reviews. I would like to comment on several issues in these reviews, and then present my own "wish list" for research in stochastics in the near future.

Issues from Past Perspectives

Bentz and Borovcnik (in press) have provided a detailed task analysis of a number of items used by other researchers in exploring students' probability concepts. In doing this, they have provided a great service to researchers, alerting us to the many possible interpretations of a task that subjects may have and suggesting alternative interpretations of subjects' responses. Their review emphasizes the complexity of some of our empirical research tasks. "We hold the view that in chasing one normative modeling of a problem, teachers [and researchers] often neglect the full complexity of it." (Bentz & Borovcnik, in press). Bentz and Borovcnik have also provided us with an excellent discussion on how simply changing the wording of tasks may produce different results.

On the other hand, however, each of the examples of tasks that Bentz and Borovcnik have chosen to analyze has been lifted out of context from its original study, without reference to the author's original purpose or methodology. Interpretation of one particular task along with a number of related tasks within a research study is an entirely different process than the out-of-context conjecturing in Bentz and Borovcnik's review. One really cannot obtain an overall picture of subjects' probability concepts by analyzing their responses to an individual task; rather, responses to a number of related tasks must be considered in order to obtain reliable information on students' thinking (Green, 1983a; Konold, 1991; Shaughnessy, 1977). If a researcher's objective is to uncover raw, naive intuitions of probability and statistics prior to instruction, it is difficult to imagine a methodology that does not present some sort of tasks to students and then attempt to interpret their responses. It is not clear what sort of a replacement Bentz and Borovcnik have suggested for empirical, process-probing methodology. One may argue—as Bentz and Borovcnik have done—that responses obtained from an empirical methodology might not represent students' potential thinking processes, because the language was misleading, the students did not know enough about probability, or there was an alternative interpretation of the data overlooked by the researcher. However, Bentz and Borovcnik have gone too far, claiming there is "no direct evidence" for the conclusions reached in the empirically based research studies they have reviewed. This is a distortion and misrepresentation of the empirical research in the literature. Perhaps the more important consideration is not "Should we be doing empirical research in stochastics?" but rather "What research questions are we interested in, and what sort of research methodology will best help us explore those questions?"

Scholz (in press) has written a tour de force of the psychological research on probability. His review is a thorough treatment, from the early behaviorist research of Bernoulli trial learning experiments, through the developmental approaches of Piaget and Fischbein, to many of the recent cognitive models of probability thinking. Models that depict humans as conservative Bayesians (Edwards, 1968) are compared to the heuristic models of Kahneman and Tversky and to information-processing models. Scholz characterizes each of these research perspectives as a "paradigm." Each theory has dominated research at one time or another in the history of psychological research in probability; each, in turn, was found to be inadequate for explaining all the evidence from subjects' responses. Scholz finds the current dominant theory in decision-making— the heuristics approach—also to be inadequate.

In Scholz' opinion, the heuristics in the field of decision-making are "fuzzily defined and show few characteristics of the kind used in cognitive psychology." In addition, researchers in the heuristics paradigm ordinarily do not gather any knowledge of "how the situation was perceived, which meaning is given to concepts, [and] which tools are available." Scholz raises these questions from the perspective of a mathematics educator, who is interested in the subjects' background knowledge of stochastics and in how subjects interpret the tasks presented to them. Psychologists may create somewhat contrived situations that mirror real life scenarios; they also seem primarily interested in "quick, intuitive responses," which are deplored by Scholz. This sort of research can be disturbing for mathematics educators, who want to get people to think and reflect on the problem situation. For example, as a result of their research on the base-rate fallacy, particularly in subjects' interpretations of the Taxi Problem, Scholz and Bentrup (1984) found a multitude of response strategies. They found that the use of heuristics was influenced both by background knowledge and by changes in

the wording of the problems. Scholz concludes that responses to the base-rate fallacy tasks were not fundamentally based on intuitive heuristics, but, rather, on analytic reasoning.

Scholz has reminded us that the task context and the subject's content knowledge are important variables that should not be ignored. When as researchers, we conduct task-based research, it is important that the subjects know what cognitive "game" we are playing today. Do we want an instant impression? Do we want a carefully reasoned explanation for their beliefs? Do we want them to try to apply normative models they may know about? What are we asking of them, and what do we expect them to use? Researchers should clarify these issues when presenting empirical tasks to subjects.

Scholz recommends a Structure-Process-Theory (SPT) model as a replacement for a simple heuristics model of analyzing thinking in stochastics. This information-processing model would require careful task administration and protocol analysis in order to capture the full spectrum of subjects' interpretations and responses to the tasks given to them. The model Scholz proposes is influenced by cognitive psychology and based on a theoretical system approach, not a computer analogy (as are many information-processing models). The model contains the basic units of most information-processing models: a sensory intake system, a working memory, long-term storage, a central processor, and an output mechanism. There are, however, four separate units within long-term storage in Scholz' model: the knowledge base, a heuristics structure, a goal system, and an evaluative system. While pointing out the limitations of the heuristics paradigm, Scholz nevertheless attempts to incorporate heuristics as a viable part of the process of making decisions under uncertainty. It is in the interaction between the knowledge base and the use of heuristics that Scholz' model has real potential.

We think that the proposed framework for the process and structure of information processing in stochastic thinking provides a basis for an integration of the classical heuristics, and furthermore, even offers a means of conceptualizing the differential status of the various heuristics (Scholz, in press).

Scholz continues by describing how the availability and representativeness heuristics are located in the initial phases of information processing, while causal schemas are more complex, involving not only the heuristics structure but also the evaluative structure in the model. Many of the details and specific examples of how Scholz' model works are not discussed in his chapter. Scholz' model shows some promise for understanding the details of an individual's task representations, which a simple heuristics model cannot reliably claim. On the other hand, the heuristics model is useful for describing general tendencies within an entire population. Whether we can infer from a subject's response to a task that he or she is using the representativeness heuristic, that he or she is aware of it and can verbalize his or her awareness, may not be so important. What may be more important is systematic attention to protocol analysis when and if we wish to go beneath surface level heuristics in exploring someone's thought processes.

In his discussion about the didactical difficulties created by our theoretical models of probability, Steinbring (in press) claims that attempts to provide ready-made, teacher-proof curriculum materials in probability and statistics have failed. This failure resulted because such curricula are unable to cope with the dynamic diversity of problems that arise within stochastics if probability theory is presented in an algorithmic manner. As a result, the experimental, relative-frequency approach may appear in direct conflict with the set-theoretic relative-proportion model. Steinbring recommends that teaching begin with "meaningful situations which permit the forming of concepts," because any attempt to teach a logical, mathematical development of probability is bound to be circular. Steinbring is also concerned with the epistemological roots of the probability concept. In his opinion, in order to introduce probability one must already have a concept of randomness. However, in order to discuss randomness, one must previously have a concept of probability. Thus, Steinbring claims we are caught in a conceptual cycle when we teach stochastics.

In my opinion, this conceptual cycle is not just reserved for probability and statistics; it is just that teaching stochastics is more recent in the history of mathematics. In geometry, for example, in order to understand properties of objects, one must have a concept of space. Triangles have interiors and are composed of line segments, so "mathematically" we need to know the fundamental notions of point, line, and plane as well as the definition of line segment and the separation postulate in order to define what we mean by "triangle." However, in order to abstract the desired properties of triangle into a mathematical definition, one must first have a good idea of what a triangle is. The idea of a triangle came first, while the logical axioms and definitions—the mathematization—came later. In teaching geometry, we try to develop *connections* between our intuitions of space, and the eventual theoretical mathematizations that we build. Sometimes we must point out how our mathematizations and our intuitions are in conflict, as when we teach that hyperbolic geometry is a logically consistent system. So, too, in teaching stochastics, we must attempt to develop connections between our intuitions of probability and our mathematical understandings of randomness, pointing out when intuitions are at odds with theory.

In their review of the psychological and pedagogical aspects of children's probability concepts, Hawkins and Kapadia (1984) refer to at least three different ways to approach instruction in probability: a priori notions, a frequentist approach, and a subjectivist approach; they reject the first two of these, and favor the subjectivist approach. From their point of view, simply telling children that each outcome is equally probable in certain special situations does not transfer smoothly to later situations with nonuniform distributions, so an a priori approach is not appropriate. Hawkins and Kapadia claim that ultimately a frequentist approach must involve some notion of infinity and that certain situations cannot be represented with repeated trials, so they reject this approach, also. Hawkins and Kapadia favor teaching probability via subjective probability because it lies on "elegant philosophical ground."

I strongly disagree with Hawkins and Kapadia. Although it is important for us to start wherever our students are, I think it would be a pedagogical nightmare to ground our teaching of probability solely in a subjective probability approach. Our job as mathematics educators is to enable our students to work with a mathematized model of probability. Hawkins and Kapadia believe that subjectivist approaches are more intuitive, because "all children are happy to make probabilistic assessments of a single unrepeatable event" (Kapadia, 1988). If we start with this point of view and stay there too long, , we may just be reinforcing Konold's (1989a) outcome approach. We need to provide another representation of probability to replace the subjective one our students already have. A frequentist approach to probability has been bolstered these days by our ability to simulate large numbers of trials of experiments on computers and to quickly examine the effects of changing parameters in our experiments. It is even possible to simulate conditional probability tasks on computers and run large numbers of trials. We can even represent probability revision tasks without having to appeal to Bayes theorem (see the discussion on Monty's Dilemma).

It is important for us to foster a notion of probability that can be shared among our students, so they can communicate effectively with one another about stochastic situations. We develop mathematical models so that students can describe their environment in a common language. If we invest too much time solely in a subjectivist approach, we may actually widen conceptual rifts among our students that exist prior to instruction. Surely this is not what Hawkins and Kapadia have in mind.

In their conclusion, Hawkins and Kapadia state that "we have little idea about the conceptions of probability that children of various ages have, and [that which we do have] is often conflicting." My point of view is just the opposite. We have learned quite a bit about children's probability conceptions; it just isn't pretty and its not complete. According to Hawkins and Kapadia, the main questions research should attempt to answer are these: (a) What conceptions of probability do children of various ages have? (b) How might these conceptions be changed? (c) Are there optimum teaching and learning techniques in stochastics? I agree these are the fundamental questions that should drive our research in stochastics. Contrary to what Hawkins and Kapadia imply, there have in fact been studies that have attempted to answer these questions, but not enough of them. I do not believe there is one true way to teach stochastics that is waiting to be discovered. Rather, there are many starting paths to introduce probability and statistics concepts to our students. For example, the relative frequency model deals with discrete data, while the relative proportion model—especially in geometric probability—deals with continuous data. If we adopt a modeling point of view for probability and statistics, the conflicts among a classical "equally likely" approach, a relative frequency approach, or a relative proportion approach do not have to become such a hurdle for our teachers and students. Rather, we can equip our students and teachers with multiple representations of probability with which to investigate Steinbring's "meaningful situations."

Future Research in Stochastics: A Wish List

Anyone who has done research in probability and statistics is aware that there are so many areas of needed research that it is hard to know where to begin. I have chosen to focus on just a few topics which I feel are important from my own perspective as a mathematics educator. This is research that I would like to see us do over the next decade.

Development of Assessment Instruments. We need to develop some standard, reliable tools to assess our students' conceptions of probability and statistics. There are currently a great deal of stochastical tasks available for researchers to modify and choose. These tasks have been generated by both psychologists and mathematics educators. They have been administered in written response settings, clinical interviews, and sometimes both. It is known that the context, framing, and wording of these tasks can all affect the responses we obtain from students. We need to develop instruments which incorporate and build on the ideas of previous research, but which also can be used and shared by many researchers and teachers. Both paper-and-pencil instruments and structured, repeatable interview scripts are needed to investigate students' conceptions of stochastics across wide ranges of grade levels and in varieties of contexts. Depending on the age group we are dealing with, the problems we use may vary, and the interview techniques may also vary slightly; however, we need to develop more consistency across age levels than we have shown up until now.

What Are Secondary Students' Conceptions and Misconceptions? With the exception of Green's study (1983a) testing large numbers of adolescents on notions of probability, there is very little large-scale information available about how secondary school students think about chance, random events, and decisions under uncertainty. Even Green's initial study did not probe students' understanding in detail. Do secondary students use heuristics similar to those exhibited by college students? Do they resort to nonstatistical or deterministic explanations of chance phenomenon? Do their conceptions change under the influence of instruction? Most of our investigations have been done with elementary school children or with college students, resulting in an age gap in our knowledge about students' conceptions of probability and statistics at the secondary level. In some countries, like the United States, the paucity of research at the secondary level is at least partly due to the lack of probability and statistics instruction at that level. However, as the new NCTM Standards are implemented, there will be an increased emphasis in probability and statistics at the secondary level. Research efforts at this level are crucial in order to help inform teachers and curriculum developers.

Absent: Cross-Cultural Studies. Most of the psychological research on decision-making under uncertainty has been done in a very few countries—principally, in the United States, Israel, the United Kingdom, and Germany. Likewise, mathematics education research on students' conceptions of stochastics has been done predominantly in a few western countries. What are the influences of culture on conceptions of probability and statistics? Are phenomena like judgmental heuristics and misconceptions of probability just artifacts of western culture, or do they appear across many cultures? It would be interesting to see if misconceptions of stochastics and probability estimates under uncertainty vary across cultures. Eisenhart (1988) discusses the advantages of conducting research in mathematics education from an educational anthropologist point of view. This area

of ethnomathematics has yet to be investigated by researchers in stochastics. We need both large scale studies like Green's across several cultures, and small-scale cross-cultural comparison studies using in-depth interviews on decision-making and probability estimation tasks.

What Are Teachers' Conceptions of Probability and Statistics? The success of NCTM's ambitious standards recommendations will ultimately depend upon teachers. What are elementary and secondary teachers' conceptions of and attitudes toward stochastics? As researchers, what can we do to change and influence their conceptions and attitudes? We need to gather information from teachers at both the pre-service and in-service levels. Teachers will be needing more and more in-service experiences in probability and statistics as the standards are implemented. At the pre-service level, we will need to develop courses which meet stochastic misconceptions and beliefs head on, and sensitize our prospective teachers to the prevalent misconceptions they can expect to encounter in their own students. The instructional experiences we design in stochastics for teachers should be informed by our research.

Needed: Teaching Experiments—What Are the Effects of Instruction? What can teachers and researchers do to mitigate against this two-fold problem (a) a lack of conceptual knowledge in stochastics, accompanied by (b) nonnormative intuitions, beliefs, and misconceptions? I believe it is essential that teachers and researchers form investigative partnerships, in which the teacher is a co-researcher and the researcher is a co-teacher. Clinical teaching experiments that carefully document changes in student's stochastic conceptions, beliefs, and attitudes over long periods of time are needed to obtain a clearer picture of the cognitive and affective development in stochastics. What cognitive and affective outcomes are evident over time among students who are using the Quantitative Literacy Series? How do students' beliefs about probability and their estimations for the likelihood of events change throughout a year when they are constantly using experimental and computer simulations? Both Soviet-type teaching experiments and more standard long-term interventions with a strong clinical methodological component are needed at all levels of instruction. It is crucial that researchers involve teachers in future research projects, because the teachers are the ultimate key to statistical literacy in our students.

What Are the Effects of Computer Software? How can we best use computers to help change students' stochastic beliefs, conceptions, and attitudes? There have been a number of computer packages developed to take advantage of the speed, graphics, and simulation possibilities of microcomputers. The possibilities for developing interactive representations of statistical concepts are exciting, as can be seen in the ELASTIC software in the Reasoning Under Uncertainty project (Rubin et al., 1988). Computers provide us with the opportunity to create whole new learning environments for our students, but little research and evaluation has been conducted on the effects these packages have on student conceptions of probability and statistics (Biehler, 1988, in press). Curriculum development projects in stochastics—particularly those developing computer environments—need to include researchers and teachers from the very beginning, as the projects of Hancock (1988) and Konold (1988) are now doing. Computer routines can easily be changed during the developmental stages using valuable input from researchers and teachers. When the software is in its later stages, it is often too late for the developers to incorporate suggestions from teaching and research.

What Is the Role of Metacognition in Decision-Making under Uncertainty? Research in stochastics has found that misconceptions of probability are difficult to remove (at least in some of our students) despite our best efforts at instruction. Similarly, research in problem-solving has determined that instruction in problem-solving heuristics and strategies is not sufficient to improve some students' problem-solving abilities. The problem-solving researchers have begun to investigate the role of metacognition. Garofalo and Lester (1985) identify two primary aspects of metacognition: knowledge of cognition and regulation of cognition. Knowledge of cognition includes knowledge of strategies and heuristics, but also includes self knowledge, such as beliefs and attitudes. Regulation of cognition includes our monitoring and decision-making mechanisms, as we mentally step outside ourselves and reflect on the processes and progress of solving a problem. We must begin to pay explicit attention to the metacognitive aspects of thinking under uncertainty, both in our teaching and in our research in stochastics.

CONCLUSION

Our journey through research in stochastics has taken us from the historical and philosophical roots of the probability concept to current research on the use of computer software to influence students' beliefs and conceptions about probability and statistics. The future for research in stochastics looks very bright. The NCTM standards should soon provide impetus for greater attention to probability and statistics in the schools, which will raise the demand and need for in-service teacher activities in stochastics and, in turn, increase the need to obtain more research information on how students learn to think and reason in stochastics. We have seen that research in stochastics comes from both the field of cognitive psychology and from mathematics education; in the future, there will be an ever greater need for these two disciplines to combine their research efforts in stochastics. Garfield and Ahlgren (1988a, 1988b) comment on the wide diversity of research endeavors in stochastics, and suggest that cooperative research endeavors between psychologists and mathematics educators will accomplish our research goals much more effectively than the isolated efforts that we have seen so far. Shared efforts in research in stochastics have been especially hard to initiate, given the great traveling distances for researchers from all over the globe who are interested in this type of research. The work of Konold (1989a) and Scholz and Bentrup (1984) are examples of cooperative research efforts between psychologists and mathematics/statistics educators that we need to encourage.

References

Ajzen, I. (1977). Intuitive theories of events and the effects of base-rate information on prediction. *Journal of Personality and Social Psychology, 35,* 303–314.

Atkinson, D. T. (1975). A Comparison of the teaching of statistical inference by Monte Carlo and analytic methods. *Dissertation Abstracts International, 36,* 5895A.

Austin, J. D. (1974). An experimental study of effects of three instructional methods in basic probability and statistics. *Journal for Research in Mathematics Education, 5,* 146–154.

Bambrough, R. (1963). *The Philosophy of Aristotle.* New York, NY: Mentor Books.

Bar-Hillel, M. (1980). The base-rate fallacy in probability judgments. *Acta Psychologica, 44,* 211–233.

Bar-Hillel, M., & Falk, R. (1982). Some teasers concerning conditional probabilities. *Cognition, 11,* 109–122.

Barnett, V. (1988). Statistical Consultancy—A basis for teaching and research. In R. Davidson, & J. Swift (Eds.), *The Proceedings of the Second International Conference on Teaching Statistics.* Victoria B.C.: University of Victoria.

Barz, T. J. (1970). *A study of two ways of presenting probability and statistics at the college level.* Unpublished Doctoral dissertation, Columbia University.

Bentz, H. J. (1983). Stochastics teaching based on common sense. In D. R. Grey, P. Holmes, V. Barnett, & G. M. Constable (Eds.), *Proceedings of the First International Conference on Teaching Statistics* (pp. 753–765). Sheffield: University of Sheffield.

Bentz, H. J., & Borovcnik, M. G. (1985). Some remarks on empirical investigations on probability and statistics concepts. In A. Bell, B. Low, & J. Kilpatrick (Eds.), *Theory, research and practice in mathe-*

matics education. (pp. 285–292). Nottingham, UK: Shell Centre for Mathematical Education.

Bentz, H. J., & Borovcnik, M. G. (in press). Empirical research on probability concepts. In R. Kapadia & M. Borovcnik (Eds.), *Chance encounters: Probability in education. A review of research and pedagogical perspectives.* Amsterdam: Kluwer.

Beyth-Marom, R., & Dekel, S. (1983). A curriculum to improve thinking under uncertainty. *Instructional Science, 12,* 67–82.

Biehler, R. (1988) Exploratory data analysis and the secondary stochastics curriculum. In R. Davidson & J. Swift (Eds.), *The Proceedings of the Second International Conference on Teaching Statistics.* Victoria, B.C.: University of Victoria.

Biehler, R. (in press). Computers in probability education. In R. Kapadia & M. Borovcnik (Eds.), *Chance encounters: Probability in education. A review of research and pedagogical perspectives.* Amsterdam: Kluwer.

Biehler, R., Rach, W., & Winkelmann, B. (1988). *Computers and mathematics teaching: the German situation and reviews of international software.* Occasional paper #103. Institute für Didaktik der Mathematik. Bielefeld, FRG: University of Bielefeld.

Borovcnik, M. (1983). Case studies for an adequate understanding and interpretation of results by statistical inference. In D. R. Grey, P. Holmes, V. Barnett & G. M. Constable (Eds.). *Proceedings of the First International Conference on Teaching Statistics* (pp. 341–353). Sheffield: University of Sheffield.

Borovcnik, M. (1984, August). *Revising probabilities according to new information.* Paper presented at the fifth meeting of the International Congress on Mathematics Education, Adelaide.

Borovcnik, M. (1986). *On "representativeness"—a fundamental statistical strategy.* Unpublished paper, University of Klagenfurt.

Borovcnik, M. (1988). Revising probabilities according to new information: A fundamental stochastic intuition. In R. Davidson, & J. Swift (Eds.), *The Proceedings of the Second International Conference on Teaching Statistics*. Victoria, B.C.: University of Victoria.

Carpenter, T. P. (1989). Teaching as problem solving. In R. Charles & E. Silver (Eds.), *The teaching and assessing of mathematical problem solving* (pp. 187–202). Reston, VA: National Council of Teachers of Mathematics.

Carpenter, T. P., Corbitt, M. K., Kepner, H. S., Lindquist, M. M., & Reys, R. E. (1981). What are the chances of your students knowing probability? *The Mathematics Teacher, 74,* 342–344.

Charles, R. I. (1989). Teacher education and mathematical problem solving: Some issues and directions. In R. Charles & E. Silver (Eds.), *The teaching and assessing of mathematical problem solving* (pp. 259–272). Reston, VA: National Council of Teachers of Mathematics.

Clement, J. (1987). Overcoming students' misconceptions in physics: The role of anchoring intuitions and analogical validity. *Proceedings of the Second International Seminar, Misconceptions and Educational Strategies in Science and Mathematics*. Ithaca, N.Y.: Cornell University.

Cobb, P. (1989). A double-edged sword. [Review of *Intuition in science and mathematics*.] *Journal for Research in Mathematics Education, 20,* 213–218.

Cohen, J. (1957). Subjective probability. *Scientific American, 197,* 128–138.

Cohen, J. (1960). *Chance, skill, and luck: The psychology of guessing and gambling*. Baltimore, MD: Penguin Books.

Copleston, F. (1962). *A history of philosophy: Greece and Rome*, Garden City, NY: Image Books.

Copleston, F. (1963). *A history of philosophy: Modern philosophy from Decartes to Leibnitz*. Garden City, NY: Image Books.

Davidson, R., and Swift, J. (Eds.). (1988). *The Proceedings of the Second International Conference on Teaching Statistics*. Victoria, BC: University of Victoria.

DeBerea, R. (1988). Statistics for college-bound students: Are the secondary schools responding? *School Science and Mathematics, 88,* 200–209.

delMas, R. C., & Bart, W. M. (1987, April). *The role of an evaluation exercise in the resolution of misconceptions of probability*. Paper presented at the Annual meeting of the American Educational Research Association.

Doherty, J. (1965). Level of four concepts of probability possessed by children of the fourth, fifth, and sixth grade before formal instruction. *Dissertation Abstracts, 27,* 1703A.

Dreyfus, T. (1986, July). *Cognitive effects of microworlds: Learning about probability*. Paper presented at the Second International Conference on LOGO and Mathematics Education, London.

Eddy, D. M. (1982). Probabilistic reasoning in clinical medicine: Problems and opportunities. In D. Kahneman, P. Slovic, & A. Tversky (Eds.), *Judgment under uncertainty: Heuristics and biases* (pp. 249–267). Cambridge, U.K.: Cambridge University Press.

Edwards, W. (1968). Conservatism in human information processing. In B. Kleinmutz (Ed.), *Formal representations of human judgment*. New York, NY: Wiley.

Eisenhart, M. (1988). The ethnographic tradition and mathematics education research. *Journal for Research in Mathematics Education, 19,* 99–114. M

Engel, A. (1975). *L'Enseignement des probabilités et de la statistique* (Vols. 1–2). Paris: Cendu.

Falk, R. (1979). Revision of probabilities and the time axis. *In Proceedings of the Third International Conference for the Psychology of Mathematics Education* (pp.64–66). Warwick, U.K.

Falk, R. (1981). The perception of randomness. In *Proceedings of the Fifth Conference of the International Group for the Psychology of Mathematics Education* (pp. 222–229). Grenoble, France.

Falk, R. (1982). Do men have more sisters than women? *Teaching Statistics, 4,* 60–62.

Falk, R. (1983). Experimental models for resolving probabilistic ambiguities. In *Proceedings of the Seventh International Conference for the Psychology of Mathematics Education* (p. 319–325). Tel-Aviv, Israel.

Falk, R. (1986). Misconceptions of statistical significance. *Journal of Structural Learning, 9,* 83–96.

Falk, R. (1988). Conditional probabilities: Insights and difficulties. In R. Davidson & J. Swift (Eds.), *The Proceedings of the Second International Conference on Teaching Statistics*. Victoria, B.C.: University of Victoria.

Falk, R. (1989). The judgment of coincidences: Mine versus yours. *American Journal of Psychology, 102,* 477–493.

Falk, R. (in press). Inference under uncertainty via conditional probabilities. In *Studies of mathematics education: Vol 7. Teaching statistics in schools*. Paris: UNESCO .

Falk, R. & Bar-Hillel, M. (1983). Probabilistic dependence between events. *Two-Year-College Mathematics Journal, 14,* 240–247.

Fischbein, E. (1975). *The intuitive sources of probabilistic thinking in children*. Dordrecht, The Netherlands: Reidel.

Fischbein, E. (1987). *Intuition in science and mathematics*. Dordrecht, The Netherlands: Reidel.

Fischbein, E., & Gazit, A. (1984). Does the teaching of probability improve probabilistic intuitions? *Educational Studies in Mathematics, 15,* 1–24.

Fischbein, E., Pampu, I., & Manzat, I. (1970a). Comparison of ratios and the chance concept in children. *Child Development, 41,* 377–389.

Fischbein, E., Pampu, I., & Manzat, I. (1970b). Effects of age and instruction on combinatory ability in children. *The British Journal of Educational Psychology, 40,* 261–270.

Fong, G. T., Krantz, D. H., & Nisbett, R. E. (1986). The effects of statistics training on thinking about everyday problems. *Cognitive Psychology, 18,* 253–292.

Garfield, J. B. (1988). *Obstacles to effective teaching of probability and statistics*. Paper presented at the Research Presession of the National Council of Teachers of Mathematics 66th Annual Meeting, Chicago.

Garfield, J. B., & Ahlgren, A. (1988a). Difficulties in learning basic concepts in probability and statistics: Implications for research. *Journal for Research in Mathematics Education, 19,* 44–63.

Garfield, J. B., & Ahlgren, A. (1988b). Difficulties in learning probability and statistics. In R. Davidson & J. Swift (Eds.), *The Proceedings of the Second International Conference on Teaching Statistics*. Victoria B.C.: University of Victoria.

Garfield, J. B., & delMas, R. (1989). Reasoning about chance events: Assessing and changing students' conception of probability. In C. Maher, G. Goldin, and B. Davis (Eds.), *The Proceedings of the Eleventh Annual Meeting of the North American Chapter of the International Group for the Psychology of Mathematics Education* (Vol II, pp. 189–195). Rutgers, NJ: Rutgers University Press.

Garofalo, J., & Lester, F. (1985). Metacognition, cognitive monitoring, and mathematical performance. *Journal for Research in Mathematics Education, 16,* 163–176.

Geeslin, W. (1974). *An analysis of content structure and cognitive structure in context of a probability unit*. (ERIC Document Reproduction Service NO. ED 090 036).

Gipson, J. H. (1971). Teaching probability in elementary school: An experimental study. *Dissertation Abstracts International, 32,* 4325A.

Gnanadesikan, M., Scheaffer, R. L., & Swift, J. (1987). *The art and techniques of simulation*. Palo Alto, CA: Dale Seymour.

Godino, J. D., Batanero, Ma. C., & Canizares, Ma. J. (1987). *Azar y Probabilidad. Fundamentos Didacticos y Propuestas Curriculares*. Madrid: Editorial Sintesis, S.A.

Goldin, G., & McClintock, E. (1984). *Task variables in mathematical problem solving*. Philadelphia, PA: Franklin Press.

Green, D. R. (1979). The chance and probability concepts project. *Teaching Statistics, 1*(3), 66–71.

Green, D. R. (1983a). A survey of probability concepts in 3,000 pupils aged 11–16 years. In D. R. Grey, P. Holmes, V. Barnett, & G. M. Constable (Eds.), *Proceedings of the First International Conference on Teaching Statistics,* (pp. 766–783) . Sheffield, UK: Teaching Statistics Trust.

Green, D. R. (1983b). School pupils' probability concepts. *Teaching Statistics, 5*(2), 34–42.

Green, D. R. (1987). Probability concepts: Putting research into practice. *Teaching Statistics, 9*(1), 8–14.

Green, D. R. (1988). Children's understanding of randomness: Report of a survey of 1600 children aged 7–11 years. In R. Davidson & J. Swift (Eds.), *The Proceedings of the Second International Conference on Teaching Statistics.* Victoria, B.C.: University of Victoria.

Hacking, I. (1975). *The emergence of probability.* Cambridge: Cambridge University Press.

Hake, R. R. (1987). Promoting student crossover to the Newtonian world. *American Journal of Physics, 55,* 878–884.

Hancock, C. (1988). *Hands on data: Direct manipulation environments for data organization and analysis.* Proposal funded by the National Science Foundation. Technical Education Research Centers, Inc. Cambridge, Massachusetts.

Hawkins, A., & Kapadia, R. (1984). Children's conceptions of probability: A psychological and pedagogical review. *Educational Studies in Mathematics, 15,* 349–377.

Heider, F. (1958). *The psychology of interpersonal relations.* New York, NY: Wiley.

Huygens, C. (1657). Ratiociniis in aleae ludo. In F. van Schooten (Ed.), *Exercitionum Mathematicorum.* Amsterdam: North Holland.

Jones, G. A. (1974). *The performance of first, second, and third grade children on five concepts of probability and the effects of grade, IQ, and embodiments on their performances.* Unpublished Doctoral dissertation, Indiana University.

Kahneman, D., & Tversky, A. (1972). Subjective probability: A judgment of representativeness. *Cognitive Psychology, 3,* 430–454.

Kahneman, D., & Tversky, A. (1973a). On the psychology of prediction. *Psychological Review, 80,* 237–251.

Kahneman, D., & Tversky, A. (1973b). Availability: A heuristic for judging frequency and probability. *Cognitive Psychology, 5,* 207–232.

Kahneman, D., Slovic, P., & Tversky, A. (1982). *Judgment under uncertainty: Heuristics and biases.* Cambridge: Cambridge University Press.

Kapadia, R. (1988). Didactical Phenomenology of Probability. In R. Davidson & J. Swift (Eds.), *The Proceedings of the Second International Conference on Teaching Statistics.* Victoria, B.C.: University of Victoria.

Kelly, I. W. & Zwiers, F. W. (1988). Mutually exclusive and independence: Unravelling basic misconceptions in probability theory. In R. Davidson & J. Swift (Eds.), *The Proceedings of the Second International Conference on Teaching Statistics.* Victoria B.C.: University of Victoria.

Kilpatrick, J., & Stannic, G. M. (1989). Historical perspectives on problem solving in the mathematics curriculum. In R. Charles & E. Silver (Eds.), *The teaching and assessing of mathematical problem solving* (pp. 1–22). Reston, VA: National Council of Teachers of Mathematics.

Kipp, W. E. (1975). An investigation of the effects of integrating topics of elementary algebra with those of elementary probability within a unit of mathematics prepared for college basic mathematics students. *Dissertation Abstracts International, 35,* 7616A.

Konold, C. (1983). Conceptions about probability: Reality between a rock and a hard place. (Doctoral disseration, University of Massachusetts, 1983). *Dissertation Abstracts International, 43,* 4179B.

Konold, C. (1989a) Informal conceptions of probability. *Cognition and Instruction, 6,* 59–98.

Konold, C. (1989b). An outbreak of belief in independence? In C. Maher, G. Goldin, & B. Davis (Eds.), *The Proceedings of the Eleventh Annual Meeting of the North American Chapter of the International Group for the Psychology of Mathematics Education* (Vol. 2, pp. 203–209). Rutgers, NJ: Rutgers University Press.

Konold, C. (1991) Understanding students' beliefs about probability. In E. von Glasersfeld (Ed.), *Radical Constructivism in Mathematics Education* (pp. 139–156). Holland: Kluwer.

Konold, C., Sutherland, M., & Lochhead, J. (1988). *A computer based curriculum for probability and statistics.* Proposal funded by the National Science Foundation. Scientific Reasoning Institute, University of Massachusetts, Amherst.

Kuhn, T. (1962). *The structure of scientific revolutions.* Chicago, IL: University of Chicago Press.

Landewehr, J. (1989). A reaction to alternative conceptions of probability. In J. Garfield (Chair), *Alternative Conceptions of Probability: Implications for Research, Teaching, and Curriculum.* Symposium conducted at the eleventh annual meeting of the North American Chapter of the International Group for the Psychology of Mathematics Education, Rutgers, New Jersey.

Landewehr, J., & Watkins, A. E. (1986). *Exploring data.* Palo Alto, CA: Dale Seymour.

Landewehr, J., Watkins, A. E., & Swift, J. (1987). *Exploring surveys: Information from samples.* Palo Alto, CA: Dale Seymour.

Leake, L. (1962). The status of three concepts of probability in children of the seventh, eighth, and ninth grades. *Dissertation Abstracts, 23,* 2010.

Leffin, W. W. (1971). *A study of three concepts of probability possessed by children in grades four–seven.* (ERIC Document Reproduction Service No. ED 070 657)

Lester, F. K. (1989). Reflections about mathematical problem-solving research. In R. Charles & E. Silver (Eds.), *The teaching and assessing of mathematical problem solving* (pp. 115–124). Reston, VA: National Council of Teachers of Mathematics.

Marshall, S. P. (1989). Assessing problem solving: A short-term remedy and a long-term solution. In R. Charles & E. Silver (Eds.), *The teaching and assessing of mathematical problem solving* (pp. 159–177). Reston, VA: National Council of Teachers of Mathematics.

McLeod, G. K. (1972). An experiment in the teaching of selected concepts of probability to elementary school children. *Dissertation Abstracts International, 32,* 1359A.

McKinley, J. E. (1960). Relationship between selected factors and achievement in a unit on probability and statistics for twelfth grade students. *Dissertation Abstracts, 21,* 561.

Mevarech, Z. (1983). A deep structure model of students' statistical misconceptions. *Educational Studies in Mathematics, 14,* 415–429.

Moyer, R. E. (1974). Effect of a unit on probability on ninth grade general mathematics students' arithmetic computation, reasoning and attitudes. *Dissertation Abstracts International, 35,* 4137A.

Mullenex, J. L. (1968). A study of the understanding of probability concepts by selected elementary school children. *Dissertation Abstracts, 29,* 3920A.

National Council of Teachers of Mathematics (1980). *An Agenda for Action.* Reston, VA: Author.

National Council of Teachers of Mathematics (1989). *Curriculum and evaluation standards for school mathematics.* Reston, VA: Author.

Newman, C. M., Obremski, T. E., & Scheaffer, R. L. (1987). *Exploring probability.* Palo Alto: Dale Seymour.

Nisbett, R. E., & Borgida, E. (1975). Attribution and the psychology of prediction. *Journal of Personality and Social Psychology, 32,* 932–943.

Nisbett, R. E., Fong, G. T., Lehman, D. R., & Cheng, P. W. (1987). Teaching reasoning. *Science, 198,* 625–631.

Nisbett, R. E., Krantz, D. H., Jepson, C., & Fong, G. T. (1982). Improving inductive inference. In D. Kahneman, P. Slovic, and A. Tversky (Eds.), *Judgement under uncertainty: Heuristics and biases* (pp. 445–459). Cambridge, U.K.: Cambridge University Press.

Nisbett, R. E., Krantz, D., Jepson, C., & Kunda, Z. (1983). The use of statistical heuristics in everyday reasoning. *Psychological Review, 90,* 339–363.

Nisbett, R. E. & Ross, L. (1980). *Human inference: Strategies and shortcomings of social judgment.* Englewood Cliffs, NJ: Prentice Hall.

Paulos, J. A. (1989) *Innumeracy: Mathematical illiteracy and its consequences.* New York: Hill & Wang Publishing.

Phillips, E., Lappan, G., Winter, M. J., & Fitzgerald, W. (1986). *Probability.* Menlo Park, CA: Addison-Wesley.

Piaget, J., & Inhelder, B. (1951). *La Genesse de l'Idée de Hazard chez l'Enfant.* Paris: Presse Universitaire de France.

Piaget, J., & Inhelder, B. (1975). *The origin of the idea of chance in children.* London: Routledge & Kegan Paul.

Piazza, T. (1988). Teaching statistics through data analysis. In R. Davidson & J. Swift (Eds.), *The Proceedings of the Second International Conference on Teaching Statistics.* Victoria, B.C.: University of Victoria.

Pollatsek, A., Konold, C., Well, A. D., & Lima, S. (1984). Beliefs underlying random sampling. *Cognition and Instruction, 12,* 395–401.

Pollatsek, A., Lima, S., & Well, A. D. (1981). Concept or computation: Students' understanding of the mean. *Educational Studies in Mathematics, 12,* 191–204.

Pollatsek, A., Well, A. D., Konold, C., & Hardiman, P. (1987). Understanding conditional probabilities. *Organizational Behavior and Human Decision Processes, 40,* 255–269.

Rade, L. (1983). Stochastics at the school level in the age of the computer. In D. R. Grey, P. Holmes, V. Barnett, & G. M. Constable (Eds.), *Proceedings of the First International Conference on Teaching Statistics* (pp. 19–33). Sheffield, U.K.: Teaching Statistics Trust.

Ross, L. (1975). The intuitive psychologist and his shortcomings: Distortions in the attribution process. In L. Berkowitz (Ed.), *Advances in experimental social psychology.* (pp. 174–177). New York: Academic Press.

Ross, L., & Anderson, C. A. (1982). Shortcomings in the attribution process: On the origins and maintenance of erroneous social assessments. In D. Kahneman, P. Slovic, & A. Tversky (Eds.), *Judgement under uncertainty: Heuristics and biases* (pp. 129–152). Cambridge, U.K.: Cambridge University Press.

Rubin, A. V., & Roseberry, A. S. (1988, August). *Teachers' misunderstandings in statistical reasoning: Evidence from a field test of innovative materials.* Paper presented at the International Statistics Round Table Conference Training Teachers to Teach Statistics, Budapest.

Rubin, A. V., Roseberry, A. S., & Bruce, B. (1988). *ELASTIC and reasoning under uncertainty* (Research report No. 6851). Boston: BBN Systems and Technologies Corporation.

Schoenfeld, A. H. (1989). Problem solving in context(s). In R. Charles & E. Silver (Eds.), *The teaching and assessing of mathematical problem solving* (pp. 82–95). Reston, VA: National Council of Teachers of Mathematics.

Scholz, R. W. (in press). Psychological research on the probability concept and its acquisition. In R. Kapadia & M. Borovcnik (Eds.), *Chance encounters: Probability in education. A review of research and pedagogical perspectives.* Amsterdam: Kluwer.

Scholz, R. W., & Bentrup, A. (1984). *Reconsidering the base rate fallacy: New data and conceptual issues.* (The Stochastic Thinking Project Report No. 1). Bielefeld, FRG: University of Bielefeld.

Schrage, G. (1983). (Mis-) Interpretation of stochastic models. In R. Scholz (Ed.), *Decision making under uncertainty* (pp. 351–361). Amsterdam: North-Holland.

Shaughnessy, J. M. (1976). A clinical investigation of college students' reliance upon the heuristics of availability and representativeness in estimating the likelihood of probabilistic events. *Dissertation Abstracts International, 37,* 5662A.

Shaughnessy, J. M. (1977). Misconceptions of probability: An experiment with a small-group, activity-based, model building approach to introductory probability at the college level. *Educational Studies in Mathematics, 8,* 285–316.

Shaughnessy, J. M. (1981). Misconceptions of probability: From systematic errors to systematic experiments and decisions. In A. Schulte (Ed.), *Teaching Statistics and Probability* (Yearbook of the National Council of Teachers of Mathematics, pp. 90–100). Reston, VA: NCTM.

Shaughnessy, J. M. (1983a). Misconceptions of probability, systematic and otherwise: Teaching probability and statistics so as to overcome some misconceptions. In D. R. Grey, P. Holmes, V. Barnett, & G. M. Constable (Eds.) *Proceedings of the First International Conference on Teaching Statistics* (pp. 784–801). Sheffield, U.K.: Teaching Statistics Trust.

Shaughnessy, J. M. (1983b). The psychology of inference and the teaching of probability and statistics: Two sides of the same coin? In R. Scholz (Ed.), *Decision making under uncertainty* (pp. 325–350). Amsterdam: North-Holland.

Shaughnessy, J. M. (1985). Problem solving derailers: The influence of misconceptions on problem-solving performance. In E. Silver (Ed.), *Teaching and learning mathematical problem-solving: Multiple research perspectives* (pp. 199–214). Hillsdale, NJ: Lawrence Erlbaum.

Shaughnessy, J. M., & Dick, T. (1991). Monty's dilemma: Should you stick or switch? *The Mathematics Teacher, 84,* 252–256.

Shepler, J. (1970). Parts of a systems approach to the development of a unit in probability and statistics for the elementary school. *Journal for Research in Mathematics Education, 1,* 197–205.

Shepler, J., & Romberg, T. (1973). Retention of probability concepts: A pilot study into the effects of mastery learning with sixth grade students. *Journal for Research in Mathematics Education, 4,* 26–32.

Shulte, A. P. (1968). Effect of a unit in probability and statistics on students and teacher of a ninth grade general mathematics class. *Dissertation Abstracts, 28,* 4962A.

Shulte, A. P. (Ed.). (1981). *Teaching statistics and probability.* (Yearbook of the National Council of Teachers of Mathematics). Reston, VA: NCTM.

Smock, C., & Belovicz, G. (1968). Understanding of the concept of probability by junior high school children. Final Report. (ERIC Document Reproduction Service No. ED 0202147).

Steinbring, H. (in press). The theoretical nature of probability and how to cope with it in the classroom. In R. Kapadia & M. Borovcnik (Eds.), *Chance encounters: Probability in education. A review of research and pedagogical perspectives.* Amsterdam: Kluwer.

Taylor, S. E. (1982). The availability bias in social perception and interaction. In D. Kahneman, P. Slovic, & A. Tversky (Eds.), *Judgement under uncertainty: Heuristics and biases* (pp. 190–200). Cambridge, U.K.: Cambridge University Press.

Thompson, A. G. (1989). Learning to teach mathematical problem solving: Changes in teachers' conceptions and beliefs. In R. Charles & E. Silver (Eds.), *The teaching and assessing of mathematical problem solving* (pp. 232–243). Reston, VA: National Council of Teachers of Mathematics.

Travers, K. J., Stout, W. F., Swift, J. H., & Sextro, J. (1985). *Using statistics.* Reading, MA: Addison-Wesley.

Tversky, A., & Kahneman, D. (1971). Belief in the law of small numbers. *Psychological Bulletin, 76,* 105–110.

Tversky, A., & Kahneman, D. (1974). Judgment under uncertainty: Heuristics and biases. *Science, 185,* 1124–1131.

Tversky, A., & Kahneman, D. (1980). Causal schemas in judgment under uncertainty. In M. Fischbein (Ed.), *Progress in social psychology.* Hillsdale, NJ: Lawrence Erlbaum.

Tversky, A., & Kahneman, D. (1982a). Judgments of and by representativeness. In D. Kahneman, P. Slovic, & A. Tversky (Eds.), *Judgement under uncertainty: Heuristics and biases* (pp. 84–100). Cambridge, U.K.: Cambridge University Press.

Tversky, A., & Kahneman, D. (1982b). Evidential impact of base rates. In D. Kahneman, P. Slovic, & A. Tversky (Eds.), *Judgement under uncertainty: Heuristics and biases* (pp. 153–162). Cambridge, U.K.: Cambridge University Press.

Tversky, A., & Kahneman, D. (1983). Extensional versus intuitive reasoning: The conjunction fallacy in probability judgment. *Psychological Review, 90*(4), 293–315.

Weissglass, J., Thies, N., & Finzer, W. (1986). *Hands-on statistics.* Belmont, CA: Wadsworth.

Well, A. D., Pollatsek, A., & Boyce, S. (1990). Understanding the effects of sample size on the mean. *Organizational Behavior and Human Decision Processes, 47,* 289–312.

Well, A. D., Pollatsek, A., & Konold, C. (1983). *Probability estimation and the use and neglect of base-rate information.* (Unpublished manuscript, The University of Massachusetts, Amherst).

White, C. W. (1974). *A study of the ability of first and eighth grade students to learn basic concepts of probability and the relationship between achievement in probability and selected factors. Dissertation Abstracts International, 35,* 1969A.

▪20▪

THE TRANSITION TO ADVANCED MATHEMATICAL THINKING: FUNCTIONS, LIMITS, INFINITY, AND PROOF

David Tall
UNIVERSITY OF WARWICK

Advanced mathematical thinking—as evidenced by publications in research journals—is characterized by two important components: precise mathematical definitions (including the statement of axioms in axiomatic theories) and logical deductions of theorems based upon them. However, the printed word is but the tip of the iceberg, the record of the final "precising phase" that is quite distinct from the creative phases of mathematical thinking in which inspirations and false turns play their part.

A major focus in mathematical education at the higher levels is not only to initiate the learner into the complete world of the professional mathematician in terms of the rigor required, but also to provide the experience on which the concepts are founded. Traditionally this has been done through a gentle introduction to the mathematical concepts and the process of mathematical proof in school before progressing to present mathematics in a more formally organized and logical framework at college and university.

The move to more advanced mathematical thinking involves a difficult transition, from a position where concepts have an intuitive basis founded on experience, to one where they are specified by formal definitions and their properties reconstructed through logical deductions. During this transition (and long after) there will exist simultaneously in the mind earlier experiences and their properties, together with the growing body of deductive knowledge. Empirical research has shown that this produces a wide variety of cognitive conflict which can act as an obstacle to learning.

In this chapter we will look at the results of research into the conceptualization of several advanced concepts, including the notions of a function, of limits and infinity, and of the process of mathematical proof, particularly during the transition phase from the later years of school to college and university. But first we must linger a little and consider the nature of our own perceptions of mathematical concepts, for even those of professional mathematicians contain idiosyncrasies dependent on personal experience.

CREASES IN THE MIND

"The human mind," wrote Antoine Lavoisier, the French chemist guillotined during the French Revolution, "gets creased into a way of seeing things." One might add that the evolving corporate mind suffers no less, since it perceives by indoctrination, from generation to generation. (Adrian Desmond, *The Hot-Blooded Dinosaurs*, p. 128)

As we look back at the historical development of mathematics, we see that successive generations develop their own corporate perception of mathematical ideas, based on mutual agreement over important concepts. The pre-Pythagorean Greeks believed that all numbers were rational, until the Pythagorean theorem revealed that the square root of two is not rational. Aristotelian dynamics suggested that the speed of a moving body is proportional to the force applied, until Newton's laws proposed that it is acceleration, not speed, that is proportional to force. For

The author gratefully acknowledges the helpful comments provided by John Harvey, University of Wisconsin, and James Schultz, Ohio State University.

two millenia, Euclidean geometry was regarded as the pinnacle of deductive logic, until 19th century mathematicians realized that there were theorems that depended on implicit assumptions (such as the fact that the diagonals of a rhombus lie inside the figure) that were not logical deductions from the axioms.

It would be a mistake to assume that at last we have "got it right" and that this generation is free of the internal conflicts and confusions of the past. On the contrary, we have our own share of corporate creases of the mind. (See, for example, Sierpińska, 1985a, 1985b, 1987.) Many of the creases that we purport to see in students are actually present in ourselves and have been passed down in varyingly modified forms from generation to generation.

For example, the idea that a function $y = f(x)$ is single-valued has become part of our mathematical culture, and we may find it strange to see students asserting that a circle $x^2 + y^2 = 1$ can be a function. Yet the term "implicit function" continues to be used in textbooks to describe such an expression. I (to my eternal shame) find that I published a computer program called the "implicit function plotter" which will draw, among other things, the graph of $x^2 + y^2 = 1$. Likewise I find myself considering the draft of a new curriculum for the 16–19 age range in Britain which says of this equation: "Strictly speaking, y is not a function of x because there is not a unique value of y for each value of x, but we might think of it as a 'double-valued' function from x to y." What are students to think? Can any of us, with hand on our heart, state that we have never indulged in any vagaries of this kind? Let him who is without sin cast the first stone.

CONCEPT DEFINITION AND CONCEPT IMAGE

What is a good definition? For the philosopher or the scientist, it is a definition which applies to all the objects to be defined, and applies only to them; it is that which satisfies the rules of logic. But in education it is not that; it is one that can be understood by the pupils. (Poincaré, 1914, p. 117)

The "new mathematics" of the 1960s was a valiant attempt to create an approach based on clear definitions of mathematical concepts, presented in a way (it was hoped) that students would understand. But it failed to achieve all its high ideals. The problem is that the individual's method of thinking about mathematical concepts depends on more than just the form of words used in a definition.

Within mathematical activity, mathematical notions are not only used according to their formal definition, but also through mental representations which may differ for different people. These "individual models" are elaborated from "spontaneous models" (models which pre-exist, before the learning of the mathematical notion and which originate, for example, in daily experience) interfering with the mathematical definition. We notice that the notion of limit denotes very often a bound you cannot cross over, which can, or cannot, be approached. It is sometimes viewed as reachable, sometimes as unreachable. (Cornu, 1981)

Thus the experience of pupils prior to meeting formal definitions profoundly affects the way in which they form men-

tal representations of those concepts. During the late 1970s and early 1980s many authors noted the mismatch between the concepts as formulated and conceived by formal mathematicians, and those as interpreted by students. For example, difficulties were noted in the understanding of the limiting process as secants tend to tangents (Orton, 1977), the meaning of infinite decimals (Tall, 1977), geometrical concepts (Vinner and Hershkowitz, 1980), the notion of function (Vinner, 1983), limits and continuity (Sierpińska, 1987; Tall & Vinner 1981), the meaning of the differential (Artigue, 1986), convergence of sequences (Robert, 1982), limits of functions (Ervynck, 1981), the tangent (Tall, 1987; Vinner, 1983), infinite series (Davis, 1982), infinite expressions (Borasi, 1985), the intuition of infinity (Fischbein, Tirosh, & Hess, 1979), and so on.

To highlight the role played by the individual's conceptual structure, the terms "concept image" and "concept definition" were introduced in Vinner and Hershkowitz (1980) and later described as follows:

We shall use the term *concept image* to describe the total cognitive structure that is associated with the concept, which includes all the mental pictures and associated properties and processes.... As the concept image develops it need not be coherent at all times.... We will refer to the portion of the concept image which is activated at a particular time the *evoked concept image*. At different times, seemingly conflicting images may be evoked. Only when conflicting aspects are evoked *simultaneously* need there be any actual sense of conflict or confusion. (Tall & Vinner, 1981, p. 152)

On the other hand, "The *concept definition* [is] a form of words used to specify that concept" (Tall & Vinner, 1981, p. 152).

The consideration of conflicts in thinking is widespread in the literature:

New knowledge often contradicts the old, and effective learning requires strategies to deal with such conflict. Sometimes the conflicting pieces of knowledge can be reconciled, sometimes one or the other must be abandoned, and sometimes the two can both be "kept around" if safely maintained in separate mental compartments. (Papert, 1980, p. 121)

In general, learning a new idea does not obliterate an earlier idea. When faced with a question or task the student now has *two* ideas, and may retrieve the new one or may retrieve the old one. What is at stake is not the possession or non-possession of the new idea; but rather the *selection* (often unconscious) of which one to retrieve. Combinations of the two ideas are also possible, often with strikingly nonsensical results. (Davis & Vinner, 1986, p. 284)

This is particularly applicable to the transition to advanced mathematical thinking when the mind simultaneously has concept images based on earlier experiences that interact with new ideas based on definitions and deductions. The very idea of *defining* a concept in a sentence, as opposed to *describing* it, is at first very difficult to comprehend, particularly when there are words in the definition that are not defined. It is impossible to make a beginning without making some assumptions, and these are based upon the individual's concept image, not on any logically formulated concept definition.

MATHEMATICAL FOUNDATIONS AND COGNITIVE ROOTS

Burrow a while and build, broad on the roots of things.
(Robert Browning, 1864, *Abt Vogler*)

In building a curriculum it is natural to attempt to start from simple ideas and move steadily to more complex concepts as the student grows in experience. What better foundations to build upon than the definitions that have evolved over many generations? The problem is that these definitions are both subtle and generative, while the experiences of students are based on the evident and particular, with the result that the generative quality of the definitions is obscured by the students' specific concept images. For example, a function may be defined as a process which assigns to each element in one set (the domain) a unique element in another (the range). It is not possible to give the full range of possibilities embedded in this definition at the outset—the sets involved may be sets of numbers, or points in n-dimensional space, or geometrical shapes, or matrices, or any other type of object, including other functions, and the method of assignment might be through a formula, an iterative or recursive process, a geometrical transformation, a list of values, or any serendipitous combination one desires, provided that it satisfies the criterion of assigning elements uniquely.

When students are first confronted with mathematical definitions, it is almost inevitable that they will meet only a restricted range of possibilities; this colors their concept images in a way that will cause future cognitive conflict.

Rather than deal initially with formal definitions that contain elements unfamiliar to the learner, it is preferable to attempt to find an approach that builds on concepts that have the dual role of being familiar to the students and providing the basis for later mathematical development. Such a concept I call a *cognitive root* (Tall, 1989). Cognitive roots are not easy to find—they require a combination of empirical research (to find out what is appropriate to the student at the current stage of development) and mathematical knowledge (to be certain of the long-term mathematical relevance). A cognitive root is different from a mathematical foundation; whereas a mathematical foundation is an appropriate starting point for a logical development of the subject, a cognitive root is more appropriate for curriculum development.

For example, the limit concept is a good example of a mathematical foundation—honed and made precise over the centuries by the combined efforts of many great mathematicians. But it proves to be difficult for students to use as a basis of their thinking and may not be a sound cognitive root for the beginning stages of calculus. On the other hand, the idea that certain graphs look less curved as they are more highly magnified is intuitively appealing and can be discovered by any student playing with a graph plotter. The fact that this can grow into the formal theory of differential manifolds that are locally like n-dimensional space suggests that "local straightness" may prove to be a suitable cognitive root for calculus. The case for local straightness is enhanced when it is realized that the solving of a (first-order) differential equation is essentially the reverse problem: to find a (locally straight) function that has a given gradient. It is possible, with software, to build a picture of an approximate solution enactively just by placing short line segments of the appropriate gradient end to end.

THE FUNCTION CONCEPT

The keynote of Western culture is the function concept, a notion not even remotely hinted at by any earlier culture. And the function concept is anything but an extension or elaboration of previous number concepts—it is rather a complete emancipation from such notions. (Schaaf, 1930, p. 500)

The function concept, according to Kleiner (1989), "goes back 4,000 years; 3,700 of these consist of anticipations" (p. 282). Its evolution has led to a complex network of conceptions: the geometric image of a graph, the algebraic expression as a formula, the relationship between dependent and independent variables, an input-output machine allowing more general relationships, and the modern set-theoretic definition (see, for example, Buck, 1970).

In the "new math," there was a valiant attempt to build the function concept from a formal definition in terms of the Cartesian product of sets A and B:

Let A and B be sets, and let $A \times B$ denote the Cartesian product of A and B. A subset f of $A \times B$ is a function if, whenever (x_1, y_1) and (x_2, y_2) are elements of f, and $x_1 = x_2$, then $y_1 = y_2$.

However, there is much empirical evidence to show that, though this definition is an excellent mathematical foundation, it may not be a good cognitive root. The "emancipation" from previous concepts suggested so eloquently by Schaaf over 60 years ago is mirrored in the total cognitive reconstruction that is necessary to use the new set-theoretic definition in place of earlier process-related notions. It is a reconstruction that students seem to find extremely difficult.

Malik (1980) highlighted the manner in which this definition represents a very different frame of thought from that experienced in traditional calculus emphasising the rule-based relationship between a dependent and independent variable.

Sierpińska focused on the latter use of the function concept and asserted:

The most fundamental conception of a function is that of a relationship between variable magnitudes. If this is not developed, representations such as equations and graphs lose their meaning and become isolated from one another. . . . Introducing functions to young students by their elaborate modern definition is a didactical error—an antididactical inversion. (Sierpińska, 1988, p. 572)

Empirical research shows that, even when students are given such a formal definition, their overwhelming experience from examples of functions with implicit common properties causes them to develop a personal concept image of a function that implicitly has these properties. For instance, if the functions

Does there exist a function whose graph is :

1.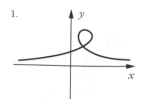

2.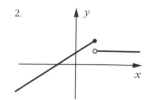

3.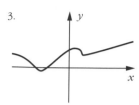

4. Does there exist a function which assigns to every number different from zero its square and to 0 it assigns 1 ?

5. What in your opinion is a function ?

FIGURE 20–1. What Do Students Think About Functions?

encountered are given mainly in terms of formulas, this causes many students to believe that the existence of a formula is essential for a function.

Dreyfus and Vinner (1982, 1989) asked a cross-section of 271 college students and 36 teachers a number of conceptual questions about functions (Figure 20.1).

The responses to the notion of function (question 5) included not only the standard definition (each value of x corresponds to precisely one value of y), but also variants such as

a correspondence between two variables
a rule of correspondence
a manipulation or operation (on one number to obtain another)
a formula, algebraic term, or equation
a graph, $y = f(x)$

However, the responses to the first four questions were not always in accord with these notions. Table 20.1 shows the percentage of students whose responses were judged correct. The percentages improve with ability and experience, but nonmathematics majors in particular have a high percentage of incorrect responses. The reasons for the responses include not only the standard definition and the variants above, but also evoked concept images such as

The graph is "continuous" or changes its character (for example, two different straight lines).

The domain of the function "splits."
There is an exceptional point.

Although the original questions are somewhat out of the ordinary, similar results have been replicated in other studies (Barnes, 1988; Markovits, Eylon, & Bruckheimer, 1986, 1988; Vinner 1983).

Markovits et al. (1986, 1988) conclude that the complexity of the modern definition causes problems because of the number of different components (domain, range, and rule), yet little emphasis is placed on domain and range at the high school level, resulting in stress being placed on the rule or relationship (which is usually given as a formula). Early emphasis on straight-line graphs seems to cause students to evoke linear graphs when asked to consider possible functions through given points (Figure 20.2).

The first graph often evoked a straight line allowing only one function because "two points can be connected by only one straight line." The second graph caused problems, perhaps because of the disposition of the points seemingly on two different lines: "If I draw a function such that all the points are on it, what will happen is for every x there will be two y, and it will not be a function." Markovits et al. observed that the "conception of functions as linear would seem to be influenced by geometry (which [students] learn simultaneously with algebra) and also by the time spent in the curriculum exclusively on linear functions" (1988, p. 54).

Barnes (1988) asked questions of grade 11 high school students and university students about different representations,

TABLE 20–1. Correct Student Responses to Function Questions

Mathematical Level	Low	Intermediate	High	Math Majors	Teachers
Question: 1	55%	66%	64%	74%	97%
2	27%	48%	67%	86%	94%
3	36%	40%	53%	72%	94%
4	9%	22%	50%	60%	75%

In the given coordinate system, draw the graph of a function such that the coordinates of each of the points A, B, (C, D, E, F) represent a pre-image and the corresponding image of the function:

The number of different such functions that can be drawn is

- 0
- 1
- 2
- more than 2 but fewer than 10
- more than 10 but not infinite
- infinite

Explain your answer.

FIGURE 20.2. More Function Questions.

for instance, whether expressions such as

$$y = 4$$

$$x^2 + y^2 = 1$$

$$y = \begin{cases} 0 & \text{if } x \leq 0 \\ x & \text{if } 0 < x \leq 1 \\ 2 - x & \text{if } x > 1 \end{cases}$$

define y as a function of x. A majority decided that the first did not, because the value of y does not depend on x. Many decided that the second *is* a function (because it is a circle, which is familiar to them), whereas the third expression presented difficulties because it appeared to define not one function but several.

When asked which graphs in Figure 20.3 represented y as a function of x, students responded in a variety of ways. The first graph evoked concept images such as "It's more like x is a function of y" or "It's a rotated function" or "That's $y = x^2$ so it's a function." The second was almost universally regarded as not being a function, not because it has vertical line segments, but because "It looks strange," or "It's not smooth and continuous," or "It's too hard to define it." The last one, in contrast to the algebraic expression $y = 4$, was regarded as being a function by all the university students, although some of the high school students were concerned that y was always the same. Some of the university students, who saw the horizontal line as a function, had asserted earlier that $y = 4$ was not a function, but now realized there was a conflict. Some, but not all, wanted to go back to the earlier question to modify their response.

At this stage, it would be of interest for the reader to look back at some of these questions to see the creases in the mind that we all share. For example, the questions related to Figure 20.1 assume that y is being considered as a (possible) function of x. The first picture could so easily be what is often described as a "parametric graph"—the image of a function from an interval to the plane. The cartoon-like blobs in the second picture are a convention to represent a discontinuity; if you think about it, you will realize that it does not truly represent the ordered pairs on the graph in the neighborhood of the discontinuity. In fact, a physical graph is only a rough representation of a function, with subtle conceptual difficulties, such as the fact that younger children see the graph as a curve and not as a set of points (Kerslake, 1977).

Could the middle graph of Figure 20.3 represent a function? It seems not, yet it could if the "vertical lines" were actually very steep but not vertical, as in the form:

$$y = -1 \quad \text{if } |x| \geq 1 + k$$

$$y = \frac{1 - |x|}{k} \quad \text{if } 1 - k < |x| < 1 + k$$

$$y = 1 \quad \text{if } |x| \leq 1 - k$$

where k is very small (say $k = \frac{1}{1,000}$).

Few students would be aware of these possibilities. However, these problem-solving possibilities indicate the implicit creases in our minds that present students with a minefield through which we trust they will choose a consistent path. Is it any wonder that so many fail?

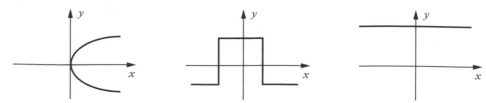

FIGURE 20–3. Are These Graphs of Functions?

Even greater difficulties with the function concept are encountered with the variety of different representations (graph, arrow diagram, formula, table, verbal description, and so on) and the relationships between them (Dorofeev, 1978; Dreyfus & Eisenberg, 1982; Janvier, 1987; Thomas, 1975). For instance, Dreyfus and Eisenberg (1987) found that students have considerable difficulties relating the algebra of transformations [such as shifts $f(x) \rightarrow f(x) + k$, $f(x) \rightarrow f(x + k)$ and stretches $f(x) \rightarrow kf(x)$, $f(x) \rightarrow f(kx)$] to their corresponding graphical representations. Of these, the transformations in the domain $f(x) \rightarrow f(x + k)$, $f(x) \rightarrow f(kx)$ naturally proved to be the more difficult.

Even (1988) studied the concept of function in prospective mathematics teachers. She found similar difficulties with student teachers in the final year of their mathematics studies.

Many of them ignored the arbitrary nature of the relationship between the two sets on which the function is defined.... Some expected functions to always be representable by an expression. Others expected all functions to be continuous. Still others accepted only "reasonable" graphs, etc. (Even, 1988, p. 216)

Can we expect teachers to be able to teach according to a modern definition of function, as it now appears in modern texts, while their conception of function is more restricted, more primitive? The participants' incomplete conception of function is problematic and may contribute to the cycle of discrepancies between concept definition and concept image of functions in students,... keeping the students' concept image of functions similar to the one from the 18th century. (Even, 1988 pp. 217–18)

Given the creases in the minds of prospective teachers, is it any wonder that it proves continually difficult to address the deep problems with the function concept in students?

In recent years the computer has been harnessed to introduce the function concept. Many of the initial moves have focused on the graphical representation of functions (Demana & Waits, 1988; Dugdale, 1982; Goldenberg et al., 1988; Schwartz, in press; Yerushalmy, in press-a, in press-b). These techniques change the conception of a function from a rule-based, pointwise process to a global visualization of overall behavior. Entirely new approaches are possible, for example, to view the qualitative shape of graphs to suggest algebraic or trigonometric relationships (Dugdale & Kibbey, 1989; Schwartz, 1990).

This brings on the one hand a great increase in potential power and, on the other, greater potential for misinterpretations of the graphical representation. For instance, the graph may look very different when drawn over different ranges (Demana & Waits, 1988), and there may be visual illusions created by changing the scale of either axis (Goldenberg et al.,

1988). The technology places enormous power in the hands of students, but serious research is necessary (and currently in progress) to gain insights into student conceptions generated by its use.

For example, most of the graph-plotting software initially available on microcomputers only accepted functions given by formulas, implicitly reinforcing the student's restricted concept image of a function as a formula. An exception is ANUGraph from the Australian National University, which allows functions to be defined by different formulas on several domains.

Only recently have graph-drawing programs appeared that allow the function notation $f(x)$. For instance, the School Mathematics Project "Function Analyser" in Britain allows functions to be typed in terms of expressions such as $g(t) = t + \sin t$, or $f(u) = e^u$, and these in turn may be used in expressions to draw graphs such as $y = g(x) + 1$, $y = g(x + 1)$, or $y = f(g(x))$. But this is still limited to functions given by formulas.

The "Triple Representation Model" (Schwarz & Bruckheimer, 1988) offers facilities to draw the graphs of functions, calculate and plot numerical values of functions, and step-search over an interval to find points that satisfy a specified equality or inequality. Here the functions are sums, differences, products, and compositions of rational, absolute value, square root, and integer part functions, defined on continuous or discrete domains. The software can be used for problem-solving activities, revealing, for example, how students use different representations to find solutions of equalities and inequalities, by "zooming in" on points where graphs cross, or using a combination of numerical evaluation and step-search strategies. Schwarz, Dreyfus, and Bruckheimer (1988) report that the software "enables students to reach higher cognitive levels in functional reasoning." For instance, the experiences with more general graphs significantly diminish the "linear graph" response to questions asking for a variety of function graphs through a given set of points. The use of the software leads to more sophisticated strategies for solving equations using the facilities provided.

There is a veritable explosion of the use of graphic calculators and graphical software on computers for the drawing of graphs of functions, some of which is indicated in a broad-ranging review of the relationships between functions and graphs by Leinhardt, Zaslavsky, and Stein (1990). Indeed, this is an excellent source of information and references for further study of empirical research into the function concept in general, indicating the depth of complexity and difficulty of the topic.

In emphasizing the many representations of the function concept—formula, graph, variable relationship, and so on—

the central idea of function as a *process* is often overlooked. For example, although graphs are often represented as an excellent way to think of a function, very few students seem to relate the graph to the underlying functional process [take a point on the *x*-axis, trace a vertical line to the graph and then a horizontal line to the *y*-axis to find the value of $y = f(x)$]. Instead, students see a graph simply as an object: a static curve (Dubinsky, 1990).

To consider the concept of function *as a process,* Dubinsky and his coworkers introduced students to the function concept via programming (Ayers, Davis, Dubinsky, and Lewin, 1988). Using the Unix operating system, a number of commands were prepared for student use, some operating on numbers and some on text. The intention was to help the students think of a function both dynamically as a process and encapsulate it statically as a mental object on which operations such as function composition may be performed. Although the number of students involved was small, there was evidence to support the hypothesis that the computer experiences were more effective for the experimental students than traditional paper and pencil exercises carried out by a control group.

From this experiment, Dubinsky progressed to the idea of programming the more general notion of function on finite sets using the language ISETL (Schwartz, Dewar, Dubinsky, & Schonberg, 1986). This programming language allows students to handle functions as arbitrary sets of ordered pairs as well as procedures, and to construct operations such as function composition in a mathematical way. More recent implementations of ISETL allow the user to graph the functions so constructed.

Empirical research shows that students can learn to think of a function as a process by programming a procedure on the computer to carry out the process (Breidenbach, Dubinsky, Hawks, & Nichols, in preparation). At a later stage, a function defined in this way can be used as an input to another procedure, hence encapsulating the process *as an object.* This suggests that the act of programming function procedures may provide a cognitive root from which the formal concept may grow. The ISETL language also provides a programming environment in which the learner may reflect on the difficult transition from function as process to function as object.

The research discussed in this section shows a wide variety of approaches to the complexity of the function concept. Some gain can be made in improving understanding and problem-solving abilities in specific areas of the function concept, but there appears to be no universal panacea. The idea of function as a process may prove to be a suitable cognitive root for the formal concept, but along the line of cognitive development there are obstacles to be overcome, including the encapsulation of the process as a single concept and the relating of this concept to its many and varied alternative representations. It remains a large and complex schema of ideas requiring a broad range of experience to grasp in any generality.

THE NOTION OF A LIMIT

Est modus in rebus, sunt certi denique fines,
Quos ultra citraque nequit consistere rectum.
[Things have their due measure; there are ultimately fixed limits, beyond which, or short of which, something must be wrong.]
(Horace, 65–8 B.C., *Satires*)

Although the function concept is central to modern mathematics, it is the concept of a limit that signifies a move to a higher plane of mathematical thinking. As Cornu observed (1983), this is the first mathematical concept that students meet where one does not find the result by a straightforward mathematical computation. Instead it is "surrounded with mystery," in which "one must arrive at one's destination by a circuitous route" (p. 151).

Limits occur in many different mathematical contexts, including the limit of a sequence, a series, a function [$f(x)$ as $x \to a$, or as $x \to \infty$], or in the notion of continuity, differentiability, or integration. In a mathematical sense, it would be appropriate to distinguish between these different types of limit—for example, the discrete limit of a sequence (a_n) as $n \to \infty$ and the continuous limit of $f(x)$ as $x \to a$. However, empirical research shows common difficulties for beginners across the various mathematical categories.

For example, the word "limit" itself has many connotations in everyday life which are at variance with the mathematical idea. An everyday limit is often something that cannot or should not to be passed, such as a "speed limit." The terminology associated with mathematical limiting processes includes phrases such as "tends to," "approaches," or "gets close to," which again have colloquial meanings differing from the mathematical meanings. For instance, when these phrases are used in relation to a sequence *approaching* a limit, they invariably carry the implication that the terms of the sequence *cannot equal* the limit (Schwarzenberger & Tall, 1977).

The problem of handling limits is exacerbated by restricted concept images of sequences and functions; for example, students are often introduced to the notion of a sequence where the terms are given as a formula. If one wished to show that some terms of a sequence might equal the limit, one might try to consider the sequence $1, 0, \frac{1}{2}, 0, \frac{1}{3}, 0, \ldots$ but students who view the terms of a sequence as a formula may insist that this is not *one* sequence, but *two*; the odd terms form a harmonic sequence $1, \frac{1}{2}, \frac{1}{3}, \ldots$ which tends down to zero, and the even terms are constants, which *are* zero (Tall, 1980b).

Davis and Vinner (1986) suggest that there are seemingly unavoidable misconception stages with the notion of a limit. One is the influence of language, mentioned earlier, in which the terms remind us of ideas that intrude into the mathematics. In addition to the *words,* there are the *ideas* that these words conjure up, which have their origins in earlier experiences. Although the authors attempted to teach a course in which the word "limit" was not used in the initial stages, they eventually concluded that "avoiding appeals to such pre-mathematical mental representation fragments may very well be futile" (p. 299). Another source of misconceptions is the sheer complexity of the ideas, which cannot appear "instantaneously in complete and mature form," so that "some parts of the idea will get adequate representations before other parts will" (p. 300). Specific examples are likely to dominate the learning; for instance Davis and Vinner found that monotonic sequences dominated the early examples, so it was not surprising that they dominated the student's concept images. This could lead to a misinterpretation of one's own experience; for instance, the fact that students dealt with many examples of sequences whose terms were given by a formula caused them to mistakenly assume that a simple algebraic formula for the nth term a_n is an essential part of the theory.

Most of the informal ideas of limit carry with them a dynamic feeling of something approaching the limiting value, for instance, as n increases, the sum

$$1 + \frac{1}{2} + \cdots + \frac{1}{2^n}$$

approaches the limit 2. This has an inevitable cognitive consequence that I term the "generic limit property": the belief that any property common to all terms of a sequence also holds of the limit (Tall, 1986). It is a belief with worthy historical precursors, for example in Leibniz's "principle of continuity" (stated in a letter to Bayle):

In any supposed transition, ending in any terminus, it is permissible to institute a general reasoning, in which the final terminus may also be included.

It permeates the history of mathematics, as in Cauchy's belief that the limit of continuous functions must again be continuous. And it remains as a crease in the mind of today, in such ideas that the limit of the sequence

$$0.9, \quad 0.99, \quad 0.999, \ldots$$

must be *less* than one—because all the terms are less than one. Thus "nought point nine recurring" is "just less than one" (Cornu, 1983; Schwarzenberger & Tall, 1977; Tall & Vinner, 1981). Cornu (1983) studied this in greater detail and found a whole array of beliefs, for instance that "0.9, 0.99, . . . *tends* to nought point nine recurring, but has *limit* one (because it "tends" to have the property of 0.99999 . . . , but cannot pass the "limit" one).

It will come as no surprise that attempting to "simplify" the limit notion by using everyday language can lead to serious conceptual problems. Orton (1980) investigated students' concepts of limit using a "staircase with treads" where extra half-size treads are inserted between each tread, then the process is repeated successively with treads half this size again (Figure 20.4).

In an interview he posed the questions:

1. If this procedure is repeated indefinitely, what is the final result?

2. How many times will extra steps have to be placed before this "final result" is reached?
3. What is the area of the final shape in terms of a, i.e. what is the area below the "final staircase"?

If a formula was given in question 3, he asked:

Can you use this formula to obtain the "final term" or limit of the sequence?

He justified the use of this terminology by stating:

The expression "final term" was again used in an attempt to help the students understand the meaning of limits. (Orton, 1980)

He is surely not alone in his attempt to help the student by an informal presentation. But a phrase such as "the final staircase" is likely to create a generic limit concept in which the student imagines a staircase with an "infinite number of steps," and this is precisely the response that it evoked.

Faced with such difficulties in the dynamic notion of a limit, it will come as no surprise that the formal definition is also fraught with cognitive problems. Even the phrase "given an epsilon greater than zero . . ." may be interpreted as taking epsilon to be "arbitrarily small," and this in turn can lead to the symbol generically representing an "arbitrarily small" number:

Everything occurs as if there exist very small numbers, smaller than "real" numbers, but nevertheless not zero. The symbol ϵ represents for many students a symbol of this type: ϵ is smaller than all real numbers, but not zero. (Cornu, 1983)

In the same way, in calculus, the introduction of symbols δx (for a small, finite increment in x) and dx (as part of the dy/dx notation) lead generically to the idea that there exist numbers that are arbitrarily, or "infinitesimally," small (Cornu, 1983; Orton, 1980; Tall, 1980a).

The introduction of the formal notion of limit does not obliterate more primitive dynamic notions; indeed, we often continue to nurture dynamic imagery in our teaching to give an intuitive flavor to rigorous proofs.

Robert (1982) studied the notion of limit of a sequence as perceived by 1,380 students at various school and university levels. She asked how the students might explain the notion of a convergent sequence to a pupil 14 or 15 years old (a question that is more likely to evoke a concept image than the

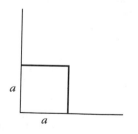
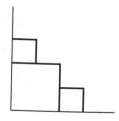
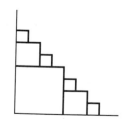

FIGURE 20–4. A Limiting Staircase.

formal definition). She classified the responses into four main categories:

1. *Monotonic and Dynamic Monotonic* (12%): "a convergent sequence is an increasing sequence bounded above (or decreasing bounded below);" "a convergent sequence is an increasing (or decreasing) sequence which approaches a limit."
2. *Dynamic* (35%): "u_n tends to l"; "u_n approaches l"; "the distance from u_n to l becomes small"; "The values approach a number more and more closely."
3. *Static* (13%): "The u_n are in an interval near l"; "The u_n are grouped round l"; "u_n is as close as you like to l."
4. *Mixed* (14%): a mixture of those above.

In addition, 4% gave the formal definition, 5% did not attempt the question, and the remainder gave incomplete or false statements such as "u_n doesn't go past l" or "u_n stays below l."

The fact that a student evokes a particular image does not mean the absence of other images:

The presentation by a student of an old (and incorrect) idea cannot be taken as evidence that the student does NOT know the correct idea. In many cases the student knows both but has retrieved the old idea. (Davis & Vinner, 1986, p. 284)

In particular, Robert's request for an explanation suitable for a 14 or 15 year old seems to exclude the formal definition because of its difficulty. This difficulty is confirmed by Tall and Vinner (1981), who asked 70 highly qualified first-year university mathematics students to write down a definition of $\lim_{x \to a} f(x) = c$ (if they knew one). They had just passed A–level examinations and would be expected to have been given a dynamic definition [$f(x)$ gets close to c as x gets close to a], although some might have been shown the formal definition. Those who replied did so as shown in Table 20.2.

Thus the majority of those who recalled the (easier) dynamic definition could state it correctly, whereas the majority of those who chose to give the formal definition were not able to recall it in a satisfactory way, misstating it in various ways such as:

$|f(x) - c| < \varepsilon$ *for all positive values of ε with x sufficiently close to a.*
As $x \to a, c - \varepsilon \leq f(x) \leq c + \varepsilon$ *for all $n > n_0$.*
$|f(n) - f(n + 1)| < \varepsilon$ *for all $n >$ given N_0.*

Teaching the notion of limit using the computer has, on the whole, fared badly. Regular computing languages, such as BASIC, Pascal or C, hold numbers in a fixed number of memory locations, which can lead to serious problems of accuracy when calculating a limit such as:

$$\lim_{h \to 0} \frac{\sin(x + h) - \sin x}{h}$$

When h is small, both numerator and denominator are tiny numbers whose quotient is likely to be highly inaccurate. For instance, for $x = \frac{\pi}{3}$, the limit as h tends to 0 should be $\frac{1}{2}$, but

TABLE 20–2. Student Definitions of the Limit Concept

	Correct	Incorrect
Formal	4	14
Dynamic	27	4

on a typical micro, taking $h = \frac{1}{10^n}$ for $n = 1$ to 10 gives the sequence:

0.455901884
0.495661539
0.499566784
0.499954913
0.499980524
0.499654561
0.500585884
0.465661287
0.232830644
0

which hardly gets close to 0.5. Numerical ideas of limits in such contexts must therefore be combined with discussion of accuracy of computer arithmetic.

Symbolic treatments of limits do not always fare better. The expression

```
((x+h)^2-x^2)/h
```

typed into Derive (Rich, Rich, and Stoutemyer, 1989) is prettily printed as

```
        2    2
(x+h)  -x
---------
    h
```

and may be automatically simplified to

```
2x+h
```

but there is no warning about the case $h = 0$. The more general expression

```
        n    n
(x+h)  -x
---------
    h
```

only simplifies to

```
        n        n
(x+h)      x
-----  -  ---  .
   h       h
```

Derive's limit option (in the first version of the software) applied to this expression, as h tends to 0, gives not nx^{n-1}, but

$$\hat{e}^n \, LN(x) - LN(x/n)$$

which may be suitable for sophisticated investigation, but is hardly appropriate for a beginner. In a later revision this was improved; however, it does illustrate the difficulties encountered when one tries to program symbol manipulation. It is hard enough to do, but far harder to get the expression into a form that may be desired.

All the evidence points once more to the fact that, although the limit concept (in a formal sense) is a good mathematical foundation, it fails to be an appropriate cognitive root. If it is difficult to start with the limit process in subjects such as calculus, what alternatives are available? Instead of introducing explicit limit ideas in differentiation, Tall (1986) begins by magnifying graphs. This builds on the thesis that a cognitive root for calculus is the idea that a differentiable function has a graph that magnifies to "look straight." To give a rich concept image, the software (Tall, Blokland, & Kok, 1990) includes not only standard functions, but also functions which are so wrinkled, that no matter how highly magnified, they never look straight. Thus in the very first lesson in calculus, it is possible to explore functions that are locally straight everywhere, functions that have different left and right gradients at certain points (because the graph magnifies to reveal a corner), and functions that are locally straight nowhere (because they are too wrinkled). This allows students to build a much richer concept image, including examples of differentiable functions, functions having different left and right derivatives, and nondifferentiable functions—providing cognitive roots on which formal theories may later be grafted.

It is not an easy path. But this is true of life itself. There is no royal road, as Euclid is said to have remarked to Ptolemy. Given the complexity of the limit concept, the road ahead is surely not to attempt to ease the student's path by trying to avoid difficulties, for the oversimplification produces inappropriate concept images that only store up problems for later. A more helpful route is to provide the rich experience necessary to enable the student to attempt to confront the difficulties and negotiate a more stable concept, mindful of the possible pitfalls.

THINKING ABOUT INFINITY

To see a World in a Grain of Sand,
And Heaven in a wild flower,
Hold Infinity in the palm of your hand
And Eternity in an hour.
(William Blake, 1757–1827, *Auguries of Innocence*)

Thoughts of infinity touch us all at some time or other as we contemplate the puny nature of our finite existence in the vastness of the universe. Research into the nature of students' concepts of infinity is probably more clouded by the creases in the minds of the researchers than by any other factor. For what do we mean by "infinity"? It would be useful for the reader to pause a moment and think what infinity means to him or her before reading on.

Historically, philosophers distinguished between *actual* infinity ("there are an infinite number of whole numbers") and *potential* infinity ("for any whole number there is always one bigger"). In modern times, actual infinity is interpreted using Cantor's theory of cardinal numbers in terms of one-one correspondences between sets. An infinite set is one that can be put in one-one correspondence with a proper subset. Thus, the natural numbers $\{1, 2, 3, \ldots n, \ldots\}$ form an infinite set because they can be put into one-one correspondence with the even numbers $\{2, 4, 6, \ldots 2n, \ldots\}$ in which n corresponds to $2n$. It is this cardinal form of infinity that is prominent in modern mathematics.

But there are properties of cardinal infinity that many find difficult to come to terms with, for instance that a set can have "as many" elements as a proper subset. In cardinal number terms, there are as many natural numbers as rationals, as many points on a unit real line segment as on a real line segment length two, or on a real line as in a square, yet there are far more real numbers than rationals. Where is the consistency?

A number of research studies are based on the inconsistency between the cardinal infinity of Cantor and our intuitions. There are creases in our minds born of our experiences comparing infinite sets that children, with their different experiences, may not share. Such research on infinity is likely to say as much about the nature of our own conceptions as it does about the conflicts in the minds of children. For this reason, it is essential that we briefly consider the nature of various conceptions of infinity before we proceed.

In Tall (1981) I suggested that experiences of infinity that children encounter rarely relate to the action of comparing sets, which means that they rarely lay the cognitive roots for the cardinal concept of infinity. For instance, when a child thinks of a "point on a line," it may be in the manner of a pencil mark (or it may be something entirely different, for example the "point" on a sword). A pencil mark has finite size. A child who views a point as having a tiny, finite size is likely to develop a generic concept of a point that has an extremely small size.

If a line segment is made up of such points, then there will be a finite number of points (100, for example) to make it up. A line segment of twice the length will require twice the number (200). The only way that the double-length line segment can have the same number of points is if the points are twice the size! In extrapolating these ideas to the infinite case, a natural generic concept would be to have a kind of infinity with an infinite number of infinitesimally small points in a unit segment and twice as many in a segment twice the length. I once suggested this to a mathematical colleague who laughed at the naiveté of it all and said, yes, there were twice as many real numbers in a line twice the length—one had $\aleph_0$ elements in it, the other had $2\aleph_0$ elements, and by cardinal arithmetic, $\aleph_0 = 2\aleph_0$! He thought it was extremely naive to think otherwise.

Let us imagine N intervals in a unit length, each of length $1/N$. If N is very large, $1/N$ is very small. A line twice the length will have $2N$ intervals of the same size, where $2N$ is even larger, and certainly not equal to N. Nonstandard analysis allows us to let N be an element in an ordered field bigger than the real numbers, so that N is larger (in the given order) than any real number. In this technical sense, N is "infinite." It follows by manipulating the order relation that $1/N$ is smaller than any positive real number and so, in this technical sense, it is "infinitesimal" (see Tall, 1980a, 1981). Thus, nonstandard analysis allows a line to be made up of an infinite number of tiny line segments of infinitesimal size. A line of twice the length will have twice the number of points of the same size. Unlike cardinal infinity, the nonstandard infinite numbers $2N$ and N are not equal—one is bigger than the other—just as in the intuition of a child. Thus, although the child's concept of infinity conflicts with cardinal infinity, it has properties that are consonant with nonstandard infinity.

As we consider the concept of infinity during the transition to advanced mathematical thinking, we now become aware of wider possibilities. *There is more than one notion of infinity.* The symbol ∞, used in phrases such as "the limit as n tends to ∞," represents the idea of *potential* infinity. Students are usually told not to think of it as a genuine number, yet they may be confused to find it used in many contexts as if it were. In mathematics, there are at least three notions of "actual infinity": *cardinal infinity* (extending the notion of counting via the comparison of sets—the favored form of infinity by mathematicians), *ordinal infinity* (the concept proposed by Cantor in terms of comparison of ordered sets), and the notion of nonstandard infinity (generalizing the notion of measuring from real numbers to a larger ordered field). For simplicity, I term this nonstandard infinity *measuring infinity.* All these kinds of infinity are logical entities appropriate for study in advanced mathematics. In judging the intuitions of a child, we should not make the mistake of considering only one kind of infinity—cardinal infinity—as the only true mathematical notion.

For example, the "infinite staircase" response to the question at the end of the last section is a perfectly reasonable nonstandard response, although it is rejected by standard analysis. Similarly reasonable is the idea that 0.999... to an infinite number of places, say N, is infinitesimally smaller than 1 (by the infinitesimal quantity $\frac{1}{10^N}$).

Tall (1980b) asked students to compute various limits, including the limits of

$$\frac{n^2}{n^2 + 1} \text{ and } \frac{n^5}{(1.1)^n}$$

as n tends to infinity. A student who wrote

$$\frac{n^2}{n^2 + 1} \to \frac{\infty}{\infty} = 1$$

was shown that a similar argument would give

$$\frac{n^5}{(1.1)^n} \to \frac{\infty}{\infty} = 1$$

but replied firmly, "No it wouldn't, because in this case the denominator is a *bigger* infinity, and the result would be zero." This sense of infinities of different sizes is not a cardinal concept—it is an extrapolation of experiences in arithmetic closer to *measuring infinity* than *cardinal* infinity.

Fischbein, Tirosh, and Hess (1979) give another clear example of measuring infinity, where

$$1 + \frac{1}{2} + \frac{1}{4} + \frac{1}{8} + \cdots$$

is stated to be

$$s = 2 - \frac{1}{\infty}$$

because there is no end to the sum of segments.

Here it is the potential infinity of the limiting process that leads to a generic concept of measuring infinity. The suggested limit is typical of all the terms: just less than 2. The arithmetic fits nicely with nonstandard analysis, but not with cardinal numbers where infinities cannot be divided.

Most experiences with limits relate to things getting *large*, or *small*, or *close* to one another. All of these extrapolate experience from *arithmetic* rather than comparisons between sets and are more likely to evoke *measuring infinity*, rather that *cardinal infinity*. It follows that the ideas of *limits* and *infinity*, which are often considered together, *relate to two different and conflicting paradigms.*

Many different ideas of infinity can occur in different students in a given class. Sierpińska (1987) analyzed the concept images of thirty-one 16-year-old precalculus mathematics and physics students and classified the students into groups which she labeled with a single name for each group.

Michael and Christopher are *unconscious infinitists* (at least at the beginning): They say "infinite," but think "very big," ... For both of them the limit should be the last value of the term; ... for Michael this last value is either plus infinity (a very big positive number) or minus infinity.... It is not so for Christopher who is more receptive to the dynamic changes of values of the terms. The last value is not always tending to infinity, it may tend to some small and known number.

George is a *conscious infinitist:* Infinity is about something metaphysical, difficult to grasp with precise definitions. If mathematics is to be an exact science then one should avoid speaking about infinity and speak about finite numbers only. In formulating general laws one can use letters denoting concrete but arbitrary finite numbers. In describing the behaviour of sequences *the most important thing is to characterize the nth term* by writing the general formula. For a given n one can then compute the exact value of the term or one can give an approximation of this value.

Paul and Robert are *kinetic infinitists:* The idea of infinity in them is connected with the idea of time.... Paul is a *potentialist:* To think of some whole, a set or a sequence, one has to run in thought through every element of it. It is impossible to think this way of an infinite number of elements. The construction of an infinite set or sequence can never be completed. Infinity exists potentially only. Robert is a *potential actualist:* It is possible [for him] to make a "jump to infinity" in thought: The infinity can potentially be ultimately actualized. For both Paul and Robert, the important thing is to see how the terms of the sequence change, if there is a tendency to approach some fixed value. For Paul, even if the terms of a sequence come closer and closer so as to differ less than any given value they will *never reach it.* Robert thinks *theoretically* the terms will *reach it in the infinity.* (pp. 74–75)

Fischbein and colleagues (1978, 1979, 1981) investigated a number of conflicts inherent between the different conceptions of infinity—for instance, the conflict between the intuition of the single potential infinity and the many infinities of cardinal number theory, or the conflict between the finite number of points that may be marked physically on a line compared with the infinite number of points that are theoretically possible. Fischbein distinguishes between "primary intuitions," which are our common heritage, and "secondary intuitions," which come from more specialized experiences. Thus, the idea of potential infinity is a primary intuition, but it takes considerable experience of cardinal infinity to develop appropriate secondary intuitions as these conflict with deeply held convictions (such as "the whole is greater than the part").

Tirosh (1985) continued the work of Fischbein and his colleagues by designing a teaching program on "finite and infinite sets" for grade 10 students, taking their intuitive background into account (for example, the fact that the students might appeal to the "part-whole" principle to declare that a set was bigger than a proper subset). The pupils were presented with quotations from mathematicians on the puzzling aspects of infinite sets, to encourage them to feel that it was legitimate to face such conflicts. It was found that it was possible to improve their understanding of the Cantorian theory by using dynamic teaching methods, including the open discussion of intuitive conflicts.

Other research has addressed the alternative paradigm of nonstandard analysis. Sullivan (1976) studied the effectiveness of teaching calculus at college level from a nonstandard viewpoint that combined axioms for the real numbers and a larger set of hyperreal numbers containing infinite and infinitesimal elements (Keisler, 1976). The approach is given a strong geometric flavor incorporating a pictorial interpretation of these elements using "microscopes" and "telescopes." She found that the students following the experimental course scored at least as well as a control group in regular analysis problems ($\varepsilon - \delta$ definitions, calculating limits, proofs, and applications), but were better at aspects of the course that presented alternative interpretations using infinitesimal arguments. The latter tend to seem easier, partly because they do not involve as many quantifiers as the standard definitions and partly because they extend intuitive experiences of "getting small" in the limit process.

Despite this empirical proof for the success of an approach using infinitesimals, the approach to calculus in higher education has hardly changed. There are genuine reasons for this, including the intrinsic sophistication of the nonstandard ideas that depend on logic at the depth of the axiom of choice. But there are also prejudices arising from traditional mathematical analysis and its links with the theory of Cantor. The creases of the mind run deep.

It is important to complement the study of student difficulties with possible sources of difficulty in the mind of the teacher. Evidence from pre-service elementary teachers enrolled in an upper-division course in mathematics methods at a large university revealed widespread inconsistencies (Wheeler & Martin, 1987, 1988). Questions asking for explanations of the symbol ∞ and the final three dots in the expression "1, 5, 25, 125, 625,..." showed that more than half the subjects were unfamiliar with the symbolism. Responses to "What is infinity?" referred either to an unending process—"The numbers go on without stopping," or to a recursive process—"No matter what

number you say, there is always one greater simply by adding one to it." In either case, potential infinity is the predominant notion of infinity evoked.

Students were asked whether the statement, "Every line segment contains an infinite number of points," was true or false. (This could evoke concepts of potential, cardinal, or measuring infinity.) There were 39 "true" responses, 24 "false," and seven did not respond. The statement, "There exists a smallest fraction greater than zero," yielded 28 "true," 29 "false," and 13 did not respond.

When cross-referenced, the responses revealed the great majority of students "holding incomplete and inconsistent concepts of infinity" and individual written responses showed a wide variety of evoked concept images riddled with conflicts and inconsistencies. However, it is also interesting to ask whether the concept of infinity provoked by asking the meaning of "..." (potential infinity) is the same kind of infinity as the number of points in a line segment (cardinal infinity). In order to research the beliefs held by students and to classify those beliefs, it is important first to analyze the concepts concerned and the kind of concept images generated by various experiences without imbuing them with a classical mathematical prejudice.

MATHEMATICAL PROOF

For nothing worthy proving can be proven,
Nor yet disproven: wherefore thou be wise,
Cleave ever to the sunnier side of doubt.
(Alfred Lord Tennyson, 1809-1892, *The Ancient Sage*)

Traditionally, the introduction to proof in school has been through Euclidean geometry. However, this disappeared from the syllabus in Britain with the arrival of the new mathematics. The NCTM *Standards* suggest a change in emphasis in the United States, with increased attention recommended for

- the development of short sequences of theorems
- deductive arguments expressed orally and in sentence or paragraph form

and decreased attention given to

- Euclidean geometry as a complete axiomatic system
- two-column proofs.

The reasons for this are not hard to find. Senk (1985) showed that only 30% of students in full-year geometry courses reach a 75% mastery on a selection of six geometric proof problems.

Not only is Euclidean proof difficult, it fails to satisfy stringent tests of modern mathematical rigor because it depends on subtle intuitive notions of space. As Hilbert put it most succinctly, "One must be able to say at all times—instead of points, straight lines and planes—tables, chairs and beer mugs"(Encyclopaedia Britannica, 1974, p. 1101).

But proof in terms of tables, chairs, and beer mugs requires a great deal of sophistication not available to younger students. Mathematical proof as a human activity requires not only an understanding of the concept definitions and the logical processes, but also insight into how and why it works. Tall (1979) asked first-year university students to comment on their preferences for the standard proof that the square root of 2 is irrational by contradiction, or for an alternative proof that showed the square of a whole number always had an even number of prime factors, hence the square of any fraction could not be $\frac{2}{1}$, because the prime 2 appeared an odd number of times. The students significantly preferred the second proof, because it gave some kind of explanation as to why the result was true (even though it was expressed in slightly loose mathematical terminology). In the transition to advanced mathematical thinking, mathematical insight in proof may be more important than mathematical precision.

Yet it does not take long before creases in the mind begin to form. Vinner (1988) gave students two proofs of the mean value theorem (if a function f is differentiable between a and b and continuous at a and b, then there is a point ξ between a and b such that $f'(\xi) = (f(b) - f(a))/(b - a)$):

1. The standard algebraic proof, applying Rolle's theorem to
$$f(x) - \frac{f(b) - f(a)}{b - a}(x - a); \text{ or}$$
2. A visual proof moving the secant AB in Figure 20.5 parallel to itself until it becomes a tangent.

Of the two proofs of the mean value theorem, 29 students found the visual proof more convincing, 28 preferred the algebraic proof, and 17 considered them equal. It would be pleasing to see some evidence that the students criticized the geometric proof on some valid ground—for instance, that it fails to give a proper algorithm to find ξ to any degree of accuracy. But almost all those preferring the visual proof mentioned that it was "clear," "evident," or "simple," while those preferring the algebraic proof tended to make general remarks that something is wrong or illegal with the visual approach. Vinner considers that students develop an "algebraic bias" not because of improved understanding of the algebra, but because of "habits, routines, convenience, and metacognitive ideas which are 'environmental,' not 'cognitive'" (p. 109). In particular, he cites the current teaching of Euclidean geometry for sowing the seeds that a visual proof is unsatisfactory.

In mathematical analysis, the need for a formal proof so often seems to arise from the fear that something might go wrong rather than from confidence that something is right. To have a good intuition of what is right, one needs appropriate experiences to give a complete range of possible concept images, and these are generally absent in undergraduates. On the other hand, students with experiences of magnifying a graph might realize that a graph may have tiny wrinkles on it— a positive reason why the smooth picture in Figure 20.5 may not tell the full story—thus leading to the need for the formal proof.

Proof is concerned "not simply with the formal presentation of arguments, but with the student's own activity of arriving at conviction, of making verification, and of communicating convictions about results to others" (Bell, 1978, p. 48). In the last decade or so, there has been a growing change of emphasis from teaching the *form* of proof to encouraging the *process*, including the earlier stages of assembling information, special-

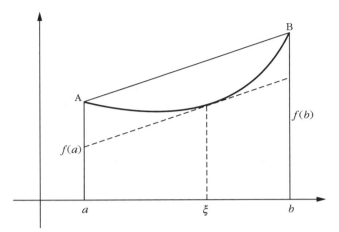

FIGURE 20–5. A visual "proof" of the mean value theorem.

izing, generalizing, and making and testing conjectures. Mason, Burton, and Stacey (1982) have developed a problem-solving approach in which the student builds up confidence through increasing levels of conviction in a conjecture they have formulated:

Convince yourself.
Convince a friend.
Convince an enemy.

The first of these requires the student to state a conjecture in a way that seems to him or her to be true; the second requires it to be articulated in a way that it can be meaningfully conveyed to someone else; and the third requires the argument to be clarified and organized in a way that will satisfy the meanest of critics. Nevertheless, this sequence of events stops short at what most professional mathematicians mean by proof: the logical deduction of theorems from carefully formulated concept definitions.

Alibert (1987, 1988a, 1988b) and his colleagues at Grenoble University have developed a course on analysis, which includes this final step through "scientific debate" in the classroom. This is based on the idea that students construct their own knowledge through "interactions, conflicts, and re-equilibrations" and that the need for proof is best emphasized through making the contradictions explicit and by involving the students in their resolution (Balacheff, 1982). Rather than simply present lectures in a logical sequence, followed by stereotyped exercises, the teachers encourage students (in a class of about 100) to make conjectures. For instance, after the introduction of the notion of integral, the teacher might say:

If I is an interval on the reals, a is a fixed element of I, and x is an element of I, we set, for f integrable over I,

$$F(x) = \int_a^x f(t) \, dt$$

Can you make some conjectures of the form "if f ... then F ... "

About 20 conjectures were formulated by the students, for example,

If f is increasing, then F is too,

(which happens to be false). The conjectures are then considered for debate. Arguments in support or against these conjectures are addressed by the students in a way that must convince everyone.

Responding to a questionnaire, 75% of students preferred the method of incorporating debates, whereas 10% rejected it as being inaccessible and not sufficiently organized. Many find debates particularly helpful when new ideas are introduced, but prefer the teacher to round off the debate by summarizing the knowledge gained.

Successful methods are thus being developed in mathematics education to improve the students' participation in the processes of mathematical thinking, including understanding the necessity for precise definitions and logical deduction. But these methods depend on radically different approaches on the part of teachers, and only time will tell if they will become more widely accepted.

REFLECTIONS

Looking back over the evidence assembled, there is a great deal of data to support the existence of serious cognitive conflict in the learning of more advanced mathematical processes and concepts such as functions, limits, infinity, and proof. What also seems to be clear is that the formal definitions of mathematics (that are such effective foundations for the logic of the subject) are less appropriate as cognitive roots for curriculum development. Their subtlety and generality are too great for the growing mind to accommodate all at once without a high risk of conflict caused by inadvertent regularities in the particular experiences encountered. There are creases of the mind everywhere: in teachers, in professional mathematicians, and in mathematics educators as well as in students. Given such a catalog of difficulties, is there a way ahead?

We should not be too downhearted. The mathematical culture of which we speak is the product of 3,000 (or is it 3,000,000) years of corporate human thought. It is asking a great deal to compress such diverse richness of experience into a decade or so of an individual's schooling. What is certain is that if we try to teach these ideas without taking account of the cognitive development of the student, we will surely fail with all but the most intelligent—and even these will have subtle creases in their minds as a result of their experiences. It is essential therefore that the expert be willing to reexamine personal beliefs about the nature of mathematical concepts and be prepared to see them also from the viewpoint of the learner.

So much of the research quoted in this chapter has been built on implicit, unspoken assumptions about the nature of the concepts being considered. The first step on a research agenda to assist students in the transition to advanced mathematical thinking must therefore be the clarification of these unspoken assumptions and the sensitizing of researchers and teachers to their existence. One source of evidence for these assumptions

is in clinical interviews with students, which requires a careful reflection on what is said, not just to see how it conflicts with formal mathematics, but also to place formal mathematics itself into perspective as a human activity that attempts to organize the complexities of human thought into a logical system. A theory of cognitive development of mathematical thought in the individual, from elementary beginnings to formal abstractions, requires a cognitive understanding of the formal abstractions themselves.

The second stage is for more detailed clinical observations of the transition process as the massive process of cognitive restructuring takes place. This transition involves a number of difficult cognitive changes:

- from the concept considered as a process (the function as a process, tending to a limit, potential infinity),
- to the concept encapsulated as a single object that is given a name (the function as an object, the limit concept, actual infinity),
- via the abstraction of properties to the concept given in terms of a definition (function as a set of ordered pairs, the epsilon-delta limit),
- to the construction of the properties of the defined object through logical deduction,
- and the relationships between various representations of the concept (including verbal, procedural, symbolic, numeric, and graphic).

These are not intended to represent a hierarchy of development. The relationships between various representations permeate the whole system in a horizontal manner, whereas the conceptual strands become more sophisticated. Empirical evidence traditionally suggests that it is necessary to become familiar with a process before encapsulating it as an object. The computer is capable of carrying out routine processes (such as drawing graphs) that now give the possibility of new learning strategies in which the objects produced by the computer are the focus of attention before the internal algorithms are studied.

The third stage is then the design and testing of learning sequences aimed at assisting the cognitive reconstruction in the transition to advanced mathematical thinking.

Research to date aimed at improving learning—as opposed to research that simply observes what is currently occurring—has a common thread. With this research, true progress in making the transition to more advanced mathematical thinking can be achieved by helping students reflect on their own thinking processes and confront the conflicts that arise in moving to a richer context where old implicit beliefs no longer hold. Such intellectual growth is stimulated by flexible environments that furnish appropriate cognitive roots and help the student build a broader concept image. Oversimplified environments designed to protect students from confusion may only serve to provide implicit regularities that students abstract, causing serious conflict at a later stage.

In taking students through the transition to advanced mathematical thinking, we should realize that the formalizing and systematizing of the mathematics is the final stage of mathematical thinking, not the total activity. As Skemp wrote in the "Psychology of Learning Mathematics" (1971):

Some reformers try to present mathematics as a logical development. This approach is laudable in that it aims to show that mathematics is sensible and not arbitrary, but it is mistaken in two ways. First it confuses the logical and the psychological approaches. The main purpose of a logical approach is to convince doubters; that of a psychological one is to bring about understanding. Second, it gives only the end-product of mathematical discovery ('this is it, all you have to do is learn it'), and fails to bring about in the learner those processes by which mathematical discoveries are made. It teaches mathematical thought, not mathematical thinking. (p. 13)

In like manner, at the advanced level, presenting definitions and theorems only by means of logical development teaches the product of advanced mathematical thought, not the process of advanced mathematical thinking.

References

Alibert, D. (1987). Alteration of didactic contract in codidactic situation. *Proceedings of the Eleventh International Conference for the Psychology of Mathematics Education* (pp. 379–385), Montréal: Université de Québec à Montréal.

Alibert, D. (1988a). Towards new customs in the classroom. *For The Learning of Mathematics, 8(2),* 31–43.

Alibert, D. (1988b). Co-didactic system in the course of mathematics: How to introduce it. *Proceedings of the Twelfth International Conference for the Psychology of Mathematics Education* (pp. 109–116). Vesprem, Hungary.

Artigue, M. (1986). The notion of differential for undergraduate students in the sciences. *Proceedings of the Tenth International Conference for the Psychology of Mathematics Education* (pp. 235–240), London: University of London.

Ayers, T., Davis, G., Dubinsky, E., & Lewin, P. (1988). Computer experiences in learning composition of functions. *Journal for Research in Mathematics Education, 19,* 248–59.

Balacheff, N. (1982). Preuve et démonstration. *Reserche en Didactique des Mathématiques, 3(3),* 261–304.

Barnes, M. (1988). Understanding the function concept. Some results of interviews with secondary and tertiary students. *Research on Mathematics Education in Australia,* 24–33.

Bell, A. W. (1978). The learning of process aspects of mathematics. *Proceedings of the Second International Conference for the Psychology of Mathematics Education* (pp. 48–78). Osnabrück, West Germany.

Blake, W. (1971). Auguries of Innocence. In W. H. Stevenson and D. V. Erdman (Eds.), *The poems of Blake.* London: Longman. Original work published in 1804.

Borasi, R. (1985). Intuition and rigor in the evaluation of infinite expressions. *Focus on Learning Problems in Mathematics, 7(3-4),* 65–75.

Breidenbach, D., Dubinsky, E., Hawks, J., & Nichols, D. (in preparation). Development of the process concept of function.

Browning, R. (1970). Abt Vogler, In *Robert Browning: Poetical works 1833–1864* (p. 808). London: Oxford University Press. Original work published in 1864.

Buck, R. C. (1970). Functions. In E. G. Begle (Ed.), *The sixty-ninth yearbook of the National Society for the Study of Education, Vol. I. Mathematics education.* Chicago, IL: National Society for the Study of Education.

Cornu, B. (1981). Apprentissage de la notion de limite: modèles spontanés et modèles propres. *Actes du Cinquième Colloque du Groupe Internationale PME,* (pp. 322–326), Grenoble: Université de Grenoble.

Cornu, B. (1983). *Apprentissage de la notion de limite: Conceptions et Obstacles.* Thèse de Doctorat, Grenoble.

Davis, R. B. (1982). *Learning mathematics: The cognitive science approach to mathematics education.* London: Croom-Helm.

Davis, R. B., & Vinner, S. (1986). The notion of limit: Some seemingly unavoidable misconception stages. *Journal of Mathematical Behaviour, 5(3),* 281–303.

Demana, F., & Waits, B. (1988). Pitfalls in graphical computation, or why a single graph isn't enough. *College Mathematics Journal, 19(2),* 177–183.

Desmond A. J. (1975) *The hot-blooded dinosaurs.* London: Blond & Briggs.

Dorofeev, G. V. (1978). The concept of function in mathematics and in school. *Mathematics in School, 2,* 10–27.

Dreyfus, T. & Eisenberg, T. (1982). Intuitive functional concepts: A baseline study on intuitions. *Journal for Research in Mathematics Education, 6,* 18–24.

Dreyfus, T., & Eisenberg, T. (1987). On the deep structure of functions. *Proceedings of the Eleventh International Conference for the Psychology of Mathematics Education,* (pp. 190–196), Montréal: Université de Québec à Montréal.

Dreyfus, T., & Vinner, S. (1982). Some aspects of the function concept in college students and junior high school teachers. *Proceedings of the Sixth International Conference for the Psychology of Mathematics Education* (pp. 12–17), Antwerp, Belgium. University of Antwerp.

Dreyfus, T., & Vinner, S (1989). Images and Definitions for the Concept of Function. *Journal for Research in Mathematics Education, 20,* 356–366.

Dubinsky, E. (in press). Computers in teaching and learning discrete mathematics and abstract algebra. In D. L. Ferguson (Ed.), *Advanced technologies in the teaching of mathematics and science.* New York, NY: Springer-Verlag.

Dugdale, S. (1982). Green globs: A microcomputer application for graphing of equations. *Mathematics Teacher, 75,* 208–214.

Dugdale, S. & Kibbey, D. (1989) Building a qualitative perspective before formalizing procedures: graphical representations as a foundation of trigonometric identities. *Proceedings of the Eleventh Annual Meeting of PME-NA.* New Brunswick, NJ.

Encyclopaedia Britannica. (1974), *Euclidean Geometry* (Vol. 7, p. 1101). Chicago, IL: University of Chicago Press, Benton.

Ervynck, G. (1981). Conceptual difficulties for first year students in the acquisition of the notion of limit of a function, *Actes du Cinquième Colloque du Groupe Internationale PME* (pp. 330–333). Grenoble: Université de Grenoble.

Even, R. (1988). *Prospective secondary mathematics teachers' knowledge and understanding about mathematical function.* Unpublished Ph.D. thesis, Michigan State University, East Lansing, MI.

Fischbein, E. (1978). Intuition and mathematical education. *Osnabrücker Schriften zür Mathematik, 1,* 148–176.

Fischbein, E., Tirosh, D., & Hess, P. (1979). The intuition of infinity. *Educational Studies in Mathematics, 10,* 3–40.

Fischbein, E., Tirosh, D., & Melamed, U. (1981). Is it possible to measure the intuitive acceptance of a mathematical statement? *Educational Studies in Mathematics, 12,* 491–512.

Goldenberg E. P. , Harvey, W., Lewis, P. G., Umiker, R. J., West, J., & Zodhiates, P. (1988). *Mathematical, technical and pedagogical challenges in the graphical representation of functions* (Tech. Rep. No.88-4).

Educational Technology Center, Harvard Graduate School of Education.

Horace, Q. F. (1968). In E. P. Morris (Ed.), *Horace: Satires and epistles* (p. 39). Norman, OK: University of Oklahoma Press. Original work published in 35 B.C.

Janvier, C. (1987). Representations and understanding: The notion of function as an example. In C. Janvier (Ed.), *Problems of representation in mathematics learning and problem-solving*. Hillsdale, NJ: Lawrence Erlbaum.

Keisler, H. J. (1976). *Elementary calculus*. Boston, MA: Prindle, Weber, & Schmidt.

Kerslake, D. (1977). The understanding of graphs. *Mathematics in School*, 6(2), 22–25.

Kleiner, I. (1989). Evolution of the function concept: A brief survey. *The College Mathematics Journal*, 20(4), 282–300.

Leinhardt, G., Zaslavsky, O., & Stein, M. (1990). Functions, graphs, and graphing: Tasks, learning and teaching. *Review of Educational Research*, 60(1), 1–64.

Malik, M. A. (1980). Historical and pedagogical aspects of the definition of a function. *International Journal of Mathematics Education in Science and Technology*, 11(4), 489–492.

Mason, J., Burton, L., & Stacey, K. (1982). *Thinking mathematically*. Reading, MA: Addison-Wesley.

Markovits, Z., Eylon, B., & Bruckheimer, M. (1986). Functions today and yesterday. *For the Learning of Mathematics*, 6, 18–24.

Markovits, Z., Eylon, B., & Bruckheimer, M. (1988). Difficulties students have with the function concept. In S. Wagner & C. Kieran (Eds.), *The Ideas of algebra, K-12: Yearbook of the National Council of Teachers of Mathematics* (pp. 43-60).

Orton, A. (1977). Chords, secants, tangents and elementary calculus. *Mathematics Teaching*, 78, 48–49.

Orton, A. (1980). *A cross-sectional study of the understanding of elementary calculus in adolescents and young adults*. Unpublished Ph.D. Thesis, Leeds University.

Papert, S. (1980). *Mindstorms*. Brighton: Harvester Press.

Poincaré, H. (1914) *Science and method* (F. Maitland, Trans.). Republished by New York, NY: Dover Publishing Inc.

Rich, A., Rich, J., & Stoutemyer, D.(1989). *Derive™—A mathematical assistant* [Computer program] Soft Warehouse. Honolulu, HI.

Robert, A. (1982). L'Acquisition de la notion de convergence des suites numériques dans l'Enseignement Supérieur. *Reserches en Didactique des Mathématiques*, 3(3), 307–341.

Schaaf, W. A. (1930). Mathematics and world history. *Mathematics Teacher*, 23, 496–503.

Schwartz, J. L. (in press). Software to think with: The case of algebra. In D. L. Ferguson (Ed.), *Advanced technologies in the teaching of mathematics and science*. New York, NY: Springer-Verlag.

Schwartz, J. T., Dewar, R. B. K., Dubinsky, E., & Schonberg, E. (1986). *Programming with sets, An introduction to SETL*. New York, NY: Springer-Verlag.

Schwarz, B., & Bruckheimer, M. (1988). Representations of functions and analogies. *Proceedings of the Twelfth International Conference for the Psychology of Mathematics Education* (pp. 552–559). Vesprem, Hungary.

Schwarz, B., Dreyfus, T., & Bruckheimer, M. (1990). A model of the function concept in a three-fold representation. *Computers and Education*, 14(3), 249–262.

Schwarzenberger, R. L. E. & Tall, D. O. (1978). Conflicts in the learning of real numbers and limits. *Mathematics Teaching*, 82, 44–49.

Senk, S. L. (1985). How well do students write geometry proofs? *Mathematics Teacher*, 78, 448–456.

Sierpińska, A. (1985a). Obstacles epistémologiques relatifs á la notion de limite, *Reserches en Didactique des Mathématiques*, 6(1), 5–67.

Sierpińska, A. (1985b). La notion d'un obstacle epistémologique dans l'enseignement. *Actes de la 37e Rencontre CIEAEM*, Leiden, Netherlands.

Sierpińska, A. (1987). Sur la relativite des erreurs. Actes de La 39e Rencontre CIEAEM, Sherbrooke. Canada.

Sierpińska, A. (1988). Epistemological remarks on functions, *Proceedings of the Twelfth International Conference for the Psychology of Mathematics Education* (pp. 568–575). Vesprem, Hungary.

Skemp, R. R. (1971). *The psychology of learning mathematics*. London: Pelican.

Sullivan, K. (1976). The teaching of elementary calculus: An approach using infinitesimals. *American Mathematical Monthly*, 83(5), 370–375.

Tall, D. O. (1977). Conflicts and catastrophes in the learning of mathematics. *Mathematical Education for Teaching*, 2(4), 2–18.

Tall, D. O. (1979). Cognitive aspects of proof, with special reference to the irrationality of $\sqrt{2}$. *Proceedings of the Third International Conference for the Psychology of Mathematics Education* (pp. 203–205). Coventry, England: University of Warwick, U.K.

Tall, D. O. (1980a). The notion of infinite measuring number and its relevance in the intuition of infinity. *Educational Studies in Mathematics, 11,* 271–284.

Tall, D. O. (1980b). Mathematical intuition, with special reference to limiting processes. *Proceedings of the Fourth International Conference for the Psychology of Mathematics Education* (pp. 170–176). Berkeley, CA: University of California-Berkeley.

Tall, D. O. (1981). Intuitions of infinity. *Mathematics in School, 10*(3), 30–33.

Tall, D. O. (1986). *Building and Testing a Cognitive Approach to the Calculus using Interactive Computer Graphics*. Ph.D. Thesis, University of Warwick.

Tall, D. O. (1987). Constructing the concept image of a tangent. *Proceedings of the Eleventh International Conference for the Psychology of Mathematics Education* (Vol. 3, pp. 69–75). Montréal: Université de Québec à Montréal.

Tall, D. O. (1989), Concept images, generic organizers, computers, and curriculum change. *For the Learning of Mathematics, 9*(3), 37–42.

Tall, D. O, Blokland, P., & Kok, D. (1990). *A graphic approach to the calculus* [Computer program]. Pleasantville, NY: Sunburst.

Tall, D. O., & Vinner, S. (1981). Concept image and concept definition in mathematics, with particular reference to limits and continuity. *Educational Studies in Mathematics, 12,* 151–169.

Tennyson, A. L. (1969). The Ancient Sage. In C. Ricks (Ed.), *The Poems of Tennyson* (pp. 1351–1352). New York, NY: Norton & Co. Inc. Original work published in 1885.

Thomas, H. L. (1975). The concept of function. In M. Rosskopf (Ed.), *Children's mathematical concepts*. New York, NY: Teachers College, Columbia University.

Tirosh, D. (1985). *The intuition of infinity and its relevance for mathematics education*. Unpublished doctoral dissertation, Tel-Aviv University.

Vinner, S. (1983). Concept definition, concept image and the notion of function. *The International Journal of Mathematical Education in Science and Technology, 14,* 293–305.

Vinner, S. (1988). Visual considerations in college calculus–students and teachers. *Theory of Mathematics Education, Proceedings of the Third International Conference* (pp. 109–116). Antwerp, Belgium: University of Antwerp.

Vinner, S., & Hershkowitz, R. (1980). Concept images and common cognitive paths in the development of some simple geometrical concepts. *Proceedings of the Fourth International Conference for the Psychology of Mathematics Education* (pp. 177–184). Berkeley, CA.

Wheeler, M. M., & Martin, W. G. (1987). Infinity Concepts among pre-service elementary school teachers. *Proceedings of the Eleventh International Conference for the Psychology of Mathematics Education* (pp. 362–368). Montréal: Université de Québec à Montréal.

Wheeler, M. M., & Martin, W. G. (1988). Explicit Knowledge of Infinity. *Proceedings of the Tenth Annual Meeting of PME-NA*. DeKalb, IL.

Yerushalmy, M. (in press-a) Students' perceptions of aspects of algebraic function using multiple representation software. *Journal of Computer Assisted Learning*. Blackwell Scientific Publications.

Yerushalmy, M. (in press-b). Understanding concepts in algebra using linked representation tools. In D. L. Ferguson (Ed.), *Advanced Technologies in the Teaching of Mathematics and Science*. New York, NY: Springer-Verlag.

CRITICAL ISSUES

·21·

TECHNOLOGY AND MATHEMATICS EDUCATION

James J. Kaput

UNIVERSITY OF MASSACHUSETTS—DARTMOUTH

Anyone who presumes to describe the roles of technology in mathematics education faces challenges akin to describing a newly active volcano—the mathematical mountain is changing before our eyes, with myriad forces operating on it and within it simultaneously. Many of these forces contain a technological component. The methods of mathematical research and application and the criteria regarding what is important to investigate are all evolving more rapidly than ever before (Steen, 1988). These changes affect decisions on what mathematics should be included in the school mathematics curricula. But the same technological forces that shape the mathematics also deeply affect the teachability and learnability of mathematics, both new and old.

Further, we must take into account the impact of accumulating experience with electronic technologies. While electronic computation has been in the hands of mathematicians for four decades, it has been in the hands of teachers and learners for at most two decades, mostly in the form of time-shared facilities. But the real breakthrough of decentralized and personalized microcomputer-based computing has been widely available for less than one decade. And it is the latter facility that has brought the greatest promise for computers in education. But even here, we must distinguish between the work of microcomputer pioneers working in laboratories, who have been at work since the early 1980s, and the end-users of computers in typical schools in typical classrooms, who were only in the late 1980s beginning to tap the resources of this class of inventions. Thus our store of reliable practical experience of computers in schools is limited, but beginning to grow rapidly, especially as access to technology increases.

Additionally, the technology itself is evolving very rapidly, especially as the economic incentives of ever wider use take effect. Most of the memory, display, and processing limitations of the popular computers of the 1980s have been pushed back by at least an order of magnitude. Exponential growth in processing power, display space, and memory capacity is reflected in the fact that these capacities more than quintupled from the Macintosh computer used to write early drafts of this chapter to the new Macintosh used to produce the last drafts. More power makes new functionality possible. Using a new outliner and employing the larger display space, I can view the entire chapter at various levels of detail and keep available a view of the entire chapter's structure while I work on details. Simultaneously, I run other mathematics learning software that I am describing in this document and import graphics from this other software into a drawing program for scaling and adjustment before bringing them into this document. Little of this was possible on the earlier computer, which was also harder to use than this more powerful one.

Since technological limitations greatly determined so many other aspects of the early computing experience (later described as analogous to the Model T Ford), we must be very cautious about extrapolating from the recent past. Major limitations of computer use in the coming decades are likely to be less a result of technological limitations than a result of limited human imagination and the constraints of old habits and social structures. Other factors, such as the role of networks in classrooms, are changing quickly. Network and human interface technologies have been among the slowest to develop but are now beginning to gain momentum. At the same time, new video and hypermedia technologies are beginning to appear in school-affordable form. And other technologies, once separate, are beginning to converge, as hand-held graphical super-calculators assume the functionality of microcomputers, as high-end personal computers assume the functionality of elaborate workstations, and as digitalization of video progresses

The author gratefully acknowledges support of portions of the work described in this chapter from the National Science Foundation grants (MDR-8850623 and MDR-8855617), and from the Apple Computer, Inc., External Research Program and the Apple Classrooms of Tomorrow Program. The views expressed are those of the author and do not necessarily reflect those of the grantors. The author also wishes to thank Roy Pea, Northwestern University, and Pat Thompson, San Diego State University, for helpful advice on the content and organization of this chapter.

to merge television and computers. For less than $500, I can purchase a device to run my television signal into this computer and display it in a separate, resizable window without affecting work as usual on the remainder of the computer screen.

Our Strategy: To Look First at Underlying Principles and Processes and to Recognize Old Questions and Issues

One approach to describing the roles of new technologies in mathematics education is to examine how they are affecting each content area, choosing some aspect of mathematical content as the basis of our analysis. For example, we can consider arithmetic, geometry, algebra, calculus, discrete mathematics, and probability and statistics, and then examine technology impacts regarding curriculum and pedagogy. Such an approach, while superficially sensible and practical, does not reach deeply enough to uncover the factors that are at the foundation of the many changes that are occurring across topics and mathematical levels. We need to identify what is different about the new electronic media and what those differences mean in terms of cognition, learning, teaching, and related matters. Nonetheless, we shall use a subject-matter breakdown later as a convenient way to summarize recent research and development activities.

Another approach is to orient our discussion toward those technologies that would be employed in a reformed practice rather than in past practice. Technologies that simply transfer the traditional curriculum from print to computer screen are at most of historical interest. The range of new technologies covered by our discussion is dominated by microcomputer technology, although we will occasionally discuss related video technologies, for example, interactive video. For the most part we will not be concerned with questions regarding particular hardware, storage media, and the like. These change too rapidly to have a place in a book such as this.

Indeed, given the rapidity and depth of ongoing changes in mathematics and technology, I have chosen an approach that attempts to reach below the surface to focus on the underlying structure of the user's interaction with mathematical notations in any medium, without considering which technologies might be involved. This will enable us to examine the ways that electronic technologies can be used to represent mathematical ideas and processes, how they differ from representation in traditional static media, and what the specific sources are of any learning efficiencies or deficiencies.

This approach based on underlying principles and processes is complemented by a practical perspective that views new technologies as frequently re-energizing age-old questions. These include questions regarding educational goals, appropriate pedagogical strategies, and underlying beliefs about the nature of the subject matter, the nature of learners and learning, and the relation between knowledge and knower. Implementation of new technologies also forces reconsideration of traditional questions about control and the social structure of classrooms and organizational structure of schools. Although we cannot dwell on all these issues, the reader is urged to reflect on them in the new circumstances that new technologies create. To assist in this reflection we will review some of the classic statements regarding computers in education before moving on to our contemporary perspective.

While I shall discuss new and emerging, mainly electronic, technologies and media, we should realize that schools have already invested heavily in technologies and media that play a large role in determining both the forms of current practice and how much that practice can be altered. Disposable workbooks and stencils, chalkboards, lined paper, and the basal text are a background for existing practice and a baseline for our expectations for the future.

Chapter Overview

The chapter begins with some context setting, first through a historical analogy and a review of the current state of affairs, and then through a review of the classic literature on technology and education, especially mathematics education. We then build a solid representational perspective on the new technologies that attempts to unify the unique phenomena that the new technologies offer. This perspective is applied to make sense of the different forms that computing environments take, and especially how they change the processes of learning and using mathematical ideas. Since one way to look at computer representations of mathematical ideas is as new forms of manipulatives (Lesh, Post, & Behr, 1987), we compare computer and physical manipulatives for a well-known manipulative, Dienes blocks, in some detail to help uncover deeper differences. The computational contribution is then traced in the major subject matter areas, with a variety of specific examples. A review and a critique of the various uses of artificial intelligence in mathematics education that draw contrasts with the field of medicine follow. The last major section considers issues of implementation, not to provide concrete tactics for dealing with particular situations, but rather to provide a general strategy. The conclusion poses a series of questions intended to focus continuing thinking about the ideas that are raised in the body of the chapter. While no claim is made that the reference list for this chapter is in any sense complete (many references were deliberately not included because they seemed dated or because their message is subsumed in more accessible references), it is intended to provide wide entry to each of the matters discussed.

WHERE ARE WE IN THE HISTORY OF EDUCATIONAL COMPUTING?

Analogy with Early Automobiles—From the Model T to the Model A

We are in the early days of the use of electronically-based technology, perhaps analogous to the days of the Model A Ford. We have progressed beyond the Model T, where one needed to be an expert to start and run the vehicle, where standards for automobile operation were not yet established, and where competition with older modes of transportation were important factors. In Model T days, one had to attend at least as much to the vehicle and its operation as one attended to where one was traveling. In computer terms this period corresponds to the days up to the early commercially available microcomputers, up to perhaps the early 1980s.

From the Model T to the Model A, operation of the transmission moved from foot control to hand control, while control of the accelerator moved in the opposite direction. Overall, the "intelligence" (both physical and cognitive) needed to operate the vehicle migrated from the user to the machine, including aids that we now take for granted, such as electric starters, the coordination of spark plug voltage, temperature, pull-load, and fuel input, automatic windshield wipers, brake lights, etc. Nowadays, of course, most of the operation of the vehicle beyond steering and acceleration is handled by the machine itself, with a considerable amount of human interface standardization across machine types. For example, on-off control of headlights is almost always found in a switch to the left of the steering wheel, and brightness control is usually on a switch attached to the steering column (although for many years it was governed by a foot switch). Such a level of standardization is not yet apparent in today's computers but is rapidly developing as graphics-oriented controls based on our visual knowledge of physical objects become widely employed. We are currently in the latter days of the Model A.

Similarly, rules of the road gradually came to be standardized and codified as experience dictated needs and refined solutions. During the transition period, a number of temporary phenomena occurred *simply because a transition was underway*—from horse and buggy to horseless carriages. Hence one needed a protocol to deal with frightened and startled horses, for example. (We see relatively few articles now about computer phobia [Jay, 1981].) And, of course, the roads themselves, reasonably well suited for earlier modes of transportation, needed rebuilding to accommodate the requirements of automobiles, which could not approach their real potential on roads not designed for them.

On the computer side of the analogy, the rebuilding of classroom facilities and especially the redesign of classroom practice has only begun. School structures of all types (physical, scheduling, curriculum, materials development and dissemination, and so on) have been developed for the use of traditional media by all parties—teachers, students, administrators. Infrastructure is slow to change. Only recently has the National Science Foundation begun funding computer laboratories for departments of mathematics, for example.

Only after sufficiently many automobile drivers had taken to the roads, creating various difficulties based on driver incompetence or inexperience, did driver credentialing processes come to be standardized (by state). Computer credentialing for teachers is in the nascent phase. An earlier, ill-formulated and aborted version of such, under the title of "computer literacy," amounted to certification on mastery of the mechanics of a Model T. It is an interesting example of education preparing for the technologically-recent past.

One last intrinsically transitional factor worth mentioning has no strong analogy with auto drivers because of the relatively short amount of time one spends learning to drive and the short apprenticeship required to become experienced drivers. In the 1980s and before, most working teachers were educated without significant experience as computer-using students, without computer use as a significant part of their teacher-preparation process, and with only marginal expectations regarding their computer use as professionals in the classroom—usually based

on Model T experience. Serious use of computers of any kind, let alone the level of computing to be available in the 1990s and beyond, is simply not part of the culture of schools and schooling. This transitional factor will be in effect for at least a generation, although recent moves by various groups, including the major teacher labor organization (NEA, 1989), are aimed directly at putting computer workstations on each teacher's desk as soon as possible. The strategy is to engage the teacher with the computer as a personal productivity tool before expecting sustained use in classroom teaching.

Consequences of Our Place in History

Let us now leave behind our analogy with automobiles and examine the current state of affairs with respect to the implementation of computer technology in schools.

1. Very few computers are available in schools, and those that do exist are primarily first-generation microcomputers that are hard to use and of very limited computational and display power. While some may regard the decrease in the number of students per computer to between 25 and 30 nationwide as reason for optimism (Becker, 1990; Office of Technology Assessment [OTA], 1988, Ch. 1.), simple arithmetic reveals that this amounts to at best an hour per week per student for mathematics. If that much time were available for paper and pencils, we would not expect much learning to take place based on paper-and-pencil technology.
2. One significant reason for the lack of computers is the lack of software in sufficient quantity and of sufficient quality to merit the investment required for large-scale classroom computer use. The bulk of commercially available software has been of the computer-assisted instruction (CAI) drill and practice genre (OTA, 1988, Ch. 6).
3. Except for a few complete, turnkey CAI systems that are pedagogically primitive and oriented to drill-and-practice learning of computation skill, that software which *is* available (a) occurs in scattered pieces that differ in user interface and underlying structure across topics and grade levels, even when produced by the same publisher, and (b) is not systematically tied to the ongoing flow of curriculum, either the traditional one or a reformed one, or (c) is based on generic tools (calculators, graphing utilities, spreadsheets, etc.) that are retrofitted as educational tools.
4. An immediate consequence of (3) is that computers are too difficult for the average teacher to use in the typical classroom on a sustained basis. The high price of entry in terms of effort, especially in the context of the historic curricular pressure toward easily measured computational skill, is likely to continue until the software publishing industry begins to produce more completely integrated systems, including teacher support systems, and systems congruent with the reform goals now widely advocated.
5. Little or no pre-service preparation for teachers includes systematic in-depth experience with technology that anticipates what those teachers could expect as little as five years into their professional lives. Schools of education mirror the

schools themselves as technological ghettos and thus help perpetuate the status quo.

6. Because of (5) and because most current teachers were educated prior to the introduction of computer use in schools, teachers' expectations regarding technological support for their daily work have been virtually nil.

7. Given all the previously cited conditions and the continuing profitability of text-based curricular material, economic incentives have favored the technological status quo rather than a move to massive computer use in mathematics classrooms—neither hardware nor software purchases are seen as worthy investments of resources (OTA, 1988, Ch. 6).

Coupled with the above factors are the traditional school inertia and the fact that, as Olds, Schwartz, and Willie (1980) said, "Educators appear to have a deep set skepticism towards anything that plugs into the wall" (p. 2; see also Cuban, 1986). Cohen (1988a, 1988b) situates the move to new technologies in the long-term historical context of a glacial move towards more ambitious constructive pedagogies, which in his view is a matter of deep cultural change requiring centuries to effect.

Given the preceding list of negatives, one might wonder why a chapter on technology and mathematics education is even needed in this book. But major changes are underway that are likely to produce substantial transformations in the 1990s, some based in the technology itself, some in the reform effort, and some in the interaction between the two.

Analogy with Bicycles

Yet another comparison may help make sense of the relation of research to practical use of computer technology in education. Let us set up an analogy between a child's use of a bicycle and a child's learning of mathematics. The question of whether a child can travel farther and faster on a bicycle than on foot is moot. It is not worth proving. The advantage of bicycles over shoes, however, depends on the terrain and even the age and physiological sophistication of the child. And the same kind of device may not be appropriate for all levels of user experience (for example, some may need extra wheels for balance purposes). And there are losses in efficiency and a certain amount of danger as one first learns how to use the new mode of transportation. Moreover, there are many circumstances where bicycles are inappropriate but where body motion is needed (for example, indoors, playing most sports, in water, on snow or ice, etc.). Despite all this, bicycles do a much better job of converting the *child's* energy to forward motion (we emphasize that it is the child's energy here, since we are not speaking of motorized vehicles). Furthermore, use of a bicycle can improve one's fundamental physiological performance off a bicycle. One can even ask, what is the maximum speed of, say a 12-year-old child on foot over a one-mile distance on a running track? Eight, maybe ten miles per hour. And on a bicycle? Perhaps 20 miles per hour. Could enough hard training, better shoes, or even careful breeding make up the difference? Not likely.

I believe that traditional versus computer modes of children's learning and doing mathematics parallel the bicycle situation. The question of whether a child can learn and do more mathematics with a computer (or other forms of electronic technology, including calculators and various video systems) versus traditional media is moot, not worth proving. That computational aids overall do a better job of converting a child's intellectual power to mathematical achievement than do traditional static media is unquestionable. *The real questions needing investigation concern the circumstances where each is appropriate.*

However, as with any analogy, one must be careful regarding its boundaries. For example, the standard bicycle, as a technology, is fairly completely evolved, whereas the computer is decidedly not. Indeed, most investigations of children's learning have involved seriously deficient software and hardware—at the level of the Model T. Furthermore and more importantly, walking as a mode of transportation is "natural" in a way that typical school use of traditional media for mathematics learning is not. While school use of traditional media has evolved into a stable state, the result may not be, and probably is not, optimal with respect to student learning—for the simple reason that the selection mechanisms underlying the evolution of school practice have not had much to do with quality, viability, and depth of student learning.

There is yet another distinction worth making here, between the evolution of traditional media-based school and classroom practice and the evolution of the languages and notations in which mathematics has traditionally been written. Again, circumstances have changed to an extent that new selection mechanisms governing the evolution of mathematical notations are coming into effect. Essentially all mathematical notations evolved in the context of static, inert media; their primary inventors of the past several centuries were the knowledge-producing elite; and their primary application beyond shopkeeper arithmetic was, again, to serve the needs of a small intellectual elite. All three underlying factors have changed: media are now dynamic and interactive; a much broader class of individuals has a hand in developing the languages in which ideas are expressed; and, most critically, virtually the entire population is now deemed to be in need of significant mathematical power (Mathematical Sciences Education Board, 1990, National Council of Teachers of Mathematics, 1989; National Research Council, 1989).

TRADITIONAL WAYS OF THINKING ABOUT COMPUTERS IN EDUCATION

By 1980, sufficient experience had been accumulated with computers in education so that underlying questions regarding their nature and appropriate use became clearly articulated. The questions, reflected in a classic paper whose title is that of the next subsection, are very much worth asking today, but their answers are likely to be more complex than when the questions were first formulated. We will review the paper from the perspective of the early 1990s, offering contemporary examples of the types of software described in the article.

"People and Computers: Who Teaches Whom?"

This paper (Olds, Schwartz, & Willie, 1980) was the result of an informal study at the Education Development Center (EDC) sponsored by Control Data Corporation; the EDC; and Robert B. Davis, David Kibbey, and others at the University of Illinois who were the pioneer developers of the Plato programs. The Plato programs, state-of-the-art CAI programs built on tutorial and games models, became available to run on mainframe computers beginning in the late 1960s and the 1970s (derivative programs continue to run on microcomputer hardware today). The paper offers a taxonomy of forms of computer use and a series of open questions regarding the optimal application of computers in education. The basic forms of computer use strictly *as an educational medium* are games, tutorials, and simulations. To the use of the computer as a medium, they add two more important uses, the computer as tool and the computer as tool-maker. The categories in this paper are not radically different from those of another classic book that helped frame early thinking about computers in education, *The Computer in the School: Tool, Tutor, Tutee* (Taylor, 1980). As we shall see, these categories have tended to run together in recent years, and will likely continue to do so.

Computer Games. Olds, Schwartz, and Willie (1980) distinguish two types of games; the goal of the first is the teaching of some particular subject matter ("content games") and of the second, the teaching of a more general cognitive or problem-solving strategy applicable across varying subject matter ("process games"). The former have more easily described and measured outcomes than the latter. Consider, for example, the classic *Green Globs,* (Dugdale, 1982). Its curricular domain is the translation between algebraic equations and coordinate graphs of those equations. The student's objective is to write equations whose graphs hit as many of a given, randomly placed set of targets as possible, and the reward for hitting the targets is exponentially related to the number hit. The families of equations can be varied, as can scaling and other parameters of the game, but its curricular target is relatively clear. A game with a slightly wider curricular range is *Guess My Rule* (Barclay, 1985). In this game students are to guess the computer's function, which is taken either randomly or in a particular order from a list which can be created by the teacher. The student inputs the domain numbers for which the computer must return the function's values. This feedback can be viewed either numerically in a table of input-output values or graphically as discrete points on a coordinate graph. When the student feels ready, a guess is made and then compared to the values of the computer's function. While no strong game goal structure is built in, the usual objective is to complete a list using as few domain values and guesses as possible. Here the curricular goal is to teach the behavior of a given set of functions or function classes, based on either numerical or graphical views, or both. Clearly, by choosing functions carefully and also by determining their order of presentation, a teacher can adjust the educational objectives of this game quite broadly to include various inference strategies. For example, it is possible to include functions containing a random component, so that students must decide whether a deterministic rule exists at all. A series of games directed toward teaching a wide variety of inference strategies based on rule-guessing (although sometimes only to predict new values of a sequence) is the *Kings Rule* series (O'Brien, 1985). Much more general and hence more difficult to specify

and measure are the skills developed in adventure games, although it is also possible to mix activities and objectives, for example, by requiring the player to perform some particular task successfully before being allowed to make the next move.

The most comprehensive study of the educational impact of educational computer games in mathematics has been offered by Bright, Harvey, and Wheeler (1985). They give a comprehensive review of research through the early 1980s and evidence from 11 studies of their own. The role of motivation has been extensively studied by the social psychologist Lepper and colleagues at Stanford University (Lepper, 1985; Malone & Lepper, 1987), although without much attention to the curricular value of the educational objectives of the games involved. The study of the particular cognitive and affective impacts of differing forms of games and reward structures is very difficult to conduct because of the large number of variables involved and the large proportion of these that cannot be adequately controlled. Most of the studies have been based on relatively trivial content games involving the learning of the syntactic transformations of a particular notation system, although Mandler has also studied the affective content of process game playing, by varying the reward structure of adventure games, in order to test his theory relating affect and cognition (Mandler, 1984, 1987).

Computer Tutorials and the Character of Early Computer-Assisted Instruction (CAI). After the choice of educational content of a tutorial has been made, simulating a tutor-tutee relationship in the computer medium involves many design decisions, some of which are based on assumptions about the structure of the domain being taught, others on assumptions about the learner, and yet others on assumptions about the appropriate form of the interaction—should it be conversational, menu-driven, or what? Besides these assumptions, the program itself is governed by the ability of the implementers to instantiate the assumptions in software, an ability that depends to a significant extent on the nature of the hardware and software systems in which the instantiation must take place. This fact was acknowledged by pioneers in applications of artificial intelligence: "Many of the instructional techniques currently in use are there because technology makes them easy to employ, not because they are educationally sound." (Shank & Edelson, 1989/90, p. 20.) A similar statement can be made about the choice of topics to which computers have been applied.

Early mathematics tutorials tended to be based on teaching syntactically-guided manipulation of the standard notation systems, both arithmetic and algebraic. Designers adopted without modification the traditional notation systems and traditional rote-teaching pedagogies and simply transferred these to the new medium using the computer's interactivity capability—the computer now indicated whether the user had provided a correct or incorrect input. This interactivity replaced the teacher's or peer's interactive role, so in a sense the computer was not a true addition to the educational environment, but rather a shift of the source of the feedback from humans to the computer medium. Early CAI thus changed very little in terms of curriculum or pedagogy, except for the impersonal nature of the feedback system. It has been argued that this impersonality can be a positive factor in student learning on the basis of

(a) the privacy of the interaction possible, and (b) the "patience" of the feedback system (Bork & Franklin, 1979).

We will review and critique the history of tutorial approaches to computer use in education as part of our discussion of the application of Artificial Intelligence (AI) in education later in this chapter.

Computer Simulations and Microworlds. The third class of educational computing identified by Olds, Schwartz, and Willie involves simulations. They distinguished two types of simulations: One is executed in parallel with the system that one is modeling and hence can be empirically checked by measurement of and comparison with the modeled system; the other does not afford easy comparative measurement. As an example of the first, they offer a computer pendulum which can be run at the same time as a real pendulum and compared with it. For example, one could use such a simulation to identify discrepancies between the model and the thing it is modeling, such as determining the role of friction in the motion of the real pendulum. One could then introduce and vary friction in the computer pendulum to match that of the real one, thereby building a more precise understanding of friction phenomena and perhaps (depending on the transparency of the underlying mathematics of the computer model) model the friction mathematically.

The second type of simulation is used when spatial or temporal scales prohibit direct checking of the simulation with what it is modeling, as might occur with a model of planetary or molecular motion. Such simulations are actually concrete models of more abstract models. Generally, simulations require more computer horsepower than do simple CAI programs, especially to the extent to which they attempt graphical depiction of the quantities and relationships in the models. With the advent of bit-mapped graphics on school-affordable computers, this constraint is considerably lessened. ("Bit-mapped graphics" means that the computer's display is controlled one pixel (or dot) at a time. In contrast, in a character-based system the computer sends messages to the screen based on a given set of displayable characters which can fit into a certain set of positions, e.g., an 80-column display. Microcomputers in the 1990s have at least bit-mapped capability.) Increased graphics capability allows for more graphic realism in the simulations, so that while early simulations might have a graphically crude pendulum with the data available for fine analysis given in tabular form, a contemporary simulation might display an ideal and a "real" pendulum side-by-side or superimposed, with the user able to see graphically subtle differences in the motion resulting from friction or other forces. The user might also be able to initiate movement directly by "picking up" the end of the pendulum with a pointing device such as a mouse.

Computer Tools—Specialized and General. Olds, Schwartz, and Willie distinguish between special purpose and general purpose tools. Given the rapid expansion in the uses to which computers have been put in recent years and the rapidly changing ways in which tools become integrated and differentiated, it is perhaps best to treat these as points on a continuum rather than as categories. Specialized computer tools abound in the world of work, such as inventory management data

systems, mailing list manipulators, airline reservation systems, stock market and weather forecasting systems. Other specialized uses, for example, automobile fuel injection systems and microwave ovens, touch the lives of most Americans daily. In contrast, the tools used in education tend to be more general, although their specific uses can vary widely. For example, symbol manipulators, spreadsheets, graphing utilities, statistics and data modeling programs, database programs, and even word processors and calculators are all relatively general-purpose tools, with origins outside of education, which can be adapted to particular educational purposes. We shall outline a number of projects below that exploit one or another such tool in the teaching of mathematics. However, as we shall discuss below, the retrofitting of generic tools into educational environments is not an easy task for the average teacher in the typical classroom, and it has not achieved widespread success. Instead, the trend seems to be toward the development of variations on generic tools designed specifically for use in educational settings (Goldenberg, 1991). We shall describe a series of such programs below.

As special purpose tools become more sophisticated, they tend to include more user-settable options and command structure. They become more like specialized programming languages rather than tools, so, as discussed further below, the distinction between programming language and tool tends to become blurred. For example, the graphically oriented spreadsheet *Wingz* (Informix software, 1989), as well as many of the more advanced database systems, allows one to create special purpose programs within it as well as to modify its structure so that it can behave according to the user's needs. A complementary trend has recently developed based on new ability of systems to support data transfer between separate applications. It is likely that one may be able to purchase modules in the future and put them together like building blocks to create customized sets of tools.

Computer as Tool-Maker and Medium-Builder. Olds, Schwartz, and Willie point out that the one aspect of computers that distinguishes them from other devices that have served as either an educational medium or tool is their capacity to serve as tool-makers. The most fundamental examples of such are the general purpose programming languages such as Logo, BASIC, Pascal, and the more narrowly purposed educationally oriented languages such as *Function Machines* (Feurzeig & Richards, 1991), and so on. The latter include the new types of authoring systems such as *Boxer* (diSessa and Abelson, 1986), *HyperCard* (©Apple; Apple Computer, Inc., 1991) and other such systems, offering easily produced on-screen animation, that are beginning to appear (Authorware, Inc., 1989; *Macromind Director Interactive,* Macromind, 1991). These systems are built within some other programming language or system and are intended more to facilitate the building of simulations, tutorials, and/or microworlds (i.e., computer media) than to build tools (although *Boxer* and *Function Machines,* both Logo derivatives, also facilitate tool-building). In fact, they usually contain means by which the student-user has access to independently constructed tools. They use a variety of techniques, such as natural language-like commands, visual depiction of program structure, and so on, to widen the field of potential authors.

HyperCard, in particular, is quite widely used, particularly in higher education. Programs built using *HyperCard* are frequently called "stacks," building on the "card" metaphor.

Such general simulation/microworld builders usually capitalize on a coherent system of metaphors rooted in some arena of activity, sometimes referred to as a "world metaphor," to help organize the user's thinking. For example, *Macromind Director Interactive* (Macromind, 1991), an animation-construction program, uses stage and choreography metaphors throughout. On the other hand, the domain-specific tool and microworld builders capitalize on the user's knowledge of that domain and its language. *Interactive Physics* (Knowledge Revolution, 1990) and STEP (Wings for Learning, 1991) (which allow the user to set up a wide variety of science laboratory explorations) or *Mathematica* (Wolfram Research, Inc., 1989; Wolfram, 1989) (a potent and general set of mathematics tools coupled with a sophisticated command language) require that the user be competent in physics or mathematics, respectively. Note that *Mathematica* is an example of a program that is both a powerful tool and a tool/microworld maker simultaneously.

The General Question of Programming Versus Using Specific Instructional Software

The traditional issue of specificity of instructional goals plays itself out in very concrete terms in the question of learning computer programming. One objective for student programming is based on the presumed general cognitive skills the student learns as a result of appropriate kinds of programming activity, for example, planning skills, problem breakdown skills, and so on. In this sense programming in a computer language substitutes for learning, say Latin, where learning the particulars of the language (which may have some degree of face validity) is subsidiary to learning more general thinking skills, as well as a whole orientation to complex problem solving (Channel & Hirsch, 1984; Papert, 1980). Another objective of programming activity is the more domain-specific learning that takes place, as one either attempts to solve, say certain mathematics problems, or uses specifically tailored commands to operate on a specific kind of mathematical object, as in a Microworld. Camp and Marchionini (1984) provide illustrations of this approach involving "PERCs" (Programming Exercises Related to Content). Shumway (1984) has argued that even the youngest children be engaged in programming. The idea that programming can act as a medium for mathematics learning goes back at least to the late 1960s (Feurzeig, Lucas, Grant, & Faflick, 1969). A particularly thoughtful analysis of one child's learning in a Logo context and how that learning relates to other learning taking place away from the computer has been provided by Lawler (1981, 1985).

Apart from teaching programming to impart the particular technical programming skills of a particular language, we see a general trend in the use of programming to focus its instructional use on domain-specific knowledge.

As already pointed out, software working and learning environments are becoming ever more flexible and user-modifiable so that the teacher and/or the student can tailor environments for particular purposes. Moreover, as the procedure-capturing capacity of systems improves, the line between "doing" in a particular domain and "programming" to accomplish tasks in that domain dissolves. (By "procedure-capturing" we mean the ability of a system to record the steps or commands that a user performs and then play them back as requested.) This is especially the case as the captured actions are represented as an editable program in some easily read language. This is sometimes called "programming by doing." (See the graphic spreadsheet *Wingz,* Informix Software, 1989, with its scripting language HyperTalk for a potent example.) The *Geometric Supposer* provides an example of procedure-capturing without the production of an editable script (although it is not hard to imagine a version for which the sequence of commands constituting a geometric construction appears as a nameable, storable, and perhaps editable program). Hence arguments regarding the relative roles of programming and use of specifically designed instructional programs become more complex as the boundaries between the two are no longer clear.

Note also that certain high level languages such as the Logo derivative *Boxer* (diSessa & Abelson, 1986) are specifically designed to make building user-modifiable tools easy to accomplish; in *Boxer* the tools that are constructed can be opened to reveal and perhaps alter their "working parts." In *Function Machines,* (Feurzeig & Richards, 1991) programs are constructed and appear on the screen as input-output machines whose workings can be made visible if the user desires.

The matter of programming versus prepackaged tools or other software presents yet another example of the way that technological changes influence what is important and how we think about it. While the programming issue was hotly debated for a decade or more, it has disappeared as an argument *in that particular form.* It remains, however, as an issue of specificity of instructional objectives (Lehrer, Guckenberg, & Sancilio, 1988). See Pea and Kurland (1984) and Kurland, Pea, Clement, and Mawby (1986) for deep examinations of the general cognitive impacts of programming as well as a study of the inconclusive impacts of Logo programming in particular, and Pea (1988) for a discussion of the whole matter of transferability and the role of technology in promoting newly defined versions of transferability.

A large Logo literature and materials supporting the use of Logo to teach a variety of subjects, including mathematics (Abelson & diSessa, 1981; Byte, 1982; Feurzeig, 1988), exist. Over the past few years, it appears that interest in Logo as a medium for learning mathematics is waning, with ever fewer instructional materials and research articles published annually. Nonetheless, important Logo work continues to appear; an example is a study by Harel (1990). She had students build fraction-learning environments for their fourth-grade peers, with strong results in several dimensions, including rational number learning (see also Harel & Papert, 1990). We review how Logo affects geometry learning later.

Reflections on the General Issue of How to Think About Computing in Mathematics Education

The previous discussion of how the programming versus instructional packages debate has become obsolete in the face of changing technology while the underlying issue of specificity of instructional goals remains, is but another example of the problem of how to think about technology and mathematics education when the technology itself changes so quickly. It is difficult to recognize what is fundamental and what is not.

Both hardware and software categories of technologies change—and even the distinctions between these two change—as programs become ever easier to instantiate in different forms of read-only memory (ROM), or "firmware." Although it has not yet been seriously discussed as a practical option, we may expect some version of a "semi-computer" that students will be able to take home just as they would a textbook after loading software from the classroom computer. They could then send back their work to the school computer on the next day. Such a semi-computer already exists at a retail price in 1991 of about $1200, perhaps five times the price of a school-affordable version. Similarly, distinctions between personal computers and workstations blur, and our expectations regarding affordable computational power are challenged by the continuing exponential growth of computing power per dollar. New genres of software appear, which in some cases can have major impact on education— *HyperCard* (Apple Computer, Inc., 1991) and "stackware," for example. The idea of a teacher workstation, once seldom discussed (Kaput, 1988), is becoming a popular notion; soon it will become part of our normal expectations for the classroom, and the issue will only be what features to include and at what cost.

Then there are the school and classroom organization issues and management issues. Not only will student computers be networked to one another and to the teacher's workstation, but teachers' workstations will be networkable to one another and to administrative computers. The long-term effects are difficult to predict, just as were the long-term effects of the automobile on population demographics and local and regional government. Among trends that seem clear is the increasing use of networks, which provide all the economy and storage convenience of time-shared facilities, but are not confined by the computational and graphics constraints of such systems. Finally, there is the matter of computationally-based intelligence and its relation with the traditional roles of the teacher—how much of the teacher's work can be automated? This will be discussed later.

The purpose of the preceding discussion was to engender a perspective so that the reader will ask how much of any issue raised in the following discussion has to do with technology and how much has to do with fundamental questions of mathematics education. It further was intended to suggest that we keep a very open mind about what is possible and appropriate in mathematics classes—after all, we are only in the age of the Model A.

FOUNDATIONS OF OUR APPROACH

As mentioned in the beginning, key strategies in coming to terms with the rapid and unpredictable changes in new technologies are to seek a point of view that reaches beyond the particulars that seem to be so susceptible to change and to create a framework in which we might make sense of the novel contributions that the new technologies seem to provide by distinguishing between the old and new media in a principled way. Historically, mathematical notation systems have been instantiated in static, inert media, but the new electronic media now afford a whole new class of dynamic, interactive notations of virtually any kind. To appreciate more fully the role of the new media and notation systems in thinking and learning processes, we will examine the interactions between mental processes and physical actions in structured physical media. Our notation-representation foundations as described here are more fully developed in a sequence of papers (Kaput, 1987a, 1989a, 1991a, 1991b).

Cognitive Structures and Notation Systems

(This section is adapted from Kaput, 1991b.) Our mental apparatus, while very limited in its processing and working memory capacity, is remarkably effective in the handling of extremely complex ideas and processes, both concrete and, even more remarkably, abstract ones. This power seems to be based in the interaction between two sources of organization of experience: (1) the structures inherent in our long-term knowledge, and (2) our ability to exploit physical means of organizing experience—in the case of mathematical experience, we utilize notation systems, mainly inherited forms for externalizing conceptual structure, but including personal, idiosyncratic marks as well. We use these in our mathematical activity just as a carpenter uses a T-square in carpentry activity. To help provide an organizational framework, the reader might think of our discussion as being about the relations between thought and language (broadly construed to include any type of mathematical language)—how we make sense of our continuing flow of experience via interactions between them.

To make sense of the interactions between processes involving mental structures and processes involving physical ones, we need a language that, in the linguists' sense, has separate "registers" for each, as well as a register for the interactions. After all, the process of making sense of our experience—in this case experience with new media and notations—requires a language for its expression, and we must adopt or define such here. Hence we posit two worlds, (i) a world of mental operations, which is always hypothetical, and (ii) a world of physical operations, which is frequently observable. As depicted in Figure 21.1, these two worlds interact in opposite directions, although in a subtle sense, each can be presumed to be cyclical.

The upward arrow in the figure depicts two types of events: (i) deliberate, active interpretation (or "reading"), and (ii) the less active, less consciously controlled and less serially organized processes of having mental phenomena evoked by physical material. The arrow pointing downward likewise depicts two types of processes: (i) the act of projecting mental structure onto existing material, and (ii) the act of producing new structures ("writing"), which includes the physical elaboration of existing ones.

Projections occur in reading and evoking as part of the underlying cyclical processes matching percepts and concepts (as described by Grossberg, 1980). I nonetheless distinguish

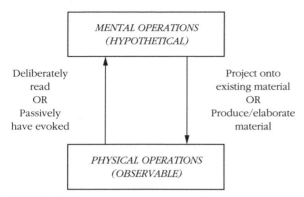

FIGURE 21–1.

downward oriented projections from upward oriented interpretive acts on the basis of the objective of the process, on where it gets its major impetus. In the downward oriented case, one has cognitive contents that one seeks to externalize for purposes of communication or testing for viability. Upward oriented processes are based on an intent to use some existing physical material to assist one's thinking.

What Do We Mean By "Notation System" and "Medium"? But now, what can exist in the bottom part of Figure 21.1? We must be able to answer this question before we know what kind of "writing" is possible. We are all familiar with a variety of traditional notation systems, including the base-ten Arabic numeration system, the algebraic notation systems for one or several variables, the systems of coordinate graphs of differing dimensions, the systems of multiple columns of tables of data, and so on. We are also familiar with special purpose educational notation systems constructed using Dienes blocks, Cuisenaire rods, fraction bars, and even bundles of sticks organized in particular ways. All these systems amount to particular organizations of physical materials based on certain syntactical rules, not always explicit, that determine the allowable objects and the allowable transformations or elaborations.

Informally then, we define a *notation system* to be a system of rules (i) for identifying or creating characters, (ii) for operating on them, and (iii) for determining relations among them (especially equivalence relations). Thus the characters need not be strings of letters or numbers, and can include graphs and diagrams or even other physical objects such as blocks, sticks, cardboard puzzle pieces, or rods. And the kinds of actions may vary depending on the particular nature of the notation system involved. For example, they might involve transformations of the form of an algebraic expression or combinations of different algebraic expressions, or putting puzzle pieces together; or they might involve trades with Dienes blocks. Thus, projections, or "writing" in the sense of Figure 21.1 need not amount to writing in the usual sense, but can include acting on virtually any kind of physical system or apparatus, where the syntax of the system itself (as well as the medium in which it is instantiated) determines the kind of "writing" that it can accept.

Sometimes the phrase "representation system" is used in place of "notation system," which emphasizes the fact that these notation systems typically are used in a representational way—

that is, one is often used to represent or stand for another, or aspects of another entity. When we use the word "representation," we mean it in this sense. However, a notation system might also be used to represent selected mathematical aspects of a non-mathematical situation, as in a mathematical model, in which case we would again be comfortable in referring to it as a "representation system." When we speak of a notation system apart from any particular representational function, we will stick with the word "notation."

But we must make a distinction between the notation system and the medium in which it is instantiated, or physically realized. Technically speaking, a notation system, as a set of rules that define the objects of the notation system and the allowable actions on them (together with some equivalence rules that tell us when actions may yield equivalent objects), is essentially an abstract thing until we decide to instantiate, or model, it using the material world. The particular aspects of the physical world that we use comprise the *medium* in which the notation system is instantiated. Such can include paper-pencil, physical objects, computer displays, sound (as in spoken language), and so on.

There is a sticky philosophical issue regarding whether we should say a notation system exists apart from some realization of it in some physical medium—it is the question of Platonism applied to notations: Do they exist in some abstract platonic world apart from physical reality? The answer does not affect our argument, so we leave the question for the reader to ponder. It has a more practical side, however, that takes the form of whether we should regard different instantiations of a notation system in different physical media as equivalent. Inevitably, different media have different "carrying dimensions" that affect the encoding of information. For example, let us consider written versus spoken language: In spoken language, there are inflections, tones, speech rates, decibel levels, and so on, all of which are naturally used by native speakers, but which do not get explicitly encoded in written text. Nonetheless, a competent reader of such text normally *projects* such additional features into the text as he/she reads it—either silently or aloud. Thus we can regard written and spoken text as different, albeit closely related, notation systems. In fact, the act of reading text aloud amounts to a translation process between the two systems. A more mathematically relevant example of this phenomenon involves the written versus the spoken versions of algebraic statements—unless one transliterates all the visible characters into spoken language, the inflections of natural spoken language usually leave some ambiguity. Consider "a times the quantity b plus c minus d." Is this a translation of "$a(b+c)-d$" or "$a(b+c-d)$"? We can't be sure without further information, which might be supplied by an extra pause after the c, for example, where the pause plays the role of the closed parenthesis. Our working approach to the issue of different instantiations of "equivalent" notation systems is to regard them as different, but similar notations so that we can regard the acts of relating two such systems as acts of translation.

Of course our key interest is the special nature of electronic media and the relations with other media. We need to expand our descriptive language a bit more to get at the heart of the matter.

Transformations Within and Across Notation Systems. Suppose one is doing typical algebra manipulation activities, operating on expressions, changing their form, and so on. Say we are (i) subtracting $6x$ from $(x^2 + 6x + 9)$, or (ii) expanding the binomial $(x + 3)^2$ to $x^2 + 6x + 9$. In either case, we are acting entirely within the same notation system, performing *transformations*. In the second case, we happen to be transforming an expression without changing its identity as a function; we refer to such as a *change in view of a fixed object*, whereas in the first case we are transforming the object into one that represents a different function. We can accomplish a change in view transformation in a coordinate system by a scale or viewing window change, for example. (See Kaput, in press-a, for a detailed discussion of change in view versus change in represented object in the particular context of graphical representations of functions.) In either case, we picture the general transformation situation as in Figure 21.2.

Although altogether too much school mathematics in arithmetic, algebra, and beyond amounts to transformations within a particular notation system, where the transformations are guided mainly by knowledge of the rules of the notation systems being used, most true mathematical activity involves the coordination of and *translations between* different notation systems. The idea of the value of multiple representations of knowledge was argued by Brown and Burton in the 1970s (Brown & Burton, 1978). Among the most common such uses involves translations between, say, a character string (or "symbolic") representation of a function and a coordinate graph of that function. Such translations are directional, depending on the intent and mental operations of the person performing them. For example, one may translate the function $y = x^2 + 6x + 9$ to the coordinate graph system—which can amount to the straightforward process of computing values of the function and plotting the resulting ordered pairs. Or one might begin with the coordinate graph and use knowledge of parabolas and quadratic functions to develop more sophisticated inferences that enable one to translate the graph to a quadratic function. We depict the general translation situation in Figure 21.3. Note that we use a filled-in double-sided arrow on the top of Figure 21.3 to indicate (i) that this is where the real action typically is (at least in traditional inert media), and (ii) that a translation in either direction is likely to involve a complex integration of cognitions associated with each system together with knowledge structures concerning the relations between them, for example, knowledge of how an ordered pair is plotted, or what the graphical y-intercept may mean in the symbolic notation. This integration and coordination of cognitions is fed by the reading/writing acts described earlier.

Fundamental Categories of Traditional Notation Systems. Some notation systems, such as coordinate graphs, tables of data, or histograms, have mainly been used to *display* information, and we refer to such as *display notations*. On the other hand, notation systems such as the algebraic systems, either equations or expressions, are used as the basis of transformations—and we refer to these as *action notations*. These are not exclusive distinctions but are based on how they are most typically used. For example, the standard horizontal notation for writing arithmetic expressions is much less typically used to transform those expressions into equivalent ones than it is used simply to display

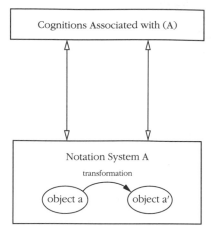

FIGURE 21–2.

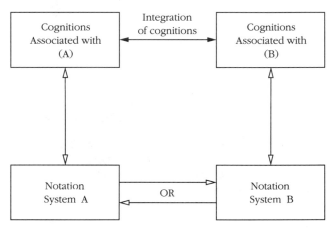

FIGURE 21–3.

the sequence of operations involved—whereas the opposite is true of the corresponding algebraic notation. However, the objects of each notation system share the characteristic of representing procedures explicitly. They are examples of *procedure-representing objects*. It happens that in the algebraic case, we often perform transformations of the procedure-representing objects. These two notation systems can be seen as different from the two-dimensional coordinate graph, which typically represents procedures (when it does so at all) only implicitly.

Four Different Classes of Mathematical Activity

We posit four types of mathematical activities in school mathematics:

1. Syntactically constrained transformations within a particular notation system, with or without reference to any external meanings,
2. Translations between notation systems, including the coordination of actions across notation systems,
3. Construction and testing of mathematical models, which amount to translation between aspects of situations and sets of notations,

4. The consolidation or crystallization of relationships and/or processes into conceptual objects or "cognitive entities" (Greeno, 1983; Harel & Kaput, in press) that can then be used in relationships or processes at a higher level of organization.

Currently the first type of activity, in the form of manipulation of symbols on paper without reference to external meanings, strongly dominates school mathematics; and early computer software tended to mirror this activity for reasons to be discussed later. Learning the arithmetic of pure number amounts to learning the syntax of a notation system, the usual base-ten Arabic system. A similar statement can be made about the bulk of standard secondary school algebra and even university calculus. The computations done with numerical calculators, symbol manipulators, traditional spreadsheets, and matrix manipulators, for example, all support transformation within a particular notation system.

Translations between graphs and algebraic expressions provide examples of Type 2 activities, while the usual modeling activities are of Type 3. Simulation software supports Type 3 activities. Type 4, as a source of mathematical-meaning building, is somewhat more subtle and has longer-term effects. It includes the act of counting as leading to whole numbers as cognitive objects (Steffe, Cobb, & von Glasersfeld, 1988), the act of taking part-of as leading to fraction numbers (Kieren, Nelson & Smith, 1985), functions as rules for transforming numbers as becoming objects that can then be further operated upon, for example, added or differentiated (Harel & Kaput, in press). Manifestation of Type 4 activities in computer media will likely take the form of procedure capturing and naming as well as the representation of the structure of sets of actions in compact form, for example, as directed graphs, to be discussed later.

We can also identify another, deeper form of translation activity that lies below the level of notations, as reflected in the translation between mathematical structures (Kaput, 1987b) or between different models of situations. An example is translating between velocity and position models of motion, or more abstractly, between a function and its derivative. This type of translation is discussed in Kaput (in press-b) in the context of MathCars, described later.

In the preceding terms, the second and third types of meaning-building activity extend the referential meaning for mathematics—either across notation systems or between mathematical structures and non-mathematical situations. It is a kind of "horizontal" growth. The first and fourth types amount to a kind of "vertical" mathematical growth by transforming actions at one level into objects and relations that can serve as inputs to and guides for actions at a higher level.

We will revisit these categories of actions repeatedly in the context of the new electronic media.

FUNDAMENTAL NEW FEATURES OF THE NEW MEDIA: THE COMPUTATIONAL CONTRIBUTION

Technologies based on dynamic interactive electronic media embody fundamental attributes that distinguish them from traditional static media in ways likely to have tremendous long-term impact on mathematics education. But we need to come to terms with why this might be the case. Given the language and framework described previously, we are now in a position to examine the particular characteristics of electronic media and contrast them with traditional media. Basically, there are three classes of distinctions that we wish to examine: (1) dynamic versus static media, (2) interactive versus inert media, and (3) procedure-capturing and executing facility in an external device versus in human memory and cognition. The first two contribute to differences in what I shall term "representational plasticity" between dynamic, interactive media and static, inert media. In the former, as opposed to the latter, one has much more freedom to create and link new notations and create variations within and across them. The third matter has to do with an entirely novel capability of interactive computational media that has yet to be fully exercised. It involves performing a sequence of actions and having the computer store those actions as a repeatable sequence, much like a program, which might be even further modified and included in other action sequences. This is related to the question of record-making and displaying. Some media inherently "overwrite" the current state whenever one creates a new state. Physical objects are usually of this type. If I have a certain configuration of Dienes blocks and rearrange them, then the prior configuration, or state, no longer exists. A similar thing happens on a simple calculator display. A computer display might allow a full record of calculator-style actions because the display space is not so limited. The simple calculator is in marked contrast to a piece of paper, where one can write or draw using the spatial organization to yield a temporally-ordered record of what one has written.

Dynamic Versus Static Media

The dynamic versus static medium distinction is the simplest. In static media, the states of notational objects cannot change as a function of time, whereas in dynamic media they can. Hence, time can become an information carrying dimension. Traditional video media are, of course, dynamic, whereas paper-pencil is not. When one writes an algebraic expression or draws a diagram, it just sits there, in a fixed state as written or drawn. Any variation needs to be projected onto it by the reader, or interpreter (in the sense of the discussion associated with Figure 21.1 above). In some cases, one may be able to facilitate the reader's projection of variation in a static medium. For example, if one wants to depict variation of a line sweeping across a geometric figure, say a line from a vertex of a triangle intersecting the opposite side, one might include several instances of the line, perhaps with an auxiliary arrow indicating the direction of the sweep. This begins to illustrate one of the major payoffs of dynamic media in mathematics learning.

One very important aspect of mathematical thinking is the abstraction of invariance. But, of course, to recognize invariance—to see what stays the same—one must have variation. *Dynamic media inherently make variation easier to achieve.* In static media, one must resort to such compensatory strategies as described with the previous diagram, which includes providing multiple instances organized spatially rather than temporally. Indeed, some notation systems, such as ordered tables of data, were designed specifically to structure multiple instances of a situation's variables. One can peruse several values of a function's input-output variables and even invest them with temporal variation by examining them in increasing order of the input variable's values. Coordinate graphs of single variable functions likewise provide simultaneous presence of multiple values (automatically ordered), allowing for the investment of temporal variation if desired. This is in contrast with a static algebraic presentation of a function, which hides all variation "inside" the letters used to denote the variables. This may be one fundamental reason why the concept of variable has historically been so difficult for students to learn (Wagner, 1981).

Physical materials such as Dienes blocks, or other "overwrite" systems, provide an interesting example of media which are weakly dynamic in the following sense: While one physically moves the elements of the system to produce a new state in a temporal process, once produced, the state remains static until changed to a new one by direct action of the user. Further, the intermediate states (for example, with some blocks in my hand and others on the desk) usually are not referentially meaningful. This is in distinction to the following situation: Given the sweeping line in the triangle described earlier, suppose one is able to "grab" with a pointing device and rotate the line continuously from one position to another so that all the intermediate states are visible and are a meaningful part of the action, as is possible with the CABRI Geometry system (Laborde, 1990). This is a stronger sense of "dynamic" than if one were to identify a new point of intersection of the line with the opposite side and have a new line drawn through the vertex and that new point, with no intermediate states shown. *Continuous transition* of intermediate states is likely to be a cognitively important feature of dynamic systems.

Interactivity: Constraints and Agents

Interactivity of the computer medium strongly distinguishes computers both from static media and from traditional video media. If you write a statement in a static medium such as pencil and paper, it merely sits there and does not interact with either the paper or the structure of anything else that may have been written on that paper. Further, the paper does not provide much in the way of help or constraint to your writing, although lined paper for writing and graph paper for coordinate graphing do provide some support and constraint. Similarly, you can sit and watch television without doing anything physical to or with the TV set, unless you decide to change the channel or adjust the picture, for example.

However, in the broad sense of Figure 21.2, essentially all notations in all media are interactive since a user must interact with them to use them productively—the user absolutely must make an interpretive contribution in the form of "reading" (in the general sense as used above). But we have a narrower definition in mind. *By interactive media, we mean interaction involving a physical contribution from the notation system and the medium in which it is instantiated.* This is depicted as the "system response" in the lower part of Figure 21.4. We characterize a medium as "inert" if the only state-change resulting

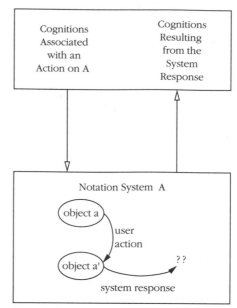

FIGURE 21–4.

from a user's input is the display of that input. Thus the key difference with notations instantiated in interactive media is the addition of something new to the result of a user's actions, something that the user must then respond to. In inert media, the user can only respond to what he or she has directly produced. Any external response to the input must be made by a third party, for example, a teacher, tutor, or peer, who happens to be observing; but it does *not* come from the notation system with which the user is interacting. Brown (1977) referred to an interactive environment as a "reactive environment."

The contributions of interactive systems are of two general types, frequently used in combination:

1. built-in constraints or supports, and
2. user-directed agents that perform actions for the user.

We will not distinguish between constraints and supports, because whether a feature is regarded as one or the other does not depend inherently on the material itself, but on the relation between the user's intentions and those of the designer of the material and the contexts for its use. For shorthand, we will refer to this as "CS structure" (constraint-support structure). Any difference may be relatively subtle, as with a constraint on the order in which information is requested for input in order to fit the logical structure of that information or its use in the system. Another example is the default presence of an autoscaling feature in a graphing utility, which automatically scales the vertical axis based on the given horizontal axis so that the graph of the inputted function extends from the bottom to the top of the coordinate graph window without any request on the part of the user. (By "default presence," we mean that the system automatically operates in this mode unless explicitly directed to do otherwise.) Or the CS structure may be blatant, whereby the system responds to an input with an admonition and perhaps an instruction: "You can't do that because.... You must do...."

User-directed agents may perform virtually any kind of actions: performing a numeric or symbolic computation, transforming within a notation system (either changing the view of a fixed object or changing the object), translating across systems, giving advice, sending messages, searching for information, checking a result, controlling other devices (e.g., a videodisc), recording actions or results for later use, displaying intermediate steps of a procedure, and so on. These are among the most obvious uses of computational resources. While most tool software is dominated by user-directed agents, it almost always embodies CS structure as well, especially tools intended to be used in educational contexts (Goldenberg, 1991).

We will now take a closer look at the complex question of how CS structure, as embodied in interactive systems, can affect learning, first in a detailed comparison of wood versus computer-based Dienes blocks, and then more generally.

Physical Manipulatives Versus Computer-Based Manipulatives: A Case Study

CS structure can exist at any level on which a learning environment can be organized. For example, a particular notation system is defined by rule systems that determine its allowable objects, allowable actions on them, and relations among objects. An example of CS structure in non-electronic media is given by the hierarchical structure of Dienes blocks, where same-numerosity sets of one size of object can be assembled to match the next size object (ten units are equivalent to a long, ten longs make a flat, and so on). Here the CS structure is at the level of objects, and of crucial importance is the *lack of inherent CS structure on actions*. Any constraints or supports on student actions with such physical manipulatives must be provided externally, by written or spoken instructions, visual templates, and so on. This lack marks an essential difference between physical materials and potential electronic instantiations of them. Cybernetic manipulatives *can* constrain actions, and they *can* define equivalence relations among objects. Patrick Thompson (1991a) has built a Dienes blocks microworld that includes constraints on actions. He and Alba Thompson compared, in a tightly designed experiment using identical teaching scripts and carefully-matched populations, a learning experience based on wooden blocks with one based on the microworld (Thompson & Thompson, 1990, in press). They found that for average- and above average-performing fourth graders the computer version led to markedly stronger understanding of the number system structure and algorithms built on it, as measured by both written tests and structured interviews. For the weakest-level students there was no difference in the results. Virtually all the attention of the weak students was focused on getting correct numerical answers, and they did not seem to attend to the constraints on their actions in either of the conditions. The Thompsons theorize that the successful students internalized the CS action structure (in the sense of Piaget) to build their own knowledge structures.

This case points to a potentially broad application of computers in mathematics education based on the internalization of CS structure embodied in a computer-instantiated learning environment. The original structural-learning conception of Dienes (1973), on which the structure of the blocks is based, is a version of the internalization of action-structure idea. However, while his imposition of structure on the *objects* of the system is strong, his constraints on *actions,* imposed via externally-provided written statements guiding activity-structure, are necessarily weak. The instantiation of the blocks system in an interactive medium changes matters substantially by radically strengthening and extending the CS structure.

But there is an even more important, higher level CS structure involved. The blocks learning system (both wood and computer medium) includes a second level of organization as previously described, since it involves maintaining translations between two notation systems—between the blocks system and the numeration system. With physical materials one can at best *point to* connections between actions on the two systems. While one can ask that students translate from one to the other, it is very difficult to cause students to maintain a translation across actions rather than merely objects. It is one thing to translate a given numeral to a blocks representation or vice versa, but it is entirely another to execute parallel sequences of actions in two systems simultaneously. It is even more difficult to attend to the respective connections between the two action sequences rather than only to their final outcomes (Resnick & Omanson, 1987). This is where the CS structure of the computer-based system is most pedagogically powerful, because it helps overcome the cognitive overload problems by handling some of the translation activities (Kaput, 1986).

In the case of the blocks system, students are constrained to act on one notation system in order to achieve a target goal in its counterpart, but with continual feedback in that counterpart provided by the computer. For example, they might be asked to add 2367 and 254 using blocks. They would then first assemble two sets of block objects to produce the two target numbers, with continuous feedback coming from the numeral representation at each step as they proceed—the result is shown in Figure 21.5. The reader will notice in the upper center of the screen that we are in base ten (which can be changed via a menu choice) and below that it indicates that we are in "Blocks Mode," which means that the user acts on blocks and gets feedback in numeral form. It also means that the command buttons (upper right) all refer to actions on blocks. The next step in working on the problem would be to "Combine" using the Combine button, which simply removes the separator between the two sets. Then the students would have a single set of blocks, whose numerosity (the "answer" to the addition problem) would be given in the usual base ten abbreviated notation. They would also have available a description in "expanded notation" using the language of blocks ("cubes," "flats," etc.), which in this case would read "2 Cubes 5 Flats 11 Longs 11 Singles." In order to put the blocks into canonical form matching the numeration system structure, they would then need to gather (10) Singles together and "Glue" them using the Glue button, which would make the glued collection into a Long. This would leave one Single and 12 Longs, 10 of which would need to be glued into a flat, and so on, until the canonical form was reached: 2 Cubes 6 Flats 2 Longs and 1 Single.

Note that for this example, as indicated in the upper left part of Figure 21.5, "A Single is 1." It is possible to change unit size so that, say, a Flat is 1. In this case, the value of a Long is .1 and

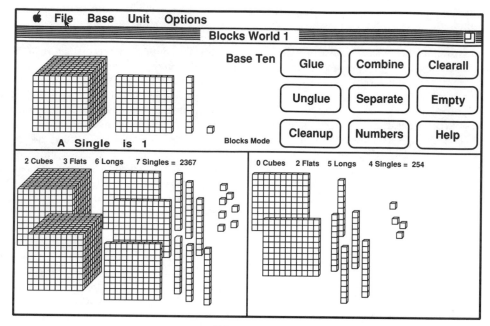

FIGURE 21–5.

a Single is .01, so we can deal with decimal fractions as well. If we were to switch to the Numbers Mode, then we would act on numerals and the blocks would perform accordingly—for example, if we typed in "36," the system would put up three Longs and six Singles. In this mode, the Glue and Unglue buttons are replaced by Carry and Borrow buttons, respectively.

A closer look at the picture reveals a third notation system, a concrete blocks version of the expanded numeration system for numerals, which appears above the blocks as a description of the current set of blocks. This is a typical situation in the use of manipulatives—an intermediate "pedagogical notation" bridges from the concrete material to the formal symbolic system. One can envision even a fourth system, not used here, consisting of the standard expanded base ten notation. The general situation is depicted in Figure 21.6, where System A is the blocks system, B is the expanded notation system, and C is the formal numeral system. In our illustration, students acted on blocks, but the results were shown in both the expanded notation and the numerals.

The Thompsons' data show that the effect of working in such linked systems is to focus attention strongly on the connections between the systems, at least for middle and higher level students. (Half of the lower level students in the study missed two of seven days' instruction, so there may have been factors other than the medium affecting outcomes.)

Reflections on Physical Versus Computer-Based Learning Environments

In general, key design questions for an environment such as we have examined above are these: Given two or more computer-based notation systems, will the learning system support translation between two notation systems, and if so, at what levels of detail and faithfulness? These questions are at their roots education questions, whose answers reflect the designer's view of the educational goals, of the underlying mathematics, and of the appropriate pedagogy.

In the case of the microworld of Thompson (1991a), the systems A and B (see Figure 21.6) were essentially locked together and overlaid, but one can easily imagine other configurations where they might be available in more separated form (Kaput & Upchurch, in preparation). Another feature of the given system is the lack of a close tie between its CS structure and the traditional algorithms for computing in the standard numeration system. This reflects the designer's view that *the operations on quantities are different from the usual procedures for putting the results of those operations in canonical form.* In this microworld, addition of blocks is simply combining disjoint sets, which in turn represents the combining of two abstract quantities. These quantities are viewed as abstractions constructed in the mind of the learner which are subject to a variety of concrete representations, including blocks and numerals. On the other hand, in the typical approach to arithmetic in schools, addition is closely tied to procedures for naming the result of the operation of addition, in particular, the traditional algorithm. Another outcome of the Thompson and Thompson (1990, in press) study was an openness to and an understanding of alternative renaming strategies among the middle and higher level computer-blocks groups not shared by the wooden blocks groups. Note that another approach to organizing the addition process could very easily parallel the standard algorithm by sequencing the order in which the blocks are combined and renamed, beginning with the Singles. This in fact is the usual way that Dienes blocks are used in instruction, to support the learning of the traditional algorithm. Ernst and Ohlsson (1989) provided a detailed analysis of the specific cognitive steps involved in carrying out the translation between blocks and numerals for some of the standard algorithms. Their

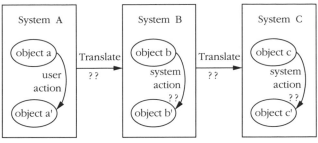

FIGURE 21–6.

analysis helps explain the neutral or even negative results of recent studies of student learning that employ such manipulatives (Resnick & Omanson, 1987).

But an even broader design question concerns the choice of or, perhaps more practically, the balance between the use of physical materials and computer analogs of those materials. Determining this balance also determines the role of ancillary written materials. Such materials usually have two different but overlapping functions. One is to provide task and activity structures, which may or may not be on-line in computer environments. This is less our concern here than the other function, to provide CS structure. We have seen that it is possible to impose much stronger CS structure using the interactive computer medium than using inert physical media, especially at the level of coordinating between notation systems, because one can transfer to the computer the mechanics of the translation process. This frees the student to focus on the connections between actions on the two systems, actions which otherwise have a tendency to consume all of the student's cognitive resources even before translation can be carried out, let alone be monitored. In addition to cognitive resource-management issues, we should mention classroom management issues. Despite the consensus that physical materials are valuable resources for mathematics learning (Friedman, 1978; Post, 1980, 1988), they are infrequently used except in the earliest grades, and even when they are used, they are seldom part of a sustained learning activity but rather are used to provide a brief illustration of a more abstract concept or operation. The two management factors, cognitive and classroom, coupled with the curricular pressure toward the development of computational skill, have effectively limited children's available representation systems to the formal symbolic ones, to be learned in static, inert media. The results have been disastrous and are one reason for the large scale reform movement now underway.

The previous paragraph, together with the previous section, reads as a call for the use of computers in education, especially the use of computer analogs of physical materials. While the disaster reflected in the status quo makes plain that fundamental change is necessary (standing still is not an option, nor is doing faster or more intensely what we have been doing), little research has been completed that would guide specific alternatives, especially in the balance between the use of traditional media and computer or related interactive video media. We can only use a common sense based analysis. For example, if we are interested in teaching the idea of function using physical dependency between physical quantities, we can ask whether the

physical material may embody the underlying causality more strongly than, say, a computer model of the physical situation, in which the dependency is likely to appear more arbitrary and indirect.

Greeno (1989) and Gurtner (1989) describe comparisons of variations in reasoning processes as students use the classic Piaget winch apparatus, or a computer model of the apparatus, or an abstract computer-based function machine. All are embodiments of linear functions expressible symbolically as $y = mx + b$. The winch apparatus involves cranking a spool to wind string tied to a block that thus slides along a marked track about a meter in length. Actually, there are two parallel winch systems whose spool axles can be either connected to the same crank or not, and whose spools can be changed to reflect different distances per single turn of the crank. Here, in terms of the formal function equation, m is the circumference of the spool, x is the number of crank-turns, b is the starting position, and y is the resulting position of the block. The computer model of the apparatus involves the same parameters and variables, but the direct tactile sense of causal dependence between number of cranks and distance traveled by the blocks cannot be duplicated in the computer model, nor can the physical experience of changing spools, or setting different starting points. The question is, how important is all that directness of experience to the learning of linear functions? It seems possible that different aspects of it may be differentially important. For example, the immediacy of the causal dependence experience—because it is an expression of the actual functional relationship—may be more important than, say, the specific actions used to change spool size.

On the other hand, Filloy and Rojano (1985) have suggested that concreteness of models in algebra may interfere with the development of adequately general understanding. However, an important factor shared by each of Greeno's three embodiments and not by the geometric embodiments studied by Filloy and Rojano, is the dynamic interactive nature of the media involved; this allows the instantiation of true variation in the input-output variables, something that formal algebraic literals do not accommodate well. Other specifics of the computer-based systems may or may not avoid confusion between the parameters and the input-output variables. For example, if the actions involved in adjusting parameter values are similar to adjusting input-output values, we should expect to see confusion.

How else might computer systems differ from either paper-pencil or physical material systems in ways that affect the balance-decision? One important difference is in how automatically records might be kept or structured. For example, in the physical winch, since the system overwrites and "forgets" its previous state whenever a crank is turned, the user must keep records in some or another form. In the computer case, a structured record in the form of an appropriately labeled table of (x, y) data is easily kept. However, the decision to keep a record, the design of the structure of records, and the act of updating the records may or may not play a significant role in teaching what we want the student to learn. Again, this is a matter of educational objectives and our informed judgment regarding the circumstances and experiences that might best lead to the achievement of those objectives. Furthermore, these decisions are likely to depend as well on the nature of the students and their prior knowledge.

Another factor involved in the balance-decision is how educationally important is the student experience in making connections between notations or between notations and models? If such connections are cognitively important to understanding some domain, then the computer's ability to instantiate live connections will probably prove advantageous. It might also be important for the students first to establish some form of those connections on their own before they are automatized by the machine—this may also convince students of the *need* for such a connection. For example, in work involving ratio reasoning (Kaput, Katz, Luke, & Pattison-Gordon, 1988; Kaput, Luke, Poholsky, & Sayer, 1987; Kaput, 1991b), we systematically led students first to establish connections between a known notation system and a new one off-line before we introduced the computer version of that connection. In each case, we initially used the new system to record a history of well-understood states of the old system. In solving problems of when (that is, after how many cranks) one block might pass another in the winch system (with different size spools connected to the same axles), it proved useful to record numbers of cranks versus the position of each block on a coordinate graph—yielding a pair of lines, the steeper one associated with the larger spool, of course. This is especially the case if the actual passing point is beyond that which can be physically realized in the winch system. Among the easily exercised opportunities in the computer version is a series of live connections between the winch model and several of the other representations, including tables, graphs, and algebraic systems—and even other models, for example, the function machine. As always, whether and how to make the connections is an education decision.

Linked Representations

Our ability to hot-link different notation, or representation, systems has become increasingly used. By a "hot-link" from system A to system B we mean the capability automatically or on command to reflect an action taken in system A in the linked system B. (Note the directionality.) It is here where the computational contribution becomes most apparent in creating a dynamic, interactive medium, as reflected in Figure 21.7. In static, inert media, there are no linkages between actions except cognitive ones, in which case cognitions that also monitor connections between representations are quite unlikely—cognitive resources are exhausted in carrying out the translations.

Two linked systems can vary greatly in the degree of correspondence between them, both at the level of correspondence of object features and at the level of correspondence of actions. Consider, for example, a formal algebraic system A of single variable expressions (hence functions) linked to a set of graphs B in such a way that for every expression *a* in A, its coordinate graph appears in B. If the actions in A include simplifications and other changes in the form of a fixed function, then the set of ordered pairs, hence points, determined by the expression is not changed. So these actions in A are not reflected in B. Similarly, if B is linked to A in the opposite direction, then there are likely to be actions in the system B

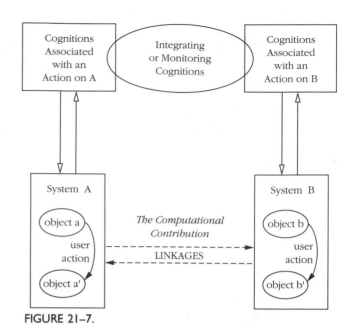

FIGURE 21–7.

not reflected in A, scale-changes, for example (Kaput, 1986, in press-a).

The use of selected feature or action translation can be a potent pedagogical aid. For example, Schwartz and Yerushalmy have developed a system (Schwartz, Yerushalmy, M., & Harvey, W., 1990; Yerushalmy & Schwartz, in press) that hot-links single variable expressions to their graphs so that a student working on algebraic manipulation of an expression can monitor, using the graphical feedback, whether or not his/her actions change the expression as a function. Thus replacing $(x + 5)^2$ by $x^2 + 10x + 25$ would not change the graph, but replacing it by $x^2 + 25$ would (see also Kaput, 1986, 1989b). This same system also allows the user to change parameters defining classes of functions easily and see the immediate effects on the graphs, or to manipulate the graphs and see the effects on the expressions. The intended impact of such systems, including more potent (but harder to use) computer algebra systems such as *Derive* (Stoutemyer, 1983) and *Maple,* (Brooks–Coles, 1990) is to help move algebra from skill-building in the syntax of character string manipulation to the study of the properties and uses of functions as represented in a variety of notation systems.

More generally, that all aspects of a complex idea cannot be adequately represented within a single notation system, and hence require multiple systems for their full expression, means that multiple, linked representations will grow in importance as an application of the new dynamic, interactive media. We examined this style of application earlier in the context of the numeration system and functions, and the author is developing a large-scale integrated system to include a variety of linked representations, including objects, number lines, tables, as well as formal representations of quantities and numbers (Kaput & Upchurch, in preparation). See Figure 21.8 for a mock-up screen illustrating some features of such a system. (Actually, for the purposes of illustrating several ideas simultaneously, this picture crowds into a single screen several windows that would typically be used one or two at a time.)

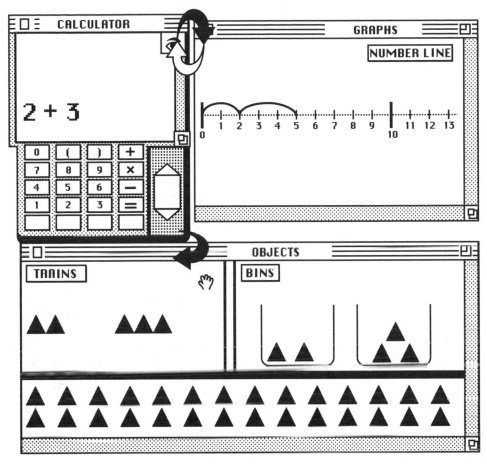

FIGURE 21–8.

An important design question to consider in relating different systems is which features to reflect across translations and how. For example, if one has a set of screen objects in system A (as in either of the Objects subwindows in Figure 21.8) whose cardinality is reflected as numerals in system B (the Calculator display), then one may not want minor movements of the objects in A to result in changes in the numerals related to the set in B. On the other hand, there is likely a strong need in A for features that allow for grouping objects into sets—one needs to know when objects are in a single set and when they are not. In the Trains subwindow this is reflected in linear contiguity, and in the Bins subwindow it is reflected in bin containment. Further, there is likely to be a need to reflect distinctions between different sets in A via distinctions between their associated numerals in B, for example, by color matching. Not shown here, but present in the software, is color coding—each of the "two-items" is blue and each of the "three-items" is green, including the (headless) arrows on the number line.

We can view simulation software in these same terms, where the key issues revolve around what is being represented in the simulation and how it is represented in the other notation systems being used. The system may employ representation-transcendent commands, for example, an incrementer that increases set size by adding elements, increases number size, and perhaps also moves points to the right along

a number line. Such representation-transcendent commands may serve to highlight mathematical structure common across the systems in question. How the designers and implementers of linked representation systems answer the complex action-connection, feature-connection and command-structure questions will largely determine their educational potential.

Control of how the connections between representations is displayed is another critical issue. Should actions in A automatically be reflected in B, or only after a user command? In work involving the web of ideas involved in multiplication, division, ratio, proportion, and linearity (Kaput, 1991b; Kaput & West, in press), we frequently found it pedagogically useful to have students predict consequences of actions taken in one representation system on one or more others. This sometimes involved whole-class activities where the teacher controlled the display. But one can expect software that will play a similar role, requesting one or more student predictions before exhibiting the corresponding representations. In Figure 21.8, the large arrows connecting windows depict different forms of data passing. The solid black arrow from one window to another means that actions in the tail window are reflected automatically and immediately in the head window. The hollow arrow (from the Numberline window to the Calculator window) is accompanied by an "eye" icon. Here, actions in the tail window are shown in the head window only when the student clicks on the eye, so display is controlled by the student. Note that the student can also be put in control of connecting windows to configure the system at the onset of use.

In summary, to the extent that different representations of ideas and actions are important and that their connections are to be internalized, computer-linked representations will encompass a growing genre of computer use in mathematics.

Stored and Captured Procedures

While in one sense, computers were literally defined by von Neuman as embodiments of stored programs, and programming itself has been an important educational use of computers, there is now another more specialized form of procedure storage that is likely to have large impact on education. The *Geometric Supposers* (Schwartz & Yerushalmy, 1985), a widely used series of programs used to help students take a more active role in learning Euclidean geometry, are based on three features: (i) easy construction of geometric shapes, which also supports easy revision, (ii) easy measurement as well as easy comparison and recording of measurements, and (iii) the capture of student constructions on particular figures as more general procedures that can be repeated on other figures that are of the same general class as the original. For example, if one constructs the three medians of a particular triangle, then the construction, stored as a generic three-median procedure, could be executed again on any other triangle. In combination these features were designed to facilitate conjecturing and more inductive approaches to the building of geometric knowledge (Schwartz & Yerushalmy, 1987a, 1987b). The procedure-capturing aspect is central, because it enables the student to reason about the relative generality of hypotheses made from particular constructions and to test these hypotheses quickly and efficiently. Continuing with the example, is the fact that the three medians intersect in a single point an accident of the particular triangle used, or is it a general consequence of the construction that applies to all triangles? Or does it apply only to some triangles?

The *Supposers* offer a form of stored procedure fundamentally different from programs explicitly written in a generic language, because it involves directly and automatically generating and storing actions on mathematical objects according to their mathematical properties and relationships. The constructive process is entirely governed by the semantics of the domain in which one is working (with the *Supposers*, for example, one is doing *geometry*), so the user works strictly in the language of the domain, with no need for learning the syntax of a general programming language.

For years other programs have embodied "macros," wherein one tells the program to "remember" what one is doing, and the exact sequence of actions (and hence their generative consequences) is stored for re-use later. While the *Supposers* show the power of this idea in an educational context, another potentially even more powerful procedure-capturing capability will soon become widely available as a feature built into operating systems and popular programs. The graphics-oriented spreadsheet *Wingz* (Informix Software, 1989) allows one to capture a sequence of actions which can then be displayed as a script in a quasi-natural language form. This script, literally an automatically generated program, can then be further edited or even incorporated into other scripts. This amounts to a hybrid between automatic procedure capturing in the language of the domain one is working in, as with the *Supposers,* and programming. The *Supposers* do not yield an explicit and *editable* sequence of commands, although another geometry program, *CABRI Geometry* (Laborde, 1990), does. This mode of computer use is likely to increase in coming years, particularly as operating systems come to support higher level languages ("higher" in the sense of farther from the binary code of the hardware and closer to the languages in which people work and communicate).

The educational potential of the ability to make a procedure into a modifiable object is not well understood since our experience with it is very limited, but we can speculate that it may help students learn about algorithms in new ways. One can envision, for example, an algorithm-builder that enables students to execute a series of actions on, say, a problem in subtraction involving a particular pair of traditionally represented numerals. The system then displays the steps as a sequence of statements that the student could edit by replacing the particular numbers by letters. The student could then attempt to run the procedure on other particular pairs of numbers to determine how robust the algorithm is and what circumstances cause it to fail. Students might also be able to compare different algorithms for efficiency, perhaps comparing their own algorithms with traditional ones. To some extent the kinds of learning available through this route are available through standard programming, but here the fundamental relationship between the particular and the general is made operative in an entirely new way. Whereas in programming one must conceive of the procedure in general terms at the outset, and then test it using particular cases later, here one *begins* with the particular before moving to the general. Indeed, one can imagine moving to the general on a step by step basis. In the hypothetical example, one might replace particular numerals with letters one at a time, testing between each stage.

Structured and Facilitated Access to Stored Information

Information storage and access have been a mainstay of computer use ever since the early days. Rapid access to large amounts of stored data has always been regarded as a primary application of computational power. While this continues to be the case, we now see much more fluency being built into the systems by which this is accomplished, particularly through the use of hypertext and hypermedia. No longer is one restricted to narrow, "one lane data highways" to get to information, but one can use varying levels of specificity and organization and in highly non-linear ways. While the ease of knowledge navigability is clearly important in education and is being applied outside mathematics in fields such as social sciences and language arts, its applications in mathematics instruction have yet to be explored. At one level, a system might employ "hot" text, which, when activated, provides further information about the item being activated, such as definitions or examples. At another level, a computer can be used to drive an external device such as a videodisc, where video helps provide a context for or illustrate a mathematical idea (more on this below).

Providing Structured and Active Records of Prior Actions

Most learning systems of the future are likely to provide at least some structured records of one's actions, a history that can be replayed, reviewed, and perhaps even modified. This is another version of the captured procedure idea, but operating at a higher, more strategic level. Of course, "actions" can involve either transformations within a notation system or translation across systems. The *Algebra Workbench* (Richards, Feurzeig, & Carter, 1991) records the actions involved in transforming algebraic expressions or equations and makes the record of such actions available in the form of a directed graph in which the nodes are the states of the object one is working with, and its edges are the transformations one has performed. The user can replay the sequence or stop it at a particular point and perform a different sequence of actions, thereby generating a new branch of the tree-graph. One can also compare one's actions with those of an "expert" (where the features of the expert can be varied, e.g., to include or not include all numerical computations explicitly). McArthur, Stasz, and Hotta, (1987) have developed an algebra system that not only provides a structured record representable as a directed graph, but it also comments on one's actions, so that the history is annotated. Their system also includes a background algebra "expert" that will perform tasks for the user and provide advice when asked.

The ability to record and conveniently display and replay a sequence of one's prior actions provides new means for reifying that most ephemeral and elusive thing called "strategy." Once reified, it can be discussed and improved. The educational implications of this computational contribution are likely to be very important, but experiences with it are currently very limited. In a sense, it is simply a dynamic enhancement of a static written record. However, the fact that it is representable in ways that expose its structure, for example as a tree-graph, and that it can be generated even while one works, are likely to be important new ingredients in the learning situation. In terms of our theoretical framework (see Figure 21.1), we are able to move not only the records of actions, but actions themselves, structured in sundry ways, from the cognitive to the physical realm, where they can serve as a resource for additional cognitions. The author is actively investigating new forms for describing the history of one's actions, for example, as sequences of icons in the form of tiny scale versions of previously produced screens that a child can activate merely by clicking. Such ways of exhibiting a history of one's prior actions and choices may also be useful to teachers in reviewing student activity at a glance or in planning sequences of activities for the future, especially if the system of history-windows can act as an editable notation system.

EDUCATIONAL PAYOFFS OF THE COMPUTATIONAL CONTRIBUTION: SOURCES OF LEARNING EFFICIENCIES

In this section, we will examine from a more subject matter-oriented perspective how the various computational contributions pay off educationally with specific examples in arithmetic, geometry, algebra, probability, and statistics.

Examples of Compacting and Enriching Experience by Off-Loading Routine Computations

Sharon Dugdale describes, in the context of graphing algebraic equations, how work with appropriate computer software can pack a large amount of graphing experience into a relatively short amount of time with the result that students deal with more graphs in a class or two playing *Green Globs* than students typically experience in an entire series of algebra courses (Dugdale, 1982). The same sort of phenomenon can be observed whenever repetitive or tedious computations are given over to technology—technology which can be relatively simple, for example, a calculator, especially one with memory (Leitzel, 1985).

However, given the resulting efficiencies in computation, one is faced with the choice of how best to utilize them, whether to do more of the same with focus on higher level objectives or to expand the activities to include less computationally convenient examples and thus deal with more realistic mathematical modeling. Given the ability, for example, to graph many functions efficiently, one can focus on *families* of functions and examine their common features, or vary parameters to study the effects of variation (Yerushalmy & Schwartz, in press). These two types of activities are obviously closely related.

A standard question is, to what extent are the actions that are being factored out needed for full understanding of the larger ideas being addressed? For example, by factoring out the numerical step in moving from an equation to a graph, how is one's understanding of the graph being changed (Kieran & Wagner, 1989)? Without its numerical component is it fundamentally incomplete in some way? Of course, most graphing systems now contain a place for the numerical representation of a function's values, so the numerical aspects of a function can be dealt with as well—although the actual computations leading to the values in a table of data are done by the computer rather than the user. Is performing the computation educationally important? Perhaps a better version of this type of question is, "*When* is performing a numerical computation important?" We will revisit this question in the different domains discussed below.

The off-loading of routine or complex computations on machines also has an experience-enriching effect that is reasonably well understood and appreciated. Extensions to more realistic examples are systematically exploited by Fey and Heid (Heid & Kunkle, 1988) and by Demana and Leitzel (1988); each offers an extended set of curricular materials using computer-based tools including symbol manipulators and graphing utilities. The range of applications that can be fruitfully dealt with is greatly enlarged, and need not be limited to the kinds of "toy" applications appearing in most texts (Fey, 1989a).

In the same vein, but with even larger implications, is the question raised by the fact that several of the new graphing and modeling systems described below do not require a closed-form algebraic description of the functions being graphed because they are based on numerical descriptions of the phenomena. One can study these phenomena in detail without using algebra. This experience compacting advantage is also applied in any geometry software that increases the efficiency of drawing and measuring, or in any situation in which routine and/or tedious actions can be factored out computationally. The factorable actions can be (either parts of or all of) transformations within a system or translations between systems.

Arithmetic and Calculators. The Hembree and Dessart (1986) meta-analysis study of the impacts of calculator use and the study by Wynands (1984) of several thousand youngsters in Sweden are convincing, at least to researchers, that heavy use of calculators in the early grades as part of instruction and assessment does not harm computational ability and frequently enhances problem-solving skill and concept development. Meissner (1987) makes the case that creativity can be fostered by good use of calculator technology. At the middle school level, new curricular materials produced by the University of Chicago School Mathematics Project (Usiskin, 1987; Usiskin et al., 1990) and the Ohio State University Computers and Calculators Project (Demana & Leitzel, 1988) systematically and successfully exploit scientific calculators.

Algebra and Symbolic Computation. Extensive studies of the impact of symbolic calculators and graphing utilities paralleling the calculator studies have not yet been published, although Lesh (1987) provides preliminary evidence that a combination of such has positive impact on students' conceptual learning without loss in symbolic computational skill; results from the other work cited above are likewise quite positive. While a few years ago symbolic manipulation systems consisted essentially of symbolic computation aids, most such systems today, despite their "computer algebra systems" title, include graphing utilities and some form of numeric approximation or root finder system in an integrated package, as does *Derive,* a new generation of *MuMath,* for example. Other well known systems include *MathCad* (Mathsoft Corporation, 1989) and *Theorist* (Prescience Corporation, 1990). *MathCad* acts as an infinitely scrollable blackboard on which one "writes" algebraic statements of virtually any kind, computes their values or relationships among them, graphs them, and so on. *Theorist* is especially intriguing because of a unique user interface that allows one to perform "natural" algebraic maneuvers even more "naturally" than one can achieve them on paper. In particular, one can isolate variables in equations by dragging them around across the equal sign, for example, while the system keeps track of their sign. One can perform substitutions simply by "picking up" a variable or a value and "dropping" it into its recipient. This is a form of a direct manipulation environment where one manipulates a traditional notation system. Both *MathCad* and *Theorist* support the creation of files (called "notebooks" in *Theorist*) which can contain a wide variety of previously constructed functions, matrices, equations, and so on, for use by others, including students. See Fey (1989b) for more examples of computer algebra systems.

Algebra and Variables: Curricular Impacts of New Technology. The issues relating algebra and technology are extremely complex and evolving as both the technology and its appropriate uses come to be better understood. I refer the reader to the *Research Issues in the Teaching and Learning of Algebra* (Wagner & Kieran, 1989) and especially chapters on AI (Lewis,

1989; Thompson, 1989), linking representations (Kaput, 1989a), and curricular choices in the face of new technologies (Fey, 1989a; Senk, 1989). Perhaps the strongest impression left by this volume is the uncertainty regarding the role of technology in defining the curriculum and its role in teaching and learning processes.

Nonetheless, we should draw attention to some larger trends. One is an increased sharing of the representational burden, once given mainly to the formal character string notation system, with graphical notations. (Of course, the geometric point of view was prominent before the character string systems were refined as tools of generalization and analysis in the 17th and 18th centuries.) This first trend is likely to continue as electronic display technologies improve and become even more ubiquitous than they are today. A second trend is to automate routine computations in ever simpler and less expensive devices, for example, graphing calculators, which are also becoming easier to use than the first generation devices intended for use by engineers and scientists. A third trend is to instantiate parametric variation in occasions for algebraic thinking, based on the ease with which numerical values can be parameterized in electronic media. This is likely to yield curricula that lead toward systems approaches to modeling and away from problems with a single set of givens and a single set of numerical answers. A fourth trend is to construct new notations to represent algebraic ideas and processes, for example, explicit and transparent function machines (Richards, Feurzig, & Carter, 1991) that can be manipulated freely as components of other machines and that can be linked to other notations such as their graphs. A fifth trend involves the increased use of dynamic media to instantiate true user-variable variables. Dynamic media are the natural "home" for variables, rather than static media, which require the user to apply much of the variation cognitively. This may lead to a deeper penetration of the key idea of variable in the curriculum, relegating the idea of variable as unknown to a much smaller role than it has in the current curriculum (Usiskin, 1988). This in turn may provide a much-needed foundation for calculus learning, which requires a true conception of variable to be meaningful (Kaput & Sims-Knight, 1983). See also the discussion of MathCars below.

The status of algebra as generalized arithmetic, coupled with the availability of computational assistance in forming and generalizing numerical patterns, is likely to move this aspect of algebra into earlier grades, as illustrated by the work of the Ohio State Pre-Algebra Project and the University of Chicago School Mathematics Project mentioned earlier. Notice that the status of algebra as generalized arithmetic is actually dual in character. On one hand, it is generalized *pure* arithmetic, based on reference-free numbers, and on the other it is a language for the generalization of *quantity* arithmetic, the arithmetic of numbers with referents, the arithmetic of modeling.

Another compaction of the curriculum may occur in conjunction with another side of algebra, namely its status as a formal system. After all, it is no great matter to tinker with the underlying axioms of a formal system and thereby create alternative systems. Given potent new environments for performing such explorations as embodied in *Mathematica* (Wolfram, 1989), we should begin to see pedagogically oriented alternative formal systems appearing soon. An excellent review of curricular impacts of graphing technologies appears in Philipp, Martin, and Richgels (1990).

The Understanding of Graphs, Including Scaling. Work in the mid-late 1980s showed that understanding of coordinate graphs comes neither automatically nor quickly (Demana & Waits, 1988; Dreyfus & Eisenberg, 1988; Goldenberg, 1987, 1988). Furthermore, to link character string and coordinate graph notations may not be sufficient to anchor students' understanding of either them or of the relations between them. Strong evidence of this has been provided by Schoenfeld, Smith, and Arcavi, (in press), who analyzed a student's attempts to use a linked system patterned after *Green Globs* to learn about slope and intercept of linear functions. This student had previous instruction in linear functions, and close analysis revealed how fragile her understandings were, how unevenly they evolved during the extended computer mediated interaction, and how unstable they were without additional semantic anchoring. Seemingly, additional cognitive anchoring is needed to stabilize learning and cognition, especially anchoring to conceptual structures based in well-understood situations being modeled.

An important aspect of this early work, especially Goldenberg's, is the barren nature of the coordinate graphs actually used by the students who were studied. In particular, there were few tick marks and labels, and no background grid was available to stabilize perceptions (Kaput, in press-c). New graphing environments by Schwartz, Yerushalmy, and Harvey (1991) provide many different views of the same graph under different scalings and support relatively easy rescaling. Similar remarks apply to recent writings on the dangers and opportunities associated with pixel-point identity confusion and computer plotting errors (e.g., the erroneous connecting of points across asymptotes [Demana & Waits, 1988]). The associated phenomena, first observed in rather primitive contexts, have not been shown to generalize across more sophisticated graphing environments, especially environments deliberately designed to confront the difficulties. New adaptive graphing techniques that track changes in the first and second derivatives of functions over a domain as they are graphed (Hausknecht & Kowalczyk, 1991; Schoenfeld, 1989) help avoid mis-connections among plotted points.

Scaling itself is in the process of moving upward in curricular priority as graphical representations of data come into greater use, both in daily life and in technical fields. Apart from the multiple scales environment just mentioned, the author knows of no other environment specifically designed to support the learning of scaling of quantitative information and/or coordinate graphs. Basically, the user must be able to adjust three display parameters: (i) the location of the base tick marks, (ii) the numerical value of those tick marks, and (iii) the overall size/shape of the display window in which the coordinate plane will appear. In earlier systems, the third parameter was seldom adjustable, but with common availability of sizable, movable and scrollable windows the student has much more display freedom than before. This freedom, as is always the case, is both a burden and an opportunity—in this case an opportunity to learn scaling, especially if accompanied by easy modification of existing scalings. New systems (Kieran, 1991) may also provide animated transitions between scalings that show all intermediate stages rather than the complete screen-redraw routines that have universally been used in the past. Work at Educational Technology Center (ETC) (Kaput et al., 1988) has indicated that scaling activity may itself be a core arena for learning about linearity and not merely an ancillary activity. See also the discussion of non-linear scaling in the section on new notation systems below. For a nontechnologically oriented but otherwise complete review of graphing issues, both as representations of algebraic relationships and modeling tools, see Leinhardt, Zaslavsky, and Stein (1990). For a more technologically oriented set of papers, see the volume edited by Romberg and colleagues on functions and graphing (Romberg, Carpenter, & Fennema, in press).

The Role of Numerical Experience in Understanding Graphs. Most function-graphing utilities operate without the display of numerical information. That is, the user types in a character string description of a function and the system provides a coordinate graph of it, with all the numerical work taking place "behind closed doors." Before computer graphing, a student could only graph a function "by hand" either after computing numerical values and plotting point by point, or after considerable practice in graphing in which the student develops various efficient techniques and knowledge of the general shape of the various families of functions. Hence computer graphing shortcuts the numerically mediated graphing experience.

The author knows of no studies directly testing the numerical understanding of functions based on the computation of particular values and the effect of shortcutting this experience. This issue parallels that of the role of the calculator in supporting or shortcutting arithmetical competence and understanding. Recent work by the author (Kaput, in press-d) has uncovered strong differences between different students' understanding of functional relationships and numerical patterns. Some students seem only to understand through the use of natural language-mediated characterizations of the perceived patterns, whereas others seem able to move to formal descriptions of those patterns. While it seems plausible that such differences may be a determinant of the amount and type of numerically based instruction in graphing that is appropriate, studies apparently have not yet been undertaken to clarify this important issue. Work by Heid and Kunkle (1988) has shown the utility of modifying the symbol manipulation package *MuMath* to support the generating of tables of numerical data in modeling situations. In Kaput (1989a) I discuss the differences among the three basic notation systems for representing functions, numerical character strings, and graphs. These differences include the ways the systems encode numerical order, variation, quantitative relationships, and so on. As already noted, each of these notations can be modified when instantiated in electronic media. In particular, a very common activity such as solving an equation can take on very different meanings when approached in the different systems.

New Relations Between Quantitative Reasoning and Algebra. As pointed out above, algebra can be regarded as a means for generalizing both arithmetic statements about pure (or abstract) numbers and quantity statements (where the numbers have dimensional referents in some situation being modeled). The *Word Problem Analyst* of Thompson (1990, 1991b) is the only computer learning environment known to the author which deliberately introduces algebra as a device for expressing relationships among quantities in modeling contexts. The program supports quantitative reasoning by having the user input quantities from the situation being modeled as well as unknown quantities (reflecting either the solution or intermediate steps) on "notecards." These can then be linked together, literally by drawing arrows from one notecard to another, to express relationships among the quantities entered. The program then does all the computing and inferring that is logically possible from the configuration that the user has entered. Its key additional feature is the ability to accept literals in place of numerical givens, which it can use to provide a literal version of the computations and inferences. Thompson (1989, 1991b) has used the program extensively with middle school students and reports substantial progress in the quality of the quantitative reasoning that resulted among the students involved.

He also reports substantial difficulties in getting students to think seriously about the process of modeling situations, and indeed, getting students to understand quantities themselves. When teachers who have been introduced to the system see that the system performs all the computations after the students have determined the relations among the quantities, they often react by suggesting that "the computer is doing all the math." This comment reveals an underlying view of what mathematics is (that is, computation) and a blindness to the key activity, which is the setting up of the quantitative relationships in the model. The large changes in student thinking associated with Thompson's program and associated curricular materials likewise reveal the shortcomings of the current calculation-oriented curriculum and the importance of underlying beliefs of teachers. Again, the analogy with the early days of automobiles is revealing.

Another program, the *Algebraic Proposer,* with a somewhat different design and aimed at older students, has been constructed by Schwartz (1988). It likewise has the students enter the quantities in a modeling situation, but in the usual form. It then displays the relationships among these quantities and their logical entailments in the form of a directed graph. It also enables a student to graph any variable which happens to be a function of any other on an appropriate coordinate graph. These modeling programs, together with more general modeling programs such as *Stella* (Richmond, 1986) will require considerable classroom and curricular exploration before their potential can be realized. Grant and colleagues have found that the representation of dynamic systems in *Stella* is difficult to understand and are developing more concrete alternatives (Grant & Borovoy, 1991). As with most other potent new technologies in schools, these modeling systems call into serious question the assumptions that drive our current curricula.

Probability, Statistics, and Data Modeling. A glance at any contemporary computer magazine or periodical will reveal the availability of dozens of increasingly powerful statistical analysis and database analysis software packages for popular microcomputers, complete with graphics capability such as

three-dimensional data viewing that was not available on any computer only a few years ago. However, most of these systems are designed for either postsecondary level educational work or professional work. Perhaps the most ambitious software development efforts in data modeling for school mathematics have been made by Rubin and colleagues at BBN Labs and Hancock at Technical Education Resource Center (TERC). The former group has produced and is now extending a statistics modeling environment called *ELASTIC* which has a number of features that exploit the dynamic interactive nature of computer media (Rubin, Bruce, Rosebery, & Horwitz, 1989; Sunburst Communications, Inc., 1991). In particular, *ELASTIC* enables a student to manipulate data represented in a histogram and watch its descriptive statistics change as a result. The actions are simple and direct: The student stretches bars of the histogram and watches markers denoting such descriptive statistics as the means move accordingly. Similarly, given a scatterplot of some data, one can drag a least squares line to fit the data and watch the sum of the squares slide up and down on a "thermometer" while trying to minimize the sum, with the equation of one's attempted regression line being updated simultaneously as one changes its position. The system also generates various population distributions which students can then sample systematically and graphically display summary data on the set of samples, including the distribution of the sample means.

In implementing *ELASTIC* in schools, certain practical bottlenecks occurred that were also observed in Russell's Used Numbers Project (Russell & Corwin, 1990), which has produced data-modeling curricular material that complements computer software. Class time limitations and logistical problems limit the use of student-acquired "real" data for analysis. In response Rubin is now leading a project at TERC to develop school-usable "Visual Databases" available on videodisc whose images can be imported into the computer which then supports various forms of measurement and counting that yield "semi-real" data (Rubin & Goodwin, 1991). The first such is a database of the states, which contains all the usual data (population and so on) in importable form, but which also supports derivable data via geometric measurements—perimeters, shore lines, or areas, for example.

A research and development project headed by Konold at the University of Massachusetts is developing software to support the learning of the theories of probability that underlie elementary descriptive and inferential statistics. Software being developed, *Chance Plus* (Konold, 1991), provides means for students to run a wide variety of probability experiments on the computer and to collect and analyze systematic data from those experiments, including sampling data.

In another research and development project, Hancock and Kaput (Hancock, 1990; Hancock & Kaput, 1990) are developing an extremely easy-to-use database system, *TableTop,* that exhibits the objects in the database as icons on the screen—see Figure 21.9. One can operate on the data, for example, by creating dynamic Venn diagrams where the objects migrate in or out of regions according to the criteria that define the sets and subsets. Similarly, one can form cross-tabulations on categorical data by simply identifying the categories on the respective axes and watching the icons move into the appropriate blocks. One

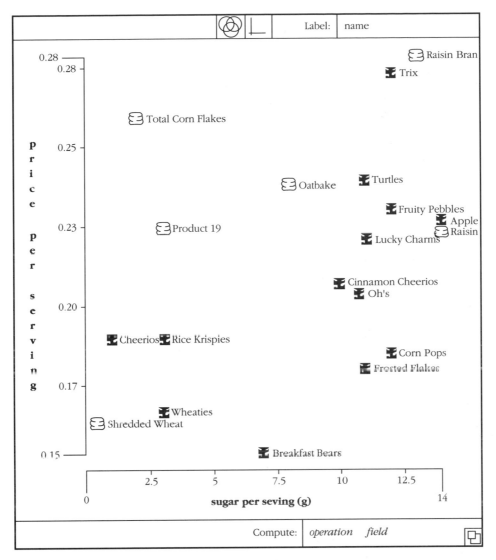

FIGURE 21-9.

can also create scatterplots for ordinal data, as well as have the system display descriptive statistics on the data. The purpose of this type of system is to render not only the manipulation of existing data, but also the creation of new data sufficiently easy so that it can be fluently performed by students beginning at the elementary grades on up. As is well known, data modeling and statistics are weak spots in the current curriculum, and computer technology is well suited to help repair the weaknesses. Data analysis software abounds at the commercial level, and a rather large amount of increasingly easy-to-use software exists for the college level. See also Shaughnessy, Chapter 19, this volume.

Geometry Learning Environments. As Fey (1984) pointed out, computers provide an ideal medium for doing geometry, another weak spot in the current curriculum. We shall look at the two basic contexts in which computer-mediated geometry instruction currently takes place, in packaged programs and in programming environments, especially Logo.

The premier example of a school geometry program, described above in the section on capturing procedures, is the *Geometric Supposer* series (Schwartz & Yerushalmy, 1985). It is perhaps the most widely used software package at the secondary level and has an enormous impact on those classrooms and laboratories where it is used as it was intended, changing the typical geometry course to a very lively exercise in conjecturing and reasoning. For detailed *Supposer* reports, see Yerushalmy and Houde (1986); Chazan (1988), and the edited collection by Schwartz, Gordon, and Butler (in press). Their section on implementation issues discusses the many challenges to using the *Supposers.* Kaput (in press-c) has also studied the substantial impact of the experience on student beliefs and attitudes. The strength of the impact differs depending on the school contexts in which the instruction occurs.

CABRI Geometry (Laborde, 1989) another geometry exploration system which is being developed in Grenoble, France, replaces the *Supposer* repeat feature by a form of animation wherein a construction can be adjusted by simple dragging actions. The movement occurs according to whether and how a given part of the construction is logically constrained; an unconstrained vertex can be moved freely, whereas those points and lines which depend on that vertex must follow accordingly, helping to reveal underlying invariance in a new way. For example, if one constructs the three medians of a triangle (and notices that they intersect in a point), then one can drag any vertex of the triangle, thereby changing its shape, but the medians remain medians and they continue to intersect in a single point. The *Supposer* equivalent of this activity is to use the repeat feature after performing the construction (on either previously built or randomly generated triangles of various types, for example, obtuse, right, etc.) and to notice that in each new case the medians intersect in a point. Another system, *GeoDraw* (Bell, 1987), greatly facilitates the construction of geometric objects, supports a geoboard, and even offers a proof checker.

While the *PreSupposer* (Schwartz & Yerushalmy, 1985) provides a somewhat simpler geometry learning environment—some of the construction primitives and the repeat feature do not exist (so that the central activity is to create and characterize useful constructions)—a widely usable and available geometry program for the elementary grades is not known to the author. This serious shortcoming may be addressed in the near future, because it is widely acknowledged that the U.S. geometry curriculum is in strong need of redefinition, especially in the earlier grades (McKnight et al., 1987).

A quadrilaterals version of the *Supposer* provides genuinely challenging geometry building at all levels including university (Schwartz & Yerushalmy, 1987a). It is not difficult to imagine versions of *Supposer*-like programs with different axioms built in, for example, non-Euclidean axioms, or programs that allow one to adjust the curvature of the space on which the constructions take place. Note that the *CABRI Geometry* program supports the usual Euclidean transformation geometry in a dynamic way, so may soon act as an alternative to the traditional approach to Euclidean geometry. Note also that generic drawing and computer-aided design (CAD) systems can be used to support geometry learning, although no literature describing such uses is known to the author. Finally, Klotz (1991) has developed the *Geometer's Sketchpad*, a system sharing many of the CABRI features. It is accompanied by interesting sets of construction explorations and challenges.

Geometry and Logo Programming. Since its development in the late 1960s by Wallace Feurzeig and Seymour Papert, Logo has been increasingly used as an environment for students to explore geometry (although as a programming language, Logo has always been more general and powerful than a "turtletalk" language). Positive impacts of Logo programming have been documented for geometric learning among children in grades 1–5 by Clements (1987). Other positive results were obtained by Olive and Lankenau (1987) regarding classification of figures and estimating angle and segment sizes, among other topics, and Noss (1987) regarding student understanding about angle size and orientation. Among older children, especially the more able students as measured in performance in other math courses, Olive (1989, 1991) has found increases in van Hiele levels and in relational thinking in the sense of Skemp (1987),

as well as in the learning of geometry specific concepts. Hoyles, Noss, and Sutherland (1989) have found improvements in student understanding of ratio and proportion, although, rather than use generally guided programming, they used relatively tightly defined tasks in a Logo microworld whose design was specifically informed by analyses of student cognitive structures prior to the instructional experience. For geometry at the elementary level, see especially the work of Lehrer, Guckenberg, and Sancilio (1988) and Lehrer, Randle, and Sancilio (1988). In response to their findings that students seem to develop separate knowledge structures for the procedural approach to geometry learned through Logo versus a kind of declarative approach to geometry learned in more traditional ways (Lehrer, Knight, Sancilio, & Love, 1989), Lehrer is developing a new learning environment that dynamically combines the two.

The approaches in most of these studies are markedly different from the earlier, more open use of microworlds as advocated by Papert (1980). Recall the earlier discussion on "how to think about technology in education."

Applying Representational Plasticity

By "representational plasticity" of a medium we mean its capability to support a variety of notational forms. Electronic media enable us not only to create any manner of new notations for mathematical objects and actions, but to create dynamic ones, and to link them so that consequences of actions in one are observable in others, and to overlay them to aid in abstraction and the imposition of structure. In this section we examine ways that this plasticity can be put to work.

Changing Display Notations Into Action Notations. The rules that define a given notation system may be instantiated in more than one medium. If we change from a static to a dynamic interactive medium, the result may be to allow new actions based on the dynamic nature of the medium, thereby enlarging the notation system. Thus a particular traditional display notation system such as the coordinate graphical system that evolved in static media may be instantiated in a dynamic and interactive medium such as a computer in a way that allows actions on its objects, thereby changing its character from a display to an action notation. For example, various systematic geometric transformations of graphs such as flips or translations are possible (Schwartz et al., 1991; Confrey, 1991) that were not among the repertoire of actions possible in static media. Similarly, the *ELASTIC* system (Rubin et al., 1989; Sunburst Communications, Inc., 1991) allows one to perform active manipulations of histograms and regression lines, again, changing them from display to action notations. Systems that allow a student to generate tables of data of arbitrary step size, form consecutive differences or ratios in columns, reorder automatically, and so on (Confrey, 1991—*Function Probe*), change the table of data notation from a display to an action notation. Indeed, *Function Probe* also changes the character of a calculator display. Not only does it eliminate the automatic over-write shortcoming of standard calculators, it allows one to store series of keystrokes as editable scripts which, as functions, can be used to generate tables of data or even coordinate graphs. Thus the calculator and computer are merged in a single system. As processing and graphical capabilities of computers increase in power, virtually any display system will be transformable into an action system.

Another example of a notation system being rendered interactive and dynamic is provided by ISETL, a computer-based mathematical language paralleling standard set-notation-based mathematical language. It was developed by Dubinsky and colleagues to enable students to interact with mathematical notation in the context of undergraduate mathematics content such as calculus, discrete mathematics, and even abstract algebra (Ayres, Davis, Dubinsky, & Lewis, 1988; Baxter, Dubinsky, & Levin, 1989; Dubinsky, in press).

A particularly strong example of the consequences of rendering a display system dynamic and interactive is provided by the *TableTop* (Hancock, 1990; Hancock & Kaput, 1990), introduced above. With a minimal amount of typing, one can define sets by giving constraints on membership, and then observe the icons move out of or into the sets one has defined. Similarly, at the click of a mouse one can form coordinate axes representing any two fields of the data at hand, and the icons then move accordingly to reflect their respective positions with respect to their values in those fields. In Figure 21.9, we see student data on cereals, collected from cereal boxes. Here we are comparing sugar per serving against price per serving, with those cereals judged to be targeted primarily towards adults represented by the whiter icons.

If one were to click on the *x*-axis, the set of previously defined fields would pop up and one could change that axis to, say, the primary grain, or the manufacturer, or the percent of recommended daily allowance of iron, and so on. The key difference in this system is that the icons would then slide to new positions reflecting the new constraints. Further, if one pulls an icon out of position, it slides back to its properly constrained position when one lets it free. Besides a variety of computational features, other dynamic motion features in this system include depicting time-series data, where, for example, time-dependent data, say the "top 40" records, literally move on the "charts" as time elapses (under the control of the user, of course). Children in elementary grades have been able to use this system to coordinate information in two dimensions, interpret Venn diagrams, and generally to create and analyze data (Hancock & Kaput, 1990). As might be imagined, children take special delight in building databases about themselves, because they can then see "themselves" moving about on the screen. The continual visual presence of the data as one manipulates them seems to be a potent factor in supporting fruitful thinking about the data.

New Notation Systems, Including Manipulable Simulations. We see these as originating from either of the two ends of the concrete-abstract spectrum, with the aim of providing linkages to a conceptually accessible middle ground. One type attempts to express abstractions in newly concrete terms, and the other expresses concrete situations in structured forms that support abstraction and generalization. Each approach can encompass many different strategies, and each is likely to employ bidirectional hot linkages between concrete and more abstract notations.

In the category of new notations we can also put the new dynamic use of traditional notations such as the directed graph in representing a record of one's actions described earlier.

One may also modify a traditional notation and use it in a new way. Goldenberg, Lewis, and O'Keefe (1990) have developed for research purposes a parallel version of horizontal coordinate axes such that, given a numerical function of a single variable, one slides a cursor along the domain number line (usually in the lower position), and observes the behavior of the corresponding point on the range axis. This offers an entirely different experience of the function's behavior. Consider, for example, what happens as one slides the domain cursor from left to right for the two cases $f(x) = 2x$ or $f(x) = -2x$; or what happens for $f(x) = x^2$. Another system designed by the author involves non-linear scaling options for orthogonal coordinate axes. The historically standard scaling is linear, although logarithmic scales have long been used in applications involving exponential functions. The new system allows one to choose the scaling of the axes to be any of the basic polynomials as well. What do quadratics and cubics look like when the vertical axis is scaled according to $f(x) = x^3$? What happens if the vertical scaling is changed to $f(x) = x^2$? It is fairly clear that these variations on traditional coordinate axes open up a new perspective on the graphical behavior of functions.

New notations often act as a bridge between a situation that mathematics is being used to model and the formal mathematical notations typically used as the modeling language. In this genre of notation, some aspects of a certain class of situations to be modeled are built into the notation and hence do not need to be abstracted. For example, the "Objects" window of Figure 21.8 includes two different but coordinated organizations of objects, "Trains" and "Bins." These serve as surrogate situations, particularly when used apart from the other systems depicted in Figure 21.8. By using the "hand" in the Objects window the student can select and move objects, in this case simple triangles, from the "reservoir" at the bottom of the screen to either the trains or bins, thereby providing a rather direct model of additive operations.

Perhaps an even stronger example of a manipulable representation of some situation, that is, a simulation, where the manipulation is done in a form natural to that situation, involves linking selected mathematical aspects of that situation to a more formal mathematical representation. Figure 21.10 depicts a scene from *MathCars*, a system being designed by the author. The basic idea is to map the motion of a vehicle onto coordinate graphical and other mathematical notations. The user controls the velocity of a simulated vehicle by controlling an accelerator (lower right side of the figure), and, depending on the user's choices, coordinated representations of time, distance traveled, or velocity appear on the simulated dashboard, continuously updated as travel proceeds. Auditory feedback takes the form of either echoes from passing posts or variable-pitch engine sounds, depending on the user's choice. Any graph generated can then be studied as an object in its own right, so that the slope of a distance graph might be examined at various points, and the slopes might even be plotted on another graph, and so on.

The particular scene pictured involves one car and a velocity versus time graph, together with a digital clock and an

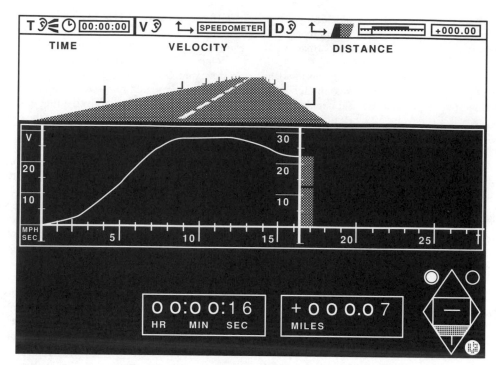

FIGURE 21–10.

odometer (which can, together with the clock, scroll to yield a table description of distance traveled). Here, the user has moved the analog speedometer to a vertical position where it slides from left to right as time progresses so that its tip leaves a trace of the vehicle's velocities as a function of time. The system can support more than one active vehicle and colors are coordinated so that all information about a vehicle is given in the color of that vehicle (the picture has been altered to accommodate the black and white medium of this book). On the top of the windshield is an array of trip representation options, several each for the basic descriptors, time, velocity, and distance traveled (actually directed distance). An enormous variety of activities is possible in such a system involving different combinations of descriptors and their different representations, some with students driving and others with students interpreting representations of their peers' driving. For example, imagine a student being asked to drive to match a given distance versus time graph while getting feedback only in terms of velocity. When the trip is completed the student can compare his/her distance graph with that of the target graph. A moment's reflection reveals that such a student is, in a very concrete sense, "enacting" the Fundamental Theorem of Calculus. This is an illustration of a different level of translation (beyond translation between notations) that is computer supportable. Here the translation is between mathematically relatable aspects of a given situation, in this case between a velocity and a distance model of the same motion event. The general point is that, as long as the relationship between models or notations can be written into a computer program, the program can then help in the translation process (Kaput, in press-b).

The aim of this system is to make accessible to students as young as middle school level some of the core ideas of the cal-

culus, such as the study of change and accumulation of quantity and the relation between change and accumulation. A key feature here is the fact that no algebra is involved. A strong but tacit assumption built into the traditional curriculum is that algebra is a prerequisite of calculus, and indeed, of most studies of quantitative relationships. The purpose of *MathCars* is to build a foundation of understanding that can then be extended to algebraic language as appropriate later.

A potent real-data version of this type of activity is based on a motion sensor connected to a computer (Mokros & Tinker, 1987), which tracks and represents graphically the velocity of objects (including students) as they move within range of the sensor. This is but one example of the general class of materials known as Microcomputer Based Laboratories (MBLs) developed at TERC that allows students to plot and analyze data in real time.

Two larger points concern:

1. The new design opportunities opened up by applying the representational plasticity of the computer and video media. Increasingly, the traditional abstract notations will be seen either as target notations accessible from a variety of concrete directions that are based in students' personal experience, or as particular options in a wider set of representational opportunities.
2. The impact of such newly possible systems on our collective sense of what curriculum may be appropriate at what grade level, and what the appropriate prerequisite structure should be.

Overlaid Notation Systems. Overlaid notations, not yet in wide use, amount to yet another application of representational plasticity. Here the idea is that one has, say, a fairly realistic depiction of some situation, perhaps a schematic map that includes the paths of some vehicles. On top of this one might superimpose some lines and measurements, in effect abstracting and schematizing the situation—for example, identifying some right triangles. Then one dissolves the underlying picture and operates on and reasons with the geometric objects that one has constructed. This technique of imposing structure on representations of situations is likely to increase in importance as digitized video becomes more widely used. Given a video that describes, say, the construction of an office complex, one can import images of the construction in progress and, utilizing specially designed drawing and measuring tools, make estimates of areas, volumes, numbers of truckloads of concrete, numbers of bricks, and so on, needed for the construction (Bardige, 1991). Furthermore, the system designer may build questions and tasks into a larger learning environment that calls for one to move among video, measurements on digitized images, and the use of traditional tools such as spreadsheets or on-line calculators.

Focusing Attention on Essentials

While I mention several ways of focusing attention on essentials, I should say at the outset that whatever features a system embodies, and however important they are to the designer or teacher, there is no guarantee that a novice to a particular mathematical topic will attend to those features simply because they are present. This conclusion was drawn especially clearly by Schoenfeld (in press) in the context of a student working with traditional notations for linear functions in a computer graphing system, and by Roschelle (1990) in the context of an elementary physics motion simulation called the *Envisioning Machine*. Particular encouragements must be present, either on- or off-line, to cause students to attend to the educationally important events or to use built-in features in an educationally appropriate way.

Translations Between Notations and Models. Logically, the educational impact of automating translations between notations falls into the compacting-of-experience category, but it involves a number of particular issues deserving separate attention. The process of relating actions in one notation A with actions or consequences in another related one B often proves cognitively overwhelming. In particular, one must become engaged in three different activities: (i) actions in A which effect state-changes in A; (ii) actions in B which effect state changes in B; and (iii) coordinations of objects, relations, and most importantly, state-changes between A and B. Furthermore, given the limits on human cognitive processing capacity, these three activities must be performed serially in some order—they cannot occur simultaneously: act in A, translate, and act in B. The matter is made even more difficult by the ephemeral nature of certain actions—they are apparent only in the doing and leave little direct trace, so evidence of an action in A may be entirely gone when one tries to effect its counterpart in B. Replayable records of actions help deal with this last difficulty, but a key computational contribution is the freeing of the translation

process from the time constraints that previously forced it to be done serially—the translation process can now be made time-independent. Actions in A can be reflected in B either instantly or when the user is ready to view them, as shown in the discussion accompanying Figure 21.8. Furthermore, the student can be put in the position of predicting in B the result of an action taken in A, and then comparing the prediction with the actual result (Kaput, 1991a, 1991b). See also Figure 21.11 for a context in which such prediction activities have been used. Requiring predictions from the student changes radically the experience of translating between different systems.

Of special interest are those cases where one of the systems depicts a familiar situation and another provides some form of mathematical representation of that situation in a model that is dynamically manipulable—either by adjusting features of the situation and examining the consequences of those changes in the mathematical representation, or by adjusting parameters in the mathematical representation and examining their consequences in the depicted situation. An early and simple but pedagogically potent example of such an environment, called "Eureka!", has been provided by Fraser, Burkhardt, and colleagues at the Shell Centre (Fraser, 1986; Phillips, 1982). The basic idea is that the student controls the flow of water in and out of a bathtub and the level of immersion of a bather. The system tracks the height of the water in a height versus time coordinate graph. Also, given a graph, the students interpret what has taken place to produce the graph. Another graph interpretation program, *Interpreting Graphs,* has been provided by Dugdale and Kibbey (1986). The system developed by Roschelle (1990) challenges the assumption that the most productive representations of phenomena are those that exactly match those that experts use.

An interesting approach to modeling being studied by Nathan, Johl, Kintch, and Lewis (1989) involves students building dynamic diagrams that are linked to algebraic representations in a program called *ANIMATE*. For example, given a standard rate/distance problem, one sets up a diagram on the computer; specifies the known rates, positions of moving vehicles, and distances; and then "runs the diagram." The system then provides quantitative output on any dependent variables that the user specifies as the animation progresses so that the user can compare results with expectations. Hall (1988) is working with related dynamic diagram systems, which he calls "qualitative diagrams." In general, this style of use amounts to a form of manipulable simulation as discussed earlier, but with very bare "simulations." We expect that directly manipulable diagrams coupled with means for collecting measurements based on the various states of the diagrams will prove to be a powerful technique for supporting the inductive reasoning steps that precede the development of formal algebraic models. One can imagine a student controlling a diagram of a ladder leaning against a vertical wall, and pulling the base away from the wall. The student would then collect data in a table relating the distance of the ladder's base from the wall and the height of the top of the ladder as it slides down the wall. Such concrete "experiments" in the context of algebra could lead to a level of quantitative understanding of such situations that far exceeds the grossly inadequate understanding now seen among calculus students as they attempt "related rate" problems in calculus courses (Monk, 1989, in press).

Kieran, Garançon, Boileau, and Pelletier (1988) are studying the role of natural language-like intermediate notations in the process of translating from a quantitative situation to an algebraic model; in their study the intermediate notation is used to drive a table of data describing the situation.

Dynamic Links Between Graphs And Character String Notations. Algebra as a component of school curriculum has been dominated by its syntactical aspects, viewed as a series of manipulations to be learned often by drill and practice, that enable one to change the form of an expression or to solve an equation. These manipulations are mainly equivalence-preserving transformations, where two expressions are equivalent if they denote the same function and two equations are equivalent if they have the same solution set. Such equivalence is not easily recognized from the character string notations, and hence a stringent adherence to the rules is required to avoid unintentional equivalence-breaking (Kaput, 1989a, 1989b). Indeed, the main rationale for the rules is given in terms of their foundation in arithmetic: $(x+5)^2 \neq x^2 + 25$, because if one substitutes (non-zero) numbers into the two sides of the equation, different numbers result. This is another way of saying that one has changed the expression as a function.

The use of linked alternate notations as pointed out earlier in the section on linked representations, especially coordinate graphs, offers promise in making the change more visible for single variable expressions. The two parabolas associated with the two expressions above are clearly different. While the specific character string representation of this difference may not be obvious, at least its existence is. A similar kind of statement can be made about single variable equation solving, where any solution is represented by the x-coordinate of the intersection of the graphs of the equation's respective sides. Thus new algebra systems that provide "hot links" between actions on expressions or equations can expose in a new way the consequences of actions taken in the formal character string realm. Research that would inform the best use of such active linkages is just beginning (Yerushalmy, 1989), and the software that would optimize those linkages likewise awaits the results of that research.

Two key purposes of multiple linked notations apply here: (i) to expose different aspects of a complex idea, and (ii) to illuminate the meanings of actions in one notation by exhibiting their consequences in another notation. The former is illustrated by the graphical interpretation of extraneous roots, which are graphically apparent as new intersections of pairs of curves resulting from extra "bends" introduced by multiplying an equation by a variable. The latter is illustrated by the kind of graphical "second opinion" of symbol manipulations described in the preceding paragraph and in Kaput (1986).

Highlighting Selected Structures and Suppressing Details. One way to view mathematics is as the study of significant structure (Steen, 1988). The representational plasticity discussed above can be applied to create explicit representations of structure that had remained tacit in traditional notations. A good example of such is provided by P. Thompson's

Expressions program (Thompson, 1988), which provides an explicit parsing tree description of any algebraic expression that a student can act upon. Actions taken in its representation-independent command structure apply to both expressions and their tree representations (for details see Thompson, 1989). The parsing structure of algebraic expressions is highly implicit and is therefore hard to learn. Its complexity is revealed when one uses a traditional symbol manipulator program such as *Mu-Math,* where much of the user's tacit knowledge of the expressions being dealt with must be brought to the surface in order to use the program (what does "simplify" really mean?) Certain software is designed to help represent the quantitative structure of situations. Two different examples of such are provided by the algebraic modeling program by Schwartz (1988), the *Algebraic Proposer,* and *Word Problem Analyst* by Thompson (1989, 1991). In the former, operations are nodes in a graph, while in the latter, quantities, represented as notecards, are nodes, with edges representing relations between quantities. Each provides an alternative to the traditional sentential quantitative description of situations. The author has used *Word Problem Analyst* for two years with academically disadvantaged college freshmen and has witnessed remarkable change in students' ability to recognize and discuss the structure of situations (see Thompson, 1989, for a description of the program—it was previously called *Word Problem Assistant*). In Figure 21.8, particularly in the two organizations of objects available, trains versus bins, we see alternative structures of discrete quantities made explicit.

Another type of structure-highlighting involves the explicit description of strategy in terms of, say, a directed graph, as with the *Algebra Workbench* described earlier.

It is in the nature of simulations of actual phenomena to suppress irrelevant detail and to provide control over those aspects of the situation being modeled that are deemed educationally relevant. However, in all the cases discussed, the actual experience of structure and invariance follows from monitoring the dynamic results of actions taken rather than from observation of a static display. Again, the dynamic and interactive nature of the medium is at the heart of its educational potential.

Varying Notation Systems to Expose Different Essentials of Complex Ideas. Complex ideas are seldom adequately represented using a single notation system. For example, differential calculus has several different facets that must be expressed in different types of notation systems. Rate of change can be seen in notations that include instantaneous change, in slopes of coordinate graphs, in formal algebraic derivatives, and in numerical difference quotients, to name a few. Similar observations apply to the complex set of ideas called rational number (Lesh, Post, & Behr, 1987), or to the idea of intensive quantity (Kaput et al., 1988). Each notation system reveals more clearly than its companions some aspect of the idea while hiding some other aspect. The ability to link different representations helps reveal the different facets of a complex idea explicitly and dynamically. In a sense, we are able to build a stereoscopic view, where the whole is much more than the sum of its parts. See Kaput (1989a, p. 180) for a discussion of the different meanings of the phrase "to solve an equation" resulting as one performs the activity using tables, graphs, symbols, and other means.

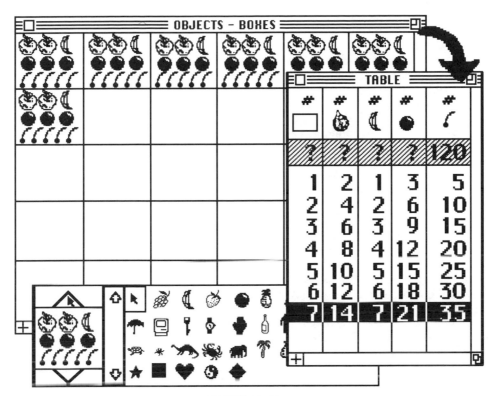

FIGURE 21-11.

In Figure 21.11, we see an illustration of a pair of representations of a "build-up" solution (Hart, 1984; Kaput & West, in press) to the following multiple ratio problem:

A fruit store owner wishes to sell fruit baskets, each of which will have 2 apples, 1 banana, 3 plums, and 5 cherries. The store owner purchased a box of 120 cherries at a special low price. How many of each of the other fruits must she buy in order to use up all the cherries in making these fruitbaskets? How many fruitbaskets will she be able to sell?

Just as with Figure 21.8, the representations are linked, as indicated by the heavy arrow in the upper right from the concrete representation to the table. Here the concrete representation, which is driven by the incrementer that creates copies of its contents in the rectangular boxes above, outputs to the table, which tracks how many of each type of fruit (together with the number of boxes used, in the left hand column). Hence, after setting up the model by copying fruit icons from the icon palette into the incrementer and creating a table column for all the information needed, the student is able to "fill baskets" until reaching 120 cherries. (This figure is bleached black and white, but the actual software color-links icons and numerical data about them.)

The student clicks on the top part of the incrementer, filling boxes and tracking the number of each type of fruit and number of boxes, as reflected in the inverse video row at the bottom of the table. The two representations of the solution-model to the problem vary in the extent to which they re-flect the structure of the situation they represent, particularly the containment relation between the baskets and the fruit, on which is based a (multilinear) functional relation between the number of baskets and the number of each type of fruit.

Reifying the Abstract or the Ephemeral

The procedure-capturing systems and the structured history systems offer new means for reflective learning through the simple device of providing perceptually concrete items on which to reflect (Collins & Brown, 1988). In a sense they map temporal events onto space and freeze them so they can be inspected and perhaps manipulated. Aspects of their structure can be explicitly represented, annotated, and hence become the subject of didactic discussion. In inert media, while a trace of one's prior actions remains, for example, a series of algebraic maneuvers in solving an equation, its global and heuristic structure is very implicit and difficult to identify, especially for a novice. Similarly, the representation of quantitative structure by P. Thompson's *Word Problem Analyst* described previously provides a concrete visual display of that structure (Thompson, 1989, 1990).

APPLYING INTELLIGENCE IN TECHNOLOGICAL LEARNING ENVIRONMENTS

The issue of employing computer intelligence in a learning context is a particular case of a much more general issue—

that of the different modes of applying research and technology in a field of intellectual work. An illuminating comparison, or rather contrast, can be drawn with another applied field, medicine. Relatively little research in medicine has ever been applied to automating the doctor, or even automating diagnosis, an exception being the GUIDON program of Clancey (1987), which is linked to the classic infectious disease diagnosis program, MYCIN (Shortliffe, 1976). Far more research has been applied to the development of medicines and the development of various diagnostic tools and therapeutic machinery to be used by health care professionals. Supporting this research is a large foundation of more basic scientific research in different areas of biology, ecology, and so on. And, increasingly, the tools and machinery have embedded within them various forms of computational intelligent support, whether visible to the user or not. A strong example is provided by CAT-scan technology, in which extremely complex computation in combination with image processing technology is used to create 3-D images from sequences of 2-D x-ray "slices" of the body part being scanned. And further processing is done to exaggerate or isolate potential abnormalities where the "abnormality" is ascertained by the tool's AI-based software. While CAT-scans are a bread-and-butter tool of physicians, their sophisticated underlying technology, including the image-creating and "intelligent" inferencing done by the software, are entirely out of sight of the user, who concentrates on the case at hand rather than the technology. All this is in strong contrast to education, where, first, a rather large research and development effort has been invested in automating the teacher (via AI-based tutors) and, second, educational technology seems to be an object of attention in its own right rather than being treated as a taken-for-granted tool of the teacher.

In this section we will first review the styles of intelligence built into CAI programs in the past and then move on to discuss the wider uses of intelligence in education—beyond its use in tutorial programs. The literature on intelligent systems in education is immense and filled with examples of preliminary systems that have not entered the mainstream of practice (see Wenger, 1987, for particulars), and we will choose among them sparingly.

A Thumbnail Historical Sketch of Artificial Intelligence in Mathematics Education: Some Facts

The earliest computer-assisted instruction (CAI) in the 1950s and 1960s consisted of "electronic page turners," where one moved from one "frame" (the term being borrowed from behaviorist programmed learning technology) to the next if one chose the correct answer from a list. The programs also often contained stored comments of one kind or another associated with each response option. These programs used stored answers and text and provided essentially the same material for all users. The only variable was the amount of time one could spend working through the material—which was the same variable available in the printed programmed instruction that the program was a transliteration of. The next step brought branching, where, depending on the student's response, the program offered one of several frames, again provided in advance. The

Intelligent CAI (ICAI) systems developed in the 1970s had a new, generative aspect, where problems and even computer responses could be generated to react to individual student input (Barr & Feigenbaum, 1982, reprinted in 1989).

Thus the first major shift was from replaying stored machine feedback and branching among a stored set of options to *generating* responses using preconstructed mechanisms. The idea of a computer tutor which would engage in dialogue with a student and make inferences regarding that student's knowledge or intentions was outlined by Carbonell (1970), who, together with Collins, built a tutor for South American geography. In the mid 1970s, particularly as reflected in the work of Brown and Goldstein (1977), the intelligence in the ICAI systems came to incorporate, ever more explicitly, teaching strategies based on hypotheses regarding student learning, including learning of heuristic knowledge. Thus these systems were no longer based merely on the deductive structure of a formal representation of the knowledge domain held by experts (Brown & R. Burton, 1978; Brown, Collins, & Harris, 1978; Brown & VanLehn, 1980).

The mathematical topics addressed were logic and formal proof (Suppes & Morningstar, 1969); algebra syntax (Sleeman, 1984, 1987); elementary arithmetic operations, particularly in an adaptation of a Plato board game (Dugdale & Kibbey, 1977) by Burton and Brown (1976) (which was given the title "How the West Was Won"); multidigit subtraction in the widely cited BUGGY programs by Brown, R. Burton, and Van Lehn described in the work cited immediately above as well as in several chapters of the key collection on AI and education edited by Sleeman and Brown (1982). These last programs were centered more in the inference of student error patterns than they were in tutorial dialogue with students. The model adopted for the tutor was essentially an observant coach who would generate appropriate diagnostic and remedial exercises based on student behavior. A recent version of a subtraction tutor based on Anderson's model of cognition (Anderson, 1987) is described by Orey and J. Burton (1989/90).

There was also considerable effort, continuing through the 1980s, in developing tutors to teach programming skills. The most popular language used to build the systems was LISP, which was also used to model student cognitions using the systems. An ambitious system for teaching the programming language Pascal (PROUST) that could diagnose about 75% of the bugs in most student Pascal programs was developed by Soloway and colleagues (Johnson & Soloway, 1985).

In the 1980s tutors were developed for high school geometry and more advanced arithmetic. Indeed, perhaps the most widely publicized tutor was Anderson's *Geometry Tutor* (Anderson, Boyle, & Reiser, 1985; Anderson, Boyle, & Yost, 1985). This tutor is based on a core expert, a deductive system that embodies the Euclidean axioms together with a variety of heuristics and a "leading questions" pedagogy. Its main goal is to teach the traditional geometry content, particularly the ability to prove theorems in the two-column style. Its pedagogy, besides providing the student with an editable figure associated with the theorem in question, provides the student with the critical missing step at any stage if the student is unable to produce it in a few tries, usually three.

In the late 1980s the *Geometry Tutor* was transferred to microcomputers and was being used in classrooms in the Pittsburgh area. Interestingly, Anderson (1989) reports differential success with both this tutor and a similar style tutor for algebra word problems, where the key factor determining their acceptance in classrooms is the alignment between their pedagogical principles and those of the teacher. Since the tutor is relatively directive, with a fixed curricular agenda, those teachers who prefer a more exploratory approach are less likely to prefer using it in their classes. Anderson has suggested that ideally a tutor should have adjustable pedagogical features, adjustable either by the teacher or by the student. Such flexibility would accommodate a greater variety of pedagogical ideologies and preferences. However pedagogically adjustable such a tutor might be, it appears that the underlying epistemology is fixed—the knowledge and the underlying authority of the tutor reside in the computer (see the discussion of implementation issues below).

Derry and Hawkes (Derry, 1989; Hawkes & Derry, 1989/90) have been developing an intelligent tutor (*TAPS*) based on fuzzy logic for arithmetic and elementary algebraic word problems as an alternative to the "buggy approach," described previously, that centered on the handling of digits in formal arithmetic that has dominated arithmetic tutors since the late 1970s (Wenger, 1987). *TAPS* also produces what amounts to a fuzzy record of student performance.

Ohlsson and colleagues are building multiple-representation tutorial systems for fractions and whole number decimals (Ohlsson, 1987, 1988a, 1988b) and are analyzing in great detail the successes and failures. In particular, their recent results and analyses call into question relatively naive assumptions regarding the educational value of concretely manipulable representations of fractions and decimals (Ohlsson & Hall, 1990). Detailed specification of the cognitive acts associated with using Dienes blocks to make sense of the standard calculation algorithms with numerals, reveals unexpected complexity and a need for intermediate, or bridging notations, even in a computer environment, as mentioned earlier in the Thompsons' case study comparing wood and computer-based blocks.

Attempts to use intelligent support in the form of tutorial assistance in the area of statistics seem to be limited, although Marcoulides (1989) reports positive results among undergraduates in the application of *ZEERA*, a tutor for elementary data analysis.

Recent developments worth noting involve the move toward coaches rather than tutors (actually begun by Brown and colleagues in the late 1970s), a study of ways to shape dialogue between two knowledgeable agents (student and software system) rather than a unidirectional information passing from system to student (Newman, 1990), attempts to incorporate multiple representations (Ohlsson, in press) and even multimedia in the tutoring environment (although mostly in domains outside of mathematics), and most importantly, an integration between tools-based approaches and tutorial, or coaching, approaches.

Some Reflections on the History of Artifical Intelligence in Mathematics Education

Generally, the use of intelligence in direct support of instruction requires the integration of three bodies of knowledge and/or hypotheses about that knowledge: (i) a theory of the subject matter, (ii) general knowledge of students' forms of learning and interaction with the subject matter, and (iii) particular knowledge of the individual learner at each state in the interaction with the machine. All three bodies of knowledge are very likely to be imperfect, as will be their implementation in a particular piece of software. In addition, systems must incorporate some teaching strategy that integrates these three bodies of knowledge and hypotheses, and finally, the integration must be accompanied by (or be incorporated in) a user interface. An important development in the 1970s and 1980s was the embedding of models of the learner into the learning systems. Increasingly, designers asked, "How would a student do this?" Their programs then attempted to compare their model's approach with that of the student at hand in order to make a decision on what to do next. There are, of course, many ways to embed intelligence into a computer environment, and the options increase with increases in the power of the hardware and software. Intelligence-embedding styles in instruction are based on two partially contrasting educational metaphors: The tutor and the collaborator. Each of these in turn aligns with a style of user interaction metaphor: The tutor aligns with the conversation metaphor (Hutchins, Hollan, & Norman, 1986), and the collaborator aligns with the collaborative version of the direct object manipulation metaphor, referred to by Hutchins (1986) as the "collaborative manipulation metaphor."

Most of the intelligent systems described above use the conversation metaphor, hence they fit the tutor metaphor. At least two major factors have been at work constraining the evolution of this vision. One significant technical constraint on early computers was simply the allowable interface and display style, which limited the user to text-based interaction. A second somewhat more subtle and more powerful constraint involved the fit between the combined tutor/conversation metaphor and the extant ability to represent the domain to be learned within a computer program. This second constraint limited the domains to which the technology would be applied to those that were capable of being directly formalized, or, rather, those domains that themselves consist of formal systems—the numeration system, formal algebra, formalized proof in logic and geometry, and programming in a computer language. This constraint was not an accident, since the evolution of computers and computer languages was directly tied to the representation of the inherited notation systems of mathematics (Papert, 1980).

Herein lies a fundamental shortcoming in educational potential: Typical ICAI systems were applied to teach the syntax of formal notations—which is the only inherent content of such notations. The ICAI systems did not, because they could not, deal with what the formalisms are used to represent and what they evolved to do in the first place. (Virtually every historically received formal system in the school mathematics curriculum evolved as a representation system [Kaput, 1987b].) They were thus primarily aimed at developing student competence in the syntax of some formal system. But this competence has come to be pedagogically inappropriate and curricularly superfluous—pedagogically inappropriate because it was experienced as meaningless and alienating by students (who were being asked to learn representation systems without anything to represent), and curricularly superfluous in the context of ever more widely available formal system manipulation tools that manipulate symbols much more effectively than humans can (Davis, 1984). Further, the objectives and methods of traditional ICAI run counter to those now commonly accepted as appropriate for mathematics students at all levels, as reflected in such recent reports as *NCTM Standards* (National Council of Teachers of Mathematics, 1989), *Everybody Counts* (National Research Council, 1989), and *Reshaping the Future of School Mathematics* (Mathematical Sciences Education, 1990).

More could be said here than we have space for regarding the narrow vision that circumscribed this style of technology use in both education and in attempts to understand and explain learning itself: Attempts to model cognitive processes and learning were living under the same constraints and, to some extent, under the same epistemological ideology. Thus formal computer-based cognitive models were used to explain phenomena arising as students attempted to learn formal systems (Gardner, 1985, esp. Ch.13; Kaput, 1985a) To see this circularity and insularity one need only recall the amount of attention once given to Lisp learning and Lisp tutor building, where the learning was modeled in, yes, Lisp. Intelligent tutors have yet to pay off in mathematics education. In the next section we will examine alternative forms for applying intelligence.

Uses of Artificial Intelligence to Support Teachers

Applications of intelligence may provide less direct support of instruction than does a tutor, by supporting the teacher's activity in some other way, or by acting in the context of a CS (constraint/support) structure for students as defined earlier. For example, a teacher might be able to obtain help in constructing materials or problems to be used by students, where the raw material for the construction is accessed from some mass storage device and assembly is guided by intelligent help, as with the Instructional Development Environment pioneered at Xerox PARC (Burton, 1989). This can allow for much more flexible individualized instruction and avoid the difficulties associated with the machine-based inferences needed in tutorial activity.

Similar intelligent systems for teacher and administrator use in assessment and report generating are under development at the Educational Testing Service (Lesh, 1990). Assessment is greatly complicated once one leaves the tidy world of calculation-oriented texts and workbooks, as indicated in the Evaluation Section of the *NTCM Standards* (1989), so intelligent help may be required to cope with the complexity. Yet another application involves computer-adaptive testing, where the computer's assessment offerings depend on the student's work up to that point.

Indeed, one can imagine computer intelligence being applied anywhere a human teacher would apply it—along the lines of a computer-based teacher's aid (Minstrell, 1988). The author has called for the systematic development of a teacher's workstation, matching that used in other professions (Kaput, 1988), and the National Education Association (NEA, 1989) has recently developed a policy to bring computers into education by first providing each teacher with a general purpose workstation, before expecting the teacher to use the computer in direct support of instruction.

Non-tutorial Uses of Artificial Intelligence for Students: Smart Notations

Since currently available computer power supports multiple, linked notations and direct actions on them, there is now fertile space for the second, collaborative manipulation metaphor in which to embed intelligence into computer learning systems for students. Systems can now be made intelligent collaborators in the elaboration and the manipulation of notations and especially in the translation across representations. Figures 21.5, 21.8, and 21.11 illustrate opportunities to provide "smart notation systems" embodying both CS structures and on-line advice as the user works. For example, if one's Dienes' blocks implementation includes "trays" for units, tens, hundreds, and so on, then the system can be programmed to query the user whenever the user moves objects across a tray's boundary—do you want to glue or unglue? If you do not have sufficient tens to glue into a hundred-flat, the system might note this and ask you where you plan to get the needed tens. Noting the problem associated with Figure 21.11, the system might contain various forms of scaffolding to help students with setting up, constraining, or interpreting the incrementing process that yields the solution. Such were built into a research prototype system (Kaput & Pattison-Gordon, 1987).

Moreover, as the *MathCars* and object-based calculation environments help illustrate (Figures 21.8, 21.10, 21.11), the notations themselves can embody both phenomenological richness and an intelligent structure. For example, the cellular structure of the four-icon calculation environment in Figure 21.11 affords a set of "natural" concretely based, ratio-reasoning strategies based on grouping and counting techniques that were found among people with little formal education (Carraher, Carraher, & Schliemann, 1985; Carraher, Schliemann, & Carraher, 1988; Saxe, 1988a, 1988b, 1990). These "natural" strategies also can be tied to the more widely applicable strategies afforded by more formal notations (Kaput, et al, 1988; Kaput and West, in press).

Non-tutorial Uses of Artificial Intelligence for Students: Smart Tools

Virtually any computer tool, almost by definition, has capability that could be called "intelligent" by the classic criterion: It does things that, if done by a human, would be considered intelligent. Consider, for example, a graphing utility. It graphs functions, and usually has an auto-scaling feature, among others. Moreover, most such utilities prompt the user for input in a certain order, while some even indicate the direction in which to search to find a graph when none of its points appear in the display window. Further, it may have a feature that is appearing with increasing frequency in various tools recently: If one selects the help utility, the cursor turns into a question mark, which the user may now use to point at a variety of objects appearing on the screen, which in turn provides advice or context-sensitive explanations. Furthermore, the explanations may be in the form of hypertext, which means that if one queries certain portions of the explanation, additional explanation appears, and so on. In other words, a whole web of context-sensitive support for the user may lie behind the surface of the tool. By this example, one sees that a range of intelligent support may be embedded in a tool. There is a tendency to include ever more user support as experience with tools accumulates.

Intelligence may also be used to structure one's use of a tool at a higher level. Hutchins, et al. (1986) describe a statistical analysis tool that helps the user decide which tests to run on a given set of data and how to follow up on a given test when it is completed. This is done in the form of an interactive dialogue that refers both to the data at hand and the array of statistical tests appearing on the screen. David McArthur and colleagues at Rand Corporation have built an intelligent tutor for algebra, but are now building means by which the underlying intelligent tutoring system can help a student do the algebra associated with exploring a mathematical microworld, acting as a smart algebra tool (in the sense of being able to carry out algebraic operations as requested) used, for example, in explorations in analytic geometry (McArthur & Stasz, 1990). Here the intelligence of the tool is experienced in the context of its use.

Adjustable Tools: Matching Tool to Educational Task

While tools become ever more powerful, the question of learning with the tool remains. An important challenge for the teacher or curriculum designer is to match the features of the tool to educational objectives. If the tool's features are fixed in advance, as with a hand-held calculator, then the challenge is greater than if the features are adjustable by the teacher, or even by software, as with a computer-based calculator. Schwartz (1988) has produced a game-like system, *What to Do When Your Calculator is Broken*, that disables certain keys on a computer-screen calculator and challenges the user to produce specified output-results using only the remaining available keys.

A development team led by the author is producing an integrated tool set for mathematics in the elementary grades in which each tool contains adjustable parameters that control its available features and representations (Kaput & Upchurch, in preparation). These parameters will be adjustable by an external agent, either a human or a software agent, to fit the educational objectives of particular tasks and explorations. In particular, the tasks and explorations are being written in the *HyperCard* authoring system (Apple Computer, Inc., 1991), which is able to adjust the tool-parameters for the tools to be used in a particular task or exploration, which we term "MathStack." For example, if the activity involves exploring geometric constructions of some kind, then any feature of the draw/measure tool that renders the construction automatic may need to be disabled, while the appropriate measuring tools may need to be activated. In effect, this is a form of "smart task." We are also designing a development system (the "MathStack Construction Kit") that will enable a non-specialist to create such smart tasks relatively easily via menu choices that involve simply checking off a list to set the tool parameters associated with that particular task.

An interesting potential consequence of easily constructible tool-task matches is a new relationship between text-dominated print media and electronic media. One reason that textbook-ancillary workbooks filled with lists of routine exercises are so widely used is that they are so easy and inexpensive to create and to use (perhaps "inflict" would be a better term). If publishers can create and teachers can use more interesting computer-mediated explorations that apply technological tools, yet maintain a curricular connection with their basal text, these materials are likely to be used on a much wider scale than they are today, offering a viable alternative to the workbook. Additionally, such materials are much more aligned with the new vision of school mathematics being widely advocated.

Summary Comments on Intelligence in Educational Technology: Some Trends

In the future, we are likely to see greater flexibility in tools designed to be used for educational purposes, enhancing the match of task to tool. In addition, the availability of context-sensitive help will increase. Further, the tools for teachers will likewise increase in sophistication, especially as teacher workstations become more prevalent and the market for such teacher support grows. While computer tutors are likely to increase in effectiveness for particular topics, the general tendency seems to be following the pattern in medicine described earlier, where technological R&D work is being applied to practice in diffuse and embedded ways rather than simply through the automation of, for example, the teaching act. This tendency also happens to fit more naturally with the vision of school mathematics being advocated in the ongoing reform movement.

Another trend worth mentioning involves a fundamental reconsideration of the nature of computer media and their relation with human cognition and cultural practice, perhaps best reflected in the work of Winograd and Flores (1986). The emerging view sees computers as a cultural medium intimately connected to our identity and our function as human beings. It sees using computers in work as a much more embedded activity than merely "using tools." Computers fundamentally alter the nature of that work, not merely its efficiency, accessibility, or scope. This reorientation, which shares some philosophic attitudes with the shift towards a view of cognition as situated in layered contexts and work (Brown, Collins, & Duguid, 1989; Lave, 1988), calls into question such basic notions as understanding, meaning, and symbolic reference (Clancey, 1989). Another, more historical, perspective in the McLuhan tradition is provided by Provenzo (1986), who sees computers as providing new forms of social organization and culture, which he refers to as the "post-typographic" culture. Cuban (1986), however, views microcomputers as merely another in a series of new technologies, such as film clips and television, to "bounce off" education in this century without changing it in important ways. This latter view may be understandable given the conditions reviewed in the history section earlier, although the author agrees with Herbert Simon that the computer is an innovation of a magnitude and importance to appear only once in several centuries. Its long term impact on mathematics is more likely to match the impact of the invention of writing than, say, the transliteration impact one might expect from the invention of tape recorders on oral mathematics. As pointed out by Provenzo, early uses of new technologies are transliterations of existing practice. Mature uses are barely hinted at in today's practice.

STRATEGIES FOR IMPLEMENTING NEW TECHNOLOGIES IN CLASSROOMS

Our approach in this section is not to offer a "practical guide," for such must repeatedly change with changing particulars—their shelf-life is especially short. Instead we will look at the issues in a way that helps the reader approach the many opportunities for practical help in trying or implementing new technologies.

The decision to implement a new technology in a classroom, school, or district is fundamentally an *educational* decision, to be guided by educational objectives. It is constrained by perceived human and monetary resource limits, but, more importantly, by the decision makers' vision and expectations regarding the likely contribution of the new technology to the achievement of their educational objectives. In a period of stable practice and relatively unquestioned objectives, the decision would be far less complex than it is in the 1990s. It would amount to determining whether new technology would improve the achievement of the accepted objectives as expressed in the existing curriculum—by hastening student progress, increasing the percentage of students succeeding, raising test scores on previously accepted assessment devices, and so on. The question to be answered is, "Will the technology help us do better what we have been trying to do?" This indeed was the criterion used through much of the history of electronic technologies in schools, and much of the research on the impact or effectiveness of computers took the form of comparative analyses on standardized tests (Becker, 1987, 1990, in press; Capper, 1988; Kulik, 1985; OTA, 1988, Ch. 3).

Readers of this volume know not only that the objectives are changing, but that they are changing in the direction of much more ambitious curriculum and pedagogy whose specification is much more problematic than the computationally oriented objectives and methods of the past. Further, they require new forms of assessment. All this radically complicates technology implementation decisions, as discussed further below.

Implementation Strategies and Challenges

Retrofitting General Application Tools as Learning Tools. Creative retrofitting of general tools, such as calculators, symbol manipulators, graphing utilities, spreadsheets, database systems, and so on, has been a widely used strategy for applying electronic technologies in mathematics classrooms. In part it is a default strategy due to the relative scarcity of more broadly useful mathematics learning software in the 1980s. As of 1988 the majority of educational software was still of the drill and practice genre (OTA, 1988, Ch. 6; Becker, in press). However, this strategy has proven to be relatively ineffective in terms of numbers of classrooms involved and proportion of learning that is computer-mediated. An underlying reason for this ineffectiveness may lie in the level of effort and expertise required to retrofit tools designed for one task to accomplish another, often one more complex, namely mathematics learning. An exploration of this issue in the case of graphing software is given by Goldenberg (1991), who outlines the specific features that generic graphing utilities do not have but that are required to support learning by novices in actual mathematics classrooms, including features that prevent the development of misunderstandings that students tend to develop otherwise (Goldenberg, 1987, 1988, 1991; Kaput, in press-a). This is not to say that excellent ideas have not been offered for the use of generic tools, for example, spreadsheets (Arganbright, 1984).

The difficulty in integrating the computer work with the ongoing flow of curriculum and instructional activity is closely related to the level of effort/expertise and to the applicability of actual educational software as it currently exists. Software that is hard to use is likely to be applied only by a small group of expert innovators and only in school circumstances that specifically support their activity.

Related to the difficulties in using generic technology in instruction is the matter of the relationship of technology with assessment, even in the simple case of calculators. For a complete review of this latter issue, see the collection edited by Kenelly (1989).

Implementing Technology Toward Reformed Objectives. Wide-scale efforts to achieve new objectives using computer technology are few, but the best known and well studied are associated with the *Geometric Supposers* described above. Using the *Supposers* as they were designed to be used, as tools for generating and examining conjectures (Schwartz & Yerushalmy, 1987a, 1987b), is a demanding task that requires a profound shift of a long-held set of assumptions, beliefs and practices about Euclidean geometry. The traditional approach to Euclidean geometry shares aspects of a museum tour, where all the knowledge resides in the text, waiting to be encountered, and, in some cases, "proved." For some of the teachers studied by Lampert (1988a), the change was akin to a religious transformation, with all the accompanying travail and exhilaration. Epistemologically, it required a radical shift in their presumed locus of epistemological reality from either text or abstract Platonic world of mathematical truth to human minds, including student minds; a revision of the role of proof from formal verification of existing truth to rational elimination of doubt about validity of hypotheses; an expansion of the forms of rational activity associated with the doing of mathematics from a narrow focus on deductive logic to include inductive reasoning based on rationally constructed and critiqued examples (Yerushalmy, 1986); and, generally, development of a sense that mathematics is something that students do and invent rather than observe. The ramifications for classroom practice are likewise profound once the museum tour is abandoned: The locus of social authority becomes more diffuse; provision must be made for students to generate, refine, and prove conjectures; the teacher must routinely negotiate between student-generated mathematics and the teacher's curricular agenda—after all, not all ideas are worth pursuing, even if time were unlimited (Chazan & Houde, 1989; Wiske, 1990; Yerushalmy et al., 1988).

An important lesson to be drawn from the *Supposer* experience is the importance of support for the teacher who must come to terms with these profound changes (Lampert, 1988b; Wiske & Houde, 1988; Wiske, 1990). While technology provoked innovation and in a sense makes it possible, the important changes are those that follow from changes in the teacher's beliefs about mathematics, teaching, learning, students, and the appropriate use of classroom time. Teachers need materials (especially, well-designed problem sets), collegial and administrative support, time, and increased understanding of the subject matter in order to have confidence and ability to make wise pedagogical decisions about how to proceed as students generate ideas and conjectures, many of which may be entirely new to the teacher. Lampert (1988b) likens the experience to leading students on a field trip to a large city where each student is equipped with a motor scooter. Teachers need to be flexible in using the software, being able to move from helping students use it in laboratory-style classes to using it as a focus of work in a teacher-centered discovery class.

They must also cope with such mundane matters as scheduling the computers and the fact that neither the school day nor the physical layout of most schools was designed for guided inquiry using computers. It takes blocks of time to run a sequence of mathematical experiments fruitfully, and targets of opportunity for specific explorations are themselves often unpredictable, so it may be difficult to predict how much computer time may be needed and when. Even the distribution of time between teacher-centered computer activity and lab-oriented activity may vary widely from topic to topic. While geometry and the *Supposers* themselves have special characteristics that limit the direct generalizability of these case studies, the broad lessons seem to merit our attention: Implementing new technologies in a reformed curriculum is not easy, and the education of teachers plays a critical role. We see no reason why the same conclusions will not apply to other mathematical topics and to other grade levels (see Kaput, 1990, for a detailed discussion of the generalizability issue).

Typical scope and sequence statements do not include sufficient flexibility for authentic guided inquiry and, most importantly, the usual assessment practices do not measure the more subtle and complex outcomes of such inquiry (Lesh, 1990). Various new technological forms for assessment are being tested, for example, electronic portfolios using hypermedia authoring tools such as *Virtual Portfolio* (Reilly, 1990), and requiring students to manipulate semantically organized information that is given in the form of a network of concepts (Niemi, 1990).

As the reform process proceeds and as individual and collective experience accumulates, the structural barriers to fruitful use of new technology may weaken. However, we recall Bernard Gifford's words, "Technology is not self implementing" (Gifford, 1986, p. 9).

Sample Pedagogical Strategies for Introducing Multiple Representations

While general strategies for the use of technology in the classroom are difficult to specify, we can report on some recent pedagogical decisions that involve the use of multiple linked notations, and how one goes about extending from one familiar notation system to another less familiar one (Kaput, 1991b). In particular, suppose students are familiar with, say, notation system A (for example, dragging and counting the concrete screen objects in Figure 21.11). We introduce notation system B (for example, a table of data as in Figure 21.11) as an efficient or useful means for recording the results obtained by acting in A. Among the useful ways of accomplishing this in a one-computer class is to print out and distribute versions of B with "blanks" that are to be filled out as "firm guesses" as the teacher generates (with class input, of course) the corresponding screens with the A portion visible, but with the B version hidden. Then B is revealed for comparison purposes.

Once B becomes familiar *as a recording and a display system*, we then introduce *actions* in B whose consequences can be viewed back in A (for example, inputting numerical values in a table where the appropriate numbers of screen objects are generated and displayed in A). We then move on to introduce a third system C as a way to represent the entities in B (for example, a coordinate graph to plot the data in the table), and repeat the process. The new notation is first used to record what one already knows, before being used as an arena for new actions. And when new actions are introduced, they are first reflected back to the familiar notation system before being shown in a new one. The intention is repeatedly to interweave the unknown with the known—not a radical idea, by any means, and yet another illustration of our recurring theme that the new technologies have a way of reasserting classical educational issues in new contexts, with new forms for applying old solutions.

Interestingly, a version of this strategy is also possible using the "smart task" system described earlier, where a student (or pair of students) working at a computer is asked to fill out a column in a table of data on-line before the computer shows what its table looks like, and so on. This kind of task can also be done on a network, where one student works with one representation and another student with a second one. Each takes input from the other in the other's representation and forwards their own (not the computer's) translated version back to the originator, who then compares it with a computer translation of their own input. In effect, the game is to predict for one's peer what a peer's computer will do. This is an example of the new kind of shared activity that can be supported in networked configurations.

OPEN QUESTIONS

Rather than summarize and thereby lengthen an already long review, we provide a list of brief questions distilled from earlier discussions. Our aim here is not to provide a research agenda, but rather to point to general areas needing further investigation as a means of assisting the reader in continuing to think about technology use in mathematics classrooms. Most questions can be regarded as having four implicit dimensions for analysis, relating to teaching, to learning, to the larger contexts in which schooling takes place, and to the technology itself.

1. How do different technologies affect the relation between procedural and conceptual knowledge, especially when the exercise of procedural knowledge is supplanted by (rather than supplemented by) machines?

2. How does one integrate multiple representations of mathematical knowledge, including the use of linked concrete and abstract notations, and action versus display notations?

3. What are the principles that would guide the design or choice of notation systems (as well as ways of linking them) for learning mathematics?

4. What criteria should guide the balance between physical materials and computer-based materials?

5. How does one balance inquiry and subject matter coverage in circumstances where the technology is deliberately designed to support inquiry?

6. What kinds of software support are needed to foster efficient inquiry?

7. What are relations between on-line and off-line instructional support, both in task setting and in learning support, and more generally, what are appropriate relations between print and electronic media?

8. How does one use computers effectively at varying levels of computer density, from one-computer classes to computer-dense, networked classes?

9. How do social patterns change in mathematics classrooms that are technologically rich?

10. What changes in school structures are needed to optimize the potential of new technologies?

11. What sorts of new technological tools and supports do teachers need in order to function effectively in the context of a more ambitious, reformed mathematics curriculum?

12. What kinds of and how much teacher support is needed in introducing teachers to new technologies?

13. What are appropriate strategies for teacher preparation and renewal, and what are appropriate technology education requirements for mathematics teacher credentials?

14. What policy steps need to be taken to bring the reform movement and technological innovation into a symbiotic relationship?

The Future

Although we have attempted to identify trends, the impossibility of accurate predictions of technology's impact on school mathematics is revealed by the following. Suppose one asked, in 1550, what the impact of the printing press might be? Or, in 1920, what the impact of the automobile might be? Or, in 1948, what the impact of television might be? Even if full and accurate answers were *possible,* they would not, indeed *could* not, be understood. The future for technology within school mathematics is promising even though we cannot completely or accurately predict or describe its eventual impact.

References

Abelson, H., & diSessa, A. (1981). *Turtle geometry: The computer as a medium for exploring mathematics.* Cambridge, MA: MIT Press.

Anderson, J. R. (1987). Skill acquisition: Compilation of weak-method problem solutions. *Psychological Review, 94*(2), 192–210.

Anderson, J. R. (1988). *The geometry tutor* [Software]. Pittsburgh, PA: Department of Psychology, Carnegie Mellon University.

Anderson, J. R. (1989, December). *Project Report to the NSF Advanced Technology Program.* Project Directors Meeting, Harvard University.

Anderson, J., Boyle, C., & Reiser, B. (1985). Intelligent tutoring systems. *Science, 228,* 456–462.

Anderson, J., Boyle, F., & Yost, G. (1985). The geometry tutor. *Proceedings of the 9th International Joint Conference on Artificial Intelligence* (Vol. 1, pp. 1–7). Los Angeles, CA.

Apple Computer, Inc. (1991). *Hypercard* [Software]. Cupertino, CA.

Arganbright, D. (1984). Mathematical applications of an electronic spreadsheet. In V. P. Hanson (Ed.), *Computers in mathematics education.* Reston, VA: National Council of Teachers of Mathematics.

Authorware, Inc. (1989). *Authorware* [Software]. Minneapolis, MN.

Ayres, T., Davis, G., Dubinsky, E., & Lewis, P. (1988). Computer experiences in learning composition of functions. *Journal for Research in Mathematics Education, 19,* 246–259.

Barclay, T. (1985). *Guess my rule* [Software]. Pleasantville, NY: HRM Software.

Bardige, A. (1991). *Copley Place* [Interactive video]. Cambridge, MA: Learningways, Inc.

Barr, A., & Feigenbaum, E. (1982, reprinted in 1989). *The handbook of artificial intelligence* (Vol. 1). Reading, MA: Addison-Wesley.

Baxter, N., Dubinsky, E., & Levin, G. (1989). *Learning discrete mathematics using ISETL.* New York, NY: Springer-Verlag.

Becker, H. J. (1987). *School uses of microcomputers: Reports from a national survey.* Baltimore, MD: Center for the Social Organization of Schools. Johns Hopkins University.

Becker, H. J. (1990, April). *Computer use in United States schools.* Paper presented at the Annual Meeting of the AERA, Boston, MA.

Bell, M. (1987). Microcomputer-based courses for school geometry. In I. Wirzup & R. Streit (Eds.), *Developments in school mathematics around the world* (pp. 604–622). Reston, VA: National Council of Teachers of Mathematics.

Bork, A., & Franklin, S. (1979). The role of personal computer systems in education, *AEDS Journal, 19,* 7–12.

Bright, G., Harvey, J., & Wheeler, M. (1985). *Learning and mathematics games* (Journal for Research in Mathematics Education Monograph, Vol. 1). Reston, VA: National Council of Teachers of Mathematics.

Brolin, H. (1987). Mathematics in Swedish schools. In I. Wirzup & R. Streit (Eds.) *Developments in school mathematics around the world* (pp. 130–155). Reston, VA: National Council of Teachers of Mathematics.

Brown, J. S. (1977). Uses of artificial intelligence and advanced computing technology in education. In R. J. Seidel & M. Rubin (Eds.), *Computers and communications: Implications for education.* New York, NY: Academic Press.

Brown, J. S., & Burton, R. B. (1978). Multiple representations of knowledge for tutorial reasoning. In D. G. Bobrow & A. Collins (Eds.), *Representation and understanding: Studies in cognitive science* (pp. 331–349). New York: Academic Press.

Brown, J. S., Collins, A., & Duguid, P. (1989). Situated cognition and the culture of learning. *Education Researcher, 18*(1), 32–42.

Brown, J. S., Collins, A., & Harris, G. (1978). Artificial intelligence and learning strategies. In H. O'Neil (Ed.), *Learning strategies.* New York, NY: Academic Press.

Brown, J. S., & Goldstein, I. (1977). *Computers in a learning society.* Testimony for the House Science and Technology Subcommittee on Domestic and International Planning, Analysis, and Cooperation.

Brown, J. S., & VanLehn, K. (1980). Repair theory: A generative theory of bugs in procedural skills. *Cognitive Science, 4,* 379–426.

Burton, R. (1989). IDE: An instructional development environment [Technical Report and Software]. Palo Alto, CA: Institute for Research on Learning.

Burton, R., & Brown, J. S. (1976). A tutor and student modeling paradigm for gaming environments. *SIGCSE Bulletin, 8,* 236–246.

Byte (1982). *Byte 7.* Logo and mathematics [Special issue].

Camp, J., & Marchionini, G. (1984). Programming and learning: Implications for mathematics education. In V. P. Hanson (Ed.), *Computers in mathematics education.* Reston, VA: National Council of Teachers of Mathematics.

Capper, J. (1988). *Computers and learning: Do they work?* (Office of Technology Assessment Contract Report). Washington, DC: Center for Research into Practice.

Carbonell, J. (1970). AI in CAI: An artificial intelligence approach to computer-assisted instruction. *IEEE Transactions on Man-Machine Systems, MMS-11,* 190–202.

Carraher, T., Carraher, D., & Schliemann, A. (1985). Mathematics in the streets and in schools. *British Journal of Developmental Psychology, 3,* 21–29.

Carraher, T. N., Schliemann, A., & Carraher, D. (1988). Mathematical concepts in everyday life. In G. Saxe (Ed.), *Children's mathematics: New directions for child development.* San Francisco, CA: Jossey Bass.

Channel, D., & Hirsch, C. (1984). Computer methods for problem solving in secondary mathematics. In V. P. Hanson (Ed.), *Computers in mathematics education.* Reston, VA: National Council of Teachers of Mathematics.

Chazan, D. (1988). Similarity: Unravelling a conceptual knot with the aid of technology. In M. Behr, C. Lacampagne, & M. M. Wheeler (Eds.), *Proceedings of the Tenth Annual Meeting of the PME NA.* DeKalb, IL.

Chazan, D., & Houde, R. (1989). *How to use conjecturing and microcomputers to teach geometry.* Reston, VA: National Council of Teachers of Mathematics.

Clancey, W. (1987). *Knowledge-based tutoring. The GUIDON program.* Cambridge, MA: MIT Press.

Clancey, W. (1989). *Symbols and computer programs.* (Technical Report). Palo Alto, CA: Institute for Research on Learning.

Clements, D. (1987). A longitudinal study of the effects of Logo programming on cognitive abilities and achievement. *Journal of Educational Computing Research, 3,* 73–94.

Cohen, D. (1988a). Educational technology and school organization. In R. Nickerson & P. Zodhiates (Eds.), *Technology in education: Looking toward 2020.* Hillsdale, NJ: Lawrence Erlbaum.

Cohen, D. (1988b). *Teaching practice: Plus ca change* (Issue Paper 88-3). East Lansing, MI: National Center for Research on Teacher Education, Michigan State University.

Collins, A., & Brown, J. S. (1988). The computer as a tool for learning through reflection. In H. Mandl & A. Lesgold (Eds.), *Learning issues for intelligent tutoring systems.* New York, NY: Springer-Verlag.

Confrey, J. (1991). *Function Probe* [Software]. Ithaca, NY: Department of Education, Cornell University.

Cuban, L. (1986). *Teachers and machines: The classroom use of technology since 1920.* New York, NY: Teachers College Press.

Davis, R. B. (1984). *Learning mathematics: The cognitive science approach to mathematics education.* Norwood, NJ: Ablex.

Demana, F., & Leitzel, J. (1988). Establishing fundamental concepts through numerical problem solving. In A. Coxford & A. Schulte (Eds.), *The Ideas of Algebra, K–12.* Reston, VA: National Council of Teachers of Mathematics.

Demana, F. & Waits, B. (1988). Pitfalls in graphical computation, or why a single graph isn't enough. *The College Mathematics Journal, 19*(2), 177–183.

Derry, S. (1989). Strategy and expertise in word problem solving. In M. Pressley, C. McCormick, & G. Miller (Eds.), *Cognitive strategy research: From basic research to educational applications.* New York, NY: Springer-Verlag.

Dienes, Z. (1973). *The six stages in the process of learning mathematics* (P. L. Seaborne, Trans.). New York, NY: Humanities Press. (Original published 1970)

diSessa, A., & Abelson, H. (1986). Boxer: A reconstructable computer medium. *Communications of the ACM, 29* (9), 859–868.

Dreyfus, T., & Eisenberg, T. (1988). *Visualizing function transformations.* Unpublished manuscript, Weizmann Institute of Science, Rehovat, Israel.

Dubinsky, E. (in press). A theory and practice of learning college mathematics. In A. Schoenfeld (Ed.), *Cognitive science and mathematics.*

Dugdale, S. (1982). Green Globs: A microcomputer application for graphing of equations. *Mathematics Teacher, 75,* 208–14.

Dugdale, S., & Kibbey, D. (1986). *Interpreting graphs* [Software]. Pleasantville, NY: Sunburst Communications.

Dugdale, S., & Kibbey, D. (1977). *Elementary mathematics with PLATO.* Urbana, IL: Computer-based Education Laboratory, University of Illinois.

Ernst, A., & Ohlsson, S. (1989). *The cognitive complexity of the regrouping and augmenting algorithms for subtraction: A theoretical analysis* (Technical Report KUL 89–6). Pittsburgh, PA: University of Pittsburgh, Learning Research and Development Center.

Feurzeig, W. (1988). Apprentice tools: Students as practitioners. In R. Nickerson & P. Zodhiates, (Eds.), *Technology in education: Looking toward 2020.* Hillsdale, NJ: Lawrence Erlbaum.

Feurzeig, W., Lucas, G., Grant, R., & Faflick, P. (1969). *Programming languages as a conceptual framework for teaching mathematics* (Technical Report). Cambridge, MA: BBN Labs.

Feurzeig, W., & Richards, J. (1991). *Function Machines* [Software]. Cambridge, MA: BBN Labs.

Fey, J. (1984). *Computing and mathematics: The impact on secondary school curricula.* Reston, VA: National Council of Teachers of Mathematics.

Fey, J. (1989a). School algebra for the year 2000. In S. Wagner & C. Kieran (Eds.), *Research issues in the learning and teaching of algebra.* Reston, VA: National Council of Teachers of Mathematics; Hillsdale, NJ: Lawrence Erlbaum.

Fey, J. (1989b). Technology and mathematics education: A survey of recent developments and important problems. *Educational Studies in Mathematics, 20,* 237–272.

Filloy, E., & Rojano, T. (1985). Obstructions to the learning of elemental algebraic concepts and teaching strategies. In S. Damarin (Ed.), *Proceedings of the Seventh Annual Meeting of the PME-NA* (pp. 154–158), Columbus, OH.

Fraser, R. (1986). *The role of computers for applications and modeling: The ITMA approach.* University of Nottingham, UK: Shell Centre for Mathematical Education.

Friedman, M. (1978). The manipulative materials strategy: The latest Pied Piper? *Journal for Research in Mathematics Education, 9,* 78–80.

Gardner, H. (1985). *The mind's new science: A history of the cognitive revolution.* New York: Basic Books.

Gifford, B. (1986). *Education and the challenge of technology.* Cupertino, CA: Apple Computer, Inc.

Goldenberg, E. P. (1987). Believing is seeing: How preconceptions influence the perception of graphs. In J. Bergeron, C. Kieran, & N. Herscovics (Eds.), *Proceedings of the 11th Annual Meeting of the PME-NA* (Vol. 1). University of Montreal.

Goldenberg, E. P. (1988). Mathematics, metaphors and human factors: Mathematical, technical, and pedagogical challenges in the educational use of graphical representation of functions. *Journal of Mathematical Behavior, 7* (2), 135–173.

Goldenberg, E. P. (1991). The difference between graphing software and *educational* graphing software. In W. Zimmerman & S. Cunningham (Eds.), *Visualization in mathematics*. Washington, DC: Mathematical Association of America.

Goldenberg, E. P., Lewis, P., & O'Keefe, J. (1990). Dynamic representation and the development of a process understanding of functions. Unpublished manuscript available from E. P. Goldenberg, Educational Development Center, Newton, MA.

Grant, W., & Borovoy, R. (1991). *Simulation environments to support the construction of mental models*. Cupertino, CA: Apple Classroom of Tomorrow Program, Apple Computer, Inc.

Greeno, J. (1983). Conceptual entities. In A. Stevens & D. Gentner (Eds.), *Mental models*. Hillsdale, NJ: Lawrence Erlbaum.

Greeno, J. (1989). Situations, mental models, and generative knowledge. In D. Klahr & K. Kotovsky (Eds.), *Complex information processing: The impact of Herbert Simon*. Hillsdale, NJ: Lawrence Erlbaum.

Greeno, J. (in press). A view of mathematical problem solving in school. In M. U. Smith (Ed.), *Problem solving in school science: Toward a unified view*. Hillsdale, NJ: Lawrence Erlbaum.

Grossberg, S. (1980). How does a brain build a cognitive code? *Psychological Review, 87,* 1–51.

Gurtner, J.-L. (1989). Understanding and discussing linear functions in situations: A developmental study. *Proceedings of the 13th International Conference on the Psychology of Mathematics Education* (Vol. 2). Paris.

Hall, R. (1988). *Qualitative diagrams: Supporting the construction of algebraic representations in problem solving* (Technical Report). Irvine: University of California, Department of Information and Computer Science.

Hancock, C. (1988). *From formal systems to symbol systems: Computer environments for a metaphor-oriented pedagogy of data structures and data analysis*. Unpublished manuscript, Technological Educational Resource Centers (TERC), Cambridge, MA.

Hancock, C. (1990). *Data modeling with the TableTop*. (Technical Report). Cambridge, MA: Technological Educational Resource Centers (TERC).

Hancock, C., & J. Kaput, (1990). Computerized tools and the process of modeling. In G. Booker, P. Cobb, & T. Mendicuti (Eds.), *Proceedings of the 14th Annual Meeting of the PME-NA*, Oaxtepec, Mexico.

Harel, G., & Kaput, J. (in press). Conceptual entities in advanced mathematical thinking: The role of notations in their formation and use. In D. Tall & E. Dubinsky (Eds.), *Advanced mathematical thinking*. London: Reidel.

Harel, I. (1990). Children as software designers: A constructionist approach to learning mathematics. *Journal of Mathematical Behavior, 9*(1), 5–95.

Harel, I., & Papert, S. (1990). Software design as a learning environment. In E. Soloway (Ed.), *Interactive learning environments*. Norwood, NJ: Ablex.

Hart, K. (1984). *Ratio: Children's strategies and errors*. Windsor, UK: The NFER-Nelson Publishing Company, Ltd.

Hausknecht, A., & Kowalczyk, R. (1991). *Tools for exploring mathematics* [Software]. New York, NY: Brooks-Cole.

Hawkes, L., & Derry, S. (1989/90). Error diagnosis and fuzzy reasoning techniques for intelligent tutoring systems. *Journal of Artificial Intelligence in Education, 1*(2), 43–56.

Heid, M. K., & Kunkle, D. (1988). Computer generated tables: Tools for concept development in elementary algebra. In A. Coxford & A. Schulte (Eds.), *The ideas of algebra, K-12*. Reston, VA: National Council of Teachers of Mathematics.

Hembree, R., & Dessart, D. (1986). Effects of hand held calculators in precollege mathematics education: A meta-analysis. *Journal for Research in Mathematics Education, 17,* 83–89.

Hoffer, A. (1987). Technology-supported learning. In I. Wirzup & R. Streit (Eds.), *Developments in school mathematics around the world* (pp. 637-646). Reston, VA: National Council of Teachers of Mathematics.

Hoyles, C., Noss, R., & Sutherland, R. (1989). A Logo-based microworld for ratio and proportion. In G. Vergnaud, J. Rogalski, & M. Artigue (Eds.), *Proceedings of the 13th International Conference on the Psychology of Mathematics Education*. University of Paris, France.

Hutchins, E. (1986). *Metaphors for interface design*. (ICS Report 8703), La Jolla, CA: University of California, Institute for Cognitive Science.

Hutchins, E., Hollan, J., & Norman, D. (1986). Direct manipulation interfaces. In D. Norman & S. Draper (Eds.), *User centered system design*. Hillsdale, NJ: Lawrence Erlbaum.

Informix Software. (1989). *Wingz* [Software]. Lawrenceville, KS.

Jay, T. (1981). Computerphobia: What to do about it. *Educational Technology*. January. Reprinted in D. Harper & J. Stewart (Eds.), *Run: Computer Education* (1983, pp.79–81). Monterey, CA: Brooks-Cole.

Johnson, W., & Soloway, E. (1985). PROUST. *Byte, 10*(4).

Kaput, J. (1985a). Representation and problem solving: Methodological issues related to modeling. In E. Silver (Ed.), *Teaching and learning mathematical problem solving: Multiple research perspectives*. Hillsdale, NJ: Lawrence Erlbaum.

Kaput, J. (1985b). *Multiplicative word problems and intensive quantities: An integrated software response* (Technical Report 85-19). Cambridge, MA: Educational Technology Center. Harvard Graduate School of Education.

Kaput, J. (1986). Information technology and mathematics: Opening new representational windows. *The Journal of Mathematical Behavior, 5,* 187–207.

Kaput, J. (1987a). Toward a theory of symbol use in mathematics. In C. Janvier (Ed.), *Problems of representation in mathematics learning and problem solving* (pp. 159–196). Hillsdale, NJ: Lawrence Erlbaum.

Kaput, J. (1987b). Representation and Mathematics. In C. Janvier (Ed.), *Problems of representation in mathematics learning and problem solving* (pp. 19–26). Hillsdale, NJ: Lawrence Erlbaum.

Kaput, J. (1988, April). *Applying technology in mathematics classrooms: Time to get serious, time to define our own technological destiny*. Paper prepared for the Annual Meeting of the American Educational Research Association, New Orleans, LA.

Kaput, J. (1989a). Linking representations in the symbol system of algebra. In C. Kieran & S. Wagner (Eds.), *A research agenda for the learning and teaching of algebra*. Reston, VA: National Council of Teachers of Mathematics; Hillsdale, NJ: Lawrence Erlbaum.

Kaput, J. (1989b). Information technologies and affect in mathematical experience. In D. McLeod & V. Adams (Eds.), *Affect and mathematical problem solving* (pp. 89–103). New York, NY: Springer-Verlag.

Kaput, J. (1990, April). *The Supposer experience: What's special and what generalizes?* Paper presented to the Annual Meeting of the AERA, Boston, MA.

Kaput, J. (1991a). Notations and representations as mediators of constructive processes. In E. von Glasersfeld (Ed.), *Radical constructivism in mathematics education*. Dordrecht, Netherlands: Kluwer.

Kaput, J. (1991b). Creating cybernetic and psychological ramps from the concrete to the abstract: Examples from multiplicative structures. Manuscript available from author.

Kaput, J. (in press-a). The urgent need for proleptic research. In T. Romberg, T. Carpenter, & E. Fennema (Eds.), *Integrating research on the graphical representation of functions*. Hillsdale, NJ: Lawrence Erlbaum.

Kaput, J. (in press-b). Democratizing access to calculus: New routes using old roots. In A. Schoenfeld (Ed.), *Cognitive science and mathematics*.

Kaput, J. (in press-c). The impact of the *Supposer* experience on students' beliefs about and attitudes towards mathematics. In J. Schwartz, M. Gordon, & P. Butler (Eds.), *The Supposer reader.* Hillsdale, NJ: Lawrence Erlbaum.

Kaput, J. (in press-d). Patterns in student construction of algebraic patterns in numerical data. In E. Dubinsky & G. Harel (Eds.), *Research in developing the concept of function: MAA Notes.* Washington, D.C.: Mathematics Association America.

Kaput, J., Katz, M. M., Luke, C., & Pattison-Gordon, L. (1988). Concrete representations for ratio reasoning. In M. Behr, C. Lacampagne, & M. M. Wheeler (Eds.), *Proceedings of the 10th Annual Meeting of the PME-NA.* DeKalb, IL.

Kaput, J., Luke, C., Poholsky, J., & Sayer, A. (1987). Multiple representations and reasoning with intensive quantities. In J. Bergeron, C. Kieran, & N. Herscovics (Eds.), *Proceedings of the 11th Annual Meeting of the PME-NA* (Vol. 1). Canada: University of Montreal.

Kaput, J., & Pattison-Gordon, L. (1987).*A concrete to abstract software ramp: Environments for learning multiplication, division, and intensive quantity* (Technical Report 87-8). Cambridge, MA: Harvard Graduate School of Education, Educational Technology Center.

Kaput, J., & Sims-Knight, J. (1983). Errors in translations to algebraic equations: Roots and implications. In M. Behr & G. Bright (Eds.), Mathematics learning problems of the post secondary student [Special issue]. *Focus on Learning Problems in Mathematics, 5,* 63–78.

Kaput, J., & Upchurch, R. (in preparation). *Integrated tools for elementary mathematics* [Software]. Scotts Valley, CA: Wings for Learning, a Sunburst Communications subsidiary.

Kaput, J., & West, M. (in press). Factors affecting informal proportional reasoning. In G. Harel and J. Confrey (Eds.), *The development of multiplicative reasoning in the learning of mathematics.* Albany, NY: SUNY Press.

Kenelly, J. (1989). *The uses of calculators in standardized testing of mathematics.* New York: The College Entrance Examination Board.

Kieran, C. (1990, April). *The graphing of functions: Integrating research on learning and instruction.* Paper given at the NCTM Research Presession on the Graphical Representation of Functions, Salt Lake City, UT. To appear in T. Romberg, T. Carpenter, & E. Fennema (Eds.), *Integrating research on the graphical representation of functions.* Hillsdale, NJ: Lawrence Erlbaum.

Kieran, C. (1991). CARAPACE [Software]. Canada: University of Quebec at Montreal.

Kieran, C., Garançon, M., Boileau, A., & Pelletier, M. (1988). Numerical approaches to algebraic problem solving in a computer environment. In M. Behr, C. Lacampagne, & M. M. Wheeler (Eds.), *Proceedings of the 10th Annual Meeting of the PME-NA.* DeKalb, IL.

Kieran, C., & Wagner, S. (1989). *A research agenda for the learning and teaching of algebra.* Reston, VA: National Council of Teachers of Mathematics; and Hillsdale, NJ: Lawrence Erlbaum.

Kieren, T., Nelson, D., & Smith, G. (1985). Graphical algorithms in partitioning tasks. *The Journal of Mathematical Behavior, 4,* 25–36.

Klotz, E. (1991). *The geometer's sketchpad* [Software]. Berkely, CA: Key Curriculum Press.

Knowledge Revolution (1990). *Interactive physics* [Software]. Chapel Hill, NC: Knowledge Revolution, Inc.

Konold, C. (1991). *Chance Plus* [Software]. Amherst, MA: Scientific Reasoning Institute, Hasbrouk Hall, University of Massachusetts.

Kulik, J. (1985). Effectiveness of computer-based instruction in elementary schools. *Computers in Human Behavior* 1, 59–74.

Kurland, M., Pea, R., Clement, C., & Mawby, R. (1986). A study of the development of programming ability and thinking skills in high school students. *Journal of Educational Computing Research, 2,* 429–458.

Laborde, J-M. (1990). *CABRI Geometry* [Software]. France: Université de Grenoble 1 (Available in the U.S. from Brooks Cole: NY).

Lampert, M. (1988a). *Teachers' thinking about students' thinking about geometry: The effects of new teaching tools* (Technical Report 88-1). Cambridge, MA: Harvard Graduate School of Education, Educational Technology Center.

Lampert, M. (1988b). *Teaching that connects students' inquiry with curricular agendas in schools* (Technical Report 88-27). Cambridge, MA: Harvard Graduate School of Education, Educational Technology Center.

Lave, J. (1988). *Cognition in practice.* New York, NY: Cambridge University Press.

Lawler, R. (1981). The progressive construction of mind. *Cognitive Science, 5,* 1–30.

Lawler, R. (1985). *Computer experience and cognitive development.* New York, NY: Wiley.

Lehrer, R., Guckenberg, T., & Sancilio, L. (1988). Influences of Logo on children's intellectual development. In R. E. Mayer (Ed.), *Teaching and learning computer programming: Multiple research perspectives* (pp. 75–110). Hillsdale, NJ: Lawrence Erlbaum.

Lehrer, R., Knight, W., Sancilio, L., & Love, M. (1989, March). *Software to link action and description.* Paper presented to the Annual Meeting of the AERA, San Francisco, CA.

Lehrer, R., Randle, L., & Sancilio, L. (1988). Learning pre-proof geometry with Logo. *Cognition and Instruction, 6*(2), 159–184.

Leinhardt, G., Zaslavsky, O., & Stein, M. (1990). Functions, graphs, and graphing: Tasks, learning, and teaching. *Review of Educational Research, 60*(1), 1–64.

Leitzel, J. (1985). Calculators do more than compute. In D. Albers, S. Rodi, & A. Watkins (Eds.), *New directions in two-year college mathematics.* New York, NY: Springer-Verlag.

Lepper, M. (1985). Microcomputers in education: Motivational and social issues. *American Psychologist, 40,* 1–18.

Lesh, R. (1987). The evolution of problem representations in the presence of powerful cultural amplifiers. In C. Janvier (Ed.), *Problems of representation in mathematics learning and problem solving* (pp. 197–206). Hillsdale, NJ: Lawrence Erlbaum.

Lesh, R. (1990). Computer-based assessment of higher-order understandings and processes in elementary mathematics. In G. Kulm (Ed.), *Assessing higher order thinking.* Washington, DC: American Association for the Advancement of Science.

Lesh, R., Post, T., & Behr, M. (1987). Dienes revisited: Multiple embodiments in computer environments. In I. Wirzup & R. Streit (Eds.), *Developments in school mathematics around the world* (pp. 647–680). Reston, VA: National Council of Teachers of Mathematics.

Lewis, M. (1989). Intelligent tutoring systems: First steps and future directions. In S. Wagner & C. Kieran (Eds.), *Research issues in the learning and teaching of algebra.* Reston, VA: National Council of Teachers of Mathematics; Hillsdale, NJ: Lawrence Erlbaum.

Macromind. (1981). *Macromind director interactive* [Software]. Macromind, Inc. Mountain View, CA: Author.

Malone, T., & Lepper, M. (1987). Making learning fun: A taxonomy of intrinsic motivations for learning. In R. E. Snow & M. J. Farr (Eds.), *Aptitude, learning and instruction: Vol. 3. Cognitive and affective process analyses* (pp. 223–253). Hillsdale, NJ: Lawrence Erlbaum.

Mandler, G. (1984). *Mind and body: Psychology of emotion and stress.* New York, NY: Norton.

Mandler, G. (1987). *Modifying reward structure in computer games.* Presentation to the meeting on mathematical problem solving and affect (D. McLeod, organizer), San Diego, CA.

Marcoulides, G. (1989). The effectiveness of an expert system and CAI in an introductory statistics course. *Journal of Artificial Intelligence in Education 1*(1), 93–101.

Mathematical Sciences Education Board. (1990). *Reshaping school mathematics: A philosophy and framework for curriculum.*

Mathsoft Corporation. (1989). *MathCad* [Software]. Cambridge, MA: David Blohm.

McArthur, D., & Stasz, C. (1990). *An intelligent tutor for algebra* (NSF Final Report R-3811-NSF). Santa Monica, CA: The Rand Corporation.

McArthur, D., Stasz, C., & Hotta, J. (1987). Learning problem-solving skills in algebra. *Journal of Educational Technology Systems, 15*(3), 303–324.

McConnell, J. (1988). Technology and algebra. In A. Coxford & A. Schulte (Eds.), *The ideas of algebra, K-12.* Reston, VA: National Council of Teachers of Mathematics.

McKnight, C., Crosswhite, F. J., Dossey, J. A., Kifer, E., Swafford, J. O., Travers, K., Cooney, T. J. (1987). *The underachieving curriculum: Assessing U.S. school mathematics from an international perspective.* Champaign, IL: Stipes Publishing Company.

Meissner, H. (1987). Schulerstrategien bei einem taschenreichnerspeil. *Journal fur Mathematik-Didaktik, 8,* 105–128.

Minstrell, J. (1988). Teachers' assistants: What could technology make feasible? In R. Nickerson & P. Zodhiates, (Eds.), *Technology in education: Looking toward 2020.* Hillsdale, NJ: Lawrence Erlbaum.

Mokros, J., & Tinker, R. (1987). The impact of microcomputer-based labs on children's ability to interpret graphs. *Journal of Research in Science Teaching, 24*(4), 369–383.

Monk, G. S. (1989, April). *A framework for understanding student understanding of functions.* Paper given at the Annual Meeting of the AERA, San Francisco. Available from the author, Department of Mathematics, University of Washington, Seattle.

Monk, G. S. (in press). Students' understanding of a function given by a physical model. In E. Dubinsky & G. Harel (Eds.), *Research in developing the concept of function: MAA Notes.* Washington, DC: Mathematics Association of America.

Nathan, M., Johl, P., Kintch, W., & Lewis, C. (1989). *An unintelligent tutoring system for solving algebra word problems* (Technical Report), Boulder, CO: University of Colorado, Institute of Cognitive Science.

National Council of Teachers of Mathematics. (1989). *Curriculum and evaluation standards for school mathematics.* Reston, VA: Author.

National Education Association. (1989). *Report of the NEA Special Committee on Technology* (E. Scott Brown, Chairperson).

National Research Council. (1989). *Everybody counts: A report to the nation on the future of mathematics education.* Washington, D.C.: National Academy Press.

Newman, D. (1990). Cognitive change by appropriation. In M. Gardner, J. Greeno, F. Reif, & A. Schoenfeld (Eds.), *Toward a scientific practice of science education.* Hillsdale, NJ: Lawrence Erlbaum.

Niemi, D. (1990). *HyperCard: New context for assessment* (Technical Report). Los Angeles: University of California, UCLA Center for Technology Assessment.

Noss, R. (1987). Children's learning of geometrical concepts through Logo. *Journal for Research in Mathematics Education, 18,* 343–362.

O'Brien, T. (1985). *The king's rule* [Software]. Pleasantville, NY: Sunburst Communications.

Office of Technology Assessment. (1988). *Power on! New tools for teaching and learning* (OTA-SET-379). U.S. Government Printing Office.

Ohlsson, S. (1987). Sense and reference in the design of interactive illustrations for rational numbers. In R. Lawler & M. Yazdani (Eds.), *Artificial intelligence and education: Learning environments and tutoring systems* (Vol. 1). Norwood, NJ: Ablex Publishing Corporation.

Ohlsson, S. (1988a), *Interactive illustrations for fractions: A progress report* (Technical Report KUL87-03). Pittsburgh, PA: University of Pittsburgh, Learning Research and Development Center.

Ohlsson, S. (1988b). *The conceptual basis of subtraction: A mathematical analysis* (Technical Report KUL87-03). Pittsburgh, PA: University of Pittsburgh, Learning Research and Development Center.

Ohlsson, S. (in press). Towards intelligent tutoring systems that teach knowledge rather than skills: Five research questions. In E. Scanlon & T. O'Shea (Eds.), *New directions in educational technology.* New York, NY: Springer-Verlag.

Ohlsson, S., & Hall, N. (1990). *The cognitive function of embodiments in mathematics education* (Technical Report KUL90-01). Pittsburgh, PA: University of Pittsburgh, Learning Research and Development Center.

Olds, H., Schwartz, J., & Willie, J. (1980). *People and computers: Who teaches whom?* Newton, MA: Educational Development Center.

Olive, J. (1989). Associations among high school students' interactions with Logo and mathematical thinking. In G. Vergnaud, J. Rogalski, & M. Artigue (Eds.) *Proceedings of the 13th Annual International Conference on the Psychology of Mathematics Education.* University of Paris, France.

Olive, J. (1991). Logo programming and geometric understanding: An in-depth study. *Journal for Research in Mathematics Education, 22,* 90–111.

Olive, J., & Lankenau, C. (1987). The effects of Logo-based learning experiences on students' non-verbal cognitive abilities. In J. Bergeron, C. Kieran, & N. Herscovics (Eds.), *Proceedings of the 11th Annual Meeting of the PME-NA* (Vol. 2). University of Montreal, Canada.

Orey, M., & Burton, J. (1989/90). POSIT: Process oriented subtraction-interface for tutoring. *Journal of Artificial Intelligence in Education, 1*(2), 77–104.

Papert, S. (1980). *Mindstorms.* New York, NY: Basic Books.

Pea, R. (1988). Putting knowledge to use. In R. Nickerson & P. Zodhiates (Eds.), *Technology in education: Looking toward 2020.* Hillsdale, NJ: Lawrence Erlbaum.

Pea, R., & Kurland, M. (1984). On the cognitive effects of learning computer programming. *New Ideas in Psychology, 2,* 137–168.

Philipp, R., Martin, W., & Richgels, G. (1990). *Curricular implications of graphical representation of functions.* Paper given at the NCTM Research Presession on the Graphical Representation of Functions, Salt Lake City, UT. To appear in T. Romberg, T. Carpenter, & E. Fennema (Eds.), *Integrating research on the graphical representation of functions.* Hillsdale, NJ: Lawrence Erlbaum.

Phillips, R. (1982). An investigation of the microcomputer as a mathematics teaching aid. *Computers and Education, 6*(1), 45–50.

Post, T. (1980). The role of manipulative materials in the learning of mathematical concepts. In M. Lindquist (Ed.), *Selected issues in mathematics education.* Reston, VA: National Council of Teachers of Mathematics.

PreScience Corporation. (1990). *Theorist* [Software]. San Francisco, CA: Alan Bonadio.

Provenzo, E. (1986). *Beyond the Gutenberg galaxy.* New York: Teachers College Press.

Reilly, B. (1990). Hypermedia software for student portfolios. Apple Classroom of Tomorrow Program, Apple Computer, Inc., Cupertino, CA.

Resnick, L. B., & Omanson, S. (1987). Learning to understand arithmetic. In R. Glaser (Ed.), *Advances in instructional psychology.* (Vol. 3, pp. 41–95). Hillsdale, NJ: Lawrence Erlbaum.

Richards, J., Feurzeig, W. & Carter, R. (1991). *The algebra workbench* [Software]. Cambridge, MA: BBN Labs.

Richmond, B. (1986). *Stella* [Software]. Lyme, NH: High Performance Systems, Inc.

Romberg, T., Carpenter, T., & Fennema, E. (in press). *Integrating research on the graphical representation of functions.* Hillsdale, NJ: Lawrence Erlbaum.

Roschelle, J. (1990, March). *Designing for conversations.* Paper presented at the Annual Meeting of the AERA, San Francisco.

Rubin, A. & Goodman, B. (1991). Tape measure: Video as data for statistical investigations. Paper presented at the Annual Meeting of the American Educational Research Association, Chicago, IL.

Rubin, A., Bruce, B., Rosebery, A., & Horwitz, P. (1989). *Reasoning under uncertainty* (BBN Report 6774). Cambridge, MA: BBN Labs.

Russel, S. & Corwin, R. (1990). *Used numbers*. Palo Alto, CA: Dale Seymour Publishing Co.

Saxe, G. (1988a). Candy selling and math learning. *Educational Researcher, 17*(6) 14–21.

Saxe, G. (1988b) The mathematics of child street vendors. *Child Development, 59*, 1415–1425.

Saxe, G. (1990). The interplay between children's learning in formal and informal social contexts. In M. Gardner, J. Greeno, F. Reif, & A. Schoenfeld (Eds.), *Toward a scientific practice of science education*. Hillsdale, NJ: Lawrence Erlbaum.

Schoenfeld, A. (1989). The curious fate of an applied problem. *The College Mathematics Journal, 20*(2), 115–123.

Schoenfeld, A., Smith, J., & Arcavi, A. (in press). Learning: The microgenetic analysis of one student's evolving understanding of a complex subject matter domain. In R. Glaser (Ed.), *Advances in instructional psychology* (Vol. 4). Hillsdale, NJ: Lawrence Erlbaum.

Schwartz, J. (1988). *The Algebraic Proposer* [Software]. Hanover, NH: True BASIC, Inc.

Schwartz, J. (1991). Shuttling between the particular and the general: Reflections on the role of conjecture and hypothesis in the generation of knowledge in science and mathematics. (Manuscript available from author, Educational Technology Center, Harvard University, Cambridge, MA 02138.)

Schwartz, J., Gordon, M., & Butler, P. (in press). *The Supposer reader*. Hillsdale, NJ: Lawrence Erlbaum.

Schwartz, J., & Yerushalmy, M. (1985). *The Geometric Supposers* [A series of four software packages] Pleasantville, NY: Sunburst Communications.

Schwartz, J., & Yerushalmy, M. (1987a). The Geometric Supposer: The computer as an intellectual prosthetic for the making of conjectures. *The College Mathematics Journal, 18*, 58–65.

Schwartz, J., & Yerushalmy, M. (1987b). Using microcomputers to restore invention to the learning of mathematics In I. Wirzup & R. Streit (Eds.), *Developments in school mathematics around the world* (pp. 623–636). Reston, VA: National Council of Teachers of Mathematics.

Schwartz, J., Yerushalmy, M., & Harvey, W. (1991). *The Algebra Toolkit* [Software]. Pleasantville, NY: Sunburst Communications, Inc.

Senk, S. (1989). Toward school algebra in the year 2000. In S. Wagner & C. Kieran (Eds.), *Research issues in the learning and teaching of algebra*. Reston, VA: National Council of Teachers of Mathematics; Hillsdale, NJ: Lawrence Erlbaum.

Shank, R., & Edelson, D. (1989/90). A role for AI in education: Using technology to reshape education. *Journal of Artificial Intelligence in Education, 1*(2), 3–20.

Shortliffe, E. (1976). *Computer-based medical consultations: MYCIN*. New York, NY: North-Holland.

Shumway, R. (1984). Young children, programming, and mathematical thinking. In V. P. Hanson (Ed.), *Computers in mathematics education, 1984 Yearbook*. Reston, VA: National Council of Teachers of Mathematics.

Skemp, R. (1987). *The psychology of learning mathematics* (Revised American edition). Hillsdale, NJ: Lawrence Erlbaum.

Sleeman, D. (1984). An attempt to understand students' understanding of basic algebra. *Cognitive Science, 8*, 387–412.

Sleeman, D. (1987). PIXIE: A shell for developing intelligent tutoring systems. In R. Lawler & M. Yazdani (Eds.), *Artificial intelligence and education: Learning environments and tutoring systems* (Vol. 1). Norwood, NJ: Ablex.

Sleeman, D., & Brown, J. S. (1982). *Intelligent tutoring systems*. New York, NY: Academic Press.

Steen, L. A. (1988). The science of patterns. *Science, 240*, 611–616.

Steffe, L. P., Cobb, P., & von Glasersfeld, E. (1988). *Construction of arithmetical meanings and strategies*. New York, NY: Springer-Verlag.

Stoutemyer, D. (1983). Nonnumeric computer applications to algebra, trigonometry, and calculus. *Two Year College Mathematics Journal, 14*, 233–239.

Sunburst Communications, Inc. (1991). ELASTIC [Software]. Pleasantville, NY. Andee Rubin.

Suppes, P., & Morningstar, J. (1969). Computer-assisted instruction. *Science, 166*, 343–350.

Taylor, R. (1980). *The computer in the school: Tutor, tool, tutee*. New York, NY: Teacher's College Press.

Thompson, P. (1988). *Expressions* [Software available from the author]. San Diego, CA: San Diego State University.

Thompson, P. (1989). Artificial intelligence, advanced technology, and learning and teaching algebra. In S. Wagner & C. Kieran (Eds.), *Research issues in the learning and teaching of algebra*. Reston, VA: National Council of Teachers of Mathematics; Hillsdale, NJ: Lawrence Erlbaum.

Thompson, P. (1990). *A theoretical model of quantity-based reasoning in arithmetic and algebra*. A revised version of a paper presented to the Annual Meeting of the AERA, San Francisco, CA. Available from the author, San Diego State University, Department of Mathematics.

Thompson, P. (1991a). *Blocks Microworld* [Software available from the author]. San Diego, CA: San Diego State University.

Thompson, P. (1991b). *Word Problem Analyst* [Software available from the author]. San Diego, CA: San Diego State University.

Thompson, P., & Thompson, A. (1990). Salient aspects of experience with concrete manipulatives In G. Booker, P. Cobb, & T. Mendicuti (Eds.), *Proceedings of the 14th International Conference for the Psychology of Mathematics Education*. Oaxtepec, Mexico.

Thompson, P., & Thompson, A. (in press). Representations, principles, and constraints: Contributions to the effective use of concrete manipulatives in elementary mathematics. *Journal for Research in Mathematics Education*.

Usiskin, Z. (1987). Lessons learned from the first eighteen months of the secondary component of the UCSMP. In I. Wirzup & R. Streit (Eds.), *Developments in school mathematics around the world* (pp. 418–429). Reston, VA: National Council of Teachers of Mathematics.

Usiskin, Z. (1988). Conceptions of algebra and uses of variables. In A. Coxford & A. Schulte (Eds.), *The ideas of algebra, K-12*. Reston, VA: National Council of Teachers of Mathematics.

Usiskin, Z., Flanders, J., Hynes, C., Polonsky, L., Porter, S., & Viktora, S. (1990). *Transition mathematics*. Glenview, IL: Scott Foresman.

Wagner, S. (1981). Conservation of equation and function under transformation of variable. *Journal for Research in Mathematics Education, 12*, 107–118.

Wagner, S., & Kieran, C. (1989). *Research issues in the learning and teaching of algebra*. Hillsdale, NJ: Lawrence Erlbaum; and Reston, VA: National Council of Teachers of Mathematics.

Wenger, E. (1987). *Artificial intelligence and tutoring systems: Computational and cognitive approaches to the communication of knowledge*. Los Altos, CA: Morgan Kaufmann.

West, M. M., Luke, C., Poholsky, J., Pattison-Gordon, L., Turner, S., & Kaput, J. (1989). Additive strategies in proportional reasoning: Results of a teaching experiment using concrete representations. In G. Goldin, C. Maher, & T. Purdy (Eds.), *Proceedings of the Eleventh Annual Meeting of the PME-NA*. Rutgers, NJ.

Wings for Learning. (1991). STEP [Software]. Scotts Valley, CA: Wings for Learning, Inc.

Winograd, T., & Flores, F. (1986). *Understanding computers and cognition*. Norwood, NJ: Ablex.

Wiske, M. S. (1990, April). *Teaching geometry through guided inquiry: A case of changing mathematics instruction with new technologies*. Paper presented to the Annual Meeting of the AERA, Boston, MA.

Wiske, M. S., & Houde, R. (1988). *From recitation to construction: Teachers change with new technologies* (Technical Report 88-28). Cambridge, MA: Harvard Graduate School of Education, Educational Technology Center.

Wolfram Research, Inc. (1989). *Mathematica* [Software]. Champaign, IL: Steven Wolfram.

Wolfram, S. (1989). *Mathematica*. Reading, MA: Addison-Wesley.

Wynands, A. (1984). Rechenfertigkeit und taschrechner—ergebnisse einer dreijarhigen untersuchung in den klassen 7–9, *Journal fur Mathematik-Didaktik, 5,* 3–32.

Yerushalmy, M. (1986). *Induction and generalization: An experiment in the teaching and learning of high school geometry.* Unpublished doctoral dissertation, Harvard Graduate School of Education.

Yerushalmy, M. (1989). *Effects of graphic feedback on the ability to transform algebraic expressions when using computers.* Unpublished manuscript, University of Haifa, Haifa, Israel.

Yerushalmy, M., Chazan, D., & Gordon, M. (1988). *Posing problems: One aspect of bringing inquiry into classrooms.* Technical Report 88-21. Cambridge, MA: Harvard Graduate School of Education, Educational Technology Center.

Yerushalmy, M., & Houde, R. A. (1986). The Geometric Supposer: Promoting thinking and learning. *Mathematics Teacher, 79,* 418–422.

Yerushalmy, M., & Schwartz, J. (in press). Seizing the opportunity to make algebra mathematically and pedagogically interesting. In T. Romberg, T. Carpenter, & E. Fennema (Eds.), *Integrating research on the graphical representation of functions.* Hillsdale, NJ: Lawrence Erlbaum.

· 22 ·

ETHNOMATHEMATICS
AND EVERYDAY COGNITION

Terezinha Nunes
THE INSTITUTE OF EDUCATION, LONDON

The purpose of this chapter is to look at current research on culture and mathematics learning and to use this research in the analysis of the acquisition of mathematical concepts and skills. The chapter is divided into five parts. The first section briefly contrasts two views of cultural influences on mathematical activities. The next three sections discuss traditional issues in mathematics education: (1) counting and measuring, (2) solving arithmetic calculations, and (3) modeling and knowledge of inverse operations. These topics do not cover all types of mathematical activities. They were chosen because of their central role in elementary mathematics education, and because current work on these topics generates an interesting picture of similarities and differences in mathematical knowledge used in distinct cultural situations. The final section turns to theoretical concepts and educational implications of research on culture and mathematics.

TWO VIEWS OF CULTURE
AND MATHEMATICS

For years mathematics educators and researchers in mathematics education have focused on the classroom as the primary setting in which mathematics learning takes place. More recently, survey studies (for example, Cockcroft, 1986) as well as analyses of children's knowledge of mathematics (for example,

Carraher, Carraher, & Schliemann, 1985, 1987; Ginsburg, 1977; Ginsburg, Posner, & Russel, 1981; Hughes, 1986; Resnick, 1984) have shown that much mathematical knowledge is acquired outside school. The realization that mathematical knowledge can be acquired outside school brings new variables into the analysis of mathematics learning and teaching. Is the mathematics learned outside school the same as that taught in school? How can teachers identify and capitalize on mathematics learned outside school?

D'Ambrosio (1984, 1985) has used the expression "ethnomathematics" to refer to forms of mathematics that vary as a consequence of being embedded in cultural activities whose purpose is other than "doing mathematics." Everyday activities such as building houses, exchanging money, weighing products, and calculating proportions for a recipe involve numbers, calculations, and precise geometrical patterns. These applications of mathematics often look different from those used in school. In the kitchen we often measure volume with spoons and cups, whereas in school activities students typically measure volume in liters or cubic meters. Everyday mathematics also varies significantly across countries, because of differences in the numeration systems used, for example, or the devices used for calculating. These differences may be perceived as deep- or surface-structure differences, depending on what views one holds of mathematical knowledge.

Two distinct approaches to the study of cultural influences on mathematical knowledge can be identified in the current

This paper was prepared while I was a visitor to the Department of Experimental Psychology, University of Oxford, supported by a grant from CAPES of the Ministry of Education, Brazil. Much of the research reported in this chapter was supported through several grants from CNPq of the Ministry of Science and Technology, Brazil. Peter Bryant discussed the ideas presented in this chapter with me and helped give them form. Frank Lester, Indiana University, and Richard Mayer, University of California, Santa Barbara, provided useful comments and suggestions on the initial chapter draft. I am grateful to them and to the institutions who have generously supported my work.

literature. One view, espoused by Stigler and Baranes (1988), holds that

Mathematics is not a universal, formal domain of knowledge...but rather an assemblage of culturally constructed symbolic representations and procedures for manipulating these representations....As children develop, they incorporate representations and procedures into their cognitive systems, a process that occurs in the context of socially constructed activities. *Mathematical skills that the child learns in school are not logically constructed on the basis of abstract cognitive structures* [italics added], but rather are forged out of a combination of previously acquired (or inherited) knowledge and skills, and new cultural input. (p. 258)

In this approach, the definition of mathematical knowledge is implicit and appears to be based on the content of knowledge. Activities that involve number, geometrical patterns, calculation, and so forth are treated as applications of mathematical knowledge. This view stresses differences rather than similarities across cultures. Within this approach, for example, variations across cultures of reciting strings of numbers are treated as a reflection of the differences in language and numeration systems. A second perspective, illustrated in the works by Gal'perin and Georgiev (1969) and D'Ambrosio (1986), suggests that the analysis of cultural influences on mathematical knowledge can demonstrate both differences and invariance in mathematical knowledge across cultures. In this view, "mathematizing" reality is representing reality in such a way that (a) more knowledge about the represented reality can be generated through inferences using mental representations, and (b) there is no need to manipulate reality further in order to verify this new knowledge. Invariant logical structures are embedded in mathematical knowledge, regardless of whether mathematical knowledge is developed in or out of school. It is the ability to make inferences using these structures—not the content of knowledge—that distinguishes mathematical knowledge. Merely reciting count words, for example, would not be considered mathematical knowledge in this perspective. Despite the fact that numbers are the content of knowledge in this case, no inferences result from mere recitation, thus no mathematical knowledge is involved. On the other hand, if count words are used to represent sets and make inferences about relationships between sets, the count words bear on mathematical knowledge. Different cultures have found distinct solutions to the organization of their count words. These differences have important effects on how quickly children learn the count words, as Miller and Stigler (1987) have shown in their comparisons of Chinese and American children. However, invariant principles underlie the activity of counting objects in both cultures (counting principles will be discussed later in this paper), and the inferences that are made by Chinese and American children using number words as representations of quantity are the same despite the differences in the words used.

In my view, understanding cultural influences on mathematics learning must involve both the analysis of the differences among particular solutions to mathematizing reality and the recognition of the logical invariants underlying these differences. Cultural differences exist not only across cultures. Within a culture, practices of mathematics vary depending on their purpose. Research on mathematics education can benefit from the analysis of various solutions to the same problem that co-exist in a single culture. In cultures where there are both written and oral representations of number, it is likely that two practices of arithmetic co-exist (Reed & Lave, 1981). Research may show that two varieties of mathematical practice within the same culture differ in some ways but are similar in others. The similarities are likely to relate to the logical invariants of mathematical knowledge. Any variations are likely to relate to the representations and the uses of the invariants in the mathematical applications. These differences are peripheral to the conceptual structure but central to the skills displayed by the subjects.

THE ETHNOMATHEMATICS OF COUNTING AND MEASURING: CONTRIBUTIONS OF CULTURE AND LOGIC

In this section, the logical invariants involved in counting will be examined, as well as cultural variations for overcoming the difficulty of applying counting principles.

Counting Principles and Cultural Contributions to the Development of Counting Systems

Counting and measuring are ways of representing selected aspects of objects and situations. In order to measure, one has to choose what dimension will be quantified—a class of objects when counting, for example, or a quality like length or weight when measuring. Like any form of representation, the initial choice of what to represent and what to ignore involves an abstraction: Everything will be ignored except the aspect that is being quantified. The activities of counting or measuring are usually carried out for some larger purpose. You count the money in your pocket to determine whether you have enough for a particular purchase. You measure a table to determine the size of the tablecloth that will cover it properly. Carrying out such activities is what makes counting and measuring meaningful.

The activities in which we use counting and measuring vary, yet an underlying logic seems present. In a thorough analysis of how children develop counting skills, Gelman and Gallistel (1978) specified four basic logical principles that must be satisfied if an activity is to be classified as counting. These principles may be summarized as (1) establishing a one-to-one correspondence between the things to be counted and the counting labels, (2) maintaining the counting labels in a fixed order, (3) recognizing the irrelevance of the order in which the objects are counted, and (4) applying the cardinality principle, that is, using the last label to represent the number of objects in the set. These four principles have such a strong logical appeal that Gelman and her collaborators are willing to treat them as culturally independent and possibly innate.

A system based only on these logical principles has limited application. In the absence of a culturally organized numeration system, one would have difficulty obeying these principles. How many labels can one remember in a fixed order (principle 2) if the labels are unrelated words and are not part of a system that makes their production easy? The human capacity to memorize ordered lists is limited in the absence of some structural support. Without a numeration system, counting would be restricted to low numbers; with a numeration system, it can go on indefinitely.

Although these counting principles may be culturally independent, specific numeration systems are culturally dependent. The important work of Lancy (1983), Saxe (1981), and others clarifies how different cultures have addressed the problem of memory load in counting. The Kewa and the Oksapmin of Papua New Guinea have developed numeration systems that help them maintain fixed order by using the names of body parts as labels in counting. For example, *thumb* indicates *one, index two, middle-finger three,* and so on. The use of body parts in counting is a cultural and conventional solution to the problem of memory load: The parts to be named and the order in which they are used must be agreed on. Some of the body parts chosen do not have clearly identifiable labels in many Western cultures, such as the three locations on the forearm and six locations between the shoulder and the neck. This use of body parts labeled systematically allows the Kewa to count to 68.

This is by no means the only or the best solution to this problem. In English another solution for the problem of memory load uses a base system in generating number labels. The count words used in English are maintained in the proper order by their generation in systematic combinations. Number words are unrelated to each other up to *twelve,* but from *thirteen* on there are cues in the labels that help generate the succeeding labels in a fixed order. These cues are even clearer after *twenty-one* when labels are used recursively to produce derivative count words. The generation of count words in this fashion is related to the introduction of a base in the counting system.

A base in a numeration system is a grouping scheme used to reorganize counting. To define a base in a numeration system is to choose a conventional unit (several conventional units for a mixed-base system) to be used in counting. According to Luria (1969), a base-numeration system involves counting natural objects, organizing them in conventional groups that become new counting units, and grasping the semantically complex structure underlying the numeration system. The number 343 expresses that there are three groups of one-hundred, four groups of ten, and three objects. In order to capture this meaning and generate number labels indefinitely, subjects must not think of the natural object as the only thing being counted; they must also understand the structure of meaning in the numeration system. Luria attributed a great importance to this distinction, arguing that the counting of natural objects without an understanding of the base system is carried out by the brain's right hemisphere, whereas understanding of the underlying base system is controlled by the left hemisphere.

In summary, counting systems rest both on a logic of invariant principles and on culturally specific devices for the implementation of these principles. Not all cultures find the same solution for the challenge of keeping number labels in a fixed order. The different solutions vary in their ability to deal with large numbers. The systematic enumeration of body parts is

one type of solution for keeping count words in a fixed order, but this device has limited range. The use of a base in a numeration system is a cultural device that can solve memory overload problems in counting. A base system allows for counting indefinitely, an impossibility with the nonbase, body-parts systems. From the psychological viewpoint, the introduction of a counting base involves the concept of counting-units. In a base system counting-units are not only the natural objects but also the conventional groups of objects indicated by the base in the system. Despite the culturally dependent nature of the resource we call a base, the use of counting-units is not simply conventional and devoid of logic. A base numeration system is supported by the concept of unit, which is common to both counting and measurement.

Cultural Variation and the Underlying Logic of the Concept of Units

There are several contexts in which the concept of units is used in everyday activities in Western cultures. In the numeration system we have counting-units of ones, tens, hundreds, and so on. For measurement we use the metric system. In monetary exchange we use coins and notes of various values. In each of these contexts, the concept of unit is present. But can it be effectively learned in everyday practice without systematic instruction?

Two types of study of everyday cognition and the use of units in counting and measuring will be discussed. The first describes the emergence of counting groups that model what is counted. The second investigates inference-making about units of different sizes.

Reinventing Counting-units for Particular Purposes. Two studies carried out with subjects from different parts of the world indicate that people can reinvent the concept of units in the context of everyday activities. These studies were conducted by Saxe (1982; 1985) among the Oksapmin and by Scribner and her colleagues (Fahrmeier, 1984; Scribner, 1984) among workers in a dairy factory in the United States.

Saxe (1982) described how a new social activity, the emergence of a money economy, influenced the Oksapmin's use of their existing numeration system. The Oksapmin had a no-money economy in the past, therefore they had little need for computation before their contact with the West. The emergence of a money economy began through their contacts with missions and farms, and marked changes were observed in the way that adults who became involved with commerce used their indigenous system. Saxe described the following changes:

Since only limited quantities can be expressed with the Oksapmin system, communicating about even small denominations of currency presents serious problems. The adaptation that has emerged is one that incorporates the base structure of the early Australian currency system into the indigenous system. With the adapted system, rather than using all 27 body parts in an enumeration, an individual counts shillings up to the inner elbow on the other side of the body (20) and calls it 1 round, or 1 pound (reflecting the organization of the first Australian currency system). If the individual needs to continue the count, he or she begins again at the thumb of the first hand (rather than progressing on the forearm [21]).... The adapted system, then, has a base structure that reflects the base structure of the early Australian currency system but nevertheless is an outgrowth of the standard indigenous system. (p. 585)

This reinvention of a grouping unit that is introduced into the indigenous system and that parallels the grouping unit of the exogenous monetary system constitutes a clear example of modeling.

The work by Scribner and her colleagues investigated the use of various methods of grouping for counting that are shaped by everyday activity. Fahrmeier (1984) described inventory-taking in a milk factory in the United States, which required assessment of quantities of some 100 products stored in a walk-in refrigerator.

Counts for each product need to be accurate within a 1% or 2% margin of error. Larger errors result either in shortages or over-production, both costly events.... Products were stored so close to each other that inventory men had limited walk room for maneuvering around arrays. They had to seek out and reach vantage points from which a count could be made, often doing this by climbing on cases to "see over" the front row of an array. By this move, they could almost always see the top cases of the stacks but not the cases underneath. For much of the time, then, they were taking counts of arrays containing invisible cases. (p. 7)

As Fahrmeier points out, these circumstances made counting different from putting items and number labels into one-to-one correspondence. Although the counting principles have to be honored for a correct count to be reached, they cannot be carried out literally. Fahrmeier describes five strategies that emerged in taking inventory under these circumstances, all of which involved the use of new units based on spatial groupings—stacks with known height in terms of number of cases. By representing the cases in stacks of known heights and using the stacks as the new unit in counting, the problem became soluble within the limits of time and accuracy required by the job. A traditional system of counting was adapted to better handle the requirements of the activity by incorporating the properties of stacks as a base for the counting procedure. Scribner (1984) describes similar uses of a new unit of grouping for product assemblers in the same factory for whom the unit was a case with 16 quarts, a number that requires awkward maneuvers (like carrying) when it is repeatedly added in the decimal numeration system.

Inference-making about Units. It is possible to go further in the study of conventional units than looking simply at counting. One can design tasks that test whether people make consistent mathematical inferences when thinking about units in these everyday practices. Four examples of inference-making about units are presented below, first in a formal description and then in the context of research.

The first example is a case of simple transitive inference: (1) If $A = B$ and $B < C$, then $A < C$.

A second example of inference-making about units is an extension of a relationship between two units and any equal number of those two units: (2) If A and B represent different units of measurement of some variable and $A > B$, then for every positive number x, $xA > xB$.

These two examples refer to the logic of units and do not involve a precise quantification of the relationship between the units. The next examples deal with the quantification of relationships between units. The first may seem trivial, but it is not simple for children (see Carraher & Schliemann, 1990):

(3) If A and B represent different units of measurement of some variable and A = xB, then A > (x − 1)B and this is true despite the greater numerosity of (x − 1)B.

The last example of inference-making using the quantification of the relationships between units involves the recognition of the additive composition of measurement:

(4) If A and B represent different units of measurement of some variable and A = xB, then any value of the variable measured in As can be expressed in terms of B values, and any B values greater than A can be expressed in terms of A values plus the remaining Bs.

Saxe and Moylan (1982) analyzed the ability of Oksapmin children and adults to make type 1 inferences and the influence of schooling on this ability when the children were using their indigenous measurement system for determining length. Their system of using body parts to represent measurements depends on the size of the body parts, which varies across subjects. Oksapmin adults and children five and older wear string bags that are measured by putting an outstretched hand into the bag and describing its expanded size as *knuckles, wrist, forearm, inner elbow, biceps,* or *shoulder.*

This measurement system creates problems in inference-making. The same bag may be measured differently by a child and an adult. Saxe and Moylan set out to investigate whether Oksapmin children and unschooled adults understood the variability of their units and whether schooling contributed to this understanding.

They gave their subjects two groups of tasks. In the first group, the unit of measurement (body part) was held constant. Subjects had to judge whether bags A and C, which they never compared directly, were the same size by measuring them against their own body part B. In these tasks, the problem structure was one of transitive inference: $A = B$; $C > B$; thus $C > A$. In the second group of tasks, the unit of measurement varied. Subjects were told that a bag had been measured on a child, and they were asked to predict whether the bag would reach up to the same body part on an adult. A represents the child's forearm, B is the adult's forearm, and C the bag; the problem structure is: $A = C$; $A < B$; thus $C < B$.

The ability to make comparisons of two bags by measuring them against a fixed unit was clearly demonstrated by school children and by unschooled adults. Unschooled children made some errors, and only about half of them produced uniformly correct responses. In summary, transitive inferences based on measurement with a fixed unit were observed among subjects with or without schooling, although schooling seemed to speed up the process of development of the ability to make transitive inferences under these circumstances.

Greater variation in performance was observed for the second group of tasks using different-sized units. Unschooled adults outperformed school children with as much as six years of schooling—a finding that indicates schooling is not the critical variable in the ability to make transitive inferences using units of different sizes. However, unschooled adults did not perform at ceiling level. Among the unschooled adults, 64% made three or four correct predictions in response to four questions, 18% made only one or two correct predictions, and the remaining 18% made no correct predictions. This study provides clear evidence for out-of-school development of reasoning and inference-making about different-sized units. It also shows that schooling speeds up the process of development in the simpler but not the more complex task.

Carraher (1985, 1989) also tested whether illiterate Brazilian adults can make type 2 and type 4 inferences in the context of a monetary system without previous instruction in writing numbers. Monetary systems involve all the inference-making about units previously described. Type 2 inferences were evaluated by asking adults a simple question: If one person had five bills of 10 *cruzados* (Brazilian currency at the time) and a second person had five bills of 100 *cruzados,* would they have the same amount of money?

Type 4 inferences were tested by asking subjects to say how they would pay imagined amounts of money (quantities were in the hundreds) using only bills of 100, 10, and 1 *cruzados* and giving out the smallest number of bills possible. All subjects responded correctly to the type 2 inference question. Some variability in performance was obtained in the second group of questions involving type 4 inferences. The results are summarized in Figure 22.1.

Despite the fact that adults did not perform at ceiling level on this second task, results indicate that a sound ability to understand the decomposition of a value into hundreds, tens, and ones can be developed through everyday practice. This task,

which was about imagined amounts of money and involved restrictions that do not apply in the monetary system (the task used only bills of 100, 10, and 1, although the system has bills of 500, 50, and 5 that could be used), was not an everyday task but a rather complex transfer task. Even so, 70% of the subjects performed at ceiling level.

In summary, studies on counting and measuring in everyday activities indicate that subjects can reinvent the concept of grouping into units for counting in ways appropriate to their activity. Further, this ability is independent of the presence of a base structure in the particular numeration system. Everyday practices also create opportunities for individuals to make inferences typical of classroom mathematical activities.

THE ETHNOMATHEMATICS OF ARITHMETIC OPERATIONS: ORAL AND WRITTEN VARIETIES OF ARITHMETIC

In the preceding section we considered only the oral representations of numbers. In this section we will look briefly at various systems for writing numbers. Having considered quantification and the use of numeration systems for measuring and making quantitative comparisons, we will turn to the use of numbers for calculating by examining the distinction between counting and solving arithmetic problems. Finally, the differences and similarities between oral and written arithmetic will be considered.

Oral and Written Numbers

Spoken numbers may be forgotten. Written systems are a culturally devised means of handling this memory load problem. As with oral counting, written numbers are uniquely represented by individual cultures. The Romans, for example, developed a written numeration system based on five ways of representing a number: (a) variation of form (*I* for *one* and *V* for *five*), (b) repetition (*III* for *three*), (c) addition (*VII* for *seven,* which is the result of adding V and II), (d) subtraction (*XL* for *forty,* which is the result of subtracting X from L), and (e) restriction rules (*V* cannot be used with repetition; *I* and *X* are used subtractively but *V* and *L* are not). This system is ingenious but not as powerful as the Hindu-Arabic system used today.

The Hindu-Arabic system is based on two major concepts: variation of form (ten digits, each with unique meaning) and place value. In a place-value system, the relative value of a digit is indicated by its position. There is no need to explicitly state its value in terms of a base. For example, one does not have to indicate that the 2 in 23 signifies the number of tens. In contrast, current Chinese written representation of numbers uses explicit reference to the grouping or base (see Figure 22.2): The representation for the number of tens is followed by the sign for 10 and the representation for the number of hundreds is followed by the sign for 100. (This explicit reference to the base or a power of the base in written representation may facilitate the understanding of the meaning of written numbers, but I know of no research that has investigated this issue.)

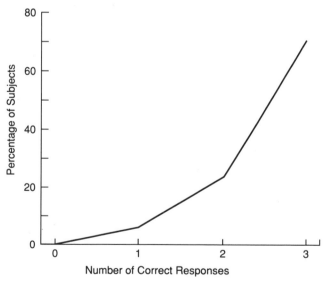

FIGURE 22–1. Percentage of unschooled Brazilian adults by number of correct responses in a three-item task analyzing numbers into hundreds, tens, and units in the context of monetary values.

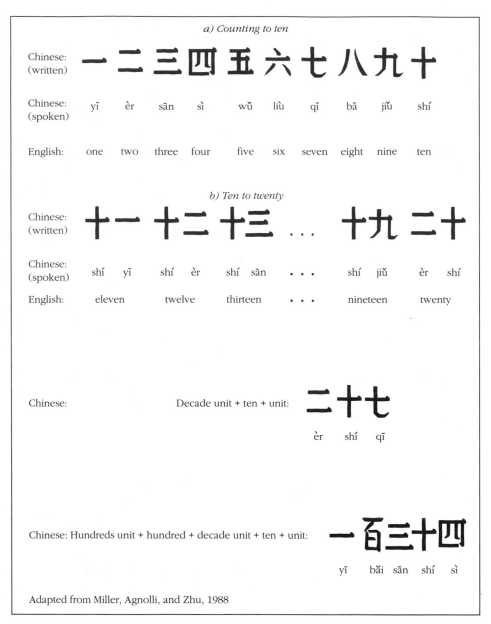

FIGURE 22–2. Oral and written numbers in English and Chinese.

Just as some oral numeration systems are more efficient for counting and measuring than others, some systems of written representation may be more efficient for calculating than others. According to Menninger (1969), Roman numbers were used to register values but not to calculate. Romans calculated by counting and using different forms of the abacus; their notation system did not facilitate calculation. In contrast, the Hindu-Arabic system, as described by Fibonacci is considered to be a calculating machine (D'Ambrosio, 1986) because it makes column arithmetic simple through positional notation and the use of zero as a place holder.

Written numeration systems vary across time and culture. Some written systems appear to make calculation easier through column arithmetic, whereas others are not useful for calculating. When the latter type of written numeration system is used, calculation tends to be carried out by counting. Across cultures, individuals differ by using concrete manipulatives to calculate or by basing calculations on written symbols. Are there common logical invariants underlying both oral and written arithmetic, or are the procedures so radically distinct that different processes underlie each?

Counting and Solving Problems

When individuals solve addition problems by putting sets together and counting the total set, are they simply counting or are they performing an arithmetic operation? If the objects were represented by fingers or pebbles, and these representations were moved about to represent the operations (for example, put together to represent addition, or separated to represent subtraction) and counted, would the subject be counting or performing an operation?

Drawing a distinction between counting and solving addition and subtraction problems is not easy. Many researchers have shown that not everyone who can count can also use counting to solve problems. The ability to use counting to solve addition and subtraction problems increases with age, varies with situation, and is influenced by schooling (see Carpenter & Moser, 1982; Fuson, 1982; Hughes, 1986; Saxe, 1985; Steffe, von Glasersfeld, Richards, & Cobb, 1983). These studies indicate that something more than counting is going on when subjects use a counting strategy in problem solving. Counting then becomes a problem-solving technique, not an activity in itself. In solving a problem by counting fingers, subjects assume (at least implicitly) that whatever results apply to fingers also apply to the original situation.

In their pioneer work on everyday mathematics among the Kpelle, Gay and Cole (1967) suggest that, despite the fact that the Kpelle solve problems by putting objects together, taking objects from sets, putting like sets of objects together, and sharing objects, "they recognize no abstract arithmetic operations as such" (p. 50). Gay and Cole were able to identify in the Kpelle language expressions for addition (for example, an expression meaning "two chickens joining three chickens is five chickens"), for subtraction ("removing two chickens from a group of three chickens leaves one chicken"), for multiplication ("three sets of two chickens are six chickens"), and for division ("putting ten bananas into five equal piles makes two bananas in each pile") Gay and Cole described the normal procedure for solving these problems to be the use of fingers or stones to represent the objects, which were then counted. These procedures were used accurately only for problems involving small numbers. With large numbers, the procedures became cumbersome and boring, and subjects often simply guessed a large number as the answer.

Gay and Cole suggested that the Kpelle "have no occasion to work with pure numerals, nor can they speak of pure numerals. All arithmetic is tied to concrete situations" (p. 50). The Kpelle can say something that translates as "two of them added onto three of them is five of them." However, they cannot say something that would be translated as "two and three is five"; their numerals act as modifiers and cannot be used abstractly.

Similarly, Hughes (1986) found that young British children treated questions such as "If there was one brick in the box and I added two more, how many would there be?" very differently from the question "What is one and two more?" Questions about this hypothetical box with small numbers of objects were answered correctly 56% of the time whereas questions strictly about numbers were answered correctly only 15% of the time. Hughes also comments on the fact that he occasionally left out (unintentionally) the nouns being quantified by the numbers in the hypothetical situations and that as long as "they were locked into a series of questions on a particular topic, they did not need the topic spelled out to them every time.... It seemed that children's difficulties arose whenever phrases such as 'two and one' *did not refer to any specific objects*" (p. 38).

Data such as those reported by Gay and Cole (1967) and by Hughes (1986) are sometimes interpreted as indicating that some people are capable of concrete but not of abstract reasoning with numbers. It is important to check this interpretation. Using fingers as representations of something concrete—like chickens or bricks—demonstrates an abstraction. This use indicates that whatever number results from combining x fingers and y fingers will also be obtained when combining x chickens and y chickens, where x and y each represent the same number in both situations. This is undoubtedly an instance of abstraction and transfer where a technique known to work in one context is used to solve a problem in another.

A counting strategy used for addition can thus be considered a form of abstraction. Yet it is distinct from the procedure used for written calculations. In problem solving, are there any logical invariants common to counting strategies and written calculation strategies? Are the differences between oral and written calculations conceptual in nature, or are they superficial in the sense that both types of calculation rest on the same underlying principles? This question is the central focus of the next section of this paper.

Oral and Written Arithmetic

The concrete versus abstract dichotomy was prevalent in the literature about arithmetic operations in diverse cultures until Reed and Lave (1981) suggested a new way of looking at this issue. Their research consisted of a detailed analysis of arithmetic problem solving among tailors in Liberia. They began with participant observation and informal interviews in the tailors' shops. The field notes were later used in the development of experimental tasks designed to sample the four arithmetic operations at five levels of difficulty, which took into account the size of the numbers in the problems and the need to carry out regroupings. Reed and Lave describe two basic strategy types observed for solving problems.

There are two classes of strategies for performing arithmetic operations, those that deal with quantities as such and those that deal with number names. Strategies that work with quantities are universal and manifest themselves as counting on fingers, manipulating pebbles and the like, or using an abacus. The tailors make extensive use of counters—either movable ones, such as pebbles, or marks on paper. Strategies using number names are typified by Western algorithmic manipulations learned in school, for example, "put down the 4 and carry the 1" litany in addition. In such treatment it is the manipulation of symbols, in a real sense divorced from reality, that carries the burden of computation. (p. 442)

Reed and Lave observed that different types of errors resulted from each of these two strategies. Tailors using the *manipulation-of-quantity* strategy were off by small amounts when calculating with small numbers but had great difficulty when large numbers were involved. (Modeling with pebbles when large numbers are involved becomes a less manageable procedure.) Tailors who had more schooling and had been exposed to Western algorithms made errors in carrying and borrowing, and thus were off by one 10 or one 100, but they showed less marked difference in their abilities to calculate with small versus large numbers. Reed and Lave conclude that the evidence does not favor a concrete versus abstract distinction but does demonstrate the existence of multiple arithmetic systems in a single culture.

Carraher et al. (1985), like Reed and Lave, were also able to document the existence of multiple arithmetic systems within a single culture. They began with the observation that children from lower-income families are often involved in the informal economy in large Brazilian cities; that is, these children carry out odd jobs such as washing cars, shining shoes, or selling candy and other low-priced items. In this setting, the children seemed competent to carry out mathematical computations. In contrast, children from lower-income families often fail in school mathematics when taught the numeration system and the arithmetic operations formally. In order to document this disparity, Carraher and colleagues bartered with five children selling fruits and vegetables in street markets, posing problems to them that involved addition, subtraction, and multiplication. The following week, the researchers presented each child with word problems and computation exercises involving the same numbers used during the vending interaction. The contrasting problem presentations identified two systems of arithmetic that seemed to function independently. Not only were the rates of correct responses significantly different across presentations, but also the procedures used by the children were clearly distinct. Carraher and colleagues' analysis of the protocols also indicated a qualitative difference between the two strategies used: As vendors, subjects solved the problems orally; acting as students, they solved them using paper and pencil.

Carraher et al. (1987) conducted a second study in which they were able to obtain better records of the strategies used in calculation. This study involved 16 third-grade children solving three types of problems: (1) problems presented in a simulated store, in which the child played the role of the vendor, and the experimenter the role of the customer; (2) word problems; and (3) computation exercises. Again, performance differed by presentation, and two distinct systems of arithmetic practice were observed.

Carraher et al. (1987) analyzed the oral practice in detail and described one general strategy used to solve addition and subtraction problems (*decomposition*) and one strategy for solving multiplication and division problems (*repeated groupings*). These two strategies for oral practice—which had already been documented by Ginsburg et al. (1981), Hunter (1977), and Plunkett (1979)—were analyzed in terms of their underlying mathematical properties by Resnick (1984) and Carraher and Schliemann (1988). A very simple example of decomposition can be seen in the following protocol, in which the child was solving the problem 200 − 35: "If it were 30, then the result would be 70. But it is 35. So, it's 65, 165." Discussing this example, Carraher and Schliemann suggested the following:

The child decomposed the problem 200 − 35 into steps which seem to be the following: (1) 200 is the same as 100 + 100; (2) 100 − 30 is 70; (3) 70 − 5 is 65; finally, (4) adding the 100 which had been "set aside," as some children say, 165.... The general principles underlying the written [algorithms] and the oral strategies seem to be the same. (pp. 182–183)

In other words, written algorithms taught in school and decomposition used in oral practices rely on the same property of addition, *associativity,* referring to the fact that the way addends are grouped does not affect the sum.

Similarly, Carraher and Schliemann (1988) propose that the strategy of repeated groupings typical of multiplication and division in oral practice and the school-taught algorithms for these operations rely on the same mathematical property, *distributivity.* The use of distributivity can be clearly identified in the following example of division, observed in a word problem in which a child tried to figure out how many marbles each of three children will get if 120 are distributed evenly.

Each one gets thirty, that will leave, three times thirty is ninety, that will leave...(E: That would leave how much?) Each one gets thirty, that leaves thirty. Then five more, that's fifteen. This leaves...fifteen. Then five more, that's fifteen. Each one gets ten and thirty, forty. (p. 184)

Carraher and Schliemann (1988) also pointed out that, despite the similarities in underlying principles, there are also striking differences between oral and written practices. Oral practices preserve the relative value of the parts of numbers that are being operated on: Hundreds are treated as hundreds; tens are treated as tens. In written algorithms, the relative values are set aside, and digits are manipulated as though they *were* units. For example, the expression *carry the one* is the same, regardless of whether what is being carried is one ten or one hundred. Oral practice is thus described as *meaning-based* in contrast to written algorithms, which are described as *rule-based.*

These characteristics of oral practice do not seem to be confined to Brazilian street vendors. They were clearly documented by Resnick (1984) in a study of an American child who discovered oral procedures for addition and subtraction before being formally instructed in written procedures. The protocol below describes this child's method when adding 152 and 149 orally.

I would have the two 100s, which equals 200. Then I would have 50 and 40, which equals 90. So I have 290. Then plus the 9 from the 49, and the 2 from the 52 equals 11. And then I add the 90 plus the 11... equals 102. (I: 102?) 101. So I put the 200 and the 101, which equals 301. (p. 6)

Finally, Carraher and Schliemann (1988) point out that written strategies have certain characteristics that make them true cultural amplifiers (see Bruner [1966] and Cole & Griffin [1980] for a discussion of cultural amplifiers). Cultural amplifiers are resources in the culture that allow for an increase in a person's ability. In the case of computation, people who know how to add large numbers may still have difficulty when calculating orally due to the memory overload that occurs when one tries to retain the numbers and the partial results of computation in memory as calculation progresses. Through the use of writing—a cultural amplifier—memory overloads are avoided; through column arithmetic the advantages of a decimal numeration system are used to their fullest power.

Saxe (1988) further pursued the analysis of multiple systems of arithmetic among Brazilian children engaged in the informal economy and was able to demonstrate the oral nature of this practice in another way. His subjects, 23 ten- to twelve-year-old candy sellers with minimal schooling, showed less than 40% correct responses in a test of the standard orthography of numbers. This low performance in written numeracy contrasts with their high rates of correct responses in bill identification and currency comparison (about 90% correct responses), in bill arithmetic (about 70% correct responses), and in comparisons of price ratios (about 70% correct responses).

Baranes, Perry, and Stigler (1989) later attempted, without success, a replication of the findings in the study by Carraher et al. (1987) with American children. They used the same tasks (simulated store, word problems, and computation exercises), the same sets of numbers, and subjects from the same grade level (third grade). However, this type of approach to the replication of cultural influences on mathematics reasoning does not take into account the need to replicate the cultural conditions that allowed for the observations. In order to replicate findings about oral arithmetic in the simulated shop in Brazil, it would have been necessary to make sure that (a) American subjects were in fact exposed to oral arithmetic practices, (b) the natural situation simulated in the experiment was one in which oral arithmetic was likely to be observed, and (c) American children's practice of written arithmetic did not differ greatly from that of Brazilian children. Their failure to replicate the results observed by Carraher et al. (1987) is not surprising but is rather instructive. First, calculating total price and making change in American stores rarely requires the use of oral arithmetic. Most stores have cash registers that do the calculations, including calculating the change in many cases. It is a setting for machine arithmetic—a third type of arithmetic practice that deserves further analysis. Second, numbers in the Brazilian study were taken from actual bills used in the culture because problems involved reference to these bills. It is unlikely that the numbers used in the Brazilian study coincided with American denominations. Further, American children may have received more instruction on written arithmetic than Brazilian children from lower-income families attending public schools, who rarely attend preschool and only start learning written arithmetic in second grade.

In contrast to Baranes et al.'s (1989) failure to replicate differences between oral and written arithmetic in the United States, Lave (1988) reports findings about oral and written calculations among American shoppers carrying out price comparisons in the supermarket that are quite similar to those of Carraher et al. (1987). The similarities involve the types of procedures used and the differential rates of success, although the actual problems are much more difficult than those presented in the Brazilian study. The similarity of the results found by Lave (1988) and Carraher et al. (1987) is best understood by comparing the type of calculations being carried out. Situations that have similar meaning and involve similar practices are comparable across cultures, whereas those that may be superficially similar but have different meanings and involve different practices must be treated as distinct. Thus, in order to find invariance both in mathematical concepts and in cultural situations, it seems necessary to look deeper than the surface features of events.

In summary, the transition from using numbers for counting to using numbers for solving arithmetic operations is not as simple as one might expect. Not everyone who can count can solve arithmetic problems, although the solution to many simple arithmetic problems can be obtained by modeling the problem situation with concrete materials and then counting.

Despite the fact that fingers or other manipulable representations are often used by children and unschooled adults in problem solving, it may be inappropriate to think of these solutions as *concrete* reasoning. They represent instances of modeling that always involves some degree of abstraction. When modeling is carried out as one-to-one correspondence and numbers are in the high tens and hundreds, counting procedures become cumbersome and boring and their efficiency in reduced. The oral practices used by Brazilian children do not use simple one-to-one correspondence and, consequently, can achieve good results with larger numbers. These practices clearly reflect an understanding of some basic properties of arithmetic operations despite the differences that result from the medium of representation used.

ETHNOMATHEMATICS OF MODELING: VARIATIONS IN FLEXIBILITY FOR PROBLEM SOLVING

I have argued that modeling of everyday situations with manipulatives involves a process of abstraction. When a situation is directly modeled, both the general number relations and some specific aspects of the situation are represented by the model. For example, to find out how many chicks will be left if nine are born and six of them run away, it is possible to represent the nine chicks by nine fingers and to represent the six chicks running away by hiding six fingers. Such direct modeling of a situation is not a particularly flexible procedure for solving problems. Suppose, for example, that one of Fahrmeier's inventory men was asked whether he could fill an order for 893 quarts of skimmed milk and, if not, to report how many quarts would be needed to fill the order. The inventory man would go into the icebox, verify that there were only 379 quarts of milk, and figure out how many quarts were needed to reach the requested 893. A direct modeling of this situation would require him to represent the 379 quarts in stock, to add some number of quarts to this until he reached the requested 893 quarts, and to count how many quarts he had added. This last count would inform him of the number of quarts needed. In formal terms, direct modeling of this problem would yield the representation $379 + x = 893$. Of course, we know that there is a quicker solution to the problem through subtraction ($893 - 379 = x$). This solution requires cognitive steps that make it different from the direct modeling procedure (see Vergnaud [1982] for an analysis of the cognitive tasks involved in inversion). The difficulty of the subtraction solution lies in the fact that it requires the subject to invert the representation of the actual situation. What he knows is that he has 379 quarts in store and he needs some more (he needs to add some) in order to come up with 893. To think of the 893 first and then to take away the 379 in stock requires an *inversion;* it requires applying an operation—subtraction—that is the inverse of what he wants to achieve—adding the number of quarts necessary to supply the 893 requested.

In many situations in everyday life, problems have a somewhat constant direction. For example, children usually know how many marbles they had at the beginning of a game and how many they won in the game. At the end of the game, they figure out how many marbles they now have. Inverting this situation, according to De Corte and Verschaffel (1987), creates difficulties from both the cognitive-relations and the social-meaning perspectives. Representations formed in everyday life refer both to the number relations and to the actual situation. A consequence of this double representation is that arithmetic problem solving learned in everyday life may not be very flexible.

Different cultural experiences may engender different degrees of flexibility for the procedures learned by the participants. In some situations, relationships may be observed consistently in one direction; in others, inversion may be practiced. Among the various cultural situations prevalent in Western cultures, schooling probably exposes individuals to inversion in the most systematic fashion. In school, pupils do not carry out actual transactions; rather, they refer to them. Numerical sentences are often written down when arithmetic problems are solved. Teachers often require children to write down the mathematical sentence that indicates the arithmetic operation leading to the correct answer. In the case of the order for 893 quarts of milk, for example, $893 - 379$ would be the expected form of the solution sentence. A consequence of this emphasis in school on solving problems through the inverse operation is that schooling may well strongly influence the development of the understanding of inversion.

This section will review research bearing on the question of flexibility in problem solving and the ability to solve problems using inversion. The first group of studies analyzes word-problem solving in school and in experimental situations. The second group analyzes problem solving by adults in imagined situations that relate to their everyday work experience.

Solving Direct and Inverse Word Problems: The Effect of Schooling

Carraher (1988) carried out two studies in which adults' ability to solve problems through the inverse operation was analyzed as a function of schooling and the size of the numbers in the problem. Bearing in mind the possible difference in difficulty between problems with small and large numbers, Carraher made two predictions about adult performance. The first was that if school instruction plays an important role in the use of inverse operations, then unschooled adults should do significantly better solving direct problems rather than inverse problems, this difference being more noticeable with larger numbers. This is not a trivial statement that some things are more difficult than others or that unschooled adults have difficulty with basic arithmetic. Unschooled Brazilian adults are expected to perform at about ceiling level using direct modeling of problems for both small and large numbers because they easily employ oral mathematics in the solution of large-number problems. Since computations per se are not a great source of difficulty among unschooled adults, their performance on direct problems is expected to be high. If schooling plays a crucial role in the use of inverse operations for problem solving, unschooled adults were expected to perform at lower levels on inverse problems even with small numbers. This poorer

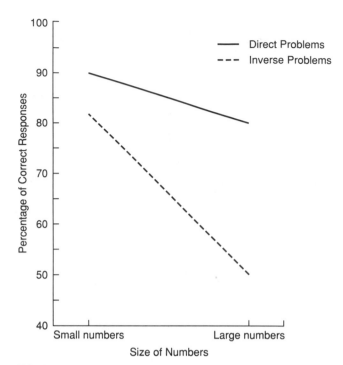

FIGURE 22–3. Percentage of correct responses by Brazilian adults enrolled in literacy programs in direct and inverse addition word problems.

performance in inverse problems would strongly contrast with the abilities displayed by American children in third grade, 80% of whom correctly solved a series of inverse addition and subtraction problems with small numbers (Riley, Greeno, & Heller, 1983). The study by Riley et al. (1983), although extremely interesting, does not offer the possibility of separating out the effects of cognitive development and schooling, which are correlated in their sample. They also used only small numbers, a methodological choice that makes it difficult to find a clear distinction between direct and inverse problems.

The second prediction in Carraher's study stated that, if schooling plays an important role in the understanding of problems to be solved through the inverse operation, then a significant effect of schooling on performance should be obtained for inverse but not for direct problems.

In order to test the first prediction, 60 adults enrolled in a literacy program (with very limited amounts of mathematics instruction) were interviewed and asked to solve six word problems, three that could be easily solved through direct modeling and three that would be more easily solved through inversion. Of the direct problems, two required addition and one required subtraction; the opposite was true of the inverse problems, which were inversions of the stories in the direct problems. Subjects solved problems with small numbers (under 20) or large numbers (above 30). A calculator was available, and all subjects were shown how to use it for each arithmetic operation before they started work on the word problems. The results of this study are summarized in Figures 22.3 and 22.4, which present the data for addition and subtraction problems, respectively. Direct problems were consistently easier than in-

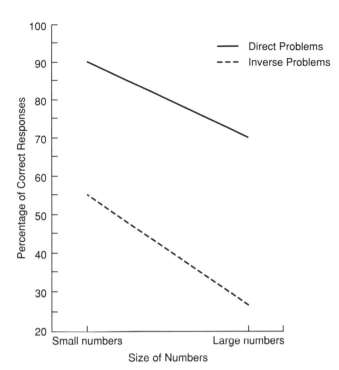

FIGURE 22-4. Percentage of correct responses by Brazilian adults enrolled in literacy programs in direct and inverse subtraction word problems.

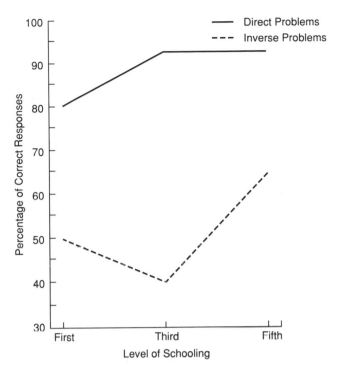

FIGURE 22-5. Percentage of correct responses by Brazilian adults enrolled in three different grade levels in night classes as a function of problem type (direct vs. inverse addition problems).

verse problems, and direct problems with large numbers were at about the same level of difficulty as inverse problems with small numbers for these subjects. The first prediction was therefore confirmed.

The second prediction, that schooling results in a significant improvement in the solution of inverse problems, was tested through a study with 90 adults enrolled in night classes. These classes were equivalent to one of three levels of schooling: first grade (basic literacy instruction), third grade, or fifth grade. Instruction on written algorithms for the four operations is completed in third grade, when students also become involved in more word-problem solving. All word problems in this study involved large numbers. A calculator was available, and its use demonstrated.

Results for the second prediction are presented in Figures 22.5 and 22.6. Direct problems remained easier than inverse problems, even with school instruction, but the effect of schooling on solving problems using the inverse operation was significant. With schooling, solving word problems becomes more flexible, and inversion is accomplished by a larger percentage of schooled than unschooled subjects.

The differences in performance just described should not lead to the conclusion that unschooled adults are concrete thinkers or that modeling is a poor resource for problem solving. The point is simply that schooling may be a source of learning about inversion, either through systematic practice using inversion in problem solving or through the introduction of symbolic systems that unify very different situations under the same symbols, like + and −. However, further research is necessary before a causal relationship between school instruc-

tion and the understanding and use of inversion can be established.

The relationship between ways of modeling situations and the ability to solve problems appears to be rather complex, and it is probably related to how the understanding of situations and solutions was initially achieved. Modeling may be helpful when there are several ways to solve a problem, some of which are formal and others unorthodox. Modeling situations is probably most helpful when the syntax for the mathematics in question is poorly understood. For example, if children are given an algebra problem to solve in the form of an algebraic sentence, like $2a + 16 = 304$, they can only solve it by manipulating the symbols according to the rules they have learned in school; there is no situation to be modeled. Had this problem been presented through words about number relations or a situational context, it might read as follows: "A number multiplied by two plus 16 equals 304; what is this number?" Under this verbal presentation, the problem can be solved by applying rules about the order in which operations must be carried out—multiplication and division are done first, addition and subtraction are only done afterward. A parallel problem can be given that does not depend exclusively on knowledge of rules. The imaginary situation could be something like the following: "The first and second graders in a school had 304 books divided between their classes, but the first graders received 16 more books than the second graders. How many books did each class receive?" This is another inverse problem; it contains a missing addend ($2a$), which when added to 16 equals 304. If students have already mastered the relationship between modeling and solving inverse problems, it would be easier for

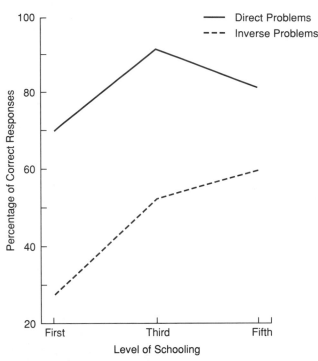

FIGURE 22–6. Percentage of correct responses by Brazilian adults enrolled in three different grade levels in night classes as a function of problem type (direct vs. inverse subtraction problems).

them to solve the problem with situational reference than with only the number sentence. Once the basic idea of modeling such situations is grasped, modeling ought to help.

This is exactly what was found by Yaroshchuk (1969), who compared the performance of a group of students solving number problems to the performance displayed by a similar group of students solving problems with the same formal mathematical description, but presented with a situational context. (The study by Yaroshchuk has a confound that should be mentioned. The numbers-only problem requires inverting two operations, addition and multiplication, whereas the problem given with situational context mentions the inverse operation of division. Since her discussion of errors indicates that the crucial step in solving the problems is the initial inversion of addition, the confound can be overlooked for the present discussion. However, a properly revised version of this study would be worth conducting.) Yaroshchuk observed 73% correctly and independently solved problems that were embedded in a situational context—a figure significantly greater than the 56% observed in number problems alone. Yaroshchuk suggested that it was *imagining* the objects that made it possible to solve the problems.

It appears that imagining real objects makes it possible for pupils to carry out certain operations which they are unable to carry out with abstract numbers (if they are not made concrete in any way). For example, in the problem cited above, one can imagine that we set aside 16 notebooks, and then imagine that we divide the remaining notebooks into two equal piles, left and right; then to one of these piles we add the 16 notebooks set aside previously. It is obvious that

being able to imagine concrete operations of transference considerably simplifies both understanding the problem's conditions and determining the mathematical operations necessary for solving it. (1969, p. 71)

Despite the fact that the study by Yaroshchuk has some flaws, it points to directions for future research that are fascinating. Problem solving that relies heavily on the learning of rules is often plagued by "buggy algorithms." This is well-known by teachers and has been clearly documented in research (Brown & Burton, 1978; Kieran, 1984; Resnick, 1982; Young & O'Shea, 1981). If students can come to understand the rules through imagining situational contexts that represent the algebraic presentation, they may be able to strengthen their understanding of the rules.

Modeling and Inversion of Problems in Everyday Life

In the preceding section, we looked at modeling and inversion in school word problems. In this section, we will look at inversion in problem situations that are close to everyday experiences. As pointed out by Lave (1988), word problems are part of the school culture and are unfamiliar to unschooled adults. The effect of schooling on the ability to solve inversion problems may not exist when situations familiar to the subjects are the context for the problem.

Two studies, one with foremen (Carraher, 1986) and one with fishermen (Schliemann & Carraher, 1990) analyzed the understanding of proportions in direct and inverse problems about everyday situations. If mathematics used in everyday situations is restricted to direct modeling, there should be an effect of inversion on problem-solving ability even for everyday problems. Furthermore, if school instruction is crucial for the understanding of inversion, level of schooling should correlate with performance in inverse problems about everyday situations.

Carraher (1986) observed that foremen use blueprints with ease in their everyday practice. Blueprints are drawn to scale so that any measurement on the blueprint is proportional to the actual size of the item represented. Foremen always know the scale they are working with because it is written on the blueprint. Although foremen may have to calculate the size of a wall when not specified on the blueprint, they never have to compute the scale used in a drawing. The direction of their practice is from a given unitary relation—the scale used in drawing—to larger values within the same scale. Carraher's research questioned whether, without indication of the scale, foreman would be able to calculate the scale used in drawing.

Carraher interviewed 17 foremen with levels of schooling ranging from 0 to 12 years (only three had studied long enough to have learned about proportions in school) on direct and inverse problems involving undisclosed familiar and unfamiliar scales. Subjects were shown the blueprints one at a time and asked to calculate the length of a specific wall *x*. The information they had at their disposal was characteristic of a proportions problem: Length of wall 1 (given) is to its length on the blueprint (given), as length of wall 2 (to be found) is to its length on the blueprint (given).

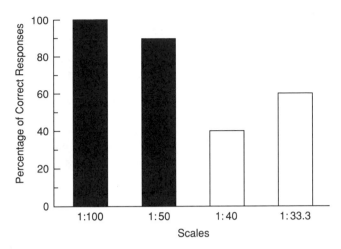

FIGURE 22–7. Percentage of correct responses by Brazilian foremen in inverse problems about scale drawings using familiar scales (1:100 and 1:50) and unfamiliar scales (1:40 and 1:33.3).

Results showed a difference in accuracy between responses for direct and inverse problems and, within inverse problems, between familiar and unfamiliar scales (see Figure 22.7). Performance on direct problems was at ceiling level. Performance on inverse problems with familiar scales was close to ceiling level, whereas only 42% and 59% of the responses to problems with unfamiliar scales were correct.

An analysis of the strategies used in the problems with unfamiliar scales showed that all the subjects who responded accurately had no trouble with inverting their procedures. These subjects would find a simplified ratio $(1/x)$, which they then used to solve the problem. An example of this strategy is illustrated in the following response by J. M., an illiterate foreman with 12 years of experience on the job (working with the unfamiliar, undisclosed scale 1:40): "This one we'll have to divide. We will take 5 centimeters here and here 2 meters (data about the first wall)....This one is hard. One meter is worth 2 and a half centimeters" (p. 535).

In contrast, all but one of the subjects who failed in problems with unfamiliar scales attempted to solve the problem through the strategy of hypothesis-testing. They worked by assuming the scale was a familiar one and tested it against the known data. When they found the scale did not fit the data, they discarded that hypothesis. After unsuccessfully testing for all the scales they knew, they gave up. An example of this follows:

L.S.: (working with the undisclosed scale 1:33.3 and the data 9 cm:3 m = 15 cm:x) Nine centimeters, 3 meters. This scale is...1 by 50, no, that would be $4\frac{1}{2}$ meters. (Pause) If you drew it like this, that is because it is correct. (Pause) Can't do it.

E: Why not? You solved all the others.

L.S.: Because it doesn't work for 1 by 50, it doesn't work for 1 by 1 [meaning 1 by 100], and it doesn't work for 1 by 20. There are three types of scale, 1 by 50, 1 by 20, and 1 by 1. The simplest scale is 1 by 1; you don't have to work on it, you look at the centimeters and you know the meters.

Now, 1 by 50 and 1 by 20 you have to calculate. Now, this one here, it shows 9 centimeters by 3 meters. I've never worked with this one. I've only worked with the other three. (p. 585)

Carraher, Schliemann, and Carraher (1988) pointed out that the hypothesis-testing procedure avoids inversion. Subjects testing hypotheses still work on the problem in the habitual direction, from the hypothesized definition of the scale to the higher value. They succeed in problems with familiar scales because there are few commonly used scales: They can eliminate those scales that do not fit the data until they find the hypothesis that works. When the scale is unfamiliar, the subject gives up after testing all of the familiar possibilities.

Unlike the previous study on addition and subtraction, no correlation between schooling and facility in solving inverse problems about proportions in everyday situations was observed. Schooling's lack of effect may be due (1) to the fact that only three subjects had stayed in school long enough to be taught proportions or (2) to their inability to apply what they had learned in an academic setting to their everyday experience with scale drawing.

Schliemann and Carraher (1990) obtained similar results in another study on direct and inverse proportion problems related to everyday situations. They asked fishermen (levels of schooling ranging from no schooling to incomplete secondary school) to solve direct and inverse problems for two types of conditions: (1) price-per-kilo relationships, which they often calculate in their everyday activities, and (2) unprocessed-to-processed seafood weight ratios, a relationship that they know to be approximately fixed and proportional but that they do not use in calculation. Direct problems proved significantly easier than inverse problems. Performance on both types of inverse problems was good, but not at ceiling level: 78% of the price-kilo problems and 77% of the food-ratio problems were solved correctly. Hypothesis-testing was observed as a strategy used to avoid the difficulties with inversion. This was generally used by subjects who had failed on the inverse problems. Some subjects succeeded by adjusting wrong hypotheses in the correct direction. No correlation was observed between schooling and performance on inverse problems. These results parallel those obtained with foremen, confirming the greater difficulty of inverse problems and the absence of a schooling effect on the ability to solve inverse problems about proportions in everyday life.

In summary, modeling can be helpful in solving mathematical problems. However, there are problems more easily solved by the operation that is the inverse of the operation suggested by the semantics of the problem or by practice in everyday situations. Schooling seems to be an important factor in facilitating inversion in word problems, but no similar effect was observed for problems related to the subjects' everyday activity.

CONCLUDING REMARKS

Ethnomathematics and everyday cognition are relatively novel topics of research, but already a substantial literature has built up based on studies that vary in style, purpose, and basic assumptions about what counts as mathematics in everyday life. Other reviews (Stigler & Baranes, 1988) have been written on culture and mathematics, and whole issues of journals (*Educational Studies in Mathematics,* Bishop, May, 1988; *Quarterly Newsletter of the Laboratory of Comparative Human Cognition,* January, 1984, and January, 1990) have been devoted to the subject. Strong criticisms of work in the area have also been written (Chevallard, 1990).

To attempt a broad coverage of such research and theoretical arguments would have forced this review to be superficial. The selection of three topics of interest in mathematics education—numeration and measurement systems, problem solving and computation, inversion and modeling—allowed for analysis of the mathematical invariants embedded in everyday situations involving these three concepts. This choice was guided by a theoretical position that identifies as mathematical those activities in which people represent objects and events in ways that allow them to extend their knowledge through mathematical inferences. It was shown here that, despite the existence of great variation in the forms of mathematical representation across cultures, the processes involved in making inferences and the types of mathematical properties used in making these inferences do not differ.

In this concluding section, we will address two general questions: (1) What sort of theories about the development of mathematical concepts can incorporate the data available on culture and mathematics learning? (2) What are the implications for mathematics education based on findings and theories about everyday mathematics?

Theoretical Generalizations

Three concepts taken from current theories about cognitive development are useful for understanding the results we have been discussing.

Cultural Amplifiers. First, we must consider the concept of cultural amplifier, initially proposed by La Barre (1954) and introduced in developmental psychology by Bruner (1966). According to this concept, the greatest changes determining man's adaptation in the last half-million years since the human brain acquired its present size and morphology are the culturally devised means to amplify human action (such as cars and hammers), sensory systems (such as radars), and reasoning skills (such as language and mathematical representation). As noted earlier, numeration systems amplify the human capacity to count and register numbers beyond the limits imposed by human memory. Place-value systems increase the range of counting and calculation beyond what is possible through manipulation of objects or oral calculation. Mathematical representational systems do not simply express meanings that people already have in mind. These systems enhance human knowledge through the use of culturally defined conventions and underlying logical structure.

Culturally defined conventions may vary across cultures and across situations in the same culture. For example, measuring systems have units of different values embedded in systems with different organizations both within and across cultures. These differences influence how people think about the measured dimension and the number relations that are familiar to them. Distances are *thought of* in terms of miles in countries that measure with miles and in terms of kilometers in countries that use kilometers.

The similarities underlying the representational systems in mathematics are perhaps more important than the differences. All measurement systems, for example, use the concept of unit and allow for inferences of the same form based on the concept of unit, regardless of the size of the unit, and its relationship with other units of measurement. Finally, what is most important is that all of these conventional symbolic systems work as cultural amplifiers for their users, amplifying their abilities more or less as a function of the systems themselves.

Socialization of Thought. A second important idea taken from the study of cognitive development that helps to analyze the relationship between culture and mathematical thinking is provided by Piaget (1962) and Vygotsky (1978), who recognized the socializing role of conventional representational systems in their analysis of natural language. Mathematical representation, like natural language, is a conventional and collective system of signifiers, the meaning of which must be provided by the subject. With respect to language, Piaget says,

Representation is thus the union of a "signifier" that allows recall, with a "signified" supplied by thought. In this respect, the collective institution of language is the main factor in both the formation and socialization of representations, but the child's ability to use verbal signs is dependent on the progress of his own thought. (1962, p. 273)

The socialization of representations in mathematical reasoning involves both the use of the same symbols (the same numeration system, the same algebraic notation, and so forth) and the use of the same boundaries for the classification of concepts (the same set of situations classified as "addition problems," for example). Luria (1966) emphasized the role of language in the establishment of common boundaries for concepts by indicating that, when the mother points at an object and says a word to refer to it, the object becomes a figure against the background. In this process, the essential functional properties of the object become more salient as it is placed in the same category as other objects designated by the same word. Similarly, mathematical representation brings, as we saw earlier, heterogeneous situations under the same designation. Children must provide the meaning for signifiers such as + (plus) or − (minus) by finding the mathematical properties common to the situations designated by the signifier. However, as Resnick (1984) pointed out, mathematical objects cannot be pointed to as the referents of natural language concepts (chair, cup, and so on) often can. You can point to three chairs and say "three" or point to the symbol 3 and say "three," but that is not pointing to a referent for the understanding of the concept of number. The task that mathematics educators have to accomplish is the socialization of mathematical reasoning. This is similar to the mother's task in the socialization of the child's representations through natural language, only more difficult.

The meaning of a mathematical concept is always abstract, and its acquisition is represented by the understanding of relations and invariants, not by the recognition of physical objects. For example, proportion problems may involve situations that vary substantially from the viewpoint of their content. These problems may involve the relationship between the number of coconuts purchased and their cost, or the relationship between raw shrimp and processed shrimp, or the relationship in a recipe between amount of flour and amount of milk. On the surface, these situations are very different, yet they share the same underlying logic. All of these problems involve at least two variables and, as one variable fluctuates, the other fluctuates proportionally in a one-to-many correspondence. In order to attribute meaning to the expression *proportion problem,* pupils must understand what is common to these situations, what is invariant in the relationship between the variables regardless of their values. This understanding of the invariants in proportional situations represents the core meaning of proportion problems. If this understanding is achieved, individuals can make changes in the values of variables without altering their relationship. For example, a shopper can buy additional coconuts without changing the unit-price relation, or a cook can change amounts of ingredients in a recipe in order to serve more guests.

Cognition and Metacognition. The third useful concept in the literature about the development of mathematical concepts relates to the distinction between cognition and metacognition. This distinction has been used in many contexts—for example, distinctions have been made between memory and metamemory (Flavell & Wellman, 1977) and between linguistic and metalinguistic knowledge (Sinclair, Jarvella, & Levelt, 1978). In general, the simple concept refers to the ability that a subject has to use a specific type of knowledge as a tool, whereas the metaconcept refers to the subject's ability to use that same piece of knowledge as an object in thinking. For example, native speakers of a language make the phonemic distinctions of their language when speaking and listening. They use phonemes as tools because phonemes are the abstract units of pronunciation. They recognize the smallest distortions in phonemes when they are, for example, pronounced by foreigners. However, the ability to isolate and count phonemes within syllables cannot be assumed for all native speakers of the language. The development of metalinguistic knowledge is closely associated not with speaking and listening but with other specific cultural practices, such as learning about rhymes and with acquiring written language. Simple concepts and metaconcepts seem to develop in association with distinct types of cultural practice.

Mathematical concepts have already been subjected to a tool-object distinction by Douady (1985). As a tool in everyday life, mathematical concepts are used in the process of structuring external situations as the subject tries to understand the mathematical relations involved in these situations. But the concepts remain in a sense transparent when they are used as tools. The subject thinks about the situation, not about the mathematical concept. From this perspective, mathematical concepts in everyday life are not restricted to concrete thinking and are not mere procedures replicated without understanding.

As the object of instruction, mathematical concepts are the focus of interest in the classroom. Pupils are expected to learn, think about, and try representing the concepts themselves. In this sense, everyday mathematics and mathematics education

are clearly distinct. Mathematical concepts are tools in everyday life and objects in the classroom. Mathematical concepts are occasionally used as tools in the classroom in word problems and other application exercises, but the focus of interest even in application exercises is the mathematical concept, not the particular situation.

Implications for Education

Does the difference between everyday and school knowledge of mathematics mean that everyday mathematics is of no use for classroom instruction? Of course not. According to developmental psychologists, in order to succeed in studying a mathematical concept as an object, the subject must already have access to the concept as a tool. Karmiloff-Smith (1988), who attributes great importance to the relationship between cognition and metacognition in cognitive development, makes this point clear:

Finally, I want to take it that the human mind not only tries to appropriate the *external* reality that is pre-set to explore and represent, but that it also tries to appropriate its own internal reality. What I mean by this is that the human mind re-represents recursively its own *internal* representations, thereby creating meta-procedures and meta-representations in general. For a long time I've argued that it is this *recursive* capacity to represent one's own representations that sets humans apart from other species. (p. 12)

If we accept for the moment the thesis that metaknowledge is the representation of one's own internal reality and that concepts must first be used in the representation of external reality in order to be transformed into metaconcepts, some directions can be extracted for mathematics education from the analysis of cultural practices embedding mathematical knowledge.

When a teacher faces a class with the aim of teaching a particular concept, the teacher must ponder whether the pupils are likely to have used the concept in everyday life. Until more research is available on this subject, teachers will have the task of looking for everyday uses of the concept and attempting to analyze the logical invariants that underlie everyday concepts and procedures. Everyday procedures will often differ from school-taught procedures, and teachers will need to look beneath the surface to recognize which invariants are being respected in everyday practice. I am hopeful that the studies reviewed here can provide a model for this enterprise.

After identifying an everyday situation in which the concept is used, the teacher will need to know how to use the situation to promote an awareness and understanding of the concept. Research on this process is scarce. The suggestions available in the literature are of two types. First, the *abstraction-generalization* model can be called into play in the learning-to-learn context (see Cole, 1977) or in the cognitive apprenticeship approach (see Brown, Collins, & Duguid, 1989). The basis of these approaches is that understanding several different situations involving the same invariant leads to the abstraction and generalization of the core concept (the invariant), and to the enrichment of the concept by extending the set of situations to which it applies. In contrast to present practice, this model of teaching proposes that pupils solve several problem situations embedding the same concept and then turn to the concept as such. For example, pupils could be given several problems about fractions without being taught how to represent fractions, and then they might connect the meanings developed in this way to the mathematical representation. This model of teaching was attempted with success by Lima (1989) in the context of fractions.

Second, the rerepresentation of the original situation through natural language can help the transition of a concept from tool to object. In this case, situations become data to be talked about rather than problems to be solved. Carraher and Ruiz (1985) attempted such a method in a study of the development of the concept of proportions. In their procedure, pupils worked in small groups to solve a problem about proportions. They were told that a second group of students would be solving an identical problem with different numbers. They were asked to write a message to this second group telling them how to solve the problem. However, since they did not know what the numbers would be in the other group's problem, they had to explain the procedure without mentioning numbers. They were told that their success in the task was not a matter of whether they had solved the problem but whether their message was useful to the other group. Under these circumstances, students realized that the message had to be a general one about the relationships between the variables, not a specific instruction for calculating, such as, "multiply 2 by 13 and divide by 3." Most students attempted to write the general message in natural language; a few attempted to devise a formula by writing a message such as "Make x your number of rings on the scale and y the number of the peg on which you put your rings." Not all the groups who could solve the problem could write general messages successfully, which showed that rerepresentation of the problem was not easy. Further studies are needed to verify what effect this type of reconstruction of the problem may have on pupils' understanding of the mathematical concept.

In summary, this chapter has attempted to show that mathematical activity can be observed as interwoven with everyday practice outside academic settings. Mathematizing objects and situations in everyday activity means representing them in ways that allow for the extraction of further information about the objects or situations on the basis of the representations without the need for further verification by returning to the represented objects. Three topics were analyzed in detail in the review of research: (a) counting and measuring, (b) problem solving and calculation, and (c) modeling and inversion. These concepts are central to the types of mathematical activity engaged in by pupils at the elementary-school level. They are also of great importance for everyday life and become interwoven in everyday activities in different forms. Some of these concepts may be acquired without schooling, but this does not mean that mathematical instruction should leave learning of these concepts for apprenticeship outside school. It is the relevance of these concepts both to mathematical activity and to everyday life that suggests the need for paying greater attention to them in mathematics teaching. By promoting their understanding in school, teachers may help children recognize how much mathematics they can learn and use outside school. It might also help children transform concepts they now use as tools into metaconcepts they can use for generalization; in the process, perhaps they will "learn to learn" mathematics and enjoy it.

References

Baranes, R., Perry, M., & Stigler, J. W. (1989). Activation of real world knowledge in the solution of word problems. *Cognition and Instruction, 6,* 287–318.

Bishop, A. J. (Ed.). (1988). Mathematics education and culture [Special issue]. *Educational Studies in Mathematics, 19*(2).

Brown, J. S., & Burton, R. R. (1978). Diagnostic models for procedural bugs in basic mathematical skills. *Cognitive Science, 2,* 155–192.

Brown, J. S., Collins, A., & Duguid, P. (1989). Situated cognition and the culture of learning. *Educational Researcher, 18*(1), 32–42.

Bruner, J. (1966). *Towards a theory of instruction.* Cambridge, MA: Harvard University Press.

Carpenter, T. P., & Moser, J. M. (1982). The development of addition and subtraction problem-solving skills. In T. P. Carpenter, J. M. Moser, & T. A. Romberg (Eds.), *Addition and subtraction: A cognitive perspective* (pp. 9–24). Hillsdale, NJ: Lawrence Erlbaum.

Carraher, T. N. (1985). The decimal system. Understanding and notation. In L. Streefland (Ed.), *Proceedings of the Ninth International Conference for the Psychology of Mathematical Education* (Vol. 1, pp. 288–303). Utrecht, Holland: University of Utrecht, Research Group on Mathematics Education and Educational Computer Center.

Carraher, T. N. (1986). From drawings to buildings. Working with mathematical scales. *International Journal of Behavioral Development, 9,* 527–544.

Carraher, T. N. (1988, April). *Adult mathematical skills. The effect of schooling.* Paper presented at the annual meeting of the American Educational Research Association, New Orleans.

Carraher, T. N. (1989, July). *Numeracy without schooling.* Paper presented at the 13th meeting of the International Group for the Study of the Psychology of Mathematics Education, Paris.

Carraher, T. N., Carraher, D. W., & Schliemann, A. D. (1985). Mathematics in the streets and in schools. *British Journal of Developmental Psychology, 3,* 21–29.

Carraher, T. N., Carraher, D. W., & Schliemann, A. D. (1987). Written and oral mathematics. *Journal for Research in Mathematics Education, 18,* 83–97.

Carraher, T. N., & Ruiz, E. R. L. (1985). *Teaching experiments on proportions.* Unpublished research report, Recife: Universidade Federal de Pernambuco.

Carraher, T. N., & Schliemann, A. D. (1988). Culture, arithmetic and mathematical models. *Cultural Dynamics, 1,* 180–194.

Carraher, T. N., & Schliemann, A. D. (1990). Knowledge of the numeration system among pre-schoolers. In L. P. Steffe & T. Wood (Eds.), *Transforming children's mathematics education. International perspectives* (pp. 135–141). Hillsdale, NJ: Lawrence Erlbaum.

Carraher, T. N., Schliemann, A. D., & Carraher, D. W. (1988). Mathematical concepts in everyday life. In G. B. Saxe & M. Gearhart (Eds.), *Children's mathematics* (pp. 71–87). San Francisco: Jossey Bass.

Chevallard, Y. (1990). On mathematics education and culture: Critical afterthoughts. *Educational Studies in Mathematics, 21*(1), 3–28.

Cockcroft, W. H. (1986). Inquiry into school teaching of mathematics in England and Wales. In M. Carss (Ed.), *Proceedings of the Fifth International Congress on Mathematical Education* (pp. 328–329). Boston: Birkhäuser.

Cole, M. (1977). An ethnographic psychology of cognition. In P. Johnson-Laird & P. Wason (Eds.), *Thinking. Readings in cognitive science* (pp. 468–482). London: Cambridge University Press.

Cole, M., & Griffin, P. (1980). Cultural amplifiers reconsidered. In D. Olson (Ed.), *The social foundations of language and thought* (pp. 343–364). New York: Norton.

D'Ambrosio, U. (1984, August). *Ethnomathematics.* Opening address to the fifth meeting of the International Conference on Mathematics Education, Adelaide, Australia.

D'Ambrosio, U. (1985). Ethnomathematics and its place in the history and pedagogy of mathematics. *For the Learning of Mathematics, 5*(1), 44–48.

D'Ambrosio, U. (1986). *Da realidade à ação Reflexões sobre educação e matemática.* Campinas: Summus Editorial.

De Corte, E., & Verschaffel, L. (1987). The effects of semantic and non-semantic factors on young children's solutions of elementary addition and subtraction word problems. In J. C. Bergeron, H. Herscovics, & C. Kieran (Eds.), *Proceedings of the Eleventh International Conference on the Psychology of Mathematics Education, 2,* 375–381. Montreal: Université de Montréal.

Douady, R. (1985). The interplay between different settings. Tool-object dialectic in the extension of mathematical ability—Examples from elementary school teaching. In L. Streefland (Ed.), *Proceedings of the Ninth International Conference on the Psychology of Mathematics Education, 1,* (pp. 33–52). Utrecht, Holland: University of Utrecht, Research Group on Mathematics Education and Educational Computer Center.

Fahrmeier, E. (1984). Taking inventory: Counting as problem solving. *The Quarterly Newsletter of the Laboratory of Comparative Human Cognition, 6*(1–2), 6–10.

Flavell, J. H., & Wellman, H. M. (1977). Metamemory. In R. V. Kail, Jr., & J. W. Hagen (Eds.), *Perspectives on the development of memory and cognition.* Hillsdale, NJ: Lawrence Erlbaum.

Fuson, K. (1982). An analysis of the counting on solution procedure in addition. In T. P. Carpenter, J. M. Moser, & T. A. Romberg (Eds.), *Addition and subtraction: A cognitive perspective* (pp. 67–82). Hillsdale, NJ: Lawrence Erlbaum.

Gal'perin, P. Ya., & Georgiev, L. S. (1969). The formation of elementary mathematical notions. In J. Kilpatrick & I. Wirszup (Eds.), *Soviet studies in the psychology of learning and teaching mathematics. Vol. I: The learning of mathematical concepts* (pp. 189–216). Chicago: University of Chicago Press.

Gay, J., & Cole, M. (1967). *The new mathematics and an old culture. A study of learning among the Kpelle in Liberia.* New York: Holt, Rinehart & Winston.

Gelman, R., & Gallistel, C. R. (1978). *The child's understanding of number.* Cambridge, MA: Harvard University Press.

Ginsburg, H. P. (1977). *Children's arithmetic. The learning process.* New York: Van Nostrand.

Ginsburg, H. P., Posner, J. K., & Russel, R. L. (1981). The development of mental addition as a function of schooling and culture. *Journal of Cross-cultural Psychology, 12*(2), 163–178.

Hughes, M. (1986). *Children and number. Difficulties in learning mathematics.* Oxford: Basil Blackwell.

Hunter, I. M. L. (1977). Mental calculation: Two additional comments. In P. N. Johnson-Laird & P. C. Wason (Eds.), *Thinking: Readings in cognitive science* (pp. 35–42). Cambridge, UK: Cambridge University Press.

Karmiloff-Smith, A. (1988). Beyond modularity: Innate constraints and developmental change. In S. Carey & R. Gelman (Eds.), *Epigenesis of the mind: Essays in biology and knowledge.* Hillsdale, NJ: Lawrence Erlbaum.

Kieran, C. (1984). A comparison between novice and more expert algebra students on tasks dealing with equivalence of equations. In J. M. Moser (Ed.), *Proceedings of the sixth annual meeting of the North American Chapter of the International Group for the Psychology of Mathematics Education* (pp. 83–91). Madison, WI: Wisconsin Center for Educational Research.

La Barre, W. (1954). *The human animal.* Chicago: University of Chicago Press.

Lancy, D. F. (1983). *Cross-cultural studies in cognition and mathematics.* New York: Academic Press.

Lave, J. (1988). *Cognition in practice. Mind, mathematics and culture in everyday life.* New York: Cambridge University Press.

Lima, M. F. (1989). Iniciação ao conceito de fração e o desenvolvimento da conservação de quantidades. In T. N. Carraher (Ed.), *Aprender Pensando. Aplicações da Psicologia Cognitiva à Educação* (4th printing, pp. 81–127). Petrópolis (RJ, Brazil): Vozes.

Luria, A. R. (1966). *El papel del lenguaje en el desarrollo de la conducta.* Buenos Aires: Ediciones Tekné.

Luria, A. R. (1969). On the pathology of computational operations. In J. Kilpatrick & I. Wirszup (Eds.), *Soviet studies in the psychology of learning and teaching mathematics. Vol. I. The learning of mathematical concepts* (pp. 37–73). Chicago: University of Chicago Press.

Menninger, K. (1969). *Number words and number symbols: A cultural history of numbers.* Cambridge, MA: The MIT Press.

Miller, K. F., & Stigler, J. W. (1987). Counting in Chinese. Cultural variation in a basic cognitive skill. *Cognitive Development, 2,* 279–305.

Piaget, J. (1962). *Play, dreams and imitation in childhood.* New York: Norton.

Plunkett, S. (1979). Decomposition and all that rot. *Mathematics in Schools, 8*(3), 2–7.

The Quarterly Newsletter of the Laboratory of Comparative Human Cognition. (1984). *6.*

The Quarterly Newsletter of the Laboratory of Comparative Human Cognition. (1990). *1–2.*

Reed, H. J., & Lave, J. (1981). Arithmetic as a tool for investigating relations between culture and cognition. In R. W. Casson (Ed.), *Language, culture and cognition: Anthropological perspectives* (pp. 437–455). New York: Macmillan.

Resnick, L. (1982). Syntax and semantics in learning to subtract. In T. Carpenter, J. Moser, & T. Romberg (Eds.), *Addition and subtraction: A cognitive perspective* (pp. 136–155). Hillsdale, NJ: Lawrence Erlbaum.

Resnick, L. (1984, October). *The development of mathematical intuition.* Paper presented at the Nineteenth Minnesota Symposium of Child Psychology, Minneapolis, MN.

Riley, M. S., Greeno, J. G., & Heller, J. I. (1983). Development of children's problem solving ability in arithmetic. In H. P. Ginsburg (Ed.), *The development of mathematical thinking* (pp. 153–196). London: Academic Press.

Saxe, G. B. (1981). Body Parts as numerals: A developmental analysis of numeration among the Oksapmin of Papua New Guinea. *Child Development, 52,* 306–316.

Saxe, G. B. (1982). Developing forms of arithmetical thought among the Oksapmin of Papua New Guinea. *Developmental Psychology, 18*(4), 583–594.

Saxe, G. B. (1985). Effects of schooling on arithmetical understandings: Studies with Oksapmin children in Papua New Guinea. *Journal of Educational Psychology, 77*(5), 503–513.

Saxe, G. B. (1988). Candy selling and mathematics learning. *Educational Researcher, 17*(6), 14–21.

Saxe, G. B., & Moylan, T. (1982). The development of measurement operations among the Oksapmin of Papua New Guinea. *Child Development, 53,* 1242–1248.

Schliemann, A. D., & Carraher, T. N. (1990). A situated schema of proportionality. *British Journal of Developmental Psychology, 8,* 259–269.

Scribner, S. (1984). Product assembly: Optimizing strategies and their acquisition. *The Quarterly Newsletter of the Laboratory of Comparative Human Cognition, 6*(1–2), 11–19.

Sinclair, A., Jarvella, R. J., & Levelt, W. J. M. (1978). *The child's conception of language.* Berlin: Springer.

Steffe, L. P., von Glasersfeld, E., Richards, J., & Cobb, P. (1983). *Children's counting types. Philosophy, theory, & applications.* New York: Praeger.

Stigler, J. W., & Baranes, R. (1988). Culture and mathematics learning. In E. Z. Rothkopf (Ed.), *Review of research in education* (vol. 15, pp. 253–306). Washington, DC: American Educational Research Association.

Vergnaud, G. (1982). A classification of cognitive tasks and operations of thought involved in addition and subtraction problems. In T. P. Carpenter, J. M. Moser, & T. A. Romberg (Eds.), *Addition and subtraction: A cognitive perspective* (pp. 39–59). Hillsdale, NJ: Lawrence Erlbaum.

Vygotsky, L. S. (1978). *Mind in society. The development of higher psychological processes.* Cambridge, MA: Harvard University Press.

Yaroshchuk, V. L. (1969). A psychological analysis of the processes involved in solving model arithmetic problems. In J. Kilpatrick & I. Wirszup (Eds.), *Soviet studies in the psychology of learning and teaching mathematics. Vol. III. Problem solving in arithmetic and algebra* (pp. 53–96). Chicago: University of Chicago Press.

Young, R. M., & O'Shea, T. (1981). Errors in children's subtraction. *Cognitive Science, 5,* 153–177.

·23·

RESEARCH ON AFFECT IN MATHEMATICS EDUCATION: A RECONCEPTUALIZATION

Douglas B. McLeod

WASHINGTON STATE UNIVERSITY AND SAN DIEGO STATE UNIVERSITY

Affective issues play a central role in mathematics learning and instruction. When teachers talk about their mathematics classes, they seem just as likely to mention their students' enthusiasm or hostility toward mathematics as to report their cognitive achievements. Similarly, inquiries of students are just as likely to produce affective as cognitive responses; comments about liking (or hating) mathematics are as common as reports of instructional activities. These informal observations support the view that affect plays a significant role in mathematics learning and instruction. Although affect is a central concern of students and teachers, research on affect in mathematics education continues to reside on the periphery of the field. If research on learning and instruction is to maximize its impact on students and teachers, affective issues need to occupy a more central position in the minds of researchers. One theme of this chapter is that all research in mathematics education can be strengthened if researchers will integrate affective issues into studies of cognition and instruction.

Current efforts to reform the mathematics curriculum place special importance on the role of affect. The National Council of Teachers of Mathematics has reaffirmed the centrality of affective issues in its recent publication of the standards for curriculum and evaluation (Commission on Standards for School Mathematics, 1989). Two of the major goals stated in this document deal with helping students understand the value of mathematics and with developing student confidence. In its Standard on mathematical disposition, the assessment of student confidence, interest, perseverance, and curiosity are all recommended. Similarly, the National Research Council's

(1989) report on the future of mathematics education (*Everybody Counts*) puts considerable emphasis on the need to change the public's beliefs and attitudes about mathematics. In the United States, there is a tendency to believe that learning mathematics is a question more of ability than effort. Adults are willing to accept poor performance in school mathematics, but they are not so willing to accept poor performance in other subjects. Both adults and children often proclaim their ignorance of mathematics without embarrassment, treating this lack of accomplishment in mathematics as a permanent state over which they have little control. The improvement of mathematics education will require changes in the affective responses of both children and adults.

As these reports show, the U. S. reform movement in mathematics education clearly takes affective factors as an important area where substantial change is needed. This emphasis on affective issues is related to the importance that the reform movement attaches to higher-order thinking. If students are going to be active learners of mathematics who willingly attack nonroutine problems, their affective responses to mathematics are going to be much more intense than if they are merely expected to achieve satisfactory levels of performance in low-level computational skills.

A variety of large-scale studies provide a substantial amount of data that indicate there is good reason to be concerned about affective factors. The Second International Mathematics Study (Robitaille & Garden, 1989) indicates that there are large differences between countries on measures of mathematical beliefs and attitudes, just as there are large differences in achie-

The assistance of Martha Nelson and Roger Parsons in reviewing the literature is gratefully acknowledged. Also, Carol Kehr Tittle, City University of New York, and Margaret R. Meyer, University of Wisconsin, made many suggestions for improving the manuscript. Preparation of this chapter was supported in part by National Science Foundation Grant No. MDR-8696142. Any opinions, conclusions, or recommendations are those of the author and do not necessarily reflect the views of the National Science Foundation.

vement. Various national assessments have also included data on affective issues. Dossey, Mullis, Lindquist, and Chambers (1988) report that students in the United States become less positive about mathematics as they proceed through school; both confidence about and enjoyment of mathematics appear to decline as students move from elementary through secondary school. Students in other countries (Foxman, Martini, & Mitchell, 1982; McLean, 1982) also show little enthusiasm for mathematics as they progress through school.

Although efforts to evaluate mathematics programs and to promote the reform and improvement of mathematics education pay considerable attention to the affective domain, these efforts usually take a very practical approach, using questionnaires to gather common-sense data on beliefs and attitudes toward mathematics. This kind of evaluation usually does not attempt to present a theoretical framework for the assessment of affect, nor does it include data from small-scale, qualitative studies that could provide a more detailed picture of students' affective responses to mathematics. The improvement of theory and the use of a variety of research methods are two additional themes that will recur throughout this chapter.

The chapter begins by considering alternative theoretical foundations for research on affect. Mandler's (1984) theory, an approach to research on affect that is based on cognitive psychology, is selected for further discussion, particularly because it illustrates how affect can be incorporated into cognitive studies of mathematics learning and teaching. The chapter then presents a framework for research on affect that reorganizes the literature into three major areas: beliefs, attitudes, and emotions. Next, research on a number of topics from the affective domain is summarized, and linked to the proposed framework. Finally, the chapter explores how qualitative as well as quantitative research methods can be used in research on affect, and discusses the implications of the chapter for future research.

TERMINOLOGY AND GENERAL BACKGROUND

For the purposes of this discussion the affective domain refers to a wide range of beliefs, feelings, and moods that are generally regarded as going beyond the domain of cognition. H. A. Simon (1982), in discussing the terminology used to describe the affective domain, suggests that we use affect as a more general term; other terms (for example, beliefs, attitudes, and emotions) will be used in this chapter as more specific descriptors of subsets of the affective domain. In the context of mathematics education, feelings and moods like anxiety, confidence, frustration, and satisfaction are all used to describe responses to mathematical tasks. Frequently these feelings are discussed in the literature as attitudes, although that term does not seem adequate to describe some of the more intense emotional reactions that occur in mathematics classrooms. For example, the "Aha!" experience in mathematical problem solving is usually recognized as a joyful event, generally of limited duration; such an event does not fit with the definitions of attitude used by most researchers.

As Hart (1989b) and H. A. Simon (1982) indicate, describing the affective domain is no easy task. Terms sometimes have dif-

ferent meanings in psychology than they do in mathematics education, and even within a given field, studies that use the same terminology are often not studying the same phenomenon. For example, Hart (1989b) notes that anxiety is sometimes described as fear, one of the more intense emotions, and in other studies as dislike or worry. Clarification of terminology for the affective domain remains a major task for researchers in both psychology and mathematics education.

Moreover, affect is generally more difficult to describe and measure than cognition. H. A. Simon (1982) notes that there may be different kinds of sadness or fear, distinctions and gradations that our ordinary language is ill-equipped to make in a dependable way. As Simon suggests, working on cognition seems relatively simple compared to the difficulties of dealing clearly with affect. As an example of these difficulties, one might consider the work on building taxonomies of educational objectives that took place over 30 years ago. The work on the cognitive domain (Bloom, Englehart, Furst, Hill, & Krathwohl, 1956) has had a major impact on curriculum and evaluation. The taxonomy for the affective domain (Krathwohl, Bloom, & Masia, 1964) has not had such an impact, and even its authors acknowledged that they found the task to be exceedingly difficult. There are, as Tittle and her colleagues point out (Tittle, Hecht, & Moore, 1989), a number of reasons for the difference in impact of these two documents, but surely one of the reasons for the differences is the complexity and difficulty of dealing adequately with the affective domain.

There have been many reviews of the literature related to affective factors and mathematics education, including those by Aiken (1970, 1976), Kulm (1980), Reyes (1980, 1984), and Leder (1987). These reviews have generally focused on attitudes toward mathematics as their major concern, rather than trying to describe and analyze all components of the affective domain. Also, they have in general worked from within the traditional paradigm of educational research, with its emphasis on quantitative methods, paper-and-pencil tests, and the positivistic perspective of behaviorist or differential psychology. This discussion will attempt to broaden the view of both the theories and the methods that might be useful in the study of affective factors in mathematics education.

PSYCHOLOGICAL THEORIES AND AFFECT

Changes in psychological theories have had a major impact on how affect is treated in mathematics education research. Frequently researchers have treated affect as an avoidable complication of modest significance; students have been viewed in rather mechanistic terms. The researcher's model of the student has a major impact on how the research is conducted, particularly in terms of the affective domain. If we believe that the learner is someone who only receives knowledge rather than someone who is actively involved in constructing knowledge, our research program could be entirely different in terms of both the affective and the cognitive domain.

The influence of behaviorism on educational psychology in this century has been an important factor in the neglect of the affective domain. Skinner (1953), for example, described the emotions as examples of imaginary constructs that were commonly used as causes of behavior. As Mandler (1984, 1985) indicates, the behaviorists have generally been unwilling to look closely at the underlying processes that are related to affective responses, particularly if data on those processes are gathered through introspection or verbal reports (Ericsson & Simon, 1980). As a result, behaviorism has much to say about stimulus and response in learning, but it has little interest in the influence of affective factors on learners.

In more recent times, experts in cognitive science and artificial intelligence have taken a serious interest in the study of mathematics learning (Schoenfeld, 1987a). However, these researchers have also tended to exclude affective factors from their considerations. As Norman (1981) has pointed out, the researcher's task would be much simpler if the emotions were superfluous; the desire to avoid complexity has been a major reason for the lack of attention to affective issues in cognitive science (Gardner, 1985). However, Norman's (1981) recommendation that cognitive science needed to focus on more than just "pure" cognition has had an impact, and current research on cognitive issues pays increased attention to affective and cultural factors.

In contrast to the behaviorists and some advocates of cognitive science, researchers in differential psychology and social psychology have given substantial attention to the notion of affective issues, especially to the study of attitudes. This work is characterized by its emphasis on definitions of terms, its preoccupation with measurement issues, and its reliance on questionnaires and quantitative methods. This approach can be characterized as the traditional paradigm for research on affect. Books by Ajzen and Fishbein (1980) and Rajecki (1982) provide extensive summaries of this work in general psychology. Research on the application of these ideas to mathematics education has been reviewed by Aiken (1970, 1976), Kulm (1980), Reyes (1984), and Leder (1987), among others. There has been considerable dissatisfaction with this traditional paradigm, both in psychology (Abelson, 1976; Berscheid, 1982; Mandler, 1972, 1989) and in mathematics education (Kulm, 1980; McLeod, 1985, 1988), particularly because of the lack of a strong theoretical foundation for the work. Nevertheless, most of what we know about affective factors in mathematics education comes from work within this traditional paradigm (Meyer & Fennema, 1988; Reyes, 1984). Fennema (1989) provides a spirited defense of the methods and contributions of this quantitative approach, noting that the traditional paradigm of differential psychology, rigorously applied, has produced a substantial amount of knowledge about affect and mathematics education. Moreover, this knowledge has been particularly useful in attacking problems of gender-related differences in participation in mathematics (Fennema, 1989; Fennema & Leder, 1990).

Although traditional quantitative approaches provide substantial information on some issues, there are many other areas (for example, emotional responses to mathematical problem solving) that are not susceptible to this approach. There are also a number of topics (for example, anxiety) where the research is confusing and contradictory. As a result, it seems useful to try to develop a new paradigm for research on affect that would be more comprehensive and more closely integrated with current research on cognition. There are always dangers in following a fad, and the new emphasis on cognitive theories in psychology does show signs of narrow-mindedness and an easy enthusiasm for new terminology rather than critical thinking. And as Messick (1987) suggests, the "hypercognitization" of affect by cognitive psychologists could result in the omission of important issues that should be central to our research agenda, rather than leading to progress in the field. In particular, there is no need to forget what has been learned in the past as we examine new approaches to old problems. Nevertheless, new approaches can lead to new progress in research, and new paradigms can lead to helpful reconceptualizations.

An alternative paradigm for research on affect has grown out of the work of developmental psychology and the rising influence of cognitive psychology in the recent past. This new paradigm for research on affect in mathematics learning can be characterized by its emphasis on theoretical issues, its interest in qualitative methods, its use of interviews and think-aloud protocols, and its attention to beliefs and emotions as well as attitudes. For examples of this kind of work in psychology, see Mandler (1984), Kagan (1978), and Ortony, Clore, and Collins (1988). Snow and Farr (1987) and Case, Hayward, Lewis, and Hurst (1988) present analyses of similar work with more direct connections to education, and Bassarear (1989), Goldin (1988), and McLeod and Adams (1989) present some applications and extensions of these ideas in mathematics education. The elaboration of both the traditional and alternative paradigms as they apply to the affective domain will be a continuing theme throughout this chapter.

COGNITIVE APPROACHES TO RESEARCH ON AFFECT

The emergence of affect as an important part of cognitive theories has been documented recently by Snow and Farr (1987). The beginning of these attempts to incorporate affective factors into cognitive theories can be traced back at least to the work of Schacter and Singer (1962) and H. A. Simon (1967); however, the leading theorists in this area now include cognitive psychologists like Lazarus (1982, 1984) and Mandler (1975, 1984). Overviews of the field (for example, Scherer & Ekman, 1984; Strongman, 1978) give substantial emphasis to the new influence of cognitive psychologists on this area. Although there are several cognitive theorists who are having a substantial impact on the study of affective issues broadly defined (including Beck & Emery, 1985; Bower, 1981; Meichenbaum, 1977; Ortony et al., 1988; Sheffler, 1977; L. R. Simon, 1986; Zajonc, 1984), the one who has done the most to apply his ideas to problems in mathematics education is Mandler (1989).

Mandler's general theory is presented in some detail in his 1984 book, and he has recently described his view of how the theory can be applied to the teaching and learning of mathematical problem solving (Mandler, 1989). At the risk of oversimplification, only a brief summary of his theory will be presented here. Most of what is presented appears to be compatible with other theorists who come from a cognitive point of view (for example, H. A. Simon, 1982).

Mandler's view is that most affective factors arise out of the emotional responses to the interruption of plans or planned behavior. In Mandler's terms, plans arise from the activation of a schema. The schema produces an action sequence, and if the anticipated sequence of actions cannot be completed as planned, the blockage or discrepancy is followed by a physiological response. This physiological arousal is typically felt as an increase in heartbeat or in muscle tension. The arousal serves as the mechanism for redirecting the individual's attention, and has obvious survival value which presumably may have played some role in evolutionary development. At the same time the arousal occurs, the individual attempts to evaluate the meaning of this unexpected or otherwise troublesome blockage. The evaluation of the interruption might classify it in one of several ways: a pleasant surprise, an unpleasant irritation, or perhaps a major catastrophe. The cognitive evaluation of the interruption provides the meaning to the arousal.

There are several important parts to the analysis of the meaning of the interruptions. First, the meaning comes out of the cognitive interpretation of the arousal. This meaning will be dependent on what the individual knows or assumes to be true. In other words, the individual's knowledge and beliefs play a significant role in the interpretation of the interruption. The role of the culture that shapes these beliefs would seem to be particularly important.

Second, the arousal that leads to the emotion is generally of limited duration. Normal individuals adjust to the unexpected event, interpret it in the context in which it occurs, and try to find some other way to carry out their plan and achieve their goal. The emotion may be intense, but it is generally transitory in normal individuals, at least initially.

Third, repeated interruptions in the same context normally result in emotions that become less intense. The individual will reduce the demand on cognitive processing by responding more and more automatically, and with less and less intensity. The responses in this situation become more stable and predictable and begin to resemble the kinds of attitudes that have been the emphasis of past research on affect in mathematics education.

To help clarify the situation, consider the affective responses of a sixth-grade student to a typical story problem. Suppose that the student believes that story problems should make sense and should have a reasonable answer that can be obtained in a minute or two. Suppose also that the student has had some success in other areas of mathematics. If the student is unable to obtain a satisfactory answer in a reasonable time, the failure to solve the problem (an interruption of the plan) is likely to lead to some arousal. The interpretation of this arousal is likely to be negative and is often reported as frustration by students who are able to verbalize their feelings. If the students are able to overcome the blockage and find a solution to the problem, they may report positive reactions to the experience. If negative reactions to story problems occur repeatedly, the response to story problems will eventually become automatic and stable. In this situation the student would have developed a negative attitude toward story problems.

In summary, there appear to be at least three major facets of the affective experience of mathematics students that are worthy of further study. First, students hold certain beliefs about mathematics and about themselves that play an important role in the development of their affective responses to mathematical situations. Second, since interruptions and blockages are an inevitable part of the learning of mathematics, students will experience both positive and negative emotions as they learn mathematics; these emotions are likely to be more noticeable when the tasks are novel. Third, students will develop positive or negative attitudes toward mathematics (or parts of the mathematics curriculum) as they encounter the same or similar mathematical situations repeatedly. These three aspects of affective experience correspond to three areas of research in mathematics education which we will now examine.

BELIEFS, ATTITUDES, AND EMOTIONS IN MATHEMATICS LEARNING: RECONCEPTUALIZING THE AFFECTIVE DOMAIN

Snow and Farr (1987), in their discussion of affect and cognition, point out that new research on affect must find ways to come to terms with the cognitive revolution in psychology. In particular, any reconceptualization of the affective domain should attempt to be compatible with cognitive-processing models of the learner. In this context, the work of Mandler (1984) should provide a useful general guide. The theoretical analyses of Mandler (1984) and the practical analyses of mathematics classrooms suggest that beliefs, attitudes, and emotions should be important factors in research on the affective domain in mathematics education. Table 23.1 provides a brief outline of these major constructs.

Beliefs, attitudes, and emotions are used to describe a wide range of affective responses to mathematics. These terms vary in the stability of the affective responses that they represent; beliefs and attitudes are generally stable, but emotions may change rapidly. They also vary in the level of intensity of the affects that they describe, increasing in intensity from "cold" beliefs about mathematics to "cool" attitudes related to liking or disliking mathematics to "hot" emotional reactions to the frustrations of solving nonroutine problems. Beliefs, attitudes, and emotions also differ in the degree to which cognition plays a role in the response, and in the time that they take to develop.

TABLE 23–1. The Affective Domain in Mathematics Education

Category	Examples
Beliefs	
About mathematics	Mathematics is based on rules
About self	I am able to solve problems
About mathematics teaching	Teaching is telling
About the social context	Learning is competitive
Attitudes	Dislike of geometric proof
	Enjoyment of problem solving
	Preference for discovery learning
Emotions	Joy (or frustration) in solving nonroutine problems
	Aesthetic responses to mathematics

For example, beliefs are largely cognitive in nature, and are developed over a relatively long period of time. Emotions, on the other hand, may involve little cognitive appraisal and may appear and disappear rather quickly, as when the frustration of trying to solve a hard problem is followed by the joy of finding a solution. Therefore, we can think of beliefs, attitudes, and emotions as representing increasing levels of affective involvement, decreasing levels of cognitive involvement, increasing levels of intensity of response, and decreasing levels of response stability. A review of some of the relevant literature provides support for the importance of these three constructs.

Beliefs

Research on beliefs in mathematics education has become an important thread linking a number of studies of both teachers and students. Thompson's (1984) work on teacher conceptions of mathematics and Schoenfeld's (1985) studies of the belief systems of problem solvers are important examples. Data have typically been gathered through observations of students and teachers, as well as through questionnaires and interviews. Researchers have generally not used a consistent framework; instead, the data have been organized in rather different ways in different studies, with each researcher trying to explain the influence of beliefs in each particular content. People who come from research on problem solving (for example, Schoenfeld, 1985; Silver, 1985) have tended to emphasize the role of student beliefs about mathematics as a discipline. For example, many students believe that problems can be solved quickly or not at all, that only geniuses can be creative in mathematics, and that proof just confirms the obvious (Schoenfeld, 1985). Other researchers, particularly those who investigate gender-related differences (for example, Fennema & Peterson, 1985) have emphasized students' beliefs about themselves as learners. In this category we find beliefs about students' ability to do mathematics or about the importance of effort to success in learning mathematics.

Although each of these approaches has contributed to our knowledge of how beliefs are important to mathematics learning and teaching, little emphasis has been given to providing an overall structure for the study of beliefs in mathematics education. There are, however, some examples of broader approaches to research on beliefs. Lester, Garofalo, and Kroll (1989) describe beliefs in terms of the subjective knowledge of students regarding mathematics, self, and problem-solving activities. Underhill (1988) discusses beliefs in terms of several dimensions, including whether mathematics is primarily rule-oriented or concept-oriented, and whether mathematics is learned by having knowledge transmitted to students or constructed by students. Fennema and Peterson (1985) suggest connections between beliefs and autonomous learning behavior, with the subsequent impact on higher-order thinking in mathematics, particularly in terms of gender-related differences in mathematics achievement.

There are a variety of ways to organize research on beliefs. Rokeach (1968), for example, organizes beliefs along a dimension of centrality to the individual. Those beliefs that are most central are those on which there is complete consensus; beliefs about which there is some disagreement would be less central. Beliefs that are imposed on individuals by authority figures would be still less central. One alternative for mathematics education would be to develop a taxonomy of beliefs like Rokeach provides for more general settings.

D'Andrade (1981) suggests that beliefs develop gradually through a process much like "guided discovery" where children respond to the situations in which they find themselves by developing beliefs that are consistent with their experience. Certainly cultural factors play an important role in this process of developing beliefs. In mathematics education most researchers seem to assume that the development of beliefs about mathematics is heavily influenced by the cultural setting of the classroom (Schoenfeld, 1989).

In this chapter beliefs related to mathematics education are discussed in terms of the experiences of students and teachers. Students' beliefs are categorized in terms of the object of the belief: beliefs about mathematics, beliefs about self, beliefs about mathematics teaching, and beliefs about the contexts in which mathematics education occurs. The discussion of these categories will also include comments on teachers' beliefs about mathematics and instruction.

Beliefs about Mathematics. Research on students' beliefs about mathematics has received considerable attention over recent years. The National Assessment of Educational Progress has included items related to beliefs about mathematics for some time. The most recent assessment (Brown et al., 1988) indicates that students believe that mathematics is important, difficult, and based on rules. These beliefs about mathematics, although not emotional in themselves, certainly would tend to generate more intense reactions to mathematical tasks than beliefs that mathematics is unimportant, easy, and based on logical reasoning. Some researchers may not see any need to include such beliefs as part of the affective domain, and it is true that these beliefs are mainly cognitive in nature. However, the role of beliefs is central in the development of attitudinal and emotional responses to mathematics, and thus beliefs are included in this description of the affective domain.

A variety of major evaluation studies have dealt with beliefs about mathematics. Dossey et al. (1988) report that students in the United States (grades 3, 7, and 11) believe that mathematics is useful, but involves mainly memorizing and following rules. McKnight et al., (1987) found similar results in the U.S. data from the Second International Mathematics Study (grades 8 and 12).

Research on beliefs has been highlighted by the results of research on problem solving. As Schoenfeld (1985) and Silver (1985) have pointed out, students' beliefs about mathematics may weaken their ability to solve nonroutine problems. If students believe that mathematical problems should always be completed in five minutes or less, they may be unwilling to persist in trying to solve problems that may take substantially longer for most students. Nevertheless, this kind of belief has been generated out of the typical classroom context in which students encounter mathematics. There is nothing wrong with the students' mechanism for developing beliefs about mathematics (D'Andrade, 1981; Schoenfeld, 1988); what needs to be changed is the curriculum (and beyond that, the culture) that encourages such beliefs.

Another important area of research on beliefs comes mainly out of the work on gender differences in mathematics education. Most of the data has come from studies that used the Fennema and Sherman (1976) scales, especially the scale on the perceived usefulness of mathematics. Fennema (1989), in summarizing this research, notes that males in general report higher perceived usefulness than females. Other scales (for example, mathematics as a male domain) also deal with beliefs about mathematics. These kinds of beliefs are important both for gender differences in mathematics achievement and for the related differences between females and males in affective responses to mathematics (see Leder, Chapter 24 on gender, this volume).

Students' beliefs about mathematics do change as students grow older. Kouba and McDonald (1987) report that students in the elementary school grades tend to think that mathematics cannot be easy. In the view of these students, if it is easy, it is not mathematics. As the children grow older, and as the material that they have learned (e.g., counting) seems increasingly elementary, they change their beliefs about what mathematics is to accommodate the notion that mathematics must be hard and unfamiliar. These students also see mathematics primarily as doing something, usually something algorithmic; the connections to the typical elementary school mathematics curriculum are relatively direct.

Stodolsky (1985) describes how beliefs about mathematics influence how students (and teachers) perform in elementary school mathematics classrooms, especially as compared to social studies classrooms. In social studies, students are much more likely to work in groups, to develop their research skills, and, in general, to work on tasks that are compatible with the development of higher-order thinking skills. In contrast, in mathematics classrooms students spend a lot of time alone doing "seatwork." Other writers have noted how students view mathematics as a skill-oriented subject, and how such limited views of the discipline lead to anxiety about mathematics (Greenwood, 1984) and more generally interfere with higher-order thinking in mathematics (Garofalo, 1989).

Most research (not including evaluation projects) on beliefs about mathematics as a discipline have relied primarily on classroom observations and interviews with students. Kloosterman and Stage (1989), however, have developed a questionnaire to measure students' beliefs about mathematics and about themselves. Data gathered through this instrument (and others) should provide a useful complement to data from qualitative studies that use interviews and observations.

Beliefs about Self. Research on self-concept, confidence, and causal attributions related to mathematics tends to focus on beliefs about the self. These beliefs about self are closely related to notions of metacognition, self-regulation, and self-awareness (Corno & Rohrkemper, 1985). Some aspects of beliefs about self have been researched quite thoroughly, especially in the area of research on gender differences. Other aspects are only beginning to be investigated. For example, a substantial amount of data has been gathered on differences between males and females in levels of confidence in doing mathematics, but very little on how children develop a belief in themselves as autonomous learners (Fennema & Peterson, 1985).

Major evaluation studies provide useful background data on some beliefs about self. National assessment data from the United States (Dossey et al., 1988) asked children in grades 3, 7, and 11 if they were good at doing mathematics. The percentage of students who responded positively dropped from 65% in grade 3 to 53% in grade 11, providing at least a rough measure of what happens to levels of confidence as students progress through school.

Research on self-concept and confidence in learning mathematics indicates that there are substantial differences between males and females in these areas. Reyes (1984) and Meyer and Fennema (1988) summarize the relevant literature. In general, males tend to be more confident than females, even when females may have better reasons, based on their performance, to feel confident. The interaction of confidence and mathematical performance, especially in the area of nonroutine problem solving, is an important research topic. It seems likely that success in problem solving will engender a belief in one's capacity for doing mathematical problems, leading to an increase in confidence, which correlates positively with achievement in mathematics (Fennema, 1989).

The notion of confidence in oneself as a learner of mathematics has also been investigated under the rubric of self-concept (Shavelson & Bolus, 1982). Literature on the notion of self-concept in relation to mathematics learning has been reviewed by Reyes (1984). The most important implications of research on self-concept will probably come from its connections to metacognition, self-regulated learning, and intrinsic motivation to learn (Corno & Rohrkemper, 1985; Kilpatrick, 1985; Schoenfeld, 1987b). Such research becomes quite complicated to sort out since the general notion of self-concept and the more specific notion of mathematical self-concept appear to be related but distinct (Marsh, 1986; Reyes, 1984).

Another set of beliefs about self has been investigated quite thoroughly in the context of causal attributions for success and failure. Although there are several antecedents of this work and many different applications of the ideas, the central themes are well explicated in a recent reformulation of the theory by Weiner (1986). The three main dimensions of the theory deal with the locus (internal or external), the stability (for example, ability versus effort), and the controllability of the causal agent. For example, a student who fails to solve a mathematics problem could say that the problem was too hard—a cause that is external, stable, and uncontrollable by the student. A student who succeeds in solving a problem might attribute that success to effort—a cause that is internal, unstable, and controllable.

The nature of the attributions of female and male students has been an important theme in recent research in mathematics education, and the results of this research provide some of the most consistent data in the literature on the affective domain. For example, males are more likely than females to attribute their success in mathematics to ability, and females are more likely than males to attribute their failures to lack of ability. In addition, females tend to attribute their successes to extra effort more than males do, and males tend to attribute their failures to lack of effort more than females do. The differences in participation in mathematics-related careers appear to reflect these gender differences in attributions (Fennema, 1989; Fennema &

Peterson, 1985; Meyer & Fennema, 1988; Reyes, 1984; Wolleat, Pedro, Becker, & Fennema, 1980).

Larger issues that are related to beliefs about self include much of the literature on motivational issues (Dweck, 1986). Much of this research is not done in a mathematical context, although there are some studies that relate the motivation and the confidence of mathematics learners (e.g., Kloosterman, 1988). The general literature, however, is full of overlapping concepts like self-efficacy (Bandura, 1977), learned helplessness (Dweck, 1986), and motivation (Covington, 1983, and many other authors). Later in the chapter the literature on each of these areas will be discussed in more detail.

Beliefs about Mathematics Teaching. So far our discussion has concentrated on students' beliefs about mathematics and about themselves. But there is a corresponding set of beliefs that students and teachers hold about mathematics teaching that are also important to the study of affect in mathematics education. There have been a number of important studies of teachers' beliefs about mathematics and mathematics teaching (for example, Barr, 1988; Grouws & Cramer, 1989; Marcilo, 1987; Peterson, Fennema, Carpenter, & Loef, 1989; Sowder, 1989; Thompson, 1984), and current recommendations for a research agenda on mathematics teaching suggest that more work be done in this area (Cooney, Grouws, & Jones, 1988). There are also more general studies of teachers' beliefs (Wittrock, 1986), including teachers' beliefs about instruction (Eisenhart, Shrum, Harding, & Cuthbert, 1988; Peterson & Barger, 1985), as well as teachers' attributions (Prawat, Byers, & Anderson, 1983); these latter investigations are more directly related to affective factors in classroom instruction. However, most of the research along these lines does not deal specifically with mathematics teaching, and we have little information on students' beliefs about mathematics instruction.

Beliefs about the Social Context. Recent research on mathematics learning has given increased attention to the social context of instruction (Cole & Griffin, 1987) and more generally to cultural issues in mathematics education (Bishop, 1988; Lave, 1988; Newman, Griffin, & Cole, 1989; Saxe, 1990). Students' beliefs about the social context appear to be another area that is closely related to affective concerns. For example, Cobb, Yackel, and Wood (1989) found that explicit teaching of social norms in a primary classroom was directly related to the kinds of affective reactions that the students expressed. Similarly, at the secondary level, Grouws and Cramer (1989) found that the classrooms of effective teachers of mathematical problem solving were characterized by a supportive classroom environment where social norms encouraged students to be enthusiastic and to enjoy mathematical problem solving. From a broader perspective, the social context provided by the school and the home can also have an effect on students' beliefs. Parsons, Adler, and Kaczala (1982), in their study of parental influences on students' attitudes and beliefs, noted that the affective reactions of students (particularly females) often reflect social norms as expressed by the parents. Research in crosscultural settings also points out the influence of the broader social context (Stevenson, 1987; Stevenson, Lee, & Stigler, 1986; Stigler & Mao, 1985).

In summary, research on beliefs and their influence on students and teachers has been an important theme in investigations of learning and instruction in mathematics. Some of this research is directly connected with affective issues (for example, confidence), but much of it is not. Since beliefs provide an important part of the context within which emotional responses to mathematics develop, we need to establish stronger connections between research on beliefs and research on emotions in the context of mathematics classrooms. More generally, research in mathematics education needs to develop a more coherent framework for research on beliefs, their relationship to attitudes and emotions, and their interaction with cognitive factors in mathematics learning and instruction.

Attitudes

Research on attitudes towards mathematics has a relatively long history. For recent reviews and analyses, see Haladyna, Shaughnessy, and Shaughnessy (1983), Kulm (1980), Leder (1987), and Reyes (1984). Many of these reviews use attitudes as a general term that includes beliefs about mathematics and about self. In this paper attitude refers to affective responses that involve positive or negative feelings of moderate intensity and reasonable stability. Examples of attitudes toward mathematics would include liking geometry, disliking story problems, being curious about topology, and being bored by algebra. As Leder (1987) and others have noted, attitudes toward mathematics are not a unidimensional factor; there are many different kinds of mathematics, as well as a variety of feelings about each type of mathematics.

Attitudes toward mathematics appear to develop in two different ways. One was referred to earlier—attitudes may result from the automatizing of a repeated emotional reaction to mathematics. For example, if a student has repeated negative experiences with geometric proofs, the emotional impact will usually lessen in intensity over time. Eventually the emotional reaction to geometric proof will become more automatic, there will be less physiological arousal, and the response will become a stable one that can probably be measured through use of a questionnaire. A second source of attitudes is the assignment of an already existing attitude to a new but related task. A student who has a negative attitude toward geometric proof may attach that same attitude to proofs in algebra. To phrase this process in cognitive terminology, the attitude from one schema is attached to a second related schema. Abelson (1976) and Marshall (1989) provide a more detailed discussion of theoretical issues related to the formation of attitudes.

There have been a large number of studies of attitudes toward mathematics over the years; the topic seems to be particularly popular among dissertation writers. The review articles listed at the beginning of this section include more extensive descriptions of these studies. Only a few selected examples of relevant research will be mentioned in this section.

Most major evaluation studies provide data on attitudes toward mathematics. National assessment data (Dossey et al., 1988) illustrate the major results: there is a positive correlation between attitude and achievement at all three grade levels assessed (grades 3, 7, and 11), but the percentage of students who say they enjoy mathematics declines from 60% in grade 3 to 50% in grade 11. Similar results appear in the Second International Mathematics Study (McKnight et al., 1987) and in studies in other countries (Foxman et al., 1982; Leder, 1987; McLean, 1982).

Research suggests that neither attitude nor achievement is dependent on the other; rather, they interact with each other in complex and unpredictable ways. For example, data from the Second International Mathematics Study indicate that Japanese students had a greater dislike for mathematics than students in other countries, even though Japanese achievement was very high (McKnight et al., 1987). Recent studies exhibit a growing appreciation for the complexity of the affective domain; the original attempts to measure attitude toward mathematics seem exceptionally primitive, given our current knowledge and experience in the area (Leder, 1987).

Some studies have assessed attitude toward various subdomains that are part of or related to mathematics. For example, McKnight et al. (1987) report data on how much students like to use calculators, as well as how they feel about checking answers and memorizing rules in mathematics. As expected, students generally like to use calculators, but dislike memorizing; checking answers falls somewhere in between. Corbitt (1984) interviewed students in the eighth grade regarding how much they liked 15 different mathematical topics. The results showed some differences from students' liking of mathematics in general; for example, students reported being bored with the typical review of computation in the eighth grade.

Research on attitude provides a broad and rather indistinct picture of a limited range of affective responses to mathematics. Researchers in the area generally limit themselves to various kinds of questionnaires (Henerson, Morris, & Fitz-Gibbon, 1978), but there are useful examples of studies that use interview data as well (Corbitt, 1984; Marshall, 1989). As the research methodology becomes more flexible and more studies use multiple research methods, including interviews rather than just questionnaires, we can expect research on attitude to make new contributions to the field of mathematics education.

In the literature it is difficult to separate research on attitudes from research on beliefs. If attitudes develop out of emotional responses, as we hypothesize, it should be possible to analyze attitudes in terms of the corresponding emotional responses. For example, if students get frustrated with computer-assisted instruction, they may develop a negative attitude toward computers; if students find it emotionally satisfying to work with their friends on mathematical problems, they may develop a positive attitude toward small-group instruction in mathematics. Further progress in research on attitudes should profit from a more careful analysis of emotional responses in mathematics education. Although Mandler's (1984) theory does not yet include any major attempt to categorize the various emotional responses to mathematics, the efforts by Ortony et al. (1988) to build a general classification system for the emotions may eventually provide some help in this area.

Emotions

The emotional reactions of students have not been major factors in research on affect in mathematics education. This lack

of attention to emotion is probably due in part to the fact that research on affective issues has mostly looked for factors that are stable and can be measured by questionnaires. To phrase this observation in another way, most research in the past has looked at products, not at processes, and at beliefs and attitudes rather than emotions. However, there have been a number of studies that have looked at the processes involved in learning mathematics, and these studies have sometimes paid attention to the emotions. Certainly the current trend toward detailed studies of a small number of subjects allows the researcher to be aware of the relationship between the emotions and cognitive processing; such an awareness was not possible in traditional large-scale studies of affective factors. In this section, we review briefly a few of these studies.

One of the early studies of problem-solving processes was conducted by Bloom and Broder (1950). In this work they noted how students' engagement in the task led them into periods of tension and frustration, especially when they felt that their attempts to reach a solution were blocked. Once the block had been overcome, the students would relax and report very positive emotions. This study was conducted before the current focus on cognition became common, and it is justifiably recognized as an early exemplar of research on cognitive processes. It also provides a useful model for integrating research on cognition and affect.

Reports of strong emotional reactions to mathematics do not appear in the research literature very often. An important exception is the work of Buxton (1981). His research deals with adults who report their emotional reaction to mathematical tasks as panic. Their reports of panic are accompanied by a high degree of physiological arousal; this arousal is so difficult to control that they find it disrupts their ability to concentrate on the task. The emotional reaction is described as fear, anxiety, and embarrassment, as well as panic. Buxton interprets these data in terms of Skemp's (1979) views of the affective domain, and suggests a number of strategies to change students' beliefs in order to reduce the intensity of the emotional response.

A number of other researchers have investigated factors that are related to the influence of emotions on cognitive processes in mathematics. Wagner, Rachlin, and Jensen (1984) report how algebra students who were stuck on a problem would sometimes get upset and grope wildly for any response that would get them past the blockage, no matter how irrational. On a more positive note, Brown and Walter (1983) discuss how making conjectures can be a source of great joy to mathematics students. In a similar way, Mason, Burton, and Stacey (1982) talk about the satisfaction of the "Aha!" experience in mathematical problem solving, and make suggestions about how students can be encouraged to savor and anticipate positive emotional experiences related to mathematics learning. Lawler (1981) also documents the positive responses that accompany that moment of insight when a child first sees the connections between two important ideas.

These observations on emotion and cognitive processing resulted from studies that were focused on cognitive rather than affective issues, and reports of students' emotional responses were frequently sidelights rather than highlights of the studies. However, there has been some research that has focused directly on the role of the emotions in mathematics learning.

McLeod, Metzger, and Craviotto (1989) report on the emotional reactions of expert and novice problem solvers, where the experts are research mathematicians and the novices are undergraduate students who are not mathematics majors. This study found that the emotional reactions to the frustrations and joys of solving problems are basically the same for each group. The experts, however, are better able to control their emotions than the novices. In another study that included emotional responses as an important component, Bassarear (1989) conducted extensive interviews with two college students over the course of a semester, observing the interaction of their beliefs, attitudes, and emotions with their performance in mathematics class. The data from these interviews suggest how emotional responses can play a significant role in students' learning of mathematics.

Although comments about emotion do appear in the research literature from time to time, it is fairly unusual for research on mathematics education to include measures of physiological changes that accompany the emotions. However, in a recent study Gentry and Underhill (1987) gathered data on muscle tension along with paper-and-pencil measures of anxiety toward mathematics. As one might expect, there was little correlation between the two measures, suggesting that traditional measures of anxiety may be quite different from the emotional responses that influence students in the classroom. Similar results were obtained by Dew, Galassi, and Galassi (1984), who compared physiological measures of heart rate and skin conductivity with data from paper-and-pencil measures of mathematics anxiety.

In summary, research on emotional responses to mathematics has been conducted, but it has never played a prominent part in research on the affective domain in mathematics. A major problem has been the lack of a theoretical framework within which to interpret the role of the emotions in the learning of mathematics. Mandler's (1984) theory should help to provide such a framework. The recent appearance of the volume by Harré (1986), with its constructivist approach to affective issues, should also provide a suitable theoretical framework for researchers who take a strong constructivist perspective. Similarly, the work of Case et al. (1988), which integrates neo-Piagetian thinking about cognition with a serious attempt to explain the development of emotion, will be of particular interest to researchers who take the perspective of developmental psychology. Although Case and his colleagues have so far concentrated on younger children and the development of jealousy and other emotions that are not specifically related to mathematics education, there is reason to expect that these ideas can be extended to older children and to a mathematical context. In conclusion, the available data from a variety of sources and a variety of theoretical perspectives suggest that careful observation of students, along with detailed interviews, should help researchers in their analysis of the emotional states of mathematics learners (McLeod, 1988)

RELATED CONCEPTS FROM THE AFFECTIVE DOMAIN

The research on beliefs, attitudes, and emotions that was outlined in the previous sections provides a general overview of research on the affective domain. There are, however, a number of other research topics—one could refer to them as mini-theories about parts of the affective domain—that have implications for mathematics education. The purpose of this section is to outline several of these areas, and to suggest how they could fit into the general rubric of beliefs, attitudes, and emotions as described above. Sometimes these topics have not been investigated in the context of mathematics education; nevertheless, they seem relevant to the field. We begin with those areas that have the most in common with traditional foci of the affective domain in mathematics education, like confidence and self-concept. Later we shall also deal with relatively new concepts from research in mathematics education, like metacognition and intuition, and suggest how those areas might be integrated more completely with research on affect.

Confidence

Confidence in learning mathematics has been studied at least since the days of the National Longitudinal Study of Mathematical Abilities (Crosswhite, 1972). Reyes (1984) provides an excellent review of the literature in this area. Although different studies have used varying methods to assess confidence, in general it is reasonable to think of confidence as a belief about one's competence in mathematics.

Confidence correlates positively with achievement in mathematics, and the relationship is generally quite strong, with correlation coefficients of greater than 0.40 appearing in studies at the secondary school level (Reyes, 1984). Confidence is also related to elective enrollment in mathematics courses, and has been used frequently in investigations of gender differences in mathematics (Fennema, 1989; Linn & Hyde, 1989). Although some data have suggested that confidence is also related to patterns of classroom interaction between students and teachers, more recent work in this area indicates that the differences are not as consistent as expected (Hart, 1989a).

Instruments to measure confidence vary greatly in complexity. In the National Assessment data, students were simply asked if they were good at mathematics (Dossey et al., 1988); on the other hand, Fennema and Sherman (1976) developed a scale to assess confidence, following approved test development procedures for validating items, assessing reliability, and so forth. Other researchers have fallen in between these two ends of the spectrum; many have developed more general scales, including items related to confidence, but have not separated confidence from other kinds of affective factors.

Some recent studies of confidence continue to produce patterns of results established in earlier research efforts. Kloosterman (1988) investigated the correlation between measures of confidence and other affective variables, like motivation and causal attributions, with grade 7 students. As predicted, these kinds of variables are correlated; however, the usefulness of these kinds of conclusions for theoretical development and for practical applications appears to be limited (Mandler, 1972). Mura (1987) investigated differences in level of confidence, career plans, and gender with college students. In line with other studies, women tended to be less confident than men, and fewer women planned to take advanced mathematics than men. In studies with students at the elementary (Newman & Wick, 1987) and secondary school levels (Newman, 1984), data indicated that boys were more confident than girls in estimation tasks at the secondary school level; however, at the elementary school level, the levels of confidence of boys and girls on these estimation tasks did not differ significantly.

Students' beliefs about their competence in mathematics are an important affective factor in mathematics classrooms. Future research on confidence needs to take into account the complete mosaic of mathematical beliefs, rather than just studying one such belief in isolation. For example, if students feel confident about doing mathematics and believe that mathematics is nothing more than doing computational exercises, their beliefs about mathematics as a discipline provide a different perspective regarding their statements of confidence. Making sense of confidence as a variable in mathematics education will require a more complete picture of the affective domain than is presently found in most studies.

Self-concept

Self-concept can be thought of as a generalization of confidence in learning mathematics (Reyes, 1984). Substantial effort has gone into research on both general self-concept and academic self-concept, and the relation of each to academic achievement (Byrne, 1984). Work by Shavelson and Bolus (1982) is representative of the area. Shavelson and his colleagues assume that students develop a general self-concept, which can be analyzed into various components, including an academic self-concept. The relationships between general and academic self-concepts, and between academic self-concept and self-concept in specific subject-matter areas like mathematics, are still being debated; however, Byrne (1984) does report some support for a hierarchical system through which more specific self-concepts are combined to make up the general self-concept. More recently, Marsh (1986) has gathered data on self-concept by discipline, and found that mathematics self-concept and verbal self-concept were not correlated, but that mathematics self-concept was correlated with achievement in mathematics, just as verbal self-concept was correlated with verbal achievement.

Since the relationship of self-concept to achievement is consistently positive, continued research on such beliefs about self seems appropriate. Since studies of self-concept have generally used only quantitative methods, there is much work that could be done in qualitative studies to further our understanding of how differences in self-concept are related to differences in mathematical performance. For example, students who have a poor self-concept in mathematics may need help in changing their beliefs about mathematics as a discipline, as well as in seeing themselves as competent learners of the subject.

Self-efficacy

A variation of self-concept is the notion of self-efficacy (Bandura, 1977). As Schunk (1984) points out, notions of self-efficacy are related to decisions about which activities students choose to participate in, how much effort they expend, and how long they persist in those activities. Norwich (1987) investigated the relationship between self-efficacy and performance in mathematics among primary school students in England, as well as the relationship between measures of self-efficacy and self-concept. The data suggest that neither self-efficacy nor self-concept were particularly significant predictors of achievement. In a study of mathematics students at the college level, Hackett and Betz (1989) found that self-efficacy was a good predictor of students' choice of major and that it also correlated positively with achievement and attitudes toward mathematics. In particular, Hackett and Betz (1989) claimed that self-efficacy was a better predictor of college major than measures of achievement. Although the data on self-efficacy are interesting, it is difficult to sort out why self-efficacy as a construct should be much more successful as a predictor than mathematical self-concept or confidence in learning mathematics. Again, a broader and more integrated view of various beliefs about self may help to make research of this type more meaningful to the field.

Mathematics Anxiety

The study of mathematics anxiety has probably received more attention than any other area that lies within the affective domain. Hembree (1990), in a meta-analysis based on 151 studies, confirmed that mathematics anxiety is related to poor performance in mathematics, and that a variety of treatments are effective in reducing mathematics anxiety and in improving performance. Treatments that involve systematic desensitization and relaxation training were found to be most effective. Efforts to change beliefs about mathematics were also of some help. From the data that are already available (Berebitsky, 1985; Gattuso & Lacasse, 1987; Hembree, 1990), it seems reasonable for researchers to propose relatively complete models of how beliefs, attitudes, and emotions are involved in the development of mathematics anxiety, and to develop treatment programs that deal with the issue in a more comprehensive way. Meanwhile, researchers are able to provide helpful suggestions to teachers (Brush, 1981), and some research has tested alternative instructional formats that may be more appropriate for students who report high levels of mathematics anxiety (Clute, 1984).

Although there has been considerable progress in investigating mathematics anxiety, the concepts underlying the research continue to be murky, and the terminology remains unclear. As Hart (1989b) points out, anxiety has sometimes been characterized as fear, a "hot" emotion, and sometimes as dislike, an attitude. Researchers have often failed to distinguish between Spielberger's notions of state and trait anxiety (Spielberger, Gonzalez, & Fletcher, 1979). The relationship of mathematics anxiety to performance in mathematics is sometimes difficult to demonstrate (Gliner, 1987; Llabre & Suarez, 1985; Mevarech & Ben-Artzi, 1987); studies that attempt to clarify the relationships between various measures of mathematics anxiety, test anxiety, and related concepts (Dew et al., 1984; Ferguson, 1986; Hendel, 1980; Richardson & Woolfolk, 1980; Rounds & Hendel, 1980) report only modest success. These kinds of studies continue to emphasize measurement issues rather than theory building. Conceptions of mathematics anxiety are often difficult to separate from test anxiety (Sarason, 1980, 1987) as it applies to mathematics; moreover, test anxiety appears to provide the main source of theoretical support for much of the research on mathematics anxiety.

Current research on mathematics anxiety, with its emphasis on statistical methodology and correlational analyses of related concepts, remains subject to the criticisms that Mandler (1972) made many years ago; significant correlations do not imply significant increases in our knowledge of the field, especially given the difficulties involved in building instruments for the affective domain and the lack of an adequate theoretical foundation for the work. Some studies have attempted to provide a stronger cognitive orientation for research on anxiety (Hunsley, 1987), to build a constructivist foundation for such work (Carter & Yackel, 1989), to investigate the influence of mathematics anxiety on the performance of elementary school teachers (Bush, 1989), and to separate more intense, emotional aspects of anxiety from the less intense, attitudinal facets (Wigfield & Meece, 1988). In spite of these significant efforts, this research area continues to make slow progress.

The difficulties involved in dealing with mathematics anxiety have led some researchers to propose alternative approaches that are based on Freudian psychology. Ginsburg and Asmussen (1988), for example, discuss the extreme anxieties that some people develop regarding mathematics, and argue that the techniques of depth psychology will be needed to treat such problems successfully. Nimier (1977), in an intriguing discussion of the impact of fears and defense mechanisms on mathematics students, provides a Freudian interpretation of certain patterns of behavior that are common in mathematics classrooms. In related work, Legault (1987) discusses how these Freudian interpretations of students' behavior have special implications for gender differences in mathematics education. Although studies of this type have not had a major impact on research in mathematics education, they may in the future yield important insights, especially regarding people who suffer from extremely negative reactions to mathematics (Buxton, 1981).

Causal Attributions

The work on causal attributions related to mathematics learning has been quite extensive. The theory, outlined earlier in this chapter, has a relatively strong theoretical foundation (Weiner, 1986). Again, as with research on confidence in learning mathematics, some of the most interesting results have dealt with gender-related differences. For example, males are more likely to attribute their success to ability than females, and females are more likely to attribute their failure to lack of ability than males. Several summaries of the research on causal attributions that is related to mathematics learning are available in the literature (Fennema, 1989; Meyer & Fennema, 1988; Reyes, 1984).

There are a number of recent studies of causal attributions among mathematics students. Kloosterman's (1988) work on relationships between attributions and confidence, discussed earlier, is a good example of the kind of research that has been done in this area. He found that students who were high in confidence were also likely to attribute success to ability and failure to effort. These interrelationships among concepts suggest new avenues for investigating relevant variables within the context of a well-developed theory. In other studies, Choroszy, Powers, Cool, and Douglas (1987) extended the work on causal attributions to community college students in American Samoa; Heckhausen (1987) analyzed various relationships between attributions and achievement; and Graham (1984) and Prawat et al. (1983) conducted research on teachers' attributions about student performance. None of these researchers provide any unexpected results, but they are all engaged in useful efforts to test and extend the theories of Weiner (1986).

Effort and Ability Attributions

Although Weiner (1986) appears to have the most complete theoretical perspective on issues related to attributions, considerable research has been done on effort and ability that is parallel to but not directly dependent upon Weiner's work. For example, Ames and Archer (1987) found that elementary school students who attributed success to effort were more likely to exhibit a mastery orientation, putting their emphasis on learning and understanding through hard work, meeting challenges, and making progress. Students who attributed success to ability were more likely to be interested in good grades than in understanding. In a review of the research on issues of effort and ability, Holloway (1988) compared data from Japan and the United States. Some of the major findings from this report are that effort is believed to be of primary importance in determining achievement in Japan, but ability is seen as the primary factor in the United States. Apparently, Japanese homes encourage task involvement in ways that promote effort attributions. Related work by Hess, Chang, and McDevitt (1987), Stevenson et al. (1986), and Stigler and Perry (1988) provides further support for the importance of cultural differences in effort and ability attributions.

Learned Helplessness

The psychological literature on learned helplessness is quite extensive, and in recent years the influence of this concept is being felt more directly in mathematics education research. Diener and Dweck (1978), in work with elementary school students, described a pattern of behavior called learned helplessness where students attributed failure to lack of ability. Such students tended to demonstrate a low level of persistence and to avoid challenges whenever possible (Dweck, 1986). The contrasting pattern, referred to as a mastery orientation, was characterized by students who made few attributions, but who concentrated on monitoring their performance. In general, mastery-oriented students saw intelligence as a growing collection of concepts and procedures that they were able to understand. Dweck and Bempechat (1983) link these two contrasting orientations (mastery versus learned helplessness) to students' beliefs about intelligence and the related attributions that the students make. Again, some of the most interesting investigations of these ideas appear in studies of gender-related differences (Parson, Meece, Adler, & Kaczala, 1982).

Motivation

Many of the studies that have been discussed in this paper have something to do with motivation. There are a great many ways to study motivation, and there has been a great deal written on the general topic (Ames & Ames, 1984; Bates, 1979; Boekaerts, 1988; Brophy, 1983; Corno & Rohrkemper, 1985; Covington, 1983; Covington & Omelich, 1985; Hatano & Inagaki, 1987; Maehr, 1984; Malone & Lepper, 1987; Morgan, 1984; Nicholls, 1984; Paris, Olson, & Stevenson, 1983; Pekrun, 1988; Stipek, 1984; Weinert, 1987). Although most of these authors do not deal with mathematics education directly, their work as a whole is certainly applicable to mathematics classrooms.

One of the difficulties with the work on motivation is the diffuse and disconnected nature of the field. In spite of the best efforts of leading researchers (Ames & Ames, 1984; Dweck, 1986; Snow & Farr, 1987), the field appears to be made up of often disconnected components dealing with such topics as achievement motivation, social motivation, extrinsic versus intrinsic motivation, fear of success, need for achievement, and so forth. According to Mandler (1989), part of the problem is that there is still no framework for research on motivation that fits comfortably into current research in cognitive psychology. Norman (1981), in his essay on cognitive science, suggested that motivation could be dealt with as a derived issue in cognitive science, where motivational factors could be explained through research on beliefs and emotions. Whether such an approach is appropriate or whether it would lead to any particular clarification in the field is yet to be determined. But clarity is obviously needed in the chaotic collection of research studies carried out under the general heading of motivation.

In summary, research areas like self-concept, causal attribution, and learned helplessness all help to complete our picture of the affective domain as it influences performance in learning and teaching. However, these research areas can be strengthened if they can be related to the complete realm of research on affect. In particular, researchers need to clarify the level of affective intensity that students are reporting. For example, if students' confidence is being assessed, then the research needs to distinguish as carefully as possible between beliefs about competence and feelings of inadequacy. Similarly, in research on anxiety, we need to distinguish between intense emotional responses (panic or fear) and other negative but less intense responses (dislike or worry). If the research in these areas can be related more directly to research on beliefs, attitudes, and emotions, the level of intensity of the affective response should be more clear, thus contributing to the overall understanding of how the affective domain is related to mathematics learning and teaching.

TOPICS RELATED TO THE COGNITIVE DOMAIN

The position of this paper—and of Mandler's (1984) theory—is that the affective and cognitive domains are intimately linked. Affective responses do not occur in the absence of cognitive evaluations, according to the theory. Nevertheless, some topics seem more closely connected to the cognitive domain than others. This section describes a number of research topics that are closely identified with research on cognition, even though they have a strong connection to the affective domain.

Autonomy

In their analysis of gender-related differences in mathematics education, Fennema and Peterson (1985) developed a model that linked performance on tasks that require higher-order thinking skills in mathematics to beliefs, social influences, and autonomous learning behaviors (ALBs). These ALBs include activities like independent thinking about a problem and willingness to persist in problem solving. Such ALBs are hypothesized to be a mediating link between students' knowledge and beliefs on the one hand and mathematical performance on the other (Fennema & Leder, 1990). Descriptions of successful teachers of mathematics (Cobb et al., 1989; Grouws & Cramer, 1989; Peterson & Fennema, 1985) place considerable emphasis on how teachers can help students develop this kind of autonomy.

Related to autonomy is Witkin's notion of field independence (Witkin & Goodenough, 1981). Although Witkin considered field independence and field dependence to be cognitive styles, rather than affective factors, the pattern of performance of field-independent students is quite similar to Fennema and Peterson's (1985) notion of autonomous learning behaviors. Given the importance of autonomy and independent thinking in the curriculum reform movement that is now so prominent in mathematics education (Commission on Standards for School Mathematics, 1989), it may be wise to return to earlier forms of these concepts and to conduct more detailed investigations of cognitive styles (Messick, 1987). One recent effort in this direction (Kelly-Benjamin, 1990) investigated differences between general and mathematical learning styles among high school seniors; results of a factor analysis suggest that students' mathematical learning styles are different from their general learning styles, and that these differences have implications for mathematics instruction.

Aesthetics

Although mathematicians often remark on the aesthetic dimensions of mathematics, the study of aesthetics has not received much attention in the mathematics education community (Dreyfus & Eisenberg, 1986). Mathematicians like Poincare and Hadamard have been quite interested in how aesthetic factors have influenced the development of mathematical thinking; it seems appropriate that the topic should receive more attention from researchers in mathematics education.

In a study of expert problem solvers who were mathematicians, Silver and Metzger (1989) concluded that expertise involved taste as well as competence. Aesthetic factors clearly influenced the problem solver's emotional reaction to the problem, which frequently involved recognition of the beauty and elegance of the problem or its solution. In addition, aesthetic factors were related to the problem solvers' monitoring of their solution strategies and cognitive processing.

This aesthetic monitoring provided an interesting link between metacognitive processing and affective responses to problem solving. The study of aesthetic influences on mathematical performance appears to be an important issue in the development of expert problem solvers, and it certainly deserves more attention in the curriculum than it currently receives (Dreyfus & Eisenberg, 1986).

Intuition

Intuition, like aesthetics, plays an important role in mathematicians' discussions of mathematical thought. Recently this topic has received considerable attention in two books (Fischbein, 1987; Noddings & Shore, 1984), both of which emphasize mathematics in their analysis. Intuitive knowledge in mathematics is knowledge which is self-evident, which carries with it characteristic feelings of certitude, and which goes beyond the facts that are available (Fischbein, 1987; Noddings & Shore, 1984). Some authors have tried to explicate the role of intuition in rational number learning (Kieren, 1988) and other arithmetic concepts (Resnick, 1986). Clearly there is much more research that could be done, particularly on the teaching and learning of mathematical problem solving, to develop an understanding of the role of intuition and of how students' intuitions could be improved.

Metacognition

Metacognition has received substantial attention in research on mathematical problem solving in recent years (Campione, Brown, & Connell, 1989; Garofalo & Lester, 1985; Schoenfeld, 1987b; Silver, 1985) and the links between metacognition and the affective domain have been duly noted as well (Brown, Bransford, Ferrara, & Campione, 1983; Garner & Alexander, 1989; Lawson, 1984; McLeod, 1988; Prawat, 1989). Lester and his colleagues (Lester et al., 1989) have been the most specific about the relationships between metacognition and affective factors like confidence and interest. In a more general setting, Weinert and Kluwe (1987) have published a volume devoted to the analysis of the relationships between metacognition and motivation. The task of specifying the ways in which metacognitive processing interacts with the affective domain is difficult; however, substantial progress is being made in understanding the area and in deriving implications for instruction, particularly in the area of problem solving (Lester et al., 1989). Some work has also been done on mathematics instruction in more general settings. For example, Newman and his colleagues (Newman, 1990; Newman & Goldin, 1990) have investigated how students regulate their own behavior in the context of seeking help in the mathematics classroom. Their results, which suggest that willingness to seek help is constrained by students' beliefs about self, provide a nice example of how to integrate research on beliefs, attitudes, and metacognition in order to analyze a specific issue related to mathematics instruction.

Social Context

The relationship of affective factors to mathematics learning and teaching is always influenced by the social context. The study of these contextual factors is receiving increased attention, particularly because of their relationship to issues of gender and ethnicity (Cole & Griffin, 1987). The role of the social context is also receiving much more attention from those in cognitive science who would like to see their research have a greater impact on real-world classrooms (Brown, Collins, & Duguid, 1988). This emphasis on the role of contextual factors has come in part from anthropology, where studies of learning in settings outside of school have led to new insights (Lave, 1988). The applicability of these insights to school settings is often unclear; for example, the notion of apprenticeships for mathematics students (as well as for future African tailors) has considerable appeal, but it is not clear how such a notion could be implemented on a broad scale in school mathematics.

The analysis of the social context has been an interest of psychologists as well as anthropologists. Saxe (1990) has studied the development of mathematical thinking with an emphasis on the interaction of culture and cognition; many of his examples are taken from investigations of how some Brazilian children develop mathematical skills through their work as candy sellers. In an example of work from another area, Magnusson (1981) provides an analysis of the characteristics of situations that are important to psychology. He includes complexity, clarity, tasks, rules, roles, physical settings, other persons, expectancies, affective tones, and emotions in his list of situational characteristics. There are many difficulties involved in making sense of a psychology of situations; nevertheless, the area presents an alternative approach to understanding the social context of teaching and learning.

Another broad approach to the issue of social context is provided by the work of Bishop (1988) on mathematical enculturation. His analysis of mathematical culture puts considerable emphasis on beliefs and attitudes, and he argues for a curriculum that puts appropriate emphasis on inducting students into the culture of our discipline. In a related effort to shed light on the role of the social context in learning, Newman et al. (1989) argue that cognitive change is as much a social as an individual process, suggesting that researchers need to focus more on the context and the social interaction among learners. From a somewhat different perspective, Cocking and Mestre (1988) and Orr (1987) look at cultural influences on learning mathematics that cause particular problems for students who are members of linguistic (or other) minority groups. Just as affective factors are particularly important to gender-related differences in mathematics performance (Fennema, 1989), it seems reasonable to hypothesize that affective factors are particularly important to differences in performance between groups that come from different cultural backgrounds.

Research that has been done in this area indicates the significant role of the family in students' beliefs and attitudes toward schooling in general and toward mathematics in particular (Parsons, Adler, & Kaczala, 1982; Stevenson et al., 1986; Stigler & Perry, 1988). The cross-cultural comparisons in these and other studies are especially interesting. Mathematics education in the United States certainly can learn much from careful analyses of classrooms in other countries, and more research on issues related to the social context of instruction should prove to be very helpful (Research Advisory Committee, 1989).

Technology

One aspect of the social context of learning and teaching is the presence of technology in the classroom. Although it is possible to utilize some kinds of technology in the classroom without changing anything in a substantive way (as computer-based drill and practice programs have demonstrated for years), other kinds of technological advances should have a significant impact on the social context and the affective environment. Kaput (1989), in a stimulating analysis of the impact that computers could have on classroom instruction, notes that students who learn in a computer environment could have quite different affective responses to learning tasks than students who do not. For example, students who use computers can discover their own errors and correct them independently, rather than being corrected by a teacher or fellow students.

The computer has become an important object in our culture, and considerable research has been done on our reactions to computers. Turkle (1984) presents an interesting analysis of how children's views of the computer develop, from the early stage where young children are still trying to figure out if the computer is alive or not, to a later stage where the computer is an object to be mastered or an object that can reflect one's own identity. Other investigators have written about attitudes toward computers (Collis, 1987; Gressard & Loyd, 1987; Swadener & Hanafin, 1987) using traditional methods and instruments. Collis and Williams (1987) investigated cross-cultural differences between Chinese and Canadian adolescents in their attitudes toward computers, noting some differences between cultures, as well as between females and males. It seems likely that technology can play an important role in changing beliefs about mathematics and possibly even in improving attitudes toward mathematics; more research and development along these lines seems appropriate, particularly studies that take affective factors into account.

Research on topics that fall in between the purely cognitive and strictly affective areas is especially important to the field of mathematics education. These topics provide a natural link between research on affect and cognition, a link that we explore further in the next section.

THEORIES AND METHODS FOR RESEARCH ON THE AFFECTIVE DOMAIN

Most research on the affective domain has followed the traditional paradigm of quantitative research, and as Fennema (1989) points out, this approach has produced valuable information on the affective domain. In recent years, however, research in the cognitive domain has made successful use of a variety of qualitative as well as quantitative techniques. Such a combination of techniques seems appropriate for the affective domain as well. Howe (1988) notes that many researchers have an ideological commitment to one set of research methods, assuming that purity of method is necessary to development of theory. However, as Howe (1988) points out, there is no convincing evidence that qualitative and quantitative methods are incompatible. Given the nature of research on the affective domain, it seems likely that a variety of reasonable research methods have a chance to make a contribution to the field, as long as the data are interpreted intelligently.

A number of researchers have discussed the strengths and weaknesses of qualitative methods in research in education. For example, Firestone (1987) and Jacob (1987) present a general analysis of the problems, and the classic work of Ericsson and Simon (1980) deals with the use of verbal reports as data. In addition to these general analyses, some authors have dealt specifically with issues related to doing qualitative research in mathematics education. Eisenhart (1988), for example, provides an analysis of ethnographic methods in the specific context of research on mathematics teaching and learning, and Ginsburg and his colleagues (Ginsburg, 1981; Swanson, Schwartz, Ginsburg, & Kossan, 1981) discuss the use of clinical interview methods in mathematics education.

If researchers are to make progress in building theory and gathering relevant data about the role of the affective domain in the learning and teaching of mathematics, they need to provide data on a wide range of issues. Some of these issues (for example, beliefs and attitudes) can be analyzed through the use of traditional quantitative techniques, but qualitative data will add substantially to the completeness of our understanding of these issues. Measures of emotional reactions to mathematics can be done quantitatively (for example, measures of heart rate), but it seems much more natural in the context of mathematics classrooms to investigate such issues through studies that use qualitative techniques. For example, college students who were asked to draw a graph that represented their emotional reactions during a problem-solving episode were able to describe the "highs" and "lows" that they felt at various points, and to specify some of the reasons for their positive and negative feelings (McLeod, Craviotto, & Ortega, 1990); presumably, ways could be found to obtain similar information from younger students as well. Having students keep journals where they write on a regular basis about mathematics can also provide data on affective responses (Adams, 1989). In addition, if research is going to help us understand the role of affect in mathematics learning and teaching, studies of affect must be integrated with studies of cognition. Most research on learning ignores affective issues even when they are quite pertinent. For example, studies of students who are attempting to solve nonroutine problems are very likely to involve fairly intense affective responses (McLeod et al., 1989), and researchers who fail to gather data on these responses are missing an important characteristic of student performance.

Integrating affect into cognitive studies of mathematics learning would improve research on both cognition and affect. The following sections describe a variety of studies of learning and teaching which attempt to use both quantitative and qualitative methods, and which integrate research on affect and cognition.

Integrating Research on Affect and Learning

Research on learning with young children often provides opportunities to include affect in studies that are designed primarily to study cognitive issues. For example, Marshall (1989) reports on the affective reactions of sixth-grade students to mathematical story problems. Although the main purpose of the research was to investigate children's development of schemas for story problems, the interviewer also encouraged students to verbalize their affective reactions to the problems. Given this opportunity to discuss their feelings in a supportive environment, many children responded quite freely. Some of the children had rather intense emotional reactions, including a few who discovered something new about mathematics during the solution of the problems and who were delighted with their new knowledge. A few others demonstrated negative reactions to the problems, including one child who reported a rapid heartbeat as well as general discomfort and fear during the interview. In this case the interviewer ended the questions and spent some time reassuring the child instead. The source of this child's difficulty appeared to be the blockage that the child experienced in attempting to solve a nonroutine problem. Most children, however, used their verbal comments to express well established attitudes about story problems; these attitudes often revealed negative views toward mathematics or toward themselves as problem solvers. In Marshall's analysis, these emotional and attitudinal responses are attached to various components of the schemas involved in solving story problems. Marshall's procedures and analysis provide a good example of how a study with cognitive objectives can be expanded to include affective issues in a natural way.

In another study involving story problems, this time at the seventh-grade level, Lester et al. (1989) focused mainly on the role of metacognition in problem solving. In order to explain the context in which metacognitive decisions were made, the researchers also gathered data on affective factors, including children's beliefs about themselves as problem solvers and their attitudes toward mathematical problem solving. The data from this study support the view that the social context and the beliefs which it engenders have an important influence on both the students' affective responses as well as their metacognitive acts.

These studies of children's learning indicate that affect plays an important role in students' (i.e., novices') mathematical performance. In another study linking research on affect and cognition, Silver and Metzger (1989) gathered related data on the performance of experts. In their study Silver and Metzger interviewed research mathematicians and asked them to solve non-routine problems while thinking aloud. These interviews provide a rich source of data on the relationship between the affective domain and expertise in mathematical problem solving. A striking result of these data is the important role played by aesthetics in the monitoring and evaluation of expert performance. Rather than viewing problems from a utilitarian perspective, these experts spoke frequently about the elegance, harmony, and coherence of various solutions (or attempted solutions) to problems. The aesthetic aspects of the problem-solving experience were clearly linked to the experts' emotional responses, including their enjoyment of the problem.

In another study of experts, Taylor (1990) investigated the attitudes of mathematicians toward mathematics. This study is interesting due to its use of qualitative methods and its observations about gender-related differences in the development of mathematicians' careers. Through the use of qualitative methods, data are gathered on such factors as the development of confidence and willingness to persist, along with information on causal attributions for the mathematical success of the participants.

A variety of studies with a cognitive orientation have included affective factors as an important part of the research. For example, Peterson and her colleagues (for example, Peterson & Swing, 1982) have completed a series of studies that have included lengthy interviews with students who were asked to comment on affective as well as cognitive matters. Ginsburg and Allardice (1984) provide an intriguing view of how beliefs and emotion can contribute to the difficulties of young children who are unsuccessful in mathematics. These and other studies suggest that the usual methods for research on cognition can be adapted to include appropriate attention to the role of affect in the learning of mathematics.

Integrating Research on Affect and Teaching

Research on teachers and teaching in mathematics education seldom focuses on the affective factors that are frequently so visible in classrooms. This section will discuss several papers that do include affect, and do so in ways that go beyond the traditional attitude measures of previous years.

Cobb et al. (1989) have provided extensive data on how a teacher in a second-grade classroom dealt with emotions in the learning of mathematics. The data were obtained through careful observation of the classroom over an entire school year. The observers were able to document how the teacher worked with the students as they developed beliefs about mathematics. For example, the teacher was very explicit about the need to justify answers to mathematical problems, and about the importance of the justification. She was also explicit in her specifications of the acceptable kinds of behavior regarding solving mathematical problems. For example, she repeatedly emphasized the satisfactions that come with solving problems independently, and instructed students not to tell the answers to those who were still working on the problems. She was very clear in letting the students know that persistence in spite of frustration was important to success in solving problems. Since these classroom norms were a change from what had been expected of the students in other contexts, the changed norms were explicitly taught and practiced, and the teacher worked hard to see that these norms were adhered to. The result was a classroom where students showed a lot of satisfaction and enthusiasm for problem solving, and viewed themselves as autonomous learners.

In a study of pre-service teachers' estimation skills, Sowder (1989) investigated the teachers' tolerance for error, their attributions of success and failure, and other beliefs about mathematics and about themselves. Through extensive interviews with a sample of teachers, Sowder was able to create a profile of the beliefs that characterized good and poor estimators. Good estimators tended to have strong self-concepts with regard to mathematics, to attribute successes to their ability rather than just to effort, and to hold the belief that estimation was important. Poor estimators were more likely to have a weak self-concept in mathematics, to attribute successes to effort, and not to value estimation. The exceptions to these general patterns were interesting cases that showed how individual beliefs about mathematics could have an important impact on individual performance on estimation tasks. The general conclusions of this study provide some indication of the difficulties that will be involved in implementing recommendations to include estimation in the elementary mathematics curriculum. Clearly many teachers who are in the field, as well as many more who are on their way, do not hold beliefs about mathematics or about themselves that are compatible with the goals of the curriculum in terms of estimation skills (Sowder, 1989).

In a third study of teachers, Grouws and Cramer (1989) observed six expert teachers of problem solving at the junior high school level. The focus of this study was to identify the affective characteristics of the classrooms of these teachers during problem-solving lessons. Each teacher was observed five to seven times over the course of a semester. The observations revealed that students enjoyed problem solving, persevered on problem-solving tasks, and worked willingly on problem-solving assignments. Interviews with teachers helped with the identification of strategies that led to this positive affective climate in the classroom. For example, teachers appeared to work hard to establish a good relationship with students. They tended to be friendly rather than formal, and to share personal anecdotes about their own problem solving that illustrated their own strengths and weaknesses as problem solvers. In addition, the teachers established a system that held students accountable for their performance in problem solving. The system itself varies, although most teachers did pay attention to more than just the answer to the problem. Also, the teachers made frequent use of cooperative groups, and noted that small-group work tended to promote independence and to reduce feelings of frustration. Although no single factor appeared to be the cause of the success of these expert teachers, further research should provide indications of how these classroom characteristics contribute to the development of positive affective environments for problem solving.

Other studies of teaching also include affective factors in some detail. Tittle (1987), for example, is investigating how data on students' affective characteristics could be provided to teachers, thus making it possible for teachers to tailor instruction to students' affective as well as cognitive characteristics. This approach could be particularly important for gender differences in mathematics education (Tittle, 1986). Brophy (1983) has also written about strategies for improving the motivational climate in classrooms through strategies for encouraging enthusiasm for learning, reducing anxiety, and inducing curiosity. Thompson and Thompson (1989) discuss how students respond to a teacher's efforts to provide a supportive atmosphere for mathematical problem solving. Further research on affect in classrooms should provide more guidance on these topics.

These studies of teachers and teaching provide useful information on how beliefs, emotions, and attitudes play a significant role in mathematics instruction. They also demonstrate how affective factors can be incorporated in cognitive studies of teaching. If studies of affect are isolated, they do not have a significant impact on researchers who are primarily interested in cognition. If research on affect can be integrated into cognitive studies of teaching and learning, our knowledge of affective factors will be more likely to have an impact on instruction.

SUMMARY

Research on affect has been voluminous, but not particularly powerful in influencing the field of mathematics education. It seems that research on instruction in most cases goes on without any particular attention to affective issues. Similarly, there is little attention to research on affect in most curriculum development efforts, and apart from the topic of gender-related differences in mathematics, research on affect appears to have little impact on curriculum development or teacher education programs in mathematics.

A major difficulty is that research on affect has not usually been grounded in a strong theoretical foundation. When such research did occur within a theoretical framework, there was little connection between that framework and the theoretical foundations of cognitive research in mathematics education. To people who work on cognition, research on affect seemed to be a collection of generally unrelated clumps of studies on issues like motivation, attitude, and causal attributions. With no overriding themes or general framework, affective studies appeared to be unconnected with each other and quite separate from the interests of most cognitive researchers.

There are several things that can be done to improve this state of affairs. For example, researchers who focus on affective issues need to be more aware of how their research can contribute to research on cognition. Similarly, those who focus on cognition need to be more aware of research on affect and to include affective issues in a meaningful way in their studies. Too often researchers who focus on affect rely on measures of achievement (like standardized tests) that would not be acceptable to cognitive researchers. On the other hand, researchers with a cognitive orientation often ignore affect or treat the issue in a cavalier manner, using inappropriate instruments that happen to be convenient. Efforts to encourage the two groups of researchers to work together are just beginning, but the results have been encouraging so far (McLeod & Adams, 1989).

There are a number of research questions where collaboration of researchers with different perspectives is needed. For example, the cognitive and affective domains intersect in the area of beliefs (Schoenfeld, 1985), and researchers need to work together to map out this area more clearly, relating various beliefs to the cognitive processes of learners and teachers. This chapter has suggested one way of organizing research on beliefs in mathematics education, but more detailed and contrasting analyses are needed. Similarly, the domains of attitudes and emotions in the context of mathematics education need to be analyzed and clarified in order for research to proceed in an orderly fashion. Recent theoretical advances (Mandler, 1984; Ortony et al., 1988) should provide some help in this effort, particularly in determining how early emotional responses may be the source of later attitudes toward mathematics.

Another research topic of special importance is the relationship of affective responses to the development of higher-order thinking skills. Current efforts at curriculum reform place special emphasis on solving nonroutine problems, on applying mathematics in new situations, and on communication regarding mathematical problems. The novelty (as well as the difficulty) of such changes in the curriculum will cause more intense affective reactions for many students and teachers; research that investigates these more intense emotional responses is particularly important if the reform movement is to succeed. Those responsible for the changes in the mathematics curriculum during the 1950s and 1960s expected students and teachers to respond as enthusiastically to mathematical abstractions as mathematicians did; those involved in the current reform movement need to know more about the affective implications of the proposed changes for students and teachers, particularly those who think of themselves as being outside the mathematics community.

Another area where research on affect is particularly needed is in studies of the uses of technology to support mathematics instruction. The rapid improvements in technological support for mathematics education are leading to changes in the organization of classrooms and the definition of mathematical tasks. The advent of graphing calculators and symbol manipulation systems, for example, should eventually result in significant changes in what mathematics we teach. These changes in the curriculum will be accompanied by changes in beliefs about mathematics, and by opportunities for more positive emotional experiences in mathematics education. Research should help guide our efforts to increase positive affective responses to mathematics through the creative use of technology.

Finally, research on affective issues in mathematics education should develop a wider variety of methods. The debate over qualitative versus quantitative research methods appears to be almost over, and the time for intelligent use of multiple research methods that fit the research problems is here. The use of clinical interviews and detailed observations should provide the field with a deeper understanding of the role of affective issues in mathematical learning and teaching.

This chapter has presented an overall theoretical framework that is consistent with current studies of cognition in mathematics education. The division of the affective domain into beliefs, attitudes, and emotions seems appropriate for the interests and views of mainline cognitive research in mathematics education. If future research on affect can be linked more closely to the study of cognitive factors in learning, the affective domain should receive more attention in curriculum development, teacher education, and research on teaching and learning in our field.

References

Abelson, R. P. (1976). Script processing in attitude formation and decision making. In J. S. Carroll & J. W. Payne (Eds.), *Cognition and social behavior* (pp. 33–45). Hillsdale, NJ: Lawrence Erlbaum.

Adams, V. M. (1989). Affective issues in teaching problem solving: A teacher's perspective. In D. B. McLeod & V. M. Adams (Eds.), *Affect and mathematical problem solving: A new perspective* (pp. 192–201). New York: Springer-Verlag.

Aiken, L. R., Jr. (1970). Attitudes toward mathematics. *Review of Educational Research, 40,* 551–596.

Aiken, L. R., Jr. (1976). Update on attitudes and other affective variables in learning mathematics. *Review of Educational Research, 46,* 293–311.

Ajzen, I., & Fishbein, M. (1980). *Understanding attitudes and predicting social behavior.* Englewood Cliffs, NJ: Prentice-Hall.

Ames, C., & Ames, R. (1984). Systems of student and teacher motivation: Toward a qualitative definition. *Journal of Educational Psychology, 76,* 535–556.

Ames, C., & Archer, J. (1987). Mothers' beliefs about the role of ability and effort in school learning. *Journal of Educational Psychology, 79,* 409–414.

Bandura, A. (1977). Self-efficacy: Toward a unifying theory of behavioral change. *Psychological Review, 84,* 191–215.

Barr, R. (1988). Conditions influencing content taught in nine fourth-grade mathematics classrooms. *Elementary School Journal, 88,* 387–411.

Bassarear, T. (1989, April). *The dynamic interaction of cognitive and non-cognitive factors in learning mathematics.* Paper presented at the annual conference of the New England Educational Research Organization, Portsmouth, NH.

Bates, J. A. (1979). Extrinsic reward and intrinsic motivation: A review with implications for the classroom. *Review of Educational Research, 49,* 557–576.

Beck, A. T., & Emery, G. (1985). *Anxiety disorders and phobias: A cognitive perspective.* New York: Basic Books.

Berebitsky, R. D. (1985). *An annotated bibliography of the literature dealing with mathematics anxiety.* (ERIC Document Reproduction Service No. ED 257 684)

Berscheid, E. (1982). Attraction and emotion in interpersonal relations. In M. S. Clark & S. T. Fiske (Eds.), *Affect and cognition* (pp. 37–54). Hillsdale, NJ: Lawrence Erlbaum.

Bishop, A. J. (1988). *Mathematical enculturation: A cultural perspective on mathematics education.* Boston: Kluwer.

Bloom, B. S., & Broder, L. J. (1950). *Problem-solving processes of college students.* Chicago: University of Chicago Press.

Bloom, B. S., Englehart, M. D., Furst, E. J., Hill, W. H., & Krathwohl, D. R. (1956). *Taxonomy of educational objectives: Handbook I. Cognitive domain.* New York: McKay.

Boekaerts, M. (1988). Introduction to emotion, motivation, and learning. *International Journal of Educational Research, 12,* 229–234.

Bower, G. H. (1981). Mood and memory. *American Psychologist, 36,* 129–148.

Brophy, J. (1983). Conceptualizing student motivation. *Educational Psychologist, 18,* 200–215.

Brown, A. L., Bransford, J. D., Ferrara, R. A., & Campione, J. C. (1983). Learning, remembering, and understanding. In J. Flavell & E. Markman (Eds.), *Mussen's handbook of child psychology* (Vol. 3, pp. 77–166). Somerset, NJ: Wiley.

Brown, C. A., Carpenter, T. P., Kouba, V. L., Lindquist, M. M., Silver, E. A., & Swafford, J. O. (1988). Secondary school results for the Fourth NAEP Mathematics Assessment: Algebra, geometry, mathematical methods, and attitudes. *Mathematics Teacher, 81,* 337–347, 397.

Brown, J. S., Collins, A., & Duguid, P. (1988). *Situated cognition and the culture of learning.* Palo Alto, CA: Institute for Research on Learning.

Brown, S. I., & Walter, M. (1983). *The art of problem posing*. Philadelphia: Franklin Institute Press.

Brush, L. R. (1981). Some thoughts for teachers on mathematics anxiety. *Arithmetic Teacher, 29*, 37–39.

Bush, W. S. (1989). Mathematics anxiety in upper elementary school teachers. *School Science and Mathematics, 89*, 499–509.

Buxton, L. (1981). *Do you panic about maths?* London: Heinemann.

Byrne, B. M. (1984). The general/academic self-concept nomological network: A review of construct validation research. *Review of Educational Research, 54*, 427–456.

Campione, J. C., Brown, A. L., & Connell, M. L. (1989). Metacognition: On the importance of understanding what you are doing. In R. I. Charles & E. A. Silver (Eds.), *The teaching and assessing of mathematical problem solving* (pp. 93–114). Reston, VA: National Council of Teachers of Mathematics, and Hillsdale, NJ: Lawrence Erlbaum.

Carter, C. S., & Yackel, E. (1989, March). *A constructivist perspective on the relationship between mathematical beliefs and emotional acts.* Paper presented at the annual meeting of the American Educational Research Association, San Francisco.

Case, R., Hayward, S., Lewis, M., & Hurst, P. (1988). Toward a neo-Piagetian theory of cognitive and emotional development. *Developmental Review, 8*, 2–51.

Choroszy, M., Powers, S., Cool, B. A., & Douglas, P. (1987). Attributions for success and failure in algebra among men and women attending American Samoa Community College. *Psychological Reports, 60*, 47–51.

Clute, P. S. (1984). Mathematics anxiety, instructional method, and achievement in a survey course in college mathematics. *Journal for Research in Mathematics Education, 15*, 50–58.

Cobb, P., Yackel, E., & Wood, T. (1989). Young children's emotional acts during mathematical problem solving. In D. B. McLeod & V. M. Adams (Eds.), *Affect and mathematical problem solving: A new perspective* (pp. 117–148). New York: Springer-Verlag.

Cocking, R. R., & Mestre, J. (Eds.). (1988). *Linguistic and cultural influences on learning mathematics.* Hillsdale, NJ: Lawrence Erlbaum.

Cole, M., & Griffin, P. (Eds.). (1987). *Contextual factors in education: Improving science and mathematics education for minorities and women.* Madison: Wisconsin Center for Education Research.

Collis, B. (1987). Sex differences in the association between secondary school students' attitudes toward mathematics and toward computers. *Journal for Research in Mathematics Education, 18*, 394–402.

Collis, B., & Williams, R. L. (1987). Cross-cultural comparison of gender differences in adolescents' attitude toward computers and selected school subjects. *Journal of Educational Research, 81*, 17–27.

Commission on Standards for School Mathematics. (1989). *Curriculum and evaluation standards for school mathematics.* Reston, VA: National Council of Teachers of Mathematics.

Cooney, T. J., Grouws, D. A., & Jones, D. (1988). An agenda for research on teaching mathematics. In D. A. Grouws & T. J. Cooney (Eds.), *Effective mathematics teaching* (pp. 253–261). Reston, VA: National Council of Teachers of Mathematics, and Hillsdale, NJ: Lawrence Erlbaum.

Corbitt, M. K. (1984). When students talk. *Arithmetic Teacher, 31*, 16–20.

Corno, L., & Rohrkemper, M. M. (1985). The intrinsic motivation to learn in classrooms. In R. Ames & C. Ames (Eds.), *Research on motivation in education* (Vol. 2, pp. 53–90). New York: Academic Press.

Covington, M. V. (1983). Motivated cognitions. In S. G. Paris, G. M. Olson, & H. W. Stevenson (Eds.), *Learning and motivation in the classroom* (pp. 139–164). Hillsdale, NJ: Lawrence Erlbaum.

Covington, M. V., & Omelich, C. L. (1985). Ability and effort valuation among failure-avoiding and failure-accepting students. *Journal of Educational Psychology, 77*, 446–459.

Crosswhite, F. J. (1972). *Correlates of attitudes toward mathematics* (National Longitudinal Study of Mathematical Abilities, Report No. 20). Palo Alto, CA: Stanford University Press.

D'Andrade, R. G. (1981). The cultural part of cognition. *Cognitive Science, 5*, 179–195.

Dew, K. M. H., Galassi, J. P., & Galassi, M. D. (1984). Math anxiety: Relation with situational test anxiety, performance, physiological arousal, and math avoidance behavior. *Journal of Counseling Psychology, 31*, 580–583.

Diener, C. I., & Dweck, C. S. (1978). An analysis of learned helplessness: Continuous changes in performance, strategy, and achievement motivation cognitions following failure. *Journal of Personality and Social Psychology, 36*, 451–462.

Dossey, J. A., Mullis, I. V. S., Lindquist, M. M., & Chambers, D. L. (1988). *The Mathematics Report Card: Trends and achievement based on the 1986 National Assessment.* Princeton: Educational Testing Service.

Dreyfus, T., & Eisenberg, T. (1986). On the aesthetics of mathematical thought. *For the Learning of Mathematics, 6*, 2–10.

Dweck, C. S. (1986). Motivational processes affecting learning. *American Psychologist, 41*, 1040–1048.

Dweck, C. S., & Bempechat, J. (1983). Children's theories of intelligence: Consequences for learning. In S. G. Paris, G. M. Olson, & H. W. Stevenson (Eds.), *Learning and motivation in the classroom* (pp. 239–256). Hillsdale, NJ: Lawrence Erlbaum.

Eisenhart, M. A. (1988). The ethnographic research tradition and mathematics education research. *Journal for Research in Mathematics Education, 19*, 99–114.

Eisenhart, M. A., Shrum, J. L., Harding, J. R., & Cuthbert, A. M. (1988). Teacher beliefs: Definitions, findings, and directions. *Educational Policy, 2*, 51–70.

Ericsson, K. A., & Simon, H. A. (1980). Verbal reports as data. *Psychological Review, 87*, 215–251.

Fennema, E. (1989). The study of affect and mathematics: A proposed generic model for research. In D. B. McLeod & V. M. Adams (Eds.), *Affect and mathematical problem solving: A new perspective* (pp. 205–219). New York: Springer-Verlag.

Fennema, E., & Leder, G. C. (Eds.). (1990). *Mathematics and gender.* New York: Teachers College Press.

Fennema, E., & Peterson, P. (1985). Autonomous learning behavior: A possible explanation of gender-related differences in mathematics. In L. C. Wilkinson & C. Marrett (Eds.), *Gender influences in classroom interaction* (pp. 17–35). Orlando: Academic Press.

Fennema, E., & Sherman, J. A. (1976). Fennema-Sherman Mathematics Attitude Scales: Instruments designed to measure attitudes toward the learning of mathematics by females and males. *Journal for Research in Mathematics Education, 7*, 324–326.

Ferguson, R. D. (1986). Abstraction anxiety: A factor of mathematics anxiety. *Journal for Research in Mathematics Education, 17*, 145–150.

Firestone, W. A. (1987). Meaning in method: The rhetoric of quantitative and qualitative research. *Educational Researcher, 16*, 16–21.

Fischbein, E. (1987). *Intuition in science and mathematics.* Boston: Reidel.

Foxman, D. D., Martini, R. M., & Mitchell, P. (1982). *Mathematical development: Secondary Survey Report No. 3.* London: Her Majesty's Stationery Office.

Gardner, H. (1985). *The mind's new science.* New York: Basic Books.

Garner, R., & Alexander, P. A. (1989). Metacognition: Answered and unanswered questions. *Educational Psychologist, 24*, 143–158.

Garofalo, J. (1989). Beliefs, responses, and mathematics education: Observations from the back of the classroom. *School Science and Mathematics, 89*, 451–455.

Garofalo, J., & Lester, F. K., Jr. (1985). Metacognition, cognitive monitoring, and mathematical performance. *Journal for Research in Mathematics Education, 16*, 163–176.

Gattuso, L., & Lacasse, R. (1987). Les mathophobes: une expérience de réinsertion au niveau collégial. In J. C. Bergeron, N. Herscovics, & C. Kieran (Eds.), *Proceedings of the Eleventh International Conference on the Psychology of Mathematics Education* (Vol. I, pp. 113–119). Montreal: University of Montreal.

Gentry, W. M., & Underhill, R. (1987). A comparison of two palliative methods of intervention for the treatment of mathematics anxiety among female college students. In J. C. Bergeron, N. Herscovics, & C. Kieran (Eds.), *Proceedings of the Eleventh International Conference on the Psychology of Mathematics Education*, (Vol. I, pp. 99–105). Montreal: University of Montreal.

Ginsburg, H. (1981). The clinical interview in psychological research on mathematical thinking: Aims, rationales, techniques. *For the Learning of Mathematics, 1*, 4–11.

Ginsburg, H. P., & Allardice, B. S. (1984). Children's difficulties with school mathematics. In B. Rogoff & J. Lave (Eds.), *Everyday cognition: Its development in social context* (pp. 194–219). Cambridge, MA: Harvard University Press.

Ginsburg, H. P., & Asmussen, K. A. (1988). Hot mathematics. In G. B. Saxe & M. Gearhart (Eds.), *Children's mathematics* (pp. 89–111). San Francisco: Jossey-Bass.

Gliner, G. S. (1987). The relationship between mathematics anxiety and achievement variables. *School Science and Mathematics, 87*, 81–87.

Goldin, G. A. (1988). Affective representation and mathematical problem solving. In M. J. Behr, C. B. Lacampagne, & M. M. Wheeler (Eds.), *Proceedings of the Tenth Annual Meeting of the International Group for the Psychology of Mathematics Education —North American Chapter* (pp. 1–7). DeKalb, IL: Northern Illinois University.

Graham, S. (1984). Teacher feelings and student thoughts: An attributional approach to affect in the classroom. *Elementary School Journal, 85*, 91–104.

Greenwood, J. (1984). My anxieties about math anxiety. *Mathematics Teacher, 77*, 662–663.

Gressard, C. P., & Loyd, B. H. (1987). An investigation of the effects of math anxiety and sex on computer attitudes. *School Science and Mathematics, 87*, 125–135.

Grouws, D. A., & Cramer, K. (1989). Teaching practices and student affect in problem-solving lessons of select junior high mathematics teachers. In D. B. McLeod & V. M. Adams (Eds.), *Affect and mathematical problem solving: A new perspective* (pp. 149–161). New York: Springer-Verlag.

Hackett, G., & Betz, N. E. (1989). An exploration of the mathematics self-efficacy/mathematics performance correspondence. *Journal for Research in Mathematics Education, 20*, 261–273.

Haladyna, T., Shaughnessy, J., & Shaughnessy, J. M. (1983). A causal analysis of attitude toward mathematics. *Journal for Research in Mathematics Education, 14*, 19–29.

Harré, R. (Ed.). (1986). *The social construction of emotion*. Oxford: Blackwell.

Hart, L. E. (1989a). Classroom processes, sex of student, and confidence in learning mathematics. *Journal for Research in Mathematics Education, 20*, 242–260.

Hart, L. E. (1989b). Describing the affective domain: Saying what we mean. In D. B. McLeod & V. M. Adams (Eds.), *Affect and mathematical problem solving: A new perspective* (pp. 37–48). New York: Springer-Verlag.

Hatano, G., & Inagaki, K. (1987). A theory of motivation for comprehension and its application to mathematics instruction. In T. A. Romberg & D. M. Stewart (Eds.), *The monitoring of school mathematics* (Vol. 2, pp. 27–46). Madison: Wisconsin Center for Education Research.

Heckhausen, H. (1987). Causal attribution patterns for achievement outcomes: Individual differences, possible types, and their origins. In F. E. Weinert & R. H. Kluwe (Eds.), *Metacognition, motivation, and understanding* (pp. 143–184). Hillsdale, NJ: Lawrence Erlbaum.

Hembree, R. (1990). The nature, effects, and relief of mathematics anxiety. *Journal for Research in Mathematics Education, 21*, 33–46.

Hendel, D. D. (1980). Experimental and affective correlates of math anxiety in adult women. *Psychology of Women Quarterly, 5*, 219–229.

Henerson, M. E., Morris, L. L., & Fitz-Gibbon, C. T. (1978). *How to measure attitudes*. Beverly Hills, CA: Sage Publications.

Hess, R. D., Chang, C.-M., & McDevitt, T. M. (1987). Cultural variations in family beliefs about children's performance in mathematics: Comparisons among People's Republic of China, Chinese-American, and Caucasian-American families. *Journal of Educational Psychology, 79*, 179–188.

Holloway, S. C. (1988). Concepts of ability and effort in Japan and the United States. *Review of Educational Research, 58*, 327–345.

Howe, K. R. (1988). Against the quantitative-qualitative incompatibility thesis, or dogmas die hard. *Educational Researcher, 11*, 10–16.

Hunsley, J. (1987). Cognitive processes in mathematics anxiety and test anxiety: The role of appraisals, internal dialogue, and attributions. *Journal of Educational Psychology, 79*, 388–392.

Jacob, E. (1987). Qualitative research traditions: A review. *Review of Educational Research, 57*, 1–50.

Kagan, J. (1978). On emotion and its development: A working paper. In M. Lewis & L. A. Rosenblum (Eds.), *The development of affect* (pp. 11–41). New York: Plenum Press.

Kaput, J. J. (1989). The role of information technologies in the affective dimension of mathematical experience. In D. B. McLeod & V. M. Adams (Eds.), *Affect and mathematical problem solving: A new perspective* (pp. 89–103). New York: Springer-Verlag.

Kelly-Benjamin, K. (1990). *Development of a scale to measure students' mathematical learning style orientations*. Unpublished manuscript, Jersey City State College, Jersey City, NJ.

Kieren, T. E. (1988). Personal knowledge of rational numbers: Its intuitive and formal development. In J. Hiebert & M. Behr (Eds.), *Number concepts and operations in the middle grades* (pp. 162–181). Reston, VA: National Council of Teachers of Mathematics and Hillsdale, NJ: Lawrence Erlbaum.

Kilpatrick, J. (1985). Reflection and recursion. *Educational Studies in Mathematics, 16*, 1–26.

Kloosterman, P. (1988). Self-confidence and motivation in mathematics. *Journal of Educational Psychology, 80*, 345–351.

Kloosterman, P., & Stage, F. K. (1989, March). *Measuring beliefs about mathematical problem solving*. Paper presented at the annual meeting of the American Educational Research Association, San Francisco.

Kouba, V., & McDonald, J. (1987). Students' perceptions of mathematics as a domain. In J. C. Bergeron, N. Herscovics, & C. Kieran (Eds.), *Proceedings of the Eleventh International Conference on the Psychology of Mathematics Education* (Vol. I, pp. 106–112). Montreal: University of Montreal.

Krathwohl, D. R., Bloom, B. S., & Masia, B. B. (1964). *Taxonomy of educational objectives: Handbook II. Affective domain*. New York: Longman.

Kulm, G. (1980). Research on mathematics attitude. In R. J. Shumway (Ed.), *Research in mathematics education* (pp. 356–387). Reston, VA: National Council of Teachers of Mathematics.

Lave, J. (1988). *Cognition in practice: Mind, mathematics and culture in everyday life*. Cambridge: Cambridge University Press.

Lawler, R. W. (1981). The progressive construction of mind. *Cognitive Science, 5*, 1–30.

Lawson, M. J. (1984). Being executive about metacognition. In J. A. Kirby (Ed.), *Cognitive strategies and educational performance* (pp. 89–109). Orlando: Academic Press.

Lazarus, R. S. (1982). Thoughts on the relations between emotion and cognition. *American Psychologist, 37*, 1019–1024.

Lazarus, R. S. (1984). On the primacy of cognition. *American Psychologist, 39,* 124–129.

Leder, G. C. (1987). Attitudes towards mathematics. In T. A. Romberg & D. M. Stewart (Eds.), *The monitoring of school mathematics* (Vol. 2, pp. 261–277). Madison: Wisconsin Center for Education Research.

Legault, L. (1987). Investigation des facteurs cognitifs et affectifs dans les blocages en mathématiques. In J. C. Bergeron, N. Herscovics, & C. Kieran (Eds.), *Proceedings of the Eleventh International Conference on the Psychology of Mathematics Education* (Vol. I, pp. 120–125). Montreal: University of Montreal.

Lester, F. K., Garofalo, J., & Kroll, D. L. (1989). Self-confidence, interest, beliefs, and metacognition: Key influences on problem-solving behavior. In D. B. McLeod & V. M. Adams (Eds.), *Affect and mathematical problem solving: A new perspective* (pp. 75–88). New York: Springer-Verlag.

Linn, M. C., & Hyde, J. S. (1989). Gender, mathematics, and science. *Educational Researcher, 18,* 17–27.

Llabre, M. M., & Suarez, E. (1985). Predicting math anxiety and course performance in college women and men. *Journal of Counseling Psychology, 32,* 283–287.

Maehr, M. L. (1984). Meaning and motivation: Toward a theory of personal investment. In R. Ames & C. Ames (Eds.), *Research on motivation in education* (Vol. 1, pp. 115–144). Orlando: Academic Press.

Magnusson, D. (Ed.). (1981). *Toward a psychology of situations: An interactional perspective.* Hillsdale, NJ: Lawrence Erlbaum.

Malone, T. W., & Lepper, M. R. (1987). Making learning fun: A taxonomy of intrinsic motivations for learning. In R. E. Snow & M. J. Farr (Eds.), *Aptitude, learning, and instruction: Volume 3: Conative and affective process analyses* (pp. 223–253). Hillsdale, NJ: Lawrence Erlbaum.

Mandler, G. (1972). Helplessness: Theory and research in anxiety. In C. D. Spielberger (Ed.), *Anxiety: Current trends in theory and research* (pp. 359–374). New York: Academic Press.

Mandler, G. (1975). *Mind and emotion.* New York: Wiley.

Mandler, G. (1984). *Mind and body: Psychology of emotion and stress.* New York: Norton.

Mandler, G. (1985). *Cognitive psychology: An essay in cognitive science.* Hillsdale, NJ: Lawrence Erlbaum.

Mandler, G. (1989). Affect and learning: Causes and consequences of emotional interactions. In D. B. McLeod & V. M. Adams (Eds.), *Affect and mathematical problem solving: A new perspective* (pp. 3–19). New York: Springer-Verlag.

Marcilo, C. (1987). *A study of implicit theories and beliefs about teaching in elementary school teachers.* (ERIC Document Reproduction Service No. ED 281 834)

Marsh, H. W. (1986). Verbal and math self-concepts: An internal/external frame of reference model. *American Educational Research Journal, 23,* 129–149.

Marshall, S. (1989). Affect in schema knowledge: Source and impact. In D. B. McLeod & V. M. Adams (Eds.), *Affect and mathematical problem solving: A new perspective* (pp. 49–58). New York: Springer-Verlag.

Mason, J., Burton, L., & Stacey, K. (1982). *Thinking mathematically.* London: Addison-Wesley.

McKnight, C. C., Crosswhite, F. J., Dossey, J. A., Kifer, E., Swafford, J. O., Travers, K. J., & Cooney, T. J. (1987). *The underachieving curriculum: Assessing U.S. school mathematics from an international perspective.* Champaign, IL: Stipes Publishing Company.

McLean, L. D. (1982). *Willing but not enthusiastic.* Toronto: Ontario Institute for Studies in Education.

McLeod, D. B. (1985). Affective issues in research on teaching mathematical problem solving. In E. A. Silver (Ed.), *Teaching and learning mathematical problem solving: Multiple research perspectives* (pp. 267–279). Hillsdale, NJ: Lawrence Erlbaum.

McLeod, D. B. (1988). Affective issues in mathematical problem solving: Some theoretical considerations. *Journal for Research in Mathematics Education, 19,* 134–141.

McLeod, D. B., & Adams, V. M. (Eds.). (1989). *Affect and mathematical problem solving: A new perspective.* New York: Springer-Verlag.

McLeod, D. B., Craviotto, C., & Ortega, M. (1990, July). *Students' affective responses to non-routine mathematical problems: An empirical study.* Paper presented at the Fourteenth International Conference for the Psychology of Mathematics Education, Mexico City.

McLeod, D. B., Metzger, W., & Craviotto, C. (1989). Comparing experts' and novices' affective reactions to mathematical problem solving: An exploratory study. In G. Vergnaud (Ed.), *Proceedings of the Thirteenth International Conference for the Psychology of Mathematics Education* (Vol. 2, pp. 296–303). Paris: Laboratoire de Psychologie du Developpement et de l'Education de l'Enfant.

Meichenbaum, D. (1977). *Cognitive behavior modification: An integrative approach.* New York: Plenum Press.

Messick, S. (1987). Structural relationships across cognition, personality, and style. In R. E. Snow & M. J. Farr (Eds.), *Aptitude, learning, and instruction: Volume 3: Conative and affective process analyses* (pp. 35–75). Hillsdale, NJ: Lawrence Erlbaum.

Mevarech, Z. R., & Ben-Artzi, S. (1987). Effects of CAI with fixed and adaptive feedback on children's mathematical anxiety and achievement. *Journal of Experimental Education, 56,* 42–46.

Meyer, M. R., & Fennema, E. (1988). Girls, boys, and mathematics. In T. R. Post (Ed.), *Teaching mathematics in grades K–8: Research-based methods* (pp. 406–425). Boston: Allyn and Bacon.

Morgan, M. (1984). Reward-induced decrements and increments in intrinsic motivation. *Review of Educational Research, 54,* 5–30.

Mura, R. (1987). Sex-related differences in expectations of success in undergraduate mathematics. *Journal for Research in Mathematics Education, 18,* 15–24.

National Research Council. (1989). *Everybody counts: A report to the nation on the future of mathematics education.* Washington, DC: National Academy Press.

Newman, D., Griffin, P., & Cole, M. (1989). *The construction zone: Working for cognitive change in school.* Cambridge: Cambridge University Press.

Newman, R. S. (1984). Children's numerical skill and judgments of confidence in estimation. *Journal of Experimental Child Psychology, 37,* 107–123.

Newman, R. S. (1990). Children's help-seeking in the classroom: The role of motivational factors and attitudes. *Journal of Educational Psychology, 82,* 71–80.

Newman, R. S., & Goldin, L. (1990). Children's reluctance to seek help with schoolwork. *Journal of Educational Psychology, 82,* 92–100.

Newman, R. S., & Wick, P. L. (1987). Effect of age, skill, and performance feedback on children's judgment of confidence. *Journal of Educational Psychology, 79,* 115–119.

Nicholls, J. G. (1984). Conceptions of ability and achievement motivation. In R. Ames & C. Ames (Eds.), *Research on motivation in education* (Vol. 1, pp. 39–73). Orlando: Academic Press.

Nimier, J. (1977). Mathématiques et affectivité. *Educational Studies in Mathematics, 8,* 241–250.

Noddings, N., & Shore, P. J. (1984). *Awakening the inner eye: Intuition in education.* New York: Teachers College Press.

Norman, D. A. (1981). Twelve issues for cognitive science. In D. A. Norman (Ed.), *Perspectives on cognitive science* (pp. 265–295). Norwood, NJ: Ablex.

Norwich, B. (1987). Self-efficacy and mathematics achievement: A study of their relation. *Journal of Educational Psychology, 79,* 384–387.

Orr, E. W. (1987). *Twice as less: Black English and the performance of black students in mathematics and science.* New York: Norton.

Ortony, A., Clore, G. L., & Collins, A. (1988). *The cognitive structure of emotions.* Cambridge: Cambridge University Press.

Paris, S. G., Olson, G. M., & Stevenson, H. W. (Eds.). (1983). *Learning and motivation in the classroom.* Hillsdale, NJ: Lawrence Erlbaum.

Parsons, J. E., Adler, T. F., & Kaczala, C. M. (1982). Socialization of achievement, attitudes, and beliefs: Parental influences. *Child Development, 53,* 310–321.

Parsons, J. E., Meece, J. L., Adler, T. F., & Kaczala, C. M. (1982). Sex differences in learned helplessness. *Sex Roles, 8,* 421–432.

Pekrun, R. (1988). Anxiety and motivation in achievement settings: Towards a systems-theoretical approach. *International Journal of Educational Research, 12,* 307–323.

Peterson, P. L., & Barger, S. A. (1985). Attribution theory and teacher expectancy. In J. B. Dusek (Ed.), *Teacher expectancies* (pp. 159–184). Hillsdale, NJ: Lawrence Erlbaum.

Peterson, P. L., & Fennema, E. (1985). Effective teaching, student engagement in classroom activities, and sex-related differences in learning mathematics. *American Educational Research Journal, 22,* 309–335.

Peterson, P. L., Fennema, E., Carpenter, T. P., & Loef, M. (1989). Teachers' pedagogical content beliefs in mathematics. *Cognition and Instruction, 6,* 1–40.

Peterson, P. L., & Swing, S. R. (1982). Beyond time on task: Students' reports of their thought processes during classroom instruction. *Elementary School Journal, 82,* 481–491.

Prawat, R. S. (1989). Promoting access to knowledge, strategy, and disposition in students: A research synthesis. *Review of Educational Research, 59,* 1–41.

Prawat, R. S., Byers, J. L., & Anderson, A. H. (1983). An attributional analysis of teachers' affective reactions to student success and failure. *American Educational Research Journal, 20,* 137–152.

Rajecki, D. W. (1982). *Attitudes. Themes and advances.* Sunderland, MA: Sinauer.

Research Advisory Committee. (1989). The mathematics education of underserved and underrepresented groups: A continuing challenge. *Journal for Research in Mathematics Education, 20,* 371–375.

Resnick, L. B. (1986). The development of mathematical intuition. In M. Perlmutter (Ed.), *Perspectives on intellectual development* (Vol. 19, pp. 159–194). Hillsdale, NJ: Lawrence Erlbaum.

Reyes, L. H. (1980). Attitudes and mathematics. In M. M. Lindquist (Ed.), *Selected issues in mathematics education* (pp. 161–184). Berkeley, CA: McCutchan.

Reyes, L. H. (1984). Affective variables and mathematics education. *Elementary School Journal, 84,* 558–581.

Richardson, F. C., & Woolfolk, R. L. (1980). Mathematics anxiety. In I. G. Sarason (Ed.), *Test anxiety: Theory, research, and applications* (pp. 271–288). Hillsdale, NJ: Lawrence Erlbaum.

Robitaille, D. F., & Garden, R. A. (Eds.). (1989). *The IEA Study of Mathematics II: Contexts and outcomes of school mathematics.* Oxford: Pergamon.

Rokeach, M. (1968). *Beliefs, attitudes, and values.* San Francisco: Jossey-Bass.

Rounds, J. B., & Hendel, D. D. (1980). Measurement and dimensionality of mathematics anxiety. *Journal of Counseling Psychology, 27,* 138–149.

Sarason, I. G. (Ed.). (1980). *Test anxiety: Theory, research, and applications.* Hillsdale, NJ: Lawrence Erlbaum.

Sarason, I. G. (1987). Test anxiety, cognitive interference, and performance. In R. E. Snow & M. J. Farr (Eds.), *Aptitude, learning, and instruction: Volume 3: Conative and affective process analyses* (pp. 131–142). Hillsdale, NJ: Lawrence Erlbaum.

Saxe, G. B. (1990). *Culture and cognitive development: Studies in mathematical understanding.* Hillsdale, NJ: Lawrence Erlbaum.

Schacter, S., & Singer, J. (1962). Cognitive, social and physiological determinants of emotional state. *Psychological Review, 69,* 379–399.

Scherer, K. S., & Ekman, P. (Eds.). (1984). *Approaches to emotion.* Hillsdale, NJ: Lawrence Erlbaum.

Schoenfeld, A. H. (1985). *Mathematical problem solving.* Orlando: Academic Press.

Schoenfeld, A. H. (1987a). Cognitive science and mathematics education: An overview. In A. H. Schoenfeld (Ed.), *Cognitive science and mathematics education* (pp. 1–31). Hillsdale, NJ: Lawrence Erlbaum.

Schoenfeld, A. H. (1987b). What's all the fuss about metacognition? In A. H. Schoenfeld (Ed.), *Cognitive science and mathematics education* (pp. 189–215). Hillsdale, NJ: Lawrence Erlbaum.

Schoenfeld, A. H. (1988). When good teaching leads to bad results: The disasters of "well-taught" mathematics courses. *Educational Psychologist, 23,* 145–166.

Schoenfeld, A. H. (1989). Explorations of students' mathematical beliefs and behavior. *Journal for Research in Mathematics Education, 20,* 238–355.

Schunk, D. H. (1984). Self-efficacy perspective on achievement behavior. *Educational Psychologist, 19,* 48–58.

Shavelson, R. & Bolus, R. (1982). Self-concept: The interplay of theory and methods. *Journal of Educational Psychology, 74,* 3–17.

Sheffler, I. (1977). In praise of cognitive emotions. *Teachers College Record, 79,* 171–186.

Silver, E. A. (1985). Research on teaching mathematical problem solving: Some underrepresented themes and needed directions. In E. A. Silver (Ed.), *Teaching and learning mathematical problem solving: Multiple research perspectives* (pp. 247–266). Hillsdale, NJ: Lawrence Erlbaum

Silver, E. A., & Metzger, W. (1989). Aesthetic influences on expert mathematical problem solving. In D. B. McLeod & V. M. Adams (Eds.), *Affect and mathematical problem solving: A new perspective* (pp. 59–74). New York: Springer-Verlag.

Simon, H. A. (1967). Motivational and emotional controls of cognition. *Psychological Review, 74,* 29–39.

Simon, H. A. (1982). Comments. In M. S. Clark & S. T. Fiske (Eds.), *Affect and cognition* (pp. 333–342). Hillsdale, NJ: Lawrence Erlbaum.

Simon, L. R. (1966). *Cognition and affect.* Buffalo: Prometheus Books.

Skemp, R. R. (1979). *Intelligence, learning, and action.* New York: Wiley.

Skinner, B. F. (1953). *Science and human behavior.* New York: Macmillan.

Snow, R. E., & Farr, M. J. (Eds.). (1987). *Aptitude, learning, and instruction: Volume 3: Conative and affective process analyses.* Hillsdale, NJ: Lawrence Erlbaum.

Sowder, J. T. (1989). Affective factors and computational estimation ability. In D. B. McLeod & V. M. Adams (Eds.), *Affect and mathematical problem solving: A new perspective* (pp. 177–191). New York: Springer-Verlag.

Spielberger, C. D., Gonzalez, H. P., & Fletcher, T. (1979). Test anxiety reduction, learning strategies, and academic performance. In H. F. O'Neil, Jr., & C. D. Spielberger (Eds.), *Cognitive and affective learning strategies* (pp. 111–132). New York: Academic Press.

Stevenson, H. W. (1987). The Asian advantage: The case of mathematics. *American Educator, 11,* 26–31.

Stevenson, H. W., Lee, S.-Y., & Stigler, J. W. (1986). Mathematics achievement of Chinese, Japanese, and American children. *Science, 231,* 693–699.

Stigler, J. W., & Mao, L.-W. (1985). The self-perception of competence by Chinese children. *Child Development, 56,* 1259–1270.

Stigler, J. W., & Perry, M. (1988). Cross-cultural studies of mathematics teaching and learning: Recent findings and new directions. In D. A. Grouws & T. J. Cooney (Eds.), *Effective mathematics teaching* (pp. 194–223). Reston, VA: National Council of Teachers of Mathematics, and Hillsdale, NJ: Lawrence Erlbaum.

Stipek, D. J. (1984). The development of achievement motivation. In R. Ames & C. Ames (Eds.), *Research on motivation in education* (Vol. 1, pp. 145–174). Orlando: Academic Press.

Stodolsky, S. S. (1985). Telling math: Origins of math aversion and anxiety. *Educational Psychologist, 20,* 125–133.

Strongman, K. T. (1978). *The psychology of emotion* (2nd ed.). New York: Wiley.

Swadener, M., & Hanafin, M. (1987). Gender similarities and differences in sixth-graders' attitude toward computers: An exploratory study. *Educational Technology, 27,* 37–42.

Swanson, D., Schwartz, R., Ginsburg, H., & Kossan, N. (1981). The clinical interview: Validity, reliability, and diagnosis. *For the Learning of Mathematics, 2,* 31–38.

Taylor, L. (1990). American female and male university professors' mathematical attitudes and life histories. In L. Burton (Ed.), *Gender and mathematics education: An international perspective* (pp. 47–59). London: Cassell.

Thompson, A. G. (1984). The relationship of teachers' conceptions of mathematics and mathematics teaching to instructional practice. *Educational Studies of Mathematics, 15,* 105–127.

Thompson, A. G., & Thompson, P. W. (1989). Affect and problem solving in an elementary school mathematics classroom. In D. B. McLeod & V. M. Adams (Eds.), *Affect and mathematical problem solving: A new perspective* (pp. 162-176). New York: Springer-Verlag.

Tittle, C. K. (1986). Gender research and education. *American Psychologist, 41,* 1161–1168.

Tittle, C. K. (1987). *A project to improve mathematics instruction for women and minorities: Comprehensive assessment and mathematics.* Unpublished manuscript, City University of New York, Graduate School and University Center, New York.

Tittle, C. K., Hecht, D., & Moore, P. (1989, March). *From taxonomy to constructing meaning in context: Revisiting the taxonomy of educational objectives, affective domain, 25 years later.* Paper presented at the annual meeting of the National Council on Measurement in Education, San Francisco.

Turkle, S. (1984). *The second self: Computers and the human spirit.* New York: Simon and Schuster.

Underhill, R. (1988). Mathematics learners' beliefs: A review. *Focus on Learning Problems in Mathematics, 10,* 55–69.

Wagner, S., Rachlin, S. L., & Jensen, R. J. (1984). *Algebra learning project: Final Report.* Athens, GA: University of Georgia.

Weiner, B. (1986). *An attributional theory of motivation and emotion.* New York: Springer-Verlag.

Weinert, F. E. (1987). Metacognition and motivation as determinants of effective learning and understanding. In F. E. Weinert & R. H. Kluwe (Eds.), *Metacognition, motivation, and understanding* (pp. 1–16). Hillsdale, NJ: Lawrence Erlbaum.

Weinert, F. E., & Kluwe, R. H. (Eds.). (1987). *Metacognition, motivation, and understanding.* Hillsdale, NJ: Lawrence Erlbaum.

Wigfield, A., & Meece, J. L. (1988). Math anxiety in elementary and secondary school students. *Journal of Educational Psychology, 80,* 210–216.

Witkin, H. A., & Goodenough, D. R. (1981). *Cognitive styles: Essence and origins.* New York: International Universities Press.

Wittrock, M. C. (Ed.). (1986). *Handbook of research on teaching* (3rd ed.). New York: Macmillan.

Wolleat, P. L., Pedro, J. D., Becker, A. D., & Fennema, E. (1980). Sex differences in high school students' causal attributions of performance in mathematics. *Journal for Research in Mathematics Education, 11,* 356–366.

Zajonc, R. B. (1984). On the primacy of affect. *American Psychologist, 39,* 117–123.

·24·

MATHEMATICS AND GENDER: CHANGING PERSPECTIVES

Gilah C. Leder

MONASH UNIVERSITY

> But how can I describe my astonishment and admiration on seeing my esteemed correspondent M. LeBlanc metamorphosed into this celebrated person, yielding a copy so brilliant it is hard to believe? The taste for the abstract sciences in general and, above all, for the mysteries of numbers is very rare: this is not surprising, since the charms of this sublime science in all their beauty reveal themselves only to those who have the courage to fathom them. But when a woman, [who] because of her sex, our customs and prejudices, encounters infinitely more obstacles than men in familiarizing herself with their knotty problems, yet overcomes these fetters and penetrates that which is most hidden, she doubtless has *the most noble courage, extraordinary talent, and superior genius.* (Gauss, April 30, 1807, emphasis added; translated in L.L. Bucciarelli & N. Dworsky, 1980)

Gender differences in mathematics learning continue to attract much research attention. An in-depth appraisal of current findings, and in particular of shifts over time of accepted beliefs and of summaries of research, requires contemplation of factors well beyond the contemporary mathematics classroom. Learning is affected by many factors: personal, situational, and cultural. To appreciate at least some of the variables which have shaped the current milieu, it is helpful to sketch the broader educational environment over the past two centuries for females and males in three different western countries whose cultures have much in common: the United States, England, and Australia. One can then evaluate more recent work on gender differences in mathematics education against a broader context.

HISTORICAL BACKGROUND

The United States

The founding of Harvard College in 1636 demonstrates that, despite their preoccupation with survival, the early settlers attached considerable importance to a liberal education. Initially, "college was beyond the reach of most men, for lack of social status, and of all women, by virtue of their sex" (Solomon, 1985, p. 2).

Differences in the educational opportunities generally available to males and females are clearly illustrated by the literacy rates of the two groups. "In 1790, perhaps only half as

The author gratefully acknowledges the helpful reviews provided by Jacqueline S. Eccles, University of Colorado, and Laurie E. Hart, University of Georgia.

many women as men were literate" (Tyack & Hansot, 1988, p. 34). There was ambivalence about the educational needs of girls. While parents realized that education could give independence to a daughter who remained single, there were fears that too much education might spoil her chances of marriage. Yet female enrollments in elementary schools grew so that by the middle of the 19th century the discrepancy in the literacy rates of females and males had largely disappeared. At the same time, coeducation had become the norm for public schools in many states. This organizational strategy was subsequently challenged.

> In the second half of the nineteenth century, for example, critics stirred up a storm of controversy by asserting that girls' attendance in mixed high schools and colleges maimed their reproductive organs and nerves. (Tyack & Hansot, 1988, p. 33)

The generally better academic performance of girls and their higher retention rate at the high school level caused much concern at the turn of this century. Despite evidence to the contrary, some educators questioned whether girls could really cope with the physical demands of secondary schooling (Brooks, 1903; Gay, 1902). Aspects of this debate as it relates to the study of mathematics have been summarized by Kroll (1985). She reported the assertion of at least one writer that "woman...is organized both bodily and mentally for dealing with an entirely different set of functions, in which mathematics plays a small part" (p. 8). Then, as now, gender differences in mathematics learning were attributed glibly by some (for example, Morrison, 1915; Thomas, 1915) to differences in innate abilities whereas others (Armstrong, 1910; Dean, 1909) pointed to differences in males' and females' interests and perceptions of future usefulness of the subject. Proposed solutions included modifying the mathematics curriculum to take account of the interests of females as well as males and a return to segregated education. The greatly increased costs of public schooling under the latter system made this an unpopular alternative. In smaller high schools the limited number of students prevented electives from being mounted so that boys and girls typically took the same subjects in the same classes. But in larger high schools, and as student numbers grew, there was experimentation with gender appropriate curricula in areas such as vocational education, physical education, and selected extracurricular activities. The relics of this policy continue to be felt to the present day.

As already indicated, for many years only a small privileged group of males was able to reap the benefits of a tertiary education. While admission of women to State Universities is reasonably long-standing—beginning with Iowa and Wisconsin in 1855 and 1867, respectively (Solomon, 1985)—various private tertiary institutions were considerably more tardy in granting equal rights to women.

> In the 1870s, a physiologist at Harvard, Dr. Edward H. Clarke, admitted that young women *could* learn rigorous subjects but argued that they *should* not. In particular Clarke opposed the admission of women to Harvard....A young woman might learn algebra, but [he argued] when the limited sum of energy flowed to the overwrought brain, it harmed the natural growth of ovaries. (Tyack & Hansot, 1988, p. 37)

TABLE 24–1. U.S. Colleges Open to Men and Women, 1870–1981

Year	No. of institutions	Percentage distribution		
		Men only	Women only	Coeducational
1870	582	59	12	29
1890	1082	37	20	43
1910	1083	27	15	58
1930	1322	15	16	69
1957*	1326	13	13	74
1976*	1849	4	5	91
1981*	1928	3	5	92

Source: Solomon, 1985, p. 44.
* excludes nondegree-granting institutions

Not until 1965 were Harvard A.B.'s granted to Radcliffe women, while female undergraduates were not admitted to Columbia until 1983. The impact of these practices must not be exaggerated. Examples such as Oberlin College, founded in Ohio in 1833, where "men and women, white and black, were to be educated together to carry out God's cause on Earth" (Solomon, 1985, p. 21), contrasted sharply with practices at other institutions. Nevertheless, there was clear evidence of the existence of institutionalized discriminatory practices in the wider society within which mathematics learning took place. The gradual shift from segregated to coeducational college education is illustrated by the data in Table 24.1.

Clearly, over the more than 100 years surveyed there has been a dramatic increase in the number of institutions offering a college education. In the vast majority of these, programs are now formally available to both female and male students. The extent to which both sets of students are in fact drawn to certain of the courses offered is discussed in a later section. Important, though beyond the scope of this chapter, is a more detailed consideration of the parallel yet sometimes more complex barriers faced by black females.

England

Because of the relatively small number of universities then in existence, the refusal of Oxford and Cambridge to admit women to full membership till 1919 and 1948, respectively (Bryant, 1979), had considerable impact on the availability of higher education to women. England's stance on this issue was decidedly more conservative than that found in several other countries which were members of the then British Empire.

At the school level, the strong private school movement produced a system segregated on the basis of class as well as sex. The custom of middle and upper class families of educating their daughters at home, but their sons in private schools, continued well into the Victorian era (Hunt, 1987). Even in the public sector single sex settings were preferred, resources permitting. Such segregation facilitated the provision of gender-differentiated curricula. It was considered appropriate that girls' schools should have staff with particular expertise in art, music, and English but not necessarily in science or mathematics. The Board of Education proclaimed in 1899 that it "was prepared to argue that girls did in fact lack the same capacity for mathematics as boys" (Hunt, 1987, p. 13).

The increased participation of working class children in schooling led to a further examination of the purpose of education.

Frequently these debates focused directly upon the assumed needs of girls and boys, respectively, and much of the curriculum was consciously shaped to be sex-specific. . . . With a growing number of subjects competing for timetable space girls spent increasingly greater periods of time in activities that separated them from boys . . . Furthermore, there is evidence that gender differences were stressed and indeed increasingly encouraged in [the] basic subjects of the curriculum. (Turnbull, 1987, p. 85)

Over the last 30 years in particular there have been various reorganizations and evaluations of the educational system. Some of these (e.g., Cockcroft, 1982) have at least formally acknowledged the disadvantages faced by girls in areas such as mathematics. In recent years initiatives aimed at achieving a more equitable educational environment have been introduced. However, political organizations and educational pressure groups have generally been more concerned with class inequalities in the educational system than with gender inequities (Thom, 1987). The legacy of the earlier notions about appropriate educational provisions for girls and boys lingers and must be taken into account by those concerned with gender differences in mathematics learning.

Australia

For much of its 200 years of formal white settlement, Australia has been greatly influenced by English beliefs and traditions. Evidence for this is still apparent in the contemporary Australian school system.

While the colonial government took considerable responsibility for the education of working class students, the more affluent sent their sons and daughters to private schools closely modeled on those in England. From the outset, sex-segregated curriculum provisions prevailed. As in England, school mathematics was both a sexist and elitist subject. Few children living in working class areas in the city or in rural areas proceeded to secondary education. Typically, only boys from wealthy families had the opportunity to study Euclid and algebra (Clements, 1979). At elementary school, girls were taught sewing and needlework in preference to arithmetic, a subject deemed more appropriate for boys (Zainu'ddin, 1975).

The emphasis on segregated education continued throughout the 19th century. A feature of a new school opened in 1854 was "the construction of a 'running bar fence' and provision for separate entrances from separate streets, to segregate all boys and girls above the age of seven" (Zainu'ddin, 1975, p. 6). Even today, many of the nation's most prestigious private schools remain single-sex, although in recent years a number have become coeducational, for financial rather than for philosophical reasons it is commonly assumed. Most public schools are coeducational, however.

As in the countries previously discussed, the path to tertiary education was more difficult for women than for men. The first woman graduate from an Australian university was admitted to the degree of Bachelor of Arts in 1883, some 30 years after the

first university in that country opened its doors. The decision to allow females to matriculate and take degree courses led to an increase in the number who studied arithmetic, algebra and Euclid, all of which were part of the matriculation examination. Differences in performance in favor of males were observed on the examination papers for Euclid and algebra in particular. While some contemporary observers attributed these results to the curricular emphases prevalent in girls' and boys' schools, others argued that these differences merely confirmed that boys were naturally inclined to mathematics, girls to literary subjects and especially modern languages (Clements, 1989). More than a century later similar arguments are still being put forward.

Much progress towards educational equity has been made in each of the three countries considered. Today the formal rights of all citizens to education, irrespective of sex, class, race or religious beliefs, tend to be taken for granted. Yet, as is shown in the sections that follow, differences remain in the extent to which females and males participate in certain facets of education. It is noteworthy that the contemporary patterns are consistent with the reported historical beliefs about appropriate educational and occupational aspirations and life styles for females and males.

RESEARCH IN MATHEMATICS EDUCATION AND GENDER ISSUES

Before reviewing the more contemporary literature on gender differences in mathematics learning, it is of interest to note the prevalence of such research. The data in Table 24.2 provide a possible measure by summarizing the frequency of work with a gender theme published in one major mathematics education journal. Shown are the number of contributions, other than book reviews, published in *The Journal for Research in Mathematics Education* between 1978 and 1990, as well as the number of articles that appeared each year with an exclusive or decided minor emphasis on gender issues. In this latter category a careful balance was maintained between acknowledging but not exaggerating references to gender effects.

Collectively, the data reveal a considerable concern with gender issues in the mathematics education research community. The overall figure of 38 out of 355 articles or approximately 10% of the total number of articles published during the period surveyed is remarkably similar to that cited by Reyes (1983) for a summary of the topics of manuscripts submitted to that journal during the two-year period 1981–1982. She reported, "The most common topics for manuscripts were problem solving, attitudes, student achievement, and sex differences—with each accounting for 8%–10% of all manuscripts" (p. 146).

A more detailed overview of the issues actually explored is given in Table 24.3. Inspection of these data indicates that the interaction between sex roles and mathematics learning has been considered from a number of different perspectives.

The majority of these articles focused on aspects of the nature and extent of differences in the mathematics achievement of females and males. Various areas of mathematics and different contributing factors were examined in these studies. For example, Swafford (1981) and Kirsher (1989) concentrated on algebra; Linn and Pulos (1983) looked at proportional reason-

TABLE 24–2. Overview of Articles Published in *Journal for Research in Mathematics Education*

Year	Articles	Number of Reports, Critiques, Forum	With Gender Theme
1978	20	12	1
1979	17	17	0
1980	22	13	3
1981	18	9	5
1982	20	4	4
1983	19	9	4
1984	19	4	2
1985	16	8	2
1986	17	5	1
1987	19	2	7
1988	20	6	2
1989	26	4	3
1990	17	12	4
Total	250	104	38

ing skills; Callahan and Clements (1984) at rote counting skills; Fennema and Tartre (1985), Ferrini-Mundy (1987), and Battista (1990) considered achievement in spatial areas; Noss (1987) concentrated on computers; Curio (1987) on graph comprehension. The role played by attitudinal and affective factors dominated the work of Wolleat, Pedro, Becker, and Fennema (1980), Fennema, Wolleat, Pedro, and Becker (1981), Armstrong and Price (1982), Leder (1982a), Perl (1982), Hackett and Betz (1989), Hart (1989), and Elliott (1990). Hembree (1990) focused on mathematics anxiety. Becker (1981) and Hart (1989) paid particular attention to the nature and quality of teacher-student interactions, Jones (1987) to the influence of previous course taking; while Marrett and Gates (1983) and Moore and Smith (1987) took account of race as an important additional variable. Many of the studies already mentioned, as well as those of Fennema and Sherman (1978), Armstrong (1981), Marshall (1983), Rubinstein (1985), Hanna (1986), and Smith and Walker (1988) used a variety of measures to explore a possible link between sex role orientation and mathematics learning.

Not all studies reported statistically significant differences in the performance of females and males. No such differences were found, for example, by Callahan and Clements (1984), Szetela and Super (1987), Curio (1987), and Smith and Walker (1988) and Ethington (1990).

The theoretical model proposed by Reyes and Stanic (1988) to explain group differences in mathematics achievement represents a useful attempt to synthesize the information gained from different studies. It illustrates not only the progress already made in this direction but also the need for continued work to enhance our understanding of this complex area.

Differential participation by males and females in post compulsory mathematics courses also attracted its share of research interest. Armstrong (1981), Olson and Kansky (1981), Perl (1982), Armstrong and Price (1982), Leder (1982a), and Mura (1987) in particular concentrated on this issue. The roles played by attitudinal factors (Armstrong & Price, 1982; Leder, 1982a; Mura, 1987; Perl, 1982) and school-based variables (Marrett & Gates, 1983; Olson & Kansky, 1981) dominated in this work.

The overview (Table 24.3) of the work published in *The Journal for Research in Mathematics Education* indicates the diversity of approaches used to explore gender differences in mathematics learning and serves as a useful introduction to a more comprehensive review of research in this area. Before presenting a more careful delineation of males' and females' patterns of performance and participation rates, the varied terminology used in the entries in Table 24.3—sex differences, sex-linked differences, and gender differences—needs to be addressed.

A Question of Terminology

There is much confusion and inconsistency in the literature in the ways the terms "sex" and "gender" are used. With time, "sex difference" is increasingly being used to refer to biological distinctions between females and males; "gender difference," to nonbiological characteristics, psychological features or social categories (Deaux, 1985; Deaux & Major, 1987; Unger, 1979). To a sensitive reader, it could be argued, sex differences in mathematics learning now convey something quite different from gender differences in mathematics learning. The former seem to focus on characteristics that are not amenable to change, whereas the latter highlight the role played by the environment, both personal and situational, in which learning occurs. Halpern (1986) challenged this interpretation:

I have decided to use the term "sex" to refer to both biological and psychological aspects of the differences between males and females because these two aspects of human existence are so closely coupled in our society. It is frequently difficult to decide if the differences that are found between females and males are due to biological (sex) differences or the psychological concomitants (gender) of biological sex. ...The preference for the term sex differences is not meant to imply a preference for biological explanations. (p. 12)

The final sentence, which indicates a need to deny that use of the term sex differences denotes a preference for biological explanations, provides one reason for the rejection in this chapter

TABLE 24–3. Articles on Gender and Mathematics in *JRME*, 1978–1988

Year	Authors	Content Summary
1978	Fennema & Sherman	The data presented are part of a larger study. Previously, sex-related differences in math achievement and cognitive and affective variables related to these differences had been explored in grades 9–12. This study follows that theme in grades 6–8. Measured were computational skills, knowledge of concepts, problem-solving ability, verbal ability, spatial visualization skills, and attitudes to mathematics. The authors concluded, "When relevant factors are controlled, sex-related differences in favor of males do not appear often, and when they do they are not large."
1980	Duval	The findings of this study indicated that teachers are not influenced unduly by the sex of the student nor that student's previous performance when they assign a grade on a geometry examination paper.
	Swafford	Changes in achievement, attitudes and applied problem-solving skills that occurred while studying first-year algebra were measured. While "females achieved as well as, and in some instances better than males" on the standardized first-year algebra test, the latter stereotyped math as a male domain more strongly than females.
	Wolleat, Pedro, Becker, & Fennema	The authors used the specially devised Mathematics Attribution Scale to examine the relevance of causal attribution theory to achievement in mathematics. They found that females, more strongly than males, attributed success and failure in mathematics according to a pattern described as learned helplessness. Females were more likely to use effort but less likely to give ability as explanations for success; ability and task were more often perceived as the reasons for failure. The attributional patterns of high achieving females were found to be particularly dysfunctional. "Sex differences in attribution to Effort in response to Success events are most pronounced at the highest level of Achievement."
1981	Fennema, Wolleat, Pedro, & Becker	An intervention program, Multiplying Options and Subtracting Bias, designed to increase females' participation in high school mathematics courses is described. The program addressed female and male high school students, their parents, math teachers, and counselors. Its main aims were to inform these groups about the importance of mathematics, its role in determining future educational and career paths, the level of bias in the presentation and delivery of mathematics; and to provide the participants with the knowledge and motivation to effect change. The results obtained indicated that the program was effective in influencing students' attitudes about math and their willingness to take more mathematics courses.
	Becker	The treatment of male and female students in high school geometry classes was the focus of this study. Factors that might negatively "affect the continued study of mathematics by young women" were of particular interest. Ten teachers were each observed 10 times. Both quantitative and qualitative data were collected. Becker reported that teachers tended to ask males more questions and to give them more attention. "The classroom environments on the whole reinforced traditional sex-typing of mathematics as male. Nothing was discerned, in teacher language and behavior or the physical environment, that could be considered as working in a positive way to stimulate young women to continue their study of mathematics." Three underlying reasons for this were suggested by the author: Teachers' expectations of students were based on sex; teachers treat students differently on the basis of sex; and students respond differently in class, on the basis of what teachers, and the wider society, expect.
	Armstrong	The findings of two national surveys (the Women in Mathematics study and the 1977–1978 National Assessment of Educational Progress data) were used to describe males' and females' achievement and participation in mathematics. Differences in mathematics achievement at the beginning and the end of high school and differences in intended and actual participation in high school mathematics courses are presented and summarized as follows: "Females enter high school with mathematical skills the same as or greater than males. Sometime during the high school years, males catch up with and even suprass the females in certain areas of achieve-

TABLE 24–3. (*Continued*)

Year	Authors	Content Summary
		ment." The differences observed are according to Armstrong, not simply a function of differences in participation, nor are they a function of sex differences in spatial visualization.
	Fennema	In this brief report Fennema reflects on the responsibilities of those engaged in, and reporting on, research that has strong social implications. She notes that secondary reporting of such work is often more extreme than is warranted by the more cautious tone of the original communication. Specific reference is made to the conclusions drawn, particularly by naive readers, from an article on sex differences in mathematics achievement originally published in *Science* by Benbow and Stanley and subsequently given much publicity in the print media.
	Olson & Kansky	Paired information was gathered on the precollege mathematics experience and the occupational aspirations of students entering the University of Wyoming. The number of mathematics courses taken, as well as the highest level of mathematics achieved, varied with the sex of the student. While males generally took more mathematics courses and reached a higher level of achievement than did females, the differences were most pronounced in the largest schools. Because of their poorer mathematics background the occupational choices available to females were more limited.
1982	Perl	A secondary analysis of data from the National Longitudinal Study of Mathematics Achievement was used to explore why more females than males study mathematics only up to the minimum level required for high school graduation or college entrance. Emphasis was placed on measures of ability, student interests and aspirations, attitudes towards mathematics, and achievement in mathematics. For both females and males, ability and achievement in mathematics, followed by positive attitudes to mathematics, differentiated most strongly between electors and nonelectors of beyond-minimal-requirement mathematics courses. As in previous research, significant differences were found in males' and females' perceptions of the usefulness of mathematics. However, negative attitude factors (confidence and anxiety) did not discriminate between electors and nonelectors for either sex.
	Armstrong & Price	A national survey of high school seniors was used to identify factors that affect women's participation in mathematics. Results are presented in three areas: the relative importance of factors that affected students' decision to take mathematics, the correlations between students' attitudes and participation in mathematics and the findings from predictive models of students' participation in mathematics. Attitudes toward mathematics, the perceived usefulness of mathematics for educational and career goals, and the positive influences of key references groups—parents, teachers, counselors, and peers—were found to be particularly important. Intervention programs should be planned accordingly.
	Leder (1982a)	The relationship between the "fear of the consequences of success" (FS) construct and continued participation in mathematics was the main focus of this study. Findings reported included the higher performance in mathematics of males, compared with females, in grades 7 and 10 (i.e. while mathematics was compulsory), the greater proportion of males intending to proceed to intensive mathematics courses, the higher FS scores of students—both males and females—intending to take a less traditional course, and the more complex relationship for females between FS and performance in mathematics. "For girls high FS is more likely to be associated with performing well in a traditional male field and not a type of post hoc rationalization for opting out of serious competition in that field." The author argued that the results obtained highlight the continuing effect of environmental pressures on gender differences in performance and participation in mathematics.
	Dreyfus & Eisenberg	This study assessed the intuitions students have on selected functional concepts presented in diagram, graph, and table format. In brief, students' intuitions grew as they progressed through school, high level students demonstrated correct

TABLE 24–3. (*Continued*)

Year	Authors	Content Summary
		intuitions more often than low level students, the setting—concrete or abstract—did not affect the intuitions, and no differences were observed in the intuitions of boys and girls in junior high school. Nevertheless, the data suggested that girls and boys developed their intuitions at different rates. "It may well be that the intuitive advantage that girls appear to have may be exploited in teaching."
1983	Linn & Pulos	The effect of aptitudes and experience on sex-related differences in proportional reasoning was investigated in this study. Less than half of the subjects (7th, 9th, and 11th graders) consistently displayed proportional reasoning. Females tended to be less successful than males. The variables used to assess aptitude clarified what proportional reasoning measured for both sexes. For both females and males the aptitude and experience measures correlated substantially with performance on proportional reasoning. "Contrary to expectations, aptitudes commonly thought to explain sex differences in mathematics performance did not account for sex differences in proportional reasoning." The authors suggest "that different mechanisms explain each observed difference."
	Schultz & Austin	The authors examined the effects of transformation type and direction on students' (first, third, and fifth graders) understanding of transformations. Slide, flip, and turn tasks were presented. No differences were found in the performance of boys and girls on the tasks.
	Marrett & Gates	Male-female differences in enrollment in mathematics courses among students in predominantly black high schools were described in this study. No male-female differences in enrollment by track were found. Most of the mathematics students were in the lower track, however. Enrollment patterns seemed to vary more by school than by sex. The authors conclude that future investigations should concentrate on factors in the school environment that either facilitate or impede student enrollment in mathematics courses.
	Marshall	Sex differences in mathematics performance were examined through the application of distractor analysis to data collected from assessment tests administered to all grade 6 students in Californian public schools between 1976 and 1979. Of prime interest were the errors made by each sex on individual test items. Differences were found in boys' and girls' selection of multiple-choice responses. Girls were more likely than boys to misuse spatial information, apply irrelevant rules, choose an incorrect operation, make errors of negative transfer and of key word association. Boys were more likely than girls to choose distractors that reflected errors of perseverance and formula interference. The author cautions that the differences reported apply only to students unable to solve the items correctly.
1984	Ethington & Wolfle	Data from the national longitudinal "High School and Beyond" study were used to examine differences in mathematics achievement between men and women. A covariance-structures causal model of mathematics achievement was used to examine the effect of selected variables (measures of general intelligence, socio-economic background, sex, spatial visualization, attitudes towards, prior coursework, and prior achievement in mathematics) on each other and on mathematics achievement. The authors reported that "women score(d) somewhat lower than men on a combined test of mathematics even after controlling for the effects of parental education, spatial and perceptual abilities, high school grades, attitudes towards mathematics, and exposure towards mathematics courses." They concluded that there is a complex interaction between sex, selected other variables, and mathematics achievement.
	Callahan & Clements	Data gathered over a five-year period on the rote-counting skills of students in first grade were presented in this study. The authors argued that the slight differences found in the counting performance of boys and girls entering first grade do not warrant differential instruction. They also cautioned that different data-gathering techniques and different statistical treatments of the data could lead to somewhat different conclusions.

TABLE 24–3. (*Continued*)

Year	Authors	Content Summary
1985	Rubinstein	Using a sample of eighth grade students, computational estimation and related mathematical skills were examined in this study. The different types of computational estimation tasks used were found to have different levels of difficulty. For example, order-of-magnitude estimation tasks were found to be considerably easier than open-ended estimation tasks, estimation with decimal numbers was more difficult than estimation with whole numbers, but estimation in a numerical setting was not more difficult than in a verbal format. Boys scored higher than girls on the total estimation test and on the order of magnitude scale, but not on the open-ended or reference number scales.
	Fennema & Tartre	This study used a longitudinal design to explore the relationships between mathematical problem-solving performance, spatial visualization, verbal skills, and sex-related differences in mathematics performance. The authors focused on students with discrepant verbal and spatial visualization skills to maximize the likelihood of gathering useful data. They summarized the findings of their complex study as follows: "Students who differed in spatial visualization skill did not differ in their ability to find correct problem solutions, but students with a higher level of spatial visualization skill tended to use spatial skills in problem solving more often than students with a lower level of skill. Girls tended to use pictures more during problem solving than boys did, but this did not enable them to get as many correct solutions. Low spatial visualization skill may be more debilitating to girls' mathematical problem solving than boys'."
1986	Hanna	Data from the Second International Mathematics Study were used to examine gender differences in the mathematics achievement of Canadian eighth grade students. Five areas were surveyed: arithmetic, algebra, probability and statistics, geometry, and measurement. No differences were found in the performance of boys and girls on the first three subtests described. However, for geometry and measurement the boys' mean percent of correct responses was somewhat higher than for the girls. "These differences, though not large, were statistically significant at the .01 level." Girls' overall performance was further affected by their much higher omission rates compared with boys (approximately 3:2) on all five topics.
1987	Mura	The performance of students enrolled in undergraduate mathematics courses at Canadian universities was compared with their predicted final grades in these courses. While for both sexes more students overestimated than underestimated their performance, males were more likely than females to overestimate, females more likely than males to underestimate, their final grades. No gender differences were found in students' expectations of successfully completing the bachelor's program, but significantly more males than females intended to proceed to doctoral studies and believed themselves capable of completing a Ph.D.
	Moore & Smith	Data gathered for the Youth Cohort of the National Longitudinal Study of Labor Force Behavior were analyzed to compare the mathematics achievement of young men and women, aged between 15 and 22, who varied in ethnic background and schooling. Males generally outperformed their female counterparts on the arithmetic reasoning and mathematics knowledge tests, that is, the mathematics tests administered. However, the size of the difference varied across ethnic groups (larger for whites and Hispanics than for blacks) and educational level (generally absent for those with less formal schooling but increasing with an increase in educational level).
	Ferrini-Mundy	The effects of spatial training on calculus achievement, spatial visualization ability, and use of visualization in solving problems on solids of revolution were investigated in this study. Students enrolled in a calculus course individually completed six spatial-training modules over an eight-week period. No differences were found in females' and males' performance on a variety of measures administered before the training sessions began. The females, however, performed better on a calculus exam administered after the training sessions. "Perhaps the most interesting finding . . . is that practice on spatial tasks enhanced women's ability and tendency to visualize while doing solid-of-revolution problems."

TABLE 24–3. (*Continued*)

Year	Authors	Content Summary
	Jones	In an earlier study, analysis of data from the High School and Beyond project had indicated that sophomore-year verbal and mathematics test scores, student socioeconomic status, and particularly the number of previous mathematics courses taken were good predictors of senior-year mathematics scores. The present study examined, and confirmed, that "the findings of the influence of mathematics courses taken pertain equally to subgroups of students classified by ethnicity (black or white) and sex."
	Szetela & Super	The main focus of the study concerned the effectiveness of a year-long program that emphasized selected problem-solving strategies. Whether sex of the student and the additional use of calculators also affected problem-solving performance was examined as well. The authors concluded from their results that sex differences in mathematics problem solving in grade 7 were "either very few or non-existent."
	Noss	Geometrical concepts (length and angle) that children learn through Logo programming were investigated in this study. While the Logo experiences generally tended to enhance the students' performance, the trend was most marked for girls. "The results of the study indicate a consistent (though largely nonsignificant) trend toward a differentially beneficial effect in favor of the girls."
	Curio	The graph comprehension of fourth and seventh graders, as well as selected factors likely to impact on this, were investigated. There was no evidence of sex-related differences in graph comprehension.
1988	Reyes & Stanic	In this article, the authors present a model to explain group differences (related to race, sex, and socioeconomic status) in mathematics achievement. The following factors are emphasized: societal influences, teacher attitudes, school mathematics curricula, classroom processes, student attitudes, and student achievement related behavior. While these researchers are able to draw on a substantial body of relevant research already carried out, much further work needs to be done to test the various elements in the model.
	Smith & Walker	An analysis of data from three 1979 New York State Regents mathematics examinations revealed that females performed slightly better than males on the ninth and eleventh grade papers, while males did better on the tenth grade paper. No substantial differences were found in the participation rates of female and male students in these examinations. The authors concluded that males and females "appear to perform as well on curriculum-specific tests" provided the two groups have a similar history of previous course taking in mathematics.
1989	Hart	Selected classroom processes were compared for high achieving 7th grade girls and boys who differed in confidence in their ability to do mathematics. Results, reported under the main headings of public and private interactions, instances of teacher-initiated behavioral praise or criticism, and time engaged in learning mathematics, revealed a number of small but consistent differences in the ways in which the girls and boys participated in mathematics classroom processes. More differences were found for the public than for the private teacher–student interactions. While some differences were also found between high-confidence and low-confidence students, "these were neither as consistent nor as pervasive as those between girls and boys." The author commented that, generally, the differences observed were not as large as those reported in earlier studies.
	Hacket and Betz	The relationship between performance and self-efficacy in mathematics, attitudes toward mathematics, and the choice of mathematics-related majors were examined in this study for a group of (volunteer) undergraduate students. The authors reported that both mathematics performance and self-efficacy (measured using an instrument previously developed by them) were positively and significantly correlated with attitudes towards mathematics and masculine sex-role orientation. They also concluded that differences in females' and males' mathematics self-efficacy ex-

TABLE 24–3. (*Continued*)

Year	Authors	Content Summary
		pectations were correlated with performance differences between the two groups; that males, and to a slightly lesser extent females, tended to overestimate their performance; and that mathematics-related self-efficacy expectations rather than past or current mathematics performance predicted mathematics-related educational and career choices.
	Kirshner	Students' syntactic knowledge representations in algebra were investigated in this study. The author argued that the visual characteristics of algebraic notation are highly correlated with propositional rules of algebraic syntax. Generally, no differences were found in the average performance of males and females on the algebra test administered. However, differences in performance in favor of males were reported on the unspaced complex nonce algebra items. The author concluded that proportionately more females than males depend on visual cues for syntactic decision making.
1990	Hembree	In this article the author used meta-analysis to synthesize the findings of 151 studies—mainly journal articles, ERIC, documents, and doctoral dissertations—concerned with the nature, effects, and relief of mathematics anxiety. Collectively the research surveyed suggested that mathematics anxiety depresses performance in mathematics, is broader than test anxiety and more appropriately designated as "a general fear of contact with mathematics, including classes, homework, and tests." At each grade level (K-12 and post secondary) females reported higher mathematics anxiety levels than males. Yet at both the junior and senior high school levels, high anxiety affected mathematics performance less for females than for males. High-anxious females were on average also more likely to continue with mathematics than high-anxious males.
	Battista	The extent to which spatial visualization and logical reasoning skills, gender, and teacher-linked differences affect performance in geometry were examined in this study. Briefly, for both males and females spatial visualization and logical reasoning were important determinants of geometry achievement, success in problem solving, and strategies used. However, spatial visualization and logical reasoning appeared to contribute differentially to the performance of females and males. While no evidence was found of gender differences in logical reasoning or use of geometry problem-solving strategies, male high school students on average scored significantly higher than their female peers on spatial visualization, geometry achievement, and geometric problem-solving tasks. The author argues that the teacher effects observed suggests that certain instructional practices may either exacerbate or minimize gender differences in geometry learning.
	Ethington	Data from eight of the 24 countries which participated in the Second International Study of Mathematics were examined in this study. The performance of eighth-grade students (seventh grade in Japan) in the different countries were compared for the content areas of fractions, ratio/proportion/percent, algebra, geometry, and measurement. No substantial gender effects were found for any of these areas. The very small differences observed favored females more frequently than males. The author concludes that the performance patterns observed are consistent with the hypothesis that cultural factors affect observed gender differences in mathematics performance.
	Elliott	Whether the relationship between selected affective variables and mathematics achievement differed for traditional and non-traditional (or mature age) college students was examined in this study. No significant differences in achievement were found between any of the groups (traditional/non-traditional males and females) on the pre- and posttests and for most of the affective variables measured. Previous achievement in mathematics was found to be a good predictor of future mathematics learning for all groups.

of the use of sex differences in favor of gender, or sex role, differences wherever possible. For those concerned with education, it is much more constructive to concentrate on factors that are at least potentially amenable to change. As is discussed in more detail later, even those (for example, Benbow, 1988) who focus on biological considerations concede that their effective contribution to gender differences in cognitive functioning is relatively small, and certainly too small to account for the differences in mathematics learning reported in the literature. Differences attributed to biological factors have less impact than the far greater pressures imposed by social and cultural stereotypes about cognitive skills, appropriate behaviors, and educational and life patterns. Thus it seems appropriate to select a terminology that emphasizes, rather than conceals, cultural pressures and socialization processes.

Participation Rates

In the United States close to 90% of the student population, both males and females, remain in school until the age of 17. A slightly higher proportion of females continue in full-time education beyond that age (Windschuttle, 1988) though somewhat more males than females complete four or more years of college (U.S. Bureau of Statistics, 1988). The trend for more females than males to remain in full-time education till the end of secondary schooling has also been noted in the United Kingdom (Annual Abstract of Statistics, 1988) and Australia (Dekkers, de Laeter, & Malone, 1986). Thus it can not be claimed that retention per se is an area of disadvantage for females. However, substantial anomalies appear when retention in specific subject areas, such as mathematics, is considered. Gender differences in participation in mathematics courses once they are no longer compulsory are in fact well documented and persistent. The National Assessment of Educational Progress (NAEP) data, for example, reveal that more American males than females take the more advanced high school mathematics courses.

The increased tendency toward taking more advanced mathematics courses was evident for both males and females. Up to the level of second-year algebra, little difference was found in the mathematics courses that male and female seventeen-year-olds reported taking. Significantly more males than females, however, reported taking precalculus or calculus courses, and advanced course enrollment by male students increased substantially from 1982 to 1986. (Silver et al., 1988, p. 724)

The same data source also reveals slightly higher enrollment rates for males, compared with females, in elective mathematics courses: probability and statistics, and a full year of computer programming (Meyer, 1989). These global data obscure possible between-States differences, however (Jacobs & Wigfield, 1989).

The comprehensive analysis of upper secondary school enrollment patterns in mathematics and science carried out in Australia by Dekkers et al. (1986) similarly indicated that "the most significant feature of the mathematically-specialised programmes in most States is that they are studied by at least twice as many males as females" (p. 53).

The pervasiveness of gender differences in participation in post-compulsory mathematics courses is confirmed by Schildkamp-Kündiger's (1982) review of international data on gender and mathematics. Participation differences were found to be most marked for high level, intensive mathematics courses and in applied fields requiring such courses as prerequisites.

The consequences of the different participation rates of males and females in mathematics and physical science subjects are far-reaching. Mathematical qualifications are commonly used as a critical entry barrier to tertiary courses, further training, and apprenticeships (Hansen, 1981; Sells, 1973). Thus those students who prematurely opt out of mathematics and related courses face more limited educational opportunities and career paths.

Performance

The somewhat contradictory findings with respect to gender differences in achievement in mathematics revealed by the overview of recent relevant research published in *The Journal for Research in Mathematics Education,* and summarized in Table 24.3, are consistent with the conclusions to be drawn from a careful reading of the wider literature.

There is much overlap in the performance in mathematics of males and females. Consistent between-gender differences are dwarfed by much larger within-group differences. Few consistent differences in performance in mathematics are reported at the early primary school level. However, substantial evidence suggests that by the beginning of secondary schooling males frequently, though certainly not invariably, outperform females on standardized tests of mathematics (see, for example, Brandon, Newton, & Hammond, 1987). Whether or not such differences are found seems to depend on the content and format of the test administered (Armstrong, 1985; Hanna, 1986 ; Hanna, Kündiger, & Larouche, 1988; Kimball, 1989; Marshall, 1983; Pattison & Grieve, 1984; Senk & Usiskin, 1985; Silver et al., 1988; Smith & Walker, 1988), the age level at which the testing takes place (Carpenter et al., 1988; Dossey, 1985; Fennema & Carpenter, 1981; Hilton & Berglund, 1974; Joffe & Foxman, 1988; Leder, 1988b), and whether classroom grades or standardized tests of achievement are considered (Kimball, 1989). The differences are particularly marked on high cognitive level questions and for high-achieving students (Benbow, 1988; Benbow & Stanley, 1980, 1983; Edwards, 1985; Fennema & Carpenter, 1981; Fox & Cohn, 1980; Hall & Hoff, 1988; Kissane, 1986; Peterson & Fennema, 1985; Wagner & Zimmerman, 1986). Yet various authors, including Becker and Hedges (1988), Braine (1988), Hyde, Fennema, and Lamon (1990), and Leder (1986c) have cautioned against the tendency to overgeneralize the occurrence of gender differences from highly selected groups to the broader student body.

In a recent article Benbow (1988) alluded to reported consistencies in the differences in performance between outstanding female and male mathematics students:

The ratios of high scoring boys to girls has remained relatively constant over the 15 years. Thus, sex differences in SAT-M scores among young adolescents are not temporary trends. They have been stable even in times of great change in attitudes towards women. (Benbow, 1988, p. 172)

Other writers (Freed, 1983; Kimball, 1989; Rosenthal, 1988) have argued that the performance gap for such students has in fact narrowed in recent years. Variations in the criteria by which children were selected for inclusion in these studies and the nature of the tasks to be performed on the tests make cross-study comparisons difficult.

Findings that have emerged from American studies (Armstrong, 1981; Ethington & Wolfe, 1984; Marshall, 1983; Moore & Smith, 1987; Silver et al., 1988) of the lower performance of females as a group, compared with males as a group, in public examinations and large scale testings have frequently been replicated in other countries. Data gathered in the two International Studies of Mathematics achievement conducted to date provide relevant information. The first study, in which 12 countries participated, revealed "clear differences in favor of boys in all populations in both verbal and computational scores" (Husen, 1967, p. 241). Yet girls in some countries performed better than boys in others. Comparable data were presented by Hanna (1989), Hanna and Kündiger (1986) and Hanna et al. (1988) in their summaries of results from the Second International Mathematics Study (SIMS). The former focused on the younger (grade 8) samples in 20 countries whereas the latter concentrated on data for grade 12 students in 15 countries. Hanna and Kündiger (1986) summarized their findings as follows:

Most of the differences did not reach statistical significance at the 1% level. Moreover, the differences that did reach statistical significance were not large, ranging from +5 to −7. Looking at each subtest separately it appears that for two of the five topics, Measurement and Geometry, the significant differences occurred consistently in the boys' favour: in 7 of the 20 countries boys had higher p-values and in 10 countries boys had higher p-values in Measurement and Geometry, respectively. (pp. 6–7)

With respect to the older students Hanna et al. (1988) reported significant differences in favor of males on at least one of the seven subtests for 14 of the 15 countries for which data were examined. In fact, significant differences, all in favor of males, were found for 62 of the 105 comparisons made. Such a generalization should again be qualified, however. The data on the Analysis subtest, for example, revealed that whereas boys scored better than girls in each of the 15 countries for which the comparisons were made, the mean score obtained by girls in some countries was considerably higher than that achieved by boys in others.

Analyses by Ethington (1990) of data for eight countries which participated in the SIMS suggest that identification of gender differences may also be influenced by the statistical techniques used to examine the data. She reported gender differences which were small, not uniformly in favor of males, and inconsistent across countries.

Over the past decade analysis of data using meta-analysis, a statistical or quantitative approach to synthesizing results from a range of studies on a common topic, has become more prevalent. Concentrating on studies listed by Maccoby and Jacklin (1974), Hyde (1981) and Rosenthal and Rubin (1982) concluded that gender differences in quantitative and visuo-spatial abilities were small and accounted for no more than 1 to 5% of the population variance. A subsequent analysis (Becker &

Hedges, 1988) controlled for the selectivity of the samples used and revealed that gender differences seemed to be larger for older studies. Friedman (1989) focused on later work, that is, published between 1974 and 1987. Pre-school and college samples were excluded from the analysis. Data on younger students predominated among the 98 studies used in the meta-analysis. Overall, no difference was found in the performance in mathematics of males and females. To allow for a comparison with the earlier meta-analysis of studies listed by Maccoby and Jacklin (1974), Friedman (1989) considered work with high school samples separately. In these, males performed slightly better than females but differences in favor of males again appeared to be decreasing over time. Hyde et al. (1990) similarly reported a decrease in the magnitude of gender differences in mathematics performance since 1974. Their meta-analysis indicated that there were no differences in various areas (for example, computation and mathematical concepts), though differences in favor of males occurred on problem-solving tests with high schools, colleges, and highly selected samples (for example, high-achieving students).

Differential Course Work Hypothesis

A number of studies have examined the effect of the differential patterns of males' and females' enrollments in mathematics courses on their achievement in mathematics. Although the findings obtained have not been unanimous, there is substantial evidence that links differences in exposure to formal mathematics courses to gender differences on standardized achievement mathematics tests (Alexander & Pallas, 1983; Ethington & Wolfle, 1984; Fennema & Sherman, 1977; Hanna, 1988; Jones, 1987; Moss, 1982; Pallas & Alexander, 1983; Rosier, 1980; Wise, Steel, & MacDonald, 1979). Controlling for comparability of in-school exposure to mathematics for females and males ignores possible differences in their participation in less formally organized and outside school mathematical experiences, however. The conclusions of Ethington and Wolfle (1984) are representative of current work:

The present research suggests once again that a great deal of the difference in mathematics achievement between men and women can be explained by differences in background, ability, attitudes, grades, and formal exposure to mathematics in the classroom. Of these, the variables measuring exposure to mathematics had the most influence on explaining variation in mathematics achievement. (p. 376)

A more detailed discussion of other factors alluded to in this excerpt is left for later sections of this chapter.

Summary

A number of consistent, though relatively small, gender differences in participation and performance in mathematics have been described in the literature. Briefly, differences in participation continue to be observed in higher level, more intensive mathematics courses and related applied fields. While generally there is much overlap in the mathematical performance of females and males, where they occur, significant differences in performance tend to favor males, particularly on higher cognitive level questions. The consequences of these data are compounding and far-reaching, for the differences observed are often accompanied by differences in the ways males and females regard themselves, and are regarded by others, as learners of mathematics. These differences further reinforce and perpetuate inequalities in a way that is highlighted by the model of academic choice proposed by Parsons et al. (1983) who argued that, when choices are available, the path selected is influenced by individuals' interpretations of reality as well as by reality itself. Students capable of continuing with mathematics but who believe that this subject is inappropriate for them are more likely to self-select out of mathematics courses and other areas for which such work is a prerequisite than students who are comfortable about taking mathematics.

POSSIBLE EXPLANATIONS

A large number of explanations, as well as various theoretical models, have been put forward to account for the observed gender differences in mathematics learning.

Theoretical Models

Fennema and Peterson (1985) have argued that external and societal influences affect internal motivational beliefs and students' autonomous learning behaviors, and that these in turn contribute to gender differences in mathematics achievement. The interactive qualities of expectancy for success and the perceived value of the task are stressed in the model proposed by Eccles et al. (1985) to explain students' decisions to enroll in mathematics courses. The cultural milieu, the behaviors, attitudes and expectations of socializers, the child's perceptions of these attitudes and expectations, the individual's goals and general self-schemata or self-image, the perceptions of the value of the task, achievement behaviors, expectancies, task-specific beliefs, past events as well as their interpretations, and the differential aptitudes of the child are all included as integral and interacting components of the model.

In their more general model of gender and social interaction Deaux and Major (1987) also focus on the individual's goals and self-schemata, the expectancies and goals of others with whom the individual interacts, and the context in which the interactions occur. The importance of different components, they argue, is not stable but will rather fluctuate with characteristics of the expectancy—its perceived social desirability, certainty, and situational context—and with the individual's concern with self-presentation or self-verification.

The model proposed by Leder (1986c) emphasized variables that are most pertinent to educators and which, with increased appreciation of their relationship to the learning of mathematics, should lead to improved classroom practices. Singled out were factors associated with the environment as well as ones associated with the learner. Included among the former were situational factors such as society, home, school, and classroom variables; personal variables, and in particular parents, peers, teachers as well as more general socializers; and curriculum variables like content areas of mathematics, types of items and

methods of assessment and instruction. Learner-related factors comprised cognitive variables, for example spatial ability, verbal ability, and mathematical ability, as well as psycho-social variables including achievement motivation, confidence, conformity, self-esteem, and independence.

An explicit emphasis on the mathematics teacher and classroom is also found in the model of Reyes and Stanic (1988) of group differences in mathematics achievement. The key components of their model—societal influences, teacher attitudes, school mathematics curricula, classroom processes, students' attitudes, and achievement-related behaviors—are hypothesized to explain not only gender but also race and socioeconomic status-related differences in mathematics learning.

The various models described share a number of common features: the emphasis on the social environment, the influence of other significant people in that environment, students' reactions to the cultural and more immediate context in which learning takes place, the cultural and personal values placed on that learning and the inclusion of learner-related affective, as well as cognitive, variables. While the proponents of the different models generally point to considerable support for their preferred representation, they also typically acknowledge the need for further testing, clarification, and substantiation of the variables identified and the links hypothesized.

Biological Variables

None of the models discussed placed particular emphasis on biological variables. The reasons for that are not difficult to trace. Those who have examined biological explanations for gender differences in mathematics learning often conclude, as did Linn and Petersen (1986):

Some have suggested that spatial ability differences might be biologically determined and provide the mechanism for a biological influence on mathematics and science. No evidence for that view can be found. (p. 94)

Furthermore, strong advocates of the notion that sex differences can be attributed at least partly to biological factors nevertheless point to the importance of personal and environmental factors.

We, therefore, cannot discount the very important interaction effects of the environment (such as home life, choice of school, adequacy of teacher, proper attitude, and family support) with genetic endowment which result in students with varying degrees of quantitative reasoning ability. (Stafford, 1972, pp. 198–199)

Even though biological factors seem to be involved in determining the sex difference in mathematical reasoning ability, this does not imply that efforts at remediation cannot make a difference....Practically speaking, one must be an environmentalist. (Benbow, 1988, p. 182)

ENVIRONMENTAL VARIABLES

In the remainder of this section the link is examined between gender differences in mathematics learning and variables such as those contained in the models described above.

Environmental variables are considered first. The use of distinct headings is convenient but somewhat artificial. An implicit thread running through the review is the link between the different components selected and the reciprocal interaction between factors in the environment and the individuals who function in it. This complex interaction between variables is consistent with the approach of the models described above.

School Variables

There are a number of ways in which schools, and teachers within the schools, differentiate between groups of male and female students. The former do so through their organizational procedures, the latter through their behaviors, expectations, and beliefs.

In recent decades, gender-segregated education has come to be viewed as an anachronism, reflecting outmoded beliefs that males and females have different educational needs. Coeducation was assumed to be the avenue through which parity of treatment could be achieved. The degree to which a community provided education in a sex-segregated setting was considered to reflect its commitment to the notion that boys and girls require different preparations for their respective adult roles. However, the extensive amount of research that highlights that coeducation per se does not achieve parity for female students or equity in policy or practice has brought about a re-examination of this assumption. Various school- and system-level interventions aimed at improving the learning climate for girls have experimented with single-sex settings (Fox & Cohn, 1980; Lee & Bryk, 1986; Lockheed, 1985; Marsh, Owens, Marsh, & Smith, 1989; Sampson, 1983, 1989; Smith, 1986, 1987; Stage, Kreinberg, Eccles, & Becker, 1985). Collectively, these studies suggest that carefully timed, organized, and implemented programs may indeed lead to qualitative (i.e., attitudinal) if not quantitative (i.e., achievement) benefits in the learning of mathematics for at least some females. As importantly, they have highlighted the many subtle factors which may disadvantage females in a coeducational setting, for example, time-tabling of courses, choice of curriculum, textbook selection and content, availability of equipment, methods of assessment, counselors' advice, administrators' implementation of certain instructional policies, as well as students' own perceptions of the learning climate.

A particularly vivid illustration of the inequities that may accompany single-sex schooling emerged from a fairly recent and well publicized test case in Australia (Shorten, 1990). A family living in the metropolitan area of Sydney sent their daughter to an all girls' school and her twin brother to a neighboring all boys' school. Concerned by the more limited resources and curriculum selection available at her school, the girl claimed discrimination, fought her case in the courts, and won on a number of points. The implications of the verdict reached are still being explored.

The benefits and disadvantages of coeducational and segregated education are worthy of further careful scrutiny. The somewhat contradictory research evidence to date does not support the adoption of long-term segregated mathematics classes. While the Assessment of Performance Unit (1982), Cockcroft (1982), Harding (1981), Ormerod (1981), Sampson

(1983, 1989), and Smith (1986) have argued that girls studying mathematics and science seem to be disadvantaged in a mixed school setting, Bone (1983), Dale (1974), Marsh, Owens, Myers, and Smith (1987), Marsh et al. (1989), Steedman (1983), and Smith (1987) reported that girls in coeducational settings and schools performed at least as well in mathematics as those in single-sex schools. The work by Smith (1986; 1987) is of particular interest. His longitudinal study of students in an English coeducational high school suggested that many of the girls monitored benefited from segregated mathematics teaching the first two years of the school, that is, when the students were in the 11- to 13-year age group, but by a coeducational setting in years 9 and 10, that is, when the students were 14 to 16. A useful review of other relevant studies can be found in Gill (1988).

Using data from the *High School and Beyond* study, Lee and Bryk (1986) found that boys performed better than girls in both mathematics and science, irrespective of type of school attended. Although females in single-sex schools showed greater performance gains from the sophomore to the senior year than those in coeducational schools in reading, writing, and science, there were no appreciable differences in the gains in the mathematics performance of females at the two types of schools. However, girls in single-sex schools were more likely to express an interest in mathematics (and English) and took slightly more mathematics courses than those attending coeducational schools. In a case study conducted in a large Victorian (Australia) coeducational high school, Rowe (1988) also reported that boys performed slightly better than girls on a specially devised test of mathematics, and that the greatest gains in mathematics achievement over the two-year period of the study were made by girls in single-sex classes, followed by boys in single-sex classes. The complexity of the issue is confirmed by the findings of another study (Eales, 1986) which revealed that students thought they were more likely to cope in class if they were with a close friend. Being in an all-girl or mixed class setting was not given the same importance.

Leder (1990a) has already pointed to the difficulty of comparing student performance across different school systems in which equipment available, staffing, and class sizes may differ substantially. The influence of these variables on mathematics achievement is confirmed by Cresswell and Gubb (1987) in their analysis of data from the SIMS. It is worth stressing that in both England and Australia where significant sections of the school population are still educated in single-sex schools, many of these cater to children from higher socio-economic homes. This point was also acknowledged by Carpenter (1985) who argued that single-sex schooling frequently operates as a subset of another social division, that of public versus private schooling. Only in recent years have studies which examine the apparent benefits or disadvantages of long-term education in a gender-segregated environment begun to control for socio-economic factors.

Teacher Variables

Teachers play a crucial role in implementing school policies, whatever the formal organization of the setting in which mathematics is taught. Indeed, the ways in which teachers interact with students have attracted much research attention. After reviewing over 130 studies on expectancy effects, Harris and Rosenthal (1985) concluded,

Teachers who hold positive expectations for a given student will tend to display a warmer socioemotional climate, express a more positive use of feedback, provide more input in terms of the amount and difficulty of the material that is taught, and increase the amount of student output by providing more response opportunities and interacting more frequently with the student. (p. 377)

Possible differences in patterns of teacher interactions with male and female students have also been examined. Brophy and Good (1974) confirmed that there were differences in the ways teachers behaved, on average, towards the two groups of students. Males, they reported, tended to receive more criticism, be praised more frequently for correct answers, have their work monitored more frequently and have more contacts with their teachers. The sex of the teacher did not seem to affect these conclusions.

More recent work has largely replicated these findings (Becker, 1981; Grieb & Easley, 1984; Hart, 1989; Koehler, 1990; Leder, 1987, 1989; Reyes, 1984; Stallings, 1979). Hart (1989) reported that while males continued to attract more attention from teachers than females for a number of the interaction categories monitored in her study, differences in the participation rates between males and females were less than described in earlier work, and were stronger for public than for private interactions between teacher and student. The conclusions of Eccles and Blumenfeld (1985) are typical.

We, like many others, have found small but fairly consistent evidence that boys and girls have different experiences in their classrooms. However, these differences seem as much a consequence of pre-existing differences in the students' behaviors as of teacher bias. Nonetheless, when differences occur, they appear to be reinforcing sex-stereotyped expectations and behaviors. (p. 12)

Differences in teacher-student interactions are not confined to American classrooms. More frequent teacher attention towards males has, for example, been reported by Galton, Simon, and Croll (1980) and Spender (1982) in English schools, by Dunkin and Doenau (1982) and Leder (1988b, 1990b) in Australian classrooms, and by Moore and Smith (1980) in Papua New Guinean classrooms. Furthermore, data gathered by Walden and Walkerdine (1985) in English mathematics classrooms revealed that teachers on average provided less encouragement and reinforcement to work independently and persistently to females than to males.

The differences in interaction patterns between teachers and certain groups of students are of more than theoretical interest. "Even if the impact on achievement scores is not immediately evident, it is quite probable that the cumulative impact over the course of the elementary school years is significant" (Leinhardt, Seewald, & Engel, 1979, p. 437). The effect on students' long-term motivational patterns has also been noted (Fennema & Peterson, 1985; Kloosterman, 1990; Koehler, 1990; Leder & Fennema, 1990; Meyer & Koehler, 1990; Weinstein, Marshall, Brattesani, & Middlestadt, 1982) and is discussed more fully in later sections.

To supplement research concerned with consistent differences in the frequency of interactions between teachers and their male and female students, a small number of studies (Gore & Roumagoux, 1983; Leder, 1987, 1990b) have examined the duration of different exchanges, as well as the wait times given by teachers, or taken by students, to reflect or to collect their thoughts before answering questions addressed to them. The former observed that teachers gave a longer wait time when interacting with males in mathematics classes. The latter reported no significant differences in the wait times given to males or females but noted that, above grade 3, females tended to receive more time than males on routine, low cognitive questions, whereas males had slightly more time than females on the more difficult and challenging high cognitive questions which serve as a useful preparation for more advanced mathematics courses. This pattern of encouragement is consistent with the patterns of gender differences in performance reported in the literature.

Even though work such as that described above has yielded useful information, the limitations of attempts to capture the complex and multifaceted classroom environment through any one classroom observation schedule are being increasingly recognized. Additional information has accumulated through studies that have focused on other aspects of teacher behavior, on small-group instruction (Barnes, 1989; Webb & Kenderski, 1985; Wilkinson, Lindow, & Chiang, 1985), and on student-student interactions and through studies which relied on detailed, intensive observations of a small sample of students (Clarke, 1989; Koehler, 1990; Lockheed, 1985; Peterson & Fennema, 1985; Stanic & Reyes, 1986; Tobin, 1984; Webb & Kenderski, 1985; Wilkinson et al., 1985). Fennema (1990) has documented a number of subtle differences in the ways in which teachers of young children explain the successes and failures in mathematics of the male and female students in their classes. Leder (1986b) used the Adjective Check List (Gough & Heilbrun, 1980) to illustrate the extent to which teachers (as well as students) stereotyped mathematics as a male domain. Case study research (Hart & Stanic, 1989; Koehler, 1990; Leder, 1991; Meyer & Koehler, 1990) in particular has highlighted a number of other subtle differences in the in-class treatment of males and females, not necessarily captured by work carried out within the broader teacher-student interaction paradigm.

The influence of the peer group is considered next; then student-related variables.

The Peer Group

The peer group acts as an important reference for childhood and adolescent socialization and further perpetuates gender-role differentiation through gender-typed leisure activities, friendship patterns, subject preferences, educational and career intentions. Peer influences operate in the classroom as well as in the broader environment. Peer values reflect, reinforce, and shape differences in the beliefs, attitudes, and behaviors of the individuals who comprise the group.

The preference of males for active games and pastimes that focus on skills and mastery of objects and the preference of females for play concerned with inter-personal relationships is well documented (Huston & Carpenter, 1985; Kelly, 1986). It has also been argued (Horner, 1968; Leder, 1986a; Stein & Bailey, 1973; Walden & Walkerdine, 1985) that females differ in the areas in which they strive or are expected to work for success. Whereas males favor achievement in the traditionally highly valued areas of intellectual expertise and leadership skills, females have been shown to be more likely to aim for excellence in areas congruent with their traditional role, that is, areas that require social skills (Boswell, 1985; Linn & Hyde, 1989). Mathematics continues to be perceived by some students as a male domain (Boswell, 1985; Fennema, Hyde, Ryan, & Frost, 1990; Fox, Brody, & Tobin, 1985; Hanna et al., 1988; Joffe & Foxman, 1986; Leder, 1986b; Sherman & Fennema, 1977). Such attributions are reflected in the subject and course choices of individual students.

Differences in leisure time activities and particularly in attitudes towards mathematics (Boswell, 1985; Fennema & Sherman, 1977; Fox et al., 1985; Joffe & Foxman, 1986; Leder, 1988b; Sherman & Fennema, 1977) are reflected in the career expectations of males and females. The occupational intentions of the two groups confirm that competence in mathematics is a more important prerequisite for the attainment of the career ambitions of males than of females. In a number of studies (Chipman & Wilson, 1985; Fennema et al., 1990; Joffe & Foxman, 1988; Kelly, 1986; Leder, 1988b; Pedro, Wolleat, Fennema, & Becker, 1981; Wolleat et al., 1980) this view is expressed by male and female students themselves. Maines (1985) has argued more generally that

The ideology of men enables them to select mathematics as a meaningful area of activity and their life structures enable them to pursue it as a career. The ideology of women deflects them away from focused attention on mathematics and their life structures militate against their pursuit of it. (p. 317)

The pervasiveness of such beliefs is highlighted by the inclusion in the above review of studies which reported American, English, and Australian data. Their self-perpetuating and reinforcing nature is described by Lockheed (1985).

Self-selected sex segregation is well documented as a widespread phenomenon among elementary and junior high school-aged children. It has been demonstrated in studies of student friendship choices and work partner preferences that utilize socioeconomic techniques, in surveys of student attitudes, and by direct observation.... Students identify same-sex but not cross-sex classmates as friends..., choose to work with same-sex but not cross-sex classmates..., sit or work in same-sex but not cross-sex age groups..., and engage in many more same-sex than cross-sex verbal exchanges. (p. 168)

The subtle ways in which students who contravene prevailing norms are disapproved by the peer group have been described by Spender and Sarah (1980), Webb and Kenderski (1985), and Wilkinson et al., (1985).

The attitudes and beliefs attributed to the peer group reflect not only the general views of the individuals in that group but also those of the broader society. Particular emphasis is placed in the next section on other social forces not yet discussed under previous headings.

The Wider Society

Many factors contribute to the environmental climate that molds community attitudes against which students measure their beliefs, aspirations, and expectations. The important role played by the media in shaping ideas and attitudes, as well as reflecting and reinforcing popular beliefs, is widely recognized. The potential for using the media as a tool for change has also been acknowledged.

The media is [sic] a powerful determinant of attitudes. Unhappily it has all too often tended to demean, trivialise, or ignore women's individuality, their contribution to social and economic life, their struggle for justice for themselves and others, their humanness.

The need to reform media portrayal of women has been recognized by international organizations, by governments of other countries, and by non-government organizations... to ensure that practices long complained of by women are avoided. (Ryan, Macphee, & Taperell, 1983, p. iii)

The impact of media reports about gender differences in mathematics achievement has been investigated by Jacobs and Eccles (1985). They examined parents' general (that is, stereotyped) as well as specific (that is, for their own children) beliefs about gender differences in mathematics learning before and after coverage of a controversial article on this topic. Many of the parents sampled became more gender-stereotyped in their beliefs after reading media excerpts of the original work and of other relevant research reports. The authors argued that "media absorption involves self-selection according to previous attitude" (Jacobs & Eccles, 1985, p. 24).

After surveying 50 print media articles that featured outstanding mathematics students, Leder (1986a) concluded that collectively the articles provided a valuable and comprehensive summary of contemporary beliefs about such students and their educational needs. However, use of sensational headlines encouraged selective reading, according to previously held beliefs and attitudes, of the wide range of views presented in the body of the text. Other examples of selective media reporting of students' achievements in mathematics continue to be found. During a week-long science conference held in Australia in 1988 only one conference paper made the front page and headline news in a national newspaper. A biased and inaccurate summary of work presented appeared under the heading "Girls don't count in maths." The authors

were dismayed at the interpretation placed on their data and at the snide use to which it was put....[T]he media attention given to any suggestion of intellectual (and sporting!) differences between males and females are likely to have a powerful impact on community perceptions of females' mathematical talent. (Willis, 1989, p. 14)

Media portrayal of males and females more generally has also been examined (Leder, 1984a; 1986a; 1988a). Articles that featured females, but not those that focused on males, frequently reported difficulties encountered by successful professionals in balancing job demands with interpersonal needs. Such accounts, as well as stereotyped portrayals in films and on television, reinforce and perpetuate the notion that females achieve success at a price and that they need to work harder to attain this success. They are also likely to reinforce individual and collective ideas about gender-appropriate activities for males and females along the lines described in the earlier section on peer values and explored further as part of the discussion on motivational variables.

Parents

In an examination of social determinants of mathematics attitudes and performance, Eccles and Jacobs (1986) identified parents as a critical force. These authors argued that parents exert a more powerful and more direct effect than teachers on children's attitudes toward mathematics. They concluded "that parents' gender-stereotyped beliefs are a key cause of sex differences in students' attitudes toward mathematics" (p. 375). Findings from the 11-year project described by Stevenson and Newman (1986) indicated not only that parents affect their children's long-term achievement and attitudes in mathematics but also that the patterns identified differed for mathematics and reading.

Armstrong and Price (1982) and Lantz and Smith (1981) found students' attitudes towards mathematics and their decisions to continue with mathematics were linked to their parents' conceptions of the educational goals of school mathematics and their perceived relevance to their children's long-term life goals. Because parents are often believed to be more encouraging of their sons' than their daughters' mathematical studies (Fennema & Sherman, 1977; Luchins & Luchins, 1980), this influence may be particularly disadvantageous for girls. This theme is explored further in the work of Hanna et al. (1988) who used the SIMS data to examine the influence of home support as a possible explanation for the gender differences in mathematics learning identified in that study. They reported that there were high levels of home support for mathematics for both sexes in countries with low levels of gender differences in mathematics performance whereas there tended to be low levels of parental support for mathematics in countries with substantial differences in achievement. In the former case, they argued, the strong home support for mathematics counterbalanced society's stronger reinforcement for participation in masculine activities, including mathematics, for males.

Gender differences in mathematics learning cannot be explained by socialization processes alone. The inadequacy of such an approach is captured by the example, cited in Maccoby and Jacklin (1974), of the four-year-old girl whose own mother was a doctor but who nevertheless insisted that girls became nurses and that only boys could become doctors. As pointed out by Kelly (1981), socialization theories that emphasize reinforcement and imitation of the behavior of others provide important explanations of gender differences in behavior but fail to explain some inconsistencies.

LEARNER-RELATED VARIABLES

Many of the issues canvassed to date are highlighted further in a review of learner-related variables. Comprehensive, useful syntheses of early research on gender differences on a range of such variables can be found in Maccoby (1967) and Maccoby and Jacklin (1974). Through their careful documentation of the differences reported in the then-available literature they facilitated objective discussion about likely explanations for their occurrence. Gender differences in quantitative skills and mathematics performance were among the areas covered. To facilitate the continued exploration here of possible explanations for the observed gender differences in mathematics learning, a distinction is made in this section between learner-related cognitive and internal belief variables. Two cognitive areas which have attracted much research attention are discussed first.

COGNITIVE VARIABLES

Intelligence. Past performance in mathematics is generally accepted as the best predictor of future achievement in the same area. Given the considerable overlap between measures of mathematical ability and tests of general intelligence, students who do well in intelligence tests tend to do well in mathematics. To what extent, then, might gender differences in intelligence explain gender differences in mathematics learning?

It is beyond the scope of this chapter to review and discuss the variety of ways in which intelligence has been defined. Of particular interest, however, is the common emphasis in the definitions and measuring instruments used on a relatively narrow range of culturally valued and needed abilities measured by "tasks favoring middle- and upper-class males" (Khan & Cataio, 1984). Through the elimination or counterbalancing of items which yield consistent differences in large-scale testings of females and males, commonly used intelligence tests such as the Stanford-Binet or Wechsler tests have been deliberately constructed to achieve a gender neutral full-scale score. Gender differences have been found, however, on some of the subtests. In particular, differences in favor of females have been found on some of the verbal subtests and in favor of males on nonverbal performance subtests that involve spatial-visualization ability (Khan & Cataio, 1984). The latter is also frequently linked to gender differences in mathematics learning.

Spatial Abilities. The relationship between mathematics achievement and visual-spatial skills continues to be a matter of debate. Connor and Serbin (1985) described a number of links between the two: significant correlations between measures of mathematical achievement and spatial visualization, a common age (adolescence) for the first occurrence of gender differences on visual-spatial tests and standardized tests of mathematics, and the "nature of visual-spatial ability and of mathematics achievement" (p. 152) per se. Linn and Petersen (1986) disputed the second assertion.

More generally, the lack of definitional precision with respect to both spatial abilities and mathematics has been used to explain an inconsistency in the literature on the expected relationship between spatial ability, mental imagery, and mathematics performance (Clements, 1981). Indeed, the way in which spatial abilities are defined and measured affects whether or not gender differences are found (Petersen & Wittig, 1979). According to Linn and Petersen (1986) gender differences in spatial ability occur on fairly specific tasks such as those that require rapid rotation of visually presented figures or tasks that require distracting information to be discounted to allow recognition of the vertical or horizontal. Many mathematical tasks, it is worth noting parenthetically, do not require these skills. Spatial skills are useful in solving certain types of problems—those which involve the perception or assimilation of patterns, the use of diagrams or graphs, for example—but are irrelevant in others.

The most common explanations for the reported gender differences on selected spatial tasks revolve around differences in the exposure of females and males to environmental factors that enhance spatial ability, the importance of a recessive gene on the X sex chromosome (see Wittig, 1979, for a review), the involvement of sex hormones (Broverman, Klaiber, Kobayashi, & Vogel, 1968; Petersen, 1979), and a possible link with differences in hemispheric specialization (Vandenberg & Kuse, 1979; Witelson, 1979). Research support for the first explanation comes from studies that have shown that spatial visualization skills can be improved through training (Ben-Chaim, 1983; Ben-Chaim, Lappan, & Houang, 1988; Brinkmann, 1966; Connor, 1977; Connor, Schackman, & Serbin, 1978; McGee, 1978). The work of those who favor the latter three explanations continues to be challenged. For example, evidence for the X-linked hypothesis has been dismissed as tenuous (Fausto-Sterling, 1985; Khan & Cataio, 1984; Sherman, 1978); evidence for the link in humans with hormones or hemispheric specialization as ambiguous, lacking, or premature (Bryden, 1988; Hines, 1988; Kimura, 1988; Sanders, 1988; Vandenberg & Kuse, 1979; Witelson, 1988). While the conflicting findings ensure that further attention will be focused on this area, the considerable body of research to date that has rejected biological considerations as adequate explanations for gender differences in spatial abilities is noteworthy.

Some insight into the link between spatial differences and gender differences in mathematics achievement is furnished by a longitudinal study reported by Fennema and Tartre (1985) and Tartre (1990). Spatial and verbal tests were administered to students in grade 6. Subjects were chosen who were high in one skill and low in the other so that four groups were formed: high-spatial visualization (SV)/low-verbal females, high-SV/low-verbal males, low-SV/high-verbal females, and low-SV/high-verbal males. The students were asked to solve a number of mathematical problems and were interviewed about their solution methods in three consecutive years (from grades 6 to 8). The group who consistently obtained the lowest scores comprised females with high verbal but low spatial skills. This group fell further behind the others over the three-year period monitored. The corresponding male group, that is, those also with high verbal but low spatial scores, each year obtained the highest number of correct solutions to the problems administered. Thus low spatial ability seemed to disadvantage females but not males with respect to mathematics achievement. The data from this longitudinal study suggest that spatial skills are more closely related to performance on certain mathematical tasks for females than for males. They also confirm that there is no simple relationship between spatial ability and mathematics achievement.

Internal Belief Variables

The gender differences in peer group values discussed earlier are reflected in students' beliefs about their own performance and long-term expectations. Particular emphasis is placed in this part of the chapter on internal belief variables—characteristics including confidence, risk-taking behavior, motivation and related constructs such as fear of success, attributional style, learned helplessness, mastery orientation, anxiety, and persistence. Once again it is worth noting that the different variables overlap and interact, both conceptually and in the ways they are typically operationalized and measured. The distinctions made are more for descriptive convenience than as an intended reflection of measures considered to be independent.

Confidence and Related Variables. The link between confidence and achievement in mathematics has been examined in a number of studies and judged to be important (Reyes, 1984). The work of Fennema and Sherman (1977, 1978) revealed that males in grades 6 to 12 consistently showed greater confidence in their ability to do mathematics than did their female classmates. Initially the differences in expressed confidence were not reflected in differences in achievement. Although this continued to be true for the males, for the older females being confident about doing mathematics was a good predictor of achievement. For these students confidence and achievement were ultimately related. The link between confidence and performance was confirmed in a further study of the same sample (Sherman, 1980). Joffe and Foxman (1986) reported that males' and females' differing attitudes towards mathematics were paralleled by differences in the test performance of the two groups. In a later publication, Joffe and Foxman (1988) hypothesized,

It seems reasonable to assume that some aspects of attitude, perhaps confidence, may be operating—positively in the case of boys, negatively in the case of girls. It also seems likely that if these feelings are already in operation at age 11 that they will become much stronger by age 15 unless some intervention is made. (pp. 27–28)

Gender differences in confidence have also been found to be predictive of gender differences in participation in noncompulsory mathematics courses (Armstrong & Price, 1982; Lantz & Smith, 1981). Highly pertinent, too, is the finding of Eccles and Jacobs (1986), who identified students' self-concept of their mathematics ability as one of three significant student factors that affected their achievement in mathematics as well as the intention to continue with further mathematics studies.

There is considerable evidence that females make more omission errors than males and are more likely to use the "I do not know" category on tests that contain these options (Annice, Atkins, Pollard, & Taylor, 1988; Hanna, 1986; Linn, De Benedictis, Delucchi, Harris, & Stage, 1987). This behavior, it has been argued (Kimball, 1989; Rowley & Leder, 1989), is not only an implicit measure of gender differences in confidence about mathematics and a preparedness to take reasonable risks in a mathematics setting, but also contributes to the differences in mathematics performance reported.

If you have a degree of partial knowledge, you would be better to guess than to omit, since your likelihood of guessing [the right answer] is better than chance. (Rowley & Leder, 1989, p. 8)

Since any penalty applied is based on chance factors, guessing is thus likely to maximize the marks obtained by those taking the test.

Also of relevance to this discussion are the findings of Hembree (1988) who used meta-analysis to integrate the results of more than 560 studies concerned with academic test anxiety. Females, he reported, consistently showed higher levels of test anxiety than males, particularly in grades 5 to 10. Females' higher test anxiety was not necessarily reflected in performance differences, however.

The importance of considering attributes such as confidence from a number of different yet related perspectives is confirmed by the multifaceted approach adopted by Hart and Stanic (1989) in their intensive monitoring of a small group of students during mathematics lessons. The richer and more comprehensive data obtained highlighted the complex interaction between students' attitudinal and achievement related behaviors and their performance in mathematics. Their interview data in particular illustrated the difficulty of judging the level of confidence displayed by students without taking into account factors such as attributions of success and failure, enjoyment, perceived usefulness, and persistence. Such variables are considered next.

Fear of Success. The motive to achieve is commonly assumed to be a major determinant of an individual's efforts to attain success. Working within the expectancy-value model of motivation, McClelland, Atkinson, Clarke, and Lowell (1953) largely confirmed that achievement motivation was influenced by three major variables: the individual's need for achievement, the expectation that the desired goal was attainable, and the value attached to the goal. However, contradictory findings, often inconsistent with those obtained with male subjects, emerged from studies with females. Cues that elicited achievement motivation and continued effort in males did not necessarily produce the same behavior in females. In an attempt to explain the gender differences observed, Horner (1968) postulated the motive to avoid success, or fear-of-success construct. She argued that Western culture's stereotyping of attainment of success in certain areas as being more congruent with the male than with the female role often meant that attainment of success by females in male-designated domains was accompanied by negative consequences such as unpopularity, guilt, abuse, or doubt about their femininity.

The prediction that success in mathematics, an area frequently stereotyped as a male domain, might evoke ambivalence or anxiety about success particularly among able, achievement-oriented females has received some research support (Butler & Nisan, 1975; Clarkson & Leder, 1984; Horner, 1968; Leder, 1982a), though not exclusively (Karabenick & Marshall, 1974; Tomlinson-Keasey, 1974). Relevant, too, are the findings of Peterson and Fennema (1985) that time spent on competitive mathematics activities was significantly related to boys', but not to girls', achievement in mathematics. For girls, mathematics achievement appeared to be enhanced more by participation in cooperative activities.

The fear of success notion is also consistent with the lower confidence expressed by females about their ability to do mathematics, their greater tentativeness about the appropriateness of their doing mathematics as discussed under the 'peer group' heading, and ultimately their lower performance in mathematics. Work carried out within the fear of success paradigm suggests that females' lower performance in mathematics is a function of internalization of, and conforming to, the expectations of others, rather than being a function of ability per se.

Attributions. Closely related to work on the fear-of-success construct, and also evolving from the expectancy-value theory of achievement motivation, is research that considers attributions of success and failure on certain tasks. A number of gender differences have been reported in this work. While both males and females considered effort to be important for the attainment of success, males were more likely than females to attribute success to a stable cause, failure to unstable factors. Females, on the other hand, were more likely to take personal responsibility for their failures, but somewhat less for their successes. It is worth noting that the higher attribution for success to effort by females, but to ability by males, reported in a number of studies is consistent with an expression of fear of success imagery (Horner, 1968; Horner, Tresemer, Berens, & Watson, 1973): a feeling that females have to work harder than males to have their achievements recognized.

Whitley, McHugh, and Frieze (1986) presented a comprehensive overview of work dealing with gender differences in attributions. They argued that publication bias towards studies in which significant findings are reported has contributed to an exaggeration of gender differences in attributions for success and failure. They concluded: "According to our analysis the achievement attributions made by men and women are really quite similar" (p. 120).

The more functional pattern of typical male attributions reported in the literature has been described as mastery oriented, the pattern more frequently associated with females as learned helplessness (Covington & Beery, 1976). A willingness to exert continued effort and to persist at difficult tasks has been identified as important distinguishing characteristics between individuals described as mastery oriented and those displaying attributional patterns considered to be less functional (Covington & Beery, 1976; Dweck & Goetz, 1978; Dweck & Licht, 1980; Dweck & Reppucci, 1973). The latter pattern of attributions is likely to lead to cognitive, motivational, and/or emotional deficits (Weiner, 1972, 1980), such as lack of persistence or a curtailment of effort (Dweck & Licht, 1980).

Studies carried out within a mathematics setting (Fennema, 1985; Gitelson, Petersen, & Tobin-Richards, 1982; Kloosterman, 1990; Leder 1982b, 1984b; Parsons, Meece, Adler, & Kaczala 1982; Pedro et al., 1981; Ryckman & Peckham, 1987; Wolleat et al., 1980) have provided qualified support for the less functional attributions of success and failure by females compared with males. When found, mean differences between the groups were generally small, however. Some evidence has also been found for the attendant negative consequences on the learning of mathematics predicted by the general attributional literature. For example, Kloosterman (1990) reported that females were more likely than males to show performance decrements following a failure experience.

It has been argued (Kloosterman, 1990; Parsons et al., 1982) that the identification of gender differences in attributions seems to be at least partly methodology-dependent. The latter, for example, elicited gender differences in attributional style with one instrument (rank-order questions) but not with another (open-ended) measure administered to the same sample. The former similarly reported that two measures of attributional style, one a paper and pencil instrument, the other defined in terms of performance decrements following failure, were found to be unrelated. A moderate, consistent, significant and positive relationship was found between achievement in mathematics and the measure of performance decrements following failure. No comparable relationship was found between achievement on a standardized mathematics test and attributions as assessed with the paper and pencil instrument. Fennema et al. (1990), noting the paucity of research on students' attributions of success and failure in mathematics and the intriguingly large effect sizes in attribution of success to ability in the studies included in their meta-analysis, argued that further research is needed in this area in particular.

Persistence. As discussed earlier, students who attribute their success to ability and their failure to effort are more likely to persevere on tasks which are not solved readily. Many higher-level mathematics tasks fall in this category. Indeed, there is some research evidence that females are less likely to work independently and to persist on more difficult mathematics tasks (Grieb & Easley, 1984; Koehler, 1990; Petersen & Fennema, 1985).

The difficulties associated with devising adequate measures of persistence have been discussed by Hart and Stanic (1989). They contrasted measures of persistence gathered with paper-and-pencil tests with information obtained from classroom observations. The former, they argued, do not necessarily capture critical behavior encouraged in the classroom.

It was easier for us to find examples of a lack of persistence than it was to find positive examples of persistence in the face of difficulty because the level and pace of instruction discouraged this kind of struggle. (Hart & Stanic, 1989, p. 14)

Given these measurement concerns, care must be taken when generalizing about gender differences in persistence. The methodological concerns expressed in this and earlier sections have distinct implications for the interpretations of the findings reported and suggest worthwhile avenues for future research.

To summarize: When gender differences are found they are typically small. Nevertheless, they indicate the extent to which students' own beliefs, expectations, and achievement behaviors reflect those of the wider society. There is some evidence that on average females are somewhat less confident about their ability to do mathematics, less certain that mathematics is an appropriate and needed area of study for them, more ambivalent about the value to them of success in mathematics; also they may be less functional than males in their attributions of success and failure and less likely to persist on the more challenging tasks. Collectively the body of research available to date suggests that there are small, subtle, interactive, and cumula-

tive links between gender differences in selected internal belief variables and gender differences in mathematics learning.

Concluding Comments. The wealth of work reviewed testifies to the considerable and continuing attention directed to gender differences in mathematics learning. The historical background presented provided a succinct illustration of the long-term antecedents of issues of current concern. This material also served to highlight the progress made to date in a number of Western countries in areas pertinent to achieving equity in mathematics learning.

The rich data base already established has not only been used to specify and delineate the range of problems encountered by certain groups of students, but, more constructively, it has also acted as an agent for change. Large numbers of females are now taking advantage of the prevailing and more accepting institutional, educational, and professional climate and have shown themselves capable of achieving well in mathematics. Yet, as the more recent work shows, certain gender differences in mathematics learning continue to be reported.

Despite the considerable overlap in the participation patterns in mathematics of females and males, somewhat more males than females enroll in intensive, high-level mathematics courses. Similarly, there is much overlap in the mathematics achievement of females and males. Yet some performance differences in favor of males continue to be found, particularly on mathematical tasks that require high cognitive level skills. Reference to cross-cultural data confirms the pervasiveness of these findings.

The importance of considering the broad context in which mathematics learning takes place has been emphasized consistently in this chapter. Possible explanations for gender differences in mathematics learning have thus comprised a diverse yet also overlapping and interactive set of variables. A number of models that have attempted to capture and synthesize the complexity of the learning environment have been described. Specifically, evidence has been presented that subtle but consistent differences can still be identified in the cultural and societal pressures exerted on females and males in a number of Western countries. These common notions about gender appropriate behavior are frequently internalized by individuals and lead to different beliefs and expectations in areas critical to the learning of mathematics.

Throughout this chapter there has been an emphasis on consistencies in the ways in which females, on average, and males, on average, behave, think, react, perform, and in particular on the ways in which these typical behavior or attitudinal patterns affect differences in mathematics learning. Indeed, the thrust of many of the studies reviewed has been on documenting the nature and scope of differences. Yet it is inappropriate to ignore or minimize the substantial variations that exist within groups of males and females. A careful examination of studies with conflicting findings often indicates that there were considerable differences in the nature of the samples, or the tasks, or the settings used in the different research. The idiosyncratic and sometimes unexpected ways in which students construct their own knowledge and make sense of the environment through a very personal framework can be gauged not only from research on gender and mathematics learning but is also confirmed by the growing body of work, discussed elsewhere in the book, that specifically addresses that issue. A more sensitive examination of apparently conflicting findings may yield more precise information about the strengths and weaknesses of smaller but more homogeneous groups of students in a range of mathematics settings. Even though gender is often a significant determinant of aspirations, expectations, and behavior, there are many other variables, including race and class for example, which have an important and interactive impact. For future research to incorporate such within-group

differences is consistent with the plea already made for a more explicit recognition of individual differences.

Supplementing the more common large-scale studies with in-depth small-sample research should provide further insights into the factors that contribute to differences in mathematics learning within as well as between groups and should lead to more constructive ways of counteracting them. Adoption of research paradigms that allow greater attention to individual differences and context-specific problems is consistent with research thrusts in other areas of mathematics education.

References

Alexander, K. L., & Pallas, A. M. (1983). Reply to Benbow and Stanley. *American Educational Research Journal, 20,* 475–477.

Annice, C., Atkins, W. J., Pollard, G. H., & Taylor, P. J. (1988, August). *Gender differences in the Australian mathematics competition.* Paper presented at the Sixth International Congress on Mathematical Education, Budapest.

Annual abstract of statistics. (1988). (No. 124). London: Her Majesty's Stationery Office.

Armstrong, J. E. (1910). Coeducation in high school: Is it a failure? *Good Housekeeping, 57,* 491–495.

Armstrong, J. M. (1981). Achievement and participation of women in mathematics: Results of two national surveys. *Journal for Research in Mathematics Education, 12,* 356–372.

Armstrong, J. M. (1985). A national assessment of participation and achievement of women in mathematics. In S. F. Chipman, L. R. Brush, & D. M. Wilson (Eds.), *Women and mathematics: Balancing the equation* (pp. 59–94). Hillsdale, NJ: Lawrence Erlbaum.

Armstrong, J. M., & Price, R. A. (1982). Correlates and predictors of women's mathematics participation. *Journal for Research in Mathematics Education, 13,* 99–109.

Assessment of Performance Unit. (1982). *Mathematical Development, Secondary Survey* (Report No. 3). London: Her Majesty's Stationery Office.

Barnes, M. (1989). A rescue operation. In G. C. Leder & S. N. Sampson (Eds.), *Educating Girls* (pp. 105–117). Sydney, Australia: Allen and Unwin.

Battista, M. T. (1990). Spatial visualization and gender differences in high school geometry. *Journal for Research in Mathematics Education, 21,* 47–60.

Becker, B. J., & Hedges, L. V. (1988). The effects of selection and variability in studies of gender differences. *Behavioral and Brain Sciences, 11,* 183–184.

Becker, J. (1981). Differential treatment of females and males in mathematics class. *Journal for Research in Mathematics Education, 12,* 40–53.

Benbow, C. P. (1988). Sex differences in mathematical reasoning ability in intellectually talented preadolescents: Their nature, effects, and possible causes. *Behavioral and Brain Sciences, 11,* 169–183.

Benbow, C. P., & Stanley, J. C. (1980). Sex differences in mathematical ability: Fact or artifact? *Science, 210,* 1262–1264.

Benbow, C. P., & Stanley, J. C. (1983). Differential course-taking revisited. *American Educational Research Journal, 20,* 469–573.

Ben-Chaim, D. (1983). Spatial visualization: Sex differences, grade-level differences and the effect of instruction on the performance of middle school boys and girls. *Dissertation Abstracts International, 43,* 2814A.

Ben-Chaim, D., Lappan, G., & Houang, R. T. (1988). The effect of instruction on spatial visualization skills of middle school boys and girls. *American Educational Research Journal, 25,* 51–71.

Bone, A. (1983). *Girls and girls only schools: A review of evidence.* London: Equal Opportunities Commission.

Boswell, S. L. (1985). The influence of sex-role stereotyping on women's attitudes and achievement in mathematics. In S. F. Chipman, L. R. Brush, & D. M. Wilson (Eds.), *Women and mathematics: Balancing the equation* (pp. 175–198). Hillsdale, NJ: Lawrence Erlbaum.

Braine, L. G. (1988). Sex differences in mathematics: Is there any news here? *Behavioral and Brain Sciences, 11,* 185–186.

Brandon, P. R., Newton, B. J., & Hammond, O. W. (1987). Children's mathematics achievement in Hawaii: Sex differences favoring girls. *American Educational Research Journal, 24,* 437–461.

Brinkmann, E. H. (1966). Programmed instruction as a technique for improving spatial visualization. *Journal of Applied Psychology, 50,* 179–184.

Brooks, S. D. (1903, November). Causes of withdrawal from school. *Educational Review,* pp. 362–393.

Brophy, J., & Good, T. L. (1974). *Teacher-student relationships: Causes and consequences.* New York: Holt, Rinehart and Winston.

Broverman, D. M., Klaiber, E. L., Kobayashi, Y., and Vogel, W. (1968). Roles of activation and inhibition in sex differences in cognitive abilities. *Psychological Review, 75,* 23–50.

Bryant, M. (1979). *The unexpected revolution.* London: University of London Institute of Education.

Bryden, M. P. (1988). Cerebral organization and mathematical ability. *Behavioral and Brain Sciences, 11,* 186–187.

Butler, R., & Nisan, M. (1975). Who is afraid of success? And why? *Journal of Youth and Adolescence, 4,* 259–270.

Callahan, L. G., & Clements, D. H. (1984). Sex differences in rote counting ability on entry to first grade: Some observations. *Journal for Research in Mathematics Education, 15,* 379–382.

Carpenter, P. (1985). Single-sex schooling and girls' academic achievements. *Australian and New Zealand Journal of Sociology, 21,* 456–472.

Carpenter, T. P., Lindquist, M. M., Brown, C. A., Kouba, V. L., Silver, E. A., & Swafford, J. O. (1988). Results of the fourth NAEP Assessment of mathematics: Trends and conclusions. *Arithmetic Teacher, 36*(1), 38–41.

Chipman, S. F., & Wilson, D. M. (1985). Understanding mathematics course enrollment and mathematics achievement: A synthesis of the research. In S. F. Chipman, L. R. Brush, & D. M. Wilson (Eds.), *Women and mathematics: Balancing the equation* (pp. 275–328). Hillsdale, NJ: Lawrence Erlbaum.

Clarke, D. J. (1989). *Mathematical behaviour and the transition from primary to secondary school.* Unpublished doctoral dissertation, Monash University, Melbourne, Australia.

Clarkson, P., & Leder, G. C. (1984). Causal attributions for success and failure in mathematics: A cross cultural perspective. *Educational Studies in Mathematics, 15,* 413–422.

Clements, M. A. (1979). *Relationships between the University of Melbourne and the secondary schools of Victoria, 1890–1912.* Unpublished Ph.D. thesis, University of Melbourne, Australia.

Clements, M. A. (1981, April). *Spatial ability, visual imagery, and mathematical learning.* Paper presented at the Annual Meeting of the American Educational Board Association, Los Angeles.

Clements, M. A. (1989). *Mathematics for the minority: Some historical perspectives of school mathematics in Victoria.* Geelong, Victoria, Australia: Deakin University Press.

Cockcroft, W. H. (1982). *Mathematics Counts.* London: Her Majesty's Stationery Office.

Connor, J. M. (1977). Sex differences in children's response to training on a visual-spatial test. *Developmental Psychology, 13,* 293–294.

Connor, J. M., Schackman, M., & Serbin, L. (1978). Sex-related differences in response to practice on a visual-spatial test and generalization to a related test. *Child Development, 49,* 24–29.

Connor, J. M., & Serbin, L. A. (1985). Visual-spatial skill: Is it important for mathematics? Can it be taught? In S. F. Chipman, L. R. Brush, D. M. Wilson (Eds.), *Women and Mathematics: Balancing the equation* (pp. 151–174). Hillsdale, NJ: Lawrence Erlbaum.

Covington, M. V., & Beery, R. (1976). *Self-worth and school learning.* New York: Holt, Rinehart and Winston.

Cresswell, M., & Gubb, J. (1987). *The Second International Mathematics Study in England and Wales.* Windsor, Berkshire: NFER-Nelson.

Curio, F. R. (1987). Comprehension of mathematical relationship expressed in graphs. *Journal for Research in Mathematics Education, 18,* 382–393.

Dale, R. (1974). *Mixed or single-sex schools?* (Vol. 3). London: Routledge and Kegan Paul.

Dean, P. R. (1909). A few algebra methods. *The Mathematics Teacher, 2,* 41–47.

Deaux, K. (1985). Sex and gender. *Annual Review of Psychology, 36,* 49–81.

Deaux, K., & Major, B. (1987). Putting gender into context: An interactive model of gender-related behaviour. *Psychological Review, 94,* 369–389.

Dekkers, J., de Laeter, J. R., & Malone, J. A. (1986). *Upper secondary school science and mathematics enrolment patterns in Australia, 1970–1985.* Bentley, Western Australia: Science and Mathematics Education Centre, Western Australian Institute of Technology.

Dossey, J. A. (1985, April). *Student/class results from the Second International Mathematics Study from United States twelfth grade classroom.* Paper presented at the annual meeting of the American Educational Research Association, Chicago.

Dreyfus, T., & Eisenberg, T. (1982). Intuitive functional concepts: A baseline study on intuitions. *Journal for Research in Mathematics Education, 13,* 360–380.

Dunkin, M. J., & Doenau, S. J. (1982). Ethnicity, classroom interaction and student achievement. *The Australian Journal of Education, 26,* 171–189.

Duval, C. M. (1980). Differential teacher grading behavior toward female students of mathematics. *Journal for Research in Mathematics Education, 11,* 203–212.

Dweck, C. S., & Goetz, T. E. (1978). Attributions and learned helplessness. In J. M. Harvey, W. Ickes, & R. F. Kidd (Eds.), *New directions in attribution research* (Vol. 2, pp. 157–179). Hillsdale, NJ: Lawrence Erlbaum.

Dweck, C. S., & Licht, B. G. (1980). Learned helplessness and intellectual achievement. In J. Garber & M. E. P. Seligman (Eds.), *Human helplessness, theory, and applications* (pp. 197–221). New York: Academic Press.

Dweck, C. S., & Reppucci, N. D. (1973). Learned helplessness and reinforcement responsibility in children. *Journal of Personality and Social Psychology, 25,* 109–116.

Eales, A. (1986). Girls and mathematics at Oadby Beauchamp College. In L. Burton (Ed.), *Girls into maths can go* (pp. 163–186). London: Holt, Rinehart and Winston.

Eccles (Parsons), J., Adler, T. F., Futterman, R., Goff, S. B., Kaczala, C. M., Meece, J. L., & Midgley, C. (1985). Self-perceptions, task perceptions, socializing influences, and the decision to enroll in mathematics. In S. F. Chipman, L. R. Brush, & D. M. Wilson (Eds.), *Women and mathematics: Balancing the equation* (pp. 95–121). Hillsdale, NJ: Lawrence Erlbaum.

Eccles, J. S., & Blumenfeld, P. (1985). Classroom experiences and student gender: Are there differences and do they matter? In L. C. Wilkinson & C. B. Marrett (Eds.), *Gender influences in classroom interaction* (pp. 79–114). New York: Academic Press.

Eccles, J. S., & Jacobs, J. E. (1986). Social forces shape math attitudes and performance. *Signs, 11,* 367–380.

Edwards, J. (1985). Boys and girls in the Australian Mathematics Competition. *Mathematics in School, 14*(5), 5–7.

Elliott, J. C. (1990). Affect and mathematics achievement of nontraditional college students. *Journal for Research in Mathematics Education, 21,* 160–165.

Ethington, C. A. (1990). Gender differnces in mathematics: An international perspective. *Journal for Research in Mathematics and Education, 21,* 74–80.

Ethington, C. A., & Wolfle, L. M. (1984). Sex differences in a causal model of mathematics achievement. *Journal for Research in Mathematics Education, 15,* 361–377.

Fausto-Sterling, A. (1985). *Myths of gender.* New York: Basic Books.

Fennema, E. (1981). Women and mathematics: Does research matter? *Journal for Research in Mathematics Education, 12,* 380–385.

Fennema, E. (1985). Attribution theory and achievement in mathematics. In S. R. Yussen (Ed.), *The growth of reflection in children* (pp. 245–265). New York: Academic Press.

Fennema, E. (1990). Teachers' beliefs and gender differences in mathematics. In E. Fennema and G. C. Leder (Eds.), *Mathematics and gender* (pp. 169–187). New York: Teachers' College Press.

Fennema, E., & Carpenter, T. (1981). The second national assessment and sex-related differences in mathematics. *Mathematics Teacher, 74,* 554–559.

Fennema, E., Hyde, J. S., Ryan, M., & Frost, L. A. (1990). Gender differences in mathematics attitudes and affect: A meta-analysis. *Psychology of Women Quarterly, 14,* 299–324.

Fennema, E., & Peterson, P. L. (1985). Autonomous learning behavior: A possible explanation of gender-related differences in mathematics. In L. C. Wilkinson, & C. B. Marrett (Eds.), *Gender-related differences in classroom interactions* (pp. 17–35). New York: Academic Press.

Fennema, E., & Sherman, J. (1977). Sex-related differences in mathematics achievement, spatial visualization, and sociocultural factors. *American Educational Research Journal, 14,* 51–71.

Fennema, E., & Sherman, J. A. (1978). Mathematics achievement and related factors: A further study. *Journal for Research in Mathematics Education, 9,* 189–203.

Fennema, E., & Tartre, L. A. (1985). The use of spatial visualization in mathematics by girls and boys. *Journal for Research in Mathematics Education, 16,* 184–206.

Fennema, E., Wolleat, P. L., Pedro, J. D., & Becker, A. D. (1981). Increasing women's participation in mathematics: An intervention study. *Journal for Research in Mathematics Education, 12,* 3–14.

Ferrini-Mundy, J. (1987). Spatial training for calculus students: Sex differences in achievement and in visualization ability. *Journal for Research in Mathematics Education, 18,* 126–140.

Fox, L. H., Brody, L., & Tobin, D. (1985). The impact of early intervention programs upon course-taking and attitudes in high school. In S. F. Chipman, L. R. Brush, & D. M. Wilson (Eds.), *Women and mathematics: Balancing the equation* (pp. 249–274). Hillsdale, NJ: Lawrence Erlbaum.

Fox, L. H., & Cohn, S. J. (1980). Sex differences in the development of precocious mathematical talent. In L. H. Fox, L. Brody, & D. Tobin (Eds.), *Women and the mathematical mystique* (pp. 94–111). Baltimore: The Johns Hopkins University Press.

Freed, N. J. (1983). Foreseeable equivalent math skill of men and women. *Psychological Reports, 52*, 334.

Friedman, L. (1989). Mathematics and the gender gap: A meta-analysis of recent studies on sex differences in mathematical tasks. *Review of Educational Research, 59*, 185–213.

Galton, M., Simon, B., & Croll, P. (1980). *Inside the primary classroom.* London: Routledge and Kegan Paul.

Gauss, C. F. (April 30, 1807), Letter translated in L. L. Bucciarelli & N. Dworsky (1980). *Sophie Germain. An essay into the history of the theory of elasticity* (p. 25). Dordrecht, Holland: D. Riedel Publishing Co.

Gay, G. E. (1902, January). Why pupils leave the high school without graduating. *Education*, pp. 300–307.

Gill, J. (1988). *Which way to school?* Canberra: Commonwealth Schools Commission.

Gitelson, I. B., Petersen, A. C., & Tobin-Richards, M. H. (1982). Adolescents' expectancies of success, self-evaluations, and attributions about performance on spatial and verbal tasks. *Sex Roles, 8*, 411–420.

Gore, D. A., & Roumagoux, D. V. (1983). Wait time as a variable in sex-related differences during fourth grade mathematics instruction. *Journal of Educational Research, 26*, 273–275.

Gough, H. G., & Heilbrun, A. B. Jr (1980). *The Adjective Check List Manual.* Palo Alto, CA: Consulting Psychologists Press, Inc.

Grieb, A., & Easley, J. (1984). A primary school impediment to mathematical equity: Case studies in rule-dependent socialization. In M. W. Steinkamp & M. Maehr (Eds.), *Advances in motivation and achievement: Vol. 2. Women in science* (pp. 317–362). Greenwich, CT: JAI.

Hacket, G., & Betz, N. E. (1989). An exploration of the mathematics self-efficacy/mathematics performance correspondence. *Journal for Research in Mathematics Education, 20*, 261–273.

Hall, C. W., & Hoff, C. (1988). Gender differences in mathematical performance. *Educational Studies in Mathematics, 19*, 395–401.

Halpern, D. F. (1986). *Sex differences in cognitive abilities.* Hillsdale, NJ: Lawrence Erlbaum.

Hanna, G. (1986). Sex differences in mathematics achievement of eighth graders in Ontario. *Journal for Research in Mathematics Education, 17*, 231–237.

Hanna, G. (1988, February). *Gender differences in mathematics achievement among eighth graders: Results from twenty countries.* Paper presented at the 154th annual meeting of the American Association for Advancement of Science, Boston.

Hanna, G. (1989, March). *Mathematics achievement of girls in grade 12: Cross cultural differences.* Paper presented at the annual meeting of the American Educational Research Association, San Francisco.

Hanna, G., & Kündiger, E. (1986, April). *Differences in mathematical achievement levels and in attitudes for girls and boys in twenty countries.* Paper delivered at the Annual Meeting of the American Educational Research Association, San Francisco.

Hanna, G., Kündiger, E., & Larouche, C. (1988, July). *Mathematical achievement of grade 12 girls in fifteen countries.* Paper delivered at the Sixth International Congress on Mathematical Education, Budapest, Hungary.

Hansen, C. (1981). Maths and jobs. *Vinculum, 18*, 8–13.

Harding, J. (1981). Sex differences in science examinations. In A. Kelly (Ed.), *The missing half* (pp. 192–204). Manchester: Manchester University Press.

Harris, M. J., & Rosenthal, R. (1985). Mediation of interpersonal expectancy effects: 31 meta-analyses. *Psychological Bulletin, 97*, 363–386.

Hart, L. E. (1989). Classroom processes, sex of student, and confidence in learning mathematics. *Journal for Research in Mathematics Education, 30*, 242–260.

Hart, L. E., & Stanic, G. M. A. (1989, April). *Attitudes and achievement related behaviour in middle school mathematics students: Views through four lenses.* Paper presented at the annual meeting of the American Educational Research Association, San Francisco.

Hayman, M. (1976). Mathematical Competitions. In J. Gibson & P. Chennels (Eds.), *Gifted children. Looking to their future.* London: Latimer New Dimensions, Ltd.

Hembree, R. (1988). Correlates, causes, effects, and treatment of test anxiety. *Review of Educational Research, 58*, 47–77.

Hembree, R. (1990). The nature, effects, and relief of mathematics anxiety. *Journal for Research in Mathematics Education, 21*, 33–46.

Hilton, T. L., & Berglund, G. W. (1974). Sex differences in mathematics achievement: A longitudinal study. *The Journal of Educational Research, 67*, 231–237.

Hines, M. (1988). Hormonal influences on human cognition. *Behavioral and Brain Sciences, 11*, 194–195.

Horner, M. S. (1968). *Sex differences in achievement motivation and performance in competitive and non-competitive situations.* Unpublished doctoral dissertation, University of Michigan.

Horner, M. S., Tresemer, D. W., Berens, A. E., & Watson, R. F., Jr. (1973). *Scoring manual for an empirically derived scoring system for motive to avoid success.* Unpublished manuscript.

Hunt, F. (Ed.) (1987). *Lessons for life. The schooling of girls and women, 1850–1950.* Oxford: Blackwell.

Husen, T. (Ed.). (1967). *International study of achievement in mathematics: A comparison in twelve countries.* Stockholm: Almquist and Wiksell.

Huston, A. C., & Carpenter, C. J. (1985). Gender differences in preschool classrooms: The effects of sex-typed activity choices. In L. C. Wilkinson & C. B. Marrett (Eds.), *Gender differences in classroom interaction* (pp. 143–165). New York: Academic Press.

Hyde, J. S. (1981). How large are cognitive gender differences? *American Psychologist, 36*, 892–901.

Hyde, J. S., Fennema, E., & Lamon, S. J. (1990). Gender differences in mathematics performance: A meta-analysis. *Psychological Bulletin. 107*, 139–155.

Jacobs, J. E., & Eccles, J. S. (1985). Gender differences in math ability: The impact of media reports on parents. *Educational Researcher, 14* (3), 20–24.

Jacobs, J. E., & Wigfield, A. (1989). Sex equity in mathematics and science education: Research-policy links. *Educational Psychology Review, 1*, 39–56.

Joffe, L., & Foxman, D. (1986). Attitudes and sex differences: Some APU findings. In L. Burton (Ed.), *Girls into maths can go* (pp. 38–50). London: Holt, Rinehart and Winston.

Joffe, L., & Foxman, D. (1988). *Attitudes and gender differences: Mathematics at age 11 and 15.* Windsor, Berkshire: NFER-Nelson.

Jones, L. V. (1987). The influence on mathematics test scores, by ethnicity and sex, of prior achievement and high school mathematics courses. *Journal for Research in Mathematics Education, 18*, 180–186.

Karabenick, S. A., & Marshall, J. M. (1974). Performance of females as a function of fear of success, fear of failure, type of opponent, and performance-contingent feedback. *Journal of Personality, 42*, 220–237.

Kelly, A. (1981). Sex differences in science achievement: Some results and hypotheses. In A. Kelly (Ed.), *The missing half* (pp. 22–41). Manchester: Manchester University Press.

Kelly, A. (1986). Gender roles at home and school. In L. Burton (Ed.), *Girls into maths can go* (pp. 90–109). London: Holt, Rinehart and Winston.

Khan, A. U., & Cataio, J. (1984). *Men and women in biological perspective.* New York: Praeger Publishers.

Kimball, M. M. (1989). A new perspective on women's math achievement. *Psychological Bulletin, 105,* 198–214.

Kimura, D. (1988). Biological influences on cognitive function. *Behavioral and Brain Sciences, 11,* 200.

Kirshner, D. (1989). The visual syntax of algebra. *Journal for Research in Mathematics Education, 20,* 274–287.

Kissane, B. V. (1986). Selection of mathematically talented students. *Educational Studies in Mathematics, 17,* 221–241.

Kloosterman, P. (1990). Attributions, performance following failure and motivation in mathematics. In E. Fennema & G. C. Leder (Eds.), *Mathematics and gender* (pp. 96–127). New York: Teachers' College Press.

Koehler, M. S. (1990). Classrooms, teachers and gender differences in mathematics. In E. Fennema & G. C. Leder (Eds.), *Mathematics and Gender* (pp. 128–148). New York: Teachers' College Press.

Kroll, D. (1985). Evidence from *The Mathematics Teacher* (1908–1920) on women and mathematics. *For the Learning of Mathematics, 5,* 7–10.

Lantz, A. E., & Smith, G. P. (1981). Factors influencing the choice of non required mathematics courses. *Journal of Educational Psychology, 73,* 825–837.

Leder, G. C. (1982a). Mathematics achievement and fear of success. *Journal for Research in Mathematics Education, 13,* 124–135.

Leder, G. C. (1982b). Learned helplessness in the classroom. A further look. *Research in Mathematics Education in Australia, 2,* 40–55.

Leder, G. C. (1984a). What price success? The view from the media. *The Exceptional Child, 31,* 223–229.

Leder, G. C. (1984b). Sex differences in attributions of success and failure. *Psychological Reports, 54,* 57–58.

Leder, G. C. (1986a). Successful females: (print) media profiles and their implications. *Journal of Psychology, 120,* 239–248.

Leder, G. C. (1986b). Mathematics: Stereotyped as a male domain? *Psychological Report, 59,* 955–958.

Leder, G. C. (1986c, April). *Gender linked differences in mathematics learning: Further explorations.* Paper presented at the Research Procession to the NCTM 64th Annual Meeting, Washington.

Leder, G. C. (1987). Teacher student interaction: A case study. *Educational Studies in Mathematics, 18,* 255–271.

Leder, G. C. (1988a). Fear of success imagery in the print media. *The Journal of Psychology, 122,* 305–306.

Leder, G. C. (1988b). Teacher-student interactions: The mathematics classroom. *Unicorn, 14,* 161–166.

Leder, G. C. (1989). Do girls count in mathematics? In G. C. Leder and S. N. Sampson (Eds.), *Educating girls: Practice and research* (pp. 84–97). Sydney: Allen and Unwin.

Leder, G. C. (1990a). Gender differences in mathematics: An overview. In E. Fennema & G. C. Leder (Eds.), *Mathematics and gender* (pp. 10–26). New York: Teachers' College Press.

Leder, G. C. (1990b). Teacher-student interactions, mathematics and gender. In E. Fennema & G. C. Leder (Eds.), *Mathematics and gender* (pp. 149–168). New York: Teachers' College Press.

Leder, G. C. (1991, April). *Early school experiences: Gender differences in mathematics learning.* Paper presented at the annual meeting of the American Educational Research Association, Chicago.

Leder, G. C., & Fennema, E. (1990). Gender differences in mathematics: A synthesis. In E. Fennema & G. C. Leder (Eds.), *Mathematics and gender.* New York: Teachers' College Press.

Lee, V. E., & Bryk, A. S. (1986). Effects of single-sex secondary schools on student achievement and attitudes. *Journal of Educational Psychology, 78,* 381–395.

Leinhardt, G., Seewald, A. M., & Engel, M. (1979). Learning what's taught: Sex differences in instruction. *Journal of Educational Psychology, 71,* 423–439.

Linn, M. C., De Benedictis, T., Delucchi, K., Harris, A., & Stage, E. (1987). Gender differences in national assessment of educational progress science items: What does "I don't know" really mean? *Journal of Research in Science Teaching, 24,* 267–278.

Linn, M. C., & Petersen, A. C. (1986). A meta-analysis of gender differences in spatial ability: Implications for mathematics and science achievement. In J. S. Hyde and M. C. Linn (Eds.), *The psychology of gender. Advances through meta-analysis* (pp. 67–101). Baltimore: The Johns Hopkins University Press.

Linn, M. C., & Pulos, S. (1983). Aptitude and experience influences on proportional reasoning during adolescence: Focus on male-female differences. *Journal for Research in Mathematics Education, 14,* 30–46.

Lockheed, M. E. (1985). Some determinants and consequences of sex segregation in the classroom. In L. C. Wilkinson & C. B. Marrett (Eds.), *Gender influences in classroom interaction* (pp. 167–184). New York: Academic Press.

Luchins, E. H., & Luchins, H. S. (1980). Female mathematicians: A contemporary appraisal. In L. H. Fox, L. Brody, & D. Tobin (Eds.), *Women and the mathematical mystique* (pp. 7–22). Baltimore: The Johns Hopkins University Press.

Maccoby, E. E. (Ed.) (1967). *The development of sex differences.* London: Tavistock Publications.

Maccoby, E. E., & Jacklin, C. N. (1974). *The psychology of sex differences.* Stanford, CA: Stanford University Press.

Maines, D. R. (1985). Preliminary notes on a theory of informal barriers for women in mathematics. *Educational Studies in Mathematics, 16,* 314–320.

Marrett, C. B., & Gates, H. (1983). Male-female enrollments across mathematics tracks in predominantly black high schools. *Journal for Research in Mathematics Education, 14,* 113–118.

Marsh, H., Owens, L., Marsh, M., & Smith, I. (1989). From single-sex to coed schools. In G. C. Leder and S. N. Sampson (Eds.), *Educating girls: Practice and research* (pp. 144–157). Sydney: Allen and Unwin.

Marsh, H. W., Owens, L., Myers, M. R., & Smith, I. D. (1987, November). *The transition from single-sex to coeducational high schools: Teacher perceptions, academic achievement, and self concept.* Paper presented at the conference, "The education of girls: Research and beyond." Monash University, Melbourne, Australia.

Marshall, S. P. (1983). Sex differences in mathematical errors: An analysis of distractor choices. *Journal for Research in Mathematics Education, 14,* 325–336.

McClelland, D., Atkinson, J. W., Clarke, R. A., & Lowell, E. L. (1953). *The Achievement Motive.* New York: Appleton, Century, Crofts.

McGee, M. G. (1978). Effects of training and practice on sex differences in mental rotation test scores. *Journal of Psychology, 100,* 87–90.

Meyer, M. R. (1989). Gender differences in mathematics. In M. M. Lindquist (Ed.), *Results from the fourth mathematics assessment of the national assessment of educational programs* (pp. 149–159). Reston, VA: National Council of Teachers of Mathematics.

Meyer, M. R., & Koehler, M. S. (1990). Internal influences on gender differences in mathematics. In E. Fennema & G. C. Leder (Eds.), *Mathematics and Gender* (pp. 60–95). New York: Teachers' College Press.

Moore, D., & Smith, P. (1980). Teacher questions: Frequency and distribution in a sample of Papua New Guinea High Schools. *Australian Journal of Education, 24,* 315–317.

Moore, E. G. J., & Smith, A. W. (1987). Sex and ethnic group differences in mathematical achievement: Results from the National Longitudinal Study. *Journal for Research in Mathematics Education, 18,* 25–36.

Morrison, H. C. (1915). Reconstructed mathematics in high school. *The Mathematics Teacher, 7,* 141–153.

Moss, J. D. (1982). *Toward equality: Progress by girls in mathematics in Australian secondary schools* (Occasional paper No. 16). Hawthorn: Australian Council for Educational Research.

Mura, R. (1987). Sex-related differences in expectations of success in undergraduate mathematics. *Journal for Research in Mathematics Education, 18,* 15–24.

Noss, R. (1987). Children's learning of geometrical concepts through Logo. *Journal for Research in Mathematics Education, 18,* 343–362.

Olson, M., & Kansky, B. (1981). Mathematical preparation versus career aspirations: Sex-related differences among college-bound Wyoming high school seniors. *Journal for Research in Mathematics Education, 12,* 375–379.

Ormerod, M. B. (1981). Factors differentially affecting the science subject preferences, choices and attitudes of girls and boys. In A. Kelly (Ed.), *The missing half* (pp. 100–112). Manchester: Manchester University Press.

Pallas, A. M., & Alexander, K. L. (1983). Sex differences in quantitative SAT performance: New evidence on the differential coursework hypothesis. *American Educational Research Journal, 20,* 165–182.

Parsons, J. E., Adler, T. F., Futterman, R., Goff, S. B., Kaczala, C. M., Meece, J. L., & Midgley, C. (1983). Expectancies, values, and academic behaviors. In J. T. Spence (Ed.), *Achievement and achievement motivation* (pp. 75–146). San Francisco: W. H. Freeman & Co.

Parsons, J. E., Meece, J. L., Adler, T. F., & Kaczala, C. M. (1982). Sex differences in attributions and learned helplessness. *Sex Roles, 8,* 421–432.

Pattison, P., & Grieve, N. (1984). Do spatial skills contribute to sex differences in different types of mathematical problems? *Journal of Educational Psychology, 76,* 678–689.

Pedro, J. D., Wolleat, P., Fennema, E., & Becker, A. D. (1981). Election of high school mathematics by females and males: Attributions and attitudes. *American Educational Research Journal, 18,* 207–218.

Perl, T. H. (1982). Discriminating factors on sex differences in electing mathematics. *Journal for Research in Mathematics Education, 13,* 66–74.

Petersen, A. C. (1979). Hormones and cognition functioning in normal development. In M. A. Wittig, and A. C. Petersen (Eds.), *Sex-related differences in cognitive functioning* (pp 289–214). New York: Academic Press.

Petersen, A. C., & Wittig, M. A. (1979). Sex-related differences in cognitive functioning: An overview. In M. A. Wittig & A. C. Petersen (Eds.), *Sex-related differences in cognitive functioning: Developmental issues* (pp. 1–17). New York: Academic Press.

Peterson, P. L., & Fennema, E. (1985). Effective teaching, student enjoyment in classroom activities, and sex-related differences in learning mathematics. *American Educational Research Journal, 22,* 309–335.

Reyes, L. H. (1983). Editorial. *Journal for Research in Mathematics Education, 14,* 146.

Reyes, L. H. (1984). Affective variables and mathematics education. *The Elementary School Journal, 84,* 558–581.

Reyes, L. H., & Stanic, G. M. (1988). Race, sex, socioeconomic status, and mathematics. *Journal for Research in Mathematics Education, 19,* 26–43.

Rosenthal, R., & Rubin, D. B. (1982). Further meta-analysis procedures for assessing cognitive gender differences. *Journal of Educational Psychology, 74,* 708–712.

Rosier, M. J. (1980). *Changes in secondary school mathematics in Australia, 1964–1978.* Hawthorn, Victoria: Australian Council for Educational Research.

Rowe, K. J. (1988). Single-sex and mixed-sex classes: The effects of class type on student achievement, confidence and participation in mathematics classes. *Australian Journal of Education, 32,* 180–202.

Rowley, G., & Leder, G. C. (1989). Mathematics competitions and the problems of not guessing. *Mathematics Competitions, 1*(3), 6–10.

Rubinstein, R. N. (1985). Computational estimation and related mathematical skills. *Journal for Research in Mathematics Education, 16,* 106–119.

Ryan, S., Macphee, I., & Taperell, K. (1983). *Fair exposure.* Canberra: Australian Government Publishing Service.

Ryckman, D. B., & Peckham, P. (1987). Gender differences in attributions for success and failure situations across subject areas. *Journal of Educational Research, 81,* 120–125.

Sampson, S. N. (1983). *Initiatives to change girls' perceptions of career opportunities. An evaluation.* Canberra: Australian Government Publishing Service.

Sampson, S. (1989). Are boys a barrier for girls in science? In G. C. Leder & S. N. Sampson (Eds.), *Educating girls. Practice and Research* (pp. 139–143). Sydney: Allen and Unwin.

Sanders, B. (1988). Mathematical ability, spatial ability and remedial training. *Behavioral and Brain Sciences, 11,* 208–209.

Schildkamp-Kündiger, E. (Ed.). (1982). *International review on gender and mathematics.* Columbus, OH: ERIC Clearinghouse for Science, Mathematics and Environmental Education.

Schultz, K. A., & Austin, J. D. (1983). Directional effects in transformation tasks. *Journal for Research in Mathematics Education, 14,* 95–101.

Sells, L. (1973). *High school mathematics as the critical filter in the job market* (ERIC No. ED080 351). Berkeley: University of California.

Senk, S., & Usiskin, Z. (1985). Geometry proof writing: A new view of sex differences in mathematics ability. *Journal of Educational Research, 91,* 187–201.

Sherman, J. (1978). *Sex-related cognitive differences: An essay on theory and evidence.* Springfield, IL: Charles C. Thomas.

Sherman, J. (1980). Mathematics, spatial visualization and related factors: Changes in girls and boys, grades 8-11. *Journal of Educational Psychology, 72,* 476–482.

Sherman, J., & Fennema, E. (1977). The study of mathematics by high school girls and boys: Related variables. *American Educational Research Journal, 14,* 159–168.

Shorten, A. (1990). Equality of eduactional opportunity in Australia. *Education Law Journal, 2,* 265–297.

Silver, E. A., Lindquist, M. M., Carpenter, T. P., Brown, C. A., Kouba, V. L., & Swafford, J. O. (1988). The fourth NAEP Mathematics Assessment: Performance trends and results and trends for instructional indicators. *Mathematics Teacher, 81,* 720–727.

Smith, S. (1986). *Separate tables? An investigation into single-sex setting in mathematics.* London: Her Majesty's Stationery Office.

Smith, S. (1987). *Separate beginnings?* Manchester Equal Opportunities Commission.

Smith, S. E., & Walker, W. J. (1988). Sex differences on New York State Regents Examinations: Support for the differential course-taking hypothesis. *Journal for Research in Mathematics Education, 19,* 81–85.

Solomon, B. M. (1985). *In the company of educated women.* New Haven: Yale University Press.

Spender, D. (1982). *Invisible women.* London: Writers and Readers Publishing Cooperative.

Spender, D., & Sarah, E. (Eds.). (1980). *Learning to lose.* London: Women's Press.

Stafford, R. E. (1972). Hereditary and environmental components of quantitative reasoning. *Review of Educational Research, 42,* 183–201.

Stage, E., Kreinberg, N., Eccles (Parsons), J., & Becker, J. (1985). Increasing participation and achievement of girls and women in mathematics, science and engineering. In S. S. Klein (Ed.), *Handbook for achieving sex equality through education* (pp. 237–269). Baltimore: The Johns Hopkins University Press.

Stallings, J. (1979). *Factors influencing women's decisions to enroll in advanced mathematics courses: Executive summary.* Menlo Park, California: SRI International.

Stanic, G. M. A., & Reyes, L. H. (1986, April). *Gender and race differences in mathematics: A case-study of a seventh-grade classroom.* Paper presented at the annual meeting of the American Educational Research Association, San Francisco.

Steedman, J. (1983). *Examination results in mixed or single sex schools: Findings from the national Child Development Study.* London: Equal Opportunities Commission.

Stein, A. H., & Bailey, M. M. (1973). The socialization of achievement orientation in females. *Psychological Bulletin, 80,* 345–366.

Stevenson, H. W., & Newman, R. S. (1986). Long-term prediction of achievement and attitudes in mathematics and reading. *Child Development, 57,* 646–659.

Swafford, J. O. (1980). Sex differences in first-year algebra. *Journal for Research in Mathematics Education, 11,* 335–346.

Szetela, W., & Super, D. (1987). Calculators and instruction in problem solving in grade 7. *Journal for Research in Mathematics Education, 18,* 215–229.

Tartre, L. A. (1990). Spatial skills, gender and mathematics. In E. Fennema and G. C. Leder (Eds.), *Mathematics and gender* (pp. 27–59). New York: Teachers' College Press.

Thom, D. (1987). Better a teacher than a hairdresser? 'A mad passion for equality', or, Keeping Molly and Betty down. In F. Hunt (Ed.), *Lessons for life. The schooling of girls and women, 1850–1950* (pp. 124–146). Oxford: Blackwell.

Thomas, F. W. (1915). What mathematics subjects should be included in the curriculum of the secondary schools? *The Mathematics Teacher, 5,* 206–213.

Tobin, K. (1984, August). *Classroom processes.* Paper presented at the Fifth International Congress on Mathematical Education, Adelaide, Australia.

Tomlinson-Keasey, C. (1974). Role variables: Their influence on female motivational constructs. *Journal of Counseling Psychology, 21,* 232–237.

Turnbull, A. (1987). Learning her womanly work: The elementary school curriculum, 1870–1914. In F. Hunt (Ed.), *Lessons for life. The schooling of girls and women, 1850–1950* (pp. 83–100). Oxford: Blackwell.

Tyack, D., & Hansot, E. (1988). Silence and policy talk. Historical puzzles about gender and education. *Educational Researcher, 17*(3), 33–41.

Unger, R. K. (1979). *Female and male: Psychological perspective.* New York: Harper and Row.

U.S. Bureau of Statistics. (1988). *Statistical Abstract of the United States* (108th ed.). Washington: U.S. Bureau of Statistics.

Vandenberg, S. G., & Kuse, A. R. (1979). Spatial ability: A critical review of the sex-linked major gene hypotheses. In M. C. Wittig & A. C. Petersen (Eds.), *Determinants of sex-related differences in cognitive functioning* (pp. 67–95). New York: Academic Press.

Wagner, H., & Zimmerman, B. (1986). Identification and fostering of mathematically gifted students. *Educational Studies in Mathematics, 17,* 243–260.

Walden, R., & Walkerdine, V. (1985). *Girls and mathematics: From primary to secondary schooling* (Bedford Way Papers 24). London: London University, Institute of Education.

Webb, N. M., & Kenderski, C. M. (1985). Gender differences in small group interaction and achievement in high and low-achieving classes. In L. C. Wilkinson & C. B. Marrett (Eds.), *Gender influences in classroom interaction* (pp. 209–236). New York: Academic Press.

Weiner, B. (1972). *Theories of motivation.* Chicago: Rand McNally.

Weiner, B. (1980). The order of affect in rational (attributions) approaches to human motivation. *Educational Research, 19,* 4–11.

Weinstein, R. H., Marshall, H., Brattesani, K., & Middlestadt, S. (1982). Student perceptions of differential treatment in open and traditional classrooms. *Journal of Educational Psychology, 74,* 678–692.

Whitley, B. E., McHugh, M. C., & Frieze, I. H. (1986). Assessing the theoretical models for sex differences in causal attributions of success and failure. In J. S. Hyde & M. C. Linn (Eds.), *The psychology of gender* (pp. 185–208). New York: Academic Press.

Wilkinson, L. C., Lindow, J., & Chiang, C. P. (1985). Sex differences and sex segregation in students' small group communication. In L. C. Wilkinson & C. B. Marrett (Eds.), *Gender influences in classroom interaction* (pp. 185–208). New York: Academic Press.

Willis, S. (1989). *Real girls don't do maths.* Geelong, Victoria, Australia: Deakin University.

Windschuttle, K. (1988). *Education in the U.S.A. Statistical comparisons with Australia.* Canberra: Australian Publishing Service.

Wise, L. K., Steel, L., & MacDonald, C. (1979). *Origins and career consequences of sex differences in high school mathematics achievement* (Report to the National Institute of Education, Grant No. NIE-G-78-001). Palo Alto, CA: American Institute for Research.

Witelson, S. F. (1979). Sex and the single hemisphere: Specialization of the right hemisphere for spatial processing. *Science, 193,* 425–427.

Witelson, S. F. (1988). Neuroanatomical sex differences: Of no consequences for cognition? *Behavioral and Brain Sciences, 11,* 215–217.

Wittig, M. A. (1979). Genetic influences on sex-related differences in intellectual performance: Theoretical and methodological issues. In M. A. Wittig & A. C. Petersen (Eds.), *Sex-related differences in cognitive functioning* (pp. 21–65). New York: Academic Press.

Wolleat, P. L., Pedro, J. D., Becker, A. D., & Fennema, E. (1980). Sex differences in high school students' causal attributions of performance in mathematics. *Journal for Research in Mathematics Education, 11,* 356–366.

Zainu'ddin, A. (1975). Reflections on the history of women's education in Australia. *Education News, 15,* 4–13.

·25·

RACE, ETHNICITY, SOCIAL CLASS, LANGUAGE, AND ACHIEVEMENT IN MATHEMATICS

Walter G. Secada

UNIVERSITY OF WISCONSIN–MADISON

For over 40 years (*Brown v. Board of Education*, 1954, footnote 11; Myrdal, 1944), we have been confronted with an ever-growing body of research documenting that the American educational system is differentially effective for students depending on their social class, race, ethnicity, language background, gender, and other demographic characteristics. This differential effectiveness has been found in mathematics as well as in many other academic subjects. Along a broad range of indicators, from initial achievement in mathematics and course taking to postsecondary degrees and later careers in mathematics-related fields, disparities can be found between Whites and Asian Americans on the one hand and African Americans, Hispanics, and American Indians on the other; between males and females; among groups based on their English language proficiency; and among groups based on social class.

There are many reasons for being concerned about disparities in the mathematics education of diverse groups. Some commentators take disparities as evidence of deep structural injustices in how the American schooling system distributes opportunities to learn mathematics and, hence, in the actual acquisition of mathematical knowledge (Oakes, 1990a, 1990b; Secada, 1989; Secada & Meyer, 1991). These writers argue that the American schooling system needs restructuring to do away with those inequities. Research helps us to understand the extent and nature of those inequities and to make reasoned conjectures about how best to go about remedying them.

Many of these same writers, as well as others, also view these disparities from a position of socially enlightened self-interest. Disparities in mathematics education are found among groups for whom demographers predict increased growth within the school-age population, both proportionately and in absolute numbers (Hodgkinson, 1985, 1989; Pallas, Natriello, & McDill, 1989). Hence, research about the nature of these disparities is needed to help schools better educate a portion of the population that is growing and for whom schools have not been successful. Consensus is developing, too, that disparities in the learning of mathematics represent a danger to our society's functioning (Johnston & Packer, 1987; National Alliance for Business [NAB], 1986a, 1986b; National Research Council [NRC], 1989; Quality Education for Minorities Project [QEMP], 1990; Secada, 1990a, 1990b, 1991a). Full participation in our most cherished democratic institutions, projections for civilian workforce and military needs, and shifts in our country's and the world's economic systems all point to the need for *everyone*—not just for a few—to possess more and different mathematical and scientific literacy than is currently made available in schools. Although we have embarked on some ambitious curriculum-reform projects for school mathematics and science (American Association for the Advancement of Science [AAAS], 1989; Mathematical Sciences Education Board [MSEB], 1990; National Council of Teachers of Mathematics [NCTM], 1989; NRC, 1989), there is a danger that mathematics education

The chapter was written, in part, with support provided by the Wisconsin Center for Education Research (WCER), School of Education, University of Wisconsin–Madison. Findings and opinions are the author's and do not necessarily reflect the views of WCER. I would like to thank Jerilyn Grignon for her insights into how being an American Indian has been constructed in our society. Thanks to Edward A. De Avila at Linguametrics, Inc., Martin Johnson at University of Maryland–College Park, and Jeannie Oakes at the University of California–Los Angeles and the RAND Corporation for their helpful comments on an earlier draft. Also, thanks to Deborah Stewart at WCER for her technical editing of this chapter.

for members of these groups will remain unreformed and unrestructured. Under such conditions, disparities in opportunities, achievement, course taking, and careers are likely to increase, resulting in the creation of a permanently unemployable underclass who will represent a threat to the United States' economic and military well-being and who will strain the country's legal and social service systems (Johnston & Packer, 1987; NAB, 1986a, 1986b; Secada, 1990b).

Perspectives like these are driven primarily by policy considerations; we seek to understand these perspectives in order to act. Many recent reviews on the mathematics education and related careers of women and minorities have been similarly grounded (Chipman & Thomas, 1987; Lockheed, Thorpe, Brooks-Gunn, Casserly, & McAloon, 1985; NRC, 1989; National Science Foundation [NSF], 1986; Oakes, 1990a, 1990b; QEMP, 1990). The intent of this review is not to cover that same ground. Rather, its purpose is to lay out, within rather broad parameters, an intellectual agenda for scholarly research.

FOCUS AND ORGANIZATION OF THIS CHAPTER

The basic assumption of this review is that research, just like mathematics, is a human activity and hence a cultural artifact. For example, Kallos and Lundgren (1975) and Sampson (1977, 1981) have argued that, despite claims of neutrality, the psychological study of people is predicated on the acceptance of specific Western norms. Sampson refers to one such norm as an excessive attention to the individual whereby constructs such as ability, learning, and social deviance are defined as individual—rather than social—constructions. Similarly, much current research on the teaching and learning of mathematics has transformed social and public events into individualistic and private states of mind or characteristics of individuals. For example, performance on an achievement test (a public event) becomes evidence of a person's ability or learning problems (private states of mind).

Elsewhere I have argued that the transformation of social events into psychological phenomena and the exclusion of diverse populations from the beginnings of the research enterprise result in the development of potentially biased models for the teaching and learning of mathematics (Secada, 1988). More generally, how we study the aforementioned group-based disparities—when that study is driven by policy concerns—is likely to be situated in culturally bound world views. For example, the literature on minorities and mathematics achievement can be seen as falling into three categories: studies that *describe* and document underachievement in mathematics by minorities, studies that *predict* that underachievement (as is the case for many policy-driven studies and studies about at-risk students), and studies that, at least tacitly, make *causal* links between minority group membership and that underachievement (for example studies on the "effects" of race, social class, and the like).

Each of these kinds of studies—description, prediction, causation—contain tacit world views about this phenomenon that may set the terms of scholarly and research-based discourse in ways that we may not want them to be set. But unless our ways of viewing the world are examined, our research may grant legitimacy to the social arrangements that originally gave rise to the disparities under study. In other words, our research is at risk; it may label without truly helping us understand the phenomenon at hand.

In this chapter, I propose to take some first steps in setting out an intellectual (as opposed to a policy-based) agenda for our understanding and subsequent scholarship. I hope to do so by examining the quantitative-research literature that links academic achievement to various social groups that have been defined along lines of race, social class, ethnicity, and language. In fashioning this review, I have tried to make sense of that literature, to ask questions such as, What does it all mean? Can the claims both explicit and tacit found in these studies be supported? In asking these questions, I hoped to explore the socially constructed nature of the categories that we use when we consider the nature of achievement disparities. From such an effort, we may be able to fashion some new understandings from which to develop new research efforts that will eventually lead to closing the achievement gap.

Elsewhere in this volume can be found a comprehensive review of the research on gender and mathematics. Hence, a discussion of gender per se lies beyond the scope of this chapter. On the other hand, gender, social class, race, and ethnicity interact in ways that research and scholarship are just beginning to explore and that create stratification of educational opportunities, processes, and outcomes (Grant & Sleeter, 1986; Reyes & Stanic, 1988; Weis, 1988). This chapter explores the extension of those insights to mathematics. Gender is included when considering more integrative issues.

Although they are important outcomes of school mathematics, careers, course taking, and affect were omitted from this review. The research on course taking, degrees, and careers has been amply reviewed elsewhere within policy-driven frameworks (Chipman & Thomas, 1987; Dix, 1987; Oakes, 1990a, 1990b; Sanders & Lubetkin, 1991). Aside from studies on gender and mathematics, I was unable to find enough studies relating affect to the mathematics education of diverse populations for more than a cursory review.

The first major section of this chapter contains a discussion of the problematic nature of how we define diverse groups in our society. The second section examines the literature on mathematics achievement by these groups. The third section is a review of efforts to close the achievement gap among various groups. The fourth section contains summary comments and suggestions for future research.

DEFINING THE POPULATIONS

Much of the renewed concern for the mathematics education of diverse learners can be linked to the realization that the school age population of the United States is becoming increasingly diverse and that growth will occur in precisely those groups for whom the educational system has not worked as well as it should. In 1976, 24% of the total student enrollment in U.S. schools was non-White; by 1984, the figure had risen to 29%; by the year 2000, between 30 and 40% of the country's school population will be minority (Center for Education Statistics [CES], 1987a, p. 64; Hodgkinson, 1985; Veltman, 1988). In the country's

20 largest school districts, the respective figures for 1976 and 1984 were 60% and 70% (CES, 1987a, p. 64). One in four students is poor (Kennedy, Jung, & Orland, 1986, p. 71); one in five students lives in a single-parent home (CES, 1987b, p. 21).

The intercorrelation among various demographic characteristics has been well-documented in the literature. For example, poverty, ethnicity and race, and family structure are correlated. Among White children living in households with an adult male present, 11.9% are poor; among African American and Hispanic children similarly situated, the respective figures rise to 23.8% and 27.3%. On the other hand, among children living in female headed households, the rates of poverty rise to 47.6, 68.5, and 70.5% for Whites, African Americans, and Hispanics, respectively (Kennedy et al., 1986, Figure 3.3, p. 36).

Increasing numbers of children from minority-language backgrounds are entering school with little or no competence in English (O'Malley, 1981; Veltman, 1988). Although Spanish is the predominant first language for such children and is likely to remain so into the next century, increasing numbers are entering school with non–Spanish-language backgrounds such as Arabic, Chinese, Hmong, Khmer, Lao, Thai, and Vietnamese (Oxford-Carpenter et al., 1984).

Hispanics are the fastest growing minority group in the United States. From 1982 to 1985, the Hispanic population grew an average of 3.0% per year; the African American population grew 1.6% per year; and the White, non-Hispanic population grew by 0.6%. Depending on assumed rate of growth, Hispanics are projected to become the largest minority group in the United States sometime between the years 2000 and 2050 (U.S. Bureau of the Census, 1986) Although Hispanics have shifted from rural to urban population centers and are found throughout the United States, over 60% of them live in California, New York, and Texas. Adding Florida and Illinois raises the total to 75% (Arias, 1986, p. 29).

Hispanics are being increasingly segregated in school (Arias, 1986; Espinosa & Ochoa, 1986). And African Americans, although they are attending nominally desegregated schools, are being grouped and treated in ways that still constrain their educational opportunities (Network of Regional Desegregation Assistance Centers, 1989).

Although information such as the preceeding is clearly warranted, it should not preclude inquiry into how the categories of ethnicity, race, social class, and language are constructed and how they are maintained in our society and, by extension, in our research. These categories not only have political and social histories; those histories also affect how we interpret research findings. Moreover, individual groups (e.g., African Americans or Hispanics) have had different histories that continue to affect how group membership is interpreted and negotiated. The relationship of, on the one hand, these histories and the negotiation of membership in various groups to, on the other hand, how mathematics education is enacted for these diverse groups has only begun to receive serious study.

Racial and Ethnic Categories

Ethnic and racial categories are neither monolithic, nor are they immutably given. Individuals often will affiliate with a given group based on one term but not on another. African Americans have been referred to in terms that could highlight race or ethnicity: Negro, Black, Afro-American, and (most recently) the term used in this chapter, African American. The use of racial terms in reference to African Americans can be traced to beliefs that people inherit a range of genetic traits that are linked to skin color (Shockley, 1971); among inheritable traits were thought to be intelligence and, consequently, aptitude for academic achievement. Exactly what is inherited has been subject to intense and often acrimonious debate (Dunn, 1987; Fernandez, 1988; Education for socially disadvantaged children, 1965, 1970), and racial categories have come to be unlinked from notions of inheritability, at least when referring to academic achievement. We have come to recognize the social and cultural meanings that are placed on racial and ethnic categories much as we have come to draw a distinction between sex and gender. Indeed, an effort to highlight the cultural and socially constructed nature of what it means to be an African American is the reason many writers prefer that label.

Accepted ways of referring to Hispanics include the term Latino. Hispanics self-identify in terms of their family's country of origin: Mexican, Puerto Rican, Cuban, and other (Nielsen & Fernandez, 1981). Individuals from Mexican ancestry will identify themselves as Mexican, Mexican American, Chicano, or Cholo (Matute-Bianchi, 1986). And New Yorkers of Puerto Rican ancestry argue that they should be seen as distinct from Puerto Ricans who have never come to the mainland.

Individuals of Asian ancestry can be differentiated in terms of country of origin, cultural, religious, and linguistic backgrounds. Chinese includes various dialects. Southeast Asian refugees and their children come from Cambodia, Laos, Vietnam, and Thailand; the Hmong were originally from the highlands across Vietnam and Laos.

The terms American Indian and Native American have been used interchangeably. The federal government has employed various criteria for determining who is an American Indian. Typically, these criteria include notions of how much American Indian blood an individual possesses or how far back one's ancestors were "pure" Indian. Originally, these criteria were developed because the federal government had to pay treaty-based annuities to the descendants of different tribes. States sometimes apply different definitions. Individual tribes have their own informal ways of monitoring people's claims to be Indian, as evidenced by the statement, "We know who we are."

Although there are cores to the constructs of race and ethnicity, that they are socially negotiated can be seen by how individuals accept or reject artifacts of those groups and define their membership accordingly. For example, Fordham (1988) and Fordham and Ogbu (1986) have described how academically successful African American students cope with the perception that they are "acting White" by hiding their abilities from others in school; by actively rejecting certain artifacts of what is typically considered to be African American culture (like music); by being successful in high-prestige nonacademic areas (like sports); or by some other means. Some individuals become extreme in their rejection of artifacts and values that are considered African American, to the point that Fordham (1988) has argued that they adopt raceless personas.

While Fordham's work has been in high schools that enroll large numbers of African American students, McCandless

(1990) studied the belief systems concerning mathematics that were held by African American students enrolled in an integrated, albeit predominantly White, high school. Enrolled in different academic tracks, these students had persisted in taking mathematics courses through their high school years. Though identified by teachers as being African American, one biracial student was enrolled in high-track mathematics courses and actively rejected any identification as being African American. The musical tastes, friendships systems, and other artifacts of adolescence discussed by this student were consistent with that rejection (and with Fordham's analysis). However, this student noted that a sibling did actively self-identify as African American. Another student who was identified by teachers as African American did self-identify as being African American, but rejected any affiliation with African American students who "don't apply themselves in school" (p. 47). For both of these students, being African American and achieving in mathematics were related, but in oppositional ways.

Matute-Bianchi (1986) has documented how individuals of Mexican ancestry self-identify with different subgroups. One group of these students, Cholos, seem to identify low achievement as a defining feature of group membership. Moreover, in a study of American Indian students' belief systems and computer-related course taking, Grignon (1991) has argued that, beyond the acceptance or rejection of cultural artifacts, many American Indian males and females define themselves in terms of their future roles as tribal members and whether specific actions (that is, course taking) will be "useful to the tribe."

These studies are beginning to document the complex ways by which membership in diverse ethnic groups is negotiated and maintained. They document how historical forces interact with the current social contexts of those groups to create what Ogbu (1978; Ogbu & Matute-Bianchi, 1986) refers to as folk theories within each group concerning a range of topics. One class of folk theories is how one gets ahead in U.S. society and how education contributes to it. These folk theories also become part of how individual members of each group interpret their educations.

These studies also raise some important issues about who defines membership in a given group and the meanings that members attach to their affiliation. For example, membership can be imposed from outside, as occurs when the government, school staff, teachers, or even researchers identify individuals as African American, Hispanic, and the like. Based on these labels, the outsiders apply certain beliefs about the individuals with whom they are interacting. Alternatively, individuals may claim (or reject) membership in a group and have their claim (or rejection) validated or questioned by other members of that group. How the negotiations, maintenance, and meanings that are applied to group membership and the group's boundaries impact mathematics classes is an area that would seem to be of critical importance.

Social Class

The *Equality of Educational Opportunity Survey* ([EEOS], also known as the *Coleman Report,* Coleman et al., 1966) ended up being as much a study of the relationship between socioeconomic status (SES) and academic achievement as it was a study of race and equal educational opportunity. Poverty and disadvantage have often been used interchangeably and as proxies for race (for example, see discussion in Cook et al., 1975). Over time, terms like race, ethnicity, poverty, and disadvantage have become differentiated from each other. Socioeconomic status refers to some combination of familial income, education, and employment. Social class carries with it an overlay of shared group values concerning a range of social issues, differentiated roles in our society, notions of oppression, and struggles among members of different social classes for power.

Interestingly, while individuals will often self-identify as belonging to a specific ethnic group or will judge their own competence in one or more languages, the research literature seems to lack examples of self-identification in terms of the students' own social class, even when it seems clear that judgments are being made about them due to their social class backgrounds or they are in conflict with individuals from other social class backgrounds. For example, although large-scale surveys contain items wherein students self-identify in terms of ethnic backgrounds, gender, and/or language proficiencies, they do not contain items wherein students directly identify themselves in terms of their own social class (for example, "What social class do you belong to: low, middle, or high?"). Rather, questions about parental education, income, books at home, and other similar indicators are used to infer students' social class.

Qualitative research studies seem equally indirect in their determinations of student social class status. For example, Grant and Sleeter (1986) described a sample of junior high school students from diverse ethnic, racial, and linguistic backgrounds. These students lived in a predominantly working-class neighborhood and attended their local school. Although they shared a common world view, these students seemed unaware of their own social class backgrounds. Grant (personal communication, October 1990) noted that he and Sleeter asked their students about their backgrounds, but that they seemed not to have well-articulated conceptions of social class. For example, students who lived just above poverty seemed to think that they were just like everyone else.

In an as yet unpublished study, Grant and Sleeter (Grant, personal communication, October 1990) found that older students in a mixed-SES high school could identify who engaged in activities that might be considered artifacts of different social classes. They knew, for example, who wore what clothes, who was on the tennis team or on the swim team. But even these older students did not seem to have a clearly articulated conception of how these activities were linked to social class backgrounds.

For this chapter, the distinctions between SES and social class membership will be glossed over, as will the distinction between poverty and disadvantage. While these terms convey specific meanings within research communities and the use of one rather than another distinguishes members of different communities, they rely on many of the same indicators— family income, parental occupation(s), and educational level of principal caretaker (usually the mother).

Language

Many students in American schools come from a variety of linguistic backgrounds. Some might be immigrants or the children of immigrants. Others might trace their familial roots in this country over generations, even to the original non-Indian settlers of an area. For example, the family trees of many people in the Southwest can be traced back to the times of the original Spanish conquest and rule. These students have in common an exposure to a language other than English—either at home or in the larger community in which they live—and they can speak and/or understand that other language. Membership in this broadly defined group is usually what is referred to as *bilingual*.

Tied to different language backgrounds, of course, are different patterns of ethnic and cultural heritages. For example, of Hispanic seniors in the High School and Beyond 1980 data-gathering wave, 79.5% self-identified as bilingual via the above criteria; in contrast, among White non-Hispanic seniors, only 5.6% were bilingual (derived from Fernandez & Nielsen, 1986, Tables 2 and 3, pp. 54–55). On the other hand, increasing numbers of students are entering our schools from other non-English-speaking backgrounds that include Arabic, Chaldean, Chinese, Hmong, Japanese, Khmer (Cambodian), Lao, Vietnamese, Thai, and other languages (O'Malley, 1981; Oxford-Carpenter et al., 1984; Veltman, 1988).

Current definitions of bilingualism include distinctions based on a person's competence in either language. Some bilingual students can speak, read, and write both languages very well, whereas others can speak but not read or not write one of their two languages, and still others have limited skills in speaking either language. Bilingual students who are limited in their mastery of English are considered limited English proficient (LEP). Although specific definitions of LEP vary from state to state and within federal policy guidelines, most definitions include evidence that there is a language other than English in the student's social milieu (usually at home, but a broader setting may be given), that the language in question has had some impact on the student (ranging from the student's having some understanding of it to being monolingual in that language), that the student is not fluent in English, and that the student's academic performance suffers as a result of that limited fluency—that is, the student has low academic achievement (Bilingual Education Act, 1988; Iowa Department of Public Instruction, 1983; Michigan Department of Education, 1977; Minnesota Education for Limited English Proficient Students Act, 1988; Wisconsin Department of Public Instruction, 1984).

There has been some debate about whether additional criteria should be placed on how to construe limited English proficiency. A common additional criterion is that a student be *dominant* in the non-English language. For example, O'Malley (1981) followed the Bilingual Education Act's then current definition of LEP and estimated that in the late 1970s there were approximately 3.6 million LEP school-age children in the United States. Barnes (1983) reanalyzed O'Malley's data, applied additional criteria, including the condition of dominance in the non-English language, and estimated a national LEP student population of 1.2 million. (Present day estimates are much higher than either.)

Beyond political argument over its definition, language proficiency would seem to be a contextually bound construct. As sociolinguists (Heath, 1986; Zentella, 1981) are quick to point out, people can be quite competent in one context but not in another. For example, monolingual students are quite proficient in their everyday uses of English, but they often fail in even the most basic communicative tasks that characterize highly academic settings like a mathematics classroom (Pimm, 1987). Similarly, people who are considered LEP in school might be perfectly competent in their use of English in the work place.

Discussion

The categories of race and ethnicity, social class, and language may have conceptual cores, but we should remember that they are also socially constructed and maintained. The social and historical processes by which groups have been created and are maintained are becoming increasingly important, since different meanings are derived from those constructions. At the very least, we should note that members of diverse groups interpret their membership in ways that differ from the interpretation of individuals who are outside of those groups. Research into how this affects the mathematics education of diverse children is just beginning.

Socially created groups exist somewhere between the individual and the larger society. It should follow that the mathematics achievement of such groups should be interpreted as a social issue rather than as a matter of individual differences.

ACHIEVEMENT

Achievement in mathematics is often used as an indicator of "how much" mathematics someone knows or possesses. Typical tests of mathematics achievement have been criticized for being dominated by low-level, basic skills items that are easily produced in a paper-and-pencil format (NCTM, 1989; Silver, in press). Despite these and other shortcomings, achievement tests have been a primary source of evidence for investigating inequality in the education of diverse groups.

Some Caveats in Using Achievement Data

Care must be taken when interpreting achievement test data as a valid indicator of any group's mathematics knowledge. For example, the National Assessment of Educational Progress (NAEP) tests only students who are in school (Burton & Jones, 1982; Moore & Smith, 1987). The mathematics achievement of school dropouts is omitted from these data; hence, the test-taking population (about whom generalizations can be made based on NAEP) changes with age. Although data for 9-year-olds may accurately represent a group, by age 13 many Hispanics—especially migrant children—have begun to drop out in large numbers (Arias, 1986). Certainly, NAEP data for 17-year-olds are not representative of all 17-year-olds.

Proportionately more African Americans, American Indians, and Hispanics drop out of school than do Whites and Asian Americans. Proportionately more students of low socioeconomic status drop out of school than do students of middle- and high-SES backgrounds. Proportionately more students of limited English proficiency drop out of school than do fluent speakers of English (Arias, 1986; Rock, Ekstrom, Goertz, & Pollack, 1986; Rumberger, 1987; Steinberg, Blinde, & Chan, 1984). Since low academic achievement is a correlate of dropping out, mathematics achievement data for these groups may be somewhat inflated in many local, state, and national assessments.

The High School and Beyond (HSB) data set (Rock et al., 1986) and the 1980 assessment of the National Longitudinal Study (NLS) of Youth Labor Force Behavior (Moore & Smith, 1987) have made efforts to include out-of-school youth in their samples.

Low-achieving groups of in-school samples are also deselected from many assessments. For example, students not considered proficient enough in English to take a test reliably and/or students enrolled in federal or local compensatory programs—such as Chapter 1, Title I, Migrant Education, and Bilingual Education— are often omitted from state and national assessments. Criteria for participation in these programs include some combination of poverty and low academic achievement. Again, the academic performance of any group involved in these programs may be overestimated by national assessments that do not test substantial segments of the group.

Racial and Ethnic Disparities

From the age of 9, African Americans and Hispanics do not perform as well as Whites on national surveys of mathematics achievement. For example, NAEP mathematics assessments were conducted in 1973, 1978, 1982, and 1986. Originally, the 1973 data were not reported in terms of demographic group membership (Carpenter, Coburn, Reys, & Wilson, 1978). But subsequently, Anick, Carpenter, and Smith (1981a, 1981b) compared the performance of African Americans and Hispanics to national norms for the 1973 and 1978 NAEP change items. (Change items are those that were administered in both the 1973 and 1978 assessments to investigate the change in student achievement across time of administration.) Overall, 9-, 13-, and 17-year-olds scored an average 38, 53, and 52%, respectively on the 1973 change items. For African Americans, the comparable statistics were 23, 32, and 34%; and for Hispanics they were 28, 40, and 38% (Anick et al., 1981a, Table 2, p. 562). For the 1978 mathematics assessment, Anick et al. (1981a) reported that, nationally, 9-, 13-, and 17-year-olds scored 52, 54, and 58% total correct; African Americans scored 41, 39, and 41%, respectively; and Hispanics scored 42, 43, and 46% (Table 1, p. 561).

These earlier surveys seemed to include African Americans and Hispanics in their national norming samples. Matthews, Carpenter, Lindquist, and Silver (1984) and Raizen and Jones (1985) disaggregated their data for reporting the results of the 1978 and 1982 NAEP mathematics change items. In general, they reported higher scores on the 1978 NAEP than did Anick

et al. (1981a, 1981b), but the differentials between Whites and African Americans and between Whites and Hispanics also were greater than reported by Anick et al. In the 1982 NAEP, White 9-, 13-, and 17-year-olds scored an average 58.8, 63.1, and 63.1% correct on the mathematics change items; African Americans scored 45.2, 48.2, and 45.0%; and Hispanics scored 47.7, 51.9, and 49.4% (Raizen & Jones, Table 25, p. 119).

Dossey, Mullis, Lindquist, and Chambers (1988) reported the 1978, 1982, and 1986 NAEP mathematics scores using 500-point scales that had been developed via item response theory (IRT). The pattern of racial and ethnic differences for the 1978 and 1982 assessments paralleled those for the 1986 NAEP mathematics assessment. In the 1986 NAEP, for 9-, 13-, and 17-year-old Whites, average proficiencies in mathematics were 226.9, 273.6 and 307.5; for African Americans, respective scores were 201.6, 249.2, and 278.6; and for Hispanics, they were 205.4, 254.3, and 283.1 (p. 138).

Hence, the general picture of racial and/or ethnic disparities in mathematics achievement that comes from the NAEP data is that Whites perform much better in mathematics than do Hispanics who, in turn, achieve slightly better than do African Americans. These cross-sectional data suggest that achievement disparities, which are great to begin with, increase over time as students grow older.

The National Longitudinal Study (NLS) and the High School and Beyond (HSB) data sets paint a similar but somewhat more detailed picture. Marrett (1987) contrasted mathematics scores for seniors in the 1972 NLS and 1980 HSB, via scores scaled to the 1972 NLS. White seniors averaged 13.95 and 12.98 on the 1972 NLS and the 1980 HSB common scales, respectively; for African Americans, the respective mean scores were 6.50 and 6.69; for Native Americans, 7.74 and 8.28; for Chicanos, 8.02 and 7.54; for Puerto Ricans, 6.33 and 7.19; for other Hispanics, 8.04 and 8.08 (Table 2, p. 10). (For the non-White groups, standard deviations range from 85% to 99% of group means.)

Hsia (1988) reported data from the NLS and HSB comparing White and Asian American students on comparable scale scores that were created via item response theory (IRT). Whereas between 1972 and 1980 White seniors dropped .14 of a standard deviation in mathematics achievement, Asian Americans dropped only half that amount (or .07 of a standard deviation; Table 4.1, p. 60). In 1972, Asian seniors scored .29 of a standard deviation higher than Whites; in 1980, they scored .37 of a standard deviation higher (Table 4.2, p. 61).

Nielsen and Fernandez (1981) reported the scores from the 1980 HSB survey for sophomores and seniors on 18 common items from the mathematics test batteries. White sophomores and seniors (non-Hispanic) averaged 10.3 and 11.6 correct, respectively; African Americans (non-Hispanic) averaged 6.7 and 7.7; Mexican Americans averaged 7.5 and 8.4; Puerto Ricans, 7.1 and 8.0; Cubans, 8.7 and 10.1; and other Latin Americans, 8.0 and 8.3 (Table 2.3, p. 29).

In a longitudinal study of the HSB 1980 sophomore cohort, Rock et al. (1986) found patterns of differential growth from 1980 to 1982, when the sophomores had become seniors. The 1980 test battery was used for the 1982 follow-up; the mathematics test consisted of two parts containing 28 and 10 items each, for a total of 38 items. Of the sophomores who stayed in school from 1980 to 1982, students of Asian American backgrounds scored the highest in 1980, and they also gained the most in total mathematics, averaging 17.94 and 20.88 for a net gain of 2.9 points. White students came next, increasing from 15.10 to 17.16 for a net gain of 2.1 points. "Other" Hispanics (non–Mexican-American, non–Puerto-Rican) and African Americans each gained 1.9 points from 1980 to 1982. (Note that, whereas Nielsen and Fernandez (1981) and Myers and Milne (1983, 1988) disaggregated Cubans from the other category, Rock et al. (1986) did not. Although all three research teams analyzed the HSB data set, this difference led to different results for Hispanics.) "Other" Hispanics, with scores of 9.56 and 11.49, ranked third in both 1980 and 1982; and African Americans, with average scores of 6.04 and 7.91, were next to last in terms of average scores. American Indians, Mexican Americans, and Puerto Rican students who stayed in school had very small gains of 1.1, 1.1, and 1.6 points in mathematics achievement to show for their efforts. American Indians averaged 8.17 and 9.25 in 1980 and 1982, respectively; Mexican Americans, 7.59 and 8.67; Puerto Ricans, 6.02 and 7.65 (Rock et al., 1986, Table 6-10, p. 177).

Similar patterns of achievement discrepancies have been reported at grades 5, 8, and 11 comparing African Americans to Whites in Pennsylvania's annual assessments from 1981 through 1984 (Kohr, Coldiron, Skiffington, Masters, & Blust, 1987; Kohr, Masters, Coldiron, Blust, & Skiffington, 1991). In California's 1984–85 assessment, the eighth-grade sample scored a mean of 54.0% items correct. In contrast, Asian eighth-grade students scored an average 63% correct; Whites, 60.3%; Filipinos, 58%; Hispanics, 45.5%; and African Americans, 42.4% (California Department of Education, 1986, Table 26, p. 80).

Achievement data can also be reported in terms of students scoring at or below a certain cutoff score (e.g., Dossey et al., 1988). At all ages reported by the NAEP mathematics assessments for 1978, 1982, and 1986, proportionately more African Americans than Hispanics, and proportionately more Hispanics than Whites, scored below predetermined cutoff scores that indicated basic skills proficiencies on arithmetic number facts and number operations, conceptual understanding, simple and multistep problem-solving, reasoning, mastery of complex procedures, and algebra (Dossey et al., 1988, pp. 139–141). Moreover, this disproportionate distribution of students scoring below given cutoff scores increased as the students grew older.

In a report that received much national press, the Montgomery County (Maryland) Public Schools found that proportionately more Asian and White students performed above grade level in the mathematics curriculum than did African American and Hispanic students. These cross-sectional differences were small for first grade, but they became progressively larger through sixth grade. Interestingly, all four groups exhibited similar rates of being on grade level from first through sixth grades; in other words, differences seem to be more a function of how students were distributed above or below grade-level performance (Gross, 1988).

In the 1985–86 California assessment, 40% of all 10th and 11th graders taking the mathematics test scored below what was considered a passing score (California Department of Education, 1987). Only 24% of African American 10th and 11th graders taking the test scored below this score. In comparison, Asian students experienced a 30% failure rate; Whites, 41%;

Filipinos, 45%; Hispanics, 48%; and American Indians, 52% (Table 7, p. 11). The very low failure rate for African American students seems a bit surprising in light of NAEP results and of California assessment results for lower grades, and it bears further investigation to determine whether it is an artifact of selection, differential dropout rates, or some other localized phenomenon or whether it is a valid and robust indicator of improving performance by African American students.

Of all 7th, 8th, and 9th graders taking the California assessment, 28% failed it. Failure rates were 10% for Asian, 24% for African Americans, 26% for Filipino students 27% for Whites; 34% for Hispanics, and 48% for American Indians (Table 9, p. 12). Of 4th, 5th, and 6th graders taking the test, the overall failure rate was 16%. For Filipino students, this rate was 9%; Asian, 11%; White, 15%; African American, 16%; Hispanic, 19%; and American Indian, 25%.

In reviewing test data for Native Americans in New Mexico, Bradley (1984) reported that just 21% of American Indian 10th graders scored at or above a 65% cutoff criterion on a test of computations. In comparison, 27% of the African American, 41% of the Hispanic, and 72% of the White 10th graders achieved this criterion. On a test of problem-solving, 57 and 58% of American Indians and African Americans, respectively, achieved 65% criterion scores; in contrast, 79% of Hispanics and 94% of Whites achieved this same criterion.

Hence, national achievement data exhibit a roughly parallel pattern: In terms of average scores and the distribution of students at or below certain cutoffs, Whites do much better than Hispanics, who do slightly better than African Americans. The California data, however, are a departure from this pattern. Mean score data from the California assessments indicate that Whites and Asian students do much better than Hispanics who do slightly better than African Americans, who in turn seem to perform slightly better than American Indians; this follows the national trends. Yet, data based on rates of students who score below a given cutoff reveal a consistent pattern in which African American students do better than Hispanics. This suggests that one of the two distributions is skewed and that there may be statewide, if not regional, variations in these disparities

Has the Mathematics Achievement Gap Been Closing? In the recent past, much attention has been given to how African Americans and Hispanics seem to have been improving their mathematics achievement scores relative to Whites. These improvements are said to be part of a pattern of general improvement in academic achievement across the country. In a report to the United States Congress concerning the underlying causes of this more general pattern of improved academic achievement, the Congressional Budget Office (CBO, 1987) stated that "Black students and probably Hispanic students have gained appreciably relative to their nonminority peers, although gaps in scores between minority and nonminority remain large. The data also suggest that relative gains were made by students in schools with high minority enrollments and in disadvantaged urban communities" (p. x).

Unfortunately, much of this initial optimism was based on reports of state- and local-district–level achievement test data (CBO, 1987). These data have become suspect due to Cannell's (1987) finding that virtually all the states and many school districts had been reporting that their students were, on average, scoring above average on standardized tests of academic achievement—much as in Lake Wobegon, where everyone is above average (Cannell, 1987, 1989; Shepard & Linn, 1989). Reasons suggested for this phenomenon have ranged from teachers teaching to the test, to students becoming more test wise, to outright cheating in test administration.

On the other hand, Burton and Jones (1982) and Jones (1984) reported a steady closing of the gap between Whites and African Americans for NAEP and SAT scores in both reading and mathematics. More recently, Mullis, Owen, and Phillips (1990) reported that, based on NAEP results from the past two decades, the achievement gaps between Whites and African Americans and between Whites and Hispanics have been closing steadily in reading, mathematics, science, and writing (see Mullis et al., 1990, pp. 35–47). Yet even these data are open to interpretation. For example, Anick et al. (1981b) reported that, between the 1973 and 1978 NAEP, "the difference between the black and national averages [at age 9] was reduced by 4 percentage points. At age 13, the difference was narrowed by 2 percentage points and at age 17, by 1 percentage point" (p. 153).

Yet even cursory examination of the table in that narrative reveals that only for 9-year-olds was there evidence of *improvement* in African American achievement scores—from 23–26%. What happened in all other cases was that national averages had dropped: for 9-year-olds, they had dropped by an average of 1 percentage point; for 13-year-olds, 2 points (that is, African American students had stood still); and for 17-year-olds, the drop was 6 points. For Hispanic 9-, 13-, and 17-year-olds, the net changes were 1, −3, and −2 percentage points from 1973 to 1978. Gains made at the expense of other groups losing more ground than a given group has lost are relative at best, and illusory at worst.

Similarly, in comparing change scores between the 1972 NLS and the 1980 HSB among seniors from various ethnic groups, Marrett (1987) reported that White seniors had dropped .98 of an NLS-based scale score. In contrast African Americans had gained .19 point; Native Americans had gained .54; Chicanos had lost .48; Puerto Rican seniors had gained .85; and other Hispanics had gained .04. Although the White loss was statistically significant, no other difference—gain or loss—was (data taken from Table 2, p. 10).

In comparing changes in average performance among Whites, African Americans, and Hispanics between the 1978 and 1982 NAEPs, Marrett (1987) and Matthews et al. (1984) reported a consistent pattern of gains for White, African American, and Hispanic 13-year-olds across all four levels of the assessment: knowledge, skills, understanding, and applications. Moreover African American and Hispanic 13-year-olds *did* gain more than did Whites—a real and meaningful closing of the gap.

Unfortunately, the results of the 1986 NAEP in mathematics, when compared to the 1978 and 1982 NAEP, revealed a pattern of gains that was limited to African Americans only. Using the 500-point scale developed for the 1986 NAEP, Dossey et al. (1988) reported that, across all age levels, African American students scored significantly better in the 1986 NAEP than they did in either 1978 or 1982. Only White 17-year-olds gained be-

tween the 1982 and 1986 NAEPs; Hispanics showed no significant gains (p. 138).

A closer reading of the 1986 NAEP data suggests that African American gains are limited to very basic skills. The percentage of 9-year-old African American students exhibiting increases in simple arithmetic facts (NAEP level 150) and beginning skills and understanding (NAEP level 200) grew between 1978 and 1986. No such growth was reported in terms of students scoring at or above higher cutoff scores. Since the higher cutoff scores indicate knowledge of more advanced topics, growth for the 9-year-old cohort was limited to simple facts and beginning skills.

Between 1978 and 1982, and between 1982 and 1986, the proportion of African American 13- and 17-year-olds at NAEP level 200 and those exhibiting mastery of basic operations and beginning problem-solving (NAEP level 250) also grew. No growth was evident for higher cutoff scores that would have indicated mastery of more advanced levels.

Over the same time period as these reported gains, there has been an increasing rate of student dropout: from 22.8% in 1972, to 27.2% in 1982, to 29.1% in 1984 (Rumberger, 1987, Table 2, p. 104). If we agree that dropping out is disproportionately high for African Americans and Hispanics and that it is correlated to lower achievement, then among older, real-world populations, the gaps may be expanding. Finally, the number of Hispanics with limited English proficiency has been increasing steadily over the same time period (O'Malley, 1981; Veltman, 1988). If schools deselect students for testing based on language proficiency, the gap between Whites and Hispanics in school may be increasing rather than staying constant.

In sum, the data on whether the gap is being narrowed are mixed. If the gap is truly narrowing, it would seem to be for African Americans only and only on items that reflect low-level and basic skills mastery. These are hard-earned gains, regardless of their causes (CBO, 1987). Indeed, if we agree that the 1970s and the 1980s were times when educational policy was predicated on mastery of basic skills, we could read these data as telling a success story: Insofar as we had set a national goal—the acquisition of basic skills—we moved in the direction of its attainment. African Americans did become more proficient in their mastery of low-level basic skills. Alternatively, we might read this as a story of incomplete success. Insofar as basic computational skills are deemed insufficient for true knowledge and mastery of mathematics, these data document how much farther we still have to go.

Similar kinds of competing interpretations of extant data have been advanced before. For example, the *Coleman Report* (Coleman et al., 1966) documented that within regions of the United States there was parity in the educational resources provided to students from diverse backgrounds. Given that the provision of equal inputs was the educational policy goal of the time, the *Report* did document a success story of sorts. Mosteller and Moynihan (1972) noted that "the country, having chosen a goal in a social problem, had come close to reaching it" (p. 11; also, see Murphy, 1981). On the other hand, given the unequal distribution of educational outcomes that were also documented by this report, the EEOS pointed out the need to set new educational goals (Coleman, 1967).

There may have been a trade-off between basic skills development and students' engagement in other mathematically worthwhile content. Compensatory programs in mathematics tend to focus on basic skills to the detriment of other content (Cole & Griffin, 1987); also, direct instruction in such programs seems to be at the expense of classroom organizational features that allow for open-ended student discussion and higher-order skills development. If so, then the purported closing of the achievement gap may have been a very costly goal to have achieved.

Integrating Gender. Research in mathematics education that seeks to integrate gender with race and ethnicity is a small, but growing, area. Care must be taken in conducting, reviewing, and interpreting that research. Very often, results are presented as indicating gender differences within ethnic and racial groups. Structurally, results could be as easily interpreted to mean that there are ethnic and racial differences within gender groups. How one negotiates these different interpretations remains problematic.

Some writers have tried to apply results from research on gender to the case for racial and/or ethnic differences. For example, in their review of literature concerning the participation of women and minorities in mathematics and related fields, Chipman and Thomas (1987) conducted a very detailed review of

the determinants of math course enrollment among young women in junior and senior high school.... To what extent can we generalize the findings of this research beyond the high school years, beyond mathematics to other fields, *and to the problem of minority representation* [italics added]? It is most likely that the overall structure of these problems is similar, but that the specifics of what is important are different. (p. 409)

Chipman and Thomas seem to be arguing that course taking is subject to the same determinants across groups, and that differential results are due to differential values of these determinants. The model for student achievement proposed by Reyes and Stanic (1988) seems to make a similar assumption as well. Others, particularly those who work within qualitative research traditions, have argued that integrative work should not result in straightforward applications of work from one area of gender, race, ethnicity, and/or social class to another (Ogbu, 1988; Sleeter & Grant, 1988). The categories used to describe student diversity have different social histories; membership in these groups is negotiated in different ways; and the scholarly research traditions within these areas define issues in different ways. Hence, straightforward generalizations across groups would seem to be highly problematic.

In their review of the literature involving middle school sex and ethnic differences in mathematics, Lockheed et al. (1985) found just six studies that considered race and gender simultaneously. Of these, only two—work by Jones and by Marshall—involved relatively large samples and provided descriptive statistics.

Marshall (in Lockheed et al., 1985) reported an analysis of the 1979 California assessment data based on gender for Hispanic, White, and Asian sixth-grade students, all of whom were fluent English speakers. Females performed better than males on computations, although Asian males and females performed equally well. Males, including Asian males, performed better than females on problem-solving.

Work by Burton and Jones (Burton & Jones, 1982; Jones, 1984, 1987; Jones, Burton, & Davenport, 1984) is noteworthy as a substantial line of research into the scale and nature of differences in African American and White mathematics achievement. They conducted analyses of large data sets—NAEP, HSB and SAT. They have also investigated how course taking and other antecedents impact achievement.

Jones et al. (1984) reported the results of a special test of mathematics exercises drawn from the 1973 mathematics assessment that was included in the 1976 NAEP. Some items had been released, most focused on basic concepts, and all were multiple choice. The sample of African American and White 13- and 17-year-old respondents was doubled to 5000 per exercise, and efforts were made to include dropouts and early graduates in the 17-year-old sample.

For the 13-year-olds, Jones and colleagues (1984) found less than 2 percentage points separating African American males and females, while less than half a point separated White males and females. In contrast, differences between African Americans and Whites were substantial—over 17 percentage points. Among 17-year-olds, African American females and males averaged 41.6 and 46% respectively, for a mean difference of 4.4. White female-to-male difference was an identical 4.4 points—66.5 versus 70.9%. Differences between African Americans and Whites averaged 25.2%. Hence, for basic skills items there were no gender differences among the 13-year-olds, and slight gender differences at age 17. On the other hand, substantial race differences were found at age 13, and they increased by age 17.

Moore and Smith (1987) reported data from the NLS 1980 assessment of young men and women ages 15 to 22, in and out of school, in and out of the military, and sampled to include African Americans, Hispanics, and members of economically disadvantaged groups. The paper-and-pencil test included scales for arithmetic reasoning and mathematics knowledge. Moore and Smith further disaggregated their data based on respondents' completed schooling level—K to grade 8, grades 9 to 11, grade 12, and grade 13 or above.

Across both scales, scores improved as a function of schooling: The more school students had, the better they performed. Whites performed better than Hispanics, who performed better than African Americans; and males performed better than females. In knowledge of mathematics, a gender-by-ethnicity interaction revealed differences favoring White and Hispanic males over females, while African American male-female differences were slight. Alternatively, one could read this interaction as indicating that, among females, differences between African Americans and Hispanics were small (on average, less than one point) when compared to White females' performance. In contrast, differences between African American and Hispanic males are significant, with even larger differences between Hispanic and White males.

In terms of arithmetic reasoning, a similar interaction study found differences favoring Hispanic males over Hispanic females, but not so for Whites or African Americans. Once again, these differences might be interpreted as within gender groups.

In sum, what little research there is regarding gender and ethnicity or race would seem to indicate that gender differences within ethnic groups are smaller for younger than for older students. These differences are differentially related to the amount of schooling completed by members of each group. The magnitude of these differences seems to be greater in terms of ethnicity and race than gender.

Social Class and Socioeconomic Disparities

For the 1978, 1982, and 1986 NAEPs, Dossey et al. (1988) provided information about mathematics achievement as a function of four levels of parental education: less than high school education, graduation from high school, some education beyond high school, and graduation from college. For students of all ages (9, 13, and 17) and for all three assessments, pair-wise differences among the first three levels of parental education ranged from 7 to 19 points along a 500-point scale. The median difference was about 10 points (data from p. 138). The difference separating 9-year-old children of parents with some college and those whose parents had graduated from college was between 1 and 3.5 scale points. For 13- and 17-year-olds whose parents had some college versus those whose parents had finished college, the difference hovered around 10 points; that is, it was similar to differences among other groups based on parental education.

Another indicator of student SES (or social class) is the type of community in which the student lives. For the 1978 NAEP, Anick et al. (1981a) reported that 9-year-old students living in Low Metro and High Metro communities averaged 43% and 60% correct, respectively; 13-year-olds similarly situated averaged 43% and 62% correct; for 17-year-olds, the respective statistics were 45% and 68%. In the 1982 NAEP, for all ages (9, 13, and 17) and across all skill levels (knowledge, skills, understanding, applications), students from rural communities outperformed students from disadvantaged urban communities, and students from advantaged urban communities performed best of all (National Center for Education Statistics [NCES], 1984, p. 54). Advantaged urban communities are defined as "cities having a population greater than 200,000 where a high proportion of residents are in professional or managerial positions" (NCES, 1984, p. 54). Disadvantaged urban communities are defined as "cities having a population greater than 200,000 where a high proportion of residents are on welfare or are not regularly employed" (NCES, 1984, p. 54).

In their longitudinal study of the HSB 1980 sophomore cohort, Rock et al. (1986) documented a consistent pattern of disparities in mathematics achievement and in growth that is related to student SES. On the total mathematics battery of 38 items, low-SES sophomores who stayed in school until the senior year (in 1982) grew from 8.47 to 9.74 correct, a gain of 1.3 points (or an effect size of 0.1 standard deviation). Middle-SES sophomores who stayed in school grew from 13.36 correct to 15.26, a gain of 1.9 points (0.2 standard deviation). High-SES sophomores who stayed in school grew from 18.41 to 21.23, a gain of 2.8 points (0.3 standard deviation).

In a meta-analysis of studies that related SES to various types of academic achievement, White (1982) found that, for 128 studies in which students were the unit of analysis, the mean correlation between SES and mathematics was .20. In contrast, for 14 studies in which schools were the unit, the mean correlation was .70 (data taken from Table 5, p. 469). This increased correlation is at least partly an artifact of the aggregation process (Myers, 1985; White, 1982). Correlations between SES and general academic achievement also varied depending on what SES measure was used. When the student was the unit of measure and when SES was measured in terms of home atmosphere (that is, parents' attitudes towards education, parents' aspirations for their children, cultural and intellectual activities of the family), White found the highest correlation between SES and achievement (.58). The next highest correlations were found when individual student SES was measured in terms of parental income and occupation (.33). When the school was the unit of analysis, parental income as an SES indicator produced the highest correlation (.77), followed by parental education (.69).

Myers (1985) investigated the relationship between (a) the concentration of poverty in a school and (b) student achievement and student learning of mathematics. In a reanalysis of the Sustaining Effects Study data, he found differences in average mathematics achievement across grades 1 through 6 in both spring and fall (Table 3, p. D-48). Using path analytic methods, Myers determined that the concentration of poverty within schools was related to student achievement even after statistically adjusting for student-level variables such as gender, racial/ethnic background, family structure, maternal work, family SES, and number of siblings. On the other hand, with the exception of first grade, the concentration of poverty in a school had little relationship to the amount of learning (that is, the change in achievement) that took place from fall to spring. In first grade, this relationship disappeared when student-level variables were held constant.

In a reanalysis of the HSB data set for the 1980 sophomore cohort, Myers (1985) essentially replicated his results for elementary students. Even after statistically adjusting for student-level variables, the concentration of poverty within schools was related to mathematics achievement but not to learning. This time, learning was measured over a two-year period, rather than six months as used in the study of the elementary school population.

Kohr and his colleagues (1987, 1991) tried to address the issue of level of analysis by stratifying schools within Pennsylvania as being of high, middle, or low SES based on the percentage of low-income families living in the area served by a school. Schools in middle-SES neighborhoods were then dropped from further analysis. Student SES scores were based on a composite of parental education, parental occupation, and amount of reading material in the home. Students were partitioned into the lowest 30 percentile, middle 40 percentile, and highest 30 percentile on these scores. Mathematics test scores for 5th-, 8th-, and 11th-grade students were converted to Z-scores for data analyses.

Across grades (5th, 8th, 11th), across school SES (high, low), and across years of testing (combined 1981/82, 1983/84), Kohr and his associates (1987, 1991) found a consistent pattern of student-based SES and racial disparities in mathematics achievement. Regardless of school SES, low-SES students achieved less than middle-class students; high-SES students scored best of all. Regardless of school SES, Whites outperformed African Americans.

Kohr and colleagues (1987, 1991) found a pattern of student-SES–by–race interactions that occurred for only the low-SES schools (that is, not for the high-SES schools), but across grade levels. Increases in White-student SES were related to greater increases in achievement than could be found for similar increases in African American SES. For example, White eighth-grade students attending a low-SES school had the following least-squares mathematics achievement estimates: Low-SES students scored an average Z score of $-.36$; middle-SES students, an average Z-score of $-.04$; and high-SES students, an average Z score of $.37$. African American eighth graders attending low-SES schools averaged $-.87$, $-.74$, and $-.61$, respectively (derived from Table 8, p. 21). In other words, not only did African American students suffer more than Whites academically from attending schools with a concentration of poverty, but also the purported benefits of increasing social class status were dampened more by attendance in such schools than they were for Whites.

Social Class Disparities in Preschool. There is evidence to suggest that many poor children enter school at an academic disadvantage to their middle class peers. Kirk, Hunt, and Volkmar (1975) compared two cohorts of five-year-old children enrolled in Head Start and in nursery school on a five-part test of number identification. Although they found no racial or gender differences, Kirk et al. (1975) did find that Head Start children performed less well than the nursery school children on some tasks involving between two and four objects.

Ginsburg and Russell (1981) administered a battery of neo-Piagetian and early number tasks to two cohorts of children. The first cohort comprised a group of poor African American children and a mixed group of middle-class African American and White preschool children. The second cohort comprised preschool and kindergarten children who were roughly evenly distributed among four groups that resulted by crossing racial (African American and White) and social-class (middle-class and poor) groups. For the latter cohort, Ginsburg and Russell (1981) included analyses based on family structure (number of parents at home) and child's age.

Ginsburg and Russell (1981) found statistically significant social-class differences and SES–by–age interactions on some, but not the majority, of their tasks. Ginsburg and Russell argued that their results point to similar robust competence across race and social class, and in a sense they do. Out of 17 tasks in their second study, Ginsburg and Russell found just four significant social-class differences. On those four tasks, performance was affected similarly across all groups by conditions under which they were administered. Age differences were found in all 17 tasks, thereby bolstering the argument that the tasks were sensitive enough to find differences if they had existed. Still, the four differences found by Ginsburg and Russell are more than one would expect to find simply by chance. Moreover, in the first part of their study, Ginsburg and Russell found nonsignificant differences that were in the direction reported by Kirk

et al. (1975). The pattern of results, therefore, supports Ginsburg and Russell's argument that these children displayed a robustness of thought regardless of social class and race. However, their findings also support the claim that there are some very specific (as opposed to general) performance disparities in preschool and that those disparities are linked to social class.

Saxe, Guberman, and Gearhart (1987) partially replicated the Ginsburg and Russell (1981) study as part of a more complex study of how parents provide metacognitive supports for the children on complex tasks. In addition to finding age-related differences, Saxe et al. (1987) found social class differences among middle- and working-class two- and four-year-olds when they worked alone on "tasks involving complex goal structures. While [they] found no differences between middle- and working-class children's counting words, their ability to read numerals, their comparison of number words, or their counting accuracy…middle class children achieved more advanced performances on tasks involving cardinality, numerical reproduction, and arithmetic" (p. 31).

Hence, it would seem that middle-class children enter school with more advanced skills in some specific areas. These are areas that are often judged to be academically important.

Integrating Social Class and Race. The simultaneous relationship of race and socioeconomic status to a wide range of outcomes—not just achievement, but also intelligence, course taking, and postschool employment and earnings—has been the subject of much research. A good deal of it seems to have pitted race and social class against each other as the explanatory causes of those outcomes. Sometimes this occurs overtly—as in the virulent nature-nurture debate of the 1960s (Light & Smith, 1969; Rohwer, 1971; Shockley, 1971), a debate that recently reappeared involving Hispanic underachievement in the United States (Dunn, 1987; Fernandez, 1988). Sometimes those themes are more subtle, as when observers argue that the roots of racial and ethnic differences in mathematics achievement can be traced to the overrepresentation of minorities in the lower socioeconomic strata of our society. For example, Jaynes and Williams (1989) have argued that

to a substantial degree, black-white differences in educational status can be traced to average social class differences between the two groups. For example…the differences in socioeconomic status explain the entire black-white difference in dropout rates and account for at least 20 percent of the difference in achievement test scores. The more significant point, however, is that black-white differences in school performance are likely to persist so long as differences in the socioeconomic status of the two groups remain. (p. 366)

In fact, poverty *is* more severely concentrated among African Americans and Hispanics than it is among Whites. In a re-analysis of the 1980 census data, for example, Kennedy, Jung, and Orland (1986) found poverty rates of 10.7, 9.2, and 7.2% among White students in grades 1–6, 7–8, and 9–12, respectively. Among African Americans, rates of poverty were roughly three to four times as great—35.8% in grades 1–6, 38.5% for grades 7–8, and 33.4% for grades 9–12. And among Hispanic students, rates of poverty were slightly lower than among African Americans—30.7% for grades 1–6, 33.7% in grades

7–8, and 25.4% for grades 9–12 (data taken from Figure 3.1, p. 34).

Not only are proportionately more African American and Hispanic students poor than are Whites; their experiences of poverty tend to be more intense. Kennedy, Jung, and Orland (1986) further reported that, of all White children between one and three years of age in 1968, 25% experienced poverty at some time during the next 15 years of their lives. For African American children similarly situated, the rate was 78%. Of White children experiencing poverty, the vast majority (79.2%) experienced poverty for less than five years. In contrast, of African American children experiencing poverty, just 41.4% got out of poverty within five years (computed from Table 3.4, p. 45).

Other characteristics denoting low socioeconomic status and educational disadvantagement are correlated with poverty, ethnicity, and race. For example, among White children living in single-parent, female-headed households, 47.5% are poor. Among African Americans and Hispanics similarly situated, the respective rates are 68.5% and 70.5% (Kennedy, Jung, & Orland, 1986, p. 36). Alternatively, among children who are not poor, less than 20% have mothers who failed to complete high school; among poor children, nearly 50% of the mothers failed to complete high school (p. 39). Among nonpoor children, 12.5% live in households with four or more children; among poor children, the rate is 31.3% (p. 37).

Researchers have typically employed one of three possible research strategies when considering race and socioeconomic status simultaneously. They have formed groups along both lines simultaneously (e.g., Ginsburg & Russell, 1981; Kirk et al., 1975; Kohr et al., 1987, 1991), or they have considered both race and SES as determinant or antecedent variables that are entered into some large-scale statistical regression analyses (for example, Welch, Anderson, & Harris, 1982), or they have grouped by race/ethnicity and considered SES as a variable *within* racial and ethnic groups (for example, Buriel & Cardoza, 1988; Hernandez, 1973; Jones, 1987; Rohwer, 1971). Although all three strategies are substantially equivalent vis-à-vis statistical modeling (see Cohen & Cohen, 1975), they seem predicated on different conceptual models and will result in different interpretations by different authors.

In preschool populations, Kirk et al. (1975) and Ginsburg and Russell (1981) found disparities based on income levels and/or family structure (single parent versus both parents at home) more often than disparities based on race. On a few of their early number tasks, Ginsburg and Russel (1981) found some interactions involving family structure (but not SES) and race wherein

middle-class or white children from intact homes performed at a somewhat higher level than middle class or white children from single-parent families. But this family status difference did not hold for lower-class or black children. Several other interactions suggest that preschool level children from intact families perform at a higher level than do children from single-parent families. (p. 50)

For all grades tested (5, 8, and 11) and for all years considered (1981/1982 and 1983/1984) Kohr et al. (1987, 1991) found consistent patterns of racial and SES group differences. Only for students enrolled in low-SES schools (*not* for students enrolled in high-SES schools) Kohr and associates found a consistent student SES-by-race interaction. As with any interaction, this one can be interpreted in two ways: (1) as indicating that discrepancies in mathematics achievement between African Americans and Whites increase with increasing SES or (2) as indicating that SES-based discrepancies in mathematics achievement are greater among Whites than they are among African Americans.

The second of these interpretations—that SES-based discrepancies are greater among Whites than among African Americans (or Hispanics)—is the more common conclusion in studies where racial/ethnic groups are formed and in which achievement is regressed on some combination of variables that includes SES as a predictor. For example, Buriel and Cardoza (1988) analyzed the relationship of a range of sociocultural variables, including SES, to academic achievement for first-, second-, and third-generation Mexican American high school seniors from the 1980 HSB survey. Out of nine step-wise regression analyses, a single SES indicator—father's education—entered just once; in all cases, student educational aspiration was the most potent predictor of academic achievement in mathematics, reading, and vocabulary. In other words, student SES was a weak predictor of achievement.

Knowledge of a second language would seem to mediate even further the relationship between SES and mathematics achievement. In an analysis of the High School and Beyond data set (1980 collection wave), Fernandez and Nielsen (1986) computed correlations between SES and mathematics achievement for each of following four groups: monolingual White seniors (.457), bilingual White seniors (.302), monolingual Hispanic seniors (.268), and bilingual Hispanic seniors (.308).

The relatively weak predictive power of student SES on academic achievement in general for African Americans and Hispanics (Jaynes & Williams, 1989; Ortiz, 1986) has been a constant source of consternation for commentators. In the background can be heard blatantly racist arguments about tracing that "unexplained variance" to genetic deficits (for example, Shockley, 1971). Alternatively, others have argued that these results support the hypothesis that the processes and outcomes of schooling are different for students from different racial and ethnic backgrounds (see discussions by Jaynes & Williams, 1989, pp. 366–373; Wolfle, 1985).

Neither of these positions may be true. Recently, both Jones (1987) and Wolfle (1985) have argued that self-reported background characteristics such as mathematics course taking or indicators of parental socioeconomic status are less reliable (or even biased) for African Americans as compared to Whites. In his study of postsecondary educational attainment based on the NLS data collected in 1972, 1973, 1974, 1976, and 1979, Wolfle (1985) statistically corrected estimates of student background variables—including parental education—and concluded that a single model could predict the postsecondary educational attainment of African Americans and Whites equally well; that is, the process of schooling does *not* differ for groups based on race.

Whether such a conclusion would hold in the case of school mathematics achievement is an open question. Wolfle created a single index of ability that was a composite of mathematics and reading scores for 1972. Father's occupation and father's education each seemed differentially related to this composite index of student ability; whereas mother's education seemed related similarly across racial groups (see Table III, p. 511). Hence, the research suggests that, although student social class may attenuate the relationship between race or ethnicity and mathematics achievement, the relationship differs across racial or ethnic groups.

More Complexity Than Meets the Eye

Much research involving mathematics achievement and other disparities among groups is unidimensional in scope. These studies consider issues of social class, race, language background, or gender in isolation from each other. Most national and statewide assessments report data only in terms of race, language group, social class, or gender; seldom, if at all, will one find large-scale assessments that report, for example, how Hispanic women from various social groups performed in mathematics as compared to Hispanic men from those same groups.

This consideration of group disparities along single dimensions would seem to find roots in research efforts that originally grew out of the Civil Rights struggles of the 1960s when African Americans, Hispanics, women, and members of other groups pressed their agendas for social change. Social and educational programs were aimed at ameliorating some of the educational disparities that gave rise to those movements. Hence, policymakers and researchers would naturally have been interested in whether those efforts had paid off. Evaluations of social innovations have typically asked the most direct question: Does this intervention work? Does Title I or Chapter 1 help poor children? Does bilingual education help students of limited English proficiency? In other words, does a particular treatment (the intervention) have main effects?

Asking whether an innovation works differentially is a more sophisticated question, and it is one that until recently seems to have been of little interest. When asked, it has often been met with antagonism. For an interesting example of how asking about the differential effects of an innovation can result in the skewering of an icon from the 1960s, see the Cook et al. (1975) reanalysis of the original *Sesame Street* evaluations (Lesser, 1974). The question of differential impact corresponds structurally to the issue of locating students within multidimensionally defined groups.

In the absence of evidence to the contrary, many researchers seem willing to assume at least tacitly that research on unidimensionally defined groups such as social class, race, or gender should generalize in a rather straightforward manner to the more complex cases that combine race, social class, gender, and other groups. Indeed, based on the achievement data reviewed up to this point, one might argue that in general we should find (1) a steady increase in achievement based on increasing SES; (2) general race and ethnic differences in which the rank ordering along lines of performance is Asians, Whites, Hispanics, and African Americans with American Indians being close to a tie; (3) a pattern of differential performance in which achievement differences that are a function of SES vary within groups according to the rank ordering above; and (4) a pattern

of gender differences within all possible groups. One may or may not accept this last premise depending on whether one accepts the findings of recent meta-analyses involving gender and mathematics (Linn & Hyde, 1989).

In an effort to see whether such a naive conjecture might be true, Del Harnisch and I ran simple descriptive statistics from the High School and Beyond data set for the 1980 sophomore cohort. We used total raw scores in mathematics (38-point maximum) and the HSB composite SES scales. The results are graphically presented in Figure 25.1.

Only for Whites and Hispanics are the patterns of mathematics achievement as predicted. For Hispanics, as noted earlier in this chapter, the relationship between achievement and SES is not as pronounced as it is for Whites. For Asian students, the steady rise based on SES is reversed at the highest SES quartile for females. Moreover, gender differences favoring Asian females are in evidence at each quartile except the top. That reversal is severe enough to result in no overall gender differences among Asian students. Gender differences among African Americans are very slight, and reversed between the highest and next highest SES quartiles; again, no overall gender differences are in evidence. Among American Indians, if there are gender differences, they favor females at the lower SES quartiles. The reversal at the highest quartile is severe enough to create a slight overall difference favoring males. Moreover, the steady improvement in mathematics that one might expect with increasing student SES is not in evidence for American Indian females; if it exists for males, it is only at the break between the highest and the three lower SES groups.

The data could be viewed in different ways. For example, American Indian males tied Hispanics in mathematics at the two lowest SES quartiles; they tied with African Americans at the next highest quartile; and they outperformed Hispanics at the highest SES quartile. In contrast, American Indian females outperformed African Americans and Hispanics at the lowest two SES quartiles; African American females outscored Hispanics at the lowest-SES grouping; Hispanics and African Americans tied and outperformed American Indian females at the hi-mid quartile.

The patterns of achievement that we might expect to occur as functions of student race, ethnicity, socioeconomic background, and gender do not materialize. There is more complexity than can be caught by unidimensional analyses of achievement patterns. Future research needs to take account of that complexity; at the very least, we should shy away from generalizations about diverse groups without first checking to see whether they are reasonable.

Language and Mathematics

Definitions of language proficiency vary by state, and often within states. Regardless of the definition of limited English proficiency (LEP), however, it is a bit redundant to note that many LEP students are low achieving in mathematics, since a student's LEP is determined by low academic achievement as well as by limited English-speaking skills. For a not very surprising example, the California Department of Education (1987) found that 42% of 12th-grade LEP students—as compared to 13% of

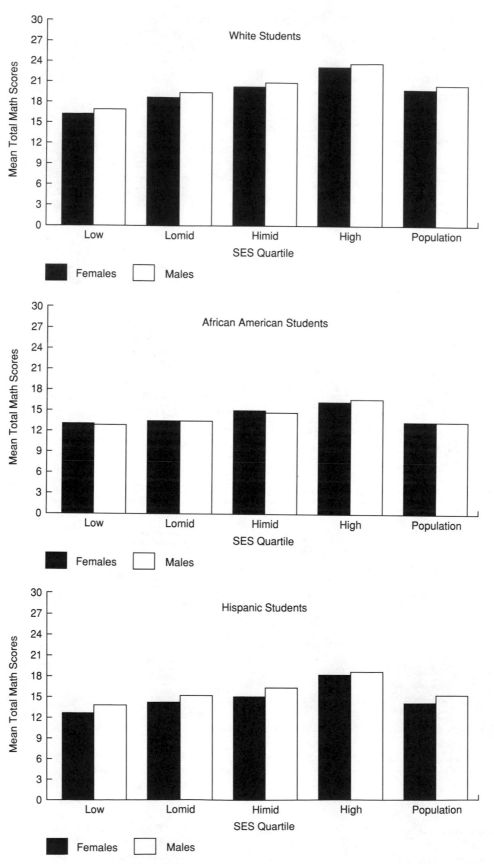

FIGURE 25–1. Math Scores by Students, High School and Beyond, 1980 Sophomore Cohort.

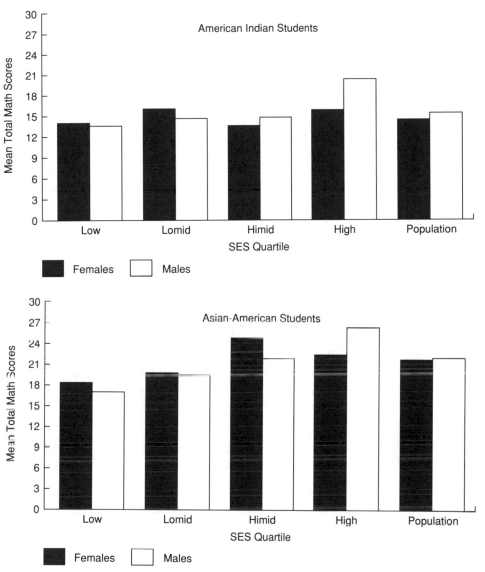

FIGURE 25–1. (*Continued*)

all other students—had failed at least one district mathematics achievement test. Also failing their district's mathematics test: 62% of LEP students in grades 10 and 11, 45% in grades 7 through 9, and 28% in grades 4 through 6. In contrast, 37, 25, and 14% of all other students in those same grades had experienced similar failure.

Proficiency and Achievement. There seems to be some relationship between degree of proficiency in a given language and mathematics achievement in that language. Fernandez and Nielsen (1986) standardized mathematics achievement scores to a mean of 50 with standard deviation of 10 for the 1980 HSB senior cohort. Also, HSB included items for which bilingual respondents judged their fluency in their respective languages. For bilingual seniors from the 1980 cohort—Whites and Hispanics—there was a significant relationship between their proficiency in English and their achievement in mathematics (Fernandez & Nielsen, 1986, Ta-

bles 4 and 5). There is some question about the validity of using self-reported language-proficiency data, especially with younger populations who might overestimate their skills. But the Fernandez and Nielsen (1986) results have been replicated in other studies.

Duran (1988) administered a battery of logical-reasoning (diagramming relationships, inferences, and logical reasoning) and reading (vocabulary, speed, level) tests in both English and Spanish to Puerto Rican bilingual adults. Duran found a single logical reasoning factor that was common across both English and Spanish administrations of the battery. In contrast, reading performance broke out into two factors, one for English and the other for Spanish. Duran found that reading in English predicted performance on the logical-reasoning factor.

In a series of studies for validating the Language Assessment Scales (LAS), De Avila and Duncan (1981) reported significant correlations between English language proficiency and mathematics achievement as measured by the CTBS for grades

1, 3, and 5 (Table 53, p. 107). Although the LAS produced the strongest correlations, De Avila and Duncan found significant correlations via many other commonly used measures of language proficiency.

Secada (1991b) investigated the ability of bilingual Hispanic first-grade children to solve arithmetic word problems in English and in Spanish. Problems were variants of those used in the Carpenter and Moser studies (1984). Language proficiency was assessed by having children retell short stories taken from the LAS that were scored along features of surface fluency and also by having children count-on and count-back. For neither language was performance on the LAS story-retelling tasks related to problem-solving in that language. In contrast, both sets of counting tasks were correlated with arithmetic problem-solving in each language.

In a study undertaken while developing an inquiry-based mathematics and science curriculum (*Finding Out/Descubrimineto,* [FO/D]) for fourth- through sixth-grade bilingual students, De Avila (1988; De Avila, Cohen, & Intili, 1981) found an overall correlation between (a) the English language LAS and (b) mathematics achievement on the CTBS as well as a specialized criterion-referenced test linked to the curriculum in question.

In sum, these studies indicate a relationship between how proficient someone is in a language and performance on measures of mathematics achievement. However, though correlations between language proficiency and mathematics achievement are significant, they tend to be in the range of .20 to .50, with most being at the lower end of the scale. Hence, there is much variance to be explained. Second, Secada's results with one of the most commonly used tests of language proficiency indicate that the relationship between language proficiency and initial arithmetic competence is complex. Certainly, judgments about that relationship should not be made on the basis of superficial indicators such as fluency.

Bilingualism and Achievement. Beyond the question of how proficiency in a single language is related to achievement in that language, researchers have been interested in whether there is a relationship between degree of bilingualism (that is, competence in both languages simultaneously) and academic achievement. This possibility has been examined in three ways. First, groups of students from similar backgrounds are classified as being either bilingual or monolingual, and their performance in mathematics is examined. For example, in the HSB survey, bilingual students were asked to make judgments concerning their relative abilities to speak, understand, read, and write English and their second languages. On their standardized mathematics achievement scales, Fernandez and Nielsen (1986) found that Hispanic bilingual seniors outperformed Hispanic monolingual seniors by more than 2.8 points (that is, 28% of a standard deviation) and that White bilingual seniors outperformed their monolingual peers by slightly more than 1.4 points (14% of a standard deviation).

Alternatively, students are classified as being bilingual (that is, strong in both languages), dominant in one but not the other of their two languages, or weak in both languages. Duncan and De Avila (1979) administered the Language Assessment Scales in English and in Spanish, the Cartoon Conservation Scales,

the Children's Embedded Figures Test, and the Draw-A-Person Test to Hispanic first and third graders. Based on performance in the LAS, children were categorized as belonging to one of five groups: late language learners (low-scoring on the LAS in both English and Spanish), limited bilinguals (high-scoring on one LAS, but not the other), monolingual, partial bilinguals, and proficient bilinguals (high-scoring in both languages). Across all three indicators of cognition, performance followed the unidimensional rank ordering just listed.

Third, the relationship between language proficiency and achievement is examined through regression analyses. Fernandez and Nielsen (1986) found that only proficiency in English was significantly related to mathematics achievement for their White bilingual population. In contrast, for the Hispanic bilingual seniors, proficiency in both English and Spanish was related to mathematics achievement. Hence, not only was there a relationship between degree of language proficiency and mathematics achievement but, for Hispanics, this relationship extended to degree of bilingualism as well.

Similarly, Duran (1988) found that reading in Spanish added to the amount of variance in logical reasoning that was explained by the regression models beyond what was explained by the English language factor. In his study of first-grade Hispanic bilingual students, Secada (1991b) found that proficiency in Spanish added to the variance in problem-solving that was explained over and above what was explained by English language proficiency, but only for problems posed in English and only when language proficiency was measured via verbal counting-on and verbal counting-back that is, not on surface fluency).

Not only does the research suggest a relationship between bilingualism and mathematics achievement, but also between degree of bilingualism and the learning of mathematics. In the development of *Finding Out/Descubrimiento,* De Avila et al. (1981) found a rank ordering of students similar to that found by De Avila and Duncan (1979) on (a) initial achievement, (b) overall growth by all students (treatment and controls), and (c) how much students in the treatment group profited relative to the controls. Secada (1982) suggested that the last finding resulted because children who were less bilingual may have been omitted from the rapid give-and-take that would have characterized the conversations that took place in the cooperative group activities of this hands-on inquiry-based program.

Language and Social Class. Many writers have criticized research involving bilingual populations for its failure to consider social class as a confounding variable when comparing bilingual populations to monolingual English speakers (Hakuta, 1986). Prior to the 1960s, recent immigrants and/or their children, who tended to be from lower-SES backgrounds, would be compared to monolingual groups from middle- and upper-class backgrounds. Recent research, including studies of mathematics achievement, has addressed this issue by creating groups that are matched along SES indices and/or through statistical adjustments of the data. An interesting twist on this theme seems to have developed, however, as many researchers seem to be asking either explicitly (A. S. Rosenthal, Baker, & Ginsburg, 1983) or implicitly (Fernandez & Nielsen, 1986; Myers & Milne, 1983, 1988) which is the primary causal agent of low mathematics achievement: SES or language proficiency. Regardless of how the issue is posed, the consistent finding has been that language proficiency, however it is measured (what is used at home, self-judged proficiency, response to questions about one's primary language), is related to mathematics achievement and to learning.

Fernandez and Nielsen (1986) found that, for monolingual students in the 1980 HSB, socioeconomic status was a stronger predictor of mathematics achievement than was any other predictor variable—length of residence in the United States, student sex (males outperformed females). In contrast, for the bilingual seniors, the relationship between SES and mathematics achievement was attenuated somewhat by students' self-reported level of language proficiency. For both White and Hispanic bilinguals, competence in English was strongly related to mathematics achievement; the relationship was almost as strong as with SES. Students' competence in their non-English language also was positively related to mathematics achievement.

Fernandez and Nielsen conducted additional analyses of their data to rule out the possibility that the attenuation of the link between SES and achievement for the bilingual populations, and especially for the Hispanics, was a statistical artifact of entering additional variables into a regression equation. They concluded that it was not.

A. S. Rosenthal, Baker, and Ginsburg (1983), in a reanalysis of data from the Sustaining Effects Study, tried to determine whether SES (a combination of parental income, education, and occupation) and race/ethnic background variables could account for the observed differences in fall mathematics achievement and in fall-to-spring mathematics learning between Hispanic bilingual students and their monolingual peers in grades 1 through 6. Rosenthal, Baker, and Ginsburg (1983) found that the less English was used at home, the worse autumn mathematics achievement was. After statistical adjustments for SES and home-language background, this relationship remained, although it became smaller. (In an earlier version of their study, A. S. Rosenthal, Milne, Ellman, Ginsburg, and Baker (1983) found a strong relationship between student SES and both mathematics achievement and learning. They failed to find any relationship between achievement or learning and children's use of Spanish and/or English at home. However, in that version of this study, A. S. Rosenthal, Milne, Ellman, Ginsburg, and Baker (1983) had excluded interactions of home language with SES and race/ethnicity from their statistical models.)

In the case of fall-to-spring learning, non-Hispanics learned more than Hispanics who spoke English at home. In turn, Hispanics who spoke English at home learned more than Hispanics who spoke both languages at home. Hispanics who spoke only Spanish at home learned more mathematics than any of the other three groups. After adjusting for SES and race/ethnic background, A. L. Rosenthal, Baker, and Ginsburg (1983) found that students who spoke only Spanish at home still learned more than would have been expected, while Hispanic students who spoke just English or a mixture of English and Spanish learned less than expected.

Using structural modeling techniques, Myers and Milne (1983, 1988) studied the mathematics achievement of seniors from various ethnic and linguistic backgrounds who had been in the 1980 HSB data-gathering wave. Myers and Milne added gender and a series of SES background variables into their

models. Intervening variables included self-reported degree of English proficiency, reading achievement, aspirations for continued education and occupation, and what turned out to be the critical intervening variable—mathematics course work.

In their first set of analyses, Myers and Milne (1983, 1988) considered the relationship between self-reported use of language at home and mathematics achievement. For all groups, social class was related to achievement, males tended to outperform females, and home language was related to mathematics achievement independently of all other variables. Students who had some exposure to a non-English language but who spoke only English at home scored on average between .24 and .30 standard deviation below their monolingual English-speaking peers. Bilingual students who spoke Chinese, Italian, French, or German at home achieved on average between .16 and .78 standard deviation higher than did their monolingual English-speaking peers. In contrast, bilingual students who spoke English and Spanish at home scored more than .20 standard deviation below their monolingual English-speaking peers; students from homes where Spanish was the *only* language spoken scored between .40 and .45 of a standard deviation below the same reference group. Across all groups, the taking of advanced mathematics courses was the single most powerful mediating variable between SES and language use, on the one hand, and mathematics achievement, on the other.

In their second set of analyses, Myers and Milne (1983, 1988) considered the relationship between self-reported primary language and mathematics achievement. Among bilingual students whose primary language was English, mathematics achievement ranged from .03 standard deviation below to .10 standard deviation above their monolingual peers, depending on ethnic group membership. Students whose primary language was Spanish scored around .30 standard deviation below monolingual English speakers. But students whose primary language was neither English nor Spanish scored even worse—an average of .78 standard deviation below the monolingual English-speaking group. Once again, course taking was a very powerful mediating variable between primary language use and mathematics achievement.

In sum, although social class is clearly related to low mathematics achievement among students from diverse language backgrounds, so, too, is language—be it competence in English, competence in Spanish, or degree of bilingualism. Despite the bad press given to being a non-English speaker, the evidence suggests that students who are primary speakers of Spanish perform better than their bilingual peers on some mathematics tasks.

Discussion

In sum, achievement disparities based on social class and racial/ethnic group membership can be detected almost as soon as students can be reliably tested. They are pervasive, evident across mathematical content domains and skill levels, and increasing over time. Over the past decade, the achievement gap between African Americans and Whites seems to have narrowed somewhat. Moreover, the research literature is beginning to consider the simultaneous relationships of multiple group membership to mathematics achievement. The literature contains many different ways of interpreting these disparities, but two specific impressions are worth mentioning.

From Description to Predication and Causation. The first impression is that the description of a social problem has been transformed into a series of predictive and causal statements. For example, it is one thing to note that Hispanics from lower-class backgrounds do not, as a group, perform as well in mathematics as do middle-class Hispanics, or even as well as Whites who are similarly situated. It is another thing, however, to argue that being Hispanic from a lower-SES background places one at risk of such an event; that is to say, membership in a group is predictive of a certain event. It is an even stronger claim to argue that membership in a group causes such an event, as is the case at least tacitly for studies that include the phrase "the effects of" in their titles, their grounding of the research question, or their discussion of results.

All three claims about what it means to belong to a socially constructed group—description, prediction, causation—are structurally the same in the various statistical analyses that are employed. In itself, this use of groups as structural variables might not be so problematic were findings to be interpreted in terms of specific group membership and nothing else. But in fact, these transformations of what it means to belong to this or that group are literally part of the fabric of the research literature.

At a minimum, causal language implies that the relationship between variables is one of prediction and control (Cook & Campbell, 1979). Through our ability to control the independent variable, we can predict occurrences of the dependent variable. I know of no studies—nor of any policy initiatives, for that matter—in which efforts were made to control the demographic characteristics of race, ethnicity, social class, or language. The only possible exception to this observation might be some of the Great Society social policy initiatives, such as the negative income tax (see Cook & Campbell, 1979). Even then, it is far from clear that the initiatives were intended to manipulate social class as much as they were intended to ameliorate the conditions in which the poor found themselves. We may select and we may stratify samples, but we cannot control who belongs to what social group as we can control treatments to which individuals are randomly assigned.

Using membership in a group for making predictive claims is also problematic. The whole point of monitoring how mathematics achievement and other educational indicators are distributed among social groups was to create interventions for improving the educational status of individuals from diverse backgrounds. The prediction that should be made, if there is one, is that, *unless we modify the educational experiences that students from these backgrounds receive, we will fail to educate them as fully as we believe they can be educated.* In and of itself, membership in these various groups cannot predict mathematics achievement later in life.

A better way of thinking of these groupings is that they help define the contexts in which students are educated and research gets conducted. Indeed, Bates and Peacock (1989) have argued that the use of groups as variables represents a fundamental flaw in how current social science research is carried out. They argued that there are major differences between (a) groups built up along the lines of demographic characteristics and (b) the structural variables (like personality traits or social roles) used for the creation of causal claims. Groups define the populations to which we can generalize claims for our results and about whom we might claim certain things. Beyond this, however, groups are not predictive variables or causal agents.

Demographic groups have had long political and social histories in the United States. Membership in them is a negotiated and complex phenomenon that is not easily expressed, nor can it be easily described as a variable. Hence, although groups may be structurally equivalent to different social and psychological variables for data analytic purposes, they are different in kind and results should be interpreted accordingly.

Social Class. The second impression from reviewing this research is a bit more difficult to describe, but it revolves around the special status that seems given to social class as a factor in educational achievement. Much of the research pits social class against some other characteristic (race, ethnicity, language background) as *the* causal variable for a given state of affairs. For the reasons just cited against regarding demographic groups as a causal variable, such arguments seem fundamentally wrongheaded. Yet SES does seem to forgive or explain away findings in ways that would be considered unacceptable were other groupings to be used.

Indicators of social class are often entered into regression analyses at some point to control for social class or to investigate the effects of social class. It is as if social class differences were inevitable or that, if we find them, the results are somehow explained. It even seems different to claim that lowered achievement is an effect of social class than to claim that it is an effect of ethnicity.

Social class differences are not as problematic in the literature as are racial, ethnic, or other disparities. For example, while the research literature and mathematics-education reform documents (for example, NCTM, 1989; NRC, 1989) at least mention women and minorities, issues of poverty and social class are absent from their discussions. Frankly, the literature does not bristle with the same sense of outrage that the poor do not do as well in mathematics as their middle-class peers as it does with similar findings along other groupings.

We should ask why. It may be that poverty and low mathematics achievement are related and that, since the 1960s (when we first undertook the challenge of improving mathematical achievement), we have been unable to break that relationship. Hence, we may have accepted it. For example, Coleman (1975) argued that equality of educational opportunity consisted in lessening the predictive power of social class on academic achievement later in life. The first generation of effective-schools research (Good & Brophy, 1986) was predicated on accepting social class differences as determinants of academic achievement; unusually effective schools were selected *within* social class groupings.

Moreover, social class does not seem to be as salient a characteristic of students as their race, ethnicity, gender, or language background. Indeed, students themselves may not be aware of social class in the same way that they are of these other characteristics.

Since 25% of school-age children experience poverty at some time in their schooling (Kennedy, Jung, & Orland, 1986) and since this rate is likely to increase with increases in single parent homes, social class must be attended to in more integrative ways than it has been in the past. Clearly, social class will become increasingly important in the future. As research becomes more integrative and sensitive to the diversity of our school age population, it will need to consider more carefully how social class enters that discourse.

IMPACT STUDIES

The preceding status studies outline a bleak picture for the mathematics education of students from diverse backgrounds. Implicit in those studies are two closely linked goals for improving the mathematics education of these students: (1) Reduce, if not eliminate, the disparities and (2) accomplish this not by lowering achievement for anyone, but by improving the achievement of students from diverse backgrounds. A third goal of social policy initiatives is even more difficult to pinpoint but is found in some of the research reported below: Once the achievement gap has been closed, keep it closed; that is, the program's impacts should be long term. Efforts to meet these goals have focused mainly on the second goal of improving achievement. These efforts have ranged from compensatory education programs to schoolwide reforms.

Compensatory Programs

Compensatory programs date back to Johnson's Great Society and its War on Poverty. The history of those programs can be read in many ways, including a basic distrust of schools, fueled in part by the *Coleman Report* (Coleman et al., 1966), which found that achievement followed lines of social class. Since schools were not effective for the poor, programs were designed to operate for students before they entered school, outside of the classroom after they entered school, or as alternatives to school. Seldom if ever were programs fashioned as integral parts of either schools or classrooms.

Compensatory programs were established under the assumption that students from diverse backgrounds suffered specific deficiencies as compared to their White, more affluent peers. Hence, the functions of programs would be to compensate for children's deficiencies. The source of those deficiencies was thought to be their upbringing and sociocultural backgrounds (Education for socially disadvantaged children, 1965, 1970).

Head Start. One of the most common policy recommendations is for children from "disadvantaged" backgrounds to enter preschools that will enable them to catch up with their more advantaged peers. One of the earliest programs was Head Start.

Current proponents of Head Start often lose sight of the fact that, initially, the program was in deep political trouble because its program evaluations could not demonstrate effects on cognitive functioning—intelligence being the primary outcome measure of the first evaluations. In fact, at no time has there been evidence that enhanced academic achievement is a long-term, sustained outcome of Head Start or similar preschool programs (McKey et al., 1985; Sarason & Klaber, 1986; Stallings & Stipek, 1986; White & Buka, 1987).

It has been primarily due to their noncognitive and out-of-school outcomes that programs like Head Start have become so highly valued. In a more recent research synthesis on the impact of Head Start, McKey et al. (1985) concluded that, while Head Start programs tend to have immediate effects on children's cognitive development, those effects fail to endure more than two years after children leave the program. Head Start programs also impact children's socioemotional development and health, community health organizations, families, and the larger community in which programs are implemented.

Key in obtaining political support for Head Start were the longitudinal studies of the Perry Preschool Program in Ypsilanti, MI. Long-term program effects included increased rates of high school completion, reduction of high school dropout rates, prevention of juvenile delinquency and of later life crime, increased college attendance, and successful later life employment (Lazar & Darlington, 1982; Stallings & Stipek, 1986). The children who attended the Perry Preschool and the comparison groups came from the same stable neighborhood; this adds to the validity of the conclusions. On the other hand, in reviewing the evidence by which Perry Preschool was judged to be successful, Stallings and Stipek (1986) suggested that the effects of the preschool on its children were likely mediated by the parents and their ongoing involvement in their children's education and that they do not represent independent and long-lasting effects of the program itself.

Sesame Street. Another program much touted for its beneficial impacts on poor children's cognitive development and hence on their eventual academic achievement is Sesame Street (Lesser, 1974), which recently celebrated its 25th anniversary. Because the program was introduced gradually into different markets, it was possible to conduct a carefully designed evaluation. Lesser (1974) demonstrated that children who had viewed the program learned about letters and acquired some other reading readiness skills earlier than children who did not view the program. In a reanalysis of those data, Cook et al. (1975) found differential effects whereby middle class children who viewed the program gained more than did the poor children who had seen the program.

The Cook et al. (1975) analysis of the Sesame Street data is particularly noteworthy because it explicitly addressed both goals at the same time: improving achievement *and* closing the achievement gap. Sesame Street succeeded in terms of the former goal while failing in terms of the latter. What makes this a problematic and hard lesson to accept is that Sesame Street was originally intended to help close the gap. Cook et al. posed the troubling question: Should public policy support innovations that widen the gap between children from different social classes?

Regardless of Cook's ongoing skepticism about the social value of Sesame Street (see Cook & Shadish, 1986), the program has achieved a special status as part of the cultural milieu from which children are thought to acquire a wide range of early childhood competencies—early number competencies among them (Fuson, 1988).

Title I/Chapter 1. The largest federal spending program supporting compensatory education is the old Title I of the Elementary and Secondary Education Act, which became Chapter 1 under the Education Consolidation and Improvement Act (ECIA) of 1981. In a far reaching assessment of that program's effectiveness, Kennedy and her colleagues (Kennedy, Birman, & Demaline, 1986; Kennedy, Jung, & Orland, 1986) synthesized the results from multiple large scale studies involving Title I (and later Chapter 1) students. Their main conclusion was that, especially in reading, the achievement of economically disadvantaged students had improved since 1965; the findings were not consistent for mathematics. Students who received Chapter 1 services gained more than comparison students, who often lost ground. Students gained the most between fall and spring of an academic year; however, disadvantaged students lost more ground over the summer than did their more advantaged peers. Summer programs did not help stem this decline, in part because summer programs did not seem designed with the same academic rigor of school-year efforts.

Students receiving Chapter 1 mathematics support gained more in mathematics relative to their peers than students enrolled in Chapter 1 reading gained in reading achievement. The greatest gains occurred for students in primary grades. When students stopped receiving services, their gains eventually dissipated.

Chapter 1 programs vary substantially in curricular-content coverage, the articulation of curriculum within and across content areas, instructional delivery models (pull-out versus in-class), and many other design features. Birman et al. (1987) tried to identify characteristics whose effectiveness could be supported by research. These features included small instructional groups, increased instructional time, well-qualified instructors, direct instruction, and lessons and materials involving higher-order academic skills (p. 65).

Chapter 1 elementary mathematics programs served a median of five students at a time; most programs served between three and eight students. Most programs met 4 to 5 times per week for between 30 and 50 minutes per meeting. Middle and secondary school mathematics programs enrolled a median of three students per meeting; most programs served between two and eight students at a time. Middle and secondary school Chapter 1 programs met 4 or 5 times per week for the length of a regular class period, that is between 40 and 55 minutes (Birman et al., 1987, p. 68).

Chapter 1 programs are typically implemented by pulling students out of their regular classes or through some other delivery mechanism whereby students forego other classroom and school activities. A common criticism of pull-out programs has been that students lose instructional time, experience disjointed lessons, and do not receive coordinated instruction on the same topic. Chapter 1 programs in elementary school mathematics seem less likely to be at the expense of other mathematics course work than Chapter 1 reading would be for other reading work. This is partly because reading takes up so much of the elementary school day that it is difficult not to interfere with it. Middle and secondary school students seem less likely to miss academic activities due to Chapter 1, since it often replaces study hall or some other free time. Hence, Chapter 1 results in increased time on task for mathematics.

The actual curriculum content that is covered in Chapter 1 programs has been subject to some study and debate. As a compensatory program, Chapter 1 focuses on basic skill development—some would argue to the virtual exclusion of any other worthwhile content (Cole & Griffin, 1987). But, since much of the elementary school curriculum is similarly focused, Chapter 1 services result in increased exposure to similar content. Much of that exposure, however, suffers from not being as coherent across settings as it might be.

Chapter 1 has limited effects on mathematics achievement (Kennedy, Birman, & Demaline, 1986; Kennedy, Jung, & Orland, 1986). Students' scores increase while they are in the program, but those gains are rapidly lost—not only over the summer but also when students leave the program. This loss may be due to the tight focus of the program on basic skills for a given grade. When students leave the program and as they progress through school, they are not tested on those narrow skills and so they no longer outperform students who did not receive Chapter 1 services. If this is the case, there may be two alternatives: (1) include content for students to learn how to learn or (2) provide support throughout students' K–12 education, even at the cost of diminishing returns for that support (Kennedy, Birman, & Demaline, 1986; Kennedy, Jung, & Orland, 1986).

We might cast the inability of sustaining Chapter 1 effects in a different light. The basic assumption for Chapter 1 (and Title I before it) was that students entered school deficient in some basic prerequisite skills. Following the medical model of illness, the remedy for these deficiencies was to remediate those students on their missing skills. Therefore, closing the achievement gap should mean that the treatment was successful. Following the medical analogy, if the deficiencies return in the failure to sustain the cure, then the original diagnosis was incorrect. To follow the medical analogy even farther, we have been treating symptoms, not the root causes of the "illness."

What might those causes be? In light of Ginsburg and Russell's (1981) findings that preschoolers from lower social class backgrounds do not suffer from massive deficits in their early number skills, it seems unlikely that we should assume that Chapter 1's failure to have sustained effects is due to the students themselves. The failure may lie in the process of schooling that Chapter 1 does not affect, or it may lie in the notion that basic skills are fundamental prerequisites for students to acquire other forms of mathematical knowledge. In either case, deficiency-based notions of Chapter 1 need rethinking.

Common criticisms of Title I and Chapter 1 have been their focus on basic skills and their reliance on notions of disadvantagement. When Title I was first enacted, poor students were thought to be culturally deprived. In current usage, *disadvantagement* has been cut off from such theoretical antecedents. Poor students are no longer culturally disadvantaged; now they are educationally disadvantaged.

These and other criticisms have been used to argue for general shifts in how Chapter 1 students are viewed, in the curriculum to which they are exposed, and in the instructional delivery of that curriculum (Knapp & Shields, 1990). Moreover, due to consistent findings that school-level variables impact student learning and that all students (even those from middle- to high-SES backgrounds) who attend schools enrolling large numbers of poor students lag behind in their mathematics achievement and learning (Kohr et al., 1991; Myers, 1985), there have been efforts to restructure Chapter 1 services so that an entire school can participate in the program, not just poor students.

Bilingual Education. Bilingual education entails the use of a language other than the politically dominant language at some points in a student's education. The organizational features of that usage have varied widely during bilingual education's long history, dating back to biblical times.

When the common school was first established in the United States, at a time when its very existence was tenuous, bilingual education was used to entice this country's large German and other northern European immigrant groups into the educational system (Perlmann, 1990). Over time, as German speakers lost control of urban school boards, and with an increasing backlash against immigrants from southern Europe, bilingual education was phased out. By the First World War, many states had passed laws prohibiting the use of non-English languages in school. By the Second World War, German had been dropped from many schools' foreign language curriculum.

After the large influx of Cuban refugees in the early 1960s and due to concerns expressed by Mexican American community groups in Texas and the Southwest, bilingual education became Title VII in the ESEA's 1966 reauthorization (Andersson & Boyer, 1978, Stein, 1986). Since that time, bilingual education's compensatory thrust has been subject to an intense and often acrimonious debate (Secada, 1990b; Stein, 1986).

In an early large scale federal study, Danoff, Coles, McLaughlin, and Reynolds (1977–1978) found no significant mathematics achievement differences between second- through sixth-grade students who were receiving program services and those who were not. This finding was true in fall to spring analysis of the entire sample, and in a fall to fall analysis of a subsample of second- and third-grade students who had been followed into their next grades. The only exception was in fourth grade where students in the program gained more than their peers who had not been in the program. Although it followed accepted practice for large scale evaluations of the time, that study was severely criticized for many methodological flaws. Among these flaws was that large numbers of comparison students had themselves been enrolled in bilingual education programs before having been mainstreamed into all-English settings. Hence, the comparison group contained large numbers of students who were likely to have been selected out of programs because they were achieving above some predetermined level (Willig, 1987).

More recently, Baker and de Kanter (1981, 1983) conducted a narrative review on the overall effectiveness of bilingual education. They concluded that evidence on the effectiveness of bilingual programs was uneven at best and that the research did not support a federal mandate for bilingual education. In a meta-analysis of a subset of the studies from Baker and de Kanter's 1981 review, Willig (1985) concluded that, along a broad range of indicators including mathematics achievement in English and in Spanish, Hispanic LEP students were better served by being enrolled in bilingual education than not. Moreover, Willig concluded that the better the technical quality of a given study—for example, if it used random assignment rather than post hoc matching of students—the greater the differences favoring the use of the students' native language.

In an effort to identify the critical features of effective bilingual instruction, Tikunoff and his colleagues conducted a longitudinal and descriptive study of 58 classrooms that were mainly in elementary schools across five states and that enrolled LEP students from Spanish-, Chinese-, and Navajo-speaking backgrounds. Time on task served as the proxy for achievement (summaries in Tikunoff, 1983, 1985). The study's categories were framed in terms of active-teaching and direct-instruction models. Teachers were surveyed and classrooms were observed to determine the incidence of these teaching behaviors. Five general features of effective classroom-based bilingual instruction were validated through two waves of data gathering: (1) the use of active teaching strategies; (2) the mediation of instruction through both languages, including alternating between languages when necessary to ensure student understanding; (3) the integration of English-language skills development with content; (4) the use of students' home cultural norms for communication and for behavior in mediating active teaching; and (5) congruence from intent to the organization and delivery of instruction, so that students understand that there are high expectations for their performance and so that teachers feel a sense of efficacy in their abilities to teach all students (Tikunoff, 1985, p. 3).

Unfortunately, Tikunoff studied only classrooms that had been designated as being effective. The features were imported from previous research on effective instruction and validated as occurring in these classrooms. Hence, as is the case for much of the effective-schools research (Good & Brophy, 1986), there is little direct evidence that these features discriminate effective from ineffective bilingual instruction.

In an analysis of classroom processes among three different kinds of elementary programs for LEP students—two bilingual and one that used almost all English—Ramirez (1986) found virtually no systematic variation in how kindergarten, first-, and third-grade teachers and their students communicated, nor in the characteristics of the instructional activities that were implemented across the programs. The sole consistent source of variation among teachers lay in how much English and/or Spanish was used during instruction. Results from the second and third year of data-gathering efforts (Ramirez, Yuen, & Ramey, 1987, 1988) have followed the first year's students and added new cohorts in the larger study. Classroom-process results have replicated the first year's findings.

The Ramirez study was intended to evaluate the effectiveness of the three program models against one another. Due to the highly politicized nature of this study, the achievement data were not to have been analyzed until the study was finished. However, first-year achievement results were leaked to the press (Secada, 1990b). During the first year, it seems that increased use of the native language during instruction led to increased achievement in mathematics. Final results of this five-year longitudinal study are expected sometime in 1991.

Burkheimer, Conger, Dunteman, Elliott, and Mowbray (1989), in the analysis of a three-year longitudinal study of federally funded programs for LEP first- and third-grade students, found that instruction in oral English (that is, the speaking of English) was negatively related to mathematics achievement.

Burkheimer et al. (1989) did not explain why this was the case. Many students receiving explicit instruction in oral English have virtually no proficiency in the English language, which is why they receive the instruction. In addition, instruction in oral English may be conducted at the expense of instruction in mathematics.

For children in second grade, instruction in English-language arts (for example, reading and writing) rather than native-language arts predicted mathematics achievement. After a certain threshold of English-language arts instruction, however, second graders achieved better in mathematics when the content was taught in Spanish than when it was taught in English.

For third-grade children, the patterns of effective instruction became increasingly complex. Within schools that enrolled large numbers of low-SES students and where ethnic heritage instruction was unavailable, instruction in Spanish resulted in better achievement than instruction in English. On the other hand, if ethnic instruction was available in such a school or if a school enrolled large numbers of high-SES students, more mathematics instruction in English than in Spanish was related to higher achievement.

In general, for students who received instruction in their ethnic heritage, oral proficiency in English was positively related to mathematics achievement; for students who did not receive this instruction, oral proficiency in English was negatively related to achievement in mathematics (Burkheimer et al., 1989, pp. 12–16). Hence, the research on bilingual education indicates that LEP students are likely to be better off receiving instruction in their native language. But we are only beginning to learn about the processes by which the use of the native language might translate into better mathematics achievement.

Teacher Expectations

Since the publication of *Pygmalion in the Classroom* (R. Rosenthal & Jacobson, 1968), teachers have borne the brunt of much of the blame for the achievement of minority and of low-SES students. The logic of this reasoning goes like this: Teachers are thought to develop unsupported preconceptions of their students' abilities based on their race, ethnicity, SES, gender, or some other demographic characteristic. Then, these expectations are communicated to students who live up (or down) to them, and/or teachers structure life in the classroom so that student opportunities and participation are constrained based on those preconceptions. Subsequently, student achievement rises or falls contingent on those initial beliefs. Teachers do establish differential patterns of interactions with their students that vary based on student demographic characteristics as well as on their expectations of student success (see reviews by Brophy, 1985a, 1985b). But questions remain: Are these expectations valid? How are expectations linked to teacher actions? Are their subsequent actions valid?

It seems that teacher expectations are built up during the first two weeks of school as they are learning about their students. At this time, expectations are most malleable and experimenters have had the most success in establishing false expectations (R. Rosenthal, 1987). Moreover, teachers create differential expectations of students as a function of race and social class, and these expectations seem to be established in

different ways (Baron, Tom, & Cooper, 1985). African American teachers working in schools that enrolled large numbers of African American students seemed to have higher expectations for their students—especially when the school itself tended to be high achieving—than did White teachers in those same settings (Beady & Hansell, 1981). Stanic and Reyes (1986) have documented how an individual middle-school mathematics teacher engaged in purposeful patterns of differential behaviors with his students on the combined basis of student race and gender.

In two large-scale studies involving the formation of elementary reading groups, Haller (1985; Haller & Davis, 1980) and Dreeben and Gamoran (1986) found that teacher assignments to reading groups tended to match student performance on achievement tests. In the Haller studies, teachers consistently assigned students to reading groups across student racial and social class characteristics. When student assignments varied from how they should have been assigned based purely on pretest scores, teachers had made the assignments based on their judgments of how much a given student would profit from being in a given group. This more generalized notion of student educability seemed based on a combination of factors that went beyond notions of student ability to include student classroom behavior, student ability to learn, and the like. Again, there were no differences along racial or social class lines on how this broader notion was used in forming reading groups.

In their study on the impact of an activities-based mathematics and science curriculum on middle-school bilingual (primarily LEP) students, De Avila and associates (De Avila et al., 1981; De Avila, Duncan, & Navarrete, 1987) asked teachers to select a group of students who would not do well with the program. Teachers were very accurate in predicting initial student performance for this group of children, who scored below their peers on the pretests. However, the teachers did not predict how well the students would actually learn, as growth for both groups of children was similar. In general mathematics, the subgroup gained an average of 15.6 points as compared to 19.3 for the larger group of children; on a special test tied to the curriculum, the "problem" subgroup gained 10.7 points as compared to a gain of 9.6 for their peers.

This result may be partly because the topics and materials were unfamiliar to the teachers. De Avila (personal communication, September 1990) has suggested that the teachers' predictions were based on notions of educability that are similar to those reported by Haller. De Avila further suggested that those notions were based on things (like student classroom behaviors) that were inconsistent with the demands of the experimental curriculum that he and his colleagues were developing. Hence, teachers were successful in predicting achievement but they were not successful in predicting learning.

Haller's and Dreeben and Gamoran's findings support the argument that teachers apply their judgments uniformly across student groups and that teachers apply their notions of ability fairly and in an effort to adapt instruction (Brophy, 1985b; Dusek, 1975, 1985). What these arguments omit, however, is the possibility that the basis of their judgments—that is, their notions of ability and educability—are constructed in a biased manner. At the very least, De Avila's observations suggest that these notions are linked to particular views of the curriculum and the learner that may be incompatible with how children actually learn. Thus, although teachers may apply notions of ability and educability uniformly across student groups, these notions may interact with other characteristics of their students and thereby limit their opportunities to learn. At the very least, notions of ability and educability can be questioned on a variety of grounds (see also Oakes, 1990b).

The link between differential expectations and actual achievement has not been documented unambiguously (Brophy, 1985b; Dusek, 1975, 1985). For example, differential patterns of student-teacher interactions or the formation and use of reading groups as an instructional strategy might also be teacher efforts to modify and to match instruction to their children's diverse abilities and interests. The problem, of course, lies in the consequences of those actions. For example, if different groups are exposed to differentiated curricula, then clearly their opportunities to learn will be constrained (see Oakes, 1990b) and their achievement will drop (see following narrative). Also, if teachers are unable or unwilling to change their judgments of student ability on the basis of new evidence concerning that ability, then the judgments become invalid even if they were valid originally.

In sum, the link between teacher expectations and student achievement is much more complex than originally thought. Possibly the best summary of this research was from Cooper (1979), who suggested that "teacher expectations often serve to sustain, rather than bias, student performance.... [But] even the *maintenance* of below-average performance through teacher-expectations effects ought to be the focus of societal concern" (pp. 392–393).

Course Taking and Curriculum

Two of the most powerful predictors of student achievement in all large-scale assessments have been increased time on mathematics and the taking of advanced coursework. In their analysis of why students enrolled in Chapter 1 mathematics gained relatively more than did students enrolled in reading, Kennedy, Birman, and Demaline (1986) suggested that the time spent in Chapter 1 mathematics was mostly in addition to what was taken in the regular mathematics program. Reading is taught throughout the school day, so instruction in reading tends to be at the expense of other reading time.

Of all subjects tested in the High School and Beyond Study — vocabulary, reading, science, and writing — mathematics was the most sensitive to (a) school completion and (b) additional course taking (Rock et al., 1986). Beyond student demographic variables, course taking was the single most powerful predictor of mathematics achievement among the linguistic groups studied by Myers and Milne (1983, 1988).

Similarly, the National Assessment of Educational Progress documents the substantial growth in achievement that is associated with advanced course taking. In a hierarchically ordered regression analysis of the 1978 NAEP achievement data for 17-year-olds, Welch et al. (1982) found that eight demographic background variables (SES and ethnicity among them) accounted for 24.5% of the variance in achievement. After these variables were first entered, the number of semesters of mathe-

matics taken by students *added* another 34.5% to the explanation of that variance.

In the 1986 NAEP (Dossey et al., 1988), 17-year-old students whose most advanced course was pre-algebra or general mathematics averaged 272 points on the 500-point scale; in contrast, students who took algebra scored 287 points; geometry, 301; algebra II, 320; and pre-calculus or calculus, 343 (Table 8.1, p. 115). Similar trends are evident for the 1978 and 1982 assessments.

Partly due to this evidence, states added to the number of mathematics courses required for high school graduation in the 1980s. Unfortunately, those requirements resulted in more low-level mathematics courses being taken by students who were enrolled in low-level or nonacademic high school tracks (Goertz, 1989; Patterson, 1991). Students enrolled in college-preparatory tracks experienced no change in their course taking patterns. These studies failed to investigate whether additional course taking, even of low level courses, resulted in enhanced achievement for the students who took those courses. Such growth may have been slight, since most students would have encountered little new content, and the greatest growth seems to be associated with exposure to new content. Hence, policy efforts related to mathematics course taking need to go beyond requiring numbers of courses; instead, they should encourage the taking of advanced courses.

Two issues arise from this link between course taking and achievement. The first concerns how tracking is related to students' mathematics education and their subsequent achievement. One of the most common secondary school practices that often begins in middle school (Oakes, 1985, 1989, 1990b), tracking incorporates three features: Students are grouped for their academic programs; student assignments to tracks are based on judgments of their ability (and more generally, educability); and tracking includes the a priori differentiation of a student's intended mathematics curriculum (see Oakes, Gamoran, & Page, in press).

The second issue is an analog to the Cook et al. (1975) finding with Sesame Street. If course taking results in the desired outcome of enhanced achievement, then is that outcome equally or unequally distributed among students (that is, does the gap close, remain the same, or become greater)?

The Antecedents of Tracking. Arguments for tracking vary but fall into two categories. First is the belief that tracking (like ability grouping) allows for programs and teaching techniques that are adapted to students' needs based on their diverse abilities. Second, tracking is thought to enable schools to better target students in terms of reasonable expectations for out-of-school outcomes later in life. For example, students who are expected to go to college are enrolled in an academic track; those who are expected to go directly from high school into the workforce are enrolled in a vocational track; students for whom such judgments are uncertain are enrolled in a general track that mixes academic and vocational courses (Oakes, 1989). The latter argument is tied to the belief that students are and should be sorted according to merit and ability.

The second of these assumptions is highly problematic because it omits things such as student career aspirations and because it places the school in the position of determining what is best for its students (Oakes, 1989). The selection of tracks seems to be negotiated between students and their schools, but much more often than not, students end up enrolling in tracks recommended by their counselors (Oakes et al., in press). Moreover, public schools do not seem to match student aspirations to track as well as do Catholic schools (Lee & Bryk, 1988), suggesting that student input into track selection is less than presumed. In addition, public schools provide students with too many options in light of the counseling that students receive for courses that prepare them for graduation, let alone for their career aspirations. Hence, students enrolled in public schools may opt out of high-track courses not because of their career aspirations but because other courses may be easier, or because their friends are taking them, or for other reasons.

Even if counselors do help students choose courses, it is far from clear that their advice is truly in the best interests of students from diverse backgrounds. In her study of four African American students enrolled in an integrated, but predominantly White and middle- to upper-class high school, McCandless (1990) found that the two students taking advanced mathematics courses were doing so at the behest of their parents or some other family member, and they did not go to the school counselor. The two students enrolled in low-level mathematics courses had received advice from their counselors that these were the best courses for them to take.

Many authors argue that the notions underlying how students are placed in tracks—ability, educability, and merit—are inherently biased and that those biases are revealed by how tracking assignments follow patterns of student variability (see Oakes et al., in press). For example, in her extensive study of the relationship of students' demographic characteristics to their opportunities to learn mathematics and science, Oakes (1990b) found that tracking patterns in schools and enrollment patterns in classes varied as functions of their respective racial and social class makeups. Regardless of social class and racial makeup, approximately 80% of all high schools surveyed engaged in tracking. However, schools that enrolled large numbers of non-Asian minorities or poor students tended to have more low-ability and fewer high-ability mathematics courses than schools enrolling large numbers of Whites or high-SES students. Similarly, African American and Hispanic students and students from low-SES backgrounds were overrepresented in low-ability mathematics classes; Whites and students from higher-SES backgrounds tended to be overrepresented in high-ability mathematics classes.

Even if we grant that schools know better than students what constitute appropriate aspirations and how to sort individuals according to merit (or ability), there is evidence that schools do not sort students strictly by merit and that they do not sort them uniformly. In a study of student mathematics and science-course enrollment across four high schools, Garet and De Lany (1988) found that high-achieving students were more likely to enroll in advanced tracks, as would be expected. However, there were substantial ethnic group differences in how students were assigned to tracks, even after adjusting for prior achievement. Asian students were more likely to take advanced mathematics, geometry, and algebra than were Whites with similar prior achievement. African American students were

also more likely to take mathematics courses than were White students of similar achievement.

Across schools, Garet and De Lany found differences in the probability that students with similar achievement would enroll in mathematics classes. For example, the likelihood of students ranked at the 25th percentile in mathematics taking no mathematics courses in each of these four schools was .26, .32, .44, and .83 (Table 5, p. 71). Garet and De Lany concluded their study by arguing that these schools had a "hierarchy of positions" (p. 75) to be filled and that students were sorted into those slots. Students with similar achievement backgrounds found themselves competing for slots in this hierarchy, and the competition was settled not solely on the basis of ability, but rather on the basis of other factors that are linked to student demographic backgrounds. This finding matches Marrett's (1986) observation that African American students who are enrolled in desegregated schools are less likely to enroll in and to successfully complete mathematics courses than are African American students of similar achievement who are enrolled in predominantly Black schools. Marrett also argued that competition for slots resulted in differential access to advanced mathematics courses.

The literature contains an anomalous finding that bears further investigation—that high-achieving African American students are more likely to be enrolled in advanced tracks and/or advanced mathematics courses than are similarly achieving White students (Gamoran & Mare, 1989; Garet & De Lany, 1988). It may be tempting to conclude from such a finding that tracking actually helps high-achieving African American students because as it provides them with enhanced opportunities to take advanced mathematics courses. However, this result may be a function of where students live. White students enroll in schools located anywhere, urban or rural. African American students, and Hispanics for that matter, tend to live in urban settings and hence enroll in urban schools. Historically, these schools have provided a range of advanced courses beyond what is available in rural settings (West, Miller, & Diodato, 1985, Table 8, p. 30). Studies that find (all other things being equal) differential enrollment by high-ability African Americans into high tracks do not take into account how the differential availability of advanced courses is linked to these patterns of enrollment. In other words, all other things are not equal; the proper comparison should be based on the availability of such courses in the first place.

The Consequences of Tracking. Oakes' (1990b) findings are particularly troublesome because she documents a broad range of other features that are also linked to social class and racial patterns in schools: access to material resources, teacher qualifications, and the kinds of instructional practices that students in different classes are exposed to. Even within schools, tracking results in an uneven distribution of resources, with advanced tracks being favored. For instance, low-track mathematics courses are often assigned to beginning teachers or to teachers who have fallen out of favor with a school administrator. General and high-track courses tend to be taught by more experienced, more powerful, or simply more favored teachers (Oakes et al., in press). Low-track mathematics courses cover less content, are paced more slowly, and seldom treat content in depth or in ways that are known to foster student engagement (Oakes, 1990b). The picture these data paint is of low-quality mathematics education programs for students in low-track courses that enroll disproportionate numbers of students who live in poverty or have non-Asian minority backgrounds. Hence, if tracking does allow for programs and teaching that are better matched to student abilities, it would seem to do so only for students enrolled in high tracks. It is hard to reconcile such a justification with what occurs in low-track mathematics courses.

Most studies on how tracking is related to achievement compare outcomes across tracks. For example, Natriello, Pallas, and Alexander (1989) compared the senior-year mathematics achievement of students who had dropped out, who were enrolled in the vocational track, who were enrolled in the general track, and who were enrolled in the academic track. Not too surprisingly, after statistically adjusting for 10th-grade achievement and various demographic characteristics, mathematics achievement differences between tracks were maintained if not increased (Natriello et al., 1989; also see Oakes, 1989).

Results based on cross-track comparison could be due simply to effects of differential course taking and not the result of some other feature of the tracking system. Moreover, as Slavin (1990) noted, the initial group differences between students assigned to different tracks are so great that whatever statistical methods are used as controls are unlikely to be adequate.

Although it is well known that there is variation in how students are assigned to tracks, most studies concerning the relative impacts of different tracks on student achievement have failed to unconfound initial biases in track assignments from the actual effects of those assignments. Gamoran and Mare (1989) incorporated specific models of how students were assigned to different tracks into their study of the differential effects of college preparatory versus other tracks on the mathematics achievement of 12th-grade students from the 1982 HSB. Based on these models of assignment and on how 10th-grade mathematics achievement predicts 12th-grade achievement, as well as on other variables, Gamoran and Mare estimated the mathematics achievement of 12th-grade students under each of two conditions: assignment to college track and assignment to noncollege track. Of course, actual data were available for only one option—the track that each student had actually been enrolled in.

As in other studies, prior levels of academic achievement, sociodemographic factors, and school factors were related to students' being placed in college tracks. For example, all other things being equal, students from higher-SES backgrounds were more likely to be in college tracks than students from lower-SES backgrounds.

Gamoran and Mare compared how SES, prior achievement, gender, race, and ethnicity were related to senior-year mathematics achievement in each of the tracks—college and noncollege. In each of the tracks they found a similar pattern of results, except for race and gender. African Americans enrolled in the college tracks were at less of an achievement disadvantage relative to Whites than were African Americans enrolled in noncollege tracks (that is, being placed in college tracks closed the gap for these students). Moreover, males in college tracks were at a greater advantage relative to females than were males in noncollege tracks (that is, tracking placements increased the gap based on gender).

In their comparisons of what would have happened under different ways of assigning students to tracks, Gamoran and Mare noted that

black-white differences are smaller in the college than in the noncollege track ($-.571$ versus $-.868$). But the race difference is even smaller under the observed system than it would be if all students were placed in the college track.... However, the means for both groups would be highest if all students enrolled in a college-preparatory program (assuming that our models would still hold under such radically changed conditions). (p. 1175)

In other words, African American and White students would, on average, achieve highest in mathematics were everyone enrolled in college tracks—the option envisioned by many who argue against tracking. However, even though the current gap between Whites and African Americans in college tracks became smaller than the gap in noncollege tracks, the alternative of placing *all* students in college tracks would result in the gap actually increasing! Gamoran and Mare argued that this was so because there were many Whites who were not enrolled in college tracks although their 10th-grade mathematics achievement was equivalent to that of many African Americans who did enroll in college tracks. Unlike the case for race, Gamoran and Mare found that

the results for the decomposition of SES effects are fundamentally different. Here, the assignment process is not compensatory, instead, tracking *increases* the gap between advantaged and disadvantaged students by almost 9%. Thus, the means for both groups would be highest, and the between-group difference would be smallest, if all students were assigned to a college-preparatory program. If all were placed in the non-college track, group differences would be reduced (because there would be no tracking to favor high-SES students), but overall mean achievement would be lower too. (pp. 1175–1176)

In other words, the case of social class disparities would seem to be the opposite of what it is for differences between African Americans and Whites. If all students were placed in the college track, the achievement gap would diminish, but at the cost of lowering achievement.

Gamoran and Mare's findings present us with a fundamental choice in terms of our values. Tracking is likely to raise overall performance by helping high-achieving students but, were all students to be placed in the college track, it is possible that we might end up exacerbating achievement differences for some groups while lowering it for others. It is worth remembering, however, that the models used by Gamoran and Mare assume that student placements were structurally similar across various possibilities—highly unlikely if everyone were placed in a college track.

Though there is ample evidence to suggest that tracking is harmful to students who are placed in low tracks, it is only relative to the quality of the education provided in high tracks. I could find no studies that compared the educational experiences of students from diverse backgrounds who were in tracked versus nontracked settings and where those settings were maintained over time. It is likely that this is due to the pervasiveness of tracking in secondary schools in the United States. On the other hand, until such studies are conducted, our evidence of the harmful effects of tracking will have been obtained by triangulation. Alternatives to tracking will be based on some idealized vision. Since the mathematics reform movement argues for the abolition of tracking (NCTM, 1989; NRC, 1989), opportunities to conduct such studies may develop. Given the history of how innovations are implemented differentially for students as functions of their backgrounds, it is far from clear that the reality will approximate such an ideal.

Differential Effects of Course Taking. Although the impact of staying in school and of taking advanced courses is substantial, results from HSB and NAEP should caution us somewhat. True, students who stayed in school gained in mathematics achievement, but gains were differential. Of the students who stayed in school,

> Whites gained .56 test score points more than Blacks and approximately .9 more than Hispanics. Males gained 1.0 of a test point more than females. The gap in gains in favor of Hispanic males was somewhat less. High SES White students gained more than low SES White students. There was no relationship between SES and gains for Blacks. (Rock et al., 1986, p. 352)

Gamoran (1989) also analyzed the relationships among a range of demographic indices, school-level experiences, and achievement for the 1982 data-gathering wave of HSB seniors. Comparing African Americans to Whites, mathematics achievement differences increased when school-based experiences—enrollment in tracks, dropping out versus staying in, number of mathematics and science courses taken—were entered into the regression analyses that he employed. Gamoran interpreted his results to indicate that schooling has differential effects for African Americans and Whites. He did not find similar results when comparing Hispanics to Whites.

Dossey et al. (1988) provided a content based analysis of 1986 NAEP results for geometry and algebra proficiency. On the geometry subscale, White, Hispanic, and African American students who had not taken a course in geometry scored 284, 274, and 264 points, respectively. Among White, Hispanic, and African American students who had taken a course in geometry, these scores were 324, 307, and 297. On the algebra subscale, White, Hispanic, and African American students not taking an algebra course scored 289, 276, and 273; respective scores for students having taken algebra were 328, 310, and 306. In all cases the gains are impressive, ranging from 33 to 40 points. On the other hand, the gains were also differentially distributed, with White students gaining about 39 to 40 points, while Hispanics and African Americans gained 33 or 34 points. Thus, the gap between students who took courses increased over the gap between students who did not.

Unlike the HSB data analyses, where Rock et al. (1986) created carefully matched groups of dropouts and students who stayed in school, NAEP data are likely to suffer from problems with sample selection. Since prior achievement predicts course taking, it is likely that students who complete a course—whether algebra or geometry—would have scored higher at some pretest than students who did not take that course. Moreover, courses themselves serve as selection devices, given that the experiences of students in courses (such as responses to content and pacing) will vary as functions of student demographic characteristics.

Using the High School and Beyond data set for the 1982 senior cohort, Jones (1987) examined how mathematics achievement during the senior year was a function of prior achievement in reading and mathematics (as measured in the 1980 wave of data gathering), mathematics courses (measured as Carnegie units), ethnicity (White versus African American), and gender (male versus female).

Jones found a strong relationship between course taking and achievement: the more high school Carnegie units in mathematics, the higher the achievement for each of the four gender-by-ethnicity groups. Also, Jones found differential growth for each of these groups, thereby partly replicating the 1986 NAEP results (Dossey et al., 1988). However, these differences could be almost completely accounted for as a function of 10th-grade achievement and student SES. Finally, 12th-grade students who had taken calculus grew more than students who had not, regardless of prior achievement and SES.

While providing suggestive insights into issues of differential growth vis-à-vis course taking, Jones's results are by no means conclusive. While the HSB mathematics test is sensitive enough to detect the differential effects of staying in school, it is unlikely to be sensitive enough to determine whether differential patterns of growth can be related to specific mathematics course work. Since HSB was a test of general mathematics achievement, it would seem unable to tell us whether taking calculus might lead to differential growth in knowledge of calculus as opposed to general mathematics achievement. Also, Jones's use of SES as an explanatory variable, while common in this research, is problematic.

Future studies of the differential effectiveness for staying in school and mathematics course taking on achievement for students from diverse backgrounds will need to disentangle true effects from selection artifacts. Of critical importance in this line of work would be the mediating role of other schooling processes. For example, students who are enrolled in different tracks are likely to experience different content and pacing even in what is nominally the same course. Often, the same course title can mean radically different things in school settings with different locales (urban, suburban, rural), student and school social class, and/or racial and ethnic makeup. The quality of instruction also is likely to be another mediating variable.

Instruction

Instruction refers here to how the curriculum gets enacted in the classroom. Instructional strategies that have been recommended for teaching mathematics to students from diverse backgrounds include direct instruction (Everston, Anderson, Anderson, & Brophy, 1980), continuous progress (continuous progress includes direct instruction as one of its features; in addition, students are to progress through a well-specified hierarchy of skills, and they should be grouped on the basis of their ongoing progress through that curriculum), individualized instruction, and, more recently, cognitively guided instruction (Brophy, 1990; McCollum, 1990; Peterson, Fennema, & Carpenter, 1990). As one reads this literature, what comes through is that instruction that works for all students can be modified to work for low-achieving students from various backgrounds (see, for example, the review by Corno & Snow, 1986). Modifications include more highly structured lessons, greater attention to basic skills, deeper coverage of less content to ensure mastery, and more teacher-led discussion as opposed to individualized student seat work in the style of active mathematics teaching (Good & Grouws, 1977).

Direct Instruction. Direct instruction is a highly structured form of teacher behaviors that are thought to support student engagement in and learning of mathematics. For example, Everston et al. (1980) found a consistent pattern of relationships between teacher behaviors and enhanced student achievement in mathematics. Effective teachers were "active, well organized, and strongly academically oriented. They emphasized whole class instruction . . . managed their classrooms effectively . . . [and] asked many questions during class discussion " (p. 58). Questions tended to engage all students rather than just a few.

The best-known example of direct instruction in mathematics education is Active Mathematics Teaching (AMT), originally developed by Good and Grouws (1977, 1979; Good, Grouws, & Ebmeier, 1983). AMT not only prescribes teaching behaviors—such as review, development, and seat work—but it also sequences those behaviors and provides guidelines for amounts of time to allocate to each of them within a given lesson.

In a study of effective instruction for low-performing high school students, Gersten, Gall, Grace, Erickson, and Stieber (1987) studied 31 classrooms of second-year Basic Algebra. Over nine months, growth from pre- to post-test was distressingly low: an average of 4.5 points on a 40-item instrument.

In their analysis of teacher behaviors, Gersten et al. (1987) followed the framework for AMT and found that increased lecture demonstration time and a high rate of questioning were correlated to enhanced achievement, while increased seatwork was correlated to lower growth. Moreover, quieter classes tended to produce greater rates of growth.

In comparing the five most to the five least effective teachers, Gersten et al. (1987) found that the most effective teachers tended to manage their classrooms so that students arrived on time, worked quietly, and remained on-task. Effective teachers tended to spend more time explaining and demonstrating material, to involve more students during questioning, to spend less time on individual seatwork, and to spend less time working with individual students.

Unfortunately, what these and similar studies fail to ask is whether such behaviors are differentially effective for different student backgrounds. For instance, students from low-SES backgrounds and African American and Hispanic students who are low-achieving are said to resist participating in classroom processes (Grant, 1989); they may not answer questions, for example. Gersten et al. provided a vignette of how the most

successful teacher slowed up the tempo of her questioning to encourage a student who had completed a homework assignment to solve a problem and to respond. Forms of contingent teacher behaviors need to be more fully documented to get a fuller picture of effective teaching for diverse learners.

In a study of the long-term effects of direct instruction on children from low-income backgrounds, Becker and Gersten (1982), followed fifth and sixth graders who had completed a three-year (grades 1–3) Follow Through Program that was based on direct instruction. (Follow Through is for low-income children who very often have graduated from a Head Start Program.) Graduates from the program outperformed a comparison group of students on mathematical problem-solving, but not on basic computational skills or mathematical concepts. Moreover, compared to national norms, these students were losing ground. Hence, as in the case of Head Start and Chapter 1, short-term academic benefits of direct instruction in a Follow Through setting dissipate over time.

Tikunoff's (1983, 1985) descriptive study of effective bilingual instruction was an effort to document characteristics of teaching in classrooms that had already been identified as successful. Effective teachers of bilingual students succeeded at four tasks that are common to all direct instruction: accurate communication, maintenance of student engagement, monitoring of progress, and provision of immediate feedback. In achieving these tasks, teachers would (1) use both languages for instruction, (2) integrate language skills development with academic skills development, and (3) respond to and use information from students' home cultures in classroom management and in the content of their lessons (Tikunoff, 1985, p. 34). These characterizations of effective instruction failed to compare ineffective with effective classroom practices, so we cannot determine whether the characteristics listed are really critical for enhancing mathematics achievement among LEP students.

Direct instruction is effective for conveying large amounts of highly structured material to students who are beginning to learn a subject (Brophy & Good, 1986; Good & Brophy, 1989). Hence, it seems to work best for basic skills development. Not surprisingly, therefore, direct instruction has been identified as a key feature in successful Chapter 1 and other basic skills programs. For content that is not easily organized, or for the teaching of higher-order thinking skills, direct instruction has yet to prove its worth (Doyle, 1983).

Cognitively Guided Instruction. A new program, known as Cognitively Guided Instruction (CGI), has been found effective for enhancing first-grade students' achievement on basic skills, problem-solving, and confidence (Carpenter, Fennema, Peterson, Chiang, & Loef, 1990). CGI does not prescribe teaching behaviors. Rather, the program is based on four interlocking principles: (1) teacher knowledge of how mathematical content is learned by their students, (2) problem solving as the focus of instruction, (3) teacher access to how students are thinking about specific problems, and (4) teacher decision-making based on teachers knowing how their students are thinking.

In a recently completed doctoral dissertation, Villasenor (1990) replicated the CGI program in the Milwaukee Public Schools. He modified the problems from Carpenter et al. (1990) to better reflect the diverse racial, ethnic, and social class backgrounds of first graders in that city. Twelve first-grade teachers from six schools volunteered to participate in this study. Their classrooms enrolled between 57.4 and 98.4% non-White students. In three classrooms, including at least one bilingual classroom taught in Spanish, the majority of the non-White students were Hispanic; in the balance, they were African American. Villasenor selected a matched sample of 12 control classrooms where teachers did not participate in the CGI workshop. Schools were comparably governed (one private school in each group), and they enrolled comparable numbers of students from diverse backgrounds. Although the treatment teachers had taught, on average, one year more than the control teachers, the controls had taught *first grade* one year more than the treatment teachers.

In his results, Villasenor documented a range of differences in which treatment teachers and their classrooms differed significantly from the control teachers and their classrooms. On average, classrooms whose teachers had participated in the CGI workshop scored one point better on an October pretest of problem solving than classrooms whose teachers had not. On this 12-item test, CGI classrooms averaged 2.3 correct versus 1.2 for the control classrooms. By the post-test in late February to early March, both classrooms had grown but the CGI-control gap had increased to nearly 7 points. CGI classrooms averaged 9.7 items correct versus 2.9 for the controls. On a 6-item problem-solving test, CGI classrooms outperformed controls 5.5 to 2.8; on a 5-item number-fact test, CGI classrooms averaged 4.8 to 2.9 for the controls. Moreover, CGI classrooms correctly used advanced problem-solving strategies more often than did controls on both problem-solving (3 to .67) and number-fact tests (3.7 to .8).

In his analysis of classroom processes, Villasenor documented how CGI teachers focused their attention on children's thought processes and engaged them in substantive conversations, while the comparison teachers focused more on drill and practice activities that were completed with little attention to student understanding.

Based on these results and on how CGI teachers from their original study adapted instruction to individual student performance and understanding, Peterson et al. (1990) have proposed that CGI shows promise for the teaching of diverse learners.

Grouping

The creation of small groups is an organizational feature of instruction that is intended to make classes more manageable and to allow teachers to fine tune their instruction to better meet the diverse abilities and needs of their students. For example, direct instruction has often been implemented in whole-class settings, consistent with Everston and colleagues' findings (1980), and AMT was developed in similar settings. Alternatively, many teachers arrange some (or even all) of their instruction by grouping students, either along lines of ability or in cooperative groups (see Slavin, 1990; Slavin & Madden, 1989a, 1989b).

Ability-Grouping. Ability-grouping refers to a range of instructional arrangements. In elementary and middle school, for instance, whole classes may be created along lines of ability (between-class ability groups) or teachers may form ability groups within their classes (see Slavin, 1987, 1989a, 1989b, for more on this distinction). In the latter case, students spend most of the day in heterogenous grouping, and the actual creation of groupings would seem to be, at least in theory, modifiable as children showed differential patterns of growth in the subject for which they are being grouped.

Beginning as early as middle school, and certainly by high school, between-class ability groups become formalized in tracking systems whereby students of similar ability are grouped together for much of their academic programs, even though they may move from classroom to classroom for instruction. Tracking, discussed earlier in this chapter, is different from between- and within-class ability-grouping in that the purpose of tracking is to assign students to differentiated course content and to differentiated amounts of courses ahead of time. Ability-grouping, on the other hand, is intended to cover the same or very similar course content, but to adjust the pacing and depth of that coverage to match the abilities of students in each group.

Few topics engender more debate than the impact of ability-grouping in all of its forms on the education of students from diverse backgrounds. On one hand, it has been well-documented that students in low-ability groups (and tracks) receive lower-quality instruction than students who are in high-ability groups (and tracks). Students in low-ability groups cover less content. They engage in small, repetitive, meaningless tasks. And the quality of their interactions with their teachers is focused more on classroom management than on academic tasks (Dreeben & Gamoran, 1986; Gamoran, 1986, 1987; Gamoran & Marc, 1989; Oakes, 1985; 1989; Sorensen & Hallinan, 1986; Veldman & Sanford, 1984). Moreover, since these low-ability groupings are overpopulated by African American, Hispanic, and low-SES students, these students are thought to receive an education inferior to what is provided for Whites, Asian Americans, and students from middle- to upper-SES backgrounds.

Sociologists of education have compared the education of students enrolled in ability-grouped contexts to that which *similar* students receive who are enrolled in mixed-ability contexts. According to them, high-ability students may be better served in homogeneous groupings and tracked contexts. However, low- and middle-ability students receive similar quality educations in either setting (Dreeben & Gamoran, 1986; Gamoran, 1986, 1987).

There are, of course, other concerns about the quality of education that is provided to students in low-ability settings. Given the likelihood that there are few or no academic benefits for being grouped, then the debate has turned on other outcomes of such grouping.

Between-Class Ability Grouping. In a review of studies where elementary students were assigned to classes based on ability, Slavin (1987) found a median effect size of 0.00. In mathematics achievement, negative effects tended to be quite large, with a couple of studies reporting effects sizes of −.52 and −.27 standard deviation against the grouped classes. In contrast, the two largest positive effect sizes were .07 and .06.

In a meta-analysis of ability grouped secondary classes, Kulik and Kulik (1982) found a small (effects size = .10 standard deviation) but statistically significant difference favoring homogeneously over heterogeneously grouped classes. However, among the 14 studies that measured mathematics achievement, the effect size was even smaller (0.05 standard deviation). The largest effect size, .33 standard deviation, was found for students in gifted and talented programs; for academically deficient and representative samples, the effect sizes were so small as to be by chance. More important, the effects were most pronounced for studies of limited duration, from 5 to 18 weeks. Studies that lasted 37 weeks or longer (that is, the same as a school year) demonstrated no differences between groups.

In a narrative best-evidence review of literature on secondary ability-grouping, Slavin (1990) found two studies in which high school students were randomly assigned to ability-grouped versus heterogenous mathematics classes. (It is likely that the ability-grouped classes were in schools where tracking took place. However, these studies are not treated as if they were tracking studies, since critical features of tracking—like differential content coverage—seem missing.) After one year, ability-grouped seventh and eighth graders in the first study were scoring .25 of a standard deviation below their non-grouped peers. In the second study, involving ninth graders from low-SES backgrounds in New York, there were no differences. In an additional five studies—four of which lasted one year and the fifth, one semester—ability-grouped classes were matched to heterogenous classes. In two studies, post test performance favored the ability grouped over heterogenous classes by .13 and .28 standard deviations; in two other studies, the reverse was true by .03 and .15 standard deviations; and in the fifth study, there were no differences. Similarly, in the correlational studies reviewed by Slavin (1990), the results were mixed. Across all the subjects reported in these studies, the median effect sizes hovered near zero.

There are two major differences between these studies and tracking studies (reviewed earlier). Tracking confounds course content with the formation of groups. In contrast, the studies reviewed by Slavin (1990) included similar content across the courses, that is, ability-grouping was a pure treatment. Second, with the exception of Gamoran and Mare's (1989) study, tracking studies compared outcomes across tracks. This not only results in confounding of grouping with content coverage; it also confounds prior achievement with student grouping. The latter problem, according to Slavin (1990), is so severe that it cannot be adjusted for via statistical methods.

Hence, the research linking between-class ability-grouping to mathematics achievement is mixed, with effect sizes showing some variation. If there are any benefits to such practices, they would seem limited to students who were initially high-achieving. These effects, however, may be more a function of differential course coverage and differential allocation of resources than of any inherent benefits to ability-grouping per se.

Within-Class Ability Grouping. In his review of the research on ability-grouping, Slavin (1987) identified five studies in which students were randomly assigned to classes that would or would not be regrouped for mathematics instruction. Effect sizes in favor of regrouping ranged from .07 to .55 of a standard deviation, with the median being .32 (Table 4, p. 318). In three out of the four studies in which effect sizes were related to initial ability, low-ability students who had been grouped gained more relative to their ungrouped cohorts than did middle- or high-ability students. This does not mean that they gained more than their middle- and high-ability peers; the effects are relative to other students of similar initial ability. Two of these studies were conducted by Slavin and his colleagues in Baltimore.

In a study of how student assignment to ability level may interact with later achievement, Veldman and Sanford (1984) found that low-ability students who had been assigned to high-ability mathematics classes gained slightly on their high-ability peers.

Most of the research on the differential effects of in-class grouping on students as a function of their demographic characteristics has focused on how reading groups are formed and their relationship to achievement (Dreeben & Gamoran, 1986; Gamoran, 1986; Haller, 1985; Haller & Davis, 1980; Sorensen & Hallinan, 1986). In these studies, the practice seems to be what Slavin refers to as in-class regrouping, as opposed to between-class grouping.

Haller (1985; Haller & Davis, 1980) studied the possibility of teacher bias—based on social class and race—in the formation of reading groups. Across both racial and student SES backgrounds, teacher assignment was consistent. There were no racial or SES differences found in the match between teacher assignment and performance at the pretest. Where there was a mismatch, teachers seemed to use the more generalized notion of educability discussed earlier in this chapter. Again, there were no differences along racial or SES lines when teachers applied this broader notion in forming ability groups. This matches Dreeben and Gamoran's (1986) findings that first-grade teachers assigned students to reading groups on the basis of initial ability.

In a study of the relationships among ability groups, student demographic characteristics, and reading achievement, Sorensen and Hallinan (1986) found no differences in achievement between students in ability-grouped versus heterogeneously grouped classrooms. Due to their pattern of results, Sorensen and Hallinan argued that students in ability-grouped classes had fewer opportunities to learn than students not enrolled in ability-grouped classes, but that there was more depth of coverage in the grouped classes. A small race-by-class grouping interaction was due to increased achievement differences between African American and other students in grouped classes, but not in the ungrouped classes.

Sorensen and Halinan's argument that there was less but deeper course coverage in the ability-grouped classes was based on how end-of-year achievement varied as different variables entered their regression analyses; they did not obtain independent indices of the actual instruction that took place in their classrooms. Dreeben and Gamoran (1986), on the other hand, did obtain indices of how much time was spent on instruction and what was covered in both ability-grouped and ungrouped first-grade classrooms.

At first, Dreeben and Gamoran found large differences in number of words learned and reading achievement that were related to race and to SES. These initial differences disappeared when instructional time and materials coverage were entered into their regression equations. Dreeben and Gamoran concluded that the real issues concerning student achievement do not lie in grouping per se, but rather in the course coverage among low-, middle-, and high-ability groups. Unfortunately, their analyses do not include interactions of race and SES with grouping variables, so it is not possible to determine whether Sorensen and Hallinan's findings would have been replicated.

What the research on in-class ability grouping seems to suggest is that, when grouping is academically intentional (it has a purpose that is acted upon by the teacher) and is clearly linked to relevant information (reading groups are not maintained for mathematics instruction, nor do groups remain fixed once established), students of diverse ability backgrounds can profit from such an arrangement. Indeed, Slavin (1987) has noted that the Joplin Plan had similar features, including between-grade grouping, and that results favored students in that plan. More recently, Slavin, Madden, Karweit, Livermon, and Dolan (1990) created flexible, cross-grade ability groups for reading among primary students in Baltimore. Students who began to fall substantially behind grade level received tutoring. First-year results indicate that students in this program are staying on grade level and are outperforming a matched group of children who are not receiving similar services.

The detrimental aspects of in-class ability grouping seem to occur because of the amount and quality of content coverage in the course. Moreover, while teacher-student interactions in low-ability groups are of lower quality than in high-ability groups, it is not clear whether such interactions would be similar in ungrouped classrooms. Finally, very little is known about how grouping affects achievement as a function of student race, ethnicity, SES, or language.

Cooperative Groups. The alternative to ability groups that is commonly proposed by authors such as Oakes (1985, 1990b) and Slavin (1990) is the use of small cooperative groups of heterogenous ability. Many studies of small groups and their processes have used direct instruction or active mathematics teaching (AMT) for the development of a lesson, but they have grouped children heterogeneously for some part of that lesson. For example, fourth graders in a study by Swing and Peterson (1982) worked in small groups of heterogeneous ability during the seatwork phase of lessons that had been patterned on AMT. Students who had received training in working cooperatively engaged in more task-related interactions than did controls. They provided and answered more high-order questions than did controls; in contrast, controls sought and provided more answers to questions than did trained students. The effects of training on small group interactions were within chance levels on tests of fractions, division, and two-week retention. However, ability group differences that were in evidence at the start of the study were maintained throughout the study.

Swing and Peterson (1982) found differential relationships between specific student behaviors and their subsequent performance on the achievement and retention tests. For all students, receiving answers (without working on them) during small group interactions was negatively related to achievement on division. For low-ability students, working on-task was re-

lated to enhanced performance on the division and retention posttests; receiving directions was related to lowered performance on division; but checking answers was related to enhanced performance on the division and retention posttests. For high-ability students, their providing conceptual responses was related to enhanced performance on division and retention; while for low-ability students, the provision of conceptual responses was related to enhanced performance on a fractions posttest. These results are consistent with an earlier study (Peterson, Janicki, & Swing, 1981) in which high- and low- but not medium-ability students profited from participation in small group instruction.

The efficacy of small cooperative groups for increasing mathematics achievement in the general population seems well established (Slavin, 1983, 1989a; Webb, 1989). In Slavin's (1989a) extensive review of studies on cooperative groups can be found research conducted in inner cities and/or involving African American, Hispanic bilingual, or low-SES students in a range of subjects, including mathematics. In a series of studies conducted in Baltimore during the early 1970s, seventh-grade students who were randomly assigned to a Teams-Games-Tournament (TGT) form of cooperative groups outperformed controls in mathematics. Third-, fourth-, and sixth-grade students, in Baltimore, randomly assigned to Student Team-Achievement Divisions (STAD) outperformed controls in mathematics. In Philadelphia, classes of low-achieving ninth graders were randomly assigned to STAD, mastery learning, combined, and controls. The STAD and the combined STAD and mastery-learning groups outperformed the controls in general mathematics. Finally, Slavin reported a study in which bilingual third and fourth graders who took part in an activity known as Jigsaw were compared to a matched group of controls; no differences in mathematics achievement were found.

Finding Out/Descubrimiento (De Avila et al., 1987) is an activity-based mathematics-and-science curriculum that uses cooperative groups. In addition to its focus on problem solving, reasoning, and the development of understanding, De Avila (personal communication, September 1990) intended for *FO/D*'s small-group processes to be an alternative to teacher-dominated direct instruction.

De Avila et al. (1981) found that Hispanic bilingual (predominantly LEP) third and fourth graders who participated in the program outperformed a similar group of students who were enrolled in transitional bilingual education programs on the CTBS total math section; third graders outperformed their control peers in math concepts.

Not only has there been interest in cooperative-group studies for developing intergroup cooperation (Johnson & Johnson, 1979); there has also been concern that children working in small groups may recreate stratification along social and demographic characteristics (Cole & Griffin, 1987; Webb, 1989). In summarizing her studies of small-group processes during mathematics instruction, Webb (1989) reported that males and females tended to give more explanations in situations where either males or females were a numerical majority, but not when groups were evenly divided. In classes of above-average, predominantly white and Asian seventh- and eighth-grade students, Webb found that females directed most of their explanations to males in both majority-female and majority-male settings. On the other hand, in similar majority-male settings, males directed most of their explanations to other males. This differential pattern of interactions among students was not found in below-average classes that enrolled African American and Hispanic children.

To guard against differential engagement and learning, Slavin (1983, 1989a) recommends a system of rewards that combines group goals and individual accountability. For example, it makes little sense to grade all students on the basis of a single worksheet (group goal, no individual accountability), because it is a relatively simple matter for the group to assign the task to its most capable member. On the other hand, if students are graded on their individual worksheets (no group goal, individual accountability), there is no incentive to work together. If all students in a group were to receive a common grade based on their individual work, then the incentive is to work together so that everyone masters the task at hand. Moreover, when the task is more complex than a simple worksheet, more refined forms of grading would seem necessary.

Discussion

The previous narrative has focused on the nominally malleable aspects of schooling that are thought to affect (1) enhanced mathematics achievement for diverse populations and (2) closing the achievement gaps between those populations. Most of these studies have concentrated on one or the other goal—primarily the improvement of achievement. The good news, of course, is that we can improve achievement for diverse students; have them stay in school; have them take mathematics courses; intervene as early as possible and provide supplementary support for as long as possible; use the student's native language for instruction; provide direct instruction where the mathematics curriculum is highly structured or is focused on basic skills; use other forms of instruction where the curriculum is more complex; replace whole class activities, especially individual seatwork, with small group work, particularly cooperative groups; do away with tracking; and use ability groups (if necessary) flexibly and for specific academic purposes.

The bad news is that we do not know whether these practices are equally effective for all students. Nor, for that matter, do we seem to know how to structure interventions whose effects are long-lasting.

Much more problematic is that, in those few instances where the two goals could be simultaneously considered, the evidence suggests that students who were situated to benefit from innovations or from specific interventions gained more than students who were not similarly situated. In the case of Sesame Street, middle-class children learned more than their poor cohorts. Whites and Asian Americans who stayed in school profited more in terms of mathematics achievement than African Americans who stayed. White students who take advanced mathematics courses seem to gain more than African Americans and Hispanics who take nominally the same courses. In other words, the rich get richer. And in the case of differences in initial ability, even cooperative groups do not seem to close the gap.

The study by Gamoran and Mare (1989) suggests that there may be hard choices to make in taking a social action—the assignment of all students to college-bound tracks and thereby abolishing all tracking—meet both goals.

A second problem with the literature revolves around how issues of race, ethnicity, social class, and language have become transformed into issues of ability. In study after study, especially those involving Chapter 1, instruction, or grouping, the differential impact of such interventions and practices tended to be recast strictly in terms of ability. In addition, ability would then be recast in terms of individual differences that were due to some combination of these various factors. Hence, questions about the relationship between demographic diversity and mathematics achievement were never addressed directly. Ability became the proxy through which such issues are considered.

This transformation is especially problematic since ability tends to be operationalized as achievement. Typically, ability would be measured at a given time by an achievement test. Ability groups would then be formed or the pretest results would be used to control for ability when achievement was measured at a later time. In either case, prior achievement became an indicator of ability (that is, it became a psychological construct), while achievement at some later time remained an indicator of program impact (that is, it referred to a socially given state of affairs). This is not only an example of how our cultural preoccupation with psychological constructs transforms social issues, it is also an example of how achievement becomes its own explanation through that transformation. Differential mathematics achievement at a given time was explained away by differential achievement at an earlier time. That the earlier achievement level was a function of schooling did not really enter the discussion, since it was now masquerading as ability. Moreover, in this vicious circle we never do address the original concern that achievement is a function of social demographic characteristics.

Future work needs to address demographic diversity directly in order to break out of this tautology. Although there have been many *status* studies that have looked explicitly at race, ethnicity, gender, social class, and the like, there were woefully few *impact* studies that did likewise. And those were primarily in the areas of tracking and ability-grouping.

Most of these intervention studies were carried out in settings that had large numbers of poor, or African American, or Hispanic students. Much time and effort went into documenting those settings, but then the data would not be presented along those lines. Many times, the studies would be cast in terms of aptitude and individual differences since the data would be used to classify students as being low, middle, or high aptitude based on some pretest, and then data would be presented in terms of differential impacts for such aptitude groups. When we read reviews of research that claim to tell us what to do vis-à-vis minorities or the poor, the research that gets cited was designed in terms of aptitude and individual difference, not in terms of ethnicity or SES.

The problems with such an approach should be obvious. Children from diverse backgrounds are not all low achievers. But the transformation of diversity into low aptitude certainly helps to legitimize such a belief. Moreover, if the goal is to close the gap among diverse groups, then we need direct evidence that specific interventions help us attain such a goal.

A third issue in the literature is in how few intervention efforts actually begin with a focus on the target groups. Indeed, many if not most of these interventions were developed from what is thought to be effective for the general student population. For instance, since direct instruction has been shown to improve mathematics achievement for "everyone," it is implemented with populations of underachieving students in an effort to see whether it will improve their achievement as well. Or efforts are made to document how extensive direct instruction is in a program that has been designed for low-achieving minority students. In either case, there was minimal effort to really grapple with the fact that many of these students are from socially diverse backgrounds and to determine what would be necessary to really adapt direct instruction to their diverse backgrounds and experiences.

With the exception of pull out compensatory programs, Sesame Street, and *Finding Out/Descubrimiento,* programs were seldom first developed within the target populations. It is worth remembering that Maria Montessori (1912) developed her preschool methods for poor children from the Italian slums; she did not import her program from what "worked" for the general population.

CONCLUDING COMMENTS

This chapter began with an argument for an intellectual agenda in scholarly research that investigates how mathematics education is unevenly distributed in our society along lines of race, ethnicity, social class, and language. The intent of this review was to move beyond simply noting what works and to grapple with underlying assumptions and issues in extant work. The result was many hours of sifting through studies to see if and how they dealt with these issues, either explicitly or tacitly. Making this task more difficult was the lack of well-developed and coherent lines of research, with the notable exceptions of Oakes, Slavin, De Avila and his colleagues, and Jones and his colleagues. Moreover, I was struck by how the social categories of student diversity were treated as unquestioned givens in much work, how often these categories were made to compete with one another, how they were transformed into other issues, and how findings in one domain were imported into others with minimal justification. Future efforts need to be cognizant of these traps and try to avoid them.

Also, what became increasingly clear as I reviewed this topic was its marginal status relative to mainstream mathematics education research. With few exceptions, work in this area was not found in mathematics education research journals, nor was it the product of mathematics educators. Such a state of affairs is both unconscionable and untenable. If the intellectual agendas of mainstream mathematics education do not (or cannot) include issues of student diversity, then the fundamental utility of mainstream efforts must be seen as suspect. Moreover, just as the differential distribution of mathematics among diverse learners can be seen as an equity issue, so, too, can the marginal status of work that deals with diversity.

It may help to develop the intellectual agenda that I am calling for if we begin with a thought experiment. Assume that the reform documents (NCTM, 1989; NRC, 1989) are right. Assume that the existing curriculum is out of balance, misaligned, full of trivial facts, structured in ways that have little to do with how real people actually learn and perform mathematics, and out of touch with the mathematics people will need to live and function in our society. Assume as well that teaching needs wholesale revamping in light of these criticisms. Then consider the logical outcomes of such a state of affairs: disengagement from the tasks that make up school mathematics, low achievement, unwillingness to take additional courses, and dropping out of school. Furthermore, consider the populations that have been characterized in these ways. Although we have labeled them in many ways, ranging from disadvantaged to disabled, they are students who are poor and ethnic minorities.

Consider, then, the possibility that their achievement patterns and the failure of so many interventions have been indicators of what the reform movement has finally realized. If so, then the label of incompetence generally applied to such students must be questioned and replaced. What should be puzzling is not their low achievement, but the social forces that coerced other students to forgive and to learn in spite of such a sorry state of affairs. If reform is to matter, it must begin with the populations for whom we have drawn these special categories. Curriculum and instruction should first be effective with these students, and then applied to other populations. Finally, the notions of disadvantage and compensatory education that are linked to these populations should be replaced by notions that acknowledge their competence.

Possibly, we may find within this alternative view the intellectual agenda I am calling for. It is a spacious enough view to encompass the concerns of those who consider themselves mainstream, since, if we are to believe what demographers tell us, the mainstream has become diverse.

References

American Association for the Advancement of Science. (1989). *Science for all Americans: Project 2061.* Washington, DC: Author.

Andersson, T., & Boyer, M. (1978). *Bilingual schooling in the United States* (2nd ed.). Austin, TX: National Educational Laboratory.

Anick, C. M., Carpenter, T. P., & Smith, C. (1981a). Minorities and mathematics: Results from the National Assessment of Educational Progress. *Mathematics Teacher, 74,* 560–566.

Anick, C. M., Carpenter, T. P., & Smith, C. (1981b). Minorities and mathematics. In T. P. Carpenter, M. K. Corbitt, H. S. Kepner, M. M. Lindquist, & Robert E. Reys (Eds.), *Results from the second mathematics assessment of the National Assessment of Educational Progress* (Appendix A, pp. 151–157). Reston, VA: National Council of Teachers of Mathematics.

Arias, M. B. (1986). The context of education for Hispanic students: An overview. *American Journal of Education, 95*(1), 26–57.

Baker, K. A., & de Kanter, A. A. (1981). *Effectiveness of bilingual education: A review of the literature* (Final draft report). Washington, DC: Office of Technical and Analytical Systems, U.S. Department of Education.

Baker, K. A., & de Kanter, A. A. (1983). Federal policy and the effectiveness of bilingual education. In K. A. Baker & A. A. de Kanter (Eds.), *Bilingual education: A reappraisal of federal policy* (pp. 33–86). Lexington, MA: Lexington Books.

Barnes, R. E. (1983). The size of the eligible language-minority population. In K. A. Baker & A. A. de Kanter (Eds.), *Bilingual education: A reappraisal of federal policy* (pp. 3–32). Lexington, MA: Lexington Books.

Baron, R. M., Tom, D. Y. H., & Cooper, H. M. (1985). Social class, race, and teacher expectations. In J. B. Dusek (Ed.), *Teacher expectancies* (pp. 251–269). Hillsdale, NJ: Lawrence Erlbaum.

Bates, F. L., & Peacock, W. G. (1989). Conceptualizing social structure: The misuse of classification in structural modeling. *American Sociological Review, 54,* 565–577.

Beady, C. H., & Hansell, S. (1981). Teacher race and expectations for student achievement. *American Educational Research Journal, 18*(2), 191–206.

Becker, W. C., & Gersten, R. (1982). A follow up of Follow Through: The later effects of the direct instruction model on children in fifth and sixth grades. *American Educational Research Journal, 19,* 75–92.

Bilingual Education Act. (1988). Public Law 100-297, Title VII. 20 U.S.C. 3281-3386. 102 Stat. 274-293.

Birman, B. F., Orland, M. E., Jung, R. K., Anson, R. J., Garcia, G. N., Moore, M. T., Funkhouser, J. E., Morrison, D. R., Turnbull, B. J., & Reisner, E. R. (1987). *The current operation of the Chapter 1 program* (Final report from the National Assessment of Chapter 1, Office of Educational Research and Improvement). Washington, DC: Government Printing Office.

Brophy, J. (1985a). Interactions of male and female students with male and female teachers. In L. C. Wilkinson & C. B. Marrett (Eds.), *Gender influences in classroom interaction* (pp. 115–142). New York: Academic Press.

Brophy, J. E. (1985b). Teacher-student interaction. In J. B. Dusek (Ed.), *Teacher expectancies* (pp. 303–328). Hillsdale, NJ: Lawrence Erlbaum.

Brophy, J. E. (1990). Effective schooling for disadvantaged students. In M. S. Knapp & P. M. Shields (Eds.), *Better schooling for the children of poverty: Alternatives to conventional wisdom* (Vol. 2: Commissioned papers and literature review. Chapter IX. Study of Academic Instruction for Disadvantaged Students, U.S. Department of Education contract no. LC88054001). Washington, DC: Government Printing Office.

Brophy, J. E., & Good, T. (1986). Teacher behavior and student achievement. In M. C. Wittrock (Ed.), *Handbook of research on teaching* (3rd ed., pp. 328–375). New York: Macmillan.

Brown v. Board of Education. 347 U.S. 483 (1954).

Buriel, R., & Cardoza, D. (1988). Sociocultural correlates of achievement among three generations of Mexican American high school seniors. *American Educational Research Journal, 25,* 177–192.

Burkheimer, G. J., Conger, A. J., Dunteman, G. H., Elliott, B. G., & Mowbray, K. A. (1989, December 13). *Effectiveness of services for language-minority limited-English-proficient students* (Executive summary). Research Triangle Park, NC: Research Triangle Institute.

Burton, N. W., & Jones, L. V. (1982). Recent trends in achievement levels of Black and White youth. *Educational Researcher, 11*(4), 10–14, 17.

California Department of Education. (1986). *Student achievement in California schools* (1984–1985 annual report). Sacramento: Author.

California Department of Education. (1987). *Statewide summary of student performance on school district proficiency assessments* (1985–1986 school year). Sacramento: Author.

Cannell, J. J. (1987). *How all fifty states are above the national average.* Albuquerque, NM: Friends for Education.

Cannell, J. J. (1989). *The "Lake Wobegone" report: How public educators cheat on standardized achievement tests.* Albuquerque, NM: Friends for Education.

Carpenter, T. P., Coburn, T. G., Reys, R. E., & Wilson, J. W. (1978). *Results from the first mathematics assessment of the National Assessment of Educational Progress.* Reston, VA: National Council of Teachers of Mathematics.

Carpenter, T. P., Fennema, E., Peterson, P. L., Chiang, C. P., & Loef, M. (1990). Using knowledge of children's mathematical thinking in classroom teaching: An experimental study. *American Educational Research Journal, 26,* 499–531.

Carpenter, T. P., & Moser, J. M. (1984). The acquisition of addition and subtraction concepts in grades one through three. *Journal for Research in Mathematics Education, 15,* 179–202.

Center for Education Statistics. (1987a). *Condition of education.* Washington, DC: Government Printing Office.

Center for Education Statistics. (1987b). *Digest of education statistics.* Washington, DC: Government Printing Office.

Chipman, S. F., & Thomas, V. G. (1987). The participation of women and minorities in mathematical, scientific, and technical fields. In E. Z. Rothkopf (Ed.), *Review of research in education* (Vol. 14, Chapter 9, pp. 387–430). Washington, DC: American Educational Research Association.

Cohen, J., & Cohen, R. (1975). *Applied multiple regression/correlation analysis for the behavioral sciences.* Hillsdale, NJ: Lawrence Erlbaum.

Cole, M., & Griffin, P. (Eds.). (1987). *Contextual factors in education: Improving science and mathematics education for minorities and women.* Madison: Wisconsin Center for Education Research, University of Wisconsin-Madison.

Coleman, J. (1967). *The concept of educational opportunity.* Baltimore, MD: Johns Hopkins University. (ERIC Document Reproduction Service No. ED 012 275)

Coleman, J. (1975). What is meant by "equal educational opportunity?" *Oxford Review of Education, 1*(1), 27–29.

Coleman, J. S., Campbell, E. Q., Hobson, C. J., McPartland, J., Mood, A. M., Weinfeld, F. D., & York, R. L. (1966). *Equality of educational opportunity.* Washington, DC: Government Printing Office. (ERIC Document Reproduction Service No. ED 012 275)

Congressional Budget Office. (1987, August). *Educational achievement: Explanations and implications of recent trends.* Washington, DC: Government Printing Office.

Cook, T. D., Appleton, H., Conner, R. F., Shaffer, A., Tamkin, G., & Weber, S. J. (1975). *Sesame Street revisited.* New York: Russell Sage Foundation.

Cook, T. D., & Campbell, D. T. (1979). *Quasi-experimentation.* Boston: Houghton-Mifflin.

Cook, T. D., & Shadish, W. R., Jr. (1986). Program evaluation: The worldly science. *Annual Review of Psychology, 37,* 193–232.

Cooper, H. M. (1979). Pygmalian grows up: A model for teacher expectation, communication, and performance influence. *Review of Educational Research, 49,* 389–410.

Corno, L., & Snow, R. E. (1986). Adapting teaching to individual differences among learners. In M. C. Wittrock (Ed.), *Handbook of research on teaching* (3rd ed., pp. 605–629). New York: Macmillan.

Danoff, M. N., Coles, G. J., McLaughlin, D. H., & Reynolds, D. J. (1977-1978). *Evaluation of the impact of ESEA Title VII Spanish/English bilingual education programs* (3 vols.). Palo Alto, CA: American Institutes for Research. (ERIC Document Reproduction Service Nos. ED 138 090, ED 138 091, & ED 154 635)

De Avila, E. A. (1988). Bilingualism, cognitive function, and language minority group membership. In R. R. Cocking & J. P. Mestre (Eds.), *Linguistic and cultural influences on learning mathematics* (pp. 101–121). Hillsdale, NJ: Lawrence Erlbaum.

De Avila, E. A., Cohen, E. G., & Intili, J. K. (1981). *Multicultural improvement of cognitive abilities* (Final Report; Contract No. 9372). Sacramento: California Department of Education. [Also in S. S. Seidner (Ed.), (1982). *Issues of language assessment: Foundations and research* (pp. 53–81). Springfield: Illinois State Board of Education.]

De Avila, E. A., & Duncan, S. E. (1979). Bilingualism and the metaset. *NABE Journal, 3*(21), 1–20.

De Avila, E. A., & Duncan, S. E. (1981). *A convergent approach to oral language assessment: Theoretical and technical specification on the Language Assessment Scales (LAS) Form A* (Stock 621). Sacramento: California Department of Education.

De Avila, E. A., Duncan, S. E., & Navarrete, C. (1987). *Finding out/Descubriminto*. Northvale, NJ: Santillana.

Dix, L. S. (Ed.). (1987). *Minorities: Their underrepresentation and career differentials in science and engineering. Proceedings of a workshop*. Washington, DC: National Academy Press.

Dossey, J. A., Mullis, I. V. S., Lindquist, M. M., & Chambers, D. L. (1988). *The mathematics report card: Are we measuring up?* (Trends and achievement based on the 1986 National Assessment, Report No. 17-M-01). Princeton, NJ: National Assessment of Educational Progress, Educational Testing Service.

Doyle, W. (1983). Academic work. *Review of Educational Research, 53,* 159–200.

Dreeben, R., & Gamoran, A. (1986). Race, instruction, and learning. *American Sociological Review, 51,* 660–669.

Duncan, S., & De Avila, E. A. (1979). Bilingualism and cognition: Some recent findings. *NABE Journal, 4*(1), 15–50.

Dunn, L. M. (1987). *Bilingual Hispanic children on the U.S. mainland: A review of research on their cognitive, linguistic, and scholastic achievement*. Circle Pines, MN: American Guidance Service.

Duran, R. P. (1988). Bilinguals' logical reasoning aptitude: A construct validity study. In R. R. Cocking & J. P. Mestre (Eds.), *Linguistic and cultural influences on learning mathematics* (pp. 241–258). Hillsdale, NJ: Lawrence Erlbaum.

Dusek, J. B. (1975). Do teachers bias student learning? *Review of Educational Research, 45,* 661–684.

Dusek, J. B. (Ed.). (1985). *Teacher expectancies*. Hillsdale, NJ: Lawrence Erlbaum.

Education for socially disadvantaged children. (1965). *Review of Educational Research, 35*(5).

Education for socially disadvantaged children. (1970). *Review of Educational Research, 40*(1).

Espinosa, R., & Ochoa, R. (1986). Concentration of California Hispanic students in schools with low achievement: A research note. *American Journal of Education, 95*(1), 77–95.

Everston, C., Anderson, C., Anderson, L., & Brophy, J. (1980). Relationships between classroom behaviors and student outcomes in junior high mathematics and English classes. *American Educational Research Journal, 17,* 43–60.

Fernandez, R. M., & Nielsen, F. (1986). Bilingualism and Hispanic scholastic achievement: Some baseline results. *Social Science Research, 15,* 43–70.

Fernandez, R. R. (Ed.). (1988). Achievement testing: Science versus ideology. *Hispanic Journal of Behavioral Sciences* [Special issue], *10*(3).

Fordham, S. (1988). Racelessness as a factor in Black students' school success: Pragmatic strategy or pyrrhic victory? *Harvard Educational Review, 58,* 54–84.

Fordham, S., & Ogbu, J. U. (1986). Black students' school success: Coping with the "burden of 'acting white.' " *The Urban Review, 18*(3), 176–206.

Fuson, K. C. (1988). *Children's counting and concepts of number*. New York, NY: Springer–Verlag.

Gamoran, A. (1986). Instructional and institutional effects of ability grouping. *Sociology of Education, 59,* 185–198.

Gamoran, A. (1987). The stratification of high school learning opportunities. *Sociology of Education, 60,* 135–165.

Gamoran, A., & Mare, R. D. (1989). Secondary school tracking and educational inequality: Compensation, reinforcement, or neutrality? *American Journal of Sociology, 94,* 1146–1183.

Garet, M. S., & De Lany, B. (1988). Students, courses, and stratification. *Sociology of Education, 61,* 61–77.

Gersten, R., Gall, M., Grace, D., Erickson, D., & Steiber, S. (1987). *Instructional correlates of achievement gains in algebra for low performing, high school students*. Paper presented at the annual meeting of the American Educational Research Association, Washington, DC.

Ginsburg, H. P., & Russell, R. L. (1981). Social class and racial influences on early mathematical thinking. *Monographs of the Society for Research in Child Development, 46,* Serial No. 6.

Goertz, M. E. (1989). *Course taking patterns in the 1980s*. New Brunswick, NJ: Center for Policy Research in Education, Rutgers University.

Good, T. L., & Brophy, J. E. (1986). School effects. In M. C. Wittrock (Ed.), *Handbook of research on teaching* (3rd ed., pp. 570–602). New York, NY: Macmillan.

Good, T. L., & Brophy, J. E. (1989). Teaching the lesson. In R. Slavin (Ed.), *School and classroom organization* (pp. 25–68). Hillsdale, NJ: Lawrence Erlbaum.

Good, T. L., & Grouws, D. (1977). Teaching effects: A process-product study in fourth-grade mathematics classrooms. *Journal of Teacher Education, 28,* 49–54.

Good, T. L., & Grouws, D. (1979). The Missouri mathematics effectiveness project: An experimental project in fourth-grade classrooms. *Journal of Educational Psychology, 71,* 355–362.

Good, T. L., Grouws, D., & Ebmeier, H. (1983). *Active mathematics teaching*. New York, NY: Longman.

Grant, C. A. (1989). Equity, equality, teachers, and classroom life. In W. G. Secada (Ed.), *Equity in education* (pp. 89–102). Lewes, England: Falmer Press.

Grant, C. A., & Sleeter, C. E. (1986). *After the school bell rings*. Lewes, England: Falmer Press.

Grignon, J. (1991). *Affect, experiences, and computer-related coursework of eighth- and twelfth-grade students in the Menominee Indian School District*. Unpublished doctoral dissertation, University of Wisconsin–Madison.

Gross, S. (1988, July). *Participation and performance of women and minorities in mathematics* (Executive summary. A project supported by National Science Foundation grant no. MDR-8470384 and the Montgomery County Public Schools). Rockville, MD: Montgomery County Public Schools.

Hakuta, K. (1986). *Mirror of language*. New York: Basic Books.

Haller, E. J. (1985). Pupil race and elementary school ability grouping: Are teachers biased against Black children? *American Educational Research Journal, 22,* 465–483.

Haller, E. J., & Davis, S. A. (1980). Does socioeconomic status bias the assignment of elementary school students to reading groups? *American Educational Research Journal, 17,* 409–418.

Heath, S. B. (1986). Sociocultural contexts of language development. In Bilingual Education Office, California Department of Education, *Beyond language: Social and cultural factors in schooling language minority students* (pp. 143–186). Los Angeles, CA: Evaluation Dissemination and Assessment Center, California State University.

Hernandez, N. G. (1973). Achievement of Mexican-American students. *Review of Educational Research, 43,* 1–40.

Hodgkinson, H. L. (1985, June). *All one system: Demographics of education, kindergarten through graduate school*. Washington, DC: Institute for Educational Leadership.

Hodgkinson, H. L. (1989, September). *The same client: The demographics of education and service delivery systems*. Washington, DC: Institute for Educational Leadership.

Hsia, J. (1988). *Asian Americans in higher education and at work.* Hillsdale, NJ: Lawrence Erlbaum.

Iowa Department of Public Instruction. (1983). *Non-English speaking students programs* (Subsection 2; Chapter 670-57 of the *Code*). Des Moines, IA: Author.

Jaynes, G. D., & Williams, R. M., Jr. (Eds.). (1989). *A common destiny: Blacks and American society.* Washington, DC: National Academy Press.

Johnson, D. W., & Johnson, R. T. (1979). Conflict in the classroom: Controversy and learning. *Review of Educational Research, 22,* 51–70.

Johnston, W. B., & Packer, A. E. (1987, June). *Workforce 2000: Work and workers for the twenty-first century.* Indianapolis: Hudson Institute.

Jones, L. V. (1984). White-black achievement differences: The narrowing gap. *American Psychologist, 39,* 1207–1213.

Jones, L. V. (1987). The influence on mathematics test scores, by ethnicity and sex, of prior achievement and mathematics courses. *Journal for Research in Mathematics Education, 18,* 180–186.

Jones, L. V., Burton, N. W., & Davenport, E. C. (1984). Monitoring the mathematics achievement of black students. *Journal for Research in Mathematics Education, 15,* 154–164.

Kallos, D., & Lundgren, U. P. (1975). Educational psychology: Its scope and its limits. *British Journal of Educational Psychology, 45,* 111–121.

Kennedy, M. M., Birman, B. F., & Demaline, R. E. (1986, July). *The effectiveness of Chapter 1 services* (Second interim report from the National Assessment of Chapter 1, Office of Educational Research and Improvement). Washington, DC: Government Printing Office.

Kennedy, M. M., Jung, R. K., & Orland, M. F. (1986, January). *Poverty, achievement, and the distribution of compensatory education services* (An interim report from the National Assessment of Chapter 1, Office of Educational Research and Improvement). Washington, DC: Government Printing Office.

Kirk, G. E., Hunt, J. M., & Volkmar, F. (1975). Social class and preschool language skill: V. Cognitive and semantic mastery of number. *Genetic Psychology Monographs, 92,* 131–153.

Knapp, M. S., & Shields, P. M. (Eds.). (1990). *Better schooling for the children of poverty: Alternatives to conventional wisdom* (2 vols. Study of Academic Instruction for Disadvantaged Children. U.S. Department of Education contract no. LC88054001). Washington, DC: Government Printing Office.

Kohr, R. L., Coldiron, J. R., Skiffington, E. W., Masters, J. R., & Blust, R. S. (1987, April). *The influence of race, class, and gender on mathematics achievement and self-esteem for fifth, eighth, and eleventh grade students in Pennsylvania schools.* Symposium presentation at the annual meeting of the American Educational Research Association, Washington, DC. (Available from the authors at the Pennsylvania Department of Education.)

Kohr, R. L., Masters, J. R., Coldiron, J. R., Blust, R. S., & Skiffington, E. W. (1991). The relationship of race, class, and gender with mathematics achievement for fifth, eighth, and eleventh grade students in Pennsylvania schools. In W. G. Secada & M. R. Meyer (Eds.), *Needed: An agenda for equity in mathematics education* [Special issue of the Peabody Journal of Education], *66* (2), 147–171.

Kulik, C. C., & Kulik, J. A. (1982). Effects of ability grouping on secondary school students: A meta-analysis of evaluation findings. *American Educational Research Journal, 19,* 415–428.

Lazar, I., & Darlington, R. (1982). Lasting effects of early education: A report from the Consortium for Longitudinal Studies. *Monographs of the Society for Research in Child Development, 47,* Serial no. 195.

Lee, V. E., & Bryk, A. S. (1988). Curriculum tracking as mediating the social distribution of high school achievement. *Sociology of Education, 68,* 78–94.

Lesser, G. S. (1974). *Children and television: Lessons from Sesame Street.* New York: Random House.

Light, R. J., & Smith, P. V. (1969). Social allocation models of intelligence. *Harvard Educational Review, 39,* 484–510.

Linn, M. C., & Hyde, J. S. (1989). Gender, mathematics, and science. *Educational Researcher, 18*(8), 17–27.

Lockheed, M. E., Thorpe, M., Brooks-Gunn, J., Casserly, P., & McAloon, A. (1985). *Sex and ethnic differences in middle mathematics, science, and computer science: What do we know?* (A report submitted to The Ford Foundation). Princeton, NJ: Educational Testing Service.

Marrett, C. B. (1986). *On minority females in precollege mathematics.* Paper presented at the annual meeting of the American Educational Research Association, San Francisco.

Marrett, C. B. (1987). Black and Native American students in precollege mathematics and science. In L. S. Dix (Ed.), *Minorities: Their underrepresentation and career differentials in science and engineering* (Proceedings of a workshop, pp. 7–31). Washington, DC: National Academy Press.

Mathematical Sciences Education Board. (1990). *Reshaping school mathematics.* Washington, DC: National Academy Press.

Matthews, W., Carpenter, T. P., Lindquist, M. M., & Silver, E. A. (1984). The third national assessment: Minorities and mathematics. *Journal for Research in Mathematics Education, 15,* 165–171.

Matute-Bianchi, M. E. (1986). Ethnic identities and patterns of school success and failure among Mexican-descent and Japanese-American students in a California high school: An ethnographic analysis. *American Journal of Education, 95,* 233–272.

McCandless, C. N. (1990). *An exploratory study on perception about mathematics as reported by four African-American high school students in a predominantly white school* (A paper submitted in partial fulfillment of the requirements for the degree of Master of Science.). Unpublished manuscript, Department of Curriculum and Instruction, University of Wisconsin–Madison.

McCollum, H. (1990). A review of research on effective instructional strategies and classroom management approaches. In M. S. Knapp & P. M. Shields (Eds.), *Better schooling for the children of poverty: Alternatives to conventional wisdom* (Vol. 2: Commissioned papers and literature review. Chapter 12. Study of Academic Instruction for Disadvantaged Students, U.S. Department of Education contract no. LC88054001). Washington, DC: Government Printing Office.

McKey, R. H., Condelli, L., Ganson, H., Barrett, B. J., McConkey, C., & Plantz, M. C. (1985). *The impact of Head Start on children, families, and communities* (Final report of the Head Start evaluation, synthesis, and utilization project. Contract no. 105-81-C-026.). Washington, DC: Government Printing Office (DHHS Publication no. OHDS 85-31193)

Michigan Department of Education. (1977). *A position statement on bilingual instruction in Michigan.* Lansing, MI: Author.

Minnesota Education for Limited English Proficient Students Act. (1988). *State of Minnesota Statutes Book,* Sections 126.261-126.269.

Moore, E. G. J., & Smith, A. W. (1987). Sex and ethnic group differences in mathematics achievement: Results from the national longitudinal study. *Journal for Research in Mathematics Education, 10,* 25–36.

Montessori, M. (1912). *The Montessori method: Scientific pedagogy as applied to children's homes.* New York, NY: Frederick A. Stokes.

Mosteller, F., & Moynihan, D. P. (1972). A pathbreaking report. In F. Mosteller & D. P. Moynihan (Eds.), *On equality of educational opportunity* (pp. 3–66). New York, NY: Vantage Books.

Mullis, J. V. S., Owen, E. H., & Phillips, G. W. (1990). *Accelerating academic achievement.* Princeton, NJ: National Assessment of Educational Progress, Educational Testing Service.

Murphy, J. (1981). Disparity and inequality in education: The crippling legacy of Coleman. *British Journal of Sociology of Education, 2*(1), 61–70.

Myers, D. E. (1985). The relationship between school poverty concentration and students' reading and math achievement and learning. In M. M. Kennedy, R. K. Jung, & M. E. Orland (Eds.), *Poverty, achievement, and the distribution of compensatory education services* (National assessment of Chapter 1, pp. D-15 to D-60). Washington, DC: Government Printing Office.

Myers, D. E., & Milne, A. M. (1983). *Mathematics achievement of language-minority students.* Paper prepared for Part C research agenda: Math and language minority student project. Washington, DC: Decision Resources. (Also available from the National Clearinghouse for Bilingual Education.)

Myers, D. E., & Milne, A. M. (1988). Effects of home language and primary language on mathematics achievement. In R. R. Cocking & J. P. Mestre (Eds.), *Linguistic and cultural influences on learning mathematics* (pp. 259–293). Hillsdale, NJ: Lawrence Erlbaum.

Myrdal, G. (1944). *An American dilemma: The Negro problem and modern democracy.* New York, NY: Harper.

National Alliance of Business. (1986a). *Employment policies: Looking to the year 2000.* Washington, DC: Author.

National Alliance of Business. (1986b). *Youth 2000. A call to action* (Report on a National Leadership Meeting, Washington, DC). Washington, DC: Author.

National Center for Education Statistics. (1984). *Condition of education* (1984 edition. A statistical report). Washington, DC: Government Printing Office.

National Council of Teachers of Mathematics. (1989). *Curriculum and evaluation standards for school mathematics.* Reston, VA: Author.

National Research Council. (1989). *Everybody counts: A report to the nation on the future of mathematics education.* Washington, DC: National Academy Press.

National Science Foundation. (1986, January). *Women and minorities in science and engineering* (NSF 86-301). Washington, DC: Author.

Natriello, G., Pallas, A., & Alexander, K. (1989). On the right track? Curriculum and academic achievement. *Sociology of Education, 62,* 109–118.

Network of Regional Desegregation Assistance Centers. (1989). *Resegregation of public schools: The third generation* (A report on the condition of desegregation in America's public schools). Portland, OR: Northwest Regional Educational Laboratory.

Nielsen, F., & Fernandez, R. M. (1981). *Hispanic students in American high schools: Background characteristics and achievement* (Contractor report to the National Center for Education Statistics). Washington, DC: Government Printing Office.

Oakes, J. (1985). *Keeping track: How schools structure inequality.* New Haven, CT: Yale University Press.

Oakes, J. (1989). Tracking in secondary schools: A contextual perspective. In R. E. Slavin (Ed.), *School and classroom organization* (pp. 173–175). Hillsdale, NJ: Lawrence Erlbaum.

Oakes, J. (1990a). Opportunities, achievement, and choice: Women and minority students in science and mathematics. In C. B. Cazden (Ed.), *Review of Research in Education, 16,* 153–222.

Oakes, J. (1990b). *Multiplying inequalities: The effects of races, social class, and tracking on opportunities to learn mathematics and science.* Santa Monica, CA: RAND.

Oakes, J., Gamoran, A., & Page, R. N. (in press). Curriculum differentiation: Opportunities, outcomes, and meanings. In P. A. Jackson (Ed.), *Handbook of research on curriculum.* New York, NY: Macmillan.

Ogbu, J. U. (1978). *Minority education and caste.* Orlando, FL: Academic Press.

Ogbu, J. U. (1988). Class stratification, racial stratification, and schooling. In L. Weis (Ed.), *Class, race, and gender in American education* (pp. 163–182). Albany, NY: SUNY Press.

Ogbu, J. U., & Matute-Bianchi, M. E. (1986). Understanding sociocultural factors: Knowledge, identity, and school adjustment. In *Beyond Language: Social and cultural factors in schooling language minority student.* (pp. 73–142). Los Angeles, CA: Evaluation, Dissemination and Assessment Center.

O'Malley, J. M. (1981). *Children's English services study: Language minority children with limited English proficiency in the United States.* Rosslyn, VA: National Clearinghouse for Bilingual Education, Inter-America Research Associates.

Ortiz, V. (1986). Reading activities and reading proficiency among Hispanic, Black, and White students. *American Journal of Education, 95,* 58–76.

Oxford-Carpenter, R., Pol, L., Lopez, D., Stupp, P., Gendell, M., & Peng, S. (1984). *Demographic projections of non-English-background and limited-English-proficient persons in the United States to the year 2000 by state, age, and language group.* Rosslyn, VA: National Clearinghouse for Bilingual Education, InterAmerica Research Associates.

Pallas, A. M., Natriello, G., & McDill, E. L. (1989). The changing nature of the disadvantaged population: Current dimensions and future trends. *Educational Researcher, 18*(5), 16–22.

Patterson, J. (1991). Minorities gain, but the gap remains. In W. G. Secada & M. R. Meyer (Eds.), *Needed: An agenda for equity in mathematics education* [Special issue of the *Peabody Journal of Education*], *66*(2), 72–94.

Perlmann, J. (1990). Historical legacies: 1840–1920. In C. B. Cazden & C. E. Snow (Eds.), *English plus: Issues in bilingual education* [Special issue of *The Annals of the American Academy of Political and Social Science,* pp. 27–37]. Newbury Park, CA: Sage.

Peterson, P. L., Fennema, E., & Carpenter, T. P. (1990). *Using children's mathematical knowledge.* Unpublished manuscript. [Available from P. L. Peterson, School of Education, Michigan State University, East Lansing, MI]

Peterson, P. L., Janicki, T. C., & Swing, S. R. (1981). Ability by treatment interaction effects on children's learning in large-group and small-group approaches. *American Educational Research Journal, 18,* 453–473.

Pimm, D. (1987). *Speaking mathematically.* New York, NY: Routledge and Kegan Paul.

Quality Education for Minorities Project. (1990). *Education that works: An action plan for the education of minorities.* Cambridge, MA: Author, MIT.

Raizen, S. A., & Jones, L. V. (Eds.). (1985). *Indicators of precollege education in science and mathematics: A preliminary review.* Washington, DC: National Academy Press.

Ramirez, J. D. (1986). Comparing structured English immersion and bilingual education: First-year results of a national study. *American Journal of Education, 95,* 122–149.

Ramirez, J. D., Yuen, S. D., & Ramey, D. R. (1987). *Second year report: Longitudinal study of immersion, early-exit, and late-exit transitional bilingual education programs for language-minority children* (SRA Report No. 386). Mountain View, CA: SRA Technologies.

Ramirez, J. D., Yuen, S. D., & Ramey, D. R. (1988). *Third year report: Longitudinal study of immersion, early-exit, and late-exit transitional bilingual education programs for language-minority children.* Mountain View, CA: Aguirre International.

Reyes, L. H., & Stanic, G. M. A. (1988). Race, sex, socioeconomic status, and mathematics. *Journal for Research in Mathematics Education, 19,* 26–43.

Rock, D. A., Ekstrom, R. B., Goertz, M. E., & Pollack, J. (1986). *Study of excellence in high school education: Longitudinal study, 1980-82 final report* (Contractor report). Washington, DC: Government Printing Office.

Rohwer, W. D., Jr. (1971). Learning, race, and school success. *Review of Educational Research, 41,* 191–210.

Rosenthal, A. S., Baker, K., & Ginsburg, A. (1983). The effect of language background on achievement level and learning among elementary school students. *Sociology of Education, 56,* 157–169.

Rosenthal, A. S., Milne, A. M., Ellman, F. M., Ginsburg, A. L., & Baker, K. A. (1983). A comparison of the effects of language background and socioeconomic status on achievement among elementary-school children. In K. A. Baker & A. A. de Kanter (Eds.), *Bilingual education: A reappraisal of federal policy* (pp. 87–111). Lexington, MA: Lexington Books.

Rosenthal, R. (1987). *Pygmalion* effect: Existence, magnitude, and social importance. *Educational Researcher, 16*(9), 37–41.

Rosenthal, R., & Jacobson, L. (1968). *Pygmalion in the classroom: Teacher expectation and pupils' intellectual development.* New York, NY: Holt, Rinehart & Winston.

Rumberger, R. W. (1987). High school dropouts: A review of issues and evidence. *Review of Educational Research, 57,* 101–121.

Sampson, E. E. (1977). Psychology and the American ideal. *Journal of Personality and Social Psychology, 35,* 767–782.

Sampson, E. E. (1981). Cognitive psychology as ideology. *American Psychologist, 36,* 730–743.

Sanders, J., & Lubetkin, R. (1991). Preparing female students for technical careers: Dealing with our own elitist biases. In W. G. Secada & M. R. Meyer (Eds.), *Needed: An agenda for equity in mathematics education* [Special issue of the *Peabody Journal of Education*], *66*(2), 113–126.

Sarason, S. B., & Klaber, M. (1986). The school as a social situation. *Annual Review of Psychology, 36,* 115–140.

Saxe, G. B., Guberman, S. R., & Gearhart, M. (1987). Social processes in early number development. *Monographs of the Society for Research in Child Development, 52,* Serial No. 2.

Secada, W. G. (1982). Improving cognition should be the concern of everyone in education. In S. S. Seidner (Ed.), *Issues of language assessment: Foundations and research* (pp. 83–85). Springfield: Illinois State Board of Education.

Secada, W. G. (1988) Equity, student diversity, and cognitivist research. In T. P. Carpenter, E. Fennema, & S. Lamon (Eds.), *Teaching and learning mathematics—Proceedings of the First Wisconsin Symposium for Research on Teaching and Learning Mathematics* (pp. 32–64). Madison, WI: Wisconsin Center for Education Research, University of Wisconsin–Madison. [Also to appear in T. P. Carpenter, E. Fennema, & S. Lamon (Eds.), *Teaching and learning mathematics.* Albany, NY: SUNY Press.]

Secada, W. G. (Ed.). (1989). *Equity in education.* London: Falmer Press.

Secada, W. G. (1990a). The challenges of a changing world for mathematics education. In T. J. Cooney (Ed.), *Teaching and learning mathematics in the 1990s.* (pp. 135–143). Reston, VA: National Council of Teachers of Mathematics.

Secada, W. G. (1990b). Student diversity and mathematics education reform. In L. Idol & B. F. Jones (Eds.), *Dimensions of cognitive instruction* (pp. 295–330). Hillsdale, NJ: Lawrence Erlbaum.

Secada, W. G. (1991a). Agenda setting, enlightened self interest, and equity in mathematics education. In W. G. Secada & M. R. Meyer (Eds.), *Needed: An agenda for equity in mathematics education* [Special issue of the *Peabody Journal of Education*], *66*(2), 22–56.

Secada, W. G. (1991b). Degree of bilingualism and arithmetic problem solving in Hispanic first graders. *Elementary School Journal, 92,* 213–231.

Secada, W. G., & Meyer, M. R. (Eds.). (1991). *Needed: An agenda for equity in mathematics education* [Special issue of the *Peabody Journal of Education*], *66*(2).

Shepard, L., & Linn, R. (1989, June 11–14). *Lake Wobegon one year later: Results from a replication of the Cannell study.* Paper presented at the Nineteenth Annual Assessment Conference sponsored by the Education Commission of the States, Boulder, CO.

Shockley, W. (1971). Negro IQ deficit: Failure of a "Malicious Coincidence" model warrants new research proposals. *Review of Educational Research, 41,* 227–248.

Silver, E. A. (in press). Assessment and mathematics education reform in the United States. In W. Secada (Ed.), *The reform of school mathematics in the United States* [Special issue of the *International Journal of Education Research*].

Slavin, R. E. (1983). *Cooperative learning.* White Plains, NY: Longman Publishing Group.

Slavin, R. E. (1987). Ability grouping: A best-evidence synthesis. *Review of Educational Research, 57,* 293–336.

Slavin, R. E. (1989a). Cooperative learning and student achievement. In R. E. Slavin (Ed.), *School and classroom organization* (pp. 129–156). Hillsdale, NJ: Lawrence Erlbaum.

Slavin, R. E. (1989b). Grouping for instruction in elementary school. In R. E. Slavin (Ed.), *School and classroom organization* (pp. 156–172). Hillsdale, NJ: Lawrence Erlbaum.

Slavin, R. E. (1990). Achievement effects of ability grouping in secondary schools: A best-evidence research synthesis. *Review of Educational Research, 60,* 471–499.

Slavin, R. E., & Madden, N. A. (1989a). What works for students at risk: A research synthesis. *Educational Leadership, 46*(5), 4–13.

Slavin, R. E., & Madden, N. A. (1989b). Effective classroom programs for students at risk. In R. E. Slavin, N. L. Karweit, & N. A. Madden (Eds.), *Effective programs for students at risk* (pp. 21–51). Boston, MA: Allyn and Bacon.

Slavin, R. E., Madden, N. A., Karweit, N. L., Livermon, B. J., & Dolan, L. (1990). Success for all: First-year outcomes of a comprehensive plan for reforming urban education. *American Educational Research Journal, 27,* 255–278.

Sleeter, C. E., & Grant, C. A. (1988). A rationale for integrating race, gender, and social class. In L. Weis (Ed.), *Class, race, and gender in American education* (pp. 144–160). Albany, NY: SUNY Press.

Sorensen, A. B., & Hallinan, M. T. (1986). Effects of grouping on growth in academic achievement. *American Educational Research Journal, 23,* 519–542.

Stallings, J. A., & Stipek, D. (1986). Research on early childhood and elementary school teaching programs. In M. C. Wittrock (Ed.), *Handbook of research on teaching* (3rd ed., pp. 727–753). New York, NY: Macmillan.

Stanic, G. M. A., & Reyes, L. H. (1986, April). *Gender and race differences in mathematics: A case study of a seventh-grade classroom.* Paper presented at the annual meeting of the American Educational Research Association, San Francisco.

Stein, C. (1986). *Sink or swim: The politics of bilingual education.* New York: Praeger.

Steinberg, L., Blinde, P. L., & Chan, K. S. (1984). Dropping out among language minority youth. *Review of Educational Research, 54,* 113–134.

Swing, S. R., & Peterson, P. L. (1982). The relationship of student ability and small-group interaction to student achievement. *American Educational Research Journal, 19,* 259–274.

Tikunoff, W. J. (Ed.). (1983, March). *Teaching in successful bilingual instructional settings* (Part I of the study report, Volume Document SBIF-81-R.6-IV). San Francisco, CA: Far West Laboratory for Educational Research and Development.

Tikunoff, W. J. (1985). *Applying significant bilingual instructional features in the classroom.* National Clearinghouse for Bilingual Education.

U.S. Bureau of the Census. (1986). *Projections of Hispanic population: 1983 to 2080* (Current population reports, Series P-25, No. 995). Washington, DC: Government Printing Office.

Veldman, D. J., & Sanford, J. P. (1984). The influence of class ability level on student achievement and classroom behavior. *American Educational Research Journal, 21,* 629–644.

Veltman, C. (1988). *The future of the Spanish language in the United States.* Washington, DC: Hispanic Policy Development Project.

Villasenor, A. (1990). *Teaching the first-grade mathematics curriculum from a problem solving perspective.* Unpublished doctoral dissertation, University of Wisconsin–Milwaukee.

Webb, N. M. (1989). Peer interaction and learning in small groups. *International Journal of Educational Research* [Special Issue], *13*(1), 21–39.

Weis, L. (Ed.) (1988). *Class, race and gender in American education.* Albany, NY: SUNY Press.

Welch, W. W., Anderson, R. E., & Harris, L. J. (1982). The effects of schooling on mathematics achievement. *American Educational Research Journal, 19,* 145–153.

West, J., Miller, W., & Diodato, L. (1985, March). *An analysis of course-taking patterns in secondary school as related to student characteristics* (Contractor report to the National Center for Education Statistics). Washington, DC: Government Printing Office.

White, K. R. (1982). The relation between socioeconomic status and academic achievement. *Psychological Bulletin, 91,* 461–481.

White, S. H., & Buka, S. L. (1987). Early education: Programs, traditions, and policies. *Review of Research in Education, 14,* 43–91.

Willig, A. C. (1985). A meta-analysis of selected studies on the effectiveness of bilingual education. *Review of Educational Research, 55,* 269–317.

Willig, A. C. (1987). Reply to Baker. *Review of Educational Research, 57,* 363–376.

Wisconsin Department of Public Instruction. (1984). *Chapter PI 13.* Madison, WI: Author.

Wolfle, L. M. (1985). Postsecondary educational attainment among Whites and Blacks. *American Educational Research Journal, 22,* 501–526.

Zentella, A. C. (1981). Ta bien: You could answer me in cualquier idioma: Puerto Rican codeswitching in bilingual classrooms. In R. Duran (Ed.), *Latino language and communication behavior* (pp. 109–131). Norwood, NJ: Ablex.

·26·

ASSESSMENT OF STUDENTS' KNOWLEDGE OF MATHEMATICS: STEPS TOWARD A THEORY

Norman L. Webb

UNIVERSITY OF WISCONSIN–MADISON

Assessment has begun to assume an increasingly important role in education. More than ever before, state and district assessments are exerting pressure on teachers and students to achieve high levels of performance. National and international assessment results have drawn the attention of policymakers, who use them as leverage for mandating change in education in hopes of bettering the nation's competitive edge. The *Nation at Risk* report (National Commission on Excellence in Education, 1983) delineated the problem by comparing American students' relatively low test scores with those of students from other industrialized nations, as well as noting the decline from 1963 to 1980 in average mathematics scores on the College Board's Scholastic Aptitude Tests (SAT). Recent developments in cognitive psychology (Brown, 1988; Carpenter & Fennema, 1988; Glaser, 1986) have resulted in the creation of new instructional programs that integrate assessment with instruction and minimize paper-and-pencil drill and exercises. These programs have influenced and are supported by recommendations included in the National Council of Teachers of Mathematics' *Curriculum and Evaluation Standards for School Mathematics* (1989).

This chapter reflects on what is known about assessment as it applies to the teaching and learning of mathematics. Relevant research and literature are reviewed to help define theoretical principles that can guide the assessment of students' mathematics knowledge. The chapter focuses on theory building through a critical evaluation of current efforts in mathematics assessment; in so doing, it challenges several existing assumptions about assessment as it relates to mathematics education. One such assumption is that assessment is organized and delineated according to certain identifiable principles and that, through understanding these principles, assessment practices can be guided to specific ends.

The chapter begins by defining mathematical assessment and explaining how it differs from other terms that have been associated with determining the achievement of educational outcomes. Then different purposes for assessment are described to indicate the breadth of the role of assessment in education. Following a short historical review of assessment in general, selected literature and research are reviewed in an effort to generate principles of mathematical assessment. This section begins by analyzing how content can be specified, based on the assumption that the approach used in specifying content, goals, and objectives will directly influence the assessment formulation and results. The main components of assessment—situation, response, analysis, and interpretation—are then discussed.

Mathematical assessment is employed for a wide range of purposes—from providing information to help a teacher work with a student so that she or he will gain a greater understanding of number sense, to plotting a national strategy that will have far-reaching implications for improving mathematics education for the nation. A comprehensive theory of mathematical assessment has to be responsive to this spectrum of purposes. Two sections are included that discuss issues at the two

I would like to thank Elizabeth Badger, Massachusetts Department of Education, Leigh Burstein, UCLA, and Vicky Kouba, State University of New York at Albany, for their thoughtful and helpful comments on an earlier draft of this chapter and Jack Bard for his assistance in doing literature searches.

extremes of this spectrum. One section reports on examples of assessment being integrated with instruction. A second section reviews the impact of large-scale assessment on mathematics teachers. These illustrative examples are used, along with information from the preceding sections, to identify the main issues that a theory of mathematical assessment would have to address.

OVERVIEW

Some issues merit initial discussion before the major themes are addressed. Is a distinct theory of mathematics assessment needed? Does the assessment of mathematics differ from the assessment of other content areas to the extent that a distinct and separate theory is meaningful? In other words, is mathematics so unique a content area that it requires content-specific means by which to assess students' proficiencies? Similarly, do existing general theories of testing invoke language and concepts that are specific enough to provide sufficient direction for assessing individual and group knowledge of mathematics? A theory of mathematical assessment can serve to describe, explain, and predict aspects and concepts within that area. One direct benefit of having a theory is to clarify terms used in association with mathematical assessment. An expanded theory would include models that give concrete structure to the application of different assessment approaches and to their interaction with the outcomes. General test theories exist such as the classical test and latent trait theories (Cronbach, 1984; Lord & Novick, 1968). These theories offer formulas for estimating parameters, true score, and error score that are couched in the tradition of psychological measurement. However, tests, as will be noted in the definition of mathematical assessment, are only one factor in assessment.

Research in cognition and learning has tended to distinguish between domain-specific knowledge and general cognitive skills (Perkins & Salomon, 1989). There still are questions regarding the interaction between general cognitive skills and domain-specific knowledge, as well as the amount of transfer across domains. It is evident that domain-specific knowledge is important and cannot be replaced by general cognitive skills. The assessment of any domain-specific knowledge, such as the assessment of mathematics, has an important role to serve in supporting the development of this specific knowledge. However, domain-specific assessment also needs to be sensitive in detecting its relationship to general cognitive skills and the facility of individuals to use this specific knowledge in conjunction with general cognitive skills.

The nature of mathematics itself and pedagogical approaches for teaching mathematics warrant consideration of specific assessment techniques in the area of mathematics. Deductive proof is prominent in mathematics in establishing truth, whereas the sciences depend heavily on observation and experimentation. The calculus, algebra, and number system are axiomatic systems that require a knowledge and deep understanding of axioms, operations, and theorems. The power of mathematics accrues when it is used to abstract a situation and then mathematical manipulations are used to gain further

knowledge of the phenomenon. Mathematics is also dynamic and ever-changing. Although calculators and computers are not unique to mathematics, their use has changed the importance of what is included in the mathematics curriculum and how it is taught. Manipulatives are assuming a greater role in mathematics classrooms to help model mathematical concepts and provide a concrete representation of students' thinking. A theory of mathematical assessment would have to take into consideration calculators, manipulatives, and the use of a variety of forms of representation. One role of such a theory would be to describe and explain differences in the assessment of mathematical knowledge with or without calculators. Such a theory would help describe why some assessment tasks are calculator sensitive while others are calculator neutral, which is not always evident for some tasks. Assessment techniques that have been used in other domains are being applied more and more to mathematics (Stenmark, 1989); for example, students writing responses to open-ended questions which are then scored, using a holistic or primary trait method. A theory of mathematical assessment would enable educational researchers to decide whether these approaches are viable in helping to identify students' knowledge of mathematics or whether they are more an indication of writing ability. Without a theory of assessment, we must resort to lengthy efforts at exploration and trial and error in learning how to adapt assessment techniques from other domains. Even though mathematics is important in many other areas, the nature of mathematics is distinctive enough and mathematics classroom practices are different enough to suggest that mathematical assessment should be distinguishable from other content area assessment.

One issue that a theory of mathematical assessment should address is how the specification of the mathematical content knowledge to be assessed is related to the responses obtained. A second issue is the priority given in assessment to validity and reliability. In testing theory, reliability is a necessary but not sufficient condition for validity. How does this rule apply in the administration of two or three lengthy mathematical problems? A third issue is the problem of aggregation of information. Is it valid to total the number of items scored correctly on a test as an indication of a student's knowledge of mathematics? How should test scores be aggregated with teacher observations and students' projects? A fourth issue is the relationship of the assessment of mathematical knowledge to the assessment of knowledge in other content areas and to the assessment of general cognitive skills. What do they have in common and how do they differ? These issues are discussed in more detail in the last section of this chapter.

Assessment

A number of terms refer to the collection of data and information for the purpose of describing an individual's or group's level of knowledge, performance, or achievement. The most frequently used terms are measurement, test, assessment, and evaluation. The term *mathematical assessment*, the focus of this paper, is used for a particular reason. In this paper, mathematical assessment refers to the comprehensive accounting of an individual's or group's functioning within mathematics or in the application of mathematics. This definition is patterned after a definition of general assessment by Wood (1987), who states that "assessment is regarded as providing a comprehensive account of an individual's functioning in the widest sense—drawing on a variety of evidence, qualitative as well as quantitative, and therefore going beyond the testing of cognitive skills by pencil-and-paper techniques which, for many people, is measurement" (p. 2). The determination of the functioning of an individual or group within mathematics or in the application of mathematics requires considering mathematics performance in a variety of contexts, including knowledge of mathematics and disposition toward mathematics. Both qualitative and quantitative approaches are needed in order to gain a comprehensive account of this functioning, with conclusions made on the basis of information from a combination of sources.

Tests are important quantitative assessment tools, but in and of themselves do not constitute the totality of assessment. A test, defined by Cronbach (1984) as a systematic procedure for observing behavior and describing it with the aid of numerical scales or fixed categories, is a very powerful tool for mathematical assessment. The same test can be used in collecting information from one person after another. A test helps standardize conditions so that inferences can be made about individuals and groups in a well-defined situation. This definition of a test is not restricted to paper-and-pencil techniques, but is broad enough to include questionnaires for obtaining reports on attitudes, procedures for observing behavior, and observation records collected systematically such as watching students using manipulatives to demonstrate a knowledge of regrouping. However, test results are generally given as a single score or a profile of scores. It is difficult, using only numerical scores, to describe how a student draws relationships between different mathematical concepts such as a proportion and similar triangles or how a student goes about solving a problem. An expanded view of mathematical assessment embracing qualitative methods, interviews, and observations, as well as teacher opinion, leads to a greater capacity to describe these important aspects of mathematical knowledge. It is this expanded view of mathematical assessment that is needed as mathematics instruction becomes directed as much toward the thinking in doing mathematics as to the doing of mathematics.

Related to testing is measurement. In education, the concept of measurement comes out of the psychological testing and behaviorist tradition. "Every important outcome of education can be measured" (Ebel, 1973, p. 10). For a more precise definition, Lord and Novick (1968) call "an observable variable a *measure* of a theoretical construct if its expected value is presumed to increase monotonically with the construct" (p. 20). Measurement in education is restricted to a quantitative description of student behavior. As with testing, the meaning given to mathematical assessment in this chapter includes measurement as one aspect, but goes beyond the provision that the object of assessment has to be quantifiable in order to be assessed.

A third term that is often associated with and sometimes used synonymously with assessment is evaluation. Included in evaluation is the systematic collection of evidence to help make decisions regarding (1) students' learning, (2) materials development, and (3) program. Assessment as defined in this chap-

ter can be a means for doing evaluation. Evaluation has also been described as the systematic investigation of the worth or merit of some objective (Joint Committee on Standards for Educational Evaluation, 1981). This definition helps somewhat if assessment is considered an accounting of what is and evaluation is the assignment of a value to the results of the assessment. This is a very cloudy distinction because what is assessed is generally considered to have value. For example, if students' computational ability is assessed, those who do better on the assessment can be considered to be more accomplished, at least in computation. The size of the group being assessed is not a factor in differentiating between mathematical assessment and evaluation. Mathematical assessment of an individual as well as of a large group, such as that carried out by the National Assessment for Education Progress, is possible. Reporting and analyzing the NAEP findings for a given year is within the realm of assessment; on the other hand, the interpretation of these findings by comparing them to previous years' results or to some projection and describing the result as "good" or "bad" is the basis of evaluation. One purpose of this paper, and one of the objectives for building a theory of mathematical assessment, is to determine what assessment is as opposed to other information-gathering enterprises. An effort will be made to clarify how terms are being used if there is any possibility for alternative interpretations.

The concept of assessment must be considered from a variety of perspectives. One purpose of assessment is for it to be used as a tool by teachers to provide evidence and feedback on what students know and are able to do. At this functional level, assessment and the observations derived from its use are viewed as inherently valid and have a direct influence on teaching and learning. A second purpose of assessment is to express what is valued regarding what students are to know, do, or believe. To the degree that assessment results are considered credible and important, the assessment instruments themselves may shape and influence the curriculum. In this way, assessment becomes a form of communication that sends a message from the teacher or others to the student about what it is important to know. A third purpose of assessment is to provide information to decision makers, including those within the educational system, governmental policymakers, and others. At this level, assessment results are used by parents, administrators, school boards, and taxpayers as the basis of judgments about the effectiveness of the educational program in general and, in some cases, about the relative skill and ability of individual teachers. In this context, assessment is a tool used to impose on teachers and schools a direct, measurable accountability for their effectiveness. In many instances, for example, policymakers exert control over the curriculum by mandating the passing of specific tests as a prerequisite to graduation.

A fourth purpose of assessment is to provide information on the effectiveness of the educational system as a whole. The results of certain forms of assessment (in the United States, the SAT and ACT) have become the nation's "bottom-line" indicators of the effectiveness of the educational system—regardless of the questionable validity of these conclusions. This was evidenced by the furor over the declining SAT scores from 1963 to 1980. "The American use of tests reflects our culture's interest in quantified and 'objective' judgments, [and is a] part of the rational management ethos" (Resnick, 1981, p. 626). Because scores on nationally prominent tests have assumed such specific cultural meaning, the public has, in large part, been resistant to reform efforts to expand assessment beyond norm-referenced testing. This remains true despite the fact that real questions have been raised about how useful normed tests are nationally for providing information to district administrators and teachers for instructional decision-making (Salmon-Cox, 1981; Sproull & Zubrow, 1981).

An analysis of the relationships between the teaching and learning of mathematics and each of the four purposes of assessment reveals the complexity of the assessment process. An evaluation of assessment must acknowledge not only its mechanics—test construction, item development, and analysis—but also the administrative, political, and cultural influences imposed upon it. In practical applications, these influences can be in conflict and create dilemmas for teaching and learning. The pressure to produce high test scores can compete with an equal but opposing pressure to identify and address individual students' instructional needs; in some cases, for example, the evaluation of students' understanding and skills may require use of an assessment instrument on which students will perform poorly. These conflicting goals and approaches make it all the more essential that educators and policymakers have a clear understanding of the different purposes for assessment and that they evaluate the approaches to be used in light of the intended purpose.

Historical Context

A historical analysis (Romberg, 1987b) of the development of assessment in the United States traces the use of written examinations to the Boston schools of 1845. Some of the first objective measures of achievement were used in 1864 by the Reverend George Fisher, who compared student work to "standard specimens" and rated spelling numerically, depending on the number of errors. In the early 1900s, the "psychometric period" brought general intelligence testing, placement testing of army recruits, aptitude tests, and achievement tests. Romberg argues that these forms of assessment marked the beginning of the modern testing movement and that they provide the bases for group testing procedures still used today: a set of questions (items), each having one unambiguous answer, with all subjects administered the same items under the same conditions. Romberg describes a variety of forms of achievement testing, including norm-referenced standardized tests, whose purpose is to identify a respondent's position in a group; profile achievement tests, whose purpose is to give a variety of scores for groups of students; and objective-referenced tests, whose purpose is to compare scores on specific objectives to an a priori criterion. Romberg concludes that such objective tests are products of an earlier era in educational thought and that present-day conditions call for the development of more relevant forms of achievement measures.

Educational measurement is premised upon a set of assumptions and a set of purposes different from psychological measurement, which continues to provide the theoretical bases of most traditional forms of educational assessment (Wood, 1986). Wood observes that educational measurement is concerned with the achievement of an individual relative to himself or herself and is designed to measure competence rather than intelligence. Educational measurement must be responsive to educators' needs and to the school cycle of planning, instruction, learning, achievement, and measurement. Criterion-referenced testing (Glaser, 1963) is an example of an educational measurement form that is a break from the psychological model that depends on homogeneity, normal distributions, and latent traits. In this form of testing, individuals are judged relative to a criterion that can result in a very skewed distribution. Because scores are not normally distributed, procedures for computing reliability that have been developed under the assumption of a normal distribution, with maximum discrimination among scores, are not appropriate. A current assessment problem is the development of good measures of reliability for criterion-referenced tests and other nontraditional forms of assessment.

This brief historical commentary indicates that our culture's assessment practices are rooted in the past and have evolved out of past practices. Mathematics assessment is deeply grounded in these traditions and is influenced by the same forces that affect testing in general; the two are shaped by the same pressures and conform to many of the same psychometric properties. Even classroom assessment is heavily dominated by a single short-answer testing format. Although not as formal and generally not using fixed-choice items, classroom tests are similar in nature to those used on a large scale. Because of traditional and general conceptions of assessment, principles that apply solely to the assessment of mathematics are difficult to isolate.

PRINCIPLES OF MATHEMATICAL ASSESSMENT

This section draws upon current literature and research to identify specific principles for assessing mathematics that can, in turn, lead to a theory of mathematics assessment. In some cases, the research reported is not specific to mathematics because of the scarcity of content-specific research. In these instances, the relevancy to mathematical assessment is projected. The use of such findings does not detract from the argument for a distinct theory of mathematical assessment, but can be used as a contrast to better highlight what may be specific to mathematics.

Specifying Content for Assessment

Fundamental to mathematical assessment is an explicit definition of the content to be assessed. This definition can be derived in many ways and depends upon the purpose for the assessment, the operative conception of mathematics involved, and the practical concerns of time, scoring, and funding. In any case, specifying content is a complex process, as illustrated by a variety of studies that have attempted it. In the final phases of the SMSG (School Mathematics Study Group) project in 1972, for example, researchers identified seven dimensions (content, cognitive level, affective dimension, verifiability, feasibility, population, and form) along which objectives for instruction and assessment can vary. In another review, Romberg (1987a) identified three major forms of achievement tests that vary by dimensions (standardized tests, profile achievement tests, and objective-referenced tests), only one of which involves specifying content. In challenging current methods of specifying content, a strong case is made by Romberg and Zarinnia (1987) that new, alternative forms of assessment need to be developed to further the aims of curriculum reform. Models such as a content-by-behavior matrix (described in more detail below), which assumes a linear ordering of knowledge and the independence of knowledge levels, are unable to represent the complexities of curricular reform that presume that knowledge is created and interrelated. Not only is specifying content complex, but the way in which content is specified affects the form of assessment as well as its results.

Assessment results will vary, depending upon how the content is specified as the assessment is designed. Content specification is derived from the purpose for assessment, a conception of mathematics, and a theory for the learning of mathematics. The purpose of assessment can range from observing a student in order to determine his or her conception of place value, to evaluating the effectiveness of a mathematics program. Conceptions of mathematics can vary from a collection of facts and skills to a hierarchical structure of concepts, procedures, and principles, to an integrated, dynamic body of knowledge that is continually changing. Learning theories could be based on constructivism, behaviorism, developmentalism, or any other applicable paradigm. This variability in purpose, mathematics conception, and learning theories necessitates a variety of approaches for specifying content. Each approach will have unique benefits and limitations.

To illustrate the interaction of conceptions of mathematics and approaches to specifying content for assessment, two extreme conceptions of mathematics will be used. In reality, both conceptions could be used to characterize mathematics depending on the purpose. The main point is that the assessment approach reflects an underlying conception of mathematics and will consequently influence the results from the assessment. One extreme view of mathematics is that it is a collection of facts, skills, and concepts. These facts, skills, and concepts can be partitioned and taught separately to students. Students who successfully master the facts, skills, and concepts are considered to have a functional level of mathematics. At the other extreme, mathematics is a structured body of knowledge of interdependent elements. This body of knowledge has a pattern of organization. For students to know mathematics, they need to know the elements—concepts, skills, properties, and principles—and the interrelationship among them. Using these two conceptions of mathematics, here are examples of assessment approaches that generate results reflective of each conception.

Approaches Reflecting Mathematics as a Collection of Facts, Skills, and Concepts.

TOPIC APPROACH. In a topic approach, a general body of content is specified without details as to how items or tasks are to be created. Topics may be as general as algebra, geome-

try, or measurement, or as specific as the content covered in a chapter, whole-number addition, or two two-digit multiplication. Although this approach makes it relatively easy to specify content, it is difficult to ensure that the full range of knowledge within the topic has been covered. An assumption that mathematics can be separated into distinct topics is compatible with this approach.

SPECIAL CASE APPROACH. On occasion, the purpose of assessment may be confined to determining or understanding one particular aspect of a student's knowledge or skills. At other times, it may be sufficient to assess a student's knowledge of a procedure by considering only those situations in which some "special trick" or skill is required. In these situations, the assessment can target special cases, such as adding two three-digit numbers with one including a zero in the tens digit, or measuring the length of an object using a scale that does not begin with zero. Diagnostic testing represents this kind of special case approach in which a relatively small number of assessment tasks are needed depending, of course, on the range of content knowledge that is targeted. This approach does not, however, provide a measure of general knowledge, but is useful in ascertaining an individual's understanding or difficulty in using a procedure or concept.

BEHAVIOR APPROACH. In a behavior approach, the activities that comprise the assessment tasks need to be precisely defined. Listing behavioral objectives or, more generally, competencies is one example of this approach. The objectives are not grouped to fit into any particular scheme other than to span the desired outcomes. An example of a competency for the third-grade level is, "The student will be able to add and subtract whole numbers" (Thoman & Moser, 1984). The behavior approach, accompanied by definite content specifications, facilitates the writing of detailed paper-and-pencil test items. The approach assumes that knowledge can be partitioned and does not acknowledge more complex situations that require an integration of knowledge. Teaching and learning that focus on what students are to do rather than on what they are to know is congruent with a behavior approach to assessment.

PERFORMANCE APPROACH. A performance approach uses direct observation to measure ability level. "Performance assessment is the process of gathering data by systematic observation for making decisions about an individual" (Berk, 1986, p. ix). A performance approach, then, defines a process by which students can be observed while performing a skill or while applying knowledge so that some judgment can be made about the level of their skill and ability. This refined behavioral approach centers on having students do exactly what is necessary to achieve the desired outcome, as compared to making an inference via some paper-and-pencil activity that only *models* the outcome. Observations of students as they use a ruler, tape measure, and scales to measure a variety of objects is one example of performance assessment. This is contrasted with having students interpret a diagram of a ruler next to a picture of a pencil and asking them to select the length of the pencil from four choices. In using a performance approach in specifying content for assessment, the emphasis is put on skills, procedures, and attitudes that are directly observable. In some cases, performance in mathematics involves paper-and-pencil computations, such as solving a quadratic equation. Performance assessment is normally directed toward an individual and is difficult to accomplish in a large-group setting. It generally requires a series of observations and a means of record keeping. The advantage of performance assessment is that it has high validity, if done properly, since the desired outcome—what is to be observed—can provide information that is directly applicable to the design of an intervention strategy where needed.

PROCESS APPROACH. A process approach identifies problem-solving processes, higher-order thinking skills, or other thinking strategies as a means of producing outcomes. Although some paper-and-pencil instruments have been used to evaluate students' ability to use processes (Charles & Lester, 1984; Wisconsin Department of Public Instruction, 1985), interviewing students or asking them to describe their thought processes have been the more accepted methods for documenting the processes being used (Carpenter & Moser, 1983; Kilpatrick, 1967; Lucas, 1972; Webb, 1975). The assessment of higher-order thinking skills is a current issue in educational practice. Most existing forms of testing do not provide for sustained reasoning (Quellmalz, 1985), and current competency tests do not provide the opportunity for students to do extended analyses, solve open-ended problems, or display a command of complex relationships (Resnick, 1987). The process approach is different from other approaches in that it draws attention to process as compared to product and gives process status as an instructional outcome. The complication in this approach is that processes are not always readily observable and may require a variety of innovative techniques to assess their use.

Approaches Reflecting Mathematics as a Structured Body of Knowledge

STATISTICAL APPROACH. Sometimes statistics are used to select assessment tasks and items using a model that assumes knowledge can be sequenced on a linear scale. When decisions regarding assessment tasks are made for statistical or psychometrical reasons that override content area concern, then the assessment approach is considered to be statistics driven. For example, assume that a general target area is specified, such as fifth-grade mathematics, and that items related to this area are judged by their characteristic of revealing distinctions among students. In this situation, a purpose of the assessment is more to develop a test that will sequence students along a scale so they can be compared against a norm rather than to determine the full extent of mathematical knowledge of the students. In short, this approach is designed to assess each student's knowledge as compared to his or her peers, rather than to evaluate students' mastery of a body of knowledge or a content area. Many norm-referenced tests are constructed in this way. Items are selected based on a scheme to match the curriculum, but the final decisions for item inclusion depend heavily on statistical parameters. Other forms of statistical approach include the application of item response theory in constructing a test. One basic assumption of this theory is that items fall along a linear scale according to difficulty and that individuals can be

sequenced on the same linear scale by ability. The theory is more complex than this, but without an initial consideration of the formation of scale from a content-area point of view, the driving rationale of this approach centers more on the statistical testing model than on the integrity of content. These approaches to assessment assume that differences in the ability to do mathematics can be sequenced on a scale so scores can be used to judge if one individual knows more or less mathematics than another individual. Rather than forming the scale from a collection of facts, skills, and concepts, statistical means are used to determine the scale.

MATRIX APPROACH. As used in assessment, this approach to specifying content generally refers to a content-by-behavior matrix that serves as a framework for planning assessment. The assessed outcomes are partitioned into distinct cells, each representing knowledge within one content area at a specific cognitive level. The dimensions of the matrix can be ordered, and the approach can be used to ensure that a range of instructional outcomes are covered in both instruction and assessment. This approach assumes that mathematics can be modeled by different dimensions and that outcomes can be sequenced and placed along these dimensions.

Even though the common form of a matrix is n-by-m, a matrix could be a vector, such as a profile achievement test, or multidimensional, n-by-m-by-p (Weaver, 1970). Bloom's Taxonomy (Bloom, 1956), or a modification such as that made for NLSMA (Romberg & Wilson, 1969) or for the National Assessment of Educational Progress (1981), has frequently been used to define cognitive levels as one dimension of the matrix. In another application of the matrix approach, the SOLO (Structure of the Learned Outcomes) Taxonomy (Biggs & Collis, 1982; Collis, 1987) has been developed to merge the cyclical nature of learning with the hierarchical nature of cognitive developments. The SOLO Taxonomy has been used to create cloze format items (Collis, Romberg, & Jurdak, 1986) by constructing an item stem with four questions, each requiring a more complex use of information that ranges from using an obvious piece of data at Level 1 to using abstract general principles at Level 4.

The matrix approach provides a scheme for assuring that a range of outcomes will be assessed, and it identifies categories for reporting results. As noted above, the approach assumes that knowledge can be partitioned into distinct compartments and that categories along the dimensions are disjoint. Another assumption of the matrix approach—this applies to the content-by-behavior matrix—is that each behavioral level is applicable to each content category.

DOMAIN APPROACH. A domain is the set of all possible tasks that will measure the desired outcomes. The domain approach defines the set of tasks by specifying the content, such as whole number operations, and the boundaries, such as size of numbers (to those numbers less than 999). Whereas a topic approach specifies a general collection of tasks, the domain approach assumes a greater structure by indicating the range in tasks that are targeted for assessment. Compared to the topic approach, the domain approach provides more specification for a possible assessment task by specifying a set of parameters indicating the limitation or boundaries of the domain. The domain approach allows weights to be assigned to different parts of the domain so that more importance can be given to one part over another. Item banking, in which items are randomly selected from a specified domain, is one example of this approach. Domain-referenced tests, which involve identifying the domain and completing those tasks that exist in the domain, are another example. The domain approach is more applicable to assessing factual procedural knowledge than problem-solving and reasoning. Since a random sample of tasks from the domain can be selected, the domain approach allows a teacher to draw an inference about a student's ability to work all of the tasks within the domain based upon his or her ability to complete the sample of problems.

CONCEPTUAL (RULE) APPROACH. This approach approximates the domain approach, but differs in the way in which the domain is defined. In a conceptual approach, the domain is generated by specifying rules which can then be used to generate a task or situation to be included in the domain; an example is the conceptual fields defined by Vergnaud (1982, 1983). Key elements of conceptual fields include problems and situations, operations of thought, and symbolic representations; they not only specify content but also consider the interrelationship between problems and situations and students' thinking in addressing them. Conceptual fields are based on the assumption that "a small number of symbols and symbolic statements can be used to represent a vast array of different problem situations" (Romberg, 1987a, p. 3). A great variety of assessment situations can be generated and identified from the few elements that define a single conceptual field, such as the additive field or the multiplicative field. Donovan and Romberg (1987) cite the work of Carpenter and Moser (1983), in which six types of addition and subtraction problems were defined based on their semantic structure, as an example of applying conceptual fields to assessment. The conceptual approach is particularly valuable in that it can be used to map what a student knows within a knowledge domain and to track the maturation of concepts within that domain. This approach requires extensive work in specifying the elements from which a field can be generated. There also is a question as to which conceptual fields can be defined.

KNOWLEDGE INTEGRATION APPROACH. This approach is based on the view that knowledge of mathematics is integrated and that assessment should involve the application of a variety of different aspects of the domain. In NCTM's *Curriculum and Evaluation Standards* (1989), for example, the student assessment and evaluation standards are based on the assumption that the determination of students' mathematical power requires assessment of all aspects of their mathematical knowledge, as well as the extent to which they have integrated this knowledge. An integration approach will draw upon situations from various content areas, requiring that students apply a variety of mathematical concepts and procedures and that they engage in reasoning and problem-solving. A set of situations that considers different forms of representations and a variety of tasks designed to assess knowledge of function and graphs has been generated at the Shell Centre for Mathematics Education in Nottingham, England (Swan, 1987). Projects or problems that require data collection and analyses exemplify an integration approach to assessment. This approach supports the goal that students are to know and do mathematics as a unified whole, not as a series of individual skills. Such assessment requires ample time, both to define appropriate assessment situations and to allow students to work through them effectively. There also is the expectation that a variety of assessment tools will be used to create a complete picture of what a student knows.

An important aspect of the knowledge integration approach is the importance of attitude, motivation, and individual conative characteristics (Snow, 1989) to knowing mathematics. The *NCTM Curriculum and Evaluation Standards* (1989) indicate the importance of students having a mathematical disposition. The *Standards* acknowledge that students' success in mathematics depends, in large part, on their motivation to do mathematics and to recognize the role mathematics plays in our culture. If developing a positive disposition towards mathematics is a goal for instruction, then disposition is a form of content and should be assessed.

Summary. The wide variety of approaches that can be used for specifying content may result in an equally diverse set of assessment tools. This discussion does not presume to have covered all of the viable approaches to identifying content, nor does it suggest that they be used independently of one another. It is suggested, however, that the approach to be adopted should depend largely on the purpose of assessment and the information that is to be gleaned from it. A premise of mathematics assessment should be that the approaches for specifying the content to be assessed should reflect a clear conception of what mathematics is and what mathematics students are to know. If one goal of a mathematics program is that students view mathematics as dynamic and understand that central concepts such as function and variable are applicable in a wide number of areas of mathematics, then a knowledge integration approach to specifying content is applicable. If mathematics is viewed as a set of skills, or if the focus of assessment is on what skills students can use, then a behavioral, domain, or performance approach in specifying content is appropriate. While all of the approaches, individually or in various combinations, are appropriate for classroom assessment, any single approach is too limited to gauge the full range and depth of what it means to know mathematics.

Formulating Mathematics Assessment

Messick (1981, p. 9) states that two questions need to be addressed whenever a new test is proposed:

1. Is the test any good as a measure of the characteristic it is interpreted to assess?
2. Should the test be used for the proposed purpose in the proposed way?

The first question is scientific, the second ethical, social, and political. Although Messick refers specifically to the testing process, these two questions apply equally well to a broader conception of what a test is, as Fredericksen (1984b) calls for, or to other forms of assessment.

Implied in the first question are the issues of validity and reliability, both essential characteristics of any form of assessment. It is these attributes that determine an assessment's value, and they are therefore of primary concern as assessment experiences are developed. The principles of validity and reliability have been discussed in great detail by Cronbach (1984), Haertel (1985), Guilford (1965) and others, and so will not be discussed here other than to note some particular concerns. Like other properties of assessment, the definitions of these two attributes depend upon certain key assumptions. Of particular importance to validating mathematics assessment are the interpretations that are made regarding a student's mathematical knowledge and what mathematics the student can do. The means of validating an assessment approach are dependent upon how the assessment will be used. Haertel (1985) notes that new uses for tests require new methods for test construction and validation. He gives as an example the lack of emergence of a unified framework for validating criterion-referenced test interpretations.

Likewise, reliability takes on new meaning when one considers the consistency of results using alternative forms of assessment of student mathematics performance that do not comply with the assumptions of classical test theory. In making educational decisions about what students know and can do, determining a student's abilities in a variety of situations is more important than obtaining a single score on a highly reliable test. For this reason, the NCTM *Standards* (1989) support the practice of basing assessment decisions on the convergence of information obtained from a variety of sources and using assessment situations that call for the integration of knowledge. The practical procedures for doing this, however, are yet to be developed.

Formulation of an assessment requires consideration of four general components of the process: the assessment situation, the response to this situation, the analysis of the response, and the interpretation of the results. The interdependent nature of these components means that specification of one affects what is possible for the other three. In addition to questions of content, the manipulation of these four aspects will determine the validity of an assessment and the consistency of the results across a variety of assessment experiences. A general principle for mathematics assessment is that the situation, response, analysis, and interpretation, as well as the mathematical knowledge being assessed, the characteristics of the individual or group who are to respond, and the purpose for assessment, must be in alignment.

Assessment Situation. In the teaching and learning of mathematics, assessment occurs when a teacher or student gathers information and makes inferences about knowledge, either for the purpose of determining the current state of that knowledge or for guiding further learning. An assessment situation can be defined as broadly as a test composed of a series of items or a project extending over a period of days, as narrowly as a stem to a question, or as unobtrusively as students working on a daily assignment.

A theory for mathematics assessment needs to be useful for generating a wide range of assessment situations. Traditionally, the most common form of assessment situation has been

a test, with psychometrics as the theoretical basis (Goetz, Hall, & Fetsco, 1990). But, as Wainer (1986) demonstrates, the traditional summary statistics of reliability, item-total biserial correlation coefficients, and mean portion correct can be misleading. This psychometric perspective of measuring knowledge does not always coincide with assessment derived from a cognitive or information-processing approach. The assessment instrument properties that are relevant to constructing a test do not apply in working with a student one-on-one to determine the student's thinking processes. In this review, the concept of the assessment situation is broadly defined. Assessment will refer to more than a test and will include a variety of situations ranging from individual assessment to large-group assessment.

In considering a broad view of assessment, some issues emerged out of the cognitive sciences that have direct implications for developing and selecting situations to be used in the assessment of mathematics learning. One is the relationship of domain-specific knowledge (declarative, procedural, or conditional knowledge an individual possesses relative to a particular field of study) to strategic or general knowledge (planned or intentionally evoked procedures used either prior to, during, or after the performance of a task, or relating attributes of a current situation to some prior encounter). In a review of research on this issue, Alexander and Judy (1988) generated several hypotheses about the interaction of domain-specific and strategic/general knowledge, but were hampered in drawing conclusions because of the lack of operational definitions for key concepts such as domains and strategies. Others (Perkins & Salomon, 1989) suggest the importance of the synthesis of general and specialized knowledge. Clarity is important in determining or specifying the relationship between mathematics-specific knowledge and general knowledge applicable across fields of study. The development of situations for assessing mathematical knowledge depends on such clarity to distinguish results that can be attributed to mathematics-specific knowledge from those attributed to more generalized knowledge applicable across many fields of study. A theory of mathematical assessment would help in making such a distinction.

Another major issue, which draws from recent cognitive research, challenges the notion that what is learned is separate from the way it is learned. If this notion is indeed correct, then assessment needs not only to consider the outcomes from instruction, but also the instruction and activities leading to the achievement of the outcomes. According to Brown, Collins, and Duguid (1989), "The activity in which knowledge is developed and deployed...is not separable from or ancillary to learning and cognition" (p. 32). The implications for assessment are clear: If knowledge and learning are situated in specific contexts, the situations used to assess them must have authenticity relative to the practice about which interpretations are being made. As important as domain, then, is the issue of assessment situation and its relationship to the situation in which the information or knowledge was acquired or learned. These issues all have implications for the specifications of situations for the purpose of assessment. In exploring assessment situations for mathematics, three major factors seem to be important to consider: the content framework, the situation characteristics, and the form of administration.

CONTENT FRAMEWORKS. Cognitive psychology and information processing have helped to define new theoretical frameworks from which to derive a deeper understanding of what it means to know mathematics. These frameworks have the potential—although not yet fully realized—for directing the construction of assessment situations based on students' mental structure of knowledge rather than on the utility of mathematics or behaviorism—a criterion that has influenced the development of tests throughout the greater part of this century.

Research in mathematics education has helped to define specific frameworks for particular topics. In the area of computation, Carpenter and Moser (1983) have developed a structure for analyzing addition and subtraction situations; Birenbaum and Shaw (1985) used a task specification chart to analyze errors in adding and subtracting fractions; and Kintsch (1986) created mental maps for children's learning from written arithmetic word problems. In algebra, Clement (1982) diagrammed the thought processes students used in solving algebra problems, and Kaput (1986) developed a framework for studying the quantity structure of algebra word problems. In geometry, Koedinger (1989) conducted a task analysis to study geometric intuitions, and Greeno (1978) constructed a computer simulation of the knowledge structures geometry students used in solving problems. In rational numbers, Janvier (1981) identified the limits imposed by a situation in which students engaged in the abstraction process of proportional reasoning. In the area of problem-solving, Collis et al. (1986) developed a structure of learned outcomes and applied it to the assessment of mathematical problem-solving ability. More general cognitive strategies have been investigated by Haylock (1987), who developed a framework for assessing mathematical creativity. These systematic efforts to understand students' mathematical thinking and the theories that articulate this understanding have generated new and important information relevant to the assessment of students' knowledge. But the general principles for guiding the construction or selection of assessment situations for particular purposes have yet to be placed in the context of recent developments in learning and teaching.

CHARACTERISTICS OF ASSESSMENT SITUATIONS. In addition to a well-grounded theory of content, other factors play a key role in the definition of assessment situations. A large body of research and literature analyzes and describes the characteristics of assessment situations that involve paper-and-pencil tests of a given set of items. Crooks (1988) has reviewed the general research on the impact of classroom evaluation practices on students, and Goldin and McClintock (1984) provide an overview of the research on mathematical problem-solving task variables (characteristics of tasks that may influence the solutions), including syntax, context, content, structure and heuristic behaviors.

Hembree (1987) performed a meta-analysis of 120 research studies that addressed non-content variables associated with mathematics assessment and their relationship to performance. His findings indicate that

- A personal form (warm and enthusiastic conditions as compared to cold and mechanical conditions) of administrating a test related to improved scores for testers of low socioeconomic status.

- Testers fared better when work space was provided adjacent to the test items.

- Testers found a multiple-choice format easier than the same test with open-response items.

- On severely and moderately speeded tests, testers performed better when items were ordered from easy to hard in terms of difficulty.

Other factors that were related to improved test performance included providing pictures in the test booklet for testers of average ability; group competition; item-by-item knowledge of results; unannounced testing; and frequent testing.

As in other areas, technology is affecting assessment and the definition of assessment situations. The use of hand-held calculators is becoming more accepted in specific assessment situations (Carter & Leinwand, 1987; Long, Reys, & Osterlind, 1989; McAloon & Robinson, 1987), but the uses that can be made of calculators in providing a deeper understanding of what mathematics a person knows have yet to be determined. Developments have been made in using computers to administer, score, and report results of mathematics assessment, but questions remain as to when an assessment situation should allow students to interact with computers as students do mathematics. Such situations might require students to create and use a spreadsheet or to develop a data base in applying mathematics. Researchers and educators have barely scratched the surface of technology's potential for improving and enhancing the assessment of mathematics learning.

Another variable in all assessment situations is the amount of information to be included and what students are expected to do with it. Fredericksen (1984a) describes "ill-structured problems" as those that approximate the "real world," exclude needed information, and lack a clear statement of what is being sought. This concept also can be used to distinguish between open-ended project situations and precisely defined word problems or exercises, where the issue of specificity is related to the interplay of mathematics-specific knowledge and strategic knowledge. Nearly all the research on characteristics of assessment situations for mathematical learning has focused on well-structured situations, which leaves many unanswered questions about more open-ended situations. Work in this area conducted in England and Wales (Pirie, 1988) and in the Netherlands (de Lange, 1987) is discussed in more detail later in this review.

ADMINISTRATION OF ASSESSMENT. A key aspect of the definition of assessment is in the mode of its administration to students. An assessment can involve the active participation of the teacher, where the teacher provides hints and probes to gain a greater understanding of the student's knowledge, or it can require the teacher to play a more passive role, such as administering a standardized test. Teachers may also ask questions in class (Biggs, 1973; Easley & Zwoyer, 1975), or administer a self-made written test (Stiggins, 1985). The assessment could be administered using a computer, which can monitor the tasks given to a student based on previous responses (McBride, 1985; Wainer & Kiely, 1987), or provide some form of error analysis (McDonald, Beal, & Ayers, 1987) or feedback. Some evidence suggests that administering tests using a computer can produce results different from the results obtained when the same tests are administered without computer use. Lee, Moreno, and Sympson (1986) administered the same test to two randomly selected groups of military recruits, with one group using paper and pencil and the second group using a computer. The computer group scored significantly lower. Reviews of research in the computer-based testing area have found the aggregation of studies to be inconclusive, with a number of studies reporting equivalence in scores between computer-based testing and conventional testing, while a number of other studies report significant mean differences between the two modes of test administration (Wise & Plake, 1989). It is clear, then, that the administration of assessment situations does interact with the assessment results. What research has not provided is an understanding of why these situational factors affect assessment in the way they do. Such an explanation is essential to a sound theory of assessment.

Response. Response in an assessment situation is what the student is expected to produce; it can vary greatly, depending on the knowledge that is required, the form of the response, and whether the response is to be made by an individual or a group. The response required is related to, but not entirely dependent upon, the situation. A test item can require the student to choose the correct answer among given options, or to produce the answer on an open-response item. The appropriateness of a particular form of response depends on the purpose of the assessment and other factors such as the time available for scoring and the number of students being tested.

One issue involved is the degree to which the required response must be authentic. Asking a student to actually measure the length of an object with a ruler is more authentic than presenting the student with a drawing of a ruler and an object to be measured. Systematic observations of students doing mathematics as they work on a project supported by their responses to probing questions are more authentic indicators of their ability to do mathematics than a test score compiled by totaling the number of correct item responses. As noted by the California Mathematics Council and EQUALS (Stenmark, 1989), students' success should be evaluated while working on worthwhile investigations or tasks. One example is this performance task:

> *There are 30 students in our class. The office has given us 144 pencils and 24 erasers as our supply for the year. How can we be sure we will still have pencils and erasers at the end of the school year?*

Are students able to make a plan?
Can they decide when to use a calculator, and then use it effectively?
Does everybody in the group participate?
Do students look at all factors of the problem, or do they jump to conclusions?
Do students use blocks or other materials appropriately?
Do they make notes or drawings to check their results?
Do they recognize and use the complexities of the problem?
(Stenmark, 1989, p. 21)

Recently, more attention is being paid to forms of assessment that deviate from the traditional forms in terms of what students are required to produce, the emphasis placed on processes, and the relationship to classroom instruction. These define three major attributes of responses—solution form, the process as the focus, and informal classroom responses.

SOLUTION FORM. Silver and Kilpatrick (1989) call for the use of open-ended problems in the assessment of mathematical problem-solving, where students generate numerous conjectures based on a set of given data or conditions. In such assessment situations, students would be asked to identify problems with similar mathematical structure and select mathematical models that can be used to represent particular problems. Students would be expected to produce conjectures, to group problems according to their mathematical structure, to offer an explanation of those structures, and to select mathematical models. Some (Collis et al., 1986; Wearne & Romberg, 1977) have recommended presenting students with questions in series, where each response indicates a higher level of thinking. In these cases, the response to the "superitem" is a set of answers rather than just one answer. Below is an example given by Collis et al. (1986, p. 212):

> *This is a machine (accompanied by a diagram of a function machine) that changes numbers. It adds the number you put in three times and then adds 2 more. So if you put in 4, it puts out 14.*

If 14 is put out, what number was put in?
If we put in a 5, what number will the machine put out?
If we got out a 41, what number was put in?
If x is the number that comes out of the machine when the number y is put in, write down the formula that will give us the value of y whatever the value of x.

Some research has investigated the efforts of varying the form of representation of an assessment response. As a means of assessing students' ability to communicate mathematically, the California Assessment Program (Pandey, 1988) has piloted mathematics situations that require them to write a paragraph or a set of directions. The Shell Centre for Mathematical Education (Swan, 1987) has directed students to write about a situation and produce other forms of representations, such as drawing a graph or diagram. Webb, Gold, and Qi (1989) systematically varied the form of presentation of problems with the form of response by presenting problems using words, algebraic equations, graphs, or pictures, and requiring high-school students to respond using numbers, words, graphs, or pictures. They found that the translation that was required when a problem was presented in a form different from the required response added a degree of difficulty to the situation. For example, students demonstrated a greater conceptual understanding in solving a problem presented using words and that required a response in words (Example 1) than they did on a problem presented using words and that required a numerical response (Example 2).

EXAMPLE 1. A problem presented in words requiring a verbal response (Webb et al., 1989, p. 19).

Tickets for a baseball game cost $10 for box seats and $5 for regular seats. 100 people came to watch the game and they paid a total of $750. Without solving the problem, explain why each of the following statements cannot be true:

1. 90 tickets for box seats were sold.
2. 60 box seats and 30 regular seats were sold.

EXAMPLE 2. A problem presented in words requiring a numerical response (Webb et al., 1989, p. 17).

John needs 120 yards of wooden planks to build a staircase. He has $420 to spend. Oak is expensive, costing $4 per yard; pine costs $3 per yard. Since he cannot afford to make an all oak staircase he would like to use as much oak as possible. This means he must spend all $420. How much wood should he buy of each type?

This finding supports the research that has been done on problem-solving and different modes of representation. Khoury and Behr (1982) observed that presenting students with conventional symbolic representations of problems is likely to produce only a limited picture of students' mathematical problem-solving processes and performance. They found an interaction between performance on ability measures (hidden figures, spatial visualization) and scores on the different parts of a mathematics retention test that varied by mode of presentation — pictorial, symbolic, or mixed. They saw promise in using alternative symbolic forms (pictorial and symbolic) of the response as a means of measuring students' conceptual understanding of mathematics. This approach, systematically varying the forms of representations on tasks, may provide a better understanding of the influence of different response-form representations. But such questions as how different response forms correspond to the situation and what information the response provides about what a student knows have not been widely explored.

PROCESS AS THE RESPONSE. As interest in the application of mathematical knowledge, strategic knowledge, higher-order thinking, and problem-solving becomes more prevalent, there is a concurrent need for assessment instruments that provide measures of these processes. Within a research environment, the think-aloud procedure has been used successfully to explore and understand students' thinking (Hart, 1983; Kilpatrick, 1967; Lucas, 1972; Webb, 1975). Although this procedure has some applicability in the classroom, it is time consuming and difficult to manage with large numbers of students. Quellmalz (1985) recommends that test items be written so as to assess students' thinking skills and sustained reasoning. She recommends that assessment tasks permit alternative interpretations or solutions and be designed with open formats that ask for explanations of reasoning.

The National Assessment of Educational Progress (Blumberg, 1986; Educational Testing Service, 1987) has piloted several approaches to assessing higher-order thinking skills in science and mathematics. These efforts included tasks that would make the students' use of higher-order skills during problem-solving more apparent through group activities, station activities, and full investigations. These tasks required students to record the findings of experiments they either observed or conducted themselves; to describe in writing the approach they would take to find the answer to questions; and to use equipment as a test administrator observed and recorded behavior on a checklist. The pilot study concluded that "hands-on" assessment was feasible and informative and that it was possible to gather descriptive information on students' skills in sorting, classifying, determining relationships, and conducting reliable experiments.

Some effort has been made to develop practical procedures for teachers to evaluate student work on open-ended problem situations. Zehavi, Bruckheimer, and Ben-Zvi (1988) conducted a series of studies with ninth-grade students working on assignment projects–open-ended mathematical problems wide in scope, but strictly related to the regular curriculum and to the core mathematics. The use of assignment projects, which included some work in class and some work at home, was found to enhance student mathematical achievement when compared to a control group. In a procedure developed for classifying student responses, the teacher used sequentially ordered hints and noted the first effective hint and its result. It was concluded that the studies produced an organized framework for qualitative evaluation of mathematical activity on open-ended problems. This research exemplifies efforts to investigate not only students' test results but also their thinking processes and the results of that thinking.

From studies such as those conducted by NAEP, from forms of assessments used in other countries and from research into higher-order thinking, there is promising evidence that viable assessment measures can be developed to produce information about students' thinking as they do mathematics. Currently, however, these assessment techniques remain in the formative stage and have not been widely used.

CLASSROOM RESPONSES. A response for assessment purposes does not have to be recorded on a formal assessment instrument, but can be a thoughtful observation of students' work during the normal operations of class. Such nonobtrusive, or embedded (Harmon, 1988) assessments may regard an activity simultaneously as a learning experience and an assessment situation. If meaningful assessment information is to be derived by this means, however, the observer needs to have a clear purpose, based on some learning or conceptual framework, in mind. Students' responses as they interact with each other in a small group are one occasion for making such observations. Asking students to prepare a portfolio (NCTM, 1989), or some other cumulative record of performance (McLean, 1987), is a variation of a classroom response that can be used for assessment. As with higher-order thinking skills, educators are beginning to express more interest in the notion of embedded assessment; even so, little is known about the procedures that need to be applied in this approach. This issue will be further discussed in the section on the integration of assessment with learning.

Analysis of the Response. A crucial step in any assessment involves the consolidation of the information collected or observed so that some meaning can be assigned for the identified purposes of the assessment. In the case of a test, items are traditionally scored as right or wrong, the number of items answered correctly is calculated, and this total is compared to a specific standard. In classroom observations, student behaviors can be observed and recorded using a coding scheme. The marks for each behavior or action can be tabulated and the frequencies reported as they relate to the other frequencies tabulated. A teacher, with some framework for depicting students' knowledge, can observe a student and then consider his or her performance in the context of this framework. This information can be used, for example, to assess whether the student is ready to learn a more complex strategy. In all of these cases, some form of analysis and abstraction of information is required in order to make judgments.

Typically, assessment results consist of a score that represents the number of correct answers. Analysis, however, involves more than simple scoring; it must consider the form of analysis, the person who does the analysis, and the level of aggregation of results. Statistical analysis is not covered in this review because it is discussed in great detail in many other sources (Guilford, 1965; Lord & Novick, 1968; Lyman, 1986). It should be noted, however, that a result can have high pedagogical validity but poor statistical qualities. For example, the National longitudinal Study of Mathematical Abilities (Wilson, Cahen, & Begle, 1968, p. 253) assessed ninth graders in the spring of the school year on this problem:

One solution of the equation $4y^2 = 1$ *is* $\frac{1}{2}$. *The other solution is*

1. $-\frac{1}{2}$
2. $-\frac{1}{4}$
3. 0
4. $\frac{1}{4}$
5. 3

On this item, 41% of the students responded correctly; however, the item had a biserial correlation of only .15 with the total score on the 16-item algebra scale. The biserial correlation is low, but the item result is important and reveals that more than 50% of the students could not find the second solution.

FORMS OF ANALYSIS. Quantification of results is one form of analysis. It may include scoring as right/wrong or using a scale. Charles and Lester (1984) provided scores of 0, 1, or 2 on each of three dimensions—understanding the problem, planning to solve the problem, and the result—in scoring the problem-solving performance of fifth and seventh grade students. In scoring the problem-solving performance of college students, Schoenfeld (1982) used a three-category scale to rate students' attempted solutions compared to plausible solutions or approaches. A student was rated 1 or 0, depending on the existence of evidence that the student used the particular strategy; 1 or 0 depending on whether the student pursued the approach; and 1 or 0 indicating the level of progress made (little, some, almost, or solved) if an approach was pursued. Schoenfeld reports that this scaling scheme was used with a high degree of consistency in its effectiveness. The state of California used for their state assessment in mathematics a scoring rubric

with six levels: 6—fully achieves the purpose of the task, while insightfully interpreting, extending beyond the task, or raising provocative questions; 5—accomplishes the purposes for the task; 4—substantially completes purposes of the task; 3—purpose of the task not fully achieved; needs elaboration; some strategies may be ineffectual or not appropriate; assumptions about the purposes may be flawed; 2—important purposes of the task not achieved; work may need redirection; approach to task may lead away from its completion; and 1—purposes of the task not accomplished (Pandey, 1991, p. 30). Other quantitative scoring schemes that have been adapted from other content areas are cited in Charles, Lester, and O'Daffer (1987).

A second form of analysis, labeled teacher's impression, is based solely on teacher judgment. This form of analysis assumes the value of teacher opinion and expertise to analyze what students know, do not know, and are able to do. Superficially, teacher impressions appear to be subjective, but in the context of a teacher's knowledge about and experience with all of his or her students, they can be considered objective since they have some consistency. This issue was explored by de Lange (1987), who asked 15 teachers to assign independent scores to the work of each of five students. Teachers had no background information on the students, on their previous work, or on how the tasks should be scored. After norming the results for each teacher by standardizing the mean and standard deviation, the rate of agreement among the teachers was judged to be very high, considering the given conditions. Although this study is somewhat impressionistic, it makes the point that teachers' judgments have consistency.

A study by Coladarci (1986) supports the efficacy of teacher judgment. Aggregated measures of teachers' judgments of how students would respond to standardized test items correlated positively and substantially with aggregated student test scores. There were some individual differences among the teachers, however, and there was variation in results according to the achievement level of students: Teachers were least accurate in judging low-performing students. In addition, there is some evidence that teachers' judgments of students' academic performance, when not related to a specific test or independent of a test, will derive results somewhat different from those obtained in a standardized test. Sharpley and Edgar (1986) concluded that teacher ratings of student performance in academic areas, including mathematics, constitute a distinct factor separate from student test performance because of the low total variance on the test scores accounted for by the teachers' ratings. The results of these two studies suggest that teachers' judgments can accurately predict student performance but, when unrestricted, focus on different factors from those measured by standardized tests.

A third form of analysis involves placing student work in a specific category or assigning it some form of classification, such as using a student's work to judge his or her developmental level. Hart et al. (1980), Osborn (1984), and Collis et al. (1986) provide examples of categorizing students based on a mental development scheme.

Descriptive analysis is another form of analysis. Rather than ranking students in some type of hierarchy or clustering them in a particular category, it may suffice to simply describe what each student has done. Schoenfeld (1982) had students answer a series of questions about a problem they had been working on for 20 minutes. Students were asked whether they had seen that problem or a related problem before and whether they had an idea of how to start the problem, as well as a variety of other questions about their approach to solving it. Information based on the answers to these questions was then used to describe differences between two groups of students. A descriptive form of analysis can also provide students with written feedback about their work, more as a means of enhancing their awareness of their progress than as a judgment about their efforts.

A fifth form of analysis involves diagnosis and prescription. Because much has been written about diagnostic teaching (for example, Denmark, 1978), the subject will not be discussed in great detail here. What distinguishes this form of analysis from the others is that it targets weaknesses rather than strengths and focuses on remedial efforts to improve performance, rather than on simply assessing and reporting what the student can currently do.

Other forms of analysis may include those created by some combination of these five. The selection of one form over another depends to a large extent on the way the information is to be used and the level of specificity that is needed. If students need to be grouped according to their preparedness to succeed in calculus, it seems reasonable that an elaborate scheme for quantifying information is not necessary. What we do not know very well is how to select the best form of analysis for a specific assessment situation and how to interpret and understand the interaction among the different forms of analyses, purposes for assessment, and other aspects of assessment.

WHO DOES THE ANALYSIS? The issue of who analyzes the assessment responses has particular implications beyond the simple provision of assessment results. Clearly, teachers are the major analyzers of assessment information collected during instruction. But, at least in the United States, teacher analyses do not have the same credibility as the more "objective" results provided by large-scale testing. This is evidenced by the number of states that have instituted testing programs to assess student competencies to make judgements regarding students rather than depend solely on teacher judgments.

In stark contrast, after 1991 England and Wales will base at least 25% of the examination marks leading to the General Certificate of Secondary Education (GCSE) on teacher assessment of pupil coursework (Pirie, 1988). One of the 17 examination objectives will include assessing students on their ability to carry out practical and investigational work and to undertake extended pieces of work. This forces the teacher into the dual role of being a promoter of learning who offers formative judgments on student work and an examiner who makes summative judgments. Such a duality of roles will be new to many teachers. Pirie considers different schemes for marking students' investigational work and then concludes with a cautionary note that unless the teacher is able to handle the dual roles, "understanding will become subordinate to assessment procedures and the feedback essential to learning will be stunted" (p. 16). Of significance to this discussion is that, depending on the weight given to the final results and how the analyses are performed, the new role of the teachers could have a critical impact on instructional emphasis.

Other sources for analyzing assessment results include such external scorers as commercial testing companies, machine scoring devices, and on-line computer analysis. While the use of machine-scoreable answer sheets has become increasingly widespread, states and districts have in recent years contracted with testing companies to holistically score writing samples of all students at certain grade levels.

Great strides have been made in the use of computers to analyze results and provide responses to students' work. McDonald et al. (1987) explored whether a computer can efficiently analyze student answers, computational error patterns, and errors related to student use of microcomputers. Their findings report the usefulness of the immediate and accurate diagnostic information computers can provide. This and other studies point to the fact that computers and other forms of technology will play an ever-increasing role in the analysis of assessment information.

LEVEL OF AGGREGATION. Information can be aggregated in many ways. In assessment, information is generally aggregated over students or over different levels of assessment—item, test, or accumulation of sources. Aggregated information helps to report and interpret data about a student, class, or larger group. A course grade is one example of aggregated information that is reduced to concise form to report one student's achievement and level of effort during the course. Procedures for aggregating information may be as simple as counting the number of correct items on a test, or as complex as applying such sophisticated statistical techniques as the item response theory to create a scale that facilitates the comparisons, across assessment years, for age groups and subpopulations (see, for example, the 1986 National Assessment by Dossey, Mullis, Lindquist, & Chambers, 1988). As with other aspects of assessment, the level of aggregation and the procedures used for aggregation will depend on the purpose for the assessment and will correspond with the assessment situation and the responses obtained.

The unit of analysis defines the group over which the aggregation will be done; it could, for example, be the student, class, district, state, or nation. In the classroom, the student is the most frequent unit of analysis. However, aggregating information across students within a class can provide useful information in considering the effectiveness of instruction. When aggregated to the class level, what may appear to be random anomalies on individual work can reveal a distinct pattern indicative of misconceptions resulting from misleading instruction. For example, in interpreting item responses from a fifth-grade mathematics test, Miller (1986) reports that at the individual student level, patterns of item response could be interpreted as a measure of carelessness or guessing. But when the information was aggregated to the class level, a poor match between test content and instructional coverage was suggested. Classroom-aggregated assessment information can provide insight into instructional practices beyond that which is revealed by results interpreted in terms of individual students.

Beyond the classroom level, aggregated information is generally used for accountability, management decision making, and policy formation. If information is needed at the district, state, or national level, but not at the student level, then techniques such as matrix sampling are sometimes used to administer subsets of the total tasks administered to randomly selected subgroups of the population. This limits the amount of data collection needed and, as a result reduces the cost of the process. The technique, however, limits the reaching of any conclusions regarding individual students. For the 1986 NAEP, a stratified multistage probability sampling design was used in which clusters of students were selected and in which certain subpopulations were sampled at higher rates than others (Johnson, 1988). For school districts, test publishers will aggregate norm-referenced standardized achievement tests scores across individual students to provide percentile ranks for students in the district. Baglin (1986) cautions that the various methods of obtaining average percentiles produce results that can vary greatly. Beyond the simplest levels, such as averaging class test scores, the aggregation problem of how to group scores quickly becomes involved with high-powered statistical issues that need to be considered in order to yield meaningful information and results. For example, two students each answered correctly 75% of the items on a test. Twenty-five percent of the items on the test are considered to be difficult items. Both students are given a score of 75 and are considered to have the same level of knowledge as measured by the test. However, one student answered all of the difficult items correctly, while the other student did not answer any of those questions correctly. A simple sum of the number of items answered correctly will misrepresent the situation.

A second form of aggregation combines results from different test items or from different sources of information to provide some score or to make some judgment about an individual student. This use of a total score as a representation of student knowledge is based on the assumption that the items whose results were summed span the content and provide a representative sampling of work within that knowledge area. Psychometric principles suggest that the greater the number of items, the more reliable the information will be. This approach seems to work well with factual knowledge and computational skills. The question remains as to whether the aggregation of independent items provides as much information about what a student knows as does considering the results of a student's investigation or project, where more open-ended responses are required.

Another method of aggregating information on an individual is using norm-referenced standardized test scores to provide a grade-equivalent score. Unfortunately, the procedures for computing grade-equivalent scores are such that their interpretations far exceed any valid meaning they may have (Lyman, 1986). In using aggregated information, it is important to understand how the information was aggregated and what conclusions are appropriate on the basis of the aggregated results.

Despite this seemingly broad range of approaches to aggregating information, we actually know very little about the means of aggregating information across a variety of sources or about the forms of information that are needed to draw conclusions regarding what a student knows. These procedures will become more important as the assessment of student reasoning and higher-order thinking becomes increasingly dependent upon exercises in which students are required to solve complex problems, make conjectures, and consider structural properties of problems. At the same time, studies such as that done by Marshall (1983) show that, using results from a sixth-grade test ad-

ministered by the California Assessment Program, even aggregating information across distractors can be useful. Marshall's analysis of distractor choices identified a sex-by-distractor interaction; from this analysis, she concluded that girls' errors, compared with boys', were more likely to relate to the misuse of spatial information, the use of irrelevant rules, or the choice of incorrect operation. Boys' errors were associated more with lack of perseverance and formula interference. Many aggregation issues exist and, as new forms of assessment are developed and put into practice, new aggregation issues will continue to arise.

Interpretations. The final aspect of assessment involves giving meaning to the assessment outcomes and the interpretation of results. The meaning derived from assessment will depend on the generalizability of the results, the form of the reports and results, and the standards that are used for comparison of the results. Interpretation is a crucial aspect of any assessment because of the potential for misapplication, such as the reporting of information by a district that goes beyond what the data or information allows. Although many forms of validity have been defined, the key question still remains: Is the assessment valid for the conclusions and interpretations that are made from the assessment?

GENERALIZABILITY. Generalizability is a field of study unto itself (Cronbach, Gleser, Nanda, & Rajaratnam, 1972). For our purposes here, it suffices to consider only the underlying principle of generality: To what level of generality will the data support conclusions? Figure 26.1, an interpretation cycle, illustrates the different levels of generality. At the most restricted level of generalization, assessment results apply only to the student responses on the tasks or items given. Based on the results of the assessment, we can conclude that the students who took the test can solve these tasks. At the next level of generality are the assessment's content specifications, which were developed as the basis of the assessment instruments. These specifications limit the domain being assessed by restricting the size of the numbers included or the form of questions asked. Approaches to specifying content have been discussed earlier in this chapter. The third level of generality includes the instructional goals, and the fourth includes the entire field of mathematics.

In addition to generalizing within subsets of mathematical knowledge, generalizations can be made for other purposes. One form of generalization produces predictions of future achievement. At the first level in this direction, assessment results are used to judge whether a student has certain prerequisites and skills for future learning. Ability grouping of students based on a test score is one example of this; levels beyond this may include predicting future success in higher education and employability. The more general the conclusions, the greater is the need for a rigorous procedure for substantiating them. The level to which the results from the assessment can be generalized will depend on factors such as the design of the assessment, how the assessment tasks were selected, how the content was specified, the theoretical basis for formulating the assessment tasks, and how representative the tasks are with respect to all possible tasks.

FORMS OF REPORTING. Reporting of assessment results could be as simple as a comment to a student or as detailed as a multi-

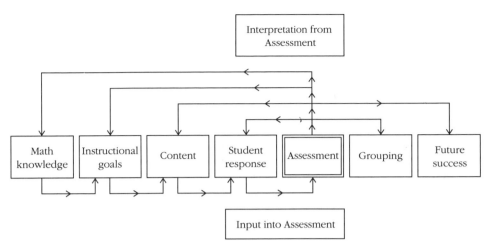

FIGURE 26–1. Interpretation cycle.

page evaluation; it could be statistical information or a narrative interpretation of the results. Whatever the form, the way in which the information is reported will influence whether and how the information will be used. The administrators of many large-scale assessment efforts, such as competency testing programs, recognize the importance of reporting the results and thus experiment with a variety of formats before selecting one that provides information in a way that is most useful to teachers.

Some research is available on the importance of providing feedback and keeping students informed of their progress (Hembree, 1987). In one study, furnishing students the correct item response after they had worked each item during a test resulted in significantly higher test performance. Similarly, positive effects have also been identified when students were provided test results during a unit of instruction, as compared to those who were not aware of their progress. Birenbaum and Tatsuoka (1987) examined three methods of providing feedback to students on the seriousness of error types committed on a post-test. Eighth-grade students used computers to take tests on adding and subtracting positive and negative whole numbers and to obtain feedback on their performance. Those students who received the most informative feedback were better able to correct their errors. However, computer feedback was ineffective in correcting serious errors. As technological development continues, computers will play an increasingly important role in assessment, and the form and frequency of feedback will become key issues. In addition, because computers provide greater facility for reporting information in a variety of formats, the form of reports and the scope of their content will become important areas of inquiry in determining effective ways of reporting information to (1) further learning in mathematics, (2) make instructional decisions, and (3) reduce the reporting of superfluous information.

STANDARDS. Interpretations on assessments are made with respect to some standard, either implied or explicit. As such, standards are an essential part of any educational assessment. A standard is a level of attainment that has some value associated with it. Livingston and Zieky (1982) describe a standard as the answer to the question, "How much is enough?". The different forms of standards (relative or absolute) are not content-specific, but apply to education assessment in general. Some critical educational decisions, such as student graduation, are based on particular standards, which may vary from one district or locale to the next. In the context of the daily classroom, standards are applied as students are assigned grades on their work. What is important to this discussion of assessment is an awareness of the various types of standards and the ways in which they relate to the other aspects of assessment.

A norm-referenced score is an example of a relative standard that is used to judge a student's score or knowledge in relation to the norm of some defined reference group. The most common use of relative standards is the comparison of an individual's score to national norms. If a student's score, and subsequently the district average score, is below the national norm, it is assumed—often falsely—that there is cause for concern. Since the reference is to some characteristic of a group, a score on a mathematics norm-referenced test does not define what mathematics a student knows or does not know. One example of a norm-referenced approach to assessment is grading students on a fixed curve, defining 93% and above as an A, 85% and above as a B, and so on. In general, these levels are not based on a sound judgment about students' knowledge, but are an artifact that assumes that a certain percentage of students will score above a certain level on all tests. Another example of a norm-referenced standard comes out of the Graded Assessment in Mathematics Project (Brown, 1986) in England, where levels were defined to represent the progress an average student should make in one year.

A second type of standard, an absolute one, is a criterion score established through some process held to represent a level of knowledge. Criterion-referenced tests are used in this way with a cutoff score that generally represents a minimally acceptable level of knowledge. Several methods are used to determine the cutoff score (Livingston & Zieky, 1982) that relate to making some judgment on how much knowledge is enough. In principle, the criterion-referenced form of standard is more compatible with educational practice that assumes all students can reach at least a certain level of knowledge. In practice, criterion-referenced decisions are based on a single score and do not take into consideration the full range of a student's knowledge of mathematics. The use of a criterion does not preclude consideration of multiple situations in which students engage in investigation, extended projects, reasoning, and problem solving. A major flaw in criterion-referenced tests involves the problem of specifying what is enough or what level of knowledge is needed if students are to derive adequate benefit from the next course.

A third standard form, a relative form, is an individual-referenced standard. With this form of standard, the student is judged on the progress that he or she makes based on what he or she knew at the beginning of instruction. Progress can be evaluated in terms of a student's acquisition of a certain body of knowledge, but the reference is made with respect to an individual rather than in comparison to some larger group. This form of standard is more appropriate in situations in which there are a variety of ways of gaining knowledge and in which it is assumed that individuals construct their own knowledge.

A fourth form of standard, the novice/expert form, has come out of the relatively new fields of cognitive theory and information processing. This format uses research in a content area to identify what it means to be an expert in that area; this description is then established as the goal to be achieved. But as Glaser (1986) indicates, a great deal of effort is being invested in understanding the cognitive structures and abilities of the skilled performer. Standards are very much related to the nature of competence and what it means to know a field such as mathematics. Increased understanding of "What is enough?"" is as important to assessment as it is to instruction. In mathematics education, we know very little about "what is enough," except that the more mathematics a person knows, the greater are his or her opportunities.

Summary of Assessment Formulation. Assessment formulation is a critical part of any assessment. Each of the distinct elements of assessment is of equal importance, as is their interaction with one another. A large body of research is now available to guide our thinking about assessment and to provide guidance as we explore assessment issues and options. An essential but elusive objective is to obtain a clear understanding of what mathematics students know and what structure of support will strengthen that knowledge and encourage its growth. Through a clear understanding of content, standards can be specified, assessment situations defined, student responses identified, and analyses performed.

ASSESSMENT AS INTEGRAL TO MATHEMATICS INSTRUCTION

This section and the next discuss two particular purposes for assessment. The first is using assessment in support of daily instruction. The second is large-scale assessment to influence what *happens* in the classroom. These two situations do not by any means constitute all the applications of assessment, but in building upon the previous section on assessment formulation,

they will help identify principles for mathematical assessment. This section explores some of the ways in which assessment and instruction can coexist and reinforce each other. However, even as assessment merges with instruction, the four features of assessment—situation, response, analysis, and interpretation—are maintained.

Viewing assessment as an integral aspect of instruction provides a framework for thinking about assessment, instruction, and their interaction. Assessment as integral to instruction means that assessment is continuous; it occurs as the teacher processes information on what students know and as he or she uses this information to guide instruction. Assessment as integral to instruction implies that assessment is more than testing; it involves the application of a variety of means for determining what students know. This section will draw upon recent research to describe some selected current assessment practices. It will then review several studies that integrate assessment with instruction. The purpose here is not simply to describe how assessment relates to instruction, but to provide evidence that the relationship between assessment and instruction is changing as the result of new developments in learning and knowledge theory and in mathematics education research.

Current Assessment Practices as Related to Instruction

According to Stiggins (1988), who draws on existing research to describe the current status of classroom assessment, teachers may spend as much as 20 to 30% of their professional time directly involved in assessment-related activities. "These activities include designing, developing, selecting, administering, scoring, recording, reporting, evaluating, and revising such items as daily assignments, tests, quizzes, observations and judgments about student performance" (p. 364). There is little evidence that teachers are given any technical assistance by the school districts in dealing with day-to-day issues of classroom assessment, Stiggins reports. Teachers are concerned about the quality of their assessments and lack confidence in them. Teacher-developed paper-and-pencil tests and many tests and quizzes provided by textbook publishers are currently dominated by questions that ask for the recall of facts and information. Many teachers lack a clear sense of their expectations about student performance and therefore depend on nonexistent or vague criteria.

Assessment practices vary by grade level and by subject matter. At the elementary level, teachers rely on diverse evaluation techniques, while secondary teachers emphasize tests (Gullickson, 1985). Teachers at all levels, however, report using a variety of assessment techniques. From the responses of 228 teachers recruited from each of four grades (2, 5, 8, and 11), Stiggins and Bridgeford (1985) found that when teaching science and mathematics, teachers tend to rely more heavily on paper-and-pencil tests than they do when teaching writing and speaking. Of the mathematics teachers responding, 85% reported using published tests. In another survey of 3,000 teachers (Linn, 1983), 74% of respondents, including persons from all content areas, said they use results from standardized achievement tests for diagnosing students' strengths and weaknesses.

Research reveals an apparent discrepancy between what K–12 teachers believe is important in assessment and what those in academia who are responsible for training teachers believe is important. Assessment is reported as one of the most complex and important tasks of teachers (Stiggins, Conklin, & Bridgeford, 1986), but there is some question about the knowledge of assessment available to those responsible for these complex tasks. Gullickson (1986) reports that teachers value nontest evaluations and formative and summative evaluations, while professors of pre-service educational measurement courses value statistics, preparation of exams, and the administration and scoring of exams. Stiggins and Bridgeford (1985) assessed the relative importance assigned to different types of tests. They found that although teacher-made objective tests and published tests have received the most attention in measurement textbooks and research articles, 47% of teachers surveyed rated performance assessment as the most important type of classroom evaluation.

Natriello (1987) reviewed research on the effects of evaluation practices on students and found that there was very little descriptive information available about actual classroom practice. Among other findings, he reports that there appears to be a curvilinear relationship between the level of standards set by teachers and student effort and performance. Students who perceived the standards for their performance as unattainable were more likely to become disinterested and disengaged from high school.

Thus, research supports the view that teachers consider assessment important; that they are engaged in assessment activities a large percentage of their working day; and that teachers impact student learning through their assessment and evaluation practices, although there may be a differential effect according to the achievement level of students. Teachers use a variety of assessment techniques, but the paper-and-pencil test is the dominant practice in upper-level mathematics classes. Little research has been done to investigate the actual practice and impact of a variety of other assessment techniques in the classroom.

Features of Assessment as Integral to Instruction

Although the literature offers scant evidence of the actual practice of classroom assessment, the dominance of paper-and-pencil and short-answer test forms and the lack of clear expectations of performance suggest that assessment in classrooms is not embedded in instruction. It is clear that teachers who use assessment to guide instruction must understand what students are expected to know and how that knowledge is structured, but short answer tests are inadequate to provide this kind of in-depth information. Instead, teachers tend to be restricted to assessing facts and are therefore limited in the extent to which they can provide complete information about students' structures of knowledge.

Glaser (1986) maintains that typical tests are inadequate to guide the specifics of instruction. Tests are primarily used as indicators to signal general rises or declines in school performance. He stresses that the measurement of achievement must rely on knowledge of learning and the course of acquisition of competence in the subject matters taught. Recent advances in two areas, information-processing analysis and the designation of experts and novices in mathematics, will move measurement toward this state. Glaser makes four key points. First, important to teaching is the ability to extrapolate from a student's performance an accurate picture of misconceptions that lead to error or an attainment that can lead to new learning. An underlying assumption of this position is that errors or misconceptions are not random but are indicators of an incomplete structure of knowledge. Second, students' comprehension and learning are based on their current beliefs; students attempt to understand and think about new information in terms of what they already know. Third, basic skills and advanced performance must be two aspects of a coordinated process. Complex activities depend upon the automation of certain lower-level tasks so that effort and attention can be concentrated on higher-level activities. Fourth, greater understanding is needed about the ways in which knowledge becomes organized and the means by which the processes that use this knowledge develop over long periods of learning and experience.

Assessment as integral to instruction is based on the assumption that teaching consists of making decisions about what students know, about processes students need to use in order to learn, about the structure of the content students are to learn, and about the means of teaching that will facilitate students' construction of their knowledge (Thompson & Briars, 1989). Teachers continually make "interactive" decisions on the average of one every two minutes (Clark & Peterson, 1986). Through a variety of means, assessment can provide information that will enable teachers to make decisions based on the best knowledge that is available about the learning process. Assessment as integral to instruction, then, embodies four features:

1. The teacher understands the structure of content knowledge and uses this structure to define expectations (standards) for learning.
2. The teacher is sensitive to the processes students use to learn, the stages of development, and the processes available to facilitate this learning.
3. Assessment is a process of first gathering information about a student's knowledge, about the structure and organization of that knowledge, and about a student's cognitive processes, then giving meaning to the information obtained.
4. Assessment is used to make informed decisions throughout instruction based on current information available about what a student knows and about what a student is striving to know.

Examples of Assessment as Integral to Instruction

Two programs that integrate assessment with instruction are described here. Each program is based on a series of studies defining the structure of knowledge and the stages of its development. As such, both programs are grounded in the theory that underpins the integration of assessment and instruction. Both programs include specific means of determining the student's current understanding, and both emphasize that instruction should be based on the information provided by this assessment.

Cognitively Guided Instruction. Cognitively Guided Instruction (CGI) (Carpenter & Fennema, 1988), which has been developed while working with students in grades 1 and 2, assumes that instructional decisions should be based on careful analyses of students' knowledge and on the goals of instruction. The program is designed to help teachers use knowledge of cognitive science to make their own instructional decisions. This approach is based on the premise that teaching itself is essentially problem solving since the teaching-learning process is too complex to predict in advance. Teachers do learn broad principles of instruction such as the form of instruction that should be appropriate for each student.

The approach used in CGI for assessing children's thinking is reported by Fennema and Carpenter (1989). The principles for assessing children's thinking that have been developed by the CGI project are based on empirical studies of children solving problems (for example, Carpenter & Moser, 1983; Carpenter et al., 1982). The approach requires teachers to make assessment a regular part of instruction by regularly asking students, as they do their individual work, to explain how they figured out an answer. Student responses are recorded or mentally noted using a framework developed from a previous analysis of students' strategies. The framework consists of skills arranged into a hierarchy of levels. A student's learning can be charted by noting progress through the levels. Because of the specification of the content, observing students as they solve a few carefully selected items provides a fairly accurate indication of the range of problems the students can solve. The major purpose of the assessment is to help the teacher decide whether a task has meaning for a student.

The Graded Assessment in Mathematics Project. The Graded Assessment in Mathematics (GAIM) project (Brown, 1988) is directed towards producing a continuous assessment scheme for recording the mathematical progress of a group of British students between the ages of 11 and 16. The levels of attainment in mathematics that provide the basis for the graded assessment scheme were developed from an extensive large-scale investigation of secondary students' understanding of mathematics (Hart, 1981; Hart et al., 1985). The scheme consists of 15 levels designed to represent the progress an average student makes during a year. Because students do not progress uniformly across all mathematical areas, teachers are encouraged to teach and assess more than one level at a time. An underlying assumption is that students will gradually construct their own mathematical knowledge in relation to the experiences they have had.

Students are assessed in terms of their knowledge of 20 to 35 criteria at each of 15 levels and for each of six topic areas. The topic criteria are profile statements, or objectives, aimed to describe the mathematics students are to know. The six topic areas are logic, measurement, number, space, statistics, and algebra and functions. They are also assessed on coursework activities designed to allow those with a broad range of abilities to approach the problem from their own level. GAIM activities include investigations and practical problems, and the project provides guidelines for describing, in general, the expected performance for each of the levels. An example of an investiga-

tion and practical problem is investigating the different symmetry patterns obtained by shading squares on a grid or planning the layout of newspaper advertisements (Brown, 1988, p. 12). Because of the number of topic criteria and because many are assessed during normal classwork, a prescribed procedure was not specified for the assessment of each activity. Teachers were responsible for devising their own assessment procedures. An external assessor did check on the procedures used and did ensure that students were assessed both orally and practically, in addition to written items. The project's fifth and final year of piloting was the 1988–89 school year.

Implications of Assessment Integrated with Instruction

Integrating assessment with instruction places strong demands on teachers; not only must they have a thorough knowledge of content structure, learning, and teaching, but they are also called upon to adopt new ways of teaching by changing their interactions with students and their use of information. This is not easy and requires teachers to have training to use assessment to inform their instructional decisions. The extensive research needed to create a cognitive framework for guiding instruction requires collaboration between researchers and teachers. Examples of such research include the work of Fredericksen (1984b) and Silver (1981) in the area of problem-solving and that of Reys et al., (1982) in the area of estimation. These and other studies demonstrate ways of assessing the organization of knowledge by considering the relationships between concepts. These studies, along with CGI and GAIM, verify that assessment can be integrated with instruction. The preliminary findings from the CGI project indicate that the approach has been effective in increasing students' knowledge of problem solving and their facility to explain the reasons behind what they do.

LARGE-SCALE ASSESSMENT

Large-scale assessment has an influence on what is taught in the classroom. Teachers are conscious of the content that is tested in large-scale assessment. The instruments used offer a model of what assessment is. Understanding more about the influence exerted on the classroom and on teaching by large-scale assessment can help illuminate more about assessment in general and the ways in which assessment results are interpreted. Recent research in this area helps clarify some of the issues.

Large-scale assessment is more prevalent now than it has ever been. Most school systems administer district-wide tests to, at minimum, a sampling of grade levels. Nearly every state as well as the Canadian provinces have some form of state assessment or mandated testing, or some combination of the two. As of June 1987, 44 of the 50 states (88%) had state mathematics testing at some grade level, through either administrating tests as a part of a state assessment program (37, or 74% of the states) or administering competency tests (26, or 54% of the states) or both (19, or 38% of the states) (Blank & Espenshade, 1988). Eighteen (36%) of the states require, or will require by 1991, a passing score on a state test for graduation. All of Canada's ten provinces have had some form of provincial mathematics testing program, although they do not test every year (Jansson, 1984). At the national level, the National Assessment for Educational Progress conducted mathematics testing in 1973, 1978, 1983, and 1986. NAEP piloted mathematics test items during 1988–89 that will allow the use of NAEP data to make state-by-state comparisons (Ferrara & Thornton, 1988). In February 1990, NAEP conducted the Trial State Assessment in 37 states, the District of Columbia, and two territories. The results of this trial, ranking the 37 participating states and territories by test score, were released in June 1991. SAT and ACT scores for each state are published annually in national newspapers as central features of the Wall Chart that the U.S. Department of Education has prepared annually since 1984 (Ginsburg, et al., 1988). The results of the Second International Mathematics Study (McKnight et al., 1987) are frequently referenced to garner support for school reform in mathematics and the sciences. This trend toward large-scale assessments seems to be growing more institutionalized as states increase high school graduation requirements (Patterson, 1989) and test scores are used as a major indicator of program quality. In 1991, the New Standards Project was initiated by the Learning Research and Development Center at the University of Pittsburgh and the National Center on Education and the Economy in Rochester, New York. The initial work of this project is to set standards for educational achievement in mathematics and literacy.

A national survey of schools (Romberg, Zarinnia, & Williams, 1989), completed for the National Center for Research in Mathematical Sciences Education (NCRMSE), provides information about how eighth grade mathematics teachers report being influenced by mandated testing. The survey sought information from eighth-grade mathematics teachers in a national sample regarding their knowledge of large-scale assessments and the impact of this knowledge on their teaching. A review of the literature, developed as background for the study, indicated that teachers spend time preparing students for large-scale assessment tests; teachers are compelled to adhere to district goals and to teach tested knowledge; teachers modify content based on what is tested; and in an attempt to raise test scores, teachers have a tendency to focus on students who exhibited lower performance. In research reported previously in this chapter (for example, Stiggins & Bridgeford, 1985), teachers used results from norm-referenced standardized achievement tests to help diagnose students' learning problems and to group students for learning; these measures, however, are of secondary importance compared to the influence of teacher-made objective tests. Based on research not necessarily related to mathematics and generally based on some restricted samples, tests do influence teachers and what they teach.

The results of the NCRMSE national survey provide information that heretofore has not been available from a national sample. Of the 701 schools sampled, responses were received from at least one teacher in 371 (53%) of the schools. A total of 552 (45%) of the 1,223 teachers who were sent surveys responded. Unfortunately, this return rate is not high enough to generalize the results to all eighth-grade teachers across the nation. However, the rate of return is sufficiently large and the frequency of schools from which replies were received is similar enough to the desired sample on variables such as district poverty level, district expenditures, and average enrollment as to be useful. Thus, the results can be interpreted to represent the conditions of testing over a wide segment of the United States.

Of the 552 teachers who responded, over 80% reported that their districts administered mandated tests in grades 7, 8, or 9. Of the total, 376 (68%) reported that the mandated tests were district tests, while 46% reported that the tests were mandated by the state. Eighteen percent of respondents indicated that state assessment programs were used in their districts in grades 7, 8 or 9. From the results of the study, it was concluded that over 80% of those who responded made some instructional changes because of district- or state-mandated tests. A large percentage (69%) of the teachers from districts with district-mandated tests reported that they examine the district tests to plan instruction or the content they will teach. Among teachers in districts with state-mandated tests, 70% reported that the state test has at least some impact on how they allocate instructional time to prepare students for the test, and 56% of the teachers reported that the test influences them to increase instructional emphases in some areas. Nearly three-quarters of the teachers reported that the district- and state-mandated tests were composed primarily of multiple-choice items with single correct answers. The content areas that teachers said they emphasize most because of the tests are basic skills (30%) and pencil-and-paper computation (25%). Areas receiving decreased instructional emphasis because of the influence of the test are extended project work (19%) and activities involving calculators (16%). The study concluded that mandated tests wield a direct influence on what a large number of eighth-grade mathematics teachers teach.

It is evident from the Romberg, Zarinnia, and Williams study, and from other studies, that the testing environment exerts a powerful influence on teaching and learning. Tests are more than a simple instrument for measuring achievement. They are interactive with the learning environment since they communicate to teachers and students society's values about what students should learn. In considering the learning and teaching of mathematics, the assessment environment—including both large-scale and classroom forms of assessment—is clearly important.

TOWARD A THEORY OF MATHEMATICS ASSESSMENT

This review of the research and literature related to the assessment of mathematics adds credence to the call for a specific theory of mathematics assessment within a general theory of educational assessment. The knowledge that is currently available through research in cognition and information processing points to the importance of knowing the structure of content in order to assess the cognitive structure of individuals related to that content. This premise, and strong arguments that educational measurement is different from psychometrics, supports the need for a general theory of educational assessment that is different from classical test theory. Existing projects in mathematics education that integrate assessment with instruction testify to the need for a clear understanding of mathematical content and learning prior to assessment. The differences between the nature of mathematics and that of other subject areas lead to the conclusion that a theory of mathematics assessment could unify the principles of assessment with the unique characteristics of the structures of mathematics. Such a unification could serve as the framework for assessing mathematical knowledge and performance, both within the classroom and across large groups.

A theory of mathematics assessment must address the general issues of validity and reliability, either by accepting these concepts for the assessment of mathematics and stipulating the conditions that meet their criteria, or by arguing for their redefinition, replacement, or elimination. When assessment includes a variety of forms of information collected from different sources, for example, the concept of the convergence of information appears more applicable than the concept of reliability. However, without a theoretical basis, the true difference or similarity between convergence and reliability is difficult to decipher.

Little is known about aggregating mathematics assessment data and what form of analyses are needed to derive maximum information from one or more assessments. A theory would provide a rationale for using different aggregation procedures and a basis for determining what meaning can be assigned to the results of the various aggregated levels. The research that has been done supports the contention that valuable information can be garnered from analyzing details such as distractor choices on multiple-choice items, as well as from the predictive value of a total score.

A theory of mathematics assessment would have to address the ways in which mathematics assessments relate to the assessment of other forms of knowledge and educational outcomes. One relationship that would need specification is how the assessment of strategic knowledge, such as general reasoning, problem-solving, and planning, interacts with the assessment of mathematics knowledge. Is it possible to separate a person's knowledge of mathematics from his or her more general knowledge? Is it possible to directly assess the thinking processes a person uses in doing mathematics? Think-aloud procedures have been used to assess process, but how relevant are the results to what a student actually does? Another issue related to assessing different forms of knowledge is the question of how the assessment of mathematical knowledge compares to the assessment of mathematics performance. From a behaviorist point of view, only performance can be assessed. But cognitive theory holds that models of mental processes and structure can be constructed and used to assess the structure of knowledge. Because both performance and knowledge are relevant to the learning and teaching of mathematics, a theory must provide the basis for assessing both.

Several approaches to specifying content have been used for assessing mathematics, depending upon the purpose for assessment. A theory would help identify the advantages of using one approach over another. It would also help identify the underlying assumptions of a content-specific approach and how these relate to the assessment results. Another issue related to content specification is the issue of generalizability and the interpretation cycle. Although a content domain has been defined for generating assessment situations, such as a conceptual field or a content-by-behavior matrix, the question remains as to whether the assessment results can be generalized to the entire domain.

Determining the domain of generalization is a critical issue in assessment in general, and for mathematics assessment in particular. Random sampling from a domain of specific test items, which has been used with item banking, seems an inefficient way to determine a student's knowledge of a mathematical domain. What is the number of assessment situations needed to understand what a student knows? Given an assessment situation, what can be generalized from a student's response? These and other questions centered on the domain of generalizability should be answered within a theory of mathematics assessment.

Such a theory must also embody a subtheory of questioning. Assessment as integral to instruction depends heavily on teachers questioning students. The type of information retrieved from students will depend on the questions that are asked and the sequence in which they are posed. Computer on-line branch testing is one response to this issue. A theory of mathematics assessment would account for this form of assessment, along with other forms of questioning, to provide guidance to teachers as they structure questions to solicit what a student knows.

An assessment involves a situation, responses to that situation, analyses of the responses, interpretations of the results, and interaction among all of these. Designing or selecting assessment situations to meet a particular purpose can be a very complicated task. A theory should help simplify this process by explaining how a situation works as students respond to it and how the different aspects of assessment interact with one another.

Many questions remain unanswered as we critically examine the state of mathematics assessment. As this and other reviews reveal, many research studies have explored different areas of assessment and thereby contributed to a better understanding of the assessment of mathematics. But there is very little mention of theory within these studies and no mention of a theory of mathematics assessment. A theory is needed to provide a structure from which to evaluate what has been done and to direct research toward what remains to be accomplished. A theory is needed to provide a language of mathematics assessment that will further discussion and establish its relationship to teaching and learning.

Time and systematic inquiry are needed to develop and test such a theory—both of which imply the need for long-range planning and directed research. In the interim, practical assessment problems exist that need answers now. Addressing these problems cannot wait for a theory to emerge. The work of developing alternative assessment measures must continue in the effort to better meet the needs of reform. Projects such as Cognitively Guided Instruction and Graded Assessment in Mathematics need to be well documented if they are to serve as a source of the vital information necessary in theory building. The paths that culminate in a viable theory of mathematics assessment will be those grounded in a diversity of practice and those searching for unifying principles.

References

Alexander, P. A., & Judy, J. E. (1988). The interaction of domain-specific and strategic knowledge in academic performance. *Review of Educational Research, 58,* 375–404.

Baglin, R. F. (1986). A problem in calculating group scores on norm-referenced tests. *Journal of Educational Measurement, 23,* 57–68.

Berk, R. A. (1986). *Performance assessment: Methods and application.* Baltimore, MD: Johns Hopkins University Press.

Biggs, E. (1973). Investigation and problem-solving in mathematical education. In A. G. Howson (Ed.), *Developments in mathematical education: Proceedings of the Second International Congress on Mathematical Education* (pp. 213–221). Cambridge, England: Cambridge University Press.

Biggs, J. B., & Collis, K. F. (1982). *Evaluating the quality of learning: The SOLO taxonomy.* New York, NY: Academic Press.

Birenbaum, M., & Shaw, D. J. (1985). Task specification chart: A key to better understanding of test results. *Journal of Educational Measurement, 22,* 219–230.

Birenbaum, M., & Tatsuoka, K. K. (1987). Effects of "on-line" test feedback on the seriousness of subsequent errors. *Journal of Educational Measurement, 24,* 145–156.

Blank, R. K., & Espenshade, P. H. (1988). 50-state analysis of education policies on science and mathematics. *Educational Evaluation and Policy Analysis, 10,* 315–324.

Bloom, B. S. (Ed.). (1956). *Taxonomy of educational objectives: The classification of educational goals: Handbook 1. Cognitive domain.* New York: Longman.

Blumberg, F. (1986). *National assessment of educational progress on higher-order thinking skills in science and mathematics: An innovative approach.* Princeton, NJ: Educational Testing Service.

Brown, J. S., Collins, A., & Duguid, P. (1989). Situated cognition and the culture of learning. *Educational Researcher, 18*(1), 32–42.

Brown, M. (1986). Developing a model to describe the mathematical progress of secondary school students (11-16 years): Findings of the graded assessment in mathematics project. In *Psychology of Mathematics Education Proceedings of the Tenth International Conference* (pp. 135-140). London, U.K.: University of London Institute of Education.

Brown, M. (1988). *The graded assessment in mathematics project.* Unpublished manuscript, King's College, London, U.K.

Carpenter, T. P., & Fennema, E. (1988). Research and cognitively guided instruction. In E. Fennema, T. P. Carpenter, & S. J. Lamon (Eds.), *Integrating research on teaching and learning mathematics* (pp. 2–17). Madison: Wisconsin Center for Education Research.

Carpenter, T. P., & Moser, J. M. (1983). The acquisition of addition and subtraction concepts. In R. Lesh & M. Landau (Eds.), *The acquisition of mathematical concepts and processes* (pp. 7–14). New York: Academic Press.

Carpenter, T. P., Moser, J. M., & Romberg, T. A. (1982). *Addition and subtraction: A cognitive perspective.* Hillsdale, NJ: Lawrence Erlbaum.

Carter, B. Y., & Leinwand, S. J. (1987). Calculators and Connecticut's eighth-grade mastery test. *Arithmetic Teacher, 34*(6), 55-56. (ERIC Document No. EJ 348 074.)

Charles, R. I., & Lester, F. K. (1984). An evaluation of a process-oriented instructional program in mathematical problem solving in grades 5 and 7. *Journal for Research in Mathematics Education, 15,* 15–34.

Charles, R., Lester, F., & O'Daffer, P. (1987). *How to evaluate progress in problem solving.* Reston, VA: National Council of Teachers of Mathematics.

Clark, C. M., & Peterson, P. L. (1986). Teachers' thought processes. In M. C. Wittrock (Ed.), *Handbook of research on teaching* (3rd ed., pp. 255–296). New York, NY: Macmillan.

Clement, J. (1982). Algebra word problem solutions: Thought processes underlying common misconceptions. *Journal for Research in Mathematics Education, 13,* 16–30.

Coladarci, T. (1986). Accuracy of teacher judgments of student responses to standardized test items. *Journal of Educational Psychology, 78,* 141–146.

Collis, K. F. (1987). Levels of reasoning and the assessment of mathematical performance. In T. Romberg & D. Stewart (Eds.), *The monitoring of school mathematics: Background papers: Vol. 2. Implications from psychology; outcomes of instruction* (pp. 203–224). Madison: Wisconsin Center for Education Research.

Collis, K. F., Romberg, T. A., & Jurdak, M. E. (1986). A technique for assessing mathematical problem solving ability. *Journal for Research in Mathematics Education, 17,* 206–221.

Cronbach, L. J. (1984). *Essentials of psychological testing.* New York, NY: Harper & Row.

Cronbach, L. J., Gleser, G. C., Nanda, H., & Rajaratnam, N. (1972). *The dependability of behavioral measurements: The theory of generalizability for scores and profiles.* New York, NY: John Wiley.

Crooks, T. J. (1988). The impact of classroom evaluation practices on students. *Review of Educational Research, 58,* 438–481.

de Lange, J. (1987). *Mathematics, insight and meaning: Teaching, learning and testing of mathematics for the life and social sciences.* Unpublished doctoral dissertation, Rijksuniversiteit Utrecht, Netherlands.

Denmark, T. (1978). *Issues for consideration by mathematics educators: Selected papers.* Kent, OH: Research Council for Diagnostic and Prescriptive Mathematics.

Donovan, B. F., & Romberg, T. A. (1987). Knowledge structures and assessment of mathematical understanding. In T. Romberg & D. Stewart (Eds.), *The monitoring of school mathematics: Background papers: Vol. 2. Implications from psychology; outcomes of instruction.* Madison: Wisconsin Center for Education Research.

Dossey, J. A., Mullis, I. V. S., Lindquist, M. M., & Chambers, D. L. (1988). *The Mathematics report card. Are we measuring up?* Princeton, NJ: Educational Testing Service.

Easley, J. A., & Zwyoer, R. E. (1975). Teaching by listening: Toward a new day in math classes. *Contemporary Education, 47,* 19–25.

Ebel, R. L. (1973). Measurement and the teacher. In L. R. Aiken, Jr. (Ed.), *Readings in psychological and educational testing.* Boston, MA: Allyn and Bacon.

Educational Testing Service. (1987). *Learning by doing: A manual for teaching and assessing higher order thinking in science and mathematics.* Princeton, NJ: Author.

Fennema, E., & Carpenter, T. (1989). Cognitively guided instruction readings: Assessing children's thinking. In E. Fennema & T. Carpenter (Eds.), *Cognitively guided instruction: A program implementation guide* (pp. 31–45 in CGI readings). Madison: Wisconsin Center for Education Research.

Ferrara, S. F., & Thornton, S. J. (1988). Using NAEP for interstate comparisons: The beginnings of a "national achievement test" and "national curriculum." *Educational Evaluation and Policy Analysis, 10,* 200–211.

Fredericksen, N. (1984a). Implications of cognitive theory for instruction in problem solving. *Review of Educational Research, 54,* 363–407.

Fredericksen, N. (1984b). The real test bias: Influences of testing on teaching and learning. *American Psychologist, 39,* 193–202.

Ginsburg, A. L., Noell, J., & Plisko, V. W. (1988). Lessons from the wall chart. *Educational Evaluation and Policy Analysis, 10,* 1–12.

Glaser, R. (1963). Instructional technology and the measurement of learning outcomes: Some questions. *American Psychologist, 18,* 519–521.

Glaser, R. (1986). The integration of instruction and testing. In E. E. Freeman (Ed.), *The redesign of testing for the 21st century: Proceedings of the 1985 ETS Invitational Conference* (pp. 45–58). Princeton, NJ: Educational Testing Service.

Goetz, E. T., Hall, R. J., & Fetsco, T. G. (1990). Implications of cognitive psychology for assessment of academic skills. In R. W. Kamphaus & C. R. Reynolds (Eds.), *Handbook of psychological and educational assessment of children* (Vol. 1). New York, NY: Guilford Press.

Goldin, G. A., & McClintock, C. E. (Eds.). (1984). *Task variables in mathematical problem solving*. Philadelphia: Franklin Institute Press.

Greeno, J. G. (1978). A study of problem solving. In R. Glaser (Ed.), *Advances in instructional psychology* (Vol. 1, pp. 13–75). Hillsdale, NJ: Lawrence Erlbaum.

Guilford, J. P. (1965). *Fundamental statistics in psychology and education*. New York, NY: McGraw-Hill.

Gullickson, A. R. (1985). Student evaluation techniques and their relationship to grade and curriculum. *Journal of Educational Research, 79*, 96–100.

Gullickson, A. R. (1986). Teacher education and teacher-perceived needs in educational measurement and evaluation. *Journal of Educational Measurement, 23*, 347–354.

Haertel, E. (1985). Construct validity and criterion-referenced testing. *Review of Educational Research, 55*, 23–46.

Harmon, M. (1988, April). *Research into assessment. New instruments, new attitudes, or new paradigms?* Paper presented at the research presession to the annual meeting of the National Council of Teachers of Mathematics, Chicago.

Hart, K. (Ed.). (1981). *Children's understanding of mathematics: 11-16*. London: John Murray.

Hart, K. (1983). Tell me what you are doing – Discussions with teachers and children. In R. Hershkowitz (Ed.), *Proceedings of the Seventh International Conference for the Psychology of Mathematics Education* (pp. 19–37). Rehovot, Israel: Wizmann Institute of Science.

Hart, K. M., Brown, M., Kerslake, D., Kuchemann, D., Johnson, D., Ruddock, G., & McCartney, M. (1980). *Secondary school children's understanding of mathematics. A report of the mathematics component of the concepts in secondary mathematics and science programme*. London: Chelsea College of Science and Technology.

Hart, K., Brown, M., & Kuchemann, D. E. (1985). *Chelsea diagnostic tests*. London: NFER-Nelson.

Haylock, D. W. (1987). A framework for assessing mathematical creativity in school children. *Educational Studies in Mathematics, 18*, 59–74.

Hembree, R. (1987). Effects of noncontent variables on mathematics test performance. *Journal for Research in Mathematics Education, 18*, 197–214.

Jansson, L.C. (1984). Mathematics assessment in Canada—An overview. *Mathematics Teacher, 77*, 382–387.

Janvier, C. (1981). Use of situations in mathematics education. *Educational Studies in Mathematics, 12*, 113–122.

Johnson, E. (1988). *Considerations and techniques for the analysis of NAEP data*. Paper presented at the annual meeting of the American Educational Research Association, New Orleans, LA.

Joint Committee on Standards for Educational Evaluation. (1981). *Standards for evaluations of educational programs, projects and materials*. New York, NY: McGraw-Hill.

Kaput, J. (1986). Quantity structure of algebra word problems: A preliminary analysis. In G. Lappan & R. Even (Eds.), *Proceedings of the Eighth Annual Meeting of the North American Chapter of the International Group for the Psychology of Mathematics Education* (pp. 115–120). East Lansing, MI: Michigan State University.

Khoury, H. A., & Behr, M. (1982). Student performance, individual differences, and modes of representation. *Journal for Research in Mathematics Education, 13*, 3–15.

Kilpatrick, J. (1967). Analyzing the solution of word problems in mathematics: An exploratory study (Doctoral dissertation, Stanford University). *Dissertation Abstracts, 28*, 4380A.

Kintsch, W. (1986). Learning from text. *Cognition and Instruction, 3*, 87–108.

Koedinger, K. R. (1989). *Tapping geometry intuitions: The role of perceptual knowledge in geometry instruction*. Paper presented at the annual meeting of the American Educational Research Association, San Francisco.

Lee, J., Moreno, K., & Sympson, J. B. (1986). The effects of mode of test administration on test performance. *Educational and Psychological Measurement, 46*, 467–474.

Linn, R. L. (1983). Testing and instruction: Links and distinctions. *Journal of Educational Measurement, 20*, 179–189.

Livingston, S. A., & Zieky, M. J. (1982). *Passing scores: A manual for setting standards of performance on educational and occupational tests*. Princeton, NJ: Educational Testing Service.

Long, V. M., Reys, B., & Osterlind, S. J. (1989). Using calculators on achievement tests. *Mathematics Teacher, 82*, 318–325.

Lord, F. M., & Novick, M. R. (1968). *Statistical theories of mental test scores*. Reading, MA: Addison-Wesley.

Lucas, J. F. (1972). An exploratory study in the diagnostic teaching of elementary calculus. *Dissertation Abstracts International, 32*, 6825A. (University Microfilms No. 72–15, 368.)

Lyman, H. B. (1986). *Test scores and what they mean*. Englewood Cliffs, NJ: Prentice-Hall.

Marshall, S. P. (1983). Sex differences in mathematical errors: An analysis of distractor choices. *Journal for Research in Mathematics Education, 14*, 325–336.

McAloon, A., & Robinson, G. E. (1987). Assessing for learning: Using calculators in assessing mathematics achievement. *Arithmetic Teacher, 35*(2), 21–23.

McBride, J. R. (1985). Computerized adaptive testing. *Educational Leadership, 43*(2), 25–28.

McDonald, J., Beal, J., & Ayers, F. (1987). Computer administered testing: Diagnosis of addition computational skills in children. *Journal of Computers in Mathematics and Science Teaching, 7*, 38–43.

McKnight, C. C., Crosswhite, F. J., Dossey, J. A., Kifer, E., Swafford, J. O., Travers, K. J., & Cooney, T. J. (1987). *The underachieving curriculum: Assessing U.S. school mathematics from an international perspective*. Champaign, IL: Stipes Publishing Company.

McLean, L. (1987). *Emerging with honor from a dilemma inherent in the validation of educational achievement measures*. Paper presented at the annual meeting of the American Educational Research Association, Washington, DC.

Messick, S. (1981). Evidence and ethics in the evaluation of tests. *Educational Researcher, 10*(9), 9–20.

Miller, M. D. (1986). Time allocation and patterns of item response. *Journal of Educational Measurement, 23*, 147–156.

National Assessment of Educational Progress. (1981). *Mathematics objectives: Assessment* (No. 13-MA 10). Princeton, NJ: Educational Testing Service.

National Commission on Excellence in Education. (1983). *A nation at risk: The imperative for educational reform*. Washington, DC: U.S. Government Printing Office.

National Council of Teachers of Mathematics. (1989). *Curriculum and evaluation standards for school mathematics*. Reston, VA: National Council of Teachers of Mathematics.

Natriello, G. (1987). The impact of evaluation processes on students. *Educational Psychologist, 22*, 155–175.

Osborn, H. H. (1984). An experimental approach to mathematics core curriculum development. *Mathematics in School, 13*(5), 12–14.

Pandey, T. (1988). *Power items and the alignment of curriculum and assessment*. Paper presented at the annual meeting of the American Educational Research Association, New Orleans, LA.

Pandey, T. (1991). *A sampler of mathematics assessment*. A document prepared in cooperation with the Mathematics Assessment Development Team and the Mathematics Assessment Advisory Committee. Sacramento, CA: California Department of Education.

Patterson, J. H. (1989). *Impact of higher academic standards on youth at risk of academic failure*. Paper presented at the annual meeting of the American Educational Research Association, San Francisco.

Perkins, D. N., & Salomon, G. (1989). Are cognitive skills context-bound? *Educational Researcher, 18*(1), 16–25.

Pirie, S. (1988). *The formal assessment of investigational work and extended projects in mathematics*. Paper presented at the International Congress of Mathematics Education, Budapest, Hungary.

Quellmalz, E. S. (1985). Needed: Better methods for testing higher-order thinking skills. *Educational Leadership, 43*(2), 29–36.

Resnick, D. P. (1981). Testing in America: A supportive environment. *Phi Delta Kappan, 62*, 625–628.

Resnick, L. B. (1987). *Education and learning to think*. Washington, DC: National Academy Press.

Reys, R. E., Rybolt, J. F., Bestgen, B. J., & Wyatt, J. W. (1982). Processes used by good computational estimators. *Journal for Research in Mathematics Education, 13*, 183–201.

Romberg, T. A. (1987a). *The domain knowledge assessment strategy* (Working Paper 87-1, Report from the School Mathematics Monitoring Center). Madison: Wisconsin Center for Education Research.

Romberg, T. A. (1987b). Measures of mathematical achievement. In T. Romberg & D. Stewart (Eds.), *The monitoring of school mathematics: Background papers: Vol. 2. Implications from psychology; outcomes of instruction*. Madison: Wisconsin Center for Education Research.

Romberg, T. A., & Wilson, J. W. (1969). *The development of tests* (NLSMA Reports No. 7). Stanford, CA: School Mathematics Study Group.

Romberg, T. A., & Zarinnia, A. (1987). Consequences of the new world view to assessment of students' knowledge of mathematics. In T. Romberg & D. Stewart (Eds.), *The monitoring of school mathematics: Background papers: Vol. 2. Implications from psychology; outcomes of instruction* (pp. 153–202). Madison: Wisconsin Center for Education Research.

Romberg, T. A., Zarinnia, E. A., & Williams, S. R. (1989). *The influence of mandated testing on mathematics instruction: Grade 8 teachers' perceptions*. Madison: National Center for Research in Mathematical Science Education, University of Wisconsin.

Salmon-Cox, L. (1981). Teachers and standardized achievement tests: What's really happening? *Phi Delta Kappan, 62*, 631–634.

Schoenfeld, A. H. (1982). Measures of problem-solving performance and of problem solving instruction. *Journal for Research in Mathematics Education, 13*, 31–49.

School Mathematics Study Group. (1972). *Section I—An SMSG statement on objectives in mathematics education. Section II—Minimum goals for mathematics education*. Stanford, CA: Stanford University.

Sharpley, C. F., & Edgar, E. (1986). Teachers' ratings vs. standardized tests: An empirical investigation of agreement between two indices of achievement. *Psychology in the Schools, 23*, 106–111.

Silver, E. A. (1981). Recall of mathematical problem information: Solving related problems. *Journal for Research in Mathematics Education, 12*, 54–64.

Silver, E. A., & Kilpatrick, J. (1989). Testing mathematical problem solving. In R. Charles & E. Silver (Eds.), *Teaching and assessing mathematical problem solving* (pp. 178–186). Hillsdale, NJ: Lawrence Erlbaum.

Snow, R. E. (1989). Toward assessment of cognitive and conative structures in learning. *Educational Researcher, 18*(9), 8–14.

Sproull, L., & Zubrow, D. (1981). Standardized testing from the administrative perspective. *Phi Delta Kappan, 62*, 628–631.

Stenmark, J. K. (1989). *Assessment alternatives in mathematics: An overview of assessment techniques that promote learning*. Prepared by the EQUALS staff and the Assessment Committee of the California Mathematics Council *Campaign for Mathematics*. Berkeley, CA: Regents, University of California.

Stiggins, R. J. (1985). Improving assessment where it means the most: In the classroom. *Educational Leadership, 43*(2), 69–74.

Stiggins, R. J. (1988). Revitalizing classroom assessment: The highest instructional priority. *Phi Delta Kappan, 69*, 363–372.

Stiggins, R. J., & Bridgeford, N. J. (1985). The ecology of classroom assessment. *Journal of Educational Measurement, 22*, 271–286.

Stiggins, R. J., Conklin, N. F., & Bridgeford, N. J. (1986). Classroom assessment: A key to effective education. *Educational Measurement: Issues and Practice, 5*(2), 5–17.

Swan, M. (1987). *The language of functions and graphs*. Nottingham, England: Shell Centre for Mathematics Education.

Thoman, J., & Moser, J. (1984). *Competencies for mathematics shelf tests. Wisconsin competency based testing program*. Madison: Wisconsin Department of Public Instruction.

Thompson, A. G., & Briars, D. J. (1989). Assessing students' learning to inform teaching: The message in the NCTM evaluation standards. *Arithmetic Teacher, 37*(4), 22–26.

Vergnaud, G. (1982). Cognitive and developmental psychology and research in mathematics education: Some theoretical and methodological issues *For the Learning of Mathematics, 3*(2), 31–41.

Vergnaud, G. (1983). Multiplicative structures. In R. Lesh & M. Landau (Eds.), *Acquisition of mathematics concepts and processes* (pp. 127–174). New York, NY: Academic Press.

Wainer, H. (1986). Can a test be too reliable? *Journal of Educational Measurement, 23*, 171–173.

Wainer, H., & Kiely, G. L. (1987). Item clusters and computerized adaptive testing. A case for testlets, *Journal of Educational Measurement, 24*, 185–201.

Wearne, D. C., & Romberg, T. A. (1977). *DMP accountability test* (Working Paper No. 217). Madison: Wisconsin Research and Development Center.

Weaver, J. F. (1970). Evaluation and the classroom teacher. In E. G. Begle (Ed.), *Mathematics education: Sixty-ninth yearbook of the National Society for the Study of Education, Part 1* (pp. 367–404). Chicago, IL: University of Chicago Press.

Webb, N. L. (1975). An exploration of mathematical problem-solving processes (Doctoral dissertation, Stanford University). *Dissertation Abstract International, 36*, 2689A.

Webb, N. M., Gold, K., & Qi, S. (1989). *Mathematical problem-solving processes and performance: Translation among symbolic representations*. Paper presented at the annual meeting of the American Educational Research Association, San Francisco.

Wilson, J. W., Cahen, L. S., & Begle, E. G. (1968). *NLSMA reports: No. 2, Part A: Y-population test batteries*. Stanford, CA: School Mathematics Study Group.

Wisconsin Department of Public Instruction. (1985). *Wisconsin pupil assessment program. Mathematics problem solving–eighth grade*. Madison: Author.

Wise, S. L., & Plake, B. S. (1989). Research on the effects of administering tests via computers. *Educational Measurement: Issues and Practice, 8*(3), 5–10.

Wood, R. (1986). The agenda for educational measurement. In D. L. Nuttall (Ed.), *Assessing educational achievement* (pp. 185–203). London, U.K.: Falmer Press.

Wood, R. (1987). *Measurement and assessment in education and psychology*. London: The Falmer Press.

Zehavi, N., Bruckheimer, M., & Ben-Zvi, R. (1988). Effect of assignment projects on students' mathematical activity. *Journal for Research in Mathematics Education, 19*, 421–438.

PERSPECTIVES

27

INTERNATIONAL STUDIES OF ACHIEVEMENT IN MATHEMATICS

David F. Robitaille
UNIVERSITY OF BRITISH COLUMBIA

Kenneth J. Travers
UNIVERSITY OF ILLINOIS AT URBANA-CHAMPAIGN

Many journal articles, papers at scholarly conferences, as well as books and monographs have considered the results of international studies in mathematics education over the past several years. The popular media have reflected this interest, as have government agencies which provide significant amounts of funding for international, comparative studies of various kinds.

More attention appears to have been given to international comparisons in mathematics than in other areas of the curriculum. There are several likely reasons for this. First, mathematics plays a prominent part in the curriculum of every country, usually second in importance only to that of the mother tongue. Second, there is a great deal of similarity of content in mathematics curricula internationally. Third, the language of mathematics is, in many ways, truly international. Whether one observes a mathematics class in Beijing or Chicago, one can usually grasp the major elements of the lesson being presented fairly easily because of the universality of mathematical symbolism and notation.

In view of the widespread interest in cross-national studies, the large expenditures involved in conducting them, and the kinds of conclusions that have been drawn from them, an examination of international studies in mathematics education seems warranted. The goal of this chapter is to provide an analysis of the major international studies of mathematics achievement and to highlight the major findings of those studies. The first part of the chapter deals with some of the arguments for and against such projects. The main body of the chapter consists of reviews of international studies, and concludes with a summary of the major findings and implications of this research along with a number of suggestions which investigators might wish to consider in future projects of this kind.

THE CASE FOR INTERNATIONAL STUDIES IN MATHEMATICS

Collaborative, international research studies in education, particularly those involving several countries, are a fairly recent phenomenon. The first International Association for the Evaluation of Educational Achievement (IEA) mathematics study, carried out in the early 1960s, was the first large-scale project of this kind. However, there are differences of opinion as to why such studies had not been conducted earlier and what made the IEA project viable.

Inkeles (1977) listed the following factors that provided an environment hospitable to the idea of comparative, international surveys: strong criticism of the American school system during the late 1950s and early 1960s, the launching of the first earth satellite by the Soviet Union, and widespread concern in the industrialized countries about the ever-escalating costs of providing free, universal public education. Education was

The authors gratefully acknowledge the helpful comments provided by M. David Miller, University of Florida, and Harold W. Stevenson, University of Michigan.

being increasingly viewed as an industry, with an increasing interest in applying principles of accountability to the field of education. Comparisons were being made with other areas of public expenditure and, more frequently, with the educational systems in other countries.

Husén(1967), writing in the introduction to the two-volume report of the first IEA mathematics study, minimizes the importance of "post-Sputnik reactions" as an enabling factor in promoting international comparative studies in education. He emphasizes, instead, the need to obtain empirical evidence of the "productivity" of educational systems, thereby agreeing with Inkeles.

Theisen, Achola, and Boakari (1983) identify five factors to explain the almost total lack of comparative international achievement data prior to the first IEA survey. They note that international studies are extremely expensive to carry out, and that cultural differences make the development of valid instruments difficult. Students in some countries might not be familiar with the testing procedures employed; political considerations might make it difficult to obtain approval for participation; and comparability might be compromised by variability in test-administration procedures.

Eckstein (1982) notes that in the 19th century many comparative studies of achievement were carried out by observers traveling abroad. These studies were "limited severely by the data available and the lack of research sophistication of their times." He identifies many of the same factors as Theisen and colleagues as accounting for the late emergence of such studies, and notes that the growth of international organizations such as the United Nations Educational, Scientific, and Cultural Organization (UNESCO) and the Organization for Economic Cooperation and Development (OECD) in the post-war period enhanced the possibility of carrying out such studies.

Before the first IEA study, the few cross-national studies that had been carried out had generally avoided focusing on comparative analyses of curricula, of performance of students, or of teaching methods because of the kinds of methodological and substantive complexities referred to above. Instead, they concentrated on system-level variables such as retention rates, teacher-pupil ratios, and per-pupil expenditures, which were more readily available and, at least superficially, more easily comparable. Husén (1967) described such studies "as a first crude stage of empiricism in comparative education" because the variables utilized consisted mainly of inputs to the educational system. He said that IEA had set as one of its goals the construction of "relevant evaluation instruments," that is, output measures.

Over the past 30 years, the number and the nature of the variables included in comparative studies of achievement have continued to expand. The Second International Mathematics Study, which was conducted by IEA in the early 1980s, included an extensive curriculum analysis, an investigation of teaching practices, and a longitudinal study of growth in students' achievement over the course of the school year.

International studies have both advocates and opponents. Some writers have directed their criticisms at particular projects, while others have commented on the more fundamental issue of the rationale for and the value to be derived from such projects in general. In the following section, some of the arguments for and against such studies are summarized.

Value of International Studies

Critics of international surveys of achievement have noted that such studies require large expenditures of time, energy, and resources on the part of participants as well as investigators. Questions have also been raised about the significance and the influence of the findings of such studies.

Given the degree of concern which exists at the present time about the need to be competitive internationally and the importance ascribed to education in that respect, there are powerful political reasons for knowing more about education in other countries. We need to know about the educational systems of other countries to ascertain their strengths and weaknesses. We need to separate fact from fiction about what school systems in other countries are able to accomplish with their students.

With respect to the practical value obtained by participating in international studies, Husén has described how IEA studies use the "world as an educational laboratory." International studies provide investigators with an opportunity to assess the importance of variables which might not be available for study in their own jurisdictions, for example, class size, single-sex schools, and out-of-school tutoring (Eckstein, 1982). Different approaches to the same goal can be compared, and their effects evaluated. Such studies provide a view of what can be accomplished, a context in which decision makers in each participating country can view their own system.

The efficacy of an educational innovation may be evaluated by examining its implementation and operation in another country. New approaches to teaching may be found by observing classroom practices of teachers from other countries. Stigler and Perry (1988) suggest that:

Cross cultural comparison also leads researchers and educators to a more explicit understanding of their own implicit theories about how children learn mathematics. Without comparison, we tend not to question our own traditional teaching practices and we may not even be aware of the choices we have made in constructing the educational process.

The IEA studies have reported on cross-national differences with regard to participation or retention rates, the extent of tracking or streaming, gender differences in achievement in mathematics, the content of the curriculum, and teaching practices. Many findings from these studies challenge the conventional wisdom which underlies educational practice in North America and elsewhere. They demonstrate that substantially different approaches to the teaching and learning of mathematics are not only viable, but demonstrably successful.

The focus of international studies in the IEA tradition is on trying to determine the effects of particular variables on teaching and learning in the context of international comparisons. For example, it is important for North American educators to learn that 13-year-old students in France and Japan are not grouped by some measure of ability for instruction in mathematics, and yet their achievement is among the highest internationally. A number of such important variables have been identified in international studies over the past 25 years. Course content, or opportunity to learn, has been shown to be an important variable in accounting for differences in achievement among students from different countries. But opportunity to learn also provides information on what students at a given age or grade level can learn.

International studies have shown that educational systems can retain very large proportions of the age cohort in the study of academic mathematics at the senior secondary level without penalizing their best students. Secondary analyses of the data from both the first and second IEA studies indicate that the highest ability students from almost all systems perform at about the same level on topics which they all have studied. Differences in achievement among the best students appear to be mainly a function of opportunity to learn.

The second IEA mathematics study included curriculum as a variable and demonstrated the central importance of that variable in accounting for differences in achievement across countries. Three levels of the mathematics curriculum—intended, implemented, and attained—were delineated, and large differences were found among them. This articulated approach to curriculum has been adopted by others, and has contributed to our understanding of what goes on in mathematics classrooms.

The entire SIMS project was curriculum-based, and emphasized the importance of viewing student success in mathematics within the contextual framework of variables such as opportunity to learn. If we are to understand differences in mathematical performance internationally, the impact of curriculum as a variable must be recognized and taken into account.

Drawing Appropriate Comparisons

International comparisons must be made judiciously. When data are aggregated to the national level, a great deal of information about sources of variation is either eliminated or obscured. For example, grade 8 students in many medium and large sized schools in the United States are tracked into four different mathematics courses each of which is labeled as grade 8 mathematics. Given the differences among these courses, it is difficult to make sense out of a statistic such as the mean achievement score in grade 8 algebra in the United States, because many of the students would have studied little or no algebra in their course. Theisen and colleagues (1983) express this concern in the following way:

National aggregations of data that focus on explaining [national] profiles of educational achievement often fall short of accounting for within-country factors that may affect standard measures of scholastic performance. Yet these are precisely the variables with the most policy relevance for enhancing school learning.

Comparing the Incomparable

Torsten Husén (1983) has said, "Comparing the outcomes of learning in different countries is in several respects an exercise in comparing the incomparable." Later he added that unless the impacts of differences in objectives and curricula, to say nothing of cultural and socioeconomic differences, are taken into account in some way, international comparisons are at best meaningless, and at worst odious.

Whatever the precise nature of the international investigation, drawing comparisons among the outcomes achieved by students in different educational systems in different countries

is a difficult task: that is to say, a task that is difficult to accomplish well. Critics such as Freudenthal (1975), commenting on the first IEA mathematics study, have been strongly critical of the selection of items used, the inappropriateness of many of the topics tested with respect to the curricula of some of the countries participating in the study, and the failure to take into account whether or not students had been taught the content required to respond to a given item. Opportunity to learn has proved to be powerful in accounting for between-country differences in achievement.

Achievement results from international studies are commonly used to rank countries and their educational systems. Such comparisons are inevitable and, to the extent that they raise questions about possible sources of variation in achievement across national systems, their importance should not be downplayed. On the other hand, such comparisons need to be accompanied by a thorough analysis of the variables which contribute to those differences.

THE IEA SURVEYS

Until recently, IEA was the only organization conducting large-scale international surveys in education. Two of those surveys have dealt exclusively with mathematics: the First International Mathematics Study (FIMS) in 1964, and the Second International Mathematics Study (SIMS) in 1980–82. Those two projects are reviewed in the following sections.

IEA as an Organization

IEA is a consortium of centers of educational research from more than 50 countries. The membership list includes institutions from a large number of industrialized countries, and some developing countries. The two main goals of IEA are to conduct cooperative international research studies in education, and to contribute to the development of research expertise internationally.

Established in 1960 (Postlethwaite, 1971), IEA first carried out a survey of mathematics achievement; the project has since become known as the First International Mathematics Study. In addition to that project, IEA carried out the Six-Subject Survey (Walker, 1976), the Second International Mathematics Study (Robitaille & Garden, 1989; Travers & Westbury, 1990), the Second International Science Study (IEA, 1988), the Classroom Environment Study (Ryan & Anderson, 1984), and the Written Composition Study (Gorman, Purves, & Degenhart, 1988). Currently, IEA has projects underway in computer applications in education, reading literacy, pre-primary education, and values in education. A third study of mathematics and science is planned for the 1990s.

The First International Mathematics Study

The choice of mathematics for IEA's first international survey was dictated, not so much by the fact that the researchers involved were particularly interested in issues surrounding the teaching and learning of mathematics, but more for convenience. The organizers believed that it would be easier to make international comparisons in mathematics than in any other area of the curriculum. In describing the study, Postlethwaite (1971) says, "Although mathematics was chosen, many of the hypotheses treated mathematics achievement as a surrogate for general school achievement." The project was designed primarily as a comparative investigation of the outcomes of schooling, with achievement in mathematics serving as the dependent variable.

Other factors also influenced the selection of mathematics as the focus of the study. First, "most countries involved in the project were concerned with improving their scientific and technical education" (Postlethwaite, 1971), and mathematics was seen to be a crucial part of such programs. Second, the "new math" movement was beginning to make its influence felt, and there was a good deal of interest in comparing what was being done to implement new curricula in different countries. Third, there was considerable interest in the report of the Royaumont Conference (O.E.E.C., 1961) and in the surveys which had been conducted in several countries in Europe and North America in preparation for that conference. That conference brought together eminent mathematicians and educators to discuss recommendations for the renewal of the mathematics curriculum. Those recommendations exerted a powerful influence on the "new math" movement of the 1960s and 1970s. See Hirstein (1980) for a summary of findings.

The first IEA mathematics study was a very ambitious project for its time. Member institutions from 12 countries agreed to participate in that project: Australia, Belgium, England, the Federal Republic of Germany, France, Finland, Israel, Japan, the Netherlands, Scotland, Sweden, and the United States. Almost all participating countries were from Europe and were more or less highly industrialized.

Design of the Study. Two populations of students were identified. The younger population consisted of 13-year olds; the older, of students in the last year of secondary school. Each country was expected to select nationally representative samples of students at each level, under the general direction of an international sampling referee.

The formal definitions of the populations were as follows (Population 2 was defined but the data were never analyzed):

Population 1a—All students who were 13:0–13:11 at the date of the testing.
Population 1b—All students at the grade level where the majority of students of age 13:0–13:11 were to be found.
Population 3—All students who were in the grades (forms) of full-time study in schools from which the universities or equivalent institutions of higher learning normally recruited their students.

Population 3 was partitioned into two sub-populations:

Population 3a—Those studying mathematics as an integral part of their course for their future training, or as part of the pre-university studies, for example, mathematicians, physicists, engineers, biologists, and so on, or all those being examined at that level.
Population 3b—Those studying mathematics as a complementary part of their studies, and the remainder. (Husén, 1967, Vol. 1, p. 46)

Achievement tests were constructed on the basis of a collaborative, international effort. Participating countries were canvassed regarding their mathematics curricula, and this information was utilized to construct a two-dimensional item-specification grid for each grade. The first dimension of the grid consisted of content topics to be included in the test items. The second dimension consisted of five cognitive behavior levels which the items would be designed to invoke (Husén, 1967, Vol. 1, p. 94):

1. Knowledge and information: recall of definitions, notation, concepts.
2. Techniques and skills: solutions.
3. Translation of data into symbols or schema and vice versa.
4. Comprehension: capacity to analyze problems, to follow reasoning.
5. Inventiveness: reasoning creatively in mathematics.

The topics selected for each grade were assigned national importance weightings, and these were used to determine the numbers of items to be included for each topic at each of the five cognitive behavior levels defined. Several hundred items were contributed to the pool, and pilot testing to select those to be included on the final forms was carried out.

The researchers considered two possible approaches to the construction of the cognitive tests. One alternative was to prepare, for each country, a set of items based on its own curriculum and taking into account variations in that curriculum within the country for different types of schools and students. This would have made the tests "fair" for all students, but would have rendered the goal of drawing comparisons among countries unreachable.

The second approach, the one utilized in the study, was to construct a pool of achievement items which students in all participating countries would take. The goal was to put together a set of items for each population that was "as free as possible from dependence on particular national organizational features, both as to grouping of students and the grade placement and sequence of mathematical topics" (Husén, 1967).

The structure of the final tests is summarized in Table 27.1. Tests A, B, and C were administered to students in Populations 1a and 1b: that is, to 13-year olds. Tests 4, 5, and 6 were administered to senior secondary students in Population 3b, the non-mathematics students. Tests 5, 7, 8, and 9 were administered to senior secondary students in Population 3a, the mathematics specialists. The report of the study makes reference to a population of 15- and 16-year olds, but no results for that population are presented.

Each test booklet included a small number, usually two or three, of open-ended items, but the vast majority were multiple choice. All items were scored on a correct-incorrect basis only. No credit was awarded for partial solutions on the open-ended items.

Today it is common to use some form of matrix sampling for administration of achievement tests in large-scale surveys, so that wide coverage of topics can be realized without requiring each student to write every item. Those techniques were not employed in FIMS. Instead, each student was required to complete either three or four test booklets, usually over a period of several days.

The fact that the number of items in the item pool was rather small means that there were very few items corresponding to

TABLE 27–1. Number of Test Items by Topic: FIMS

Topic	Test Form									Total
	A	B	C	4	5	6	7	8	9	
Basic arithmetic	6	6	3							15
Advanced arithmetic	6	7	5	2	3	4				27
Elementary algebra	5	4	3	1	1					14
Intermediate algebra	1	2	1	7	7	4	8	4		34
Euclidean geometry	5	2	6	6	5	2				26
Analytical geometry		1			3	1	1	1	3	10
Sets		2	2	1		2	3	1		11
Trigonometry					1	2		2		5
Analysis						1	1	4	3	9
Calculus									9	9
Probability						1		1		2
Logic				1	1		4	3		9
Affine geometry			3							3
TOTAL	23	24	23	18	21	17	17	16	15	174

Adapted from Husén, 1967, Vol. 1, p. 105.

any specific topic. Each student in Populations 1a and 1b wrote a total of 70 items. Students in Population 3a wrote 69 items, and those in Population 3b wrote 58 items. A major implication of this is that interpretation of the results for levels other than the whole test or for anything other than specific items is somewhat problematic, except for a number of subtests on broadly defined topics such as those listed in the first column of Table 27.1.

In addition to the achievement tests, several descriptive and attitude scales were developed for the study, in keeping with IEA's view of attitude development as one of the important outcomes of schooling. Five attitude scales were developed for use in the study. They were titled Mathematics as a Process, Difficulties of Learning Mathematics, The Place of Mathematics in Society, School and School Learning, and Man and His Environment.

Several questionnaires were also developed for use in the project. These were completed by students, by teachers, by school principals, and by "an expert on the educational system" (Postlethwaite, 1971) of the country. Students were asked for basic demographic and personal information, about their school work, their career aspirations, and their parents. Teachers were asked to supply information about their qualifications, experience, and education: both pre-service and in-service. They were also asked about their familiarity with the new mathematics curriculum, and about how well they expected their students to perform on each test item. This last item was designed to learn how differences in curricula among the participating countries would affect the performance of students.

Major Findings. The results summarized in Table 27.2 indicate that all groups of students from all of the participating countries found the tests difficult. The mean scores (percent correct) for Populations 1a and 1b were all below 50%; in fact, the majority were below 40%. Similarly low scores occurred at the senior level, although three of them did exceed 50% by narrow margins.

The study produced a number of findings of particular interest to mathematics educators, such as:

1. Males outperformed females in both age groups, and this was in line with findings from other achievement surveys in the 1960s and 1970s. Husén (1967, Volume 2, pp. 240–241) summarized these findings by saying that, "Sex was related to mathematics achievement in almost all countries, the boys scoring higher than the girls at all levels and especially in the terminal non-mathematics population (3b)." More recent studies have shown that these achievement differences between males and females have narrowed significantly in many countries over the last 20 years.

2. For Populations 1a and 1b, parents' level of education was found to be positively correlated with students' achievement. Among older students there was much less variability with regard to this variable: Parents of students at this level tended to be much more homogeneous with regard to level of education and occupational category. This finding is similar to results from many studies which have found a significant correlation between students' achievement and their parents' socioeconomic status.

3. Data on students' attitudes toward mathematics indicated that 13-year olds in all of the participating countries had a more positive view of mathematics as a process than did the senior students. The authors suggested that the more positive views among younger students might have been due to the influence of the "new math" curricula which had recently been introduced.

Similar trends were noted in the second IEA mathematics study, and more research is needed to investigate the reasons for this decline with age. On the one hand, the results may simply reflect a more "realistic" view on the part of older students of the importance of mathematics to their current and future plans or toward the nature of school mathematics itself. On the other hand, and this is more troubling, these findings may be an accurate reflection of

TABLE 27–2. Achievement Test Scores — FIMS

| Country | Mean of Raw Scores (Standard Deviation) | | | |
| | Pop. 1a | Pop. 1b | Pop. 3a | Pop. 3b |
	70 items	70 items	69 items	58 items
Australia	20 (14)	19 (12)	22 (10)	—[a]
Belgium	28 (15)	30 (14)	35 (13)	24 (10)
England	19 (17)	24 (18)	35 (13)	21 (10)
Finland	—[b]	—[b]	25 (10)	22 (8)
France	18 (12)	21 (13)	33 (11)	26 (10)
Germany, Federal Republic	—[a]	25 (12)	29 (10)	28 (8)
Israel	—[a]	32 (15)	36 (9)	—[a]
Japan	31 (17)	31 (17)	31 (15)	25 (14)
Netherlands	24 (16)	21 (12)	32 (8)	25 (10)
Scotland	19 (15)	22 (16)	26 (10)	21 (10)
Sweden	16 (11)	15 (11)	27 (12)	13 (6)
United States	16 (13)	18 (13)	14 (13)	8 (9)
Overall	20 (15)	23 (15)	26 (14)	21 (13)

Adapted from Husén (1967) Volume II, pp. 22–25.

[a] Country did not supply data for this population.

[b] Note that the Finnish data for Populations 1a and 1b have been omitted from this table because they were found to be in error as the international reports were being printed.

a decline in students' interest in and attitude toward the continued study of mathematics.

4. Senior students tended to rate the importance of the role of mathematics in contemporary society less highly than did the 13-year olds, and this effect was most pronounced in English-speaking countries. While this finding is surprising and somewhat disturbing at first glance, it may be that senior students were indicating that they did not see very much in the way of applications of the mathematics they were studying in school to everyday life. That would seem to be a realistic view considering the nature of many of the topics included in the curriculum at the time.

5. Results on the relationship between class size and achievement were conflicting. At the Population 3 level, smaller classes tended to have higher achievement. This pattern was reversed for Population 1. Postlethwaite (1971) suggested, "There were so many complicating factors in this study that it was almost impossible to separate out the effect of class size from the others, especially since most of them were not under control." He concluded that it was unlikely that merely reducing class size would result in an improvement in students' level of performance. Similar results have been reported from other international surveys.

6. Many of the European countries participating in FIMS had highly selective school systems which retained only small proportions of students until the completion of secondary school, and even fewer in pre-university mathematics courses. Belgium, England, Finland, and Scotland all reported having less than 20% of the age cohort in school at the Population 3 level. Retention in Sweden was the highest among European participants at 23%. In the United States, on the other hand, 70% of the age cohort was reported to be still in school.

7. The importance of retentivity as a factor in accounting for differences in achievement between countries was underscored by Peaker (1969). On the basis of his analysis he

concluded, "...retentivity accounts for the major part of the variation between the countries, on all items. When the [Population] 3a scores are regressed against the [Population] 1a scores...retentivity accounts for most of the variation in total scores."

8. Analysis of the achievement data indicated that the most able students from all countries performed at similarly high levels. Good students from countries with comprehensive school systems and high rates of retention did as well as their peers from countries with much more selective systems. A related finding showed that "the higher the proportion of students retained in the school system, the higher the proportion of the age group making high scores is likely to be" (Postlethwaite, 1971). Similar analyses were conducted on the SIMS data, and those findings are discussed later.

9. A positive relationship was found between students' achievement on a given item and their opportunity to learn the mathematics needed to respond to that item correctly. The opportunity to learn variable was an IEA innovation and, although there may be room for criticism for the way in which it was operationalized in FIMS, the fact that its importance as an explanatory variable was recognized by the researchers is worthy of note.

10. Results on the value of homework were difficult to interpret, as they often are in surveys of this kind. The term "homework" itself is open to a wide variety of interpretations, and the amount of time spent on homework activities must surely be related to students' ability.

Comments. The two-volume international report of IEA's first mathematics study was published in 1967, three years after the data were collected. The first comprehensive critique of the study and its findings appeared in March, 1971, as a special issue of the *Journal for Research in Mathematics Education* (Wilson & Peaker, 1971). Other papers were published over the next several years, most of them appearing after the publication of the reports of IEA's Six-Subject Survey (Walker, 1976).

The First International Mathematics Study was the first of its kind. The findings contributed to the development of a better understanding of the immense variability between countries on a large number of variables that are important to the teaching and learning of mathematics. They also demonstrated that there are significant differences in the levels at which students in different countries achieve.

The achievement results made headlines in the popular press and caused considerable controversy. Several of the papers which appeared in the special issue of JRME refer to the achievement results and raise concerns about the validity of comparisons between systems with such widely different curricula and retention rates, particularly at the senior secondary level. The researchers themselves seemed to be somewhat ambivalent about how to handle this aspect of the study. Although Husén (1967) states very clearly that IEA was not intending to conduct a "horse race," the achievement results are presented at the very beginning of Volume II without any mitigating evidence about the effects of variables such as opportunity to learn and participation rates. Postlethwaite (1971), on the other hand, in his JRME paper which summarizes the study, never refers to the achievement results at all. He confines his commentary to the relational analyses of the links between achievement and a number of explanatory variables.

Kilpatrick (1971) makes the very important point that, from a mathematics educator's viewpoint, the most interesting aspects of the study are to be found in the item-level results. Unfortunately, while the items and the results are included in Volume II, they are relegated to an appendix unaccompanied by any commentary. The study might have generated more interest and secondary analyses by mathematics educators if item-level statistics had been accorded more prominence.

The Second International Mathematics Study

After completion of FIMS and the subsequent Six-Subject Survey, IEA went through a period of several years in which no major new projects were undertaken. In the late 1970s, mainly at the instigation of Roy Phillipps of the New Zealand Department of Education, a decision was made to launch a second study of mathematics. This project has come to be known as the Second International Mathematics Study (SIMS).

The goals established for SIMS were more ambitious, and its composition was fundamentally different from that of FIMS. Unlike the first study, SIMS was intended to be a study of mathematics education internationally, with significant input and guidance at every stage from the mathematics education community. Garden (1987) says that the overall goal of the Second International Mathematics Study was "to produce an international portrait of mathematics education with a particular focus on the mathematics classroom."

Design of the Study. SIMS was designed to be an in-depth study of the mathematics curriculum at three levels, as shown in Figure 27.1: the Intended Curriculum, the Implemented Curriculum, and the Attained Curriculum (Travers, Garden, & Rosier, 1989). The Intended Curriculum is the curriculum as

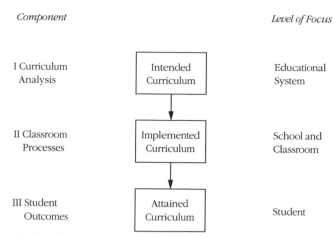

FIGURE 27–1. Framework for SIMS.

defined at the national or system level, as codified in curriculum guides, and as operationalized in textbooks approved for teachers' use. The Implemented Curriculum is the curriculum as it is taught by teachers in their own classrooms. The Attained Curriculum is what is learned by students and manifested in their achievement and attitudes.

Two populations of students, roughly comparable to those in the first study, were targeted for participation in SIMS. They were designated Populations A and B and defined as follows (Robitaille & Garden, 1989, pp. 6–7):

Population A: All students in the grade (year, level) where the majority have attained the age of 13.00 to 13.11 years by the middle of the school year.

Population B: All students who are in the normally accepted terminal grade of the secondary education system and who are studying mathematics as a substantial part of their academic program.

The widely different makeup of Population B in participating countries caused a number of difficulties, and made interpretation of the achievement results at that level difficult. As in the case of the first study, the educational systems in a number of countries were found to be much more selective than others, and to retain relatively few students in the study of pre-university mathematics. Some places, such as Canada (Ontario) and Scotland, have two grade levels which can be regarded as terminal grades for secondary education. Population B in Hungary was not comparable with those in other systems because all students at this grade level are expected to take "substantial" mathematics courses, including calculus. Basic data on the makeup of Population B internationally are presented in Table 27.3.

A commonly expressed concern about survey research methodology (for example, Walberg, Harnisch, & Tsai, 1986) is that surveys are typically cross-sectional. That is, data are collected on one occasion, and it is not possible to make causal attributions or comparisons about the extent or rate of change. In a cross-sectional survey of mathematics such as FIMS, one can talk about students' achievement after a given number of years, but one cannot compare rates of growth or link teachers' pedagogical practices to the achievement of their students.

TABLE 27–3. Makeup of Population B: SIMS

	Mid-year modal age (years)	Population B as percentage of age group	Population B as percentage of grade group	Percentage of age group in school
Belgium (Flemish)	17	10	30	65
Belgium (French)	17	10	30	65
Canada (B.C.)	17	30	38	82
Canada (Ontario)	18	19	55	33
England & Wales	17	6	35	17
Finland	18	15	38	59
Hong Kong	18	6	—[a]	—[a]
Hungary	17	50	100	50
Israel	17	6	10	60
Japan	17	12	13	92
New Zealand	17	11	67	17
Scotland	16	18	42	43
Sweden	18	12	50	24
Thailand	18	—[a]	—[a]	—[a]
United States	17	12	15	82

[a] Data not available.

From Robitaille and Garden (1989, p. 126).

In order to make it possible for attributional analyses to be carried out, and for growth in mathematics achievement to be charted, the full design of the Second International Mathematics Study was longitudinal. Pretests were administered at the beginning of the school year and post-tests at the end. Monitoring of teaching methods during the school year was accomplished through the administration of questionnaires to teachers about the practices they employed in teaching particular topics. The basic or reduced version of the study was cross-sectional, with almost all of the data being collected toward the end of the school year.

Students, teachers, and school administrators from a wide range of countries participated in one or more aspects of the study. In most cases, countries were represented by a single, national probability sample; however, there were several exceptions. Belgium was represented by separate Flemish and French samples; Scotland participated independently of England and Wales; and the Canadian provinces of British Columbia and Ontario participated independently of one another. Table 27.4 lists participants in either the cross-sectional or the longitudinal versions of the study at both population levels.

Based on the results of a preliminary survey of mathematics curricula in several countries, a content-by-cognitive-behavior grid was developed for each population. The cognitive behavior dimension of the grids was partitioned into four categories after a scheme proposed by Wilson (1971): Computation, Comprehension, Application, and Analysis. The second dimension of the Population A grid, the content dimension, was subdivided into five strands: Arithmetic, Algebra, Geometry, Descriptive Statistics, and Measurement. Population B content was subdivided into nine strands: Sets and Relations, Number Systems, Algebra, Geometry, Elementary Functions and Calculus, Probability and Statistics, Finite Mathematics, Computer Science, and Logic.

Items were collected from a variety of sources in several countries, some were developed especially for use in this study, and a number of so-called "anchor items" from the first IEA study were used to chart changes over time in those 11 countries that took part in both the first and second studies. Population A students, in both the longitudinal and cross-sectional versions of the study, were administered a core form and one of four rotated forms. Population B students wrote either one or two of eight rotated forms constructed at that level. The content of the achievement tests is summarized in Tables 27.5 and 27.6.

Major Findings. The SIMS data were collected over a two-year period from 1980 to 1982, and the international reports appeared several years later. The findings summarized here are

TABLE 27–4. Countries Participating in SIMS

Country	Cross-Sectional		Longitudinal	
	Pop. A	Pop. B	Pop. A	Pop. B
Belgium (Flemish)	X	X	X	
Belgium (French)	X	X		
Canada (B.C.)	X	X	X	X
Canada (Ontario)	X	X	X	X
England and Wales	X	X		
Finland	X	X		
France	X		X	
Hong Kong	X	X		
Hungary	X	X		
Israel	X	X		
Japan	X	X	X	
Luxembourg	X			
Netherlands	X			
New Zealand	X	X	X	
Nigeria	X			
Scotland	X	X		
Swaziland	X			
Sweden	X	X		
Thailand	X	X	X	
United States	X	X	X	X

TABLE 27–5. Structure of the SIMS Item Pool — Population A

Major Strands	Numbers of Items (Percent)		
	Longitudinal Version	Cross-Sectional Version	Common to both Versions
Arithmetic	62 (34%)	46 (25%)	46 (29%)
Algebra	32 (18%)	40 (23%)	30 (18%)
Geometry	42 (23%)	48 (27%)	39 (25%)
Measurement	26 (14%)	24 (14%)	24 (15%)
Descriptive Statistics	18 (10%)	18 (10%)	18 (11%)
TOTAL	180	176	157

subdivided into three groups corresponding to the three major levels of the study: the national level, the school or teacher level, and the student level.

FINDINGS AT THE COUNTRY LEVEL. One of the major national-level findings is that wide discrepancies between countries continue to exist in the degree of opportunity provided to students to complete secondary school to the grade 12 level or equivalent. These differences are to be found not only between developing and developed countries, but within the group of developed countries.

Some countries have seen dramatic growth in this measure of educational opportunity whereas in others the growth has been much more modest. For example, the proportion of the age cohort completing secondary school in Belgium went from 13% to 65%, and in Finland from 14% to 59%, between the two IEA mathematics studies. In the United States, the proportion of students completing secondary school increased from 70% to 82%. On the other hand, the proportion of students included in Population B dropped significantly during that same period: from 18% of the age cohort in FIMS, to 12% in SIMS (Robitaille & Taylor, 1989).

During the period between the first and second IEA studies, substantial changes occurred in the intended mathematics curricula of most countries. Questionnaires concerning curricular emphases were administered in conjunction with both IEA studies, and it is possible to draw some comparisons from them even though the questionnaire items used in the two studies,

TABLE 27–6. Structure of the SIMS Item Pool — Population B

Major Strands	Number of Items (Percent)
Sets, Relations, Functions	7 (5%)
Number Systems	17 (13%)
Algebra	26 (19%)
Geometry	26 (19%)
Analysis	46 (34%)
Probability and Statistics	7 (5%)
Other	7 (5%)
TOTAL	136

as well as the specific content topics, were not identical. More detailed information on the exact wording of the items used in each study and the mathematical topics included may be found in Robitaille and Taylor (1989).

Overall, the results indicate that the importance of geometry at the Population A level declined between the two studies, whereas the importance of arithmetic and algebra increased. The United States had the highest importance rating for arithmetic in both studies, and almost the lowest for geometry.

At the Population B level, the importance of algebra and elementary functions and calculus showed an increase in most countries, while geometry scores decreased rather dramatically. All of these trends were evident in the results for the United States, although the change in the importance score for elementary functions and calculus was very small there. The importance scores for elementary functions and calculus was lower in the United States than in any of the other countries in both studies. This last finding is not surprising since Population B students in most other countries take a full course in calculus.

There are also large between-country differences at the senior secondary school level with respect to the proportions of either the age or the grade cohort who are studying Population B mathematics, that is, those students who are completing the requirements needed to pursue the study of mathematics at the post-secondary level. Such differences have important implications for the curriculum, since a course designed to be offered to, say, the top 10% of the age cohort would likely be either too advanced or too rigorous for a much wider cross-section of the population. In the majority of countries taking part in the second study—including Japan, the United States, and most of the European countries—Population B accounted for no more than 12% of the age cohort. Notable exceptions to this trend were found in Hungary, where 50% of the age cohort was included in Population B, and in the Canadian province of British Columbia, with 30%.

The rate of retention of students in the study of mathematics has implications not only for the content of the curriculum but also for the scientific and economic progress of the countries involved. As more and more fields require students to have a firm grounding in mathematics, the need for students well-qualified in mathematics will continue to grow. Of particular concern in this regard is the fact that in some countries, the United States and Sweden, for example, the proportion of students enrolled in mathematics programs at the Population B level has apparently declined in the period since the first study, although this may be an artifact of differences in the sampling procedures employed in the two studies. In the case of the United States, the decline in the size of Population B was from 18% of the age cohort in the first study to 12% in the second.

At the Population A level, except perhaps for geometry, there are many similarities among countries insofar as the content of the mathematics curriculum is concerned, at least in a cumulative sense. That is to say, students have studied many of the same topics although not necessarily in the same order or in the same depth. The vast majority of achievement items used in the second study were considered to be appropriate in all of the participating countries, and teachers agreed that most of the items concerned topics which were part of the curriculum implemented in their classrooms. No such agreement was evident for Population B. Instead, the situation at that level is characterized by large differences in content between countries.

FINDINGS AT THE SCHOOL OR TEACHER LEVEL. The majority of mathematics teachers at both the Population A and Population B levels were experienced and apparently well-qualified to teach mathematics. The vast majority of Population B teachers reported that they specialized in the teaching of mathematics. Specialization was also widespread at the Population A level except in Canada (Ontario), where Population A classes are usually housed in elementary schools and where teachers are expected to teach several subjects.

Analysis of teachers' responses to questionnaire items concerning their teaching practices revealed many similarities across countries. According to these results, the teaching of mathematics appears to be largely a "chalk-and-talk" affair, with teachers using whole-class instructional techniques, relying heavily on prescribed textbooks, and rarely giving differentiated instruction or assignments. In some countries, but not in the United States, many teachers of mathematics have had no academic preparation in either the pedagogy of mathematics or in general pedagogy, and this may account for their reluctance to use alternative teaching strategies, or to make more use of teaching aids and resources.

Data obtained from questionnaires administered to teachers in the longitudinal version of the study indicate that the teaching of mathematics, at least at the grade 8 level, is characterized by an enormous amount of review, especially in the United States and Canada. Based on his analysis of the content of several contemporary series of North American mathematics textbooks, Flanders (1988) reported that large portions of those books were given over to review of material covered in previous grades.

Teachers in France and Japan, on the other hand, apparently do not devote as much time to review. Repeatedly, their questionnaire results indicated that a given topic was neither reviewed nor taught in the Population A year because it was presumed to have been taught and mastered in an earlier grade. This was rarely the case for North American teachers.

These findings are summarized in the stacked bar graph displayed in Figure 27.2. Each bar is divided into three sections to depict, respectively, content which was not taught prior to or during the Population A year, content which was taught or reviewed during the Population A year, and content which was taught during an earlier year and assumed as prerequisite for the Population A year.

The graphs for the United States and the two Canadian provinces indicate that teachers do not assume that students have mastered much of this content in earlier grades. The areas corresponding to "taught previous year" are very small. This stands in marked contrast to the graphs for Belgium, France, and Japan, which show that teachers assume that students have learned a good deal of this content in earlier grades. The graphs also show that the curriculum in those three countries concentrates on fewer areas than is the case in North America.

Teachers' responses to certain items on the questionnaires indicated that they tend to use more abstract teaching practices with topics which are being reviewed than they do with topics being presented for the first time. While this seems quite appropriate at first glance, the pattern does have at least one unsettling aspect. Teachers also report that they are more likely to use these more abstract approaches with lower ability students than with higher ability classes (Dirks, 1986). Of course, it is difficult to know precisely the extent to which teachers' responses to questionnaires such as these accurately reflect their teaching practices. In the case of the SIMS data, some validation work has been undertaken and the results are encouraging. However, self-report data must always be interpreted with caution.

Questionnaire results also indicate that Population A teachers did not expect or encourage their students to use calculators, and the somewhat fragmentary data available on this matter indicated that they were not in widespread use in mathematics classrooms. Students' mean performance in arithmetic showed some evidence of decline between the first and second studies; and this fact, taken in combination with teachers' reluctance to permit students to use calculators in school even though they use them at home, is disturbing. It may be, however, given the rapid growth of access to calculators and the corresponding decrease in costs, that this situation is very different now than it was when these data were collected between 1980 and 1982.

There were substantial differences among systems with regard to average class sizes (Robitaille & Garden, 1989). The range was from 19 to 43 students at the Population A level, and from 14 to 43 students at Population B. Some of the largest class sizes were reported from countries where performance levels of students were among the highest in the study, notably in Hong Kong and Japan. Students in Hong Kong were not only enrolled in some of the largest classes reported in the study, but they were also among the youngest participants and among the most likely to have teachers who were less than fully qualified to teach mathematics. The performance levels attained by Hong Kong students given these kinds of handicaps were outstanding, and they underscore the important role played by factors such as motivation and parental or societal encouragement.

Class sizes and student-teacher ratios in the United States and the two Canadian provinces were larger than those in most of the other industrialized countries participating in the second study. Moreover, North American teachers reported teaching more hours per week. Taken together, these results indicate that North American mathematics teachers have comparatively heavy workloads.

Some form of intra-class grouping or tracking of students for mathematics classes is practiced in every country at some grade level. At the senior secondary level, many students are, for one reason or another, no longer taking any mathematics courses, while the remainder are usually divided among two or more alternative programs. At the junior secondary level, it is much more common for all students to be taking either the same course or very similar courses. However, in some countries, widespread tracking of students into different classes on the basis of some measure of mathematical or overall ability at that level is the rule. Such tracking is particularly prevalent in Canada (British Columbia) and the United States, where over 70% of teachers reported that tracking of Population A stu-

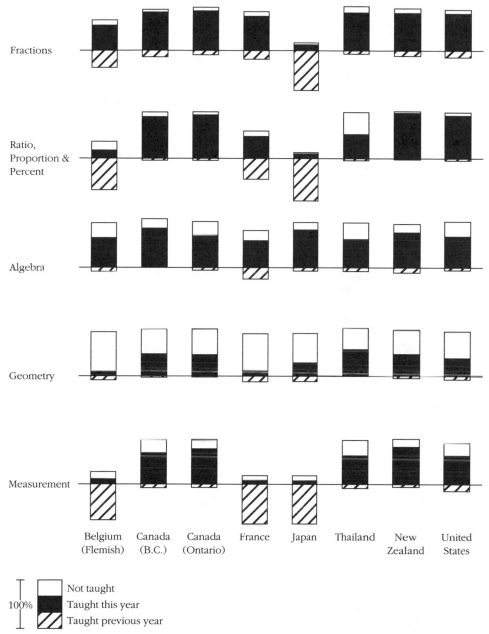

Fractions

Ratio,
Proportion &
Percent

Algebra

Geometry

Measurement

Belgium   Canada   Canada   France   Japan   Thailand   New   United
(Flemish) (B.C.)   (Ontario)   Zealand States

100%   Not taught
        Taught this year
        Taught previous year

FIGURE 27–2. Comparison of content for Population A taught as new, reviewed, or not taught.

dents by ability is widely practiced. At the opposite extreme, over 70% of Population A teachers in France and all teachers in Japan and Hungary reported that they did not place students into different classes or programs for mathematics at that level.

The data from the achievement tests do not provide much support for grouping students into different courses or programs in this way. Students from countries where little or no tracking takes place did very well on the achievement tests, particularly those from France and Japan. Moreover, it is interesting to note that, whether or not some form of tracking is widely practiced in a system, most teachers still report that the range of abilities in their class is either wide or very wide.

Achievement results show that French and Japanese students perform very well in mixed-ability groups at the Population A level, but the results do not provide an explanation. Some hints may be found in data from other parts of the study, such as the attitude questionnaires administered to participating teachers. For example, Japanese teachers were much more prone to attribute causes for students' lack of satisfactory progress to their own lack of prowess as teachers than to some weakness on the part of the students. They also rated teaching mathematics as being much more difficult than did teachers from most other countries.

These findings need to be augmented by comparative data collected over a substantial period of time, such as a complete school year or more. Perhaps interviews with teachers and students would help us develop a more complete depiction of the important variables at work in mathematics classrooms.

Gamoran (1987) points out that, in research on ability grouping, attention must be paid not only to how classes are organized and what content is taught but also to instructional treatment. The findings of Dirks (1986) about the inverse relationship between students' assumed level of ability and degree of abstraction employed by teachers are a case in point. So too are the dramatically widening gaps in achievement between classes of different ability levels reported in *The Underachieving Curriculum* (McKnight et al., 1987).

At the Population B level, although countries vary significantly in the proportion of the age or grade cohort retained in the study of pre-university mathematics, the evidence indicates that the very best students in almost every country do about as well as their counterparts in other countries (Husén, 1967; Miller & Linn, 1989). In other words, a policy of maintaining a high rate of retention of students in an academic mathematics program, such as in Hungary and Canada (British Columbia), does not necessarily affect the performance of the very best students adversely. The best students in such systems do as well as their counterparts elsewhere on topics which both have studied. One would expect, however, that students in the more selective systems will have been exposed to more topics than those in systems with high rates of retention.

The stacked bar-graph in Figure 27.3 summarizes data on achievement levels in algebra attained by all students in each country (solid black), the best 5% of students (shaded), and

the best 1%. There are substantial differences in the heights of the solid black bars, indicating wide disparities in achievement levels among countries when the entire Population B cohort is included. However, the differences are generally smaller when the comparisons are restricted to the best 5% of students; and, even smaller when restricted to the best 1%. This is true in all of these countries except the United States, where the performance level attained by even the best students remains lower than that of the best students in other countries, although the size of the achievement gap is narrower.

FINDINGS AT THE STUDENT LEVEL. In most countries, as was shown in Figure 27.2 for the countries participating in the longitudinal study, the Population A mathematics curriculum is a combination of topics from arithmetic, algebra, and geometry. In Canada and the United States, the curriculum tends to include more arithmetic than anything else, with a sprinkling of algebra and geometry for some students. In the United States a complete algebra course is offered to only a small minority of students who are less than 14 years old.

Similarly large curricular differences occur at the Population B level. In most countries, a significant amount of selection of students has taken place by this time, and the Population B cohort tends to be relatively small, usually between 4% and 20%. The curriculum in these countries is likely to include both calculus and algebra as important strands, with less emphasis on geometry and trigonometry because they have been covered in earlier years. In Canada, where there is a higher degree of retention of students in the study of pre-university mathe-

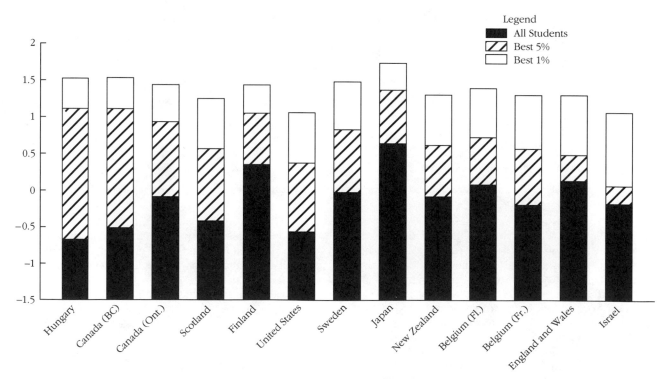

FIGURE 27–3. Mean achievement in algebra (z-scores) for all
Population B students, for the best 5%, and for the best 1%.
(Countries listed in order of their retention rates.)

matics, the Population B curriculum consists mainly of algebra and trigonometry, as it does for most students in the United States. Few students take a full course in calculus. However, in Hungary teachers report both a high retention rate and a large proportion of students studying calculus.

Most students at the Population A and Population B levels indicate that they believe that mathematics is important, that they want to do well in mathematics, and that a good knowledge of mathematics will be important to them in their careers. They also indicate that their parents share these opinions, and that their parents encourage or exhort them to do well in mathematics. While their opinions about mathematics cannot be construed as negative, students are not overly enthusiastic either. It is particularly interesting to note that Japanese students, who performed very well on the achievement tests, were much more likely than students elsewhere to feel that mathematics was difficult and not enjoyable. Japanese teachers expressed similar opinions about teaching mathematics, unlike teachers from Canada and the United States, for example, who tended to find teaching mathematics easier and more enjoyable.

Population A students found the achievement test items fairly difficult on the whole. The mean percentage correct across countries for all items was 47%. Performance on items involving computation with whole numbers and other straightforward applications of basic concepts was generally good; however, performance fell off sharply on items calling for the use of higher order thinking skills. Performance levels on items involving rational numbers, whether expressed as common or decimal fractions, were generally disappointing. Even simple computational items and basic applications of percent and ratio and proportion were found to be difficult. Possible causes of students' difficulties in these areas are explored by Behr, Harel, Post, and Leah, Chapter 14, this volume.

Results from the longitudinal version of SIMS indicate that growth in students' achievement from pretest to post-test was modest. For the United States sample, the median gain in achievement on the 180 items used in the longitudinal study was only 7 percentage points. That is, on a typical item, the percentage of students who obtained the correct answer at the end of the school year was only 7 percentage points higher than it was at the beginning of the school year. In the United States, growth scores exceeded 15 percentage points on only 21 of the 180 items.

In France and Japan, where teachers' responses to questionnaire items indicated that they spent significantly less time on review than in North America, the proportion of items for which there were large growth scores tended to be higher. Unfortunately, only a few of the items which produced high growth scores in the United States were included in the Japanese pretests. Japan used only 60 items on their pretest instead of the 180 items used in France and the United States. In France, 40 of the 180 items produced growth scores of 15 percentage points or more; in Japan, 19 of 60 did.

Figure 27.4 depicts the changes in achievement from pretest to post-test on 6 of the 21 high-growth items in the United States which were also used in France and Japan. The length of each bar shows the amount of growth in achievement during the

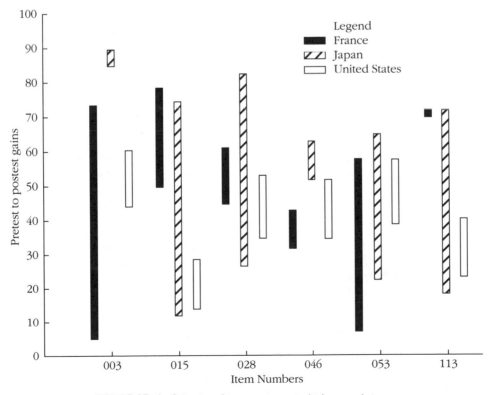

FIGURE 27–4. Gains in achievement on six high-growth items.

Item 003: $\frac{2}{5} + \frac{3}{8}$ **is equal to**

A. $\frac{5}{13}$
B. $\frac{5}{40}$
C. $\frac{6}{40}$
D. $\frac{16}{15}$
E. $\frac{31}{40}$

Item 015: Simplify: $5x + 3y + 2x - 4y$

A. $7x + 7y$
B. $8x - 2y$
C. $6xy$
D. $7x - y$
E. $7x + y$

Item 028: What are the coordinates of Point P?

A. $(-3, 4)$
B. $(-4, -3)$
C. $(3, 4)$
D. $(4, -3)$
E. $(-4, 3)$

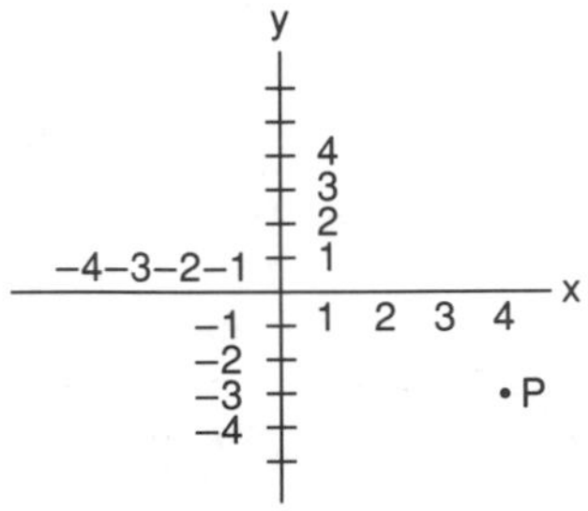

Item 046: 20% of 125 is equal to

A. 6.25
B. 12.50
C. 15
D. 25
E. 50

Item 053: When $x = 2$, $\dfrac{7x + 4}{5x - 4} =$

A. 11
B. 3
C. $\frac{11}{5}$
D. $\frac{9}{5}$
E. $\frac{7}{5}$

Item 113: $(-6) - (-8)$ is equal to

A. 14
B. 2
C. -2
D. -10
E. 14

FIGURE 27–5. High-growth items.

school year. Thus, the mean percent correct on Item 003 in France grew from 5% on the pretest to 73% on the post-test. The corresponding results for Japan and the United States were from 84% to 89%, and from 44% to 60%, respectively.

The items themselves are presented in Figure 27.5 in a somewhat reduced format to enable all six to be printed on the same page. Note that all six items deal with fairly low-level cognitive tasks, and this trend was evident across all countries participating in the longitudinal version of the study. High growth was much more likely to occur on a low-level cognitive task than on a high-level one.

All of the high growth items produced interesting results, particularly when the achievement results are examined in the light of data from other aspects of the study. Consider Item 003, for example. In North America, this content is typically encountered for the first time by students in grade 5. By the time they finish grade 8, they have been taught this algorithm four years in a row. And, after all that exposure, 40% failed to select the correct response.

In Japan this topic is taught in an earlier grade, and their pretest score of 84% correct indicates that it was mastered there. In France, addition of rational numbers is apparently taught for the first time during the Population A year, and the size of the gain in the growth score on this item—68 percentage points—is extremely large.

Of the six items shown in Figure 27.5, at least four deal with content which almost all grade 8 students in North America should have seen in an earlier grade or grades. (The two possible exceptions are Items 015 and 053.) This is reflected in the pretest scores where the graph indicates that pretest scores in the United States are frequently higher than those in Japan or in France, depending on the item. The main difference is that the performance of American students did not increase to nearly the same degree as it did in France or Japan.

It would be a mistake to read too much into these results since they are based on relatively few items, and there is not a great deal of detailed information available concerning the teaching practices employed. However, they do suggest that a re-examination of the way mathematics is taught in North America may be warranted, particularly with regard to the effects of repeated review on students' achievement in and attitudes toward mathematics.

On the subject of gender differences in achievement at the Population A level, girls tended to outperform boys in the use of computational skills and in algebra. Boys tended to outperform girls in geometry and measurement. These latter differences, which were the most marked and consistent, may be related to differences in boys' and girls' spatial visualization abilities. It should be noted, in connection with these gender differences, that substantial between-country differences also exist. Girls outperformed boys fairly consistently in some countries, and the reverse was true in others. No investigation of country-level variables which might account for variations like those found in this study have been carried out as yet. However, Leder, Chapter 24, this volume, includes additional information about variables associated with gender differences.

Girls are a distinct minority in Population B in most countries. This fact, along with the already noted differences between countries in rates of retention of students in mathematics, complicates the investigation of gender differences in achievement at this level. The analyses conducted thus far indicate that Population B boys outperformed girls on every subtest in almost every country, and that many of the differences were quite large.

Comparisons of performance levels between countries are extremely complex and problematic, particularly at the Population B level where, in addition to differences in retention rates, significant between-country differences in curriculum also exist. But even at the lower grade level, where countries appear to have much more similar curricula, the situation is complicated. Thus, countries such as Japan, New Zealand, and England and Wales have similar curricula for Population A students, but achievement levels are very different. Large differences in performance between countries with comparable curricula exist elsewhere. This underscores the importance of taking into account the influence of sociocultural factors which may affect students' and teachers' attitudes and behaviors in highly significant ways.

Comments. One of the weaknesses of SIMS was the length of time it took for the study to be completed and for the international report to be published: about 12 years from the date of the first planning meeting to the publication of the third volume of the international report. This delay has resulted in the piecemeal publication and announcement of many results and some "garbling" of the findings.

The longitudinal version of SIMS has proved to be very fertile ground for investigating links between growth in students' achievement and aspects of the way in which mathematics is taught. Certainly, from the point of view of persons whose primary interest is mathematics education—as opposed, say, to the study of variables that might have general relevance for policy making in education—the information on teaching practices and student growth is a valuable resource.

Although twice as many items were utilized in the second IEA mathematics study as in the first, there still were not enough to provide thorough coverage of the curriculum across countries. Nor was there a sufficiently close correspondence between the content of the items and the classroom process questionnaires. A third IEA mathematics study might well use several hundred items for each population, and these should be closely linked with the variables of interest in the instruments developed for use in the study. The SIMS data showed that there were indeed differences between countries in the way that certain topics were taught or in the degree to which they were emphasized. Unfortunately, there were not enough items which related specifically to those topics to enable relational analyses of the comparative effects of these differences to be carried out.

Designers of future international studies should give serious consideration to including a younger, elementary-school-age population of students. Also, attention needs to be paid to means for increasing the usefulness of comparative results at the senior secondary level, where large differences in curriculum make comparisons difficult. One possibility might be to conduct a survey of participation rates, attitudes, curriculum, teaching practices, and the like, but to exclude any kind of achievement survey at that level.

Inclusion of an elementary school population could provide very important information. For example, it would be useful to find out whether the kinds of achievement differences which are apparent by the time students reach the age of 13 are present as early as age 9 or 10. Some indications that significant cross-national achievement differences are evident early have been shown in a number of smaller-scale studies (for example, Stevenson, Lee, & Stigler, 1986; Stigler & Perry, 1988).

OTHER INTERNATIONAL STUDIES

Both IEA mathematics studies involved fairly large numbers of participating countries. The first mathematics study had 12 participants, and the second had 20. In the past few years, as increasing attention has been paid to international comparisons, a number of smaller scale cross-national studies have been carried out. Among these are the recently completed international survey conducted by the Educational Testing Service (ETS) and the Stevenson-Lee-Stigler studies of students' achievement in the United States, Japan, and Taiwan.

The International Assessment of Educational Progress

The International Assessment of Educational Progress (IAEP) constituted a significant departure for the Educational Testing Services (ETS) from previous surveys. That organization has for many years been very active in numerous aspects of educational testing in the United States, and recently assumed responsibility for the National Assessment of Educational Progress (NAEP) program. The ETS publication, *The Mathematics Report Card: Are We Measuring Up?* (Dossey, Mullis, Lindquist, & Chambers, 1988) summarizes the major findings and trends from the four NAEP assessments of mathematics conducted between 1973 and 1986.

The president of ETS states that improvement in the measurement of student's achievement is an important part of the mission of that organization (Lapointe, Mead, & Phillips, 1989). He describes the IAEP project in the following way:

Working with colleagues from five other countries, ETS's measurement specialists have translated and adapted the techniques perfected in the United States by the National Assessment of Educational Progress, and together they have conducted mini-assessments in five different countries. (Lapointe et al., 1989, p. 5)

Design of the Study. Students from six countries participated in the IAEP survey in early 1988: Canada, Ireland, South Korea, Spain, the United Kingdom, and the United States. In the case of Canada, four provinces—British Columbia, Ontario, Quebec, and New Brunswick—took part independently of one another, so that it was not possible to estimate one set of results for the country as a whole. Moreover, the three provinces other than British Columbia each selected two samples of students for participation: one from English-speaking schools, and the other from French-speaking ones.

The mathematics items used in the IAEP were taken from the pool of items used in the 1986 NAEP assessment of mathematics. Representatives from each of the participating countries reviewed 281 NAEP items and agreed on a set of 63 items which were considered most appropriate internationally.

In addition to the achievement items, students were asked a number of questions about themselves, about their attitudes toward mathematics and science, about instructional activities, and about homework and amount of television watched. Schools were asked to supply information about students' opportunity to learn by having a teacher or coordinator estimate "the percentages of the seventh- and eighth-grade students who had already had an opportunity to learn the concepts tested by each item in the assessment" (Lapointe et al., 1989).

Major Findings. Results from the mathematics testing are summarized in Figure 27.6. The graph shows the 12 participants ranked in descending order and partitioned into four subgroups. Differences within each of the four subgroups are not statistically significant, but differences between the subgroups are.

The vertical dimension of the graph, the Mathematics Proficiency Scale, was developed using item-response theory (IRT):

more specifically, a three-parameter logistic model. Five points on the scale—300, 400, 500, 600, and 700—were selected as "anchor" points to describe "what students *know* or *can do*," and the scale was used to estimate the percentage of students in each participating system who were functioning at each of the five "anchor" points (Lapointe et al., 1989). These were defined as follows:

Level 300: Perform simple addition and subtraction.
Level 400: Use basic operations to solve simple problems.
Level 500: Use intermediate level mathematics skills to solve two-step problems.
Level 600: Understand measurement and geometry concepts and solve more complex problems.
Level 700: Understand and apply more advanced mathematical concepts.

The authors report that there are striking differences between countries as regards the percentage of students at each level of the scale. For example, while 78% of Korean 13-year olds are estimated to be at or above the 500 level on the scale, only 40% of American and Franco-Ontarian students have attained that level. At the 600 level, the corresponding percentages are 40% and 10%, respectively. The difficulties that arise in interpreting these results are considered in the section following this.

Analysis of student responses to questions about instructional activities reveals there are a great many similarities

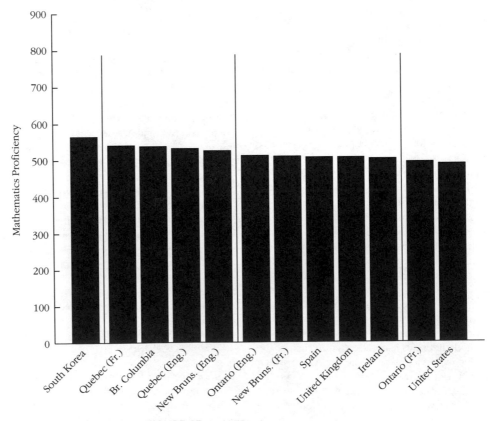

FIGURE 27–6. IAEP achievement results.

in the way their teachers teach, regardless of which country they come from. The most frequent activities are listening to teachers lecture or present the material for the lesson and individual seatwork. Relatively few students reported that they worked in small groups on a regular basis.

Approximately two-thirds of American 13-year olds agreed with a statement to the effect that they were "good at" mathematics, in spite of their comparatively weak performance overall. By way of contrast, less than 25% of Korean students expressed this opinion. Differences like this may have more to do with cultural influences than with students' actual beliefs, although it would be difficult to substantiate that claim from the survey data.

Comments. Less than a year elapsed between the collection of the achievement data in February 1988 and the publication of the final report (Lapointe et al., 1989). This is a significant accomplishment given the amount of work involved in coordinating the collection and analysis of data from several countries with the concomitant complications of language differences.

The goals of IAEP, compared with those of the second IEA mathematics study, were limited almost entirely to achievement comparisons and to one age-level of students. Moreover, the items employed in the achievement survey had already been used, thereby shortening, by a considerable margin, the development time required. On the other hand, this approach meant that the other participants were invited to compare their students' performance with that of students in the United States and other countries on items that had been developed for use in the United States. The report contains no information about the degree to which the final set of 63 items "covered" the major portion of the curriculum in any of the participating countries, including the United States.

The fact that there were so few items in the survey is a serious limitation of the usefulness and interpretability of the results. For example, the topics of relations, functions, and algebraic expressions were represented by a total of only six items. Another serious limitation is that none of the items will be released so that mathematics educators can make an analysis of differences in the ways that students from different countries responded to them.

The use of item-scaling and other IRT techniques has raised a number of concerns and made the interpretation of results problematic. First, the theory as applied in this study suggests that achievement in mathematics is a uni-dimensional trait, even though the items covered topics from a variety of strands in the curriculum, including arithmetic, algebra, geometry, logic, and problem solving. In addition, measurement of the trait is based on only 63 achievement items, a small number on which to base an assessment of mathematical proficiency.

The scaled scores shown in Figure 27.6 are difficult to interpret. That is, it is not possible to say what the numbers mean. Korean students attained a mathematics proficiency score of 568; Spanish students, 512. The report indicates that such a difference is statistically significant at the 0.05 level, but that by itself is not very informative. What does a score of 568 mean? Is it a good score or a bad score? Is the difference between 568 and 512 significant from an educational viewpoint? Should the administrators of the school system in Spain be concerned

about the performance of their students? In the absence of much more information than is available from the report, these questions are unanswerable.

The validity of the opportunity-to-learn data in this study is also open to question. The teachers of the students being tested were *not* asked to supply the data. Instead, a teacher or coordinator in each school was asked what proportion of the students of that age in the school had been taught the content needed to respond to the item. It is, therefore, not surprising to learn that, "In about half of the comparisons between average percents correct and opportunity-to-learn ratings, students perform better in various mathematics topics than their teachers' opportunity to learn ratings would suggest" (Lapointe et al., 1989, p. 33).

Smaller-Scale Studies

In the past several years, a number of researchers have conducted somewhat more focused international studies on a more modest scale than either the IEA or the IAEP projects. Several of these are reviewed here.

Dallas Times Herald Survey. In late 1983, the *Dallas Times–Herald,* perhaps motivated by the widespread attention given in the United States and many other countries around the world to the publication of the report *A Nation at Risk* (National Commission on Excellence in Education, 1983), carried out its own international survey in order to compare the educational achievement of students in Dallas with those of students from seven other countries: namely Australia, Canada, England, France, Japan, Sweden, and Switzerland. The tests—in mathematics, science, and geography—were developed by American educators for 12-year-old students and were administered in schools that were selected by educational officials in each country to "mirror as closely as possible the national average" in achievement.

Results showed the United States as lowest in mathematics, in the bottom third in science, and at the international median in geography. While Japan scored highest in mathematics (a mean score of 50% correct compared with 25% for the United States), they scored only about 2 percentage points higher than the United States in science. The reported scores for science were 45.3% for Japan and 43.7% for the United States. Commentary from leading American educators on the findings of the Dallas study included a statement by Ernest Boyer of the Carnegie Foundation for the Advancement of Teaching attributing the relatively poor American performance to an "unwillingness to establish priorities" around a core of academic subjects.

Japan/Illinois Study. Harnisch and his colleagues (1985) explored patterns of mathematics achievement and, for background, measures of achievement motivation for samples of high school students in the state of Illinois and in Japan. The Illinois sample consisted of approximately 10,000 students tested by the State Board of Education. The Japanese sample included 1700 students randomly selected from across the country by the Nippon Electric Company (NEC) in a collaborative project with the University of Illinois. The mathematics test included 60 items on algebra, geometry, modern mathematics, data interpretation, and probability. (The term "modern mathematics" was not defined in the report, but referred to topics which first appeared in curricula in the 1960s: for example, sets and non-decimal numeration systems.) The student background questionnaire assessed the level of mathematics completed, frequency of leisure reading, and student perception of the importance of the test.

The mathematics achievement data showed that, for all three age groups (15-, 16- and 17-year olds), Japanese students outperformed their Illinois counterparts by more than two standard deviations. Japanese males outscored Japanese females by about one standard deviation, whereas Illinois male and female scores differed very little. Japanese females far outscored both sexes in the Illinois sample.

For the Japanese students, the largest correlates of mathematics achievement were age, being male, and exposure to the content of the test items. The largest correlates for the Illinois students included the number of mathematics courses taken, confidence as a reader, and the importance which the student attributed to the test. The researchers commented that, although the study confirmed others in showing a strong link between courses completed and achievement, it should not be taken to imply that simply requiring more courses or more extensive coverage would improve achievement. They noted, for example, that "insufficient teachers may be available to teach rigorous courses; many American high school students may lack the ability and motivation to study calculus or physics; and it is unclear that American parents are willing to sacrifice the time and energy required to support their children's personal and educational growth to Japanese standards."

Development of Mathematical Thinking. Song and Ginsburg (1987) used the *Test of Early Mathematical Ability* (Ginsburg & Baroody, 1983) to assess mathematical thinking in Korean and American children at ages 4, 5, 6, 7, and 8. They found that Korean children between the ages of 4 and 6 did not do as well as American children on informal mathematics items: relative magnitude, counting, and so on. No significant differences between the groups were found for formal mathematics measures such as number facts, calculation, and base ten concepts.

At ages 7 and 8, Korean children showed superior achievement on both informal and formal mathematics tasks. The researchers concluded that Korean children do not begin with a "head start" in mathematics. They attributed the relative success of Korean children and the correspondingly poor performance of American children in early school mathematics to such environmental and cultural factors as classroom practices, attitudes of teachers, and parental values.

Attributions of Success in Mathematics. The superior performance of Japanese students in international surveys of mathematics achievement gave rise to parallel longitudinal studies in Japan and the United States of family influences on school readiness and achievement. The Japanese component was directed by Hiroshi Azuma of the University of Tokyo, while the United States' portion was directed by Robert D. Hess of Stanford University. The study began as an investigation of preschool children, with a view to identifying family variables associated with school readiness and to describing the socialization processes involved (Azuma, Kashigawa, & Hess, 1981). Follow-up studies were carried out when the children were in grade 5 in Japan and grade 6 in the United States.

During the preschool phase of the study, which lasted three years, mothers were interviewed about their child-rearing beliefs and behavior, and about their children's social and cognitive development. Tests of cognitive ability were administered to the children at ages 4, 5, and 6. During the follow-up phase, information about the children was obtained from tests and teacher ratings. The mothers were interviewed again to determine their assessment of their children's performance in mathematics and their attributions of causes for their children's relative success in mathematics. Interviews with the children were also conducted and were essentially identical to those with the mothers, with appropriate wording changes (Hess et al., 1986).

The research revealed that mothers in both countries had high aspirations for their children. It also found that observed maternal behavior in both countries was correlated with school readiness and subsequent school-related performance. For example, communication efficiency was significantly related to school readiness (at ages 5 and 6) and achievement (in grades 5 and 6) for Japan and the United States, respectively. There were, however, rather striking differences between the countries. Contrasts were noted, for example, in how a mother responded to her child when he or she resisted the mother's requests. Japanese mothers tended to yield to the child, giving in rather than injuring the relationship: Protecting feelings of closeness with the child is an important goal of Japanese mothers.

This contrasted with views commonly held by parents in the United States: "Be consistent, don't lose a battle with the child, be firm..." (Hess et al., 1986). Japanese mothers interacted with their children in ways that promoted internalization of adult norms and standards, whereas American mothers were more oriented toward applications of external authority and direction. Furthermore, American mothers relied more on verbal explanations and the use of techniques encouraging verbalization by children, while Japanese mothers were more prone to emphasize following correct procedures as a route to understanding.

In the same study, Holloway, Kashiwagi, Hess, and Azuma (1986) reported on dramatic differences founded in mothers' beliefs about the causes of success and failure in school. Mothers were asked to identify sources of success of their children in grade 5 mathematics. In Japan, poor performance was attributed to lack of effort. They were less likely to blame the school, the child's ability, or luck; nor did they see the difficulty of the task as a significant element in the child's level of performance. In the United States, mothers cited lack of effort as the most significant factor, but they also cited lack of ability and poor classroom instruction. Generally, American mothers assigned more blame to the school for low performance than did Japanese mothers.

Mathematics Achievement in the Elementary Grades: The Michigan Studies. The well-documented superior mathematics achievement of Japanese students at the secondary school level provided the background for two major studies at the University of Michigan designed to explore cross-national achievement in mathematics at the elementary school level. The research was carried out by Harold Stevenson and his colleagues at various universities in the United States and in Asia, and a summary of this research is reported by Stigler and Baranes (1988).

The first cycle of data collection, carried out in 1979–80, involved large samples of grade 1 and grade 5 children, their teachers, principals, and parents in Sendai (Japan), in Taipei (Taiwan), and in metropolitan Minneapolis. The second investigation, in 1985–1986, involved the same two Asian cities, but metropolitan Chicago was chosen for the United States sample. Grades 1 and 5 were used once again. Stigler and Baranes (1988) summarized the findings of these studies as follows:

DIFFERENCES IN MATHEMATICAL KNOWLEDGE. A primary objective of the Michigan studies was to identify the grade level at which Japanese superiority in mathematics achievement first emerged. (It has already been noted that Song and Ginsburg reported the appearance of superior achievement of Korean children, as compared with American children, by ages 7 and 8.) The Michigan studies clearly showed the pronounced superiority of Japanese and Taiwanese children during the first grade and even greater differences at the fifth grade.

The first Michigan study used what is described in the report as a standard measure of mathematics achievement. However, the second Michigan study took a very careful look at what the term "mathematics achievement" meant. In particular, there was interest in assessing the extent to which the achievement of the Japanese represented mainly computational proficiency as contrasted with "the creative mathematical problem-solving skills that American mathematics educators care about most" (p. 292). Therefore, the second Michigan study used tests carefully designed to measure a wide variety of concepts and skills, including computation, estimation, graphing, and visualization. At the time of the preparation of the Stigler and Baranes report (1988), all of the analyses had not been completed. However, based on the results they had, the researchers concluded that at both the first and fifth grades, Japanese children outscored the American children on almost every test. The Taiwanese results were similar to the Japanese data for the fifth grade but mixed at the first grade.

OBSERVATION OF MATHEMATICS CLASSROOMS. Both of the Michigan studies made use of classroom observations, although the methodologies were rather different. In the first, a structured coding scheme was used with an elaborate time-sampling plan. Consequently, it was possible to obtain estimates of the percentage of time devoted to various classroom activities. In the second study, in the words of the researchers, they "decided to trade the greater reliability of an objective coding scheme for the inherent richness of detailed narrative descriptions of mathematics lessons in three cultures" (Stigler & Baranes, 1988, p. 294). Results of preliminary analyses of the classroom data indicated dramatic differences in the classrooms. Highlights are as follows:

1. Time devoted to mathematics instruction. The school year is much longer in Japan and Taiwan than in the United States:

approximately 240 days compared to 180 for the United States. The school days are also longer, with a greater percentage of time in school devoted to academic activities. In the first grade, all three countries spend more time on reading than on mathematics. In the United States, the differential in favor of reading is even greater. However, by the fifth grade, Japan and Taiwan devote about as much time to mathematics as to reading while teachers in the United States reported spending twice as much time on reading as on mathematics.

2. Patterns of classroom organization. Classroom organization in the Asian and American samples was rather different. The Asian teachers were found to work with the whole class about 80% of the time, while the United States teachers did so less than 50% of the time. American teachers spent the remainder of the time working with small groups or individual students. The researchers commented that, because of this practice in Asian classrooms (which typically contained about 40 students compared with about 22 in the United States), the children were spending close to 80% of their time in the mathematics class under the direct supervision of the teacher. In the United States, on the other hand, students were under the direct supervision of the teacher less than 50% of the time. The authors summarized the situation by saying that, "American students ...spend significantly more time engaged in off-task, inappropriate behavior than do their counterparts in Japan and Taiwan" (p. 295).

3. Verbal discussion and reflection. It was this aspect of classroom life that most clearly distinguished Japanese from both Taiwanese and American classes. Nearly 50% of all instruction in Japanese first-grade classrooms consisted of verbal explanations by either the teacher or students. This contrasted with only about 20% in Taiwan and the United States. The researchers report that, "By the fifth grade, the gap had narrowed somewhat, but there still was more verbal explanation in Japanese classrooms than in classrooms of the other two cultures" (p. 296).

Verbal interaction was evident in other ways in Japanese classrooms. For example, the most common way of evaluating student work in Japan was for the teacher to ask students who got incorrect answers to put their work on the board and then discuss with the entire class the process that led to the error. By contrast, in American classrooms, the most common form of evaluation was simply to praise students who obtained correct answers. The researchers concluded that, "By focusing on errors, Japanese teachers have a natural basis on which to build a discussion; praise, on the other hand, functions as a conversation stopper" (p. 296). The emphasis on verbal discussion and reflection in Japanese classrooms was also evident in the pace of the lessons. It was only in Japan that teachers were observed to devote an entire 40-minute lesson to one or two problems. For example, Stigler and Perry (in press) analyzed the number of problems covered in instructional segments which lasted five minutes, and found that only one problem was dealt with in 75% of the segments in Japan and in only 17% of the segments in American classrooms. As the researchers comment, "Japanese teachers seem not to rush through the material, but rather are constantly pausing to discuss and explain" (p. 296).

The role of manipulatives in elementary mathematics classrooms was also studied. It was noted that, even though Japanese teachers placed a lot of emphasis on verbal explanations, they (and the Chinese) used more manipulatives and real-world problem-solving scenarios than American teachers. Furthermore, the manipulatives were used in different ways in Asian classrooms. While American teachers were found to use a great variety of objects (e.g. Popsicle sticks, poker chips, bags of oranges, Dienes' blocks, and Cuisenaire rods), the Japanese (and, to some extent, the Chinese) teachers used a very limited number of materials (for example, objects similar to Dienes' blocks) and used them repeatedly for different instructional purposes. It was also found that, in Japan, while the manipulatives were being used, the frequency of verbal explanation increased. In the United States, on the other hand, the frequency remained the same. As the researchers concluded, "... in the presence of concrete referents, Japanese teachers use the objects as a topic of discussion, whereas American teachers tend to use the objects as a substitute for discussion" (p. 297).

The Michigan Studies have made an important contribution to our knowledge of the teaching and learning of mathematics across cultures. They have demonstrated that remarkable differences in achievement between Asian and American children appear almost immediately after the children enter the school system. The studies are also important in that they have applied innovative and informative techniques for providing data on the contexts in which teaching and learning take place in different cultures.

Correlates of Achievement. Stevenson et al. (1986) surveyed the mathematics achievement of American, Chinese, and Japanese children in kindergarten and grades 1 and 5. Participants, who were selected from one city in each country, were administered a battery of tests. The children, their mothers, and their teachers were interviewed and classrooms were observed. The achievement of the American children at all three grade levels was found to be lower than that of the Japanese, and was lower than that of the Chinese children at grades 1 and 5.

American children in grades 1 and 5 spend less than 20% of class time on mathematics, less than the corresponding percentages for Japan and China. Moreover, there is much more variability among the American classrooms as regards the amount of time devoted to academic subjects than among Japanese or Taiwanese classrooms.

The classroom observations revealed "strikingly different" patterns of leadership in the classrooms. Asian classrooms were led by the teacher more than 70% of the observed time, while American classrooms were teacher-led less than 50% of the time. Interview data showed that more than 40% of the American mothers were "very satisfied" with their children's academic performance in school. Fewer than 6% of Asian mothers were as positive.

More findings from this study are reported in Lee, Ishikawa, & Stevenson (1987). The hypothesis that the superior achieve-

ment of Japanese children was attributable to superior intellectual ability could not be supported on the basis of the children's performance on cognitive tasks. The investigators therefore focused on relationships between children's beliefs and their achievement in school. Children were asked to rate how well they liked mathematics, language arts, school, and homework. American children, whose achievement was the lowest of the three groups, had the most positive attitudes toward mathematics. Chinese children expressed the greatest degree of liking for reading, and also gave the highest ratings as to how well they liked school.

Similar results were found with respect to liking for homework. The researchers commented that this finding was obtained, "despite the fact that Chinese fifth graders spend approximately 40 hours a week at school, Japanese children, 37 hours, and American children, only 30 hours." The average amount of homework (apparently, for all school subjects) done each weekday night, as estimated by the mothers, was 115 minutes for the Chinese, 57 minutes for the Japanese, and 46 minutes for the Americans.

Textbook Analyses. Teachers of mathematics in all countries rely very heavily on textbooks in their day-to-day teaching, and this is perhaps more characteristic of the teaching of mathematics than of any other subject in the curriculum. Teachers decide what to teach, how to teach it, and what sorts of exercises to assign to their students largely on the basis of what is contained in the textbook authorized for their course.

Concerns have been expressed in the United States and in Canada about the quality of textbooks, about the way they are written, and about their pervasive influence. There is no doubt that the content of textbooks is a significant factor in determining students' opportunity to learn and their achievement.

ELEMENTARY SCHOOL (GRADES K-8) TEXTBOOKS. Fuson, Stigler, & Bartsch (1988) studied the grade placement of topics in addition and subtraction in textbooks from several countries, including the United States. The investigators reported a high degree of uniformity in the grade placement of these topics in Japan, China, the Soviet Union, and Taiwan. Both the simplest and the most difficult multi-digit addition and subtraction problems appeared from one to three years earlier in the textbooks of those countries than in the textbooks of the United States.

Stevenson and Bartsch (in press) surveyed the content of American and Japanese textbooks and found that the following topics were introduced earlier in Japan: addition, subtraction, decimals and fractions, dimensions of solids, time, money, relation of circles and spheres, triangles, squares, rectangles, polygons, rhombus, parallelograms, and calculation of areas. The only topics which appeared earlier in American textbooks were ratio and proportion, problem solving, fractions, and weight.

Sugiyama (1987) analyzed approaches to problem solving in selected Japanese and United States textbooks for grades 1 through 8. He found, for example, that the American textbooks for grades 1 and 2 which he examined contained no word problems suggesting the meaning of addition and subtraction— only figures or pictures. By contrast, the Japanese textbooks for grades 1 and 2 included many problems to be solved using addition and subtraction. At this grade level Sugiyama also noted that, in Japan, teachers combine the meaning of operations with the development of computational skill. In the United States, as he inferred from the textbooks, "Writers of the... textbooks plan to teach how to compute first, and application comes later."

Sugiyama concluded that word problems in Japanese textbooks for grades 7 and 8 were more difficult than those found in American texts. Problems in American textbooks for grades 7 and 8 were found in grade 5 in Japan. Strategies for solving problems are presented in American textbooks whereas, in Japan, this material is not in the textbook. It is, however, taught by the teachers. A richer variety of problems, dealing with more situations from the life of the students (for example, foods, sports, and pets), was found in the American books than in those from Japan.

SECONDARY SCHOOL TEXTBOOKS. The Stevenson and Bartsch study (in press) also included an examination of the content of secondary school textbooks, using the same general approach as that used for elementary books. A general finding was that concepts tended to be introduced earlier in Japan than in the United States. There was no tendency for more concepts to be introduced in the textbooks of one country than the other. However, concepts did tend to be introduced up to a year earlier in Japan. Furthermore, a great deal of repetition of concepts was found in American books, and this was interpreted as reflecting "the American spiral curriculum." Over 70% of concepts were repeated at least once in American books after their initial introduction. Almost 25% were repeated twice, and 10% were repeated three times. By contrast, in Japan, 38% of the topics were reviewed once, and only 6% were repeated more than once.

American textbooks tend to be much longer than the Japanese ones examined. The average number of pages in the American textbooks examined was 540, with a range from 400 to 856 pages. Japanese textbooks had a mean length of 178 pages, with none having more than 230 pages. As the investigators stated, "Japanese texts tend to be tersely written, while the American textbooks contain long, sometimes repetitive presentations of the basic arguments." The researchers also found that more of the problems in the Japanese textbooks tended to be complex. Simple problems appear at the beginning of a set of exercises, but the problems rapidly become difficult in the Japanese books. Again, the investigators report, "As (is) the case with the elementary school textbooks, many of the elementary steps in the development of concepts are omitted in the Japanese textbooks. American textbooks tend to represent the information step-by-step and all details are specified."

CONCLUSION

Comments about what is to be learned from international studies can be made at two levels. On the one hand, at what we shall call the substantive level, there are findings which pertain to the improvement of the teaching and learning of mathematics. On the other hand, there are methodological lessons to be learned about how future studies of mathematics should be designed and implemented to avoid weaknesses found in

earlier studies and to do a better job of operationalizing and assessing the crucial variables.

Improving Mathematics Education

International studies of the teaching and learning of mathematics can serve as valuable sources of data and information against which educators in a given country can compare and contrast the curriculum, the teaching practices, and the outcomes attained by students in their own system. The possible impact of alternative curricular offerings, teaching strategies, administrative arrangements, and the like can be estimated efficiently by examining their implementation in other jurisdictions, even when the countries involved are quite dissimilar culturally or economically.

A number of the findings from the two IEA mathematics studies demonstrate this point clearly. For example, we know from both studies that a system can retain a large proportion of students in the study of mathematics until the end of secondary school without jeopardizing the opportunity for success of the most able students. We also know that it is possible to delay streaming of students into alternative programs in mathematics at the secondary school level for a year or two longer than is common practice in many parts of the United States and Canada. A corollary of this finding is that all students at the junior high level can follow the same mathematics curriculum.

Achievement comparisons can also provide indications about what is possible. Although growth in level of achievement from pretest to post-test in the longitudinal version of SIMS was generally quite modest, the fact that eighth grade students in France and Japan had extremely high growth scores on items directly related to topics emphasized in their curricula demonstrates that much more significant rates of growth are possible. Similarly, the consistently high levels of achievement attained by students from some countries illustrate what can be accomplished. They should serve as a spur and an incentive for improvement.

Methodological Considerations

The importance of opportunity to learn as an explanatory variable for differences in students' achievement has been underscored in several of the studies reviewed in this chapter. Attempts should be made to develop methods of measuring the influence of this variable on students' achievement in such a way that students' raw scores could be moderated to reflect its impact. Even if such an approach proves unfeasible in the short term, much can be done to improve the way in which opportunity to learn is measured so as to maximize the validity of the data collected.

The longitudinal version of the Second International Mathematics Study has demonstrated the value of having several data-collection points rather than just one. Information about changes in students' achievement and attitudes, and their relationship to the teaching practices employed by teachers, are crucial to the development of an adequate understanding of what happens in mathematics classrooms. That understanding is, in turn, fundamental to the successful design and implementation of revised curricula or alternative teaching practices.

Virtually all of the information gathered in the studies included in this review was collected through the administration of test items and questionnaires presented in multiple-choice format. This is a cost-effective and efficient technique for collecting large amounts of data from many respondents, but its limitations are well known. Information about how students approach the solution of a given problem is, in the final analysis, more important than whether or not they are able to recognize the correct solution. Similarly, to know what really goes on in mathematics classrooms, we need to observe what teachers actually do there, and not rely solely on what they say they do.

International, comparative studies of achievement have a significant contribution to make to the ultimate goal of improving the teaching and learning of mathematics. Properly conducted and appropriately reported studies of this kind will help increase our understanding of the role played by key variables in the process.

References

Azuma, H., Kashigawa, K., & Hess, R. D. (Undated). *Maternal attitudes, behaviors, and children's cognitive development: A cross-national survey between Japan and the United States.* Tokyo: University of Tokyo Press. (In Japanese)

Dirks, M. K. (1986). *The operational curricula of Mathematics 8 teachers in British Columbia.* Unpublished doctoral dissertation, University of British Columbia.

Dossey, J. A., Mullis, I. V. S., Lindquist, M. M., & Chambers, D. L. (1988). *The mathematics report card. Are we measuring up?* Princeton: Educational Testing Service.

Eckstein, M.A. (1982). Comparative school achievement. In H. E. Mitzel (Ed.), *Encyclopedia of educational research* (Vol. 1). New York, NY: The Free Press.

Flanders, J. R. (1987). How much of the content in mathematics textbooks is new? *Arithmetic Teacher, 35* (1), 18–23.

Freudenthal, H. (1975). Pupils' achievements internationally compared. *Educational Studies in Mathematics, 6,* 127–186.

Fuson, K., Stigler, J., & Bartsch, K. (1988). Grade placement of addition and subtraction topics in Japan, China, the USSR, Taiwan, and the U.S. *Journal for Research in Mathematics Education, 19,* 449–456.

Gamoran, A. (1987). Organization, instruction, and the effects of ability grouping. *Review of Educational Research, 57,* 341–346.

Garden, R. A. (1987). The second IEA mathematics study. *Comparative Education Review, 31*(1), 47–68.

Ginsburg, H. P., & Baroody, A. J. (1983). *TEMA Test of Early Math Ability.* Austin, TX: Pro Ed.

Gorman, T. P., Purves, A. & Degenhart, R. E. (1988). *The IEA study of written composition: Vol. 1. The international writing tasks and scoring scales.* Oxford: Pergamon Press.

Harnisch, D. L., Walberg, H. J., Shiow-Ling, T., Takahiro, S., & Fyans, L. J., Jr. (1985). Mathematics productivity in Japan and Illinois. *Evaluation in Education: An International Review Series, 9* (3), 277–284.

Hess, D., Azuma, H., Kashigawa, I., Dickson, W., Nagano, S., Holloway, S., Miyake, K., Price, G., Hatano, G. & McDevitt, T. (1986). Family influences on school readiness and achievement in Japan and the United States: An overview of a longitudinal study. In H. Stevenson, H. Azuma, and K. Hakuta (Eds.), *Child development and education in Japan.* New York, NY: Freeman.

Hirstein, J. (1980). From Royaumont to Bielefeld: A twenty year cross-national survey of the content of school mathematics. In H. G. Steiner (Ed.), *Comparative studies of mathematics curricula—Change and stability, 1960–1980* (pp. 55–65). Bielefeld FRG: Institut für Didaktik der Mathematik.

Holloway, S., Kashiwagi, K., Hess, R. D., & Azuma, H. (1986). Causal attributions by Japanese and American mothers and children about performance in mathematics. *International Journal of Psychology, 21,* 269–286.

Husén, T. (Ed.). (1967). *International study of achievement in mathematics* (Vols. I and II). Stockholm: Almqvist & Wiksell.

Husén, T. (1983, March). Are standards in U.S. schools really lagging behind those in other countries? *Phi Delta Kappan,* pp. 455–461.

Inkeles, A. (1977). A Review of "International Studies in Education." In *Proceedings of the National Academy of Education, 4,* 139–200.

International Association for the Evaluation of Educational Achievement. (1988). *Science achievement in seventeen countries: A preliminary report.* Oxford: Pergamon Press.

Kilpatrick, J. (1971). Some implications of the international study of achievement in mathematics for mathematics educators. *Journal For Research in Mathematics Education, 2,* 164–171.

Lapointe, A. E., Mead, N. A., & Phillips, G. W. (1989). *A world of difference: An international assessment of mathematics and science.* Princeton, NJ: Educational Testing Service.

Lee, S., Ishikawa, V., & Stevenson, H. W. (1987). Beliefs and achievement in mathematics and reading: A cross-national study of Chinese, Japanese and American children and their mothers. *Advances in Motivation and Achievement: Enhancing Motivation, 5,* 149–179.

McKnight, C. C., Crosswhite, F. J., Dossey, J. A., Kifer, E., Swafford, J. O., Travers, K. J., & Cooney, T. J. (1987). *The underachieving curriculum.* Champaign, IL: Stipes Publishing Company.

Miller, D., & Linn, R. (1989). Cross-national achievement with differential retention rates. *Journal for Research in Mathematics Education, 20,* 28–40.

Organization for European Economic Co-operation (1961). *New thinking in school mathematics.* Paris: O.E.E.C. Publications.

Peaker, G. F., (1969). The international study of achievement in mathematics: A few examples of special interest to teachers. *International Review of Education, 15,* 222–229.

Postlethwaite, T. N. (1971). International association for the evaluation of educational achievement (IEA)—The mathematics study. *Journal for Research in Mathematics Education, 2,* 69–103.

Robitaille, D. F., & Garden, R. A. (1989). *The IEA study of mathematics II: Contexts & outcomes of school mathematics.* Oxford: Pergamon Press.

Robitaille, D. F., & Taylor, A. R. (1989). Changes in patterns of achievement between the first and second mathematics studies. In D. F. Robitaille & R. A. Garden (Eds.), *The IEA study of mathematics 11: Contexts & outcomes of school mathematics.* Oxford: Pergamon Press.

Ryan, D. W., & Anderson, L. W. (Eds.). (1984). Rethinking research on teaching: Lessons learned from an international study. *Evaluation in Education, 8.*

Song, M. J., & Ginsburg, H. (1987). The development of informal and formal mathematical thinking in Korean and U.S. children. *Child Development, 58*, 1286–1296.

Special Report: Dallas Times Herald (1983, December 11–12). American Education, The ABCs of Failure. Author.

Stevenson, H., & Bartsch, K. (in press). An analysis of Japanese and American textbooks in mathematics. In R. Leetsma & H. Walberg (Eds.), *Japanese education.* Greenwich, CT: JAI Press.

Stevenson, H. W., Lee, S. Y., & Stigler, J. (1986). Mathematics achievement of Chinese, Japanese and American children. *Science, 231,* 693–699.

Stigler, J. W. & Baranes, R. (1988). Culture and mathematics learning. In E. Z. Rothkopf (Ed.), *Review of research in education* (pp. 253–306). Washington, D.C.: American Educational Research Association.

Stigler, J. W., & Perry, M. (1988). Cross-cultural studies of mathematics teaching and learning: Recent finding and new directions. In D. A. Grouws, T. J. Cooney, & D. Jones (Eds.), *Effective mathematics teaching.* Reston, VA: National Council of Teachers of Mathematics.

Stigler, J. W. & Perry, M. (in press). Mathematics learning in Japanese, Chinese and American classrooms. In G. Saxe & M. Gearhart (Eds.), *Children's Mathematics.* San Francisco, CA: Jossey-Bass.

Sugiyama, Y. (1987). Comparison of word problems in textbooks between Japan and the U.S. In J. P. Becker & T. Miwa (Eds.), *Proceedings of U.S.-Japan Seminar on Problem Solving.* Carbondale, IL: Board of Trustees, Southern Illinois University.

Theisen, G. L., Achola, P. W., & Boakari, F. M. (1983). The underachievement of cross-national studies of achievement. *Comparative Education Review, 27,* 46–68.

Travers, K. J., Garden, R. A., & Rosier, M. (1989). Introduction to the study. In D. F. Robitaille & R. A. Garden (Eds.), *The IEA study of mathematics II: Contexts and outcomes of school mathematics* (pp. 1–16). Oxford: Pergamon Press.

Travers, K. J., & Westbury, I. (1990). *The IEA study of mathematics I: Analysis of mathematics curricula.* Oxford: Pergamon Press.

Walberg, H. J., Harnisch, D., & Tsai, S. L. (1986). Elementary school mathematics productivity in twelve countries. *British Educational Research Journal, 12,* 237–248.

Walker, D. A. (1976). *The IEA six subject survey: An empirical study of education in twenty-one countries.* New York, NY: John Wiley.

Wilson, J. W. (1971). Evaluation of learning in secondary school mathematics. In B. S. Bloom, J. T. Hastings, & G. F. Madaus (Eds.), *Handbook on Formative and Summative Evaluation of Student Learning* (pp. 643–696). New York, NY: McGraw-Hill.

Wilson, J. W., & Peaker, G. F. (Eds.). (1971). International study of achievement in mathematics [Special Issue]. *Journal for Research in Mathematics Education, 2*(2).

28

INTERNATIONAL PERSPECTIVES ON
RESEARCH IN MATHEMATICS EDUCATION

Alan J. Bishop
UNIVERSITY OF CAMBRIDGE

The last 20 years have witnessed a remarkable growth in the awareness of the international dimension in mathematics education. This book is being written 20 years after the first International Congress on Mathematics Education took place in Lyons, France, when nearly 1,000 participants from many countries came together and began seriously to try and understand each others' approaches to issues in mathematics education. At that Congress there was apparently only one plenary address which considered research directly—Begle's "The role of research in the improvement of mathematics education." In that address he made the plea:

I see little hope for any further substantial improvement in mathematics education until we turn mathematics education into an experimental science, until we abandon our reliance on philosophical discussion based on dubious assumptions and instead follow a carefully correlated pattern of observation and speculation, the pattern so successfully employed by the physical and natural scientist. (Begle, 1969, p.110)

Nearly 20 years later, at the 1988 Congress in Budapest, with 2,500 participants, there was a whole Theme Group concerned with "The Practice of Teaching and Research in Didactics," which was, itself, organized into nine working groups with titles such as "On the relations between teachers and research," "Observation of teaching practice," and "Theory-practice relations," and with key contributors from about 20 different countries (Hirst & Hirst, 1989).

As the amount of research has increased in that period, together with the number of people from different countries engaged in research, so has the range of perspectives on research. Several of these perspectives are, of course, reflected in the ear-

lier chapters of this book, but it is the intention in this chapter to focus on international perspectives, to describe them more formally, and to reflect on their significance.

Why are some perspectives being referred to as "international?" Are there not just different perspectives because individual researchers have different preferences? If there are, indeed, any national characteristics of such research, as is implied by the word "international," should not the chapter have been entitled "National perspectives on...?" Why, then, are we considering international perspectives?

It is my contention that our research field has grown from a situation where, even if it was once possible to talk of national perspectives, given the recency of organized mathematics education research in most countries it is now not only more difficult, but also, in my view, relatively unproductive to continue to do that. As a result of an increase in international meetings and conferences, internationally cooperative research projects, and books, reports, and journals coming from different countries, there is an obvious increase in the range of international influences to which researchers are subjecting themselves. As the number of people engaged in research in mathematics education has increased, and as they have made their ideas more public and shareable, so there has been a growing mutual international influencing of ideas, methods, practices, and expectations (see Bauersfeld, 1977).

This mutual influencing has not, however, led to a unification and conformity of research, although there are similarities in approach to be seen in different countries. Such a unification might lead to what might be termed *the* international perspective on research if it did exist, and as one might well be able to describe in a more unified field like mathematics, itself. It has,

The author gratefully acknowledges the helpful comments provided by David Wheeler, Concordia University.

however, increased researchers' awareness of a number of key issues arising from the historically, culturally, and socially different approaches to mathematics education seen around the world.

This heightened awareness has led both to positive effects, such as the development of a richer understanding of the value of different approaches to enquiry, and also to negative effects such as tedious, verbal, power struggles over what constitutes "proper" research. The objective of this chapter, then, is not just to document and delineate this heightened awareness but also to try to indicate productive directions for its development. What can we learn from research as it is practiced by researchers in different countries? How can we gain more access to knowledge of their procedures? How can we productively use the contrasts which exist between what any one of us might do and what anyone else does? What aspects of our research activity become heightened by these comparisons?

HOW TO CONSTRUE THE WORD RESEARCH IN THIS CHAPTER

Let us begin by returning to what I said in the first paragraph, that at the Lyons ICME there was *apparently* only one address which focused on research in mathematics education. There were however several other speakers who, while not perhaps focusing on the role of the enquiry process, indicated first that research was an important part of their work and second that the kind of research procedures they used were not necessarily the same. For example, B. Christiansen (1969) from Denmark talked about "Induction and deduction in the learning of mathematics and in mathematical instruction" in which he described some experimental teaching done with 65 classes, averaging 25 students each, in the sixth and seventh grade. This experimental teaching was evaluated by getting "informal reports from the teachers on the experiences collected in connection with the rather extensive change of curriculum and of educational approach" (p.24).

In a similar vein, R. Gauthier (1969) from France in "Essai d'individualisation de l'enseignement" described some experiments with individualized teaching carried out by l'Institut Pedagogique National with 30 teachers and all their classes in the Lyons area of France. This study involved the extensive use of *fiches de travail* (work cards) and although the details of the research procedure were not spelled out in the paper it is clear that "collaborative observation" was used.

A. Revuz (1969) also reported on some French research in "Les premiers pas en analyze" concerned with revising the teaching of elementary analysis in the French secondary schools. He talked about some studies carried out by teachers in the Institut de Recherche sur l'Enseignement des Mathématiques (I.R.E.M.) at Paris into the effects of introducing the new programs.

On an even larger scale, A. Markouchevitch (1969) reported on "Certains problèmes de l'enseignement des mathématiques à l'école" in the USSR, including the involvement of 500 researchers, academics, and teachers in evaluating academic teaching prior to the introduction of compulsory secondary

schooling from 7 to 17 years throughout the USSR. Kolmogorov was in charge of revising the mathematics part of the curriculum.

At the other end of the "size of the sample" scale, E. Fischbein (1969), then in Rumania, talked about "Enseignement mathématique et developpement intellectuel" in which he described various research studies in the "psycho-pedagogique" style concerned with aspects of intuition, particularly focusing on the Piaget/Bruner controversy. Also, E. Castelnuova (1969) (Italy) and F. Papy (1969) (Belgium) each reported on teaching experiments they had personally carried out using some new practical aids—Castelnuova's being concerned with "différents représentations utilisant la notion de barycentre" (center of gravity) and Papy with a device she called a "minicomputer."

As another example, Z. Krygowska (Poland) in "La texte mathématique dans l'enseignement" (1969) focused on the role of the mathematics textbook, analyzed various problems with its use and referred to some observations obtained from classes of 14–15-year-olds.

Already, at this first ICME, it was clear that serious and systematic enquiry in mathematics education was an important part of several mathematics educators' armory but it was also clear that there were several differences in approach, particularly over questions like. What should be the use of classroom evidence? How necessary is any theory to what evidence is collected? What is the research aimed at? How should the research relate to existing practice? How big should the sample of evidence be?

It is not my task in this chapter to define, describe and analyse the different research paradigms used in our field. It is necessary, though, for me to make clear what I will consider under the heading of research in this chapter, for two principal reasons. First, it is necessary because any author's orientation towards research will inevitably influence what perspectives emerge in their chapter. Second, it is necessary because of the particular task of this chapter to review the international context of research, and no one can be a neutral observer of the international scene.

I take the view that to qualify as research in mathematics education, a study in our field needs three components:

- Enquiry, which concerns the reason for the research activity. It represents the systematic quest for knowledge, the search for understanding, and gives the dynamism to the activity. Research must be *intentional* enquiry.

- Evidence, which is necessary in order to keep the research related to the reality of the mathematical education situation under study, be it classrooms, syllabuses, textbooks, or historical documents. Evidence samples the reality on which the theorizing is focused.

- Theory, which recognizes the existence of values, assumptions, and generalized relationships. It is the way in which we represent the knowledge and understanding that comes from any particular research study. Theory is the essential product of the research activity, and theorizing is, therefore, its essential goal.

All three components are necessary in my view, although there can be many arguments over what constitutes enquiry, theory and evidence, and over the relationships between the three. Because of my task in this chapter, I accept broad definitions of all three components.

However, I would already argue that some activities, such as the mere accumulation of data, the reporting of accidentally occurring incidents, the detailing of abstract analyses, the offering of "armchair" speculations, the planning of a curriculum or a lesson, the designing of some teaching materials, or the setting of an examination, do not, of themselves, constitute research activities, although each may make a contribution to the research process at some stage. Indeed, much research has developed from deliberately considering as problematic the various activities associated with doing mathematics education (for example, writing materials, planning curricula, teaching lessons, and assessing children). If the object of research is the improvement of mathematics teaching then it clearly makes sense to examine the normal activities associated with mathematics teaching to see if they are creating obstacles to improvement; that is, to what extent are they part of the solution, rather than just part of the problem? Those activities are not, themselves, *research* activities.

How, then, should we understand three other plenary contributions to the first ICME congress, which also had a highly scholarly and "researchy" appearance? I am referring here to the contributions of Servais, Steiner and Rosenbloom. This is not to denigrate or ignore the worth of the remaining addresses; it is just that they do not shed as much light on the issues of this chapter.

These three studies were all concerned with experimental approaches to teaching the mathematical content, but their emphasis was particularly on making the mathematical *content* accessible to the learners. W. Servais (1969) (Belgium) in his paper "Logique et enseignement mathématique" talked about the use of switching circuits to help the teaching of logical relations. H. G. Steiner (1969) (W. Germany) discussed "Magnitudes and rational numbers—a didactical analysis" while P. C. Rosenbloom (1969) (USA) focussed on "Vectors and Symmetry." All three speakers presented extended mathematical analyses of their topics, from an educationally theoretical point of view. They were clearly not mere "armchair" speculations. However they only briefly discussed their actual teaching.

What, then, are we to make of these scholarly offerings? Are they to be considered as research, as preliminaries to research, as alternatives to research, or as applications of research? Where do they fit into a schema of research on mathematics education relevant to this chapter and this book?

If we are to try to make sense of the international research scene, then I suggest a profitable way to understand these studies is as research, but from a rather different tradition. The 1969 ICME was the first occasion that such a large international group of mathematics educators had congregated to reflect on issues in mathematics education, and for most members it was the first time they had the opportunity to survey different approaches to the serious study of our field. Speakers came from different countries, but more significantly, they came from rather different research traditions.

A research tradition is very different from a research paradigm. One doesn't choose a research tradition as one might a research paradigm; one is part of it, as a result of upbringing,

education, cultural background, and research training. From my reading and my analyses, there are three very different research traditions that I think have informed international perceptions about research in mathematics education, and that are illustrated in part by the contributions to the first ICME. I will refer to them as the Pedagogue tradition, the Empirical-scientist tradition, and the Scholastic-philosopher tradition. (This analysis has several connections with Habermas' 1972 analysis, extended into educational research by Carr, 1985, who in his paper explores the Historical/Hermaneutic, Empirical-analytic, and Critical sciences traditions, and their value frameworks.)

The Pedagogue tradition is probably the oldest one. The names of Socrates, Comenius, Pestalozzi, and Froebel would be associated with this tradition, as would more recently, and more mathematically, researchers such as Polya, Beberman, Gattegno, Dienes, and Freudenthal. This tradition celebrates the role of the gifted thinking, reflective teacher. The teacher is perceived as a craft practitioner continually endeavoring to improve his or her practice. Research is an integral part of being a good teacher in this tradition, with experiment and observation being the key components of the research. There is a strong humanistic and personal value element in this tradition, with a tendency towards a liberal-progressive educational ideal.

Research reporting in this tradition is largely a matter of passing on both the wisdom of the expert practitioner and their teaching methods. The ideas of Polya (1954), Beberman (1958), Gattegno (1971), Dienes (1973), and Freudenthal (1973) are considered to be highly significant; theory in this tradition concerns the developed ideas of these expert teachers, with the criteria for their acceptability being their intuitive validity for all teachers. The evidence presented is usually highly selective and exemplary, demonstrating perhaps what a gifted child has produced or what even the slowest learner has managed to do. The research method is deceptively simple: be ingenious and inventive as a teacher, be brave and try out ideas with your pupils, and observe, observe, observe. "Learn from the children as they try to learn with you," seems to be the message. (See for example, Brousseau, Davis, & Werner, 1986.)

At the ICME in 1969 the papers of Emil Castelnuova and of Frederique Papy were strictly in that tradition but others were influenced by it. Both of them reported on aspects of creative teaching illuminated by examples of significant contributions from their pupils. Their reports aimed to persuade by means of demonstration. The ideas were intended to be intuitively reasonable, the observed responses from the children to convince, and the methodology of the enquiry was clear, as was the goal of the liberal-progressive improvement of mathematics teaching. This tradition is still alive and well, and probably best illustrated today by the work of the International Commission for the Study and Improvement of Mathematics Teaching (CIEAEM) at its annual conferences.

The second tradition, that of the Empirical-scientist, is a much more recent phenomenon in education, and is well described in Begle's paper at the 1969 ICME meeting and later in Begle (1979). In this tradition it is evidence which is the key to knowledge, and the research process focuses attention on the methods of obtaining that evidence and of analyzing it, often quantitatively. A radical Empiricist would probably judge that much of the research reported at that ICME would not have been acceptable, with the possible exception of Fischbein's. A less radical view would have recognized several of the already mentioned studies as having been influenced by the tradition. However the data of Castelnuova and of Papy already referred to would not be accepted by this tradition, as the teacher is arguably in the worst position to be the evidence gatherer. The evidence, as far as is possible, should be objective and indisputable. The analysis would often be prescribed and standardized, so that the results obtained from the analysis would be the same across researchers.

In this tradition the research methodology is very important. Indeed to a large extent, one tries to minimize the influence of the teacher's individuality. Improvement in mathematics education is an ultimate goal of this tradition, but it comes about by the gradual accumulation of relevant evidence. Reports of such research are judged by the quality of the evidence and research procedure used. Assuming that acceptable procedures have been followed, the evidence must then be explained by reference to theory.

As Begle argued, the radical Empiricist would turn mathematics education into an experimental science, like natural science, with disputes between competing theories being settled by reference to the critical evidence. As I have said, the only research paper presented at the 1969 ICME which would have come anywhere near these strong criteria would have been Fischbein's, but Begle's call was heeded by many mathematics educators and the Empirical-scientist tradition is a much more dominant one today in mathematics education than it was then. The *Journal for Research in Mathematics Education* has long been associated with this tradition, but we can also see it influencing much of the work of the International Group for the Psychology of Mathematics Education (PME) at its annual conferences. (See, for example, Bergeron, Herscovics, & Kieran, 1987.)

Third, there is the Scholastic-philosopher tradition. This also has a long history, and in this tradition education is taken to be one of the humanities. Here it is analysis, rational theorizing, and criticism which are important, with the real teacher and real classrooms being somewhat imperfect manifestations of the theoretic educational situation. This tradition has long had an appeal for mathematicians and for certain mathematics educators, perhaps because of its Platonic-like approach, with educational abstractions and critical theorizing as its principal components.

This, I would argue, is the tradition into which the 1969 ICME presentations of Servais, Steiner, and Rosenbloom fit. The hallmark is extensively theoretical and didactical analysis, with only minimal reference to evidence. Servais reported some limited examples of pupils' activity, only to demonstrate what could be achieved with his approach and as befits someone who was also a strong Pedagogue. Steiner makes this tradition's view of the role of evidence clearer with his concluding remarks:

Much of the complex mentioned above is left open for further development and experimentation. This is especially true for the use of systems of quantities in school mathematics. On the basis of a precise mathematical understanding, I am sure it will be possible to find adequate ways of teaching it. (p.259)

Rosenbloom was less sanguine in his comments. He said:

When vector geometry is introduced in high schools in an experimental program, there may be difficulty in assuring the continuity of the participation of the students from one year to the next. Otherwise one will not be able to build on the intuitive preparation provided in the earlier years. Of course, in most countries there will be a serious problem in teacher education. (p.274)

Improvement in mathematics education under this tradition implies the imperfect reality coming closer to the perfect theory. Improvement here would be recognized in the development of a greater critical and reflective awareness, but the real research goal is to develop theoretical knowledge by analysis and criticism, with evidence assumed to be known. This tradition also exists today and is perhaps best illustrated by the work of the international group on the Theory of Mathematics Education (TME) (see Steiner, 1984)

As we have seen, the roles of evidence and theory vary between the three traditions, as does the goal of enquiry, as the summary table below shows. However, the three components of research are all recognizably present in some role.

Theory	Goal of Enquiry	Role of Evidence	Role of Theory
Pedagogue tradition	Direct improvement of teaching	Providing selective and exemplary children's behavior	Accumulated and shareable wisdom of expert teachers
Empirical scientist tradition	Explanation of educational reality	Objective data, offering facts to be explained	Explanatory, tested against the data
Scholastic philosopher tradition	Establishment of rigorously argued theoretical position	Assumed to be known. Otherwise remains to be developed	Idealized situation to which educational reality should aim

For the purposes of this chapter I have characterized these three traditions rather starkly and I apologize to any colleagues who feel unhappy with my characterization insofar as it applies to them. It is very much a preliminary structuring for the rest of my analysis.

I see it as preliminary in two senses. First, I feel it represents where research in our field was 20 years ago. I have chosen the 1969 ICME deliberately as my starting point because of its uniqueness in the recent history of our field. Things were never quite the same again. Although as I have indicated one can still find research groups, conferences, and papers which mainly follow one of the three traditions outlined here, it is more the case that research studies nowadays are influenced to some degree by all three traditions.

Second, it is a preliminary attempt at a classification from an international perspective, and one which, in my view, provides a better introduction than would a national classification to the issues important to address in the rest of this chapter.

This choice needs justifying further, by asking the question "To what extent can we perceive national perspectives on research in mathematics education?" Is there a typically American research study, a typically British approach to research, or a typically German tradition of research in mathematics education? There are clearly phenomena which those living and working outside a country will recognize as a significant feature. For example, the British are assumed to prefer a pragmatic approach to everything, including research, the Dutch research seems to be closely related to curriculum development, the Germans apparently value didactical analysis, the French are developing what appears to be a new tradition with their *didactique* school of research, the Russians created their teaching experiment, and empiricism seems to epitomize North American research.

But does this convenient labeling really describe *national* characterizations? Are there no British theoretical developments? Is there no one in Holland who is not pursuing curriculum development research? Are there no German or French empiricists? Is all Russian work experimental? Is there no North American research not of an empiricist style? It seems to me that national descriptors are likely to be unproductive as analytic tools.

However, we can begin to understand some of those developments in different countries in terms of the three traditions just outlined. For example, I would argue that Britain has been strongly associated with the Pedagogue tradition, which as we have seen does have a strong realistic and pragmatic feel about it, while Holland and Scandinavia balance this tradition with that of the Scholastic-philosopher. German and French researchers are also strongly influenced by the latter tradition, while with their different socio-political styles, the Russians and Americans are strongly associated with the Empirical-science tradition.

The extent to which the reader agrees or disagrees with these assessments will undoubtedly vary, but my hope is that any researcher will be able to use the description of the three traditions to reflect on their own research activities. Moreover, I want a preliminary orientation and structure which is recognizable everywhere. The Pedagogue tradition is well known, for example, throughout Asia and Africa, the Empirical-scientist tradition has its roots in the systematic observation of phenomena, a practice known by scientists on every continent, and the Scholastic-philosopher tradition has its roots in the language of dispute, also known everywhere. These traditions, therefore, exist to some degree in all countries, and they will provide us with a more satisfactory basis for international analysis than simple national descriptors.

CRITICAL RELATIONSHIPS IN RESEARCH IN MATHEMATICS EDUCATION

From an international perspective certain relationships stand out as being very important when conducting any research in mathematics education; they are important because they delineate the ways researchers in different societies and countries approach their tasks. One value of a chapter such as this is to move beyond mere comparison of approaches, and to help those who do research to understand their work from the wider

perspective offered by the international context. Furthermore, I would maintain that nobody in the position of determining what research is done and how it is done should avoid consideration of these relationships. This is why I offer them as critical relationships, and the international perspective highlights them more than could any within-country analysis.

The critical relationships I will consider in the rest of this chapter are those between the following

1. "What is" and "what might be"
2. Mathematics and education
3. The problem and the research method
4. The teacher and the researcher
5. The researcher and the educational system

The Relationship Between "What Is" and "What Might Be"

I have placed this relationship first because it is one which has been present ever since enquiry began, and because for a researcher it probably determines more of the balances in the subsequent relationships than does any other. To what extent is one concerned with researching the educational situation as it currently exists, rather than researching educational possibilities for the future? For someone schooled in the Pedagogue tradition there is an immediate concern for the improvement of mathematics teaching (however unfocused the criteria might be), with no need to uncover "what is," except as a starting point. For the Scholastic- philosopher one can see more of the "what should be," imperative present, while for the Empirical-scientist, the immediate concern is with "what is" and with how to explain it, and improvement must be far more gradual.

The contrast in this relationship is well illustrated by what are referred to as status studies on the one hand, essentially looking at "what is," and exploratory studies on the other, focusing more on "what might be." There are two main reasons for the existence of many status studies. First, there is the clear influence of the Empirical-scientist tradition—expressing the sentiment that natural science's main research concern is with understanding the situation rather than with changing it. That is another's task, not the scientist's. The second reason has more to do with the individual country's organization of education. In noncentralized systems such as exist in most parts of Europe, America, and Australasia, there is often such a wide diversity of practice of mathematics education that there is a great need to document what is actually happening. This is the main reason why survey research methods are so often taught to and used by researchers in these countries. In the more centralized educational systems of socialist countries and much of Asia and Africa, the need to discover what is happening in the system has not been so important a stimulant for research. This is not to say that in centralized systems there is no knowledge of what is happening in those systems, nor that everyone knows what is happening, but rather that in decentralized systems the variety of practice is accepted and is a powerful stimulant of research. When a researcher moves beyond mere data acquisition of those practices, the diversity and variety of practice which exists in many systems can certainly be valued as a source for ideas.

Surveys, however, are not the only way of finding out "what is" in the educational process. We have seen more and more researchers in several countries using case studies to give a finer focus on individual situations, observing either learners, teachers, or specific classrooms. (See for example Booth, 1981; Brousseau, 1984; Krutetskii, 1976; and Thompson, 1984.) Often these studies are undertaken from a psychological perspective but more recently anthropological approaches have been employed (Carraher, Carraher, & Schliemann, 1985; Cobb, Yackel, & Wood, 1988). Perhaps indeed the "what is" approach is more aptly demonstrated by these approaches, since it is clearly felt by both psychologists and anthropologists that they are not attempting to change situations but merely to study them, thus reflecting their empiricist position strongly.

At the opposite end of the scale is exploratory research, which has the express aim of pursuing "what might be" in mathematics education. Such studies can often be informed by the Pedagogue tradition, with at times a strongly innovative and unconventional character (for example, Papert, 1980), but they can also be strongly influenced by the Scholastic tradition (see Mellin-Olsen, 1987). Good exploratory research needs to be strongly theorized and argued research because of the social and political investment it entails. To engage in exploratory research is to challenge the status quo, and to be able to do this a researcher needs the full armory of theory analysis and logic allied to good craft credentials. I am not, of course, referring to practices which are easily accepted within the educational norms, but to the development of ideas which will challenge whatever the educational norms are for that society. In the case of mathematics education, we certainly have seen examples of strong theoretical argument which develops the new ideas, and experimental, innovative practice follows. In the past these experiments were both small-scale, as in the case of experiments with structured apparatus, and much larger, as was seen in the Modern Mathematics era. (See for example the more recent studies of Bednarz & Janvier, 1988; Fielker, 1987; Gerdes, 1988; Hillel, Kieran, & Gurtner, 1989; and Hoyles & Noss, 1986.) Nowadays much of the pioneering work with microcomputers is of the small-scale exploratory variety, challenging many of the educational and mathematical norms of present society.

The tensions which exist between these two research extremes are sometimes voiced but rarely written about seriously; the "what is" researcher will accuse the "what might be" researcher of ignoring the real situation, of not knowing where to start from, of proselytizing rather than doing proper research, and perhaps of exhibiting intellectual arrogance. The "what is" researcher can equally be accused of upholding unstated educational values, of an obsession with gathering so-called objective facts, of ivory-tower remoteness, and of a lack of committment to any educational ideals. (The Brophy/Confrey exchanges in the November 1986 issue of the *Journal for Research in Mathematics Education* illustrate the contrasts well.) Both can learn from the other, of course, and the appropriate balance must be struck by every researcher, depending on their social and political situation. We will have more to say about this aspect in the discussion of the final relationship.

The Relationship between Mathematics and Education

This relationship is critical for any researcher in mathematics education because of the fact that each of these aspects can drive the research in very different directions. Both the Pedagogue and the Empirical-scientist traditions have focused on education predominantly while the Scholastic-philosopher tradition has a strongly mathematical emphasis, illustrated by such concerns as the "elementarizing" of mathematics for education purposes (see Kirsch, 1977).

For those researchers who belong to the education camp, it is obvious that problems in mathematics education are a subset of problems in education, while for the mathematics-oriented researchers, it is crucial to recognize the uniqueness of mathematics in the research. For those following the scientific tradition today, the issue is one of the appropriateness of any generalization; is the object of the study to be able to generalize to other educational situations or to other mathematical situations? For the modern Pedagogue, the difficulty with this particular relationship is that it doesn't focus essentially on children; if "education" implies a focus on children, then there is no doubt that the Pedagogue's sympathies will lie there. For the Scholastic-philosopher, it appears to be theorising about mathematical issues which provides the basis for theorising about mathematics education.

To some extent the researcher's position may well be determined by the kind of institution in which the researcher works. In a few countries there are now specialist research centers in mathematics education, but in most countries researchers either work in the context of education or of mathematics (for example, in the respective University, College, Departments, or Governmental Institutes). This situation inevitably influences them and will determine whether the research is predominantly educational, and where mathematics operates as a kind of place-holder, able to be replaced by science or language; or whether the research is predominantly mathematical, which in most cases, almost views mathematics education as mathematical training in the sense that the education of the whole child will tend to be ignored in the process.

Educationally driven research into mathematics education will tend to consider mathematics as a generalizable, often undifferentiated concept. Since the focus is on educational problems in the research, such as the influence of teachers' attitudes, it is possible to explore this in the field of mathematics education by attending to, for example, teachers' and pupils' attitudes towards mathematics together with measures of mathematical attainment. As long as a representative sample of mathematical concepts and skills are being used, the argument goes, one is able to infer about mathematical aspects, in general. Standardized mathematical measures which sample the mathematical field will often be used, and emphasis will be given to the educational problems involved. (See Brophy, 1986; Clarkson & Leder, 1984; Fennema, 1985; Good & Grouws, 1977; and Reyes & Stanic, 1988.) The relevant theory will also be educational, and this area of research will continue to be strongly influenced by research and theoretical developments from other educational fields, and from those disciplines which impact on education, notably psychology, anthropology, sociology and philosophy.

More mathematically-oriented researchers in mathematics education will, however, not be content to broadly sample the area of mathematics; the uniqueness of various mathematical ideas needs, in their view, to be recognized and represented. Geometry is very different from algebra, and arithmetic is very different from each of the other two. But finer distinctions than those are made, since each mathematical field can be found to provoke its own unique educational problems. The area of research known as mathematical didactics, which has a strong base in continental Europe, concentrates in this way on detailed analyses of mathematical concepts from an educational standpoint and is clearly in the best traditions of the Scholastic-philosophers (see Freudenthal, 1983; Krygowska, 1988; and Wittmann, 1984). Mathematics curriculum research is also heavily dependent on this tradition and on this approach to research (see for example, Howson, Keitel, & Kilpatrick, 1981) as are many psychological studies (see Vergnaud, 1988). The dangers of over-compartmentalization are great, however, and exciting developments in educational aspects of science, language, the arts, and technology can tend to be ignored. It is clear, though, that attention to mathematical detail in research has had a strong influence in our field. Indeed the growth of mathematics education as a unique discipline has come about largely by the increasing interrelationships between mathematical and educational concerns in research.

The Relationship between the Problem and the Research Method

This relationship is always critical in any field; for researchers in mathematics education, it is no different. The situation in the three traditions is clear: The Pedagogues were concerned with the development of new procedures and practices, the Empirical-scientists were concerned to develop standardized evidential procedures in order to explain reality, and the Scholastic-philosophers were occupied with theoretical analyses and qualities of argument.

It is the Empirical-scientists' tradition which appears to be most conservative in terms of methodology employed, with the two other traditions being much more concerned with the quality of the ideas and insights generated rather than with the method used. Certainly in terms of the criteria of quality of research, the scientist's tradition appears to have the most emphasis on method. In that tradition, for example, it is assumed that the research methods are teachable—the courses available and the books demonstrate this. For both of the other traditions the method is far less important than the insights obtained, and one has great difficulty, for example, in locating books on "how to be a good Pedagogue" or "how to be a good Scholastic-philosopher in education."

Clearly, a research procedure so closely linked with evidence must have ways to verify that evidence, and the Empirical-scientists' concerns with ideas of randomness, control procedures, sample quality, and other statistical concepts and methods are geared to that end. However, unless the experimental situation is well chosen and theoretically important, all that will be generated is yet more evidence, which as already has been pointed out is of little value by itself. Conversely, the value and power of this method is that, when used on interesting, well-theorized problems, it can generate important, interesting data which can advance our understanding. Perhaps there has been less attention to theory than is warranted, if this approach is to achieve its goal of explanation.

One peculiarity about Empirical-scientific research into mathematics education is that there appears to be little tradition of actual replication of significant research findings. When one thinks of the replications of Piaget's research around the world, it is remarkable how little of that happens with any other research developments. Certainly it is in the natural science tradition to replicate important studies, so one wonders why it is, when so many canons of the natural scientist have been assumed by the educational scientists, that the critical one of replicability has been ignored. It is not just a matter of *verifying* a particular theory such as Piaget's, it is that by attempting replication in particularly interesting situations one can help to *develop* the theoretical position. (See Lin's 1988 research in Taiwan which replicates and extends the CSMS research of Hart, 1981.) Perhaps journal editors have not welcomed such replication studies.

It is not the case of course that the only method-led research is in the Empirical- scientific tradition. There is method in the scholastic-philosopher's approach and in that of the pedagogue; thus, the logical analysis and the quality of the theorized argument will be important as a consideration in scholastic work. If the assumptions are well founded and the logical development sound, then the consequences must be correct. Once again, however, whether they are of any interest and relevance will also depend, as with scientific research, on the starting point and on the problem addressed.

For the researcher in the Pedagogue tradition, the key method is observation allied to the ability to draw insightful and meaningful conclusions. The sharpness of the observation will be an important criterion therefore in the quality of the research, together with the sensibility of the conclusions. The reports of the pedagogue researcher are judged by the plausibility, and by their perceived significance by the reader. Few people will recall the *methods* of Dienes, Gattegno, Beberman, Polya or Freudenthal, but their observations, ideas and theories have had a profound influence on our thought.

There are therefore both drawbacks and benefits from a method orientation to research. What of the problem-orientation? A truly problem-led approach has tended not to be possible in our field because of the limited range of research procedures assumed to be legitimate. However, now that more sociological, anthropological, and ethnographical research methods are available and their value is recognized, it is far more possible to undertake true problem-led research (see Bauersfeld, 1977; and Eisenhart, 1988). What criteria, though, would apply to this research? How can it be judged? The method-researcher would argue that, without methodological criteria, such research is worthless. The problem-researcher would judge research quality by what light is shed by the research on the problem under consideration: What can the research tell us that we didn't know already about that problem?

The real problem will not, of course, actually be solved by the researcher but the intention is to use whatever research procedures seem appropriate and useful for shedding light on the problem. This means that it may well be interesting and feasible to use for example some of the scientific methodology on this problem; no methodology is ruled out by a problem-led approach. Several researchers were quite surprised when reading Krutetskii's (1976) work to meet first a vitriolic attack on another's use of factor analysis, and then to see Krutetskii himself using factor analysis. The point, of course, is not that this particular method is right or wrong, but that it was considered by Krutetskii to be inappropriate for the first problem yet appropriate for the second; for a problem-led researcher like Krutetskii, no method is right or wrong, merely more or less appropriate. Indeed, the problem-led approach can be seen to cut across all three traditions in research, although I personally feel it seems to be an extension of the pedagogue approach. The improvement stance of the Pedagogue tradition posits the idea of pedagogical problems to be solved, and what I suspect has happened has been that other methods and techniques have been encompassed in that approach.

The other issue which the Krutetskii story highlights concerns whether the researcher should play a strong role in choosing and defining the problem, and determining the research procedures to use, as well as in analyzing and theorizing the findings. The method-led research approach plays down the role of the individual researcher and tends to deny the humanness of the whole activity, thus emphasizing the assumed 'objectivity' of the method. Problem-led research on the other hand, depends for its success on very personal qualities of the researcher. Somewhere the researcher, like the artist, the craft practitioner, or, indeed, the teacher, needs to strike their own personal balance between technique and task, between problem and method, within the constraints laid down by their educational system.

The Relationship between Researcher and Teacher

This is a particularly significant relationship in mathematics education, a field clearly concerned with practicalities and so clearly involving other professionals, such as teachers. In the Pedagogue tradition, the teacher was the researcher and that tradition still continues, with teachers doing action-research and researchers often being reflective teachers, trying ideas out with children in laboratory schools, for example. For the other two traditions the separation is clear. The Scholastic-philosopher had no pretentions to be the teacher but was content to develop the appropriate theoretical construction of the mathematics educational situation. The Empirical-scientist was equally separated from the teacher; indeed it makes no sense for the teacher to be the researcher in this tradition. The scientist studies the teaching/learning situation and, to the extent that situation includes the teacher, the teacher must also be an object of study.

Indeed for both of these latter traditions the roles of teacher and researcher are incompatible: The teacher must act and must interact with the learners in an immediate and developmental manner, but the researcher needs to reflect and be analytic in one tradition and be observant and concerned with data collection and analysis in the other. Moreover, it is not just the incompatibility of roles which is important. The teacher is the practitioner whose practice, it is felt, needs to be informed by the research of the researcher. So, we have a clear hierarchy involved, with the researcher informing the teacher, but not necessarily vice versa. In a sense, then, both the scientist and the

scholastic-researcher have the "expert" role, with the teacher either trying to put into operation the researcher's recommendations personally or using whatever teaching materials have been developed as a result of the researcher's work.

At another level, the analysis and study of mathematics teaching from both these perspectives can make the teacher an object—not a subject—in the research. The individuality and humanness of the teacher can become at best an irrelevance and at worst a confounding influence. In much scientific research, for example, the aim will be to eliminate the influence of the individual teacher; in philosophical analysis, the teacher will tend to be an undifferentiated object, if considered at all. The same criticism can apply to these traditions' treatments of the children. They can be mere examples of stages or of particular types, demonstrators of certain errors or skills, or possessors of certain abilities and attitudes. They are rarely considered to be young people.

If it is the case that researchers believe their task is to help teachers to change their practice, it would seem much more sensible for these researchers and teachers to engage together in joint research. This is, indeed, happening more and more, but perhaps still not enough. For example, in the research on gender in mathematics education, we can still find status studies which, in their results, accuse teachers of bad practice, and not enough of the more powerful exploratory studies in which researchers and teachers collaborate with investigating procedures designed to help girls make better progress in mathematics. Of course it may still be the case that researchers believe their role is merely to develop human knowledge and understanding, and that it is someone else's role to determine what should be done about the use of that knowledge, (for example, government education officials, curriculum developers, or teacher educators).

In particular, it is clear that in some systems in the world there is much more investment in developing teaching materials and aids rather than in developing teacher education. Although this kind of statement is difficult to support with evidence and could easily be challenged, it is interesting to see how much effort goes into developing teaching materials in, for example, both Russia and the U.S. As would be expected in a highly centralized system, such as that in Russia, the control exerted by the textbooks is paramount, and thus the research in mathematics education is more generally directed at developing the textbook, rather than at developing the teachers. In the U.S., for other reasons perhaps to do with the shortages of well-qualified mathematics teachers, there is again a heavy investment in research which can feed ideas into textbook development. This tendency also is influenced by the Empirical-scientist's tradition, in which the results of research tend to be instrumentalized into technological products.

In countries which have had more of the Pedagogue tradition, as in parts of Western Europe and Asia, there has developed more research which recognizes the centrality of the teachers in the teaching situation (see the work of the BACOMET group, in Christiansen, Howson, & Otte, 1986). Consequently, in countries such as the United Kingdom, France, and the Netherlands, research on the teaching of mathematics is not only carried out in cooperation by teachers, but is often teacher-directed as well. The teacher as an action-researcher is a clear Pedagogue-related model. In teachers' centers in the United Kingdom, in the IREM in France, and in the Institute for the Development of Mathematics Education (IOWO) in the Netherlands, research and teacher development were almost synonymous, to the extent that researchers from other traditions would criticize these activities as teacher in-service education and not proper research.

Criteria of quality are difficult to describe in the Pedagogue tradition. By its nature, the quality of the research relies on the personal influence and standing of the particular pedagogue who is, therefore, open to charges of merely developing educational fashions or bandwagons. However, good quality Pedagogue research will document and elaborate the goals being sought, the social context of the innovative educational encounter, the observations of the pupils, the actions and feelings of the teacher, and the post-experience reflections preferably of all participants. The accounts are intended for teachers and other researchers-as-teachers, and need to be read in detail. Pedagogue-related research is not appropriate for reduction or encapsulation; its power lies in its richness rather than in any succinct theoretical construct. The pedagogical wisdom of Polya or Freudenthal cannot be understood by terms such as "discovery learning" or "learner-centered instruction." The wisdom of the pedagogue is revealed in the well-theorized and articulated innovative practice, in the empathetic elaborations of the educational situation, and in the insightful analysis of the experience.

Complex though all this seems, the more empathetic view of this kind of teacher-researcher relationship appears to be spreading because it is clear from other research that, no matter how good or bad the textbook or teaching materials may be, the role of the actual teacher is critical and cannot be ignored (see the account by Laborde, 1989, about the French research on "Didactiques des mathematiques"; and also Verstappen, 1986).

Another interesting aspect of this relationship is whether the research focuses on teaching or on learning in mathematics. There is a long history of research in mathematics education related to children and their struggles to come to grips with mathematical ideas. This research clearly has links with the Pedagogue tradition. One can also see a respect for the teacher's role and position with research containing a high element of child observation. The further one moves away from that approach and towards the scientists' use of sophisticated test materials reliant for their understanding on detailed statistical analysis, for example, the more the ideas are likely to be of value to text writers and, perhaps, to curriculum developers.

Equally with research concerned more with *teaching* mathematics, one can find examples which take an empathetic stance towards teachers and their human tasks (for example, Bishop, 1976; Bishop & Nickson, 1983; Thompson, 1984), as well as examples where the teacher is an object of study and is characterized as one example of a set of variables (for example, Merrill & Wood, 1974). Once again the former kind of research is likely to be of more value to those who work in the field of teacher development, whereas the latter may be of more use to textbook and material designers.

In countries with a strong Scholastic-philosopher tradition in research, one can find more concern for the issues of the curriculum. Indeed, where the teacher and the pupils are the central factors for the Pedagogue, and where the Empirical-scientist prefers to focus on situational conditions—and thus feed ideas into learning materials, institutional features, and textbooks—the Scholastic-philosopher appears to look to the mathematics curriculum and teaching schemes for solutions to the major problems of mathematics education (see Christiansen & Walther, 1986). With respect to the relationship under discussion, it is interesting to note the division of concerns into macro-curricular research (those analyses leading to the development of better overall school curricula and schemes) and micro-curricular research (those relating more to the local concerns of the teacher and to the smaller scale of the classroom group). The recognition of the mathematics teacher as a micro-curriculum developer is not yet present in many countries, and even if it is recognized, it may not necessarily be seen to be something to be valued and developed. Where it is recognized, there tends to be an importance in teacher education for developing the teacher's own mathematical knowledge and understanding.

This relationship, then, has many important aspects to it. Perhaps finally one might reflect on how people become researchers. I would suspect that the majority of people doing research in mathematics education are or have been teachers at either school or University level. This experience and the researchers' perceptions of this experience will surely influence, to some degree, the personal balance struck by any researcher in both their theoretical and practical stance with their teacher colleagues.

The Relationship between the Researcher and the Educational System

We have so far considered four relationships of critical concern to every researcher in mathematics education in any country. It cannot be claimed that there is any unique linear ordering of these, yet the reader should have sensed we are moving away from a within-researcher focus to a gradual consideration of other people with whom the researcher needs to interact. In this final relationship, the aim is to focus specifically on the balance to be sought by every mathematics education researcher in relation to the educational system in which they work.

It may seem to some that this relationship ought to have been considered first, because for many researchers their work situation—in relation to the educational system—appears to determine their space for research. However, as this chapter is intended to be of help to researchers in different countries, I feel it was vitally important *not* to start from that standpoint. That is why I began by considering the relationship of "what is" to "what might be." That is why I now approach this fifth relationship as the last of the group, because at stake is the autonomy and integrity of the individual researcher.

As we saw in the last relationship, the balance the researcher strikes with the teacher can vary enormously; this is also the case in relation to other people in the educational system. In many countries researchers in mathematics education exist in both research institutes and educational institutions. The former are either funded independently of the government or directly by the government, and their research is, therefore, influenced by this funding. Can anyone, then, be said to be a free researcher, in that sense? Certainly it is rare to find government funded researchers free to investigate whatever they wish. It is much more likely that researchers who are working in independently funded institutes and those who are also paid to do another job will perceive their research situation as being freer. The nature of the employer will, therefore, have an influence on the researcher's activities.

Traditionally, the assumption has been that it is in the Universities that academic freedom has been enjoyed, and certainly they are the institutions in which a great deal of research has developed and flourished. That assumption can be challenged in two ways: First, it is no longer the case that it is *only* in the Universities that such academic freedoms can exist. In educational systems which are relatively open and locally controlled, there is a great deal of local research activity between teachers, teacher educators, curriculum developers, and material developers, none of whom may work at a University. Equally, and perhaps more soberly, one can note that even the academic freedoms of the University researchers can be curtailed by the need to secure research contracts, to maintain satisfactory lists of publications, and to satisfy various working constraints and contracts.

Researcher freedom is, of course, entirely relative and controlled by the social and political norms of the society as reflected in the educational system of that country. Even in a socalled free society, as has already been noted, the researcher is subject to many social controls, particularly in the field of educational research. Society may happily tolerate a technological inventor researching a lost cause such as perpetual motion, but will not so easily tolerate an educational inventor seeking to challenge fundamental assumptions about how children should be educated. Paradoxically for some, there is no necessary reason why, in a more centrally controlled educational system and in an overtly less free society, researchers cannot generate interesting new ideas. The West, for example, has no exclusive rights on the development of knowledge, nor on research approaches in mathematics education, and one must certainly not confuse mathematics educational practice with mathematics education research. Equally, what researchers do and what they are allowed to research is one thing, and what the educational system does with what the researchers produce is quite another.

If there is a concern about the researcher's freedom, then there must also be a concern about the researcher's responsibility. To whom is the mathematics education researcher responsible and accountable? Rather, since these relationships are to be determined by any researcher, to whom *should* the researcher be accountable? We have already seen the differences between educational systems where research ideas feed into teacher education programs, text and material development, or curriculum development, and it is likely that such differences will also relate to the issue at hand.

If a researcher works in a way to help teacher development, then to a large extent the accountability should lie with the teachers, as we have seen. However, what appears to be happening is that, as educational systems develop their own complexities, it is becoming more difficult for researchers to talk directly with teachers. There are now many people working in very different ways that affect the quality of children's mathematical education, and, increasingly, researchers need to be able to communicate with these groups. Also, as research has developed our knowledge, we understand much more about the effects of institutional, curricular, examination, parental, text, media, and teacher influences. Researchers need to communicate with a wide variety of people about these ideas, and potentially many more people are interested in the results of their research.

Initially, it might seem that researchers are responding to this challenge. Research journals are growing, and more and more books and publications exist which contain reports of research and summaries of research. More and more conferences have research as their principal focus.

Yet, in reality, it seems as if researchers are not talking *directly* to the other people who are key aspects of the education system, but are increasingly talking to each other. Research journals are edited by researchers, with editorial boards of researchers, and they are increasingly read only by researchers. Books on research are expensive because the publishers know they only have a very limited market—only researchers buy them. Researchers are concentrating on communicating with other researchers and it is becoming increasingly difficult for any one else to hear what is being said, let alone to understand it (see Crosswhite's 1987 critique). Perhaps this situation is due to the method-led research taking over from the problem-led research. Perhaps it is that as research has grown, so theory has developed its own momentum, which means that it has lost touch with the educational reality it is intended to inform. Perhaps it is that obsessions with data gathering have meant that theoretical and generalizable interpretation has been relatively ignored, or the researchers' concerns about their academic freedoms have made them lose sight of their educational responsibility.

Accountability and responsibility concerns like these are not just about where, how, and what to publish. It is not sufficient for researchers to write their reports and think they have finished their work. As with the previous relationship, the researcher should not be the educational expert, telling those in the rest of the system what to do. Researchers should be expert at research and can advise, inform, educate, and generally democratize the research process, but they cannot be the educational expert. Many more people within the educational system have significant expertise that cannot be ignored.

Equally, those other experts in the educational system need to be in a position to tell researchers what they need to be advised and informed about. The situation which exists at present in some systems is essentially that researchers can determine both what problems they will research and how they will research them. This situation has come about partly because of the powerful positions in the educational hierarchy occupied by many of those who control research, and partly because of the desire for the maintenance of academic freedom by the researchers, particularly when this is felt to be under threat.

Perhaps the greatest difficulty for the researcher in striking the appropriate balance between freedom and responsibility is that the process of enquiry has its own satisfactions, it generates its own momentum, and it is so clearly, for anyone who engages with it, a neverending process. Most researchers are capable of generating research agendas stretching way beyond their lifespans, and being able to pursue their own agenda is one of the researcher's cherished goals. Even more frustrating, it always appears that researchers in other fields, such as literature, philosophy, ancient civilizations, and nuclear physics, are much more able to pursue their own agendas than are those in mathematics education. That, if one talks to such researchers, is largely a myth. The social and academic controls in other fields are just as great and the tensions between freedom and responsibility are as present as they are in mathematics education.

For the researcher in mathematics education, their responsibility is to the other people who work in the educational field, particularly to those whose expertise influences children's mathematical education. There is a great need for the development of team and group approaches to research in mathematics education, not just interdisciplinary teams (though they can be useful) but also inter-colleague teams, involving teachers, teacher educators, curriculum developers, text writers, and government advisers, as well as researchers. The responsibility must be demonstrated *before* the research begins, *during* it, and *after* it is over. Researchers who deny that that responsibility exists must suffer the consequences when their freedoms are curtailed.

ON SOURCES

In this brief survey of an immensely disparate and complex field I have tried to structure the key issues which I feel underlie research in mathematics education everywhere. In particular, I have approached the chapter from the viewpoint of researchers who are seeking to understand more about their field and how to research it. We can only begin to understand our own problems, research methods, approaches, and values by getting outside our own system, so to speak, and by contrasting those aspects with those of researchers in other national contexts. Researchers from other systems offer us a rich source of contrast if we are prepared and willing to understand them.

This chapter could have stopped here, but I felt that there is a final issue raised by my last exhortation: It concerns the sources of information. How can we learn about these other approaches? Essentially, there are two sources of evidence which have been used for this chapter and which can be used by other researchers. The first is written documentation, and the second is personal experience gained through discussions at conferences, meetings, and cooperative research activities. The main problems associated with the latter are fairly obvious: personal bias, accidental data, self-selected samples, and selective memory. However, there are also many advantages. Concerning the written documentation, there is clearly a great deal of research reporting in journals, books, reports, conference proceedings, and the like, all of which I shall generally refer to as *research reports*. From the final perspective of this chapter, though, we should consider these reports to be problematic. For example, if we are trying to get closer to what research is actually *done*,

then research reports are a less than perfect source due to factors such as competitive publications and marketing decisions.

Individually presented papers at conferences are another public source of information, although not every researcher with something important to tell the research community can attend international conferences because of costs and travel and political restrictions. Furthermore, the limited number of languages used for international communication also controls the accessibility of reports on research. For example, I would expect that perhaps 95% of the research reports used as data for this book were written in English, with perhaps another 4% written in French or German. There are, however, many reports presented or written, in local languages for the information of local people, but never known outside that situation.

Thus it is likely that only *some* research studies will be allowed to take place, because of the socio-political controls which apply in any society, only *some* reports of that research will be published, and only *some* of those reports will be accessible to a global public by virtue of being written in an internationally accessible language.

However, of value in writing this chapter have also been the various papers, articles, and books which have surveyed research in mathematics education either nationally or internationally (Bell, Costello, & Kuchemann, 1983; Easley, 1980; Galbraith, 1988; Jones, 1988; Lorenz, 1982; Romberg & Carpenter, 1986; Rouchier & Steinbring, 1988), together with the annual JRME research listings. In particular the ICME summary reports have been valuable, (Athen & Kunle, 1977; Carss, 1986; Editorial Board, 1969; Hirst & Hirst, 1989; Howson, 1973; Zweng, Green, Kilpatrick, Pollack, & Suydam, 1983) as have reports from national and international conferences (see Blane & Leder, 1988; Fennema, Carpenter, & Lamon, 1988; Howson & Wilson, 1986; Laborde, 1989; Nisbet, Megarry, & Nisbet, 1985; Popkewitz, 1984a; Yates, 1971; Steiner, 1984; Verstappen, 1986). These international reports and summaries can be considered to be just as problematic as the primary sources, if not more so in view of the complex socio-political contexts in which they operate.

In particular, these publications, with what I see as their tendency to proselytize, reveal to us that our research field appears in general to be neither a politically open nor a critically aware field. Having recognized the socio-political controls on research it is remarkable how little is actually written about this aspect. Of course it is only within the last twenty years that mathematics educators generally have become seriously aware of the socio-political dimension of their roles (see Griffiths & Howson, 1974; and Swetz, 1978), and it was only at the last ICME Congress in Budapest, 1988, that there was due recognition of this aspect in the official program (in the Fifth Day Special, Bishop, Damerow, Gerdes, & Keitel, 1989). Missing from even that part of the program was any discussion of the socio-political controls on research, although many of those present who carry out research surely know what those controls feel like and what they do to one's research agenda. (Sierpinska, 1989, refers to this aspect.)

There appear to be few internationally agreed rules about doing research with which one can criticize and challenge the research procedures and ideas of others or against which one can measure one's own research situation and context (see

Freudenthal, 1978; Popkewitz, 1984(b); and Kilpatrick, 1985 for different perspectives on this question). One wonders whether this is a passing state-of-affairs or whether it is endemic to the systematic study of education, itself an activity so subject to socio-political controls.

What I am certain about is that researchers, more than any other group in the mathematics education field, ought to be leading this field in fostering international understanding in developing shared criticisms, because at the heart of the researchers' activity is the admission of ignorance. Not knowing, or not understanding, is the root of our motivation for research; researchers everywhere share this motivation. Once we can recognize our internationally mutual state of ignorance, we can be freer to listen to alternative views, and we will undoubtedly find that our understanding benefits from this constructive alternativism.

References

Athen, H., & Kunle, H. (Eds.) (1977). *Proceedings of the Third International Congress on Mathematical Education.* Karlsruhe, Germany: University of Karlsruhe.

Bauersfeld, H. (1977). Research related to the mathematical learning process. (Survey lecture to group B4). In H. Athen and H. Kunle (Eds.), *Proceedings of the Third International Congress on Mathematical Education,* (pp. 231–243). Karlsruhe, Germany: University of Karlsruhe.

Beberman, M. (1958), *An emerging program of secondary school mathematics.* Harvard, MA: Harvard University Press.

Bednarz, N., & Janvier, B. (1988). A constructivist approach to numeration in primary school: Results of a three year intervention with the same group of children. *Educational Studies in Mathematics, 19,* 299–331.

Begle, E. G. (1969). The role of research in the improvement of mathematics education. In Editorial Board of Educational Studies in Mathematics (Eds.), *Proceedings of the First International Congress of Mathematical Education* (pp. 100–112). Dordrecht, The Netherlands: Reidel.

Begle, E. G. (1979). *Critical variables in mathematics education.* Washington: Mathematics Association of America and the National Council of Teachers of Mathematics.

Bell, A. W., Costello, J., & Kuchemann, D. E. (1983). *Research on learning and teaching mathematics.* Slough, U.K.: NFER-Nelson.

Bergeron, J. C., Herscovics, N., & Kieran, C. (Eds.). (1987). *Proceedings of the Eleventh International Conference on the Psychology of Mathematics Education.* Montreal, Canada.

Bishop, A. J. (1976). Decision-making the intervening variable. *Educational Studies in Mathematics, 7,* 41–47.

Bishop, A. J., Damerow, P., Gerdes, P., & Keitel, C. (1989). Mathematics education and society. In A. Hirst and K. Hirst (Eds.), *Proceedings of the Sixth International Congress on Mathematical Education,* (pp. 311–325). Budapest, Hungary: ICMI Secretariat and Janos Bolyai Mathematical Society.

Bishop, A. J. & Nickson, M. (1983). *Research on the social context of mathematics education.* Slough, U.K.: NFER-Nelson.

Blane, D. C., & Leder, G. C. (1988). *Mathematics education in Australia.* Melbourne, Australia: MERGA and the Mathematics Education Centre, Monash University.

Booth, L. R. (1981). Child-methods in secondary mathematics. *Educational Studies in Mathematics, 12,* 29–41.

Brophy, J. (1986). Teaching and learning mathematics: Where research should be going. *Journal for Research in Mathematics Education, 17,* 323–346.

Brousseau, G. (1984). Les obstacles epistémologiques et les problèmes en mathématiques. *Recherches en Didactique des Mathematiques, 4,*(2).

Brousseau, G., Davis, R. B., & Werner, T. (1986) Observing students at work. In B. Christiansen, A. G. Howson and M. Otte (Eds.). *Perspectives on mathematics education,* Dordrecht, The Netherlands: Kluwer.

Carr, W. (1985). Philosophy, values and educational science. *Journal of Curriculum Studies, 17,* 119–132.

Carraher, T. N., Carraher, D. W., & Schliemann, A. D. (1985). Mathematics in the streets and in schools. *British Journal of Developmental Psychology, 3,* 21–29.

Carss, M. (Ed.). (1986). *Proceedings of the Fifth International Congress on Mathematical Education.* Boston, MA: Birkhäuser.

Castelnuova, E. (1969). Différentes représentations utilisant la notion de barycentre. In Editorial Board of Educational Studies in Mathematics (Eds.), *Proceedings of the First International Congress on Mathematical Education* (pp. 175–200). Dordrecht, The Netherlands: Reidel.

Christiansen, B. (1969). Induction and deduction in the learning of mathematics and in mathematical instruction. In Editorial Board of Educational Studies in Mathematics (Eds.) *Proceedings of the First International Congress on Mathematical Education* (pp. 7–27). Dordrecht, The Netherlands: Reidel.

Christiansen, B., Howson, A. G., & Otte, M. (1986). *Perspectives on mathematics education.* Dordrecht, The Netherlands: Reidel.

Christiansen, B., and Walther, G. (1986). Task and activity. In B. Christiansen, A. G. Howson and M. Otte (Eds.), *Perspectives in mathematics education* (pp. 243–307). Dordrecht, The Netherlands: Reidel.

Clarkson, P., & Leder, G. C. (1984). Causal attributions for success and failure in mathematics: A cross cultural perspective. *Educational Studies in Mathematics, 15,* 413–422.

Cobb, P., Yackel, E., & Wood, T. (1988). Curriculum and teacher development: Psychological and anthropological perspectives. In E. Fennema, T. P. Carpenter, & S. J. Lamon (Eds.), *Integrating research on teaching and learning mathematics* (pp. 92–130). University of Wisconsin–Madison.

Crosswhite, F. J. (1987). Cognitive science and mathematics education: A mathematics educator's perspective. In A. H. Schoenfeld (Ed.), *Cognitive science and mathematics education* (pp. 265–277). Hillsdale, NJ: Lawrence Erlbaum.

Dienes, Z. P. (1973). *The six stages in the process of learning mathematics.* Slough, U.K.: NFER-Nelson.

Easley, J. (1980). Alternative research metaphors and the social context of mathematics teaching and learning. *For the Learning of Mathematics, 1,* 32–40.

Editorial Board of Educational Studies in Mathematics (Ed.). (1969). *Proceedings of the First International Congress on Mathematical Education.* Dordrecht, The Netherlands: Reidel.

Eisenhart, M. A. (1988). The ethnographic research tradition and mathematics education research. *Journal for Research in Mathematics Education, 19,* 99–114.

Fennema, E. (1985). Explaining sex-related differences in mathematics: Theoretical models. *Educational Studies in Mathematics, 16,* 303–320.

Fennema, E., Carpenter, T. P., & Lamon, S. J. (Eds.). (1988). *Integrating research on teaching and learning mathematics.* University of Wisconsin–Madison.

Fielker, D. S. (1987). A calculator, a tape recorder and thou. *Educational Studies in Mathematics, 18,* 417–437.

Fischbein, E. (1969). Enseignement mathématique et developpement intellectuel. In Editorial Board of Educational Studies in Mathematics (Eds.), *Proceedings of the First International Congress on Mathematics Education* (pp. 158–174). Dordrecht, The Netherlands: Reidel.

Freudenthal, H. (1973). *Mathematics as an educational task.* Dordrecht, The Netherlands: Reidel.

Freudenthal, H. (1978). *Weeding and sowing.* Dordrecht, The Netherlands: Reidel.

Freudenthal, H. (1983). *Didactical phenomenology of mathematical structures.* Dordrecht, The Netherlands: Reidel.

Galbraith, P. (1988). Mathematics education and the future: A long wave view of change. *For the Learning of Mathematics, 8*(3), 27–33.

Gattegno, C. (1971). *What we owe children.* London: Routledge and Kegan Paul.

Gauthier, R. (1969). Essai d'individualisation de l'enseignement (enfants de dix a quatorze ans). In Editorial Board of Educational Studies in Mathematics (Eds.), *Proceedings of the First International Congress on Mathematics Education* (pp. 57–68). Dordrecht, The Netherlands: Reidel.

Gerdes, P. (1988). On possible uses of traditional Angolan sand drawings in the mathematics classroom. *Educational Studies in Mathematics, 19,* 3–22.

Good, T., & Grouws, D. (1977). Teaching effects: Process-product study in fourth-grade mathematics classrooms. *Journal of Teacher Education, 28,* 49–54.

Griffiths, H. B., & Howson, A. G. (1974). *Mathematics, society and curricula.* Cambridge, U.K.: Cambridge University Press.

Habermas, J. (1972). *Knowledge and human interests.* London: Heinemann.

Hart, K. M. (1981). *Children's understanding of mathematics*: 11–16. London: John Murray.

Hillel, J., Kieran, C., and Gurtner, J-L. (1989). Solving structured geometric tasks on the computer: The role of feedback in generating strategies. *Educational Studies in Mathematics, 20,* 1–39.

Hirst, A., & Hirst, K. (1989). *Proceedings of the Sixth International Congress on mathematical education.* Budapest, Hungary: ICMI Secretariat and Janos Bolyai Mathematical Society.

Howson, A. G. (Ed.) (1973). Developments in mathematical education. *Proceedings of the Second International Congress on Mathematical Education.* Cambridge: Cambridge University Press.

Howson, A. G., & Wilson, B. J. (1986). *School mathematics in the 1990s.* ICMI Study Series. Cambridge: Cambridge University Press.

Howson, A. G., Keitel, C., & Kilpatrick, J. (1981). *Curriculum development in mathematics.* Cambridge: Cambridge University Press.

Hoyles, C., & Noss, R. (1986). How does the computer enlarge the scope of do-able mathematics. In C. Hoyles, R. Noss, & R. Sutherland (Eds.), *Proceedings of the Second International conference for LOGO and Mathematics Education* (pp. 142–153). London: University of London Institute of Education.

Jones, G. A. (1988). Critical variables in mathematics education revisited. In D. C. Blane and G. C. Leder (Eds.), *mathematics education in Australia* (pp. 3–21). Melbourne, Australia: MERGA and the Mathematics Education Centre, Monash University.

Kilpatrick, J. (1985). Reflection on and recursion in mathematics education. *Educational Studies in Mathematics, 16,* 1–26.

Kirsch, A. (1977). Aspects of simplification in mathematics teaching. In H. Athen & H. Kunle (Eds.), *Proceedings of the Third International Congress on Mathematical Education,* (pp. 98–120). Karlsruhe, Germany: University of Karlsruhe.

Krutetskii, V. A. (1976). *The psychology of mathematical abilities in schoolchildren.* Chicago, IL: University of Chicago Press.

Krygowska, Z. (1969). Le texte mathématique dans l'enseignement. In Editorial Board of Educational Studies in Mathematics (Eds.), *Proceedings of the First International Congress on Mathematical Education* (pp. 228–238). Dordrecht, The Netherlands: Reidel.

Krygowska, Z. (1988). Composants de l'activité mathématique qui devraient jouer le rolé essentiel dans la mathématique pour tous. *Educational Studies in Mathematics, 19,* 423–433.

Laborde, C. (1989). Hardiesse et raison des recherches francaises en didactique des mathématiques. In *Proceedings of the 13th Conference on the Psychology of Mathematics Education* (pp. 46–61). Paris, France.

Lin, F. L. (1988). Societal differences and their influences on children's mathematics understanding. *Educational Studies in Mathematics, 19,* 471–497.

Lorenz, J. H. (1982). Research since 1976 on affective student characteristics. *For the Learning of Mathematics. 3*(1), 24–29.

Markouchevitch, A. (1969). Certains problèmes de l'enseignement des mathématiques à l'école. In Editorial Board of Educational Studies in Mathematics (Eds.), *Proceedings of the First International Congress on Mathematical Education* (pp. 147–157). Dordrecht, The Netherlands: Reidel.

Mellin-Olsen, S. (1987). *The politics of mathematics education.* Dordrecht, The Netherlands: Reidel.

Merrill, M. D., & Wood, N. D. (1974). *Instructional strategies: A preliminary taxonomy.* Columbus, OH: ERIC.

Nisbet, J., Megarry, J., & Nisbet, S. (1985). *World yearbook of education, research, policy and practice.* London: Kogan Page.

Papert, S. (1980). *Mindstorms: Children, computers and powerful ideas.* Great Britain: The Harvester Press, Ltd.

Papy, F. (1969). Minicomputer. In Editorial Board of Educational Studies in Mathematics (Eds.), *Proceedings of the First International Congress on Mathematical Education* (pp. 201–213). Dordrecht, The Netherlands: Reidel.

Polya, G. (1954). *Mathematics and plausible reasoning* (2 vols). Princeton, NJ: Princeton University Press.

Popkewitz, T. S. (1984a). Soviet pedagogical science: Visions and contradictions. *Journal of Curriculum Studies, 16,* 111–130.

Popkewitz, T. S. (1984b). *Paradigm and ideology in educational research: The social functions of the intellectual.* London: The Falmer Press.

Revuz, A. (1969). Les premiers pas en analyse. In Editorial Board of Educational Studies in Mathematics (Eds.), *Proceedings of the First International Congress on Mathematical Education* (pp. 138–146). Dordrecht, The Netherlands: Reidel.

Reyes, L. H., & Stanic, G. M. A. (1988). Race, sex, socioeconomic status and mathematics. *Journal for Research in Mathematics Education, 19,* 26–43.

Romberg, T. A., & Carpenter, T. P. (1986). Research on teaching and learning mathematics: Two disciplines of scientific enquiry. In M. C. Wittrock (Ed.), *Handbook of Research on Teaching* (3rd ed., pp. 850–873). New York, NY: MacMillan.

Rosenbloom, P. C. (1969). Vectors and Symmetry. In Editorial Board of Educational Studies in Mathematics (Eds.), *Proceedings of the First International Congress on Mathematical Education* (pp. 273–283). Dordrecht, The Netherlands: Reidel.

Rouchier, A., & Steinbring, H. (1988). *The practice of teaching and research in didactics.* Survey Lecture presented at Theme Group 5, ICME 6, Budapest, Hungary (unpublished).

Servais, W. (1969). Logique et enseignement mathématique. In Editorial Board of Educational Studies in Mathematics (Eds.), *Proceedings of the First International Congress on Mathematical Education* (pp. 28–47). Dordrecht, The Netherlands: Reidel.

Sierpinska, A. (1989). Reaction to work of Theme Group 5. In A. Hirst and K. Hirst (Eds.), *Proceedings of the Sixth International Congress on Mathematics Education* (pp. 273–274). Budapest, Hungary: ICMI Secretariat and Janos Bolyai Mathematical Society.

Steiner, H. G. (1969). Magnitudes and real numbers—a didactical analysis. In Editorial Board of Educational Studies in Mathematics (Eds.), *Proceedings of the First International Congress on Mathematical Education* (pp. 239–260). Dordrecht, The Netherlands: Reidel.

Steiner, H. G. (Ed.), (1984). *Theory of mathematics education.* Bielefeld, W. Germany: I.D.M., Paper 54, University of Bielefeld.

Swetz, F. J. (Ed.) (1978). *Socialist mathematics education.* Southampton, PA: Burgundy Press.

Thompson, A. G. (1984). The relationship of teachers' conception of mathematics and mathematics teaching to instructional practice. *Educational Studies in Mathematics, 15,* 105–127.

Vergnaud, G. (1988). Theoretical frameworks and empirical facts in the psychology of mathematics education. In A. Hirst and K. Hirst, *Proceedings of the Sixth International Congress on Mathematical Education* (pp. 29–47). Budapest, Hungary: ICMI Secretariat and Janos Bolyai Mathematical Society.

Verstappen, P. F. L. (Ed.). (1986). *Proceedings of second conference on systematic cooperation between theory and practice in mathematics education.* National Institute for Curriculum Development, Enschede, The Netherlands.

Wittmann, E. (1984). Teaching units as the integrating core of mathematics education. *Educational Studies in Mathematics, 15,* 1, 25–36.

Yates, A. (Ed.). (1971). *The role of research in educational change.* UNESCO Institute for Education, Hamburg. Palo Alto, CA: Pacific Books.

Zweng, M., Green, T., Kilpatrick, J., Pollak, H., & Suydam, M. (Eds.). (1983). *Proceedings of the Fourth International Congress on Mathematical Education.* Boston, MA: Birkhäuser.

·29·

REFLECTIONS ON WHERE MATHEMATICS EDUCATION NOW STANDS AND ON WHERE IT MAY BE GOING

Robert B. Davis

RUTGERS UNIVERSITY

One might hope, at this point, for a final chapter that would predict the future, but the fact is that prediction does not work. Who would have predicted the current importance of personal computers? Not the presidents of two of the largest computer manufacturers, who ruled against corporate efforts in this direction so that it was left to two young men, Steve Jobs and Steve Wozniak, to develop the idea and to build the company that introduced the microcomputer. Who would have predicted high school geometry courses based on chaos and fractals? Yet three separate research and development teams have created three such courses (for one of these, see Goldenberg, 1989; the two other teams are directed by Wallace Feurzeig, at Bolt, Beranek and Newman, and by Eugene Stanley, at Boston University). Who would have predicted our present unemployment situation: that the national unemployment rate of 5.3% would be "misleading" because "definitions of the work force excluded a growing number of Americans," and that, by the summer of 1988, 45.3% of New York City residents over the age of 16 could not be counted as labor force participants because they were "either unemployed or unemployable" and were in fact lost to the system of counting (Phillips, 1990, p.20). Almost half of the people of New York City unemployed. Who predicted that?

On a more positive note, who could have predicted the appearance of major publications as far-sighted, educationally sound, and influential as the NCTM *Standards* (NCTM, 1989) and the National Research Council's *Everybody Counts* (NRC, 1989)? It is the best of times, it is the worst of times. How does it all add up? Where are we heading? Who can begin to say?

If predicting the future is too difficult, there may be a useful alternative: We can look at what seem to be some of the most important challenges and opportunities now facing research and development in mathematics education, which is what this chapter attempts to do. I first present a list of 15 important current challenges or opportunities and then discuss some of them more extensively.

CHALLENGES AND OPPORTUNITIES

WARNING: At first this list seems made up of disconnected items that have little in common and therefore carry no recognizable message. Please do not give up. Deeper scrutiny will reveal the common thread. The exciting prospect is that the *unity* of these items will give us some reason for hope.

1. Build on the ways that students actually think. Traditional instruction has typically started from a notion of how people ought to think (about some specific mathematical situation) and has sought to impart this method to students. In the last few years an alternative approach has gained ground: Observe carefully how students *actually* think about the topic under discussion, and build upon this process. (See, for example, Maher & Alston, 1989.)

2. Improve the situation with the "underclass" of urban poor. Phillips's figure of 45.3% "unemployed or unemployable" resi-

The author gratefully acknowledges the helpful comments provided by James Fey, University of Maryland, and Douglas B. McLeod, San Diego State University.

dents of New York City over the age of 16 is clearly a disaster, and it is clearly unacceptable. That is coming close to *half of the people* in our largest city. (Nor is this true only of New York; Phillips points out: "similar circumstances were reported in Detroit and Baltimore, while the ratio...for the nation as a whole is 34.5%" [Phillips, 1990, p. 20].) An almost unbelievable number of Americans are unable to fit into the productive part of our society. (As some demographers point out, the people in New York or elsewhere who are not listed as "employed" in any usual sense may in fact *be* employed, but they are not reporting their income for tax purposes. This could apply to any plumber, baby-sitter, typist, lawn mower, or dog walker; it also applies to anyone who makes a living selling stolen goods and to everyone in the drug trade—whoever deals only in the cash economy, as many people seem to do nowadays.)

3. Create, implement, and evaluate more diverse school mathematics programs. In the 1960s it was widely assumed that part of the solution to many of our most pressing educational problems would involve more diverse school mathematics programs, such as David Page's *Illinois Arithmetic Project* (Lockard, 1967) and Earle Loman's *Unified Science and Mathematics for Elementary Schools* (USMES) (Lockard, 1977). The era of "back to basics" parried this thrust in most school systems, but the *Standards* and *Everybody Counts* may signify a new interest in this approach. If school systems are going to move far in this direction, however, research and development must create some reasonably explicit programs and demonstrate that they are capable of producing acceptable test scores.

4. Develop evaluation procedures that put more emphasis on creativity, and less on repeating simple rote procedures.

5. Pay more attention to some of the implicit messages that tests and school programs send to students. A recent letter to the editor in the *New York Times* (July 26, 1990), "As Johnny Learns to Shop, Hiroko Learns to Think" by Richard Askey, the Gabor Szego Professor of Mathematics at the University of Wisconsin, compared test items on Japanese and U.S. mathematics tests. Here is an item from a Japanese test at the ninth-grade level:

The numbers a, b, and c are integers such that a is greater than 1 and less than 9, and b and c are each greater than 0 and less than 9. Find all triples (a, b, c) that satisfy the inequality

$$100a + 10b + c < 5(12a + 5b + c).$$

By contrast, Professor Askey reports that, on the U.S. tests he examined, "Almost all [of the problems] are one-step problems." Askey goes on to make an important point: Perhaps 75 one-step problems on a test will produce about the same ranking of students as will 6 multistep problems that require serious thought (and perhaps some originality). *But the message that they send to students is entirely different.* The one-step problems say to students, "You do not have to do much hard thinking in mathematics, nor must you be very creative; all you have to do is pay attention in class, memorize dutifully, practice diligently, and you will get no surprises on the tests." The Japanese tests send

a different message—rather more in the spirit of the contest problems that a very few U.S. students encounter—where it is clear from the outset that, if you have developed nothing more than routine skills, you will be hopelessly ineffective. You *must* strive for ingenuity and originality.

Of course, comparing Japanese and U.S. school programs is no simple matter, and the comparison made here involves some tricky questions that are not easy to resolve. Nonetheless, Askey's basic point is an important one: We must look at textbooks, tests, and classroom practices not merely in terms of their obvious *explicit* purposes, but also in terms of the *implicit* or "partly hidden" messages that they send to students. In particular, what kind of behavior do we seem to be asking students to display? What sort of things must a student do in order to be successful?

6. Study the effect of assuming that we must "tell" students how to do mathematics. There is one special case that deserves much more study: Many teachers believe that they *must* tell students how to deal with mathematics problems. Others disagree. In the 1960s this disagreement became the basis for numerous controversies over "discovery learning"; see, for example, Shulman and Keislar (1966). Those who defend the assumption that we must *tell* students how to solve mathematics problems usually argue that if we do not do so, some (or many) students will be lost and will quickly become demoralized. By contrast, those who argue against this assumption usually claim that teaching based on it sends students the message that nobody can solve a math problem unless someone tells them how to do it. Hence, students quickly give up the habit of trying to think for themselves, and they adopt the strategy of merely trying to remember what the teacher has said. Anyone who interviews students extensively will find many who say, at least in effect, "I couldn't possibly do this problem, because you haven't told me how to do it." The opponents of "telling" argue, first, that this attitude on the part of students renders them almost incapable of making good progress in mathematics, and, second, that the attitude is not inevitable, but rather was taught to the students by implicit messages repeatedly sent by teachers and textbooks.

7. Study mental representations. The topic mentioned first, "build on the way students actually think," has a corollary of particular importance, namely, the study of the mental representations of students. (See, for example, Janvier, 1987.) It is surely no coincidence that one of the liveliest working groups at the recent meetings of the International Group for the Psychology of Mathematics Education, in Oaxtapec, Mexico (July, 1990) was the Working Group on Representations, led by Gerald Goldin. One knows very little about how someone thinks if one has no knowledge at all of his or her mental representations. It is particularly interesting to study teachers, and determine (where possible) whether the teacher has made a correct, or incorrect, assessment of the representations that individual students are using.

8. Find out what really happens in classrooms. A number of researchers are accumulating videotaped records of what actually happens in classrooms. This sheds light on several of the topics already listed and also appears to show how common it is for a teacher to talk about mathematical situations in imprecise and misleading language.

9. Help schools make more use of excellent existing materials. In recent years a number of very high-quality courses or instructional sequences have been created, including the Feurzeig & Roberts sixth-grade algebra materials (Roberts, Carter, Davis, & Feurzeig, 1989), the three high school courses on fractals and chaos mentioned earlier (Goldenberg, 1989), the high school statistics course developed by Andee Rubin (Rosebery & Rubin, 1989), John Anderson's geometry tutor (Anderson, Boyle, & Yost, 1986), and many others. These courses have been tested and are known to work well, but they are not appearing in schools. In part, this can be attributed to the cost of the computers that are needed to implement the programs, but this is probably not the whole story. Schools do not see where this material fits, and teachers are unsure how to make use of it. There is a large research and development job here that needs to be done, if U.S. students are to benefit substantially from this potentially valuable material.

10. Develop new lessons in discrete mathematics. It is clear that there is now so much interest and activity in the area of discrete mathematics that it deserves special consideration. As one example, the Rutgers Center for Discrete Mathematics and Theoretical Computer Science has received approximately 10 million dollars from the National Science Foundation for a program that includes the development of instructional approaches at the precollege level. Within the United States, discrete mathematics has usually been seen as an alternative to the usual calculus-based sequence intended for physical scientists and engineers (but also appropriate for economists, biologists, and the like). In the Netherlands, it has been seen in a different light; thanks to Freudenthal, de Lange, and others, the Dutch approach has been to recognize that many things have gone seriously wrong with the existing school mathematics sequence and that setting things right (or even changing much of anything at all) has turned out to be difficult, if not impossible. The Dutch have seen discrete mathematics as an opportunity to make a fresh beginning (de Lange, 1987). But to carry this off, one needed a rather precise list of the main things that are wrong with the existing school mathematics sequence. The United States will not take full advantage of this opportunity without serious guidance from the mathematics education community, particularly in identifying weaknesses of the existing mathematics sequence that could be avoided in the new sequence that may emerge under the name of "discrete mathematics." One should also be concerned to avoid introducing *new* weaknesses into the pathway through beginning mathematics.

11. Delineate alternatives. At present it is by no means easy to describe the major alternative approaches to school mathematics. A few years ago we saw attempts to contrast and compare "guided discovery" versus "telling and showing," but this discussion often went astray (Shulman & Keislar, 1966; for a more recent discussion, together with some experimental results, see McDaniel & Schlager, 1990). We saw interest in "open education" (Devaney, 1974; Kohl, 1969), but here, too, the alter-

natives were often poorly posed and seriously misunderstood. How is "constructivism" (Cobb, 1987) distinctive at the classroom level? What defines the approaches of Cobb, or Maher, or Saxon? How do they differ from one another? Two of the best discussions of the ways in which various alternative approaches are different can be found in Howson, Keitel, and Kilpatrick (1981) and in Solomon (1986), but a careful delineation of alternative approaches remains a largely uncompleted task. It is urgently needed; without it we are unable to identify the approaches we may be observing (or attempting to implement) and what outcomes we should be concerned with. Gross misinterpretations are common, as we surely saw in much of the work on "open education" and in the 1960s work on "new mathematics."

A later section, "Looking More Deeply," will consider three major current alternatives, but by the end of this chapter the reader will surely conclude that the full job of defining, comparing, and contrasting alternative approaches to school mathematics still lies ahead of us and will probably not be easy.

12. Establish an epistemology. Closely related to the teaching practice of trying to build on a student's own ways of thinking has come a new notion concerning the *kind* of knowledge one wants students to acquire. The new approach puts less emphasis on learning rules (for example, knowing the rules for adding fractions or solving quadratic equations) and being able to use them; instead, the focus is on the acquisition of an appropriate and powerful collection of basic metaphors (Davis, 1990; Papert, 1980). Support for this change in epistemology also derives from efforts to get computers to process information somewhat more as human beings seem to (Allman, 1989).

13. Determine how the curriculum should respond to the new technological capabilities. It was once the case that there was no choice; people *had* to be able to add up columns of numbers by using paper-and-pencil algorithms. They *had* to be able to carry out the other familiar algorithms of arithmetic. Otherwise there simply was no effective way to obtain answers that were needed from time to time.

Clearly, this is not true today. Hand-held calculators are rapidly approaching ubiquity. We can get answers by means other than paper-and-pencil algorithms, and most of the time we do.

Somewhat the same situation is being created for more advanced portions of mathematics, thanks to Maple and Reduce and Mathematica and other so-called "computer algebra systems" (Hosack, 1988; Karian, 1990; Wilf, 1982; Wilf & Zeilberger, 1990; Wolfram, 1984; Zorn, 1985). This surely ought to be reflected somehow in changes in the school mathematics curriculum. Far too little attention has been paid to this question, although the most notable exception is the work of James Fey and his colleagues (Fey, 1984; see also Child, 1990; Devlin, 1990). The conservatism of education is clear, but it is also clear that many of the algorithms we teach are no longer necessary, and we need to plan our curriculum and our lessons to recognize this reality. (See also Judson, 1990.)

Of course, mathematics education could choose to ignore the curricular implications of new technological capabilities. Perhaps the strongest case for looking at them carefully is that they offer a possibility of making some changes, and there are very few such opportunities.

14. Upgrade public policy debate. Various unsatisfactory aspects of U.S. schooling have been appearing prominently in recent media discussions. If any real improvements in U.S. education are going to be made, they will indeed require public discussion, but that discussion needs to be more carefully reasoned and better informed than most of it has been. Will the mathematics education research and development community play a role in this? Do we have anything to contribute?

15. Revise methodology. The change in the *kind* of mathematical knowledge that students are expected to develop has a precise parallel in a change in the kind of pedagogical knowledge that teachers need. In both cases there is a shift *away from* "rule-based" knowledge of word and formula and algorithm and *toward* construction of one's collection of basic metaphors or assimilation paradigms.

In fact, it is precisely this shift that underlies nearly every item on the preceding list. It is so important that it warrants three examples:

EXAMPLE 1. You have a pile of three blue poker chips, and a pile of five red chips. How many chips do you have all together?

We all know (and most students do, too) that this real situation is properly modeled by the abstract notion of *addition*. At this point we could use a calculator, keying in the problem

$$3 + 5 =$$

and getting from the calculator the answer "8."

EXAMPLE 2 (From Marilyn Burns, 1987). You have a large bowl and a small measuring cup. As quantities to measure, you have some rather large pebbles and some fine-grained sand. You repeatedly fill the measuring cup with the pebbles, transferring them to the large bowl until it is filled. By counting you have determined that it required 27 cups of pebbles to fill the bowl. You now empty the bowl, and get ready for the next part of this task.

You are now asked to *calculate* how many cups of sand will be needed in order to fill the bowl with sand. You look at the rather large pebbles and at the quite small size of the grains of sand, and at first you are uncertain about which calculation (if there is one) that you ought to perform. You finally decide that there should be the same ratio of sizes in either case, and therefore probably the same ratio of numbers of cups. Hence, you answer that it will require 27 cups of sand to fill the bowl. You now carry out the physical task of doing this, and you find that you were in fact correct. It *does* require 27 cups of sand to fill the bowl to the top. You again empty the bowl and prepare for the next step.

Finally, you are asked to put 20 cups of pebbles into the empty bowl, and are asked to *calculate* how many cups *of sand* will be needed to fill the bowl to the top. You have to consider whether this situation calls for *addition* (in which case I suppose you might *add* 27 and 20 and get the answer 47, although this does not seem promising), or for *subtraction* (so

that you subtract 20 from 27 and conclude that it will require 7 cups of sand, in addition to the 20 cups of pebbles that are already in the bowl, to fill the bowl to the top), or for *some other arithmetical calculation*—or perhaps you will conclude that *there is no arithmetical operation* that will give a correct answer. After you have made your choice, you carry out the appropriate calculation (again, let us suppose that you use a hand-held calculator). You now carry out the physical operations, count how many cups of sand are required to fill the bowl to the top, and determine whether you chose the correct calculation.

EXAMPLE 3. You are given this problem:

In a certain town, $\frac{2}{3}$ of the men are married to $\frac{3}{5}$ of the women [we do not count children]. What fraction of the adults in the town are married?

You again decide which arithmetical operations, if any, are appropriate. Having made your decision, you use a calculator (perhaps one that will calculate with fractions and return a fractional answer).

Now, here is the main point behind these three examples: Most people who have not had an opportunity to think seriously about such matters would claim that the *mathematics is that part of the problem that the calculator did*. They might find the decisions about particle size in the bowl problem, or the choice of arithmetical operations in the marriage problem, to be thought provoking, but they would probably *not* consider them an essential part of the *mathematics*. In the case of the red and blue poker chips, they might *not even notice that there was any thinking involved* other than the computation that the calculator carried out.

I would argue that such observers are precisely wrong. There is very little *mathematics* in the actual carrying out of the computations, which is why the "stupid" hand-held machine is able to do this part of the job so reliably. The mathematics lies mainly in analyzing the real situation and *deciding how to represent it in an appropriate abstract symbolic form*.

Because until recently machines were not available to do calculations, it has long been necessary to teach human beings some slow and unreliable paper-and-pencil methods of carrying out these computations. In the process, we have come (unwisely) to focus nearly *all* of our pedagogical efforts and our testing programs on these computations—*and we have come to take for granted the truly mathematical task of modeling the reality in an appropriate way.*

An interview with a fifth-grade girl makes the distinction clear (Maher & Alston, 1989). Ling Chen, the student, is given this problem:

Jane had $\frac{1}{3}$ of a candy bar. She gives $\frac{1}{2}$ of what she has to Mike. How much of the candy bar does she give to Mike?

Ling Chen is given her choice of a method for solving the problem. She has available to her paper and pencil, a set of wooden Pattern Blocks, and various other manipulatable materials. Although she has never received instruction in the use of physically manipulatable materials, she nonetheless chooses the Pattern Blocks. She selects a yellow hexagon to represent the entire original candy bar, a blue parallelogram to represent the "one-third" that Jane had, and a small green triangle to represent the part that Jane gave to Mike. These are all good choices, since three blue parallelograms will just cover one large yellow hexagon, and two small green triangles will just cover one blue parallelogram. Using this physical translation of the problem, Ling Chen gets the correct answer: Mike has received $\frac{1}{6}$ of a candy bar.

Ling Chen is then asked if she could solve the same problem "by using numbers." She writes

$$\frac{1}{3} \text{ divided by } \frac{1}{2} = \frac{2}{3}$$

She has, of course, carried out the calculation correctly, *but she chose the wrong calculation to model this problem.*

In a long sequence of studies extending over 6 years, Maher and her colleagues have videotaped many students at various grade levels, often following the same students for several years. The kind of difficulty that Ling Chen demonstrated is more the rule than the exception. These students have learned to do calculations with a high rate of success, but they cannot solve problems correctly *because they do not know how to model problems with appropriate abstract representations.*

But here is the second important point: *How do people think about real situations, such as stacks of poker chips, or men marrying women, or filling a bowl with a mixture of sand and pebbles?*

We are coming in fact to know quite a bit about this kind of information processing, in part because now we are trying to get computers to do it. It isn't easy but, as we all know, humans learn to do it well and at present, most computer programs do it very badly, or not at all.

This distinction is central to the tasks of mathematics education. I need to emphasize that this is *not* the "procedure" versus "concept" distinction, which it somewhat resembles. It is a distinction between two quite different modes of information processing, and it is easier to approach it by thinking of computers, although it becomes more important when we consider human methods of thinking.

Until recently, most computers were so-called "von Neumann machines," defined by their ability to carry out one command at a time and requiring quite explicit identification of each command. The result, as William Allman points out (Allman, 1989), was an information-processing machine that was very different from the human mind. For a human being, "it's easy to recognize your mother, but hard to do logic" (and Allman has some good examples of how "hard" it is for humans "to do logic"). By contrast, with a von Neumann machine, "it's hard to recognize your mother, and easy to do logic. " (Allman, p. 20).

Now this is a truly fundamental difference. Von Neumann computers are very *explicit* machines. Every step is well-defined, their total command set is well-defined, the systematic way in which they perform can be plotted, goals and subgoals can be identified, and so on. Everything is basically very orderly, which does not mean that it is always simple—as anyone who studies sorting algorithms is well aware—but it does make possible one definite way of describing what is going on.

It has been known for some time that the human brain is not this kind of machine. The brain has around 100 billion neurons, and each neuron may be connected to as many as 100 thousand other neurons. So the brain is vast, even compared against the kind of huge numbers one often encounters nowadays.

It is also very slow—at least its constituent parts are. A single neuron is about 100 thousand times slower than a typical computer switch. Yet somehow this vast array of slow switches regularly performs information-processing feats that computers cannot come close to achieving. Driving an automobile is one good example: Most humans can get to be very good at this; no computer has been able to learn even a low-level version of this skill. Nor is it an accident that Allman uses "recognizing your mother" as a test case; again, people do it effortlessly, whereas computers are usually unable to do it.

In recent years a few pioneers have taken up this challenge more directly, by building so-called "non-von" or "connection" machines, where large numbers of very simple processors interact, often in ways that cannot be programmed or described (just as one cannot now describe how the human mind recognizes your mother). On the biological side, great progress has been made in neuroscience; although we do not yet understand how the mind works its wonders, we are rapidly getting to know more and more about at least a few pieces of this vast puzzle.

This becomes relevant to school mathematics when we notice how *explicit* typical instruction usually is and how much it works in a simple, step-by-step way. It also focuses on the "algorithm written down on paper" aspect of the task of creating and manipulating abstract models of reality, or, when it does attempt to deal with visualization and conceptualization tasks, it often resorts to "systematic" procedures such as "looking for key 'clue' words." It is an important sign of progress that a volume as important as the *Standards* takes up this matter explicitly, as when it recommends "less attention" to "use of clue words to determine which operation to do" (p. 21).

If the human mind is the non-von machine that it certainly appears to be, then the common British educational practice has much to recommend it: Have children play with relevant materials, talk about them, make up their own problems with them, and so on. British practice is often unsystematic. As one observer wrote, about British instruction in reading, "You can watch classes all day long. You never see anyone teach reading. Yet all of the children learn to read" (personal communication). This should not be seen as an accident.

The truly remarkable thing about modeling reality is being able to think about two stacks of poker chips, being able to *visualize* them, being able to *imagine* the act of "putting the two stacks together," and so on. The calculation of *2 + 3* is the trivial part. Yet it is this trivial part on which most school programs have chosen to concentrate. It is the *visualizing* and the *imagining* that is the hard part, and the part that we must make sure that students can do. Speaking of this same distinction, Marvin Minsky points out that early programs in artificial intelligence "could do college-level calculus," but when programmers began to work on getting computers to deal with ninth-grade algebra problems, they found that this task was much harder. Things become much harder still when programmers turned to elementary school arithmetic—and harder even than that when they tried to get computers to play with blocks, like a pre-school child might do. (Allman, 1989, p. 29)

To use the language of George Lakoff and his colleagues (Johnson, 1987; Lakoff, 1987; Lakoff & Johnson, 1980), instruction and testing have focused on abstract calculations and neglected the basic repertoire of *metaphors* that play so large a role in determining how we think. To be good at arithmetic means primarily to be good at visualizing stacks of poker chips: visualizing how they can be moved around, combined, split apart into a larger number of smaller piles, and so on. It means being able to visualize Cuisenaire® rods being placed end to end. It means being able to visualize shapes made by rubber bands on geoboards or to determine how many interior cubes must be inside a $4 \times 4 \times 4$ Dienes MAB block (Davis, 1984).

It is again no accident that many British programs involve tasks such as asking a child to imagine that small cubes, 1 cm along each edge, are temporarily glued together to make a larger cube, say 5 cm along each edge. The outside is now painted red, after which the small cubes are pulled apart as individuals once again. How many cubes are red on all six faces? On exactly five faces? On exactly four faces? On exactly three faces? On exactly two faces? On exactly one face? On no faces?

But if these visualizations form the basis for the large collection of metaphors that defines the true nature of the knowledge we want children to acquire about mathematical models of reality, it is equally true that researchers and teachers need to acquire the same sort of knowledge about teaching and learning mathematics. Mathematics education has focused too much on the kind of abstract general law that has proved useful in physics—things like F = ma, or pV = nRT. What is most needed is an enlarging of our basic collection of metaphors, or "assimilation paradigms," to use a more Piagetian formulation. I have argued this case elsewhere at considerable length (see Davis, 1990), and there is hardly space to reproduce those arguments here, but I think that they need careful consideration, especially in an age when modern videotaping equipment allows us to *show* people directly the behavior that we could once only describe in words. When people have *seen* children invent mathematics, they are less disposed to imagine that children *must be told* what to do and how to do it. Arguments about "pure discovery" versus "guided discovery" and all its other variants would probably never have happened if all of the participants had been able to *watch* exactly how the interactions between teachers and students took place as they can today.

Perhaps the central argument has been best stated by Squire (1986):

Memory is stored as changes in the same neural systems that ordinarily participate in perception, analysis, and processing of the information to be learned....[For example, the inferotemporal cortex] has been proposed to be not only a higher order visual processing region, *but also a repository of the visual memories that result from this processing* [emphasis added]. (p. 1612; see also Davis, 1986)

If, indeed, Squire and his colleagues are correct, then there can be little long-term gain in telling students how to model reality; *they must experience it themselves.*

It seems clear that humans use both kinds of information processing. Sometimes they follow explicit directions, and sometimes they carry out "pattern-recognition," gestalt-like processing that they have not learned in any rule-like way, but rather from personal exploration and experience. Both school mathematics programs and mathematics education research have focused almost exclusively on the former learning process and largely neglected the latter. Nearly every other topic in this chapter stems in some degree from this excessive concentration on only one of two possibilities. In our own work in preservice teacher education, we have time and again taught courses where we presented the best descriptions we could give to let prospective teachers know about possible ways in which children and teachers could interact during mathematics lessons; yet when we subsequently showed them some films of actual classroom lessons, they were amazed. "Oh," to quote a typical comment, "is *this* what you meant!" They did not possess, among their basic assimilation paradigms, any exemplars of many of the kinds of teaching that we were attempting to describe, and the best descriptions we could create were not powerful enough to put these ideas into their minds.

This aspect of knowledge has not received nearly enough attention. What did Pasteur and his colleagues give us as "the germ theory of disease"? Hardly a true abstract generalization. As Herbert Simon has put it, their "generalization" seems to amount to this: "If you want to know why someone is ill, look for some kind of bug!" That sounds a good deal more like friendly advice than like a "scientific" generalization. In fact, we have missed the true nature of this important contribution to the advancement of science. What it really represents is an *increase in our basic collection of metaphors or assimilation paradigms.* Many of the most important advances in science have exactly this form, and are not abstract generalizations like F = ma. (See Davis, 1990.) This will probably also prove to be true of most advances in our knowledge of the teaching and learning of mathematics. In the past we have overlooked much of this kind of knowledge because we were looking for the wrong kind of knowledge.

Summary of the Epistemological Argument

The child who knows arithmetic is one who can visualize rearrangements of stacks of poker chips or rearrangements of keys on a key ring. This may entail some use of matters like 5 times 4 times 3 times 2 times 1 divided by 2, but it is the correct modeling of the real situation that is central, not the part of the task that could be performed by a three-dollar calculator. To be good at arithmetic you must possess a large collection of basic metaphors or assimilation paradigms, like "stacks of poker chips" or "ways of arranging keys on a key ring."

The teacher (or educational researcher) who has a deep knowledge of what is involved in teaching and learning mathematics must possess a large repertoire of basic metaphors or assimilation paradigms—he or she must know a great many alternatives and possibilities. The teacher may also know something about the frequencies of these possibilities and so on, but the collection of basic assimilation paradigms is the source of the greatest power in the teacher's accumulated knowledge.

Relatively little of the most powerful knowledge will take the form of abstract generalizations of the kind that we have traditionally sought.

LOOKING MORE DEEPLY

The list of challenges and opportunities deserves to be looked at more deeply, and probably the best way to do that is to view matters from the perspective of some typical classes in some typical schools. We can start with a classroom I observed in a school in the northeastern United States.

The students in this class have failed a statewide examination on "basic mathematics" that they must pass if they are to receive high school diplomas. The class we are observing is intended to prepare these students for their next encounter with this test.

Some of the students are in this class for the first time. For some, it is the second time. For a few it is their *third* time. This is what is currently called an urban class. Most—perhaps all—of the students are from low-income homes, and most are members of various minorities.

This particular class is a quintessential example of how things can go wrong in contemporary urban education. The topic of the day is percent. The teacher has tried to insert at least some interest into the class by asking students to bring in advertisements or newspaper stories that make reference to percent in one way or another. Few of the students have done so.

The classroom is noisy—not because of noise made by the students, but because of a very noisy steam heating system and other mechanical noises from the building itself. This would make thoughtful discussion impossible for anyone, but most of these students make no pretense of trying to follow the class discussion or of participating in any way. Some students put their heads down on their desks and are unresponsive—one observer wonders if they are on drugs (which would not be unusual in this school).

The question in the textbook asks, "Express 3.50 as a percent." No student in the class is able to do this. There are so many things wrong in this situation that one hardly knows where to start. Perhaps that specific question is a good starting point. Not only are the students unable to "express 3.50 as a percent," but so am I. However, in the days that follow, I decide to try it out on others. I ask several of the most highly respected mathematicians in the world to do this. Not one of them can. I ask the treasurer of one of the Fortune 500 companies. He can't either. We all believe that 3.50 is 10% of 35, or 1% of 350, or 50% of 7, or 700% of 0.5.

Now, actually I do know what the textbook authors had in mind (see Davis, 1988), but the authors have managed to come up with something that appears only in school mathematics. This is not percent as it is known to successful mathematicians, nor to successful money managers in major industries. Here, as nearly always, one sees surface details and must deal with them, when what lies beneath the surface is much more important.

With the best of intentions, we have created a curriculum of mathematics that has been severed from the real world. It consists of meaningless bits and pieces, and we ask students to learn it as a large collection of meaningless bits and pieces.

It has been our hope that this would somehow make things easier. *In the very short run it does.* Students can learn a few bits, pass the test on them, forget them, learn a few more small bits, pass the test on them, forget them, and so on. This does not add up to anything very important, and students know it. For a student whose parents see the larger meanings, uses, and patterns, this can be endurable because the student can learn a meaningful context into which the pieces fit. But for young people who feel alienated from our society (and it seems clear that most of these students do) there is no compelling intrinsic interest in what they are being asked to do, and they see little reason for doing it.

One can try to analyze this problem by looking at the students, the previous schooling of these students, the course we are presently observing, or the system of planning and evaluation that society has created. On the matter of the students, for the most part I defer to those who understand it better, even though I have been working with minority urban students since the 1950s and continue to do so today. I am convinced that the situation is worse now than it was in 1958, and a number of other observers agree, but I am unsure of the explanation and I certainly don't see a cure.

Something may be gleaned from a series of books recently published by the State University of New York Press, including Lomotey (1990). Perhaps the most interesting analyses have been made by Robert Moses, but he has thus far presented these only in workshops and at professional meetings (including the Humanistic Mathematics Network meeting in Louisville, Kentucky, January 15, 1990), and they are not yet available in print.

Another approach to trying to understand the plight of these young people, and their world view, is sometimes labeled the study of "affect" (see, for example, McLeod, 1990). One could argue about the precise boundary between "affect" and "cognition"—if I think that I will be unable to succeed, that *idea* would probably be called a "cognitive" matter, but the *feelings* that it may engender would probably be called an "affective" matter.

Yet one more approach that should be taken seriously is the work of Urie Treisman, at the University of California–Berkeley and elsewhere. This is another valuable activity that is not adequately documented in print; one has to observe it directly or talk with its creator. A major aspect of Treisman's work is expecting *more*, not less, from minority students. Watch Reggie Jackson or Leon Durham before you dismiss this approach. (See also Newman, 1990.)

On the matter of the previous schooling that such students received and how one would wish it to be different in the future, mathematics education has something to offer. Here we see at least three clear alternatives. One is a highly structured program, largely paper and pencil, but including as much meaning and experience as possible. This has its perils because it easily slips into dry pedantic classes that may fail to give as much inspiration as many of these students need. Unfortunately, this failure, like some forms of cancer, does not show up until years later, by which time the students may have come to put too much emphasis on following directions and on getting the right answers, with far too little time and energy spent on reflective thought about alternative ways of dealing with problems.

A second choice is the kind of small-group work, using intrinsically interesting tasks and making extensive essential use of manipulatable materials, that has been developed by Marilyn Burns, by Carolyn Maher, and others (Burns, 1987; Burns & McLaughlin, 1990; Burns & Tank, 1988; Maher, 1987).

The third alternative goes further by introducing a very large array of quite diverse activities that together relate mathematics to almost every aspect of living. This is most familiar in its British variations, as in Schools Council, 1969; Nuffield Mathematics Teaching Project, 1965; Biggs & McLean, 1969; Biggs, 1987. It involves making graphs of all sorts of things: color preferences, pet ownership, who has the longest hair, and so on. It includes measuring all sorts of things–such as pulse rate when resting versus pulse rate after exercising, weight and height—mapping the playground, finding the height of trees and flagpoles, and so on. Children use "trundle wheels" to find how far each of them walks in getting to school in the morning. Children study the rate of growth of plants with varying amounts of sunlight. Children play games of chance and work out probabilities.

All three of these alternative curricula can work in valuable ways. What research and development has not done is describe the strengths of each. Instead, past studies have put all three kinds of curricula up against the same yardstick, one that tests only the performance of written algorithms. Moreover, because the programs themselves have never been carefully described, how would you know whether you were using any one of them in a form that was consistent with its own proper goals and methods?

Nor do these three alternatives exhaust the list of important possibilities for enhancing education. A quite different, but very creative, approach has been developed by Constance Kamii and presented in a series of videotapes showing actual lessons. (See, for example, Kamii, undated videotape.) This focuses on paper-and-pencil algorithms but *leaves it up to the students to invent these algorithms*, and it seeks a great diversity of them, not merely those that are customarily taught in U.S. schools.

Yet another approach has been developed by Paul Cobb, Erna Yackel, and Terry Wood, at Purdue University. In the Cobb-Yackel-Wood program, second-grade children engage in problem-centered instructional activities that were created using the cognitive models of children's learning developed by Steffe and colleagues (Steffe, Cobb, & von Glasersfeld, 1988; Steffe, von Glasersfeld, Richards, & Cobb, 1983) to encourage both conceptual and procedural development. In these classes, children work on their own original solutions to problems in small-group and whole-class settings. Evaluation studies have indicated that students in this program get scores on the computational part of standardized tests that are similar to the scores of students in traditional programs, but that they do significantly better on conceptual and applications subtests (Cobb, et al., in press). Beyond this, the students in the Cobb-Yackel-Wood program believe that mathematics should make sense and that understanding and collaborating are important for success (Nicholls, Cobb, Wood, Yackel, & Patashnick, 1990). The approach of Cobb and colleagues places far more emphasis on helping children to think creatively about real situations, and it makes little if any use of traditional teacher demonstrations of the usual algorithms. In this they may be getting closer to the famous work of L. P. Benezet, who introduced no arithmetic at all until grade 6—but he prepared for it with years of

thinking mathematically about real situations. (See Benezet, 1935; Benezet, 1936; Whitney, 1986.)

As yet we do not have really basic descriptions of any of these programs that can serve to define them unequivocally, nor do we know how they differ from one another. We also do not have satisfactory methods of student evaluation that are consistent with the goals of most of them, although at least two states, Vermont and California, are now working on the evaluation question.

These programs seek to create a *culture* of mathematics, rather than allow it to remain merely a subject that is "taught." Recent work of George Miller and his colleagues (Miller & Gildea, 1987), Brown, Collins, & Duguid (1989), and others (Davis, 1989) can be summarized by saying: For serious long-term learning, one does not learn facts, *one acquires a culture*. This theme has also been stressed by Seymour Papert (1980).

Returning to the high school remedial course that we were observing, one could try to develop a course with far greater intrinsic interest—perhaps one based on carpentry or on art or on operating a small commercial enterprise—and with greater emphasis on small-group cooperative work and effective communication. Of course, one *cannot* do this if the students are going to be tested with problems like "Express 3.50 as a percent." They will not learn such things, because such things are not really part of the genuine culture of mathematics—not in "pure" mathematics, not in applied mathematics, not in everyday uses of mathematics, and not anywhere else that math plays a genuine role in a viable culture.

If one analyzes the remedial class in terms of how our society is dealing with the needs of those young people, an interesting set of questions arises. Is this class really only a warehouse for these students? It certainly looks that way. Some of these students have far deeper needs than learning about percent, and these needs are not being identified or addressed. What is going on in the life of a student who comes to class high on drugs or who is so sleepy he cannot stay awake? One could say that these are properly family concerns, but apparently some of these families are not dealing with the problems successfully. Isn't this a problem that we could solve, if we really tried?

One could consider other classes (given far more space and time, I would love to discuss some classes in wealthy suburbs that we have been studying) and ask how research and development work in mathematics education is helping them, but further analysis in this direction will be left as an exercise for the reader. It is worth doing.

The Ideas of Mathematics

A particularly important series of studies have been started by Carolyn Maher and her colleagues, using videotapes of classroom lessons and teacher diaries, to identify how students think about problems (especially what representations they use), how teachers believe the students think about these problems, and how close an agreement there is between the student's actual thoughts and the teacher's notion of what these thoughts are.

In one study (Davis & Maher, 1990; Maher & Davis, 1990), a fifth-grade teacher is teaching a lesson on improper fractions. She has made up a problem about children eating some slices of two pizzas. The class is working in small groups; Brian and

Scott are working together, and a videotape camera is watching both boys. Brian models the problem with wooden shapes known as Pattern Blocks. He gets the correct answer (which happens to be $\frac{14}{24}$). Meanwhile, the teacher is circulating around the rest of the class and has not had an opportunity to observe what Brian and Scott have been doing. She does finally come to their table and looks at their work.

Of course, the teacher here is at a great disadvantage compared to a researcher who can study the tapes at leisure, study the teacher's diary, conduct further task-based interviews, and so on. The teacher did not have all of this data available and had to make an instant decision in order to make an effort to help the students.

The teacher makes an error herself by concluding that the answer should be $\frac{14}{12}$. In her diary she describes Brian as "adding denominators." In fact, she is thrice wrong. The answer is $\frac{14}{24}$; Brian has made no error. Second, Brian has not added denominators—in fact, he has had no denominators. He did not solve the problem by any paper-and-pencil calculation, and he did not write any fractions except for his final answer. He solved it, as many children do, by using his concrete model of the situation. Finally, the source of the disagreement in answers hinges on the choice of the *unit* to be used. Brian has used the unit that the problem statement actually called for; the teacher has not. The teacher did not realize that this was the source of the disagreement.

Research of this type gets to the true heart of mathematics. Mathematics is about *ideas*, not about symbols written on paper. Yet very little research in mathematics education has focused on the actual ideas in students' minds or on how well teachers are able to identify these ideas, interact with them, and help students improve on them. The recent attention to "constructivism" and to "mental representations" may indicate a growing recognition of what is most central to human mathematical thought and of the need to study it more closely. Incidentally, these tapes of actual classroom lessons show that both teachers and students have incorrect ideas a very large percentage of the time, and that they misunderstand one another frequently—far more than most of us would ever have imagined (Maher & Davis, 1990).

Public Discussion

The problems we saw in the remedial class for urban youngsters will not be solved without public involvement and public discussion. Superintendents have told us that they hardly dare make any changes that might result in lower test scores—even though there are good reasons why such changes should be made, and good reasons why lower scores might not be signals of a poorer situation. For one thing, the tests often deal with matters that the students do not really need to know or, as in the case of 3.50 as a percent, might be better off *not* knowing. These improvements will simply not take place unless mathematics educators play a role in helping the public understand why current tests are often not good measures of the success of an educational program. (For an interesting discussion, see Koretz, 1988.)

As one small aspect of a large problem of designing an appropriate curriculum, consider a discussion that some Rutgers University faculty recently had with top management from two Fortune 500 companies. Texas Instruments had been trying to find out why they had succeeded in getting calculators into schools in Europe but not in the United States. This question arose during the discussion with the leaders of two other corporations. The businessmen felt strongly that they did not expect their employees to do paper-and-pencil arithmetic; using calculators, they were convinced, produced answers faster and with greater accuracy, both of which were important in their businesses. However, in determining which job applicants they would invite in for interviews, the personnel departments of both corporations made use of tests that involved arithmetical calculations. The arithmetical skills being tested were not job related; they were tested merely as a way of reducing the number of people who had to be interviewed. In schools in our area, there is considerable resistance to using calculators, especially from low-SES and blue-collar parents. In light of the conversation with the businessmen, it became clearer why this might be the case. Many of these parents had probably been asked to take tests that required paper-and-pencil arithmetical calculations, and foresaw their children as disadvantaged when applying for jobs if they needed calculators to do arithmetic.

Here is a very paradoxical situation. Corporations that do not particularly want paper-and-pencil skills in their employees are nonetheless behaving in a way that makes it necessary for schools to spend a large amount of time developing these skills. A curriculum that was more powerfully oriented to the actual desires of these companies could be installed in schools—but only if some of the companies' own practices could be changed—and even then only if this change could be communicated effectively to the general public. For all such matters, serious public discussion is an absolute necessity. Many other examples could also be given. Indeed, one wealthy suburban school system in New Jersey recently *stopped using computers that they already owned* because a majority of the new board of education thought that "computers were a fad" and had no large role to play in quality education so this appeared to be a good place to try to save some money.

Concluding Remarks

When one looks at the tasks that need to be worked on, the problems that need to be solved, and the far more powerful resources that are available today, it seems quite reasonable to argue that mathematics education has a major research and development job ahead of it. Carrying out this work successfully will require use of many resources that have been little used in the past. We need to look at ways of expanding our individual and collective repertoires of basic assimilation paradigms. We need to pay more attention to problems of describing learning environments and curricula, and to recognizing differences among them. We need far better measures of learning outcomes, especially long-term outcomes and outcomes that look more like personal attributes than like task-specific skills. But serious needs, important opportunities, and powerful resources are the ingredients of significant research and development. The ingredients seem to be in place. Who can predict whether mathematics education will rise to the occasion?

References

Allman, W. F. (1989). *Apprentices of wonder: Inside the neural network revolution*. New York, NY: Bantam Books.

Anderson, J. R., Boyle, C. F., & Yost, G. (1986). The geometry tutor. *Journal of Mathematical Behavior, 5,* 5–20.

Askey, R. (1990). "As Johnny learns to shop, Hiroko learns to think." Letter to the Editor, *New York Times,* July 26, 1990.

Benezet, L. P. (1935). The story of an experiment, Part I. *Journal of the National Education Association, 24,* 241–244, 301–303.

Benezet, L. P. (1936). The story of an experiment, Part II. *Journal of the National Education Association, 25,* 7–8.

Biggs, E. (1987). Understanding area. *Journal of Mathematical Behavior, 6,* 183–190.

Biggs, E., & MacLean, J. R. (1969). *Freedom to learn: An active approach to mathematics*. Don Mills, Ontario: Addison-Wesley (Canada).

Brown, J. S., Collins, A., & Duguid, P. (1989). Situated cognition and the culture of learning. *Educational Researcher, 18*(1), 32–42.

Burns, M. (1987). Explorations with raisins. *Journal of Mathematical Behavior, 6,* 135–148.

Burns, M., & McLaughlin, C. (1990). *A collection of math lessons from grades 6 through 8*. New Rochelle, NY: Cuisenaire.

Burns, M., & Tank, B. (1988). *A collection of math lessons from grades 1 through 3*. New Rochelle, NY: Cuisenaire.

Child, J. D. (1990). Almost no stuff in, wrong stuff out. *Notices of the American Mathematical Society, 37,* 425–426.

Cobb, P. (1987). Information-processing psychology and mathematics education—A constructivist perspective. *Journal of Mathematical Behavior, 6,* 3–40.

Cobb, P., Wood, T., Yackel, E., Nicholls, J., Wheatley, G., Trigatti, B., & Perlwitz, M. (1991). Assessment of a problem-centered second grade mathematics project. *Journal for Research in Mathematics Education, 22,* 3–29.

Davis, R. B. (1984). *Learning mathematics: The cognitive science approach to mathematics education*. Norwood, NJ: Ablex Publishing Corp.

Davis, R. B. (1986). The convergence of cognitive science and mathematics education. *Journal of Mathematical Behavior, 5,* 321–333.

Davis, R. B. (1988). Is percent a number? *Journal of Mathematical Behavior, 7,* 299–302.

Davis, R. B. (1989). The culture of mathematics and the culture of schools. *Journal of Mathematical Behavior, 8,* 143–160.

Davis, R. B. (1990). The knowledge of cats: Epistemological foundations of mathematics education. *Proceedings of the Fourteenth International Conference for the Psychology of Mathematics Education, 1,* PI.1 – PI.24.

Davis, R. B., & Maher, C. A. (1990). The nature of mathematics: What do we do when we "do mathematics"? In R. B. Davis, C. A. Maher, & N. Noddings (Eds.), *Constructivist views on the teaching and learning of mathematics* (pp. 65–78). Reston, VA: National Council of Teachers of Mathematics.

de Lange, J. (1987). *Mathematics: Insight and meaning*. Utrecht, The Netherlands: OW & OC.

Devaney, K. (1974). *Developing open education in America. A review of theory and practice in public schools*. Washington, DC: National Association for the Education of Young Children.

Devlin, K. (1990). The right stuff. *Notices of the American Mathematical Society, 37,* 417–425.

Fey, J. (1984). *Computing and mathematics: The impact on the secondary school curriculum*. Reston, VA: National Council of Teachers of Mathematics.

Goldenberg, E. P. (1989) Seeing beauty in mathematics: Using fractal geometry to build a spirit of mathematical inquiry. *Journal of Mathematical Behavior, 8,* 169–204.

Hosack, J. (1988). Computer algebra systems. In D. A. Smith, G. J. Porter, L. C. Leinbach, & R. H. Wenger (Eds.), *Computers and mathematics* (pp. 35–41). Washington, DC: Mathematical Association of America.

Howson, G., Keitel, C., & Kilpatrick, J. (1981). *Curriculum development in mathematics*. Cambridge, England: Cambridge University Press.

Janvier, C. (Ed.). (1987). *Problems of representation in the teaching and learning of mathematics*. Hillsdale, NJ: Lawrence Erlbaum.

Johnson, M. (1987). *The body in the mind: The bodily basis of reason and imagination*. Chicago: University of Chicago Press.

Judson, P. (1990) Using computer algebra systems in mathematics courses. *Journal of Mathematical Behavior, 9,* 153–158.

Kamii, C. (undated). *Double-column addition: A teacher uses Piaget's theory* [Video]. Birmingham, Alabama: Promethean Films South.

Karian, Z. A. (1990). Symbolic computation: A revolutionary force. *UME Trends, 2*(3), 3.

Kohl, H. R. (1969). *The open classroom*. New York: Random House.

Koretz, D. (1988). Arriving in Lake Wobegon: Are standardized tests exaggerating achievement and distorting instruction? *American Educator, 12*(2), 8–15; 46–52.

Lakoff, G. (1987). *Women, fire, and dangerous things*. Chicago, IL: University of Chicago Press.

Lakoff, G., & Johnson, M. (1980). *Metaphors we live by*. Chicago, IL: University of Chicago Press.

Lockard, J. D. (Ed.). (1967). *Report of the International Clearinghouse on Science and Mathematics Curricular Developments*. Washington, DC: American Association for the Advancement of Science.

Lockard, J. D. (Ed.). (1977). *Twenty years of science and mathematics curriculum development. The Tenth Report of the International Clearinghouse on Science and Mathematics Curricular Developments*. Washington, DC: American Association for the Advancement of Science.

Lomotey, K. (Ed.). (1990). *Going to school: The African-American experience*. Albany, NY: State University of New York Press.

Maher, C. A. (1987). The teacher as designer, implementer, and evaluator of children's mathematical learning environments. *Journal of Mathematical Behavior, 6,* 295–303.

Maher, C. A., & Alston, A. (1989). Is meaning connected to symbols? An interview with Ling Chen. *Journal of Mathematical Behavior, 8,* 241–248.

Maher, C. A., & Davis, R. B. (1990). Teacher's learning: Building representations of children's meanings. In R. B. Davis, C. A. Maher, & N. Noddings (Eds.), *Constructivist views on the teaching and learning of mathematics* (pp. 79–92). Reston, VA: National Council of Teachers of Mathematics.

McDaniel, M. A., & Schlager, M. S. (1990). Discovery learning and transfer of problem-solving skills. *Cognition and Instruction, 7,* 129–159.

McLeod, D. B. (1990). Information-processing theories and mathematics learning: The role of affect. *International Journal of Educational Research, 14*(1), 13–29.

Miller, G. A., & Gildea, P. M. (1987). How children learn words. *Scientific American, 257*(3), 94–99.

National Council of Teachers of Mathematics. (1989). *Curriculum and evaluation standards for school mathematics*. Reston, VA: National Council of Teachers of Mathematics.

National Research Council. (1989). *Everybody counts: A report to the nation on the future of mathematics education*. Washington, DC: National Academy Press.

Newman, R. J. (1990). Expect more from minority students. *UME Trends, 2*(3), 4.

Nicholls, J., Cobb, P., Wood, T., Yackel, E., & Patashnick, M. (1990). Dimensions of success in mathematics: Individual and classroom differences. *Journal for Research in Mathematics Education, 21,* 109–122.

Nuffield Mathematics Teaching Project. (1965). *I do and I understand.* London, England: The Nuffield Foundation.

Papert, S. (1980). *Mindstorms: Children, computers, and powerful ideas.* New York, NY: Basic Books.

Phillips, K. (1990). *The politics of rich and poor.* New York, NY: Random House.

Roberts, N., Carter, R., Davis, F., & Feurzeig, W. (1989). Power tools for algebra problem solving. *Journal of Mathematical Behavior, 8,* 251–265.

Rosebery, A. S., & Rubin, A. (1989). Reasoning under uncertainty: Developing statistical reasoning. *Journal of Mathematical Behavior, 8,* 205–219.

Schools Council. (1969). *Mathematics in primary schools.* London, England: Her Majesty's Stationery Office.

Shulman, L. S., & Keislar, E. R. (1966). *Learning by discovery: A critical appraisal.* Chicago, IL: Rand McNally.

Solomon, C. (1986). *Computer environments for children: A reflection on theories of learning and education.* Cambridge, MA: Massachusetts Institute of Technology Press.

Squire, L. R. (1986). Mechanisms of memory. *Science, 23,* 1612–1619.

Steffe, L. P., Cobb, P., & von Glasersfeld, E. (1988) *Construction of arithmetical meanings and strategies.* New York, NY: Springer-Verlag.

Steffe, L. P., von Glasersfeld, E., Richards, J., & Cobb, P. (1983). *Children's counting types: Philosophy, theory, and application.* New York, NY: Praeger Scientific.

Whitney, H. (1986). Coming alive in school math and beyond. *Journal of Mathematical Behavior, 5,* 129–140.

Wilf, H. S. (1982). The disk with the college education. *American Mathematical Monthly, 89,* 4–8.

Wilf, H. S., & Zeilberger, D. (1990). Towards computerized proofs of identities. *Bulletin (New Series) of the American Mathematical Society, 23,* 77–83.

Wolfram, S. (1984). Computer software in science and mathematics. *Scientific American, 251*(3), 188–203.

Zorn, P. (1985, August). Calculus from a discrete viewpoint: A course with computer symbolic manipulation. Paper presented at the Mathematics Association of America meeting, in Laramie, Wyoming.

ABOUT THE CONTRIBUTORS

Michael T. Battista is professor of mathematics education at Kent State University. He is a member of the editorial panel of the *Journal for Research in Mathematics Education* and has authored numerous articles on the teaching and learning of mathematics, focusing on geometry, spatial visualization, and the use of computers in instruction. He is currently exploring how the use of computer microworlds can enhance the learning of elementary school geometry.

Merlyn J. Behr is professor of mathematical sciences and education at Northern Illinois University. He received his Ph.D. in mathematics education from Florida State University. Professor Behr's research interests are in the area of teaching, knowing, and learning number concepts among children and teachers, especially rational number concepts and concepts of proportionality. He has taught courses in mathematics and mathematics education.

Alan J. Bishop is a university lecturer in the department of education at Cambridge University. He has been researching in the field of mathematics education for the last 30 years and has recently developed a particular interest in social and cultural issues. For 22 years he was the editor of *Educational Studies in Mathematics*, one of the foremost international journals on research in mathematics education, and he has been published widely. His most recent book is *Mathematical Enculturation—A Cultural Perspective on Mathematics Education*.

Hilda Borko is associate professor of curriculum and instruction at the University of Colorado–Boulder. During the writing of this chapter, she was on the faculty of the University of Maryland. Her current research interests include the process of learning to teach and expert-novice differences in teacher cognition and action. She is presently co-principal investigator with Catherine Brown and Robert Underhill of a longitudinal study of learning to teach mathematics, funded by the National Science Foundation. She is also editor of the teaching, learning, and human development section of the *American Educational Research Journal*.

Catherine A. Brown is associate professor of mathematics education in the departments of mathematics and curriculum and instruction at Virginia Polytechnic Institute and State University. She teaches undergraduate and graduate courses in both mathematics and pedagogy. Her research focuses on the process of becoming a mathematics teacher. In particular, she is interested in the role university teacher education programs play in the development of teachers' knowledge, beliefs, thinking, and actions. She is active in a number of research-related activities of the National Council of Teachers of Mathematics.

Thomas P. Carpenter is professor of curriculum and instruction at the University of Wisconsin-Madison. He is currently serving as editor of the *Journal for Research in Mathematics Education* and is associate director of the National Center for Research in Mathematical Sciences Education. His research has focused on the development of quantitative concepts in primary school children. He is currently investigating how teachers use knowledge about children's thinking in planning and implementing instruction.

Douglas H. Clements is associate professor of mathematics and computer education at the State University of New York–Buffalo. Previously a kindergarten teacher for five years, he has conducted research and has been published widely in computer applications in mathematics education, the early development of mathematical ideas, and the learning and teaching of geometry. Through a National Science Foundation (NSF) grant he has developed an elementary geometry curriculum based on Logo, *Logo Geometry,* with Michael T. Battista. He is currently working with colleagues on a new NSF project to develop a K–6 mathematics curriculum. His most recent book is *Computers in Elementary Mathematics Education*. He is active in the National Council of Teachers of Mathematics (NCTM), is editor and author of the forthcoming NCTM addenda materials, and was chair of the editorial panel of NCTM's research journal, the *Journal for Research in Mathematics Education*.

Robert B. Davis is professor of mathematics education at Rutgers University and editor of the *Journal of Mathematical Behavior*. He studied mathematics and psychology at MIT and taught there, as well as at Syracuse University, at Cornell University, and elsewhere. Since 1957 he has been concerned with the design of experiences that help students learn mathematics at all levels, from kindergarten through university-level mathematics. He created the Madison Project for curriculum improvement, the NSF-supported Big Cities in-service teacher education project, the mathematics curriculum at University High School in Urbana, Illinois, and directed the team that produced courseware for the NSF-sponsored trials of the PLATO computer system at the University of Illinois.

John A. Dossey is distinguished university professor of mathematics at Illinois State University. His research focuses on national and international studies of student achievement and curriculum in mathematics. During the period from 1986 to 1988, he served as president of the National Council of Teachers of Mathematics. He also served on the Commission on Standards for School Mathematics, the Mathematical Sciences Education Board, and the board of governors of the Mathematical Association of America. Active in teacher education, he served on the mathematics panel for the National Board for Professional Teaching Standards. He has authored or coauthored 25 books and nearly 100 other publications.

Elizabeth Fennema holds a joint appointment as professor in the department of curriculum and instruction and the women's studies program at the University of Wisconsin–Madison. She also is a co-director of the National Center for Research in Mathematical Sciences Education. Well known for the Fennema-Sherman studies on gender-related differences in mathematics learning, the Fennema-Sherman Mathematics Attitude Scales, her research on gender issues in education, and her commitment to equity issues, Dr. Fennema is equally respected for her contributions to the study of teaching. Her current research in cognitively guided instruction concerns teacher knowledge and beliefs about children's cognitions in addition and subtraction and their effect on learning.

Megan Loef Franke is associate researcher with the cognitively guided instruction project at the Wisconsin Center for Educational Research. Her research interests focus on understanding and measuring teacher knowledge as it affects the teaching and learning environment.

Karen C. Fuson is professor of education and social policy at Northwestern University. She has been involved in preservice and in-service teacher education, undergraduate and graduate teaching, and research. Her research has traced the development of children's understanding of numbers and number words. This work has included basic developmental and cognitive research as well as instructional studies. This research is reported in a book, *Children's Counting and Concepts of Number,* and in many research articles and reviews. Her most recent work addresses how cultures affect understanding of various mathematical topics, with the ultimate aim of improving the mathematical experiences of the many cultural subgroups in the United States.

Thomas L. Good is professor of education in the department of curriculum and instruction and research associate in the Center for Research in Social Behavior at the University of Missouri–Columbia. He is editor of the *Elementary School Journal* and author of numerous articles and books, including *Looking in Classrooms* with Jere Brophy. His research interests focus upon the study of teacher-student communication and learning in classroom settings.

Brian Greer is a reader in the school of psychology at Queen's University in Belfast and has qualifications in mathematics, education, and psychology. His general research area is children's mathematical thinking and learning, with a specific interest in multiplicative concepts. He coauthored *Analysis of Structural Learning* with M. A. Jeeves and co-edited *New Directions in Mathematics Education* with G. A. Mulhern and has been guest editor for recent issues of the *Journal of Mathematical Behavior* and the *International Journal of Educational Research.* He is currently a co-director of Oxford Mathematics, a secondary level mathematics curriculum project.

Douglas A. Grouws is professor of mathematics education and research associate in the Center for Research in Social Behavior at the University of Missouri–Columbia. His research interests include the relationships between teaching practices and student mathematical learning and the development of problem-solving ability in mathematics. He is coauthor of *Active Mathematics Teaching,* co-editor of *Effective Mathematics Teaching,* and has been published in numerous journals. He has served on the executive committee of the American Educational Research Association's Special Interest Group for Research in Mathematics Education, has been chair of the editorial board of the *Journal for Research in Mathematics Education* (JRME), and is editor of JRME Research Monographs. He teaches undergraduate and graduate mathematics education courses and is a recipient of the William T. Kemper Fellowship for Teaching Excellence.

Guershon Harel is associate professor at Purdue University with a joint appointment in the department of curriculum and instruction and the department of mathematics. For 10 years, while completing his undergraduate and graduate studies in mathematics at Ben-Gurion University, he was a mathematics teacher in a middle school and high school. He began his research in mathematics education with a five-year study on the learning and teaching of linear algebra to secondary students. His current research is in the multiplicative structures domain, where he has conducted theoretical analyses and empirical investigations. He also studies the nature and growth of students' conceptions of proof and function.

James Hiebert is professor of mathematics education at the University of Delaware. His research focuses on how elementary school students acquire understandings of mathematics and how alternative instructional approaches influence these understandings. He is especially interested in the ways in which children develop meaning for written mathematical symbols. He is editor of *Conceptual and Procedural Knowledge: The Case of Mathematics* and co-editor of *Number Concepts and Operations in the Middle Grades.*

James J. Kaput is professor of mathematics at the University of Massachusetts at Dartmouth. He is also associate director of the National Center for Research in Mathematical Sciences Education, and research associate at both the Educational

Technology Center at the Harvard Graduate School of Education and the Technical Educational Resource Center (TERC) in Cambridge. At each of these locations, he is engaged both in research and in curriculum and software development projects. These involve the application of new technologies in mathematics curricula at levels ranging from kindergarten through college. His key interest is in empowering students and teachers with the help of appropriate technologies.

Carolyn Kieran is professor of mathematics education in the department of mathematics and computer science at the Université du Québec à Montréal. Her research interests include the learning and teaching of algebra, the use of technology in school mathematics, and the application of historical and psychological models to mathematics education research. Recent publications include volume 4 of the National Council of Teachers of Mathematics' *Research Agenda for Mathematics Education: Research Issues in the Learning and Teaching of Algebra* with S. Wagner, *Perspectives on Mathematical Literacy*, and *Cognitive Processes Involved in Learning School Algebra*. She was a past chair of the editorial panel of the *Journal for Research in Mathematics Education* and a former member of the international committee of the *Psychology of Mathematics Education* group. She is currently on the executive committee of the Seventh International Congress on Mathematical Education.

Jeremy Kilpatrick is professor of mathematics education at the University of Georgia. He is vice-president of the International Commission on Mathematical Instruction, a former member of the Mathematical Sciences Education Board, and former editor of the *Journal for Research in Mathematics Education*. His research and writing primarily concern problem solving, research, curriculum, and assessment in mathematics education. He is coauthor of *Curriculum Development in Mathematics* with Geoffrey Howson and Christina Keitel and co-editor of *Mathematics and Cognition: A Research Synthesis* with Pearla Nesher by the International Group for the Psychology of Mathematics Education.

Mary Schatz Koehler is associate professor of mathematics education in the department of mathematical sciences at San Diego State University and is affiliated with the center for research in mathematics and science education. Her research interests lie in the area of mathematics equity, focusing particularly on gender-related issues. The way in which teacher behaviors and classroom processes may differentially affect females and males in terms of achievement, participation, and affect is a focus of her research. In addition, she teaches undergraduate courses for pre-service elementary and secondary teachers, as well as graduate courses for community college instructors. She has been involved in curriculum development to incorporate research on the learning of mathematics into the pre-service teacher preparation program.

Gilah Leder is associate professor of education at Monash University in Clayton, Victoria, in Australia. Her research interests include the impact of affective variables on mathematics learning, gender differences, exceptionality, and the interaction between teaching, assessment, and learning in mathematics. Dr. Leder recently co-edited *Educating Girls: Practice*

and Research and *Mathematics and Gender* and served as joint guest editor of the special bicentennial issue of the *Australian Journal of Education*.

Richard Lesh is a principal research scientist and director of the Educational Testing Service's (ETS) program on technology and assessment. He joined ETS in 1989, after being dean of education, director of teacher education, and professor in the departments of mathematics and education at Northwestern University. For five years, Professor Lesh was also the director of math/science software development at WICAT—which produced IBM's current library of educational software in mathematics—including the national award-winning Mathematics Education Toolkit. Professor Lesh graduated from Indiana University in 1971 and has been director or principal investigator for a series of NSF-funded research projects focusing on topics ranging from teacher education to mathematical concept development to students' abilities to use mathematics in everyday problem-solving situations. In addition to many chapters and books dealing with his research in mathematics teaching and learning, he has authored books ranging from Scott Foresman's kindergarten textbook to books for mathematics teacher education.

Mary M. McCaslin is associate professor in the department of curriculum and instruction in the college of education at the University of Missouri–Columbia. She received her Ph.D. in educational psychology from Michigan State University, where she was an intern in the Institute for Research on Teaching. Prior to moving to Missouri, Professor McCaslin was an associate professor in the department of human development at Bryn Mawr College. For the past decade, her research has examined socialization processes in the classroom, specifically the enhancement of students' self-regulated, adaptive learning.

Douglas B. McLeod, professor of mathematics and education, formerly served as chair of the department of elementary and secondary education at Washington State University. His research on affective issues grew out of his continuing interest in the general topic of individual differences among learners of mathematics. Recent publications include *Affect and Mathematical Problem Solving: A New Perspective* with V. M. Adams. He currently serves as professor of mathematics at San Diego State University.

Catherine M. Mulryan is a lecturer in education at St. Patrick's College of Education in Dublin, Ireland. She has taught pre-service and in-service teacher education courses in elementary mathematics, language studies, and research on teaching for several years. She recently completed her doctoral degree in curriculum and instruction at the University of Missouri–Columbia. Her present academic interests are in cooperative small-group instruction in mathematics, mathematical problem solving, student mediation of instruction, and the professional development of practicing teachers. Among her most recent publications is "Teacher Ratings: A Call for Teacher Control and Self-evaluation" (with Thomas L. Good) in *The New Handbook of Teacher Evaluation*.

Marilyn Nickson is research officer in the Council for Examination Development at the University of Cambridge Local Examination Syndicate in Cambridge, England. She formerly served as head of mathematics in the department of education at Anglia Polytechnic. Her research interests are in the philosophical and social areas of mathematics education. She is coauthor of *The Social Context of Mathematics Education: A Review of Research in Mathematical Education* with Alan Bishop, and her work related to perceptions of mathematics and their effects upon classroom practice appears in several other books in the United Kingdom and the United States. She is presently engaged in a project funded by the School Examinations and Assessment Council that is concerned with the differentiation of mathematics syllabi according to pupil ability and related assessment procedures.

Nel Noddings is professor of education and associate dean at Stanford. Her areas of special interest are feminist ethics, moral education, and mathematical problem solving. She is president of the Philosophy of Education Society. She was a Phi Beta Kappa visiting scholar for the school year from 1989 to 1990. In addition to three books—*Caring: A Feminine Approach to Ethics and Moral Education, Women and Evil*, and *Awakening the Inner Eye: Intuition in Education* with Paul Shore—and two co-edited volumes, she is the author of more than 70 articles and chapters on various topics ranging from the ethics of caring to mathematical problem solving.

Terezinha Nunes is associate professor at the department of psychology at the Universidade Federal de Pernambuco in Recife, Brazil, and presently a research fellow at The Open University, Milton Keynes, United Kingdom. She has conducted research on cultural influences on the development of mathematical concepts in and out of school and on reading development. She was responsible for the development of the *Learning Through Thinking* project, which forged new relations between the university and the state educational authority. Her books (published in Portuguese) include *The Clinical Method, Society and Intelligence, Learning Through Thinking* (as an editor), and *Mathematics in the Streets and in Schools* with A. D. Schliemann and D. W. Carraher.

Thomas R. Post is currently professor of mathematics education at the University of Minnesota in Minneapolis. He has taught mathematics in public schools in the state of New York and a wide variety of undergraduate and graduate courses in mathematics education at Minnesota. He has conducted research dealing with mathematical learning and concept development and is interested in the implications that psychologically related findings have for the development of instructional activities, particularly those utilizing manipulative materials. Dr. Post has served as national co-chairperson of the Special Interest Group for Research in Mathematics Education of the American Educational Research Association (AERA), as chairperson of the North American chapter of the International Group for the Psychology of Mathematics Education (NA-PME), and on the editorial board of the *Journal for Research in Mathematics Education*. He has been co-principal investigator of the National Science Foundation–supported Rational Number Project. Dr. Post has coauthored three college-level

texts, and his work has appeared in over 100 professional publications.

David F. Robitaille is professor of mathematics education and head of the department of mathematics and science education at the University of British Columbia in Vancouver, Canada. His research has focused on large-scale assessment of student outcomes and instructional practices in mathematics, and his work has been published in a wide variety of journals in the field. He co-edited one of three volumes on the results of the Second International Mathematics Study with Robert Garden of the New Zealand Ministry of Education and is currently serving as international coordinator for the International Association for Educational Achievement's (IEA) Third International Mathematics and Science Study. Professor Robitaille is an author of elementary school mathematics textbooks with Scott, Foresman and Company.

Thomas A. Romberg is the Sears Roebuck Foundation—Bascom professor in education at the University of Wisconsin–Madison, and the director of the National Center for Research in Mathematical Sciences Education for the U.S. Department of Education. He has a long history of involvement with mathematics curriculum reform, including work with the School Mathematics Study Group and Developing Mathematical Processes Curriculum Group. He served as chair of the project School Mathematics: Options for the 1990s for the U.S. Department of Education and of *Curriculum and Evaluation Standards for School Mathematics* for the National Council of Teachers of Mathematics, for which he received the American Educational Research Association's (AERA) Interpretive Scholarship/Professional Service Award in 1991. His publications include *Learning to Add and Subtract* and *Toward Effective Schooling: The IGE Experience*. His article, "Research on teaching and learning mathematics: Two disciplines of scientific inquiry," was selected by the AERA as the best review of research in 1987. He is well known for his study and involvement with mathematics curriculum reform efforts internationally and has had Fulbright Fellowships to both Australia and the USSR.

Alan H. Schoenfeld is professor of education and mathematics at the University of California-Berkeley, where he chairs the division of education in mathematics, science, and technology. His research is on mathematical thinking and problem solving. Professor Schoenfeld is author of *Mathematical Problem Solving* and editor of *Cognitive Science and Mathematics Education* and *A Source Book for College Mathematics Teaching*. He recently finished a term on the California mathematics framework panel and now serves on the mathematics panel of the National Board for Professional Teaching Standards.

Walter G. Secada is associate professor of curriculum and instruction at the University of Wisconsin–Madison. A senior researcher at the Wisconsin Center for Education Research and a principal investigator for the Center on Organization and Restructuring of Schools, he also directs a technical assistance and resource center funded by the Office of Bilingual Education and Minority Languages Affairs of the U.S. Department of Education. His scholarly and research interests include equity in education, mathematics education, and bilingual education. Among his recent works are a longitudinal study on the relationship between degree of bilingualism and arithmetic problem solving by Hispanic primary school children; a special issue of the *Peabody Journal of Education* with Margaret Meyer entitled "Needed: An agenda for equity in mathematics education;" a special issue of the *International Journal of Educational Research* entitled "The reform of school mathematics in the United States;" a monograph, *New directions for equity in mathematics education* with Elizabeth Fennema; and an in-press book on the teaching of mathematics to bilingual students.

Mike Shaughnessy has worked in the mathematics department at Oregon State University since 1976, when he received his Ph.D. from Michigan State. Throughout that time he has nurtured an interest in the teaching and learning of probability and statistics, with particular interest in misconceptions of probability and the interaction and conflict between psychological and mathematical construction in probability. Professor Shaughnessy has done research on the van Hiele levels of geometric thinking with William Burger. He teaches mathematics content courses for pre-service and in-service mathematics teachers and has been published extensively in the areas of probability and geometry.

Judith T. Sowder is professor of mathematical sciences at San Diego State University and is a member of the university's Center for Research in Mathematics and Science Education. She has recently served as chair of the Research Advisory Committee for the National Council of Teachers of Mathematics (NCTM), as director of the NCTM Research Agenda Project, and as co-chair of the American Educational Research Association's (AERA) Special Interest Group for Research in Mathematics Education. Her own research focuses on the development of number sense and estimation skills. She is the author of several book chapters and journal articles.

David Tall is reader in mathematics education and director of the Mathematics Education Research Centre at the University of Warwick, England. He received his D.Phil. in mathematics at Oxford and Ph.D. in education at Warwick. He is author of *Complex Analysis, Foundations of Mathematics,* and *Algebraic Number Theory* with Ian Stewart, writer of much computer software, including *A Graphic Approach to the Calculus, Real Functions and Graphs, and Numerical Solutions of Equations*, and author of over 100 articles on mathematics and mathematics education. He was the founder and chairman of the Advanced Mathematical Thinking Group of the Psychology of Mathematics Education (PME), chairman of the Mathematical Association Committee on Computers in the secondary mathematics curriculum, and editor of *Mathematics Review*. He was recently awarded the Bronze Medallion of the International Percy Grainger Society for his distinguished contribution to research and promotion of the music of Percy Grainger. His education research interests include cognitive development of mathematical concepts, visualization, the use of computers in concept development, problem solving, and the nature of advanced mathematical thinking.

Alba G. Thompson is associate professor of mathematics education in the department of mathematical sciences at San Diego State University. Her research and writing are primarily in the area of teachers' conceptions and philosophies of mathematics and mathematics teaching and the relationship of these to instructional practices. Her work has been published in various journals and books. She was a member of the author team for the National Council of Teachers of Mathematics' *Curriculum and Evaluation Standards* and has served on the editorial board of the *Journal for Research in Mathematics Education*. She has been the recipient of two outstanding teaching awards from the Associated Students at San Diego State University. Currently, she is conducting research with Patrick W. Thompson funded by the National Science Foundation that focuses on middle-grade children's development of quantitative reasoning and the necessary cognitive and affective transformations teachers must undergo in order to help bring about such development in their students.

Kenneth Travers is professor of mathematics education at the University of Illinois in Urbana-Champaign. He has taught mathematics at the secondary school and college levels. He currently teaches courses in curriculum research and evaluation at the university level. His research interests are in international assessment and secondary school curriculum development. Currently he is on leave to the National Science Foundation as head of the office for studies, evaluation, and dissemination. Professor Travers is a past chair of the International Mathematics Committee for the Second International Mathematics Study and was director of the coordinating center for U.S. participation in the study.

Norman L. Webb is senior research scientist for the Wisconsin Center for Education Research at the University of Wisconsin–Madison. His work is in the area of assessment and evaluation, and he is the issue editor for the National Council of Teachers of Mathematics' (NCTM) 1993 yearbook on assessment. He documented the Ford Foundation's Urban Mathematics Collaborative Project and chaired the working group that wrote the evaluation standards for the NCTM *Curriculum and Evaluation Standards for School Mathematics*. He consults for the Woodrow Wilson Fellowship Foundation to evaluate the history, science, and mathematics leadership program of one-week summer institutes. He has served as the program evaluator of other programs and supervised the competency-based testing program for the Wisconsin Department of Public Instruction.

NAME INDEX

SUBJECT INDEX